AF334085

COMPREHENSIVE DEVELOPMENTAL NEUROSCIENCE:
NEURAL CIRCUIT DEVELOPMENT AND FUNCTION IN THE HEALTHY AND DISEASED BRAIN

COMPREHENSIVE DEVELOPMENTAL NEUROSCIENCE: NEURAL CIRCUIT DEVELOPMENT AND FUNCTION IN THE HEALTHY AND DISEASED BRAIN

Editors-in-Chief

Professor JOHN L. R. RUBENSTEIN
Department of Psychiatry,
University of California at San Francisco, San Francisco, CA, USA

Professor PASKO RAKIC
Duberg Professor of Neurobiology and Neurology,
Director Kavli Institute for Neuroscience,
Yale University School of Medicine, New Haven, CT, USA

AMSTERDAM · BOSTON · HEIDELBERG · LONDON · NEW YORK · OXFORD
PARIS · SAN DIEGO · SAN FRANCISCO · SINGAPORE · SYDNEY · TOKYO
Academic Press is an imprint of Elsevier

Academic Press

Academic Press is an imprint of Elsevier

525 B Street, Suite 1900, San Diego, CA 92101-4495, USA

32 Jamestown Road, London NW1 7BY, UK

225 Wyman Street, Waltham, MA 02451, USA

First edition 2013

Copyright © 2013 Elsevier Inc. All rights reserved.

No part of this publication may be reproduced, stored in a retrieval system, or transmitted in any form or by any means electronic, mechanical, photocopying, recording or otherwise without the prior written permission of the publisher.

Permissions may be sought directly from Elsevier's Science & Technology Rights, Department in Oxford, UK: phone (+44) (0) 1865 843830; fax (+44) (0) 1865 853333; email: permissions@elsevier.com. Alternatively, visit the Science and Technology Books website at www.elsevierdirect.com/rights for further information.

Notice

No responsibility is assumed by the publisher for any injury and/or damage to persons, or property as a matter of products liability, negligence or otherwise, or from any use or, operation of any methods, products, instructions or ideas contained in the material herein. Because of rapid advances in the medical sciences, in particular, independent verification of diagnoses and drug dosages should be made.

British Library Cataloguing-in-Publication Data

A catalogue record for this book is available from the British Library

Library of Congress Cataloging-in-Publication Data

A catalog record for this book is available from the Library of Congress

ISBN: 978-0-12-397267-5

For information on all Academic Press publications visit our website at elsevierdirect.com

Typeset by SPi Global
www.spi-global.com

Editors-in-Chief

Professor John L. R. Rubenstein Department of Psychiatry, University of California at San Francisco, San Francisco, CA, USA

Professor Pasko Rakic Duberg Professor of Neurobiology and Neurology, Director Kavli Institute for Neuroscience, Yale University School of Medicine, New Haven, CT, USA

Section Editors

Professor Arturo Alvarez-Buylla Department of Neurological Surgery and The Eli and Edythe Broad Center of Regeneration Medicine and Stem Cell Research University of California, San Francisco, School of Medicine, San Francisco, CA, USA

Professor Yehezkel Ben-Ari Institute of Neurobiology of the Mediterranean Sea (INMED) AND CEO of Neurochlore Company, INSERM (the French Institute of Health and Medical Research), Marseille, Department of the "Bouches du Rhone", France

Professor Kenneth Campbell Divisions of Developmental Biology and Neurosurgery Cincinnati Children's Hospital Medical Center University of Cincinnati College of Medicine Cincinnati, OH, USA

Professor Hollis T. Cline Department of Molecular and Cellular Neuroscience, The Scripps Research Institute, La Jolla, CA, USA

Dr. François Guillemot Division of Molecular Neurobiology MRC National Institute for Medical Research, London, UK

Professor Takao Hensch Department of Molecular and Cellular Biology Harvard University, Cambridge, MA, USA

Professor Pat Levitt Zilkha Neurogenetic Institute and Department of Cell and Neurobiology, Keck School of Medicine of University of Southern California, Los Angeles, CA, USA

Dr Oscar Marín Instituto de Neurociencias, CSIC and Universidad Miguel Hernández, Alicante, Spain

Professor Dennis D.M. O'Leary Vincent J. Coates Chair in Molecular Neurobiology, Molecular Neurobiology Laboratory, The Salk Institute, La Jolla, CA, USA

Professor Franck Polleux The Scripps Research Institute, Dorris Neuroscience Center, La Jolla, CA, USA

Dr David H. Rowitch University of California, San Francisco, CA, USA

Dr Gordon M. Shepherd Department of Neurobiology, Yale School of Medicine, New Haven, CT, USA

Professor Helen Tager-Flusberg Department of Psychology and Department of Anatomy & Neurobiology, Boston University, Boston, MA, USA

Contents

Contributors

N. Akshoomoff   University of California San Diego, La Jolla, CA, USA

P.J. Bauer   Emory University, Atlanta, GA, USA

A.M. Beltz   The Pennsylvania State University, University Park, PA, USA

S.A. Berenbaum   The Pennsylvania State University, University Park, PA, USA

J.E.O. Blakemore   Indiana University – Purdue University Fort Wayne, Fort Wayne, IN, USA

A.B. Bowman   Vanderbilt University, Nashville, TN, USA

J.E. Bramen   University of California at Los Angeles (UCLA), Los Angeles, CA, USA

A.R. Brooks-Kayal   University of Colorado, Aurora, CO, USA

B.J. Casey   Sackler Institute for Developmental Psychobiology, New York, NY, USA

M.M. Cohen   Sackler Institute for Developmental Psychobiology, New York, NY, USA

J.B. Colby   University of California at Los Angeles (UCLA), Los Angeles, CA, USA

M. Colonnese   INSERM U901, University Aix-Marseille II, Marseille, France

C. Cordeaux   Yale University, New Haven, CT, USA

J.F. Cubells   Emory University School of Medicine, Atlanta, GA, USA

E.P. Davis   University of Denver, Denver, CO, USA; University of California, Irvine, Orange, CA, USA

J. Decety   University of Chicago, Chicago, IL, USA

K.C. Ess   Vanderbilt University, Nashville, TN, USA

D.E. Feldman   University of California Berkeley, Berkeley, CA, USA

L. Fernandez   University of California, San Francisco, CA, USA

R.H. Fitch   University of Connecticut, Storrs, CT, USA

N.A. Fox   University of Maryland, College Park, MD, USA

N. Franklin   Sackler Institute for Developmental Psychobiology, New York, NY, USA

D.C. Gillespie   McMaster University, Hamilton, ON, Canada

M.R. Gunnar   University of Minnesota, Minneapolis, MN, USA

H. Gweon   Massachusetts Institute of Technology, Cambridge, MA, USA

R.J. Hagerman   University of California at Davis, M.I.N.D. Institute, Sacramento, CA, USA

F. Haist   University of California San Diego, La Jolla, CA, USA

C. Hughes   University of Cambridge, Cambridge, UK

M.H. Johnson   Birkbeck College, University of London, London, UK

S.P. Johnson   University of California, Los Angeles, CA, USA

M. Kano   The University of Tokyo, Tokyo, Japan

R. Khazipov   INSERM U901, University Aix-Marseille II, Marseille, France; Kazan Federal University, Kazan, Russia

S.-J. Kim   University of Washington, Seattle, WA, USA; Center for Integrative Brain Research, Seattle Children's Research Institute, Seattle, WA, USA

A.J. King   University of Oxford, Oxford, UK

K.K. Kumar   Vanderbilt University, Nashville, TN, USA

A. Lahat   McMaster University, Hamilton, ON, Canada

J. Lany   University of Notre Dame, Notre Dame, IN, USA

M.J. Leigh   University of California at Davis, M.I.N.D. Institute, Sacramento, CA, USA

P.J. Lein   University of California at Davis School of Veterinary Medicine, Davis, CA, USA

G. Lepousez   Institut Pasteur, Paris, France; Centre National de la Recherche Scientifique, Paris, France

M. Lewis   University of Florida, Gainesville, FL, USA

P.-M. Lledo   Institut Pasteur, Paris, France; Centre National de la Recherche Scientifique, Paris, France

K.J. Michalska   University of Chicago, Chicago, IL, USA

M.W. Miller   State University of New York Upstate Medical University, Syracuse, NY, USA; Veterans Affairs Medical Center, Syracuse, NY, USA

M. Minlebaev   INSERM U901, University Aix-Marseille II, Marseille, France; Kazan Federal University, Kazan, Russia

Z. Molnár   University of Oxford, Oxford, UK

S.M. Mooney   University of Maryland, Baltimore, MD, USA; State University of New York Upstate Medical University, Syracuse, NY, USA

D. Moreno-De-Luca   Emory University School of Medicine, Atlanta, GA, USA

D. Muller   University of Geneva, Geneva, Switzerland

C.A. Nelson III   Harvard Medical School, Boston, MA, USA

I. Nikonenko   University of Geneva, Geneva, Switzerland

L. Niswander   University of Colorado Denver, Aurora, CO, USA; Howard Hughes Medical Institute, Aurora, CO, USA

E.D. O'Hare   University of California at Berkeley, Berkeley, CA, USA

K. Pelphrey   Yale University, New Haven, CT, USA

A.M. Persico   University Campus Bio-Medico, Rome, Italy; I.R.C.C.S. Fondazione Santa Lucia, Rome, Italy

D.B. Polley   Eaton-Peabody Laboratory, Massachusetts Eye and Ear Infirmary & Dept. of Otology and Laryngology, Harvard Medical School, Boston, MA, USA

M.I. Posner   University of Oregon, Eugene, OR, USA

C. Pyrgaki   University of Colorado Denver, Aurora, CO, USA

L.T. Reiter   University of Tennessee Health Science Center, Memphis, TN, USA

G. Righi   Boston Children's Hospital, Boston, MA, USA

M.K. Rothbart   University of Oregon, Eugene, OR, USA

E. Rubenstein   Stanford University School of Medicine, Stanford, CA, USA

M.R. Rueda   University of Granada, Granada, Spain

J.R. Saffran   University of Wisconsin-Madison, Madison, WI, USA

J.T. Sanchez   Virginia Merrill Bloedel Hearing Research Center & Dept. of Otolaryngology, Head and Neck Surgery, University of Washington, Seattle, WA, USA

R. Saxe   Massachusetts Institute of Technology, Cambridge, MA, USA

A.M. Seery   Boston University, Boston, MA, USA

A.H. Seidl   Virginia Merrill Bloedel Hearing Research Center & Dept. of Otolaryngology, Head and Neck Surgery, University of Washington, Seattle, WA, USA

E.H. Sherr   University of California, San Francisco, CA, USA

D.E. Shulz   Centre National de la Recherche Scientifique, Gif sur Yvette, France

E.R. Sowell   University of Southern California (USC), Los Angeles, CA, USA

T.R. Stanford   Wake Forest School of Medicine, Winston-Salem, NC, USA

B.E. Stein   Wake Forest School of Medicine, Winston-Salem, NC, USA

J. Stiles   University of California San Diego, La Jolla, CA, USA

K.L. Summar   George Washington University, Washington, DC, USA

H. Tager-Flusberg   Boston University, Boston, MA, USA

A.X. Thomas   University of Colorado, Aurora, CO, USA

J. Tirrell   Yale University, New Haven, CT, USA

N. Urraca   University of Tennessee Health Science Center, Memphis, TN, USA

J. Veenstra-VanderWeele   Vanderbilt University, Nashville, TN, USA

A. Voos   Yale University, New Haven, CT, USA

Y. Wang   Virginia Merrill Bloedel Hearing Research Center & Dept. of Otolaryngology, Head and Neck Surgery, University of Washington, Seattle, WA, USA

M. Watanabe   Hokkaido University, Sapporo, Japan

Introduction to Comprehensive Developmental Neuroscience

It is broadly accepted that understanding the genetic, molecular, and cellular mechanisms of neural development is essential for understanding evolution and disorders of neural systems. Recent advances in genetic, molecular, and cell biological methods have generated a massive increase in new information. By contrast, there is a paucity of comprehensive and up-to-date syntheses, references, and historical perspectives on this important subject. Therefore, we embarked on the formidable task of assembling a novel resource entitled 'Comprehensive Developmental Neuroscience.' We hope that the books in this series will serve as valuable references for basic and translational neuroscientists, clinicians, and students.

To help with this enormous task, we invited leading experts in various subfields to select the subjects and invite appropriate authors. We were gratified by the number of busy scientists who accepted the invitation to write their articles. All the chapters have been peer reviewed by the Section Editors to ensure accuracy, thoroughness, and scholarship.

In the resulting three volumes, we cover a broad array of subjects on neural development. We organized the volumes chronologically according to the ordered steps in neural development. In addition, each volume is subdivided into three to four sections, each edited by world experts in these areas. The sections have 10–20 chapters that are written and illustrated by leading scientists.

This Volume in the series has 40 chapters devoted to the anatomical and functional development of neural circuits and neural systems, as well as chapters that address neurodevelopmental disorders in humans and experimental organisms. This volume is subdivided into three sections. The first is on the mechanisms that control the assembly of neural circuits in specific regions of the nervous system, and as a function of neural activity and critical periods. The second section concentrates on multiple aspects of cognitive development, particularly in humans. The final section addresses disorders of the nervous system that arise through defects in neural development, building on the principles that are addressed in earlier sections of the book.

Volume 1 has 48 chapters devoted mainly to patterning and cell type specification in the developing central and peripheral nervous systems (CNS and PNS). This volume is subdivided into three sections. The first is on the mechanisms that control regional specification, which generate subdivisions of the nervous system. The second is on mechanisms that regulate the proliferation of neuronal progenitors and that control differentiation and survival of specific neuronal subtypes. The third section addresses the mechanisms controlling development of non-neural cells: astrocytes, oligodendroyctes, Schwann cells, microglia, meninges, blood vessels, ependyma, and choroid plexus.

Volume 2 in the series has 56 chapters devoted to migration (cell and axonal), the formation of neuronal connections, and the maturation of neural functions. This volume is subdivided into four sections. The first is on mechanisms that control the formation of axons and dendrites. The second is on the mechanisms that regulate cell migration that disperses specific subtypes of cells along highly defined pathways to specific destinations. The third section is on the regulation of synapse formation and maintenance during development; in addition, it has chapters on synaptogenesis in the mature nervous system in response to neurogenesis, neural activity, and neural trauma. The final section is on the developmental sequences that regulate neural activity, from cell-intrinsic maturation to early correlated patterns of activity.

John L.R. Rubenstein
Pasko Rakic

CIRCUIT DEVELOPMENT

The Form and Functions of Neural Circuits in the Olfactory Bulb

G. Lepousez, P.-M. Lledo

Institut Pasteur, Paris, France; Centre National de la Recherche Scientifique, Paris, France

OUTLINE

1.1 INTRODUCTION

Sensory systems are specialized biological devices by which organisms perceive the external sensory space. Sensory perception allows the deconstruction of this external world and the subsequent emergence of an internal percept. Following this Aristotelian principle, animals can discriminate and recognize an enormous range of physical and chemical signals in their environment, which may profoundly influence their behavior and provide them with vital information for reproduction and survival.

Several sophisticated sensory channels are available for that purpose, but all of them rely on a specific set of rules by which information is transposed from one dimension to another. For the chemical senses, this transposition concerns the ways in which chemical information gives rise to specific neuronal responses in a dedicated sensory organ (Ache and Young, 2005). In comparison to other sensory stimuli, odorant molecules, that is, volatile molecules that are perceived as odors, are infinite in terms of molecular formulae and cannot be classified with only objective dimensions such as frequency in the case of auditory stimuli (Laurent, 2002). The task of perceiving an odorant is made even more complex because a single odor usually is composed of many types of molecules (e.g. chocolate contains more

© 2013 Elsevier Inc. All rights reserved.

than 600 types) and is still perceived as one unique object, distinct from its components (a process called pattern completion). In addition, odor intensity (i.e., molecular concentration) can vary without changes in the perceived quality (a process called perceptual stability; Gottfried, 2010).

The origin of chemical detection (also called chemosensation) certainly dates back to the beginning of prokaryotes. It has evolved into distinct modalities in vertebrates to meet crucial needs such as locating potential food sources, detecting dangers such as predators, and mediating social and sexual interactions. Despite these functions, which are highly conserved throughout the evolution of species, interest in audition and vision has largely dominated the sensory research field in neuroscience, leaving behind the understanding of the more primitive chemical sense. Nevertheless, during the last two decades, neuroscience has made considerable progress in understanding how the brain perceives, discriminates, and recognizes odorant molecules so precisely. In 1991, the description of a multigene family of olfactory G-protein-coupled receptors provided a molecular basis for odor recognition (Buck and Axel, 1991). This discovery was of great significance to the neurobiology of olfaction and later was recognized with the Nobel Prize in 2004 (Mombaerts, 2004a). Since then, olfaction has attracted the attention of neuroscientists who started to investigate the different stages of chemosensory systems. As a result, converging approaches including anatomy, neurogenetics, biochemistry, cellular biology, neurophysiology, psychology, and computational neuroscience have contributed to provide a picture of how chemical information is processed in the olfactory system, starting from the periphery to higher brain regions.

Although still incomplete, today we have a fairly comprehensive picture about how chemicals interact with their cognate receptors to initiate signal transduction in the sensory receptor cells. We also know how the olfactory sensory information is first transduced in the sensory neurons located in dedicated olfactory sensory organs (Mombaerts, 2004b). Among the different elements along the olfactory pathway, local circuits in the second- and third-order central areas are key elements that process the simple monophasic sensory signal conveyed by the sensory neurons and convert it into a multidimensional content (Figure 1.1) (Gottfried, 2010; Wilson and Sullivan, 2011).

Below, we discuss how the chemical information is encoded and processed at the first central relay of the olfactory system, the main olfactory bulb, as well as the functions of the bulbar neural circuits that are relevant for triggering specific behavioral responses. In addition to the main olfactory bulb, which relays odorant information detected by sensory neurons of the olfactory epithelium, a similar structure called the accessory olfactory bulb, located in the caudal part of the olfactory bulb, relays pheromonal information detected in the vomeronasal organ. Because of space constraints, this chapter focuses exclusively on the main olfactory bulb and odor coding in rodents.

1.2 SYNAPTIC ORGANIZATION OF THE MAMMALIAN OLFACTORY BULB

The olfactory system is responsible for correctly coding sensory information from thousands of odorous stimuli. To accomplish this complex task, odor information is

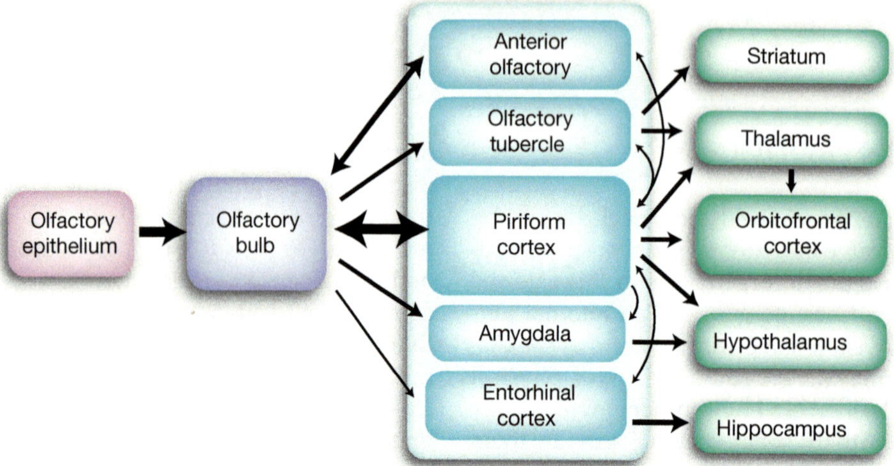

FIGURE 1.1 Schematic representation of the olfactory system. The olfactory system is composed of a sensory organ, the olfactory epithelium, followed by the olfactory bulb, the first central relay of odor information. The olfactory bulb then projects to different olfactory cortical areas involved in olfactory processing and behavior. Structures in blue that receive direct inputs from the output neurons of the olfactory bulb composed together the so-called 'olfactory cortex.'

processed through distinct units. At each of these units, a modified representation of the odor stimulus is generated. Following a bottom–up approach, we will start our description from the olfactory sensory organ located in the nasal cavity.

1.2.1 Organization of Sensory Inputs

The ability to process correctly the physical and statistical properties of chemical information is constrained by the functional architecture and the synaptic organization of some dedicated neural circuits. As our knowledge about the neurobiology of olfaction grows, it is becoming evident that the main olfactory systems of animals in disparate phyla share many striking features regarding the functional architecture and the synaptic connectivity (Bargmann, 2006; Hildebrand and Shepherd, 1997). For instance, mammals and invertebrate olfactory systems display common organizational features and functional characteristics (Wilson and Mainen, 2006). In all cases, the initial event shared by virtually all odorant detection systems is the requirement of specific interaction of odorant molecules with specific receptors expressed on the cilia of sensory olfactory neurons before conveying information to central structures (Ache and Young, 2005; Bargmann, 2006; Mombaerts, 2004b). Essentially, there are four different features that are common to all olfactory systems: (1) The presence of soluble odorant binding proteins in the mucus overlying the receptor cell dendrite, (2) the existence of G-protein-coupled receptor acting as specific odorant receptors, (3) the use of a two-step signaling cascade in odorant transduction, (4) the presence of functional anatomical structures at the first central target in the olfactory pathway called glomeruli (reviewed by Bargmann, 2006; Firestein, 2001; Mombaerts, 2004b). If these common features represent adaptive mechanisms that have evolved independently, their study might bring valuable knowledge about the way the nervous system extracts information from olfactory space.

In mammals, olfactory transduction starts with the activation of some of the thousand different types of odorant receptors located on the cilia of bipolar olfactory sensory neurons that comprise the olfactory neuroepithelium (1000–1300 genes in the mouse; Buck and Axel, 1991; Mombaerts, 2004b). It is believed that these sensory neurons recognize more than a thousand airborne volatile molecules called odorants. A general principle is that these receptors exist as separate information channels, because most sensory neurons express only one type of olfactory receptor. Because of the broad chemical tuning of these receptors (Araneda et al., 2000), the general coding principle is "one odor = one set of receptor activations" (Firestein, 2001; Mombaerts, 2004b). The olfactory epithelium then projects to the first central relay of the olfactory system: the main olfactory bulb (referred to below as the olfactory bulb). As advances in understanding olfactory transduction progressed, interest in deciphering some of the olfactory bulb functions grew concomitantly. This heightened interest has been spurred on, at least in part, by the discovery of the way in which the sensory organ connects to the olfactory bulb. We know now that the sensory neurons that express the same odorant receptor project their axons centrally to one or two spherical modules in the olfactory bulb called glomeruli (Zou et al., 2009). Here, sensory information is transmitted to higher brain centers via output neurons called mitral/tufted cells. The discrete and spherical glomeruli are morphological units made of distinctive bundles of neuropil (Figure 1.2). This terminology reflects both the homogeneity of the sensory inputs (they all express only one odorant receptor) and the degree to which the olfactory bulb neurons connected to the same glomerular unit display a similar receptive field (Dhawale et al., 2010; Tan et al., 2010). Recently, dramatic progress has been made in understanding of how olfactory sensory neurons develop and how they express only one odorant receptor. This discovery prompted the use of genetic tools to probe and manipulate specific populations of sensory neurons, resulting in several major breakthroughs

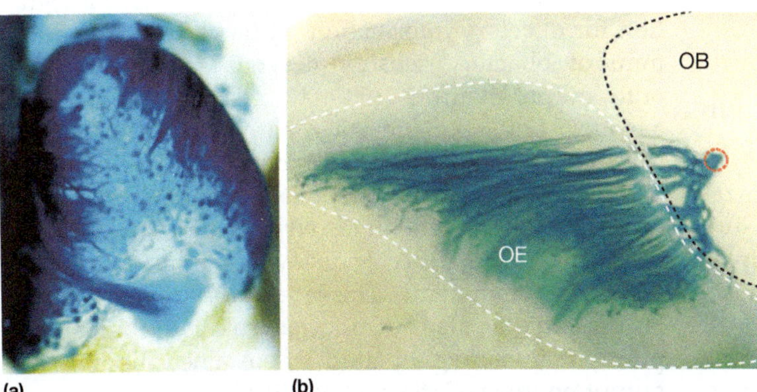

FIGURE 1.2　Convergence of the olfactory sensory neuron axons into glomeruli. (a) Olfactory sensory neurons labeled in blue project their axons on the surface of the olfactory bulb (note the single glomeruli). (b) Whole-mount view of the olfactory epithelium (OE) and the olfactory bulb (OB). Olfactory sensory neurons expressing the olfactory receptor P2 in the olfactory epithelium (OE) project the axons towards a single glomerulus (red dashed circle) on the surface of the olfactory bulb (OB). *Adapted from Mombaerts P, Wang F, Dulac C, et al. (1996) Visualizing an olfactory sensory map.* Cell *87: 675–686, with permission.*

regarding axon guidance, axon sorting and axon positioning (Sakano, 2010; Zou et al., 2009).

In different species, each glomerular structure results from the convergence of 5 to 40 thousand axon terminals of sensory neurons that express the same odorant receptor (Figure 1.2). Therefore, the glomerular layer can be considered as a two-dimensional anatomical representation of the olfactory receptors' repertoire (also called 'chemotopic map'; Johnson and Leon, 2007; Wachowiak and Shipley, 2006). Because one odor can activate several olfactory neurons, odor information is first encoded by the combinatorial patterns of glomerular activation. Odorants activate a specific array of olfactory sensory neurons that lead to a chemotopically fragmented map of activated glomeruli on the surface of the olfactory bulb (Meister and Bonhoeffer, 2001). Remarkably, the precise projection pattern can be reproduced from one animal to another and even between different rodent species (Soucy et al., 2009). Distinct odorants activate different combinations of glomeruli and two odors are more difficult to discriminate when these show a greater overlap in this glomerular chemotopic map (Linster et al., 2001). Nevertheless, if such a spatial pattern coding scheme were only applicable to several odors, it would not be able to provide a sufficiently large coding space to discern between the millions of potential odors or mixtures of odor present in our environment (Laurent, 2002).

The sensory neurons project to paired olfactory glomeruli on both the medial and lateral aspects of the olfactory bulb, thus creating two mirror-symmetric maps (Mombaerts et al., 1996). As each group of glomerulus-specific output neurons is odorant receptor-specific, glomeruli form a morphological as well as functionally defined network somewhat analogous to barrels in the somatosensory cortex (Johnson and Leon, 2007). In mammals, the convergence ratio of sensory neurons to olfactory bulb output neuron is very large: about 1000:1 (Zou et al., 2009). A bulbar output neuron thus forms its responses to odors from very large numbers of converging inputs, ensuring detection of faint signals, increase signal-to-noise ratios and temporal noise average.

1.2.2 Synaptic Processing Within Olfactory Bulb Microcircuits

Because of its relatively simple anatomical organization and easy access, the olfactory bulb has been a privileged model system for deciphering the principles underlying network processing of sensory information. There, odors elicit a well-organized pattern of glomeruli activation across the surface of the olfactory bulb, but the mechanisms by which this chemotopic map is processed into an odor code by the bulbar circuitry has recently attracted more attention. With the advance in recent years of *in vitro* brain slice preparation, as well as *in vivo* recording techniques that were applied on behaving animals, the complex processing of the olfactory information is starting to be revealed. Since Cajal's pioneering studies, it has been known that the main output neurons of the bulb, the so-called mitral cells, are located in a single lamina, the mitral cell layer (Figure 1.3). A second population of output neurons, namely tufted cells, are scattered in the external plexiform layer (EPL). The primary (or apical) dendrite of mitral and tufted cells, extending vertically from its soma, contacts one glomerulus, whereas their multiple long secondary dendrites spread in the EPL.

About 50–100 output neurons (mitral/tufted cells) emanate from each glomerulus and project to a number of higher centers that compose the olfactory cortex (see Figure 1.1). Output neurons are the backbone of two serial intrabulbar microcircuits: one between primary apical dendrites and juxtaglomerular cells and the other between secondary dendrites and granule cells. The main difference between juxtaglomerular and granule cells is that the former mediate mostly interactions between cells affiliated with the same glomerulus, while granule cells mostly mediate interactions between output neurons projecting to many different glomeruli (Figure 1.3). Therefore, two potential distinct sites of odor processing can be distinguished according to the topographical organization of the bulbar circuit. The first one resides in the glomeruli where local interneurons shape excitatory inputs coming from sensory neurons. The second one lies in the EPL, where reciprocal dendrodendritic synapses are heavily distributed between dendritic spines of local interneurons and the dendrites of output neurons. These two inhibitory microcircuits are also under the control of centrifugal fibers coming from cortical and neuromodulatory area.

1.2.2.1 *Synaptic Transmission at the First Synapses*

The glomerular layer constitutes a first site of integration for olfactory information. In this layer, axonal termini of olfactory sensory neurons synapse directly onto output neurons (50–100 cells per glomerulus) and also onto local neurons, namely juxtaglomerular cells (1000–2000 cells per glomerulus). The olfactory nerve-evoked excitatory responses of both neuron types comprise fast amino-3-hydroxy-5-methyl-4-isoaxozole-propionic acid (AMPA) and slow N-methyl-D-aspartate (NMDA) components. The latter is particularly long-lasting and may play an important role in the bulbar output by maintaining a pattern of sustained discharge of output neurons (Carlson et al., 2000).

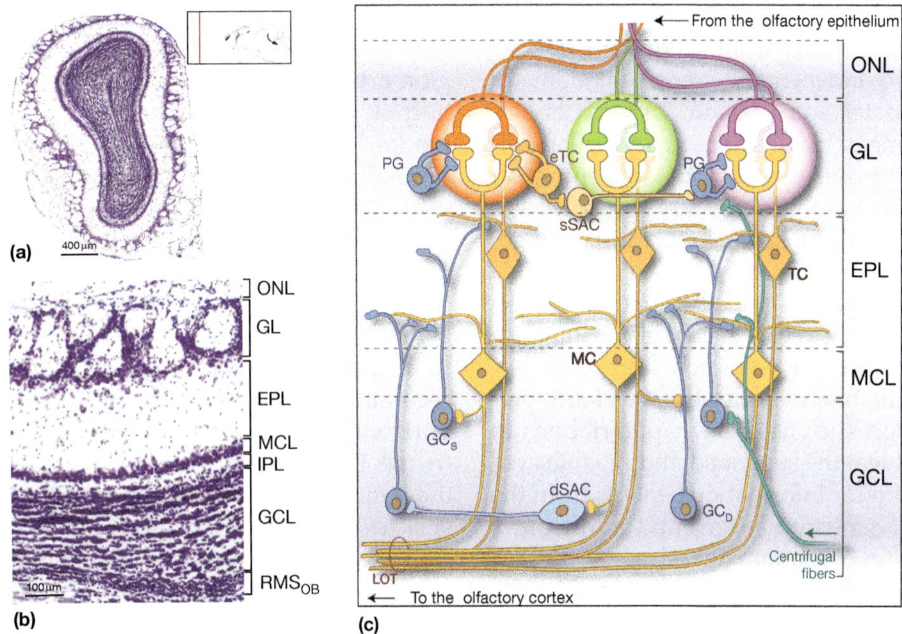

FIGURE 1.3 Anatomical organization of the main olfactory bulb. (a) Nissl-stained coronal section of the mouse olfactory bulb showing the different concentric layers. Inset, a sagittal section showing the rostrocaudal level of the coronal section. (b) Higher magnification view. ONL, olfactory nerve layer; GL, glomerular layer; EPL, external plexiform layer; MCL, mitral cell layer; IPL, internal plexiform layer; GCL, granule cell layer; RMS OB, rostral migratory stream of the olfactory bulb. (a) Olfactory sensory neurons (OSNs) in the olfactory epithelium (OE) that express the same odorant receptors project their axons to the main olfactory bulb into one of the glomeruli that form the glomerular layer (GL). In the GL, sensory neuron terminals synapse onto the apical dendrites of output neurons, namely the mitral cell (MC) and the tufted cells (TC). In addition, periglomerular cells (PGC), superficial short-axon cells (sSAC) and external tufted cells (eTC) act on glomerular synaptic transmission exerting diverse functional effects. In the external plexiform layer (EPL), the lateral dendrites of mitral and tufted cells interact with the dendrites of granule cells (GC). Granule cells can also be subdivided into distinct subpopulations: superficial granule cells (GC_S), which target the superficial lamina of the external plexiform layer and synapse with tufted cells, and deep granule cells (GC_D), which target the deep lamina of the external plexiform layer are connected to mitral cells. The somas of mitral cells are aligned and delineate the mitral cell layer (MCL), and the somas of tufted cells are scattered in the EPL. Granule cell somas and also some deep short-axon cells (dSAC) compose the granule cell layer (GCL). Centrifugal fibers from other brains regions innervate specific layers of the olfactory bulb depending on their brain origin. Lastly, output neuron axons fasciculate to form the lateral olfactory tract (LOT). All of the cell types colored in orange are glutamatergic; GABAergic cells are in blue. ONL, olfactory nerve layer.

Juxtaglomerular cells have dendrites restricted to one glomerulus and impinge onto olfactory nerve terminals or primary dendrites of mitral/tufted cells. Juxtaglomerular cells can be subdivided into three categories (Figure 1.3). The first category is comprised by periglomerular cells, small axonless interneurons that are the main source of γ-aminobutyric acid (GABA) in the glomerular layer and which provide feedback and feedforward inhibition to output neurons (Wachowiak and Shipley, 2006). It is noteworthy that a subpopulation of these GABAergic cells is also dopaminergic (Kosaka and Kosaka, 2005). Their inhibitory action alters the strength of sensory signals as they pass out of the bulb, and it is thought to provide a mechanism of filtering of weak sensory inputs (Gire and Schoppa, 2009). In addition, periglomerular cells also exert a presynaptic regulation of sensory inputs thanks to the expression of $GABA_B$ and D_2 dopamine receptors on olfactory sensory terminals (Aroniadou-Anderjaska et al., 2000; Hsia et al., 1999). The second category concerns the superficial short-axon cells. This heterogeneous cell population, comprising GABAergic and glutamatergic neurons, displays dendritic and axonal branching in several glomeruli and is thought to mediate lateral inhibition between glomeruli (Aungst et al., 2003; Kiyokage et al., 2010). The third group is composed of external tufted cells. These local glutamatergic cells, distinct from the output tufted cells, innervate periglomerular cells, short-axon cells, and mitral cells. Their spontaneous intrinsic rhythmic activity, synchronized by gap junctions and with the breathing rhythm, coordinates the activity of the cells located within each glomerulus (De Saint Jan et al., 2009; Hayar et al., 2004). In addition to direct fast excitation from OSN, external tufted cells provide a significant slow feed-forward excitation onto mitral cells (Najac et al., 2011). Moreover, activation of external tufted cells and glutamatergic superficial short-axon cells triggers lateral excitation of distant periglomerular cells, thus providing feedforward inhibition onto mitral/tufted cells of remote glomeruli (Aungst et al., 2003; Wachowiak and Shipley, 2006). This distributed inhibition mediates a globally averaged level of feedforward inhibition onto distant mitral/tufted cells and contributes to normalization of the intensity of sensory input, a key function for constructing intensity-independent representations of stimulus quality (Linster and Cleland, 2009). It is also known that several major classes of periglomerular cells are generated in the adult through the subventricular zone (SVZ) and the rostral migratory stream (RMS) and that their respective proportions remain constant throughout life (Lledo et al., 2008).

Since extremely faint signals can be detected by the olfactory system, there may be some mechanisms that promote reliable information transmission from olfactory sensory neurons to the brain. In the absence of any presynaptic specialization such as the synaptic ribbons in the retina or the cochlea, the sustained and reliable synaptic transmission of odor information benefits from the high convergence of sensory neurons and also from a very high probability of glutamate release, reflected by a marked paired-pulse depression (Hsia et al., 1999;

Murphy et al., 2004). In addition to the synaptic mechanisms described above, glutamate spillover within glomeruli and the presence of gap junctions between output neurons connected to the same glomerulus would help in increasing the signal-to-noise ratio and normalizing the activity of output neurons (Christie et al., 2005; Linster and Cleland, 2009).

1.2.2.2 The Dendrodendritic Synapse Provides the Major Source of Inhibitory Contact to Output Neurons

In contrast to the primary dendrite, mitral/tufted cell secondary (or basal) dendrites radiate up to 1000 μm horizontally, spanning almost the entire olfactory bulb. In the EPL, mitral/tufted cell dendrites interact with the dendrites of granule cells and both contain synaptic vesicles (Figure 1.4; Price and Powell, 1970). As granule cells are the largest group of bulbar interneurons (there are about 100 granule cells for one output neuron), and

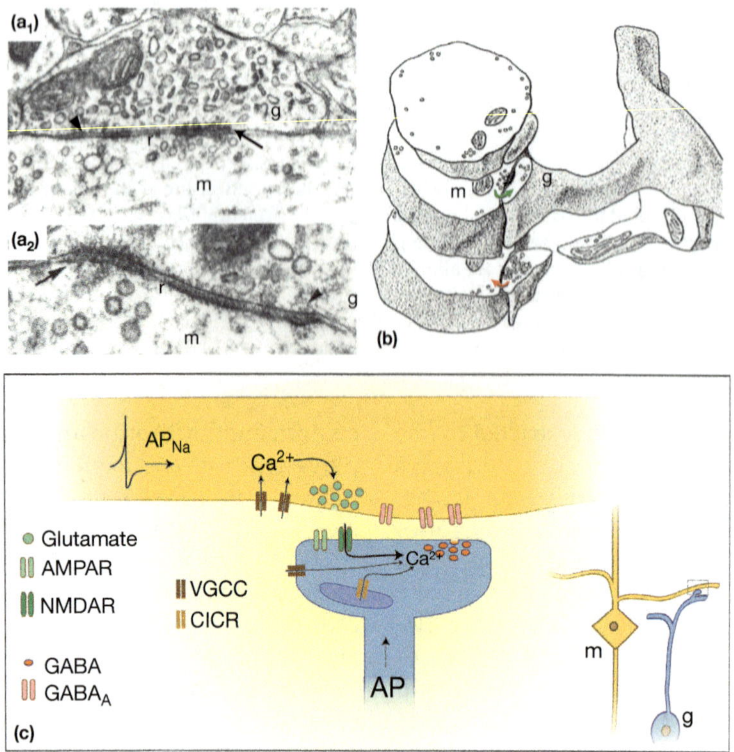

FIGURE 1.4   The dendrodendritic reciprocal synapse in the main olfactory bulb. (a) Two electron micrographs of a mitral cell dendrite (m) making a mitral-to-granule asymmetric synapse (arrow) onto a granule cell spine (g), which in turn makes a granule-to-mitral symmetric synapse (arrowhead) back onto the same mitral cell dendrite (adapted from Price JL and Powell TP (1970). The synaptology of the granule cells of the olfactory bulb. *Journal of Cell Science* 7: 125–155, with permission). (b) Serial section reconstitution of a reciprocal synapse between a lateral dendrite of mitral cell (m) and a granule cell spine (g). Green arrow, glutamatergic synapse; red arrow, GABAergic synapse (adapted from Rall W, Shepherd GM, Reese TS, and Brightman MW (1966) Dendrodendritic synaptic pathway for inhibition in the olfactory bulb. *Experimental Neurology* 14: 44–56, with permission). (c) Schematic representation of the dendrodendritic synapse. Action potentials (AP) propagating in mitral cell secondary dendrites (m, represented in orange) activate voltage-gated calcium channels (VGCC), which trigger the release of glutamate at reciprocal synapses. This glutamate locally depolarizes granule cell spines (g, represented in blue) via AMPA receptors (AMPAR) and NMDA receptors (NMDAR), which triggers calcium entry from NMDA receptors and also from voltage-gated calcium channels and calcium-induced calcium release from internal stores (CICR). This in turn causes GABA release back onto the mitral cell dendrite that activates GABA_A receptors (GABA_A). In addition, action potentials generated in the soma of the granule cell can propagate in the dendritic tree and trigger a global release of GABA.

as they are axonless neurons, synaptic transmission between dendrites remains the dominant mode of neuronal interaction in the olfactory bulb (Schoppa and Urban, 2003; Shepherd et al., 2007).

Granule cells can also be subdivided into distinct subpopulations: superficial granule cells that target the superficial lamina of the EPL and are believed to establish synapses mostly with tufted cells and deep granule cells targeting the deep lamina of the EPL which are connected to the mitral cells (Figure 1.3; Mori et al., 1983; Shepherd et al., 2007). These two microcircuits (tufted/superficial granule cells versus mitral/deep granule cells) are thought to maintain different functions in odor processing. Because of their short lateral dendrites and the intrabulbar connections between mirror-symmetric glomeruli they support, tufted cells may be important for intensity perception of odorants (Lodovichi et al., 2003; Nagayama et al., 2004). By contrast, the mitral-granule cell circuit is thought to mediate complex inhibitory functions that are important for odor discrimination (Lledo and Lagier, 2006).

The synaptic interactions that play a key role in the mitral/tufted-granule cell circuits have some simple yet unusual features. Output cells and granule cells communicate via reciprocal dendrodendritic synapses (Figure 1.4; Jahr and Nicoll, 1980; Price and Powell, 1970; Rall et al., 1966; review in Schoppa and Urban, 2003). This feature also extends to periglomerular cells that synapse with mitral/tufted apical dendrites in the glomerulus. The reciprocal circuit provides inhibition that forms the basis for a reliable, spatially localized, and recurrent inhibition. Synaptic depolarization in mitral/tufted cells driven by the long-lasting excitatory inputs from sensory neurons triggers the generation of an action potential. This action potential invades the lateral dendrites (Xiong and Chen, 2002) and triggers glutamate release from output neuron dendrites onto granule cell spines. The activation of AMPA and NMDA receptors present in the spines of granule cells leads to a depolarization and a local calcium entry in interneuron dendrites and spines. This, in turn, elicits the synchronous and asynchronous release of GABA directly back onto output neurons. The release of GABA from granule cell spines is driven by calcium entry, mainly from NMDA receptors, but also from voltage-gated calcium channels and internal stores (Figure 1.4; Egger et al., 2005; Isaacson and Strowbridge, 1998; Schoppa et al., 1998). Thus, dendritic GABA release from granule cell spines might occur (1) through an action potential-independent manner, providing a local and graded form of inhibition, or (2) through an action potential-dependent manner, triggered by somatic excitatory inputs and supporting a global form of inhibition (Lledo and Lagier, 2006). Additionally, interneurons of the olfactory bulb also receive GABAergic inputs, as

classically reported in several other brain areas. Recent studies have revealed that this inhibition is mediated by inframitral interneurons called 'deep short-axon cells' (Figure 1.3), which represent an unexpectedly large and diversified population of interneurons in the olfactory bulb (Eyre et al., 2008; Pressler and Strowbridge, 2006). These deep short-axon cells receive inputs from output neurons and possibly from centrifugal fibers and provide a feedback inhibition onto granule cells (and to a lesser extent periglomerular cells) thanks to a large axonal arbor.

Since secondary dendrites have large projection fields and extensive reciprocal connections with interneurons, each local bulbar interneuron may contact the dendrites of numerous output neurons. This suggests not only that dendrodendritic interactions provide a fast and graded feedback inhibition, but that they also underlie lateral inhibition between output neurons that innervate different glomeruli. In this sense, granule cells would provide both local (recurrent) and integrated (lateral and global) inhibition to output neurons (Lledo and Lagier, 2006). How do inhibition and local microcircuits shape and encode the spatial representation of sensory inputs?

1.2.3 Sensory Processing in the Output Layer

The mitral/tufted-granule cell circuit is thought to perform three main functions: sharpening the tuning of output neurons, decorrelating sensory inputs into time-varying temporal patterns, and synchronizing activated output neuron activity. These computations are crucial for odor coding and especially for odor discrimination: reducing or facilitating inhibition between granule cells and output neurons impairs or improves odor-discrimination performance, respectively (Abraham et al., 2010).

First, olfactory information is transmitted not only vertically across the glomerular relay, between sensory neurons and output neurons, but also horizontally, through local interneuron connections that are activated in odor-specific patterns. Such a model based on lateral inhibition, originally introduced in the 1950s to explain visual contrast enhancement in the retina (see Shepherd, 2010) has been extensively characterized mathematically. Anatomical and functional characterizations of the olfactory bulb circuit have revealed the importance of lateral inhibition. Thanks to this inhibitory mechanism, activity in few stimulated output neurons may lead to the inhibition of other neighboring neurons innervating distinct glomeruli (Egger and Urban, 2006; Yokoi et al., 1995). This inhibition, which critically relies on dendrodendritic synapses, was proposed as a mechanism for sharpening the tuning of output cells compared to their inputs (e.g., Koulakov and Rinberg, 2011). For instance, examination of the responses of individual mitral cells to inhalation of aliphatic aldehydes reveals that many individual cells are excited by one subset of these odorants, unaffected by a second subset, and inhibited by a third subset (Figure 1.5; Tan et al., 2010; Yokoi et al., 1995). Interestingly, the magnitude of lateral inhibition received by neighboring output neurons through the granule cell circuit is dynamically regulated, since it depends on the activity level of the postsynaptic output neuron (Arevian et al., 2008). Such interactions may underlie the limited mitral cell activation observed in animals that are awake (Koulakov and Rinberg, 2011). Because odor maps generated by different odors are extensive all over the surface of the olfactory bulb and often overlap, the propagation of action potentials into the lateral dendrites, and the possible spread of excitation through granule cell dendrites, contributes to a contrast enhancement mechanism that sharpens the tuning of widely dispersed output neuron odorant-receptive fields. Rather than a center-surround receptive field like that in the retina, mitral cells have a more distributed and sparse receptive field (Fantana et al., 2008). Lastly, the interglomerular lateral inhibition provided by superficial short-axon cells in the glomerular layer described above could, to a lesser extent, provide an additional form of lateral inhibition (Aungst et al., 2003; Linster and Cleland, 2009).

Second, the long-range projections of secondary dendrites underlie fundamental computations characteristic of the nonlinear dynamical system (Laurent, 2002). Indeed, output cells respond to a constant odor exposure with odor-specific temporal spike patterns that may constitute a "temporal code" of odor identity for downstream regions (Figure 1.6(a)). The spatially distributed pattern of activated output neurons – sometimes ambiguous for similar odorants – evolves progressively over time and diverges from the initial pattern, following a trajectory in the coding space that is characteristic of a given odorant and a given intensity (Figure 1.6(b); Bathellier et al., 2008; Laurent, 2002; Mazor and Laurent, 2005; Stopfer et al., 2003). Within this

framework, the main function of bulbar microcircuits would be to decorrelate similar input patterns into divergent odor-specific temporal dynamics of output cell activity. This computation, called decorrelation, would reformat combinatorial representations by reducing the overlap between clustered input activity patterns, thus making odor representation more distinct and facilitating their discrimination by downstream olfactory centers (Friedrich and Laurent, 2001). This temporal coding scheme may also be generalized in the case of terrestrial mammals, in which active breathing introduces an additional temporal dynamic to the stimulus (Wachowiak, 2011; see respiratory traces in Figure 1.5). Odor inhalation triggers precise and reliable cell- and odor-specific temporal spike patterns, which are tightly time-locked to the sniff phase and carry sufficient information for the discrimination of odors (Cury and Uchida, 2010; Dhawale et al., 2010; Shusterman et al., 2011). The timing of mitral cell activation relative to the sniff phase has been recently highlighted as another possible mechanism for the rapid encoding of odor (Smear et al., 2011). This additional extrinsic rhythmicity imposed by the breathing cycle provides an external "clock" for the entire olfactory system that may promote spike-timing dependent coding relative to the phase of the underlying respiratory cycle (Wachowiak, 2011).

Third, the long-range projection of secondary dendrites and recurrent inhibition provided by granule cells shape the temporal patterns of output neuron activity and lead to the synchronization of the network. This synchronization gives rise to oscillations of the local field potentials (LFP) in the gamma frequency band (30–100 Hz; see LFP traces in Figure 1.6(a)), a phenomenon already described by Sir Adrian in 1942 (Adrian, 1942; see also Shepherd, 2010). Since the pioneering theoretical work of Rall and Shepherd (Rall and Shepherd, 1968; Rall et al., 1966), experimental and computational studies have confirmed that recurrent inhibition provided by the dendrodendritic reciprocal synapse is

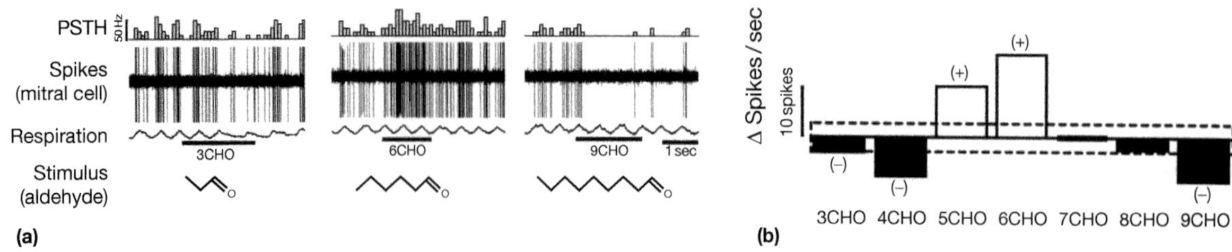

FIGURE 1.5   Molecular receptive field property of mitral cell. (a) *In vivo* recording of mitral cell spiking activity in anesthetized rat exposed to a homologous series of aliphatic aldehydes. Top, peristimulus time histogram (PSTH), which represents the average of five single-trial responses. The bottom traces indicate the respiratory cycle. The black bar marks the period of odor presentation. Note the respiratory-driven spiking activity of mitral cell. (b) Histogram of the change in spike rates (white column = increase; black column = decrease) during odorant-stimulation compared with the spike rate during prestimulation period. Stimulus odorants are propylaldehyde (3CHO), butylaldehyde (4CHO), valeraldehyde (5CHO), hexylaldehyde (6CHO), heptylaldehyde (7CHO), octylaldehyde (8CHO), and nonylaldehyde (9CHO). *Adapted from Nagayama S, Takahashi YK, Yoshihara Y, et al. (2004) Mitral and tufted cells differ with respect to the manner in which odor maps are decoded in the rat olfactory bulb.* Journal of Neurophysiology *91: 2532–2540, with permission.*

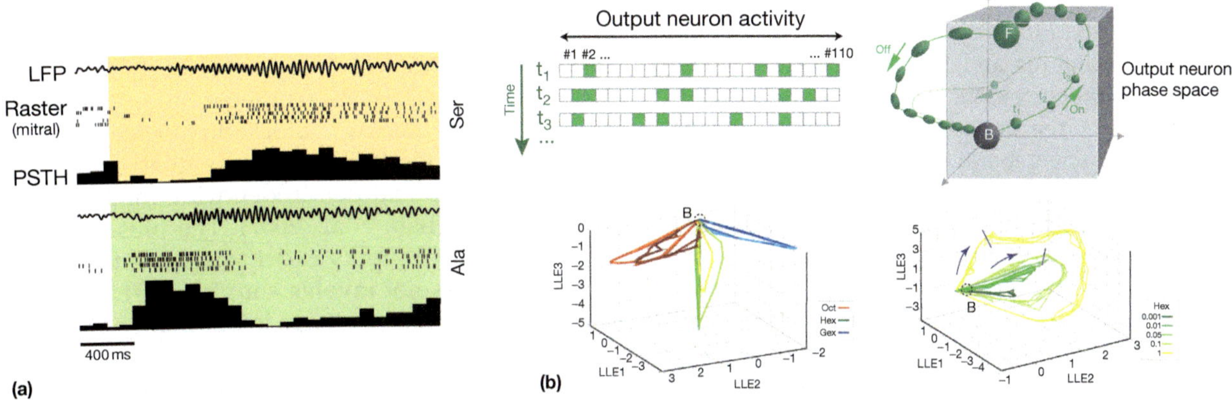

FIGURE 1.6 Temporal pattern and population coding of odor responses. (a) Five single-trial responses of one zebrafish mitral cell to two odor stimuli (Ala, alanine; Ser, serine). LFP, local field potential (bandpass filtered 5–50 Hz); middle, spike raster in which each tick represents the timing of one action potential from the mitral cell; bottom, peristimulus time histogram (PSTH), which represents the average of ten single-trial responses. Color shading indicates odor presentation. Note the odor-specific slow temporal response patterns. (b) The activity of the population of 99 recorded output neurons at a given time is represented as a point in a 110-dimensional space, in which each dimension represents the firing rate of one of the 110 PNs in the corresponding time bin. Each odor representation is thus represented by a multidimensional vector of output neuron states that evolves with time in a stimulus-specific manner and represents odor-evoked trajectories. Then the 110-dimensional data were analyzed with a dimensionality reduction method (namely principal component analysis) and projected in the space of the first three principal components to allow visualization. From an initial resting state (origin of the coordinate system, b), the population vector evolves through an extended trajectory to an odor-specific fixed point (F). When the pulse terminates, the population vector returns from the fixed point back to baseline via a different trajectory. Different odors evoke different trajectories for the population vector. These trajectories are highly reproducible and specific of the odor and of the odor concentration. *Adapted from Friedrich RW and Stopfer M (2001) Recent dynamics in olfactory population coding. Current Opinion in Neurobiology 11: 468–474; Stopfer M, Jayaraman V, and Laurent G (2003) Intensity versus identity coding in an olfactory system. Neuron 39: 991–1004; Mazor O and Laurent G (2005) Transient dynamics versus fixed points in odor representations by locust antennal lobe projection neurons. Neuron 48: 661–673, with permission.*

crucial for establishing this network synchrony (Kay et al., 2009; Laurent, 2002; Lledo and Lagier, 2006). For example, specific modification of GABA$_A$ receptors expressed in the lateral dendrites of mitral cells disrupts the properties of gamma oscillations (Lagier et al., 2007). Such synchrony may play an important role in odor processing, especially for odor discrimination (Stopfer et al., 1997). As proposed in other brain regions, synchronization to a specific frequency would help in "binding together" remotely distributed mitral/tufted cells that participate in the encoding for the same odor, creating synchronized cell assemblies (Laurent, 2002). This synchronization also promotes temporal coincidence of spatially segregated inputs. Interestingly, downstream regions in the olfactory cortex can act as coincidence detectors (Franks and Isaacson, 2006; Franks et al., 2011).

Taking these data into account, it is clear that the interplay between local interneurons and output neurons provides a multidimensional device for the representation of olfactory information. Sparse and distributed connectivity between local interneurons and mitral output neurons may contribute to a global reformatting of odor representations in the form of an odor-specific spatiotemporal redistribution of activity across the olfactory bulb. Within this framework, two high-dimensional encoders involved in information coding within the bulbar network should be distinguished. The first is composed of the olfactory receptor repertoire expressed by the olfactory sensory neurons, which transduces receptor activation patterns into glomerular odor maps throughout a highly reliable synaptic transmission. The secondary encoder lies in the mitral-granule cell network that extracts higher-order features from the odor maps in order to process them into temporal spiking patterns across sparse ensembles of activated output neurons. Thus, spatiotemporal pattern coding of odors provides a large coding space that scales to the millions of discernible odors of our environment.

An important issue concerning odor coding theories relates to how downstream regions can read odor representations built in the olfactory bulb neuronal circuit. One important particularity of the olfactory system is that it is the only sensory pathway through which peripheral olfactory information can propagate towards cortical regions without any relay via the thalamus (Figure 1.1). The axons of bulbar output neurons project directly to different cerebral cortical structures that compose the so-called olfactory cortex (Wilson et al., 2006). The olfactory cortex includes the anterior olfactory cortex (also called the anterior olfactory nucleus), which connects the two olfactory bulbs through a portion of the anterior commissure (Kikuta et al., 2010), the piriform cortex (this region receives the major part of olfactory bulb projection and is considered to be the primary olfactory cortex), the olfactory tubercle, the cortical nucleus of the amygdala, and the entorhinal cortex, which in turn projects to the hippocampus (Figure 1.1). From the olfactory cortex, olfactory information is ultimately relayed to the thalamus and to the neocortex, notably to the orbitofrontal cortex, the region of cortex thought to be involved in the conscious perception of smell (Gottfried, 2010). The general topographic organization of the olfactory bulb is not conserved in the downstream structures. Mitral cells from the same glomerulus project broadly in the entire piriform cortex without any apparent spatial correlation (Ghosh et al., 2011; Sosulski et al., 2011). Pyramidal cells of the piriform cortex receive convergent inputs from distinct glomeruli (Apicella et al., 2010; Davison and Ehlers, 2011; Miyamichi et al., 2011) and respond to multiple odorants (Poo and Isaacson, 2009; Stettler and Axel, 2009). This high degree of overlap between distributed bulbar projections to higher olfactory centers may constitute the anatomical basis for a combinatorial coding and a crosstalk between information strands emanating from different odorant receptors. This characteristic probably helps to integrate multiple modules of olfactory information into a synthetic representation of a particular scent made of numerous chemical compounds. Alternatively, the synaptic organization of olfactory bulb inputs to the olfactory cortex and the local inhibition within the olfactory cortex that enforces coincidence detection in pyramidal cells suggest that spike-timing-dependent coding of odor representation is an important element of the processing carried out by the olfactory cortex (Poo and Isaacson, 2009; Stokes and Isaacson, 2010).

1.2.4 Centrifugal Fibers from Higher Brain Structures Profusely Innervate the Olfactory Bulb

In addition to sensory inputs from the olfactory epithelium, the olfactory bulb receives a large number of centrifugal inputs from a variety of cortical and noncortical structures (Figure 1.3). Thus, rather than a simple bottom–up odor coding of olfactory signals, the olfactory bulb also deeply integrates top–down information, which modulates the odor representation depending on the internal state or experience of the animal. These direct centrifugal inputs include (1) glutamatergic fibers from olfactory cortex, mainly from the piriform cortex and the anterior olfactory cortex, and (2) modulatory projections from the locus coeruleus (noradrenaline), the horizontal limb of the diagonal band of Broca (acetylcholine and GABA), and the dorsal raphe nucleus (serotonin).

Most of the glutamatergic centrifugal fibers synapse onto granule cell somas, creating a trisynaptic loop between the olfactory cortex and the olfactory bulb. These excitatory inputs tune the excitability of granule cells, modulating the mitral cell–granule cell microcircuit

and the associated processes mentioned above, such as synchronization (Mouret et al., 2009; Strowbridge, 2009). For instance, repetitive excitatory inputs can produce a large depolarization in the granule cell soma sufficient to relieve the Mg^{2+} blockade of NMDA receptors at distal dendrodendritic synapses, thereby promoting recurrent and lateral dendrodendritic inhibition in the olfactory bulb (Balu et al., 2007). The presence of these dense centrifugal inputs clearly suggests that olfaction processing does not involve simple feed-forward pathways. Rather, feedback loops involving long-range axonal projections from downstream regions of the olfactory cortex to the olfactory bulb continually reset the network and provide a dynamic processing of odor.

In addition to the massive glutamatergic innervation from the olfactory cortex, the bulb receives diffuse inputs from neuromodulatory regions. It has been difficult to correlate cellular results and behavioral experiments, especially because these neuromodulators exert multiple and sometimes opposite effects on olfactory bulb neurons depending on the receptor and ligand concentration.

Acetylcholine is released in all the layers of the olfactory bulb by cholinergic fibers originating from the horizontal limb of the diagonal band of Broca and exerts multiple effects on the activity of both output neurons and local interneurons via the expression of different nicotinic and muscarinic receptors (Castillo et al., 1999). At the functional level, changes of behavioral state are partly mediated by changes in the level of acetylcholine tone (Tsuno et al., 2008). Disruption of this cholinergic tone affects output cell receptive fields, gamma oscillations, and olfactory discrimination (Chaudhury et al., 2009; Tsuno et al., 2008).

Serotoninergic fibers extend from the dorsal raphe nuclei and densely innervate the glomerular layer. Recently, serotonin has been shown to regulate sensory inputs to the olfactory bulb indirectly by promoting GABA release from periglomerular cells (Petzold et al., 2009). Noradrenergic fibers from the locus coeruleus are present in the deeper layers, in the granule cell layer, and also in the EPL. Noradrenaline exerts multiple effects on both partners of the dendrodendritic synapse, which may result in the modulation of output cell activity (Jahr and Nicoll, 1982; Nai et al., 2009). Both serotonin and noradrenaline levels increase after odor presentation, and both have been involved in odor preference in young animals and odor discrimination in adults (review in Mandairon and Linster, 2009).

Thus, neuromodulatory innervation is thought to promote flexible and context-dependent changes in the information-processing mode of local neuronal circuits. As will be discussed at greater length later in this chapter, neuromodulators play a major role in the plasticity of the olfactory bulb cells and circuits.

1.3 CIRCUIT DEVELOPMENT: A LESSON FROM ADULT NEUROGENESIS

In contrast to other sensory modalities, the olfactory system exhibits lifelong turnover of both peripheral sensory neurons and central interneurons of the olfactory bulb. In these two places, the addition of new neurons represents another means, in addition to molecular, synaptic or morphological alterations within individual cells, by which circuits can change their own functional organization. This cell-level renovation is neither static nor merely restorative. The process of neuron production during adulthood (called hereafter adult neurogenesis) constitutes an adaptive response to challenges imposed by an animal's environment and/or by its internal state (hormones, stress). Adult neurogenesis in the sensory organ and in the olfactory bulb also raises a number of important questions concerning the role of neurogenesis in olfaction.

1.3.1 Neurogenesis of Sensory Neurons in the Adult Olfactory Epithelium

Facing continuous environmental assaults due to their relatively unprotected position in the nasal cavity, olfactory sensory neurons of the olfactory epithelium are continuously renewed throughout adult life. This site of neurogenesis is made possible by the presence of multipotent progenitors found deep in the olfactory epithelium, near the basal lamina that separates the epithelium from the underlying lamina propria. This progenitor population is mainly composed of globose basal cells and, to a lesser extent, horizontal basal cells. Globose basal cells include transient amplifying precursors and immediate neuronal precursors that express specific markers such as Mash1 and neurogenin-1 and give rise to all the different cell types of the epithelium, including sensory neurons (Cau et al., 1997). During their journey towards the bulb, sensory neuron axons are enveloped by a class of glial cells called ensheathing glial cells, which may act as an extrinsic orientation cue (Sakano, 2010). The progressive generation of mature neurons is relatively rapid and takes only 8–10 days after lesion or toxin exposure. The turnover of sensory neurons, and by extension, the rate of neurogenesis in the olfactory epithelium, is normally regulated by environmental factors. Indeed, the mitotic rates of sensory neurons can be bidirectionally regulated: Naris occlusion reduces sensory neuron turnover and progenitor division, whereas naris reopening, chemical lesions of the epithelium, or olfactory bulb ablation stimulates progenitor activity, restoring the sensory neuron population (review in Schwob, 2002). It is noteworthy that differentiated neurons send back regulatory signals to inform

progenitor cells about the number of new neurons that need to be produced to maintain cell population homeostasis. Thus, neurogenesis of sensory neurons depends on a proper balance of positive regulatory factors that stimulate proliferation and differentiation and negative regulatory molecules produced by mature sensory neurons to inhibit additional neuron production. There has been considerable interest in growth factors that control neurogenesis, differentiation, and apoptosis, such as transforming growth factor, fibroblast growth factor, and bone morphogenetic protein (Schwob, 2002). This adult neurogenesis in a peripheral organ, coupled with the fact that there are a limited number of cell classes in the olfactory epithelium, makes this area attractive for studying mechanisms that control the rate of formation of neurons and their death throughout adulthood.

What could be the functional meaning of this never-ending rejuvenation of the sensory organ? Mature sensory neurons that have been damaged by exposure or by pathogens and immature sensory neurons that cannot find adequate synaptic targets in the olfactory bulb are two obvious candidates to support the existence of this peripheral neurogenesis. Once mature, sensory neurons must extend along a long route to the correct glomerulus. As odor quality remains constant throughout life, the glomerular array must remain constant to a certain degree. Axon guiding through preexisting axons and guidance cues present in the olfactory bulb allow for correct targeting to the olfactory bulb (Sakano, 2010; Zou et al., 2009). Moreover, apoptotic cell death has been observed in cells throughout all stages of regeneration, implying apoptotic regulation of neuron numbers and targeting at all levels of the neuronal lineage. It remains to be determined whether the newly generated sensory neurons indeed bring unique features (such as high probability of glutamate release, lack of adaptation, etc.) to the targeted olfactory bulb circuits.

1.3.2 Adult-Born Interneurons in the Olfactory Bulb

The second adult region where ongoing neuronal addition/replacement takes place is the olfactory bulb (Figure 1.7). In this case, the ongoing neuronal production of olfactory bulb neurons occurs mainly in the SVZ under normal conditions. There, astrocytes in the adult SVZ, which line the border between the striatum and the lateral ventricle, act as slow-dividing adult primary neural stem cells, capable of generating a progeny of neuroblast precursors. Stem cell astrocytes (also called type-B cells) divide and generate rapidly dividing type-C transit-amplifying cells that in turn give rise to type-A migrating neuroblasts (reviewed in Doetsch, 2003; Figure 1.7). Once generated, these neuroblasts proceed

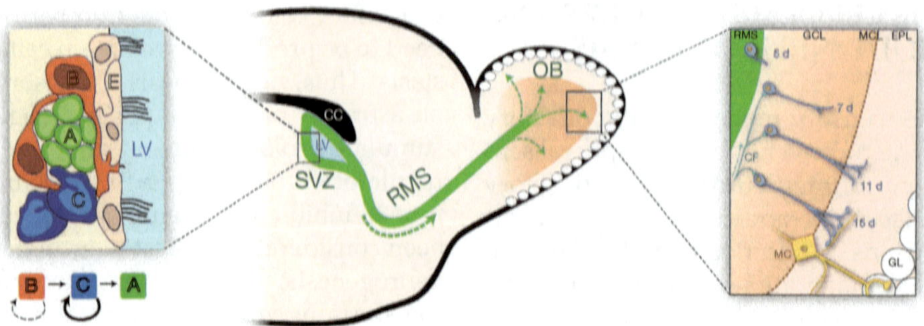

FIGURE 1.7   The subventricular zone – olfactory bulb neurogenic pathway in the adult rodent. The center panel is a schematic diagram of a sagittal section of the rodent forebrain. The subventricular zone (SVZ) lies in the walls of the lateral ventricle (LV), below the corpus calosum (CC). After their generation in the SVZ, neuroblasts migrate tangentially along the RMS to their final destination in the olfactory bulb (OB). Left panel shows the neurogenic niche. Separated from the lateral ventricle by a monolayer of ependymal cells (E), slow-dividing astrocytic stem cells (red, type-B cells) divide to generate transit-amplifying cells (blue, type-C cells), which in turn give rise to neuroblasts (in green, type-A cells) that start to migrate in chain to the rostral migratory stream (RMS). Right panel illustrates the sequence of morphological maturation, from the migrating neuroblast in the RMS (5 days after birth) to their final differentiation into interneurons (mainly granule cells, here represented in blue) and their integration into the network. Adult-born granule cells first receive somatic glutamatergic inputs from centrifugal fibers (CF) before establishing dendrodendritic synapse with mitral cells (MC).

towards the olfactory bulb along an intricate path of migration, up to 5 mm long in rodents, called the RMS (Lois and Alvarez-Buylla, 1994). Along the RMS, another population of astrocytes forms a glial tunnel that guides the chain migration of neuroblasts from the SVZ to the olfactory bulb. More than 30 000 neuroblasts exit the rodent SVZ for the RMS each day and reach the center of the olfactory bulb within 6 days (Doetsch, 2003). In the olfactory bulb, neuroblasts detach from these chains and migrate radially from the RMS. Finally, after 1–3 weeks, neuroblasts mature into olfactory inhibitory interneurons of two main types, granule cells (90% of the adult-born cells) and juxtaglomerular cells (5% of the adult-born cells), found in their respective olfactory bulb layers (Lledo et al., 2008; Petreanu and Alvarez-Buylla, 2002. Interneuron identity is specified at birth and is not specified within the olfactory bulb. Progenitors within the various domains of the SVZ are heterogeneous and are preprogrammed at birth to generate different subsets of olfactory bulb interneurons (Lledo et al., 2008; Merkle et al., 2007).

Using replication-incompetent viral vectors to transduce newly generated neurons in the SVZ and label them with GFP, the morphological and functional properties of these newborn bulbar neurons during their migration and differentiation have been characterized (Lledo et al., 2006). Morphologically, newly generated cells become more complex within the first week after their birth and they become fully mature morphologically as early as 4 weeks of age (Petreanu and Alvarez-Buylla, 2002; Whitman and Greer, 2007; Figure 1.7). After starting migration, neuroblasts express functional GABA and glutamate receptors (Platel et al., 2007). Upon reaching the olfactory bulb, they receive GABAergic and glutamatergic synaptic contacts, notably from centrifugal fibers (Figure 1.7). Then, as they progressively become excitable,

they become spiking neurons and start to release GABA onto output cell dendrites (Bardy et al., 2010; Belluzzi et al., 2003; Carleton et al., 2003). Thus, the formation of synaptic contacts before their activation and GABA release would prevent any network perturbations due to aberrant contacts. Interestingly, of the neurons that successfully mature, only 50% survive to the first month (Petreanu and Alvarez-Buylla, 2002). Therefore, new neurons are intensively selected early in their life.

Although the ongoing bulbar neurogenesis has been extensively documented at the cellular level, the functional consequences are not yet clear. Ablation of adult neurogenesis using genetic, pharmacological, or radiation methods results in faint olfactory phenotype alterations, notably in short-term and long-term olfactory memory (reviewed in Lazarini and Lledo, 2010); however, all previous behavioral analysis has indicated that adult-born neurons might be recruited in short-term or long-term odor memory, two phenomena that may involve neural plasticity (see Section 1.4.2).

1.4 STRUCTURAL AND EXPERIENCE-INDUCED PLASTICITY IN THE OLFACTORY BULB

1.4.1 Activity-Dependent Plasticity in the Olfactory Bulb: Cell Properties and Transmitters

The mammalian olfactory system is known for undergoing experience-dependent plasticity. In the olfactory bulb, granule cells are the major sites of synaptic plasticity. In contrast to many axodendritic synapses described in the brain, the major synaptic interaction in the olfactory bulb, the dendrodendritic transmission between output neurons and granule cells, does not exhibit a strong activity-dependent form of synaptic plasticity, but rather seems to be functionally hard-wired (although a form of short-term plasticity has been described, see Dietz and Murthy, 2005). On the contrary, plasticity at the dendro-dendritic synapse is under the control of neuromodulatory inputs originating from brain regions known to be involved in attention and learning processes. As mentioned above, neuromodulators such as acetylcholine and noradrenaline are known to modulate the synaptic properties of the dendrodendritic synapse (Tsuno et al., 2008). From a general point of view, neuromodulators can have important and long-lasting effects on odor discrimination, learning, and memory, providing the olfactory system with a high degree of plasticity. The neuromodulators act at two distinct, yet complementary functional levels: cell excitability and synaptic activity. By acting on both inhibitory local interneurons and mitral/tufted output neurons (reviewed in Mandairon and Linster, 2009), they have profound effects on both odor processing at the first central relay and odor representation.

In addition to neuromodulatory top–down projections, feedback projections from the olfactory cortex to granule cells are the major source of synaptic plasticity in the olfactory bulb (Gao and Strowbridge, 2009; Nissant et al., 2009; Stripling and Patneau, 1999). In addition to different forms of short-term plasticity, tetanic stimulation of centrifugal excitatory inputs onto granule cells produces a long-term potentiation (LTP) of synaptic strength. Moreover, the same tetanic stimulation that triggers LTP in granule cells also produces a long-lasting enhancement of inhibition onto mitral cells (Gao and Strowbridge, 2009). This form of plasticity is thought to play a role in the long-term modification of the mitral–granule cell network and may shape the spatial and temporal firing patterns of output cell activity. Indeed, the activity of individual output cells can change dramatically depending on the odor context or during learning (Doucette and Restrepo, 2008; Fletcher and Wilson, 2003; Kay and Laurent, 1999; Pager, 1983).

The most thoroughly documented region of the olfactory system exhibiting synaptic plasticity is the olfactory cortex (Wilson et al., 2006). Synaptic contact between the olfactory bulb and the olfactory cortex occurs within the piriform cortex and through recurrent associative connections (Best and Wilson, 2004), as well as from distant and higher order areas, such as the orbitofrontal cortex (Cohen et al., 2008). In the piriform cortex, the massive recurrent connections as well as inputs from distant and high-order areas, such as the orbitofrontal cortex, provide the anatomical structure of an associative network (Best and Wilson, 2004; Cohen et al., 2008; Wilson and Sullivan, 2011).

Lastly, thanks to the accessibility of the olfactory epithelium, responses to naris closure or destruction of olfactory epithelium have highlighted how general decreases in sensory inputs can induce long-term activity-dependent structural plasticity in a sensory network. These experiments reveal how sensory deafferentation mainly affects interneuron populations, interneuron number and phenotype (notably dopamine expression) in particular, as well as the reorganization of the dendritic arbor and synapse density (Leo et al., 2000; Saghatelyan et al., 2005).

1.4.2 Adult-Born Neurons Are Substrates for Experience-Induced Plasticity

The continuous arrival of new interneurons provides another major source of plasticity in the bulbar network. In addition to bringing new building blocks and new connections into the network, adult-generated neurons have unique functional properties compared to neurons generated during early life. These attributes increase their functional impact in the network relative to more mature neurons. By using viral labeling approaches to distinguish adult-born granule cells from preexisting ones, LTP of glutamatergic synapses was found specifically in adult-born cells shortly after their arrival in the olfactory bulb, but this property progressively faded after several weeks (Nissant et al., 2009). Thus, adult-born granule cells may be particularly sensitive to synaptic plasticity but may also potentiate the LTP of cortical glutamatergic inputs to the olfactory bulb. LTP of cortical inputs – specifically to adult-born granule cells – provides an intriguing mechanism to regulate the spatial and temporal firing patterns of output neurons. A role for this synaptic plasticity in olfactory learning remains to be found. Interestingly, olfactory activity influences the maturation and the survival of newborn neurons. Anosmic mice exhibit a strong decrease in the survival of newly formed granule cells (Petreanu and Alvarez-Buylla, 2002), whereas enrichment of the olfactory environment potentiates the survival and accelerates the formation of glutamatergic synapses onto newborn cells (Kelsch et al., 2009; Mouret et al., 2008; Rochefort et al., 2002). Postprandial sleep has been recently shown to promote newborn cell death, possibly though the action of top–down inputs from the olfactory cortex (Yokoyama et al., 2011). Together, these results suggest that postnatal neurogenesis in the olfactory bulb is part of a plasticity mechanism coupled to sensory experience. Moreover, the link between the rate of neuronal survival and olfactory activity suggests that it constitutes a way for the system to adapt to the olfactory environment.

It takes about 2–3 weeks for an adult-generated neuron to become part of the existing olfactory bulb circuit

(Lledo et al., 2006). In this network, the presence of young evolving neurons brings special functions that the preexisting bulbar neurons cannot achieve. Since it takes time for new neurons to mature and become synaptically integrated, adult neurogenesis may contribute to slow, long-term adjustments of the olfactory bulb circuitry, rather than to fast and acute plastic changes. In contrast, the action of centrifugal fibers into the bulb may mediate the faster adaptive changes that adult neurogenesis cannot support. Interestingly, adult-born survival is also controlled by the concerted action of neuromodulators and feedback excitatory projections (Mouret et al., 2009). Therefore, the structural plasticity achieved through adult neurogenesis can be seen as a very long-term form of metaplasticity in the olfactory bulb network: Synaptic plasticity at centrifugal inputs facilitates further integration of long-lasting plastic elements provided by adult neurogenesis.

1.5 CONCLUDING REMARKS

Studies of architectural and functional organization of bulbar circuits have both revealed a wide range of distinct neuronal functioning. This diversity reflects the complex task that neuronal networks have to fulfill in order to process a high-dimensional sensory space. Obviously, information is encoded across neuron assemblies in the olfactory bulb that cannot be extracted by simply averaging the firing frequency. GABAergic inhibition is crucial for olfactory coding, but the functional architecture of dendrodendritic inhibitory microcircuits differ from conventional networks described in other sensory systems. In addition, odor representation is dynamic and highly complex, therefore requiring a unique mechanism of neuronal plasticity. Adult neurogenesis and the actions of centrifugal projections to the olfactory bulb are among the most prominent processes that allow for adaptive mechanisms of plasticity.

Acknowledgments

The authors are grateful to Andrew Sideris for valuable comments on the manuscript. This work was supported by the Fondation pour la Recherche Médicale 'Equipe FRM' and the Medical Insurance Company 'Arpège.'

References

Abraham, N.M., Egger, V., Shimshek, D.R., et al., 2010. Synaptic inhibition in the olfactory bulb accelerates odor discrimination in mice. Neuron 65, 399–411.

Ache, B.W., Young, J.M., 2005. Olfaction: Diverse species, conserved principles. Neuron 48, 417–430.

Adrian, E.D., 1942. Olfactory reactions in the brain of the hedgehog. Journal of Physiology (London) 100, 459–473.

Apicella, A., Yuan, Q., Scanziani, M., Isaacson, J.S., 2010. Pyramidal cells in piriform cortex receive convergent input from distinct olfactory bulb glomeruli. Journal of Neuroscience 30, 14255–14260.

Araneda, R.C., Kini, A.D., Firestein, S., 2000. The molecular receptive range of an odorant receptor. Nature Neuroscience 3, 1248–1255.

Arevian, A.C., Kapoor, V., Urban, N.N., 2008. Activity-dependent gating of lateral inhibition in the mouse olfactory bulb. Nature Neuroscience 11, 80–87.

Aroniadou-Anderjaska, V., Zhou, F.M., Priest, C.A., Ennis, M., Shipley, M.T., 2000. Tonic and synaptically evoked presynaptic inhibition of sensory input to the rat olfactory bulb via GABA(B) heteroreceptors. Journal of Neurophysiology 84, 1194–1203.

Aungst, J.L., Heyward, P.M., Puche, A.C., et al., 2003. Centre-surround inhibition among olfactory bulb glomeruli. Nature 426 (6967), 623–629.

Balu, R., Pressler, R.T., Strowbridge, B.W., 2007. Multiple modes of synaptic excitation of olfactory bulb granule cells. Journal of Neuroscience 27, 5621–5632.

Bardy, C., Alonso, M., Bouthout, W., Lledo, P.M., 2010. How, when, and where new inhibitory neurons release neurotransmitters in the adult olfactory bulb. Journal of Neuroscience 30 (50), 17023–17034.

Bargmann, C.I., 2006. Comparative chemosensation from receptors to ecology. Nature 444 (7117), 295–301.

Bathellier, B., Buhl, D., Accolla, R., Carleton, A., 2008. Dynamic ensemble odor coding in the mammalian olfactory bulb: Sensory information at different timescales. Neuron 57, 586–598.

Belluzzi, O., Benedusi, M., Ackman, J., Loturco, J.J., 2003. Electrophysiological differentiation of new neurons in the olfactory bulb. Journal of Neuroscience 23, 10411–10418.

Best, A.R., Wilson, D.A., 2004. Coordinate synaptic mechanisms contributing to olfactory cortical adaptation. Journal of Neuroscience 24, 652–660.

Buck, L., Axel, R., 1991. A novel multigene family may encode odorant receptors: A molecular basis for odor recognition. Cell 65, 175–187.

Carleton, A., Petreanu, L.T., Lansford, R., Alvarez-Buylla, A., Lledo, P.-M., 2003. Becoming a new neuron in the adult olfactory bulb. Nature Neuroscience 6, 507–518.

Carlson, G.C., Shipley, M.T., Keller, A., 2000. Long-lasting depolarizations in mitral cells of the rat olfactory bulb. Journal of Neuroscience 20, 2011–2021.

Castillo, P.E., Carleton, A., Vincent, J.D., Lledo, P.M., 1999. Multiple and opposing roles of cholinergic transmission in the main olfactory bulb. Journal of Neuroscience 19, 9180–9191.

Cau, E., Gradwohl, G., Fode, C., Guillemot, F., 1997. Mash1 activates a cascade of bHLH regulators in olfactory neuron progenitors. Development 124, 1611–1621.

Chaudhury, D., Escanilla, O., Linster, C., 2009. Bulbar acetylcholine enhances neural and perceptual odor discrimination. Journal of Neuroscience 29, 52–60.

Christie, J.M., Bark, C., Hormuzdi, S.G., Helbig, I., Monyer, H., Westbrook, G.L., 2005. Connexin36 mediates spike synchrony in olfactory bulb glomeruli. Neuron 46, 761–772.

Cohen, Y., Reuveni, I., Barkai, E., Maroun, M., 2008. Olfactory learning-induced long-lasting enhancement of descending and ascending synaptic transmission to the piriform cortex. Journal of Neuroscience 28, 6664–6669.

Cury, K.M., Uchida, N., 2010. Robust odor coding via inhalation-coupled transient activity in the mammalian olfactory bulb. Neuron 68, 570–585.

Davison, I.G., Ehlers, M.D., 2011. Neural circuit mechanisms for pattern detection and feature combination in olfactory cortex. Neuron 70, 82–94.

De Saint Jan, D., Hirnet, D., Westbrook, G.L., Charpak, S., 2009. External tufted cells drive the output of olfactory bulb glomeruli. Journal of Neuroscience 29 (7), 2043–2052.

Dhawale, A.K., Hagiwara, A., Bhalla, U.S., Murthy, V.N., Albeanu, D.F., 2010. Non-redundant odor coding by sister mitral cells revealed by light addressable glomeruli in the mouse. Nature Neuroscience 13, 1404–1412.

Dietz, S.B., Murthy, V.N., 2005. Contrasting short-term plasticity at two sides of the mitral-granule reciprocal synapse in the mammalian olfactory bulb. Journal of Physiology (London) 569, 475–488.

Doetsch, F., 2003. The glial identity of neural stem cells. Nature Neuroscience 6, 1127–1134.

Doucette, W., Restrepo, D., 2008. Profound context-dependent plasticity of mitral cell responses in olfactory bulb. PLoS Biology 6, e258.

Egger, V., Urban, N.N., 2006. Dynamic connectivity in the mitral cell-granule cell microcircuit. Seminars in Cell & Developmental Biology 17, 424–432.

Egger, V., Svoboda, K., Mainen, Z.F., 2005. Dendrodendritic synaptic signals in olfactory bulb granule cells: Local spine boost and global low-threshold spike. Journal of Neuroscience 25, 3521–3530.

Eyre, M.D., Antal, M., Nusser, Z., 2008. Distinct deep short-axon cell subtypes of the main olfactory bulb provide novel intrabulbar and extrabulbar GABAergic connections. Journal of Neuroscience 28, 8217–8229.

Fantana, A.L., Soucy, E.R., Meister, M., 2008. Rat olfactory bulb mitral cells receive sparse glomerular inputs. Neuron 59 (5), 802–814.

Firestein, S., 2001. How the olfactory system makes sense of scents. Nature 413 (6852), 211–218.

Fletcher, M.L., Wilson, D.A., 2003. Olfactory bulb mitral-tufted cell plasticity: Odorant-specific tuning reflects previous odorant exposure. Journal of Neuroscience 23, 6946–6955.

Franks, K.M., Isaacson, J.S., 2006. Strong single-fiber sensory inputs to olfactory cortex: Implications for olfactory coding. Neuron 49, 357–363.

Franks, K.M., Russo, M.J., Sosulski, D.L., Mulligan, A.A., Siegelbaum, S.A., Axel, R., 2011. Recurrent circuitry dynamically shapes the activation of piriform cortex. Neuron 72, 49–56.

Friedrich, R.W., Laurent, G., 2001. Dynamic optimization of odor representations by slow temporal patterning of mitral cell activity. Science 291, 889–894.

Friedrich, R.W., Stopfer, M., 2001. Recent dynamics in olfactory population coding. Current Opinion in Neurobiology 11, 468–474.

Gao, Y., Strowbridge, B.W., 2009. Long-term plasticity of excitatory inputs to granule cells in the rat olfactory bulb. Nature Neuroscience 12, 731–733.

Ghosh, S., Larson, S.D., Hefzi, H., et al., 2011. Sensory maps in the olfactory cortex defined by long-range viral tracing of single neurons. Nature 472, 217–220.

Gire, D.H., Schoppa, N.E., 2009. Control of on/off glomerular signaling by a local GABAergic microcircuit in the olfactory bulb. Journal of Neuroscience 29, 13454–13464.

Gottfried, J.A., 2010. Central mechanisms of odour object perception. Nature Reviews Neuroscience 11 (9), 628–641.

Hayar, A., Karnup, S., Shipley, M.T., Ennis, M., 2004. Olfactory bulb glomeruli: External tufted cells intrinsically burst at theta frequency and are entrained by patterned olfactory input. Journal of Neuroscience 24, 1190–1199.

Hildebrand, J.G., Shepherd, G.M., 1997. Mechanisms of olfactory discrimination: Converging evidence for common principles across phyla. Annual Review of Neuroscience 20, 595–631.

Hsia, A.Y., Vincent, J.D., Lledo, P.M., 1999. Dopamine depresses synaptic inputs into the olfactory bulb. Journal of Neurophysiology 82, 1082–1085.

Isaacson, J.S., Strowbridge, B.W., 1998. Olfactory reciprocal synapses: Dendritic signaling in the CNS. Neuron 20, 749–761.

Jahr, C.E., Nicoll, R.A., 1980. Dendrodendritic inhibition: Demonstration with intracellular recording. Science 207, 1473–1475.

Jahr, C.E., Nicoll, R.A., 1982. Noradrenergic modulation of dendrodendritic inhibition in the olfactory bulb. Nature 297, 227–229.

Johnson, B.A., Leon, M., 2007. Chemotopic odorant coding in a mammalian olfactory system. The Journal of Comparative Neurology 503, 1–34.

Kay, L.M., Laurent, G., 1999. Odor- and context-dependent modulation of mitral cell activity in behaving rats. Nature Neuroscience 2, 1003–1009.

Kay, L.M., Beshel, J., Brea, J., Martin, C., Rojas-Líbano, D., Kopell, N., 2009. Olfactory oscillations: The what, how and what for. Trends in Neurosciences 32, 207–214.

Kelsch, W., Lin, C.-W., Mosley, C.P., Lois, C., 2009. A critical period for activity-dependent synaptic development during olfactory bulb adult neurogenesis. Journal of Neuroscience 29, 11852–11858.

Kikuta, S., Sato, K., Kashiwadani, H., Tsunoda, K., Yamasoba, T., Mori, K., 2010. From the cover: Neurons in the anterior olfactory nucleus pars externa detect right or left localization of odor sources. Proceedings of the National Academy of Science of the United States of America 107, 12363–12368.

Kiyokage, E., Pan, Y.-Z., Shao, Z., et al., 2010. Molecular identity of periglomerular and short axon cells. Journal of Neuroscience 30, 1185–1196.

Kosaka, K., Kosaka, T., 2005. Synaptic organization of the glomerulus in the main olfactory bulb: Compartments of the glomerulus and heterogeneity of the periglomerular cells. Anatomical Science International/Japanese Association of Anatomists 80, 80–90.

Koulakov, A.A., Rinberg, D., 2011. Sparse incomplete representations: A potential role of olfactory granule cells. Neuron 72, 124–136.

Lagier, S., Panzanelli, P., Russo, R.E., et al., 2007. GABAergic inhibition at dendrodendritic synapses tunes gamma oscillations in the olfactory bulb. Proceedings of the National Academy of Science of the United States of America 104, 7259–7264.

Laurent, G., 2002. Olfactory network dynamics and the coding of multidimensional signals. Nature Reviews Neuroscience 3, 884–895.

Lazarini, F., Lledo, P.-M., 2010. Is adult neurogenesis essential for olfaction? Trends in Neurosciences 34 (1), 20–30.

Leo, J.M., Devine, A.H., Brunjes, P.C., 2000. Focal denervation alters cellular phenotypes and survival in the developing rat olfactory bulb. The Journal of Comparative Neurology 417, 325–336.

Linster, C., Cleland, T.A., 2009. Glomerular microcircuits in the olfactory bulb. Neural Networks 22, 1169–1173.

Linster, C., Johnson, B.A., Yue, E., et al., 2001. Perceptual correlates of neural representations evoked by odorant enantiomers. Journal of Neuroscience 21, 9837–9843.

Lledo, P.-M., Lagier, S., 2006. Adjusting neurophysiological computations in the adult olfactory bulb. Seminars in Cell & Developmental Biology 17, 443–453.

Lledo, P.-M., Alonso, M., Grubb, M.S., 2006. Adult neurogenesis and functional plasticity in neuronal circuits. Nature Reviews Neuroscience 7, 179–193.

Lledo, P.-M., Merkle, F.T., Alvarez-Buylla, A., 2008. Origin and function of olfactory bulb interneuron diversity. Trends in Neurosciences 31, 392–400.

Lodovichi, C., Belluscio, L., Katz, L.C., 2003. Functional topography of connections linking mirror-symmetric maps in the mouse olfactory bulb. Neuron 38, 265–276.

Lois, C., Alvarez-Buylla, A., 1994. Long-distance neuronal migration in the adult mammalian brain. Science 264, 1145–1148.

Mandairon, N., Linster, C., 2009. Odor perception and olfactory bulb plasticity in adult mammals. Journal of Neurophysiology 101 (5), 2204–2209.

Mazor, O., Laurent, G., 2005. Transient dynamics versus fixed points in odor representations by locust antennal lobe projection neurons. Neuron 48, 661–673.

Meister, M., Bonhoeffer, T., 2001. Tuning and topography in an odor map on the rat olfactory bulb. Journal of Neuroscience 21, 1351–1360.

Merkle, F.T., Mirzadeh, Z., Alvarez-Buylla, A., 2007. Mosaic organization of neural stem cells in the adult brain. Science 317, 381–384.

Miyamichi, K., Amat, F., Moussavi, F., et al., 2011. Cortical representations of olfactory input by trans-synaptic tracing. Nature 472, 191–196.

Mombaerts, P., 2004a. Love at first smell – The 2004 Nobel Prize in Physiology or Medicine. The New England Journal of Medicine 351, 2579–2580.

Mombaerts, P., 2004b. Genes and ligands for odorant, vomeronasal and taste receptors. Nature Reviews Neuroscience 5, 263–278.

Mombaerts, P., Wang, F., Dulac, C., et al., 1996. Visualizing an olfactory sensory map. Cell 87, 675–686.

Mori, K., Kishi, K., Ojima, H., 1983. Distribution of dendrites of mitral, displaced mitral, tufted, and granule cells in the rabbit olfactory bulb. The Journal of Comparative Neurology 219, 339–355.

Mouret, A., Gheusi, G., Gabellec, M.-M., de Chaumont, F., Olivo-Marin, J.-C., Lledo, P.-M., 2008. Learning and survival of newly generated neurons: When time matters. Journal of Neuroscience 28, 11511–11516.

Mouret, A., Murray, K., Lledo, P.-M., 2009. Centrifugal drive onto local inhibitory interneurons of the olfactory bulb. Annals of the New York Academy of Sciences 1170, 239–254.

Murphy, G.J., Glickfeld, L.L., Balsen, Z., Isaacson, J.S., 2004. Sensory neuron signaling to the brain: Properties of transmitter release from olfactory nerve terminals. Journal of Neuroscience 24 (12), 3023–3030.

Nagayama, S., Takahashi, Y.K., Yoshihara, Y., Mori, K., 2004. Mitral and tufted cells differ in the decoding manner of odor maps in the rat olfactory bulb. Journal of Neurophysiology 91, 2532–2540.

Nai, Q., Dong, H.-W., Hayar, A., Linster, C., Ennis, M., 2009. Noradrenergic regulation of GABAergic inhibition of main olfactory bulb mitral cells varies as a function of concentration and receptor subtype. Journal of Neurophysiology 101, 2472–2484.

Najac, M., de Saint Jan, D., Reguero, L., Grandes, P., Charpak, S., 2011. Monosynaptic and polysynaptic feed-forward inputs to mitral cells from olfactory sensory neurons. Journal of Neuroscience 31, 8722–8729.

Nissant, A., Bardy, C., Katagiri, H., Murray, K., Lledo, P.-M., 2009. Adult neurogenesis promotes synaptic plasticity in the olfactory bulb. Nature Neuroscience 12, 728–730.

Pager, J., 1983. Unit responses changing with behavioral outcome in the olfactory bulb of unrestrained rats. Brain Research 289, 87–98.

Petreanu, L., Alvarez-Buylla, A., 2002. Maturation and death of adultborn olfactory bulb granule neurons: Role of olfaction. Journal of Neuroscience 22, 6106–6113.

Petzold, G.C., Hagiwara, A., Murthy, V.N., 2009. Serotonergic modulation of odor input to the mammalian olfactory bulb. Nature Neuroscience 12, 784–791.

Platel, J.-C., Lacar, B., Bordey, A., 2007. GABA and glutamate signaling: Homeostatic control of adult forebrain neurogenesis. Journal of Molecular Histology 38, 303–311.

Poo, C., Isaacson, J.S., 2009. Odor representations in olfactory cortex: 'Sparse' coding, global inhibition, and oscillations. Neuron 62, 850–861.

Pressler, R.T., Strowbridge, B.W., 2006. Blanes cells mediate persistent feedforward inhibition onto granule cells in the olfactory bulb. Neuron 49, 889–904.

Price, J.L., Powell, T.P., 1970. The synaptology of the granule cells of the olfactory bulb. Journal of Cell Science 7, 125–155.

Rall, W., Shepherd, G.M., 1968. Theoretical reconstruction of field potentials and dendrodendritic synaptic interactions in olfactory bulb. Journal of Neurophysiology 31, 884–915.

Rall, W., Shepherd, G.M., Reese, T.S., Brightman, M.W., 1966. Dendrodendritic synaptic pathway for inhibition in the olfactory bulb. Experimental Neurology 14, 44–56.

Rochefort, C., Gheusi, G., Vincent, J.-D., Lledo, P.-M., 2002. Enriched odor exposure increases the number of newborn neurons in the adult olfactory bulb and improves odor memory. Journal of Neuroscience 22, 2679–2689.

Saghatelyan, A., Roux, P., Migliore, M., et al., 2005. Activity-dependent adjustments of the inhibitory network in the olfactory bulb following early postnatal deprivation. Neuron 46, 103–116.

Sakano, H., 2010. Neural map formation in the mouse olfactory system. Neuron 67, 530–542.

Schoppa, N.E., Kinzie, J.M., Sahara, Y., Segerson, T.P., Westbrook, G.L., 1998. Dendrodendritic inhibition in the olfactory bulb is driven by NMDA receptors. Journal of Neuroscience 18, 6790–6802.

Schoppa, N.E., Urban, N.N., 2003. Dendritic processing within olfactory bulb circuits. Trends in Neurosciences 26, 501–506.

Schwob, J.E., 2002. Neural regeneration and the peripheral olfactory system. Anatomical Record 269, 33–49.

Shepherd, G.M., 2010 Creating Modern Neuroscience. The Revolutionary 1950s. Oxford University Press, New York.

Shepherd, G.M., Chen, W.R., Willhite, D., Migliore, M., Greer, C.A., 2007. The olfactory granule cell: From classical enigma to central role in olfactory processing. Brain Research. Brain Research Reviews 55, 373–382.

Shusterman, R., Smear, M.C., Koulakov, A.A., Rinberg, D., 2011. Precise olfactory responses tile the sniff cycle. Nature Neuroscience 1–8.

Smear, M., Shustermann, R., O'Connor, R., Bozza, T., Rinberg, D., 2011. Perception of sniff phase in mouse olfaction. Nature 479, 397–400.

Sosulski, D.L., Bloom, M.L., Cutforth, T., Axel, R., Datta, S.R., 2011. Distinct representations of olfactory information in different cortical centres. Nature 472, 213–216.

Soucy, E.R., Albeanu, D.F., Fantana, A.L., Murthy, V.N., Meister, M., 2009. Precision and diversity in an odor map on the olfactory bulb. Nature Neuroscience 12, 210–220.

Stettler, D.D., Axel, R., 2009. Representations of odor in the piriform cortex. Neuron 63, 854–864.

Stokes, C.C.A., Isaacson, J.S., 2010. From dendrite to soma: Dynamic routing of inhibition by complementary interneuron microcircuits in olfactory cortex. Neuron 67, 452–465.

Stopfer, M., Bhagavan, S., Smith, B.H., Laurent, G., 1997. Impaired odour discrimination on desynchronization of odour-encoding neural assemblies. Nature 390, 70–74.

Stopfer, M., Jayaraman, V., Laurent, G., 2003. Intensity versus identity coding in an olfactory system. Neuron 39, 991–1004.

Stripling, J.S., Patneau, D.K., 1999. Potentiation of late components in olfactory bulb and piriform cortex requires activation of cortical association fibers. Brain Research 841, 27–42.

Strowbridge, B.W., 2009. Role of cortical feedback in regulating inhibitory microcircuits. Annals of the New York Academy of Sciences 1170, 270–274.

Tan, J., Savigner, A., Ma, M., Luo, M., 2010. Odor information processing by the olfactory bulb analyzed in gene-targeted mice. Neuron 65, 912–926.

Tsuno, Y., Kashiwadani, H., Mori, K., 2008. Behavioral state regulation of dendrodendritic synaptic inhibition in the olfactory bulb. Journal of Neuroscience 28, 9227–9238.

Wachowiak, M., 2011. All in a sniff: Olfaction as a model for active sensing. Neuron 71, 962–973.

Wachowiak, M., Shipley, M.T., 2006. Coding and synaptic processing of sensory information in the glomerular layer of the olfactory bulb. Seminars in Cell & Developmental Biology 17, 411–423.

Whitman, M.C., Greer, C.A., 2007. Synaptic integration of adultgenerated olfactory bulb granule cells: Basal axodendritic centrifugal input precedes apical dendrodendritic local circuits. Journal of Neuroscience 27, 9951–9961.

Wilson, D.A., Sullivan, R.M., 2011. Cortical processing of odor objects. Neuron 72 (4), 506–519.

Wilson, D.A., Kadohisa, M., Fletcher, M.L., 2006. Cortical contributions to olfaction: Plasticity and perception. Seminars in Cell & Developmental Biology 17, 462–470.

Wilson, R.I., Mainen, Z.F., 2006. Early events in olfactory processing. Annual Review of Neuroscience 29, 163–201.

Xiong, W., Chen, W.R., 2002. Dynamic gating of spike propagation in the mitral cell lateral dendrites. Neuron 34, 115–126.

Yokoi, M., Mori, K., Nakanishi, S., 1995. Refinement of odor molecule tuning by dendrodendritic synaptic inhibition in the olfactory bulb. Proceedings of the National Academy of Sciences of the United States of America 92 (8), 3371–3375.

Yokoyama, T.K., Mochimaru, D., Murata, K., et al., 2011. Elimination of adult-born neurons in the olfactory bulb is promoted during the postprandial period. Neuron 71, 883–897.

Zou, D.-J., Chesler, A., Firestein, S., 2009. How the olfactory bulb got its glomeruli: A just so story? Nature Reviews Neuroscience 10, 611–618.

Functional Circuit Development in the Auditory System

D.B. Polley[1], A.H. Seidl[2], Y. Wang[2], J.T. Sanchez[2]

[1]Eaton-Peabody Laboratory, Massachusetts Eye and Ear Infirmary & Dept. of Otology and Laryngology, Harvard Medical School, Boston, MA, USA; [2]Virginia Merrill Bloedel Hearing Research Center & Dept. of Otolaryngology, Head and Neck Surgery, University of Washington, Seattle, WA, USA

OUTLINE

© 2013 Elsevier Inc. All rights reserved.

2.1 INTRODUCTION TO AUDITORY SYSTEM DEVELOPMENT

2.1.1 A Neurobiological Approach to Studying Auditory System Development

The auditory system undergoes a series of profound changes from the time neural circuits begin forming in the fetal brain to the day, years later, when a child first comprehends a complete sentence. The processes unfolding during this period are a fascinating mixture of intrinsic molecular orchestration and activity-dependent refinement. In humans, auditory perceptual development is a protracted process that begins late in the second trimester when the fetus first shows discriminative changes in heart rate to variations in specific sound features. Progressive improvements in the ability to resolve variations in frequency, temporal patterns and spatial positions of sounds are observed throughout infancy and early childhood, and these changes often parallel an increasing capacity to discriminate phonological units of a child's native language (Werker, 2005). While a greater understanding of the processes at work in human auditory development is of paramount importance, these efforts are often complicated by an inability to isolate the contributions of sensory system development from other cognitive and physical factors.

The use of model systems such as birds and rodents has provided researchers with direct access to central and peripheral auditory circuits, and has elucidated many of the basic mechanisms that underlie their changes during ontogeny. Songbirds and chickens have long been popular model systems due to the greater ease of studying and manipulating the embryo in the egg (*in ovo*) rather than *in utero*. As a result, we can now draw information from an extensive corpus of work detailing the organization of brainstem and forebrain pathways, and the similarities by which songbirds acquire their song and humans acquire speech, particularly in their dependence on auditory feedback during sensitive periods of development. Interestingly, birds begin hearing approximately eight days before hatching, around embryonic (E) day 11, while altricial mammals such as mice, rats and gerbils do not respond to airborne sound until the second postnatal (P) week of life. The relatively late onset of hearing in rodents has aided research into the cellular and molecular changes that underlie abrupt changes in circuit development in the days before and after the onset of hearing.

Recent studies suggest that molecular cues, whose expression is genetically controlled, play an essential role in the formation of topographically ordered connections in the auditory system. It is also evident that factors linked to the flux of ions across the cell membrane during and following the action potential also support neuron survival and regulate the growth and topographic specificity of axons and dendrites within auditory brain areas. In some brain areas, instructive electrical signals generated through the "closed-loop" spontaneous activity is sufficient to promote normal neural circuit development, whereas higher levels of the auditory pathway require structured activity patterns arising from acoustic signals to guide their final stages of assembly. The influence of each activity-dependent and activity-independent factor waxes and wanes within defined windows of development, and piecing together the chronology and mechanisms behind each epoch of auditory system development represents one of the fundamental challenges for researchers in this field. The chapter describes the current state of knowledge concerning the interplay of these factors in the establishment of functional circuits from the cochlea to the cerebral cortex.

2.1.2 Basic Concepts of Cochlear Transduction

The encoding of auditory information begins with a highly specialized receptor organ known as the cochlea in mammals and the basilar papilla in birds. In mammals, pressure waves are delivered into the fluid filled cochlea via the middle ear bones, setting the cochlear partition into motion. The cochlear partition consists of the basilar membrane (BM), the organ of Corti, and the tectorial membrane (Figure 2.1(a)). As a result of this motion and the mechanical properties of these structures, a relatively crude spectral analysis is performed by a vibratory pattern that occurs along the partition, a passive process referred to as the traveling wave (Von Békésy, 1960). This traveling wave results in a vibratory pattern such that high frequency sounds produce maximum displacement near the beginning (base) of the partition, while low frequency sounds vibrate maximally near the end (apex) of the partition.

The sensory receptors in mammals responsible for transducing these hydro-mechanical vibrations are located in the organ of Corti and are known as inner and outer hair cells (IHC and OHC, respectively). Both types of receptors are polarized epithelial cells that contain mechano-sensitive organelles located on their apical surface. These actin-filled hair-like processes are termed stereocilia and contain mechanotransducer channels near the tips of the ciliary bundles. The bundles are deflected as a result of partition movement, opening the channels and depolarizing the hair cell due to an influx of potassium (K^+). This unconventional form of depolarization activates voltage gated calcium (Ca^{2+}) channels, which triggers the release of the excitatory neurotransmitter glutamate. Glutamate binds to postsynaptic receptors located on first-order spiral

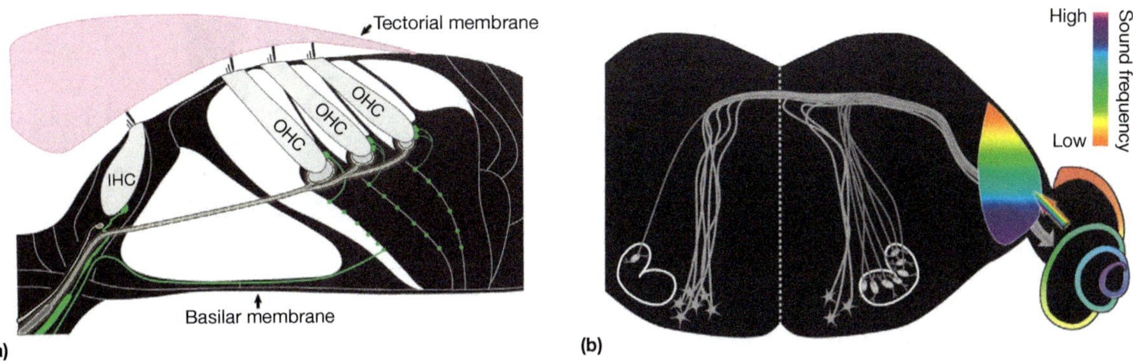

FIGURE 2.1 **Connections within the adult auditory system.** Schematics illustrate the position of cell bodies and their interconnections in mammalian Organ of Corti (a) and initial components of the auditory pathway (b). (a) Two classes of mechanical sensory receptors, inner hair cells (IHC) and outer hair cells (OHC) are innervated by dendrites that carry afferent signals to the brain (green) and efferent axons that convey signals to the brain (gray). (*B*) Sounds of varying frequencies maximally excite particular regions along the basilar membrane (BM). Maximal sensitiveness are mapped systematically across the length of the BM, such that high frequency vibrations excite the base of the BM and low frequencies excite the apex. This tonotopic organization of frequency preference is preserved through topographically ordered projections to the central auditory system. Distinct populations of cells on both sides of the brainstem send efferent projections to the cochlea.

ganglion neurons (green dendrites, Figure 2.1(a)), initiating saltatory conduction of action potentials to auditory nuclei in the brainstem.

The generic description above for sensory transduction holds true for all hair cells in both birds and mammals. However, the separation into IHCs and OHCs, and their differences, are remarkable and unique to mammals. The IHCs are the true sensory receptor, receiving approximately 95% of afferent innervation. In contrast, OHCs receive only around 5% of afferent contacts, suggesting a minor role in sensory transduction. Despite this dichotomy, OHCs are thought to have highly-specialized "motor" functions. One such active function is to enhance the bundle displacement caused by the vibration of the cochlear partition, acting as a positive feedback on the bundle and the consequent amplification of its displacement. A second active function involves an elongation and contraction of the OHC itself. As the OHC changes length, it feeds back mechanical force onto its surrounding environment, a process termed electromotility. This results in a highly significant and spatially segregated enhancement of the basilar membrane vibratory amplitude. Accordingly, the relatively crude sensitivity and frequency selectivity of the basilar membrane that arises through its passive biomechanical properties is substantially refined through hair cell active amplification. Not only do hair cells transmit signals to the central nervous system, they are also innervated by efferent axons from the brain, which communicate with OHCs to extend the dynamic range of hair cell signaling and protect the Organ of Corti from acoustic overexposure (Figure 2.1(b)). For detailed reviews of auditory transduction and the distinct mechanisms that create active cochlear amplification in mammals versus birds, readers are referred to (Hudspeth, 2008; Dallos, 2006)

2.1.3 Scope of this Chapter

A functional circuit can be defined as the connection between one or more cells or nuclei that transmit – and often transform – a signal. As such, topics relating to the proliferation, delamination and migration of cells and the development of their connecting projections in the auditory system are touched upon only briefly. Rather, this chapter is primarily concerned with describing the relative contributions of intrinsic molecular events, spontaneous action potentials and sound-evoked action potentials in the assembly of functional circuits within the peripheral and central auditory pathways.

This chapter is divided into three principal sections. The first section covers the ontogeny of local circuits within the cochlea and as well as the development of afferent and efferent circuits connecting the cochlea to the brain. The second section describes the establishment of circuits within the developing auditory brainstem and the influence of signaling from the auditory periphery. The final section addresses the formation of functional circuits in the auditory midbrain and cortex. When appropriate, we have cited notable seminal research papers, breakthrough findings and comprehensive review materials so that the interested reader may avail themselves of these more focused sources of information.

2.2 DEVELOPMENT OF PERIPHERAL CIRCUITS

2.2.1 Developing Networks Within the Cochlea

Cochlear hair cells initiate the process of hearing by converting mechanical deflections of their stereocilia bundles into electrochemical signals that are distributed throughout the rest of the auditory system. Before mature and normal transduction can occur, a number of critical developmental events take place between hair cells and non-sensory cells within the cochlea. The specialized function of IHCs and OHCs depends in part upon developing networks of non-sensory supporting cells within the organ of Corti and the lateral wall of the cochlea.

The precision of mechanoelectrical transduction can be attributed, in part, to the unusual electrical potential and ionic milieu in the endolymphatic space surrounding the apical surface of the hair cell. Prior to and during the first week of hearing, an endocochlear potential is established between the endolymph and surrounding perilymph, which increases from 0 mV to +80 mV. The ramping up of the electrical potential is complemented by the accumulation of high levels of K^+ in the endolymphatic space, which further exaggerates the electrical gradient across the negative resting potential of the hair cell membrane. The combination of high extracellular K^+ and the positive endocochlear potential work synergistically to effectively drive ionic currents through open mechanotransducer channels, creating the large and rapid receptor potential changes that mediate glutamate release at the synapse between the hair cell and the auditory nerve. The endocochlear potential is established through the development of tight cellular junctions between local networks of epithelial cells, connective tissue and supporting cells that completely partition the endolymph from the surrounding perilymph. These tightly bound networks also efficiently recycle K^+ from the hair cell back into the endolymphatic space where they can once again be used in sensory transduction.

The spontaneous generation of action potentials from sensory receptors is considered essential for normal neural circuit development throughout the brain. In the developing auditory system, the mechanisms responsible for spontaneous action potential activity are still unresolved but recent reports suggest that this spontaneous activity is generated by IHCs of the cochlea. The cartoon of the IHC region in the immature Organ of Corti represents one proposed set of developmental changes that occur in cochlear circuitry (Figure 2.1(a)). Compared to IHCs in mature animals, which are surrounded by one or two supporting cells (see Figure 2.1(a)), the pre-hearing Organ of Corti features a structure known as the greater epithelial ridge, or Kollikers organ (Ko). This structure consists of non-neuronal inner supporting cells (ISCs) that are present up to the onset of hearing. However, by the time of hearing onset, Ko undergoes programmed cell death and subsequent removal of the majority of ISCs. Despite this dramatic change in the structure of the organ of Corti, recent studies have identified a potential role for ISCs in the initiation of electrical signaling within the auditory nerve (Tritsch, 2007). One to two weeks prior to the onset of hearing, the elongated

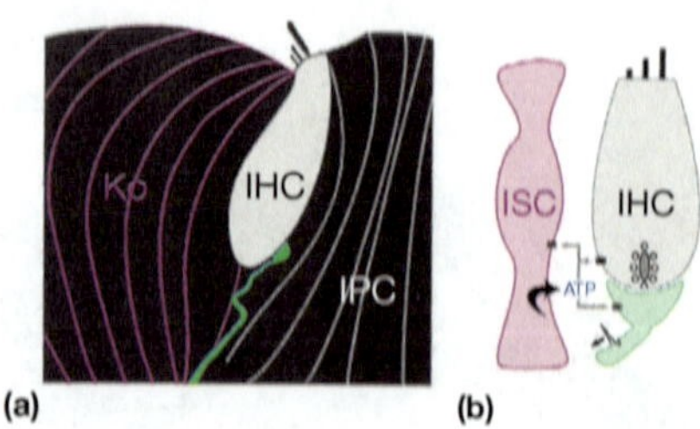

FIGURE 2.2 **Transient microcircuits in the developing Organ of Corti.** (a) Schematic of the region surrounding the inner hair cell in the pre-hearing rodent. Compared to the same region in the adult cochlea (Fig. 2.1(a)), the immature cochlea features a proliferation of elongated inner supporting cells called Kölliker's organ (Ko, in purple) and inner pillar cells (IPC). (b) Inner supporting cells in Ko (ISC, purple) release ATP prior to hearing onset. ATP binds to purinergic receptors (black ellipses) to promote Ca^{2+}-dependent glutamate release from ribbon synapses within the inner hair cell (IHC, gray) and action potentials in auditory nerve fibers (green).

ISCs within Ko begin to spontaneously release adenosine tri-phosphate (ATP) into the extracellular space (Figure 2.2(b)). ATP activates purinergic receptors on neighboring IHCs, peripheral processes of the auditory nerve and on the ISCs themselves. Binding of ATP on the IHC depolarizes the membrane potential, inducing Ca^{2+}-dependent glutamate release and bursts of action potentials in auditory nerve fibers. ATP release is local and desynchronized along the length of the cochlea. In this manner, spatially and temporally independent volleys of electrical signals initiated by non-sensory neurons entrain the firing patterns of SG and, ultimately, central auditory neurons. This process is thought to play a role in the strengthening of functional circuits prior to the onset of hearing.

Despite this role of ATP release from Ko, it remains uncertain how early action potential activity is patterned and whether ATP binding drives IHC membrane voltage or provides weaker modulatory control. More recent data suggests that during the first postnatal week of life, developing IHCs intrinsically generate the voltage changes that elicit action potentials in SG neurons. The frequency and pattern of this spontaneous action potential activity varies between regions of the cochlea (i.e., high-frequency versus low-frequency) and are modulated in multiple ways by the release of acetylcholine (ACh) and ATP near the IHCs (Johnson, 2011). It has been proposed that this pattern of action potential activity, along with ACh and ATP modulation, could be important for guiding tonotopic organization and the refinement of sensory information along the central auditory pathways before the occurrence of experience-drive information becomes relevant.

2.2.2 Development of the Place Code

The basilar membrane acts as a spectral analyzer that translates vibration frequencies within the cochlear fluid pressure waves into positions of maximal displacement along its length. In mature animals, the BM is relatively narrow and taut at its base (violet region, Figure 2.1(b)) compared to the apex, which is wider and more mobile (red region, Figure 2.1(b)). As previously mentioned, this structural gradient confers a smooth shift of preferred vibration frequency along its length, with high frequencies maximally activating basal regions of the BM and low frequencies maximally activating apical areas. As with the visual and somatosensory pathways, the spatial organization of the receptor organ is maintained through topographic connections between the receptor epithelia and successive levels of the central sensory pathways. In the auditory system, the one-dimensional tonotopic arrangement of preferred frequency along the length of the BM is preserved in tonotopic maps of preferred frequency within the central auditory nuclei.

In nearly every respect, the development of basal (high frequency) regions of the cochlea occurs before apical (low frequency) regions; apical hair cells are the last to differentiate and the last to be innervated by afferent and efferent nerve fibers that convey signals to and from the brain. Therefore, one would predict that sensitivity to high frequency sounds would emerge before low frequencies in development. Interestingly, behavioral and neurophysiological hearing assessments in dozens of avian and mammalian species show just the opposite: high frequency sensitivity is the last to mature. This developmental mismatch implies that either tonotopic connections between the periphery and central auditory system are undergoing large-scale rewiring, or that developmental changes within the cochlea cause a given position along the BM to vibrate at progressively higher frequencies during the early period of hearing. Direct neurophysiological recordings from first-order auditory ganglion neurons that innervate a fixed point within the basal cochlea provide clear support for this latter scenario (Echteler, 1989). Basal cochlear regions were found to be maximally sensitive to low frequencies at the onset of the hearing and then become gradually responsive to higher frequencies. This developmental shift in the cochlear place code has been traced to a progressive maturation of OHC mechanics (Norton, 1991).

2.2.3 Development of Afferent and Efferent Circuits

The mature functional circuit linking the auditory periphery to the brain has four essential processing stations: 1) sensory hair cells in the auditory periphery, 2) first-order spiral ganglion cells (SG) that send a peripheral dendrite to the IHCs and OHCs and a central axon to the brain (i.e., the auditory nerve), 3) the cochlear nucleus (CN), a second-order auditory brain nucleus that is heavily innervated by auditory nerve fibers, and 4)

brainstem neurons whose efferent projections innervate IHCs, OHCs and neurons of the CN (Figure 2.3(a)). The two types of sensory receptors in the mammalian cochlea (IHCs and OHCs) are each innervated by two types of SG neurons: Type I and Type II. Type I afferents comprise approximately 95% of all SG neurons and each contact a single IHC. A single IHC, in turn, can be innervated by 10–20 Type I afferents, providing parallel and topographically specific information transfer from a single IHC to the CN. By contrast, a single Type II SG neuron contacts 30–60 OHCs, providing a weaker, spatially integrated signal to CN neurons. Although the OHCs contribute comparatively little afferent input to the brain, they are the predominant targets of efferent axons from the medial olivocochlear neurons (MOC) in the brainstem. In mature animals, the central processes of the auditory nerve terminate in the CN. In some cases, one or two auditory nerve fibers contact a single CN neuron via a massive axosomatic synapse called the endbulb of Held. The process through which this circuit achieves its mature form reflects an interplay of molecular processes and intrinsic activity-dependent processes that can be broken down into three phases.

Phase 1: Well Before Hearing Onset

In rodents, the peripheral processes of SG neurons innervate basal regions of the cochlea approximately five days before birth, or approximately 17 days prior to hearing onset (Figure 2.3(b)). Within a day of growing into the peripheral epithelia, afferents can be sorted into Type I or Type II morphologies. It is generally agreed that there are no gross errors or widely exuberant connections between SG dendrites and sensory hair cells. However, the exact precision of longitudinal (basal to apical) and radial (IHC to OHC) innervation patterns by Type I afferents is not entirely understood. Genetic fate mapping studies have described highly precise and rapid initial targeting (Koundakjian, 2007), while ultrastructural and histochemical studies find that Type I afferents initially innervate multiple IHCs and OHCs (Echteler, 1992). The weight of evidence favors this latter characterization. ACh-releasing efferent fibers grow into the sensory epithelia at the same time or slightly before afferent fibers. Compared to afferent innervation, however, efferent fibers show a clear developmental shift in their spatial targeting, as olivocochlear efferent fiber innervation is initially biased towards IHCs rather than OHCs (Pujol, 1978).

The central projections of SG neurons reach their targets in the CN approximately one day before their peripheral dendrites innervate the sensory epithelia. Despite the fact that the cochlear nucleus is still forming at this stage, projections from the SG are remarkably precise and demonstrate a clear spatial organization long before intrinsic or sensory-evoked action potential signaling begins. Although little is known about the

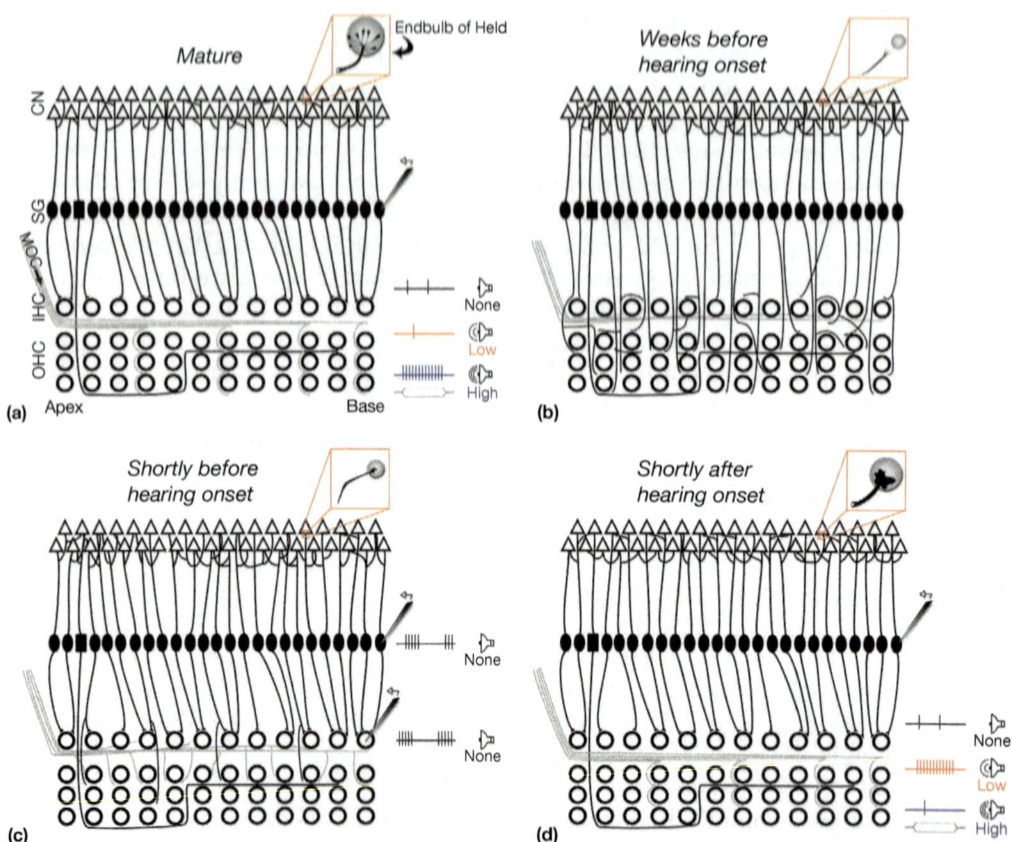

FIGURE 2.3   **Development of afferent and efferent circuits linking the periphery and the brain**. (a) In mature mammals, Type I (ellipse) and Type II (rectangle) spiral ganglion neurons (SG) extend a peripheral processes to a single row of inner hair cells (IHC) or three rows of outer hair cells (OHC), respectively. The central projection of SG neurons form elaborate axosomatic synapses called endbulbs of Held on specific types of neurons within the cochlear nucleus (CN, triangle). Efferent axons from medial olivocochlear (MOC) neurons innervate OHCs. Recording from a SG neuron innervating to the base of the cochlea would reveal occasional action potentials in the absence of sound or in response to low frequency sounds but elevated firing rates in response to high frequency sound (sound onset and offset represented by the trapezoidal stimulus). (b) Weeks before hearing onset, patterns of peripheral, central and efferent connections have been established, but lack the precise organization seen in mature animals. (c) In the days leading up to hearing onset, Type I/ II and MOC connections are sorted into their correct hair cell targets and rhythmic trains of Ca^{2+} spikes initiated in IHCs entrain the firing patterns of Type I SG neurons. Endbulb synapses are beginning to form. (d) In the days following hearing onset, peripheral innervation patterns are largely mature and central projection are approaching the topographic specificity and synaptic structure observed in mature animals. SG neurons innervating basal regions of the cochlea are responsive to low frequency tones, rather than high, due to immature basilar membrane and stereocilia mechanics.

molecules or cellular interactions participating in the establishment of auditory nerve fiber topography, several growth factors and receptors are expressed at approximately the time that connections are being established. In particular, differential distribution of the Eph/ephrin family of receptor tyrosine kinases in the auditory nerve and CN suggests an axon guidance mechanisms that shapes the formation of topographic connectivity within the auditory brainstem (Cramer, 2005).

Phase 2: Shortly Before Hearing Onset

Following the initial period of exuberant connectivity, the dendrites from Type I ganglion cells begin to coalesce around individual IHCs (Figure 2.3(c)). Conversely, efferent MOC fibers are in a transitional state during which immature connections to IHCs are maintained alongside newly formed connections with OHCs. IHC

mediated spontaneous spiking is robust during this period and electrophysiological recordings reveal temporally patterned IHC Ca^{2+} spikes that are mimicked by the firing patterns of SG neurons (Tritsch, 2010). In addition, functional synapses between SG and CN neurons are established, but the shape and overall size of endbulbs in the CN are still quite immature.

Phase 3: Shortly After Hearing Onset

At this stage, peripheral innervation of the receptor epithelia is essentially mature with Type I and Type II afferent endings making appropriate contacts with IHCs and OHCs, respectively. MOC efferents drop their IHC connection and make nearly exclusive contact with OHCs. Rhythmic spontaneous bursting gives way to stochastic spontaneous action potentials intermingled with temporall and topographically structured sound-evoked responses. However, recordings from SG neurons innervating basal IHCs reveal an immature preference for low frequency sounds, as described in the section on place code development above. The topography from the SG neurons to the CN is initially very precise. However, more recent findings in mammals show that subtle topographic refinement continues around the onset of hearing and continues for several months, suggesting a combination of intrinsic and extrinsic activity-dependent mechanisms (Leake, 2002). Additionally, endbulb synapses continue to undergo clear structural modifications. Though they do not yet reach the state of a fully developed calyx that engulfs most of the CN cell body in mature animals, the axon terminals increase in diameter by an order of magnitude relative to the pre-hearing period (Jhaveri, 1982; Ryugo, 1982).

2.2.4 Conclusions

In a matter of weeks, what began as an outpocketing of epithelial cells near the embryonic hindbrain becomes an exquisite functional circuit, capable of encoding mechanical vibrations spanning three orders of magnitude in frequency, a 120 dB dynamic range for amplitude encoding (a million-million-fold change in signal energy), and sensitivity to sub-atomic stereocilia displacements with microsecond mechanical response times. The physical attributes of sound are captured by a tonotopically-organized array of sensory hair cells, which form topographic connections with neurons in the CN to initiate the psychological experience of hearing. This complex circuitry arises through the interaction of sensory and non-sensory supporting cells within the developing cochlea in addition to afferent and efferent connections that link the cochlea to the brain. Other than a modest topographic refinement in the spatial distribution of central projections to the CN and a fairly dramatic transformation of the endbulb synaptic terminal shape and overall size, these circuits reach a mature form independent of sensory input. Instead, maturation appears to reflect the dominant influence of genetic programming and molecular guidance cues with a subordinate role for cell-cell interactions that may include - but are not limited to – internally generated action potential patterns.

In humans, the development of peripheral circuitry is paralleled by enormous changes in infant phonological perception. The earliest stages of perceptual refinement can be attributed, at least in part, to the physical maturation of the outer ear, middle ear and cochlea. However, the scope of perceptual processes that continue to come online after peripheral circuits have matured as well as their dependence upon normally patterned sensory input both point toward the essential role of central auditory circuits in the development of hearing.

2.3 DEVELOPMENT OF BRAINSTEM CIRCUITS

2.3.1 Functional Circuit Assembly in the Brainstem

The assembly of functional circuits within the auditory system has been the subject of intense study over the past 30 years. Auditory brainstem nuclei are derived from progenitor cells within the hindbrain that migrate to their appropriate positions shortly before hair cells in basal regions of the cochlea are born. Like the central projection of SG neurons to the CN, guidance of embryonic brainstem neurons and assembly of their interconnections are thought to be mediated by the Eph/ephrin signaling, although the detailed signaling pathways have yet to be defined (Cramer, 2005). Axonal connections into second- and third-order nuclei are fully formed several days before hearing onset in birds and rodents. Direct electrical stimulation of the afferent axons in pre-hearing animals reveal that these connections form functional synapses shortly after they innervate their target nuclei (Jackson, 1982). The tonotopically organized connection between the auditory spiral ganglion neurons and the CN is preserved in projections to higher-order nuclei. That this topography is initially present from the time connections are formed further supports the influence of activity-dependent mechanisms on circuit formation in the auditory system.

Functional circuits in the auditory brainstem of mammals (Figure 2.4(a)) and birds (Figure 2.4(b)) can be separated into three functional divisions: 1) an excitatory projection from the second-order auditory nucleus (green), 2) inhibitory inputs (red), and 3) third-order nuclei that establish binaural sensitivity by integrating these excitatory and inhibitory inputs (white).

In mammals, CN neurons receive afferent input from the SG and project ipsilaterally to the lateral superior olive (LSO) and contralaterally to the medial nucleus of the trapezoid body (MNTB). The medial superior olive (MSO) gets binaural excitatory inputs from both the ipsilateral and contralateral CN. The MNTB extends a glycineric (inhibitory) projection ipsilaterally to the MSO and LSO.

In birds, cochlear ganglion neurons extend a central process into two second-order brainstem nuclei, nucleus magnocellularis (NM) and nucleus angularis (NA). NM neurons make bilateral projections to nucleus laminaris (NL), a third-order nucleus analogous to the MSO in mammals. Whereas the mammalian MSO receives a powerful direct inhibitory input from MNTB, the inhibitory brainstem nucleus in birds, SON, modulates the convergent excitatory strengths of bilateral inputs to NA, NM, and NL.

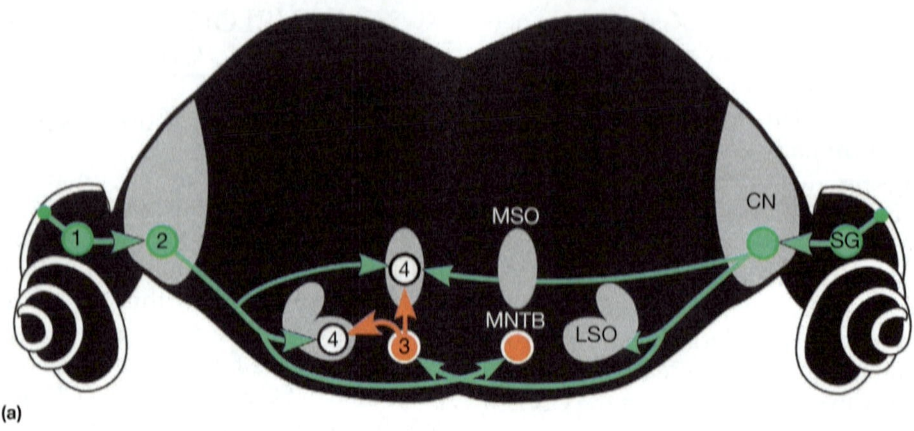

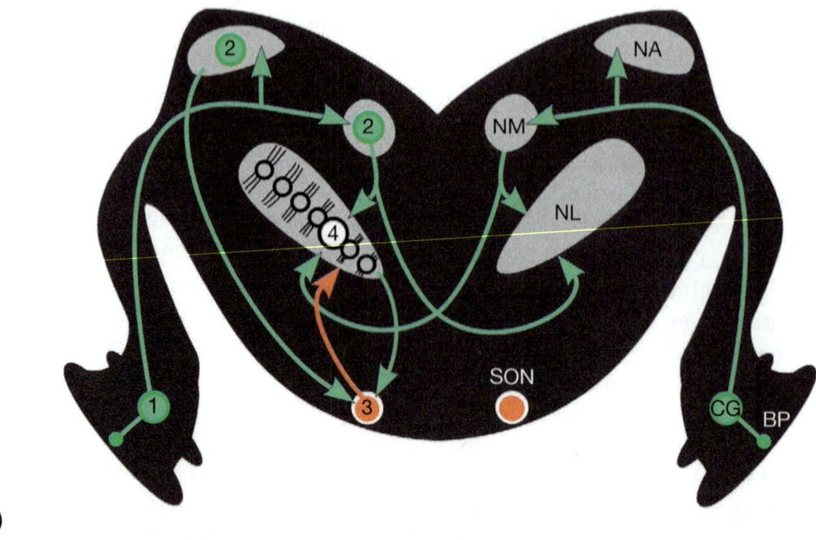

FIGURE 2.4   **Organization of mammalian and avian brainstem circuitry**. Schematic coronal sections of the auditory brainstem nuclei in rodents (A) and chicken (B). Brainstem circuits feature central projections from the primary sensory ganglion neurons (1) to 2nd order auditory nuclei in the brainstem (2). 3rd order brainstem nuclei (4) integrate inputs from 2nd order nuclei and local inhibitory nuclei (3). Excitatory projections are shown in green, inhibitory in red, nuclei in grey. For mammals, SG = spiral ganglion, CN = cochlear nucleus, LSO = lateral superior olive, MSO = medial superior olive, MNTB = medial nucleus of the trapezoid body. For birds, BP = basilar papilla, CG = cochlear ganglion, NA = nucleus angularis, NM = nucleus magnocellularis, NL = nucleus laminaris, SON = superior olivary nucleus. For both schematics, the dorsal surface of the brainstem is facing up.

2.3.2 Development of Fine-Scale Connectivity in the MSO

The spatial position of a visual or tactile stimulus can be encoded according to where, along the two-dimensional layout of the retina or skin, activity is greatest. In the auditory system, the functional layout of the cochlea is reduced to a single dimension mapped to sound frequency. The horizontal position of a sound source in space must be computed centrally, by neurons sensitive to differences in the loudness or timing of sounds arriving to each ear. Neurons in the MSO of mammals (NL in birds) are specialized to extract microsecond differences in the timing of excitatory inputs from the CN (or NM) associated with each ear. In mammals with low frequency hearing, such as gerbils, this precise tuning to interaural time differences (ITDs) appears to be enhanced by developmental changes in the subcellular positioning of inhibitory glycinergic inputs from the MNTB (Werthat, 2008). Glycinergic terminals from MNTB are evenly distributed across the dendrites and cell bodies of MSO neurons round the time of hearing onset. In the days and weeks that follow, the less effective inhibitory synapses on the distal ends of the dendritic tree are eliminated, sparing the proximal axosomatic inhibitory synapses (Figure 2.5). Alterations at the subcellular level are paralleled by a progressive refinement in the overall topographic breadth of presynaptic axon innervation across the topographic map in MSO.. These anatomical changes are associated with increased temporal precision and efficacy of synaptic inhibition onto MSO neurons as well as sharpening of neural ITD tuning functions in the auditory brainstem, suggesting a direct link between physiological maturation and its structural underpinnings. Taken together, the developmental shifts in the topographic and subcellular distribution of inhibitory inputs to the MSO work synergistically to sharpen the spatial and temporal sensitivity of MSO neurons to the excitatory CN inputs arriving from each side of the brain.

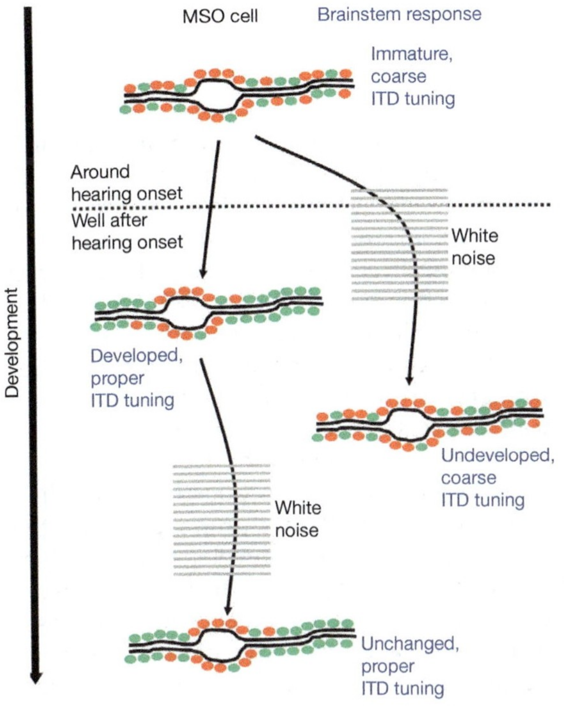

FIGURE 2.5 **Correlation between development of inhibitory inputs and ITD tuning**. Before hearing onset, inhibitory glycinergic synapses (red dots) and excitatory glutamatergic synapse (green dots) on MSO neurons are distributed over soma and dendrites. After hearing onset, the glycinergic synaptic inputs get refined to the cell body. This refinement depends on meaningful acoustic experience and can be interrupted by the exposure of omnidirectional white noise during hearing onset. White noise exposure at an adult stage has no effect on glycine receptor distribution. ITD tuning responses in the auditory brainstem correlate with the development of inhibitory inputs to MSO. When animals were exposed to white noise between P10 and P25, ITD tuning responses recorded were found to have the same characteristics as right after hearing onset, when neuronal responses to ITDs are immature and coarse. Under control conditions, i.e. with normal exposure to sound, ITD sensitivity developed normally. White noise exposure in adult animals has no effect on ITD tuning.

2.3.3 Development of Fine-Scale Connectivity in the LSO

LSO neurons are sensitive to interaural sound level differences via tuning to the relative strength of excitatory inputs from the ipsilateral ear and inhibitory inputs from the contralateral ear. Unlike the MSO, contralateral inputs are only expressed indirectly, via a local inhibitory connection from the MNTB. The sharp binaural tuning of LSO neurons is thought to arise from the integration of tonotopically matched excitatory inputs from the ipsilateral CN and MNTB. By contrast to MSO synapses, which are predominantly remodeled after hearing onset, topographic plasticity in the LSO proceeds in two phases: a functional silencing of MNTB inhibitory inputs prior to hearing onset followed by a structural pruning of MNTB axon terminals after the onset of hearing.

The pre-hearing functional refinement phase takes place contemporaneously with the period of ATP-induced Ca^{2+} spikes in IHCs. Accordingly, recordings from MNTB neurons at this age reveal spontaneous action potential bursts with the same rhythmic patterns observed in SG neurons (Tritsch, 2010). These activity patterns also depend upon an intact connection with IHCs, which may indicate that intrinsic spontaneous "test patterns" generated in the Organ of Corti prior to hearing onset may play a role in both peripheral and central circuit formation. During this 3–4 day period, the spatial spread of MNTB-derived inhibition shrinks by approximately 75%, considerably sharpening the breadth of inhibition along the tonotopic axis (Kim, 2003).

The onset of hearing ushers in a phase of structural refinement. In the week following hearing onset, dendritic arbors of postsynaptic LSO neurons increase in branching complexity yet are culled to more confined space within the topographic map. These postsynaptic modifications are accompanied by the physical elimination of tonotopically misaligned presynaptic axon terminals from the MNTB. For a more detailed description of inhibitory circuit development the reader is referred to Chapter 131 of this volume by K. Kandler et. al.

2.3.4 Afferent Regulation of Cochlear Nucleus Development

These functional and structural changes in the brainstem synaptic networks during the weeks surrounding hearing onset provide correlational evidence that spontaneous and sensory-evoked events are needed to achieve the elegant organization of neurons and connectivity observed in mature animals. The strong test of this hypothesis has been carried out by dozens of studies over the last sixty years that examine age-dependent deviations from normative development following disruption or complete elimination of afferent input from the cochlea. The seminal study by Rita Levi-Montalcini (Levi-Montalcini, 1949) removed the otocyst (the precursor of hair cells and ganglion neurons) of chicks at an early stage of embryonic development, thereby depriving auditory brainstem nuclei of cochlear input. She noted that neurons in NA and NM developed normally until E11 (the approximate time of hearing onset in the chick), at which point the size and overall number of surviving neurons rapidly declined. This observation suggested a dependence of CN neuron survival upon an intact connection with the periphery and has inspired a number of researchers to delve deeper into the role of afferent signaling in the formation and maintenance of brainstem circuitry.

Subsequent studies in the chick and rodents have further characterized the nature and timing of CN degeneration following deafferentation. Unilateral otocyst removal at E3 or cochlea removal prior to sexual maturity in chicks or cochlear destruction shortly after birth in rodents produces a variety of changes, including: 1) the CN on the ablated side has significantly fewer neurons than the intact side and the surviving neurons have smaller soma and neuropil area; 2) an ectopic – yet tonotopically aligned – projection from the normally innervated CN grows into the CN on the deafferented side; and 3) in mammals, the endbulb of Held development described in Figure 2.3 never reaches the fully mature state (reviewed in Harris, 2006).

In most animals, these plasticity effects are strictly limited to a developmental critical period. As noted above, the effects of otocyst removal in the E3 chick are not apparent for another eight days, when active connections with the periphery are first established. The critical period for CN neuron survival ends as abruptly as it begins, as cochlear destruction in gerbils during the first postnatal week causes 45-90% of CN cells to die, yet has no effect on the number of surviving neurons when the same manipulation is performed at P9, just prior to the onset of hearing (Tierney, 1997). The remarkably rapid changes in susceptibility of CN neurons to deprivation appears to reflect a differential weighting of factors that promote versus inhibit cell death.

Collectively, studies of CN development demonstrate that the proliferation, migration and formation of appropriate topographic connections are complete before action potentials begin to appear in auditory nerve fibers. The arrival of normal spontaneous and sound-evoked afferent action potentials has little effect on cellular morphology or topographic specificity. However, pathological deviations from the normal developmental trajectory (e.g., cochlear removal) that occur within a defined critical period window radically alter CN organization, leading to pronounced cell death and atrophy of surviving cells. The cellular mechanisms that close the critical period of CN vulnerability are not yet fully established, but gene array analyses suggest that glial proliferation and up-regulation of immunity-related genes may play an important role.

In addition to deafferentation-induced cell death and atrophy, hearing loss also induces a long-term enhancement of neuronal excitability. *In vitro* measurements of neurons in acute slices of the CN demonstrate a shift towards enhanced excitation and diminished inhibition. The loss of balanced excitation and inhibition arises from an abnormal sorting of membrane-bound ligand-gated neurotransmitter receptors and voltage-gated ion channels. Ongoing research is exploring the hypothesis that this pathological over-excitability in the brainstem and other stations of the central auditory pathways may be the source of tinnitus, the perception of phantom sounds that can accompany hearing loss.

2.3.5 Afferent Regulation of 3rd-Order Brainstem Nuclei

All auditory-evoked signals in the brain are initially routed through the CN. Therefore, alterations in CN morphology resulting from cochlear ablation could also impact the organization of downstream nuclei. Indeed, the post-hearing structural refinement of axon terminals from the MNTB to the MSO and LSO described previously is substantially diminished without a connection to an intact cochlea. Moreover, unilateral cochlea removal before hearing onset enhanced an otherwise weak physical connection between the normally innervated CN and the opposing side of the MSO.

Unlike the inhibitory projections from the MNTB to the LSO and MSO, excitatory projections from the CN undergo considerably less developmental refinement and are largely insensitive to cochlear removal. The CN, for example, forms a giant excitatory synapse onto MNTB neurons called the calyx of Held. This calyx is the largest synapse in the mammalian brain and features glutamatergic CN terminals that almost completely envelop the MNTB neuron, ensuring high-fidelity transmission necessary for sharp interaural time- and level-dependent tuning observed in the MSO and LSO, respectively. In mature brains, one calyx innervates a single MNTB neuron. Although the calyx undergoes substantial changes in shape during the first weeks of postnatal development, the one-to-one connectivity is present from the time CN projections initially arrive in the MNTB and is established with or without spontaneous or evoked action potential from the auditory nerve (Hoffpauir, 2006).

The effects of deafferentation have been extensively studied in NL of the chicken. The intrinsic organization and connectivity patterns within NL lend themselves to studies of afferent regulation of individual dendrites. First, NL exhibits a clear gradient of dendritic geometry that varies from small, short and stubby to large, long and elaborate across the high-to-low tonotopic map. This gradient appears to begin to form around the time of hearing onset. Second, NL neurons have two sets of symmetrical dendrites, one set oriented dorsally to contact glutamatergic inputs from the ipsilateral NM, the other ventrally to contact glutamatergic inputs from the contralateral NM. This organization permits researchers to directly compare the effects of unilateral manipulations (which would affect the dorsal dendrites on the ipsilateral NL and the ventral dendrites on the contralateral NL) to transection of the NM axons at midline (which would affect the ventral dendrites on both sides of the brain).

Using both of these approaches, researchers have discovered an interesting dichotomy in the afferent regulation of intrinsic features versus functional circuit properties. Removal of synaptic input to one side of NL induces a progressive retraction of dendrites on the corresponding side on a surprisingly short timescale (Figure 2.6). Tracking these changes over time reveals that retraction begins within 1 hour, can last over 2 weeks, and can amount to as much as a 60% reduction in length, demonstrating that NL neurons can rapidly regulate significant amounts of membrane surface devoted to specific excitatory inputs (Deitch, 1984). By contrast to the rapid calibration of dendritic length allocated to the dorsal versus ventral NM inputs, the short-to-long gradient of relative dendrite length across the tonotopic axis of NL is largely unaffected by cochlear removal.

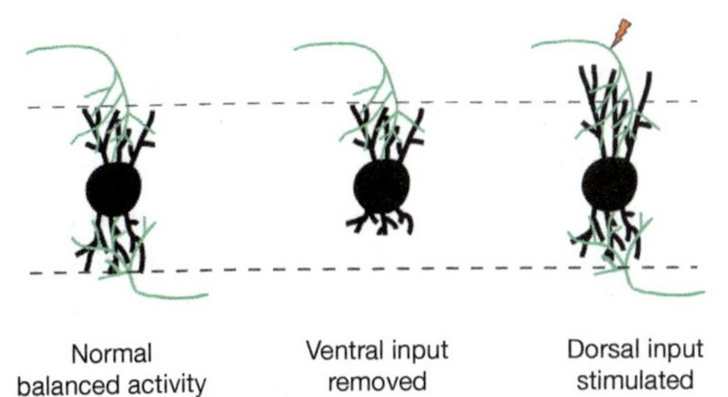

FIGURE 2.6 **Compartment-specific regulation of afferent activity on dendritic structure in the chick nucleus laminaris (NL).** NL neurons have bipolar dorsal and ventral dendrites (black), receiving highly segregated excitatory inputs by an axon from either the ipsilateral or contralateral ear via the nucleus magnocellularis (NM). (*Left*), Under normal physiological conditions, the two dendritic domains receive balanced inputs (green), each elicited by one ear, and their dendrites are of similar length. (*Middle*), Following the deafferentation of the inputs to one domain, e.g. when axons from one side of the brain get cut, deprived dendritic branches rapidly retract, while the length of the other dendrite remains unchanged. (*Right*) Physiological stimulation of the axon inputs to one set of dendrites (red lightning bolt), but not the other, leads to a growth of the stimulated dendritic branches and a retraction of unstimulated dendritic domain.

2.3.6 Influence of the Source and Pattern of Afferent Activity on Brainstem Circuits

The effects of cochlear removal on brainstem circuits point towards a panoply of developmental events that depend upon afferent signals from the periphery. Cochlear removal eliminates both spontaneous and sound-evoked action potentials in addition to the physical degeneration of SG neurons. Because the dependence on afferent activity is most commonly observed after hearing onset, one might assume that sound-evoked activity provides important signals for the fine-tuning of brainstem circuits. An alternative explanation holds that any afferent action potential signaling, be it spontaneous or evoked, could be sufficient for normal assembly of brainstem circuits.

To isolate the relative contributions of spontaneous action potentials versus sound-evoked activity, researchers have compared the effect of pharmacologically silencing all afferent action potentials versus simply blocking the transduction of acoustic signals. The effects of tetrodotoxin infusion into the inner ear at the time electrical signaling first begins were indistinguishable from the effects of cochlear removal: approximately 40% of the neurons died with widespread neuronal atrophy in the survivors (Born, 1988). On the other hand, simply blocking sound-evoked activity (by disrupting the sound transmission mechanisms of the outer or middle ear) without eliminating spontaneous activity in auditory nerve did not cause atrophy or cell death in the CN (Tucci, 1985). These results suggest that action potentials, or more probably the voltage-gated changes in glutamate and calcium signaling, provide a necessary source of trophic support during the critical period of CN development. Although the particular patterning of sound-evoked action potentials was not necessary for the normal cellular maturation of this second-order nucleus, it had a significant impact on stimulus selectivity and circuit formation in downstream nuclei. For instance, rearing animals in omnidirectional noise interferes with low frequency signals necessary to calculate interaural time differences. Absent these instructive environmental signals, the MSO fails to develop at a normal pace and features widely branching axodendritic MNTB synapses rather than the topographically focused axosomatic synapses found in normally reared animals (Seidl, 2005); Figure 2.5. Similarly, interaural level difference tuning in brainstem neurons is significantly altered when owls are reared with an earplug that deprives them of normally calibrated binaural cues (Mogdans, 1994).

2.3.7 Conclusions

In summary, functional circuit development and refinement in the auditory brainstem is thought to reflect an interplay between molecular cues coupled with spontaneous and evoked action potential activity. Factors such as connectional topography in the LSO and spatial gradients of dendritic length in NL appear to be governed by intrinsic, activity-independent mechanisms and form prior to the onset of hearing. Cochlear action potentials before and after the onset of hearing predominantly regulate the size and subcellular positioning of synaptic contacts. The development of precise connectivity is less categorical: the proper development of axonal projections from the MNTB to MSO and LSO is dependent upon afferent signaling after hearing onset, the topographic specificity of SG connections into the CN are subtly modified by afferent signaling, while excitatory projections from the CN to their ipsilateral and contralateral brainstem targets develop independently of cochlear signaling (although they make aberrant connections should one cochlea be removed).

Compared to peripheral circuits, which have matured in many respects by hearing onset, brainstem circuits undergo additional refinement after sound-evoked activity is introduced to the system. These phenomena are likely to contribute to the progressive improvement in infant auditory perceptual acuity in humans and animals. For example, human infants become increasingly capable of resolving the fine positioning of sounds along a horizontal plane, which may reflect the developmental calibration of synaptic properties within the MSO and LSO. Other aspects of phonological development, including the neural specializations that underlie the acquisition of language, can only be understood by examining the development of functional circuits in higher levels of the central auditory system.

2.4 DEVELOPMENT OF AUDITORY MIDBRAIN AND FOREBRAIN CIRCUITS

Higher auditory circuits are assembled from a diverse set of excitatory and inhibitory connections arising from three critical sources: 1) the inferior colliculus (IC), a midbrain auditory brain structure, 2) the medial geniculate (MG) and reticular (Rt) divisions of the thalamus, and 3) the auditory cortex (Actx) (Figure 2.7). These nuclei are interconnected through a complex array of feedforward, feedback and intrinsic connections. Inputs from the two ears are heavily intermixed, given the feedforward binaural inputs from the MSO and LSO as well as interhemispheric connections between each IC and Actx. Both the IC and MG are obligatory relays for auditory signals reaching the Actx and, consistent with brainstem circuits, a tonotopic organization within these connections is evident 5-7 days before birth in rodents, long before the initiation of afferent signaling from the periphery.

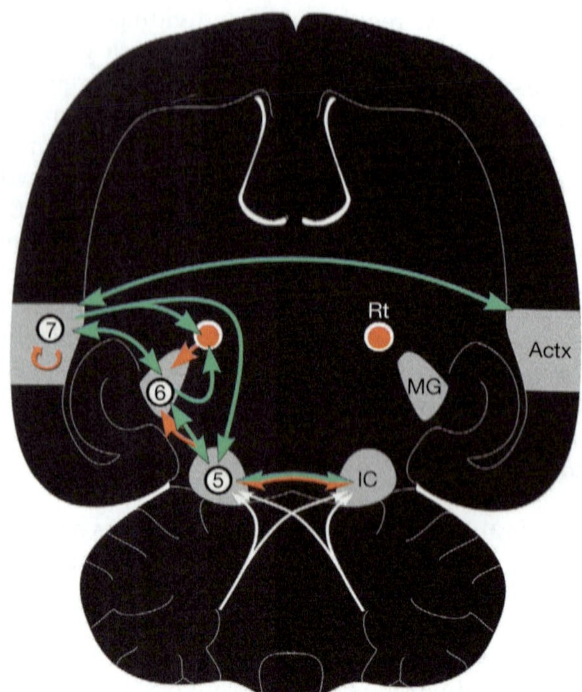

FIGURE 2.7　**Organization of higher auditory circuits.** Schematic of a horizontal section through a rodent brain. Convergent bilateral input from the brainstem (white arrows) projects to the inferior colliculus (IC), an auditory midbrain structure (5). The IC connects to the auditory thalamus (6) and cortex (7) through a complex chain of feedforward, feedback and intrinsic excitatory (green) and inhibitory (red) connections. MG = medial geniculate body, Rt = reticular nucleus of the thalamus, Actx = auditory cortex. Top of the figure is rostral (i.e., anterior), bottom is caudal (i.e., posterior).

2.4.1 Development of Thalamocortical Subplate Circuitry

Compared to the wealth of studies in NL, LSO and MSO, functional circuit developments in higher auditory brain areas have yet to be characterized in detail at the cellular level. One emerging exception is the observation of a transient microcircuit linking the neonatal thalamus and cortex. Cortical neurons originate from the ventricular zone and reach their final positions within the six-layered cortical plate via local radial migration or, in the case of several classes of inhibitory interneurons, via long-distance horizontal migration along the rostral stream (additional information on patterning and neural migration within the cerebral cortex can be found in Rubenstein and Rakic, 2013). Subplate neurons (SPN) reside in the white matter beneath the cortical plate, physically interposed between the ascending MG axons and the incipient circuits forming in the more superficial layers of the cerebral cortex (for a comprehensive review of subplate neurons, see (Kanold, 2010)).

SPN development occurs at an accelerated pace compared to the other cortical neurons; SPNs are among the first neurons to appear in the cerebral cortex, the first cortical neurons to fire action potentials, and are almost completely eliminated around the time of hearing onset. SPNs have elaborate dendritic trees that integrate excitatory inputs from the MG and local inhibitory inputs. SPN axon terminals ramify extensively in the same layers of the cerebral cortex that will receive the bulk of direct axonal input from the MG in subsequent stages of development. SPNs have high input resistances, relatively depolarized resting membrane potentials and are electrically coupled to one another via gap junctions, making them a sensitive and potent source of excitatory input to Actx prior to hearing onset.

Their precocious morphological and biophysical development combined with their physical position between MG and Actx make SPNs ideal interlocutors in the postnatal assembly of thalamocortical circuits. Indeed, as represented in Figure 2.8, MG axons innervate the subplate days before birth and wait there for days or weeks (depending on the species) before innervating the middle cortical layers. During this waiting period, SPNs make excitatory glutamatergic projections to excitatory and inhibitory neuron subtypes within Actx, potentially providing a source of activity-dependent synaptic refinement before connections are established with the MG.

2.4.2 Postnatal Development of Local Cortical Circuits

Cortical circuits undergo a subsequent wave of refinement during the second postnatal week, after the majority of SPNs have been eliminated. At the onset of hearing, sound-evoked responses in Actx are restricted to frequencies at the center of the hearing range presented at high sound levels. Over the following 2–3 days, response thresholds decrease and sensitivity to a broader range of sound frequencies begins to emerge. Following this brief period of change, which almost certainly reflects peripheral maturation, recordings from Actx neurons reveal a gradual improvement in their ability to synchronize action potential timing to rapid modulations of an incoming sound source. This improvement in cortical temporal processing is a hallmark feature of central auditory development and thought to be essential for the accurate encoding of important environmental sounds such as speech.

Through targeted recordings of excitatory and inhibitory neuron subtypes in Actx using an *in vitro* brain slice preparation, researchers have discovered that the progressive elimination of temporal "sluggishness" towards the end of the second postnatal week is associated with a confluence of synaptic and intrinsic changes in Actx neurons. In addition to afferent input from MG, acetylcholine (ACh)-positive axon terminals from neurons located in the basal forebrain begin innervating the Actx in postnatal week 1 and reach adult levels by postnatal week 2. The maturation of cholinergic terminals coincides with significant changes in the levels or composition of nicotinic ACh receptors and NMDA receptors (nAChRs and NMDARs, respectively) in Actx (Metherate, 2003). Although ACh does not excite Actx neurons directly, it can enhance excitatory transmission indirectly by binding to nAChRs, which in turn modulate glutamatergic NMDARs. The decline of nAChRs

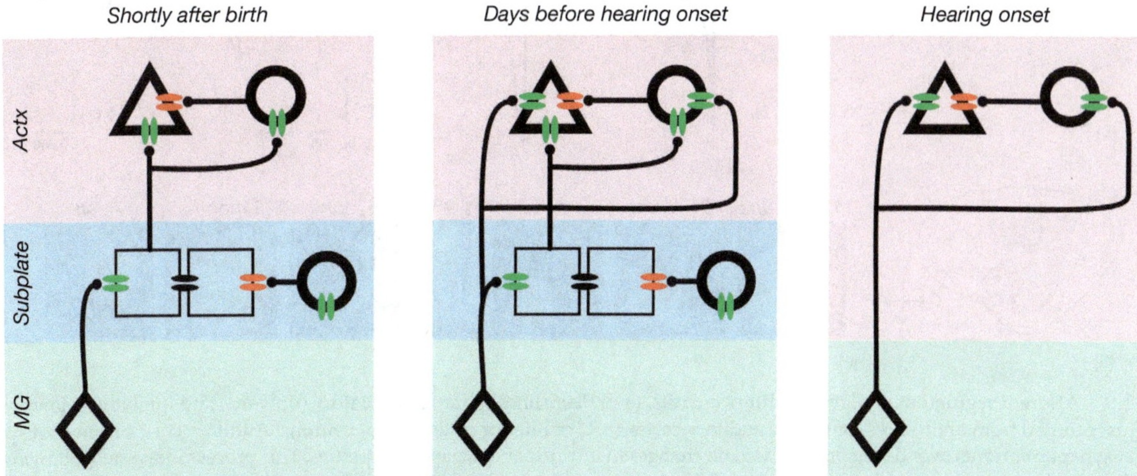

FIGURE 2.8 **Subplate neurons form a transient microcircuit linking the auditory thalamus and cortex.** Subplate neurons (SPNs, squares) mediate excitatory signaling from the medial geniculate nucleus of the thalamus (MG, diamond) to excitatory (triangle) and inhibitory (circle) cells in the auditory cortex (Actx, triangle). Shortly after birth (left), SPNs are believed to be the exclusive source of subcortical input to Actx. In the days hearing onset (middle), SPNs mediate direct excitatory projections to Actx before they are eliminated, shortly thereafter (right). Green ellipses=glutamate receptor, red ellipses=GABA receptor, black ellipses=electrical synapse.

in Actx by postnatal week 3 works synergistically with changes in AMPA and NMDA glutamate receptor subunit composition to reduce the duration and increase the amplitude of excitatory postsynaptic currents in Actx (Figure 2.7(a)). Thus, the transient appearance of nAChRs during postnatal week 2 represents a critical period during which ACh can prolong the time course of NMDA-dependent synaptic excitation around the period of hearing onset.

The sharpening of synaptic excitation in Actx around the time of hearing onset is mirrored by a progressive enhancement of synaptic inhibition. Presynaptically, the voltage-gated K^+ channel subtypes present in GABAergic interneurons change significantly over the first three weeks of postnatal development to adjust the resting membrane potential and shorten the refractory period following an action potential. These biophysical changes are complemented by the progressive loss of $GABA_B$ receptors on presynaptic interneurons, enabling higher sustained firing rates without fatiguing. Postsynaptically, the $GABA_A$ receptor subunit composition in excitatory Actx neurons changes over the first weeks of development to eliminate the α3 subunit in favor of the α1 and β2/3, both of which are associated with faster inhibitory synaptic current rise times (Figure 2.9(a)). Collectively, these intrinsic and synaptic changes in the time course of excitatory and inhibitory synaptic

transmission endows Actx neurons with an improved ability to track rapid temporal fluctuations in sound signals with high fidelity (for review see Sanes, 2009).

2.4.3 Afferent Regulation of Higher Auditory Circuit Development

Detailed characterizations of auditory brainstem circuit development in the absence of cochlear signaling have revealed a combination of activity-dependent and activity-independent processes. To assess the generality of these principles throughout the central auditory pathways, researchers have also examined the role of an intact periphery on the maturation of IC and Actx circuits. The potentially confounding influence of CN degeneration has been avoided in this type of experiment by bilaterally removing the cochleae after the critical period for CN cell death, but before the onset of hearing, such that higher auditory circuits have the potential to be shaped by spontaneous – but not sound-evoked – action potentials.

Recordings made in the acute brain slice preparation weeks after cochlear ablation reveal dramatic alterations in the strength and time course of excitatory and inhibitory synaptic currents in the IC (Figure 2.9(b)) and Actx (Figure 2.9(c)). Compared to control slices taken

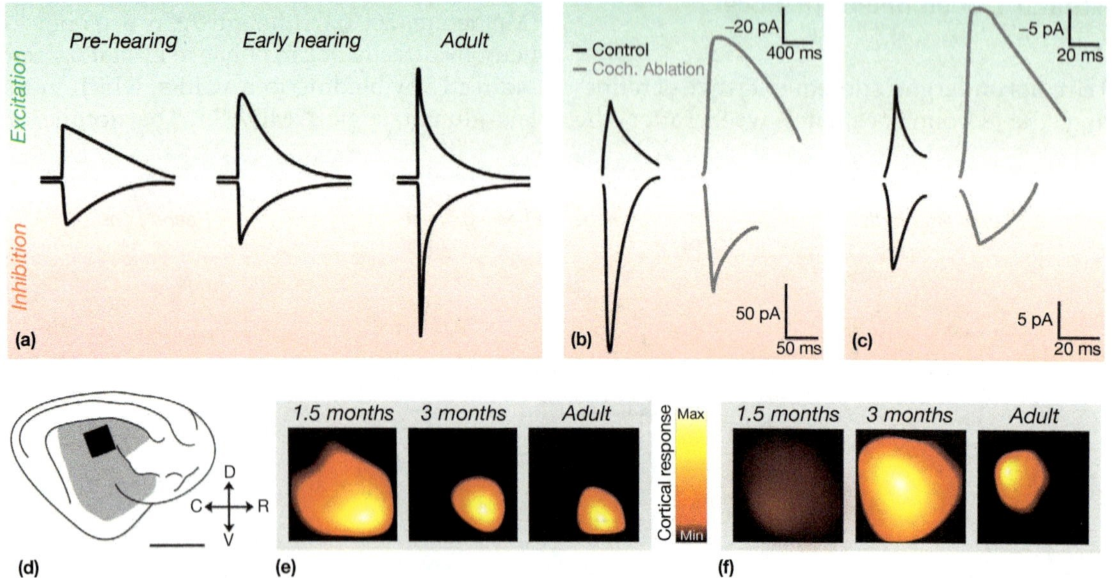

FIGURE 2.9 **Afferent regulation of higher auditory circuits.** (a-c) Black lines represent excitatory (upward) and inhibitory (downward) synaptic currents recorded from brain slices containing auditory cortex (a, c) or inferior colliculus (b) neurons. Auditory cortex neurons display faster and stronger synaptic currents over development based on changes in intrinsic and synaptic properties. This process is arrested following bilateral cochlear ablation, wherein neurons are hyperexcitable and display immature temporal dynamics. Current amplitude is plotted on the vertical axis. (d) Lateral surface of the cat brain. Gray denotes total area of auditory cortex in normal cats. Black square denotes area from which neurophysiological signals are measured in panels e and f. D=dorsal, R=rostral, V=ventral C=caudal. Horizontal bar=1 cm. Inward negative EPSCs are plotted upwards, and outward positive IPSCs are plotted downwards, by convention. (e-f) Areal extent of neural responses in auditory cortex evoked from activation of a stimulating electrode implanted proximal to the auditory nerve fibers in acutely deafened (e) or congenitally deaf (f) cats. Current traces in (b) and (c) are adapted from data presented in Sanes and Bao, 2009.

from normally hearing animals, neurons in cochlear-ablated slices areas show weaker, prolonged inhibition that is qualitatively similar to activity found in normal animal prior to hearing onset. Synaptic excitation is also prolonged in a similar fashion to pre-hearing animals, yet the amplitudes are far greater than those observed in the course of normal development, suggesting that deafferentation tips the homeostatic balance between excitation and inhibition towards greatly enhanced excitability, in a similar fashion to the CN (Kotak, 2005).

Recent studies have identified a combination of intrinsic and synaptic factors that may explain the failure of synaptic inhibition to mature normally. In terms of intrinsic biophysical mechanisms, the first postnatal weeks are marked by a substitution of voltage-gated K^+ channels that mediate faster membrane kinetics as well as the appearance of a K^+-dependent intracellular chloride transporter, KCC2. During the first week of postnatal development, when KCC2 expression levels are low, intracellular chloride concentrations are higher than the electrochemical equilibrium potential and GABA release from inhibitory neurons induces a depolarization of the membrane potential in postsynaptic neurons, rather than the expected hyperpolarization. As levels of membrane-bound KCC2 increase in the second week of postnatal development, greater amounts of intracellular chloride are extruded into the extracellular space, thereby lowering the equilibrium potential and establishing the normal hyperpolarizing influence of GABA. This developmental process is arrested in the IC of cochlea-ablated animals, where the inhibitory reversal potential can be elevated by as much as 24 mV above normally hearing age-matched controls, reducing the hyperpolarizing effect of GABA binding. Although overall expression levels of KCC2 are not affected, pharmacological experiments reveal that cochlear ablation arrests the normal age-dependent maturation of KCC2 function, contributing to reduced levels of inhibitory signaling (Vale, 2003).

A synaptic piece of the puzzle was discovered through a comparison of $GABA_A$ receptor subunit composition in the Actx of normal and deafferented animals. Recall from the description above that $GABA_A$ receptors composed of the $\alpha1$ and $\beta2/3$ subunits appear during the second postnatal week and are partially responsible for the transition to shorter, larger-amplitude inhibitory synaptic currents. Following bilateral cochlear ablation, juvenile Actx neurons fail to express the mature form of the membrane-bound $GABA_A$ receptor, thereby preventing the sharpening of the inhibitory postsynaptic currents observed in age-matched controls (Kotak, 2008). Taken together, these results demonstrate a broad spectrum of intrinsic and synaptic events that fail to develop in the absence of sound evoked-activity. The sum total of these events render IC and Actx neurons

incapable of tracking rapid fluctuations in sound properties, an essential characteristic of normal hearing.

2.4.4 Developmental Regulation over Reinstating Hearing in the Deaf

Synaptic transmission studies in the brain slice preparation shed some light on the molecular targets of afferent signaling and help to identify the complications and possibilities associated with reinstating hearing in deaf individuals. Unlike birds, and other non-mammalian vertebrates, which can regrow hair cells throughout life, mammals are born with all the cochlear hair cells they will ever have. Nevertheless, hearing is a possibility for profoundly deaf individuals through the use of the cochlear implant, a neural prosthetic device that bypasses the dysfunctional transduction machinery within the cochlea and reinstates afferent signals through direct electrical stimulation of auditory nerve fibers. Approximately 200,000 individuals have been fitted with cochlear implants over the past 40 years and it was discovered early on that the age of surgical implantation plays a crucial role in the quality of hearing experienced by cochlear implant users. While post-lingually deaf individuals often recover acceptable hearing and speech recognition whether they are implanted as children or adults, congenitally deaf individuals stand the best chance of experiencing the full benefit of the cochlear implant if they undergo the implantation procedure at an early age, typically by the time they are 7 years old (Dorman, 2007).

Through careful study of cochlear implants in a special breed of congenitally deaf cats, researchers have begun to understand how auditory brain areas represent signals delivered through the cochlear implant and the manner by which these representations are shaped through development and experience. As an experimental control, normally hearing cats are acutely deafened with an ototoxic drug and immediately fit with a cochlear implant. Neural recordings are made from the Actx of acutely or congenitally deaf cats at various ages in response to brief electrical pulses delivered the auditory nerve (Figure 2.9(d)). A comparison of activation patterns across development in acutely deafened cats reveals an exuberant spatial spread of neural activity across the Actx in young kittens that is culled to a topographically restricted activation area by 3 months of age (Figure 2.9(e)). By contrast, activating the implant in congenitally deaf kittens at 1.5 months evokes a weak cortical response (Figure 2.9(f)). At 3 months postnatal, deaf cats show the exuberant activation patterns comparable to normally hearing kittens at 1.5 months and these activation areas are not consolidated until early adulthood (Kral, 2005). The hyper-excitability of the Actx in congenitally deaf cats at three months may stem from the diminished synaptic inhibition and augmented synaptic excitation observed in brain slices of rodents that undergo cochlear ablation in infancy (Figure 2.9(c)).

Although the mechanisms governing the recovery of hearing in cochlear implant users remain unclear, one possible clue has been found through analyzing the endbulb of Held synapse in deaf cats that began hearing through chronic use of the cochlear implant at an young age. As described previously, the endbulb synapse fails to develop normally in the absence of cochlear signaling. Strikingly, the endbulb synapse from cats using the cochlear implant for 3 months beginning at an early age was largely indistinguishable from normally hearing cats (Ryugo, 2005). Therefore, reinstating afferent activity to the central auditory system in early life can rescue the progressive synaptic degradation observed at several levels of the central auditory system.

2.4.5 Experience-Dependent Influences on Functional Circuit Development

For the most part, the influence of afferent signaling on neural circuit development has been described in the context of all-or-nothing manipulations; comparisons to normally hearing animals are made through cochlea removal, genetic deafness, or pharmacological silencing of the auditory nerve. However, many facets of auditory perceptual development depend upon the specific patterns of auditory experience rather than its presence or absence. For example, before a child utters a first word in its native language, it will have been exposed to hundreds of thousands of words that bear the phonemic structure specific to that language. One school of developmental cognitive psychology has argued that this repeated auditory exposure to speech sounds, when appropriately timed during childhood development, "primes" the auditory and vocal motor areas of the brain to specialize in the nuances of a given language. A half century of inspired neurobiology and neuroethology research has shown that the processes through which songbirds learn their vocal repertoire offers a striking parallel to human language acquisition. Song and speech learning both involve a complex interplay between innate predispositions and experience, both forms of learning are shaped by developmental critical periods, both require skilled control over the motor vocal apparatus, and both depend upon precisely calibrated auditory feedback (for a review of song learning, the reader is referred to (Brainard, 2002).

Although the circuitry underlying song acquisition is more closely linked to sensorimotor integration than auditory processing *per se*, auditory experience has also been found to exert a profound influence over the development of dedicated auditory circuitry within IC and Actx. In rodents, a tonotopic map of preferred sound frequency tuning can be readily delineated from the primary field of the auditory cortex (AI), which is visible along the lateral surface of the brain (Figure 2.10(a)). In mature animals, the tonotopic map features an orderly low-to-high gradient of frequency tuning across the posterior-to-anterior extent of AI, with similar tuning organized into orthogonal iso-frequency contours (Figure 2.10(b)). Most recording sites feature closely matched frequency tuning for tones presented to each ear, albeit tuning evoked from contralateral stimuli is more robust than the weaker, high-threshold tuning for tones presented to the ipsilateral ear. By contrast to the adult map, AI recordings made shortly after the onset of hearing (P11-14) reveal narrowly tuned, high threshold responses restricted to the middle frequency regions of the tonotopic map (Figure 2.10(c)).

A series of studies over the past decade demonstrate that experimentally modifying the ambient acoustic environment experienced by young animals during critical periods of higher auditory circuit development can dramatically alter the normal developmental sequence described in Figures 2.10(b) and 2.10(c). For example, rearing litters of rodents in a sound environment dominated by the repeated presentation of a single tone at a fixed frequency more than doubles the area of the tonotopic map tuned to that exposure frequency (Figure 2.10(d)). As illustrated in Figure 2.10(d), exposure to a repetitive middle frequency tone for just 96 hours beginning at the onset of hearing is associated with an expansion of the corresponding area within the tonotopic map at the expense of flanking frequency representations. The tonotopic map retains this distorted organization into adulthood despite the fact that animals hear normal acoustic stimuli from P16 onwards. Conversely, rearing animals older than P16 to the repeated single frequency has no effect on the map regardless of the length of exposure time (de Villers-Sidani, 2007). Similar effects of tone rearing have been reported in the IC.

Whereas rearing animals in fixed frequency tone environments induces a region-specific reallocation of territory within the cortical map, rearing animals in broad-spectrum noise is associated with a degeneration of Actx tonotopy and an abundance of recording sites with abnormally broad tuning (Figure 2.10(e)). Similar to the extended visual cortex critical periods observed in animals reared in total darkness, the critical period for the effects of single tone exposure is abnormally prolonged in animals reared in white noise. Thus, deprivation of patterned acoustic input postpones the flip of a molecular switch that normally limits the critical period, and the effects of singe tone exposure can distort the map many weeks after the critical period would normally have closed (Chang, 2003). That a critical period normally observed in infancy can be delayed

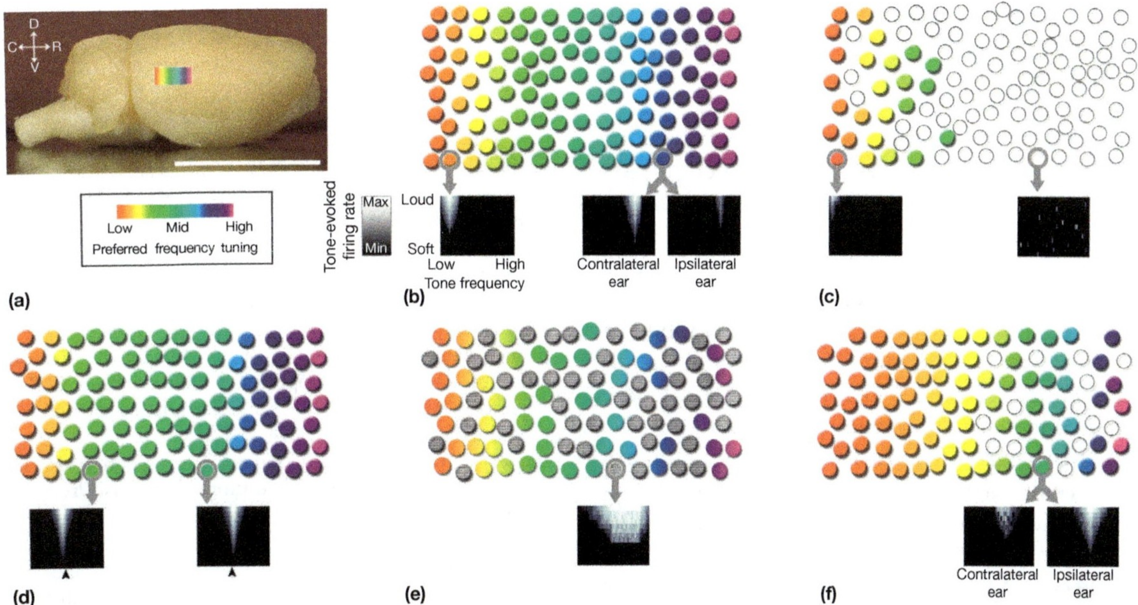

FIGURE 2.10   **Experience-dependent map reorganization in the developing auditory cortex.** (a) Photograph of the lateral surface of the rat-brain. Color gradient denotes the position and orientation of the tonotopic map within the primary auditory cortex (AI). Horizontal line = 1 cm. (b-f) Schematics of tonotopic maps reconstructed from microelectrode recordings of tone-evoked action potentials at approximately 100 different points within AI. At each recording site, the frequency response area (grayscale surface plots) is determined by presenting tones at various frequencies and levels and identifying the frequency to which the neuron is tuned at its threshold intensity (indicated by the color of each dot). (B) Normal adult animals display a smooth and orderly tonotopic gradient in which neurons at the caudal and rostral boundaries of AI prefer low and high frequency tones, respectively. Neurons are tuned to the same frequency for sounds presented to each ear, yet responses to the contralateral ear are more robust than the ipsilateral ear. (c) Tonotopy at the onset of hearing is narrow, high-threshold and restricted to low frequency areas of the map. Open circles denote recordings sites with no evoked response. (d) Rearing rodents in acoustic environments dominated by presentation of a single middle frequency tone (represented by green colors and vertical arrows on surface plots) show an over-representation of that tone frequency within the tonotopic map such that recording sites at distant points within the map have similar tuning. (e) Rearing rodents in continous white noise is associated with a disrupted tonotopic map featuring numerous recording sites with abnormally broad frequency tuning (represented by gray dots). (f) Reversibly closing the contralateral ear during critical periods of development is associated with an over-representation of low frequency tones, the loss of contralateral tuning at several points (open circles) and an enhancement of tuning strength to tones delivered to the developmentally unobstructed ipsilateral ear.

into adulthood supports the possibility that progressive development of auditory feature representation in Actx may be more closely linked to an experiential timeline, rather than a strict chronological timeline.

A parallel can also be drawn between the reversible lid suture method used extensively in studies of visual cortex development and the effects of reversible ear canal ligation on Actx and IC development. Ligating the ear canal temporarily interferes with the transmission of acoustic signals to the middle ear, and ultimately the brain, particularly at high frequencies. By reopening the ear canal prior to recording from the contralateral AI, researchers have observed a degraded tonotopic map populated by weak high-threshold responses to contralateral tones and the occasional absence of contralateral tuning in higher frequency regions of the map (Popescu, 2010). By contrast to the contralateral bias observed in normal animals (Figure 2.(10)b), the quality of tuning for stimuli delivered to the ipsilateral ear is often superior to the normally dominant contralateral ear (Figure 2.10 (f)). Interestingly, ipsilateral inputs were only enhanced

when the ear canal was ligated in infant and juvenile animals, but not in adulthood. Collectively, these results indicate that the allocation of representational resources within the Actx and IC is quite dynamic and can be adaptively reassigned to the inputs that are most prominent during critical periods of early postnatal development.

2.4.6 Conclusions and Directions for Future Research

Compared to the periphery and brainstem, the assembly of circuits in higher auditory brain areas appears to be particularly sensitive to sound-evoked patterns of afferent action potentials. Eliminating hearing or even perturbing the normal balance of signals between the ears or across various frequency channels can dramatically alter auditory signal processing. This plasticity has been documented at multiple levels of analysis, ranging from the synaptic transmission between two neurons up to the coordinated arrangement of frequency tuning across hundreds of thousands of neurons. Although circuit reorganization at higher levels of the auditory system is striking, additional information on the normative course of development in the IC, MG and Actx as well as a deeper mechanistic understanding about the contributions of transient microcircuits such as the SPN to circuit assembly prior to hearing onset will greatly aid efforts to understand how the stable networks that underlie normal hearing are constructed. Absent data that bridge the gap between synapses and neural networks in normally developing animals, it is difficult to know whether activity- or experience-dependent modifications derail normal development or introduce an abnormal outcome following the conclusion of normal maturation.

The past fifty years of auditory developmental research have revealed a great deal about the interplay between intrinsic molecular events and dynamic electrical signaling in functional circuit maturation. The scope of research in the coming years may widen to include other biological factors that help to shape functional circuits. Although neurons are the central players in functional circuits, they do not operate in isolation. Networks of developing non-sensory glial cells, cells that form the developing vasculature and the molecules that form the extracellular matrix play may key roles in modulating neural signaling and providing a physical substrate for growth. As we have seen in the cochlea, the non-sensory cells that make up Kölliker's organ may act as the catalyst for electrical signaling throughout the pre-hearing auditory system and it is probable that exciting new connections between neural development and the non-neural cells and molecules that support them remain to be discovered. On the other end of the spectrum, it will be important to link the observations made at a neurobiological level to their potential behavioral consequences measured at the level of the whole organism. Demonstrations of abnormal circuit connectivity or map organization are worthwhile and interesting in their own right, but take on additional importance when they can be associated with changes in perceptual abilities. Studies that bridge the gaps between these various levels of analysis and stations of processing along the auditory pathway promise to reveal more about the etiology and therapeutic possibilities for treating hearing impairment, and may teach us a great deal about the fundamental principles of neural development.

References

Born, D.E., Rubel, E.W., 1988. Afferent influences on brain stem auditory nuclei of the chicken: Presynaptic action potentials regulate protein synthesis in nucleus magnocellularis neurons. The Journal of Neuroscience 8, 901–919.

Brainard, M.S., Doupe, A.J., 2002. What songbirds teach us about learning. Nature 417, 351–358.

Chang, E.F., Merzenich, M.M., 2003. Environmental noise retards auditory cortical development. Science 300, 498–502.

Cramer, K.S., 2005. Eph proteins and the assembly of auditory circuits. Hearing Research 206, 42–51.

Dallos, P., Zheng, J., Cheatham, M.A., 2006. Prestin and the cochlear amplifier. The Journal of Physiology 576, 37–42.

de Villers-Sidani, E., Chang, E.F., Bao, S., Merzenich, M.M., 2007. Critical period window for spectral tuning defined in the primary auditory cortex (A1) in the rat. The Journal of Neuroscience 27, 180–189.

Deitch, J.S., Rubel, E.W., 1984. Afferent influences on brain stem auditory nuclei of the chicken: Time course and specificity of dendritic atrophy following deafferentation. The Journal of Comparative Neurology 229, 66–79.

Dorman, M.F., Sharma, A., Gilley, P., Martin, K., Roland, P., 2007. Central auditory development: Evidence from CAEP measurements in children fit with cochlear implants. Journal of Communication Disorders 40, 284–294.

Echteler, S.M., 1992. Developmental segregation in the afferent projections to mammalian auditory hair cells. Proceedings of the National Academy of Sciences of the United States of America 89, 6324–6327.

Echteler, S.M., Arjmand, E., Dallos, P., 1989. Developmental alterations in the frequency map of the mammalian cochlea. Nature 341, 147–149.

Harris, J.A., Rubel, E.W., 2006. Afferent regulation of neuron number in the cochlear nucleus: Cellular and molecular analyses of a critical period. Hearing Research 216–217, 127–137.

Hoffpauir, B.K., Grimes, J.L., Mathers, P.H., Spirou, G.A., 2006. Synaptogenesis of the calyx of Held: Rapid onset of function and one-to-one morphological innervation. The Journal of Neuroscience 26, 5511–5523.

Hudspeth, A.J., 2008. Making an effort to listen: Mechanical amplification in the ear. Neuron 59, 530–545.

Jackson, H., Hackett, J.T., Rubel, E.W., 1982. Organization and development of brain stem auditory nuclei in the chick: Ontogeny of postsynaptic responses. The Journal of Comparative Neurology 210, 80–86.

Jhaveri, S., Morest, D.K., 1982. Sequential alterations of neuronal architecture in nucleus magnocellularis of the developing chicken: a Golgi study. Neuroscience 7, 837–853.

Johnson, S.L., Eckrich, T., Kuhn, S., et al., 2011. Position-dependent patterning of spontaneous action potentials in immature cochlear inner hair cells. Nature Neuroscience 14, 711–717.

Kanold, P.O., Luhmann, H.J., 2010. The subplate and early cortical circuits. Annual Review of Neuroscience 33, 23–48.

Kim, G., Kandler, K., 2003. Elimination and strengthening of glycinergic/GABAergic connections during tonotopic map formation. Nature Neuroscience 6, 282–290.

Kotak, V.C., Takesian, A.E., Sanes, D.H., 2008. Hearing loss prevents the maturation of GABAergic transmission in the auditory cortex. Cerebral Cortex 18, 2098–2108.

Kotak, V.C., Fujisawa, S., Lee, F.A., Karthikeyan, O., Aoki, C., Sanes, D.H., 2005. Hearing loss raises excitability in the auditory cortex. The Journal of Neuroscience 25, 3908–3918.

Koundakjian, E.J., Appler, J.L., Goodrich, L.V., 2007. Auditory neurons make stereotyped wiring decisions before maturation of their targets. The Journal of Neuroscience 27, 14078–14088.

Kral, A., Tillein, J., Heid, S., Hartmann, R., Klinke, R., 2005. Postnatal cortical development in congenital auditory deprivation. Cerebral Cortex 15, 552–562.

Leake, P.A., Snyder, R.L., Hradek, G.T., 2002. Postnatal refinement of auditory nerve projections to the cochlear nucleus in cats. The Journal of Comparative Neurology 448, 6–27.

Levi-Montalcini, R., 1949. Development of the acousticovestibular centers in the chick embryo in the absence of the afferent root fibers and of descending fiber tracks. The Journal of Comparative Neurology 91, 209–242.

Metherate, R., Hsieh, C.Y., 2003. Regulation of glutamate synapses by nicotinic acetylcholine receptors in auditory cortex. Neurobiology of Learning and Memory 80, 285–290.

Mogdans, J., Knudsen, E.I., 1994. Site of auditory plasticity in the brain stem (VLVp) of the owl revealed by early monaural occlusion. Journal of Neurophysiology 72, 2875–2891.

Norton, S.J., Bargones, J.Y., Rubel, E.W., 1991. Development of otoacoustic emissions in gerbil: Evidence for micromechanical changes underlying development of the place code. Hearing Research 51, 73–91.

Popescu, M.V., Polley, D.B., 2010. Monaural deprivation disrupts development of binaural selectivity in auditory midbrain and cortex. Neuron 65, 718–731.

Pujol, R., Carlier, E., Devigne, C., 1978. Different patterns of cochlear innervation during the development of the kitten. The Journal of Comparative Neurology 177, 529–536.

Rubenstein, J.L.R., Rakic, P., 2013. Patterning and Cell Types Specification in the Developing CNS and PNS.

Ryugo, D.K., Fekete, D.M., 1982. Morphology of primary axosomatic endings in the anteroventral cochlear nucleus of the cat: A study of the endbulbs of Held. The Journal of Comparative Neurology 210, 239–257.

Ryugo, D.K., Kretzmer, E.A., Niparko, J.K., 2005. Restoration of auditory nerve synapses in cats by cochlear implants. Science 310, 1490–1492.

Sanes, D.H., Bao, S., 2009. Tuning up the developing auditory CNS. Current Opinion in Neurobiology 19, 188–199.

Seidl, A.H., Grothe, B., 2005. Development of sound localization mechanisms in the mongolian gerbil is shaped by early acoustic experience. Journal of Neurophysiology 94, 1028–1036.

Tierney, T.S., Russell, F.A., Moore, D.R., 1997. Susceptibility of developing cochlear nucleus neurons to deafferentation-induced death abruptly ends just before the onset of hearing. The Journal of Comparative Neurology 378, 295–306.

Tritsch, N.X., Yi, E., Gale, J.E., Glowatzki, E., Bergles, D.E., 2007. The origin of spontaneous activity in the developing auditory system. Nature 450, 50–55.

Tritsch, N.X., Rodriguez-Contreras, A., Crins, T.T., Wang, H.C., Borst, J.G., Bergles, D.E., 2010. Calcium action potentials in hair cells pattern auditory neuron activity before hearing onset. Nature Neuroscience 13, 1050–1052.

Tucci, D.L., Rubel, E.W., 1985. Afferent influences on brain stem auditory nuclei of the chicken: Effects of conductive and sensorineural hearing loss on n. magnocellularis. The Journal of Comparative Neurology 238, 371–381.

Vale, C., Schoorlemmer, J., Sanes, D.H., 2003. Deafness disrupts chloride transporter function and inhibitory synaptic transmission. The Journal of Neuroscience 23, 7516–7524.

Von Békésy, G., 1960. Experiments in Hearing. McGraw-Hill, New York.

Werker, J.F., Yeung, H.H., 2005. Infant speech perception bootstraps word learning. Trends in Cognitive Science 9, 519–527.

Werthat, F., Alexandrova, O., Grothe, B., Koch, U., 2008. Experience-dependent refinement of the inhibitory axons projecting to the medial superior olive. Developmental Neurobiology 68, 1454–1462.

Development of the Superior Colliculus/Optic Tectum

B.E. Stein, T.R. Stanford

Wake Forest School of Medicine, Winston-Salem, NC, USA

3.1 NOMENCLATURE

Anatomically, the *tectum* is that portion of the mesencephalon, or midbrain, sitting between the hindbrain and the forebrain. The name is the Latin word for *roof* and reflects the view of early anatomists that the tectum formed a roof over the fluid-filled cerebral aqueduct and the tegmentum. In mammals, it sometimes is referred to as the *tectal plate* (or *quadrigeminal plate*) and is composed of two pairs of bumps or colliculi (*collis* means hill in Latin), one on each side of the midbrain. The more rostral pair is referred to as the *superior colliculi*, and the more caudal pair is known as the *inferior colliculi*.

However, the reader should be aware that there is some confusion in the nomenclature. The term *optic tectum* (OT) properly refers to the nonmammalian homolog of the superior colliculus (SC), and its name reflects its preeminent visual role in these animals. Thus, discussions of the OT are generally specific to the major central nervous system visual nucleus in birds, reptiles, amphibians, and fish, animals from which much has been learned about the function of the OT. But while the mammalian homolog of the OT is called the *superior colliculus*, projections to and from it are called *tectopetal* (e.g., *retinotectal*, *corticotectal*) and *tectofugal* (e.g., *tectoreticular*, *tectospinal*), respectively. The confusing nature of this

*Neural Circuit Development and Function in the Brain: Comprehensive Developmental
Neuroscience, Volume 3* http://dx.doi.org/10.1016/B978-0-12-397267-5.00150-3

© 2013 Elsevier Inc. All rights reserved.

nomenclature is compounded by the tendency of some to refer to the SC as the OT and of others to use *colliculo(ar)* for *tecto(al)*.

The primary subject of the following discussion is the SC; however, the OT and SC have fundamental functional similarities, and research on the OT will be referred to when it is instructive for understanding SC organization, function, or both.

3.2 FUNCTIONAL ROLE

Although the SC, like its nonmammalian counterpart, is clearly visually dominant, it is not strictly visual, and neither is the OT. Furthermore, the role of the SC is not restricted to visuomotor behavior (nor is the OT). Rather, it is a multisensory structure that contains visual, auditory, and somatosensory representations, all of which contribute to its role in initiating orientation and localization behaviors that involve multiple sensory organs. Thus, although the visuomotor role in gaze shifts (i.e., movement of the eyes with or without corresponding movement of the head) is well known (Sparks, 1986), it also is involved in orientation of the ears (Stein and Clamann, 1981) and limbs (Stein and Gaither, 1981; Stuphorn et al., 2000). It is best to think of the SC as involved in the transformation of sensory signals (visual, auditory, and somatosensory) into motor commands.

Its ability to represent salient stimuli and engage in sensorimotor transduction so that the organism can initiate rapid orientation to the initiating event is critical to survival. The rapid maturation of the SC compared with that of the cortex reflects its importance in minimizing early ecological vulnerabilities as neonates develop greater independence. This is especially evident in altricial species, whose birth at an early stage of maturation makes observation of neural development far easier than

in their precocial counterparts. However, in the adult stage of both altricial and precocial species, the communication between the cortex and SC reaches its apex, and their complementary and interdependent functions become linked through their reciprocal connections.

Before discussing the developmental features that render the SC capable of performing its adult role or how its inherent plasticity is reflected in changes resulting from sensory experience, trauma, or both, it first is necessary to describe its overall structure and function in the mature brain. A discussion of the major features of this remarkable structure follows.

3.3 GENERAL ANATOMICAL ORGANIZATION OF THE SUPERIOR COLLICULUS

The SC is a laminated structure, composed of seven alternating fibrous and cellular layers (Kanaseki and Sprague, 1974), as shown in the illustration of a cat in Figure 3.1. Operationally, however, it is divided into two broad regions: superficial (I–III) and deep (IV–VII) layers (Edwards, 1980; Harting et al., 1973; Stein, 1984). The former is strictly visual. It receives a substantial direct projection from the retina, as well as many visual projections indirectly from a host of other subcortical and cortical structures (Edwards et al., 1979). Its outputs ascend the neuraxis via the sensory thalamus and are relayed from there to the extrastriate cortex. Its descending projections pass through the deeper layers to target another visual site, the parabigeminal nucleus.

In contrast, the deep layers of the SC contain a much more heterogeneous group of neurons, which includes recipients not only of visual inputs (few come directly from the retina, but many are relayed from other brain regions, especially the cortex) but also of inputs from

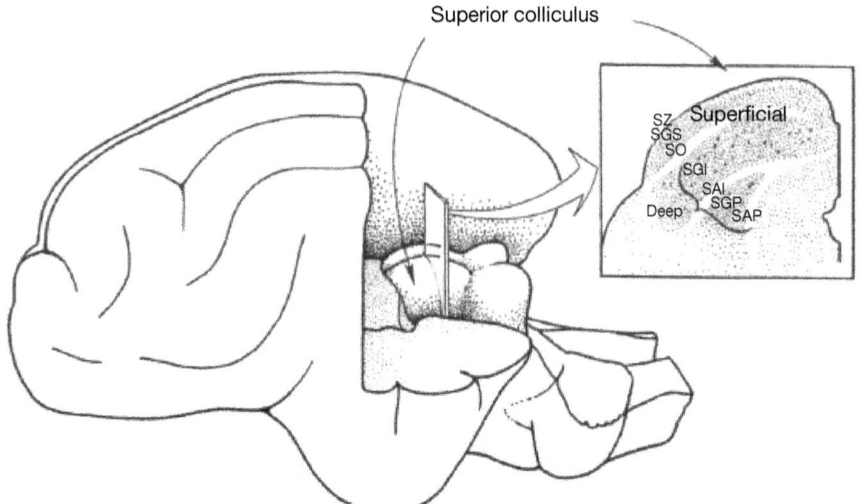

FIGURE 3.1   Lateral view of the cat brain and superior colliculus (SC). A portion of cortex has been removed to reveal the SC and IC (unlabeled). A coronal section of the SC to the right shows its layers: SZ, stratum zonale; SGS, stratum griseum superficiale; SO, stratum opticum; SGI, stratum griseum intermediale; SAI, stratum album intermediale; SGP, stratum griseum profundum; SAP, stratum album profundum. *Reproduced from Stein BE, Meredith MA (1993) The Merging of the Senses. Cambridge, MA: MIT Press.*

the auditory, somatosensory, and motor systems (Stein and Meredith, 1993). In keeping with their role in sensorimotor transformation, the deeper layers send heavy descending projections to the regions of the brainstem and spinal cord that control movements of the eyes, ears, head, and limbs, as well as outputs to nonspecific sensory and motor areas of the thalamus (Edwards, 1980; Edwards and Henkel, 1978; Edwards et al., 1979; Harting, 1977; Harting et al., 1973, 1980; Moschovakis and Karabelas, 1985; Redgrave et al., 1985, 1986a,b; Rhoades et al., 1987; Sommer and Wurtz, 1998, 2004a,b; Stein and Edwards, 1979; Weber et al., 1979). It is the deep layers that are most closely associated with the sensorimotor roles for which the SC is best known; because they contain neurons responsive to multiple sensory modalities, they are referred to as the multisensory layers.

3.4 SPATIAL TOPOGRAPHIES, MULTISENSORY INTEGRATION, AND MOTOR OUTPUT

3.4.1 Visuotopy

Physically, the SC is aligned with the general axis of the brain, so its rostral pole points forward and its caudal pole points backward. Neurons in its superficial visual layers are arranged in a visuotopic (i.e., retinotopic) map of contralateral visual space and the map in the cat SC (e.g., see Feldon et al., 1970) is generally representative of those in animals having forward-facing eyes (see Cynader and Berman, 1972 for a description in monkeys). SC neurons having receptive fields in central visual space are located rostrally in the structure and are the smallest. Thus, this portion of the map has the greatest spatial resolution. Neurons having receptive fields progressively more peripheral (i.e., temporal) in space are located progressively more caudal in the structure and are the largest. This portion of the map has the lowest spatial fidelity. In short, the horizontal meridian from central to peripheral visual space runs roughly rostrocaudal. The vertical meridian is roughly orthogonal to the horizontal meridian, with superior visual space represented medial and inferior space represented lateral. The central visual point (i.e., the fovea in primates and area centralis in carnivores) is the point at which the horizontal and vertical meridians cross.

The central representation of visual space is expanded in the SC. Thus, for example, the central 10° of space occupies more than a third of the tissue devoted to the map in the cat SC and more than half the tissue in monkey SC. In some species, such as the cat, neurons located rostral to the vertical meridian extend the map to represent up to 10° of the ipsilateral hemifield. This nasal representation is essentially nonexistent in monkeys (Cynader and

Berman, 1972), but is much larger in animals such as rats, which have eyes on the sides of their head (e.g., see Siminoff et al., 1966).

A retinotopic organization is also characteristic of the OT, although in some nonmammalian species, the vertical and horizontal meridians are not always as well aligned with the brain's rostrocaudal axis as they are in mammals (e.g., see Gaither and Stein, 1979). The ubiquity of a retinotopy likely represents the ease of using a map to determine the location of a visual event and of transforming the visual cues it provides into motor coordinates for orientation responses. However, this sensorimotor transduction is primarily a function of the deep (i.e., multisensory) layers of the SC, which also contain a map of the visual space.

The retinotopy in the deep layers is similar to that of the overlying superficial layer map; however, the source of visual afferents differs somewhat. Receptive fields are considerably larger than those in the overlying layers, and as a result, there is lower spatial resolution in this visual map (McIlwain, 1975, 1991; Meredith and Stein, 1990). Yet, it has a better representation of the far periphery and also extends a bit further into the ipsilateral hemifield than does the superficial layer map (Meredith and Stein, 1990). Shifts in receptive field size follow the same trend as those in superficial layers, with central visual fields being the smallest and far peripheral receptive fields being the largest. But the visual map in the multisensory layers is only one of three overlapping spatiotopic sensory representations in this region of the structure.

3.4.2 Somatotopy

The somatosensory representation in the multisensory SC is largely of the cutaneous surface and its maplike representation is called 'somatotopic.' Its relationship to the visuotopic representation has also been studied most extensively in cats (see Meredith et al., 1991; Stein et al., 1975; but see also Benedetti and Ferro, 1995; Dräger and Hubel, 1975; McHaffie et al., 1989). Like the visual representation, it is formed from comparatively large receptive fields that are organized into a map of the body, with a geometric expansion of the representation of the face and head. The face representation is made up of the smallest receptive fields, is located rostrally, and roughly overlaps the representation of central visual space. The body and rump are represented more caudal and the limbs extend laterally, so that the upper body space is represented medially to coincide with the representation of upper visual space and the lower body space is represented laterally to coincide with the representation of lower visual space. Given that the SC also receives inputs from pathways carrying information about potentially harmful stimuli, it is not surprising to find that it also has neurons responsive to noxious stimuli (Stein and Dixon, 1978, 1979). These neurons have many of the same properties found in structures better known for dealing with nociceptive information (Larson et al., 1987; Rhoades et al., 1983). Although this representation has been studied only in rodents and is largely restricted to the face and forelimb (McHaffie et al., 1989; see also Auroy et al., 1991), it incorporates the same general topographic features employed by the other sensory maps (Stein and Meredith, 1993).

3.4.3 Audiotopy

Unlike the visual and somatotopic maps, which are formed directly via afferents from different regions of the retina or skin (albeit with some geometrical distortions), the spatiotopic nature of the auditory representation must be derived via neural computation based on comparisons of the timing, intensity, and frequency of the sound signals at the two ears. Although the term 'audiotopic' is not commonly used and hence seems a bit awkward, it is as appropriate as 'retinotopic' and 'somatotopic' in this context. Its properties have also been examined closely in cats and follow the same general organizational features of the maps that have been outlined earlier. Forward auditory space is represented in the rostral aspect of the structure, temporal auditory space in its caudal aspect, and superior–inferior space is laid out along its mediolateral axis (Middlebrooks and Knudsen, 1984; see also Palmer and King, 1982; King and Palmer, 1983).

3.4.4 Multisensory Integration

The alignment of SC sensory maps is evident in the overlap among modality-specific receptive fields of multisensory SC and OT neurons. Indeed, the individual maps are largely a reflection of the receptive field properties of multisensory neurons that, in cats at least, comprise the majority of sensory responsive neurons in the deep layers. Each multisensory neuron has at least two receptive fields, one for each of the sensory modalities to which it responds (e.g., see Stein and Meredith, 1993 for a review). For example, a visual–auditory neuron with a visual receptive field in central visual space will have an auditory receptive field in an overlapping region of central visual space, so that only visual and auditory cues from roughly the same locations will affect that neuron.

A key element of SC function is derived from the ability of multisensory neurons to integrate the influences of the different sensory modalities and the nature of such integration hinges strongly on map alignment. While

much of the information about multisensory integration comes from studies of single neurons in cat SC, similar observations have also been made in the SC and OT of other species (see, e.g., Bell et al., 2001, 2005; Gaither and Stein, 1979; Hartline et al., 1978; King and Palmer, 1978; Stein and Gaither, 1981; Van Wanrooij et al., 2009; Wallace et al., 1996; Zahar et al., 2009). Operationally defined, 'multisensory integration' is the process by which stimuli from different senses combine (i.e., *cross-modal*) to produce a response that differs from those produced by the component stimuli individually. At the level of the single neuron, integration corresponds to *a statistically significant difference between the number of impulses evoked by a cross-modal combination of stimuli and the number evoked by the most effective of these stimuli individually* (see Stein and Stanford, 2008 for discussion). This difference can be manifested as either a response increase or decrease, depending on how the stimuli are configured in space. Cross-modal stimuli that signal a common event (they are in the same place at the same time) impinge on overlapping modality-specific receptive fields, and the integration of their influences yields response enhancement. Conversely, modality-specific cues emanating from disparate locations affect nonoverlapping regions of their respective sensory maps; disparate cues thus fail to produce enhancement and, in some instances, produce response depression (Kadunce et al., 1997). Examples of multisensory integration that yield response enhancement and response depression are shown in Figure 3.2.

Remarkably, convergence of inputs from different senses alone is not sufficient for multisensory integration in the SC. A matching set of converging unisensory inputs from association cortex is required (primarily from the anterior ectosylvian sulcus (AES); Alvarado et al., 2009; Fuentes-Santamaria et al., 2008a, 2009; Stein et al., 1983 but also inputs from the rostral–lateral suprasylvian sulcus (rLS) play a role – see Jiang et al., 2001). Without these cortical inputs, SC neurons respond to individual cues from different senses but do not integrate them to yield (Alvarado et al., 2007, 2008, 2009; Jiang et al., 2001; Wallace and Stein, 1994). Rather, the neural response to cross-modal stimulation is no greater than that to the best of the stimulus components alone, a finding that is mirrored by an absence of the multisensory advantages that are typical of SC-mediated behaviors (Jiang et al., 2002, 2006, 2007; Wilkinson et al., 1996). The requirement for cortical input has implications for the development of SC multisensory integration and the behaviors it supports, as will be discussed later in this chapter.

Insofar as the magnitude of activity within SC sensory maps corresponds to the physiological salience of an external stimulus, and thus the likelihood that it will generate a motor response, the implications of multisensory integration for behavior are quite clear. Enhanced

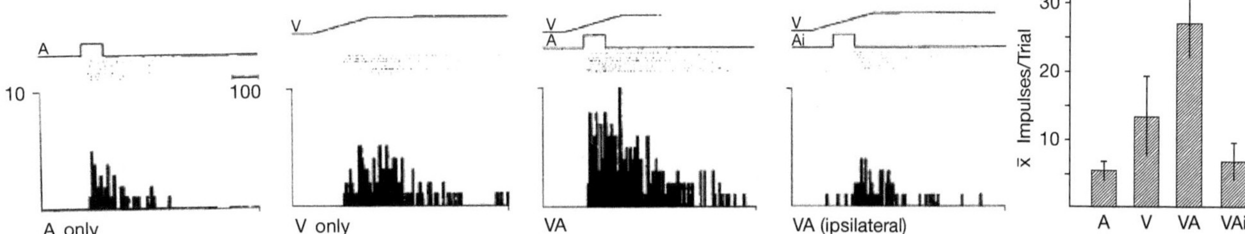

FIGURE 3.2   Multisensory enhancement. Left: A square-wave broadband auditory stimulus (A, 1st panel) and a moving visual stimulus (ramp labeled V, 2nd panel) evoked unisensory responses from this neuron (illustrated below the stimulus by rasters and peristimulus time histograms). Each dot in the raster represents a single impulse; each row, a single trial. Trials are ordered from bottom to top. The 3rd panel shows the response to auditory and visual stimulus presentation at the same time and location, which is much more robust than either unisensory response. Yet, when the auditory stimulus was moved into ipsilateral auditory space (Ai) and out of the receptive field, its combination with the visual stimulus elicited fewer impulses than did the visual stimulus individually. This 'response depression' is illustrated within the 4th panel. Right: The mean number of impulses/trial elicited by each of the four stimulus configurations. Note the difference between multisensory enhancement (VA) and multisensory depression (VAi). *Reproduced from Stein and Rowland, 2011, Progress in Brain Research.*

activity for spatially concordant cues corresponds to an increased likelihood of stimulus detection and an associated motor response to orient to that stimulus. In contrast, the multisensory depression that results from stimuli at competing locations would have an opposing effect (Calvert et al., 2004; Gillmeister and Eimer, 2007; Spence et al., 2004; Stein and Stanford, 2008). In either case, sensory-related activation must be transformed into motor commands, which, like their sensory counterparts, assume the form of a topographically organized map.

3.4.5 Mototopic Representation

Many of the neurons in the deep/multisensory layers are output neurons that project to one or more brainstem and spinal cord regions responsible for moving the eyes, ears, whiskers, head, and limbs (Coulter et al., 1979; Edwards and Henkel, 1978; Grantyn and Grantyn, 1982; Harting, 1977; Holcombe and Hall, 1981a,b; Huerta and Harting, 1982a,b; Moschovakis et al., 1998; Weber et al., 1979; see Gandhi and Katnani, 2011; Hall and Moschovakis, 2004; Sparks, 1986; Sparks and Mays, 1990 for reviews). The term 'mototopic,' like 'audiotopic,' is not in common usage but, once again, seems equally appropriate in this context, given the more commonly used terms 'visuotopic' and 'somatotopic.' By far, the most is known about the motor topography for producing gaze shifts and for movements of eyes (or the eyes and head) to place stimuli of interest into the line of sight. Accordingly, the motor map for gaze shifts is in register with the visual, auditory, and somatosensory topographies; therefore, the sensory evoked activity represents the distance and direction of a stimulus from the current line of sight, and the motor activity at that site represents a command for shifting gaze to the corresponding distance and in the corresponding direction.

The SC motor map is two-dimensional, with gaze amplitude (from small to large) represented along its rostrocaudal axis and gaze direction (from upward to downward) represented along its mediolateral axis (du Lac and Knudsen, 1990; Goldberg and Wurtz, 1972; Pare et al., 1994; Robinson, 1972; Schiller and Stryker, 1972; Wurtz and Goldberg, 1972). In a manner analogous to sensory receptive fields, constituent neurons of the SC motor map have *movement fields*, such that they respond most vigorously in association with gaze shifts within a particular range of amplitude and direction, and gaze shifts are produced consequent to the activity of their premotor activity. Figure 3.3 illustrates a typical SC movement field for a neuron recorded from the right SC of a monkey. This neuron discharges most vigorously (peak of the 3D plot) for saccades having a direction of approximately 190° (leftward and slightly downward) and having an amplitude of approximately 7° of visual angle (i.e., deviation of eye from the straight ahead position). Considered in the context of the motor map of the right SC, this neuron would be located toward the rostral pole (small movements) and slightly lateral (downward). Thus, like the SC sensory maps, which constitute 'place codes' for stimulus location in sensory space, the SC motor map is a 'place code' for movement vector, such that the locus of activation determines the amplitude and direction of an impending movement.

3.4.6 Maintaining Sensory and Motor Map Alignment

The overlap among the sensory and motor maps is unlikely to be serendipitous, as it appears to be the most straightforward way to coordinate the cross-modal sensory cues and movements of the various sensory organs toward a salient event. Nevertheless, given the ability to move the different sensory organs independently, retaining the registry of their midbrain maps presents

(Metzger et al., 2004). Maintaining intermap registry ensures that any visual, auditory, or somatosensory cue (and any combination of them) derived from the same location in space activates neurons in approximately the same SC location. This, in turn, accesses the same point in the motor map to produce coordinated orientation of all the sensory organs toward the location of the initiating event. This not only helps determine the nature of that event but also puts the animal in an appropriate position to evaluate it.

3.5 THE MATURATION OF THE SUPERIOR COLLICULUS

3.5.1 The Neonate

In altricial species, such as cats, the ability of the SC to use sensory information to initiate overt behaviors is poorly developed at birth. Unlike primates, ungulates, and other precocial species, the newborn carnivore (rodent, lagomorph, marsupial, etc.) is poorly equipped to deal with its sensory environment. Its eyes are closed, there is a vascular network around the lens that impedes the transmission of light (Bonds and Freeman, 1978; Thorn et al., 1976), and it is functionally blind. Similarly, its ear canals are still sealed, and it is deaf. The only sensory inputs that activate SC neurons at this time are tactile. These inputs are already functional prenatally but are only weakly effective and stay that way for some time after birth (Stein et al., 1973a,b). The newborn carnivore's motor capabilities are also poorly developed, and it depends heavily on its mother for things like warmth and protection, and even the initiation of feeding (Larson and Stein, 1984; Rosenblatt, 1971). The immaturity of the newborn cat's SC makes it a good model for exploring how sensory responses develop, how the different sensory representations become established, and how multisensory integration develops so that SC neurons can use the available sensory information synergistically to optimize SC-mediated behavior.

3.5.2 Sensory Chronology

As noted, somatosensory responsiveness in some cat SC neurons precedes birth and provides the structure with its first source of information about the world. Auditory responses begin to appear in some SC neurons at 5 days postnatally, and neurons develop visual responsiveness last (Stein et al., 1973a,b). Interestingly, visual responses develop in two sequential phases, first appearing from top to bottom in the most superficial (i.e., visual) layers of the structure and then in a similar manner in neurons in its multisensory (deeper) layers. These phases are widely separated in time, with neurons in the

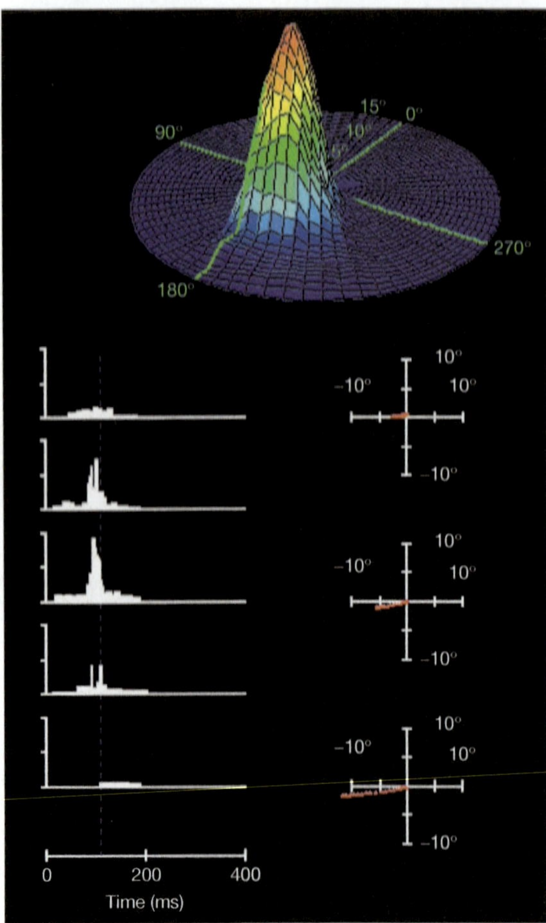

FIGURE 3.3   Superior colliculus movement field. Neurons in the SC have movement fields and discharge most for movements within particular ranges of amplitude and direction. The 3D polar plot (above) illustrates the movement field of an SC neuron that discharges for relatively small and slightly downward saccadic eye movements. Below left, the premotor activity profiles are shown aligned on the onsets of saccades (blue dotted line) of five different amplitudes (top to bottom) and representing a 'slice' through the peak of the movement field along the line of best direction. Representative saccade trajectories are shown to the right of the activity profiles. Note that maximum activity corresponds to a slightly downward saccade having amplitude corresponding to roughly 7° of visual angle. *Reproduced from Stanford and Sparks, unpublished observations.*

a nontrivial problem. The problem is partially solved by the nonstatic nature of these maps. Studies in various species show that shifting the eyes voluntarily or inducing long-term shifts in the optical axis via surgical or prismatic means produces a corresponding shift in the SC auditory map (Brainard and Knudsen, 1998; Hartline et al., 1995; Jay and Sparks, 1984; King et al., 1988; Peck et al., 1995). A similar shift may also be initiated in the somatosensory map (Groh and Sparks, 1996). These observations suggest that the different senses are linked to an oculocentric coordinate system, though compensatory shifts in the nonvisual maps do not completely compensate for large ocular misalignments most superficial strata of the superficial layers beginning to respond to visual stimuli at about 6 postnatal days, and those in the multisensory layers not showing responsiveness to visual stimuli for several weeks (Kao et al., 1994).

3.5.2.1 Retinotectal Inputs and the Development of a Superficial Layer Visuotopy

Because of the visually dominant role of the SC, the development of its visual inputs (especially those coming directly from the retina) has received a great deal of attention. Once again, cats have been one of the primary sources of this information, along with rodents, opossums, and monkeys. Given the functional immaturity of the newborn cat's SC (there is no visual responsiveness yet), it is surprising to note the advanced state of its retinotectal topography during late embryonic development (Graybiel, 1975; Williams and Chalupa, 1982). Although, at this point, the retinotectal projections appear to be restricted to the superficial layers of the SC, segregation of inputs from the contralateral and ipsilateral eyes is already apparent several days before birth (Figure 3.4(b); embryonic day 56), and by embryonic day 61 (Figure 3.4(c)) further refinement leads to an almost adult-like patterning of inputs. This advanced pattern of retinal projections is achieved by sculpting it from a far more widespread projection that is evident at approximately embryonic day 38 (Figure 3.4(a)). At that point, the projections from the two retinas are intermingled and distributed across the entire rostrocaudal and mediolateral extent of the SC. This pattern is

progressively altered so that by embryonic day 61 (4–7 days before parturition) the pattern of ipsilateral and contralateral retinotectal inputs closely resembles the pattern characteristic of the adult. Although the timing is different, similar developmental processes have been noted in monkeys (Rakic, 1977), hamsters (Frost et al., 1979), rats (Land and Lund, 1979), and opossums (Cavalcante and Rocha-Miranda, 1978).

Data from a variety of nonmammalian and mammalian models have suggested that chemoaffinity cues provide the basis for the fundamental topographical order of retinotectal projections and that this coarse representation is subsequently refined by activity-dependent mechanisms (Drescher et al., 1997; Fraser, 1992; Ruthazer and Cline, 2004; Ruthazer et al., 2003; Sperry, 1963; Walter et al., 1987). There is good reason to assume that similar processes are involved in determining the adult-like pattern of sensory afferents to the multisensory layers of the SC, but observations of the development of its retinotectal projections are hampered by weak input, and less attention has been directed toward the maturation of nonvisual inputs. Nevertheless, it has been noted that somatosensory tectopetal afferents are already widespread in the multisensory layers of the SC at birth, and they too appear to have an adult-like configuration (McHaffie et al., 1986, 1988).

The apparent maturity of the representation of tectopetal sensory afferents (inputs) at birth stands in contrast to the significant immaturity in at least one of its tectofugal efferent (output) sensory projections, namely, those to the lateral geniculate nucleus (LGN). This transient

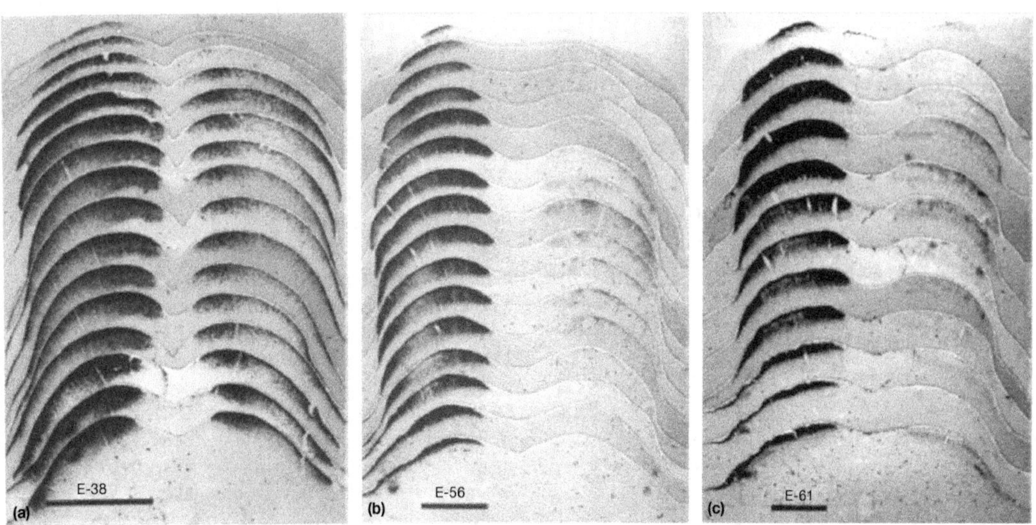

FIGURE 3.4 Retinotectal projection patterns to the superficial superior colliculus are highly developed in cats before birth. Shown are photographic illustrations of the distribution of a tracer (horseradish peroxidase shown as dark regions) from one eye to both colliculi approximately 17 ((a), at E38) and 9 days before birth ((b), at E56). Sections are ordered from rostral (bottom) to caudal (top), with the contralateral SC on the left. The calibration bar is 1 mm. Note that the projection has largely withdrawn from the ipsilateral SC and changed its distribution in the contralateral SC by E56. Several days later, ((c), at E61), an almost adult-like pattern input is apparent. *Adapted from Williams RW, Chalupa LM (1982) Prenatal development of retinocollicular projections in the cat: An anterograde tracer transport study.* Journal of Neuroscience 2:604–622.

tectogeniculate projection is topographic and is part of the ascending wing of the SC's visual projection that eventually reaches the cortex via polysynaptic pathways. In adults, the tectogeniculate projection is confined to the ventral C layers, but in neonates it extends across all layers of the LGN as well as into the medial interlaminar nucleus. It is retained for approximately 3 postnatal weeks, long after visual function has been initiated, and is eliminated by mechanisms that are not yet fully understood (Stein et al., 1985). Its immaturity contrasts with the more rapidly maturing tectofugal pathways involved in motor functions (Stein et al., 1982, 1984).

Despite the adult-like pattern of retinotectal projections at birth, the visual system is not yet functional. Visual responsiveness begins in superficial layer SC neurons at 6 days of age. At this time, the eyelids are still closed (they open naturally at 7–11 days postnatal) and visually responsive neurons are rare, which are clustered together between ineffective loci (Kao et al., 1994; Stein et al., 1973a). This reflects a random onset of visual activity via random functional coupling of retinotectal afferents and their SC target neurons and/or the random openings in the vascular networks around the lens that begin at about the time of eye opening (Freeman and Lai, 1978). Most active sites are located in the middle portion of the structure and in its most superficial aspect. Nevertheless, the general topographic organization established by the afferent projections is apparent at this time (Kao et al., 1994; Figure 3.5).

Curiously, the initial visual activity is restricted not only across the horizontal aspect of the SC, but also, as noted earlier, in its vertical aspect. Neurons in the most superficial portion of the SC, where inputs from retinal W cells terminate, begin responding earliest. Neurons deeper in the superficial layers, where Y-cell inputs dominate, develop later, and visual responsiveness in the subjacent multisensory layers develops last (Kao et al., 1994; Wallace and Stein, 1997). The bases for this particular pattern of functional development remain obscure, but it occurs at the same time that the number of active loci across the horizontal aspect of the structure increases to yield a continuous retinotopy that underlies its visuotopic organization (Figure 3.6).

These early responsive neurons are functionally immature and have many properties that typify functionally immature neurons elsewhere in the nervous system. Their receptive fields are very large and require either long-duration flashed stationary targets or very slowly moving stimuli for activation. They fatigue readily with repeated stimulation, require very long interstimulus intervals to respond to sequential stimuli, and have very long latencies (Kao et al., 1994; Stein et al., 1973b). They also lack binocularity and direction selectivity, properties that characterize these neurons when

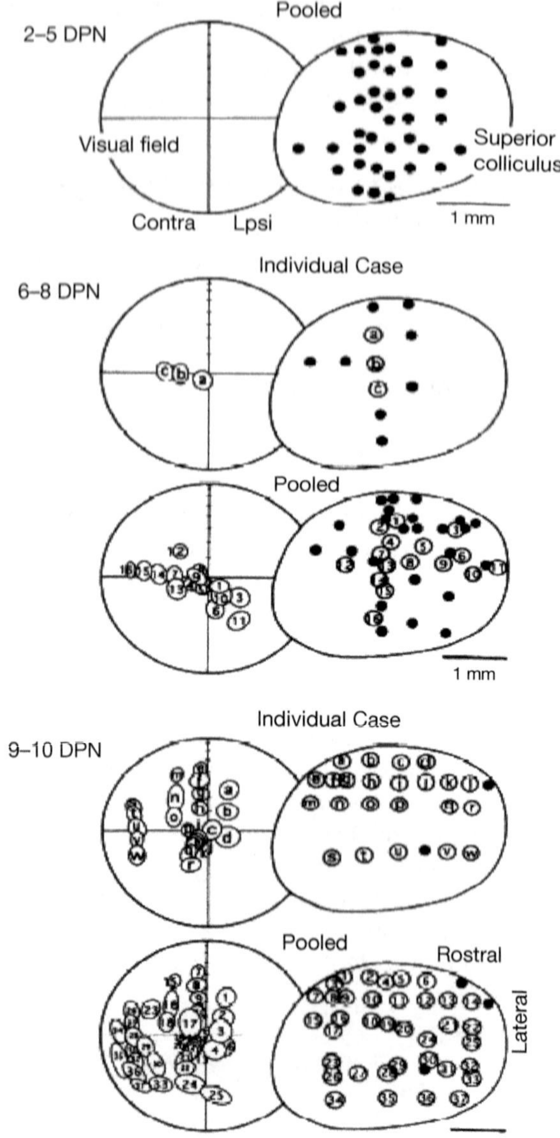

FIGURE 3.5 The maturation of the visuotopic map in the superficial superior colliculus. Data from three age groups are shown. Each contains a diagram of the visual field on the left and a schematic of the dorsal surface of the SC on the right. Circles on the SC represent electrode penetrations (filled, no visual activity; open, visual activity). Correspondence between the electrode penetrations and visual receptive fields are shown by numbers and letters (only the 'best area' of a receptive field was mapped). No visual activity was encountered before 6 days postnatal (dpn) but was already represented in a maplike pattern when first encountered. Adjacent electrode penetrations in the SC yielded visual activity at adjacent sites in visual space. By 9–10 dpn, most SC locations had become responsive to visual stimulation. *Reproduced from Kao CQ, Stein BE, Coulter DA (1994) Postnatal development of excitatory synaptic function in deep layers of SC. Abstracts – Society for Neuroscience 20: 1186.*

mature and are believed to facilitate orientation to moving targets. These properties develop over a 6- to 8-week period. It is not clear what, if any, visual behaviors these early responding neurons can support, as overt visual

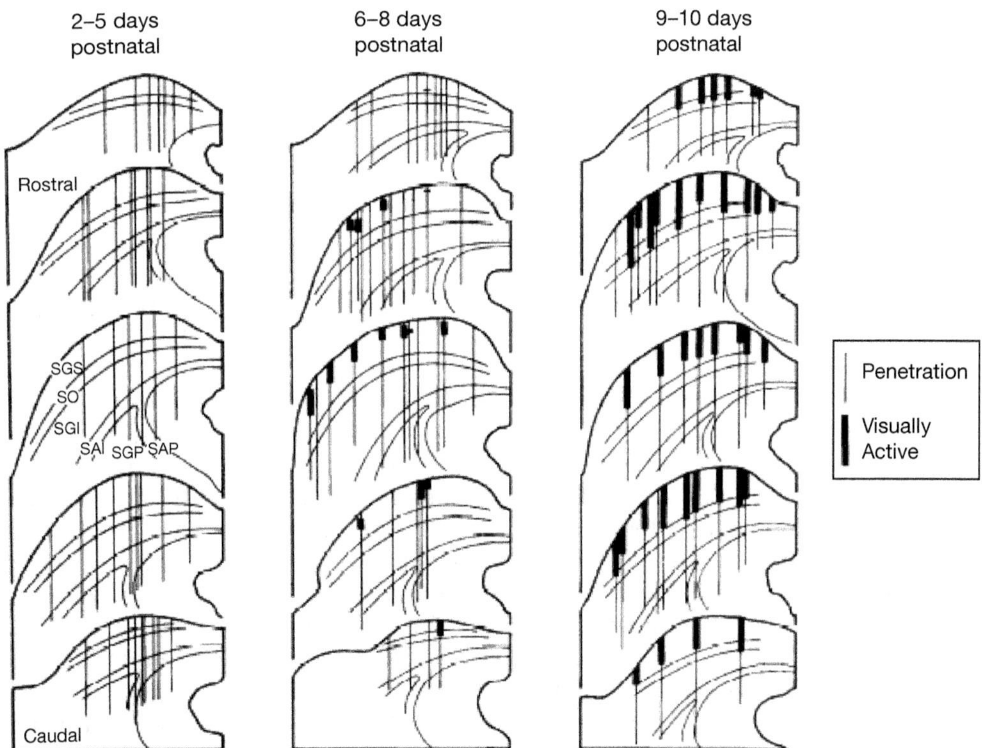

FIGURE 3.6 During maturation, superior colliculus visual activity progresses from the top to the bottom of the superficial layers. Vertical lines illustrate electrode penetrations through coronal sections of the SC in each age group. Sections are arranged rostrocaudally. Each region of visual activity is represented by a thick vertical black line spanning the distance between the first and last location of visually active neurons. Note that the visually active spans increase with age so that by 9–10 dpn, visual activity is evident throughout the depths of the superficial layers. No deep-layer visual activity was present at this time. *Reproduced from Kao CQ, Stein BE, Coulter DA (1994) Postnatal development of excitatory synaptic function in deep layers of SC. Abstracts – Society for Neuroscience 20: 1186.*

function is not observed in cats until 2–3 weeks after birth, by which point they have already matured considerably (Fox et al., 1978; Stein et al., 1973a). Nevertheless, the physiological maturation of the receptive field properties of these superficial layer neurons takes 2 months or more to reach their adult status. This reflects, in part, the physiological maturation of visual tectopetal afferents, especially those from the cortex (e.g., see Stein and Gallagher, 1981).

Although sensory experience is critical for the formation of many neural properties in the central nervous system, the fundamental features of SC visuotopy and the responsiveness of its constituent neurons appear to progress in a manner independent of visual experience. Superficial layer and deep (multisensory) layer neurons in dark-reared cats show robust visual responses and adult-like resistance to fatigue with repeated stimulation, although their receptive fields remain comparatively large and their more sophisticated properties develop more gradually. Similarly, in the newborn monkey, the ocular properties are far more mature than those of the cat, and SC neurons already show a well-ordered and continuous superficial layer visuotopy. Although the receptive fields of its neurons are also larger than

those in the adult and their visual latencies are longer, the neurons respond robustly to visual stimuli and already have many adult-like response properties (Wallace et al., 1997). The maturational differences in the newborn monkey and cat SC likely reflect both the shorter gestational period in the cat and its higher dependence on corticotectal inputs for construction of its neuronal response properties (Stein, 1984).

3.5.2.2 Development of Deep Layer Sensory Topographies

Attention to the maturation of neurons in the deep layers has been focused on their functional properties rather than their topographies. Thus, much of what is known about this feature of their development has to be inferred. The somatosensory and auditory receptive fields of neonatal neurons are, like their visual counterparts, extremely large. Neonatal somatosensory receptive fields cover much of the contralateral body and early auditory receptive fields are 'omnidirectional,' having receptive fields that encompass the whole of the contralateral auditory space. They gradually shrink in size over the first few months of life, and in the absence of evidence that neurons have receptive fields that are 'mislocated,' it is assumed that the resolution of their maps gradually increases as individual receptive fields contract. The functional changes that ensue in these layers as a consequence of development are discussed in the section on the maturation of multisensory integration. Examples of receptive field development are also provided.

3.5.3 The Development of Multisensory Neurons

Although a great deal of effort has been expended on understanding the maturation of the superficial layer visuotopy to understand the maturation of SC-mediated orientation behaviors, as noted earlier, it is actually the deep (multisensory) layer visual responsiveness that is most closely linked to this function. Also, these layers are slower to develop. Because the maturation of these neurons has been most closely studied in cats, unless otherwise stated, the descriptions in what follows relate to this species.

Because neurons in the multisensory layers become responsive to visual cues only after 3 postnatal weeks (Kao et al., 1994; Wallace and Stein, 1997), it is obvious that neither the weak direct retinal projection to these layers (Beckstead and Frankfurter, 1983; Berson and McIlwain, 1982) nor the relay of visual input from active neurons in the superficial layers (e.g., Behan and Appell, 1992; Grantyn et al., 1984; Moschovakis and Karabelas, 1985) are capable of activating these neurons before this time. Yet somatosensory- and auditory-evoked neuronal activity is already apparent, though the responsive neurons also have the large receptive fields and tendency toward fatigue that are characteristic of neonatal neurons. This developmental chronology (somatosensory first, auditory second, visual last) parallels the development of sensory-evoked orientation behaviors (Fox et al., 1978; Norton, 1974; Villablanca and Olmstead, 1979).

At this time, there is a gradual increase in the number of SC neurons that can respond to multiple sensory inputs. Thus, each of the different possible convergence patterns of responsiveness to visual, auditory, and/or somatosensory inputs becomes evident. One of the hallmark features of multisensory SC neurons, however, is their ability to use information from the different senses synergistically. This process of integrating cross-modal inputs markedly facilitates SC-mediated behavior and begins at about the fourth postnatal week.

3.5.4 Superficial Layer and Deep (Multisensory) Layer Maturational Delay

As noted earlier, visual responsiveness occurs comparatively early in superficial layer neurons, beginning before the end of the first week, and comparatively late in the multisensory layers, where it begins in the third or fourth postnatal week. This maturational delay relative to superficial layer visual activity underscores the significant maturational difference between the sensory-specific properties of superficial-layer neurons and their multisensory counterparts in the deep layers. Synthesizing information from multiple senses is surely a more complex process than responding to any one of them individually and therefore requires greater maturational time. These superficial–deep-layer differences also extend to the functional roles to which their information processing contributes. Although both neuronal populations probably share functions related to perception and overt behavior, superficial layer visual neurons are believed to contribute more to the former and deep-layer multisensory neurons to the latter.

Developing the ability to integrate cross-modal inputs is a multistep process. Neurons first develop responsiveness to a single sensory input, then to at least two different (i.e., cross-modal) inputs, and can finally develop the ability to integrate the information carried in multiple sensory channels.

The maturation of the various possible multisensory convergence patterns follows closely the chronology of unisensory maturation. Thus, somatosensory–auditory neurons are the first multisensory neurons to appear. They become evident 10–12 days post parturition, and visual–nonvisual neurons become evident at approximately the third postnatal week, as soon as visual responsiveness begins in the multisensory layers. Yet, it still takes many weeks of maturation before the adult-like incidence of the various modality-convergence patterns is reached and the adult-like incidence of neurons capable of multisensory integration is achieved. As might be suspected from their unisensory counterparts, the receptive fields of the initial multisensory neurons are very large. They contract over several months, thereby progressively increasing the spatial resolution of the individual sensory maps to which they contribute and enhancing their spatial register with one another (Figure 3.7). The changes in receptive field size are accompanied by an increase in response vigor and reliability, as well as by a decrease in response latency. These functional changes reflect a combination of developmental factors taking place in afferent systems, as well as within the internal circuitry of the SC.

3.5.5 The Development of Multisensory Integration

The ability to integrate cross-modal inputs is delayed until at least 1 month after birth (Wallace and Stein, 1997), and before it develops, the cross-modal responses of multisensory neurons look like the responses elicited

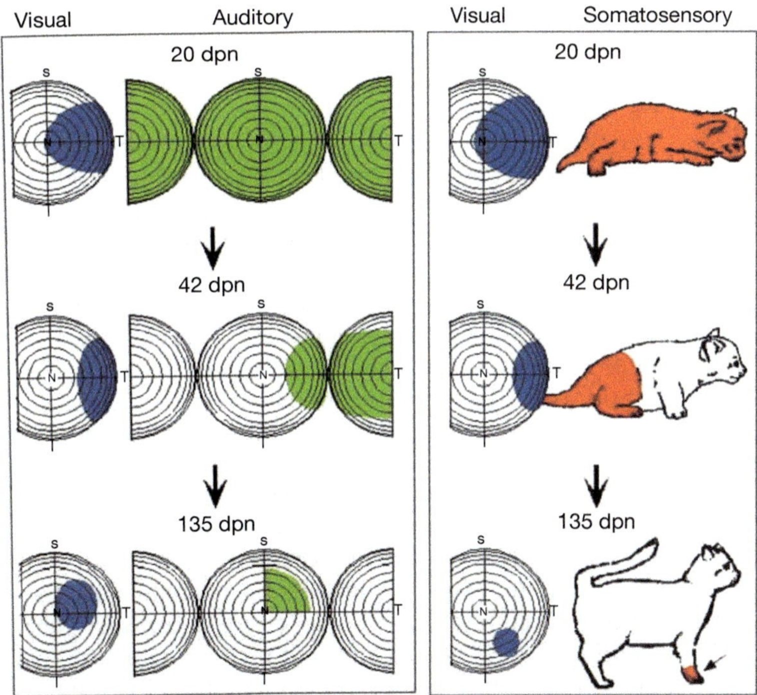

FIGURE 3.7 The maturation of multisensory receptive fields. Shown are exemplar SC receptive fields from visual–auditory (left) and visual–somatosensory (right) neurons that were mapped in animals aged 22–135 dpn. Note that receptive fields become progressively smaller at older ages, increasing their spatial resolution and their register with one another. *Adapted from Wallace MT and Stein BE (1997) Development of multisensory neurons and multisensory integration in cat superior colliculus.* Journal of Neuroscience 17: 2429–2444 *(see also Stein and Rowland, 2011).*

by one (the more effective) of the component stimuli alone. At this age, however, some neurons respond to cross-modal stimuli with responses that significantly exceed the presumptive unisensory responses, but such neurons are rare at this time. The incidence of these integrative neurons progressively increases with age, but the mature condition is not reached until the animal is several months old and has had a great deal of sensory experience.

The absence of an ability to integrate cross-modal inputs in the neonatal cat's SC is not due to the general immaturity of these neurons. The macaque monkey is born much later in development than the cat. Its eyes and ears are open at birth and it sees and hears quite well, and unlike the cat, it already has multisensory SC neurons in its SC. But they cannot integrate their cross-modal inputs to produce response enhancement (Wallace and Stein, 2001) and, therefore, respond to these stimuli very differently than do their adult counterparts (Wallace et al., 1996). The likely reason is that they have not yet had the necessary sensory experience. Apparently, postnatal sensory experience is not critical for the appearance of multisensory neurons, but these observations strongly suggest that it is necessary for them to integrate their different sensory inputs. Observations from experiments with human subjects are consistent with this hypothesis (e.g., Gori et al., 2008; Neil et al., 2006; Putzar et al., 2007), though critical direct observations showing that the newborn cannot yet use cross-sensory cues synergistically are not yet available. Nevertheless, newborn human infants

are capable of engaging in a host of multisensory tasks, the best known of which is cross-modal matching (see Stein et al., 2010 for further discussion).

3.5.6 The Impact of Sensory Experience on the Maturation of Multisensory Integration

As noted earlier, when discussing multisensory integration, inputs from association cortex (e.g., primarily from the anterior ectosylvian sulcus, or AES, in cats) have been found to be essential for SC multisensory integration. Thus, one might expect that the functional coupling of this input with multisensory SC neurons is a necessary precondition for the development of SC multisensory integration, and this appears to be the case. Anatomically, the corticotectal projections from AES have already grown into the multisensory layers of the SC prior to birth and long before SC neurons are responsive to multiple sensory inputs (McHaffie et al., 1988). However, they are unlikely to be functional at this time. Their critical contribution to multisensory integration becomes obvious as soon as a neuron in the SC exhibits this capability. As noted earlier, this first happens for some rare neurons at about 1 month of age. Deactivating AES at this stage eliminates that capacity as effectively as it does in adulthood. The neurons' responses to the cross-modal stimulus combination are now no longer significantly different from those to the most effective component stimulus (Wallace and Stein, 2000; Figure 3.8).

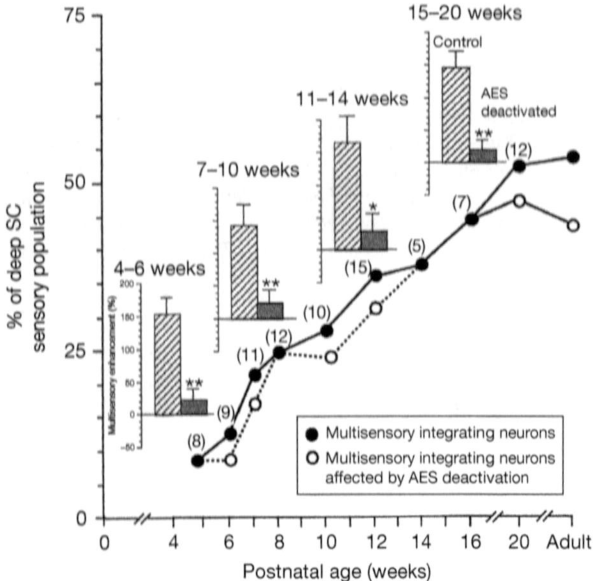

FIGURE 3.8　The maturation of multisensory integration parallels the functional coupling of AES-SC projections. The developmental increase in the incidence of SC neurons capable of multisensory integration is paralleled by the effectiveness of AES deactivation in blocking the expression of this capability. Nearly all SC neurons lost this ability during AES deactivation, regardless of age (the number of neurons tested at every time point is shown in parentheses). The small proportion of SC neurons whose multisensory integration capability was unaffected by AES deactivation likely depended on an adjacent area of association cortex (i.e., the rostrolateral suprasylvian cortex, see Jiang W, Wallace MT, Jiang H, Vaughan JW, Stein BE (2001) Two cortical areas mediate multisensory integration in superior colliculus neurons. *Journal of Neurophysiology* 85: 506–522). *Reproduced from Wallace MT and Stein BE (2000) Onset of cross-modal synthesis in the neonatal superior colliculus is gated by the development of cortical influences.* Journal of Neurophysiology *83:3578–3582.*

Although the association cortex component of the SC circuit appears to be critical for multisensory integration, it is not sufficient. Appropriate sensory experience is also a key factor.

During this early phase of life, the brain is learning about external events and the statistical relationships among stimuli that are derived from the same event. Cross-modal stimuli that are produced by the same event are temporally and spatially coincident, or at least proximate in space and time. These relationships must be learned (e.g., via Hebbian learning rules) and somehow represented in the underlying circuitry responsible for multisensory integration. Given the arbitrary nature of many cross-modal relationships, it is difficult to conceive of an effective scheme for incorporating this information without actual experience. Experience with cross-modal stimuli establishes a principled way of interpreting and interacting with external events so that only a select group of cross-modal stimuli will produce multisensory integration. For example, the brain learns to expect that some visual and auditory cues are linked to the same event, based on their location and timing. This information is used to establish principles for categorizing cross-modal stimuli derived from the same, or different, events.

To test the hypothesis that sensory experience plays a key role in the development of multisensory integration capabilities, cats were raised in darkness in order to preclude visual–nonvisual experiences. Normally, the physiology of SC multisensory integration is reached at 3–4 months of age (Wallace and Stein, 1997), and these dark-reared animals were not studied until at least 6 months of age. Visual experience was not essential for the appearance of visually responsive neurons, and such neurons were common in these dark-reared animals, as well as in newborn monkeys (Wallace and Stein, 2001). Each of the modality-convergence patterns was also well represented among the neurons encountered (Wallace et al., 2004). However, their receptive fields were still quite large and more typical of the neonate than the mature animal (Figure 3.9). The failure to contract their receptive fields was indicative of their physiological immaturity and this was most apparent in their inability to integrate cross-modal cues. As shown in Figure 3.9, their multisensory response to a cross-modal visual–nonvisual stimulus was approximately equal to their response to the visual component stimulus alone. Multisensory responses that approximate the response to the most effective component stimulus are typical of neonatal neurons that have not yet developed their capacity for multisensory integration and of adult multisensory neurons deprived of the critical input from association cortex.

Human subjects who have had their early vision compromised by congenital cataracts also show persistent visual–nonvisual integration deficits even many years after the cataracts have been removed (Putzar et al., 2007). These observations support the hypothesis that cross-modal experience is a critical factor in the maturation of the capacity to integrate multisensory cues (Figure 3.10), but they do not provide information about whether the nature of that early experience determines the principles that govern this process. To test this possibility, animals were reared from birth to 6 months of age in a dark room in which pairs of visual and auditory stimuli were periodically presented, but these stimuli were always spatially disparate (Wallace and Stein, 2007). The properties of their multisensory SC neurons were then assessed.

Once again, visually responsive neurons were common, as were visual–nonvisual neurons. Yet, the majority of visual–auditory neurons appeared to have been unaffected by the experience with the disparate visual–auditory cues and had properties characteristic of neonates. Their receptive fields were very large and were unable to integrate these cross-modal cues. But there

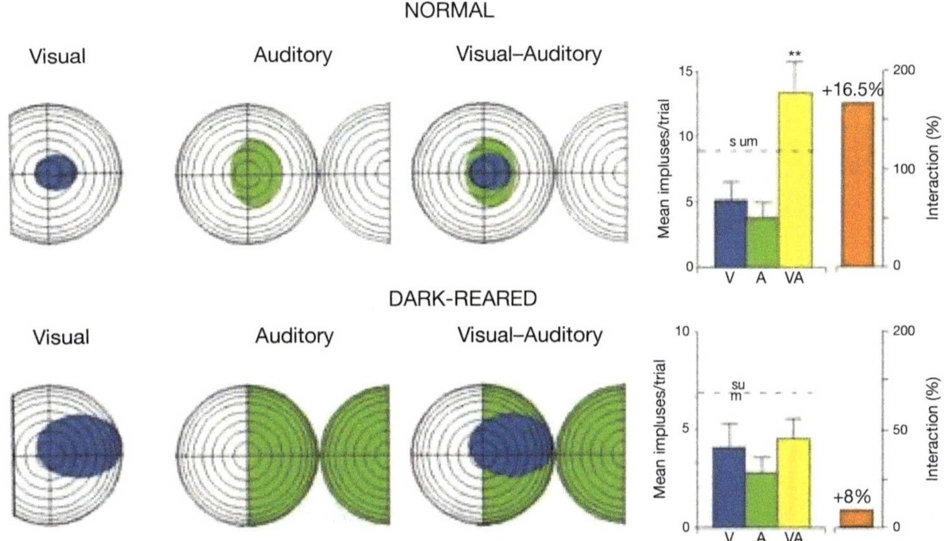

FIGURE 3.9   Multisensory neurons have large receptive fields and lack multisensory integration capabilities after dark rearing. The visual and auditory receptive fields and the multisensory responses of a typical normal SC multisensory neuron are shown at the top. The bar graph to the right shows the characteristic enhanced response to the cross-modal pair of stimuli in spatial and temporal register. In contrast, the neuron below, from the SC of a dark-reared animal, has much larger receptive fields and shows no evidence of multisensory response enhancement to the cross-modal stimulus. V, visual stimulus; A, auditory stimulus; VA, cross-modal stimulus. *Adapted from Wallace MT, Perrault TJ, Jr., Hairston WD, Stein BE (2004) Visual experience is necessary for the development of multisensory integration 1.* Journal of Neuroscience 24: 9580–9584.

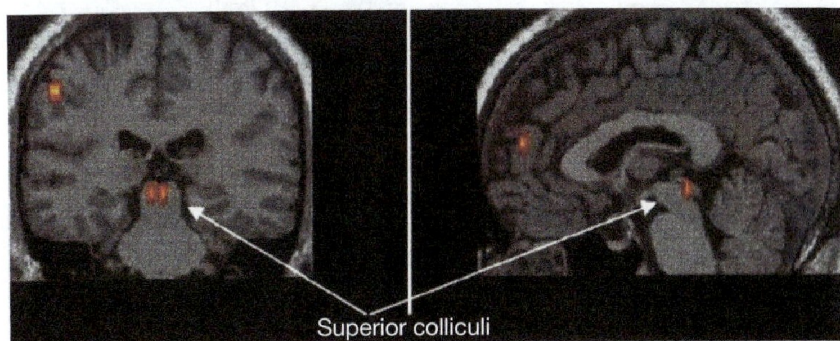

FIGURE 3.10   Multisensory enhancement is evident in the human superior colliculus using fMRI. Coronal (left) and sagittal (right) sections reveal high degrees of enhanced SC multisensory activity (red indicates elevated BOLD responses). Elevated activity is in response to visual–auditory as opposed to the best of these stimuli (i.e., visual) individually. *Adapted from Calvert GA, Hansen PC, Iversen SD, Brammer MJ (2001) Detection of audio-visual integration sites in humans by application of electrophysiological criteria to the BOLD effect.* Neuroimage 14: 427–438.

was a substantial minority of such neurons that did appear to have incorporated the visual–auditory experience. Their receptive fields were somewhat smaller (but still quite large) and were in poor spatial register with one another. Some neurons had visual and auditory receptive fields that were elongated along their horizontal axes and had only small portions overlapping one another. In other cases, there was no receptive field overlap, a configuration that is highly unusual in normal animals, but one that was consistent with the early visual–auditory experience of these animals. Most important in the present context was that they could integrate visual–auditory stimuli. But the cross-modal stimuli had to be spatially disparate in order to fall within their respective receptive fields simultaneously. An example of such a neuron is shown in Figure 3.11.

Collectively, the data reveal that experience is essential for the development of multisensory integration and that the nature of the experience directs the formation of the underlying neural circuit through which this integration is achieved. Although the specifics of that circuit remain to be fully explored (e.g., see Fuentes-Santamaria et al., 2006, 2008a,b, 2009), the cortex is known to play a critical role (e.g., see Alvarado et al., 2009; Jiang et al., 2001; Stein, 2005; Wallace and Stein, 1994) and its ablation early in life precludes the maturation of SC multisensory integration (Jiang et al., 2006, 2007). It appears that early experience is essential for the brain to learn the statistics of cross-modal events, and there is good reason to suspect that this experience exerts its critical impact on the AES-SC (anterior ectosylvian sulcus-superior colliculus) projection.

This possibility was explored by Rowland et al. (2005) and Stein et al. (2008), who deactivated AES and adjacent association cortex ipsilaterally during the period in which SC multisensory integration normally develops. This deprived the cortex of cross-modal experience but did not compromise the responsiveness of SC

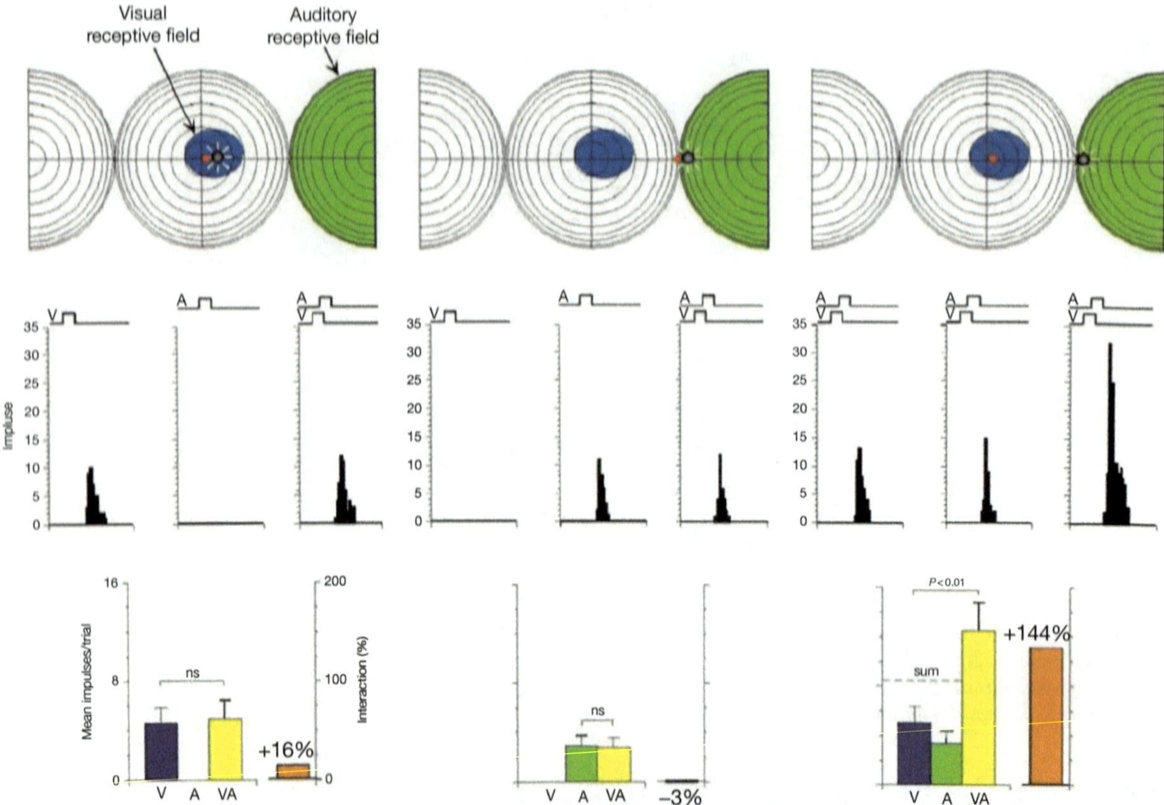

FIGURE 3.11　Early experience with spatially disparate visual–auditory stimuli results in atypical requirements for their integration. Shown is an SC neuron from an animal reared from birth to 6 months of age with simultaneous but spatially disparate visual–auditory cues. The neuron's visual and auditory receptive fields were atypical by being nonoverlapping. When visual–auditory stimuli were spatiotemporally coincident and within the visual (left) or auditory (center) receptive fields, the multisensory response was no greater than that evoked by the most effective component stimulus. But, when the two stimuli were disparate in space and simultaneously presented within their respective receptive fields, they elicited a significantly enhanced multisensory response. Thus, the neuron integrated spatially disparate visual–auditory stimuli as normal animals integrate spatially concordant visual–auditory stimuli, a seeming 'reversal' of the spatial principle. *Adapted from Wallace MT and Stein BE (2007) Early experience determines how the senses will interact.* Journal of Neurophysiology 97: 921–926.

neurons to these stimuli via many other input channels. Chronic deactivation was accomplished by implanting muscimol-infused pledgets of Elvax (a polymer) over the association cortex. The GABAa agonist was slowly released from the polymer, deactivating underlying neurons. When the polymer was depleted or removed, cortical activity returned rapidly and was once again responsive to environmental stimuli. Behavioral and physiological studies were then conducted when the animals had reached 1 year of age. Although experiments are still ongoing, the preliminary data reveal that the animals appeared normal in their ability to respond to visual stimuli in both visual fields. They also benefited from visual–auditory cues in the ipsilateral hemifield as much as normal animals. However, they were severely compromised in their multisensory responses to crossmodal stimuli in their contralateral hemifield. Crossmodal stimuli in this hemifield were no more effective in facilitating SC-mediated behavior than was the visual stimulus alone. Thus, as expected, the inability of association cortex to monitor the statistics of visual–auditory events compromised the maturation of the circuit necessary for SC multisensory integration. This was also evident in the inability of multisensory neurons in the ipsilateral SC to integrate visual–auditory information.

That these deficits in multisensory integration were apparent long after the cortex was once again active (a period far longer than that required for its normal acquisition in early life) could be interpreted as reflecting a 'critical' or 'sensitive' period for instantiating this process. If the former were the case, this capacity would never develop. This possibility was examined in some of these same animals that were retained for several years. These animals were then tested as before. Their behavior appeared to be normal, indicating that they acquired the ability to integrate visual–auditory cues later in life. That this acquisition likely involved multisensory SC neurons was indicated by the finding that the multisensory responses of ipsilateral SC neurons, while not completely normal, did exhibit multisensory response enhancement to spatiotemporally coincident visual–auditory stimuli.

3.5.6.1 *Motor Development*

Much less is known about the development of the motor than the sensory properties of SC neurons. Although there is little doubt that the fundamental properties involved in gaze shifts are not species-specific, most of the available information comes from studies in one preparation: the alert monkey trained to make gaze shifts. For obvious reasons, technical limitations make conducting such studies in neonates quite difficult, especially making any direct correlations with the maturation of sensory properties in the primary sensory developmental model, the cat. Nevertheless, for neonatal SC sensory responses to have any direct impact on behavior, their efferents to the brainstem and/or spinal cord must be in place and capable of carrying tectofugal signals.

These pathways can be seen exiting the SC and contacting targets involved in eye movements within hours of birth in the cat. The target areas include the central gray overlying the oculomotor nucleus, an area into which oculomotor dendrites project, as well as segments of the pontine and medullary reticular formation that connect to the abducens nucleus (Stein et al., 1982). Tectofugal projections also reach the cervical spinal cord. This region is involved in controlling head and limb movement, and these projections are detailed in adults by Huerta and Harting (1982a,b).

Using electrical stimulation of the SC, this motor pathway has already been demonstrated to have functional capabilities in 2-day-old cats (Stein et al., 1980), days before natural auditory stimuli can activate multisensory layer neurons and weeks before visual stimuli are effective. Such electrical stimulation elicits eye, ear, neck, whisker, and limb movements, albeit with a higher threshold and lower reliability than in adults. Furthermore, the motor topography in the SC is already evident. Stimulation of homotopic loci in each SC produces mirror-image eye movements. Nevertheless, many SC stimulation sites proved to be ineffective at this stage of maturation. Presumably, this is due to the immaturity of the SC, as direct stimulation of the oculomotor nucleus evoked reliable eye movements. It is not yet known, however, whether tactile stimuli, which can already activate SC neurons in preterm kittens, can initiate SC-mediated movements in neonates. If, on the other hand, the SC is organized in motor coordinates, as has been suggested by some investigators (e.g., see Sparks, 1986; Sparks and Porter, 1983), it could indicate that early motor function could influence later auditory and visual organization. It would also be consistent with the idea (e.g., see Hein et al., 1979) that eye-movement-generated movement of an image across the retina is necessary for interpreting that visual image and that during development, the former precedes the latter (though spontaneous or vestibular cues could produce the necessary eye movements in neonatal cats; Fish and Windle, 1932; Windle and Fish, 1932).

3.6 SUMMARY

It is important to acknowledge that this chapter has focused on certain anatomical and physiological aspects of SC development for which functional implications are readily apparent. Thus, for example, the formation of topographic representations and the development of multisensory integrative capabilities have obvious implications for producing behavioral output and behavioral correlates in developing animals. In doing so, however, voluminous bodies of literature on, for example, the chronology of synapse formation and the appearance of histochemical markers that parallel SC functional development have been ignored. Likewise, with few exceptions, the focus has been primarily on the SC, but there is also a wealth of comparative data pertaining to the OT from studies of a wide array of nonmammalian species. While the same general principles of development apply, maturational differences dictated by the varied and specialized environmental niches that these animals occupy would require a discussion that is beyond the scope of this chapter. Lastly, it is important to point out what already may be obvious: knowledge of SC development disproportionately favors aspects relevant to its sensory capabilities. This weighting belies the truly integrated sensorimotor function of the SC and highlights a rather substantial gap in the knowledge of how the SC develops to play its adult role. It can only be hoped that future studies will address this imbalance to provide a more complete picture of the SC developmental chronology.

Acknowledgments

The research described here was supported in part by NIH grants NS36916 and EY016716 and EY12389.

References

Alvarado, J.C., Stanford, T.R., Vaughan, J.W., Stein, B.E., 2007. Cortex mediates multisensory but not unisensory integration in superior colliculus. Journal of Neuroscience 27, 12775–12786.

Alvarado, J.C., Rowland, B.A., Stanford, T.R., Stein, B.E., 2008. A neural network model of multisensory integration also accounts for unisensory integration in superior colliculus. Brain Research 1242, 13–23.

Alvarado, J.C., Stanford, T.R., Rowland, B.A., Vaughan, J.W., Stein, B.E., 2009. Multisensory integration in the superior colliculus requires synergy among corticocollicular inputs. Journal of Neuroscience 29, 6580–6592.

Auroy, P., Irthum, B., Woda, A., 1991. Oral nociceptive activity in the rat superior colliculus. Brain Research 549, 275–284.

Beckstead, R.M., Frankfurter, A., 1983. A direct projection from the retina to the intermediate gray layer of the superior colliculus demonstrated by anterograde transport of horseradish peroxidase in monkey, cat and rat. Experimental Brain Research 52, 261–268.

Behan, M., Appell, P.P., 1992. Intrinsic circuitry in the cat superior colliculus: Projections from the superficial layers. The Journal of Comparative Neurology 315, 230–243.

Bell, A.H., Corneil, B.D., Meredith, M.A., Munoz, D.P., 2001. The influence of stimulus properties on multisensory processing in the awake primate superior colliculus. Canadian Journal of Experimental Psychology 55, 123–132.

Bell, A.H., Meredith, M.A., Van Opstal, A.J., Munoz, D.P., 2005. Cross-modal integration in the primate superior colliculus underlying the preparation and initiation of saccadic eye movements. Journal of Neurophysiology 93, 3659–3673.

Benedetti, F., Ferro, I., 1995. The effects of early postnatal modification of body shape on the somatosensory-visual organization in mouse superior colliculus. The European Journal of Neuroscience 7, 412–418.

Berson, D.M., McIlwain, J.T., 1982. Retinal Y-cell activation of deep-layer cells in superior colliculus of the cat. Journal of Neurophysiology 47, 700–714.

Bonds, A.B., Freeman, R.D., 1978. Development of optical quality in the kitten eye. Vision Research 18, 391–398.

Brainard, M.S., Knudsen, E.I., 1998. Sensitive periods for visual calibration of the auditory space map in the barn owl optic tectum. Journal of Neuroscience 18, 3929–3942.

Calvert, G.A., Hansen, P.C., Iversen, S.D., Brammer, M.J., 2001. Detection of audio-visual integration sites in humans by application of electrophysiological criteria to the BOLD effect. NeuroImage 14, 427–438.

Calvert, G.A., Spence, C., Stein, B.E., 2004. The Handbook of Multisensory Processes. MIT Press, Cambridge, MA.

Cavalcante, L.A., Rocha-Miranda, C.E., 1978. Postnatal development of retinogeniculate, retinopretectal and retinotectal projections in the opossum. Brain Research 146, 231–248.

Coulter, J.D., Bowker, R.M., Wise, S.P., Murray, E.A., Castiglioni, A.J., Westlund, K.N., 1979. Cortical, tectal and medullary descending pathways to the cervical spinal cord. Progress in Brain Research 50, 263–279.

Cynader, M., Berman, N., 1972. Receptive-field organization of monkey superior colliculus. Journal of Neurophysiology 35, 187–201.

Dräger, U.C., Hubel, D.H., 1975. Responses to visual stimulation and relationahip between visual, auditory, and somatosensory inputs in mouse superior colliculus. Journal of Neurophysiology 38, 690–713.

Drescher, U., Bonhoeffer, F., Müller, B.K., 1997. The Eph family in retinal axon guidance. Current Opinion in Neurobiology 7, 75–80.

du Lac, S., Knudsen, E.I., 1990. Neural maps of head movement vector and speed in the optic tectum of the barn owl. Journal of Neurophysiology 63, 131–146.

Edwards, S.B., 1980. The deep layers of the superior colliculus: Their reticular characteristics and structural organization. In: Hobson, J. A., Brazier, M.A. (Eds.), The Reticular Formation Revisited. Raven Press, New York, pp. 193–209.

Edwards, S.B., Henkel, C.K., 1978. Superior colliculus connections with the extraocular motor nuclei in the cat. The Journal of Comparative Neurology 179, 451–467.

Edwards, S.B., Ginsburgh, C.L., Henkel, C.K., Stein, B.E., 1979. Sources of subcortical projections to the superior colliculus in the cat. The Journal of Comparative Neurology 184, 309–329.

Feldon, S., Feldon, P., Kruger, L., 1970. Topography of the retinal projection upon the superior colliculus of the cat. Vision Research 10, 135–143.

Fish, M.W., Windle, W.F., 1932. The effect of rotatory stimulation on the movements of the head and eyes in newborn and young kittens. The Journal of Comparative Neurology 54, 103–107.

Fox, P., Chow, K.L., Kelly, A.S., 1978. Effects of monocular lid closure on development of receptive-field characteristics of neurons in rabbit superior colliculus. Journal of Neurophysiology 41, 1359–1372.

Fraser, S.E., 1992. Patterning of retinotectal connections in the vertebrate visual system. Current Opinion in Neurobiology 2, 83–87.

Freeman, R.D., Lai, C.E., 1978. Development of the optical surfaces of the kitten eye. Vision Research 18, 399–407.

Frost, D.O., So, K.F., Schneider, G.E., 1979. Postnatal development of retinal projections in Syrian hamsters: A study using autoradioagraphic and anterograde techniques. Neuroscience 4, 1649–1677.

Fuentes-Santamaria, V., Stein, B.E., McHaffie, J.G., 2006. Neurofilament proteins are preferentially expressed in descending output neurons of the cat superior colliculus: A study using SMI-32. Neuroscience 138, 55–68.

Fuentes-Santamaria, V., Alvarado, J.C., Stein, B.E., McHaffie, J.G., 2008a. Cortex contacts both output neurons and nitrergic interneurons in the superior colliculus: Direct and indirect routes for multisensory integration. Cerebral Cortex 18, 1640–1652.

Fuentes-Santamaria, V., McHaffie, J.G., Stein, B.E., 2008b. Maturation of multisensory integration in the superior colliculus: Expression if nitric oxide synthase and neurofilament SMI-32. Brain Research 1242, 45–53.

Fuentes-Santamaria, V., Alvarado, J.C., McHaffie, J.G., Stein, B.E., 2009. Axon morphologies and convergence patterns of projections from different sensory-specific cortices of the anterior ectosylvian sulcus onto multisensory neurons in the cat superior colliculus. Cerebral Cortex 19, 2902–2915.

Gaither, N.S., Stein, B.E., 1979. Reptiles and mammals use similar sensory organizations in the midbrain. Science 205, 595–597.

Gandhi, N.J., Katnani, H.A., 2011. Motor functions of the superior colliculus. Annual Review of Neuroscience 34, 205–231.

Gillmeister, H., Eimer, M., 2007. Tactile enhancement of auditory detection and perceived loudness. Brain Research 1160, 58–68.

Goldberg, M.E., Wurtz, R.H., 1972. Activity of superior colliculus in behaving monkey. I. Visual receptive fields of single neurons. Journal of Neurophysiology 35 (4), 542–559.

Gori, M., Del, V.M., Sandini, G., Burr, D.C., 2008. Young children do not integrate visual and haptic form information. Current Biology 18, 694–698.

Grantyn, A., Grantyn, R., 1982. Axonal patterns and sites of termination of cat superior colliculus neurons projecting in the tecto-bulbo-spinal tract. Experimental Brain Research 46, 243–256.

Grantyn, R., Ludwig, R., Eberhardt, W., 1984. Neurons of the superficial tectal gray. An intracellular HRP-study on the kitten superior colliculus in vitro.

Graybiel, A.M., 1975. Anatomical organization of retinotectal afferents in the cat: An autoradiographic study. Brain Research 96, 1–23.

Groh, J.M., Sparks, D.L., 1996. Saccades to somatosensory targets. III. Eye-position-dependent somatosensory activity in primate superior colliculus. Journal of Neurophysiology 75, 439–453.

Hall, W.C., Moschovakis, A., 2004. The Superior Colliculus: New Approaches for Studying Sensorimotor Integration. CRC Press, Boca Raton, FL.

Harting, J.K., 1977. Descending pathways from the superior colliculus: An autoradiographic analysis in the rhesus monkey (Macaca mulatta). The Journal of Comparative Neurology 173, 583–612.

Harting, J.K., Hall, W.C., Diamond, I.T., Martin, G.F., 1973. Anterograde degeneration study of the superior colliculus in Tupaia glis: Evidence for a subdivision between superficial and deep layers. The Journal of Comparative Neurology 148, 361–386.

Harting, J.K., Huerta, M.F., Frankfurter, A.J., Strominger, N.L., Royce, G.J., 1980. Ascending pathways from the monkey superior colliculus: An autoradiographic analysis. The Journal of Comparative Neurology 192, 853–882.

Hartline, P.H., Kass, L., Loop, M.S., 1978. Merging of modalities in the optic tectum: Infrared and visual integration in rattlesnakes. Science 199, 1225–1229.

Hartline, P.H., Vimal, R.L., King, A.J., Kurylo, D.D., Northmore, D.P., 1995. Effects of eye position on auditory localization and neural representation of space in superior colliculus of cats. Experimental Brain Research 104, 402–408.

Hein, A., Vital-Durand, F., Salinger, W., Diamond, R., 1979. Eye movements initiate visual-motor development in the cat. Science 204, 1321–1322.

Holcombe, V., Hall, W.C., 1981a. Laminar origin of ipsilateral tectpontine pathways. Neuroscience 6, 255–260.

Holcombe, V., Hall, W.C., 1981b. The laminar origin and distribution of the crossed tectoreticular pathways. Journal of Neuroscience 1, 1103–1112.

Huerta, M.F., Harting, J.K., 1982a. Tectal control of spinal cord activity: Neuroanatomical demonstration of pathways connecting the superior colliculus with the cervical spinal cord grey. Progress in Brain Research 57, 293.

Huerta, M.F., Harting, J.K., 1982b. Projections of the superior colliculus to the supraspinal nucleus and the cervical spinal cord grey of the cat. Brain Research 242, 326.

Jay, M.F., Sparks, D.L., 1984. Auditory receptive fields in primate superior colliculus shift with changes in eye position. Nature 309, 345–347.

Jiang, W., Wallace, M.T., Jiang, H., Vaughan, J.W., Stein, B.E., 2001. Two cortical areas mediate multisensory integration in superior colliculus neurons. Journal of Neurophysiology 85, 506–522.

Jiang, W., Jiang, H., Stein, B.E., 2002. Two corticotectal areas facilitate multisensory orientation behavior. Journal of Cognitive Neuroscience 14, 1240–1255.

Jiang, W., Jiang, H., Stein, B.E., 2006. Neonatal cortical ablation disrupts multisensory development in superior colliculus. Journal of Neurophysiology 95, 1380–1396.

Jiang, W., Jiang, H., Rowland, B.A., Stein, B.E., 2007. Multisensory orientation behavior is disrupted by neonatal cortical ablation. Journal of Neurophysiology 97, 557–562.

Kadunce, D.C., Vaughan, J.W., Wallace, M.T., Benedek, G., Stein, B.E., 1997. Mechanisms of within- and cross-modality suppression in the superior colliculus 2. Journal of Neurophysiology 78, 2834–2847.

Kanaseki, T., Sprague, J.M., 1974. Anatomical organization of pretectal nuclei and tectal laminae in the cat. The Journal of Comparative Neurology 158, 319–337.

Kao, C.Q., Stein, B.E., Coulter, D.A., 1994. Postnatal development of excitatory synaptic function in deep layers of SC. Abstracts – Society for Neuroscience 20, 1186.

King, A.J., Palmer, A.R., 1983. Cells responsive to free-field auditory stimuli in guinea-pig superior colliculus: Distribution and response properties. The Journal of Physiology 342, 361–381.

King, A.J., Palmer, A.R., 1985. Integration of visual and auditory information in bimodal neurones in the guinea-pig superior colliculus. Experimental Brain Research 60, 492–500.

King, A.J., Hutchings, M.E., Moore, D.R., Blakemore, C., 1988. Developmental plasticity in the visual and auditory representations in the mammalian superior colliculus. Nature 332, 72–86.

Land, P.W., Lund, R.D., 1979. Development of the rat's uncrossed retinotectal pathway and its relation to plasticity studies. Science 205, 698–700.

Larson, M.A., Stein, B.E., 1984. The use of tactile and olfactory cues in neonatal orientation and localization of the nipple. Developmental Psychobiology 17, 423–436.

Larson, M.A., McHaffie, J.G., Stein, B.E., 1987. Response properties of nociceptive and low-threshold mechanoreceptive neurons in the hamster superior colliculus. Journal of Neuroscience 7, 547–564.

McHaffie, J.G., Ogasawara, K., Stein, B.E., 1986. Trigeminotectal and other trigeminofugal projections in neonatal kittens: An anatomical demonstration with horseradish peroxidase and tritiated leucine. The Journal of Comparative Neurology 249, 411–427.

McHaffie, J.G., Kruger, L., Clemo, H.R., Stein, B.E., 1988. Corticothalamic and corticotectal somatosensory projections from the anterior ectosylvian sulcus (SIV cortex) in neonatal cats: An anatomical demonstration with HRP and 3H-leucine. The Journal of Comparative Neurology 274, 115–126.

McHaffie, J.G., Kao, C.Q., Stein, B.E., 1989. Nociceptive neurons in rat superior colliculus: Response properties, topography, and functional implications. Journal of Neurophysiology 62, 510–525.

McIlwain, J.T., 1975. Visual receptive fields and their images in superior colliculus of the cat. Journal of Neurophysiology 38, 219–230.

McIlwain, J.T., 1991. Distributed spatial coding in the superior colliculus: A review. Visual Neuroscience 6, 3–13.

Meredith, M.A., Stein, B.E., 1990. The visuotopic component of the multisensory map in the deep laminae of the cat superior colliculus. Journal of Neuroscience 10, 3727–3742.

Meredith, M.A., Clemo, H.R., Stein, B.E., 1991. Somatotopic component of the multisensory map in the deep laminae of the cat superior colliculus. The Journal of Comparative Neurology 312, 353–370.

Metzger, R.R., Mullette-Gillman, O.A., Underhill, A.M., Cohen, Y.E., Groh, J.M., 2004. Auditory saccades from different eye positions in the monkey: Implications for coordinate transformations. Journal of Neurophysiology 92, 2622–2627.

Middlebrooks, J.C., Knudsen, E.I., 1984. A neural code for auditory space in the cat's superior colliculus. Journal of Neuroscience 4, 2621–2634.

Moschovakis, A.K., Karabelas, A.B., 1985. Observations on the somatodendritic morphology and axonal trajectory of intracellularly HRP-labeled efferent neurons located in the deeper layers of the superior colliculus of the cat. The Journal of Comparative Neurology 239, 276–308.

Moschovakis, A.K., Kitama, T., Dalezios, Y., Petit, J., Brandi, A.M., Grantyn, A.A., 1998. An anatomical substrate for the spatiotemporal transformation. Journal of Neuroscience 18, 10219–10229.

Neil, P.A., Chee-Ruiter, C., Scheier, C., Lewkowicz, D.J., Shimojo, S., 2006. Development of multisensory spatial integration and perception in humans. Developmental Science 9, 454–464.

Norton, T.T., 1974. Receptive-field properties of superior colliculus cells and development of visual behavior in kittens. Journal of Neurophysiology 37, 674–690.

Palmer, A.R., King, A.J., 1982. The representation of auditory space in the mammalian superior colliculus. Nature 299 (5880), 248–249.

Pare, M., Crommelinck, M., Guitton, D., 1994. Gaze shifts evoked by stimulation of the superior colliculus in the head-free cat conform to the motor map but also depend on stimulus strength and fixation activity. Experimental Brain Research 101, 123–139.

Peck, C.K., Baro, J.A., Warder, S.M., 1995. Effects of eye position on saccadic eye movements and on the neuronal responses to auditory and visual stimuli in cat superior colliculus. Experimental Brain Research 103, 227–242.

Putzar, L., Goerendt, I., Lange, K., Rosler, F., Roder, B., 2007. Early visual deprivation impairs multisensory interactions in humans. Nature Neuroscience 10, 1243–1245.

Rakic, P., 1977. Prenatal development of the visual system in rhesus monkey. Philosophical Transactions of the Royal Society of London. Series B, Biological Sciences 278, 245–260.

Redgrave, P., Odekunle, A., Dean, P., 1985. Tectal cells of origin of predorsal bundle in rat: Location, and segregation from ipsilateral descending pathway. Neuroscience Letters. Supplement 22, S320.

Redgrave, P., Mitchell, I.J., Dean, P., 1986a. Stereotaxic mapping of efferent descending pathways of the superior colliculus in rat: An anterograde WGA-HRP study. Neuroscience Letters. Supplement 26, S349.

Redgrave, P., Odekunle, A., Dean, P., 1986b. Tectal cells of origin of predorsal bundle in rat: Location and segregation from ipsilateral descending pathway. Experimental Brain Research 63, 279–293.

Rhoades, R.W., Mooney, R.D., Jacquin, M.F., 1983. Complex somatosensory receptive fields of cells in the deep laminae of the hamster's superior colliculus. Journal of Neuroscience 3, 1342–1354.

Rhoades, R.W., Mooney, R.D., Klein, B.G., Jacuin, M.F., Szczepanik, A.M., Chiaia, N.L., 1987. The structural and functional characteristics of tectospinal neurons in the golden hamster. The Journal of Comparative Neurology 255, 451–465.

Robinson, D.A., 1972. Eye movements evoked by collicular stimulation in the alert monkey. Vision Research 12 (11), 1795–1808.

Rosenblatt, J.S., 1971. Suckling and home orientation in the kitten: A comparative developmental study. In: Tobach, E., Aronson, L., Shaw, E. (Eds.), The Biopsychology of Development. Academic Press, New York, pp. 345–410.

Rowland, B.A., Jiang, W., Stein, B.E., 2005. Long-term plasticity in multisensory integration. Program No. 505.8 In: 2005 Neuroscience Meeting Planner. Washington, DC: Society for Neuroscience Online.

Ruthazer, E.S., Cline, H.T., 2004. Insights into activity-dependent map formation from the retinotectal system: A middle-of-the-brain perspective. Journal of Neurobiology 59, 134–146.

Ruthazer, E.S., Akeman, C.J., Cline, H.T., 2003. Control of axon branch dynamics by correlated activity in vivo. Science 301, 66–70.

Schiller, P.H., Stryker, M., 1972. Single-unit recording and stimulation in superior colliculus of the alert rhesus monkey. Journal of Neurophysiology 35 (6), 915–924.

Siminoff, R., Schwassmann, H.O., Kruger, L., 1966. An electrophysiological study of the visual projection to the superior colliculus of the rat. The Journal of Comparative Neurology 127, 435–444.

Sommer, M.A., Wurtz, R.H., 1998. Frontal eye field neurons orthodromically activated from the superior colliculus. Journal of Neurophysiology 80, 3331–3335.

Sommer, M.A., Wurtz, R.H., 2004a. What the brain stem tells the frontal cortex. I. Oculomotor signals sent from superior colliculus to frontal eye field via mediodorsal thalamus. Journal of Neurophysiology 91, 1381–1402.

Sommer, M.A., Wurtz, R.H., 2004b. What the brain stem tells the frontal cortex. II. Role of the SC-MD-FEF pathway in corollary discharge. Journal of Neurophysiology 91, 1403–1423.

Sparks, D.L., 1986. Translation of sensory signals into commands for control of saccadic eye movements: Role of primate superior colliculus. Physiological Reviews 66, 118–171.

Sparks, D.L., Mays, L.E., 1990. Signal transformations required for the generation of saccadic eye movements. Annual Review of Neuroscience 13, 309–336.

Sparks, D.L., Porter, J.D., 1983. Spatial localization of saccade targets. II. Activity of superior colliculus neurons preceding compensatory saccades. Journal of Neurophysiology 49, 64–74.

Spence, C., Pavani, F., Driver, J., 2004. Spatial constraints on visual-tactile cross-modal distractor congruency effects. Cognitive, Affective, & Behavioral Neuroscience 4, 148–169.

Sperry, R.W., 1963. Chemoaffinity in the orderly growth of nerve fiber patterns and connections. Proceedings of the National Academy of Sciences of the United States of America 50, 703–710.

Stein, B.E., 1984. Development of the superior colliculus. Annual Review of Neuroscience 7, 95–125.

Stein, B.E., 2005. The development of a dialogue between cortex and midbrain to integrate multisensory information. Experimental Brain Research 166, 305–315.

Stein, B.E., Clamann, H.P., 1981. Control of pinna movements and sensorimotor register in cat superior colliculus. Brain, Behavior and Evolution 19, 180–192.

Stein, B.E., Dixon, J.P., 1978. Superior colliculus cells respond to noxious stimuli. Brain Research 158, 65–73.

Stein, B.E., Dixon, J.P., 1979. Properties of superior colliculus neurons in the golden hamster. The Journal of Comparative Neurology 183, 269–284.

Stein, B.E., Edwards, S.B., 1979. Corticotectal and other corticofugal projections in neonatal cat. Brain Research 161, 399–409.

Stein, B.E., Gaither, N.S., 1981. Sensory representation in reptilian optic tectum: Some comparisons with mammals. The Journal of Comparative Neurology 202, 69–87.

Stein, B.E., Gallagher, H.L., 1981. Maturation of cortical control over superior colliculus cells in cat. Brain Research 223, 429–435.

Stein, B.E., Meredith, M.A., 1993. The Merging of the Senses. MIT Press, Cambridge, Mass.

Stein, B.E., Rowland, B.A., 2011. Organization and plasticity in multisensory integration: early and late experience affects its governing principles. Progress in Brain Research 191, 145–163.

Stein, B.E., Stanford, T.R., 2008. Multisensory integration: Current issues from the perspective of the single neuron. Nature Reviews. Neuroscience 9, 255–266.

Stein, B.E., Labos, E., Kruger, L., 1973a. Sequence of changes in properties of neurons of superior colliculus of the kitten during maturation. Journal of Neurophysiology 36, 667–679.

Stein, B.E., Labos, E., Kruger, L., 1973b. Determinants of response latency of neurons in superior colliculus in kittens. Journal of Neurophysiology 36, 680–689.

Stein, B.E., Magalhaes-Castro, B., Kruger, L., 1975. Superior colliculus: Visuotopic-somatotopic overlap. Science 189, 224–226.

Stein, B.E., Clamann, H.P., Goldberg, S.J., 1980. Superior colliculus: Control of eye movements in neonatal kittens. Science 210, 78–80.

Stein, B.E., Spencer, R.F., Edwards, S.B., 1982. Efferent projections of the neonatal superior colliculus: Extraoculomotor-related brain stem structures. Brain Research 239, 17–28.

Stein, B.E., Spencer, R.F., Edwards, S.B., 1983. Corticotectal and corticothalamic efferent projections of SIV somatosensory cortex in cat. Journal of Neurophysiology 50, 896–909.

Stein, B.E., Spencer, R.F., Edwards, S.B., 1984. Efferent projections of the neonatal cat superior colliculus: Facial and cerebellum-related brainstem structures. The Journal of Comparative Neurology 230, 47–54.

Stein, B.E., McHaffie, J.G., Harting, J.K., Huerta, M.F., Hashikawa, T., 1985. Transient tectogeniculate projections in neonatal kittens: An autoradiographic study. The Journal of Comparative Neurology 239, 402–412.

Stein, B.E., Price, D.D., Gazzaniga, M.S., 1989. Pain perception in a man with total corpus callosum transection. Pain 38, 51–56.

Stein, B.E., Perrault Jr, T.J., Vaughan, J.W., Rowland, B.A., 2008. Long term plasticity of multisensory neurons in the superior colliculus. Program No. 457.14 In: 2008 Neuroscience Meeting Planner. Washington, DC: Society for Neuroscience, 2008; Online.

Stein, B.E., Burr, D., Constantinidis, C., et al., 2010. Semantic confusion regarding the development of multisensory integration: A practical solution. The European Journal of Neuroscience 31, 1713–1720.

Stuphorn, V., Bauswein, E., Hoffmann, K.P., 2000. Neurons in the primate superior colliculus coding for arm movements in gaze-related coordinates. Journal of Neurophysiology 83, 1283–1299.

Thorn, F., Gollender, M., Erickson, P., 1976. The development of the kittens visual optics. Vision Research 16, 1145–1149.

Van Wanrooij, M.M., Bell, A.H., Munoz, D.P., Van Opstal, A.J., 2009. The effect of spatial-temporal audiovisual disparities on saccades in a complex scene. Experimental Brain Research 198, 425–437.

Villablanca, J.R., Olmstead, C.E., 1979. Neurological development of kittens. Developmental Psychobiology 12, 101–127.

Wallace, M.T., Stein, B.E., 1994. Cross-modal synthesis in the midbrain depends on input from cortex. Journal of Neurophysiology 71, 429–432.

Wallace, M.T., Stein, B.E., 1997. Development of multisensory neurons and multisensory integration in cat superior colliculus. Journal of Neuroscience 17, 2429–2444.

Wallace, M.T., Stein, B.E., 2000. Onset of cross-modal synthesis in the neonatal superior colliculus is gated by the development of cortical influences. Journal of Neurophysiology 83, 3578–3582.

Wallace, M.T., Stein, B.E., 2001. Sensory and multisensory responses in the newborn monkey superior colliculus. Journal of Neuroscience 21, 8886–8894.

Wallace, M.T., Stein, B.E., 2007. Early experience determines how the senses will interact. Journal of Neurophysiology 97, 921–926.

Wallace, M.T., Wilkinson, L.K., Stein, B.E., 1996. Representation and integration of multiple sensory inputs in primate superior colliculus. Journal of Neurophysiology 76, 1246–1266.

Wallace, M.T., McHaffie, J.G., Stein, B.E., 1997. Visual response properties and visuotopic representation in the newborn monkey superior colliculus. Journal of Neurophysiology 78, 2732–2741.

Wallace, M.T., Perrault Jr., T.J., Hairston, W.D., Stein, B.E., 2004. Visual experience is necessary for the development of multisensory integration 1. Journal of Neuroscience 24, 9580–9584.

Walter, J., Henke-Fahle, S., Bonhoeffer, F., 1987. Avoidance of posterior tectal membranes by temporal retinal axons. Development 101, 909–913.

Weber, J.T., Martin, G.F., Behan, M., Huerta, M.F., Harting, J.K., 1979. The precise origin of the tectospinal pathway in three common laboratory animals: A study using the horseradish peroxidase method. Neuroscience Letters 11, 121–127.

Wilkinson, L.K., Meredith, M.A., Stein, B.E., 1996. The role of anterior ectosylvian cortex in cross-modality orientation and approach behavior. Experimental Brain Research 112, 1–10.

Williams, R.W., Chalupa, L.M., 1982. Prenatal development of retinocollicular projections in the cat: An anterograde tracer transport study. Journal of Neuroscience 2, 604–622.

Windle, W.F., Fish, M.W., 1932. The development of the vestibular righting reflex in the cat. The Journal of Comparative Neurology 54, 85–96.

Wurtz, R.H., Goldberg, M.E., 1972. Activity of superior colliculus in behaving monkey. 3. Cells discharging before eye movements. Journal of Neurophysiology 35 (4), 575–586.

Zahar, Y., Reches, A., Gutfreund, Y., 2009. Multisensory enhancement in the optic tectum of the barn owl: Spike count and spike timing. Journal of Neurophysiology 101, 2380–2394.

Multisensory Circuits

A.J. King

University of Oxford, Oxford, UK

4.1 INTRODUCTION: MULTISENSORY PERCEPTION AND BEHAVIOR

As they interact with their surroundings, humans and other animal species typically experience multiple stimuli that are registered by different sense organs. These signals may provide complementary information about the same object or event. This is likely to be the case if the stimuli occur at more or less the same time and originate from the same location in space or – in the case of more complex signals, such as the auditory and visual components of speech – if they are semantically related. The capacity of the central nervous system to merge and integrate input from different sensory modalities can have profound effects on perception and behavior. This can be seen as an improvement in an animal's ability to detect or discriminate objects, or determine where they are located, and a reduction in the time required to react to them (reviewed by Alais et al., 2010; Calvert et al., 2004).

In addition to enhancing perceptual judgments, particularly when the stimulus information supplied by one or more sensory modalities is weak or ambiguous, crossmodal interactions can induce various illusions if the different sensory inputs provide conflicting cues. These illusions provide further evidence of how the brain attempts to merge multisensory inputs so that they are perceived as if they originate from the same external event. One of the most striking examples involves the effect of vision on speech perception. Being able to view the speaker's face improves speech comprehension so long as the audiovisual cues are congruent (Ma et al., 2009; Sumby and Pollack, 1954), but watching the speaker articulate a different speech sound from the one that is being presented results in the perception of a third sound that represents a combination of what was seen and heard (the "McGurk effect"; McGurk and MacDonald, 1976). Another well-known example of the influence of vision on auditory perception is provided by the "ventriloquism effect." Although spatially aligned visual cues can improve the accuracy of auditory localization (Shelton and Searle, 1980; Stein et al., 1988), misaligned visual cues can bias or capture the perceived location of a sound source (Howard and Templeton, 1966; Slutsky and Recanzone, 2001).

In addition to their effects on the perceived identity and location of stimuli, interactions between the senses can determine when those stimuli are perceived to occur. For example, if a single light flash is accompanied by multiple auditory beeps, subjects report seeing multiple

Neural Circuit Development and Function in the Brain: Comprehensive Developmental Neuroscience, Volume 3 http://dx.doi.org/10.1016/B978-0-12-397267-5.00144-8

© 2013 Elsevier Inc. All rights reserved.

flashes (Shams et al., 2000). Similarly, sounds can alter the perceived timing of visual events (Fendrich and Corballis, 2001; Recanzone, 2003). Thus, whereas vision tends to dominate audition in the spatial domain, the reverse is true for temporal tasks (Welch and Warren, 1980). This is because in many species, including humans, the spatial resolution of the visual system is superior to that of the auditory system, whereas temporal sensitivity is much greater for audition (King and Nelken, 2009). However, this relationship between the senses is not fixed because the ventriloquism effect can work in reverse, with audition capturing vision, if visual signals are degraded so that they become harder to localize (Alais and Burr, 2004). It seems, therefore, that the integration of different sensory cues in the brain is statistically optimal, in the sense that each cue is weighted in a task-specific way according to how reliable it is (Alais et al., 2010; Ernst and Banks, 2002).

These examples illustrate how interactions between the senses can have a considerable impact on how the world is perceived. Circuits within the brain must therefore be able to merge multisensory signals that are linked in space and time, as well as more complex cues, such as facial expressions and their associated vocalizations. Although certain brain regions such as the superior colliculus (SC) in the midbrain and the posterior parietal cortex, prefrontal cortex, and superior temporal sulcus have long been associated with specific multisensory functions, it is now clear that convergence of different modality inputs is much more widespread than previously realized. This is particularly the case in the cortex, where even those areas typically associated with modality-specific processing, such as the auditory or visual cortex, are now thought to be part of an interactive network of brain regions that are influenced by multiple sensory systems (reviewed by Alais et al., 2010; Ghazanfar and Schroeder, 2006). Inputs from other modalities to primary and secondary cortical areas may serve to amplify responses to related input from the dominant modality (Schroeder et al., 2008), but there is also evidence that they can alter the sensitivity of cortical neurons in more specific ways (e.g., Bizley and King, 2008; Ghazanfar et al., 2005). Moreover, the finding that early cortical sensory areas are engaged when crossmodal illusions, including the ventriloquism effect (Bonath et al., 2007) and the sound-induced perception of illusory light flashes (Mishra et al., 2007), are experienced indicates that these brain regions as well as the more traditional association areas are likely to contribute to perception in a multisensory world.

In spite of the widespread intermixing of sensory modalities in the cerebral cortex, surprisingly little is known about the development of those circuits. Instead, the SC has, for many years, served as a model system for investigating both the principles of multisensory integration by neurons in the adult brain and the developmental processes that allow different sensory inputs to be combined and coordinated in the first place. This chapter therefore focuses primarily on the developmental mechanisms involved in merging multisensory spatial information in this midbrain structure. Although relatively few attempts have been made to examine the maturation of multisensory processing in the cortex, valuable insights into how the different senses interact at this level during development have been obtained by investigating the crossmodal consequences of early sensory deprivation. These studies, which are based predominantly on humans and other species that have been deprived of their vision or hearing, are considered in the second part of the chapter.

4.2 MULTISENSORY PROCESSING IN THE SC

The mammalian SC receives visual, auditory, and somatosensory afferents from a large number of subcortical and cortical brain areas (e.g., Edwards et al., 1979; King et al., 1998a; Wallace et al., 1993). As far as we know, these inputs are all unisensory and converge, often on the same neurons, in the deeper layers of the SC. In contrast, the superficial layers contain only visually responsive neurons. Nevertheless, the sensory representations found in both the superficial and deeper layers are arranged topographically with respect to their receptive field locations (Cynader and Berman, 1972; King and Hutchings, 1987; King and Palmer, 1983; Middlebrooks and Knudsen, 1984; Stein et al., 1976; Wallace and Stein, 1996). Moreover, these maps lie in spatial register with one another so that different modality signals arising from a particular direction in space, and therefore, potentially from the same source, are represented in a corresponding region of the SC. For those deep-layer neurons that receive converging visual, auditory, and somatosensory afferents, this means that their unisensory receptive fields overlap and covary with the anatomical location of the neurons within the SC.

One of the most important and well-studied properties of SC neurons is that the number of spikes evoked by a combination of two or more stimulus modalities is often higher than that of the response to each stimulus presented by itself and may even exceed the sum of those responses. Multisensory enhancement is typically most pronounced when the individual stimuli are weakly effective in driving the neurons (Meredith and Stein, 1986) and is observed when those stimuli are presented in close temporal and spatial proximity (King and Palmer, 1985; Meredith and Stein, 1996; Meredith et al., 1987). In contrast, pairing multisensory signals that are widely separated in time or space does not enhance, and may even depress, the response to unisensory stimulation. These interactions between the different sensory inputs reaching the SC should therefore serve to amplify the neuronal responses to a multisensory target, such as an object that can be seen and heard.

Neurons in the deeper layers also discharge prior to eye and head movements, and while the SC is implicated in different motor behaviors, its principal function is to shift an animal's gaze toward sensory stimuli located on the contralateral side of the body (reviewed by Gandhi and Katnani, 2011). Superimposing the different sensory representations in the SC therefore allows each of the spatial cues associated with a multisensory target to trigger, via a common motor output map, a gaze shift in that direction. The principles of multisensory integration exhibited by SC neurons suggest that spatiotemporally coincident stimuli will, by evoking stronger responses in SC neurons, promote more accurate orienting behavior, whereas pairing unisensory stimuli from different locations or at different times should have the opposite effect. Similarly, integration of inputs from different sensory modalities would be expected to have the greatest effect on orienting behavior when the individual cues are weak. Studies in cats (Stein et al., 1988) and humans (Corneil et al., 2002) have indeed confirmed that this is the case. Moreover, in cats, deactivation of descending projections from two association areas in the cortex, the anterior ectosylvian sulcus (AES) and the rostrolateral suprasylvian sulcus (rLS), eliminates both the multisensory integrative processes displayed by SC neurons (Jiang et al., 2001) and the behavioral benefits of synthesizing inputs from different sensory modalities (Jiang et al., 2002).

Because multisensory enhancement of the responses of SC neurons occurs only when each stimulus falls within its excitatory receptive field, the spatial registration of the sensory maps in the SC would appear to be essential if these interactions are to signal the different modality cues associated with a common source. However, maintaining map registration is problematic because spatial information is specified by each sensory modality using a different frame of reference. The visual map is coded in retinal coordinates, and so the receptive fields of SC neurons will shift relative to the head as the eyes move and therefore potentially become misaligned with the auditory receptive fields, which are centered on the head and ears, and the body-centered somatosensory receptive fields. However, electrophysiological recordings from SC neurons in awake animals have shown that somatosensory (Groh and Sparks, 1996) and auditory (Hartline et al., 1995; Jay and Sparks, 1984; Populin et al., 2004) responses are modified by changes in eye position, suggesting that these signals are partially transformed into a common eye-centered reference frame that matches the coordinates of the visual map.

Combining appropriate inputs from different sensory modalities during development is also challenging, particularly since the spatial relationship between the sense organs can change as the body grows. The following section examines the role of sensory experience in aligning the different modality maps and in the maturation of multisensory integration in the SC.

4.3 DEVELOPMENT OF MULTISENSORY RESPONSES IN THE SC

Each of the sensory systems appears to begin functioning at a different stage of development, with somatosensory sensitivity emerging first and visual last (Gottlieb, 1971) (see Chapter 3). This sequence has also been observed in the SC, where recordings from newborn kittens have shown that the earliest sensory responsive neurons are activated only by tactile stimuli, followed a few days later by the first acoustically responsive neurons and subsequently by responses to visual stimuli (Figure 4.1; Stein et al., 1973). The maturation of multisensory neurons follows the same order, with neurons sensitive to both tactile and auditory stimuli appearing during the second postnatal week, almost 2 weeks before the first visually responsive multisensory neurons (Figure 4.1; Wallace and Stein, 1997). However, in more precocial species, including humans and other primates, most sensory systems become functional before birth, and the SC of newborn monkeys has been shown to contain many multisensory neurons (Wallace and Stein, 2001).

Despite these differences in the postnatal age at which multisensory neurons develop, a common feature of all the species, both precocial and altricial, that have been studied is that the sensory response properties of deep SC neurons initially differ in a number of ways from those seen in older animals (Campbell et al., 2008; Stein et al., 1973; Wallace and Stein, 1997, 2001; Withington-Wray et al., 1990a). In common with other brain areas, sensory responses recorded from SC neurons in very young animals typically have longer latencies, are more sluggish, and show greater habituation to repeated stimulation than is the case in later life. They also lack at least some of the stimulus selectivity that is characteristic of the adult. The details of these age-dependent differences vary to some extent between species and stimulus modalities, but it has consistently been observed that the spatial receptive fields of SC neurons in infant animals are larger than those found in adults and that the earliest multisensory neurons lack the capacity to integrate different modality cues in ways that either enhance or depress their responses.

The key events in the maturation of the sensory representations in the SC therefore comprise a gradual contraction of receptive fields and the appearance of

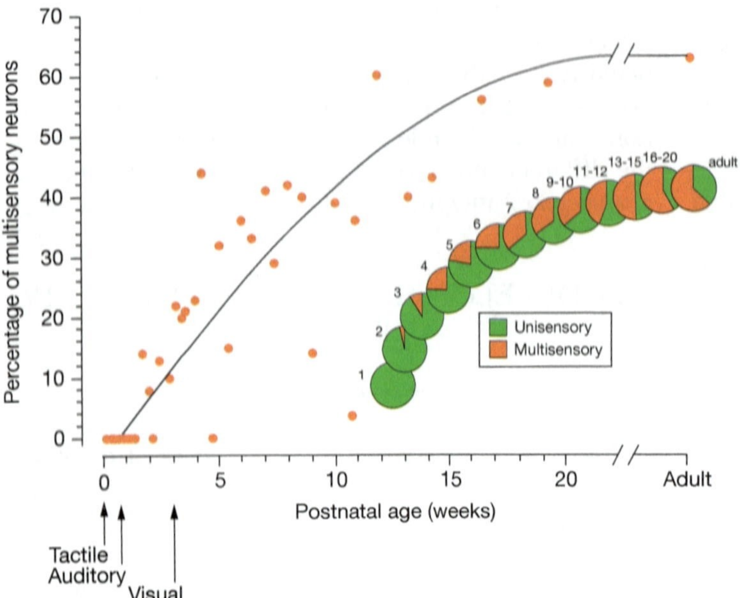

FIGURE 4.1   Postnatal development of multisensory neurons in the deeper layers of the cat superior colliculus (SC). The percentage of sensory neurons responding to stimuli in two or more modalities is plotted as a function of age. The arrows indicate the ages at which the first unisensory responses have been recorded in the deep SC layers of this species. The pie charts show how the incidence of multisensory neurons increases as development proceeds. *Adapted from Wallace MT and Stein BE (1997) Development of multisensory neurons and multisensory integration in cat superior colliculus.* Journal of Neuroscience 17: 2429–2444, *with permission.*

multisensory integration. Indeed, Wallace and Stein (1997) observed that there is a clear correlation between these developmental changes, with the size of the deep-layer receptive fields in cats being a good predictor of whether the neurons will display the capacity to synthesize multisensory inputs. Although this suggests that these events may be causally related, perhaps involving maturation of the same or closely related cellular mechanisms, other aspects of sensory processing also need to be considered.

Foremost among the other factors is the sometimes overlooked fact that the development of sensory representations in the brain depends on the maturation of the relevant sense organs as well as the neural circuits involved (See Rubenstein and Rakic, 2013. Readers are also encouraged to read Chapters 2, 3, 14 within this book). Determining the relative contributions of these factors is generally not straightforward, but studies of the developing auditory representation have provided unique insights into this. Because sound frequency is mapped along the length of the cochlea in the inner ear, sound source location has to be derived by comparing the amplitude and timing of sounds reaching each ear and by sensing the spectral-shape cues produced by the direction-dependent way in which the external ears filter the incoming sound (King et al., 2001). Individual variations in the size and shape of the head and external ears are matched by differences in the binaural and monaural localization cue values corresponding to each direction in space. Consequently, the localization cues available in infancy, when the auditory receptive fields are very large and lack topographic order, are different from those provided by adult ears. But, if virtual acoustic space stimuli are used to allow infant ferrets to hear through adult ears, the auditory receptive fields immediately shrink and are no longer any different in size from those found in adult animals (Campbell et al., 2008).

This result would seem to indicate that growth-related changes in the physical dimensions of the head and external ears can account for the contraction of auditory spatial receptive fields that occurs during postnatal development. However, whether a similar process contributes to the maturation of spatial tuning in other modalities is less clear. In the visual system, for example, the maturation of receptive fields seems to be determined more by the specificity of the afferent connections than by the resolution of the optics or the photoreceptors (Jacobs and Blakemore, 1988; Kiorpes and Movshon, 1990; Tavazoie and Reid, 2000) (see Chapter 14). Moreover, even though providing infant ferrets with adult ears removes age-dependent differences in spatial tuning, no improvement is seen in the topography of the auditory representation in the SC, which gradually emerges over the course of several weeks following hearing onset (Figure 4.2; Campbell et al., 2008). Thus, central mechanisms must also be involved in auditory map development (see Chapter 2).

As noted in the previous section, descending signals from association areas of the cerebral cortex must be available if multisensory integration is to be observed in adult cats. Those inputs also appear to be critically involved in the maturation of the multisensory integrative properties of SC neurons. Evidence for this comes from the observation that the development in these neurons of a capacity to integrate their different sensory inputs coincides with the functional maturation of inputs from the AES (Wallace and Stein, 2002). Moreover, in cats in which the AES and rLS were lesioned in infancy, no difference is seen either physiologically (Jiang et al., 2006) or behaviorally (Jiang et al., 2007) between the

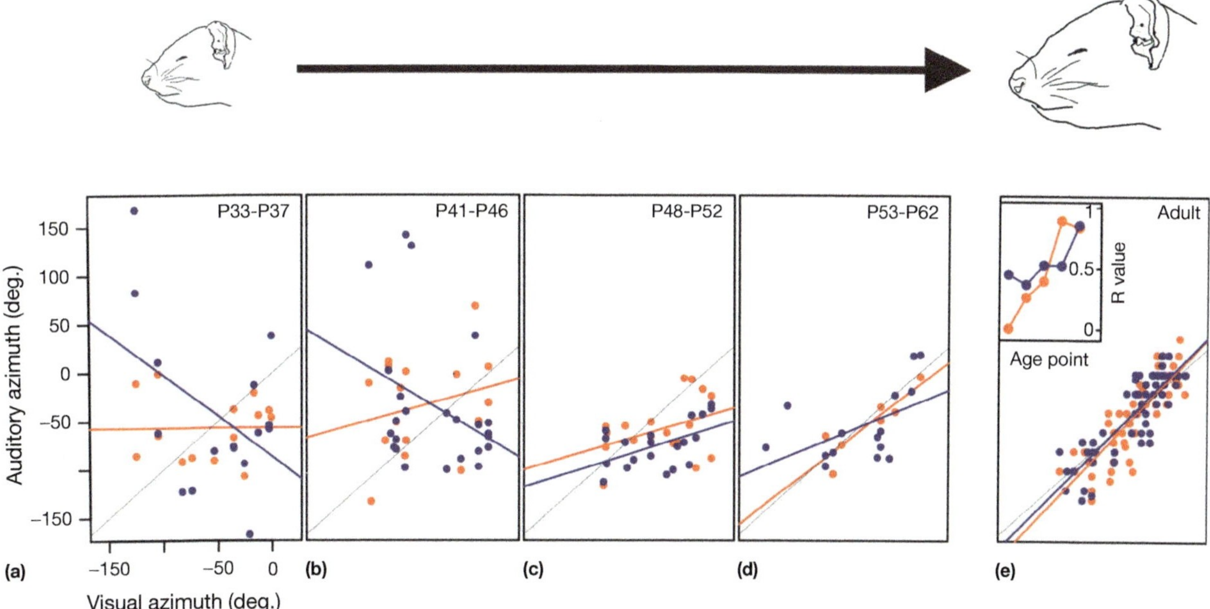

FIGURE 4.2 Maturation of auditory spatial topography in the ferret SC. (a–e) Each panel plots the auditory best azimuth of deep-layer SC neurons as a function of the visual best azimuth of neurons recorded in the overlying superficial layers. The data were obtained at the postnatal ages indicated at the top of each panel using free-field stimuli. Because adultlike visual topography is present in the superficial SC layers throughout this developmental period, the topographic order of the auditory map is reflected in the degree to which it is in register with the visual map. Auditory responses recorded at near-threshold sound levels ($\sim$10 dB > unit threshold) are shown in red; suprathreshold levels ($\sim$25 dB > unit threshold), in blue. A linear regression was fitted to the data from each sound level at each age group (red and blue lines; the black line is the 45° diagonal indicating perfect alignment of the visual and auditory data). The inset panel in (e) plots the correlation coefficient (R) of each regression slope as a function of age. At both sound levels, there is a steady increase in the R value during development, indicating an improvement in topographic order in the auditory representation. *Adapted from Campbell RA, King AJ, Nodal FR, Schnupp JWH, Carlile S, and Doubell TP (2008) Virtual adult ears reveal the roles of acoustical factors and experience in auditory space map development.* Journal of Neuroscience 28: 11557–11570, *with permission.*

responses to spatiotemporally matched visual and auditory stimuli and those made to the most effective unisensory stimuli. What signals the cortex provides that are apparently so critical for multisensory integration in the midbrain have yet to be identified.

Collectively, these studies suggest that infant animals should not be able to benefit from combining inputs across the different senses. Unfortunately, little is known about the multisensory perceptual abilities in developing animals because of the difficulty of conducting such experiments. In contrast, a great deal of research has been carried out on human infants. Although these studies have demonstrated that certain multisensory processing skills are present at birth, the ability to match or integrate particular combinations of sensory cues continues to mature over the following months and years. Of particular relevance to the maturation of multisensory spatial processing is the finding that infants only start to integrate auditory and visual localization cues toward the end of the first year of life (Neil et al., 2006), which is clearly consistent with the gradual appearance of the multisensory integrative properties of SC neurons. But while many studies point to a progressive expansion in multisensory abilities, such as the ability to perceive

more complex crossmodal relationships based on gender or affect, recent work suggests that a developmental narrowing also takes place, leading to infants losing their initial ability to match the visual and auditory attributes of nonnative social signals (reviewed by Lewkowicz and Ghazanfar, 2009). What appears to be critical for these changes in multisensory perception is exposure to appropriate sensory inputs during early life. How experience influences the maturation of multisensory circuits in the brain is considered in the following sections.

4.3.1 Role of Experience in Aligning the Sensory Maps in the SC

The importance of experience in merging the spatial cues provided by different sense systems has been demonstrated in a number of studies in which sensory inputs have been altered in particular ways. For example, if barn owls (Knudsen, 1985; Mogdans and Knudsen, 1992) or ferrets (King et al., 1988, 2000) are raised with one ear blocked, the auditory receptive fields in the SC (or its homolog in birds, the optic tectum) develop relatively normally and follow the topography of the visual representation, even though the sound localization cues available are highly abnormal. A compensatory change in the developing auditory representation can also be induced by manipulating the visual inputs. This has been achieved in barn owls by mounting prisms in front of the animal's eyes, which, because these animals have very limited eye movements, results in a corresponding displacement of the visual map in the optic tectum. Prism rearing produces a corresponding shift in both the tectal map of auditory space (Figure 4.3(a) and 4.3(b); Knudsen and Brainard, 1991) and in the accuracy of sound-evoked head-orienting responses (Knudsen and Knudsen, 1990). A shift in the auditory map in the SC has also been observed in ferrets in which the orbital position of the contralateral eye was changed by removing the medial rectus muscle in infancy (King et al., 1988). As discussed,

auditory responses in the mammalian SC are modulated whenever an animal alters its direction of gaze. However, the shift in auditory spatial tuning observed in this study appeared to be caused by visual experience rather than the change in eye position itself (King and Carlile, 1995).

In barn owls, prism-induced plasticity of auditory spatial tuning has also been described in another midbrain nucleus, the external nucleus of the inferior colliculus (ICX), where a map of auditory space is first generated and then relayed to the optic tectum (Brainard and Knudsen, 1993; Figure 4.3(c) and 4.3(d)). The shift in the auditory map in the ICX is brought about by a rewiring of connections from the central nucleus of the inferior colliculus (ICC; DeBello et al., 2001) and appears to be triggered by visual signals that are transmitted from the upper layers of the optic tectum to the ICX (Hyde and

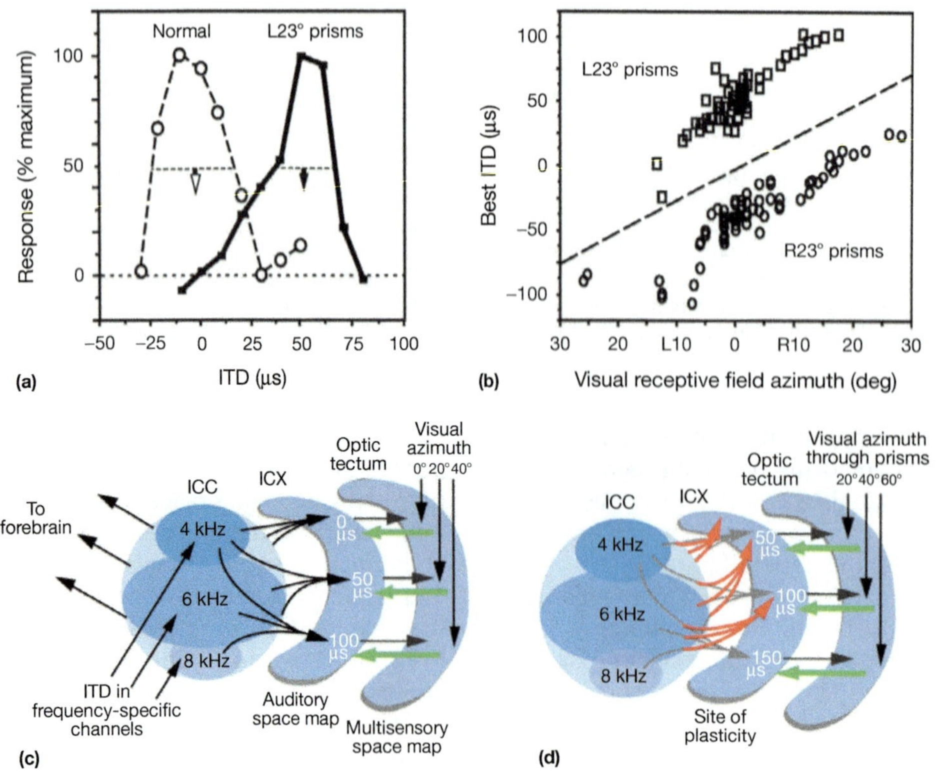

FIGURE 4.3   Prism adaptation in the barn owl. (a) The effects of prism experience on the spatial tuning of neurons in the optic tectum. These neurons derive their azimuth sensitivity through their tuning to interaural time differences (ITDs). Before prism experience (Normal), the neuron shown is tuned to about 0° µs ITD, matching its visual receptive field, which is centered at 0° azimuth. After a young owl has experienced prisms that displace the visual field to the left by 23° (L23°) for more than 8 weeks, the ITD tuning at this site has shifted toward the values produced by sounds at the locations of the neuron's optically displaced visual receptive field (in this case, 50 µs right-ear leading). Down arrows indicate the best ITD in each case. Negative ITD values indicate left-ear leading. (b) Best ITDs plotted as a function of visual receptive field azimuth for neurons recorded in the optic tecta of owls raised with either L23° prisms (boxes) or R23° prisms (circles). The expected relationship found in normal owls is indicated by the dashed line. (c) The midbrain pathway in a normal barn owl. ITDs are mapped in frequency-specific channels in the central nucleus of the inferior colliculus (ICC). Information across frequencies is then integrated by neurons in the external nucleus of the inferior colliculus (ICX) to create space-specific auditory neurons that are organized to form a map of contralateral space. The map is conveyed to the optic tectum, where it merges with a visual map of space from the retina and forebrain. Topographic projections back from the optic tectum to the ICX (green arrows) are thought to carry visual spatial information to instruct auditory plasticity in the ICX. (d) Following a period of several weeks of prism experience during early life, a systematic change occurs in the ICC-ICX projection (red arrows), which shifts the auditory maps in both the ICX and the OT, so that they realign with the shifted visual map. *Adapted from Keuroghlian AS and Knudsen EI (2007) Adaptive auditory plasticity in developing and adult animals.* Progress in Neurobiology *82: 109–121, with permission.*

Knudsen, 2002). A similar circuit is likely to exist in mammals, where the exclusively visual superficial layers of the SC project topographically both to the deeper multisensory layers and to the nucleus of the brachium of the inferior colliculus (IC; Doubell et al., 2000, 2003), which is the principal source of auditory input to the SC (Figure 4.4; King et al., 1998a). The finding that partial lesions of the superficial SC layers in neonatal ferrets disrupt the developing topography of the underlying auditory representation suggests that they might provide a template for guiding the development and plasticity of the auditory responses (King et al., 1998b).

Therefore, vision plays an instructive role in merging multisensory spatial information during development through its influence on the maturation of the auditory space map. This resembles the visual dominance seen in studies of stimulus localization by adult humans described in the first section of this chapter and is consistent with the fact that vision generally provides more precise and reliable spatial information. Moreover, as previously discussed, a map of auditory space has to be computed in the brain by tuning neurons to spatial cues

whose values depend on the physical dimensions of the head and external ears, and those dimensions often change considerably during development. The retinotopic visual map most likely guides the emergence of auditory spatial selectivity by helping to overcome the uncertainty and variability in the relationship between auditory localization cues and directions in space and thus promotes the alignment of sensory maps that are constructed in different ways (Gutfreund and King, 2012).

Early loss of vision therefore results in a partial breakdown in visual–auditory map alignment (King and Carlile, 1993; Knudsen et al., 1991; Wallace et al., 2004). Some studies have also reported that auditory neurons in the SC of visually deprived animals can have abnormally large receptive fields (Knudsen et al., 1991; Wallace et al., 2004; Withington, 1992; Withington-Wray et al., 1990b) or are ambiguously tuned to multiple sound directions (King and Carlile, 1993). Although confirming that visual inputs play a pivotal role in guiding the formation of a normal map of auditory space in the brain, accurate sound localization can develop in the absence of vision. Indeed, as discussed in more detail later

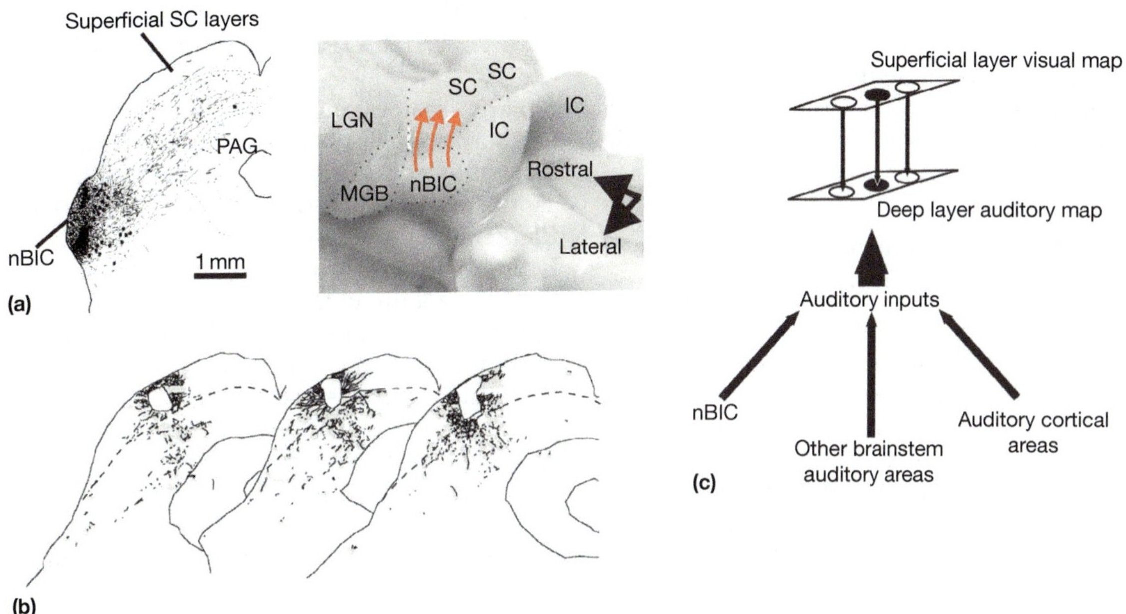

FIGURE 4.4 Circuits underlying auditory map formation in the mammalian superior colliculus (SC). (a) The left panel shows in a camera lucida drawing of a coronal section through the ferret midbrain the distribution of retrograde and anterograde labeling following a single injection of biotinylated dextran amine conjugated with fluorescein in the nucleus of the brachium of the inferior colliculus (nBIC). Fibers and terminals are represented by lines and labeled somata by circles. The right panel shows a lateral view of the ferret midbrain, with the arrows illustrating the topographic organization of the projection from the nBIC to the deeper SC layers. (b) Series of sections from a ferret in which a single injection of biocytin had been made in the superficial visual layers of the SC. The light gray shading depicts areas containing boutons, and black stippling indicates axons. (c) Summary of visual and auditory inputs to the deeper layers of the SC. This part of the SC receives converging inputs from several auditory brainstem and cortical areas, including a spatially ordered projection from the nBIC. Neurons in the superficial layers of the SC project topographically both to the underlying deeper layers and also to the nBIC (not shown). The map of visual space in the superficial layers appears to provide an activity template that guides the development and plasticity of the auditory representation. Other abbreviations: IC, inferior colliculus; LGN, lateral geniculate nucleus; MGB, medial geniculate body; PAG, periaqueductal gray. Anatomical data from Doubell et al. (2003) and Nodal et al. (2005).

in this chapter, visually impaired or blind individuals can localize sound just as well as or even better than subjects with normal sight (Collignon et al., 2009; King and Parsons, 1999; Lessard et al., 1998; Rauschecker and Kniepert, 1994; Röder et al., 1999). Such improvements in perceptual abilities could, of course, be based on changes in other regions of the brain, particularly in the cortex. But, where multisensory spatial representations are brought together, as in the SC, concurrently available visual inputs are used to constrain the developing auditory map so that it matches the visual field representation.

Although these studies emphasize the importance of vision in coordinating multisensory inputs in the developing brain, it is not always vision that dominates intersensory spatial relations. If the somatosensory receptive fields of SC neurons are shifted by modifying the position of the vibrissae in newborn mice, a compensatory reorganization of the visual map takes place (Benedetti and Ferro, 1995). Indeed, while there is strong evidence that plastic auditory inputs are guided by a stable visual template, it has also been proposed that each of the senses is weighted during development according to the reliability of the spatial information they provide (Witten et al., 2008) in much the same way that visual and auditory spatial signals are now thought to interact when humans make localization judgments (Alais et al., 2010). In this scenario, auditory receptive fields are plastic simply because they are larger than the visual receptive fields and therefore convey less precise spatial information.

4.3.2 Role of Experience in the Development of Multisensory Integration in the SC

In addition to its importance in aligning the different maps in the SC, appropriate sensory experience is critical for the emergence of the multisensory integrative abilities of SC neurons. Thus, multisensory neurons in the SC of dark-reared cats have abnormally large spatial receptive fields and lack the capacity to integrate different modality cues (Wallace et al., 2004). In other words, combining spatiotemporally congruent visual and auditory stimuli did not produce the enhancement of unisensory responses that is characteristic of animals that have been raised normally.

However, while experience with objects that can be both seen and heard promotes intersensory map alignment, this is not a prerequisite for multisensory integration. Wallace and Stein (2007) showed this by raising cats in the dark and periodically presenting them with simultaneous light and sound at a fixed spatial disparity. Although most of the SC neurons recorded when the animals were fully grown again had receptive fields that had failed to contract and lacked multisensory

integration, a few had relatively small visual and auditory receptive fields whose relative locations matched the spatial configuration of the bisensory stimuli to which they had been exposed (Wallace and Stein, 2007). As in normally raised control cats, these neurons did exhibit multisensory enhancement when visual and auditory stimuli occurred together within their respective receptive fields, but this now meant presenting those stimuli from different locations in space. Provision of abnormal multisensory experience can therefore lead to neurons developing the capacity to integrate visual and auditory cues from seemingly unrelated sources. This illustrates the importance of aligning the different sensory maps in the SC, as it is presumably only then that behaviorally relevant crossmodal interactions can take place.

Behavioral observations in visually deprived humans also highlight the importance of experience in acquiring the ability to synthesize inputs from different modalities. Like the dark-reared cats, humans lacking early patterned visual experience as a result of congenital cataracts show reduced auditory–visual interactions, whereas their performance in unisensory tasks was normal (Putzar et al., 2007).

4.4 DEVELOPMENT OF MULTISENSORY CIRCUITS IN THE CORTEX

Early anatomical studies reported the presence in newborn animals of transient inputs from other sensory systems to areas of the cortex that are traditionally viewed as modality specific (Dehay et al., 1988; Innocenti and Clarke, 1984). However, since multisensory convergence is observed to a greater or lesser extent throughout the fully mature cerebral cortex, it is unlikely that such connections disappear altogether. Little is known about the maturation of the multisensory response properties of cortical neurons. However, Wallace et al. (2006) recorded in the AES of cats at different postnatal ages and found that responsiveness to somatosensory, auditory, and visual stimulation emerges in that order and that the capacity of the multisensory neurons to integrate crossmodal cues takes several months to mature. This sequence therefore parallels the ontogeny of sensory function in the SC.

By raising barn owls with prismatic spectacles, Miller and Knudsen (1999) found that, as in the space-mapped regions of the midbrain, visual experience shapes the developing auditory spatial receptive fields of neurons in the forebrain. However, the great majority of evidence for multisensory interactions in the cortex during development has come from studies in mammals in which experience of one modality has been restricted or eliminated altogether. In the cat AES, for example, early visual deprivation has been reported to cause an expansion in the territory of auditory neurons (Rauschecker and Korte, 1993) and a sharpening of their spatial sensitivity (Korte and Rauschecker, 1993), as well as a change in multisensory integrative properties, which become dominated by response depression rather than enhancement (Carriere et al., 2007). These results therefore differ from those described previously for the SC, in which visual deprivation tends to either broaden or leave auditory spatial tuning unchanged and prevents the appearance of multisensory integration. Nevertheless, they show that, as in the midbrain, sensory experience plays a critical role in shaping the functional organization of cortical multisensory areas.

Many other studies have also examined the crossmodal consequences of temporary or permanent loss of one of the senses. The enhanced perceptual abilities that are often reported in those studies could result from altered processing within the neural pathways of the intact modalities (Elbert et al., 2002; Korte and Rauschecker, 1993), perhaps reflecting the greater attention paid to those modalities in the absence of the sense that has been lost. But the functional identity of an area of sensory cortex can also change if its primary input is removed, to be replaced by inputs from other modalities (Figure 4.5). Remarkably, this can happen even if existing multisensory inputs are temporarily unmasked in adulthood. Thus, wearing a blindfold for a few days leads to primary visual cortex becoming recruited for tactile processing (Merabet et al., 2008). However, the most dramatic examples of sensory substitution have been reported in humans who are congenitally deaf or blind. For example, the visual cortex appears to contribute to the superior auditory localization abilities of blind individuals (Collignon et al., 2007; Gougoux et al., 2005; Renier et al., 2010). This presumably results either from the formation of novel functional connections from auditory to visual brain areas or from the unmasking of connections that are now known to exist normally.

The capacity of a cortical area to take on the functions of a different sensory modality has been illustrated in a series of studies in animals in which visual inputs were surgically redirected into the auditory thalamus and cortex. Neonatal lesions of the midbrain (SC and IC) partially remove normal retinal targets and the main source of auditory input to the medial geniculate body. As a result, retinal afferents form connections that allow visual information to be conveyed to the auditory cortex (Kalil and Schneider, 1975; Sur et al., 1988). In 'rewired' ferrets, neurons in the auditory cortex have been shown to display receptive field properties, including orientation and direction selectivity, characteristic of the visual cortex (Roe et al., 1992), and even support limited visually guided behavior (von Melchner et al., 2000). Similar results have been obtained in hamsters in which visual afferents were redirected into either the auditory (Frost et al., 2000) or somatosensory cortex (Métin and Frost, 1989).

These studies show that the modality specificity of a given area of the cortex is determined by its inputs and can change in behaviorally relevant ways if normal inputs are replaced by those from another sense. Nevertheless, the functional specificity of different cortical areas seems to be retained following crossmodal reorganization. In early blind humans, part of the occipital cortex normally involved in visual spatial processing is more strongly activated when subjects perform auditory or tactile localization rather than identification tasks (Renier et al., 2010). Similarly, visual localization deficits are produced in deaf cats by deactivating auditory cortical areas that are known to be involved in sound localization in animals with normal hearing (Lomber et al., 2010; Meredith et al., 2011). Such crossmodal plasticity could potentially arise from an expansion or unmasking of connections that selectively link cortical areas sharing the same or related functions in different modalities. Alternatively, this reallocation of resources might be mediated via top–down pathways to different sensory areas involved in stimulus localization.

4.5 SENSITIVE PERIODS IN THE MATURATION OF MULTISENSORY PROCESSING

The finding that human multisensory perceptual abilities narrow during infancy (Lewkowicz and Ghazanfar, 2009) suggests that a sensitive period of development may exist during which experience is particularly effective in shaping the functional architecture of the brain. Several studies have indicated that the guiding influence of vision on the maturation of auditory space maps in the midbrain is

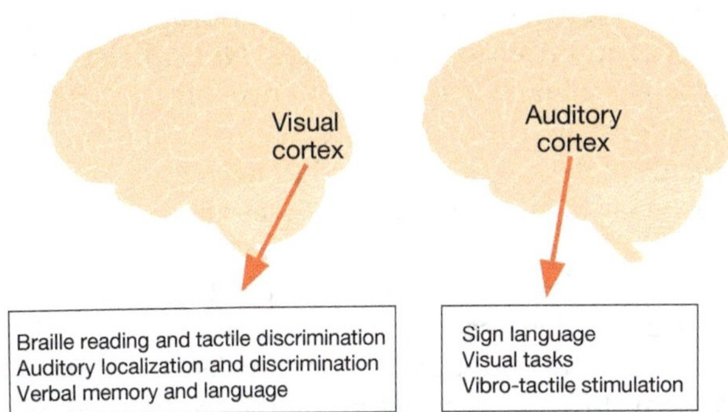

FIGURE 4.5 Crossmodal plasticity in the cortex following sensory deprivation. (a) Recruitment of visual cortex in the blind for tactile processing (e.g., Braille reading, sound localization, and verbal memory). (b) Recruitment of auditory and language-related areas in deaf people for viewing sign language, peripheral visual processing, and vibrotactile stimulation. *Reproduced from Merabet LB and Pascual-Leone A (2010) Neural reorganization following sensory loss: The opportunity of change.* Nature Reviews Neuroscience 11: 44–52.

developmentally regulated (King et al., 1998b; Knudsen and Brainard, 1991), and similar findings have been obtained from behavioral studies (Knudsen and Knudsen, 1990). However, more recent experiments have shown that adaptive crossmodal plasticity is possible throughout life. To some extent, plasticity in later life is constrained by changes in neural circuitry that take place during the sensitive period. For example, visual experience can adjust auditory spatial tuning in the midbrain of adult barn owls if the animals were previously exposed to the same prismatic displacement, but not if adult birds are fitted with prisms that displace the visual field in the opposite direction (Knudsen, 1998). This experience-specific plasticity in adulthood can be accounted for by changes in midbrain connectivity induced by prism rearing (Figure 4.3), which leave a 'trace' that can be reactivated by appropriate experience in later life. A similar explanation likely applies to other forms of sensory plasticity that were previously thought to be restricted to a sensitive period of development (Hofer et al., 2009).

The question of whether sensitive periods exist can also be addressed in deprivation studies either by determining whether the age at which the sensory loss occurs influences the nature and extent of the changes that take place in the brain or by examining the potential for recovery if the missing function is restored. That a sensitive period is present for the acquisition of multisensory integration is suggested by studies of patients who have had congenital cataracts removed. Even after several years of multisensory experience, they lack the ability to synthesize different modality cues normally (Putzar et al., 2007). Integration of auditory and visual cues is possible, however, if sensory function is restored early enough. Congenitally deaf children fitted with cochlear implants within the first 2 ½ years of life exhibit the McGurk effect, whereas, after this age, auditory and visual speech cues can no longer be fused (Schorr et al., 2005). In contrast, recent animal studies suggest that sensitivity to crossmodal experience might be maintained in later life. Following repeated presentation of spatiotemporally congruent visual and auditory cues to anesthetized adult cats that had been reared in the dark, SC neurons start to acquire their characteristic multisensory integrative properties (Yu et al., 2010).

Most of the evidence for compensatory crossmodal changes in neural processing and behavior following sensory deprivation has come from subjects who have been blind or deaf from birth or early in life. But while functional and anatomical differences do exist between early- and late-onset blind individuals (e.g., Cohen et al., 1999; Jiang et al., 2009; Stevens and Weaver, 2009), enhanced nonvisual abilities have also been reported following late-onset blindness (Fieger et al., 2006; King and Parsons, 1999; Occelli et al., 2008; Voss et al., 2004).

Together, these studies show that the mature brain retains some capacity for crossmodal plasticity, although to a lesser extent than that seen during development. As with other aspects of sensory processing (e.g., Kacelnik et al., 2006), behavioral training is likely to be an important factor in determining the extent to which the mature brain can adapt to changes in input. Similarly, adult barn owls exhibit much greater visually induced auditory plasticity if the stimuli acquire behavioral significance for the animals (Bergan et al., 2005). This suggests that top–down regulation of sensory responses is part of the circuitry involved in integrating multisensory information in the brain.

4.6 CONCLUDING REMARKS

The studies reviewed in this article show that sensory experience plays a vital role during development in establishing and maintaining the neural circuits responsible for synthesizing information across different sensory modalities. Much of this evidence has come from research in animals, which provide an opportunity both to alter sensory inputs in a controlled and reproducible manner in ways that are not possible in humans and to investigate how different senses interact at the level of individual neurons. Nevertheless, many of the animal studies have examined the crossmodal consequences of depriving inputs in one sensory modality and therefore offer valuable insights into the changes that likely take place in the human brain following blindness or deafness.

The emphasis in most of the animal work has been on the development of multisensory spatial interactions, particularly at the level of the SC, where vision has been shown to play a pivotal role in aligning the different modality maps and in the maturation of multisensory integration. The SC has provided a model system for research in this area, leading to the formulation of a number of general principles that also apply to other aspects of multisensory processing. Nevertheless, most multisensory functions, at least in primates, rely on cortical activity, and it will clearly be important to extend the range of developmental studies to include other processes, such as the integration of faces and voices during infancy.

The considerable crossmodal plasticity that takes place during development, and to some extent in later life too, has clear adaptive value. This enables information from different sensory modalities to be merged in behaviorally relevant ways and allows for crossmodal compensatory changes to take place following sensory deprivation. However, these changes can potentially be maladaptive in that they may limit the capacity of the brain to utilize sensory inputs that are restored as a result, for example, of cataract surgery or cochlear implantation. The growing evidence for adult crossmodal plasticity nevertheless holds out the promise of developing effective rehabilitation strategies in such cases for patients who are recovering from early sensory loss.

Acknowledgment

The author is supported by a Wellcome Trust Principal Research Fellowship (WT076508A1A).

References

Alais, D., Burr, D., 2004. The ventriloquist effect results from near-optimal bimodal integration. Current Biology 14, 257–262.

Alais, D., Newell, F.N., Mamassian, P., 2010. Multisensory processing in review: From physiology to behaviour. Seeing and Perceiving 23, 3–38.

Benedetti, F., Ferro, I., 1995. The effects of early postnatal modification of body shape on the somatosensory-visual organization in mouse superior colliculus. European Journal of Neuroscience 7, 412–418.

Bergan, J.F., Ro, P., Ro, D., Knudsen, E.I., 2005. Hunting increases adaptive auditory map plasticity in adult barn owls. Journal of Neuroscience 25, 9816–9820.

Bizley, J.K., King, A.J., 2008. Visual–auditory spatial processing in auditory cortical neurons. Brain Research 1242, 24–36.

Bonath, B., Noesselt, T., Martinez, A., et al., 2007. Neural basis of the ventriloquist illusion. Current Biology 17, 1697–1703.

Brainard, M.S., Knudsen, E.I., 1993. Experience-dependent plasticity in the inferior colliculus: A site for visual calibration of the neural representation of auditory space in the barn owl. Journal of Neuroscience 13, 4589–4608.

Calvert, G., Spence, C., Stein, B.E. (Eds.), 2004. The Handbook of Multisensory Processes. MIT Press, Cambridge, MA.

Campbell, R.A., King, A.J., Nodal, F.R., Schnupp, J.W.H., Carlile, S., Doubell, T.P., 2008. Virtual adult ears reveal the roles of acoustical factors and experience in auditory space map development. Journal of Neuroscience 28, 11557–11570.

Carriere, B.N., Royal, D.W., Perrault, T.J., et al., 2007. Visual deprivation alters the development of cortical multisensory integration. Journal of Neurophysiology 98, 2858–2867.

Cohen, L.G., Weeks, R.A., Sadato, N., Celnik, P., Ishii, K., Hallett, M., 1999. Period of susceptibility for cross-modal plasticity in the blind. Annals of Neurology 45, 451–460.

Collignon, O., Lassonde, M., Lepore, F., Bastien, D., Veraart, C., 2007. Functional cerebral reorganization for auditory spatial processing and auditory substitution of vision in early blind subjects. Cerebral Cortex 17, 457–465.

Collignon, O., Voss, P., Lassonde, M., Lepore, F., 2009. Cross-modal plasticity for the spatial processing of sounds in visually deprived subjects. Experimental Brain Research 192, 343–358.

Corneil, B.D., Van Wanrooij, M., Munoz, D.P., Van Opstal, A.J., 2002. Auditory–visual interactions subserving goal-directed saccades in a complex scene. Journal of Neurophysiology 88, 438–454.

Cynader, M., Berman, N., 1972. Receptive-field organization of monkey superior colliculus. Journal of Neurophysiology 35, 187–201.

DeBello, W.M., Feldman, D.E., Knudsen, E.I., 2001. Adaptive axonal remodeling in the midbrain auditory space map. Journal of Neuroscience 21, 3161–3174.

Dehay, C., Kennedy, H., Bullier, J., 1988. Characterization of transient cortical projections from auditory, somatosensory, and motor cortices to visual areas 17, 18, and 19 in the kitten. The Journal of Comparative Neurology 272, 68–89.

Doubell, T.P., Baron, J., Skaliora, I., King, A.J., 2000. Topographic projection from the superior colliculus to the nucleus of the brachium of the inferior colliculus: Evidence for convergence of visual and auditory information. European Journal of Neuroscience 12, 4290–4308.

Doubell, T.P., Skaliora, I., Baron, J., King, A.J., 2003. Functional connectivity between the superficial and deeper layers of the superior colliculus: An anatomical substrate for sensorimotor integration. Journal of Neuroscience 23, 6596–6607.

Edwards, S.B., Ginsburgh, C.L., Henkel, C.K., Stein, B.E., 1979. Sources of subcortical projections to the superior colliculus in the cat. The Journal of Comparative Neurology 184, 309–329.

Elbert, T., Sterr, A., Rockstroh, B., Pantev, C., Müller, M.M., Taub, E., 2002. Expansion of the tonotopic area in the auditory cortex of the blind. Journal of Neuroscience 22, 9941–9944.

Ernst, M.O., Banks, M.S., 2002. Humans integrate visual and haptic information in a statistically optimal fashion. Nature 415, 429–433.

Fendrich, R., Corballis, P.M., 2001. The temporal cross-capture of audition and vision. Perception & Psychophysics 63, 719–725.

Fieger, A., Röder, B., Teder-Sälejärvi, W., Hillyard, S.A., Neville, H.J., 2006. Auditory spatial tuning in late-onset blindness in humans. Journal of Cognitive Neuroscience 18, 149–157.

Frost, D.O., Boire, D., Gingras, G., Ptito, M., 2000. Surgically created neural pathways mediate visual pattern discrimination. Proceedings of the National Academy of Sciences of the United States of America 97, 11068–11073.

Gandhi, N.J., Katnani, H.A., 2011. Motor functions of the superior colliculus. Annual Review of Neuroscience 34, 205–231.

Ghazanfar, A.A., Schroeder, C.E., 2006. Is neocortex essentially multisensory? Trends in Cognitive Sciences 10, 278–285.

Ghazanfar, A.A., Maier, J.X., Hoffman, K.L., Logothetis, N.K., 2005. Multisensory integration of dynamic faces and voices in rhesus monkey auditory cortex. Journal of Neuroscience 25, 5004–5012.

Gottlieb, G., 1971. Ontogenesis of sensory function in birds and mammals. In: Tobach, E., Aronson, L.R., Shaw, E. (Eds.), The Biopsychology of Development. Academic Press, New York, pp. 67–128.

Gougoux, F., Zatorre, R.J., Lassonde, M., Voss, P., Lepore, F., 2005. A functional neuroimaging study of sound localization: Visual cortex activity predicts performance in early-blind individuals. PLoS Biology 3, e27.

Groh, J.M., Sparks, D.L., 1996. Saccades to somatosensory targets. III. Eye-position-dependent somatosensory activity in primate superior colliculus. Journal of Neurophysiology 75, 439–453.

Gutfreund, Y., King, A.J., 2012. Role of visual input in the development of the auditory space map in the midbrain. In: Stein, B.E. (Ed.), The New Handbook of Multisensory Processes. MIT Press, Cambridge, MA.

Hartline, P.H., Vimal, R.L., King, A.J., Kurylo, D.D., Northmore, D.P., 1995. Effects of eye position on auditory localization and neural representation of space in superior colliculus of cats. Experimental Brain Research 104, 402–408.

Hofer, S.B., Mrsic-Flogel, T.D., Bonhoeffer, T., Hübener, M., 2009. Experience leaves a lasting structural trace in cortical circuits. Nature 457, 313–317.

Howard, I.P., Templeton, W.B., 1966. Human Spatial Orientation. Wiley, New York.

Hyde, P.S., Knudsen, E.I., 2002. The optic tectum controls visually guided adaptive plasticity in the owl's auditory space map. Nature 415, 73–76.

Innocenti, G., Clarke, S., 1984. Bilateral transitory projection to visual areas from auditory cortex in kittens. Brain Research 14, 143–148.

Jacobs, D.S., Blakemore, C., 1988. Factors limiting the postnatal development of visual acuity in the monkey. Vision Research 28, 947–958.

Jay, M.F., Sparks, D.L., 1984. Auditory receptive fields in primate superior colliculus shift with changes in eye position. Nature 309, 345–347.

Jiang, W., Wallace, M.T., Jiang, H., Vaughan, J.W., Stein, B.E., 2001. Two cortical areas mediate multisensory integration in superior colliculus neurons. Journal of Neurophysiology 85, 506–522.

Jiang, W., Jiang, H., Stein, B.E., 2002. Two corticotectal areas facilitate multisensory orientation behavior. Journal of Cognitive Neuroscience 14, 1240–1255.

Jiang, W., Jiang, H., Stein, B.E., 2006. Neonatal cortical ablation disrupts multisensory development in superior colliculus. Journal of Neurophysiology 95, 1380–1396.

Jiang, W., Jiang, H., Rowland, B.A., Stein, B.E., 2007. Multisensory orientation behavior is disrupted by neonatal cortical ablation. Journal of Neurophysiology 97, 557–562.

Jiang, J., Zhu, W., Shi, F., et al., 2009. Thick visual cortex in the early blind. Journal of Neuroscience 29, 2205–2211.

Kacelnik, O., Nodal, F.R., Parsons, C.H., King, A.J., 2006. Training-induced plasticity of auditory localization in adult mammals. PLoS Biology 4, 627–638.

Kalil, R.E., Schneider, G.E., 1975. Abnormal synaptic connections of the optic tract in the thalamus after midbrain lesions in newborn hamsters. Brain Research 100, 690–698.

King, A.J., Carlile, S., 1993. Changes induced in the representation of auditory space in the superior colliculus by rearing ferrets with binocular eyelid suture. Experimental Brain Research 94, 444–455.

King, A.J., Carlile, S., 1995. Neural coding for auditory space. In: Gazzaniga, M.S. (Ed.), The Cognitive Neurosciences. MIT Press, Cambridge, MA, pp. 279–293.

King, A.J., Hutchings, M.E., 1987. Spatial response properties of acoustically-responsive neurons in the superior colliculus of the ferret: A map of auditory space. Journal of Neurophysiology 57, 596–624.

King, A.J., Nelken, I., 2009. Unraveling the principles of auditory cortical processing: Can we learn from the visual system? Nature Neuroscience 12, 698–701.

King, A.J., Palmer, A.R., 1983. Cells responsive to free-field auditory stimuli in guinea-pig superior colliculus: Distribution and response properties. The Journal of Physiology 342, 361–381.

King, A.J., Palmer, A.R., 1985. Integration of visual and auditory information in bimodal neurones in the guinea-pig superior colliculus. Experimental Brain Research 60, 492–500.

King, A.J., Parsons, C.H., 1999. Improved auditory spatial acuity in visually deprived ferrets. European Journal of Neuroscience 11, 3945–3956.

King, A.J., Hutchings, M.E., Moore, D.R., Blakemore, C., 1988. Developmental plasticity in the visual and auditory representations in the mammalian superior colliculus. Nature 332, 73–76.

King, A.J., Jiang, Z.D., Moore, D.R., 1998. Auditory brainstem projections to the ferret superior colliculus: Anatomical contribution to the neural coding of sound azimuth. The Journal of Comparative Neurology 390, 342–365.

King, A.J., Schnupp, J.W.H., Thompson, I.D., 1998. Signals from the superficial layers of the superior colliculus enable the development of the auditory space map in the deeper layers. Journal of Neuroscience 18, 9394–9408.

King, A.J., Parsons, C.H., Moore, D.R., 2000. Plasticity in the neural coding of auditory space in the mammalian brain. Proceedings of the National Academy of Sciences of the United States of America 97, 11821–11828.

King, A.J., Schnupp, J.W.H., Doubell, T.P., 2001. The shape of ears to come: Dynamic coding of auditory space. Trends in Cognitive Sciences 5, 261–270.

Kiorpes, L., Movshon, J.A., 1990. Behavioral analysis of visual development. In: Coleman, J.R. (Ed.), Development of Sensory Systems in Mammals. John Wiley & Sons, New York, pp. 125–154.

Knudsen, E.I., 1985. Experience alters the spatial tuning of auditory units in the optic tectum during a sensitive period in the barn owl. Journal of Neuroscience 5, 3094–3109.

Knudsen, E.I., 1998. Capacity for plasticity in the adult owl auditory system expanded by juvenile experience. Science 279, 1531–1533.

Knudsen, E.I., Brainard, M.S., 1991. Visual instruction of the neural map of auditory space in the developing optic tectum. Science 253, 85–87.

Knudsen, E.I., Knudsen, P.F., 1990. Sensitive and critical periods for visual calibration of sound localization by barn owls. Journal of Neuroscience 10, 222–232.

Knudsen, E.I., Esterly, S.D., du Lac, S., 1991. Stretched and upside-down maps of auditory space in the optic tectum of blind-reared owls; acoustic basis and behavioral correlates. Journal of Neuroscience 11, 1727–1747.

Korte, M., Rauschecker, J.P., 1993. Auditory spatial tuning of cortical neurons is sharpened in cats with early blindness. Journal of Neurophysiology 70, 1717–1721.

Lessard, N., Paré, M., Lepore, F., Lassonde, M., 1998. Early-blind human subjects localize sound sources better than sighted subjects. Nature 395, 278–280.

Lewkowicz, D.J., Ghazanfar, A.A., 2009. The emergence of multisensory systems through perceptual narrowing. Trends in Cognitive Sciences 13, 470–478.

Lomber, S.G., Meredith, M.A., Kral, A., 2010. Cross-modal plasticity in specific auditory cortices underlies visual compensations in the deaf. Nature Neuroscience 13, 1421–1427.

Ma, W.J., Zhou, X., Ross, L.A., Foxe, J.J., Parra, L.C., 2009. Lip-reading aids word recognition most in moderate noise: A Bayesian explanation using high-dimensional feature space. PLoS One 4, e4638.

McGurk, H., MacDonald, J., 1976. Hearing lips and seeing voices. Nature 264, 746–748.

Merabet, L.B., Hamilton, R., Schlaug, G., et al., 2008. Rapid and reversible recruitment of early visual cortex for touch. PLoS One 3, e3046.

Meredith, M.A., Stein, B.E., 1986. Visual, auditory, and somatosensory convergence on cells in superior colliculus results in multisensory integration. Journal of Neurophysiology 56, 640–662.

Meredith, M.A., Stein, B.E., 1996. Spatial determinants of multisensory integration in cat superior colliculus neurons. Journal of Neurophysiology 75, 1843–1857.

Meredith, M.A., Nemitz, J.W., Stein, B.E., 1987. Determinants of multisensory integration in superior colliculus neurons. I. Temporal factors. Journal of Neuroscience 7, 3215–3229.

Meredith, M.A., Kryklywy, J., McMillan, A.J., Malhotra, S., Lum-Tai, R., Lomber, S.G., 2011. Crossmodal reorganization in the early deaf switches sensory, but not behavioral roles of auditory cortex. Proceedings of the National Academy of Sciences of the United States of America 108, 8856–8861.

Métin, C., Frost, D.O., 1989. Visual responses of neurons in somatosensory cortex of hamsters with experimentally induced retinal projections to somatosensory thalamus. Proceedings of the National Academy of Sciences of the United States of America 86, 357–361.

Middlebrooks, J.C., Knudsen, E.I., 1984. A neural code for auditory space in the cat's superior colliculus. Journal of Neuroscience 4, 2621–2634.

Miller, G.L., Knudsen, E.I., 1999. Early visual experience shapes the representation of auditory space in the forebrain gaze fields of the barn owl. Journal of Neuroscience 19, 2326–2336.

Mishra, J., Martinez, A., Sejnowski, T.J., Hillyard, S.A., 2007. Early cross-modal interactions in auditory and visual cortex underlie a sound-induced visual illusion. Journal of Neuroscience 27, 4120–4131.

Mogdans, J., Knudsen, E.I., 1992. Adaptive adjustment of unit tuning to sound localization cues in response to monaural occlusion in developing owl optic tectum. Journal of Neuroscience 12, 3473–3484.

Neil, P.A., Chee-Ruiter, C., Scheier, C., Lewkowicz, D.J., Shimojo, S., 2006. Development of multisensory spatial integration and perception in humans. Developmental Science 9, 454–464.

Nodal, F.R., Doubell, T.P., Jiang, Z.D., Thompson, I.D., King, A.J., 2005. Development of the projection from the nucleus of the brachium of the inferior colliculus to the superior colliculus in the ferret. The Journal of Comparative Neurology 485, 202–217.

Occelli, V., Spence, C., Zampini, M., 2008. Audiotactile temporal order judgments in sighted and blind individuals. Neuropsychologia 46, 2845–2850.

Populin, L.C., Tollin, D.J., Yin, T.C.T., 2004. Effect of eye position on saccades and neuronal responses to acoustic stimuli in the superior colliculus of the behaving cat. Journal of Neurophysiology 92, 2151–2167.

Putzar, L., Goerendt, I., Lange, K., Rösler, F., Röder, B., 2007. Early visual deprivation impairs multisensory interactions in humans. Nature Neuroscience 10, 1243–1245.

Rauschecker, J.P., Kniepert, U., 1994. Auditory localization behaviour in visually deprived cats. European Journal of Neuroscience 6, 149–160.

Rauschecker, J.P., Korte, M., 1993. Auditory compensation for early blindness in cat cerebral cortex. Journal of Neuroscience 13, 4538–4548.

Recanzone, G.H., 2003. Auditory influences on visual temporal rate perception. Journal of Neurophysiology 89, 1078–1093.

Renier, L.A., Anurova, I., De Volder, A.G., Carlson, S., VanMeter, J., Rauschecker, J.P., 2010. Preserved functional specialization for spatial processing in the middle occipital gyrus of the early blind. Neuron 68, 138–148.

Röder, B., Teder-Sälejärvi, W., Sterr, A., Rösler, F., Hillyard, S.A., Neville, H.J., 1999. Improved auditory spatial tuning in blind humans. Nature 400, 162–166.

Roe, A.W., Pallas, S.L., Kwon, Y.H., Sur, M., 1992. Visual projections routed to the auditory pathway in ferrets: Receptive fields of visual neurons in primary auditory cortex. Journal of Neuroscience 12, 3651–3664.

Rubenstein, J.L.R., Rakic, P., 2013. Patterning and Cell Types Specification in the Developing CNS and PNS.

Schorr, E.A., Fox, N.A., van Wassenhove, V., Knudsen, E.I., 2005. Auditory–visual fusion in speech perception in children with cochlear implants. Proceedings of the National Academy of Sciences of the United States of America 102, 18748–18750.

Schroeder, C.E., Lakatos, P., Kajikawa, Y., Partan, S., Puce, A., 2008. Neuronal oscillations and visual amplification of speech. Trends in Cognitive Sciences 12, 106–113.

Shams, L., Kamitani, Y., Shimojo, S., 2000. Illusions: What you see is what you hear. Nature 408, 788.

Shelton, B.R., Searle, C.L., 1980. The influence of vision on the absolute identification of sound-source position. Perception & Psychophysics 28, 589–596.

Slutsky, D.A., Recanzone, G.H., 2001. Temporal and spatial dependency of the ventriloquism effect. Neuroreport 12, 7–10.

Stein, B.E., Labos, E., Kruger, L., 1973. Sequence of changes in properties of neurons of superior colliculus of the kitten during maturation. Journal of Neurophysiology 36, 667–679.

Stein, B.E., Magalhães-Castro, B., Kruger, L., 1976. Relationship between visual and tactile representations in cat superior colliculus. Journal of Neurophysiology 39, 401–419.

Stein, B.E., Huneycutt, W.S., Meredith, M.A., 1988. Neurons and behavior: The same rules of multisensory integration apply. Brain Research 448, 355–358.

Stevens, A.A., Weaver, K.E., 2009. Functional characteristics of auditory cortex in the blind. Behavioural Brain Research 196, 134–138.

Sumby, W., Pollack, I., 1954. Visual contribution to speech intelligibility in noise. Journal of the Acoustical Society of America 26, 212–215.

Sur, M., Garraghty, P.E., Roe, A.W., 1988. Experimentally induced visual projections into auditory thalamus and cortex. Science 242, 1437–1441.

Tavazoie, S.F., Reid, R.C., 2000. Diverse receptive fields in the lateral geniculate nucleus during thalamocortical development. Nature Neuroscience 3, 608–616.

von Melchner, L., Pallas, S.L., Sur, M., 2000. Visual behaviour mediated by retinal projections directed to the auditory pathway. Nature 404, 871–876.

Voss, P., Lassonde, M., Gougoux, F., Fortin, M., Guillemot, J.P., Lepore, F., 2004. Early- and late-onset blind individuals show supra-normal auditory abilities in far-space. Current Biology 14, 1734–1738.

Wallace, M.T., Stein, B.E., 1996. Sensory organization of the superior colliculus in cat and monkey. Progress in Brain Research 112, 301–311.

Wallace, M.T., Stein, B.E., 1997. Development of multisensory neurons and multisensory integration in cat superior colliculus. Journal of Neuroscience 17, 2429–2444.

Wallace, M.T., Stein, B.E., 2001. Sensory and multisensory responses in the newborn monkey superior colliculus. Journal of Neuroscience 21, 8886–8894.

Wallace, M.T., Stein, B.E., 2002. Onset of cross-modal synthesis in the neonatal superior colliculus is gated by the development of cortical influences. Journal of Neurophysiology 83, 3578–3582.

Wallace, M.T., Stein, B.E., 2007. Early experience determines how the senses will interact. Journal of Neurophysiology 97, 921–926.

Wallace, M.T., Meredith, M.A., Stein, B.E., 1993. Converging influences from visual, auditory, and somatosensory cortices onto output neurons of the superior colliculus. Journal of Neurophysiology 69, 1797–1809.

Wallace, M.T., Perrault Jr., T.J., Hairston, W.D., Stein, B.E., 2004. Visual experience is necessary for the development of multisensory integration. Journal of Neuroscience 24, 9580–9584.

Wallace, M.T., Carriere, B.N., Perrault Jr., T.J., Vaughan, J.W., Stein, B.E., 2006. The development of cortical multisensory integration. Journal of Neuroscience 26, 11844–11849.

Welch, R.B., Warren, D.H., 1980. Immediate perceptual response to intersensory discrepancy. Psychological Bulletin 88, 638–667.

Withington, D.J., 1992. The effect of binocular lid suture on auditory responses in the guinea-pig superior colliculus. Neuroscience Letters 136, 153–156.

Withington-Wray, D.J., Binns, K.E., Keating, M.J., 1990a. The developmental emergence of a map of auditory space in the superior colliculus of the guinea pig. Brain Research. Developmental Brain Research 51, 225–236.

Withington-Wray, D.J., Binns, K.E., Keating, M.J., 1990b. The maturation of the superior collicular map of auditory space in the guinea pig is disrupted by developmental visual deprivation. European Journal of Neuroscience 2, 682–692.

Witten, I.B., Knudsen, E.I., Sompolinsky, H., 2008. A Hebbian learning rule mediates asymmetric plasticity in aligning sensory representations. Journal of Neurophysiology 100, 1067–1079.

Yu, L., Stein, B.E., Rowland, B.A., 2010. Initiating the development of multisensory integration by manipulating sensory experience. Journal of Neuroscience 30, 4904–4913.

Cerebellar Circuits

M. Kano[1], M. Watanabe[2]

[1]The University of Tokyo, Tokyo, Japan; [2]Hokkaido University, Sapporo, Japan

5.1 OVERVIEW OF THE MICROCIRCUIT IN THE CEREBELLAR CORTEX

5.1.1 Cell Types and Afferent Fibers

The cerebellum consists of the cortex and the centrally located deep cerebellar nuclei (DCN). The cerebellar cortex exhibits a characteristic trilaminar structure composed of the molecular layer, Purkinje cell (PC) layer, and granular layer. In its mediolateral extent, the cerebellar cortex is divided into three longitudinal regions: vermis (medial cerebellum), paravermis (intermediate cerebellum or pars intermedia), and hemisphere (lateral cerebellum). Each of these regions is folded into lobules. The DCN also have three divisions: the medial (fastigial),

interpositus (globose and emboliform), and lateral (dentate) nuclei, each of which is connected topographically with the vermis, paravermis, and hemisphere, respectively. Cerebellar neurons with distinct cytological and neurochemical properties reside in specific layers and sites of the cerebellum. They are connected with each other and also with specific brain regions outside the cerebellum (Figure 5.1).

PCs are the sole output neurons of the cerebellar cortex. The somata of PCs are aligned in the PC layer. PCs extend well-arborized dendrites in the molecular layer and project γ-aminobutyric acid (GABA)ergic axons to DCN and vestibular nuclei. There are two distinct excitatory afferents to the cerebellum, that is, climbing fibers (CFs) and mossy fibers (MFs; Palay and Chan-Palay, 1974). In adulthood, each PC is innervated by a single

© 2013 Elsevier Inc. All rights reserved.

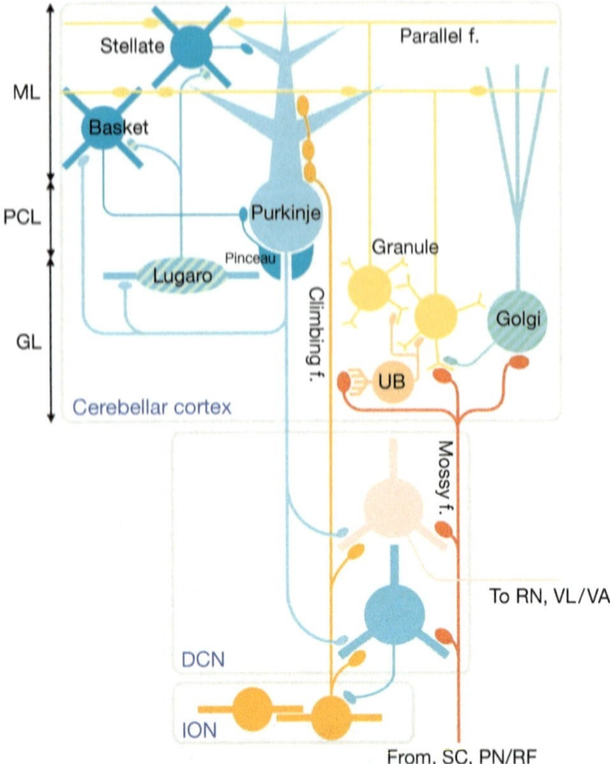

FIGURE 5.1   Neuronal component and synaptic wiring diagram of the cerebellum. Neurons painted with warm colors (yellow, orange, and pink) represent excitatory neurons, while those with cold colors (blue, light blue, and green) indicate inhibitory neurons. DCN, deep cerebellar nuclei; GL, granular layer; ION, inferior olivary nucleus; ML, molecular layer; PCL, Purkinje cell layer; UB, unipolar brush cell. PN, pontine nuclei; RF, reticular formation; RN, red nucleus; SC, spinal cord; VL/VA, ventral lateral/anterior thalamic nucleus.

(Barbour, 1993). PFs also excite two types of GABAergic interneurons in the molecular layer, basket cells, and stellate cells. Although these two types of interneuron share similar features (Sultan and Bower, 1998), they differ in short-term plasticity (Bao et al., 2010), gene expression (Schilling and Oberdick, 2009), and most importantly, the target of projection. Basket cell axons innervate the soma and surround the axon initial segment (AIS) of PCs, while stellate cells innervate dendrites of PCs (Palay and Chan-Palay, 1974). The specialized conical structure formed by basket cell axons around the AIS is called the pinceau formation (Ango et al., 2004).

In the granular layer, there are several types of interneurons, which do not project to PCs directly. Golgi cells, Lugaro cells, and globular cells are inhibitory interneurons with a dual glycinergic/GABAergic phenotype (Ottersen et al., 1988). Golgi cells extend dendritic trees radiating through the cerebellar cortex. They receive excitatory inputs from PFs in the molecular layer and from MFs and presumably also from CFs in the granular layer. Golgi cells, in turn, innervate GC dendrites in the cerebellar glomeruli, thus providing both feedforward and feedback inhibition to GCs. Golgi cells also receive inhibitory inputs presumably from other Golgi cells, basket cells, stellate cells, and Lugaro cells. Lugaro cells are fusiform inhibitory interneurons lying just beneath the PC layer. They receive GABAergic inputs from axon collaterals of PCs and project axons to the molecular layer to innervate basket and stellate cells, thus providing feedback inhibition to PCs through these molecular layer interneurons (Laine and Axelrad, 2002). Globular cells have globular somata at variable depths in the granular layer and are thought to be a subtype of Lugaro cells (Laine and Axelrad, 2002). Lugaro cells and globular cells are neurochemically distinguished from Golgi cells; the former express calretinin, while most of the latter express mGluR2 and neurogranin (Simat et al., 2007). Unipolar brush cells are excitatory interneurons enriched in the granular layer of the vestibulocerebellum. They are characterized by single short dendrites terminating with a brush of dendrites, which engulf one or two rosettes of glutamatergic and cholinergic MFs. They innervate dendrites of other unipolar brush cells and GCs, thus regarded as an intermediate component that amplify excitatory drives of MFs on to GCs (Mugnaini and Floris, 1994).

CF originating from the inferior olive of the contralateral medulla oblongata (Eccles et al., 1966). Each CF forms hundreds of synapses by twisting around the proximal dendritic compartment. Therefore, activation of CFs causes strong depolarization of PC dendrites, triggers regenerative 'Ca^{2+} spikes' due to activation of voltage-gated Ca^{2+} channels (VDCCs) in PC dendrites (Miyakawa et al., 1992), and generates characteristic 'complex spikes' in the PC soma (Eccles et al., 1966) that are composed of a fast somatic Na$^+$ action potential followed by slow dendritic Ca^{2+} spikes. In contrast, MFs originating from various extracerebellar regions, such as the spinal cord, pontine nuclei, and reticular formation, convey motor and sensory information to the distal dendritic compartment of PCs through parallel fibers (PFs), the axons of granule cells (GCs; Ito, 1984). Approximately, 10^5–10^6 PFs innervate a given PC, while each PF forms only one or two synapses onto individual PCs (Napper and Harvey, 1988). Thus, excitation of a single PF only weakly depolarizes PC dendrites, and about 50 GCs should fire synchronously to generate a single Na$^+$ action potential (called 'simple spike') in the PC soma

DCN neurons receive excitatory inputs from collaterals of MFs and CFs, and inhibitory inputs from PC axons. Thus, information from the cerebellar cortex is integrated with direct inputs from MFs and CFs in DCN neurons. In the DCN, GABAergic neurons are either local interneurons or projection neurons targeting their axons to the inferior olive (De Zeeuw et al., 1988), while non-GABAergic neurons (presumably glutamatergic) project to the rest of the brain, including the red nucleus and the thalamus. Thus, the DCN is in the key position of the cerebellar system and the source of cerebellar output.

5.1.2 Generation of Neurons that Constitute Microcircuits Related to PCs

All GABAergic neurons in the cerebellum originate from Ptf1a-expressing cells in the ventricular zone (Hoshino et al., 2005). Of these, projection neurons, that is, PCs and DCN neurons projecting to the inferior olive, are specified within the ventricular zone at the onset of cerebellar neurogenesis in early prenatal life (Altman and Bayer, 1997; Miale and Sidman, 1961). GABAergic interneurons also derive from the ventricular zone, but the progenitors continue to proliferate in the prospective white matter up to postnatal development (Zhang and Goldman, 1996). Phenotypic specification and diversity of GABAergic interneurons appear to be created by instructive cues provided by the microenvironment of the prospective white matter (Leto et al., 2009). On the other hand, the rhombic lip generates all progenitors of glutamatergic cerebellar neurons which then migrate via different pathways. Glutamatergic neurons in the DCN migrate to the nuclear transitory zone before descending to the prospective DCN (Fink et al., 2006). GC precursors first migrate to the cerebellar surface and form the external granular layer. There, they continue to proliferate during the postnatal period, and then descend to the internal granular layer (Rakic, 1971). Unipolar brush cells migrate to their destination through developing white matter (Englund et al., 2006).

5.1.3 Compartmentalization of the Cerebellum

Although the basic cellular composition and wiring diagram are uniform across the cerebellum, longitudinal organization of the cerebellar cortex has been demonstrated using anatomical, physiological, and molecular mapping techniques (Apps and Hawkes, 2009). Longitudinal cerebellar zones have been defined anatomically by cholinesterase labeling in the white matter and topographic projection of CFs and PC axons (Voogd and Ruigrok, 2004). Longitudinal zones are further divided into smaller units called microzones, based on high synchrony of complex spike activity (Llinas and Sasaki, 1989; Sugihara et al., 1993) and Ca^{2+} spikes (Mukamel et al., 2009; Schultz et al., 2009). Each microzone is ~500 μm in width, stable across behavioral states, and has a sharp boundary with the neighboring microzones (Mukamel et al., 2009). This synchrony is based on electrical coupling of nearby olivary neurons through dendrodendritic gap junctions (Llinas et al., 1974; Sotelo et al., 1974) and topographical olivocerebellar projections, that is, from given subregions of the inferior olive to specific longitudinal cortical zones (Sugihara et al., 2001).

Elaborate and fine compartmentalization can be recognized as cerebellar stripes by histochemistry for various molecules expressed in PCs. The best example of 'late' markers, which reveal the adult topography (postnatal day 15 onward), is aldolase C or zebrin II antigen (Hawkes and Leclerc, 1987), while 'early' markers of topography, such as calbindin and L7, typically reveal zones and stripes during perinatal development (embryonic day 13 (E13) to postnatal day 5(P5); Wassef et al., 1985)). These stripes are reproducible between individuals and conserved across species. Using a novel molecular marker, phospholipase Cβ4 (PLCβ4), which is continuously expressed in zebrin II-negative PCs from embryonic stage to adulthood (Nakamura et al., 2004; Watanabe et al., 1998), some stripes in the adult cerebellum have been shown to derive from two or more distinct embryonic clusters (Marzban et al., 2007). Importantly, studies with small tracer injections have shown that the topography of zebrin II expression pattern corresponds to that of the olivary projection to the cerebellar cortex and further to that of the olivary projection to the DCN and cerebellar cortical projection to DCN (Pijpers et al., 2005; Sugihara and Shinoda, 2004, 2007). These lines of evidence support the idea that the entire cerebellar system is formed by parallel assembly of an olivo–cortico–nuclear microcomplex (Ito, 1984).

5.2 DEVELOPMENT OF CF–PC SYNAPSES

5.2.1 Multiple Innervation of PCs by CFs in Early Postnatal Period

PCs undergo drastic changes in their morphology from late embryonic to early postnatal days (Armengol and Sotelo, 1991). PCs are born at the cerebellar ventricular zone at E10–E13 and migrate to form a multilayer below the molecular layer. At the end of radial migration, PCs exhibit bipolar shapes at E19–P0 ('simple and fusiform cell' by Armengol and Sotelo, 1991). Then, at P1–P3, new stem dendrites emerge from all aspects of the cell bodies, which confer the complex shapes of PCs ('complex-fusiform cell' by Armengol and Sotelo, 1991). From P3 to P6, such stem dendrites disappear by retraction of the long dendritic branches ('regressive–atrophic dendrites' by Armengol and Sotelo, 1991). In the meantime, PCs line up in a monolayer, which is completed by P5. PCs undergo explosive outgrowth of perisomatic protrusions that emerge in all directions from the cell bodies (the stage of 'stellate cells' by

Armengol and Sotelo, 1991). Upto P10, PCs extend single or double stem dendrites into the molecular layer, which grow and split into many dendritic branches, with simultaneous withdrawal of long somatic processes. The polarity of PCs is determined during this developmental period. From P10 to P15, the growth of the dendritic arbor occurs mainly in its lateral domain, whereas from P15 on, the dendritic growth occurs in the vertical plane and the height of the dendritic field reaches the adult level at P30.

After reaching the primitive cerebellum around E18, axons of inferior olivary neurons give rise to thick and thin collaterals (Wassef et al., 1992). This stage is called the 'creeper stage' (Chedotal and Sotelo, 1993) at which the 'CFs,' the thick collaterals of olivary axons, creep between immature PCs. At this stage, PCs have just completed their migration and are organized in a multilayer of 'simple and complex-fusiform cells'. Initially, each olivocerebellar axon forms about 100 'creeper' CFs (Sugihara, 2005). Then, the CF to PC synapse undergoes three distinct phases of postnatal development, which are described by Ramón y Cajal in his pioneering studies (Cajal, 1911): the 'pericellular nest' stage, the 'capuchon' stage, and the 'dendritic' stage. At the 'pericellular nest' stage, CFs surround the cell bodies of PCs and establish contacts with the abundant pseudopodia stemming from the soma and form a plexus on the lower part of the PC soma. At this stage, PCs have many perisomatic protrusions that emerge in all directions from the cell bodies, and therefore, this stage is called the phase of 'stellate cells' (Armengol and Sotelo, 1991). Among the 100 'creeper' CFs of each olivocerebellar axon, only around 10 can develop 'pericellular nests.' The 'capuchon' stage is characterized by the displacement of the plexus of CF collaterals to the apical portion of PC somata and main dendrites. Finally, at the 'dendritic' stage, CFs undergo translocation to growing PC dendrites and expanding their innervation territories.

Earlier electrophysiological studies on juvenile rats *in vivo* showed that stimulation to the inferior olive after P3 elicits CF-mediated responses in PCs (Crepel, 1971). However, in contrast to the all-or-none nature of CF responses in the adult animals, the responses of juvenile PCs are graded in parallel with the increase in the stimulus strength (Crepel et al., 1976). This was the first evidence that PCs are innervated by multiple CFs in early postnatal development. Later extensive studies *in vivo* revealed that both the percentage of PCs innervated by multiple CFs and the average number of CFs innervating individual PCs decrease with postnatal development and that most PCs become singly innervated by CFs (Crepel et al., 1981; Mariani and Changeux, 1981). These results clearly indicate that developmental elimination of redundant CF inputs occurs during postnatal development.

5.2.2 Functional Differentiation of Multiple CFs

Multiple CFs initially form functional synapses on the PC soma at around P3 (Chedotal and Sotelo, 1993; Morando et al., 2001). When recorded from PCs in cerebellar slices at this developmental stage, excitatory postsynaptic currents (EPSCs) elicited by stimulating multiply-innervating CFs are much smaller than those of mature CFs (Bosman et al., 2008; Hashimoto and Kano, 2003, 2005; Ohtsuki and Hirano, 2008; Scelfo and Strata, 2005). Therefore, CF inputs become stronger, while redundant CFs are eliminated during postnatal development.

Changes in the relative synaptic strengths of multiple CFs innervating the same PC have been systematically studied during postnatal development (Hashimoto and Kano, 2003) by recording CF-mediated EPSCs in PCs from cerebellar slices of mice aged P2–P21. This systematic study showed that more than five discrete CF-EPSCs with similar amplitudes are recorded in PCs from neonatal mice around P3 (Figure 5.2(a), ~P3). In contrast, in the second postnatal week, PCs with multiple CF-EPSCs have one large CF-EPSC and a few small CF-EPSCs (Figure 5.2(a), ~P7 and ~P12). These results indicate that synaptic strengths of multiply-innervating CFs are relatively uniform in neonatal mice, and one CF is selectively strengthened during postnatal development (Bosman et al., 2008; Hashimoto and Kano, 2003, 2005). Quantitative assessments of the disparity among the amplitudes of multiple CF-EPSCs in individual PCs demonstrate that one CF is selectively strengthened among multiple CFs innervating the same PC from P3 to P7 (Figure 5.2(b); Hashimoto and Kano, 2003). These electrophysiological data are consistent with the morphological observation that the innervation pattern of CFs over PCs drastically changes during this postnatal period in rats (Sugihara, 2005). At P4, CFs have many creeping terminals in the PC layer and their swellings do not aggregate at particular PC somata (creeper type). Then, from P4 to P7, CFs surround several specific PC somata and form aggregated terminals on them (nest type; Sugihara, 2005).

There are clear differences in electrophysiological properties between EPSCs elicited by the strongest CF input and those by other weaker inputs. Transient rises of glutamate concentration in the synaptic cleft are significantly higher after stimulation of the strongest CF than the weaker CFs (Hashimoto and Kano, 2003). This is thought to result from the fact that the probability of multivesicular release (i.e., more than one synaptic vesicle released simultaneously to a given postsynaptic site from the corresponding presynaptic release site) is higher for the strongest CF than for the weaker CFs. Further electrophysiological examination suggested that

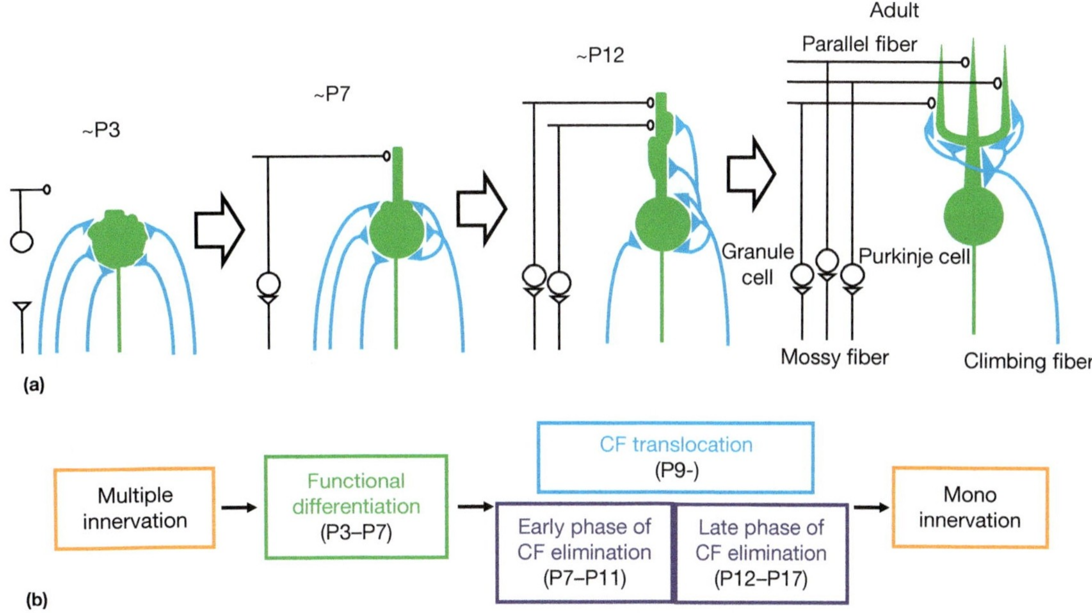

FIGURE 5.2 Postnatal development of CF–PC synapses. (a) Diagrams of CF–PC synapses at four representative stages of postnatal development in mice. (b) Four distinct phases in postnatal development of CF–PC synapses. *Reproduced with permission from* European Journal of Neuroscience *34, 1697–1710, with permission.*

the number of release sites facing a narrow postsynaptic PC region is larger in the strongest CF than in weaker CFs (Hashimoto and Kano, 2003).

5.2.3 Dendritic Translocation of Single CFs

Morphological evidence indicates that the site of CF innervation of PC changes from soma to dendrite during early postnatal development, a phenomenon known as 'CF translocation' (Altman and Bayer, 1997). The relationship between the selective strengthening of single CFs and CF translocation was investigated by using both electrophysiological and morphological techniques (Hashimoto et al., 2009a). The location of synapses along the somatodendritic domains of PCs can be estimated by analyzing the kinetics of quantal EPSCs (qEPSCs) arising from single synaptic vesicles in CF terminals.

At P7–P8 when the selective strengthening of single CFs on each PC has just been completed, there is no significant difference in the distribution of the rise times (i.e., time from the onset to the peak) of qEPSCs for the strongest compared with the weaker CFs. Since the rise time of qEPSCs is proportional to the distance from the synaptic sites to the somatic recording site (Roth and Hausser, 2001), synapses of the strongest and weaker CFs are thought to be located on the soma at around P7 (Figure 5.2(a), ~P7). At P9–P10, the incidence of qEPSCs with slow rise times was more frequent for the strongest than for the weaker CFs, suggesting the initiation of CF translocation (Figure 5.2(b)). The difference in the distribution of qEPSC rise times for the strongest

compared with the weaker CFs becomes larger from P11 to P14. While the incidence of qEPSCs with slow rise times becomes more frequent for the strongest CFs with age, the qEPSC rise times for weaker CFs remain almost unchanged from P9 to P14. These electrophysiological data collectively indicate that (1) synaptic competition among multiple CFs occurs on the soma before P7 (Figure 5.2(a), ~P3 and ~P7, Figure 5.2(b)), (2) only the strongest CF ('winner' CF) starts to translocate to dendrites at P9 and the translocation continues thereafter (Figure 5.2(a), ~P12, Figure 5.2(b)), and (3) synapses of the weaker CFs ('loser' CFs) remain around the soma (Figure 5.2(a), ~P12).

Morphological data are consistent with these electrophysiological observations. When subsets of CFs are labeled by an anterograde tracer, biotinylated dextran amine (BDA), injected into the inferior olive, pericellular nests with extensive branching of CFs are observed at P7, P9, and P12. At P7, in spite of the presence of immature stem dendrites in PCs, CFs innervate the soma and spare the dendrites (Figure 5.3(a)). Dendritic innervation of CFs starts at P9 (Figure 5.3(b)) and at P12, and thereafter, the territory of innervation extends progressively along the PC dendrites (Figure 5.3(c) and 5.3(d)).

5.2.4 Early Phase of CF Synapse Elimination

Detailed assessment of the postnatal development of CF innervations in mouse cerebellar slices demonstrates that there is no significant reduction in the average number of CFs per PC from P3 to P6 when functional

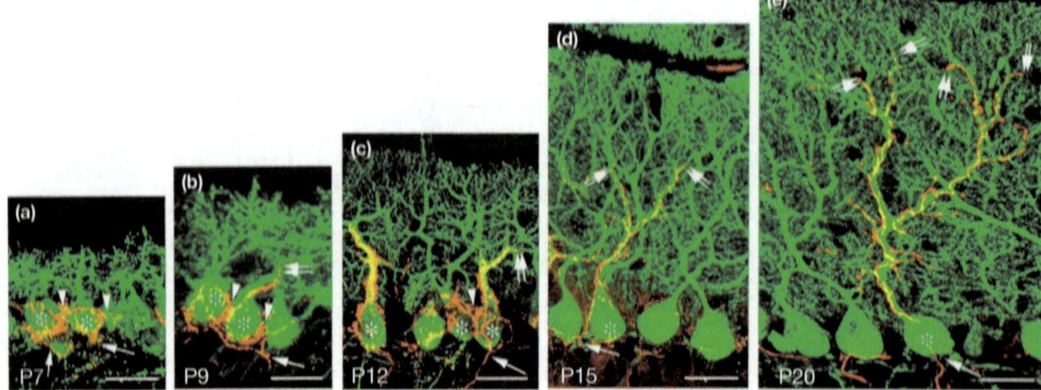

FIGURE 5.3   Developmental profile of CF innervations from perisomatic nest stage to peridendritic stage. (a–e) Fluorescent labeling of CFs with BDA (red) and PCs with calbindin antibody (green) at respective postnatal days. *Reproduced from Hashimoto K, Ichikawa R, Kitamura K, Watanabe M, and Kano M (2009) Translocation of a 'winner' climbing fiber to the Purkinje cell dendrite and subsequent elimination of 'losers' from the soma in developing cerebellum.* Neuron 63: 106–118, with permission.

differentiation of multiple CFs occurs (Hashimoto et al., 2009b). The value then decreases progressively from P6 to around P15 (Hashimoto et al., 2009b; Scelfo and Strata, 2005). CF synapse elimination therefore does not proceed in parallel with functional differentiation of multiple CFs but starts after the strengthening of single CFs to individual PCs. Crepel et al. (1981) showed that elimination of surplus CFs consisted of two distinct phases, the early phase up to around P8 and the late phase from around P9 to P17 (Crepel et al., 1981). The early phase occurs normally in animals with mild X-irradiation to the cerebellum during the early postnatal period, which causes selective loss of GCs and PFs while leaving PCs intact. In marked contrast, the late phase is severely impaired by inhibiting GC production by X-irradiation. This study indicates that the early phase of CF synapse elimination is independent of PF–PC synapse formation, whereas the late phase is critically dependent on it. However, since the animal models with 'hypogranular' or 'agranular' cerebella often have abnormalities of cerebellar development other than GC genesis and PF–PC synapse formation, there remains a possibility that CF synapse elimination might be influenced by such developmental defects.

The analysis of mutant mice deficient in the glutamate receptor δ2 subunit (GluRδ2 or GluD2) demonstrates that there are two distinct phases of CF synapse elimination. GluRδ2 is richly expressed in PCs and its deletion causes impairment of PF–PC synapse formation leading to reduction of PF–PC synapse number to about half of that in wild-type mice. In spite of the severe impairment of PF synapse formation, GluRδ2 deletion does not significantly affect the laminar structure of the cerebellum and morphology of the PC and its dendritic tree (Kashiwabuchi et al., 1995; Kurihara et al., 1997). In GluRδ2 knockout mice, the average number of CFs innervating each PC is similar to that of control mice from P5 to P11. However, the value is significantly larger than that of control mice from P12 to P14. Thus, CF synapse elimination in mice can be classified into two distinct phases, namely, the 'early phase' from P6 to around P11 which is independent of PF–PC synapse formation and the 'late phase' from around P12 and thereafter which requires normal PF–PC synapse formation (Figure 5.2(b); Hashimoto et al., 2009b).

Molecular mechanisms of the early phase of CF synapse elimination remain largely unknown. However, patterns of CF activity have been reported to influence the early phase of CF synapse elimination. Andjus et al. disrupted the normal activity pattern of CFs in rat at P9–P12 by administration of harmaline, which induced synchronous activation of inferior olive neurons (Andjus et al., 2003). This treatment caused persistent multiple CF innervations of PCs in rats at P15–P87. Furthermore, a recent report strongly suggests that PC activity is crucial for CF synapse elimination. Lorenzetto et al. (2009) generated transgenic mice that expressed a chloride channel-YFP fusion protein specifically in PCs to suppress their excitabilities. In these mice, the expression of chloride channel was observed in PCs during the 'early phase' at P9, and multiple CF innervations persisted up to P90. Therefore, perturbation of PC activity is considered to cause impairment of the 'early phase' of CF synapse elimination. In addition, insulin-like growth factor I (IGF-1) is reported to be involved in CF synapse elimination from P8 to P12 (Kakizawa et al., 2003). IGF-1 is thought to enhance the strengths of CF synapses and promote their survival, whereas the shortage of IGF-1 appears to impair the development of CF synapses (Kakizawa et al., 2003). In addition, Sherrard et al. reported recently that the active, phosphorylated form of full-length TrkB, a receptor for brain-derived neurotrophic factor (BDNF), fell around the onset of the early phase of CF synapse elimination. In contrast, the expression of the truncated form, which acts as a negative regulator of TrkB signaling, rose at the same developmental stage (Sherrard et al., 2009). This finding suggests that decrease in TrkB signaling might permit the elimination of surplus CF synapses, although TrkB signaling appears to be involved in the 'late phase' of CF synapse elimination presumably through the maturation of GABAergic inhibitory synapses (see Section 5.2.5).

Morphological data indicate that CFs that undergo dendritic translocation keep their synapses on the PC soma during the second postnatal week. In contrast, synaptic terminals of the weaker CFs are confined to the soma and the basal part of the primary dendrite. The characteristic pericellular nest consists of somatic synapses originating from collaterals of a single predominant CF and from weaker CFs and thus represents multiple CF innervation of PCs (Hashimoto et al., 2009a). Therefore, CF synapse elimination is thought to be a process of nonselective pruning of perisomatic synapses, which spares dendritic synapses of a single predominant CF and leads to monoinnervations of that CF (Hashimoto et al., 2009a).

Hashimoto et al. (2011) have recently reported that the P/Q-type VDCC expressed in PCs drives the early phase of CF synapse elimination (Hashimoto et al., 2011). They generated mice with PC-selective deletion of $Ca_V2.1$, a pore-forming subunit of the P/Q-type VDCC that constitutes the major Ca^{2+} current component in PCs (PC-$Ca_V2.1$ KO mice; Mintz et al., 1992; Stea et al., 1994). Although initial CF to PC synapse formation appears normal at around P4 in PC-$Ca_V2.1$ KO mice, subsequent CF synapse development and elimination are severely impaired. First, biased strengthening of a single CF input in each PC from P5 to P7 is absent in PC-$Ca_V2.1$ KO mice, and multiple CF inputs equally become larger by about fourfold. Second, more than one CF undergoes translocation to dendrites in PC-$Ca_V2.1$ KO mice. Third, CF synapse elimination is severely impaired in PC-$Ca_V2.1$ KO mice until around P12 (Hashimoto et al., 2011). Global $Ca_V2.1$ KO mice have essentially the same defects in CF synapse development and elimination as PC-$Ca_V2.1$ KO mice (Hashimoto et al., 2011; Miyazaki et al., 2004). These results indicate that Ca^{2+} influx through P/Q-type VDCC into PCs is essential for selective strengthening of a single CF input in each PC, dendritic translocation of the strengthened CF, and the early phase of CF synapse elimination.

5.2.5 Late Phase of CF Synapse Elimination

In GluRδ2 knockout mice, CFs invade into the distal dendrites and form ectopic synapses there (Hashimoto et al., 2001; Ichikawa et al., 2002). These ectopic CF synapses appear around P10 when PF synapse formation and PC dendritic arborization occur most vigorously. The similar type of multiple CF innervation is also found in a mutant mouse deficient in cbln1 in which PF to PC synapse formation is severely impaired (Hirai et al., 2005). These results indicate that PFs compete for the innervation territory with CFs during development and play a role in restricting CF innervation to proximal dendrites (Figure 5.4). The details of this phenomenon will be described in Section 5.3.3.

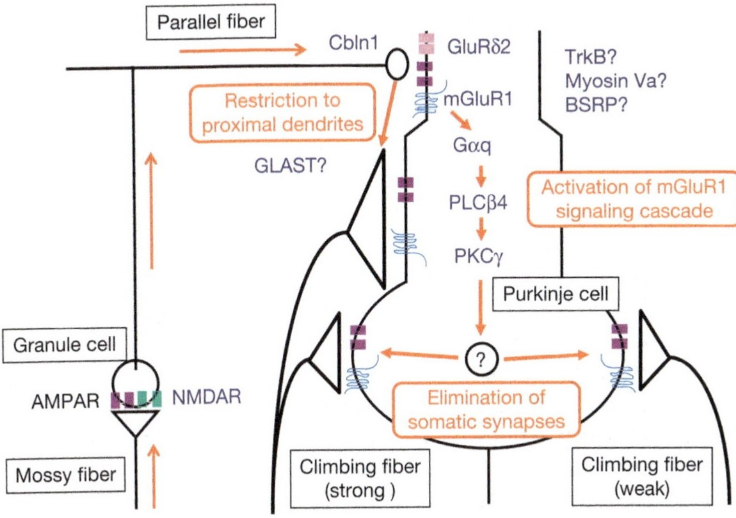

FIGURE 5.4 Mechanisms underlying the late phase of CF synapse elimination. PF–PC synapses play two distinct roles in the late phase. PF synapses are formed and maintained on distal dendrites of PCs through the interaction of Cbln1 and GluRδ2. First, PF synapses occupy the postsynaptic sites on the distal dendrites and confine the CF innervation sites to the proximal dendrites (restriction to proximal dendrites). Second, neural activity along the pathway of mossy fiber, GC, and PF involving N-methyl-D-aspartate (NMDA) receptor at MF to GC synapses drives mGluR1 to PKCγ signaling cascades in PCs (activation of mGluR1 signaling cascade). The signaling molecules downstream of PKCγ are currently unknown, but this eventually leads to elimination of somatic CF synapses including those of weak CFs and of somatic collaterals of strong CF (elimination of somatic synapses). TrkB, myosin Va, brain-specific receptor-like protein (BSRP), and glutamate-aspartate transporter (GLAST) are also involved in the late phase of CF synapse elimination, but details of the mechanisms of their actions in CF synapse elimination are unknown. *Modified from Kano, M., Hashimoto, K., Tabata, T., 2008. Type-1 metabotropic glutamate receptor in cerebellar Purkinje cells: a key molecule responsible for long-term depression, endocannabinoid signalling and synapse elimination. Philosophical Transactions of the Royal Society B 363, 2173–2186.*

Another role of PF synapses is to activate type 1 metabotropic glutamate receptor (mGluR1) and its downstream signaling cascades in PCs to drive the process of CF synapse elimination. It is shown that the mutant mouse deficient in mGluR1 is impaired in CF synapse elimination (Kano et al., 1997; Levenes et al., 1997). Mice deficient in signaling molecules downstream of mGluR1, that is, $G_{\alpha q}$, PLCβ4, and PKCγ, are also impaired in CF synapse elimination (Hashimoto et al., 2000; Kano et al., 1995, 1998; Offermanns et al., 1997). Electrophysiological examination of CF innervation following postnatal development demonstrates that the regression of CF synapse normally occurs during the first and second postnatal weeks in all of the four mouse strains. However, these mice display abnormality in CF synapse elimination during the third postnatal week. These results suggest that the signaling cascade from mGluR1 to protein kinase C (PKC)γ is essential for the late phase of CF synapse elimination, but it is dispensable for the early phase of CF synapse elimination (Figure 5.2(b)). Importantly, the formation and function of PF to PC synapses are normal in these mutant mice. Therefore, the impaired CF synapse elimination is not caused secondarily by the defect in PF synaptogenesis.

The defect in the CF synapse elimination in the mGluR1-deficient mouse is restored in the mGluR1-rescue mice in which mGluR1a has been introduced specifically into PCs (Ichise et al., 2000). Regression of CF synapses is impaired in mice by PC-specific expression of a PKC inhibitor peptide (De Zeeuw et al., 1998). Furthermore, the distribution of multiply-innervated PCs in the cerebellum of PLCβ4 knockout mice exactly matches that of the PCs with predominant expression of PLCβ4 in the wild-type mouse cerebellum (Kano et al., 1998). These lines of evidence clearly indicate that the signaling from mGluR1 to PKCγ in PCs but not other cell types plays a central role in CF synapse elimination.

The mGluR1 signaling required for the late phase of synapse elimination is thought to be driven by PF activity, since mGluR1 can readily be activated by PF inputs (Batchelor et al., 1994; Finch and Augustine, 1998; Takechi et al., 1998). Furthermore, chronic blockade of N-methyl-D-aspartate (NMDA) receptors within the cerebellum results in the impairment of CF synapse elimination (Rabacchi et al., 1992) specifically in its later phase (Kakizawa et al., 2000). NMDA receptors are not present at either PF or CF synapses onto PCs during second and third postnatal weeks, but they are abundantly expressed at MF to GC synapses (Kakizawa et al., 2000). Therefore, the chronic blockade of NMDA receptors within the cerebellum should affect MF to GC transmission. These results suggest that neural activity along MF–GC–PF–PC pathway and subsequent activation of mGluR1 are prerequisite for the late phase of CF synapse elimination (Figure 5.4; Kakizawa et al., 2000).

Besides the mGluR1 signaling in PCs, a neurotrophin receptor, TrkB (Bosman et al., 2006; Johnson et al., 2007), a motor protein, myosin, Va (Takagishi et al., 2007), a glutamate transporter, glutamate-aspartate transporter (GLAST) (Watase et al., 1998), and a novel brain-specific receptor-like protein family, brain-specific receptor-like protein (Miyazaki et al., 2006), are also involved in CF synapse elimination. Since genetic or pharmacological deletion of these molecules in mice impairs CF synapse elimination in the second postnatal week, these signaling cascades are thought to be involved in the 'late phase' of CF synapse elimination. The involvement of a neurotrophin receptor, TrkB, is especially interesting, because TrkB signaling is required for normal development of GABAergic innervations of PCs. In TrkB knockout mice, the number of GABAergic synapses is reduced and the inhibitory postsynaptic currents are prolonged (Bosman et al., 2006), which suggest the GABA$_A$ receptors do not undergo the normal α3 to α1 subunit switching (Takayama and Inoue, 2004). It is therefore possible that normal development of GABAergic inhibitory synapses onto PCs is prerequisite for the late phase of CF synapse elimination (Nakayama et al., 2012). As for other molecules potentially involved in the late phase of CF synapse elimination, null mutant mice deficient in Ca^{2+}/calmodulin-dependent protein kinase IV (CaMKIV) are reported to have persistent multiple CF innervations, but it is unclear at what stage of postnatal development the impairment occurs (Ribar et al., 2000). It is also reported that null mutant mice deficient in α-calcium/calmodulin-dependent protein kinase II (CaMKIIα) display multiple CF innervations at P21–P28, but this phenotype disappears in adulthood (Hansel et al., 2006), suggesting that CaMKIIα deficiency delays but does not prevent CF synapse elimination.

During synapse elimination in the neuromuscular junction, bulb-shaped tips of retreating motor axons and the axon fragments ('axozomes') are engulfed by Schwann cells (Bishop et al., 2004). These axon bulbs, axozomes, and Schwann cell cytoplasm are often positively stained with Lysotracker Red, a marker for the lysosomes and late endosomes of living cells, suggesting axonal digestion through autophagy and subsequent heterophagy by Schwann cells (Song et al., 2008). It is also reported that Lysotracker-positive structures surrounding PCs, which are presumed to be within Bergmann glia, are abundant during the second and third postnatal weeks (Song et al., 2008). This result suggests that retreating CF axons might be digested in a manner similar to the retreating motor axons at neuromuscular junction.

The current model for the mechanisms underlying the late phase of CF synapse elimination is illustrated in Figure 5.4. First, PF synapses confine the CF innervation territories to proximal dendrites of PCs (Figure 5.4). Second, PF activity involving NMDA receptor at MF to

GC synapses drives mGluR1 to PKCγ signaling cascades in PCs (Figure 5.4), which leads to nonselective elimination of somatic CF synapses (Figure 5.4). The signaling molecules downstream of PKCγ are currently unknown. The molecular mechanisms underlying morphological elimination of the weaker CFs are also unknown. Some mechanisms must convey transsynaptic retrograde signaling from PCs to weaker CFs.

A recent study using organotypic slice culture shows that CF synapse elimination occurs only during the critical period that depends on the maturation stage of postsynaptic PCs but not on presynaptic olivary neurons (Letellier et al., 2009). The authors cocultured immature or mature medulla containing the inferior olive with naïve or non-naïve PCs (i.e., PCs that have not undergone synapse elimination or those that have experienced synapse elimination, respectively). They found that multiple CF innervation was observed when either the PCs were naïve or CFs were immature. Interestingly, non-naïve PCs could not eliminate multiple CFs. These results suggest that CF synapse elimination during the critical period leaves indelible trace in PCs that prevents the elimination process from occurring in the later stage (Letellier et al., 2009).

5.3 DEVELOPMENT OF PF–PC SYNAPSES

5.3.1 Formation of PF–PC Synapses

PF–PC synapses increase in number and mature in the early postnatal period, concomitant with differentiation of GCs and growth of PC dendrites. During the first 10 days of a rodent's life, production and migration of GCs as well as growth of PC dendrites are slow (Altman and Bayer, 1997). In this period, the proliferative or outer zone of the external granular layer has constant thickness of four to five cells, while the depth of the premigratory or inner zone increases progressively (Altman and Bayer, 1997). In the premigratory zone, postmitotic GCs extend future PFs in the transverse plane, and then migrate downward along Bergmann fibers in the molecular layer (Altman and Bayer, 1997; Rakic, 1971). As a consequence, T-shaped axon of GCs differentiates, and horizontal beams of PFs become a part of the upper zone of the molecular layer. PC dendrites are immature, particularly in the upper zone of the molecular layer, where dendrites extend filopodium-like protrusions, PF–PC synapses are few in number, spines are often free of innervation, and coverage by Bergmann glia is incomplete (Kurihara et al., 1997; Yamada et al., 2000).

In the next ten postnatal days, PC dendrites grow dynamically, the bulk of GCs come into existence, and PF–PC synapses explosively increase in number (Takacs and Hamori, 1994). Moreover, almost all spines form synaptic contact with PF terminals, and PF–PC synapses are equipped with well-developed postsynaptic density and complete astroglial coverage (Kurihara et al., 1997; Spacek, 1985; Yamada et al., 2000). Analyses using agranular and hypogranular animal models, where PF–PC synaptogenesis is hindered by spontaneous gene mutation or postnatal X-ray irradiation, have demonstrated that PF synapse formation in PCs plays a critical role in the elongation, branching, and planar development of dendritic trees (Sotelo, 2004).

5.3.2 Stabilization and Maintenance of PF–PC Synapses

Our understanding of the molecular mechanism of PF synapse formation in PCs has been greatly advanced by molecular identification of GluRδ2 (Araki et al., 1993; Lomeli et al., 1993) and analyses of mutant mice defective in this gene *grid2* (GluRδ2-knockout mice and spontaneous *hotfoot* mutant mice; Guastavino et al., 1990; Kashiwabuchi et al., 1995). Like other ionotropic GluR subunits, GluRδ2 preserves three transmembrane domains (TM1, TM3, and TM4), reentrant hairpin loop (TM2) surrounding a channel pore, ligand-binding domains in the N-terminal region, and protein–protein interaction sites in the C-terminal region (Uemura et al., 2004; Yuzaki, 2004). However, GluRδ2 does not function as a glutamate-gated ion channel (Kakegawa et al., 2007a,b).

In the brain, GluRδ2 is expressed almost exclusively in PCs (Araki et al., 1993; Lomeli et al., 1993) and selectively localized at PF but not CF synapses (Landsend et al., 1997; Takayama et al., 1995). PCs in GluRδ2-defective mice display characteristic phenotypes mostly related to PF synapse structure and function, including reduction in the number of PF synapses per PC to about half of the wild-type mice (54% in control PCs; Kurihara et al., 1997), emergence of free spines lacking synaptic contact in the distal dendritic domain (37% of the total spines; (Ichikawa et al., 2002; Kurihara et al., 1997)), mismatching of pre- and postsynaptic specialization at PF synapses (Guastavino et al., 1990; Takeuchi et al., 2005), impaired long-term depression (LTD) at PF synapses (Kakegawa et al., 2008; Kashiwabuchi et al., 1995; Uemura et al., 2007), impaired motor learning (Kakegawa et al., 2008; Kishimoto et al., 2001), and severe ataxia (Guastavino et al., 1990; Kashiwabuchi et al., 1995). In a drug-inducible, PC-specific GluRδ2-knockout mouse strain, mismatched PF–PC synapses, free spines, and motor discoordination are induced and exacerbated in the adult cerebellum, concomitant with a decrease in GluRδ2 protein (Takeuchi et al., 2005). Furthermore, expression of the N-terminal domain of GluRδ2 in human embryonic kidney cells induces presynaptic differentiation (Uemura and Mishina, 2008) and its viral transfer to adult GluRδ2-knockout mice rapidly restores PF synapse formation and motor discoordination (Kakegawa et al., 2009). Thus, GluRδ2 strengthens and regulates the connectivity of PF–PC synapses in both the developing and adult cerebella. On the other hand, the last seven amino acids known as the T-site, which binds to various PDZ domain-containing proteins including postsynaptic density (PSD)-93, PTPMEG, delphilin, nPIST, and synaptic scaffolding molecule are essential for cerebellar LTD and motor learning (Kakegawa et al., 2008; Uemura et al., 2007).

Cbln1 or precerebellin was originally identified as a precursor of PC-specific peptide cerebellin (Slemmon et al., 1984). However, C-terminal two-thirds of Cbln1 shares significant structural similarity with the globular domain of complement C1q chain (Urade et al., 1991), and the full-length Cbln1 is released into the culture medium as a hexameric complex (Bao et al., 2005). Thus, Cbln1 now belongs to the C1q/tumor necrosis factor superfamily. Of four members (Cbln1–4), Cbln1 is highly expressed in cerebellar GCs together with Cbln3 (Hirai et al., 2005; Miura et al., 2006), exists as Cbln1 homomeric and Cbln1/3 heteromeric complexes (Iijima et al., 2007; Pang et al., 2000), and selectively accumulates in the synaptic cleft facing PF terminals, but not CF terminals, in PCs (Iijima et al., 2007; Miura et al., 2009). Cbln1-knockout mice show characteristic phenotypes similar to, or even severer than, GluRδ2-knockout mice (Hirai

et al., 2005). In PCs of Cbln1-knockout mice, 78% of the spines are free of innervation and 14% have mismatching in pre- and postsynaptic differentiation (Figure 5.5). Hence, only 8% of spines in Cbln1-knockout mice establish normal matched synapses with PF terminals. LTD at PF–PC synapses and motor coordination is also impaired. When recombinant Cbln1 is applied to the subarachnoid space of adult Cbln1-knockout mice, only a single injection can rapidly restore PF–PC synapse structure and function, and cerebellar ataxia (Ito-Ishida et al., 2008). These common phenotypes in GluRδ2- and Cbln1-knockout mice and their rapid rescue by genetic or molecular supplementation are explained by the fact that transsynaptic interaction of postsynaptic GluRδ2 and presynaptic neurexins is mediated by Cbln1. Thus, through this unique interaction, Cbln1 acts as a bidirectional synaptic organizer for both pre- and postsynaptic components at PF–PC synapses (Matsuda et al., 2010; Uemura et al., 2010).

The strengths of PF synapses, but not CF synapses, in PCs are selectively decreased when postsynaptic mGluR1 or inositol 1,4,5-trisphosphate (IP$_3$) signaling is chronically inhibited, when PF activity is inhibited by suppressing NMDA receptor-mediated inputs to GCs or when antibody against BDNF is applied *in vivo* (Furutani et al., 2006). The weakening is due to reduction of glutamate release probability from PFs. The weakening of PF–PC synaptic strength is reversed by

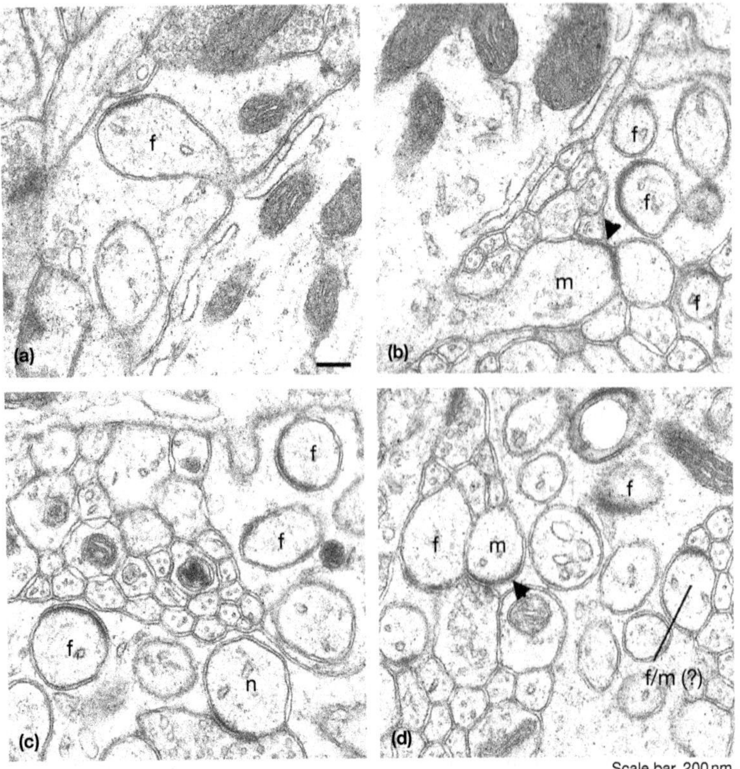

FIGURE 5.5   Electron micrographs of free spines (f) and mismatched synapses (m) in Cbln1-knockout mice. Free spines are thoroughly enwrapped by the lamellate processes of Bergmann glia. *Arrowheads* indicate the portion of the PSD that is not opposed by the presynaptic active zone. Normal matched synapses (n) are very rare in this mutant. In (d), the presence of postsynaptic density with no presyanptic terminal differentiation suggests that the spine with such postsynaptic density is either free or mismatched spine (hence labeled "f/m(?)." Scale bar = 0.2 μm. *Reproduced from Watanabe, M., Molecular mechanisms governing competitive synaptic wiring in cerebellar Purkinje cells.* Tohoku Journal of Experimental Medicine 214, 2008:175–190, *with permission.*

in vivo application of BDNF. These results suggest that postsynaptic mGluR1 activation and the following IP$_3$ signaling maintain presynaptic function through BDNF at PF–PC synapses (Furutani et al., 2006). In this regard, it is interesting to note that free spines on PC dendrites emerge in the spontaneous ataxic mutant rigoletto (*rig*) (also known as waddles; *wdl*), which is caused by a 19-bp deletion in the exon 8 of carbonic anhydrase-related protein *Car8* (Hirasawa et al., 2007). Car8 is known to bind to IP$_3$ receptor and reduce its affinity for IP$_3$ (Hirota et al., 2003). In future studies, it is important to investigate how Car8 modulates the mGluR1-IP$_3$ signaling and whether Car8 is involved in the GluRδ2–Cbln1–neurexin interaction, both of which regulate the connectivity of PF–PC synapses.

5.3.3 Heterosynaptic Competition Between PF and CF Inputs

The proximal compartment of PC dendrite appears smooth in contour due to low spine density and is innervated by single CFs. On the other hand, the distal compartment is made up of spiny branchlets studded with numerous spines and innervated by PFs (Figure 5.6, middle). Accumulated experimental evidence indicates that the construction of such excitatory synaptic organization stands on competitive equilibrium between CFs and PFs whose expansions are promoted by distinct mechanisms. Surgical, pharmacological, and genetic manipulations that shift this equilibrium can alter the organized innervations pattern.

Disruption of GluRδ2 gene in mice not only causes abnormal structure and function of PF synapses but also affects the mode of CF innervations (Ichikawa et al., 2002). In the molecular layer, CF branches are distributed in the inner four-fifths (84% of the molecular layer thickness) in

control mice, whereas their distribution almost reaches the pial surface (95%) in GluRδ2-knockout mice. When the tracer-labeled CFs were followed from the soma to the tips of PC dendrites by serial electron microscopy, CF branches in GluRδ2-knockout mice extend distally and take over the free spines on the distal dendrites. Such aberrant extension occurs toward not only distal dendrites of the same PCs but also those of the neighboring PCs. The latter type of spine takeover results in multiple innervation of a PC by CFs of different neuronal origins. This anatomical evidence for multiple CF innervation is consistent with electrophysiological recording combined with Ca^{2+} imaging. In GluRδ2-knockout mice, a single strong CF elicits large EPSCs with a fast rise time and large Ca^{2+} transients over the entire dendritic tree, whereas weak CFs elicit small EPSCs with a slow rise time and small Ca^{2+} elevation that is confined to distal dendrites (Hashimoto et al., 2001). These findings indicate that GluRδ2 is essential for restricting CF innervation to the proximal dendritic compartment and thereby preventing multiple CF innervation at the distal dendritic compartment (Figures 5.4 and 5.6, right).

This mechanism is also active in the adult cerebellum. The ablation of GluRδ2 in adulthood also leads to progressive distal extension of ascending branches of CFs, and they aberrantly innervate distal dendrites of the target and neighboring PCs (Miyazaki et al., 2010). Furthermore, transverse branches of CFs, which are short motile collaterals forming no synapses in wild-type animals (Nishiyama et al., 2007), display aberrant mediolateral extension and innervate distal dendrites of neighboring and remote PCs. Consequently, many PCs are connected by single main CFs and surplus CFs that innervate small parts of the distal dendrites. Surplus CF-EPSCs with slow rise time and small amplitude also emerge progressively after GluRδ2 ablation. Therefore,

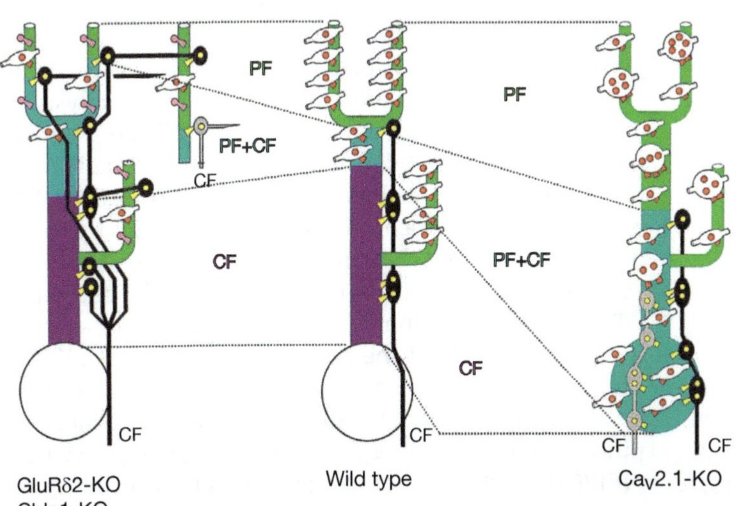

FIGURE 5.6 Summary diagram of molecular mechanisms for competitive synaptic wiring in PCs. Note that the climbing fiber (CF) and parallel fiber (PF) territories are reversed in mutant mice defective in GluRδ2/Cbln1 (left) and Ca$_V$2.1 (right). With both mechanisms, CF and PF territories are sharply segregated, and CF mono-innervation is established in wild-type animals (middle). *Reproduced from Watanabe, M., 2008. Molecular mechanisms governing competitive synaptic wiring in cerebellar Purkinje cells.* Tohoku Journal of Experimental Medicine *214, 2008:175–190, with permission.*

GluRδ2 is essential to keep CF mono-innervation in the adult cerebellum by suppressing aberrant invasion of CF branches to the territory of PF innervation.

In contrast, CF innervation is regressed but PF innervation expands to the proximal compartment, when surgical lesion to olivocerebellar projections is made in adult animals or activities in the cerebellar cortical neurons are blocked with the sodium channel blocker tetrodotoxin or with the synaptic scaffolding molecule AMPA) receptor antagonist 2,3-Dioxo-6-nitro-1,2,3,4-tetrahydrobenzo [f]quinoxaline-7-sulfonamide (Bravin et al., 1995; Cesa et al., 2007; Kakizawa et al., 2005). The latter change often accompanies hyperspiny transformation at the proximal dendritic compartment (Bravin et al., 1995; Cesa et al., 2007). Similar changes are reproduced in global knockout mice of $Ca_V2.1$, a pore-forming subunit of the P/Q-type VDCC (Miyazaki et al., 2004). In $Ca_V2.1$-knockout mice, hyperspiny transformation is induced at proximal dendrites and somata of PCs, and many of these ectopic spines are innervated by PF terminals. Conversely, the distribution of CFs is regressed to lower portions of the molecular layer, and they innervate spines from somata and basal dendrites. Furthermore, in more than 90% of $Ca_V2.1$-knockout PCs, their basal dendrites and somata are innervated by CFs of different neuronal origins. As a result, the proximal somatodendritic compartment in $Ca_V2.1$-lacking PCs receives chaotic innervation by numerous PFs and multiple CFs (Figure 5.6, right). Thus, CF activities leading to AMPA receptor activation and subsequent Ca^{2+} influx through P/Q-type Ca^{2+} channels are essential for monopolizing the proximal dendritic compartment by a single main CF and for expelling other excitatory inputs from that compartment.

Taken altogether, excitatory synaptic wiring in PCs is formed and maintained through homosynaptic competition among CFs and heterosynaptic competition between PFs and CFs. GluRδ2 and Cbln1 fuel heterosynaptic competition in favor of PF innervation, whereas P/Q-type VDCCs facilitate both heterosynaptic and homosynaptic competitions in favor of single main CFs. Based on these molecular mechanisms, PCs establish territorial innervations by PF and CF and mono-innervation by CF.

5.4 DEVELOPMENT OF INHIBITORY SYNAPSES FROM STELLATE CELLS AND BASKET CELLS TO PCS

5.4.1 Formation of Basket Cell–PC Synapses

Basket cells innervate the AIS of PC and constract characteristic pinceau formation in the mature cerebellum (Ito, 1984; Palay and Chan-Palay, 1974). The innervation by basket cell axons of PCs seems to begin when basket cells migrate across the PC layer at the

end of the first postnatal week (Ango et al., 2004). The basket cell axons initially form synaptic contacts on the somata of PCs. Then, they appear to move directly to AIS without searching for other possible targets. Upon reaching the AIS, basket cell axons extend multiple terminal branches and eatablish pinceau formation. This behavior strongly suggests that there are guidance cues for basket cell axons along the surface of the PC with gradient from soma to AIS.

Several molecules have been identified that accumulate at AISs of PCs. These include membrane-associated adaptor protein ankyrin-G and one of its binding partner, neurofascin 186 (NF186), a splice variant of neurofascin which belongs to the L1 subgroup of the Ig superfamily (Brummendorf et al., 1998). It is thought that ankyrin-G is stabilized at the AIS partly through its interaction with β4-spectrin tetramers which bind to actin network (Davis et al., 1996). Ankyrins and β-spectrins are known as intracellular adaptor proteins that recruit ion channels, transporters, and cell adhesion molecules to subcellular domains. They are thought to constitute microdomains for intercellular contact and signaling (Bennett and Baines, 2001; Bennett and Chen, 2001). On the other hand, neurofascin is known as a cell-surface glycoprotein that is shown to mediate axon–axon interactions *in vitro* (Rathjen et al., 1987). The distribution of NF186 on PC surface is particularly intriguing. NF186 exhibits subcellular concentration gradient in PCs from AIS toward the soma and dendrites, being highest at AIS and very low at the top of the soma and in the dendrites (Figure 5.7(a); Ango et al., 2004). This gradient is already formed at the end of the first postnatal week when the basket cell axons first contact the somata of PCs. Therefore, it is highly likely that the gradient of NF186 plays an important role in guiding basket cell axons to AIS (Figure 5.7(a)). Since ankyrin-G is expressed exclusively at AISs in PCs, there may be two forms of NF186, the ankyrin-G bound form that is restricted to AIS and the ankyrin-G free form that is distributed to the surface of the PC soma.

Evidence for the requirement of ankyrin-G and NF186 in basket cell axon targeting and pinceau synapse formation at AIS of PC has been presented by the analysis of ankyrin-G knockout mice in which the NF186 gradient in PCs is abolished (Figure 5.7(b); Ango et al., 2004). In these mice, basket cell axons are not restricted to AIS but instead are present on the soma and slightly more distal portion of PC axons (Figure 5.7(b)). Although some basket cell axon bundles successfully reach the AIS, they are very thin, extend along PC axons abnormally, and followed the ectopic localization of NF186. The synaptic contacts visualized by a GABA synthesizing enzyme, GAD65, were greatly reduced at pinceau synapses. These results strongly suggest that NF186 is a substrate for the growth of basket cell axons and its gradient functions as a guidance cue. The results

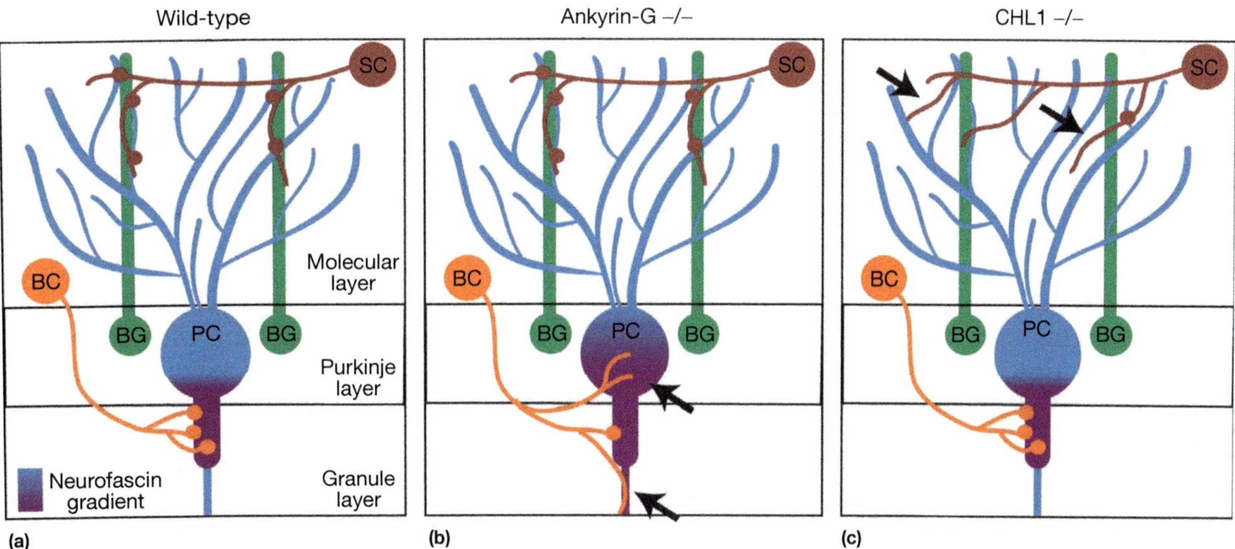

FIGURE 5.7 Subcellular specificity of basket cell and stellate cell inhibitory connections to PCs. Schematics showing the inhibitory synaptic connectivity patterns of the wild-type (a), ankyrin-G-deficient (b), and CHL1-deficient (c) mice. In wild-type mice, a sharp gradient of neurofascin is present from AIS toward the soma. In ankyrin-G-deficient mice, this gradient is no longer restricted to the AIS, which causes mistargeting of basket cell axons and reduced synapse formation. In CHL1-deficient mice, stellate cell axons are not properly guided by Bergmann glia fibers, and synapse formation is decreased. SC, stellate cell; BC, basket cell; PC, Purkinje cell; BG, Bergmann glia. *Reproduced from Williams ME, de Wit J, and Ghosh A (2010) Molecular mechanisms of synaptic specificity in developing neural circuits. Neuron 68: 9–18, with permission.*

also suggest that NF186 bound to ankyrin-G at AIS is required for the stabilization of pinceau synapses. Further supporting evidence of this notion has been obtained from the experiments to disrupt NF186 function directly in PCs (Ango et al., 2004). Expression of the dominant-negative form of NF186 into PCs during the process of establishing pinceau does not affect AIS-restricted ankyrin-G distribution or guidance of basket cell axons to AIS. However, GAD65-labeled pinceau formation is decreased in the infected PCs to the same extent as in ankyrin-G-deficient PCs. Therefore, either remaining endogenous NF186 can guide the basket cell axons to AIS or NF186 is required specifically for pinceau synapse stabilization while other molecules interacting with ankyrin-G guide the basket cell axons to AIS (Huang et al., 2007; Williams et al., 2010).

5.4.2 Formation of Stellate Cell–PC Synapses

Stellate cells are GABAergic inhibitory interneurons whose somata are located in the molecular layer. In the mature cerebellum, their axons innervate dendrites of PCs with ascending and descending collaterals, and with a plexus of finer branches and terminals. Similarly to basket cells, stellate cells are derived from dividing progenitors in the white matter of postnatal cerebellum (Zhang and Goldman, 1996). The stellate cell precursors migrate into the molecular layer a few days later than the basket cell precursors, with a peak between P8 and P11 but continuing till P14 (Yamanaka et al., 2004). By using a green fluorescent protein- bacterial artificial chromosome transgenic reporter mouse line in which GFP was mainly expressed in stellate cells and basket cells, Huang and colleagues demonstrated how stellate cell axons establish their innervations of PC dendrites after they reach the molecular layer (Ango et al., 2008). Between P12 and P16, stellate cells become bipolar and extend neuritis in horizontal orientation. Then, at P16–P18, stellate cell axons send ascending and descending collaterals, which are further elaborated with appearance of plexus of finer branches up to P40. Importantly, both ascending and descending collaterals of stellate cell axons are strictly associated with the fibers of Bergmann glia that are visualized by the staining of the glia-specific cytoskeleton protein glial fibrillary acidic protein (GFAP). During the third and fourth postnatal weeks, Bergmann glia fibers are known to extend lateral varicoses and fine processes and form an extensive reticular meshwork. Radial fibers from neighboring Bergmann glia are aligned to form palisades in parlobular plane, which are perpendicular to PC dendrites (Altman and Bayer, 1997). In contrast to the close association to Bergmann glia fibers, stellate cell axons do not follow PC dendrites. Bergmann glia fibers enwrap segments of PC dendrites in a patchy, en passant pattern, which is in contrast to the close association to stellate cell axons. Triple immunolabeling of stellate cell axon terminals, Bergmann glia fibers and PC dendrites indicates that stellate cell boutons are formed at the intersection between Bergmann glia fibers and PC dendrites. Thus, Bergmann glia fibers may function as an intermediate scaffold to guide stellate cell axons along the characteristic trajectories toward multiple PC dendrites and to form synaptic contacts (Figure 5.7(a); Ango et al., 2008).

Since neurofascin, a member of the L1 cell adhesion molecule (L1CAM) subfamily, is crucial for targeting of basket cell axons to the AIS of PC, it is possible that other members of L1CAM might be important for the targeting of stellate cell axons to PC dendrites. A systematic survey of the expression patterns of L1CAMs during postnatal cerebellar development revealed that close homologue of L1 (CHL1) is distributed in a radial stripe pattern that exactly matches the expression of the Bergmann glia marker GFAP, but not the PC marker calbindin (Ango et al., 2008). CHL1 expression in Bergmann glia fibers is prominent as early as P8, reaching higher levels around P18, and then declining in adulthood. CHL1 is also expressed in stellate cells, but the expression is delayed, being undetectable at P8, becoming obvious around P14, and remaining in adulthood (Ango et al., 2008). These expression patterns suggest the involvement of CHL1 in the stellate cell axon targeting to PC dendrites.

A series of experiments using CHL1-deficient mice have clarified the importance of Bergmann glia fibers and CHL1 in organizing stellate cell innervations of PC dendrites (Figure 5.7(c); Ango et al., 2008). In CHL1 knockout mice, stellate cell axons exhibit abnormal trajectories and orientation, and aberrant innervations of PC dendrites (Figure 5.7(c)). In addition, there is a clear reduction in the staining of GAD65, a marker for GABAergic inhibitory synaptic terminal, and the density of stellate cell synapses along PC dendrites. Such aberrant stellate cell axons can form morphologically normal synapses onto PC dendrites albeit with reduced efficacy and density, but many of them are not maintained and the stellate cell axon terminals become atrophic with age. In contrast, there is no change in the GAD65 staining in the AIS of PC (Figure 5.7(c)). Furthermore, there is no change in excitatory synapses from PFs and CFs in CHL1-deficient mice. Importantly, the selective defect in stellate cell innervations and synapse formation on PC dendrites is observed in mice with conditional deletion of CHL1 in Bergmann glia. These results demonstrate that Bergmann glia fibers function as guiding scaffolds, and CHL1 is a molecular signal for organization of stellate cell axon arbors and directing their innervations of PC dendrites (Huang et al., 2007; Williams et al., 2010).

5.4.3 Activity-Dependent Remodeling of Inhibitory Synapses

Several lines of evidence indicate that $GABA_A$ receptor-mediated signaling coordinates pre- and postsynaptic maturation during activity-dependent development of inhibitory synapses (Huang, 2009; Huang and Scheiffele, 2008; Huang et al., 2007). For example, altering GABA synthesis by manipulating the expression of GAD67 greatly influences inhibitory synaptic innervation in the visual cortex (Chattopadhyaya et al., 2007). Acute suppression of the $\gamma 2$ subunit of $GABA_A$ receptor not only disrupts $GABA_A$ receptor clustering but also reduces innervations of the $\gamma 2$-deficient neurons by GABAergic terminals (Li et al., 2005). In cerebellar PCs, genetic deletion of the $\alpha 1$ subunit of $GABA_A$ receptor in mice causes complete loss of functional $GABA_A$ receptors and synaptic inhibition in PCs by P18 (Fritschy et al., 2006). Morphologically, GABAergic synaptic terminals from stellate cells are reduced by 75%, whereas basket cell synapses on PC soma are not affected. During postnatal development, GABAergic terminals from stellate cells are initially formed normally onto PC dendritic shafts. PCs of the $\alpha 1$ knockout mice transiently express $\alpha 3$ subunit and have functional $GABA_A$ receptors during early postnatal development (Patrizi et al., 2008). However, subsequent down-regulation of $\alpha 3$ results in complete loss of GABAergic currents and a decreased rate of GABAergic synaptogenesis (Patrizi et al., 2008). Simultaneously, ectopic mismatched synapses begin to be formed between GABAergic terminals and PC dendritic spines (Fritschy et al., 2006; Patrizi et al., 2008) on which normally glutamatergic excitatory terminals make synaptic contacts. Interestingly, the postsynaptic adhesion molecule neuroligin-2 is correctly targeted to inhibitory synapses lacking $GABA_A$ receptors, whereas neuroligin-2 is absent from the mismatched synapses albeit the presence of GABAergic terminals (Patrizi et al., 2008). These results indicate that $GABA_A$ receptors are not required for the formation of synapses, but they appear to be crucial for activity-dependent regulation of synaptic density, presumably through promoting the stabilization of transient axodendritic contact into mature inhibitory synapses.

5.5 SUMMARY AND CONCLUSIONS

The cerebellum provides a good system to study how microcircuits are formed during peri- and postnatal development. The cerebellar cortex consists of only seven types of neurons, that is, PC, GC, basket cell, stellate cell, Golgi cell, Lugaro cell, and unipolar brush cell, and there are two glutamatergic excitatory afferents. The PC is the sole output neuron of the cerebellar cortex and inhibits neurons in the DCN. Bergmann glia, the characteristic astrocyte in the cerebellar cortex, plays multiple roles in neural circuit formation and synaptic transmission such as migration of GCs and guidance of stellate cell axons. The neural circuits made by these cell types and afferents are basically the same throughout the cerebellum.

In the first section of this chapter, we briefly described the cell types and the synaptic organization of the cerebellum and how these cells are generated and migrate to their final positions. We also mentioned the mediolateral compartmentalization based on olivocerebellar projection and some molecular background of the compartmentalization.

In the second section, we made an overview of postnatal development of CF–PC synapses, which is one of the best-studied examples of synapse elimination in the brain. Shortly after birth, each PC is innervated by multiple CFs with similar synaptic strengths on the soma. Subsequently, a single CF is selectively strengthened during the first postnatal week. Then, at around P9, only the strongest CF ('winner' CF) starts to extend its innervation to PC dendrites. In contrast, synapses of the weaker CFs ('loser' CFs) remain on the soma and the most proximal portion of the dendrite, and they are eliminated progressively during the second and third postnatal weeks. From P6 to P11, the elimination proceeds independently of PF–PC synapse formation. From P12 and thereafter, the elimination of weaker CFs requires normal PF–PC synapse formation and is dependent on the PF synaptic inputs that activate mGluR1 and its downstream signaling in PCs.

In the third section, we described how PF synapses are formed and maintained on dendritic spines of PCs. We introduced a recent hot topic that Cbln1 interacts with GluRδ2 on PC dendritic spines and neurexin on PF terminals and that the GluRδ2–Cbln1–neurexin system stabilizes and maintains PF–PC synapses. We also mentioned that mGluR1-mediated calcium signaling in PC dendrites causes release of BDNF and maintains presynaptic function of PFs. Furthermore, we showed that innervation territories of PFs and CFs on PC dendrites stand on the equilibrium caused by heterosynaptic competition between PFs and CFs and homosynaptic competition between multiple CFs.

In the fourth section, we summarized how GABAergic inhibitory synapses from basket and stellate cells are targeted to the PC's AIS and dendrites, respectively. Basket cell axons seem to be guided to PC's AIS following the gradient of neurofascin, a member of L1CAM. This gradient is caused by cross-linking of neurofascin to ankyrin-G that is localized to AISs. On the other hand, stellate cells direct their axons along Bergmann glia fibers to PC dendrites. The association of stellate cell axons to Bergmann glia fibers is mediated by CHL1, another member of L1CAM. We also mentioned that GABA$_A$ receptors appear to be crucial for activity-dependent regulation of the density of inhibitory synapses on PCs.

The cerebellum has been attracting many neuroscientists who pursue the mechanisms of synapse formation, synapse elimination, and synapse remodeling. The small number of cell types and little regional variation in the layer structure and synaptic organization in the cerebellum enable us to perform quantitative and detailed morphological and electrophysiological analyses, when compared to other brain areas. An example is the study of developmental synapse elimination. While this process is intensively studied in peripheral synapses such as neuromuscular junction and autonomic ganglia, it is generally very difficult to do detailed analyses of synapse elimination in the CNS because of small synapse size, heterogeneity and abundance of synaptic inputs to each neuron, and the complexity of synaptic organization. The CF to PC synapse is one of a few examples in the CNS in which developmental synapse elimination can be studied quantitatively by electrophysiological and morphological techniques. Synapses from retinal ganglion cells to the lateral geniculate nucleus and those from the lateral lemniscus to the ventral basal thalamus are also known to undergo massive elimination during postnatal development and have been studied intensively. Formation of inhibitory topographic map in the auditory brainstem is another example in which developmental refinement occurs through synapse elimination. As for the molecular and cellular mechanisms, developmental synapse elimination at CF to PC connection is best characterized among the four types of synapses, as detailed in this review. Thus, CF synapse elimination is an excellent model system for the study of developmental synapse refinement, which is comparable to that of the visual cortex.

The topics introduced in this chapter, that is, CF synapse elimination, maintenance of PF synapses through GluD2-cbln1-neurexin interaction, and targeting of basket cell and stellate cell axons to PCs, are breakthroughs in this field of neuroscience. Continuing researches on cerebellar microcircuits will elucidate fundamental mechanisms of the formation, elimination, maturation, and maintenance of neural circuits in developing CNS.

Acknowledgments

This work has been supported in part by the Strategic Research Program for Brain Sciences (Development of Biomarker Candidates for Social Behavior), Grants-in-Aid for Scientific Research 21220006 (M.K.) and 19100005 (M.W.), and the Global Centers of Excellence Program (Integrative Life Science Based on the Study of Biosignaling Mechanisms) from MEXT, Japan.

References

Altman, J., Bayer, S.A., 1997. Development of the Cerebellar System: In Relation to its Evolution, Structure, and Functions. CRC Press, Boca Raton, FL.

Andjus, P.R., Zhu, L., Cesa, R., Carulli, D., Strata, P., 2003. A change in the pattern of activity affects the developmental regression of the Purkinje cell polyinnervation by climbing fibers in the rat cerebellum. Neuroscience 121, 563–572.

Ango, F., di Cristo, G., Higashiyama, H., Bennett, V., Wu, P., Huang, Z.J., 2004. Ankyrin-based subcellular gradient of neurofascin, an immunoglobulin family protein, directs GABAergic innervation at Purkinje axon initial segment. Cell 119, 257–272.

Ango, F., Wu, C., Van der Want, J.J., Wu, P., Schachner, M., Huang, Z.J., 2008. Bergmann glia and the recognition molecule CHL1 organize GABAergic axons and direct innervation of Purkinje cell dendrites. PLoS Biology 6, e103.

Apps, R., Hawkes, R., 2009. Cerebellar cortical organization: a one-map hypothesis. Nature Reviews Neuroscience 10, 670–681.

Araki, K., Meguro, H., Kushiya, E., Takayama, C., Inoue, Y., Mishina, M., 1993. Selective expression of the glutamate receptor channel $\delta 2$ subunit in cerebellar Purkinje cells. Biochemical and Biophysical Research Communications 197, 1267–1276.

Armengol, J.A., Sotelo, C., 1991. Early dendritic development of Purkinje cells in the rat cerebellum. A light and electron microscopic study using axonal tracing in 'in vitro' slices. Brain Research. Developmental Brain Research 64, 95–114.

Bao, D., Pang, Z., Morgan, J.I., 2005. The structure and proteolytic processing of Cbln1 complexes. Journal of Neurochemistry 95, 618–629.

Bao, J., Reim, K., Sakaba, T., 2010. Target-dependent feedforward inhibition mediated by short-term synaptic plasticity in the cerebellum. Journal of Neuroscience 30, 8171–8179.

Barbour, B., 1993. Synaptic currents evoked in Purkinje cells by stimulating individual granule cells. Neuron 11, 759–769.

Batchelor, A.M., Madge, D.J., Garthwaite, J., 1994. Synaptic activation of metabotropic glutamate receptors in the parallel fibre-Purkinje cell pathway in rat cerebellar slices. Neuroscience 63, 911–915.

Bennett, V., Baines, A.J., 2001. Spectrin and ankyrin-based pathways: metazoan inventions for integrating cells into tissues. Physiological Reviews 81, 1353–1392.

Bennett, V., Chen, L., 2001. Ankyrins and cellular targeting of diverse membrane proteins to physiological sites. Current Opinion in Cell Biology 13, 61–67.

Bishop, D.L., Misgeld, T., Walsh, M.K., Gan, W.B., Lichtman, J.W., 2004. Axon branch removal at developing synapses by axosome shedding. Neuron 44, 651–661.

Bosman, L.W., Hartmann, J., Barski, J.J., et al., 2006. Requirement of TrkB for synapse elimination in developing cerebellar Purkinje cells. Brain Cell Biology 35, 87–101.

Bosman, L.W., Takechi, H., Hartmann, J., Eilers, J., Konnerth, A., 2008. Homosynaptic long-term synaptic potentiation of the 'winner' climbing fiber synapse in developing Purkinje cells. Journal of Neuroscience 28, 798–807.

Bravin, M., Rossi, F., Strata, P., 1995. Different climbing fibres innervate separate dendritic regions of the same Purkinje cell in hypogranular cerebellum. The Journal of Comparative Neurology 357, 395–407.

Brummendorf, T., Kenwrick, S., Rathjen, F.G., 1998. Neural cell recognition molecule L1: from cell biology to human hereditary brain malformations. Current Opinion in Neurobiology 8, 87–97.

Cajal, S.R., 1911. Histologie du systeme nerveux de l'homme et des vertebres, vol. II. Maloine, Paris.

Cesa, R., Scelfo, B., Strata, P., 2007. Activity-dependent presynaptic and postsynaptic structural plasticity in the mature cerebellum. Journal of Neuroscience 27, 4603–4611.

Chattopadhyaya, B., Di Cristo, G., Wu, C.Z., et al., 2007. GAD67-mediated GABA synthesis and signaling regulate inhibitory synaptic innervation in the visual cortex. Neuron 54, 889–903.

Chedotal, A., Sotelo, C., 1993. The 'creeper stage' in cerebellar climbing fiber synaptogenesis precedes the 'pericellular nest' – ultrastructural evidence with parvalbumin immunocytochemistry. Brain Research. Developmental Brain Research 76, 207–220.

Crepel, F., 1971. Maturation of climbing fiber responses in the rat. Brain Research 35, 272–276.

Crepel, F., Mariani, J., Delhaye-Bouchaud, N., 1976. Evidence for a multiple innervation of Purkinje cells by climbing fibers in the immature rat cerebellum. Journal of Neurobiology 7, 567–578.

Crepel, F., Delhaye-Bouchaud, N., Dupont, J.L., 1981. Fate of the multiple innervation of cerebellar Purkinje cells by climbing fibers in immature control, X-irradiated and hypothyroid rats. Brain Research 227, 59–71.

Davis, J.Q., Lambert, S., Bennett, V., 1996. Molecular composition of the node of Ranvier: identification of ankyrin-binding cell adhesion molecules neurofascin (mucin+/third FNIII domain-) and NrCAM at nodal axon segments. The Journal of Cell Biology 135, 1355–1367.

De Zeeuw, C.I., Holstege, J.C., Calkoen, F., Ruigrok, T.J., Voogd, J., 1988. A new combination of WGA-HRP anterograde tracing and GABA immunocytochemistry applied to afferents of the cat inferior olive at the ultrastructural level. Brain Research 447, 369–375.

De Zeeuw, C.I., Hansel, C., Bian, F., et al., 1998. Expression of a protein kinase C inhibitor in Purkinje cells blocks cerebellar LTD and adaptation of the vestibulo-ocular reflex. Neuron 20, 495–508.

Eccles, J.C., Llinas, R., Sasaki, K., 1966. The excitatory synaptic action of climbing fibres on the Purkinje cells of the cerebellum. The Journal of Physiology 182, 268–296.

Englund, C., Kowalczyk, T., Daza, R.A., et al., 2006. Unipolar brush cells of the cerebellum are produced in the rhombic lip and migrate through developing white matter. Journal of Neuroscience 26, 9184–9195.

Finch, E.A., Augustine, G.J., 1998. Local calcium signalling by inositol-1,4,5-trisphosphate in Purkinje cell dendrites. Nature 396, 753–756.

Fink, A.J., Englund, C., Daza, R.A., et al., 2006. Development of the deep cerebellar nuclei: transcription factors and cell migration from the rhombic lip. Journal of Neuroscience 26, 3066–3076.

Fritschy, J.M., Panzanelli, P., Kralic, J.E., Vogt, K.E., Sassoe-Pognetto, M., 2006. Differential dependence of axo-dendritic and axo-somatic GABAergic synapses on GABA$_A$ receptors containing the $\alpha 1$ subunit in Purkinje cells. Journal of Neuroscience 26, 3245–3255.

Furutani, K., Okubo, Y., Kakizawa, S., Iino, M., 2006. Postsynaptic inositol 1,4,5-trisphosphate signaling maintains presynaptic function of parallel fiber-Purkinje cell synapses via BDNF. Proceedings of the National Academy of Sciences of the United States of America 103, 8528–8533.

Guastavino, J.M., Sotelo, C., Damez-Kinselle, I., 1990. Hot-foot murine mutation: behavioral effects and neuroanatomical alterations. Brain Research 523, 199–210.

Hansel, C., de Jeu, M., Belmeguenai, A., et al., 2006. αCaMKII Is essential for cerebellar LTD and motor learning. Neuron 51, 835–843.

Hashimoto, K., Kano, M., 2003. Functional differentiation of multiple climbing fiber inputs during synapse elimination in the developing cerebellum. Neuron 38, 785–796.

Hashimoto, K., Kano, M., 2005. Postnatal development and synapse elimination of climbing fiber to Purkinje cell projection in the cerebellum. Neuroscience Research 53, 221–228.

Hashimoto, K., Watanabe, M., Kurihara, H., et al., 2000. Climbing fiber synapse elimination during postnatal cerebellar development requires signal transduction involving Gαq and phospholipase C $\beta 4$. Progress in Brain Research 124, 31–48.

Hashimoto, K., Ichikawa, R., Takechi, H., et al., 2001. Roles of glutamate receptor $\delta 2$ subunit (GluR$\delta 2$) and metabotropic glutamate receptor subtype 1 (mGluR1) in climbing fiber synapse elimination during postnatal cerebellar development. Journal of Neuroscience 21, 9701–9712.

Hashimoto, K., Ichikawa, R., Kitamura, K., Watanabe, M., Kano, M., 2009a. Translocation of a 'winner' climbing fiber to the Purkinje cell dendrite and subsequent elimination of 'losers' from the soma in developing cerebellum. Neuron 63, 106–118.

Hashimoto, K., Yoshida, T., Sakimura, K., Mishina, M., Watanabe, M., Kano, M., 2009b. Influence of parallel fiber-Purkinje cell synapse formation on postnatal development of climbing fiber-Purkinje cell synapses in the cerebellum. Neuroscience 162, 601–611.

Hashimoto, K., Tsujita, M., Miyazaki, T., et al., 2011. Postsynaptic P/Q-type Ca^{2+} channel in Purkinje cell mediates synaptic competition and elimination in developing cerebellum. Proceedings of the National Academy of Sciences of the United States of America 108, 9987–9992.

Hawkes, R., Leclerc, N., 1987. Antigenic map of the rat cerebellar cortex: the distribution of parasagittal bands as revealed by monoclonal anti-Purkinje cell antibody mabQ113. The Journal of Comparative Neurology 256, 29–41.

Hirai, H., Pang, Z., Bao, D., et al., 2005. Cbln1 is essential for synaptic integrity and plasticity in the cerebellum. Nature Neuroscience 8, 1534–1541.

Hirasawa, M., Xu, X., Trask, R.B., et al., 2007. Carbonic anhydrase related protein 8 mutation results in aberrant synaptic morphology and excitatory synaptic function in the cerebellum. Molecular and Cellular Neuroscience 35, 161–170.

Hirota, J., Ando, H., Hamada, K., Mikoshiba, K., 2003. Carbonic anhydrase-related protein is a novel binding protein for inositol 1,4,5-trisphosphate receptor type 1. The Biochemical Journal 372, 435–441.

Hoshino, M., Nakamura, S., Mori, K., et al., 2005. Ptf1a, a bHLH transcriptional gene, defines GABAergic neuronal fates in cerebellum. Neuron 47, 201–213.

Huang, Z.J., 2009. Activity-dependent development of inhibitory synapses and innervation pattern: role of GABA signalling and beyond. The Journal of Physiology 587, 1881–1888.

Huang, Z.J., Di Cristo, G., Ango, F., 2007. Development of GABA innervation in the cerebral and cerebellar cortices. Nature Reviews Neuroscience 8, 673–686.

Huang, Z.J., Scheiffele, P., 2008. GABA and neuroligin signaling: linking synaptic activity and adhesion in inhibitory synapse development. Current Opinion in Neurobiology 18, 77–83.

Ichikawa, R., Miyazaki, T., Kano, M., et al., 2002. Distal extension of climbing fiber territory and multiple innervation caused by aberrant wiring to adjacent spiny branchlets in cerebellar Purkinje cells lacking glutamate receptor δ2. Journal of Neuroscience 22, 8487–8503.

Ichise, T., Kano, M., Hashimoto, K., et al., 2000. mGluR1 in cerebellar Purkinje cells essential for long-term depression, synapse elimination, and motor coordination. Science 288, 1832–1835.

Iijima, T., Miura, E., Matsuda, K., Kamekawa, Y., Watanabe, M., Yuzaki, M., 2007. Characterization of a transneuronal cytokine family Cbln – regulation of secretion by heteromeric assembly. European Journal of Neuroscience 25, 1049–1057.

Ito, M., 1984. The Cerebellum and Neural Control. Raven Press, New York.

Ito-Ishida, A., Miura, E., Emi, K., et al., 2008. Cbln1 regulates rapid formation and maintenance of excitatory synapses in mature cerebellar Purkinje cells *in vitro* and *in vivo*. Journal of Neuroscience 28, 5920–5930.

Johnson, E.M., Craig, E.T., Yeh, H.H., 2007. TrkB is necessary for pruning at the climbing fibre-Purkinje cell synapse in the developing murine cerebellum. The Journal of Physiology 582, 629–646.

Kakegawa, W., Kohda, K., Yuzaki, M., 2007a. The δ2 'ionotropic' glutamate receptor functions as a non-ionotropic receptor to control cerebellar synaptic plasticity. The Journal of Physiology 584, 89–96.

Kakegawa, W., Miyazaki, T., Hirai, H., et al., 2007b. Ca^{2+} permeability of the channel pore is not essential for the δ2 glutamate receptor to regulate synaptic plasticity and motor coordination. The Journal of Physiology 579, 729–735.

Kakegawa, W., Miyazaki, T., Emi, K., et al., 2008. Differential regulation of synaptic plasticity and cerebellar motor learning by the C-terminal PDZ-binding motif of GluRδ2. Journal of Neuroscience 28, 1460–1468.

Kakegawa, W., Miyazaki, T., Kohda, K., et al., 2009. The N-terminal domain of GluD2 (GluRδ2) recruits presynaptic terminals and regulates synaptogenesis in the cerebellum *in vivo*. Journal of Neuroscience 29, 5738–5748.

Kakizawa, S., Yamasaki, M., Watanabe, M., Kano, M., 2000. Critical period for activity-dependent synapse elimination in developing cerebellum. Journal of Neuroscience 20, 4954–4961.

Kakizawa, S., Yamada, K., Iino, M., Watanabe, M., Kano, M., 2003. Effects of insulin-like growth factor I on climbing fibre synapse elimination during cerebellar development. European Journal of Neuroscience 17, 545–554.

Kakizawa, S., Miyazaki, T., Yanagihara, D., Iino, M., Watanabe, M., Kano, M., 2005. Maintenance of presynaptic function by AMPA receptor-mediated excitatory postsynaptic activity in adult brain. Proceedings of the National Academy of Sciences of the United States of America 102, 19180–19185.

Kano, M., Hashimoto, K., Chen, C., et al., 1995. Impaired synapse elimination during cerebellar development in PKCγ mutant mice. Cell 83, 1223–1231.

Kano, M., Hashimoto, K., Kurihara, H., et al., 1997. Persistent multiple climbing fiber innervation of cerebellar Purkinje cells in mice lacking mGluR1. Neuron 18, 71–79.

Kano, M., Hashimoto, K., Watanabe, M., et al., 1998. Phospholipase Cβ4 is specifically involved in climbing fiber synapse elimination in the developing cerebellum. Proceedings of the National Academy of Sciences of the United States of America 95, 15724–15729.

Kano, M., Hashimoto, K., Tabata, T., 2008. Type-1 metabotropic glutamate receptor in cerebellar Purkinje cells: a key molecule responsible for long-term depression, endocannabinoid signalling and synapse elimination. Philosophical Transactions of the Royal Society B 363, 2173–2186.

Kashiwabuchi, N., Ikeda, K., Araki, K., et al., 1995. Impairment of motor coordination, Purkinje cell synapse formation, and cerebellar long-term depression in GluR δ2 mutant mice. Cell 81, 245–252.

Kishimoto, Y., Kawahara, S., Fujimichi, R., Mori, H., Mishina, M., Kirino, Y., 2001. Impairment of eyeblink conditioning in GluRδ2-mutant mice depends on the temporal overlap between conditioned and unconditioned stimuli. European Journal of Neuroscience 14, 1515–1521.

Kurihara, H., Hashimoto, K., Kano, M., et al., 1997. Impaired parallel fiber − > Purkinje cell synapse stabilization during cerebellar development of mutant mice lacking the glutamate receptor δ2 subunit. Journal of Neuroscience 17, 9613–9623.

Laine, J., Axelrad, H., 2002. Extending the cerebellar Lugaro cell class. Neuroscience 115, 363–374.

Landsend, A.S., Amiry-Moghaddam, M., Matsubara, A., et al., 1997. Differential localization of δ glutamate receptors in the rat cerebellum: coexpression with AMPA receptors in parallel fiber-spine synapses and absence from climbing fiber-spine synapses. Journal of Neuroscience 17, 834–842.

Letellier, M., Wehrle, R., Mariani, J., Lohof, A.M., 2009. Synapse elimination in olivo-cerebellar explants occurs during a critical period and leaves an indelible trace in Purkinje cells. Proceedings of the National Academy of Sciences of the United States of America 106, 14102–14107.

Leto, K., Bartolini, A., Yanagawa, Y., et al., 2009. Laminar fate and phenotype specification of cerebellar GABAergic interneurons. Journal of Neuroscience 29, 7079–7091.

Levenes, C., Daniel, H., Jaillard, D., Conquet, F., Crepel, F., 1997. Incomplete regression of multiple climbing fibre innervation of cerebellar Purkinje cells in mGluR1 mutant mice. Neuroreport 8, 571–574.

Li, R.W., Yu, W., Christie, S., et al., 2005. Disruption of postsynaptic GABA receptor clusters leads to decreased GABAergic innervation of pyramidal neurons. Journal of Neurochemistry 95, 756–770.

Llinas, R., Sasaki, K., 1989. The Functional organization of the olivo-cerebellar system as examined by multiple Purkinje cell recordings. European Journal of Neuroscience 1, 587–602.

Llinas, R., Baker, R., Sotelo, C., 1974. Electrotonic coupling between neurons in cat inferior olive. Journal of Neurophysiology 37, 560–571.

Lomeli, H., Sprengel, R., Laurie, D.J., et al., 1993. The rat δ-1 and δ-2 subunits extend the excitatory amino acid receptor family. FEBS Letters 315, 318–322.

Lorenzetto, E., Caselli, L., Feng, G., et al., 2009. Genetic perturbation of postsynaptic activity regulates synapse elimination in developing cerebellum. Proceedings of the National Academy of Sciences of the United States of America 106, 16475–16480.

Mariani, J., Changeux, J.P., 1981. Ontogenesis of olivocerebellar relationships. I. Studies by intracellular recordings of the multiple innervation of Purkinje cells by climbing fibers in the developing rat cerebellum. Journal of Neuroscience 1, 696–702.

Marzban, H., Chung, S., Watanabe, M., Hawkes, R., 2007. Phospholipase Cβ4 expression reveals the continuity of cerebellar topography through development. The Journal of Comparative Neurology 502, 857–871.

Matsuda, K., Miura, E., Miyazaki, T., et al., 2010. Cbln1 is a ligand for an orphan glutamate receptor δ2, a bidirectional synapse organizer. Science 328, 363–368.

Miale, I.L., Sidman, R.L., 1961. An autoradiographic analysis of histogenesis in the mouse cerebellum. Experimental Neurology 4, 277–296.

Mintz, I.M., Adams, M.E., Bean, B.P., 1992. P-type calcium channels in rat central and peripheral neurons. Neuron 9, 85–95.

Miura, E., Iijima, T., Yuzaki, M., Watanabe, M., 2006. Distinct expression of Cbln family mRNAs in developing and adult mouse brains. European Journal of Neuroscience 24, 750–760.

Miura, E., Matsuda, K., Morgan, J.I., Yuzaki, M., Watanabe, M., 2009. Cbln1 accumulates and colocalizes with Cbln3 and GluRδ2 at parallel fiber-Purkinje cell synapses in the mouse cerebellum. European Journal of Neuroscience 29, 693–706.

Miyakawa, H., Lev-Ram, V., Lasser-Ross, N., Ross, W.N., 1992. Calcium transients evoked by climbing fiber and parallel fiber synaptic inputs in guinea pig cerebellar Purkinje neurons. Journal of Neurophysiology 68, 1178–1189.

Miyazaki, T., Hashimoto, K., Shin, H.S., Kano, M., Watanabe, M., 2004. P/Q-type Ca²⁺ channel α1A regulates synaptic competition on developing cerebellar Purkinje cells. Journal of Neuroscience 24, 1734–1743.

Miyazaki, T., Hashimoto, K., Uda, A., et al., 2006. Disturbance of cerebellar synaptic maturation in mutant mice lacking BSRPs, a novel brain-specific receptor-like protein family. FEBS Letters 580, 4057–4064.

Miyazaki, T., Yamasaki, M., Takeuchi, T., Sakimura, K., Mishina, M., Watanabe, M., 2010. Ablation of glutamate receptor GluRδ2 in adult Purkinje cells causes multiple innervation of climbing fibers by inducing aberrant invasion to parallel fiber innervation territory. Journal of Neuroscience 30, 15196–15209.

Morando, L., Cesa, R., Rasetti, R., Harvey, R., Strata, P., 2001. Role of glutamate δ-2 receptors in activity-dependent competition between heterologous afferent fibers. Proceedings of the National Academy of Sciences of the United States of America 98, 9954–9959.

Mugnaini, E., Floris, A., 1994. The unipolar brush cell: a neglected neuron of the mammalian cerebellar cortex. The Journal of Comparative Neurology 339, 174–180.

Mukamel, E.A., Nimmerjahn, A., Schnitzer, M.J., 2009. Automated analysis of cellular signals from large-scale calcium imaging data. Neuron 63, 747–760.

Nakamura, M., Sato, K., Fukaya, M., et al., 2004. Signaling complex formation of phospholipase Cβ4 with metabotropic glutamate receptor type 1α and 1,4,5-trisphosphate receptor at the perisynapse and endoplasmic reticulum in the mouse brain. European Journal of Neuroscience 20, 2929–2944.

Nakayama, H., Miyazaki, T., Kitamura, K., et al., 2012. GABAergic inhibition regulates developmental synapse elimination in the cerebellum. Neuron 74, 384–396.

Napper, R.M., Harvey, R.J., 1988. Number of parallel fiber synapses on an individual Purkinje cell in the cerebellum of the rat. The Journal of Comparative Neurology 274, 168–177.

Nishiyama, H., Fukaya, M., Watanabe, M., Linden, D.J., 2007. Axonal motility and its modulation by activity are branch-type specific in the intact adult cerebellum. Neuron 56, 472–487.

Offermanns, S., Hashimoto, K., Watanabe, M., et al., 1997. Impaired motor coordination and persistent multiple climbing fiber innervation of cerebellar Purkinje cells in mice lacking Gaq. Proceedings of the National Academy of Sciences of the United States of America 94, 14089–14094.

Ohtsuki, G., Hirano, T., 2008. Bidirectional plasticity at developing climbing fiber-Purkinje neuron synapses. European Journal of Neuroscience 28, 2393–2400.

Ottersen, O.P., Storm-Mathisen, J., Somogyi, P., 1988. Colocalization of glycine-like and GABA-like immunoreactivities in Golgi cell terminals in the rat cerebellum: a postembedding light and electron microscopic study. Brain Research 450, 342–353.

Palay, S.L., Chan-Palay, V., 1974. Cerebellar Cortex. Springer-Verlag, New York.

Pang, Z., Zuo, J., Morgan, J.I., 2000. Cbln3, a novel member of the precerebellin family that binds specifically to Cbln1. Journal of Neuroscience 20, 6333–6339.

Patrizi, A., Scelfo, B., Viltono, L., et al., 2008. Synapse formation and clustering of neuroligin-2 in the absence of GABA_A receptors. Proceedings of the National Academy of Sciences of the United States of America 105, 13151–13156.

Pijpers, A., Voogd, J., Ruigrok, T.J., 2005. Topography of olivo-cortico-nuclear modules in the intermediate cerebellum of the rat. The Journal of Comparative Neurology 492, 193–213.

Rabacchi, S., Bailly, Y., Delhaye-Bouchaud, N., Mariani, J., 1992. Involvement of the N-methyl D-aspartate (NMDA) receptor in synapse elimination during cerebellar development. Science 256, 1823–1825.

Rakic, P., 1971. Neuron-glia relationship during granule cell migration in developing cerebellar cortex. A Golgi and electronmicroscopic study in *Macacus rhesus*. The Journal of Comparative Neurology 141, 283–312.

Rathjen, F.G., Wolff, J.M., Chang, S., Bonhoeffer, F., Raper, J.A., 1987. Neurofascin: a novel chick cell-surface glycoprotein involved in neurite-neurite interactions. Cell 51, 841–849.

Ribar, T.J., Rodriguiz, R.M., Khiroug, L., Wetsel, W.C., Augustine, G.J., Means, A.R., 2000. Cerebellar defects in Ca²⁺/calmodulin kinase IV-deficient mice. Journal of Neuroscience 20, RC107.

Roth, A., Hausser, M., 2001. Compartmental models of rat cerebellar Purkinje cells based on simultaneous somatic and dendritic patch-clamp recordings. The Journal of Physiology 535, 445–472.

Scelfo, B., Strata, P., 2005. Correlation between multiple climbing fibre regression and parallel fibre response development in the postnatal mouse cerebellum. European Journal of Neuroscience 21, 971–978.

Schilling, K., Oberdick, J., 2009. The treasury of the commons: making use of public gene expression resources to better characterize the molecular diversity of inhibitory interneurons in the cerebellar cortex. Cerebellum 8, 477–489.

Schultz, S.R., Kitamura, K., Post-Uiterweer, A., Krupic, J., Hausser, M., 2009. Spatial pattern coding of sensory information by climbing fiber-evoked calcium signals in networks of neighboring cerebellar Purkinje cells. Journal of Neuroscience 29, 8005–8015.

Sherrard, R.M., Dixon, K.J., Bakouche, J., Rodger, J., Lemaigre-Dubreuil, Y., Mariani, J., 2009. Differential expression of TrkB isoforms switches climbing fiber-Purkinje cell synaptogenesis to selective synapse elimination. Developmental Neurobiology 69, 647–662.

Simat, M., Parpan, F., Fritschy, J.M., 2007. Heterogeneity of glycinergic and gabaergic interneurons in the granule cell layer of mouse cerebellum. The Journal of Comparative Neurology 500, 71–83.

Slemmon, J.R., Blacher, R., Danho, W., Hempstead, J.L., Morgan, J.I., 1984. Isolation and sequencing of two cerebellum-specific peptides. Proceedings of the National Academy of Sciences of the United States of America 81, 6866–6870.

Song, J.W., Misgeld, T., Kang, H., et al., 2008. Lysosomal activity associated with developmental axon pruning. Journal of Neuroscience 28, 8993–9001.

Sotelo, C., 2004. Cellular and genetic regulation of the development of the cerebellar system. Progress in Neurobiology 72, 295–339.

Sotelo, C., Llinas, R., Baker, R., 1974. Structural study of inferior olivary nucleus of the cat: morphological correlates of electrotonic coupling. Journal of Neurophysiology 37, 541–559.

Spacek, J., 1985. Three-dimensional analysis of dendritic spines. III. Glial sheath. Anatomy and Embryology (Berlin) 171, 245–252.

Stea, A., Tomlinson, W.J., Soong, T.W., et al., 1994. Localization and functional properties of a rat brain α1A calcium channel reflect similarities to neuronal Q- and P-type channels. Proceedings of the National Academy of Sciences of the United States of America 91, 10576–10580.

Sugihara, I., 2005. Microzonal projection and climbing fiber remodeling in single olivocerebellar axons of newborn rats at postnatal days 4–7. The Journal of Comparative Neurology 487, 93–106.

Sugihara, I., Shinoda, Y., 2004. Molecular, topographic, and functional organization of the cerebellar cortex: a study with combined aldolase C and olivocerebellar labeling. Journal of Neuroscience 24, 8771–8785.

Sugihara, I., Shinoda, Y., 2007. Molecular, topographic, and functional organization of the cerebellar nuclei: analysis by three-dimensional mapping of the olivonuclear projection and aldolase C labeling. Journal of Neuroscience 27, 9696–9710.

Sugihara, I., Lang, E.J., Llinas, R., 1993. Uniform olivocerebellar conduction time underlies Purkinje cell complex spike synchronicity in the rat cerebellum. The Journal of Physiology 470, 243–271.

Sugihara, I., Wu, H.S., Shinoda, Y., 2001. The entire trajectories of single olivocerebellar axons in the cerebellar cortex and their contribution to cerebellar compartmentalization. Journal of Neuroscience 21, 7715–7723.

Sultan, F., Bower, J.M., 1998. Quantitative Golgi study of the rat cerebellar molecular layer interneurons using principal component analysis. The Journal of Comparative Neurology 393, 353–373.

Takacs, J., Hamori, J., 1994. Developmental dynamics of Purkinje cells and dendritic spines in rat cerebellar cortex. Journal of Neuroscience Research 38, 515–530.

Takagishi, Y., Hashimoto, K., Kayahara, T., et al., 2007. Diminished climbing fiber innervation of Purkinje cells in the cerebellum of myosin Va mutant mice and rats. Developmental Neurobiology 67, 909–923.

Takayama, C., Inoue, Y., 2004. Transient expression of GABA$_A$ receptor α2 and α3 subunits in differentiating cerebellar neurons. Brain Research. Developmental Brain Research 148, 169–177.

Takayama, C., Nakagawa, S., Watanabe, M., Mishina, M., Inoue, Y., 1995. Light- and electron-microscopic localization of the glutamate receptor channel δ2 subunit in the mouse Purkinje cell. Neuroscience Letters 188, 89–92.

Takechi, H., Eilers, J., Konnerth, A., 1998. A new class of synaptic response involving calcium release in dendritic spines. Nature 396, 757–760.

Takeuchi, T., Miyazaki, T., Watanabe, M., Mori, H., Sakimura, K., Mishina, M., 2005. Control of synaptic connection by glutamate receptor δ2 in the adult cerebellum. Journal of Neuroscience 25, 2146–2156.

Uemura, T., Mishina, M., 2008. The amino-terminal domain of glutamate receptor δ2 triggers presynaptic differentiation. Biochemical and Biophysical Research Communications 377, 1315–1319.

Uemura, T., Mori, H., Mishina, M., 2004. Direct interaction of GluRδ2 with Shank scaffold proteins in cerebellar Purkinje cells. Molecular and Cellular Neuroscience 26, 330–341.

Uemura, T., Kakizawa, S., Yamasaki, M., et al., 2007. Regulation of long-term depression and climbing fiber territory by glutamate receptor δ2 at parallel fiber synapses through its C-terminal domain in cerebellar Purkinje cells. Journal of Neuroscience 27, 12096–12108.

Uemura, T., Lee, S.J., Yasumura, M., et al., 2010. Trans-synaptic interaction of GluRδ2 and Neurexin through Cbln1 mediates synapse formation in the cerebellum. Cell 141, 1068–1079.

Urade, Y., Oberdick, J., Molinar-Rode, R., Morgan, J.I., 1991. Precerebellin is a cerebellum-specific protein with similarity to the globular domain of complement C1q B chain. Proceedings of the National Academy of Sciences of the United States of America 88, 1069–1073.

Voogd, J., Ruigrok, T.J., 2004. The organization of the corticonuclear and olivocerebellar climbing fiber projections to the rat cerebellar vermis: the congruence of projection zones and the zebrin pattern. Journal of Neurocytology 33, 5–21.

Wassef, M., Zanetta, J.P., Brehier, A., Sotelo, C., 1985. Transient biochemical compartmentalization of Purkinje cells during early cerebellar development. Developmental Biology 111, 129–137.

Wassef, M., Chedotal, A., Cholley, B., Thomasset, M., Heizmann, C.W., Sotelo, C., 1992. Development of the olivocerebellar projection in the rat: I. Transient biochemical compartmentation of the inferior olive. The Journal of Comparative Neurology 323, 519–536.

Watanabe, M., 2008. Molecular mechanisms governing competitive synaptic wiring in cerebellar Purkinje cells. Tohoku Journal of Experimental Medicine 214, 175–190.

Watanabe, M., Kano, M., 2011. Climbing fiber synapse elimination in cerebellar Purkinje cells. European Journal of Neuroscience 34, 1697–1710.

Watanabe, M., Nakamura, M., Sato, K., Kano, M., Simon, M.I., Inoue, Y., 1998. Patterns of expression for the mRNA corresponding to the four isoforms of phospholipase Cβ in mouse brain. European Journal of Neuroscience 10, 2016–2025.

Watase, K., Hashimoto, K., Kano, M., et al., 1998. Motor discoordination and increased susceptibility to cerebellar injury in GLAST mutant mice. European Journal of Neuroscience 10, 976–988.

Williams, M.E., de Wit, J., Ghosh, A., 2010. Molecular mechanisms of synaptic specificity in developing neural circuits. Neuron 68, 9–18.

Yamada, K., Fukaya, M., Shibata, T., et al., 2000. Dynamic transformation of Bergmann glial fibers proceeds in correlation with dendritic outgrowth and synapse formation of cerebellar Purkinje cells. The Journal of Comparative Neurology 418, 106–120.

Yamanaka, H., Yanagawa, Y., Obata, K., 2004. Development of stellate and basket cells and their apoptosis in mouse cerebellar cortex. Neuroscience Research 50, 13–22.

Yuzaki, M., 2004. The δ2 glutamate receptor: a key molecule controlling synaptic plasticity and structure in Purkinje cells. Cerebellum 3, 89–93.

Zhang, L., Goldman, J.E., 1996. Generation of cerebellar interneurons from dividing progenitors in white matter. Neuron 16, 47–54.

6

Dendritic Spines

D. Muller, I. Nikonenko
University of Geneva, Geneva, Switzerland

Abbreviations

AMPA α-Amino-3-hydroxyl-5-methyl-4-isoxazole-propionate
AMPAR α-Amino-3-hydroxyl-5-methyl-4-isoxazole-propionate receptor
Cdc42 Cell division control protein 42
DISC1 Disrupted-in-schizophrenia 1
EM Electron microscopy
FMRP Fragile X mental retardation protein
GKAP Guanilate cyclase associated protein
GluR Glutamate receptor subunit
LTD Long-term depression
LTP Long-term potentiation
NMDA N-Methyl-D-aspartate
NO Nitric oxide
NMDAR N-Methyl-D-aspartate receptor
PAK P21 activated kinase
PSD Postsynaptic density
PSD-95 Postsynaptic density 95
SAP97 Synapse associated protein 97
SER Smooth endoplasmic reticulum
mGluR Metabotropic glutamate receptor

6.1 INTRODUCTION

Dendritic spines are the principal site for excitatory transmission in the brain. Since their description by Ramon y Cajal and Tanzi at the end of the nineteenth century, dendritic spines have been at the focus of intense research aimed at understanding their function in the processing of neuronal activity. The initial belief that they could play a role in learning and memory processes actually has received strong support from numerous studies that have analyzed their morphological and functional properties. First, dendritic spines exhibit important activity-dependent forms of plasticity that affect both their strength and structural organization, contributing in this way to information processing. Second, recent developments in confocal imaging techniques have provided evidence that spines are dynamic structures that can grow and be eliminated throughout life and thus make possible a continuous adaptation of brain circuits to experience. Third, at the molecular level, dendritic spines, and particularly their postsynaptic density, the region where receptors are located, form a highly complex structure in terms of protein composition, diversity, and implicated signaling mechanisms, illustrating the key role played by dendritic spines in signal integration. Finally, progress in the genetic identification of molecular defects underlying human diseases has provided strong evidence that alterations of a wide array of synaptic proteins lead to important

Neural Circuit Development and Function in the Brain: Comprehensive Developmental
Neuroscience, Volume 3 http://dx.doi.org/10.1016/B978-0-12-397267-5.00145-X

© 2013 Elsevier Inc. All rights reserved.

cognitive and behavioral disorders, establishing a direct link between dendritic spine synapses and higher brain functions.

6.2 GENERAL MORPHOLOGICAL CHARACTERISTICS OF DENDRITIC SPINES

Dendritic spines are mostly found on cortical excitatory neurons. They can however also be found on inhibitory interneurons such as cerebellar Purkinje cells, spiny stellate cells of basal ganglia or olfactory granule cells. Spines are also found in invertebrates, for example in mushroom bodies in flies or arthropods, and many of the properties described here also apply to these other types of spines.

A major characteristic of dendritic spines is their high level of morphological variability (Figure 6.1(a)). To make sense of the functional implications of this variability, researchers have tried to classify them according to various morphological criteria, including size, shape, organization of the spine and/or postsynaptic density (PSD), or presence of specific organelles revealed in electron microscopic (EM) studies. In the rat hippocampus, where most studies of dendritic spines have been performed, the size of the spine head determined by the length of the largest diameter varies between 0.2 and 1.3 μm, which roughly corresponds to volumes of 0.004–1.2 μm^3; thus, corresponding to variations differed by a factor of about 300. These variations in spine volume usually also are correlated with variations in the size of the PSD, size of the presynaptic terminal, and presence of various organelles in the spine: large spines are more likely to contain a spine apparatus or ribosomes.

Dendritic spine synapses are composed of several different elements. The PSD is a highly organized structure characterized by a high density of receptors and channels, associated signaling proteins, adhesion molecules, and cytoskeletal elements assembled together by a variety of scaffold proteins. It represents the contact zone, where synaptic transmission occurs. PSD size, measured with three-dimensional (3D) EM reconstruction, can vary widely, between 0.008 and 0.54 μm^2 at hippocampal CA1 excitatory synapses (Harris and Stevens, 1989). The shape of the PSD is also quite variable and in most cases, it appears on 3D EM reconstruction as a single, macular area, but more complex shapes are not uncommon and particularly have been associated with increased synaptic remodeling and possibly receptor content and turnover (Ganeshina et al., 2004; Geinisman et al., 1993; Toni et al., 1999, 2001). These include PSDs interrupted in the middle, often referred to as perforated PSDs, or PSDs composed of two or multiple individual parts that may correspond to separate transmission zones, referred to as segmented PSDs (Figure 6.1(b)). Synapses with complex PSDs are mostly present on large mushroom-type spines and may represent, depending on conditions, between 5% and 25% of all PSDs.

The postsynaptic membrane of dendritic spines is separated from the presynaptic terminal by a synaptic cleft (Figure 6.1(a)) that is usually 10–20 nm wide and contains dense material binding the two membranes together. The exact content of the synaptic cleft material is not known, but it likely consists of transsynaptic fibrils that often are regularly spaced (Zuber et al., 2005).

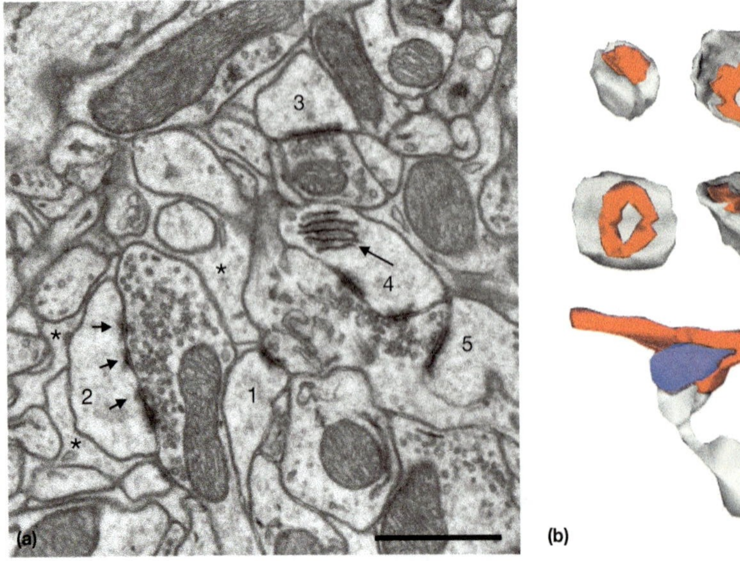

FIGURE 6.1　EM illustrations of the variability of spine and PSD size and shape. (a) Spine heads of different sizes and shapes (nos 1–5). Spine no. 1 is of a thin type, spine nos. 2–5 are of mushroom type. PSDs in spines 2 and 4 are of a complex shape. Arrow points at spine apparatus in spine no. 4. Spine nos. 1, 4, and 5 contact the same presynaptic bouton (multi-synaptic bouton). Arrows in spine no. 2 point at a complex PSD. Astrocytic processes surrounding this synapse are marked with asterisks. Scale bar 0.5 μm. (b) 3D reconstructions of different types of PSDs (macular, U-shaped, perforated, segmented, and multi-innervated spine with two separate PSDs, shown with contacting presynaptic boutons).

Dendritic spines are almost always in contact with a presynaptic bouton formed by an enlargement of the axonal shaft and filled with synaptic vesicles containing neurotransmitter that is released at the presynaptic active zone facing a PSD (Figure 6.1(a)). In addition, spine synapses are often surrounded by fine astrocytic processes (Figure 6.1(a)) that may contribute to synaptic network functions by controlling the local synaptic environment and coordinating synaptic activity and plasticity (Genoud et al., 2006; Haydon, 2001; Lushnikova et al., 2009; Volterra and Meldolesi, 2005; Witcher et al., 2007).

Dendritic spines may also contain several intracellular organelles. About half of them have what is referred to as the spine apparatus composed of smooth endoplasmic reticulum (ER) (Spacek and Harris, 1997). Smooth ER forms a continuous network throughout dendritic segments and spines (Cooney et al., 2002) and is likely to be important for regulating calcium content, local protein synthesis, or both. Confocal imaging has shown a continuity of spine ER with dendritic ER and a surprisingly high degree of turnover in individual spines (Toresson and Grant, 2005). Ribosomes are often found at the base of dendritic spines. They are either associated with ER or free in the cytoplasm and serve local protein synthesis. Free polyribosomes showed redistribution from dendritic shafts into enlarged spines after induction of long-term potentiation (LTP), pointing at the importance of local synthesis for synaptic plasticity (Bourne et al., 2007; Ostroff et al., 2002). Spines may contain endosomal systems consisting of clathrin-coated vesicles and pits, large uncoated vesicles, tubular compartments, multivesicular bodies, and multivesicular bodies–tubule complexes. The recycling endosomes, the part of the system that transports membrane-bound proteins onto and off the cell surface, is of particular importance for activity-related delivery of glutamate receptors to the synapse, as well as for activity-induced structural plasticity of spines (Park et al., 2006; Racz et al., 2004). Another important component of dendritic spines is actin filaments (F-actin). They constitute a major element of dendritic spine cytoskeleton and as such play an essential role in many aspects of spine formation, plasticity and dynamics (Hotulainen and Hoogenraad, 2010). Finally, microtubules, a major cytoskeletal element of dendrites and small unmyelinated axons, are generally absent from dendritic spines. However, recent studies have provided evidence for the presence of microtubules in a small portion of dendritic spines. Notably, growing microtubule ends can transiently enter dendritic spines, interact with F-actin binding protein cortactin, and presumably modulate actin dynamics in the spine in an activity-dependent fashion (Hu et al., 2011; Jaworski et al., 2009). A recent EM study, using a special microtubule-stabilizing fixative, demonstrated the presence of microtubules in CA1

dendritic spines after strong LTP induction and suggested that they could be implicated in AMPA receptors (AMPAR) trafficking to stimulated PSDs (Mitsuyama et al., 2008).

6.3 VARIATIONS IN SPINE SYNAPSE ORGANIZATION

Although the significance of the variability in spine size and shape is still largely unclear, one distinction that has emerged as potentially meaningful is between spines with a large head, a constricted neck and a large PSD (usually referred to as mushroom type spines: Figures 6.1(a) and 6.2(e)), and long and thin spines, with a small head and a filopodia-like shape (often referred to as thin spines: Figures 6.1(a) and 6.2(d)). There is now evidence that these two classes of spines may have different functional properties and could correspond to different phases of the life of a spine: thin spines representing immature, newly formed protrusions that are less likely to remain stable, while mushroom type spines would represent mature, stable spines that have undergone a process of activity- or plasticity-mediated enlargement and remodeling (Bourne and Harris, 2007). Along with mushroom spines and thin spines, which represent more than 80% of all excitatory spines in mature tissue, several other protrusion morphologies or spine structures have been reported under particular developmental or activity conditions.

A first type of protrusion, mainly seen during the early phases of synaptic network development, is comprised by filopodia (Figure 6.2(c)). Filopodia are typically referred to as long, very motile protrusions devoid of an enlargement at the tip. They can grow and retract within minutes and have been shown to be able to transform into dendritic spines. They are therefore believed to represent precursors of dendritic spines (Ziv and Smith, 1996). Filopodia are mainly seen during early stages of development, where they can represent up to 20% of all protrusions, but their proportion then decreases with maturation, so that in mature tissue they constitute only a few percent depending upon criteria used to identify them. A clear distinction between filopodia and thin spines is indeed often difficult. At the EM level, filopodia are often defined by the absence of a PSD at the tip, while with confocal imaging it is the absence of enlargement at the tip and their motility that best characterize them. Filopodia may sometimes express a PSD, which is then usually located at the base or along the length of the protrusion. Formation and stability of filopodia appears to be controlled by calcium transients present in the developing dendrite (Lohmann et al., 2005) and their growth seems to be targeted, as they are able to form transient contacts specifically with excitatory axons and not with inhibitory ones (Lohmann and Bonhoeffer, 2008).

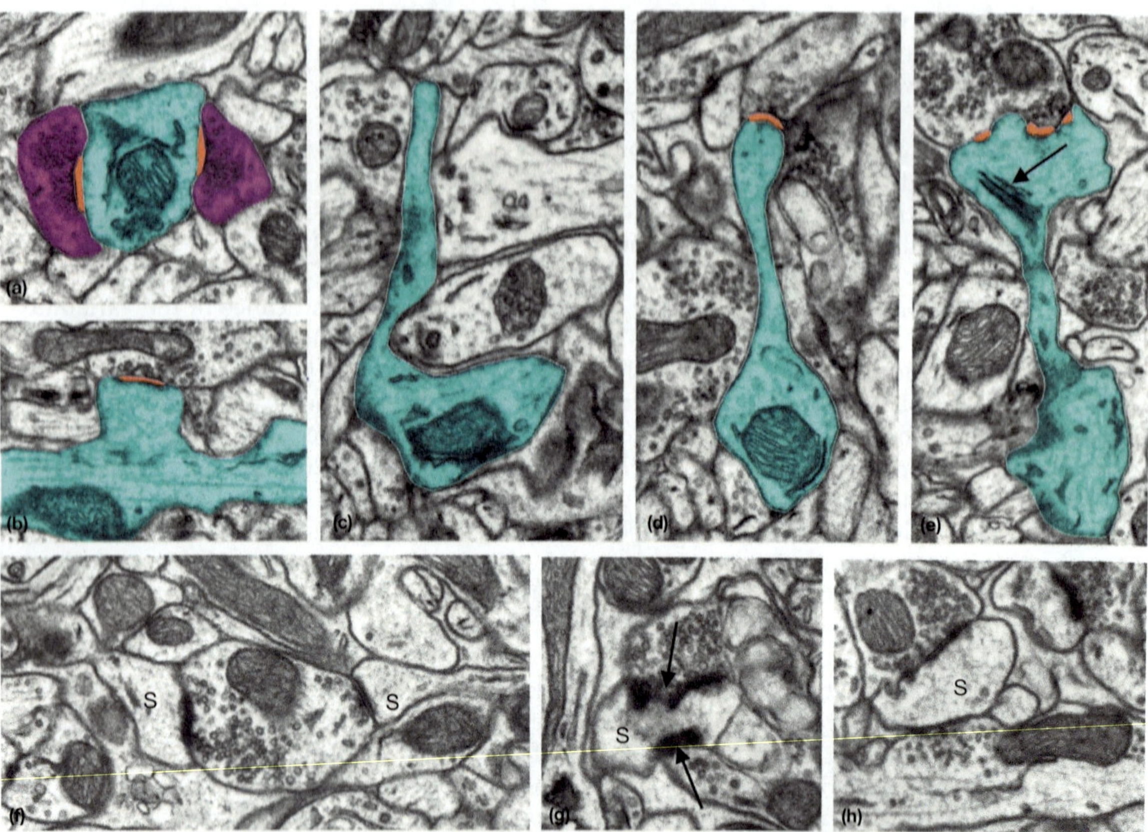

FIGURE 6.2   EM illustrations of different types of excitatory synapses. (a) Two presynaptic terminals (magenta) making shaft synapses (red) on the dendritic shaft (blue), cut transversally, with a mitochondria in the center surrounded with endoplasmic reticulum. (b) Stubby spine on a dendritic segment (blue) with a macular PSD (red). (c) Filopodia. (d) Thin spine with a macular PSD. (e) Mushroom spine with a complex PSD and spine apparatus (arrow). (f) Two spines (s) sharing a common presynaptic bouton (multi-synaptic bouton). (g) Multi-innervated spine (s) with two different PSDs (arrows) contacting two different presynaptic boutons. (h) Cortical spine (s) with excitatory (top) and inhibitory (bottom) synaptic contacts. Scale bars for the top and bottom rows: 0.5 Mm.

While filopodia grow directly from dendrites, thin and motile protrusions extending from dendritic spines can also be observed. These have been referred to as spinules, and they are usually associated with spines having large heads and complex PSDs. Their occurrence appears to be promoted by glutamate application and seems to require AMPAR transmission (Richards et al., 2005). They could therefore result from a reorganization of the spine actin cytoskeleton mediated by synaptic activity and thus be associated with activity- or plasticity-mediated spine remodeling.

Other particular categories of synaptic contacts are made up of shaft and sessile synapses. Shaft synapses are contacts made by an axonal bouton directly on a dendritic shaft (Figure 6.2(a)). In adults, most shaft synapses seen on major apical dendrites are actually formed by inhibitory contacts, but a fraction of them may also be excitatory. Overall, excitatory shaft synapses represent less than 10% of all excitatory synapses in mature tissue. This proportion, however, is higher during development and may also vary as a function of neuronal type and brain region. It has been shown that shaft synapses can give rise to dendritic spine synapses through the extension of a protrusion (Marrs et al., 2001). Alternatively, shaft synapses could also result from the retraction of dendritic spines. Finally, they might have specific functions, as their contribution to signal integration in dendrites is more important than for spine synapses. Consistent with this interpretation, their formation and number appear to be specifically regulated by a signaling system implicating ephrin B3 (Aoto et al., 2007).

Synaptic contacts can also be made on short dendritic protrusions devoid of a neck; these are referred to as stubby or sessile spines (Figure 6.2(b)). They probably represent an intermediate stage between excitatory shaft synapses and spine synapses with a neck. At the EM level, they usually constitute only a very small proportion (in the range of a few percent) of all protrusions in mature tissue. Reportedly, their number is modulated by activity and induction of LTP (Chang and Greenough, 1984).

While most spine synapses are formed between individual axonal boutons and dendritic spines, there are examples where a single axonal bouton contacts multiple dendritic spines (also referred to as multiple spine bouton, Figure 6.2(f)) and cases where a single spine is contacted by multiple axonal boutons (forming multi-innervated spines, Figures 6.1(b) and 6.2(g)). These cases are usually quite rare, representing only a few percent of all synaptic contacts in mature tissue, but their occurrence has been shown to increase under developmental or plasticity conditions. Notably, an increase in multiple spine boutons has been reported following induction of LTP, with the additional interesting feature that in many cases the two spines contacted by the same axonal boutons arose from the same dendrite (Toni et al., 1999). This observation was therefore interpreted as indicating a process of new spine growth and synapse formation triggered by induction of plasticity. Similar images of multiple spine boutons have also been reported when analyzing the development and integration of new neurons in the hippocampal dentate gyrus (Toni et al., 2007) or under conditions of spine growth promoted by whisker trimming in the somatosensory cortex (Knott et al., 2006). These results therefore suggest that images of multiple spine boutons could reflect a transient competition for the same presynaptic boutons between a pre-existing spine and a newly formed one, and thus translate an ongoing process of synaptogenesis.

Multi-innervated spines are similarly rather infrequent in mature tissue, reaching 2–3% of all protrusions. Recent analysis of the mechanisms underlying this particular synaptic configuration has revealed the role played by the transsynaptic release of nitric oxide (NO) by the growing spine. Under conditions where this release is enhanced, it promotes the differentiation of nearby axonal shafts into presynaptic boutons and the formation of additional synaptic contacts on the same spine. Conversely, blockade of NO production during development is associated with a reduction in the formation of spines and synapses (Nikonenko et al., 2008). As the production of NO at synapses is linked to the level of expression of scaffold proteins such as PSD-95 or SAP97 (Poglia et al., 2010), one might consider the possibility that the formation of a multi-innervated spine occurs as a result of enhanced growth of the PSD. Also, as the strength of a synapse correlates with the size of the PSD, multi-innervated spines could reflect the existence of highly potent synaptic contacts.

Finally, there are also cases of excitatory dendritic spines that bear at the same time an inhibitory synaptic contact (Figure 6.2(h)). This situation is not seen in all tissues and has been mainly reported at spine synapses in different cortical areas. In the somatosensory cortex for example, the proportion of these spines with excitatory/inhibitory synapses is low under control situations, but can markedly increase as a result of strong, lasting sensory stimulation (Knott et al., 2002). The mechanisms underlying the formation of these particular spine synapses are currently unknown, but could represent a means for controlling the level and specificity of excitation in cortical regions.

6.4 MOLECULAR COMPOSITION AND SIGNALING MECHANISMS

Excitatory dendritic spine synapses represent one of the most complex biological structures in terms of their molecular composition and diversity. Despite their small size, dendritic spines contain and express more than 1000 different proteins often organized in multi-protein signaling complexes (Emes et al., 2008). Additionally, synaptic proteins very often exist in different subtypes that confer subtle functional properties.

A central mechanism targeted by many synaptic regulators appears to involve the control of the trafficking of synaptic proteins, among which are receptors and PSD-associated proteins. At excitatory synapses, multiple neurotransmitter receptor subtypes (NMDA, AMPA and metabotropic receptors) are usually expressed and further characterized by various subunit compositions and distinct biophysical properties. These receptors are embedded in a matrix of scaffolding proteins, which together with adhesion and signaling molecules constitute the PSD. Structurally, the PSD is a large network of interconnected protein assemblies, one important function of which might be to regulate the number and subunit composition of the receptors present at the synapse.

Regulation of these mechanisms likely occurs in a number of different ways, but a key role therein is certainly played by synaptic activity. At many excitatory synapses, high-frequency trains of stimulation are able to trigger specific signal transduction pathways that serve as important mechanisms for the regulation of synaptic strength. Changes in synaptic efficacy are currently viewed as the most prevalent molecular model for mechanisms of information processing, learning and memory (Cooke and Bliss, 2006). Numerous forms of synaptic plasticity and changes in synaptic strength have been reported at central synapses depending on brain region, cell type, and age. Some of these are short-lasting such as facilitation or post-tetanic potentiation, but the most interesting forms result in long-lasting changes in synaptic efficacy (Bliss and Collingridge, 1993). These may lead either to increases or to decreases in synaptic strength (Dudek and Bear, 1993) and involve pre- or postsynaptic mechanisms

(Nicoll and Malenka, 1995). The best-known example of synaptic plasticity is probably the form of LTP that is expressed at CA1 hippocampal synapses. In this type of plasticity, the change in synaptic strength is triggered by an activation of postsynaptic NMDA type of glutamate receptors (NMDAR) leading to calcium fluxes in dendritic spines. This in turn activates signal transduction pathways, among which protein kinases – calcium/calmodulin-dependent protein kinase II (CaMKII) in particular – play an important role (Lisman et al., 2002). While the specific sequence of molecular events involved in the lasting increase in synaptic strength is still a matter of debate, much recent evidence suggests that synaptic potentiation depends heavily upon the number of receptors expressed at synapses and thus upon local trafficking of receptors within individual dendritic spines (Malenka and Nicoll, 1999; Malinow and Malenka, 2002).

One interesting hypothesis to account for the lasting changes in synaptic strength is the idea that receptors and scaffolding proteins interact to form receptor-binding slots controlling synaptic strength (Bats et al., 2007; Lisman et al., 2002). Biochemical and imaging studies indicate that the main scaffold proteins (e.g. PSD-95, SAP97, GKAP, Shank or Homer) are more densely expressed at synapses (60–400 molecules of each per synapse) than receptors and thus outnumber glutamate receptors (0–200 receptors per synapse). They have therefore been proposed to provide a structural basis for the formation of a pool of receptor-confining domains likely to account for modifications of synaptic strength. Glutamate receptors, especially AMPARs, exhibit a high level of lateral mobility within membranes and can exchange rapidly at synapses in a way that is highly sensitive to activity (Triller and Choquet, 2008). Synaptic stimulation or calcium uncaging reduce AMPAR diffusion, so that active synapses tend to trap AMPARs more efficiently than inactive synapses (Borgdorff and Choquet, 2002). Additionally, large PSDs capture and retain more PSD-95 than smaller PSDs and thus may favor trapping of AMPARs (Gray et al., 2006). Synaptic activity is therefore able to regulate glutamate receptor and scaffold protein dynamics at synapses, thereby providing a mechanism for activity-dependent targeting and retention of receptors at individual synapses.

Numerous proteins participate to the regulation of receptor expression at synapses. In addition to the major scaffold proteins, auxiliary transmembrane regulatory proteins (TARPs), PDZ proteins and protein kinases also play an important role (Nicoll et al., 2006). These proteins may affect not only the lateral mobility of receptors at synapses (Opazo et al., 2010) but also their intracellular trafficking, which is crucial for controlling the abundance of glutamate receptors at synapses. Within dendritic spines, AMPAR internalization is thought to occur at endocytotic zones, defined as clathrin-coated membrane domains located adjacent to the PSD (Racz et al., 2004). These internalized receptors are then transferred to recycling endosomes, which may function as a supply mechanism for providing AMPARs to synapses. Upon induction of plasticity, AMPARs are rapidly inserted at synapses, although the precise location at which this exocytotic process might take place remains unclear. Evidence suggests that AMPAR incorporation at synapses following LTP induction could occur through lateral diffusion from a pool driven to the surface primarily on dendrites (Makino and Malinow, 2009).

In addition to activity, several other signaling pathways also contribute to regulate receptor trafficking. Among these, small GTPases play a particularly important role. GTPases form a large family of proteins characterized by their ability to bind and hydrolyze GTP. They generally act as molecular switches affecting various biological activities regulating growth and migration of many cell types. Their action is therefore not restricted to neurons and synapses. GTPases cycle between an active GTP-bound and inactive GDP-bound state and are tightly regulated by a variety of modulators, many of which are either activity-dependent or linked to signaling from synaptic adhesion molecules. At synapses, GTPases appear to control receptor trafficking in different ways, either directly or through their action on the actin cytoskeleton or the local translation machinery (Boda et al., 2010). Ras and Rap, for example, are two GTPases that exert opposing effects on the regulation of AMPAR delivery to synapses. Ras activation promotes the trafficking of GluA1-containing AMPARs from a deliverable pool located near the PSD to the synapse, while stimulation of Rap signaling promotes spine shrinkage, removal of GluA2/3 containing AMPARs and depression of synaptic transmission (Kielland et al., 2009).

It is interesting that changes in synaptic strength are also associated with a remodeling of potentiated synapses. Repetitive confocal imaging of identified synapses upon induction of plasticity has shown that the changes in synaptic strength are accompanied by an enlargement of the dendritic spine head which correlates with the increased sensitivity to glutamate (Matsuzaki et al., 2004). This effect is rapid and can be long-lasting. Although its significance is still unclear, these results indicate that remodeling of the spine architecture through modifications of the actin cytoskeleton is part of the processes that underlie plasticity. Regulation of the actin cytoskeleton is a complex process that involves different signaling pathways, among which are the small GTPases Rac and Cdc42. Much recent evidence suggests that this cascade could participate in the regulation of spine size. By integrating activity and transsynaptic signals through adhesion molecules such as N-cadherin or EphrinB receptors, this pathway could tightly control the function of the actin cytoskeleton through PAK, LIM kinase, or cofilin-dependent mechanisms and in this way regulate the growth, size, and morphology of spines (Penzes and Jones, 2008; Tolias et al., 2007). Another pathway implicated in the regulation of the spine actin cytoskeleton could integrate signals from the extracellular matrix (Gundelfinger et al., 2010). The importance of this signaling is particularly well illustrated by studies of matrix metalloproteases (MMPs), which degrade the extracellular matrix and markedly affect synaptic structure and function. One of them, MMP9, is associated with enhanced activity, is required for LTP at hippocampal synapses, and mediates through integrin signaling spine enlargement and synaptic potentiation (Wang et al., 2008). The extracellular matrix additionally plays an important role in controlling the lateral diffusion of membrane proteins at synapses by forming compartments that might affect plasticity properties (Frischknecht et al., 2009).

Another mechanism likely to play an important role in the functional maturation and plasticity properties of excitatory spine synapses is the local regulation of protein synthesis. Pharmacological inhibition of protein synthesis interferes with induction of plasticity and much recent evidence indicates that activity is able to promote a local translation of messenger RNAs in dendritic spines (Kelleher et al., 2004). This is associated with a redistribution of polyribosomes into large dendritic spines upon induction of LTP (Ostroff et al., 2002). At the molecular level, evidence points to an activity-mediated translation of proteins such as CaMKII or Arc (Bramham, 2008) that could play a role in mechanisms of synaptic plasticity by regulating actin dynamics, spine size, and the organization of the postsynaptic density.

6.5 MECHANISMS OF SPINE FORMATION

There appear to be at least two different processes through which new excitatory spine synapses can be formed. One includes as an initial step the growth of a very motile protrusion, the filopodium, which extends from the dendrite up to several microns and then makes contact with potential partners before eventually retracting into the form of a dendritic spine. Filopodia are mainly seen during early phases of synaptic development, when the distance to potential partners is greater (Ziv and Smith, 1996). Due to their high motility, they are ideally suited to explore the space around the dendrites searching for appropriate binding partners.

How exactly they contribute to synapse formation remains poorly understood. Filopodia have been shown to make repeated, transient contacts with axons, which in some cases can be stabilized, over a range of minutes, through the generation of calcium transients (Lohmann and Bonhoeffer, 2008). This implies a capacity

of filopodia to sense their environment. It does not seem that this signaling is mediated through the neurotransmitter glutamate, since glutamate receptor antagonists do not affect contact formation by filopodia. However, specific recognition mechanisms likely play a role, since filopodia are able to discriminate between partners and never make stabilized contacts with inhibitory axons (Lohmann and Bonhoeffer, 2008). Once a contact has been made, filopodia may eventually be transformed into spines (De Roo et al., 2008a; Maletic-Savatic et al., 1999; Marrs et al., 2001; Trachtenberg et al., 2002; Zuo et al., 2005). The success rate of this process, however, seems to be variable. Imaging studies in young mice or in hippocampal slice cultures suggest that only 10–20% of filopodia may actually be transformed into spines and that most of these spines still disappear within subsequent days (De Roo et al., 2008a; Zuo et al., 2005). Filopodia might however play a more important role during early phases of development, when the probability to reach a partner is lower and the motility of the filopodium is an advantage. Recent work in fact showed that decreasing filopodia motility by interfering with EphrinB signaling reduced the rate of synaptogenesis. This effect was notably prominent in early, but not late, development (Kayser et al., 2008). This is consistent with the notion that the contribution of filopodia to synapse formation may be restricted to early development.

Several different molecules have been identified as regulating the motility and behavior of filopodia. These can be classified in two categories: "accelerators" and "brakes". The accelerators include calcium/calmodulin-dependent protein kinase II (CaMKII) (Jourdain et al., 2003), syndecan-2 (Ethell and Yamaguchi, 1999; Lin et al., 2007), and paralemmin-1 (Arstikaitis et al., 2008), which enhance filopodia formation and further accelerate spine maturation. In striking contrast, the brakes are molecules that not only induce but also maintain dendritic filopodia, thus slowing spine maturation and sometimes even causing spine-to-filopodia reversion (Furutani et al., 2007; Kumar et al., 2005; Matsuno et al., 2006; Pak and Sheng, 2003; Vazquez et al., 2004). A particularly good example of a brake molecule is telancephalin (Yoshihara et al., 2009).

In more mature tissue, time-lapse imaging has shown that new protrusions may also directly appear as spines (Engert and Bonhoeffer, 1999; Trachtenberg et al., 2002). This process, which occurs within minutes, probably accounts for about half of all protrusions formed in young hippocampal tissue (De Roo et al., 2008a; Engert and Bonhoeffer, 1999; Jourdain et al., 2003). Typically, these new spines have long necks and small heads, which sometimes makes them difficult to distinguish from filopodia, except that they are less motile. In young neurons, new spines and filopodia are produced at a high rate and seemingly in a random fashion (Lendvai et al., 2000).

Also, a significant proportion are essentially transient and tend to disappear within hours (Cruz-Martin et al., 2010; De Roo et al., 2008a; Holtmaat et al., 2005). The reasons for this are still unknown, but could be linked to a failure to find a proper partner, to establish a contact, or to sense or relay activity. There might also be specific conditions in very young tissue that do not favor the induction of plasticity and thus reduce the probability that these new spines will become stabilized (Ehrlich et al., 2007).

What are the different steps that lead to formation of a synaptic contact at nascent synapses? 3D EM reconstruction of newly formed spines *in vitro* and *in vivo* or following LTP-inducing protocols in slices has revealed that they do not seem initially to express a PSD (De Roo et al., 2008a; Knott et al., 2006; Nagerl et al., 2007). Consistent with this, EM analyses in the cortex or hippocampus have shown the existence of a small population of spine-like protrusions devoid of PSD or even without presynaptic partners, suggesting that spine growth could precede synapse formation (Arellano et al., 2007; De Roo et al., 2008a; Knott et al., 2006). The speed of this process appears to be quite variable. Based on morphological criteria (presence of a postsynaptic density on 3D EM reconstruction or expression of PSD-95-EGFP), the formation of mature synapses appears to require between 5 and 24 h. However, functional analyses of spine sensitivity to glutamate uncaging suggest that a new protrusion could respond to released transmitter much faster, within tens of minutes (Zito et al., 2009). This suggests that receptors may be present or sense transmitter before a morphologically mature PSD is apparent. One principal takeaway, however, is that formation of new, functional synapses can be a fast process, taking just minutes to hours.

Another interesting set of questions pertains to the respective roles of the pre- and postsynaptic partners in regulating the formation of the new synapse. The observation that new protrusions initially grow without expressing a PSD indicates that the growth of the protrusion is probably initiated postsynaptically. Similarly, recent evidence indicating that the control of the number of PSDs and contacts made on a spine is regulated by the postsynaptic release of nitric oxide (NO) further suggests a primary role for the postsynaptic side in the control of the formation of a contact with a given partner (Nikonenko et al., 2008). However, there is also evidence indicating a role for active axon terminals in the promotion of spine growth. Notably, presynaptic activity and glutamate have been shown to promote spine growth (Maletic-Savatic et al., 1999; Richards et al., 2005). In addition, new protrusions seem to prefer to grow towards axonal boutons with active synapses (Knott et al., 2006; Toni et al., 2007); conversely, activated presynaptic boutons may grow filopodia-like protrusions that establish synaptic contacts (Nikonenko et al., 2003). These observations therefore suggest that glutamate may also act as a trophic factor that could guide new protrusions to active boutons. Additionally, a great variety of adhesion and signaling molecules expressed either on pre- or postsynaptic structures have been shown to promote synapse formation even between non-neuronal partners (Dalva et al., 2007; Han and Kim, 2008). The process of synapse formation is therefore very likely to depend upon complex signaling mechanisms linked to numerous secreted or membrane-bound molecules present on both pre- and postsynaptic structures.

6.6 SPINE DYNAMICS AND DEVELOPMENT OF SYNAPTIC NETWORKS

In vivo analyses of spine behavior over periods ranging from minutes to days and months has revealed that spines are very dynamic structures that exhibit various forms of motility and morphological remodeling. A first type of motility initially reported in cultures, but also described *in vivo*, relates to fast (seconds to minutes) twitching of spine heads that has been linked to a continuous rearrangement of the actin cytoskeleton (Fischer et al., 1998; Majewska et al., 2006). Although this form of motility has been associated with activity and experience-dependent plasticity (Oray et al., 2004), its significance in terms of synapse function and properties remains unclear. One possibility is that this spine twitching relates to the mechanisms of protein turnover, which is particularly high in dendritic spines. Proteins such as PSD-95, Ras, or Shank3 have been shown to redistribute between synaptic and dendritic pools within a few minutes to a few hours (Gray et al., 2006; Tsuriel et al., 2006). These trafficking mechanisms might possibly require a continuous rearrangement of the actin cytoskeleton.

On a different timescale, spines have also been shown to undergo a continuous turnover through a process of formation and elimination (Figure 6.3(a)). The rate of spine growth and disappearance is particularly high during development and critical periods when the main organization of the cortical synaptic network is established (Cruz-Martin et al., 2010; Lendvai et al., 2000; Zuo et al., 2005). Spine turnover then decreases with brain maturation but remains effective at a low level even in adult mice (Grutzendler et al., 2002; Trachtenberg et al., 2002). Quantification of spine turnover in various brain regions has been difficult and shows great variability. In mice, turnover rate can affect as many as 5–20% of total spines per hour during the first weeks after birth, decreasing to about 5–10% per week at 1 month of age and 1–2% per month in adult

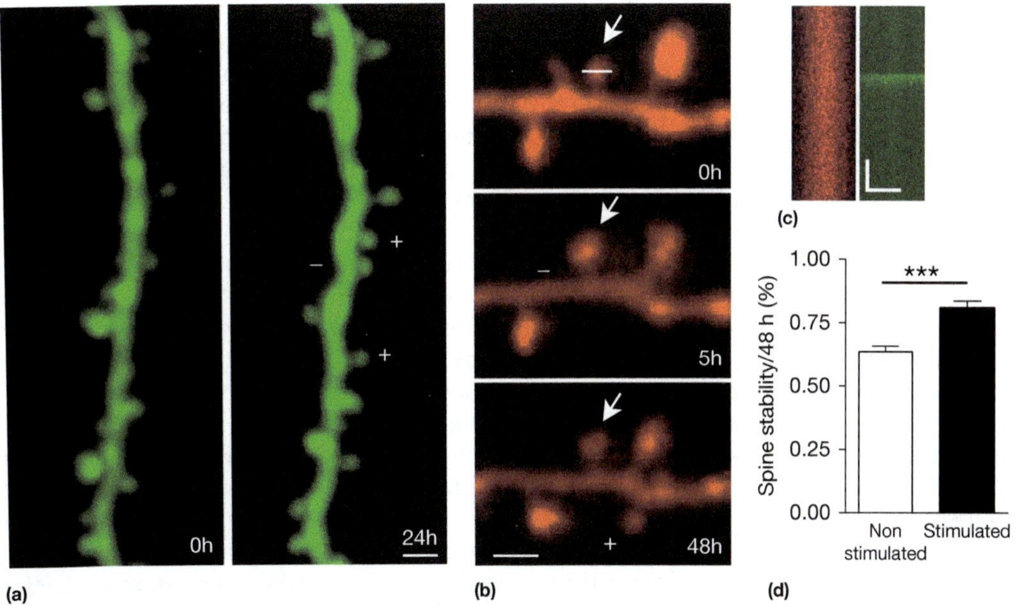

FIGURE 6.3 Confocal images illustrating spine dynamics in hippocampal slice cultures. (a) Dentritic segment of a CA1 pyramidal neuron expressing eGFP and imaged at 24 h interval. Plus and minus signs indicate newly formed and lost spines. Scale bar: 1 μm. (b) Enlargement and stabilization of a stimulated spine in a cell expressing mRFP and loaded with the calcium indicator Fluo4-AM. Line scan analysis of the spine observed at time 0 h showed an increased calcium transient upon electrical stimulation of a group of CA3 cells. Theta burst activation of this spine resulted in an enlargement of the spine head 5 h later and promoted the persistence of the spine over the next 48 h. Scale bar: 1 μm. (c) Line scan analyses through the spine illustrated in (b) and showing the red (mRFP) and green (Fluo4-AM) signals. Scale bars: 0.5 μm; 0.5 s. (d) Stimulated spines show a significantly increased stability over the next 48 h than non-stimulated spines (n=37 and 44 spines analyzed).

tissue (Cruz-Martin et al., 2010; Holtmaat and Svoboda, 2009); however, these values vary quite substantially depending on the cortical region or cell type analyzed, the criteria used for analysis, and the experimental approach. For example, the possible effects of anesthesia must be considered (De Roo et al., 2009). Despite this vartiability, the important message delivered by these studies is that remodeling of synaptic connections is very substantial during development and remains a significant phenomenon throughout life. Further interesting information provided by these experiments concerns the lifetime of a synapse. Analyses of spine stability over weeks or months have shown that not all spines have the same lifespan. While many new spines appear to be very transient, others may persist for prolonged periods of time, up to years under *in vivo* conditions (Grutzendler et al., 2002; Holtmaat et al., 2005). Consistent with the changes in spine dynamics, the fraction of persistent spines also increases as a function of development, with more than 95% of spines surviving months in adult tissue.

An important issue regarding the mechanisms of spine dynamics pertains to their contribution to the function and development of brain networks. Repetitive imaging approaches have shown that stimulation protocols that are used to induce LTP are associated with the growth of new protrusions (Engert and Bonhoeffer, 1999; Maletic-Savatic et al., 1999; Nagerl et al., 2007;

Toni et al., 1999). This growth process can, however, also be associated with an increased elimination of spines and thus involves a more subtle control of spine dynamics (De Roo et al., 2008b; Nagerl et al., 2004). Sensory stimulation and motor learning have both been reported to stimulate an increase in spine turnover, leading to modifications of network organization (Holtmaat et al., 2006; Wilbrecht et al., 2010; Xu et al., 2009; Yang et al., 2009). Similarly, in the visual cortex, monocular deprivation and restoration of binocular vision are associated with marked changes in spine dynamics and spine density (Hofer et al., 2009). All of these results suggest a strong link between functional plasticity and specific synaptic rearrangements.

The question, however, is to understand the specificity of these synaptic rearrangements and how they may affect brain networks at a functional level. An important hint is provided by studies of spine dynamics in relation to induction of synaptic plasticity. When LTP is induced at a group of synapses, several structural rearrangements can be observed (Figure 6.3(b–c)). First, stimulated synapses enlarge (De Roo et al., 2008b; Ehrlich et al., 2007; Harvey and Svoboda, 2007; Kopec et al., 2007; Matsuzaki et al., 2004), a process closely correlated with the accumulation of AMPAR at the synapse (Zito et al., 2009) and the reorganization of the actin cytoskeleton (Honkura et al., 2008). Spine size and synaptic efficacy are thus tightly linked.

A second important characteristic of potentiated synapses is that they are switched to a stable state (De Roo et al., 2008b). This effect is very specific for potentiated synapses since unstimulated synapses tend to undergo elimination (Figure 6.3(d)). How this activity-dependent structural remodeling of spines confers stability on them is still unclear. The close correlation existing between spine head size, PSD size, receptor number, and stability suggests that the phenomenon could be non-specifically related to the amount of receptors and proteins accumulated at the synapse: larger spines with larger PSDs express more adhesion and cross-linking molecules and are more stable. An important aspect to consider in this interpretation is that the size of spine heads shows continuous fluctuations over time (De Roo et al., 2008b; Holtmaat et al., 2005; Yasumatsu et al., 2008). Small spines can enlarge and retract over periods of days. This would suggest that stability, together with PSD size and strength, could continuously fluctuate according to the activity history of the synapse. Another interpretation, however, might be that there exist specific molecular events able to reinforce spine stability. For example, it has been suggested by ultrastructural analyses that stabilization could be provided through the acquisition by the potentiated spines of the machinery for mRNA translation. This would allow local regulation of protein synthesis and trafficking and thus confer independence and stability to the spine (Bramham, 2008; Ostroff et al., 2002; Steward and Falk, 1985). Additionally, this could be associated with the expression at the synapse of specific proteins able to stabilize the cytoskeleton or anchor pre- and postsynaptic structures. Recent evidence suggests that N-cadherin could play such a role (Mendez et al., 2010a). N-cadherin is expressed on both sides of the synaptic cleft and is involved in the formation of homophilic binding domains, which might account for the periodic protein complexes observed across the cleft (Zuber et al., 2005). Expression of dominant-negative mutant N-cadherin or knockdown of N-cadherin expression leads to the formation of unstable synapses. Conversely, expression of N-cadherin is associated with a synapse-specific increase in stability. Additionally, N-cadherin expression is regulated by synaptic activity, and it is selectively expressed in potentiated synapses upon induction of LTP, leading to a switch in stability of synapses. A main function of LTP during development might therefore be to promote stabilization through a change in the composition of the PSD and, importantly, an increase in the expression of the adhesion molecule N-cadherin.

The third important structural change induced by induction of LTP is an increase in spine dynamics. This process also shows specificity. Following induction of plasticity, growth predominantly takes place at the vicinity of stimulated synapses, promoting in this way the formation of clusters of activated synapses (De Roo et al., 2008b). This might have important consequences, particularly for spines on remote dendrites, where spatiotemporal clustering of synaptic activity may play an important role for information processing (Nevian et al., 2007). Additionally, spine elimination also occurs in a very selective manner, since the process concerns mainly unstimulated spines. Thus, the changes in spine dynamics triggered by patterns of activity will affect differentially synaptic connections by favoring clustered activity with new partners, while eliminating inactive synapses. Together, these studies indicate that during development, plasticity mechanisms not only change synaptic strength, but also act as a major process organizing the wiring of synaptic networks. By switching stimulated synapses to a persistent state and additionally promoting the replacement of unstimulated synapses by new ones, LTP operates as a selection mechanism that maintains connections mediating coherent, synchronized activity, while, at the same time, favoring adaptation of the network. This fine-tuning process could be particularly important during critical periods, ensuring the specificity of network development, and may serve as an efficient learning and memory system. In keeping with this interpretation, *in vivo* work analyzing spine dynamics associated with a learning task in motor cortex found that memory of the task was associated with an increase in spine turnover and a selective stabilization of newly formed spines over the course of months (Xu et al., 2009; Yang et al., 2009).

In addition to activity patterns inducing synaptic plasticity, spine growth or elimination appears to be also regulated by a number of other homeostatic processes. One factor that seems to be very important for the regulation of spine dynamics during critical periods of development is the balance between excitation and inhibition (Hensch et al., 1998; Morishita and Hensch, 2008). In the visual cortex, alteration of this balance markedly affects the onset, formation, and plasticity of ocular dominance columns. This process, which is directly related to mechanisms of spine formation and pruning (Hofer et al., 2009), is regulated by GABAergic inhibition. Recent work has identified an important functional, bidirectional recruitment of fast-spiking interneurons during experience-dependent plasticity (Yazaki-Sugiyama et al., 2009). Also, in a manner consistent with this, the application of anesthetics to developing cortical tissue promotes a rapid enhancement of spine and synapse formation leading to a marked increase in spine density on pyramidal neurons (De Roo et al., 2009). Interestingly, this control by the excitation/inhibition balance could still persist even in the adult visual cortex. Recent experiments show that the removal of a molecular brake responsible for the age-dependent loss of visual plasticity demonstrates the persistence of a control by the excitation/inhibition balance and suggests that the mechanisms which reduce adult structural plasticity could work by adjusting this balance (Morishita et al., 2010).

In addition, growth factors and hormones are also very likely to participate in the regulation of spine dynamics. For example, there is strong evidence suggesting an important role for estrogens in the regulation of spine formation and density (McEwen, 2010). The effect is rapid and affects spine growth but not spine elimination (Mendez et al., 2010b); it may thus represent a homeostatic mechanism for adapting spine density and the complexity of synaptic networks through changes in spine dynamics.

6.7 SPINE ALTERATIONS AND BRAIN DISEASE

Alterations of dendritic spine morphology or function represent key features of various developmental psychiatric disorders and neurodegenerative diseases. In the last two decades, progress concerning the identification of genetic alterations associated with mental retardation and autism spectrum disorders has revealed that, in many instances, these defects concern genes coding for synaptic proteins or proteins involved in the regulation of synaptic properties (Bourgeron, 2009; Laumonnier et al., 2007; Ropers, 2006). Analyses of the underlying mechanisms using gain and loss of function approaches have revealed various kinds of alterations, especially defects affecting spine morphology, spine density, or synaptic plasticity. One of the best examples of intellectual disability is probably the fragile X syndrome linked to the silencing of the fmr1 gene (Bagni and Greenough, 2005). The protein coded by the fmr1 gene (FMRP) is associated with polyribosomes in dendrites and spines, and can be found in dendritic RNA granules that travel on microtubules in dendrites. FMRP is thus believed to play a role in the regulation of local translation at individual synapses, acting both as translational suppressor and/or activator. The main defects reported in Fmr1 knockout mice include an increased spine density with more long, thin, tortuous spines reminiscent of immature filopodia-like spines, as well as defects in synaptic plasticity characterized by impaired cortical LTP and enhanced metabotropic glutamate receptor (mGluR)-mediated long-term depression (LTD). This has led to the hypothesis that FMRP could function as a negative feedback mechanism to suppress mGluR-stimulated translation implicated in LTD and the endocytosis of AMPARs (Huber et al., 2002). As such, knockdown of FMRP is believed to result in an increased mGluR-dependent removal of surface AMPARs, and evidence suggests that this defect could be corrected by antagonists of metabotropic receptors (Dolen et al., 2007).

Several other mutations of synaptic proteins identified as associated with cognitive disabilities, autism spectrum disorders, or even schizophrenia also result in dendritic spine abnormalities. This is the case for oligophrenin, PAK3, ARGHEF6, SynGAP, Shank, and EPAC2, but defects have also been reported in a mouse model of the 22q11 syndrome, a condition at high risk for schizophrenia, or following interference with disrupted-in-schizophrenia 1 (DISC1) (Boda et al., 2004; Govek et al., 2004; Node-Langlois et al., 2006; Sala et al., 2001; Vazquez et al., 2004; Woolfrey et al., 2009). In many of these cases, the synaptic defects include an increase in the fraction of long, thin and immature spines, and this is also often associated with alteration of glutamate receptor recycling. These results have therefore led to the hypothesis that cognitive disability could primarily result from alterations of dendritic spine plasticity, either because of defects in receptor recycling mechanisms that affect synaptic strength or because of alterations in spine formation, maturation, or dynamics that could interfere with the development and specificity of synaptic networks (Boda et al., 2010).

In addition to these developmental psychiatric disorders, spine pathology has also been observed in association with degenerative diseases such as Alzheimer's disease (Selkoe, 2002; Spires et al., 2005), Parkinson's disease (Day et al., 2006), or prion disease (Fuhrmann et al., 2007). In mouse models of Alzheimer's disease, the vicinity of amyloid plaques is characterized by highly dismorphic neurites and spine loss, a phenotype that could be caused by the inhibition by amyloid oligomers of LTP mechanisms and the promotion of LTD (Wei et al., 2010). These results thus also suggest that changes in spine dynamics may provide an important contribution to the alterations and dysfunctions of neuronal networks in degenerative and memory disorders.

6.8 CONCLUSION

Dendritic spines are remarkable structures that exhibit a high degree of structural and functional specialization. They are tightly controlled by activity and signaling mechanisms that determine their properties of plasticity and their long-term maintenance in brain networks. The complexity of these regulations and their impact on the formation and function of brain circuits clearly highlight the key role played by dendritic spines in the processing of information by neuronal networks.

References

Aoto, J., Ting, P., Maghsoodi, B., Xu, N., Henkemeyer, M., Chen, L., 2007. Postsynaptic ephrinB3 promotes shaft glutamatergic synapse formation. Journal of Neuroscience 27, 7508–7519.

Arellano, J.I., Espinosa, A., Fairen, A., Yuste, R., DeFelipe, J., 2007. Non-synaptic dendritic spines in neocortex. Neuroscience 145, 464–469.

Arstikaitis, P., Gauthier-Campbell, C., Carolina Gutierrez Herrera, R., et al., 2008. Paralemmin-1, a modulator of filopodia induction is required for spine maturation. Molecular Biology of the Cell 19, 2026–2038.

Bagni, C., Greenough, W.T., 2005. From mRNP trafficking to spine dysmorphogenesis: The roots of fragile X syndrome. Nature Reviews Neuroscience 6, 376–387.

Bats, C., Groc, L., Choquet, D., 2007. The interaction between Stargazin and PSD-95 regulates AMPA receptor surface trafficking. Neuron 53, 719–734.

Bliss, T.V., Collingridge, G.L., 1993. A synaptic model of memory: Long-term potentiation in the hippocampus. Nature 361, 31–39.

Boda, B., Alberi, S., Nikonenko, I., et al., 2004. The mental retardation protein PAK3 contributes to synapse formation and plasticity in hippocampus. Journal of Neuroscience 24, 10816–10825.

Boda, B., Dubos, A., Muller, D., 2010. Signaling mechanisms regulating synapse formation and function in mental retardation. Current Opinion in Neurobiology 20, 519–527.

Borgdorff, A.J., Choquet, D., 2002. Regulation of AMPA receptor lateral movements. Nature 417, 649–653.

Bourgeron, T., 2009. A synaptic trek to autism. Current Opinion in Neurobiology 19, 231–234.

Bourne, J., Harris, K.M., 2007. Do thin spines learn to be mushroom spines that remember? Current Opinion in Neurobiology 17, 1–6.

Bourne, J.N., Sorra, K.E., Hurlburt, J., Harris, K.M., 2007. Polyribosomes are increased in spines of CA1 dendrites 2 h after the induction of LTP in mature rat hippocampal slices. Hippocampus 17, 1–4.

Bramham, C.R., 2008. Local protein synthesis, actin dynamics, and LTP consolidation. Current Opinion in Neurobiology 18, 524–531.

Chang, F.L., Greenough, W.T., 1984. Transient and enduring morphological correlates of synaptic activity and efficacy change in the rat hippocampal slice. Brain Research 309, 35–46.

Cooke, S.F., Bliss, T.V., 2006. Plasticity in the human central nervous system. Brain 129, 1659–1673.

Cooney, J.R., Hurlburt, J.L., Selig, D.K., Harris, K.M., Fiala, J.C., 2002. Endosomal compartments serve multiple hippocampal dendritic spines from a widespread rather than a local store of recycling membrane. Journal of Neuroscience 22, 2215–2224.

Cruz-Martin, A., Crespo, M., Portera-Cailliau, C., 2010. Delayed stabilization of dendritic spines in fragile X mice. Journal of Neuroscience 30, 7793–7803.

Dalva, M.B., McClelland, A.C., Kayser, M.S., 2007. Cell adhesion molecules: Signalling functions at the synapse. Nature Reviews Neuroscience 8, 206–220.

Day, M., Wang, Z., Ding, J., et al., 2006. Selective elimination of glutamatergic synapses on striatopallidal neurons in Parkinson disease models. Nature Neuroscience 9, 251–259.

De Roo, M., Klauser, P., Mendez, P., Poglia, L., Muller, D., 2008a. Activity-dependent PSD formation and stabilization of newly formed spines in hippocampal slice cultures. Cerebral Cortex 18, 151–161.

De Roo, M., Klauser, P., Muller, D., 2008b. LTP promotes a selective long-term stabilization and clustering of dendritic spines. PLoS Biology 6, e219.

De Roo, M., Klauser, P., Briner, A., et al., 2009. Anesthetics rapidly promote synaptogenesis during a critical period of brain development. PLoS One 4, e7043.

Dolen, G., Osterweil, E., Rao, B.S., et al., 2007. Correction of fragile X syndrome in mice. Neuron 56, 955–962.

Dudek, S.M., Bear, M.F., 1993. Bidirectional long-term modification of synaptic effectiveness in the adult and immature hippocampus. Journal of Neuroscience 13, 2910–2918.

Ehrlich, I., Klein, M., Rumpel, S., Malinow, R., 2007. PSD-95 is required for activity-driven synapse stabilization. Proceedings of the National Academy of Sciences of the United States of America 104, 4176–4181.

Emes, R.D., Pocklington, A.J., Anderson, C.N., et al., 2008. Evolutionary expansion and anatomical specialization of synapse proteome complexity. Nature Neuroscience 11, 799–806.

Engert, F., Bonhoeffer, T., 1999. Dendritic spine changes associated with hippocampal long-term synaptic plasticity. Nature 399, 66–70.

Ethell, I.M., Yamaguchi, Y., 1999. Cell surface heparan sulfate proteoglycan syndecan-2 induces the maturation of dendritic spines in rat hippocampal neurons. The Journal of Cell Biology 144, 575–586.

Fischer, M., Kaech, S., Knutti, D., Matus, A., 1998. Rapid actin-based plasticity in dendritic spines. Neuron 20, 847–854.

Frischknecht, R., Heine, M., Perrais, D., Seidenbecher, C.I., Choquet, D., Gundelfinger, E.D., 2009. Brain extracellular matrix affects AMPA receptor lateral mobility and short-term synaptic plasticity. Nature Neuroscience 12, 897–904.

Fuhrmann, M., Mitteregger, G., Kretzschmar, H., Herms, J., 2007. Dendritic pathology in prion disease starts at the synaptic spine. Journal of Neuroscience 27, 6224–6233.

Furutani, Y., Matsuno, H., Kawasaki, M., Sasaki, T., Mori, K., Yoshihara, Y., 2007. Interaction between telencephalin and ERM family proteins mediates dendritic filopodia formation. Journal of Neuroscience 27, 8866–8876.

Ganeshina, O., Berry, R.W., Petralia, R.S., Nicholson, D.A., Geinisman, Y., 2004. Differences in the expression of AMPA and NMDA receptors between axospinous perforated and nonperforated synapses are related to the configuration and size of postsynaptic densities. The Journal of Comparative Neurology 468, 86–95.

Geinisman, Y., de Toledo-Morrell, L., Morrell, F., Heller, R.E., Rossi, M., Parshall, R.F., 1993. Structural synaptic correlate of long-term potentiation: Formation of axospinous synapses with multiple, completely partitioned transmission zones. Hippocampus 3, 435–445.

Genoud, C., Quairiaux, C., Steiner, P., Hirling, H., Welker, E., Knott, G.W., 2006. Plasticity of astrocytic coverage and glutamate transporter expression in adult mouse cortex. PLoS Biology 4, e343.

Govek, E.E., Newey, S.E., Akerman, C.J., Cross, J.R., Van der Veken, L., Van Aelst, L., 2004. The X-linked mental retardation protein oligophrenin-1 is required for dendritic spine morphogenesis. Nature Neuroscience 7, 364–372.

Gray, N.W., Weimer, R.M., Bureau, I., Svoboda, K., 2006. Rapid redistribution of synaptic PSD-95 in the neocortex in vivo. PLoS Biology 4, e370.

Grutzendler, J., Kasthuri, N., Gan, W.B., 2002. Long-term dendritic spine stability in the adult cortex. Nature 420, 812–816.

Gundelfinger, E.D., Frischknecht, R., Choquet, D., Heine, M., 2010. Converting juvenile into adult plasticity: A role for the brain's extracellular matrix. European Journal of Neuroscience 31, 2156–2165.

Han, K., Kim, E., 2008. Synaptic adhesion molecules and PSD-95. Progress in Neurobiology 84, 263–283.

Harris, K.M., Stevens, J.K., 1989. Dendritic spines of CA 1 pyramidal cells in the rat hippocampus: Serial electron microscopy with reference to their biophysical characteristics. Journal of Neuroscience 9, 2982–2997.

Harvey, C.D., Svoboda, K., 2007. Locally dynamic synaptic learning rules in pyramidal neuron dendrites. Nature 450, 1195–1200.

Haydon, P.G., 2001. GLIA: Listening and talking to the synapse. Nature Reviews Neuroscience 2, 185–193.

Hensch, T.K., Fagiolini, M., Mataga, N., Stryker, M.P., Baekkeskov, S., Kash, S.F., 1998. Local GABA circuit control of experience-dependent plasticity in developing visual cortex. Science 282, 1504–1508.

Hofer, S.B., Mrsic-Flogel, T.D., Bonhoeffer, T., Hubener, M., 2009. Experience leaves a lasting structural trace in cortical circuits. Nature 457, 313–317.

Holtmaat, A., Svoboda, K., 2009. Experience-dependent structural synaptic plasticity in the mammalian brain. Nature Reviews Neuroscience 10, 647–658.

Holtmaat, A.J., Trachtenberg, J.T., Wilbrecht, L., et al., 2005. Transient and persistent dendritic spines in the neocortex in vivo. Neuron 45, 279–291.

Holtmaat, A., Wilbrecht, L., Knott, G.W., Welker, E., Svoboda, K., 2006. Experience-dependent and cell-type-specific spine growth in the neocortex. Nature 441, 979–983.

Honkura, N., Matsuzaki, M., Noguchi, J., Ellis-Davies, G.C., Kasai, H., 2008. The subspine organization of actin fibers regulates the structure and plasticity of dendritic spines. Neuron 57, 719–729.

Hotulainen, P., Hoogenraad, C.C., 2010. Actin in dendritic spines: Connecting dynamics to function. The Journal of Cell Biology 189, 619–629.

Hu, X., Ballo, L., Pietila, L., et al., 2011. BDNF-induced increase of PSD-95 in dendritic spines requires dynamic microtubule invasions. Journal of Neuroscience 31, 15597–15603.

Huber, K.M., Gallagher, S.M., Warren, S.T., Bear, M.F., 2002. Altered synaptic plasticity in a mouse model of fragile X mental retardation. Proceedings of the National Academy of Sciences of the United States of America 99, 7746–7750.

Jaworski, J., Kapitein, L.C., Gouveia, S.M., et al., 2009. Dynamic microtubules regulate dendritic space morphology and synaptic plasticity.

Jourdain, P., Fukunaga, K., Muller, D., 2003. Calcium/calmodulin-dependent protein kinase II contributes to activity-dependent filopodia growth and spine formation. Journal of Neuroscience 23, 10645–10649.

Kayser, M.S., Nolt, M.J., Dalva, M.B., 2008. EphB receptors couple dendritic filopodia motility to synapse formation. Neuron 59, 56–69.

Kelleher Jr., R., Govindarajan, A., Jung, H.Y., Kang, H., Tonegawa, S., 2004. Translational control by MAPK signaling in long-term synaptic plasticity and memory. Cell 116, 467–479.

Kielland, A., Bochorishvili, G., Corson, J., et al., 2009. Activity patterns govern synapse-specific AMPA receptor trafficking between deliverable and synaptic pools. Neuron 62, 84–101.

Knott, G.W., Quairiaux, C., Genoud, C., Welker, E., 2002. Formation of dendritic spines with GABAergic synapses induced by whisker stimulation in adult mice. Neuron 34, 265–273.

Knott, G.W., Holtmaat, A., Wilbrecht, L., Welker, E., Svoboda, K., 2006. Spine growth precedes synapse formation in the adult neocortex in vivo. Nature Neuroscience 9, 1117–1124.

Kopec, C.D., Real, E., Kessels, H.W., Malinow, R., 2007. GluR1 links structural and functional plasticity at excitatory synapses. Journal of Neuroscience 27, 13706–13718.

Kumar, V., Zhang, M.X., Swank, M.W., Kunz, J., Wu, G.Y., 2005. Regulation of dendritic morphogenesis by Ras-PI3K-Akt-mTOR and Ras-MAPK signaling pathways. Journal of Neuroscience 25, 11288–11299.

Laumonnier, F., Cuthbert, P.C., Grant, S.G., 2007. The role of neuronal complexes in human x-linked brain diseases. American Journal of Human Genetics 80, 205–220.

Lendvai, B., Stern, E.A., Chen, B., Svoboda, K., 2000. Experience-dependent plasticity of dendritic spines in the developing rat barrel cortex in vivo. Nature 404, 876–881.

Lin, Y.L., Lei, Y.T., Hong, C.J., Hsueh, Y.P., 2007. Syndecan-2 induces filopodia and dendritic spine formation via the neurofibromin-PKA-Ena/VASP pathway. The Journal of Cell Biology 177, 829–841.

Lisman, J., Schulman, H., Cline, H., 2002. The molecular basis of CaMKII function in synaptic and behavioural memory. Nature Reviews Neuroscience 3, 175–190.

Lohmann, C., Bonhoeffer, T., 2008. A role for local calcium signaling in rapid synaptic partner selection by dendritic filopodia. Neuron 59, 253–260.

Lohmann, C., Finski, A., Bonhoeffer, T., 2005. Local calcium transients regulate the spontaneous motility of dendritic filopodia. Nature Neuroscience 8, 305–312.

Lushnikova, I., Skibo, G., Muller, D., Nikonenko, I., 2009. Synaptic potentiation induces increased glial coverage of excitatory synapses in CA1 hippocampus. Hippocampus 19, 753–762.

Majewska, A.K., Newton, J.R., Sur, M., 2006. Remodeling of synaptic structure in sensory cortical areas in vivo. Journal of Neuroscience 26, 3021–3029.

Makino, H., Malinow, R., 2009. AMPA receptor incorporation into synapses during LTP: The role of lateral movement and exocytosis. Neuron 64, 381–390.

Malenka, R.C., Nicoll, R.A., 1999. Long-term potentiation–a decade of progress? Science 285, 1870–1874.

Maletic-Savatic, M., Malinow, R., Svoboda, K., 1999. Rapid dendritic morphogenesis in CA1 hippocampal dendrites induced by synaptic activity. Science 283, 1923–1927.

Malinow, R., Malenka, R.C., 2002. AMPA receptor trafficking and synaptic plasticity. Annual Review of Neuroscience 25, 103–126.

Marrs, G.S., Green, S.H., Dailey, M.E., 2001. Rapid formation and remodeling of postsynaptic densities in developing dendrites. Nature Neuroscience 4, 1006–1013.

Matsuno, H., Okabe, S., Mishina, M., Yanagida, T., Mori, K., Yoshihara, Y., 2006. Telencephalin slows spine maturation. Journal of Neuroscience 26, 1776–1786.

Matsuzaki, M., Honkura, N., Ellis-Davies, G.C., Kasai, H., 2004. Structural basis of long-term potentiation in single dendritic spines. Nature 429, 761–766.

McEwen, B.S., 2010. Stress, sex, and neural adaptation to a changing environment: Mechanisms of neuronal remodeling. Annals of the New York Academy of Sciences 1204 (Suppl), E38–E59.

Mendez, P., De Roo, M., Poglia, L., Klauser, P., Muller, D., 2010. N-cadherin mediates plasticity-induced long-term spine stabilization. The Journal of Cell Biology 189, 589–600.

Mendez, P., Garcia-Segura, L.M., Muller, D., 2010. Estradiol promotes spine growth and synapse formation without affecting pre-established networks. Hippocampus 28, 8.

Mitsuyama, F., Niimi, G., Kato, K., 2008. Redistribution of microtubules in dendrites of hippocampal CA1 neurons after tetanic stimulation during long-term potentiation. Ital J Anat Embryol 113, 17–27.

Morishita, H., Hensch, T.K., 2008. Critical period revisited: Impact on vision. Current Opinion in Neurobiology 18, 101–107.

Morishita, H., Miwa, J.M., Heintz, N., Hensch, T.K., 2010. Lynx1, a cholinergic brake, limits plasticity in adult visual cortex. Science 330, 1238–1240.

Nagerl, U.V., Eberhorn, N., Cambridge, S.B., Bonhoeffer, T., 2004. Bidirectional activity-dependent morphological plasticity in hippocampal neurons. Neuron 44, 759–767.

Nagerl, U.V., Kostinger, G., Anderson, J.C., Martin, K.A., Bonhoeffer, T., 2007. Protracted synaptogenesis after activity-dependent spinogenesis in hippocampal neurons. Journal of Neuroscience 27, 8149–8156.

Nevian, T., Larkum, M.E., Polsky, A., Schiller, J., 2007. Properties of basal dendrites of layer 5 pyramidal neurons: A direct patch-clamp recording study. Nature Neuroscience 10, 206–214.

Nicoll, R.A., Malenka, R.C., 1995. Contrasting properties of two forms of long-term potentiation in the hippocampus. Nature 377, 115–118.

Nicoll, R.A., Tomita, S., Bredt, D.S., 2006. Auxiliary subunits assist AMPA-type glutamate receptors. Science 311, 1253–1256.

Nikonenko, I., Jourdain, P., Muller, D., 2003. Presynaptic remodeling contributes to activity-dependent synaptogenesis. Journal of Neuroscience 23, 8498–8505.

Nikonenko, I., Boda, B., Steen, S., Knott, G., Welker, E., Muller, D., 2008. PSD-95 promotes synaptogenesis and multiinnervated spine formation through nitric oxide signaling. The Journal of Cell Biology 183, 1115–1127.

Node-Langlois, R., Muller, D., Boda, B., 2006. Sequential implication of the mental retardation proteins ARHGEF6 and PAK3 in spine morphogenesis. Journal of Cell Science 119, 4986–4993.

Opazo, P., Labrecque, S., Tigaret, C.M., et al., 2010. CaMKII triggers the diffusional trapping of surface AMPARs through phosphorylation of stargazin. Neuron 67, 239–252.

Oray, S., Majewska, A., Sur, M., 2004. Dendritic spine dynamics are regulated by monocular deprivation and extracellular matrix degradation. Neuron 44, 1021–1030.

Ostroff, L.E., Fiala, J.C., Allwardt, B., Harris, K.M., 2002. Polyribosomes redistribute from dendritic shafts into spines with enlarged synapses during LTP in developing rat hippocampal slices. Neuron 35, 535–545.

Pak, D.T., Sheng, M., 2003. Targeted protein degradation and synapse remodeling by an inducible protein kinase. Science 302, 1368–1373.

Park, M., Salgado, J.M., Ostroff, L., et al., 2006. Plasticity-induced growth of dendritic spines by exocytic trafficking from recycling endosomes. Neuron 52, 817–830.

Penzes, P., Jones, K.A., 2008. Dendritic spine dynamics–a key role for kalirin-7. Trends in Neurosciences 31, 419–427.

Poglia, L., Muller, D., Nikonenko, I., 2010. Ultrastructural modifications of spine and synapse morphology by SAP97. Hippocampus 21, 990–998.

Racz, B., Blanpied, T.A., Ehlers, M.D., Weinberg, R.J., 2004. Lateral organization of endocytic machinery in dendritic spines. Nature Neuroscience 7, 917–918.

Richards, D.A., Mateos, J.M., Hugel, S., et al., 2005. Glutamate induces the rapid formation of spine head protrusions in hippocampal slice cultures. Proceedings of the National Academy of Sciences of the United States of America 102, 6166–6171.

Ropers, H.H., 2006. X-linked mental retardation: Many genes for a complex disorder. Current Opinion in Genetics and Development 16, 260–269.

Sala, C., Piech, V., Wilson, N.R., Passafaro, M., Liu, G., Sheng, M., 2001. Regulation of dendritic spine morphology and synaptic function by Shank and Homer. Neuron 31, 115–130.

Selkoe, D.J., 2002. Alzheimer's disease is a synaptic failure. Science 298, 789–791.

Spacek, J., Harris, K.M., 1997. Three-dimensional organization of smooth endoplasmic reticulum in hippocampal CA1 dendrites and dendritic spines of the immature and mature rat. Journal of Neuroscience 17, 190–203.

Spires, T.L., Meyer-Luehmann, M., Stern, E.A., et al., 2005. Dendritic spine abnormalities in amyloid precursor protein transgenic mice demonstrated by gene transfer and intravital multiphoton microscopy. Journal of Neuroscience 25, 7278–7287.

Steward, O., Falk, P.M., 1985. Polyribosomes under developing spine synapses: Growth specializations of dendrites at sites of synaptogenesis. Journal of Neuroscience Research 13, 75–88.

Tolias, K.F., Bikoff, J.B., Kane, C.G., Tolias, C.S., Hu, L., Greenberg, M.E., 2007. The Rac1 guanine nucleotide exchange factor Tiam1 mediates EphB receptor-dependent dendritic spine development. Proceedings of the National Academy of Sciences of the United States of America 104, 7265–7270.

Toni, N., Buchs, P.A., Nikonenko, I., Bron, C.R., Muller, D., 1999. LTP promotes formation of multiple spine synapses between a single axon terminal and a dendrite. Nature 402, 421–425.

Toni, N., Buchs, P.A., Nikonenko, I., Povilaitite, P., Parisi, L., Muller, D., 2001. Remodeling of synaptic membranes after induction of long-term potentiation. Journal of Neuroscience 21, 6245–6251.

Toni, N., Teng, E.M., Bushong, E.A., et al., 2007. Synapse formation on neurons born in the adult hippocampus. Nature Neuroscience 10, 727–734.

Toresson, H., Grant, S.G., 2005. Dynamic distribution of endoplasmic reticulum in hippocampal neuron dendritic spines. European Journal of Neuroscience 22, 1793–1798.

Trachtenberg, J.T., Chen, B.E., Knott, G.W., et al., 2002. Long-term in vivo imaging of experience-dependent synaptic plasticity in adult cortex. Nature 420, 788–794.

Triller, A., Choquet, D., 2008. New concepts in synaptic biology derived from single-molecule imaging. Neuron 59, 359–374.

Tsuriel, S., Geva, R., Zamorano, P., et al., 2006. Local sharing as a predominant determinant of synaptic matrix molecular dynamics. PLoS Biology 4, e271.

Vazquez, L.E., Chen, H.J., Sokolova, I., Knuesel, I., Kennedy, M.B., 2004. SynGAP regulates spine formation. Journal of Neuroscience 24, 8862–8872.

Volterra, A., Meldolesi, J., 2005. Astrocytes, from brain glue to communication elements: The revolution continues. Nature Reviews Neuroscience 6, 626–640.

Wang, X.B., Bozdagi, O., Nikitczuk, J.S., Zhai, Z.W., Zhou, Q., Huntley, G.W., 2008. Extracellular proteolysis by matrix metalloproteinase-9 drives dendritic spine enlargement and long-term potentiation coordinately. Proceedings of the National Academy of Sciences of the United States of America 105, 19520–19525.

Wei, W., Nguyen, L.N., Kessels, H.W., Hagiwara, H., Sisodia, S., Malinow, R., 2010. Amyloid beta from axons and dendrites reduces local spine number and plasticity. Nature Neuroscience 13, 190–196.

Wilbrecht, L., Holtmaat, A., Wright, N., Fox, K., Svoboda, K., 2010. Structural plasticity underlies experience-dependent functional plasticity of cortical circuits. Journal of Neuroscience 30, 4927–4932.

Witcher, M.R., Kirov, S.A., Harris, K.M., 2007. Plasticity of perisynaptic astroglia during synaptogenesis in the mature rat hippocampus. Glia 55, 13–23.

Woolfrey, K.M., Srivastava, D.P., Photowala, H., et al., 2009. Epac2 induces synapse remodeling and depression and its disease-associated forms alter spines. Nature Neuroscience 12, 1275–1284.

Xu, T., Yu, X., Perlik, A.J., et al., 2009. Rapid formation and selective stabilization of synapses for enduring motor memories. Nature 462, 915–919.

Yang, G., Pan, F., Gan, W.B., 2009. Stably maintained dendritic spines are associated with lifelong memories. Nature 462, 920–924.

Yasumatsu, N., Matsuzaki, M., Miyazaki, T., Noguchi, J., Kasai, H., 2008. Principles of Long-Term Dynamics of Dendritic Spines. Journal of Neuroscience 28, 13592–13608.

Yazaki-Sugiyama, Y., Kang, S., Cateau, H., Fukai, T., Hensch, T.K., 2009. Bidirectional plasticity in fast-spiking GABA circuits by visual experience. Nature 462, 218–221.

Yoshihara, Y., De Roo, M., Muller, D., 2009. Dendritic spine formation and stabilization. Current Opinion in Neurobiology 19, 146–153.

Zito, K., Scheuss, V., Knott, G., Hill, T., Svoboda, K., 2009. Rapid functional maturation of nascent dendritic spines. Neuron 61, 247–258.

Ziv, N.E., Smith, S.J., 1996. Evidence for a role of dendritic filopodia in synaptogenesis and spine formation. Neuron 17, 91–102.

Zuber, B., Nikonenko, I., Klauser, P., Muller, D., Dubochet, J., 2005. The mammalian central nervous synaptic cleft contains a high density of periodically organized complexes. Proceedings of the National Academy of Sciences of the United States of America 102, 19192–19197.

Zuo, Y., Lin, A., Chang, P., Gan, W.B., 2005. Development of long-term dendritic spine stability in diverse regions of cerebral cortex. Neuron 46, 181–189.

Cortical Columns

Z. Molnár

University of Oxford, Oxford, UK

Neural Circuit Development and Function in the Brain: Comprehensive Developmental Neuroscience, Volume 3 http://dx.doi.org/10.1016/B978-0-12-397267-5.00137-0

© 2013 Elsevier Inc. All rights reserved.

7.1 INTRODUCTION

Modularity is a common organizational principle in all parts of the brain. The cerebral cortex exhibits a laminar and a radial organization. Cortical columns are sometimes regarded as the basic functional units in cerebral cortical processing, development, and evolution. *Cortical column* is a historic term that can refer to a vertically arranged cell constellation, a pattern of connectivity, myelin distribution, metabolic characteristics, staining property, vasculature, magnitude of specific gene expression, embryonic origin, or functional properties. The columnar organization reflects the intermittently recursive mapping of several variables under the two-dimensional surface of the neocortex in a variably independent or combined (e.g., hypercolumn) manner. The term *cortical column* is constantly evolving with the improved understanding of the organization and function of the various radially oriented groups of cells that share certain functional or anatomical properties.

Columns are ubiquitous in the brain but are in no way obligatory, and comprehensive descriptions of the various specific forms of "columns" in the brain are still evolving. To date, comparative studies have failed to identify a specific sensory, motor, or cognitive function that is specifically associated with the presence or absence of a particular form of cortical column. The developing cortex contains numerous radial determinants. Pyramidal neurons are generated in "ontogenic units" and subsequently disperse radially. It has been shown that sibling cells have a stronger tendency to establish synaptic connections with each other in the cortical plate. However, the link between the embryonic and adult columnar constellations is currently not known.

There are, therefore, several problems related to the term and concept of cortical column. (1) The often loose, general, and uncritical use of the term can be confusing. (2) Related to the lack of a universal presence of certain types of column within some cortical areas, brains, or species undermines the idea that similar building blocks comprise all cortical circuits. (3) The concept that the cortical column (or even just an arbitrary columnar unit along the depth of the cortex) has a universally constant number of neurons associated with it, with only the primate visual cortex showing a difference. (4) The lack of correlation between the absence or presence of particular types of column and specific mode of sensory or cognitive processing capacities (across the same brain or across close or more distant species). (5) Although there is evidence for the overall radial disposition of the pyramidal neuron clones and a higher probability of synapse formation between sibling cells, there is a lack of correlation between the columnar development of the brain and columns in the adult cortex.

Knowledge of the laminar and columnar organization of the cerebral cortex is continuously advancing, and with this the conceptual details of the columnar organization also is changing. The time may arrive when both the concept and the nomenclature will have to adapt to these changes.

The hypothesis of the column as the fundamental processing unit of the cerebral cortex was formulated by Mountcastle (1957) from studies of cells responding to a single modality of tactile stimuli (cutaneous or deep joint receptors) in the somatosensory cortex of cats. The concept emerged from Mountcastle's work and was developed further over five decades; he claimed that the cerebral cortex can be further subdivided into "complex processing and distributing units that link a number of inputs to a number of outputs via overlapping internal processing chains" (Mountcastle, 1957).

By exploring the physiological, anatomical, genetic, and developmental properties of the cerebral cortex, more details of its organization were revealed, and many of these new entities were referred to as 'columns.' The emerging concept in cerebellar circuits by Eccles et al. (1967) fueled the quest for a fundamental cortical processing unit, an archetypical cortical column, which was intensified in the hope of identifying modules that are general for all cortical areas. There are references to functional columns, minicolumns, hypercolumns, ontogenetic or embryonic columns, ocular dominance (OD) columns, orientation columns, and barrel columns. The only common theme linking these terms is that they refer to a structural, physiological, or developmental organization that transcends the laminar pattern and is perpendicular to it. None of these several types of columns are general to all cortical areas, and several are restricted to the primary sensory areas. Table 7.1 gives three definitions and Table 7.2 gives a list of some of the terms that refer to cortical columnar structures. There are so many varieties of cortical columns defined by different criteria and by different authors that it is difficult to define, relate, or compare these columns. The concept of an archetypical cortical column is no better defined now than when it was first introduced.

In this chapter, the following points are discussed and examined: (1) the problems associated with the current nomenclature; (2) the evidence for and against the idea that columns are the common building blocks of the cortex; (3) the question of how constant the cell numbers are within a column and how homogeneous is the structure of the various columns; (4) the possible functions of the columns; and (5) the current knowledge of the columnar development in the cortex.

TABLE 7.1 Examples for definitions of cortical columns

1. Mountcastle (1997) "The modular organization of nervous systems is a widely documented principle of design for both vertebrate and invertebrate brains of which the columnar organization of the neocortex is an example. The classical cytoarchitectural areas of the neocortex are composed of smaller units, local neural circuits repeated iteratively within each area. Modules may vary in cell type and number, in internal and external connectivity, and in mode of neuronal processing between different large entities; within any single large entity they have a basic similarity of internal design and operation. Modules are most commonly grouped into entities by sets of dominating external connections. This unifying factor is most obvious for the heterotypical sensory and motor areas of the neocortex. Columnar defining factors in homotypical areas are generated, in part, within the cortex itself. The set of all modules composing such an entity may be fractionated into different modular subsets by different extrinsic connections. Linkages between them and subsets in other large entities form distributed systems. The neighborhood relations between connected subsets of modules in different entities result in nested distributed systems that serve distributed functions. A cortical area defined in classical cytoarchitectural terms may belong to more than one and sometimes to several distributed systems. Columns in cytoarchitectural areas located at some distance from one another, but with some common properties, may be linked by long-range, intracortical connections."

2. http://en.wikipedia.org/wiki/Cortical_minicolumn "A cortical column, also called hypercolumn or sometimes cortical module, [1] is a group of neurons in the brain cortex, which can be successively penetrated by a probe inserted perpendicularly to the cortical surface and which have nearly identical receptive fields. Neurons within a minicolumn encode similar features, whereas a hypercolumn 'denotes a unit containing a full set of values for any given set of receptive field parameters.' [2] A cortical module is defined as either synonymous with a hypercolumn (Mountcastle) or as a tissue block of multiple overlapping hypercolumns (Hubel & Wiesel).

   Various estimates suggest there are 50–100 cortical minicolumns in a hypercolumn, each comprising around 80 neurons. An important distinction is that the columnar organization is functional by definition, and reflects the local connectivity of the cerebral cortex. Connections 'up' and 'down' within the thickness of the cortex are much denser than connections that spread from side to side."

3. Boucsein et al. (2011) http://www.ncbi.nlm.nih.gov/pmc/articles/PMC3072165/ "This slightly ambiguous term loosely describes the concept of vertically arranged groups of cells that share certain functional and/or anatomical properties and could represent a 'basic functional unit' in cortical processing. They are ubiquitous in the brain but in no way obligatory, and a comprehensive description of the various forms of 'columns' in the brain is still lacking."

[1] Towards cortex sized artificial neural systems, Johansson C and Lansner A (2007) *Neural Networks*, vol. 20(1), pp. 48–61. Elsevier.
[2] The columnar organization of the neocortex, Mountcastle VB (1997) *Brain*, vol. 20(4), pp. 701–722. Oxford University Press.
Boucsein C, Nawrot MP, Schnepel P, Aertsen A (2011) Beyond the cortical column: Abundance and physiology of horizontal connections imply a strong role for inputs from the surround. Frontiers in Neuroscience 5: 32.

7.2 THE CORTICAL COLUMN NOMENCLATURE REFLECTS THE HISTORY OF CHANGING CONCEPTS OF CORTICAL ANATOMY, FUNCTIONAL REPRESENTATION, AND DEVELOPMENT

7.2.1 Is the Isocortex Hexalaminar?

The cortex (hereafter referring to the forebrain 'neocortex' as distinct from the olfactory and hippocampal three-layered cortex) is organized horizontally into six laminae and vertically into groups of cells linked synaptically across and along the horizontal layers. The definitions of both laminae and radial modules are a historic convention rather than a biologically or functionally related demonstrable reality. All mammalian cerebral neocortices have a uniform laminar structure (therefore also called the isocortex) that has been historically and arbitrarily divided into six layers (Brodmann, 1909; Lorente de Nó, 1922, 1949; von Economo and Koskinas, 2008). This basic master plan is modified according to variations. These variations expose the problem of fitting the hexalaminar universal model to all mammalian cortices. Subdivisions of layers III, IV, V, and VI in some species, missing layers (IV) in some cortical areas, or fused layers (II/III) in other areas indicate that there is uncertainty and a random element to current laminar nomenclature (Molnár, 2011; Molnár and Belgard, 2012).

The radially oriented apical dendrites and processes and the general radial orientation within the cortex have been widely known since the cerebral cortex was first examined microscopically (see, e.g., Cajal, 1909). Also, the vertical connectivity linking neurons across cortical layers was described by Lorente de Nó (1938) in the primary somatosensory cortex of the mouse. The concept of a column emerged from the functional properties of the cortex and received attention after Vernon Mountcastle discovered that the neurons are arranged vertically (or radially in the convoluted cerebrum) in the form of columns spanning the width of the primate somatosensory cortex with cells in each column responding with distinct receptive field properties (superficial as compared to deep skin receptors) to a single receptive field at the periphery (Mountcastle et al., 1957). Although it is now known that these functionally distinct 'columns' were separate, distinct cortical areas and not functional units within a cortical area (Kaas et al., 2011), these observations drove further discoveries of an array of iterative neuronal groups (also called modules) that extend radially across cellular layers VI to II with layer I at the top. Subsequently, Hubel and Wiesel (1968) revealed the orientation and OD columns in the primary visual cortex, and this was followed by the observations of Abeles and Goldstein (1970) in the primary auditory cortex (Table 7.2). These physiological observations led to the

TABLE 7.2 Examples for terms that refer to columnar structures in the cortex

Module	Cortical area	Definition	Dimension	References
Cortical column/module	S1	Penetrations parallel to the pial surface and crossing the vertical axis of the cortex pass through 300–500-μm blocks of tissue in each of which neurons with identical properties are encountered. Sharp transitions are observed from a block with one set of properties to the adjacent block with different properties. The defining property for place is the peripheral receptive field, the zone on the body surface within which an adequate stimulus evokes a response of cortical cells	300–500-μm	Mountcastle (1957), Powell and Mountcastle (1959)
Ocular dominance column	V1	Ocular dominance (OD) columns or OD stripes are regions of neurons in the visual cortex that respond to the stimulation from either the left or right eye, and they can be defined both anatomically and physiologically	Variable	Hubel and Wiesel (1969)
Orientation columns	V1	Form orientation slabs that measure 0.5–1.0 mm in the iso-orientation direction and in which a full 180° rotation of orientation preference is repeated	560 μm	Hubel and Wiesel (1968)
Blobs	V1	Metabolic activity 2DG or cytochrome oxidase staining Blob cells respond differentially at low spatial frequencies (1.1 cycles per degree), interblob cells at higher frequencies (3.8 cycles per degree)	150-μm diameter, most prominent in layers II and III. Repeat intervals of 500–550 μm; the parallel rows are 350-μm apart	Hendrickson and Wilson (1979), Wong-Riley (1979), Livingstone and Hubel (1984)
Isofrequency bands	A1	Neurons of similar frequency preference are arranged in isofrequency bands (IFBs) across the primary auditory cortex (AI) of many mammals	No wider than 200 μm and 5–7 mm in length extending across the gyrus	Tunturi (1950), Brugge and Merzenich (1973)
Binaural summation columns	A1	Most neurons arrayed in a column perpendicular to the cortical surface display the same aural dominance and binaural interaction Summation columns occupy about two-thirds of the area sampled; suppression columns, about one-third. Within most suppression columns, the contralateral ear was dominant. Within summation columns, aural dominance varied. Summation columns appear to be composed of smaller columns differing in aural dominance	The sizes of binaural interaction columns vary considerably; some occupy several square millimeters of cortical surface. At least some binaural interaction columns occupy strips of cortex oriented orthogonal to isofrequency contours	Imig and Adrián (1977)
Motor columnar aggregates	Motor cortex	Pyramidal and nonpyramidal cells are clustered into columnar aggregates	300 μm wide, separated by 100-μm cell-sparse zones	Meyer (1987)
Motion columns	MT	Neurons in monkey MT with similar axes of motion preference are arranged in vertical columns, and these columns are themselves arranged in slabs in which a full rotation of 180° of axis of motion is represented	400–500 μm	Albright et al. (1984)

TABLE 7.2 Examples for terms that refer to columnar structures in the cortex—cont'd

Module	Cortical area	Definition	Dimension	References
Shape and face recognition columns	Homotypical inferotemporal cortex	Require moderately complex features (shapes and faces) for their activation		Gross et al. (1972), Perrett et al. (1992)
Microcolumns	All cortical areas	The dendrites of 3–20 large pyramidal cells of layer V form clusters that ascend together through layer IV	These modules are −30 μm in diameter and coccur with center-to-center spacing that varies from 20 to 80 μm; the wider spacing occurs in the larger brains of the macaque monkey and man	Fleischhauer et al. (1972), Peters and Walsh (1972)
Barrels	Some rodent S1	Cytoarchitectonic patterning of the layer IV neurons forming a ring-like structure on tangential sections	Variable 300 μm	Woolsey and van der Loos (1970)
Synaptic ZN	Monkey primary visual cortex	Synaptic Zn patches correspond to a subset of corticocortical terminations		Dyck et al. (2003)
VGLUT-2 columns	Rat and mouse barrel field	VGLUT-2 marker for thalamocortical termination	Variable 300 μm	Liguz-Lecznar and Skangiel-Kramska (2007)
Ontogenic units/columns	Monkey	The progenitor cells that generate the minicolumn	Each proliferative unit in the ventricular zone of the monkey consists of 3–5 stem cells, a number that gradually increases to 10–12 stem cells during development; the units are separated by glial septa (Rakic, 1988)	Rakic (1988)
Domains	Early postnatal rat cortex	Domains of spontaneously coactive neurons using optical recordings of brain slices labeled with the fluorescent calcium indicator fura-2 in early postnatal rat cortex	The functional domains were 50–120 μm in diameter on tangential slices; they spanned several cortical layers and resembled columns found in the adult cortex in coronal slices	Yuste et al. (1992)

concept that "neurons within a given column are stereotypically interconnected in the vertical dimension, share extrinsic connectivity, and hence act as basic functional units subserving a set of common static and dynamic cortical operations that include not only sensory and motor areas but also association areas subserving the highest cognitive functions" (Jones and Rakic, 2010). The inclusion of the highest cognitive functions was, of course, an extrapolation that lacked evidence. The concept that the cortex comprises similarly structured units is an attractive one, but it seems that there are far too many variations and individual units that can be highly specialized and vary within certain cortical areas, or sectors within areas.

7.2.2 The Loose and Uncritical Use of the Term in Ways That Are So Generalized as to Be Unhelpful and Even Confusing

Although the anatomical and functional columnarity of the neocortex has never been in doubt, over time and with more discoveries of radial arrangements in the cortex, the term 'cortical column' became looser as columns were defined by cell constellation, pattern of connectivity, myelin content, staining property, magnitude of gene expression, or functional properties (Rockland, 2010; Table 7.2). Although the term column is used by some to refer only to: 'interconnected neurons, with common input, common output, and common response properties extending through the thickness of the cortex,' others do not use these criteria, and the term 'column' evolved into a loose and somewhat ambiguous term referring to some aspect of the vertical organization of the cortex (Table 7.1, third definition).

Montcastle's definition of a column in 1997 is different from the one formulated in 1957, and it includes references to physiological, anatomical, and embryological aspects: "The basic unit of the mature neocortex is the *minicolumn*, a narrow chain of neurons extending vertically across the cellular layers II–VI, perpendicular to the pial surface. Each minicolumn in primates contains 80–100 neurons, except for the striate cortex where the number is 2.5 times larger. Minicolumns contain all the major cortical neural cell phenotypes, interconnected in the vertical dimension. The minicolumn is produced by the iterative division of a small cluster of progenitor cells, a polyclone, in the neuroepithelium, via the intervening ontogenetic unit in the cortical plate of the developing neocortex" (Mountcastle, 1997).

This excerpt shows how, over the decades, the increasingly protean imagery evoked by the term 'column' now obliges investigators to acknowledge its conceptual and linguistic shortcomings (Rockland, 2010). Structural, functional, and embryological definitions are used loosely, and there is a lack of proper and unequivocal definitions. Therefore, it is difficult to define what constitutes a particular 'cortical column.' Moreover, the use of the terminology is not stringent. Most columns have no definable 'solid' borders; some of the structures referred to as a column do not extend across the entire thickness of the cortex from the pial surface to the white matter (e.g., barrels, microcolumns). The term 'column' has become too general. To convey the complex aspects of cortical organization adequately, additional adjectives are required to specify a particular entity. Mountcastle used the terms column and module interchangeably, but nowadays the term module is used more loosely. Other terms, such as 'patch' or 'domain,' suffer, to varying degrees, from the same problem (Rockland, 2010).

7.3 A LACK OF UNIVERSAL PRESENCE OF CERTAIN COLUMNS WITHIN CORTICAL AREAS, BRAINS, AND SPECIES IS UNDERMINING THE IDEA THAT SIMILAR BUILDING BLOCKS COMPRISE ALL CORTICAL CIRCUITS

7.3.1 The Use of Physiological Methods to Reveal Columns

Evidence for neocortical columnar organization was initially obtained in electrophysiological studies of single neurons in the somatic sensory cortex in anesthetized cats and monkeys (Mountcastle, 1957; Powell and Mountcastle, 1959). Microelectrode penetrations made normal to the pial surface encounter neurons in each cellular layer with similar properties of place and modality. Penetrations parallel to the pial surface and crossing the vertical axis of the cortex pass through 300–500-µm-sized blocks of tissue in each of which neurons with identical properties are encountered. Sharp transitions are observed from a block with one set of properties to the adjacent block with different properties. The defining property for place is the peripheral receptive field, the zone on the body surface within which an adequate stimulus evokes a response of cortical cells. To reveal the functional modularity in the cortex, 2-deoxyglucose,

optical recordings of intrinsic signals, voltage- and calcium-sensitive dyes, and expression of immediate early genes have also been used in both somatosensory and visual cortical areas (Horton and Adams, 2005).

7.3.2 Columnar Organization of Some Afferent and Efferent Projections

Both intrinsic and extrinsic cortical connections are often patchy and appear columnar in cross sections of the cerebral cortex. Various anterograde tracers injected *in vivo* are often dramatically patterned in cross section, especially in layer IV and adjacent layers. The problem here is that, very commonly, a patchy distribution of label that may involve only one or two laminae at most is interpreted as columnar, whereas the label can be a stripe, area, or spot, and the 'column' is projected to the structure because of the investigator's interpretation (Figure 7.1). By contrast, cortical or thalamic terminations in layer I are, in fact, transcolumnar, typically extending over several millimeters. "Bundles of axons from cells of thalamic modules project to columnar zones of termination in layers IV and IIb of the postcentral cortex, forming clusters separated by zones in which terminals are much less dense. Clustering obtains also for the ipsilateral corticocortical and transcallosal systems" (Mountcastle, 1997).

It is widely considered that the effective unit of operation in such a distributed system is not the single neuron and its axon, but groups of cells with similar functional properties and anatomical connections (Jones, 1999, 2010). This modular arrangement might allow a large number of neurons to be connected without a significant increase in cortical volume. Mitchison (1992) estimated that fusing 100 cortical columns would lead to a tenfold increase in cortical volume. The explanation for this surprising estimate is that within a column only restricted subsets of neurons are involved in long-distance connections, whereas the majority is only connected locally within the columns. Consequently, the length of axons that interconnect neurons is shortened, also reducing the cortical volume. The hypothesis requires that nerve cells in the middle layers of the cortex, in which most thalamic afferents terminate, should be joined by narrow vertical connections to cells in layers lying superficial and deep to them, so that all cells in the column are excited by incoming stimuli with only small latency differences. Experiments in the monkey, however, did not show such homogeneous arrangement and revealed that terminal arbors of individual thalamocortical axons are often smaller than the cross-sectional width of the region that showed a response revealed by optical recording in the so-called activity columns (Blasdel and Lund, 1983; Freund et al., 1989). Thus, 'activity columns' are

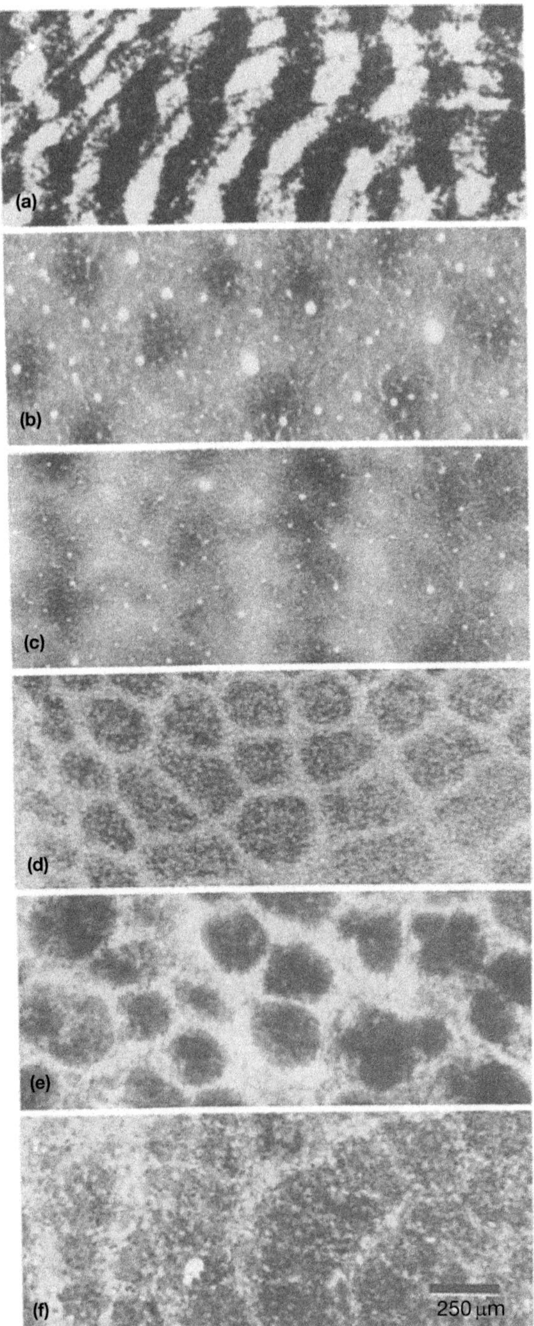

FIGURE 7.1 Examples of modular circuitry in the mammalian brain. (a) Ocular dominance columns in layer IV in primary visual cortex (V1) of a rhesus monkey (autoradiograph after injection of radioactive proline into one eye). (b) Blobs in layers II–III in V1 of a squirrel monkey (cytochrome oxidase (CO) histochemistry). (c) Stripes in layers II–III in V2 of a squirrel monkey (CO histochemistry). (d) Barrels in layer IV in the primary sensory cortex (S1) of a rat (succininc dehydrogenase histochemistry). (e) Barreloids in the ventrobasal nucleus of the thalamus in a rat (succinic dehydrogenase histochemistry). (f) Glomeruli in the olfactory bulb of a mouse (Sudan Black staining). *Reproduced from Purves D, Riddle DR, and LaMantia AS (1992) Iterated patterns of brain circuitry (or how the cortex gets its spots). Trends in Neuroscience 15(10): 362–368.*

assembled from the convergence of smaller units defined by arbors in a 300–500-μm wide space and not merely by activity-related or molecular factors (Inan and Crair, 2007). Moreover, analysis of a large data set of recordings has revealed that, within a cortical column, connectivity is highly nonuniform (Song et al., 2005).

7.3.3 Modules of the Visual Cortex

The visual *cortex* processes information concerning form, pattern, and motion within functional maps that reflect the layout of neuronal circuits. The columnar organization in the primate V-1 is defined by the neuronal properties of ocularity and place imposed by geniculate input and by orientation specificity generated by intracortical processing. The neurons studied in tangential penetrations vary systematically in ocularity (Figure 7.1(a)) and orientation selectivity (Hubel and Wiesel, 1969). As the primary visual cortex has been studied in great detail, this is the area where the cortical columns have been identified most methodically: OD columns, orientation columns, hypercolumns, and alternating callosal and ipsilateral columns (see Table 7.2).

7.3.3.1 OD Columns/Stripes

OD columns or OD stripes are regions of neurons in the visual cortex that respond to stimulation from either the left or right eye, and they can be defined both anatomically and physiologically (Hubel and Wiesel, 1969). Thalamocortical projections carrying signals from one eye or the other synapse mostly within layer IV. In a normally developed visual system, the area of dominance columns for each eye is the same, and each cortical cell responds to visual input predominantly according to its column. OD columns were revealed by single unit recording and transneuronal transport across the lateral geniculate synapse of radioactive amino acids (Hubel and Wiesel, 1969). OD columns are slab-like domains; columnar width is variable as a function of the visual field; that is, they are larger in the foveal representation (Hubel and Wiesel, 1977). In the peripheral visual field representation, the slab-like confirmation breaks up into patches (Adams et al., 2007). Monocular deprivation during early life prevents this balance from developing, and the nondeprived eye assumes control of nearly all cortical cells. These effects were largely identified by Wiesel and Hubel through studies on cats and monkeys (Hubel and Wiesel, 1969).

Similarly, for OD columns of the primate visual cortex, classical anatomical and physiological studies identified core and edge regions functionally distinguished by different degrees of monocular bias (LeVay et al., 1975). More recently, different conditions of visual deprivation have revealed functional subcompartments within OD columns, visualized by either changes in

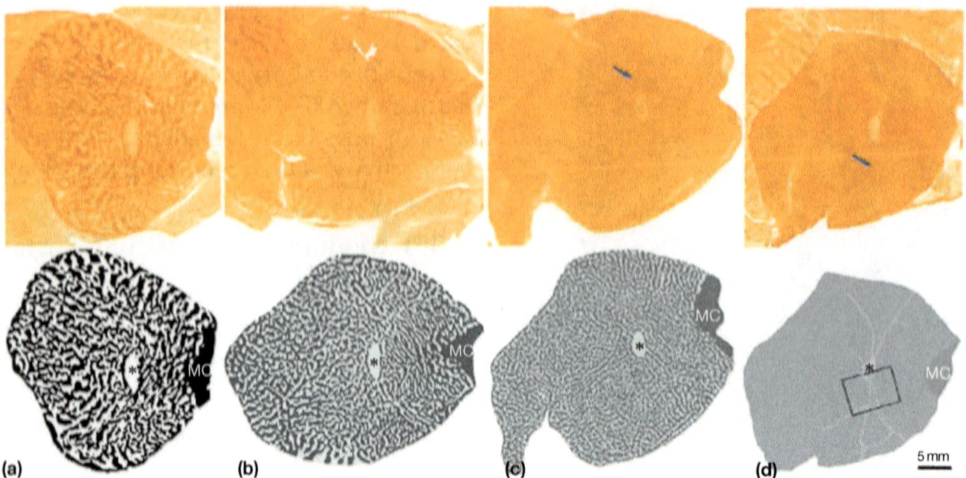

FIGURE 7.2   Variable appearance of ocular dominance columns in normal squirrel monkeys (*reproduced from Adams DL and Horton JC (2003).* *Capricious expression of cortical columns in the primate brain.* Nature Neuroscience *6: 113–114*). Shown are photographic montages of layer 4 C from the left striate cortex of four squirrel monkeys with normal vision 10 days after left eye enucleation. Flatmounts were reacted for CO, and examples for the individual variations are presented: (a) Large, crisply segregated columns; (b) intermediate columns; (c) fine, indistinct columns; (d) rudimentary columns. The columns are essentially absent, although hints were visible in a few peripheral regions of cortex. Note thin profiles (arrows) radiating from the blind spot in (c) and (d), which represent shadows (angioscotomas) from retinal blood vessels. Columns were virtually absent in one-third of the animals (d). MC, monocular crescent; *, blind spot.

cytochrome oxidase (CO) activity (Horton and Hocking, 1998) or differential expression of immediate early genes (Takahata et al., 2009). Species variation in columnar structure is hard to reconcile with ideas about the fundamental importance of columns. Some members of single species, for example, squirrel monkeys, show enormous variability in the expression of OD columns (Adams and Horton, 2003). Some individuals have normal OD columns throughout the visual cortex; others have them only in parts of the visual cortex, while in some, they are nearly absent (Figure 7.2).

7.3.3.2 *Orientation Columns*

Within an orientation column, neurons throughout the vertical thickness of the cortex respond to stimuli oriented at the same angle (Hubel and Wiesel, 1968; Hubel et al., 1977, 1978). A neighboring column will then have neurons responding to a slightly different orientation from the one next to it. The functional maps of orientation preference in the ferret, tree shrew, and galago – three species separated since the basal radiation of placental mammals more than 65 Mya ago – share this common organizing principle (Kaschube et al., 2010). Maps of orientation tuning as viewed from the cortical surface (not in sections) contain singularities where orientation columns converge (Blasdel and Salama, 1986), also called pinwheels (Bonhoeffer and Grinvald, 1991).

7.3.3.3 *Gene Expression in the Cortex in 'Columnar' Fashion*

Zones of heightened CO levels can reveal metabolic zones arranged in a modular fashion across the cortex (Livingstone and Hubel, 1984; Figure 7.1(b)), but these do not extend through all cortical layers. In the monkey, primary visual cortex CO staining reveals metabolic activity in a nonuniform fashion (Wong-Riley, 1979). The patches of CO activity correspond to thalamocortical terminations (Livingstone and Hubel, 1982). Adjacent sections, reactive for synaptic zinc, show patches that correspond to another subset of corticocortical terminations (e.g., Ichinohe and Rockland, 2004). There are numerous other differences that distinguish cortical modules; some of these are associated with thalamic inputs (e.g., Vglut2; 5HTT).

7.3.4 Overlap Between Columnar Entities Within the Same Structures; Combining Physiological and Anatomical Definitions

Individual columns are embedded within distributed networks, and cortical modules are composed of groups of minicolumns. The same column can be part of different networks (e.g., OD and orientation columns or hypercolumns). Despite decades of work, the organization of these modules and their connections, singly or in relation to each other, is only poorly understood. Anatomical observations are often linked to modular patterns of increased metabolic activity, blood flow studies, 2DG uptake, or expression of immediate early genes depending on the stimulus, and may be limited to a single layer, a few layers, or a whole cortical thickness. However, some of these studies revealed no discrete anatomical arrangements that would explain modularity of function, or perceived anatomical arrangements were not in line with detected physiological changes, raising the question: how can an 'imperfect' anatomical arrangement generate functionally distinct modules? At the cellular level, there is growing evidence that cortical columns contain multiple, highly specific, fine-scale subcircuits (Otsuka and Kawaguchi, 2008; Yoshimura et al., 2005) and that within columns, there are locally heterogeneous response properties (Sato et al., 2007).

7.4 NUMBER OF NEURONS IN A CORTICAL COLUMN

The various cortical columns that have been described by different anatomical or physiological methods have very different sizes, shapes, and diameters (examples shown in Figure 7.1). As discussed earlier, the term cortical 'column' is ambiguous – it can refer to small-scale minicolumns (diameter 50 µm), to larger scale macrocolumns (diameter – 300–500 µm), or to multiple different structures within both categories (Jones, 2000; Rakic, 2007; Rockland, 2010).

7.4.1 General Concept that the Cortical Column (Even Just an Arbitrary Unit Column that Includes the Full Depth of the Cortex) Has a Universal Constant Number of Neurons Associated with It

Although there is very little quantitative work on the number of neurons in anatomically or physiologically identified cortical columns, it is expected that they are different. It is also generally accepted that the cortical surface areas vary much more than the radial thickness of the cortex. Powell, after returning to Oxford from Mountcastle's laboratory, was influenced by the concept of the column and set out to quantify parameters within a cortical segment that had roughly the same dimensions as the physiological columns that were estimated (Jones, 1999). After quantification in selected species, it has been proposed that regardless of the thickness of the cortex within an arbitrary (30-µm-wide, 25-µm-deep) vertical 'column' between the pial surface and the white matter of the cortex, the number of neurons is 110 in all cytoarchitectonic areas (Rockel et al., 1974, 1980). Under such conditions, the neuronal number was claimed constant in all mammalian species (mouse, rat, cat, Old World monkey, and human) and for all cortical thicknesses, the numbers of cells in these arbitrary columns in prefrontal, primary motor, somatosensory, (posterior) parietal, and temporal neocortex (except the primary visual cortex in primates) all being the same. There was only one area in the cortex that showed a difference from this constant number; in all primates studied (galago, marmoset, squirrel monkey, macaque monkey, baboon, and human), the number per 30-µm-wide, 25-µm-deep column of the visual cortex was increased to about 260–270. This increase is a reflection of the much higher packing density of cells in the true striate cortex. In a later study with Anita Hendrickson, Powell found that the neuronal number remained constant across both the monocular and binocular segments of the macaque visual cortex (Powell and Hendrickson, 1981). The changes in packing density of neurons in the arbitrary unit columns were inversely related to the volume of neuropil.

Using similar methods in marsupials, it has been established that the neuronal numbers are half the ones observed in the mouse in a similar arbitrary unit column (Cheung et al., 2007, 2010; Figure 7.3). Using a more recent 'unbiased' stereology method, Herculano-Houzel and her colleagues showed that the density of neurons in the neocortex varies as much as three times even among the highly related primate species (Collins et al., 2010; Herculano-Houzel et al., 2008; Lent et al., 2012). In spite of these observations, it is still true that cortical expansion in evolution is achieved by expanding the cortical surface area, with relatively little change in the thickness. The ratio for the cerebral cortical surface areas in the mouse, macaque, and human is 1:100:1000, whereas for cortical thickness, it is more like 1:1:1 as it is in the range of 2–4 mm in all three species (Rakic, 2009). However, the idea that all mammalian cortices in most areas have a very similar numerical constancy has to be abandoned. In fact, the differences noted earlier might hold a key to understanding cortical specializations for specific functions.

7.5 LACK OF CORRELATION BETWEEN THE ABSENCE OR PRESENCE OF PARTICULAR COLUMNS AND A SPECIFIC SENSORY OR COGNITIVE PROCESSING NETWORK (COMPARISONS ACROSS THE SAME BRAIN AND ACROSS CLOSE AND MORE DISTANT SPECIES)

7.5.1 Microscopic and Macroscopic Cell Patterning Defining Cortical Modules

Most cortical columns can be related to some forms of cellular patterning in the cortex. These can be from subtle microscopic patterns to macroscopically identifiable features. Some distinctive body attachments with characteristic shapes are even recognizable in the somatosensory cortex. Examples include the barrel cortex of the mouse (Woolsey and Van der Loos, 1970), representation of the nasal probosci of the star-faced mole (Catania and Kaas, 1995), and representation of the raccoon hand (Welker et al., 1964) or primate hand (Jain et al., 1998). These visible columns received special status in neurobiology because they helped investigators to understand questions related to synaptic plasticity and map

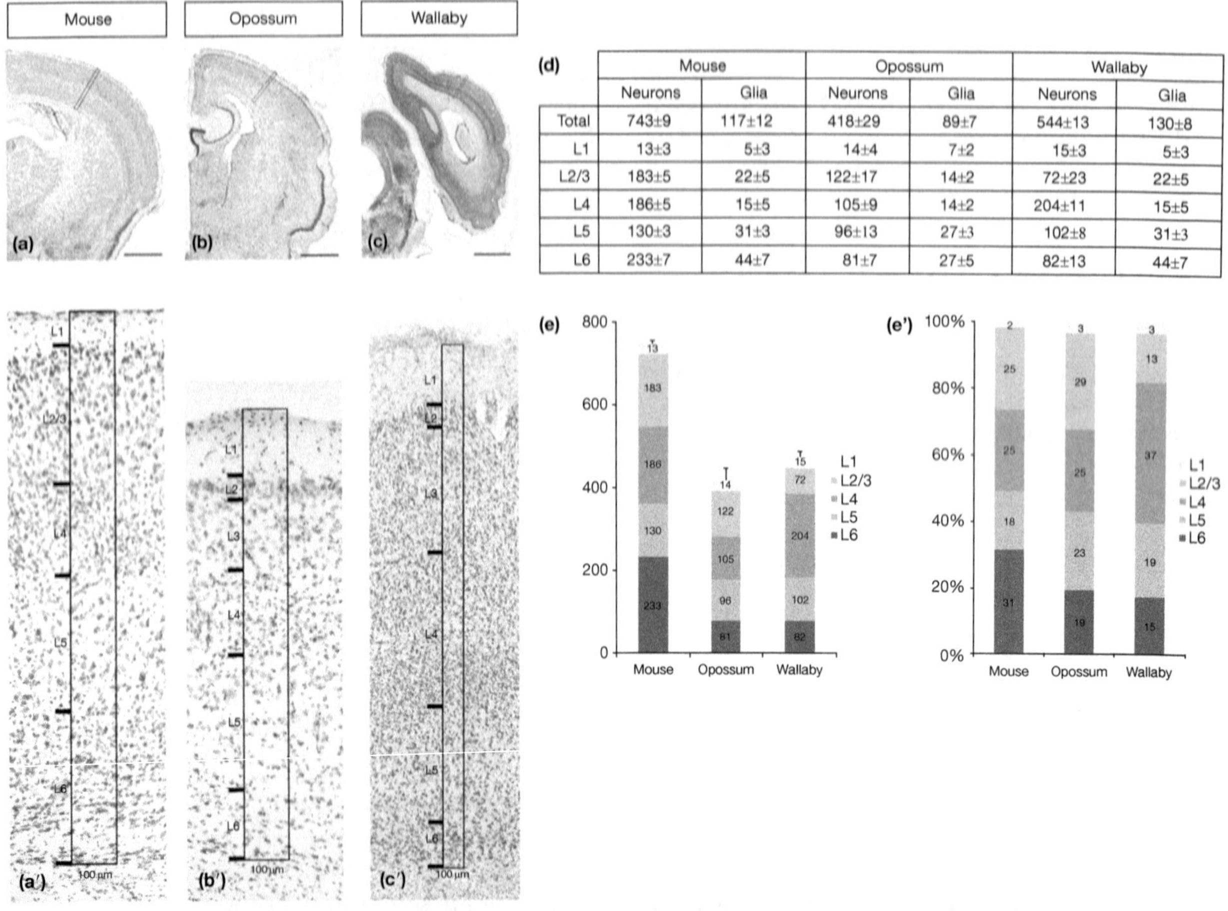

(d)

	Mouse		Opossum		Wallaby	
	Neurons	Glia	Neurons	Glia	Neurons	Glia
Total	743±9	117±12	418±29	89±7	544±13	130±8
L1	13±3	5±3	14±4	7±2	15±3	5±3
L2/3	183±5	22±5	122±17	14±2	72±23	22±5
L4	186±5	15±5	105±9	14±2	204±11	15±5
L5	130±3	31±3	96±13	27±3	102±8	31±3
L6	233±7	44±7	81±7	27±5	82±13	44±7

FIGURE 7.3   Quantification of the number of neurons in the mouse, opossum, and wallaby. (a–c) Cresyl violet-stained sections of adult (a) mouse, (b) opossum, and (c) tammar wallaby. An arbitrary 'unit column' (a 100-µm wide) spanning from layer 1 to 6 was marked in the primary somatosensory/visual area (boxed areas in a–c, higher magnification in a'–c'). The number of neurons and glia was quantified in each layer and expressed as mean±SEM in (d). (e) The mean number of neurons present in each cortical layer, showing that the number of neurons in a unit column is not constant between different infraclasses within mammals. (e') The proportion of neurons in each cortical layer. Scale bar: a–b = 500 µm, c = 1 mm. *Reproduced from Cheung AF, Kondo S, Abdel-Mannan O, et al. (2010) The subventricular zone is the developmental milestone of a 6-layered neocortex: comparisons in metatherian and eutherian mammals. Cerebral Cortex 20(5): 1071–1081, with permission.*

formation. The barrel field is one of the best studied model systems in the mouse cortex; yet, its functional significance has still not been comprehended. Are they structures without any particular function (Horton and Adams, 2005)?

7.5.2 Are Barrels Cortical Columns?

It is puzzling that barrel fields are present in rats, mice, squirrels, rabbits, possums, and porcupines, but not in raccoons, beavers, or cats (Woolsey et al., 1975). The presence or absence of barrels is not related to the presence of actively mobile whiskers (whisking behavior), as guinea pigs do not whisk but nonetheless have a barrel field. The peripheral somatic sensory input is relayed through the brainstem and the ventrobasal complex of the thalamus before it is transmitted to layer IV, the gateway of the sensory cortical circuitry. Thalamic axons form arbors and establish synapses in a periphery-related pattern in layer IV. This pattern formed by thalamocortical axons can be present in the absence of the cytoarchitectonic pattern that was originally termed barrels by Woolsey and Van der Loos (1970) (see López-Bendito and Molnár, 2003). The individual thalamocortical axon clusters that form periphery-related patterns are first surrounded by densely packed layer IV cells that form the walls of the actual cytoarchitectural 'barrels.' In the middle of each barrel is a plexus of thalamic fibers carrying signals from one corresponding whisker (Figure 7.4). In the barrel field of the rodent somatosensory cortex, dendritic bundles are mostly located in the barrel walls and septa, avoiding the hollows (mouse, Escobar et al., 1986).

In the rodent barrel cortex, dendrites of neurons in layer IV conform to barrel limits (Harris and Woolsey,

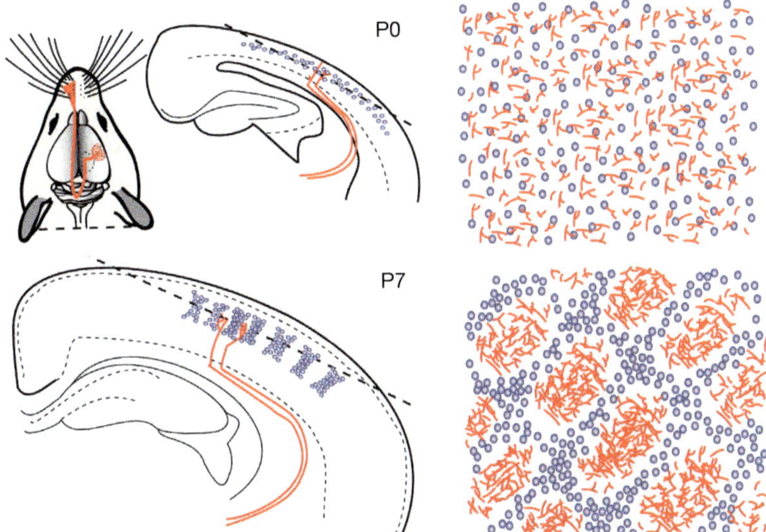

FIGURE 7.4 Schematic overview of the development of the periphery-related thalamocortical patterning and the cytoarchitectonic barrel formation in the mouse. Thalamocortical fiber clusters arise from an initially uniform distribution of thalamocortical arbors (red), and they impose the characteristic patterning of layer IV neurons (blue). Left column represents coronal, right column tangential sections. *Reproduced from Molnár Z and Molnár E (2006) Calcium and NeuroD2 control the development of thalamocortical communication.* Neuron 49(5): 639–642.

1979), but this seems to be an exceptional case. The best documented example of two thalamic systems in the same layer is from rodent barrel cortex (Alloway, 2008). In layer IV, thalamocortical projections from the ventral posterior medial (VPM) and the posterior nuclei target the barrels (lemniscal pathway) and their intervening septa (paralemniscal pathway), respectively. A similar segregation occurs more generally, but with segregation in different layers. The functional significance of the thalamocortical–corticocortical patterning is unknown, but could be related to differential processing by distinct postsynaptic populations. Barrels show variation in size and shape across S1. There have been few quantitative studies of the differences and their relevance.

Sakmann's group has presented a set of papers in which they define a 'standard' column in the rat somatosensory cortex as based on the topographically specific input from the large bundle of thalamocortical axons emanating from a single 'barreloid' in the VPM nucleus of the thalamus and terminating in one of the 'barrels' in the layer IV aggregations of neurons (Helmstaedter et al., 2007; Meyer et al., 2010; Wimmer et al., 2010). In this case, then, their column is of the kind defined originally by Mountcastle and not a minicolumn, although it may contain minicolumns as defined above. On the basis of measurements of concentrations of thalamocortical axon terminals labeled by green fluorescent protein expressed in their parent cells and extending the width of the periodic densities of terminations (which in layer IV are ~300 μm wide) across the depth of the cortex, this column has a cross-sectional area of about 121 000 square microns and a depth of ~1840 μm from the pia to white matter. A second kind of column defined by the authors has its basis in the terminations of axons arriving from the posterior medial (Pom) nucleus of the thalamus and ending deep and superficial to the barrels and especially in the zones of reduced cell density or 'septa' lying between them. This column, as measured from septum to septum and across the intervening barrel, is thus a little wider than the column defined by inputs to the barrels; it has a cross-sectional area of approximately 124 000 square microns, but when projected across the depth of the cortex, it has the same length as the VPM-based column. The measurement of the Pom-based barrel might be rather arbitrary, as the authors describe the axons of Pom neurons as spreading horizontally for seemingly wider extents than those from VPM (Jones and Rakic, 2010).

The barrels in the rodent somatosensory cortex are not stereotyped. Hollow barrels, with cell-sparse cores, are typical of mice, young rats, and the anterolateral subfield of mature rats, but solid columns, with cell-dense cores, are typical of the main posteromedial field in rats (Rice, 1995). Variability is not reported for other columnar systems of connections, but this is likely because many of the systems are harder to visualize globally or require specialized tissue processing.

The function of the barrels in the rodent primary somatosensory cortex is not known. The cortical architecture can be missing or significantly disrupted and yet apparently remain functionally intact. For example, the disrupted barrel cortex in the reeler and in other mutant or transgenic mice is not associated with marked somatosensory deficits (López-Bendito and Molnár, 2003; Rakic and Caviness, 1995).

The degree to which the cortex is modifiable, and by what mechanisms, has been extensively investigated under various environmental manipulations. Although it is not known what the functional relevance (if any) of the barrel arrangements may be, this system helped the understanding of various aspects of cortical circuit formation and plasticity. Study of the barrel field in various mouse mutants proved to be instrumental in the understanding of the molecular mechanisms of these interactions (Erzurumlu and Kind, 2001). The development of the periphery-related patterning of the thalamocortical projections and the induction of the cytoarchitectonic barrels require both pre- and postsynaptic interactions. During the first days of postnatal development, thalamic projections assume a periphery-related pattern within layer IV precisely mirroring the arrangements of the whiskers. Thalamocortical axon segregation is soon followed by the relocation of layer IV cells from an initially homogeneous distribution to the walls of the barrels surrounding the clustered thalamic projections (Molnár and Molnár, 2006; Figure 7.4).

Van der Loos and Woolsey (1973) provided evidence for the environmental influence on cortical cytoarchitectonic differentiation by demonstrating that changing or blocking the flow of sensory input from specific whiskers during the early stages of development results in a cascade of events that will change the arrangements and somatodendritic morphology of layer IV cells. With the development of finer techniques of clonal analysis and neuronal cell-type specification, one can anticipate a new generation of genetic and molecular manipulations that will help us elucidate the underlying mechanisms of barrel formation. Overexpression of NT3 is reported to result in an enhanced expression of dendritic bundles ('minicolumns') in the rat barrel cortex (Miyashita et al., 2010). However, it is not clear to what extent the barrels represent a general and valid model for cortical columns.

7.5.3 Microcolumns and Apical Dendritic Bundles

There are a number of examples of repeating microarrays of intracortical elements that are interpreted as conforming to a microcolumnar pattern of vertical connections. The observations on patterning of apical dendrites of pyramidal cells with somata located in layers II, III, and V have led, as noted above, to the introduction of the term minicolumns or microcolumns (Fleischhauer et al., 1972; Peters and Walsh, 1972). Innocenti and Vercelli (2010) distinguished minicolumns and bundles, whereas some investigators have used these terms interchangeably. Minicolumns of radially aligned cell bodies can be demonstrated by regular Nissl preparations or other histological methods that reveal cell bodies. Bundles are composed of the apical dendrites of pyramidal neurons whose cell bodies are in different layers and can be seen in material prepared by the Golgi technique, stained with osmium for electron microscopical analysis, or with markers of somatodendritic morphology (e.g., microtubule-associated protein 2 or SMI32) (Peters and Walsh, 1972; Figure 7.5). Innocenti and Vercelli demonstrated bundles using retrograde transport of lipophilic tracers or intracellular injection of neurons in slice preparations (Innocenti and Vercelli, 2010). Myelinated axons are also organized in bundles; these bundles course close to those of the dendrites, and at least some of them originate from neurons whose apical dendrites are in a bundle (Peters and Sethares, 1996). Depending also on tangential location and depth, the minicolumns and bundles can be more or less sharp.

An average bundle is composed of the dendrites of 3–20 large pyramidal cells of layer V that form clusters that ascend together through layer IV. They are joined in the supragranular layers by the successive addition of the apical dendrites of pyramidal cells of layers II and III, and all ascend further, many sending their terminal arrays to layer I (Figure 7.5). Reconstructions have revealed that individual dendrites change their neighborhood relations along a bundle, superficial dendrites can be added between the dendrites from deeper layers, and individual dendrites can bifurcate to sending branches to neighboring bundles (Massing and Fleischhauer, 1973). In the monkey visual cortex, the microcolumns are estimated to consist of the dendrites of ≈ 142 pyramidal neurons. These modules are 30 µm in diameter and occur with center-to-center spacing that varies from 20 to 80 µm, the wider spacing occurring in the larger brains of the macaque monkey and man. Their estimated density is ≈ 1270 per mm^2 in the monkey visual cortex. In the visual cortex, the mean spacing between modules was found to be 60 µm in the rat, 56 µm in the cat, and 23 µm in the rhesus monkey (Peters, 1997). Not all apical dendrites from layer V enter into the composition of dendritic bundles (Rockland and Ichinohe, 2004). The presence of layer V is not considered the prerequisite to consider a bundle (Innocenti and Vercelli, 2010); however, this issue has not been investigated in mutant mouse cortex that lacks a particular subtype of layer V.

7.5.4 Complex Relationship Relations Between Minicolumns and Dendritic Bundles

Cell bodies of neurons in a minicolumn can be seen to orient obliquely to engage their apical dendrite into the neighboring dendritic bundles already in layer V and more so in layer III (Gabbott, 2003; Peters and Kara, 1987; Peters and Walsh, 1972; Figure 7.5). The progressive addition of dendrites to the bundle from depth to surface in the cortex ('like onions held by their stem';

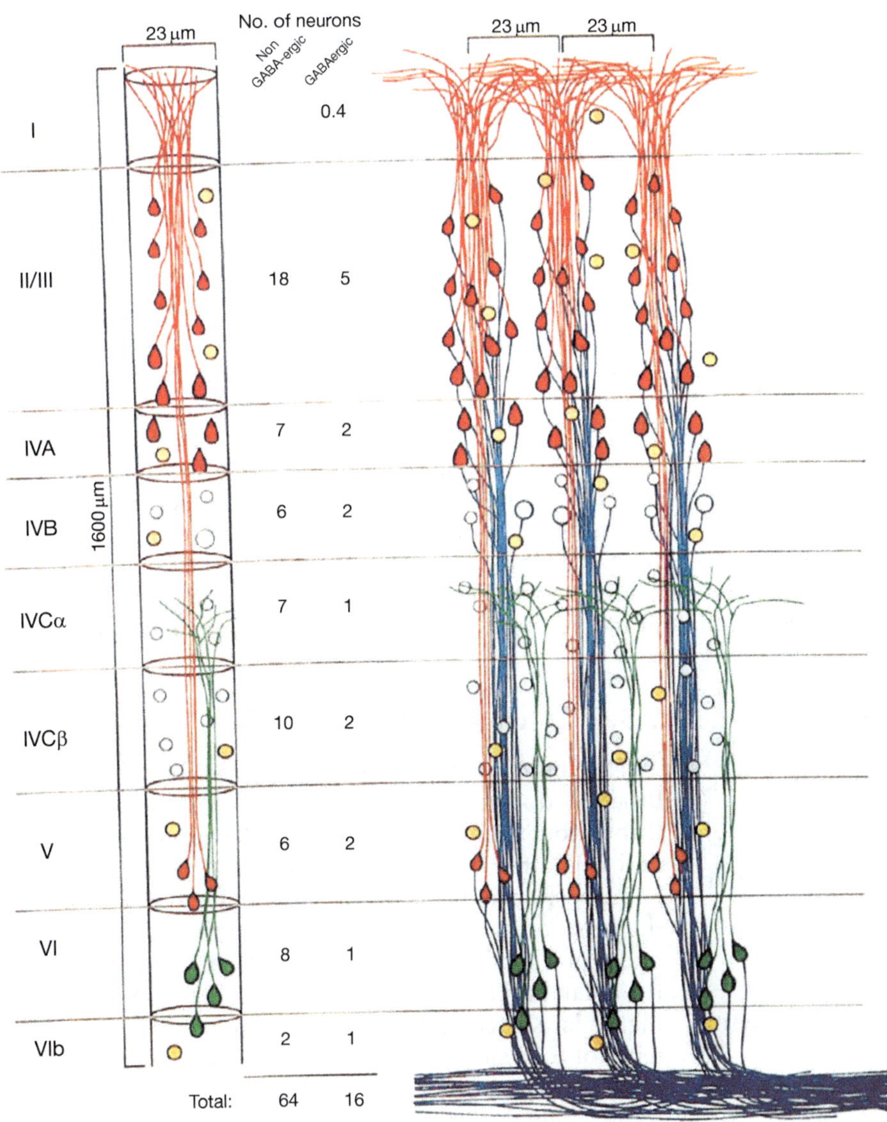

FIGURE 7.5 Bundles and microcolumns in the primary visual cortex of the macaque monkey. (*Reproduced from Peters A and Sethares C (1996) Myelinated axons and the pyramidal cell modules in monkey primary visual cortex.* Journal of Comparative Neurology 365: 232–255). Left panel: Schematic representation of the arrangements of the apical dendrites of pyramidal cells. Layer II/III, IVA, and V pyramidal cells are shown in red, layer VI in green. Gray represents neurons of IVB and IVC without dendrites; azure represents GABAergic neurons. Right panel: Pyramidal minicolumns are represented adjacent to the dendritic bundles. Axons of pyramidal cells are depicted in blue.

Peters and Kara, 1987) also indicates that the bundles collect dendrites from more than one minicolumn of cell bodies (Massing and Fleischhauer, 1973; Peters and Kara, 1987; Vercelli et al., 2004).

Peters and Sethares (1996) observed that those neurons with apical dendrites in the same bundle also bundle their myelinated axons, indicating that neurons in a dendritic bundle might send their axons to the same target. Subsequently, Lev and White (1997) showed that, in the mouse area MsI, following injection of horseradish peroxidase in the contralateral hemisphere, all dendrites in a labeled bundle belonged to callosally projecting neurons, thus suggesting that dendritic bundles are target specific. This issue has been further investigated by Vercelli et al. (2004) in the visual cortex of the rat for different targets (ipsilateral cortex, superior colliculus (SC), pons, lateral geniculate nucleus, and striatum). This study strongly supports the concept that dendritic bundles are target specific. Moreover, the composition of dendritic bundles does not seem to depend on the age of the animal and is already established at P3. It is an interesting possibility that some transient elements join the bundles at early developmental stages from the early-generated neurons of the subplate. Thus, it has recently been demonstrated that subplate apical dendrites have a close association with layer V apical dendrites, having the same targets in the early postnatal barrel cortex (Hoerder-Suabedissen and Molnár, 2012). However, it is not known whether dendrodendritic synapses occur in particular dendritic bundles or not.

Unlike what might have been expected, neurons of the same bundle are not more interconnected than neurons of different bundles (Krieger et al., 2007). There are preferential connections between certain output neurons, interestingly between corticocortical and corticotectal neurons, whose apical dendrites lie in separate dendritic bundles in mouse (Brown and Hestrin, 2009).

Innocenti and Vercelli (2010) proposed that neurons in the different layers of one minicolumn, projecting to different targets, send their apical dendrites to separate dendritic bundles, where they join apical dendrites of neurons from neighboring minicolumns projecting to the same target or combination of targets. They propose that, in a given cortical locale (area or part of an area), an assembly of apical dendritic bundles, which includes each of the outputs to distant cortical or subcortical structures, their parent somata and basal dendrites, and the portion of the neuropil that pertains to them, constitutes a cortical output unit.

Dendritic bundles and microcolumns can be identified in all cortical areas in the cerebral cortex of different mammalian species, such as rodents, carnivores, and primates including humans. The dendritic bundling seems to offer two important advantages. It might minimize the length of the axonal arbors that contact specific neuronal classes, and in development it might simplify the axonal search and recognition of targets. The link, if any, between the minicolumn and apical dendritic bundles and functional cortical units is not yet established. The link between the anatomical organization and the physiological units is not clear.

7.5.5 Columns Outside the Mammalian Isocortex

Columnar arrangements are present in numerous structures in mammals. Iterated circuitry is present in the olfactory bulb glomeruli (Figure 7.1(f)), and in the barreloids (Figure 7.1(e)) and barrelettes in the ventrobasal thalamic nucleus and brainstem, respectively. Columnar structures are also present in the laminated structure of the SC (Harting et al., 1992; Illing and Graybiel, 1986). These are revealed by acetylcholinesterase reactivity as 200–600-μm-wide patches (Harting et al., 1992; Mana and Chevalier, 2001).

Rockland (2010) gives an overview of these patterns: "The periaqueductal gray contains longitudinal columns of afferent inputs, output neurons, and intrinsic interneurons thought to coordinate different strategies for coping with different types of aversive stimuli' (Bandler and Shipley, 1994; Keay and Bandler, 2001). The lateral septal nucleus is reported to have a complex system of chemically and connectionally distinct zones of transverse sheets (Risold and Swanson, 1998). Some thalamic nuclei have distinct domains, which are neurochemically and connectionally distinguishable (Rausell and Jones, 1991). The basal ganglia are organized into

neurochemically and connectionally distinct striosomes and matrix (Graybiel and Ragsdale, 1978)."

In the cerebellar cortex, an elaborate array of modular subdivisions is revealed by histochemical markers, the topography of afferent projections and some efferent projections, and gene expression in subpopulations of Purkinje cells (PCs) (Sillitoe and Joyner, 2007; Voogd and Glickstein, 1998). Zebrin II expression in PC reveals a parasagittal stripe pattern, each stripe consisting of a few hundred to a few thousand PCs, that is highly reproducible, activity independent, and conserved across species. Other molecular and connectivity markers have an orderly relation to zebrin+ or zebrin− stripes (Larouche and Hawkes, 2006). The functional importance of this striking organization remains to be elucidated, but, as compared to the mosaicism of the SC, it has been suggested to subserve a massively parallel architecture with a high number of processing channels (Larouche and Hawkes, 2006). Minicolumn-like dendritic bundles can also be found in numerous noncortical structures (e.g., Roney et al., 1979).

7.5.6 Columns in Nonmammals

The columnar organization of the cerebral cortex is a broadly documented principle of design preserved throughout mammalian evolution (Mountcastle, 1997), and it is often considered unique to mammals. Karten has questioned the assumption of the uniqueness of the neocortical cells and circuits in mammals and has argued for a similar laminar and columnar organization in the avian brain (Wang et al., 2010). Using contemporary methods, Karten and colleagues demonstrated the existence of comparable columnar functional modules in the laminated auditory telencephalon of an avian species (*Gallus gallus*). Tracer placed into individual layers of the telencephalon within the cortical region, which is considered similar to the mammalian auditory cortex by Karten and colleagues, revealed extensive interconnections across layers and between neurons within narrow radial columns perpendicular to the laminae (Wang et al., 2010). This columnar organization was further confirmed by visualization of radially oriented axonal collaterals of individual intracellularly filled neurons. These findings indicate that laminar and columnar properties of the neocortex are not unique to mammals and may have evolved from cells and circuits found in more ancient vertebrates (Shepherd and Grillner, 2010; Montiel et al., 2012).

7.5.7 What Is the Function of a Cortical Column?

The functional rationales for the columnar organization of the cerebral cortex include arguments for 'augmenting of cortical surface area during speciation;

modular segregation of inputs; and facilitation of computation by enhancing information processing' (Purves et al., 1992). Modular clustering is believed to be important to allow a large number of neurons to be connected without a significant increase in cortical volume (Mitchison, 1992).

However, if modules are essential for information processing, why is it that they are present in some species but not in others without any noticeable perceptual differences (Horton and Adams, 2005; Purves et al., 1992)? Or, if they are essential in enhanced computation, why are they not present in higher motor and association areas (Purves et al., 1992)? The criteria for the identification for modules/columns are so diverse that it is possible that some variables that might define patchy or modular arrangements might be identified in the future.

The experimental paradigms provided by the barrel and the OD column tend to influence the way the cerebral cortex is looked at as a whole, but neither is clearly built up from microcolumnar units of cells or connections. None of these cortical arrangements is associated to a particular cognitive or perceptual ability in species where they are present or absent. Horton and Adams argue that the lack of correlation across species with or without OD columns and vision, and with or without cortical barrel field and whisker function, strongly argues for the 'lack of particular function of these striking but inconstantly expressed anatomical features' (Horton and Adams, 2005).

Dendritic bundles have been found throughout the mammalian brain and are believed to serve fundamental roles in the brain's functioning. However, no physiological experiments to determine their function have been performed on these well-established anatomical units. The function of the anatomical microcolumns as fundamental units of organization is also not clear. Much more comprehensive analysis of the intricacies of intracortical connectivity and the anatomy and physiology of microcolumns in all cortical areas of several species is needed. Combining these approaches could clarify several issues (see Bock et al., 2011). Microcolumns might represent fine-grain functional modularity of the cortex. The radial dispersion of these clusters in some columns is about 400 µm in all species, similar to the spread of some dendritic arbors.

7.5.8 Columns in Neuropathology

Cortical modules have drawn the attention of neuropathologists not because their function is known, but because they can be visualized, quantified, and compared between subjects. They might be epiphenomena, but they are detectable entities and therefore can be used as diagnostic signs for abnormal cortical organization.

The periodicities of microcolumnar structures (that contain about 11 neurons and have a periodicity of about 80 µm) were disrupted in two examples of neurodegenerative disease in human (Jones, 2000). Some alterations of the microcolumnar structures have been described in the brains of the more elderly (Peters, 1997).

Currently, it is not known what degree of radial allocation and lateral neuronal dispersion is essential for the proper radial delivery and intermixing of neuronal types in cortical columns. Alterations in the clonal dispersion of neurons have been linked to neuropsychiatric disorders associated with abnormal columnar organization (Torii et al., 2009).

7.6 WHAT IS THE CORRELATION BETWEEN THE COLUMNAR DEVELOPMENT OF THE BRAIN AND FUTURE COLUMNS?

7.6.1 Cortical Columns During Development

Mountcastle emphasized that the mode of generation of the cortex already reflects its basic columnar organization (Mountcastle, 1997). From Golgi preparations and from Nissl-stained material, the radial orientation of neurons within the developing cerebral cortex is apparent from the very beginning (Cajal, 1909). Neurons assume a radial orientation and dendritic polarity shortly after their generation. These observations triggered theories that much of the anatomical substrate for a columnar organization would already be specified at early developmental stages before activity-dependent mechanisms could be activated (Rakic, 1988).

7.6.2 Ontogenic Units/Columns – The Fundamental Building Blocks in the Developing Neocortex

There is strong evidence for the overall radial migration of pyramidal neurons in all mammalian cortices (Rakic, 2009). Clonally related postmitotic pyramidal neurons are initially deployed in a geometrically columnar pattern in the embryonic primate cerebrum. Rakic proposed that the location of the cohorts of cortical neurons from a single neuronal progenitor is not random but is largely predictable (Rakic, 1988). Each neuron in mature cortical 'minicolumns' is derived from one of a small group of progenitors forming a polyclonal group in the ventricular zone (VZ) (Kornack and Rakic, 1995).

The progenitor cells that generate the minicolumn were termed ontogenic columns (Rakic, 1988). Rakic estimated that "each proliferative unit in the ventricular zone of the monkey consists of 3–5 stem cells, a number that gradually increases to 10–12 stem cells during development; the units are separated by glial septa" (Rakic, 1988). According to this theory the surface area, and thus the size of the neocortex, is determined by the number of ontogenetic units set by the number of symmetric divisions of progenitor cells in the neural epithelium before migration begins (Rakic, 1988, 2009). It has been suggested that one important phenomenon for the increased cerebral complexity during evolution may be the multiplication of neuronal columns throughout the cerebral cortex (Rakic and Caviness, 1995).

According to this theory, functional columns in the adult cerebral cortex must consist of several ontogenic columns (polyclones). The visual display of these ontogenetic units has not been achieved, and it is still unclear to what extent and how gene expression in the VZ could play a role in the development of discrete functional units, such as minicolumns or columns. Cell lineage experiments using replication-incompetent retroviral vectors have shown that the pyramidal neuronal progeny of a single neuroepithelial/radial glial cell in the dorsal telencephalon is organized into discrete radial clusters of sibling excitatory neurons (Kornack and Rakic, 1995; Noctor et al., 2001). Costa et al. (2009) noted that most neuronal clones derived from E13 progenitors span 150–250 µm in the horizontal axis and contribute to all cortical layers generated after that embryonic stage. The same authors performed mathematical extrapolations for injections at the onset of neurogenesis in the cerebral cortex (E10–11) and suggested that neuronal siblings would not disperse by more than 400–500 µm. Thus, both the radial and horizontal dispersions of excitatory neuronal clones fit well with the possibility that they could help to create a structural basis for the future specification of columns.

How these developmental neural clusters relate to adult anatomical and physiological columns has not been addressed. Neurons from different clones intermix with the adjacent columns as they migrate across the intermediate zone. In addition to the radial allocation of clonally related neurons, short lateral shifts and transfers from their parental to the neighboring radial glial fibers have been described (Kornack and Rakic, 1995; Noctor et al., 2001; Tan and Breen, 1993). These dispersed neurons intermix with neurons originating from neighboring proliferative units. The molecular mechanisms, their role, and the significance of this lateral dispersion for cortical development are not understood. A recent study revealed that the lateral dispersion depends on the expression levels of Eph receptor As (EphAs) and ephrin-As during neuronal migration. Torii et al. (2009) demonstrated that an EphA and ephrin-A (Efna) signaling-dependent shift in the allocation of clonally related neurons is essential for the proper assembly of cortical columns in the mouse cerebral cortex (Figure 7.6). Currently, it is not known what degree of radial allocation and lateral neuronal dispersion seems to be essential or optimal for the proper radial delivery and intermixing of neuronal types in the cortical columns. The degrees of mixing of derivatives of different progenitors have not been estimated in different species.

7.6.3 Sibling-Neuron Circuits in the Developing Columns

Thus, both the developing cortex and the adult cortical columns have overwhelmingly radial arrangements. In the developing brain, the clonally related neurons have higher chances of being situated within the same radial volume of cortical tissue (Tan and Breen, 1993). It has been proposed that the initial columnar organization may act as a seed to establish the primary information-processing unit in the cortex.

This raises the question as to whether neurons from the same clone develop connectivity preferentially. Are

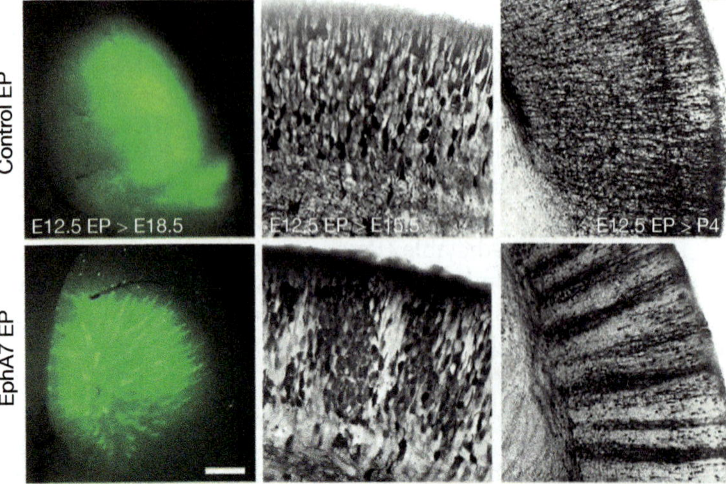

FIGURE 7.6 Eph receptor A (EphA) and ephrin-A (Efna) signaling-dependent shift in the allocation of clonally related neurons changes the level of lateral dispersion in the embryonic brain. Torii and colleagues demonstrated that lateral dispersion depends on the expression levels of EphAs and ephrin-As. Increased EphAs and ephrins during neuronal migration leads to increased tangential sorting of cortical neurons. The figure represents the distribution of EYFP-labeled neurons (green or black) during development in the cortex in which control or EphA7 expression plasmid was delivered using electroporation (EP) with EYFP plasmid. Scale bars, 600 mm (for E18.8), 50 mm (for E15.5), and 200 mm (for P4). *Reproduced from Torii M, Hashimoto-Torii K, Levitt P and Rakic P (2009) Integration of neuronal clones in the radial cortical columns by EphA and ephrin-A signalling.* Nature 461(7263): 524–528.

they more likely to develop chemical synapses with each other rather than with neighboring nonsiblings? Cell lineage or clonal analysis studies have been combined with recording experiments to study the possibility that the cell lineage of single neuroepithelial/radial glia cells could form radial columns of sibling, interconnected neurons. Yu and colleagues (2009) identified individual clones of cortical pyramidal neurons by injecting enhanced green fluorescent protein (EGFP)-expressing retroviruses into the lateral ventricle of mouse embryos at early stages of neurogenesis. They made simultaneous whole-cell recordings on two EGFP-expressing sister neurons and observed that these cells displayed unidirectional synaptic connections in 35% of pairs. In contrast, less than 7% of radially situated nonsister excitatory neurons were connected (Yu et al., 2009). This experiment provides strong support for the idea that excitatory neurons generated from the same progenitor keep spatial relationships and display (mutual) connectional preferences, but it stops short of relating the clones to adult cortical columns. There is a distinction between the idea that sibling neurons have predictable arrangements and connectivity and the idea that adult columns are 'preformed' and prespecified in ontogenic units in the VZ. More work is needed to clarify these issues.

The potential molecular mechanisms involved in the establishment of sibling-neuron circuits are not known. It has been hypothesized that neurons derived from the same progenitor are more likely to display similar chemical and physical properties because of their genetic inheritance (Costa and Hedin-Pereira, 2010). Sister neurons might share more of the combinatorial transcription code that has been present in the common cortical progenitors, and therefore the sister neurons might share a similar set of surface molecules that are important for cell–cell recognition or for molecular guidance cues. The molecular determinants of the cell-intrinsic properties for cell–cell recognition between sibling neurons in the cortex remain a largely uncharted territory.

7.6.4 Transient Columnar Domains During Development

The coordinated calcium fluctuation patterns underlying gap junction-mediated communication were suggested as a possible basis for the formation of initial functional cell assemblies in the postnatal cerebral cortex. Yuste et al. (1992) observed distinct domains of spontaneously coactive neurons using optical recordings of brain slices labeled with the fluorescent calcium indicator fura-2 in early postnatal rat cortex. Their observations emphasized the discrete multicellular patterns that are mediated through communication via gap junctions. The functional domains were 50–120 μm in diameter on tangential slices; they spanned several cortical layers and resembled columns found in the adult cortex in coronal slices. In the developing somatosensory cortex, domains were smaller than, and distinct from, the barrels. Gap junctions coupled the neurons within each domain. Gap junction domains persisted after blockade of sodium- and calcium-dependent action potentials, suggesting that they may promote metabolic rather than activity-related assemblies (Kandler and Katz, 1998).

There are modules and columns within the developing cortex that are present only transiently during development, but not in the adult. Numerous stains are transient during barrel development (Erzurumlu et al., 1990; Mitrovic and Schachner, 1996). OD columns are more apparent during development or upon visual deprivation than in normal adults in the marmoset, and the presence or absence of OD columns has been debated in the adult marmoset (Chappert-Piquemal et al., 2001; Spatz, 1989). There are transient circuits that show a pattern that is transiently related to the (sensory) periphery of the vibrissae in the barrel cortex in the mouse, and this involves the neurites of the early generated and largely transient subplate neurons (Piñon et al., 2009; Hoerder-Suabedissen and Molnár, 2012). These changes in cortical patterning may reflect the development of synaptic integration that will provide coherent activity among groups of target cells, but it has been questioned whether the observed patterns themselves have any functional relevance (Purves et al., 1992). Induction of visible and distinguishable barrel patterns or of OD columns has not been linked to a particular sensory or motor capacity (Horton and Adams, 2005; Purves et al., 1992).

Prior to birth, monocular transduction pathways are already established through a process known as Hebbian learning. Spontaneous retinal activity in one eye of the developing fetus leads to neuronal depolarization (Galli and Maffei, 1988) that can propagate through the thalamus (Mooney et al., 1996). Synapses that receive multiple inputs are more likely to propagate the signal, whereas errant connections will not be sufficient to trigger another action potential. If glutamate has been released by the presynaptic axon terminal, postsynaptic neurons that depolarize become permeable to calcium ions. Calcium entry leads to a chemical process that strengthens the synapse, making it more likely to survive than other connections.

Although orientation columns can develop without any externally elicited sensory visual input (before birth), their maintenance relies on postnatal sensory-driven visual activity (Crair et al., 1998).

7.7 SUMMARY AND THE WAY FORWARD

1. There are several problems associated with the current nomenclature of columns. The concept of a 'universal cortical column' is very captivating in anatomical, physiological, and developmental models of the cerebral cortex, and this is reflected in the current terminology that aims to gloss over differences rather than expose them. There is overwhelming evidence for various forms of radial organization, and some of the modular (columnar) organizations are striking (Douglas and Martin, 2004). However, there are various types of cortical columns, and there is a need to define them more clearly as a better understanding of their properties is gained. The term 'column' might be modified or abandoned altogether when there is more information about the types of cortical circuits and the range of their operations. The term column is still used because of its captivating concept. For the time being, there is no easy alternative to 'column.' It is necessary to establish more specific terminology that will allow specific reference to particular entities. As the cell types of the cerebral cortex become better characterized morphologically, chemically, and physiologically, the details of the types of connections and circuits that they establish with one another within the cortex are becoming understood.

2. It is now known that columns/modules are characteristic of the neocortex, but there is no single structure or function that is *the* common building block of all cortical areas in all mammals. The observations made on the barrels in S1 and the OD columns in V1 have had an enormous influence on the way the cerebral cortex is looked at as a whole; however, it is increasingly apparent that this cannot be generalized to all regions. Horton and Adams write: "At some point, one must abandon the idea that columns are the basic functional entity of the cortex. It now seems doubtful that any single, transcendent principle endows the cerebral cortex with a modular structure. Each individual area is constructed differently, and each will need to be taken apart cell by cell, layer by layer, circuit by circuit, and projection by projection to describe fully the architecture of the cortex" (Horton and Adams, 2005). Under the influence of the new data, the concept will also gradually change and the elusive idea of a 'universal cortical unit' that extends radially as a homogeneous building block in all areas in all mammals may finally have to be rejected.

3. The previous finding of constant cell numbers within an arbitrary unit column and the homogeneous structure of the column is not supported by recent observations. On the contrary, it is apparent that cortical areas exhibit huge differences in cell composition, cell numbers, and connectivity (Lent et al., 2012).

4. The possible functions specifically associated with the presence or absence of a particular column (e.g., OD columns, barrels) is not clear. Comparative studies have yet to identify a specific sensory, motor, or cognitive function that is specifically associated with a particular form of cortical column. Purves and colleagues postulated that the columnar patterns arise because they are an "incidental consequence of the rules of synapse formation" (Purves et al., 1992).

5. There is overwhelming evidence for early columnar allocation of the developing pyramidal neurons. It is also evident that at early developmental stages an early organization has already been specified before activity-dependent mechanisms could take place. The link between the clonally related neuronal assemblies and future modules of the cortex is not yet clear, but with the current repertoire of methodologies, these issues can now be addressed.

Acknowledgments

The author is very grateful to Ray Guillery, Tony Rockel, Gordon Shepherd, and Anna Hoerder-Suabedissen for the lively discussions and comments on an earlier version of this chapter. The work of the author's laboratory is funded by Medical Research Council UK, The Wellcome Trust, BBSRC UK, and the FP7-HEALTH-2009.

References

Abeles, M., Goldstein Jr., M.H., 1970. Functional architecture in cat primary auditory cortex: Columnar organization and organization according to depth. Journal of Neurophysiology 33 (1), 172–187.

Adams, D.L., Horton, J.C., 2003. Capricious expression of cortical columns in the primate brain. Nature Neuroscience 6, 113–114.

Adams, D.L., Sincich, L.C., Horton, J.C., 2007. Complete pattern of ocular dominance columns in human primary visual cortex. Journal of Neuroscience 27 (39), 10391–10403.

Albright, T.D., Desimone, R., Gross, C.G., 1984. Columnar organization of directionally selective cells in visual area MT of the macaque. Journal of Neurophysiology 51 (1), 16–31.

Alloway, K.D., 2008. Information processing streams in rodent barrel cortex: The differential functions of barrel and septal circuits. Cerebral Cortex 18, 979–989.

Bandler, R., Shipley, M.T., 1994. Columnar organization in the midbrain periaqueductal gray: Modules for emotional expression? Trends in Neurosciences 17, 379–389.

Blasdel, G.G., Lund, J.S., 1983. Termination of afferent axons in macaque striate cortex. Journal of Neuroscience 3, 1389–1413.

Blasdel, G.G., Salama, G., 1986. Voltage-sensitive dyes reveal a modular organization in monkey striate cortex. Nature 321 (6070), 579–585.

Bock, D.D., Lee, W.C., Kerlin, A.M., et al., 2011. Network anatomy and *in vivo* physiology of visual cortical neurons. Nature 471 (7337), 177–182.

Bonhoeffer, T., Grinvald, A., 1991. Iso-orientation domains in cat visual cortex are arranged in pinwheel-like patterns. Nature 353 (6343), 429–431.

Brodmann, K., 1909. Vergleichende Lokalizationslehre der Grosshirnrinde in ihren Prinzipien dargestellt auf Grund des Zellenbaues. Barth, Leipzig.

Brown, S.P., Hestrin, S., 2009. Intracortical circuits of pyramidal neurons reflect their long-range axonal targets. Nature 457, 1133–1137.

Brugge, J.F., Merzenich, M.M., 1973. Responses of neurons in auditory cortex of the macaque monkey to monaural and binaural stimulation. Journal of Neurophysiology 36 (6), 1138–1158.

Cajal, S, Ry, (1909), Cajal SRy (1909–1911) *Histologie du système nerveux de l'homme et des vertébrés*, 2 vols. Translated by Azoulay L and Maloine A, Paris. English translation by Swanson N, Swanson LW (1995) *Histology of the Nervous System of Man and Vertebrates*, 2 vols. New York: Oxford University Press..

Catania, K.C., Kaas, J.H., 1995. Organization of the somatosensory cortex of the star-nosed mole. Journal of Comparative Neurology 351 (4), 549–567.

Chappert-Piquemal, C., Fonta, C., Malecaze, F., Imbert, M., 2001. Ocular dominance columns in the adult New World Monkey *Callithrix jacchus*. Visual Neuroscience 18 (3), 407–412.

Cheung, A.F., Pollen, A.A., Tavare, A., DeProto, J., Molnár, Z., 2007. Comparative aspects of cortical neurogenesis in vertebrates. Journal of Anatomy 211 (2), 164–176.

Cheung, A.F., Kondo, S., Abdel-Mannan, O., et al., 2010. The subventricular zone is the developmental milestone of a 6-layered neocortex: Comparisons in metatherian and eutherian mammals. Cerebral Cortex 20 (5), 1071–1081.

Collins, C.E., Airey, D.C., Young, N.A., Leitch, D.B., Kaas, J.H., 2010. Neuron densities vary across and within cortical areas in primates. Proceedings of the National Academy of Sciences of the United States of America 107 (36), 15927–15932.

Costa, M.R., Hedin-Pereira, C., 2010. Does cell lineage in the developing cerebral cortex contribute to its columnar organization? Frontiers in Neuroanatomy 4, 26.

Costa, M.R., Bucholz, O., Schroeder, T., Götz, M., 2009. Late origin of glia-restricted progenitors in the developing mouse cerebral cortex. Cerebral Cortex supplement 1, i135–i143.

Crair, M.C., Gillespie, D.C., Stryker, M.P., 1998. The role of visual experience in the development of columns in cat visual cortex. Science 279 (5350), 566–570.

Douglas, R.J., Martin, K.A., 2004. Neuronal circuits of the neocortex. Annual Review of Neuroscience 27, 419–451.

Dyck, R.H., Chaudhuri, A., Cynader, M.S., 2003. Experience-dependent regulation of the zincergic innervation of visual cortex in adult monkeys. Cerebral Cortex 13 (10), 1094–1109.

Eccles, J.C., Ito, M., Szentágothai, J., 1967. The Cerebellum as a Neuronal Machine. Springer-Verlag, Berlin.

Erzurumlu, R.S., Kind, P.C., 2001. Neural activity: Sculptor of 'barrels' in the neocortex. Trends in Neurosciences 24 (10), 589–595.

Erzurumlu, R.S., Jhaveri, S., Benowitz, L.I., 1990. Transient patterns of GAP-43 expression during the formation of barrels in the rat somatosensory cortex. Journal of Comparative Neurology 292 (3), 443–456.

Escobar, M.I., Pimienta, H., Caviness Jr., V.S., Jacobson, M., Crandall, J.E., Kosik, K.S., 1986. Architecture of apical dendrites in the murine neocortex: Dual apical dendritic systems. Neuroscience 17, 975–989.

Fleischhauer, K., Petsche, H., Wittkowski, W., 1972. Vertical bundles of dendrites in the neocortex. Zeitschrift für Anatomie und Entwicklungsgeschichte 136, 213–223.

Freund, T.F., Martin, K.A., Soltesz, I., Somogyi, P., Whitteridge, D., 1989. Arborisation pattern and postsynaptic targets of physiologically identified thalamocortical afferents in striate cortex of the macaque monkey. Journal of Comparative Neurology 289, 315–336.

Gabbott, P.L., 2003. Radial organisation of neurons and dendrites in human cortical areas 25, 32, and 32'. Brain Research 992, 298–304.

Galli, L., Maffei, L., 1988. Spontaneous impulse activity of rat retinal ganglion cells in prenatal life. Science 242 (4875), 90–91.

Graybiel, A.M., Ragsdale, C.W., 1978. Histochemically distinct compartments in the striatum of humans, monkeys, and cat demonstrated by acetylthiocholinesterase staining. Proceedings of the National Academy of Sciences of the United States of America 75, 5723–5726.

Gross, C.G., Rocha-Miranda, C.E., Bender, D.B., 1972. Visual properties of neurons in inferotemporal cortex of the Macaque. Journal of Neurophysiology 35 (1), 96–111.

Harris, R.M., Woolsey, T.A., 1979. Morphology of Golgi-impregnated neurons in mouse cortical barrels following vibrissae damage at different post-natal ages. Brain Research 161, 143–149.

Harting, J.K., Updyke, B.V., Lieshout, D.P.V., 1992. Corticotectal projections in the cat: Anterograde transport studies of twenty-five cortical areas. Journal of Comparative Neurology 324, 379–414.

Helmstaedter, M., de Kock, C.P.J., Feldmeyer, D., Bruno, R.M., Sakmann, B., 2007. Reconstruction of an average cortical column in silico. Brain Research Reviews 55, 193–203.

Hendrickson, A.E., Wilson, J.R., 1979. A difference in [14C]deoxyglucose autoradiographic patterns in striate cortex between Macaca and Saimiri monkeys following monocular stimulation. Brain Res 170 (2), 353–358.

Herculano-Houzel, S., Collins, C.E., Wong, P., Kaas, J.H., Lent, R., 2008. The basic nonuniformity of the cerebral cortex. Proceedings of the National Academy of Sciences of the United States of America 105 (34), 12593–12598.

Hoerder-Suabedissen, A., Molnár, Z., 2012. Morphology of mouse subplate cells with identified projection targets changes with age. Journal of Comparative Neurology 520 (1), 174–185.

Horton, J.C., Adams, D.L., 2005. The cortical column: A structure without a function. Philosophical Transactions of the Royal Society of London. Series B, Biological Sciences 360 (1456), 837–862.

Horton, J.C., Hocking, D.R., 1998. Monocular core zones and binocular border stripes in primate striate cortex revealed by contrasting effects of enucleation, eyelid suture, and retinal laser lesions on cytochrome oxidase activity. Journal of Neuroscience 15, 5433–5455.

Hubel, D.H., Wiesel, T.N., 1968. Receptive fields and functional architecture of monkey striate cortex. Journal of Physiology 195 (1), 215–243.

Hubel, D.H., Wiesel, T.N., 1969. Anatomical demonstration of columns in the monkey striate cortex. Nature 221, 747–750.

Hubel, D.H., Wiesel, T.N., 1977. Functional architecture of macaque monkey visual cortex. Proceedings of the Royal Society of London, Series B: Biological Sciences 198, 1–59.

Hubel, D.H., Wiesel, T.N., Stryker, M.P., 1977. Orientation columns in macaque monkey visual cortex demonstrated by the 2-deoxyglucose autoradiographic technique. Nature 269 (5626), 328–330.

Hubel, D.H., Wiesel, T.N., Stryker, M.P., 1978. Anatomical demonstration of orientation columns in macaque monkey. Journal of Comparative Neurology 177 (3), 361–380.

Ichinohe, N., Rockland, K.S., 2004. Region specific micromodularity in the uppermost layers in primate cerebral cortex. Cerebral Cortex 14, 1173–1184.

Illing, R.-B., Graybiel, A.M., 1986. Complementary and non-matching afferent compartments in the cat's superior colliculus: Innervation of the acetycholinesterase-poor domain of the intermediate zone. Neuroscience 18, 373–394.

Imig, T.J., Adrián, H.O., 1977. Binaural columns in the primary field (A1) of cat auditory cortex. Brain Res 138 (2), 241–257.

Inan, M., Crair, M.C., 2007. Development of cortical maps: Perspectives from the barrel cortex. The Neuroscientist 13, 49–60.

Innocenti, G.M., Vercelli, A., 2010. Dendritic bundles, minicolumns, columns, and cortical output units. Frontiers in Neuroanatomy 4, 11.

Jain, N., Catania, K.C., Kaas, J.H., 1998. A histologically visible representation of the fingers and palm in primate area 3b and its immutability following long-term deafferentations. Cerebral Cortex 8 (3), 227–236.

Jones, E.G., 1999. Making brain connections: Neuroanatomy and the work of TPS Powell, 1923–1996. Annual Review of Neuroscience 22, 49–103.

Jones, E.G., 2000. Microcolumns in the cerebral cortex. Proceedings of the National Academy of Sciences of the United States of America 97, 5019–5021.

Jones, E.G., 2010. Cellular structure of the human cerebral cortex (Book review). Brain 113, 945–946.

Jones, E.G., Rakic, P., 2010. Radial columns in cortical architecture: It is the composition that counts. Cerebral Cortex 20 (10), 2261–2264 Epub 28 July 2010.

Kaas, J.H., Gharbawie, O.A., Stepniewska, I., 2011. The organization and evolution of dorsal stream multisensory motor pathways in primates. Frontiers in Neuroanatomy 5, 34.

Kandler, K., Katz, L.C., 1998. Coordination of neuronal activity in developing visual cortex by gap junction-mediated biochemical communication. Journal of Neuroscience 18 (4), 1419–1427.

Kaschube, M., Schnabel, M., Löwel, S., Coppola, D.M., White, L.E., Wolf, F., 2010. Universality in the evolution of orientation columns in the visual cortex. Science 330 (6007), 1113–1116.

Keay, K.A., Bandler, R., 2001. Parallel circuits mediating distinct emotional coping reactions to different kinds of stress. Neuroscience and Biobehavioral Reviews 25, 669–678.

Kornack, D.R., Rakic, P., 1995. Radial and horizontal deployment of clonally related cells in the primate neocortex: Relationship to distinct mitotic lineages. Neuron 15, 311–321.

Krieger, P., Kuner, T., Sakmann, B., 2007. Synaptic connections between layer 5B pyramidal neurons in mouse somatosensory cortex are independent of apical dendrite bundling. Journal of Neuroscience 27, 11473–11482.

Larouche, M., Hawkes, R., 2006. From clusters to stripes: The developmental origins of adult cerebellar compartmentation. Cerebellum 5, 77–88.

Lent, R., Azevedo, F.A., Andrade-Moraes, C.H., Pinto, A.V., 2012. How many neurons do you have? Some dogmas of quantitative neuroscience under revision. European Journal of Neuroscience 35 (1), 1–9.

Lev, D.L., White, E.L., 1997. Organization of pyramidal cell apical dendrites and composition of dendritic clusters in the mouse: Emphasis on primary motor cortex. European Journal of Neuroscience 9, 280–290.

LeVay, S., Hubel, D.H., Wiesel, T.N., 1975. The pattern of ocular dominance columns in macaque visual cortex revealed by a reduced silver stain. Journal of Comparative Neurology 159 (4), 559–576.

Liguz-Lecznar, M., Skangiel-Kramska, J., 2007. Vesicular glutamate transporters VGLUT1 and VGLUT2 in the developing mouse barrel cortex. International Journal of Developmental Neuroscience 25 (2), 107–114.

Livingstone, M.S., Hubel, D.H., 1982. Thalamic inputs to cytochrome oxidase-rich regions in monkey visual cortex. Proceedings of the National Academy of Sciences of the United States of America 79 (19), 6098–6101.

Livingstone, M.S., Hubel, D.H., 1984. Specificity of intrinsic connections in primate primary visual cortex. Journal of Neuroscience 4 (11), 2830–2835.

López-Bendito, G., Molnár, Z., 2003. Thalamocortical development: How are we going to get there? Nature Reviews Neuroscience 4 (4), 276–289.

Lorente de Nó, R., 1922. Corteza cerebral del ratón. I. La corteza acústica. Trabajos del Laboratorio de Investigaciones Biológicas 20, 41–78.

Lorente de Nó, R., 1938. Architectonics and structure of the cerebral cortex. In: Fulton, J.F. (Ed.), Physiology of the Nervous System. Oxford University Press, New York, pp. 291–330.

Lorente de Nó, R., 1949. Cerebral cortex: Architecture, intracortical connections, motor projection. In: Fulton, J.F. (Ed.), Physiology of the Nervous System. Oxford University Press, Oxford, pp. 274–313 (first edition was 1938).

Mana, S., Chevalier, G., 2001. Honeycomb-like structure of the intermediate layers of the rat superior colliculus: Afferent and efferent connections. Neuroscience 103, 673–693.

Massing, W., Fleischhauer, K., 1973. Further observations on vertical bundles of dendrites in the cerebral cortex of the rabbit. Zeitschrift für Anatomie und Entwicklungsgeschichte 141, 115–123.

Meyer, G., 1987. Forms and spatial arrangement of neurons in the primary motor cortex of man. Journal of Comparative Neurology 262 (3), 402–428.

Meyer, H.S., Wimmer, V.C., Oberlaender, M., de Kock, C.P.J., Sakmann, B., Helmstaedter, M., 2010. Number and laminar distribution of neurons in a thalamocortical projection column of rat vibrissal cortex. Cerebral Cortex 20 (10), 2277–2286.

Mitchison, G., 1992. Axonal trees and cortical architecture. Trends in Neurosciences 15 (4), 122–126.

Mitrovic, N., Schachner, M., 1996. Transient expression of NADPH diaphorase activity in the mouse whisker to barrel field pathway. Journal of Neurocytology 25 (8), 429–437.

Miyashita, T., Wintzer, M., Kurotani, T., Konishi, T., Ichinohe, N., Rockland, K.S., 2010. Neurotrophin-3 is involved in the formation of apical dendrite bundles in cortical layer 2 of the rat. Cerebral Cortex 20, 229–240.

Molnár, Z., 2011. The need for research on human brain development. Brain 34 (7), 2177–2185.

Molnár, Z., Belgard, T.G., 2012. Transcriptional profiling of layers of the primate cerebral cortex. Neuron 73 (6), 1053–1055.

Molnár, Z., Molnár, E., 2006. Calcium and NeuroD2 control the development of thalamocortical communication. Neuron 49 (5), 639–642.

Montiel, J.F., Wang, W.Z., Oeschger, F.M., et al., 2011. Hypothesis on the dual origin of the Mammalian subplate. Front Neuroanatomy 5, 25.

Mooney, R., Penn, A.A., Gallego, R., Shatz, C.J., 1996. Thalamic relay of spontaneous retinal activity prior to vision. Neuron 17 (5), 863–874.

Mountcastle, V.B., 1957. Modality and topographic properties of single neurons of cat's somatic sensory cortex. Journal of Neurophysiology 20 (4), 408–434.

Mountcastle, V.B., 1997. The columnar organization of the neocortex. Brain 120 (Pt 4), 701–722.

Mountcastle, V.B., Davies, P.W., Berman, A.L., 1957. Response properties of neurons of cat's somatic sensory cortex to peripheral stimuli. Journal of Neurophysiology 20 (4), 374–407.

Noctor, S.C., et al., 2001. Neurons derived from radial glial cells establish radial units in neocortex. Nature 409, 714–720.

Otsuka, T., Kawaguchi, Y., 2008. Firing-pattern-dependent specificity of cortical excitatory feed-forward subnetworks. Journal of Neuroscience 28 (44), 11186–11195.

Perrett, D.I., Hietanen, J.K., Oram, M.W., Benson, P.J., 1992. Organization and functions of cells responsive to faces in the temporal cortex. Philosophical Transactions of the Royal Society of London: B Biology Sciences 335 (1273), 23–30.

Peters, A., 1997. Pyramidal cell modules in cerebral cortex. In: Motta, P. M. (Ed.), Recent Advances in Microscopy of Cells, Tissues and Organs. A. Delfino, Rome, pp. 187–192.

Peters, A., Kara, D.A., 1987. The neuronal composition of area 17 of rat visual cortex. IV. The organization of pyramidal cells. Journal of Comparative Neurology 260, 573–590.

Peters, A., Sethares, C., 1996. Myelinated axons and the pyramidal cell modules in monkey primary visual cortex. Journal of Comparative Neurology 365, 232–255.

Peters, A., Walsh, M.A., 1972. Study of the organization of apical dendrites in the somatic sensory cortex of the rat. Journal of Comparative Neurology 144, 253–268.

Piñon, M.C., Jethwa, A., Jacobs, E., Campagnoni, A., Molnár, Z., 2009. Dynamic integration of subplate neurons into the cortical barrel field circuitry during postnatal development in the Golli-tau-eGFP (GTE) mouse. Journal of Physiology 587 (Pt 9), 1903–1915.

Powell, T.P., Hendrickson, A.E., 1981. Similarity in number of neurons through the depth of the cortex in the binocular and monocular parts of area 17 of the monkey. Brain Research 216 (2), 409–413.

Powell, T.P., Mountcastle, V.B., 1959. Some aspects of the functional organization of the cortex of the postcentral gyrus of the monkey: A correlation of findings obtained in a single unit analysis with cytoarchitecture. Bulletin of the Johns Hopkins Hospital 105, 133–162.

Purves, D., Riddle, D.R., LaMantia, A.S., 1992. Iterated patterns of brain circuitry (or how the cortex gets its spots). Trends in Neurosciences 15 (10), 362–368.

Rakic, P., 1988. Specification of cerebral cortical areas. Science 241, 170–176.

Rakic, P., 2007. The radial edifice of cortical architecture: From neuronal silhouettes to genetic engineering. Brain Research Reviews 55, 204–219.

Rakic, P., 2009. Evolution of the neocortex: A perspective from developmental biology. Nature Reviews Neuroscience 10 (10), 724–735.

Rakic, P., Caviness Jr., V.S., 1995. Cortical development: View from neurological mutants two decades later. Neuron 14 (6), 1101–1104.

Rausell, E., Jones, E.G., 1991. Chemically distinct compartments of the thalamic VPM nucleus in monkeys relay principal and spinal trigeminal pathways to different layers of the somatosensory cortex. Journal of Neuroscience 11, 226–237.

Rice, F.L., 1995. Comparative aspects of barrel structure and development. In: Peters, A., Jones, E.G., Diamond, I.T. (Eds.), Cerebral Cortex: The Barrel Cortex of Rodents, vol. 11. Plenum Press, New York, pp. 1–76.

Risold, P.Y., Swanson, L.W., 1998. Connections of the rat lateral septal complex. Brain Research Reviews 24, 115–195.

Rockel, A.J., Hiorns, R.W., Powell, T.P., 1974. Numbers of neurons through full depth of neocortex. Journal of Anatomy 118 (Pt 2), 371.

Rockel, A.J., Hiorns, R.W., Powell, T.P., 1980. The basic uniformity in structure of the neocortex. Brain 103 (2), 221–244.

Rockland, K.S., 2010. Five points on columns. Frontiers in Neuroanatomy 4, 22.

Rockland, K.S., Ichinohe, N., 2004. Some thoughts on cortical minicolumns. Experimental Brain Research 158, 265–277.

Roney, K.J., Scheibel, A.B., Shaw, G.L., 1979. Dendritic bundles: Survey of anatomical experiments and physiological theories. Brain Research 180 (2), 225–271.

Sato, T.R., Gray, N.W., Mainen, Z.F., Svoboda, K., 2007. The functional microarchitecture of the mouse barrel cortex. PLoS Biology 5, e189. http://dx.doi.org/10.1371/journal.pbio.0050189.

Shepherd, G., Grillner, S., 2010. Handbook of Brain Microcircuits. Oxford University Press, New York pp. 544.

Sillitoe, R.V., Joyner, A.L., 2007. Morphology, molecular codes, and circuitry produce the three-dimensional complexity of the cerebellum. Annual Review of Cell and Developmental Biology 23, 549–577.

Song, S., Sjöström, P.J., Reigl, M., Nelson, S., Chklovskii, D.B., 2005. Highly nonrandom features of synaptic connectivity in local cortical circuits. PLoS Biology 3 (3), e68 Epub 1 March 2005.

Spatz, W.B., 1989. Loss of ocular dominance columns with maturity in the monkey, *Callithrix jacchus*. Brain Research 488 (1–2), 376–380.

Takahata, T., Higo, N., Kaas, J.H., Yamamori, T., 2009. Expression of immediate-early genes reveals functional compartments within ocular dominance columns after brief monocular inactivation. Proceedings of the National Academy of Sciences of the United States of America 106, 12151–12155.

Tan, S.S., Breen, S., 1993. Radial mosaicism and tangential cell dispersion both contribute to mouse neocortical development. Nature 362, 638–640.

Torii, M., Hashimoto-Torii, K., Levitt, P., Rakic, P., 2009. Integration of neuronal clones in the radial cortical columns by EphA and ephrin-A signalling. Nature 461 (7263), 524–528.

Tunturi, A.R., 1950. Physiological determination of the arrangement of the afferent connections to the middle ectosylvian auditory area in the dog. American Journal of Physiology 162 (3), 489–502.

Van der Loos, H., Woolsey, T.A., 1973. Somatosensory cortex: Structural alterations following early injury to sense organs. Science 179 (71), 395–398.

Vercelli, A., Garbossa, D., Curtetti, R., Innocenti, G.M., 2004. Somatodendritic minicolumns of output neurons in the rat visual cortex. European Journal of Neuroscience 20, 495–502.

von Economo, C., Koskinas, G.N., 2008. Atlas of the Cytoarchitectonics of the Adult Human Cerebral Cortex. Karger, Basel (translated, revised and edited by L. C. Triarhou).

Voogd, J., Glickstein, M., 1998. The anatomy of the cerebellum. Trends in Neurosciences 21, 370–375.

Wang, Y., Brzozowska-Prechtl, A., Karten, H.J., 2010. Laminar and columnar auditory cortex in avian brain. Proceedings of the National Academy of Sciences of the United States of America 107 (28), 12676–12681.

Welker, W.I., Johnson Jr., J.I., Pubols Jr., B.H., 1964. Some morphological and physiological characteristics of the somatic sensory system in raccoons. American Zoologist 4, 75–94.

Wimmer, V.C., Bruno, R.M., de Kock, C.P.J., Kuner, T., Sakmann, B., 2010. Dimensions of a projection column and architecture of VPM and POm axons in rat vibrissal cortex. Cerebral Cortex 20 (10), 2265–2276.

Wong-Riley, M., 1979. Changes in the visual system of monocularly sutured or enucleated cats demonstrable with cytochrome oxidase histochemistry. Brain Res 171 (1), 11–28.

Woolsey, T.A., Van der Loos, H., 1970. The structural organization of layer IV in the somatosensory region (SI) of mouse cerebral cortex. The description of a cortical field composed of discrete cytoarchitectonic units. Brain Research 17 (2), 205–242.

Woolsey, T.A., Welker, C., Schwartz, R.H., 1975. Comparative anatomical studies of the SmL face cortex with special reference to the occurrence of 'barrels' in layer IV. Journal of Comparative Neurology 164 (1), 79–94.

Yoshimura, Y., Dantzker, J.L., Callaway, E.M., 2005. Excitatory cortical neurons form fine-scale functional networks. Nature 433 (7028), 868–873.

Yu, Y.C., Bultje, R.S., Wang, X., Shi, S.H., 2009. Specific synapses develop preferentially among sister excitatory neurons in the neocortex. Nature 458, 501–504.

Yuste, R., Peinado, A., Katz, L.C., 1992. Neuronal domains in developing neocortex. Science 257 (5070), 665–669.

Neonatal Cortical Rhythms

R. Khazipov[1,2], M. Colonnese[1], M. Minlebaev[1,2]

[1]INSERM U901, University Aix-Marseille II, Marseille, France; [2]Kazan Federal University, Kazan, Russia

Nomenclature

AMPA α-Amino-3-hydroxyl-5-methyl-4-isoxazole-propionate
DC Direct coupled
EEG Electroencephalography
ENO Early network oscillations
GABA γ-Aminobutyric acid
GDP Giant depolarizing potential
LGN Lateral geniculate nucleus
MEG Magnetoencephalography
NMDA *N*-Methyl-D-aspartic acid
ODCs Ocular dominance columns
P Postnatal day
R Receptor
S1 Primary somatosensory cortex
SAT Slow activity transient
STDP Spike-time-dependent plasticity
V1 Primary visual cortex
μV Microvolt

8.1 INTRODUCTION

The fetal period in humans is characterized by a number of fundamental events in the construction of the nervous system, such that at birth, many of the primary circuits already have been formed and display remarkable functional performance, although development evidently continues after birth until full maturity is reached at around age 30. Considerable evidence indicates that electrical activity expressed in the human fetal brain – and in lower mammals at corresponding developmental stages – controls a number of developmental processes, including neuronal differentiation, migration, synaptogenesis, and synaptic plasticity (for review, see Ben-Ari et al., 1997; Blankenship and Feller, 2010; Feldman et al., 1999; Feller and Scanziani, 2005; Fox, 2002; Henley and Poo, 2004; Katz and Crowley, 2002; Katz and Shatz, 1996; Rakic and Komuro, 1995; and Zhou and Poo, 2004a). Probably the most thoroughly elaborated evidence has been generated by studying sensory cortices, in which development of sensory maps is critically influenced by activity from the sensory periphery. However, the physiology of the fetal central nervous system, and notably the electrical patterns of organized neuronal activity that underlie map formation, has remained obscure for a long time. This is mainly a result of technical limitations in recording electrical activity from the fetal brain in utero. An important and almost paradoxical aspect of the problem is that the fetus develops

© 2013 Elsevier Inc. All rights reserved.

in utero under conditions of virtually complete sensory deprivation. Knowing the importance of input from the sensory periphery for development of the nervous system raises an important question: What are the mechanisms that provide sensory stimulation to the developing sensory systems *in utero*? This question is accompanied by a number of related issues. For example, how are the early sensory inputs processed in developing circuits, and what are the specific activity patterns associated with the activity-dependent formation of the cortical maps? How are these early activity patterns generated, and how are they transformed to the mature mode of sensory processing necessary to support behavioral function?

In this chapter, we attempt to answer these questions by reviewing experimental evidence obtained in the premature human neonate and in the postnatal rodent, which seems to be an excellent model for studying the processes that occur during the human fetal development. In both systems, endogenous mechanisms for the activation of sensory pathways exist in two developing sensory systems: the somatosensory and the visual. In the somatosensory system, sensory input is generated by sensory feedback resulting from spontaneous movements. In the visual system, it is provided by spontaneous retinal waves generated in the initially light-insensitive retina. In both systems, this endogenously activated "sensory" input drives oscillatory bursts in thalamocortical networks. These central oscillations occur in a topographic manner and thus provide binding between the aligned elements of sensory circuits to create conditions for the activity-dependent formation of cortical maps. The generation of the early oscillatory patterns, which primarily include oscillations in alpha–beta and gamma frequency ranges, involves glutamatergic synapses including an input from the thalamus. The fast rhythmic components of early activities are generated primarily by α-amino-3-hydroxyl-5-methyl-4-isoxazole-propionate (AMPA) receptors at glutamatergic synapses, whereas *N*-methyl-D-aspartic acid (NMDA) receptors, also activated during the early oscillatory bursts due to summation of rhythmic input, provide conditions for NMDA-R-dependent synaptic plasticity. The local nature of early activity is generated as a result of the topographic organization of the thalamic input. Spread of early activity is limited by the delayed maturation of the long-range horizontal connections in the cortex and by surround inhibition provided by GABAergic interneurons even at a very early age. We propose a model in which the early oscillatory patterns shape early circuits according to spike-time-dependent plasticity (STDP) in localized regions.

We will finally discuss how early cortical activity patterns are related to the development of sensory signal processing and explorative functions. The main function of our brain consists of the online processing of the input from the external world and the body and the generation of an output according to previous experience and prediction. Although the early oscillatory bursts are reliably activated by endogenous mechanisms, these activity patterns are poorly suited for the exploration of the environment. Early in development, external sensory stimuli evoke all-or-none oscillatory bursts similar to those triggered by endogenously activated sensory input; however, this paradigm enables only a primitive form of sensation, not the high-frequency graded sensory processing required for explorative functions. With maturation, early oscillatory bursts disappear in association with a rapid developmental switch in the mode of sensory processing from bursting to "acuity." In the visual system, this switch occurs shortly before the onset of patterned sensory input – birth in the human and eye opening in the rat – in association with an emergence of the active cortical state. A similar switch also occurs in the somatosensory whisker-related rat barrel cortex just before the onset of active whisking. Thus, the early bursting mode of sensory signal processing is related to the development of the sensory cortex, but not to the exploration of the external world. This suggests an inside-out development of the sensory cortex, which initially is tuned to the internally generated activity at the sensory interface and serves to embed it into topographically organized sensory cortical maps and, once this is achieved and the maps are well tuned, switches to the exploration mode, enabling reliable high-frequency processing of sensory signals from the environment.

8.2 NEOCORTICAL PATTERNS IN PREMATURE HUMAN NEONATES

Determining the patterns of activity expressed in the fetal brain is a challenging task. While the standard way to characterize electrical brain activity is to record the electrical signals from the head surface using scalp electroencephalography (EEG), such recordings cannot be performed in the fetus (although some attempts have been made to record the EEG and the magnetoencephalograph (MEG) through the mother's abdomen). At present, the main approach to fetal brain activity consists of scalp EEG recordings from premature neonates. In this group of patients surveyed in specialized intensive care units, neurophysiological monitoring is a part of the standard patient examination. Evidently, as the environment is very different and the functions of many systems (cardiovascular, respiratory, and digestive) undergo significant development after birth, a question could be raised as to whether cortical activity in premature neonates is the same as in the fetus. For the most part, the answer is 'yes,' because EEG development is largely age-dependent, and the EEG patterns expressed in premature neonates match the gestational age, but not the actual postnatal age. Moreover, studies using magneto-encephalography in the fetus *in utero* document a remarkable similarity with the temporal organization of activity and electrographic patterns observed in the gestational age-matching premature neonates. Therefore, scalp EEG from premature neonates is today considered a reliable approach to measuring fetal brain function.

Characteristic adult EEG patterns emerge essentially during the postnatal period and undergo pronounced changes in the amplitude and distribution of oscillations in different frequency bands until age 30 (Niedermeyer and Da Silva, 2005; Uhlhaas et al., 2010). Thus, the premature infant EEG displays a number of unique activity patterns. These include a number of distinct transient periods of rhythmic activity and intermittent sharp events that are expressed during certain periods of development (Anderson et al., 1985; Lamblin et al., 1999; Scher, 2006; Stockard-Pope et al., 1992). At mid-gestation, activity is dominated by intermittent delta waves from 0.3 to 2 Hz. By the 7th month, slow oscillations become intermixed with rapid rhythms. During the second half of gestation, the dominant EEG pattern in central, temporal, and rostral regions is the delta-brush (Anderson et al., 1985; Lamblin et al., 1999; Scher, 2006; Stockard-Pope et al., 1992) (Figure 8.1). Different terms have also been used in the literature to describe this pattern including spindle-shaped bursts of fast activity (Ellingson, 1958), rapid rhythm (Dreyfus-Brisac, 1962; Nolte et al., 1969; Parmelee et al., 1969), rapid bursts (Dreyfus-Brisac, 1962), spindle-like fast (Watanabe and Iwase, 1972), fast activity at 14–24 Hz (Goldie et al., 1971), and ripples of prematurity (Engel, 1975). A delta-brush consists of 8–25 Hz spindle-like, rhythmic activity superimposed on a delta wave. Often, delta-brushes occur in a sequence, and these grouped delta-brush activities may stay for up to 10 s, giving rise to so-called slow activity transients, or SATs, that, in direct coupled (DC) recordings, attain unusually large amplitudes of up to 800 μV (Vanhatalo et al., 2002, 2005). Such DC shifts are filtered and therefore not observed using conventional high-pass (>0.5–1 Hz) EEG recordings. Delta brushes are expressed in all cortical areas and fade near term. While resembling sleep spindles in some ways, delta brushes and sleep spindles appear to be distinct patterns. Sleep spindles emerge only during the second postnatal month.

Besides delta-brushes, other patterns expressed in the premature brain include the neonatal 'delta crest' (isolated frontopolar delta waves with superimposed fast activity), midline frontal theta–alpha burst, EEG spikes and sharp transients, anterior slow dysrhythmia, and temporal sawtooth or temporal theta bursts (Anderson et al., 1985; Lamblin et al., 1999; Scher, 2006; Stockard-Pope et al., 1992). Because nearly all information about the earliest cortical patterns in humans derives from scalp-recorded EEGs of premature infants, these

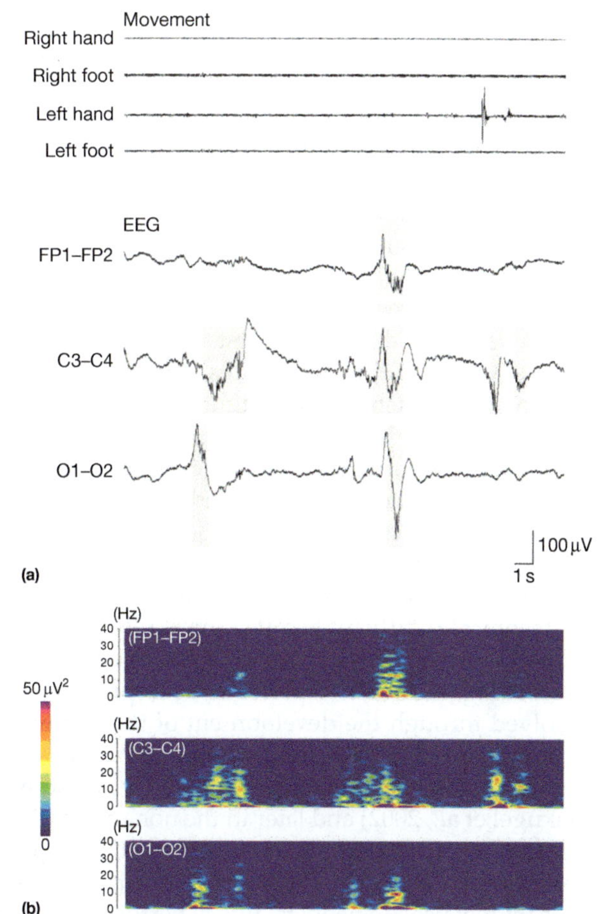

FIGURE 8.1   Delta-brushes in the human preterm neonate. (a) Representative example of three simultaneous EEG traces recorded in bipolar transversal montage (frontal FP1–FP2, central C3–C4, and occipital O1–O2) during quiet sleep in a 30-week, postconceptional-age neonate. Bursts of delta waves alternate with periods of hypoactivity. Delta-brushes are characterized by alpha–beta oscillations superimposed on delta waves (gray squares). Traces above show concomitant hand- and foot-movement recordings. (b) Wavelet analysis of bipolar EEG recordings shown in (a). *Adapted with permission from Milh M, Kaminska A, Huon C, Lapillonne A, Ben Ari Y, Khazipov R (2007) Rapid cortical oscillations and early motor activity in premature human neonate. Cerebral Cortex 17: 1582–1594.*

observations alone do not inform us whether they represent pathological activity of the immature brain or the normal physiological patterns of developing neuronal networks. Addressing these issues requires simultaneous recording of neuronal spiking and EEG activity in intact developing tissue.

8.3 THE FIRST ORGANIZED CORTICAL NETWORK PATTERNS IN THE NEONATAL RODENT

Offspring of rats and mice are altricial, that is, they are born in an immature state (Clancy et al., 2001). Although it is difficult to provide exact comparisons between humans and rodents, the level of rat brain development at the day of birth (P0) can be roughly compared to the state of human cortex at mid-gestation, and term in humans roughly corresponds to the postnatal day P12 in the rat or mouse. Therefore, postnatal rodents could be an excellent model to study the developmental events that occur during the human fetal period. However, despite a rich repertoire of activity patterns observed in preterm human neonates during the second half of gestational age, until recently no organized brain activity had been reported during the first ten postnatal days in rodents. Based on EEG recordings, organized cortical activity in infant rodents was thought to commence with the emergence of delta waves, starting from P11 (Frank and Heller, 1997; Gramsbergen, 1976; Jouvet-Mounier et al., 1970). This stands in contrast with a number of patterns of correlated activity found in the neonatal cortical slices *in vitro* (for review, see Allene and Cossart, 2010). It should be noted that recording activity in neonatal rats is technically difficult because the skull is cartilaginous, making mechanically stable, movement-artifact-free recordings difficult to perform. This problem has been solved through the development of the technique of recordings from head-restrained animals with a dental cement-enforced head cup, first under anesthesia (Leinekugel et al., 2002) and later in the nonanesthetized (Khazipov et al., 2004b) and decerebrated rat pups (Karlsson and Blumberg, 2005). At present, this technique is widely used in neonatal rats to perform intracortical recordings of the local field potential and multiple units using electrode arrays, patch-clamp recordings from individual neurons, and imaging (Colonnese et al., 2010; Mohns and Blumberg, 2008; Sipila et al., 2006; Yang et al., 2009). Using this technique, several activity patterns had been described in neonatal rodents, revealing a remarkable similarity to the activities seen in human premature neonates and enabling investigation of their underlying mechanisms. These patterns so far have been thoroughly investigated in only the sensory cortices (notably somatosensory and visual areas) and the hippocampus.

8.3.1 Spindle- and Gamma-Bursts in Somatosensory Cortex

In the neonatal rat somatosensory cortex, formation of the body map representation occurs during the first postnatal week. Because modulation of sensory-driven experience strongly influences this process during this time, this period is also known as the critical period of the somatosensory map development. During this period, thalamocortical axons grow into the neocortex and form input-specific patterns of synapses with neocortical neurons (e.g., the whisker-specific-barrel pattern

in the barrel cortex) (Erzurumlu and Jhaveri, 1990; Higashi et al., 2002; Molnar et al., 2003; Petersen, 2007; Price et al., 2006; Woolsey and Van Der Loos, 1970). Manipulations of the peripheral receptors, or of cortical activity, during this critical period can disrupt the formation of thalamocortical synapses (Cases et al., 1996; Catalano and Shatz, 1998; Feldman et al., 1999; Foeller and Feldman, 2004; Fox, 1992, 2002; Fox and Wong, 2005; Lu et al., 2006; O'Leary et al., 1994; Persico et al., 2001; Van der Loos and Woolsey, 1973; Woolsey and Wann, 1976). Pharmacological or genetic manipulations associated with a loss of function of NMDA-Rs result in malformations of barrel cortex development and functional deficits (Dagnew et al., 2003; Fox, 2002; Fox et al., 1996; Iwasato et al., 2000; Lee et al., 2005a,b; Schlaggar et al., 1993). During the critical period, thalamocortical synapses display an enhanced NMDA-R contribution and increased NMDA-R-dependent synaptic plasticity, including the conversion of 'silent' pure NMDA-R-based synapses to fully functional mixed AMPA/NMDA-R synapses, as well as switching to fast AMPA-receptor-mediated synaptic transmission from slow kainate-mediated transmission (Bannister et al., 2005; Barth and Malenka, 2001; Carmignoto and Vicini, 1992; Crair and Malenka, 1995; Daw et al., 2006; Feldman et al., 1998, 1999; Hestrin, 1992; Isaac et al., 1995, 1997; LoTurco et al., 1991; Monyer et al., 1994).

What are the patterns of cortical activity that underlie this activity-dependent plasticity during this critical period? Extracellular and patch-clamp recordings from rats during the first postnatal week revealed two predominant organized patterns of activity in the somatosensory neocortex: so-called spindle-bursts and gamma-bursts (Khazipov et al., 2004b; Marcano-Reik and Blumberg, 2008; Marcano-Reik et al., 2010; Minlebaev et al., 2007, 2009; Seelke and Blumberg, 2010; Yang et al., 2009) (Figure 8.2). Both are transient local oscillatory events, with a difference in the dominant frequency of oscillation and in the size of the recruited network. A spindle-burst is a transient burst of rhythmic, 5–25 Hz activity with a duration of approximately 1 s and a recruiting cortical zone of approximately a half millimeter. Gamma-bursts (40–50 Hz) are typically shorter in duration (150–300 ms) and are more local (~200 μm) events. Spindle bursts and gamma bursts may also intermingle. Recordings in the DC mode, without high-pass filtering of the signal, revealed that spindle bursts are associated with relatively large (up to hundreds of μV) negative delta waves. A tight temporal and quantitative correlation between the power of the alpha–beta oscillations and the time course and amplitude of the delta wave, a similar depth profile and location of the major current sinks in the dense cortical plate, and involvement (though differential, see below) of ionotropic glutamate receptors in the generation of

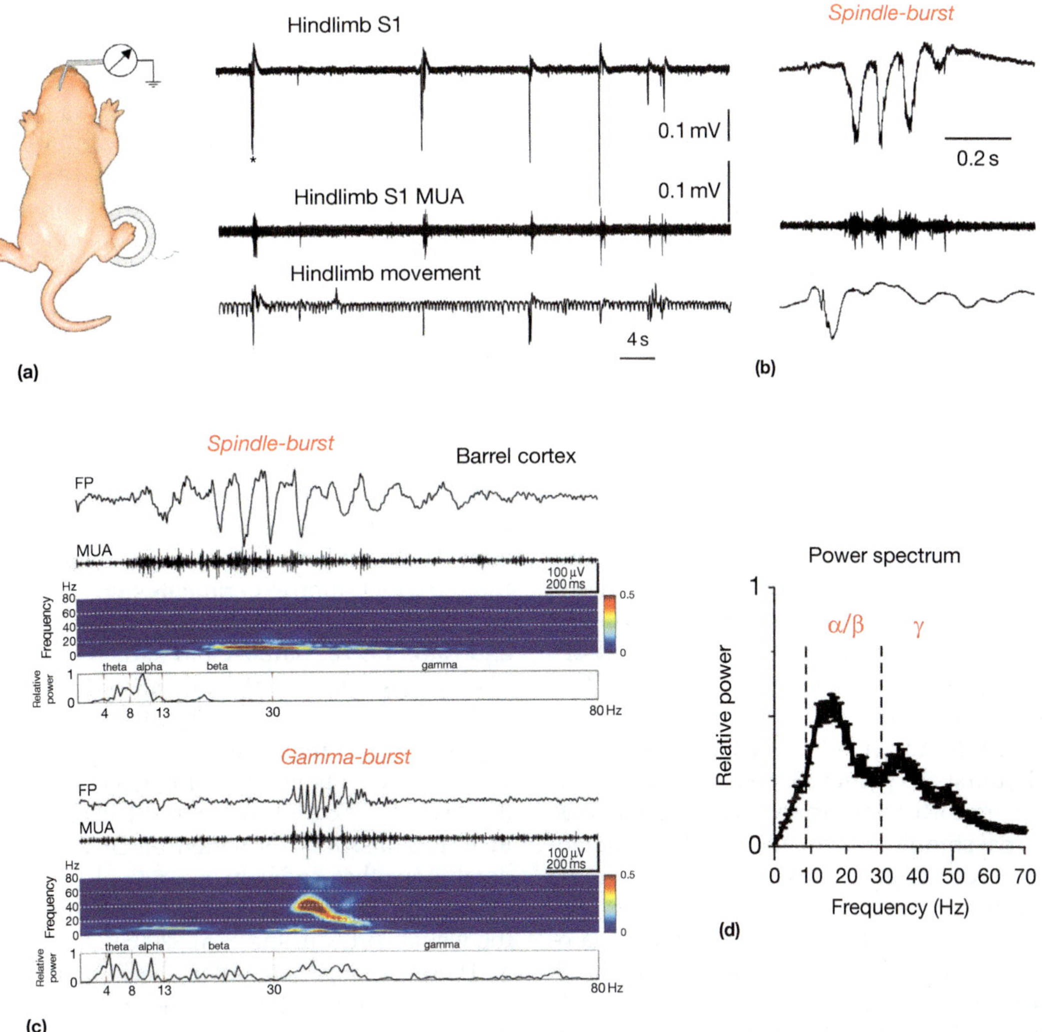

FIGURE 8.2   Spindle- and gamma-bursts in primary somatosensory cortex (S1) of the newborn rat. (a) Wide-band recordings of extracellular activity and filtered (0.3–5 kHz) MUA in S1 hindlimb area of a P2 rat. Positivity is up. Bottom, movement of the contralateral hindlimb. Continuous rhythm reflects respiration. Note that field events and synchronized unit bursts are associated with movements. The event marked by * is shown at an expanded time scale in (b). (c) Examples of spindle-bursts (top panel, P1) and gamma-bursts (bottom panel, P3) in the neonatal rat barrel cortex. Below the traces are shown color-coded wavelet spectra and relative powers in different frequency domains. (d) Average power spectrum of the bursts recorded in neonatal rat barrel cortex (pooled data from 14 P2–7 rates, total 499 bursts). Note two peaks at alpha–beta (8–30 Hz; spindle-bursts) and gamma (30–80 Hz, gamma-bursts) frequency ranges. *Adapted with permission from (a, b): Khazipov R, Sirota A, Leinekugel X, Holmes GL, Ben Ari Y, Buzsaki G (2004b) Early motor activity drives spindle bursts in the developing somatosensory cortex. Nature 432: 758–761; (c) Yang JW, Hanganu-Opatz IL, Sun JJ, Luhmann HJ (2009) Three patterns of oscillatory activity differentially synchronize developing neocortical networks in vivo. Journal of Neuroscience 29: 9011–9025; and (d) Minlebaev M, Ben-Ari Y, Khazipov R (2007) Network mechanisms of spindle-burst oscillations in the neonatal rat barrel cortex in vivo. Journal of Neurophysiology 97: 692–700.*

both components, indicate that the high-frequency oscillations and the delta waves are two components of a single activity (Minlebaev et al., 2009). The identification of the slow component of spindle-bursts supports their homology with the human electrographic pattern of delta brushes, which are also expressed in somatosensory cortical areas of human premature neonates during the second half of gestation. The remarkable similarities between these two patterns indicate that they are the same physiological phenomenon (Khazipov and Luhmann, 2006). This also confirms that

the neonatal rodent can be a useful model for studying the mechanism and physiological roles of this activity pattern in cortical development.

Spindle bursts share some similar electrographic characteristics with adult sleep spindles (Steriade, 2001). However, in contrast to sleep spindles, neonatal spindle bursts are local events with a limited tendency to spread. Furthermore, spindle bursts are present in the waking pup even during walking and feeding and are typically triggered by myoclonic twitches of isolated muscles or whole-body startles. Myoclonic twitches are one of the most remarkable developmental motor phenomena in the neonatal rat (Blumberg and Lucas, 1994; O'Donovan, 1999; Petersson et al., 2003), human fetus, and premature neonate (Cioni and Prechtl, 1990; de Vries et al., 1982; Hamburger, 1975; Prechtl, 1997; and is discussed extensively in Rubenstein and Rakic, 2013.). This particular type of motor activity results from the stochastic bursts of activity generated in the spinal cord under brainstem control (Blumberg and Lucas, 1994; Karlsson et al., 2005; Kreider and Blumberg, 2000). Delay between the movements and cortical spindle bursts, and the observation that spindle bursts can also be induced by direct sensory stimulation, indicated that spindle bursts are triggered by sensory feedback initiated by spontaneous movements (Figure 8.2(a)). At present, whether gamma bursts are initiated by sensory feedback is unknown, but both spindle bursts and gamma bursts are efficiently triggered by external stimulation (Khazipov et al., 2004b; Marcano-Reik and Blumberg, 2008; Marcano-Reik et al., 2010; Minlebaev et al., 2007; Yang et al., 2009). Importantly, spindle bursts and gamma bursts persist after sensory deafferentation (e.g., spinal cord transection or application of local anesthetics), although at a reduced frequency (Khazipov et al., 2004b; Yang et al., 2009). These results suggest that spindle bursts and gamma bursts are endogenous – probably thalamocortical or intracortical – oscillations. However, external stimuli brought about by the thalamocortical afferents can trigger these oscillations in the somatosensory cortex in a somatotopic manner.

In addition to gamma and spindle bursts, the somatosensory cortex of rodents during the first postnatal week also displays sharp wave transients (Khazipov et al., 2004b; Seelke and Blumberg, 2010) that may reflect sensory feedback events resulting from brief movements that failed to initiate oscillatory bursts. Long oscillations lasting for >40 s and reminiscent of SATs have also been described, although these events are rare in the somatosensory cortex (Yang et al., 2009).

8.3.2 Spindle Bursts and SATs in Visual Cortex

Despite the similarities between the functional organization of somatosensory and visual systems during the first postnatal week, there is an important difference. In the somatosensory system, sensory stimulation reliably evokes cortical responses starting from near birth – as soon as thalamic axons enter the cortex; in the visual system, however, the retina is insensitive to light during the first postnatal week. During this developmental period, when the visual cortical map is formed, the retina generates spontaneous waves of activity (see Rubenstein and Rakic, 2013). These waves are generated in the network of retinal ganglion and amacrine cells, and locally synchronize retinal activity local domains (Galli

and Maffei, 1988; Meister et al., 1991; Torborg and Feller, 2005; Wong et al., 1993). Using an original *in vitro* preparation of the neonatal mouse intact retina – lateral geniculate nucleus (LGN) thalamic nucleus, it was demonstrated that spontaneous retinal activity is transmitted via the optic nerve to the LGN, where it drives bursts of activity (Mooney et al., 1996). Modulation of retinal waves during the first postnatal week results in alteration of retinal projections to their subcortical targets, suggesting an instructive role for retinal waves in the development of retinogeniculate connectivity (Chandrasekaran et al., 2005; Grubb et al., 2003; McLaughlin et al., 2003; Mrsic-Flogel et al., 2005; Muir-Robinson et al., 2002; Nicol et al., 2007; Penn et al., 1998; Shatz and Stryker, 1988; Stellwagen and Shatz, 2002). Evidence also exists for the contribution of retinal waves to cortical development. In monkeys, ocular dominance columns (ODCs) are formed already *in utero* before visual experience (Rakic, 1976). Although enucleation experiments suggest that retinal input may not be required for the formation of ODCs (Crowley and Katz, 1999), complete blockade of retinal activity can disturb segregation of thalamocortical connections in ODCs (Stryker and Harris, 1986). In neonatal mice, suppression of retinal waves during the first postnatal week also results in imprecise geniculocortical mapping (Cang et al., 2005). Furthermore, blockade of retinal waves in ferrets disrupts formation of ODCs. Together, these data show that spontaneous retinal waves are also involved in the development of thalamic connections to the visual cortex (Cang et al., 2005).

These findings suggested the hypothesis that retinal waves are transmitted to and trigger activity in the developing visual cortex. This hypothesis has been tested in neonatal rats (Colonnese and Khazipov, 2010; Hanganu et al., 2006). Using extracellular and whole-cell recordings from the visual cortex of neonatal rats *in vivo*, it was shown that, as in the somatosensory cortex, the dominant pattern of activity in the visual cortex during the first postnatal week is a spindle burst (Figure 8.3). Simultaneous recordings from the retina and the primary visual (V1) cortex revealed a strong correlation between spindle bursts in the visual cortex and spontaneous retinal waves. In addition, V1 spindle bursts could be reliably evoked by direct stimulation of the optic nerve. Pharmacological modulation of retinal activity affected the rate of occurrence of V1 spindle bursts; for instance, intraocular forskolin injection, known to increase the frequency and amplitude of retinal waves (Tsai et al., 1987), greatly increased the rate of occurrence of cortical spindle bursts in the contralateral V1 cortex. On the other hand, blocking the propagation of retinal activity with local application of tetrodotoxin or removing the retina resulted in a twofold reduction of V1 spindle-burst frequency, analogous to the reduction of somatosensory spindles after spinal cord transections.

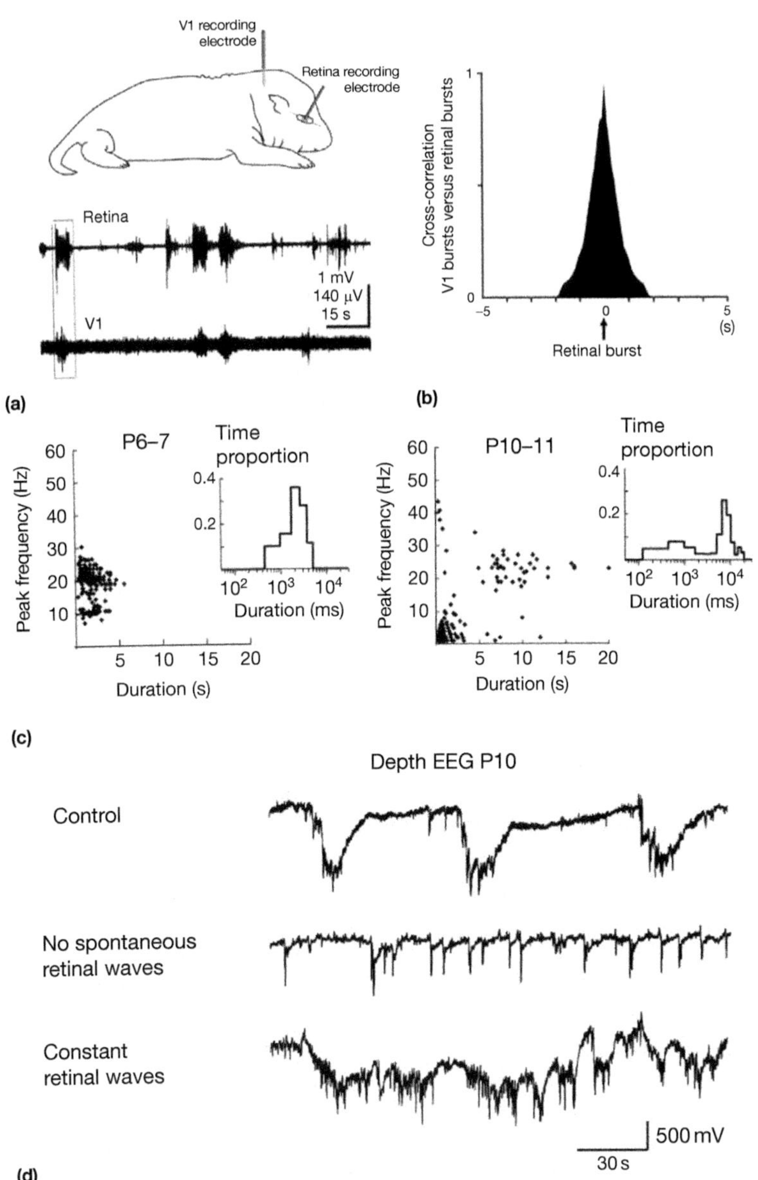

FIGURE 8.3 Retinal drive of visual cortex activity in neonatal rats. (a) Simultaneous recordings of retinal and cortical activity during the first postnatal week reveal synchrony of retinal bursts and cortical spindle-bursts. (b) Cross-correlation of cortical and retinal activity bursts. (c) Diversification of activity patterns during the second postnatal week. Spontaneous activity P6–7 forms a single distribution of duration and presence of rapid oscillations (spindles). By P10–11, elongation of bursts containing rapid oscillations (SATs) and the emergence of short-activity bursts with no oscillatory component lead to two clearly separable activity patterns. (d) Manipulation of spontaneous retinal waves specifically affects SATs, either eliminating them when retinal waves are blocked (enucleation or intra-ocular urethane injection) or increasing their occurrence when retinal activity is increased (retinal inhibition blockade). *Adapted with permission from Hanganu IL, Ben Ari Y, Khazipov R (2006) Retinal waves trigger spindle bursts in the neonatal rat visual cortex.* Journal of Neuroscience 26: 6728–6736 (a, b) *and Colonnese MT and Khazipov R (2010) "Slow activity transients" in infant rat visual cortex: A spreading synchronous oscillation patterned by retinal waves.* Journal of Neuroscience 30: 4325–4337 (c, d).

During the second postnatal week, when the retina becomes responsive to light, spontaneous activity in the visual cortex starts to diversify. Spindle bursts start to group in long-lasting events associated with large amplitude, slow (5–10 s long) DC shifts that are highly reminiscent of SATs in the human premature neonate (Figures 8.3 and 8.4). SATs in the rat are completely eliminated by enucleation. In addition, their main characteristics (duration, interevent intervals, and subburst structure) exactly match those of phase III retinal waves recorded *in vitro*. This indicates that SATs are also driven by retinal waves, as are spindle bursts during the first postnatal week. In parallel, starting from postnatal day P9 a separate class of events of short duration (200 ms) emerges, reminiscent of cortical up-states, which also start to be seen at around the same age in cortical slices *in vitro* (Allene et al., 2008; Rheims et al., 2008). Similarly, in postnatal day P22–39 ferrets, V1 multiple unit activity is organized in bursts with about 10-Hz intraburst multi-unit activity (MUA). Studied with multielectrode arrays, this bursting activity exhibited a patchy structure that reflects ODCs in the visual cortex (Chiu and Weliky, 2001, 2002).

In parallel with the electrophysiological discovery of the early oscillatory patterns, a related pattern, termed early network oscillations (ENOs), has been described using calcium imaging of large neuronal populations *in vivo* (Adelsberger et al., 2005). ENOs are characterized by synchronous intracellular calcium increases that last for about 1 s and recur at about 10-s intervals in P3–4 rats, similar to spindle-bursts. Recorded in the temporal cortex, ENOs mainly occurred during movement-free

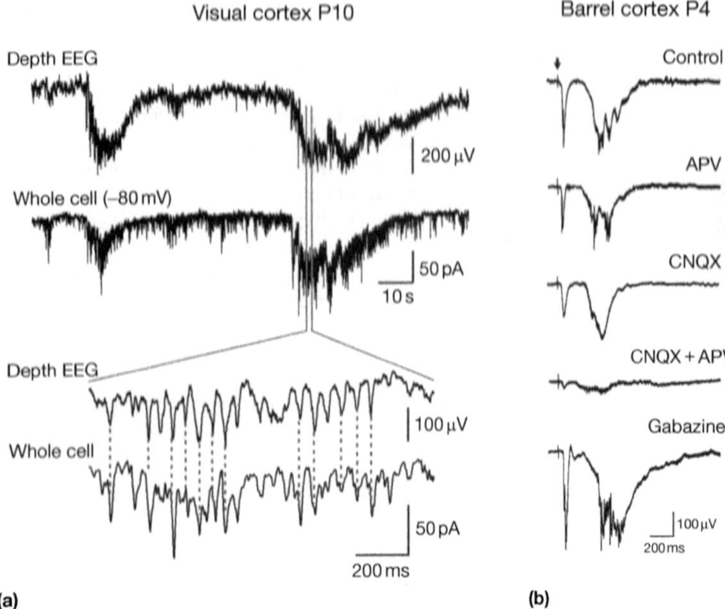

FIGURE 8.4 The mechanisms of the early oscillatory bursts. (a) Simultaneous depth EEG (field-potential, top trace) and whole-cell (400 µm depth, bottom trace) recordings from P10 rat visual cortex show that the infra-slow wave of SATs is composed of a similar long-duration depolarizing current. At expanded time base, rapid oscillations in the field potential are closely correlated with excitatory synaptic currents. (b) Pharmacological analysis of spindle-bursts evoked by electrical whisker pad stimulation in the P4 rat barrel cortex. Note that rapid oscillatory component (brush) is suppressed by the cortical application of the AMPA/kainate receptors (CNQX), whereas the delta-component is blockaded by combined application of AMPA/kainate and NMDA-receptor antagonists (CNQX and APV). Blockade of GABA(A) receptors with gabazine enhances the response. *Adapted with permission from: (a) Colonnese MT and Khazipov R (2010) "Slow activity transients" in infant rat visual cortex: A spreading synchronous oscillation patterned by retinal waves.* Journal of Neuroscience 30: 4325–4337; *(b) Minlebaev M, Ben Ari Y, Khazipov R (2009) NMDA receptors pattern early activity in the developing barrel cortex in vivo.* Cerebral Cortex 19: 688–696.

resting periods, and it remains unknown whether ENOs in the somatosensory cortex are associated with movements. Nevertheless, optical ENOs appear similar to the electrographically defined spindle-bursts, including duration of events and inter-event intervals. Although the dynamics of intracellular calcium increases during ENOs were rather smooth and no high-frequency component characteristic oscillation was evident, this may be due to slow dynamics of intracellular calcium and the limited temporal resolution of calcium imaging methods compared to electrophysiological recordings. Further experiments using simultaneous electrophysiological and imaging approaches are needed to determine whether or not spindle bursts and ENOs reflect the same fundamental physiological patterns. Before speculation about the functional role of the early spontaneous patterns, we discuss the available observations that provide insights into their physiological mechanisms.

8.4 MECHANISMS OF EARLY NETWORK PATTERNS

The synaptic basis of the generation of the early activity patterns was explored using whole-cell patch-clamp recordings from neonatal rat somatosensory (Khazipov et al., 2004b; Minlebaev et al., 2007, 2009) and visual (Colonnese and Khazipov, 2010; Colonnese et al., 2010; Hanganu et al., 2006) cortex. These studies have revealed a pivotal role for glutamatergic and GABAergic synapses in the generation of the early activity patterns.

The mechanisms of spindle-bursts were investigated in detail in the neonatal rat barrel cortex (Minlebaev

et al., 2007, 2009) using a superfused cortex preparation *in vivo*, enabling the application of drugs directly to the cortex. Pharmacological analysis revealed that generation of spindle-bursts in the neonatal barrel cortex primarily involves glutamatergic mechanisms (Figure 8.4). Interestingly, the relative contribution of AMPA/kainate and NMDA-Rs to the generation of the delta and alpha–beta components is different. Rapid alpha–beta oscillations mainly require AMPA/kainate receptors, and blockade of NMDA-Rs has no significant effect on this rapid oscillatory component. On the other hand, delta waves are generated by both types of glutamate receptors acting in concert, though with a primary contribution of NMDA-Rs. The differential contribution of the AMPA/kainate and the NMDA-Rs to the two components of the spindle-burst probably reflects a difference in the kinetics of the synaptic currents mediated by these receptors. AMPA-R-mediated synaptic currents have fast rise and decay times in the millisecond range (see, e.g., Crair and Malenka, 1995; Kidd and Isaac, 1999; Khazipov et al., 2004b) and therefore are ideally suited for synchronization of the rapid activities, such as alpha–beta oscillations. NMDA-R-mediated synaptic currents have rise times in the range of tens of milliseconds and decay times of hundreds of milliseconds; they are particularly slow at the immature synapse (Carmignoto and Vicini, 1992; Chittajallu and Isaac, 2010; Hestrin, 1992; Khazipov et al., 1995; Monyer et al., 1994). The slow kinetics of NMDA-R-mediated synaptic currents enables their powerful summation during rhythmic activation of synaptic inputs during spindle-bursts. The high NMDA/AMPA ratio at the immature synapses is another important factor contributing to the increased contribution of NMDA-Rs to the delta wave (Chittajallu and Isaac, 2010; Crair and Malenka, 1995; Durand et al., 1996; Gasparini et al., 2000; Isaac et al., 1997; Voronin et al., 2004).

NMDA-R-dependent patterns of activity in developing cortical networks have also been described in cortical tissue *in vitro*, including hippocampal giant depolarizing potentials (GDPs) (Ben-Ari et al., 1989) and associated calcium oscillations (Allene and Cossart, 2010; Allene et al., 2008; Crepel et al., 2007; Leinekugel et al., 1997) and neocortical bursting and oscillatory activity (Arumugam et al., 2005; Dupont et al., 2006; Garaschuk et al., 2000; Kandler and Thiels, 2005; LoTurco et al., 1991). Interestingly, in the case of hippocampal GDPs, depolarizing γ-aminobutyric acid (GABA) may facilitate activity of NMDA-Rs by attenuation of their voltage-dependent magnesium block (Khazipov et al., 1997; Leinekugel et al., 1997). Activation of NMDA-Rs during spindle-bursts may be directly linked to the plasticity mediated by these receptors in the developing cortex. Indeed, pharmacological or genetic manipulations associated with a loss of function in NMDA-Rs results in malformations of barrel cortex development and functional deficits (Dagnew et al., 2003; Fox, 2002; Fox et al., 1996; Iwasato et al., 2000; Lee et al., 2005a,b; Schlaggar et al., 1993). During the critical period, thalamocortical synapses display an enhanced NMDA-R contribution and increased NMDA-R-dependent synaptic plasticity, including the conversion of 'silent' pure NMDA-R-based synapses to fully functional mixed AMPA/NMDA-R synapses, as well as a switch to fast AMPA-receptor-mediated synaptic transmission from slow kainate-mediated transmission (Bannister et al., 2005; Barth and Malenka, 2001; Carmignoto and Vicini, 1992; Crair and Malenka, 1995; Daw et al., 2006; Feldman et al., 1998, Hestrin, 1999, 1992; Isaac et al., 1995, LoTurco et al., 1997, 1991; Monyer et al., 1994). Activation of NMDA-Rs achieved by massive summation of thalamocortical input during spindle bursts provides a long time window for co-incident activation of cortical neurons by the thalamocortical cells. This physiological paradigm may underlie the NMDA-R-dependent plasticity in developing thalamocortical synapses, including both potentiation/maintenance of the activity during spindle-burst synapses and depression/elimination of those synapses that are less recruited by spindle-bursts. Plasticity in developing synapses is spike-time-dependent, and the synaptic strength can be bi-directionally modified by correlated pre-/postsynaptic spiking within a narrow time window on the order of 10 ms (Mu and Poo, 2006; Chapter 9). While recent findings indicated that the early oscillatory patterns may induce bi-directional plasticity at thalamocortical synapses (Minlebaev et al., 2011), whether STDP occurs during early oscillatory patterns is an important question for the further research.

While a considerable amount of data has now accumulated about network mechanisms of spindle-bursts, the generation of gamma-bursts is less well understood. In the adult brain, neuronal synchronization by gamma oscillations is a fundamental process in cortical computation (Buzsaki, 2006; Fries, 2009). Synchronized by inhibition (Bartos et al., 2007), gamma oscillations subserve perceptual binding (Gray and Singer, 1989) and support synaptic plasticity (Wespatat et al., 2004). The ontogeny and role of gamma oscillations in developing networks, however, remain controversial. It was generally accepted that gamma oscillations emerge relatively late in development (Uhlhaas et al., 2010) as associative cortical layers and large-scale connections (Bureau et al., 2004; Luhmann et al., 1986), which are required for perceptual binding, and GABAergic inhibition (Daw et al., 2007; Doischer et al., 2008; Luhmann and Prince, 1991), which is pivotal for gamma rhythmogenesis, both develop on a delayed timescale. Yet, several studies have now indicated the existence of transient, local, gamma-burst oscillations in the neonatal rat somatosensory (Yang et al., 2009) and visual (Colonnese et al., 2010) cortices. Moreover, studies in the neonatal rat barrel cortex revealed that the early gamma oscillations (EGOs) are specifically evoked in a single cortical barrel by stimulation of the corresponding (principal) whisker (Minlebaev et al., 2011). Simultaneous recordings from the ventro-posterior-medial (VPM) barreloids and cortical barrels have shown that the EGOs are primarily driven by a thalamic oscillator and synchronize neurons in a single thalamic barreloid and the corresponding cortical barrel. Basing on these findings and the results of whole-cell recordings from L4 neurons, the following network EGOs model has been suggested: (1) Sensory input from a whisker activates the gamma oscillator in the thalamic barreloid, which provides topographic feedforward synchronization in the corresponding cortical barrel; (2) cortical interneurons become involved in EGOs in an age-dependent manner: until ~P5, EGOs are independent of cortical inhibition, but starting from P5, along with the development of feedforward inhibition, interneurons are recruited and support EGOs by controlling runaway recurrent cortical excitation. Thus, during the first postnatal week, EGOs undergo evolution from a primitive form of cortical activity passively following a thalamic oscillator to a more complex interactive model in which an active cortical oscillator, by virtue of emerging inhibition, starts to support gamma oscillations. Interestingly, artificial EGOs mimicked by pairing subthreshold gamma-rhythmic thalamic input with action potentials in L4 neurons in thalamocortical slices resulted in long-lasting potentiation of thalamocortical EPSPs. It has been suggested that in contrast to the inhibition-based "adult" gamma oscillations, which emerge at the end of the second postnatal week and enable horizontal synchronization, EGOs are primarily driven by gamma-rhythmic excitatory thalamic inputs and provide vertical synchronization between topographically aligned thalamic and cortical neurons. Multiple replay of the sensory input in the thalamocortical synapses during EGOs ("repetitio est mater studiorum!") may allow thalamic and cortical neurons to be woven into vertical topographic functional units prior to the development of horizontal binding and other integrative cortical functions subserved by "adult" gamma oscillations in the mature brain.

Another intriguing question concerns the contributions of depolarizing and excitatory GABA actions to the early cortical-activity patterns. GABA is the main inhibitory neurotransmitter in the adult brain. Synchronous inhibition by hyperpolarization and a shunt provided by GABAergic interneurons are instrumental for the generation of various activity patterns in the adult brain (Bartos et al., 2007; Buzsaki, 2006; Freund and Buzsaki, 1996; Wang, 2010). However, slice preparations suggest that early in development – including the embryonic period and the first postnatal week in rodents – GABA, acting via chloride-permeable GABA (A) receptors, exerts a depolarizing and excitatory action on immature neurons as a result of their elevated intracellular chloride concentration (LoTurco et al., 1995; Luhmann and Prince, 1991; Owens et al., 1996; Tyzio et al., 2006; Yamada et al., 2004; Yuste and Katz, 1991; reviewed in Ben Ari et al., 2007; and is discussed extensively in Rubenstein and Rakic, 2013). Elevated intracellular chloride in the immature neurons is a result of high activity of the chloride-loading co-transporter NKCC1 and delayed expression of the chloride extruder KCC2 (Rivera et al., 1999; Yamada et al., 2004). Depolarizing GABA is involved in the generation of the primitive pattern of neuronal network activity in the immature hippocampus and cortex – so-called GDPs (Allene et al., 2008; Ben-Ari et al., 1989; Dzhala et al., 2005; Khazipov et al., 2004a; Rheims et al., 2008; Sipila et al., 2005). During GDPs, both pyramidal cells and interneurons fire randomly within a very large time window of few hundreds of milliseconds. Excitation of pyramidal cells and interneurons during GDPs is brought about by synergistic excitatory actions of GABA and glutamate (Bolea et al., 1999; De la Prida et al., 1998; Khazipov et al., 1997; Lamsa et al., 2000; Leinekugel et al., 1997; Menendez et al., 1996; Valeeva et al., 2010).

Does GABA play a similar excitatory action in the generation of the early oscillatory patterns of cortical activity *in vivo*? Unfortunately, at present there is surprisingly little experimental evidence that GABA exerts excitatory actions on immature cortical neurons *in vivo*, as it does *in vitro*. At present, the only information available about the roles of GABAergic interneurons *in vivo* is based on the effects of pharmacological manipulations of GABAergic synaptic transmission on network-driven activities.

In the neonatal rat hippocampus, the dominant neuronal network patterns of activity (GDPs *in vitro* and sharp waves *in vivo*) are blocked by the NKCC1 antagonist, bumetanide, which shifts the reversal potential of the GABA(A)-receptor-mediated responses toward negative values (Dzhala et al., 2005; Sipila et al., 2006; Tyzio et al., 2006). Although a similar effect of bumetanide on the GABA(A) reversal potential was found in neocortical neonatal neurons (Tyzio et al., 2006; Yamada et al., 2004), bumetanide did not significantly affect spindle-bursts in the barrel cortex *in vivo* (Minlebaev et al., 2007). Therefore, it appears that early hippocampal patterns of sharp waves *in vivo* and GDPs *in vitro* are more dependent on the depolarizing actions of GABA than the neocortical pattern of spindle-bursts, consistent with observations *in vitro* (Garaschuk et al., 2000; Rheims et al., 2008).

However, GABAergic interneurons do clearly participate in the generation of spindle-bursts and gamma-bursts in an age-dependent manner, as evidenced by the effects of the GABA(A)-receptor antagonists and positive allosteric GABA(A)-receptor modulators. Although blockade of GABA(A) receptors does not significantly affect the frequency of oscillations in the alpha–beta frequency domain, it does increase the power of these oscillations, as well as increase the amplitude of the delta component and the occurrence of spindle-bursts. Gamma oscillations are little affected by blockade of cortical inhibition until P5, but they are suppressed in older animals. The opposite manipulation, enhancement of GABA(A)-receptor-mediated currents by diazepam, reduces by twofold the occurrence of spindle-bursts. These results suggest that GABAergic interneurons play an inhibitory role in spindle-burst generation. In keeping with these findings, blockade of GABA(A) receptors strongly increases the size of cortical areas activated during spindle-bursts. Spindle-bursts are local events, and the diameter of the cortical zones activated during spindle-bursts usually does not exceed 0.5 mm. After blockade of GABA(A) receptors, spindle-bursts can be recorded at distances up to 1–2 mm. Thus, the spread of spindle-bursts is determined not only by the topographic thalamocortical excitatory input (Agmon et al., 1996; Bureau et al., 2004; Ferezou et al., 2006; Higashi et al., 2002; Khazipov et al., 2004b; Kidd and Isaac, 1999; Petersen and Sakmann, 2001), but also by surround GABAergic inhibition that prevents horizontal spread of the activity via long-range glutamatergic cortical connections, a pattern also observed in the adult neocortex (Chagnac-Amitai and Connors, 1989; Fox et al., 2003; Sun et al., 2006). The inhibitory action of GABA at the network level does not necessarily imply a hyperpolarizing action, as even depolarizing GABA may produce strong inhibition via shunting mechanisms amplified by activation of the voltage-gated potassium channels and inactivation of sodium channels (Borg-Graham et al., 1998; Gao et al., 1998; Gulledge and Stuart, 2003; Lu and Trussell, 2001). These results, which suggest an inhibitory role of GABA during generation of spindle-bursts, are in general agreement with the finding that GABA(A) antagonists induce hypersynchronous seizure-like activity in the neocortex *in vivo* by P3 (Baram and Snead, 1990) and *in vitro* by P2 (Wells et al., 2000). These results are also in keeping with the plasticity changes induced in the developing barrel cortex by chronic treatment with the GABA(A)-receptor agonist, muscimol (Wallace et al., 2001).

8.5 DISCONTINUOUS TEMPORAL ORGANIZATION OF THE EARLY ACTIVITY

The uniqueness of immature cortical activity consists not only of developmentally restricted patterns of activity, but also stems from its *discontinuous temporal organization* (Figures 8.5 and 8.6). The first reports of the discontinuous nature of early cortical activity were made in human preterm neonates using scalp electrographic recordings by Dreyfus-Brisac, Monod, and their colleagues. Analyzing EEGs from preterm neonates during the second half of gestation (Dreyfus-Brisac, 1962; Dreyfus-Brisac and Larroche, 1971; Dreyfus-Brisac et al., 1956), they noted that the cortical EEG was organized in intermittent bursts separated by periods of isoelectric EEG that could last for tens of seconds. This temporal organization was named *tracé discontinu*. With maturation, flat periods between the bursts became shortened, and starting from about 30 weeks of postconceptional age, *tracé discontinu* evolves to *tracé alternant* with low-voltage activity between the bursts, though

even at term, some discontinuity is still evident (Lamblin et al., 1999; Stockard-Pope et al., 1992).

Such extreme discontinuity of cortical activity is striking. Indeed, showing EEG recordings obtained from a healthy premature neonate to an "adult" neurophysiologist without revealing the age of the patient would result in a diagnosis such as severe posthypoxic encephalopathy or barbiturate coma. Yet, this discontinuity is a normal feature of the immature cortex, and this raises several questions: Does discontinuity simply reflect immaturity of the nervous system, which is incapable of maintaining continuous activity, or, in keeping with Mozart's claim that the most meaningful element of a musical masterpiece is a pause, has it physiological roles for development?

The first question about human discontinuity is whether it results from true network silence as observed during slow-wave sleep in adults, or is it the result of temporally uncoordinated activity of neurons? This question has been examined in neonatal rats. Extracellular recordings of multiple unit activity showed that neurons fire virtually no action potentials during isoelectric EEG epochs. In agreement with this result, patch-clamp recordings revealed that neurons stay relaxed at their resting membrane potential $\sim 80\,mV$ and receive very little synaptic input during isoelectric EEG periods. This means that isoelectric EEG epochs reflect synchronous neuronal silence. Such a state of collective silence, called the down-state, also exists in the adult EEG. Considerable evidence suggests that isoelectric epochs are by nature the same phenomenon as down-states of long duration. In adults, down-states with durations from tens of milliseconds to a few hundred milliseconds occur during deep sleep in alternation with up-states (Buzsaki, 2006; Steriade, 2001; Steriade et al., 1993). Oscillation

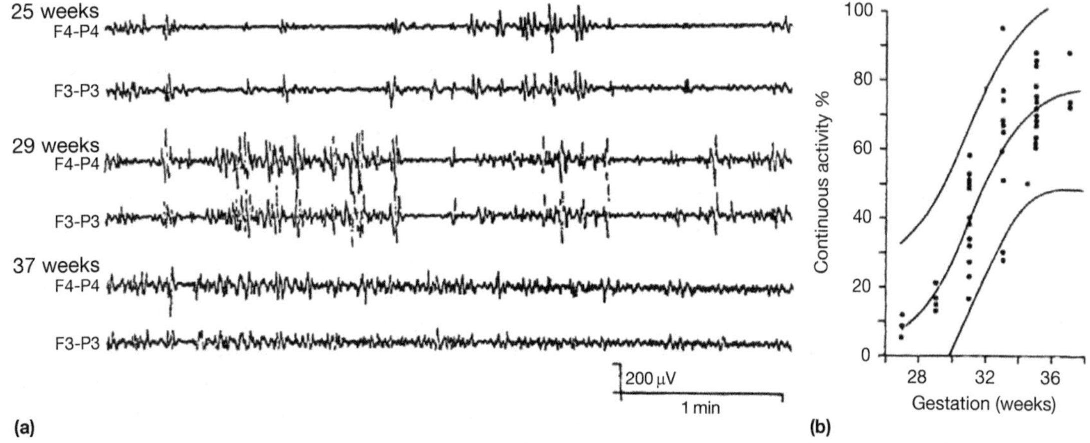

FIGURE 8.5   Developmental changes in continuity of human cortical activity. (a) EEG of preterm infants at 25, 29, and 37 gestational weeks. The duration of silent periods decreases, while the length of periods of continuous activity increases with age. (b) Quantification of the proportion of EEG containing continuous activity (continuity defined as >80% continuous activity in 5-min epochs with attenuated activity 10–20 µV lasting no longer than 5 s). *Adapted with permission from Connell JA, Oozeer R, Dubowitz V (1987) Continuous 4-channel EEG monitoring: A guide to interpretation, with normal values, in preterm infants.* Neuropediatrics 18: 138–145.

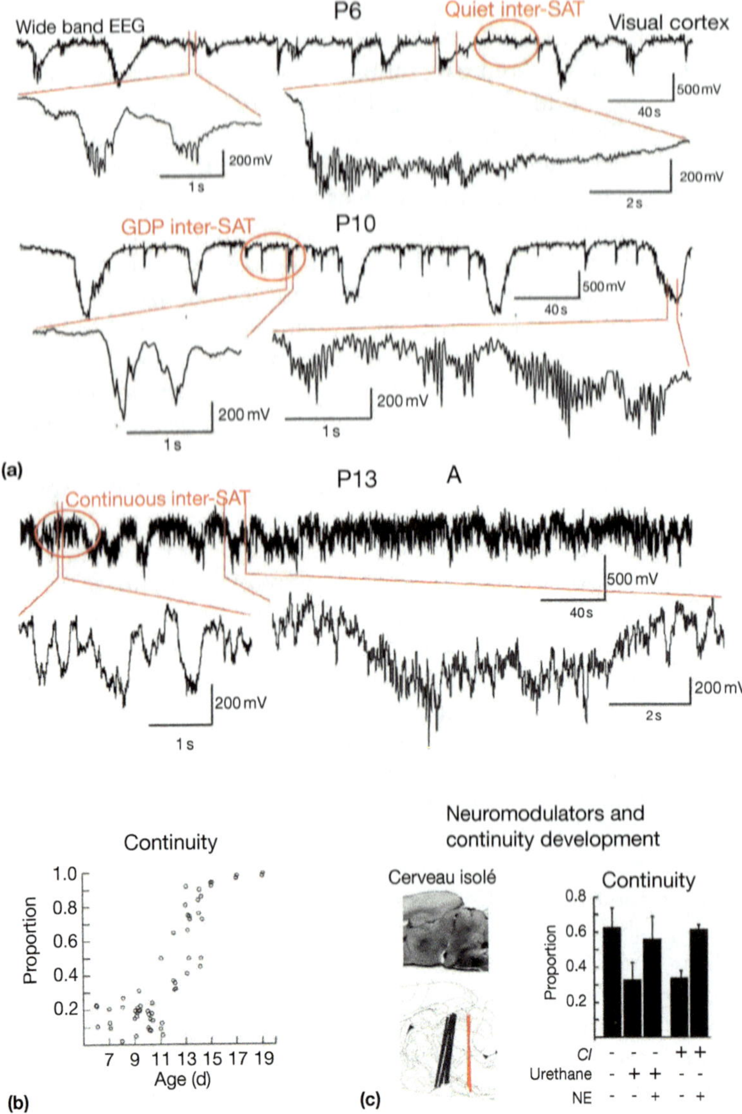

FIGURE 8.6 Developmental changes in diversity and continuity of rodent cortical activity. (a) Wide-band extracellular recording of spontaneous activity in layer 4 of the visual cortex in awake rats at P6, 10, and 13. At P6, all events consist of retinal-wave driven (see Figure 8.3), slow negative-activity transients (SATs) of varying length that contain faster rhythmic oscillations (left and right example). Inter-SAT activity is rare. During the second postnatal weeks, SATs become elongated (left example), while inter-SAT activity consists of occasional 200-ms–1-s bursts of action potentials (giant-depolarizing potentials (GDPs), right example). At the end of the second week, SATs become reduced and irregular (left example), while inter-SAT activity becomes continuous and diverse (right example). See Figure 8.3(c) for quantification. (b) Continuity of activity by post-natal age. Continuity of activity as a function of post-natal age was measured as the proportion of 500-ms epochs that contain at least one action potential in multi-unit recordings. (c) Role of ascending neuromodulators in the development of continuous activity. Surgical isolation at the rostral midbrain border (right image, CI) or urethane anesthesia reduces the continuity of spontaneous activity in P13–15 rats. Application of norephinephrine (NE) to the cortical surface increases continuity in lesioned and anesthetized animals. Error bars SEM. *Adapted with permission from Colonnese MT and Khazipov R (2010) "Slow activity transients" in infant rat visual cortex: a spreading synchronous oscillation patterned by retinal waves.* Journal of Neuroscience 30: 4325–4337 (a) and Colonnese MT, Kaminska A, Minlebaev M, Milh M, Bloem B, Lescure S, Moriette G, Chiron C, Ben-Ari Y, Khazipov R (2010) *A conserved switch in sensory processing prepares developing neocortex for vision.* Neuron 67: 480–498 (b, c).

between up- and down-states generates slow waves in the EEG, and therefore these deep stages of sleep are also called slow-wave sleep oscillations. Intracellular studies indicate that when the level of mutual excitation in the cortical network falls below the critical level maintaining the up-state, which is partly due to the activity-dependent depression in glutamatergic synapses, neurons return to their resting membrane potential during down-states (Contreras et al., 1996; Sanchez-Vives and McCormick, 2000; Shu et al., 2003).

Immature neurons do just the same, but for much longer periods of time; however, one important difference between immature cortical activity and adult sleep is that the immature cortex maintains these long-duration down-states even when the animal is awake. In adults, the awake state is characterized by a continuous (a.k.a. activated or desynchronized) mode of cortical function.

Intracellular recordings show little difference between continuous cortical activity during awake, REM sleep, and up-states during slow-wave sleep. The activated cortical state is not present in neonatal rats or young premature neonates and starts to emerge in the rat during the second postnatal week and a month before term in humans, coinciding with the emergence of clearly differentiable sleep states in both species (Jouvet-Mounier et al., 1970; Lamblin et al., 1999). Interestingly, the active state does not emerge simultaneously in the entire cortex. Increase in the amount and continuity of spontaneous activity occurs first in the somatosensory and then in the visual cortex, both in the rat and humans, which also coincides with elimination of delta-brushes and spindle-bursts in these areas (Colonnese et al., 2010; Curzi-Dascalova et al., 1993; Dreyfus-Brisac and Larroche, 1971; Lamblin et al., 1999).

At least two developmental factors are likely involved in the emergence of the activated cortical state: development of arousal systems and intracortical connections. In the rat, maturation of sleep/wake transitions depends on noradrenergic development (Gall et al., 2009) and maturation of cholinergic (Mechawar and Descarries, 2001) and noradrenergic afferents and receptor distributions (Latsari et al., 2002; Venkatesan et al., 1996) around this time. Isolating the forebrain from ascending neuromodulatory arousal inputs in P13–15 rats reverses the developmental increase in spontaneous cortical activity, reinstating an immature mode of discontinuity (Figure 8.6). Continuity can then be reinstated by cortical application of norepinephrine (Colonnese et al., 2010), a key initiator of the activated cortical state (Berridge and Waterhouse, 2003; Foote and Morrison, 1987a,b). However, this is not to say that arousal systems are not functional before this time. Actually, many aspects of brainstem and forebrain arousal systems operate well before the emergence of the active cortical state; for example, infant rats and humans give evidence of behavioral sleep–wake cycles, state-dependent firing of hindbrain sleep control neurons (Karlsson et al., 2005), and modulation of cortical EEG patterns by the sleep state (Seelke and Blumberg, 2010). Furthermore, functional monaminergic and cholinergic connections are in place quite early (Hanganu et al., 2007; Johnston and Coyle, 1981). Yet, arousal systems appear incapable of inducing a continuous active state in the immature cortex. Thus, the acquisition of brainstem control of cortical states likely requires more than a simple engagement of neuromodulatory systems, and also thalamocortical changes. A prominent candidate is the strengthening of long- and short-range intracortical connectivity, required for maintenance of the activated cortical state (Sanchez-Vives and McCormick, 2000; Shu et al., 2003).

A number of observations are consistent with an increase in functional cortical connectivity over this time period. First, while k-complexes and sleep-spindles, stimulated by light in adults, are synchronized throughout the cortex (Amzica et al., 1998) via horizontal cortical and corticothalamic connections (Contreras et al., 1996, 1997), delta-brushes are local events (Khazipov et al., 2004b; Yang et al., 2009). Second, as noted earlier, in adults, episodes of network silence (down-states) during sleep do not exceed 500 ms (Steriade, 2001), whereas the silent periods in the young rat pups or preterm infants can run to tens of seconds. Such long silent periods are observed in cortical slabs and slices, that is, under conditions of markedly reduced cortical connectivity (Sanchez-Vives et al., 2000; Steriade et al., 2005; Timofeev et al., 2000). Finally, the network of horizontal connections between pyramidal neurons emerges shortly before eye opening in cats (Callaway and Katz, 1990; Galuske and Singer, 1996) and ferrets

(Durack and Katz, 1996; Ruthazer and Stryker, 1996), independently of visual input. In humans, dense horizontal connections are also first observed at GW37 (Burkhalter et al., 1993). Therefore, emergence of the continuous mode of cortical activity likely results from a coincidence of two developmentally regulated factors: (1) maturation of the brainstem arousal input to the cortex, which provides tonic neuronal depolarization, and (2) formation of excitatory synaptic connections between cortical neurons, which are needed for maintenance of the activated cortical state.

In conclusion, it is tempting to propose a hypothesis whereby discontinuity and long silent periods are important for circuit development. Growing evidence indicates that developing synapses display very high levels of plasticity, as governed by the Hebbian principle "neurons that fire together wire together" (Hebb, 1949), which in a more elaborated way is formulated as STDP (Dan and Poo, 2006; Mu and Poo, 2006). STDP implies that the strength of synaptic connections can be bi-directionally modified – either potentiated or depressed – depending on spike timing in pre- and postsynaptic neurons. If the spike in a presynaptic neuron precedes the spike in the postsynaptic neuron (as in the case the postsynaptic neuron was driven by the presynaptic neuron), the synapse will be potentiated; if the opposite occurs, the synapse is depressed. The most efficient changes in both directions occur when correlated spiking occurs within a time window of ten of milliseconds. These functional changes in the strength of synaptic connections were proposed as precursors of further anatomic changes – synapse stabilization and elimination in the cases of potentiation and depression, respectively.

Let us put STDP into the context of temporal organization, that is continuity/discontinuity, and examine how this is expected to impact plasticity in a thalamocortical synapse. We know that a sensory input triggers topographic delta-brushes, providing pre-before-post firing in the appropriate thalamocortical synapse and thus conditions for potentiation/stabilization. From the plasticity standpoint, the ideal case for potentiation would be if two conditions are met: (1) each pre-spike is followed by a post-spike, and (2) each post-spike is preceded by a pre-spike. These conditions will be achieved if topographic thalamocortical synapses provide a necessary and sufficient excitatory drive to cortical cells. Excitation of cortical neurons by intracortical synapses during intracortically generated spontaneous activity would introduce noise to the system, as random – in relation to thalamic input – firing of cortical neurons will create variable conditions for STDP in the thalamocortical synapse. In the developing cortex, this noise is eliminated by delayed introduction of horizontal connections, which strongly increases the plasticity impact of sensory-driven thalamocortical bursts. This hypothesis is supported by studies in the developing *Xenopus* retinotectal system, in which it was shown that activity-induced synaptic modifications are quickly reversed either by subsequent spontaneous activity in the tectum or by exposure to random visual inputs (Zhou and Poo, 2004b; Zhou et al., 2003). This reversal depended on the burst spiking and activation of the N-methyl-D-aspartate subtype of glutamate receptors. Stabilization of synaptic modifications could be achieved by an appropriately spaced pattern of induction stimuli. These findings underscore the vulnerable nature of activity-induced synaptic modifications *in vivo* and suggest a temporal constraint on the pattern of visual inputs for effective induction of stable synaptic modifications. These findings are also linked to the developmental changes in the sensory-evoked activity patterns that will be discussed in the next section.

8.6 EARLY ACTIVITY PATTERNS AND THE DEVELOPMENT OF PERCEPTION

In this chapter, we have identified two characteristics of neocortical activity during the time of cortical layer formation and thalamic in-growth and map formation: the presence of "slow-activity transients" containing rapid oscillations (including spindle-bursts and delta-brushes) and a discontinuous pattern of activity, that is, the presence of long periods of neuronal silence. We argue that these features are the result of a unique network state specific to early development that is optimized to drive synaptic plasticity in thalamocortical networks in response to spontaneous activity generated in the sense organs. Because early spontaneous activity at the sensory periphery is infrequent, the early cortex must remain mostly quiet. Secondly, because cortical circuits are concerned primarily with the presence and absence of activity within topographic space, they produce bursts of highly correlated firing containing little additional information.

The final question that concerns us is the transition to mature patterns of cortical activity: when do they occur and what are their behavioral correlates and mechanisms? This issue has been studied most extensively in the visual system. Here the activity patterns in the sense organ are best understood, and relative roles of neuronal activity and the environment in maturation can be examined. Cortical visual development can be roughly divided up into four periods: (1) the completely previsual period, when phase 2 retinal waves are transmitted to the cortex; (2) an obscured visual phase, when phase 3 (glutamatergic) retinal waves are prominent but coexist with initial photo-receptor-driven light responses (Huberman et al., 2008) that allow cortical visual responses to be driven through the closed eye-lids

(Akerman et al., 2004; Chapman and Stryker, 1993; Ohshiro and Weliky, 2006); (3) a pre-critical period (Feller and Scanziani, 2005) that follows eye-opening, during which new visual experience drives development of binocularity, orientation, and direction selectivity (Smith and Trachtenberg, 2007; White et al., 2001); and (4) the classical critical period for monocular deprivation plasticity (Hubel and Wiesel, 1970).

As reviewed above, the early period of spontaneous cortical activity, consisting of delta-brushes, SATs, and discontinuity occurs throughout the first and second visual development periods, rapidly maturing around the time of eye-opening and the transition to the third period. Colonnese et al. (2010) examined the role of this transition of thalamocortical networks in determining visual-response properties in rats and human preterm infants. In the somatosensory system, both spontaneous muscular twitches and direct touch are effective in eliciting cortical delta-brushes. Perhaps this is not surprising, given that the ultimate effect on the sense organ was similar. However, the activation of retinal ganglion cells by light is weaker and patterned differently than during retinal waves (Blankenship et al., 2009; Kerschensteiner and Wong, 2008; Tian and Copenhagen, 2003). This casts doubt on a possible interaction between these transmission systems for spontaneous and evoked visual activity. Furthermore, primate retinas have weaker and less-organized spontaneous activity during the later gestational period (Warland et al., 2006), further questioning a potential interaction between spontaneous and evoked activity patterns in humans.

Despite these facts, whole-field light flashes directed to the closed eyelids of rats evoked a complex of oscillatory patterns in the cortex typical of the spontaneous activity patterns observed at this age (Figure 8.7(a)). Despite weak or absent retinal visual responses *in vitro*, robust light responses could be evoked as early as P8. Cortical light responses consisted of a complex of described activity patterns, including an initial response that included a gamma-burst and a GDP, followed by an evoked delta-brush. Electrical stimulation of the retina showed that these patterns did not result from simply the triggering of retinal waves by light, but emerged in the thalamocortical circuit following even brief inputs. Similar visual stimuli evoked a similar complex of activity patterns, including delta-brushes, in the occipital EEGs of sleeping preterm infants from as early as they could be recorded (27 gestational weeks). Thus, the network properties dedicated to transmitting retinal waves are also recruited during visual stimulation, even though the importance of such stimuli to thalamocortical activity at these ages is questionable. Together, these data show that the early activity bursts are a fundamental response to thalamic input, regardless of source, that shapes sensory processing.

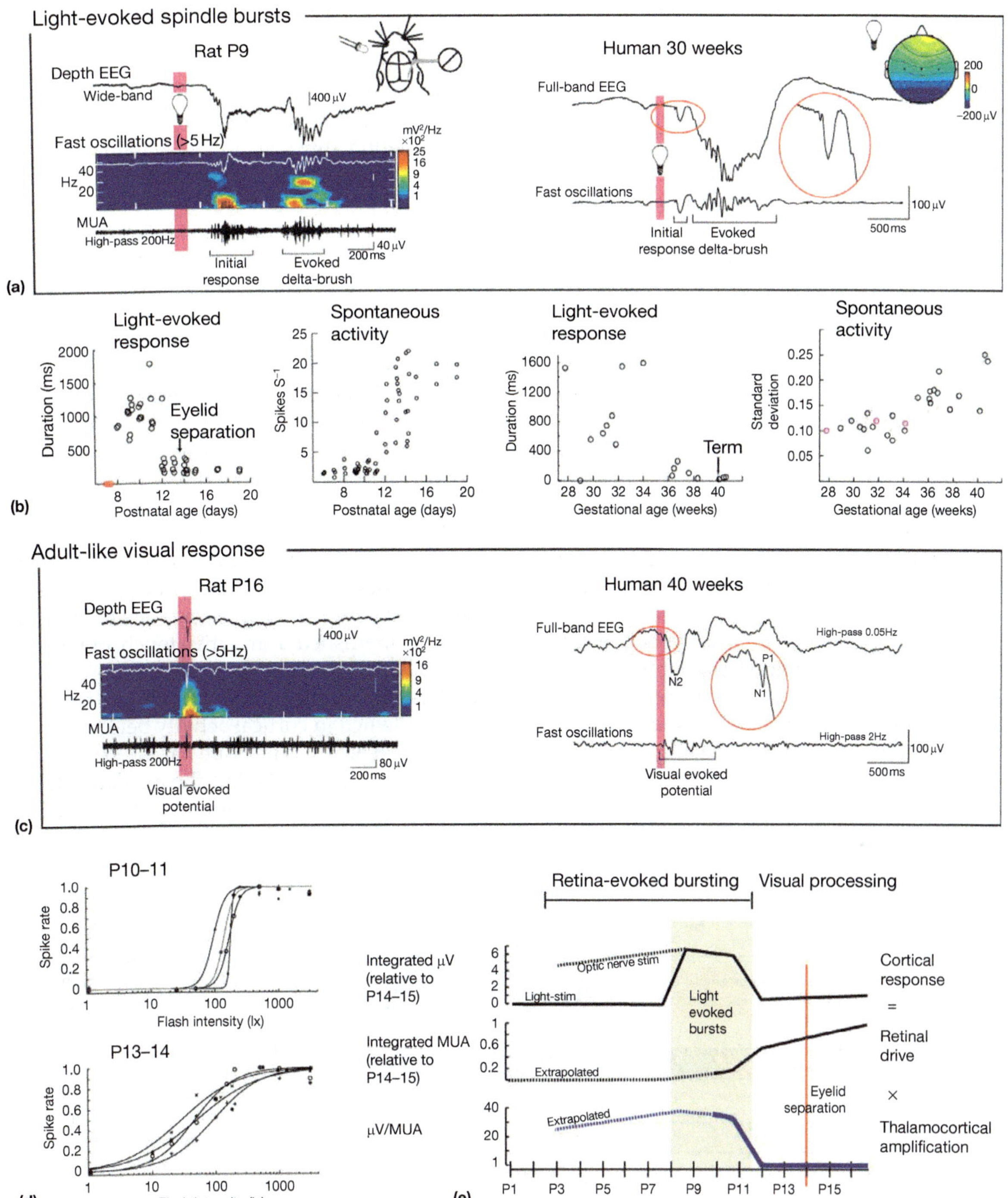

FIGURE 8.7 Role of thalamocortical-network changes in the development of visual processing. (a) Cortical visual responses during early development in rats (left) and preterm infants (right). During the period of discontinuous cortical activity, full-field light flashes trigger bursts of endogenous delta-brush oscillations resulting in a large amplitude potential and long-lasting response. Example wide-band traces from depth EEG (rat) or surface EEG (human) are shown along with a time-series spectrogram and MUA for the rat and a high-pass filtered trace to show rapid oscillations for the human. Electrode map in the top-right corner shows localization of the negative slow wave to occipital electrodes. (b) Linked development of the mature pattern of visual processing and continuous activity. In both species, short visual responses and continuous cortical activity rapidly emerged just before active visual exploratory behavior at eye-opening (rats) or birth (humans). (c) Example traces of visual-evoked responses after eye-opening (rats) or birth (humans). (d) Bursting behavior of early cortical circuits prevents graded visual responses. Spike rates in layer 4 of the rat visual cortex during the initial response show all-or-nothing bursting to graded light intensities during the period of discontinuity. Graded visual responses are observed rapidly following the switch in cortical activities. (e) Early cortical oscillations amplify thalamic input during the early period. This amplification is downregulated to allow for graded visual processing just before pattern vision. Thalamocortical amplification was estimated by dividing the evoked cortical potential by the *in vitro* retinal spiking response in age-matched rats. *Adapted with permission from Colonnese MT, Kaminska A, Minlebaev M, Milh M, Bloem B, Lescure S, Moriette G, Chiron C, Ben-Ari Y, Khazipov R (2010) A conserved switch in sensory processing prepares developing neocortex for vision.* Neuron 67: 480–498.

Using this controlled stimulation to assay the development of the thalamocortical response properties revealed that the developmental transition to mature patterns of visual response occurs as a rapid switch 1–2 days before eye-opening in rats (Figure 8.7(b) and 8.7(c)). In addition to this switch in visual-response patterns, the amount and continuity of spontaneous activity began to increase (and would continue to do so over the next weeks) on the same day. A similar switch in evoked and spontaneous activity occurred in the occipital cortex of human infants between 34 and 36 gestational weeks, that is, just before natural term (human fetuses open their eyes at the end of the second trimester). In both species, this switch consisted of a loss of the early cortical burst oscillations and their replacement by the visual evoked potential (VEP). As such, it results in a surprising decrease in the size and duration of visual responses.

Functionally, why would visual responses need to become reduced in size just before the onset of visual processing? One clue comes from examining the response curves to varying light intensities before and after the switch (Figure 8.7(d)). Before the switch, visual responses were 'all-or-none' bursts, effectively registering the presence of a minimal stimulation, but not its intensity. Only after the reduction of visual response with the switch could graded responses be attained. Measurement of the spiking responses of retinal ganglion cells in acutely excised retina allowed further quantification of the input–output relationships before and after the switch (Figure 8.7(e)). These revealed an inverse relationship between retinal and cortical light responses, with the reliability of retinal spiking to light flashes rapidly increasing just as the amplitude of the cortical response decreased. This had the effect of massively decreasing the amplification in thalamocortical circuits. Thus, we suggest that the primary role of early activity patterns, in addition to precisely synchronizing activity, is to amplify weak, young inputs. Such an amplification is incompatible with visual processing and thus is downregulated before vision is initiated at eye-opening or birth.

This switch in cortical-activity patterns, from a discontinuous mode with early bursting oscillations to continuous baseline activity with more restricted, graded response, appears to be universal for all sensory systems studied so far (Figure 8.8). Intriguingly, while the developmental template remains the same, the timing of the switch appears to differ between sensation and species, with the common feature being that it occurs shortly before the onset of active sensation for that system. This is best illustrated in the somatosensory whisker system, where passive activation (i.e., external touch of the whisker) is distinguished from active touch, or whisking. Rats begin active whisking 2–3 days before eye-opening (Landers and Philip, 2006). Similarly, they experience the same switch in cortical activity before the visual cortex (Figure 8.8, top). The issue has not been examined in depth in human infants, but the available data suggest less heterogeneity in the timing of the switch between

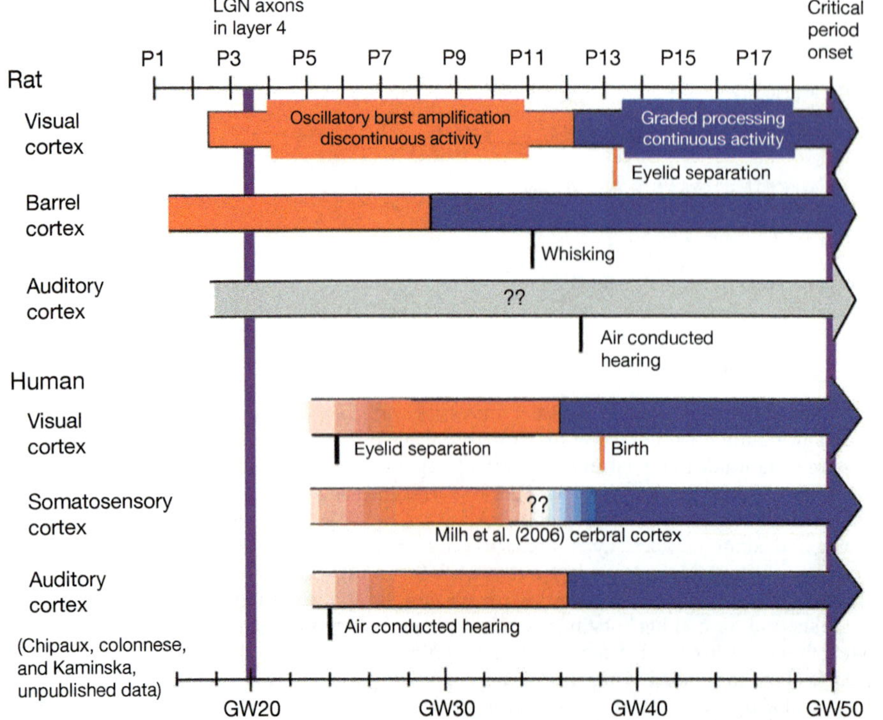

FIGURE 8.8 Two-state model of cortical-activity development. Known transition times from early bursting responses (evoked delta-brushes) to mature responses for multiple sensory systems in the human and the rat. Developmental timelines are aligned to the independent time points shown by purple lines. The timing of the switch in cortical circuits is correlated to the onset of exploratory sensation (e.g., whisking or eye-opening), which occurs at different times for the rat, but simultaneously at birth for humans, and this heterochrony is matched in the cortical-activity pattern development. ?? signifies unexplored time points. *Adapted with permission from Colonnese MT, Kaminska A, Minlebaev M, Milh M, Bloem B, Lescure S, Moriette G, Chiron C, Ben-Ari Y, Khazipov R (2010) A conserved switch in sensory processing prepares developing neocortex for vision. Neuron 67: 480–498.*

sensory systems. This is consistent with our notion that timing is tied to active exploration, which in humans occurs for all systems around the same time, namely birth.

The mechanisms responsible for the timing of this switch are poorly understood. The close correlation with the acquisition of continuous activity implicates the regulation of spontaneous activity in the switch. This is confirmed by observations that surgical isolation of mid/hindbrain ascending neuromodulatory systems from the neocortex reinstates early bursting as well as increases the discontinuity of activity (Figure 8.6). Such a regulation of sensory processing by the level of synaptic background activity is consistent with observations in adult animals (Destexhe et al., 2003). Neocortical neurons, during waking or 'up-states' observed during slow-wave sleep, receive constant synaptic activity, which is maintained through local network dynamics (Haider and McCormick, 2009) that can modulate gain (Chance et al., 2002) and adjust the sensitivity to inputs (Arieli et al., 1996; Borg-Graham et al., 1998). The timing of this switch is closely linked to the surge in synaptic density observed in the cerebral cortex that begins just before birth in monkeys and humans (Bourgeois and Rakic, 1996; Huttenlocher and Dabholkar, 1997) and before nest-exit/eye-opening in rats (Blue and Parnavelas, 1983). Interestingly, this is one of the sole features of neocortical development that violates the conserved timing of neurogenic events across mammalian species (Clancy et al., 2001), suggesting it is independently regulated to support the active exploration of the sensory world that will follow. Despite this violation of neurogenic timing and heterogeneity between sensory systems, the timing of the switch does not appear to be experience-dependent, as neither dark rearing nor early eye-opening could modify its timing in rats (Colonnese et al., 2010).

The evidence reviewed so far in this chapter is consistent with the hypothesis that early circuit development is characterized by a unique state of cortical-network dynamics in which spontaneous activity is maintained in a discontinuous mode and inputs are amplified by cortical-burst oscillations. This period appears tightly regulated to coincide with the sensory isolation experienced in the womb (or maintained by eye- or ear-canal closure in rodents), and to end when sensory processing is initiated. From this point on, many of the gross characteristics of adult cortical activity can be discerned, including clear regulation of cortical EEG by sleep state (Jouvet-Mounier et al., 1970), increased frequency of delta-waves in slow-wave sleep (Seelke and Blumberg, 2008, 2010), and the reduction and eventual elimination of SATs in counterpoint to continuous activity (Vanhatalo and Kaila, 2006). However, the development of cortical activity is by no means complete by this point. Again, this has been examined most deeply in the visual

system. Immediately following the change from immature patterns in the visual cortex, the pre-critical period is initiated and appears to exist in a transitional time, after the elimination of early activity patterns but before the acquisition of completely adult patterns. For example, cortical activity in ferrets freely viewing natural scenes just after eye-opening was dominated by highly correlated spontaneous activity bursts, an effect that was reduced by the time of the critical period (Fiser et al., 2004). These continuing changes are likely driven by many of the same developmental processes underlying the early switch in cortical activities such as increasing synaptic density and changes in neuromodulatory systems. However, several unique processes are also occurring over this time to mature cortical dynamics, including but not limited to the increasing sparcification of cortical circuits (Rochefort et al., 2009) driven by changes in intrinsic excitability (Golshani et al., 2009), increased reliability of the synaptic drive of layer 2/3 cells (Stern et al., 2001), the development of perisomatic inhibition (Hensch et al., 1998; Huang et al., 1999), and the final acquisition of strong sleep-associated rhythms (Miyamoto et al., 2003) that are necessary for the opening of the critical period for ocular-dominance plasticity.

Thus, throughout the development of the sensory systems, the changing network dynamics of the thalamocortical circuits are closely tied to the changing processes of synaptic plasticity and circuit formation that depend on them.

Acknowledgments

Financial support was provided by the Agence Nationale de la Recherche (ANR_09MNPS006), the Fondation pour la Recherche Médicale en France (FRM_DEq. 20110421301), and the Government of the Russian Federation (Grant 11.G34.31.0075).

References

Adelsberger, H., Garaschuk, O., Konnerth, A., 2005. Cortical calcium waves in resting newborn mice. Nature Neuroscience 8, 988–990.

Agmon, A., Hollrigel, G., O'dowd, D.K., 1996. Functional GABAergic synaptic connection in neonatal mouse barrel cortex. Journal of Neuroscience 16, 4684–4695.

Akerman, C.J., Grubb, M.S., Thompson, I.D., 2004. Spatial and temporal properties of visual responses in the thalamus of the developing ferret. Journal of Neuroscience 24, 170–182.

Allene, C., Cossart, R., 2010. Early NMDA receptor-driven waves of activity in the developing neocortex: Physiological or pathological network oscillations? Journal de Physiologie 588, 83–91.

Allene, C., Cattani, A., Ackman, J.B., et al., 2008. Sequential generation of two distinct synapse-driven network patterns in developing neocortex. Journal of Neuroscience 28, 12851–12863.

Amzica, F., Steriade, M., 1998. Cellular substrates and laminar profile of sleep K-complex. Neuroscience 82, 671–686.

Anderson, C.M., Torres, F., Faoro, A., 1985. The EEG of the early premature. Electroencephalography and Clinical Neurophysiology 60, 95–105.

Arieli, A., Sterkin, A., Grinvald, A., Aertsen, A., 1996. Dynamics of ongoing activity: Explanation of the large variability in evoked cortical responses. Science 273, 1868–1871.

Arumugam, H., Liu, X., Colombo, P.J., Corriveau, R.A., Belousov, A.B., 2005. NMDA receptors regulate developmental gap junction uncoupling via CREB signaling. Nature Neuroscience 8, 1720–1726.

Bannister, N.J., Benke, T.A., Mellor, J., et al., 2005. Developmental changes in AMPA and kainate receptor-mediated quantal transmission at thalamocortical synapses in the barrel cortex. Journal of Neuroscience 25, 5259–5271.

Baram, T.Z., Snead, O.C., 1990. Bicuculline induced seizures in infant rats: Ontogeny of behavioral and electrocortical phenomena. Brain Research. Developmental Brain Research 57, 291–295.

Barth, A.L., Malenka, R.C., 2001. NMDAR EPSC kinetics do not regulate the critical period for LTP at thalamocortical synapses. Nature Neuroscience 4, 235–236.

Bartos, M., Vida, I., Jonas, P., 2007. Synaptic mechanisms of synchronized gamma oscillations in inhibitory interneuron networks. Nature Reviews Neuroscience 8, 45–56.

Ben-Ari, Y., Cherubini, E., Corradetti, R., Gaïarsa, J.-L., 1989. Giant synaptic potentials in immature rat CA3 hippocampal neurones. The Journal of Physiology (London) 416, 303–325.

Ben-Ari, Y., Khazipov, R., Leinekugel, X., Caillard, O., Gaïarsa, J.-L., 1997. GABA$_a$, NMDA and AMPA receptors: A developmentally regulated 'Ménage A Trois'. Trends in Neurosciences 20, 523–529.

Ben Ari, Y., Gaiarsa, J.L., Tyzio, R., Khazipov, R., 2007. GABA: A pioneer transmitter that excites immature neurons and generates primitive oscillations. Physiological Reviews 87, 1215–1284.

Berridge, C.W., Waterhouse, B.D., 2003. The locus coeruleus-noradrenergic system: Modulation of behavioral state and state-dependent cognitive processes. Brain Research. Brain Research Reviews 42, 33–84.

Blankenship, A.G., Feller, M.B., 2010. Mechanisms underlying spontaneous patterned activity in developing neural circuits. Nature Reviews Neuroscience 11, 18–29.

Blankenship, A.G., Ford, K.J., Johnson, J., et al., 2009. Synaptic and extrasynaptic factors governing glutamatergic retinal waves. Neuron 62, 230–241.

Blue, M.E., Parnavelas, J.G., 1983. The formation and maturation of synapses in the visual cortex of the rat. I. Qualitative analysis. Journal of Neurocytology 12, 599–616.

Blumberg, M.S., Lucas, D.E., 1994. Dual mechanisms of twitching during sleep in neonatal rats. Behavioral Neuroscience 108, 1196–1202.

Bolea, S., Avignone, E., Berretta, N., Sanchez-Andres, J.V., Cherubini, E., 1999. Glutamate controls the induction of GABA-mediated giant depolarizing potentials through AMPA receptors in neonatal Rat Hippocampal slices. Journal of Neurophysiology 81, 2095–2102.

Borg-Graham, L.J., Monier, C., Fregnac, Y., 1998. Visual input evokes transient and strong shunting inhibition in visual cortical neurons. Nature 393, 369–373.

Bourgeois, J.P., Rakic, P., 1996. Synaptogenesis in the occipital cortex of macaque monkey devoid of retinal input from early embryonic stages. The European Journal of Neuroscience 8, 942–950.

Bureau, I., Shepherd, G.M., Svoboda, K., 2004. Precise development of functional and anatomical columns in the neocortex. Neuron 42, 789–801.

Burkhalter, A., Bernardo, K.L., Charles, V., 1993. Development of local circuits in human visual cortex. Journal of Neuroscience 13, 1916–1931.

Buzsaki, G., 2006. Rhythms of the Brain. Oxford University Press, New York.

Callaway, E.M., Katz, L.C., 1990. Emergence and refinement of clustered horizontal connections in cat striate cortex. Journal of Neuroscience 10, 1134–1153.

Cang, J., Renteria, R.C., Kaneko, M., Liu, X., Copenhagen, D.R., Stryker, M.P., 2005. Development of precise maps in visual cortex requires patterned spontaneous activity in the retina. Neuron 48, 797–809.

Carmignoto, G., Vicini, S., 1992. Activity-dependent decrease in NMDA receptor responses during development of the visual cortex. Science 258, 1007–1011.

Cases, O., Vitalis, T., Seif, I., De Maeyer, E., Sotelo, C., Gaspar, P., 1996. Lack of barrels in the somatosensory cortex of monoamine oxidase A-deficient mice: Role of a serotonin excess during the critical period. Neuron 16, 297–307.

Catalano, S.M., Shatz, C.J., 1998. Activity-dependent cortical target selection by thalamic axons. Science 281, 559–562.

Chagnac-Amitai, Y., Connors, B.W., 1989. Horizontal spread of synchronized activity in neocortex and its control by GABA-mediated inhibition. Journal of Neurophysiology 61, 747–758.

Chance, F.S., Abbott, L.F., Reyes, A.D., 2002. Gain modulation from background synaptic input. Neuron 35, 773–782.

Chandrasekaran, A.R., Plas, D.T., Gonzalez, E., Crair, M.C., 2005. Evidence for an instructive role of retinal activity in retinotopic map refinement in the superior colliculus of the mouse. Journal of Neuroscience 25, 6929–6938.

Chapman, B., Stryker, M.P., 1993. Development of orientation selectivity in ferret visual cortex and effects of deprivation. Journal of Neuroscience 13, 5251–5262.

Chittajallu, R., Isaac, J.T., 2010. Emergence of cortical inhibition by coordinated sensory-driven plasticity at distinct synaptic loci. Nature Neuroscience 13, 1240–1248.

Chiu, C., Weliky, M., 2001. Spontaneous activity in developing ferret visual cortex in vivo. Journal of Neuroscience 21, 8906–8914.

Chiu, C., Weliky, M., 2002. Relationship of correlated spontaneous activity to functional ocular dominance columns in the developing visual cortex. Neuron 35, 1123–1134.

Cioni, G., Prechtl, H.F., 1990. Preterm and early postterm motor behaviour in low-risk premature infants. Early Human Development 23, 159–191.

Clancy, B., Darlington, R.B., Finlay, B.L., 2001. Translating developmental time across mammalian species. Neuroscience 105, 7–17.

Colonnese, M.T., Khazipov, R., 2010. "Slow activity transients" in infant rat visual cortex: A spreading synchronous oscillation patterned by retinal waves. Journal of Neuroscience 30, 4325–4337.

Colonnese, M.T., Kaminska, A., Minlebaev, M., et al., 2010. A conserved switch in sensory processing prepares developing neocortex for vision. Neuron 67, 480–498.

Connell, J.A., Oozeer, R., Dubowitz, V., 1987. Continuous 4-channel EEG monitoring: A guide to interpretation, with normal values, in preterm infants. Neuropediatrics 18, 138–145.

Contreras, D., Timofeev, I., Steriade, M., 1996. Mechanisms of long-lasting hyperpolarizations underlying slow sleep oscillations in cat corticothalamic networks. Journal de Physiologie 494 (Pt 1), 251–264.

Contreras, D., Destexhe, A., Sejnowski, T.J., Steriade, M., 1997. Spatiotemporal patterns of spindle oscillations in cortex and thalamus. Journal of Neuroscience 17, 1179–1196.

Crair, M.C., Malenka, R.C., 1995. A critical period for long-term potentiation at thalamocortical synapses [see comments]. Nature 375, 325–328.

Crepel, V., Aronov, D., Jorquera, I., Represa, A., Ben Ari, Y., Cossart, R., 2007. A parturition-associated nonsynaptic coherent activity pattern in the developing hippocampus. Neuron 54, 105–120.

Crowley, J.C., Katz, L.C., 1999. Development of ocular dominance columns in the absence of retinal input. Nature Neuroscience 2, 1125–1130.

Curzi-Dascalova, L., Figueroa, J.M., Eiselt, M., et al., 1993. Sleep state organization in premature infants of less than 35 weeks' gestational age. Pediatric Research 34, 624–628.

Dagnew, E., Latchamsetty, K., Erinjeri, J.P., Miller, B., Fox, K., Woolsey, T.A., 2003. Glutamate receptor blockade alters the development of intracortical connections in rat barrel cortex. Somatosensory and Motor Research 20, 77–84.

Dan, Y., Poo, M.M., 2006. Spike timing-dependent plasticity: From synapse to perception. Physiological Reviews 86, 1033–1048.

Daw, M.I., Bannister, N.V., Isaac, J.T.R., 2006. Rapid, activity-dependent plasticity in timing precision in neonatal barrel cortex. Journal of Neuroscience 26, 4178–4187.

Daw, M.I., Ashby, M.C., Isaac, J.T., 2007. Coordinated developmental recruitment of latent fast spiking interneurons in layer IV barrel cortex. Nature Neuroscience 10, 453–461.

De La Prida, L.M., Bolea, S., Sanchez-Andres, J.V., 1998. Origin of the synchronized network activity in the rabbit developing hippocampus. European Journal of Neuroscience 10, 899–906.

De Vries, J.I., Visser, G.H., Prechtl, H.F., 1982. The emergence of fetal behaviour. I. Qualitative aspects. Early Human Development 7, 301–322.

Destexhe, A., Rudolph, M., Pare, D., 2003. The high-conductance state of neocortical neurons in vivo. Nature Reviews Neuroscience 4, 739–751.

Doischer, D., Hosp, J.A., Yanagawa, Y., et al., 2008. Postnatal differentiation of basket cells from slow to fast signaling devices. Journal of Neuroscience 28, 12956–12968.

Dreyfus-Brisac, C., 1962. The electroencephalogram of the premature infant. World Neurology 3 (5–15), 5–15.

Dreyfus-Brisac, C., Larroche, J.C., 1971. Discontinuous electroencephalograms in the premature newborn and at term. electro-anatomo-clinical correlations. Revue d'Électroencéphalographie et de Neurophysiologie Clinique 1, 95–99.

Dreyfus-Brisac, C., Fischgold, H., Samson-Dollfus, D., et al., 1956. Veille sommeil et reactivite sensorielle chez le premature et le nouveau-ne. Electroencephalography and Clinical Neurophysiology 6, 418–440.

Dupont, E., Hanganu, I.L., Kilb, W., Hirsch, S., Luhmann, H.J., 2006. Rapid developmental switch in the mechanisms driving early cortical columnar networks. Nature 439, 79–83.

Durack, J.C., Katz, L.C., 1996. Development of horizontal projections in layer 2/3 of ferret visual cortex. Cerebral Cortex 6, 178–183.

Durand, G.M., Kovalchuk, Y., Konnerth, A., 1996. Long-term potentiation and functional synapse induction in developing hippocampus. Nature 381, 71–75.

Dzhala, V.I., Talos, D.M., Sdrulla, D.A., et al., 2005. NKCC1 transporter facilitates seizures in the developing brain. Nature Medicine 11, 1205–1213.

Ellingson, R.J., 1958. Electroencephalograms of normal, full-term newborns immediately after birth with observations on arousal and visual evoked responses. Electroencephalography and Clinical Neurophysiology. Supplement 10, 31–50.

Engel, R., 1975. Abnormal Electroencephalograms in the Neonatal Period. Charles C Thomas, Springfield, IL.

Erzurumlu, R.S., Jhaveri, S., 1990. Thalamic axons confer a blueprint of the sensory periphery onto the developing rat somatosensory cortex. Brain Research. Developmental Brain Research 56, 229–234.

Feldman, D.E., Nicoll, R.A., Malenka, R.C., Isaac, J.T., 1998. Long-term depression at thalamocortical synapses in developing rat somatosensory cortex. Neuron 21, 347–357.

Feldman, D.E., Nicoll, R.A., Malenka, R.C., 1999. Synaptic plasticity at thalamocortical synapses in developing rat somatosensory cortex: LTP, LTD, and silent synapses. Journal of Neurobiology 41, 92–101.

Feller, M.B., Scanziani, M., 2005. A precritical period for plasticity in visual cortex. Current Opinion in Neurobiology 15, 94–100.

Ferezou, I., Bolea, S., Petersen, C.C.H., 2006. Visualizing the cortical representation of whisker touch: Voltage-sensitive dye imaging in freely moving mice. Neuron 50, 617–629.

Fiser, J., Chiu, C., Weliky, M., 2004. Small modulation of ongoing cortical dynamics by sensory input during natural vision. Nature 431, 573–578.

Foeller, E., Feldman, D.E., 2004. Synaptic basis for developmental plasticity in somatosensory cortex. Current Opinion in Neurobiology 14, 89–95.

Foote, S.L., Morrison, J.H., 1987a. Development of the noradrenergic, serotonergic, and dopaminergic innervation of neocortex. Current Topics in Developmental Biology 21, 391–423.

Foote, S.L., Morrison, J.H., 1987b. Extrathalamic modulation of cortical function. Annual Review of Neuroscience 10, 67–95.

Fox, K., 1992. A critical period for experience-dependent synaptic plasticity in rat barrel cortex. Journal of Neuroscience 12, 1826–1838.

Fox, K., 2002. Anatomical pathways and molecular mechanisms for plasticity in the barrel cortex. Neuroscience 111, 799–814.

Fox, K., Wong, R.O.L., 2005. A comparison of experience-dependent plasticity in the visual and somatosensory systems. Neuron 48, 465–477.

Fox, K., Schlaggar, B.L., Glazewski, S., O'leary, D.D., 1996. Glutamate receptor blockade at cortical synapses disrupts development of thalamocortical and columnar organization in somatosensory cortex. Proceedings of the National Academy of Sciences of the United States of America 93, 5584–5589.

Fox, K., Wright, N., Wallace, H., Glazewski, S., 2003. The origin of cortical surround receptive fields studied in the barrel cortex. Journal of Neuroscience 23, 8380–8391.

Frank, M.G., Heller, H.C., 1997. Development of REM and slow wave sleep in the rat. American Journal of Physiology 272, R1792–R1799.

Freund, T., Buzsaki, G., 1996. Interneurons of the hippocampus. Hippocampus 6, 345–470.

Fries, P., 2009. Neuronal gamma-band synchronization as a fundamental process in cortical computation. Annual Review of Neuroscience 32, 209–224.

Gall, A.J., Joshi, B., Best, J., Florang, V.R., Doorn, J.A., Blumberg, M.S., 2009. Developmental emergence of power-law wake behavior depends upon the functional integrity of the locus coeruleus. Sleep 32, 920–926.

Galli, L., Maffei, L., 1988. Spontaneous impulse activity of rat retinal ganglion cells in prenatal life. Science 242, 90–91.

Galuske, R.A., Singer, W., 1996. The origin and topography of long-range intrinsic projections in cat visual cortex: A developmental study. Cerebral Cortex 6, 417–430.

Gao, X.B., Chen, G., Van Den Pol, A.N., 1998. GABA-dependent firing of glutamate-evoked action potentials at AMPA/kainate receptors in developing hypothalamic neurons. Journal of Neurophysiology 79, 716–726.

Garaschuk, O., Linn, J., Eilers, J., Konnerth, A., 2000. Large-scale oscillatory calcium waves in the immature cortex. Nature 3, 452–459.

Gasparini, S., Saviane, C., Voronin, L.L., Cherubini, E., 2000. Silent synapses in the developing hippocampus: Lack of functional AMPA receptors or low probability of glutamate release? Proceedings of the National Academy of Sciences of the United States of America 97, 9741–9746.

Goldie, L., Svedsen-Rhodes, U., Easton, J., Roberton, N.R., 1971. The development of innate sleep rhythms in short gestation infants. Developmental Medicine and Child Neurology 13, 40–50.

Golshani, P., Goncalves, J.T., Khoshkhoo, S., Mostany, R., Smirnakis, S., Portera-Cailliau, C., 2009. Internally mediated developmental desynchronization of neocortical network activity. Journal of Neuroscience 29, 10890–10899.

Gramsbergen, A., 1976. The development of the EEG in the rat. Developmental Psychobiology 9, 501–515.

Gray, C.M., Singer, W., 1989. Stimulus-specific neuronal oscillations in orientation columns of cat visual cortex. Proceedings of the National Academy of Sciences of the United States of America 86, 1698–1702.

Grubb, M.S., Rossi, F.M., Changeux, J.P., Thompson, I.D., 2003. Abnormal functional organization in the dorsal lateral geniculate nucleus of mice lacking the beta 2 subunit of the nicotinic acetylcholine receptor. Neuron 40, 1161–1172.

Gulledge, A.T., Stuart, G.J., 2003. Excitatory actions of GABA in the cortex. Neuron 37, 299–309.

Haider, B., Mccormick, D.A., 2009. Rapid neocortical dynamics: Cellular and network mechanisms. Neuron 62, 171–189.

Hamburger, V., 1975. Fetal behavior. In: Hafez, E.S. (Ed.), The Mammalian Fetus: Comparitive Biology and Methodology. Charles C Thomas, Springfield, pp. 69–81.

Hanganu, I.L., Ben Ari, Y., Khazipov, R., 2006. Retinal waves trigger spindle bursts in the neonatal rat visual cortex. Journal of Neuroscience 26, 6728–6736.

Hanganu, I.L., Staiger, J.F., Ben Ari, Y., Khazipov, R., 2007. Cholinergic modulation of spindle bursts in the neonatal rat visual cortex in vivo. Journal of Neuroscience 27, 5694–5705.

Hebb, D.O., 1949. The Organization of Behaviour. John Wiley & Sons, New York.

Henley, J., Poo, M.M., 2004. Guiding neuronal growth cones using Ca^{2+} signals. Trends in Cell Biology 14, 320–330.

Hensch, T.K., Fagiolini, M., Mataga, N., Stryker, M.P., Baekkeskov, S., Kash, S.F., 1998. Local GABA circuit control of experience-dependent plasticity in developing visual cortex. Science 282, 1504–1508.

Hestrin, S., 1992. Developmental regulation of NMDA receptor-mediated synaptic currents at a central synapse. Nature 357, 686–689.

Higashi, S., Molnar, Z., Kurotani, T., Toyama, K., 2002. Prenatal development of neural excitation in rat thalamocortical projections studied by optical recording. Neuroscience 115, 1231–1246.

Huang, Z.J., Kirkwood, A., Pizzorusso, T., et al., 1999. BDNF regulates the maturation of inhibition and the critical period of plasticity in mouse visual cortex. Cell 98, 739–755.

Hubel, D.H., Wiesel, T.N., 1970. The period of susceptibility to the physiological effects of unilateral eye closure in kittens. The Journal of Physiology (London) 206, 419–436.

Huberman, A.D., Feller, M.B., Chapman, B., 2008. Mechanisms underlying development of visual maps and receptive fields. Annual Review of Neuroscience 31, 479–509.

Huttenlocher, P.R., Dabholkar, A.S., 1997. Regional differences in synaptogenesis in human cerebral cortex. The Journal of Comparative Neurology 387, 167–178.

Isaac, J.T.R., Nicoll, R.A., Malenka, R.C., 1995. Evidence for silent synapses: Implications for the expression of LTP. Neuron 15, 427–434.

Isaac, J.T.R., Crair, M.C., Nicoll, R.A., Malenka, R.C., 1997. Silent synapses during development of thalamocortical inputs. Neuron 18, 269–280.

Iwasato, T., Datwani, A., Wolf, A.M., et al., 2000. Cortex-restricted disruption of NMDAR1 impairs neuronal patterns in the barrel cortex. Nature 406, 726–731.

Johnston, M.V., Coyle, J.T., 1981. Development of central neurotransmitter systems. Ciba Foundation Symposium 86, 251–270.

Jouvet-Mounier, D., Astic, L., Lacote, D., 1970. Ontogenesis of the states of sleep in rat, cat, and guinea pig during the first postnatal month. Developmental Psychobiology 2, 216–239.

Kandler, K., Thiels, E., 2005. Flipping the switch from electrical to chemical communication. Nature Neuroscience 8, 1633–1634.

Karlsson, K.A., Blumberg, M.S., 2005. Active medullary control of atonia in week-old rats. Neuroscience 130, 275–283.

Karlsson, K.A., Gall, A.J., Mohns, E.J., Seelke, A.M., Blumberg, M.S., 2005. The neural substrates of infant sleep in rats. PLoS Biology 3, E143.

Katz, L.C., Crowley, J.C., 2002. Development of cortical circuits: Lessons from ocular dominance columns. Nature Reviews Neuroscience 3, 34–42.

Katz, L.C., Shatz, C.J., 1996. Synaptic activity and the construction of cortical circuits. Science 274, 1133–1138.

Kerschensteiner, D., Wong, R.O., 2008. A precisely timed asynchronous pattern of on and off retinal ganglion cell activity during propagation of retinal waves. Neuron 58, 851–858.

Khazipov, R., Ragozzino, D., Bregestovski, P., 1995. Kinetics and Mg^{2+} block of N-methyl-D-aspartate receptor channels during postnatal development of hippocampal CA3 pyramidal neurons. Neuroscience 69, 1057–1065.

Khazipov, R., Leinekugel, X., Khalilov, I., Gaïarsa, J.-L., Ben-Ari, Y., 1997. Synchronization of GABAergic interneuronal network in CA3 subfield of neonatal rat hippocampal slices. The Journal of Physiology (London) 498, 763–772.

Khazipov, R., Khalilov, I., Tyzio, R., Morozova, E., Ben Ari, Y., Holmes, G.L., 2004a. Developmental changes in GABAergic actions and seizure susceptibility in the rat hippocampus. European Journal of Neuroscience 19, 590–600.

Khazipov, R., Luhmann, H.J., 2006. Early patterns of electrical activity in the developing cerebral cortex of humans and rodents. Trends in Neuroscience 29, 414–418.

Khazipov, R., Sirota, A., Leinekugel, X., Holmes, G.L., Ben Ari, Y., Buzsaki, G., 2004b. Early motor activity drives spindle bursts in the developing somatosensory cortex. Nature 432, 758–761.

Kidd, F.L., Isaac, J.T., 1999. Developmental and activity-dependent regulation of kainate receptors at thalamocortical synapses. Nature 400, 569–573.

Kreider, J.C., Blumberg, M.S., 2000. Mesopontine contribution to the expression of active 'Twitch' sleep in decerebrate week-old rats. Brain Research 872, 149–159.

Lamblin, M.D., Andre, M., Challamel, M.J., et al., 1999. Electroencephalography of the premature and term newborn. Maturational aspects and glossary. Neurophysiologie Clinique 29, 123–219.

Lamsa, K., Palva, J.M., Ruusuvuori, E., Kaila, K., Taira, T., 2000. Synaptic GABA(A) activation inhibits AMPA-kainate receptor-mediated bursting in the newborn (P0-P2) rat hippocampus. Journal of Neurophysiology 83, 359–366.

Landers, M., Philip, Z.H., 2006. Development of rodent whisking: Trigeminal input and central pattern generation. Somatosensory and Motor Research 23, 1–10.

Latsari, M., Dori, I., Antonopoulos, J., Chiotelli, M., Dinopoulos, A., 2002. Noradrenergic innervation of the developing and mature visual and motor cortex of the rat brain: A light and electron microscopic immunocytochemical analysis. The Journal of Comparative Neurology 445, 145–158.

Lee, L.J., Iwasato, T., Itohara, S., Erzurumlu, R.S., 2005a. Exuberant thalamocortical axon arborization in cortex-specific NMDAR1 knockout mice. Journal of Comparative Neurology 485, 280–292.

Lee, L.J., Lo, F.S., Erzurumlu, R.S., 2005b. Nmda receptor-dependent regulation of axonal and dendritic branching. Journal of Neuroscience 25, 2304–2311.

Leinekugel, X., Medina, I., Khalilov, I., Ben-Ari, Y., Khazipov, R., 1997. Ca^{2+} oscillations mediated by the synergistic excitatory actions of $GABA_a$ and NMDA receptors in the neonatal hippocampus. Neuron 18, 243–255.

Leinekugel, X., Khazipov, R., Cannon, R., Hirase, H., Ben Ari, Y., Buzsaki, G., 2002. Correlated bursts of activity in the neonatal hippocampus in vivo. Science 296, 2049–2052.

Loturco, J.J., Blanton, M.G., Kriegstein, A.R., 1991. Initial expression and endogenous activation of NMDA channels in early neocortical development. Journal of Neuroscience 11, 792–799.

Loturco, J.J., Owens, D.F., Heath, M.J., Davis, M.B., Kriegstein, A.R., 1995. GABA and glutamate depolarize cortical progenitor cells and inhibit DNA synthesis. Neuron 15, 1287–1298.

Lu, T., Trussell, L.O., 2001. Mixed excitatory and inhibitory GABA-mediated transmission in chick cochlear nucleus. The Journal of Physiology 535, 125–131.

Lu, H.C., Butts, D.A., Kaeser, P.S., She, W.C., Janz, R., Crair, M.C., 2006. Role of efficient neurotransmitter release in barrel map development. Journal of Neuroscience 26, 2692–2703.

Luhmann, H.J., Prince, D.A., 1991. Postnatal maturation of the GABAergic system in rat neocortex. Journal of Neurophysiology 65, 247–263.

Luhmann, H.J., Martinez, M.L., Singer, W., 1986. Development of horizontal intrinsic connections in cat striate cortex. Experimental Brain Research 63, 443–448.

Marcano-Reik, A.J., Blumberg, M.S., 2008. The corpus callosum modulates spindle-burst activity within homotopic regions of somatosensory cortex in newborn rats. European Journal of Neuroscience 28, 1457–1466.

Marcano-Reik, A.J., Prasad, T., Weiner, J.A., Blumberg, M.S., 2010. An abrupt developmental shift in callosal modulation of sleep-related spindle bursts coincides with the emergence of excitatory-inhibitory balance and a reduction of somatosensory cortical plasticity. Behavioral Neuroscience 124, 600–611.

Mclaughlin, T., Torborg, C.L., Feller, M.B., O'leary, D.D., 2003. Retinotopic Map refinement requires spontaneous retinal waves during a brief critical period of development. Neuron 40, 1147–1160.

Mechawar, N., Descarries, L., 2001. The cholinergic innervation develops early and rapidly in the rat cerebral cortex: A quantitative immunocytochemical study. Neuroscience 108, 555–567.

Meister, M., Wong, R.O., Baylor, D.A., Shatz, C.J., 1991. Synchronous bursts of action potentials in ganglion cells of the developing mammalian retina. Science 252, 939–943.

Menendez, D.L.P., Bolea, S., Sanchez-Andres, J.V., 1996. Analytical characterization of spontaneous activity evolution during hippocampal development in the rabbit. Neuroscience Letters 218, 185–187.

Milh, M., Kaminska, A., Huon, C., Lapillonne, A., Ben Ari, Y., Khazipov, R., 2007. Rapid cortical oscillations and early motor activity in premature human neonate. Cerebral Cortex 17, 1582–1594.

Minlebaev, M., Ben-Ari, Y., Khazipov, R., 2007. Network mechanisms of spindle-burst oscillations in the neonatal rat barrel cortex in vivo. Journal of Neurophysiology 97, 692–700.

Minlebaev, M., Ben Ari, Y., Khazipov, R., 2009. Nmda receptors pattern early activity in the developing barrel cortex in vivo. Cerebral Cortex 19, 688–696.

Minlebaev, M., Colonnese, M., Tsintsadze, T., Sirota, A., Khazipov, R., 2011. Early gamma oscillations synchronize developing thalamus and cortex. Science 334, 226–229.

Miyamoto, H., Katagiri, H., Hensch, T., 2003. Experience-dependent slow-wave sleep development. Nature Neuroscience 6, 553–554.

Mohns, E.J., Blumberg, M.S., 2008. Synchronous bursts of neuronal activity in the developing hippocampus: Modulation by active sleep and association with emerging gamma and theta rhythms. Journal of Neuroscience 28, 10134–10144.

Molnar, Z., Higashi, S., Lopez-Bendito, G., 2003. Choreography of early thalamocortical development. Cerebral Cortex 13, 661–669.

Monyer, H., Burnashev, N., Laurie, D.J., Sakmann, B., Seeburg, P.H., 1994. Developmental and regional expression in the rat brain and functional properties of four NMDA receptors. Neuron 12, 529–540.

Mooney, R., Penn, A.A., Gallego, R., Shatz, C.J., 1996. Thalamic relay of spontaneous retinal activity prior to vision. Neuron 17, 863–874.

Mrsic-Flogel, T.D., Hofer, S.B., Creutzfeldt, C., et al., 2005. Altered map of visual space in the superior colliculus of mice lacking early retinal waves. Journal of Neuroscience 25, 6921–6928.

Mu, Y., Poo, M.M., 2006. Spike timing-dependent LTP/LTD mediates visual experience-dependent plasticity in a developing retinotectal system. Neuron 50, 115–125.

Muir-Robinson, G., Hwang, B.J., Feller, M.B., 2002. Retinogeniculate axons undergo eye-specific segregation in the absence of eye-specific layers. Journal of Neuroscience 22, 5259–5264.

Nicol, X., Voyatzis, S., Muzerelle, A., et al., 2007. Camp oscillations and retinal activity are permissive for ephrin signaling during the establishment of the retinotopic Map. Nature Neuroscience 10, 340–347.

Niedermeyer, E., (2005). Maturation of the EEG: Development of waking and sleep patterns. In: Niedermeyer, E. Lopez Da Silva, F. (Eds.), Electroencephalography: Basic Principles, Clinical Applications and Related Fields, pp. 209–234. Philadelphia: Lippincott Williams and Wilkins.

Nolte, R., Schulte, F.J., Michaelis, R., Weisse, U., Gruson, R., 1969. Bioelectric brain maturation in small-for-dates infants. Developmental Medicine and Child Neurology 11, 83–93.

O'Donovan, M.J., 1999. The origin of spontaneous activity in developing networks of the vertebrate nervous system. Current Opinion in Neurobiology 9, 94–104.

Ohshiro, T., Weliky, M., 2006. Simple fall-off pattern of correlated neural activity in the developing lateral geniculate nucleus. Nature Neuroscience 9, 1541–1548.

O'leary, D.D., Ruff, N.L., Dyck, R.H., 1994. Development, critical period plasticity, and adult reorganizations of mammalian somatosensory systems. Current Opinion in Neurobiology 4, 535–544.

Owens, D.F., Boyce, L.H., Davis, M.B., Kriegstein, A.R., 1996. Excitatory GABA responses in embryonic and neonatal cortical slices demonstrated by gramicidin perforated-patch recordings and calcium imaging. Journal of Neuroscience 16, 6414–6423.

Parmelee, A.H., Akiyama, Y., Stern, E., Harris, M.A., 1969. A periodic cerebral rhythm in newborn infants. Experimental Neurology 25, 575–584.

Penn, A.A., Riquelme, P.A., Feller, M.B., Shatz, C.J., 1998. Competition in retinogeniculate patterning driven by spontaneous activity. Science 279, 2108–2112.

Persico, A.M., Mengual, E., Moessner, R., et al., 2001. Barrel pattern formation requires serotonin uptake by thalamocortical afferents, and not vesicular monoamine release. Journal of Neuroscience 21, 6862–6873.

Petersen, C., 2007. The functional organization of the barrel cortex. Neuron 56, 339–355.

Petersen, C.C.H., Sakmann, B., 2001. Functionally independent columns of rat somatosensory barrel cortex revealed with voltage-sensitive dye imaging. Journal of Neuroscience 21, 8435–8446.

Petersson, P., Waldenstrom, A., Fahraeus, C., Schouenborg, J., 2003. Spontaneous muscle twitches during sleep guide spinal self-organization. Nature 424, 72–75.

Prechtl, H.F., 1997. State of the art of a new functional assessment of the young nervous system. An early predictor of cerebral palsy. Early Human Development 50, 1–11.

Price, D.J., Kennedy, H., Dehay, C., et al., 2006. The development of cortical connections. European Journal of Neuroscience 23, 910–920.

Rakic, P., 1976. Prenatal genesis of connections subserving ocular dominance in the rhesus monkey. Nature 261, 467–471.

Rakic, P., Komuro, H., 1995. The role of receptor/channel activity in neuronal cell migration. Journal of Neurobiology 26, 299–315.

Rheims, S., Minlebaev, M., Ivanov, A., et al., 2008. Excitatory GABA in rodent developing neocortex in vitro. Journal of Neurophysiology 100, 609–619.

Rivera, C., Voipio, J., Payne, J.A., et al., 1999. The K^+/Cl^- Cotransporter KCC2 renders GABA hyperpolarizing during neuronal maturation. Nature 397, 251–255.

Rochefort, N.L., Garaschuk, O., Milos, R.I., et al., 2009. Sparsification of neuronal activity in the visual cortex at eye-opening. Proceedings of the National Academy of Sciences of the United States of America 106, 15049–15054.

Rubenstein, J.L.R., Rakic, P., 2013. Patterning and Cell Types Specification in the Developing CNS and PNS.

Ruthazer, E.S., Stryker, M.P., 1996. The role of activity in the development of long-range horizontal connections in area 17 of the ferret. Journal of Neuroscience 16, 7253–7269.

Sanchez-Vives, M.V., McCormick, D.A., 2000. Cellular and network mechanisms of rhythmic recurrent activity in neocortex. Nature Neuroscience 3, 1027–1034.

Scher, M.S., 2006. Electroencephalography of the newborn: Normal features. In: Holmes, G.L., Moshe, S., Jones, R.H. (Eds.), Clinical Neurophysiology of Infancy, Childhood and Adolescence. Elsevier, St. Louis, MO, pp. 46–69.

Schlaggar, B.L., Fox, K., O'leary, D.M., 1993. Postsynaptic control of plasticity in developing somatosensory cortex. Nature 364, 623–626.

Seelke, A.M., Blumberg, M.S., 2008. The microstructure of active and quiet sleep as cortical delta activity emerges in infant rats. Sleep 31, 691–699.

Seelke, A.M., Blumberg, M.S., 2010. Developmental appearance and disappearance of cortical events and oscillations in infant rats. Brain Research 1324, 34–42.

Shatz, C.J., Stryker, M.P., 1988. Prenatal tetrodotoxin infusion blocks segregation of retinogeniculate afferents. Science 242, 87–89.

Shu, Y., Hasenstaub, A., Mccormick, D.A., 2003. Turning on and off recurrent balanced cortical activity. Nature 423, 288–293.

Sipila, S.T., Huttu, K., Soltesz, I., Voipio, J., Kaila, K., 2005. Depolarizing GABA acts on intrinsically bursting pyramidal neurons to drive giant depolarizing potentials in the immature hippocampus. Journal of Neuroscience 25, 5280–5289.

Sipila, S.T., Schuchmann, S., Voipio, J., Yamada, J., Kaila, K., 2006. The Na-K-Cl cotransporter (NKCC1) promotes sharp waves in the neonatal rat hippocampus. Journal of Physiology 573, 765–773.

Smith, S.L., Trachtenberg, J.T., 2007. Experience-dependent binocular competition in the visual cortex begins at eye opening. Nature Neuroscience 10, 370–375.

Stellwagen, D., Shatz, C.J., 2002. An instructive role for retinal waves in the development of retinogeniculate connectivity. Neuron 33, 357–367.

Steriade, M., 2001. Impact of network activities on neuronal properties in corticothalamic systems. Journal of Neurophysiology 86, 1–39.

Steriade, M., Mccormick, D.A., Sejnowski, T.J., 1993. Thalamocortical oscillations in the sleeping and aroused brain. Science 262, 679–685.

Stern, E.A., Maravall, M., Svoboda, K., 2001. Rapid development and plasticity of layer 2/3 maps in rat barrel cortex in vivo. Neuron 31, 305–315.

Stockard-Pope, J.E., Werner, S.S., Bickford, R.G., 1992. Atlas of Neonatal Electroencelography, 2nd edn. Raven Press, New York.

Stryker, M.P., Harris, W.A., 1986. Binocular impulse blockade prevents the formation of ocular dominance columns in cat visual cortex. Journal of Neuroscience 6, 2117–2133.

Sun, Q.Q., Huguenard, J.R., Prince, D.A., 2006. Barrel cortex microcircuits: Thalamocortical feedforward inhibition in spiny stellate cells is mediated by a small number of fast-spiking interneurons. Journal of Neuroscience 26, 1219–1230.

Tian, N., Copenhagen, D.R., 2003. Visual stimulation is required for refinement of on and off pathways in postnatal retina. Neuron 39, 85–96.

Timofeev, I., Grenier, F., Bazhenov, M., Sejnowski, T.J., Steriade, M., 2000. Origin of slow cortical oscillations in deafferented cortical slabs. Cerebral Cortex 10, 1185–1199.

Torborg, C.L., Feller, M.B., 2005. Spontaneous patterned retinal activity and the refinement of retinal projections. Progress in Neurobiology 76, 213–235.

Tsai, W.H., Koh, S.W., Puro, D.G., 1987. Epinephrine regulates cholinergic transmission mediated by rat retinal neurons in culture. Neuroscience 22, 675–680.

Tyzio, R., Cossart, R., Khalilov, I., et al., 2006. Maternal oxytocin triggers a transient inhibitory switch in GABA signaling in the fetal brain during delivery. Science 314, 1788–1792.

Uhlhaas, P.J., Roux, F., Rodriguez, E., Rotarska-Jagiela, A., Singer, W., 2010. Neural synchrony and the development of cortical networks. Trends in Cognitive Science 14, 72–80.

Valeeva, G., Abdullin, A., Tyzio, R., et al., 2010. Temporal coding at the immature depolarizing GABAergic synapse. Frontiers in Cellular Neuroscience 4, 17.

Van Der Loos, H., Woolsey, T.A., 1973. Somatosensory cortex: Structural alterations following early injury to sense organs. Science 179, 395–398.

Vanhatalo, S., Kaila, K., 2006. Development of neonatal EEG activity: From phenomenology to physiology. Seminars in Fetal & Neonatal Medicine 11, 471–478.

Vanhatalo, S., Tallgren, P., Andersson, S., Sainio, K., Voipio, J., Kaila, K., 2002. DC-EEG discloses prominent, very slow activity patterns during sleep in preterm infants. Clinical Neurophysiology 113, 1822–1825.

Vanhatalo, S., Palva, J.M., Andersson, S., Rivera, C., Voipio, J., Kaila, K., 2005. Slow endogenous activity transients and developmental expression of K+-Cl- cotransporter 2 in the immature human cortex. European Journal of Neuroscience 22, 2799–2804.

Venkatesan, C., Song, X.Z., Go, C.G., Kurose, H., Aoki, C., 1996. Cellular and subcellular distribution of alpha 2a-adrenergic receptors in the visual cortex of neonatal and adult rats. The Journal of Comparative Neurology 365, 79–95.

Voronin, L.L., Altinbaev, R.S., Bayazitov, I.T., et al., 2004. Postsynaptic depolarisation enhances transmitter release and causes the appearance of responses at "silent" synapses in rat hippocampus. Neuroscience 126, 45–59.

Wallace, H., Glazewski, S., Liming, K., Fox, K., 2001. The role of cortical activity in experience-dependent potentiation and depression of sensory responses in rat barrel cortex. Journal of Neuroscience 21, 3881–3894.

Wang, X.J., 2010. Neurophysiological and computational principles of cortical rhythms in cognition. Physiological Reviews 90, 1195–1268.

Warland, D.K., Huberman, A.D., Chalupa, L.M., 2006. Dynamics of spontaneous activity in the fetal macaque retina during development of retinogeniculate pathways. Journal of Neuroscience 26, 5190–5197.

Watanabe, K., Iwase, K., 1972. Spindle-like fast rhythms in the EEGs of low-birth weight infants. Developmental Medicine and Child Neurology 14, 373–381.

Wells, J.E., Porter, J.T., Agmon, A., 2000. GABAergic inhibition suppresses paroxysmal network activity in the neonatal rodent hippocampus and neocortex. Journal of Neuroscience 20, 8822–8830.

Wespatat, V., Tennigkeit, F., Singer, W., 2004. Phase sensitivity of synaptic modifications in oscillating cells of rat visual cortex. Journal of Neuroscience 24, 9067–9075.

White, L.E., Coppola, D.M., Fitzpatrick, D., 2001. The contribution of sensory experience to the maturation of orientation selectivity in ferret visual cortex. Nature 411, 1049–1052.

Wong, R.O., Meister, M., Shatz, C.J., 1993. Transient period of correlated bursting activity during development of the mammalian retina. Neuron 11, 923–938.

Woolsey, T.A., Van Der Loos, H., 1970. The structural organization of layer IV in the somatosensory region (Si) of mouse cerebral cortex. The description of a cortical field composed of discrete cytoarchitectonic units. Brain Research 17, 205–242.

Woolsey, T.A., Wann, J.R., 1976. Areal changes in mouse cortical barrels following vibrissal damage at different postnatal ages. Journal of Comparative Neurology 170, 53–66.

Yamada, J., Okabe, A., Toyoda, H., Kilb, W., Luhmann, H.J., Fukuda, A., 2004. Cl$^-$ uptake promoting depolarizing GABA actions in immature rat neocortical neurones is mediated by NKCC1. The Journal of Physiology Online 557, 829–841.

Yang, J.W., Hanganu-Opatz, I.L., Sun, J.J., Luhmann, H.J., 2009. Three patterns of oscillatory activity differentially synchronize developing neocortical networks in vivo. Journal of Neuroscience 29, 9011–9025.

Yuste, R., Katz, L.C., 1991. Control of postsynaptic Ca^{2+} influx in developing neocortex by excitatory and inhibitory neurotransmitters. Neuron 6, 333–344.

Zhou, Q., Poo, M.M., 2004a. Reversal and consolidation of activity-induced synaptic modifications. Trends in Neurosciences 27, 378–383.

Zhou, Q., Tao, H.W., Poo, M.M., 2003. Reversal and stabilization of synaptic modifications in a developing visual system. Science 300, 1953–1957.

9

Spike Timing-Dependent Plasticity

D.E. Shulz[1], D.E. Feldman[2]

[1]Centre National de la Recherche Scientifique, Gif sur Yvette, France; [2]University of California Berkeley, Berkeley, CA, USA

OUTLINE

9.1 INTRODUCTION

Since its discovery 15 years ago, spike timing-dependent plasticity (STDP) has emerged as a leading candidate mechanism for activity-dependent synaptic plasticity during neural circuit development and for driving neural plasticity and learning in adults. This is due to its physiologically realistic induction requirements, powerful Hebbian properties, and prevalence across many cell types and synapses from insects to mammals. However, STDP is a more complex and less understood phenomenon than apparent from initial studies in *in vitro* brain slices. Though STDP is widely observed *in vitro*, whether STDP is a plausible learning mechanism under natural conditions *in vivo* has been questioned. A substantial number of studies have now characterized STDP from the molecular to the circuit and systems levels. The purpose of this review is to critically analyze the data on diversity of STDP, predicted functional properties, cellular mechanisms, and its role in circuit development and adult plasticity *in vivo*. Several excellent reviews of specific aspects of STDP

Neural Circuit Development and Function in the Brain: Comprehensive Developmental Neuroscience, Volume 3 http://dx.doi.org/10.1016/B978-0-12-397267-5.00029-7

© 2013 Elsevier Inc. All rights reserved.

have been published (Abbott and Nelson, 2000; Caporale and Dan, 2008; Dan and Poo, 2006; Fino and Venance, 2011; Letzkus et al., 2007; Lisman and Spruston, 2005; Paulsen and Sejnowski, 2000). We conclude that STDP occurs broadly and with diverse cellular mechanisms; that it is powerfully modulated by dendritic excitability, network activity, and neuromodulation, which both enriches its computational role and will constrain its relevance *in vivo*; that STDP does occur *in vivo* at least under controlled experimental conditions; and that STDP may contribute to several basic features of neural circuit development and to several forms of adult learning.

9.2 DISCOVERY OF STDP

Activity-dependent long-term modification of synapse strength is critical for the development of neural circuits and underlies learning and memory in young and adult brains. The seminal theoretical description of activity-dependent plasticity was by Donald Hebb, who proposed that when cell A reliably contributes to spiking of postsynaptic cell B, synapse strength from A to B is functionally enhanced (Hebb, 1949). This rule was later amended to include weakening of ineffective synapses (Bienenstock et al., 1982; Sejnowski, 1977; Stent, 1973; von der Malsburg, 1973). In the period between the 1970s and 1990s, long-term synaptic potentiation and depression (LTP and LTD) were discovered and shown to implement Hebbian synaptic plasticity. These forms of synaptic plasticity have since been observed in almost every area of the brain. In classical LTP and LTD, the change in synaptic weight depends on temporally correlated pre- and postsynaptic activity (reviewed in Fregnac and Shulz, 1999). This is termed correlation-dependent plasticity (CDP). The temporal requirement for pre- and postsynaptic coactivation in CDP was not considered to be very stringent (±50 ms or more), and plasticity was thought to be largely independent of the precise timing and order of action potentials.

Some studies, however, noted that the order of activation of the presynaptic and postsynaptic neurons was crucial in determining the sign of plasticity (LTP or LTD) (Levy and Steward, 1983). Gustafsson et al. (1987) showed that LTP was induced if an excitatory postsynaptic potential (EPSP) preceded a postsynaptic spike burst by <100 ms and was not induced if the EPSP immediately followed the burst. This order of pre- and postsynaptic activity was later shown to induce LTD (Debanne et al., 1994, 1996). In 1997, several groups discovered that induction of LTP and LTD by pairing pre- and postsynaptic action potentials was critically dependent on the order and relative timing of single spikes, down to the millisecond scale (Bell et al., 1997; Bi and

Poo, 1998; Debanne et al., 1994, 1997; Markram et al., 1997; Figure 9.1). Confirmation at many other synapses followed. This form of temporally precise bidirectional Hebbian plasticity was termed STDP (Abbott and Nelson, 2000). STDP has been proposed on theoretical grounds to be a major mechanism for the induction of *in vivo* synaptic plasticity (Abbott and Nelson, 2000; Gerstner et al., 1996; Song et al., 2000; van Rossum et al., 2000). In the canonical form of STDP, as occurs in cortical pyramidal cells, when a presynaptic spike (and the resulting EPSP) leads a postsynaptic spike by up to 10–20 ms, an increase in synapse strength (LTP) is induced. Conversely, LTD is observed when a postsynaptic spike precedes a presynaptic spike and EPSP by short (0–20 ms) or long (0–100 ms) intervals, depending on the synapse being studied (Feldman, 2000; Froemke and Dan, 2002; Markram et al., 1997; Nishiyama et al., 2000; Sjostrom and Nelson, 2002; Sjostrom et al., 2001). Functionally, this rule causes synaptic inputs that contribute to postsynaptic firing being potentiated, while uncorrelated inputs onto otherwise active postsynaptic cells are depressed. This implements Hebb's learning rule (Abbott and Nelson, 2000; Paulsen and Sejnowski, 2000).

STDP has been extensively studied both experimentally and computationally. The mechanisms and functional consequences of STDP have been characterized in many different systems both *in vivo* and *in vitro*. Here, recent progress in the relevance of STDP in the intact brain is reviewed, with a particular emphasis on its functional properties, its multiple cellular mechanisms, and its implication in development, adult plasticity, and learning.

9.3 PREVALENCE AND DIVERSITY OF STDP

STDP is highly prevalent in the nervous system, from insects to mammals. It has been observed at a wide variety of excitatory and inhibitory synapses in many brain areas and in multiple cell types and preparations, including neuronal cell cultures, *in vitro* brain slices, and *in vivo*. While all STDP depends on precise timing and order of pre- and postsynaptic spikes, the precise temporal windows, magnitude, and sign of plasticity are remarkably diverse across synapses (Figure 9.2) (reviewed in Caporale and Dan, 2008; Shulz and Jacob, 2010). This diversity of STDP rules reflects both the existence of distinct cell type-specific forms of STDP, and the fact that nonassociative factors (i.e., factors besides pre–post spike timing) strongly modulate STDP induction. These factors include the number and frequency of postsynaptic spikes during induction (Froemke and Dan, 2002; Froemke et al., 2005; Sjostrom et al., 2001), the level

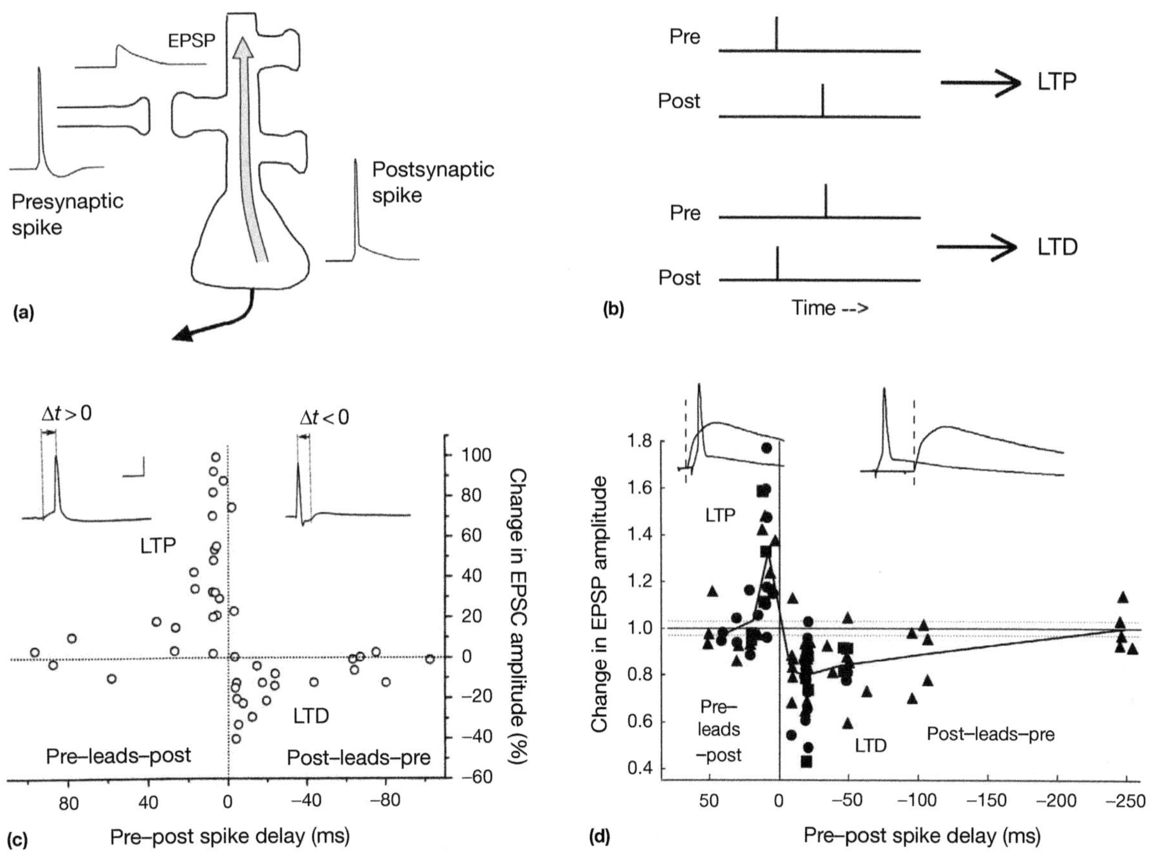

FIGURE 9.1 Spike timing-dependent plasticity. (a) Induction of STDP by pairing presynaptic spikes, which evoke EPSPs in dendrites, with postsynaptic spikes that backpropagate from the soma through the dendritic tree. (b) Temporal order dependence of STDP. (c) STDP learning rule reported by Bi and Poo (1998) for synapses in hippocampal cell culture. Each symbol is one neuron, which was presented with pre–post spike pairing at one delay. (d) STDP learning rule reported by Feldman (2000) for putative L4 synapses onto L2/3 pyramidal cells in acute slices of somatosensory cortex.

of postsynaptic depolarization evoked by presynaptic spikes (Sjostrom and Hausser, 2006; Sjostrom et al., 2001, 2004, 2008), synapse location in the dendritic tree (Froemke et al., 2005; Letzkus et al., 2006; Sjostrom and Hausser, 2006), and regulation by neuromodulators (Pawlak and Kerr, 2008; Reynolds and Wickens, 2002; Seol et al., 2007).

Three major classes of STDP can be distinguished within the substantial diversity of STDP forms. These are (a) canonical Hebbian STDP in which pre-leading-post and post-leading-pre spike pairings lead to LTP and LTD, respectively; (b) anti-Hebbian STDP in which both LTP and LTD occur but with an inverse relationship to spike timing compared to Hebbian STDP; and (c) all-LTD STDP in which LTD of synaptic transmission occurs irrespective of pre–post temporal order but only for pre–post intervals within a defined time window. This latter case is correlation-dependent, but with the opposite sign to that predicted by Hebb, and is often termed anti-Hebbian LTD. Variation within these classes is substantial and may reflect the existence of cell type-

specific subforms of STDP or the effects of nonassociative factors that vary across experiments.

In canonical Hebbian STDP (e.g., Bi and Poo, 1998; Feldman, 2000; Markram et al., 1997), induction of LTP occurs when presynaptic spikes occur up to ~10–20 ms before postsynaptic spikes. LTD is induced by the reverse order, when postsynaptic spikes lead presynaptic spikes by up to 20–100 ms, depending on the synapse. This form of STDP is prevalent at excitatory synapses onto cortical pyramidal neurons (Figure 9.2(j); Bi and Poo, 1998; Feldman, 2000; Markram et al., 1997; Nevian and Sakmann, 2006; Nishiyama et al., 2000; Sjostrom et al., 2001; Wittenberg and Wang, 2006) and nonpyramidal excitatory neurons in the auditory brainstem (Figure 9.2(g)) (Tzounopoulos et al., 2004). It also occurs at excitatory synapses onto some striatal interneurons (Figure 9.2(a) and 9.2(d); Fino et al., 2008, 2009). In anti-Hebbian STDP, spike order and timing trigger LTP versus LTD induction but with a time dependence opposite to canonical Hebbian STDP: post-leading-pre spiking drives LTP, while pre-leading-post spike order drives LTD. This

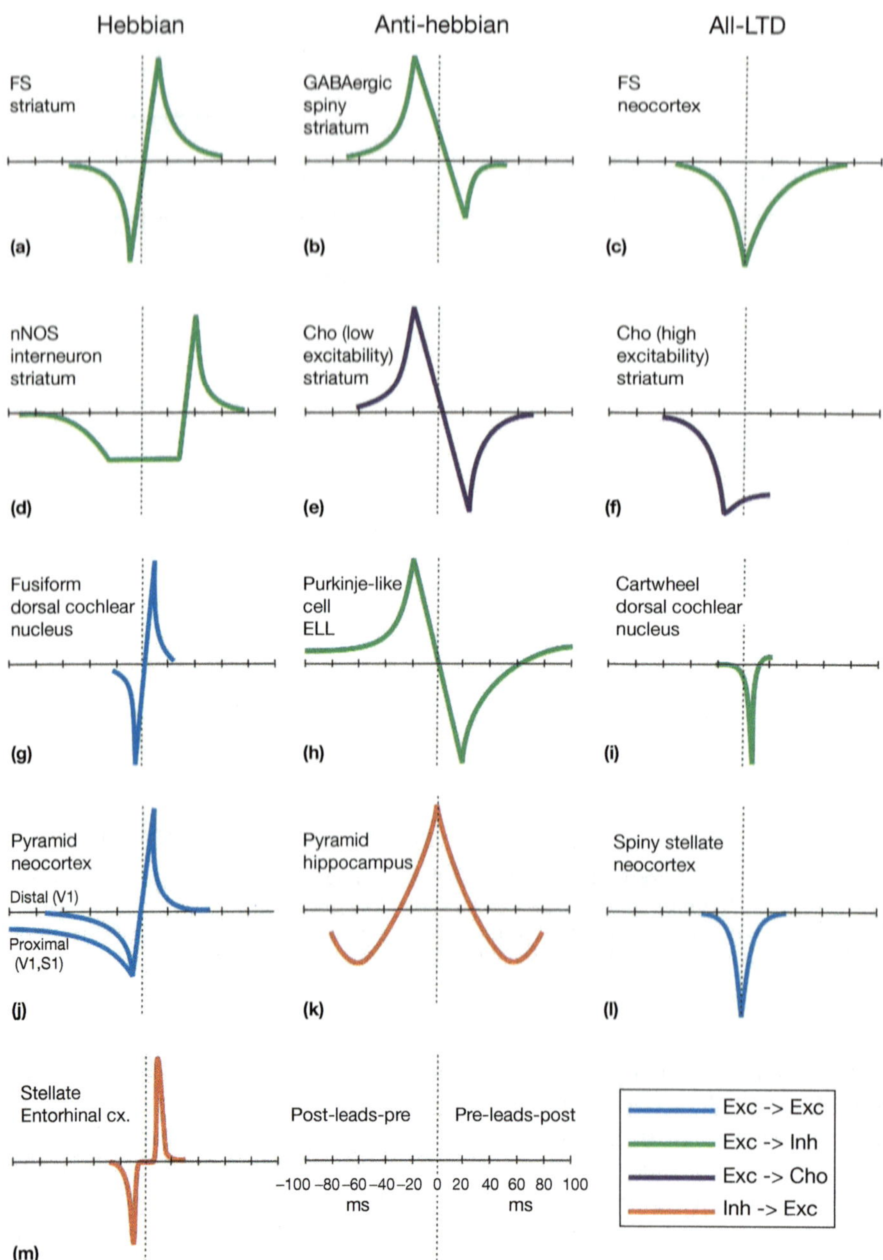

FIGURE 9.2   Three classes of STDP rules. Schematic STDP learning curves for selected examples of STDP. Light blue, excitatory synapses onto excitatory postsynaptic neurons. Green, excitatory synapses onto inhibitory interneurons. Blue, cholinergic (Cho) interneurons. Red, inhibitory synapses onto excitatory neurons. All curves are normalized to the maximal effect. For truncated curves, no data are available for longer intervals. (a, e, f) Adapted from Fino E, Deniau JM, Venance L (2008) Cell-specific spike-timing-dependent plasticity in GABAergic and cholinergic interneurons in corticostriatal rat brain slices. *The Journal of Physiology* 586: 265–282. (b) Adapted from Fino E, Glowinski J, Venance L (2005) Bidirectional activity-dependent plasticity at corticostriatal synapses. *Journal of Neuroscience* 25: 11279–11287. (c) Adapted from Lu JT, Li CY, Zhao JP, Poo MM, Zhang XH (2007) Spike-timing-dependent plasticity of neocortical excitatory synapses on inhibitory interneurons depends on target cell type. *Journal of Neuroscience* 27: 9711–9720. (d) Adapted from Fino E, Paille V, Deniau JM, Venance L (2009) Asymmetric spike-timing dependent plasticity of striatal nitric oxide-synthase interneurons. *Neuroscience* 160: 744–754. (g, i) Adapted from Tzounopoulos T, Kim Y, Oertel D, Trussell LO (2004) Cell-specific, spike timing-dependent plasticities in the dorsal cochlear nucleus. *Nature Neuroscience* 7: 719–725. (h) Adapted from Bell CC, Han VZ, Sugawara Y, Grant K (1997) Synaptic plasticity in a cerebellum-like structure depends on temporal order. *Nature* 387: 278–281. (j) Adapted from Feldman DE (2000) Timing-based LTP and LTD at vertical inputs to layer II/III pyramidal cells in rat barrel cortex. *Neuron* 27: 45–56; Froemke RC, Dan Y (2002) Spike-timing-dependent synaptic modification induced by natural spike trains. *Nature* 416: 433–438. (k) Adapted from Woodin MA, Ganguly K, Poo MM (2003) Coincident pre- and postsynaptic activity modifies GABAergic synapses by postsynaptic changes in Cl^- transporter activity. *Neuron* 39: 807–820. (l) Adapted from Egger V, Feldmeyer D, Sakmann B (1999) Coincidence detection and changes of synaptic efficacy in spiny stellate neurons in rat barrel cortex. *Nature Neuroscience* 2: 1098–1105. (m) Adapted from Haas JS, Nowotny T, Abarbanel HD (2006) Spike-timing-dependent plasticity of inhibitory synapses in the entorhinal cortex. *Journal of Neurophysiology* 96: 3305–3313. ELL, electrosensory lobe; FS, fast-spiking; S1, primary somatosensory cortex; V1, primary visual cortex.

has been observed at parallel fiber synapses onto GABAergic Purkinje-like neurons in the electrosensory lobe of the electric fish (a structure analogous to the mammalian cerebellum) (Bell et al., 1997; Figure 9.2(h)) and at excitatory synapses onto medium-sized spiny neurons of the striatum, which are GABAergic output neurons (Fino et al., 2005; Figure 9.2(b)), as well as excitatory inputs onto cholinergic striatal interneurons (Fino et al., 2008; Figure 9.2(e)). In all-LTD STDP, synapses undergo LTD irrespective of the temporal order, as long as the pre–post interval is less than ∼20–50 ms, depending on cell type. All-LTD STDP occurs at excitatory inputs onto several excitatory and inhibitory neurons (Figure 9.2(c), 9.2(i), and 9.2(l); Birtoli and Ulrich, 2004; Egger et al., 1999; Tzounopoulos et al., 2004), including fast-spiking inhibitory neurons in neocortex (Birtoli and Ulrich, 2004; Egger et al., 1999; Lu et al., 2007; Tzounopoulos et al., 2004). Classical cerebellar LTD at parallel fiber inputs onto Purkinje neurons is also a form of all-LTD STDP (Wang et al., 2000).

These distinct STDP forms occur in the same brain structure and even at synapses made by single axons on two distinct target cell types. For example, parallel fiber excitatory input onto fusiform principal neurons of the dorsal cochlear nucleus shows classical Hebbian STDP, while parallel fiber input onto glycinergic cartwheel neurons shows an all-LTD form of STDP triggered by pre–post spike intervals (Tzounopoulos et al., 2004; Figure 9.2(g) and 9.2(i)). This difference is likely to result from the interaction of different transmitter systems. An even stronger example is the excitatory cortical input to the striatum (Fino and Venance, 2011). Excitatory synapses from cortical pyramidal cells onto striatal parvalbumin-positive fast-spiking neurons show Hebbian STDP (Fino et al., 2008; Figure 9.2(a)). Cortical inputs onto neuronal nitric oxide synthase (nNOS)-expressing interneurons also show Hebbian STDP but with a peculiar time course in which the LTD window extends to pre-leading-post intervals up to +20 ms (Fino et al., 2009; Figure 9.2(d)). Excitatory inputs onto cholinergic interneurons show either anti-Hebbian STDP (Figure 9.2(e)) or LTD-only STDP (Figure 9.2(f)), depending on the excitability state of the postsynaptic neuron. Thus, multiple postsynaptic cell type-specific forms of STDP coexist in the striatum. This diversity of STDP forms may affect striatal output in ways that are not yet fully explored (Fino and Venance, 2011).

STDP of inhibitory synapses onto excitatory neurons has been much less studied but seems distinct from excitatory synapses. In the entorhinal cortex, inhibitory inputs from layer two onto stellate cells exhibit a modified Hebbian STDP with a temporal range between −5 and +5 ms in which no plasticity is induced (Haas et al., 2006; Figure 9.2(m)). In the hippocampus, inhibitory inputs to CA1 pyramids exhibit a symmetrical curve, with potentiation induced by both pre–post and post–pre pairings within ±20 ms and depression induced by longer negative or positive intervals (Woodin et al., 2003; Figure 9.2(k)).

In conclusion, STDP has diverse forms, though three main classes can be distinguished. The wide variety of cell type-specific STDP rules may extend the computational power of neuronal circuits and enable differential control of processing and storage of information by subpopulations of neurons. Despite this diversity, most of our understanding of STDP in neural development and learning focuses on canonical Hebbian STDP. Additional regional variation in STDP may result from neuromodulatory gradients (see Reynolds and Wickens, 2002 for dopamine).

9.4 FUNCTIONAL PROPERTIES OF STDP

STDP differs from classical CDP in its requirement for postsynaptic somatic action potentials, its dependence on pre- versus postsynaptic spike order, and its 10-ms-scale spike timing dependence. Historically, correlation-dependent LTP and LTD were first discovered in response to sustained high-frequency or low-frequency presynaptic spiking, which drive strong postsynaptic depolarization and LTP or weak postsynaptic depolarization and LTD, respectively (Bliss and Lomo, 1973; Dudek and Bear, 1992; Mulkey and Malenka, 1992). The critical induction requirement at most synapses is temporally correlated presynaptic spiking and postsynaptic depolarization, as shown by pairing presynaptic activation with direct intracellular depolarization of the postsynaptic neuron (Wigstrom et al., 1986). The molecular coincidence detector is the postsynaptic N-methyl-D-aspartate (NMDA) receptor, which fluxes calcium in response to simultaneous glutamate and postsynaptic depolarization (Wigstrom and Gustafsson, 1986). Correlation-dependent, NMDA receptor (NMDAR)-dependent LTP and LTD became regarded as the canonical form of LTP and LTD at excitatory synapses onto excitatory neurons, though other, distinct forms of LTP and LTD are plentiful (e.g., Chevaleyre et al., 2006; Nicoll and Malenka, 1995).

In CDP, postsynaptic depolarization can arise from any source including subthreshold dendritic depolarization, local dendritic spiking, or somatic action potentials, and the sign of plasticity is determined by the magnitude of postsynaptic depolarization, not the precise order or timing of pre- and postsynaptic spikes (Lisman and Spruston, 2005; Paulsen and Sejnowski, 2000). In contrast, STDP explicitly depends on postsynaptic somatic action potentials, which backpropagate through the dendrites to synapses, and the sign and magnitude of plasticity depend on the sequential order and precise

(10-ms scale) timing of pre- versus postsynaptic spikes (Bi and Poo, 1998; Magee and Johnston, 1997; Markram et al., 1997). The existence of STDP therefore reveals the critical importance of precise spike timing for LTP and LTD.

A common function of Hebbian STDP and CDP is that they both implement bidirectional Hebbian synaptic plasticity at excitatory synapses, which is the basis for most modern theories of activity-dependent synapse development and associative learning (Miller, 1994). In bidirectional Hebbian plasticity, when neuron A consistently participates in driving spikes in neuron B, the A → B synapse is strengthened (Hebb, 1949), while ineffective inputs that do not drive postsynaptic spikes are weakened (Bienenstock et al., 1982; Sejnowski, 1977; Stent, 1973; von der Malsburg, 1973). CDP approximates this rule by assuming that when pre- and postsynaptic activities are strongly correlated, effective synapses exist that causally drive postsynaptic spikes and therefore should be potentiated. In contrast, synapses with weak firing correlations are assumed to be ineffective and should be depressed. Hebbian STDP implements this rule by virtue of precise spike timing: synapses at which presynaptic spikes lead postsynaptic spikes by a brief 10–20 ms interval must help drive postsynaptic spikes and are potentiated, while synapses at which postsynaptic spikes lead presynaptic spikes are ineffective synapses onto otherwise active neurons and are depressed (Song and Abbott, 2001; Song et al., 2000; van Rossum et al., 2000). Moreover, Hebbian STDP at many synapses is biased toward LTD (Figures 9.1(d) and 9.2(j)) (e.g., Debanne et al., 1998; Feldman, 2000; Froemke et al., 2005; Sjostrom et al., 2001). This LTD-biased STDP drives depression of presynaptic inputs that are uncorrelated with postsynaptic spiking, thus providing an additional means of weakening ineffective inputs (Feldman, 2000). Thus, Hebbian STDP implements bidirectional Hebbian plasticity (Abbott and Nelson, 2000; Paulsen and Sejnowski, 2000). Because STDP drives Hebbian potentiation at low, physiologically realistic firing rates, STDP may be the natural means of LTP induction in low firing rate brain regions, neuron classes, or brain states (Paulsen and Sejnowski, 2000).

Hebbian STDP has other functional properties that are distinct from CDP, and arise from the spike order dependence and overall shape of the STDP learning rule. (1) LTD-biased STDP promotes stable firing rates in neuronal networks because LTP at effective synapses (which promotes higher firing rates) is counterbalanced by synapse weakening by spontaneous uncorrelated spiking (which reduces firing rates). This results in a stable state in which postsynaptic spikes occur relatively rarely and spike trains are irregular and physiologically realistic (Song et al., 2000; van Rossum et al., 2000). In contrast, CDP is unstable due to positive feedback between firing rate, firing correlation, and synapse strengthening (unless negative correlation is taken into account or a homeostatic form of metaplasticity is added). (2) STDP prevents formation of strong bidirectional recurrent excitation, which forms with correlation-based learning rules and helps drive runaway network activity (Abbott and Nelson, 2000; Clopath et al., 2010). (3) STDP implements competition between convergent inputs to a postsynaptic cell, which is a key feature of developmental plasticity (Katz and Shatz, 1996). Competition arises because synapses that fire first tend to generate pre-leading-post spike order and are strengthened, while later inputs exhibit post-leading-pre spike order and are weakened. As a result, strong early inputs competitively weaken other synapses with later or less effective input (Abbott and Nelson, 2000; Kempter et al., 1999; Song et al., 2000; Zhang et al., 1998). Correlation-based learning rules do not generate competition between inputs without constraints on total synapse strength or other normalization rules like history-dependent metaplasticity (Bienenstock et al., 1982). (4) STDP helps maintain synchronous spiking during signal propagation in feedforward networks, which is critical for network function (Bruno and Sakmann, 2006; Swadlow and Gusev, 2000; Usrey and Reid, 1999). Consider a feedforward network in which neurons exhibit a wide range of spike latency to a synchronous network input. With STDP, feedforward synapses onto postsynaptic cells that spike earliest will be weakened, thereby increasing spike latency, while synapses onto those cells that spike later will be strengthened, reducing their spike latency (Cassenaer and Laurent, 2007; Suri and Sejnowski, 2002; Zhigulin et al., 2003). (5) STDP is a powerful mechanism for learning temporal sequence information because sequential activation of connected neurons drives LTP at synapses from the first to the second neuron, but drives LTD at synapses in the reverse direction. The result is emergence of directional connectivity, tuning for learned sequences, and the ability to predict future events from past stimuli (Blum and Abbott, 1996; Engert et al., 2002; Fiete et al., 2010; Rao and Sejnowski, 2001). In contrast, CDP cannot, on its own, learn rapid sequence information.

Because of these theoretical properties, Hebbian STDP is widely proposed to store learned associations and temporal sequence information in excitatory networks and to contribute to use-dependent developmental maturation of excitatory synapses. In contrast, anti-Hebbian STDP is appropriate to reduce the representation of inputs that are temporally associated with a strong, spike-eliciting input (e.g., as occurs in cerebellum during storage of negative images of predicted sensory input) (Abbott and Nelson, 2000; Bell, 2001).

A major, unresolved debate is the relationship between Hebbian STDP and CDP. Section 9.5 discusses whether these are mechanistically distinct forms of plasticity versus different operating regimes of the same fundamental plasticity mechanism. Here, functional evidence that STDP and CDP are strongly interrelated is reviewed. Despite the widespread notion that STDP depends only on spike timing, and not on firing rate, substantial evidence shows that STDP depends on pre- and postsynaptic firing rate and burst structure (Froemke and Dan, 2002; Markram et al., 1997; Nevian and Sakmann, 2006; Pfister and Gerstner, 2006; Sjostrom et al., 2001; Tzounopoulos et al., 2004; Wang et al., 2005; Wittenberg and Wang, 2006; Zilberter et al., 2009). For example, neocortical L5 and CA1 pyramidal cell synapses exhibit Hebbian STDP at moderate firing rates, but high pre- and postsynaptic firing rates (>10–50 Hz) induce LTP independent of spike timing, and low firing rates generate LTD with post-leading-pre firing order and no plasticity with pre-leading-post order (Markram et al., 1997; Sjostrom et al., 2001; Wittenberg and Wang, 2006). Unitary L2/3–L2/3 synapses similarly exhibit only LTD at low postsynaptic firing rates but exhibit Hebbian STDP during high-frequency postsynaptic bursts (Zilberter et al., 2009). These observations suggest that STDP operates primarily in a permissive middle range of firing frequency, while high firing rates drive correlation-dependent LTP and low firing rates produce a strong bias toward LTD (Figure 9.3(a)). STDP also requires a critical level of

subthreshold dendritic depolarization prior to spiking, which is generated by summation of EPSPs across multiple synaptic inputs and is necessary to allow effective backpropagation of spikes (Delgado et al., 2010; Sjostrom and Hausser, 2006; Sjostrom et al., 2001; Stuart and Hausser, 2001). Thus, STDP is interdependent on spike timing, firing rate, and subthreshold dendritic depolarization.

These experimental data suggest that STDP and CDP represent two functional modes of a single, more general plasticity mechanism. Some groups have argued that STDP is the fundamental 'kernel' of long-term plasticity – that STDP is driven by individual pairs of pre- and postsynaptic spikes and that nonlinear summation of STDP across spike pairs explains firing rate-dependent plasticity driven by sustained or complex spike trains (Froemke and Dan, 2002; Froemke et al., 2006; Pfister and Gerstner, 2006; Wang et al., 2005; Wittenberg and Wang, 2006). Other groups have argued that STDP and CDP represent distinct fundamental modes of a single, unified plasticity process that is sensitive to spike timing, firing rate, and postsynaptic voltage. Indeed, a unified biochemical model based solely on calcium and NMDA receptor dynamics successfully explains major properties of both STDP and firing rate-dependent LTP and LTD (Badoual et al., 2006; Shouval et al., 2002, 2010). Similarly, a phenomenological plasticity model based on interaction of presynaptic spikes with time-filtered postsynaptic membrane potential predicts all

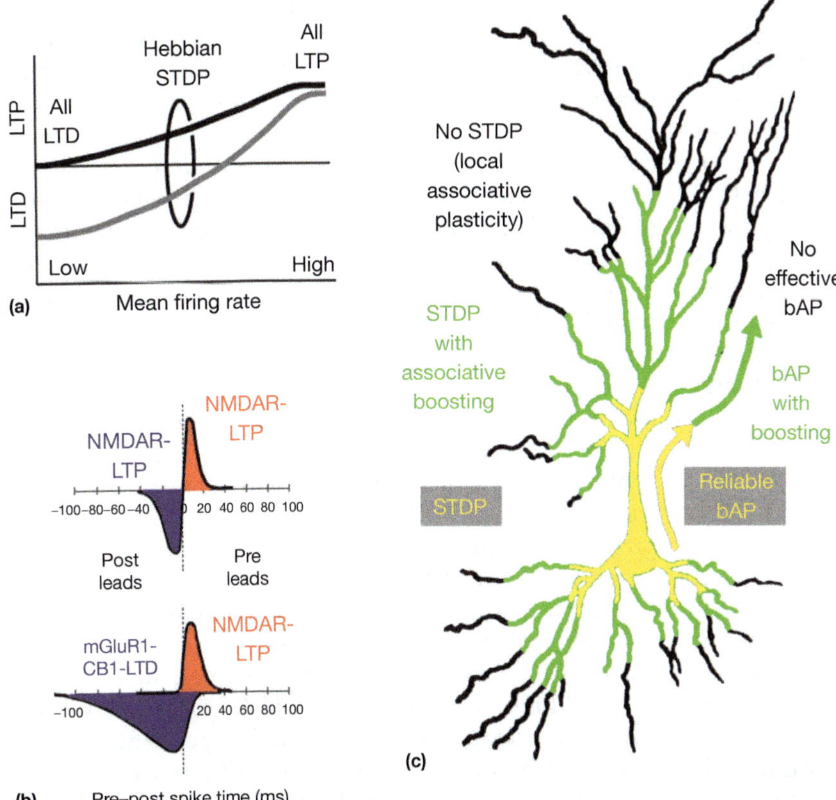

FIGURE 9.3 Cellular mechanisms for STDP. (a) Firing rate dependence of STDP. Curves show magnitude of LTP or LTD that results from pre-leading-post (black) or post-leading-pre (gray) spike pairings, as a function of mean pre- or postsynaptic firing rate. Bidirectional Hebbian STDP occurs with moderate firing rates. (b) Hebbian STDP can be composed of NMDAR-LTP and NMDAR-LTD (top) or NMDAR-LTP and mGluR1–CB1-LTD (bottom). (c) Proposed dendritic zones for STDP, based on efficiency of action potential backpropagation through the dendrites.

major features of Hebbian STDP (Clopath et al., 2010; Pfister and Gerstner, 2006). If this unified view is correct, the term 'STDP' should be taken to describe the spike time dependence of the general hybrid time- and rate-dependent plasticity process, rather than a biochemically distinct form of plasticity.

The hybrid time- and rate-dependent model of plasticity powerfully predicts major features of circuit development in a cortical column-like network containing feedforward and recurrent synapses (Clopath et al., 2010). When input to the model consists of long-duration spike trains mimicking rate coding of sensory stimuli, feedforward synapses modify to generate small receptive fields, and recurrent synapses develop strong bidirectional connections between neurons with similar receptive fields (due to timing-independent LTP driven by high firing rates of coactive neurons). This behavior is classically expected from correlation-dependent Hebbian plasticity. In contrast, when model input consists of spatiotemporal sequences of brief spiking responses, feedforward synapses still undergo Hebbian plasticity to generate receptive fields, but recurrent synapses develop strong unidirectional connections that reflect the temporal sequence in the learned pattern, as predicted by STDP. Thus, the hybrid time- and rate-dependent model can explain activity-dependent development of key circuit features according to input statistics to different brain regions (Clopath et al., 2010).

In summary, Hebbian STDP has robust Hebbian properties and appears well suited to drive realistic circuit development for neurons, brain regions, and activity states with low firing rates. Despite its name, STDP depends not only on spike timing but also on firing rate and postsynaptic voltage, suggesting a close relationship to CDP. One view, supported by recent modeling studies, suggests that STDP and CDP are different operating regimes of a hybrid spike timing-, rate-, and depolarization-dependent plasticity mechanism. Because neuromodulation can also powerfully regulate STDP (see Section 9.5), three-factor plasticity rules that include presynaptic spiking, postsynaptic activity, and neuromodulatory signals are capturing increasing interest computationally (e.g., Fremaux et al., 2010) and experimentally.

9.5 CELLULAR MECHANISMS OF STDP

The distinct forms of STDP (Section 9.3) are mediated by distinct biochemical and molecular plasticity mechanisms. In addition, even nominally identical forms of STDP (e.g., Hebbian STDP at excitatory synapses) can be mediated by different biochemical pathways in different neurons or synapses. Thus, STDP is mechanistically heterogeneous across synapses. Here, a discussion is presented on STDP mechanisms in comparison with

mechanisms for classical correlation-dependent LTP and LTD, which are triggered by joint activation of glutamate receptors (NMDA receptors or metabotropic glutamate receptor – mGluRs) and postsynaptic depolarization. STDP appears to involve identical biochemical pathways, but the primary source of postsynaptic depolarization is somatic action potentials that backpropagate through the dendrites to active synapses (backpropagating APs, bAPs).

Three major, biochemically distinct forms of LTP and LTD have been shown to underlie both CDP and STDP: (1) NMDAR-dependent LTP, in which NMDARs detect correlated presynaptic glutamate release and postsynaptic depolarization by their dual requirement for glutamate binding and voltage-dependent relief of Mg^{2+} blockade. NMDARs flux calcium, which is the postsynaptic second messenger for plasticity. Strong correlated activity generates strong NMDA currents and high postsynaptic calcium, which activates protein kinases including CaMKII. CaMKII phosphorylates AMPA receptors (AMPARs), increasing single-channel conductance and triggering AMPAR delivery to the postsynaptic membrane, which is the primary expression mechanism for LTP (Malinow and Malenka, 2002). Presynaptic components of expression can also occur but are less understood (Enoki et al., 2009). (2) NMDAR-dependent LTD, in which weaker pre- and postsynaptic activity correlations evoke less NMDAR current than for LTP, leading to lower dendritic calcium. This activates protein phosphatases including PP1 and calcineurin, which trigger LTD by trafficking of AMPARs away from the postsynaptic membrane (Malinow and Malenka, 2002). In NMDAR-dependent LTP and LTD, the NMDAR is the sole coincidence detector for plasticity, and the sign of plasticity is determined by the magnitude and time course of NMDAR-mediated calcium flux, with high calcium generating LTP, moderate calcium LTD, and low calcium no plasticity (Artola and Singer, 1993; Lisman, 1989; Yang et al., 1999). (3) mGluR-dependent and/or cannabinoid type 1 (CB1) receptor-dependent LTD, which usually leads to LTD via a decrease in presynaptic transmitter release probability. Postsynaptic NMDARs are not required for this LTD. In CB1-dependent LTD, the retrograde signal is an endocannabinoid (eCB), a phospholipid transmitter synthesized by postsynaptic dendrites in response to calcium and/or mGluR activation, which diffuses retrogradely to activate CB1 receptors on the presynaptic terminal. CB1 receptors are G protein-coupled receptors that drive a long-lasting decrease in release probability via mechanisms that are still not completely understood (Chevaleyre et al., 2006; Wilson and Nicoll, 2002). While mGluR-dependent LTD typically involves CB1 signaling, some mGluR-dependent, CB1-independent forms of LTD also exist; however, they are not discussed separately here.

The large majority of STDP is composed of combinations of these three forms of LTP and LTD. Other forms of LTP and LTD exist but are less studied or not yet linked to STDP and are generally not considered here (Malenka and Bear, 2004).

9.5.1 Biochemical Signaling Pathways for Hebbian STDP

At least two mechanistically distinct forms of Hebbian STDP exist. The first form is composed of NMDA-dependent LTP and NMDA-dependent LTD and occurs at CA3–CA1 hippocampal synapses and some synapses on neocortical L2/3 pyramidal cells (Figure 9.3(b)). In this form of STDP, the NMDAR is the sole coincidence detector for appropriate pre- and postsynaptic spike timing. Thus, both LTP and LTD components of STDP require NMDARs (Froemke et al., 2005; Nishiyama et al., 2000). When a presynaptic spike occurs just before a postsynaptic spike (pre-leads-post), a strong supralinear calcium signal is produced via dendritic NMDARs, while post-leading-pre spike order triggers a weaker, sublinear calcium signal (Koester and Sakmann, 1998; Magee and Johnston, 1997). The magnitude of calcium signal is thought to be the sole trigger for plasticity, with high calcium driving LTP and low calcium driving LTD (Lisman, 1989). Spike timing is thought to control NMDAR-mediated calcium signals by several mechanisms. Brief pre-leading-post spike intervals drive maximal calcium signals because (i) the noninstantaneous kinetics of Mg^{2+} unblock of NMDA receptors causes maximal NMDA current to occur when glutamate release leads postsynaptic depolarization by a short interval (Kampa et al., 2004) and (ii) presynaptically evoked EPSPs inactivate A-type K^+ channels and activate voltage-gated sodium channels, generating a brief temporal window in which bAPs are boosted. This boosting of the bAP promotes greater NMDA current (Hoffman et al., 1997; Holbro et al., 2010; Stuart and Hausser, 2001). Post-leading-pre spike order generates weaker calcium signals because (i) glutamate release coincides with the modest depolarization following the bAP, generating NMDA currents only modestly greater than would occur at V_{rest} (Karmarkar and Buonomano, 2002; Shouval et al., 2002), and (ii) at some synapses, calcium influx during the bAP causes calcium-dependent inactivation of NMDA receptors so that presynaptic release evokes even less NMDA current (Froemke et al., 2005; Rosenmund et al., 1995; Tong et al., 1995).

This single coincidence detector model for STDP predicts that as pre-leading-post spike interval increases beyond the brief window for LTP, a second temporal window for LTD will occur because calcium will be reduced compared to its high value at short pre-leading-post intervals (Abarbanel et al., 2003; Karmarkar and Buonomano, 2002; Shouval et al., 2002). This second LTD window has been observed at CA3–CA1 synapses (Wittenberg and Wang, 2006) but, surprisingly, not at L2/3 synapses that are proposed to also use the single coincidence detector mechanism (Froemke and Dan, 2002).

The second major form of Hebbian STDP involves a combination of NMDA-dependent LTP and mGluR- and/or CB1-dependent LTD (Figure 9.3(b)). This form occurs at several synapses in L2/3 and L5 of the somatosensory and visual cortex and at cortical synapses onto striatal medium spiny neurons under conditions of GABAergic blockade. Here, postsynaptic NMDA receptors are required for spike timing-dependent LTP, but not LTD (Bender et al., 2006b; Corlew et al., 2007; Fino et al., 2010; Nevian and Sakmann, 2006; Rodriguez-Moreno and Paulsen, 2008; Sjostrom et al., 2003). LTD instead requires postsynaptic group I mGluRs; its effector phospholipase C; low-threshold T-, R-, or L-type voltage-sensitive calcium channels (VSCCs); and calcium release from inositol trisphosphate (IP3) receptor-gated internal stores (Bender et al., 2006b; Bi and Poo, 1998; Fino et al., 2010; Nevian and Sakmann, 2006; Nishiyama et al., 2000; Seol et al., 2007). These are components of a major NMDAR-independent pathway for synaptically evoked postsynaptic calcium release (Berridge, 1993). In addition, mGluRs and postsynaptic calcium synergistically drive eCB synthesis and release (Nakamura et al., 1999), leading to activation of presynaptic CB1 receptors and a reduction in release probability (Chevaleyre et al., 2006; Wilson and Nicoll, 2002). Thus, this form of LTD requires retrograde eCB signaling to CB1 receptors, and expression occurs by a decrease in presynaptic transmitter release probability (Bender et al., 2006b; Fino et al., 2010; Nevian and Sakmann, 2006; Rodriguez-Moreno and Paulsen, 2008; Sjostrom et al., 2003). This STDP involves two separate coincidence detectors: NMDA receptors detect pre-leading-post spike intervals and exclusively trigger LTP, whereas a separate mechanism within the mGluR–VSCC–PLC–IP3R–CB1 pathway detects post-leading-pre spike intervals and exclusively triggers LTD (Bender et al., 2006b; Fino et al., 2010; Nevian and Sakmann, 2006). As a result, dendritic calcium concentration is not strictly correlated with the sign of plasticity (Nevian and Sakmann, 2006), and unlike for the single coincidence detector form of STDP, no second LTD window exists (Bender et al., 2006b; Feldman, 2000; Froemke and Dan, 2002; Sjostrom et al., 2001). Remarkably similar mechanisms have been observed during non-Hebbian STDP LTD at the parallel fiber–cartwheel cell synapse in the dorsal cochlear nucleus (Tzounopoulos et al., 2007).

Though independent of postsynaptic NMDARs, the mGluR–CB1-dependent form of LTD often depends on presynaptic NMDARs (preNMDARs) (Bender et al., 2006b; Casado et al., 2002; Corlew et al., 2007; Rodriguez-Moreno and Paulsen, 2008; Sjostrom et al., 2003). Extensive anatomical and physiological evidence supports the existence of preNMDARs at some synapses, which act as autoreceptors to boost release probability during baseline synaptic transmission (Berretta and Jones, 1996; Brasier and Feldman, 2008; Corlew et al., 2008; McGuinness et al., 2010). At synapses with this form of STDP, intracellularly loading the NMDAR blocker MK-801 into the presynaptic neuron blocks only LTD, while loading MK-801 into the postsynaptic neuron blocks only LTP (Rodriguez-Moreno and Paulsen, 2008). PreNMDARs contain NR2B, NR2C/D, and/or NR3A subunits, and STDP LTD is selectively blocked by NR2B and NR2C/D antagonists and in NR3 knockouts (Banerjee et al., 2009; Bender et al., 2006b; Corlew et al., 2008; Larsen et al., 2011; McGuinness et al., 2010; Sjostrom et al., 2003; Woodhall et al., 2001). PreNMDAR function is most prominent in early postnatal development, and spike timing-dependent LTD that is presynaptic and requires preNMDARs in juvenile cortex can become postsynaptic and independent of preNMDARs in older animals (Corlew et al., 2007) or can disappear altogether (Banerjee et al., 2009).

How the mGluR–VSCC–CB1–preNMDA signaling pathway detects appropriate post-leading-pre spike intervals for LTD is not known. In one model, each postsynaptic spike releases eCB to activate presynaptic CB1Rs, each presynaptic spike provides depolarization (and likely glutamate) to activate preNMDARs, and coincident CB1 and preNMDAR activation is required to drive LTD (Duguid and Sjostrom, 2006; Sjostrom et al., 2003). In this model, the post-leading-pre window for LTD reflects the delay and duration of CB1 activation following each postsynaptic spike, and preNMDAR activation restricts LTD to temporally coactive synapses. Consistent with this model, increasing eCB signal duration by inhibiting eCB catalysis broadens the LTD window, and pairing presynaptic spikes with exogenous CB1 agonists drives LTD (Sjostrom et al., 2003). In a second model, the spike order and timing detector for LTD is either phospholipase C (PLC), which is a known molecular coincidence detector that detects joint mGluR activation and VSCC-derived cytosolic calcium, or the IP3 receptor, which is activated by synergism between PLC-produced IP3 and cytosolic calcium (Berridge et al., 2003; Manita and Ross, 2009; Nakamura et al., 1999; Sarkisov and Wang, 2008). Acting through these or other mechanisms, mGluRs and calcium synergistically facilitate eCB synthesis, release, and synaptic plasticity (Chevaleyre et al., 2006; Hashimotodani et al., 2005, 2007). In this model, presynaptic spikes activate mGluRs, postsynaptic spikes drive VSCC calcium entry, and appropriately timed activation of both pathways is required for sufficient eCB release. This eCB signal then instructs LTD

at presynaptic terminals at which preNMDARs have been recently active (Bender et al., 2006b).

Recent evidence suggests that Hebbian STDP can also involve two other forms of LTP. Retinotectal synapses in *Xenopus laevis* and immature hippocampal mossy fibers exhibit spike timing-dependent LTD that is presynaptically expressed and is thought to use brain-derived neurotrophic factor (BDNF) as a retrograde signal (Mu and Poo, 2006; Sivakumaran et al., 2009; Zhou et al., 2003). BDNF has also been implicated in postsynaptic structural changes associated with STDP LTP (Tanaka et al., 2008). STDP LTP at L4–L2/3 synapses in the mature somatosensory cortex appears to require retrograde signaling by nitric oxide (Hardingham and Fox, 2006).

9.5.2 Biochemical Signaling Pathways for Anti-Hebbian and All-LTD STDP

Excitatory synapses onto inhibitory cartwheel cells in dorsal cochlear nucleus exhibit STDP that consists only of presynaptic CB1-mediated LTD at pre-leading-post time intervals (Figure 9.2(i)). Strikingly, at higher stimulation frequencies, these cells express postsynaptic NMDAR-dependent LTP, echoing the coexistence of these mechanisms in Hebbian STDP (Tzounopoulos et al., 2007). All-LTD STDP at parallel fiber–Purkinje cell synapses also involves postsynaptic mGluRs, VSCCs, IP3Rs, and presynaptic CB1 receptor activation but is expressed postsynaptically by AMPAR internalization (Safo and Regehr, 2005; Steinberg et al., 2006). Strong evidence suggests that the IP3 receptor is the order-dependent coincidence detector for this form of STDP (Berridge, 1993; Sarkisov and Wang, 2008; Wang et al., 2000). At other synapses, all-LTD STDP involves postsynaptic mGluR signaling (Birtoli and Ulrich, 2004; Egger et al., 1999; Lu et al., 2007) and sometimes IP3R signaling (Lu et al., 2007).

Overall, no unique cellular mechanisms have been discovered that differentiate STDP from previously known, non-spike timing-dependent forms of LTP and LTD. Instead, STDP at a given synapse seems to be a combination of a classical LTP mechanism (primarily NMDA-dependent, postsynaptic LTP) plus a known LTD mechanism (either postsynaptic NMDA-dependent LTD or mGluR/CB1-dependent presynaptic LTD). Therefore, the time-dependent features of STDP must result from previously unknown, short-timescale temporal dependence of these signaling pathways.

9.5.3 Dendritic Excitability and STDP

For postsynaptic somatic spikes to control STDP, they must backpropagate from the axonal initiation site to the synapse, where they relieve Mg^{2+} blockade of glutamate-bound NMDA receptors at active synapses or open nearby VSCCs. Backpropagation is governed by voltage-dependent sodium, potassium, and hyperpolarization-activated cyclic nucleotide-gated (HCN) channels in dendrites (Spruston, 2008). Dendritic excitability therefore critically governs STDP. A major feature is that bAPs backpropagate decrementally so that in distal dendrites, they do not provide sufficient depolarization, when paired with a small EPSP for STDP. In the most distal branches, bAPs fail completely (Spruston, 2008). As a result, STDP in distal dendrites requires additional dendritic depolarization in addition to the bAP: either summated EPSPs from nearby convergent inputs (Sjostrom and Hausser, 2006; Sjostrom et al., 2001) or a brief burst of postsynaptic spikes at sufficiently high frequency (Froemke and Dan, 2002; Markram et al., 1997; Nevian and Sakmann, 2006; Sjostrom et al., 2001; Tzounopoulos et al., 2004; Wittenberg and Wang, 2006). These supralinearly boost the level of depolarization at the synapse and can activate dendritic sodium channels to allow bAPs to propagate more effectively to the synapse (Sjostrom and Hausser, 2006; Stuart and Hausser, 2001). This additional depolarization can also evoke local sodium or calcium spikes in the dendrite, which strongly promote STDP and other forms of plasticity (Golding et al., 2002; Kampa et al., 2006; Zhou et al., 2005). The dependence of STDP at many synapses on postsynaptic firing rate and postsynaptic bursts likely reflects this requirement for additional dendritic depolarization in addition to the bAP.

Dendritic filtering of bAPs appears to establish dendritic zones for STDP in pyramidal neurons (Kampa et al., 2007; Spruston, 2008; Figure 9.3(c)). Synapses on proximal dendrites, which experience strong bAPs, may exhibit STDP in response to single presynaptic and postsynaptic spikes. Somewhat more distal synapses require bAPs plus large EPSPs from multiple nearby synapses, or postsynaptic spike bursts, to drive STDP, perhaps with local dendritic spikes as intermediaries. Synapses on the most distal dendrites are likely to be outside the influence of bAPs so that STDP governed by somatic spikes is absent. These synapses instead exhibit plasticity based on strong local EPSPs that elicit dendritic sodium or calcium spikes or regenerative NMDA spikes (Golding et al., 2002; Gordon et al., 2006). The different magnitude and kinetics of depolarization in proximal versus distal dendrites can result in dramatically different STDP rules (Froemke et al., 2005; Letzkus et al., 2006). Thus, dendritic filtering of bAPs, plus other specializations (e.g., gradients of voltage-activated K^+ channels), leads to distinct dendritic 'plasticity zones' in which different rules for synapse modification exist. These may contribute to activity-dependent stabilization of different functional classes of synapses within these regions (Froemke et al., 2005).

As a result, dynamic modulation of dendritic excitability is predicted to strongly influence the magnitude and spatial extent of STDP. Local network states (e.g., UP states) and recent somatic depolarization can increase bAP amplitude (Tsubokawa et al., 2000; Waters and Helmchen, 2004), which may enhance STDP. Neuromodulators that alter bAP propagation (e.g., muscarinic acetylcholine receptors) could similarly increase or decrease the power of somatic spikes in controlling synaptic plasticity (Sourdet and Debanne, 1999; Tsubokawa and Ross, 1997). GABAergic inhibition increases the threshold for bAP backpropagation, which may suppress STDP (van den Burg et al., 2007), although STDP clearly occurs in networks with inhibition intact (Jacob et al., 2007; Meliza and Dan, 2006). Background activity, neuromodulation, and inhibition represent global factors that when added to mathematical STDP rules are likely to generate new adaptive capabilities in large recurrent networks of neurons (Clopath et al., 2010; Legenstein et al., 2008; Pfister and Gerstner, 2006).

9.5.4 Neuromodulatory Control of STDP

Many forms of plasticity are strongly regulated by behavioral state, including attention, vigilance, and reinforcement (e.g., Ahissar et al., 1992; Bao et al., 2001). These behavioral variables are mediated in the brain by acetylcholine, dopamine, noradrenaline, and other neuromodulators (Aston-Jones et al., 1991; Devauges and Sara, 1990; Sarter and Bruno, 2000; Sarter et al., 2005; Schultz, 2002). Substantial evidence shows that neuromodulation (particularly by acetylcholine) is required for use-dependent sensory plasticity in primary sensory cortex *in vivo*, both in juveniles (Bear and Singer, 1986; Kasamatsu and Pettigrew, 1976; Weinberger, 2003) and adults (Bakin and Weinberger, 1996; Delacour et al., 1990; Edeline et al., 1994a,b; Hars et al., 1993; Juliano et al., 1991; Metherate et al., 1988; Molina-Luna et al., 2009; Rasmusson and Dykes, 1988; Webster et al., 1991). Correspondingly, pairing sensory stimulation with microstimulation of cholinergic or dopaminergic afferents or focal application of acetylcholine induces robust long-term plasticity of sensory responses (Ego-Stengel et al., 2001; Kilgard and Merzenich, 1998; Shulz et al., 2000, 2003). These pairing effects can be temporally asymmetric: during pairing of auditory stimuli with microstimulation of the dopaminergic ventral tegmental area (VTA), responses to auditory stimuli that preceded VTA activity are enhanced in auditory cortex, while responses to stimuli that followed VTA activity are reduced (Bao et al., 2001). Thus, neuromodulators powerfully regulate cortical plasticity and can selectively reinforce sensory responses depending on their timing relative to neuromodulatory supervising signals. This suggests that neuromodulation may be a third parameter (besides pre- and postsynaptic spiking) that governs the outcome of synaptic plasticity, including STDP (Ahissar et al., 1996; Crow, 1968; Kety, 1970; Pawlak et al., 2010; Reynolds and Wickens, 2002).

Neuromodulatory signals may affect STDP induction via a number of cellular and network parameters. On the network level, attention-related modulatory signals alter the sparseness of cortical activity (Vinje and Gallant, 2002) potentially rendering the system more sensitive to STDP induction. For instance, noradrenaline release in the visual cortex produces a reduction in the level of spontaneous and evoked activity (Ego-Stengel et al., 2002) which may lead the system into an optimized range of activity for STDP induction. However, no direct experimental evidence for this hypothesis is available yet.

On the cellular level, acetylcholine could dynamically regulate STDP by modifying the biophysical properties of dendrites and the amplitude and extent of bAPs (Sandler and Ross, 1999; Tsubokawa and Ross, 1997), whose backpropagation into the dendritic tree is required for STDP (Engelmann et al., 2008; Sjostrom et al., 2008). bAP backpropagation is well known to be modulated by the network state (Waters and Helmchen, 2004) and dendritic depolarization (Sjostrom and Hausser, 2006), both of which are modulated by ascending cholinergic signals (Colbert and Johnston, 1998; Hoffman and Johnston, 1998). Direct evidence for neuromodulatory control of STDP was shown by Seol et al. (2007), who discovered that acetylcholine and noradrenaline synergistically regulate STDP in the visual cortex in vitro: combined application of a muscarinic M1 agonist and a beta-adrenergic agonist enabled bidirectional STDP, while separate application of these agonists enabled spike timing-dependent LTD only and LTP only. Strong evidence also shows that dopamine (DA) modulates STDP in several brain structures (Bissiere et al., 2003; Couey et al., 2007; Lin et al., 2008; Pawlak and Kerr, 2008; Seol et al., 2007; Shen et al., 2008; Zhang et al., 2009; reviewed in Pawlak et al., 2010). Dramatic changes in the shape of the STDP curve were observed, for example, in CA1 hippocampal neurons in the presence of DA (Zhang et al., 2009) with a widening of the LTP side of the rule for pre–post pairings and an inversion of plasticity for the post–pre pairings.

Most, if not all, of the studies exploring the effect of neuromodulatory agents on STDP have been done in vitro (but see Cassenaer and Laurent, 2012 for an in vivo example of neuromodulation of the STDP rule by octopamine in the olfactory system of the locust). One cannot exclude that the timing of drug application relative to the conditioning stimuli and/or the local concentration of the neuromodulator at relevant synapses, both parameters not under control in the in vitro studies, are of particular importance in vivo (Ahissar et al., 1996). Further in vivo experiments combining STDP induction protocols and selective activation of neuromodulatory ascending systems are needed to explore how local rules of synaptic plasticity are regulated by global factors acting on several spatial and temporal scales.

9.6 IS STDP RELEVANT IN VIVO?

Neuronal networks in the intact brain show an activity regime radically different from that in the in vitro quiescent slice or cell culture. This includes differences in the levels of neuromodulation and inhibition and the presence of spontaneous network activity that generates strong ongoing synaptic bombardment in vivo. Because all these factors powerfully regulate STDP induction, the relevance of STDP in vivo has been questioned. Of particular importance is whether backpropagating somatic action potentials are the major source of dendritic depolarization for plasticity in vivo (Lisman and Spruston, 2005). Strong inhibition and intense background synaptic activity reduce the ability of backpropagating action potentials to invade the dendritic tree (Sjostrom and Hausser, 2006), possibly preventing STDP induction (van den Burg et al., 2007). Instead, it has been proposed that the primary drivers of synaptic plasticity in vivo are locally generated dendritic spikes, which implement local associative plasticity within each dendritic branch or compartment (Golding et al., 2002; Gordon et al., 2006). Such plasticity would be computationally distinct from STDP because it is driven purely by local synaptic associations, rather than associations between synaptic input and somatic spiking.

To evaluate whether STDP is relevant in vivo, the evidence that STDP is induced in vivo under experimental conditions tailored to elicit it is discussed first and, second, whether STDP is likely to be a prominent learning rule during natural (nonexperimental) conditions. The current evidence is suggestive, but not yet compelling, that STDP does occur in vivo under experimental conditions tailored to elicit STDP. This section summarizes this evidence, which is discussed in more detail in Sections 9.7 and 9.8.

The strongest evidence comes from studies in which sensory stimuli are carefully timed with respect to evoked spikes in single recorded neurons, and STDP is assessed at the level of subthreshold synaptic responses (Bell et al., 1997; Cassenaer and Laurent, 2007; Engert et al., 2002; Jacob et al., 2007; Levy and Steward, 1983; Meliza and Dan, 2006; Mu and Poo, 2006; Zhang et al., 1998). The first evidence was from Levy and Steward (Levy and Steward, 1983), who electrically stimulated pre- and postsynaptic neurons in the hippocampus in the anesthetized rat and showed that associative induction of potentiation and depression depended on the temporal order of stimulation. A series of studies in the retinotectal system of *X. laevis* tadpoles showed that natural visual motion stimuli elicit STDP at tectal synapses, measured from changes in visually evoked synaptic currents in tectal neurons (Engert et al., 2002; Mu and Poo, 2006; Zhang et al., 1998). In visual and somatosensory cortex of anesthetized rats, pairing sensory stimuli with postsynaptic spiking induced by intracellular current injection causes STDP of sensory-evoked postsynaptic potentials (Jacob et al., 2007; Meliza and Dan, 2006), although this plasticity is of substantially lower amplitude and more variable than in cortical slices (Feldman, 2000; Froemke and Dan, 2002).

Additional evidence comes from studies that infer STDP indirectly from changes in extracellularly recorded spiking and sensory perception following precisely timed presentation of sensory stimuli. In visual cortex of young and adult cats, pairing visual and/or electrical stimulation at precise time intervals induces changes in neural tuning (e.g., in orientation selectivity and receptive field location) that are compatible with STDP at horizontal, cross-columnar synapses (Fu et al., 2002; Yao and Dan, 2001; Yao et al., 2004) and the corresponding reorganization of cortical maps (Schuett et al., 2001). These same sensory stimulation protocols drive shifts in perception of orientation and position in humans with order and interval dependence consistent with STDP. This strongly suggests that STDP or STDP-like plasticity occurs in the intact brain under attentive conditions (Fu et al., 2002; Yao and Dan, 2001). Similar neurophysiological results were observed in primary auditory cortex of adult ferrets (Dahmen et al., 2008). However, these effects are much smaller and variable than STDP in brain slices, despite large numbers of pairings (Fu et al., 2002; Jacob et al., 2007; Meliza and Dan, 2006; Yao and Dan, 2001). In somatosensory cortex of adult rats, pairing spontaneous action potentials with subsequent whisker deflection drives selective depression of neural responses to the paired whisker consistent with STDP (Jacob et al., 2007). The magnitude of plasticity is again rather small, which may reflect the complex neural network activity that occurs in response to sensory stimuli and which may affect the probability of STDP induction (see Frégnac et al., 2010 for a critical review of these data). Nevertheless, the temporal specificity and the sign of plasticity observed in these studies are in agreement with STDP.

Some properties of plasticity induced *in vivo* are not identical to STDP *in vitro*. For example, STDP *in vivo*, in addition to being smaller and more variable than in brain slices, persists for just 10–15 min before being reversed by ongoing spontaneous activity (Yao and Dan, 2001; Zhou et al., 2003). In addition, the range of synaptic delays that drives synaptic depression *in vivo* is often narrower than *in vitro* (e.g., Cassenaer and Laurent, 2007; Dahmen et al., 2008; Fu et al., 2002; Jacob et al., 2007; Yao and Dan, 2001). Importantly, evidence for spike timing-dependent potentiation in mammalian cortex is weaker than for depression. Only a few studies have attempted to measure potentiation separately from depression in mammalian cortex, but these have found depression to be consistently induced, while potentiation is rarer and may be absent on average (Jacob et al., 2007; Meliza and Dan, 2006). In contrast, spike timing-dependent potentiation does clearly occur at developing retinotectal synapses (Engert et al., 2002). Thus, spike timing-dependent depression may be more robust in the cortex *in vivo* than potentiation, although more studies are needed to evaluate this. One difficulty in comparing STDP between *in vivo* and *in vitro* models is the heterogeneity in experimental protocols applied to induce STDP (Shulz, 2010; Shulz and Jacob, 2010). These include pairing sensory or synaptic stimulation with intracellular current injection to elicit one postsynaptic spike, pairing stimulation with a vigorous postsynaptic spike burst, pairing sensory–sensory stimulation, and differences in the number of pairings and anesthetic state. Neural networks react radically differently under these conditions, making comparisons of plasticity difficult.

While these studies suggest that STDP or STDP-like plasticity can occur *in vivo*, the functional importance of STDP relative to other synaptic learning rules during natural sensory input is unclear (reviewed in Caporale and Dan, 2008; Dan and Poo, 2006; Shulz and Jacob, 2010). That is, is STDP a major learning rule under natural conditions *in vivo*? This question is almost entirely untested and depends critically on both the availability of STDP mechanisms at the cellular level *in vivo* and the temporal patterns of spiking induced by natural sensory stimulation. In the primary visual cortex of the anesthetized adult cat, imposed covariance of pre- and postsynaptic spiking drives plasticity more robustly than presynaptic theta burst stimulation, which is thought to evoke postsynaptic spikes with a pre-leading-post spike order. This has been interpreted to suggest that firing correlations drive plasticity more efficiently than STDP (Frégnac et al., 2010). In contrast, evidence strongly suggests that spike order drives STDP during natural visual stimulation in the developing *Xenopus* optic tectum (Section 9.7.3). A major complication in all *in vivo* tests of STDP is the use of anesthesia, except for demonstration of small STDP-like effects on visual perception in humans (Fu et al., 2002; Yao and Dan, 2001). Additional work in awake animals will be required to assess the prominence of STDP under natural conditions.

9.7 FUNCTIONS OF STDP IN DEVELOPMENT

STDP is an increasingly prominent candidate mechanism to mediate activity- and experience-dependent components of neural circuit development. However, proof that it plays a causal role in circuit development is limited to a few brain regions. Here, evidence for this developmental role for STDP is summarized. The focus is on three questions: (1) Are developing synapses capable of STDP? (2) What are the predicted functions of STDP in development, based on theory and simulation? (3) What developmental functions have been empirically demonstrated to result from STDP *in vivo*?

9.7.1 Are Developing Synapses Capable of STDP?

Most studies of STDP at the synaptic and mechanistic levels have been performed in brain slices from juvenile, 2–3-week-old rats and mice or from developing *Xenopus* or chicks (Egger et al., 1999; Feldman, 2000; Froemke and Dan, 2002; Lu et al., 2007; Markram et al., 1997; Nevian and Sakmann, 2006; Sjostrom et al., 2001; Tzounopoulos et al., 2004; Wittenberg and Wang, 2006; Zhang et al., 1998). This is a period of robust circuit development and activity- and experience-dependent synapse refinement. Thus, STDP operates during activity-dependent development. Indeed, Hebbian STDP has been demonstrated *in vivo* at these ages, by pairing sensory stimulation with postsynaptic spikes evoked by extracellular or intracellular current injection or by sensory stimulation (discussed in detail below) (Engert et al., 2002; Jacob et al., 2007; Meliza and Dan, 2006; Mu and Poo, 2006; Schuett et al., 2001; Vislay-Meltzer et al., 2006; Zhang et al., 2000).

At older ages, STDP rules may change, though the evidence is somewhat conflicting. In brain slice experiments in L2/3 pyramidal cells in mouse S1, STDP LTD cannot be induced after P25, while STDP LTP persists into adulthood (Banerjee et al., 2009; Hardingham and Fox, 2006). In L2/3 of V1, STDP LTD similarly becomes more difficult to elicit after P23 but can be rescued if gamma-aminobutyric acid receptor type A (GABA-A) receptors are blocked during LTD induction (Corlew et al., 2007). These findings support a prevalent view that LTD is primarily a developmental phenomenon, but LTP is robust throughout life (Bear and Abraham, 1996). At odds with this view, STDP LTD can occur in L2/3 and L5 of adult S1 *in vivo* by pairing spontaneous postsynaptic spikes with whisker stimulation (Jacob et al., 2007), though this effect is weaker than in juveniles. It remains possible that LTD is robust in adults with appropriate neuromodulation (Seol et al., 2007).

Despite its prevalence in developing circuits, STDP is not universal, and is confined to specific synapses and dendritic locations, and requires specific spike train patterns to be activated (as reviewed in Sections 9.3 and 9.4). As a result, it is an empirical question whether STDP is a dominant force shaping neural circuit development, relative to other activity-dependent forms of synaptic plasticity.

9.7.2 What Are the Predicted Functions of STDP in Development, Based on Theory and Simulation?

In the activity-dependent phase of neural circuit development, coarse initial circuits that were specified by innate molecular cues are refined and optimized by sensory-driven and spontaneous neural activity. This occurs prominently during early postnatal life. It has been studied extensively in developing sensory maps, where early sensory experience shapes neuronal sensory responses, stimulus selectivity (sensory tuning), and microcircuit topography (Feldman and Brecht, 2005; Hensch, 2005; Ruthazer, 2005; White and Fitzpatrick, 2007). Computational studies over the past 25 years have shown that key features of activity-dependent development and plasticity can be explained by classical CDP, working together with additional mechanisms that implement activity-dependent competition between inputs (Miller, 1994). More recent computational studies show that Hebbian STDP drives realistic circuit development and plasticity, including features not predicted by CDP.

STDP has been shown in computational models to drive six common features of network development and experience-dependent plasticity (reviewed in detail in Abbott and Nelson, 2000; Gilson et al., 2010a). These include: (1) Basic Hebbian strengthening of coactive inputs that associatively evoke postsynaptic spikes and weakening of inputs with later or uncorrelated firing that fails to evoke postsynaptic spikes. This is illustrated in Figure 9.4(A). (2) Segregation of inputs onto target neurons based on temporal correlations of input spiking (Clopath et al., 2010; Gilson et al., 2010b; Gutig et al., 2003; Song and Abbott, 2001; Song et al., 2000). This property can explain both experience-dependent development of sensory receptive fields and gross segregation of inputs into distinct zones within a target region. Correlation-driven input segregation occurs in both networks with initially random connection strength and networks with coarse preexisting structure, corresponding to *de novo* emergence of circuit structure and plasticity of early innate circuits (Song and Abbott, 2001). (3) Selectivity for experienced spatiotemporal patterns (sequences) of input, including direction-selective responses in vision and spatial trajectories in the

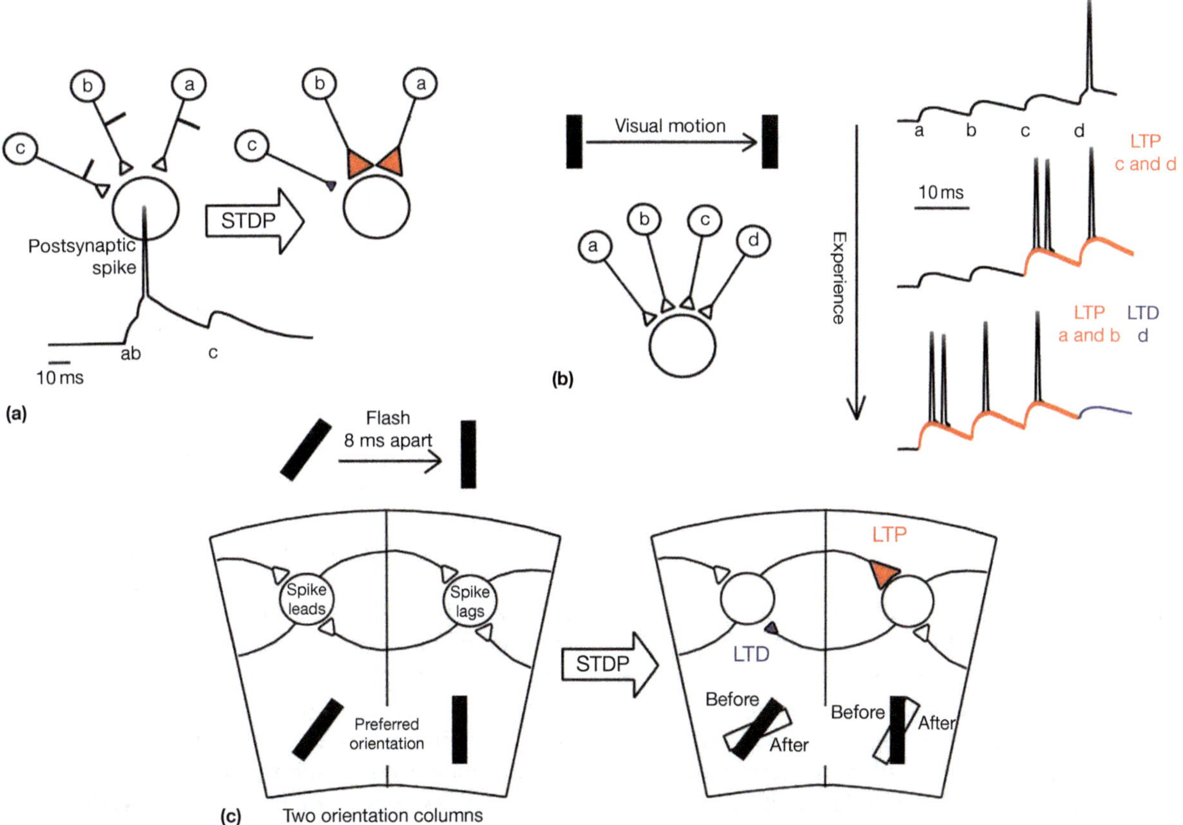

FIGURE 9.4 Some roles of STDP in synapse development and adult plasticity. (A) STDP strengthens initially weak synapses that show correlated firing (a, b) and weakens synapses with later or uncorrelated firing (c). Trace shows EPSPs mediated by each synapse and an evoked postsynaptic spike, prior to STDP. LTP is in red; LTD is in blue. (B) Development of motion direction-selective responses by STDP. Four presynaptic cells (a–d) are imagined to be driven sequentially by a rightward moving visual stimulus. Traces show visual motion-evoked EPSPs and spikes in the postsynaptic cell prior to visual experience (top) and with increasing experience. Synapses are initially weak (black EPSPs). Visual experience causes synapses active before postsynaptic spikes to strengthen (red) and synapses active after postsynaptic spikes to weaken (blue). This increases responses to motion stimuli and shifts the receptive field to 'upstream' locations. (C) Hypothesis for plasticity of orientation tuning induced by flashing two oriented bars at a precise time delay. When an oblique bar is flashed before a vertical bar, cells in the oblique orientation column spike before the vertical column. This induces STDP at intracolumnar synapses between columns. These changes in intracolumnar projection efficacy cause orientation tuning to shift toward the first orientation and away from the second.

hippocampal place system (Blum and Abbott, 1996; Buchs and Senn, 2002; Mehta et al., 2000; Rao and Sejnowski, 2001). This reflects emergence of directional connections within recurrent networks, which does not occur by CDP (Clopath et al., 2010). In motor networks, this sequence learning mechanism causes the emergence of spontaneous repeated spike train sequences, which are useful in motor patterning (Fiete et al., 2010). (4) Competition between inputs, in which strong correlated activity in one set of inputs weakens other inputs (Kempter et al., 1999; Song et al., 2000; van Rossum et al., 2000; Zhang et al., 1998). Competition is a common feature of developmental plasticity but is not inherent in CDP (Miller, 1994). Competition emerges in STDP because correlated inputs summate to drive short-latency postsynaptic spikes, which in turn weaken subsequent, noncorrelated inputs. (5) Emergence of stable, physiologically realistic firing rates and naturalistic irregular spike trains (Kempter et al., 2001; Song et al., 2000). See Section 9.3 for explanation. (6) Establishment of coincidence detection by neurons (Fontaine and Peremans, 2007; Gerstner et al., 1996) and enhancement of temporal synchrony across neurons (Masuda and Kori, 2007). While submillisecond coincidence detection is a unique property of some auditory neurons, ~10-ms timescale coincidence detection is a basic feature of sensory neocortex and could be tuned, in part, by STDP (Azouz and Gray, 2003; Roy and Alloway, 2001).

Because of these findings, Hebbian STDP has emerged as a strong candidate mechanism for activity-dependent development of neural circuits. A detailed comparison between STDP and specific CDP models is beyond the scope of this review, but one major competing model is the Bienenstock, Cooper, and Munro (BCM) model that incorporates firing rate-dependent LTP and LTD with activity-dependent metaplasticity (Bienenstock et al.,

1982). This model explains many features of activity-dependent development and plasticity in sensory cortex (Smith et al., 2009). Remarkably, several implementations of STDP have been shown to be functionally equivalent to BCM, suggesting that they may reflect equivalent cellular processes (Clopath et al., 2010; Izhikevich and Desai, 2003; Senn et al., 2001).

While STDP may play a significant role in development, it cannot be universal. In some early developing networks, postsynaptic neurons are too immature to generate somatic sodium spikes (e.g., in L2/3 of S1 cortex in the first few days of whisker use) (Stern et al., 2001). STDP cannot guide development of synapses onto these neurons, but may function a few days later, as sensory-evoked spikes increase (Bureau et al., 2004). Likewise, brain regions that generate relatively few, precisely timed spikes in response to sensory input may be most likely to utilize STDP, whereas regions with high sensory-driven or spontaneous firing rates that are modulated at slow timescales are likely to use other forms of plasticity. For example, ON and OFF retinal ganglion cells exhibit prolonged spontaneous spike bursts (retinal waves) with ~1 s temporal offset between these cell populations. Development of segregated ON and OFF input zones in the lateral geniculate nucleus requires a burst-dependent learning rule that is sensitive to longer timescale correlations, rather than STDP, which is sensitive to ms-scale correlations (Gjorgjieva et al., 2009).

9.7.3 What Developmental Functions Have Been Empirically Demonstrated to Result from STDP *In Vivo*?

Experimental studies testing how STDP contributes to circuit development have focused primarily on three areas: use-dependent development of visual response properties in the optic tectum of *X. laevis*, use-dependent development of visual response properties in the mammalian visual cortex (V1), and deprivation-induced plasticity in the mammalian sensory cortex, primarily the rodent somatosensory cortex (S1). These are reviewed here.

9.7.3.1 *STDP in Emergence of Direction Selectivity in* Xenopus *tectum*

STDP can store information about spatiotemporal patterns of input activity (Blum and Abbott, 1996; Clopath et al., 2010; Mehta et al., 2000; Rao and Sejnowski, 2001). In vision, a highly relevant spatiotemporal pattern is visual motion, and many neurons in adults are selective (tuned) for visual motion direction. In the retinotectal system of *X. laevis*, strong evidence indicates that early visual experience with moving stimuli causes neurons to develop motion direction tuning via STDP.

In young *Xenopus* tadpoles, tectal neurons initially lack selectivity for visual motion direction. When a bar is repeatedly moved in a consistent direction across a young neuron's receptive field, excitatory synaptic responses evoked by the trained movement direction are selectively increased, causing tectal neurons to become tuned for the trained direction (Engert et al., 2002). Several lines of evidence show that this is due to STDP at retinotectal synapses. First, retinotectal synapses exhibit robust Hebbian STDP *in vivo*, by pairing either electrically or visually evoked presynaptic spikes with postsynaptic spikes (Zhang et al., 1998, 2000). Second, successful motion training occurs only when visual motion stimuli elicit postsynaptic spikes, and training causes retinal inputs active before evoked tectal spikes to be potentiated, while inputs active after tectal spikes are depressed. This is the hallmark of Hebbian STDP (Engert et al., 2002; Mu and Poo, 2006). The mechanics of this process have been determined using three sequentially flashed bars at different spatial positions to simulate visual motion. When sequentially flashed bars are paired with postsynaptic spikes that occur just after the center bar stimulus (either evoked by this stimulus or by current injection), responses to the first and second bars are increased, while responses to the third bar are decreased, as predicted by Hebbian STDP. Moreover, training with both real and simulated motion increases visual responses to flashed stimuli at spatial locations that are active prior to the receptive field center (i.e., locations that are 'upstream' in the trained movement direction). This asymmetrically expands the receptive field toward earlier-activated spatial locations (Engert et al., 2002), as predicted by computational models of Hebbian STDP driven by moving stimuli (Blum and Abbott, 1996; Mehta et al., 2000). In a recent computational model, STDP at retinotectal synapses was shown to explain all these findings (Honda et al., 2011). These results strongly suggest that natural motion stimuli drive emergence of motion direction tuning via STDP. This phenomenon is illustrated in Figure 9.4(B).

Several studies show that nonmoving visual stimuli also shape tectal visual receptive field properties via STDP. Pairing a small flashed bar or spot with a postsynaptic spike evoked by intracellular current injection induces LTP or LTD of visually evoked synaptic currents according to Hebbian STDP rules (Mu and Poo, 2006; Vislay-Meltzer et al., 2006). This increases or decreases visual responses to stimuli within the trained subregion of the tectal cell's visual receptive field, causing a systematic shift in receptive field location (Vislay-Meltzer et al., 2006). When repetitively flashed stationary stimuli are strong enough to evoke tectal spikes on their own, visual responses are enhanced. This is likely to reflect STDP LTP because strengthening occurs only when stimuli successfully evoke postsynaptic spikes, which imposes the pre-leading-post spiking order within the

20-ms temporal window for STDP LTP at this synapse (Zhang et al., 2000). This suggests that visually driven STDP mediates the activity-dependent increase in visual response strength during normal tectal development and may also contribute to the normal developmental decrease in receptive field size (Tao and Poo, 2005).

A final feature of tectal development that may be driven by STDP is the development of synchronous spiking mediated by recurrent excitation between tectal neurons. Visual stimulation elicits rapid, direct retinotectal excitation of tectal neurons, plus slower, longer-lasting excitation via recurrent synapses, which mediate sustained spiking to visual stimuli. Recurrent excitation is long lasting and temporally variable early in development and becomes rapid and synchronous with experience (Pratt and Aizenman, 2007). Sensory training with optic nerve stimulation or visual flashes strengthens recurrent inputs that are active prior to the mean spike time of tectal cells and weakens recurrent inputs active after this time. As a result, recurrent inputs become more rapid and temporally precise (Pratt and Aizenman, 2007), as predicted by Hebbian STDP (Cassenaer and Laurent, 2007; Suri and Sejnowski, 2002; Zhigulin et al., 2003). However, this behavior has not been proven to reflect STDP directly.

9.7.3.2 STDP in Experience-Dependent Development of Sensory Tuning in V1 Cortex

In mammalian V1, retinotopic and ocular dominance maps are already well developed at eye opening but are strongly modified by visual experience in the first weeks of life. In contrast, tuning for orientation and motion direction are weak or absent at eye opening and are induced by early visual experience (White and Fitzpatrick, 2007). Some evidence suggests that STDP contributes to each of these aspects of development. Arguably, the weakest evidence is for retinotopy. In developing kitten V1, pairing a focal visual stimulus with postsynaptic spiking elicited by current injection during whole-cell recording strengthens or weakens visually evoked responses, with temporal dependence consistent with Hebbian STDP. While this 'stimulus timing-dependent plasticity' modulates response strength, it does not shift receptive field location (Meliza and Dan, 2006). Stimulus timing-dependent plasticity of retinotopy has been observed in adult cats, where repeated sequential presentation of two neighboring retinotopic stimuli (with <20-ms delay) shifts the spatial location of V1 receptive fields toward the retinotopic location activated first. The direction and timing dependence of this plasticity is consistent with Hebbian STDP at intracortical connections (Fu et al., 2002). Retinotopic map plasticity also occurs in adults in response to focal retinal lesions that binocularly deprive a small V1 region of visual input. Postlesion visual experience causes neurons in the deprived V1 region to acquire novel receptive fields outside the deprived region of visual space

(Gilbert and Wiesel, 1992). This reflects functional and anatomical reorganization of intracortical horizontal connections (Yamahachi et al., 2009). A computational study found that the spatial pattern of acquired receptive fields was consistent with intracortical reorganization via STDP, but not with classical CDP (Young et al., 2007). However, whether these mechanisms contribute to retinotopic refinement during development is unknown.

Stronger evidence links STDP to development of motion direction selectivity. Like in *Xenopus*, motion direction tuning is absent at eye opening and develops soon thereafter as a result of visual experience (White and Fitzpatrick, 2007). Training with visual motion stimuli immediately after eye opening induces motion direction tuning in V1 of young ferrets (Li et al., 2008). This is consistent with STDP driven by spatiotemporal spike patterns evoked in V1 inputs (Buchs and Senn, 2002). However, whether STDP is the causal mechanism, and whether it explains the emergence of direction selectivity during normal development, is unclear. Some support for this hypothesis derives from a careful analysis of motion-selective properties of receptive fields in V1 complex cells in adult cats (Fu et al., 2004). Fu et al. found that complex cells received stronger rightward (leftward) motion input from visual field locations to the left (right) of receptive field center. This anisotropy in intracortical circuits is exactly as predicted by STDP driven by natural visual motion and suggests that STDP was active during the development of the circuits for motion direction tuning (Fu et al., 2004). However, cellular studies that demonstrate that STDP is the causal process for development of motion selectivity are lacking.

Similar evidence exists for plasticity of orientation tuning. Orientation tuning exists at eye opening but is robustly plastic in response to early postnatal sensory experience, with experienced orientations gaining representation in the orientation map (Hirsch and Spinelli, 1970; Sengpiel et al., 1999). To test whether orientation tuning could by altered via STDP, Schuett et al. paired oriented visual stimuli with extracellular electrical stimulation to elicit postsynaptic spikes in V1 in anesthetized kittens. When visually evoked responses preceded electrical stimulation, cortical neurons shifted their orientation toward the presented orientation; when visually evoked responses followed electrical stimulation, orientation shifted away from the presented orientation, characteristic of Hebbian STDP. The effect, which required several hundreds of pairings to be induced, occurred predominantly in L2/3 and L5–6, suggesting a locus in intracortical connections (Schuett et al., 2001). In a similar study in adults, stimulus timing-dependent plasticity was induced by flashing a conditioned oriented stimulus <20 ms before or after the presentation of the preferred orientation. This training caused the peak of the orientation tuning to shift toward or away from the conditioned orientation, respectively. The temporal order and timing dependence was consistent with Hebbian STDP at horizontal projections between neurons tuned to the trained orientations (Yao and Dan, 2001; Yao et al., 2004). This effect is illustrated in Figure 9.4(C). Thus, carefully controlled sensory experience can alter orientation tuning in a timing-dependent manner consistent with STDP at intracortical connections. However, whether natural visual experience uses this method to refine or maintain orientation tuning during development is unknown.

STDP may contribute to experience-dependent refinement of whisker receptive fields in developing rodent S1, but evidence is weaker than for V1. In anesthetized juvenile rats, STDP LTD occurs in L2/3 pyramidal cells in response to pairing whisker deflection with intracellularly evoked postsynaptic spikes: postsynaptic spikes that precede whisker-evoked subthreshold potentials ($\Delta t < 30$ ms) cause weakening of whisker-evoked responses, which lasts for 5–10 min. Conversely, when postsynaptic spikes follow whisker-evoked potentials, no depression, or sometimes potentiation, is observed, suggestive of Hebbian STDP (Jacob et al., 2007). In adults, spike timing-dependent depression of whisker-evoked spiking responses was observed in L2/3 and L5–6 pyramidal cells (Jacob et al., 2007). However, whether STDP is engaged by natural whisker stimuli to drive receptive field plasticity during normal development is not known. One possible role for STDP is to generate whisker direction tuning in L2/3 of S1 from natural wave fronts of whisker deflection (Andermann and Moore, 2006; Leger et al., 2009).

9.7.3.3 STDP in Deprivation-Induced Plasticity in S1 and V1 Cortex

During postnatal development, sensory deprivation drives rapid depression of cortical sensory responses to deprived inputs, followed more slowly by increased responses to spared inputs. The overall effect is to bias neural selectivity toward the most active inputs. This deprivation-induced plasticity can be explained by Hebbian weakening of synapses mediating deprived inputs, coupled with some form of competition that strengthens synapses mediating spared inputs (Feldman, 2009). It is commonly hypothesized that such plasticity utilizes the same synaptic mechanisms that drive emergence or refinement of sensory response properties during normal development. Substantial evidence indicates that deprivation-induced plasticity involves LTP and LTD at cortical synapses, coupled with synapse formation, remodeling, and removal (Feldman, 2009).

STDP appears to be one of the mechanisms driving deprivation-induced weakening of sensory responses in rodent somatosensory (S1 or barrel) cortex. Rodent S1 contains a somatotopic map of the whiskers, each represented by a cortical column. Deflection of a single whisker drives spikes in L4, followed by L2/3, of the corresponding column, due to feedforward, column-specific

excitatory projections from thalamus to L4 to L2/3. In addition, whisker deflection drives weaker responses in neighboring columns via horizontal cross-columnar projections (Lubke and Feldmeyer, 2007). In juvenile rats, trimming or plucking a subset of whiskers weakens and shrinks the representation of deprived whiskers in L2/3, mediated in part by weakening L4–L2/3 excitatory synapses (Feldman and Brecht, 2005). This deprivation-induced weakening appears to represent CB1-LTD induced *in vivo* by sensory deprivation because it occludes subsequent CB1-LTD, is expressed presynaptically by reduced release probability, and is prevented by CB1 antagonist treatment *in vivo* during whisker deprivation (Allen et al., 2003; Bender et al., 2006a; Feldman, 2009; Li et al., 2009).

L4–L2/3 synapses in rat S1 exhibit LTD-biased STDP consisting of NMDAR-dependent LTP and CB1-LTD (Bender et al., 2006b; Feldman, 2000; Nevian and Sakmann, 2006). This STDP rule drives net LTD in response to either uncorrelated spiking or systematic post-leading-pre spiking (Feldman, 2000). Deprivation is likely to drive LTD *in vivo* via STDP because whisker deprivation acutely alters mean L4 and L2/3 firing rate in S1 of awake rats only modestly but powerfully alters L4–L2/3 spike timing. This was shown in anesthetized animals, where simultaneous deflection of all whiskers (to mimic normal whisking) evokes L4 spikes reliably before L2/3 spikes, whereas deflection of all but one whisker (to mimic acute whisker deprivation) immediately causes L4–L2/3 firing in the deprived column to decorrelate and firing order to reverse. These spike timing changes are quantitatively appropriate to drive spike timing-dependent LTD (Celikel et al., 2004). These findings suggest that spike timing, not spike rate, may be the key parameter driving synapse weakening in response to whisker deprivation.

STDP has also been hypothesized to drive ocular dominance plasticity in developing V1, but there is currently little direct evidence for this hypothesis. V1 neurons exhibit characteristic ocular dominance, which is a measure of the relative response to stimuli in the right versus left eye. Ocular dominance is already mature at eye opening but is highly plastic to visual experience in a defined developmental critical period (19–32 days of age in mice). During this period, closure of one eye (monocular deprivation) causes a rapid loss of responses to the deprived eye, followed by a slower gain of responses to the open eye, thus shifting ocular dominance (Wiesel and Hubel, 1963). Ocular dominance plasticity involves both rapid physiological changes in excitatory synaptic strength (e.g., LTP and LTD) and structural rearrangement of V1 synapses (Hensch, 2005; Hofer et al., 2006). While CDP can explain the basic features of ocular dominance plasticity, an STDP model has been proposed. In this model, monocular deprivation alters the precise temporal patterning of V1 spikes, thus inducing STDP in deprived-eye or open-eye pathways (Hensch, 2005; Hofer et al., 2006). Direct evidence that STDP causally drives ocular dominance plasticity is lacking, but the dynamics of plasticity in one cell class (fast-spiking interneurons) may be consistent with STDP (Yazaki-Sugiyama et al., 2009). In addition, the STDP model may explain why, during development, the critical period does not begin until inhibitory basket cells mature sufficiently to provide an 'optimal' balance between inhibition and excitation (Katagiri et al., 2007). Basket cells make dense perisomatic synapses on pyramidal cells that potently control spike timing (Huang et al., 2007; Pouille and Scanziani, 2001). Sufficient inhibition may be required to enable the precise timing of visually evoked spikes so that experience-dependent changes in spike timing can engage STDP (Hofer et al., 2006; Kubota and Kitajima, 2010; Kuhlman et al., 2010). Similar inhibitory gating of plasticity occurs in *Xenopus*, where tectal inhibitory circuits are required to ensure that visual motion stimuli evoke precise spatiotemporal patterns of spiking in the tectum. When GABAergic transmission is blocked, precise encoding of motion stimuli is lost and spikes become highly correlated between neurons. Under these conditions, training with visual motion stimuli does not cause development of motion direction tuning (Richards et al., 2010).

Thus, STDP is a strong candidate for driving deprivation-induced weakening of synapses in S1 and may play a similar role in V1, but for this, substantially less evidence exists. Inhibitory gating of plasticity may reflect the need for optimal inhibitory–excitatory balance to precisely time spikes and enable STDP.

In summary, STDP is well suited to explain activity-dependent development of network connectivity and stimulus selectivity during initial circuit formation. Experimentally, the best evidence that STDP is involved in circuit development is in experience-dependent development of direction selectivity in the *Xenopus* retinotectal system. STDP may also contribute to deprivation-induced plasticity in developing S1 and possibly to experience-dependent development of stimulus selectivity in V1. However, direct evidence that STDP is the relevant synaptic plasticity rule for circuit development or developmental plasticity outside of the retinotectal system remains largely lacking. In contrast, stronger evidence exists for STDP in adult circuit plasticity (see Section 9.8).

9.8 FUNCTIONS OF STDP IN ADULT PLASTICITY AND LEARNING

Theoretical work suggests that STDP could mediate several forms of learning, including shaping of neuronal selectivity (Guyonneau et al., 2005), coordinating transformations of multimodal information (Davison and Fregnac, 2006), tuning of auditory response delays (Gerstner et al., 1996), reinforcement learning (Farries and Fairhall, 2007), temporal difference learning (Rao and Sejnowski, 2003), input pattern detection (Masquelier et al., 2009), and learning of temporal sequences (Fiete et al., 2010). However, whether STDP plays these roles *in vivo* in the behaving animal is not yet clear (Letzkus et al., 2007). Evidence for the involvement of STDP in sensory learning in the primary sensory cortex and in the electrosensory lobe of electric fish is strong; for other forms of learning, however, its involvement remains primarily theoretical.

9.8.1 Sensory Learning and Primary Sensory Cortex Plasticity

The occurrence of STDP has been indirectly studied in the visual cortex *in vivo* by sequentially presenting two visual stimuli at time intervals suitable for inducing STDP (Fu et al., 2002; Yao and Dan, 2001; Yao et al., 2004). The two stimuli differ in spatial location or orientation. Sensory stimulation increases the firing probability of neurons within a defined window of time, and thus the pairing of two stimuli increases the occurrence of the imposed spike timing interactions. In these studies of stimulus timing-dependent plasticity using sensory–sensory associations, pairing causes modifications of neuronal tuning that are rather small but have a temporal specificity and sign expected from STDP (see also Dahmen et al. (2008) for a similar study on the auditory cortex). These results support the idea that STDP could mediate experience-dependent modulation of receptive fields in the visual cortex *in vivo*. Parallel psychophysical experiments using similar plasticity protocols show perceptual changes that are compatible with the induced neurophysiological effects, indicating that sensory-induced STDP may drive plasticity of human visual perception (Fu et al., 2002; Yao and Dan, 2001).

In the *in vivo* somatosensory cortex of the rat, cortical map reorganization can be induced by whisker deprivation. This procedure modifies the relative timing of thalamic and cortical action potentials within a range compatible with STDP (Allen et al., 2003; Celikel et al., 2004). Thus, STDP could underlie modifications of cellular responses during experience-driven network reorganizations, although these observations should be confirmed in the adult rat. Evidence for STDP in the somatosensory cortex of adult animals *in vivo* is still scarce but Jacob et al. (2007) have shown that pairing spontaneously emitted postsynaptic spiking with subsequent whisker deflections within a brief time window leads to synaptic and functional depression specific to the paired whisker, consistent with spike timing-dependent LTD.

9.8.2 Sensory Image Cancellation in Electric Fish

Anti-Hebbian forms of STDP have been described in cerebellum-like structures containing comparable cell types to mammalian cerebellum (Bell et al., 1997; Tzounopoulos et al., 2004) and in some corticostriatal connections (Fino et al., 2005). In the electrosensory lobe of three distinct groups of electric fish, anti-Hebbian STDP at parallel fiber synapses on Purkinje-like cells has been proposed to generate a representation of predictable electrosensory input arising from motor commands. The comparison of this 'negative' image with the actual sensory inflow suppresses the expected sensory consequences of a motor act, facilitating the detection of unexpected stimuli (reviewed in Bell, 2001; Bell et al., 1999).

9.8.3 Hippocampus and Memory

The hippocampus shows prominent STDP (Bi and Poo, 1998; Wittenberg and Wang, 2006), which has been proposed to be involved in the modification of hippocampal spatial receptive fields (place fields) (O'Keefe and Dostrovsky, 1971) during exploration of novel environments (Wilson and McNaughton, 1993). Place fields of hippocampal CA1 pyramidal cells are spatially skewed such that firing is asymmetric across the spatial extent of the place field, with lower firing rates when the animal enters the field and higher firing rates when it exits. In addition, the center of gravity of place fields expands backward as an animal repetitively explores a track. These features of place fields are experience-dependent (Mehta et al., 2000). As proposed on the basis of computational models of CA3 to CA1 plasticity, these modifications of CA1 receptive fields could result from NMDA-dependent STDP of synaptic inputs to CA1 cells (Ekstrom et al., 2001; Mehta et al., 2000; Shouval et al., 2002, 2010) (see also Yu et al., 2008 for alternative biophysical models). These receptive field shifts are similar to those observed in the primary visual cortex during stimulus timing-dependent plasticity (Fu et al., 2002; Yao and Dan, 2001).

9.8.4 Olfactory Learning

Olfactory learning in the insect represents one of the most robust examples of STDP induction. In the beta-lobe of the mushroom body, a central structure of the locust olfactory system, spiking activity elicited by odor presentation associated with postsynaptic spiking of intrinsic neurons (known as the Kenyon cells) triggers STDP. This form of STDP facilitates the transmission of odor-specific information through the olfactory system by synchronizing the target neurons of the Kenyon cells, thus improving the readout of the sparse olfactory code in Kenyon cells (Cassenaer and Laurent, 2007). In moths, an appetitive associative procedure induces a conditioned response to an odor. This conditioned response is reduced, however, if the reward delivery overlaps with Kenyon cell activity induced by the odor (Ito et al., 2008). Thus, changing the temporal interval between odor and reward modifies the probability of induction of the conditioned response. This was considered as evidence that STDP in Kenyon cells alone cannot underlie the olfactory learning. However, STDP has not been directly measured in this study. An alternative possibility is to reinterpret these results within the theoretical framework of reinforcement learning where in addition to pre- and postsynaptic activity, a third element (here, an appetitive reward) provides a behavioral validation of the network state during the presentation of the odor but does not in itself induce postsynaptic action potentials (see Cassenaer and Laurent, 2012). Its permissive action is rather mediated through activation of metabotropic receptors and second messenger cascades that might interact with a sustained response (Drew and Abbott, 2006) or with some intracellular signature left by the sensory input (Izhikevich, 2007), but not with the early sensory-driven activity.

9.8.5 STDP in Human Cortex

In humans, paired association of transcranial magnetic stimulation over the somatosensory cortex (S1) and median nerve stimulation induces bidirectional changes of the median nerve somatosensory evoked potential (SSEP) (Wolters et al., 2005). See Wolters et al. (2003) for a similar study on the motor cortex. The changes were confined to the P25 component of the SSEP, which is believed to originate in the upper cortical layers of S1. Interestingly, the direction of the changes depended on the timing of the stimuli. Enhancement of the P25 component was induced by a pre–post arrangement of stimulation-induced events, while a depression was noted with a reversal of events. These observations may constitute a signature of STDP in human S1. Litvak et al. (2007) further confirmed the regional and laminar location of neuroplastic changes induced by the paired associative stimulation and reported congruent behavioral consequences of the STDP-like plasticity in human S1.

In summary, theory indicates that STDP could mediate multiple features of learning. Currently, the strongest experimental evidence for Hebbian STDP in natural learning in adults is during sensory perceptual learning in the primary sensory cortex, where it may store associations and temporal sequences in response to precisely timed sequential sensory stimuli. For anti-Hebbian STDP, evidence strongly supports a role in learning and cancelling expected sensory patterns in the electrosensory lobe of electric fish. Additional work is needed to evaluate the biological role of STDP in other forms of adult learning and plasticity.

Acknowledgments

The authors thank Yves Fregnac for insightful comments. This work was supported by grants to DS from the European Union Seventh Framework Programme (FP7/2007–2013; grant numbers 243914, Brain-i-Nets, and 269921, BrainScaleS), and to DF from the National Science Foundation Temporal Dynamics of Learning Center (SBE-0542013).

Glossary

AMPAR Alpha-amino-3-hydroxy-5-methyl-4-isoxazolepropionic acid subtype of ionotropic glutamate receptor.

bAP Backpropagating action potential, that is, action potential that propagates from soma to dendrites.

CB1 Cannabinoid receptor type 1.

Correlation-dependent plasticity Form of LTP and LTD in which plasticity is determined by magnitude of correlated pre- and post-synaptic activity, but not precise pre–post timing and order.

eCB Endocannabinoid.

GABA-A receptor Gamma-aminobutyric acid receptor type A, the primary fast inhibitory receptor in the central nervous system.

IP3 Inositol trisphosphate.

LTD Long-term depression, or activity-dependent, long-term decrease in functional synapse strength.

LTP Long-term potentiation, or activity-dependent, long-term increase in functional synapse strength.

mGluR Metabotropic glutamate receptor.

NMDAR N-Methyl-D-aspartate subtype of ionotropic glutamate receptor.

PLC Phospholipase C.

S1 Primary somatosensory cortex.

Spike timing-dependent plasticity Form of LTP and LTD in which the magnitude and sign of plasticity are determined by the precise (10–100-ms scale) timing of pre- and postsynaptic action potentials.

V1 Primary visual cortex.

VSCC Voltage-sensitive calcium channel.

References

Abarbanel, H.D., Gibb, L., Huerta, R., Rabinovich, M.I., 2003. Biophysical model of synaptic plasticity dynamics. Biological Cybernetics 89, 214–226.

Abbott, L.F., Nelson, S.B., 2000. Synaptic plasticity: Taming the beast. Nature Neuroscience 3 (supplement), 1178–1183.

Ahissar, E., Vaadia, E., Ahissar, M., Bergman, H., Arieli, A., Abeles, M., 1992. Dependence of cortical plasticity on correlated activity of single neurons and on behavioral context. Science 257, 1412–1415.

Ahissar, E., Haidarliu, S., Shulz, D.E., 1996. Possible involvement of neuromodulatory systems in cortical Hebbian-like plasticity. Journal of Physiology, Paris 90, 353–360.

Allen, C.B., Celikel, T., Feldman, D.E., 2003. Long-term depression induced by sensory deprivation during cortical map plasticity in vivo. Nature Neuroscience 6, 291–299.

Andermann, M.L., Moore, C.I., 2006. A somatotopic map of vibrissa motion direction within a barrel column. Nature Neuroscience 9, 543–551.

Artola, A., Singer, W., 1993. Long-term depression of excitatory synaptic transmission and its relationship to long-term potentiation. Trends in Neurosciences 16, 480–487.

Aston-Jones, G., Chiang, C., Alexinsky, T., 1991. Discharge of noradrenergic locus coeruleus neurons in behaving rats and monkeys suggests a role in vigilance. Progress in Brain Research 88, 501–520.

Azouz, R., Gray, C.M., 2003. Adaptive coincidence detection and dynamic gain control in visual cortical neurons in vivo. Neuron 37, 513–523.

Badoual, M., Zou, Q., Davison, A.P., et al., 2006. Biophysical and phenomenological models of multiple spike interactions in spike-timing dependent plasticity. International Journal of Neural Systems 16, 79–97.

Bakin, J.S., Weinberger, N.M., 1996. Induction of a physiological memory in the cerebral cortex by stimulation of the nucleus basalis. Proceedings of the National Academy of Sciences of the United States of America 93, 11219–11224.

Banerjee, A., Meredith, R.M., Rodriguez-Moreno, A., Mierau, S.B., Auberson, Y.P., Paulsen, O., 2009. Double dissociation of spike timing-dependent potentiation and depression by subunit-preferring NMDA receptor antagonists in mouse barrel cortex. Cerebral Cortex 19, 2959–2969.

Bao, S., Chan, V.T., Merzenich, M.M., 2001. Cortical remodelling induced by activity of ventral tegmental dopamine neurons. Nature 412, 79–83.

Bear, M.F., Abraham, W.C., 1996. Long-term depression in hippocampus. Annual Review of Neuroscience 19, 437–462.

Bear, M.F., Singer, W., 1986. Modulation of visual cortical plasticity by acetylcholine and noradrenaline. Nature 320, 172–176.

Bell, C.C., 2001. Memory-based expectations in electrosensory systems. Current Opinion in Neurobiology 11, 481–487.

Bell, C.C., Han, V.Z., Sugawara, Y., Grant, K., 1997. Synaptic plasticity in a cerebellum-like structure depends on temporal order. Nature 387, 278–281.

Bell, C.C., Han, V.Z., Sugawara, Y., Grant, K., 1999. Synaptic plasticity in the mormyrid electrosensory lobe. Journal of Experimental Biology 202, 1339–1347.

Bender, K.J., Allen, C.B., Bender, V.A., Feldman, D.E., 2006a. Synaptic basis for whisker deprivation-induced synaptic depression in rat somatosensory cortex. Journal of Neuroscience 26, 4155–4165.

Bender, V.A., Bender, K.J., Brasier, D.J., Feldman, D.E., 2006b. Two coincidence detectors for spike timing-dependent plasticity in somatosensory cortex. Journal of Neuroscience 26, 4166–4177.

Berretta, N., Jones, R.S., 1996. Tonic facilitation of glutamate release by presynaptic N-methyl-D-aspartate autoreceptors in the entorhinal cortex. Neuroscience 75, 339–344.

Berridge, M.J., 1993. Inositol trisphosphate and calcium signalling. Nature 361, 315–325.

Berridge, M.J., Bootman, M.D., Roderick, H.L., 2003. Calcium signalling: Dynamics, homeostasis and remodelling. Nature Reviews. Molecular Cell Biology 4, 517–529.

Bi, G.Q., Poo, M.M., 1998. Synaptic modifications in cultured hippocampal neurons: Dependence on spike timing, synaptic strength, and postsynaptic cell type. Journal of Neuroscience 18, 10464–10472.

Bienenstock, E.L., Cooper, L.N., Munro, P.W., 1982. Theory for the development of neuron selectivity: Orientation specificity and binocular interaction in visual cortex. Journal of Neuroscience 2, 32–48.

Birtoli, B., Ulrich, D., 2004. Firing mode-dependent synaptic plasticity in rat neocortical pyramidal neurons. Journal of Neuroscience 24, 4935–4940.

Bissiere, S., Humeau, Y., Luthi, A., 2003. Dopamine gates LTP induction in lateral amygdala by suppressing feedforward inhibition. Nature Neuroscience 6, 587–592.

Bliss, T.V., Lomo, T., 1973. Long-lasting potentiation of synaptic transmission in the dentate area of the anaesthetized rabbit following stimulation of the perforant path. The Journal of Physiology 232, 331–356.

Blum, K.I., Abbott, L.F., 1996. A model of spatial map formation in the hippocampus of the rat. Neural Computation 8, 85–93.

Brasier, D.J., Feldman, D.E., 2008. Synapse-specific expression of functional presynaptic NMDA receptors in rat somatosensory cortex. Journal of Neuroscience 28, 2199–2211.

Bruno, R.M., Sakmann, B., 2006. Cortex is driven by weak but synchronously active thalamocortical synapses. Science 312, 1622–1627.

Buchs, N.J., Senn, W., 2002. Spike-based synaptic plasticity and the emergence of direction selective simple cells: Simulation results. Journal of Computational Neuroscience 13, 167–186.

Bureau, I., Shepherd, G.M., Svoboda, K., 2004. Precise development of functional and anatomical columns in the neocortex. Neuron 42, 789–801.

Caporale, N., Dan, Y., 2008. Spike timing-dependent plasticity: A Hebbian learning rule. Annual Review of Neuroscience 31, 25–46.

Casado, M., Isope, P., Ascher, P., 2002. Involvement of presynaptic N-methyl-D-aspartate receptors in cerebellar long-term depression. Neuron 33, 123–130.

Cassenaer, S., Laurent, G., 2007. Hebbian STDP in mushroom bodies facilitates the synchronous flow of olfactory information in locusts. Nature 448, 709–713.

Cassenaer, S., Laurent, G., 2012. Conditional modulation of spike-timing-dependent plasticity for olfactory learning. Nature 482, 47–52.

Celikel, T., Szostak, V.A., Feldman, D.E., 2004. Modulation of spike timing by sensory deprivation during induction of cortical map plasticity. Nature Neuroscience 7, 534–541.

Chevaleyre, V., Takahashi, K.A., Castillo, P.E., 2006. Endocannabinoid-mediated synaptic plasticity in the CNS. Annual Review of Neuroscience 29, 37–76.

Clopath, C., Busing, L., Vasilaki, E., Gerstner, W., 2010. Connectivity reflects coding: A model of voltage-based STDP with homeostasis. Nature Neuroscience 13, 344–352.

Colbert, C.M., Johnston, D., 1998. Protein kinase C activation decreases activity-dependent attenuation of dendritic Na + current in hippocampal CA1 pyramidal neurons. Journal of Neurophysiology 79, 491–495.

Corlew, R., Wang, Y., Ghermazien, H., Erisir, A., Philpot, B.D., 2007. Developmental switch in the contribution of presynaptic and postsynaptic NMDA receptors to long-term depression. Journal of Neuroscience 27, 9835–9845.

Corlew, R., Brasier, D.J., Feldman, D.E., Philpot, B.D., 2008. Presynaptic NMDA receptors: Newly appreciated roles in cortical synaptic function and plasticity. The Neuroscientist 14, 609–625.

Couey, J.J., Meredith, R.M., Spijker, S., et al., 2007. Distributed network actions by nicotine increase the threshold for spike-timing-dependent plasticity in prefrontal cortex. Neuron 54, 73–87.

Crow, T.J., 1968. Cortical synapses and reinforcement: A hypothesis. Nature 219, 736–737.

Dahmen, J.C., Hartley, D.E., King, A.J., 2008. Stimulus-timing-dependent plasticity of cortical frequency representation. Journal of Neuroscience 28, 13629–13639.

Dan, Y., Poo, M.M., 2006. Spike timing-dependent plasticity: From synapse to perception. Physiological Reviews 86, 1033–1048.

Davison, A.P., Fregnac, Y., 2006. Learning cross-modal spatial transformations through spike timing-dependent plasticity. Journal of Neuroscience 26, 5604–5615.

Debanne, D., Gahwiler, B.H., Thompson, S.M., 1994. Asynchronous pre- and postsynaptic activity induces associative long-term depression in area CA1 of the rat hippocampus in vitro. Proceedings of the National Academy of Sciences of the United States of America 91, 1148–1152.

Debanne, D., Gahwiler, B.H., Thompson, S.M., 1996. Cooperative interactions in the induction of long-term potentiation and depression of synaptic excitation between hippocampal CA3–CA1 cell pairs in vitro. Proc. Natl. Acad. Sci. USA 93, 11225–11230.

Debanne, D., Gahwiler, B.H., Thompson, S.M., 1997. Bidirectional associative plasticity of unitary CA3–CA1 EPSPs in the rat hippocampus in vitro. Journal of Neurophysiology 77, 2851–2855.

Debanne, D., Gahwiler, B.H., Thompson, S.M., 1998. Long-term synaptic plasticity between pairs of individual CA3 pyramidal cells in rat hippocampal slice cultures. The Journal of Physiology 507 (Pt 1), 237–247.

Delacour, J., Houcine, O., Costa, J.C., 1990. Evidence for a cholinergic mechanism of "learned" changes in the responses of barrel field neurons of the awake and undrugged rat. Neuroscience 34, 1–8.

Delgado, J.Y., Gomez-Gonzalez, J.F., Desai, N.S., 2010. Pyramidal neuron conductance state gates spike-timing-dependent plasticity. Journal of Neuroscience 30, 15713–15725.

Devauges, V., Sara, S.J., 1990. Activation of the noradrenergic system facilitates an attentional shift in the rat. Behavioural Brain Research 39, 19–28.

Drew, P.J., Abbott, L.F., 2006. Extending the effects of spike-timing-dependent plasticity to behavioral timescales. Proceedings of the National Academy of Sciences of the United States of America 103, 8876–8881.

Dudek, S.M., Bear, M.F., 1992. Homosynaptic long-term depression in area CA1 of hippocampus and effects of N-methyl-D-aspartate receptor blockade. Proceedings of the National Academy of Sciences of the United States of America 89, 4363–4367.

Duguid, I., Sjostrom, P.J., 2006. Novel presynaptic mechanisms for coincidence detection in synaptic plasticity. Current Opinion in Neurobiology 16, 312–322.

Edeline, J.M., Hars, B., Maho, C., Hennevin, E., 1994a. Transient and prolonged facilitation of tone-evoked responses induced by basal forebrain stimulations in the rat auditory cortex. Experimental Brain Research 97, 373–386.

Edeline, J.M., Maho, C., Hars, B., Hennevin, E., 1994b. Non-awaking basal forebrain stimulation enhances auditory cortex responsiveness during slow-wave sleep. Brain Research 636, 333–337.

Egger, V., Feldmeyer, D., Sakmann, B., 1999. Coincidence detection and changes of synaptic efficacy in spiny stellate neurons in rat barrel cortex. Nature Neuroscience 2, 1098–1105.

Ego-Stengel, V., Shulz, D.E., Haidarliu, S., Sosnik, R., Ahissar, E., 2001. Acetylcholine dependent induction and expression of functional plasticity in the barrel cortex of the adult rat. Journal of Neurophysiology 86, 422–437.

Ego-Stengel, V., Bringuier, V., Shulz, D.E., 2002. Noradrenergic modulation of functional selectivity in the cat visual cortex: An in vivo extracellular and intracellular study. Neuroscience 111, 275–289.

Ekstrom, A.D., Meltzer, J., McNaughton, B.L., Barnes, C.A., 2001. NMDA receptor antagonism blocks experience-dependent expansion of hippocampal "place fields" Neuron 31, 631–638.

Engelmann, J., van den Burg, E., Bacelo, J., et al., 2008. Dendritic back-propagation and synaptic plasticity in the mormyrid electrosensory lobe. Journal of Physiology, Paris 102, 233–245.

Engert, F., Tao, H.W., Zhang, L.I., Poo, M.M., 2002. Moving visual stimuli rapidly induce direction sensitivity of developing tectal neurons. Nature 419, 470–475.

Enoki, R., Hu, Y.L., Hamilton, D., Fine, A., 2009. Expression of long-term plasticity at individual synapses in hippocampus is graded, bidirectional, and mainly presynaptic: Optical quantal analysis. Neuron 62, 242–253.

Farries, M.A., Fairhall, A.L., 2007. Reinforcement learning with modulated spike timing dependent synaptic plasticity. Journal of Neurophysiology 98, 3648–3665.

Feldman, D.E., 2000. Timing-based LTP and LTD at vertical inputs to layer II/III pyramidal cells in rat barrel cortex. Neuron 27, 45–56.

Feldman, D.E., 2009. Synaptic mechanisms for plasticity in neocortex. Annual Review of Neuroscience 32, 33–55.

Feldman, D.E., Brecht, M., 2005. Map plasticity in somatosensory cortex. Science 310, 810–815.

Fiete, I.R., Senn, W., Wang, C.Z., Hahnloser, R.H., 2010. Spike-time-dependent plasticity and heterosynaptic competition organize networks to produce long scale-free sequences of neural activity. Neuron 65, 563–576.

Fino, E., Venance, L., 2011. Spike-timing dependent plasticity in striatal interneurons. Neuropharmacology 60, 780–788.

Fino, E., Glowinski, J., Venance, L., 2005. Bidirectional activity-dependent plasticity at corticostriatal synapses. Journal of Neuroscience 25, 11279–11287.

Fino, E., Deniau, J.M., Venance, L., 2008. Cell-specific spike-timing-dependent plasticity in GABAergic and cholinergic interneurons in corticostriatal rat brain slices. The Journal of Physiology 586, 265–282.

Fino, E., Paille, V., Deniau, J.M., Venance, L., 2009. Asymmetric spike-timing dependent plasticity of striatal nitric oxide-synthase interneurons. Neuroscience 160, 744–754.

Fino, E., Paille, V., Cui, Y., Morera-Herreras, T., Deniau, J.M., Venance, L., 2010. Distinct coincidence detectors govern the corticostriatal spike timing-dependent plasticity. The Journal of Physiology 588, 3045–3062.

Fontaine, B., Peremans, H., 2007. Tuning bat LSO neurons to interaural intensity differences through spike-timing dependent plasticity. Biological Cybernetics 97, 261–267.

Frégnac, Y., Pananceau, M., René, A., et al., 2010. A re-examination of Hebbian covariance rules and spike timing-dependent plasticity in cat visual cortex in vivo. Frontiers in Synaptic Neuroscience 2, 147.

Fregnac, Y., Shulz, D.E., 1999. Activity-dependent regulation of receptive field properties of cat area 17 by supervised Hebbian learning. Journal of Neurobiology 41, 69–82.

Fremaux, N., Sprekeler, H., Gerstner, W., 2010. Functional requirements for reward-modulated spike-timing-dependent plasticity. Journal of Neuroscience 30, 13326–13337.

Froemke, R.C., Dan, Y., 2002. Spike-timing-dependent synaptic modification induced by natural spike trains. Nature 416, 433–438.

Froemke, R.C., Poo, M.M., Dan, Y., 2005. Spike-timing-dependent synaptic plasticity depends on dendritic location. Nature 434, 221–225.

Froemke, R.C., Tsay, I.A., Raad, M., Long, J.D., Dan, Y., 2006. Contribution of individual spikes in burst-induced long-term synaptic modification. Journal of Neurophysiology 95, 1620–1629.

Fu, Y.X., Djupsund, K., Gao, H., Hayden, B., Shen, K., Dan, Y., 2002. Temporal specificity in the cortical plasticity of visual space representation. Science 296, 1999–2003.

Fu, Y.X., Shen, Y., Gao, H., Dan, Y., 2004. Asymmetry in visual cortical circuits underlying motion-induced perceptual mislocalization. Journal of Neuroscience 24, 2165–2171.

Gerstner, W., Kempter, R., van Hemmen, J.L., Wagner, H., 1996. A neuronal learning rule for sub-millisecond temporal coding. Nature 383, 76–81.

Gilbert, C.D., Wiesel, T.N., 1992. Receptive field dynamics in adult primary visual cortex. Nature 356, 150–152.

Gilson, M., Burkitt, A.N., Grayden, D.B., Thomas, D.A., van Hemmen, J.L., 2010a. Emergence of network structure due to spike-timing-dependent plasticity in recurrent neuronal networks V: Self-organization schemes and weight dependence. Biological Cybernetics 103, 365–386.

Gilson, M., Burkitt, A., van Hemmen, L.J., 2010b. STDP in recurrent neuronal networks. Frontiers in Computational Neuroscience 4, 23.

Gjorgjieva, J., Toyoizumi, T., Eglen, S.J., 2009. Burst-time-dependent plasticity robustly guides ON/OFF segregation in the lateral geniculate nucleus. PLoS Computational Biology 5, e1000618.

Golding, N.L., Staff, N.P., Spruston, N., 2002. Dendritic spikes as a mechanism for cooperative long-term potentiation. Nature 418, 326–331.

Gordon, U., Polsky, A., Schiller, J., 2006. Plasticity compartments in basal dendrites of neocortical pyramidal neurons. Journal of Neuroscience 26, 12717–12726.

Gustafsson, B., Wigström, H., Abraham, W.C., Huang, Y.Y., 1987. Long-term potentiation in the hippocampus using depolarizing current pulses as the conditioning stimulus to single volley synaptic potentials. Journal of Neuroscience 7, 774–780.

Gutig, R., Aharonov, R., Rotter, S., Sompolinsky, H., 2003. Learning input correlations through nonlinear temporally asymmetric Hebbian plasticity. Journal of Neuroscience 23, 3697–3714.

Guyonneau, R., VanRullen, R., Thorpe, S.J., 2005. Neurons tune to the earliest spikes through STDP. Neural Computation 17, 859–879.

Haas, J.S., Nowotny, T., Abarbanel, H.D., 2006. Spike-timing-dependent plasticity of inhibitory synapses in the entorhinal cortex. Journal of Neurophysiology 96, 3305–3313.

Hardingham, N., Fox, K., 2006. The role of nitric oxide and GluR1 in presynaptic and postsynaptic components of neocortical potentiation. Journal of Neuroscience 26, 7395–7404.

Hars, B., Maho, C., Edeline, J.M., Hennevin, E., 1993. Basal forebrain stimulation facilitates tone-evoked responses in the auditory cortex of awake rat. Neuroscience 56, 61–74.

Hashimotodani, Y., Ohno-Shosaku, T., Tsubokawa, H., et al., 2005. Phospholipase Cbeta serves as a coincidence detector through its Ca^{2+} dependency for triggering retrograde endocannabinoid signal. Neuron 45, 257–268.

Hashimotodani, Y., Ohno-Shosaku, T., Kano, M., 2007. Endocannabinoids and synaptic function in the CNS. The Neuroscientist 13, 127–137.

Hebb, D.O., 1949. The Organization of Behavior. Wiley, New York.

Hensch, T.K., 2005. Critical period plasticity in local cortical circuits. Nature Reviews Neuroscience 6, 877–888.

Hirsch, H.V., Spinelli, D.N., 1970. Visual experience modifies distribution of horizontally and vertically oriented receptive fields in cats. Science 168, 869–871.

Hofer, S.B., Mrsic-Flogel, T.D., Bonhoeffer, T., Hubener, M., 2006. Lifelong learning: Ocular dominance plasticity in mouse visual cortex. Current Opinion in Neurobiology 16, 451–459.

Hoffman, D.A., Johnston, D., 1998. Downregulation of transient K+ channels in dendrites of hippocampal CA1 pyramidal neurons by activation of PKA and PKC. Journal of Neuroscience 18, 3521–3528.

Hoffman, D.A., Magee, J.C., Colbert, C.M., Johnston, D., 1997. K+ channel regulation of signal propagation in dendrites of hippocampal pyramidal neurons. Nature 387, 869–875.

Holbro, N., Grunditz, A., Wiegert, J.S., Oertner, T.G., 2010. AMPA receptors gate spine Ca(2+) transients and spike-timing-dependent potentiation. Proceedings of the National Academy of Sciences of the United States of America 107, 15975–15980.

Honda, M., Urakubo, H., Tanaka, K., Kuroda, S., 2011. Analysis of development of direction selectivity in retinotectum by a neural circuit model with spike timing-dependent plasticity. Journal of Neuroscience 31, 1516–1527.

Huang, Z.J., Di Cristo, G., Ango, F., 2007. Development of GABA innervation in the cerebral and cerebellar cortices. Nature Reviews Neuroscience 8, 673–686.

Ito, I., Ong, R.C., Raman, B., Stopfer, M., 2008. Olfactory learning and spike timing dependent plasticity. Communicative and Integrative Biology 1, 170–171.

Izhikevich, E.M., 2007. Solving the distal reward problem through linkage of STDP and dopamine signaling. Cerebral Cortex 17, 2443–2452.

Izhikevich, E.M., Desai, N.S., 2003. Relating STDP to BCM. Neural Computation 15, 1511–1523.

Jacob, V., Brasier, D.J., Erchova, I., Feldman, D., Shulz, D.E., 2007. Spike timing-dependent synaptic depression in the in vivo barrel cortex of the rat. Journal of Neuroscience 27, 1271–1284.

Juliano, S.I., Ma, W., Eslin, D., 1991. Cholinergic depletion prevents expansion of topographic maps in somatosensory cortex. Proceedings of the National Academy of Sciences of the United States of America 88, 780–784.

Kampa, B.M., Clements, J., Jonas, P., Stuart, G.J., 2004. Kinetics of Mg^{2+} unblock of NMDA receptors: Implications for spike-timing dependent synaptic plasticity. The Journal of Physiology 556, 337–345.

Kampa, B.M., Letzkus, J.J., Stuart, G.J., 2006. Requirement of dendritic calcium spikes for induction of spike-timing-dependent synaptic plasticity. The Journal of Physiology 574, 283–290.

Kampa, B.M., Letzkus, J.J., Stuart, G.J., 2007. Dendritic mechanisms controlling spike-timing-dependent synaptic plasticity. Trends in Neurosciences 30, 456–463.

Karmarkar, U.R., Buonomano, D.V., 2002. A model of spike-timing dependent plasticity: One or two coincidence detectors? Journal of Neurophysiology 88, 507–513.

Kasamatsu, T., Pettigrew, J.D., 1976. Depletion of brain catecholamines: Failure of ocular dominance shift after monocular occlusion in kittens. Science 194, 206–209.

Katagiri, H., Fagiolini, M., Hensch, T.K., 2007. Optimization of somatic inhibition at critical period onset in mouse visual cortex. Neuron 53, 805–812.

Katz, L.C., Shatz, C.J., 1996. Synaptic activity and the construction of cortical circuits. Science 274, 1133–1138.

Kempter, R., Gerstner, W., Van Hemmen, J.L., 1999. Hebbian learning and spiking neurons. Physical Review E – Statistical, Nonlinear, and Soft Matter Physics 59, 4498–4514.

Kempter, R., Gerstner, W., van Hemmen, J.L., 2001. Intrinsic stabilization of output rates by spike-based Hebbian learning. Neural Computation 13, 2709–2741.

Kety, S.S., 1970. Biogenic amines in the central nervous system: Their possible roles in arousal, emotion and learning. In: Schmitt, F.O. (Ed.), The Neurosciences: Second Study Program. Rockefeller University Press, New York, pp. 324–336.

Kilgard, M.P., Merzenich, M.M., 1998. Cortical map reorganization enabled by nucleus basalis activity. Science 279, 1714–1718.

Koester, H.J., Sakmann, B., 1998. Calcium dynamics in single spines during coincident pre- and postsynaptic activity depend on relative timing of back-propagating action potentials and subthreshold excitatory postsynaptic potentials. Proceedings of the National Academy of Sciences of the United States of America 95, 9596–9601.

Kubota, S., Kitajima, T., 2010. Possible role of cooperative action of NMDA receptor and GABA function in developmental plasticity. Journal of Computational Neuroscience 28, 347–359.

Kuhlman, S.J., Lu, J., Lazarus, M.S., Huang, Z.J., 2010. Maturation of GABAergic inhibition promotes strengthening of temporally coherent inputs among convergent pathways. PLoS Computational Biology 6, e1000797.

Larsen, R.S., Corlew, R.J., Henson, M.A., et al., 2011. NR3A-containing NMDARs promote neurotransmitter release and spike timing-dependent plasticity. Nature Neuroscience 14, 338–344.

Legenstein, R., Pecevski, D., Maass, W., 2008. A learning theory for reward-modulated spike-timing-dependent plasticity with application to biofeedback. PLoS Computational Biology 4, e1000180.

Leger, J.-F., Kremer, Y., Bourdieu, L., 2009. Two photon calcium imaging reveals a whisker direction tuning map in a single barrel of adult rats. Society for Neuroscience Abstracts 174, 13.

Letzkus, J.J., Kampa, B.M., Stuart, G.J., 2006. Learning rules for spike timing-dependent plasticity depend on dendritic synapse location. Journal of Neuroscience 26, 10420–10429.

Letzkus, J.J., Kampa, B.M., Stuart, G.J., 2007. Does spike timing-dependent synaptic plasticity underlie memory formation? Clinical and Experimental Pharmacology and Physiology 34, 1070–1076.

Levy, W.B., Steward, O., 1983. Temporal contiguity requirements for long-term associative potentiation/depression in the hippocampus. Neuroscience 8, 791–797.

Li, L., Bender, K.J., Drew, P.J., Jadhav, S.P., Sylwestrak, E., Feldman, D.E., 2009. Endocannabinoid signaling is required for development and critical period plasticity of the whisker map in somatosensory cortex. Neuron 64, 537–549.

Li, Y., Van Hooser, S.D., Mazurek, M., White, L.E., Fitzpatrick, D., 2008. Experience with moving visual stimuli drives the early development of cortical direction selectivity. Nature 456, 952–956.

Lin, Y., Yang, H., Min, M., Chiu, T., 2008. Inhibition of associative long-term depression by activation of beta-adrenergic receptors in rat hippocampal CA1 synapses. Journal of Biomedical Science 15, 123–131.

Lisman, J., 1989. A mechanism for the Hebb and the anti-Hebb processes underlying learning and memory. Proceedings of the National Academy of Sciences of the United States of America 86, 9574–9578.

Lisman, J., Spruston, N., 2005. Postsynaptic depolarization requirements for LTP and LTD: A critique of spike timing-dependent plasticity. Nature Neuroscience 8, 839–841.

Litvak, V., Zeller, D., Oostenveld, R., et al., 2007. LTP-like changes induced by paired associative stimulation of the primary somatosensory cortex in humans: Source analysis and associated changes in behaviour. European Journal of Neuroscience 25, 2862–2874.

Lu, J.T., Li, C.Y., Zhao, J.P., Poo, M.M., Zhang, X.H., 2007. Spike-timing-dependent plasticity of neocortical excitatory synapses on inhibitory interneurons depends on target cell type. Journal of Neuroscience 27, 9711–9720.

Lubke, J., Feldmeyer, D., 2007. Excitatory signal flow and connectivity in a cortical column: Focus on barrel cortex. Brain Structure & Function 212, 3–17.

Magee, J.C., Johnston, D., 1997. A synaptically controlled, associative signal for Hebbian plasticity in hippocampal neurons. Science 275, 209–213.

Malenka, R.C., Bear, M.F., 2004. LTP and LTD: An embarrassment of riches. Neuron 44, 5–21.

Malinow, R., Malenka, R.C., 2002. AMPA receptor trafficking and synaptic plasticity. Annual Review of Neuroscience 25, 103–126.

Manita, S., Ross, W.N., 2009. IP(3) mobilization and diffusion determine the timing window of $Ca(2+)$ release by synaptic stimulation and a spike in rat CA1 pyramidal cells. Hippocampus 20, 524–539.

Markram, H., Lubke, J., Frotscher, M., Sakmann, B., 1997. Regulation of synaptic efficacy by coincidence of postsynaptic APs and EPSPs. Science 275, 213–215.

Masquelier, T., Guyonneau, R., Thorpe, S.J., 2009. Competitive STDP-based spike pattern learning. Neural Computation 21, 1259–1276.

Masuda, N., Kori, H., 2007. Formation of feedforward networks and frequency synchrony by spike-timing-dependent plasticity. Journal of Computational Neuroscience 22, 327–345.

McGuinness, L., Taylor, C., Taylor, R.D., et al., 2010. Presynaptic NMDARs in the hippocampus facilitate transmitter release at theta frequency. Neuron 68, 1109–1127.

Mehta, M.R., Quirk, M.C., Wilson, M.A., 2000. Experience-dependent asymmetric shape of hippocampal receptive fields. Neuron 25, 707–715.

Meliza, C.D., Dan, Y., 2006. Receptive-field modification in rat visual cortex induced by paired visual stimulation and single-cell spiking. Neuron 49, 183–189.

Metherate, R., Tremblay, N., Dykes, R.W., 1988. Transient and prolonged effects of acetylcholine on responsiveness of cat somatosensory cortical neurons. Journal of Neurophysiology 59, 1253–1276.

Miller, K.D., 1994. Models of activity-dependent neural development. Progress in Brain Research 102, 303–318.

Molina-Luna, K., Pekanovic, A., Rohrich, S., et al., 2009. Dopamine in motor cortex is necessary for skill learning and synaptic plasticity. PLoS One 4, e7082.

Mu, Y., Poo, M.M., 2006. Spike timing-dependent LTP/LTD mediates visual experience-dependent plasticity in a developing retinotectal system. Neuron 50, 115–125.

Mulkey, R.M., Malenka, R.C., 1992. Mechanisms underlying induction of homosynaptic long-term depression in area CA1 of the hippocampus. Neuron 9, 967–975.

Nakamura, T., Barbara, J.G., Nakamura, K., Ross, W.N., 1999. Synergistic release of Ca^{2+} from IP3-sensitive stores evoked by synaptic activation of mGluRs paired with backpropagating action potentials. Neuron 24, 727–737.

Nevian, T., Sakmann, B., 2006. Spine Ca^{2+} signaling in spike-timing-dependent plasticity. Journal of Neuroscience 26, 11001–11013.

Nicoll, R.A., Malenka, R.C., 1995. Contrasting properties of two forms of long-term potentiation in the hippocampus. Nature 377, 115–118.

Nishiyama, M., Hong, K., Mikoshiba, K., Poo, M.M., Kato, K., 2000. Calcium stores regulate the polarity and input specificity of synaptic modification. Nature 408, 584–588.

O'Keefe, J., Dostrovsky, J., 1971. The hippocampus as a spatial map. Preliminary evidence from unit activity in the freely-moving rat. Brain Research 34, 171–175.

Paulsen, O., Sejnowski, T.J., 2000. Natural patterns of activity and long-term synaptic plasticity. Current Opinion in Neurobiology 10, 172–179.

Pawlak, V., Kerr, J.N., 2008. Dopamine receptor activation is required for corticostriatal spike-timing-dependent plasticity. Journal of Neuroscience 28, 2435–2446.

Pawlak, V., Wickens, J.R., Kirkwood, A., Kerr, J.N., 2010. Timing is not everything: Neuromodulation opens the STDP gate. Frontiers in Synaptic Neuroscience 2, 1–14.

Pfister, J.P., Gerstner, W., 2006. Triplets of spikes in a model of spike timing-dependent plasticity. Journal of Neuroscience 26, 9673–9682.

Pouille, F., Scanziani, M., 2001. Enforcement of temporal fidelity in pyramidal cells by somatic feed-forward inhibition. Science 293, 1159–1163.

Pratt, K.G., Aizenman, C.D., 2007. Homeostatic regulation of intrinsic excitability and synaptic transmission in a developing visual circuit. Journal of Neuroscience 27, 8268–8277.

Rao, R.P., Sejnowski, T.J., 2001. Predictive learning of temporal sequences in recurrent neocortical circuits. Novartis Foundation Symposium 239, 208–229 discussion 29–40.

Rao, R.P., Sejnowski, T.J., 2003. Self-organizing neural systems based on predictive learning. Philosophical Transactions. Series A, Mathematical, Physical, and Engineering Sciences 361, 1149–1175.

Rasmusson, D.D., Dykes, R.W., 1988. Long-term enhancement of evoked potentials in cat somatosensory cortex produced by co-activation of the basal forebrain and cutaneous receptor. Experimental Brain Research 70, 276–286.

Reynolds, J.N., Wickens, J.R., 2002. Dopamine-dependent plasticity of corticostriatal synapses. Neural Networks 15, 507–521.

Richards, B.A., Voss, O.P., Akerman, C.J., 2010. GABAergic circuits control stimulus-instructed receptive field development in the optic tectum. Nature Neuroscience 13, 1098–1106.

Rodriguez-Moreno, A., Paulsen, O., 2008. Spike timing-dependent long-term depression requires presynaptic NMDA receptors. Nature Neuroscience 11, 744–745.

Rosenmund, C., Feltz, A., Westbrook, G.L., 1995. Calcium-dependent inactivation of synaptic NMDA receptors in hippocampal neurons. Journal of Neurophysiology 73, 427–430.

Roy, S.A., Alloway, K.D., 2001. Coincidence detection or temporal integration? What the neurons in somatosensory cortex are doing. Journal of Neuroscience 21, 2462–2473.

Ruthazer, E.S., 2005. You're perfect, now change – Redefining the role of developmental plasticity. Neuron 45, 825–828.

Safo, P.K., Regehr, W.G., 2005. Endocannabinoids control the induction of cerebellar LTD. Neuron 48, 647–659.

Sandler, V.M., Ross, W.N., 1999. Serotonin modulates spike backpropagation and associated $[Ca^{2+}]_i$ changes in the apical dendrites of hippocampal CA1 pyramidal neurons. Journal of Neurophysiology 81, 216–224.

Sarkisov, D.V., Wang, S.S., 2008. Order-dependent coincidence detection in cerebellar Purkinje neurons at the inositol trisphosphate receptor. Journal of Neuroscience 28, 133–142.

Sarter, M., Bruno, J.P., 2000. Cortical cholinergic inputs mediating arousal, attentional processing and dreaming: Differential afferent regulation of the basal forebrain by telencephalic and brainstem afferents. Neuroscience 95, 933–952.

Sarter, M., Hasselmo, M.E., Bruno, J.P., Givens, B., 2005. Unraveling the attentional functions of cortical cholinergic inputs: Interactions between signal-driven and cognitive modulation of signal detection. Brain Research. Brain Research Reviews 48, 98–111.

Schuett, S., Bonhoeffer, T., Hubener, M., 2001. Pairing-induced changes of orientation maps in cat visual cortex. Neuron 32, 325–337.

Schultz, W., 2002. Getting formal with dopamine and reward. Neuron 36, 241–263.

Sejnowski, T.J., 1977. Storing covariance with nonlinearly interacting neurons. Journal of Mathematical Biology 4, 303–321.

Sengpiel, F., Stawinski, P., Bonhoeffer, T., 1999. Influence of experience on orientation maps in cat visual cortex. Nature Neuroscience 2, 727–732.

Senn, W., Markram, H., Tsodyks, M., 2001. An algorithm for modifying neurotransmitter release probability based on pre- and postsynaptic spike timing. Neural Computation 13, 35–67.

Seol, G.H., Ziburkus, J., Huang, S., et al., 2007. Neuromodulators control the polarity of spike-timing-dependent synaptic plasticity. Neuron 55, 919–929.

Shen, W., Flajolet, M., Greengard, P., Surmeier, D.J., 2008. Dichotomous dopaminergic control of striatal synaptic plasticity. Science 321, 848–851.

Shouval, H.Z., Bear, M.F., Cooper, L.N., 2002. A unified model of NMDA receptor-dependent bidirectional synaptic plasticity. Proceedings of the National Academy of Sciences of the United States of America 99, 10831–10836.

Shouval, H.Z., Wang, S.S., Wittenberg, G.M., 2010. Spike timing dependent plasticity: A consequence of more fundamental learning rules. Frontiers in Computational Neuroscience 4, 19.

Shulz, D.E., 2010. Synaptic plasticity in vivo: More than just spike timing? Frontiers in Synaptic Neuroscience 2, 137.

Shulz, D.E., Jacob, V., 2010. Spike-timing-dependent plasticity in the intact brain: Counteracting spurious spike coincidences. Frontiers in Synaptic Neuroscience 2, 1–10.

Shulz, D.E., Ego-Stengel, V., Ahissar, E., 2003. Acetylcholine-dependent potentiation of temporal frequency representation in the barrel cortex does not depend on response magnitude during conditioning. Journal of Physiology, Paris 97, 431 439.

Shulz, D.E., Sosnik, R., Ego, V., Haidarliu, S., Ahissar, E., 2000. A neuronal analogue of state-dependent learning. Nature 403, 549–553.

Sivakumaran, S., Mohajerani, M.H., Cherubini, E., 2009. At immature mossy-fiber-CA3 synapses, correlated presynaptic and postsynaptic activity persistently enhances GABA release and network excitability via BDNF and cAMP-dependent PKA. Journal of Neuroscience 29, 2637–2647.

Sjostrom, P.J., Hausser, M., 2006. A cooperative switch determines the sign of synaptic plasticity in distal dendrites of neocortical pyramidal neurons. Neuron 51, 227–238.

Sjostrom, P.J., Nelson, S.B., 2002. Spike timing, calcium signals and synaptic plasticity. Current Opinion in Neurobiology 12, 305–314.

Sjostrom, P.J., Turrigiano, G.G., Nelson, S.B., 2001. Rate, timing, and cooperativity jointly determine cortical synaptic plasticity. Neuron 32, 1149–1164.

Sjostrom, P.J., Turrigiano, G.G., Nelson, S.B., 2003. Neocortical LTD via coincident activation of presynaptic NMDA and cannabinoid receptors. Neuron 39, 641–654.

Sjostrom, P.J., Turrigiano, G.G., Nelson, S.B., 2004. Endocannabinoid-dependent neocortical layer-5 LTD in the absence of postsynaptic spiking. Journal of Neurophysiology 92, 3338–3343.

Sjostrom, P.J., Rancz, E.A., Roth, A., Hausser, M., 2008. Dendritic excitability and synaptic plasticity. Physiological Reviews 88, 769–840.

Smith, G.B., Heynen, A.J., Bear, M.F., 2009. Bidirectional synaptic mechanisms of ocular dominance plasticity in visual cortex. Philosophical Transactions of the Royal Society of London. Series B, Biological Sciences 364, 357–367.

Song, S., Abbott, L.F., 2001. Cortical development and remapping through spike timing-dependent plasticity. Neuron 32, 339–350.

Song, S., Miller, K.D., Abbott, L.F., 2000. Competitive Hebbian learning through spike-timing-dependent synaptic plasticity. Nature Neuroscience 3, 919–926.

Sourdet, V., Debanne, D., 1999. The role of dendritic filtering in associative long-term synaptic plasticity. Learning and Memory 6, 422–447.

Spruston, N., 2008. Pyramidal neurons: Dendritic structure and synaptic integration. Nature Reviews Neuroscience 9, 206–221.

Steinberg, J.P., Takamiya, K., Shen, Y., et al., 2006. Targeted in vivo mutations of the AMPA receptor subunit GluR2 and its interacting protein PICK1 eliminate cerebellar long-term depression. Neuron 49, 845–860.

Stent, G.S., 1973. A physiological mechanism for Hebb's postulate of learning. Proceedings of the National Academy of Sciences of the United States of America 70, 997–1001.

Stern, E.A., Maravall, M., Svoboda, K., 2001. Rapid development and plasticity of layer 2/3 maps in rat barrel cortex in vivo. Neuron 31, 305–315.

Stuart, G.J., Hausser, M., 2001. Dendritic coincidence detection of EPSPs and action potentials. Nature Neuroscience 4, 63–71.

Suri, R.E., Sejnowski, T.J., 2002. Spike propagation synchronized by temporally asymmetric Hebbian learning. Biological Cybernetics 87, 440–445.

Swadlow, H.A., Gusev, A.G., 2000. The influence of single VB thalamocortical impulses on barrel columns of rabbit somatosensory cortex. Journal of Neurophysiology 83, 2802–2813.

Tanaka, J., Horiike, Y., Matsuzaki, M., Miyazaki, T., Ellis-Davies, G.C., Kasai, H., 2008. Protein synthesis and neurotrophin-dependent structural plasticity of single dendritic spines. Science 319, 1683–1687.

Tao, H.W., Poo, M.M., 2005. Activity-dependent matching of excitatory and inhibitory inputs during refinement of visual receptive fields. Neuron 45, 829–836.

Tong, G., Shepherd, D., Jahr, C.E., 1995. Synaptic desensitization of NMDA receptors by calcineurin. Science 267, 1510–1512.

Tsubokawa, H., Ross, W.N., 1997. Muscarinic modulation of spike backpropagation in the apical dendrites of hippocampal CA1 pyramidal neurons. Journal of Neuroscience 17, 5782–5791.

Tsubokawa, H., Offermanns, S., Simon, M., Kano, M., 2000. Calcium-dependent persistent facilitation of spike backpropagation in the CA1 pyramidal neurons. Journal of Neuroscience 20, 4878–4884.

Tzounopoulos, T., Kim, Y., Oertel, D., Trussell, L.O., 2004. Cell-specific, spike timing-dependent plasticities in the dorsal cochlear nucleus. Nature Neuroscience 7, 719–725.

Tzounopoulos, T., Rubio, M.E., Keen, J.E., Trussell, L.O., 2007. Coactivation of pre- and postsynaptic signaling mechanisms determines cell-specific spike-timing-dependent plasticity. Neuron 54, 291–301.

Usrey, W.M., Reid, R.C., 1999. Synchronous activity in the visual system. Annual Review of Physiology 61, 435–456.

van den Burg, E.H., Engelmann, J., Bacelo, J., Gomez, L., Grant, K., 2007. Etomidate reduces initiation of backpropagating dendritic action potentials: Implications for sensory processing and synaptic plasticity during anesthesia. Journal of Neurophysiology 97, 2373–2384.

van Rossum, M.C., Bi, G.Q., Turrigiano, G.G., 2000. Stable Hebbian learning from spike timing-dependent plasticity. Journal of Neuroscience 20, 8812–8821.

Vinje, W.E., Gallant, J.L., 2002. Natural stimulation of the nonclassical receptive field increases information transmission efficiency in V1. Journal of Neuroscience 22, 2904–2915.

Vislay-Meltzer, R.L., Kampff, A.R., Engert, F., 2006. Spatiotemporal specificity of neuronal activity directs the modification of receptive fields in the developing retinotectal system. Neuron 50, 101–114.

von der Malsburg, C., 1973. Self-organization of orientation sensitive cells in the striate cortex. Kybernetik 14, 85–100.

Wang, S.S., Denk, W., Hausser, M., 2000. Coincidence detection in single dendritic spines mediated by calcium release. Nature Neuroscience 3, 1266–1273.

Wang, H.X., Gerkin, R.C., Nauen, D.W., Bi, G.Q., 2005. Coactivation and timing-dependent integration of synaptic potentiation and depression. Nature Neuroscience 8, 187–193.

Waters, J., Helmchen, F., 2004. Boosting of action potential backpropagation by neocortical network activity in vivo. Journal of Neuroscience 24, 11127–11136.

Webster, H.H., Rasmusson, D.D., Dykes, R.W., et al., 1991. Long-term enhancement of evoked potentials in raccoon somatosensory cortex following co-activation of the nucleus basalis of Meynert complex and cutaneous receptors. Brain Research 545, 292–296.

Weinberger, N.M., 2003. The nucleus basalis and memory codes: Auditory cortical plasticity and the induction of specific, associative behavioral memory. Neurobiology of Learning and Memory 80, 268–284.

White, L.E., Fitzpatrick, D., 2007. Vision and cortical map development. Neuron 56, 327–338.

Wiesel, T.N., Hubel, D.H., 1963. Single-cell responses in striate cortex of kittens deprived of vision in one eye. Journal of Neurophysiology 26, 1003–1017.

Wigstrom, H., Gustafsson, B., 1986. Postsynaptic control of hippocampal long-term potentiation. Journal of Physiology, Paris 81, 228–236.

Wigstrom, H., Gustafsson, B., Huang, Y.Y., Abraham, W.C., 1986. Hippocampal long-term potentiation is induced by pairing single afferent volleys with intracellularly injected depolarizing current pulses. Acta Physiologica Scandinavica 126, 317–319.

Wilson, M.A., McNaughton, B.L., 1993. Dynamics of the hippocampal ensemble code for space. Science 261, 1055–1058.

Wilson, R.I., Nicoll, R.A., 2002. Endocannabinoid signaling in the brain. Science 296, 678–682.

Wittenberg, G.M., Wang, S.S., 2006. Malleability of spike-timing-dependent plasticity at the CA3–CA1 synapse. Journal of Neuroscience 26, 6610–6617.

Wolters, A., Sandbrink, F., Schlottmann, A., et al., 2003. A temporally asymmetric Hebbian rule governing plasticity in the human motor cortex. Journal of Neurophysiology 89, 2339–2345.

Wolters, A., Schmidt, A., Schramm, A., et al., 2005. Timing-dependent plasticity in human primary somatosensory cortex. The Journal of Physiology 565, 1039–1052.

Woodhall, G., Evans, D.I., Cunningham, M.O., Jones, R.S., 2001. NR2B-containing NMDA autoreceptors at synapses on entorhinal cortical neurons. Journal of Neurophysiology 86, 1644–1651.

Woodin, M.A., Ganguly, K., Poo, M.M., 2003. Coincident pre- and postsynaptic activity modifies GABAergic synapses by postsynaptic changes in Cl^- transporter activity. Neuron 39, 807–820.

Yamahachi, H., Marik, S.A., McManus, J.N., Denk, W., Gilbert, C.D., 2009. Rapid axonal sprouting and pruning accompany functional reorganization in primary visual cortex. Neuron 64, 719–729.

Yang, S.N., Tang, Y.G., Zucker, R.S., 1999. Selective induction of LTP and LTD by postsynaptic $[Ca^{2+}]_i$ elevation. Journal of Neurophysiology 81, 781–787.

Yao, H., Dan, Y., 2001. Stimulus timing-dependent plasticity in cortical processing of orientation. Neuron 32, 315–323.

Yao, H., Shen, Y., Dan, Y., 2004. Intracortical mechanism of stimulus-timing-dependent plasticity in visual cortical orientation tuning.

Proceedings of the National Academy of Sciences of the United States of America 101, 5081–5086.

Yazaki-Sugiyama, Y., Kang, S., Cateau, H., Fukai, T., Hensch, T.K., 2009. Bidirectional plasticity in fast-spiking GABA circuits by visual experience. Nature 462, 218–221.

Young, J.M., Waleszczyk, W.J., Wang, C., Calford, M.B., Dreher, B., Obermayer, K., 2007. Cortical reorganization consistent with spike timing-but not correlation-dependent plasticity. Nature Neuroscience 10, 887–895.

Yu, X., Shouval, H.Z., Knierim, J.J., 2008. A biophysical model of synaptic plasticity and metaplasticity can account for the dynamics of the backward shift of hippocampal place fields. Journal of Neurophysiology 100, 983–992.

Zhang, L.I., Tao, H.W., Holt, C.E., Harris, W.A., Poo, M., 1998. A critical window for cooperation and competition among developing retinotectal synapses. Nature 395, 37–44.

Zhang, L.I., Tao, H.W., Poo, M., 2000. Visual input induces long-term potentiation of developing retinotectal synapses. Nature Neuroscience 3, 708–715.

Zhang, J.C., Lau, P.M., Bi, G.Q., 2009. Gain in sensitivity and loss in temporal contrast of STDP by dopaminergic modulation at hippocampal synapses. Proceedings of the National Academy of Sciences of the United States of America 106, 13028–13033.

Zhigulin, V.P., Rabinovich, M.I., Huerta, R., Abarbanel, H.D., 2003. Robustness and enhancement of neural synchronization by activity-dependent coupling. Physical Review E: Statistical, Nonlinear, and Soft Matter Physics 67, 021901.

Zhou, Q., Tao, H.W., Poo, M.M., 2003. Reversal and stabilization of synaptic modifications in a developing visual system. Science 300, 1953–1957.

Zhou, Y.D., Acker, C.D., Netoff, T.I., Sen, K., White, J.A., 2005. Increasing Ca^{2+} transients by broadening postsynaptic action potentials enhances timing-dependent synaptic depression. Proceedings of the National Academy of Sciences of the United States of America 102, 19121–19125.

Zilberter, M., Holmgren, C., Shemer, I., et al., 2009. Input specificity and dependence of spike timing-dependent plasticity on preceding postsynaptic activity at unitary connections between neocortical layer 2/3 pyramidal cells. Cerebral Cortex 19, 2308–2320.

COGNITIVE DEVELOPMENT

Introduction to Cognitive Development from a Neuroscience Perspective

H. Tager-Flusberg

Boston University, Boston, MA, USA

10.1 INTRODUCTION

Newborns depend for their survival on caregivers, almost always mothers, who provide them with food, warmth, and comfort. They come into this world equipped with a set of reflexes, basic sensory capacities and preferences, and primitive means for signaling their needs. Over the first 2 years of life, infants undergo rapid developments in cognition and behavior, and their brains undergo exponential growth and change. During these early years, infants are transformed into toddlers capable of goal-directed actions and independent mobility; they develop sophisticated concepts of the world around them, acquire the basic use and understanding of language, and become equipped to navigate their social environment with ease. Nevertheless, at the age of 2, toddlers still have a great deal to learn, and developmental changes continue, at a steadier rate, over the next two decades.

In this volume, the authors follow this remarkable journey in human development with particular emphasis on the early years. Studies of child development began almost a century ago; however, the conceptual frameworks and information about what children know as they grow have shifted over the years, driven largely by the introduction of new methods and technology that

is now opening up the possibility of asking key questions about the neural and cognitive mechanisms that drive developmental change. This chapter serves as a roadmap to the basic frameworks and methods that have guided the field of child development over the past few decades and then introduces the chapters in this volume.

10.2 FRAMEWORKS AND METHODS

10.2.1 Conceptual Frameworks

The scientific field of cognitive development began with the seminal work of Jean Piaget, though its intellectual roots go back to philosophical questions about the origins of knowledge and mind raised by Plato, Aristotle, and others in ancient Greece. Piaget, who was a biologist by training, was the first to develop a comprehensive theoretical framework for how children acquire core concepts that are the foundation of human thought. His 'constructivist' theory claimed that fundamental concepts, such as space, time, and causality, are acquired by the baby operating on the environment. Through very basic processes, complex abstract concepts are built up through *actions* and *interactions* with objects

© 2013 Elsevier Inc. All rights reserved.

and people. In Piaget's theory, children develop through a series of qualitatively distinct stages, each defined by a new form of mental representation. He was committed to a strong view on the biological contributions to cognitive development by arguing for the importance of evolution, maturation, and adaptation for understanding change over time; however, his empirical work focused exclusively on observable behavior, which clearly contributed to the significant role he attributed to action, especially as the foundation of representational knowledge (Flavell, 1996).

Piaget's work lies between the extremes of nativism and empiricism. He began writing in the early part of the twentieth century, but his work did not become well known in the English-speaking scientific community until the 1970s, when psychologists became open to ideas that were grounded in a more complex view of how biology intersects with experience. Nevertheless, even though Piaget firmly argued that developmental changes took place in children's cognition, he was not able to articulate how such changes were accomplished; instead, he focused more on generating the important questions to be addressed by developmental scientists that remain central to the field today.

Over the last few decades, developmental science has blossomed as it has taken on questions about what the starting point is for newborns, what the child knows at different ages, how this knowledge is organized and changes over time, and which factors and basic learning mechanisms contribute to these changes. These questions are being addressed using new tools that provide a more direct window into the minds and brains of babies and young children than simply observations of their behavior. While developmental scientists are informed by parallel research on other species, the unique social and linguistic abilities of humans transform the child's conceptions of the world in ways that require different frameworks and approaches for investigating developmental processes (cf. Vygotsky, 1978).

10.2.2 Eye Gaze

For Piaget, infants' actions on the world revealed their underlying knowledge; however, motor development is a slow process in humans and thus provides a very indirect assessment of their cognitive capacities. Beginning with the seminal work of Fantz (1958), researchers have been using eye gaze patterns to provide a more direct approach to what infants see, discriminate, prefer, remember, and expect. Because they have relatively good ocular-motor control, babies' eye movements are a reliable and valid way of revealing mental processing. Studies using eye gaze patterns, including measures of first fixations, time spent looking at images, and anticipatory looking patterns, have demonstrated that even

newborns are not just a bundle of reflexes with basic sensory capacities. Instead, people now know that infants have perceptual preferences that are biased toward attending to particular events and entities in their environment, and rapid changes over the first few months of life consolidate and expand on these initial biases (see Chapter 37).

For many years, researchers relied on either manual on-line coding of eye gaze patterns or videotaping of infants' looking patterns and later laborious coding their eye movements by hand. It is the most common behavioral method in use today for studies on infant perception, cognitive, social, and language development (Aslin, 2007). The recent advent of automated eye trackers, particularly ones that do not require head-mounted cameras or complex calibration procedures, has led to changes in the ability to capture the microstructure of eye gaze patterns in infants including much finer temporal and spatial resolutions, without requiring manual coding that could potentially introduce error or bias. Nevertheless, despite the advantages of using automated eye trackers to capture eye movement patterns, there are still significant challenges about how data collected from these devices should be analyzed and interpreted (Aslin, 2012).

10.2.3 Electrophysiology

If eye gaze is the method of choice for investigating the minds of babies, then electrophysiology is the method of choice for investigating their brain function. The broader field of cognitive neuroscience has relied on a range of technologies to probe brain structure and function in people; however, not all of them are easily adapted for use with infants. Electrophysiological recordings, including electroencephalography (EEG) and event-related potentials (ERPs), provide relatively good measures of temporal processing and dynamics of cognitive processes. Their primary advantages for their use in studying infants and young children include safety, ease of use, and tolerance of some movement.

The EEG signals collected from multiple electrodes placed over the scalp reflect ongoing electrical activity in the brain; ERPs are signals that occur in response to a specific stimulus and are most widely used as a neural assessment for a range of cognitive processes. The analysis of EEG activity is generally not time locked, but can be decomposed into constituent frequencies to quantify power in specific bandwidths, each of which reflects different aspects of neural processing related to cognition. ERPs are thought to detect postsynaptic potentials from pyramidal cells summed over a large number of neurons. Invariant stimulus-related electrical activity is extracted through an averaging process over a large number of trials in order to reduce the noise due to random components. The average ERP is then analyzed temporally, as a series of positive and negative components characterized by their polarity, peak latency, amplitude, and distribution over the scalp (Csibra et al., 2008). The introduction of high-density arrays, which include a large number of electrodes embedded in a net that is quickly and easily fitted over the scalp, provides more complete spatial coverage and has led to new methods of analysis with better spatial and temporal resolutions.

Different components in the ERP signal have been linked to specific behavioral tasks including, for example, attention, face processing, and a range of linguistic processes from speech to syntax (see, for example, Chapter 32 and Rubenstein and Rakic, 2013). Much of what is known about the cognitive processes associated with specific ERP components comes from studies of adults. There are significant developmental changes in the latency, morphology, and topography in the known ERP components, and while the precise roots of these changes are not yet fully understood, they are thought to reflect a combination of cognitive advances and more efficient neural processing. While electrophysiological methods are now widely used in human developmental neuroscience, they do have several limitations including their poor spatial resolution and lack of sensitivity to processing in subcortical brain areas.

10.2.4 Magnetic Resonance Imaging and Other Imaging Methods

The growing field of cognitive neuroscience has relied most extensively on magnetic resonance imaging (MRI) as the single most effective, noninvasive, and safe method for examining *in vivo* human brain structure and function. MRI provides high-resolution spatial information, and its application in developmental science has given people detailed information about volumetric changes in regional gray and white matter. The advent of diffusion tensor imaging (DTI), a variant of conventional MRI, has led to advances in the ability to identify and characterize developmental changes in white matter pathways (Wozniak et al., 2008).

Functional aspects of brain development, particularly in older children, have been tracked using functional MRI (fMRI), which measures changes in blood oxygenation levels (BOLD response), an indirect measure of increases in regional neuronal activity. fMRI provides excellent spatial resolution when local differences in the BOLD response are analyzed between tasks or groups. Advances in analytic methods such as connectivity analyses have led to new ways of tracking developmental changes in systems-level cortical representations.

Despite their importance in developmental cognitive neuroscience, there are several challenges in using MRI with young children, not least of which is the requirement to lie completely still in a noisy enclosed tunnel for relatively long time periods while the scan data are collected. This involves a good deal of cooperation from an awake child, which is possible in children over the age of 5 who have been carefully prepared, or the use of alternative approaches such as scanning during sleep or sedation. It is also not clear how developmental changes in brain morphology and metabolism may influence the BOLD response, and there are other technical concerns that need to be carefully considered when interpreting findings from fMRI studies conducted on young children or children with neurodevelopmental disorders (Casey et al., 2005).

The newest technology that has been introduced to investigate functional brain development, particularly in infants, is near-infrared spectroscopy (NIRS), which, like fMRI, measures changes in hemodynamic responses that are assumed to be related to regional brain activity. Optical probes placed on the scalp measure changes in blood oxygenation levels that reflect surface cortical activity, offering moderately good spatial resolution, depending on the number and location of the probes. Considerable methodological and technological advances have been made, and the number of studies using NIRS to investigate functional activity in the brains of very young infants is increasing each year (cf. Gervain et al., 2011). It has been used in studies of infant perception, cognition, and language; however, it cannot detect neural responses generated in structures that lie below the cortical surface, which limits its use in studying key aspects of memory, emotion, or social perception.

10.2.5 Summary

Important advances in developmental cognitive neuroscience have been made in recent years based on the introduction of new conceptual frameworks and methods for probing cognition and brain processes. People are now beginning to be able to link behavioral and brain changes in ways that allow them to test theoretically grounded hypotheses about the neural bases of cognitive development. Yet, progress can only be made if their methods and technologies are used in the context of well-designed experiments and an appreciation of the limitations in the application and interpretation of findings from each available method. As noted, there are real challenges in using all the methods surveyed here with pediatric populations: they all require a considerable amount of cooperation and minimal movement. While young infants and older children can tolerate the requirements for most of these methodologies, toddlers and preschoolers are far more active than infants and far less compliant than school-aged children, and therefore, relatively less is known about development during these critical years. It is expected that technological innovations in the coming years will help to fill in these gaps to provide a more complete picture of cognitive development from birth through adolescence.

10.3 OVERVIEW OF CHAPTERS

This volume provides comprehensive and detailed coverage of the current state of research on cognitive development including both behavioral and neuroscience perspectives on the field. Each chapter highlights key developmental questions and illustrates the primary methods that have been used to address them. Throughout this volume, the emphasis is on typical development, but applications to atypical populations are discussed, particularly in those areas where more significant work has been conducted on children with neurodevelopmental disorders. The field is still young, and studies that attempt to address the developmental relationship between cognitive and neural processes have really only begun in earnest during the past decade. Each chapter concludes with a discussion of the future directions that people can expect to see in the coming decades.

In the opening chapter, Mark Johnson (see Rubenstein and Rakic, 2013) discusses the theories that have dominated the newly emerging field of developmental cognitive neuroscience. He highlights the importance of having theoretically driven research and evaluates the predictions and evidence from each theory using research from different cognitive domains. He concludes that the theory of interactive specialization, which has its roots in the Piagetian theory, offers the most promise of a framework that captures biological developmental change that is grounded in experience.

The next two chapters summarize the foundations of brain and behavioral development. Colby and his colleagues provide a detailed summary of what is currently known about postnatal structural changes in the brain based primarily on studies using MRI and DTI. They describe the very different trajectories in gray and white matter development, emphasizing the significance of developmental timing and key genetic and experience-dependent factors driving the dynamic processes of brain development that continue through late adolescence. Lamy and Saffran (see Rubenstein and Rakic, 2013) take on the central question about how infants learn, in particular, how they can acquire abstract and complex cognitive structure based on inputs from the environment. They focus on infants' capacities to extract different types of statistical regularities from the perceptual information that are central to the formation of linguistic and object categories. Evidence from studies conducted over the past decade

suggests that infants actively use probability distributions, sequential structure, correlations, and associations in auditory and visual inputs, beginning early in the first year of life. These active learning mechanisms, which are presumably basic capacities that evolve via experience, are the building blocks that drive cognitive development.

For humans, vision is arguably the most important perceptual system that drives conceptual development, particularly for growth in understanding objects and events. It is also the system that is most well studied and understood across different species. In his chapter, Scott Johnson (Chapter 37) discusses how infants come to experience a world of stable objects beginning with limited but organized vision at birth that undergoes rapid functional changes during the first year of life driven by brain maturation coupled with learning from experience and manual explorations. More advanced aspects of visuospatial abilities are taken up by Stiles and her colleagues who discuss the development of ventral and dorsal neurocognitive processes including global and local pattern perception, spatial construction, localization, attention, and manipulation. The last part of their chapter describes research on these processes, particularly the vulnerability of the dorsal stream in children who have experienced brain injury or with neurogenetic syndromes.

Human memory serves a range of important functions that are carried out by different systems. In her chapter, Bauer describes the major forms of memory, each of which follows distinct developmental trajectories. She summarizes recent studies that document the emergence of both declarative and nondeclarative memory systems during the first 2 years of life, which is far earlier than has previously been thought. At the same time, Bauer argues that some aspects of declarative memory have a more protracted course not reaching maturity until middle childhood, paralleling what is known about the development of the underlying neurobiology of memory processing systems as demonstrated in studies of ERPs in children.

The earliest autobiographical memories are sparse and usually for events that took place during the preschool years. One reason that has been offered for this phenomenon is that encoding and retrieving these memories depend on language. By the time most children are 3 years old, they have mastered the fundamentals of language from sounds to words to grammar, which allow them to form narrative memories of their lives. In the next chapter, Tager-Flusberg and Seery (Chapter 32) describe the development of language covering the major milestones, as well as how language intersects with aspects of motor, conceptual, and social cognitive development. Much is known about the importance of left-hemisphere frontal/temporal cortical systems in adult language processing. These left lateralized systems emerge during the first year of life in typically developing children according to studies using ERP and NIRS with infants. Failure to lateralize language functions to the left hemisphere during this early sensitive period appears to be one hallmark finding among children with developmental language disorders.

The next four chapters are concerned with different aspects of social–emotional development. This has become a very active area of research in developmental science, in part because of the complexity and richness of human social lives and in part because of the important consequences when development in the social domain is impaired in children with genetic syndromes or specific forms of psychopathology. Righi and Nelson (see Rubenstein and Rakic, 2013) focus on the development of face-processing skills. They describe the initial rapid developments that take place during the first year of life in the foundational cognitive processes for identifying faces and facial expressions that depend on dedicated neural architecture in the fusiform region of the temporal cortex. At the same time, these early acquired abilities are followed by a more protracted period of development during which time more subtle behavioral advances are accompanied by volumetric changes in the fusiform 'face area' as well as temporal and morphological changes in the signature ERP signal elicited by faces, the N170. The development of the neural systems underlying more complex aspects of social perception, including the evaluation of a person's intentions, communicative signals, and psychological disposition from eye gaze and body motion cues, is addressed by Voos and his colleagues (see Rubenstein and Rakic, 2013) in the next chapter, drawing heavily on fMRI studies of school-aged children. This is followed by Gweon and Saxe's chapter (see Rubenstein and Rakic, 2013), which focuses on the development of children's theory of mind: the ability to reason about people's actions based on mental states such as thoughts or beliefs. The classic studies in this area concluded that the theory of mind emerges at about the age of 4 based on behavioral and ERP studies that used a task evaluating children's understanding of false beliefs. Gweon and Saxe review more recent behavioral studies suggesting that this understanding may already be in place at least in an implicit form by around 18 months. At the same time, fMRI studies suggest that the brain region that is crucially involved in the theory of mind processing, the right temporal–parietal junction, continues to show functional developmental changes through middle childhood, providing another example of the early emergence of a cognitive achievement followed by a more protracted developmental period into middle childhood. Decety and Michalska (see Rubenstein and Rakic, 2013) focus more on the affective components of social behavior: the development of the capacity to respond to another person's distress, or empathy. Their chapter summarizes the normal course of development in the psychological processes and neurobiological mechanisms that drive empathic behavior, and then how these might go awry in children with conduct disorder or related forms of psychopathy.

The next three chapters focus on the aspects of executive processing that cut across different developmental domains. Posner and his colleagues (see Rubenstein and Rakic, 2013) review the development of the complex set of networks that is involved in executive attentional processes and self-regulation from early infancy through middle childhood. An exciting recent advance in this area is a research that finds associations between common genetic variants with individual differences in specific attentional components; in turn, these differences influence environmental experiences of children as mediated, for example, by parenting behaviors. Lahat and Fox explore the development of two aspects of cognitive control that play important roles in decision-making and social behavior: inhibitory control and self-monitoring. The neural substrates for these advanced cognitive systems, as indexed by fMRI and ERP measures, depend on areas in the prefrontal cortex that do not reach the end point of development until late adolescence. Hughes (see Rubenstein and Rakic, 2013) summarizes the research on classic executive function measures in both typical and atypical children. She argues that because the executive functions and their neural substrates, which encompass multiple brain regions, develop incrementally over the entire period of childhood and adolescence, they are more susceptible to environmental influences. Her chapter describes examples of such influences including studies of training, parent–child interaction, and clinical populations.

The final two chapters in this volume are concerned with primary influences on behavioral and brain development. Gunnar and Davis (see Rubenstein and Rakic, 2013) focus on the effects of stress, drawing heavily on what is known from animal models to investigate whether the findings from that body of literature can be extended to current work on prenatal and postnatal stress responses in human development. Then, Beltz and her colleagues (see Rubenstein and Rakic, 2013) take up the issue of sex differences in development, which has important implications for understanding many neurodevelopmental disorders that differentially affect males and females for reasons that are still not well understood.

The cognitive neuroscience of human development is still in its formative years. The exponential growth of this field, as evidenced by the numbers of papers, journals, and books published over the past decade, has largely been driven by methodological advances that allow people to view more directly the minds and brains of babies and children and observe how they change over time. As these methods develop further, it is expected that greater progress will be made by asking broader questions about the significance of the developmental trajectories and timing within and across cognitive domains, the constraints that operate on individual variation and developmental plasticity, and the precise ways in which biological and nonbiological factors influence the developmental course of brain and cognitive development.

References

Aslin, R., 2007. What's in a look? Developmental Science 10, 48–53.

Aslin, R., 2012. Infant eyes: A window on cognitive development. Infancy 17, 126–140.

Casey, B.J., Tottenham, N., Liston, C., Durston, S., 2005. Imaging the developing brain: What have we learned about cognitive development? Trends in Cognitive Sciences 9, 104–110.

Csibra, G., Kushnerenko, E., Grossmann, T., 2008. Electrophysiological methods in studying infant cognitive development. In: Nelson, C.A., Luciana, M. (Eds.), Handbook of Developmental Cognitive Neuroscience. 2nd edn. MIT Press, Cambridge, MA, pp. 247–262.

Fantz, R.L., 1958. Pattern vision in young infants. Psychological Record 8, 43–47.

Flavell, J., 1996. Piaget's legacy. Psychological Science 7, 200–203.

Gervain, J., Mehler, J., Werker, J.F., et al., 2011. Near-infrared spectroscopy: A report from the McDonnell infant methodology consortium. Developmental Cognitive Neuroscience 1, 22–46.

Rubenstein, J.L.R., Rakic, P., 2013. Patterning and Cell Types Specification in the Developing CNS and PNS.

Vygotsky, L., 1978. Mind in Society. Harvard University Press, Cambridge, MA.

Wozniak, J., Mueller, B., Lim, K.O., 2008. Diffusion tensor imaging. In: Nelson, C.A., Luciana, M. (Eds.), Handbook of Developmental Cognitive Neuroscience. 2nd edn. MIT Press, Cambridge, MA, pp. 301–310.

Theories in Developmental Cognitive Neuroscience

M.H. Johnson

Birkbeck College, University of London, London, UK

11.1 INTRODUCTION

Recently a new field of science has emerged at the interface between developmental neuroscience and developmental psychology. This field has come to be known as developmental cognitive neuroscience (DCN) (Johnson, 2010). The exciting mix of these previously separate fields has led to a burgeoning of new observations, methods, and ideas. However, a characteristic shared with most newly emerging interdisciplinary areas in the biological sciences is the feeling that the knowledge acquired to date is rather fragmentary, with many intuitively surprising observations remaining unexplained. How can one understand these phenomena, and interpret and explain them within a broader context of other findings with different methods and populations? To do this, one needs to develop theories, and, the author will argue that these theories need to be of a particular kind to be useful.

As a scientific discipline, DCN sits at the convergence of two of the oldest philosophical and scientific debates of mankind. The first of these questions concerns the relation between mind and body, and specifically between the physical substrate of the brain and the mental processes it supports. This issue is fundamental to the scientific discipline of cognitive neuroscience. The second debate concerns the origin of organized biological structures, such as the highly complex structure of the adult human brain. This issue is fundamental to the study of development. Light can be shed on these two fundamental issues by tackling them both simultaneously, and specifically by focusing on the relation between the postnatal development of the human brain and the cognitive processes it supports.

The second of the two debates above, that of the origins of organized biological structure, can be posed in terms of phylogeny or ontogeny. The phylogenetic

Neural Circuit Development and Function in the Brain: Comprehensive Developmental Neuroscience, Volume 3 http://dx.doi.org/10.1016/B978-0-12-397267-5.00050-9

© 2013 Elsevier Inc. All rights reserved.

(evolutionary) version of this question concerns the origin of the characteristics of species, and has been addressed by Charles Darwin and many others since. The ontogenetic version of this question concerns individual development within a life span. The ontogeny question has been somewhat neglected relative to phylogeny since some influential scientists have held the view that once a particular set of genes has been selected by evolution, ontogeny is simply a process of executing the instructions coded for by those genes. By this view, the ontogenetic question essentially reduces to phylogeny (a position sometimes termed 'nativism'). In contrast to this view, many now agree that ontogenetic development is an active process through which biological structure is constructed afresh in each individual by means of complex and variable interactions between genes and their environments. The information is not in the genes, but emerges from the constructive interaction between genes and their environment (see also Oyama, 2000).

What actually is ontogenetic development? Many introductory biology textbooks define development in terms of an increasing restriction of fate. This refers to the basic observation that as the biological development of an individual organism (ontogeny) proceeds, the range of options for further specification or specialization available to the organism at that stage decreases. Structural or functional specialization is an end state in which there are few or no options left to the organism. By this view, plasticity can be defined as a developmental stage in which there are still options available for alternative developmental pathways (Thomas and Johnson, 2008). Another dimension of ontogenetic development is that it involves the construction of increasingly complex hierarchical levels of biological organization, including the brain and the cognitive processes it supports. As is seen later in this chapter, organizational processes at one level, such as cellular interactions, can establish new emergent functions at a higher level, such as that associated with overall brain structure. This characteristic of ontogeny means that a full picture of developmental change requires different levels of analysis to be investigated simultaneously. The author argues that the developmentalist needs to go beyond statements such as a psychological change being due to maturation, and actually provide an account of the processes causing the change at cellular and molecular levels. Thus, in contrast to most other areas of psychology and cognitive science, a complete account of developmental change specifically requires an interdisciplinary approach.

Given the above considerations, it is perhaps surprising that only recently has there been renewed interest in examining relations between brain and cognitive development. Although the field of developmental psychology was originally founded by biologists (such as Darwin and Piaget), biological approaches to human behavioral development fell out of favor in the 1970s and 1980s for a variety of reasons, including the widely held belief among cognitive psychologists in that period that the software of the mind is best studied without reference to the hardware of the brain (see section 11.2 for the details of this argument). However, the recent explosion of basic knowledge on mammalian brain development makes the task of relating brain to behavioral changes considerably more viable than previously. In parallel, molecular and cellular methods, along with theories of self-organizing dynamic networks, have led to great advances in our understanding of how vertebrate brains are constructed during ontogeny. These advances, along with those in noninvasive structural and functional neuroimaging, have led to the recent emergence of DCN.

11.2 WHY DO WE NEED THEORIES?

Some might argue that, as a subfield of biology, investigators in DCN should proceed with their empirical investigations of human development unbiased by any perspective or theory, or simply wait until enough data are in before speculating on its significance. However, even a cursory read of philosophy of science tells you that this view is, at best, somewhat naïve. While Victorian naturalists simply made observations about the animals and plants they studied, when one moves to a modern science led by experiments, the rules change. The experiments choosen to conduct (out of the many millions of potential experiments that could be done) are inevitably guided by implicit assumptions and biases. For example, many of the earliest functional MRI experiments with children professed to be neutral and exploratory while also assuming that new brain regions becoming on-line with increased development would be seen. As will be seen later, in many cases this basic starting assumption was violated. In addition, theories in science do not just attempt to explain sets of data *post hoc*, but they should actually generate predictions and direct whole lines of empirical investigation.

Having emphasized the importance of theories, we must also be sure to achieve an appropriate balance with data gathered from a variety of different sources and methods. In one of the parent disciplines of DCN, cognitive development, several grand theoretical castles have been built on the shaky sands of just one particular type of behavioral test. In most of the biological sciences, confidence in an observation or conclusion is greatly increased by seeking multiple different sources of converging evidence.

Interdisciplinary fields such as DCN face a formidable challenge in the development of adequate theories since scientists are required to develop theories that not only cross different levels of observation (such as genetic, neural, and behavioral), but also relate those different levels together in some way. Ideally, DCN theories should relate evidence from different levels of observation in terms of one level of explanation. As mentioned earlier, for several decades in the field of cognitive development, it was generally considered inappropriate to attempt to relate different levels of explanation. Rather, the aim was to explain one level of observation (change in behavior) in terms of one level of explanation (cognitive). This widespread view was taken for a variety of reasons, but one influential source was the work of Marr (1982). Marr argued that because the same computation can, in principle, be implemented on different computer or neural architectures, a computational account of cognition could, and should, be constructed independently of the details of its implementation on hardware. This influential argument led to the view that considering the role of the brain in cognitive development was reductionist in the sense that molecular and cellular processes could never provide an adequate explanation of cognitive processes.

While the case against a simple reductionist view is clearly correct, as stated earlier, theories of biological development critically need to explain the reverse process (to reductionism) of the emergence of higher-order structures of organization. Thus, while not denying Marr's anti-reductionist point, constructing a specific type of neural computer hardware will constrain the range of possible computations that could be supported. With these considerations in mind, Mareschal et al. (2007), among others, argue that there are constraints on computation imposed by its detailed implementation. Further, when attempting to bridge levels of explanation, mechanistic accounts of processes of computation and developmental change should be consistent across different levels, that is, there is a need for isomorphism between levels of description. Since the goal of DCN is to relate the genetic, neural, cognitive, and behavioral accounts of human development, devising theories that relate the different levels of observation seems crucial. Mareschal et al. (2007, p. 209) thus describe it "We would argue that strong pragmatic considerations mean that what is achievable in real time at one level of description strongly constrains the appropriate theories of what is going on at other levels. What is more easily achievable in the brain will strongly constrain the character of cognition. More specifically, the brain can implement much more readily certain representational states and transformations more than others. It is these primitives that the researcher should initially use to construct theories of cognition."

Theories come in different shapes and sizes. Specifically, the amount and range of DCN data accounted for can independently vary along at least two dimensions: (a) how domain specific or domain general (domain here is used in a general sense to refer to an aspect of cognition) a theory is, and (b) how many levels of explanation are incorporated or integrated. It is a

defining feature of DCN, as opposed to traditional cognitive development, that multiple levels of observation are considered and related in terms of a single process or causal mechanism. One reason for this is that the parent discipline of cognitive development had been built on the strategy of explaining changes in behavior during development in terms of cognition – a level of explanation that is not itself directly observable. While the scientific strategy of theorizing at a level that is not directly observable is not unique to cognitive psychology, constraining theories of this kind by only one level of observation is of high risk because of the lack of constraints it imposes. In other words, a very wide variety of theories can successfully account for data at one level of observable only. It is proposed that a better strategy is to sandwich a nonobservable level of explanation (such as cognition) between two levels of observable, such as those of brain and behavior.

Theories in DCN could potentially vary enormously in the scope of data that they account for, from a single cognitive domain in a single population to an account that crosses domains of cognition and populations (typical and atypical development). Often, in biology, the broader the scope of a theory, the less clearly it makes detailed domain-specific predictions. Thus, some have referred to such broad-scope theories as frameworks (Kuhn, 1996). Put simply, frameworks are ways of thinking about, or viewpoints on, a large body of data. Frameworks may have some testable elements but primarily serve as a coherent set of assumptions that, taken together, offer an account of a wide range of phenomena. In addition, within a framework, more specific and detailed theories (and thence hypotheses) can be constructed. Further, they guide lines of research and the kinds of hypotheses that are explored. In a young and newly emerged field, it is suggested that the first priority should be to develop appropriate and useful frameworks, since adopting the wrong framework could be an expensive diversion in terms of both time and money.

11.3 THREE FRAMEWORKS FOR UNDERSTANDING HUMAN FUNCTIONAL BRAIN DEVELOPMENT

A review of the literature reveals that three different frameworks on human postnatal functional brain development are currently commonly adopted and explored (Johnson, 2001).

11.3.1 Maturational Viewpoint

According to the maturational viewpoint, newly emerging sensory, motor, and cognitive functions are related to the maturation of particular areas of the brain, usually regions of cerebral cortex. Much of the research to date attempting to relate brain to behavioral development in humans has taken this approach. Evidence concerning the differential neuroanatomical development of brain regions should then predict the age when a particular region is likely to become functional. Conversely, success in a new behavioral task at given age is attributed to the maturation of a new brain region. Functional brain development is, in this sense, depicted as the reverse of adult neuropsychological studies of patients with brain damage, with specific brain regions being added-in during development (with the converse effects from being depleted by damage).

Despite the intuitive appeal and attractive simplicity of the maturational approach, it does not successfully explain many observations on human functional brain development. For example, recent evidence (discussed later) suggests that some of the regions that are slowest to develop by neuroanatomical criteria can be activated shortly after birth and appear to mediate cognitive functions even before they would be considered anatomically mature. Thus, the emergence of new behaviors is not necessarily linked to a previously immature, silent neural region becoming active when it matures. In fact, where functional activity has been assessed by fMRI during a behavioral transition, multiple cortical and subcortical areas appear to change their response pattern (e.g., Luna et al., 2001; Supekar et al., 2009) rather than a few specific areas becoming active. Another difficulty for the maturational viewpoint is that associations between neural and cognitive changes based on age of onset are theoretically weak because of the great variety of neuroanatomical and neurochemical measures that change at different times in different regions of the brain. Thus, as the brain is continuously developing until the teenage years, it is nearly always possible to find a potential neural correlate for any behavioral change in development.

11.3.2 Interactive Specialization

In contrast to the maturational viewpoint, the interactive specialization (IS) viewpoint assumes that postnatal functional brain development, at least within cerebral cortex, involves a process of organizing patterns of interregional interactions (Johnson, 2000, 2001, 2010). According to this view, the response properties of a specific cortical region are partly determined by its patterns of connectivity to other regions, and their patterns of activity. During postnatal development, changes in the response properties of cortical regions occur as they interact and compete with each other to acquire their role in new computational abilities. From this perspective, some cortical regions may begin with poorly defined broad functions, and consequently are partially

activated in a wide range of different stimuli and task contexts. During development, activity-dependent interactions between regions sharpen the functions and response properties of cortical regions such that their activity becomes restricted to a narrower set of circumstances (e.g., a region originally activated by a wide variety of visual objects may come to confine its response to upright human faces). The onset of new behavioral competencies during infancy will, therefore, be associated with changes in activity over several regions, and not just by the onset of activity in one or more additional region(s).

11.3.3 Skill Learning

The third perspective on human functional brain development, skill learning, involves the proposal that the brain regions active in infants during the onset of new perceptual or motor abilities are similar, or even identical, to those involved in complex skill acquisition in adults. For example, Gauthier and colleagues have shown that extensive training of adults to identify individual artificial objects (called greebles) eventually results in activation of a cortical region previously preferentially activated by faces, the fusiform face area (Gauthier et al., 1999). This indicates that this region is normally activated by faces in adults, not because it is prespecified to do so, but because of our extensive expertise with that class of stimulus. Extended to development, this view would argue that development of face processing during infancy and childhood could proceed in a similar manner to acquisition of perceptual expertise for a novel visual category in adults (see Gauthier and Nelson, 2001). While the degree to which parallels can be drawn between adult expertise and infant development remains unclear, to the extent that the skill-learning hypothesis is correct it presents a clear view of a continuity of mechanisms throughout the life span.

11.4 ASSUMPTIONS UNDERLYING THE THREE FRAMEWORKS

Underlying these frameworks are differing sets of key assumptions.

11.4.1 Deterministic Versus Probabilistic Epigenesis

Gottlieb (1992) distinguished between two approaches to the study of development; deterministic epigenesis in which it is assumed that there is a unidirectional causal path from genes to structural brain changes and then to psychological function, and probabilistic epigenesis in which interactions between genes, structural brain changes, and psychological function are viewed as bidirectional, dynamic, and emergent. In many ways, it is a defining feature of the maturational approach that it assumes deterministic epigenesis; region-specific gene expression is assumed to effect changes in intraregional connectivity that, in turn, allows new functions to emerge. A related assumption commonly made within the maturational approach is that there is a one-to-one mapping between brain and cortical regions and particular cognitive functions, such that specific computations come on-line following that maturation of circuitry intrinsic to the corresponding cortical region. In some respect, this view parallels mosaic development at the cellular level in which simple organisms (such as *Caenorhabditis elegans*) are constructed through cell lineages that are largely independent of each other (see Elman et al., 1996 for discussion). Similarly, different cortical regions are assumed to have different maturational timetables, thus enabling new cognitive functions to emerge at different ages.

In contrast to the maturational approach, IS has a number of different underlying assumptions. Specifically, a probabilistic epigenesis assumption is coupled with the view that cognitive functions are the emergent product of interactions between different brain regions. In this respect, IS follows current trends in adult functional neuroimaging. For example, Friston and Price (2001) point out that it may be an error to assume that particular functions can be localized within a certain cortical region. Rather, they suggest that the response properties of a region are determined by its patterns of connectivity to other regions as well as by their current activity states. By this view, "the cortical infrastructure supporting a single function may involve many specialized areas whose union is mediated by the functional integration among them" (p. 276). Similarly, in discussing the design and interpretation of adult functional MRI studies, Carpenter and collaborators have argued that "In contrast to a localist assumption of a one-to-one mapping between cortical regions and cognitive operations, an alternative view is that cognitive task performance is subserved by large-scale cortical networks that consist of spatially separate computational components, each with its own set of relative specializations, that collaborate extensively to accomplish cognitive functions" (Carpenter et al., 2001, p. 360). Extending these ideas to development, the IS approach emphasizes changes in interregional connectivity as opposed to the maturation of intraregional connectivity. While the maturational approach may be analogous to mosaic cellular development, the IS view corresponds to the regulatory development seen in higher organisms, in which cell–cell interactions are critical in determining developmental fate. While mosaic development can be faster than regulatory, the latter has several advantages. Namely, regulatory development

is more flexible and better able to respond to damage, and it is more efficient in terms of genetic coding. In regulatory development, genes do not code directly, but need only orchestrate cellular-level interactions to yield more complex structures (see Elman et al., 1996).

11.4.2 Static Versus Dynamic Mapping

As well as the mapping between structure and function at one age, it can also be considered how this mapping might change during development. When discussing functional imaging of developmental disorders, Johnson et al. (2002) point out that many laboratories have assumed that the relation between brain structure and cognitive function is unchanging during development. Specifically, in accord with the maturational view, when new structures come on line, the existing (already mature) regions continue to support the same functions they did at earlier developmental stages. The static assumption is partly why it is sometimes considered acceptable to study developmental disorders in adulthood and then extrapolate back in time to early development. Contrary to this view, the IS approach suggests that when a new computation or skill is acquired, there is a reorganization of interactions between different brain structures and regions. This reorganization process could even change how previously acquired cognitive functions are represented in the brain. Thus, the same behavior could be supported by different neural substrates at different ages during development.

Stating that structure–function relations can change with development is all very well, but it lacks the specificity required to make all but the most general predictions. Fortunately, the view that there is competitive specialization of regions during development gives rise to expectations about the types of changes in structure–function relations that should be observed. Specifically, as regions become increasingly selective in their response, properties during development patterns of cortical activation during behavioral tasks may therefore be more extensive than those observed in adults, and involve different patterns of activation. Additionally, within broad constraints, successful behavior in the same tasks can be supported by different patterns of cortical activation in infants and adults. Evidence in support of this view will be discussed later.

The basic assumption underlying the skill-learning approach is that there is a continuity of the circuitry underlying skill acquisition from birth through to adulthood (see Poldrack, 2002; Ungerleider, et al., 2002 for review of the neural systems involved in perceptual and motor skill learning in adults). These circuits are likely to involve a network of structures that retains the same basic function across developmental time (a static brain–cognition mapping). However, other brain regions may respond to learning with dynamic changes in functionality similar or identical to those hypothesized within the IS framework. For example, neuroimaging studies of adults acquiring the skill of mirror reading show both increases and decreases in cortical activity over widespread regions during learning: unskilled performance is associated with activation in bilateral occipital, parietal, and temporal lobes and cerebellum, with acquisition of skill leading to decreases in bilateral occipital and right parietal activation and to increases in inferior temporal lobe and caudate nucleus activation (Poldrack, 2002). According to the skill-learning view, similar dynamic changes in brain activation would occur as skills emerge during development.

11.4.3 Plasticity

Another way in which the three perspectives differ is with regard to the concept of, and assumptions about, plasticity. Plasticity in brain development is a phenomenon that has generated much controversy, with several different conceptions and definitions having been presented. According to the maturational framework, plasticity is a specialized mechanism that is activated following brain injury. According to the IS approach, plasticity is simply the state of having a region's function not yet fully specialized. That is, there is still remaining scope for developing more finely tuned responses. As mentioned earlier, this definition corresponds well with the view of developmental biologists that development involves the increasing restriction of fate. Finally, according to the skill-learning hypothesis view, the functional plasticity present in early development may share many characteristics with the plasticity underlying acquisition and retention of skills in adults (Karni and Bertini, 1997). Thus, unlike the IS approach, plasticity does not necessarily reduce during development.

11.4.4 Summary

To summarize, the maturational view is characterized by the interrelated assumptions that (1) cortical areas are a mosaic of regions with independent developmental timetables; (2) deterministic epigenesis means that inherent structural development in a region causes or allows functional changes; (3) there is fixed regional structure/function mapping; and (4) plasticity involves specialized mechanisms triggered by injury. In contrast, the IS approach is based on the assumptions that (1) cortical areas are inextricably linked through dense patterns of interconnections that contribute to coordinated sequences of development; (2) probabilistic epigenesis gives a vital role to intrinsically and extrinsically

generated activity in sculpting anatomical development; (3) combinations of cortical regions may support similar or identical behaviors in different ways during the course of development; and (4) plasticity is the inherent state of an unspecialized neural system. The third perspective, skill learning, is based on the assumptions that (1) specific cortical areas or networks are specialized for perceptual and motor skill acquisition from early on; (2) the acquisition of these skills shapes the response functions of the same or other cortical regions; (3) dynamic changes occur in the neural substrate of a skill as the brain becomes expert; and (4) plasticity is retained at fairly constant level throughout development.

11.5 PREDICTIONS AND EVIDENCE

Frameworks are useful for a variety of reasons, but particularly so when they help to generate predictions that direct research (albeit that they will not always make opposing predictions), and when they offer coherent explanations of previously puzzling observations. This section presents, three of the sources of evidence that an adequate account of human functional brain development will need to address. Before this, however, the author reviews the types of predictions that arise from the three approaches. The first set of predictions concerns the neural correlates of the onset of new abilities during development. According to a maturational view, new behavioral abilities are mediated by new components of cognition that are, in turn, enabled or allowed by the maturation of one or more brain regions. A consequence of this is a general increase in the number of structures that can be activated in tasks with development. In contrast to these predictions, in the IS approach, it is anticipated that widespread networks of brain regions will change their patterns of activation in association with the onset of new behavioral abilities. Specifically, regions become more specialized (finely tuned) in their response properties with experience. A consequence of this specialization is that on at least some occasions, fewer brain regions will be activated with specific stimulus or task contexts. Turning to the skill-learning approach, here it is anticipated that the onset of new abilities will often be associated with activation of a network of skill acquisition areas. As the new behavioral ability is acquired, a different network of brain regions may become involved.

With regard to the issue of how functions are mapped on to patterns of brain activity, according to the maturational approach, if two age groups are compared in a task for which they show identical behavioral performance, it should also be expected to see identical patterns of brain activation. This is not necessarily the case for the IS approach since the exact patterns of brain activation that support a function will change according to the degree of specialization of component regions within the supporting network. Indeed, the IS approach predicts that the patterns of regional brain activation supporting a function will change during development. While the exact nature of this change will depend on the degree of specialization achieved in different component structures, the IS view predicts a general trend for a decrease in the extent of cortical activation with increasing development/experience. The skill-learning view invokes the reactivation of one or more skill-learning circuits at the onset of a task, followed by a different pattern of activation after the skill is acquired. In this case, in many comparisons between age groups, the younger group will have acquired the skill in less depth than the older group, giving rise to different patterns of underlying brain activation. Interestingly, however, patterns of changing brain activation while adults acquire new skills should mirror the changes seen during development as infants and children acquire simpler skills.

11.6 FUNCTIONAL BRAIN IMAGING

Elsewhere, evidence from functional imaging pertaining to the development of face perception and social cognition (Cohen-Kadosh and Johnson, 2007; Johnson, 2010; Johnson et al., 2009) has been reviewed in detail. In this section, evidence is reviewed from the functional neuroimaging of normal development that pertains to the three perspectives on functional brain development described above. The author suggests that the evidence currently available does not offer much support to the maturational view, at least not without substantial modification. Instead, behavioral change often seems to be accompanied by large-scale dynamic changes in the interactions between regions, and different cortical regions become more specialized for functions as a consequence of development.

A maturational approach to human postnatal functional brain development predicts that a neural correlate of increasing behavioral abilities is an increasing number of active cortical areas. In functional imaging paradigms, therefore, infants and children should show fewer regions active in tasks where they show poorer behavioral performance than adults. In contrast, if new behaviors require changes in interregional interaction, a greater or equal extent of cortical activation can be predicted, and different patterns of activation might be found early in development even in task domains where behavioral performance is similar to that of adults. Which of the three approaches is adopted not only has theoretical implications, but also practical implications for data collection. For example, if one adopts the maturational approach and is expecting a particular area to become

active during development for a particular task, then brain imaging may be focused on that region rather than on the whole brain. As a consequence, any possible changes in more distant brain structures would not be detected. By contrast, if one adopts the IS approach, then the importance of whole-brain imaging is clearly apparent.

A number of authors have described developmental changes in the spatial extent of cortical activation in a given situation during postnatal life. Event-related potential experiments with infants have indicated that for both word learning (Neville et al., 1992) and face processing (de Haan et al., 2002), there is increasing localization of processing with age and experience of a stimulus class. That is, electrophysiological recordings reveal a wider area of processing for words or faces in younger infants than in older ones whose processing has become more specialized and localized. From the IS framework, such developmental changes are accounted for in terms of more pathways being partially activated in younger infants before experience with a class of stimuli. With increasing experience, the specialization of one or more of those pathways occurs over time. Taking the example of word recognition, ERP activity differentiating between comprehended and noncomprehended words is initially found over widespread cortical areas. This narrows to left temporal leads after children's vocabularies have reached a certain level, irrespective of maturational age (Mills et al., 1993). Changes in the extent of localization can be viewed as a direct consequence of specialization. Initially, multiple pathways are activated for most stimuli. With increasing experience, fewer pathways become activated by each specific class of stimulus. Pathways become tuned to specific functions and are therefore no longer engaged by the broad range of stimuli, as was the case earlier in development. Additionally, there may be inhibition from pathways that are becoming increasingly specialized for that function. In this sense, then, there is competition between pathways to recruit functions, with the pathway best suited for the function (by virtue of its initial biases) usually winning out.

Further evidence to support the IS view comes from functional MRI studies in children. For example, Luna et al. (2001) tested participants aged 8–30 years in an occulomotor response-suppression task. Their behavioral results showed that the adult level of ability to inhibit prepotent responses developed gradually through childhood and adolescence. The difference between prosaccade and antisaccade conditions was investigated with functional MRI, and revealed changing patterns of brain activation during development. Both children and adolescents had less activation than adults in a few cortical areas (superior frontal eye fields, intraparietal sulcus), and several subcortical areas, a finding broadly consistent with maturational hypotheses. However, both children and adolescents also had differential activation in regions not found to show differences in adults. Children displayed increased relative activation in the supramarginal gyrus compared to the other age groups, and the adolescents showed greater differential activity in the dorsolateral prefrontal cortex than children or adults. These findings illustrate that the neural basis of behavior can change over developmental time, with different patterns of activation being evident at different ages, a pattern consistent with the IS and the skill-learning viewpoints.

A similar conclusion can be reached after examination of the developmental fMRI data produced by Casey and colleagues. These authors (Casey et al., 1997; Thomas et al., 1999) administered a go/no-go task to assess inhibitory control and frontal lobe function to healthy volunteers from 7 years of age to adult. The task involved participants responding to a number of letters, but withholding their response to a rarely occurring 'X.' More than twice the volume of prefrontal cortex activity (dorsolateral prefrontal cortex) was observed in children compared to adults. One explanation of this finding is that children found the task more difficult and demanding than adults. However, children with error rates similar to those in adults showed some of the largest volumes of prefrontal activity suggesting that task difficulty was not the important factor. It is difficult to account for these decreases in the extent of cortical activation in terms of the progressive maturation of prefrontal cortical areas. The finding that children and adults appeared to show different patterns of activation even when performance was equated is inconsistent with the skill-learning viewpoint, but is in line with expectations from the IS viewpoint.

The third example of the use of fMRI to study the development of cortical activation patterns during childhood comes from studies of verbal fluency tasks in which participants are asked to generate words in response to a cue (e.g., to generate examples of a target category, or generate a verb that relates to a cued noun). Several studies have shown that adults (Lehéricy et al., 2000) and school-age children (Gaillard et al., 2000; Hertz-Pannier et al., 1997, 1999) typically activate left hemisphere frontal cortical networks including Broca's area, premotor, prefrontal, and supplementary motor areas as well as, less consistently, temporal cortical areas including superior temporal, middle temporal, and supramarginal gyri. In addition, some degree of activation in homologous right frontal regions is almost always found both in adults (Pujol et al., 1999; Springer et al., 1999) and in children (Gaillard et al., 2000; Hertz-Pannier et al., 1997, 1999). Two studies have found that both the degree to which activation is bilateral (rather than left dominant) and the extent of this activation is greater in children than in adults (Gaillard et al., 2000; Holland et al., 2001). Thus, as in the other examples discussed above, typical development is associated with a reduction in the extent of activation of cortical areas and, as a consequence, an increased lateralization of activation to the left hemisphere with age (Holland et al., 2001).

In sum, the balance of evidence to date suggests that (1) new behavioral skills are accompanied by widespread changes across many regions of cortex, and (2) functional brain development involves the twin process of increasing localization and increasing specialization.

11.7 CRITICAL OR SENSITIVE PERIODS

The three perspectives on human functional brain development differ in their views as to the effects of early atypical experience. By the maturational view, differences in experience might influence the speed at which a function matures or the ultimate level of performance; in the skill-learning view, any atypical early experiences can potentially be compensated for later in development as the mechanisms for learning remain in operation in the same way; in the IS view, atypical early experiences could have long-lasting effects because they could affect the specialization and localization of function, which may not be able to be altered later in life when there is less scope for plasticity.

There have only been a limited number of studies examining the effects of atypical early experience in humans. Some studies have investigated the perception of facial information in children who experienced deprivation of patterned visual input in the early months of life due to bilateral, congenital cataracts. These patients were tested years after their cataracts were removed and they were fitted with contact lenses (i.e., years after visual input had been restored), thus any effects of the few months of deprivation following birth would likely be absent or very minimal according to the maturation or skill-learning views. However, investigation of these patients reveals persistent deficits in selective aspects of face processing. One study found that patients showed impairments in matching facial identity over changes in viewpoint (and tended to show an impairment in recognizing identity over changes in emotional expression), but performed normally on tests of lip-reading, perception of eye gaze, and matching of emotional expressions (Geldart et al., 2002). The second study demonstrated that this difficulty in processing facial identity may be due to deficits in processing the spacing among facial features since patients performed normally in discrimination of faces that differed only on individual features (e.g., mouth) but they were impaired in discrimination of faces that differ only in the spacing of the features (Le Grand et al., 2001). This was not due to a general impairment in perception of spacing of features, as they performed normally in discriminating nonface patterns whether they differed by the shape of the features or the spacing of the features. The fact that these impairments persisted even after years of visual input to compensate for the early deprivation is not consistent with the maturational or skill-learning views, but is consistent with the IS view that early atypical experience may have long-lasting consequences.

The three perspectives that have been discussed yield different types of predictions about the consequences of perinatal brain damage. According to the maturational view additional mechanisms of plasticity are activated following early damage. Specific additional explanations are then required to account for incidents of recovery of function. In addition, it is not obvious why the extent of plasticity is greater earlier in life. From the IS perspective, there is a parsimonious explanation of recovery of function following perinatal damage since the regional specialization of the remaining brain regions will be altered to compensate, particularly, the corresponding regions in the other hemisphere. In cases of bilateral or extensive damage, recovery is less likely. From the skill-learning perspective, plasticity is a life-long feature of the brain. Damage to the general circuits critical for skill acquisition will have long-lasting and widespread consequences, whereas damage to circuits specific to acquisition of particular skills or their retention may result in more isolated impairments. Of the three approaches, IS gives the simplest account of sensitive periods for plasticity since plasticity is reduced when specialization of corresponding regions is achieved.

In sum, with regard to the long-term effects of atypical early experience, or even variations of experience within the normal range, once again the three frameworks lead to different sets of expectations. From the skill-learning perspective variations in early experience will determine the extent of skills acquired. Early deprivation will be potentially reversible since the same mechanisms of skill acquisition are available later in life. From the IS perspective, long-term effects of atypical early experience can result from atypical patterns of regional specialization arising early in life. Such atypical patterns of specialization may be difficult to reverse once established. Finally, under the maturational view, a primary variable influenced by the environment is the speed of maturation that may affect the level or maintenance of a skill. It is sometimes argued that early sensory deprivation may have a general slowing effect on the sequence of maturation.

11.8 ATYPICAL DEVELOPMENT: FROM GENETICS TO BEHAVIOR IN DCN

It has been seen in earlier sections that IS is a promising framework for understanding human functional brain development. While this level of explanation may be appropriate for characterizing the proximal causes of atypical cognition or behavior in developmental disorders such as autism or Williams's syndrome, when development goes awry, a satisfactory explanation of the disorder also requires an account of the causal mechanisms that initiate the atypical development trajectory. This usually entails hypotheses about how an atypicality at one level (such as genetics) causes or induces the onset of atypical development at other levels (brain, cognitive, behavior). An initial attempt to provide a framework for unraveling these complex causal pathways in developmental disorders came with the causal modeling approach of Morton and Frith (1995). According to its originators, causal modeling provided a theory-neutral system for modeling different theories about the paths of causation from a biological level to cognitive and behavioral levels. The models were represented in a graphical notation that allowed for easy comparison between competing explanations, but the framework itself did not involve the construction of computational or neural network models (although a theory represented in the notation could potentially be implemented this way).

Causal modeling is a useful way to compare different theories where these theories are based on the assumption of predetermined epigenesis (a one-way causal pathway from genetics to behavior), but it is a less natural format for capturing the complexities of probabilistic epigenesis (Gottlieb, 2007) in which cause can also run in reverse – for example, sensory experience or internal states such as stress are known to affect gene expression profiles. In addition, causal modeling is intended as a notation for comparing different theories, and as such does not provide an explanation or generate testable hypotheses itself.

A different perspective on identifying causal factors in development involves using implemented computational models that can capture complex nonlinear interactions that are hard to conceptualize with just schematic illustrations or verbal descriptions. Here, the assumption is that multiple factors can interact in complex ways to determine outcome. In one of several initial attempts to apply neural networks to developmental disorders, Oliver and colleagues charted the different ways in which the 'normal' formation of structured representations in the cerebral cortex can go wrong (Oliver et al., 2000). These models were originally designed to investigate the mechanisms underlying the IS process discussed earlier. Several groups have used simple cortical matrix models to investigate the factors and mechanisms responsible for cortical specialization (e.g., Kerszberg et al., 1992; Oliver et al., 2000; Shrager and Johnson, 1995). In these artificial neural networks, connections between nodes are pruned according to variations of Hebbian learning: links between nodes that are often active together are strengthened, whereas links between nodes that are not often coactive get weaker and are pruned. In some of these models, the degree of pruning of connections during learning approximately matches that seen during the course of human brain development. During exposure to patterned input (roughly equivalent to sensory stimulation), nodes become more selective in their response properties, and under certain conditions, clusters of nodes with similar response properties emerge. Thus, in these computational models, selective pruning plays a role in the emergence of clusters of nodes (localization) that share common specific response properties (specialization).

In order to explore developmental disorders, Oliver and colleagues made simulations with a simple cortical matrix model in which one or other of the parameters known to be important for the emergence of structured representations was deliberately changed. In this case, when the authors manipulated an aspect of the intrinsic structure of the network, the relative length of excitatory and inhibitory links, this initial state change totally disrupted the formation of structured representations. In other simulations, structured representations emerged, but were distorted in different ways relative to the typical development case. Oliver et al. (2000) aimed to generate taxonomy of the ways that structured cortical representations could go awry in development, with the long-term aim that some of these artificial developmental disorders could potentially map onto those that occur in the real world.

A criticism of the Oliver et al. (2000) approach, and several related models, is that while they could potentially provide an explanation at the level of brain structure and function, they do not capture the genetic, cognitive, and behavioral levels. Very recent models have begun the ambitious task of simulating from the genetic to the behavioral levels, albeit within a restricted domain of cognition (past tense acquisition) (Thomas et al., submitted).

11.9 IS: FUTURE CHALLENGES

To recap on the basis of the IS view, it argues that early in postnatal development, many areas begin with poorly defined functions, and consequently can be partially activated by a wide variety of sensory inputs and tasks. During development, activity-dependent interactions between regions result in modifications of the intraregional connectivity such that the activity of a given area becomes restricted to a narrower range of circumstances. As a result of becoming more finely tuned, small-scale functional areas become increasingly distinct from their surrounding cortical tissue, and this will be evident in functional imaging studies as increasing localization of function. In summary, according to the IS view, small-scale areas of cortex become tuned for certain functions as a result of a combination of factors, including (i) the suitability or otherwise of the biases within the large-scale region (e.g., transmitter types and levels, synaptic density, etc.); (ii) the information within the sensory inputs (sometimes partly determined by other brain systems); and (iii) competitive interactions with neighboring regions (so that functions are not duplicated).

To date, the majority of the research on the emergence of specialized functions in human cortex has focused on specific regions. However, it is clear from the IS viewpoint that the next step is to understand how networks involving different regions, each with their own different specializations, emerge. In other words, while understanding the functional brain development at the level of individual cortical regions is beginning, still the knowledge about how the larger scale of cortical function in terms of networks of regions develops (Johnson and Munakata, 2005) is lacking. This section reviews, some initial evidence and theory that may begin to address this intriguing issue.

However, before considering the empirical evidence, one needs to consider what makes a network of functional nodes more or less successful. A branch of mathematics called graph theory concerns itself with the relative efficiency of different kinds of networks. While it may seem at first that a lattice or grid pattern is the optimal design for a network, formal analysis of measures of local network connectivity and the average path length from one node to another show that so-called small-world networks are the most efficient (Bassett and Bullmore, 2006). In contrast to the grid pattern of streets found in many American cities, small-world networks are more like the clusters of small streets in a village that is then linked to other such villages by fast highways. Although the overall balance of the small local streets and highways can vary, most biological systems (and even the World Wide Web) are small-world networks. Several studies have shown the regional interconnectivity of the adult brain is a highly efficient small-world network, but how does this efficient network emerge?

The first piece of the jigsaw comes from recent work by Fair et al. (2007, 2009) who used functional connectivity analyses in fMRI to study resting state control networks in school-age children and adults. Their analysis allows them to infer the nature and strength of functional connections between 39 different cortical regions. They found that development entailed both segregation (i.e., decreased short-range connectivity) and integration (i.e., increased long-range connectivity) of brain regions that contribute to a network. In a similar study, the general developmental transition from more local connectivity to greater and stronger long-range network connectivity was confirmed using slightly different methods and 90 different cortical and subcortical regions (Supekar et al., 2009; but see Power et al., 2012; Van Dijk et al., 2012 for technical limitations of these analyses).

The decrease in short-range interregional functional connectivity is readily explicable in terms of the IS view. As neighboring regions of cortical tissue become increasingly specialized for different functions (e.g., objects vs faces), they will less commonly be coactivated. This process may also involve synaptic pruning and, as heard in the last section, has been simulated in neural network models of cortex in which nodes with similar response properties cluster together spatially distinct from nodes with other response properties (Oliver et al., 1996). Thus, decreasing functional connectivity between neighboring areas of cortex is readily predicted by models implementing the IS view. More challenging from the current perspective is to account for the increase in long-range functional connections.

A maturational explanation of the increase in long-range functional connectivity would suggest that this increase is due to the establishment or strengthening of the relevant fiber bundles. However, the increase in functional connectivity during development may occur after the relevant long-range fiber bundles are in place (see Fair et al., 2009; Supekar et al., 2009 for discussion). While increased myelination is likely to be a contributory factor, (1) myelination itself can be a product of the activity/usage of a connection (Markham and Greenough, 2004); and (2) a general increase in myelin does not in itself account for the specificity of interregional activity into functional networks that support particular computations (but see Nagy et al., 2004). Thus, the strengthening and maintenance of long-range brain connections is also likely to be an activity-dependent aspect of brain development. This raises the question of why and how do particular anatomically distant brain regions begin to cooperate in a functional network?

A key to answering this question may lie in scaling up the basic mechanisms of Hebbian learning. Instead of cells that fire together wiring together, regions that tend to be coactivated in a given task context strengthen or maintain the neural pathways between them. While each region is becoming individually specialized for a particular function, this intraregion change in tuning is modulated and influenced by its presence within an emerging network of coactivated structures. For example, in a task that requires visually guided action, a variety of visual and motor areas will be coactivated along with multimodal integration areas. If the task is repeated sufficiently often then these patterns of coactivation will be strengthened, and specialization of individual regions will proceed within this context of overall patterns of activation.

The second source of coactivation in the developing human brain is commonly overlooked – spontaneous activity during the resting state (with no current task demands). Although there has been great interest in the resting state or default network in adults, only recently has this been studied using fMRI in children (although there is a long history of studying resting EEG in children). It seems likely that the oscillatory resting activity of the brain, which probably occupies more waking hours than those when the child is engaged in any specific tasks, may play a key role in strengthening and pruning the basic architecture of long-range connections.

The third reason why anatomically distant regions may strengthen and maintain their connectivity relates to the fact that most of the long-range functional connections studied by Fair et al. (2007) involved links to parts of the prefrontal cortex (PFC). This part of the cortex is generally considered to have a special role during development in childhood and skill acquisition in adults (Gilbert and Sigman, 2007; Thatcher, 1992). Indeed, Prefrontal Cortex (PFC) may play a role in orchestrating the collective functional organization of other cortical regions during development. While there are several neural network models of PFC functioning in adults (e.g., O'Reilly, 2006), few if any of these have addressed development. However, another class of model intended to simulate aspects of development may be relevant both to PFC and to the issue of how networks of specialized regions come to coordinate their activity to support cognition. Knowledge-based cascade correlation (Shultz et al., 2007) involves an algorithm and architecture that recruits previously learned functional networks when required during learning. Computationally, this dynamic neural network architecture has a number of advantages over other learning systems. Put simply, it can learn many tasks faster, or learn tasks that other networks cannot, because it can recruit the knowledge (computational abilities) of other self-contained networks as and when required. In a sense, it selects from a library of available computational systems to orchestrate the best combination for the learning problem at hand. While this class of model is not intended to be a detailed model of brain circuits (Shultz and Rivest, 2001; Shultz et al., 2007), it has been used to characterize frontal systems (Thivierge et al., 2005) and may capture important elements of the emerging interactions between PFC and other cortical regions at an abstract level. In addition, it offers initially attractive accounts of (1) why PFC is required for the acquisition of new skills; (2) why PFC is active from early in development, but also shows prolonged developmental change; and (3) why early damage to PFC can have widespread effects over many domains.

Although much work remains to be done to understand in more detail, the factors that lead to the emergence of long-range networks, the graph theory analyses of changes during the school-age years are generating important insights. While, as described earlier, there are differences in the balance of short and long connections between children and adults, it is important to note that the network organization of children's brains is as efficient as that of adults. In other words, while children's brains are wired differently from those of adults, they are still optimally geared for the rapid and high-fidelity transmission of information. Whether the same is true in infancy and early childhood remains unknown.

Aside from the shift from local to long-range connectivity, another change in network structure observed using graph theory analysis during development is in the hierarchical structure. Adult networks have a more hierarchical structure that is optimally connected to support top-down relations between one part of the network and another (Supekar et al., 2009). While hierarchical networks have a number of computational advantages to be discussed below, they are known to be less plastic and more vulnerable to damage or noise in the particular nodes at the top of the hierarchy. Thus, the network arrangement of children may be more flexible and plastic in response to unusual or atypical sensory input or environmental context. Further, the response to focal brain damage, particularly in the prefrontal cortex, may be more clearly understood in the light of these different network structures.

One of the features of a hierarchical network is the capacity for one region to feedback highly processed sensory or motor input to the earlier stages of processing. In much the same way as we hypothesized that lateral interregional interactions help shape the intrinsic connectivity of areas to result in functional specialization, interactions between regions connected by feedback and feed-forward connections may also help shape the specialization of the areas involved. Top-down effects play an important role in sensory information processing in the adult brain (e.g., Siegel et al., 2000). For example, during perception, information propagates through the visual processing hierarchy from primary sensory areas to higher cortical regions while feedback connections convey information in the reverse direction. In a neuro-computational model of feedback in visual processing in the adult brain, Spratling and Johnson (2004) demonstrated that a number of different phenomena associated with visual attention, figure/ground segmentation, and contextual cueing could all be accounted for by a common mechanism underlying cortical feedback. Extending these ideas to development, there are potentially two important implications of feedback that will benefit from future exploration. The first of these will be to examine how the specialization of early sensory areas is shaped by top-down feedback, and vice versa, during development. The second topic for investigation will be to examine the consequences of relatively poor or diffuse cortical feedback in the immature cortex.

Top-down feedback from PFC may also have a direct role in shaping the functional response properties of posterior cortical areas. In cellular recording studies from both humans and animals, evidence has accrued that the selectivity of response of neurons in areas such as the fusiform cortex may increase in real time following the presentation of a stimulus. For example, McCarthy et al. (1997) measured local field potentials in face-selective regions of lateral fusiform cortex in human adults and found that responses of these neurons go from being face selective at around 200 ms after stimulus presentation, to being face identity or emotion selective at later temporal windows. This suggests that top-down cortical feedback pathways, in addition to their importance in attention and object processing (Spratling and Johnson, 2004, 2006), may increase the degree of specialization and localization in real time, as well as in developmental time. Thus, some of the changes in functional specialization and localization seen in face-sensitive regions may reflect the increasing influence of interregional coordination with other regions, including the PFC.

A final aspect of the transition from child brain network to the adult one is the greater connectivity between cortical and subcortical structures seen at younger ages (Supekar et al., 2009). This observation may be fundamental for our understanding of the emergence of the social brain and memory systems as it implies that the specialization of some cortical areas may be initially more dominated by structures such as the amygdala and hippocampus. As adulthood is approached, more networks become intrinsic to the cortex and develop a complex hierarchical structure more dominated by PFC.

11.10 SUMMARY, CONCLUSIONS, AND RECOMMENDATIONS

Now that DCN has become established as an interdisciplinary field in its own right, it has become time to evaluate and question the directions one is going in. One of the most common criticisms leveled at the newly emerging field is that it is primarily being driven forward by the powerful new methods for imaging brain structure and function in an infant- and child-friendly way (as well as new techniques for genetic analyses), and that it lacks the theory-driven approach that characterizes much of the best work in cognitive development. Similar concerns are expressed, albeit less directly, by students who can be daunted by the somewhat fragmentary islands of data that have been acquired to date about human functional brain development. Where is the overarching theory or framework within which they can make sense of disparate observations? A related concern sometimes expressed by those in cognitive science is that the hypotheses which are presented in DCN are reductionist, or otherwise impoverished as a cognitive explanation of infant or child behavior. In other words, this criticism is that what hypotheses and theories there are in the field are of the wrong type, and do not offer a satisfactory explanation of behavioral change in development.

Starting with the criticism of a relative lack of theories in DCN, acknowledgement has to be made that, at least compared to the parent discipline of cognitive development, work in DCN is generally less theory driven (albeit with the exceptions discussed in this chapter). Why is this? A large part of the explanation is believed to be due to the sudden increase in the volume and diversity of data available because of the new methods that have become available. Many theories that successfully accounted for sets of behavior observations in child development founder on the rocks when attempts are made to account for neuroscience data relating to the same behavioral tasks. For one thing, when one more than doubles the quantity of data to be accounted for, then many previously successful theories will no longer offer a satisfactory explanation, simply because the chance of observing refuting evidence is much higher. Bringing powerful new methods into a field is analogous to a catastrophic environmental change during evolution – the majority of species (theories) simply cannot adapt and, therefore, die off. It takes generations for the better-adapted species to emerge.

The second issue is that of accommodating to new types of data. When one begins to study brain function directly, the first thing that strikes is the complexity of the processes involved. For example, neuroscience evidence indicates that the brain has at least three partially independent routes for executing eye movements. While these routes may have slightly different attributes, duplication of computations and (apparent) redundancy seems to be a basic feature of how the brain does things. Thus, at a sweep, simple single-route cognitive models appear less plausible. Add to this the complexity of feedback routes interacting with sensory-driven information, and the undoubted importance of temporal synchrony, and many existing theories of cognitive development look hopelessly simplistic. Of course, a common reaction to this is that theories of cognitive development are not intended to account for neuroscience data – that is, merely a matter of implementation. However, if you accept this argument, I contend that you are not doing DCN (and I would argue that satisfactory explanations of development necessitate bridging between levels of observation (see Johnson, 2010)).

This leads us to the second common criticism of theory in developmental science; the theories are of the wrong type to be of relevance for explaining the development of human behavior. Commonly, the view is expressed that theories in DCN are reductionist and, therefore, do not offer good explanations of cognitive change. As discussed earlier, it has been argued following Marr (1982), that cognition is a level of explanation independent from the underlying neuroscience. Recent directions in neuroscience suggest that, to the contrary, there is a large degree of interdependence between levels in real complex biological systems such as the brain. This has led to the proposal that theories that are consistent between different levels of explanation should be sought (see Mareschal et al., 2007, for a detailed discussion of this point). Ultimately, theories that are consistent with both behavioral and brain development evidence will have greater explanatory power than those confined to one level of observation.

In considering the issues above, the current dearth of plausible theories in DCN seems unsurprising. After all, new fields in the biological sciences (in contrast to some physical sciences) often go through a natural history phase in which collection of basic data is the priority. However, in this chapter, it was argued that one needs to strive to bring more adequate and appropriate theories into the field. Thus, three positive suggestions for hallmarks of a good theory in DCN are offered.

1. The theory advanced should genuinely relate neural observations to behavioral ones, and can be equally well tested (and refuted) by either neural or behavioral level observations. I suspect that a variety of different types of theories will emerge to serve this bridging function, but that they are unlikely to look like many existing cognitive development theories. Theories that have been developed purely on the basis of behavioral data are unlikely to naturally map on to brain imaging data, and there is a danger in seeking only confirmatory data. Ideally, theories of functional brain development that are equally compatible with brain and behavioral observations should be developed.

2. Theories in developmental science should involve mechanisms of change. This suggestion is not new (e.g., Karmiloff-Smith, 1998; Mareschal and Thomas, 2007), but it is still surprisingly common to see theories that explain the state of affairs before and after a developmental transition, but that do not specify the mechanisms of the transition itself (other than using the terms such as maturation or learning). Theories of development need to be theories focused on change.

3. Given that theories in DCN are accounting for several levels of observation, and that they also need to be compatible with undoubtedly complex and dynamic aspects of neural processing, we need to find ways to elucidate and present those theories so that they are both comprehensible and clarifying. This is the attraction and importance of formal computational modeling, be it symbolic, connectionist or hybrid (see Mareschal et al., 2007). While theories may initially develop as informal ideas, ultimately we should aim to implement them as computational models.

Finally, the author cautions against being too prescriptive. In the long term it is probably good for the field to have a heterogeneous mix of different types of theories and let the data, and time, select those with the best fit to reality. After all, despite their prolonged domination, the dinosaurs did not inherit the globe.

Acknowledgments

The author thanks several members of the Centre for Brain and Cognitive Development, Birkbeck, for their comments on earlier versions of this chapter. Financial support was provided by the UK Medical Research Council PG071484.

References

Bassett, D.S., Bullmore, E., 2006. Small-world brain networks. The Neuroscientist 12, 512–523.

Carpenter, P.A., Just, M.A., Keller, T.A., Cherkassky, V., Roth, J.K., Minshew, N.J., 2001. Dynamic cortical systems subserving cognition: fMRI studies with typical and atypical individuals. In: McClelland, J.L., Siegler, R.S. (Eds.), Mechanisms of Cognitive Development: Behavioral and Neural Perspectives. Erlbaum, Mahwah, NJ, pp. 353–383.

Casey, B.J., Trainor, R.J., Orendi, J.L., et al., 1997. A developmental functional MRI study of prefrontal activation during performance of a go-no-go task. Journal of Cognitive Neuroscience 9 (6), 835–847.

Cohen-Kadosh, K., Johnson, M.H., 2007. Developing a cortex specialized for face perception. Trends in Cognitive Neuroscience 11, 367–369.

de Haan, M., Humphries, K., Johnson, M.H., 2002. Developing a brain specialized for face processing: A converging methods approach. Developmental Psychobiology 40, 200–212.

Elman, J., Bates, E., Johnson, M.H., Karmiloff-Smith, A., Parisi, D., Plunkett, K., 1996. Rethinking Innateness: A Connectionist Perspective on Development. MIT Press, Cambridge, MA.

Fair, D.A., Dosenbach, N.U.F., Church, J.A., et al., 2007. Development of distinct control networks through segregation and integration. Proceedings of the National Academy of Sciences of the United States of America 104, 13507–13512.

Fair, D.A., Cohen, A.L., Power, J.D., et al., 2009. Functional brain networks develop from a 'local to distributed' organization. PLoS Computational Biology 5, e1000381.

Friston, K.J., Price, C.J., 2001. Dynamic representations and generative models of brain function. Brain Research Bulletin 54, 275–285.

Gaillard, W.D., Hertz-Pannier, L., Mott, S.H., Barnett, A.S., LeBihan, D., Theodore, W.H., 2000. Functional anatomy of cognitive development: fMRI of verbal fluency in children and adults. Neurology 54, 180–185.

Gauthier, I., Nelson, C., 2001. The development of face expertise. Current Opinion in Neurobiology 11, 219–224.

Gauthier, I., Tarr, M.J., Anderson, A.W., Skudlarski, P., Gore, J.C., 1999. Activation of the middle fusiform 'face area' increases with expertise recognizing novel objects. Nature Neuroscience 2 (6), 568–573.

Geldart, S., Mondlach, C.J., Maurer, D., de Schonen, S., Brent, H.P., 2002. The effect of early visual deprivation on the development of face processing. Developmental Science 5, 490–501.

Gilbert, C.D., Sigman, M., 2007. Brain state: Top-down influences in sensory processing. Neuron 54, 677–696.

Gottlieb, G., 1992. Individual Development and Evolution. Oxford University Press, New York.

Gottlieb, G., 2007. Probabilistic epigenesis. Developmental Science 10, 1–11.

Hertz-Pannier, L., Gaillard, W.D., Mott, S., et al., 1997. Assessment of language hemispheric dominance in children with epilepsy using functional MRI. Neurology 48, 1003–1012.

Hertz-Pannier, L., Chiron, C., van de Moortele, P., Bourgeois, M., Fohlen, M., Dulac, O., 1999. Multi-task fMRI presurgical language mapping in children with cognitive impairment. NeuroImage 9, S691.

Holland, S.K., Plante, E., Byars, A.W., Strawburg, R.H., Schmithorst, V.J., Ball Jr., W.S., 2001. Normal fMRI brain activation patterns in children performing a verb generation task. NeuroImage 14, 837–843.

Johnson, M.H., 2000. Functional brain development in infants: Elements of an interactive specialization framework. Child Development 71, 75–81.

Johnson, M.H., 2001. Functional brain development in humans. Nature Reviews Neuroscience 2, 475–483.

Johnson, M.H., 2010. Developmental Cognitive Neuroscience, 3rd edn. Wiley-Blackwell, Chichester, West Sussex.

Johnson, M.H., Munakata, Y., 2005. Processes of change in brain and cognitive development. Trends in Cognitive Neuroscience 9, 152–158.

Johnson, M.H., Grossmann, T., Cohen-Kadosh, K., 2009. Mapping functional brain development: Building a social brain through interactive specialization. Developmental Psychology 45, 151–159.

Johnson, M.H., Halit, H., Grice, S., Karmiloff-Smith, A., 2002. Neuroimaging of typical and atypical development: A perspective from multiple levels of analysis. Development and Psychopathology 14, 521–536.

Karmiloff-Smith, A., 1998. Development itself is the key to understanding developmental disorders. Trends in Cognitive Neuroscience 2, 389–398.

Karni, A., Bertini, G., 1997. Learning perceptual skills: Behavioral probes into adult cortical plasticity. Current Opinion in Neurobiology 4, 530–535.

Kerszberg, M., Dehaene, S., Changeux, J., 1992. Stabilization of complex input-output functions in neural clusters formed by synapse selection. Neural Networks (5), 403–413.

Kuhn, T.S., 1996. The Structure of Scientific Revolutions, 3rd edn. University of Chicago Press, Chicago, IL.

Le Grand, R., Mondloch, C.J., Maurer, D., Brent, H., 2001. Early visual experience and face processing. Nature 410, 890.

Lehéricy, S., Cohen, L., Bazin, B., et al., 2000. Functional MRI evaluation of temporal and frontal language dominance compared with the Wada test. Neurology 54, 1625–1633.

Luna, B., Thulborn, K.R., Munoz, P.D., et al., 2001. Maturation of widely distributed brain function subserves cognitive development. NeuroImage 13 (5), 786–793.

Mareschal, D., Thomas, M.S.C., 2007. Computational modeling in developmental psychology. IEEE Transactions on Evolutionary Computation 11, 137–150.

Mareschal, D., Johnson, M.H., Sirois, S., Spratling, M., Thomas, M., Westermann, G., 2007. Neuroconstructivism, Vol. I: How the Brain Constructs Cognition. Oxford University Press, Oxford.

Markham, J.A., Greenough, W.T., 2004. Experience-driven brain plasticity: Beyond the synapse. Neuron Glia Biology 1, 351–363.

Marr, D., 1982. Vision. W.H. Freeman, San Francisco, CA.

McCarthy, G., Puce, A., Gore, J.C., Allison, T., 1997. Face-specific processing in the human fusiform gyrus. Journal of Cognitive Neuroscience 9 (5), 605–610.

Mills, D.L., Coffey, S.A., Neville, H.J., 1993. Language acquisition and cerebral specialization in 20-month-old infants. Journal of Cognitive Neurosciences 5, 317–334.

Morton, J., Frith, U., 1995. Causal modelling: A structural approach to developmental psychopathology. In: Cicchetti, D., Cohen, D. (Eds.), Manual of Developmental Psychopathology, vol. 1. Wiley, New York, pp. 357–390.

Nagy, Z., Westerberg, H., Klingberg, T., 2004. Maturation of white matter is associated with the development of cognitive functions during childhood. Journal of Cognitive Neuroscience 16, 1227–1233.

Neville, H.J., Mills, D.L., Lawson, D.S., 1992. Fractionating language: Different neural subsystems with different sensitive periods. Cerebral Cortex 2 (3), 244–258.

O'Reilly, R.C., 2006. Biologically based computational models of high-level cognition. Science 314, 91–94.

Oliver, A., Johnson, M.H., Shrager, J., 1996. The emergence of hierarchical clustered representations in a Hebbian neural, model that simulates development in the neocortex. Network: Computation in Neural Systems 7, 291–299.

Oliver, A., Johnson, M.H., Karmiloff-Smith, A., Pennington, B., 2000. Deviations in the emergence of representations: A neuroconstructivist framework for analysing developmental disorders. Developmental Science 3, 1–23.

Oyama, S., 2000. The Ontogeny of Information: Developmental Systems and Evolution, 2nd edn. Duke University Press, Durham.

Poldrack, R.A., 2002. Neural systems for perceptual skill learning. Behavioral and Cognitive Neuroscience Reviews 1 (1), 76–83.

Power, J.D., Barnes, K.A., Snyder, A.Z., Schlaggar, B.L., Peterson, S.E., 2012. Spurious but systematic correlations in functional connectivity MRI networks arise from subject motion. NeuroImage 59, 2142–2154.

Pujol, J., Deus, J., Losilla, J.M., Capdevila, A., 1999. Cerebral lateralization of language in normal left-handed people studied by functional MRI. Neurology 52, 1038–1043.

Shrager, J., Johnson, M.H., 1995. Waves of growth in the development of cortical function: A computational model. In: Julesz, B., Kovacs, I. (Eds.), Maturational Windows and Adult Cortical Plasticity. The Santa Fe Institute Studies in the Sciences of Complexity, vol. XXIII. Addison-Wesley, New York, pp. 31–44.

Shultz, T.R., Rivest, F., 2001. Knowledge-based cascade-correlation: Using knowledge to speed learning. Connection Science 13, 43–72.

Shultz, T.R., Rivest, F., Egri, L., Thivierge, J.P., Dandurand, F., 2007. Could knowledge based neural learning be useful in developmental robotics? The case of KBCC. International Journal of Humanoid Robotics 4, 245–279.

Siegel, M., Körding, K.P., König, P., 2000. Integrating top-down and bottom-up sensory processing by somato-dendritic interactions. Journal of Computational Neuroscience 8, 161–173.

Spratling, M., Johnson, M.H., 2004. A feedback model of visual attention. Journal of Cognitive Neuroscience 16, 219–237.

Spratling, M., Johnson, M.H., 2006. A feedback model of perceptual learning and categorization. Visual Cognition 13, 129–165.

Springer, J.A., Binder, J.R., Hammeke, T.A., et al., 1999. Language dominance in neurologically normal and epilepsy subjects: A functional MRI study. Brain 122, 2033–2046.

Supekar, K.S., Musen, M.A., Menon, V., 2009. Development of large-scale functional brain networks in children. PLoS Biology 7 (7), e1000157. http://dx.doi.org/10.1371/journal.pbio.1000157.

Thatcher, R.W., 1992. Cyclic cortical reorganization during early childhood. Special issue: The role of frontal lobe maturation in cognitive and social development. Brain and Cognition 20, 24–50.

Thivierge JP, Titone D, and Shultz TR (2005) Simulating frontotemporal pathways involved in lexical ambiguity resolution. *Poster Proceedings of the Cognitive Science Society*, 2005.

Thomas, M., Johnson, M.H., 2008. New advances in understanding sensitive periods in brain development. Current Directions in Psychological Science 17, 1–5.

Thomas MSC, Ronald A, and Forrester NA (submitted). Modelling the mechanisms underlying population variability across development: Simulating genetic and environmental effects on cognition.

Thomas, K.M., et al., 1999. A developmental functional MRI study of spatial working memory. NeuroImage 10 (3), 327–338.

Ungerleider, L.G., Doyon, J., Karni, A., 2002. Imaging brain plasticity during motor skill learning. Neurobiology of Learning and Memory 78 (3), 553–564.

Van Dijk, K.R., Sabuncu, M.R., Buckner, R.L., 2012. The influence of head motion on intrinsic functional connectivity MRI. NeuroImage 59, 431–438.

12

Structural Brain Development
Birth Through Adolescence

J.B. Colby[1], E.D. O'Hare[2], J.E. Bramen[1], E.R. Sowell[3]

[1]University of California at Los Angeles (UCLA), Los Angeles, CA, USA; [2]University of California at Berkeley, Berkeley, CA, USA; [3]University of Southern California (USC), Los Angeles, CA, USA

OUTLINE

Abbreviations

AD Axial diffusivity
CT Computed tomography
DTI Diffusion tensor imaging
FA Fractional anisotropy
fMRI Functional MRI
IQ Intelligence quotient
MD Mean diffusivity
MRI Magnetic resonance imaging
PET Positron emission tomography
RD Radial diffusivity
ROI Region of interest
TBSS Tract-based spatial statistics
VBM Voxel-based morphometry

12.1 INTRODUCTION

Human brain development is a dynamic process that begins *in utero* and continues prominently through childhood, adolescence, and young adulthood. While strongly influenced by genetic factors, the environment also prominently affects brain maturation by acting on the cellular and macroscopic levels. This experiential learning impacts both brain structure and function through forms of neuronal plasticity that continue throughout our lifetimes. However, despite the fact that investigating brain development is undoubtedly one of

Neural Circuit Development and Function in the Brain: Comprehensive Developmental Neuroscience, Volume 3 http://dx.doi.org/10.1016/B978-0-12-397267-5.00051-0

© 2013 Elsevier Inc. All rights reserved.

the keys to appreciating how we emerge as unique human beings and how this process can go awry in disease, our understanding of this important period has historically been hindered by two main factors. First, there has been a lack of reliable postmortem data as, thankfully, children are generally healthy during development. Second, technological limitations of past methods such as positron emission tomography (PET) and computed tomography (CT) often imposed some modest risk of harm to the subject (e.g., ionizing radiation), which made the study of healthy, typically developing children ethically questionable. The situation changed dramatically with the dissemination of magnetic resonance imaging (MRI) technology during the 1980s, which not only offers higher-quality images of the brain parenchyma than ultrasound, x-ray, CT, or PET, but also does so in a way that is remarkably safe for the subject (see Chapter 34).

This discussion will begin with a review of the historical postmortem and histological literature, and will then move on to the groundbreaking neuroimaging investigations of the 1990s, that first examined brain development with this MRI technology. A collection of more detailed phenomena will then be examined, which have been uncovered with advanced brain mapping techniques and have come together as a set of classic features that characterize typical brain development. Finally, we will conclude with a discussion of the cutting-edge efforts being made to integrate these diverse observations within a more generalized 'multimodal' imaging framework and to relate them to advancements in cognitive development. A focus on prominent sex-specific, regional and temporal variations will be continually threaded throughout this discussion.

12.2 POSTMORTEM STUDIES AND HISTOLOGY

12.2.1 Comparison to MRI

Although datasets were sparse, postmortem and histological studies were able to provide key insight into normal brain structure and development, as well as pathology, decades before the introduction of neuroimaging methods like PET and MRI. Further, the rich literature that developed from this early effort has provided a strong foundation of data against which newer imaging modalities can be validated. Compared to a modern imaging method like MRI, there are several distinct advantages and disadvantages of these postmortem studies. Not only are datasets relatively small in postmortem samples, as mentioned above, but longitudinal studies – valued for their statistical power to detect changes over time within individuals among the highly variable population – are also impossible to conduct. Conversely, because postmortem methods can directly visualize the

brain tissue, spatial resolution far exceeds even the best neuroimaging protocol, and there is less validation needed to ensure that the raw signal being measured faithfully represents the underlying neuronal architecture. Artifacts, though, are an important concern for either method. While postmortem methods may introduce artifacts due to cell death, fixation/staining procedures, and morphological changes due to osmotic pressure and mechanical damage, MRI data suffer artifacts from other sources like magnetic susceptibility effects (signal loss in regions near large caverns of air), local image distortions caused by magnetic field inhomogeneities, and partial volume effects that occur when different structures fall within the same voxel. Many of these issues present less of a problem for the interpretation of larger data sets obtained with MRI, relative to postmortem data, as effects of artifacts generally become small as the number of samples becomes large. However, it should still be remembered that MRI only offers a wide-angle indirect view of tissue, which cannot reach down to the cellular level and must be observed through the complex lens of magnetic resonance.

12.2.2 Synaptogenesis and Pruning

By the time an infant is born, the human brain already contains on the order of 100 billion neurons (Kandel et al., 2000). The period of rapid overall brain growth that began *in utero* continues after birth through the first years of life. Surprisingly, however, postmortem studies during the early part of the twentieth century showed that total brain volume and weight actually plateau early and reach approximately 90% of their adult values by age 5 (Dekaban, 1978; Riddle et al., 2010; Vignaud, 1966; and is discussed extensively in Rubenstein and Rakic, 2013).

Even during this early period of pronounced overall growth, brain development is characterized as a dynamic process with both progressive and regressive changes that are influenced by complex genetic influences as well as experience-dependent plasticity due to environmental influences. As the infant brain grows in size, it also grows in complexity. Neurons undergo dendritic branching, forming an arbor of neural connections through synaptogenesis and then ultimately refine this global brain network through the processes of myelination and synaptic pruning. Much of our understanding of the complex balance between synaptogenesis and synaptic pruning has evolved from the seminal histological work performed by Huttenlocher and colleagues, who mapped synaptic density in different areas of the brain throughout childhood. Overall synaptic density is comparable to the adult level at birth. It then rises even further through the first year of life to its peak at 12–18 months and then decreases during late childhood and young adulthood towards a stable adult plateau of ~1 billion synapses*mm^{-3} (Huttenlocher, 1979). This has helped to form the theory that the flexible groundwork laid through an initial overabundance of connections gives way to a reduced – but more targeted and efficient – network through experience-dependent synaptic pruning. Interesting regional variations were also observed during these studies, with primary visual and auditory cortex reaching their peak synaptic densities earlier than prefrontal cortex (Huttenlocher, 1979; Huttenlocher and Dabholkar, 1997; Huttenlocher et al., 1982). The extended period of synapse elimination also has regional variations, with pruning ending by age 12 in the auditory cortex but continuing through mid-adolescence in the prefrontal cortex. This temporal pattern parallels concurrent gains in the cognitive domains that are thought to relate to these regions (Luna et al., 2004; Spear, 2000).

12.2.3 Myelination

Myelination of axonal projections by oligodendroglia is also a prominent component of early brain development. This process begins *in utero*, continues rapidly through the first 5 years of life, and remarkably extends – although at a slower rate – through young adulthood. Intracortical histological preparations by Kaes in 1907 were some of the first to demonstrate this prolonged trajectory of myelination, along with its striking regional variability in timing (Kaes, 1907; Kemper, 1994). His work not only demonstrated earlier trajectories in some areas (posterior temporal, precentral, and postcentral cortex) than others (superior parietal, anterior temporal, and anterior frontal cortex), but also showed that regions with a more protracted developmental trajectory have more pronounced changes during older age. This has helped to form the 'first-in-last-out' theory of aging (Davis et al., 2009), which suggests that higher-order cognitive manifestations (e.g., problem solving and logical reasoning) – some of the last to develop (Luna et al., 2004) – are some of the first to degenerate in old age. Furthermore, the visible spread of myelin outwards into the cortex results in an apparent cortical thinning, which suggests that normal developmental decreases in cortical thickness (discussed in Section 12.4.3) may be due, in part, to this progressive increase in myelin and not simply to regressive changes like synaptic pruning and cell loss (see Rubenstein and Rakic, 2013).

These initial observations in intracortical tissue were extended to the white matter in the pioneering work performed by Yakovlev and Lecours in the 1960s. They demonstrated that white matter myelination begins *in utero* during the second trimester of pregnancy and continues throughout young adulthood (Yakovlev and Lecours, 1967). Additionally, they extended the earlier observations of regional variations in the timing of myelination and described a general posterior-to-anterior trend in the timing of white matter myelination during development that has also been replicated in other samples (Kinney, 1988). Later independent research, targeting the hippocampal formation, has also noted striking increases in myelination, with a 95% increase observed in the extent of myelination relative to brain weight during the first two decades of life. Surprisingly, the authors noted that expanding myelination continued even through the fourth to sixth decades of life (Benes et al., 1994). Taken together, these observations suggest that structural white matter development, in the form of advancing myelination, proceeds in tune with overall cognitive development – with areas involved in lower-order sensory and motor function myelinating earlier than areas involved with higher-order executive function. This correlated timing implies there may be some relationship between advancing brain function and increased myelination; however, postmortem studies are limited from investigating this directly.

12.2.4 Sex-Specific Differences

A pronounced sexual dimorphism in overall brain size emerges during the first 5 years of human brain development, with males having brains that are, on average, approximately 10% larger than females at their adult plateau (Dekaban, 1978). This simple and widely reproducible observation has served as a catalyst for continued interest in the study of sex-specific differences during brain development in order to (1) map other detailed components of brain development that may also show sex-specific differences, (2) determine if there are any cognitive correlates with these findings (Kimura, 1996), (3) establish what – if not total volume of brain matter – are the driving structural contributors to individual cognitive differences in areas like language skills and overall intelligence, and, perhaps most importantly, (4) better understand and clinically address the range of neuropsychiatric disorders that tend to emerge during adolescence with prominent sex-specific affinities (Marsh et al., 2008). Interestingly, while some of this sex-specific variance in brain size can be attributed to height, which is consistent with broader trends across different mammalian species, there remains a significant sex-specific effect on brain size even when differences in body size are taken into account (Peters et al., 1998). Although the brains of adult males tend to be larger than adult females, this increase is actually smaller than what would be predicted based on differences in adult height alone. Histological findings indicating a 15% higher neuronal density in males than females are consistent with this (Rabinowicz et al., 2002), although conflicting reports from other studies prohibit firm conclusions on this point (Haug, 1987; Pakkenberg and Gundersen, 1997). A consideration of the fact that females actually tend to be taller than males during late childhood, perhaps due to faster pubertal maturation in girls, further weakens the idea of such a simple allometric relationship when age-matched males and females are compared (Giedd et al., 2006). These discrepancies highlight the diversity that exists among the postmortem literature on the topic of sex differences in brain development, which is also likely to be influenced by a variety of confounds (including cohort effects and observational bias) that have made interpretation challenging (Peters et al., 1998). Additionally, these reports are limited to either simple global measures, like total brain volume or weight, or very local measures, like neuronal density, and generally do not account for regional variations in measures like cortical thickness and folding complexity (Luders et al., 2004; Sowell et al., 2007) (see Chapters 35 and 38).

12.2.5 Summary

The central theme that emerges from this early postmortem work is that brain development from birth through adolescence is a uniquely dynamic process, encompassing both progressive and regressive events, with varying magnitudes and timing across different regions of the brain. In particular, the concurrent decrease in synaptic density and increase in white matter myelination is consistent with the principle of selective specialization, which has been postulated to be the driving force behind the creation of cognitive networks and thought to form the foundation for higher cognitive processes (Fuster, 2002; Post and Weiss, 1997; Tsujimoto, 2008). The initial overabundance of neurons and synapses during infancy is thought to provide a flexible substrate through which activity-dependent plasticity can fine-tune neural network activity via processes like synaptic pruning, which continue robustly through adolescence and, in some form, throughout life.

12.3 *IN VIVO* VOLUME ANALYSES

With the development of MRI, not only were clinicians provided with a superior technology for the diagnosis of brain injury and disease (Barkovich, 2006; Panigrahy and Blüml, 2009; Prager and Roychowdhury, 2007), but researchers were also provided with an unparalleled technology for the study of typical brain development *in vivo*. This, together with the expansion of computing technology during the 1980s, led to the first wave of

structural neuroimaging studies aimed at extending previous postmortem results. Much of this early work utilized volumetric parcellation methods, whereby brain images are segmented according to different anatomical landmarks, and the volumes and tissue content (gray matter, white matter, cerebrospinal fluid) of these different regions are computed and compared between subject groups or throughout development. This parcellation step has been performed with a variety of methods, including the use of stereotactic coordinates (Jernigan et al., 1991a,b; Reiss, et al., 1996), manually drawn regions of interest (ROIs) (Giedd et al., 1996c; Sowell et al., 2002b), and automated protocols (Giedd et al., 1996a, 1999a) (see Chapter 28).

12.3.1 Gray Matter Decreases in Development

Given the previous postmortem observations of regional and temporal variations in synaptic density and myelination throughout the brain, the gray and white matter volume estimates extracted through these volumetric parcellation methods would be expected to show similar age-related developmental trajectories and regional differences. This was first demonstrated by Jernigan and Tallal, who observed that a group of children aged 8–10 had significantly more cortical gray matter than a group of young adults, as well as a higher gray matter to white matter ratio (Jernigan and Tallal, 1990). A subsequent study extended these findings to confirm that the group differences were due to continuous age-related decreases in gray matter volume with time – independent of brain size – and localized these effects to superior frontal and parietal cortices (Jernigan et al., 1991a,b). These studies marked the first *in vivo* morphological evidence in support of the earlier postmortem histological work by Huttenlocher and colleagues. While not a direct measure of synaptic density, the volumetric MRI finding of decreased gray matter volume is consistent with the regressive synaptic pruning changes previously described and aligns with the theory that evolutionarily more complex regions like the frontal lobe show more protracted timing in their development than evolutionarily simpler regions like primary motor/visual cortex. Even this early on, Jernigan and colleagues were also aware of the possible relationship between their *in vivo* MRI findings and the postmortem white matter myelination studies of Yakovlev and Lecours and suggested that an 'apparent' cortical thinning could be due, in part, to progressing myelination. Thus, a component of these observed changes might not be a gray matter loss, *per se*, but a transition of unmyelinated 'gray' matter into white matter, which, on MRI, would appear as a gray matter volume 'loss' during the childhood and adolescent years.

12.3.2 Regional and Temporal Dynamics

Since these initial observations of childhood and adolescent gray matter volume loss, other investigations have confirmed the general trend (Caviness et al., 1996; Giedd et al., 1999a; Ostby et al., 2009; Pfefferbaum et al., 1994; Reiss et al., 1996; Sowell et al., 2002b; Wilke et al., 2007) and extended these observations in several important ways. In a large cross-sectional sample of 161 subjects aged 3 months to 70 years, Pfefferbaum and others were able to demonstrate the early rise and plateau in total brain volume by approximately age 5, as well as the late childhood and adolescent decline in cortical gray matter volume. The extended age range of the sample allowed them to characterize the trajectory of the gray matter volume decline as curvilinear, which suggested an overall inverted 'U'-shaped curve consisting of early childhood gray matter increases followed by a relatively early peak and then late childhood and adolescent reductions (Pfefferbaum et al., 1994). This general time course of cortical development is a feature that has gone on to become one of the hallmarks of structural brain development (Courchesne et al., 2000; Paus et al., 2001; Sowell et al., 2003).

Other studies have investigated the relative volume changes (controlling for global increases in total brain volume) more closely in broader age samples and in specific cortical and subcortical structures. In doing so, this work has demonstrated further heterogeneity in maturational timing and trajectory complexity across the brain. Importantly, the relative gray matter volume reduction during adolescence was confirmed to be most concentrated in the frontal and parietal lobes (Sowell et al., 2002b). Meanwhile, subcortical gray matter structures like the basal ganglia also generally showed a relative volume reduction, although with a simpler linear trajectory than the cortex over the age range of late childhood to young adulthood (Ostby et al., 2009; Sowell et al., 2002b). Tzarouchi et al. recently applied some of the modern spatial normalization methods to align each individual's structural MRI brain data to a group average of all individuals studied and conducted an analysis in the vein of these classical volumetric studies by examining volume change over time in over 100 regions throughout the cortex. Using an exponential nonlinear function to model the upstroke portion of the developmental curve, they were able to estimate the age at which these different gray matter regions reached full development. In doing so, they provided a compelling demonstration of the previously theorized maturational sequence, with primary somatosensory and visual cortices maturing the earliest and then posterior-to-anterior and inferior-to-superior trends in the developmental timing of the remaining temporal, parietal, and frontal lobes (Tzarouchi et al., 2009).

12.3.3 White Matter Increases in Development

Interestingly, while gray matter volume was observed to peak early, researchers began to consistently observe that white matter volume continues to steadily increase roughly linearly from birth through adolescence and young adulthood (Caviness et al., 1996; Paus et al., 2001; Pfefferbaum et al., 1994; Sowell et al., 2002b; Wilke et al., 2007). The timing of these changes shows a posterior-to-anterior gradient, which generally parallels the overlying gray matter and has led to continued investigation into the interaction between these processes (Barkovich et al., 1988). The white matter volume increase is consistent with the widespread reports of relative gray matter reductions during later childhood and adolescence, as a protracted increase in underlying white matter volume (due, in part, to increased oligodendroglial wrapping of axonal fibers) will increase total brain volume and, therefore, decrease the relative gray matter volumes of specific structures compared to this total. The midsagittal corpus callosum was one of the first white matter areas to be examined in more detail, with volumetric analyses showing robust increases in total area throughout adolescence (Bellis et al., 2001; Giedd et al., 1999b) and a surprising anterior-to-posterior trend in the timing of the growth curve when the corpus callosum was subdivided into seven distinct segments (Giedd et al., 1996a). This protracted nature of white matter development is a thread that we will see repeated in the following sections, as imaging modalities and analysis techniques have advanced (Giedd et al., 1999a; Lebel et al., 2008b; Sowell et al., 2003), and as one that has gained increasing interest as more attention is being focused on the network properties of the brain as a potential, important mediator for the late cognitive development seen in domains like risk/reward processing, cognitive control, and working memory (Spear, 2000).

12.3.4 Sex Differences

Sex-specific differences in brain structure were also extended with these structural imaging techniques. Total brain volume was confirmed to be approximately 10% larger in males than females at the plateau of overall brain volume that is reached during childhood (Caviness et al., 1996; Courchesne, et al., 2000; Durston et al., 2001; Gur et al., 2002; Lenroot and Giedd, 2010), and the significant sex-specific effect remains even when height and weight are covaried (Giedd et al., 1996b). However, more detailed regional volumetric observations of different cortical regions have been inconsistent – with varying reports of increased or decreased volumes in males and females that are further complicated by whether or not absolute or relative changes (to total brain volume differences) were reported (Sowell et al., 2007).

Despite this variation, strong evidence suggests that there may be sexual dimorphism in the timing of the developmental trajectory in the cortex, such that males and females have similar overall trajectories of regional brain maturation (both with an inverted 'U'-shaped curve) but differing gender effects with time because of a difference in the timing of this trajectory (Giedd et al., 1999a). Specifically, there appears to be approximately a 1–2 years' phase difference between girls and boys, with peaks in gray matter occurring earlier in girls than boys and regional variations in both phase and the actual differences between the sexes (Lenroot et al., 2007). Because this is a temporally dynamic period of development (and not a static one, e.g., like comparing fully mature adults), assessing sex differences during childhood and adolescence has become a more complicated problem, which requires the dissociation of phase differences (particularly those caused by differences in age of pubertal onset) from sex differences in the maturational trajectories. An additional challenge is identifying those differences that persist into adulthood and actually have functional relevance.

Observations of sex differences in subcortical regions have also been somewhat more reproducible. In particular, over the course of development, the amygdala seems to increase in volume more in males and the hippocampus more in females (Giedd et al., 1996b,c, 1997; Wilke et al., 2007). This is in line with animal studies that have shown high densities of steroid hormone receptors in the medial temporal lobe (Sarkey et al., 2008) and also that sex steroids exert trophic effects on these structures (Cooke, 2006; Galea et al., 2006; Zhang et al., 2008). In one recent study specifically targeted to investigate the degree to which the rise of gonadal hormones during puberty contributes to the emergence of these sex-specific differences, we examined subcortical volume measures among a group of 80 adolescent boys and girls matched on sexual maturity within a relatively narrow age range (Bramen et al., 2011). This focused analysis revealed an interaction between sex and the effect of puberty in predicting amygdala and hippocampal volumes: While females actually had larger left amygdala and right hippocampal volumes than boys during early puberty (relative to total brain size), by late puberty, the amygdala volume had increased in males but stayed relatively stable in females. In the right hippocampus, the effect of puberty was also to increase the volume in males but, surprisingly, to decrease the volume in females. This reiterates the importance of considering the timing in the interpretation of developmental phenomena like sexual dimorphisms. Furthermore, these results suggest that the sex-specific differences in amygdala volume previously observed across a broader age sample (Giedd et al., 1996c) are likely due, in large part, to the effects of puberty, but that the previous observation of relatively larger hippocampal volumes in females may be due to nonpubertal influences, as the direct contribution of puberty, demonstrated in our recent observations, would be to blunt this effect. These findings are particularly important in the context of adolescent brain development, as maturation of these processing centers and their connections to areas like the prefrontal cortex may contribute to the dramatic changes seen in social and emotional domains during this period of development (Dahl, 2004; Steinberg, 2005). The caudate nucleus has also been shown to be relatively larger, controlling for total brain volume, in females across several distinct samples (Giedd et al., 1996b, 1997; Sowell et al., 2002b; Wilke et al., 2007). Put another way, the caudate is spared the reduction in volume that is typically shown by other structures in female brains. This observation of sexual dimorphism in the basal ganglia is also important, as it may relate to the emergence of similar sex differences in the incidence of several neuropsychiatric disorders (e.g., attention deficit hyperactivity disorder, Tourette syndrome) that are thought to involve these structures (Marsh et al., 2008) (see Chapters 35 and 38).

12.4 BRAIN MAPPING APPROACHES

12.4.1 Advantages

The early volumetric MRI imaging observations by Jernigan and others helped lay the foundation for the next wave of neuroimaging studies designed to further characterize the anatomical changes that occur during normal development. While the volumetric protocols were able to validate much of the classical postmortem literature, as well as provide further evidence for gray matter loss, white matter gain, and regional/temporal dynamics, they are unable to precisely localize where these maturational changes are taking place within the relatively large ROIs studied. Instead, these methods collapse entire regions of the brain down into one or several summary descriptive statistics that may not be characteristic of all functional and structural brain circuits within these large lobar regions. In contrast, newer methods like voxel-based morphometry (VBM) and cortical thickness analysis are distinct in that they allow for statistical analysis at many points throughout the entire brain volume or at many points across the entire cortical surface and allow the creation of whole-brain 'maps' to visualize these data. These enhanced analysis modalities, together with the traditional methods discussed in Sections 12.2 and 12.3, have contributed greatly to our understanding of normal brain development and provide an important context for the further study of neurodevelopmental and psychiatric disease (Eliez and Reiss, 2000; Marsh et al., 2008) (see Chapters 30, 35 and 38).

12.4.2 Voxel-Based Strategies

In VBM, the local fractional gray matter volume is analyzed in the neighborhood around each voxel in the brain to generate whole-brain maps of gray matter 'density' or 'concentration' (Ashburner and Friston, 2000). Spatial normalization algorithms are applied to align the brains of individual subjects so that each voxel then can be compared throughout development or between groups. Consistent with the previous volumetric studies and postmortem examples, whole-brain mapping strategies utilizing VBM show decreasing gray matter density during later development. In line with the coarse frontal and parietal lobar localizations of the earlier volumetric reports, the regions showing the most protracted changes in these new analyses include clusters in the dorsal frontal and parietal cortices during the transition from childhood to adolescence (Sowell et al., 1999a), as well as a distinct grouping of dorsal, medial, and orbital frontal cortical areas during the later transition from adolescence into young adulthood (see Figure 12.1; Sowell et al., 1999b). The relative specificity of these later changes to the frontal lobes is consistent with the similarly protracted time course of cognitive development in executive function domains, which are also typically thought to involve these frontal regions (Casey et al., 2005; Luna et al., 2004, 2010; Spear, 2000).

This notion of gray matter density was extended to allow for analysis on the cortical surface through the method of cortical pattern matching, where sulcal landmarks are manually identified and used to drive accurate nonlinear spatial normalization into a common template, while helping to account for regional, gender, and individual variability (Ashburner et al., 2003; Luders et al., 2004). Using this technique, protracted postadolescent gray matter density decreases were again demonstrated in dorsal frontal cortex (Gogtay et al., 2004; Sowell et al., 2001b), and, for the first time, shown to correlate significantly with underlying relative brain growth in these regions (Sowell et al., 2001b). This suggests the combined influences of both regressive processes like synaptic pruning, as well as progressive processes, like myelination, are acting in these areas. Using similar gray matter density measurement techniques, and a powerful longitudinal design that tracked individuals prospectively for 8–10 years, Gogtay et al. provided further evidence that the lower-order somatosensory and visual areas develop earlier than the higher-order association cortices that integrate these processes and also that phylogenetically older areas develop earlier than younger areas (Gogtay et al., 2004). Surprisingly, however, gray matter density increases were actually observed in bilateral perisylvian regions

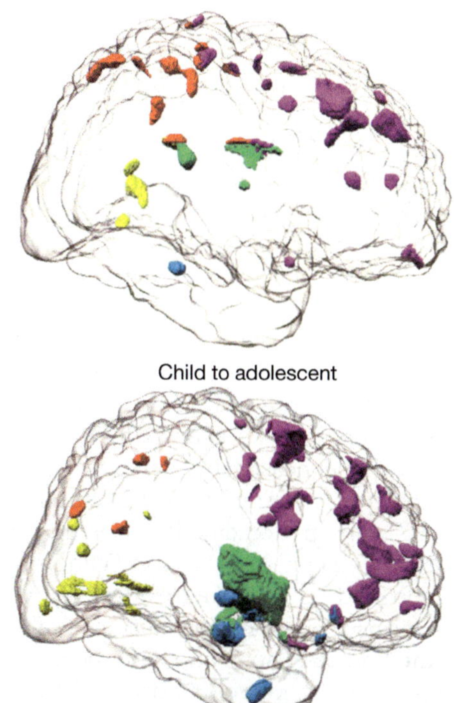

FIGURE 12.1 Gray matter density maturation. Voxel-based morphometry (VBM) measurements of fractional gray matter density/concentration show typical decreases during development. Colored volumes within a transparent cortical surface rendering represent the extent of significant decreases in gray matter density during the transition from childhood to adolescence (top panel) and adolescence to adulthood (bottom panel). Color-coding indicates which changes occurred in the frontal lobe (purple), parietal lobe (red), occipital lobe (yellow), temporal lobe (blue), and subcortical regions (green). *Sowell ER, Thompson PM, Holmes CJ, Batth R, Jernigan TL, and Toga AW (1999a) Localizing age-related changes in brain structure between childhood and adolescence using statistical parametric mapping. Neuroimage 9: 587–597; Sowell ER, Thompson PM, Holmes CJ, Jernigan TL, and Toga AW (1999b) in vivo evidence for post-adolescent brain maturation in frontal and striatal regions.* Nature Neuroscience 2: 859–861.

during the transition from adolescence to adulthood and shown to correlate with both lateralized differences in sylvian fissure morphology and concomitant local brain growth (Sowell et al., 2001b, 2002a). This suggests the presence of a particularly extended developmental trajectory in these gray matter regions beyond that in the dorsal frontal lobe and perhaps implies a unique position for these canonical language areas in the developmental landscape – with the typical inverse correlation between density and volume (decreasing density, increasing volume) reversed to give a direct relationship (increasing density, decreasing volume) in these areas during this age range (Sowell et al., 2003). Taken together, these findings again highlight both regional as well as temporal complexities to the normal developmental sequence of brain structure.

12.4.3 Cortical Thickness

The investigation of apparent cortical gray matter decreases during development reached full stride with the development of MRI cortical thickness measurement techniques. These automated algorithms extract mesh models of the white-matter/gray-matter boundary surface and the pial (i.e., cortical) surface and then directly calculate the cortical thickness at many points throughout the cortical sheet (see Figure 12.2; Fischl and Dale, 2000). Unlike the rather abstract interpretation of fractional gray matter 'density' estimates, cortical thickness estimates are in physical units of millimeters and validate exceptionally well against the historical postmortem cortical thickness maps – with average measurements in children ranging from 1.5 mm in occipital cortex to 5.5 mm in dorsomedial frontal cortex (see Figure 12.2; Sowell et al., 2004a; Von Economo, 1929).

In addition to their strong agreement with postmortem data in terms of absolute thickness estimates, reports using this method are also in line with the mounting evidence from postmortem, volumetric, and VBM density measurements, which support the picture that gray matter thickness peaks early and then declines due to a combination of progressive events, like enhanced myelin penetration into the cortical neuropil, and regressive events, like continued synaptic pruning (O'Donnell et al., 2005; Shaw et al., 2008; Sowell et al., 2004a; Tamnes et al., 2010). In a longitudinal study of 45 typically developing children aged 5–11 years, who were scanned 2 years apart, these techniques were able to demonstrate gray matter thinning of ~0.15–0.30 mm year^{-1} coupled to relative brain volume increases in right frontal and bilateral parietooccipital regions (see Figure 12.3). This study was also able to reproduce the surprising earlier findings of gray matter increases in bilateral perisylvian language areas (Wernike's area) and extended these observations to the left inferior frontal gyrus – another language area (Broca's area) (Sowell et al., 2004a). Cortical thickening was estimated to be at a rate of 0.10–0.15 mm year^{-1} in these areas. This unique pattern of cortical thickening in the canonical language regions could be related to parallel gains in language processing made during this period of development. In another large longitudinal study of 375 children and young adults, changes in cortical thickness were modeled with a low-order polynomial basis set in order to investigate regional differences in the complexity of the developmental trajectory (Shaw et al., 2008). Patterns of varying complexity were found to parallel the established histological maps of cytoarchitectonic complexity and agree with the previous literature (Gogtay et al., 2004; Sowell et al., 2004a) – with simpler laminar areas, like the limbic cortex, having simpler trajectories and more complex laminar areas, like association cortex, having more complex trajectories. Although cortical thinning during adolescence reflects developmental processes like myelination and synaptic

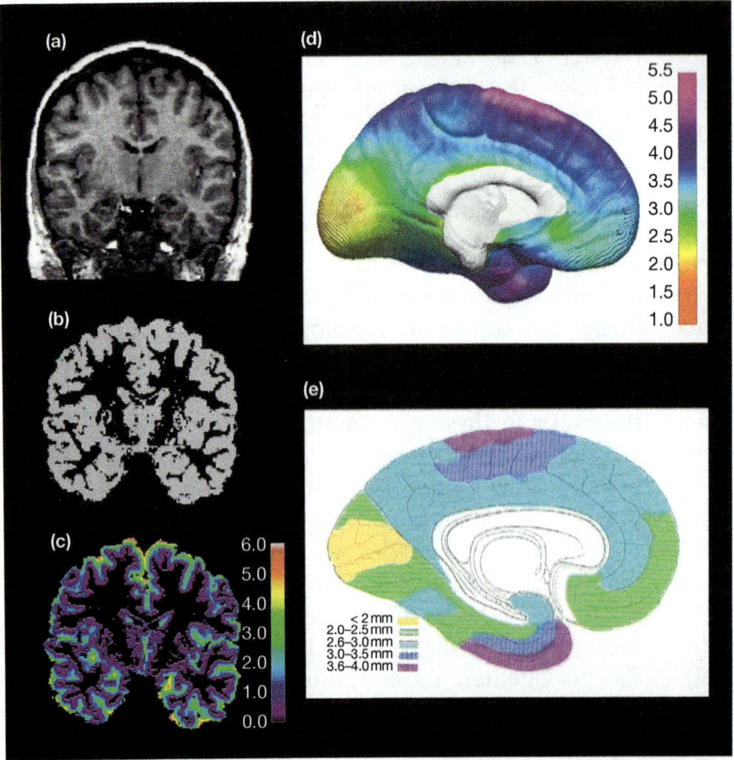

FIGURE 12.2   Cortical thickness analysis. (a–c) show a single representative slice for one subject: (a) Raw T1-weighted anatomical MRI scan. (b) Gray/white matter tissue segmentation. (c) Gray matter thickness image, with thickness (mm) coded by color (warmer colors overlie the areas with the thickest cortex). (d) shows an *in vivo* average cortical thickness map generated by performing this analysis on a cross-sectional sample of 45 subjects. The brain surface rendering is color-coded according to the underlying cortical thickness (mm) and the color bar on the right. The regional variations in cortical thickness can be compared to an adapted version of the classical Von Economo postmortem cortical thickness map (e), which has been color-coded in a similar manner over the original stippling pattern to highlight the similarity between the two maps. *Sowell ER, Thompson PM, Leonard CM, Welcome SE, Kan E, and Toga A W (2004a) Longitudinal mapping of cortical thickness and brain growth in normal children.* Journal of Neuroscience 24: 8223–8231; *Von Economo CV (1929) The Cytoarchitectonics of the Human Cerebral Cortex. London: Oxford Medical Publications.*

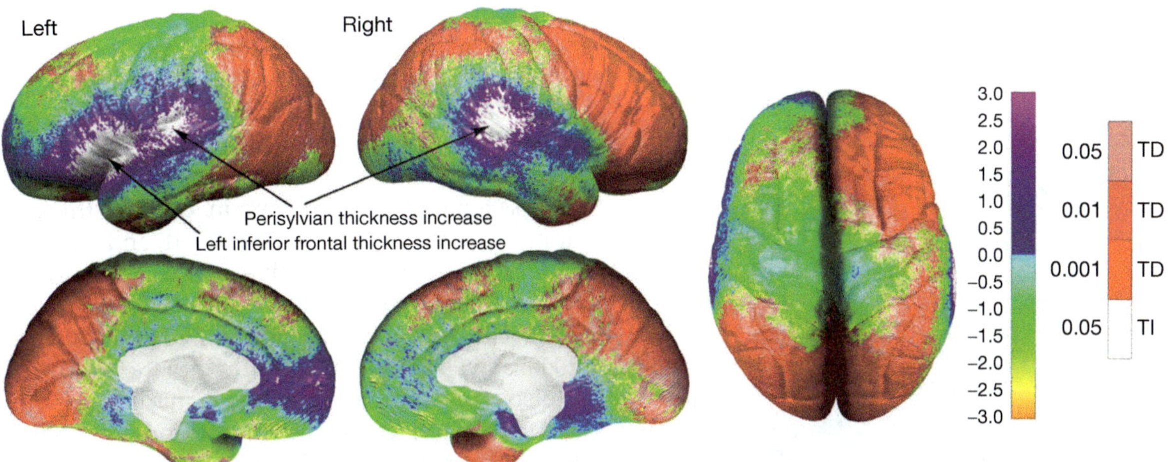

FIGURE 12.3   Gray matter thickness maturation. Statistical maps showing the significance of cortical thickness change in a longitudinal sample of 45 children scanned twice between the ages of 5 and 11. Areas showing significant thickness decrease (TD) are displayed in red, and areas showing significant thickness increase (TI) are displayed in white (see color bar and significance thresholds at right). Nonsignificant areas are coded by their t-statistic according to the left rainbow color bar. Arrows highlight the relative specificity of thickness increases during this age range to canonical language areas in the left inferior frontal gyrus (Broca's area) and perisylvian region (Wernike's area). *Sowell ER, Thompson PM, Leonard CM, Welcome SE, Kan E, and Toga AW (2004a) Longitudinal mapping of cortical thickness and brain growth in normal children.* Journal of Neuroscience 24: 8223–8231.

pruning, it is important to note that cortical thinning also continues in some form throughout the rest of the lifespan (Sowell et al., 2003). This likely belies a shift in etiology to degenerative changes associated with aging (Sowell et al., 2004b), and recent work has sought to delineate this inflection point more precisely. By analyzing local gray and white matter signal intensities in the context of cortical thinning, the timing of the developmental peak was found to range from 8 to 30 years of age in different regions of the cortex, with the regional pattern following the general posterior-to-anterior gradient discussed in Section 12.3.2 (Westlye et al., 2010).

12.4.4 White Matter

Even before the widespread adoption of diffusion imaging, which will be discussed in Section 12.5, researchers were able to adapt traditional anatomical MRI analysis techniques to study white matter development (Wozniak and Lim, 2006). Magnetization transfer ratio imaging is sensitive to the 'bound' protons found on the phospholipids of myelin (Wolff and Balaban, 1989) and reflects the increasing myelination during early development (Engelbrecht et al., 1998) as well as the posterior-to-anterior trend in the timing of this process (Buchem et al., 2001). T2 relaxometry, which estimates the fraction of water in the brain, that is associated with the phospholipid bilayer of myelin (MacKay et al., 1994), has also been used to demonstrate the caudal-to-rostral wave of myelination (Lancaster et al., 2003). In an application of the VBM technology to the white matter, Paus and colleagues were able to powerfully interrogate the rather general 'global white

matter increases' observation previously described to obtain a much richer localization of the precise anatomical regions involved. In an 88-subject sample of children aged 4–17 years, they observed a prominent increase in white matter density in the internal capsule bilaterally, as well as the in left arcuate fasciculus, suggesting continued maturation of corticospinal and frontotemporal fibers through this age range (Paus et al., 1999). This work agrees with the postmortem data from Yakovlev and Lecours and demonstrates the unique progressive changes that occur in the white matter while the cortex shifts to undergo predominantly regressive events. Confirming the surprising corpus callosum results of the classical volumetric study by Giedd et al., which was discussed in Section 12.3.3, Thompson and colleagues applied a continuum mechanics approach to obtain maps of local tissue deformation in the corpus callosum during development. Their longitudinal design studied 6 children aged 3–11 with a follow-up interval of up to 4 years and again demonstrated an anterior-to-posterior wave in the timing of maximal local growth (see Figure 12.4; Thompson et al., 2000). This contrasts with the general posterior-to-anterior trend that has been observed in gray matter cortical regions and suggests a unique pattern of development in this region of interhemispheric fiber connectivity.

12.4.5 Sex Differences

Continuing the trend from volumetric results, VBM gray matter density and cortical thickness observations of sex-specific effects during development have also been variable (Wilke et al., 2007). However, this topic

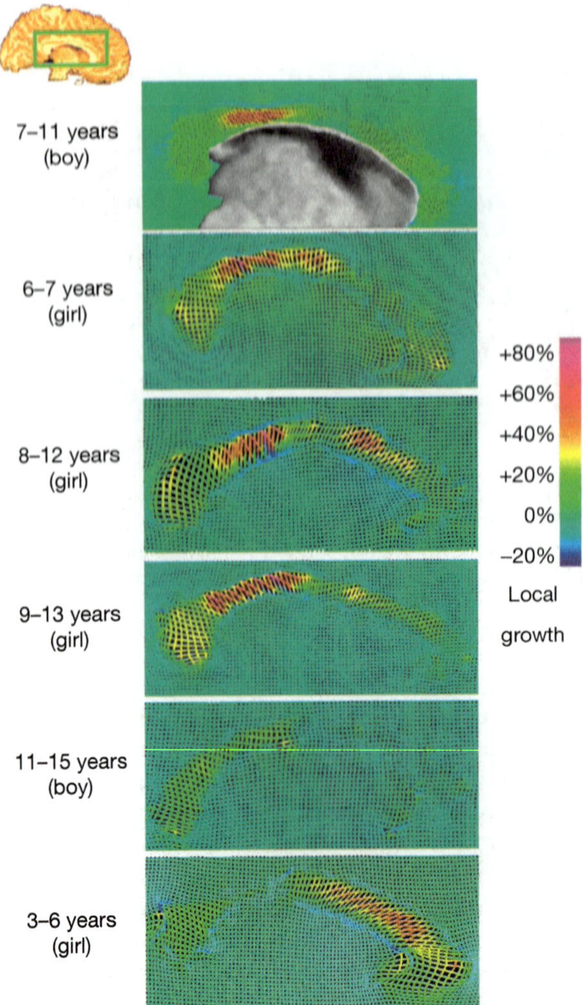

FIGURE 12.4   Corpus callosum maturation. Maps of the local volume changes in the corpus callosum are shown for six individuals aged 3–15 years, who were scanned twice longitudinally with an interval of up to 4 years. Maturation includes outward tissue expansion (warmer colors), with a dynamic wave in timing such that more frontal regions show prominent change early, and more posterior regions show prominent change later. *Thompson PM, Giedd JN, Woods RP, MacDonald D, Evans AC, and Toga AW (2000) Growth patterns in the developing brain detected by using continuum mechanical tensor maps. Nature 404: 190–193.*

remains a critical issue, as sex-specific differences in brain development are likely to contribute to the sexually dimorphic susceptibilities of a variety of psychiatric disorders – like schizophrenia and major depression – that emerge during adolescence (Durston et al., 2001; Lenroot and Giedd, 2010). Returning to the issue of gender differences in development, Sowell et al. analyzed cortical thickness and local brain size (taken as the distance from the center of the brain) in a large sample of 176 healthy subjects aged 7–87 years (Sowell et al., 2007). In line with previous studies, male brains were larger than females, at all locations. Strikingly, however, absolute cortical thickness was greater in females in right inferior parietal and posterior temporal regions even without accounting for the smaller overall size of female

brains. This finding was not significantly modulated by age and was demonstrated even more robustly across broad right temporal and parietal regions when an age-matched and brain-volume-matched subset of 18 males and 18 females was evaluated (see Figure 12.5). These findings suggest that there are both regional- and sex-specific differences in cortical thickness that appear relatively early in childhood, and support earlier reports of selective relative increases in gray matter volumes in females (Allen et al., 2003; Goldstein et al., 2001; Gur et al., 2002; Im et al., 2006; Nopoulos et al., 2000; Sowell et al., 2002b). Although the corpus callosum is also a frequent target for brain mapping research into sex-specific effects on brain development, no consensus has been reached and the topic remains frequently debated (Giedd et al., 2006) (see Chapters 35 and 38).

Because of the overall smaller brain volume in females, it has also been proposed that there may be evolutionary pressure to develop other compensatory mechanisms. Through sulcal delineation and cortical-pattern-matching techniques, it has been shown that females tend to develop a greater degree of cortical 'complexity' by young adulthood (Luders et al., 2004). This suggests that there is more cortical surface per unit volume in females and may be one mechanism through which female brains have become optimized for their smaller size.

An increasing focus is also being shifted away from sex-specific differences, *per se*, to the known differences in pubertal timing and sex steroid levels that are likely to be major contributors to observed sex-specific effects and their frequently observed modulation by age (Giedd et al., 2006; Lenroot et al., 2007). The emerging picture suggests that puberty and sex steroids do indeed have organizing effects on brain development (Bramen et al., 2011; Neufang et al., 2009; Peper et al., 2009a; Witte et al., 2010). One recent study of 107, 9-year-old monozygotic and dizygotic twin pairs noted strong overall heritability in regional brain volumes but also demonstrated decreased frontal and parietal gray matter density among the subgroup of individuals who had begun to develop secondary sexual characteristics of puberty (Peper et al., 2009b). Further investigation among the same cohort revealed that the serum level of luteinizing hormone, one of the first indicators of puberty, is associated with both increased overall white matter volume and increased white matter density in the cingulum, middle temporal gyrus, and splenium of the corpus callosum (Peper et al., 2008). The splenium observation is particularly intriguing, as this is the same region shown to have maximal growth over the 9–13 age range in a different study (Thompson et al., 2000). These results – observed between otherwise very well-matched groups – suggest that the onset of puberty and sex steroid levels may directly contribute to the decreases in gray matter and increases in white matter that are prominent features of normal brain development during late childhood and adolescence.

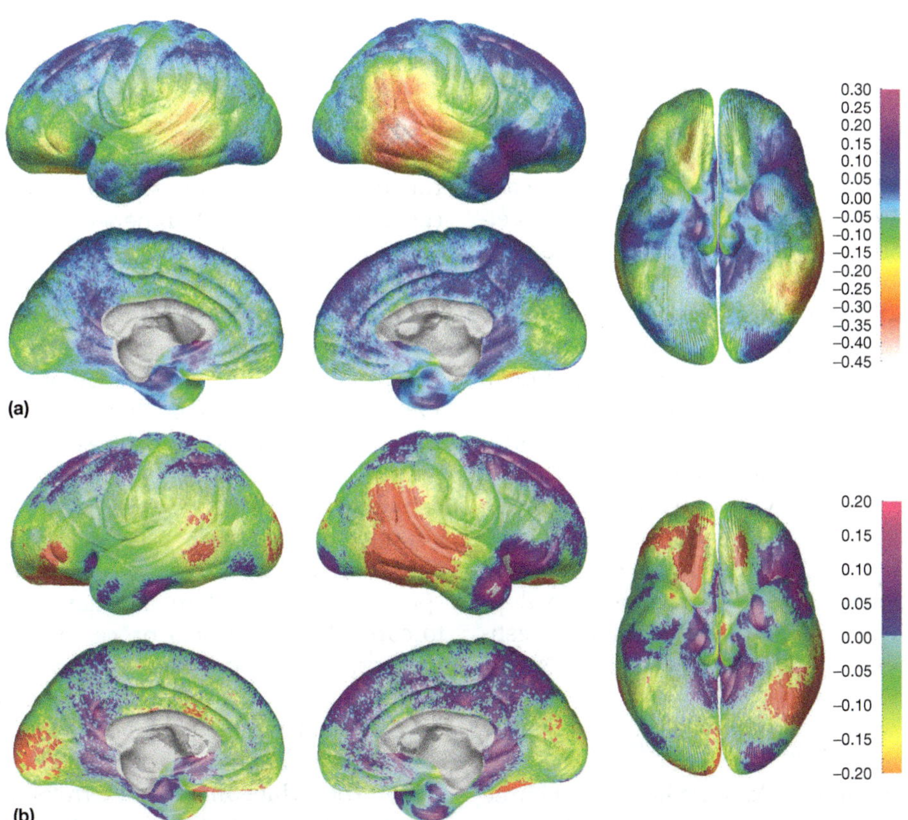

FIGURE 12.5   Sex-specific differences in cortical thickness. (a) Sex differences in cortical thickness (mm) among an age- and brain volume-matched sample of 18 males and 18 females. Warmer colors (<0 on the color bar at right) are regions where females have thicker cortex, and cooler colors (>0 on the color bar at right) are regions where females have thinner cortex, relative to males. (b) Statistical maps showing the significance of these sex differences. Areas where the cortex is significantly thicker in females are shown in red, and include right inferior parietal and posterior temporal, and left posterior temporal and ventral frontal regions. Areas where the cortex is significantly thinner in females are shown in white, and are limited to small regions in the right temporal pole and orbitofrontal cortex. The correlation coefficient is mapped for nonsignificant regions according to the color bar at right. *Sowell ER, Peterson BS, Kan E, et al. (2007) Sex differences in cortical thickness mapped in 176 healthy individuals between 7 and 87 years of age.* Cerebral Cortex 17: 1550–1560.

12.4.6 Summary

Taken together, these structural imaging studies represent a powerful evolution in our understanding of brain development during childhood and adolescence. The overall picture remains one of early overall brain growth, followed by a transition around age 5 to gray matter decreases coupled with persistent white matter increases. These processes continue through adolescence but relatively balance each other in magnitude. Thus, while overall net brain volume changes relatively little past the age of 5, adolescence remains a period of dynamic change beneath the pial surface.

12.5 DIFFUSION MRI

One of the remarkable discoveries to emerge from these developmental neuroimaging studies is the continued expansion of white matter volume well into adulthood (Giedd et al., 1999a; Sowell et al., 2003). This robust and protracted increase has rewritten the age range associated with brain development (Pujol et al. 1993) and has driven an increasing focus on the white matter and its network connectivity as a possible mediator for the late cognitive gains seen in executive function domains during typical development (Liston et al., 2006), as well as a possible mechanism for neuropathology (Le Bihan, 2003) and training-induced increases in performance (Bengtsson et al., 2005; Carreiras et al., 2009).

12.5.1 Diffusion Tensor Imaging Theory

Simultaneously with this growing interest in studying the white matter, as it relates to connectivity between still-maturing brain regions and cognitive function, diffusion imaging was maturing as an MRI variation specifically tuned to examine the white matter (Basser et al., 1994; Bihan et al., 1986; Pierpaoli et al., 1996). Since the diffusion properties of water within neural tissue are affected by the geometry of the neuronal microenvironment, it is intuitive that diffusion imaging can provide a sensitive lens through which the microstructural properties of the white matter can be investigated. Specifically, differences in microstructural properties like fiber coherence, axon packing, and myelination have all been shown to manifest as changes in the diffusion MRI signal (Beaulieu, 2002). By viewing this diffusion landscape within the brain from multiple angles, a more complete 'tensor' model of diffusion can be generated for each voxel (Basser et al., 1994). This can be thought of geometrically as a diffusion ellipsoid, with diffusion components in the radial (RD, radial diffusivity) and axial (AD, axial diffusivity) directions (see Figure 12.6). The size of this ellipsoid corresponds to the overall mean diffusivity (MD). The shape of the ellipsoid corresponds to the directionality of diffusion and is termed fractional anisotropy (FA). It can vary from 0, for perfectly isotropic diffusion, to 1, for perfectly anisotropic diffusion (e.g., the ventricles have low FA, while the corpus callosum has high FA). Because it has been shown to be sensitive to myelination, this FA metric has received considerable attention as a way to track the developmental maturation within the white matter and investigate disease. See Le Bihan (2003) for an excellent review.

12.5.2 Diffusion Parameters in Development

Using this unique framework, there has been a surge in research aimed at more deeply characterizing the normal developmental processes in these important regions

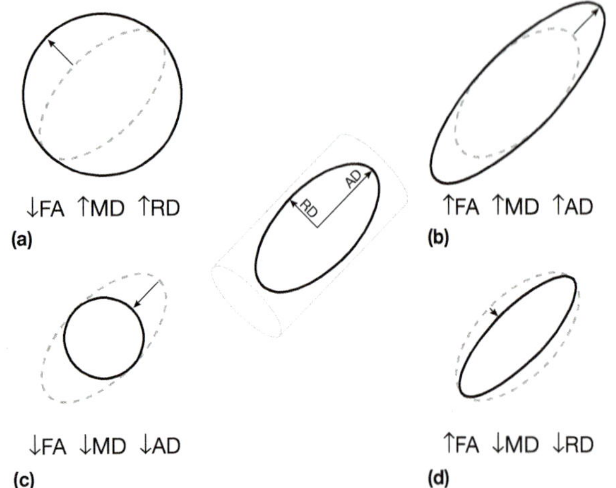

FIGURE 12.6   Diffusion tensor imaging (DTI) metrics. DTI metrics include fractional anisotropy (FA), which is a unitless measure of the directionality of diffusion, and mean diffusivity (MD), which is the overall magnitude of diffusion. The center panel shows a cross-section of the DTI ellipsoid model of diffusion, which is assumed to be oriented along the fiber axis (shown here as a cylinder). Individual diffusion components along the axial (AD) and radial (RD) directions contribute to the FA and MD values at each point in the brain. (a–d) show different changes in the individual diffusion parameters, and their varying effects on FA and MD. Note that changes in different diffusion components (AD or RD) can lead to the same effect on one diffusion metric, but have opposite effects on the other. (d) represents the prevailing regime during development, where decreasing RD – due, in part, to advancing myelination – leads to increasing FA (a more pointed ellipsoid) and decreasing MD (a smaller ellipsoid).

of connectivity that were previously obscured by low contrast within the white matter on traditional T1-weighted anatomical MRI. Similar to the general description in the overlying gray matter, the developmental trajectory within the white matter is both a nonlinear function of time and has prominent regional variations (Lebel et al., 2008b; Mukherjee et al., 2001; Snook et al., 2005). From birth, there is a rapid rise in diffusion directionality (FA; see Figure 12.7), coupled with a decrease in overall diffusivity (MD) (Bava et al., 2010; Engelbrecht et al., 2002; Hüppi et al., 1998; Löbel et al., 2009; Morriss et al., 1999; Mukherjee et al., 2001; Neil et al., 1998; Schmithorst and Yuan, 2010; Schneider et al., 2004). In an interesting contrast to this general pattern within the white matter, gray matter cortical regions actually have been observed to have decreasing FA in a sample of preterm infants (McKinstry et al., 2002). This could reflect the fact that changes in FA are not highly specific for myelination and may also occur in response to cortical maturational processes, like synaptogenesis. Further, these observations may be related to the perinatal period of selective vulnerability in neural tissue, which has been demonstrated in animal studies and confirmed in humans through MRI (Miller and Ferriero, 2009). The white matter pattern of increasing FA and decreasing overall diffusion, although not universally reported in later development (Schneiderman et al., 2007), generally continues in a decelerating fashion throughout childhood, adolescence, and, in some areas, into adulthood (Bonekamp et al., 2007; Klingberg et al., 1999; Schmithorst et al., 2002; Zhang et al., 2007). There is a relatively stable plateau of these parameters during adulthood, which then eventual declines later in life (Davis et al., 2009; Salat et al., 2005). Accordingly, the developmental rising portion of this arc has been modeled as a linear (Snook et al., 2005), polynomial (Hsu et al., 2010), or exponential (Lebel et al., 2008b; Mukherjee et al., 2001; Schneider et al., 2004) function. The earliest reports utilized an ROI approach to look at diffusion properties averaged across specific anatomical locations and were able to reproduce the 'increasing FA, decreasing MD' pattern across a broad variety of regions within the brain and during different periods of development. In one example, Suzuki and colleagues examined ROIs placed bilaterally in the frontal and parietal white matter of 16 children and young adults. They observed increased FA and decreased overall diffusivity with age but went on to make the important determination that the etiology of these changes in FA and MD was a primary decrease in both radial (RD) and AD diffusion components, with a larger decrease along the radial direction (Suzuki et al., 2003). This explains the overall decreased diffusivity that was observed (both components decreased) but also the increased diffusion directionality (one component decreased more than the other). The dominance of changes in radial diffusivity (RD) during development is an important phenomenon that has been broadly replicated (Giorgio

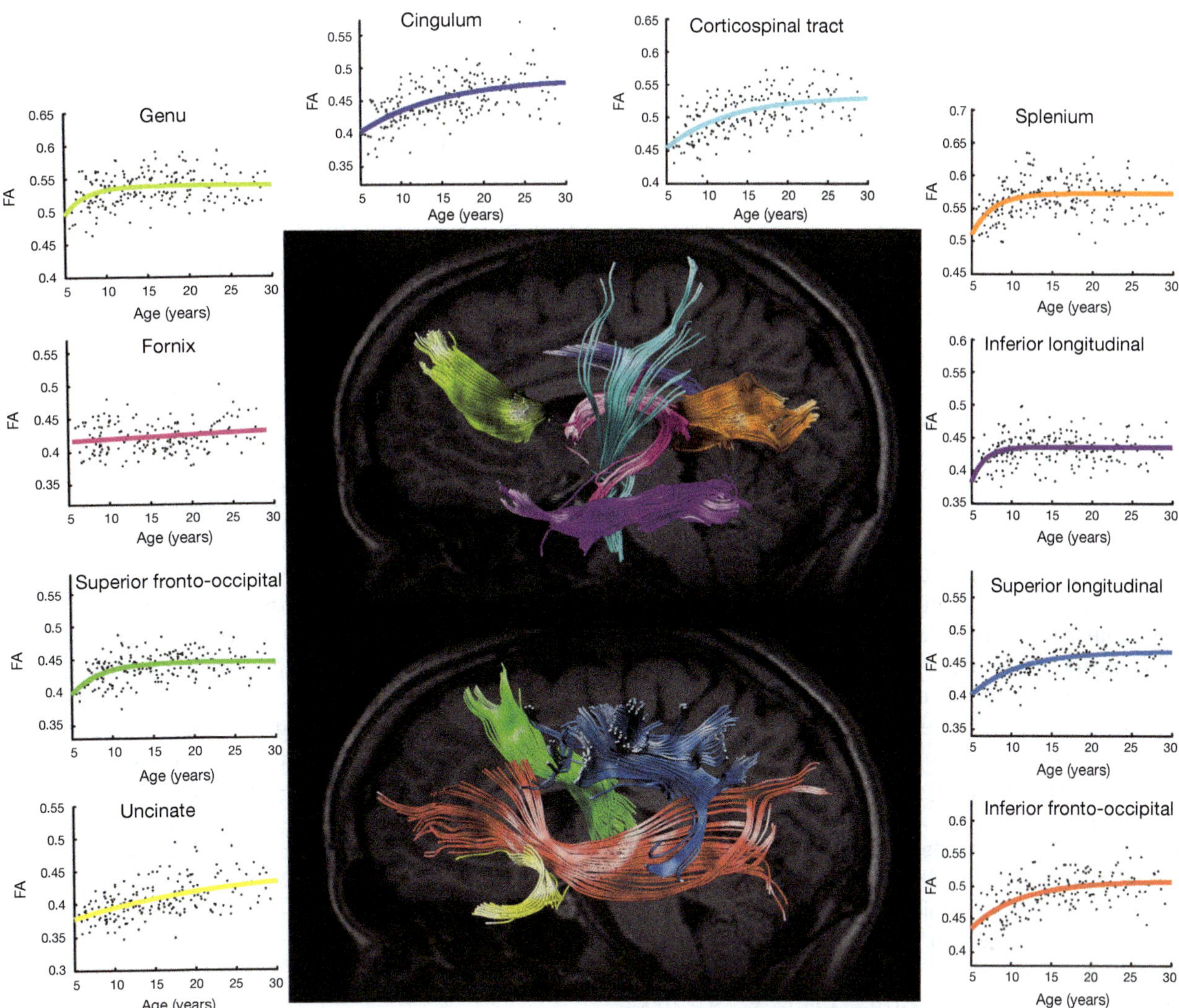

FIGURE 12.7 White matter maturation. Diffusion tensor imaging (DTI) tractography was used to identify ten major white matter tracts in 202 individuals aged 5–30 years (center panels show the extracted tracts for a representative subject). Broad age-related increases in fractional anisotropy (FA), a DTI index of white matter maturation that is sensitive to myelination, were observed across all tracts. Maturational trajectories generally followed an exponential rise, with regional variations in mean FA as well as developmental timing. The surrounding scatterplots demonstrate these relationships, and are color-coded according to the tracts in the center panels. *Lebel C, Walker L, Leemans A, Phillips L, and Beaulieu C (2008b) Microstructural maturation of the human brain from childhood to adulthood.* Neuroimage 40: 1044–1055.

et al., 2008; Lebel et al., 2008b; Löbel et al., 2009; Qiu et al., 2008), although not universally (Ashtari et al., 2007; Giorgio et al., 2010), and is thought to relate to the primary role that extended myelination plays during this age range (Song et al., 2002).

Paralleling the advancements made in the analysis of the cortex, methods have quickly adapted to include whole-brain mapping techniques that are able to examine the brain in a spatially continuous manner and better localize developmental changes. In general, these later efforts using VBM and similar techniques have both confirmed and extended the earlier ROI findings of broadly increasing FA and decreasing MD (Snook et al., 2007). Tract-based spatial statistics (TBSS) is an evolution of these methods that is tailored specifically to the analysis of diffusion tensor imaging (DTI) data and has been used successfully to demonstrate age-related changes in diffusion imaging parameters (Bava et al., 2010; Burzynska et al., 2010; Giorgio et al., 2008, 2010). By projecting the imaging data onto a tract 'skeleton' consisting of the cores of the white matter tracts, TBSS avoids some of the alignment problems that arise when the high-contrast FA maps are compared using traditional voxel-by-voxel techniques (Smith et al., 2006, 2007). In a sample of 75 children through young adults that were analyzed using this approach, widespread FA increases and diffusivity decreases were again demonstrated spanning the frontal, temporal, and parietal lobes and the cerebellum (Qiu et al., 2008). Recognizing the need to synthesize these reports into a normative reference standard against which to judge clinical abnormalities, effort has also been directed towards generating developmental brain atlases that integrate this diverse set of information (Hermoye et al., 2006; Löbel et al., 2009; Mori et al., 2008; Verhoeven et al., 2010).

12.5.3 Fiber Tractography

By making the assumption that the direction of the diffusion ellipsoid (i.e., the direction of principal diffusion) is pointing in the same direction as the neuronal fiber axis, streamlines can be generated passing from voxel to voxel along the path of principal diffusion. In this manner, the DTI technology has been extended to allow for *in vivo* fiber tractography (Behrens et al., 2003; Catani et al., 2002; Conturo et al., 1999; Mori et al., 1999). This allows for individualized measurements to be made that are tailored to each subject's anatomy, which circumvents many of the problems associated with attempting to register a diverse set of brains to a single template. Although these algorithms have generally validated well against postmortem dissections for many major white matter tracts, specific limitations related to issues, like partial volume averaging and complex fiber geometries, must be considered (Pierpaoli et al., 1996). Using this technology, together with standardized protocols for delineating the major white matter tracts of interest (Wakana et al., 2007), researchers have mapped the development of white matter fiber connectivity from before birth (Huang, 2010; Huang et al., 2006, 2009), through childhood, adolescence, and adulthood (Behrens et al., 2003; Liu et al., 2010; Wakana et al., 2004) and even through evolution (Rilling et al., 2008). Like other developmental neuroimaging efforts, these data provide important insight into human brain development in their own right and, additionally, serve as important normative markers against which pathology can be judged (Adams et al., 2010; Lebel et al., 2008a; Thomas et al., 2009). In a seminal report on the typical developmental trajectories within 10 major white matter tracts in a large sample of 202 subjects aged 5–30 years, Lebel et al. observed continually increasing FA in all regions (generally approximated well by an exponential function), but regional variations in timing such that the time to reach 90% of the adult plateau varied from approximately 7 years old in the inferior longitudinal fasciculus to beyond 25 years old in the cingulum and uncinate fasciculus (see Figure 12.7; Lebel et al., 2008b). Overall, they note that frontotemporal connections were the slowest to develop. In a representative example of the degree of intersubject diversity that exists even within tracts, DTI tractography has been used to demonstrate lateralization of different white matter tracts (Bonekamp et al., 2007). In one particular study, left lateralization was shown for the arcuate fasciculus (temporoparietal part of the superior longitudinal fasciculus), with higher FA and more streamlines in the left hemisphere (Lebel and Beaulieu, 2009). These findings are in line with previous observations of left lateralization of perisylvian regions (Geschwind and Levitsky, 1968; Pujol et al., 2002) and are thought to relate to the left hemisphere language dominance that exists in the majority of the population. Interestingly, this same pattern has been demonstrated even in neonates, suggesting that the structural basis of left hemisphere language dominance is present long before the development of speech (Liu et al., 2010). Previous morphometric findings of local volume increases within the corpus callosum (Giedd et al., 1996a; Thompson et al., 2000) have also been explored with tractography. In a large sample of 315 subjects aged 5–59 years, Lebel and others demonstrated the typical trajectory of increasing FA and decreasing MD in the fiber tracts leading from all midsagittal sections of the corpus callosum (Lebel et al., 2010). They also observed an 'outer-to-inner' trend in the timing of these maturational arcs, which contrasts with the anterior-to-posterior volumetric trend observed on T1-weighted MRI (Thompson et al., 2000) and highlights the additional insight that can be uncovered when the full extent of a tract is considered.

12.5.4 Sex Differences

Diffusion imaging also reveals sex-specific structural differences within the white matter (Lenroot and Giedd, 2010; Schmithorst et al., 2008). In one tractography study of 114 children, adolescents, and young adults, Asato et al. found generally decreasing radial diffusivity (RD) and protracted maturation past adolescence, in projection and association fibers that included connections between the prefrontal cortex and the striatum. Furthermore, they observed that white matter microstructural maturation proceeded in parallel with pubertal changes, with females having overall earlier maturation of white matter tracts than males (Asato et al., 2010). This suggests that there may be hormonal influences on white matter maturation and that by considering these aspects, one may obtain a more appropriate estimate of developmental progress than by only considering chronological age. This notion is supported by concurrent findings with structural MRI that demonstrate that white matter volume increases during adolescence, especially in boys, are affected by testosterone levels and androgen receptor genes (Paus et al., 2010; Perrin et al., 2008).

12.5.5 Summary

Taken together, diffusion imaging studies generally show increasing diffusion directionality (FA) and decreasing overall diffusion (MD) during development. These changes are predominantly due to decreasing diffusivity in the radial direction (i.e. radial diffusivity; RD) from the fiber axis, which suggests a primary role for myelination in this process. These changes progress rapidly from birth through childhood and, eventually, level off to a relatively stable adult plateau. Paralleling what has been observed in the cortex and through volumetric observations, there are regional variations in the timing of this developmental trajectory that follow a roughly posterior-to-anterior trend. Sexual dimorphism is also present, with females exhibiting earlier white matter maturation than males – a trend that mimics their differences in pubertal timing.

12.6 CONNECTING DIFFERENT TECHNIQUES

12.6.1 Multimodal Imaging

Although the development of cortical gray matter and the development of white matter microstructure have been investigated independently, one needs to consider their dynamics jointly in order to determine what relationships exist between them. This challenge returns to one of the original questions that stemmed from the postmortem histological findings – that is, 'To what degree do myelination and synaptic pruning (and other cellular processes) contribute to the decreasing gray matter and increasing white matter that is found during brain development?' While these phenomena are undoubtedly linked, it remains unclear which is dominant and exactly how they interact. The maturation of DTI and structural MRI analysis techniques has now made it possible to investigate these questions using *in vivo* imaging data; however, in the end, it will likely be necessary to complete the circle and validate these observations back in histological preparations.

In a study focusing on adolescence, Giorgio et al. began by using the TBSS method, discussed in Section 12.5.2, to demonstrate broad increases in FA that were driven predominantly by decreases in radial diffusivity (RD). They then made an important and innovative step by incorporating both DTI tractography and gray matter VBM to show that the putative fibers leading from the white matter regions, with the strongest developmental effects, connect with regions showing significantly decreased gray matter density in the cortex. Further, they observed that the gray matter density decreases were significantly correlated with the FA increases in the connected white matter (Giorgio et al., 2008). By following the structural connectivity present in the actual data, and using these patterns to guide their comparisons, this protocol links the concurrent phenomena of white matter FA increases and gray matter density decreases more convincingly than was possible with previous qualitative visual inspections. Tamnes et al. investigated this same general question in a different manner by integrating cortical thickness, volumetric, and DTI measurements derived from a single sample of 168 participants, aged 8–30 years (see Figure 12.8; Tamnes et al., 2010). As expected, they were able to demonstrate a combination of the phenomena seen in earlier individual studies, including broad cortical thickness decreases, white matter volume increases, FA increases (predominantly decreases in radial diffusion), and MD decreases. Most importantly, however, they were able to go on to demonstrate that, of the three measures, cortical thickness had the strongest relationship with age. Further, although the DTI and volume measures explained some of the variance in cortical thickness and each other, none of the measures were redundant. This implies that each may be sensitive to different microstructural processes and that all are useful indicators of brain development and microstructural integrity (Fjell et al., 2008). This reiterates the likely mixed regime of both synaptic pruning within the cortex and advancing myelination at the gray–white cortical interface, which contributes to the brain morphological changes seen during adolescence.

12.6.2 Brain–Behavior Relationships

While important neuroanatomical insights can be gleaned from these structural brain mapping observations, perhaps the most significant outgrowth of this research has been an expanded understanding of the cognitive and behavioral changes that accompany this underlying maturation of brain structure. There has been a long tradition of investigation into the cognitive correlates of brain structure, but unfortunately many of the early findings – which commonly focused on differences between ethnic or social groups – are unreliable because of data collection and analysis bias (Gould, 1978, 1981). With the advent of MRI, however, volumetric measurements of total brain size have shown a modest but reproducible correlation with general intelligence that emerges over the course of development (Peters et al., 1998; Reiss et al., 1996; Willerman et al., 1991; Witelson et al., 2006). However, the correlational nature of these findings does not at all suggest that groups with different brain sizes, like males and females, will have different intelligence. Indeed, independent of the possible relationships with neuroanatomy, it remains exceptionally controversial whether there is even any overall

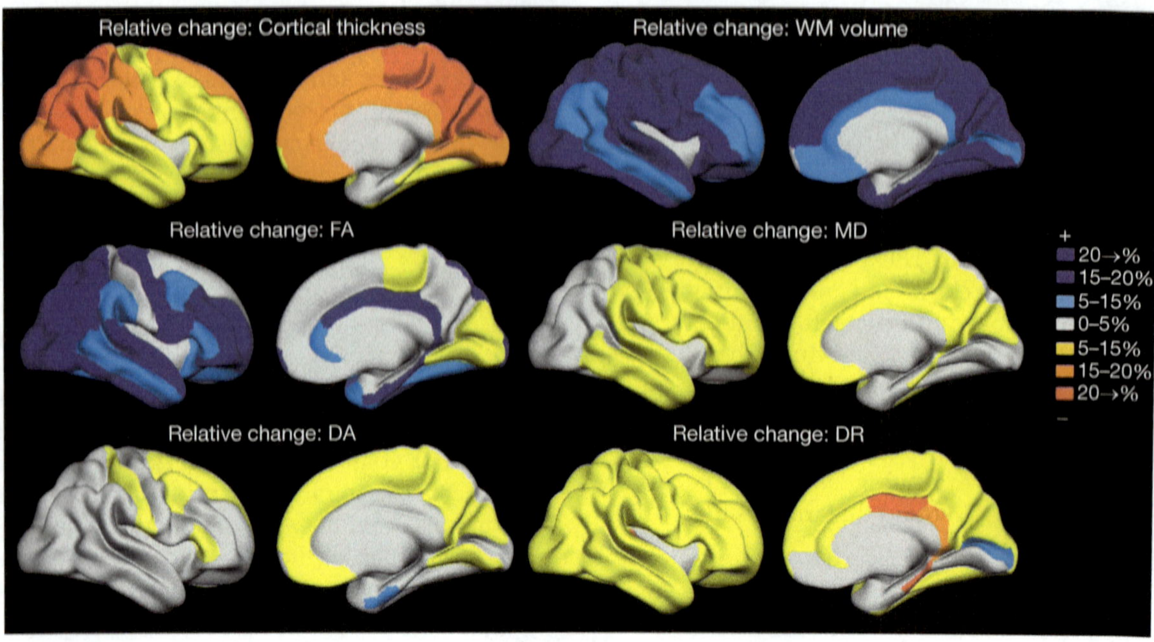

FIGURE 12.8   Multimodal imaging: volumes, cortical thickness, and DTI. Concurrent volumetric, cortical thickness, and diffusion tensor imaging (DTI) analyses were performed in the same sample of 168 participants aged 8–30. The percent changes in cortical thickness, white matter volume, fractional anisotropy (FA), mean diffusivity (MD), axial diffusion component (AD), and radial diffusion component (RD) are mapped by region and color coded according to the color bar on the right. Medial structures and corpus callosum are masked out. *Tamnes C, Ostby Y, Fjell A, Westlye L, Due-Tønnessen P, and Walhovd K (2010) Brain maturation in adolescence and young adulthood: Regional age-related changes in cortical thickness and white matter volume and microstructure.* Cerebral Cortex 20: 534–548.

gender effect on intelligence (Blinkhorn, 2005; Hedges and Nowell, 1995; Irwing and Lynn, 2006; Jorm et al., 2004; Lynn and Irwing, 2004; Neisser et al., 1996) and, if so, whether the small effect magnitudes that have been reported are relevant, given the possible biases that may have contributed. An important additional phenomenon to consider is that both brain structure and intelligence are highly heritable (Shaw, 2007; Thompson et al., 2001). Both are further impacted by environmental influences in a process that begins *in utero*, continues throughout life, and contributes to individual variations in structural brain development and cognitive function that exist even among monozygotic twins. Although not exclusive, the orchestration of structural brain development by these genetic and environmental factors is one way in which they can converge to influence cognitive development (Toga and Thompson, 2005) (see Chapter 26).

Since there is evidence that brain development takes place through selective elimination and connectivity optimization, with prominent regional and temporal variability, it is not surprising that a global measure like total brain volume may not be the optimal choice for investigating the structural basis of cognitive development. Fortunately, the brain mapping strategies, discussed in Sections 12.3–12.5, have had more success examining brain-region-specific relationships between structure and function. This work has supported many

of the classical structure–function relationships discovered through lesion studies – for example, that the prefrontal cortex is related to cognitive control (Damasio et al., 1994) – and also has extended these findings by (1) providing more detail, (2) including more normative subjects without pathology, and (3) allowing for broader investigation in the pediatric population. In this way, these modern neuroanatomical imaging studies, together with complementary results from functional neuroimaging methods that can measure task-dependent blood flow response within the brain (Casey et al., 1995; Luna et al., 2010), have formed a powerful framework to investigate how brain development relates to cognitive function during childhood and adolescence. In this vein, continued investigation into the structural basis of general intelligence has revealed age-variable relationships between intelligence quotient (IQ) and regional brain structure. In line with the total brain volume results, a correlation between IQ and gray matter volume develops by adulthood (Wilke et al., 2003). However, regional relationships between IQ and gray matter structural measures appear earlier and have been reported to include the anterior cingulate during childhood (Wilke et al., 2003), the orbitofrontal cortex during adolescence (Frangou et al., 2004), and the frontal lobe – particularly the prefrontal cortex – by adulthood (Haier et al., 2004; Reiss et al., 1996; Thompson et al., 2001). Interestingly, these regional relationships between gray matter development and IQ appear to be modulated by sex, although the specific regions reported to be most associated with IQ for each sex have been variable (Haier et al., 2005; Narr et al., 2007). In one important study, which investigated the relationship between cortical thickness maturation and IQ in a large longitudinal sample of 307 children and adolescents, IQ was observed to correlate most closely not with cortical thickness, *per se*, but rather with the shape of the developmental trajectory in cortical thickness change (see Figure 12.9; Shaw et al., 2006). The subjects that had the highest IQs tended to have the most dynamic cortical maturation, with more rapid cortical thickening during early childhood and more rapid cortical thinning during late childhood and adolescence. However, in terms of absolute thickness, the superior intelligence group actually had thinner cortex at the start of the age range studied (approximately age 7), peaked later, and then had relatively equal thickness to the others by the end of the age range (approximately age 19). This observation highlights the notion that, like the pattern of structural maturation itself, the relationships between brain structure and cognitive ability are complicated by their dependency on age during the course of development. In another example, Choi et al. took a multimodal approach, as described in Section 12.6.1, and investigated correlations between intelligence and both cortical thickness *and* functional MRI (fMRI) blood flow response during a reasoning task. Because both sets of scans were performed on the same sample of subjects, the authors were able to go a step further than single modal studies and examine if different intelligence subcomponents correlated more with one imaging modality or the other. Their findings quite eloquently demonstrated that the *crystallized* component of intelligence (related to our ability to utilize previously acquired knowledge and past experiences) correlated more strongly with cortical thickness, while the *fluid* component of intelligence (related to our problem-solving and critical-thinking ability in novel situations, independent of past experience) was more strongly related to functional blood flow response (Choi et al., 2008). While the specific pattern and methodologies of these studies have varied widely, the common pattern that has emerged is a relationship between frontal lobe structural brain development and general intellectual ability.

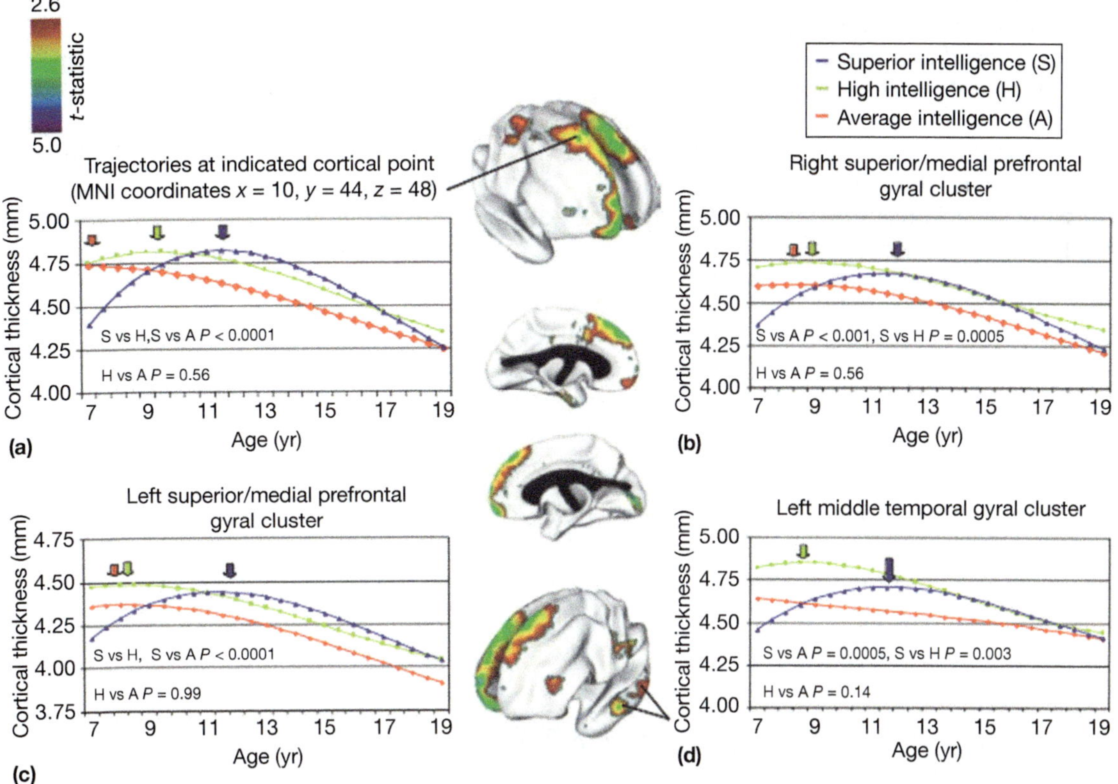

FIGURE 12.9 Trajectory of cortical thickness change versus IQ. Higher IQ was associated with a more dynamic trajectory (more rapid thickening and thinning) in cortical thickness maturation among a sample of 307 children and adolescents scanned longitudinally. The center panel shows regions where there was a significant interaction between IQ group (superior, high, or average) and a cubic age^3 term in the regression analysis, which implies a varying trajectory shape in these regions. These individual trajectories are plotted in panels (a–d), and are color-coded according to intelligence group. Arrows indicate the age at peak cortical thickness for each trajectory. *Shaw P, Greenstein D, Lerch J, et al. (2006) Intellectual ability and cortical development in children and adolescents. Nature 440: 676–679.*

Other studies have investigated more specific cognitive functions and their relation to gray matter structure. In the same longitudinal sample of 45 typically developing children, that was described previously, we observed inverse correlations between performance on the vocabulary subtest of the Wechsler Intelligence Scale for Children (Wechsler, 2003) – a test of general verbal intellectual functioning – and gray matter thickness in left dorsolateral frontal and lateral parietal regions (see Figure 12.10; Sowell et al., 2004a). This is consistent with the language dominance of the left hemisphere and suggests a possible relationship between these concurrent structural and cognitive developmental processes. While originally interpreted as possibly relating to developmental cortical thinning, the results of the Shaw et al. (2006) study suggest that the individuals with the greatest verbal intellectual function here may still have been on the upstroke of their developmental arc in our much younger sample (age 5–11 years) and simply had thinner cortex at the time sampled. This nuance is also reflected in another study, which had an older sample (age 6–18 years), during the later period of development where increased cortical thickness is associated with higher IQ (Karama et al., 2009). Further studies, again in the young sample of 5–11-year-olds, have investigated even more targeted cognitive subtests, including phonological processing and motor speed and dexterity (Lu et al., 2007). Structural development in the inferior frontal gyrus (a phylogenetically more complex area that matures slower and is still on the upward stroke of cortical thickening) was expected to relate to advances in phonological processing, which has been shown to involve this area on functional imaging studies (Bookheimer, 2002) but not to relate to advances in motor processing.

Conversely, structural development in the hand motor region (a phylogenetically simpler area that matures earlier and is already experiencing cortical thinning) was expected to relate to advances in motor processing but not phonological processing. This predicted double dissociation was demonstrated as expected, which not only illustrates a specific alignment between language development and structural development in the inferior frontal gyrus but also reiterates the regionally specific definition of 'structural development' during childhood – with some cortical regions thinning but some relatively specific language areas still exhibiting thickening. A similar analysis has also revealed relationships between cortical thinning and both delayed verbal recall functioning and visuospatial memory ability, which is again consistent with the functional neuroimaging literature that suggests the dorsolateral prefrontal cortex is involved with memory recall (Casey et al., 1995; Sowell et al., 2001a). The relationship between cognitive development and structural brain development is further supported by intervention/training studies, which suggest even relatively short periods of cognitive or motor training can be associated with, at least, short-term morphological changes in brain structure (Draganski et al., 2004).

Diffusion imaging indicators of white matter development also relate to cognitive function. In a sample of 23 children and adolescents, there was a significant direct relationship between diffusion characteristics (FA) and working memory ability in inferior frontal and temporooccipital regions and the genu of the corpus callosum (Nagy et al., 2004). This relationship existed above and beyond the correlation of each individual measure with age, which suggests that the maturation

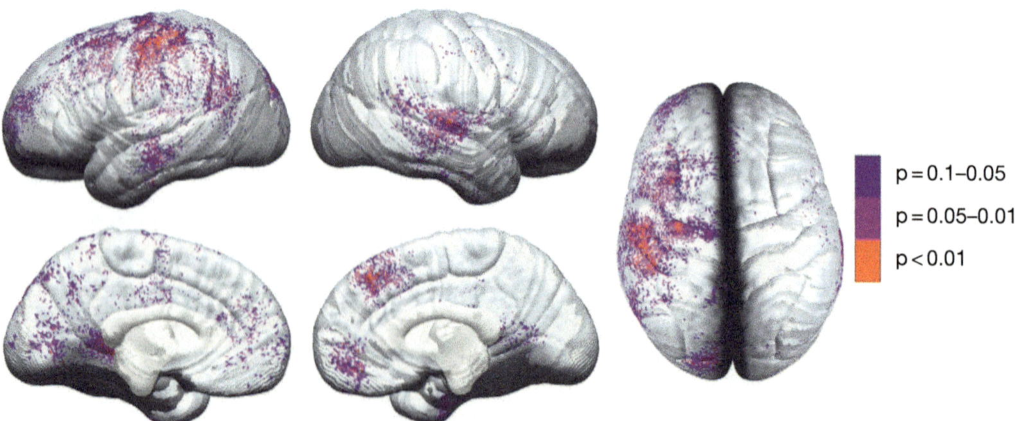

FIGURE 12.10   Cortical thickness versus language functioning. Statistical maps showing the significance of the relationship between changes in cortical thickness and changes in vocabulary scores in a longitudinal sample of 45 children scanned twice between the ages of 5 and 11. Areas with a significant negative relationship (cortical thinning was associated with improved language performance) are color-coded according to their *P* value, with the significance thresholds shown in the color bar on the right. No positive correlations were observed. *Sowell ER, Thompson PM, Leonard CM, Welcome SE, Kan E, and Toga AW (2004a) Longitudinal mapping of cortical thickness and brain growth in normal children.* Journal of Neuroscience 24: 8223–8231.

of the white matter in specific areas – as indexed by FA – may play a role in the development of (or simply reflect the development of) specific cognitive domains. In another study related to structure–function specialization within the brain, others have shown correlations between Chinese reading score and FA in the anterior limb of the left internal capsule and English reading score and FA in the corona radiata (Qiu et al., 2008). While a preliminary interpretation, this dissociation could represent that distinct brain networks are more or less involved with language development, depending on the specific language or mode of acquisition. For instance, it is quite reasonable to hypothesize that learning written Chinese, with symbolic characters representing whole words that are generally independent of pronunciation, could drive development (or reflect development) of different brain regions compared to learning written English, which uses an alphabetic system to describe how words sound. As a final example, in the arcuate fasciculus lateralization tractography study discussed in Section 12.5.3, greater leftward lateralization was associated with better performance on cognitive tests of receptive vocabulary and phonological processing (Lebel and Beaulieu, 2009). These studies suggest that diffusion imaging is not only a useful technique for tracking normal anatomical maturation within the white matter, but also that regional DTI metrics can provide reflections of cognitive development in specific domains.

12.7 CONCLUSIONS AND FUTURE DIRECTIONS

Our understanding of human brain development has accelerated over the last 20 years through the use of MRI and *in vivo* human brain mapping. Postmortem and histological studies have demonstrated that brain maturation, on the cellular level, encompasses both progressive and regressive events. These include synaptic pruning and protracted myelination, which continue to shape the underlying neural microstructure and regional brain morphology long after overall brain volume begins to plateau, around age 5. Brain development, in general, can be characterized as both nonlinear with respect to time, and also variable with respect to region. The hallmark of structural brain development during childhood is a striking change in the relative proportions of gray and white matter – with a peak and then decline in gray matter volume and cortical thickness but a relatively sustained increase in white matter beyond adolescence. Across these different regions, there is a general posterior-to-anterior and inferior-to-superior trend in the timing of maturation, such that primary somatosensory and phylogenetically older areas of the brain tend to

mature earlier than higher-order association cortices – particularly areas in the frontal lobe. Within the white matter, diffusion imaging indicators show decreasing diffusivity (MD) and increasing directionality (FA), which suggests that myelination continues through young adulthood and perhaps even beyond. Performance across a variety of cognitive domains has also been shown to relate to these structural changes, with the specificity of these relationships generally in line with classic functional neuroanatomical localizations.

Although the complexity of the regional and temporal patterns of structural brain development makes investigating and interpreting these brain–behavior relationships challenging, future work should continue to focus on the possible functional manifestations of structural brain development. Particularly, by integrating different structural and functional imaging modalities with thorough cognitive assessments, we can investigate the ways in which these processes interact with each other within a more inclusive framework that more realistically encompasses the full developmental landscape. With the increasingly broad array of radiological features of development that have been characterized, there is additionally a growing need to reintegrate a firm neurobiological understanding of the cellular mechanisms that facilitate these changes. Finally, effort should continue to be directed towards uncovering the ways in which this basic neuroscientific knowledge concerning human brain development can be translated into a better context for the understanding and clinical treatment of neurodevelopmental disorders.

References

Adams, E., Chau, V., Poskitt, K.J., Grunau, R.E., Synnes, A., Miller, S.P., 2010. Tractography-based quantitation of corticospinal tract development in premature newborns. Journal of Pediatrics 156, 882–888.

Allen, J.S., Damasio, H., Grabowski, T.J., Bruss, J., Zhang, W., 2003. Sexual dimorphism and asymmetries in the gray–white composition of the human cerebrum. NeuroImage 18, 880–894.

Asato, M.R., Terwilliger, R., Woo, J., Luna, B., 2010. White matter development in adolescence: A DTI study. Cerebral Cortex 20, 2122–2131.

Ashburner, J., Csernansky, J.G., Davatzikos, C., Fox, N.C., Frisoni, G.B., Thompson, P.M., 2003. Computer-assisted imaging to assess brain structure in healthy and diseased brains. Lancet Neurology 2, 79–88.

Ashburner, J., Friston, K.J., 2000. Voxel-based morphometry – The methods. NeuroImage 11, 805–821.

Ashtari, M., Cervellione, K.L., Hasan, K.M., et al., 2007. White matter development during late adolescence in healthy males: A cross-sectional diffusion tensor imaging study. NeuroImage 35, 501–510.

Barkovich, A.J., 2006. MR imaging of the neonatal brain. Neuroimaging Clinics of North America 16, 117–135 viii–ix.

Barkovich, A.J., Kjos, B.O., Jackson, D.E., Norman, D., 1988. Normal maturation of the neonatal and infant brain: MR imaging at 1.5 T. Radiology 166, 173–180.

Basser, P.J., Mattiello, J., LeBihan, D., 1994. MR diffusion tensor spectroscopy and imaging. Biophysical Journal 66, 259–267.

Bava, S., Thayer, R., Jacobus, J., Ward, M., Jernigan, T.L., Tapert, S.F., 2010. Longitudinal characterization of white matter maturation during adolescence. Brain Research 1327, 38–46.

Beaulieu, C., 2002. The basis of anisotropic water diffusion in the nervous system – A technical review. NMR in Biomedicine 15, 435–455.

Behrens, T.E.J., Johansen-Berg, H., Woolrich, M.W., et al., 2003. Non-invasive mapping of connections between human thalamus and cortex using diffusion imaging. Nature Neuroscience 6, 750–757.

Bellis, M.D.D., Keshavan, M.S., Beers, S.R., et al., 2001. Sex differences in brain maturation during childhood and adolescence. Cerebral Cortex 11, 552–557.

Benes, F.M., Turtle, M., Khan, Y., Farol, P., 1994. Myelination of a key relay zone in the hippocampal formation occurs in the human brain during childhood, adolescence, and adulthood. Archives of General Psychiatry 51, 477–484.

Bengtsson, S.L., Nagy, Z., Skare, S., Forsman, L., Forssberg, H., Ullén, F., 2005. Extensive piano practicing has regionally specific effects on white matter development. Nature Neuroscience 8, 1148–1150.

Bihan, D.L., Breton, E., Lallemand, D., Grenier, P., Cabanis, E., Laval-Jeantet, M., 1986. MR imaging of intravoxel incoherent motions: Application to diffusion and perfusion in neurologic disorders. Radiology 161, 401–407.

Blinkhorn, S., 2005. Intelligence: A gender bender. Nature 438, 31–32.

Bonekamp, D., Nagae, L.M., Degaonkar, M., et al., 2007. Diffusion tensor imaging in children and adolescents: Reproducibility, hemispheric, and age-related differences. NeuroImage 34, 733–742.

Bookheimer, S., 2002. Functional MRI of language: New approaches to understanding the cortical organization of semantic processing. Annual Review of Neuroscience 25, 151–188.

Bramen, J.E., Hranilovich, J.A., Dahl, R.E., et al., 2011. Puberty influences medial temporal lobe and cortical gray matter maturation differently in boys than girls matched for sexual maturity. Cerebral Cortex 21, 636–646.

Buchem, M.A.V., Steens, S.C., Vrooman, H.A., et al., 2001. Global estimation of myelination in the developing brain on the basis of magnetization transfer imaging: A preliminary study. American Journal of Neuroradiology 22, 762–766.

Burzynska, A., Preuschhof, C., Bäckman, L., et al., 2010. Age-related differences in white-matter microstructure: Region-specific patterns of diffusivity. NeuroImage 49, 2104–2112.

Carreiras, M., Seghier, M.L., Baquero, S., et al., 2009. An anatomical signature for literacy. Nature 461, 983–986.

Casey, B.J., Cohen, J.D., Jezzard, P., et al., 1995. Activation of prefrontal cortex in children during a nonspatial working memory task with functional MRI. NeuroImage 2, 221–229.

Casey, B.J., Tottenham, N., Liston, C., Durston, S., 2005. Imaging the developing brain: What have we learned about cognitive development? Trends in Cognitive Sciences 9, 104–110.

Catani, M., Howard, R.J., Pajevic, S., Jones, D.K., 2002. Virtual *in vivo* interactive dissection of white matter fasciculi in the human brain. NeuroImage 17, 77–94.

Caviness, V.S., Kennedy, D.N., Richelme, C., Rademacher, J., Filipek, P.A., 1996. The human brain age 7–11 years: A volumetric analysis based on magnetic resonance images. Cerebral Cortex 6, 726–736.

Choi, Y.Y., Shamosh, N.A., Cho, S.H., et al., 2008. Multiple bases of human intelligence revealed by cortical thickness and neural activation. Journal of Neuroscience 28, 10323–10329.

Conturo, T.E., Lori, N.F., Cull, T.S., et al., 1999. Tracking neuronal fiber pathways in the living human brain. Proceedings of the National Academy of Sciences of the United States of America 96, 10422–10427.

Cooke, B.M., 2006. Steroid-dependent plasticity in the medial amygdala. Neuroscience 138, 997–1005.

Courchesne, E., Chisum, H.J., Townsend, J., et al., 2000. Normal brain development and aging: Quantitative analysis at in vivo MR imaging in healthy volunteers. Radiology 216, 672–682.

Dahl, R.E., 2004. Adolescent brain development: A period of vulnerabilities and opportunities. Keynote address. Annals of the New York Academy of Sciences 1021, 1–22.

Damasio, H., Grabowski, T., Frank, R., Galaburda, A.M., Damasio, A.R., 1994. The return of Phineas Gage: Clues about the brain from the skull of a famous patient. Science 264, 1102–1105.

Davis, S.W., Dennis, N.A., Buchler, N.G., White, L.E., Madden, D.J., Cabeza, R., 2009. Assessing the effects of age on long white matter tracts using diffusion tensor tractography. NeuroImage 46, 530–541.

Dekaban, A.S., 1978. Changes in brain weights during the span of human life: Relation of brain weights to body heights and body weights. Annals of Neurology 4, 345–356.

Draganski, B., Gaser, C., Busch, V., Schuierer, G., Bogdahn, U., May, A., 2004. Neuroplasticity: Changes in grey matter induced by training. Nature 427, 311–312.

Durston, S., Pol, H.E.H., Casey, B.J., Giedd, J.N., Buitelaar, J.K., Engeland, H.V., 2001. Anatomical MRI of the developing human brain: What have we learned? Journal of the American Academy of Child and Adolescent Psychiatry 40, 1012–1020.

Eliez, S., Reiss, A.L., 2000. MRI neuroimaging of childhood psychiatric disorders: A selective review. Journal of Child Psychology and Psychiatry 41, 679–694.

Engelbrecht, V., Rassek, M., Preiss, S., Wald, C., Mödder, U., 1998. Age-dependent changes in magnetization transfer contrast of white matter in the pediatric brain. American Journal of Neuroradiology 19, 1923–1929.

Engelbrecht, V., Scherer, A., Rassek, M., Witsack, H.J., Mödder, U., 2002. Diffusion-weighted MR imaging in the brain in children: Findings in the normal brain and in the brain with white matter diseases. Radiology 222, 410–418.

Fischl, B., Dale, A.M., 2000. Measuring the thickness of the human cerebral cortex from magnetic resonance images. Proceedings of the National Academy of Sciences of the United States of America 97, 11050–11055.

Fjell, A.M., Westlye, L.T., Greve, D.N., et al., 2008. The relationship between diffusion tensor imaging and volumetry as measures of white matter properties. NeuroImage 42, 1654–1668.

Frangou, S., Chitins, X., Williams, S.C.R., 2004. Mapping IQ and gray matter density in healthy young people. NeuroImage 23, 800–805.

Fuster, J.M., 2002. The Prefrontal Cortex: Anatomy, Physiology, and Neuropsychology of the Frontal Lobe. Lippincott-Raven, New York.

Galea, L.A.M., Spritzer, M.D., Barker, J.M., Pawluski, J.L., 2006. Gonadal hormone modulation of hippocampal neurogenesis in the adult. Hippocampus 16, 225–232.

Geschwind, N., Levitsky, W., 1968. Human brain: Left-right asymmetries in temporal speech region. Science 161, 186–187.

Giedd, J.N., Rumsey, J.M., Castellanos, F.X., et al., 1996a. A quantitative MRI study of the corpus callosum in children and adolescents. Brain Research. Developmental Brain Research 91, 274–280.

Giedd, J.N., Snell, J.W., Lange, N., et al., 1996b. Quantitative magnetic resonance imaging of human brain development: Ages 4–18. Cerebral Cortex 6, 551–560.

Giedd, J.N., Vaituzis, A.C., Hamburger, S.D., et al., 1996c. Quantitative MRI of the temporal lobe, amygdala, and hippocampus in normal human development: Ages 4–18 years. Journal of Comparative Neurology 366, 223–230.

Giedd, J.N., Castellanos, F.X., Rajapakse, J.C., Vaituzis, A.C., Rapoport, J.L., 1997. Sexual dimorphism of the developing human brain. Progress in Neuro-Psychopharmacology and Biological Psychiatry 21, 1185–1201.

Giedd, J.N., Blumenthal, J., Jeffries, N.O., et al., 1999a. Brain development during childhood and adolescence: A longitudinal MRI study. Nature Neuroscience 2, 861–863.

Giedd, J.N., Blumenthal, J., Jeffries, N.O., et al., 1999b. Development of the human corpus callosum during childhood and adolescence: A longitudinal MRI study. Progress in Neuro-Psychopharmacology and Biological Psychiatry 23, 571–588.

Giedd, J.N., Clasen, L.S., Lenroot, R., et al., 2006. Puberty-related influences on brain development. Molecular and Cellular Endocrinology 254–255, 154–162.

Giorgio, A., Watkins, K.E., Douaud, G., et al., 2008. Changes in white matter microstructure during adolescence. NeuroImage 39, 52–61.

Giorgio, A., Watkins, K.E., Chadwick, M., et al., 2010. Longitudinal changes in grey and white matter during adolescence. NeuroImage 49, 94–103.

Gogtay, N., Giedd, J.N., Lusk, L., et al., 2004. Dynamic mapping of human cortical development during childhood through early adulthood. Proceedings of the National Academy of Sciences of the United States of America 101, 8174–8179.

Goldstein, J.M., Seidman, L.J., Horton, N.J., et al., 2001. Normal sexual dimorphism of the adult human brain assessed by in vivo magnetic resonance imaging. Cerebral Cortex 11, 490–497.

Gould, S.J., 1978. Morton's ranking of races by cranial capacity: Unconscious manipulation of data may be a scientific norm. Science 200, 503–509.

Gould, S.J., 1981. The Mismeasure of Man. W. W. Norton, New York.

Gur, R.C., Gunning-Dixon, F., Bilker, W.B., Gur, R.E., 2002. Sex differences in temporo-limbic and frontal brain volumes of healthy adults. Cerebral Cortex 12, 998–1003.

Haier, R.J., Jung, R.E., Yeo, R.A., Head, K., Alkire, M.T., 2004. Structural brain variation and general intelligence. NeuroImage 23, 425–433.

Haier, R.J., Jung, R.E., Yeo, R.A., Head, K., Alkire, M.T., 2005. The neuroanatomy of general intelligence: Sex matters. NeuroImage 25, 320–327.

Haug, H., 1987. Brain sizes, surfaces, and neuronal sizes of the cortex cerebri: A stereological investigation of man and his variability and a comparison with some mammals (primates, whales, marsupials, insectivores, and one elephant). American Journal of Anatomy 180, 126–142.

Hedges, L.V., Nowell, A., 1995. Sex differences in mental test scores, variability, and numbers of high-scoring individuals. Science 269, 41–45.

Hermoye, L., Saint-Martin, C., Cosnard, G., et al., 2006. Pediatric diffusion tensor imaging: Normal database and observation of the white matter maturation in early childhood. NeuroImage 29, 493–504.

Hsu, J., Hecke, W.V., Bai, C., et al., 2010. Microstructural white matter changes in normal aging: A diffusion tensor imaging study with higher-order polynomial regression models. NeuroImage 49, 32–43.

Huang, H., 2010. Structure of the fetal brain: What we are learning from diffusion tensor imaging. The Neuroscientist 16, 634–649.

Huang, H., Zhang, J., Wakana, S., et al., 2006. White and gray matter development in human fetal, newborn and pediatric brains. NeuroImage 33, 27–38.

Huang, H., Xue, R., Zhang, J., et al., 2009. Anatomical characterization of human fetal brain development with diffusion tensor magnetic resonance imaging. Journal of Neuroscience 29, 4263–4273.

Hüppi, P.S., Maier, S.E., Peled, S., et al., 1998. Microstructural development of human newborn cerebral white matter assessed in vivo by diffusion tensor magnetic resonance imaging. Pediatric Research 44, 584–590.

Huttenlocher, P.R., 1979. Synaptic density in human frontal cortex – Developmental changes and effects of aging. Brain Research 163, 195–205.

Huttenlocher, P.R., Dabholkar, A.S., 1997. Regional differences in synaptogenesis in human cerebral cortex. Journal of Comparative Neurology 387, 167–178.

Huttenlocher, P.R., De Courten, C., Garey, L.J., 1982. Synaptic development in human cerebral cortex. International Journal of Neurology 17, 144–154.

Im, K., Lee, J., Lee, J., et al., 2006. Gender difference analysis of cortical thickness in healthy young adults with surface-based methods. NeuroImage 31, 31–38.

Irwing, P., Lynn, R., 2006. Intelligence: Is there a sex difference in IQ scores? Nature 442, E1 discussion E1–2.

Jernigan, T.L., Tallal, P., 1990. Late childhood changes in brain morphology observable with MRI. Developmental Medicine and Child Neurology 32, 379–385.

Jernigan, T.L., Archibald, S.L., Berhow, M.T., Sowell, E.R., Foster, D.S., Hesselink, J.R., 1991a. Cerebral structure on MRI. Part I: Localization of age-related changes. Biological Psychiatry 29, 55–67.

Jernigan, T.L., Trauner, D.A., Hesselink, J.R., Tallal, P.A., 1991b. Maturation of human cerebrum observed in vivo during adolescence. Brain 114, 2037–2049.

Jorm, A.F., Anstey, K.J., Christensen, H., Rodgers, B., 2004. Gender differences in cognitive abilities: The mediating role of health state and health habits. Intelligence 32, 7–23.

Kaes, T., 1907. Die Grosshirnrinde des Menschen in ihern Massen und in iherm Fasergehalt. Gustav Fischer, Jena.

Kandel, E.R., Schwartz, J.H., Jessell, T.M., 2000. Principles of Neural Science. McGraw-Hill, New York.

Karama, S., Ad-Dab'bagh, Y., Haier, R., et al., 2009. Positive association between cognitive ability and cortical thickness in a representative US sample of healthy 6 to 18 year-olds. Intelligence 37, 145–155.

Kemper, T.L., 1994. Neuroanatomical and neuropathological changes during aging. In: Albert, M.L., Knoefel, J.E. (Eds.), Clinical Neurology of Aging. Oxford University Press, Oxford, pp. 8–9.

Kimura, D., 1996. Sex, sexual orientation and sex hormones influence human cognitive function. Current Opinion in Neurobiology 6, 259–263.

Kinney, H.C., Brody, B.A., Kloman, A.S., Gilles, F.H., 1988. Sequence of central nervous system myelination in human infancy. II. Patterns of myelination in autopsied infants. Journal of Neuropathology and Experimental Neurology 47, 217–234.

Klingberg, T., Vaidya, C.J., Gabrieli, J.D., Moseley, M.E., Hedehus, M., 1999. Myelination and organization of the frontal white matter in children: A diffusion tensor MRI study. Neuroreport 10, 2817–2821.

Lancaster, J.L., Andrews, T., Hardies, L.J., Dodd, S., Fox, P.T., 2003. Three-pool model of white matter. Journal of Magnetic Resonance Imaging 17, 1–10.

Le Bihan, D., 2003. Looking into the functional architecture of the brain with diffusion MRI. Nature Reviews Neuroscience 4, 469–480.

Lebel, C., Beaulieu, C., 2009. Lateralization of the arcuate fasciculus from childhood to adulthood and its relation to cognitive abilities in children. Human Brain Mapping 30, 3563–3573.

Lebel, C., Rasmussen, C., Wyper, K., et al., 2008a. Brain diffusion abnormalities in children with fetal alcohol spectrum disorder. Alcoholism, Clinical and Experimental Research 32, 1732–1740.

Lebel, C., Walker, L., Leemans, A., Phillips, L., Beaulieu, C., 2008b. Microstructural maturation of the human brain from childhood to adulthood. NeuroImage 40, 1044–1055.

Lebel, C., Caverhill-Godkewitsch, S., Beaulieu, C., 2010. Age-related regional variations of the corpus callosum identified by diffusion tensor tractography. NeuroImage 52, 20–31.

Lenroot, R.K., Giedd, J.N., 2010. Sex differences in the adolescent brain. Brain and Cognition 72, 46–55.

Lenroot, R.K., Gogtay, N., Greenstein, D.K., et al., 2007. Sexual dimorphism of brain developmental trajectories during childhood and adolescence. NeuroImage 36, 1065–1073.

Liston, C., Watts, R., Tottenham, N., et al., 2006. Frontostriatal microstructure modulates efficient recruitment of cognitive control. Cerebral Cortex 16, 553–560.

Liu, Y., Balériaux, D., Kavec, M., et al., 2010. Structural asymmetries in motor and language networks in a population of healthy preterm neonates at term equivalent age: A diffusion tensor imaging and probabilistic tractography study. NeuroImage 51, 783–788.

Löbel, U., Sedlacik, J., Güllmar, D., Kaiser, W.A., Reichenbach, J.R., Mentzel, H., 2009. Diffusion tensor imaging: The normal evolution of ADC, RA, FA, and eigenvalues studied in multiple anatomical regions of the brain. Neuroradiology 51, 253–263.

Lu, L., Leonard, C., Thompson, P., et al., 2007. Normal developmental changes in inferior frontal gray matter are associated with improvement in phonological processing: A longitudinal MRI analysis. Cerebral Cortex 17, 1092–1099.

Luders, E., Narr, K.L., Thompson, P.M., et al., 2004. Gender differences in cortical complexity. Nature Neuroscience 7, 799–800.

Luna, B., Garver, K.E., Urban, T.A., Lazar, N.A., Sweeney, J.A., 2004. Maturation of cognitive processes from late childhood to adulthood. Child Development 75, 1357–1372.

Luna, B., Padmanabhan, A., O'Hearn, K., 2010. What has fMRI told us about the development of cognitive control through adolescence? Brain and Cognition 72, 101–113.

Lynn, R., Irwing, P., 2004. Sex differences on the progressive matrices: A meta-analysis. Intelligence 32, 481–498.

MacKay, A., Whittall, K., Adler, J., Li, D., Paty, D., Graeb, D., 1994. In vivo visualization of myelin water in brain by magnetic resonance. Magnetic Resonance in Medicine 31, 673–677.

Marsh, R., Gerber, A.J., Peterson, B.S., 2008. Neuroimaging studies of normal brain development and their relevance for understanding childhood neuropsychiatric disorders. Journal of the American Academy of Child and Adolescent Psychiatry 47, 1233–1251.

McKinstry, R.C., Mathur, A., Miller, J.H., et al., 2002. Radial organization of developing preterm human cerebral cortex revealed by noninvasive water diffusion anisotropy MRI. Cerebral Cortex 12, 1237–1243.

Miller, S.P., Ferriero, D.M., 2009. From selective vulnerability to connectivity: Insights from newborn brain imaging. Trends in Neurosciences 32, 496–505.

Mori, S., Crain, B.J., Chacko, V.P., Zijl, P.C.V., 1999. Three-dimensional tracking of axonal projections in the brain by magnetic resonance imaging. Annals of Neurology 45, 265–269.

Mori, S., Oishi, K., Jiang, H., et al., 2008. Stereotaxic white matter atlas based on diffusion tensor imaging in an ICBM template. NeuroImage 40, 570–582.

Morriss, M.C., Zimmerman, R.A., Bilaniuk, L.T., Hunter, J.V., Haselgrove, J.C., 1999. Changes in brain water diffusion during childhood. Neuroradiology 41, 929–934.

Mukherjee, P., Miller, J.H., Shimony, J.S., et al., 2001. Normal brain maturation during childhood: Developmental trends characterized with diffusion-tensor MR imaging. Radiology 221, 349–358.

Nagy, Z., Westerberg, H., Klingberg, T., 2004. Maturation of white matter is associated with the development of cognitive functions during childhood. Journal of Cognitive Neuroscience 16, 1227–1233.

Narr, K.L., Woods, R.P., Thompson, P.M., et al., 2007. Relationships between IQ and regional cortical gray matter thickness in healthy adults. Cerebral Cortex 17, 2163–2171.

Neil, J.J., Shiran, S.I., McKinstry, R.C., et al., 1998. Normal brain in human newborns: Apparent diffusion coefficient and diffusion anisotropy measured by using diffusion tensor MR imaging. Radiology 209, 57–66.

Neisser, U., Boodoo, G., Bouchard Jr., T., et al., 1996. Intelligence: Knowns and unknowns. American Psychologist 51, 77–101.

Neufang, S., Specht, K., Hausmann, M., et al., 2009. Sex differences and the impact of steroid hormones on the developing human brain. Cerebral Cortex 19, 464–473.

Nopoulos, P., Flaum, M., O'Leary, D., Andreasen, N.C., 2000. Sexual dimorphism in the human brain: Evaluation of tissue volume, tissue composition and surface anatomy using magnetic resonance imaging. Psychiatry Research 98, 1–13.

O'Donnell, S., Noseworthy, M.D., Levine, B., Dennis, M., 2005. Cortical thickness of the frontopolar area in typically developing children and adolescents. NeuroImage 24, 948–954.

Ostby, Y., Tamnes, C.K., Fjell, A.M., Westlye, L.T., Due-Tønnessen, P., Walhovd, K.B., 2009. Heterogeneity in subcortical brain development: A structural magnetic resonance imaging study of brain maturation from 8 to 30 years. Journal of Neuroscience 29, 11772–11782.

Pakkenberg, B., Gundersen, H.J., 1997. Neocortical neuron number in humans: Effect of sex and age. Journal of Comparative Neurology 384, 312–320.

Panigrahy, A., Blüml, S., 2009. Neuroimaging of pediatric brain tumors: From basic to advanced magnetic resonance imaging (MRI). Journal of Child Neurology 24, 1343–1365.

Paus, T., Zijdenbos, A., Worsley, K., et al., 1999. Structural maturation of neural pathways in children and adolescents: in vivo study. Science 283, 1908–1911.

Paus, T., Collins, D.L., Evans, A.C., Leonard, G., Pike, B., Zijdenbos, A., 2001. Maturation of white matter in the human brain: A review of magnetic resonance studies. Brain Research Bulletin 54, 255–266.

Paus, T., Nawaz-Khan, I., Leonard, G., et al., 2010. Sexual dimorphism in the adolescent brain: Role of testosterone and androgen receptor in global and local volumes of grey and white matter. Hormones and Behavior 57, 63–75.

Peper, J.S., Brouwer, R.M., Schnack, H.G., et al., 2008. Cerebral white matter in early puberty is associated with luteinizing hormone concentrations. Psychoneuroendocrinology 33, 909–915.

Peper, J.S., Brouwer, R.M., Schnack, H.G., et al., 2009a. Sex steroids and brain structure in pubertal boys and girls. Psychoneuroendocrinology 34, 332–342.

Peper, J.S., Schnack, H.G., Brouwer, R.M., et al., 2009b. Heritability of regional and global brain structure at the onset of puberty: A magnetic resonance imaging study in 9-year-old twin pairs. Human Brain Mapping 30, 2184–2196.

Perrin, J.S., Hervé, P., Leonard, G., et al., 2008. Growth of white matter in the adolescent brain: Role of testosterone and androgen receptor. Journal of Neuroscience 28, 9519–9524.

Peters, M., Jäncke, L., Staiger, J.F., Schlaug, G., Huang, Y., Steinmetz, H., 1998. Unsolved problems in comparing brain sizes in Homo sapiens. Brain and Cognition 37, 254–285.

Pfefferbaum, A., Mathalon, D.H., Sullivan, E.V., Rawles, J.M., Zipursky, R.B., Lim, K.O., 1994. A quantitative magnetic resonance imaging study of changes in brain morphology from infancy to late adulthood. Archives of Neurology 51, 874–887.

Pierpaoli, C., Jezzard, P., Basser, P.J., Barnett, A., Chiro, G.D., 1996. Diffusion tensor MR imaging of the human brain. Radiology 201, 637–648.

Post, R.M., Weiss, S.R.B., 1997. Emergent properties of neural systems: How focal molecular neurobiological alterations can affect behavior. Development and Psychopathology 9, 907–929.

Prager, A., Roychowdhury, S., 2007. Magnetic resonance imaging of the neonatal brain. Indian Journal of Pediatrics 74, 173–184.

Pujol, J., Vendrell, P., Junqué, C., Martí-Vilalta, J.L., Capdevila, A., 1993. When does human brain development end? Evidence of corpus callosum growth up to adulthood. Annals of Neurology 34, 71–75.

Pujol, J., López-Sala, A., Deus, J., et al., 2002. The lateral asymmetry of the human brain studied by volumetric magnetic resonance imaging. NeuroImage 17, 670–679.

Qiu, D., Tan, L., Zhou, K., Khong, P., 2008. Diffusion tensor imaging of normal white matter maturation from late childhood to young adulthood: Voxel-wise evaluation of mean diffusivity, fractional anisotropy, radial and axial diffusivities, and correlation with reading development. NeuroImage 41, 223–232.

Rabinowicz, T., Petetot, J.M., Gartside, P.S., Sheyn, D., Sheyn, T., de, C., 2002. Structure of the cerebral cortex in men and women. Journal of Neuropathology and Experimental Neurology 61, 46–57.

Reiss, A.L., Abrams, M.T., Singer, H.S., Ross, J.L., Denckla, M.B., 1996. Brain development, gender and IQ in children: A volumetric imaging study. Brain 119, 1763–1774.

Riddle, W.R., Donlevy, S.C., Lee, H., 2010. Modeling brain tissue volumes over the lifespan: Quantitative analysis of postmortem weights and in vivo MR images. Magnetic Resonance Imaging 28, 716–720.

Rilling, J.K., Glasser, M.F., Preuss, T.M., et al., 2008. The evolution of the arcuate fasciculus revealed with comparative DTI. Nature Neuroscience 11, 426–428.

Rubenstein, J.L.R., Rakic, P., 2013. Patterning and Cell Types Specification in the Developing CNS and PNS.

Salat, D.H., Tuch, D.S., Hevelone, N.D., et al., 2005. Age-related changes in prefrontal white matter measured by diffusion tensor imaging. Annals of the New York Academy of Sciences 1064, 37–49.

Sarkey, S., Azcoitia, I.N., Garcia-Segura, L.M., Garcia-Ovejero, D., DonCarlos, L.L., 2008. Classical androgen receptors in non-classical sites in the brain. Hormones and Behavior 53, 753–764.

Schmithorst, V., Yuan, W., 2010. White matter development during adolescence as shown by diffusion MRI. Brain and Cognition 72, 16–25.

Schmithorst, V.J., Wilke, M., Dardzinski, B.J., Holland, S.K., 2002. Correlation of white matter diffusivity and anisotropy with age during childhood and adolescence: A cross-sectional diffusion-tensor MR imaging study. Radiology 222, 212–218.

Schmithorst, V.J., Holland, S.K., Dardzinski, B.J., 2008. Developmental differences in white matter architecture between boys and girls. Human Brain Mapping 29, 696–710.

Schneider, J.F.L., Il'yasov, K.A., Hennig, J., Martin, E., 2004. Fast quantitative diffusion-tensor imaging of cerebral white matter from the neonatal period to adolescence. Neuroradiology 46, 258–266.

Schneiderman, J.S., Buchsbaum, M.S., Haznedar, M.M., et al., 2007. Diffusion tensor anisotropy in adolescents and adults. Neuropsychobiology 55, 96–111.

Shaw, P., 2007. Intelligence and the developing human brain. BioEssays 29, 962–973.

Shaw, P., Greenstein, D., Lerch, J., et al., 2006. Intellectual ability and cortical development in children and adolescents. Nature 440, 676–679.

Shaw, P., Kabani, N.J., Lerch, J.P., et al., 2008. Neurodevelopmental trajectories of the human cerebral cortex. Journal of Neuroscience 28, 3586–3594.

Smith, S.M., Jenkinson, M., Johansen-Berg, H., et al., 2006. Tract-based spatial statistics: Voxelwise analysis of multi-subject diffusion data. NeuroImage 31, 1487–1505.

Smith, S.M., Johansen-Berg, H., Jenkinson, M., et al., 2007. Acquisition and voxelwise analysis of multi-subject diffusion data with tract-based spatial statistics. Nature Protocols 2, 499–503.

Snook, L., Paulson, L., Roy, D., Phillips, L., Beaulieu, C., 2005. Diffusion tensor imaging of neurodevelopment in children and young adults. NeuroImage 26, 1164–1173.

Snook, L., Plewes, C., Beaulieu, C., 2007. Voxel based versus region of interest analysis in diffusion tensor imaging of neurodevelopment. NeuroImage 34, 243–252.

Song, S., Sun, S., Ramsbottom, M.J., Chang, C., Russell, J., Cross, A.H., 2002. Dysmyelination revealed through MRI as increased radial (but unchanged axial) diffusion of water. NeuroImage 17, 1429–1436.

Sowell, E.R., Thompson, P.M., Holmes, C.J., Batth, R., Jernigan, T.L., Toga, A.W., 1999a. Localizing age-related changes in brain structure between childhood and adolescence using statistical parametric mapping. NeuroImage 9, 587–597.

Sowell, E.R., Thompson, P.M., Holmes, C.J., Jernigan, T.L., Toga, A.W., 1999b. In vivo evidence for post-adolescent brain maturation in frontal and striatal regions. Nature Neuroscience 2, 859–861.

Sowell, E.R., Delis, D., Stiles, J., Jernigan, T.L., 2001a. Improved memory functioning and frontal lobe maturation between childhood and adolescence: A structural MRI study. Journal of the International Neuropsychological Society 7, 312–322.

Sowell, E.R., Thompson, P.M., Tessner, K.D., 2001b. Mapping continued brain growth and gray matter density reduction in dorsal frontal cortex: Inverse relationships during postadolescent brain maturation. Journal of Neuroscience 21.

Sowell, E.R., Thompson, P.M., Rex, D., et al., 2002a. Mapping sulcal pattern asymmetry and local cortical surface gray matter distribution in vivo: Maturation in perisylvian cortices. Cerebral Cortex 12, 17–26.

Sowell, E.R., Trauner, D.A., Gamst, A., Jernigan, T.L., 2002b. Development of cortical and subcortical brain structures in childhood and adolescence: A structural MRI study. Developmental Medicine and Child Neurology 44, 4–16.

Sowell, E.R., Peterson, B.S., Thompson, P.M., Welcome, S.E., Henkenius, A.L., Toga, A.W., 2003. Mapping cortical change across the human life span. Nature Neuroscience 6, 309–315.

Sowell, E.R., Thompson, P.M., Leonard, C.M., Welcome, S.E., Kan, E., Toga, A.W., 2004a. Longitudinal mapping of cortical thickness and brain growth in normal children. Journal of Neuroscience 24, 8223–8231.

Sowell, E.R., Thompson, P.M., Toga, A.W., 2004b. Mapping changes in the human cortex throughout the span of life. The Neuroscientist 10, 372–392.

Sowell, E.R., Peterson, B.S., Kan, E., et al., 2007. Sex differences in cortical thickness mapped in 176 healthy individuals between 7 and 87 years of age. Cerebral Cortex 17, 1550–1560.

Spear, L.P., 2000. The adolescent brain and age-related behavioral manifestations. Neuroscience and Biobehavioral Reviews 24, 417–463.

Steinberg, L., 2005. Cognitive and affective development in adolescence. Trends in Cognitive Sciences 9, 69–74.

Suzuki, Y., Matsuzawa, H., Kwee, I.L., Nakada, T., 2003. Absolute eigenvalue diffusion tensor analysis for human brain maturation. NMR in Biomedicine 16, 257–260.

Tamnes, C., Ostby, Y., Fjell, A., Westlye, L., Due-Tønnessen, P., Walhovd, K., 2010. Brain maturation in adolescence and young adulthood: Regional age-related changes in cortical thickness and white matter volume and microstructure. Cerebral Cortex 20, 534–548.

Thomas, C., Avidan, G., Humphreys, K., Jung, K., Gao, F., Behrmann, M., 2009. Reduced structural connectivity in ventral visual cortex in congenital prosopagnosia. Nature Neuroscience 12, 29–31.

Thompson, P.M., Giedd, J.N., Woods, R.P., MacDonald, D., Evans, A.C., Toga, A.W., 2000. Growth patterns in the developing brain detected by using continuum mechanical tensor maps. Nature 404, 190–193.

Thompson, P.M., Cannon, T.D., Narr, K.L., et al., 2001. Genetic influences on brain structure. Nature Neuroscience 4, 1253–1258.

Toga, A.W., Thompson, P.M., 2005. Genetics of brain structure and intelligence. Annual Review of Neuroscience 28, 1–23.

Tsujimoto, S., 2008. The prefrontal cortex: Functional neural development during early childhood. The Neuroscientist 14, 345–358.

Tzarouchi, L.C., Astrakas, L.G., Xydis, V., et al., 2009. Age-related grey matter changes in preterm infants: An MRI study. NeuroImage 47, 1148–1153.

Verhoeven, J.S., Sage, C.A., Leemans, A., et al., 2010. Construction of a stereotaxic DTI atlas with full diffusion tensor information for studying white matter maturation from childhood to adolescence using tractography-based segmentations. Human Brain Mapping 31, 470–486.

Vignaud, J., 1966. Development of the nervous system in early life. Part III. Radiological study of the normal skull in premature and newborn infants. In: Falkner, F.T. (Ed.), Human Development. W. B. Saunders, Philadelphia, PA.

Von Economo, C.V., 1929. The Cytoarchitectonics of the Human Cerebral Cortex. Oxford Medical Publications, London.

Wakana, S., Jiang, H., Nagae-Poetscher, L.M., Zijl, P.C.M.V., Mori, S., 2004. Fiber tract-based atlas of human white matter anatomy. Radiology 230, 77–87.

Wakana, S., Caprihan, A., Panzenboeck, M.M., et al., 2007. Reproducibility of quantitative tractography methods applied to cerebral white matter. NeuroImage 36, 630–644.

Wechsler, D., 2003. Wechsler Intelligence Scale for Children, 4th edn The Psychological Corporation, San Antonio, TX.

Westlye, L.T., Walhovd, K.B., Dale, A.M., et al., 2010. Differentiating maturational and aging-related changes of the cerebral cortex by use of thickness and signal intensity. NeuroImage 52, 172–185.

Wilke, M., Sohn, J., Byars, A.W., Holland, S.K., 2003. Bright spots: Correlations of gray matter volume with IQ in a normal pediatric population. NeuroImage 20, 202–215.

Wilke, M., Krägeloh-Mann, I., Holland, S.K., 2007. Global and local development of gray and white matter volume in normal children and adolescents. Experimental Brain Research 178, 296–307.

Willerman, L., Schultz, R., Rutledge, J.N., Bigler, E.D., 1991. In vivo brain size and intelligence. Intelligence 15, 223–228.

Witelson, S.F., Beresh, H., Kigar, D.L., 2006. Intelligence and brain size in 100 postmortem brains: Sex, lateralization and age factors. Brain 129, 386–398.

Witte, A.V., Savli, M., Holik, A., Kasper, S., Lanzenberger, R., 2010. Regional sex differences in grey matter volume are associated with sex hormones in the young adult human brain. NeuroImage 49, 1205–1212.

Wolff, S.D., Balaban, R.S., 1989. Magnetization transfer contrast (MTC) and tissue water proton relaxation in vivo. Magnetic Resonance in Medicine 10, 135–144.

Wozniak, J.R., Lim, K.O., 2006. Advances in white matter imaging: A review of in vivo magnetic resonance methodologies and their applicability to the study of development and aging. Neuroscience and Biobehavioral Reviews 30, 762–774.

Yakovlev, P.I., Lecours, A.R., 1967. The myelogenetic cycles of regional maturation of the brain. In: Minkowski, A. (Ed.), Regional Development of the Brain Early in Life. Blackwell Scientific, Oxford, pp. 3–70.

Zhang, J., Evans, A., Hermoye, L., et al., 2007. Evidence of slow maturation of the superior longitudinal fasciculus in early childhood by diffusion tensor imaging. NeuroImage 38, 239–247.

Zhang, J., Konkle, A.T.M., Zup, S.L., McCarthy, M.M., 2008. Impact of sex and hormones on new cells in the developing rat hippocampus: A novel source of sex dimorphism? European Journal of Neuroscience 27, 791–800.

Statistical Learning Mechanisms in Infancy

J. Lany[1], J.R. Saffran[2]

[1]University of Notre Dame, Notre Dame, IN, USA; [2]University of Wisconsin-Madison, Madison, WI, USA

While psychologists long imagined that infants experience the world as a bewildering array of sights, sounds, and sensations (James, 1890), research since the 1950s has shown that infants are born with more advanced perceptual and cognitive abilities than was once thought. For example, infants' visual, auditory, tactile, and vestibular systems are functional at birth, and their discrimination abilities in these areas are already remarkably acute (see Kellmann and Arterberry, 2000 for a comprehensive discussion). They show evidence of processing and remembering stimuli in their environment; even very young infants differentially attend to repeated stimuli versus novel stimuli (e.g., Fagan, 1970). Infants also exhibit differential attention to stimulus features. For example, newborns prefer complex, patterned stimuli to homogeneous ones (Olson and Sherman, 1983), facelike stimuli to nonfacelike ones (Valenza

et al., 1996), and speech to other acoustic stimuli (Vouloumanos et al., 2010).

In addition to advances in our understanding of infants' perceptual abilities, there has been great progress in the last two decades in the study of the mechanisms by which infants learn about their environment. Scientists began studying whether infants can learn via classical and operant conditioning, and by the 1970s, researchers had made substantial progress in discovering the conditions under which such learning occurred (Rovee-Collier, 1986; Sameroff, 1971). In the decades since, researchers have continued to make exciting discoveries about the nature of the learning mechanisms by which infants become sensitive to structure in their environment. In this chapter, the focus is on recent research on infants' ability to learn four different kinds of statistical structure: probability distributions, sequential

© 2013 Elsevier Inc. All rights reserved.

structure, correlations between stimulus dimensions, and associations between different forms of information. These types of statistical structures appear to be foundational to many aspects of development, including auditory and visual perception, language development, early cognition, and event processing. Much of our discussion is focused on the role of these learning mechanisms in language acquisition, though connections are also made between language learning and learning in other domains. Whenever possible, research investigating the relationships between behavioral findings and underlying neural processes has also been presented. The studies reviewed in this chapter reveal the power, nuance, and complexity of infant statistical learning mechanisms and contribute to an understanding of their role in early cognitive development.

13.1 LEARNING PROBABILITY DISTRIBUTIONS

13.1.1 Tracking Probability Distributions in Speech Sounds

A basic form of learning available to infants is sensitivity to the frequencies with which events occur in the environment. Early research on infant perceptual development revealed that infants can differentiate between frequent and infrequent events: when presented with successive trials of two pictures side by side in which one is repeated while the other changes, infants spend more time looking toward the novel picture (Fantz, 1964). Recent studies revealed that by the time they are about 8 months old, infants can do something potentially much more powerful than tracking frequencies of individual events: they can track the relative frequencies of items within a set, or probability distributions.

Differences between speech sounds, or phonemes, that correspond to differences in the meanings of words are called 'phonetic contrasts.' For example, the English sounds /r/ and /l/ are considered to be different phonemes, representing a phonetic contrast, because a switch from one to the other changes the meaning of a word (i.e., 'right' vs. 'light'). However, while /r/ and /l/ are perceived differently by adult speakers of English, they are not by speakers of Japanese, as these speech sounds do not correspond to a meaningful difference in their language. Thus, adults' ability to discriminate among speech sounds reflects the phonetic organization of their native language; adults are best able to discriminate between sounds that embody phonetic contrasts.

While adults' phonetic perception differs according to their native language, landmark research in the 1970s revealed that early in development, infants across languages readily discriminate among almost all of the speech sounds of the world (e.g., Eimas et al., 1971; for review, see Aslin et al., 1998a). Within the first year of life, however, infants' phonetic discrimination patterns become closely matched to the phonetic contrasts relevant in their native language (Kuhl et al., 1992; Werker and Tees, 1984). Within a given language, the distribution of speech sounds along a particular continuum appears to reflect phonetic category information (e.g., Lisker and Abramson, 1964). Specifically, every production of a speech sound differs on a variety of acoustic dimensions, influenced by the characteristics of individual speakers' vocal tract, speech rate, coarticulation, and phrase or sentence-level prosody. However, for acoustic dimensions that are critical in the perceived differences between phonemes, the values will tend to form nonuniform distributions, in which some values along a dimension are much more frequent than others (i.e., forming a bimodal or trimodal distribution). Interestingly, the differences relevant for highlighting phonemic contrasts may be exaggerated in infant-directed speech (Burnham et al., 2002; Kuhl et al., 1997; Werker et al., 2007). Adults can capitalize on nonuniform variation in speech sounds to discriminate between phonemes (Kluender et al., 1998), and computational modeling studies suggest that sensitivity to probability distributions may play an important role in learning phonetic contrasts (McMurray et al., 2009; Valhaba et al., 2007).

Maye et al. (2002) tested whether experience with unevenly distributed sound profiles influences phonetic perception in English learning in infants at 6 and 8 months of age. They created a set of eight speech sounds forming a continuum between [da] and [ta]. The endpoints of the continuum differed in the amount of voicing in the initial aspiration, from [da] (voiced) to [ta] (unvoiced), with six intermediate values. Infants in the unimodal condition heard a distribution in which the intermediate values (tokens 4 and 5) occurred most frequently, with decreasing frequencies toward the tails of the distribution. For infants in the bimodal condition, tokens 2 and 7 occurred with high frequency, while the tokens at the midpoint and endpoints were relatively less frequent.

After the infants were familiarized to the unimodal or bimodal distribution, they were tested to determine whether they discriminated between sounds at the opposite endpoints of the [da] – [ta] continuum, using a preferential listening task. In this procedure, the presentation of an auditory stimulus is contingent on infants' attention to a visual stimulus, such as a flashing light or a checkerboard. As long as the infants continue to look toward the visual stimulus, the auditory stimulus continues to play, but once they look away, it stops and a new trial begins. If infants' phonetic discrimination is affected by the distributional properties of the speech sounds, they should continue to discriminate between stimuli at the endpoints when the distributional information supports the maintenance of two phonetic categories. However, when the values of the tokens form a unimodal distribution, infants should treat the sounds as belonging to a single phonetic category, ignoring variation in voicing. Only infants in the bimodal condition showed discrimination of the endpoints, suggesting that experience with unimodal distributions along an acoustic continuum may play a role in the loss of sensitivity to phonetic contrasts not relevant in their language.

Recent research suggests that experience with distributional variation in speech input can also result in an enhancement or sharpening of discrimination. For example, Kuhl et al. (2006) found that at 6–8 months of age, both English- and Japanese-learning infants discriminate between /r/ and /l/, a phonetic contrast present only in English. By 10–12 months of age, the Japanese-learning infants showed reduced discrimination, consistent with other studies showing a loss of discrimination of nonnative language contrasts at this age. the English-learning infants, on the other hand, showed better discrimination between /r/ and /l/ at 10–12 months than at 6–8 months. Moreover, infants who hear speech in which acoustic differences in phonemic contrasts are exaggerated also show enhanced speech discrimination skills (Liu et al., 2003), suggesting that they capitalize on acoustic information differentiating speech sounds.

Maye et al. (2008) tested the role of distributional learning in the enhancement of speech-sound discrimination using a contrast that is difficult for infants to discriminate: the contrast between pre-voiced and short-lag stop consonants, such as [da] versus [ka] (Aslin et al., 1981). They found that 8-month-old infants learning English failed to discriminate the endpoints when given no additional exposure to speech sounds on the continuum, confirming that this is a difficult distinction for infants. Infants exposed to a unimodal distribution of tokens in the lab also failed to show discrimination. However, infants exposed to a bimodal distribution showed evidence of discrimination, suggesting that sensitivity to distributional information can lead to enhanced discrimination of difficult speech-sound contrasts. Interestingly, at 10–12 months of age, just after the period of perceptual reorganization, infants' speech-sound discrimination abilities show diminishing effects of experience with distributional information. Specifically, they need more extensive experience with a bimodal distribution to show evidence of discrimination of speech sounds that are not contrastive in their native language (Yoshida et al., 2010).

In sum, developmental changes in infants' phonetic perception appear to result, at least in part, from their ability to track distributional regularities in their native language. Kuhl (2004) has proposed that infants' experience with acoustic and distributional cues works in concert with maturational changes in neural development to produce such changes. Data from neurophysiological recordings suggest that there are neural changes in infants' processing of nonnative phonetic contrasts. Cheour et al. (1998) used an oddball paradigm to test Finnish infants' discrimination of native and nonnative vowel contrasts at both 6 and 12 months of age. In these studies, a vowel present in the infants' native language was repeated, with another 'oddball' vowel presented on about 1 in 10 trials. The oddball was either another native-language vowel, which differed in the frequency of the second formant, or a vowel that does not occur in Finnish. Evoked-response potential (ERP) studies using the oddball paradigm have shown that when infants detect a change to the repeating stimulus, the ERP wave shows a negative deflection beginning at about 200 ms, called a mismatch negativity, or MMN. At 6 months of age, Finnish infants showed a similar MMN to both the native and nonnative oddball vowels, but by 12 months, they showed a much greater MMN response to the native vowel. Using a similar paradigm in a longitudinal study with English-learning infants, Rivera-Gaxiola et al. (2005b) found an enhancement in the neural response to native-language contrasts between 7 and 11 months. They also found that infants continued to show a neural response indicative of discrimination for nonnative vowels, though the neural manifestation of the discrimination took two different forms. In one group of infants, the oddball elicited a negative component, similar to their response at 6 months of age, while in a second group, it elicited a greater positive component. Interestingly, Rivera-Gaxiola et al. (2005a) found that infants who continued to show a negative component in response to nonnative oddball phonemes later had smaller native-language vocabularies, while those that showed a positive response had larger vocabularies. The authors hypothesized that the development of the positive response in infants who later developed more advanced language skills reflects neural reorganization in response to native-language patterns. These data are in line with Kuhl's (2004) hypothesis that changes in neural organization may play a role in changes in phoneme perception during infancy.

In sum, experience with distributional information plays an important role in the dramatic changes in speech-sound discrimination that take place between 8 and 10 months of age. While infants remain sensitive to distributional cues marking nonnative phonemic contrasts after the period of reorganization, these findings suggest that neural changes, increased experience with the distributional properties of one's native language, or a combination of these factors, reduces their impact on discrimination. It is important to note that the social context in which infants experience nonnative language distributional patterns may also modulate the effects of such patterns, but by increasing sensitivity to novel phonetic contrasts – infants show greater sensitivity to difference in nonnative contrasts after listening to a native speaker in person than after watching and listening to a native speaker on a video tape (Kuhl et al., 2003). These findings suggest that although reorganization takes place early and has dramatic effects on speech perception, the capacity to learn novel contrasts is maintained and can be facilitated through social interaction.

13.1.2　Tracking Probabilities in Sampling Events

The studies described so far suggest that distributional information may facilitate perceptual reorganization in the auditory realm over the first year. Another line of recent research suggests that sensitivity to the distributional regularities plays a pervasive role in learning and cognition across diverse areas of development. Inductive learning involves encountering a finite set of examples (a sample), and using that information to extrapolate beyond the sample to behave flexibly when encountering new examples. To do so, learners must use information about the distribution of samples to learn about the underlying properties of the population from which the samples were drawn.

In a recent series of studies, Xu and Garcia (2008) explicitly tested infants' sensitivity to the relationship between the distribution of individual samples and the underlying populations they belong to. In one experiment, 8-month-old infants watched an experimenter bring out a covered box and remove 5 balls of mostly one color, either 4 red and 1 white, or 4 white and 1 red ball with her eyes closed. After she had removed the 5 balls, the experimenter removed the box's cover. Thus, the infants could not see the composition of balls in the box until after the sample balls were removed. The infants looked at the box longer on trials when the sample was unlikely under conditions of random sampling (i.e., when the container contained mostly red balls after the experimenter had just drawn mostly white balls from it). This suggests that infants' knowledge of the properties of the sample influenced their expectations about the properties of the population (i.e., the covered box). Interestingly, when the same experiment was repeated, but the experimenter drew the ping-pong balls from her pocket, infants showed no differences in looking times toward boxes containing mostly red versus mostly white balls. In other words, when the test balls were not sampled from the box, infants did not respond differentially to the contents of the box as a function of the properties of the sample. This suggests that in the

initial experiment, infants were responding to the relation between a sample and the population they were drawn from, rather than to more superficial perceptual relationships between the sample balls and the balls in the box.

Teglas et al. (2007) also investigated whether infants' processing of sampling events is affected by population information. Twelve-month-olds viewed a movie depicting a container with a small opening at its base, with four objects bouncing inside. Critically, three of the objects were identical, and one was unique. Next, an occluder covered the container, and one of the objects exited the container through the opening. On half of the trials, the infants saw events with probable outcomes (one of the three identical objects exited), and on the other trials, they saw the less probable outcome (the unique object exited the container). The infants looked longer at the improbable than the probable outcome, suggesting that the composition of objects in the container influenced their processing of the outcomes. However, when the container was partitioned such that only the unique object was physically capable of exiting, the infants looked longer when one of the three identical objects exited the container. This suggests that they were not simply responding to surface information about object frequency, looking longer toward a less common or less familiar object, but rather were responding to the relationship between the population of objects in the container and the outcome of the sampling event.

These studies also hint that infants are sensitive to information about the sampling process itself. For example, when the sampling process is random, as when an experimenter with her eyes closed picked balls out of a box in Xu and Garcia (2008), infants expect the sample to reflect the global properties of the population. However, when the sampling was not random, infants showed evidence of adjusting their expectations about the relations between a sample and the population. For example, when an occluder made some sampling events impossible in Teglas et al. (2007), infants did not appear to expect that the more frequent elements would be sampled. Moreover, Xu and Denison (2009) found that infants use information about an individual's preferences (e.g., for red balls) to guide their sampling behavior, and they are not surprised when she looks into a box containing mostly white balls and picks out mostly red ones.

Taken together, these findings suggest that across domains, infants are highly sensitive to the probabilities of individual events and can integrate this information in order to track the probabilities of sets of items related to one another in an underlying distribution or population. In the domain of speech perception, this process appears to facilitate differentiation between random noise and meaningful variations in similar sounds. In the domain of event perception, sensitivity to distributional information allows infants to generalize from individual experiences to anticipate future scenarios.

13.2 LEARNING CO-OCCURRENCE STATISTICS

Within a given domain, the reliable co-occurrence of elements is typically a surface manifestation of a meaningful relationship between those elements. Co-occurrence information thus has the potential to play a pervasive role in learning, as long as learners are sensitive to it. Despite the seemingly simple nature of such associative relationships, tracking sequential associations in most environmental patterns can be quite challenging. It entails encoding a stimulus, as well as those that precede and follow it, and maintaining memory representations across multiple occurrences of each stimulus. This might not seem difficult to do for one item that occurs in a reliable context, but consider tracking this information for a set of 100, or even 1000 different items that occur in highly variable contexts.

Natural language provides a compelling illustration of this difficult problem. Given that the sounds comprising words and sentences unfold over time, as opposed to being expressed simultaneously, spoken language is rich with sequential structure. Language is hierarchically organized, consisting of patterns at both very fine-grained and larger-grained levels, and there are sequential regularities at each level of structure. In part because of the high demands that tracking such complex sequential structure would place on infants, the potential role of co-occurrence learning mechanisms in language acquisition has been the object of a great deal of interest. Numerous studies have tested two related hypotheses: (1) tracking sequential associations could lead to the discovery of language structure, both within words as well as across words and in simple grammatical patterns, and (2) infants have sufficient computational resources to track such complex information.

13.2.1 Sequential Learning: Phonotactics

At a very fine-grained level, languages are organized according to phonotactic patterns: regularities in how sounds are structured within words. What makes these regularities interesting is that they differ across languages, and thus must be learned. For example, English syllables can begin with some consonant clusters (e.g., /st/ and /fr/), but not others (e.g., /fs/ and /ng/). However, none of these sequences can appear in syllable onsets in Japanese. Infants appear to be sensitive to phonotactic patterns by 9 months of age (Friederici and

Wessels, 1993; Jusczyk et al., 1993, 1994). In addition, 9-month-old infants can learn novel phonotactic patterns given brief laboratory experience with novel languages (e.g., Saffran and Thiessen, 2003). Following exposure to lists of novel words in which the initial consonant within a syllable was always voiced (e.g., /g/, /d/, /b/) and the final consonant was unvoiced (e.g., /k/, /t/, /p/), infants were able to distinguish between novel words consistent with those patterns and ones that followed the opposite pattern. However, when voicing was not a consistent feature of syllable onsets and offsets (i.e., /g/, /b/, and /t/ occur in syllable-initial position, while /k/, /d/, and /p/ occur in syllable-final position), infants failed to show evidence of learning the phonotactic regularities. This suggests that infants can learn positional regularities when they follow a consistent pattern, but that learning restrictions on the positions of individual elements is taxing for infants at this age.

Interestingly, by 16.5 months, infants successfully learn individual restrictions on sound locations within syllables (Chambers et al., 2003). These results suggest that with age and linguistic experience, infants' ability to learn more specific regularities is enhanced. Because learning an abstract pattern typically results from the ability to track regularities or similarities across individual exemplars, it is often more demanding than tracking information about specific exemplars. However, when exemplars share a salient feature, the need for tracking exemplar-specific information can be reduced and may lessen the computational demands on learning. This may be why younger infants were able to track phonological regularities only when they conformed to a phonological generalization.

13.2.2 Sequential Structure: Word Segmentation

Another source of sequential regularity in language pertains to how syllables are combined to form words. Syllables that reliably co-occur often belong to the same word, while syllables that rarely co-occur are more likely to span word boundaries (e.g., Swingley, 2005). Because of this feature of language, transitional probabilities between syllables tend to be higher within words than between words. The transitional probability (TP) of a co-occurrence relationship between two elements, X and Y, is computed by dividing the frequency of XY by the frequency of X. This yields the probability that if X occurs, Y will also occur. Saffran et al. (1996) tested whether 8-month-old infants, who are just beginning to learn their first words, can capitalize on TP cues. Infants listened to a stream of synthesized speech in which the only potential cue to word boundaries was the TPs between adjacent syllables (1.0 within words; 0.33 at word boundaries). The infants listened to the stream for about 2 min and were then tested on their ability to discriminate between the syllable sequences with high TPs ('words') and novel combinations of the familiar syllables ('nonwords'). Infants listened longer to the strings containing nonwords, suggesting they distinguished between syllable sequences that were attested in their input (TPs = 1.0) and sequences that did not occur (TPs = 0). A second group of infants was familiarized in the same manner and tested on sequences with high TPs 'words' (all internal TPs were 1.0), and 'part-words': sequences of syllables that had occurred across word boundaries (e.g., the final syllable of one word and the first two syllables of another word), which had lower TPs of 0.33. Infants again displayed a novelty preference, suggesting that they can also discriminate sequences that contain highly reliable transitions from those that contain less reliable ones.

Subsequent studies more specifically probed the kinds of sequential statistics that infants are sensitive to in these studies, in particular, distinguishing between the frequency of a sequence and its transitional probability. The frequency of a sequence is simply the number of times it occurs, while the transitional probability of a co-occurrence relationship provides a measure of how tightly linked or connected X and Y are, controlling for the raw frequency of X. If X occurs many times without Y, then no matter how many times it occurs with Y, the conditional probability will be relatively low. Conversely, a sequence can have low frequency but a high conditional probability.

In the Saffran et al. (1996) study, the syllable transitions within words were both more frequent and had higher TPs than the syllable transitions in part-words. Aslin et al. (1998b) thus investigated whether infants can track TPs, or just frequencies. Specifically, they modified the design of Saffran et al. (1996) such that two of the words occurred twice as often as the other words. This manipulation permitted a design in which the four test items – two words and two part-words – were equally frequent; however, the TPs within words were 1.0, and the TPs spanning the word boundaries were still 0.33. Infants showed discrimination between the words and part-words, despite the fact that the syllable sequences occurred equally often in both types of test items. This suggests that by 8 months of age, infants have a powerful mechanism for tracking co-occurrence relationships and for distinguishing potentially spurious co-occurrences from ones in which there is a very tight connection.

Another recent study further specified the nature of the regularities that infants can use to learn reliable co-occurrence relationships. The TPs in these studies have primarily been described as prospective relationships, or the probability that, given the occurrence of a syllable, another syllable will follow (i.e., the probability that the

syllable 'by' will follow the syllable 'ba,' as in the word baby). However, infants could also be sensitive to the probabilities of retrospective relationships (i.e., the probability that the syllable 'ba' will precede the syllable 'by'). In the studies of sensitivity to TPs reviewed so far, the words contained forward TPs (FTPs) and backward TPs (BTPs) of 1.0, and thus infants could have used either or both to discriminate words from part-words. Moreover, in natural languages, FTPs and BTPs are both likely to be reliable cues to word boundaries, and thus tracking both would potentially be advantageous to infants. Pelucchi et al. (2009a) thus familiarized 8-month-old infants with a corpus in which HTP and LTP words differed in their BTPs, but both had FTPs of 1.0. Thus, infants could only use BTPs to distinguish between the HTP and LTP words. Infants showed significant discrimination between HTP and LTP words, suggesting that they were sensitive to the BTPs of the syllable sequences. While FTPs may facilitate anticipating upcoming stimuli, infants' sensitivity to BTPs may help them to remember what came before a particular element. Thus, infants' sensitivity to both types of relationships potentially enhances word segmentation and other tasks that require sequential learning and processing.

In most of the studies described so far, infants were given limited exposure to artificial languages, produced as synthetic speech, which is not as engaging as naturally spoken infant-directed speech. More recent studies have investigated whether infants also use these cues to segment speech when there are many more competing regularities to attend to, and many more individual segments over which to compute TPs. For example, Pelucchi et al. (2009b) tested whether infants can track forward TPs in a natural language. They presented English-learning 8-month-olds with Italian sentences spoken with infant-directed prosody. Embedded in these sentences were two words with high internal TPs (1.0) and two words with low internal TPs (0.33). Over the course of the 2-min familiarization phase, infants heard each of these words presented just 18 times. When tested, infants discriminated between high TP words and low TP words, suggesting that they were able to track TPs despite the complexity and variability of the materials. Pelucchi et al. (2009a), described above, also used naturally spoken Italian in their work showing that infants can track backward TPs.

In another study examining how statistical learning mechanisms scale up to the complexity of natural language, Johnson and Tyler (2010) tested infants' ability to use TPs in a corpus in which the words varied in syllable number. They familiarized 8- and 5.5-month-old infants to a synthesized speech stream in which two bisyllabic words and two trisyllabic words were concatenated in a continuous stream. Interestingly, infants failed to discriminate words from part-words under these conditions. This suggests that infants benefit from the presence of other regularities beyond TPs (i.e., consistent syllable length). Indeed, learners also appear to benefit from rhythmic patterns that point to word boundaries, though the degree to which this occurs depends on the infants' age, presumably reflecting the amount of native-language knowledge the infant has acquired (e.g., Johnson and Jusczyk, 2001; Theissen and Saffran, 2003, 2007).

Altogether, these results provide strong evidence that infants are sensitive to sequential relationships between phonemes and syllables. The extent to which such cues interact with other potential cues to word boundaries, such as stress patterns, sentential prosody, and the occurrence of words in isolation, are areas particularly in need of further study.

13.2.3 Sequential Learning: Grammatical Patterns

At yet another level up in the hierarchy of language structure, sensitivity to sequential relationships appears to play a role in learning grammatical patterns, such as learning how words can be combined into phrases and sentences. For example, Saffran and Wilson (2003) asked whether sensitivity to word boundaries arising from tracking TPs between syllables provides a foundation for tracking word combinations. They played 12-month-old infants synthesized speech strings in which the TPs between syllables served as a reliable cue to word boundaries (i.e., TPs within words were 1.0, while TPs of syllables spanning word boundaries were 0.25). The strings also contained word-order patterns that were not directly cued by relationships between syllables, but could only be detected by tracking the TPs between the words. Infants were then tested on their ability to discriminate novel grammatical strings from ungrammatical ones. Despite the fact that the TPs of adjacent syllables could not be used to distinguish between the grammatical and ungrammatical test strings, infants showed significant discrimination. These findings suggest that sensitivity to sequential relationships resulting in segmentation may bolster subsequent learning by helping infants to learn sequential relationships between words.

Gómez and Gerken (1999) also tested whether 12-month-old infants can learn TPs within multiword 'sentences.' They created an artificial language in which nonsense words were combined to form sentences in which there were probabilistic regularities in the ordering of words. Infants were then tested on novel grammatical and ungrammatical strings that contained familiar words. Infants successfully discriminated between grammatical and ungrammatical strings, providing evidence that they learned the probabilistic co-occurrence relationships between words. In a subsequent experiment, infants were also able to distinguish between grammatical and ungrammatical strings when the strings were instantiated in a novel vocabulary. This finding suggests that learning sequential statistics can lead not just to knowledge of individual sequences but also potentially to a more abstract level of structure.

13.2.4 Learning Nonadjacent Co-occurrence Probabilities

The studies on sequential learning described thus far suggest that infants readily track predictive relationships between adjacent segments such as syllables and words. However, in natural languages, predictive relationships can also occur between nonadjacent elements. For example, grammatical dependencies marking tense are often nonadjacent, as in the relationship between auxiliaries such as 'is' and the progressive inflection 'ing,' as they are necessarily separated by a verb (e.g., 'is running,' 'is eating,' 'is talking'). Likewise, a plural noun predicts plural marking on the subsequent verb, but the noun and verb can be separated by modifiers, as in 'The kids who were late to school are in trouble.' Tracking nonadjacent dependencies places greater demands on memory than tracking adjacent dependencies, as elements must be remembered long enough to be linked to other elements occurring later in time. In addition, because nonadjacent dependencies can be separated by several word elements, there are many potentially irrelevant relationships for the learner to track, presenting a considerable computational burden.

Given the demands involved in detecting nonadjacent regularities, it is perhaps unsurprising that both infants and adults have substantial difficulty learning them. For example, while infants can track the relationships between adjacent elements well before they turn a year old, infants start showing sensitivity to grammatical relationships involving nonadjacent elements in their native language only at about 18 months of age (Santelmann and Jusczyk, 1998), suggesting that some combination of language experience and maturation of neural substrates for memory is needed to facilitate nonadjacent dependency learning.

Gomez (2002) investigated the conditions that promote sensitivity to nonadjacent relationships. In particular, she hypothesized that there is a relationship between the presence of salient adjacent structure and the tendency to track nonadjacent structure. Because tracking adjacent relationships appears to be relatively easy for infants, they are likely to focus on adjacent structure as long as it is reliable. However, when the variability in adjacent relationships is high (i.e., when adjacent TPs are low), infants may be less likely to track those relationships and more likely to track reliable nonadjacent regularities.

To test this hypothesis, Gomez (2002) exposed 18-month-olds to artificial language strings containing nonadjacent dependencies with TPs of 1.0. Critically, these dependencies were separated by an intervening element drawn from a set of 3, 12, or 24 different elements, and thus the adjacent TPs between words varied across conditions: relatively high (0.33, set size 3), medium (0.08, set size 12), or very low (less than 0.01, set size 24). Only infants in the set size 24 condition, in which the predictability of adjacent sequences was very low, successfully discriminated between familiar grammatical strings and strings that violated the nonadjacent relationships.

In subsequent research, Gómez and Maye (2005) found that 15-month-old infants track nonadjacent dependencies even under conditions of high variability but that 12-month-olds fail to do so. This developmental pattern suggests that increases in memory capacity over the second year facilitate nonadjacent dependency tracking. In addition to developments in memory, another factor in the development of nonadjacent dependency learning is prior language experience. Just as segmentation facilitates learning higher-order patterns in which those elements are combined, Lany and Gómez (2008) found that prior learning about adjacent relationships facilitates infants' detection of nonadjacent patterns. In particular, when 12-month-old infants were first given experience with adjacent relationships between word categories, they then successfully detected novel nonadjacent relationships between words from those categories.

13.2.5 Learning Co-occurrence Statistics in Other Domains

The studies described so far suggest that infants' sensitivity to sequential regularities could play a role in detecting multiple layers of language structure, from phonotactics to grammar. Far from being specific to learning in the auditory domain, infants' sensitivity to sequential structure plays an important role in learning across domains. Sequential relationships are important dimensions of event structure, such as in brushing one's teeth (e.g., picking up the toothbrush, applying toothpaste, wetting the brush, and brushing), making coffee, or opening a toy container. The subcomponents of the action tend to co-occur in reliable sequences, and thus tracking sequential structure in the visual domain may play a role in both detecting event boundaries and in learning their internal structure.

Indeed, Baldwin et al. (2008) found that adults can use statistical information to learn regularities in dynamic action sequences. Employing a design similar to studies investigating word segmentation (e.g., Saffran et al., 1996), adults viewed a series of discrete actions spliced together. There were action triplets embedded within the stream such that within a sequence of individual actions (e.g., poke, scrub, drink), TPs were 1.0, and across action sequences, the TPs between actions were 0.33. The only cues to the action sequence boundaries were the TPs. Adults were able to discriminate trained action sequences (TPs of 1.0) from unattested sequences of familiar actions (TPs of 0.0), as well as from sequences spanning boundaries (TPs of 0.33). They also showed evidence of detecting sequential regularities in action sequences when the individual actions (e.g., the action of drinking) varied considerably in their perceptual features across training.

While similar studies with action sequences have not yet been done with infants, numerous studies suggest that infants track sequential statistics in other domains of visual processing. For example, Kirkham et al. (2002) tested infants' ability to detect sequential relationships in a series of shapes. In these experiments, 2-, 5-, and 8-month-old infants were habituated to a series of six shapes that appeared sequentially on a screen. The objects were grouped into three sets of sequential pairs. Within a pair, one shape was reliably followed by another shape; the TPs between objects within a pair were 1.0, while the TPs at pair boundaries were 0.33. At all of the ages tested, infants discriminated between strings that preserved the statistical regularities and sequences in which the sequential regularities were violated. While infants in this experiment could use either frequency of co-occurrence or TPs to distinguish the test trials, Marcovitch and Lewkowicz (2009) found that by 4 months of age, infants can separately track frequency of shape co-occurrence as well as the conditional probability of shape pairs. These findings suggest that the ability to track sequential co-occurrence relationships in the visual domain emerges quite early in development.

Other studies have focused on infants' ability to learn sequential regularities in dynamic events. Kirkham et al. (2007) showed 8- and 11-month-old infants a spatiotemporal sequence in which a red dot appeared consecutively in the six positions of a 2×3 grid. There were reliable 'location pairs' in the sequence: for example, if the dot first appeared in the top left, it always subsequently appeared in the top middle. After the second element in a pair, the dot could appear in one of three locations. Thus, within the continuous sequence, the TPs within a location pair were 1.0, and TPs spanning pairs were 0.33. Only the 11-month-olds distinguished between sequences in which location pairs were preserved and sequences containing the unattested transitions. However, when each location in the habituation sequence was occupied by a different object, 8-month-olds distinguished between test sequences in which the location pairings were preserved and those in which they were violated. Eye tracking data revealed that infants were also faster to reach the second element in a location pair, which was predictable given the previous element, than to initial elements, which were not predictable given the last object/location, providing additional evidence that they had learned the spatiotemporal regularities. These results suggest that learning location-based co-occurrence relationships is challenging, developing between 8 and 11 months of age, despite robust abilities to track purely temporal co-occurrence relationships that develop many months earlier. However, infants can succeed at younger ages when the sequential relationships are also marked by featural cues, or when the occurrence of the object at one location predicts not just a subsequent location but also the identity of the object that will appear next.

13.2.6 Neural Correlates of Learning Sequential Structure

Given the behavioral results amassed over the past decade, researchers have become extremely interested in the neurophysiological processes involved in tracking sequential regularities. Cunillera et al. (2006, 2009) recorded ERPs as adults listened to a continuous stream of 3-syllable words cued by TPs, similar to the materials used by Saffran et al. (1996). They observed an increased negativity (an N400) occurring just after the onset of syllables in word-initial position appearing over the course of familiarization with the stream. Abla et al. (2008) found a similar pattern when learners were presented with continuous streams of 'tone words,' an effect that was more pronounced in participants who showed better discrimination in a subsequent forced-choice test. This is similar to behavioral studies by Saffran et al. (1999), which showed successful segmentation of tone sequences using TPs. Sanders et al. (2002) measured ERPs to trisyllabic words presented in isolation before and after hearing those words presented in continuous strings with TPs as the only cues to word boundary locations. Despite the difference in their methods, they also found that words evoked a greater N400 after training, as well as an N100 at syllable onset. Moreover, ERP recordings of sleeping newborn infants who were played similar streams of continuous syllables show a negative deflection of the ERP wave during the first syllable of a word (Teinonen et al., 2009). These findings suggest that there is a reliable neural signature related to tracking TPs in the auditory domain.

Interestingly, similar ERP signatures have been observed in adults learning sequential co-occurrence relationships in *visually* presented stimuli. For example, Abla and Okanoya (2009) found that triplet onsets within a visually presented stream of shapes also evoked an N400. In other words, just as in the case of syllable and tone streams, the first element of a statistically defined unit within a stream of shapes evoked a distinctive response. There was also overlap in the areas in which the increased N400 was observed across these studies – specifically middle frontal and central sites – across both auditory and visual modalities.

Together, the findings suggest that the N400 reflects something about learning reliable sequences across domains. In many studies of language processing, N400 components reflect the occurrence of an unexpected word, and thus its appearance at the first syllable of statistically defined words could reflect the fact that its occurrence was less predictable than the internal syllables. It could also reflect a search for the representation of the sequence as the first syllable is heard.

13.3 LEARNING CORRELATIONAL STRUCTURE

Another form of learning involves tracking correlations among properties of objects and events. Similar to tracking a sequence of events such as syllables or objects, tracking correlational structure entails learning associations, and thus there is potential overlap in the learning mechanisms involved. In the case of learning correlational structure, however, the relevant associations lie between features or dimensions of an object or event (such as an object or its label) rather than sequential components as they unfold over time. First, recent research suggesting that tracking correlations facilitates word segmentation and learning grammatical categories is described. Research suggesting that sensitivity to correlational structure plays an important role in segmenting the visual world, as well as in forming object categories (such as learning to distinguish between cats and dogs or between plates and spoons) is also discussed.

13.3.1 Sensitivity to Correlated Cues in Language Acquisition

In the previous section, evidence that tracking sequential structure facilitates finding word boundaries in fluent speech (Aslin et al., 1998a,b; Saffran et al., 1996), as well as some evidence suggesting that sensitivity to sequential TPs may interact with other phonological regularities in word forms was reviewed. Specifically, Johnson and Tyler (2010) found that when word length is consistent, even 5.5-month-old infants can track TPs in continuous speech. However, when word length varies, infants have a harder time using TPs to segment the stream. This suggests that infants' ability to use TPs as a segmentation cue may be facilitated by the presence of correlated cues such as regular word length.

Sahni et al. (2010) investigated whether sensitivity to TPs marking word boundaries can facilitate learning other correlated cues relevant to segmentation. They exposed 9-month-old infants to a continuous stream of bisyllabic words in which TPs within words were 1.0 and TPs across word boundaries were 0.25. Critically, all the 'words' began with /t/ (e.g., *tohsigh, teemay, tiepu*). The question of interest was whether infants could use TP information to learn enough about the speech stream to discover the /t/-initial cue. Infants successfully discriminated between novel words that had /t/ onsets and words in which /t/ occupied a medial location, suggesting they had learned that words begin with /t/. Critically, infants failed this same discrimination test when they had no prior familiarization to the speech stream, or when they were familiarized with a stream in which every other syllable began with /t/ but there were no TPs indicating word boundaries. These results suggest that infants can learn correlations between word properties to aid segmentation – in this case, using a known cue (TPs) to discover a novel correlated cue to word boundaries (/t/-onsets).

Evidence that infants can learn sequential relationships between words by 12 months of age has already been reviewed (Gómez and Gerken, 1999; Saffran and Wilson, 2003). However, a critical property of grammatical structure is that it pertains not just to how individual words are combined (e.g., '*the cat*' is grammatical, but '*cat the*' is not) but also to how words from different grammatical categories can be combined (e.g., determiners like '*the*' and '*a*' precede nouns rather than follow them). While tracking distributional information is useful for learning word-order patterns, learning grammatical categories with distributional cues alone can be very difficult (Braine, 1987; Smith, 1969). Importantly, in natural languages, words from different syntactic categories are distinguished both by distributional (Cartwright and Brent, 1997; Mintz, 2003; Mintz et al., 2002; Monaghan et al., 2005; Redington et al., 1998) and phonological cues (Farmer et al., 2006; Kelly, 1992; Monaghan et al., 2005). In other words, words from the same grammatical category share correlated distributional and phonological features. For example, nouns tend to occur after both '*a*' and '*the*,' and to have a strong–weak stress pattern. Critically, Braine (1987) found that the presence of correlated cues facilitates learning the category-level co-occurrence relationships in adults. Braine hypothesized that the presence of the additional cues reduces the computational demands involved in tracking a large number of individual co-occurrence relationships between specific words. Rather than tracking and accurately remembering each individual co-occurrence relationship, the presence of correlated cues may allow learners to detect the associations between the phonological features shared by words within a category.

To test whether infants can also use correlated cues to form grammatical categories, Gerken et al. (2005) exposed 17-month-old English-learning infants to Russian words drawn from two grammatical categories. In Russian, words have complex morphological structure, often consisting of a stem plus multiple grammatical morphemes. The familiarization set consisted of six feminine and six masculine words, and all of the words contained additional grammatical morphemes: feminine words ended in the case markers '*oj*' and '*u*' and masculine words ended in the case markings '*ya*' and '*em*.' The case markings provided distributional cues to the feminine and masculine categories. An additional phonological cue marking the category distinction was present on half of the words: three of the feminine words contained a derivational suffix '*k*,' and three of the masculine words contained the suffix '*tel*.' Thus, in many feminine words, '*k*' was followed by '*oj*' and '*us*' (e.g., '*polku*' and '*polkoj*'), while in many masculine words, '*tel*' was followed by '*ya*' and '*em*' (e.g., '*zhitelya*' and '*zhitelyem*'). Infants familiarized with these words were subsequently able to distinguish between novel grammatical words containing those relationships and ungrammatical ones: for example, even if they had not heard '*zhitelyem*,' they were able to distinguish it from the ungrammatical '*zhitelu*.' They were also able to distinguish between '*vannoj*' (grammatical combination) and '*vannya*' or '*vannyem*' (ungrammatical combinations). In this case, they could not have been using the co-occurrence relationships between the case markers and derivational morphemes (the '*telya*' and '*koj*' sequences), but were generalizing based on distributional information.

Thus, when word categories are marked by correlated cues sharing phonological properties and distributional characteristics, infants successfully learn and generalize the category relationships. While 17-month-olds show evidence of generalizing based on distributional information, Gerken et al. (2005) found that 12-month-olds failed to do so. However, using a similar category-learning paradigm, Gómez and Lakusta (2004) found that 12-month-olds can learn correlations between distributional and phonological features. These findings are consistent with the hypothesis that category learning may initially involve tracking the correlations between distributional and phonological features of words.

13.3.2 Learning Correlational Structure in the Visual Domain

Infants' visual environments are exceptionally complex. Nevertheless, elements that reliably co-occur across time and space tend to belong to the same object and thus provide information about object boundaries. Thus, just as in the case of segmenting words from continuous speech, the co-occurrence of features in the visual field is potentially a powerful cue that could be used to learn what clusters form objects. To test whether infants can track such conditional probabilities in complex visual displays, Fiser and Aslin (2002) showed 9-month-old infants scenes containing three elements. In each scene, three discrete shapes were presented simultaneously in a 2×2 grid. Two of the elements formed a 'base pair': those objects always occurred together in a consistent spatial orientation (the probability of co-occurrence, or TP, was 1.0). The two elements of each base pair also occurred with a third element, but its position relative to the base pair varied (TP = 0.25). Infants were then tested on scenes containing two elements – either base pair elements in their proper configuration or two elements that did not form a base pair. Infants were able to discriminate between base pairs and nonbase pairs when they differed in the frequency with which they appeared during habituation, and also when they were equally frequent but differed in their TPs. This suggests that sensitivity to statistical information facilitates object perception and raises the question of whether such learning can support object perception in younger infants.

Category learning is another domain in which sensitivity to correlations between object attributes might facilitate learning. Object categories (e.g., cups, dogs, birds, and trees) are structured such that properties characteristic of the category tend to co-occur within individual instances of the category. For example, the presence of one feature of a tree (branches, leaves, bark, trunk) is correlated with the presence of the others, but less strongly correlated with the presence of features associated with objects from other categories (e.g., ceramic handles, fur, and beaks).

Younger (1985) asked whether infants' sensitivity to such correlations plays a role in category learning. Ten-month-old infants viewed drawings of a set of animals that were composed of several continuously varying features (i.e., leg length, tail width, neck length, and ear separation). In the 'broad' condition, infants were exposed to a set of animals with features that were uniformly distributed (e.g., animals with long necks had both long and short legs), forming one category with a broad distribution of feature values. In the 'narrow' condition, the feature values formed two correlated clusters: for example, long-necked animals all had short legs, while short-necked animals had long legs. In this condition, the animals clustered together into two potential categories, each with a narrow range of features.

Infants were habituated to a set of these animals and then tested with novel animals. One animal contained feature values that were the average of all the animals

in the familiarization set for the broad condition, but falling in between the average of the two narrow categories. The second animal contained a set of features that were closer to the average of one of the categories in the narrow condition, but farther from the average of the broad category. Infants in the narrow condition dishabituated to the broad stimulus, while infants in the broad condition dishabituated to the test animal closer to the average of one of the narrow categories. This suggests that when feature values formed two clusters, infants formed categories corresponding to correlations between values on these dimensions. However, when feature values were randomly distributed, infants did not group the animals into separate categories.

In similar studies, Younger and Cohen (1986) found that 7-month-olds, but not 4-month-olds, are sensitive to correlations among object features. However, Mareschal et al. (2005) noted that because 4-month-old infants appear to have learned some perceptually defined categories, for example, the difference between squares and diamonds (Bomba and Siqueland, 1983), and cats and dogs (Eimas and Quinn, 1994), the studies of Younger and colleagues may have underestimated infants' categorization abilities. They hypothesized that for young infants with limited memory, longer exposures to individual animals may make it more difficult to keep all the animals in memory. In particular, if infants can only retain information about one animal at a time, they would be unable to detect the correlations between object features across a set of animals. Thus, they modified the habituation phase such that the individual animals were presented for briefer durations. Under these conditions, 4-month-olds showed evidence of forming categories based on correlations between features.

These findings suggest that tracking correlated cues plays an important role in learning perceptually based object categories. There is also a potentially important connection between infants' ability to track correlations between features and their sensitivity to distributional information. Specifically, studies of distributional learning, such as Maye et al. (2002, 2008), suggest that infants can use the frequencies across a single dimension of a complex stimulus to form groupings. In these studies, infants show evidence of tracking correlations among the values of several distributions.

Quinn et al. (2006) began to investigate the neural underpinnings of categorization in infants. In particular, 6-month-old infants were presented with a set of images of cats and then tested on their looking behavior toward a set of pictures of dogs interspersed with novel cat pictures while ERPs were recorded. When tested, infants looked longer to the novel dogs than to the novel cats, suggesting they treated the within-category images (the novel cats) as more similar to the images from familiarization, despite the fact that all images were novel.

The ERPs to the pictures of cats and dogs also suggested that infants were sensitive to category information. In particular, the ERPs to the cats viewed early on during familiarization, as well as to the set of novel dog images, showed a negative slow wave (or a negative deflection of the ERP wave) between 1 and 1.5 s after the picture appeared. The decrease of these components to cats during the latter part of familiarization indicates that infants were beginning to respond to novel cats as though they were familiar. In other words, infants were responding to category-level information, not just item-level features. In addition, the dog pictures, but not the novel cat pictures, elicited a negative central component between 300 and 750, a component thought to reflect attentional allocation to novel stimuli in infants (Nelson, 1994). Quinn et al. suggest that this component may be related to the behavioral differences in looking to the dog and cat pictures, specifically the novelty preference for pictures of dogs observed after infants have been viewing only pictures of cats. Noting that the negative slow wave indexing categorization occurs after the component reflecting novelty detection, they suggest that recognizing similarity between within-category objects is a more complex process than recognition of novelty.

Using a similar procedure, Grossmann et al. (2009) also investigated the ERP signatures involved in infant categorization. The ERP recordings revealed a negative component at 300–500 ms in anterior cortical regions when infants were shown an image from a novel category, similar to the findings of Quinn et al. (2006). Grossmann et al. (2009) also found that novel exemplars of the familiarized category evoked a late positive component relative to the response to repeated items from the category. The authors suggest that this component may reflect updating of the category representation.

In sum, infants can track correlated properties of elements by the time they are 4 months of age. This ability plays a role in object perception, both in object segmentation and categorization, as well as in forming grammatical categories in the auditory domain. The data from the ERP recordings add to our understanding of the behavioral findings by showing real-time learning, helping to isolate different aspects of the process (responding to familiarity, novelty, and memory updating), and by shedding light on the neural processes underlying discrimination observed at test. In future studies, it will be important to test the neural correlates of learning novel categories, as current evidence pertains to categories that infants are familiar with prior to the experiment. In addition, it will be interesting to test whether there are parallels between the neural correlates observed in the visual domain and in grammatical category formation, as well as whether the signatures of category learning differ from those of learning sequential structure.

13.4 LEARNING ASSOCIATIONS BETWEEN WORDS AND REFERENTS

A central problem facing infant language learners is that of forming associations between words and their referents. One possibility is that this process begins no differently than the process of learning correlated object properties, as described in the previous section. Indeed, one might think of a label as simply another feature that is shared by objects within a category (Sloutsky and Robinson, 2008). Others have suggested that words are not simply a feature like any other but that they have a privileged role because they are inherently symbolic (Waxman and Gelman, 2009). On this view, words have a different relation to objects than do other features of objects, such as their appearance, from the very beginning of word learning. In either case, it is clear that learning the associations between words and their referents is a highly complex process, requiring a powerful but selective associative learning mechanism. Even the occurrence of a word with a very concrete and observable meaning, such as 'rabbit,' will coincide with many objects and events in the environment, sometimes, but not always, including an actual rabbit (e.g., Quine, 1960). Thus, establishing the referent of a novel word poses a formidable challenge. Even if the infant is fixating on a rabbit, she still must determine that the label refers to the animal itself, as opposed to its floppy ears, or to its color. Thus, the challenge facing the infant is to form an association between the word and some aspect of the environment that contains the referent, but that is neither overly broad nor too narrow.

While the demands of learning words are substantial, recent research suggests that infants' ability to form associations is both remarkably powerful and highly selective. Infants' sensitivity to the reliability of co-occurrence relationships between words and particular aspects of the environment may facilitate learning in such complex environments. Smith and Yu (2008) tested whether infants in the early stages of word learning can capitalize on these relationships using a cross-situational learning paradigm. They presented 12- and 14-month-old infants with six words whose referents were embedded in complex scenes. On each trial, infants heard a label (e.g., 'bosa') while viewing a scene consisting of the referent along with a distractor object. Given just a single trial of this nature, infants would be unable to determine which object was the referent. However, on other trials in which 'bosa' was presented, one of the same objects was presented along with a *different* distractor. On other trials, the label 'manu' was presented, and the object that consistently occurred with it served as a distractor on other trials. Thus across trials, each label consistently occurred with one object. After infants were familiarized with the label–object training trials, they were tested using a preferential looking procedure. On each trial, infants were shown two objects simultaneously while the label for one of them was repeated. Infants looked significantly longer toward the object that matched the label at both ages, with stronger effects for the 14-month-olds than for the 12-month-olds.

These findings suggest that infants are sensitive to the reliability of label–object pairings across occurrences and can use information gathered across trials to settle on the most probable referent. However, it is important to note that words often occur in the absence of their referent and vice versa. Voloumanos and Werker (2009) thus tested whether infants learn label–object associations under more stochastic conditions. They presented 18-month-old infants with three word–object pairings, with each object labeled 10 times. One of the objects was labeled by the same word all 10 times, but the other two objects were labeled stochastically: most of the time, they occurred with one label (8 times), but on two occurrences, they were paired with a label that occurred predominantly with the other object. Infants were then tested on how well they learned the associations using a preferential looking task. The results suggest that infants were able to find the referent of words labeled stochastically: performance for word labeled 10 times or 8 times was the same when the correct referent was paired with a distractor object that had never occurred with the label. However, when a word labeled 8 times was tested with the correct referent (the object it had co-occurred with 8 times) paired with a distractor object that it had co-occurred with 2 times, infants' performance was at chance. Thus, experiencing the distractor object paired with the label just twice over the course of training disrupted recognition of the more frequent word/object pairing. Together, these findings suggest that probabilistic co-occurrences between labels and objects are harder for infants to learn than deterministic pairings, but that infants nonetheless develop some sensitivity to the association.

Recent work has shed light on the neurophysiological correlates of detecting reliable word–object associations during infancy. Friedrich and Friederici (2008) recorded ERPs while 14-month-olds were presented with two novel nonsense words, each occurring 8 times. One word was always presented with the same novel object (consistent referent), while the other was presented with a new object on each occurrence (inconsistent referent). After the first four exposures to each word, they found an early fronto-laterally localized negative component (N200-500) that was greater for the word with a consistent referent than for the word that had inconsistent referents. Over the course of the second set of four exposures, this component diminished, and the words with consistent referents began to evoke an N400 in parietal regions. Infants were also tested on their memory for the consistent pairings 24 h later. On these trials, those words were paired with either the correct or incorrect referent. A larger N400 was observed when the words were presented with an incorrect referent.

The authors suggest that the N200-500 reflects early-occurring associative processes that begin to link together the auditory and visual features of the consistent events, and that the N400 reflects semantic integration: a stronger encoding of and memory for the word-referent relation. This interpretation is consistent with the finding that when 14-month-olds are presented with familiar words from their own language, there is a larger N400 to words presented with incongruous labels than congruous ones (Friedrich and Friederici, 2005). Taken together, these findings suggest that infants are sensitive to consistent co-occurrence relationships between words and objects, and that distinctive neural processes may underlie the formation of these referential associations.

13.4.1 Sensitivity to the Statistics of Objects Taking the Same Label

Another way infants' associations become more refined is in terms of the kinds of referents they associate with novel words. While infants are initially flexible in the kinds of referents they consider, recent studies suggest that this process rapidly becomes constrained by prior learning. Smith and colleagues (Colunga and Smith, 2003, 2005; Jones and Smith, 2002; Smith et al., 2002) suggest while learning novel words may initially require repeated associations between objects and labels, once a critical mass of labels has been acquired, they detect statistical regularities that characterize those associations. For example, object labels such as *'cup,' 'ball,'* and *'phone'* refer to a set of objects that have a common shape, and many of the words in children's early vocabularies are objects that are organized by shape (Samuelson and Smith, 1999). However, as children learn more words, a higher-order abstraction emerges from these specific associations: object labels tend to pick out groups of things with similar shapes. Such a generalization would allow children to extend object labels on the basis of shape as opposed to other features, such as color or texture. As a result, upon hearing the label *'crayon'* referring to a red crayon, they can use it to refer to new crayon-shaped objects regardless of their color.

Based on these considerations, Smith et al. (2002) hypothesized that teaching children words from shape-based categories should result in increased word learning. They tested this hypothesis by bringing

17-month-old infants, who do not yet show a systematic shape bias, into the lab several times over 2 months, and teaching them names for novel objects. In one condition, the object categories were shape based: objects given the same names had similar shapes. In a control condition, objects that shared names did not have shape in common, but rather had some other property, such as their material, in common. Infants taught shape-based categories extended trained words to novel items based on shape, while infants in the nonshape-based conditions, or given no training, did not. Moreover, infants in the shape-based category training rapidly extended novel labels based on shape as well and showed a greater increase in their overall vocabularies over the 2-month training period than infants in the other conditions. These findings suggest that infants' experience with well-structured label–object pairings facilitates the rapid formation of new label–object associations, and the appropriate extension of those labels. Critically, this process results in rapid gains in vocabulary size and word-learning skill.

Just as increases in vocabulary size lead to more efficient word learning (Smith et al., 2002), the neurophysiological processes underlying word learning also appear to change as infants' vocabulary size increases. Torkildsen et al. (2008b) investigated the ERP signatures of 20-month-old infants in a novel word-learning task as a function of whether infants had fewer or more than 75 words in their productive vocabularies, a period that typically precedes a period of rapid vocabulary growth. In this task, infants were exposed to training blocks in which they viewed three familiar pictures, each labeled by the correct referent. They also saw three pictures of imaginary animals, each paired with a nonsense word. After each word had been presented with its referent five times, the pictures were presented with incongruous labels. All infants showed a larger N400 to the pairings of real words with incongruous objects at central and parietal recording sites, suggesting that the infants recognized the mismatch between label and referent. For the newly trained words, only the high-vocabulary infants showed an N400 when they were paired with an incongruent referent. This suggests that the low-vocabulary infants either did not learn the words or were able to form only a very weak association. The high-vocabulary infants showed more robust evidence of learning the novel mappings, though the N400 response was more broadly distributed than for the familiar words, suggesting that their sensitivity to the novel words differed from that to more established word–referent pairings.

Torkildsen et al. (2008a) also examined the ERPs during the five training trials. Consistent with the fact that only high-vocabulary infants showed an N400 indicative of successful learning at test, they found that only the high-vocabulary infants showed an increase in an early

negative component (N200-400) as the novel words repeated during training. This component has previously been associated with the early stages of learning the meaning of a word (see Friedrich and Friederici (2008)). While these findings do not suggest a neural mechanism by which infants narrow in on specific features as they gain skill in word learning, they do confirm that the process becomes more efficient as infants' vocabulary grows.

13.4.2 Sensitivity to the Internal Statistics of Labels in Word Learning

Additional constraints on the associations that infants are likely to form during word learning come from the properties of potential labels themselves. Graf Estes et al. (2007) tested the hypothesis that experience with statistical cues to word boundaries influences infants' ability to learn word–object pairings. In this study, 16-month-old infants were first familiarized with a stream of syllables in which the only cues to word boundaries were the TPs between adjacent syllables. Infants were then habituated to two label–object pairs. For some infants, the labels corresponded to the words (TP = 1.0), and for the other infants, the labels were equally frequent part-words (TP = 0.33) from the speech stream. After habituation to the label–object pairings, infants were tested using a Same-Switch procedure (Werker et al., 1998). The Same trials preserved the label–object pairings present during habituation, while the label–object pairings were switched on Switch trials. Learning is measured by the degree of increased looking to Switch trials, which is evidence of dishabituation. Only infants for whom the labels corresponded to words showed longer looking to the Switch trials, suggesting that learning words composed of syllables with high TPs is easier than learning labels that had internal statistics suggesting that the syllables did not cohere into a word. Likewise, Graf Estes et al. (2010) found that infants learn word–object associations better when the labels conform to the predominant phonotactic patterns of the native language. Experience with regularities in fluent speech may thus facilitate word learning by providing infants with candidate sound sequences to map to referents, and by promoting more accurate processing of those sound sequences, facilitating subsequent recall.

Data from ERP recordings in word-learning tasks also suggest that semantic processing is influenced by phonotactic information. As previously discussed, between 12 and 14 months of age, infants show a larger N400 when familiar words are presented with incongruous referents. Friedrich and Friederici (2005) found that both 12- and 19-month-old infants showed an early negative component that differed between real and phonotactically legal nonsense words and phonotactically illegal nonsense words presented without referents, which likely reflects sensitivity to the relative familiarity of the words. At 19 months, infants also show an N400 response when novel words that are phonotactically legal in their language are presented with objects that have known labels, suggesting they recognize the label–object mismatch. However, phonotactically illegal words did not elicit an N400, suggesting that they may not have been perceived as potential labels. These findings are consistent with the hypothesis that the N400 response reflects an advance or change in word-learning skill, and specifically that phonotactic information plays an important role by the end of the second year.

Another potential source of information guiding word learning is the overlap between words' statistical and semantic properties at the level of grammatical categories. Words from different lexical categories are correlated with different semantic regularities (e.g., nouns tend to refer to objects and people, adjectives to properties such as color or texture, and verbs to actions or events). As previously discussed, words from different lexical categories can also be distinguished by a constellation of statistical cues, including distributional and phonological regularities (e.g., Christiansen et al., 2009; Kelly, 1992; Mintz et al., 2002; Monaghan et al., 2005).

A recent study tested whether infants can capitalize on experience with such cues in a word-learning task (Lany and Saffran, 2010). In this study, 22-month-old infants first listened to an artificial language that contained two word categories, disyllabic *X-words* and monosyllabic *Y-words*. For infants in the experimental group, phrases took the form aX and bY, and thus words from the X and Y categories were reliably marked by correlated distributional and phonological cues. Infants in the control group also heard aX and bY phrases, but in addition, they heard an equal number of aY and bX phrases. Thus, for the control infants, the word categories were not marked by correlated cues. Infants were then trained on pairings between phrases from the language and pictures of unfamiliar animals and vehicles. For both experimental and control groups, familiar aX phrases were paired with animal pictures, and familiar bY phrases were paired with vehicle pictures. Infants were then tested using a preferential looking procedure.

Interestingly, only the experimental infants were able to learn the trained associations between phrases and pictures, despite the fact that control infants had the same amount of experience with them. Moreover, only the experimental infants successfully generalized to novel pairings: when hearing a word with distributional and phonological properties of other words referring to animals, they mapped the word to a novel animal over a novel vehicle. These findings suggest that infants' experience with statistical cues marking word categories lays

an important foundation for learning the meanings of those words.

In sum, infants' ability to form associations is remarkably powerful, but it is also selective. Infants do not simply form an association between a word and object that happen to co-occur, but rather they form an association only when that object reliably occurs across other presentations of that word. And, while infants are initially relatively flexible in the kinds of associations they form between words and referents, recent studies suggest that this process rapidly becomes constrained by prior learning. For example, by 17 months of age, words with good sequential statistics are more readily associated with referents than sound sequences that do not have the characteristics of typical words in that language (Graf Estes et al., 2007). Moreover, once infants have begun to learn associations between words and referents, this knowledge influences the associations that infants will subsequently form. For example, infants rapidly begin to associate novel nouns with objects' shape over objects' color (e.g., Smith et al., 2002). Likewise, words that have statistical properties of a particular category are readily mapped to new referents from that category (Lany and Saffran, 2010).

13.5 CONCLUSIONS

The research reviewed in this chapter suggests that infants' sensitivity to statistical structure is powerful and nuanced and plays a role in diverse aspects of learning over the first several months of life. We have discussed infants' sensitivity to four different kinds of statistical structure: probability distributions, sequential structure, correlations between stimulus dimensions, and associations between different forms of information. While we have discussed these types of structure separately, these are likely not independent learning processes. For example, there are important similarities between learning sequential structure in word segmentation and in learning correlational structure in object segmentation. Tracking sequential structure within words may share important properties with learning sequential regularities between words and word categories. Likewise, the processes involved in learning word–label associations may be similar to those in learning feature co-occurrences in object categorization. Furthermore, real-world learning processes tap a combination of these learning mechanisms – for example, learning word–object associations critically relies on a complex interaction of sensitivity to phonotactic regularities, word segmentation, object categorization, word-order patterns, and grammatical categories.

A related point is that statistical learning mechanisms are tuned through experience. For example, in the auditory domain, infants' sensitivity to reliable TPs between syllables facilitates segmenting words from fluent speech. In turn, this makes it possible to learn phonological regularities correlated with TPs, which may further facilitate segmentation in natural speech (Sahni et al., 2010). Likewise, segmentation facilitates learning word-order patterns, and tracking correlations between such distributional properties of words and their phonological properties can facilitate learning grammatical structure. In the domain of visual perception, tracking the co-occurrence of features may facilitate segmenting objects within complex scenes (Fiser and Aslin, 2002), and tracking how these features co-occur across instances facilitates the formation of object categories (Younger, 1985). In other words, as learners track simple structure, they build a foundation for tracking more complex, higher-order structure. They also appear to highlight relevant structure, which can facilitate learning more computationally challenging regularities (Lany and Gómez, 2008).

These studies also raise many intriguing questions for future research. For example, many studies manipulate the statistical properties of infants' input and show that this influences learning, but this does not suggest that infants are tracking input statistics veridically. Indeed, studies with older children suggest that they, unlike adults, tend to distort input statistics, maximizing probabilities rather than matching them (e.g., Hudson Kam and Newport, 2005). It also leads to questions of how infants encode these statistical regularities and how fine-grained their sensitivity is. Also, learning inherently entails perceiving, remembering, acting on the environment, and interacting with others, and thus in future research, it will be important to relate infants' statistical learning abilities to other developmental processes, such as perception, memory, and social interaction.

Another important area for future research is the underlying neurophysiology of statistical learning. Researchers have just begun to investigate the neurophysiological substrates of the behavioral findings, namely, how ERPs to stimuli consistent with familiarized patterns differ from ERPs to stimuli that are inconsistent. Given the challenges inherent in ERP research in infants, much of this work has tested learning in adults. However, recent advances in neurophysiological methodology in infants suggest that these techniques hold much promise for investigating learning mechanisms directly. Indeed, while the discrimination tasks typically used in infant behavioral studies can only probe the endpoint (or other static snapshot) of the learning process, psychophysiological techniques are particularly well suited to observing the process of learning.

In sum, infants' environments are rich with meaningful statistical structure, and infants readily track it. Far from being limited to simple associative learning, such as the stimulus–response learning of classical learning theory, the studies reviewed in this chapter reveal a sensitivity to statistical structure that is powerful and nuanced, capable of not only tracking fine-grained detail but also forming generalizations. Advances in the study of statistical learning mechanisms have shed new light on many aspects of development, such as auditory and visual perception, language development, and event processing. Continued study of infants' statistical learning mechanisms holds great promise for advancing the study of early cognitive development.

Acknowledgments

Preparation of this chapter was supported by National Institute of Child Health and Human Development Grants F32 HD057698 to J Lany and R01HD37466 to JR Saffran, by a James S. McDonnell Foundation Scholar Award to JR Saffran. The content is solely the responsibility of the authors and does not necessarily represent the official views of the Eunice Kennedy Shriver National Institute of Child Health and Human Development or the National Institutes of Health. The authors thank Jessica Payne for helpful comments on a previous draft.

References

Abla, D., Okanoya, K., 2009. Visual statistical learning of shape sequences: An ERP study. Neuroscience Research 64, 185–190.

Abla, D., Katahira, K., Okanoya, K., 2008. Online assessment of statistical learning by event-related potentials. Journal of Cognitive Neuroscience 20, 952–964.

Aslin, R.N., Pisoni, D.B., Hennessy, B.I., Perey, A.J., 1981. Discrimination of voice onset time by human infants: New findings and implications for the effects of early experience. Child Development 52, 1135–1145.

Aslin, R.N., Jusczyk, P.W., Pisoni, D.B., 1998a. Speech and auditory processing during infancy: Constraints on and precursors to language. In: Kuhn, D., Siegler, R. (Eds.), Handbook of Child Psychology: Cognition, Perception, and Language, vol. 2. Wiley, New York, pp. 147–254.

Aslin, R.N., Saffran, J.R., Newport, E.L., 1998b. Computation of conditional probability statistics by 8-month-old infants. Psychological Science 9, 321–324.

Baldwin, D., Andersson, A., Saffran, J.R., Meyer, M., 2008. Segmenting dynamic human action via statistical structure. Cognition 106, 1382–1407.

Bomba, P.C., Siqueland, E.R., 1983. The nature and structure of infant form categories. Journal of Experimental Child Psychology 35, 609–636.

Braine, M.D.S., 1987. What is learned in acquiring words classes: A step toward acquisition theory. In: MacWhinney, B. (Ed.), Mechanisms of Language Acquisition. Erlbaum, Hillsdale, NJ, pp. 65–87.

Burnham, D., Kitamura, C., Vollmer-Conna, U., 2002. What's new pussycat? On talking to babies and animals. Science 296, 1435.

Cartwright, T.A., Brent, M.R., 1997. Syntactic categorization in early language acquisition: Formalizing the role of distributional analysis. Cognition 63, 121–170.

Chambers, K.E., Onishi, K.H., Fisher, C., 2003. Infants learn phonotactic regularities form brief auditory experience. Cognition 87, B69–B77.

Cheour, M., Ceponiene, R., Lehtokoski, A., et al., 1998. Development of language-specific phoneme representations in the infant brain. Nature Neuroscience 1, 351–353.

Christiansen, M.H., Onnis, L., Hockema, S.A., 2009. The secret is in the sound: From unsegmented speech to lexical categories. Developmental Science 12, 388–395.

Colunga, E., Smith, L.B., 2003. The emergence of abstract ideas: Evidence from networks and babies. Philosophical Transactions by the Royal Society 358, 1205–1214.

Colunga, E., Smith, L.B., 2005. From the Lexicon to expectations about kinds: A role for associative learning. Psychological Review 112, 347–382.

Cunillera, T., Toro, J.M., Sebastian-Galles, N., Rodriguez-Fornells, A., 2006. The effects of stress and statistical cues on continuous speech segmentation: An event-related brain potential study. Brain Research 1123, 168–178.

Cunillera, T., Càmara, E., Toro, J.M., et al., 2009. Time course and functional neuroanatomy of speech segmentation in adults. NeuroImage 48, 541–553.

Eimas, P.D., Quinn, P.C., 1994. Studies on the formation of perceptually based basic-level categories in young infants. Child Development 65, 903–917.

Eimas, P.D., Siqueland, E.R., Jusczyk, P., Vigorito, J., 1971. Speech perception in infants. Science 171, 303–306.

Fagan, J.F., 1970. Memory in the infant. Journal of Experimental Child Psychology 9, 217–226.

Fantz, R.L., 1964. Visual experience in infants: Decreased attention to familiar patterns relative to novel ones. Science 146, 668–670.

Farmer, T.H., Christiansen, M.H., Monaghan, P., 2006. Phonological typicality influences on-line sentence comprehension. Proceedings of the National Academy of Sciences of the United States of America 103, 12203–12208.

Fiser, J., Aslin, R.N., 2002. Statistical learning of new visual feature combinations by infants. Proceedings of the National Academy of Sciences of the United States of America 99, 15822–15826.

Friederici, A.F., Wessels, J.M.I., 1993. Phonotactic knowledge of word boundaries and its use in infant speech perception. Perception & Psychophysics 54, 287–295.

Friedrich, M., Friederici, A.D., 2005. Phonotactic knowledge and lexical–semantic processing in one-year-olds: Brain responses to words and nonsense words in picture contexts. Journal of Cognitive Neuroscience 17, 1785–1802.

Friedrich, M., Friederici, A.D., 2008. Neurophysiological correlates of online word learning in 14-month-old infants. Cognitive Neuroscience and Neuropsychology 19, 1757–1761.

Gerken, L.A., Wilson, R., Lewis, W., 2005. Seventeen-month-olds can use distributional cues to form syntactic categories. Journal of Child Language 32, 249–268.

Gomez, R.L., 2002. Variability and detection of invariant structure. Psychological Science 13, 431–436.

Gómez, R.L., Gerken, L.A., 1999. Artificial grammar learning by one-year-olds leads to specific and abstract knowledge. Cognition 70, 109–135.

Gómez, R.L., Lakusta, L., 2004. A first step in form-based category abstraction in 12-month-old infants. Developmental Science 7, 567–580.

Gómez, R.L., Maye, J., 2005. The developmental trajectory of nonadjacent dependency learning. Infancy 7, 183–206.

Graf Estes, K., Evans, J.L., Alibali, M.W., Saffran, J.R., 2007. Can infants map meaning to newly segmented words? Psychological Science 18, 254–260.

Graf Estes, K., Edwards, J., Saffran, J.R., 2010. Phonotactic constraints on infant word learning. Infancy 16, 180–197.

Grossmann, T., Gliga, T., Johnson, M.H., Mareschal, D., 2009. The neural basis of perceptual category learning in human infants. Journal of Cognitive Neuroscience 21, 2276–2286.

Hudson Kam, C.L., Newport, E.L., 2005. Regularizing unpredictable variation: The roles of adult and child learners in language formation and change. Language Learning and Development 1, 151–195.

James, W., 1890. The Principles of Psychology. Holt, New York.

Johnson, E.K., Jusczyk, P.W., 2001. Word segmentation by 8-month-olds: When speech cues count more than statistics. Journal of Memory and Language 44, 548–567.

Johnson, E.K., Tyler, M.D., 2010. Testing the limits of statistical learning for word segmentation. Developmental Science 13, 339–345.

Jones, S.S., Smith, L.B., 2002. How children know the relevant properties for generalizing object names. Developmental Science 5, 219–232.

Jusczyk, P.W., Friederici, A.D., Wessels, J.M., Svenkerud, V.Y., Jusczyk, A.M., 1993. Infants' sensitivity to the sound patterns of native language words. Journal of Memory and Language 32, 402–420.

Jusczyk, P.W., Luce, P.A., Charles-Luce, J., 1994. Infants' sensitivity to phonotactic patterns in their native language. Journal of Memory and Language 33, 630–645.

Kellmann, P.J., Arterberry, M.E., 2000. The Cradle of Knowledge. The MIT Press, Cambridge, MA.

Kelly, M.H., 1992. Using sound to solve syntactic problems: The role of phonology in grammatical category assignments. Psychological Review 99, 349–364.

Kirkham, N.Z., Slemmer, J.A., Johnson, S.P., 2002. Visual statistical learning in infancy: Evidence for a domain general learning mechanism. Cognition 83, B35–B42.

Kirkham, N.Z., Slemmer, J.A., Richardson, D.C., Johnson, S.P., 2007. Location, location, location: Development of spatiotemporal sequence learning in infancy. Child Development 78, 1559–1571.

Kluender, K.R., Lotto, A.J., Holt, L.L., Bloedel, S.L., 1998. Role of experience for language-specific functional mappings of vowel sounds. Journal of the Acoustical Society of America 104, 3568–3582.

Kuhl, P.K., 2004. Early language acquisition: Cracking the speech code. Nature Reviews Neuroscience 5, 831–843.

Kuhl, P.K., Williams, K.A., Lacerda, F., Stevens, K.N., Lindblom, B., 1992. Linguistic experience alters phonetic perception in infants by 6 months of age. Science 255, 606–608.

Kuhl, P.K., Andruski, J.E., Chistovich, I.A., et al., 1997. Cross-language analysis of phonetic units in language addressed to infants. Science 277, 684–686.

Kuhl, P.K., Tsao, F.M., Liu, H.M., 2003. Foreign-language experience in infancy: Effects of short-term exposure and social interaction on phonetic learning. Proceedings of the National Academy of Sciences of the United States of America 100, 9096–9101.

Kuhl, P.K., Stevents, E., Hayashi, A., Deguchi, T., Kiritani, S., Iverson, P., 2006. Infants show a facilitation effect for native language phonetic perception between 6 and 12 months. Developmental Science 9, F13–F21.

Lany, J., Gómez, R.L., 2008. Twelve-month-old infants benefit from prior experience in statistical learning. Psychological Science 19, 1247–1252.

Lany, J., Saffran, J.R., 2010. From statistics to meaning: Infant acquisition of lexical categories. Psychological Science 21, 284–291.

Lisker, L., Abramson, A.S., 1964. A cross-language study of voicing in initial stops: Acoustical measurements. Word 20, 384–482.

Liu, H.M., Kuhl, P.K., Tsao, F.M., 2003. An association between mothers' speech clarity and infants' speech discrimination skills. Developmental Science 6, F1–F10.

Marcovitch, S., Lewkowicz, D.L., 2009. Sequence learning in infancy: The independent contributions of conditional probability and pair frequency information. Developmental Science 12, 1020–1025.

Mareschal, D., Powell, D., Westermann, G., Volein, A., 2005. Evidence of rapid correlation-based perceptual category learning by 4-month-olds. Infant and Child Development 14, 445–457.

Maye, J., Werker, J.F., Gerken, L., 2002. Infant sensitivity to distributional information can affect phonetic discrimination. Cognition 82, B101–B111.

Maye, J., Weiss, D.J., Aslin, R.N., 2008. Statistical phonetic learning in infants: Facilitation and feature generalization. Developmental Science 11, 122–134.

McMurray, B., Aslin, R.N., Toscano, J.C., 2009. Statistical learning of phonetic categories: Insights from a computational approach. Developmental Science 12, 369–378.

Mintz, T.H., 2003. Frequent frames as a cue for grammatical categories in child directed speech. Cognition 90, 91–117.

Mintz, T.H., Newport, E.L., Bever, T.G., 2002. The distributional structure of grammatical categories in speech to young children. Cognitive Science 26, 393–424.

Monaghan, P., Chater, N., Christiansen, M.H., 2005. The differential role of phonological and distributional cues in grammatical categorization. Cognition 96, 143–182.

Nelson, C.A., 1994. Neural correlates of recognition memory in the first postnatal year of life. In: Dawson, G., Fischer, K. (Eds.), Human Behavior and the Developing Brain. Guilford Press, New York, pp. 269–313.

Olson, G.M., Sherman, T., 1983. Attention, learning and memory in infants. In: Haith, M.M., Campos, J.J. (Eds.), Handbook of Child Psychology. Infancy and Developmental Psychobiology, vol. 2. Wiley, New York, pp. 1001–1080.

Pelucchi, B., Hay, J.F., Saffran, J.R., 2009a. Learning in reverse: Eight-month-old infants track backwards transitional probabilities. Cognition 11, 244–247.

Pelucchi, B., Hay, J.F., Saffran, J.R., 2009b. Statistical learning in a natural language by 8-month-old infants. Child Development 80, 674–685.

Quine, W.V.O., 1960. Word and Object. MIT Press, Cambridge, MA.

Quinn, P.C., Westerlund, A., Nelson, C.A., 2006. Neural markers of categorization in 6-month-old infants. Psychological Science 17, 59–66.

Redington, M., Chater, N., Finch, S., 1998. Distributional information: A powerful cue for acquiring syntactic categories. Cognitive Science 22, 425–469.

Rivera-Gaxiola, M., Klarman, L., Garcia-Sierra, A., Kuhl, P.K., 2005a. Neural patterns to speech and vocabulary growth in American infants. NeuroReport 16, 495–498.

Rivera-Gaxiola, M., Silva-Pereyra, J., Kuhl, P.K., 2005b. Brain potentials to native and non-native speech contrasts in 7- and 11-month-old American infants. Developmental Science 8, 162–172.

Rovee-Collier, C., 1986. The rise and fall of infant classical conditioning research: Its promise for the study of early development. Advances in Infancy Research 4, 139–159.

Saffran, J.R., Thiessen, E.D., 2003. Pattern induction by infant language learners. Developmental Psychology 39, 484–494.

Saffran, J.R., Wilson, D.P., 2003. From syllables to syntax: Multi-level statistical learning by 12-month-old infants. Infancy 4, 273–284.

Saffran, J.R., Aslin, R.N., Newport, E.L., 1996. Statistical learning by 8-month-old infants. Science 274, 1926–1928.

Saffran, J.R., Johnson, E.K., Aslin, R.N., Newport, E.L., 1999. Statistical learning of tone sequences by human infants and adults. Cognition 70, 27–52.

Sahni, S.D., Seidenberg, M.S., Saffran, J.R., 2010. Connecting cues: Overlapping regularities support cue discovery in infancy. Child Development 81, 727–736.

Sameroff, A.J., 1971. Can conditioned responses be established in the newborn infant: 1971? Developmental Psychology 5, 1–12.

Samuelson, L.K., Smith, L.B., 1999. Early noun vocabularies: Do ontology, category structure and syntax correspond? Cognition 73, 1–33.

Sanders, L.D., Newport, E.L., Neville, H.J., 2002. Segmenting nonsense: An event-related potential index of perceived onsets in continuous speech. Nature Neuroscience 5, 700–703.

Santelmann, L.M., Jusczyk, P.W., 1998. Sensitivity to discontinuous dependencies in language learners: Evidence for limitations in processing space. Cognition 69, 105–134.

Sloutsky, V.M., Robinson, C.W., 2008. The role of words and sounds in infants' visual processing: From overshadowing to attentional tuning. Cognitive Science 32, 354–377.

Smith, K.H., 1969. Learning co-occurrence restrictions: Rule learning or rote learning? Journal of Verbal Behavior 8, 319–321.

Smith, L.B., Yu, C., 2008. Infants rapidly learn word-referent mappings via cross-situational statistics. Cognition 106, 1558–1568.

Smith, L.B., Jones, S.S., Landau, B., Gershkoff-Stowe, L., Samuelson, L., 2002. Object name learning provides on-the-job training for attention. Psychological Science 13, 13–19.

Swingley, D., 2005. Statistical clustering and the contents of the infant vocabulary. Cognitive Psychology 50, 86–132.

Teglas, E., Girotto, V., Gonzales, M., Bonatti, L.L., 2007. Intuitions of probabilities shape expectations of the future at 12 months and beyond. Proceedings of the National Academy of Sciences of the United States of America 104, 19156–19159.

Teinonen, T., Fellman, V., Näätänen, R., Alku, P., Huotilainen, M., 2009. Statistical language learning in neonates revealed by event-related brain potentials. BMC Neuroscience 10, 21.

Theissen, E.D., Saffran, J.R., 2003. When cues collide: Use of stress and statistical cues to word boundaries by 7- to 9-month-old infants. Developmental Psychology 39, 706–716.

Theissen, E.D., Saffran, J.R., 2007. Learning to learn: Infants' acquisition of stress-based strategies for word segmentation. Language Learning and Development 3, 73–100.

Torkildsen, J.V.K., Hansen, H.F., Svangstu, J.M., 2008a. Brain dynamics of word familiarization in 20-month-olds: Effects of productive vocabulary size. Brain and Language 108, 73–88.

Torkildsen, J.V.K., Svangstu, J.M., Hansen, H.F., 2008b. Productive vocabulary size predicts event-related potential correlates of fast-mapping in 20-month-olds. Journal of Cognitive Neuroscience 20, 1266–1282.

Valenza, E., Simion, F., Macchi Cassia, V., Umilta, C., 1996. Face preference at birth. Journal of Experimental Psychology. Human Perception and Performance 22, 892–903.

Valhaba, G.K., McClelland, J.L., Pons, F., Werker, J.F., Amano, S., 2007. Unsupervised learning of vowel categories from infant-directed speech. Proceedings of the National Academy of Sciences of the United States of America 104, 13273–13278.

Voloumanos, A., Werker, J.F., 2009. Infants' learning of novel words in a stochastic environment. Developmental Psychology 45, 1611–1617.

Vouloumanos, A., Hauser, M., Werker, J.F., Martin, A., 2010. The tuning of human neonates' preference for speech. Child Development 81, 517–527.

Waxman, S.R., Gelman, S.A., 2009. Early word-learning entails reference not merely associations. Trends in Cognitive Sciences 13, 258–263.

Werker, J.F., Tees, R.C., 1984. Cross-language speech perception: Evidence for perceptual reorganization during the first year of life. Infant Behaviour and Development 7, 49–63.

Werker, J.F., Cohen, L.B., Lloyd, V.L., Casasola, M., Stager, C.L., 1998. Acquisition of word-object associations by 14-month-old infants. Developmental Psychology 34, 1289–1309.

Werker, J.F., Pons, F., Dietrich, C., Kajikawa, S., Fais, L., Amano, S., 2007. Infant-directed speech supports phonetic category learning in English and Japanese. Cognition 103, 147–162.

Xu, F., Denison, S., 2009. Statistical inference and sensitivity to sampling in 11-month-old infants. Cognition 112, 97–104.

Xu, F., Garcia, V., 2008. Intuitive statistics by 8-month-old infants. Proceedings of the National Academy of Sciences of the United States of America 105, 5012–5015.

Yoshida, K.A., Pons, F., Maye, J., Werker, J.F., 2010. Distributional phonetic learning at 10 months of age. Infancy.

Younger, B.A., 1985. The segregation of items into categories by ten-month-old infants. Child Development 56, 1574–1583.

Younger, B.A., Cohen, L.B., 1986. Developmental changes in infants' perception of correlation among attributes. Child Development 57, 803–815.

Development of the Visual System

S.P. Johnson

University of California, Los Angeles, CA, USA

O U T L I N E

The purpose of vision is to obtain information about the surrounding environment so that we may plan appropriate actions. Consider, for example, a stroll on the beach vs. a hike in the Grand Canyon (Figure 14.1). Both activities involve locomotion, but each places very different demands on the perceptual and action systems, including the visual system. In the case of the stroll, the beach is wide, there are few obstacles and little risk. In the case of the hike, in contrast, the path is narrow; rocks, vegetation, and abrupt precipices must be avoided. To remain safe, the hiker must know what the risks are, and this invariably involves knowing what objects there are in the visual scene. The importance of accurate perception of our surroundings is attested by the allotment of cortical tissue devoted to vision: By some estimates, over 50% of the cortex of the macaque monkey (a phylogenetically close cousin to *Homo sapiens*) is involved in visual perception, and there are perhaps 30 distinct cortical areas that participate in visual or visuomotor processing (Felleman and Van Essen, 1991; Van Essen et al., 1992).

This chapter reviews theory and data concerning development of the human visual system with an emphasis on object perception. As will be seen, infants are prepared to see objects and understand many of their properties (e.g., permanence, coherence) well in advance of locomotion, so that by the time infants begin to crawl and walk, they have a good sense of what and where obstacles might be, even if the hazards these objects pose remain unknown.

Preparation of this chapter was supported by NIH grants R01-HD40432 and R01-HD48733.

© 2013 Elsevier Inc. All rights reserved.

FIGURE 14.1   Two visual scenes.

There is much else to learn. Visual scenes, for example, tend to be very complex: a multitude of overlapping and adjacent surfaces with distinct shapes, colors, textures, and depths relative to the observer. Yet our visual experience as adults is not one of incomplete fragments of surfaces, but instead one of objects, most of which have a shape that can be inferred from partial views and incomplete information. Is the infant's visual system sufficiently functional and organized to make sense of the world from the onset of visual experience at birth, able to bind shapes, colors, and textures into coherent forms, and to perceive objects as regular and predictable and complete across space and time? Or does the infant's visual system require a period of maturation and experience within which to observe and learn about the world?

These 'nature versus nurture' questions begin to lose their steam when the details of visual development are examined and explained, because visual development stems from growth, maturation, and experience from learning and from action; all happen simultaneously and all influence one another. Infants free of disability or developmental delay are born with a functional visual system that is prepared to contribute in important ways to learning, but incapable of perceiving objects in an adult-like fashion. Developmental processes that lead to mature perception and interpretation of the visual world as coherent, stable, and predictable are an area of active investigation and are only beginning to be understood.

14.1 CLASSIC THEORETICAL ACCOUNTS

Discussions of the nature versus nurture of cognitive development are entrenched and persistent. Such discussions are particularly vigorous when concerning infant cognition and have tended to be long on rhetoric but short on evidence, in part because the evidence has been, until recently, relatively sparse. Research on visual development, in contrast, has tended to focus on developmental changes in neural mechanisms, with much of the evidence coming from animal models (Kiorpes and Movshon, 2004; Teller and Movhson, 1986). Research on human infants' visual development has often been motivated by two theoretical accounts, each of which considers seriously both the starting point for postnatal development and the mechanisms of change that yield stable, mature object perception: Piagetian theory and Gestalt theory.

14.1.1 Piagetian Theory

The first systematic study of infants' perception and knowledge of objects was conducted by Jean Piaget in the 1920s and 1930s (Piaget, 1952/1936, 1954/1937). According to Piaget, knowledge of objects and space developed in parallel, and were interdependent: One cannot perceive or act on objects accurately without awareness of their position in space relative to other objects and to the observer. Knowledge of the self and of external objects as distinct, coherent, and permanent entities grew from active manual search, initiated by the child. When the child experiences her own movements, she comes to understand them as movements of objects through space and applies the same knowledge to movements of other objects.

Initially, prior to any manual action experience, infants understand the world as a 'sensory tableaux' in which images shift unpredictably and lack permanence or substance; in an important sense, the world of objects that we take for granted does not yet exist. Active search behavior emerges only after 4 months and marks the beginnings of 'true' object knowledge. Over the next few months, infants reveal this knowledge, for example, by following the trajectory of thrown or dropped objects, and by retrieval of a desired object from under a cover where it had been seen previously. Later in infancy, infants are able to search accurately for objects even when there are multiple potential hiding places, marking the advent of full 'object permanence.'

Piaget placed more emphasis on the importance of manual search for developmental changes in object perception than visual skills, yet the lessons from his theory for questions of development of the visual system could not be more relevant. Upon the infant's first exposure to patterned visual input, he does not inhabit a world of objects, but rather a world of disconnected images devoid of depth, coherence, and permanence. Building coherent things from these disconnected images comes from action and experience with objects over time.

14.1.2 Gestalt Theory

Piagetian theory can be contrasted with a coeval, competing account. The Gestalt psychologists, unlike Piaget, were not strictly developmentalists, but they did have much to say about how visual experience might be structured in the immature visual system. They suggested that subjective experience corresponds to the simplest and most regular interpretation of a particular visual array in accord with a general 'minimum principle,' or Prägnanz (Koffka, 1935). The relatively basic shapes of most objects are more coherent, regular, and simple than disconnected and disorganized forms. The minimum principle and Prägnanz were thought to be rooted in the tendency of neural activity toward minimum work and minimum energy, which impel the visual system toward simplicity (Koffka, 1935).

The minimum principle is a predisposition inherent in the visual system, and so it follows that young infants should experience the visual environment as do adults. In one of the few sections of Gestalt writings to focus on development, a 'primitive mentality' was attributed to the human infant (Koffka, 1959/1928; Köhler, 1947), implying that one's perceptual experience is never one of disorganized chaos, no matter what one's position in the lifespan. Hebb (1949) noted, in addition, the neonate's electroencephalogram was organized and somewhat predictable, perhaps reflecting organized sensory systems at birth and serving as a stable foundation for subsequent perceptual development. Gibson (1950) suggested that visual experience begins with 'embryonic meanings,' a position echoed by Zuckerman and Rock (1957), who argued that an organized world could not arise from experience in the form of memory for previously encountered scenes and objects, because experience cannot operate in an organized fashion over inherently disorganized inputs. Necessarily, therefore, the starting point of visual organization is inherently organized. Like Piaget, Gestalt psychologists proposed

that development of object perception *per se* involved active manual exploration, which imparts additional information about specific object kinds (Koffka, 1959), but the starting point for visual experience is necessarily quite different on the two accounts. On the Gestalt view, perceptual organization precedes object knowledge; on the Piagetian view, object knowledge and perceptual organization develop in tandem.

Piagetian and Gestalt accounts specify a starting point for postnatal development, and each has particular views about how development of the infant's visual world might proceed. Neither account is wholly on one side of the nature–nurture issue, and both accounts have offered testable predictions that have guided subsequent research, and as will be seen later in this chapter, both accounts have influenced important research on object perception in infants. Yet neither can be taken as complete, in part because neither took a sufficiently comprehensive approach to vision. A quote from Gibson (1979) helps explain why this is so: The visual system comprises "the eyes in the head on a body supported by the ground, the brain being only the central organ of a complete visual system. When no constraints are put on the visual system, we look around, walk up to something interesting and move around it so as to see it from all sides, and go from one vista to another" (p. 1). Vision is not passive, even in infancy; at no point in development are infants simply inactive recipients of visual stimulation. Instead, they are active perceivers, and active participants in their own development, from the beginning of postnatal life (von Hofsten, 2004). Young infants do not have all the action systems implied by Gibson's quote at their disposal, but eye movements are a notable exception, and as will be seen in a later section, there are strong reasons to suspect a critical role for oculomotor behavior in cognitive development.

14.2 PRENATAL DEVELOPMENT OF THE VISUAL SYSTEM

The mammalian visual system, like other sensory and cortical systems, begins to take shape early in prenatal development. For example, in humans, the retina starts to form around 40 days postconception and is thought to have a relatively complete set of cells by 160 days, though the growth of individual cells and their lattice-like organization characteristic of mature structure continue to develop well past birth (Finlay et al., 2003). The distinction between foveal and extrafoveal regions (viz., what will become thalamus and cortex) is present early; like the retina, the topology and patterning of receptors and neurons continue to change throughout prenatal development and the first year after birth. Foveal receptors are overrepresented in the cortical visual system, and detailed information about different parts of the scene is enabled by moving the eyes to different points (see Section 14.4.3). The musculature responsible for eye movements develops before birth in humans, as do subcortical systems (e.g., superior colliculus and brainstem) to control these muscles (Johnson, 2001; Prechtl, 2001).

Many developmental mechanisms are common across mammalian species, including humans, though the timing of developmental events varies (Clancy et al., 2000; Finlay and Darlington, 1995). Data from humans are sparse, but the few cases where deceased embryos and fetuses are available demonstrate that many major structures (neurons, areas, and layers) in visual cortical and subcortical areas are in place by the end of the second trimester *in utero* (e.g., Zilles et al., 1986). Later developments consist of the physical growth of neurons and the proliferation and pruning of synapses, which is, in part, activity dependent (Greenough et al., 1987; Huttenlocher et al., 1986).

14.2.1 Development of Structure in the Visual System

The visual system consists of a richly interconnected yet functionally segregated network of areas specializing in processing different aspects of visual scenes and visually guided behavior: contours, motion, luminance, color, objects, faces, approach versus avoidance, and so forth. Areal patterns are present in a rudimentary form during the first trimester, but the final forms continue to take shape well after birth; like synaptic pruning, developmental processes are partly the result of experience. Some kinds of experience are intrinsic to the visual system, as opposed to outside stimulation. Spontaneous prenatal activity in visual pathways contributes to retinotopic mapping (Sperry, 1963) and the preservation of sensory structure, beginning in the retina and extending through the thalamus, primary visual cortex, and higher visual areas. Waves of coordinated, spontaneous firing of retinal cells have been observed in chicks and ferrets (Wong, 1999). Waves travel across the retinal surface and are then systematically propagated through to the higher areas. This might be one way by which correlated inputs remain coupled and dissimilar inputs become dissociated, prior to exposure to light.

As soon as neurons are formed, find their place in cortex, and grow, they begin to connect to other neurons. There is a surge in synaptogenesis in visual areas around the time of birth and then a more protracted period in which synapses are eliminated, reaching adult-like levels at puberty (Bourgeois et al., 2000). This process is activity dependent: Synapses are preserved in active cortical circuits and lost in inactive circuits. Auditory

cortex, in contrast, experiences a synaptogenesis surge several months earlier, which corresponds to its earlier functionality relative to visual cortex (viz., prenatally). Here, too, pruning of synapses extends across the next several years. (In other cortical areas, such as frontal cortex, there is a more gradual accrual of synapses without extensive pruning.) For the visual system, the addition and elimination of synapses, the onset of which coincides with the start of visual experience, provides an important mechanism by which the cortex tunes itself to environmental demands and the structure of sensory input.

14.3 VISUAL PERCEPTION IN THE NEWBORN

Human infants are born with a functional visual system. The eye of the newborn is sensitive to light, and if motivated (i.e., awake and alert), the baby may react to visual stimulation with head and eye movements. Vision is relatively poor, however: acuity (detection of fine detail), contrast sensitivity (detection of different shades of luminance), color sensitivity, and sensitivity to direction of motion all undergo improvements after birth (Banks and Salapatek, 1983). The field of view is also relatively small, so that newborns often fail to detect targets too far distant or too far in the periphery. In addition, as far as we know, neonates lack stereopsis, perception of depth from binocular disparity (differences in the input to the two eyes). Maturation of the eye and cortical structures (see previous section) supports developments in these visual functions, and learning plays an important role as well, as discussed in greater detail in Section 14.5.

14.3.1 Visual Organization at Birth

Testing newborn infants is not for the faint of heart. Success is entirely dependent on the baby's mood; this is at its most capricious early in postnatal life, and there is no predicting neonate behavior. Having said this, a number of patient scientists have conducted careful experiments with neonates; these experiments have revealed that despite relatively poor vision, neonates actively scan the visual environment. Early studies, summarized by Haith (1980), revealed systematic oculomotor behaviors that provided clear evidence of visual organization at birth. Newborns, for example, will search for patterned visual stimulation, tending to scan broadly until encountering an edge, at which point scanning narrows so that the edge can be explored. Such behaviors are clearly adaptive for investigating and learning about the visual world.

In addition, newborn infants show consistent visual preferences. Fantz (1961) presented newborns with pairs of pictures and other two-dimensional (2D) patterns and recorded member of the pair which attracted the infant's visual attention, which he scored as proportion of fixation times per exposure. Infants typically showed longer looking at one member of the pair: bull's-eyes versus stripes, or checkerboards versus solid forms, for example. Visual preferences have served as a method of choice ever since, in older infants as well as neonates. Slater (1995) described a number of newborns' preferences: patterned versus unpatterned stimuli, curvature versus rectilinear patterns, moving versus static patterns, 2D versus 3D forms, and high- versus low-contrast patterns, among others. In addition, perhaps due to the relatively poor visual acuity of the newborn visual system, there is a preference for 'global' form versus 'local' detail in newborns (Macchi Cassia et al., 2002).

14.3.2 Visual Behaviors at Birth

Fantz (1964) reported that repeated exposure to a single stimulus led to a decline of visual attention, and increased attention to a new stimulus, in 2- to 6-month-olds. A substantial number of subsequent investigations examined infants' preferences for familiar and novel stimuli as a function of increasing exposure, and these in turn led to standardized methods for testing infant perception and cognition, such as habituation paradigms (Cohen, 1976), as well as a deeper understanding of infants' information processing (Aslin, 2007; Hunter and Ames, 1989; Sirois and Mareschal, 2002).

Neonates (and older infants) will habituate to repeated presentations of a single stimulus; habituation is operationalized as a decrement of visual attention across multiple exposures according to a predetermined criterion. Following habituation, infants generally show preferences for novel versus familiar stimuli, implying both discrimination of novel and familiar stimuli and memory for the stimulus shown during habituation. Neonates and older infants also recognize visual constancies or invariants, the identification of common features of a stimulus across some transformation, for instance, shape, size, slant, and form (Slater et al., 1983). Recognition of invariants forms the basis for categorization.

14.3.3 Faces and Objects

Newborns prefer faces and face-like forms relative to other visual stimuli and are thus well prepared to begin engaging in social interactions with conspecifics. Some have speculated that there is an innate representation

Stimuli	Total fixation time	Number of discrete looks
	53.86 s vs. 37.62 s $p < 0.03$	10 vs. 8.09 $p < 0.05$
	34.70 s vs. 41.08 s $p > 0.20$	7.6 vs. 8.3 $p > 0.30$
	44.15 s vs. 22.89 s $p < 0.003$	10.43 vs. 6.5 $p < 0.01$

FIGURE 14.2 Face-like stimuli from experiments on neonates' preferences. *Reproduced from Turati C, Simion F, Milani I, and Umiltà C (2002) Newborns' preference for faces: What is crucial? Developmental Psychology 38: 875–882.*

for facial structure (Morton and Johnson, 1991); others have suggested that the preference stems from general-purpose visual biases that guide attention toward stimuli of a particular spatial frequency, with a prevalence of stimulus elements in the top portion, as seen in Figure 14.2 (Turati et al., 2002; Valenza et al., 1996).

Newborns' object perception is not so precocious. Neonates perceive segregation of figure and ground (i.e., seeing objects as distinct from backgrounds), but there are limits in the ability to perceive object occlusion, as seen in Figure 14.3(a). Adults and older infants perceive this display as consisting of two objects, one moving back and forth behind the other (Kellman and Spelke, 1983). Neonates, however, seem to perceive this display as consisting of three disconnected parts (Slater et al., 1990). In these experiments, infants were habituated with the partly occluded rod display, followed by two test displays. One test display (Figure 14.3(b)) consisted of the whole rod (no occluder), and the other consisted of two rod parts, separated by a gap in the space where the occluder was seen, corresponding to the visible rod portions in the habituation stimulus (Figure 14.3(c)). For 4-month-olds, longer looking at the broken rod is taken as evidence that they perceived unity of the rod parts as unified behind the box, but for neonates longer looking at the complete rod implies perception of disjoint surfaces in similar displays. The developmental processes underlying this shift in perceptual abilities are discussed in Section 14.5.1. See also Chapters 18 and 19.

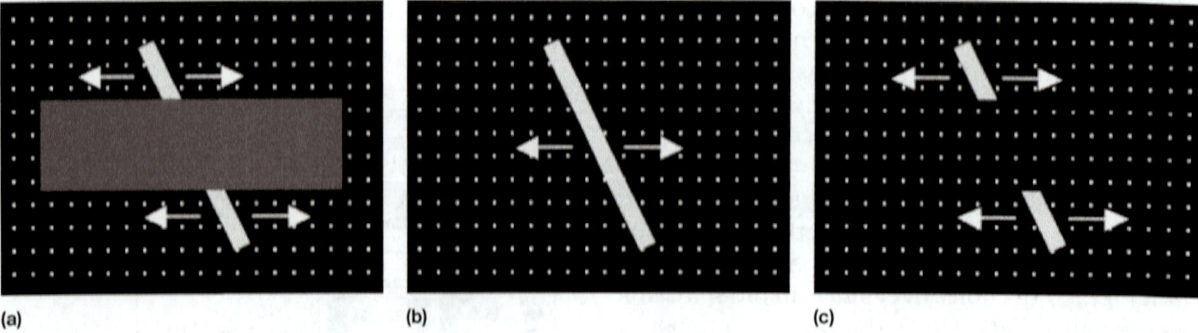

FIGURE 14.3   Rod-and-box displays from experiments on infants' perception of partly occluded objects: (a) habituation stimulus and (b and c) test stimuli.

14.4 POSTNATAL VISUAL DEVELOPMENT

As noted in Section 14.2, visual development begins prenatally; in this section, some of the ways in which it continues after birth are described. Both infants and adults scan visual scenes actively – on the order of 2–4 eye movements per second in general (Johnson et al., 2004; Melcher and Kowler, 2001) – but visual function is relatively poor at birth in terms of processing and analyzing visual information. Functional visual development has been explained in terms of visual maturation (Atkinson, 2000; Johnson, 1990, 2005). Acuity, for example, improves in infancy with a number of developments, all taking place in parallel: migration of receptor cells in the retina toward the center of the eye, elongation of the receptors to catch more incoming light, growth of the eyeball to augment the resolving power of the lens, myelination of the optic nerve and cortical neurons, and synaptogenesis and pruning. See also Chapter 12.

14.4.1 Visual Physiology

The visual system, like the rest of the brain, is organized modularly and hierarchically. Incoming light is transduced into neural signals by the retina, which passes information to the lateral geniculate nucleus (part of the thalamus), and then to primary visual area (V1) in cortex and higher visual areas. Successively higher visual areas code for visual attributes in larger portions of visual field and participate in more complex visual functions (see Figure 14.4). For example, visual pathways extending from V5 (also known as area MT, or medial temporal) through parietal cortex are largely responsible for coding motion. Infants younger than 2 months appear unable to discriminate different directions of motion until maturation of pathways extending to and originating in V5 (Johnson, 1990). For motion

processing, therefore, development centers on a limited number of visual areas and a relatively small number of mechanisms (e.g., myelination, synaptic growth, and pruning). Object perception, in contrast, is far more complex, involving many areas, each of which is responsible for processing one or more of the many visual attributes that defines edges, surfaces, and objects.

14.4.2 Critical Periods

A critical period refers to a time in an individual's ontogeny when some function or ability must be stimulated or it will be lost permanently (see Daw, 1995). This notion can be contrasted with a sensitive period, similar in concept but generally referring to scenarios in which effects of deprivation are not so severe. The formal study of critical periods was initiated by Wiesel and Hubel (1963), who covered or sutured one eye in kittens from birth for a period of 1–4 months and examined the effects of visual deprivation by patching the unaffected eye and observing visual function of the affected eye alone. The deprived eye was effectively blind, as revealed by both behavioral and neural effects. Behavioral effects included an inability to navigate visually or respond to objects introduced by the experimenters, though the animals behaved normally under these circumstances when permitted to use the unaffected eye. Neural effects were examined by recording from single cells in visual cortex; in general, few cortical cells could be driven by the deprived eye in cortical regions normally responsive to input from both eyes, such as the postlateral gyrus. Wiesel and Hubel also reported the effects of eye closure in animals that were allowed some visual experience prior to deprivation, highlighting the distinction between critical and sensitive periods. The unaffected eye dominated activity of cells in the visual cortex but depended on both the extent of visual experience prior to deprivation and the duration of deprivation.

Stereopsis, the detection of distance differences in near space (e.g., threading a needle), seems to emerge

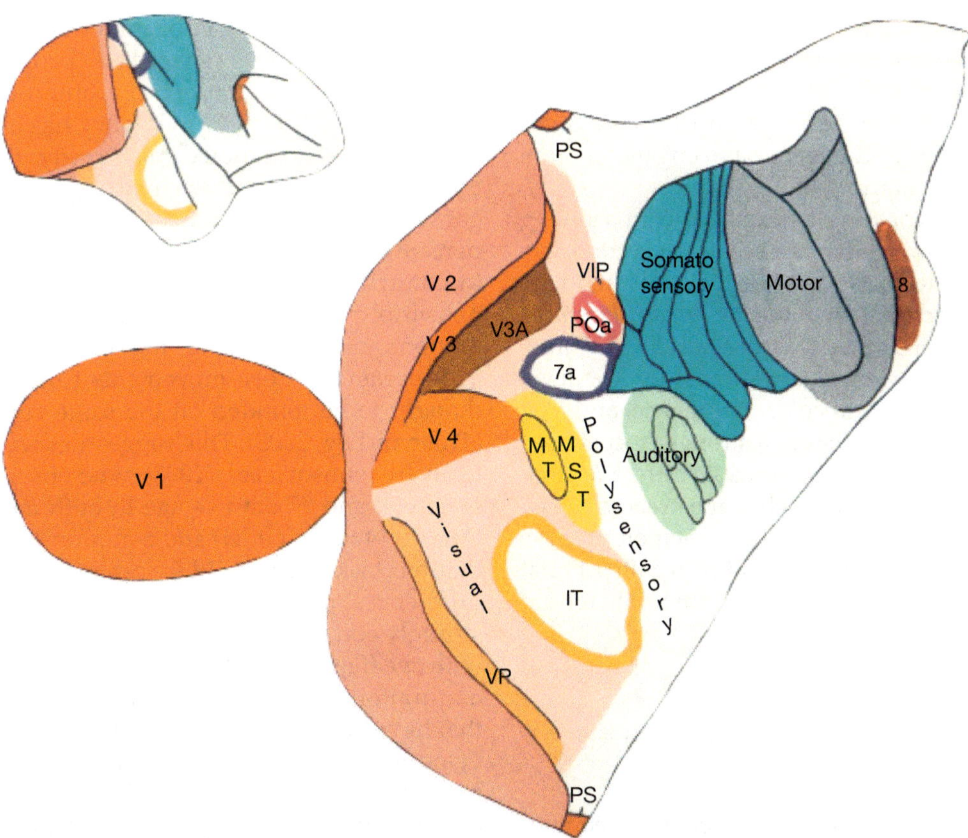

FIGURE 14.4 Cortical areas in the macaque monkey showing an outer view of the left hemisphere (upper left) and a flattened representation of sensory and motor regions. Visual areas are depicted in red, orange, and yellow. PS, prestriate area; VIP, ventral intraparietal area; MST, medial superior temporal area; POa, parieto-occipital area; IT, inferotemporal area; VP, ventral posterior area. *Reproduced from Van Essen DC and Maunsell JHR (1983) Hierarchical organization and the functional streams in the visual cortex.* Trends in Neurosciences 6: 370–375.

during a critical period. Stereopsis relies on slight differences in the inputs to the two eyes when they are directed to the same point, also known as disparity. Cells in the primary visual cortex are organized into 'ocular dominance' columns that receive inputs from the two eyes and register the amount of disparity between them. These require binocular function early in life – the two eyes must be directed consistently at the same points and focus on them. This can be disrupted by amblyopia (poor vision in one eye) or strabismus (misalignment of the eyes). Normally, mature visual cortex contains cells responsive to both eyes, and a few to only one eye. Abnormal visual experience can produce a preponderance of cells responsive to only one or the other eye, but not to both. In typically developing infants, stereopsis emerges at about 4 months, as inputs from the two eyes into the ocular dominance columns become segregated (Held, 1985). (Prior to this time, the inputs are more likely to be superimposed, which may result in frequent diplopia, or double vision, early in life.) The critical period for development of stereopsis in humans is estimated to be 1–3 years (Banks et al., 1975).

14.4.3 Development of Visual Attention

Visual attention – eye movements – is a combination of saccades and fixations. During a saccade, the point of movements sweeps rapidly across the scene, and during a fixation, the point of gaze is stationary. Analysis of the scene is performed during fixations. Eye movements can also be smooth rather than saccadic, as when the head translates or rotates as the point of gaze remains stabilized on a single point in space (the eyes move to compensate for head movement), or when following a moving target.

Visual attention in infancy has attracted a great deal of interest, because it is a behavior that is relatively mature, even at birth, and because it is relatively easy to observe (Johnson, 2005; Richards, 1998). Oculomotor behaviors that have been examined include detection of targets in the periphery, saccade planning, oculomotor anticipations, sustained versus transient attention, effects of spatial cuing, and eye/head movement integration; other tasks examined inhibition of eye movements, such as disengagement of attention, inhibition of return, and spatial negative priming. Bronson (1990, 1994) examined scanning patterns as infants viewed simple geometric forms, and reported changes with development in attention to distributed visual features, including a greater tendency to scan between features, to direct saccades with greater accuracy, and in general to engage in more 'volitional' scanning, starting at 2–3 months.

There are important developments also in viewing complex scenes. In my lab, we recently recorded eye movements of infants and adults as they watched segments of an animated cartoon, *A Charlie Brown Christmas*, that was rich in social content (Frank et al., 2009). Three-month-olds' attention was captured most by low-level image salience (variations in color, luminance, and motion), and by 9 months, there was a stronger focusing of attention on faces. There were no reliable differences between age groups in measures such as mean saccade distance and fixation duration. One interpretation of these results is a developmental transition toward attentional capture by semantic content – the 'meaning' inherent in social stimuli. See also Chapter 22.

14.4.4 Cortical Maturation and Oculomotor Development

Gaze control in mature primates is accomplished with a coordinated system of both subcortical and cortical brain areas, as seen in Figure 14.5. Control of eye movements originates in areas with outputs that are connected to the brainstem, which sends signals to the oculomotor musculature. Development of visual attention has often been interpreted as revealing development of cortical systems that control it. Visual attention has been suggested to be largely under subcortical control until the first few months after birth, after which there is increasing cortical control (Atkinson, 1984; Colombo, 2001; Johnson, 1990).

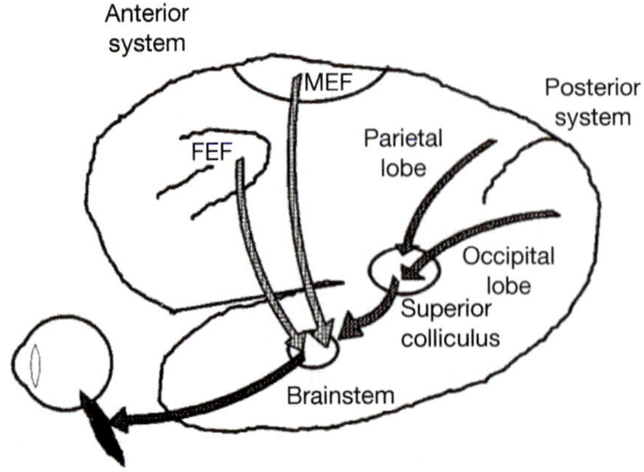

FIGURE 14.5 Subcortical and cortical structures involved in oculomotor control. FEF, frontal eye fields; MEF, medial eye fields. *Reproduced from Schiller and Tehovnik (2001).*

For example, oculomotor smooth pursuit and perception of motion direction have been proposed to rely on a common cortical region, area V5, and their development of these visual functions in infancy has been tied to maturation of V5, as noted previously (Johnson, 1990). Smooth pursuit is maintenance of gaze on a moving target with nonsaccadic, eye movements; motion direction perception is often tested with random-dot displays to control for the possibility that motion following is not simply a detection of change in position. Perceiving motion and performing the computations involved in programming eye movements to follow motion are thought to be founded on the same cortical structures (Thier and Ilg, 2005). This suggestion was tested empirically by Johnson et al. (2008), who observed infants between 58 and 97 days of age in both a smooth pursuit (Figure 14.6, top panel) and a motion direction discrimination task (Figure 14.6, center panel). Individual differences in performance on the two tasks were strongly correlated and were also positively correlated with age (Figure 14.6, bottom panel), consistent with the maturation theory. Other visual functions in infancy that have been linked to cortical maturation include development of form and motion perception, stemming from maturation of parvocellular and magnocellular processing streams, respectively (Atkinson, 2000), and development of visual memory for object features and object locations, stemming from maturation of ventral and dorsal processing streams (Mareschal and Johnson, 2003).

14.4.5 Development of Visual Memory

Memory for events, object features, and locations improves over the first several postnatal months (Rose et al., 2004). As noted previously, newborns will habituate to repeated presentations of a visual stimulus and recover interest to a novel one, clear evidence for a functional short-term visual memory store available at birth. Visual short-term memory in older infants has been examined with a 'change-detection' task in which infants viewed a pair of displays side by side, each of which contained one or more shapes. On one side, the object or objects underwent color changes every 250 ms (Ross-Sheehy et al., 2003). When there was one object per side, 4-month-olds looked longer toward the side with color changes, implying a short-term store of the color information across the 250-ms temporal gap. Visual short-term memory develops rapidly: 10-month-olds retained color information across a set size of four, all different colors (Ross-Sheehy et al., 2003), and 7.5-month-olds retained information about color-location combinations (set size of three) across a 300-ms delay (Oakes et al., 2006).

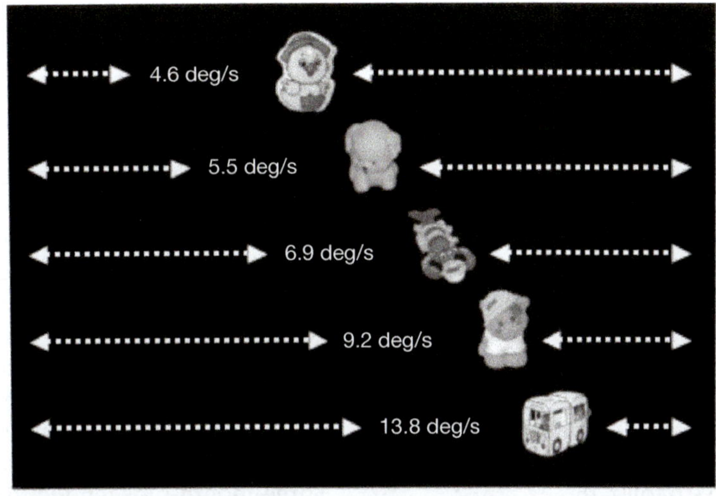

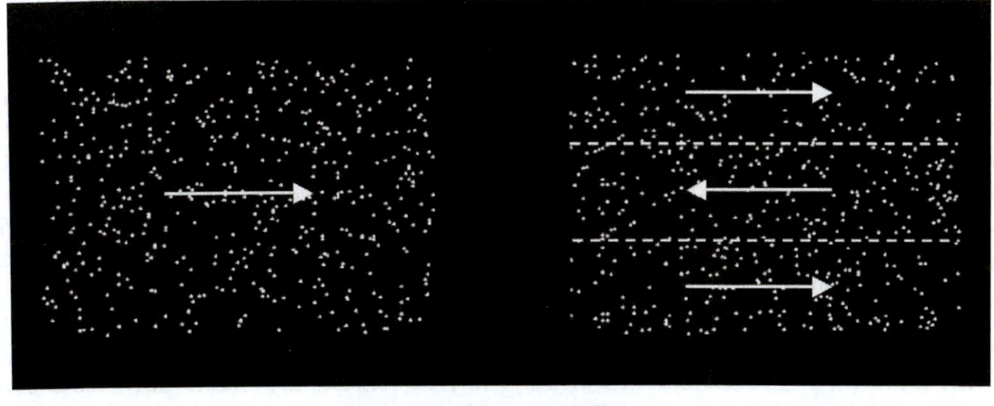

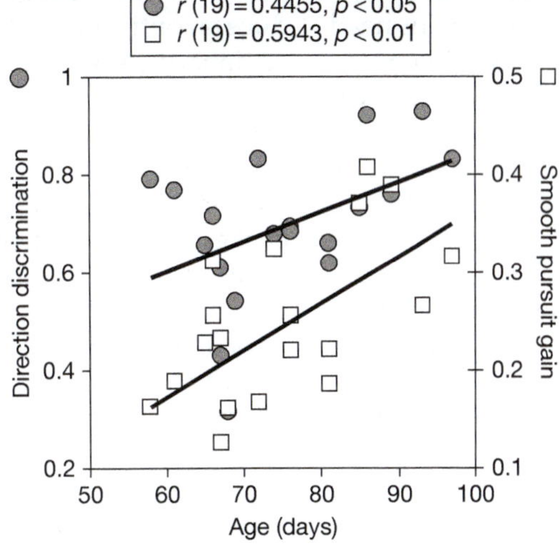

FIGURE 14.6 Top: schematic depiction of stimuli used to examine smooth pursuit in young infants. A toy moved laterally at one of five speeds in one of five vertical positions on the screen. Only one toy was shown at a time. Center: Random-dot kinematograms used to examine motion direction discrimination in young infants. Dotted lines and dots, shown here to demarcate regions of motion, were not present in the stimulus. Bottom: Individual infant's performance in smooth pursuit and direction discrimination tasks were correlated with age. *Adapted from Johnson SP, Davidow J, Hall-Haro C, and Frank MC (2008) Development of perceptual completion originates in information acquisition.* Developmental Psychology 44: 1214–1224.

Studies of infant memory employing operant conditioning paradigms, in which infants are trained to kick their legs to move a mobile, have demonstrated long-term visual recognition stores that are available from at least 2 months under some conditions; the memories formed can last for several days or even weeks given sufficient training with the mobile and reminders (Rovee-Collier, 1999). Infants at 6 months can imitate observed behaviors after 24 h, and the retention interval is considerably longer in older infants (Barr et al., 1996). Developments in visual memory, like many other visual functions, have been proposed to stem from cortical development, in particular areas of the medial temporal lobe such as hippocampus, perirhinal and entorhinal cortices, and amygdala (Bauer, 2004; Nelson, 1995; Rose et al., 2004). See also Chapter 16.

14.4.6 Development of Visual Stability

Our gaze moves frequently from point to point in the visual scene, and our bodies move from place to place. Despite these continual disruptions and interruptions in visual input, we experience the visual world as an inherently stable place. Consider, for example, the difference in your visual experience when you read this page while shaking your head back and forth (as if you wanted to signify 'no' to someone). Reading is not much compromised. Now shake the page back and forth while holding your head steady. You will discover reading to be more difficult, yet the spatial relation between your head and the page in the two situations is similar. When you rotate your head, compensatory eye movements known as the vestibulo-ocular response (VOR) allow the point of gaze to remain fixed or to continue moving volitionally as desired (as when reading). When the page moves, there is no such compensatory mechanism.

Evidence from three paradigms suggests that visual stability emerges gradually across the first year after birth. First, young infants have difficulty discriminating optic flow patterns that simulate different directions of self-motion (Gilmore et al., 2004). Infants viewed a pair of random-dot displays in which the dots repeatedly expanded and contracted around a central point to simulate the effect of moving forward and backward under real-world conditions. On one side, the location of this point shifted periodically, which for adults specifies a change in heading direction; the location on the other side remained stationary. Under these circumstances, adults detected a shift simulating a 5° change in heading, but infants were insensitive to all shifts below 22°, and sensitivity was unchanged between 3 and 6 months. Gilmore et al. speculated that optic flow sensitivity may be improved by self-produced locomotion after 6 months of age, or by maturation of the ventral visual stream.

Second, young infants' saccade patterns tend to be retinocentric, rather than body centered, in a 'double-step' tracking paradigm (Gilmore and Johnson, 1997). Retinocentric saccades are programmed without taking into account previous eye movements. Body-centered eye movements, in contrast, are programmed while updating the spatial frame of reference or coordinate system in which the behaviors occur. Infants first viewed a fixation point that then disappeared, followed in succession by the appearance and extinguishing of two targets on either side of the display. The fixation point was located at the top center of the display, and targets were located below it at the extreme left and right sides. As the infant viewed the fixation point and targets in sequence, there was an age-related transition in saccade patterns. Three-month-olds tended to direct their gaze downward from the first target, as if directed toward a target below the current point of gaze. In reality, the second target was below the first location – the original fixation point – not the current point of gaze. Seven-month-old infants, in contrast, were more likely to direct gaze directly toward the second target. These findings imply that young infants' visual–spatial coordinate system, necessary to support perception of a stable visual world, may be insensitive to extraretinal information, such as eye and head position, in planning eye movements.

Third, there are limits in the ability of infants younger than 2 months to switch attention flexibly and volitionally to consistently maintain a stable gaze. Movement of one's body through the visual environment can produce an optic flow pattern, as can head movement while stationary (recall the head-shaking example). The two scenarios may produce similar visual inputs from optic flow, yet we readily distinguish between them. In addition, adult observers can generally direct attention to either moving or stationary targets, nearby or in the background, as desired. These are key features of visual stability, and four eye movement systems work in concert to produce it. Optokinetic nystagmus (OKN) stabilizes the visual field on the retina as the observer moves through the environment. OKN is triggered by a large moving field, as when gazing out the window of a train: The eyes catch a feature, follow it with a smooth movement, and saccade in the opposite direction to catch another feature, repeating the cycle. The VOR, described previously, helps maintain a stable gaze to compensate for head movement. (OKN and the VOR are present and functional at birth, largely reflexive or obligatory, and are likely mediated by subcortical pathways; Atkinson and Braddick, 1981; Preston and Finocchio, 1983.) The others are the saccadic eye movement system and smooth pursuit, to compensate for or cancel the VOR or OKN as appropriate. Aslin and Johnson (1994) observed suppression (cancellation) of the VOR to fixate a small moving target in 2- and

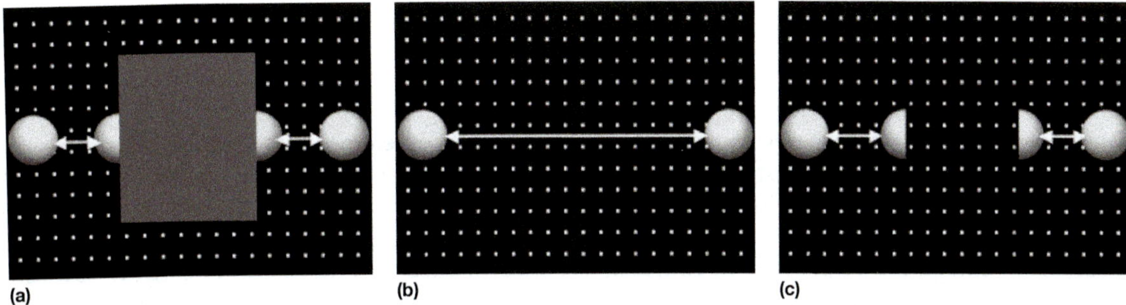

FIGURE 14.7 Ball-and-box displays from experiments on infants' perception of existence constancy: (a) habituation stimulus and (b and c) test stimuli.

4-month-olds, but not 1-month-olds, and Aslin and Johnson (1996) observed suppression of OKN to fixate a stationary target in 2-month-olds, but not in a younger group.

14.4.7 Object Perception

As noted previously in Section 14.3.3, 'piecemeal' or fragmented perception of the visual environment extends from birth through the first several months afterward under some conditions, implying a fundamental shift in the infant's perceptual experience. Because neonates and 4-month-olds appear to construe rod-and-box displays differently – as disjoint surfaces and as occluded objects, respectively – an important step in understanding development of perceptual completion is investigations of performance in 2-month-olds. In the first such investigation, 2-month-olds were found to show an 'intermediate' pattern of performance (no reliable posthabituation preference), consistent with the possibility that spatial completion is developing at this point but not yet in final form (Johnson and Náñez, 1995). A follow-up study examined the hypothesis that 2-month-olds may perceive unity if given additional perceptual support. We simply increased the amount of visible rod surface revealed behind the occluder by reducing box height and by adding gaps in it, and under these conditions, 2-month-olds provided evidence of unity perception (Johnson and Aslin, 1995). Adopting this approach with newborns, however, failed to reveal similar evidence: Even in 'enhanced' displays, newborns seemed to perceive disjoint rather than unified rod parts (Slater et al., 1994, 1996).

A number of studies have shown that young infants can maintain representations of the solidity and location of fully hidden objects across brief delays (e.g., Aguiar and Baillargeon, 1999; Spelke et al., 1992). Yet newborns provide little evidence of perceiving partly occluded objects, begging the question of how perception of complete occlusion, or existence constancy, emerges during the first few months after birth. To address this question, experiments have examined infants' responses to objects that move forward on a trajectory, disappear behind an occluder, reappear on the far side, and reverse direction, repeating the cycle (Figure 14.7(a)). Following habituation to this display, infants viewed test displays consisting of continuous and discontinuous trajectories (Figures 14.7(b) and 14.7(c)), analogous to the broken and complete test stimuli described previously. Four-month-olds appeared to treat the ball-and-box display depicted in Figure 14.7(a) as consisting of two disconnected trajectories, rather than a single, partly hidden path (Johnson et al., 2003a,b), but by 6 months, infants perceived this trajectory as unitary. When occluder size was reduced, however, 4-month-olds' posthabituation preferences (and thus, by inference, their percepts of spatiotemporal completion) were shifted, partway by an intermediate width, and fully by a narrow width, so narrow as to be only slightly larger than the ball itself. Reducing the spatial gap, therefore, supported perception of a complete trajectory in 4-month-olds. In 2-month-olds, this manipulation appeared to have no effect, implying a lower age limit for trajectory completion in infants, just as there may be for spatial completion.

14.4.8 Face Perception

As noted in Section 14.3.3, infants are better prepared to perceive faces than objects at birth, showing preferences for faces and face-like structures. Research on face perception in infants provides additional insights on mechanisms of recognition. In adults, face recognition is near ceiling when faces are upright, but when faces are inverted, performance is relatively poor – the inversion effect (Yin, 1969). This appears to be specific to faces; other visual configurations normally seen upright, such as houses, are not vulnerable to the effect. These findings are thought to reflect a difference in processing 'strategies' when viewing upright versus inverted faces. When faces are upright, they are processed in terms of both the individual features and the spatial relations among features (viz., both piecemeal and holistic processing), but when inverted, these relations are more difficult to access, forcing greater reliance on only a single source of information for recognition – the features – and thus impairing performance.

Carey and Diamond (1977) reported that children younger than 10 years of age do not show the inversion effect. This led to the suggestion that young children process faces according to features only, and that piecemeal-to-holistic processing develops during childhood, perhaps from experience viewing faces or maturation of the right cerebral hemisphere implicated in complex visual–spatial tasks. Consistent with these findings, children's discrimination of faces was impaired more by a mismatch in the spacing of features than by a mismatch in the features themselves (eyes, nose, and mouth) or faces' outer contours, as seen in Figure 14.8 (Mondloch et al., 2002), and there were dramatic improvements in performance from 6 years through adulthood in matching identity of faces across changes in facial expression,

orientation, and 'lip reading' (mouthing different vowels), all of which require sensitivity to spatial relations among features (Mondloch et al., 2003).

Other reports, however, provide evidence for a much earlier piecemeal-to-holistic shift in processing faces. First, Younger (1992) found that 10-month-old infants were sensitive to correlations among facial attributes in a face discrimination task; 7-month-olds provided evidence of discrimination from featural variations only. Second, evidence from a 'switch' paradigm showed that 7-month-olds processed configurations of facial features that were disrupted by inversion (Cohen and Cashon, 2001). In the switch design, infants are habituated to a pair of distinct stimuli (in this case, faces); at the test, selected features are switched from one stimulus to the other, and infants are observed for recovery of interest to the new configuration. A more recent study using this

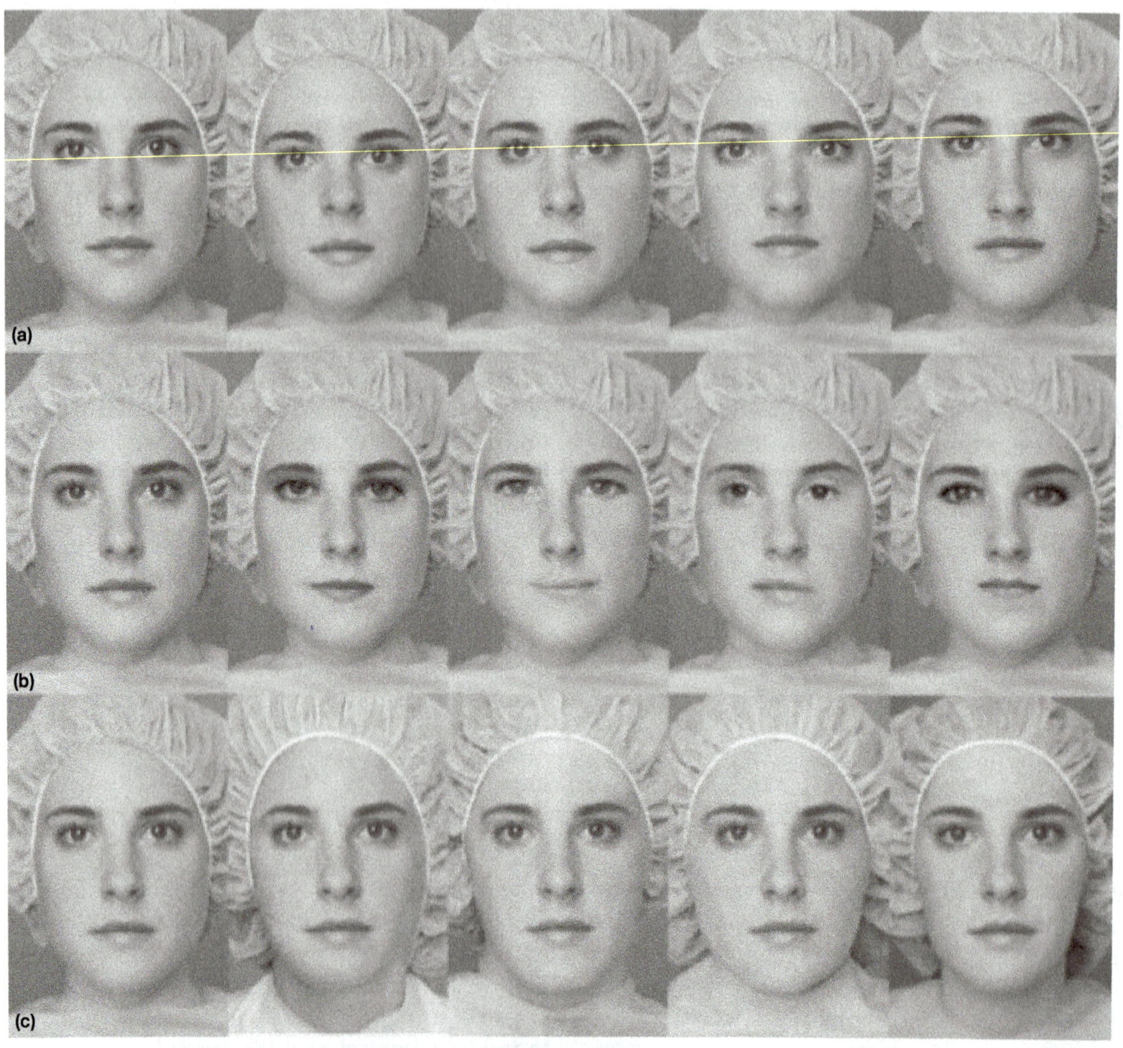

FIGURE 14.8   Stimuli used to test recognition of faces in which the spacing of features is varied (top row), the features (but not their spacing) are varied (center row), or the outer contours (but not features or spacing) are varied (bottom row). The faces in the leftmost positions of each row are identical; other faces in each row are variations of it. *Reproduced from Mondloch CJ, Le Grand R, Maurer D (2002) Configural face processing develops more slowly than featural face processing.* Perception 31: 553–566.

design found a developmental progression toward processing configurations between 4 and 10 months (Schwarzer et al., 2007). Third, the inversion effect was found in face recognition tasks with 5-, 7-, and 9-month-olds, but when outer contours and inner facial features were inverted in separate experiments, only the 2 older age groups showed impairment from inversion, suggesting a greater flexibility in their processing – utilizing either internal or external features to recognize the faces (Rose et al., 2008).

14.4.9 Critical Period for Development of Holistic Perception

Evidence for a critical period for holistic face perception comes from a study of individuals born with cataracts who underwent surgery to correct the problem (Le Grand et al., 2001). Each individual had at least 9 years of visual experience after surgery. The individuals were tested with face recognition tasks as described in the previous section, including tests of inversion effects, using some of the stimuli shown in Figure 14.8. There was a specific deficit in recognition from configurational information – the spacing of features – but not from featural information, where performance was not reliably different from controls. A particularly striking characteristic of these findings concerns the timing of cataract replacement, which for every patient was less than 7 months of age – and in a few cases, as little as 2–3 months. The critical period for development of holistic processing, therefore, appears to be exceedingly brief. Interestingly, infants at 2–3 months show no signs of the inversion effect (Cashon and Cohen, 2003), and sensitivity to some kinds of holistic information in faces is not adultlike until several years after this time, as noted previously.

Some kinds of holistic object perception appear to be comprised by visual deprivation, but the evidence is complex. On the one hand, patients treated for cataracts showed no deficits, relative to controls, in identifying pictures of houses on the basis of both featural and configurational information, in contrast to face recognition (Robbins et al., 2008). And a case study of SRD, a woman who had cataracts removed at age 12, revealed few obvious deficits in object perception when tested 22 years later on shape matching, visual memory, and image segmentation tasks (Ostrovsky et al., 2006). Her performance at face recognition was impaired relative to controls, as expected from the Le Grand et al. (2001) study, but she was not tested explicitly for holistic object perception.

On the other hand, a case study of MM, a man who lost his vision at 3.5 years and had cataract replacement nearly 40 years later, revealed marked deficits in object perception skills (Fine et al., 2003). Five months after surgery, MM was unable to detect transparency in overlapping forms, to see depth from perspective in a Necker cube, or to identify a shape defined by illusory contours (a Kanizsa square) – the latter a paradigmatic instance of holistic processing, the binding of visual features across a spatial gap. He was also limited in recognition of everyday objects and had difficulty discriminating faces and identifying emotional expression, reporting to rely on individual features rather than a 'Gestalt' for these purposes. Cortical areas that give strong responses in normally sighted observers when viewing faces and objects (lingual and fusiform gyri) were largely inactive in MM. (Other visual functions were well preserved, such as contrast sensitivity, color perception, and motion perception, implying that they may have been more established and consequently robust to deprivation by the time MM was blinded in childhood.)

A recent study of illusory contour perception in cataract replacement patients provides additional evidence for severe compromise in feature binding from early visual deprivation (Putzar et al., 2007). Patients were divided into two groups, one with cataract replacement prior to 6 months and the second after this time, and their performance was compared to controls. The patient group treated after 6 months showed elevated reaction times and greater miss rates when searching for illusory shapes among distracters, relative to real shapes; the other groups showed reliably less of a difference on these measures. Interviews conducted after testing revealed that the post-6-month patient group did not perceive the illusory figures at all, but rather adopted a strategy of finding regions in the scenes where the inducing elements pointed inward. Consistent with experiments on face perception described previously, these results point to the first several months after birth as a critical period for spatial integration of visual information.

14.5 HOW INFANTS LEARN ABOUT OBJECTS

14.5.1 Learning from Targeted Visual Exploration

Infants in the transition toward spatial completion in rod-and-box displays – 2–3 months of age – have been observed for evidence that scanning patterns are associated with unity perception. These links are clear. Amso and Johnson (2006) and Johnson et al. (2004) observed 3-month-old infants in a perceptual completion task using the habituation paradigm described previously. Infants' eye movements were recorded with a corneal reflection eye tracker during the habituation phase of the experiment. We found systematic differences in scanning patterns between infants whose posthabituation test display preferences indicated unity perception and infants who provided evidence of perception of disjoint surfaces: 'Perceivers' tended to scan more in the vicinity of the two visible rod segments and to scan back and forth between them. In a younger sample (58–97 days), Johnson et al. (2008) found a reliable correlation between posthabituation preference (viz., our index of spatial completion) and targeted visual exploration, operationalized as the proportion of saccadic eye movements directed toward the moving rod parts, obviously the most relevant aspect of the stimulus for perception of completion. Spatial completion was not predicted by other measures of oculomotor performance, including mean number of fixations per second, mean saccade distance (to assess overall scanning activity), mean vertical position of each infant's fixations (to assess a bias for the upper portion of the stimulus), and mean dispersion of visual attention (to assess scanning of limited portions of the stimulus vs. scanning more broadly). Nor was spatial completion associated with another measure of oculomotor control, smooth pursuit. Rather, spatial completion was best predicted by saccades directed toward the vicinity of the moving rod parts. This can be a challenge for a developing oculomotor system, attested by the fact that targeted scans almost always followed the rod as it moved, rarely anticipating its position.

Targeted visual exploration develops with time, stems from increasing endogeous control of oculomotor behavior, and consists of both selection of desired visual targets and inhibition of everything else in the visual scene. Evidence for development of selection comes from studies of orienting, discussed previously in Section 14.4.3. Evidence for development of its complement, inhibition, is relatively scarce. Newborns exhibit inhibition of return of the point of gaze to recently visited locations (Valenza et al., 1994), but inhibition of eye movements to covertly attended locations develops more slowly across the first year (Amso and Johnson, 2005, 2008). How selection and inhibition work together to maximize effective uptake of visual information is not yet known, but the experiments on spatial completion and eye movements begin to provide important insights. Very young infants' ability to perceive occlusion may be precluded by insufficient access to visual information for unity: alignment, common motion, and other Gestalt cues such as similarity and interposition. An alternate view stressing developmental mechanisms that are independent of learning and experience might posit that emergence of spatial completion stems exclusively from maturation of neural structures responsible for object perception, and, as infants begin to perceive occlusion, their eye movement patterns support or confirm this percept. Amso and Johnson (2006) found that both spatial

completion and scanning patterns were strongly related to performance in an independent visual search task in which targets were selected from among distracters. This finding is inconsistent with the possibility that scanning patterns were tailored specifically to perceptual completion, and instead suggests that a general facility with targeted visual behavior leads to improvements across multiple tasks – precisely the pattern of performance that was observed.

How might developing object perception systems benefit from targeted scans? Eye movements may serve as a vital binding mechanism due to the relatively restricted visual field and poor acuity characteristic of infant vision. Visual information in the periphery is more difficult to access with a single glance, increasing the need to scan between features to ascertain their relations to one another. The developmental timing of targeted visual exploration in infants seems just right for another reason: the critical period for development of holistic object processing. It may be that motor feedback from scanning eye movements serves as a trigger for consolidation of neural circuits in areas that represent the stimulus, enabling association of the separate parts of an object seen on sequential fixations (Rodman, 2003). As an observer views salient object features, the point of gaze falls in rapid succession on components that will later be perceived as part of a coherent whole. Motor feedback signaling a series of sequential fixations within the central visual field could thus be a powerful cue to bind features together, a possibility consistent with close relations in adults between scan paths and pattern recognition (Noton and Stark, 1971; Rizzo et al., 1987) and scene perception (Henderson, 2003).

14.5.2 Learning from Associations Between Visible and Occluded Objects

By 6 months, infants' short-term representations of unseen objects are sufficiently robust to guide reaching and oculomotor systems prospectively to intercept objects on hidden trajectories (Clifton et al., 1991; Johnson et al., 2003a; von Hofsten et al., 1998). At 4 months, prospective behavior – anticipations from eye and head movements to the place of reappearance of an object seen to move behind an occluder – is adapted to variations in occluder width and object speed, implying that under some conditions, infants may track with their 'mind's eye' (von Hofsten et al., 2007). Yet, under other circumstances, 4-month-olds process partly occluded trajectories in terms of visible components only, not complete paths (Figure 14.7). Representations of occluded objects in 4-month-olds, therefore, appear to be rather fragile and not completely established.

To examine the possibility that learning can facilitate spatiotemporal completion, my colleagues and I presented ball-and-box displays to 4- and 6-month-olds as we recorded their eye movements (Johnson et al., 2003a). We reasoned that a representation of the object and its trajectory under occlusion would be reflected in a consistent pattern of anticipatory eye movements toward the place of reemergence, before the object's appearance. The stimulus was identical to the displays used by Johnson et al. (2003b) to investigate spatiotemporal completion (Figure 14.7(a)). Because 6-month-olds provided evidence of spatiotemporal completion in these displays when tested with an habituation paradigm, we predicted that oculomotor anticipations would be more frequent in the older age group. This prediction was supported. A higher proportion of 6-month-olds' object-directed eye movements was classified as anticipatory (i.e., initiated prior to the ball's emergence from behind the occluder; Figure 14.9, top panel) relative to 4-month-olds (Figure 14.9, center panel), corroborating the likelihood that spatiotemporal completion strengthens between 4 and 6 months.

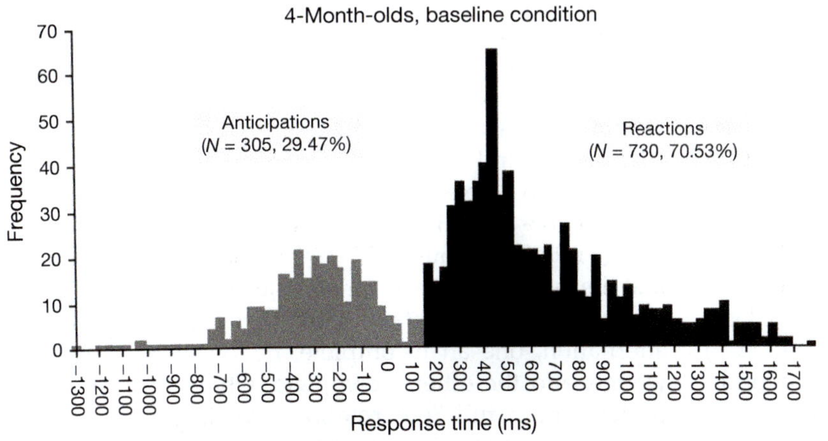

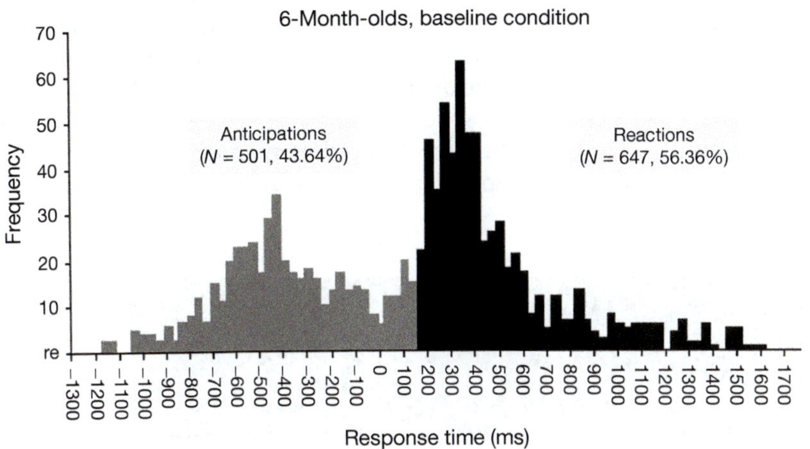

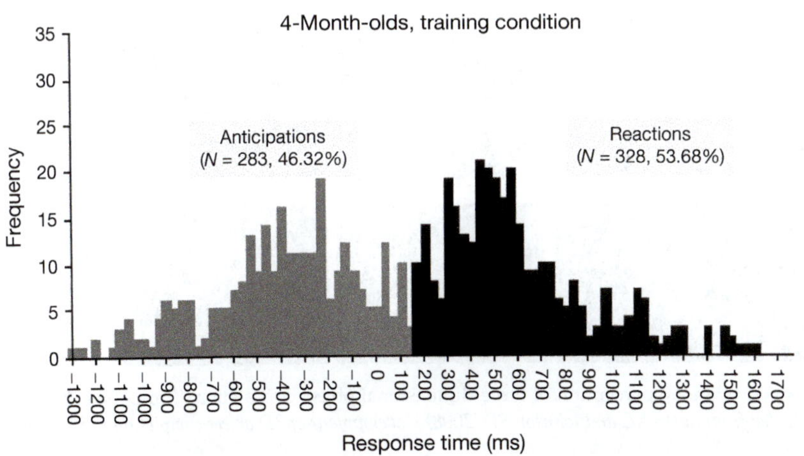

FIGURE 14.9   Histograms showing oculomotor anticipations (gray bars) versus reactions (black bars) as infants view ball-and-box displays. Each eye movement is coded for latency with respect to the emergence of the ball from behind the box, time 0. Eye movements initiated prior to this time are anticipations, and eye movements initiated after this time are reactions. Top panel: 4-month-olds. Center panel: 6-month-olds. Bottom panel: 4-month-olds after 'training' with a fully visible trajectory. *Adapted from Johnson SP, Amso D, Slemmer JA (2003) Development of object concepts in infancy: Evidence for early learning in an eye tracking paradigm.* Proceedings of the National Academy of Sciences of the United States of America *100: 10568–10573.*

Evidence for learning as an important contributor to this developmental change came from a new group of 4-month-olds in a 'training' condition. These infants were first presented with an unoccluded, fully visible ball trajectory (no occluder) for 2 min followed by the ball-and-box display as per the other conditions, and their eye movements were recorded. Here, the proportion of anticipations was reliably greater than that observed in the 'baseline' conditions with untrained 4-month-olds, but not reliably different from that of untrained 6-month-olds (Figure 14.9, bottom panel). In other words, 2 min of exposure led to behaviors characteristic of infants who are 2 months older. This rapid learning may stem from the ability to form associations between fully visible to partly or fully hidden objects.

In the real world, infants are exposed to many different objects moving in different ways, presenting multiple opportunities for learning. For associative learning about occlusion to be a viable means of dealing with real-world events, associations between visible and partly occluded paths must be committed to memory. How long do such rapidly acquired associations last? To address this question, the Johnson et al. (2003b) methods were replicated with new groups of 4-month-olds and a nearly identical pattern of anticipatory behaviors in baseline and training conditions were observed (Johnson and Shuwairi, 2009). A third group received a half hour break between training and test, and performance reverted to baseline, implying that memory for the association was lost during the delay. But a fourth group, provided with a single 'reminder' trial after an identical delay, showed a recovery of oculomotor anticipations equivalent to the no-delay training condition. (A fifth group, provided only a single training trial, showed no benefit in the form of anticipatory looking.) These findings suggest that accumulated exposure to occlusion events may be an important means by which existence constancy arises in infancy.

14.5.3 Learning from Visual–Manual Exploration

Spatial and spatiotemporal completions involve occlusion of far objects by nearer ones. Solid objects also occlude parts of themselves, meaning we cannot see the opposite surfaces from our present vantage point. Perceiving objects as solid in 3D space constitutes 3D object completion, and we recently asked whether young infants perceive objects in this way (Soska and Johnson, 2008). Four- and 6-month-olds were habituated to a wedge rotating through 15° around the vertical axis such that the far sides were never revealed (Figure 14.10). Following habituation infants viewed two test displays in alternation, one an incomplete, hollow version of the wedge, and the other a complete, whole version, both undergoing a full 360° rotation revealing the entirety of the object shape. Four-month-olds showed no consistent posthabituation preference, but 6-month-olds looked longer at the hollow stimulus, indicating perception of the wedge during habituation as a solid, volumetric object in 3D space.

How does 3D object completion arise? One possibility is that developmental changes in infants' motor skills

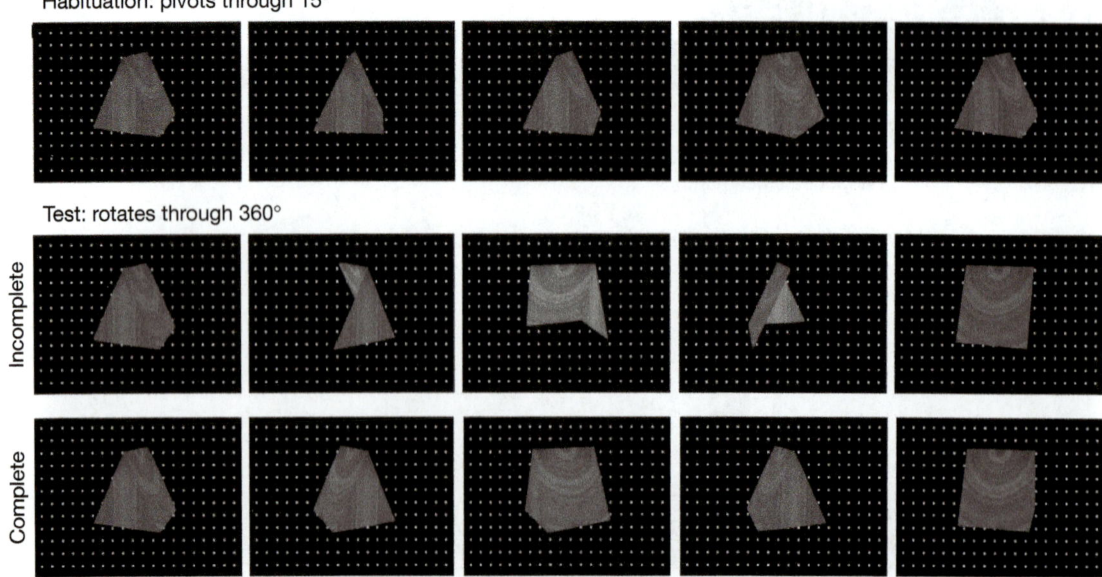

FIGURE 14.10   Rotating object displays from experiments on infants' perception of three-dimensional object completion. Top panels: habituation stimulus. Center and bottom panels: test stimuli. *Adapted from Soska KC and Johnson SP (2008) Development of 3D object completion in infancy. Child Development 79: 1230–1236.*

might underlie the ability to perceive the unseen parts of objects. Two types of motor skill, self-sitting and coordinated visual–manual object exploration, seem particularly important, because independent sitting frees the hands for play and promotes gaze stabilization during manual actions (Rochat and Goubet, 1995). Thus, self-sitting might spur improvements in coordinating object manipulation (e.g., rotating and transferring hand to hand) with visual inspection, providing infants with multiple views of objects. We tested these hypotheses in a group of 4.5- to 7.5-month-olds by replicating the Soska and Johnson (2008) methods and evaluating the infants' motor skills (self-sitting and manipulation of different objects) on the same day (Soska et al., 2010). Strong and significant relations were found between both self-sitting and visual–manual coordination (from the motor skills assessment) and the measure of 3D object completion (from the habituation paradigm). (Other motor skills we recorded, such as holding skill and manual exploration without visual attention to the objects, did not predict 3D object completion.) These results provide evidence for a cascade of developmental events following from the advent of visual–motor coordination, including learning from self-produced experiences.

Evidence from spatial completion experiments reveals that newborns perceive surface segregation even under conditions in which older infants and adults see the identical surfaces as unified (Slater et al., 1990), yet under other circumstances, say, when stationary surfaces are directly adjacent, their connectivity or segregation may be ambiguous (Needham, 1997). This was demonstrated by Needham and Baillargeon (1998) for 4.5-month-olds' interpretation of stimulus displays containing two dissimilar but adjacent, stationary objects (Figure 14.11). After viewing these objects during a familiarization trial, infants were presented with test events in which a hand pulled the cylinder; the box either remained stationary or moved with the cylinder.

The authors reasoned that infants would look longer at the event that was unexpected (e.g., the 'move-apart' event if the objects were perceived as connected), a result found with 8-month-olds (Needham and Baillargeon, 1997), but the 4.5-month-old infants looked about equally at the two test events, providing no evidence for either interpretation on the infants' part.

Needham and Baillargeon (1998) asked whether 4.5-month-olds would learn from a brief prior exposure to either object in isolation and subsequently perceive the two as segregated. Their hypothesis was confirmed: Either a 5-s exposure to the box or a 15-s exposure to the cylinder alone supported segregation of the adjacent cylinder-and-box display into two separate units when infants were tested immediately afterward. Some effects of such training last as long as 72 h (Dueker et al., 2003). This learning effect has been extended in a number of important ways. For example, the effect generalizes from exposure to objects in different orientations (Needham, 2001), but not to objects with distinct features, unless infants are introduced to the different objects in a variety of settings or contexts prior to testing, prompting formation of a perceptual category for the objects (Dueker and Needham, 2005). Categorization is facilitated as well by increasing the number or variety of exemplars during the learning phase of the experiment (Needham et al., 2005).

14.6 SUMMARY AND CONCLUSIONS

From its prenatal origins to its postnatal refinement, learning to see is a mixture of developmental mechanisms, some of which operate outside of experience, and some of which are dependent on it. Although newborn infants can see fairly well upon their first exposure to patterned visual stimulation, as best we can tell the initial inputs are not bound into a stable, predictable,

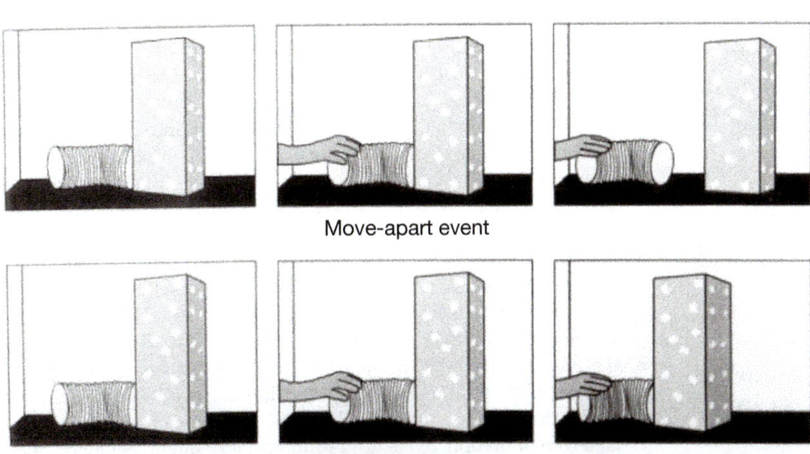

FIGURE 14.11   Schematic depictions of stimuli used to assess object segregation in infants. *Reproduced from Needham (2001).*

coherent visual world. The visual world as we adults know it emerges across the first year after birth. Developmental mechanisms are not limited only to cortical maturation, or to experience, or to learning, but instead comprise all of these and their interactions.

The theoretical views most relevant to questions of visual cognitive development described previously – the views of Piaget and the Gestalt theorists – forecasted some of the research described here. As Piaget proposed, an infant's experience of the visual world begins with a limited capacity to detect object boundaries, particularly under occlusion, and develops in part as a result of the infant's interactions with the environment. And as the Gestalt theorists proposed, visual perception is organized at birth and elaborated with experience; many of the organizational principles characteristic of adult vision appear to be operational in infants (if not at birth). These theories have proven prescient and have given direction to many investigations of infant perception and cognition, yet neither theory is fully adequate to explain the foundations of vision and its development.

Although our understanding of visual cognitive development continues to grow, the current state of knowledge is substantial, and the outlines of a comprehensive account can now be realized. This account, described in this chapter, can be summarized as follows:

- To understand how infants come to experience a stable and predictable world of substantial, volumetric objects, overlapping and extending in depth – the visual world that we adults experience – we must look to experiments that elucidate visual development.
- Visual development begins well before birth. The visual system begins to develop within weeks after conception and continues to develop rapidly prior to the onset of patterned visual stimulation.
- Vision is partially organized at birth. Neonates show systematic scanning patterns and visual preferences, in particular preferences for areas of high contrast and motion. These preferences are well suited for directing attention to features of the visual world relevant to learning about objects. But neonates do not perceive objects as do adults – as solid and substantial entities.
- Newborns' experience of the visual world is fragmented and unstable. Visual and motor systems that yield an experience of coherent objects and the position of the observer relative to a stable environment emerge across the first postnatal year.
- There is a critical period for development of face and object perception. Normal visual experience during this time is essential to their development, as are patterns of eye movements, and other action systems, in binding features into wholes.

- Developmental mechanisms include cortical maturation, visual experience, and learning, and the interplay between these developmental events.
- Developments in some visual functions have been linked directly to maturation of specific cortical regions and visual pathways. Development of smooth pursuit eye movements and motion direction discrimination are thought to stem from maturation of cortical area V5 (also known as MT), form and motion perception from parvocellular and magnocellular processing streams, respectively, and visual memory from structures in the medial temporal lobe. These developments occur between birth and 6 months of age.
- Infants have multiple means of learning at their disposal, and learning is an indispensable part of understanding the visual world. Infants learn from their own behavior as well as by observing relevant events in the environment.

References

Aguiar, A., Baillargeon, R., 1999. 2.5-month-old infants' reasoning about when objects should and should not be occluded. Cognitive Psychology 39, 116–157.

Amso, D., Johnson, S.P., 2005. Selection and inhibition in infancy: Evidence from the spatial negative priming paradigm. Cognition 95, B27–B36.

Amso, D., Johnson, S.P., 2006. Learning by selection: Visual search and object perception in young infants. Developmental Psychology 6, 1236–1245.

Amso, D., Johnson, S.P., 2008. Development of visual selection in 3- to 9-month-olds: Evidence from saccades to previously ignored locations. Infancy 13, 675–686.

Aslin, R.N., 2007. What's in a look? Developmental Science 10, 48–53.

Aslin, R.N., Johnson, S.P., 1994. Suppression of VOR by human infants. In: Poster presented at the Association for Research in Vision and Ophthalmology Conference Sarasota, FL, USA, May 1994.

Aslin, R.N., Johnson, S.P., 1996. Suppression of the optokinetic reflex in human infants: Implications for stable fixation and shifts of attention. Infant Behavior and Development 19, 233–240.

Atkinson, J., 1984. Human visual development over the first 6 months of life: A review and hypothesis. Human Neurobiology 3, 61–74.

Atkinson, J., 2000. The Developing Visual Brain. Oxford University Press, New York.

Atkinson, J., Braddick, O., 1981. Development of optokinetic nystagmus in infants: An indicator of cortical binocularity? In: Fisher, D.E., Monty, R.A., Senders, J.W. (Eds.), Eye Movements: Cognition and Visual Perception. Erlbaum, Hillsdale, NJ, pp. 53–64.

Banks, M., Salapatek, P., 1983. Infant visual perception. (series ed.), (vol. eds.) In: Mussen, P.H., Haith, M.M., Campos, J.J. (Eds.), Handbook of Child Psychology, Vol. 2: Infancy and Developmental Psychobiology. 4th edn Wiley, New York, pp. 435–572.

Banks, M.S., Aslin, R.N., Letson, R.D., 1975. Sensitive period for the development of human binocular vision. Science 190, 675–677.

Barr, R., Dowden, A., Hayne, H., 1996. Developmental changes in deferred imitation by 6- to 24-month-olds. Infant Behavior and Development 19, 159–170.

Bauer, P.J., 2004. Getting explicit memory off the ground: Steps toward construction of a neuro-developmental account of changes in the first two years of life. Developmental Review 24, 347–373.

Bourgeois, J.P., Goldman-Rakic, P.S., Rakic, P., 2000. Formation, elimination, and stabilization of synapses in the primate cerebral cortex. In: Gazzaniga, M.S. (Ed.), The new cognitive neurosciences. MIT Press, Cambridge, MA, pp. 45–53.

Bronson, G., 1990. Changes in infants' visual scanning across the 2- to 14-week age period. Journal of Experimental Child Psychology 49, 101–125.

Bronson, G., 1994. Infants' transitions toward adult-like scanning. Child Development 65, 1243–1261.

Carey, S., Diamond, R., 1977. From piecemeal to configurational representation of faces. Science 195, 312–314.

Cashon, C.H., Cohen, L.B., 2003. Beyond U-shaped development in infants' processing of faces: An information-processing account. Journal of Cognition and Development 5, 59–80.

Clancy, B., Darlington, R.B., Finlay, B.L., 2000. The course of human events: Predicting the timing of primate neural development. Developmental Science 3, 57–66.

Clifton, R.K., Rochat, P., Litovsky, R.Y., Perris, E.E., 1991. Object representation guides infants' reaching in the dark. Journal of Experimental Psychology: Human Perception and Performance 17, 323–329.

Cohen, L.B., 1976. Habituation of infant visual attention. In: Tighe, R.J., Leaton, R.N. (Eds.), Habituation: Perspectives from Child Development, Animal Behavior, and Neurophysiology. Erlbaum, Hillsdale, NJ, pp. 207–238.

Cohen, L.B., Cashon, C.H., 2001. Do 7-month-old infants process independent features or facial configurations? Infant and Child Development 10, 83–92.

Colombo, J., 2001. The development of visual attention in infancy. Annual Review of Psychology 52, 337–367.

Daw, N.W., 1995. Visual Development. Plenum Press, New York.

Dueker, G.L., Needham, A., 2005. Infants' object category formation and use: Real-world context effects on cateory use in object processing. Visual Cognition 12, 1177–1198.

Dueker, G., Modi, A., Needham, A., 2003. 4.5-month-old infants' learning, retention, and use of object boundary information. Infant Behavior and Development 26, 588–605.

Fantz, R.L., 1961. The origin of form perception. Scientific American 204, 66–72.

Fantz, R.L., 1964. Visual experience in infants: Decreased attention to familiar patterns relative to novel ones. Science 146, 668–670.

Felleman, D.J., Van Essen, D.C., 1991. Distributed hierarchical processing in the primate cerebral cortex. Cerebral Cortex 1, 1–47.

Fine, I., Wade, A.R., Brewer, A.A., et al., 2003. Long-term deprivation affects visual perception and cortex. Nature Neuroscience 6, 915–916.

Finlay, B.L., Darlington, R.B., 1995. Linked regularities in the development and evolution of mammalian brains. Science 268, 1578–1584.

Finlay, B.L., Clancy, B., Kingsbury, M.A., 2003. The developmental neurobiology of early vision. In: Hopkins, B., Johnson, S.P. (Eds.), Neurobiology of Infant Vision. Praeger Publishers, Westport, CT, pp. 1–41.

Frank, M.C., Vul, E., Johnson, S.P., 2009. Development of infants' attention to faces during the first year. Cognition 110, 160–170.

Gibson, J.J., 1950. The Perception of the Visual World. Greenwood Press, Westport, CT.

Gibson, J.J., 1979. The Ecological Approach to Visual Perception. Erlbaum, Hillsdale, NJ.

Gilmore, R.O., Johnson, M.H., 1997. Body-centered representations for visually-guided action emerge during early infancy. Cognition 65, B1–B9.

Gilmore, R.O., Baker, T.J., Grobman, K.H., 2004. Stability in young infants' discrimination of optic flow. Developmental Psychology 40, 259–270.

Greenough, W.T., Black, J.E., Wallace, C.S., 1987. Experience and brain development. Child Development 58, 539–559.

Haith, M.M., 1980. Rules that Babies Look by: The Organization of Newborn Visual Activity. Erlbaum, Hillsdale, NJ.

Hebb, D.O., 1949. The Organization of Behavior: A Neuropsychological Theory. Erlbaum, Mahwah, NJ.

Held, R., 1985. Binocular vision: Behavioral and neural development. In: Mehler, J., Fox, R. (Eds.), Neonate Cognition: Beyond the Blooming Buzzing Confusion. Erlbaum, Hillsdale, NJ, pp. 37–44.

Henderson, J.M., 2003. Human gaze control during real-world scene perception. Trends in Cognitive Sciences 7, 498–504.

Hunter, M.A., Ames, E.W., 1989. A multifactor model of infant preferences for novel and familiar stimuli. In: In: Rovee-Collier, C., Lipsitt, L.P. (Eds.), Advances in Child Development and Behavior, vol. 5. Ablex, Norwood, NJ, pp. 69–93.

Huttenlocher, P.R., de Courten, C., Garey, L.J., Van Der Loos, H., 1986. Synaptogenesis in human visual cortex – Evidence for synapse elimination during normal development. Neuroscience Letters 33, 247–252.

Johnson, M.H., 1990. Cortical maturation and the development of visual attention in early infancy. Journal of Cognitive Neuroscience 2, 81–95.

Johnson, S.P., 2001. Neurophysiological and psychophysical approaches to visual development. In: Kalverboer, A.F., Gramsbergen, A., Hopkins, J.B. (Eds.), Handbook of Brain and Behaviour in Human Development. Elsevier, Amsterdam, pp. 653–675 (series eds.), (section ed.).

Johnson, M.H., 2005. Developmental Cognitive Neuroscience, 2nd edn. Blackwell, Oxford, UK.

Johnson, S.P., Aslin, R.N., 1995. Perception of object unity in 2-month-old infants. Developmental Psychology 31, 739–745.

Johnson, S.P., Náñez, J.E., 1995. Young infants' perception of object unity in two-dimensional displays. Infant Behavior and Development 18, 133–143.

Johnson, S.P., Shuwairi, S.M., 2009. Learning and memory facilitate predictive tracking in 4-month-olds. Journal of Experimental Child Psychology 102, 122–130.

Johnson, S.P., Amso, D., Slemmer, J.A., 2003. Development of object concepts in infancy: Evidence for early learning in an eye tracking paradigm. Proceedings of the National Academy of Sciences of the United States of America 100, 10568–10573.

Johnson, S.P., Bremner, J.G., Slater, A., Mason, U., Foster, K., Cheshire, A., 2003. Infants' perception of object trajectories. Child Development 74, 94–108.

Johnson, S.P., Slemmer, J.A., Amso, D., 2004. Where infants look determines how they see: Eye movements and object perception performance in 3-month-olds. Infancy 6, 185–201.

Johnson, S.P., Davidow, J., Hall-Haro, C., Frank, M.C., 2008. Development of perceptual completion originates in information acquisition. Developmental Psychology 44, 1214–1224.

Kellman, P.J., Spelke, E.S., 1983. Perception of partly occluded objects in infancy. Cognitive Psychology 15, 483–524.

Kiorpes, L., Movshon, J.A., 2004. Neural limitations on visual development in primates. In: Chalupa, L.M., Werner, J.S. (Eds.), The Visual Neurosciences. MIT Press, Cambridge, MA, pp. 159–188.

Koffka, K., 1935. Principles of Gestalt Psychology. Routledge & Kegan Paul, London.

Koffka, K., 1959. The Growth of the Mind: An Introduction to Child Psychology. (Translated by Ogden RM) Littlefield, Adams, Patterson, NJ (Original work published 1928).

Köhler, W., 1947. Gestalt Psychology: An Introduction to New Concepts in Modern Psychology. Liveright, New York.

Le Grand, R., Mondloch, C.J., Maurer, D., Brent, H.P., 2001. Early visual experience and face processing. Nature 410, 890.

Macchi Cassia, V., Simion, F., Milani, I., Umiltà, C., 2002. Dominance of global visual properties at birth. Journal of Experimental Psychology. General 131, 398–411.

Mareschal, D., Johnson, M.H., 2003. The 'what' and 'where' of object representations in infancy. Cognition 88, 259–276.

Melcher, D., Kowler, E., 2001. Visual scene memory and the guidance of saccadic eye movements. Vision Research 41, 3597–3611.

Mondloch, C.J., Le Grand, R., Maurer, D., 2002. Configural face processing develops more slowly than featural face processing. Perception 31, 553–566.

Mondloch, C.J., Geldart, S., Maurer, D., Le Grand, R., 2003. Developmental changes in face processing skills. Journal of Experimental Child Psychology 86, 67–84.

Morton, J., Johnson, M.H., 1991. CONSPEC and CONLERN: A two-process theory of infant face recognition. Psychological Review 98, 164–181.

Needham, A., 1997. Factors affecting infants' use of featural information in object segregation. Current Directions in Psychological Science 6, 26–33.

Needham, A., 2001. Object recognition and object segregation in 4.5-month-old infants. Journal of Experimental Child Psychology 78, 3–24.

Needham, A., Baillargeon, R., 1997. Object segregation in 8-month-old infants. Cognition 62, 121–149.

Needham, A., Baillargeon, R., 1998. Effects of prior experience in 4.5-month-old infants' object segregation. Infant Behavior and Development 21, 1–24.

Needham, A., Dueker, G.L., Lockhead, G., 2005. Infants' formation and use of categories to segregate objects. Cognition 94, 215–240.

Nelson, C.A., 1995. The ontogeny of human memory: A cognitive neuroscience perspective. Developmental Psychology 31, 723–738.

Noton, D., Stark, L., 1971. Scan paths in eye movements during pattern perception. Science 171, 308–331.

Oakes, L.M., Ross-Sheehy, S., Luck, S.J., 2006. Rapid development of feature binding in visual short-term memory. Psychological Science 17, 781–787.

Ostrovsky, Y., Andalman, A., Sinha, P., 2006. Vision following extended congenital blindness. Psychological Science 17, 1009–1014.

Piaget, J., 1952. The Origins of Intelligence in Children. (Translated by Cook M) International Universities Press, New York (Original work published 1936).

Piaget, J., 1954. The Construction of Reality in the Child. (Translated by Cook M) Basic Books, New York.

Prechtl, H.F.R., 2001. Prenatal and early postnatal development of human motor behavior. In: Kalverboer, A.F., Gramsbergen, A. (Eds.), Handbook of Brain and Behaviour in Human Development. Kluwer Academic Publishers, Norwell, MA, pp. 415–427.

Preston, K.L., Finocchio, D.V., 1983. Development of the vestibulo-ocular and optokinetic reflexes. In: Simons, K. (Ed.), Early Visual Development: Normal and Abnormal. Oxford University Press, New York, pp. 80–88.

Putzar, L., Hötting, K., Rösler, F., Röder, B., 2007. The development of visual feature binding processes after visual deprivation in early infancy. Vision Research 47, 2616–2626.

Richards, J.E. (Ed.), 1998. Cognitive Neuroscience of Attention: A Developmental Perspective. Erlbaum, Mahwah, NJ.

Rizzo, M., Hurtig, R., Damasio, A.R., 1987. The role of scanpaths in facial recognition and learning. Annals of Neurology 22, 41–45.

Robbins, R.A., Maurer, D., Mondloch, C.J., Nishimura, M., Lewis, T.L., 2008. A house is not a face: The effects of early deprivation on the later discrimination of spacing and featural changes in a non-face object. Poster presented at the Annual Meeting of the Cognitive Neuroscience Society. San Francisco, April 2008.

Rochat, P., Goubet, N., 1995. Development of sitting and reaching in 5- to 6-month-old infants. Infant Behavior and Development 18, 53–68.

Rodman, H.R., 2003. Development of temporal lobe circuits for object recognition: Data and theoretical perspectives from nonhuman primates. In: Hopkins, B., Johnson, S.P. (Eds.), Neurobiology of Infant Vision. Praeger, Westport, CT, pp. 105–145.

Rose, S.A., Feldman, J.F., Jankowski, J.J., 2004. Infant visual recognition memory. Developmental Review 24, 74–100.

Rose, S.A., Jankowski, J.J., Feldman, J.F., 2008. The inversion effect in infancy: The role of internal and external features. Infant Behavior and Development 31, 470–480.

Ross-Sheehy, S., Oakes, L.M., Luck, S.J., 2003. The development of visual short-term memory capacity in infants. Child Development 74, 1807–1822.

Rovee-Collier, C., 1999. The development of infant memory. Current Directions in Psychological Science 8, 80–85.

Schiller, P.H., Tehovnik, E.J., 2001. Look and see: How the brain moves your eyes about. Progress in Brain Research 134, 127–142.

Schwarzer, G., Zauner, N., Jovanovic, B., 2007. Evidence of a shift from featural to configural face processing in infancy. Developmental Science 10, 452–463.

Sirois, S., Mareschal, D., 2002. Models of habituation in infancy. Trends in Cognitive Sciences 6, 293–298.

Slater, A., 1995. Visual perception and memory at birth. In: In: Rovee-Collier, C., Lipsitt, L.P. (Eds.), Advances in Infancy Research, vol. 9. Ablex, Norwood, NJ, pp. 107–162.

Slater, A.M., Morison, V., Rose, D.H., 1983. Perception of shape by the new-born baby. British Journal of Developmental Psychology 1, 135–142.

Slater, A., Morison, V., Somers, M., Mattock, A., Brown, E., Taylor, D., 1990. Newborn and older infants' perception of partly occluded objects. Infant Behavior and Development 13, 33–49.

Slater, A., Johnson, S.P., Kellman, P.J., Spelke, E.S., 1994. The role of three-dimensional depth cues in infants' perception of partly occluded objects. Early Development and Parenting 3, 187–191.

Slater, A., Johnson, S.P., Brown, E., Badenoch, M., 1996. Newborn infants' perception of partly occluded objects. Infant Behavior and Development 19, 145–148.

Soska, K.C., Johnson, S.P., 2008. Development of 3D object completion in infancy. Child Development 79, 1230–1236.

Soska, K.C., Adolph, K.A., Johnson, S.P., 2010. Systems in development: Motor skill acquisition facilitates 3D object completion. Developmental Psychology 46, 129–138.

Spelke, E.S., Breinlinger, K., Macomber, J., Jacobson, K., 1992. Origins of knowledge. Psychological Review 99, 605–632.

Sperry, R.W., 1963. Chemoaffinity in the orderly growth of nerve fiber patterns and their connections. Proceedings of the National Academy of Sciences of the United States of America 50, 703–710.

Teller, D.Y., Movhson, J.A., 1986. Visual development. Vision Research 26, 1483–1506.

Thier, P., Ilg, U.J., 2005. The neural basis of smooth-pursuit eye movements. Current Opinion in Neurobiology 15, 645–652.

Turati, C., Simion, F., Milani, I., Umiltà, C., 2002. Newborns' preference for faces: What is crucial? Developmental Psychology 38, 875–882.

Valenza, E., Simion, F., Umiltà, C., 1994. Inhibition of return in newborn infants. Infant Behavior and Development 17, 293–302.

Valenza, E., Simion, F., Macchi Cassia, V., Umiltà, C., 1996. Face preference at birth. Journal of Experimental Psychology: Human Perception and Performance 22, 892–903.

Van Essen, D.C., Maunsell, J.H.R., 1983. Hierarchical organization and the functional streams in the visual cortex. Trends in Neurosciences 6, 370–375.

Van Essen, D.C., Anderson, C.H., Felleman, D.J., 1992. Information processing in the primate visual system: An integrated systems perspective. Science 255, 419–423.

von Hofsten, C., 2004. An action perspective on motor development. Trends in Cognitive Sciences 8, 266–272.

von Hofsten, C., Vishton, P., Spelke, E.S., Feng, Q., Rosander, K., 1998. Predictive action in infancy: Tracking and reaching for moving objects. Cognition 67, 255–285.

von Hofsten, C., Kochukhova, O., Rosander, K., 2007. Predictive tracking over occlusions by 4-month-old infants. Developmental Science 10, 625–640.

Wiesel, T.N., Hubel, D.H., 1963. Single-cell responses in striate cortex of kittens deprived of vision in one eye. Journal of Neurophysiology 26, 1003–1017.

Wong, R.O.L., 1999. Retinal waves and visual system development. Annual Review of Neuroscience 22, 29–47.

Yin, R.K., 1969. Looking at upside-down faces. Journal of Experimental Psychology 81, 141–145.

Younger, B., 1992. Developmental change in infant categorization: The perception of correlations among facial features. Child Development 63, 1526–1535.

Zilles, K., Werners, R., Busching, U., Schleicher, A., 1986. Ontogenesis of the laminar structure in areas 17 and 18 of the human visual cortex: A quantitative study. Anatomy and Embryology 174, 339–353.

Zuckerman, C.B., Rock, I., 1957. A re-appraisal of the roles of past experience and innate organizing processes in visual perception. Psychological Bulletin 54, 269–296.

The Development of Visuospatial Processing

J. Stiles, N. Akshoomoff, F. Haist
University of California San Diego, La Jolla, CA, USA

OUTLINE

Visual input is a critical source of knowledge about the organization and structure of the spatial world. It provides information about everything from the structure of objects and scenes to their location or movement in space. Visuospatial processing encompasses a wide variety of neurocognitive abilities ranging from the basic ability to analyze how parts or features of an object combine to form an organized whole, to the dynamic and interactive spatial processes required to track moving objects, to visualize displacement, and to localize, attend, or reach for objects or visual targets in a spatial array. These varied processes work in concert to provide a seamless and immediate perception of the intricacies of the visual world. This perception provides an essential basis for precise and effective action in the world as well as a rich source of input for cognitive functions across many domains.

A complex neural architecture involving dozens of interrelated visual areas in the posterior cortices supports visuospatial processing (Van Essen et al., 1992). Ungerleider and Mishkin (1982) first proposed a model for understanding the organization of this complex set of cortical areas and functions in the early 1980s (Ungerleider, 1995). In their model, the cortical visual system is anatomically and functionally subdivided into the ventral and dorsal processing pathways or streams (see Figure 15.1). The ventral stream is dominant for processing information about patterns and objects, while the dorsal stream mediates spatial processing associated with attention to movement and location. Subsequent models describe dorsal stream functions as specialized for supporting visual processing related to action (e.g., Andersen et al., 1997; Goodale and Milner, 1992; Goodale and Westwood, 2004; Rizzolatti and Matelli, 2003).

This chapter begins with a summary of the neuroarchitecture of the ventral and dorsal visual streams. The summary focuses on the flow of visual information beginning, for both streams, in the primary visual cortex and then extending to the temporal and parietal lobes for the ventral and dorsal streams, respectively. Connections between the two major visual pathways as well as connections with the frontal lobes are also considered.

Neural Circuit Development and Function in the Brain: Comprehensive Developmental Neuroscience, Volume 3 http://dx.doi.org/10.1016/B978-0-12-397267-5.00058-3

© 2013 Elsevier Inc. All rights reserved.

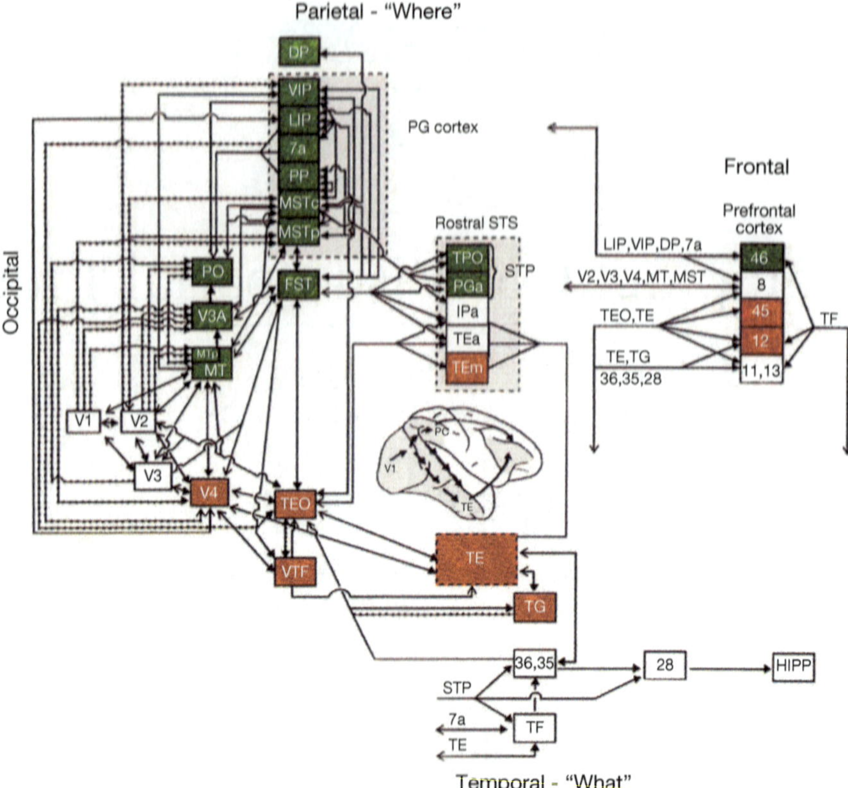

FIGURE 15.1　Dorsal and ventral visual-processing pathways in monkey. Solid lines indicate connections arising from both central and peripheral visual field representations; dotted lines indicate connections restricted to peripheral field representations. Red boxes indicate ventral stream areas related primarily to object vision; green boxes indicate dorsal stream areas related primarily to spatial vision; and white boxes indicate areas not clearly allied with either stream. Shaded region on the lateral view of the brain represents the extent of the cortex included in the diagram. Abbreviations: DP, dorsal prelunate area; FST, fundus of superior temporal area; HIPP, hippocampus; LIP, lateral intraparietal area; MSTc, medial superior temporal area, central visual field representation; MSTp, medial superior temporal area, peripheral visual field representation; MT, middle temporal area; MTp, middle temporal area, peripheral visual field representation; PO, parietal–occipital area; PP, posterior parietal sulcal zone; STP, superior temporal polysensory area; V1, primary visual cortex; V2, visual area 2; V3, visual area 3; V3A, visual area 3, part A; V4, visual area 4; and VIP, ventral intraparietal area. Inferior parietal area 7a; prefrontal areas 8, 11–13, 45, and 46, perirhinal areas 35 and 36; and entorhinal area 28 are from Brodmann (1909). Inferior temporal areas TEO and TE, parahippocampal area TF, temporal pole area TG, and inferior parietal area PG are from Von Bonin and Bailey (1947). Rostral superior temporal sulcal (STS) areas are from Seltzer and Pandya (1978), and VTF is the visually responsive portion of area TF (Boussaoud et al., 1991). *Reproduced from Ungerleider LG (1995) Functional brain imaging studies of cortical mechanisms for memory.* Science 270(5237): 769–775, *with permission.*

The next two sections consider the development of cognitive processes associated with the two principal brain visual systems. The section on ventral stream processing examines the development of visual pattern processing from infancy through adolescence focusing on changes in the perception of visual patterns and faces, and in the ability to construct spatial arrays. The section on dorsal stream processing examines the development of spatial attention, location processing, and mental rotation. The final section of the chapter turns the focus to neurodevelopmental disorders where visuospatial processing is a primary feature. It examines both the effects of frank neural insult on the development of spatial processes and the effects of specific genetic abnormalities on the development of the neural systems that underlie the development of spatial processes. The original descriptions of ventral and dorsal stream organization came from studies of adults with injury to various subsystems within the cortical visual pathways. Data from children with neurodevelopmental disorders provide insight into the emergence of visual system organization following early pathology, and can address questions about how specific neural compromise and neural plasticity interact to affect the developmental trajectories of basic visuospatial functions and the neural systems that mediate them.

15.1　ANATOMICAL ORGANIZATION OF THE PRIMARY VISUAL SYSTEMS

The organization of the primary visual pathways has been most fully described for rhesus macaque monkeys; thus, the description presented here uses the nomenclature typically used for nonhuman primates.

However, the basic pathways in humans and monkeys appear to be largely homologous (Brewer et al., 2002; Orban et al., 2004). The ventral visual pathway begins at the retina and projects via the lateral geniculate nucleus (LGN) of the thalamus to the primary visual cortex, area V1. From there, the pathway proceeds to extrastriate visual areas V2 and V4, and then projects ventrally to the posterior (PIT) and anterior (AIT) regions of the inferior temporal lobe. Input to the ventral pathway is derived principally, though not exclusively, from P-type retinal ganglion cells that project to the parvocellular layers of the LGN and then to layer 4C beta of V1. Parvocellular input to V1 organizes into distinct areas called the blob and interblob regions (Kaas and Collins, 2004; Livingstone and Hubel, 1984; Wong-Riley, 1979). Cells in the blob regions are maximally sensitive to form, while cells in the interblob regions respond principally to color. The ventral stream processes information about visual properties of objects and patterns, and has been described as the 'what' pathway.

The dorsal visual pathway also begins at the retina and projects via the LGN to area V1. From there, the pathway proceeds to extrastriate areas V2 and V3, then projects dorsally to the medial (MT/V5) and medial superior (MST) regions of the temporal lobe, and then to the ventral inferior parietal (IP) lobe. Input to the dorsal pathway is derived principally, though not exclusively, from the large M-type retinal ganglion cells that project to the magnocellular layers of LGN and then to layer 4C alpha of V1. Cells in this pathway are maximally sensitive to movement and direction and are less responsive to color or form. The original functions identified for the dorsal stream involved processing of information about spatial location, optic flow, and motion, and allocation and maintenance of spatial attention. It was thus described as the 'where' pathway. More recently, work examining the dorsal stream's role in visually guided movements suggests that the pathway is and also involved in the integration of visual and motor functions. It has thus been called the 'how' system (e.g., Andersen et al., 1997; Goodale, 2011; Goodale and Milner, 1992; Rizzolatti and Matelli, 2003).

The dorsal and ventral pathways project rostrally to common and distinct, albeit adjacent, areas of the prefrontal cortex. Imaging studies suggest that these prefrontal networks are involved in a variety of dorsal and ventral stream functions (Farivar, 2009). For example, spatial working memory and attention rely on networks connecting the dorsolateral prefrontal cortex (DLPFC) and posterior parietal cortex (Awh and Jonides, 2001; Corbetta et al., 2002; Curtis, 2006), whereas object memory relies on systems connecting the prefrontal cortex with inferior temporal areas (Ranganath, 2006; Ranganath and D'Esposito, 2005; Ranganath et al., 2004). At least three principal projection pathways from the parietal lobe have been described: a parietal–prefrontal pathway mediating eye movement and spatial working memory, a parietal premotor pathway mediating visually guided movement (eye movement, reach, and grasp), and a parietal medial temporal pathway that processes complex spatial information for navigation (Kravitz et al., 2011). There is substantial evidence that the dorsal and ventral pathways are richly interconnected and at least partially overlapping in the mature (e.g., Dobkins and Albright, 1994, 1995, 1998; Marangolo et al., 1998; Merigan and Maunsell, 1993; Rosa et al., 2009; Sincich and Horton, 2005; Thiele et al., 2001; Zanon et al., 2010) and the developing (Dobkins and Anderson, 2002; Dobkins and Teller, 1996a,b) visual system. The dissociation of function across the two pathways may be less complete than originally thought. Subregions within each system may respond to functions typically associated with the other pathway (Husain and Nachev, 2007; Kawasaki et al., 2008; Konen and Kastner, 2008; Lehky and Sereno, 2007). For example, regions in the parietal lobe may respond to color and shape features (Kawasaki et al., 2008), and area MT/V5 in extrastriate visual cortex may show object-selective responses (Konen and Kastner, 2008).

15.2 VENTRAL STREAM PROCESSES

A major function of the ventral visual stream is the analysis of pattern information. Here, findings from three specific functions within the ventral stream are summarized: global–local processing, face processing, and spatial construction. Behaviorally, visuospatial analysis is defined as the ability to specify the parts and the overall configuration of a visually presented pattern, and to understand how the parts are related to form an organized whole (e.g., Delis et al., 1986, 1988; Palmer, 1977, 1980; Palmer and Bucher, 1981; Robertson and Delis, 1986; Smith and Kemler, 1977; Vurpillot, 1976). Thus, it involves the ability to segment a pattern into a set of constituent parts (referred to as featural or local-level processing), and integrate those parts into a coherent whole (referred to as configural or global-level processing). Different approaches to the study of spatial analysis have focused on level of processing and type of input. Perceptual processing studies focus largely on issues of global versus local or configural versus featural processing. Much of the data on perceptual processing of global and local aspects of objects and patterns come from hierarchical form-processing tasks (e.g., see Figure 15.2). Perceptual processing of faces is a related but generally distinct line of study. Faces constitute a critically important class of social stimuli for which most individuals acquire considerable processing expertise. Because of the importance of the information faces provide to typical social interaction and communication, faces may be

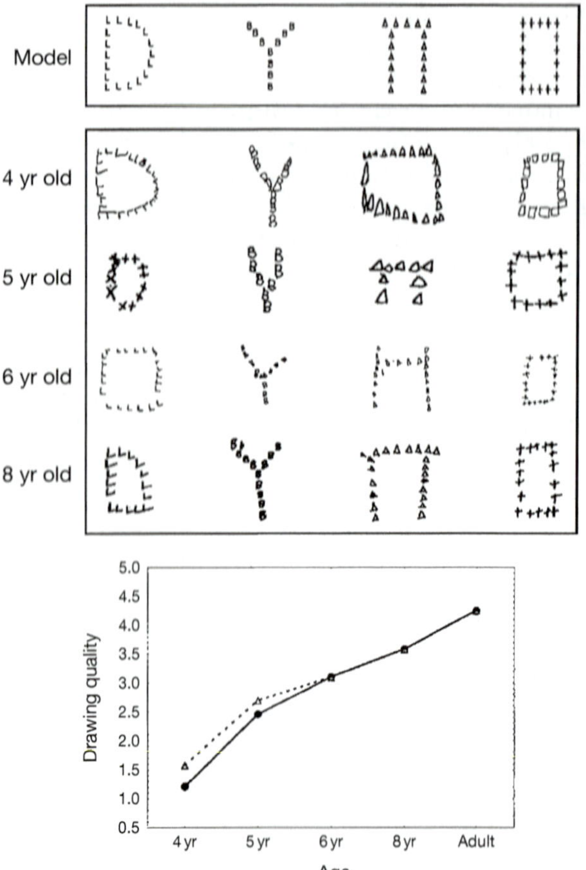

FIGURE 15.2   The model forms (top) provide examples of hierarchical form stimuli. Hierarchical form stimuli have two levels of organization: a large global/configural level comprised of appropriately arranged smaller forms constituting the local/featural level. A series of hierarchical form stimuli (the models) were presented one at a time and children were given 10 s to study the form. After a 30 s delay, they were asked to reproduce the forms from memory. The graph illustrates that systematic improvement in the accuracy of reproductions was observed among typically developing 4–8-year-old children, and also that there are no differences at any age in the relative accuracy of reproducing the global and local levels of the forms. At each age, children are equally accurate in their reproduction of global and local pattern information. *Reproduced from Dukette D and Stiles J (2001) The effects of stimulus density on children's analysis of hierarchical patterns. Developmental Science 4(2): 233–251, with permission.*

processed differently compared to other classes of visual objects. Construction tasks provide a window to children's conceptual organization of spatial arrays. The processes and strategies children use to recreate spatial scenes can provide insight into their understanding of their spatial world.

15.2.1 Perception of the Global and Local Levels of Visual Pattern Structure

Differential laterality for global and local processing is well documented for adults with right hemisphere (RH) dominance for global and left hemisphere (LH)

dominance for local processing (e.g., Han et al., 2002; Martin, 1979; Martinez et al., 1997; Sergent, 1982; Volberg and Hubner, 2004; Yovel et al., 2001). Sergent (1982) suggested that these differences arise from preferential processing of lower spatial frequencies in the RH and higher spatial frequencies in the LH. Several experiments have presented sinusoidal gratings containing a single spatial frequency presented to the right visual field (RVF) or left visual field (LVF) to evaluate this hypothesis with generally positive results. Low spatial frequencies elicit faster responses when presented to the LVF-RH than the RVF-LH, while high spatial frequencies elicit the opposite pattern (Kitterle and Selig, 1991; Kitterle et al., 1990, 1992). Event-related potential (ERP) and other functional imaging studies have supported these basic patterns of lateralization (e.g., Fink et al., 1997; Heinze et al., 1998; Martinez et al., 1997).

In addition to the evidence for the laterality of global- and local-level processing, there is also strong evidence of a global–local processing asymmetry. Specifically, inconsistent or competing information at the global level interferes with local processing, but inconsistent local information does not affect global processing. These two findings led to the postulation of a global precedence effect in visual pattern processing, which states that global-level information is processed prior to local-level information (Navon, 1977). Although many factors may mitigate the global precedence effect in adults, it remains a robust finding within the standard task (Ivry and Robertson, 1998; Kimchi, 1992; Navon, 2003; Robertson and Delis, 1986; Robertson and Lamb, 1991; Robertson et al., 1993).

The ability to analyze spatial patterns begins to emerge in the first year of life. Newborns exhibit configural preferences and rudimentary part–whole processing (Cassia et al., 2002; Farroni et al., 2000; Quinn et al., 1993; Slater et al., 1991). There are dramatic changes in the complexity of visual pattern processing reflecting a systematic improvement in the infant's ability to process global- and local-level pattern information across the first year of life (Cohen and Younger, 1984). These patterns of change appear to reflect early hemispheric differences in processing. Infants as young as 4 months exhibit lateralized processing differences on global and local-processing tasks similar to those observed in adult neuroimaging studies (Deruelle and de Schonen, 1991, 1998).

Studies using the standard hierarchical form stimuli (see Figure 15.2) have also consistently documented a protracted period of developmental change in global–local processing that extends well into adolescence (Dukette and Stiles, 1996, 2001; Harrison and Stiles, 2009; Mondloch et al., 2003; Moses et al., 2002; Porporino et al., 2004; Vinter et al., 2010). The classic global precedence effect emerges slowly over the course of development. Although some studies of children report a global processing bias (Cassia et al., 2002; Mondloch et al., 2003; Moses et al., 2002; Porporino et al., 2004), others report only modest effects that are modulated by altering task and stimulus demands. For example, increased task demands (Harrison and Stiles, 2009) and selective degradation of the global-level stimulus induce a shift in processing bias from the global to the local level that is much more pronounced in children than in adults (Dukette and Stiles, 1996). The combined data from studies of hierarchical form processing show that children are clearly able to engage in global- and local-level processing from a very early age. However, stable and mature levels of visuospatial processing emerge slowly over a protracted period of development. Variations in stimulus and task demands play an important role in modulating the dominant level of processing. Thus, the functional role of a global or local-processing bias or advantage may be different during development than it is later in life, and may reflect growing expertise and facility in processing complex visuospatial patterns.

Imaging studies of typical children confirm the behavioral findings and suggest that the neural systems associated with spatial analytic processing undergo a protracted period of development. Moses et al. (2002) tested children between 11 and 15 years of age using a hemifield reaction time (RT) task and functional magnetic resonance imaging (fMRI) protocols identical to those used by Martinez et al. (1997) with adult subjects. The pattern of RT data obtained from children across this age range differed from that of adults. Similar to the findings from the Mondloch et al. (2003) study, children were faster with global than with local targets and did not manifest the kinds of hemifield RT differences observed among adults. Importantly, children's profiles of activation in the fMRI study differed from those of the adults. For the global and local tasks, children showed statistically greater activation in the RH than in the LH. Overall activation among children was greater than among adults, and children showed considerably more bilateral activation particularly on the local-processing tasks than adults. Thus, at least for these perceptually demanding tasks, children showed a global processing advantage and overall RH dominance.

Anatomical changes are shown to be associated with the shift from local to global processing biases in children. In one study, 6-year-old children were assigned to one of two groups depending on performance on a behavioral global–local processing task (Poirel et al., 2011). One group of children exhibited the mature profile of global-level bias, and the other the more immature local-level bias profile. The investigators used voxel-based morphology to assess group gray matter density differences in brain regions implicated in global processing, specifically the calcarine sulcus, the inferior occipital gyrus, the RH occipital lingual gyrus, the right parietal precuneus, and the precentral gyrus. The group of children exhibiting the behaviorally more mature 'global bias' showed reduced gray matter density in all of the brain regions associated with global-level processing. These findings suggest a link between brain maturation and performance on this important spatial-processing task.

15.2.2 Perception of Faces

The ability to recognize a face is essential for everyday social exchange. Although such recognition depends on the discrimination of subtle differences among faces, the task of identifying a face is effortless for adults, suggesting considerable expertise with this important class of stimuli (see also Chapter 18 for a more extended discussion of the development of face processing). Face processing is thought to rely disproportionately on configural cues such as the spacing between features. For example, unlike other objects, face recognition is significantly impaired when the stimuli are turned upside down (Rossion and Gauthier, 2002). It has been suggested that face inversion selectively disrupts facial configural information processing (Yin, 1969). Consistent with this interpretation, neuroimaging studies with adults find a strong RH bias for face activation within what has been described as the core brain network for face processing (Epstein et al., 2006; Gauthier et al., 2005; Grill-Spector et al., 2004; Kanwisher et al., 1997, 1999; Mazard et al., 2006; Rhodes et al., 2004; Wojciulik et al., 1998; Xu, 2005; Yovel and Kanwisher, 2004, 2005). The core face network is a ventral–occipital–temporal (VOT) system that includes the middle aspects of the lateral fusiform gyrus, often referred to as the fusiform face area (FFA), the inferior occipital gyrus in Brodmann's area 18, often referred to as the occipital face area (OFA), and the posterior superior temporal sulcus (Haxby et al., 2000a).

Preference for face stimuli has been documented from the first hours of life (Johnson et al., 1991). Infants as young as 2–3 months show selective cortical responses to faces (Halit et al., 2004; Tzourio-Mazoyer et al., 2002). Some studies suggest that infants show an RH bias for faces (de Schonen and Deruelle, 1991; de Schonen and Mathivet, 1990; De Schonen et al., 1996). Despite these early competences, there is overwhelming evidence for developmental change in face processing that extends at least through the school-age period (Chung and Thomson, 1995; Taylor et al., 2001). Early studies suggested that changes in face processing might reflect a shift from a more feature-based to a more configural or analytic strategy (Carey and Diamond, 1977; Diamond and Carey, 1986; Tanaka and Farah, 1993). However, accumulating evidence supports a pattern of slower, quantitative age-related change (Itier and Taylor, 2004; Taylor et al., 1999) and increasingly more effective use of the same types of cues used by adults (Baenninger, 1994; Freire and Lee, 2001). These kinds of change may be associated with the acquisition of greater expertise in processing faces and other visual objects (Carey, 1996; Diamond and Carey, 1986; Gauthier and Nelson, 2001).

Recent developmental fMRI studies of face processing suggest that the core brain network for face processing undergoes a protracted change that extends through the school-age period into adolescence (Aylward et al., 2005; Gathers et al., 2004; Golarai et al., 2007; Grill-Spector et al., 2008; Passarotti et al., 2003). Most developmental studies have focused on individual components within the core VOT network, particularly the FFA. The preponderance of evidence indicates that school-age children may produce reliable FFA activation, but the patterns of activation within the fusiform gyrus region vary considerably from those observed among adults. Systematic increases in fMRI blood oxygen level dependent (BOLD) signal activation both in terms of the extent (Brambati et al., 2010; Golarai et al., 2007; Peelen and Kastner, 2009) and intensity of activation (Brambati et al., 2010; Cohen Kadosh et al., 2011; Golarai et al., 2007; Joseph et al., 2011) have been reported from the early school-age period through adulthood. These developmental activation changes correlate with improvement in recognition memory for faces (Golarai et al., 2007, 2010) and with task demand (Scherf et al., 2011). A small number of studies have looked at changes in the organization of brain networks for face processing, and report that young children appear to nonselectively recruit much more extensive networks. With age and growing expertise, these networks become more focused and task specific (Joseph et al., 2011).

15.2.3 Spatial Construction

Spatial construction tasks such as drawing or block assembly provide insight into an individual's conceptualization of the organization of spatial arrays. They can reveal how the participant construes both the parts of an array and the relations among parts that combine to form the overall configuration. Studies of adults with unilateral brain injury use construction tasks extensively. These studies consistently report lateralized differences in the kinds of construction errors produced. Specifically, adults with injury to right posterior brain regions are able to identify, or segment, the parts of spatial forms but have difficulty organizing these parts into integrated spatial configurations. In contrast, adults with injury to left posterior brain regions are able to reproduce the overall pattern configuration, but fail to

incorporate pattern detail and tend to simplify the spatial arrays (Akshoomoff et al., 1989; Delis et al., 1986, 1988; Piercy et al., 1960b; Shorr et al., 1992).

Studies of children's spatial construction activities suggest that before 12 months children engage in very little systematic organization of objects (Forman, 1982; Gesell, 1925; Guanella, 1934; Langer, 1980). In block construction tasks, stacking begins at about 12 months, and by 18 months, children begin to arrange blocks in lines by placing the blocks next to one another (Bayley, 1969; Forman, 1982; Gesell, 1925; Stiles-Davis, 1988). It is not until 3–4 years that children regularly build both vertical and horizontal components within a single spatial construction (Guanella, 1934; Stiles-Davis, 1988). There is also systematic change in processes used to generate block constructions (Stiles and Stern, 2001; Stiles-Davis, 1988). At 24 months, children rely upon a simple repetitive process with a single relation (e.g., stacking). By 36 months, they can use more than one relation (e.g., including a stack and a line in the same construction), but they typically generate them in sequence (e.g., completing the stack, and then the line). By 48 months, children are able to produce multicomponent constructions, with multiple spatial relations, extending in multiple directions in space, and with multiple points of contact between components (e.g., shifting between the line and the stack while building; creating multicomponent construction such as a bridge and several roads). Similar changes are observed in studies of children's drawings. For example, Prather and Bacon (1986) showed that children can attend to either the parts or the whole of a spatial pattern, but their performance can be influenced by specific task and stimulus manipulations. Data from a large series of studies using different measures with children ranging in age from 3 to 12 years show that initially children segment out well-formed, independent parts and use simple combinatorial rules to integrate the parts into the overall configuration (Akshoomoff and Stiles, 1995a,b; Feeney and Stiles, 1996; Stiles and Stern, 2001; Tada and Stiles, 1996). Across the preschool and school-age period, change is observed in both the nature of the parts and the relations children use to organize the parts. Further, pattern complexity affects how children approach the problem of analysis. In a study using the Rey-Osterrieth Complex Figure and simplified variants (see Figure 15.3), it was shown that simplification of the pattern induced more advanced reproduction strategies (Akshoomoff and Stiles, 1995a,b).

15.3 DORSAL STREAM PROCESSES

A variety of spatial processes have been associated with dorsal visual stream processing. We consider three of these processes in this section: spatial localization, spatial

Rey-Osterreith complex form

Representative copies

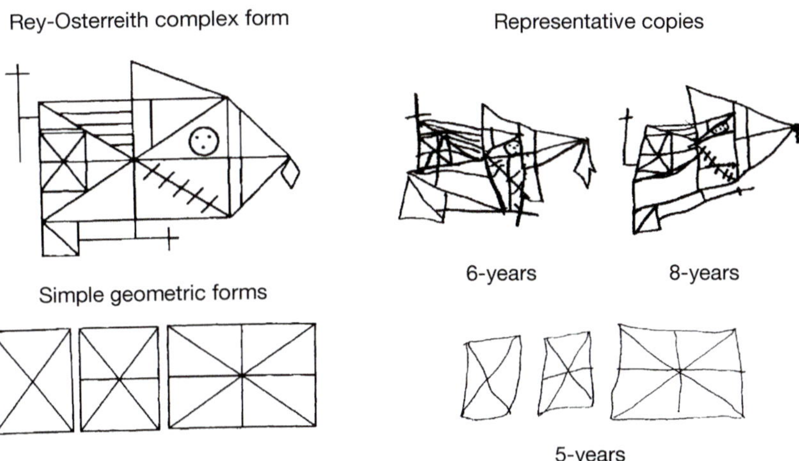

FIGURE 15.3 Models of the Rey-Osterrieth Complex Figure (ROCF) and the three simple geometric forms are shown on the left. Children were given unlimited time to copy each model form. On the right side are examples from 6- and 8-year-old children's copies of the ROCF, and a representative example of a 5-year-old child's copies of the simple geometric forms. Although the simpler forms are contained within the ROCF, children are more accurate in reproducing them in isolation than in the context of the more complex form. *Reproduced from Akshoomoff NA and Stiles J (1995) Developmental trends in visuospatial analysis and planning: I. Copying a complex figure.* Neuropsychology 9(3): 364–377, *with permission.*

attention, and mental rotation. The characterization of these basic dorsal system processes as independent and distinct from those of the ventral stream is somewhat artificial in that, for example, localization of an object may also require a shift in spatial attention, and mental translation of an object must involve both localization and attention in space. Nonetheless, there is substantial evidence for functional and anatomical independence of key features of each process.

15.3.1 Spatial Localization

Evidence from human and animal studies shows that the dorsal stream plays a critical role in perceptual localization (Belger et al., 1998; Chiba et al., 2002). In a series of studies using positron emission tomography (PET) imaging, Haxby and colleagues examined profiles of posterior brain activation using tasks that required adults to compare the location of objects in two visually presented arrays (Haxby et al., 1991, 1994). In addition to activation in bilateral extrastriate cortex presumed to be involved in early visual processing, there was also robust activation of bilateral regions of parietal lobe, including posterior superior parietal areas extending rostrally to the intraparietal sulcus (Brodmann's area 7). These human brain activation findings on location processing are consistent with animal studies (Colby and Duhamel, 1996; Colby and Goldberg, 1999; Rizzolatti and Matelli, 2003), and have been largely replicated in subsequent fMRI, PET, and transcranial magnetic stimulation studies (Belger et al., 1998; Casey et al., 1998; Ellison and Cowey, 2006; Jonides et al., 1993; Nelson et al., 2000; Oliveri et al., 2001; Smith et al., 1995, 1996). Moreover, subsequent studies identified the IP lobe as important in perceptual processing of location (Colby and Duhamel, 1996; Courtney et al., 1996). A large number of functional neuroimaging studies have demonstrated the importance of frontal regions in spatial working memory for locations. Two regions that appear

to be particularly important for spatial working memory in humans include the superior frontal cortex (Courtney et al., 1998; Curtis, 2006; Haxby et al., 2000b; Sala et al., 2003) and DLPFC (Curtis, 2006; Postle et al., 2000).

The task of looking or reaching to a spatial location involves a complex network of neural areas within the dorsal frontoparietal system (Colby and Duhamel, 1996; Colby and Goldberg, 1999; Johnson et al., 1996; Pierrot-Deseilligny et al., 2004; Rizzolatti and Matelli, 2003; Wise et al., 1997). Prefrontal motor areas mediate planning and preparation for motor action; activation of these areas typically precedes the actual motor event. There is considerable evidence for superior parietal input to dorsal premotor and motor cortices; activation in frontal and superior parietal areas is concordant, suggesting a network of spatial-motor control (Rizzolatti and Matelli, 2003; Rizzolatti and Sinigaglia, 2010). In addition, IP areas connect to frontal premotor areas and play an important modulatory role in spatial-motor activity (Andersen et al., 1997). Rizzolatti and Matelli (2003) have suggested that the dorsal system may comprise two separate but interrelated systems: an IP system dominated by visual perceptual inputs and a superior parietal system governed by somatosensory information that is used to guide action.

Location processing is postulated to rely on the computation of two distinct types of relations: categorical and coordinate (Kosslyn, 1987, 2006; Kosslyn et al., 1989, 1992). Categorical relations provide generalized abstract positional information about the relative location of two elements, such as above/below or right/left. Coordinate relations provide precise metric information about spatial relations. Neuroimaging studies have implicated posterior parietal regions for both categorical and coordinate relational processing (Kosslyn et al., 1989, 1995a, 1998; Trojano et al., 2002) but the laterality of the two processes appears to differ. Specifically, categorical processing is LH dominant, while coordinate processing is RH dominant (Kosslyn, 2006; Kosslyn et al., 1989, 1995a).

One of the largest bodies of data on the early development of visuospatial processing comes from a simple, spatial hiding task, originally introduced by Piaget (1952). Infants watch as a toy is hidden under one of two screens (A or B) and are then encouraged to retrieve it. Eight-month-olds easily retrieve the object hidden under A (but also see Smith et al., 1999), but when the object is then hidden under B, they continue to search at A committing what has been termed the A not B error (AB error). This error has been widely conceptualized as an index of object permanence, that is, of the infant's knowledge that objects exist independently over space and time. A wide range of factors have been shown to affect the likelihood of making the AB error. For example, the beginning of self-locomotion reduces the likelihood of AB error (Bertenthal and Campos, 1990; Horobin and Acredolo, 1986; Kermoian and Campos, 1988). In addition, healthy preterm infants are more advanced on the AB search task compared to full-term peers matched for gestational age, suggesting that extra experience in the world offers the healthy preterm infants a developmental advantage (Matthews et al., 1996). Altering task demands effects AB task performance. Some factors, such as the use of salient landmarks, distinctive screens, or increased distance between the screens, improve performance (Butterworth et al., 1982; Wellman et al., 1987). By contrast, increasing task demands by increasing the delay between hiding and search negatively impacts performance. Introduction of a delay between hiding and retrieval increases error frequency among children as old as 12 months (Diamond, 1985; Spencer et al., 2001).

Although neuropsychological data on AB task performance are limited, several studies implicate the DLPFC. In adult rhesus monkeys, bilateral lesions of the DLPFC disrupt AB search task performance (Diamond, 1991; Diamond et al., 1994). Studies using near-infrared spectroscopy to measure localized brain activation in infants provide converging evidence for the association between frontal lobe development and successful search performance (Baird et al., 2002). Electroencephalography (EEG) data have been used to examine potential markers of object representation. Gamma-band activity has been associated with maintenance of mental representations of objects among adults (Tallon-Baudry et al., 1998). Studies measuring gamma-band activity in the EEG of 6-month-old infants during object processing and object occlusion tasks suggest that the neural signature of object representation can be detected by the middle of the first year of life (Csibra et al., 2000; Kaufman et al., 2003, 2005). In summary, these data suggest that a complex network of neural systems emerge across the first year of life to support performance on this seemingly simple task. The data point to changes in both frontal and parietal regions within the dorsal stream, and suggest comparable changes within temporal and frontal regions of the ventral stream. As Johnson noted, changes within these neural regions are unlikely to be unitary events; rather, neural development likely reflects a more gradual 'coming online' of the different components of the complex neural system that progressively comes to support the range of behaviors involved in the visual search task (Johnson et al., 2001).

Although studies of location coding in toddlers suggest that they can make use of fine-grained distance information when searching for hidden objects, the tendency to subdivide space (hierarchical coding) to facilitate remembering an object's location does not emerge until approximately age 4 (Huttenlocher et al., 1994). Further, it is not until age 10 that children show reliable, adult-like spatial coding of fine-grained, multidimensional categorical information (Sandberg et al., 1996). This is consistent with other studies demonstrating improvements in location memory through mid to late childhood (Bell, 2002; Luciana et al., 2005; Orsini et al., 1987; Zald and Iacono, 1998). Increasing task demands by requiring that multiple spatial positions be recalled in a certain order extends the period of immature performance into early adolescence (Farrell Pagulayan et al., 2006; Gathercole et al., 2004; Luciana et al., 2005). Fine-tuning of location information encoded in memory is reported to extend through late adolescence (Luna et al., 2004).

Visual hemifield tasks are used to examine hemispheric specialization of categorical and coordinate image generation (Kosslyn et al., 1995b). In these studies, participants decide whether probe marks (X) presented on a blank grid (categorical task) or bracketed square (coordinate task) appeared on a previously studied letter (see Figure 15.4). Target grids or brackets are presented to either the right (RVF) or left visual hemifield (LVF). For adults, the grid task elicits an LH 'categorical' advantage, whereas the bracket task elicits an RH 'coordinate' advantage. This profile of lateralization appears to emerge gradually during middle childhood. Specifically, 8-year-olds show an RH advantage for both categorical and coordinate tasks, but 10-year-olds begin to show the profile of lateralized differences characteristic of adults (Reese and Stiles, 2005). It is notable that the overall performance of 8-year-olds is considerably poorer than that of 10-year-olds, and the RH advantage is evident primarily on less challenging trials when the probe appears in a salient location marked by global spatial cues (so-called early probes). The finding of a developmentally early RH advantage on these location-processing tasks is consistent with the finding of an RH-mediated, global-processing advantage for children on a global–local processing task discussed earlier (Moses and Stiles, 2002). It appears that the more detailed LH-mediated processing required for both local-level and coordinate spatial processing are later emerging aspects of neural specialization.

Categorical task

Study stimulus   Test stimulus

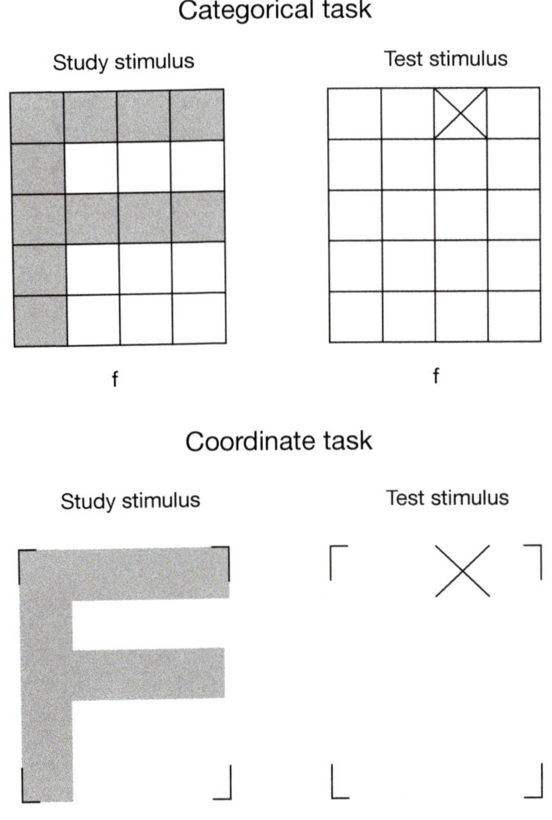

Coordinate task

Study stimulus   Test stimulus

FIGURE 15.4 Examples of the categorical (grids, above) and coordinate (brackets, below) stimuli (Kosslyn et al., 1995a). During the test phase, subjects were asked to read the lowercase letter beneath the stimulus and decide whether the corresponding block letter would cover the X mark in the grid (categorical task) or brackets (coordinate task) if it were present. *Reproduced from Kosslyn SM, Maljkovic V, Hamilton SE, Horwitz G, and Thompson WL (1995) Two types of image generation: Evidence for left and right hemisphere processes. Neuropsychologia 33(11): 1485–1510, with permission.*

15.3.2 Spatial Attention

A closely related line of investigation focuses on the neural systems associated with the ability to shift attention to different spatial locations. In contrast to work examining profiles of brain activity when subjects are required to directly perceive or remember the location of an object, spatial attention tasks investigate the brain systems engaged when attention must shift to a new location. There is considerable clinical and experimental evidence that the posterior parietal lobes play a crucial role in the ability to shift attention (Heilman and Valenstein, 1993; Hillyard and Anllo-Vento, 1998; Ivry and Robertson, 1998; Posner, 1980; Posner et al., 1984; Rafal and Robertson, 1995; Robertson, 1992). Posner's influential model of the attention system involves an interconnected network of structures that modulate and control different aspects of attention (Posner, 1980;

Posner and Petersen, 1990). The posterior parietal network plays an essential role in disengaging attention from one location and allowing a shift of attention to another location.

In the standard task used to test covert shifts of attention (Posner and Cohen, 1980), subjects are required to fixate on a point located centrally between two identical, flanking squares. After a fixed period, a visual cue is presented either centrally (e.g., an arrow) or peripherally (e.g., one box brightens), and soon after a target appears briefly in one box. The subject responds as soon as the target is detected. The critical variable is the validity of the cue. On most trials (75–80%), the cue is 'valid' and the target appears in the cued box. On the remaining trials, the cue is 'invalid,' and the target appears in the opposite box. If cueing serves to covertly shift attention, it should take less time to detect the target when the cue is valid than when it is invalid. One additional, well-established finding concerns response differences associated with the length of the interval between the valid cue and target, or stimulus onset asynchrony (SOA). With short SOAs (<200 ms), the classic facilitation of RT is observed. However, at longer SOAs (300–1300 ms), responses to the cued target are slowed (e.g., Posner et al., 1985). This phenomenon, which has been called inhibition of return (IOR), is thought to reflect the suppression of responses to an already attended location.

To examine patterns of brain activation associated with shifting attention, Corbetta used a variant of the attentional cueing task (Corbetta, 1998; Corbetta et al., 1993). The results of this study confirmed earlier reports from human and animal work on the role of the parietal lobes in shifting spatial attention. Significant foci of brain activation were observed in left and right superior parietal regions. However, the patterns of activation to stimuli presented to the RVF and LVF were not symmetrical across the hemispheres. Presentation of targets to the LVF produced significantly more activation in the RH than the LH, whereas presentation of targets to the RVF produced significant levels of activation in both the RH and LH. Furthermore, distinct activation sites for RVF and LVF targets were identified within the right superior parietal region, suggesting that different brain regions within the RH are responsible for processing information from the two sides of space.

There is a small, but growing literature on infants' ability to shift attention in the visual field (also see Colombo, 2001). A number of studies have shown that by 6 months, infants show both facilitation and IOR (Clohessy et al., 1991; Harman et al., 1994; Hood, 1993; Johnson et al., 1994; Johnson and Tucker, 1996; Varga et al., 2010). Attempts to evoke these responses from younger infants have been mixed. However, control of factors such as SOA duration and cue/target eccentricity appears to be critical for eliciting the responses. Using 200 and 700 SOAs, Johnson and Tucker (1996) demonstrated reliable facilitation and IOR among 4-month-olds, but not among 6-month-olds. However, when a 133 ms SOA was introduced, 6-month-olds showed strong facilitation. This finding suggests that while the basic attentional responses may be robust as early as 4 months, the timing parameters that elicit the response may change with development. Consistent with this, Varga et al. (2010) reported a developmental shift from facilitation to inhibition between 4.5 and 6 months with 300 ms SOAs. Similarly, Harman et al. (1994) found no IOR response among 3-month-old children when stimuli were presented at 30° eccentricity, but a strong response at 10°. Thus, distribution of attention across the visual field may also change with development. Few studies have examined facilitation and IOR in children under 2 months. Johnson and Tucker (1996) reported only weak facilitation effects and no IOR effects among 2-month-old infants. However, Valenza et al.'s (1994) study of newborns suggests that IOR may be present in the first days of life. Further, the child's prior experience in the world, as indexed by familiarity responses, has been shown to affect individual components of the EEG response. Specifically, a negative ERP component that is measured over frontal and central electrodes (the so-called negative central or Nc component) has been shown to increase in response to novelty. Reynolds and Richards (2005) reported a smaller Nc response to familiar compared with novel stimuli in children as young as 4.5 months, suggesting that memory may have a modulatory effect on attention from very early in life. Finally, social cues can direct covert shifts of attention in children as young as 4 months of age (Reid et al., 2004). Direction of eye gaze is a potent cue for shared attention. When infants observed an adult shift gaze toward (cued) or away (uncued) from a target object, ERP responses to subsequent presentations of the cued or uncued object differed in frontotemporal brain regions. These findings suggest that a social cue can induce covert shifts in the infants' attention.

The existing literature on the development of spatial attention in the school-age period is limited, and the findings somewhat inconsistent (Brodeur and Enns, 1997; Enns and Brodeur, 1989; Nougier et al., 1992; Pearson and Lane, 1990). Schul et al. (2003) studied a large sample of children ranging in age from 7 to 17 years on a classic covert orienting task. They reported systematic developmental improvement in aspects of visual attention, including orienting, disengaging, and attending an uncued location. Across the 10-year age window, there was a systematic linear decline in response time, coupled with a linear increase in accuracy (see Figure 15.5).

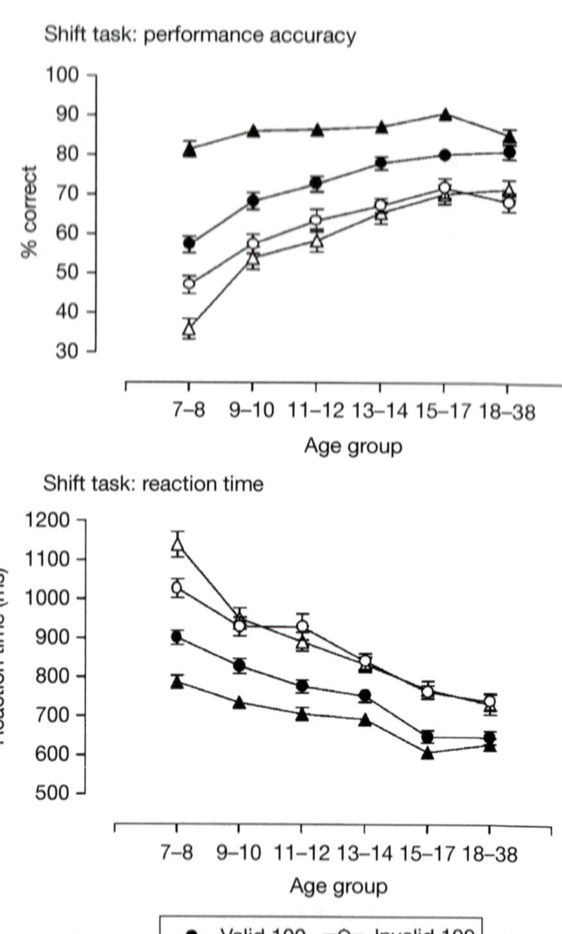

FIGURE 15.5　The accuracy and reaction time performance of typically developing 7–20-year-olds on a shift attention task show a regular pattern of improvement that emerges gradually through the school-age and adolescent periods. *Reproduced from Schul R, Townsend J, and Stiles J (2003) The development of attentional orienting during the school-age years.* Developmental Science 6(3): 262–272, *with permission.*

15.3.3 Mental Rotation

Mental rotation is an important spatial operation that involves the ability to mentally transpose the orientation of an object in space, thus allowing for the computation of a canonical mental representation of a noncanonically presented stimulus, for example, constructing a canonical side-view image of a dog or cat when presented a noncanonical view from behind. Mental rotation requires a host of visuospatial skills including visual pattern processing, visuospatial attention, and visuospatial working memory. A common method used to study mental rotation is to present two objects, one upright and one rotated off vertical, and ask participants if the objects are the same or mirror images. The robust result is that response times vary as a linear, monotonically increasing function of angular disparity between the two objects. This linear response time has become the hallmark characteristic of mental rotation (see Shepard and Cooper, 1986). Over the past decade, a number of neuroimaging studies have documented a neural network associated with mental rotation that includes the right superior parietal lobe, higher-order visual areas (such as MT), and the premotor area (Richter et al., 1997; Riecansky, 2004).

Children as young as age 5 can perform mental rotation (Kosslyn et al., 1990; Marmor, 1975). In an early study, Marmor (1975) found that the RT regression slopes of 5-year-olds were similar to adults and concluded that children are able to perform mental spatial transformations. Subsequent studies have replicated this basic finding (Kosslyn et al., 1990) and confirmed through verbal reports that children use mental rotation to make judgments even under conditions where no explicit instructions to mentally rotate are given (Estes, 1998). Although young children can engage in mental rotation, developmental differences are observed in speed and efficiency of processing (Hale, 1990; Kail et al., 1980; Merriman et al., 1985; Snow, 1990). Further, a wide range of factors including IQ, gender, socioeconomic status, videogame playing, stimulus type employed, and practice on the mental rotation task all impact performance (Cai and Chen, 2000; De Lisi and Wolford, 2002; Okagaki and Frensch, 1994; Waber et al., 1982; Willis and Schaie, 1988). These data suggest that the observed developmental changes could reflect improvement of initially rudimentary mental rotation skills, or they reflect changing strategies for solving the matching problems presented within the context of the standard mental rotation tasks.

Only a few studies have reported data on the neural systems that underlie mental rotation during development. In general, children show patterns of parietal activation similar to adults, but their activation appears more diffuse (Booth et al., 1999, 2000; Roberts and Bell, 2002). The distinctive patterns of activation, including greater superior parietal activation, reported for adults compared to 9- to 10-year-olds may be an index of increasing functional specialization (Booth et al., 2000). Further, an ERP study of 8-year-old children reported bilateral, but asymmetrical parietal activation (right < left) that could reflect emerging hemispheric specialization for mental rotation (Roberts and Bell, 2002). More recently, Ark (Ark, 2005) tested 9–10-year-old children using behavioral and fMRI measures of mental rotation with challenging 3D stimuli. The activation data suggested differences between adults and children in two important brain areas related to mental rotation: the parietal area and MT. Consistent with earlier studies, children produced greater bilateral and widespread activation in the parietal lobe than adults. In addition,

adults produced greater activation in MT than the low-performing children, but not the high-performing children. MT is thought to play a role in imagining the movement of the figures in the mental rotation task. Based on the behavioral data, the low-performing children did not activate MT because they were not performing mental rotation efficiently.

15.4 TRAJECTORIES OF DORSAL AND VENTRAL STREAM DEVELOPMENT

Although many studies have examined the development of ventral or dorsal stream functions separately, work comparing the developmental trajectories of these two systems is limited. The available data present contradictory views of the relative rates of maturation of the two visual systems. One body of data that draws largely from studies of infants younger than a year suggests that dorsal stream functions involved in motion and location processing emerge earlier than ventral stream functions involved in feature processing. In contrast to these findings, studies of older children tend to support the view that the ventral stream matures earlier than dorsal stream.

Much of the evidence for the early maturation of the dorsal stream comes from infant studies of object individuation in which spatiotemporal cues involving motion and location processing are pitted against featural cues. The violation-of-expectation paradigm studies contrast infant responses when spatiotemporal or featural violations are introduced during the test. Infants under about 12 months of age (8 months with simplified tasks) recognize spatiotemporal violations, but fail to notice featural changes (Bonatti et al., 2002; Feigenson and Carey, 2003; Krojgaard, 2000, 2003, 2007; Van de Walle et al., 2000; Xu and Carey, 1996; Xu et al., 2004). This has led to the suggestion that the dorsal system develops earlier than the ventral stream. Other evidence indicates that ventral stream information, such as color, is incorporated into object processing only toward the end of the first year of life (Kaldy and Leslie, 2003; Leslie et al., 1998).

In contrast to the infant work, studies of older preschool and school children generally report that the development of dorsal stream lags behind the ventral stream. In their ERP studies, Neville and colleagues reported significant effects of response latency for the motion but not color stimuli that extended across the age span (Armstrong et al., 2002; Coch et al., 2005; Mitchell and Neville, 2004). However, visual evoked potential (VEP) studies contrasting spatial frequency (Gordon and McCulloch, 1999) and chromaticity (Madrid and Crognale, 2000) found evidence of a lag in ventral stream functioning. Other studies testing thresholds for motion and form coherence reported that ventral stream-mediated form coherence matures significantly ahead of dorsal stream-mediated motion coherence (Atkinson et al., 2005; Braddick et al., 2003; Gunn et al., 2002). Behavioral studies of children using the dual-stream framework comparing 'what' versus 'how' (Milner and Goodale, 1995) are rare. Atkinson (1998) reported data from a small sample of 4–7-year-old children using Milner and Goodale's (1995) 'postbox' task, which requires manual posting of a letter into a slot at a particular angle (dorsal) or visual matching of the perceived angle of the slot (ventral). They reported a significantly better performance with the visual-matching task. A second study using the same task but focused on 3–4-year-old children suggested a ventral stream advantage based on a similar pattern of the results (Dilks et al., 2008).

The mixed and often contradictory results from studies comparing the relative rates of dorsal and ventral stream development have led a number of investigators to suggest that it may be misleading to treat the development of either visual system as a unitary event. Quinn and Bhatt (2006) have suggested that the global dichotomy likely overlooks subtler changes that occur within each stream across development. In addition, Johnson, Mareschal, and colleagues suggest that the inconsistency

in the data may also arise from another factor related to the immaturity of the visual system (Johnson et al., 2001; Kaufman et al., 2003; Mareschal and Johnson, 2003). They note that while there is good evidence that both streams are functional from very early in development, there is little support indicating that information from the two streams is integrated until late in the first year. This lack of integration may account for some of the findings of the early dominance of spatiotemporal information in the infant literature.

Recent neuroimaging studies of older children examining structural connectivity and functional networks provide relevant data for the developmental trajectory of the dorsal and ventral streams. In a cross-sectional study of development, Lebel et al. (2008) evaluated fractional anisotropy (FA) from diffusion tensor imaging (DTI) in multiple major white matter tracts in typically developing people aged 5–30 years. Figure 15.6 shows an example from their findings comparing FA measures from a major dorsal pathway, the superior longitudinal fasciculus, and a major ventral pathway, the inferior longitudinal fasciculus. While the inferior longitudinal fasciculus appears mature in the late childhood period, the superior longitudinal fasciculus does not reach full maturity until mid to late adolescence. An extended trajectory for dorsal stream processing was also evident in a study of functional network structure in a developmental study of face processing. Haist et al.

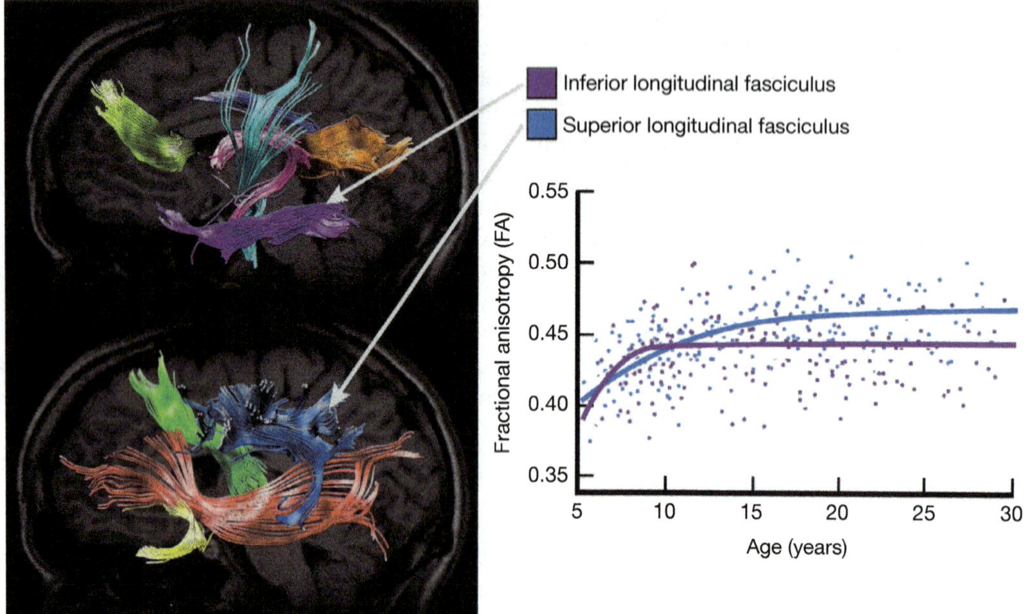

FIGURE 15.6   Evidence for developmental differences in dorsal and ventral stream white matter tracts from diffusion tensor imaging (DTI). Fractional anisotropy (FA) was measured in 202 healthy people ranging from 5 to 29 years in ten major white matter pathways. The tractography in these pathways from a representative adult participant is shown in the left panel. The right panel displays the FA findings across development in the superior longitudinal fasciculus (SLF; dorsal stream) and inferior longitudinal fasciculus (ILF; ventral stream), indicating an extended developmental trajectory for the dorsal stream relative to the ventral stream. Specifically, the SLF does not reach maturity until adolescence, whereas the ILF reaches maturity in late childhood. *Reproduced from Lebel C, Walker L, Leemans A, Phillips L, and Beaulieu C (2008) Microstructural maturation of the human brain from childhood to adulthood.* Neuroimage 40(3): 1044–1055, *with permission.*

(2011) evaluated functional activation in children (ages 7–12), adolescents (ages 13–17), and adults (ages 18–40) during a simple face and object viewing task (i.e., no explicit memory or other cognitive task requirements). Adults produced a modestly greater extent of activation of the fusiform gyrus (FFA) relative to children but not adolescents, and there was a positive correlation with age and activation in the OFA. The striking finding was that children produced hyperactivation relative to adults in regions in the so-called extended face network (Haxby et al., 2000a), including regions in the ventral stream such as the anterior temporal pole (superior temporal gyrus) and amygdala, and dorsal stream such as the IP lobule and inferior frontal gyrus. Adolescents produced hyperactivation relative to adults only in the dorsal stream regions.

15.5 NEURODEVELOPMENTAL DISORDERS OF VISUOSPATIAL PROCESSING

Patient data have been an important source of information on the functional organization of the two major visual pathways. Studies of adult patients with frank injury to either the dorsal or ventral stream networks have been an important source of data on functional organization within the human brain. These studies rely on the logic of subtraction, looking for associations between site of lesion and specific functional loss. The study of developmental disorders requires a somewhat different perspective. Rather than simple associations, the central questions concern the multiple, alternative patterns of brain organization that can arise following early injury to the developing brain or disruption of molecular signaling pathways at critical points in early brain development. These studies thus address issues concerning neural plasticity and compromise, and their effects on the development of basic functions. This section reviews a few of the neurodevelopmental disorders that affect visuospatial functions. It examines both the effects of frank neural insult on the development of spatial processes and the effects of specific genetic abnormalities.

15.5.1 Perinatal Stroke

Perinatal stroke (PS) is a cerebrovascular event that occurs in the period just before birth or immediately after and is usually observed among infants born at term (Lynch and Nelson, 2001). The incidence rate of PS is estimated at 1 in 4000, but it is widely believed that the estimates are low reflecting only those cases that present with identifiable symptoms (Nelson and Lynch, 2004). PS most commonly involves the middle cerebral artery distribution, creating large lesions that

compromise much of one cerebral hemisphere. In adults, such lesions result in significant cognitive deficits, the specific patterns of which differ depending on the side and site of the injury. However, children with such large lesions often achieve considerably better functional outcomes. They typically have normal or corrected-to-normal sensory functions, and intellectual functioning that falls within the normal range on standardized IQ tests (Aram and Ekelman, 1986; Ballantyne et al., 1994; Bates, 1999; Levine et al., 2005; Nass et al., 1989; Stiles et al., 2012).

There is, however, evidence that different neural systems and functions may vary in their capacity for adaptive reorganization, even when injury is early. While basic sensory and motor systems are capable of considerable reorganization, the residual effects on function are often greater than for other domains (Himmelmann et al., 2006; Nelson and Ellenberg, 1982; Van Heest et al., 1993; Wu et al., 2006; Yekutiel et al., 1994). Within cognitive domains, level of function is consistently superior to that of adults with comparable injury, but varies by skill domain. Early developing functions such as those associated with visuospatial processing appear to be more vulnerable than later-developing functions such as language (Reilly et al., 2008; Stiles et al., 2009, 2012). Similarly, functions such as visuospatial processing that have a long evolutionary history and are closely linked to a specific sensory system exhibit somewhat less functional plasticity.

Studies of ventral stream processing among adults with unilateral injury have shown that different patterns of spatial deficit are associated with LH and RH injury (Arena and Gainotti, 1978; Gainotti and Tiacci, 1970; McFie and Zangwill, 1960; Piercy et al., 1960a; Ratcliff, 1982; Swindell et al., 1988; Warrington et al., 1966). Injury to LH brain regions results in disorders involving difficulty defining the parts of a spatial array. For example, patients with LH injury tend to oversimplify spatial patterns and omit details when drawing. On perceptual judgment tasks, they rely upon overall configural cues and ignore specific elements. In studies of global versus local processing, LH injury is associated with local-processing deficits. By contrast, patients with RH lesions have difficulty with the configural aspects of spatial analysis. In drawing, they include details, but fail to maintain a coherent organization among the elements. In perceptual judgment tasks, they focus on the parts of the pattern without attending to the overall form. In studies of global versus local processing, RH injury results in global-level deficits (Delis et al., 1986, 1988).

Data from children with PS suggest that the basic organization of the ventral stream is established early, but is capable of at least limited adaptive organization. Children with RH and LH injury show similar patterns of impairment as adults, but their deficits are milder and performance improves with development (Stiles et al., 2008, 2012). For example, reproduction accuracy for the global-level forms, but not local-level forms, is significantly lower than controls in children with RH injury; the reverse pattern is observed in children with LH injury (see Figure 15.7(a)). Further, while accuracy improves in all groups with development, the pattern of deficit persists for both of the groups with PS (see Figure 15.7(b) for examples at two developmental time points for one child with LH and one with RH injury). These performance differences reflect alternative patterns of brain organization that can arise following early injury. In fMRI studies of global–local processing, adolescents with PS do not show a typical profile of right posterior activation for global- and left posterior activation for local-level processing (Moses et al., 2002). Rather, regardless of the side of lesion, activation for both tasks is confined to the ventral–temporal regions of the contralesional hemisphere. As shown in Figure 15.6(c), LH (left hemisphere lesion) shows

extensive activation on the right and little or no activation on the left on both the global- and local-processing tasks, while RH (right hemisphere lesion) shows extensive activation of the LH and very little activation of the RH. These findings suggest that an alternative, lateralized pattern of brain organization emerges in the wake of early injury. While functional, this alternative pattern of activation is not optimal as reflected in the behavioral performance profiles. Similar lateralized differences in global/configural versus local/featural deficits are observed across a range of tasks including block construction (Stiles et al., 1996; Vicari et al., 1998), copying and drawing (Akshoomoff and Stiles, 2003; Akshoomoff et al., 2002), and face perception (de Schonen et al., 2005; Mancini et al., 1994; Stiles et al., 2006).

Consistent with data from ventral stream functioning, studies of children with PS report patterns of deficit for

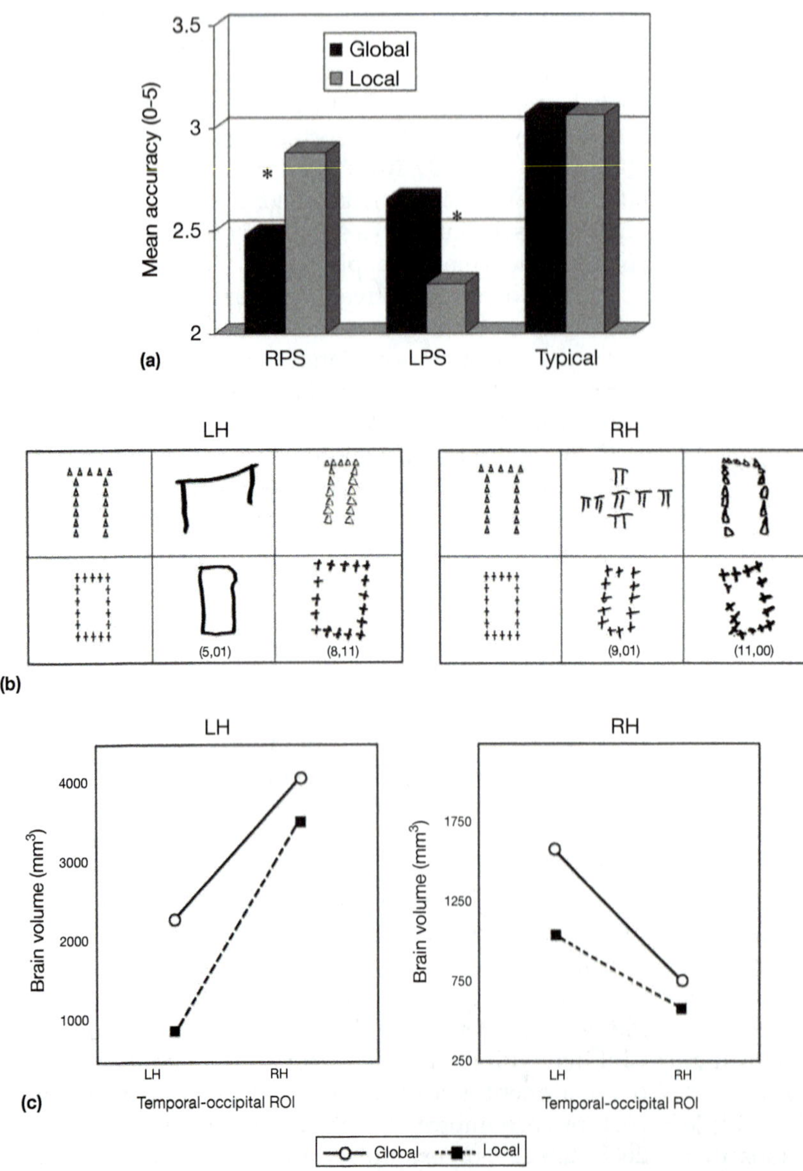

FIGURE 15.7 Behavioral and functional brain activation data from hierarchical form-processing tasks. (a) In contrast to age- and IQ-matched controls, who are equally accurate in reproducing the global and local pattern levels, 5–12-year-old children with RPS injury were more accurate in reproducing the local pattern level, and children with LPS injury were more accurate in reproducing the global pattern level. *Reproduced from Stiles J, Stern C, Appelbaum M, Nass R, Trauner D, and Hesselink J (2008) Effects of early focal brain injury on memory for visuospatial patterns: Selective deficits of global–local processing.* Neuropsychology 22(1): 61–73, with permission. (b) Examples of reproductions from two children, one with LH injury (left) and one with RH injury (right) at two developmental time points. While performance improves with age, subtle levels of specific deficits persist. (c) Brain activation to a perceptual hierarchical-form-processing task is lateralized to the contralesional hemisphere for both global and local processing. Activation is based on an ROI analysis in a region of the posterior lateral temporal–occipital lobe. *Activation data are from the same two children whose earlier drawings are shown in panel b. Reproduced from Stiles, J., Moses, P., Passarotti, A., Dick, F. and Buxton, R. 2003. Exploring developmental change in the neural bases of higher cognitive functions: The promise of functional magnetic resonance imaging.* Developmental Neuropsychology, 24(2&3), 641–668 with permission.

dorsal stream function that are consistent with those observed among adults, though the severity of deficit is less pronounced. Here again, there is clear evidence for performance improvements from the preschool to the late adolescent period, suggesting that children may be better able to compensate for their spatial-processing deficits than adults with comparable injury. Studies of adults suggest that LH injury interferes with categorical processing of spatial locations, while RH injury interferes with coordinate processing (Laeng, 1994; Palermo et al., 2008; Postma et al., 2008; van Asselen et al., 2008). Very similar patterns of results have been reported for children with PS. In an object-retrieval task, 3-year-olds with RH injury were impaired in their use of coordinate relations, performing below the level of typical 18-month-olds (Lourenco and Levine, 2009). Performance improved with development such that by age 5 children showed mastery of this task. However, evidence of persistent subtle deficit emerges among older children on more challenging tasks. Reese et al. (in preparation) tested 10–16-year-olds on Kosslyn's visual hemifield categorical and coordinate processing task (Kosslyn et al., 1995a). Similar to findings from adults with unilateral injury, children with RH injury showed subtle deficits in coordinate processing, and children with LH injury in categorical processing. These studies focused on dorsal stream processing are consistent with those focused on ventral stream processing in that they show evidence of developmental improvement in spatial processing against a backdrop of subtle, persistent lateralized deficit.

15.5.2 Spina Bifida

Spina bifida meningomyelocele is a major neurodevelopmental disorder caused by an open lesion in the spinal cord through which the meninges protrude into a fluid-filled sac (Fletcher and Dennis, 2009; Mitchell et al., 2004). It is usually associated with a malformation of the cerebellum and hindbrain (Chiari II malformation), which in turn causes hydrocephalus. Callosal dysgenesis is also common. Each of the primary CNS abnormalities associated with spina bifida can directly affect neurobehavioral outcomes. Eye movement disorders are common. The ability to perform visually guided hand and arm movements is often affected, particularly in children with higher spinal cord lesions (Fletcher et al., 2005). These are likely a result of insult to the midbrain and tectum as well as impact to the cerebellum.

In addition to hydrocephalus, spina bifida can also be characterized by hypoplasia of the corpus callosum, cortical thinning, and/or white matter loss. The secondary effects of these abnormalities can also impact neuropsychological functioning. Although outcomes are variable, children with spina bifida generally have a relatively stronger language performance and weaker perceptual and motor skills, particularly as demonstrated by comparing their verbal and performance IQ scores (Fletcher et al., 1992).

In order to examine the visual perceptual deficits found in children with spina bifida, Dennis et al. (2005) compared the results across a series of studies on measures emphasizing object-based perception or ventral processing and those emphasizing action-based or dorsal processing. Among children with IQs at or above 70, performance was relatively better on ventral stream tasks (face recognition and visual illusions) and poorer on dorsal stream tasks, particularly those requiring action-based movement (visual pursuit, drawing, route finding, and route planning). Poorest performance relative to control subjects was found on stereopsis and visual figure-ground tasks. Results from a study of object-based visual processing also provide further support for sparing of the ventral visual stream in spina bifida, despite damage to posterior brain regions (Vinter et al., 2010). Weaknesses in both visual–spatial ability and phonological processing are related to poor math performance in preschoolers with spina bifida (Barnes et al., 2011).

The early disruption of brain development associated with spina bifida appears to lead to relatively more disruption of functions associated with the dorsal visual stream. Spatial test performance is correlated with corpus callosum measures in children with hydrocephalus, including those with spina bifida (Fletcher et al., 1996). Recent MRI and DTI results indicate disruption of the white matter and reorganization of cortical regions, with the greatest impact on posterior regions (Hasan et al., 2008; Juranek et al., 2008). Direct comparisons between neuropsychological test performance and imaging results are needed to examine specific aspects of visuospatial processing in more detail.

15.5.3 Neurogenetic Syndromes

Three neurogenetic syndromes are associated with deficits in the development of visuospatial skills, particularly those skills associated with the dorsal visual stream. Neuroimaging data from patients with these syndromes also implicate greater neurodevelopmental abnormalities and perhaps greater early neurodevelopmental vulnerability within the dorsal visual stream.

15.5.3.1 Williams Syndrome

Williams syndrome (WS) is caused by a hemizygous microdeletion of approximately 25 genes on chromosome 7q11.23. Most individuals with WS have mild to moderate intellectual impairment (IQs range from the high 50s to the low 70s). There is an unusual cognitive profile associated with WS, with relative sparing of language and relative impairment in visuospatial and visuomotor task performance (Bellugi et al., 2000; Mervis et al., 2000; Meyer-Lindenberg et al., 2006; Sarpal et al., 2008). Difficulties with visuospatial construction tasks, particularly drawing and block construction tasks, are hallmark deficits in WS. Deficits in location-processing tasks are also found. In contrast, face processing is an area of remarkable strength in WS. These results led to the suggestion of a clear deficit in the dorsal stream with relative sparing of the ventral stream in WS (Galaburda and Bellugi, 2001; Mills et al., 2000; Sarpal et al., 2008). However, task demand differences inherent in these tasks left open the possibility that there were other explanations for this phenomenon. This was examined directly in a study of face and place processing in children and adults with WS (Paul et al., 2002). The perceptual tasks were precisely matched. Individuals with WS did not differ significantly in performance from controls in the face-processing task but were significantly worse in the place (location)-processing task, providing further evidence of a dorsal stream deficit.

Deficits on construction tasks in individuals with WS are consistent across different ages, paradigms, and samples (Bellugi et al., 2001; Donnai and Karmiloff-Smith, 2000). These deficits may be related to the considerable motor demands of such tasks in contrast to tasks that tap other ventral stream skills, such as face processing, with minimal motor demands. However, the reliance on both ventral and dorsal stream processing is an alternative explanation for these discrepancies. In order to more closely examine the dorsal stream deficit hypothesis without the possible impact of intellectual impairment, a group of adults with WS with normal intelligence participated in a series of fMRI experiments (Meyer-Lindenberg et al., 2004). One task used identical stimuli in two conditions. The object-based condition required participants to indicate if two shapes match while the visuospatial decision condition required participants to indicate if two shapes could be constructed to make a square (motor). Both control and WS participants showed comparable activation in the ventral stream during these conditions. While control participants activated bilateral regions of the parietal portion of the dorsal stream in the match minus motor contrast, individuals with WS showed no significant activation. Similar results were found in the attention to object versus attention to location condition. The authors concluded that this hypoactivation reflects a persistent functional deficit in WS that becomes rate limiting when higher demands are placed on the dorsal stream during construction tasks. A subsequent fMRI study demonstrated dorsal stream deficits as well as results that suggest that the medial ventral stream is affected more than the lateral ventral stream (O'Hearn et al., 2011).

Gray matter volume reduction was found in the adjoining parietal–occipital and intraparietal sulcus in WS. Subsequent studies have reported smaller superior parietal lobe volumes in WS (Eckert et al., 2005) as well as DTI abnormalities that suggest that the underlying white matter tracts subserving the dorsal stream of visual processing may be aberrant in WS (Hoeft et al., 2007).

15.5.3.2 Fragile X Syndrome

Fragile X syndrome is a single-gene disorder caused by an expansion of CGG repeats in the promoter region upstream of the *FMR1* gene on the long arm of the X chromosome (Walter et al., 2009). It is the most common known cause of autism, with 15–30% of males with fragile X meeting DSM-IV criteria for autistic disorder and a higher percentage showing more autistic behaviors than expected based on their developmental level. The cognitive deficits can range from mild learning disabilities to severe intellectual disability, with males more likely to be severely affected than females (Schneider et al., 2009). In addition to difficulties in social abilities and executive functions, individuals with fragile X have impaired visuospatial skills (Kwon et al., 2001).

Males and females with fragile X have difficulty with mental manipulation of the spatial relationships between objects and visuomotor coordination, as well as construction tasks and visuospatial working memory (Reiss and Dant, 2003). Children with fragile X also have difficulty with arithmetic skills, with the appearance of poor math achievement scores in early childhood (Mazzocco, 2001). In a study of the relationship between visuospatial skills and math performance, girls with fragile X showed impairment on some aspects of 'where' spatial processing, but no apparent difficulty in identifying objects (Mazzocco et al., 2006). Specifically, girls had difficulty with processing global-level information during a memory for location task and integrating parts during the visual closure task. Significant correlations between math tasks and visual perception tasks were also found. These visuospatial deficits may be tied to the dorsal stream, which in turn implicates dysfunction in the posterior parietal cortex. Evidence from both structural and functional neuroimaging supports the dorsal stream deficit hypothesis in fragile X. As shown in Figure 15.8, DTI neuroimaging has shown that dorsal stream white matter tracts, including the frontal–striatal and parietal sensory–motor tracts, are altered in fragile X (Barnea-Goraly et al., 2003). Functional MRI indicates that fragile X is associated with decreased activation in the frontal–parietal areas during math computation and working memory tasks (Reiss and Dant, 2003).

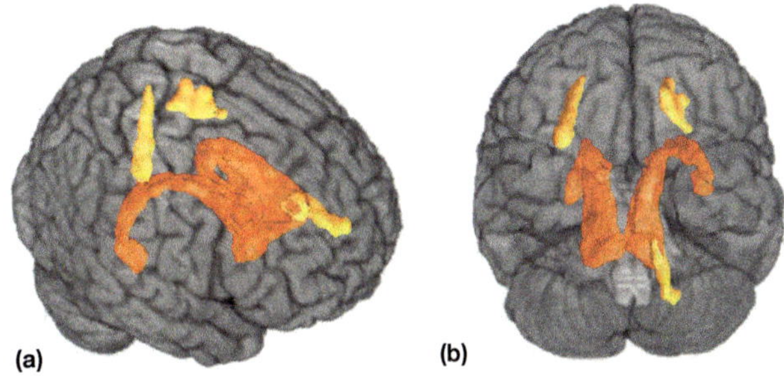

FIGURE 15.8 Example of neuroimaging evidence of a dorsal stream deficit in a neurogenetic disorder. Ten females with fragile X (mean age = 16.7 years) and ten typically developed females (age = 17.0) were examined using diffusion tensor imaging (DTI). The figure shows regions of reduced DTI fractional anisotropy (FA; colored in yellow), a measure of white matter integrity, in fragile X syndrome compared to the control participants. Specifically, abnormal FA was found in the frontal–parietal and parietal sensory–motor pathways. The caudate nucleus (colored in red) is shown for reference. Findings are shown in 3D perspective viewed from (a) superior lateral perspective of the right hemisphere and (b) inferior perspective viewed from under the frontal lobe. Rendered T1 images are shown representing the orientation of the viewing perspective. *Adapted from Barnea-Goraly N, Eliez S, Hedeus M, et al. (2003) White matter tract alterations in fragile X syndrome: Preliminary evidence from diffusion tensor imaging.* American Journal of Medical Genetics Part B: Neuropsychiatric Genetics *118B(1): 81–88, with permission.*

15.5.3.3 Turner Syndrome

Turner syndrome is caused by the partial or complete loss of one female X chromosome. Females with this condition typically have IQ scores in the average range, with performance IQ scores lower than verbal IQ scores, and specific difficulties on visuospatial and math tasks and visuomotor control (Mazzocco, 2001; Walter et al., 2009). There is decreased volume of the parietal and occipital cortices, particularly the superior parietal lobe and postcentral gyri, in Turner syndrome (Brown et al., 2004). Although full-scale IQ is correlated with volume of the postcentral gyri, the relationship between the volume of these posterior brain regions and visuospatial deficits has not been investigated.

Girls with Turner syndrome demonstrate a different pattern of performance across a battery of visuospatial tasks than females with fragile X syndrome (Mazzocco et al., 2006). Girls with Turner syndrome have difficulty with tests of both the 'what' and 'where' systems of visuospatial processing. Specifically, performance on a 'where' task was correlated with performance on a counting task, although it was not clear whether this reflected the visuospatial or the working memory demands of the enumeration task.

Reduced activation of the frontal–parietal network is associated with reduced performance on a visuospatial working memory task (Tamm et al., 2003). Reduced activation in the parietal–occipital regions of the cortex was found on a functional imaging version of the

judgment of line orientation task (Kesler et al., 2004). DTI data also implicate abnormalities within the frontal–parietal network (Walter et al., 2009). There is additional evidence for abnormalities within the temporal lobe that may help to explain recent evidence of ventral stream visuospatial deficits in Turner syndrome (Kesler et al., 2003; Rae et al., 2004).

The data from these three distinct neurogenetic syndromes indicate relative deficits in visuospatial skills, particularly the skills associated with the dorsal stream or 'where' visual system. The results from available neuroimaging studies implicate greater involvement of posterior brain regions, but it is not clear how these generally similar behavioral and neurobiological findings result from such different genetic abnormalities. Walter et al. (2009) observed that disruption of the *FMR1* gene associated with fragile X and disruption of the LIMK1 gene associated with WS affect early dendritic morphology. This disruption may have a relatively greater impact on the dorsal visual stream due to its reliance on larger dendritic fields.

15.6 SUMMARY AND CONCLUSIONS

The ultimate product of typical visuospatial processing is a rich and fine-grained understanding of the visual world around us that provides the basis for efficient behavioral and cognitive interactions with the environment. This ability arises from the intricate coordinated activity of multiple brain regions organized into two generally construed neurofunctional brain systems architecturally defined as the ventral and dorsal visual-processing streams, and functionally distinguished for pattern and object processing (i.e., 'what') and spatial or action processing (i.e., 'where' or 'how'), respectively.

The basic neural systems mediating both of these aspects of spatial functioning appear to be specified in a rudimentary fashion early in development. Infants are able to track and retrieve hidden objects by the middle of the first year of life. Basic markers for control of spatial attention can be documented by 4 months and may be available earlier. Dissociable patterns of spatial analytic deficit can be documented in children with perinatal brain injury. All of these data are indicative of the early emergence of basic neural systems that are specified for processing certain types of information. However, early specification of a neural system does not imply full or optimal functioning early in development. There is ample evidence for protracted change extending throughout childhood, with some abilities not reaching maturity until adolescence. Therefore, the developmental perspective using both typical and atypical development is imperative to provide insights into the functional organization of visuospatial processing, and to understand the plasticity and cognitive–behavioral consequences of early pathology.

Consideration of the dorsal and ventral processing systems as independent is widely believed to be too simplistic, as there is clear evidence for interaction or potential interaction between the systems at multiple points of anatomy and function. Nevertheless, the major developmental hypothesis regarding the trajectory posits that the dorsal system-dependent visuospatial processing matures after the ventral system pattern and object processing abilities. As summarized, this hypothesis is not entirely consistent with findings from infants, but is generally supported by findings in typically developing preschool to school-age children. This has profound implications in developmental disorders. Specifically, dorsal stream functions may be particularly vulnerable to insult at multiple points during development (Atkinson and Braddick, 2007). Findings from several developmental disorders, including frank neurological insult from perinatal stroke and spina bifida, and neurogenetic syndromes, including WS, fragile X syndrome, and Turner syndrome, were summarized that clearly demonstrate disproportionate deficits in dorsal-stream visuospatial functions relative to ventral stream functions. Together, the preponderance of evidence from developmental disorders converges to suggest that visuospatial dorsal-stream functions are specifically vulnerable and less capable of compensation.

There is considerable speculation regarding the neurofunctional basis for dorsal-stream vulnerability in development (Atkinson and Braddick, 2011; Grinter et al., 2010). From an anatomical perspective, evidence points to the possibility of greater vulnerability to insult in magnocellular pathways, the dominant visual cell type in the dorsal stream, relative to the parvocelluar pathway, the dominant visual cell type in the ventral stream. Thus, there is good reason to suspect that dorsal-stream vulnerability results from deficits originating from the earliest visual-processing stages and cascading to advanced spatial processing within the parietal cortex. From a neurocognitive systems perspective, there are additional reasons to suspect greater vulnerability within dorsal stream processing. As summarized, recent models consider the primary role for dorsal stream function to be in the service of behavioral and cognitive 'action,' and there is compelling evidence that typical visuospatial functions depend on the tightly integrated activity between the parietal and frontal cortex networks. Recent neuroimaging findings investigating structural (DTI) and functional connectivity (Fair et al., 2007, 2008, 2009) converge to suggest that the parietal–frontal networks have a protracted development period that does not reach maturity until adolescence. Thus, disruption of typical visual development that impacts parietal lobe-related visuospatial functions would be

compounded later in childhood when advanced visuospatial abilities require the contributions from frontal lobe-dependent processes in addition to the posterior visuospatial abilities in the ventral occipital–temporal and parietal networks.

As this chapter suggests, the present state of the field requires a much greater emphasis on coordinated studies of the development of dorsal and ventral stream visual-processing networks. In the real world, there is no dissociation between these important processing systems; they work seamlessly and in concert to guide visual perception and action. Across development, it is likely that change in one system affects change in the other. The challenge is to better define the emergence in developmental time of each of these systems and to understand how their separate activities become coordinated. Studies of functional connectivity across development hold considerable promise for defining the emergence and integration of these systems at the neural level. At the behavioral level, studies designed to directly compare and contrast the developmental trajectories can inform our understanding. It is essential that studies link spatial perception and cognition to address the understudied problem of how spatial action systems emerge.

References

Akshoomoff, N.A., Stiles, J., 1995a. Developmental trends in visuospatial analysis and planning: I. Copying a complex figure. Neuropsychology 9 (3), 364–377.

Akshoomoff, N.A., Stiles, J., 1995b. Developmental trends in visuospatial analysis and planning: II. Memory for a complex figure. Neuropsychology 9 (3), 378–389.

Akshoomoff, N.A., Stiles, J., 2003. Children's performance on the ROCF and the development of spatial analysis. In: Knight, J.A., Kaplan, E. (Eds.), The Handbook of Rey–Osterrieth Complex Figure Usage: Clinical and Research Applications. Psychological Assessment Resources, Inc., Lutz, FL, pp. 393–409.

Akshoomoff, N.A., Delis, D.C., Kiefner, M.G., 1989. Block constructions of chronic alcoholic and unilateral brain-damaged patients: A test of the right hemisphere vulnerability hypothesis of alcoholism. Archives of Clinical Neuropsychology 4 (3), 275–281.

Akshoomoff, N.A., Feroleto, C.C., Doyle, R.E., Stiles, J., 2002. The impact of early unilateral brain injury on perceptual organization and visual memory. Neuropsychologia 40 (5), 539–561.

Andersen, R.A., Snyder, L.H., Bradley, D.C., Xing, J., 1997. Multimodal representation of space in the posterior parietal cortex and its use in planning movements. Annual Review of Neuroscience 20, 303–330.

Aram, D.M., Ekelman, B.L., 1986. Cognitive profiles of children with early onset of unilateral lesions. Developmental Neuropsychology 2 (3), 155–172.

Arena, R., Gainotti, G., 1978. Constructional apraxia and visuoperceptive disabilities in relation to laterality of cerebral lesions. Cortex 14 (4), 463–473.

Ark, W.S., 2005. Comparing Mental Rotation and Feature Matching Strategies in Adults and Children with Behavioral and Neuroimaging Techniques. Escholarship, University of California, California Digital Library.

Armstrong, B.A., Neville, H.J., Hillyard, S.A., Mitchell, T.V., 2002. Auditory deprivation affects processing of motion, but not color. Brain Research. Cognitive Brain Research 14 (3), 422–434.

Atkinson, J., 1998. The 'where and what' or 'who and how' of visual development. In: Simion, F., Butterworth, G. (Eds.), The Development of Sensory, Motor and Cognitive Capacities in Early Infancy: From Perception to Cognition. Psychology Press/Erlbaum (UK) Taylor & Francis, Hove, England, pp. 3–24.

Atkinson, J., Braddick, O., 2007. Visual and visuocognitive development in children born very prematurely. Progress in Brain Research 164, 123–149.

Atkinson, J., Braddick, O., 2011. From genes to brain development to phenotypic behavior: 'Dorsal-stream vulnerability' in relation to spatial cognition, attention, and planning of actions in Williams syndrome (WS) and other developmental disorders. Progress in Brain Research 189, 261–283.

Atkinson, J., Nardini, M., Anker, S., Braddick, O., Hughes, C., Rae, S., 2005. Refractive errors in infancy predict reduced performance on the movement assessment battery for children at 3 1/2 and 5 1/2 years. Developmental Medicine and Child Neurology 47 (4), 243–251.

Awh, E., Jonides, J., 2001. Overlapping mechanisms of attention and spatial working memory. Trends in Cognitive Science 5 (3), 119–126.

Aylward, E.H., Park, J.E., Field, K.M., et al., 2005. Brain activation during face perception: Evidence of a developmental change. Journal of Cognitive Neuroscience 17 (2), 308–319.

Baenninger, M., 1994. The development of face recognition: Featural or configurational processing? Journal of Experimental Child Psychology 57 (3), 377–396.

Baird, A.A., Kagan, J., Gaudette, T., Walz, K.A., Hershlag, N., Boas, D.A., 2002. Frontal lobe activation during object permanence: Data from near-infrared spectroscopy. NeuroImage 16 (4), 1120–1125.

Ballantyne, A.O., Scarvie, K.M., Trauner, D.A., 1994. Verbal and performance IQ patterns in children after perinatal stroke. Developmental Neuropsychology 10 (1), 39–50.

Barnea-Goraly, N., Eliez, S., Hedeus, M., et al., 2003. White matter tract alterations in fragile X syndrome: Preliminary evidence from diffusion tensor imaging. American Journal of Medical Genetics. Part B, Neuropsychiatric Genetics 118B (1), 81–88.

Barnes, M.A., Stubbs, A., Raghubar, K.P., et al., 2011. Mathematical skills in 3- and 5-year-olds with spina bifida and their typically developing peers: A longitudinal approach. Journal of International Neuropsychological Society 1–14.

Bates, E., 1999. Plasticity, localization, and language development. In: Broman, S.H., Fletcher, J.M. (Eds.), The Changing Nervous System: Neurobehavioral Consequences of Early Brain Disorders. Oxford University Press, New York, NY, pp. 214–253.

Bayley, N., 1969. Bayley Scales of Infant Development. Psychological Corporation, New York, NY.

Belger, A., Puce, A., Krystal, J.H., Gore, J.C., Goldman-Rakic, P., McCarthy, G., 1998. Dissociation of mnemonic and perceptual processes during spatial and nonspatial working memory using fMRI. Human Brain Mapping 6 (1), 14–32.

Bell, S., 2002. Spatial cognition and scale: A child's perspective. Journal of Environmental Psychology 22 (1–2), 9–27 Special Issue: Children and the Environment.

Bellugi, U., Lichtenberger, L., Jones, W., Lai, Z., St George, M., 2000. I. The neurocognitive profile of Williams Syndrome: A complex pattern of strengths and weaknesses. Journal of Cognitive Neuroscience 12 (supplement 1), 7–29.

Bellugi, U., Lichtenberger, L., Jones, W., Lai, Z., St. George, M., 2001. The neurocognitive profile of Williams syndrome: A complex pattern of strengths and weaknesses. In: Bellugi, U., George, M.S. (Eds.), Journey from Cognition to Brain to Gene: Perspectives from Williams Syndrome. The MIT Press, Cambridge, MA, pp. 1–41.

Bertenthal, B.I., Campos, J.J., 1990. A systems approach to the organizing effects of self-produced locomotion during infancy. Advances in Infancy Research 6, 1–60.

Bonatti, L., Frot, E., Zangl, R., Mehler, J., 2002. The human first hypothesis: Identification of conspecifics and individuation of objects in the young infant. Cognitive Psychology 44, 388–426.

Booth, J.R., Macwhinney, B., Thulborn, K.R., Sacco, K., Voyvodic, J., Feldman, H.M., 1999. Functional organization of activation patterns in children: Whole brain fMRI imaging during three different cognitive tasks. Progress in Neuro-Psychopharmacology & Biological Psychiatry 23 (4), 669–682.

Booth, J.R., MacWhinney, B., Thulborn, K.R., Sacco, K., Voyvodic, J.T., Feldman, H.M., 2000. Developmental and lesion effects in brain activation during sentence comprehension and mental rotation. Developmental Neuropsychology 18 (2), 139–169.

Boussaoud, D., Desimone, R., Ungerleider, L.G., 1991. Visual topography of area TEO in the macaque. The Journal of Comparative Neurology 306 (4), 554–575.

Braddick, O., Atkinson, J., Wattam-Bell, J., 2003. Normal and anomalous development of visual motion processing: Motion coherence and 'dorsal-stream vulnerability'. Neuropsychologia 41 (13), 1769–1784.

Brambati, S.M., Benoit, S., Monetta, L., Belleville, S., Joubert, S., 2010. The role of the left anterior temporal lobe in the semantic processing of famous faces. NeuroImage 53 (2), 674–681.

Brewer, A.A., Press, W.A., Logothetis, N.K., Wandell, B.A., 2002. Visual areas in macaque cortex measured using functional magnetic resonance imaging. Journal of Neuroscience 22 (23), 10416–10426.

Brodeur, D.A., Enns, J.T., 1997. Covert visual orienting across the lifespan. Canadian Journal of Experimental Psychology 51 (1), 20–35.

Brodmann, K., 1909. Vergleichende Lokalisationslehre der Grosshirnrinde in ihren Prinzipien Dargestellt auf Grund des Zellenbaues. J.A. Barth, Leipzig.

Brown, W.E., Kesler, S.R., Eliez, S., Warsofsky, I.S., Haberecht, M., Reiss, A.L., 2004. A volumetric study of parietal lobe subregions in Turner syndrome. Developmental Medicine and Child Neurology 46 (9), 607–609.

Butterworth, G., Jarrett, N., Hicks, L., 1982. Spatiotemporal identity in infancy: Perceptual competence or conceptual deficit? Developmental Psychology 18 (3), 435–449.

Cai, H., Chen, C., 2000. The development of mental rotation ability and its correlationship with intelligence. Psychological Science (China) 23 (3), 363–365.

Carey, S., 1996. Perceptual classification and expertise. In: Gelman, R., Kit-Fong, T. (Eds.), Perceptual and Cognitive Development. Academic Press, Inc., San Diego, CA, pp. 49–69.

Carey, S., Diamond, R., 1977. From piecemeal to configurational representation of faces. Science 195 (4275), 312–314.

Casey, B.J., Cohen, J.D., O'Craven, K., et al., 1998. Reproducibility of fMRI results across four institutions using a spatial working memory task. NeuroImage 8 (3), 249–261.

Cassia, V.M., Simion, F., Milani, I., Umilta, C., 2002. Dominance of global visual properties at birth. Journal of Experimental Psychology. General 131 (3), 398–411.

Chiba, A.A., Kesner, R.P., Jackson, P.A., 2002. Two forms of spatial memory: A double dissociation between the parietal cortex and the hippocampus in the rat. Behavioral Neuroscience 116 (5), 874–883.

Chung, M.-S., Thomson, D.M., 1995. Development of face recognition. British Journal of Psychology 86 (1), 55–87.

Clohessy, A.B., Posner, M.I., Rothbart, M.K., Vecera, S.P., 1991. The development of inhibition of return in early infancy. Journal of Cognitive Neuroscience 3 (4), 345–350.

Coch, D., Skendzel, W., Grossi, G., Neville, H., 2005. Motion and color processing in school-age children and adults: An ERP study. Developmental Science 8 (4), 372–386.

Cohen, L.B., Younger, B.A., 1984. Infant perception of angular relations. Infant Behavior & Development 7 (1), 37–47.

Cohen Kadosh, K., Cohen Kadosh, R., Dick, F., Johnson, M.H., 2011. Developmental changes in effective connectivity in the emerging core face network. Cerebral Cortex 21 (6), 1389–1394.

Colby, C.L., Duhamel, J.R., 1996. Spatial representations for action in parietal cortex. Brain Research. Cognitive Brain Research 5 (1–2), 105–115.

Colby, C.L., Goldberg, M.E., 1999. Space and attention in parietal cortex. Annual Review of Neuroscience 22, 319–349.

Colombo, J., 2001. The development of visual attention in infancy. Annual Review of Psychology 52, 337–367.

Corbetta, M., 1998. Frontoparietal cortical networks for directing attention and the eye to visual locations: Identical, independent, or overlapping neural systems? Proceedings of the National Academy of Sciences of the United States of America 95 (3), 831–838.

Corbetta, M., Miezin, F.M., Shulman, G.L., Petersen, S.E., 1993. A PET study of visuospatial attention. Journal of Neuroscience 13 (3), 1202–1226.

Corbetta, M., Kincade, J.M., Shulman, G.L., 2002. Neural systems for visual orienting and their relationships to spatial working memory. Journal of Cognitive Neuroscience 14 (3), 508–523.

Courtney, S.M., Ungerleider, L.G., Keil, K., Haxby, J.V., 1996. Object and spatial visual working memory activate separate neural systems in human cortex. Cerebral Cortex 6 (1), 39–49.

Courtney, S.M., Petit, L., Maisog, J.M., Ungerleider, L.G., Haxby, J.V., 1998. An area specialized for spatial working memory in human frontal cortex. Science 279 (5355), 1347–1351.

Csibra, G., Davis, G., Spratling, M.W., Johnson, M.H., 2000. Gamma oscillations and object processing in the infant brain. Science 290 (5496), 1582–1585.

Curtis, C.E., 2006. Prefrontal and parietal contributions to spatial working memory. Neuroscience 139 (1), 173–180.

De Lisi, R., Wolford, J.L., 2002. Improving children's mental rotation accuracy with computer game playing. Journal of Genetic Psychology 163 (3), 272–282.

de Schonen, S., Deruelle, C., 1991. Hemispheric specialization and recognition of shapes and faces by the baby/Spécialisation hémisphérique et reconnaissance des formes et des visages chez le nourrisson. L'Année Psychologique 91 (1), 15–46.

de Schonen, S., Mathivet, E., 1990. Hemispheric asymmetry in a face discrimination task in infants. Child Development 61 (4), 1192–1205.

De Schonen, S., Deruelle, C., Mancini, J., 1996. Pattern processing in infancy: Hemispheric differences and brain maturation. In: Vital-Durand, F., Atkinson, J. et al., (Eds.), Infant Vision. Oxford University Press, Oxford, pp. 327–344.

de Schonen, S., Mancini, J., Camps, R., Maes, E., Laurent, A., 2005. Early brain lesions and face-processing development. Developmental Psychobiology 46 (3), 184–208.

Delis, D.C., Robertson, L.C., Efron, R., 1986. Hemispheric specialization of memory for visual hierarchical stimuli. Neuropsychologia 24 (2), 205–214.

Delis, D.C., Kiefner, M.G., Fridlund, A.J., 1988. Visuospatial dysfunction following unilateral brain damage: Dissociations in hierarchical hemispatial analysis. Journal of Clinical and Experimental Neuropsychology 10 (4), 421–431.

Dennis, M., Edelstein, K., Copeland, K., et al., 2005. Covert orienting to exogenous and endogenous cues in children with spina bifida. Neuropsychologia 43 (6), 976–987.

Deruelle, C., de Schonen, S., 1991. Hemispheric asymmetries in visual pattern processing in infancy. Brain and Cognition 16 (2), 151–179.

Deruelle, C., de Schonen, S., 1998. Do the right and left hemispheres attend to the same visuospatial information within a face in infancy? Developmental Neuropsychology 14 (4), 535–554.

Diamond, A., 1985. Development of the ability to use recall to guide action, as indicated by infants' performance on AB. Child Development 56 (4), 868–883.

Diamond, A., 1991. Frontal lobe involvement in cognitive changes during the first year of life. In: Gibson, K.R., Petersen, A.C. et al., (Eds.), Brain Maturation and Cognitive Development: Comparative and Cross-Cultural Perspectives. Aldine De Gruyter, New York, NY, pp. 127–180.

Diamond, R., Carey, S., 1986. Why faces are and are not special: An effect of expertise. Journal of Experimental Psychology. General 115 (2), 107–117.

Diamond, A., Werker, J.F., Lalonde, C., 1994. Toward understanding commonalities in the development of object search, detour navigation, categorization, and speech perception. In: Dawson, G., Fischer, K.W. et al., (Eds.), Human Behavior and the Developing Brain. The Guilford Press, New York, NY, pp. 380–426.

Dilks, D.D., Hoffman, J.E., Landau, B., 2008. Vision for perception and vision for action: Normal and unusual development. Developmental Science 11 (4), 474–486.

Dobkins, K.R., Albright, T.D., 1994. What happens if it changes color when it moves? The nature of chromatic input to macaque visual area MT. Journal of Neuroscience 14 (8), 4854–4870.

Dobkins, K.R., Albright, T.D., 1995. Behavioral and neural effects of chromatic isoluminance in the primate visual motion system. Visual Neuroscience 12 (2), 321–332.

Dobkins, K.R., Albright, T.D., 1998. The influence of chromatic information on visual motion processing in the primate visual system. In: Watanabe, T. et al., (Ed.), High-Level Motion Processing: Computational, Neurobiological, and Psychophysical Perspectives. The MIT Press, Cambridge, MA, pp. 53–94.

Dobkins, K.R., Anderson, C.M., 2002. Color-based motion processing is stronger in infants than in adults. Psychological Science 13 (1), 76–80.

Dobkins, K.R., Teller, D.Y., 1996a. Infant contrast detectors are selective for direction of motion. Vision Research 36 (2), 281–294.

Dobkins, K.R., Teller, D.Y., 1996b. Infant motion: Detection (M:D) ratios for chromatically defined and luminance-defined moving stimuli. Vision Research 36 (20), 3293–3310.

Donnai, D., Karmiloff-Smith, A., 2000. Williams syndrome: From genotype through to the cognitive phenotype. American Journal of Medical Genetics 97 (2), 164–171.

Dukette, D., Stiles, J., 1996. Children's analysis of hierarchical patterns: Evidence from a similarity judgment task. Journal of Experimental Child Psychology 63 (1), 103–140.

Dukette, D., Stiles, J., 2001. The effects of stimulus density on children's analysis of hierarchical patterns. Developmental Science 4 (2), 233–251.

Eckert, M.A., Hu, D., Eliez, S., et al., 2005. Evidence for superior parietal impairment in Williams syndrome. Neurology 64, 152–153.

Ellison, A., Cowey, A., 2006. TMS can reveal contrasting functions of the dorsal and ventral visual processing streams. Experimental Brain Research 175 (4), 618–625.

Enns, J.T., Brodeur, D.A., 1989. A developmental study of covert orienting to peripheral visual cues. Journal of Experimental Child Psychology 48 (2), 171–189.

Epstein, R.A., Higgins, J.S., Parker, W., Aguirre, G.K., Cooperman, S., 2006. Cortical correlates of face and scene inversion: A comparison. Neuropsychologia 44 (7), 1145–1158.

Estes, D., 1998. Young children's awareness of their mental activity: The case of mental rotation. Child Development 69 (5), 1345–1360.

Fair, D.A., Cohen, A.L., Dosenbach, N.U., et al., 2008. The maturing architecture of the brain's default network. Proceedings of the National Academy of Sciences of the United States of America 105 (10), 4028–4032.

Fair, D.A., Dosenbach, N.U., Church, J.A., et al., 2007. Development of distinct control networks through segregation and integration. Proceedings of the National Academy of Sciences of the United States of America 104 (33), 13507–13512.

Fair, D.A., Cohen, A.L., Power, J.D., et al., 2009. Functional brain networks develop from a 'local to distributed' organization. PLoS Computational Biology 5 (5), e1000381.

Farivar, R., 2009. Dorsal-ventral integration in object recognition. Brain Research Reviews 61, 144–153.

Farrell Pagulayan, K., Busch, R.M., Medina, K.L., Bartok, J.A., Krikorian, R., 2006. Developmental normative data for the Corsi Block-tapping task. Journal of Clinical and Experimental Neuropsychology 28 (6), 1043–1052.

Farroni, T., Valenza, E., Simion, F., Umilta, C., 2000. Configural processing at birth: Evidence for perceptual organisation. Perception 29 (3), 355–372.

Feeney, S.M., Stiles, J., 1996. Spatial analysis: An examination of preschoolers' perception and construction of geometric patterns. Developmental Psychology 32 (5), 933–941.

Feigenson, L., Carey, S., 2003. Tracking individuals via object files: Evidence from infants' manual search. Developmental Science 6, 568–584.

Fink, G.R., Halligan, P.W., Marshall, J.C., Frith, C.D., Frackowiak, R.S.J., Dolan, R.J., 1997. Neural mechanisms involved in the processing of global and local aspects of hierarchically organized visual stimuli. Brain 120 (10), 1779–1791.

Fletcher, J.M., Dennis, M., 2009. Spina bifida and hydrocephalus. In: Yeates, K.O., Ris, M.D., Taylor, H.G., Pennington, B. (Eds.), Pediatric Neuropsychology: Research, Theory, and Practice. 2nd edn. The Guilford Press, New York.

Fletcher, J.M., Francis, D.J., Thompson, N.M., Davidson, K.C., Miner, M.E., 1992. Verbal and nonverbal skill discrepancies in hydrocephalic children. Journal of Clinical and Experimental Neuropsychology 14 (4), 593–609.

Fletcher, J.M., Bohan, T.P., Brandt, M.E., et al., 1996. Morphometric evaluation of the hydrocephalic brain: Relationships with cognitive development. Childs Nervous System 12, 192–199.

Fletcher, J.M., Copeland, K., Frederick, J.A., et al., 2005. Spinal lesion level in spina bifida: A source of neural and cognitive heterogeneity. Journal of Neurosurgery 102 (3 supplement), 268–279.

Forman, G., 1982. A search for the origins of equivalence concepts through a microanalysis of block play. In: Forman, G. (Ed.), Action and Thought: From Sensorimotor Schemes to Symbolic Operations. Academic Press, New York, NY, pp. 97–135.

Freire, A., Lee, K., 2001. Face recognition in 4- to 7-year-olds: Processing of configural, featural, and paraphernalia information. Journal of Experimental Child Psychology 80 (4), 347–371.

Gainotti, G., Tiacci, C., 1970. Patterns of drawing disability in right and left hemispheric patients. Neuropsychologia 8 (3), 379–384.

Galaburda, A.M., Bellugi, U., 2001. Cellular and molecular cortical neuroanatomy in Williams syndrome. In: Bellugi, U., George, M.S. (Eds.), Journey from Cognition to Brain to Gene: Perspectives from Williams Syndrome. The MIT Press, Cambridge, MA, pp. 123–145.

Gathercole, S.E., Pickering, S.J., Ambridge, B., Wearing, H., 2004. The structure of working memory from 4 to 15 years of age. Developmental Psychology 40 (2), 177–190.

Gathers, A.D., Bhatt, R., Corbly, C.R., Farley, A.B., Joseph, J.E., 2004. Developmental shifts in cortical loci for face and object recognition. NeuroReport 15 (10), 1549–1553.

Gauthier, I., Nelson, C.A., 2001. The development of face expertise. Current Opinion in Neurobiology 11 (2), 219–224.

Gauthier, I., Curby, K.M., Skudlarski, P., Epstein, R.A., 2005. Individual differences in FFA activity suggest independent processing at different spatial scales. Cognitive, Affective, & Behavioral Neuroscience 5 (2), 222–234.

Gesell, A., 1925. Mental Growth of the Preschool Child: A Psychological Outline of Normal Development from Birth to the Sixth Year. Macmillan, New York, NY.

Golarai, G., Ghahremani, D.G., Whitfield-Gabrieli, S., et al., 2007. Differential development of high-level visual cortex correlates with category-specific recognition memory. Nature Neuroscience 10 (4), 512–522.

Golarai, G., Liberman, A., Yoon, J.M., Grill-Spector, K., 2010. Differential development of the ventral visual cortex extends through adolescence. Frontiers in Human Neuroscience 3, 80.

Goodale, M.A., 2011. Transforming vision into action. Vision Research 51 (13), 1567–1587.

Goodale, M.A., Milner, A.D., 1992. Separate visual pathways for perception and action. Trends in Neurosciences 15 (1), 20–25.

Goodale, M.A., Westwood, D.A., 2004. An evolving view of duplex vision: Separate but interacting cortical pathways for perception and action. Current Opinion in Neurobiology 14 (2), 203–211.

Gordon, G.E., McCulloch, D.L., 1999. A VEP investigation of parallel visual pathway development in primary school age children. Documenta Ophthalmologica 99 (1), 1–10.

Grill-Spector, K., Knouf, N., Kanwisher, N., 2004. The fusiform face area subserves face perception, not generic within-category identification. Nature Neuroscience 7 (5), 555–562.

Grill-Spector, K., Golarai, G., Gabrieli, J., 2008. Developmental neuroimaging of the human ventral visual cortex. Trends in Cognitive Science 12 (4), 152–162.

Grinter, E.J., Maybery, M.T., Badcock, D.R., 2010. Vision in developmental disorders: Is there a dorsal stream deficit? Brain Research Bulletin 82 (3–4), 147–160 (Research Support, Non-U.S. Gov't Review).

Guanella, F.M., 1934. Block building activities of young children. Archives of Psychology 174, 5–92.

Gunn, A., Cory, E., Atkinson, J., et al., 2002. Dorsal and ventral stream sensitivity in normal development and hemiplegia. NeuroReport 13 (6), 843–847.

Haist, F., Adamo, M., Han, J., Lee, K., Stiles, J., 2011. On the development of human face-processing abilities: Evidence for hyperactivation of the extended face system in children. Journal of Vision 11 (11), 461. http://dx.doi.org/10.1167/11.11.461

Hale, S., 1990. A global developmental trend in cognitive processing speed. Child Development 61 (3), 653–663.

Halit, H., Csibra, G., Volein, A., Johnson, M.H., 2004. Face-sensitive cortical processing in early infancy. Journal of Child Psychology and Psychiatry 45 (7), 1228–1234.

Han, S., Weaver, J.A., Murray, S.O., Kang, X., Yund, E.W., Woods, D.L., 2002. Hemispheric asymmetry in global/local processing: Effects of stimulus position and spatial frequency. NeuroImage 17 (3), 1290–1299.

Harman, C., Posner, M.I., Rothbart, M.K., Thomas-Thrapp, L., 1994. Development of orienting to locations and objects in human infants. Canadian Journal of Experimental Psychology 48 (2), 301–318.

Harrison, T.B., Stiles, J., 2009. Hierarchical forms processing in adults and children. Journal of Experimental Child Psychology 103 (2), 222–240.

Hasan, K.M., Eluvathingal, T.J., Kramer, L.A., Ewing-Cobbs, L., Dennis, M., Fletcher, J.M., 2008. White matter microstructural abnormalities in children with spina bifida myelomeningocele and hydrocephalus: A diffusion tensor tractography study of the association pathways. Journal of Magnetic Resonance Imaging 27, 700–709.

Haxby, J.V., Grady, C.L., Horwitz, B., et al., 1991. Dissociation of object and spatial visual processing pathways in human extrastriate cortex. Proceedings of the National Academy of Sciences of the United States of America 88 (5), 1621–1625.

Haxby, J.V., Horwitz, B., Ungerleider, L.G., Maisog, J.M., Pietrini, P., Grady, C.L., 1994. The functional organization of human extrastriate cortex: A PET-rCBF study of selective attention to faces and locations. Journal of Neuroscience 14 (11 Pt 1), 6336–6353.

Haxby, J.V., Hoffman, E.A., Gobbini, M.I., 2000a. The distributed human neural system for face perception. Trends in Cognitive Sciences 4 (6), 223–233.

Haxby, J.V., Petit, L., Ungerleider, L.G., Courtney, S.M., 2000b. Distinguishing the functional roles of multiple regions in distributed neural systems for visual working memory. NeuroImage 11 (2), 145–156.

Heilman, K.M., Valenstein, E. (Eds.), 1993. Clinical Neuropsychology. 3rd edn. Oxford University Press, New York, NY.

Heinze, H.J., Hinrichs, H., Scholz, M., Burchert, W., Mangun, G.R., 1998. Neural mechanisms of global and local processing. A combined PET and ERP study. Journal of Cognitive Neuroscience 10 (4), 485–498.

Hillyard, S.A., Anllo-Vento, L., 1998. Event-related brain potentials in the study of visual selective attention. Proceedings of the National Academy of Sciences of the United States of America 95 (3), 781–787.

Himmelmann, K., Beckung, E., Hagberg, G., Uvebrant, P., 2006. Gross and fine motor function and accompanying impairments in cerebral palsy. Developmental Medicine and Child Neurology 48 (6), 417–423.

Hoeft, F., Barnea-Goraly, N., Haas, B.W., et al., 2007. More is not always better: Increased fractional anisotropy of superior longitudinal fasciculus associated with poor visuospatial abilities in Williams syndrome. Journal of Neuroscience 27 (44), 11960–11965.

Hood, B.M., 1993. Inhibition of return produced by covert shifts of visual attention in 6-month-old infants. Infant Behavior & Development 16 (2), 245–254.

Horobin, K., Acredolo, L., 1986. The role of attentiveness, mobility history, and separation of hiding sites on Stage IV search behavior. Journal of Experimental Child Psychology 41 (1), 114–127.

Husain, M., Nachev, P., 2007. Space and the parietal cortex. Trends in Cognitive Science 11 (1), 30–36.

Huttenlocher, J., Newcombe, N., Sandberg, E.H., 1994. The coding of spatial location in young children. Cognitive Psychology 27 (2), 115–148.

Itier, R.J., Taylor, M.J., 2004. Face recognition memory and configural processing: A developmental ERP study using upright, inverted, and contrast-reversed faces. Journal of Cognitive Neuroscience 16 (3), 487–502.

Ivry, R.B., Robertson, L.C., 1998. The Two Sides of Perception. The MIT Press, Cambridge, MA.

Johnson, M.H., Tucker, L.A., 1996. The development and temporal dynamics of spatial orienting in infants. Journal of Experimental Child Psychology 63 (1), 171–188.

Johnson, M.H., Dziurawiec, S., Ellis, H., Morton, J., 1991. Newborns' preferential tracking of face-like stimuli and its subsequent decline. Cognition 40 (1–2), 1–19.

Johnson, M.H., Posner, M.I., Rothbart, M.K., 1994. Facilitation of saccades toward a covertly attended location in early infancy. Psychological Science 5 (2), 90–93.

Johnson, P.B., Farraina, S., Bianchi, L., Caminiti, R., 1996. Cortical networks for visual reaching: Physiological and anatomical organization of frontal and parietal lobe arm regions. Cerebral Cortex 6 (2), 102–119.

Johnson, M.H., Mareschal, D., Csibra, G., 2001. The functional development and integration of the dorsal and ventral visual pathways: A neurocomputational approach. In: Nelson, C.A., Luciana, M. (Eds.), Handbook of Developmental Cognitive Neuroscience. The MIT Press, Cambridge, MA, pp. 339–351.

Jonides, J., Smith, E.E., Koeppe, R.A., Awh, E., Minoshima, S., Mintun, M.A., 1993. Spatial working memory in humans as revealed by PET. Nature 363 (6430), 623–625.

Joseph, J.E., Gathers, A.D., Bhatt, R.S., 2011. Progressive and regressive developmental changes in neural substrates for face processing: Testing specific predictions of the Interactive Specialization account. Developmental Science 14 (2), 227–241.

Juranek, J., Fletcher, J.M., Hasan, K.M., et al., 2008. Neocortical reorganization in spina bifida. NeuroImage 40, 1516–1522.

Kaas, J.H., Collins, C.E. (Eds.), 2004. The Primate Visual System. CRC Press, Boca Raton, FL.

Kail, R., Pellegrino, J., Carter, P., 1980. Developmental changes in mental rotation. Journal of Experimental Child Psychology 29 (1), 102–116.

Kaldy, Z., Leslie, A.M., 2003. Identification of objects in 9-month-old infants: Integrating 'what' and 'where' information. Developmental Science 6 (3), 360–373.

Kanwisher, N., McDermott, J., Chun, M.M., 1997. The fusiform face area: A module in human extrastriate cortex specialized for face perception. Journal of Neuroscience 17 (11), 4302–4311.

Kanwisher, N., Stanley, D., Harris, A., 1999. The fusiform face area is selective for faces not animals. NeuroReport 10 (1), 183–187.

Kaufman, J., Csibra, G., Johnson, M.H., 2003. Representing occluded objects in the human infant brain. Proceedings of the Royal Society B: Biological Sciences 270 (supplement 2), S140–S143.

Kaufman, J., Csibra, G., Johnson, M.H., 2005. Oscillatory activity in the infant brain reflects object maintenance. Proceedings of the National Academy of Sciences of the United States of America 102 (42), 15271–15274.

Kawasaki, M., Watanabe, M., Okuda, J., Sakagami, M., Aihara, K., 2008. Human posterior parietal cortex maintains color, shape and motion in visual short-term memory. Brain Research 1213, 91–97.

Kermoian, R., Campos, J.J., 1988. Locomotor experience: A facilitator of spatial cognitive development. Child Development 59 (4), 908–917.

Kesler, S.R., Blasey, C.M., Brown, W.E., et al., 2003. Effects of X-monosomy and X-linked imprinting on superior temporal gyrus morphology in Turner syndrome. Biological Psychiatry 54 (6), 636–646.

Kesler, S.R., Haberecht, M.F., Menon, V., et al., 2004. Functional neuroanatomy of spatial orientation processing in Turner syndrome. Cerebral Cortex 14 (2), 174–180.

Kimchi, R., 1992. Primacy of wholistic processing and global/local paradigm: A critical review. Psychological Bulletin 112 (1), 24–38.

Kitterle, F.L., Selig, L.M., 1991. Visual field effects in the discrimination of sine-wave gratings. Perception & Psychophysics 50 (1), 15–18.

Kitterle, F.L., Christman, S., Hellige, J.B., 1990. Hemispheric differences are found in the identification, but not the detection, of low versus high spatial frequencies. Perception & Psychophysics 48 (4), 297–306.

Kitterle, F.L., Hellige, J.B., Christman, S., 1992. Visual hemispheric asymmetries depend on which spatial frequencies are task relevant. Brain and Cognition 20 (2), 308–314.

Konen, C.S., Kastner, S., 2008. Two hierarchically organized neural systems for object information in human visual cortex. Nature Neuroscience 11 (2), 224–231.

Kosslyn, S.M., 1987. Seeing and imagining in the cerebral hemispheres: A computational approach. Psychological Review 94 (2), 148–175.

Kosslyn, S.M., 2006. You can play 20 questions with nature and win: Categorical versus coordinate spatial relations as a case study. Neuropsychologia 44 (9), 1519–1523.

Kosslyn, S.M., Koenig, O., Barrett, A., Cave, C.B., Tang, J., Gabrieli, J.D., 1989. Evidence for two types of spatial representations: Hemispheric specialization for categorical and coordinate relations. Journal of Experimental Psychology. Human Perception and Performance 15 (4), 723–735.

Kosslyn, S.M., Margolis, J.A., Barrett, A.M., et al., 1990. Age differences in imagery abilities. Child Development 61 (4), 995–1010.

Kosslyn, S.M., Chabris, C.F., Marsolek, C.J., Koenig, O., 1992. Categorical versus coordinate spatial relations: Computational analyses and computer simulations. Journal of Experimental Psychology. Human Perception and Performance 18 (2), 562–577.

Kosslyn, S.M., Maljkovic, V., Hamilton, S.E., Horwitz, G., Thompson, W.L., 1995a. Two types of image generation: Evidence for left and right hemisphere processes. Neuropsychologia 33 (11), 1485–1510.

Kosslyn, S.M., Thompson, W.L., Kim, I.J., Alpert, N.M., 1995b. Topographical representations of mental images in primary visual cortex. Nature 378 (6556), 496–498.

Kosslyn, S.M., Thompson, W.L., Gitelman, D.R., Alpert, N.M., 1998. Neural systems that encode categorical versus coordinate spatial relations: PET investigations. Psychobiology 26 (4), 333–347.

Kravitz, D.J., Saleem, K.S., Baker, C.I., Mishkin, M., 2011. A new neural framework for visuospatial processing. Nature Reviews Neuroscience 12 (4), 217–230.

Krojgaard, P., 2000. Object individuation in 10-month-old infants: Do significant objects make a difference. Cognitive Development 15, 169–184.

Krojgaard, P., 2003. Object individuation in 10-month-old infants: Manipulating the amount of introduction. British Journal of Developmental Psychology 21, 447–463.

Krojgaard, P., 2007. Comparing infants' use of featural and spatiotemporal information in an object individuation task using a new event-monitoring design. Developmental Science 10 (6), 892–909.

Kwon, H., Menon, V., Eliez, S., et al., 2001. Functional neuroanatomy of visuospatial working memory in fragile X syndrome: Relation to behavioral and molecular measures. The American Journal of Psychiatry 158 (7), 1040–1051.

Laeng, B., 1994. Lateralization of categorical and coordinate spatial functions: A study of unilateral stroke patients. Journal of Cognitive Neuroscience 6 (3), 189–203.

Langer, J.S., 1980. The Origins of Logic: Six to Twelve Months. Academic Press, New York, NY.

Lebel, C., Walker, L., Leemans, A., Phillips, L., Beaulieu, C., 2008. Microstructural maturation of the human brain from childhood to adulthood. NeuroImage 40 (3), 1044–1055.

Lehky, S.R., Sereno, A.B., 2007. Comparison of shape encoding in primate dorsal and ventral visual pathways. Journal of Neurophysiology 97 (1), 307–319.

Leslie, A.M., Xu, F., Tremoulet, P.D., Scholl, B.J., 1998. Indexing and the object concept: Developing 'what' and 'where' systems. Trends in Cognitive Sciences 2 (1), 10–18.

Levine, S.C., Kraus, R., Alexander, E., Suriyakham, L.W., Huttenlocher, P.R., 2005. IQ decline following early unilateral brain injury: A longitudinal study. Brain and Cognition 59 (2), 114–123.

Livingstone, M.S., Hubel, D.H., 1984. Anatomy and physiology of a color system in the primate visual cortex. Journal of Neuroscience 4 (1), 309–356.

Lourenco, S.F., Levine, S.C., 2009. Location representation following early unilateral brain injury: Evidence of distinct deficits and degrees of plasticity. In: Paper Presented at the Society for Research in Child Development. Denver, CO.

Luciana, M., Conklin, H.M., Hooper, C.J., Yarger, R.S., 2005. The development of nonverbal working memory and executive control processes in adolescents. Child Development 76 (3), 697–712.

Luna, B., Garver, K.E., Urban, T.A., Lazar, N.A., Sweeney, J.A., 2004. Maturation of cognitive processes from late childhood to adulthood. Child Development 75 (5), 1357–1372.

Lynch, J.K., Nelson, K.B., 2001. Epidemiology of perinatal stroke. Current Opinion in Pediatrics 13 (6), 499–505.

Madrid, M., Crognale, M.A., 2000. Long-term maturation of visual pathways. Visual Neuroscience 17 (6), 831–837.

Mancini, J., de Schonen, S., Deruelle, C., Massoulier, A., 1994. Face recognition in children with early right or left brain damage. Developmental Medicine and Child Neurology 36 (2), 156–166.

Marangolo, P., Di Pace, E., Rafal, R., Scabini, D., 1998. Effects of parietal lesions in humans on color and location priming. Journal of Cognitive Neuroscience 10 (6), 704–716.

Mareschal, D., Johnson, M.H., 2003. The 'what' and 'where' of object representations in infancy. Cognition 88 (3), 259–276.

Marmor, G.S., 1975. Development of kinetic images: When does the child first represent movement in mental images? Cognitive Psychology 7 (4), 548–559.

Martin, M., 1979. Hemispheric specialization for local and global processing. Neuropsychologia 17 (1), 33–40.

Martinez, A., Moses, P., Frank, L., Buxton, R., Wong, E., Stiles, J., 1997. Hemispheric asymmetries in global and local processing: Evidence from fMRI. NeuroReport 8 (7), 1685–1689.

Matthews, A., Ellis, A.E., Nelson, C.A., 1996. Development of preterm and full-term infant ability on AB, recall memory, transparent Barrier Detour, and Means-End tasks. Child Development 67 (6), 2658–2676.

Mazard, A., Schiltz, C., Rossion, B., 2006. Recovery from adaptation to facial identity is larger for upright than inverted faces in the human occipito-temporal cortex. Neuropsychologia 44 (6), 912–922.

Mazzocco, M.M., 2001. Math learning disability and math LD subtypes: Evidence from studies of Turner syndrome, fragile X syndrome, and neurofibromatosis type 1. Journal of Learning Disabilities 34 (6), 520–533.

Mazzocco, M.M., Singh Bhatia, N., Lesniak-Karpiak, K., 2006. Visuospatial skills and their association with math performance in girls with fragile X or Turner syndrome. Child Neuropsychology 12 (2), 87–110.

McFie, J., Zangwill, O.L., 1960. Visual-constructive disabilities associated with lesions of the left cerebral hemisphere. Brain 83, 243–259.

Merigan, W.H., Maunsell, J.H., 1993. How parallel are the primate visual pathways? Annual Review of Neuroscience 16, 369–402.

Merriman, W.E., Keating, D.P., List, J.A., 1985. Mental rotation of facial profiles: Age-, sex-, and ability-related differences. Developmental Psychology 21 (5), 888–900.

Mervis, C.B., Robinson, B.F., Bertrand, J., Morris, C.A., Klein-Tasman, B.P., Armstrong, S.C., 2000. The Williams syndrome cognitive profile. Brain and Cognition 44 (3), 604–628.

Meyer-Lindenberg, A., Kohn, P., Mervis, C.B., et al., 2004. Neural basis of genetically determined visuospatial construction deficit in Williams syndrome. Neuron 43 (5), 623–631.

Meyer-Lindenberg, A., Mervis, C.B., Berman, K.F., 2006. Neural mechanisms in Williams syndrome: A unique window to genetic influences on cognition and behaviour. Nature Reviews Neuroscience 7 (5), 380–393.

Mills, D.L., Alvarez, T.D., St George, M., Appelbaum, L.G., Bellugi, U., Neville, H., 2000. III. Electrophysiological studies of face processing in Williams syndrome. Journal of Cognitive Neuroscience 12 (supplement 1), 47–64.

Milner, A.D., Goodale, M.A., 1995. The Visual Brain in Action. Oxford University Press, New York, NY.

Mitchell, T.V., Neville, H.J., 2004. Asynchronies in the development of electrophysiological responses to motion and color. Journal of Cognitive Neuroscience 16 (8), 1363–1374.

Mitchell, L.E., Adzick, N.S., Melchionne, J., Pasquariello, P.S., Sutton, L.N., Whitehead, A.S., 2004. Spina bifida. Lancet 364 (9448), 1885–1895.

Mondloch, C.J., Geldart, S., Maurer, D., de Schonen, S., 2003. Developmental changes in the processing of hierarchical shapes continue into adolescence. Journal of Experimental Child Psychology 84 (1), 20–40.

Moses, P., Stiles, J., 2002. The lesion methodology: Contrasting views from adult and child studies. Developmental Psychobiology 40 (3), 266–277.

Moses, P., Roe, K., Buxton, R.B., Wong, E.C., Frank, L.R., Stiles, J., 2002. Functional MRI of global and local processing in children. NeuroImage 16 (2), 415–424.

Nass, R., deCoudres Peterson, H., Koch, D., 1989. Differential effects of congenital left and right brain injury on intelligence. Brain and Cognition 9 (2), 258–266.

Navon, D., 1977. Forest before trees: The precedence of global features in visual perception. Cognitive Psychology 9 (3), 353–383.

Navon, D., 2003. What does a compound letter tell the psychologist's mind? Acta Psychologica 114 (3), 273–309.

Nelson, K.B., Ellenberg, J.H., 1982. Children who 'outgrew' cerebral palsy. Pediatrics 69 (5), 529–536.

Nelson, K.B., Lynch, J.K., 2004. Stroke in newborn infants. Lancet Neurology 3 (3), 150–158.

Nelson, C.A., Monk, C.S., Lin, J., Carver, L.J., Thomas, K.M., Truwit, C.L., 2000. Functional neuroanatomy of spatial working memory in children. Developmental Psychology 36 (1), 109–116.

Nougier, V., Azemar, G., Stein, J.F., Ripoll, H., 1992. Covert orienting to central visual cues and sport practice relations in the development of visual attention. Journal of Experimental Child Psychology 54 (3), 315–333.

O'Hearn, K., Roth, J.K., Courtney, S.M., et al., 2011. Object recognition in Williams syndrome: Uneven ventral stream activation. Developmental Science 14 (3), 549–565.

Okagaki, L., Frensch, P.A., 1994. Effects of video game playing on measures of spatial performance: Gender effects in late adolescence. Journal of Applied Developmental Psychology 15 (1), 33–58. Special Issue: Effects of Interactive Entertainment Technologies on Development.

Oliveri, M., Turriziani, P., Carlesimo, G.A., et al., 2001. Parieto-frontal interactions in visual-object and visual-spatial working memory: Evidence from transcranial magnetic stimulation. Cerebral Cortex 11 (7), 606–618.

Orban, G.A., Van Essen, D., Vanduffel, W., 2004. Comparative mapping of higher visual areas in monkeys and humans. Trends in Cognitive Science 8 (7), 315–324.

Orsini, A., Grossi, D., Capitani, E., Laiacona, M., Papagno, C., Vallar, G., 1987. Verbal and spatial immediate memory span: Normative data from 1355 adults and 1112 children. Italian Journal of Neurological Sciences 8 (6), 539–548.

Palermo, L., Bureca, I., Matano, A., Guariglia, C., 2008. Hemispheric contribution to categorical and coordinate representational processes: A study on brain-damaged patients. Neuropsychologia 46 (11), 2802–2807.

Palmer, S.E., 1977. Hierarchical structure in perceptual representation. Cognitive Psychology 9 (4), 441–474.

Palmer, S.E., 1980. What makes triangles point: Local and global effects in configurations of ambiguous triangles. Cognitive Psychology 12 (3), 285–305.

Palmer, S.E., Bucher, N.M., 1981. Configural effects in perceived pointing of ambiguous triangles. Journal of Experimental Psychology. Human Perception and Performance 7 (1), 88–114.

Passarotti, A.M., Paul, B.M., Bussiere, J.R., Buxton, R.B., Wong, E.C., Stiles, J., 2003. The development of face and location processing: An fMRI study. Developmental Science 6 (1), 100–117.

Paul, B.M., Stiles, J., Passarotti, A., Bavar, N., Bellugi, U., 2002. Face and place processing in Williams syndrome: Evidence for a dorsal-ventral dissociation. NeuroReport 13 (9), 1115–1119.

Pearson, D.A., Lane, D.M., 1990. Visual attention movements: A developmental study. Child Development 61 (6), 1779–1795.

Peelen, M.V., Kastner, S., 2009. A nonvisual look at the functional organization of visual cortex. Neuron 63 (3), 284–286.

Piaget, J., 1952. The Origins of Intelligence in Children. International Universities Press, New York.

Piercy, M., Hecaen, H., de, A., 1960a. Constructional apraxia associated with unilateral cerebral lesions-left and right sided cases compared. Brain 83, 225–242.

Piercy, M., Hecaen, H., De Ajuriaguerra, J., 1960b. Constructional apraxia associated with unilateral cerebral lesions: Left and right sided cases compared. Brain 83, 225–242.

Pierrot-Deseilligny, C., Milea, D., Muri, R.M., 2004. Eye movement control by the cerebral cortex. Current Opinion in Neurology 17 (1), 17–25.

Poirel, N., Simon, G., Cassotti, M., et al., 2011. The shift from local to global visual processing in 6-year-old children is associated with grey matter loss. PLoS One 6 (6), e20879.

Porporino, M., Shore, D.I., Larocci, G., Burack, J.A., 2004. A developmental change in selective attention and global form perception. International Journal of Behavioral Development 28 (4), 358–364.

Posner, M.I., 1980. Orienting of attention. Quarterly Journal of Experimental Psychology 32 (1), 3–25.

Posner, M.I., Cohen, Y., 1980. Attention and the control of movements. In: Stelmach, G.E., Requin, J. (Eds.), Tutorials in Motor Behavior. North Holland Publishing, Amsterdam, pp. 243–258.

Posner, M.I., Petersen, S.E., 1990. The attention system of the human brain. Annual Review of Neuroscience 13, 25–42.

Posner, M.I., Walker, J.A., Friedrich, F.J., Rafal, R.D., 1984. Effects of parietal injury on covert orienting of attention. Journal of Neuroscience 4 (7), 1863–1874.

Posner, M.I., Rafal, R.D., Choate, L.S., Vaughan, J., 1985. Inhibition of return: Neural basis and function. Cognitive Neuropsychology 2 (3), 211–228.

Postle, B.R., Berger, J.S., Taich, A.M., D'Esposito, M., 2000. Activity in human frontal cortex associated with spatial working memory and saccadic behavior. Journal of Cognitive Neuroscience 12 (supplement 2), 2–14.

Postma, A., Kessels, R.P., van Asselen, M., 2008. How the brain remembers and forgets where things are: The neurocognition of object-location memory. Neuroscience and Biobehavioral Reviews 32 (8), 1339–1345.

Prather, P.A., Bacon, J., 1986. Developmental differences in part/whole identification. Child Development 57 (3), 549–558.

Quinn, P.C., Bhatt, R.S., 2006. Are some gestalt principles deployed more readily than others during early development? The case of lightness versus form similarity. Journal of Experimental Psychology. Human Perception and Performance 32 (5), 1221–1230.

Quinn, P.C., Burke, S., Rush, A., 1993. Part-whole perception in early infancy: Evidence for perceptual grouping produced by lightness similarity. Infant Behavior & Development 16 (1), 19–42.

Rae, C., Joy, P., Harasty, J., et al., 2004. Enlarged temporal lobes in Turner syndrome: An X-chromosome effect? Cerebral Cortex 14 (2), 156–164.

Rafal, R., Robertson, L., 1995. The neurology of visual attention. In: Gazzaniga, M.S. et al., (Ed.), The Cognitive Neurosciences. The MIT Press, Cambridge, MA, pp. 625–648.

Ranganath, C., 2006. Working memory for visual objects: Complementary roles of inferior temporal, medial temporal, and prefrontal cortex. Neuroscience 139 (1), 277–289.

Ranganath, C., D'Esposito, M., 2005. Directing the mind's eye: Prefrontal, inferior and medial temporal mechanisms for visual working memory. Current Opinion in Neurobiology 15 (2), 175–182.

Ranganath, C., Cohen, M.X., Dam, C., D'Esposito, M., 2004. Inferior temporal, prefrontal, and hippocampal contributions to visual working memory maintenance and associative memory retrieval. Journal of Neuroscience 24 (16), 3917–3925.

Ratcliff, G., 1982. Disturbances of spatial orientation associated with cerebral lesions. In: Potegal, M. (Ed.), Spatial Abilities: Development and Physiological Foundations. Academic Press, New York, pp. 301–331.

Reese, C.J., Stiles, J., 2005. Hemispheric specialization for categorical and coordinate spatial relations during an image generation task: Evidence from children and adults. Neuropsychologia 43 (4), 517–529.

Reese, C. J., Nass, R., Trauner, D., Stiles, J. (in preparation). Categorical and coordinate image generation in children with early focal brain injury.

Reid, V.M., Striano, T., Kaufman, J., Johnson, M.H., 2004. Eye gaze cueing facilitates neural processing of objects in 4-month-old infants. NeuroReport 15 (16), 2553–2555.

Reilly, J.S., Levine, S.C., Nass, R.D., Stiles, J., 2008. Brain plasticity: Evidence from children with perinatal brain injury. In: Reed, J., Warner, J. (Eds.), Child Neuropsychology. Blackwell Publishing, Oxford, pp. 58–91.

Reiss, A.L., Dant, C.C., 2003. The behavioral neurogenetics of fragile X syndrome: Analyzing gene-brain-behavior relationships in child developmental psychopathologies. Development and Psychopathology 15 (4), 927–968.

Reynolds, G.D., Richards, J.E., 2005. Familiarization, attention, and recognition memory in infancy: An event-related potential and cortical source localization study. Developmental Psychology 41 (4), 598–615.

Rhodes, G., Byatt, G., Michie, P.T., Puce, A., 2004. Is the fusiform face area specialized for faces, individuation, or expert individuation? Journal of Cognitive Neuroscience 16 (2), 189–203.

Richter, W., Ugurbil, K., Georgopoulos, A., Kim, S.G., 1997. Time-resolved fMRI of mental rotation. NeuroReport 8 (17), 3697–3702.

Riecansky, I., 2004. Extrastriate area V5 (MT) and its role in the processing of visual motion. Ceskoslovenská Fysiologie 53 (1), 17–22.

Rizzolatti, G., Matelli, M., 2003. Two different streams form the dorsal visual system: Anatomy and functions. Experimental Brain Research 153 (2), 146–157.

Rizzolatti, G., Sinigaglia, C., 2010. The functional role of the parieto-frontal mirror circuit: Interpretations and misinterpretations. Nature Reviews Neuroscience 11 (4), 264–274 (Research Support, Non-U.S. Gov't Review).

Roberts, J.E., Bell, M.A., 2002. The effects of age and sex on mental rotation performance, verbal performance, and brain electrical activity. Developmental Psychobiology 40 (4), 391–407.

Robertson, L.C., 1992. Perceptual organization and attentional search in cognitive deficits. In: Margolin, D.I. et al., (Ed.), Cognitive Neuropsychology in Clinical Practice. Oxford University Press, New York, NY, pp. 70–95.

Robertson, L.C., Delis, D.C., 1986. 'Part-whole' processing in unilateral brain-damaged patients: Dysfunction of hierarchical organization. Neuropsychologia 24 (3), 363–370.

Robertson, L.C., Lamb, M.R., 1991. Neuropsychological contributions to theories of part/whole organization. Cognitive Psychology 23 (2), 299–330.

Robertson, L.C., Lamb, M.R., Zaidel, E., 1993. Interhemispheric relations in processing hierarchical patterns: Evidence from normal and commissurotomized subjects. Neuropsychology 7 (3), 325–342.

Rosa, M.G., Palmer, S.M., Gamberini, M., et al., 2009. Connections of the dorsomedial visual area: Pathways for early integration of dorsal and ventral streams in extrastriate cortex. Journal of Neuroscience 29 (14), 4548–4563.

Rossion, B., Gauthier, I., 2002. How does the brain process upright and inverted faces? Behavioral and Cognitive Neuroscience Reviews 1 (1), 63–75.

Sala, J.B., Rama, P., Courtney, S.M., 2003. Functional topography of a distributed neural system for spatial and nonspatial information maintenance in working memory. Neuropsychologia 41 (3), 341–356 (Special Issue: Functional Neuroimaging of Memory).

Sandberg, E.H., Huttenlocher, J., Newcombe, N., 1996. The development of hierarchical representation of two-dimensional space. Child Development 67 (3), 721–739.

Sarpal, D., Buchsbaum, B.R., Kohn, P.D., et al., 2008. A genetic model for understanding higher order visual processing: Functional interactions of the ventral visual stream in Williams syndrome. Cerebral Cortex 18 (10), 2402–2409.

Scherf, K.S., Luna, B., Avidan, G., Behrmann, M., 2011. 'What' precedes 'which': Developmental neural tuning in face- and place-related cortex. Cerebral Cortex 21 (9), 1963–1980.

Schneider, A., Hagerman, R.J., Hessl, D., 2009. Fragile X syndrome – From genes to cognition. Developmental Disabilities Research Reviews 15 (4), 333–342.

Schul, R., Townsend, J., Stiles, J., 2003. The development of attentional orienting during the school-age years. Developmental Science 6 (3), 262–272.

Seltzer, B., Pandya, D.N., 1978. Afferent cortical connections and architectonics of the superior temporal sulcus and surrounding cortex in the rhesus monkey. Brain Research 149 (1), 1–24.

Sergent, J., 1982. The cerebral balance of power: Confrontation or cooperation? Journal of Experimental Psychology. Human Perception and Performance 8 (2), 253–272.

Shepard, R.N., Cooper, L.A., 1986. Mental Images and Their Transformations. The MIT Press, Cambridge, MA.

Shorr, J.S., Delis, D.C., Massman, P.J., 1992. Memory for the Rey-Osterrieth Figure: Perceptual clustering, encoding, and storage. Neuropsychology 6, 43–50.

Sincich, L.C., Horton, J.C., 2005. The circuitry of V1 and V2: Integration of color, form, and motion. Annual Review of Neuroscience 28, 303–326.

Slater, A., Mattock, A., Brown, E., Bremner, J.G., 1991. Form perception at birth: Cohen and Younger (1984) revisited. Journal of Experimental Child Psychology 51 (3), 395–406.

Smith, L.B., Kemler, D.G., 1977. Developmental trends in free classification: Evidence for a new conceptualization of perceptual development. Journal of Experimental Child Psychology 24 (2), 279–298.

Smith, E.E., Jonides, J., Koeppe, R.A., Awh, E., Schumacher, E.H., Minoshima, S., 1995. Spatial versus object working memory: PET investigations. Journal of Cognitive Neuroscience 7 (3), 337–356.

Smith, E.E., Jonides, J., Koeppe, R.A., 1996. Dissociating verbal and spatial working memory using PET. Cerebral Cortex 6 (1), 11–20.

Smith, L.B., Thelen, E., Titzer, R., McLin, D., 1999. Knowing in the context of acting: The task dynamics of the A-not-B error. Psychological Review 106 (2), 235–260.

Snow, J.H., 1990. Investigation of a mental rotation task with school-age children. Journal of Psychoeducational Assessment 8 (4), 538–549.

Spencer, J.P., Smith, L.B., Thelen, E., 2001. Tests of a dynamic systems account of the A-not-B error: The influence of prior experience on the spatial memory abilities of two-year-olds. Child Development 72 (5), 1327–1346.

Stiles, J., Stern, C., 2001. Developmental change in spatial cognitive processing: Complexity effects and block construction performance in preschool children. Journal of Cognition and Development 2 (2), 157–187.

Stiles, J., Stern, C., Trauner, D., Nass, R., 1996. Developmental change in spatial grouping activity among children with early focal brain injury: Evidence from a modeling task. Brain and Cognition 31 (1), 46–62.

Stiles, J., Moses, P., Passarotti, A., Dick, F., Buxton, R., 2003. Exploring developmental change in the neural bases of higher cognitive functions: The promise of functional magnetic resonance imaging. Developmental Neuropsychology 24 (2&3), 641–668.

Stiles, J., Paul, B., Hesselink, J., 2006. Spatial cognitive development following early focal brain injury: Evidence for adaptive change in brain and cognition. In: Munakata, Y., Johnson, M.H. (Eds.), Process of Change in Brain and Cognitive Development. Attention and Performance XXI. Oxford University Press, Oxford, pp. 535–561.

Stiles, J., Stern, C., Appelbaum, M., Nass, R., Trauner, D., Hesselink, J., 2008. Effects of early focal brain injury on memory for visuospatial patterns: Selective deficits of global–local processing. Neuropsychology 22 (1), 61–73.

Stiles, J., Nass, R.D., Levine, S.C., Moses, P., Reilly, J.S., 2009. Perinatal stroke: Effects and outcomes. In: Yeates, K.O., Ris, M.D., Taylor, H.G., Pennington, B. (Eds.), Pediatric Neuropsychology: Research, Theory, and Practice. 2nd edn. The Guilford Press, New York, pp. 181–210.

Stiles, J., Reilly, J.S., Levine, S.C., Trauner, D.A., Nass, R., 2012. Neural Plasticity and Cognitive Development: Insights from Children with Perinatal Brain Injury. Oxford University Press, New York.

Stiles-Davis, J., 1988. Developmental change in young children's spatial grouping activity. Developmental Psychology 24 (4), 522–531.

Swindell, C.S., Holland, A.L., Fromm, D., Greenhouse, J.B., 1988. Characteristics of recovery of drawing ability in left and right brain-damaged patients. Brain and Cognition 7 (1), 16–30.

Tada, W.L., Stiles, J., 1996. Developmental change in children's analysis of spatial patterns. Developmental Psychology 32 (5), 951–970.

Tallon-Baudry, C., Bertrand, O., Peronnet, F., Pernier, J., 1998. Induced gamma-band activity during the delay of a visual short-term memory task in humans. Journal of Neuroscience 18 (11), 4244–4254.

Tamm, L., Menon, V., Reiss, A.L., 2003. Abnormal prefrontal cortex function during response inhibition in Turner syndrome: Functional magnetic resonance imaging evidence. Biological Psychiatry 53 (2), 107–111.

Tanaka, J.W., Farah, M.J., 1993. Parts and wholes in face recognition. Quarterly Journal of Experimental Psychology: Human Experimental Psychology 46A (2), 225–245.

Taylor, M.J., McCarthy, G., Saliba, E., Degiovanni, E., 1999. ERP evidence of developmental changes in processing of faces. Clinical Neurophysiology 110 (5), 910–915.

Taylor, M.J., Edmonds, G.E., McCarthy, G., Allison, T., 2001. Eyes first! Eye processing develops before face processing in children. NeuroReport 12 (8), 1671–1676.

Thiele, A., Dobkins, K.R., Albright, T.D., 2001. Neural correlates of chromatic motion perception. Neuron 32 (2), 351–358.

Trojano, L., Grossi, D., Linden, D.E., et al., 2002. Coordinate and categorical judgements in spatial imagery. An fMRI study. Neuropsychologia 40 (10), 1666–1674.

Tzourio-Mazoyer, N., De Schonen, S., Crivello, F., Reutter, B., Aujard, Y., Mazoyer, B., 2002. Neural correlates of woman face processing by 2-month-old infants. NeuroImage 15 (2), 454–461.

Ungerleider, L.G., 1995. Functional brain imaging studies of cortical mechanisms for memory. Science 270 (5237), 769–775.

Ungerleider, L.G., Mishkin, M., 1982. Two cortical visual systems. In: Ingle, D.J., Goodale, M.A., Mansfield, R.J.W. (Eds.), Analysis of Visual Behavior. The MIT Press, Cambridge, MA, pp. 549–586.

Valenza, E., Simion, F., Umilta, C., 1994. Inhibition of return in newborn infants. Infant Behavior & Development 17 (3), 293–302.

van Asselen, M., Kessels, R.P., Kappelle, L.J., Postma, A., 2008. Categorical and coordinate spatial representations within object-location memory. Cortex 44 (3), 249–256.

Van de Walle, G.A., Carey, S., Prevor, M., 2000. Bases for object individuation in infancy: Evidence from manual search. Journal of Cognition and Development 1 (3), 249–280.

Van Essen, D.C., Anderson, C.H., Felleman, D.J., 1992. Information processing in the primate visual system: An integrated systems perspective. Science 255 (5043), 419–423.

Van Heest, A.E., House, J., Putnam, M., 1993. Sensibility deficiencies in the hands of children with spastic hemiplegia. The Journal of Hand Surgery 18 (2), 278–281.

Varga, K., Frick, J.E., Kapa, L.L., Dengler, M.J., 2010. Developmental changes in inhibition of return from 3 to 6 months of age. Infant Behavior & Development 33 (2), 245–249.

Vicari, S., Stiles, J., Stern, C., Resca, A., 1998. Spatial grouping activity in children with early cortical and subcortical lesions. Developmental Medicine and Child Neurology 40 (2), 90–94.

Vinter, A., Puspitawati, I., Witt, A., 2010. Children's spatial analysis of hierarchical patterns: Construction and perception. Developmental Psychology 46 (6), 1621–1631.

Volberg, G., Hubner, R., 2004. On the role of response conflicts and stimulus position for hemispheric differences in global/local processing: An ERP study. Neuropsychologia 42 (13), 1805–1813.

Von Bonin, G., Bailey, P., 1947. The Neocortex of Macaca mulatta. In: Illinois Monographs in the Medical Sciences, vol. 5., No. 4. University of Illinois Press, Urbana, IL.

Vurpillot, E., 1976. The Visual World of the Child. International Universities Press, New York, NY.

Waber, D.P., Carlson, D., Mann, M., 1982. Developmental and differential aspects of mental rotation in early adolescence. Child Development 53 (6), 1614–1621.

Walter, E., Mazaika, P.K., Reiss, A.L., 2009. Insights into brain development from neurogenetic syndromes: Evidence from fragile X syndrome, Williams syndrome, Turner syndrome and velocardiofacial syndrome. Neuroscience 164 (1), 257–271.

Warrington, E.K., James, M., Kinsbourne, M., 1966. Drawing disability in relation to laterality of cerebral lesion. Brain 89 (1), 53–82.

Wellman, H.M., Cross, D., Bartsch, K., 1987. Infant search and object permanence: A meta-analysis of the A-not-B error. Monographs of the Society for Research in Child Development 51 (3), 1–51.

Willis, S.L., Schaie, K.W., 1988. Gender differences in spatial ability in old age: Longitudinal and intervention findings. Sex Roles 18 (3–4), 189–203.

Wise, S.P., Boussaoud, D., Johnson, P.B., Caminiti, R., 1997. Premotor and parietal cortex: Corticocortical connectivity and combinatorial computations. Annual Review of Neuroscience 20 (10), 25–42.

Wojciulik, E., Kanwisher, N., Driver, J., 1998. Covert visual attention modulates face-specific activity in the human fusiform gyrus: fMRI study. Journal of Neurophysiology 79 (3), 1574–1578.

Wong-Riley, M., 1979. Changes in the visual system of monocularly sutured or enucleated cats demonstrable with cytochrome oxidase histochemistry. Brain Research 171 (1), 11–28.

Wu, Y.W., Lindan, C.E., Henning, L.H., et al., 2006. Neuroimaging abnormalities in infants with congenital hemiparesis. Pediatric Neurology 35 (3), 191–196.

Xu, Y., 2005. Revisiting the role of the fusiform face area in visual expertise. Cerebral Cortex 15 (8), 1234–1242.

Xu, F., Carey, S., 1996. Infants' metaphysics: The case of numerical identity. Cognitive Psychology 30 (2), 111–153.

Xu, F., Carey, S., Quint, N., 2004. The emergence of kind-based object individuation in infancy. Cognitive Psychology 49 (2), 155–190.

Yekutiel, M., Jariwala, M., Stretch, P., 1994. Sensory deficit in the hands of children with cerebral palsy: A new look at assessment and prevalence. Developmental Medicine and Child Neurology 36 (7), 619–624.

Yin, R.K., 1969. Looking at upide-down faces. Journal of Experimental Psychology 81 (1), 141–145.

Yovel, G., Kanwisher, N., 2004. Face perception: Domain specific, not process specific. Neuron 44 (5), 889–898.

Yovel, G., Kanwisher, N., 2005. The neural basis of the behavioral face-inversion effect. Current Biology 15 (24), 2256–2262.

Yovel, G., Levy, J., Yovel, I., 2001. Hemispheric asymmetries for global and local visual perception: Effects of stimulus and task factors. Journal of Experimental Psychology. Human Perception and Performance 27 (6), 1369–1385.

Zald, D.H., Iacono, W.G., 1998. The development of spatial working memory abilities. Developmental Neuropsychology 14 (4), 563–578.

Zanon, M., Busan, P., Monti, F., Pizzolato, G., Battaglini, P.P., 2010. Cortical connections between dorsal and ventral visual streams in humans: Evidence by TMS/EEG co-registration. Brain Topography 22 (4), 307–317 (Research Support, Non-U.S. Gov't).

16

Memory Development

P.J. Bauer

Emory University, Atlanta, GA, USA

16.1 INTRODUCTION

The title of this chapter – *Memory Development* – creates the impression that a single entity – *memory* – has a single course of development. Instead, there are several different types of memory, each with its own characteristics and course of development. For example, one type of memory is short-term, lasting only seconds or minutes before fading away, whereas another type is long-term, lasting as long as a lifetime. Some types of memory have a limited capacity, whereas others are, for all practical purposes, limitless. Memory sometimes is context-specific and other times highly flexible. In fact, it sometimes seems that the only thing all memories have in common is that they are about the past. Yet even this characterization is not universally true, in that remembering to do something *in the future* also qualifies as 'memory' of the prospective type.

One primary goal of this chapter is to characterize the major forms of memory. This step is necessary in order to accomplish the second goal, which is to describe the course of development of different types of memory over the first years of life. As will become apparent, different types of memory have different courses of development. Differences in the timing and course of development shed light on some of the mechanisms of age-related change, consideration of which is the third major goal of the chapter.

© 2013 Elsevier Inc. All rights reserved.

16.2 DIFFERENT FORMS OF MEMORY

The suggestion that memory is not a unitary construct is an old one. Maine de Biran (1804/1929) is credited as the first to suggest that there might be different forms of memory (Schacter et al., 2000). The notion was furthered at the beginning of the twentieth century with studies of wounded veterans from World War I. Kleist (1934), a German physician, examined veterans who had received head wounds from gun shots or shrapnel and the behavioral patterns that seemed to result from them. He observed that there were systematic relations between the site of the wound (and resulting brain lesion) and the type of mental impairment experienced by the veteran. The notion that different parts of the brain subserve different cognitive functions, including different types of memory, received especially strong impetus from the famous case of H. M. who, at the age of 27, underwent experimental surgery to treat intractable seizures. To treat the seizures, his surgeon removed large portions of the temporal lobes on both sides of the brain (Scoville and Milner, 1957). Subsequent to the surgery, H. M. suffered impairment of some forms of memory, yet not all of his memory capacities were disrupted. His case in particular led to the notion that there are different types or forms of memory. Whereas some distinctions within the domain are readily accepted, others are sources of debate.

16.2.1 Short- and Long-Term Memory

Characterizing memory along a temporal dimension is relatively uncontroversial. Whereas some memories are short-term, lasting only seconds, others last much longer, on the order of days, months, and even years. Short-term memory generally is recognized as capacity-limited. It is commonly thought to hold 7 'units' of information – such as digits in a phone number – plus or minus 2 (Miller, 1956), though estimates of capacity vary. In contrast, long-term memory is virtually limitless in its capacity. There seemingly is no upper limit on the number of items, pieces of information, or personal experiences that one can maintain in long-term memory stores (more on the temporal dimension and capacity limitations of memory below).

16.2.2 Declarative and Nondeclarative Memory

Within long-term memory, many recognize a division based on the contents of memory, its function, its rules of operation, and the neural substrates that support it. Although the precise distinctions captured by the different labels given to the divisions are not identical, there is substantial overlap in contemporary conceptualizations of declarative or explicit memory versus nondeclarative, procedural, or implicit memory. Each is discussed in turn.

16.2.3 Declarative or Explicit Memory

Declarative or explicit memory is devoted to processing of names, dates, places, facts, events, and so forth. These are entities that are thought of as being encoded symbolically and that thus can be described with language. In terms of function, declarative memory is specialized for fast processing and learning. New information can be entered into the declarative memory system on the basis of a single trial or experience. In terms of rules of operation, declarative is fallible: one forgets names, dates, places, and so forth. Although there are compelling demonstrations of long-term remembering of lessons learned in high school and college (e.g., foreign language vocabulary: Bahrick, 2000), a great deal of forgetting from declarative memory occurs literally minutes, hours, and days after an experience. Declarative memory also has a specific neural substrate. Current conceptualizations suggest that the formation, maintenance, and subsequent retrieval of declarative or explicit memories depend on a multicomponent network involving cortical structures (including posterior–parietal, anterior–prefrontal, and limbic–temporal association areas) as well as medial temporal structures (including the hippocampus and entorhinal, perirhinal, and parahippocampal cortices: e.g., Eichenbaum and Cohen, 2001; Murray and Mishkin, 1998; Zola and Squire, 2000). The medial temporal structures may be considered 'primary' in the sense that without them, whether measured by recall or recognition, declarative memory is impaired (Moscovitch, 2000).

Declarative memory is itself subdivided into the categories of semantic and episodic memory (e.g., Schacter and Tulving, 1994), with a finer distinction between episodic and autobiographical memory. Semantic memory supports general knowledge about the world (Tulving, 1972, 1983). People are consulting semantic memory when they retrieve the facts that the capital of the United States is Washington, DC, that the United States has 50 states, and that with over 660 000 square miles, Alaska is the largest state in terms of land mass. For practical purposes, both the capacity of semantic memory and the longevity of the information stored in it seem infinite. Semantic memory also is not tied to a particular time or place. That is, people know facts and figures, names and dates, yet in most cases, they do not know when and where they learned this information. People might be able to reconstruct how old they were or what grade they were in when they learned some tidbits of information, but unless there was something unique about the experience surrounding the acquisition of this information, they carry it around without address or reference to a specific episode.

In contrast to semantic memory, episodic memory supports retention of information about unique events (Tulving, 1972, 1983), such as a specific visit to

Washington, DC, or the fact that Alaska was one of the states on a list of state names studied in a memory experiment. Some episodic memories, such as whether a specific state was included in a word list, may not stay with one for very long and are not especially personally relevant or significant. Yet other episodic memories are personally significant and even self-defining. These so-called *autobiographical* memories are episodic memories that are infused with a sense of personal involvement or ownership (Bauer, 2007). They are the episodes on which people reflect when they consider who they are and how their previous experiences have shaped them.

16.2.4 Nondeclarative, Procedural, or Implicit Memory

Nondeclarative, procedural, or implicit memory is devoted to perceptual and motor skills and procedures. Skilled motor behavior, such as dancing or swinging a tennis racket, is not a name, date, place, fact, or event but a collection of finely tuned motor patterns, behaviors, and perceptual skills that one cannot verbally describe. Most types of nondeclarative memory function to support gradual, incremental learning. That is, behavior is modified through practice, experience, or multiple trials. Perhaps as a result of its slow function, a rule of the operation of nondeclarative memory is that the learning is relatively infallible (see next paragraph). A typical example is riding a bicycle – one may not have ridden a bicycle in many years but when he or she rides one again, he or she 'just knows how.' Indeed, breaking an old nondeclarative pattern can be quite difficult, as exemplified by the tendency to, in a moment of panic, slam the pedals backward (as one does on his or her old 1-speed bike) instead of squeezing the hand brakes (as one should on his or her adult, multi-speed bike). As might be expected, given the diversity of behaviors supported by nondeclarative memory, different types of nondeclarative memory depend on different neural structures. For example, motor skill learning and many types of conditioning seemingly are dependent on cerebellum and subcortical structures such as the basal ganglia and priming seems dependent on extrastriate cortex.

Just as declarative memory is subdivided, so too is nondeclarative memory. The most common categories of nondeclarative memory are (1) motor skill learning and (2) priming, and (3) classical conditioning. The range of motor skills that people acquire is limitless. What they have in common is that although there might be a declarative component to them (to stop the bike, put on the brakes), their fluid execution is not accomplished via learning and frequent repetition of the declarative 'rules,' but motor practice. Hours and hours and hours of practice lie behind the fluid swish of a tennis

champion's racket or the dance steps of Ginger Rogers and Fred Astaire. A tennis or dance instructor can tell one how to hold the racket or how to position one's foot to produce certain steps, but it is not this knowledge on which accomplished athletes depend. Closer to home, people all know that to stop cars that they are driving, they put their feet on the brakes. Yet few if any people know how many pounds of pressure they must put on the brake pedal to stop quickly versus more gradually. The information is encoded in their muscles and joints; it is not accessible to conscious access or description.

Another form of nondeclarative memory is priming, which is facilitated processing of a stimulus following prior exposure to the stimulus. Perceptual priming occurs when subsequent processing of a stimulus is facilitated by prior perceptual exposure to it. Conceptual priming occurs when there is overlap between related concepts in memory. A person who had recently studied a list that included the word 'Alaska' would be more likely to include it in a list of states than an individual who had not studied the item. With regard to the distinction between conscious (declarative) and unconscious (nondeclarative) memory, an important point to remember about priming is that it can occur without conscious awareness that the item had been studied earlier, or that subsequent processing was facilitated by earlier processing. That is, facilitated processing occurs even in the absence of recognition or recall of the originally studied item (see Lloyd and Newcombe, 2009 and Roediger and McDermott, 1993, for discussions).

Finally, classical conditioning occurs when two stimuli that naturally do not co-occur become associated with one another through repeated pairing. Typically, one of the stimuli – the unconditioned stimulus – evokes a high-probability or even reflexive response termed the unconditioned response. For example, a puff of air to the eye (unconditioned stimulus) produces a blink (unconditioned response). The other stimulus – the conditioned stimulus – is behaviorally neutral, such as a tone. Classical conditioning occurs when, over time and repeated pairing of the two stimuli, the conditioned stimulus (the tone) takes on the same significance as the unconditioned stimulus, such that it alone is sufficient to produce the response: anticipatory eye closure at the sound of the tone (see Woodruff-Pak and Disterhoft, 2008, for a review). This simple form of learning occurs across phyla (e.g., rodents, nonhuman primates, and humans), making it one of the clearest examples of learning and memory without awareness.

16.2.5 Relations Between Different Forms of Memory

An important point to keep in mind with regard to the distinction between declarative and nondeclarative forms of memory is that, in most cases, one derives both declarative and nondeclarative memories from the same experience. Learning to drive a car is a good example. In driver's education class, students learn that to go, they step on the accelerator and to stop, they step on the brake. Tests to obtain a license to drive probe the student's memory of appropriate following and stopping distance and the steps involved in parallel parking, for example. But one would never become an expert driver based on this type of memory alone. To become a good driver, one must practice executing the motor movements. Although one may be consciously aware that vehicle stopping distance is 15 feet per second, it is not this knowledge that permits him or her to bring the vehicle to a gentle stop at a stop sign. That skill is acquired through practice, practice, and more practice at driving and braking.

Another important point is that, in most cases, declarative and nondeclarative memories not only are acquired in parallel but also continue to coexist, even after execution of the behavior no longer seems to require conscious awareness. Many skills, such as driving, start out very demanding of attentional resources yet eventually can be performed almost 'automatically,' that is, without conscious attention paid to them. Such changes tempt the conclusion that once these skills no longer require conscious attention to execute, they no longer are declarative. Yet such a conclusion is not valid: If declarative memories were to become nondeclarative, that would mean that they were no longer accessible to consciousness or declaration. The fact that one can tell that vehicle stopping distance is 15 feet per second proves that memory of the procedure is accessible to consciousness, even though one does not have to think about feet per second in order to brake. In other words, one still has the declarative memory, even though it is not that memory on which he or she depends to execute the behavior. In the intact organism, declarative and nondeclarative memory coexist. Many behaviors are executed based on nondeclarative memory alone, but that does not mean that one type of memory has 'turned into' another. Rather, at the time of learning, both types of memory were acquired; skilled performance of the motor behavior may be supported by one (nondeclarative), yet the other (declarative) continues to exist.

The observations that different forms of memory are acquired in parallel and coexist, possibly throughout the life of the memory, beg the question of why organisms would have multiple types of memory. It is likely that either these different forms of memory evolved in order to deal with competing demands for different kinds of information storage or nature took fair advantage of structures that had evolved for other reasons, in order to deal with the demands. Either way, one memory system seems specialized for rapid encoding of information that is subject to equally rapid forgetting

whereas the other seems specialized for acquisition of information at a slower rate yet seemingly permits more robust retention. Why could not a single system accomplish both tasks? Although this question cannot be answered definitively, computer simulations have revealed that a system that can change rapidly to accept new inputs has difficulty maintaining old inputs. Conversely, a system that is good at maintaining old inputs has difficulty 'learning' new things (e.g., McClelland et al., 1995). This analysis suggests that complementary memory systems work in concert in order to avoid interference with existing knowledge yet still maintain flexibility.

16.3 DEVELOPMENTAL CHANGES IN NONDECLARATIVE AND DECLARATIVE MEMORY

Discussion of developmental changes in all of the different forms of memory is well beyond the scope of this chapter. Instead, exemplars of each major type of memory are discussed as follows: priming, as an exemplar of nondeclarative memory, and episodic and autobiographical memory, as exemplars of declarative memory.

16.3.1 Priming

In the modern cognitive science and neuroscience literatures, priming is perhaps the most thoroughly studied form of nondeclarative memory. As defined above, priming involves facilitated processing of a stimulus as a result of prior exposure to it. Perceptual priming occurs when facilitated processing is based on the surface features of the stimulus; conceptual priming occurs when the facilitation is based on the meaning of the stimuli. In both cases, the facilitation is observed independent of explicit recognition of the stimuli as having been seen before (thus leading to characterization of the effect as nondeclarative).

Perceptual priming is apparent from an early age and is thought to show only small developmental changes. In a typical paradigm, subjects encounter pictures of common objects early in an experimental session. Later in the session they are shown degraded pictures that are slowly made sharper and clearer. Some of the pictures are of the objects studied earlier whereas others are of new objects. Children as young as 3 are faster to name the pictures (i.e., they name them at lower levels of clarity) when they have seen them previously than when they have not. Age-related differences in performance on these types of tasks are not pronounced (e.g., Hayes and Hennessey, 1996; Parkin and Streete, 1988), even when children are compared with adults (e.g., Drummey and Newcombe, 1995). For this reason, perceptual priming is thought to be relatively developmentally invariant (see Lloyd and Newcombe, 2009, and Parkin, 1998 for reviews).

Age-related differences are apparent in conceptual priming paradigms. In a representative conceptual priming paradigm, subjects are presented with words early in a session. Later they are challenged to list members of target categories. For example, in response to the instruction to list as many states as they can, subjects are more likely to nominate 'Alaska' if they saw the word earlier in the session than if they did not. Effects of this nature – effects that seemingly depend on activation of related concepts in memory – are more robust in older relative to younger children (e.g., Perez et al., 1998). Data such as these have been used to argue that conceptual priming shows salient improvements with age. Lloyd and Newcombe (2009) sounded a cautionary note in their review of this literature, however. They noted the distinct possibility that rather than differences in priming, these paradigms may reflect age-related improvements in conceptual knowledge. That is, the mechanism of priming may be age-invariant, and what accounts for improvements with age is category knowledge. Testing of this possibility is an area for future research.

16.3.2 Episodic and Autobiographical Memory

Episodic and autobiographical memory have a protracted course of development, beginning in the first year of life and extending into early adulthood. As discussed above, episodic memory supports retention of information about unique events that can be located in a particular place and time (Tulving, 1972, 1983). Autobiographical memories are episodic memories that are especially personally relevant. They are what is thought to make up the life story or personal past.

16.3.3 Questioning the Existence of Episodic Memory in Young Children

Until the middle 1980s, it was widely believed that infants and young children were incapable of remembering specific episodes and thus of forming autobiographical memories. As discussed in Bauer (2006b, 2007), the pessimism as to children's mnemonic abilities stemmed from a number of sources. One salient source of the perspective was the literature on *infantile* or *childhood amnesia*. Most adults remember few if any memories from the first 3–4 years of life, and from the ages of 3.5–7, they have a smaller number of memories than would be expected based on forgetting alone. As reviewed in Bauer (2007, 2008), the phenomenon is strikingly robust and consistent across time, population, and method (e.g., free recall,

response to cue words, and questionnaire). Although a variety of explanations for the amnesia have been advanced (see Bauer, 2007, 2008, for reviews), one of the most common was also the simplest: adults lacked memories from early in life because children failed to create them.

The suggestion that children younger than age 3 years did not form episodic memories was consistent with the dominant theoretical perspective at the time. A central tenant of Jean Piaget's *genetic epistemology* (see Flavell, 1963, for an introduction to the perspective) was that for the first 18–24 months of life, infants lacked symbolic capacity and, thus, the ability to mentally represent objects and events (e.g., Piaget, 1952). Instead, they were thought to live in a 'here and now' world that included physically present entities, yet the entities had no past and no future. In other words, infants were described as living an 'out of sight, out of mind' existence. Piaget hypothesized that by 18–24 months of age, children had constructed the capacity for mental representation. However, even then, they were thought to be without the cognitive structures that would permit them to organize events along coherent dimensions that would make the events memorable. Consistent with this suggestion, in retelling fairy tales, children as old as 7 years made errors in temporal sequencing (Piaget, 1926, 1969). Piaget attributed their poor performance to the lack of reversible thought. Without it, children could not organize information temporally and thus could not tell a story from beginning, to middle, to end. Without this ability, they could not be expected to retain coherent episodic memories.

16.3.4 A New Perspective on Young Children's Memory Abilities

The perspective on infants' and young children's mnemonic abilities began to change in the middle 1980s as a result of recognition of the importance to memory of meaningful and familiar stimuli and development of a means of assessing event memory in pre- and early-verbal children. In an influential series of studies, Mandler and her colleagues (e.g., Mandler and DeForest, 1979; Mandler and Johnson, 1977) demonstrated that even young children are sensitive to the structure inherent in story materials. When the stories were well-organized, children had high levels of recall. They also tended to 'correct' poorly organized stories, to conform to the hierarchical structure (see Mandler, 1984, for a review). At roughly the same time, Nelson and her colleagues (e.g., Nelson, 1986) demonstrated that when children were asked to recall 'what happens' in the context of everyday events and routines, such as going to McDonald's, their performance was qualitatively similar to that of older children and even adults. Moreover, it became apparent that children did not require multiple experiences of events in order to remember them. For example, Fivush (1984) interviewed kindergarten children after only a single day of school. Although the children had experienced the school-day routine just once, they nevertheless provided well-organized reports of the experience.

The change in perspective on memory ability extended to infants as well, with the development of nonverbal means of assessing memories for specific episodes, namely, elicited and deferred imitation. Elicited and deferred imitation entail use of objects to demonstrate an action or sequence of actions that, either immediately (elicited imitation), after some delay (deferred imitation), or both, infants are invited to imitate. Piaget (1952) himself had identified deferred imitation as one of the hallmarks of the development of symbolic thought. Meltzoff (1985) and Bauer and her colleagues (Bauer and Mandler, 1989 and Bauer and Shore, 1987) brought the technique under experimental control. Meltzoff (1988) demonstrated that infants as young as 9 months of age were able to defer imitation of an action for 24 hours. Bauer and Shore (1987) published findings that over a 6-week delay, infants 17–23 months of age remembered not only individual actions but temporally ordered sequences of action. That is, even after 6 weeks, they were able to reproduce in the correct temporal order the steps of putting a ball into a cup, covering it with another cup, and shaking the cups to make a rattle.

Whereas originally, the argument that imitation-based paradigms provide a means of testing declarative memory was based in Piaget's (1952) observations, there are a number of characteristics that support the claim. Because the argument has been developed in detail elsewhere (e.g., Bauer, 2006b, 2007 and Carver and Bauer, 2001), only two components of it are presented. First, once children acquire the requisite language, they talk about events that they experienced as preverbal infants, in the context of imitation tasks (e.g., Bauer et al., 2002b; Cheatham and Bauer, 2005). This is strong evidence that the format in which the memories are encoded is declarative, as opposed to nondeclarative or implicit (formats inaccessible to language). Second, the paradigm passes the 'amnesia test.' Whereas intact adults accurately imitate sequences after a delay, patients with amnesia due to hippocampal lesions perform no better than naïve controls (McDonough et al., 1995). Adolescents and young adults who sustained hippocampal damage early in life also exhibit deficits in performance on the task (Adlam et al., 2005). This suggests that the paradigm taps the type of memory that gives rise to verbal report. For these reasons, the task has come to be widely accepted as a nonverbal analogue to verbal report (e.g., Bauer, 2002; Mandler, 1990; Meltzoff, 1990; Nelson and Fivush, 2000; Rovee-Collier and Hayne, 2000; Schneider and

Bjorklund, 1998; Squire et al., 1993) and is widely used to examine developments in memory for specific episodes in the first two years of life.

16.3.5 Memory for Specific Episodes in the First Two Years of Life

Over the first two years of life, there are developmental changes in memory for specific episodes along a number of dimensions. Perhaps the most salient change is in the length of time over which memory is apparent. Importantly, because like any complex behavior, the length of time an episode is remembered is multiply-determined, there is no 'growth chart' function that specifies that children of X age should remember for Y long. Nonetheless, across numerous studies, there has emerged evidence that with increasing age, infants tolerate lengthier retention intervals. For example, at 6 months of age, infants remember an average of one action of a three-step sequence for 24 hours (Barr et al., 1996; see also Collie and Hayne, 1999). Nine-month-olds remember individual actions over delays from 24 hours (Meltzoff, 1988) to 5 weeks (Carver and Bauer, 1999, 2001). By 10–11 months of age, infants remember over delays as long as 3 months (Carver and Bauer, 2001 and Mandler and McDonough, 1995). Thirteen- to fourteen-month-olds remember actions over delays of 4–6 months (Bauer et al., 2000 and Meltzoff, 1995). By 20 months of age, children remember the actions of event sequences over as many as 12 months (Bauer et al., 2000).

Infants also recall the temporal order of actions in multistep sequences, though retaining order information presents a cognitive challenge to young infants, in particular, as evidenced by low levels of ordered recall and substantial within-age-group variability in the first year. Although 67% of Barr et al.'s (1996) 6-month-olds remembered individual actions over 24 hours, only 25% of them remembered actions in the correct temporal order. Among 9-month-olds, approximately 50% of infants exhibit ordered reproduction of sequences after a 5-week delay (Bauer et al., 2001, 2003; Carver and Bauer, 1999). By 13 months of age, the substantial individual variability in ordered recall has resolved: 78% of 13-month-olds exhibit ordered recall after 1 month. Nevertheless, throughout the second year of life, there are age-related differences in children's recall of the order in which actions of multistep sequences unfolded. The differences are especially apparent under conditions of greater cognitive demand, such as when less support for recall is provided, and after longer delays (Bauer et al., 2000).

The first two years of life also are marked by changes in the robustness of memory for specific episodes. For instance, there are changes in the number of experiences that seem to be required in order for infants to remember. In Barr et al. (1996), at 6 months, infants required six exposures to events in order to remember them 24 hours later. If instead they saw the actions demonstrated only three times, they showed no memory after 24 hours (i.e., performance of infants who had experienced the puppet sequence did not differ from that of naïve control infants). By 9 months of age, the number of times actions need to be demonstrated to support recall after 24 hours has reduced to three (e.g., Meltzoff, 1988). Indeed, 9-month-olds who see sequences modeled as few as two times within a single session recall individual actions of them 1 week later (Bauer et al., 2001). However, over the same delay, ordered recall was observed only among infants who had seen the sequences modeled a total of six times, distributed over three exposure sessions. Three exposure sessions also support ordered recall over the longer delay of 1 month. By the time infants are 14 months of age, a single exposure session is all that is necessary to support recall of multiple different single actions over 4 months (Meltzoff, 1995). Ordered recall of multistep sequences is apparent after as many as 6 months for infants who received a single exposure to the events at the age of 20 months (Bauer and Leventon, in press).

Another index of the robustness of memory is the extent to which it is disrupted by interference. One form of interference that has been studied in infancy is changes in context between encoding and retrieval. Reports on infants' sensitivity to contextual changes are mixed. There are some suggestions that recall is disrupted if between exposure and test, the appearance of the test materials is changed (e.g., Hayne et al., 1997, 2000; and Herbert and Hayne, 2000). However, there also are reports of robust generalization from encoding to test by infants across a wide age range. Infants have been shown to generalize imitative responses across changes in (1) the size, shape, color, and/or material composition of the objects used in demonstration versus test (e.g., Bauer and Dow, 1994; Bauer and Fivush, 1992; Bauer and Lukowski, 2010; and Lechuga et al., 2001), (2) the appearance of the room at the time of the demonstration of modeled actions and at the time of the memory test (e.g., Barnat et al., 1996 and Klein and Meltzoff, 1999), (3) the setting for the demonstration of the modeled actions and the test of memory for them (e.g., Hanna and Meltzoff, 1993; Klein and Meltzoff, 1999), and (4) the individual who demonstrated the actions and the individual who tested for memory of the actions (e.g., Hanna and Meltzoff, 1993). Evidence of flexible memory extends to infants as young as 9–11 months of age (e.g., Baldwin et al., 1993; Lukowski et al., 2009; McDonough and Mandler, 1998). In summary, whereas there is evidence that with age, infants' memories as tested in imitation-based paradigms become more generalizable (e.g., Herbert and Hayne, 2000), there is substantial evidence that from an early age, infants' memories survive changes in context and stimuli.

16.3.6 Developments in the Preschool Years and Beyond

Beginning in the third year of life, verbal assessments become a viable means for testing episodic memory. This opens up new possibilities: children can be tested not only for memory for controlled laboratory events but for events from their lives outside the laboratory as well. This combination of approaches has yielded a wealth of data about children's memories for the routine events that make up their everyday lives, and about their memories for unique events. Some of the events are highly personally significant and contribute to an emerging autobiography or personal past. Major findings from each of these categories are reviewed.

Early studies of young children's memories for the events of their own lives focused on everyday, routine events. The children's reports included actions common to the activities, and almost invariably, the actions were mentioned in the temporal order in which they typically occurred. Representative of the findings was the answer provided by a 3-year-old child to the question, "What happens when you have a birthday party?": "You cook a cake and eat it" (K. Nelson and Gruendel, 1986, p. 27). This early research revealed 'minimalist,' yet nevertheless accurate, reports by children as young as 3 years of age (see also K. Nelson, 1986, 1997). Subsequent studies revealed that with development, children's reports included more information. For example, in addition to mention of cooking a cake and then eating it, 6- and 8-year-old children told of putting up balloons, receiving and then opening presents from party guests, eating birthday cake, and playing games. Second, relative to younger children, older children more frequently mentioned alternative actions: "...and then you have lunch *or whatever you have.*" Third, with age, children include in their reports more optional activities, such as "*Sometimes* then they have three games...then *sometimes* they open up the other presents...." Finally, with increasing age, children mentioned more conditional activities, such as "*If you're like at Foote Park or something, then* it's time to go home...." (Nelson and Gruendel, 1986, p. 27). Whereas some of the differences in younger and older children's reports might be due to the greater number of experiences of events such as birthday parties that older children have, relative to younger children, experience alone does not account for the developmental differences. In laboratory research in which children of different ages are given the same amount of experience with a novel event, older children produce more elaborate reports relative to younger children (e.g., Fivush et al., 1992; Price and Goodman, 1990).

Young children also form memories of unique events. In an early study, Fivush et al. (1987) found that all of the children in a sample of 2.5–3-year-olds recalled at least one event that had happened 6 or more months in the past. The children reported the same amount of information about events that had taken place more than 3 months ago as they did about events that had taken place within 3 months. In Hamond and Fivush (1991), 3- and 4-year-old children recalled a trip to Disney World they had taken either 6 months previously or 18 months previously. The amount they remembered did not differ as a function of the delay. Moreover, the older and younger children did not differ in the amount of information they reported about the event. Yet the age groups did differ in how elaborate their reports were. Whereas the younger children tended to provide the minimum required response to a question, the older children tended to provide more elaborate responses.

With development, there are changes in what children include in their reports about events. For example, young children seemingly focus on what is common or routine across experiences whereas older children and adults focus on what is unique or distinctive. This trend is illustrated in Fivush and Hamond (1990). In response to an interviewer's invitation to talk about going camping, after providing the interviewer with the distinctive information that the family had slept in a tent, a 2.5-year-old child went on to report on the more typical features of the camping experience. In total, 48% of the information that the children reported was judged to be distinctive, implying that 52% of it was not. By 4 years of age, children report about three times more distinctive information than typical information (Fivush and Hamond, 1990). One consequence of focus on what is common across experience is that a unique event such as camping gets 'fused' into the daily routine of eating and sleeping. In the process, the features that distinguish events from one another may fade into the background and be lost. The result would be fewer memories of episodes that are truly unique. Conversely, with increasing focus on the more distinctive features of events, there is a resulting increase in the number of memories that are truly unique.

With age, children not only include different types of information in their narratives but also include more information. For example, in research by Fivush and Haden (1997), from 3.5 to 6 years of age, the number of propositions children included in the average narrative increased more than twofold, from 10 to 23. Young children's narratives include basic information about what actions occurred in the event; they feature intensifiers, qualifiers, and internal evaluations, and the actions in the narrative are joined by simple temporal and causal connections (e.g., *then*, *before*, and *after*, and *because*, *so*, *in order to*, respectively). What accounts for the increase in narrative length over this age period is that with age, children provide (1) more information about who was involved and when and where the event occurred, (2) more information about optional or variable actions (e.g., *"When it turned red light*, we stopped"; Fivush and Haden, 1997, p. 186), and (3) more elaborations (Fivush and Haden, 1997). As a result, relative to younger children's, older children's stories are more complete, easier to follow, and engaging.

The dramatic increases with age in the amount of information that children *report* tempt the conclusion that there also are age-related increases in the amount of information that children *remember* about events. This is not a 'safe' conclusion, however, in light of evidence that perhaps especially for younger children, verbal reports underestimate the richness of memories (e.g., Fivush et al., 2004; see Bauer, 1993, in press and Mandler, 1990, for discussions). Indeed, because of the inevitable confounding between increases in age and increases in narrative skills, whereas it is clear that children report more with age, it is not clear whether they also remember more.

16.3.7 Autobiographical Memory

Over the course of the preschool years, autobiographical or personal memory becomes increasingly apparent. Autobiographical memories are the memories of events and experiences that make up one's life story or personal past. They are the stories that people tell about themselves that reveal who they are and how their experiences have shaped their characters. As implied by this description, autobiographical memories differ from 'run-of-the-mill' episodic memories in that autobiographical memories are infused with a sense of personal involvement or ownership in the event. They are memories of events that happened to one's self, in which one participated, and about which one had emotions, thoughts, reactions, and reflections. It is this feature that puts the 'auto' in '*auto*biographical.'

Throughout the preschool years, children's memories take on more and more autobiographical features (see Bauer, 2007, for discussion). From a very young age, children include references to themselves in their narratives: "*I* fell down." With age, they increasingly pepper their narratives with the subjective perspective that indicates the significance of the event for the child (Fivush, 2001). For example, they go beyond comment on the objective reality of 'falling down' to convey how they felt about the fall: "I fell down and *was so embarrassed* because everybody was watching!" It is this subjective perspective that provides the explanation for why events are funny, or sad, for instance, and thus of significance to one's self.

There also are changes in the marking of events as having taken place at a specific place and time.

For instance, children increasingly include specific references to time, such as "on my birthday," "at Christmas," or "last summer" (Nelson and Fivush, 2004). Markings such as these not only establish that an event happened at a time different from the present but also begin to establish a time line along which an organized historical record of when events occurred can be constructed. Children also include in their narratives more orienting information, including where events took place and who participated in them (e.g., Fivush and Haden, 1997). These changes serve to distinguish events from one another, thereby making them more distinctive. Children also include more descriptive detail in their reports, suggestive of a sense of reliving the experience. For example, they include more intensifiers ("Cause she was *very* naughty"), qualifiers ("I *didn't like* her video tape"), elements of suspense ("And *you know what*?"; examples from Fivush and Haden, 1997), and even repetition of the dialogue spoken in the event ("...I said, 'I hope my Nintendo my Super Nintendo is still here'.", from Ackil et al., 2003). The result is a much more elaborate narrative that brings both the storyteller and the listener to the brink of reliving the experience. It is tempting to conclude that these changes account for the finding among adults of a steadily increasing number of memories of events that took place from the ages of 3–7 years (Bauer, 2007).

Relative to those in the preschool years, developments in autobiographical or personal memory in later childhood and adolescence have been relatively neglected. Yet age-related changes in autobiographical reports continue throughout the elementary school years and beyond. An illustration of the types of changes that occur during this period is the breadth or completeness of children's narratives. Like a good newspaper story, a 'good' autobiographical narrative includes a number of elements, including information about the *who, what, where, when, why,* and *how* of the experience. The average 7-year-old includes only half the number of these narrative elements than the average 11-year-old (see Bauer et al., 2007).

16.4 MECHANISMS OF DEVELOPMENTAL CHANGE

Given that memory is not a single entity or unity construct, it is not surprising that there is not a single answer to the question of 'what develops' in the development of memory. Rather, like all complex behaviors, memory is multiply-determined. An adequate explanation of why it develops as it does will entail multiple levels of analysis, ranging from the cellular and molecular events that allow for the storage of information to the cultural influences that shape the expression of memory (see Bauer, 2007, for discussion). Because the focus of this volume is on basic neural and cognitive processes, the author focuses on these two categories of explanation. The reader is referred to other sources (e.g., Bauer, 2007, in press and Nelson and Fivush, 2004) for elaboration of the roles of other aspects of development, including conceptual change and social influences on remembering.

16.4.1 Neural Structures and Processes

A thorough review of the neural substrates that support the different types of memory, and the courses of development of each, is well beyond the scope of this chapter (see Bauer, 2007, 2009 and Nelson et al., 2006 for reviews of subsets of this large literature). Yet a brief review is essential to the goal of identifying possible mechanisms of developmental change. Studies of patients with specific types of lesions and disease and animal models thereof, as well as neuroimaging studies, have made clear that registration of experience and formation of memory traces to represent it involve multistage processes that depend on networks of neural structures. For example, encoding, storage, and later retrieval of declarative memories depend on a multicomponent network involving temporal (hippocampus, and entorhinal, parahippocampal, and perirhinal cortices) and cortical (including prefrontal and other association areas) structures (e.g., Eichenbaum and Cohen, 2001; Markowitsch, 2000; Zola and Squire, 2000). In the review to follow, the author focuses primarily on this network in large part because its course of development is better worked out, relative to the networks supporting nondeclarative memory.

16.4.2 The Neural Substrate of Declarative Memory

Formation of a declarative memory begins as the elements that constitute an experience register across primary sensory areas (e.g., visual and auditory). Inputs from primary cortices are projected to unimodal association areas where they are integrated into whole percepts of what objects look, feel, and sound like. Unimodal association areas in turn project to polymodal prefrontal, posterior, and limbic association cortices where inputs from the different sense modalities are integrated and maintained over brief delays (see, e.g., Petrides, 1995). Prefrontal structures not only are involved in the initial processing or *encoding* of experiences into long-term traces but also are implicated in the temporary maintenance of material in short-term and working memory. As well, the neocortical regions involved in the initial perception and registration of experience are thought to be responsible for both perceptual and conceptual priming (e.g., Gabrieli, 1998; see Toth, 2000 for a review).

For maintenance of traces of experience over delays of longer than a few seconds or minutes, the inputs to the association areas must be stabilized or *consolidated*, a task attributed to medial–temporal structures, in concert with cortical areas (McGaugh, 2000). Specifically, information from association areas converges on perirhinal and para-hippocampal structures from which it is projected to the entorhinal cortex and in turn to the hippocampus. Within the hippocampus, conjunctions and relations among the elements of experience are linked into a single event. Association areas share the burden of consolida-tion by relating new memories to episodes already in storage: information processed in the hippocampus is projected back through the temporal cortices which in turn project to the association areas that gave rise to their inputs. Eventually, traces are stabilized such that the hippocampus is no longer required to maintain them; consolidated traces are *stored* in neocortex (although, whether memories are ever wholly indepen-dent of the hippocampus is debated: see, e.g., Moscovitch and Nadel, 1998 and Reed and Squire, 1998, for opposing views).

Finally, behavioral and neuroimaging data implicate prefrontal cortex in the *retrieval* of memories from long-term storage (e.g., Cabeza et al., 1997, 2004; Maguire, 2001; Markowitsch, 1995). For example, dam-age to prefrontal cortex disrupts retrieval of facts and ep-isodes. Deficits are especially apparent (1) in free recall versus recognition, (2) for temporal information versus items, (3) for specific event features, and (4) for source of information. Imaging studies have revealed high levels of activation in prefrontal cortex during retrieval of episodic memories from long-term stores (reviewed in Gilboa, 2004). Activations in medial prefrontal cortex are observed during retrieval of internally generated in-formation, such as the thoughts and feelings that put the *auto* in autobiographical memories (Cabeza et al., 2004). Lateral posterior parietal and precuneus also are implicated in retrieval of autobiographical memories. The activations are greater when subjects report retriev-ing more details about the memory (reviewed in Gilboa, 2004).

16.4.3 Development of the Neural-Substrate-Supporting Declarative Memory

Developments in the neural-substrate-supporting de-clarative memory are summarized in a number of sources (e.g., Bauer, 2007, 2009, in press; Nelson, 2000; Nelson et al., 2006; Richman and Nelson, 2008). In terms of brain development in general, there are changes in both gray and white matter from infancy well into adolescence (e.g., Caviness et al., 1996; Giedd et al., 1999; Gogtay et al., 2004; and Sowell et al., 2004). Reflect-ing changes in vasculature, glia, neurons, and neuronal processes, gray matter increases until puberty. Beyond puberty, as a result of pruning and other regressive events (i.e., loss of neurons and axonal branches), the thickness of the cortical mantle actually declines (e.g., Giedd et al., 1999; Gogtay et al., 2004; Sowell et al., 2001; Van Petten, 2004). In contrast to curvilinear change in gray matter volume, white matter volume increases linearly with age (Giedd et al., 1999). Increases in white matter are associated with greater connectivity between brain regions and with myelination processes that con-tinue into young adulthood (e.g., Johnson, 1997; Klingberg, 2008; Schneider et al., 2004).

In terms of the temporal–cortical episodic-memory network, there are a number of indicators that in the hu-man, many components of the medial temporal lobe de-velop early. For instance, as reviewed by Seress and Abraham (2008), the cells that make up most of the hip-pocampus are formed in the first half of gestation and virtually all are in their adult locations by the end of the prenatal period. The neurons in most of the hippo-campus also begin to connect early in development: syn-apses are present as early as 15 weeks gestational age. The number and density of synapses both increase rap-idly after birth and reach adult levels by approximately 6 postnatal months. Perhaps as a consequence, glucose uti-lization in the temporal cortex reaches adult levels at the same time (i.e., by about 6 months: Chugani, 1994; Chugani and Phelps, 1986). Thus, there are numerous in-dices of early maturity of major portions of the medial temporal components of the network.

In contrast to early maturation of most of the hippocampus, development of the dentate gyrus of the hippocampus is protracted (Seress and Abraham, 2008). At birth, the dentate gyrus includes only about 70% of the adult number of cells. Thus, roughly 30% of the cells are produced postnatally. Indeed, neurogenesis in the dentate gyrus of the hippocampus continues throughout childhood and adulthood (Tanapat et al., 2001). It is not until 12–15 postnatal months that the mor-phology of the structure appears adultlike. Maximum density of synaptic connections in the dentate gyrus also is delayed, relative to that in the other regions of the hip-pocampus. In humans, synaptic density increases dra-matically (to well above adult levels) beginning at 8–12 postnatal months and reaches its peak at 16–20 months. After a period of relative stability, excess synapses are pruned until adult levels are reached at about 4–5 years of age (Eckenhoff and Rakic, 1991).

Although the functional significance of later develop-ment of the dentate gyrus is not clear, there is reason to speculate that it impacts behavior. As already noted, upon experience of an event, information from distributed regions of cortex converges on the entorhinal cortex. From there, it makes its way into the hippocampus in one of two ways: via a 'long route' or a 'short route.' The long route involves projections from entorhinal cortex into the hippocampus, by way of the dentate gyrus; the short route bypasses the dentate gyrus. Whereas the short route may support some forms of memory (Nelson, 1995, 1997a,b), based on data from rodents, it seems that adultlike memory behavior depends on passage of information through the dentate gyrus (Czurkó et al., 1997; Nadel and Willner, 1989). This implies that maturation of the dentate gyrus of the hippocampus may be a rate-limiting variable in the development of episodic memory early in life (e.g., Bauer, 2007, 2009; Nelson, 1995, 1997a,b, 2000). Finally, hippocampal volume continues to increase gradually throughout childhood and into adolescence (e.g., Gogtay et al., 2004; Pfluger et al., 1999; Utsunomiya et al., 1999). Myelination in the hippocampal region continues throughout adolescence (Arnold and Trojanowski, 1996; Benes et al., 1994; Schneider et al., 2004).

The association areas also undergo a protracted course of development (Bachevalier, 2001). For example, it is not until the seventh prenatal month that all six cortical layers are apparent. The density of synapses in the prefrontal cortex increases dramatically at 8 postnatal months and peaks between 15 and 24 months. Pruning to adult levels does not begin until late childhood; adult levels are not reached until late adolescence or early adulthood (Huttenlocher, 1979; Huttenlocher and Dabholkar, 1997; see Bourgeois, 2001, for discussion). In the years between, in some cortical layers, there are changes in the size of cells and the lengths and branching of dendrites (Benes, 2001). Although the maximum density of synapses may be reached as early as 15 postnatal months, it is not until 24 months that synapses develop adult morphology (Huttenlocher, 1979). There also are changes in glucose utilization and blood flow over the second half of the first year and into the second year: blood flow and glucose utilization increase above adult levels by 8–12 and 13–14 months of age, respectively (Chugani et al., 1987). Other maturational changes in prefrontal cortex, such as myelination, continue into adolescence, and adult levels of some neurotransmitters are not seen until the second and third decades of life (Benes, 2001). It is not until adolescence that neurotransmitters such as acetylcholine reach adult levels (discussed in Benes, 2001).

Although much of the attention to developmental changes has focused on the medial–temporal and prefrontal regions, there also are age-related changes in the lateral temporal and parietal cortices. Cortical gray-matter changes occur earlier in the frontal and occipital poles, relative to the rest of the cortex, which matures in a parietal-to-frontal direction. The superior temporal cortex is last to mature (though the temporal poles mature early; Gogtay et al., 2004). The late development of this portion of the cortex is potentially significant for memory as it is one of the polymodal association areas that plays a role in integration of information across sense modalities.

16.4.4 Functional Consequences of Development of the Temporal–Cortical Network

What are the consequences for behavior of the slow course of development of the neural network that supports declarative memory? At a general level, one may expect concomitant behavioral development: As the neural substrate develops, so does behavior (and vice versa, of course). But precisely how do changes in the medial temporal and cortical structures, and their interconnections, produce changes in behavior? In other words, how do they affect memory representations? To address this question, one must consider the basic processes involved in memory-trace formation, storage, and retrieval and how the 'recipe' for a memory might be affected by changes in the underlying neural substrate. In other words, one must consider how developmental changes in the substrate for memory relate to changes in the efficacy and efficiency with which information is maintained over the short-term, encoded and stabilized for long-term storage, in the reliability and ease with which it is retrieved.

16.4.5 Basic Cognitive and Mnemonic Processes

With developmental changes in the temporal–cortical network, one may expect changes in basic cognitive processes and in behavior. The basic processes involved in memory are encoding, consolidation, and retrieval of memory traces. Although the processes are difficult to cleanly separate from one another (e.g., when the encoding ends and the consolidation begins is a challenging question to address), they do build on one another and thus are described in the nominal order in which they occur: short-term maintenance and encoding, consolidation and storage, and retrieval.

16.4.6 Encoding

Association cortices are involved in the initial registration and temporary maintenance of experience. Because prefrontal cortex in particular undergoes considerable postnatal development, it is reasonable to expect that neurodevelopmental changes in it relate to age-related changes in the speed and efficiency with which information is encoded into long-term storage. Consistent with this suggestion, in a longitudinal study,

Bauer and her colleagues (Bauer et al., 2006) found differences in the amplitudes of event-related potential (ERP) responses to familiar stimuli between 9 and 10 months of age that correlated with age-related improvements in recall after a 1-month delay. Behavioral data also indicate developments in encoding throughout the second year of life. For example, relative to 15-month-olds, 12-month-olds require more trials to learn multistep events to a criterion (learning to a criterion indicates that the material was fully encoded). In turn, 15-month-olds are slower to achieve criterion, relative to 18-month-olds (Howe and Courage, 1997). Indeed, across development, older children learn more rapidly than younger children (Howe and Brainerd, 1989). Changes in the temporary registration of information are apparent throughout the preschool years and school years (e.g., Cowan and Alloway, 2009). The net result of these changes is that children become increasingly adept not only at maintaining information in temporary registration but also in initiating the type of organizational processing that promotes consolidation of it.

16.4.7 Consolidation and Storage

As reviewed earlier, medial temporal structures are implicated in the processes by which new memories become 'fixed' for long-term storage; cortical association areas are the presumed repositories for long-term memories. In a fully mature, intact adult, the changes in synaptic connectivity associated with memory-trace consolidation continue for hours, weeks, and even months, after an event. Memory traces are vulnerable throughout this time, as evidenced by the fact that lesions inflicted during the period of consolidation result in deficits in memory whereas lesions inflicted after a trace has been consolidated do not (e.g., Kim and Fanselow, 1992; Takehara et al., 2003). For the developing organism, the period of consolidation may be one of greater vulnerability for a memory trace, relative to the adult. Not only are some of the implicated neural structures relatively undeveloped (i.e., the dentate gyrus and prefrontal cortex) but also the connections between them are still being sculpted and thus are less than fully effective and efficient. As a consequence, even once children have successfully encoded an event, they remain vulnerable to forgetting. Younger children may be more vulnerable to forgetting, relative to older children (Bauer, 2004, 2006a).

To examine the role of consolidation and storage processes in long-term memory in 9-month-old infants, Bauer et al. (2003) combined ERP measures of immediate recognition (as an index of encoding), ERP measures of 1-week delayed recognition (as an index of consolidation and storage), and deferred imitation measures of recall after 1 month. After the delay, 46% of the infants evidenced ordered recall of the sequences, and 54% did not. At the immediate ERP test, regardless of whether they subsequently recalled the events, the infants evidenced recognition: Their ERP responses were different to the old and new stimuli. This strongly implies that the infants had encoded the events. Nevertheless, 1 week later, at the delayed-recognition test, the infants who would go on to recall the events recognized the props, whereas infants who would not evidence ordered recall did not. Thus, in spite of having encoded the events, a subset of 9-month-olds failed to recognize them after 1 week and subsequently failed to recall them after 1 month. Moreover, the size of the difference in delayed-recognition response predicted recall performance 1 month later. Thus, infants who had stronger memory representations after a 1-week delay exhibited higher levels of recall 1 month later (see also Carver et al., 2000). These data strongly imply that at 9 months of age, consolidation and/or storage processes are a source of individual differences in mnemonic performance.

In the second year of life, there are behavioral suggestions of between-age group differences in consolidation and/or storage processes as well as a replication of the finding among 9-month-olds that intermediate-term consolidation and/or storage failure relates to recall over the long-term. In Bauer et al. (2002a), 16- and 20-month-olds were exposed to multistep events and tested for recall immediately (as a measure of encoding) and after 24 hours. Over the delay, the younger children forgot a substantial amount of the information they had encoded: they produced only 65% of the target actions and only 57% of the ordered pairs of actions that they had learned just 24 hours earlier. For the older children, the amount of forgetting over the delay was not statistically reliable. It is not until 48 hours that children 20 months of age exhibit significant forgetting (Bauer et al., 1999). These observations suggest age-related differences in the vulnerability of memory traces during the initial period of consolidation.

The vulnerability of memory traces during the initial period of consolidation is related to the robustness of recall after 1 month. This is apparent from another of the experiments in Bauer et al. (2002a), this one involving 20-month-olds only. The children were exposed to multistep events and then tested for memory for some of the events immediately, some of the events after 48 hours (a delay after which, based on Bauer et al., 1999, some forgetting was expected), and some of the events after 1 month. Although the children exhibited high levels of initial encoding (as measured by immediate recall), they nevertheless exhibited significant forgetting after both 48 hours and 1 month. The robustness of memory after 48 hours predicted 25% of the variance in recall

1 month later; variability in level of encoding did not predict significant variance. This effect is a conceptual replication of that observed with 9-month-olds in Bauer et al. (2003) (see Bauer, 2005 and Howe and Courage, 1997, for additional evidence of a role for post-encoding processes in long-term recall). The findings that infants who are 'good consolidators' have high levels of long-term recall are reminiscent of Bosshardt et al. (2005) with adults: fMRI activations 1 day after learning were predictive of forgetting 1 month later.

Changes in the processes by which memory representations are consolidated and stored can be expected to continue throughout the preschool years. However, although neuroimaging techniques such as ERPs could be brought to bear on the question, as they are in the infancy period, such studies have not been conducted with preschool-age children. Neither is there a plethora of behavioral studies to address the question. A major reason is that few studies include the requisite type or number of tests. Frequently, studies of long-term memory fail to include a measure of initial encoding (e.g., Liston and Kagan, 2002), thus making it impossible to determine the variance associated with encoding processes. They also tend to measure recall only once, at the end of the retention interval, thus making it impossible to determine the variance associated with post-encoding processes during the period of consolidation. An exception to this approach is described in the next section.

16.4.8 Retrieval

Prefrontal cortex is implicated in retrieval of memories from long-term storage sites. Prefrontal cortex undergoes a long period of postnatal development, making it a likely candidate source of age-related differences in long-term recall. Surprisingly, although retrieval processes are a compelling candidate source of developmental differences in long-term recall, there are few data with which to evaluate their contribution. A major reason is that most studies do not allow for assignment of relative roles of the processes that take place before retrieval, namely, encoding and consolidation. As discussed in the section on encoding, older children learn more rapidly than younger children. Yet age-related differences in encoding effectiveness rarely are taken into account. In fact, as just noted, in many studies, no measures of encoding or initial learning are obtained. In addition, with standard testing procedures, it is difficult to know whether a memory representation has lost its integrity and become unavailable (consolidation or storage failure) or whether the memory trace remains intact but has become inaccessible with the cues provided (retrieval failure). Implication of retrieval processes as a source of developmental change requires that encoding

be controlled and that memory be tested under conditions of high support for retrieval.

In the infancy period, one of the studies that permits assessment of the contributions of consolidation and/or storage relative to retrieval processes is Bauer et al. (2000) (see also Bauer et al., 2003, described earlier). The study provided data on children of multiple ages (13, 16, and 20 months) tested over delays of 1–12 months. Immediate recall of half of the events was tested, thus providing a measure of encoding. Because the children were given what amounted to multiple test trials, without intervening study trials, there were multiple opportunities for retrieval. As discussed by Howe and his colleagues (e.g., Howe and Brainerd, 1989; Howe and O'Sullivan, 1997), for intact memory traces, retrieval attempts strengthen the trace and route to retrieval of it, thereby increasing accessibility on each test trial. Conversely, lack of improvement across test trials implies that the trace was no longer available (although see Howe and O'Sullivan, 1997, for multiple nuances of this argument). Third, immediately after the recall tests, relearning was tested. That is, after the second test trial, the experimenter demonstrated each event once and allowed the children to imitate. Since Ebbinghaus (1885), relearning has been used to distinguish between an intact but inaccessible memory trace and a trace that has disintegrated. Specifically, if the number of trials required to relearn a stimulus was smaller than the number required to learn it initially, savings in relearning were said to have occurred. Savings presumably accrue because the products of relearning are integrated with an existing (though not necessarily accessible) memory trace. Conversely, the absence of savings is attributed to storage failure: there is no residual trace upon which to build. In developmental studies, age-related differences in relearning would suggest that the residual memory traces available to children of different ages are differentially intact.

To eliminate encoding processes as a potential source of developmental differences in long-term recall, in a reanalysis of the data from Bauer et al. (2000), subsets of 13- and 16-month-olds and subsets of 16- and 20-month-olds were matched for levels of encoding (as measured by immediate recall; Bauer, 2005). The amount of information the children forgot over the delays then was examined. For both comparisons, even though they were matched for levels of encoding, younger children exhibited more forgetting relative to older children. The age effect was apparent on both test trials. Moreover, in both cases, for older children, levels of performance after the single relearning trial were as high as those at initial learning. In contrast, for younger children, performance after the relearning trial was lower than at initial learning. Together, the findings of age-related differential loss of information over time and of age effects in relearning strongly implicate storage processes, as opposed to retrieval processes, as the major source of age-related differences in long-term recall.

The conclusions from the infancy literature are consistent with the results of research with older children conducted within the trace-integrity framework (Brainerd et al., 1990) and conceptually related fuzzy-trace theory (Brainerd and Reyna, 1990). In this tradition, to eliminate encoding differences as a source of age-related effects, participants are brought to a criterion level of learning prior to imposition of a delay. To permit evaluation of the contributions of storage processes versus retrieval processes, participants are provided multiple test trials, without intervening study trials. In one such study, 4- and 6-year-old children learned and then recalled eight-item picture lists (Howe, 1995). In this study, as in virtually every other study conducted within this tradition (reviewed in Howe and O'Sullivan, 1997), the largest proportion of age-related variance in children's recall was accounted for by memory failure at the level of consolidation and/or storage, as opposed to retrieval. Whereas consolidation and/or storage failure rates decline throughout childhood, retrieval failure rates remain at relatively constant levels (Howe and O'Sullivan, 1997). The apparent lack of change in retrieval failure rates throughout childhood undermines the suggestion that retrieval processes are a major source of developmental change during this period.

If not by affecting retrieval, by what means do developmental changes in prefrontal cortex influence memory development in infancy and childhood? As discussed in Bauer (2006a, 2007), rather than on retrieval processes, a major effect of developments in prefrontal structures may be on consolidation and/or storage processes. Consolidation is an interactive process between medial temporal and cortical structures. As such, changes in cortical structures may be as important to developments in consolidation processes as are changes in medial temporal structures. Moreover, the ultimate storage sites for long-term memories are the association cortices. Thus, developmental changes in prefrontal cortex may play their primary role in supporting more efficient consolidation and more effective storage.

16.5 CONCLUSION

The title of this chapter – *Memory Development* – gives the impression that its subject will be singular: a singular system with a singular course of development. On the contrary, there are many different forms of memory, each with its own characteristics and developmental course. Although some broad generalizations apply, most of what one can be said to know about memory is relevant within a limited frame and for a subset of the types of memory. Continued progress in understanding memory and its development requires that appropriate distinctions be maintained.

Historically, most types of memory were thought to be relatively late to develop. This expectation was perhaps nowhere more apparent than in reference to episodic and autobiographical memory. Research in the last decades of the twentieth century made clear that the assumption was unwarranted. When they are tested with structured stimuli and personally relevant events and materials, even young children show evidence of mnemonic competence. Thus, in sharp contrast to expectations of developmental discontinuities in event memory, there is ample evidence that the capacity to remember past events develops early. The research also made clear, however, that development is a protracted event, beginning in infancy and continuing into late adolescence. A full accounting of the development of 'memory' thus requires a long-term perspective.

Finally, it is a truism to say that complex behaviors are multiply determined. In the field of the development of memory, this dictum must be embraced wholeheartedly. Within the same space of time as the mnemonic capabilities of even young children were chronicled, progress in explaining the timing and course of development was made at a variety of different levels. Although much remains to be discovered, understanding of the cellular and molecular events that permit the storage and later retrieval of information is now within reach. Similarly, people are on the verge of understanding how basic memory processes determine the life course of a memory at different points in developmental time. People may look forward to the day when multiple levels of explanation come together into a comprehensive account of the processes and determinants of the capacities called *memory*.

SEE ALSO

Cognitive Development: Structural Brain Development: Birth Through Adolescence; Developing Attention and Self Regulation in Infancy and Childhood; A Neuroscience perspective on empathy and its development; Executive Function: Development, Individual Differences and Clinical Insights.

References

Ackil, J.K., Van Abbema, D.L., Bauer, P.J., 2003. After the storm: Enduring differences in mother-child recollections of traumatic and non-traumataic events. Journal of Experimental Child Psychology 84, 286–309.

Adlam, A.-L.R., Vargha-Khadem, F., Mishkin, M., de Haan, M., 2005. Deferred imitation of action sequences in developmental amnesia. Journal of Cognitive Neuroscience 17 (2), 240–248.

Arnold, S.E., Trojanowski, J.Q., 1996. Human fetal hippocampal development: I. Cytoarchitecture, myeloarchitecture, and neuronal morphologic features. The Journal of Comparative Neurology 367, 274–292.

Bachevalier, J., 2001. Neural bases of memory development: Insights from neuropsychological studies in primates. In: Nelson, C.A., Luciana, M. (Eds.), Handbook of Developmental Cognitive Neuroscience. The MIT Press, Cambridge, MA, pp. 365–379.

Bahrick, H.P., 2000. Long-term maintenance of knowledge. In: Tulving, E., Craik, F.I.M. (Eds.), The Oxford Handbook of Memory. Oxford University Press, New York, pp. 347–362.

Baldwin, D.A., Markman, E.M., Melartin, R.L., 1993. Infants' ability to draw inferences about nonobvious object properties: Evidence from exploratory play. Child Development 64, 711–728.

Barnat, S.B., Klein, P.J., Meltzoff, A.N., 1996. Deferred imitation across changes in context and object: Memory and generalization in 14-month-old children. Infant Behavior & Development 19, 241–251.

Barr, R., Dowden, A., Hayne, H., 1996. Developmental change in deferred imitation by 6- to 24-month-old infants. Infant Behavior & Development 19, 159–170.

Bauer, P.J., 1993. Identifying subsystems of autobiographical memory: Commentary on Nelson. In: Nelson, C.A. (Ed.), Memory and Affect in Development. Erlbaum, Hillsdale, NJ, pp. 25–37.

Bauer, P.J., 2002. Long-term recall memory: Behavioral and neurodevelopmental changes in the first 2 years of life. Current Directions in Psychological Science 11, 137–141.

Bauer, P.J., 2004. New developments in the study of infant memory. In: Teti, D.M. (Ed.), Blackwell Handbook of Research Methods in Developmental Science. Blackwell Publishing, Oxford, pp. 467–488.

Bauer, P.J., 2005. Developments in declarative memory: Decreasing susceptibility to storage failure over the second year of life. Psychological Science 16, 41–47.

Bauer, P.J., 2006a. Constructing a past in infancy: A neurodevelopmental account. Trends in Cognitive Sciences 10, 175–181.

Bauer, P.J., 2006b. Event memory. In: Damon, W., Lerner, R.M. (Eds.), Handbook of Child Psychology. 6th edn. In: Kuhn, D., Siegler, R. (Eds.), Cognition, Perception, vol. 2. John Wiley & Sons, Inc, Hoboken, NJ, pp. 373–425.

Bauer, P.J., 2007. Remembering the Times of Our Lives: Memory in Infancy and Beyond. Erlbaum, Mahwah, NJ.

Bauer, P.J., 2008. Infantile amnesia. In: Haith, M.M., Benson, J.B. (Eds.), Encyclopedia of Infant and Early Childhood Development. Academic Press, San Diego, CA, pp. 51–61.

Bauer, P.J., 2009. The cognitive neuroscience of the development of memory. In: Courage, M.L., Cowan, N. (Eds.), The Development of Memory in Infancy and Childhood. 2nd edn. Psychology Press, New York, NY, pp. 115–144.

Bauer, P.J., (in press). Memory. In: Zelazo PD (ed.) *Oxford Handbook of Developmental Psychology*. New York, NY: Oxford University Press.

Bauer, P.J., Dow, G.A.A., 1994. Episodic memory in 16- and 20-month-old children: Specifics are generalized, but not forgotten. Developmental Psychology 30, 403–417.

Bauer, P.J., Fivush, R., 1992. Constructing event representations: Building on a foundation of variation and enabling relations. Cognitive Development 7, 381–401.

Bauer, P.J., Leventon, J.S. (in press). Memory for one time experiences in the second year of life: Implications for the status of episodic memory. Infancy.

Bauer, P.J., Lukowski, A.F., 2010. The memory is in the details: Relations between memory for the specific features of events and long-term recall in infancy. Journal of Experimental Child Psychology 107 (1), 1–14.

Bauer, P.J., Mandler, J.M., 1989. One thing follows another: Effects of temporal structure on one- to two-year-olds' recall of events. Developmental Psychology 25, 197–206.

Bauer, P.J., Shore, C.M., 1987. Making a memorable event: Effects of familiarity and organization on young children's recall of action sequences. Cognitive Development 2, 327–338.

Bauer, P.J., Van Abbema, D.L., de Haan, M., 1999. In for the short haul: Immediate and short-term remembering and forgetting by 20-month-old children. Infant Behavior & Development 22, 321–343.

Bauer, P.J., Wenner, J.A., Dropik, P.L., Wewerka, S.S., 2000. Parameters of remembering and forgetting in the transition from infancy to early childhood. Monographs of the Society for Research in Child Development , vol. 65. (4, Serial No. 263).

Bauer, P.J., Wiebe, S.A., Waters, J.M., Bangston, S.K., 2001. Reexposure breeds recall: Effects of experience on 9-month-olds' ordered recall. Journal of Experimental Child Psychology 80, 174–200.

Bauer, P.J., Cheatham, C.L., Cary, M.S., Van Abbema, D.L., 2002a. Short-term forgetting: Charting its course and its implications for long-term remembering. In: In: Shohov, S.P. (Ed.), Advances in Psychology Research, vol. 9. Nova Science Publishers, Hungtington, NY, pp. 53–74.

Bauer, P.J., Wenner, J.A., Kroupina, M.G., 2002b. Making the past present: Later verbal accessibility of early memories. Journal of Cognition and Development 3, 21–47.

Bauer, P.J., Wiebe, S.A., Carver, L.J., Waters, J.M., Nelson, C.A., 2003. Developments in long-term explicit memory late in the first year of life: Behavioral and electrophysiological indices. Psychological Science 14, 629–635.

Bauer, P.J., Wiebe, S.A., Carver, L.J., et al., 2006. Electrophysiological indices of encoding and behavioral indices of recall: Examining relations and developmental change late in the first year of life. Developmental Neuropsychology 29, 293–320.

Bauer, P.J., Burch, M.M., Scholin, S.E., Güler, O.E., 2007. Using cue words to investigate the distribution of autobiographical memories in childhood. Psychological Science 18, 910–916.

Benes, F.M., 2001. The development of prefrontal cortex: The maturation of neurotransmitter systems and their interaction. In: Nelson, C.A., Luciana, M. (Eds.), Handbook of Developmental Cognitive Neuroscience. The MIT Press, Cambridge, MA, pp. 79–92.

Benes, F.M., Turtle, M., Khan, Y., Farol, P., 1994. Myelination of a key relay zone in the hippocampal formation occurs in the human brain during childhood, adolescence, and adulthood. Archives of General Psychiatry 51, 477–484.

Bosshardt, S., Degonda, N., Schmidt, C.F., et al., 2005. One month of human memory consolidation enhances retrieval-related hippocampal activity. Hippocampus 15, 1026–1040.

Bourgeois, J.-P., 2001. Synaptogenesis in the neocortex of the newborn: The ultimate frontier for individuation? In: Nelson, C.A., Luciana, M. (Eds.), Handbook of Developmental Cognitive Neuroscience. The MIT Press, Cambridge, MA, pp. 23–34.

Brainerd, C.J., Reyna, V.F., 1990. Gist is the grist: Fuzzy-trace theory and the new intuitionism. Developmental Review 10, 3–47.

Brainerd, C.J., Reyna, V.F., Howe, M.L., Kingma, J, 1990. The development of forgetting and reminiscence. Monographs of the Society for Research in Child Development , vol. 55. (3–4, Serial No. 222).

Cabeza, R., McIntosh, A.R., Tulving, E., Nyberg, L., Grady, C.L., 1997. Age-related differences in effective neural connectivity during encoding and recall. NeuroReport 8, 3479–3483.

Cabeza, R., Prince, S.E., Daselaar, S.M., et al., 2004. Brain activity during episodic retrieval of autobiographical and laboratory events: An fMRI study using a novel photo paradigm. Journal of Cognitive Neuroscience 16, 1583–1594.

Carver, L.J., Bauer, P.J., 1999. When the event is more than the sum of its parts: Nine-month-olds' long-term ordered recall. Memory 7, 147–174.

Carver, L.J., Bauer, P.J., 2001. The dawning of a past: The emergence of long-term explicit memory in infancy. Journal of Experimental Psychology. General 130, 726–745.

Carver, L.J., Bauer, P.J., Nelson, C.A., 2000. Associations between infant brain activity and recall memory. Developmental Science 3, 234–246.

Caviness, V.S., Kennedy, D.N., Richelme, C., Rademacher, J., Filipek, P.A., 1996. The human brain age 7–11 years: A volumetric analysis based on magnetic resonance images. Cerebral Cortex 6, 726–736.

Cheatham, C.L., Bauer, P.J., 2005. Construction of a more coherent story: Prior verbal recall predicts later verbal accessibility of early memories. Memory 13, 516–532.

Chugani, H.T., 1994. Development of regional blood glucose metabolism in relation to behavior and plasticity. In: Dawson, G., Fischer, K. (Eds.), Human behavior and the developing brain. Guilford, New York, pp. 153–175.

Chugani, H.T., Phelps, M.E., 1986. Maturational changes in cerebral function determined by 18FDG positron emission tomography. Science 231, 840–843.

Chugani, H.T., Phelps, M., Mazziotta, J., 1987. Positron emission tomography study of human brain functional development. Annals of Neurology 22, 487–497.

Collie, R., Hayne, H., 1999. Deferred imitation by 6- and 9-month-old infants: More evidence of declarative memory. Developmental Psychobiology 35, 83–90.

Cowan, N., Alloway, T., 2009. Development of working memory in childhood. In: Courage, M.L., Cowan, N. (Eds.), The Development of Memory in Infancy and Childhood. Taylor & Francis, New York, NY, pp. 303–342.

Czurkó, A., Czéh, B., Seress, L., Nadel, L., Bures, J., 1997. Severe spatial navigation deficit in the Morris water maze after single high dose of neonatal X-ray irradiation in the rat. Proceedings of the National Academy of Science 94, 2766–2771.

de Biran, M., 1804. The Influence of Habit on the Faculty of Thinking. Williams & Wilkins, Baltimore, MD.

Drummey, A.B., Newcombe, N.S., 1995. Remembering versus knowing the past: Children's explicit and implicit memoris for pictures. Journal of Experimental Child Psychology 59, 549–565.

Ebbinghaus H (1885) On Memory (Ruger HA and Bussenius CE, Translators). New York: Teachers' College, 1913. Paperback edition, New York: Dover, 1964.

Eckenhoff, M., Rakic, P., 1991. A quantitative analysis of synaptogenesis in the molecular layer of the dentate gyrus in the rhesus monkey. Developmental Brain Research 64, 129–135.

Eichenbaum, H., Cohen, N.J., 2001. From conditioning to conscious recollection: Memory systems of the brain. Oxford University Press, New York.

Fivush, R., 1984. Learning about school: The development of kindergarteners' school scripts. Child Development 55, 1697–1709.

Fivush, R., 2001. Owning experience: Developing subjective perspective in autobiographical narratives. In: Moore, C., Lemmon, K. (Eds.), The Self in Time: Developmental Perspectives. Erlbaum, Mahwah, NJ, pp. 35–52.

Fivush, R., Haden, C.A., 1997. Narrating and representing experience: Preschoolers' developing autobiographical accounts. In: van den Broek, P., Bauer, P.J., Bourg, T. (Eds.), Developmental Spans in Event Representation and Comprehension: Bridging Fictional and Actual Events. Erlbaum, Mahwah, NJ, pp. 169–198.

Fivush, R., Hamond, N.R., 1990. Autobiographical memory across the preschool years: Toward reconceptualizing childhood amnesia. In: Fivush, R., Hudson, J.A. (Eds.), Knowing and Remembering in Young Children. Cambridge University Press, New York, pp. 223–248.

Fivush, R., Gray, J.T., Fromhoff, F.A., 1987. Two-year-olds talk about the past. Cognitive Development 2, 393–409.

Fivush, R., Keubli, J., Clubb, P.A., 1992. The structure of events and event representations: Developmental analysis. Child Development 63, 188–201.

Fivush, R., Sales, J.M., Goldberg, A., Bahrick, L., Parker, J.F., 2004. Weathering the storm: Children's long-term recall of Hurricane Andrew. Memory 12, 104–118.

Flavell, J.H., 1963. The Developmental Psychology of Jean Piaget. Van Nostrand, Princeton, NJ.

Gabrieli, J.D.E., 1998. Cognitive neuroscience of human memory. Annual Review of Psychology 49, 87–115.

Giedd, J.N., Blumenthal, J., Jeffries, N.O., et al., 1999. Brain development during childhood and adolescence: A longitudinal MRI study. Nature Neuroscience 2, 861–863.

Gilboa, A., 2004. Autobiographical and episodic memory – One and the same? Evidence from prefrontal activation in neuroimaging studies. Neuropsychologia 42, 1336–1349.

Gogtay, N., Giedd, J.N., Lusk, L., et al., 2004. Dynamic mapping of human cortical development during childhood through early adulthood. Proceedings of the National Academy of Sciences 101, 8174–8179.

Hamond, N.R., Fivush, R., 1991. Memories of Mickey Mouse: Young children recount their trip to Disneyworld. Cognitive Development 6, 433–448.

Hanna, E., Meltzoff, A.N., 1993. Peer imitation by toddlers in laboratory, home, and day-care contexts: Implications for social learning and memory. Developmental Psychology 29, 702–710.

Hayes, B.K., Hennessey, R., 1996. The nature and development of nonverbal implicit memory. Journal of Experimental Child Psychology 63, 22–43.

Hayne, H., MacDonald, S., Barr, R., 1997. Developmental changes in the specificity of memory over the second year of lie. Infant Behavior & Development 20, 233–245.

Hayne, H., Boniface, J., Barr, R., 2000. The development of declarative memory in human infants: Age-related changes in deferred imitation. Behavioral Neuroscience 114, 77–83.

Herbert, J., Hayne, H., 2000. Memory retrieval by 18–30-month-olds: Age-related changes in representational flexibility. Developmental Psychology 36, 473–484.

Howe, M.L., 1995. Interference effects in young children's long-term retention. Developmental Psychology 31, 579–596.

Howe, M.L., Brainerd, C.J., 1989. Development of children's long-term retention. Developmental Review 9, 301–340.

Howe, M.L., Courage, M.L., 1997. Independent paths in the development of infant learning and forgetting. Journal of Experimental Child Psychology 67, 131–163.

Howe, M.L., O'Sullivan, J.T., 1997. What children's memories tell us about recalling our childhoods: A review of storage and retrieval processes in the development of long-term retention. Developmental Review 17, 148–204.

Huttenlocher, P.R., 1979. Synaptic density in human frontal cortex: Developmental changes and effects of aging. Brain Research 163, 195–205.

Huttenlocher, P.R., Dabholkar, A.S., 1997. Regional differences in synaptogenesis in human cerebral cortex. The Journal of Comparative Neurology 387, 167–178.

Johnson, M.H., 1997. Developmental Cognitive Neuroscience. Blackwell Publishers, Inc., Oxford.

Kim, J.J., Fanselow, M.S., 1992. Modality-specific retrograde amnesia of fear. Science 256, 675–677.

Klein, P.J., Meltzoff, A.N., 1999. Long-term memory, forgetting, and deferred imitation in 12-month-old infants. Developmental Science 2, 102–113.

Kleist, K., 1934. Kriegsverletzungen des Gehirns in inhrer Bedeutung fur die Hirnlokalisation und Hirnpathologie. In: Bonhoeffer, K. (Ed.), Handbuch der Aerztlichen Erfahrungen im Weltkriege 1914/1918. Geistes- und Nervenkrankheiten4, Barth, Leipzig, pp. 343–1360.

Klingberg, T., 2008. White matter maturation and cognitive development during childhood. In: Nelson, C.A., Luciana, M. (Eds.), Handbook of Developmental Cognitive Neuroscience. 2nd edn. The MIT Press, Cambridge, MA, pp. 237–243.

Lechuga, M.T., Marcos-Ruiz, R., Bauer, P.J., 2001. Episodic recall of specifics and generalisation coexist in 25-month-old children. Memory 9, 117–132.

Liston, C., Kagan, J., 2002. Memory enhancement in early childhood. Nature 419, 896.

Lloyd, M.E., Newcombe, N.S., 2009. Implicit memory in childhood: Reassessing developmental invariance. In: Courage, M.L., Cowan, N. (Eds.), The Development of Memory in Infancy and Childhood. Taylor & Francis, New York, NY, pp. 93–113.

Lukowski, A.F., Wiebe, S.A., Bauer, P.J., 2009. Going beyond the specifics: Generalization of single actions, but not temporal order, at nine months. Infant Behavior & Development 32 (3), 331–335.

Maguire, E.A., 2001. Neuroimaging studies of autobiographical event memory. Philosophical Transactions of the Royal Society of London 356, 1441–1451.

Mandler, J.M., 1984. Stories, Scripts and Scenes: Aspects of Schema Theory. Lawrence Erlbaum Associates, Hillsdale, NJ.

Mandler, J.M., 1990. Recall of events by preverbal children. In: Diamond, A. (Ed.), The Development and Neural Bases of Higher Cognitive Functions. New York Academy of Science, New York, pp. 485–516.

Mandler, J.M., DeForest, M., 1979. Is there more than one way to recall a story? Child Development 50, 886–889.

Mandler, J.M., Johnson, N.S., 1977. Remembrance of things parsed: Story structure and recall. Cognitive Psychology 9, 111–151.

Mandler, J.M., McDonough, L., 1995. Long-term recall of event sequences in infancy. Journal of Experimental Child Psychology 59, 457–474.

Markowitsch, H.J., 1995. Which brain regions are critically involved in the retrieval of old episodic memory? Brain Research Reviews 21, 117–127.

Markowitsch, H.J., 2000. Neuroanatomy of memory. In: Tulving, E., Craik, F.I.M. (Eds.), The Oxford Handbook of Memory. Oxford University Press, New York, pp. 465–484.

McClelland, J.L., McNaughton, B.L., O'Reilly, R.C., 1995. Why there are complementary learning systems in the hippocampus and neocortex: Insights from the successes and failures of connectionist models of learning and memory. Psychological Review 102, 419–457.

McDonough, L., Mandler, J.M., 1998. Inductive generalization in 9- and 11-month-olds. Developmental Science 1, 227–232.

McDonough, L., Mandler, J.M., McKee, R.D., Squire, L.R., 1995. The deferred imitation task as a nonverbal measure of declarative memory. Proceedings of the National Academy of Sciences 92, 7580–7584.

McGaugh, J.L., 2000. Memory – A century of consolidation. Science 287, 248–251.

Meltzoff, A.N., 1985. Immediate and deferred imitation in fourteen- and twenty-four-month-old infants. Child Development 56, 62–72.

Meltzoff, A.N., 1988. Infant imitation and memory: Nine-month-olds in immediate and deferred tests. Child Development 59, 217–225.

Meltzoff, A.N., 1990. The implications of cross-modal matching and imitation for the development of representation and memory in infants. In: Diamond, A. (Ed.), The Development and Neural Bases of Higher Cognitive Functions. New York Academy of Science, New York, pp. 1–31.

Meltzoff, A.N., 1995. What infant memory tells us about infantile amnesia: Long-term recall and deferred imitation. Journal of Experimental Child Psychology 59, 497–515.

Miller, G.A., 1956. The magical number seven, plus or minus two: Some limits on our capacity for processing information. Psychological Review 63, 81–97.

Moscovitch, M., 2000. Theories of memory and consciousness. In: Tulving, E., Craik, F.I.M. (Eds.), The Oxford Handbook of Memory. Oxford University Press, New York, pp. 609–625.

Moscovitch, M., Nadel, L., 1998. Consolidation and the hippocampal complex revisited: In defense of the multiple-trace model. Current Opinion in Neurobiology 8, 297–300.

Murray, E.A., Mishkin, M., 1998. Object recognition and location memory in monkeys with excitotoxic lesions of the amygdala and hippocampus. Journal of Neuroscience 18, 6568–6582.

Nadel, L., Willner, J., 1989. Some implications of postnatal maturation of the hippocampus. In: Chan-Palay, V., Köhler, C. (Eds.), The hippocampus – New Vistas. Alan R. Liss, New York, pp. 17–31.

Nelson, K., 1986. Event Knowledge: Structure and Function in Development. Erlbaum, Hillsdale, NJ.

Nelson, C.A., 1995. The ontogeny of human memory: A cognitive neuroscience perspective. Developmental Psychology 31, 723–738.

Nelson, C.A., 1997. The neurobiological basis of early memory development. In: Cowan, N. (Ed.), The Development of Memory in Childhood. Psychology Press, Hove East Sussex, pp. 41–82.

Nelson, K., 1997. Event representations then, now, and next. In: van den Broek, P., Bauer, P.J., Bourg, T. (Eds.), Developmental Spans in Event Representation and Comprehension: Bridging Fictional and Actual Events. Erlbaum, Mahwah, NJ, pp. 1–26.

Nelson, C.A., 2000. Neural plasticity and human development: The role of early experience in sculpting memory systems. Developmental Science 3, 115–136.

Nelson, C.A., de Haan, M., Thomas, K., 2006. Neural bases of cognitive development. In: Damon, W., Lerner, R.M. (Eds.), Handbook of Child Psychology. 6th edn. In: Kuhn, D., Siegler, R. (Eds.), Cognition, Perception, and Language, vol. 2. John Wiley and Sons Inc, Hoboken, NJ, pp. 3–57.

Nelson, K., Fivush, R., 2000. Socialization of memory. In: Tulving, E., Craik, F.I.M. (Eds.), The Oxford Handbook of Memory. Oxford University Press, New York, pp. 283–295.

Nelson, K., Fivush, R., 2004. The emergence of autobiographical memory: A social cultural developmental theory. Psychological Review 111, 486–511.

Nelson, K., Gruendel, J., 1986. Children's scripts. In: Nelson, K. (Ed.), Event Knowledge: Structure and Function in Development. Erlbaum, Hillsdale, NJ, pp. 21–46.

Parkin, A.J., 1998. The development of procedural and declarative memory. In: Cowan, N. (Ed.), The Development of Memory in Childhood. Psychology Press, Hove, pp. 113–137.

Parkin, A.J., Streete, S., 1988. Implicit and explicit memory in young children and adults. British Journal of Psychology 79, 361–369.

Perez, L.A., Peynircioglu, Z.F., Blaxton, T.A., 1998. Developmental differences in implicit and explicit memory performance. Journal of Experimental Child Psychology 70, 167–185.

Petrides, M., 1995. Impairments on nonspatial self-ordered and externally ordered working memory tasks after lesions of the mid-dorsal part of the lateral frontal cortex in monkeys. Journal of Neuroscience 15, 359–375.

Pfluger, T., Weil, S., Wies, S., et al., 1999. Normative volumetric data of the developing hippocampus in children based on magnetic resonance imaging. Epilepsia 40, 414–423.

Piaget, J., 1926. The Language and Thought of the Child. Hartcourt, Brace, New York.

Piaget, J., 1952. The Origins of Intelligence in Children. International Universities Press, New York.

Piaget, J., 1969. The Child's Conception of Time. Routledge & Kegan Paul, London.

Price, D.W.W., Goodman, G.S., 1990. Visiting the wizard: Children's memory for a recurring event. Child Development 61, 664–680.

Reed, J.M., Squire, L.R., 1998. Retrograde amnesia for facts and events: Findings from four new cases. Journal of Neuroscience 18, 3943–3954.

Richman, J., Nelson, C.A., 2008. Mechanisms of change: A cognitive neuroscience approach to declarative memory development. In: Nelson, C.A., Luciana, M. (Eds.), Handbook of Developmental Cognitive Neuroscience. 2nd edn. The MIT Press, Cambridge, MA, pp. 541–552.

Roediger, H.L., McDermott, K.B., 1993. Implicit memory in normal human subjects. In: In: Boller, F., Grafman, J. (Eds.), Handbook of Neuropsychology, vol. 8. Elsevier, Amsterdam, pp. 63–131.

Rovee-Collier, C., Hayne, H., 2000. Memory in infancy and early childhood. In: Tulving, E., Craik, F.I.M. (Eds.), The Oxford Handbook of Memory. Oxford University Press, New York, pp. 267–282.

Schacter, D.L., Tulving, E., 1994. What are the memory systems of 1994? In: Schacter, D.L., Tulving, E. (Eds.), Memory Systems. MIT Press, Cambridge, MA, pp. 1–38.

Schacter, D.L., Wagner, A.D., Buckner, R.L., 2000. Memory systems of 1999. In: Tulving, E., Craik, F.I.M. (Eds.), The Oxford Handbook of Memory. Oxford University Press, New York, pp. 627–643.

Schneider, W., Bjorklund, D.F., 1998. Damon, W. (Ed.), Handbook of Child Psychology. 5th edn. In: Kuhn, D., Siegler, R.S. (Eds.), Cognition, Perception, and Language, vol. 2. John Wiley and Sons, New York, pp. 467–521.

Schneider, J.F.L., Il'yasov, K.A., Hennig, J., Martin, E., 2004. Fast quantitative diffusion-tensor imaging of cerebral white matter from the neonatal period to adolescence. Neuroradiology 46, 258–266.

Scoville, W.B., Milner, B., 1957. Loss of recent memory after bilateral hippocampal lesions. Journal of Neurological and Neurosurgical Psychiatry 20, 11–12.

Seress, L., Abraham, H., 2008. Pre- and postnatal morphological development of the human hippocampal formation. In: Nelson, C.A., Luciana, M. (Eds.), Handbook of Developmental Cognitive Neuroscience. 2nd edn. The MIT Press, Cambridge, MA, pp. 187–212.

Sowell, E.R., Delis, D., Stiles, J., Jernigan, T.L., 2001. Improved memory functioning and frontal lobe maturation between childhood and adolescence: A structural MRI study. Journal of International Neuropsychological Society 7, 312–322.

Sowell, E.R., Thompson, P.M., Leonard, C.M., Welcome, S.E., Kan, E., Toga, A.W., 2004. Longitudinal mapping of cortical thickness and brain growth in normal children. Journal of Neuroscience 24, 8223–8231.

Squire, L.R., Knowlton, B., Musen, G., 1993. The structure and organization of memory. Annual Review of Psychology 44, 453–495.

Takehara, K., Kawahara, S., Kirino, Y., 2003. Time-dependent reorganization of the brain components underlying memory retention in trace eyeblink conditioning. Journal of Neuroscience 23, 9897–9905.

Tanapat, P., Hastings, N.B., Gould, E., 2001. Adult neurogenesis in the hippocampal formation. In: Nelson, C.A., Luciana, M. (Eds.), Handbook of Developmental Cognitive Neuroscience. The MIT Press, Cambridge, MA, pp. 93–105.

Toth, J.P., 2000. Nonconscious forms of human memory. In: Tulving, E., Craik, F.I.M. (Eds.), The Oxford Handbook of Memory. Oxford University Press, New York, pp. 245–261.

Tulving, E., 1972. Episodic and semantic memory. In: Tulving, E., Donaldson, W. (Eds.), Organization of Memory. Academic Press, New York, pp. 381–403.

Tulving, E., 1983. Elements of Episodic Memory. Oxford University Press, Oxford.

Utsunomiya, H., Takano, K., Okazaki, M., Mistudome, A., 1999. Development of the temporal lobe in infants and children: Analysis by MR-based volumetry. American Journal of Neuroradiology 20, 717–723.

Van Petten, C., 2004. Relationship between hippocampal volume and memory ability in healthy individuals across the lifespan: Review and meta-analysis. Neuropsychologia 42, 1394–1413.

Woodruff-Pak, D.S., Disterhoft, J.F., 2008. Where is the trace in trace conditioning? Trends in Neurosciences 31, 105–112.

Zola, S.M., Squire, L.R., 2000. The medial temporal lobe and the hippocampus. In: Tulving, E., Craik, F.I.M. (Eds.), The Oxford Handbook of Memory. Oxford University Press, New York, pp. 485–500.

Early Development of Speech and Language
Cognitive, Behavioral, and Neural Systems

H. Tager-Flusberg, A.M. Seery

Boston University, Boston, MA, USA

OUTLINE

17.1 INTRODUCTION

The acquisition of spoken language is one of the most remarkable and uniquely human accomplishments. At birth, newborns are capable of producing only cries and other vegetative sounds, have no control over their oral-vocal tract, and understand no meaningful speech. Within 3 years, without overt instruction, children have mastered the complex speech sound system of their language and acquired a rich and varied vocabulary. They have sufficient grammatical knowledge to allow them to talk about a wide range of topics, referring to events, people, and abstract ideas that are not immediately present in their environment and may exist only in their imaginations, thus going well beyond the communicative abilities of any other species. Human languages are highly complex, hierarchically organized abstract systems, composed of a phonological rule system specifying how speech sounds are combined, lexicon, syntax, morphology, and pragmatic and discourse rules. These systems depend on the development of a range of domain-specific and domain-general cognitive and

© 2013 Elsevier Inc. All rights reserved.

neural mechanisms that interact and become integrated into organized functional networks over the course of the first few years of life. This chapter summarizes what is currently known about the early development of speech and language skills in typically developing children, beginning with the prenatal period and then addresses how these developmental processes may go awry in children with language disorders.

17.1.1 What Is Language?

Spoken languages are composed of sounds; phonemes are the minimal linguistic units of a language and include vowels and consonants. The rapidly changing acoustic properties of phonemes are free to vary across speakers, time, and context, which makes the problem of speech perception particularly challenging. On the production side, speech sounds involve highly complex and intricately timed sequences of movements from the lower to the upper respiratory tract, through the larynx and oral and nasal cavities, and involving the lips, tongue, teeth, and glottis to produce phonemes. Within each language, there are constraints, or rules, which specify the permissible sequences of phonemes, referred to as phonotactic constraints. In addition, speech involves prosodic features that characterize the stress patterns of words and sentences that may be influenced by lexical, grammatical, or pragmatic factors.

Phonemes are combined to create the minimal unit of a language that carries meaning (semantic function), referred to as morphemes. Words (or lexical items) make up the vast majority of a language's morphemes. Most languages also include a smaller set of morphemes called 'bound morphemes' because they are affixed to other morphemes but still carry meaning such as tense, mood, number, and gender. Examples in English include past tense (*-ed*), the negative *un-*, or the plural *-s*. In turn, words and morphemes are combined to create sentences. The rules governing the order and hierarchical organization of sentence structure make up the syntax of a language. Together, the phonology, morphology, semantics, and syntax form the structural components of language. The functional uses of language, or pragmatics, include a variety of speech acts (e.g., directive, question) as well as the social rules governing the use of language in a range of discourse contexts (e.g., conversation, narrative).

The neural bases of language perception and production are relatively well known (e.g., Hickock, 2009). Large networks of cortical and subcortical regions are involved, spanning basic sensory and motor systems to higher-order, relatively specialized cortical areas. In the majority of adults, in whom most *in vivo* studies have been carried out, these specialized areas, devoted primarily to linguistic processing, are functionally organized asymmetrically in the left hemisphere of the brain; the two most significant areas include Broca's area in the inferior frontal cortex and Wernicke's area in the temporal lobe. These regions are connected via white matter fiber bundles, including the primary dorsal pathway, the *arcuate fasciculus*, as well as secondary dorsal and ventral pathways (Friederici, 2009). Although less is known about the development of these neural systems because of limitations to the use of neuroimaging methods in infants and young children, there is now an emerging body of research focusing primarily on perceptual and language processing using electrophysiological methods, which are the most easily adapted for use with infants (Kuhl, 2010).

17.2 SPEECH PERCEPTION

Speech perception and language comprehension rely on complex neural networks that, in most adults, heavily involve the perisylvian region of the left hemisphere. Research has explored the early behavioral responses and neural correlates of exposure to language in neonates and infants to investigate the development of the language networks.

17.2.1 Prenatal Responses to Speech and Language

The auditory system is almost fully developed by the third trimester of pregnancy, so the developing fetus can process auditory and speech stimuli, although the rapid temporal features of phonemic stimuli are, for the most part, filtered out by the mother's body and uterine environment. Near-term fetuses show a more sustained heart rate deceleration to sentences than to music but do not differ in their response to sentences and chimeras of these sentences with all phonetic information removed (Granier-Deferre et al., 2010). Most prenatal exposure to phonemic stimuli is provided by the mother's voice, the vibrations of which are transmitted directly to the fetus.

Infants learn about many aspects of language during the last weeks of the prenatal period, including the identity of their mother's voice, stress patterns of the language, as well as specific segments of language (such as repeated passages from storybooks) to which they have been exposed. Beginning around 32 weeks' gestation, fetuses exhibit heart rate deceleration in response to recordings of their mother's voice (Kisilevsky and Hains, 2011), and near-term infants show differential patterns of heart rate response to recordings of the mother's voice and a stranger's voice (Kisilevsky et al.,

2003). At birth, infants distinguish and prefer their own mother's voice over the voice of a female stranger (DeCasper and Fifer, 1980), speech passages they were exposed to prenatally over novel passages (DeCasper and Spence, 1986), and their native language over other rhythmically dissimilar languages (Byers-Heinlein et al., 2010; Mehler et al., 1988; Nazzi et al., 1998).

17.2.2 Newborn Speech Perception

From birth onward, infants display a strong preference for listening to speech over other auditory stimuli. Neonates exhibit more nonnutritive, high-amplitude sucking when rewarded with linguistic stimuli than with white noise (Butterfield and Siperstein, 1970), tones, or scrambled speech sounds (Vouloumanos and Werker, 2007). Even very young infants possess sophisticated speech perception abilities, which is similar to adults in a variety of ways.

Adults perceive phonetic information categorically. For instance, the speech sounds /da/ and /ta/ differ with respect to the phonetic feature of voice onset time (VOT), and when presented with syllables with VOTs varying continuously, adults perceive only two sounds: /da/ or /ta/. Furthermore, adults are easily able to extract phonetic information despite the fact that the acoustic properties of speech vary drastically depending on the individual speaker's voice. Behavioral studies have found that 2-month-old infants detect phonetic changes regardless of whether this change is accompanied by a change in pitch (Kuhl and Miller, 1982) or speaker (Jusczyk et al., 1992). In newborns, electrophysiological studies have shown that the neonatal brain responds as quickly to a change in phonemic category whether or not this is accompanied by a change of speaker (Dehaene-Lambertz and Pena, 2001). Neonates extract linguistically relevant phonemic information from the extraneous acoustic information related to specific voices and thus normalize across speakers. Notably, however, this is not due to an inability of infants to discern the different voices as neonates have been shown to discriminate at least between male and female voices (Floccia et al., 2000). In addition to normalizing across speakers, very young infants, like adults, perceive speech sounds categorically rather than continuously. Infants will dishabituate more to a change in syllable when this change crosses adult categorical boundaries than when the change occurs within a given phonetic category (Eimas et al., 1971). Newborns are sensitive to linguistic or auditory 'gestalts,' such as syllable repetition. Specifically, they show increased response over the temporal and left frontal areas to grammars containing a consecutive repetition (ABB sequences, e.g., 'mubaba'), suggesting the possibility of automatic

perceptual detection of auditory repetition (Gervain et al., 2008).

17.2.3 Development of Speech Perception

After birth, infants continue to learn about the language that they are exposed to, and by 4 months, infants discriminate their own language even from other very similar languages, such as Spanish and Catalan (Bosch and Sebastián-Gallés, 2001). Infants also undergo a drastic transition where their perception of speech becomes shaped by exposure to a specific language. Each language contrastively uses only subsets of the available universal phonetic categories, and adults cannot easily discriminate between phonetic categories not used in their native language.

Infants are prepared to perceive any language they might encounter but gradually restrict this ability to their own language over the first year of life through a process of perceptual narrowing. Infants stop discriminating consonant contrasts not used in their native language by around 10 months of age (Werker and Tees, 1984). Measuring electrophysiological response to a phonetic change after a repeated syllable has shown that this perceptual narrowing involves decreased response when the change involves nonnative phonemes (Cheour et al., 1998; Rivera-Gaxiola et al., 2005) coupled with increased response when the change involves native-language phonemes (Rivera-Gaxiola et al., 2005), suggesting a process of neural commitment to native-language phonemes (Kuhl, 2004). A similar study using near-infrared spectroscopy, which is more sensitive than electrophysiology to spatial localization of neural processing, found that increased response to native phoneme changes did not stabilize until around 10 months of age, and a left-lateralized response to native-language phonemes emerged around 13 months of age (Minagawa-Kawai et al., 2007).

Using magnetoencephalography, Imada and colleagues found that a wide range of sounds, including speech, elicited responses in left hemisphere superior temporal areas in infants from birth through 12 months of age (Imada et al., 2006). In contrast, inferior frontal areas of the left hemisphere were not responsive to speech until after age 6 months; this response was correlated with temporal region responses to speech by 12 months of age, suggesting the basic neural network for processing language is in place by the end of the first year. It may be that there are structural predispositions for these functional networks to develop in the left hemisphere (Dubois et al., 2009).

In sum, it is evident that the infant brain possesses some degree of functional organization of the language networks present in human adults. The superior temporal region, in particular, responds to language at birth, although this response is not specific to speech. To some extent, left-lateralized responses may be driven by acoustic characteristics of the auditory stimulus in addition to truly linguistic properties. Although there are some inconsistencies in the current infant literature, which may be the result of different experimental paradigms and stimuli as well as imaging modality, by the end of the first year, the neural and cognitive bases for speech perception appear to be firmly in place in perisylvian regions of the left hemisphere.

17.3 SPEECH PRODUCTION

In contrast to the perception of speech, at birth, infants are not able to produce any speech-like sounds. This capacity develops over the course of the first year of life and continues as the child's developing phonological system is acquired alongside other aspects of language development.

17.3.1 Infant Speech Production

The developmental progression in speech production follows universal pathways. During the first 2 months of life, infants produce a range of reflexive vocalizations that appear to be automatic responses to their physical state. These include crying, fussing, and vegetative sounds. The range of sounds produced by very young infants is constrained by the size of their oral cavity and the position of the larynx (Lieberman et al., 1972). Between 2 and 4 months, there are still significant limitations to the infant's ability to articulate, but the vocal repertoire expands to include pleasure-related cooing sounds. These sounds, which may include some vowels or consonants, are produced in the back of the mouth with articulation limited to movements of the jaw (Kent, 1999).

Beginning at around 4 months, as developmental changes take place in the morphology of the vocal tract, infants engage in more vocal play that includes both nonspeech and speech sounds such as rudimentary vowels and consonants. At 6 months, babbling begins, with the infant producing well-formed consonant–vowel (CV) syllables that now extend to those involving movements of the lips and tongue. Canonical babbling (repeated sequences of CV syllables) is viewed as the most significant milestone in speech production (Oller, 2000). During the latter half of the first year, the infant's sound combinations become increasingly more variable and frequent, though there appears to be a relatively small set of consonants produced across infants acquiring many different languages (Locke, 1983). As infants move closer to the onset of meaningful speech, their babbles increase in length and incorporate varied stress and intonation patterns, often referred to as 'jargon' or conversational babble.

The development of speech motor control plays an important role in speech sound production. A computational model of this process was introduced by Guenther (1995). According to this self-organizing model, the initiation of speech is the product of the perception of intended target sounds. Thus, auditory feedback from both adult and self-produced speech is critical in developing the mapping between target sounds and developmental changes in the vocal tract. On this view, there is a close link between the development of speech perception and production, which is consistent with the finding that although deaf infants produce early vocalizations, they rarely engage in canonical babbling and their sound production lessens over time (Stoel-Gammon, 1998).

17.3.2 Relationship Between Speech and Motor Developments

During the first year of life, infants not only develop the ability to articulate complex speech sounds, they also make significant advances in other aspects of motor development. While there are parallels between speech and motor developments, the relationship between them is both complex and quite specific. Gross motor milestones are not closely tied to developmental stages in babbling; however, it has been found that rhythmic arm banging, with or without objects, consistently emerges just prior to the onset of canonical babbling (Iverson and Thelen, 1999). Iverson (2010) argues that these repetitive hand movements provide opportunities for practicing skills required for canonical babbling as they both involve rhythmically organized motor stereotypies.

Babbling begins as a behavior tied closely to the speech motor system, offering the opportunity to practice complex articulatory movements in the context of proprioceptive and auditory feedback. Over time, babbling rapidly becomes integrated with other developmental changes and events in the environment to emerge as an early linguistic skill (Iverson, 2010). The close neural links between brain regions involved in language and motor behavior provide support for the view that there are reciprocal influences between these systems over the course of development, with the onset of intentional communication in infants evident in manual and other body gestures (Iverson et al., 2007).

17.3.3 Phonological Development

It is generally agreed that there is essential continuity between prelinguistic babbling and the earliest stages of phonological development evidenced in the child's first words (Stoel-Gammon, 1998). The majority of sounds produced in the earliest words of children are the same as those preferred in their babbles. Initially, children's words are composed of simple CV syllable structures, using a relatively small inventory of sounds. Gradually, over time and with growth in vocabulary, there is an expansion in the range of sounds produced by children. Although there is no universal order in the acquisition of phonological features, certain regularities have been found in the phonological sounds that are used across children, reflecting developmental changes in the vocal tract (Stoel-Gammon and Sosa, 2007). Mastery over vowels occurs before that over consonants. The main consonant classes that are used earlier in development include stops (e.g., *b, d*), nasals (e.g., *m, n*), and glides (e.g., *w*). Later developing consonants include fricatives (e.g., *v*) and liquids (e.g., *l, r*).

As children begin producing more elaborate syllable structures and a wider range of sounds in words, they begin making speech sound errors. The most striking feature of these errors is that while there are individual differences in the particular kinds of errors made, the errors are not random, but instead fall into common patterns, reflecting developmental changes in the child's representation of the speech sound system (Grunwell, 1981; Ingram, 1976). In children with more severe articulation difficulties, words may be produced that involve combinations of different error patterns, but in most children, speech sound errors do not persist beyond the preschool years.

Young children will often omit syllables or specific sounds as they attempt to reproduce more complex adult words (Menn and Stoel-Gammon, 2009). Typically, unstressed syllables will be omitted; those occurring at the beginning of words (e.g., *mato* for *tomato*) or in the unstressed medial position (e.g., *e'phant* for *elephant*), and consonant clusters often lead to omissions (e.g., *top* for *stop*). Another class of error patterns is to change sounds at the level of individual articulatory features. For example, voiced consonants may be changed to unvoiced consonants (e.g., *bot* for *pot*). Place changes also may be found in some children, with back consonants becoming more frontal (e.g., *dame* for *game*). These kinds of errors demonstrate the significance of features in children's phonological representations. A third class of errors illustrates how the child's representation of the target word may influence the kinds of sound substitutions that are made. Assimilation errors entail the change in one sound in the target word to make it more similar to another sound in that word. Such errors may involve assimilation in different feature classes such as voicing (e.g., *doad* for *toad*) or place (*gog* for *dog*).

Several theoretical frameworks have been proposed to account for the acquisition of phonology, but the most promising current formal model is optimality theory

(Archangeli and Langendoen, 1997). On this model, representations of phonological inputs are evaluated against a set of finite and universal constraints to determine an optimal output phonological form. Here, the notion of constraints replaces earlier theories that defined absolute, serially applied abstract rules. Within optimality theory, constraints are not absolute but operate in parallel in a language-specific order. Optimality theory has been shown to be highly successful in capturing children's phonological error patterns as well as variation found between, and within, individual children acquiring different languages (Barlow and Geirut, 1999).

17.4 SOCIAL–COGNITIVE FOUNDATIONS OF LANGUAGE

Language is a cultural system, acquired within the context of specific environments and embedded in the emergence of developments in the child's cognitive, emotional, and social capacities. Thus, language is the vehicle that satisfies the child's motivation to engage and communicate with social partners. In the absence of any social partners, for example, in children raised in complete isolation, language is not acquired (cf. Curtiss, 1977).

17.4.1 Social Engagement

From birth, the social niche for language is clearly established in the infant's preference for human voices and faces. Over the first few months of life, rapid changes take place. Mothers and their infants begin to interact in finely tuned ways with one another. They synchronize their eye gaze, movements, and facial expressions of affect in patterns that resemble turn-taking patterns in conversations (Snow, 1977). By the age of 4 months, there is a marked increase in vocal turn-taking during these rich interactions between infants and their caretakers (Ratner and Bruner, 1978). Thus, engagement with other people provides the context that lays the groundwork for infants' motivation to communicate. Interestingly, the neural foundations for these social capacities are available very early in the first year of life (Grossmann and Farroni, 2009).

17.4.2 Infant-Directed Talk

In their interactions with infants, adults in most cultures speak in a special register, including a higher pitch, more variable intonation patterns, and shorter utterances that focus on objects and people in the immediate environment (Fernald, 1989). This register, now referred to as 'infant-directed talk' (IDT), is strongly preferred by infants (e.g., Werker and McLeod, 1989). Its properties capture infants' attention and facilitate the analysis of the phonological properties and statistical regularities of their native language (Burnham et al., 2002; Thiessen et al., 2005). Together, these features serve to facilitate the acquisition of the formal features of language as well as the child's ability to learn the meanings of words toward the end of the first year (Waxman, 2003).

IDT also incorporates rich emotional content captured in the prosodic contours that convey different emotional states. Younger infants are very sensitive to this component of IDT and respond to IDT statements of approval and praise with higher rates of smiling and attention even when they hear adults speaking in a different language (Fernald, 1989). Together, the features of IDT ensure that infants attend to the speech and language in their environment and facilitate the processing of its essential features and characteristics.

17.4.3 Intentional Communication

Toward the end of the first year of life, vocalizations as well as other nonvocal behaviors become genuinely integrated into social interaction as infants' developing social–cognitive capacities lead to the onset of intentional communication (Carpenter et al., 1998). At this point, infants become capable of coordinating their attention to objects or events with other people through eye gaze patterns (joint attention), gestures, and vocalizations. This developmental achievement is generally viewed as a critical step in language acquisition, with the onset of communicative intent. Infants at this stage are able to communicate a variety of meanings, including *protodeclaratives*, which involve pointing or other gestures to draw another person's attention to objects of interest, and *protoimperatives*, gestures or vocalizations to express requests for objects or actions. The significance of these communicative attempts is that they indicate the infant's capacity to understand the intentions of others (the beginning of a theory of mind), at least in a rudimentary or implicit form (Tomasello, 1999).

Infants' preverbal communications go beyond pointing to encompass other conventional gestures such as nodding or shaking their head to signify acceptance or rejection, waving as a salutation, or invented gestures that may incorporate symbolic features of an object or event, such as wiggling a finger to refer to a dog (Carter, 1979). Gestural communication is an important predictor of spoken language development in children and is another example of the close developmental links between the motor and linguistic systems (Rowe and Goldin-Meadow, 2009).

17.4.4 Pragmatic Development

As children begin to speak, their utterances express the same functions as their early preverbal gestural forms. By the time children are three, new functions emerge, including the use of language to describe objects or events, or to assert an opinion, for example, and they employ a range of conversational devices. At this point, children are able to express each of these functions using a variety of linguistic forms. The development of functional and communicative aspects of language is closely tied to developments in theory of mind and related social–cognitive achievements (Bartsch and Wellman, 1995).

There is a more protracted period of development for expressing functions using more indirect forms, such as indirect requests. Although 2-year-olds use terms such as *want* or *need* as a way of asking for something (e.g., *I need new ball*), genuine indirect requests do not emerge until around the age of three (e.g., *Where is the truck?*). By three years of age, children can use polite forms to make their request (e.g., *Would you give me a cookie?*), but hints or oblique indirect requests are not used until the early school years (Bryant, 2009).

Research has focused on children's developing awareness of the effect of their speech on a listener by taking into account the listener's knowledge. For example, 4-year-olds change the way they speak depending on their audience; thus, they use simpler language if they are talking to 2-year-olds compared to their language addressed to older children or adults. These modifications in children's discourse show some awareness of the distinct needs of a very young conversational partner (Shatz and Gelman, 1973).

Communicative competence entails knowing how to engage in conversations in appropriate and informative ways. Ultimately, this depends on appreciating both literal and nonliteral uses of language (e.g., metaphor, irony, and white lies), but these developments continue well into middle childhood (Siegal and Surian, 2007). Children also must master a range of different forms of discourse from conversation, to personal narratives to storytelling. The use of language in various contexts provides the interactive, communicative framework within which children acquire knowledge of the linguistic structures available in their native language so that they can express more fully the ideas that are generated by their developing cognitive and social systems.

17.5 LEXICAL DEVELOPMENT

By the end of the first year, infants already understand a number of words and phrases and soon begin to produce words. This is a critical milestone that is often used as an objective proxy for the onset of language. Long before this milestone is reached, infants have acquired the skill of segmenting words within streams of continuous speech where there are often no clear boundaries, using a variety of cues, including sequential statistical information (Saffran et al., 1996) and consistent stress patterns (Jusczyk et al., 1999).

17.5.1 Stages of Lexical Development

Lexical development can be divided into three broad periods. The first covers the acquisition of the initial 50 words or so, during which children are learning what words do. At this stage, some words appear to be tied to particular contexts and serve primarily social or pragmatic purposes. Word learning during this initial phase is relatively slow and uneven (Nelson, 1981). The child's vocabulary at this stage, especially in Western middle-class children, is dominated by names for objects, including animals, people, toys, and familiar household things. Other early words include social terms (e.g., *hi*, *bye*), modifiers (e.g., *more*, *wet*), and relational terms that express success, failure, recurrence, direction, and so forth (Bates et al., 1994).

By the middle of the second year, there is a significant increase in the rate at which children acquire new words. This new period is usually referred to as the vocabulary spurt and may be punctuated by requests from children for adults to label things in the world around them. Words are learned very quickly, often after only a single exposure that may take place without any explicit instruction (Tomasello, 2003). This process of rapid word learning is referred to as *fast mapping* (Houston-Price et al., 2005).

By the time children reach their third birthday, they begin to develop a more organized lexicon, in which the meaning relations among groups of words are discovered. For example, at this time, children begin to learn words from a semantic domain, such as kinship, and they are able to organize the words according to their similarities and differences in dimensions of meanings (Nagy and Scott, 2000). For nouns labeling concrete objects, children begin to organize taxonomies, now also learning words at the superordinate and subordinate levels and understanding the hierarchical relations among terms such as dachshund, dog, and animal. Semantic developments at this stage will often lead to reorganizational processes as these kinds of relationships among words are realized by the child (Bowerman, 1978). The rate of word learning continues to be very rapid with estimates suggesting that children acquire about 15–20 new words a day during the preschool years and beyond (Bloom, 2000).

17.5.2 Developmental Processes

How do children accomplish the challenge of acquiring arbitrary symbols and their associated sound forms (words) to communicate concepts so rapidly? Research has identified several factors that facilitate the task of word learning in young children. At the most transparent level, the input to children provides one critical constraint on the words they acquire. The number and frequency of specific words in parents' speech to children correlates significantly with the frequency of those words in the children's vocabulary (Huttenlocher et al., 1991), and these variables correlate with socioeconomic status, reflecting the fact that more highly educated mothers speak far more to their children than mothers with little education (Hoff, 2006). The social context of word learning also helps young children who, because of their ability to infer intentions, can map word meanings by observing where their mother is looking or what she is pointing at when she labels objects (Baldwin and Meyer, 2007).

The child's developing conceptual knowledge influences the meanings they map onto words (Bloom, 2000). The relationship between language and conceptual development, or more generally between language and thought, is highly complex with each system placing constraints on the other, and with both being dependent on the social environment for their elaboration in development. Other general cognitive processes, including attention and memory, are also clearly important in word learning (Samuelson and Smith, 2000).

Children bring to the task of word learning several constraints that guide their hypotheses about the possible meanings of words. Markman (1989) has proposed that young children rely on three primary constraints including the *mutual exclusivity constraint*, which leads the child to assume that each object only has a single name and that a name can only refer to one category of objects; the *whole-object constraint*, which states that new words refer to whole objects rather than parts of objects; and the *taxonomic constraint*, which states that words refer to categories of objects (not specific exemplars). While some view these kinds of constraints as principles that are specific to lexical development, others view them as more general biases that may be an aspect of broader pragmatic or cognitive processes (Diesendruck, 2007).

Lexical development takes place in parallel with other aspects of language, particularly the acquisition of syntactic knowledge. Children use syntactic information to facilitate word learning, especially when other cues are not available, as is often the case for learning the meanings of related verbs such as *look* and *see*. In cases such as this, children can use information about the number and kind of arguments that occur with particular verbs to work out their meanings (Naigles and

Swensen, 2007). Thus, transitive verbs take object arguments while intransitive verbs do not. Syntactic information is also useful for figuring out the distinction between mass nouns (e.g., *spaghetti*) from count nouns (e.g., *a potato*), or between common nouns and proper names, and very young children have been shown to use this information when they hear new words in ambiguous contexts (Hall and Waxman, 2004).

Children depend on a wide range of cognitive, social, and linguistic mechanisms to learn the meanings of new words, and the input provided by their conversational partners helps to guide and constrain the learning process. Several integrative theoretical models have been proposed to account for how children might attend to different types of cues at different developmental stages and weigh the information carried by competing cues to lexical meaning (Bloom, 2000; Hollich et al., 2000). Such models have the potential to account for individual differences in, for example, children learning different languages or raised in different kinds of learning environments. A number of computational models of early word learning have been developed (e.g., Frank et al., 2009), though thus far, none has incorporated the full range of processes or constraints that are central to these integrative developmental models.

17.5.3 Neural Bases of Word Learning

By 5 months of age, infants' neural response shows sensitivity to the word stress patterns used in their language and will show a mismatch event related potential (ERP) response to deviations from this stress pattern (Weber et al., 2004). By 10 months, infants show distinct patterns of brain activity in response to familiar two-syllable words embedded in continuous speech, even when they have previously heard the word only in isolation (Kooijman et al., 2005).

As infants become more verbal, experience with language influences the organization of their neural response, resulting in distinct patterns of brain activity in response to words that an infant understands by 11 months (Thierry et al., 2003). Between 13 and 17 months, infants exhibit a broadly distributed bilateral response to words in general and a stronger N200 response to words that are understood than to words that are not (Mills et al., 1997). By 20 months, this pattern of response is localized to temporal and parietal electrode sites over the left hemisphere (Mills et al., 1993). The lateralized and focal response to known versus unknown words is linked more closely to linguistic ability rather than simply developmental maturation of the brain or brain connectivity. Thus, 13- to 17-month-old infants with more advanced receptive language skills showed a more localized neural response, whereas infants with low receptive language showed a less mature broader response pattern. One study with 20-month-old infants used a training paradigm to teach new words to label pictures of objects. After the training, there was an increased N200 response to the trained words over anterior electrode sites. This change in N200 response was independent of expressive vocabulary size, highlighting the influence of experience with a particular word and its meaning, not simply the infants' maturational level or overall language ability (Mills et al., 2005). Still, developmental maturation of language areas may influence the latency of response to new words as it decreases with age independent of language level (Mills et al., 1997).

In infants, semantic knowledge of word meaning is captured by a later negative response similar to the adult N400, which processes the integration of word meaning in context and involves the middle and superior temporal gyri (Kutas and Federmeier, 2011). In 19-month-old infants, the N400 response is elicited when shown pictures of objects paired with incongruent verbal labels (Friedrich and Friederici, 2004). Response to incongruent picture–word pairings is present early in the waveform over lateral frontal electrode sites, but is largest later in the waveform over both posterior and frontal sites. This later N400 was stronger, earlier, and more similar to that of adults in infants with more advanced receptive vocabularies. At 19 months, robust N400 responses are also elicited in priming contexts in the majority of infants. At 12 months, infants with advanced productive vocabularies also show an N400 response to priming, though at this early age, it is slightly delayed in onset and smaller in duration than in older children or adults (Friedrich and Friederici, 2010). The early functioning of N400 neural mechanisms highlights their significance for word learning, though the precise nature of this relationship is not yet known.

17.6 SYNTACTIC DEVELOPMENT

By the time they are 2 years old, toddlers begin combining words to form two or three word phrases or primitive sentences. From the earliest stages, children demonstrate their sensitivity to some of the basic rule-governed properties of their target language, for example, word order rules in English, though the precise nature of the knowledge this represents is still hotly debated.

17.6.1 Stages of Syntactic Development

Children must segment the sound stream they hear into morphemes, phrases, and sentences (Morgan, 1986) and discover word classes (nouns, verbs, determiners, etc.) because grammatical rules operate on these more abstract categories to create hierarchically organized sentences. They must also figure out if their target language encodes tense, person, number, etc., and if so, whether these are marked by grammatical morphemes or other linguistic means. Different theories have been proposed for how children acquire this knowledge, ranging from distributional statistical (e.g., Mintz et al., 2002), to functional (Tomasello, 2002), to semantic bootstrapping (Pinker, 1984), to linguistic nativist approaches (Chomsky, 1995).

Across different languages, children begin by building phrasal units (e.g., noun phrase), move on to simple sentence structures and different sentence modalities (e.g., statements, questions), and finally master complex sentences such as coordinations and embeddings. As children's grammatical development proceeds, the length of their utterances increases, reflecting both the acquisition of linguistic knowledge (Brown, 1973) and growth in general cognitive capacities, particularly verbal working memory.

17.6.2 Early Sentences

As early as 18 months of age, the hierarchical nature of phrasal structures (e.g., *the red barn*) is available to children (Lidz et al., 2003). At this stage, English-speaking children adhere closely to using word order in their productive speech but omit most grammatical morphemes that mark, for example, tense or number. In contrast, children acquiring inflectionally rich languages such as Italian or Hebrew use the morphemes marking these features even in their very early productions (Caselli et al., 1999; Levy, 1988).

Children's early 'sentences' express a universal set of meanings: they talk about objects and people present in the environment, their locations, attributes, and interrelationships (Brown, 1973). Often their sentences will omit the subject, and while children acquiring languages where subjects are obligatory do so less than children acquiring languages in which so-called null subjects can be grammatical (Valian, 1990), these omissions reflect the limited cognitive capacity of young children and the fact that sentence subjects may be inferred from the context and are therefore pragmatically not necessary (O'Grady et al., 1989).

17.6.3 Grammatical Morphology

In contrast to many other languages, English has a relatively impoverished set of affixes and lexical terms that comprise its grammatical morphology to mark tense, number, person, etc. The course of development is thus a gradual process that is quite protracted, though across

children, there is a similar order in which different morphemes are acquired reflecting their linguistic complexity. The developmental process is more rapid in languages such as Italian, though within each language, there is still a constrained order in which different morphemes are mastered (Brown, 1973).

In English, there are both regular and irregular forms to mark, for example, tense (e.g., *walk–walked*; *run–ran*) or number (e.g., *boy–boys*; *mouse–mice*). As children acquire these morphemes, they go through a stage of making errors, particularly by providing regular affixes to irregular verbs or nouns (e.g., *runned*; *mouses*). Pinker and his colleagues (Pinker, 1999; Ullman, 2001) argue that the acquisition and use of these different forms reflect two distinct mechanisms: a rule-learning system for regular forms and a lexical-memory system for irregular forms. In contrast, some computational models have been developed that can acquire the different forms based on a unitary mechanism that operates on the distributional properties of lexical items in the input (McClelland and Patterson, 2002).

17.6.4 Later Grammatical Development

Children's early utterances may be used to express a range of functions including statements, questions, and negations; however, they rely on single words or intonation to do so. By the preschool years, children acquire auxiliary verbs (do, be), and their negations and questions include the requisite verb morphology and inversion needed for questions (de Villiers et al., 1990). Sentences are now combined to create longer and more complex coordinated structures using 'and' and later, 'if,' 'because,' and other coordinating conjunctions. Rarer complex linguistic constructions, for example, relative clauses, or the passive form are not mastered until the early school-age years. The order and speed of syntactic development reflect both the frequency of forms in the input and the complexity of the abstract rules that underlie them (Tager-Flusberg and Zukowski, 2009).

17.6.5 Neural Bases of Grammatical Development

In comparison to research on the neural bases of speech perception and word learning, few studies have been conducted on early grammatical development, perhaps reflecting the relative difficulty of obtaining reliable data from preschoolers who are less tolerant of electrode caps or sitting still. In adults, the most widely studied ERP components are an early left anterior negativity (ELAN), which is related to automatic sentence parsing based on word-category information or morphological violations and is thought to be generated in the left inferior frontal gyrus and anterior superior temporal gyrus (STG; Friederici et al., 2000). Later, and more controlled, syntactic processing occurs with a later positivity, the P600, during which syntax is integrated with other information and sentence reanalysis occurs in the context of syntactic anomalies or ambiguities (Kaan and Swaab, 2003).

By around 30 months, the ERP responses of toddlers to syntactic violations show a large, late positivity similar to the P600 (Bernal et al., 2010; Silva Pereyra et al., 2005) and, in some cases, display an earlier left-lateralized negativity similar to the adult ELAN (Bernal, et al., 2010). This similarity in waveform topography and morphology suggests that toddlers have already developed or constructed a relatively adult-like system for processing the syntax of simple sentences, including to some extent a system of automatic processing. Passive sentences, on the other hand, involve higher processing demands than do active sentences and do not evoke adult-like automatic components until late childhood or adolescence (Hahne et al., 2004).

Nevertheless, the neural response to syntactic violations is immature in children in that it is not yet functionally specified the way it is in adults. For example, functional magnetic resonance imaging (MRI) studies in adults show that distinct areas of the inferior frontal gyrus and STG are activated in response to syntactic violations versus semantic violations, particularly as processing demands increase. At the age of five or six, children still show substantial overlap in their fMRI activation to syntactic and semantic violations, particularly in the STG (Brauer and Friederici, 2007). These findings provide some support for the view that the acquisition of syntactic structure may be closely linked to semantics in young children.

17.7 LANGUAGE DISORDERS

The vast majority of children follow the developmental pathway to language described in the previous sections. But for some children, the onset of language is quite delayed; thereafter, the rate of development may be slowed, and the normal synchrony between the structural and pragmatic components of language may not hold. Children whose language is significantly delayed may never reach the same endpoints as typically developing children; this is especially true for those who also have intellectual disability, for example, children with known genetic syndromes including Down syndrome or fragile X syndrome (Tager-Flusberg, 2007). Limits in the acquisition of language beyond the early school years, specifically in the domains of syntax and morphology, have been taken as evidence for a critical period (e.g., Lenneberg, 1967).

Two complex neurodevelopmental disorders are defined, in part, on the basis of primary impairments in language and communication: specific language impairment (SLI) and autism spectrum disorder (ASD). SLI is diagnosed on the basis of delays and slowed rate of development of language in the absence of hearing impairment, frank neurological damage, social deprivation, or other neurodevelopmental disorders. ASD is diagnosed on the basis of impairments in communication, social reciprocity, and restricted or repetitive behaviors and interests. The majority of children with ASD also have comorbid impairments in structural aspects of language, and there is a growing interest in the overlap between ASD and SLI in their language phenotypes as well as genetic risk factors (e.g., Abrahams and Geschwind, 2008; Tager-Flusberg, 2004; Tager-Flusberg et al., 2008). A selective review of what is currently known about the developmental course and neurocognitive bases of language impairment in young children with these disorders, both of which are marked by significant heterogeneity in the severity and characterization of core and associated symptoms, is presented here.

17.7.1 Speech Perception

SLI and ASD are not diagnosed until the preschool years; however, in recent years, research has focused on infants at risk for these disorders, defined usually on the basis of family risk (e.g., presence of older sibling or other first-degree relative with the disorder), in order to capture the earliest emergence of deviations from patterns of normal development. Studies of 2-month-old infants at risk for SLI show significant differences in ERP responses for discriminating between long versus short syllable lengths (Friederici, 2006), and by 4–5 months, they show a reduced ERP response for discriminating the stress pattern of two-syllable words (Weber et al., 2004), suggesting that deficits in processing speech duration very early in life may be a marker for later language impairment. At 6 months of age, infants at risk for SLI are less able to discriminate tones presented in rapid succession, and the amplitude of the N250 mismatch ERP response is smaller and delayed in onset when listening to tones with brief interstimulus intervals (Benasich et al., 2006). These ERP measures predicted to language outcomes at the age of two. Taken together, these studies suggest that differences, which may affect auditory as well as speech perception, are present very early in life for infants at risk for SLI. Similar studies have not yet been carried out with infants at risk for ASD.

17.7.2 Speech Production

Retrospective reports highlight delays in the onset of babbling in children with SLI and ASD (Norbury et al., 2008; Tager-Flusberg et al., 2005). Children with ASD who have more severe deficits in acquiring spoken language also have a history of deficits in a range of other oral-motor skills (Gernsbacher et al., 2008). Prospective studies of infants at risk for ASD have confirmed that babbling is delayed in this group: between 6 and 12 months, infants at risk have fewer speech vocalizations and produce fewer consonants and less canonical babble (Paul et al., 2010). In addition, there is not as close a connection between the onset of babbling and rhythmic arm banging in high-risk infants as in typical infants (Iverson and Wozniak, 2007).

When children with SLI begin speaking, they are at higher risk for articulation disorders: though these are distinct syndromes, there is some overlap between SLI and speech sound disorders (Sices et al., 2007). Most children with ASD do not have impaired articulation (Kjelgaard and Tager-Flusberg, 2001), and their order of acquiring consonants follows the normal pattern (McCleery et al., 2006). Nevertheless, there is a group of children with more severe ASD (and often comorbid intellectual disability) who remain without spoken language and minimal vocal and speech repertoires despite years of intervention. It is not known what mechanisms are impaired in these children.

17.7.3 Social–Cognitive Foundations of Language

Relatively little research has focused on the early social development of children with SLI. Given that these children do not have primary social impairments, communicative and pragmatic aspects of language are less likely to be affected. There is some evidence that children with SLI may use more nonverbal communicative signals, particularly gestures, to compensate for their limited expressive language skills (Iverson and Braddock, 2011).

In contrast, given their primary impairments in social reciprocity, children with ASD have been extensively studied for the social–cognitive foundations of language. Infants at risk for ASD who later receive a diagnosis show significant declines in social engagement (eye contact, social smiling) beginning around 12 months (Ozonoff et al., 2010); they also produce significantly fewer gestures and do not respond to their name (Landa et al., 2007; Nadig et al., 2007). Early indicators of social cognition, including joint attention, imitation, and gestural communication, are all strongly correlated with language development in toddlers with ASD (Luyster et al., 2008), highlighting the ties between these domains.

Unlike typically developing children, most preschoolers with ASD show no preference for IDT, even their own mothers' IDT, over nonspeech analogs or other environmental sounds (Klin, 1991; Kuhl et al., 2005).

A stronger preference for nonspeech sounds was associated with a lack of ERP response to differentiating speech sounds in a mismatch negativity paradigm and with more severe impairments in language development (Kuhl et al., 2005).

When they do begin speaking, children with ASD use a restricted range of functions, failing to use language in prosocial ways, such as directing attention to an object of interest, or exchanging thoughts and experiences (Wetherby et al., 2007). They have poor conversational skills, which is associated with deficits in theory of mind (Hale and Tager-Flusberg, 2005), and other discourse genres, including narration and storytelling, are also impaired when compared to children at the same language level (e.g., Diehl et al., 2006; Losh and Capps, 2003). Even children with ASD who have intact structural language skills continue to struggle with interpreting nonliteral uses of language, such as lies or metaphors, and show subtle deficits in integrating verbal and nonverbal (prosody, gesture, and eye gaze) modalities when communicating in face-to-face interactions (Tager-Flusberg et al., 2011).

17.7.4 Lexical Development

Children with SLI and ASD do acquire words, though usually at a slower rate. Among children with SLI, vocabulary size has been linked to the ability to accurately repeat multisyllabic nonsense words (e.g., Ellis Weismer et al., 2000). Nonword repetition provides a sensitive measure of phonological working memory, which is an important component of learning to map novel sound sequences to meaning. Children with ASD and comorbid language impairment also show impairments in nonword repetition (Kjelgaard and Tager-Flusberg, 2001).

Researchers have investigated whether children with language disorders are able to use the full range of processes and constraints to guide their acquisition of new words. Young children with SLI are less able to acquire new words using fast mapping (Rice et al., 1994) and are especially vulnerable when using syntactic cues (Rice et al., 2000). Studies of children with ASD have shown that they are able to use the mutual exclusivity constraint (e.g., Preissler and Carey, 2005), but they are less sensitive to social cues for word learning such as direction of eye gaze that signals a speaker's intention (Parish-Morris et al., 2007).

Toddlers who were later significantly delayed in language milestones showing signs of SLI did not exhibit the expected N400 ERP response for lexical-semantic processing in a picture–word matching paradigm at either 14 or 19 months (Friedrich and Friederici, 2005). Even older children with SLI only show a P600 response (but no N400) in response to semantic violations in sentence contexts. These differences in neural response suggest weaker lexical-semantic representations (Friederici, 2006).

17.7.5 Syntactic Development

There is general agreement that language disorders involve more serious impairments in syntax and morphology, compared to phonological or lexical development (e.g., Tager-Flusberg, 2007; Tomblin and Zhang, 1999). Children with SLI begin combining words to form simple utterances at least 2 years later than typical children, though they show the same general developmental trajectory in the growth of sentences over time (Rice, 2007). Children with ASD are also delayed, though there is far greater heterogeneity in both the onset of word combinations and rate of development in this population (Tager-Flusberg et al., 1990).

Rice (2007) has argued that SLI is characterized by primary deficits in the acquisition of particular aspects of grammar: those related to the morphology and syntax of tense, agreement, and case marking. For example, children with SLI go through a more protracted developmental period of omitting past or present tense morphemes (Rice et al., 1998). Children with ASD who have impairments in structural language also omit tense morphemes (Roberts et al., 2004), underscoring the parallels between these disorders. While Rice and her colleagues argue that these tense-marking deficits can be explained on the basis of disruptions in the acquisition of linguistic knowledge, others argue that they reflect domain-general processing limitations (e.g., Leonard et al., 1997).

A number of studies have investigated neural responses to different aspects of syntax in children with SLI. An early study found that lexicalized grammatical morphemes (function words), but not content words, elicited a more bilateral or even right-lateralized negativity, in contrast to the LAN evoked in typical children (Neville et al., 1993). Van der Leley (2005) investigated ELAN responses in SLI children with severe grammatical deficits. In this subgroup of children with SLI, syntactic violations did not elicit any ELAN response. Similar studies have not been conducted on children with ASD. While these preliminary findings are interesting, there are still only limited investigations of the mechanisms that underlie impairments in syntactic development in children with language disorders.

17.7.6 Neural Bases of Language Disorders

As noted earlier, in most individuals, regions that comprise the language network are asymmetric and larger in the left hemisphere. Functionally, the left hemisphere assumes a primary role in processing phonological, semantic, and grammatical aspects of language. Developmental studies have found that while the left hemisphere assumes language processing functions by the end of the first year, there is continuing developmental growth in the structure and asymmetry of language-related cortices into late adolescence (e.g., Sowell et al., 2004).

In children with SLI, both structural and functional differences have been found in studies using MRI. In general, these children have reduced volumes in the primary cortical language areas and reduced asymmetry in the frontal areas, and they are functionally less left lateralized (Tager-Flusberg et al., 2008). In ASD, similar atypical structural patterns have been found, although findings across different studies vary, depending on the methods used, and the age and characteristics of the participants. The majority of studies, which almost always include only boys with ASD, report reduced volumes of the pars triangularis in Broca's area and posterior language regions (e.g., McAlonan et al., 2005; but see also, Knaus et al., 2009). Interestingly, a recent study suggested that reduced volumes in these regions may be associated with later onset of language (McAlonan et al., 2008). Reduction in left hemisphere asymmetry of frontal language regions has also been found in several studies and is associated with greater impairment in language abilities (de Fossé et al., 2004). Almost all functional studies have been conducted with older children or adults. The most consistent findings are reduced left hemisphere asymmetry of language processing, especially in frontal areas, and reduced correlations in activation patterns across language regions (see Tager-Flusberg et al., 2008). One study found atypical activation to speech in toddlers with ASD (Redcay and Courchesne, 2008). In contrast to the control children, who activated the left hemisphere language network, toddlers with ASD primarily activated corresponding regions in the right hemisphere, suggesting that these atypical patterns begin early in development. It is not known, however, the extent to which these differences in functional organization of language in ASD are related to fundamental differences in neuroanatomy that are present at birth, to developing language skills, or to compensatory mechanisms. Nevertheless, the consistent pattern across a range of neuroimaging studies is that children with SLI or ASD are less likely than nondisordered children to depend on language areas in the left hemisphere for processing language.

17.8 CONCLUSIONS

The development of language is a remarkable accomplishment that depends on critical interactions between infants and their environment. Already at birth, attention and motivation bias newborns toward social stimuli, especially human language, and their brains seem tuned to selectively respond to linguistic stimuli. Over the first year of life, experience with language embedded in supportive social contexts leads to greater neural specialization and behavioral changes, and by their first birthday, most infants are already well on their way to acquiring the phonology, lexicon, and grammar of their target language. The course of development involves complex interactions between language-specific and other general cognitive, social, and physiological mechanisms, and the normal pathway depends on exquisite timing and synchrony across multiple domains. Future research will unravel further the neurocognitive systems that contribute to successful language acquisition, and as progress is made in discovering the genetic bases of language-related neurodevelopmental disorders, we will begin to understand how genes help to build a brain that is capable of learning how to speak and understand an abstract symbolic system in just a few short years (see Rubenstein and Rakic, 2013).

Acknowledgments

Support for preparation of this chapter was provided by grants from the National Institutes of Health (RO1 DC 10290) to H Tager-Flusberg and an Autism Speaks Weatherstone Predoctoral Fellowship to AM Seery.

References

Abrahams, B., Geschwind, D., 2008. Advances in autism genetics: On the threshold of a new neurobiology. Nature Reviews Genetics 9, 341–355.

Archangeli, D., Langendoen, D. (Eds.), 1997. Optimality Theory: An Overview. Blackwell, Malden, MA.

Baldwin, D., Meyer, M., 2007. How inherently social is language? In: Hoff, E., Shatz, M. (Eds.), Blackwell Handbook of Language Development. Blackwell, Oxford, pp. 87–106.

Barlow, J., Geirut, J., 1999. Optimality theory in phonological acquisition. Journal of Speech, Language, and Hearing Research 42, 1482–1498.

Bartsch, K., Wellman, H.M., 1995. Children Talk About the Mind. Oxford University Press, Oxford.

Bates, E., Marchman, V., Thal, D., Fenson, L., Dale, P., Reznick, S., 1994. Developmental and stylistic variation in the composition of early vocabulary. Journal of Child Language 21, 85–123.

Benasich, A., Choudhury, N., Friedman, J., Realpe Bonilla, T., Chojnowlska, C., Gou, Z., 2006. Infants as a prelinguistic model for language learning impairments: Predicting from event-related potentials to behavior. Neuropsychologia 44, 396–411.

Bernal, S., Dehaene-Lambertz, G., Millotte, S., Christophe, A., 2010. Two-year-olds compute syntactic structure on-line. Developmental Science 13, 69–76.

Bloom, P., 2000. How Children Learn the Meanings of Words. MIT Press, Cambridge, MA.

Bosch, L., Sebastián-Gallés, N., 2001. Evidence of early language discrimination abilities in infants from bilingual environments. Infancy 2, 29–49.

Bowerman, M., 1978. Systematizing semantic knowledge: Changes over time in the child's organization of word meaning. Child Development 49, 977–987.

Brauer, J., Friederici, A.D., 2007. Functional neural networks of semantic and syntactic processes in the developing brain. Journal of Cognitive Neuroscience 19, 1609–1623.

Brown, R., 1973. A First Language. Harvard University Press, Cambridge, MA.

Bryant, J.B., 2009. Language in social contexts: Communicative competence in the preschool years. In: Gleason, J.B., Ratner, N.B. (Eds.), The Development of Language. Pearson, New York, pp. 192–226.

Burnham, D., Kitamura, C., Vollmer-Conna, U., 2002. What's new, pussycat? On talking to babies and animals. Science 296, 1435.

Butterfield, E., Siperstein, G., 1970. Influence of contingent auditory stimulation upon non-nutritional suckle. In: Bosma, J.F. (Ed.), Third Symposium on Oral Sensation and Perception: The Mouth of the Infant. Charles C. Thomas, Springfield, IL, pp. 313–334.

Byers-Heinlein, K., Burns, T., Werker, J., 2010. The roots of bilingualism in newborns. Psychological Science 21, 343–348.

Carpenter, M., Nagell, K., Tomasello, M., 1998. Social cognition, joint attention, and communicative competence from 9 to 15 months of age. Monographs of the Society for Research in Child Development 63 (4, Serial No. 255), pp. 1–176.

Carter, A., 1979. Prespeech meaning relations: An outline of one infant's sensorimotor morpheme development. In: Fletcher, P., Garman, M. (Eds.), Language Acquisition. Cambridge University Press, Cambridge, pp. 71–92.

Caselli, M., Casadio, P., Bates, E., 1999. A comparison of the transition from first words to grammar in English and Italian. Journal of Child Language 26, 69–111.

Cheour, M., Ceponiene, R., Lehtokoski, A., et al., 1998. Development of language-specific phoneme representations in the infant brain. Nature Neuroscience 1, 351–353.

Chomsky, N., 1995. The Minimalist Program. MIT Press, Cambridge, MA.

Curtiss, S., 1977. Genie: A Psycholinguistic Study of a Modern Day 'Wild' Child. Academic, New York.

de Fossé, L., Hodge, S.M., Makris, N., et al., 2004. Language-association cortex asymmetry in autism and specific language impairment. Annals of Neurology 56, 757–766.

de Villiers, J., Roeper, T., Vainikka, A., 1990. The acquisition of long distance rules. In: Frazier, L., de Villiers, J. (Eds.), Language Processing and Acquisition. Kluwer, Dordrecht.

DeCasper, A.J., Fifer, W.P., 1980. Of human bonding: Newborns prefer their mothers' voices. Science 208, 1174–1176.

DeCasper, A., Spence, M., 1986. Prenatal maternal speech influences newborns' perception of speech sounds. Infant Behavior and Development 9, 133–150.

Dehaene-Lambertz, G., Pena, M., 2001. Electrophysiological evidence for automatic phonetic processing in neonates. Neuroreport 12, 3155–3158.

Diehl, J.J., Bennetto, L., Young, E.C., 2006. Story recall and narrative coherence of high-functioning children with autism spectrum disorders. Journal of Abnormal Child Psychology 34, 87–102.

Diesendruck, G., 2007. Mechanisms of word learning. In: Hoff, E., Shatz, M. (Eds.), Blackwell Handbook of Language Development. Blackwell, Oxford, pp. 257–276.

Dubois, J., Hertz-Pannier, L., Cachia, A., Mangin, J.F., Le Bihan, D., Dehaene-Lambertz, G., 2009. Structural asymmetries in the infant language and sensori-motor networks. Cerebral Cortex 19, 414–423.

Eimas, P.D., Siqueland, E.R., Jusczyk, P., Vigorito, J., 1971. Speech perception in infants. Science 171, 303–306.

Ellis Weismer, S., Tomblin, J.B., Zhang, X., Buckwalter, P., Chynoweth, J.G., Jones, M., 2000. Nonword repetition performance in school-age children with and without language impairment. Journal of Speech, Language, and Hearing Research 43, 865–878.

Fernald, A., 1989. Intonation and communicative intent in mothers' speech to infants: Is the melody the message? Child Development 60, 1497–1510.

Floccia, C., Nazzi, T., Bertoncini, J., 2000. Unfamiliar voice discrimination for short stimuli in newborns. Developmental Science 3, 333–343.

Frank, M.C., Goodman, N.D., Tenenbaum, J., 2009. Using speakers' referential intentions to model early cross-situational word learning. Psychological Science 20, 579–585.

Friederici, A., 2006. The neural basis of language development and its impairment. Neuron 52, 941–952.

Friederici, A., 2009. Pathways to language: Fiber tracts in the human brain. Trends in Cognitive Science 13, 175–181.

Friederici, A.D., Wang, Y.H., Herrmann, C.S., Maess, B., Oertel, U., 2000. Localization of early syntactic processes in frontal and temporal cortical areas: A magnetoencephalographic study. Human Brain Mapping 11, 1–11.

Friedrich, M., Friederici, A.D., 2004. N400-like semantic incongruity effect in 19-month-olds: Processing known words in picture contexts. Journal of Cognitive Neuroscience 16, 1465–1477.

Friedrich, M., Friederici, A., 2005. Lexical priming and semantic integration reflected in the ERP of 14 month-olds. Neuroreport 16, 653–656.

Friedrich, M., Friederici, A., 2010. Maturing brain mechanisms and developing behavioral language skills. Brain and Language 114, 66–71.

Gernsbacher, M., Sauer, E., Geye, H., Schweigert, E., Goldsmith, H., 2008. Infant and toddler oral- and manual-motor skills predict later speech fluency in autism. Journal of Child Psychology and Psychiatry 49, 43–50.

Gervain, J., Macagno, F., Cogoi, S., Pena, M., Mehler, J., 2008. The neonate brain detects speech structure. Proceedings of the National Academy of Sciences of the United States of America 105, 14222–14227.

Granier-Deferre, C., Ribeiro, A., Jacquet, A.Y., Bassereau, S., 2010. Near-term fetuses process temporal features of speech. Developmental Science 13, 1–17.

Grossmann, T., Farroni, T., 2009. Decoding social signals in the infant brain: A look at eye gaze perception. In: de Haan, M., Gunnar, M. (Eds.), Handbook of Developmental Social Neuroscience. Guilford, New York, pp. 87–106.

Grunwell, P., 1981. The development of phonology: A descriptive profile. First Language 3, 161–191.

Guenther, F., 1995. Speech sound acquisition, coarticulation, and rate effects in a neural network model of speech production. Psychological Review 102, 594–621.

Hahne, A., Eckstein, K., Friederici, A.D., 2004. Brain signatures of syntactic and semantic processes during children's language development. Journal of Cognitive Neuroscience 16, 1302–1318.

Hale, C., Tager-Flusberg, H., 2005. Social communication in children with autism: The relationship between theory of mind in discourse development. Autism 9, 157–178.

Hall, G., Waxman, S. (Eds.), 2004. Weaving a Lexicon. MIT Press, Cambridge, MA.

Hickock, G., 2009. The functional neuroanatomy of language. Physics of Life Reviews 6, 121–143.

Hoff, E., 2006. How social contexts support and shape language development. Developmental Review 26, 55–88.

Hollich, G., Hirsh-Pasek, K., Golinkoff, R., 2000. Breaking the language barrier: An emergentist coalition model for the origins of word learning. Monographs of the Society for Research in Child Development 65 (3, Serial No. 262), pp. 1–123.

Houston-Price, C., Plunkett, K., Harris, P., 2005. 'Word learning wizardry' at I;6. Journal of Child Language 32, 175–189.

Huttenlocher, J., Haight, W., Bryk, A., Seltzer, M., Lyons, T., 1991. Early vocabulary growth: Relation to language input and gender. Developmental Psychology 27, 236–248.

Imada, T., Zhang, Y., Cheour, M., Taulu, S., Ahonen, A., Kuhl, P., 2006. Infant speech perception activates Broca's area: A developmental magnetoencephalography study. Neuroreport 17, 957–962.

Ingram, D., 1976. Phonological Disability in Children. Edward Arnold, London.

Iverson, J., 2010. Developing language in a developing body: The relationship between motor development and language development. Journal of Child Language 37, 229–261.

Iverson, J., Braddock, B., 2011. Gesture and motor skill in relation to language in children with language impairment. Journal of Speech, Language, and Hearing Research 54, 72–86.

Iverson, J., Thelen, E., 1999. Hand, mouth, and brain: The dynamic emergence of speech and gesture. Journal of Consciousness Studies 6, 19–40.

Iverson, J., Wozniak, R., 2007. Variation in vocal-motor development in infant siblings of children with autism. Journal of Autism and Developmental Disorders 37, 158–170.

Iverson, J., Hall, A., Nickel, L., Wozniak, R., 2007. The relationship between reduplicated babble onset and laterality biases in infant rhythmic arm movements. Brain and Language 101, 198–207.

Jusczyk, P.W., Pisoni, D.B., Mullennix, J., 1992. Some consequences of stimulus variability on speech processing by 2-month-old infants. Cognition 43, 253–291.

Jusczyk, P., Houston, D., Newsome, M., 1999. The beginnings of word segmentation in English-learning infants. Cognitive Psychology 39, 159–207.

Kaan, E., Swaab, T., 2003. Repair, revision, and complexity in syntactic analysis: An electrophysiological differentiation. Journal of Cognitive Neuroscience 15, 98–110.

Kent, R., 1999. Motor control: Neurophysiology and functional development. In: Caruso, A., Strand, E. (Eds.), Clinical Management of Speech Disorders in Children. Thieme, New York.

Kisilevsky, B., Hains, S., 2011. Onset and maturation of fetal heart rate response to the mother's voice over late gestation. Developmental Science 14, 214–223.

Kisilevsky, B., Hains, S., Lee, K., et al., 2003. Effects of experience on fetal voice recognition. Psychological Science 14, 220.

Kjelgaard, M.M., Tager-Flusberg, H., 2001. An investigation of language impairment in autism: Implications for genetic subgroups. Language and Cognitive Processes 16, 287–308.

Klin, A., 1991. Young autistic children's listening preferences in regard to speech: A possible characterization of the symptom of social withdrawal. Journal of Autism and Developmental Disorders 21, 29–42.

Knaus, T.A., Silver, A.M., Dominick, K.C., et al., 2009. Age-related changes in the anatomy of language regions in autism spectrum disorder. Brain Imaging and Behavior 3, 51–63.

Kooijman, V., Hagoort, P., Cutler, A., 2005. Electrophysiological evidence for prelinguistic infants' word recognition in continuous speech. Cognitive Brain Research 24, 109–116.

Kuhl, P., 2004. Early language acquisition: Cracking the speech code. Nature Reviews Neuroscience 5, 831–843.

Kuhl, P., 2010. Brain mechanisms in early language acquisition. Neuron 67, 713–727.

Kuhl, P.K., Miller, J.D., 1982. Discrimination of auditory target dimensions in the presence or absence of variation in a second dimension by infants. Perception and Psychophysics 31, 279–292.

Kuhl, P.K., Coffey-Corina, S., Padden, D., Dawson, G., 2005. Links between social and linguistic processing of speech in preschool children with autism: Behavioral and electrophysiological measures. Developmental Science 8, F1–F12.

Kutas, M., Federmeier, K., 2011. Thirty years and counting: Finding meaning in the N400 component of the event-relation brain potential (ERP). Annual Review of Psychology 62, 621–647.

Landa, R., Holman, K., Garrett-Mayer, E., 2007. Social and communication development in toddlers with early and later diagnosis of autism spectrum disorders. Archives of General Psychiatry 64, 853–864.

Lenneberg, E., 1967. The Biological Foundations of Language. Wiley, New York.

Leonard, L., Eyer, J., Bedore, L., Grela, B., 1997. Three accounts of the grammatical morpheme difficulties in English-speaking children with specific language impairment. Journal of Speech, Language, and Hearing Research 40, 741–753.

Levy, Y., 1988. On the early learning of formal grammatical systems: Evidence from studies of the acquisition of gender and countability. Journal of Child Language 15, 179–187.

Lidz, J., Waxman, S., Freedman, J., 2003. What infants know about syntax but couldn't have learned. Cognition 89, B65–B73.

Lieberman, P., Crelin, E., Klatt, D., 1972. Phonetic ability and related anatomy of the newborn, adult human, neanderthal man and the chimpanzee. American Anthropologist 74, 287–307.

Locke, J., 1983. Phonological Acquisition and Change. Academic, New York.

Losh, M., Capps, L., 2003. Narrative ability in high-functioning children with autism or Asperger's syndrome. Journal of Autism and Developmental Disorders 33, 239–251.

Luyster, R.J., Kadlec, M.B., Carter, A., Tager-Flusberg, H., 2008. Language assessment and development in toddlers with autism spectrum disorders. Journal of Autism and Developmental Disorders 38, 1426–1438.

Markman, E., 1989. Categorization and Naming in Young Children. MIT Press, Cambridge, MA.

McAlonan, G.M., Cheung, V., Cheung, C., et al., 2005. Mapping the brain in autism. A voxel-based MRI study of volumetric differences and intercorrelations in autism. Brain 128, 268–276.

McAlonan, G.M., Suckling, J., Wong, N., et al., 2008. Distinct patterns of grey matter abnormality in high-functioning autism and Asperger's syndrome. Journal of Child Psychology and Psychiatry 49, 1287–1295.

McCleery, J., Tully, L., Slevc, L., Schreibman, L., 2006. Consonant production patterns of young severely language-delayed children with autism. Journal of Communication Disorders 39, 217–231.

McClelland, J., Patterson, K., 2002. Rules or connections in past-tense inflections: What does the evidence rule out? Trends in Cognitive Sciences 6, 465–472.

Mehler, J., Jusczyk, P., Lambertz, G., Halsted, N., Bertoncini, J., Amiel-Tison, C., 1988. A precursor of language acquisition in young infants. Cognition 29, 143–178.

Menn, L., Stoel-Gammon, C., 2009. Phonological development: Learning sounds and sound patterns. In: Gleason, J.B., Ratner, N.B. (Eds.), The Development of Language. Pearson, New York, pp. 58–103.

Mills, D., Coffey-Corina, S., Neville, H., 1993. Language acquisition and cerebral specialization in 20-month-old infants. Journal of Cognitive Neuroscience 5, 317–334.

Mills, D., Coffey-Corina, S., Neville, H., 1997. Language comprehension and cerebral specialization from 13 to 20 months. Developmental Neuropsychology 13, 397–445.

Mills, D., Plunkett, K., Prat, C., Schafer, G., 2005. Watching the infant brain learn words: Effects of vocabulary size and experience. Cognitive Development 20, 19–31.

Minagawa-Kawai, Y., Mori, K., Naoi, N., Kojima, S., 2007. Neural attunement processes in infants during the acquisition of a language-specific phonemic contrast. Journal of Neuroscience 27, 315–321.

Mintz, T., Newport, E., Bever, T., 2002. The distributional structure of grammatical categories in speech to young children. Cognitive Science 26, 393–424.

Morgan, J., 1986. From simple input to complex grammar. MIT Press, Cambridge, MA.

Nadig, A., Ozonoff, S., Young, G., Rozga, A., Sigman, M., Rogers, S., 2007. A prospective study of response to name in infants at risk for autism. Archives of Pediatrics and Adolescent Medicine 161, 378–383.

Nagy, W., Scott, J., 2000. Vocabulary processes. In: In: Kamil, M., Mosenthal, P., Pearson, P., Barr, R. (Eds.), Handbook of Reading Research, vol. 3. Erlbaum, Mahwah, NJ, pp. 269–284.

Naigles, L., Swensen, L., 2007. Syntactic supports for word learning. In: Hoff, E., Shatz, M. (Eds.), Blackwell Handbook of Language Development. Blackwell, Oxford, pp. 212–231.

Nazzi, T., Bertoncini, J., Mehler, J., 1998. Language discrimination by newborns: Toward an understanding of the role of rhythm. Language 24, 756–766.

Nelson, K., 1981. Acquisition of words by first-language learners. Annals of the New York Academy of Sciences 379, 148–159.

Neville, H., Coffey, S., Holcomb, P., Tallal, P., 1993. The neurobiology of sensory and language processing in language-impaired children. Journal of Cognitive Neuroscience 5, 235–253.

Norbury, C., Tomblin, J.B., Bishop, D.V.M. (Eds.), 2008. Understanding Developmental Language Disorders: From Theory to Practice. Psychology Press, Hove.

O'Grady, W., Peters, A.M., Masterson, D., 1989. The transition from optional to required subjects. Journal of Child Language 16, 513–529.

Oller, K., 2000. The Emergence of Speech Capacity. Erlbaum, Mahwah, NJ.

Ozonoff, S., Iosif, A.-M., Baguio, F., et al., 2010. A prospective study of the emergence of early behavioral signs of autism. Journal of the American Academy of Child and Adolescent Psychiatry 49, 256–266.

Parish-Morris, J., Hennon, E., Hirsh-Pasek, K., Golinkoff, R., Tager-Flusberg, H., 2007. Children with autism illuminate the role of social intention in word learning. Child Development 78, 1265–1287.

Paul, R., Fuerst, Y., Ramsay, G., Chawarska, K., Klin, A., 2010. Out of the mouths of babes: Vocal production in infant siblings of children with ASD. Journal of Child Psychology and Psychiatry 52, 588–598.

Pinker, S., 1984. Language Learnability and Language Development. Harvard University Press, Cambridge, MA.

Pinker, S., 1999. Words and Rules. Basic Books, New York.

Preissler, M., Carey, S., 2005. The role of inferences about referential intent in word learning: Evidence from autism. Cognition 97, B13–B23.

Ratner, N., Bruner, J., 1978. Games, social exchange and the acquisition of language. Journal of Child Language 5, 391–401.

Redcay, E., Courchesne, E., 2008. Deviant functional magnetic resonance imaging patterns of brain activity to speech in 2–3-year-old children with autism spectrum disorder. Biological Psychiatry 64, 589–598.

Rice, M., 2007. Children with specific language impairment: Bridging the genetic and developmental perspectives. In: Hoff, E., Shatz, M. (Eds.), Blackwell Handbook of Language Development. Oxford, Blackwell, pp. 411–431.

Rice, M., Oetting, J., Maruis, J., Bode, J., Pae, S., 1994. Frequency of input effects on word comprehension of children with specific language impairment. Journal of Speech, Language, and Hearing Research 37, 106–122.

Rice, M., Wexler, K., Hershberger, S., 1998. Tense over time: The longitudinal course of tense acquisition in children with specific language impairment. Journal of Speech, Language, and Hearing Research 41, 1412–1431.

Rice, M., Cleave, P., Oetting, J., 2000. The use of syntactic cues in lexical acquisition by children with SLI. Journal of Speech, Language, and Hearing Research 43, 582–594.

Rivera-Gaxiola, M., Silva-Pereyra, J., Kuhl, P.K., 2005. Brain potentials to native and non-native speech contrasts in 7- and 11-month-old American infants. Developmental Science 8, 162–172.

Roberts, J., Rice, M., Tager-Flusberg, H., 2004. Tense marking in children with autism. Applied Psycholinguistics 25, 429–448.

Rowe, M.L., Goldin-Meadow, S., 2009. Early gesture selectively predicts later language learning. Developmental Science 12, 182–187.

Rubenstein, J.L.R., Rakic, P., 2013. Patterning and Cell Types Specification in the Developing CNS and PNS.

Saffran, J., Aslin, R., Newport, E., 1996. Statistical learning by 8 month-old infants. Science 274, 1926–1928.

Samuelson, S., Smith, L.B., 2000. Grounding development in cognitive processes. Child Development 71, 98–106.

Shatz, M., Gelman, R., 1973. The development of communication skills: Modifications in the speech of young children as a function of listener. Monographs of the Society for Research in Child Development 38 (5, Serial No. 152), pp. 1–37.

Sices, L., Taylor, H.G., Freebairn, L., Hansen, A., Lewis, B., 2007. The relationship between speech sound disorders and early literacy skills in preschool-aged children: Impact of co-morbid language impairment. Journal of Developmental and Behavioral Pediatrics 28, 1–10.

Siegal, M., Surian, L., 2007. Conversational understanding in young children. In: Hoff, E., Shatz, M. (Eds.), Blackwell Handbook of Language Development. Blackwell, Oxford, pp. 304–323.

Silva Pereyra, J.F., Klarman, L., Lin, L.J., Kuhl, P.K., 2005. Sentence processing in 30-month-old children: An event-related potential study. Neuroreport 16, 645–648.

Snow, C., 1977. The development of conversation between mothers and babies. Journal of Child Language 4, 1–22.

Sowell, E.R., Thompson, P.M., Leonard, C.M., Welcome, S.E., Kan, E., Toga, A.W., 2004. Longitudinal mapping of cortical thickness and brain growth in normal children. Journal of Neuroscience 24, 8223–8231.

Stoel-Gammon, C., 1998. The role of babbling and phonology in early linguistic development. In: Wetherby, A., Warren, S., Reichle, J. (Eds.), Transitions in Prelinguistic Communication: Preintentional to Intentional and Presymbolic to Symbolic. Brookes, Baltimore, MD, pp. 87–110.

Stoel-Gammon, C., Sosa, A., 2007. Phonological development. In: Hoff, E., Shatz, M. (Eds.), Blackwell Handbook of Language Development. Blackwell, Oxford, pp. 238–256.

Tager-Flusberg, H., 2004. Do autism and specific language impairment represent overlapping language disorders? In: Rice, M.L., Warren, S. (Eds.), Developmental Language Disorders: From Phenotypes to Etiologies. Lawrence Erlbaum Associates, Mahwah, NJ, pp. 31–52.

Tager-Flusberg, H., 2007. Atypical language development: Autism and other neurodevelopmental disorders. In: Hoff, E., Shatz, M. (Eds.), Blackwell Handbook of Language Development. Oxford, Blackwell, pp. 432–453.

Tager-Flusberg, H., Zukowski, A., 2009. Putting words together: Morphology and syntax in the preschool years. In: Gleason, J.B., Ratner, N.B. (Eds.), The Development of Language. Pearson, New York, pp. 139–191.

Tager-Flusberg, H., Calkins, S., Noin, I., Baumberger, T., Anderson, M., Chadwick-Denis, A., 1990. A longitudinal study of language acquisition in autistic and Down syndrome children. Journal of Autism and Developmental Disorders 20, 1–22.

Tager-Flusberg, H., Paul, R., Lord, C.E., 2005. Language and communication in autism. In: Volkmar, F., Paul, R., Klin, A., Cohen, D.J. (Eds.), 3rd edn. Handbook of Autism and Pervasive Developmental Disorder, vol. 1. Wiley, New York, pp. 335–364.

Tager-Flusberg, H., Lindgren, K., Mody, M., 2008. Structural and functional imaging research on language disorders: Specific language impairment and autism spectrum disorders. In: Wolf, L., Schreiber, H., Wasserstein, J. (Eds.), Adult Learning Disorders: Contemporary Issues. Psychology Press, New York, pp. 127–157.

Tager-Flusberg, H., Edelson, L., Luyster, R., 2011. Language and communication in autism spectrum disorders. In: Amaral, D., Dawson, G., Geschwind, D. (Eds.), Autism Spectrum Disorders. Oxford University Press, Oxford.

Thierry, G., Vihman, M., Roberts, M., 2003. Familiar words capture the attention of 11-month-olds in less than 250 ms. Neuroreport 14, 2307–2310.

Thiessen, E., Hill, E., Saffran, J., 2005. Infant-direct speech facilitates word segmentation. Infancy 7, 53–71.

Tomasello, M., 1999. The Cultural Origins of Human Cognition. Harvard University Press, Cambridge, MA.

Tomasello, M., 2002. Do young children have adult syntactic competence? Cognition 74, 209–253.

Tomasello, M., 2003. Constructing a Language. Harvard University Press, Cambridge, MA.

Tomblin, J.B., Zhang, X., 1999. Language patterns and etiology in children with specific language impairment. In: Tager-Flusberg, H. (Ed.), Neurodevelopmental Disorders. MIT Press, Cambridge, MA, pp. 361–382.

Ullman, M., 2001. A neurocognitive perspective on language: The declarative/procedural model. Nature Reviews Neuroscience 2, 717–726.

Valian, V., 1990. Null subjects: A problem for parameter-setting models of language acquisition. Cognition 35, 105–122.

Van der Leley, H., 2005. Domain-specific cognitive systems: Insights from grammatical-SLI. Trends in Cognitive Science 9, 53–59.

Vouloumanos, A., Werker, J.F., 2007. Listening to language at birth: Evidence for a bias for speech in neonates. Developmental Science 10, 159–164.

Waxman, S., 2003. Links between object categorization and naming: Origins and emergence in human infants. In: Rakison, D., Oakes, L. (Eds.), Early Category and Concept Development: Making Sense of the Blooming Buzzing Confusion. Oxford University Press, New York, pp. 213–241.

Weber, C., Hahne, A., Friedrich, M., Friederici, A., 2004. Discrimination of word stress in early infant perception: Electrophysiological evidence. Cognitive Brain Research 18 (2), 149–161.

Werker, J., McLeod, P., 1989. Infant preference for both male and female infant-directed talk: A developmental study of attentional and affective responsiveness. Canadian Journal of Psychology 43, 230–246.

Werker, J., Tees, R., 1984. Cross-language speech perception: Evidence for perceptual reorganization during the first year of life. Infant Behavior and Development 7, 49–63.

Wetherby, A.M., Watt, N., Morgan, L., Shumway, S., 2007. Social communication profiles of children with autism spectrum disorders late in the second year of life. Journal of Autism and Developmental Disorders 3, 960–975.

The Neural Architecture and Developmental Course of Face Processing

G. Righi[1], C.A. Nelson III[2]

[1]Boston Children's Hospital, Boston, MA, USA; [2]Harvard Medical School, Boston, MA, USA

O U T L I N E

18.1 INTRODUCTION

Faces are ubiquitous in our environment and convey information that is used extensively in all social interactions. For example, human adults can easily classify faces on the bases of their gender, race, and identity and thus recognize quickly whether they have come across a friend or a foe (Bruce and Young, 1986). Moreover, faces also convey information that can assist an observer in gaining insight into the internal state of another person; for example, slight changes in facial features are related to different emotional states, from happiness to sadness, from anger to fear, from surprise to disgust (Ekman, 1993). Even though for most adult

© 2013 Elsevier Inc. All rights reserved.

observers, gathering information from a face is a relatively seamless process, this skill undergoes a lengthy developmental trajectory that has its origins in infancy and continues well into adolescence. Throughout development, both the behaviors that accompany face processing and the neural underpinning of these behaviors undergo substantial modification. The bias brought to this chapter is that we cannot fully understand the mature face-processing system unless we make a parallel effort to understand the development of this system. The goal of this chapter is to provide an extensive overview of the research on face processing from infancy to adulthood. In each of the sections, the results of both behavioral and neurophysiological research that examines different aspects of face processing, including facial identity, facial categorization (e.g., of gender, race), and the discrimination and recognition of facial expressions are reviewed. The various sections that follow are organized chronologically with the exception of the first section. In the first section, face processing in adults is discussed in order to provide the reader with a framework on which to understand the results of the developmental work, which is presented in chronological order from infancy through adolescence. The last section of this chapter discusses impairments in face development that result from either traumatic brain injury or neurological syndromes.

18.2 FACE PROCESSING IN ADULTS

Despite the chronological organization of the majority of this chapter, the discussion of face processing in adults is dealt with first. This will provide the reader with an illustration of the mature face-processing system, including its neural architecture, which in turn will serve as a template against which to compare the development of this system. In this first section, primarily experimental evidence is reviewed, but the methodological tasks most commonly used to measure face processing are also discussed, as they apply to both the study of adults and the study of developmental populations.

18.2.1 How Adults Process Faces

Most people can recognize hundreds of individual faces by the time they reach adulthood, and their ability to memorize facial identities remains more or less intact throughout their lifespan. Moreover, face recognition is surprisingly robust. For example, it has been shown that recognition performance is relatively impervious to physical transformations including blurring (Harmon, 1973; Yip and Sinha, 2002), changes in lighting conditions (Braje, 2003; Braje et al., 1998), and changes in

viewing angle (Hill et al., 1997; O'Toole et al., 1998). The resilience to these transformations is something that any observer has experienced as these variations make the recognition of identity more challenging but nevertheless successful. One of the reasons behind the robustness of face processing is that faces are not identified by focusing on specific features in isolation; rather, human observers rely on both specific facial features and on the relations among them and perform what has been termed 'holistic processing' (Tanaka and Farah, 1993; Young et al., 1987). More recently, several authors have suggested that there are three different types of holistic processing that all come into play during face processing but that can be specifically tested using different experimental manipulations (Gauthier and Tarr, 2002; Maurer et al., 2002). Aside from differences in terminology, both Gauthier and Tarr (2002) and Maurer and colleagues (2002) suggest that when viewing faces, observers process (1) the specific special arrangement of features in a face (i.e., two eyes above nose and mouth), (2) the specific spatial relations among features in a face (i.e., distance between the eyes, length of the forehead, etc.), and (3) a face gestalt in which all facial features are integrated into a representation. Healthy adults are also very proficient at detecting facial expressions, regardless of whether they are familiar or unfamiliar with a specific face. There are several well-established behavioral demonstrations of these specific types of processing strategies. The sensitivity to facelike configurations is such that adults excel at detecting the presence of faces even in very degraded situations such as in extremely blurred images or even when they are presented with highly schematic faces (Diamond and Carey, 1986). Adults are also very good at detecting variations in the spacing of features within a face, such that they can recognize different identities even when they vary only in spacing among facial features (Le Grand et al., 2001). This sensitivity to the 'configural' properties of a face has also been shown by using the 'face inversion' paradigm (Yin, 1969), which has revealed that the ability to recognize faces is greatly hindered by picture plane inversions and even more so when the distinctions between the experimental stimuli are created by spacing manipulations (Goffaux and Rossion, 2007). Lastly, the bias to produce 'holistic' face representations has been shown using the 'face composite' paradigm (Young et al., 1987). In this paradigm, subjects are presented with a face created by combining tops and bottoms of different individuals, and they are asked to pay attention to only one of the two specific parts; what has been shown is that when the two parts are presented in alignment with one another, subjects make more errors because they get distracted by the irrelevant face part. These experimental effects have been extensively replicated in the adult literature and are considered the hallmarks of expert face processing.

18.2.2 Models of Face Processing

Going beyond face recognition, adults are also very quick at determining many other different types of information from faces, including gender, race, direction of gaze, and emotional expressions. In order to understand fully the nuances of face processing, it is worth discussing in brief theoretical models that have been proposed in the literature. The most influential model of face perception remains Bruce and Young's dual-route model (1986) formulated to provide a theoretical framework that can explain how perceivers extrapolate and process different types of information from faces, such as identity, emotional expression, and facial speech. The core assumption of this model is that independent modules and processing streams support these different tasks and they work in parallel with no cross talk of information between them. Face processing begins with 'structural encoding,' which produces a number of descriptions of a face, some of which are dependent on the specific instance of the face that is presented (view-centered), while others can be more general and contain knowledge about the global structure of that face in a more invariant manner (expression-independent), meaning that they are not related to a specific expression or viewing condition. The segregation of the processing streams takes place after the initial structural encoding phase.

Regardless of whether the face is novel or familiar, expression and facial speech-processing modules receive only the information contained in the view-centered descriptions, while expression-independent descriptions are connected to a processing module that works exclusively for identity recognition. A module separate from the recognition module, which receives both the view-centered and the expression-independent information, processes the physical structure of a novel face. This module is also responsible for extracting information about the physical configurations that can be used to judge gender, race, age, etc. To summarize, after a common initial visual analysis, the information is sent to one or more of the four different modules depending on the objective of the task at hand. In turn, the outputs of each of the modules are sent to a common 'cognitive system.' This system is responsible for directing attention and decision processes, but most importantly it contains memory information about all the semantic knowledge that can be associated with a face. Bruce and Young (1986) divide the stored semantic information on the basis of which processing module provides the physical analysis needed to retrieve it. They suggest four different types of codes: expression codes, facial speech codes, visually derived semantic codes, and identity-derived semantic codes. Expression codes contain the labels for facial expressions. Facial speech codes contain representations for mouth movements connected to speech. Visually

derived semantic codes contain the information relative to judgments that can be based on the physical structure alone and are independent of identity, emotional expression, and facial speech, such as gender, race, age, and some social attributions. Identity-derived semantic codes contain all the information one has acquired about a specific person that one knows.

Another class of face-processing models is represented by the prototype-based or face space model (Valentine, 1991, 1999). This type of model has been formulated with the goal of providing a unitary account of various phenomena of face processing including recognition and identification of race and gender. The basic assumption that these models make is that any face can be represented within a multidimensional space. The number of variables needed to discriminate between faces determines the dimensions of the space. The center of the space is assumed to represent the average value of a population on that specific dimension. What makes a face more or less recognizable among other faces is the distance between the target face and neighboring faces in the space. The creation of average faces gives rise to prototypes, which are presumably stored in memory. It has been suggested that observers have prototypes not only for identity but also for gender, race, and possibly even age because despite sharing the basic shape and features, individuals classified in these categories differ on the basis of specific featural and configural information (O'Toole et al., 1997).

18.2.3 Neural Substrates of Face Processing

The neural substrates of adult face processing have been extensively studied over the last decade. The advent of functional magnetic resonance imaging (fMRI) has opened the way to look at the neural networks recruited by face processing. One of the most significant findings of this literature is the identification of a portion of the cortex located within the medial fusiform gyrus that is found to be primarily responsive to faces, compared to most common objects; this functional area has been termed the fusiform face area or FFA (Figure 18.1; Kanwisher et al., 1997).

Subsequent to Kanwisher's important contribution regarding the role of the fusiform gyrus in face processing, other researchers have identified a rich network of areas that are recruited by a variety of face-processing tasks (Haxby et al., 2000; Ishai et al., 2005; Tsao et al., 2008). For example, the findings from a variety of neuroimaging studies have been integrated by Haxby and colleagues (Haxby et al., 2002) into a model of the brain network of face perception. Their proposal is strongly influenced by Bruce and Young's (1986) dual-route model, as they suggest that the recognition of identity

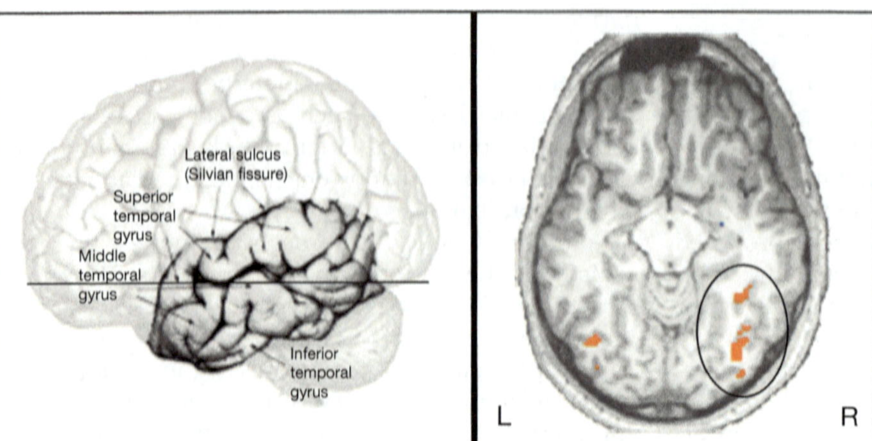

FIGURE 18.1   The fusiform face area in adult subjects.

and the processing of social information from a face are dependent on different variables that give rise to independent cognitive representations, which in turn produce distinct neural representations (Haxby et al., 2002). In line with Bruce and Young's model (1986), they suggest that while face identification is based on those features that remain invariant over time, other tasks, such as eye gaze perception, expression identification, and lip movement recognition, are derived from the analysis of the changeable features of a face. After an initial visual analysis carried out in the inferior occipital gyri, the two classes of features mentioned above are directed to either the superior temporal sulcus (STS) (changeable features) or to the lateral fusiform gyrus (invariant features). These two areas are in agreement with studies that have reported the STS to be activated by different types of biological movement including those of the eyes and mouth, while the fusiform gyrus has been consistently found to be activated when subjects are asked to judge facial identity, but less so when subjects are asked to detect eye gaze direction. From the STS, information is segregated further such that eye gaze processing is directed to the intraparietal sulcus, lip reading information is directed to auditory cortex, and emotion-related information is directed to the amygdala, insula, and other limbic system structures. The identification of the neural substrates for these three different processes is inspired by considering each process a face-specific instance of a more generic cognitive process. That is, these specific tasks recruit areas that perform the more general operations and become part of this network only when face stimuli are the input of the system. The direction of gaze can be used to direct the subject's attention toward what another individual might be looking at and as such can be considered a special case of attentional mechanisms. Emotional expressions provide information about the emotion that an individual might be experiencing and as such are

processed by the same areas that are involved in the experiencing of emotions. Lip movements are the face-specific component of the multimodal speech signal and as such are processed by areas involved in speech perception. The information that reaches the fusiform gyrus is used for judgments of identity and other types of classifications that are independent of dynamic changes in facial structure (possibly gender, race, age, etc.) and as such is suggested to be relayed to areas that contain task-relevant stored knowledge, such as biographical or semantic information. In support of this idea, studies have shown that the recognition of a face is associated with activity in the anterior temporal cortex (Gobbini and Haxby, 2007; Leveroni et al., 2000), suggesting that these areas might contain semantic knowledge that is associated with known faces.

Beyond the individuation of anatomical substrates involved in face processing, many researchers have been interested in identifying functional signatures using methods such as event-related potentials (ERPs) in order to shed light on the time course of different aspects of face processing. A review of the findings using this methodology is also important in the context of development because ERPs are the most widely used technique to study the brain's response to faces in infants, toddlers, and children. Studies using ERP have identified a series of electrical potentials that respond to different aspects of face processing. The most widely studied of these components is the N170, a negative deflection in the ongoing electroencephalography (EEG) that occurs on average around 170 ms after the presentation of a face and is measured over posterior scalp electrodes, which is suggested to reflect the activity of the occipitotemporal face-selective neurons and is specifically responsive to the structural encoding of faces (Bentin et al., 1996; Carmel and Bentin, 2002; Eimer, 2000; Rossion et al., 1999). Moreover, the N170 has also been associated with configural processing as it shows both amplitude and latency modulations in response to inverted faces, when compared to upright faces (Bentin et al., 1999; Jacques and Rossion, 2007). The recognition of facial identity has been suggested to take place later in the processing stream and has been associated with other components such as the N250 (Rossion et al., 1999; Rossion and Gauthier, 2002), the N400, and the P600 (Eimer, 2000).

18.2.4 Conclusions

In summary, with a few exceptions, adults are very good at recognizing and processing the information present in faces. However, it should be evident from the data reviewed in this section that these abilities are dependent on complex processing strategies and rely on the interactions of a rich network of brain regions. Moreover, adult expertise is also dependent on the constant interactions with faces that we all experience in everyday life. The next sections in this chapter will examine how face expertise arises throughout development, with the hope to provide the reader with an understanding on both the behavioral and neural specialization for face processing.

18.3 FACE PROCESSING IN THE FIRST YEAR OF LIFE

18.3.1 How Infants Learn to See Faces

Despite a very immature visual system, infants develop the ability to process facial identities and facial expressions surprisingly quickly over the first year of life (Figure 18.2) (see Chapter 14).

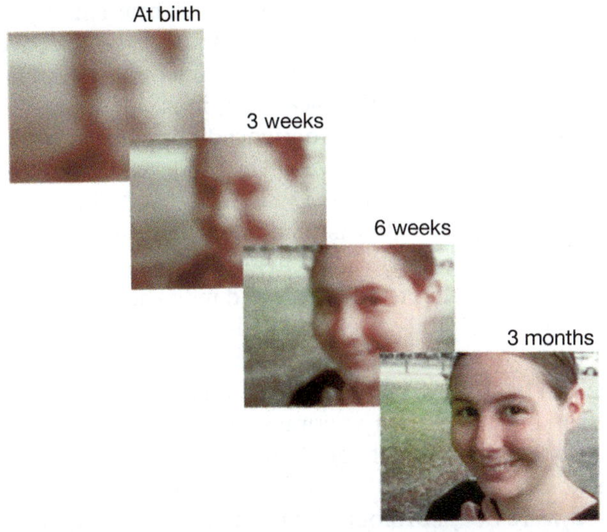

FIGURE 18.2 Development of the infant visual system within the first year of life.

Their experience with faces begins at birth, and one of the most striking findings of the literature looking at face processing in infants is that since birth they show a preference to pay attention to faces and facelike stimuli, compared to other visual objects (Johnson and Morton, 1991; Maurer, 1983; Mondloch, et al., 1999; Valenza et al., 1996). The ontogeny of their face preference is still under debate. On one hand, it has been hypothesized that infants are 'innately' attracted to stimuli that resemble faces (e.g., an oval with three dots in it) and that from birth there is a subcortical brain system, possibly including the superior colliculus and the pulvinar, that causes infants to orient toward facelike patterns (Johnson, 2005). This mechanism is supposed to be replaced at around 2 months of age by a cortical mechanism (Conlearn) that is already somewhat specialized for faces (Johnson, 2005). On the other hand, it has been hypothesized that faces happen to be optimal stimuli for a developing visual system in terms of their physical characteristics (Banks and Salapatek, 1976; Kleiner, 1987). Following this initial orienting toward faces, it is through exposure and active experience that the infant brain starts to develop what become face-sensitive processes and neural structures (Nelson, 2001).

Over the first months of life, infants rapidly learn to perform several types of facial discriminations. Several studies have shown that infants as young as 4 days old show signs of discriminating their mother's face from that of a stranger, although this ability is not yet robust enough to withstand, for example, the presence of a headscarf (Pascalis et al., 1995). However, infants rapidly progress in their ability to recognize their mother's faces such that at 3 months of age, they are able to do so even across variations in viewpoint (Pascalis et al., 1998). It is around 3 months that infants also show the ability to discriminate between male and female faces (Quinn et al., 2002) and also between faces of different races (Bar-Haim et al., 2006; Kelly et al., 2005; Quinn et al., 2008).

Related to the development of the ability to discriminate between faces of different identities, different gender, and different races is a phenomenon called perceptual narrowing (see Nelson, 2001 for discussion). While at 3 months infants seem to be learning to differentiate between different categories of faces, it has been suggested that by 9 months they develop the ability to distinguish between exemplars within a category but only if they have experience with that specific category. Perceptual narrowing refers to the fact that between 3 and 9 months, on average around 6 months, infants are able to discriminate between individual exemplars of many different visual categories. However, by 9 months, this flexibility appears lost in most categories, except for the ones with which the infant has the most experience.

In behavioral studies, it has been shown that 9-month-olds do not successfully differentiate between different individuals of a race other than their own (Kelly et al., 2007). Moreover, by 9 months of age, infants are not discriminating among a pair of monkey faces, something that 6-month-olds are capable of doing, while successfully differentiating between human faces (Pascalis et al., 2002, 2005).

During the first year of life, infants also experience changes in the strategies they use when they process faces. One of the hallmarks of face processing in adults is their reliance on the configural properties of a face, which is at the root of their expertise with faces. Infants do not seem to show the ability to use configural cues at birth, but some investigators have suggested that this ability emerges within the first year of life, even though it may not be reaching adultlike levels until later in development. Recent studies have shown that by 7 months of age, infants are sensitive to prototypical face configurations (Thompson et al., 2001), such that they can discriminate between faces that contain atypical distances between their features and more prototypical exemplars. A recent study has suggested that infants as young as 5 months are able to discriminate between faces that differ only in spacing between the eyes and spacing between the nose and mouth while maintaining identical features (Bhatt et al., 2005).

18.3.2 How Infants Process Facial Expressions

Within the first year of life, infants are also able to discriminate some facial expressions (Barrera and Maurer, 1981; de Haan et al., 1998; Nelson and Dolgin, 1985; Nelson and Salapatek, 1986; Nelson et al., 1979; Young-Browne et al., 1977), and within their first 12 months, they undergo a series of developmental changes. Starting as early as 36 h after birth, infants show evidence of being able to discriminate among happy, sad, and surprised faces (Field et al., 1982). Experimental evidence from infants in their first 6 months of life shows that they continue to be able to discriminate at least some facial expressions from one another, with some variability depending on the intensity of the exemplars used on their specific characteristics. Overall, it seems that happy faces are successfully differentiated from other expressions and neutral faces (Farroni et al., 2007; LaBarbera et al., 1976; Young-Browne et al., 1977), while negative expressions are more difficult to differentiate for young infants (LaBarbera et al., 1976; Nelson and Dolgin, 1985; Young-Browne et al., 1977). It has been hypothesized that during the first few months of life, infants may learn to discriminate the expressions that they are most often exposed to, which are usually positive, and are not yet capable of discriminating among ones that they do not

see often. It is between 7 and 12 months that, perhaps, emotion processing experiences the most dramatic changes and infants start showing the ability to discriminate expression with which they do not have much experience. For example, it is around 7 months that infants first show the ability to consistently discriminate fearful faces (Nelson and Dolgin, 1985) and start showing longer fixations toward this expression (Leppänen et al., 2007; Peltola et al., 2009). Another important shift that takes place starting at around 7 months is the perception of facial expression in a categorical manner, which is a signature of emotion processing present in adults. Thus, it is around this age that infants demonstrate the ability to (a) generalize their discrimination of facial expressions across multiple exemplars (e.g., Nelson et al., 1979) and (b) discriminate qualitatively between two facial expressions even though the stimuli used are created using a quantitative morph continuum. This effect has been using faces that are morphed in a continuous way between happiness to sadness (Leppänen et al., 2009) and fear and happiness (Kotsoni et al., 2001).

18.3.3 Neural Substrates of Face Processing in Infants

The study of the neural substrates of face processing in infants has received much attention in recent years. One of the questions that is still debated is whether faces are processed by specialized neural structures since birth (Johnson, 2005; Johnson and Morton, 1991) or whether the neural structures that are found to be face-selective in the adult brain become specialized for face processing through active experience over the course of development (Gauthier and Nelson, 2001; Nelson, 2001). Because of obvious methodological difficulties, there are very few studies that have used functional brain imaging methods to test whether infants show face-selective brain activations. To our knowledge, the only investigation of this type used PET scanning in six 2-month-old infants, who at birth had experienced hypoxic–ischemic encephalopathy (Tzourio-Mazoyer et al., 2002). In this study, infants were presented with faces of adult females and schematic dot patterns. The infants' brain showed a surprisingly rich network of clusters of activation in response to faces akin to what is found in the adult brain in the inferior occipital cortex, but it also showed activation for faces in parietal and frontal regions (Tzourio-Mazoyer et al., 2002). These results are important as they confirm the presence of a neural system sensitive to faces as early as 2 months of age. However, there are two issues to consider with this study. First of all, the infants tested were not neurologically normal. Second, the comparison stimulus used was far less attractive and complex compared to faces, and as such it is difficult to determine whether any complex object would have elicited similar patterns of activation.

18.3.4 Neural Signatures of Face Processing in Infants

The most widely used method to assess infants' brain responses to faces is ERPs, and the adaptability of this methodology to infant studies has given rise to an extensive literature that has investigated the functional signatures of face processing in infants. ERP components that are sensitive to faces have been shown to emerge as early as 3 months of age (Halit et al., 2004). There are two components that are considered to be the antecedents of the adult N170: the N290 and the P400. The N290 shows an adultlike effect of inversion starting around 6 months of age (Halit et al., 2004), but as pointed out by de Haan and colleagues (de Haan et al., 2003), the selectivity for faces of this component has yet to be tested in infants. The P400 shows an adultlike latency difference between faces and objects and by 12 months also shows an effect of inversion (Halit et al., 2004). Similar to the adult literature, these two components have been linked to the structural analysis of faces, while another set of components has been linked to the recognition of familiar faces: the Nc and the positive slow wave. The Nc appears to show a difference between the mother's face and a stranger's face starting at 6 months of age as long as the two faces are different enough from one another (Figure 18.3; de Haan and Nelson, 1997).

ERP components elicited in response to faces have also been shown to be modulated by the presence of emotional expressions. Among face-sensitive components, the Nc component appears to be sensitive to several facial expressions in infants. In one of the first studies using ERP to look at facial expression processing, Nelson and de Haan (1996) showed that fearful faces elicited an increased amplitude of the Nc component, compared to happy faces. Leppänen and colleagues (Leppänen et al., 2009) also found differential sensitivity to happy and sad faces in the Nc component in 7-month-olds, but no differences were found in the N290 and P400. Grossman and colleagues found amplitude modulations in this component in response to angry and happy faces in 7-month-olds (Grossmann et al., 2007). Interestingly, it is only around 12 months that infants start showing modulations of emotional expressions on more posterior components (Grossmann et al., 2007).

18.3.5 Conclusions

During the first year of life, face processing undergoes rapid development both behaviorally and neurally. In the first 12 months of life, infants start to show many of the face-processing skills that will gain strength throughout development and will reach full maturity in adolescence and adulthood. Moreover, their neural responses to faces also start to show signs of face specialization, although it is apparent that much anatomical and functional maturation will take place with

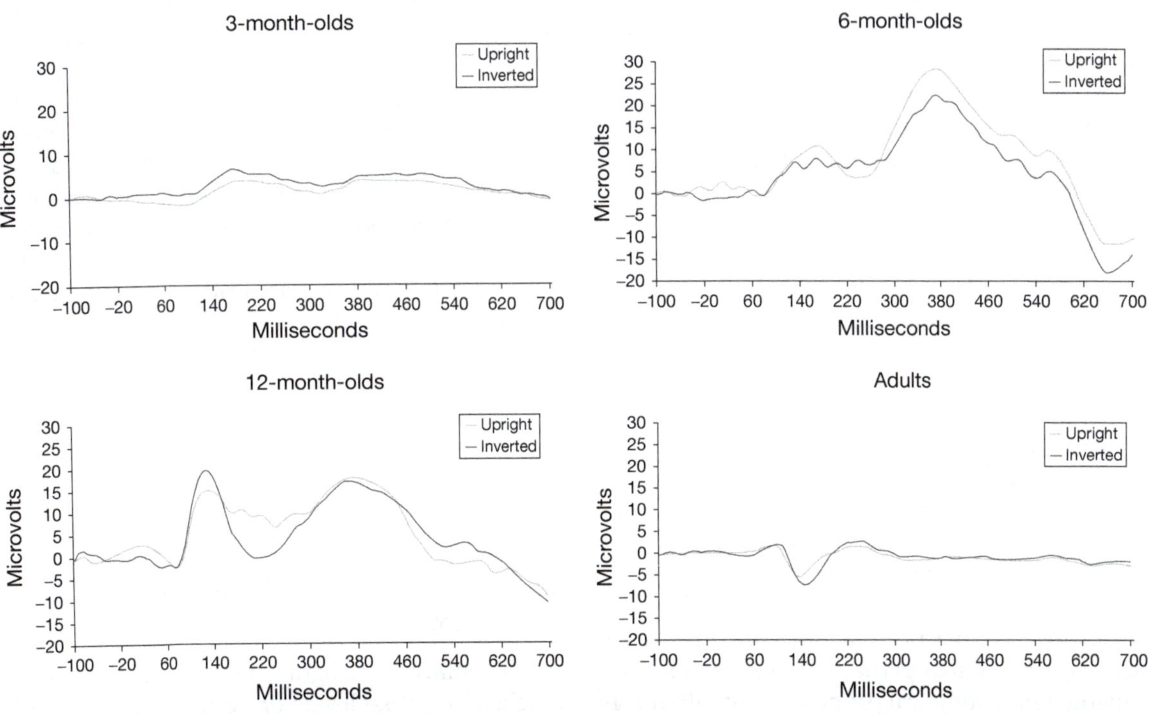

FIGURE 18.3 Functional signatures of face processing as measured using ERPs in the first year of life and adulthood.

development in order to reach adultlike neural signatures of face processing. Even though issues regarding the ontogeny of face processing have yet to be resolved in the literature, it is clear that experience with faces since birth is very important for infants in order to lay the groundwork to develop expert face processing.

18.4 FACE PROCESSING IN TODDLERS AND PRESCHOOLERS

18.4.1 How Young Children Process Faces

Even though infants make incredible strides in the first year of life, their face-processing abilities are far from fully developed by the time they turn one. As discussed below, a great deal of development takes place after a child's first birthday. However, while infants' face-processing abilities are studied extensively, there are fewer investigations available of toddlers and preschoolers, with very few studies documenting the performance of 2- and 3-year-olds and relatively more data available for 4- and 5-year-olds.

There are few studies that have quantified the behavioral performance of toddlers and preschoolers in facial recognition and facial expression recognition. Bruce and colleagues (Bruce et al., 2000) have carried out an extensive behavioral investigation of how children as young as 4 and 5 perform on a variety of measures of face processing including the processing of facial identity, the processing of direction of gaze, and the processing of facial expressions. In this investigation, young children's performance was above chance across all tasks but still significantly below the proficiency attained by adults. Preschoolers also had trouble matching facial expression and direction of gaze with a specific target face, while they were easily able to discriminate along these dimensions (Bruce et al., 2000).

The results of Bruce and colleagues' investigation of face processing in young children suggest that they have yet to reach adultlike face-processing proficiency. There are two main hypotheses as to how these children differ from adults. On one hand, these differences have been hypothesized to be due to the fact that toddlers and preschoolers have yet to develop the face-specific processing strategies found in adults, namely, configural and holistic processing, and rely more heavily on feature-based strategies (see Mondloch et al., 2003a,b for a review). This hypothesis has been supported by studies showing that preschoolers experience difficulty in recognition when faces are 'disguised' (i.e., presented with a hat) between the original presentation and the subsequent ones, suggesting that perhaps they are not encoding faces using their configural properties but rather rely on specific isolated features (Carey and Diamond, 1977; Freire and Lee, 2001). Further support for this hypothesis

has come from studies using artificial face sets that are created such that different faces can vary exclusively either in terms of individual facial features, in terms of the spacing between facial features, or in terms of the facial contours (Mondloch et al., 2002). These studies have shown that toddlers and preschoolers fail to discriminate different identities when they have to rely exclusively on configural information but perform above chance when they can use featural and contour information (Mondloch et al., 2002, 2003a,b, 2006).

Alternatively, some researchers have proposed that the differences found in performance between young children, older children, and adults are produced by the protracted development of generic cognitive functions that can support the employment of more complex processing strategies (i.e., configural and holistic processing) more effectively (see McKone and Boyer, 2006 for a review). That is, it has been suggested that toddlers and preschoolers show a quantitative shift in their processing strategies and not a qualitative one. In support of this hypothesis, several researchers have used tasks such as the classic inversion effect task (Yin, 1969) and the composite effect task (Young et al., 1987) and measured young children's performance using faces and nonface objects such as cars or shoes. These investigators have found positive evidence of a difference in performance between upright and inverted faces in children as young as 3 (Macchi Cassia et al., 2006; Picozzi et al., 2009; Sangrigoli and de Schonen, 2004) and of holistic processing in children as young as 3½ years of age (de Heering et al., 2007; Macchi Cassia et al., 2009; Pellicano and Rhodes, 2003), even though across all these studies, overall performance in young children was usually worse compared to older children and adults.

Even though these two hypotheses point to diverging developmental trajectories, it may also be the case that their findings are not diametrically opposed. First of all, there are marked methodological differences among these studies, and different types of stimulus manipulations are used to measure configural and holistic processing, which could in part lead to diverging behavioral effects. Moreover, the tasks used vary in the relative difficulty of the different manipulations, and as such it is difficult to compare their findings objectively. Perhaps, results from electrophysiological investigations, which will be discussed below in this section, may shed more light on this controversy (Box 18.1).

18.4.2 How Young Children Process Facial Expressions

The ability to recognize facial expression has also not reached adultlike levels of performance in toddlers and preschoolers. Interestingly, there appear to be different developmental trajectories for different facial expressions

---

BOX 18.1

Electrophysiological and neuroimaging methods used in face processing

There are two primary techniques that are most widely used in the study of face perception: related potentials (ERP) and functional magnetic resonance imaging (fMRI). Traditionally, ERP has been used to study face processing in developmental populations, while both techniques have been used with adults. However, in recent years, fMRI has been used successfully with children as young as 7 years of age.

Event-related potentials measure the synchronous activity of large populations of neurons in response to a specific event. In a typical ERP experiment, stimuli of various types will be repeated many times, in order to be able to compute an average response by collapsing across repeated presentations of the same stimulus or category of stimuli. ERPs are collected using arrays of electrodes. In recent years, the number of recording electrodes has grown from 32 to 64, 128, and even 256, available in caps or nets that can be used with infants, children, and adults. Researchers use several characteristics of the ERP signal in a diagnostic

manner: latency of a specific waveform, amplitude of a specific waveform, and the location of potentials across the scalp at specific time points (scalp topography).

fMRI measures the hemodynamic response in the brain related to the presentation of specific experimental stimuli. It has been shown that when nerve cells are active, they increase their oxygen consumption. In response to the increased need for oxygen, there is an increase in blood flow in local capillaries. As a consequence, there will be an imbalance in the relative concentration of oxyhemoglobin and deoxyhemoglobin. Because the magnetic properties of blood vary depending on the level of oxygenation, pulse sequences can be used to detect these imbalances and measure the blood oxygen level-dependent (BOLD) contrast. In turn, the BOLD response can be used as an indirect measure of localized brain activity. Studies of face processing using fMRI will present subjects with faces and other types of objects and then measure the BOLD response to the different stimuli across the entire brain, in order to understand whether there are specific regions that modulate their response on the basis of the stimuli that they see.

---

with happiness being the most readily recognized expression followed by sadness, anger, and fear (see Herba and Phillips, 2004 for a review). Moreover, it has also been shown that preschoolers need more intense expressions in order to successfully recognize emotions, compared to older children (Gao and Maurer, 2009). These findings may initially seem at odds with those described in the previous section suggesting that infants are already capable of discriminating basic facial expression. However, it is important to point out that the nature of the tasks used between infants and toddlers can be very different and, as such, it is difficult to compare directly the results across age groups (McClure, 2000), and as a consequence researchers have wondered whether the different tasks are measuring similar processing abilities. Moreover, given that the neural substrates supporting emotion recognition are far from fully mature at these ages, it is not surprising that young children show behavioral differences in emotion processing compared to older children and adults.

18.4.3 Functional Signatures of Face Processing in Young Children

If infants present a challenge for the study of the neural underpinnings of face processing, toddlers and preschoolers may be even more difficult to test. Thus, there are relatively few studies investigating how the

brain of young children responds to faces, and for the most part, these studies employed ERPs. Nevertheless, these investigations can be particularly informative to help us understand the ability of young children to discriminate among different faces and different facial expressions, as the behavioral investigations alone have produced somewhat conflicting evidence. Overall, one would not expect major changes in the types of components elicited by faces, such as the P1 and the N170, and the nature of the changes is expected to be primarily in terms of latency (i.e., decreased latencies), morphology (i.e., more defined waveforms), and topography (i.e., hemispheric specialization) of these components, as toddlers' and preschoolers' brains undergo structural changes such as increased myelination, increased functional specialization, and changes in the underlying neural generators (de Haan et al., 2003; Gauthier and Nelson, 2001). There are, however, other changes that are taking place in these components that are more specific to the stimuli used, as the specific faces used acquire different significance with age. For example, Carver and colleagues (Carver et al., 2003) found that relative amplitude differences in the Nc and P400 components in the mother's face and a stranger's face switched between 2–3-year-olds and 4–5-year-olds with the younger children showing increased amplitude to the mother's face and the older children showing the opposite pattern.

Overall, while the morphology and topography of face-sensitive ERP components is changing toward adultlike brain signatures, their development is far from over within this age group (Taylor et al., 1999, 2004), possibly suggesting that even though young children show some evidence of adultlike face processing, they are still undergoing maturational changes both behaviorally and neurally.

Components reflecting differences in responding to emotional faces also show developmental changes, and these studies can further the understanding of how young children perceive facial expressions. Despite the fact that from a behavioral perspective, toddlers and preschoolers have difficulty recognizing fear in a face, Dawson and colleagues have shown that fearful faces elicit faster and larger P200 and N300 components, compared to neutral faces (Dawson et al., 2004). Batty and Taylor (2006) found that facial expressions also affected early components such as the P1. Toddlers and preschoolers had faster P1 latencies across all six emotions compared to neutral faces and more specifically showed the fastest latencies for positive emotions and, in particular, for happiness (Batty and Taylor 2006).

18.4.4 Conclusions

Taken together, the results of behavioral investigations and ERP studies looking at various aspects of face processing and face-processing strategies have led the majority of investigators to believe that toddlers and preschoolers are still on a developmental trajectory toward adultlike face processing. Nevertheless, it is important to point out that in very few years of life, young children demonstrate an uncanny proficiency in extracting information from faces.

18.5 FACE PROCESSING IN SCHOOL-AGE CHILDREN AND ADOLESCENTS

18.5.1 How Children and Adolescents Process Faces

The behavioral performance of school-age children and adolescents is relatively well documented, compared to that of toddlers and preschoolers. The majority of studies performed with these age groups have used experimental paradigms that are commonly used with adults (i.e., recognition memory paradigms, inversion face task, composite face task, etc.) in order to examine when and how children achieve adultlike levels of performance in face recognition and also when they show evidence of using configural and holistic processing strategies.

It is primarily within this age group that the scientific debate over children experiencing quantitative or qualitative changes in face-processing strategies is contested. This debate ensues because experimental results are heterogeneous. On the one hand, there are several studies pointing to the fact that school-age children, and in some cases even adolescents, differ in the type of information they use within a face (see Mondloch et al., 2003a,b for a review). For example, it has been demonstrated that children up to the age of 15 are not yet able to rely on the internal facial features of a face without contour information in order to successfully identify unfamiliar faces but rather perform better when they are provided only with the external contour information, compared to the internal features alone (Campbell and Tuck, 1995; Campbell et al., 1999; Want et al., 2003), which is opposite to the pattern that is observed in adults. Using a set of faces created in the laboratory that can differ only in terms of the external contour, or internal features, or the spacing among the internal features, Mondloch and colleagues found that, unlike adults, 6-, 8-, and 10-year-olds produced the most errors when discriminating faces differing in feature spacing (Mondloch et al., 2002). Moreover, Mondloch and colleagues have shown that 6-, 8-, and 10-year-olds failed to show an increased cost of face inversion in the feature-spacing set, compared to the other two stimuli manipulations, which is also different from how adult subjects perform (Mondloch et al., 2002).

Although these studies and others (Freire and Lee, 2001; Mondloch et al., 2006) suggest that children are less sensitive than adults to configural information, other researchers have produced results providing support for the hypothesis that children apply adultlike processing strategies from a young age. Children as young as 4 have demonstrated evidence of holistic processing as demonstrated using the composite face task (Mondloch et al., 2007; Pellicano and Rhodes, 2003), and Tanaka and colleagues (Tanaka et al., 1998) have found an adultlike whole-part advantage in 6-year-olds. Moreover, children as young as 6 have performed in a manner similar to that of adults on an inversion effect task, producing comparable differences between performance on upright and inverted faces (Mondloch et al., 2002).

In order to examine the controversy mentioned in the previous paragraphs regarding the nature of the development of face-processing strategies, there are some methodological considerations to be made. First, one possible explanation for the divergent results is the methodological differences employed across studies. For example, in some cases, memory-encoding paradigms are used, which require children to learn a specific set of faces prior to the testing phase, while in other cases the children are presented with simultaneous matching paradigms, in which there is very little memory load. Second, depending on the specific nuances of each study, the children are sometimes compared to the adults in terms of their quantitative performance, while in other cases their behavioral trends are compared to those found in adults. Thus, this makes it difficult to examine directly all these results and draw unified conclusions. In order to aid the resolution of this debate, it may be useful to consider it in the context of neurophysiological findings, which will be discussed below in this section.

Identity recognition, however, appears to be a process that does not reach full maturation until adolescence. In an extensive study looking at performance both on face and emotion recognition, Bruce and colleagues (Bruce et al., 2000) showed that performance on face recognition and emotion recognition increased steadily from 5 years of age until about 11 years of age. Moreover, they showed that when the experimental faces used were highly dissimilar, 11-years-olds performed as well as adolescents, but this was not the case when more similar faces were used, suggesting that more challenging face recognition follows a slow progression. Similarly, using a recognition memory paradigm, Golarai and colleagues have shown that adultlike levels of recognition memory performance for faces are not reached until at least 14 years of age (Golarai et al., 2007), while these authors did not find any age effects in the recognition memory performance for common objects. Moreover, school-age children tend to make more errors than adults when asked to recognize faces across viewpoints, lighting conditions, and changes in facial expression (Bruce et al., 2000; Mondloch et al., 2002, 2003a,b). While in the majority of studies it has been found that children perform worse than adolescents in recognition memory tasks, it is important to acknowledge that this may not be specific to faces, but it may be an issue of task difficulty. Why this issue is not easily solved is because, to our knowledge, there are no extensive studies comparing face processing to the processing of other complex visual objects.

18.5.2 How Children and Adolescents Process Facial Expressions

The recognition of emotional facial expression is also still undergoing development from childhood through adolescence, and the specific pattern of improvement appears dependent on the types of emotional faces used, on the intensity of the facial expression, and on the specific expression itself (Durand et al., 2007; Gao and Maurer, 2009; Herba et al., 2006; Kolb et al., 1992; Thomas et al., 2007; Vicari et al., 2000). Overall, behavioral studies of emotion recognition with children and adolescents show that children as young as 5 recognize happy faces accurately (Gao and Maurer, 2009; Vicari et al., 2000) but also show that sad and fearful faces remain more difficult to recognize especially if the face

stimuli used depict less intense emotion (Gao and Maurer, 2009; Thomas et al., 2007; Vicari et al., 2000). The speed of processing of facial expression also undergoes a maturation change between the ages of 7 and 10 (De Sonneville et al., 2002). There are nevertheless some differences in studies, in terms of the specific age ranges and their performance, and these differences are likely due to the different methodologies employed and the types of stimuli used. However, a protracted pattern of development for the recognition of facial expression makes sense even in the context of the development of the neural regions recruited when children and adolescents look at emotional faces.

The behavioral studies of development provide somewhat heterogeneous evidence, but by and large, they suggest that by 10–11 years of age, children perform in a manner comparable to adults across a variety of tasks that involve facial recognition, the use of face-specific processing strategies (i.e., configural processing, holistic processing), and recognition of emotional expressions. Given the heterogeneity of behavioral findings, it is particularly important to study the neural substrates of face processing in different age groups in order to understand whether by the time performance has reached adultlike levels, the underlying neural substrates and processes have also stabilized to what is found in adults or whether brain and behavior follow two different trajectories.

18.5.3 Neural Substrates of Processing in Children and Adolescents

In contrast to the relatively few studies examining the neural bases of face processing in toddlers and pre-schoolers, this topic has been more heavily addressed in older children. Of particular interest are recent investigations using fMRI to look not only at the functional signatures of face processing but also at the specificity of the neural substrates recruited. The question of interest in this literature is whether neural selectivity for faces has emerged by the time children are of school-age, that is, whether there are already measurable clusters of neurons that respond more strongly to faces than to most other objects. Several recent studies have been particularly focused on whether category selectivity for faces is present in the brains of children as young as 7 years of age (Aylward et al., 2005; Golarai et al., 2007, 2010; Passarotti et al., 2003; Scherf et al., 2007). Overall, these studies suggest that category selectivity in the right fusiform gyrus emerges around 7 years of age. More specifically, Golarai et al. (2007) have shown that the volume of these regions in the adult brain is larger compared to that in both children and adolescents and that it gets progressively larger between children and adolescents. More evidence of slow maturation is provided also by a recent study comparing the neural activations in response to upright and inverted faces in fMRI. Pasarrotti and colleagues (Passarotti et al., 2007) found that children ages 8–11 did not show showed increased activation for inverted faces, compared to upright faces, while older children ages 13–15 showed a trend in the opposite direction, which is what is commonly reported in adult studies.

Further evidence of protracted neural face selectivity is also provided by studies using ERP. Investigations using this technique have demonstrated that in children and adolescents, there are still differences in the latency and topography of face-selective neural signatures, compared to adult data. Adultlike characteristics of the latency of N170, for example, are not reached until adolescence (Taylor et al., 1999). Moreover, children do not show an adultlike latency difference between upright and inverted faces in the N170 until about 11 years of age (Itier and Taylor, 2004; Taylor et al., 2004), and the differences in amplitude with the inverted faces eliciting larger amplitudes, compared to upright faces, are observed only in adolescents (Taylor et al., 2004). Moreover, adultlike topography with higher amplitudes in the right hemisphere compared to the left for the N170 does not appear to emerge until adolescence (Taylor et al., 1999).

Although the neural responses to emotional faces are not widely studied within this age group, the few studies available on the topic have shown differences between adults and school-age children, and adolescents have also been reported in the context of emotional face processing. Passarotti and colleagues (Passarotti et al., 2009) tested adolescents with happy and sad faces in fMRI and found that while there were no differences in the areas that were recruited when adolescents and adults looked at emotional faces, there were differences that emerged in the relative activations in different brain regions, such that the adolescents activated paralimbic regions more strongly than the adults, but the adults activated prefrontal regions more strongly than the adolescent subjects. Similarly, Guyer and colleagues found more activation of the amygdala in response to fearful faces in adolescents, compared to adults, who in turn showed stronger functional connectivity between the amygdala and the hippocampus.

Differences in facial expression processing across children, adolescents, and adults are also observed using ERPs. Batty and Taylor (2006) have shown that from the age of 5 to the age of 15, there are changes in the sensitivity of the P1 and the N170 to facial expressions. More specifically, while children show latency modulations in the P1, older children show latency modulations only in the N170. Amplitude changes also vary across components within this age group, showing the same patterns as the latencies. Overall, what does not seem to be changing across development is the fact that positive emotions are processed with shorter latencies compared to negative emotions (Batty and Taylor, 2003, 2006).

18.5.4 Conclusions

Taken together, both the majority of behavioral findings and the results of neurophysiological investigations seem to suggest that face processing is still undergoing maturational changes through childhood and adolescence. Thus, even though children and adolescents already show some of the hallmarks of adultlike face processing, experimental evidence points to the fact that throughout childhood and adolescence, there are quantitative changes that take place in how these observers process faces.

18.6 IMPAIRMENTS IN FACE PROCESSING

While most of us rarely forget a face we have seen before, or have trouble identifying facial expressions, there is a subset of children and adults that experience great difficulties with these seemingly effortless operations. Dysfunctions in different aspects of face processing can arise from structural abnormalities within the brain and visual organs such as brain lesions and cataracts. Difficulties in processing faces can also be found in populations with neurological disorders such as autism spectrum disorder (see Chapter 34) and Williams syndrome (WS).

18.6.1 Prosopagnosia

In the literature, the majority of cases describing face-processing impairments are linked to brain injury to the occipital and temporal cortices, occurring either in the right hemisphere or bilaterally. The impairment that results from this kind of injury and produces primarily a deficit in recognizing familiar people has been termed prosopagnosia (Bodamer, 1947). Prosopagnosics not only show great difficulty in recognizing and learning new faces but also display atypical face-processing strategies (Barton, 2003). More specifically, they appear to be overly dependent on specific facial features, to be insensitive to the configural properties of faces, and to be unable to use holistic processing as reliably as healthy controls (Barton, 2003; Bukach et al., 2006, 2008; Wilkinson et al., 2009). However, prosopagnosics do not typically show difficulty identifying facial expressions (Young, 1992).

While prosopagnosia usually occurs in adults, there are documented cases of prosopagnosia in children following brain injury (Dutton, 2003; Dutton et al., 2006; Young and Ellis, 1989). Similarly to the adults, these children have great difficulty in recognizing individuals by their face, and they also show differences with configural/holistic processing compared to age-matched controls. However, as brain injury in children is rarely localized to the occipitotemporal cortex, these patients manifest neurological and perceptual problems beyond face perception (Dutton et al., 2006). Moreover, in contrast with adult prosopagnosics, children with this type of brain damage also tend to have difficulty recognizing facial expressions (Dutton, 2003).

Prosopagnosia can be also diagnosed in people who have not suffered any brain trauma (Behrmann et al., 2005a; Duchaine and Nakayama, 2006; Kress and Daum, 2003). Recent case studies documenting this type of prosopagnosia have termed it 'congenital,' to distinguish it from the form that follows brain injury ('acquired' prosopagnosia). Congenital prosopagnosics self-report difficulties in recognizing familiar faces that date to childhood and usually have learned to rely on cues other than facial information to recognize people; thus, they appear less impaired than acquired prosopagnosics. In recent years, this population has received much attention, and several studies have been aimed to categorize the nature of these patients' deficits. These patients can reliably detect a face among nonface objects and perform judgments of gender and age on faces successfully. However, they have difficulty in both matching and recognition tasks when they are under time pressure and/or the stimuli used are impoverished (i.e., face ovals with no hair; Behrmann et al., 2005b). Similar to acquired prosopagnosia, congenital prosopagnosics show a different manner of processing of face stimuli, demonstrating a heavier reliance on featural strategy in place of configural and holistic processing (Behrmann et al., 2005b). So far in the literature, this syndrome has primarily been studied in adults. However, a few cases of this syndrome have also been reported in children, and even the adult studies suggest an early onset of these problems (Ariel and Sadeh, 1996; Grueter et al., 2007).

The neural bases of congenital prosopagnosia are yet to be fully understood. Studies using ERPs have shown a reduced brain response to faces in congenital prosopagnosics as measured by the amplitude of face-sensitive components (Bentin et al., 1999). Studies using fMRI have produced discordant results, such that in few instances congenital prosopagnosics did not show brain activation in the right fusiform gyrus in response to faces (Hadjikhani and de Gelder, 2002), while others report typical right fusiform gyrus activation in response to faces (Avidan et al., 2005; Hasson et al., 2003). Brain activations alone have not been a diagnostic measure with these patients, as the relationship between these neural responses and their impairment is not clear. Thus, it has been suggested that congenital prosopagnosia could be caused by anatomical abnormalities in the temporal lobes, the same structures that when damaged produce acquired prosopagnosia. For example, Behrmann and colleagues have measured a reduction in the volume of the anterior fusiform gyrus in congenital prosopagnosics, compared to control subjects (Behrmann et al., 2007). Recent data using diffusion tensor imaging (DTI) have suggested that the recognition impairments may be due to the thinning of white matter fiber tracts connecting the ventral temporal cortex to the anterior temporal and prefrontal cortices (Thomas et al., 2009).

18.6.2 Congenital Cataract Patients

Given how rapidly face processing develops in the first year of life, researchers have wondered whether congenital visual impairments such as cataracts, which are present at birth but are usually surgically repaired within the first 2–6 months of life, have an impact on the ability to process faces later in life. Le Grand and colleagues have conducted a series of studies with children and young adults, aged 9–29, who were congenitally blind at birth because of bilateral cataracts but gained vision between 2 and 6 months of age following corrective surgery. These studies have shown that, despite the extensive experience with faces that these subjects have acquired since their cataracts were removed, they are not able to achieve the same level of performance as age-matched controls on tasks that tap complex face-processing strategies. More specifically, these patients had difficulty recognizing different faces that varied in terms of the distance between internal features (Le Grand et al., 2001), and they also failed to show the traditional 'composite effect' (Le Grand et al., 2004), which is used as an indirect measure of holistic processing (Maurer et al., 2002). Moreover, they experience particular difficulty with recognition when faces change orientation or facial expression across multiple presentations (Geldart et al., 2002). However, these patients are able to recognize faces fairly well in the real world, which is likely due to the fact that they can identify specific faces using the distinctive shape of the internal features and contour information (Mondloch et al., 2003a,b). No studies to date have looked at the functional responses that these patients' brains show when they are viewing faces. However, the dissociation found in this population between featural and configural/holistic processing abilities suggests that their neural network supporting face processing may have developed differently compared to healthy adults (Mondloch et al., 2003a,b).

18.6.3 Autism and WS

18.6.3.1 How Subjects with Autism Process Faces

Atypical face processing has also been reported in certain populations of individuals diagnosed with autism spectrum disorder (ASD) and WS. ASD is a neurological disorder that is diagnosed early in childhood, usually around the age of 3, and characterized by impairments in social interactions and language development; it is usually accompanied by repetitive behaviors and restricted interests (DSM-IV). In recent years, great interest has been devoted to understanding the nature of face-processing difficulties experienced by children and adults diagnosed with ASD because face-processing deficits are suggested to be strongly related to the social impairments experienced by these individuals. Moreover, some investigators have suggested that face-processing deficits may be one of the earliest indicators for the presence of autism (Dawson et al., 2005; Schultz, 2005; Schultz et al., 2000). The impairments found in ASD individuals since childhood span many different face-processing tasks with studies reporting deficits in recognizing facial identity (Boucher and Lewis, 1992; Boucher et al., 1998; Klin et al., 1999), less reliance on holistic and configurational processing (Behrmann et al., 2006; Gauthier et al., 2009; Joseph and Tanaka, 2003; Schultz et al., 2000; Teunisse and de Gelder, 2003) and reduced visual attention to internal facial features (Chawarska and Shic, 2009; Klin et al., 2002).

18.6.3.2 How Subjects with ASD Process Facial Expressions

Because of the social nature of ASD impairments, much attention has also been devoted to studying whether these patients have difficulties recognizing facial expressions. Overall, the majority of studies conducted on this topic would argue that ASD individuals have difficulty, beginning in childhood, recognizing, identifying, and classifying facial expressions (Braverman et al., 1989; Celani et al., 1999; Gross, 2004; Hobson, 1986). In several studies, it has been reported that these patients experience particular difficulties in recognizing negative emotions (Humphreys et al., 2007; Pelphrey et al., 2002) possibly because of ineffective use of information from the eye region of a face (Baron-Cohen et al., 1997, 2001; Gross, 2004). Moreover, recent studies using facial morphs that vary parametrically in the strength of visible facial expressions have shown that most ASD individuals have difficulty identifying and categorizing more subtle facial expressions (Rump et al., 2009; Teunisse and de Gelder, 2001).

18.6.3.3 Neural Substrates of Face Processing in ASD

The behavioral differences found between ASD individuals and typically developing controls have been connected to a variety of differences found in brain responses to faces between these two subject groups. Studies using ERPs have shown that both children and adults with ASD show shorter latencies for objects compared to faces for the N170 component (Webb et al., 2006). Moreover, the scalp topography of face-sensitive ERP components is different between these two subject groups, suggesting a smaller degree of hemispheric lateralization for face processing in ASD subjects (Dawson et al., 2005). Atypical ERP responses have also been demonstrated when ASD individuals are shown emotional faces (Dawson et al., 2004). Differences between ASD individuals and healthy controls have also been found using fMRI, with several studies showing hypoactivity not only in brain regions associated with processing of facial identity, such as the right fusiform gyrus (Dawson et al., 2002; Grelotti et al., 2002; Pierce et al., 2001; Schultz et al., 2000), but also in brain regions associated with the processing of facial expressions and gaze, such as the STS (Dalton et al., 2005; Pelphrey et al., 2005) and the amygdala (Adolphs et al., 2001; Hadjikhani et al., 2007; Schultz, 2005).

While the majority of studies of face processing in ASD individuals point to atypical development of this skill, this population shows a great deal of variability both in terms of behavioral patterns and neural responses. This heterogeneity of results has produced several debates in the literature regarding the true origins of these deficits. Nevertheless, atypical face processing, both at the behavioral and neural levels, remains one of the hallmarks of ASD.

18.6.3.4 Face Processing in WS

Another neurological disorder that has received attention in association with face processing is WS. WS is a rare genetic disorder characterized by a series of physical deformities, language and motor delays, and atypical cognitive functions in a variety of domains, including atypical social functioning (see Bellugi et al., 2000 for a review). Unlike ASD individuals, children and adults diagnosed with WS show face recognition abilities comparable to healthy controls (Bellugi et al., 1994; Tager-Flusberg, and Joseph, 2003). However, it has been suggested that they do not process faces configurally and holistically but rather rely more heavily on featural processing (Deruelle et al., 1999); but not all researchers agree on this interpretation (Karmiloff-Smith et al., 2004; Tager-Flusberg et al., 2006). Patients with WS are akin to ASD patients in that they also show deficits in emotion recognition from facial expressions (Tager-Flusberg et al., 2003).

These two subject populations also show divergences and similarities in the context of neural responses to faces, which map the dissociations found at the behavioral level. Studies using fMRI in adults diagnosed with WS have found activation in the right fusiform gyrus in response to faces that is comparable to that of healthy controls (Meyer-Lindenberg et al., 2004; Schultz et al., 2000). However, WS patients show hypoactivation in brain regions recruited when processing facial expressions, specifically the amygdala and the orbitofrontal cortex (Meyer-Lindenberg et al., 2005).

18.7 CONCLUSIONS

The goal of this chapter was to provide the reader with a comprehensive overview of the literature concerning face processing and its development from birth to adulthood.

Faces are very important stimuli for humans, as they convey a multitude of information about the people that we encounter in the world. While some debate remains regarding the initial mechanisms that orient infants toward faces, it is most likely that the development of face processing relies on both experience-independent early specificity and an experience-dependent sensitive period (Sugita, 2008). That is, while the initial orienting toward faces may be driven by a genetically specified primitive architecture (Johnson, 2005), the development of the ability to extract information from a face is heavily dependent on experience, and it is through exposure and interactions with specific types of faces that we learn to become 'face experts,' which is a lengthy process that requires the first 10–15 years of life (Nelson, 2001).

Despite the many things that are known in the face-processing literature, it is the authors' wish to conclude this review with some of the questions that are still unanswered. First of all, while it has been demonstrated that experience with faces is very important for one to become an 'expert,' relatively little is known about the specific nature of this experience. Secondly, it is unclear whether the effect of experience is dependent on a sensitive period. Although the data produced by investigations with cataract patients and congenital prosopagnosics would suggest that any derailment from typical development affects face processing, more work will be necessary to establish the precise nature of a sensitive period and its relations to different aspects of face processing. Lastly, we would like to highlight a real conundrum of face processing, which is its apparent lack of plasticity. It has been demonstrated that once a portion of the face-processing system becomes compromised, the ability to recognize faces appears gone for good. This is a very troublesome observation for this literature because it is difficult to reconcile how an operation of such

evolutionary importance that relies on a large network of neural substrates can show so little ability to reorganize itself following an insult, especially given the fact that our brains are overall quite plastic. In our view, this remains an essential issue to resolve in the coming years.

Acknowledgments

The writing of this chapter was made possible by a grant from the National Institute of Mental Health to the second author MH078829. For inquiries about this chapter, contact the second author at the Laboratories of Cognitive Neuroscience, 1 Autumn Street, 6th Floor, Boston, MA 02215, and Charles.nelson@childrens.harvard.edu.

References

Adolphs, R., Sears, L., Piven, J., 2001. Abnormal processing of social information from faces in autism. Journal of Cognitive Neuroscience 13, 232–240.

Ariel, R., Sadeh, M., 1996. Congenital visual agnosia and prosopagnosia in a child: A case report. Cortex 32, 221–240.

Avidan, G., Hasson, U., Malach, R., Behrmann, M., 2005. Detailed exploration of face-related processing in congenital prosopagnosia: Functional neuroimaging findings. Journal of Cognitive Neuroscience 17, 1150–1167.

Aylward, E.H., Park, J.E., Field, K.M., et al., 2005. Brain activation during face perception: Evidence of a developmental change. Journal of Cognitive Neuroscience 17, 308–319.

Banks, M.S., Salapatek, P., 1976. Contrast sensitivity function of the infant visual system. Vision Research 16, 867–869.

Bar-Haim, Y., Ziv, T., Lamy, D., Hodes, R.M., 2006. Nature and nurture in own-race face processing. Psychological Science 17, 159–163.

Baron-Cohen, S., Jolliffe, T., Mortimore, C., Robertson, M., 1997. Another advanced test of theory of mind: Evidence from very high functioning adults with autism or Asperger syndrome. Journal of Child Psychology and Psychiatry 38, 813–822.

Baron-Cohen, S., Wheelwright, S., Hill, J., Raste, Y., Plumb, I., 2001. The 'Reading the Mind in the Eyes' Test revised version: A study with normal adults, and adults with Asperger syndrome or high-functioning autism. Journal of Child Psychology and Psychiatry 42, 241–251.

Barrera, M.E., Maurer, D., 1981. The perception of facial expressions by the three-month-old. Child Development 52, 203–206.

Barton, J.J., 2003. Disorders of face perception and recognition. Neurologic Clinics 21, 521–548.

Batty, M., Taylor, M.J., 2003. Early processing of the six basic facial emotional expressions. Cognitive Brain Research 17, 613–620.

Batty, M., Taylor, M.J., 2006. The development of emotional face processing during childhood. Developmental Science 9, 207.

Behrmann, M., Avidan, G., Marotta, J.J., Kimchi, R., 2005a. Detailed exploration of face-related processing in congenital prosopagnosia: Behavioral findings. Journal of Cognitive Neuroscience 17, 1130–1149.

Behrmann, M., Marotta, J., Gauthier, I., Tarr, M.J., McKeeff, T.J., 2005b. Behavioral change and its neural correlates in visual agnosia after expertise training. Journal of Cognitive Neuroscience 17, 554–568.

Behrmann, M., Thomas, C., Humphrey, K., 2006. Seeing it differently: Visual processing in autism. Trends in Cognitive Sciences 10, 258–264.

Behrmann, M., Avidan, G., Gao, F., Black, S., 2007. Structural imaging reveals anatomical alterations in inferotemporal cortex in congenital prosopagnosia. Cerebral Cortex 17, 2354–2363.

Bellugi, U., Wang, P.P., Jerrigan, T.J., 1994. Williams Syndrome: An unusual neuropsychological profile. In: Broman, S., Grafman, J. (Eds.), Atypical Cognitive Deficits in Developmental Disorders: Implications for Brain Function. Lawrence Erlbaum Associates, Hillsdale, NJ.

Bellugi, U., Lichtenberger, L., Jones, W., Lai, Z., St. George, M., 2000. The neurocognitive profile of Williams Syndrome: A complex patterns of strengths and weaknesses. Journal of Cognitive Neuroscience 12, 7–29.

Bentin, S., Puce, A.T., Perez, A., McCarthy, G., 1996. Electrophysiological studies of face perception in humans. Journal of Cognitive Neuroscience 8, 551–565.

Bentin, S., Deouell, L.Y., Soroker, N., 1999. Selective visual streaming in face recognition: Evidence from developmental prosopagnosia. NeuroReport 10, 823–827.

Bhatt, R.S., Bertin, E., Hayden, A., Reed, A., 2005. Face processing in infancy: Developmental changes in the use of different kinds of relational information. Child Development 76, 169–181.

Bodamer, J., 1947. Prosopagnosia. Archiv Der Psychiatrichen Nervenkrankheiten 179, 6–54.

Boucher, J., Lewis, V., 1992. Unfamiliar face recognition is relatively able in Autistic children. Journal of Child Psychology and Psychiatry 33, 843–859.

Boucher, J., Lewis, V., Collis, G., 1998. Familiar face and voice matching and recognition in children with autism. Journal of Child Psychology and Psychiatry, and Allied Disciplines 39, 171–181.

Braje, W.L., 2003. Illumination encoding in face recognition: Effect of position shift. Journal of Vision 3, 161–170.

Braje, W.L., Kersten, D., Tarr, M.J., Troje, N.F., 1998. Illumination effects in face recognition. Psychobiology 26, 371–380.

Braverman, M., Fein, D., Lucci, D., Waterhouse, L., 1989. Affect comprehension in children with pervasive developmental disorders. Journal of Autism and Developmental Disorders 19, 301–320.

Bruce, V., Young, A., 1986. Understanding face recognition. British Journal of Psychology 77, 305–327.

Bruce, V., Campbell, R.N., Doherty-Sneddon, G., et al., 2000. Testing face processing skills in children. British Journal of Developmental Psychology 18, 319–333.

Bukach, C.M., Bub, D.N., Gauthier, I., Tarr, M.J., 2006. Perceptual expertise effects are not all or none: Spatially limited perceptual expertise for faces in a case of prosopagnosia. Journal of Cognitive Neuroscience 18, 48–63.

Bukach, C.M., Le Grand, R., Kaiser, M.D., Bub, D.N., Tanaka, J.W., 2008. Preservation of mouth region processing in two cases of prosopagnosia. Journal of Neuropsychology 2, 227–244.

Campbell, R., Tuck, M., 1995. Recognition of parts of famous-face photographs by children: An experimental note. Perception 24, 451–456.

Campbell, R., Coleman, M., Walker, J., et al., 1999. When does the inner-face advantage in familiar face recognition arise and why? Visual Cognition 6, 197–215.

Carey, S., Diamond, R., 1977. From piecemeal to configurational representation of faces. Science 195, 312–314.

Carmel, D., Bentin, S., 2002. Domain specificity versus expertise: Factors influencing distinct processing of faces. Cognition 83, 1–29.

Carver, L.J., Dawson, G., Panagiotides, H., et al., 2003. Age-related differences in neural correlates of face recognition during the toddler and preschool years. Developmental Psychobiology 42, 148–159.

Celani, G., Battacchi, M.W., Arcidiacono, L., 1999. The understanding of the emotional meaning of facial expressions in people with autism. Journal of Autism and Developmental Disorders 29, 57–66.

Chawarska, K., Shic, F., 2009. Looking but not seeing: Atypical visual scanning and recognition of faces in 2 and 4-year-old children with Autism Spectrum Disorder. Journal of Autism and Developmental Disorders 39, 1663–1672.

Dalton, K.M., Nacewitz, B.M., Johnstone, T., et al., 2005. Gaze fixation and the neural circuitry of face processing in autism. Nature Neuroscience 8, 519–526.

Dawson, G., Carver, L., Meltzoff, A., Panagiotides, H., McPartland, J., Webb, S.J., 2002. Neural correlates of face and object recognition in young children with Autism Spectrum disorder, developmental delay and typical development. Child Development 73, 700–717.

Dawson, G., Webb, S.J., Carver, L., Panagiotides, H., McPartland, J., 2004. Young children with autism show atypical brain responses to fearful versus neutral facial expressions of emotion. Developmental Science 7, 340–359.

Dawson, G., Webb, S.J., McPartland, J., 2005. Understanding the nature of face processing impairment in autism: Insights from behavioral and electrophysiological studies. Developmental Neuropsychology 27, 403–424.

de Haan, M., Nelson, C.A., 1997. Recognition of the mother's face by six-month-old infants: A neurobehavioral study. Child Development 68, 187–210.

de Haan, M., Nelson, C.A., Gunnar, M.R., Tout, K.A., 1998. Hemispheric differences in brain activity related to the recognition of emotional expressions by 5-year-old children. Developmental Neuropsychology 14, 495–518.

de Haan, M., Johnson, M.H., Halit, H., 2003. Development of face-sensitive event-related potentials during infancy: A review. International Journal of Psychophysiology 51, 45–58.

de Heering, A., Houthuys, S., Rossion, B., 2007. Holistic face processing is mature at 4 years of age: Evidence from the composite face effect. Journal of Experimental Child Psychology 96, 57–70.

De Sonneville, L.M., Verschoor, C.A., Njiokiktjien, C., Op het Veld, V., Toorenaar, N., Vranken, M., 2002. Facial identity and facial emotions: Speed, accuracy, and processing strategies in children and adults. Journal of Clinical and Experimental Neuropsychology 24, 200–213.

Deruelle, C., Mancini, J., Livet, M.O., Cassé-Perrot, C., de Schonen, S., 1999. Configural and local processing of faces in children with Williams Syndrome. Brain and Cognition 41, 276–298.

Diamond, R., Carey, S., 1986. Why faces are and are not special: An effect of expertise. Journal of Experimental Psychology. General 115, 107–117.

Duchaine, B., Nakayama, K., 2006. The Cambridge face memory test: Results for neurologically intact individuals and an investigation of its validity using inverted face stimuli and prosopagnosic participants. Neuropsychologia 44, 576–585.

Durand, K., Gallay, M., Seigneuric, A., Robichon, F., Baudouin, J.Y., 2007. The development of facial emotion recognition: The role of configural information. Journal of Experimental Child Psychology 97, 14–27.

Dutton, G.N., 2003. Cognitive vision, its disorders and differential diagnosis in adults and children: Knowing where and what things are. Eye 17, 289–304.

Dutton, G.N., McKillop, E.C.A., Saldkasimova, S., 2006. Visual problems as a result of brain damage in children. British Journal of Ophthalmology 90, 931–932.

Eimer, M., 2000. The face-specific N170 component reflects late stages in the structural encoding of faces. NeuroReport 11, 2319–2324.

Ekman, P., 1993. Facial expression and emotion. American Psychologist 48, 376–379.

Farroni, T., Menon, E., Rigato, S., Johnson, M.H., 2007. The perception of facial expressions in newborns. The European Journal of Developmental Psychology 4, 2–13.

Field, T.M., Woodson, R., Greenberg, R., Cohen, D., 1982. Discrimination and imitation of facial expression by neonates. Science 218, 179–181.

Freire, A., Lee, K., 2001. Face recognition in 4- to 7-year-olds: Processing of configural, featural, and paraphernalia information. Journal of Experimental Child Psychology 80, 347–371.

Gao, X., Maurer, D., 2009. Influence of intensity on children's sensitivity to happy, sad, and fearful facial expressions. Journal of Experimental Child Psychology 102, 503–521.

Gauthier, I., Nelson, C.A., 2001. The development of face expertise. Current Opinion in Neurobiology 11, 219–224.

Gauthier, I., Tarr, M.J., 2002. Unraveling mechanisms for expert object recognition: Bridging brain activity and behavior. Journal of Experimental Psychology. Human Perception and Performance 28, 431–446.

Gauthier, I., Klaiman, C., Schultz, R.T., 2009. Face composite effects reveal abnormal face processing in autism spectrum disorders. Vision Research 49, 470–478.

Geldart, S., Mondloch, C.J., Maurer, D., de Schonen, S., Brent, H.P., 2002. The effect of early visual deprivation on the development of face processing. Developmental Science 5, 490–501.

Gobbini, M.I., Haxby, J.V., 2007. Neural systems for recognition of familiar faces. Neuropsychologia 45, 32–41.

Goffaux, V., Rossion, B., 2007. Face inversion disproportionately impairs the perception of vertical but not horizontal relations between features. Journal of Experimental Psychology. Human Perception and Performance 33, 995–1002.

Golarai, G., Ghahremani, D.G., Whitfield-Gabrieli, S., et al., 2007. Differential development of high-level visual cortex correlates with category-specific recognition memory. Nature Neuroscience 10, 512–522.

Golarai, G., Liberman, A., Yoon, J.M., Grill-Spector, K., 2010. Differential development of the ventral visual cortex extends through adolescence. Frontiers in Human Neuroscience 3, 80.

Grelotti, D.J., Gauthier, I., Schultz, R.T., 2002. Social interest and the development of cortical face specialization: What autism teaches us about face processing. Developmental Psychobiology 40, 213–225.

Gross, T.F., 2004. The perception of four basic emotions in human and nonhuman faces by children with autism and other developmental disabilities. Journal of Abnormal Child Psychology 32, 469–480.

Grossmann, T., Striano, T., Friederici, A.D., 2007. Developmental changes in infants' processing of happy and angry facial expressions: A neurobehavioral study. Brain and Cognition 64, 30–41.

Grueter, M., Grueter, T., Bell, V., et al., 2007. Hereditary prosopagnosia: The first case series. Cortex 43, 734–749.

Guyer, A.E., Monk, C.S., McClure-Tone, E.B., et al., 2008. A developmental examination of amygdala response to facial expressions. Journal of Cognitive Neuroscience 20 (9), 1565–1582.

Hadjikhani, N., de Gelder, B., 2002. Neural basis of prosopagnosia: An fMRI study. Human Brain Mapping 16, 176–182.

Hadjikhani, N., Joseph, R.M., Snyder, J., Tager-Flusberg, H., 2007. Abnormal activation of the social brain during face perception in autism. Human Brain Mapping 28, 441–449.

Halit, H., Csibra, G., Volein, A., Johnson, M.H., 2004. Face-sensitive cortical processing in early infancy. Journal of Child Psychology and Psychiatry, and Allied Disciplines 45, 1228–1234.

Harmon, L.D., 1973. The recognition of faces. Scientific American 229, 71–82.

Hasson, U., Avidan, G., Deouell, L.Y., Bentin, S., Malach, R., 2003. Face-selective activation in a congenital prosopagnosic subject. Journal of Cognitive Neuroscience 15, 419–431.

Haxby, J.V., Ishai, I.I., Chao, L.L., Ungerleider, L.G., Martin, I.I., 2000. Object-form topology in the ventral temporal lobe: Response to I. Gauthier (2000). Trends in Cognitive Sciences 4, 3–4.

Haxby, J.V., Hoffman, E.A., Gobbini, M.I., 2002. Human neural systems for face recognition and social communication. Biological Psychiatry 51, 59–67.

Herba, C., Phillips, M., 2004. Annotation: Development of facial expression recognition from childhood to adolescence: Behavioural and neurological perspectives. Journal of Child Psychology and Psychiatry, and Allied Disciplines 45, 1185–1198.

Herba, C.M., Landau, S., Russell, T., Ecker, C., Phillips, M.L., 2006. The development of emotion-processing in children: Effects of age, emotion, and intensity. Journal of Child Psychology and Psychiatry, and Allied Disciplines 47, 1098–1106.

Hill, H., Schyns, P.G., Akamatsu, S., 1997. Information and viewpoint dependence in face recognition. Cognition 62, 201–222.

Hobson, R.P., 1986. The autistic child's appraisal of expressions of emotion. Journal of Child Psychology and Psychiatry, and Allied Disciplines 27, 321–342.

Humphreys, K., Avidan, G., Behrmann, M., 2007. A detailed investigation of facial expression processing in congenital prosopagnosia as compared to acquired prosopagnosia. Experimental Brain Research 176, 356–373.

Ishai, A., Schmidt, C.F., Boesiger, P., 2005. Face perception is mediated by a distributed cortical network. Brain Research Bulletin 67, 87–93.

Itier, R.J., Taylor, M.J., 2004. Face inversion and contrast-reversal effects across development: In contrast to the expertise theory. Developmental Science 7, 246–260.

Jacques, C., Rossion, B., 2007. Early electrophysiological responses to multiple face orientations correlate with individual discrimination performance in humans. NeuroImage 36, 863–876.

Johnson, M.H., 2005. Subcortical face processing. Nature Reviews Neuroscience 6, 766–774.

Johnson, M.H., Morton, H., 1991. Biology and Cognitive Development: The Case of Face Recognition. Blackwell, Oxford.

Joseph, R.M., Tanaka, J., 2003. Holistic and part-based face recognition in children with autism. Journal of Child Psychology and Psychiatry 44, 529–542.

Kanwisher, N., McDermott, J., Chun, M.M., 1997. The fusiform face area: A module in human extrastriate cortex specialized for face perception. Journal of Neuroscience 17, 4302–4311.

Karmiloff-Smith, A., Thomas, M., Annaz, D., et al., 2004. Exploring the Williams Syndrome face-processing debate: The importance of building developmental trajectories. Journal of Child Psychology and Psychiatry, and Allied Disciplines 45, 1258–1274.

Kelly, D.J., Quinn, P.C., Slater, A.M., et al., 2005. Three-month-olds, but not newborns, prefer own-race faces. Developmental Science 8, 431–436.

Kelly, D.J., Quinn, P.C., Slater, A.M., Lee, K., Ge, L., Pascalis, O., 2007. The other-race effect develops during infancy: Evidence of perceptual narrowing. Psychological Science 18, 1084–1089.

Kleiner, K.A., 1987. Amplitude and phase spectra as indices of infants' pattern preference. Infant Behavior & Development 10, 49–59.

Klin, A., Jones, W., Schultz, R., Volkmar, F., Cohen, D., 2002. Visual fixation patterns during viewing of naturalistic social situations as predictors of social competence in individuals with autism. Archives of General Psychiatry 59, 809–816.

Klin, A., Sparrow, S.S., de Bildt, A., Cicchetti, D.V., Cohen, D.J., Volkmar, F.R., 1999. A normed study of face recognition in autism and related disorders. Journal of Autism and Developmental Disorders 29, 499–508.

Kolb, B., Wilson, B., Taylor, L., 1992. Developmental changes in the recognition and comprehension of facial expression: Implications for frontal lobe function. Brain and Cognition 20, 74–84.

Kotsoni, E., de Haan, M., Johnson, M.H., 2001. Categorical perception of facial expressions by 7-month-old infants. Perception 30, 1115–1125.

Kress, T., Daum, I., 2003. Developmental prosopagnosia: A review. Behavioral Neurology 14, 109–121.

LaBarbera, J.D., Izard, C.E., Vietze, P., Parisi, S.A., 1976. Four- and six-month-old infants' visual responses to joy, anger, and neutral expressions. Child Development 47, 535–538.

Le Grand, R., Mondloch, C.J., Maurer, D., Brent, H.P., 2001. Neuroperception: Early visual experience and face processing. Nature 410, 890.

Le Grand, R., Mondloch, C.J., Maurer, D., Brent, H.P., 2004. Impairment in holistic face processing following early visual deprivation. Psychological Science 15, 762–768.

Leppänen, J.M., Moulson, M.C., Vogel-Farley, V.K., Nelson, C.A., 2007. An ERP study of emotional face processing in the adult and infant brain. Child Development 78, 232–245.

Leppänen, J.M., Richmond, J., Vogel-Farley, V.K., Moulson, M.C., Nelson, C.A., 2009. Categorical representation of facial expressions in the infant brain. Infancy 14, 346–362.

Leveroni, C.L., Seidenberg, M., Mayer, A.R., Mead, L.A., Binder, J.R., Rao, S.M., 2000. Neural systems underlying the recognition of familiar and newly learned faces. Journal of Neuroscience 20, 878–886.

Macchi Cassia, V., Kuefner, D., Westerlund, A., Nelson, C.A., 2006. Modulation of face-sensitive event-related potentials by canonical and distorted human faces: The role of vertical symmetry and up-down featural arrangement. Journal of Cognitive Neuroscience 18, 1343–1358.

Macchi Cassia, V.M., Picozzi, M., Kuefner, D., Bricolo, E., Turati, C., 2009. Holistic processing for faces and cars in preschool-aged children and adults: Evidence from the composite effect. Developmental Science 12 (2), 236–248.

Maurer, D., 1983. The scanning of compound figures by young infants. Journal of Experimental Child Psychology 35, 437–448.

Maurer, D., Le Grand, R., Mondloch, C.J., 2002. The many faces of configural processing. Trends in Cognitive Sciences 6, 255–260.

McClure, E.B., 2000. A meta-analytic review of sex differences in facial expression processing and their development in infants, children, and adolescents. Psychological Bulletin 126, 424–453.

McKone, E., Boyer, B.L., 2006. Sensitivity of 4-year-olds to featural and second-order relational changes in face distinctiveness. Journal of Experimental Child Psychology 94, 134–162.

Meyer-Lindenberg, A., Kohn, P., Mervis, C.B., et al., 2004. Neural basis of genetically determined visuospatial construction deficit in Williams Syndrome. Neuron 43, 623–631.

Meyer-Lindenberg, A., Hariri, A.R., Munoz, K.E., et al., 2005. Neural correlates of genetically abnormal social cognition in Williams Syndrome. Nature Neuroscience 8, 991–993.

Mondloch, C.J., Lewis, T.L., Budreau, D.R., et al., 1999. Face perception during early infancy. Psychological Science 10, 419–422.

Mondloch, C.J., Le Grand, R., Maurer, D., 2002. Configural face processing develops more slowly than featural face processing. Perception 31, 553–566.

Mondloch, C.J., Geldart, S., Maurer, D., Le Grand, R., 2003a. Developmental changes in face processing skills. Journal of Experimental Child Psychology 86, 67–84.

Mondloch, C.J., Le Grand, R., Maurer, D., 2003b. Early visual experience is necessary for the development of some-but not all- aspects of face processing. In: Pascalis, O., Slater, A. (Eds.), The Development of Face Processing in Infancy and Early Childhood. Nova Science Publisher Inc., Hauppauge, New York, pp. 99–117.

Mondloch, C.J., Dobson, K.S., Parsons, J., Maurer, D., 2004. Why 8-year-olds cannot tell the difference between Steve Martin and Paul Newman: Factors contributing to the slow development of sensitivity to the spacing of facial features. Journal of Experimental Child Psychology 89, 159–181.

Mondloch, C.J., Maurer, D., Ahola, S., 2006. Becoming a face expert. Psychological Science 17, 930–934.

Mondloch, C.J., Pathman, T., Maurer, D., Grand, R.L., de Schonen, S., 2007. The composite face effect in six-year-old children: Evidence of adult-like holistic face processing. Visual Cognition 15, 564–577.

Nelson, C.A., Morse, P.A., Leavitt, L.A., 1979. Recognition of facial expressions by seven-month olds infants. Child Development 50, 1239–1242.

Nelson, C.A., Dolgin, K.G., 1985. The generalized discrimination of facial expressions by seven-month-old infants. Child Development 56, 58–61.

Nelson, C.A., Salapatek, P., 1986. Electrophysiological correlates of infant recognition memory. Child Development 57, 1483–1497.

Nelson, C.A., De Haan, M., 1996. Neural correlates of infants' visual responsiveness to facial expressions of emotion. Developmental Psychobiology 29 (7), 577–595.

Nelson, C.A., 2001. The development and neural bases of face recognition. Infant and Child Development 10, 3–18.

O'Toole, A., Deffenbacher, K.A., Valentin, D., McKee, K., Huff, D., Abdi, H., 1997. The perception of face gender: The role of stimulus structure in recognition and classification. Memory and Cognition 25, 1–20.

O'Toole, A.J., Edelman, S., Bülthoff, H.H., 1998. Stimulus-specific effects in face recognition over changes in viewpoint. Vision Research 38, 2351–2363.

Pascalis, O., De Schonen, S., Morton, J., Deruelle, C., Fabre-Grenet, M., 1995. Mother's face recognition by neonates: A replication and an extension. Infant Behavior & Development 18, 79–85.

Pascalis, O., de Haan, M., Nelson, C.A., de Schonen, S., 1998. Long-term recognition memory for faces assessed by visual paired comparison in 3- and 6-month-old infants. Journal of Experimental Psychology 24, 249–260.

Pascalis, O., de Haan, M., Nelson, C.A., 2002. Is face processing species-specific during the first year of life? Science 296, 1321–1323.

Passarotti, A.M., Paul, B.M., Bussiere, J.R., Buxton, R.B., Wong, E.C., Stiles, J., 2003. The development of face and location processing: An fMRI study. Developmental Science 6, 100–117.

Pascalis, O., Scott, L.S., Kelly, D.J., et al., 2005. Plasticity of face processing in infancy. Proceedings of the National Academy of Sciences of the United States of America 102, 5297–5300.

Passarotti, A.M., Smith, J., DeLano, M., Huang, J., 2007. Developmental differences in the neural bases of the face inversion effect show progressive tuning of face-selective regions to the upright orientation. NeuroImage 34, 1708–1722.

Passarotti, A.M., Sweeney, J.A., Pavuluri, M.N., 2009. Neural correlates of incidental and directed facial emotion processing in adolescents and adults. Social Cognitive and Affective Neuroscience 4, 387–398.

Pellicano, E., Rhodes, G., 2003. Holistic processing of faces in preschool children and adults. Psychological Science 14, 618–622.

Pelphrey, K.A., Sasson, N.J., Reznick, J.S., Goldman, P.G., Piven, J., 2002. Visual scanning of faces in autism. Journal of Autism and Developmental Disorders 32, 249–261.

Pelphrey, K.A., Morris, J.P., McCarthy, G., 2005. Neural basis of eye gaze processing deficits in autism. Brain 128, 1038–1048.

Peltola, M.J., Leppänen, J.M., Mäki, S., Hietanen, J.K., 2009. Emergence of enhanced attention to fearful faces between 5 and 7 months of age. Social Cognitive and Affective Neuroscience 4, 134–142.

Picozzi, M., Cassia, V.M., Turati, C., Vescovo, E., 2009. The effect of inversion on 3- to 5-year-olds' recognition of face and nonface visual objects. Journal of Experimental Child Psychology 102, 487–502.

Pierce, K., Muller, R.A., Ambrose, J., Allen, G., Courchesne, E., 2001. Face processing occurs outside the fusiform face area in autism: Evidence from functional MRI. Brain 124, 2059–2073.

Quinn, P.C., Yahr, J., Kuhn, A., Slater, A.M., Pascalils, O., 2002. Representation of the gender of human faces by infants: A preference for female. Perception 31, 1109–1121.

Quinn, P.C., Uttley, L., Lee, K., et al., 2008. Infant preference for female faces occurs for same- but not other-race faces. Journal of Neuropsychology 2, 15–26.

Rossion, B., Campanella, S., Gomez, C.M., et al., 1999. Task modulation of brain activity related to familiar and unfamiliar face processing: An ERP study. Clinical Neurophysiology 110, 449–462.

Rossion, B., Gauthier, I., 2002. How does the brain process upright and inverted faces? Behavioral and Cognitive Neuroscience Reviews 1, 63–75.

Rump, K.M., Giovannelli, J.L., Minshew, N.J., Strauss, M.S., 2009. The development of emotion recognition in individuals with autism. Child Development 80, 1434–1447.

Sangrigoli, S., de Schonen, S., 2004. Effect of visual experience on face processing: A developmental study of inversion and non-native effects. Developmental Science 7, 74–87.

Scherf, K.S., Behrmann, M., Humphreys, K., Luna, B., 2007. Visual category-selectivity for faces, places and objects emerges along different developmental trajectories. Developmental Science 10, F15–F30.

Schultz, R.T., 2005. Developmental deficits in social perception in autism: The role of the amygdala and fusiform face area. International Journal of Developmental Neuroscience 23, 125–141.

Schultz, R.T., Gauthier, I., Klin, A., et al., 2000. Abnormal ventral temporal cortical activity during face discrimination among individuals with autism and Asperger syndrome. Archives of General Psychiatry 57, 331–340.

Sugita, Y., 2008. Face perception in monkeys reared with no exposure to faces. Proceedings of the National Academy of Sciences of the United States of America 105, 394–398.

Tager-Flusberg, H., Joseph, R.M., 2003. Identifying neurocognitive phenotypes in autism. Philosophical Transactions of the Royal Society of London. Series B, Biological Sciences 358, 303–314.

Tager-Flusberg, H., Plesa-Skwerer, D., Faja, S., Joseph, R.M., 2003. People with Williams Syndrome process faces holistically. Cognition 89, 11–24.

Tager-Flusberg, H., Skwerer, D.P., Joseph, R.M., 2006. Model syndromes for investigating social cognitive and affective neuroscience: A comparison of autism and Williams Syndrome. Social Cognitive and Affective Neuroscience 1, 175–182.

Tanaka, J.W., Farah, M.J., 1993. Parts and wholes in face recognition. Quarterly Journal of Experimental Psychology 46, 225–245.

Tanaka, J.W., Kay, J.B., Grinnell, E., Stansfield, B., Szechter, L., 1998. Face recognition in young children: When the whole is greater than the sum of its parts. Visual Cognition 5, 479–496.

Taylor, M.J., McCarthy, G., Saliba, E., Degiovanni, E., 1999. ERP evidence of developmental changes in processing of faces. Clinical Neurophysiology 110, 910–915.

Taylor, M.J., Batty, M., Itier, R.J., 2004. The faces of development: A review of early face processing over childhood. Journal of Cognitive Neuroscience 16, 1426–1442.

Teunisse, J.P., de Gelder, B., 2001. Impaired categorical perception of facial expressions in high-functioning adolescents with autism. Child Neuropsychology 7, 1–14.

Teunisse, J.P., de Gelder, B., 2003. Face processing in adolescents with autistic disorder: The inversion and composite effects. Brain and Cognition 52, 285–294.

Thomas, L.A., De Bellis, M.D., Graham, R., LaBar, K.S., 2007. Development of emotional facial recognition in late childhood and adolescence. Developmental Science 10, 547–558.

Thomas, C., Avidan, G., Humphreys, K., Jung, K.J., Gao, F., Behrmann, M., 2009. Reduced structural connectivity in ventral visual cortex in congenital prosopagnosia. Nature Neuroscience 12, 29–31.

Thompson, L.A., Madrid, V., Westbrook, S., Johnston, V., 2001. Infants attend to second-order relational properties of faces. Psychonomic Bulletin & Review 8, 769–777.

Tsao, D.Y., Moeller, S., Freiwald, W.A., 2008. Comparing face patch systems in macaques and humans. Proceedings of the National Academy of Sciences of the United States of America 105, 19514–19519.

Tzourio-Mazoyer, N., De Schonen, S., Crivello, F., Reutter, B., Aujard, Y., Mazoyer, B., 2002. Neural correlates of woman face processing by 2-month-old infants. NeuroImage 15, 454–461.

Valentine, T., 1991. A unified account of the effects of distinctiveness, inversion, and race in face recognition. Quarterly Journal of Experimental Psychology Section A: Human Experimental Psychology 43, 161–204.

Valentine, T., 1999. Face-space models of face recognition. In: Wenger, M.J., Townsend, J.T. (Eds.), Computational, Geometric, and Process Perspectives on Facial Cognition: Context and Challenges. Lawrence Erlbaum Associates Inc., Mahwah, New Jersey.

Valenza, E., Simion, F., Macchi Cassia, V., Umilta, C., 1996. Face preference at birth. Journal of Experimental Psychology 22, 892–903.

Vicari, S., Reilly, J.S., Pasqualetti, P., Vizzotto, A., Caltagirone, C., 2000. Recognition of facial expressions of emotions in school-age children: The intersection of perceptual and semantic categories. Acta Paediatrica 89, 836–845.

Want, S.C., Pascalis, O., Coleman, M., Blades, M., 2003. Recognizing people from the inner and outer parts of their faces: Developmental data concerning unfamiliar faces. British Journal of Developmental Psychology 21, 125–135.

Webb, S.J., Dawson, G., Bernier, R., Panagiotides, H., 2006. ERP evidence of atypical face processing in young children with autism. Journal of Autism and Developmental Disorders 36, 881–890.

Wilkinson, D., Ko, P., Wiriadjaja, A., Kilduff, P., McGlinchey, R., Milberg, W., 2009. Unilateral damage to the right cerebral hemisphere disrupts the apprehension of whole faces and their component parts. Neuropsychologia 47, 1701–1711.

Yin, R.K., 1969. Looking at upside down faces. Journal of Experimental Psychology 81, 141–145.

Yip, A.W., Sinha, P., 2002. Contribution of color to face recognition. Perception 31, 995–1003.

Young, A.W., 1992. Face recognition impairments. Philosophical Transactions of the Royal Society of London. Series B, Biological Sciences 335, 47–53.

Young, A.W., Ellis, H.D., 1989. Childhood prosopagnosia. Brain and Cognition 9, 16–47.

Young, A.W., Hellawell, D., Hay, D.C., 1987. Configurational information in face perception. Perception 16, 747–759.

Young-Browne, G., Rosenfeld, H.M., Degen Horowitz, F., 1977. Infant discrimination of facial expressions. Child Development 48, 555–562.

Developmental Neuroscience of Social Perception

A. Voos, C. Cordeaux, J. Tirrell, K. Pelphrey
Yale University, New Haven, CT, USA

19.1 INTRODUCTION

Within adult social psychology, the study of social cognition encompasses a range of phenomena including moral reasoning, attitude formation, stereotyping, and related topics. This definition includes the processes underlying the perception, memory, and judgment of social stimuli; the effects of social, cultural, or affective factors on the processing of information; and even the behavioral and interpersonal consequences of those cognitive processes. Social psychology has provided an arsenal of measurement techniques and theoretical models to inform efforts within social neuroscience to study the mature human brain. However, neuroscience has traditionally defined social cognition more narrowly as the ability to perceive the intentions, actions, and psychological dispositions of others (Brothers, 1990). Within developmental psychology, the study of social cognition and its development has focused most intensely upon 'theory of mind,' or a person's awareness that other individuals maintain thoughts, beliefs, and desires that are different from his or her own, and that the actions of others can be best explained with reference to their

individual mental states (Frith and Frith, 1999; Premack and Woodruff, 1978). Developmental psychologists have characterized the stages of development in theory of mind abilities but have yet to provide a mechanistic account of 'how' theory of mind develops in children. Such questions necessitate studies that bridge the gap between cognitive accounts of theory of mind and the underlying brain mechanisms and, thus, require measurement at both the behavioral and neural system levels. One definition that cuts across disciplines and levels of analysis dictates that *social cognition* refers to the fundamental abilities to perceive, categorize, remember, analyze, reason with, and behave toward conspecifics (Adolphs, 2001; Pelphrey et al., 2004a). Currently, very little is known about the neural correlates of social cognition in children or about the changes in brain structure, function, and connectivity that underlie normative development in this domain. Providing an understanding of brain development in relation to changes in social cognition is important to the field, as it will allow for the construction of normative 'growth charts' for the function of circuits supporting various aspects of social cognition. Additionally, this knowledge

Neural Circuit Development and Function in the Brain: Comprehensive Developmental Neuroscience, Volume 3 http://dx.doi.org/10.1016/B978-0-12-397267-5.00056-X

© 2013 Elsevier Inc. All rights reserved.

may aid researchers in their search for genetic and environmental factors (and gene × environment interactions) related to suboptimal social cognitive development as observed in autism and related neurodevelopmental disorders.

This chapter focuses on the development of a critical aspect of social cognition – *social perception*. Social perception is the initial stage of evaluating intentions and psychological dispositions of others by analysis of gaze direction, body movement, and other types of biological motion (Allison et al., 2000). It is closely linked to *action understanding*: the ability to appreciate other people's actions in terms of the mental states that drive behavior. It has been argued that social perception is an ontogenetic and phylogenetic prequel to more sophisticated social cognition abilities, particularly theory of mind (e.g., Baron-Cohen et al., 1995; Frith and Frith, 1999; Pelphrey and Morris, 2006; Premack and Woodruff, 1978; Saxe et al., 2004a). A task analysis supports this argument and reveals that successful social perception involves a set of three distinct but interrelated social cognition abilities: (1) individuating and recognizing other people, (2) perceiving their emotional states, and (3) analyzing their intentions and motivations. Successful social perception, in turn, facilitates a fourth and more sophisticated aspect of social cognition: (4) representing another person's perceptions and beliefs, a core component of theory of mind. In this chapter, the five neuroanatomically and functionally dissociable neural circuits that underlie these aspects of social perception are elaborated on and what is known about the functional development of each circuit in infants, children, and adolescents discussed. The focus is on functional magnetic resonance imaging (fMRI) and developmental studies of social perception involving other neuroimaging techniques, for example, electrophysiology are not included. This focus was selected because of the interest in identifying the exact neural systems involved in distinct and dissociable (at the neural systems level) aspects of social perception. It must be noted, however, that fMRI is limited in that it is not yet practical to study awake, behaving infants and toddlers with this method. In contrast, electrophysiology can be readily applied to infants and toddlers. However, in even the best, most controlled conditions with adults, it is a challenge to make precise statements about the neuroanatomical origins of scalp-recorded electrical signals. See, for example, the ongoing debate concerning the origins of the N170 in response to faces (Halgren et al., 2000; Itier and Taylor, 2004). This inferential problem is compounded when one wishes to study the development of a brain system with electrophysiology because then one has to identify a developmental precursor to the well-characterized adult signals and then make inferences about the underlying neural generator. Following the review of the available

literature, a model of social cognitive development has been put forward that aims to explain how an early developing social perception network can give rise to more sophisticated aspects of social cognition, including theory of mind.

19.2 NEUROANATOMICAL SUBSTRATES OF SOCIAL PERCEPTION

19.2.1 Ventral Occipital Temporal Cortex and the Recognition of Other People

Several areas located in the ventral occipitotemporal cortex (VOTC) support the basic representation and recognition of other people. These include the lateral fusiform gyrus (FFG), which contains the 'fusiform face area' (FFA), and the 'extrastriate body area.' The former has a clear role in face perception and recognition (e.g., Kanwisher et al., 1997; Puce et al., 1996), while the latter has been implicated in the visual perception of human bodies (e.g., Downing et al., 2001). These brain regions provide the basic representation of their specific visual category that supports both bottom-up visual processing of social information and top-down imagery of those categories.

In adults, there are well-defined expectations of how brain activity related to visual object processing should be organized (Dehaene and Cohen, 2007; Martin, 2007). Faces and letter strings are two visual categories that have been studied extensively and elicit distinct responses in VOTC. Faces evoke selective activity in the mid-FFG relative to other objects (Allison et al., 1994; Kanwisher, 2000; Kanwisher et al., 1997). In contrast, the lateral mid-fusiform/inferior temporal gyrus shows a bias for processing words, letters, and letter strings over digits and other objects (Allison et al., 1994; Baker et al., 2001; Cohen and Dehaene, 2004; Dehaene et al., 2002; Hashimoto and Sakai, 2004; Polk and Farah, 1998; Polk et al., 2002). Both categories are hypothesized to recruit specialized neural processes that best suit the features that define the category (Kanwisher, 2000; Polk and Farah, 1998). The developmental origins of their specializations, however, are presumably quite different.

To date, the most is known about the development of the face-sensitive area of the FFG. Current evidence indicates that face-selective neural processing is already present in the FFG by 6 years of age and becomes increasingly robust throughout adolescence (Aylward et al., 2005; Golarai et al., 2007; Grill-Spector et al., 2008; Libertus et al., 2009; Scherf et al., 2007; Tzourio-Mazoyer et al., 2002). Interestingly, the gradual refinement of face-related neural processing over development

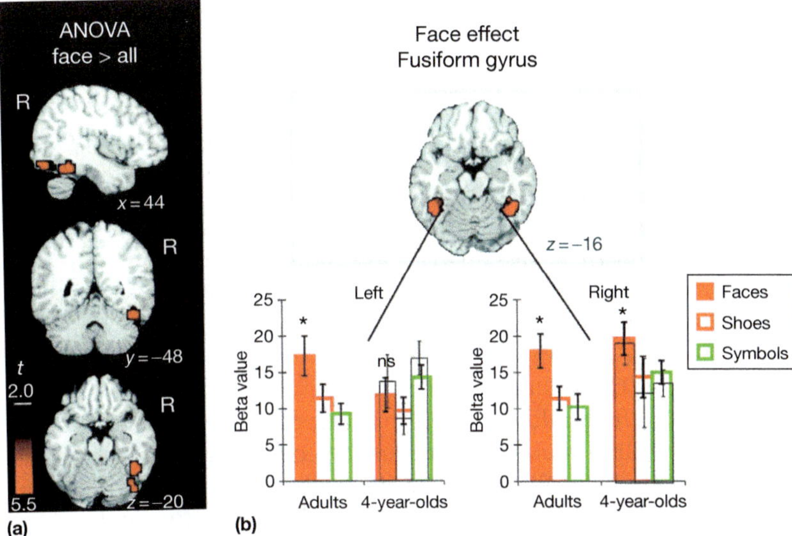

FIGURE 19.1 The right fusiform face effect. The right fusiform gyrus face effect for adults and children was evident in (a) a main effect of faces > all categories (shoes, letters, and numbers) in an ANOVA ($p < .05$, corrected) and (b) an ROI analysis in which voxels were selected based on their average response across all categories (faces, shoes, letters, and numbers) > scrambled. The light gray bars within the data from 4-year-olds represent children with motion estimates equal to those of adults (these data are equivalent to the full 4-year-old sample). The asterisks indicate significantly different from all categories at $p < 0.05$, corrected for multiple comparisons.

parallels the well-known changes in children's face recognition abilities (Carey and Diamond, 1994).

In order to differentiate the developmental origins of category-specific areas in VOTC, our group recently used fMRI to examine the organization of the ventral visual pathway for basic processing of symbols, faces, and nonface objects in 4-year-old children (Cantlon et al., 2011). We tested 4-year-old children's and adults' responses to faces, letters, numbers, and shoes in VOTC. As illustrated in Figure 19.1, children and adults exhibited a common pattern of face selectivity in the right mid-FFG at a locus consistent with previously reported face-selective sites in adults. This indicates that certain features of adult brain organization have already taken formed by 4 years of age (Cantlon et al., 2006; Grill-Spector et al., 2008; Mahon et al., 2009; Polk et al., 2007). This biased activity to faces in the right mid-FFG in overlapping regions previously reported as face selective in older children and adults (Grill-Spector et al., 2008) represents an early developing visual specialization for social information in VOTC.

Some evidence indicates that face selectivity in the FFG is related to face recognition memory in children (Golarai et al., 2007), where children who perform better on face recognition tasks exhibit a greater spatial extent of face-related activity. However, it is unknown whether the refinement of face-related processing hinges on 'increases' in the FF response to faces, 'decreases' in the responses to nonfaces, or 'both.' This question is critical for understanding the nature of the developmental process underlying category selectivity in the brain. In fact, a long-standing debate in the developmental literature concerns the question of whether neural development is driven by building up or pruning back representations in the brain (Bourgeois and Rakic, 2003; Changeux, 1985; Changeux and Danchin, 1976; Changeux and Dehaene,

1989; Dehaene-Lambertz and Dehaene, 1997; Purves et al., 1996; Quartz, 1999; Quartz and Sejnowski, 1997).

Faces have an evolutionary significance and spontaneously attract children's attention from birth, with research indicating that infants who are just days old prefer to look at images of faces instead of nonfaces, and familiar rather than novel faces (Johnson et al., 1991; Nelson, 2001). In the constructivist view, experience-dependent input specifies the connectivity and functions of cortical regions and thereby gradually builds up specialized cortical functions (Quartz, 1999). Selectionism, in contrast, postulates that redundant and irrelevant neuronal connections exist from birth and are gradually eliminated on the basis of experience-evoked activity in order to define specialized cortical functions (Changeux and Danchin, 1976). In principle, these two developmental processes can be distinguished in the category-selective brain responses of young children. For instance, in the case of face representation, a constructivist pattern of activity would predict an increase in face-related activity with increasing face recognition ability, whereas a selectionist pattern would predict a decrease in nonface activity with increasing face recognition ability. Naturally, both patterns are nonexclusive and may jointly occur, either simultaneously or at different ages.

In order to test these two theories, the relationship between children's developing category knowledge and their category-related brain activity was examined in the aforementioned study (Cantlon et al., 2011). The children were tested on a series of identification and naming tests with the same stimuli presented in the fMRI session. As illustrated in Figure 19.2, children's accuracy on the face-matching task was not correlated with an increase in the neural response to faces in the right mid-fusiform face-selective region of interest (ROI), as

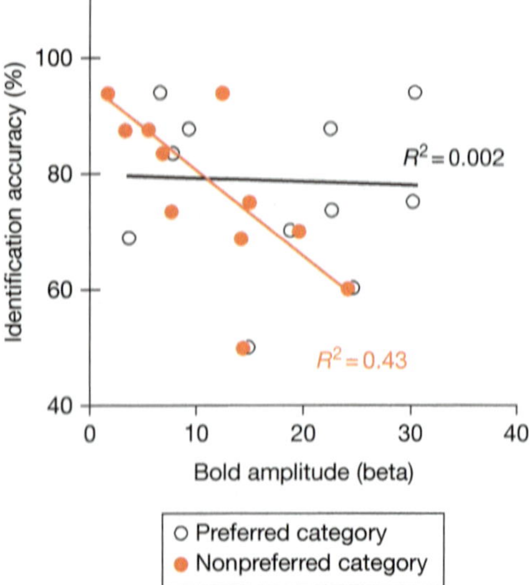

FIGURE 19.2　Scatterplot illustrating negative correlations between behavioral performance and the magnitude of the brain response to the nonpreferred category. In the brain region that showed the face effect, children's accuracy on a face identification task was correlated with a decrease in activity to the nonpreferred category (shoes) rather than an increase in activity to the preferred category (faces). Chance on the face identification task was 50%.

might be expected. Instead, accuracy on the face-matching task correlated with a decrease in the response to shoes in the face-selective ROI. This finding suggests that as children in this age range become more proficient at recognizing faces, the mid-fusiform face-selective region reduces its response to other visual classes (as opposed to increasing its response to faces). These data suggest that brain regions that will eventually become selective for a particular category already produce a relatively strong response to their preferred category in early childhood, but they gradually decrease their responses to nonpreferred categories as knowledge is acquired. While adult category-selective brain regions show similar types of category biases in early childhood, there are undoubtedly developmental changes in the structure and function of those regions. For example, Golarai et al. (2007) reported that the spatial extent of activation in face-selective fusiform regions expands with increased memory for faces between 7 years of age and adulthood. The degree to which face-selective cortex expands over the course of early childhood (i.e., 0–7 years) remains in question. Our own findings suggest that one catalyst of those functional changes could be a decreasing representation of nonpreferred (i.e., nonface) entities in face-selective regions.

Many questions remain about exactly how the development of face specialization unfolds. One aspect of this process, evident in our fMRI results, is that neural responses in category-selective regions to nonpreferred

categories need to be pruned away. This is supported by the notion that learning proceeds by 'selection,' 'attrition,' or a 'use it or lose it' principle, which has long been proposed at the theoretical level (Bourgeois and Rakic, 2003; Changeux, 1985; Changeux and Danchin, 1976; Changeux and Dehaene, 1989; Dehaene-Lambertz and Dehaene, 1997; Purves et al., 1996) and has received empirical support in domains such as bird song acquisition or speech perception, both of which exhibit perceptual narrowing over development (Kuhl et al., 1992; Werker and Lalonde, 1988). Our data suggest a similar selectionist principle in the development of face-selective VOTC in human children.

It should be noted that our data are not inconsistent with a moderate form of the constructivist view (Sirois et al., 2008), which posits a selection process via lateral inhibition following category learning (for evidence of this phenomenon in adults, see Allison et al., 2002; Pelphrey et al., 2003). Under that hypothesis, however, one might expect preferred and nonpreferred category-related responses in VOTC to become increasingly anti-correlated over development. Although this pattern did not emerge in our data, longitudinal fMRI data on children's category-related visual responses at different points in their acquisition of category knowledge will help to further adjudicate among these hypotheses. Such data could reveal developmental periods during which representations are being constructed (with increasing responses to preferred categories) as well as periods during which a selection mechanism is engaged (with decreasing responses to the nonpreferred category). The degree to which selection or construction is observed at a given point in early childhood likely will depend on children's experience with the specific categories examined. At a single point in development, some categories could exhibit a pattern of increasing responses to preferred stimuli, while other categories exhibit decreasing responses to nonpreferred stimuli. Such a proposal is consistent with our data and with previously reported studies of VOTC activation in older children (Golarai et al., 2007; Grill-Spector et al., 2008; Libertus et al., 2009; Scherf et al., 2007).

A tentative biological mechanism for the reduction in high-level visual activity to nonpreferred categories over development may be the known reduction in synaptic density between 2 and 11 years of age (Chugani et al., 1987; Giedd et al., 1999; Huttenlocher and Dabholkar, 1997; Huttenlocher et al., 1982; Shaw et al. 2008). Synaptic density in visual areas steadily increases between birth and 1–2 years of age, reaches levels that are approximately 50–60% greater than adult levels, and then gradually decreases over the next several years. Some evidence indicates that vascular density parallels synaptic density in primary visual areas and thus blood supply might be related, at least in sensory areas, to neural plasticity and synapse formation/elimination (Duvernoy et al., 1981; Logothetis and Wandell, 2004).

19.2.2 Limbic Circuitry and the Perception and Experience of Emotions

The limbic system is composed of a set of brain structures that support a variety of functions, including emotion. The amygdala is one such structure, and several fMRI studies have focused on its response to emotional (especially fearful) faces. Baird et al. (1999) first identified amygdala activation to fearful faces in adolescents aged 12–17 years. Thomas et al. (2001) later reported that adults demonstrated greater amygdala activation to fearful facial expressions compared to other expressions, whereas 11-year-old children showed greater amygdala activation to neutral faces compared to other expressions. One explanation offered by the authors is that the neutral faces were seen as more ambiguous than fearful facial expressions, resulting in increases in amygdala activation. In a cross-sectional study, Killgore et al. (2001) reported sex differences in amygdala responses in children and adolescents. They found that the left amygdala responded to fearful facial expressions in all children, although its activity decreased over the adolescent period in females but not in males. In a follow-up study, Killgore and Yurgelun-Todd (2004) compared children, adolescents, and adults during the perception of fearful faces. They reported that males and females differed in the right hemisphere – left hemisphere asymmetry of activation of the amygdala and prefrontal cortex (PFC) and this interacted with age. For boys, activation within the dorsolateral PFC was bilateral in childhood, right lateralized in adolescence, and bilateral in adulthood, whereas females showed a monotonic relationship with age, such that older females showed more bilateral activation than younger ones, but with significant bilateral activation at all ages. In contrast, amygdala activation was similar for both sexes, with bilateral activation in childhood, right-lateralized activation in adolescence, and bilateral activation in adulthood.

Lobaugh et al. (2006) reported that fear, disgust, and sadness recruit distinct neural systems both in 10-year-old children and adults. Two recent cross-sectional studies reported that adolescents display more amygdala activity in response to affective faces than either children or adults (Guyer et al., 2008; Hare et al., 2008). Further, adolescents also show less response to emotions in ventromedial PFC (vmPFC), a region whose functional connectivity with the amygdala is associated with habituation to emotional stimuli (Etkin et al., 2006; Hare et al., 2008). This suggests that teenagers may be more emotionally reactive and also less capable of relying on PFC for affect regulation (see also Grosbras et al., 2007; Lévesque et al., 2004).

In a recent fMRI study, we examined the development of the neural circuitry supporting emotion regulation in school-age children (Pitskel et al., 2011). We focused on the circuitry supporting cognitive reappraisal, a particular approach to emotion regulation frequently utilized in behavioral psychotherapies. Despite a wealth of research on cognitive reappraisal in adults, little is known about the developmental trajectory of brain mechanisms subserving this form of emotion regulation in children. We asked children and adolescents to increase and decrease their emotional response to disgusting images (e.g., a picture of a person with mucus hanging from her nose) by either pretending it was real and right there in front of them (for the increase condition) or by pretending it was just make believe (for the decrease condition). Distinct patterns of brain activation were identified during successful up- and downregulation of emotion, as well as an inverse correlation between activity in vmPFC and limbic structures, particularly the amygdala, during downregulation, indicative of the regulatory role for vmPFC. Further, as illustrated in Figure 19.3, age-related effects on activity in the vmPFC and amygdala were discovered. Of particular interest, during downregulation of emotion, significant negative correlations with age in the amygdala were observed, consistent with more effective emotion regulation.

While this review highlights a number of findings from cross-sectional studies of age-related changes in brain function related to the perception and regulation of emotion, only one study has reported valuable longitudinal data on the neural circuitry underlying emotion processing. In a groundbreaking study, Pfeifer et al. (2011) reported longitudinal data from 38 neurotypical participants who underwent two fMRI sessions across the transition from late childhood (10 years) to early adolescence (13 years). Strikingly, responses to affective facial displays exhibited a combination of general and emotion-specific changes in ventral striatum (VS), vmPFC, amygdala, and temporal pole. Furthermore, VS activity increases correlated with decreases in susceptibility to peer influence and risky behavior. VS and amygdala responses were also significantly more negatively coupled in early adolescence than in late childhood while processing sad and happy versus neutral faces. Together, these results suggest that VS responses to viewing emotions may play a regulatory role that is critical to adolescent interpersonal functioning.

19.2.3 Lateral Temporal Cortex and the Perception of Biological Motion

Biological motion perception refers to the visual perception of a biological entity engaged in a recognizable activity. This definition includes the observation of

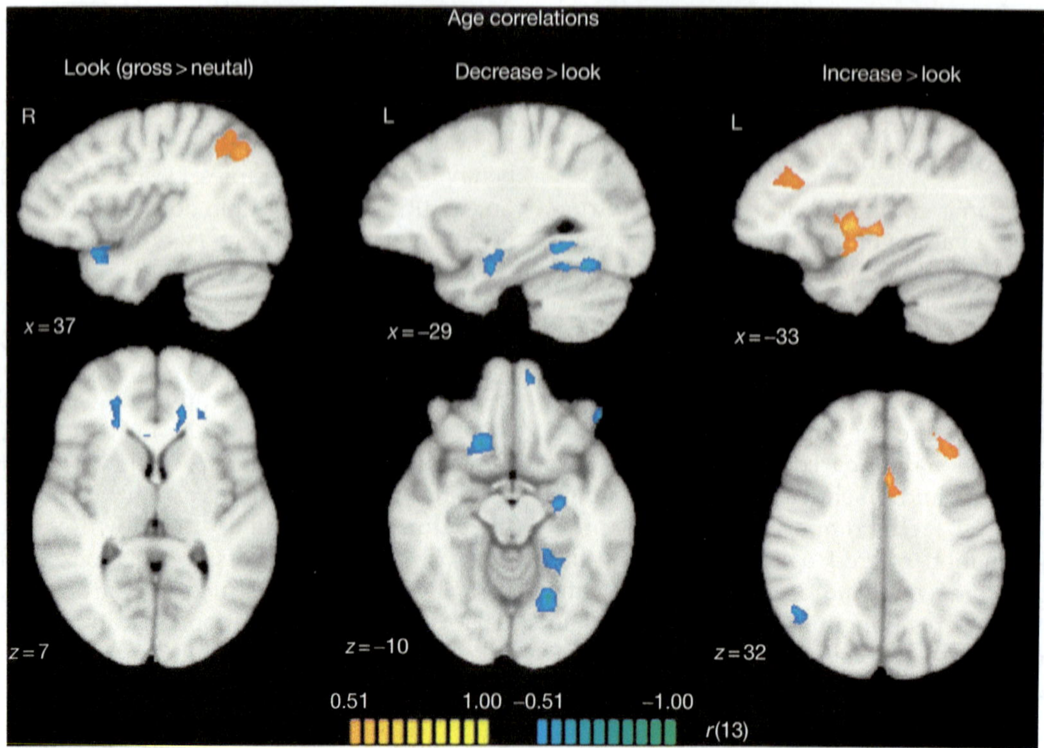

FIGURE 19.3   Activation correlated with age in each of three key contrasts, look-gross > look-neutral (reflecting the perception of disgust), decrease-gross > look-gross (reflecting efforts to downregulate emotion), and increase-gross > look-gross (reflecting efforts to increase emotion responses). Areas in orange are those that were positively correlated with age; areas in blue exhibited negative correlations with age. All activations are shown at threshold of $p < .05$. Images are displayed in radiologic convention.

humans walking and making eye and mouth movements, but the term can also refer to the visual system's ability to recover information about another's motion from sparse input. The latter is well illustrated by the discovery that point-light displays (moving images created by placing lights on the major joints of a walking person and filming them in the dark), while being relatively impoverished stimuli, contain the information necessary to identify the agent of motion and the kind of motion produced by the agent (Johansson, 1973). Biological motion is integral to social perception. The superior temporal sulcus (STS) region, particularly the posterior STS, has been implicated in the perception of biological motion including eye, hand, and whole-body movements (e.g., Bonda et al., 1996; Pelphrey et al., 2003). Specifically, it exhibits differences in activation to actions that are congruent with contextual factors relative to actions that are incongruent with the context. It is hypothesized that the STS plays such a role because intentions are used, chiefly, for predicting what people will do in the future. Since future actions are intrinsically dynamic biological motion, it seems reasonable that the STS, whose job it is to represent such motion, could be used for this role. However, it is currently unknown whether regions that represent intention for anticipation and prediction

are distinct from areas that simply represent the current state of perceived biological motion.

Frith and Frith (1999) first suggested that the ability to distinguish between biological and nonbiological figures and their actions is one of the likely evolutionary and developmental precursors to theory of mind. They also noted that the STS region previously implicated in biological motion processing is adjacent to regions of the brain used for other, higher level aspects of 'mentalizing,' a concept closely related to theory of mind. The ability to infer goal states using the actions of an agent has been called an 'intentionality detector' and proposed as a component of the human mentalizing system (Baron-Cohen et al., 1994). By the age of 4 months, infants can detect biological motion from impoverished stimuli, 'point-light walkers,' and prefer these movies to those of nonbiological motion (Fox and McDaniel, 1982). A study of 8-month-old infants using ERPs suggested that there was activity in the right hemisphere in response to biological motion (Hirai and Hiraki, 2005). Lloyd-Fox et al. (2011) recently identified a more precise neurobiological basis for this very early developing social perception ability using functional near-infrared spectroscopy. In their pioneering study, 5-month-old infants watched videos of adult actors moving their hands, their mouth, or their eyes, all in contrast to nonbiological mechanical movements. They observed that different regions of the frontal and temporal cortex responded to these biological movements and that different patterns of cortical activation emerged according to the type of movement watched. These findings demonstrate that from an early age, our brains selectively respond to biologically relevant movements, and further, selective patterns of regional specification to different cues occur within the lateral temporal cortex, including the STS region.

Only a handful of fMRI studies have examined the brain mechanisms for the perception of biological motion compared to nonbiological motion in children. In one fMRI study (Carter and Pelphrey, 2006), the functional development of the STS region was explored using a paradigm that was previously employed with a sample of adults (Pelphrey et al., 2003). In order to determine whether the STS region responded to biological motion more than to other types of complex, meaningful motion or to random motion, adult participants viewed a virtual scene that included four types of carefully matched, animated stimuli: a walking human (human), a walking robot (robot), a grandfather clock (clock), and a disjointed mechanical figure (mechanical) (see Figure 19.4). While the figures differed markedly in form, their movements were nearly identical. These figures allowed for the exploration of whether a biological motion pattern (walking) would be processed differently if made by a biological (human) versus non-biological (robot) figure and whether there were differences in brain activation patterns for organized versus disorganized mechanical motion.

In adults, there were no significant differences between the human and the robot or between the clock and the disjointed mechanical figure in the posterior STS region. The posterior STS (right hemisphere > left hemisphere) responded more strongly to biological motion (robot and human) than to nonmeaningful but complex nonbiological motion (mechanical) or complex and meaningful nonbiological motion (clock) (Pelphrey et al., 2003). Importantly, not every brain region showed this pattern of effects. We observed a dissociation of function between the STS region and an area posterior and inferior to the posterior STS region corresponding to the motion-sensitive visual area MT or V5 (MT/V5, e.g., McCarthy et al., 1995; Tootell et al., 1995; Zeki et al., 1991). The STS region responded selectively to biological motion, whereas MT/V5 responded equally to all four types of motion.

As illustrated in Figure 19.5, 7- to 10-year-old children exhibited robust biological > nonbiological activity in the middle and posterior portions of the STS region. To examine possible age-related changes in these responses, it was assumed that higher levels of biological > nonbiological activity (i.e., greater biological−nonbiological difference scores) were indicative of a more maturely functioning system. Consequently, a positive correlation between the specificity of the STS region for biological motion and age was anticipated. Consistent with this prediction, the magnitude of the biological−nonbiological difference score was positively correlated with age in the right posterior STS region ($r=0.64$, $p=0.03$, two-tailed), with age accounting for approximately 41% of the variance in the biological−nonbiological difference scores. This finding highlights developmental changes in the functioning of the STS region during biological motion perception across middle childhood.

Much of what develops in regard to social cognitive abilities in childhood likely involves changes in connections among the various brain regions involved in social information processing. Increasing connections allow for increasingly sophisticated social perception and mentalizing abilities. Given the available research on the STS region, future research should aim to explore the development of structural and functional connectivity to and from this area. It could be the case that what appears to be a relative increase in differentiation in the STS region could represent changes in its inputs and outputs and

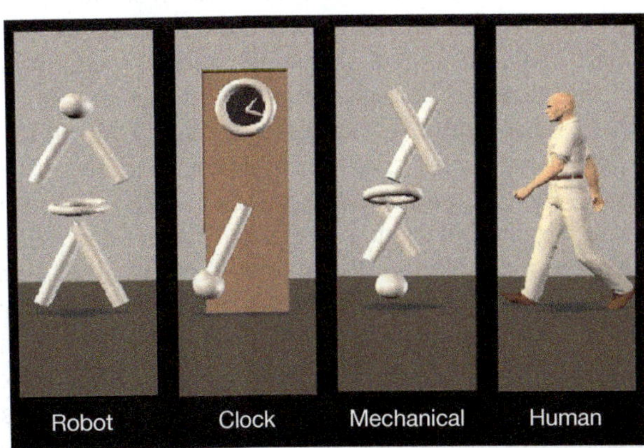

FIGURE 19.4 Biological (robot and human) and nonbiological (clock and mechanical) motion stimuli.

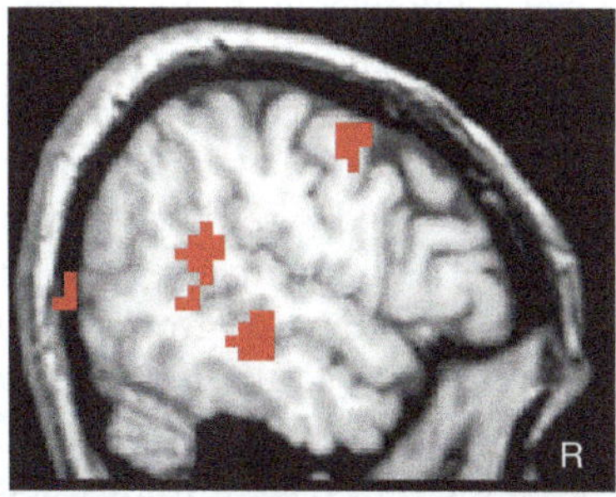

FIGURE 19.5 The red-colored map indicates regions of significant biological > nonbiological motion activity in school-age children.

could thus either be a function of tuning of the STS response *per se*, or a development in the input by top-down regions to the STS. This could be examined using functional connectivity (to explore when regions work together) and diffusion tensor imaging (to determine the structural connections between these regions).

19.2.4 Medial PFC and the Representation of the Self and the Other

In adults, the medial PFC (mPFC) has been implicated in a wide range of social cognitive tasks, including making inferences about other people's intentions and mental states (Castelli et al., 2002; Gregory et al., 2002), the attribution of emotion to the self and others (Ochsner et al., 2004), self-reflection (Heatherton et al., 2006; Kelley et al., 2002; Northoff et al., 2006), and representing semantic knowledge about the psychological aspects of other people (e.g., Mitchell et al., 2005a,b). Additionally, the precuneus and posterior cingulate in the medial posterior parietal cortex (mPPC) are more active during self-knowledge retrieval than during other types of social or semantic tasks (e.g., D'Argembeau et al., 2010; Kelley et al., 2002). The capacity to perceive and reflect upon oneself as an individuated, unified, and stable entity with lasting qualities and traits (e.g., shy, intelligent, and kind) is a critically important and perhaps unique aspect of the human experience.

Many theoretical accounts of social cognition make self-referential processing a central component of processing others (Mitchell et al., 2005a,b; Spengler et al., 2009; Uddin et al., 2007), positing that one strategy for predicting the mental states of others is by referencing one's own thoughts, feelings, or behaviors in a similar situation. Recent studies have shown developmental changes from childhood to adulthood in activity in the frontoparietal network during self-perception tasks. To illustrate, Pfeifer et al. (2011) studied a sample of 12 children and 12 adults (average age = 10.2 and 26.1 years, respectively) using fMRI. The participants reported whether short phrases described themselves or a highly familiar other (Harry Potter). In both children and adults, the mPFC was relatively more active during self – than social – knowledge retrieval, and the mPPC was relatively more active during social – than self-knowledge retrieval. Direct comparisons between children and adults indicated that children activated the mPFC during self-knowledge retrieval to a much greater extent than adults. The particular regions of the mPPC involved varied between the two groups, with the posterior precuneus engaged by adults, but the anterior precuneus and posterior cingulate engaged by children. Only children activated the mPFC significantly above baseline during self-knowledge retrieval. This last finding suggests that the presence of a cortical, prefrontal-mediated,

neural system supporting self-referential processing in children, is perhaps no longer needed in adults. Longitudinal data will be needed to more fully understand this pattern.

19.2.5 Posterior Parietal Cortex and Thinking About the Thoughts of Others

Thinking about the thoughts of others, or 'theory of mind,' has been studied most intensely using false belief tasks. In the typical design, a child watches while a puppet places an object in location A. The puppet leaves the scene and the object is transferred to location B. The puppet returns and the child is asked to predict where the puppet will look for the object. Three-year-olds think the puppet will look in location B, where the object actually is; older children think the puppet will look in location A, where the puppet last saw the object (Wellman et al., 2001). The 3-year-olds who fail the false belief task are not performing at chance, nor are they confused by the questions, but instead make systematically below-chance predictions with high confidence (Ruffman et al., 2001). The standard interpretation of these results is that 3-year-olds lack a representational theory of mind. That is, 3-year-olds fail to understand how the contents of thoughts can differ from reality (Gopnik and Astington, 1988; Wellman et al., 2001).

Although the false belief task has been used in literally hundreds of studies, it remains controversial whether success on this task depends on the deployment of a 'special' domain-specific mechanism for reasoning about other minds. As many researchers have noted (Bloom and German, 2000; Leslie, 2000; Roth and Leslie, 1998), children might pass or fail the false belief task for reasons having nothing to do with deficits in understanding other minds. In particular, the false belief task requires a high level of executive control – that is, the ability to plan and carry out a sequence of thoughts or actions, while inhibiting distracting alternatives. Thus, researchers have suggested that the false belief task underestimates children's ability to think about mental states (Bloom and German, 2000). Alleged shifts in children's theory of mind might reflect changes only in children's executive function – especially the abilities to select from among competing responses and to inhibit the tendency to respond based on reality (Carlson et al., 2004). Recently, an even bigger obstacle has arisen for the standard view: multiple reports that infants can make correct predictions on false belief tasks, when measured by violation-of-expectation looking time measured at 12–15 months (Onishi and Baillargeon, 2005; Surian et al., 2007) or predictive looking at 24 months (Southgate et al., 2007). These results have been taken as evidence for very early emerging, or even innate, cognitive mechanisms for theory of mind (Leslie, 2005).

Cognitive neuroscience provides a complementary route to address the same theoretical concerns. Are there cognitive and neural mechanisms selectively implicated in theory of mind, independent of executive demands? If so, do these brain regions' response profiles mature around age 4, the age when children reliably pass explicit false belief tasks, or early in childhood or even infancy?

Recent neuroimaging of adult brains has revealed a small but remarkably consistent set of cortical regions associated with thinking about other people's thoughts, or 'theory of mind' (Frith and Frith, 2003; Gallagher et al., 2000; Saxe and Kanwisher, 2003): bilateral temporoparietal junction (TPJ), mPFC, and posterior cingulate. The mPFC is recruited when processing many kinds of information about people (Amodio and Frith, 2006; Mitchell et al., 2005a; Ochsner et al., 2005), whereas the right TPJ is recruited selectively for thinking about thoughts (Saxe and Kanwisher, 2003; Saxe and Powell, 2006). Many functional neuroimaging studies have borrowed paradigms from the rich, older tradition of studying theory of mind in children, though few have directly investigated the development of these neural mechanisms in childhood (Kobayashi et al., 2006, 2007). In two initial studies, Kobayashi et al. (2006, 2007) reported that unlike adults, 9-year-old children did not show activation in the right or left TPJ during belief-reasoning tasks. These results suggested the tantalizing possibility of surprisingly late developmental changes in the neural mechanisms for theory of mind. However, many important questions remain opened. For example, if the TPJ, bilaterally, is not recruited for theory of mind in 9-year-olds, when do these regions develop adult-like selectivity? Are these brain regions involved in some other social cognitive function in the younger children? Or are these brain regions perhaps involved in domain-general functions in younger children and acquire a social role only later in life?

Another important question concerned the developmental relation between theory of mind and the perception of human body actions. Basic perception and understanding of human action are very early emerging, with preverbal infants demonstrating their ability to attend to human action and interpret human body movements in terms of pursuit of goals (Gergely et al., 1995; Johnson, 2003; Meltzoff and Brooks, 2001; Woodward, 1998; Woodward et al., 2001). A longitudinal study found that infants' early action understanding predicts their later success at age 4 years on explicit false belief tasks, suggesting that early perceiving and later reasoning about other people may rely on common cognitive mechanisms (Wellman et al., 2004). But how are the neural mechanisms for action perception and theory of mind related?

The neuroimaging findings reviewed above indicate that action perception (including watching hand, body, and head movements) recruits a region near the right TPJ, in the right posterior STS (Allison et al., 2000; Pelphrey et al., 2003, 2005; Puce et al., 1998). Critically, the posterior STS response depends not only on the pattern of biological motion but on its relation to the environmental context, suggesting that these regions are involved in interpreting human behavior in terms of intentions and goals (Brass et al., 2007; Pelphrey et al., 2004b,c; Saxe et al., 2004b). Early reviews of the adult neuroimaging literature proposed the existence of a single neural substrate (sometimes called posterior STS/TPJ) 'for detection of the behavior of agents and analysis of the goals and outcomes of this behavior' (Frith and Frith, 1999). However, more recent research has revealed that, at least in adults, these two regions are functionally distinct (Gobbini et al., 2007). The posterior STS shows a high response during action observation, and the TPJ shows a high response during reasoning about beliefs, but not vice versa. The possibility remained that theory of mind and action perception initially depend on a single region in posterior STS and TPJ, which then differentiates into two separate regions with distinct functions, later in development. In order to test this hypothesis, Saxe et al. (2009) compared the patterns of brain activation associated with perceiving biological motion (see Figure 19.4) and thinking about thoughts, in the same children.

Children's brains differentiated, within ongoing stories, sections that described the characters' thoughts from sections describing the physical context. Regions in precuneus and bilateral TPJ showed significantly higher responses during the mental rather than physical sections, as did mPFC, but at a lower threshold. To investigate selectivity for thinking about thoughts, the responses to people versus physical and mental versus people subsections in each ROI were compared. All the regions showed a significantly higher response for people rather than physical sections as well as a higher response for mental rather than people sections, except the mPFC, in which this latter difference did not reach any significance. That is, we did not find evidence that the mPFC reliably differentiated information about characters' thoughts from any other facts about people, although the average response in the mPFC was also not significantly different from that observed in the other regions in a direct comparison. These results suggest that in children, unlike previous results in adults, the right TPJ is not significantly more selective for mental state facts, relative to other social facts, than the mPFC. In order to determine whether this difference reflected a developmental change, changes in response patterns with age were examined. A selectivity index was calculated for each brain region, used to measure the difference (in units of percent signal change from rest) between the mental and people sections, relative to the difference between the mental and physical sections, for each individual: $100 \times (\text{mental} - \text{people})/(\text{mental} - \text{physical})$.

Only one brain region showed a significant correlation between age and the selectivity index: the right TPJ. Critically, the brain regions implicated in theory of mind did not overlap with those recruited during perception of biological motion.

This finding of age-related change in brain activation raises as many questions as it answers. There is a broad consensus among developmental psychologists that theory of mind is largely mature well before age 6 years (Onishi and Baillargeon, 2005; Southgate et al., 2007; Wellman et al., 2001). The above neuroimaging results, in contrast, suggest that a key component of the neural organization underlying theory of mind is still changing 3 years later, around age 9 years. So what are the cognitive (and behavioral) correlates of the increased neural specialization? The observed changes are hypothesized to reflect changes in neural organization, specifically in the selectivity of the right TPJ neural response, and are consistent with the pattern of anatomical development of human cortex as revealed in longitudinal MRI studies (e.g., Gogtay et al., 2004). Additionally, a longitudinal study found that gray matter does not reach mature density in 'higher order association areas,' including regions near the TPJ, until early adolescence (Gogtay et al., 2004). The aforementioned findings suggest that the strong selectivity observed in adult brain regions for social perception is not innate but emerges gradually over many years in childhood. In particular, regions showing the most selective response profiles in adulthood show late developmental change (the right TPJ in this study), whereas regions with more general response profiles in adulthood show less developmental change (the mPFC in this study). For both perceiving and reasoning about other people, these results suggest that the basic cognitive signatures of domain specificity may be in place long before the brain systems underlying these processes have reached an adult-like state. The implications of this conclusion are as yet unclear. But one conclusion does seem clear: the finding of late-emerging cortical selectivity undermines the interpretation of category-selective brain regions in adults as evidence for innate and early-maturing domain-specific cognitive or perceptual modules. In particular, our results in the right TPJ are challenging for theories of cognitive development that emphasize an innate and early-maturing domain-specific module for theory of mind.

19.3 A DEVELOPMENTAL MODEL OF SOCIAL PERCEPTION

To date, the field of developmental social neuroscience has begun to dissect some of the neurobiological mechanisms underlying key aspects of social perception and social cognition in infants, children, and adolescents.

This initial work and associated methodological developments have provided the exciting opportunity to address a number of fascinating theoretical questions including: (1) Do later-developing social cognition abilities colonize the systems engaged by earlier-developing abilities? Or, as development proceeds, are new regions recruited into a more basic underlying network? (2) How do the components of the social brain come to interact with each other and with other brain circuits involved in such activities as executive function and language? (3) What are the constraints for specific regions to take on aspects of social perception and social cognition (e.g., proximity and connectivity to motion and speech decoding)?

As our review has illustrated, there is a diverse set of brain regions that contribute to the complex set of social information processing tasks that we call 'social perception.' Given principles of division of labor in the brain, we would expect that each of these regions would show functional specialization. However, social perception emerges from the joint activity and connectivity of each brain region as it contributes to information processing. Early in development, the main source of social input comes from looking at faces and actions of others. Panel (a) of Figure 19.6 depicts this early social network. External input is processed for its basic social components – emotionality in the limbic areas, static faces and body information in the VOTC, and dynamic biological motion in the STS. This information can be used to anticipate other actions and guide social exploration, but only if it is integrated across these individual domains and sensory modalities. We know that in school-age children and adults, the integration of information for the understanding of action takes place in the STS, but we do not know when this begins to happen (Mosconi et al., 2005).

This early circuit relies on sensory input available to even young infants and probably does not require sophistication on the part of the child in terms of their ability to represent complex environmental knowledge or knowledge about others' mental states (Senju et al., 2006). However, it is the very simplicity that provides the child with their first abilities to perceive and anticipate the actions of others and to reap the positive rewards of being able to do so. With this early success, the child (and their social perception system) would be motivated to dedicate more resources to the problem of predicting and understanding others.

Indeed, this simple circuit, while probably remaining critically important throughout life, can only make very temporally limited predictions. For example, while smiling and reaching may provide information enough to anticipate actions over the next few seconds, it probably cannot be used to predict what a person will be doing an hour from now. To predict this longer scale of actions,

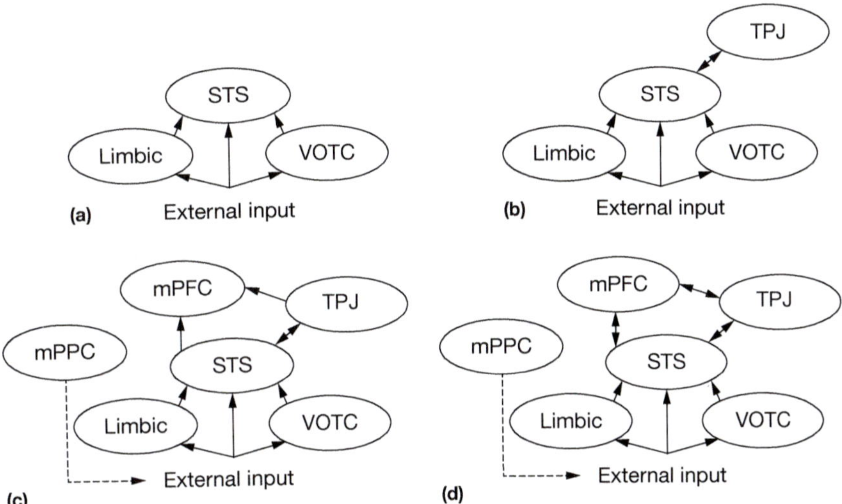

FIGURE 19.6 A developmental model of social perception development.

more sophisticated and enduring constructs about others need to be built. At this point, the TPJ is recruited (panel (b)) to allow the attribution of more abstract mental states, most likely by communicating with other areas representing language and semantic knowledge (not shown). It also must communicate with the STS, with initial projections from the STS to the TPJ needed to keep the latter informed about the current state of action, as it is perceived. Reciprocal connections from the TPJ back to the STS would allow the system to take advantage of the knowledge about how intentions (and more sophisticated representations) can unfold into action.

Further development of the system (panel (c)) would bring online prefrontal areas. It is well known that prefrontal regions and connections to the prefrontal regions from the posterior portion of the brain continue to develop well into adulthood. Thus, in early stages, the prefrontal contribution to social perception is probably limited and impoverished. However, as these regions develop, children gain the ability to reason and simulate social interactions, albeit with limited capacity. Additionally, mPPC regions such as the precuneus may begin to allow the child to switch between internal and external foci, instead of being driven completely by external stimuli. Behaviorally, this likely manifests in and is strengthened by pretense (the capability to make believe or to act 'as if'), which could be thought of as a precursor to adult-like abilities to deliberately ruminate and ponder. Indeed, pretense may be critical for adult-level performance for two reasons. First, pretense allows the child to generate scenarios, including social situations, which can be reprocessed for additional learning. Second, by reducing the dominance of external factors, the frontal-executive control of posterior areas, which may be relatively weak, can be developed.

In the final stage (panel (d)), the child develops adult-like control over posterior areas representing social information. These regions can still provide bottom-up processing of social information but can now also be activated by top-down mechanisms. One consequence of this is that social events that are separate in time can be brought together for the purposes of deliberate social reasoning. For example, we might be able to recollect an expression we perceived our friend make earlier in the day (via top-down reactivation of limbic and FFA regions) and couple that with body language we are perceiving now (via the STS) and reach a conclusion about her mental state.

19.4 FUTURE DIRECTIONS

The work reviewed in this chapter has set the stage for critical research to inform our neural systems level understanding of autism spectrum disorder (ASD). ASD is a common, early-onset neurodevelopmental disorder characterized by difficulties in social interaction and communication and repetitive or restricted interests and behaviors. ASD displays great phenotypic heterogeneity and etiological diversity, but social dysfunction is its hallmark and unifying feature. This social dysfunction is revealed by abnormalities in both simple behaviors, such as sharing gaze, and more complex social behaviors, such as triadic attention sharing. Anomalies of social perception, unlike communication problems or repetitive behaviors that are present in numerous disorders (such as anxiety or expressive language impairment), are unique to ASD and are documented across sensory modalities. Autism is a developmental disorder; early deficits derail subsequent experiences, thereby canalizing development toward more severe dysfunction and creating sequelae in additional domains of function. Consequently, the lack of reliable predictors of the condition during the first year of life has been a major impediment to the effective treatment of ASD. Without early predictors and in the absence of a firm diagnosis until behavioral symptoms emerge, treatment is often delayed for 2 or more years.

A significant body of research has already informed our understanding of the brain basis of ASD via research on the development of systems for social perception in children and adults with and without autism (for a recent review of this research, please see McPartland et al., 2011). There is, however, a noticeable lack of information about the very early development (e.g., the first 2–3 years) of brain systems for social perception in infants and toddlers with or without ASD. This is because of the enormous challenge involved in successfully conducting this research. However, a recent study by Elsabbagh et al. (2012) illustrates the great potential for neuroimaging to contribute to our understanding and early diagnosis of ASD. They tested the hypothesis that neural sensitivity to eye gaze in early infancy would predict development of ASD in toddlerhood. The study involved a prospective longitudinal sample of infants at high familial risk for ASD and a comparison group of infants at low risk. The researchers recorded electrophysiological brain responses (event-related potentials; ERPs) while 6–10-month-old infants viewed faces with dynamic eye gaze directed either toward them or away from them. Approximately 18–30 months later, these children were clinically evaluated for the presence of an ASD. Strikingly, neural responses to dynamic eye gaze shifts during the first year predicted clinical outcomes at 36 months, despite similar patterns of gaze as measured by eye tracking. The authors conclude that ERP responses to eye gaze in the first year of life reflect developmental processes leading to the later emergence of ASD.

As the field strives for earlier methods of detecting autistic development, these remarkable findings offer hope for future clinical practice, suggesting the possibility of noninvasive, brain-based screening methods that could detect differences prior to behavioral emergence. Of course, prior to realization of such clinical benefits, it will be critical to investigate the specificity of this biomarker to autism, its presence in an unselected, population-based sample, and, most importantly, its viability in individual patient data. Given historical difficulty parsing heterogeneity in ASD, these findings suggest the potential power of systems neuroscience approaches to identify meaningful subtypes of ASD to inform treatment and predicting outcome. For the future, a strategy of deep behavior and brain phenotyping over longitudinal development is envisioned to offer a detailed profile of brain–behavior performance for a given individual for the purpose of detection of atypical development, subcategorization (e.g., for genetic analysis), treatment selection, and prediction of treatment response (see Rubenstein and Rakic, 2013).

References

Adolphs, R., 2001. The neurobiology of social cognition. Current Opinion in Neurobiology 11 (2), 231.

Allison, T., McCarthy, G., Nobre, A., Puce, A., Belger, A., 1994. Human extrastriate visual cortex and the perception of faces, words, numbers, and colors. Cerebral Cortex 4 (5), 544.

Allison, T., Puce, A., McCarthy, G., 2000. Social perception from visual cues: Role of the STS region. Trends in Neurosciences 4 (7), 267.

Allison, T., Puce, A., McCarthy, G., 2002. Category-sensitive excitatory and inhibitory processes in human extrastriate cortex. Journal of Neurophysiology 88 (5), 2864.

Amodio, D.M., Frith, C.D., 2006. Meeting of minds: The medial frontal cortex and social cognition. Nature Reviews Neuroscience 7 (4), 268.

Aylward, E.H., Park, J.E., Field, K.M., et al., 2005. Brain activation during face perception: Evidence of a developmental change. Journal of Cognitive Neuroscience 17 (2), 308.

Baird, A.A., Gruber, S.A., Fein, D.A., et al., 1999. Functional magnetic resonance imaging of facial affect recognition in children and adolescents. Journal of the American Academy of Child and Adolescent Psychiatry 38 (2), 195.

Baker, J.T., Sanders, A.L., Maccotta, L., Buckner, R.L., 2001. Neural correlates of verbal memory encoding during semantic and structural processing tasks. NeuroReport 12 (6), 1251.

Baron-Cohen, S., Ring, H., Moriarty, J., Schmitz, B., Costa, D., Ell, P., 1994. Recognition of mental state terms clinical findings in children with autism and a functional neuroimaging study of normal adults. British Journal of Psychiatry 165 (5), 640.

Bloom, P., German, T., 2000. Two reasons to abandon the false belief task as a test of theory of mind. Cognition 77 (1), B25.

Bonda, E., Petrides, M., Ostry, D., Evans, A., 1996. Specific involvement of human parietal systems and the amygdala in the perception of biological motion. Journal of Neuroscience 16 (11), 3737.

Bourgeois, J.P., Rakic, P., 2003. Changes of synaptic density in the primary visual cortex of the macaque monkey from fetal to adult stage. Journal of Neuroscience 13 (7), 2801–2820.

Brass, M., Schmitt, R.M., Spengler, S., Gergely, G., 2007. Investigating action understanding: Inferential processes versus action simulation. Current Biology 17 (24), 2117.

Brothers, L., 1990. The neural basis of primate social communication. Motivation and Emotion 14 (2), 81.

Cantlon, J.F., Brannon, E.M., Carter, E.J., Pelphrey, K.A., 2006. Functional imaging of numerical processing in adults and 4-y-old children. PLoS Biology 4 (5), e125.

Cantlon, J.F., Pinel, P., Dehaene, S., Pelphrey, K.A., 2011. Cortical representations of symbols, objects, and faces are pruned back during early childhood. Cerebral Cortex 21 (1), 191.

Carey, S., Diamond, R., 1994. Are faces perceived as configurations more by adults than by children? Visual Cognition 1, 253–274.

Carlson, S.M., Moses, L.J., Claxton, L.J., 2004. Individual differences in executive functioning and theory of mind: An investigation of inhibitory control and planning ability. Journal of Experimental Child Psychology 87 (4), 299.

Carter, E.J., Pelphrey, K.A., 2006. School-aged children exhibit domain-specific responses to biological motion. Social Neuroscience 1 (3–4), 396.

Castelli, F., Frith, C., Happé, F., Frith, U., 2002. Autism, Asperger syndrome and brain mechanisms for the attribution of mental states to animated shapes. Brain 125 (8), 1839.

Changeux, J.P., 1985. Neuronal Man. Pantheon Books, New York, NY.

Changeux, J.P., Danchin, A., 1976. Selective stabilisation of developing synapses as a mechanism for the specification of neuronal networks. Nature 264 (5588), 705–712.

Changeux, J.P., Dehaene, S., 1989. Neuronal models of cognitive functions. Cognition 33 (1–2), 63–109.

Chugani, H.T., Phelps, M.E., Mazziotta, J.C., 1987. Positron emission tomography study of human brain functional development. Annals of Neurology 22 (4), 487.

Cohen, L., Dehaene, S., 2004. Specialization within the ventral stream: The case for the visual word form area. NeuroImage 22 (1), 466.

D'Argembeau, A., Stawarczyk, D., Majerus, S., Collette, F., Van der Linden, M., Salmon, E., 2010. Modulation of medial prefrontal and inferior parietal cortices when thinking about past, present, and future selves. Social Neuroscience 5, 187–200.

Dehaene, S., Cohen, L., 2007. Cultural recycling of cortical maps. Neuron 56, 384–398.

Dehaene, S., Le Clec, H.G., Poline, J.B., Le Bihan, D., Cohen, L., 2002. The visual word form area: A prelexical representation of visual words in the fusiform gyrus. NeuroReport 13, 321–325.

Dehaene-Lambertz, G., Dehaene, S., 1997. In defense of learning by selection: Neurobiological and behavioral evidence revisited. Behavioral Brain Science 20, 560–561.

Downing, P., Jiang, Y., Shuman, M., Kanwisher, N., 2001. A cortical area selective for visual processing of the human body. Science 293 (5539), 2470.

Duvernoy, H.M., Delon, S., Vannson, J.L., 1981. Cortical blood vessels of the human brain. Brain Research Bulletin 7 (5), 519–579.

Elsabbagh, M., Mercure, E., Hudry, K., et al., 2012. Infant neural sensitivity to eye gaze is associated with later emerging autism. Current Biology 22 (4), 338–342.

Etkin, A., Egner, T., Peraza, D.M., Kandel, E.R., Hirsch, J., 2006. Resolving emotional conflict: A role for the rostral anterior cingulate cortex in modulating activity in the amygdala. Neuron 51, 871–882.

Fox, R., McDaniel, C., 1982. The perception of biological motion by human infants. Science 218, 486–487.

Frith, C.D., Frith, U., 1999. Interacting minds – A biological basis. Science 286, 1692–1695.

Frith, U., Frith, C.D., 2003. Development and neurophysiology of mentalizing. Philosophical Transactions of the Royal Society B 358, 459–473.

Gallagher, H.L., Happé, F., Brunswick, N., Fletcher, P.C., Frith, U., Frith, C.D., 2000. Reading the mind in cartoons and stories: An fMRI study of 'theory of mind' in verbal and nonverbal tasks. Neuropsychologia 38, 11–21.

Gergely, G., Nadasdy, Z., Csibra, G., Biro, S., 1995. Taking the intentional stance at 12 months of age. Cognition 56, 165–193.

Giedd, J.N., Blumenthal, J., Jeffries, N.O., et al., 1999. Brain development during childhood and adolescence: A longitudinal MRI study. Nature Neuroscience 2 (10), 861–863.

Gobbini, M.I., Koralek, A.C., Bryan, R.E., Montgomery, K.J., Haxby, J.V., 2007. Two takes on the social brain: A comparison of theory of mind tasks. Journal of Cognitive Neuroscience 19, 1803–1814.

Gogtay, N., Giedd, J.N., Lusk, L., et al., 2004. Dynamic mapping of human cortical development during childhood through early adulthood. Proceedings of the National Academy of Sciences 101, 8174–8179.

Golarai, G., Ghahremani, D.G., Whitfield-Gabrieli, S., et al., 2007. Differential development of high-level visual cortex correlates with category-specific recognition memory. Nature Neuroscience 10 (4), 512.

Gopnik, A., Astington, J.W., 1988. Children's understanding of representational change and its relation to the understanding of false belief and the appearance-reality distinction. Child Development 59, 26–37.

Gregory, C., Lough, S., Stone, V., et al., 2002. Theory of mind in patients with frontal variant frontotemporal dementia and Alzheimer's disease: Theoretical and practical implications. Brain 125, 752–764.

Grill-Spector, K., Golarai, G., Gabrieli, J., 2008. Developmental neuroimaging of the human ventral visual cortex. Trends in Cognitive Science 12, 152–162.

Grosbras, M.H., Jansen, M., Leonard, G., et al., 2007. Neural mechanisms of resistance to peer influence in early adolescence. Journal of Neuroscience 27 (30), 8040.

Guyer, A.E., Monk, C.S., McClure-Tone, E.B., et al., 2008. A developmental examination of amygdala response to facial expressions. Journal of Cognitive Neuroscience 20 (9), 1565.

Halgren, E., Raij, T., Marinkovic, K., Jousmaki, V., Hari, R., 2000. Cognitive response profile of the human fusiform face area as determined by MEG. Cerebral Cortex 10, 69–81.

Hare, T.A., Tottenham, N., Galvan, A., Voss, H.U., Glover, G.H., Casey, B.J., 2008. Biological substrates of emotional reactivity and regulation in adolescence during an emotional go-nogo task. Biological Psychiatry 63, 927–934.

Hashimoto, R., Sakai, K.L., 2004. Learning letters in adulthood: Direct visualization of cortical plasticity for forming a new link between orthography and phonology. Neuron 42 (2), 311.

Heatherton, T.F., Wyland, C.L., Macrae, C.N., Demos, K.E., Denny, B.T., Kelley, W.M., 2006. Medial prefrontal activity differentiates self from close others. Social Cognitive and Affective Neuroscience 1 (1), 18–25.

Hirai, M., Hiraki, K., 2005. An event-related potentials study of biological motion perception in human infants. Cognitive Brain Research 22 (2), 301.

Huttenlocher, P.R., Dabholkar, A.S., 1997. Regional differences in synaptogenesis in human cerebral cortex. Journal of Comparative Neurology 387 (2), 167–178.

Huttenlocher, P.R., de Courten, C., Garey, L.J., Van der Loos, H., 1982. Synaptogenesis in human visual cortex-evidence for synapse elimination during normal development. Neuroscience Letters 3 (3), 247–252.

Itier, R.J., Taylor, M.J., 2004. Source analysis of the N170 to faces and objects. NeuroReport 15 (8), 1261–1265.

Johansson, G., 1973. Visual perception of biological motion and a model for its analysis. Perception and Psychophysics 14, 201–211.

Johnson, S.C., 2003. Detecting agents. Philosophical Transactions of the Royal Society of London, Series B 358, 549–559.

Johnson, M.H., Dziurawiec, S., Ellis, H.D., Morton, J., 1991. Newborns' preferential tracking of face-like stimuli and its subsequent decline. Cognition 40, 1–19.

Kanwisher, N., 2000. Domain specificity in face perception. Nature Neuroscience 3, 759–763.

Kanwisher, N., McDermott, J., Chun, M.M., 1997. The fusiform face area: A module in human extrastriate cortex specialized for face perception. Journal of Neuroscience 17 (11), 4302–4311.

Kelley, W., Macrae, C., Wyland, C., Caglar, S., Inati, S., Heatherton, T., 2002. Finding the self? An event-related fMRI study. Journal of Cognitive Neuroscience 14, 785–794.

Killgore, W.D., Oki, M., Yurgelun-Todd, D.A., 2001. Sex-specific developmental changes in amygdala responses to affective faces. NeuroReport 12, 427–433.

Killgore, W.D., Yurgelun-Todd, D.A., 2004. Sex-related developmental differences in the lateralized activation of the prefrontal cortex and amygdala during perception of facial affect. Perceptual and Motor Skills 99 (2), 371–391.

Kobayashi, C., Glover, G., Temple, E., 2006. Cultural and linguistic influence on neural bases of 'theory of mind': An fMRI study with Japanese bilinguals. Brain and Language 98, 210–220.

Kobayashi, C., Glover, G.H., Temple, E., 2007. Cultural and linguistic effects on neural bases of 'theory of mind' in American and Japanese children. Brain Research 11 (64), 95–107.

Kuhl, P.K., Williams, K.A., Lacerda, F., Stevens, K.N., Lindblom, B., 1992. Linguistic experience alters phonetic perception in infants by 6 months of age. Science 255 (5044), 606.

Leslie, A.M., 2000. How to acquire a 'representational theory of mind'. In: Sperber, D. (Ed.), Metarepresentations: A Multidisciplinary Perspective. Oxford University Press, New York, NY, pp. 197–223.

Leslie, A.M., 2005. Developmental parallels in understanding minds and bodies. Trends in Cognitive Sciences 9, 459–462.

Lévesque, J., Joanette, Y., Mensour, B., et al., 2004. Neural basis of emotional self-regulation in childhood. Neuroscience 129, 361–369.

Libertus, M.E., Brannon, E.M., Pelphrey, K.A., 2009. Developmental changes in category-specific brain responses to numbers and letters in a working memory task. NeuroImage 44 (4), 1404–1414.

Lloyd-Fox, S., Blasi, A., Everdell, N., Elwell, C.E., Johnson, M.H., 2011. Selective cortical mapping of biological motion processing in young infants. Journal of Cognitive Neuroscience 23 (9), 2521–2532.

Lobaugh, N.J., Gibson, E., Taylor, M.J., 2006. Children recruit distinct neural systems for implicit emotional face processing. NeuroReport 17 (2), 215–219.

Logothetis, N.K., Wandell, B.A., 2004. Interpreting the BOLD signal. Annual Review of Physiology 66, 735–769.

Mahon, B.Z., Anzellotti, S., Schwarzbach, J., Zampini, M., Caramazza, A., 2009. Category-specific organization in the human brain does not require visual experience. Neuron 63 (3), 397–405.

Martin, A., 2007. The representation of object concepts in the brain. Annual Review of Psychology 58 (1), 25–45.

McCarthy, G., Spicer, M., Adrignolo, A., Luby, M., Gore, J., Allison, T., 1995. Brain activation associated with visual motion studied by functional magnetic resonance imaging in humans. Human Brain Mapping 2, 234–243.

McPartland, J.C., Coffman, M., Pelphrey, K.A., 2011. Recent advances in understanding the neural bases of autism spectrum disorder. Current Opinion in Pediatrics 23 (6), 628–632.

Meltzoff, A.N., Brooks, R., 2001. 'Like me' as a building block for understanding other minds: Bodily acts, attention, and intention. In: Malle, B.F., Moses, L.J., Baldwin, D.A. (Eds.), Intentions and Intentionality: Foundations of Social Cognition. MIT Press, Cambridge, MA, pp. 171–191.

Mitchell, J.P., Banaji, M.R., Macrae, C.N., 2005a. The link between social cognition and self-referential thought in the medial prefrontal cortex. Journal of Cognitive Neuroscience 17 (8), 1306.

Mitchell, J.P., Banaji, M.R., Macrae, C.N., 2005b. General and specific contributions of the medial prefrontal cortex to knowledge about mental states. NeuroImage 28 (4), 757–762.

Mosconi, M.W., Mack, P.B., McCarthy, G., Pelphrey, K.A., 2005. Taking an 'intentional stance' on eye-gaze shifts: A functional neuroimaging study of social perception in children. NeuroImage 27 (1), 247–252.

Nelson, C.A., 2001. The development and neural bases of face recognition. Infant and Child Development 10 (1–2), 3–18.

Northoff, G., Heinzel, A., de Greck, M., Bermpohl, F., Dobrowolny, H., Panksepp, J., 2006. Self-referential processing in our brain – A meta-analysis of imaging studies on the self. NeuroImage 31 (1), 440–457.

Ochsner, K.N., Knierim, K., Ludlow, D.H., et al., 2004. Reflecting upon feelings: An fMRI study of neural systems supporting the attribution of emotion to self and other. Journal of Cognitive Neuroscience 16 (10), 1746–1772.

Ochsner, K.N., Beer, J.S., Robertson, E.R., et al., 2005. The neural correlates of direct and reflected self-knowledge. NeuroImage 28 (4), 797–814.

Onishi, K.H., Baillargeon, R., 2005. Do 15-month-old infants understand false beliefs? Science 308 (5719), 255–258.

Pelphrey, K.A., Morris, J.P., 2006. Brain mechanisms for interpreting the actions of others from biological-motion cues. Current Directions in Psychological Science 15 (3), 136–140.

Pelphrey, K.A., Adolphs, R., Morris, J.P., 2004. Neuroanatomical substrates of social cognition dysfunction in autism. Mental Retardation and Developmental Disabilities Research Reviews 10 (4), 259–271.

Pelphrey, K.A., Mitchell, T.V., McKeown, M.J., Goldstein, J., Allison, T., McCarthy, G., 2003. Brain activity evoked by the perception of human walking: Controlling for meaningful coherent motion. Journal of Neuroscience 23 (17), 6819–6825.

Pelphrey, K.A., Morris, J.P., McCarthy, G., 2004. Grasping the intentions of others: The perceived intentionality of an action influences activity in the superior temporal sulcus during social perception. Journal of Cognitive Neuroscience 16 (10), 1706–1716.

Pelphrey, K.A., Viola, R.J., McCarthy, G., 2004. When strangers pass: Processing of mutual and averted social gaze in the superior temporal sulcus. Psychological Science 15 (9), 598–603.

Pelphrey, K.A., Morris, J.P., Michelich, C.R., Allison, T., McCarthy, G., 2005. Functional anatomy of biological motion perception in posterior temporal cortex: An fMRI study of eye, mouth and hand movements. Cerebral Cortex 15 (12), 1866–1876.

Pfeifer, J.H., Masten, C.L., Moore 3rd, W.E., et al., 2011. Entering adolescence: Resistance to peer influence, risky behavior, and neural changes in emotion reactivity. Neuron 69 (5), 1029–1036.

Pitskel, N.B., Bolling, D.Z., Kaiser, M.D., Crowley, M.J., Pelphrey, K.A., 2011. How grossed out are you? The neural bases of emotion regulation from childhood to adolescence. Developmental Cognitive Neuroscience 1 (3), 324–337.

Polk, T.A., Farah, M.J., 1998. The neural development and organization of letter recognition: Evidence from functional neuroimaging, computational modeling, and behavioral studies. Proceedings of the National Academy of Sciences of the United States of America 95 (3), 847–852.

Polk, T.A., Stallcup, M., Aguirre, G.K., et al., 2002. Neural specialization for letter recognition. Journal of Cognitive Neuroscience 14 (2), 145–159.

Polk, T.A., Park, J., Smith, M.R., Park, D.C., 2007. Nature versus nurture in ventral visual cortex: A functional magnetic resonance imaging study of twins. Journal of Neuroscience 27 (51), 13921–13925.

Premack, D., Woodruff, G., 1978. Chimpanzee problem-solving: A test for comprehension. Science 202 (4367), 532–535.

Puce, A., Allison, T., Asgari, M., Gore, J.C., McCarthy, G., 1996. Differential sensitivity of human visual cortex to faces, letterstrings, and textures: A functional magnetic resonance imaging study. Journal of Neuroscience 16 (16), 5205–5215.

Puce, A., Allison, T., Bentin, S., Gore, J.C., McCarthy, G., 1998. Temporal cortex activation in humans viewing eye and mouth movements. Journal of Neuroscience 18 (6), 2188–2199.

Purves, D., White, L.E., Riddle, D.R., 1996. Is neural development Darwinian? Trends in Neurosciences 19 (11), 460–464.

Quartz, S.R., 1999. The constructivist brain. Trends in Cognitive Sciences 3 (2), 48–57.

Quartz, S.R., Sejnowski, T.J., 1997. The neural basis of cognitive development: A constructivist manifesto. Behavioral and Brain Sciences 20 (4), 537–556.

Roth, D., Leslie, A.M., 1998. Solving belief problems: Toward a task analysis. Cognition 66 (1), 1–31.

Rubenstein, J.L.R., Rakic, P., 2013. Patterning and Cell Types Specification in the Developing CNS and PNS.

Ruffman, T., Garnham, W., Import, A., Connolly, D., 2001. Does eye gaze indicate implicit knowledge of false belief? Charting transitions in knowledge. Journal of Experimental Child Psychology 80 (3), 201–224.

Saxe, R., Kanwisher, N., 2003. People thinking about thinking people. The role of the temporo-parietal junction in 'theory of mind'. NeuroImage 19 (4), 1835–1842.

Saxe, R., Powell, L.J., 2006. It's the thought that counts: Specific brain regions for one component of theory of mind. Psychological Science 17 (8), 692–699.

Saxe, R., Carey, S., Kanwisher, N., 2004. Understanding other minds: Linking developmental psychology and functional neuroimaging. Annual Review of Psychology 55, 87–124.

Saxe, R., Xiao, D.K., Kovacs, G., Perrett, D.I., Kanwisher, N., 2004. A region of right posterior superior temporal sulcus responds to observes intentional actions. Neuropsychologia 42 (11), 1435–1446.

Saxe, R.R., Whitfield-Gabrieli, S., Scholz, J., Pelphrey, K.A., 2009. Brain regions for perceiving and reasoning about other people in school-aged children. Child Development 80 (4), 1197–1209.

Scherf, K.S., Behrmann, M., Humphreys, K., Luna, B., 2007. Visual category-selectivity for faces, places and objects emerges along different developmental trajectories. Developmental Science 10 (4), F15–F30.

Senju, A., Johnson, M.H., Csibra, G., 2006. The development and neural basis of referential gaze perception. Social Neuroscience 1 (3–4), 220–234.

Shaw, P., Kabani, N.J., Lerch, J.P., et al., 2008. Neurodevelopmental trajectories of the human cerebral cortex. Journal of Neuroscience 28 (14), 3586–3594.

Sirois, S., Sprattling, M., Thomas, M.S., Westermann, G., Mareschal, D., Johnson, M.H., 2008. Précis of neuroconstructivism: How the brain constructs cognition. Behavioral and Brain Sciences 31 (3), 321–331.

Southgate, V., Senju, A., Csibra, G., 2007. Action anticipation through attribution of false belief by 2-year-olds. Psychological Science 18 (7), 587–592.

Spengler, S., von Cramon, D.Y., Brass, M., 2009. Control of shared representations relies on key processes involved in mental state attribution. Human Brain Mapping 30 (11), 3704–3718.

Surian, L., Caldi, S., Sperber, D., 2007. Attribution of beliefs by 13-month-old infants. Psychological Science 18 (7), 580–586.

Thomas, K.M., Drevets, W.C., Whalen, P.J., et al., 2001. Amygdala response to facial expressions in children and adults. Biological Psychiatry 49 (4), 309–316.

Tootell, R.B., Reppas, J.B., Kwong, K.K., et al., 1995. Functional analysis of human MT and related visual cortical areas using magnetic resonance imaging. Journal of Neuroscience 15 (4), 3215–3230.

Tzourio-Mazoyer, N., De Schonen, S., Crivello, F., Reutter, B., Aujard, Y., Mazoyer, B., 2002. Neural correlates of woman face processing by 2-month-old infants. NeuroImage 15 (2), 454–461.

Uddin, L.Q., Iacoboni, M., Lange, C., Keenan, J.P., 2007. The self and social cognition: The role of cortical midline structures and mirror neurons. Trends in Cognitive Science 11 (4), 153–157.

Wellman, H.M., Cross, D., Watson, J., 2001. Meta-analysis of theory-of-mind development: The truth about false belief. Child Development 72 (3), 655–684.

Wellman, H.M., Phillips, A.T., Dunphy-Lelii, S., LaLonde, N., 2004. Infant social attention predicts preschool social cognition. Developmental Science 7 (3), 283–288.

Werker, J.F., Lalonde, C.E., 1988. Cross-language speech perception: Initial capabilities and developmental change. Developmental Psychology 24 (5), 672–683.

Woodward, A.L., 1998. Infants selectively encode the goal object of an actor's reach. Cognition 69, 1–34.

Woodward, A.L., Sommerville, J.A., Guajardo, J.J., 2001. How infants make sense of intentional action. In: Malle, B., Moses, L., Baldwin, D. (Eds.), Intentions and Intentionality: Foundations of Social Cognition. MIT Press, Cambridge, MA, pp. 149–169.

Zeki, S., Watson, J.D., Lueck, C.J., Friston, K.J., Kennard, C., Frackowiak, R.S., 1991. A direct demonstration of functional specialization in human visual cortex. Journal of Neuroscience 11 (3), 641–649.

Developmental Cognitive Neuroscience of Theory of Mind

H. Gweon, R. Saxe

Massachusetts Institute of Technology, Cambridge, MA, USA

It's the best possible time to be alive, when almost everything you thought you knew is wrong.

Arcadia, by Tom Stoppard

Imagine you arrive back at the laundromat and see a stranger take your clothes out of the dryer and start to fold them. What is going on? With only seconds of perceptual data about this stranger, you can immediately conjure up multiple plausible explanations: maybe he intends to steal your clothes, maybe he is feeling amazingly generous and wants to help someone out, or maybe he just falsely believes that those are his own clothes. In this example and in countless other brief and extended social interactions every day, we do not just describe people's actions as movements through space and time. Instead we seek to explain and judge and predict their actions, and we do so by appealing to a rich but invisible causal structure of thoughts, beliefs, desires, emotions, and intentions inside their heads. This capacity to reason about people's actions in terms of their mental states is called a 'theory of mind' (ToM).

This chapter is about what we know, and what we do not know, about how the human brain acquires its amazing capacity for ToM. In the past few decades, ToM has been studied intensively in childhood development (using behavioral measures) and in the adult human brain (using functional neuroimaging). Converging evidence from these two approaches provides insight into the cognitive and neural basis of this key human cognitive capacity. However, as we highlight later, we are especially excited about the future of ToM in developmental cognitive neuroscience: studies that combine both methods, using neuroimaging methods to directly study cognitive and neural development in childhood.

We start by describing an account of ToM, in development and in neuroscience, that we shall call the 'Standard' view. Next, we describe some recent challenges that shake the foundations of the Standard view. Finally, we point to the open questions, and especially the key contributions that developmental cognitive neuroscience can make in the next generation of studies of ToM.

20.1 WHAT WE THOUGHT WE KNEW: THE STANDARD VIEW

20.1.1 Development

The laundromat example makes clear a central feature of ToM: it is especially useful when other people have false beliefs. When strangers fold their own laundry,

Neural Circuit Development and Function in the Brain: Comprehensive Developmental Neuroscience, Volume 3 **http://dx.doi.org/10.1016/B978-0-12-397267-5.00057-1**

© 2013 Elsevier Inc. All rights reserved.

no special explanation seems warranted. We can describe their actions in behavioral terms: folding laundry is a thing people frequently do in laundromats. However, when a stranger starts to fold your laundry, then it becomes important to figure out: what are they thinking? The understanding that they may have a false belief makes an otherwise highly unlikely action suddenly predictable. If they believe it is their own laundry, then of course they will start folding it.

Because false belief scenarios are so diagnostic of ToM inferences, understanding of false beliefs has been considered a key milestone in ToM development. Children's ability to predict and explain actions based on a false belief is typically assessed in a 'false belief task' (see Figure 20.1). In a typical example, children hear a story like this one: Sally sees her dog run to hide behind the sofa; then, while Sally is out of the room, the dog moves to behind the TV. Children are then asked to predict where Sally will look first for her dog when she returns to the room (Baron-Cohen et al., 1985; Wimmer and Perner, 1983).

Adults immediately recognize that Sally will look for her dog behind the sofa, where she thinks it is. Surprisingly, 3-year-olds systematically make the opposite prediction; they confidently insist that Sally will look behind the TV, where the dog really is (Wellman et al., 2001). Moreover, if 3-year-olds actually see Sally looking behind the sofa, they still do not appeal to Sally's false belief to explain her action, but instead appeal to changed desires (e.g., 'she must not want the dog,' Goodman et al., 2006; Moses and Flavell, 1990). In contrast, typical 5-year-old children correctly predict and explain Sally's action, by appealing to her false belief.

This pattern of children's judgments, over development, is incredibly robust; it has been replicated in hundreds of studies conducted over four decades. The same shift in understanding false beliefs and the ability to use beliefs to explain actions occurs between 3 and 5 years in children from rural and urban societies, in Peru, India, Samoa, Thailand, and Canada (Callaghan et al., 2005) and even in children of a group of hunter-gatherers in Cameroon (Avis and Harris, 1991).

Developmental psychologists do not disagree about these data; they disagree about the interpretation. To start with, we will articulate two claims we call the 'Standard' interpretation of these results. This is the view that most informed, and converged with, the first neuroimaging studies of ToM in the adult brain. It was never, however, a consensus opinion; in whole and in part, every aspect of the Standard view has been hotly debated all along.

First, the Standard view of development on the false belief task proposes that children undergo a key conceptual change in their ToM between ages 3 and 5 years. They

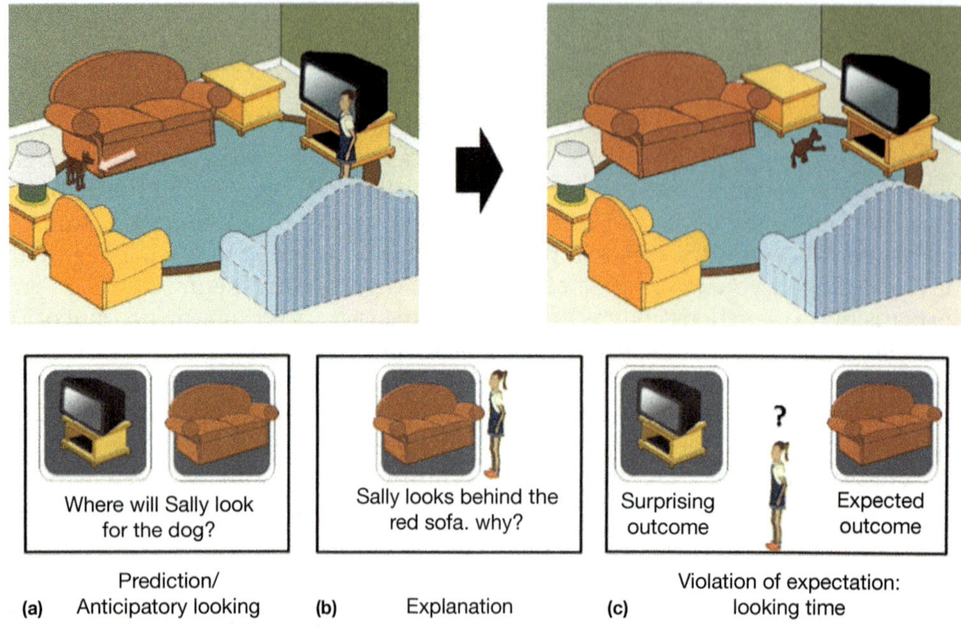

FIGURE 20.1   An example of a standard false belief scenario. Sally sees her dog hide behind the red sofa, but the dog moves to behind the TV while she is out of the room. Children's understanding of false beliefs at different developmental stages can be assessed using various methods: (a) by asking children to predict where Sally will look, or measuring their anticipatory look when Sally comes back to find her dog; (b) by asking children to explain Sally's action; or (c) by comparing infants' looking times to observing Sally go toward the sofa (the 'expected' outcome, if they understand false beliefs) or the TV ('unexpected' outcome). In spite of the logical similarity between these tasks, children make the correct prediction/explanation (Sally will look behind the sofa) around age 4 years, but make correct anticipatory looks/longer looking to the unexpected outcome around age 15–24 months. Art thanks to Steven Green.

are slowly acquiring a fully 'metarepresentational' ToM (Perner, 1991), which lets them understand people's beliefs and thoughts as representations of the world. Representations are designed to accurately reflect the real world, but sometimes fail to do so. So a representational ToM allows children to understand when and how the content of a person's belief can be false (Gopnik and Astington, 1988; Wellman et al., 2001), and that in these cases, people's actions will depend on what they believe, not on what's really true.

Note that learning to understand false beliefs involves change within a child's ToM, not the acquisition of a ToM. Even very young infants understand that people's actions depend on what they want (e.g., Phillips and Wellman, 2005; Woodward, 1998; see Gergely and Csibra, 2003 for a review), and what they can see (Meltzoff and Broks, 2008). That is already a sophisticated ToM, and it makes the right predictions for people's actions in many, if not most, circumstances. Specifically, when someone has a false belief, though, a ToM based mainly on understanding intentions makes the wrong prediction: if Sally wants her dog, she will go get her dog, so she will go where it is, behind the TV. Thus, on the Standard view, young children do have a ToM founded upon very early developing concepts of intention and perception. Nevertheless, their ToM changes substantially between ages 3 and 5 years by the addition of a full concept of 'belief.'

Second, the Standard view proposes that changes in false belief understanding reflect maturation of a 'domain-specific' mechanism for ToM. Between the ages of 3 and 5 years, children change and mature in many ways: they come to have a richer vocabulary and a better memory and a larger shoe size. However, on the Standard view, ToM develops separately: ToM task performance is not just a matter of getting smarter or faster in general, but specifically of conceptual change within ToM.

One way to assess domain specificity is to compare children's development of reasoning about false beliefs with their ability to solve very similar puzzles about other false representations: outdated photographs. For example, children might see Sally take a photograph of the dog behind the sofa; after the dog moves to the TV, the children are asked where the dog is in the photograph. The false photograph task is logically very similar to the false belief task, requiring very similar capacities for language, memory, and inhibitory control (e.g., the ability to choose between two competing response alternatives). Nevertheless, young children are significantly better at the false belief task (Zaitchik, 1990), possibly because they have a 'special mechanism' for ToM which gives their performance a boost.

Stronger evidence that ToM is separate from other parts of cognition comes from studies of children with neurodevelopmental disorders. Children diagnosed with autism spectrum disorders (ASDs) are significantly delayed in passing false belief tasks, compared with typically developing children or children with other developmental disorders like Down syndrome (Baron-Cohen, 1997; Baron-Cohen et al., 1985). On a larger set of tasks, tapping multiple different aspects of ToM (e.g., understanding desires, beliefs, knowledge, and emotions), children with ASD show both delayed development and also disorganized development. That is, typically developing children pass these tasks in a stable order (e.g., understanding false beliefs is easier than understanding false emotions), but children with autism pass the tasks in a scrambled order, as if they were passing for different reasons (Peterson et al., 2005). Furthermore, the difficulty seems to be specific to ToM: compared with typically developing children matched for IQ and verbal abilities, children with autism show comparable (or even better) performance on nonsocial tasks that require similar logical and executive capacities, like the false photograph task (Charman and Baron-Cohen, 1992; Leekam and Perner, 1991; Leslie and Thaiss, 1992).

The observation that a neurobiological developmental disorder, ASD, could disproportionately affect development of ToM supports the Standard view that ToM development depends on a distinct neural mechanism. That is, some brain region, chemical, or pattern of connectivity might be specifically necessary for ToM, and disproportionately targeted by the mechanism of ASD. This hypothesis was difficult to test, though, until the advent of neuroimaging allowed researchers to investigate the brain regions underlying high-level cognitive functions like ToM.

20.1.2 Neuroimaging

The Standard view proposes that children undergo a conceptual change in their ToM between the ages of 3–5 years, from a conception involving actions and goals to one involving beliefs, and that this development of ToM is supported by a domain-specific mechanism. Early neuroimaging studies of ToM in adults appeared to converge nicely with both of these predictions.

Following the tradition in developmental psychology, the early neuroimaging studies of ToM required participants to attribute false beliefs to characters in stories or cartoons. Meanwhile, the scientists measured the oxygen in the blood of the participant's brain, either using radioactive labels (positron emission tomography, PET; Happe et al., 1996) or by measuring intrinsic differences in the magnetic response of oxygenated blood (functional magnetic resonance imaging, fMRI; Brunet et al., 2000; Fletcher et al., 1995; Gallagher et al., 2000; Goel et al., 1995; Saxe and Kanwisher, 2003; Vogeley et al., 2001). At that point, something remarkable happened.

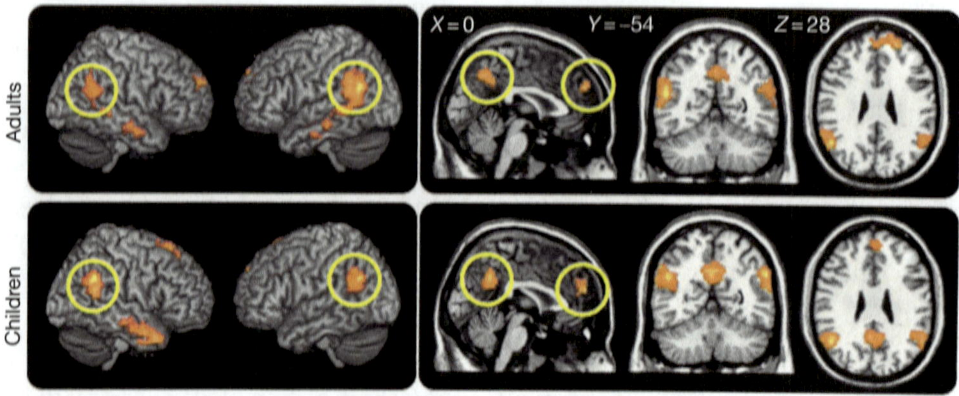

FIGURE 20.2   ToM brain regions in adults (top) and children aged 5–12 years (bottom): the right temporoparietal junction (RTPJ), left temporoparietal junction (LTPJ), precuneus, and dorsal medial prefrontal cortex (yellow circle, from left to right). The bilateral TPJ and precuneus are shown on the coronal section, and all four regions (except for the precuneus in adults) are shown on the horizontal section. *Data shown are from Gweon et al. (2012).*

Across different labs, countries, tasks, stimuli, and scanners, every lab that asked 'which brain regions are involved in ToM?' got basically the same answer: a group of brain regions in the right and left temporoparietal junction (TPJ), right anterior superior temporal sulcus (STS) and temporal pole, the medial precuneus and posterior cingulate (PC), and the medial prefrontal cortex (MPFC; Castelli et al., 2000; Fletcher et al., 1995; Gallagher et al., 2000; German et al., 2004; Goel et al., 1995; Saxe and Kanwisher, 2003; Vogeley et al., 2001; see Figure 20.2). These studies provided initial evidence for a distinct mechanism for ToM, which converged with the predictions of the Standard view.

First, these regions did not respond to other tasks that require similar logical and executive capacities as the false belief task. As described earlier, the 'false photograph' task requires participants to answer questions based on a physical, tangible representation of the past (i.e., a photograph) that used to be true but is currently false. Similar to the false belief task is the false sign task (Parkin, 1994). Understanding a false directional sign involves the use of a symbolic representation that misrepresents the current reality: for example, a signpost indicates that an object is in location A, but the object is then moved to location B. The false photograph and false sign task provide a good test of 'domain specificity' for brain regions; these tasks are very similar to the false belief task in most cognitive demands, and differ mainly in whether they require reasoning about a belief. Using that logic, we can claim that at least some brain regions in human adults are specific for ToM. The bilateral TPJ, MPFC, and PC all respond much more during false belief compared with false photograph stories (Saxe and Kanwisher, 2003). Of these regions, the right TPJ (RTPJ) in particular responds more during stories about false beliefs compared with very closely matched stories

about false signs (Perner et al., 2006). Also, identical nonverbal cartoon stimuli elicit RTPJ activity when participants construe the cartoons in terms of a character's false beliefs, but not when participants produce the exact same responses using a nonsocial 'algorithm' (Saxe et al., 2006). So at least the RTPJ, and possibly the other regions in this group, are plausible candidates for the domain-specific mechanism predicted by the Standard view.

Second, brain regions for thinking about beliefs and desires are near, but distinct from, brain regions involved in understanding actions and goals. In the right temporal lobe are brain regions involved in perceiving human bodies and body postures (extrastriate body area, EBA; Downing et al., 2001), movements (MT/V5; Grossman et al., 2000; Tootell et al., 1995), and in particular, people's facial expressions and bodily movements (right posterior STS, pSTS; Howard et al., 1996; see Chapter 19). Interestingly, the activity in the right pSTS depends not just on the action itself; it responds more to actions that are unexpected or incongruent in context (Brass et al., 2007; Pelphrey et al., 2003, 2004). For example, Brass et al. (2007) showed that the pSTS response is enhanced when participants look at a man using his knee to push an elevator button with nothing in his hands, compared with a man performing the same action but with his arms full of books (i.e., when using his knee is rational, in context).

The right pSTS response to intentional actions is impressive, but it would not support your inference about the stranger folding your clothes in the laundromat: in addition to detecting his action as intentional, you specifically need to infer his beliefs and desires (does he know those are your clothes? does he want to help or harm you?) in order to explain why he is doing so. These inferences appear to depend especially on the RTPJ.

The RTPJ is adjacent to the right pSTS (Gobbini et al., 2007), but has a different response profile (see

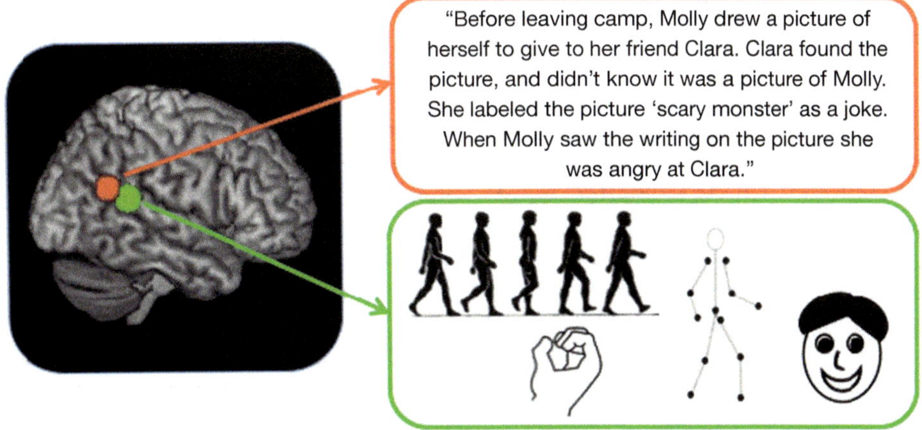

FIGURE 20.3   The RTPJ (red) and the right posterior superior temporal sulcus (pSTS) (green). The RTPJ responds to stimuli that invoke reasoning about thoughts and beliefs, whereas the right pSTS responds to human actions such as walking, grasping, eye gazes, and mouth movements.

Figure 20.3). The RTPJ is recruited during stories or cartoons that depict a person's thoughts, but not in general for depictions of human actions, whereas the right pSTS is recruited when watching movies of simple actions, like reaching for a cup, but not for verbal descriptions of actions. The RTPJ also does not respond to photographs of people (Saxe and Kanwisher, 2003) or to descriptions of people's physical appearance (Saxe and Kanwisher, 2003; Saxe and Powell, 2006). Activity in the RTPJ is low during descriptions of people's physical sensations like hunger, thirst, or tiredness (Bedny et al., 2009; Saxe and Powell, 2006). Even within a single story about a person, the timing of the response in the RTPJ is predicted by the timing of sentences describing the character's thoughts. The response in the RTPJ shows a peak just at the time someone's thoughts are described (Saxe & Wexler, 2005; Saxe et al., 2009).

In sum, neuroimaging studies on ToM with adults provide compelling evidence for a domain-specific mechanism for reasoning about thoughts. Moreover, these studies show that the brain regions involved in ToM are distinct from regions that respond to the perception of goal-directed motion. These results converge nicely with the Standard view, which proposes that children gradually shift from an earlier understanding of goals and intentions (perhaps relying on the pSTS) to the concept of 'beliefs' as reasons for people's actions (relying on the RTPJ).

20.1.3 Developmental Cognitive Neuroscience

The Standard view thus makes a set of testable predictions for the future of ToM studies in developmental cognitive neuroscience (Saxe et al., 2004a).

First, there should be qualitative changes in the anatomy of ToM brain regions, especially including the TPJ and dorsal medial prefrontal cortex (DMPFC), around age 4 years, correlated with children's acquisition of the concept of false belief.

Second, prior to age 4 years, children should rely on early-maturing neural mechanisms for social cognition, especially including the pSTS representation of action, that support recognition of goal-directed actions, but not a theory of mental states like beliefs.

Third, after age 4 years, continued development is more likely to occur outside of the ToM regions, in brain systems that support difficult task performance more generally. That is, children might get generally faster or better at resolving conflicting representations, or more sophisticated in their use of language to describe mental states. But the most momentous shift in the neural mechanisms for ToM would already be complete.

The first prediction of the Standard view is that around age 4, the emergence of children's concept of 'false belief' would be supported by qualitative maturational changes in the ToM brain regions. And, in an initial test, this prediction received impressive support. Sabbagh et al. (2009) first tested a large group of 4-year-old children on standard false belief tasks. As predicted, false belief task performance in this group of children was liminal: some reliably passed the false belief tasks, some reliably failed, and some were intermediate. Next, Sabbagh and colleagues used electroencephalograms (EEG) to measure the amplitude and coherence of alpha waves in the same children's brains, while they were just sitting quietly at rest. These measures are thought to reflect anatomical maturation in a cortical region (Thatcher, 1992). Through an analysis technique called standardized low-resolution brain electromagnetic tomography (sLORETA; Pascual-Marqui et al.,

2002), researchers can use these measures to estimate the current density of alpha waves independently in every region. So Sabbagh and colleagues could then ask: where in the child's brain is maturational change specifically predictive of performance on false belief tasks (controlling for both age, and performance on other demanding tasks)? That is, in which brain regions is the density of the alpha signal best correlated with children's ToM development? The answer was: in the RTPJ and DMPFC – the same two regions most commonly associated with ToM in functional neuroimaging studies of adults!

Sabbagh et al.'s (2009) results are exciting because (1) they offer converging evidence implicating the RTPJ and DMPFC in ToM development, using a completely different method and experimental design, and (2) they support the a priori prediction of an association between anatomical maturation in these brain regions, and performance on standard false belief tasks, when children are around 4 years old.

While source localization methods such as LORETA are shown to render reliable results in estimating the sources of EEG signals (Pascual-Marqui et al., 2002; Wagner et al., 2004), EEG methods still offer much lower spatial resolution compared with other neuroimaging techniques such as functional magnetic resonance imaging (fMRI). So the three basic predictions of the Standard view remain to be thoroughly tested.

But what's most exciting is that the most recent research suggests that each of these predictions is at least partly wrong.

20.2 EVERYTHING WE THOUGHT WE KNEW WAS WRONG

20.2.1 Infants

The biggest challenge for the Standard view comes from a rapidly growing body of research showing that even 15–18-month-old infants understand that people can have, and act on, false beliefs. If so, there is a big lacuna in the foundation of the Standard view. Four-year-old children cannot be acquiring a concept of 'false belief' if that concept is already available to one-and-a-half year olds!

The first report that infants understand false beliefs was a looking-time study by Onishi and Baillargeon (2005). The basic logic of a looking-time study with infants is that first one shows the infants a partial event, setting up an expectation of what will happen next. Then, one shows the infants two possible 'completions' for that event. One of the completions is designed to fit the infants' expectations, and the other completion is designed to violate those expectations. Even preverbal infants can then show which completion they 'expected,'

by looking longer at the unexpected completion. Onishi and Baillargeon (2005) made very elegant use of this logic to test what infants expect a person to do when she is acting based on a false belief (see Figure 20.1).

In the original study, 15-month-olds watched an experimenter repeatedly hide, and then retrieve, a toy in one of two boxes (a yellow box and a green box). Then, on the critical trial, the experimenter hid the toy in the yellow box. After a short pause, the toy then moved by itself to the green box. The key manipulation was that either the experimenter watched the toy move ('true belief') or the experimenter turned away, and did not see the toy move ('false belief'). Finally, the experimenter reached either into the green box (where the toy really was) or into the yellow box (where she last put the toy). Which reaching action did the infants expect? Consistent with previous evidence that infants understand goal-directed actions, when the experimenter had a true belief, infants looked longer when she reached into the yellow box, than when she reached into the green box, where the toy was. Amazingly, though, when the experimenter had not seen the toy move, infants looked longer when she reached into the green box than into the yellow box – as if these 15-month-old infants expected her to reach for the toy where she falsely believed it to be.

Initially, these results were met with some skepticism, as they seem to contradict so much evidence that children do not understand false beliefs until many years later. However, in the intervening years, more and more studies of false belief understanding in infants and toddlers have accumulated. Baillargeon and colleagues have conducted a whole series of elegant studies, expanding on their original results. In these experiments, 1-year-olds (12–24 months) have demonstrated systematic expectations about actions based on false beliefs about contents as well as locations, based on indirect inferences as well as direct perception, and based on beliefs updated from other people's verbal reports as well as from observation (Onishi and Baillargeon, 2005; Scott and Baillargeon, 2009; Song et al., 2008; Surian et al., 2007).

Evidence that toddlers understand false beliefs is not restricted to studies using violation of expectation looking-time measures. Using anticipatory looking, Southgate et al. (2007) measured infants' predictions about where the experimenter would reach for her toy. If the experimenter had a false belief about her toy, 24-month-olds (but not 18-month-olds) anticipated that she would reach for her toy in the location where she saw it last, on the very first trial of the experiment.

Apparently, by their second year of life, children are already able to use inferred false beliefs to correctly predict others' actions. Why, then, is there a dramatic change in false belief task performance, when children are 4 years old? Baillargeon and colleagues suggest that 2- and 3-year-old children have a fully mature understanding of representational mental states, but the demanding format of standard false belief tasks masks their abilities, rendering them incapable of expressing their knowledge (Baillargeon et al., 2010).

If so, the predictions for developmental cognitive neuroscience studies of ToM should be very different (Scott and Baillargeon, 2009). Brain regions specifically involved in ToM, like the TPJ and DMPFC, should not show any distinct qualitative developmental change around age 4 years, when children pass standard explicit false belief tasks. Instead, passing explicit false belief tasks should be associated with development in brain regions for executive function and language (see Botvinick et al., 2004; Caplan, 2007, for reviews). The developmental changes in ToM regions, by contrast, should occur much earlier, perhaps just after a child's first birthday.

Another interpretation of the infants' performance might be that while infants make systematic action predictions based on false beliefs, they do not actually have the same 'concept' of false beliefs that adults do. For example, infants may have a restricted, implicit understanding of beliefs, while 3–5-year-old children struggle to acquire a richer, more flexible, explicit concept of beliefs (Apperly and Butterfill, 2009). This view is plausible in part because a similar process occurs in other domains of cognition. The best-studied example is children's numerical concepts. Preverbal infants have representations with numerical content that allow them, for example, to track, differentiate, and even add and subtract small numbers (e.g., $1+1=2$; Feigenson et al., 2002; Wynn, 1992). Preverbal infants can also distinguish large numbers, when they differ by a large enough ratio (e.g., $8 > 4$, $16 > 8$; Xu and Spelke, 2000). However, these infants cannot form exact representations of large numbers that would let them, for example, distinguish between seven and eight objects (Carey, 2009; Xu and Spelke, 2000). Between the ages of 2 and 4 years, children then slowly and effortfully construct the concepts of the natural numbers, using as a necessary scaffold a culturally constructed list of names for numbers (Le Corre and Carey, 2007; Sarnecka and Gelman, 2004). The new concepts thus constructed are vastly more powerful than the infant's initial numerical conceptions.

It is thus tempting to believe that a similar process differentiates the infant's implicit conception of beliefs from the 4-year-old's explicit concept. Perhaps infants have an efficient, but limited, system for representing a person's belief about simple perceptual experiences (e.g., where she thinks the watermelon is). As they grow older, children gradually acquire an independent system that supports much richer representations that can be integrated with other processes to allow sophisticated explanations for others' behaviors, flexible revisions of the beliefs, and even moral judgments about others based on their beliefs. However, adherents of this view need

to provide a characterization of the implicit versus explicit systems of ToM. What kind of competence does the implicit system support, and what are the key restrictions or limitations on the implicit conception? How does the explicit system overcome the limitations of the implicit one? In order to address this question, we need to go beyond the format of the tasks (i.e., looking-time or anticipatory looking versus explicit pointing or verbal responses) and focus on the nature of beliefs entertained by the two systems. Apperly and Butterfill (2009), for example, proposed that an implicit ToM might be limited to 'tracking attitudes to object's locations,' so infants might not be able to 'track beliefs involving both the features and the location of an object (e.g., 'the red ball is in the cupboard').' Since then, Baillargeon and colleagues have shown that infants can track this kind of belief too (e.g., the belief that 'the disassembled toy penguin is in the opaque box'; Scott and Baillargeon, 2009). However, this kind of proposal is precisely what is needed to give substance to the idea of two independent 'systems' for ToM.

Developmental cognitive neuroscience could make a key impact here. If implicit ToM and explicit ToM are truly distinct systems, they might have distinct neural mechanisms. The distinction might appear in the recruitment of different brain regions for implicit versus explicit ToM tasks. Alternatively, there might be two distinct temporal phases in the development of a single brain region (or group of regions): an early change corresponding to the initial development of an implicit ToM, and a later change when the same region is co-opted for a qualitatively different explicit ToM. All of these possibilities remain to be tested.

20.2.2 Neuroimaging

Functional neuroimaging of children is a new research program, so very few fMRI studies have looked at the development of ToM brain regions. However, these few studies also pose challenges to the predictions of the Standard view. Specifically, the Standard view predicted that very young children rely on the pSTS representation of intentional actions, and that sometime around age 4 there is qualitative change in the neural representations of beliefs and desires. What the Standard model certainly did not predict was that the neural representations of both intentional action and ToM would show qualitative functional change in 5–10-year-old children. But that is exactly what the recent fMRI studies are finding.

The first demonstration of a neural response to intentional action in children was provided by Mosconi et al. (2005), who showed that the right pSTS in 7–10-year-old children is sensitive to the intentionality of action as in adults: while any gaze shift evokes activity in the right pSTS, the activity is higher when the gaze shift is directed to unexpected locations than when the shift predictively follows a moving target. Interestingly, Carter and Pelphrey (2006) found a developmental change in the functional profile of this region much later than what the Standard view would predict. They measured brain activity in 7–10-year-old children while they viewed animated clips of biological (e.g., a person walking) and nonbiological motion (e.g., a grandfather clock moving). While biological motion induced higher activity than nonbiological motion in general, they found that this difference grows larger with age. These results suggest that although the right pSTS is already sensitive to intentional action before age 7, there are still ongoing changes in the specificity of response in this region well beyond this age (see Chapter 19). Therefore, the prediction of the Standard view that the right pSTS develops early in childhood before age 4 seems at least partly wrong.

Similarly, the few existing developmental fMRI studies on ToM (Kobayashi et al., 2007a, 2007b; Saxe et al., 2009) have all reported activations in similar brain regions in children that are shown to be recruited for ToM in adults: these areas include the bilateral TPJ, PC, and the MPFC. However, these studies also found developmental changes in the neural basis of ToM between school-age children and adults (Kobayashi et al., 2007a), or among school-age children (Saxe et al., 2009) in some of these regions.

In a recent study, Gweon et al. (2012) asked children between 5 and 12 years of age and adults to listen to stories inside the scanner and measured their brain responses to these stories. The stories described people's thoughts, beliefs, and feelings (mental condition), people's appearance and their social relationships (social condition), or purely physical events involving objects that do not involve people or their mental states (physical condition). The first question was whether children, just like adults, show higher responses to mental stories compared with physical stories. The answer was yes: the regions that were significantly more active to mental versus physical condition in children were strikingly similar to those found in adults (see Figure 20.1). However, there was an age-related change in how some these regions respond to the social stories. In the older half of the participants (8.5–12 years), the bilateral TPJ and PC did not respond to the social stories but only showed heightened activity to mental stories just like in adults. In contrast, the same brain regions in the younger half of the participants (5–8.5 years) responded to both mental and social stories, and did not discriminate stories with and without mental state content. That is, the selectivity of these regions to mental states increased with age. Saxe et al. (2009) found the same pattern (see Figure 20.4). In line with the studies on developmental change in the right pSTS, these results suggest that although the neural basis for ToM seems to be already in place before age 5, there are qualitative changes in the response selectivity of these brain areas that occur well past age 5. Furthermore, the selectivity in the RTPJ was correlated with children's performance outside the scanner on tasks designed to tap into later-developing aspects of ToM, such as making moral decisions based on mental states or understanding nonliteral utterances in context. The role of maturational factors and experience in the neural and behavioral development remains to be tested. However, these results provide initial evidence for a link between neural and cognitive development in ToM.

Rather than resolving the controversy between the Standard view and its more recent opponent, these results challenge both views and add another puzzle. The Standard view predicts that ToM brain regions

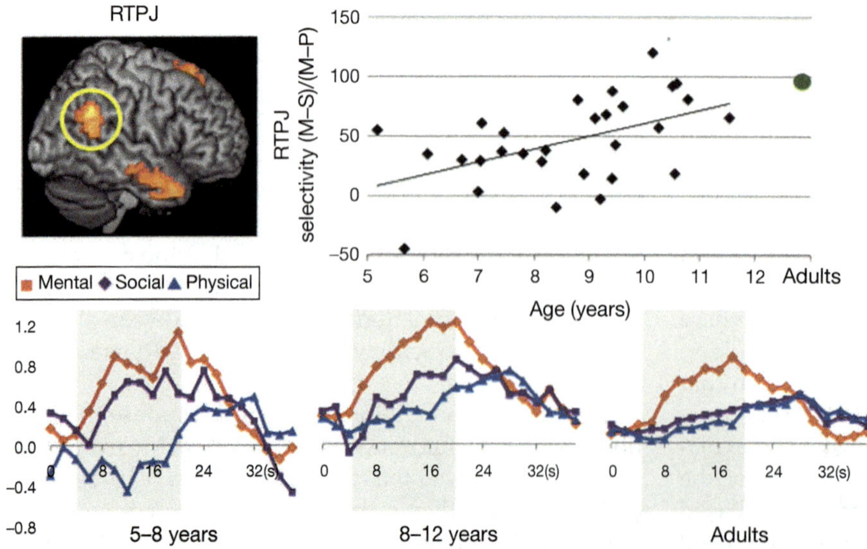

FIGURE 20.4 Developmental change in the RTPJ. The three graphs (bottom) show the time course of activation in the RTPJ in mental, social, and physical conditions in three different age groups (Gweon et al., 2012). The X-axis shows the time (seconds), and the Y-axis is the percent signal change in the activation relative to baseline. The stories were presented from 4 to 24 s, indicated by the shaded box. While the RTPJ in younger children responds to both mental and social conditions, the response to social conditions gets gradually lower as children grow older. That is, the RTPJ response becomes more 'selective' for mental stories. The scatterplot (upper right) shows that the selectivity of the RTPJ for mental, versus social, information becomes higher with age, in a cross-sectional sample of 32 children (combined data from Saxe et al., 2009; Gweon et al., 2012).

should show functional changes between the ages of 3 and 5 years. More recent data predict that significant changes in ToM brain regions should occur around the child's first birthday, when infants start to show signs of false belief understanding, with changes around age 4 years in regions supporting executive function. Therefore, the recent heated debates concerning the time course of development in the ToM mechanism have focused on these two age ranges (see Leslie, 2005; Perner and Ruffman, 2005; Scott and Baillargeon, 2009); neither predicted functional change in the neural basis of ToM well past 5 years of age.

Of course, these results do not provide direct evidence against either view, and these possibilities are not mutually exclusive. It remains possible that major changes occur in the ToM brain regions either around 12–15 months, or age 4 years, or both, supporting the acquisition of a concept of implicit and explicit false belief. However, if so, those hypothesized changes do not produce a brain region with a highly selective role in attributing mental states before 5 years of age: before age 8 years, children show normal brain regions involved in ToM (bilateral TPJ, PC, and regions in MPFC), but none of them are selectively recruited just for thinking about thoughts.

20.3 CONCLUSIONS

In sum, cognitive scientists have made foundational discoveries about the development of ToM in childhood, and about the neural basis of ToM in adulthood. Initially, these two independent methods of inquiry seemed to provide converging evidence of the structure of ToM: a domain-specific, human-unique mechanism that undergoes qualitative change in early childhood. Just recently, however, this unified view has begun to fracture, and much of what we thought we knew seems to be wrong.

The current evidence from behavioral studies of ToM suggests that infants have some understanding of false beliefs and how they affect actions, while 3-year-olds struggle to apply the same knowledge in similar tasks. Furthermore, while behavioral studies push the critical age for ToM development younger, neuroimaging studies suggest that the whole ToM system in the brain undergoes significant functional change much later, around age 8 years.

Of course, these puzzles may be partially due to the limitations in methods and the difficulty in using the same paradigm across different age groups. For example, standard false belief tasks have been mostly used for children past 3 years of age, whereas nonverbal false belief tasks using looking-time methods have been mostly used with infants and toddlers (but see Scott et al., 2012; Senju et al, 2009). This makes it difficult to directly test the differences in the nature of belief representations in infants and older children. Similarly, the available methods for measuring neural responses to these tasks vary by age. fMRI, which has been most extensively used for studies of ToM in adults and older children, is currently not useful for children younger than 4–5 years of age. In the future, it will be necessary to establish a clear relationship between fMRI results and those from other neuroimaging methods such as near infrared spectroscopy (NIRS), EEG, or magnetoencephalography (MEG), available for younger children.

The impact of a new understanding of the neural development of ToM could be tremendous. ToM deficit is a central issue in ASD. Knowing how brain regions for ToM emerge during typical development will provide the foundation for understanding how this development can go awry.

We are confident that the current uncertainty provides an important window of opportunity for new methods to make key theoretical contributions. We look forward to the emerging developmental cognitive neuroscience of ToM.

SEE ALSO

Cognitive Development: A Neuroscience perspective on empathy and its development; Developmental Neuroscience of Social Perception; Early Development of Speech and Language: Cognitive, Behavioral and Neural Systems; The Neural Architecture and Developmental Course of Face Processing; Theories in Developmental Cognitive Neuroscience. **Diseases:** Autisms.

Acknowledgments

The authors thank Jorie Koster-Hale for comments on the manuscript and Shawn Green for help with the figures. R Saxe was supported by the John Merck Scholars Program, the Simons Foundation, the Packard Foundation, and the Ellison Medical Foundation.

References

Apperly, I., Butterfill, S., 2009. Do humans have two systems to track beliefs and belief-like states. Psychological Review 116, 953–970.

Avis, J., Harris, P., 1991. Belief-desire reasoning among Baka children: Evidence for a universal conception of mind. Child Development 62, 460–467.

Baillargeon, R., Scott, R., He, Z., 2010. False-belief understanding in infants. Trends in Cognitive Sciences 14, 110–118.

Baron-Cohen, S., 1997. Mindblindness: An Essay on Autism and Theory of Mind. The MIT Press, Cambridge, MA.

Baron-Cohen, S., Leslie, A.M., Frith, U., 1985. Does the autistic child have a "theory of mind"? Cognition 21, 37–46.

Bedny, M., Pascual-Leone, A., Saxe, R., 2009. Growing up blind does not change the neural bases of theory of mind. Proceedings of the National Academy of Sciences 106, 11312–11317.

Botvinick, M., Cohen, J., Carter, C., 2004. Conflict monitoring and anterior cingulate cortex: An update. Trends in Cognitive Sciences 8, 539–546.

Brass, M., Schmitt, R.M., Spengler, S., Gergely, G., 2007. Investigating action understanding: Inferential processes versus action simulation. Current Biology 17, 2117–2121.

Brunet, E., Sarfati, Y., Hardy-Baylè, M., Decety, J., 2000. A PET investigation of the attribution of intentions with a nonverbal task. NeuroImage 11, 157–166.

Callaghan, T., Rochat, P., Lillard, A., et al., 2005. Synchrony in the onset of mental-state reasoning. Psychological Science 16, 378–384.

Caplan, D., 2007. Functional neuroimaging studies of syntactic processing in sentence comprehension: A critical selective review. Language and Linguistics Compass 1, 32–47.

Carey, S., 2009. The Origin of Concepts. Oxford University Press Inc., New York.

Carter, E., Pelphrey, K., 2006. School-aged children exhibit domain-specific responses to biological motion. Social Neuroscience 1, 396–411.

Castelli, F., Happe, F., Frith, U., Frith, C., 2000. Movement and mind: A functional imaging study of perception and interpretation of complex intentional movement patterns. NeuroImage 12, 314–325.

Charman, T., Baron-Cohen, S., 1992. Understanding drawings and beliefs: A further test of the metarepresentation theory of autism: A research note. Journal of Child Psychology and Psychiatry 33, 1105–1112.

Downing, P.E., Jiang, Y., Shuman, M., Kanwisher, N., 2001. A cortical area selective for visual processing of the human body. Science 293, 2470–2473.

Feigenson, L., Carey, S., Hauser, M., 2002. The representations underlying infants' choice of more: Object files versus analog magnitudes. Psychological Science 13, 150–156.

Fletcher, P.C., Happe, F., Frith, U., et al., 1995. Other minds in the brain: A functional imaging study of "theory of mind" in story comprehension. Cognition 57, 109–128.

Gallagher, H.L., Happe, F., Brunswick, N., Fletcher, P.C., Frith, U., Frith, C.D., 2000. Reading the mind in cartoons and stories: An fMRI study of 'theory of mind' in verbal and nonverbal tasks. Neuropsychologia 38, 11–21.

Gergely, G., Csibra, G., 2003. Teleological reasoning in infancy: The naive theory of rational action. Trends in Cognitive Sciences 7, 287–292.

German, T.P., Niehaus, J.L., Roarty, M.P., Giesbrecht, B., Miller, M.B., 2004. Neural correlates of detecting pretense: Automatic engagement of the intentional stance under covert conditions. Journal of Cognitive Neuroscience 16, 1805–1817.

Gobbini, M.I., Koralek, A.C., Bryan, R.E., Montgomery, K.J., Haxby, J.V., 2007. Two takes on the social brain: A comparison of theory of mind tasks. Journal of Cognitive Neuroscience 19:11, 1803–1814.

Goel, V., Grafman, J., Sadato, N., Hallett, M., 1995. Modeling other minds. Neuroreport 6, 1741–1746.

Goodman, N.D., Baker, C.L., Bonawitz, E.B., et al., 2006. Intuitive theories of mind: A rational approach to false belief. In: Proceedings of the Twenty-Eighth Annual Conference of the Cognitive Science Society, Vancouver, Canada.

Gopnik, A., Astington, J.W., 1988. Children's understanding of representational change and its relation to the understanding of false belief and the appearance–reality distinction. Child Development 59, 26–37.

Grossman, E., Donnelly, M., Price, R., et al., 2000. Brain areas involved in perception of biological motion. Journal of Cognitive Neuroscience 12, 711–720.

Gweon, H., Dodell-Feder, D., Bedny, M., and Saxe, R., 2012. Theory of mind performance in children correlates with functional specialization of a brain region for thinking about thoughts. *Child Development* doi:10.1111/j.1467-8624.2012.01829.x.

Happe, F., Ehlers, S., Fletcher, P., et al., 1996. 'Theory of mind' in the brain. Evidence from a PET scan study of Asperger syndrome. Neuroreport 8, 197.

Howard, R., Brammer, M., Wright, I., Woodruff, P., Bullmore, E., Zeki, S., 1996. A direct demonstration of functional specialization within motion-related visual and auditory cortex of the human brain. Current Biology 6, 1015–1019.

Kobayashi, C., Glover, G.H., Temple, E., 2007a. Children's and adults' neural bases of verbal and nonverbal 'theory of mind'. Neuropsychologia 45, 1522–1532.

Kobayashi, C., Glover, G.H., Temple, E., 2007b. Cultural and linguistic effects on neural bases of 'theory of mind' in American and Japanese children. Brain Research 1164, 95–107.

Le Corre, M., Carey, S., 2007. One, two, three, four, nothing more: An investigation of the conceptual sources of the verbal counting principles. Cognition 105, 395–438.

Leekam, S.R., Perner, J., 1991. Does the autistic child have a metarepresentational deficit? Cognition 40, 203–218.

Leslie, A.M., 2005. Developmental parallels in understanding minds and bodies. Trends in Cognitive Sciences 9, 459–462.

Leslie, A.M., Thaiss, L., 1992. Domain specificity in conceptual development: Neuropsychological evidence from autism. Cognition 43, 225–251.

Meltzoff, A.N., Broks, R., 2008. Self-experience as a mechanism for learning about others: A training study in social cognition. Developmental Psychology 44 (5), 1257–1265.

Mosconi, M., Mack, P., McCarthy, G., Pelphrey, K., 2005. Taking an "intentional stance" on eye-gaze shifts: A functional neuroimaging study of social perception in children. NeuroImage 27, 247.

Moses, L., Flavell, J., 1990. Inferring false beliefs from actions and reactions. Child Development 61, 929–945.

Onishi, K.H., Baillargeon, R., 2005. Do 15-month-old infants understand false beliefs? Science 308, 255–258.

Parkin, L., 1994. Children's Understanding of Misrepresentation. University of Sussex, UK (unpublished manuscript).

Pascual-Marqui, R.D., Esslen, M., Kochi, K., Lehmann, D., 2002. Functional imaging with low-resolution brain electromagnetic tomography (LORETA): A review. Methods and Findings in Experimental and Clinical Pharmacology 24, 91–95.

Pelphrey, K., Singerman, J., Allison, T., McCarthy, G., 2003. Brain activation evoked by perception of gaze shifts: The influence of context. Neuropsychologia 41, 156–170.

Pelphrey, K.A., Morris, J.P., McCarthy, G., 2004. Grasping the intentions of others: The perceived intentionality of an action influences activity in the superior temporal sulcus during social perception. Journal of Cognitive Neuroscience 16, 1706–1716.

Perner, J., 1991. Understanding the Representational Mind. MIT Press, Cambridge, MA.

Perner, J., Ruffman, T., 2005. Infants' insight into the mind: How deep? Science 308, 214.

Perner, J., Aichorn, M., Kronblicher, M., Staffen, W., Ladurner, G., 2006. Thinking of mental and other representations: The roles of right and left temporo-parietal junction. Social Neuroscience 1, 245–258.

Peterson, C., Wellman, H., Liu, D., 2005. Steps in theory-of-mind development for children with deafness or autism. Child Development 76, 502–517.

Phillips, A.T., Wellman, H.M., 2005. Infants' understanding of object-directed action. Cognition 98, 137–155.

Sabbagh, M., Bowman, L., Evraire, L., Ito, J., 2009. Neurodevelopmental correlates of theory of mind in preschool children. Child Development 80, 1147–1162.

Sarnecka, B., Gelman, S., 2004. Six does not just mean a lot: Preschoolers see number words as specific. Cognition 92, 329–352.

Saxe, R., Kanwisher, N., 2003. People thinking about thinking people. The role of the temporo-parietal junction in "theory of mind" NeuroImage 19, 1835–1842.

Saxe, R., Powell, L.J., 2006. It's the thought that counts: Specific brain regions for one component of theory of mind. Psychological Science 17, 692–699.

Saxe, R., Wexler, A., 2005. Making sense of another mind: The role of the right temporo-parietal junction. Neuropsychologia 43, 1391–1399.

Saxe, R., Carey, S., Kanwisher, N., 2004a. Understanding other minds: Linking developmental psychology and functional neuroimaging. Annual Review of Psychology 55, 87–124.

Saxe, R., Xiao, D.K., Kovacs, G., Perrett, D.I., Kanwisher, N., 2004b. A region of right posterior superior temporal sulcus responds to observed intentional actions. Neuropsychologia 42, 1435–1446.

Saxe, R., Schulz, L.E., Jiang, Y.V., 2006. Reading minds versus following rules: Dissociating theory of mind and executive control in the brain. Social Neuroscience 1, 284–298.

Saxe, R., Whitfield-Gabrieli, S., Scholz, J., Pelphrey, K.A., 2009. Brain regions for perceiving and reasoning about other people in school-aged children. Child Development 80, 1197–1209.

Scott, R., Baillargeon, R., 2009. Which penguin is this? Attributing false beliefs about object identity at 18 months. Child Development 80, 1172–1196.

Scott, R., He, Z., Baillargeon, R., Cummins, D., 2012. False-belief understanding in 2.5-year-olds: Evidence from two novel verbal spontaneous-response tasks. Developmental Science 15, 181–193.

Senju, A., Southgate, V., White, S., Frith, U., 2009. Mindblind eyes: An absence of spontaneous theory of mind in Asperger syndrome. Science 325 (5942), 883–885.

Song, H., Onishi, K., Baillargeon, R., Fisher, C., 2008. Can an actor's false belief be corrected by an appropriate communication? Psychological reasoning in 18.5-month-old infants. Cognition 109, 295–315.

Southgate, V., Senju, A., Csibra, G., 2007. Action anticipation through attribution of false belief by 2-year-olds. Psychological Science 18, 587–592.

Surian, L., Caldi, S., Sperber, D., 2007. Attribution of beliefs by 13-month-old infants. Psychological Science 18, 580–586.

Thatcher, R., 1992. Cyclic cortical reorganization during early childhood. Brain and Cognition 20, 24–50.

Tootell, R., Reppas, J., Kwong, K., et al., 1995. Functional analysis of human MT and related visual cortical areas using magnetic resonance imaging. Journal of Neuroscience 15, 3215–3230.

Vogeley, K., Bussfeld, P., Newen, A., et al., 2001. Mind reading: Neural mechanisms of theory of mind and self-perspective. NeuroImage 14, 170–181.

Wagner, M., Fuchs, M., Kastner, J., 2004. Evaluation of sLORETA in the presence of noise and multiple sources. Brain Topography 16, 277–280.

Wellman, H.M., Cross, D., Watson, J., 2001. Meta-analysis of theory-of-mind development: The truth about false belief. Child Development 72, 655–684.

Wimmer, H., Perner, J., 1983. Beliefs about beliefs: Representation and constraining function of wrong beliefs in young children's understanding of deception. Cognition 13, 103–128.

Woodward, A.L., 1998. Infants selectively encode the goal object of an actor's reach. Cognition 69, 1–34.

Wynn, K., 1992. Addition and subtraction by human infants. Nature 358, 749–750.

Xu, F., Spelke, E., 2000. Large number discrimination in 6-month-old infants. Cognition 74, B1–B11.

Zaitchik, D., 1990. When representations conflict with reality: The preschooler's problem with false beliefs and "false" photographs. Cognition 35, 41–68.

A Neuroscience Perspective on Empathy and Its Development

J. Decety, K.J. Michalska

University of Chicago, Chicago, IL, USA

Abbreviations

aMCC Anterior medial cingulate cortex
CADS Child and adolescent dispositions scale
CD Conduct disorder
dACC Dorsal anterior cingulate cortex
dlPFC Dorsolateral prefrontal cortex
fMRI Functional magnetic resonance imaging
IFG Inferior frontal gyrus
mPFC Medial prefrontal cortex
OFC Orbitofrontal cortex
PAG Periaqueductal gray
SMA Supplementary motor area
SPL Superior parietal lobule
TPJ Temporoparietal junction
vmPFC Ventromedial prefrontal cortex

21.1 INTRODUCTION

Developmental affective neuroscience is an emerging area of research that has the potential to improve one's understanding of human social behavior by integrating theory and research into psychology, neuroscience, clinical psychology, and neuropsychiatry. Among the psychological processes that are the basis for much of social perception and smooth social interaction, empathy and sympathy play key roles. Empathy-related responding, including caring and sympathetic concern, motivates prosocial behavior, inhibits aggression, and paves the way to moral conduct. On the other hand, certain developmental disorders, such as conduct disorder (CD), are marked by psychopathic tendencies and empathy deficits, which likely influence antisocial responses to others' distress, for example, active aggression. Understanding how these behaviors are implemented in the brain, in both typically developing children and children with aggressive tendencies can help elucidate why empathy does or does not automatically lead to prosocial behavior.

This chapter critically examines one's current knowledge about the development of the mechanisms that support the experience of empathy and associated behavioral responses such as prosocial behavior. The affective and cognitive components that give way to empathy are reviewed, starting first with the automatic proclivity to share emotions with others, and then the cognitive processes of perspective taking and executive control, which allow individuals to be aware of the intentions

Neural Circuit Development and Function in the Brain: Comprehensive Developmental Neuroscience, Volume 3 **http://dx.doi.org/10.1016/B978-0-12-397267-5.00026-1**

© 2013 Elsevier Inc. All rights reserved.

and feelings of others and keep separate self and other perspectives. The goal is to address the underlying cognitive and affective neural architecture that instantiates empathy and to examine the dysfunction of these processes in developmental disorders marked by social-cognitive impairments.

Based on the theoretical and empirical evidences from developmental psychology, cognitive neuroscience, and lesion studies, it can be argued that a number of distinct and interacting components contribute to the experience of empathy: (1) affective sharing, a bottom-up process grounded in affective arousal; (2) understanding emotion that relies on a sense of agency, self-awareness, and other awareness and critically involves the medial and ventromedial prefrontal cortex (vmPFC) and temporoparietal junction; and (3) executive functions instantiated in the prefrontal cortex, which operate as a top-down mediator, helping to regulate emotions, appraise social context, and yield mental flexibility (Decety, 2011; Decety and Jackson, 2004; Decety and Meyer, 2008).

Drawing from multiple sources of data can help paint a more complete picture of the phenomenological experience of empathy, as well as an understanding of the development and interaction between the respective mechanisms that drive the phenomenon. Furthermore, studying subcomponents of more complex psychological constructs such as empathy can be particularly useful from a developmental perspective because only some of its components or precursors may be observable. Developmental studies can provide unique opportunities to see how the components of the system interact in ways that are not possible in adults – where all the components are fully mature and operational (De Haan and Gunnar, 2009). Until quite recently, research on the development of empathy-related responding from a neurobiological level of analysis has been relatively sparse. It is believed that integrating this perspective with behavioral work can shed light into the neurobiological mechanisms underpinning the basic building blocks of empathy and sympathy and their age-related functional changes. Such integration can help to understand the neural processes that underpin prosocial behavior including pathological altruism while also benefiting interventions for individuals with atypical development, such as antisocial behavior problems.

21.2 CLEARING UP DEFINITIONAL ISSUES

The construct of empathy is applied to various phenomena that cover a broad spectrum. This spectrum ranges from feeling concern for others (creating a motivation to help them), experiencing emotions that match another individual's emotions, knowing what another is thinking or feeling, to blurring the line between self and other (Hodges and Klein, 2001).

Key concepts
- Empathy is a construct that can be decomposed into a model that includes bottom-up processing of affective sharing and emotion awareness and top-down processing in which the perceiver's motivation, memories, intentions, and self-regulation influence the extent of an empathic experience
- The experience of empathy can lead to sympathy, which refers to feelings of concern for the wellbeing of another. It includes another-oriented motivation or an egoistic motivation to reduce stress by withdrawing from the stressor in the case of personal distress
- Empathy and sympathy play key roles in motivating prosocial behavior and provide the affective and motivational foundation for moral development
- Empathy is implemented by a network of distributed, often recursively connected, interacting neural regions, including the brainstem, hypothalamus, superior temporal sulcus, insula, medial and orbitofrontal cortices, amygdala, and anterior cingulate cortex (ACC), as well as autonomic and neuroendocrine processes implicated in social behaviors and emotional states
- Neurodevelopmental studies provide unique opportunities to explore how the components of empathic responding interact in ways that are not possible in adults
- Investigating dysfunction of the components of empathy provides important clues for understanding deviations that can lead to the lack of empathy and concern for others

In developmental psychology and social psychology (the two academic disciplines that have produced most of the research on this subject), empathy is generally defined as an affective response stemming from the understanding of another's emotional state or a condition similar to what the other person is feeling or would be expected to feel in the given situation (Eisenberg et al., 1991). Other theorists more narrowly characterize empathy as one specific set of congruent emotions – those feelings that are more other-focused than self-focused (Batson, 2009; Batson et al., 1987). Very often, empathy and sympathy are conflated. Here, the authors distinguish between empathy (the ability to appreciate the emotions and feelings of others with a minimal distinction between self and other) and sympathy (feelings of concern about the welfare of others). While empathy and sympathy are often confused, the two can be dissociated. Although sympathy may stem from the apprehension of another's emotional state, it does not have to be congruent with the affective state of the other. The experience of empathy can lead to sympathy (which includes another-oriented motivation) or personal distress (an egoistic motivation to reduce stress by withdrawing from the stressor, thereby decreasing the likelihood of prosocial behavior). Emotion regulation is a critical component of empathy because the modulation of emotional experience allows the person to remain aware of a situation without being overwhelmed or

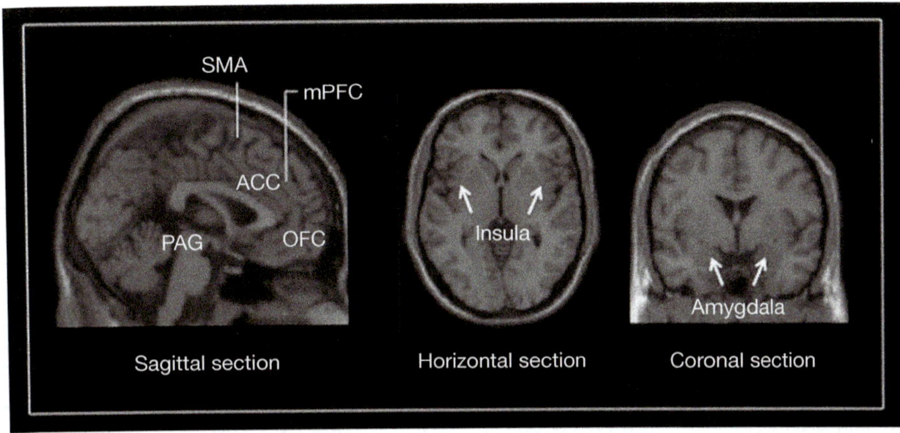

FIGURE 21.1 Brain regions that play crucial roles in the experience of empathy and associated phenomena such as sympathy labeled on sagittal, horizontal, and coronal sections of a structural MRI scan. Functional imaging studies reveal that the anterior insula and the anterior cingulate cortices are conjointly activated during the experience of emotion and during the perception of emotion in others. The insula provides a foundation for the representation of subjective bodily feelings, which substantiates emotional awareness. The ACC can be divided anatomically based on cognitive (dorsal) and emotional (ventral) components: the dorsal part is connected with the prefrontal cortex and parietal cortex as well as the motor system, making it a central station for processing top-down and bottom-up stimuli and assigning appropriate control to other areas in the brain; by contrast, the ventral part of the ACC is connected with amygdala (a structure involved in assigning affective significance to stimuli), ventral striatum, hypothalamus, and anterior insula and is involved in assessing the salience of emotion and motivational information. Many functions are attributed to ACC, such as error detection, anticipation of tasks, motivation, and modulation of emotional responses. The medial prefrontal cortex (mPFC) is critically associated with theory of mind processes and emotion understanding. The orbitofrontal cortex (OFC) is involved in sensory integration, in representing the affective value of reinforcers, and in decision making and expectation. In particular, the OFC is thought to regulate planning behavior associated with sensitivity to reward and punishment and is closely connected to the anterior insula and amygdala.

numbed by it. This is especially important in the case of negative arousal (Decety and Lamm, 2009). Developmental research also indicates that children who can regulate their emotions and emotion-related behavior should be relatively likely to experience sympathy rather than personal distress (Eisenberg and Eggum, 2009).

The complex construct of empathy can be decomposed into a model that includes bottom-up processing of affective sharing and top-down processing in which the perceiver's motivation, intentions, and self-regulation influence the extent of an empathic experience. Shared neural circuits, self-awareness, emotion understanding, and self-regulation constitute the basic macrocomponents of empathy, which are mediated by specific and interacting neural circuits, including aspects of the prefrontal cortex, ACC, medial prefrontal cortex (mPFC), orbitofrontal cortex (OFC), insula, limbic system, and frontoparietal attention networks (Decety, 2011). Consequently, this model assumes and predicts that dysfunction in either of these macrocomponents may lead to an alteration of the experience of empathy and corresponds with selective social cognitive disorders depending on which aspect is disrupted (Blair, 2005; Decety and Moriguchi, 2007; Figure 21.1).

It is also important to keep in mind that both interpersonal as well as contextual factors impact a person's subjective experience of empathy. For instance, mood states, relationship to the person, and the social context in which the interaction occurs influence the way and the extent to which the observer will react.

Recent affective neuroscience research with children and adult participants indicates that the affective, cognitive, and regulatory aspects of empathy involve interacting yet partially nonoverlapping neural circuits. Furthermore, there is now evidence for age-related changes in these neural circuits, which together with behavioral measures reflects how brain maturation influences the reaction to the distress of others (Decety and Michalska, 2010).

21.3 THE DEVELOPMENT OF EMPATHY

Empathy is one of the higher-order emotions that typically emerges as a child comes to a greater awareness of the experience of others, during the second and third years of life, and that arises in the context of someone else's emotional experience (Robinson, 2008). Each of the components of empathy (affective sharing, cognitive understanding, and emotion regulation) will be considered separately from both a developmental and neuroscience perspective. These components are indeed dissociable as documented in neurological patients (Strum et al., 2006), yet mature empathic sensitivity and concern depend on their functional integration in the service of goal-directed social behavior. In addition, both genetic and environmental factors contribute to the development of empathy and prosociality (Knafo et al., 2008).

21.3.1 Affective Sharing

While there is some controversy concerning the nature of empathic responding in very young children, there is ample behavioral evidence demonstrating that the affective component of empathy develops earlier than the cognitive component. Affective responsiveness is known to be present at an early age, is involuntary, and relies on somato-sensorimotor resonance between other and self (Decety and Meyer, 2008). The general consensus of work in this area is that infants and toddlers are sensitive and responsive to others' emotional cues and that some of the basic building blocks of empathy, such as emotion sharing, are present in the first days of life.

Discrete facial expressions of emotion have been identified in newborns, including joy, interest, disgust, and distress (Izard, 1982), suggesting that subcomponents of full emotional experience and expression are present at birth and supporting the possibility that these processes are hardwired in the brain. Very young infants are able to send emotional signals and to receive and detect the emotional signals sent by others. Human newborns are capable of imitating expressions of fear, sadness, and surprise (Field et al., 1982; Haviland and Lewica, 1987), preparing individuals for later empathic connections through affective interactions with others. Moreover, both infant humans and nonhuman primates are known to respond to others' distress with distress. Infants exposed to newborn cries cry significantly more often than those exposed to silence and those exposed to a synthetic newborn cry of the same intensity (Dondi et al., 1999; Sagi and Hoffman, 1976). A study by Martin and Clark (1987), which examined infants' reactions to audiotapes of neonatal crying, showed not only that 1-day-old babies cry in response to other infant cries but also that newborns do not respond to the sound of their own cries. Together, these findings demonstrate that infants' auditory perception of another's aversive affective state elicits the same distressful emotional state in the self and suggests a neurobiologically based predisposition for humans to be connected to others. The latter results also indicate that there is some self–other distinction already functioning at birth.

Research using measures of facial electromyography (EMG) in adult participants demonstrates that viewing facial expressions triggers distinctive patterns of facial muscle activity similar to that of the observed expression, even in the absence of conscious recognition of the stimulus (Dimberg et al., 2000). In one such study,

participants were exposed very briefly (56 ms) to pictures of happy or angry facial expressions and EMG was recorded from their faces (Sonnby-Borgstrom et al., 2003). Results demonstrated facial mimicry despite the fact that the participants were unaware of the stimuli. A study conducted with school-age boys demonstrated that angry and happy facial stimuli spontaneously elicit different EMG response patterns (de Wied et al., 2006). Angry faces evoked a stronger increase in corrugator activity than happy faces, while happy faces evoked a stronger increase in zygomaticus activity than angry faces. Such a mimicry mechanism may be driven by the so-called mirror neuron system (somatosensory motor neurons localized in the premotor, motor, and posterior parietal cortices) that directly links perception and action.

This automatic emotional resonance between other and self provides the basic mechanism on which intersubjective feelings develop. Infant arousal in response to the affects and emotions signaled by others can serve as an instrument for social learning, reinforcing the significance of the social exchange, which then becomes associated with the infant's own emotional experience. As a consequence, infants come to experience emotions as shared states and learn to differentiate their own states partly by witnessing the resonant responses that they elicit in others.

21.3.2 Emotion Understanding

Thus, it is clear that from very early on in development, infants are capable of emotional resonance, which is one important precursor of empathy (Hoffman, 2000). Although the capacity for two people to resonate with each other emotionally before any cognitive understanding is the basis for developing shared emotional meanings, it is not enough for mature empathic understanding. Such an understanding requires forming an explicit representation of the feelings of another person, an intentional agent, which necessitates additional computational mechanisms beyond the emotion-sharing level, as well as self-regulation to modulate negative arousal in the observer (Decety and Moriguchi, 2007; Decety et al., 2008). The cognitive components that give way to empathic understanding have a more protracted course of development than the affective components, even though many precursors are already in place very early in life.

Emotion understanding refers to conscious knowledge about emotion processes or beliefs about how emotions work. Such understanding includes the recognition of emotion expression and knowledge about one's own and others' emotions, the detection of cues for others' feelings, as well as ways of intentionally using emotion expression to communicate to others (or vice versa; e.g., hiding emotions). Children's development of gradually more sophisticated understanding of emotion fosters many adaptive processes, such as social functioning and coping. Consequently, delayed or limited emotion understanding may place youth at risk for disorders.

As discussed in the previous section, children recognize facial expressions associated with emotions at an early age (Haviland and Lewica, 1987). Some questions remain as to whether these early reactions represent recognition of emotion in another or simple mimicry. Either way, most children are using emotion labels for facial expressions and are talking about emotion topics by the age of 2 years (Gross and Ballif, 1991). Recent work has also documented that even very young children (18–25 months old) can sympathize with a victim even in the absence of overt emotional cues (Vaish et al., 2009), which suggests some early form of affective perspective taking that does not rely on emotion contagion or mimicry. Prior work by Zahn-Waxler et al. (1992) demonstrated that prosocial behaviors (help, sharing, and provision of comfort) emerge between the ages of 1 and 2 and that these behaviors are linked to expressions of concern as well as efforts to understand the other's plight. Regarding the causes and effects of emotion and the cues used in inferring emotion, developmental research has detailed a progression from situation-bound, behavioral explanations of emotion to broader, more mentalistic understandings (Harris et al., 1981). For example, children's early explanations of emotion are largely based on the external world (e.g., 'I am sad because someone took my toy.'), whereas as children develop, their explanations of emotions focus more on internal causes (e.g., 'I am sad because that toy was important to me.'). As children develop, their emotional inferences contain a more complex and differentiated use of several types of information, such as moral variables (Nunner-Winkler and Sodian, 1988), relational and contextual factors, and the target child's goals or beliefs (Harris, 1994). This development appears to be somewhat slower for complex emotions such as pride, shame, or embarrassment (Lewis, 2000). Children also develop an understanding of multiple emotions, comprehending that a person can feel more than one emotion at a time. Development of this understanding proceeds from lack of acknowledgment of multiple emotions in younger children, to acknowledgment, and to an appreciation of different variables, such as emotion valence and emotion intensity (e.g., one very strong and one very weak emotion are easier to understand than two strong ones; Carroll and Steward, 1984).

Understanding that appraisal can modulate a person's emotional experience to a given situation develops from being desire-based to being belief-based. At first, 2- and 3-year-old children understand the role that desires or goals play in determining a person's appraisal and ensuing emotion (Repacholi and Gopnik, 1997). By 18 months, infants can not only infer that another person can hold a desire that may be different from their own but also recognize how desires are related to emotions and understand something about the subjectivity of these desires. By 4 and 5 years of age, this desire-based concept of emotion develops to include beliefs and expectations. Children at this age begin to understand that an emotion is not necessarily triggered by whether or not a desire and an outcome match but rather whether a desire and an expected outcome match. The shift from a desire to a belief–desire conception of mind and emotion is well established. For example, Bartsch and Wellman (1995) have examined children's references to other people's mental states and demonstrated that children talk systematically about desires and goals throughout their third year, but that beginning at about their third birthday, children also begin to make reference to beliefs. At about 5 years of age, talk about beliefs becomes as frequent as talk about desires.

Young children's understanding of emotions has been shown to be related to a number of factors, including children's language ability, family discourse about feelings, and the quality of family relationships. Families vary in the linguistic environment that they offer to children for the interpretation and regulation of emotion. A child with a parent who frequently discusses emotions will have a different kind of conversation partner than a child with a parent who is much more controlled in talking about emotions. Research has established that there is indeed substantial variation among the extent to which emotions are discussed in the home. The frequency with which children engage in discussion about emotions and their causes is correlated with their later ability to identify how someone feels (Brown and Dunn, 1996). While it is possible that such a correlation reflects some disposition of the child that manifests itself in both talking about emotions and sensitivity to emotions, it is more likely that frequent family discussion may prompt children to talk about emotion and increase their understanding and perspective-taking abilities (Harris, 2000). Consistent with this idea, Lewis (2000) demonstrated that 3-year-olds who normally do poorly on a standard test of psychological understanding (the false belief task) perform better if they are prompted to structure the events leading to the false belief into a coherent narrative. Allowing children to express emotion and report on current and past emotions also provides them with an opportunity to share, explain, and regulate emotional experience. Conversation helps to develop empathy, for it is often here that people learn of shared experiences and feelings.

With cognitive empathy, an individual is thought to use perspective-taking processes to imagine or project into the place of the other in order to understand what she/he is feeling. This aspect of empathy is closely related to processes involved in mental state attribution (also known as theory of mind), which requires executive functions such as cognitive flexibility, inhibitory control, and working memory (Decety and Jackson, 2004; Stone and Gerrans, 2006). While relatively little is known about whether children who have a strong grasp of mental states also are advanced in their understanding of emotions, emotion understanding and mental state understanding seem to engage common, as well as distinct, computational processes. When seeing another child who is upset, a child has to hold two different perspectives in mind in order to correctly identify what that child is feeling and comfort them and their own perspective (and emotions), which may not be congruent with that of the other child, and the point of view of the other child. There is indeed some evidence for a link between understanding of mental state reasoning and emotion. Several studies have shown that by around 4 years of age, children can appreciate that the emotion a person feels about a given event depends on that person's perception of the event and their beliefs and desires about it. Hughes and Dunn (1998) conducted a longitudinal study of 50 children aged 47–60 months, examining developmental changes in understanding of false belief and emotion and mental state in their conversation with friends. They found that individual differences in understanding of both false belief and emotion were stable over this time period and were significantly related to each other.

Emotion recognition continues to develop into later adolescence (Tonks et al., 2007), which might improve social cognition performance. Social performance as a whole continues to improve into adulthood, but it is unclear if this reflects changes in social cognition *per se* or the continued development of cognitive functions that are not specific to social behavior, such as declarative knowledge, metacognition, speed of processing, and working memory.

Neuroimaging studies that have examined the neural systems engaged during mental state understanding in adults consistently identify a neural network involving the mPFC, the posterior temporal cortex at the junction of the parietal cortex (TPJ), and the temporal poles (e.g., Brunet et al., 2000; Frith and Frith, 2003). These regions were activated in children aged 6–11 years, while they listened to sections of a story describing a character's thoughts compared to sections of the same story that described a physical context (Saxe et al., 2009). Furthermore, change in response selectivity with age was observed in the right TPJ, which was recruited equally for mental and physical facts about people in younger children, but only for mental facts in older children. Further support for age-related changes in brain activity

associated with metacognition is provided by a neuroimaging investigation of theory of mind in participants whose age ranged between 9 and 16 years (Moriguchi et al., 2007). Both children and adolescents demonstrated significant activation in the neural circuits associated with mentalizing tasks, including the TPJ, the temporal poles, and the mPFC. Furthermore, the authors found a positive correlation between age and the degree of activation in the dorsal part of the mPFC. Direct evidence for the implication of these regions during accurate identification of interpersonal emotional states was documented in a recent functional magnetic resonance imaging (fMRI) study in which participants were requested to rate how they believe target persons felt while talking about autobiographical emotional events (Zaki et al., 2009).

In sum, the neural circuits implicated in emotion understanding largely overlaps with those involved in ToM processing, especially the mPFC and right TPJ. They continue to undergo maturation until late adolescence.

21.4 EMOTION REGULATION

The regulation of internal emotional states and processes is particularly relevant to the modulation of vicarious emotions and the experience of empathy. Given the shared nature of the representations of one's own emotional states and others' emotional states, it would seem difficult not to experience emotional distress while viewing another's distressed state (while personal distress does not contribute to the empathic concern and prosocial behavior). Personal distress can actually deter one's inclination to soothe another person's distress. It is, therefore, adaptive for this automatic sharing mechanism between self and other to be modulated by cognitive control. To this end, executive functions (i.e., the processes that serve to monitor and control thoughts and actions, including effortful control, planning, cognitive flexibility, and response inhibition; Russell, 1996), work in a top-down fashion to regulate one's inclinations to be biased in one's self-perspective while judging another person's emotional state and promoting a sympathetic regard for the other, rather than a desire to escape aversive arousal (Decety, 2005). Difficulty in the ability to regulate emotions can result in deleterious emotional arousal, thereby hindering the ability to socially function adaptively and appropriately.

Although research on emotion regulation has increased rapidly in the past decade, definitions of emotion regulation have typically been implied and not stated. While researchers agree that the construct of emotion regulation involves both emotion as a *behavior regulator* and emotion as a *regulated phenomenon*, the authors emphasize the latter – that is, how one attempts to regulate emotion – and examine various regulatory processes. As an example, Thomson (1994) defined emotion regulation as the extrinsic and intrinsic processes responsible for monitoring, evaluating, and modifying emotional reactions especially their intensive and temporal features, to accomplish one's goals.

Sympathy is strongly related to effortful control, with children high in effortful control showing greater empathic concern (Rothbart et al., 1994). A number of developmental studies conducted by Eisenberg and her colleagues (1994) found that individual differences in the tendency to experience sympathy versus personal distress vary as a function of dispositional differences in individuals' abilities to regulate their emotions. Well-regulated children who have control over their ability to focus and shift attention have been found to be relatively prone to sympathy regardless of their emotional reactivity. This is because they can modulate their negative vicarious emotions to maintain an optimal level of emotional arousal. In contrast, children who are unable to regulate their emotions, especially if they are dispositionally prone to intense negative emotions, are found to be low in dispositional sympathy and prone to personal distress (Eisenberg et al., 1994).

Interestingly, the development of executive functions is functionally linked to the development of mental state understanding. There is now increasing evidence of a specific link between the development of mentalizing and improved self-control at around the age of 4 years (Carlson and Moses, 2001). Improvement in inhibitory control corresponds with increasing metacognitive abilities (Zelazo et al., 2004), as well as with maturation of brain regions that underlie working memory and inhibitory control (Tamm et al., 2002). A series of studies by Posner and Rothbart (2000) strongly suggest that executive regulation undergoes dramatic change during the third year of life.

Emotion regulation taps into executive function resources implemented in the prefrontal cortex (Zelazo et al., 2008), with different regions subserving distinct functions. Ventral and dorsal regions of the prefrontal cortex have been associated with response inhibition and self-control, which are both key components of emotion regulation (Ochsner et al., 2002). Support for this hypothesis in the domain of empathy comes from a study that compared the neurohemodynamic response in a group of physicians and a group of matched control participants when they were exposed to short video clips depicting hands and feet being pricked by a needle (painful situations) or being touched by a Q-tip (non-painful situations; Cheng et al., 2007). Unlike control participants, physicians showed a significantly reduced neurohemodynamic empathic response in the anterior insula, anterior medial cingulate cortex (aMCC), and dorsal anterior cingulate cortex (dACC) and no activation of the periaqueductal gray (PAG; a mediator of the flight-or-fight response) when shown video clips of body parts being pricked by a needle. Instead, cortical regions underpinning executive functions and self-regulation (dorsolateral and mPFC) and executive attention (precentral, superior parietal, and temporoparietal junction) were found to be activated. Connectivity analysis further demonstrated that activation in the medial and dorsolateral prefrontal cortices subserving executive control and self-regulation (Ochsner and Gross, 2005) was inversely correlated with activity in the insular cortex in the physicians, indicating executive suppression of the emotional response to the others' pain.

It is well documented that the prefrontal cortex and its functions follow an extremely protracted developmental course and age-related changes continue well into adolescence (Bunge et al., 2002; Casey et al., 2005; Sowell et al., 1999). Frontal lobe maturation is associated with an increase in a child's ability to activate areas involved in emotional control and exercise inhibitory control over their thoughts, attention, and action. The maturation of the prefrontal cortex also allows children to use verbalizations to achieve self-regulation of their feelings (Diamond, 2002). It is, therefore, likely that different parts of the brain may be differentially involved in empathy at different ages. For example, Killgore et al. (2001) and Killgore and Yurgelun-Todd (2007) provided evidence that as a child matures into adolescence there is a shift in response to emotional events from using more limbic-related anatomic structures, such as the amygdala, to using more frontal lobe regions to control emotional responses. Thus, not only may there be less neural activity related to the regulation of cognition and emotion in younger individuals but the neural pattern itself is likely to differ.

21.5 PERCEIVING OTHER PEOPLE IN DISTRESS

Pain evolved as a protective function by not only warning a person suffering that something is awry but also impelling expressive behaviors that attract the attention of others. It has been argued that the long history of mammalian evolution has shaped maternal brains to be sensitive to signs of suffering in one's own offspring (Haidt and Graham, 2007). In many primates and other social animals, this sensitivity extends beyond the mother–child relationship, so that all typically developed individuals dislike seeing others suffering.

A growing number of functional MRI studies have demonstrated that the same neural circuits involved in the experience of physical pain are also involved in the perception or the imagination of another individual in pain (Jackson et al., 2006). This neural network includes the supplementary motor area (SMA), cerebellum, dACC, aMCC, and anterior insular cortex. In addition, studies using different modalities of neuroimaging including transcranial magnetic stimulation (Avenanti et al., 2005), somatosensory-evoked potentials (Bufalari et al., 2007), functional MRI (Lamm et al., 2007), and magnetoencephalography (Cheng et al., 2008) indicate that areas processing the sensory dimension of pain (the somatosensory cortex and posterior insula) are also activated by the mere visual perception of others' pain. It is worth mentioning, however, that activation of these regions reflects a general aversive response not specific to nociception (i.e., the neural processing of noxious stimuli). This network of regions underpins a physiological mechanism that mobilizes the organism to react – with heightened arousal and attention – to threatening situations. The dACC plays a key role in conflict monitoring; the aMCC is involved in autonomic regulation associated with processing of fear and anxiety; the anterior insula processes visceral bodily sensations; the PAG integrates physiological changes in response to stress, and in the context of danger, the SMA as a result of feedback from the limbic system represents one anatomical substrate for activating motor responses associated with danger and threats (Decety, 2011; Yamada and Decety, 2009).

Distress, of course, often does not occur in a social vacuum. The social context in which pain occurs influences cognitive appraisal and the neural mechanisms underpinning its perception in the observer. To evaluate whether the neural processing in the pain matrix is modulated by social context, a study that compared patterns of brain activation while adult participants observed painful situations occurring by accident and painful situations intentionally caused by another individual (Akitsuki and Decety, 2009). Since pain is the result of a type of social interaction in the latter situation, its recognition is likely to involve not only the perception of pain but also the cognitive evaluation of the social interaction.

Results show that attending to painful situations caused by accident is associated with activation of the pain matrix, including the aMCC, insula, PAG, and somatosensory cortex. Interestingly, when watching another person intentionally inflicting pain onto another, regions that are consistently engaged in mental state understanding and affective evaluation (mPFC, TPJ, OFC, and amygdala) were additionally recruited. Furthermore, stronger connectivity between the left amygdala and the vmPFC was found when participants perceived painful situations caused by another individual relative to situations where pain occurred accidentally. These data demonstrate the impact of social context on the neural response to the perception of others' pain. Similar data were obtained with a group of young children (Decety et al., 2008) and adolescents (Decety et al., 2009). When watching sympathy-eliciting stimuli, increased effective connectivity was found between regions of the mPFC and the frontoparietal supramodal attention network, as well as between the right TPJ and the mPFC.

21.6 NEURODEVELOPMENTAL CHANGES IN EMPATHIC RESPONDING

One limitation of these above-mentioned studies is that they cannot capture any continuous functional changes across age. This is unfortunate because among areas of the brain undergoing considerable remodeling from childhood to adolescence is the prefrontal cortex, both dorsal and ventromedial, which plays a key role in understanding and experiencing social emotions (Kringelbach and Rolls, 2004; Shamay-Tsoory et al., 2006). Furthermore, both the insula and the amygdala may differentially contribute to the experience of interpersonal sensitivity during development. While human neuroimaging studies using pain empathy paradigms all report activations in the insula, no systematic attention has been paid to the anatomical subdivisions of the insula, particularly regarding their respective functional contribution across age.

To examine age-related changes associated with empathy and sympathy, functional MRI and behavioral data were collected from a group of 57 participants ranging from 7 to 40 years of age (Decety and Michalska, 2010). Results at the whole group level showed that attending to accidentally caused painful situations was associated with activation of the pain matrix, including the aMCC, insula, PAG, and somatosensory cortex. Interestingly, when watching one person intentionally inflicting pain onto another, regions that are consistently engaged in mental state understanding and affective evaluation (mPFC, TPJ, and OFC) were also recruited. The younger the participants, the more strongly the amygdala, posterior insula, and SMA were recruited when they watched painful situations that were accidentally caused. While participants' subjective ratings of the painful situations decreased with age and were significantly correlated with hemodynamic response in the mPFC, increases in pain ratings were correlated with bilateral amygdala activation (Figure 21.2).

A significant negative correlation between age and degree of activation was found in the posterior insula. In contrast, a positive correlation was found in the anterior portion of the insula. A posterior-to-anterior progression of increasingly complex rerepresentations in the human insula is thought to provide a foundation for the sequential integration of the individual homeostatic

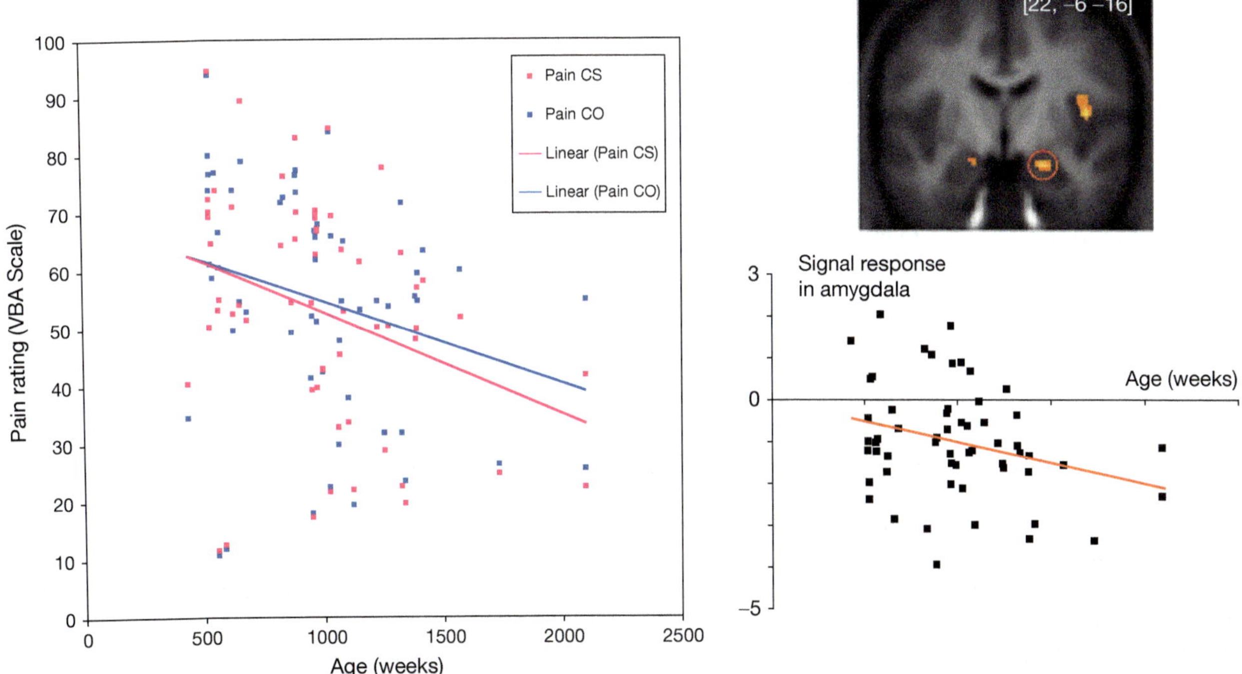

FIGURE 21.2   Left: Subjective ratings to dynamic visual stimuli depicting painful situations accidentally caused by self (Pain CS) and painful situations intentionally caused by another individual (Pain CO) across age (in weeks) in 57 participants (from 7 to 40 years old). A gradual decrease in the subjective evaluation of pain intensity for both painful conditions was found across age, with younger participants rating them significantly more painful than older participants (for Pain CS $r = -0.327$, $p < 0.01$; for Pain CO $r = -0.267$, $p < 0.05$). Further, while on an average participants rated the pain caused intentionally (Pain CO) conditions as significantly more painful than when pain was accidentally caused by self (Pain CS; $t(56) = 2.581$, $p < 0.01$), this effect was not driven by age. Right: Significant negative correlation between age and degree of activation in the amygdala when the participants are exposed to painful situations accidentally caused. *Adapted from Decety J and Michalska KJ (2010) Neurodevelopmental changes in the circuits underlying empathy and sympathy from childhood to adulthood.* Developmental Science 13: 886–899.

condition with one's sensory environment and motivational condition (Craig, 2004). The posterior insula receives inputs from the ventromedial nucleus of the thalamus that is highly specialized to convey emotional and homeostatic information such as pain, temperature, hunger, thirst, itch, and cardiorespiratory activity. It serves as a primary sensory cortex for each of these distinct interoceptive feelings from the body. The posterior part has been shown to be associated with interoception, due to its intimate connections with amygdala, hypothalamus, and cingulate and orbitofrontal cortices (Jackson et al., 2006). It has been proposed that the right anterior insula serves to compute a higher-order metarepresentation of the primary interoceptive activity, which is related to the feeling of pain and its emotional awareness (Craig, 2003).

Results from this study also showed that activation in the OFC in response to sympathy-eliciting stimuli shifts from the engagement of the medial portion in young participants to the lateral portion in older participants. The medial OFC appears integral in guiding visceral and motor responses, whereas lateral OFC integrates the external sensory features of a stimulus with its impact on the

homeostatic state of the body (Hurliman et al., 2005). Greater signal change with increasing age was associated with prefrontal regions that are responsible for cognitive control and response inhibition, such as the dorsolateral prefrontal cortex (dlPFC) and the inferior frontal gyrus (IFG). Indeed, the older the participants the greater the activity in the dlPFC and IFG, which are involved in cognitive control and response inhibition (Kawashima et al., 1996; Swick et al., 2008). This is in line with evidence that regulatory mechanisms continue into late adolescence and early adulthood.

Overall, this pattern in the amygdala and insula can be interpreted in terms of the frontalization of inhibitory capacity, hypothesized to provide a greater top-down modulation of activity within more primitive emotion-processing regions (Yurgelun-Todd, 2007). This finding provides neurophysiological support for developmental studies showing that emotion regulation is an important aspect of empathy and sympathy, especially in relation with prosocial behavior. Indeed, this pattern of developmental change in the OFC appears to reflect a gradual shift between the monitoring of somatovisceral responses in young children, mediated by the medial aspect of the

OFC, and the executive control of emotion processing implemented by its lateral portion in older participants.

In light of these neurophysiological considerations, the authors posited that the anatomical progression (from sensory to emotional awareness) parallels the neurodevelopmental response to seeing people in pain or distress. In other words, a visceral response to painful stimuli associated with danger and negative affect is less likely to occur with increasing age and such a response may be replaced by a more detached appraisal of the stimulus. This was indeed confirmed by having participants rate the pain of the people in the clips outside of the scanner. It was found that on an average, all participants rated the pain caused intentionally conditions as significantly more painful than when pain was caused accidentally by the self. Interestingly, the results also indicated a gradual decrease in the subjective evaluation of pain intensity for both these conditions across age, with younger participants rating them as significantly more painful than older participants. Moreover, while decreases in pain evaluation were significantly correlated with hemodynamic response in the mPFC, increases in pain ratings were correlated with bilateral amygdala and somatosensory cortex activation. These behavioral results are important for interpreting region-specific differences in activation with age as reflecting functional maturation and not simply differences in performance.

In sum, the behavioral evaluations of pain intensity and the pattern of brain activation from childhood to adulthood reflect a gradual change from a visceral emotional response critical for the analysis of the affective significance of stimuli to a more evaluative function.

21.7 ATYPICAL EMPATHIC PROCESSING IN CHILDREN WITH ANTISOCIAL BEHAVIORAL DISORDERS

Children and adolescents diagnosed with antisocial behavioral disorders are marked by lack of regard for others, inability to feel remorse, and even a derivation of pleasure from the distress of others. It has been hypothesized that empathy and sympathetic concern for others are essential factors inhibiting aggression toward others (Eisenberg, 2005; Zahn-Waxler et al., 1995). That is, if a person vicariously experiences the distress that they have caused to others because of their aggression, they will be less likely to continue to hurt others and more likely to help them. Conversely, lack of sympathy is an important risk factor for antisocial behavior problems such as CD (Lahey and Waldman, 2003). The propensity for aggressive behavior has been thought to reflect a blunted empathic response to the suffering of others (Blair and Blair, 2009) and may be a consequence of a failure to be aroused by others' distress (Raine, 1997).

It has also been suggested that aggressive behavior can arise from abnormal processing of affective information, resulting in a deficiency in experiencing fear, empathy, and guilt, which in normally developing individuals, inhibits acting out violent impulses (Davidson et al., 2000).

Another hypothesis regarding the relation between affect and aggression can be drawn from the research with animal models and psychiatric populations, which indicates that there may in fact be no blunting of the emotional response toward the other. Rather, heightened emotional reactivity, potentially coupled with diminished regulatory processes may trigger aggressive impulses (Coccaro et al., 2007). This emotional reaction can have either a negative or a positive valence. Previous work has shown that negative affect is generally positively associated with aggression, suggesting that empathic mimicry in conjunction with poor emotion regulation might produce negative affect that increases aggression. For instance, there are numerous empirical studies that document that physical pain often instigates aggressive inclinations (Berkowitz, 1993). The role of heightened negative emotional arousal in antisocial boys, including callous disregard for victims in distress (either caused or observed by aggressive youth) has been examined in young children with normative, subclinical, or clinical levels of behavior problems (Hastings et al., 2000). Results showed that this process can develop early. Boys (but not girls) with clinical levels of aggressive/disruptive problems showed more callous disregard toward victims during simulated distresses than typically developing children (at 4–5 years and again at 7 years). Despite this, 4- to 5-year-old children with clinical problems also showed as much empathic concern as controls at ages of 4–5 years. But by 7 years of age, the clinical group was less empathic than controls according to parent, teacher, and child reports, as well as when observed in the laboratory. Thus, early emotional negativity/callous disregard may have long-term adverse effects.

A recent study by Cheng et al. (2012) found that incarcerated juvenile psychopaths who scored high callous–unemotional traits, compared to participants with low callous–unemotional traits exhibited a reduced frontal N120 (measured with ERPs) in response to visual stimuli depicting people in pain, indicating an absence of early affective arousal. However, there was no deficit in sensorimotor resonance in both groups, as assessed by measures of the *mu* suppression over the sensorimotor cortex, which is considered as a reliable index of the mirror neuron system. This finding indicates that affective arousal is not mediated by the mirror neuron system.

In an fMRI study by Decety et al. (2009), two groups of 16–18-year-olds (matched on age, sex, and race–ethnicity) were scanned using the same animated stimuli and procedures as used with normally developing

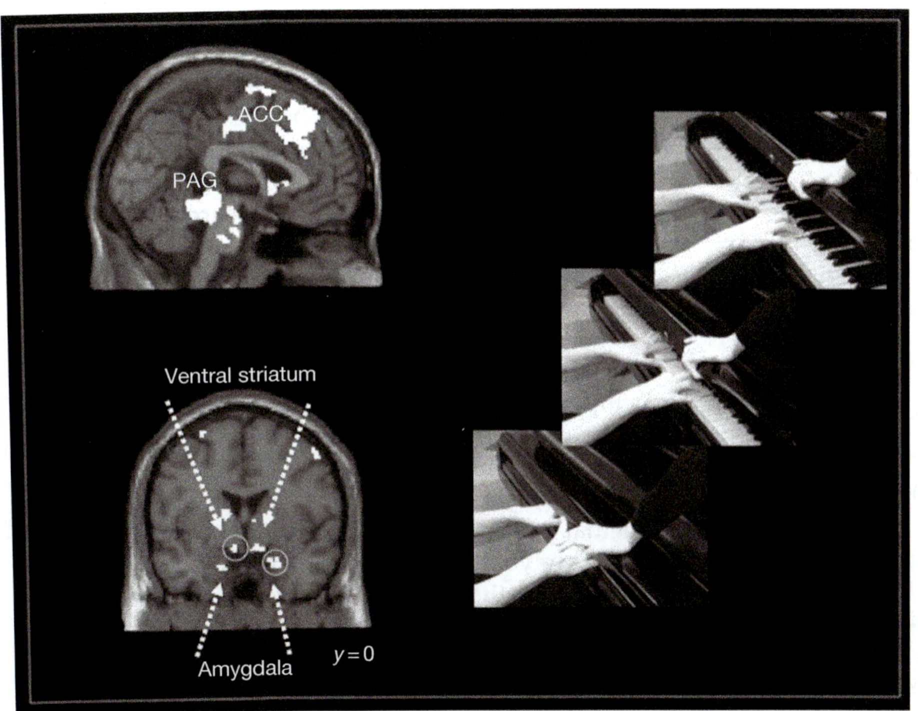

FIGURE 21.3 When youth with aggressive conduct disorder (CD) watch an individual intentionally hurting another (such as closing a piano lid), regions of the brain that process nociceptive information were activated (anterior insula, dorsal and ventral ACC, somatosensory cortex, and PAG), as well as the amygdala and ventral striatum which are part of the neural circuit involved in reward processing. These latter regions were not engaged in healthy control adolescents. Dispositional ratings of daring and sadism in youth with CD correlated with amygdala and striatal response ($p < 0.005$). These adolescents seem to enjoy seeing people in pain, and this may be rewarding and lead to repeated aggression. *Adapted from Decety J, Michalska KJ, Akitsuki Y, and Lahey B (2009) Atypical empathic responses in adolescents with aggressive conduct disorder: A functional MRI investigation.* Biological Psychology *80: 203–211.*

children (Decety et al., 2008, 2010). Results showed that the pain matrix (anterior insula, aMCC, and PAG) was significantly activated in both groups when viewing others in pain, but there was significantly greater activation of the pain matrix and somatosensory cortex in adolescents with CD than in non-CD controls (Figure 21.3).

This suggests that youth with CD do not fail to respond to viewing pain in others. Indeed, this somatic sensorimotor resonance was significantly greater in participants with CD than those without CD. In addition, there was significantly greater activation of the temporal pole in youth with CD than in controls. The temporal pole is part of a system that modulates visceral responses to emotionally evocative stimuli, and electrical stimulation of the TP produces increases in heart rate, respiration, and blood pressure (Gloor et al., 1982). In addition, the temporal poles are tightly connected with the limbic system and project to the hypothalamus, a neuromodulatory region important for autonomic regulation. There is evidence that the temporal pole is involved in the processing of both positive and negative affects (Olson et al., 2007).

Furthermore, adolescents with CD exhibited significantly greater activation of both the amygdala and the ventral striatum when viewing others in pain versus others not in pain. It was also observed that in participants with CD, the extent of amygdala activation to viewing pain in others was positively correlated with their number of aggressive acts and their ratings of daring and sadism scores on the child and adolescent dispositions scale (CADS). This is consistent with greater affective response in youth with CD to viewing others in pain.

What these findings suggest is that individuals with CD actually react to viewing pain in others at least similarly or possibly to a greater extent in some important brain regions than youth without CD, especially in the insula, amygdala, and aMCC. This stands in contrast with one previous fMRI study that reported reduced amygdala response in youth with CD during the viewing of pictures with negative emotional valence (Marsh et al., 2008). The study by Decety and colleagues suggests that youth with CD do not exhibit reduced amygdala response to all negatively valenced stimuli; indeed, they appear to exhibit enhanced response to images of people in pain, including a specific activation of the ventral striatum.

Different patterns of response were detected in the OFC across the two groups. While the lateral OFC was selectively activated in control participants when observing pain inflicted by another, activation of the medial OFC was found in the participants with CD. A direct comparison between the groups confirmed this finding. The OFC/mPFC has been specifically implicated in a variety of areas relevant to CD and aggression, including the regulation of negative affect (Phan et al., 2005). Importantly, functional connectivity analyses demonstrated significantly greater amygdala/PFC coupling when watching pain being intentionally caused by another individual in the control group than those in the CD group. In particular, in healthy adolescents, left amygdala activity covaried with activity in the prefrontal cortex to a greater extent while watching situations of pain being intentionally caused compared to viewing accidentally caused pain. Adolescents with CD showed no significant functional connectivity between frontal regions and the amygdala.

These findings are consistent with at least two hypotheses that are currently being investigated with younger children who meet diagnosis for CD:

- The first hypothesis is that highly aggressive antisocial youth enjoy seeing their victims in pain and, because of their diminished PFC/amygdala connectivity, may not effectively regulate positively reinforced aggressive behavior. The amygdala is involved in the processing of more than just negative affect. Several studies point to a role for the amygdala in positive affect, and its coupling with the striatum enables a general arousing effect of reward (Murray, 2007). It is possible, therefore, that the robust hemodynamic response in the amygdala to viewing others in pain in youth with aggressive CD reflects a positive affective response (e.g., 'enjoyment' or 'excitement'). The finding that CADS ratings of daring items (which reflect enjoyment of novel and risky situations) and sadism items (which reflect enjoyment of hurting others or viewing people or animals being hurt) correlated positively with amygdala response in youth with CD is in line with this hypothesis. In addition, this interpretation is consistent with the significant activation among youth with aggressive CD of the ventral striatum, which is part of the system implicated in reward and pleasure, among other things.
- The second hypothesis is that youth with CD have a lower threshold for responding to many situations with negative affect, including viewing pain in others, and are less able to regulate these negative emotions through cortical processes. Many studies indicate that individuals with CD tend to respond to aversive stimuli with greater negative affect than most other youth (Lahey and Waldman, 2003). This is potentially important as their negative affect may increase the likelihood of aggression, especially in the absence of effective emotion regulation (Berkowitz, 2003). This interpretation fits well with the hypothesis of a dysfunction in the neural circuitry of emotion regulation (Davidson et al., 2000) and is consistent with the analyses of effective PFC/amygdala connectivity. Aggression may be related to instability of negative affect and poor impulse control (Raine, 2002). Children with aggressive behavior problems have difficulties regulating negative emotions, which may result in harmful patterns of interpersonal

behavior. Often triggered by hypersensitivity to specific stimuli, aggressive adults experience escalating agitation followed by an abrupt outburst of aggressive and threatening behavior (Gollan et al., 2005). Failure to discriminate between pain to others and to oneself may further lead to negative emotion when viewing others in pain. Research with nonhumans demonstrates that physical pain often elicits aggression (Berkowitz, 2003). It has been hypothesized that aggressive persons are disposed to experience negative affect (Lahey and Waldman, 2003). This suggests that in certain situations, empathic mimicry might produce high levels of distress in youth predisposed to be aggressive that, ironically, increases their aggression. It is possible that strong activation of neural circuits that underpin actual pain processing is associated with negative affect in youth with CD. This, in conjunction with reduced activation in areas associated with emotion regulation, could result in a dysregulated negative affective state, which may instigate aggression under some circumstances. For example, youth with CD who see an injured friend (or fellow member of a gang) may be more likely to respond aggressively than other youth for this reason. Finally, the strong and specific activation of the amygdala and ventral striatum in the aggressive adolescents with CD during the perception of pain in others is an important and intriguing finding, which necessitates additional research in order to understand the relative roles of negative and positive affect when viewing others in pain in aggression and empathic dysfunction.

Overall, youth with aggressive CD show atypical patterns of hemodynamic response and effective connectivity in regions that regulate emotion when exposed to the distress of others, as well as reward-related activation. More work is needed to disambiguate how such abnormal emotional processing is associated with callous disregard for others, insensitivity to their distress, or even enjoyment that may lead to rewarding offending behavior.

21.8 CONCLUSION

Empathy and sympathy develop as a result of complex biological and psychological processes involving emotion sharing, emotional regulation, mental state understanding, and cognitive abilities that are continuously interactive between the individual and the social environment. Empathy can be viewed as both intrapersonal and interpersonal processes. Breaking down empathy and related phenomena into components and examining their neurodevelopment can contribute to a more complete model of interpersonal sensitivity. Likewise, drawing from multiple sources of data can improve one's understanding of the nature and causes of empathy deficits in individuals with antisocial behavior disorders. Recent advances in cognitive neuroscience indicate that distinct but interacting brain circuits underpin the different components of empathy, each having their own developmental trajectory.

One important direction for future research is investigating the functional link between empathy, prosocial behavior, and reward. Several studies have documented that the frontomesolimbic reward network is engaged to the same extent when individuals receive monetary rewards and when they freely choose to donate money to charity, and even more so when they are observed by others, suggesting that altruism draws on general mammalian neural systems of reward and social attachment (e.g., Izuma et al., 2010; Moll et al., 2006). Thus, prosocial behaviors may stem from a plurality of motives and intertwined social and motivational contingencies. For instance, witnessing people in pain or distress triggers a neural response associated with aversion. This response, in turn, may initiate helping or soothing behaviors motivated to both reduce one's own discomfort and to feel good about oneself, as well as to lessen another's distress. These behaviors may be reinforced by both endogenous reward (dopamine system) and positive social feedback from others.

Given the importance of empathy for healthy social interaction, it is clear that a developmental approach using functional neuroimaging to elucidate the computational mechanisms underlying affective reactivity, regulation and behavioral outcomes is essential to complement traditional behavioral methods and gain a better understanding of how deficits may arise in the context of development.

Acknowledgment

The writing of this paper was supported by a grant (# BCS-0718480) from the National Science Foundation and a grant from NIH/NIMH (# MH84934-01A1) to J Decety.

References

Akitsuki, Y., Decety, J., 2009. Social context and perceived agency affects empathy for pain: An event-related fMRI investigation. NeuroImage 47, 722–734.

Avenanti, A., Bueti, D., Galati, G., Aglioti, S.M., 2005. Transcranial magnetic stimulation highlights the sensorimotor side of empathy for pain. Nature Neuroscience 8, 955–960.

Bartsch, K., Wellman, H.M., 1995. Children Talk About the Mind. Oxford University Press, New York.

Batson, C.D., 2009. These things called empathy: Eight related but distinct phenomena. In: Decety, J., Ickes, W. (Eds.), The Social Neuroscience of Empathy. MIT Press, Cambridge, MA, pp. 3–15.

Batson, C.D., Fultz, J., Schoenrade, P., 1987. Distress and empathy: Two qualitatively distinct vicarious emotions with different motivational consequences. Journal of Personality 55, 19–39.

Berkowitz, L., 1993. Pain and aggression: Some findings and implications. Motivation and Emotion 17, 277–293.

Berkowitz, L., 2003. Affect, aggression, and antisocial behavior. In: Davidson, R.J., Scherer, K.R., Goldsmith, H.H. (Eds.), Handbook of Affective Sciences. Oxford University Press, New York, pp. 804–823.

Blair, R.J.R., 2005. Responding to the emotions of others: Dissociating forms of empathy through the study of typical and psychiatric populations. Consciousness and Cognition 14, 698–718.

Blair, R.J.R., Blair, K.S., 2009. Empathy, morality, and social convention: Evidence from the study of psychopathy and other psychiatric disorders. In: Decety, J., Ickes, W. (Eds.), The Social Neuroscience of Empathy. MIT Press, Cambridge, MA, pp. 139–152.

Brown, J.R., Dunn, J., 1996. Continuities in emotion understanding from three to six years. Child Development 67, 789–802.

Brunet, E., Sarfati, Y., Hardy-Bayle, M.C., Decety, J., 2000. A PET investigation of attribution of intentions to others with a non-verbal task. NeuroImage 11, 157–166.

Bufalari, I., Aprile, T., Avenanti, A., DiRusso, F., Aglioti, S.M., 2007. Empathy for pain and touch in the human somatosensory cortex. Cerebral Cortex 17, 2553–2561.

Bunge, S.A., Dudukovic, N.M., Thomasson, M.E., Vaidya, C.J., Gabrieli, J.D.E., 2002. Immature frontal lobe contributions to cognitive control in children: Evidence from fMRI. Neuron 33, 301–311.

Carlson, S.M., Moses, L.J., 2001. Individual differences in inhibitory control and children's theory of mind. Child Development 72, 1032–1053.

Carroll, J.J., Steward, M.S., 1984. The role of cognitive development in children's understandings of their own feelings. Developmental Psychology 55, 1486–1492.

Casey, B.J., Tottenham, N., Liston, C., Durston, S., 2005. Imaging the developing brain: What have we learned about cognitive development? Trends in Cognitive Sciences 9, 104–110.

Cheng, Y., Lin, C., Liu, H.L., et al., 2007. Expertise modulates the perception of pain in others. Current Biology 17, 1708–1713.

Cheng, Y., Yang, C.Y., Ching-Po, L., Lee, P.L., Decety, J., 2008. The perception of pain in others suppresses somatosensory oscillations: A magnetoencephalography study. NeuroImage 40, 1833–1840.

Cheng, Y., Hung, A., Hung, D., Decety, J., 2012. Dissociation between affective sharing and emotion understanding in juvenile psychopaths. Development and Psychopathology 24, 623–636.

Coccaro, E., McCloskey, M., Fitzgerald, D., Phan, K., 2007. Amygdala and orbitofrontal reactivity to social threat in individuals with impulsive aggression. Biological Psychiatry 62, 168–178.

Craig, A.D., 2003. Interoception: The sense of the physiological condition of the body. Current Opinion in Neurobiology 13, 500–505.

Craig, A.D., 2004. Human feelings: Why are some more aware than others? Trends in Cognitive Sciences 8, 239–241.

Davidson, R.J., Putnam, K.M., Larson, C.L., 2000. Dysfunction in the neural circuitry of emotion regulation – A possible prelude to violence. Science 289, 591–594.

De Haan, M., Gunnar, M.R., 2009. The brain in a social environment. Why study development? In: De Haan, M., Gunnar, M.R. (Eds.), Handbook of Developmental Social Neuroscience. The Guilford Press, New York, pp. 3–10.

de Wied, M., Van Boxtel, A., Zaalberg, R., Goudena, P.P., Matthys, M., 2006. Facial EMG responses to dynamic emotional facial expressions in boys with disruptive behavior disorders. Journal of Psychiatric Research 40, 112–121.

Decety, J., 2005. Perspective taking as the royal avenue to empathy. In: Malle, B.F., Hodges, S.D. (Eds.), Other Minds: How Humans Bridge the Divide Between Self and Others. Guilford Publications, New York, pp. 135–149.

Decety, J., 2011. Dissecting the neural mechanisms mediating empathy. Emotion Review 3, 92–108.

Decety, J., Jackson, P.L., 2004. The functional architecture of human empathy. Behavioral and Cognitive Neuroscience Reviews 3, 71–100.

Decety, J., Lamm, C., 2009. Empathy versus personal distress – Recent evidence from social neuroscience. In: Decety, J., Ickes, W. (Eds.), The Social Neuroscience of Empathy. MIT Press, Cambridge, MA, pp. 199–213.

Decety, J., Meyer, M., 2008. From emotion resonance to empathic understanding: A social developmental neuroscience account. Development and Psychopathology 20, 1053–1080.

Decety, J., Michalska, K.J., 2010. Neurodevelopmental changes in the circuits underlying empathy and sympathy from childhood to adulthood. Developmental Science 13, 886–899.

Decety, J., Moriguchi, Y., 2007. The empathic brain and its dysfunction in psychiatric populations: Implications for intervention across different clinical conditions. BioPsychoSocial Medicine 1, 22–65.

Decety, J., Michalska, K.J., Akitsuki, Y., 2008. Who caused the pain? A functional MRI investigation of empathy and intentionality in children. Neuropsychologia 46, 2607–2614.

Decety, J., Michalska, K.J., Akitsuki, Y., Lahey, B., 2009. Atypical empathic responses in adolescents with aggressive conduct disorder: A functional MRI investigation. Biological Psychology 80, 203–211.

Diamond, A., 2002. Normal development of prefrontal cortex from birth to young adulthood: Cognitive functions, anatomy, and biochemistry. In: Stuss, D.T., Knight, R.T. (Eds.), Principles of Frontal Lobe Function. Oxford University Press, New York, pp. 446–503.

Dimberg, U., Thunberg, M., Elmehed, K., 2000. Unconscious facial reactions to emotional facial expressions. Psychological Science 11, 86–89.

Dondi, M., Simion, F., Caltran, G., 1999. Can newborns discriminate between their own cry and the cry of another newborn infant? Developmental Psychology 35, 418–426.

Eisenberg, N., 2005. Age changes in prosocial responding and moral reasoning in adolescence and early adulthood. Journal of Research on Adolescence 15, 235–260.

Eisenberg, N., Eggum, N.D., 2009. Empathic responding: Sympathy and personal distress. In: Decety, J., Ickes, W. (Eds.), The Social Neuroscience of Empathy. MIT Press, Cambridge, MA, pp. 71–83.

Eisenberg, N., Shea, C.L., Carlo, G., Knight, G.P., 1991. Empathy-related responding and cognition: A chicken and the egg dilemma. In: Kurtines, W.M. (Ed.), Handbook of Moral Behavior and Development. Research 2, Erlbaum, Hillsdale, NJ, pp. 63–88.

Eisenberg, N., Fabes, R.A., Murphy, B., et al., 1994. The relations of emotionality and regulation to dispositional and situational empathy-related responding. Journal of Personality and Social Psychology 66, 776–797.

Field, T.M., Woodson, R., Greenberg, R., Cohen, D., 1982. Discrimination and imitation of facial expression by neonates. Science 219, 179–181.

Frith, U., Frith, C.D., 2003. Development and neurophysiology of mentalizing. Philosophical Transactions of the Royal Society of London B358, 459–473.

Gloor, P., Olivier, A., Quesney, L.F., Andermann, F., Horowitz, S., 1982. The role of the limbic system in experiential phenomena of temporal lobe epilepsy. Annals of Neurology 12, 129–144.

Gollan, J.K., Lee, R., Coccaro, E.F., 2005. Developmental psychopathology and neurobiology of aggression. Development and Psychopathology 17, 1151–1171.

Gross, A.L., Ballif, B., 1991. Children's understanding of emotion from facial expressions and situations: A review. Developmental Review 11, 368–398.

Haidt, J., Graham, J., 2007. When morality opposes justice: Conservatives have moral intuitions that liberals may not recognize. Social Justice Research 20, 98–116.

Harris, P.L., 1994. The child's understanding of emotion: Developmental change and the family environment. Journal of Child Psychology and Psychiatry 35, 3–28.

Harris, P.L., 2000. Understanding emotion. In: Lewis, M., Haviland-Jones, J.M. (Eds.), Handbook of Emotions. Guilford Press, New York, pp. 281–292.

Harris, P.L., Olthof, T., Meerum-Terwogt, M., 1981. Children's knowledge of emotion. Journal of Child Psychology and Psychiatry 22, 247–261.

Hastings, P.D., Zahn-Waxler, C., Robinson, J., Usher, B., Bridges, D., 2000. The development of concern for others in children with behavior problems. Developmental Psychology 36, 531–546.

Haviland, J.M., Lewica, M., 1987. The induced affect response: Ten-week-old infants' responses to three emotion expressions. Developmental Psychology 23, 97–104.

Hodges, S.D., Klein, K.J.K., 2001. Regulating the costs of empathy: The price of being human. Journal of Socio-Economics 30, 437–452.

Hoffman, M.L., 2000. Empathy and Moral Development: Implications for Caring and Justice. Cambridge University Press, Cambridge.

Hughes, C., Dunn, J., 1998. Understanding mind and emotion: Longitudinal associations with mental state talk between young friends. Developmental Psychology 34, 1026–1037.

Hurliman, E., Nagode, J.C., Pardo, J.V., 2005. Double dissociation of exteroceptive and interoceptive feedback systems in the orbital and ventromedial prefrontal cortex of humans. Journal of Neuroscience 25, 4641–4648.

Izard, C.E., 1982. Measuring Emotions in Infants and Young Children. Cambridge Press, New York.

Izuma, K., Saito, D.N., Sadato, N., 2010. Processing of the incentive for social approval in the ventral striatum during charitable donation. Journal of Cognitive Neuroscience 22, 621–631.

Jackson, P.L., Rainville, P., Decety, J., 2006. To what extent do we share the pain of others? Insight from the neural bases of pain empathy. Pain 125, 5–9.

Kawashima, R., Sato, K., Itoh, H., et al., 1996. Functional anatomy of go/no-go discrimination and response selection – A PET study in man. Brain Research 728, 79–89.

Killgore, W.D.S., Yurgelun-Todd, D.A., 2007. Unconscious processing of facial affect in children and adolescents. Social Neuroscience 2, 28–47.

Killgore, W.D.S., Oki, M., Yurgelun-Todd, D.A., 2001. Sex-specific developmental changes in amygdala responses to affective faces. NeuroReport 12, 427–433.

Knafo, A., Zahn-Waxler, C., Van Hulle, C., Robinson, J.L., Rhee, S.H., 2008. The developmental origins of a disposition toward empathy: Genetic and environmental contributions. Emotion 8, 737–752.

Kringelbach, M.L., Rolls, E.T., 2004. The functional neuroanatomy of the human orbitofrontal cortex: Evidence from neuroimaging and neuropsychology. Progress in Neurobiology 72, 341–372.

Lahey, B.B., Waldman, I.D., 2003. A developmental propensity model of the origins of conduct problems during childhood and adolescence. In: Lahey, B.B., Moffitt, T.E., Caspi, A. (Eds.), Causes of Conduct Disorder and Juvenile Delinquency. Guilford Press, New York, pp. 76–117.

Lamm, C., Batson, C.D., Decety, J., 2007. The neural substrate of human empathy: Effects of perspective-taking and cognitive appraisal. Journal of Cognitive Neuroscience 19, 42–58.

Lewis, M., 2000. Self-conscious emotions: Embarrassment, pride, shame, and guilt. In: Lewis, M., Haviland, J.M. (Eds.), Handbook of Emotions. Guilford Press, New York, pp. 623–636.

Marsh, A.A., Finger, E.C., Mitchell, D.G.V., et al., 2008. Reduced amygdala response to fearful expressions in children and adolescents with callous unemotional traits and disruptive behavior disorders. American Journal of Psychiatry 165, 712–720.

Martin, G.B., Clark, R.D., 1987. Distress crying in neonates: Species and peer specificity. Developmental Psychology 18, 3–9.

Moll, J., Krueger, F., Zahn, R., Pardini, M., de Oliveira-Souza, R., Grafman, J., 2006. Human fronto–mesolimbic networks guide decisions about charitable donation. Proceedings of the National Academy of Sciences 103, 15623–15628.

Moriguchi, Y., Ohnishi, T., Mori, T., Matsuda, H., Komaki, G., 2007. Changes of brain activity in the neural substrates for theory of mind in childhood and adolescence. Psychiatry and Clinical Neurosciences 61, 355–363.

Murray, E.A., 2007. The amygdala, reward and emotion. Trends in Cognitive Sciences 11, 489–497.

Nunner-Winkler, G., Sodian, B., 1988. Children's understanding of moral emotions. Child Development 59, 1323–1338.

Ochsner, K.N., Gross, J.J., 2005. The cognitive control of emotion. Trends in Cognitive Sciences 9, 242–249.

Ochsner, K.N., Bunge, S.A., Gross, J.J., Gabrieli, J.D.E., 2002. Rethinking feelings: An fMRI study of cognitive regulation of emotion. Journal of Cognitive Neuroscience 14, 1215–1229.

Olson, I.R., Plotzker, A., Ezzyat, Y., 2007. The enigmatic temporal pole: A review of social and emotional processing. Brain 130, 1718–1731.

Phan, K.L., Fitzgerald, D.A., Nathan, P.J., Moore, G.J., Uhde, T.W., Tancer, M.E., 2005. Neural substrates for voluntary suppression of negative affect: A functional magnetic resonance imaging study. Biological Psychiatry 57, 210–219.

Posner, M.I., Rothbart, M.K., 2000. Developing mechanisms of self-regulation. Development and Psychopathology 12, 427–441.

Raine, A., 1997. Antisocial behavior and psychophysiology: A biosocial perspective and a prefrontal dysfunction hypothesis. In: Stoff, D., Breiling, J., Maser, J.D. (Eds.), Handbook of Antisocial Behavior. New York, Wiley, pp. 289–304.

Raine, A., 2002. Biosocial studies of antisocial and violent behavior in children and adults: A review. Journal of Abnormal Child Psychology 30, 311–326.

Repacholi, B.M., Gopnik, A., 1997. Early reasoning about desires: Evidence from 14- and 18-month-olds. Developmental Psychology 33, 12–21.

Robinson, J., 2008. Empathy and prosocial behavior. Encyclopedia of Infant and Early Childhood Development 1, 441–450.

Rothbart, M.K., Ahadi, S.A., Hershey, K.L., 1994. Temperament and social behavior in childhood. Merrill-Palmer Quarterly 40, 21–39.

Russell, J., 1996. Agency and its Role in Mental Development. Psychology Press, Hove.

Sagi, A., Hoffman, M.L., 1976. Empathic distress in the newborn. Developmental Psychology 12, 175–176.

Saxe, R.R., Whitfield-Gabrieli, S., Pelphrey, K.A., Scholz, J., 2009. Brain regions for perceiving and reasoning about other people in school-aged children. Child Development 80, 1197–1209.

Shamay-Tsoory, S.G., Tibi-Elhanany, Y., Aharon-Peretz, J., 2006. The ventromedial prefrontal cortex is involved in understanding affective but not cognitive theory of mind stories. Social Neuroscience 1, 149–166.

Sonnby-Borgstrom, M., Jonsson, P., Svensson, O., 2003. Emotional empathy as related to mimicry reactions at different levels of information processing. Journal of Nonverbal Behavior 27, 3–23.

Sowell, R.E., Thompson, P.M., Holmes, C.J., Jernigan, T.L., Toga, A.W., 1999. In vivo evidence for post-adolescent brain maturation in frontal and striatal regions. Nature Neuroscience 2, 859–861.

Stone, V.E., Gerrans, P., 2006. What's domain-specific about theory of mind? Social Neuroscience 1, 309–319.

Strum, V.E., Rosen, H.J., Allison, S., Miller, B.L., Levenson, R.W., 2006. Self-conscious emotion deficits in frontotemporal lobar degeneration. Brain 129, 2508–2516.

Swick, D., Ashley, V., Turken, A.U., 2008. Left inferior frontal gyrus is critical for response inhibition. BMC Neuroscience 9, 102e.

Tamm, L., Menon, V., Reiss, A.L., 2002. Maturation of brain function associated with response inhibition. Journal of the American Academy of Child and Adolescent Psychiatry 41, 1231–1238.

Thomson, R.A., 1994. Emotion regulation: A theme in search of definition. Monographs of the Society for Research in Child Development 59, 24–52.

Tonks, J., Williams, H., Frampton, I., Yates, P., Slater, A., 2007. Assessing emotion recognition in 9- to 15-years olds: Preliminary analysis of abilities in reading emotion from faces, voices and eyes. Brain Injury 21, 623–629.

Vaish, A., Carpenter, M., Tomasello, M., 2009. Sympathy through affective perspective-taking, and its relation to prosocial behavior in toddlers. Developmental Psychology 45, 534–543.

Yamada, M., Decety, J., 2009. Unconscious affective processing and empathy: An investigation of subliminal priming on the detection of painful facial expressions. Pain 143, 71–75.

Yurgelun-Todd, D., 2007. Emotional and cognitive changes during adolescence. Current Opinion in Neurobiology 17, 251–257.

Zahn-Waxler, C., Radke-Yarrow, M., Wagner, E., Chapman, M., 1992. Development of concern for others. Developmental Psychology 28, 126–136.

Zahn-Waxler, C., Cole, P.M., Welsh, J.D., Fox, N.A., 1995. Psychophysiological correlates of empathy and prosocial behaviors in preschool children with behavior problems. Development and Psychopathology 7, 27–48.

Zaki, J., Weber, J., Bolger, N., Ochsner, K., 2009. The neural bases of empathic accuracy. Proceedings of the National Academy of Sciences 106, 11382–11387.

Zelazo, P.D., Craik, F.I., Booth, L., 2004. Executive function across the life span. Acta Psychologica 115, 167–183.

Zelazo, P., Carlson, S., Kesek, A., 2008. The development of executive function in childhood. In: Nelson, C.A., Luciana, M. (Eds.), Handbook of Developmental Cognitive Neuroscience. MIT Press, Cambridge, MA, pp. 553–574.

Developing Attention and Self-Regulation in Infancy and Childhood

M.I. Posner[1], M.K. Rothbart[1], M.R. Rueda[2]

[1]University of Oregon, Eugene, OR, USA; [2]University of Granada, Granada, Spain

OUTLINE

22.1 INTRODUCTION

William James (1890) defined attention as follows:

> Attention is the taking possession of the mind in clear and vivid form of one out of what seem several simultaneous objects or trains of thought.

James' definition, however, provides little perspective toward an understanding of the normal development of attention or its pathologies. The theme of this chapter is that it is now possible to view attention much more concretely as an organ system. The authors follow the Webster's Dictionary definition of an organ system:

> An organ system may be defined as differentiated structures in animals and plants made up of various cells and tissues and adapted for the performance of some specific function and grouped with other structures into a system.

The authors believe that viewing attention as an organ system aids in answering many perplexing issues raised in developmental psychology, psychiatry, and neurology. Neuroimaging studies have systemically shown that a wide variety of cognitive tasks activate a distributed set of neural areas, each of which can be identified with specific mental operations (Posner and Raichle, 1994, 1998). These areas of activation may be more consistent for the study of attention than for any other cognitive system. Attention can be viewed as involving

Neural Circuit Development and Function in the Brain: Comprehensive Developmental Neuroscience, Volume 3 http://dx.doi.org/10.1016/B978-0-12-397267-5.00059-5

© 2013 Elsevier Inc. All rights reserved.

specialized networks to carry out functions such as achieving and maintaining the alert state, orienting to sensory events, and controlling thoughts and feelings.

The goals of this chapter are first to delineate the attentional networks of the human brain (Section 22.2), mostly from adult studies. There are obvious changes in behavior between infancy and adulthood, and in the next two sections (Sections 22.3 and 22.4), the authors describe brain changes at these ages and seek to link them to developmental differences in behavior. Section 22.5 examines changes in the attention networks themselves during child development and discusses how adult control relates to these changes. The authors use the connection between attention networks and neuromodulators to discuss some of the genes that influence individual differences and network development (Section 22.6). Finally, some likely new areas for research are outlined (Section 22.7).

22.2 BRAIN NETWORKS

22.2.1 Taxonomies of Attention

Although many efforts have been made to develop taxonomies of attention, imaging studies have suggested that at least three somewhat independent networks are involved in different aspects of attention, carrying out the functions of alerting, orienting, and executive attention (Posner and Fan, 2008) (Figure 22.1). Alerting refers to achieving and maintaining a state of high sensitivity to incoming stimuli, orienting refers to the selection of information from sensory input, and executive attention includes mechanisms for monitoring and resolving conflict between thoughts, feelings, and responses.

The brain network involved in achieving and maintaining the alert state is represented by squares in Figure 22.1. Alertness is an important prerequisite for other attentional operations. Although the alert state is often contrasted with sleep, it is probably better to consider it in relation to the default state. The default state is defined in studies using functional magnetic resonance imaging (fMRI) with an instruction not to actively process anything (Raichle, 2009). The default state is characterized by a slow oscillation between two large-scale networks. A warning signal moves the brain away from the default state toward a high level of alertness. This change involves widespread variation in autonomic signals such as heart rate (Kahneman, 1973) and cortical changes, for example, a negative shift in the scalp recorded an electroencephalogram called the contingent negative variation (CNV) (Walter, 1964).

Much of attention research involves orienting to sensory events. The ease with which the experimenter can control the presentation of visual and auditory stimuli probably accounts for the popularity of studying orienting in different organisms and ages. Precisely controlling the presentation of stimulation also allows the study of overt and covert forms of orientation. Overt orienting involves eye movements, head movements, or both toward the source of stimulation and usually occurs when sufficient time is allowed between the presentation of an orienting cue and the target. Covert orienting involves only orientation of attention and can be studied when not enough time is allowed to move the eyes, head, or both or by instructing individuals to attend to the cue

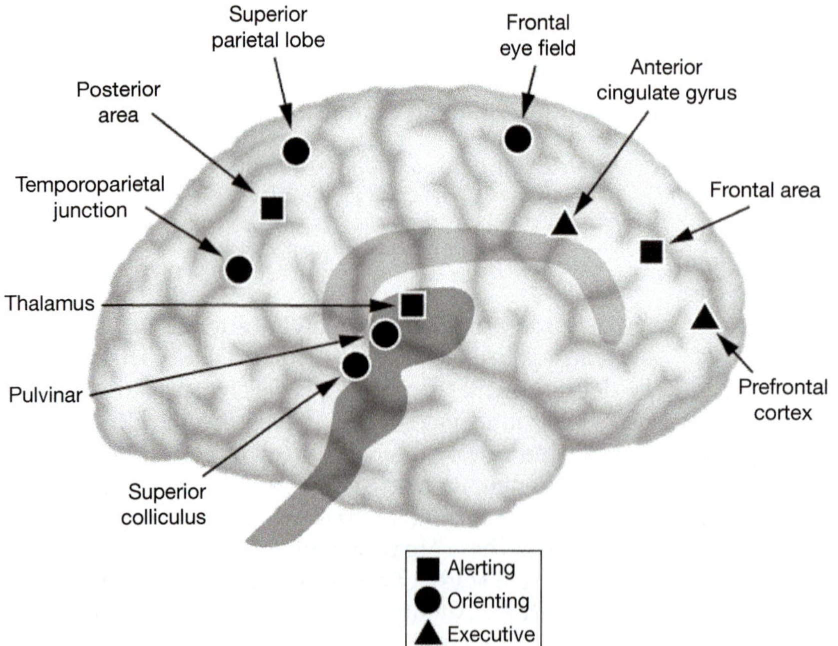

FIGURE 22.1   The alerting, orienting, and executive attention networks. The triangles, circles, and squares indicate nodes of activation of each network.

without looking at it. The close relation of eye movements and covert shifts of attention has led to studies examining their relationship. In general, the same brain areas active before saccades also are active before covert shifts of attention (Corbetta and Shulman, 2002). However, cellular studies of alert monkeys have found that within the frontal eye fields there exist two different but overlapping populations of cells. One is active before saccades and the other before attention shifts not leading to saccades (Thompson et al., 2005). These findings suggest that eye, head, and covert attention movements become coordinated during early development so that in adults it becomes difficult, but still possible, to separate them.

Every sensory signal provides input to both sensory-specific cortical pathways and brain stem arousal systems related to alerting. Together, these signals may yield changes in a brain network of parietal and frontal areas (see circles in Figure 22.1) that orchestrate covert shifts of attention. The orienting network acts to boost the strength of input signals in sensory-specific pathways in comparison with nonattended signals.

The executive attention network is involved in the regulation of feelings (emotions), thoughts (cognitions), and actions (Posner and Rothbart, 2007a,b). The anterior cingulate gyrus, one of the main nodes of the executive attention network, has been linked to a variety of specific self-regulating functions. These include the monitoring of conflict (Botvinick et al., 2001), control of working memory (Duncan et al., 2000), regulation of emotion (Bush et al., 2000), and response to error (Holroyd and Coles, 2002). In emotional studies, the cingulate often is seen as part of a network involving the orbital frontal cortex and the amygdala that regulates our emotional response to input. Activation of the anterior cingulate is observed when people are asked to control their natural reactions to strong positive (Beauregard et al., 2001) or negative emotions (Ochsner et al., 2002). Analysis of the functional connectivity between brain areas has shown that when emotionally neutral sensory information is involved, there is strong connectivity between the dorsal anterior cingulate cortex (ACC) and the relevant sensory area (Crottaz-Herbette and Mennon, 2006); when emotional control is involved, there is functional connectivity between the ventral ACC and the amygdala (Etkin et al., 2006).

22.2.2 Sites and Sources of Attention

Normally, all sensory events contribute to both a state of alertness and an orienting of attention. In order to distinguish the brain areas involved in alerting and orienting (sources) from the sites at which they operate, it is useful to separate the presentation of a cue indicating where a target will occur from the presentation of the target requiring a response (Corbetta and Shulman, 2002; Posner, 1978). This methodology has been used in behavioral studies with normal individuals (Posner, 1978), patients (Posner and Fan, 2008), and monkeys (Marrocco and Davidson, 1998), and in studies using scalp electrical recording and event-related neuroimaging (Corbetta and Shulman, 2002). Two types of cue are of interest. Some cues provide information only on when the target will occur. These warning signals lead to changes in a network of brain areas related to alerting. Other cues provide information on aspects of the target, such as where it will occur, and lead to changes in the orienting network.

Studies using event-related fMRI have shown that following the presentation of the cue and before the target is presented, a network of brain areas becomes active (Corbetta and Shulman, 2002). There is widespread agreement on the identity of these areas, but a considerable amount of work remains to be done to understand the function of each area.

When a target is presented in isolation at the cued location, the subsequent target is processed more efficiently than when no cue to its location has been presented (see Posner and Fan, 2008, for a review). The brain areas influenced by orienting will be those that would normally be used to process the target. For example, in the visual system, orienting can influence sites of processing in the primary visual cortex or in a variety of extrastriate visual areas where the computations related to the target are performed. Orienting to target motion influences area MT (V5) while orienting to target color influences area V4. This principle of activation of brain areas also extends to higher-level visual input. For example, attention to faces modifies activity in the face-sensitive area of the fusiform gyrus. The finding that attention can modify activity in primary visual areas has been of particular importance because this brain area has been more extensively studied than any other. When multiple targets are presented, they tend to suppress the normal level of activity that they would have produced if presented in isolation. One important role of orienting to a particular location is to provide a relative enhancement of the target at that location in comparison with other items presented in the visual field.

22.3 BRAIN CHANGES IN HUMAN DEVELOPMENT

Much of this handbook reviews the developmental brain changes that have been traced in nonhuman animals. They include material on migration of cells into cortical areas and the changes that take place in synaptic density and myelination with age. Also well

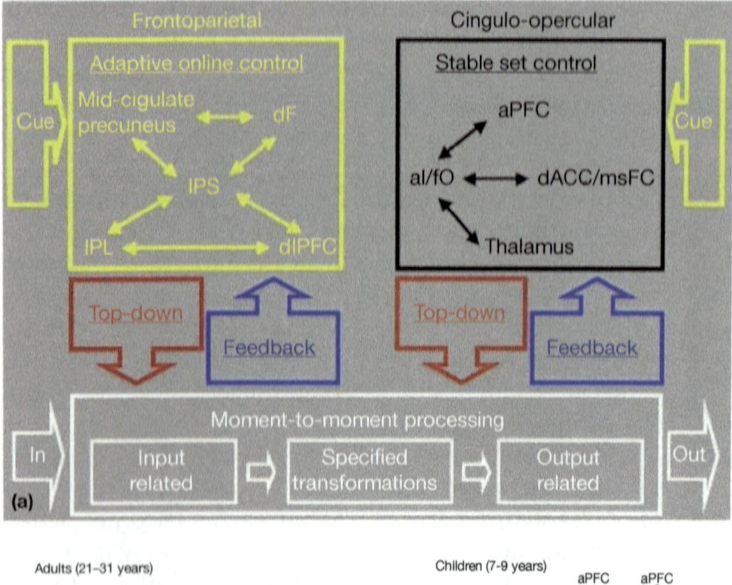

FIGURE 22.2 The frontoparietal (orienting) and cingulo-opercular (executive) networks following the work of Dosenbach et al. (2007) (a) Greater long connections and more separation of networks in adults (b) than children (c).

documented in this volume is the role of experience in helping to sculpt gray and white matter development. This chapter deals with efforts to trace changes in development in resting connectivity and in brain activation during cognitive tasks in humans using fMRI.

An important finding was that even at rest, there is a common set of brain areas that appear to be active together (default state). While evidence for connection between brain areas related to attention is found even in infancy (Gao et al., 2009), studies suggest that the connectivity between these areas changes over the course of development. Figure 22.2 illustrates two sets of attention-related brain networks that are active at rest: they are a network of frontoparietal brain areas (related to orienting) and a cingulo-opercular (related to executive attention) network. The frontoparietal network (see top left panel of Figure 22.2) in adults is involved in short-term control operations common when orienting to sensory signals. The cinguloparietal network (see top right panel of Figure 22.2) is involved in longer, more strategic control that fits well with an executive system (Dosenbach et al., 2007).

Connections change over the life span. The bottom panels of Figure 22.2 show that adults (left panel) have separate networks related to orienting and executive attention, while these are more integrated in children. Children at age 9 years show many shorter connections.

Adults show more segregation of the two networks and longer connections (Dosenbach et al., 2007; Fair et al., 2007, 2008). Because resting connectivity analysis requires no task, it has been studied during infancy (Gao et al., 2009). During the first year of life, the anterior cingulate shows little or no connectivity to other areas. After the first year, infants begin the slow process of developing the long-range connectivity that is typical of adults.

Some of the same brain areas found active during rest change when the person is given a task. For example, while the organization of anatomical areas in alerting and orienting is not fully known, some promising beginnings have taken place (see Posner, 2008, for a review). In alerting, the source of attention appears to be the locus coeruleus (lc). Cells in the lc have two modes of processing. One mode is sustained and is perhaps related to the tonic level of alertness over long time intervals. This function is known to involve the right cerebral hemisphere more strongly than the left. Alertness is influenced by sensory events and by the diurnal rhythm. However, its voluntary maintenance during task performance may be orchestrated from the anterior cingulate. More phasic shifts of alerting can result from presenting any environmental signal. However, if the signal is likely to warn of an impending target, this shift results in a characteristic suppression of the intrinsic brain rhythms (e.g., alpha) within a few tens of milliseconds and a strong negative wave (CNV) recorded from surface electrodes that moves from a frontal generator toward the sensory areas of the hemisphere opposite the expected target.

Another aspect of attention, especially prominent in the transition between infancy and early childhood, is self-regulation. Effortful and voluntary control is central to a broad range of abilities in executive functioning, such as error detection, planning, memory, and problem solving. There is evidence that effortful control (EC) is linked to the executive attention network (see triangles in Figure 22.1).

It has been possible to show that infants carry out at least one of the important functions related to the anterior cingulate, the recognition of error. In one study, 7-month-old infants were found to look longer at erroneous rather than correct events (Wynn, 1992). Error events appear to cause a scalp negative electrical potential over a set of electrodes that have been localized in the anterior cingulate (Berger et al., 2006; Dehaene et al., 1994). However, the typical regulation of behavior found in adults, that is, to slow down following an error, does not seem to emerge until about the age of 3 years (Jones et al., 2003). Together, the connectivity and task data fit well with the idea that infants can process information and use executive attention to a limited degree, but they are unable to show regulation of their behavior by this network.

22.4 BEHAVIORAL DEVELOPMENT IN INFANCY

Attention in infancy is less developed than it is later in life, and the functions of alerting, orienting, and, particularly, executive control are less independent during infancy. Alerting and orienting are examined first, and then, executive attention in relation to self-regulation is considered. The measurement of these variables must be different in infancy than later when voluntary responses can be directed by the experimenter. Efforts have been made to design tasks that can be performed by infants that tap the same networks of brain areas shown in Figure 22.1.

22.4.1 Alerting and Orienting

The early life of the infant is concerned with changes in state. Sleep dominates at birth, and the waking state is relatively rare at first. The newborn infant spends nearly three-quarters of the time sleeping at birth (Colombo and Horowitz, 1987). Many of the changes in the alert state depend on external input. Arousal of the central nervous system involves input from brain stem systems that modulate activation of the cortex. As in adults, primary

among these is the lc, which is the source of the brain's norepinephrine. It has been demonstrated that the influence of warning signals operates via this brain system, since drugs that block it also prevent the changes in the alert state that lead to improved performance after a warning signal has been provided (Marrocco and Davidson, 1998). It is likely that the endogenous changes during waking that take place without external input also involve this system.

There is a dramatic change in the percentage of time in the waking state over the first 3 months of life. By the 12th postnatal week, the infant has become capable of maintaining the alert state during much of the daytime. This ability still depends heavily on external stimulation, much of it provided by the caregiver.

Much of the response to external stimuli involves orienting toward a stimulus. Newborns show head and eye movements toward novel stimuli. Eye movements are preferentially directed toward moving stimuli and have been shown to involve properties of the stimulus, for example, how much they resemble human faces (Johnson and Morton, 1991). It has also been demonstrated that newborns can make imitative responses. When shown, for example, a face with a protruding tongue, they respond with a similar movement (Meltzoff and Moore, 1977). However, the reliability and complexity of these responses to sensory input change dramatically over the first months of life.

The most frequent method of studying orienting in infancy involves tracking of saccadic eye movements. As in adults, there is a close relation, but not identity, between the direction of gaze and the infants' attention. The attention system can be driven by external input from birth (Richards and Hunter, 1998); however, the system continues to improve in precision over many years. Infant eye movements often fall short of the target, and peripheral targets are often foveated by a series of head and eye movements. Although not as easy to track, the covert system likely follows a similar trajectory. Studies that have attempted to examine the covert system by the use of brief cues that do not produce an eye movement, followed by targets that do, show that the speed of the eye movement to the target is enhanced by the cue and this enhancement improves over the first year of life (Butcher, 2000). In more complex situations, for example, when there are competing targets, the improvement may continue for longer periods.

Orienting to sensory input is a major mechanism for regulating distress. Infants often have a hard time disengaging from high spatial frequency targets and may become distressed before they are able to move away from the target. Caregivers provide a hint of how attention is used to regulate the state of the infant when they attempt to distract the infants by bringing their attention to other stimuli. As infants orient, they are often quieted, and their distress appears to diminish. In one study, 3- to 6-month-old infants were first shown a sound and light display; about 50% of the infants became distressed by the stimulation, but then they strongly oriented to interesting visual and auditory soothing events when these were presented (Harman et al., 1997). While the children oriented, facial and vocal signs of distress disappeared. However, as soon as the orienting stopped, for example, when the object was removed, the infants' distress returned to almost the same levels as before the presentation of the soothing object. An internal system, which was termed the *distress keeper*, and which the authors believe involves the amygdala, appears to hold a computation of the initial level of distress so that it returns if the infant's orientation to the novel event is lost. Interestingly, infants were quieted by distraction for as long as 1 min, without changing the eventual level of distress reached once the orienting ended (Harman et al., 1997).

For newborn infants, the control of orienting is initially largely in the hands of caregiver presentations. By 4 months, however, infants have gained considerable control over disengaging their gaze from one visual location and moving it to another, and greater orienting skill in the laboratory is associated with lower temperamental negative emotion and greater soothability as reported by parents (Johnson et al., 1991). Late infancy is the time when self-regulation develops. Increasingly, infants are able to gain control of their emotions and other behaviors. This transition marks the development of the executive attention system.

22.4.2 Executive Attention in Infancy

It is difficult to assess executive attention in infants because, as outlined earlier, caregivers provide most of the regulation of infant behavior. Effortful control (EC) is a high-level factor from parental reports on children's temperament (Rothbart and Rueda, 2005). This factor is defined as the ability to withhold a dominant response to carry out a nondominant one. Parents observing their children's behavior in particular daily-life situations (e.g., putting away toys on command) can readily respond to questions that relate to this factor. This can be done for children about 2 years of age and older. Below this age, temperament questionnaires (Rothbart et al., 1994) are confined to factors such as orienting and positive and negative affect. Older children receive scores on EC. Moreover, children older than 2 years can be scored on tasks that involve voluntary responding, such as pressing keys to visual input.

As mentioned earlier, there is evidence of the presence of the executive attention network at about 7 months for the detection of error. Later in the first year of life, there is evidence of further development of executive attention. One example is Diamond's work using the 'A not B' task and the reaching task. These two marker tasks involve inhibition of an action that is strongly elicited by the situation. In the 'A not B' task, the experimenter shifts the location of a hidden object from location A to location B after the infant's retrieving from location A had been reinforced as correct in the previous trials (Diamond, 1991). In the reaching task, visual information about the correct route to a toy is put in conflict with the cues that normally guide reaching. A toy is placed under a transparent box. The opening of the box is on the side (it can be the front side, the back side, etc.), and the infant can reach it only if the tendency to reach directly along the line of sight through the transparent top of the box is inhibited. Important changes in performance on these tasks are observed from 6 to 12 months. Comparison of performance between monkeys with brain lesions and human infants on the same marker tasks suggests that the tasks are sensitive to the development of the prefrontal cortex, and maturation of this brain area seems to be critical for the development of this form of inhibition.

Another task that reflects the executive system involves anticipatory looking in a visual sequence task (Clohessy et al., 2001; Haith et al., 1988). In the visual sequence task, stimuli are presented to the infant in a fixed and predictable sequence of locations. The infant's eyes are drawn reflexively to the stimuli because they are designed to be attractive and interesting. After a few trials, some infants will begin to anticipate the location of the next target by correctly moving their eyes in anticipation of the target. It has been shown that anticipatory looking occurs with infants as young as 3.5–4 months (Clohessy et al., 2001; Haith et al., 1988). Learning more complex sequences of stimuli, such as sequences in which a location is followed by one or two or more different locations, the particular location depending on the location of the previous stimulus within the sequence (e.g., 121312...), requires the monitoring of context and, in adult studies, has been shown to depend on the lateral prefrontal cortex (Keele et al., 2003). It was found that infants of 4 months do not learn to go to locations where there is conflict as to which location is the correct one. The ability to respond when such conflict occurs is not present until about 18–24 months of age (Clohessy et al., 2001). At 3 years, the ability to respond correctly when there is conflict in the sequential looking task correlates with the ability to resolve conflict in a spatial conflict task (SCT) (Rothbart et al., 2003). These findings support the slow development of the executive attention network during the first and second years of life.

The visual sequence task is related to other features that reflect control. One of these is the cautious reach toward novel toys. Rothbart and colleagues found that the slow cautious reach of infants at 10 months predicted higher levels of EC as measured by parent report at 7 years of age (Rothbart et al., 2001). Infants of 7 months who show higher levels of correct anticipatory looking in the visual sequence task also show longer inspection times before reaching toward novel objects and slower reaching toward the objects (Sheese et al., 2008). This suggests that successful anticipatory looking at 7 months is one feature of self-regulation. In addition, infants with higher levels of correct anticipatory looking also showed evidence of higher levels of emotionality in a distressing task and more evidence of efforts to self-regulate their emotional reactions. Even at 7 months, the executive attention system is showing some properties of self-regulation, even though it is not yet sufficiently developed to resolve conflict in the visual sequence task (Clohessy et al., 2001) or the task of reaching away from the line of sight in the transparent box task (Diamond, 1991).

An important question about early development of executive attention is its relationship to the orienting network (Figure 22.1). Recall that the orienting network develops very early and has a critical role in regulation of emotion by the caregiver as early as 4 months. It was also found that orienting as measured from the Infant Behavior Questionnaire at 7 months was not correlated with EC as measured in the same infants aged 2 years (Sheese et al., 2009). However, orienting did show some early regulation of emotional responding in the infants. Orienting was positively related to positive affect and negatively related to negative affect. It was expected that orienting would be positively related to positive affect because previous work had shown that the duration of orienting as reported by caregivers was longer for children who smiled and laughed more. It was also expected that orienting would be negatively related to negative affect during infancy, given its use as a tool for soothing infants (Harman et al., 1997). It was also found that when the same infants were run in a conflict task (child Attention Network Test (ANT), see Section 22.5.1) at age 4 years, the regulatory functions found at 7 months were correlated more with the orienting network than with the executive network (Posner et al., 2012). As it was also found in parent reports that at the age of 2 years, the orienting network was related to control of emotion as mentioned earlier, it was concluded that in infancy and very early childhood, control is exercised by the orienting network, and only later as it becomes more connected does the executive network exert the major control.

By 2 years of age, parents can give reports that lead to a measure of EC. Two-year-old children, however, did not show the usual pattern of negative relationships between EC and negative affect that has been repeatedly found at other ages (Rothbart and Rueda, 2005). This

was unexpected, and it is possible that toddlers may be at a transition stage between emotional control by orienting and control by executive attention.

Colombo (2001) presented a summary of attentional functions in infancy, which included alertness, spatial orienting, object-oriented attention, and endogenous attention. This division is similar to the network approach but divides orienting into space and features and includes the functions of interstimulus shifts and sustained attention as part of endogenous attention. The researcher argues that alerting reaches the mature state at about 4 months, orienting by 6–7 months, and endogenous attention by 4–5 years. This schedule is similar to the order of development described earlier, but as discussed in the next major section, all these functions continue to develop during childhood. The genetic findings discussed below are new since the Colombo summary and add additional substance to the distinctions between functions and their integration in the achievement of self-regulation.

22.4.3 Summary

The findings to date suggest that orienting plays some of the regulatory roles in early infancy that are later exercised by the executive network. This fits with the more integrated networks shown in childhood at the bottom of Figure 22.2. Parenting may play an important role in the development of the executive attention network, perhaps partly through the presentation of novel objects that have been shown to activate the executive network in adults (Shulman et al., 2009). Further evidence for the role of parenting is presented in Section 22.6.

22.5 ATTENTIONAL NETWORKS AND CHILD DEVELOPMENT

22.5.1 Alerting Network

As discussed earlier, the state of alertness can be elicited by external stimulation, but it also varies in a regular rhythm over the course of the day and can be attained in a voluntary endogenously generated way. Young infants are able to attain the alert state when elicited by external stimulation, and they show a progressive increase in the frequency and duration of alert periods during the first year of life, whereas the ability to voluntarily deploy attention seems to emerge later and shows a steadier developmental course during childhood (Colombo, 2001).

Preparation from warning cues (phasic alertness) can be measured by comparing the speed and accuracy of response to stimulation with and without warning signals (Posner, 2008). Presentation of warning cues prior to targets allows the individual to get ready to respond by increasing the state of alertness. This commonly results in increased response speed, although it may also cause declines in the accuracy of the response, particularly at short intervals between the warning cue and target (Posner, 1978).

The difficulty of using reaction time (RT) tasks with very young children makes studying developmental differences in preparation from alerting cues more challenging; yet several studies have examined developmental changes in phasic alertness between preschoolers, older children, and adults. In the authors' work, the ANT has been used to examine the efficiency of the three brain networks underlying attention: alerting, orienting, and executive attention (Fan et al., 2002). The task requires the person to press one key if the central arrow points to the left and another if it points to the right. Conflict is introduced by having flankers surrounding the target point in either the same (congruent) or the opposite (incongruent) direction as the target. Cues presented prior to the target provide information on where or when the target will occur. RTs for the separate conditions are subtracted, providing three measures that represent the efficiency of the individual in alerting, orienting, and executive networks. The child ANT is the same as described previously, but instead of arrows, fish are used and the child presses the key to feed or catch the central fish. Using the child ANT, Mezzacappa (2004) observed a trend to larger alerting scores (difference between RT in trials with and without warning cues) with age in a sample of 5- to 7-year-old children. Increasing age was associated with larger reductions in RT in response to warning cues. Older children also showed lower rates of omissions overall, indicating greater ability to remain vigilant during the task period. Young children (aged 5 years) also appear to need more time than older children (aged 8 years) and adults to get full benefit from a warning cue, and they also seem to be less able to sustain the optimal level of alertness over time (Morrison, 1982). The difficulty of maintaining the alert state without a cue is also observed in older children (aged 10 years) when compared to adults (Rueda et al., 2004a), suggesting that tonic or sustained attention continues to develop through late childhood.

Sustained attention is frequently measured by examining variations in performance on a task over a relatively extended period, as in the so-called Continuous Performance Tasks (CPT). Variations in the level of alertness can be observed by examining the percentage of correct and/or omitted responses to targets or through indexes of perceptual sensitivity (d') over time. With young children, the percentage of individuals able to complete the task can also indicate maturational differences in the ability to sustain attention. In a study conducted with preschoolers, only 30–50% of those aged 3–4 years were able to complete the task, whereas the percentages rose to 70% for those aged 4–4½ years and close to 100% for those older than 4½ years (Levy, 1980). Even though the largest development of vigilance seems to occur during the preschool period, children continue to show larger declines in performance in CPT over time than adults through middle and late childhood, especially under more difficult task conditions, reaching the adult level by approximately 13 years of age (Curtindale et al., 2007; Lin et al., 1999).

Developmental changes in alertness during childhood and early adolescence appear to relate to continuous maturation of frontal systems during this period. One way to examine brain mechanisms underlying changes in alertness is by registering patterns of brain-generated electrical activation through electrodes placed on the scalp while warning cues are processed. Typically, several hundred milliseconds after a cue predicting the upcoming occurrence of a target stimulus is presented, a negative variation of brain activity is generated up until the target appears (Walter, 1964). This electrophysiological index is called the CNV, and it appears to be related to a source of activation in the right ventral and medial frontal (ACC) brain areas (Segalowitz and Davies, 2004). The CNV is related to performance on various measures of intelligence and executive functions as well as to the functional capacity of the frontal cortex (Segalowitz et al., 1992). The CNV and other slow waves have been related to changes in activation as studied by fMRI (Raichle, 2009). Various studies have shown that the amplitude of the CNV increases with age, especially during middle childhood. For instance, Jonkman found that the CNV amplitude is significantly smaller in 6- to 7-year-old children than in adults, but no differences were observed between 9- and 10-year-old children and adults (Jonkman, 2006). The difference in CNV amplitude between children and adults seems to also be restricted to early components of the CNV observed over the right frontocentral channels (Jonkman et al., 2003), suggesting a role of maturation of the frontal alerting network.

22.5.2 Orienting Network

Infants are able to orient attention to external stimulation from early in their life. Nonetheless, aspects of the attention system that increase precision and voluntary control of orienting continue developing throughout childhood and adolescence. Most infant studies examine overt forms of orienting. By the time children are able to follow instructions and respond to stimulation by pressing keys, both overt and covert orienting can be measured with this method. The cuing task has been widely used to study the development of visual orienting over the life span. In this task, a cue is displayed prior to the presentation of a target to which a response, usually a key press, is required. The cue is aimed to induce orientation of attention to a particular location. Then, the target may appear at the cued location or at an uncued one. When the target appears at the cued location, benefits of orienting attention to that location in the RT and the accuracy of response to the target can be measured. When the cue is presented at a location different from that of the target, a decrease in RT is observed that is thought to be due to operations of disengagement of attention from the cued location and reorientation to the location occupied by the target. Imaging research and studies on patients have provided information on the brain anatomy related to each of these orienting operations (Posner and Fan, 2008). For example, endogenous (voluntary) orientation of attention is associated with structures of the superior parietal lobe and the frontal eye fields, whereas exogenous (automatic) orienting seems to be the function of a network comprised by the temporoparietal junction and ventral frontal cortex, largely lateralized to the right hemisphere (Corbetta and Shulman, 2002). Activation of these cortical areas is required for disengaging from the current focus of attention. Moving attention from one location to another involves the superior colliculus, whereas engaging attention requires thalamic areas such as the pulvinar nucleus.

Mostly using a cuing paradigm, several studies have examined the development of orienting during childhood. Despite a progressive increase in orienting speed to valid cues during childhood (Schul et al., 2003), data generally show no age differences in the orienting benefit effect among young children (5–6 years of age), older children (aged 8–10 years), and adults (Enns and Brodeur, 1989), regardless of whether the effect is measured in covert or overt orienting conditions (Wainwright and Bryson, 2002). However, there seems to be an age-related decrease in the orienting cost (Enns and Brodeur, 1989; Schul et al., 2003; Wainwright and Bryson, 2002). In addition, the effect of age when disengaging and reorienting to an uncued location appears to be larger under endogenous orienting conditions (e.g., longer intervals between cue and target) (Schul et al., 2003; Wainwright and Bryson, 2005). This suggests that mostly aspects of orienting related to the control of disengagement and voluntary orientation, which depend on cortical regions of the parietal and temporal lobes, improve with age during childhood. In a study in which endogenous orienting was examined in children aged 6–14 years and in adults, all groups but the youngest children showed larger orienting effects, calculated as the difference in RT to targets appearing at cued and uncued locations, with longer cue–target

intervals (Wainwright and Bryson, 2005). This indicates that young children seem to have problems endogenously adjusting the scope of their attentional focus. This idea was also suggested by Enns and Girgus (1985), who found that attentional focusing as well as the ability to effectively divide or switch attention between stimuli improves with age, between ages 5, 8, and 10 years and adulthood.

22.5.3 Executive Attention Network

Self-regulation of cognition and action can be measured in the laboratory by registering responses to tasks that involve conflict. Common conflict tasks such as the classic Stroop task require the participant to avoid paying attention to aspects of the stimulation (e.g., the word *WHITE*) that may be dominant (i.e., semantic information contained in the word) while responding to nondominant features (i.e., the color in which the word is written). From 2 years of age and older, children are able to perform simple conflict tasks in which their RT can be measured. The SCT (Gerardi-Caulton, 2000) induces conflict between the identity and the location of an object. In this task, pictures of the houses of two animals (e.g., a duck and a cat) are presented at the bottom left and bottom right sides of the screen; then one of the two animals appears either on the left or right side of the screen in each trial and the child is required to show the animal what its house is by touching it. Location is the dominant aspect of the stimulus, although instructions require responding according to its identity. Thus, conflict trials in which the animal appears on the side of the screen opposite to its house usually result in slower responses and larger error rates than nonconflict (when the animal appears on the same side as its house) trials. Between 2 and 4 years of age, children progressed from an almost complete inability to carry out the task to a relatively good performance. Although 2-year-old children tended to perseverate on a single response, 3-year-old children performed at high accuracy levels, although, like adults, they responded more slowly and with reduced accuracy to conflict trials (Gerardi-Caulton, 2000; Rothbart et al., 2003).

The detection and correction of errors is another form of action monitoring. While performing the SCT, 30- and 36-month-old children showed longer RTs following erroneous trials than those following correct ones, indicating that children were noticing their errors and using them to guide their performance on the next trial. However, no evidence of slowing following an error was found at 24 months of age (Rothbart et al., 2003). A similar result with a different time frame was found when using a version of the Simple Simon game. In this task, children are asked to execute a response when a command is given by a stuffed animal, while inhibiting responses that are commanded by a second animal (Jones et al., 2003). Children of age 36–38 months were unable to inhibit their response and showed no slowing following an error, but at 39–41 months of age, children showed both an ability to inhibit action and a slowing of RT following an error. These results suggest that between 30 and 39 months, children greatly develop their ability to detect and correct erroneous responses and that this ability may be related to the development of inhibitory control.

The development of executive attention has also been traced into the primary school period (Rueda et al., 2004a) using the child version of the Attention Networks Test (ANT). Overall, children's RTs were much longer than those of adults, but considerable development in the speed of resolving conflict from age 4 to about 7 years was observed. However, the ability to resolve conflict on the flanker task, as measured by increases in RT and percentage of errors produced by the presence of incompatible compared to compatible flankers, remained about the same from age 7 years to adulthood. Nonetheless, studies in which the difficulty of the conflict task is increased by other demands, such as switching rules or holding more information in working memory, have shown further development of conflict resolution between late childhood and adulthood (Davidson et al., 2006).

To study the brain mechanisms that underlie the development of executive attention, some developmental studies have been carried out using event-related potentials (ERPs) and conflict tasks. In one of these studies, a flanker task was used to compare conflict resolution in three groups of children aged 5–6, 7–9, and 10–12 years, and a group of adults (Ridderinkhof and van der Molen, 1995). In this study, developmental differences were examined in two ERP components, one related to response preparation (lateralized readiness potential (LRP)) and the other related to stimulus evaluation (P3). The authors found differences between children and adults in the latency of the LRP peak, but not in that of the P3 peak, suggesting that developmental differences in the ability to resist interference are mainly related to response competition and inhibition, but not to stimulus evaluation.

Brain responses to errors are also informative of the function of the executive attention system. The error-related negativity (ERN) is a potential with a frontocentral scalp distribution that appears some time (usually between 60 and 120 ms) after an error response (Gehring et al., 1993) and is thought to be generated by the ACC (van Veen and Carter, 2002). The amplitude of the ERN seems to reflect detection of an error as well as salience of the error for a particular individual in the context of the task, and it is therefore subject to individual differences in affective style or motivation. Generally, larger ERN amplitudes are associated with

greater engagement in the task and/or greater efficiency of the error detection system (Santesso et al., 2005; Tucker et al., 1999). Developmentally, the amplitude of the ERN shows a progressive increase during childhood into late adolescence (Segalowitz and Davies, 2004), with younger children (aged 7–8 years) less likely to show the ERN to errors than older children and adults. This is likely to reflect the progressive maturation of the brain system for action monitoring and regulation.

Another evoked potential, the N2, has been related to situations that require executive control (Koop et al., 1996) and has been directly associated with activation coming from the ACC (van Veen and Carter, 2002). An ERP study was conducted in which the fish flanker task of the child ANT was used on 4-year-old children and on adults (Rueda et al., 2004b). Adults showed larger N2 for incongruent trials than for congruent trials over the midfrontal leads. Four-year-old children also showed a larger negative deflection for the incongruent condition at the midfrontal electrodes that, compared to adults, had greater amplitude and was extended over a longer period. While the frontal effect was evident in adults at around 300 ms post target, children did not show any effect until approximately 550 ms after the target. In addition, the effect was sustained over a period of 500 ms before the children's responses, in contrast with only 50 ms in the case of adults. Another important difference between 4-year-old children and adults was the distribution of effects over the scalp. In adults, the frontal effects appear to be focalized on the midline, whereas in children, the effects were observed mostly at prefrontal sites and in a broader number of channels, including the midline and lateral areas.

The focalization of signals in adults as compared to children is consistent with neuroimaging studies conducted with older children in which children appeared to activate the same network of areas as adults when performing similar tasks, but the average volume of activation appeared to be remarkably greater in children than in adults (Casey et al., 2002; Durston et al., 2002). Altogether, these data suggest that the brain circuitry underlying executive functions becomes more focal and refined as it gains in efficiency. This maturational process not only involves greater anatomical specialization but also reduces the time these systems need to resolve each of the processes implicated in the task. This is consistent with the recent findings that the network of brain areas involved in attentional control shows increased segregation of short-range connections but ncreased integration of long-range connections with maturation (Fair et al., 2007). Segregation of short-range connectivity may be responsible for greater local specialization, whereas integration of long-range connectivity likely increases efficiency by improving coordinated responses between different processing modules.

22.5.4 Summary

The attention network shows substantial development during childhood. Rates of development appear to differ among the three networks, with alerting showing a very slow developmental time course. While the ANT shows most of the development in the executive network to occur before the age of 7 years, the connectivity data (Figure 22.2) indicate increased segregation between orienting and executive networks beyond the age of 9 years.

22.6 GENES AND EXPERIENCE BUILD NETWORKS

The finding that common brain networks are involved in self-regulation provides an important approach to evolution by looking at commonalities and differences in nonhuman organisms. Another approach of equal importance involves an examination of individual differences in the efficiency of this network. Such differences could rest in part on the genetic variation known to exist among individuals and in part on differences in cultural or individual experiences. The study of temperament examines individual differences in reactivity and self-regulation that are biologically based (Rothbart and Bates, 2006). One of the most important individual differences has been called 'effortful control.' It is a higher order factor consisting of a number of subscales measuring attentional and behavioral control.

22.6.1 Effortful Control

EC is derived from parents and self-reports of behavior in dimensions such as inhibitory control, attentional control, and low-intensity pleasure.

22.6.2 Attention Network Test

The ANT was used in a sample of 40 normal adults (Fan et al., 2002) where each of these measures was found to be reliable over repeated presentations. In addition, no correlation was found among the measures. An analysis of RTs in this task showed large main effects for cue type and target type. There were only two small interactions indicating some lack of independence among the cue conditions. One of these interactions was found when a cue directed orienting to the correct target location and the influence of the surrounding flankers was reduced. In addition, omitting a cue, which produces relatively long RTs, also reduced the size of the flanker interference. Presumably, this is because some of conflict is resolved in parallel with alerting.

Subsequent work has confirmed the relative independence among networks, while showing that they can interact when conditions are made more difficult or otherwise changed (Fan et al., 2009). A study using fMRI showed that the anatomy of these three networks was for the most part independent (Fan et al., 2005). In addition, each of the networks has a dominant neuromodulator arising from the subcortical brain areas. The alerting network is modulated by norepinephrine produced in the lc; the orienting network, by acetylcholine from the basal forebrain; and the executive network, by dopamine from the ventral tegmental area (Posner and Fan, 2008).

Scores on the conflict network of the ANT have been shown to correlate with the temperament factor of EC at several ages during childhood. Gerardi-Caulton (2000) carried out some of the first research linking EC to underlying brain networks of executive attention using spatial conflict as a laboratory marker task. Similar findings linking parent-reported EC to performance on laboratory attention tasks have been shown for 24-, 30-, and 36-month-old children (Rothbart et al., 2003); 3- and 5-year-old children (Chang and Burns, 2005); and 7-year-old children (Gonzalez et al., 2001). Some adult studies have also found a correlation between conflict resolution ability and EC (Kanske, 2008), and some disorders involve both executive attention (Fernandez-Duque and Black, 2006) and EC.

22.6.3 Daily Life

The correlation between conflict scores in this simple and easily administered cognitive task and parental reports of EC forms the basis for the association between self-regulation and executive attention.

As discussed, ANT executive attention scores and EC have been related to many aspects of child development. EC is related to the empathy that children show toward others, their ability to delay an action, and their ability to avoid behaviors such as lying or cheating when given the opportunity (Posner and Rothbart, 2007b; Rothbart and Rueda, 2005). High levels of EC and the ability to resolve conflict are related to fewer antisocial behaviors such as truancy in adolescents (Ellis et al., 2004). These findings have convinced the authors that self-regulation, a psychological function crucial for child socialization, can also be studied in terms of specific anatomical areas and their connections.

The common nature of brain networks such as those in Figure 22.1 argues strongly for the role of genes in their construction. This has led cognitive neuroscience to incorporate data into the growing field of human genetics (Green et al., 2008; Posner et al., 2007). One method for doing this relates individual variations in genes (genetic alleles) to aspects of human behavior. Brain activity can serve as an intermediate level for relating genes to behavior. As one example, the ANT has been used to examine individual differences in the efficiency of executive attention. Strong heritability of the executive network (Fan et al., 2002) supported the search for genes related to individual differences in network efficiency.

22.6.4 Neuromodulators

The association of the executive network with the neuromodulator dopamine provides a way of searching for candidate genes that might relate to the efficiency of the network (Fossella et al., 2002; Posner and Fan, 2008). For example, several studies using conflict-related tasks have found that alleles of the *catechol-O-methyl transferase* (COMT) gene were related to the ability to resolve conflict (Blasi et al., 2005; Diamond et al., 2004). A number of other dopamine genes have also been found to be related to this form of attention, and research has suggested that genes related to serotonin transmission also influence executive attention (see Posner et al., 2007, for a summary). It has also been possible to show in brain imaging studies that some of these genetic differences are related to the degree to which the anterior cingulate is activated during task performance (Fan et al., 2003). In the future, it may be possible to relate genes to specific nodes within neural networks, allowing a much more detailed understanding of the origins of brain networks underlying attention.

22.6.5 Longitudinal Study

One goal of our longitudinal study was to see if the genes that were shown to influence attention in adults (Posner et al., 2007) would have specific roles in the development of self-regulation during infancy and childhood. The study began when the infants were 7 months old. The children were retested and genotyped at age 2 years and were tested a final time at age 4 years when they are able perform the ANT as a measure of executive attention.

Rothbart and Derryberry (1981) proposed a distinction between reactive and self-regulatory aspects of child temperament. They argued that early in life, negative affect, particularly fear and orienting of attention, served as regulatory mechanisms that were supplemented by parental regulation. Rothbart and Bates (2006) argued for developmental changes in which EC arose at about 3–4 years when parents could first report on their children's self-regulatory ability.

The longitudinal study has not only confirmed but also revised and extended this analysis. A negative correlation was found at 7 months between parental reports of infant orienting of attention and negative affect.

Orienting also correlates positively with reports of positive affect. By 2 years, orienting is no longer related to affect, but EC shows modest nonsignificant negative correlations with both positive and negative affect. There is substantial evidence that for children, adolescents, and adults, EC shows a negative correlation with negative affect in Western countries, although Ahadi et al. (1993) found that in China, EC was negatively correlated with positive affect in children. They argued that culture shaped the direction of the interaction between EC and emotion.

As discussed previously, the results of the longitudinal study suggest that early in life, orienting serves as a regulatory system, and it both reduces negative affect (Harman et al., 1997) and increases positive affect. In this view, both orienting and executive networks serve parallel regulatory functions during infancy. Later on, executive attention appears to dominate in regulating emotions and thoughts, but orienting still serves as a control system. This parallel use of the two networks fits with the findings of Dosenbach et al. (2007) showing that in adults their frontoparietal network controls behavior at short time intervals while their cinguloopercular network exercises strategic control over long intervals. These changes in regulation during early childhood served as a framework to examine continuities and discontinuities in the influence of genetic variations.

22.6.5.1 CHRNA4

The nicotinic cholinergic receptor modulates the release of dopamine in the mesolimbic system. Polymorphisms in the CHRNA4 gene have been associated with nicotine dependence in humans and with cognitive performance (Rigbi et al., 2008). The CHRNA4 C1545T polymorphism (rs1044396) has been associated with variation in performance of shifts of attention during visual search (Parasuraman et al., 2005) and in brain activity when performing visual attention tasks (Winterer et al., 2007).

During infancy, the T/T allele of CHRNA4 is related to a better performance in anticipatory looking, but at 18 months, the C/C homozygotes have the higher scores on EC (Voelker et al., 2009). The finding that alleles associated with higher attention during infancy (as measured by anticipation) have lower parent-reported EC at age 2 years may indicate that infants who exercise control through orienting are slower to transition to control via the executive network. As the measurement at 18–20 months occurs well before regulation by executive attention is complete, infants with strong orienting may not yet have made the transition.

As visuospatial attention requires orienting, the CHRNA4 polymorphism was expected to influence orienting in the subjects, and it was thought that it might also influence higher-order attention. The authors' genetic findings provided support for the parallel model of regulation discussed earlier. At 7 months, CHRNA4 is related to aspects of anticipatory looking (Sheese et al., 2008). At about 2 years, the main influence of this gene appears to be on EC, which depends on executive attention. In adults, CHRNA4 seems to be related to tasks that clearly involve the orienting network (Parasuraman et al., 2005), but these tasks may involve executive attention as well. These findings support the idea that gene expression during development may vary as different brain networks change in importance.

22.6.5.2 Catechol-O-methyl Transferase

COMT plays an important role in dopamine metabolism by modulating extracellular levels of dopamine. The functional Val/Met polymorphism of COMT has a measurable effect on COMT enzyme activity, with the Val allele degrading extracellular dopamine more quickly than the less enzymatically active Met allele. The Met allele has been associated with anxiety and negative mood states (Drabant et al., 2006), affective disorders (Karayiorgou et al., 1997), and decreased novelty seeking (Reuter and Hennig, 2005).

A finding from the authors' current longitudinal study is that the COMT gene, which has consistently been shown to be related to executive attention in adults and older children (Blasi et al., 2005; Diamond et al., 2004), is also related to aspects of attention in toddlers (Voelker et al., 2009). It was found that haplotypes of the COMT gene (Diatchenko et al., 2005) influenced both anticipatory looking and nesting cup activity at 18–20 months, both tasks thought to be at least partly related to executive attention. At 7 months, COMT was also related to positive affect as reported by parents. The finding of a relation of COMT to positive affect together with the influence of this gene on executive attention at 18–20 months could provide a genetic link between reactive emotion and emotional regulation during early development. However, it is also possible that COMT's relation to positive affect in infancy is mediated by regulatory aspects of executive attention. Evidence for this is mixed in the authors' current study in that positive affect in infancy was unrelated to later EC, but other studies have shown such a connection (Rothbart and Rueda, 2005).

22.6.5.3 DRD4

The 7-repeat allele of the DRD4 gene has been linked to attention deficit hyperactivity disorder (ADHD) and to the temperamental dimension of risk taking. Adults and children with the 7-repeat allele have been shown to be higher in the temperamental quality of risk-taking and at greater risk for attention deficit disorder than those with fewer repeats (Auerbach et al., 1999). In one series of studies (Auerbach et al., 1999), it was found that the orienting of 2-month-old infants as rated by parents and as observed

during inspection of toys was related to the presence of the 7-repeat allele of the dopamine 4 receptor gene. This allele appears to interact with a gene related to serotonin transmission (5-HTT) to influence orienting.

Evidence that the environment can have a strong influence in the presence of the 7-repeat alleles has been reported by other investigators (Bakermans-Kranenburg and van Ijzendoorn, 2006; van Ijzendoorn and Bakermans-Kranenburg, 2006). The same group (Bakersmans-Kranenburg et al., 2008) carried out a parenting training intervention and found that the training decreased externalizing behavior, but only for those children with the DRD4 7-repeat allele. This finding is important because assignment to the training group was random, ensuring that the result was not due to something about the parents other than the training.

In the longitudinal study, the authors added an observation of caregiver–child interaction in which the children played with toys in the presence of one of their parents. Raters observed the caregiver–child interaction and rated the parents on five dimensions of parental quality according to a schedule developed by NICHD (1993): the ratings included support, autonomy, stimulation, lack of hostility, and confidence in the child. Although all the parents were likely concerned and caring, they did differ in their scores, and the combined scores were divided at the median into two groups. One of the groups was considered to show a higher quality of parenting and the other a lower quality.

A strong interaction was found between genes and parenting. For children without the 7-repeat polymorphism, variations in parenting within the range examined were unrelated to the children's scores on impulsivity and risk taking. For children carrying the 7-repeat gene variant, however, variations in parenting quality mattered. Children with this allele and high-quality parenting showed normal levels of risk taking, but those with lower-quality parenting showed very high values for risk taking. It seems paradoxical that the 7-repeat allele associated with developmental psychopathology (ADHD) is also under positive selective pressure in recent human evolution (Ding et al., 2002). Why should an allele related to ADHD be positively selected? It is the authors' view that positive selection of the 7-repeat allele could well arise from its sensitivity to environmental influences.

Parenting provides training for children in the values favored by the culture in which they live. For example, Rothbart and colleagues (Ahadi et al., 1993) found that in Western culture, EC appears to regulate negative affect (sadness and anger), while in China (at least in the 1980s), it was found to regulate positive affect (outgoingness and enthusiasm). In recent years, the genetic part of the nature by nurture interaction has been given a lot of emphasis, but if genetic variations are selected according to the sensitivity they give children to cultural influences, this could support a greater balance between genes and environment. Theories of positive selection in the DRD4 gene have stressed the role of sensation seeking in human evolution (Wang et al., 2006). The authors' findings do not contradict this emphasis but suggest an interpretation that could have even wider significance. It remains to be seen whether the other 300 genes estimated to show positive selection would also increase an individual's sensitivity to variations in rearing environments.

How could variation in genetic alleles lead to enhanced influence of cultural factors such as parenting? The anterior cingulate receives input on both reward value and pain or punishment, and this information is clearly important in regulating thoughts and feelings. Dopamine is the most important neuromodulator in these 'reward' and 'punishment' pathways. Thus, changes in the availability of dopamine could enhance the influence of signals from parents related to reward and punishment. Another interaction has been reported between the serotonin transporter gene and parental social support on the temperamental dimension of behavioral inhibition or social fear (Fox et al., 2005). To explain this interaction, Fox et al. (2007) argue that children with a short form of the serotonin transporter gene who also have lower social support from parents show enhanced attention to threat and greater social fear. In the authors' study of the DRD4, however, attention did not appear to be the mechanism by which the genetic variation influenced the child's behavior. In the authors' study, there was no influence of the 7-repeat allele on executive attention; instead, gene and environment interacted to influence the child's behavior as observed by the caregiver. However, an interaction between the DRD4 7-repeat allele and EC as rated by parents did appear at age 4 years and older (Sheese et al., 2012). This finding lends support to the idea that control shifts to the executive network during childhood. The authors also found that the COMT genotype showed an interaction between attention and parenting quality, and unlike the DRD4, it did operate through attention even at age 2 years. However, in the case of COMT, attention involved anticipatory looking, which is at least partly influenced by orienting at this age.

Overall, the authors found both continuity and discontinuity in the influence of genetic variation in early childhood. The DRD4 influence on executive attention emerged only at age 4 years and continued into adulthood, while the COMT influence on tasks involving attention shifting was exhibited as early as 2 years. The CHRNA4 genetic variation was most variable for attention changes between infancy and adulthood. These variations appear related to the changes in dominant control networks described in this chapter.

22.6.6 Summary

While a few candidate genes have been related to individual differences between infants and young children, these account for only a small portion of the differences in temperament and behavior. One reason for these small effects is that parenting and other forms of cultural variation interact with genes in determining behavior. However, it is likely that the genes whose variation is related to individual differences in attention are also those important in building the common attentional networks. Combined human and animal studies may be helpful in further explicating connections between genes and control networks. See also Chapter 38 and Rubenstein and Rakic, 2013.

22.7 FUTURE RESEARCH

In this chapter, the authors have examined some of the tools used in studies of the human brain and mind during development. These tools may allow for a deeper understanding of how the developing brain makes possible the changes in attention and self-regulation that occur in behavior early in life. Future research should enable the use of these tools to understand how developmental changes in functional activation and connectivity relate to the specific behavioral markers at the same age. Research can help one understand how changes in activation relate to differences found in functional connectivity and in white and gray matter. Is there a fixed order of these changes, or does their rate and order of occurrence depend on whether they take place as the result of development or from practice on a task? Better coordination of human and nonhuman animal work may allow one to determine the relationship of changes found with noninvasive imaging to those seen in studies of the microanatomy and circuitry of brain areas in animal research.

It is also difficult to know how brain changes relate to behavioral changes found at the same time. Longitudinal studies will allow one to better define this relationship. Doing so may require the use of tasks and methods that remain relatively stable across ages. The use of resting fMRI may be the most important of these since it allows testing of different ages without the need to develop comparable tasks (Fair et al., 2007). The discovery that the electroencephalogram signal for error detection involves similar brain areas at 7 months as it does for adults provides another means of examining an event that may be comparable across wide differences in age. The use of more analytic behavioral observations (e.g., anticipatory eye movement, the ANT, and parent reports of temperament) may allow for the mapping of changes in mental operations to brain changes in development.

The growing knowledge of genetic and epigenetic methods has only just begun to influence research in human development. Mainly genetic variation has been related to individual differences in behavior (Posner et al., 2007). However, it seems likely that the same genes related to individual differences are involved in building the common networks underlying attention. Thus, studies designed to relate the expression of genes at particular nodes in neural networks to key aspects of behavior will be of great importance in realizing the goal of understanding how neural networks become organized in development.

Although it is known that some genetic variants interact with environmental experience, the actual mechanisms involved are not yet known, as genetic variations are expressed at numerous places in the brain and the rest of the body. As the mechanisms by which genes can be altered by the environment are enlarged in the field of epigenetics, it is possible to learn more about how training influences development.

Acknowledgment

The research for this chapter was supported by NICHD grant HD060563 to Georgia State University.

References

Ahadi, S.A., Rothbart, M.K., Ye, R., 1993. Children's temperament in the U.S. and China: Similarities and differences. European Journal of Personality 7, 359–378.

Auerbach, J., Geller, V., Letzer, S., et al., 1999. Dopamine D4 receptor (D4DR) and serotonin transporter promoter (5-HTTLPR) polymorphisms in the determination of temperament in two month old infants. Molecular Psychiatry 4, 369–374.

Bakermans-Kranenburg, M.J., van Ijzendoorn, M.H., 2006. Gene-environment interaction of the dopamine D4 receptor (DRD4) and observed maternal insensitivity predicting externalizing behavior in preschoolers. Developmental Psychobiology 48, 406–409.

Bakersmans-Kranenburg, M.J., Van Ijzendoorn, M.H., Pijlman, F.T.A., Mesman, J., Juffer, F., 2008. Experimental evidence for differential susceptibility: Dopamine D4 receptor polymorphism (DRD4 VNTR) moderates intervention effects on toddlers externalizing behavior in a randomized controlled trial. Developmental Psychology 44, 293–300.

Beauregard, M., Levesque, J., Bourgouin, P., 2001. Neural correlates of conscious self-regulation of emotion. Journal of Neuroscience 21, RC165.

Berger, A., Tzur, G., Posner, M.I., 2006. Infant babies detect arithmetic error. Proceeding of the National Academy of Science of the United States of America 103, 12649–12653.

Blasi, G., Mattay, G.S., Bertolino, A., et al., 2005. Effect of catechol-O-methyltransferase val^{158}met genotype on attentional control. Journal of Neuroscience 25 (20), 5038–5045.

Botvinick, M.M., Braver, T.S., Barch, D.M., Carter, C.S., Cohen, J.D., 2001. Conflict monitoring and cognitive control. Psychological Review 108, 624–652.

Bush, G., Luu, P., Posner, M.I., 2000. Cognitive and emotional influences in the anterior cingulate cortex. Trends in Cognitive Science 4 (6), 215–222.

Butcher, P.R., 2000. Longitudinal Studies of Visual Attention in Infants: The Early Development of Disengagement and Inhibition of Return. Aton, Meppel.

Casey, B., Thomas, K.M., Davidson, M.C., Kunz, K., Franzen, P.L., 2002. Dissociating striatal and hippocampal function developmentally with a stimulus–response compatibility task. Journal of Neuroscience 22 (19), 8647–8652.

Chang, F., Burns, B.M., 2005. Attention in preschoolers: Associations with effortful control and motivation. Child Development 76, 247–263.

Clohessy, A.B., Posner, M.I., Rothbart, M.K., 2001. Development of the functional visual field. Acta Psychologica 106, 51–68.

Colombo, J., 2001. The development of visual attention in infancy. Annual Review of Psychology 52, 337–367.

Colombo, J., Horowitz, F.D., 1987. Behavioral state as a lead variable in neonatal research. Merrill Palmer Quarterly 33, 423–438.

Corbetta, M., Shulman, G.L., 2002. Control of goal-directed and stimulus-driven attention in the brain. Nature Reviews Neuroscience 3 (3), 201–215.

Crottaz-Herbette, S., Mennon, V., 2006. Where and when the anterior cingulate cortex modulates attentional response: Combined fMRI and ERP evidence. Journal of Cognitive Neuroscience 18, 766–780.

Curtindale, L., Laurie-Rose, C., Bennett-Murphy, L., Hull, S., 2007. Sensory modality, temperament, and the development of sustained attention: A vigilance study in children and adults. Developmental Psychology 43 (3), 576–589.

Davidson, M.C., Amso, D., Anderson, L.C., Diamond, A., 2006. Development of cognitive control and executive functions from 4 to 13 years: Evidence from manipulations of memory, inhibition, and task switching. Neuropsychologia 44 (11), 2037–2078.

Dehaene, S., Posner, M.I., Tucker, D.M., 1994. Localization of a neural system for error detection and compensation. Psychological Science 5, 303–305.

Diamond, A., 1991. Neuropsychological insights into the meaning of object concept development. In: Carey, S., Gelman, R. (Eds.), The Epigenesis of Mind: Essays on Biology and Cognition. Erlbaum, Hillsdale, NJ, pp. 67–110.

Diamond, A., Briand, L., Fossella, J., Gehlbach, L., 2004. Genetic and neurochemical modulation of prefrontal cognitive functions in children. American Journal of Psychiatry 161, 125–132.

Diatchenko, L., Slade, G.D., et al., 2005. Genetic basis for individual variations in pain perception and the development of a chronic pain condition. Human Molecular Genetics 14 (1), 135–143.

Ding, Y.C., Chi, H.C., Grady, D.L., et al., 2002. Evidence of positive selection acting at the human dopamine receptor D4 gene locus. Proceedings of the National Academy of Sciences of the United States of America 99 (1), 309–314.

Dosenbach, N.U.F., Fair, D.A., Miezin, F.M., et al., 2007. Distinct brain networks for adaptive and stable task control in humans. Proceedings of the National Academy of Sciences of the United States of America 104, 1073–1978.

Drabant, E.M., Hariri, A.R., Meyer-Lindenberg, A., et al., 2006. Catechol O-methyltransferase val^{158}met genotype and neural mechanisms related to affective arousal and regulation. Archives of General Psychiatry 63, 1396–1406.

Drevets, W.C., Raichle, M.E., 1998. Reciprocal suppression of regional cerebral blood flow during emotional versus higher cognitive processes: Implications for interactions between emotion and cognition. Cognition and Emotion 12, 353–385.

Duncan, J., Seitz, R.J., Kolodny, J., et al., 2000. A neural basis for general intelligence. Science 289, 457–460.

Durston, S., Thomas, K.M., Yang, Y., Ulug, A.M., Zimmerman, R.D., Casey, B., 2002. A neural basis for the development of inhibitory control. Developmental Science 5 (4), F9–F16.

Ellis, E., Rothbart, M.K., Posner, M.I., 2004. Individual differences in executive attention predict self-regulation and adolescent psychosocial behaviors. Annals of the New York Academy of Sciences 1031, 337–340.

Enns, J.T., Brodeur, D.A., 1989. A developmental study of covert orienting to peripheral visual cues. Journal of Experimental Child Psychology 48 (2), 171–189.

Enns, J.T., Girgus, J.S., 1985. Developmental changes in selective and integrative visual attention. Journal of Experimental Child Psychology 40, 319–337.

Etkin, A., Egner, T., Peraza, D.M., Kandel, E.R., Hirsch, J., 2006. Resolving emotional conflict: A role for the rostral anterior cingulate cortex in modulating activity in the amygdala. Neuron 51, 871–882.

Fair, D.A., Dosenbach, N.U.F., Church, J.A., et al., 2007. Development of distinct control networks through segregation and integration. Proceedings of the National Academy of Sciences of the United States of America 104 (33), 13507–13512.

Fair, D., Cohen, A.L., Dosenbach, A.U.F., et al., 2008. The maturing architecture of the brain's default network. Proceedings of the National Academy of Sciences of the United States of America 105, 4028–4032.

Fan, J., McCandliss, B.D., Sommer, T., Raz, M., Posner, M.I., 2002. Testing the efficiency and independence of attentional networks. Journal of Cognitive Neuroscience 3 (14), 340–347.

Fan, J., Fossella, J.A., Summer, T., Wu, Y., Posner, M.I., 2003. Mapping the genetic variation of executive attention onto brain activity. Proceedings of the National Academy of Sciences of the United States of America 100, 7406–7411.

Fan, J., McCandliss, B.D., Fossella, J., Flombaum, J.I., Posner, M.I., 2005. The activation of attentional networks. NeuroImage 26, 471–479.

Fan, J., Gu, X., Guise, K.G., et al., 2009. Testing the behavior interaction and integration of attentional networks. Brain and Cognition 70, 209–220.

Fernandez-Duque, D., Black, S.E., 2006. Attentional networks in normal aging and Alzheimer's disease. Neuropsychology 20, 133–143.

Fossella, J., Posner, M.I., Fan, J., Swanson, J.M., Pfaff, D.M., 2002. Attentional phenotypes for the analysis of higher mental function. Scientific World Journal 2, 217–223.

Fox, N.A., Nichols, K.E., Henderson, H.A., et al., 2005. Evidence for a gene-environment interaction in predicting behavioral inhibition in middle school children. Psychological Science 16 (12), 921–926.

Fox, N.A., Hane, A.A., Pine, D.S., 2007. Plasticity for affective neurocircuitry – How the environment affects gene expression. Current Directions in Psychological Science 16 (1), 1–5.

Gao, W., Zhu, H., Giovanello, K.S., et al., 2009. Evidence on the emergence of the brain's default network from 2 week-old to 2-year old healthy pediatric subjects. Proceedings of the National Academy of Sciences of the United States of America 106, 6790–6795.

Gehring, W.J., Gross, B., Coles, M.G.H., Meyer, D.E., Donchin, E., 1993. A neural system for error detection and compensation. Psychological Science 4, 385–390.

Gerardi-Caulton, G., 2000. Sensitivity to spatial conflict and the development of self-regulation in children 24–36 months of age. Developmental Science 3 (4), 397–404.

Gonzalez, C., Fuentes, L.J., Carranza, J.A., Estevez, A.F., 2001. Temperament and attention in the self-regulation of 7-year-old children. Personality and Individual Differences 30, 931–946.

Green, A.E., Munafo, M.R., DeYoung, C.G., Fossella, J.A., Fan, J., Gray, J.R., 2008. Using genetic data in cognitive neuroscience: From growing pains to genuine insights. Nature Reviews Neuroscience 9, 710–720.

Haith, M.M., Hazan, C., Goodman, G.S., 1988. Expectation and anticipation of dynamic visual events by 3.5 month old babies. Child Development 59, 467–469.

Harman, C., Rothbart, M.K., Posner, M.I., 1997. Distress and attention interactions in early infancy. Motivation and Emotion 21, 27–43.

Holroyd, C.B., Coles, M.G.H., 2002. The neural basis of human error processing: Reinforcement learning, dopamine and error-related negativity. Psychological Review 109, 679–709.

James, W., 1890. Principles of Psychology. Holt, New York.

Johnson, M.H., Morton, J., 1991. Biology and Cognitive Development: The Case of Face Recognition. Blackwell, Oxford.

Johnson, M.H., Posner, M.I., Rothbart, M.K., 1991. Components of visual orienting in early infancy: Contingency learning, anticipatory looking and disengaging. Journal of Cognitive Neuroscience 3 (4), 335–344.

Jones, L.B., Rothbart, M.K., Posner, M.I., 2003. Development of executive attention in preschool children. Developmental Science 6 (5), 498–504.

Jonkman, L.M., 2006. The development of preparation, conflict monitoring and inhibition from early childhood to young adulthood: A Go/Nogo ERP study. Brain Research 1097 (1), 181–193.

Jonkman, L.M., Lansbergen, M., Stauder, J.E.A., 2003. Developmental differences in behavioral and event-related brain responses associated with response preparation and inhibition in a Go/NoGo task. Psychophysiology 40 (5), 752–761.

Kahneman, D., 1973. Attention and Effort. Prentice Hall, Englewood Cliffs, NJ.

Kanske, P., 2008. Exploring Executive Attention in Emotion. Sachsisches Digitaldruckzentrum, Dresden.

Karayiorgou, M., Altemus, M., Galke, B.L., et al., 1997. Genotype determining low catechol-O-methyltransferase activity as a risk factor for obsessive-compulsive disorder. Proceedings of the National Academy of Sciences of the United States of America 94, 4572–4575.

Keele, S.W., Ivry, R., Mayr, U., Hazeltine, E., Heuer, H., 2003. The cognitive and neural architecture of sequence representation. Psychological Review 110, 316–339.

Koop, B., Rist, F., Mattler, U., 1996. N200 in the flanker task as a neurobehavioral tool for investigating executive control. Psychophysiology 33, 282–294.

Levy, F., 1980. The development of sustained attention (vigilance) in children: Some normative data. Journal of Child Psychology and Psychiatry 21 (1), 77–84.

Lin, C.C.H., Hsiao, C.K., Chen, W.J., 1999. Development of sustained attention assessed using the continuous performance test among children 6–15 years of age. Journal of Abnormal Child Psychology 27 (5), 403–412.

Marrocco, R.T., Davidson, M.C., 1998. Neurochemistry of attention. In: Parasuraman, R. (Ed.), The Attentive Brain. MIT Press, Cambridge, MA, pp. 35–50.

Meltzoff, A.N., Moore, M.K., 1977. Imitation of facial and manual gestures by human neonates. Science 198, 74–78.

Mezzacappa, E., 2004. Alerting, orienting, and executive attention: Developmental properties and sociodemographic correlates in an epidemiological sample of young, urban children. Child Development 75 (5), 1373–1386.

Morrison, F.J., 1982. The development of alertness. Journal of Experimental Child Psychology 34 (2), 187–199.

NICHD Early Child Care Research Network (1993) The NICHD Study of Early Child Care: A comprehensive longitudinal study of young children's lives. ERIC Document Reproduction Service No. ED3530870.

Ochsner, K.N., Bunge, S.A., Gross, J.J., Gabrieli, J.D.E., 2002. Rethinking feelings: An fMRI study of the cognitive regulation of emotion. Journal of Cognitive Neuroscience 14, 1215–1229.

Parasuraman, R., Greenwood, P.M., Kumar, R., Fossella, J., 2005. Beyond heritability: Neurotransmitter genes differentially modulate visuospatial attention and working memory. Psychological Science 16 (3), 200–207.

Pidoplichko, V.I., De Biasi, M., Williams, J.T., Dani, J.A., 1997. Nicotine activates and desensitizes midbrain dopamine neurons. Nature 390, 401–404.

Posner, M.I., 1978. Chronometric Explorations of Mind. Lawrence Erlbaum Associates, Hillsdale, NJ.

Posner, M.I., 2008. Measuring alertness. Annals of the New York Academy of Sciences 1129, 193–199 Molecular and Biophysical Mechanisms of Arousal, Alertness, and Attention.

Posner, M.I., Fan, J., 2008. Attention as an organ system. In: Pomerantz, J. R. (Ed.), Topics in Integrative Neuroscience. Cambridge University Press, New York, pp. 31–61 ch. 2.

Posner, M.I., Raichle, M.E., 1994. Images of Mind. Scientific American Books, Washington, DC.

Posner, M.I., Raichle, M.E. (Eds.), 1998. Neuroimaging of cognitive processes. Proceedings of the National Academy of Sciences of the United States of America 95, 763–764.

Posner, M.I., Rothbart, M.K., 2007a. Research on attention networks as a model for the integration of psychological science. Annual Review of Psychology 58, 1–23.

Posner, M.I., Rothbart, M.K., 2007b. Educating the Human Brain. APA Books, Washington, DC.

Posner, M.I., Rothbart, M.K., Sheese, B.E., 2007. Attention genes. Developmental Science 10, 24–29.

Posner, M.I., Rothbart, M.K., Sheese, E., Voelker, P., 2012. Control networks and neuromodulators of early development. Developmental Psychology 48 (3), 827–835. http://dx.doi.org/10.1037/a0025530.

Raichle, M.E., 2009. A paradigm shift in functional brain imaging. Journal of Neuroscience 29, 12729–12734.

Reuter, M., Hennig, J., 2005. Association of the functional catechol-O-methyltransferase Val^{158}Met polymorphism with the personality trait of extraversion. NeuroReport 16, 1135–1138.

Richards, J.E., Hunter, S.K., 1998. Attention and eye movements in young infants: Neural control and development. In: Richards, J.E. (Ed.), Cognitive Neuroscience of Attention. LEA, Mahwah, NJ.

Ridderinkhof, K.R., van der Molen, M.W., 1995. A psychophysiological analysis of developmental differences in the ability to resist interference. Child Development 66 (4), 1040–1056.

Rigbi, A., Kanyas, K., Yakir, A., et al., 2008. Why do young women smoke? V. Role of direct and interactive effects of nicotinic cholinergic receptor gene variation on neurocognitive function. Genes, Brain, and Behavior 7, 164–172.

Rothbart, M.K., Bates, J.E., 2006. Temperament in children's development. In: Damon, W., Lerner, R. (Eds.), Handbook of Child Psychology. 6th edn. In: Eisenberg, N. (Ed.), Social, Emotional, and Personality DevelopmentVol. 3. Wiley, New York, pp. 99–166.

Rothbart, M.K., Derryberry, D., 1981. Development of individual differences in temperament. In: Lamb, M.E., Brown, A.L. (Eds.), Advances in Developmental Psychology. Erlbaum, Hillsdale, NJ, pp. 37–86.

Rothbart, M.K., Rueda, M.R., 2005. The development of effortful control. In: Mayr, U., Awh, E., Keele, S.W. (Eds.), Developing Individuality in the Human Brain: A Festschrift Honoring Michael I Posner. American Psychological Association, Washington, DC.

Rothbart, M.K., Ahadi, S.A., Hershey, K.L., 1994. Temperament and social behavior in childhood. Merrill-Palmer Quarterly 40, 21–39.

Rothbart, M.K., Ahadi, S.A., Hershey, K.L., Fisher, P., 2001. Investigations of temperament at three to seven years: The Children's Behavior Questionnaire. Child Development 72, 1394–1408.

Rothbart, M.K., Ellis, L.K., Rueda, M.R., Posner, M.I., 2003. Developing mechanisms of effortful control. Journal of Personality 71, 1113–1143.

Rubenstein, J.L.R., Rakic, P., 2013. Patterning and Cell Types Specification in the Developing CNS and PNS.

Rueda, M., Fan, J., McCandliss, B.D., et al., 2004. Development of attentional networks in childhood. Neuropsychologia 42 (8), 1029–1040.

Rueda, M.R., Posner, M.I., Rothbart, M.K., Davis-Stober, C.P., 2004. Development of the time course for processing conflict: An event-related potentials study with 4 year olds and adults. BMC Neuroscience 5, 39.

Santesso, D.L., Segalowitz, S.J., Schmidt, L.A., 2005. ERP correlates of error monitoring in 10-year olds are related to socialization. Biological Psychology 70, 79–87.

Schul, R., Townsend, J., Stiles, J., 2003. The development of attentional orienting during the school-age years. Developmental Science 6 (3), 262–272.

Segalowitz, S.J., Davies, P.L., 2004. Charting the maturation of the frontal lobe: An electrophysiological strategy. Brain and Cognition 55 (1), 116–133.

Segalowitz, S.J., Unsal, A., Dywan, J., 1992. Cleverness and wisdom in 12-year-olds: Electrophysiological evidence for late maturation of the frontal lobe. Developmental Neuropsychology 8, 279–298.

Sheese, B.E., Rothbart, M.K., Posner, M.I., White, L., Fraundorf, S., 2008. Executive attention in infancy. Infant Behavior and Development 31, 501–510.

Sheese, B.E., Voelker, P., Posner, M.I., Rothbart, M.K., 2009. Genetic variation influences on the early development of reactive emotions and their regulation by attention. Cognitive Neuropsychiatry 14 (4), 332–355.

Sheese, B.E., Rothbart, M.K., Voelker P., Posner, M.I., 2012. The dopamine receptor D4 gene 7 repeat allele interacts with parenting quality to predict Effortful Control in four-year-old children. Child Development Research 2112, ID 863242.; http://dx.doi.org/10.1155/2012/863242.

Shulman, G.L., Astafiev, S.V., Franke, D., et al., 2009. Interaction of stimulus-driven reorienting and expectation in ventral and dorsal frontoparietal and basal ganglia-cortical networks. Journal of Neuroscience 29, 4392–4407.

Thompson, K.G., Biscoe, K.L., Sato, T.R., 2005. Neuronal basis of covert spatial attention in the frontal eye fields. Journal of Neuroscience 25, 9479–9487.

Tucker, D.M., Hartry-Speiser, A., McDougal, L., Luu, P., deGrandpre, D., 1999. Mood and spatial memory: Emotion and right hemisphere contribution to spatial cognition. Biological Psychology 50, 103–125.

van Ijzendoorn, M.H., Bakermans-Kranenburg, M.J., 2006. DRD4 7-repeat polymorphism moderates the association between maternal unresolved loss or trauma and infant disorganization. Attachment and Human Development 8, 291–307.

van Veen, V., Carter, C.S., 2002. The timing of action-monitoring processes in the anterior cingulate cortex. Journal of Cognitive Neuroscience 14, 593–602.

Voelker, P., Sheese, B.E., Rothbart, M.K., Posner, M.I., 2009. Variations in COMT gene interact with parenting to influence attention in early development. Neuroscience 164 (1), 121–130.

Wainwright, A., Bryson, S.E., 2002. The development of exogenous orienting: Mechanisms of control. Journal of Experimental Child Psychology 82 (2), 141–155.

Wainwright, A., Bryson, S.E., 2005. The development of endogenous orienting: Control over the scope of attention and lateral asymmetries. Developmental Neuropsychology 27 (2), 237–255.

Walter, G., 1964. The convergence and interaction of visual, auditory and tactile responses in human non-specific cortex. Annals of the New York Academy of Sciences 112, 320–361.

Wang, E.T., Kodama, G., Baldi, P., Moyzis, R.K., 2006. Global landscape of recent inferred Darwinan selection for Homo sapiens. Proceedings of the National Academy of Sciences of the United States of America 103, 135–140.

Winterer, G., Musso, F., Konrad, A., et al., 2007. Association of attentional network function with exon 5 variations of the CHRNA4 gene. Human Molecular Genetics 16, 2165–2174.

Wynn, K., 1992. Addition and subtraction by human infants. Nature 358, 749–750.

The Neural Correlates of Cognitive Control and the Development of Social Behavior

A. Lahat[1], N.A. Fox[2]

[1]McMaster University, Hamilton, ON, Canada; [2]University of Maryland, College Park, MD, USA

23.1 THE DEVELOPMENT OF COGNITIVE CONTROL AND ITS NEURAL BASIS

With development, children and adolescents improve in their ability to control their thoughts, behavior, and impulses and create long-term goals while ignoring distracting information (Casey et al., 2005; Diamond, 2002; Zelazo, 2004). This ability has been termed *cognitive control* or *executive function* (EF) (Bunge and Crone, 2009). According to Bunge and Crone (2009), cognitive control is not a single mechanism, but rather includes several specific functions. This chapter focuses on two of these specific functions. The first function is inhibitory control, or the ability to inhibit dominant responses in favor of more appropriate responses (Rothbart et al., 2003). The second function includes two processes thought to reflect self-monitoring: conflict monitoring, which is indicative of interference and interactions between different information processing pathways (Braver et al., 2001), and error/feedback monitoring, or updating a response based on feedback and error detection (Bunge and Crone, 2009).

The chapter examines these cognitive control functions and their relation to the development of social behavior. First, the development of cognitive control and its neural correlates is reviewed, focusing on the specific functions of inhibitory control and self-monitoring (including conflict and error monitoring). Then, individual differences associated with the development of these functions and their neural correlates are reviewed. Specifically, the relationship between temperament and inhibitory control/self-monitoring is examined. Temperament refers to stable behavioral and psychological profiles in infancy that shape patterns of behavior across contexts early in development (Kagan, 2001). It is important to examine how intrinsic factors, such as cognitive control processes, are related to temperament because they may influence developmental outcomes. Data suggesting cross-cultural variations in these abilities also are presented. Finally, the role of these functions in the

Neural Circuit Development and Function in the Brain: Comprehensive Developmental Neuroscience, Volume 3 http://dx.doi.org/10.1016/B978-0-12-397267-5.00060-1

© 2013 Elsevier Inc. All rights reserved.

neurodevelopment of decision making and motivation, including moral and social decisions, as well as moral behavior, is reviewed. This chapter focuses mostly on childhood and adolescence because the neural correlates of cognitive control develop markedly during this period.

23.1.1 Inhibitory Control

Inhibitory control, as defined earlier, is the ability to inhibit a dominant response in favor of a more appropriate response (Rothbart et al., 2003). This function includes the ability to restrain a prepotent response or choosing an alternative response pattern. It has a well-characterized developmental course, with improvements in children's ability to inhibit prepotent responses detected during toddlerhood (Gerardi-Caulton, 2000) and improvements in inhibitory control occurring during the preschool years (e.g., Gerstadt et al., 1994; Rueda et al., 2004; Zelazo, 2006), with marked improvements in inhibitory control continuing during middle childhood (e.g., Simonds et al., 2007) and adolescence (e.g., Anderson, 2002; Huizinga et al., 2006).

Research has pointed to several changes that occur with development in neural activation during inhibitory control tasks. For example, several studies have indicated that activation in the prefrontal cortex (PFC) becomes more focal with age (e.g., Bunge et al., 2002; Durston et al., 2006; Luna et al., 2001). These studies show that activation in certain brain regions increases with age while, at the same time, activation in other brain regions decreases. For example, in a study examining inhibitory control (Durston et al., 2006), decreased activation with development was found in prefrontal regions that were not relevant for task performance. However, an increase in activation, which was associated with improved task performance, was found in the ventrolateral PFC (VLPFC).

In a study examining developmental changes in inhibitory control, Bunge et al. (2002) administered to children between the ages of 8 and 12, as well as adults, a task that combined elements of the go/no-go (response inhibition) and the flanker (interference suppression) paradigms. In a typical go/no-go paradigm, participants are asked to respond to the majority of trials (go trials) and to withhold their response on a minority of the trials (no-go trials). In a typical flanker paradigm, participants are asked to respond on the basis of a central stimulus while ignoring flanking stimuli (flankers). In the task administered by Bunge et al. (2002), participants were asked to press a left button when a central arrow pointed to the left and press a right button when it pointed to the right. The flanker stimuli were congruent with the central target (go trials), incongruent with the target (go trials), or Xs, in which case participants were asked to refrain from responding (no-go trials).

The findings indicated that for both interference suppression (flanker) and response inhibition (go/no-go), children failed to engage a region in the right VLPFC that young adults recruited (Bunge et al., 2002). The findings also revealed decreased activation with age in the left VLPFC for interference suppression (flanker) and in the parietal and temporal cortices for response inhibition (go/no-go). According to Bunge et al. (2002), these data show that for different types of inhibitory control, children recruit different brain regions to adults. Specifically, children in this study failed to recruit a region in the right VLPFC, which was most robustly activated by both tasks in adults. This can be a result of immature brain structures in children or due to a shift in cognitive strategy between childhood and adulthood (Bunge et al., 2002).

Patterns of increasing and decreasing activation with development have been observed in other studies using the go/no-go paradigm. For example, Tamm et al. (2002) found age-related decreases in activation in the superior and middle frontal gyri as well as increased activation in the left inferior frontal gyrus between ages 8 and 20. In a study by Rubia et al. (2006), functional brain activation was compared between adolescents and adults during three different executive tasks measuring inhibitory control (go/no-go), cognitive interference inhibition (Simon task, also called the directional or motor Stroop task, in which participants are presented with congruent and incongruent stimuli on either side of a computer screen indicating a left or right button press), and attentional set shifting (switch task, during which interference from a previous stimulus–response association has to be inhibited). In all three tasks, adults recruited portions of the PFC, the anterior cingulate cortex (ACC), and the striatum more strongly than adolescents. Additionally, adults engaged the inferior parietal cortex more strongly than adolescents on the Simon and switch tasks, a finding similar to that reported for the go/no-go task by Bunge et al. (2002).

Additional evidence for focalization comes from a study by Luna et al. (2001), who administered an antisaccade task to children (7–12 years), adolescents (13–17 years), and young adults (18–22 years). This task requires participants to suppress looking at a visual target that appears suddenly in the peripheral visual field and instead look away from the target in the opposite direction. The findings indicate that, with age, activation decreased in the superior middle gyrus but increased in the striatum, intraparietal sulcus, frontal eye field, and lateral cerebellum. Additionally, activation in the dorsolateral PFC (DLPFC) was greatest for adolescents as compared with children and young adults.

In a functional magnetic resonance imaging (fMRI) study using the Stroop task (a classic task tapping inhibitory control), Adleman et al. (2002) studied children (ages 7–11 years), adolescents (ages 12–16 years), and young adults (ages 18–22 years). In this task, participants are asked to name the color of ink in which a color word is displayed. Adleman et al. (2002) found that young adults showed increased activation compared to adolescents in the left frontal gyrus. Additionally, young adults showed greater activation than children in the ACC and left parietal and parietooccipital regions as well as in the left frontal gyrus. Compared to children, adolescents showed greater activation in the parietal cortex. Adult and adolescent groups, however, did not differ in activation for this region. These findings suggest that when performing the Stroop task, young subjects are able to recruit parietal structures while PFC function contributing to performance on this task continues to develop into young adulthood. These age-related increases in activation occurred in conjunction with improvements in behavioral performance. These findings suggest that with development, resource recruitment becomes more focal.

Electrophysiological studies have pointed to increased cortical efficiency with age during tasks that require inhibitory control (Lewis et al., 2006). For example, Lewis et al. (2006) used dense-array electroencephalography (EEG) to measure event-related potentials (ERPs) as children and adolescents performed a go/no-go task. The N2 component of the ERP, an index of inhibitory control, was found to decrease in amplitude and latency with age. Additionally, source modeling of the N2 indicated a developmental decline in posterior midline activity, which was paralleled by increasing activity in frontal midline regions suggestive of the ACC. This pattern of frontalization is in line with other studies that observed reliance on more anterior regions of PFC with improvement of cognitive control (e.g., Lamm et al., 2006; Rubia et al., 2000). For example, Lamm et al. (2006) found that the N2 in a go/no-go task was source-localized to the cingulate cortex and orbitofrontal cortex (OFC). However, the source of the N2 in older children and in children who showed improved inhibitory control (regardless of age) was estimated to be in the ACC compared to younger children and children who performed poorly, whose N2 source was estimated in the posterior cingulate cortex.

In sum, research examining the neural correlates of the development of inhibitory control suggests that, with age, improvements in this ability are associated with increased focalization and efficiency in recruitment of the PFC. Thus, with development, brain regions that are not relevant for task performance decrease in their activation, while regions relevant for the task increase in their activation. This focalization of the PFC has been consistently found with various tasks that tap inhibitory control, including flanker, go/no-go, and Stroop tasks.

These changes in functional activation of inhibitory control that occur with development may reflect disengagement from immature neural circuits used by children and adolescents and recruitment of more mature alternative neural networks. However, the differences in activation patterns between children and adults may also reflect children's reliance on different cognitive strategies compared with adults. It is also possible that children engage in ongoing strategy-learning processes or differ from adults in their efforts to complete the task (Ernst and Mueller, 2008).

23.1.2 Self-Monitoring: Conflict and Error Monitoring

Self-monitoring is often measured through cognitive tasks (e.g., Stoop, flanker, and go/no-go paradigms) by examining response times on trials following an error as compared to response times following correct trials. If inaccurate performance is particularly salient to an individual, more controlled and slower responding in the trial following an error is typically exhibited (Davies et al., 2004; Luu et al., 2000). This form of self-monitoring can be viewed as a compensatory strategy where subjects slow their reaction time (RT) after an error in order to maximize accurate performance on the upcoming trial.

Studies have examined whether children, after making an error, respond more slowly on subsequent trials. Backen Jones et al. (2003) found that as 3- to 4-year-old children's performance on a Simon Says-type task improved, they tended to show posterror slowing. Using a more fast-paced task-switching paradigm, however, Davidson et al. (2006) failed to find evidence of posterror slowing in 4- to 5-year-old children. Typically, younger children make more errors than adults and may be less aware of them. Some studies have shown that children are able to detect errors, but are unable to correct them. For example, Bullock and Lutkenhaus (1984) found that 18- and 24-month-old children could distinguish between correctly and incorrectly built towers, even when they themselves failed to build the towers correctly. In a study designed to examine this question with preschool children, Jacques et al. (1999) presented 3-year-old children with the dimensional change card sort, in which children are shown two target cards (e.g., a blue rabbit and a red boat) and are asked to sort a series of test cards (e.g., red rabbits and blue boats) first according to one dimension (e.g., color) and then according to the other (e.g., shape). Most 3-year-olds perseverate during the postswitch phase, continuing to sort test cards by the first dimension (e.g., Zelazo et al., 2003). In order to assess error detection, Jacques et al. (1999) asked children to evaluate the sorting of a puppet. When 3-year-olds watched the puppet perseverate, they judged the puppet to be correct. When they saw the puppet sort correctly, they judged the puppet to be wrong, which suggests that 3-year-olds' perseverating performance and error detection are closely linked in this task.

Behavioral measures used to study self-monitoring can also be supplemented with psychophysiological methods to obtain a more comprehensive picture of ACC and PFC involvement in monitoring processes. A good deal of work studying these functions has focused on an error-related potential known as the error-related negativity (ERN). The ERN is a specific neural activity pattern associated with cognitive monitoring that is produced after the commission of an error and has a centromedial scalp distribution (Falkenstein et al., 1991; Gehring et al., 1993; van Veen and Carter, 2002). It is usually observed between 50 and 150 ms postresponse when participants make errors (Falkenstein et al., 1991; Scheffers et al., 1996). Due to its localization and involvement with monitoring, the ERN is seen as part of a neural system that interacts with the PFC to contribute to more efficient cognitive processing (Luu and Tucker, 2001; van Veen and Carter, 2002). The ERN is followed by a positive deflection (Pe) peaking between 200 and 500 ms after the ERN (Davies et al., 2004) and has been associated with awareness of an error (Nieuwenhuis et al., 2001). Source localization models have indicated a generator for the ERN in ACC (e.g., van Veen and Carter, 2002).

Although debate remains as to whether the ERN is a result of error detection (Falkenstein et al., 1991; Gehring et al., 1993) or response conflict processes (Carter et al., 1998; Gehring and Fencsik, 2001), it is generally hypothesized that the ERN is part of a larger performance monitoring system that is influential in the development of self-regulatory skills. As such, the ERN is suggested to serve as a feedforward control mechanism by which self-monitoring can influence future cognitive strategies and overall behavioral performance (Rodriguez-Fornells et al., 2002).

Davies et al. (2004) examined error detection in children using a flanker task and found smaller (i.e., more positive) ERN amplitudes for 7- to 8-year-old children as compared to adolescents. Although ERN amplitudes increase slowly with age, these young children showed behavioral evidence of error detection because, as found in adults, their responses slowed down following an erroneous response. Furthermore, even the youngest children in the study showed evidence of the Pe component, further demonstrating error detection (Davies et al., 2004).

Another type of self-monitoring is known as *conflict monitoring*. As mentioned above, conflict monitoring can be defined as interference and interactions between different information processing pathways (Braver et al.,

2001). Conflict monitoring includes the ability to identify discrepancies in self-performance, and some consider error monitoring as detection of postresponse conflict as opposed to preresponse conflict. Indeed, research has shown an overlap in brain regions associated with both conflict and error monitoring (Ridderinkhof et al., 2004). Conflict monitoring has been found to have a neural basis in ACC. Studies have shown activation in this region for both the conflict elicited by the external environment and that elicited by the individual's own behavior or thoughts (e.g., Carter and van Veen, 2007; Casey et al., 1997a; Posner and Fan, 2004). Once conflict has been detected, the ACC signals for the recruitment of higher-order cognitive processes associated with PFC activation in order to alleviate this conflict (Kerns et al., 2004).

While the majority of neuroimaging work demonstrating ACC activation during conflict monitoring has been found with adults, research shows that this structure also underlies conflict monitoring in children and adolescents (Rubia et al., 2007). For example, in a neuroimaging study with 5- to 16-year-old children, performance on a visual discrimination task that required attention to conflict was positively correlated with the size of a child's right ACC (Casey et al., 1997a). Additionally, in a study with 7- to 12-year-old children, behavioral performance on a go/no-go task was correlated with the degree of ACC activation (Casey et al., 1997b).

In sum, review of studies examining self-monitoring and its neural correlates in children suggests that even young children are capable of self-monitoring, as reflected in error and conflict monitoring. Behaviorally, children have been found to be capable of detecting errors in simple tasks, but incapable of correcting them or identifying them in more complex situations. In addition, for certain tasks, children slow down their responses following an error. Although evidence has been found for self-monitoring in children, this function continues to develop with age as evidenced by increasing ERN amplitude. Thus, the research reviewed suggests that, to a certain degree, children recruit similar neural functions associated with self-monitoring as adults. Specifically, children have been found to recruit the ACC, which has been associated with both error and conflict monitoring.

23.2 INDIVIDUAL DIFFERENCES IN COGNITIVE CONTROL

23.2.1 The Role of Temperament in Cognitive Control

Individual differences in the development of cognitive control have been reported (e.g., Gehring et al., 2000; Henderson, 2010; Lewis et al., 2006; McDermott et al., 2009; Stieben et al., 2007; White et al., 2011) and have been linked to social outcomes in children. For example, behavioral and physiological measures of cognitive control have been linked to negative mood induction conditions (Lewis et al., 2006, 2007) and to heightened levels of trait anxiety and internalizing symptoms (e.g., Gehring et al., 2000; Stieben et al., 2007). Differences among individuals in cognitive control have also been associated with different types of temperament. For example, a number of studies from different laboratories have found that specific cognitive control functions moderate the developmental trajectory of behavioral inhibition (BI) in children (Henderson, 2010; Thorell et al., 2004; White et al., 2011). BI is a temperament identified early in childhood that predicts heightened emotional reactivity throughout childhood (Fox et al., 2005) and increased risk for anxiety in adolescence (Chronis-Tuscano et al., 2009).

For example, Thorell et al. (2004) examined how BI and inhibitory control assessed at 5 years of age were associated with socioemotional functioning at 9 years of age. The results indicated that behaviorally inhibited children with high levels of inhibitory control were reported as being more socially anxious than behaviorally inhibited children with low levels of inhibitory control. Similarly, Fox and Henderson (2000) found that behaviorally inhibited 4-year-olds with high inhibitory control were less socially competent and more socially withdrawn than behaviorally inhibited children with low inhibitory control.

These findings about the role of inhibitory control for temperamental BI are the opposite of results regarding another EF, attention shifting (Eisenberg et al., 1998). For example, Eisenberg et al. (1998) found that children low on attention shifting and high on parental reports of negative emotions, such as fear, sadness, and anxiety, were rated by their parents and teachers as more shy 2 years later.

White et al. (2011) examined how attention shifting and inhibitory control, which were tested at 48 months of age, moderated the association between BI assessed at 24 months of age and anxiety problems in the preschool years. The results indicated that high levels of inhibitory control increased the risk for anxiety disorders among behaviorally inhibited children whereas high levels of attention shifting decreased the risk for anxiety problems in these children.

In a different study, Lahat et al. (2012a) examined how these two EFs – attention shifting and inhibitory control – moderated the relation between exuberant temperament in infancy and propensity for risk taking in childhood. Temperamental exuberance has been defined by positive reactivity to novelty, approach behavior, and sociability (Putnam and Stifter, 2005). Children with an exuberant temperament are also characterized by impulsivity, sensitivity to reward, fearlessness, and risk taking (Fox et al., 2001; Polak-Toste and Gunnar, 2006; Rothbart and Bates, 2006). EF was assessed at 48 months of age. Risk-taking propensity was measured when children were 60 months of age. The results indicated that exuberance was positively associated with risk-taking propensity among children relatively low in attention shifting but unrelated for children high in attention shifting. Inhibitory control did not significantly moderate the link between exuberance and risk-taking. Taken together, the findings from these sets of studies on temperament and different types of EF demonstrate that attention shifting and inhibitory control have differential influences on levels of risk or adaptation. Furthermore, these two studies (Lahat et al., 2012a; White et al., 2011) suggest that high levels of attention shifting may serve as a protective factor in the link between temperament and negative outcomes. This conclusion may have important implications for prevention and intervention efforts in the form of training in order to improve attention-shifting skills.

McDermott et al. (2009) provided neurophysiological evidence for the role of cognitive control among individuals with different temperaments. These authors studied adolescents who were part of a larger longitudinal study and were assessed during infancy and early childhood for BI. These adolescents were administered a flanker task, during which ERPs were recorded. The study focused on the ERN component, which is a negative deflection in the ERP waveform produced after the commission of an error. The ERN was found to be larger for adolescents with high childhood BI than for adolescents low on childhood BI. This finding suggests increased error monitoring among highly behaviorally inhibited individuals. In addition, adolescents and their parents completed semistructured diagnostic interviews to assess lifetime presence or absence of anxiety disorders. The results indicated that the ERN moderated the relation between early BI and later anxiety disorders such that, for the participants high on BI, larger ERNs were related to a higher risk of anxiety disorders (McDermott et al., 2009).

In a recent study using dense-array ERPs and source analyses, Lamm et al. (2012) showed that, during an inhibitory control task, children high in temperamental fearfulness showed modeled source activation in areas suggestive of VLPFC across both emotional and nonemotional conditions of the task. However, children low in temperamental fearfulness only showed this pattern of activation during the emotional condition. Results from this study suggest that while children low in temperamental fearfulness recruited increased inhibitory control only during the emotional conditions, or those conditions in which more cognitive control recourses were likely needed, children with high fearful temperaments sustain this increased level of inhibitory control across both neutral and emotional contexts (Lamm et al., 2012). These findings suggest that temperamentally fearful individuals show increased vigilance not only in emotional situations but also in nonemotional ones. Thus, these fearful individuals may refrain even from social situations that do not induce negative emotions.

Other neurophysiological evidence for the moderating role of inhibitory control has been found for the association between shyness and social–emotional maladjustment (Henderson, 2010). Henderson (2010) examined the associations between shyness, N2 component of ERP, and social adjustment in 9- to 13-year-old children. Participants were administered a flanker task while ERPs were being recorded. In addition, they completed questionnaires assessing temperament, social anxiety, attribution style, and self-perceptions. The results indicated that shyness was associated with poor social outcomes primarily among children with relatively large N2 amplitudes. The results point to the role of conflict in shy children's social adjustment (Henderson, 2010).

Individual differences in cognitive control and their association with social and cognitive adjustment have also been studied using the Stroop task. Perez-Edgar and Fox (2003) examined 11-year-old children who were administered an emotional Stroop task in which emotionally charged words substitute the traditional color words. Based on their RT performance on this task, participants were divided into either an interference or a facilitation group. Children in the interference group were slower to respond to either negative or positive words, whereas children in the facilitation group were faster to respond to these words. The two groups were assessed on cognitive, emotional, and social measures collected at ages 4, 7, and 11. The interference group showed greater socioemotional (but not cognitive) maladjustment over time. In a second study, ERPs were collected during this task. The findings reveal that negative words, as compared with positive words, involve ERP components that are considered to tap attentional processes. Additionally, larger positive slow-wave amplitudes were observed for the facilitation group.

In a different study (Warren et al., 2010), using the emotional Stroop during fMRI, participants were also administered the Attachment Script Assessment in which they were asked to generate stories in response to attachment-related word prompts. This measure assesses secure-base-script knowledge or the degree to which an individual is able to generate narratives in a hypothetical situation in which attachment-related threats are recognized and resolved. The findings indicated that individuals with lower levels of secure-base-script knowledge showed increased activity in brain regions associated with emotion regulation, such as the right OFC, as well as activity in regions associated with inhibitory control, such as the left DLPFC, ACC,

and superior frontal gyrus. These findings suggest that insecure attachment is associated with a greater need in inhibitory control in order to attend to task-relevant nonemotional information (Warren et al., 2010).

In sum, studies examining individual differences in cognitive control and their social–emotional outcomes have shown links with negative mood inductions (Lewis et al, 2006) as well as anxiety disorders and other social problems (e.g., Fox and Henderson, 2000; McDermott et al., 2009; White et al., 2011). For example, highly behaviorally inhibited children, who are also high in inhibitory control, have been found to show negative social outcomes as well as increased anxiety disorders. Children with this temperament have also been found to show increased self-monitoring, as reflected by larger ERNs relative to children with low BI (McDermott et al., 2009). Additionally, children with a fearful temperament showed increased VLPFC source activation even in response to nonemotional stimuli (Lamm et al., 2012). Taken together, these studies suggest that high inhibitory control, as well as high self-monitoring, may result in negative social outcomes for children who are shy or have a more fearful temperament.

23.2.2 Cross-cultural Differences in the Development of Cognitive Control

Research on the development of cognitive control has mostly been conducted in Western cultures. However, an emerging body of cross-cultural studies suggests that Asian children may outperform Western children on measures of cognitive control (e.g., Lahat et al., 2010; Sabbagh et al., 2006).

Research comparing children from Western and Asian cultures has shown that Chinese children perform better on behavioral and neurophysiological measures of EFs than North American children (e.g., Chen et al., 1998; Ho, 1994; Lahat et al., 2010; Sabbagh et al., 2006; Wu, 1996). For example, Sabbagh et al. (2006) administered a battery of EF and theory of mind (ToM) tasks to preschoolers from China and the United States. ToM is the ability to attribute mental states – beliefs, intents, and desires – to oneself and others and to recognize that mental representations can differ across individuals (Premack and Woodruff, 1978). The Chinese preschoolers showed better performance than the US preschoolers on all measures of EF, but not on measures of ToM. However, individual differences in EF predicted ToM performance in both groups.

Chinese children's advanced performance on EF tasks may stem from the opportunities to exercise and practice these abilities that they encounter within their culture. For example, Chinese parents expect children to master impulse control at a much younger age than North American parents (Chen et al., 1998; Ho, 1994;

Wu, 1996). Compared with Western parents, Chinese parents are more controlling and protective in child rearing. For example, they often encourage their young children to stay close to and to be dependent on them. Indeed, most Chinese infants and toddlers sleep in the same bed or in the same room as their parents (Chen et al., 1998). In addition, impulse control is more highly valued in Chinese daycare settings than in North American daycare settings (Tobin et al., 1989).

Another possibility for Chinese children's superior cognitive control was suggested by Lahat et al. (2010) in a study that focused on the N2 component of ERP. Cultural differences in the importance that parents place on impulse control could affect children's motivation to succeed on a task such as the go/no-go task, resulting in greater PFC activation. The study compared 5-year-old children from a Chinese-Canadian ethnic background with children from a European-Canadian background on a go/no-go task. No behavioral differences between the two cultural groups were observed, but robust N2 amplitude differences were found. Chinese-Canadian children showed larger (i.e., more negative) N2 amplitudes than European-Canadian children on the right side of the scalp on no-go trials as well as on the left side of the scalp on go trials. Source analyses of the N2 showed greater modeled source activation for Chinese-Canadian children in dorsomedial, ventromedial, and (bilateral) ventrolateral PFC. These findings reveal that Chinese-Canadian children show greater hemispheric differentiation than European-Canadian children, suggesting more advanced cognitive control. Moreover, the asymmetrically lateralized N2 may be a reliable marker of both effortful inhibition (on the right) and effortful approach (on the left).

Behavioral cultural differences were not observed in the Lahat et al. cross-cultural study, despite cultural differences found in other research (e.g., Sabbagh et al., 2006). This discrepancy can stem from the different cognitive control tasks that were used or from the different age groups that were studied. Furthermore, the Chinese-Canadians in the Lahat et al. study grew up in Canada, whereas previous research examined Asian children living in Asia. In any case, this study demonstrates that neurophysiological techniques can provide a measure of neurocognitive function that is more sensitive than behavioral data alone.

In sum, individual differences in cognitive control have also been observed in cross-cultural variations. Specifically, an emerging body of research comparing children from Asian and Western cultures has shown behavioral (e.g., Sabbagh et al., 2006) as well as neurophysiological (Lahat et al., 2010) evidence, suggesting advanced cognitive control abilities among children from a Chinese cultural background compared with children from a Western cultural background. Although the reasons for these differences are not clear yet, it is possible that differences in socialization between the two cultures play a major role.

23.3 THE ROLE OF COGNITIVE CONTROL IN DECISION MAKING AND MOTIVATION

23.3.1 Motivation-related Decisions

Cognitive control has also been implicated in decision-making processes. Studies with adults support the idea of a frontostriatal network linking cognitive control and motivation during decision making (see Somerville and Casey, 2010). Fewer studies have examined the role of the development of cognitive control as it affects the emergence of decision-making skills.

A simple paradigm that has been used to assess basic decision making with young children is delay of gratification, which measures children's ability to give up an immediate reward in favor of a larger reward later. For example, participants are seated in front of a piece of candy while an experimenter leaves the room. If they wait for the experimenter to return, they get two treats; otherwise they get only one (Mischel et al., 1972). Variations of the standard delay of gratification paradigm have shown developmental differences throughout the preschool years. For example, in one study (Kochanska et al., 1996), children were asked to hold candy in their mouth without eating it until they were told to do so. In a second task, children were asked not to peek while they could hear that the experimenter was wrapping a gift for them. Children's ability to delay gratification increased significantly from 3 to 4 years of age. In a study linking delay of gratification and decision making, Prencipe and Zelazo (2005) examined children's delay of gratification for self and other (the experimenter). Three-year-olds typically chose an immediate reward for themselves and a delayed reward for other (the experimenter). According to Prencipe and Zelazo, these findings suggest that 3-year-olds are capable of adaptive decision making but still have difficulty regulating their own approach behavior in motivationally salient situations. According to these authors, it is possible that their behavior is driven by the relatively automatic processes rather than by more deliberate prefrontal networks. The findings can be explained in light of Barresi and Moore's (1996) model of the development of perspective taking. It is possible that 3-year-olds may have made impulsive choices in the self condition because they took an exclusively first-person perspective on their own behavior and had difficulty adopting a more objective, third-person perspective according to which delay would be preferred. In contrast, 3-year-olds may have chosen delayed rewards in the other condition because they took an exclusively third-person perspective and had difficulty appreciating the experimenter's subjective perspective (i.e., his or her own desire for immediate gratification).

In more complex situations, individuals are often required to make approach-avoidance decisions in the face of uncertainty. One common measure of complex decision making in adults is the Iowa gambling task (IGT; Bechara et al., 1994). In this task, participants are asked to choose cards from four decks that contain a different number of cards that could lead to their winning or losing money. During the task, the participants learn that some decks are more advantageous than others. Kerr and Zelazo (2004) modified the IGT to create a version for children that included two decks of cards, one advantageous and one disadvantageous. Feedback on participants' decisions was provided in the form of happy (reward) and sad (loss) faces. Choosing cards from the disadvantageous deck resulted in more rewards on every trial but also with occasional (unpredictable) large losses. Three-year-olds failed to develop a preference for the advantageous deck. However, 4- and 5-year-olds were able to make advantageous decisions (Kerr and Zelazo, 2004).

Although these studies suggest a steady improvement in cognitive control and decision making with development, according to Somerville and Casey (2010), cognitive control and decision making can often be impaired in light of emotionally charged interactions. This impaired cognitive control is especially pronounced during adolescence when rates of risky behavior, such as drug use and risky sexual conduct, increase (Casey et al., 2005; Steinberg, 2008). Thus, although adolescents show improvements in cognitive control, their goal-oriented behavior can be diminished in light of motivational cues of potential reward (Cauffman et al., 2010; Figner et al., 2009; Steinberg et al., 2009). For example, Figner et al. (2009) used a gambling task in which reward feedback was given either during decision making or after decision making. The findings show that adolescents made riskier choices compared to adults, but only in the condition in which the reward was given during the decision. This condition is more emotionally charged and elicits greater arousal.

Another study examined participants between the ages of 10 and 30 years and used a delay discounting task (Steinberg et al., 2009) in which participants were asked to choose between an immediate reward of less value (e.g., $400 today) and a variety of delayed rewards of more value (e.g., $700 1 month from now or $800 6 months from now). The findings indicate that, before age 16, children showed a greater willingness to accept a smaller reward immediately than a large reward that was delayed. Cauffman et al. (2010) obtained similar results examining the same age range with a modified version of the IGT. Results indicate that approach behaviors

(a tendency to choose from the advantageous decks) display an inverted U-shape relation to age, peaking in mid- to late adolescence. In contrast, avoidance behaviors (a tendency to refrain from choosing from the disadvantageous decks) increase linearly with age, with adults avoiding disadvantageous decks at higher rates than both preadolescents and adolescents. Taken together, these studies show that risky choices tend to peak between 14 and 16 years of age, followed by a decline in risky behavior.

Casey et al. (2008) proposed a model describing the neurocircuitry of the development of control and motivational processes. According to this model, top-down prefrontal regions involved in cognitive control develop linearly with age, whereas bottom-up striatal regions, which are involved in processing of salient cues in the environment, develop in an inverted U-shaped function. Evidence for such a frontostriatal circuitry comes from studies using diffusion tensor imaging and fMRI. Casey and colleagues (Casey et al., 2007; Liston et al., 2006) have found that the strength of the connection between frontal and striatal regions is associated with effective cognitive control in typically and atypically developing individuals.

fMRI studies examining the role of the striatum in salient and motivational contexts support the idea that adolescents show enhanced sensitivity to incentives relative to children and adults (Ernst et al., 2005; Galvan et al., 2006; Geier et al., 2010; May et al., 2004). For example, Ernst et al. (2005) used a monetary reward task and found stronger activation among adolescents than adults in the left nucleus accumbens, a structure in the striatum thought to be involved in reward processing. In addition, a reduction in the fMRI blood oxygen level dependent (BOLD) signal in the amygdala in response to reward omission was larger for adults than for adolescents.

Few studies (e.g., Hardin et al., 2009; Jazbec et al., 2006) have examined the link between cognitive control and motivational processes. Recent research has shown that adolescents' cognitive control can be enhanced in light of incentives. For example, when participants were promised a financial reward for accurate performance on certain trials of an antisaccade task, cognitive control was improved for adolescents more than adults (Jazbec et al., 2006). This finding has also been obtained with social rewards, such as a happy face (Kohls et al., 2009). Geier et al. (2010) studied the neural underpinnings of reward processing and its influence on cognitive control in adolescence using a modified version of an antisaccade task. The results indicate that faster correct inhibitory responses were provided on reward trials than on neutral trials by both adolescents and adults. Additionally, fewer inhibitory errors were made by adolescents. For reward trials, the BOLD signal was attenuated in ventral striatum in adolescents during cue assessment.

This was followed by overactivation in adolescents during response preparation (i.e., during fixation after the reward cue) in the ventral striatum, as well as the precentral sulcus, which is important for controlling eye movements. These findings suggest enhanced activation in adolescents in control regions as a result of reward anticipation (Geier et al., 2010).

In sum, studies examining the development of cognitive control in decision making have pointed to an inverted U-shaped trend in development (see Somerville and Casey, 2010). Studies with young children show steady improvements in the ability to delay gratification (e.g., Prencipe and Zelazo, 2005) as well as more effective decision making (e.g., Kerr and Zelazo, 2004). However, studies have shown a greater propensity for risky decisions during adolescence in light of motivationally salient situations (e.g., Cauffman et al., 2010). fMRI studies examining the link between cognitive control processes and motivation have found enhanced activation during adolescence in brain regions associated with cognitive control during anticipation of reward (Geier et al., 2010).

23.3.2 Social and Moral Decisions and Behavior

Evidence for the role of cognitive control in influencing more specific types of decisions, such as social and moral decisions comes from behavioral, fMRI, and ERP studies (e.g., Greene et al., 2001, 2004; Lahat et al., in press, 2012b; Lahat and Zelazo, 2010). In these studies, participants are typically asked to make a decision about whether a social or moral violation is acceptable or unacceptable to perform. Moral violations involve intrinsic negative consequences for others, such as physical harm or issues of fairness (Nucci, 1981; Smetana, 2006; Turiel, 1983).

For example, Greene et al. (2001, 2004) found evidence for the role of cognitive conflict and inhibitory control in the moral judgments of adult participants. These authors distinguish between impersonal moral dilemmas and personal moral dilemmas. An example of an impersonal dilemma is the *trolley* dilemma (Thomson, 1986) in which one has to decide whether to allow an out-of-control trolley to continue down a track where it will kill five people or whether to push a switch diverting it to a track where it will kill only one person. However, in a variation of this dilemma, the *footbridge* dilemma, the only way to save the five people is to push a large person in front of the trolley, killing him but saving the others. This latter dilemma is a personal dilemma as it is introduced in an 'up-close-and-personal manner' (Greene et al., 2001) and the bystander witnessing the event now becomes a moral agent. Most individuals assert that it is acceptable to sacrifice one person in order to save five in the case of the trolley dilemma, but not in the case of the footbridge dilemma (Greene et al., 2001). Greene et al. (2001) found that RTs were longer when participants judged personal moral violations as acceptable than when they judged personal moral violations as unacceptable. This effect was not found for impersonal moral judgments. According to Greene et al. (2001), personal moral violations elicit a negative social–emotional prepotent response, which results in judging the moral violation as unacceptable. An individual must overcome this prepotent response in order to judge a personal moral violation as acceptable. Thus, according to Greene, making a utilitarian judgment to serve the greater good involves cognitive control in order to overcome prepotent responses elicited by personal dilemmas.

In a different study (Greene et al., 2004), the authors focused on personal moral dilemmas only and explored whether different patterns of neural activity in response to these dilemmas are correlated with differences in moral decision-making behavior. For this purpose, Greene et al. (2004) made a further distinction within the class of personal dilemmas. They differentiated between difficult personal moral dilemmas, which elicit high conflict, and easy personal moral dilemmas, which elicit low conflict. Difficult personal dilemmas are a class of dilemmas that create cognitive and emotional tension compared to easy dilemmas. In response to these dilemmas, participants tend to answer slowly, and they exhibit no consensus in their judgments. An example of a difficult dilemma is the *crying-baby* dilemma in which an individual and other townspeople have sought refuge in a cellar to escape from enemy soldiers who have taken over the village. The protagonist's baby begins to cry loudly, and this could summon the attention of the soldiers, who might kill the protagonist, his child, and the others hiding out in the cellar. To save himself and the others, the protagonist must smother his child to death. This case contrasts with easy, low-conflict, personal moral dilemmas that receive relatively rapid and uniform judgments. One such example is the *infanticide* dilemma, in which a teenage mother must decide whether or not to kill her unwanted newborn infant. The latter dilemma is considered easy in the sense that most participants tend to agree that an unwanted infant should not be killed.

Greene et al. (2004) examined whether different brain regions are involved in judgments of difficult versus easy moral dilemmas. The dilemmas were classified as either difficult or easy according to their RTs, with difficult dilemmas having longer RTs. The findings indicate that judgments of difficult dilemmas, as compared to easy dilemmas, involved increased activity bilaterally in both the DLPFC and the ACC. This contrast also revealed activity in the inferior parietal lobes and the posterior cingulate cortex. Greene et al. (2004) also compared the neural activity associated with utilitarian judgments (judging a personal moral violation as acceptable in favor of the greater good, such as smothering the baby in the crying-baby dilemma) to that associated with non-utilitarian judgments (prohibiting a personal moral violation despite its utilitarian value, such as allowing the baby to live in the crying-baby dilemma). The authors found increased activity for utilitarian, as compared with nonutilitarian, moral judgments bilaterally in the anterior DLPFC and in the right inferior parietal lobe. In addition, they found increased activity for utilitarian moral judgments in the more anterior region of the posterior cingulate (Greene et al., 2004). These findings show that judgments of difficult dilemmas engage brain areas associated with the detection of conflict and the operation of inhibitory control. According to Greene et al. (2004), when participants respond in a utilitarian manner, such responses reflect not only the involvement of abstract reasoning but also the engagement of cognitive control in order to overcome prepotent social–emotional responses elicited by these dilemmas.

Evidence for the role of cognitive control in moral judgments has been found in children as young as 10 years of age (Lahat et al., 2012b, in press; Lahat and Zelazo, 2010). These authors obtained behavioral and neurophysiological data suggesting that the detection of cognitive conflict distinguishes between moral decisions and social conventional decisions. Social conventions are behavioral uniformities that serve to coordinate individuals' interactions in a social system (Nucci, 1981; Smetana, 1981, 2006; Turiel, 1983). These conventions, such as forms of address and modes of greeting, are symbolic elements of social organization. Social conventions, such as eating with one's fingers or wearing pyjamas to school, can vary across different social systems, are contingent on societal rules, and can be altered by authority or social consensus. This is in contrast to moral acts, such as hitting, lying, and stealing, which are considered universal, independent from rules and authority, and unalterable.

In a set of studies, Lahat et al. (2012b, in press) measured RTs and focused on the N2 component of ERP, which, as mentioned earlier, has been considered to be an index of cognitive control (e.g., Nieuwenhuis et al., 2003). Adults, children (10 years), and adolescents (12–14 years) were administered a moral–conventional judgments task while ERPs were being recorded and RTs were measured. The task included scenarios describing a social interaction and had three possible endings: a moral violation, a conventional violation, and a neutral act. Participants were to judge whether the act was acceptable or unacceptable to perform in a situation where a social rule was assumed or removed (i.e., "imagine that there is no rule against the act; is it okay or not okay to perform the act?"). The findings revealed that, at all ages, RTs were faster for moral rather than conventional violations when a rule was assumed (Lahat et al., 2012b).

ERP data indicated that adults', but not adolescents', N2 amplitudes were larger (i.e., more negative) for conventional rather than moral violations when a rule was assumed (Lahat and Zelazo, 2010). Taken together, these results suggest that judgments of conventional violations involve increased conflict detection compared to moral violations, and these two domains are processed differently across development. The findings were explained by the idea that judgments of conventional violations are more explicitly dependent on rules, and thus a violation of the rule results in increased conflict detection. However, judgments of moral violations are based more directly on the intrinsic negative consequences of the act and thus less cognitive conflict is detected in these trials. These findings inform theories of moral development and suggest that cognitive processing of moral judgments changes with age.

The role of cognitive control has also been found to be crucial for moral behavior, specifically that of lie-telling behavior in children and adolescents (Evans and Lee, 2011; Evans et al., 2011; Talwar and Lee, 2008). For example, Talwar and Lee (2008) studied children between the ages of 3 and 8 years. In this study, participants were told not to peek at a toy. However, most children could not resist the temptation and peeked and later lied about it. Children's conceptual moral understanding of lies, executive functioning, and theory-of-mind understanding were also assessed. Results indicated that children's initial denials about peeking were related to inhibitory control measured with a Stroop and whisper task (in which children are required to whisper their responses rather than say them aloud). Denials about peeking were also related to moral understanding and first-order false belief understanding; that is, telling a lie successfully required deliberately creating a false belief in the mind of another. Additionally, children's ability to maintain their lies was related to their second-order belief understanding (e.g., Ann knows that Sally thinks her toy is in the box), a finding in line with research showing that second-order beliefs begin to emerge around 6 years of age and undergo steady development well into adolescence (Sullivan et al., 1994). These findings suggest that social and cognitive factors may play an important role in children's lie-telling abilities.

In a similar paradigm, designed for older participants, Evans and Lee (2011) examined 8- to 16-year-old children's tendency to lie, the sophistication of their lies, and related cognitive factors. Participants were left alone and asked not to look at the answers to a test, but the majority of children peeked. The researcher then asked a series of questions to examine whether the participants would lie about their cheating and, if they did lie, evaluate the sophistication of their lies. Additionally, participants completed measures of EF, including inhibitory control measured with a Stroop task. Results revealed that the sophistication of 8- to 16-year-old children's lies, but not their decision to lie, was significantly related to executive functioning skills and performance on the Stroop task. In these studies, participants with a better performance on the Stroop were better able to conceal incriminating knowledge they ought not to know. According to these authors, the Stroop task is thought to measure both inhibitory control and working memory. In the situation created in these studies, participants were required to hold in working memory their incriminating knowledge in order to create answers different from this knowledge. At the same time, they must inhibit reporting the truth. Taken together, these findings suggest that whether a person is a good liar who is able to, or a bad liar who is not able to, maintain his/her lie, is dependent on working memory in conjunction with inhibitory control.

In a different study (Evans et al., 2011), Chinese preschoolers' ability to tell a strategic lie by making it consistent with the physical evidence of their transgression was investigated. Participants in this study were left alone in a room and asked not to lift a cup to see its contents. If children lifted up the cup, the contents would be spilled and evidence of their transgression would be left behind. Upon returning to the room, the experimenter asked children whether they peeked and how the contents of the cup ended up on the table. The findings revealed that young children were able to tell strategic lies to be consistent with the physical evidence by about 4 or 5 years of age, and this ability increases in sophistication with age. Furthermore, the study revealed that children's theory-of-mind understanding and inhibitory control skills were significantly related to their ability to tell strategic lies in the face of physical evidence. Although this study was conducted in a non-Western culture, no cross-cultural comparison was made and, although research suggests that Chinese children are more advanced than Western children in cognitive control (e.g., Lahat et al., 2010; Sabbagh et al., 2006), it is not known if Chinese children are more advanced at telling strategic lies than Western children.

In sum, studies examining the neural correlates of moral judgments with adults have shown that certain classes of moral dilemmas (i.e., difficult dilemmas) involve inhibitory control and detection of conflict. These studies show increased activation in ACC and DLPFC during judgments of these dilemmas (Greene et al., 2004). The role of cognitive control in moral and social judgments has also been studied with a behavioral/ERP paradigm in children, adolescents, and adults (Lahat et al., 2012b; Lahat and Zelazo, 2010). Findings reveal larger N2 amplitudes for adults, but not children, in response to conventional rather than moral violations. This finding suggests that the morality–convention distinction continues to develop well into early adolescence. Finally, the role of inhibitory control has also been demonstrated in moral behavior (e.g., Evans and Lee, 2011).

23.4 CHAPTER SUMMARY AND FUTURE DIRECTIONS

In this chapter, the role of cognitive control across different social behaviors was reviewed. The chapter focused on the development of the neural correlates of two specific cognitive control functions: inhibitory control and self-monitoring (including conflict and error monitoring). As described earlier, these specific functions have been found to develop with the maturation of the PFC and patterns of activation increase with age (Zelazo et al., 2008).

Studies examining the neural correlates of the development of inhibitory control have shown that PFC activation not only increases with age, but also becomes more focalized (e.g., Bunge et al., 2002; Durston et al., 2006; Luna et al., 2001). As discussed, with development, brain regions not associated with task performance decrease in activation, while regions relevant to task performance increase in activation. With development, children also show improvements in self-monitoring abilities. For example, they are better able to detect and correct errors (e.g., Backen Jones et al., 2003; Bullock and Lutkenhaus, 1984). These behavioral data are in line with findings from neurophysiological studies, which show an increase in ERN amplitude with age (Davies et al., 2004). An additional form of self-monitoring that has been studied is conflict monitoring. Some argue that error monitoring is a specific type of conflict monitoring, which occurs postresponse and, although these abilities are distinct, there is an overlap in their functional activation, namely the ACC (Ridderinkhof et al., 2004). Even young children have been found to recruit the ACC in tasks that require self-monitoring (Casey et al., 1997a,b).

Inhibitory control and self-monitoring have been found to show associations with individual differences and their social–emotional outcomes (e.g., Fox and Henderson, 2000; Lamm et al., 2012; McDermott et al., 2009; White et al., 2011). This body of research has shown that children with a fearful temperament who are also high in inhibitory control or self-monitoring are at risk of developing negative social–emotional outcomes, such as social withdrawal and anxiety disorders. Evidence on individual differences associated with cognitive control come mostly from behavioral and electrophysiological studies. Future research should examine the association between temperament and cognitive control using neuroimaging in order to better localize differences in brain activation.

Individual differences relating to cultural variations have also been found, particularly when comparing Chinese children with children from Western cultures. These studies provide behavioral (Sabbagh et al., 2006) as well as neurophysiological (Lahat et al., 2010) evidence for advanced performance on cognitive control tasks among Chinese children. It is possible that socialization processes that differ across the two cultures contribute to these differences in cognitive control. The cross-cultural neurophysiological work has been conducted with Asian children who grew up in North America. Future research should compare Western children with Asian children who are raised in Asia. This would provide a more pure cross-cultural comparison.

In this chapter, the role of cognitive control in the development of decision making and motivation was also reviewed. Studies have shown that the development of decision making takes an inverted U-shaped form, with steady improvements during childhood (e.g., Kerr and Zelazo, 2004) and a propensity for risk taking during adolescence (Somerville and Casey, 2010). fMRI studies with adolescents (e.g., Geier et al., 2010) have shown increased activation in brain regions associated with cognitive control during anticipation of reward. Most of the work has focused on monetary or other types of non-social rewards. Future research should examine the relation between cognitive control and social reward, such as social acceptance.

Research has shown that cognitive control also plays a role in the development of moral and social decisions, as well as moral behavior. Specifically, the role of cognitive conflict has been demonstrated in studies with adults, in which increased activation has been found in the ACC and DLPFC (e.g., Greene et al., 2001, 2004). Cognitive conflict has also been found to play a role in adults', but not children's, distinction between moral and conventional judgments, as reflected in N2 amplitudes (Lahat and Zelazo, 2010). Finally, a role for the development of inhibitory control has been identified in children's and adolescents' moral behavior (Evans and Lee, 2011). Future research on the role of cognitive control in the development of moral judgment should examine direct relations between neurophysiological correlates of moral judgments and various types of cognitive control. This would allow to better pinpoint which cognitive control functions are involved in moral development.

SEE ALSO

Diseases: The Developmental Neurobiology of Repetitive Behavior; **Induction and Patterning of the CNS and PNS:** Area Patterning of the Mammalian Cortex; The Formation and Maturation of Neuromuscular Junctions. **Migration:** Cell Polarity and Initiation of Migration.

References

Adleman, N.E., Menon, V., Blasey, C.M., et al., 2002. A developmental fMRI study of the Stroop color-word task. NeuroImage 16, 61–75.

Anderson, P., 2002. Assessment and development of executive function (EF) during childhood. Child Neuropsychology 8, 71–82.

Backen Jones, L., Rothbart, M., Posner, M., 2003. Development of executive attention in preschool children. Developmental Science 6, 498–504.

Barresi, J., Moore, C., 1996. Intentional relations and social understanding. The Behavioral and Brain Sciences 19, 107–154.

Bechara, A., Damasio, A.R., Damasio, H., Anderson, S.W., 1994. Insensitivity to future consequences following damage to human prefrontal cortex. Cognition 50, 7–15.

Braver, T.S., Barch, D.M., Gray, J.R., Molfese, D.L., Snyder, A., 2001. Anterior cingulate cortex and response conflict: Effects of frequency, inhibition and errors. Cerebral Cortex 11, 825–836.

Bullock, M., Lutkenhaus, P., 1984. The development of volitional behavior in the toddler years. Child Development 59, 664–674.

Bunge, S.A., Crone, E.A., 2009. Neural correlates of the development of cognitive control. In: Rumsey, J., Ernst, M. (Eds.), Neuroimaging in Developmental Clinical Neuroscience. Cambridge University Press, pp. 22–37.

Bunge, S.A., Dudukovic, N.M., Thomason, M.E., Vaidya, C.J., Gabrieli, J.D.E., 2002. Development of frontal lobe contributions to cognitive control in children: Evidence from fMRI. Neuron 33, 301–311.

Carter, C.S., van Veen, V., 2007. Anterior cingulate cortex and conflict detection: An update of theory and data. Cognitive, Affective, & Behavioral Neuroscience 7, 367–379.

Carter, C.S., Braver, T.S., Barch, D.M., Botvinick, M.M., Noll, D., Cohen, J.D., 1998. Anterior cingulate cortex, error detection, and the online monitoring of performance. Science 280, 747–749.

Casey, B.J., Trainor, R., Giedd, J., et al., 1997a. The role of the anterior cingulate in automatic and controlled processes: A developmental neuroanatomical study. Developmental Psychobiology 30, 61–69.

Casey, B.J., Trainor, R.J., Orendi, J.L., et al., 1997b. A developmental functional MRI study of prefrontal activation during performance of a go-no-go task. Journal of Cognitive Neuroscience 9, 835–847.

Casey, B.J., Galvan, A., Hare, T.A., 2005. Changes in cerebral functional organization during cognitive development. Current Opinion in Neurobiology 15, 239–244.

Casey, B.J., Epstein, J.N., Buhle, J., et al., 2007. Frontostriatal connectivity and its role in cognitive control in parent–child dyads with ADHD. The American Journal of Psychology 164, 1729–1736.

Casey, B.J., Getz, S., Galvan, A., 2008. The adolescent brain. Developmental Review 28, 62–77.

Cauffman, E., Shulman, E.P., Steinberg, L., et al., 2010. Age differences in affective decision making as indexed by performance on the Iowa gambling task. Developmental Psychology 46, 193–207.

Chen, X., Hastings, P.D., Rubin, K.H., Chen, H., Cen, G., Stewart, S.L., 1998. Child rearing attitudes and behavioral inhibition in Chinese and Canadian toddlers: A cross cultural study. Developmental Psychology 34, 677–686.

Chronis-Tuscano, A., Degnan, K.A., Pine, D.S., et al., 2009. Stable early maternal report of behavioral inhibition predicts lifetime social anxiety disorder in adolescence. Journal of the American Academy of Child and Adolescent Psychiatry 48, 928–935.

Davidson, M.C., Amso, D., Cruess-Anderson, L., Diamond, A., 2006. Development of cognitive control and executive functions from

4–13 years: Evidence from manipulations of memory, inhibition and task switching. Neuropsychologia 44, 2037–2078.

Davies, P.L., Segalowitz, S.J., Gavin, W.J., 2004. Development of response-monitoring ERPs in 7- to 25-year-olds. Developmental Neuropsychology 25, 355–376.

Diamond, A., 2002. Normal development of prefrontal cortex from birth to young adulthood: Cognitive functions, anatomy and biochemistry. In: Stuss, D.T., Knight, R.T. (Eds.), Principles of Frontal Lobe Function. Oxford University Press, London, pp. 466–503.

Durston, S., Davidson, M.C., Tottenham, N., et al., 2006. A shift from diffuse to focal cortical activity with development. Developmental Science 9, 1–20.

Eisenberg, N., Shepard, S.A., Fabes, R.A., Murphy, B.C., Guthrie, I.K., 1998. Shyness and children's emotionality, regulation, and coping: Contemporaneous, longitudinal, and across-context relations. Child Development 69, 767–790.

Ernst, M., Mueller, S.C., 2008. The adolescent brain: Insights from functional neuroimaging research. Developmental Neurobiology 68, 729–743.

Ernst, M., Nelson, E.E., Jazbec, S., et al., 2005. Amygdala and nucleus accumbens in responses to receipt and omission of gains in adults and adolescents. NeuroImage 25, 1279–1291.

Evans, A.D., Lee, K., 2011. Verbal deception from late childhood to middle adolescence and its relation to executive functioning skills. Developmental Psychology 47, 1108–1116.

Evans, A.D., Xu, F., Lee, K., 2011. When all signs point to you: Lies told in the face of evidence. Developmental Psychology 47, 39–49.

Falkenstein, M., Hohnsbein, J., Hoormann, J., Blanke, L., 1991. Effects of crossmodal divided attention on late ERP components II. Error processing in choice reaction tasks. Electroencephalography and Clinical Neurophysiology 78, 447–455.

Figner, B., Mackinlay, R.J., Wilkening, F., Weber, E.U., 2009. Affective and deliberative processes in risky choice: Age differences in risk taking in the Columbia Card Task. Journal of Experimental Psychology: Learning, Memory, and Cognition 35, 709–730.

Fox NA and Henderson HA (2000) Temperament, emotion, and executive function: Influences on the development of self-regulation. *Paper Presented at the Annual Meeting of the Cognitive Neuroscience Society*. San Francisco.

Fox, N.A., Henderson, H.A., Rubin, K.H., Calkins, S.D., Schmidt, L.A., 2001. Continuity and discontinuity of behavioral inhibition and exuberance: Psychophysiological and behavioral influences across the first four years of life. Child Development 72, 1–21.

Fox, N.A., Henderson, H.A., Marshall, P.J., Nichols, K.E., Ghera, M.M., 2005. Behavioral inhibition: Linking biology and behavior within a developmental framework. Annual Review of Psychology 56, 235–262.

Galvan, A., Hare, T.A., Parra, C.E., et al., 2006. Earlier development of the accumbens relative to orbitofrontal cortex might underlie risk-taking behavior in adolescents. Journal of Neuroscience 26, 6885–6892.

Gehring, W.J., Fencsik, D.E., 2001. Functions of the medial frontal cortex in the processing of conflict and errors. Journal of Neuroscience 21, 9430–9437.

Gehring, W.J., Goss, B., Coles, M.G.H., Meyer, D.E., Donchin, E., 1993. A neural system for error detection and compensation. Psychological Science 4, 385–390.

Gehring, W.J., Himle, J., Nisenson, L.G., 2000. Action-monitoring dysfunction in obsessive-compulsive disorder. Psychological Science 11, 1–6.

Geier, C.F., Terwilliger, R., Teslovich, T., Velanova, K., Luna, B., 2010. Immaturities in reward processing and its influence on inhibitory control in adolescence. Cerebral Cortex 20, 1613–1629.

Gerardi-Caulton, G., 2000. Sensitivity to spatial conflict and the development of self-regulation in children 24–26 months of age. Developmental Science 3, 397–404.

Gerstadt, C.L., Hong, Y.J., Diamond, A., 1994. The relationship between cognition and action: Performance of children 3 1/2-7 years old on a Stroop-like day-night test. Cognition 53, 129–153.

Greene, J.D., Sommerville, R.B., Nystrom, L.E., Darley, J.M., Cohen, J.D., 2001. An fMRI investigation of emotional engagement in moral judgment. Science 293, 2105–2108.

Greene, J.D., Nystrom, L.E., Engell, A.D., Darley, J.M., Cohen, J.D., 2004. The neural bases of cognitive conflict and control in moral judgment. Neuron 44, 389–400.

Hardin, M.G., Mandell, D., Mueller, S.C., Dahl, R.E., Pine, D.S., Ernst, M., 2009. Inhibitory control in anxious and healthy adolescents is modulated by incentive and incidental affective stimuli. Child Psychological Psychiatry 50, 1550–1558.

Henderson, H.A., 2010. Neural correlates of cognitive control and the regulation of shyness in children. Developmental Neuropsychology 35, 177–193.

Ho, D.Y.F., 1994. Cognitive socialization in Confucian heritage cultures. In: Greenfield, P.M., Cocking, R.R. (Eds.), Cross-Cultural Roots of Minority Development. Erlbaum, Hillsdale, NJ, pp. 285–313.

Huizinga, M., Dolan, C.V., van der Molen, M.W., 2006. Age-related change in executive function: Developmental trends and a latent variable analysis. Neuropsychologia 44, 2017–2036.

Jacques, S., Zelazo, P.D., Kirkham, N.Z., Semcesen, T.K., 1999. Rule selection versus rule execution in preschoolers: An error-detection approach. Developmental Psychology 35, 770–780.

Jazbec, S., Hardin, M.G., Schroth, E., McClure, E., Pine, D.S., Ernst, M., 2006. Age-related influence of contingencies on a saccade task. Experimental Brain Research 174, 754–762.

Kagan, J., 2001. Temperamental contributions to affective and behavioral profiles in childhood. In: Hoffmann, S.G., Dibartolo, P.M. (Eds.), From Social Anxiety to Social Phobia: Multiple Perspectives. Allyn & Bacon, Needham Heights, MA, pp. 216–234.

Kerns, J.G., Cohen, J.D., MacDonald III, A.W., Cho, R.Y., Stenger, V.A., Carter, C.S., 2004. Anterior cingulate conflict monitoring and adjustments in control. Science 303, 1023–1026.

Kerr, A., Zelazo, P.D., 2004. Development of 'hot' executive function: The Children's Gambling Task. Brain and Cognition 55, 148–157.

Kochanska, G., Murray, K., Jacques, T.Y., Koenig, A., Vandegeest, K.A., 1996. Inhibitory control in young children and its role in emerging internalization. Child Development 67, 490–507.

Kohls, G., Petzler, J., Hepertz-Dahlmann, B., Konrad, K., 2009. Differential effects of social and nonsocial reward on response inhibition in children and adolescents. Developmental Science 12, 614–625.

Lahat A and Zelazo PD (2010) The neurophysiological correlates of moral development: The distinction between moral and social conventional violations. Paper presented at the Annual Meeting of the Jean Piaget Society, St. Louis, MI, USA, 3–5 June.

Lahat, A., Todd, R.M., Mahy, C.E.V., Lau, K., Zelazo, P.D., 2010. Neurophysiological correlates of executive function: A comparison of European-Canadian and Chinese-Canadian 5-yearold children. Frontiers in Human Neuroscience 3, 72. http://dx.doi.org/10.3389/neuro.09.072.2009.

Lahat, A., Degnan, K.A., White, L.K., et al., 2012a. Temperamental exuberance and executive function predict propensity for risk-taking in childhood. Development & Psychopathology 24 (3), 847–856.

Lahat, A., Helwig, C.C., Zelazo, P.D., 2012b. Age-related changes in cognitive processing of moral and social conventional violations. Cognitive Development 27, 181–194.

Lahat, A., Helwig, C.C., and Zelazo, P.D., in press. An ERP study of adolescents' and young adults' judgments of moral and social conventional violations. Child Development.

Lamm, C., Zelazo, P.D., Lewis, M.D., 2006. Neural correlates of cognitive control in childhood and adolescence: Disentangling the contributions of age and executive function. Neuropsychologia 44, 2139–2148.

Lamm, C., White, L.K., McDermett, J.M., Fox, N.A., 2012. Neural activation underlying cognitive control in the context of neutral and affectively charged pictures in children. Brain and Cognition 79, 181–187.

Lewis, M.D., Lamm, C., Segalowitz, S.J., Steiben, J., Zelazo, P.D., 2006. Neurophysiological correlates of emotion regulation in children and adolescents. Journal of Cognitive Neuroscience 18, 430–443.

Lewis, M.D., Todd, R.M., Honsberger, M.J.M., 2007. Event-related potential measures of emotion regulation in early childhood. NeuroReport 18, 61–65.

Liston, C., Watts, R., Tottenham, N., et al., 2006. Frontostriatal microstructure modulates efficient recruitment of cognitive control. Cerebal Cortex 16, 553–560.

Luna, B., Thulborn, K.R., Munoz, D.P., et al., 2001. Maturation of widely distributed brain function subserves cognitive development. NeuroImage 13, 786–793.

Luu, P., Tucker, D.M., 2001. Regulating action: Alternating activation of midline frontal and motor cortical networks. Clinical Neurophysiology 112, 1295–1306.

Luu, P., Collins, P., Tucker, D.M., 2000. Mood, personality and self-monitoring: Negative affect and emotionality are related to frontal lobe mechanisms of error detection. Journal of Experimental Psychology. General 129, 1–18.

May, J.C., Delgado, M.R., Dahl, R.E., et al., 2004. Event-related functional magnetic resonance imaging of reward-related brain circuitry in children and adolescents. Biological Psychiatry 55, 359–366.

McDermott, J.M., Perez-Edgar, K., Henderson, H.A., Chronis-Tuscano, A., Pine, D.S., Fox, N.A., 2009. A History of childhood behavioral inhibition and enhanced response monitoring in adolescence are linked to clinical anxiety. Biological Psychiatry 65, 445–448.

Mischel, W., Ebbesen, E.B., Zeiss, A.R., 1972. Cognitive and attentional mechanisms in delay of gratification. Journal of Personality and Social Psychology 21, 204–218.

Nieuwenhuis, S., Ridderinkhof, K.R., Blom, J., Band, G.P., Kok, A., 2001. Error-related brain potentials are differentially related to awareness of response errors: Evidence from an antisaccade task. Psychophysiology 38, 752–760.

Nieuwenhuis, S., Yeung, N., van den Wildenberg, W., Ridderinkhof, R., 2003. Electrophysiological correlates of anterior cingulate function in a go/no-go task: Effects of response conflict and trial type frequency. Cognitive, Affective, & Behavioral Neuroscience 3, 17–26.

Nucci, L.P., 1981. Conceptions of personal issues: A domain distinct from moral or societal concepts. Child Development 52, 114–121.

Perez-Edgar, K., Fox, N.A., 2003. Individual differences in children's performance during an emotional stroop task: A behavioral and electrophysiological study. Brain and Cognition 52, 33–51.

Polak-Toste, C.P., Gunnar, M.R., 2006. Temperamental exuberance: Correlates and consequences. In: Marshall, P.J., Fox, N.A. (Eds) The development of social engagement: Neurobiological perspectives. Series in affective science. Oxford University Press, New York, NY, pp. 19–45.

Posner, M.I., Fan, J., 2004. Attention as an Organ System. Cambridge University Press, Cambridge.

Premack, D., Woodruff, G., 1978. Does the chimpanzee have a theory of mind? The Behavioral and Brain Sciences 1, 516–526.

Prencipe, A., Zelazo, P.D., 2005. Development of affective decision-making for self and other: Evidence for the integration of first- and third-person perspectives. Psychological Science 16, 501–505.

Putnam, S.P., Stifter, C.A., 2005. Behavioral approach-inhibition in toddlers: Prediction from infancy, positive and negative affective components, and relations with behavior problems. Child Development 76, 212–226.

Ridderinkhof, K.R., Ullsperger, M., Crone, E.A., Nieuwenhuis, S., 2004. The role of the medial frontal cortex in cognitive control. Science 306, 443–447.

Rodriguez-Fornells, A., Kurzbuch, A.R., Munte, T.F., 2002. Time course of error detection and correction in humans: Neurophysiological evidence. Journal of Neuroscience 22, 9990–9996.

Rothbart, M.K., Bates, J.E., 2006. Temperament. In: Damon, W., Eisenberg, N. (Eds.), Handbook of Child Psychology. 6th edn. Social, Emotional, and Personality Development, vol. 3. Wiley, New York, pp. 99–166.

Rothbart, M.K., Ellis, L.K., Rueda, M.R., Posner, M.I., 2003. Developing mechanisms of temperamental effortful control. Journal of Personality 71, 1113–1143.

Rubia, K., Overmeyer, S., Taylor, E., et al., 2000. Functional frontalisation with age: Mapping neurodevelopmental trajectories with fMRI. Neuroscience and Biobehavioral Reviews 24, 13–19.

Rubia, K., Smith, A.B., Woolley, J., et al., 2006. Progressive increase of frontostriatal brain activation from childhood to adulthood during event-related tasks of cognitive control. Human Brain Mapping 27, 973–993.

Rubia, K., Smith, A.B., Taylor, E., Brammer, M., 2007. Linear age-correlated functional development of right inferior fronto-striato-cerebellar networks during response inhibition and anterior cingulate during error-related processes. Human Brain Mapping 28, 1163–1177.

Rueda, M.R., Fan, J., McCandliss, B.D., et al., 2004. Development of attentional networks in childhood. Neuropsychologia 42, 1029–1040.

Sabbagh, M.A., Xu, F., Carlson, S.M., Moses, L.J., Lee, K., 2006. The development of executive functioning and theory of mind: A comparison of Chinese and U.S. preschoolers. Psychological Science 17, 74–81.

Scheffers, M.K., Coles, M.G., Bernstein, P., Gehring, W.J., Donchin, E., 1996. Event-related brain potentials and error-related processing: An analysis of incorrect responses to go and no-go stimuli. Psychophysiology 33, 42–53.

Simonds, J., Kieras, J.E., Rueda, M.R., Rothbart, M.K., 2007. Effortful control, executive attention, and emotional regulation in 7–10-year-old children. Cognitive Development 22, 474–488.

Smetana, J.G., 1981. Preschool children's conceptions of moral and social rules. Child Development 52, 1333–1336.

Smetana, J.G., 2006. Social-cognitive domain theory: Consistencies and variations in children's moral and social judgments. In: Killen, M., Smetana, J.G. (Eds.), Handbook of Moral Development. Lawrence Erlbaum Associates, Mahwah, NJ, pp. 119–153.

Somerville, L.H., Casey, B.J., 2010. Developmental neurobiology of cognitive control and motivational systems. Current Opinion in Neurobiology 20, 236–241.

Steinberg, L., 2008. A social neuroscience perspective on adolescent risk taking. Developmental Review 28, 78–106.

Steinberg, L., Graham, S.J., O'Brien, L., Woolard, J., Cauffman, E., Banich, M., 2009. Age differences in future orientation and delay discounting. Child Development 80, 28–44.

Stieben, J., Lewis, M.D., Granic, I., Zelazo, P.D., Segalowitz, S., Pepler, D., 2007. Neurophysiological mechanisms of emotion regulation for subtypes of externalizing children. Development and Psychopathology 19, 455–480.

Sullivan, K., Zaitchik, D., Tager-Flusberg, H., 1994. Preschoolers can attribute second-order beliefs. Developmental Psychology 30, 395–402.

Talwar, V., Lee, K., 2008. Social and cognitive correlates of children's lying behavior. Child Development 79, 866–881.

Tamm, L., Menon, V., Reiss, A.L., 2002. Maturation of brain function associated with response inhibition. Journal of the American Academy of Child and Adolescent Psychiatry 41, 1231–1238.

Thomson, J.J., 1986. Rights, Restitution, and Risk: Essays in Moral Theory. Harvard University Press.

Thorell, L., Bohlin, G., Rydell, A., 2004. Two types of inhibitory control: Predictive relations to social functioning. International Journal of Behavioral Development 28, 193–203.

Tobin, J.J., Wu, D.Y.H., Davidson, D.H., 1989. Preschool in Three Cultures: Japan China and the United States. Yale University Press, London.

Turiel, E., 1983. The development of social knowledge: Morality and convention. Cambridge University Press, Cambridge.

Van Veen, V., Carter, C.S., 2002. The timing of action-monitoring processes in the anterior cingulate cortex. Journal of Cognitive Neuroscience 14, 593–602.

Warren, S.L., Bost, K.K., Roisman, G.I., et al., 2010. Effect of adult attachment and emotional distractors on brain mechanisms of cognitive control. Psychological Science 21, 1818–1826.

White, L.K., McDermott, J.M., Degnan, K.A., Henderson, H.A., Fox, N.A., 2011. Behavioral inhibition and anxiety: The moderating roles of inhibitory control and attention shifting. Journal of Abnormal Child Psychology 39, 735–747.

Wu, D.Y.H., 1996. Chinese childhood socialization. In: Bond, M.H. (Ed.), The Handbook of Chinese Psychology. Oxford University Press, New York, pp. 143–151.

Zelazo, P.D., 2004. The development of conscious control in childhood. Trends in Cognitive Science 8, 12–17.

Zelazo, P.D., 2006. The dimensional change card sort (DCCS): A method of assessing executive function in children. Nature Protocols 1, 297–301.

Zelazo, P.D., Müller, U., Frye, D., Marcovitch, S., 2003. The development of executive function in early childhood. Monographs of the Society for Research in Child Development 68, vii-137.

Zelazo, P.D., Carlson, S.M., Kesek, A., 2008. Development of executive function in childhood. In: Nelson, C.A., Luciana, M. (Eds.), Handbook of Developmental Cognitive Neuroscience. 2nd edn. MIT Press, Cambridge, MA, pp. 553–574.

Executive Function
Development, Individual Differences, and Clinical Insights

C. Hughes

University of Cambridge, Cambridge, UK

OUTLINE

24.1 INTRODUCTION

Executive function (EF) is an umbrella term that encompasses the set of higher-order processes (such as inhibitory control, working memory, and attentional flexibility) that govern goal-directed action and adaptive responses to novel, complex, or ambiguous situations (Hughes et al., 2005). Research into the neural substrate for EF (Golden, 1981) has long focused on the prefrontal cortex (PFC), but this traditional view recently has been challenged for two reasons. First, positron emission tomography studies demonstrate that EF tasks activate parietal areas involved in basic attentional processes more strongly than the PFC (for a review, see Jurado and Rosselli, 2007); similar findings also have been reported in a recent magnetic resonance imaging (MRI) study of EF in children and adolescents (Tamnes et al., 2010). Second, clinical studies show that early pathology in any brain region leads to executive deficits, such that, for children at least, intact EF depends upon the integrity of the entire brain, not just the frontal regions (Anderson and Catroppa, 2005).

Research into EF in children has grown exponentially over the past few decades. A recent Scopus search using the terms *executive functions* and *children* showed just 5 studies prior to 1980; 26 studies between 1980 and 1990; 216 studies between 1990 and 2000; and 1092 studies between 2000 and 2010. This massive expansion of research reflects the growth of interest in childhood clinical groups, and EF deficits are apparent in a number of different developmental disorders – especially attention deficit hyperactivity disorder (ADHD) and autism spectrum disorder (ASD) (Pennington and Ozonoff, 1996; Sergeant et al., 2002). However, recent years also have seen a rapid growth in studies of normative age-related changes in EF and its neural substrates, enabling a valuable crossover between research on typical and atypical development. As Tau and Peterson (2010, p. 162) put it, documenting normative pathways of brain development "provides the Archimedean point from which to interpret and understand the aberrant pathways of brain development that produce disease," whereas studying atypical groups "informs our knowledge of normal brain

Neural Circuit Development and Function in the Brain: Comprehensive Developmental Neuroscience, Volume 3 http://dx.doi.org/10.1016/B978-0-12-397267-5.00062-5

© 2013 Elsevier Inc. All rights reserved.

development by throwing into relief the developmental pathways that are most sensitive to perturbation."

Both the pervasiveness of EF deficits in childhood disorders and the salience of EF for studies of normal brain development can be understood in terms of the protracted nature of EF development. Thus, EF skills begin to emerge in infancy (Diamond, 1988), show marked improvements across toddlerhood and the preschool period (Carlson et al., 2004; Hughes and Ensor, 2007; Hughes et al., 2010), and continue to improve during the school-age years (e.g., Huizinga et al., 2006), with some aspects of EF continuing to develop throughout adolescence (Luciana et al., 2005; Luna et al., 2004). Interestingly, although several reviews of EF development are available (e.g., Anderson, 2002; Best et al., 2009; Blair et al., 2005; Blakemore and Choudhury, 2006; Garon et al., 2008; Hughes and Graham, 2002), few span the full period of development (for an exception, see Diamond, 2002). To address this gap, the first section of this chapter provides a synopsis of research findings on developmental trajectories for EF from infancy to adolescence in both typical and atypical populations.

Having a protracted developmental course also makes EF a focus of interest for researchers interested in exploring environmental influences. As noted by Garon (2008) and Hughes and Ensor (2009), this topic of environmental influence on EF is a promising new direction for research because the slow maturation of the frontal cortex and its networks makes it heavily dependent on the environment (e.g., Noble et al., 2005). However, until recently, the processes underpinning environmental influences on EF have been, in research terms, largely *terra incognita*. To rectify this gap, the second section of this chapter provides an overview of three strands of recent research that investigate (1) effects of training or intervention programs on children's emerging EF skills, (2) influences of parent–child interactions on EF, and (3) studies of environmental influence on EF in clinical groups.

Interwoven with perspectives on EF that emphasize typical or atypical development, the expansion of research in this field also mirrors the intensity of research into children's understanding of mind; numerous studies have reported a robust association between EF and children's theory-of-mind skills that is evident across a broad age range, independent of covariant effects of language ability and intelligence quotient, and applies to both typical and atypical populations. In addition, recent research has demonstrated that early emerging skills in both EF and theory of mind are strong predictors of children's readiness for school and their performance in academic subjects. The third section of this chapter aims to bring together the findings from research, tracing links between EF and children's academic, sociocognitive, and social success at school.

24.2 NORMATIVE DEVELOPMENTAL TRAJECTORIES FOR EF FROM INFANCY TO ADOLESCENCE

The past two decades have seen a massive increase in the availability of child-friendly EF tasks (for a useful summary, see Garon et al., 2008), leading to dramatic improvements in the understanding of the development of EF. For instance, it is now known that EF is a unitary construct with partially dissociable components (Garon et al., 2008) that begins to emerge in the first few years of life (e.g., Diamond, 1991) and continues to develop through to adulthood (Huizinga et al., 2006). However, most studies in this field have been cross-sectional in design and have not controlled for peripheral task demands. As a result, both cohort effects and developmental changes in how children cope with peripheral task demands may contribute to age-related contrasts in EF performance. In addition, simplifying adult tasks to make them age-appropriate for young children carries the danger of losing the critical EF component (Garon et al., 2008). These notes of caution need to be remembered when considering the summary review of findings from infants to adolescents given below.

24.2.1 Infancy

The first evidence that EF emerges in the first year of life (much earlier than previously believed) came from studies using Piaget's object permanence task in which babies are repeatedly allowed to retrieve an attractive object from one location (A) before seeing it hidden at a new location (B). Early studies indicated that while 8-month-olds and older babies typically search correctly at location B, 5-month-olds persist in searching for the object at location A (e.g., Harris, 1975). Although Piaget interpreted this 'A not B' error as a failure to recognize that objects have an independent existence in the world (i.e., a lack of 'object permanence'), later studies showed that when looking times rather than physical reaches are used to assess perceptual understanding, even 4-month-olds do well (e.g., Baillargeon et al., 1985). Thus, young babies who make the A not B error in their reaching responses may know that the object has been moved, but fail to inhibit their previously successful (and therefore prepotent) reach to A. Thus, success on this task can be seen as reflecting infants' growing cognitive flexibility and volitional control (Diamond and Goldman-Rakic, 1989). Other tasks that also demonstrate early executive skills in infancy include a detour-reaching task, in which infants are invited to retrieve an object that is visible behind a Perspex screen; success on this task depends on making a 'detour reach' around the side of the screen (Diamond et al., 1989). In short, a rudimentary ability to inhibit prepotent responses is clearly evident by 7–12 months (Diamond, 2002).

More recent studies of infants also indicate an early emergence of other aspects of EF. In particular, building on the finding by Rothbart et al. (2003) that 2-year-olds' anticipatory looking (i.e., looking to the location of a target prior to its appearance) is associated with parental ratings of self-regulation, Sheese and colleagues (2008) have reported that 6- and 7-month-olds who display high levels of anticipatory looking also show more signs of self-regulation in their approach toward novel toys. Together, these two studies indicate that executive attention (indexed by anticipatory looking) is also evident very early in life. Interestingly, Sheese et al. (2008) found that anticipatory looking was also associated with looking away from disturbing stimuli (facemasks), supporting proposals (e.g., Aksan and Kochanska, 2004; Rothbart et al., 2000) for a link between early systems of emotional control (e.g., fear and caution) and later systems of cognitive control.

24.2.2 Preschoolers

Research into the preschool years accounts for the lion's share of studies of EF in childhood. As will be discussed in the third section of this chapter, this focus on preschoolers partly reflects the intensity of research into children's understanding of mind, which shows dramatic improvements in the preschool years (e.g., Wellman et al., 2001). Another reason for a research focus on preschoolers is that a wide variety of age-appropriate tasks have been developed in response to the previous dearth of tasks suitable for young children (e.g., Carlson, 2005; Diamond et al., 1997; Espy et al., 1999; Hughes, 1998; Zelazo and Muller, 2002).

However, even when these conceptual and methodological factors are taken into account, the growth of research on preschool EF remains remarkable and is difficult to summarize briefly. Fortunately, a recent comprehensive review by Garon et al. (2008) is helpful in at least three respects. First, adopting the 'unity and diversity' model proposed by Miyake and colleagues (2000) to account for the coherent yet fractionated nature of EF in adults, Garon et al. (2008) demonstrate that the three key components of EF (working memory, response inhibition, and set shifting) in adults are all evident before the age of 3 years. Second, Garon et al. (2008) address the question of how developmental changes in EF across the preschool years should be characterized, and, consistent with proposals from other theorists (e.g., Rothbart and Posner, 2005), highlight the importance of age-related improvements in attention and coordination of distinct EF components. As they observe, attention also appears central to theoretical accounts that

characterize EF development in terms of increased ability to overcome prepotent thoughts/acts (Diamond, 2002) or latent representations (Munakata, 2001) or to integrate conflicting rules (Zelazo et al., 2003).

Third, Garon et al. (2008) offer a number of suggestions for future research. One suggested direction (discussed more fully in the next section of this chapter) concerns the importance of studying environmental influences, as the slow maturation of the PFC and its networks make it heavily dependent on the environment (Noble et al., 2005). Another suggested direction is the use of longitudinal designs in order to examine developmental changes in EF. This is an important point as research into preschool EF (like so many fields within developmental psychology) is heavily dominated by cross-sectional studies that do not allow researchers to control for individual differences. In one exceptional longitudinal study, Hughes et al. (2010) applied latent-variable analyses to demonstrate that: (1) latent factors for EF in preschool and early school age show measurement invariance (supporting the validity of across-time comparisons of average EF performance); (2) individual differences in the preschool latent factor for EF are positively related to both maternal education and early verbal ability; and (3) rates of growth in EF are inversely related to verbal ability (such that preschoolers with low verbal ability begin to catch up with their peers following the transition to school) but are unrelated to maternal education (i.e., there is no independent catch-up effect for children with less-educated mothers). As these findings illustrate, adopting longitudinal study designs to examine developmental trajectories in early EF is valuable, both theoretically and for the development of educational policy.

24.2.3 School Age

Studies of EF in school-aged children often involve computerized tasks, which allow researchers to standardize administration, to include large numbers of trials, and to collect information on reaction times. One widely used battery of computerized EF tasks is the CANTAB (CAmbridge Neuropsychological Tests – Automated Battery), which was developed for work with adult clinical groups (Luciana, 2003) and was first administered to children in the Hughes et al. (1994) study of children with ASD; subsequent studies have confirmed that the CANTAB is sensitive to EF deficits in ASD across a wide range of ages and ability levels (Ozonoff et al., 2004). Normative data on age-related improvements in EF in typically developing children (screened to exclude children with learning or behavioral difficulties) have also been gathered (Luciana, 2003).

A key finding to emerge from studies of EF across the school years is that step-wise improvements are evident at different ages for different aspects of EF. Specifically (as noted earlier), the preschool years are characterized by dramatic improvements in inhibitory control, but studies of young school-aged children highlight improvements in cognitive flexibility and studies of preadolescents emphasize improvements in working memory and planning ability. For example, using the intradimensional/extradimensional (IDED) shift task from the CANTAB battery, Luciana (2003) reported a marked improvement in children's ability to shift their mental set around the age of 6 or 7 years. The IDED task is a multistage task that is based on the Wisconsin Card Sorting Task, which requires participants to work out a rule for sorting cards (e.g., sort by color, number, or shape of stimuli) and, then, when they receive negative feedback indicating that the rule has changed, to shift strategy in order to sort by a new rule. In the same study, Luciana (2003) showed that clear improvements on the Tower of London planning task or on a self-ordered search test of working memory were often not evident before the age of 11 or 12 years. Other studies using different tasks have reported a similar contrast in the developmental trajectories of different aspects of EF. For example, several studies report improvements in mental flexibility around age 8 years (e.g., Anderson, 2002; Anderson et al., 2001c), while planning, organizing, and strategic thinking are typically reported to emerge later and to show age-related improvements throughout adolescence (Anderson et al., 1996, 2001a; De Luca et al., 2003; Krikorian et al., 1994; Welsh et al., 1991).

At odds with this general pattern, however, are findings from another study that suggest a long developmental progression for cognitive flexibility, with 13-year-olds still not at adult levels (Davidson et al., 2006). A closer look at the specific tasks suggests that the contrast in these findings may be explained by Diamond's (2009) 'all or none' theory. According to this theory, the brain and mind work effortlessly (or under difficult conditions) at a gross level, but require effort (or more optimal conditions) to work in a more selective manner. Thus, it is easier to inhibit a dominant response all the time than only some of the time. As a result, even older children are likely to show frequent errors on task-switching paradigms, such as those used in the Davidson et al. (2006) study. This study also revealed several other interesting developmental contrasts. For example, adults slowed down on difficult trials to preserve accuracy, but children (and especially young children) were impulsive and so made errors on difficult trials. These contrasting speed accuracy trade-offs highlight the value of using computerized tasks to assess EF. At the same time, these advantages may be offset by a reduction in sensitivity. For example, children with ASD have been found to perform significantly better on a computerized set-shifting task than on a manual version (Ozonoff, 1995). In addition, computerized tasks are likely to have lower ecological validity, such that there is still a need for manual tasks that mimic everyday EF demands. Here, one task battery that deserves mention is the behavioral assessment of dysexecutive syndrome (Norris and Tate, 2000), which includes tasks that tap into abilities for multitasking, problem solving, and strategic thinking.

24.2.4 Adolescence

Consistent with the findings from younger samples described above, different aspects of EF have been reported to show distinct trajectories across the adolescent years. For example, in a study of Australian 11- to 17-year-olds that included a variety of EF tasks, Anderson et al. (2001b) found clear linear age-related improvements on tests of selective attention, working memory, and problem solving, but no age-related difference in planning performance.

Also echoing findings from studies of younger samples is an emerging theme from the adolescent literature regarding the importance of considering EF alongside other key brain systems. For example, Tau and Peterson (2010) note that adolescents and adults differ not just in the maturity of their EF functions, but also in the extent to which they avoid risk and respond to reward/peer influence (adolescents are less risk-averse, more driven by reward and more easily influenced by peers). As a result, accounts of developmental change in everyday behavior should consider not only top-down EF systems but also bottom-up motivational and emotional responses to situations of risk and reward.

Similar conclusions emerge from a recent community-based study of relations between EF, problem behaviors, and risk taking in 10- to 12-year-olds (Romer et al., 2009). In this study, the children's self-reported impulsivity was inversely related to both working memory and reversal learning and explained individual differences in both externalizing problems and performance on a risk-taking task. Noting that interventions to improve children's working memory have led to reductions in impulsive behaviors (e.g., Klingberg et al., 2005), the authors of this study concluded that young people who have difficulties in considering multiple (and potentially conflicting) goals will be less likely to either 'look before they leap' or temper their interest in novel or exciting experiences.

24.3 CLINICAL INSIGHTS, FROM INFANCY TO ADOLESCENCE

In parallel with, and perhaps fueling the growth of, research on normative development in EF, studies of EF in atypical child populations have grown dramatically and yielded several interesting findings that together emphasize the extent to which EF can be impaired by brain abnormalities or insults, especially if these occur early in life.

24.3.1 Infancy

The first clinical study of EF in infancy highlighted the importance of dopamine for EF. Specifically, Diamond et al. (1997) conducted a 4-year longitudinal study of children treated early and continuously for phenylketonuria (PKU), a metabolic disorder characterized by a failure to convert phenylalanine to tyrosine (the precursor of dopamine) that, if untreated, is the most common biochemical cause of intellectual disability. On tests of working memory and inhibitory control, children with PKU with high plasma levels of phenylalanine performed more poorly than all other control groups (including siblings and PKU children with lower levels of phenylalanine). Moreover, this impairment was specific to performance on EF tasks, directly related to levels of phenylalanine, and evident across the age groups, from the youngest (6–12 months) to the oldest (3½–7 years).

Long-term deficits in EF are also found in infants born prematurely and infants exposed prenatally to high levels of alcohol. In a recent meta-analysis, Mulder et al. (2009) showed that children born prematurely displayed a variety of EF impairments, with the degree of impairment predicted by: (1) gestational age (i.e., extremely premature infants show greater deficits), (2) age at test (group differences attenuate with age), and (3) aspect of EF under test (this 'catch-up' effect is evident for selective attention skills but not for attentional set-shifting skills). No corresponding systematic review of EF in children with prenatal alcohol exposure is yet available, but individual studies consistently report deficits in EF that are independent of IQ (e.g., Connor et al., 2000; Green et al., 2009; Mattson et al., 1999; Schonfeld et al., 2001) and somewhat distinct from EF deficits in other groups, such as children with ADHD (e.g., Vaurio et al., 2008). Moreover, a recent friendship training study involving primary-school-aged children with prenatal alcohol exposure showed that parental ratings of EF predicted gains in social skills, even when effects of IQ were taken into account (Schonfeld et al., 2009). In short, prenatal alcohol exposure has long-term adverse effects on children's emerging EF skills, which in turn not only underpin academic achievement (see later in this chapter) but also play a key role in these children's social competencies.

24.3.2 Preschoolers

From a clinical perspective, preschool milestones in EF have attracted considerable attention from researchers studying the cognitive profiles associated with ASD. For example, Hughes and Russell (1993) demonstrated that children with ASD (with a verbal mental age of at least 4 years) displayed significant difficulties on two tasks that most 4-year-olds passed with ease. The first of these was the 'Windows' task in which, for each of 20 trials, children could win a treat (visible through a window in a box) by choosing a visibly empty box. Children with ASD (and many 3-year-olds) chose the baited box, and often persisted in this incorrect response across all 20 trials. Moreover, administering the task in four different conditions (verbal/nonverbal; competitor present/absent) led to no change in the results. In the second task, to retrieve a large and attractive marble from inside a metal box, children needed to perform a simple but arbitrary means–end action (flicking a switch at the side of the box that turned off a simple circuit with an infrared beam that, if tripped, activated a small trapdoor on which the marble was resting). Once again, most older children with ASD and typically developing 3-year-olds (but not 4-year-olds) persisted in making a direct and unsuccessful attempt to reach into the box.

These findings added weight to reports of impaired performance by children with ASD on more traditional (and complex) EF tasks (e.g., Hughes et al., 1994; Ozonoff et al., 1991; Prior, 1979) and sparked a sustained program of research on EF deficits in ASD (for reviews, see Hill, 2004; Pennington and Ozonoff, 1996). The breadth and sophistication of current research in this field are nicely illustrated by a handful of recent findings that include: (1) evidence for associations between impaired inhibitory control and high-level repetitive behaviors (e.g., compulsions and preoccupations) in children with ASD (Mosconi et al., 2009); (2) imaging results that indicate reduced functional connectivity and network integration between the frontal, parietal, and occipital regions among individuals with ASD in completing EF tasks (Solomon et al., 2009); (3) longitudinal evidence for the importance of early EF in shaping the developmental trajectory of theory-of-mind skills in children with ASD (Pellicano, 2010); and (4) evidence for age-related improvements in EF from childhood to adolescence in ASD, indicating the presence of plasticity and suggesting a prolonged window for effective treatment (for a review, see O'Hearn et al., 2008).

24.3.3 School Age

Earlier, in the section on EF in typically developing school-aged children, concern was raised about the ecological validity of the computerized tasks that are often used with this age group. This issue of ecological validity is also salient from a clinical perspective, as EF deficits in children with ADHD (who, by definition, show marked problems of impulsivity/inattention/disorganization in their everyday lives) are often less pervasive and severe than the EF deficits shown by children with ASD (e.g., Geurts et al., 2004; Goldberg et al., 2005; but see also Happé et al., 2006), such that contemporary causal accounts of ADHD also include additional deficits in the signaling of delayed rewards (e.g., Sonuga-Barke, 2005).

In his review, Sonuga-Barke (2005) also argues for the importance of considering EF deficits alongside environmental influences, a view that is supported by the peak in diagnosis for ADHD at age 12 (Mandell et al., 2005); that is, just following the transition to secondary (or 'middle') school, which brings a significant increase in demands for self-controlled and planful behavior. Indeed, a recent longitudinal study that modeled trajectories for parent-rated ADHD symptoms confirms that the age-related decline in symptoms is at least transiently reversed following the transition to secondary school (Langberg et al., 2008). Likewise, in a study that involved carefully matched samples, Happé et al. (2006) found that children with ASD, but not children with ADHD, showed age-related improvements in EF performance. Together, these findings highlight the importance of adopting a developmental perspective when examining EF deficits in atypical groups.

24.3.4 Adolescence

Much of the clinical work on EF in adolescence has concerned risk taking and substance abuse. In particular, several researchers have reported a robust predictive association between poor EF and impulsivity in preadolescence and high levels of drug use in late adolescence (e.g., Aytaclar et al., 1999). Similarly, in a review of alcohol addiction in adolescence, Wiers et al. (2007) proposed a model in which repeated alcohol use leads to compromised EF development coupled with sensitization of a reward-based approach system. Note that this model adopts Tau and Peterson's (2010) recommended dual focus on top-down and bottom-up processes.

In the dual-focus study by Fairchild et al. (2009), there is a comparison between how adolescent boys with early- versus late-onset conduct disorder (CD) or controls perform on tests of EF (Wisconsin Card Sort Test) and decision making (Risky Choice task) under conditions of varying motivation and stress. Controlling for effects of IQ, they found nonsignificant group differences in EF, but more risky choices in both CD groups than in controls; adolescent boys with early-onset CD made risky choices even when the gains were relatively small. These findings suggest a shift in the balance between sensitivity to reward and punishment among boys with CD (particularly the early-onset form) that is similar to the imbalance proposed by Wiers et al. (2007) in top-down and bottom-up processes in addictive adolescents. Perhaps unsurprisingly then, CD is, as Fairchild et al. (2009) note, associated with a significantly increased susceptibility to substance-use disorders.

24.4 FROM BIOLOGICAL TO ENVIRONMENTAL PREDICTORS OF INDIVIDUAL DIFFERENCES IN EF

Biological studies have been hugely influential in research on EF in childhood. For example, the impact of Diamond and colleagues' (1989) EF account of age-related improvements in infants' performance on the A not B task was greatly increased by the convergent findings from parallel studies of: (1) rhesus monkeys (Diamond et al., 1989), which highlighted the dorsolateral PFC as pivotal to success on this task, and (2) children with PKU, which demonstrated the importance of dopamine for success on a wide battery of simple EF tasks (Diamond et al., 1997). More recently, in their review of studies using functional MRI, Tau and Peterson (2010) concluded that age-related improvements in EF in childhood are associated with increased activation of (dopamine-rich) frontal and striatal circuits.

Other seminal findings include the demonstration of Golden (1981) that myelination of the prefrontal cortex is associated with age-related improvements in EF in children. More recent technological advances have led to significant progress in documenting parallels between milestones in EF development and changes in brain myelination (e.g., Nagy et al., 2004). This progress is particularly evident in Giedd et al.'s (1999) large-scale longitudinal MRI work on adolescence, which has shown that this period is characterized by both a linear increase in white matter and a nonlinear decrease in gray matter. Gains in white matter have clear functional consequences, which include faster and more efficient sharing of information within the frontostriatal circuits and smoother communication between the frontal cortex and other brain regions (Paus, 2010). Peak periods of reduction in gray matter occur just after puberty and at the transition from adolescence to adulthood; although typically attributed to synaptic pruning, this 'loss' of gray matter may simply reflect gains in white matter (Paus, 2010). As noted in a recent review (Blakemore and Choudhury, 2006), structural changes in the adolescent brain are particularly evident in the frontal cortex and are linked to age-related improvements in inhibitory control (Luna et al., 2004), working memory (e.g., Luciana et al., 2005), and decision making (e.g., Hooper et al., 2004).

The studies of EF in atypical groups summarized in the previous section provide a third strand of research with a clear biological perspective. In particular, impairments in EF are most pronounced among children with ADHD or ASD (Pennington and Ozonoff, 1996), two disorders that show substantial genetic influence (Kuntsi et al., 2004; Ronald et al., 2006). Indeed, EF has recently been implicated in the genetic basis for ADHD (e.g., Bidwell et al., 2007; Gau and Shang, 2010). More direct evidence for strong genetic influences on early EF comes from studies that include genotyping (e.g., Fossella et al., 2002; Rueda et al., 2005), which demonstrate that children with the homozygous long allele for the DAT1 gene (associated with high effortful control and low extroversion) outperform those with the heterozygous (long/ short) allele on EF tests of conflict resolution (for a recent review of molecular genetic studies of EF in children, see Brocki et al., 2009).

However, none of the above findings precludes environmental factors also contributing to either developmental change or individual differences in EF. Indeed, genetic factors often show substantial interactions with environmental influences, such that genetic vulnerability is expressed only among individuals exposed to environmental stressors, such as harsh parenting or family chaos (Asbury et al., 2003, 2005). Moreover, as noted earlier, the development of EF has a very protracted course, which makes it particularly sensitive to environmental influence (Farah et al., 2006; Noble et al., 2005). In addition, current cognitive models of EF (e.g., Duncan, 2001) highlight the fluidity of relations between the prefrontal cortex and EF performance. Specifically, neurons within the PFC show a rapid adaptation to changing task demands (Freedman et al., 2001), making it difficult to map between behavioral and neuronal functions. In comparison with research on biological influences, studies of environmental influences on EF are largely *terra incognita*. Nevertheless, there is growing recognition that early experiences contribute to children's neurocognitive development. Specifically, unfavorable early environmental experiences adversely affect both brain structure (e.g., De Bellis, 2005) and function (e.g., Rutter and O'Connor, 2004); conversely, positive experiences (especially with caregivers) appear to have a positive impact on brain development (e.g., Nelson and Bloom, 1997; Schore, 2001). As Bierman et al. (2008, p. 823) put it, 'EF development depends, in part, upon sensitive responsive caregiving and opportunities for guided exploration of the social and physical environment, fostering sustained joint attention, emotional understanding, planning, and problem solving.'

Direct evidence for environmental effects on EF comes from intervention studies, which can be considered in two sets. The first set of studies involves direct training on task analogs. In the first of these studies, Kloo and Perner (2003) gave a simplified version of the Wisconsin Card Sort task to typically developing preschoolers and found that both card-sorting training and false-belief training (each delivered in two sessions on separate days) led to positive effects on EF at posttest. More recently, Rueda et al. (2005) gave 5 days of attention training to groups of 4- and 6-year-olds, and reported that children who performed poorly at pretest showed the greatest gains from this training program, with the overall improvements in executive attention and IQ performance being equivalent to half the difference between older and younger participants (i.e., to gains expected from 1 year's difference in age). Similarly, in a recent review, Klingberg (2010) reported that, across a wide variety of age groups, training leads to significant improvements in working memory. However, not all aspects of EF appear so malleable to training. For example, a 5-week preschool training program has been shown to produce significant improvements in working memory but not in inhibitory control (Thorell et al., 2009), raising interesting questions as to whether the distinct processes that underpin different aspects of EF lead to contrasts in the extent to which performance can be improved by training. The final study in this set of training interventions was conducted by Karbach and Kray (2009) and involved training on task switching, administered to both school-aged children (aged 8–10 years) and two groups of adults (aged 18–26 and 62–76 years). All three groups showed positive effects that transferred to other EF tasks and to tests of fluid intelligence (e.g., tests of abstract thinking and reasoning); however, when the training tasks were variable, improvements were reduced for children but increased for adults. This interaction effect highlights the importance of adopting a developmentally sensitive approach to the development of interventions.

The second set of intervention studies adopts a broader and more naturalistic approach and is thus perhaps of greater relevance for theories of how everyday social environments might impinge on children's EF development. At least three such interventions have been assessed using randomized control trials (RCTs). The first of these, Head Start REDI (REsearch based, Developmentally Informed), is integrated into the Head Start prekindergarten program for disadvantaged children and involves brief lessons, 'hands-on' extension activities, and specific teaching strategies linked empirically with the promotion of social–emotional competencies, language development, and emergent literacy skills. In their RCT, Bierman et al. (2008) reported that the REDI intervention led to significant improvements in children's abilities to stay on task, coupled with marginally significant gains in set-shifting performance. Another well-recognized intervention is the Vygotskian 'Tools in the Mind' preschool curriculum (Bodrova and Leong, 1996; Diamond et al., 2007), which includes a variety of specially designed activities (e.g., sociodramatic play and shared reading) that enable children to progress from external to shared to self-regulation; teachers are also trained to foster early skills in literacy and mathematics by encouraging reflective thinking and metacognition. Interestingly, although language plays a pivotal role in Vygotskian accounts of cognitive development, the Tools curriculum appears to have positive effects on EF (as indexed by low problem behavior scores), but no significant impact on language development (Barnett et al., 2008). The third RCT focused on an 8-week school-based intervention for older children (7- to 9-year-olds) that aims to promote mindful awareness practices through twice-weekly half-hour sessions. Comparing teachers' and parents' pre- and post-program ratings of EF skills, Flook et al. (2010) reported that less well-regulated children showed particularly clear treatment-related gains (evident both at school and at home); note that this finding echoes earlier results from preschoolers (e.g., Rueda et al., 2005).

Together, the findings from these three RCTs support two of the three dimensions of adult–child interactions that Carlson (2003) has proposed as likely to favor EF in children: *scaffolding* (which provides children with successful experiences of problem-based learning) and *mind-mindedness* (which provides children with verbal tools for progressing from external to internal forms of self-regulation). Positive effects on EF of the third dimension of *sensitivity* (which provides infants with successful experiences of impacting on the environment) have been reported by Bernier et al. (2010) in their recent longitudinal study of 80 infants in which maternal sensitivity, mind-mindedness, and scaffolding (or 'autonomy support') were evaluated at 12–15 months of age, and EF was assessed at 18 and 26 months. All three measures of parenting predicted child EF; however, once the effects of maternal education and general cognitive ability were taken into account, scaffolding was the strongest predictor of EF at each age. Another recent study, conducted by Bibok ct al. (2009), offers a refinement of how parental scaffolding may promote young children's EF skills (indexed by performance on an attention-switching task). Specifically, observations of parents scaffolding children's goal-directed activities revealed that the *timing* of parents' elaborative utterances was a key predictor of children's attention-switching skills.

In short, there is now good evidence that parents' deliberate efforts to scaffold children's goal-directed activities do indeed foster the development of early EF skills. However, as highlighted by research on children's developing social understanding, family influences are often incidental rather than deliberate. For example, young children are extremely acute observers of family life, paying particularly close attention to injustices such as parents' differential treatment of siblings (Dunn, 1993). This point can be applied to young children's observational skills more generally; most parents will be able to recall episodes in which their actions have been mimicked with uncanny accuracy. This attention to detail favors children's rapid mastery of complex action plans: while simple mimicry of adults' planful behavior does not constitute executive control, acquiring a repertoire of 'goal-directed' acts is likely to promote EF skills. Support for this view comes from a recent longitudinal observational study by Hughes and Ensor (2009) involving a socially diverse sample of 125 children. In this study, both maternal scaffolding (in structured play with jigsaws) and opportunities for observational learning (indexed by maternal strategic behavior in a multitasking paradigm and in a shared tidy-up task) predicted improvements between the ages of 2 and 4 in children's EF scores, even when effects of verbal ability were controlled. Note that including EF assessments at both time points enabled Hughes and Ensor (2009) to take the temporal stability of individual differences in EF into account and so minimize the confounding effects of genetic factors (Kovas et al., 2007).

Reinforcing the importance of incidental effects, Hughes and Ensor (2009) also found that EF development from age 2 to age 4 was *negatively* correlated with parental ratings of disorganized and unpredictable family life, suggesting that families can hinder as well as help young children's emerging EF skills. Here it is worth noting that the social diversity of Hughes and Ensor's (2009) study sample is important: just as socioeconomic (SES) effects appear strongest at the bottom end of the scale (e.g., Scarr, 2000), adverse environmental measures (e.g., ratings of family chaos) may only show significant variance if low SES families (who are most at risk of exposure to multiple stressors) are included. Using an expanded sample from the same longitudinal study (i.e., the target children, plus friends recruited at age 4; $N=191$), Hughes et al. (in preparation) have applied latent-variable analyses to demonstrate that each of three distinct measures of maternal well-being (specifically, depression, parental efficacy, and satisfaction) showed independent associations with individual differences in EF. Interestingly, of these three factors, only maternal depression was also (negatively) related to individual differences in children's verbal ability; that is, while positive effects of favorable family environments appear specific to EF, unfavorable family environments appear to adversely affect cognitive development in general. Note also that the significant inverse relation between maternal depression and preschool EF reported by Hughes et al. (in preparation) appears developmentally specific, as research on both older children and adolescents with depressed mothers has led to null findings (Klimes-Dougan et al., 2006; Micco et al., 2009). That said, the significant inverse relation between maternal depression and poor EF in preschoolers reported by Hughes et al. (in preparation) is supported by findings from a recent study of internationally adopted children, which demonstrate that, despite good catch-up in many specific areas of development, these young children (who were exposed to severe adverse early environments) showed persistent difficulties in EF and attentional regulation (Jacobs et al., 2010).

Clinical studies have also highlighted the importance of environmental influences. For example, studies of children who have experienced maltreatment or severe neglect highlight the impact of such extreme adverse environments on neuroendocrine and autonomic stress reactivity, which in turn leads to increased demands on EF systems of regulatory control (e.g., Bierman et al., 2008; Cicchetti, 2002). Moreover, findings from children with traumatic brain injury indicate that higher-order brain functions (such as EF) are particularly vulnerable while they are still emerging. Specifically, in their recent review of the effects of early brain injury on EF, Anderson et al. (2010) examined findings from children with focal brain pathology evident on MRI scans; they compared EF performance in late childhood/adolescence (assessed across a variety of domains) in children who sustained early brain injury at each of six developmental periods (congenital/perinatal/infancy/preschool/mid-childhood/late childhood), with these six groups being matched for gender, SES, lesion size, location, or laterality. Their findings clearly supported theoretical perspectives that emphasize vulnerability rather than plasticity in brain function. Specifically, children who experienced brain injury very early in life displayed markedly more severe deficits in EF (and IQ). In other words, while the development of EF can be disrupted, with either transient or more permanent consequences, EF skills, once established, are relatively robust (Johnson, 2005; Thomas and Johnson, 2008).

24.4.1 Early EF Predicts Academic, Sociocognitive, and Social Success at School

This third section aims to bring together the findings from research tracing links between EF and children's academic, sociocognitive, and social success at school. Interestingly, each of these research areas highlights different aspects of EF: working memory appears central in accounts of EF and academic performance; cognitive flexibility is highlighted by more than one account of EF and social cognition; and inhibitory control is central to accounts of EF and social success. At the same time, overlapping associations are likely, as there is very close interplay between children's academic ability, social understanding, and social behavior at school.

24.4.1.1 EF and Academic Performance

Over the past two decades, research into the neurocognitive underpinnings of children's competence in core academic domains such as literacy and numeracy has expanded rapidly; interestingly, several accounts give EF (especially working memory) a prominent role (e.g., Blair and Razza, 2007; Gathercole and Pickering, 2000; Geary, 1990). Empirical research confirms the importance of working memory for academic achievement. For example, in a recent longitudinal study, Alloway and Alloway (2010) found that working memory at age 5 (i.e., at the start of formal education) eclipsed IQ as a predictor of academic success 6 years later. Likewise, accounts of academic failure among children and adolescents with ADHD highlight the importance of deficits in EF (e.g., Alloway et al., 2010; Clark et al., 2000), while Dahlin (2010) found that primary school children with special needs who completed a cognitive training program designed to enhance working memory also displayed accelerated reading development. That said, the role of working memory in mathematical competence is more complex than previously thought (Geary, 2010). In particular, echoing the finding (discussed in the first section of this chapter) that developmental timing is a stronger predictor of EF impairment than location of brain injury (Anderson et al., 2010), recent studies have shown significant changes in brain–mathematics relations as children develop and mature (Ansari, 2010; Meyer et al., 2010).

Working with younger children, and adopting a rather different theoretical perspective (in which early EF rather than academic performance is center stage), Blair and colleagues have shown that individual differences in EF in preschool predict both school readiness and children's success in numeracy and literacy (Blair and Diamond, 2008; Blair and Peters, 2003; Blair and Razza, 2007; Razza and Blair, 2009). In their review of this field, Blair and Diamond (2008) argue that early self-regulation reflects an emerging balance between emotional arousal and cognitive regulation, such that self-regulation (and hence children's school readiness) is likely to be enhanced by school interventions that link emotional/motivational arousal with activities designed to promote EF.

More recently, Hughes and Ensor (in press) have extended this research field in two ways: adopting a developmentally dynamic approach to examine *growth* of EF across the transition to school and considering individual differences in EF trajectories in relation to children's own *perceptions* of their (academic and social) success at school. Research on children's self-perceived academic abilities has burgeoned over recent years, fueled by the finding that individual differences in IQ fail to account for up to 50% of the variance in academic performance (e.g., Chamorro-Premuzic and Furnham, 2005; Rhode and Thompson, 2007). Indeed, a meta-analytic review has shown that, even controlling for previous achievement, self-perceived abilities exert small but consistent effects on later achievement (Valentine et al., 2004). Conversely, poor self-perceptions in early childhood are associated with loneliness, withdrawal, and peer exclusion (Coplan et al., 2004). Together, these findings suggest that self-perceptions may be a key aspect of children's psychological (as opposed to practical) school readiness. In their study (in which 191 children were followed from ages 4 to 6), Hughes and Ensor (in press) found that, even with effects of concurrent verbal ability and EF controlled, children who had made rapid gains in EF across the transition to school reported higher levels of academic competence at age 6. Given that this is the first study to report an association between preschool EF and school-children's self-perceived abilities, it is worth noting that similar findings have been obtained from adult samples. In particular, Tangney and colleagues (2004) have reported that, among adults, individual differences in self-control (a construct that is closely related to EF) show robust associations with individual differences in self-esteem.

24.4.1.2 EF and Social Cognition

The finding that individual differences in EF trajectories predict children's self-perceived academic competence brings us to another hot topic, namely the robust association between variation in preschoolers' EF skills and in their understanding of mind. Interestingly, this link between EF and understanding of mind is evident at several different periods of development, from toddlerhood (e.g., Carlson et al., 2004; Hughes and Ensor, 2005) to adolescence (Dumontheil et al., 2010). Equally remarkable, the association between EF and theory of mind has been reported for a variety of clinical groups, including children with ASD (e.g., Pellicano, 2007), hyperactivity or conduct problems (Hughes et al., 1998), traumatic brain injuries (Dennis et al., 2009), and fetal alcohol syndrome (Rasmussen et al., 2009). Thus, the link between EF and theory of mind appears pervasive. How then should it be explained?

In an early review of the evidence for associations between EF and theory of mind, Perner and Lang (1999) offered five possible explanations: (1) *theory-of-mind skills are necessary for children to pass EF tasks* (e.g., in order to inhibit a particular response, children have to be able

to represent it as maladaptive); (2) *EF is needed for children to develop their understanding of mental states* (e.g., experience of goal-directed action improves children's conceptual understanding of intentional states); (3) the relevant theory-of-mind tasks require EF (e.g., standard false-belief tasks place heavy demands on children's inhibitory control and/or working memory); (4) *tests of both EF and theory of mind require the same kind of embedded conditional reasoning*; and (5) the two systems are not functionally related, but have overlapping or neighboring *neural substrates*.

Having reviewed the evidence, Perner and Lang (1999) concluded that only a third of these proposals could be ruled out with confidence; in a subsequent study (see also Perner et al., 2002), the fourth account was also ruled out (as false-belief tasks appear as difficult as card-sorting tasks, despite requiring less complex conditional reasoning). Since then, evidence that similar brain regions are implicated in EF and theory of mind has grown (for reviews, see Perner and Aichhorn, 2008; Perner et al., 2006); thus, the fifth account (neuroanatomical proximity may well contribute to the association between the two domains) remains plausible. Indeed, very recent research demonstrates that dopamine, long recognized as pivotal to EF (e.g., see studies of PKU described in the first section of this chapter), also plays a key role in children's growing understanding of mind. For example, findings from a recent EEG study indicate that the dorsal medial PFC (which is rich in dopamine receptors and lies at the end of the mesocortical dopamine pathway) is a specific neurodevelopmental correlate of preschoolers' theory-of-mind development (Sabbagh et al., 2009). Likewise, using an archive of preschoolers' EEG recordings, Lackner et al. (2010) have reported that individual differences in eye-blink rate (an indirect but reliable measure of dopamine function) predicted theory-of-mind performance even controlling for several other related factors, including age, verbal ability, gender, and performance on a Stroop task (which taps the ability to inhibit a maladaptive response). Interestingly, as Lackner et al. (2010) note, dopamine provides a mechanism that may explain *both* neurobiological and experiential influences on theory-of-mind development. Specifically, dopamine promotes the neural plasticity needed to respond flexibly to environmental feedback by changing goals and expectations (e.g., Montague et al., 2004) and so may mediate the impact of family factors known to predict theory-of-mind development (e.g., frequencies of family conversations about mental states) (Ensor and Hughes, 2008) or of interactions with siblings (Perner et al., 1994) that depend on children's ability to reflect on (and revise) their own concepts of mind in order to accommodate new information from the environment.

Another area of progress in this research field has been the growth of longitudinal studies, including microgenetic studies (e.g., Flynn, 2006) and studies of toddlers (Carlson et al., 2004; Hughes and Ensor, 2007). One consistent finding to emerge from these studies is that early EF predicts later mental-state awareness more strongly than early mental-state awareness predicts later EF. This asymmetry in predictive relationships challenges Perner and Lang's first account (namely that theory of mind provides a foundation for EF). Instead, without going as far as stipulating that EF is, in some sense, necessary for the emergence of mental-state awareness, it seems reasonable to argue that EF improvements in the preschool years help explain how children *make use of* their early intuitive understanding of mind.

It is also worth noting that the relationship between EF and theory of mind may well be developmentally dynamic. For example, in a critique of the original theory-of-mind account of ASD (which is often diagnosed long before children are expected to pass false-belief tasks), Tager-Flusberg (2001) proposed that early-onset 'socioperceptual' skills (or intuitive mentalizing) depend on modular cognitive processes, whereas later-onset 'sociocognitive' skills (or off-line mental-state reasoning) depend on other aspects of cognition, such as language and EF. This model of dual processes (e.g., Apperly et al., 2009; de Vignemont, 2009) also goes some way to explaining why typically developing young children can show quite sophisticated mentalizing skills in their everyday interactions and yet fail experimental false-belief tasks.

Note also that other models of theory-of-mind development also suggest a developmentally dynamic relationship with EF. For example, Wellman and colleagues (Wellman and Liu, 2004) have proposed that improvements in children's understanding of mind involve a series of distinct achievements, such that different aspects of EF may be particularly important for specific milestones. Conversely, the stage-like nature of development in EF outlined in the first section of this chapter also suggests the need for a more differentiated approach. For example, in their study of children with traumatic brain injuries, Dennis et al. (2009) used path analysis to elucidate the nature of relations between EF and theory of mind and argued that their findings support models of EF in which inhibitory control serves as a foundation for other aspects of EF, in particular working memory (Barkley, 1997). Specifically, Dennis et al. (2009) found that the relationship between impaired inhibitory control and theory of mind was mediated, at least among children with traumatic brain injury, by impairments in working memory.

24.4.1.3 EF and Social Competence

For adults, there is robust evidence that deficits in EF are associated with problems of antisocial behavior: in a meta-analytic review of this literature, Morgan and Lilienfeld (2000) reported that the average EF performance of antisocial groups fell 0.62 standard deviations below that of control groups. Building on this work, Raine (2002) reviewed evidence from neuropsychological, neurological, and brain imaging studies and concluded that prefrontal structural and functional deficits are implicated in antisocial or aggressive behavior throughout the lifespan. Most recently, Beauchamp and Anderson (2010) have conducted a theoretical review of the cognitive underpinnings of children's developing social skills and noted that, within EF, attentional control (i.e., self-monitoring, response inhibition, and self-regulation) is especially critical.

In one of the earliest studies to link EF to young children's social competence (and the first to use direct observational methods), Hughes et al. (2000) found that, in a socially diverse sample of preschoolers (half of whom had been rated by parents as 'hard to manage'), poor performance on a battery of EF tasks (but not on theory-of-mind tasks) was associated with higher frequencies of anger and antisocial behavior toward friends. In other words, the interpersonal problems of these hard-to-manage preschoolers appear not to reflect difficulties in social understanding *per se*, but rather failure of behavioral regulation. In a follow-up study with the same sample, Hughes et al. (2001) showed that poor EF at age 4 predicted negative behavior at age 5 and that this group of hard-to-manage children continued to show rule violations and perseverative errors at age 7 (Brophy et al., 2002).

In a further longitudinal observational study (for more details, see Hughes, 2011), Hughes and colleagues followed a socially diverse sample of 140 children from toddlerhood through to school age, recruiting best friends for each target child at age 4, such that their findings are best presented in two parts. In the first, Hughes and Ensor (2008) examined children's EF, verbal ability, and theory-of-mind scores at ages 2, 3, and 4 in relation to aggregate (multi-informant and multi-setting) measures of problem behaviors at each time point and made the following findings: (1) poor EF at age 2 predicted worsening problem behaviors from ages 2 to 4; (2) individual differences in EF at age 3 fully mediated the influence of age 2 language deficits upon age 4 problem behaviors; and (3) by age 4, individual differences in problem behaviors showed specific associations with individual differences in EF (but not theory of mind or verbal ability). Capitalizing on the expanded sample ($N = 191$), Hughes and Ensor applied latent growth models, which showed that high EF gains across the transition to school predicted low levels of teacher-rated emotional symptoms, hyperactivity, conduct problems, and peer problems at age 6 (as well as higher self-reported academic competence, as noted earlier). Together, these findings highlight the importance of preschool EF for early social adjustment and demonstrate the value of examining developmental change in EF (gains in EF across the transition to school predicted social outcomes even when individual differences in concurrent EF were controlled).

Links between EF and social competence may also be indirect. For example, Razza and Blair (2009) have reported that false-belief understanding mediates the association between early individual differences in children's EF and later teacher ratings of children's social competence. On a related note, Maszk et al. (1999) found that 4- to 6-year-olds rated by peers and teachers as high in behavioral and emotional self-control became increasingly popular over the school year and so argued that individual differences in self-control may be meaningful for how children are viewed by others and hence for how they view themselves. This raises the interesting possibility with regard to Hughes and Ensor's findings. Specifically, children's awareness of their own gains in EF across the transition to school may shape both social behavior and self-concepts. If so, interventions to improve children's social adjustment might aim beyond increasing children's EF to ensuring that children are aware of their own progress in regulating and organizing their thoughts and behaviors.

At this point, it is worth noting that the findings reviewed in this chapter and elsewhere suggest that EF can act as a moderator, mediator, and outcome of interventions. For example, studies of both school-aged children (e.g., Flook et al., 2010) and preschoolers (e.g., Rueda et al., 2005) indicate that the effects of interventions are often particularly impressive for children with poor EF. Likewise, improvements in children's inhibitory control have been shown to at least partially mediate the positive effects of the PATHS curriculum on children's behavior (Riggs et al., 2006). Similarly, the three RCTs reviewed earlier in this chapter all demonstrated that positive effects on EF can be expected from enriched and structured curricula that promote scaffolding (and hence lead to experiences of successful problem-based learning) and adult mind-mindedness (enabling a progression from external to internal forms of self-regulation). These multiple roles for EF in interventions to promote social behavior highlight the importance of adopting a broad and contextualized approach to identifying underlying mechanisms. The next step is to place research on interventions within a developmental perspective to identify whether different processes are pivotal for different age groups.

24.5 CONCLUSIONS

This chapter on EF in childhood has covered considerable ground, including development from infancy to adolescence in typically developing and atypical groups; positive and negative effects of environmental influences (from training on specific tasks to exposure to enriched and predictable environments); and academic, sociocognitive, and social outcomes associated with individual differences in early EF (or EF trajectories). Perhaps the main pair of conclusions to emerge from this review is that both continuities and contrasts in EF in children of different ages are striking. One striking commonality across studies of typically developing children of different ages is a close interplay between top-down systems of EF and bottom-up reward-oriented systems, such that, from infancy through to adolescence, poor EF appears associated with risk taking and sensation seeking. Indeed, research on several atypical groups, including children with ADHD, CD, or problems of substance abuse, also highlights this interplay between top-down and bottom-up processes. Although not yet evident in research on ASD, it is worth noting that the amygdala, which is a key substrate involved in reward processing, is central to at least one prominent account of ASD (Baron-Cohen et al., 2000). Thus, extending this dual focus on EF and reward processing to children with ASD would appear a fruitful direction for future research.

A second notable developmental continuity is that, across a wide age range, typically developing individuals with good EF are more likely than their peers to do well on tests of theory of mind and show positive self-concepts and less likely to display antisocial behaviors. Perhaps related to these stable correlates of EF, longitudinal studies support EF as a predictor of later academic achievement in both young children and adolescents. Finally, across a wide variety of ages, at least some aspects of EF (e.g., working memory) appear malleable to training effects.

Examples of age-related contrasts include differences in the nature of EF: improvements in some aspects of EF (such as inhibitory control) can be seen from a very early age, while other aspects (e.g., planning) do not show marked improvements until much later on in development. Another important contrast concerns the extent to which EF can be associated with a localized neural base: age-related improvements in EF appear hand in hand with an increase in frontostriatal activation, such that development is characterized by a progression from diffuse to specific neural substrate. Several age-related functional changes in children's performance on EF tasks suggest that this progressive localization of neural substrate may reflect increases in how strategic children and adolescents are when completing EF tasks. For example, adults and children differ markedly in how they respond to more challenging situations; while adults can reduce their speed of response to remain accurate, young children typically show a drop in accuracy. Similarly, young children are particularly likely to show an 'all or none' effect, in that they can inhibit a response if this is consistently required of them, but find it much harder to cope with situations that place varying demands on this system of inhibitory control. Finally, related to these contrasts in strategy use, training studies indicate an age-related contrast in the optimal format of the training tasks, with task variability increasing training benefits in adults but reducing training benefits in children.

Together, the above age-related contrasts lead to a third key conclusion, namely the need to take developmental issues seriously when examining a construct such as EF that shows such a protracted developmental course. For example, if the differences noted above do indeed reflect an age-related contrast in strategy use on EF tasks, then the validity of across-age comparisons is in question, as different sets of skills may well underpin performance on the same task for children of different ages. An important first step in addressing this issue is to establish measurement invariance before comparing EF skills across different age groups (cf. Hughes et al., 2010). Developmental issues are also raised by findings from studies of atypical groups. Thus, studies comparing different clinical groups (e.g., children with ASD and children with ADHD) should be designed so that contrasts in developmental *trajectories* can be elucidated. The few existing studies that adopt a developmental perspective indicate that children with ASD may show greater progress than children with ADHD, but the reasons for this are not yet known.

The final conclusions to emerge from this review concern the interplay between EF and children's environments. First, although individual differences in EF have been viewed as almost entirely genetic in origin (e.g., Friedman et al., 2008), there is growing evidence that, for young children at least, environmental influences can be substantial. Thus, detailed longitudinal studies highlight the importance of family factors (e.g., maternal well-being, sensitivity, and consistency of parenting). In addition, at least three RCTs show that early educational interventions have positive effects on EF. These positive effects may be: (1) strongest for children with low levels of EF (i.e., EF moderates the impact of interventions); (2) pivotal to explaining the substantial improvement in children's behavior as a result of such interventions (i.e., EF is a mediator of intervention effects); and (3) achieved indirectly via improvements in children's theory-of-mind skills, or in how children are viewed by others, or how they view themselves. Clearly then, tracing out the mechanisms that underpin associations between family environments and children's growing EF skills as well as between interventions and children's social and cognitive achievements is an important challenge for future research.

References

Aksan, N., Kochanska, G., 2004. Links between systems of inhibition from infancy to preschool years. Child Development 75, 1477–1490.

Alloway, T., Alloway, R., 2010. Investigating the predictive roles of working memory and IQ in academic attainment. Journal of Experimental Child Psychology 106, 20–29.

Alloway, T., Gathercole, S., Elliott, J., 2010. Examining the link between working memory behaviour and academic attainment in children with ADHD. Developmental Medicine and Child Neurology 52, 632–636.

Anderson, P., 2002. Assessment and development of executive function (EF) during childhood. Child Neuropsychology 8, 71–82.

Anderson, V., Catroppa, C., 2005. Recovery of executive skills following pediatric traumatic brain injury (TBI): A two year follow-up. Brain Injury 19, 459–470.

Anderson, P., Anderson, V., Lajoie, G., 1996. The Tower of London test: Validation and standardization for pediatric populations. The Clinical Neuropsychologist 10, 54–65.

Anderson, P., Anderson, V., Garth, J., 2001a. Assessment and development of organisational ability: The Rey Complex Figure organisational strategy score (RCF-OSS). The Clinical Neuropsychologist 15, 81–94.

Anderson, V., Anderson, P., Northam, E., Jacobs, R., Catroppa, C., 2001b. Development of executive functions through late childhood and adolescence in an Australian sample. Developmental Neuropsychology 20, 385–406.

Anderson, W.W., Damasio, H., Tranel, D., Damasio, A.R., 2001c. Long-term sequelae of prefrontal cortex damage acquired in early childhood. Developmental Neuropsychology 18, 281–296.

Anderson, V., Spencer-Smith, M., Coleman, L., et al., 2010. Children's executive functions: Are they poorer after very early brain insult. Neuropsychologia 48, 2041–2050.

Ansari, D., 2010. Neurocognitive approaches to developmental disorders of numerical and mathematical cognition: The perils of neglecting the role of development. Learning and Individual Differences 20, 123–129.

Apperly, I., Samson, D., Humphreys, G., 2009. Studies of adults can inform accounts of theory of mind development. Developmental Psychology 45, 190–201.

Asbury, K., Dunn, J., Pike, A., Plomin, R., 2003. Nonshared environmental influences on individual differences in early behavioral development: A monozygotic twin differences study. Child Development 74, 933–943.

Asbury, K., Wachs, T., Plomin, R., 2005. Environmental moderators of genetic influence on verbal and nonverbal abilities in early childhood. Intelligence 33, 643–661.

Aytaclar, S., Tarter, R., Kirisci, L., Sandy, L., 1999. Association between hyperactivity and executive cognitive functioning in childhood and substance use in early adolescence. Journal of the American Academy of Child and Adolescent Psychiatry 38, 172–178.

Baillargeon, R., Spelke, E., Wasserman, S., 1985. Object permanence in five-month-old infants. Cognition 20, 191–208.

Barkley, R.A., 1997. Behavioral inhibition, sustained attention and executive functions: Constructing a unified theory of ADHD. Psychological Bulletin 121, 65–94.

Barnett, W., Jung, K., Yarosz, D., et al., 2008. Educational effects of the Tools of the Mind curriculum: A randomized trial. Early Childhood Research Quarterly 23, 299–313.

Baron-Cohen, S., Ring, H., Bullmore, E., Wheelwright, S., Ashwin, C., Williams, S., 2000. The amygdala theory of autism. Neuroscience and Biobehavioral Reviews 24, 355–364.

Beauchamp, M., Anderson, V., 2010. SOCIAL: An integrative framework for the development of social skills. Psychological Bulletin 136, 39–64.

Bernier, A., Carlson, S., Whipple, N., 2010. From external regulation to self-regulation: Early parenting precursors of young children's executive functioning. Child Development 81, 326–339.

Best, J., Miller, P., Jones, L., 2009. Executive functions after age 5: Changes and correlates. Developmental Review 29, 180–200.

Bibok, M., Carpendale, J., Muller, U., 2009. Parental scaffolding and the development of executive function. New Directions in Child and Adolescent Psychiatry: Special Issue on Social Interaction and the Development of Executive Function 123, 17–34.

Bidwell, L., Willcutt, E., DeFries, J., Pennington, B., 2007. Testing for neuropsychological endophenotypes in siblings discordant for attention-deficit/hyperactivity disorder. Biological Psychiatry 62, 991–998.

Bierman, K., Nix, R., Greenberg, M., Blair, C., Domitrovich, C., 2008. Executive functions and school readiness intervention: Impact, moderation, and mediation in the Head Start REDI program. Development and Psychopathology 20, 821–843.

Blair, C., Diamond, A., 2008. Biological processes in prevention and intervention: The promotion of self-regulation as a means of preventing school failure. Development and Psychopathology 20, 899–911.

Blair, C., Peters, R., 2003. Physiological and neurocognitive correlates of adaptive behavior in preschool among children in Head Start. Developmental Neuropsychology 24, 479–497.

Blair, C., Razza, R., 2007. Relating effortful control, executive function, and false belief understanding to emerging math and literacy ability in kindergarten. Child Development 78, 647–663.

Blair, C., Zelazo, P., Greenberg, M., 2005. The measurement of executive function in early childhood. Developmental Neuropsychology 28, 561–571.

Blakemore, S.-J., Choudhury, S., 2006. Development of the adolescent brain: Implications for executive function and social cognition. Journal of Child Psychology and Psychiatry, and Allied Disciplines 47, 296–312.

Bodrova, E., Leong, D., 1996. Tools of the Mind: The Vygotskian Approach to Early Childhood Education. Merrill/Prentice Hall, Englewood Cliffs, NJ.

Brocki, K., Clerkin, S., Guise, K., Fan, J., Fossella, J., 2009. Assessing the molecular genetics of the development of executive attention in children: Focus on genetic pathways related to the anterior cingulate cortex and dopamine. Neuroscience 164, 241–246.

Brophy, M., Taylor, E., Hughes, C., 2002. To go or not to go: Inhibitory control in 'hard to manage' children. Infant and Child Development: Special Issue on Executive Functions and Development 11, 125–140.

Carlson, S., 2003. Executive function in context: Development, measurement, theory, and experience. Monographs of the Society for Research in Child Development 68, 138–151.

Carlson, S., 2005. Developmentally sensitive measures of executive function in preschool children. Developmental Neuropsychology 28, 595–616.

Carlson, S., Mandell, D., Williams, L., 2004. Executive function and theory of mind: Stability and prediction from ages 2 to 3. Developmental Psychology 40, 1105–1122.

Chamorro-Premuzic, T., Furnham, A., 2005. Personality and Intellectual Competence. Erlbaum, Hillsdale, NJ.

Cicchetti, D., 2002. The impact of social experience on neurobiological systems: Illustration from a constructivist view of child maltreatment. Cognitive Development 17, 1407–1428.

Clark, C., Prior, M., Kinsella, G., 2000. Do executive function deficits differentiate between adolescents with AD/HD and oppositional defiant/conduct disorder? A neuropsychological study using the six elements test and Hayling Sentence Completion Test. Journal of Abnormal Child Psychology 28, 403–414.

Connor, P., Sampson, P., Bookstein, F., Barr, H., Streissguth, A., 2000. Direct and indirect effects of prenatal alcohol damage on executive function. Developmental Neuropsychology 18, 331–354.

Coplan, R., Findlay, L., Nelson, L., 2004. Characteristics of preschoolers with lower perceived competence. Journal of Abnormal Child Psychology 32, 399–408.

Dahlin, K., 2010. Effects of working memory training on reading in children with special needs. Reading and Writing 23, 1–13.

Davidson, M., Amso, D., Anderson, L., Diamond, A., 2006. Development of cognitive control and executive functions from 4 to 13 years: Evidence from manipulations of memory, inhibition, and task switching. Neuropsychologia 44, 2037–2078.

De Bellis, M., 2005. The psychobiology of neglect. Child Maltreatment 10, 150–172.

De Luca, C., Wood, S., Anderson, V., et al., 2003. Normative data from the CANTAB: Development of executive function over the lifespan. Journal of Clinical and Experimental Neuropsychology 25, 242–254.

de Vignemont, F., 2009. Drawing the boundary between low-level and high-level mindreading. Philosophical Studies 144, 457–466.

Dennis, M., Agostino, A., Roncadin, C., Levin, H., 2009. Theory of mind depends on domain-general executive functions of working memory and cognitive inhibition in children with traumatic brain injury. Journal of Clinical and Experimental Neuropsychology 31, 835–847.

Diamond, A., 1988. Abilities and neural mechanisms underlying A not B performance. Child Development 59, 523–527.

Diamond, A., 1991. Developmental time course in human infants and infant monkeys, and the neural bases of inhibitory control in reaching. Annals of the New York Academy of Sciences: Part VII. Inhibition and Executive Control 608, 637–704.

Diamond, A., 2002. Normal development of prefrontal cortex from birth to young adulthood: Cognitive functions, anatomy, and biochemistry. In: Stuss, D., Knight, R. (Eds.), Principles of Frontal Lobe Function. Oxford University Press, London, pp. 466–503.

Diamond, A., 2009. All or none hypothesis: A global-default mode that characterizes the brain and mind. Developmental Psychology 45, 130–138.

Diamond, A., Goldman-Rakic, P., 1989. Comparison of human infants and rhesus monkeys on Piaget's A-not-B task: Evidence for dependence on dorsolateral prefrontal cortex. Experimental Brain Research 74, 24–40.

Diamond, A., Zola-Morgan, S., Squire, L., 1989. Successful performance by monkeys with lesions of the hippocampal formation on AB and object retrieval, two tasks that mark developmental changes in human infants. Behavioral Neuroscience 103, 526–537.

Diamond, A., Prevor, M., Callender, G., Druin, D., 1997. Prefrontal cortex cognitive deficits in children treated early and continuously for PKU. Monographs of the Society for Research in Child Development 62, 1–208.

Diamond, A., Barnett, W., Thomas, J., Munro, S., 2007. Preschool program improves cognitive control. Science 318, 1387–1388.

Dumontheil, I., Apperly, I., Blakemore, S.-J., 2010. Online usage of theory of mind continues to develop in late adolescence. Developmental Science 13, 331–338.

Duncan, J., 2001. An adaptive coding model of neural function in prefrontal cortex. Nature Reviews Neuroscience 2, 820–829.

Dunn, J., 1993. Young Children's Close Relationships: Beyond Attachment, vol. 4. Sage Publications, Newbury Park, CA.

Ensor, R., Hughes, C., 2008. Content or connectedness? Early family talk and theory of mind in the toddler and preschool years. Child Development 79, 201–216.

Espy, K., Kaufmann, P., McDiarmid, M., Glisky, M., 1999. Executive functioning in preschool children: Performance on A-not-B and other delayed response format tasks. Brain and Cognition 41, 178–199.

Fairchild, G., van Goozen, S., Stollery, S., et al., 2009. Decision making and executive function in male adolescents with early-onset or adolescence-onset conduct disorder and control subjects. Biological Psychiatry 66, 162–168.

Farah, M., Shera, D., Savage, J., et al., 2006. Childhood poverty: Specific associations with neurocognitive development. Brain Research 1110, 166–174.

Flook, L., Smalley, S., Kitil, M., et al., 2010. Effects of mindful awareness practices on executive functions in elementary school children. Journal of Applied School Psychology 26, 70–95.

Flynn, E., 2006. A microgenetic investigation of stability and continuity in theory of mind development. British Journal of Developmental Psychology 24, 631–654.

Fossella, J., Sommer, T., Fan, J., Wu, Y., Swanson, J., Pfaff, D., 2002. Assessing the molecular genetics of attention networks. BMC Neuroscience [Electronic Resource] 3, 14.

Freedman, D.J., Riesenhuber, M., Poggio, T., Miller, E.K., 2001. Categorical representation of visual stimuli in the primate prefrontal cortex. Science 291, 312–316.

Friedman, N., Miyake, A., Young, S., DeFries, J., Corley, R., Hewitt, J., 2008. Individual differences in executive functions are almost entirely genetic in origin. Journal of Experimental Psychology. General 137, 201–225.

Garon, N., Bryson, S., Smith, I., 2008. Executive function in preschoolers: A review using an integrative framework. Psychological Bulletin 134, 31–60.

Gathercole, S., Pickering, S., 2000. Working memory deficits in children with low achievements in the national curriculum at 7 years of age. British Journal of Educational Psychology 70, 177–194.

Gau, S.-F., Shang, C.-Y., 2010. Executive functions as endophenotypes in ADHD: Evidence from the Cambridge Neuropsychological Test Battery (CANTAB). Journal of Child Psychology and Psychiatry, and Allied Disciplines 51, 838–849.

Geary, D., 1990. A componential analysis of an early learning deficit in mathematics. Journal of Experimental Child Psychology 49, 363–383.

Geary, D., 2010. Mathematical disabilities: Reflections on cognitive, neuropsychological, and genetic components. Learning and Individual Differences 20, 130–133.

Geurts, H., Verté, S., Oosterlaan, J., Roeyers, H., Sergeant, J., 2004. How specific are executive functioning deficits in attention deficit hyperactivity disorders and autism? Journal of Child Psychology and Psychiatry, and Allied Disciplines 45, 836–854.

Giedd, J., Blumenthal, J., Jeffries, N., Castellanos, F., Liu, H., Zijdenbos, A., 1999. Brain development during childhood and adolescence: A longitudinal MRI study. Nature Neuroscience 2, 861–863.

Goldberg, M., Mostofsky, S., Cutting, L., et al., 2005. Subtle executive impairment in children with autism and children with ADHD. Journal of Autism and Developmental Disorders 35, 279–293.

Golden, C., 1981. Diagnosis and Rehabilitation in Clinical Neuropsychology. Charles C. Thomas, Springfield, IL.

Green, C., Mihic, A., Nikkel, S., et al., 2009. Executive function deficits in children with fetal alcohol spectrum disorders (FASD) measured using the Cambridge Neuropsychological Tests Automated Battery (CANTAB). Journal of Child Psychology and Psychiatry, and Allied Disciplines 50, 688–697.

Happé, F., Hughes, C., Booth, R., Charlton, R., 2006. Executive dysfunction in autism spectrum disorders and attention deficit/hyperactivity disorder: Developmental profiles. Brain and Cognition (Special Issue on Asperger's Syndrome) 61, 25–39.

Harris, P., 1975. Development of search and object permanence during infancy. Psychological Bulletin 82, 332–344.

Hill, E., 2004. Executive function in autism. Trends in Cognitive Sciences 8, 26–32.

Hooper, C., Luciana, M., Conklin, H., Yarger, R., 2004. Adolescents' performance on the development of decision making and ventromedial prefrontal cortex. Developmental Psychology 40, 1148–1158.

Hughes, C., 1998. Executive function in preschoolers: Links with theory of mind and verbal ability. British Journal of Developmental Psychology 16, 233–253.

Hughes, C., 2011. Social Understanding, Social Lives: From Toddlerhood Through to the Transition to School. Psychology Press, London.

Hughes, C., Ensor, R., 2005. Theory of mind and executive function in 2-year-olds: A family affair? Developmental Neuropsychology 28, 645–668.

Hughes, C., Ensor, R., 2007. Executive function and theory of mind: Predictive relations from ages 2- to 4-years. Developmental Psychology 43, 1447–1459.

Hughes, C., Ensor, R., 2008. Does executive function matter for preschoolers' problem behaviors? Journal of Abnormal Child Psychology 36, 1–14.

Hughes, C., Ensor, R., 2009. How do families help or hinder the development of executive function? New Directions in Child and Adolescent Psychiatry: Special Issue on Social Interaction and the Development of Executive Function 123, 35–50.

Hughes C and Ensor R (in press) Executive function trajectories across the transition to school predict externalizing and internalizing behaviors and children's self-perceived academic success at age 6. Journal of Experimental Child Psychology (Special Issue on Executive Functions).

Hughes, C., Graham, A., 2002. Measuring executive functions in childhood: Problems and solutions? Child and Adolescent Mental Health 7, 131–142.

Hughes, C., Russell, J., 1993. Autistic children's difficulty with mental disengagement from an object: Its implications for theories of autism. Developmental Psychology 29, 498–510.

Hughes, C., Russell, J., Robbins, T., 1994. Evidence for central executive dysfunction in autism. Neuropsychologia 32, 477–492.

Hughes, C., Dunn, J., White, A., 1998. Trick or treat? Uneven understanding of mind and emotion and executive function among 'hard to manage' preschoolers. Journal of Child Psychology and Psychiatry, and Allied Disciplines 39, 981–994.

Hughes, C., White, A., Sharpen, J., Dunn, J., 2000. Antisocial, angry and unsympathetic: 'Hard to manage' preschoolers' peer problems, and possible social and cognitive influences. Journal of Child Psychology and Psychiatry 41, 169–179.

Hughes, C., Cutting, A., Dunn, J., 2001. Acting nasty in the face of failure? Longitudinal observations of 'hard to manage' children playing a rigged competitive game with a friend. Journal of Abnormal Child Psychology 29, 403–416.

Hughes, C., Graham, A., Grayson, A., 2005. Executive function in childhood: Development and disorder. In: Oates, J. (Ed.), Cognitive Development. Open University Press, Oxford, pp. 205–230.

Hughes C, Roman G, and Ensor R (in preparation) Links between maternal depression and parental self-efficacy and satisfaction and preschool executive function.

Huizinga, M., Dolan, C., van der Molen, M., 2006. Age-related change in executive function: Developmental trends and a latent variable analysis. Neuropsychologia 44, 2017–2036.

Hughes, C., Ensor, R., Wilson, A., Graham, A., 2010. Tracking executive function across the transition to school: A latent variable approach. Developmental Neuropsychology 35, 20–36.

Jacobs, E., Miller, L., Tirella, L., 2010. Developmental and behavioral performance of internationally adopted preschoolers: A pilot study. Child Psychiatry and Human Development 41, 15–29.

Johnson, M., 2005. Sensitive periods in functional brain development: Problems and prospects. Developmental Psychobiology 46, 287–292.

Jurado, M., Rosselli, M., 2007. The elusive nature of executive functions: A review of our current understanding. Neuropsychology Review 17, 213–233.

Karbach, J., Kray, J., 2009. How useful is executive control training? Age differences in near and far transfer of task-switching training. Developmental Science 12, 978–990.

Klimes-Dougan, B., Ronsaville, D., Wiggs, E., Martinez, P., 2006. Neuropsychological functioning in adolescent children of mothers with a history of bipolar or major depressive disorders. Biological Psychiatry 60, 957–965.

Klingberg, T., 2010. Training and plasticity of working memory. Trends in Cognitive Sciences 14, 317–324.

Klingberg, T., Olesen, P., Johnson, M., et al., 2005. Computerized training of working memory in children with ADHD – A randomized, controlled trial. Journal of the American Academy of Child and Adolescent Psychiatry 44, 177–186.

Kloo, D., Perner, J., 2003. Training transfer between card sorting and false belief understanding: Helping children apply conflicting descriptions. Child Development 74, 1823–1839.

Kobayashi, C., Glover, G., Temple, E., 2007. Cultural and linguistic effects on neural bases of 'Theory of Mind' in American and Japanese children. Brain Research 1164, 95–107.

Kovas, Y., Haworth, C., Dale, P., Plomin, R., 2007. The Genetic and Environmental Origins of Learning Abilities and Disabilities in the Early School Years, vol. 72. Blackwell, Boston, MA.

Krikorian, R., Bartok, J., Gay, N., 1994. Tower of London procedure: A standard method and developmental data. Journal of Clinical and Experimental Neuropsychology 16, 840–850.

Kuntsi, J., Eley, T.C., Taylor, A., et al., 2004. Co-occurrence of ADHD and low IQ has genetic origins. American Journal of Medical Genetics. Part B, Neuropsychiatric Genetics 124, 41–47.

Lackner, C., Bowman, L., Sabbagh, M., 2010. Dopaminergic functioning and preschoolers' theory of mind. Neuropsychologia 48, 1767–1774.

Langberg, J., Epstein, J., Altaye, M., Molina, B., Arnold, L., Vitiello, B., 2008. The transition to middle school is associated with changes in the developmental trajectory of ADHD symptomatology in young adolescents with ADHD. Journal of Clinical Child and Adolescent Psychology 37, 651–663.

Luciana, M., 2003. Practitioner review: Computerized assessment of neuropsychological function in children: Clinical and research applications of the Cambridge Neuropsychological Testing Automated Battery (CANTAB). Journal of Child Psychology and Psychiatry 45, 649–663.

Luciana, M., Conklin, H., Cooper, C., Yarger, R., 2005. The development of nonverbal working memory and executive control processes in adolescents. Child Development 76, 697–712.

Luna, B., Garver, K., Urban, T., Lazar, N., Sweeney, J., 2004. Maturation of cognitive processes from late childhood to adulthood. Child Development 75, 1357–1372.

Mandell, D., Thompson, W., Weintraub, E., DeStefano, F., Blank, M., 2005. Trends in diagnosis rates for autism and ADHD at hospital discharge in the context of other psychiatric diagnoses. Psychiatric Services 56, 56–62.

Maszk, P., Eisenberg, N., Guthrie, I., 1999. Relations of children's social status to their emotionality and regulation: A short-term longitudinal study. Merrill-Palmer Quarterly 45, 468–492.

Mattson, S., Goodman, A., Caine, C., Delis, D., Riley, E., 1999. Executive functioning in children with heavy prenatal alcohol exposure. Alcoholism, Clinical and Experimental Research 23, 1808–1815.

Meyer, M., Salimpoor, V., Wu, S., Geary, D., Menon, V., 2010. Differential contribution of specific working memory components to mathematics achievement in 2nd and 3rd graders. Learning and Individual Differences 20, 101–109.

Micco, J., Henin, A., Biederman, J., et al., 2009. Executive functioning in offspring at risk for depression and anxiety. Depression and Anxiety 26, 780–790.

Miyake, A., Friedman, N., Emerson, M., Witzki, A., Howerter, A., Wager, T., 2000. The unity and diversity of executive functions and their contributions to complex 'frontal lobe' tasks: A latent variable analysis. Cognitive Psychology 41, 49–100.

Montague, P., Hyman, S., Cohen, J., 2004. Computational roles for dopamine in behavioural control. Nature 431, 760–767.

Morgan, A., Lilienfeld, S., 2000. A meta-analytic review of the relation between antisocial behavior and neuropsychological measures of executive function. Clinical Psychology Review 20, 113–156.

Mosconi, M., Kay, M., D'Cruz, A.-M., et al., 2009. Impaired inhibitory control is associated with higher-order repetitive behaviors in autism spectrum disorders. Psychological Medicine 39, 1559–1566.

Mulder, H., Pitchford, N., Hagger, M., Marlow, N., 2009. Development of executive function and attention in preterm children: A systematic review. Developmental Neuropsychology 34, 393–421.

Munakata, Y., 2001. Graded representations in behavioral dissociations. Trends in Cognitive Sciences 5, 309–315.

Nagy, Z., Westerberg, H., Klingberg, T., 2004. Maturation of white matter is associated with the development of cognitive functions during childhood. Journal of Cognitive Neuroscience 16, 1227–1233.

Nelson, C., Bloom, F., 1997. Child development and neuroscience. Child Development 68, 970–987.

Noble, K., Norman, M.F., Farah, M., 2005. Neurocognitive correlates of socioeconomic status in kindergarten children. Developmental Science 8, 74–87.

Norris, G., Tate, R.L., 2000. The behavioural assessment of the dysexecutive syndrome (BADS): Ecological, concurrent and construct validity. Neuropsychological Rehabilitation 10 (1), 33–45.

O'Hearn, K., Asato, M., Ordaz, S., Luna, B., 2008. Neurodevelopment and executive function in autism. Development and Psychopathology 20, 1103–1132.

Ozonoff, S., 1995. Reliability and validity of the Wisconsin Card Sorting Test in studies of autism. Neuropsychology 9, 491–500.

Ozonoff, S., Pennington, B.F., Rogers, S.J., 1991. Executive function deficits in high functioning autistic children: Relationship to theory of mind. Journal of Child Psychology and Psychiatry 32, 1081–1105.

Ozonoff, S., Cook, I., Coon, H., et al., 2004. Performance on Cambridge neuropsychological test automated battery subtests sensitive to frontal lobe function in people with autistic disorder: Evidence from the collaborative programs of excellence in autism network. Journal of Autism and Developmental Disorders 34, 139–150.

Paus, T., 2010. Growth of white matter in the adolescent brain: Myelin or axon? Brain and Cognition 72, 26–35.

Pellicano, E., 2007. Links between theory of mind and executive function in young children with autism: Clues to developmental primacy. Developmental Psychology 43, 974–990.

Pellicano, E., 2010. Individual differences in executive function and central coherence predict developmental changes in theory of mind in autism. Developmental Psychology 46, 530–544.

Pennington, B., Ozonoff, S., 1996. Executive function and developmental psychopathology. Journal of Child Psychology and Psychiatry 37, 51–87.

Perner, J., Aichhorn, M., 2008. Theory of mind, language and the temporoparietal junction mystery. Trends in Cognitive Sciences 12, 123–126.

Perner, J., Lang, B., 1999. Development of theory of mind and executive control. Trends in Cognitive Sciences 3, 337–344.

Perner, J., Ruffman, T., Leekam, S., 1994. Theory of mind is contagious: You catch it from your sibs. Child Development 65, 1228–1238.

Perner, J., Lang, B., Kloo, D., 2002. Theory of mind and self-control: More than a common problem of inhibition. Child Development 73 (3), 752–767.

Perner, J., Aichhorn, M., Kronbichler, M., Staffen, W., Ladurner, G., 2006. Thinking of mental and other representations: The roles of left and right temporo-parietal junction. Social Neuroscience 1, 245–258.

Prior, M.R., 1979. Cognitive abilities and disabilities in infantile autism: A review. Journal of Abnormal Child Psychology 7, 357–380.

Raine, A., 2002. Annotation: The role of prefrontal deficits, low autonomic arousal, and early health factors in the development of antisocial and aggressive behavior in children. Journal of Child Psychology and Psychiatry, and Allied Disciplines 43, 417–434.

Rasmussen, C., Wyper, K., Talwar, V., 2009. The relation between theory of mind and executive functions in children with fetal alcohol spectrum disorders. The Canadian Journal of Clinical Pharmacology 16, 370–380.

Razza, R., Blair, C., 2009. Associations among false-belief understanding, executive function, and social competence: A longitudinal analysis. Journal of Applied Developmental Psychology 30, 332–343.

Rhode, T., Thompson, L., 2007. Predicting academic achievement with cognitive ability. Intelligence 35, 83–92.

Riggs, N., Greenberg, M., Kusché, C., Pentz, M., 2006. The mediational role of neurocognition in the behavioral outcomes of a social-emotional prevention program in elementary school students: Effects of the PATHS curriculum. Prevention Science 7, 91–102.

Romer, D., Betancourt, L., Giannetta, J., Brodsky, N., Farah, M., Hurt, H., 2009. Executive cognitive functions and impulsivity as correlates of risk taking and problem behavior in preadolescents. Neuropsychologia 47, 2916–2926.

Ronald, A., Happe, F., Bolton, P., et al., 2006. Genetic heterogeneity between the three components of the autism spectrum: A twin study. Journal of the American Academy of Child and Adolescent Psychiatry 45, 691–699.

Rothbart, M., Posner, M., 2005. Genes and experience in the development of executive attention and effortful control. New Directions for Child and Adolescent Development 109, 101–108.

Rothbart, M., Derryberry, D., Hershey, K., 2000. Stability of temperament in childhood: Laboratory infant assessment to parent report at seven years. In: Molfese, V., Molfese, D. (Eds.), Temperament and Personality Development Across the Life Span. Erlbaum, Hillsdale, NJ, pp. 85–119.

Rothbart, M., Ellis, L., Rueda, M., Posner, M., 2003. Developing mechanisms of temperamental effortful control. Journal of Personality 71, 1113–1144.

Rueda, M., Rothbart, M., McCandliss, B., Saccomanno, L., Posner, M., 2005. Training, maturation, and genetic influences on the development of executive attention. Proceedings of the National Academy of Sciences of the United States of America 102, 14931–14936.

Rutter, M., O'Connor, T., 2004. Are there biological programming effects for psychological development? Findings from a study of Romanian adoptees. Developmental Psychology 40, 81–94.

Sabbagh, M., Bowman, L., Evraire, L., Ito, J., 2009. Neurodevelopmental correlates of theory of mind in preschool children. Child Development 80, 1147–1162.

Scarr, S., 2000. American childcare today. In: Slater, A., Muir, D. (Eds.), Blackwell Reader in Developmental Psychology. Blackwell, Oxford.

Schonfeld, A., Mattson, S., Lang, A., Delis, D., Riley, E., 2001. Verbal and nonverbal fluency in children with heavy prenatal alcohol exposure. Journal of Studies on Alcohol 62, 239–246.

Schonfeld, A., Paley, B., Frankel, F., O'Connor, M., 2009. Behavioral regulation as a predictor of response to children's friendship training in children with fetal alcohol spectrum disorders. The Clinical Neuropsychologist 23, 428–445.

Schore, A., 2001. Effects of a secure attachment relationship on right brain development, affect regulation, and infant mental health. Infant Mental Health Journal 22, 7–66.

Sergeant, J., Geurts, H., Oosterlaan, J., 2002. How specific is a deficit of executive functioning for attention-deficit/hyperactivity disorder?. Behavioural Brain Research 130, 3–28.

Sheese, B., Rothbart, M., Posner, M., White, L., Fraundorf, S., 2008. Executive attention and self-regulation in infancy. Infant Behavior and Development 31, 501–510.

Solomon, M., Ozonoff, S., Ursu, S., et al., 2009. The neural substrates of cognitive control deficits in autism spectrum disorders. Neuropsychologia 47, 2515–2526.

Sonuga-Barke, E., 2005. Causal models of attention-deficit/hyperactivity disorder: From common simple deficits to multiple developmental pathways. Biological Psychiatry 57, 1231–1238.

Tager-Flusberg, H., 2001. A re-examination of the theory of mind hypothesis of autism. In: Burack, J., Charman, T., Yirmiya, N., Zelazo, P. (Eds.), The Development of Autism: Perspectives from Theory and Research. Lawrence Erlbaum Associates, Mahwah, NJ, pp. 173–194.

Tamnes, C., Østby, Y., Walhovd, K., Westlye, L., Due-Tønnessen, P., Fjell, A., 2010. Neuroanatomical correlates of executive functions in children and adolescents: A magnetic resonance imaging (MRI) study of cortical thickness. Neuropsychologia 48, 2496–2508.

Tangney, J., Baumeister, R., Boone, A., 2004. High self-control predicts good adjustment, less pathology, better grades, and interpersonal success. Journal of Personality and Social Psychology 72, 271–324.

Tau, G., Peterson, B., 2010. Normal development of brain circuits. Neuropsychopharmacology 35, 147–168.

Thomas, M., Johnson, M., 2008. New advances in understanding sensitive periods in brain development. Current Directions in Psychological Science 17, 1–5.

Thorell, L., Lindqvist, S., Nutley, S., Bohlin, G., Klingberg, T., 2009. Training and transfer effects of executive functions in preschool children. Developmental Science 12, 106–113.

Valentine, J., DuBois, D., Cooper, H., 2004. The relation between self-beliefs and academic achievement: A meta-analytic review. Educational Psychologist 39, 111–133.

Vaurio, L., Riley, E., Mattson, S., 2008. Differences in executive functioning in children with heavy prenatal alcohol exposure or attention-deficit/hyperactivity disorder. Journal of International Neuropsychological Society 14, 119–129.

Wellman, H., Liu, D., 2004. Scaling of theory-of-mind tasks. Child Development 75, 523–541.

Wellman, H., Cross, D., Watson, J., 2001. Meta-analysis of theory of mind development: The truth about false belief. Child Development 72, 655–684.

Welsh, M.C., Pennington, B.F., Groisser, D.B., 1991. A normative-developmental study of executive function: A window on prefrontal function in children. Developmental Neuropsychology 7, 131–149.

Wiers, R., Bartholow, B., van den Wildenberg, E., et al., 2007. Automatic and controlled processes and the development of addictive behaviors in adolescents: A review and a model. Pharmacology Biochemistry and Behavior 86, 263–283.

Zelazo, P., Muller, U., 2002. Executive function in typical and atypical development. In: Goswami, U. (Ed.), Handbook of Childhood Cognitive Development. Blackwell, Oxford.

Zelazo, P., Muller, U., Frye, D., Marcovitch, S., 2003. The development of executive function: Cognitive complexity and control – Revised. Monographs of the Society for Research in Child Development 68, 93–119.

The Effects of Stress on Early Brain and Behavioral Development

M.R. Gunnar[1], E.P. Davis[2,3]

[1]University of Minnesota, Minneapolis, MN, USA; [2]University of Denver, Denver, CO, USA; [3]University of California, Irvine, Orange, CA, USA

25.1 INTRODUCTION

Over a half century of research using animal models has documented the impact of early-life stress on neurobehavioral development (Sanchez et al., 2001). Both stress to the mother or more directly to the fetus during prenatal development and stressors that affect mother and infant during postnatal development have been shown to impact circuits that are developing during the period of stressor exposure, including the development of stress-mediating systems. Alterations in stress-mediating systems, in turn, influence how the organism responds to stressors throughout development, producing cascading effects that may result in significant physical and mental health problems later in life. Research on the neurobiological sequelae of stress during human pre- and postnatal development has a much shorter history. However, inroads are being made in understanding how exposure to stress early in life influences neurobehavioral development and lifelong health (Shonkoff et al., 2009).

Neural Circuit Development and Function in the Brain: Comprehensive Developmental Neuroscience, Volume 3 http://dx.doi.org/10.1016/B978-0-12-397267-5.00039-X

© 2013 Elsevier Inc. All rights reserved.

Activity of the hypothalamic–pituitary–adrenocortical (HPA) axis, a stress-sensitive neuroendocrine system, has figured prominently in animal studies of early-life stress ever since it was noted in the 1950s that early experiences permanently altered its reactivity and regulation (Levine, 1957). Because the HPA axis produces hormones that function as gene transcription factors in numerous organs and tissues and because experience alters its activity as well as the activity of its receptors, research on early-life stress has continued to include a focus on the HPA axis. Attention to activity of this system in studies of human development has been promoted by the availability of assays that allow noninvasive measurement of cortisol, its end hormone, in small samples of saliva (Gunnar and Donzella, 2002). Consistent with the history of research in this area, a focus is maintained, although not exclusively, on the HPA axis in reviewing the research on early-life stress and human neurobehavioral development.

This chapter consists of four parts. First, it contains a brief review of the anatomy and physiology of the mature HPA axis and related stress-mediating systems and second, a discussion of prenatal stress and fetal programming. Third, there is a discussion on the postnatal development of the HPA system, the importance of social regulation of the HPA axis in early human development, and what is currently known about long-term impacts of early-life stress on later physical and mental health. Finally, issues that need to be addressed are considered as this field moves forward.

25.2 THE ANATOMY AND PHYSIOLOGY OF STRESS

Stressors are real or perceived threats to psychological or physical viability that are responded to by stressor-specific release of molecules termed stress mediators. These molecules bind to their receptor targets and orchestrate integrated responses that have evolved to increase survival in the immediate face of threat (Joels and Baram, 2009). Glucocorticoids (cortisol in humans) are steroid hormones that serve as a major mediator of the mammalian stress response. Glucocorticoids are produced by the cortex of the adrenal glands; the medulla of the adrenals produces adrenaline, a hormone that is central to the fight/flight response. Glucocorticoids serve multiple roles in defensive responding (Sapolsky et al., 2000). At basal levels they are permissive, in the sense that they maintain organs and tissues in a state that permits rapid and sustained mobilization by other neurotransmitters or hormones. At elevated levels they suppress the actions of other stress-mediating systems and, through negative feedback, return the HPA system to basal levels of activity. Via effects on gene transcription,

glucocorticoids also can have long-term effects on neural systems mediating perception and response to threat, both up- and down-regulating reactions to subsequent stressors. Critically, the effects of acute activations of the HPA system and those of chronic activation are markedly different, with chronic activation resulting in progressive changes in the expression of stress-mediating genes, alteration in neuronal systems that process signals of threat, and changes in neuronal firing patterns throughout the brain.

The cascade of events that produce changes in cortisol release by the adrenals begins with the release of corticotropin-releasing hormone (CRH) and arginine vasopressin (AVP) by cells in the paraventricular nuclei of the hypothalamus (see Figure 25.1, reviewed in Gunnar and Vazquez (2006)). CRH and AVP are released through small blood vessels to the anterior pituitary where they stimulate the release of adrenocorticotropic hormone (ACTH) into the bloodstream. Cells on the cortex of the adrenal glands respond to ACTH and start a cascade of enzymatic actions that convert cholesterol to cortisol (corticosterone in rodents). Activation of the adrenal cortex by ACTH also results in the production of dehydroepiandrosterone (DHEA), an adrenal androgen that because of its anabolic effects has antistress properties. Once released into the circulation, because of its lipid solubility, cortisol enters the cytoplasm of cells throughout the body and brain where it interacts with its receptors if they are present.

Cortisol has affinity to two receptors, mineralocorticoid receptors (MRs) and glucocorticoid receptors (GRs). Its affinity with MRs is many times greater than to GRs; hence, if both are present MRs will be occupied and activated first, followed by GRs. In most areas of the body, however, cortisol cannot access MRs because an enzyme is present (11 beta hydroxysteroid dehydrogenase or 11β-HSD) that converts cortisol to a form with low MR affinity. As discussed later, this enzyme is also present in the placenta where it serves to regulate impacts of maternal cortisol on the placenta and fetus. In the brain, however, the enzyme is not present, allowing the levels of cortisol in circulation to determine the balance between MR and GR activation. Under basal levels MRs tend to be almost wholly occupied; while when cortisol rises to stress levels and also at the peak of the diurnal rhythm, GRs become occupied as well. MRs tend to mediate many of the permissive effects of cortisol, while GRs mediate many of the more catabolic stress effects. GRs are also involved in negative feedback of the axis, functioning at the level of the pituitary, hypothalamus, hippocampus, and likely also the medial frontal cortex, to contain the HPA response and help return the axis to basal levels of activity.

While there is increasing evidence that under conditions of stress, rapid cell-membrane-mediated effects

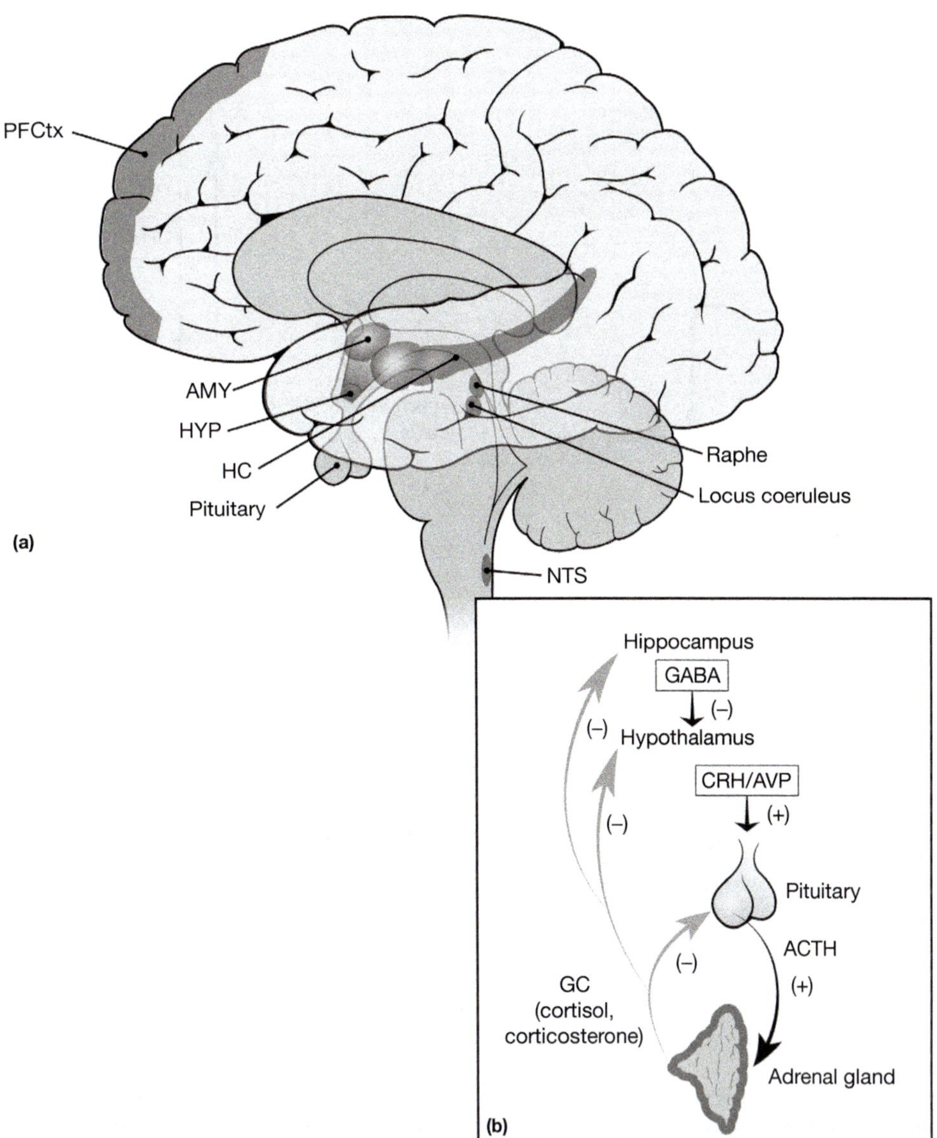

FIGURE 25.1 The HPA system. (a) depicts the anatomy of the HPA system and structures important in its regulation. PFCtx, prefrontal cortex; AMY, amygdala; HYP, hypothalamus; HC, hippocampus; NTS, nucleus of the tractus solitarius. (b) depicts the activation (+) and negative feedback inhibition (−) pathways of the HPA system. Increases in GCs are initiated by the release of CRH/AVP from the medial parvocellular region of the paraventricular nucleus (mpPVN) in the hypothalamus. Negative feedback inhibition operates through GCs acting at the level of the pituitary, hypothalamus, and hippocampus. GABA, gamma animobutyric acid; CRH, corticotropin releasing hormone; AVP, arginine vasopressin; ACTH, adrenocorticotropic hormone. *Reproduced from Gunnar MR and Vazquez D (2006) Stress neurobiology and developmental psychopathology. In: Cicchetti D and Cohen D (eds.)* Developmental Psychopathology:Developmental Neuroscience, *2nd edn., vol. 2, pp. 533–577. New York: Wiley, with permission.*

of cortisol occur, most effects of cortisol involve translocation of the cortisol–receptor complex from the cytoplasm to the cell nucleus where cortisol interacts with glucocorticoid receptive elements (GREs) in the promotor region of many genes. Activation of GREs increases or decreases gene transcription in interaction with other gene transcription factors. Because many of cortisol's effects are produced via gene transcription, they take minutes to hours to be produced. This means that while acute threat may stimulate increases in cortisol production, cortisol itself is not a major factor in fight/flight responses that proceed on the basis of seconds to minutes.

Activation of the HPA axis is regulated by complex signals derived across a number of pathways that carry information about the state of the internal and external milieu (see Figure 25.2). Activation of the system in response to threats to internal homeostasis (e.g., blood volume loss), travel to the CRH-producing cells in the hypothalamus through brainstem pathways. Psychological threats that require integration of information about

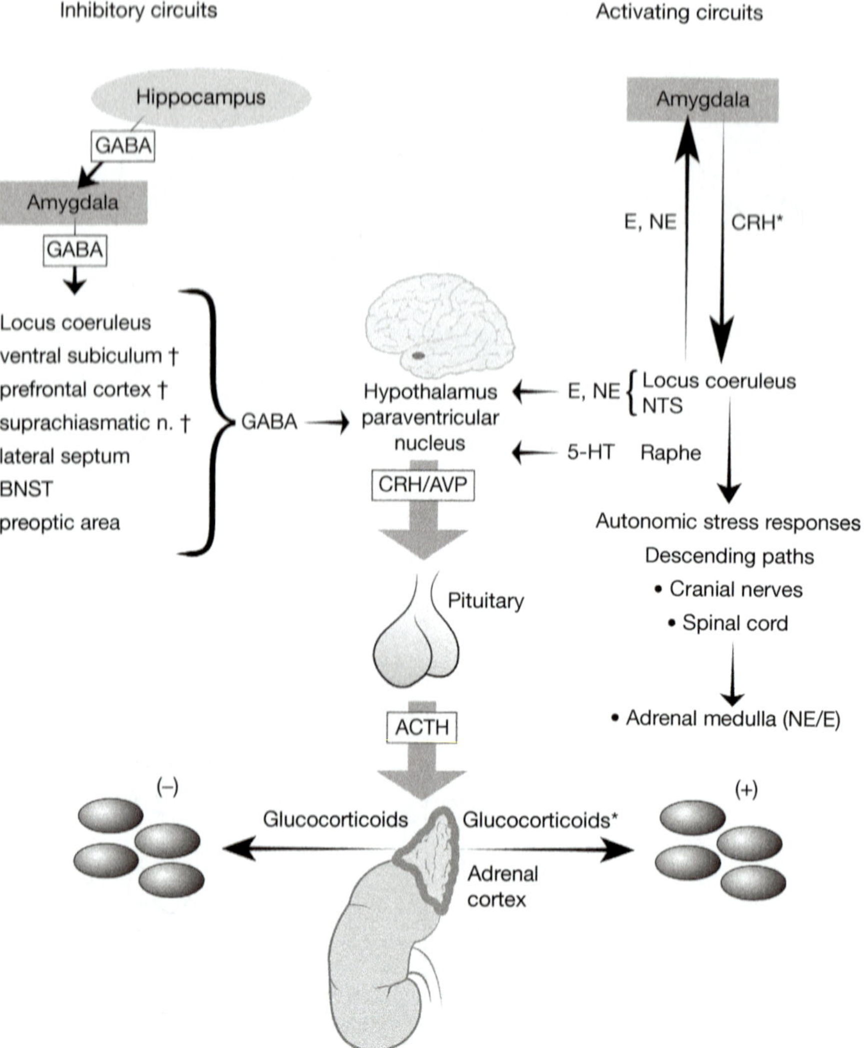

FIGURE 25.2   Schematic representation of the activating (right side) and inhibiting (left side) circuits that contribute to the regulation of the HPA system. Catecholamines, *norepinephrine (NE), and epinephrine (E)* arising from medullary nuclei of the brainstem are the primary neurotransmitters providing activation of CRH synthesis and release from the mpPVN. *Serotonin* originating from dorsal raphe is weakly activating; it acts both directly on mpPVN and indirectly through *excitatory glutamate neurons* or *inhibitory gamma aminobutyric acid (GABA)* inputs. Paradoxically, the inhibitory GABA neurotransmitter activates the mpPVN to secrete CRH, as two GABA neurons activated in series leads to excitation and not inhibition. *Extrahypothalamic CRH also acts as a neurotransmitter to initiate autonomic and behavioral responses to stress.* The activation of the extrahypothalamic CRH system is initiated by *rising* glucocorticoid levels that operate on the amygdala to secrete CRH that, in turn, impacts on the locus ceruleus (LC). Through the activation of catecholaminergic brainstem nuclei there is also stimulation of descending pathways leading to NE/E release from the adrenal medulla that facilitates cardiovascular autonomic responses to stress. Inhibition of the HPA axis seen in the left side is provided by *glucocorticoids* acting on glucocorticoid receptors (GR) in the hypothalamus and pituitary where CRH and ACTH release is halted. The hippocampus serves to inhibit the stress response via multiple circuits, some of which are direct inhibitory GABA inputs; others are indirect through glutamate excitatory inputs to GABA neurons converging in the mpPVN. GABA neurons located in each of the structures further modify the stress reactivity and inhibition from other brain regions such as the thalamus, association cortex, cortical and limbic afferents. *Glucocorticoids provide positive stimulation to the amygdala for the synthesis and release of CRH, but negative to the pituitary, hypothalamus, and hippocampus. †Interaction is through glutamate outflow from these regions that synapse on local GABAergic neurons, producing inhibition of mpPVN. *Reproduced from Gunnar MR and Vazquez D (2006) Stress neurobiology and developmental psychopathology. In: Cicchetti D and Cohen D (eds.) Developmental Psychopathology: Developmental Neuroscience, 2nd edn., vol. 2, pp. 533–577. New York: Wiley, with permission.*

external events are mediated through pathways involving the amygdala and bed nucleus of the stria terminalis (BNST). Notably, neural systems involved in activating the axis in response to psychosocial threats either produce CRH or have receptors for CRH. The central nucleus of the amygdala is one region rich in CRH-producing neurons, activation of which plays a role not only on activating the HPA axis but also in stimulating increases in central norepinephrine and peripheral activation of the sympathetic nervous system. The extrahypothalamic CRH system is part of the fight/flight system and a key orchestrator of fear behavior (Rosen and Schulkin, 1998).

The expression of CRH and its receptors in the various brain regions involved in emotion and cognition is age-dependent and regulated by stress throughout the life span. Recent evidence indicates that effects of stress on neurodevelopment are mediated by CRH, as well as cortisol (Korosi and Baram, 2008).

25.3 PRENATAL STRESS AND NEUROBEHAVIORAL DEVELOPMENT

25.3.1 Fetal Programming

Fetal development proceeds at a more rapid pace than any later developmental stage (Barker, 1998a,b). For this reason, the human fetus is particularly vulnerable to both organizing and disorganizing influences, which have been described as programming. Programming is the process by which a stimulus or insult during a vulnerable developmental period has a long-lasting or permanent effect (Barker, 1998a,b; Kuzawa, 2005). The effects of programming are dependent on the timing (i.e., the developmental stage of organ systems and the changes in maternal and placental physiology) and the duration of exposure (Davis and Sandman, 2010; Nathanielsz, 1999). There is convincing support for fetal programming of adult health outcomes including heart disease, diabetes, and obesity; however, the evidence comes primarily from retrospective studies that rely on birth phenotype (e.g., small size at birth or preterm delivery) as an index of fetal development (e.g., Barker, 1998a,b). It is unlikely that birth phenotype alone is the cause of subsequent health outcomes. Birth phenotype, instead, reflects fetal adaptation to exposures that shape the structure and function of physiological systems that underlie health and disease risk (Gluckman and Hanson, 2004). Prenatal exposure to maternal stress signals is one of the primary pathways for prenatal programming of later health and development. In the first section, the role that prenatal maternal stress signals might play in preparing the fetus for adaptation to the postnatal world is discussed.

25.3.2 Stress Regulation and Pregnancy

During the prenatal period, signals from the maternal host environment influence the fetal developmental trajectory. Some of these effects may have evolved to help prepare the infant for the postnatal environment. The HPA axis participates in a surveillance and response system that is present in many species, from the desert-dwelling Western Spadefoot tadpole to the human fetus, and allows for the detection of threat so that development can be adjusted accordingly. For instance, rapidly evaporating pools of desert water result in the elevation of CRH in the tadpole, accelerating metamorphosis, and increasing the likelihood of survival. If the CRH response is blocked during environmental desiccation, then development is not accelerated and the tadpole's survival is compromised. There are long-term consequences for the tadpole that survives this early-life stress because its growth is stunted and it is at a disadvantage in the competition for food and reproduction (Denver, 1997). It has been argued that a similar signaling pathway participates in the regulation of human fetal development. Detection by the fetal/placental unit of stress signals from the maternal environment (e.g., cortisol) informs the fetus that there may be a threat to survival. This information may prime or advance the placental clock (McLean et al., 1995) by activating the promoter region of the CRH gene and increasing the placental synthesis of CRH (Sandman et al., 2006). The rapid increase in circulating CRH begins the cascade of events resulting in myometrial activation and in extreme cases, preterm birth. Early departure (i.e., preterm birth) from the inhospitable host environment may be essential for survival, but it also may have long-term consequences for the human fetus just as it does for the tadpole. The developmental trajectory of the fetus, whether born early or at term, is influenced by the maternal environment and adaptation of the developmental program to maternal stress signals may prepare the fetus for postnatal survival.

25.3.2.1 Changes in the Maternal HPA and Placental Axis Over the Course of Pregnancy

Regulation of the HPA axis is altered dramatically during pregnancy because the placenta expresses the genes for CRH (hCRHmRNA) and the precursor for ACTH and betaendorphin (proopiomelanocortin; see Figure 25.3). All of these stress-responsive hormones increase as pregnancy advances, but the exponential increase in placental CRH (pCRH) over the course of gestation in maternal plasma is remarkable, as by the latter half of gestation it reaches levels observed only in the hypothalamic portal system during physiological stress. pCRH is identical to hypothalamic CRH in structure, immunoreactivity, and bioactivity. There is, however, one crucial difference in its regulation. In contrast to the negative feedback regulation of hypothalamic CRH, cortisol stimulates the expression of hCRHmRNA in the placenta, establishing a positive feedback loop that allows for the simultaneous increase of pCRH, ACTH, and cortisol over the course of gestation (see for review, Sandman and Davis, 2010). The normative increase in stress-responsive hormones such as cortisol and pCRH plays an important role in the regulation of pregnancy as well as facilitating maturation of the fetus. However, because of the positive feedback between cortisol and pCRH, the effects of maternal stress on the fetus may

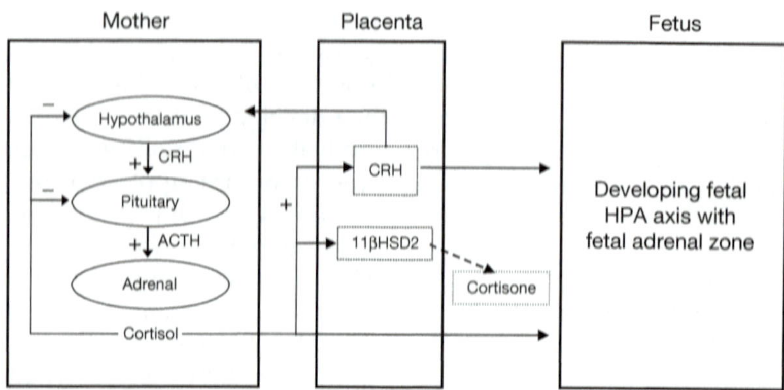

FIGURE 25.3　The regulation of the maternal HPA axis changes dramatically over the course of gestation with profound implications for the mother and the fetus. One of the most significant changes during pregnancy is the development of the placenta, a fetal organ with significant endocrine properties. In nonpregnant women, exposure to stress activates a cascade of events including the release of CRH, ACTH, and cortisol. This stress system is regulated by a negative feedback loop in which cortisol 'turns off' the HPA axis. During pregnancy, CRH is released from the placenta into both the maternal and fetal compartments. In contrast to the negative feedback regulation of hypothalamic CRH, cortisol *increases* the production of CRH from the placenta. Placental CRH (pCRH) concentrations rise exponentially over the course of gestation. Because of the positive feedback between cortisol and pCRH, the effects of maternal stress on the fetus may be amplified representing one pathway by which stress may exert influences on the fetus. In addition to its effects on pCRH, maternal cortisol passes through the placenta. However, the effects of maternal cortisol on the fetus are modulated by the presence of a placental enzyme 11βHSD2 which oxidizes it into an inactive form, cortisone. Activity of this enzyme increases as pregnancy advances, and then drops precipitously so that maternal cortisol is available to promote maturation of the fetal lungs and central nervous system as well as other organ systems. Structures of the HPA axis begin their development early in gestation and become increasingly functional with the progression toward term. See the text for description.

be amplified with potentially negative consequences for the developing fetus.

The effects of CRH and cortisol are modulated by the activities of binding proteins and enzymes. For example, in conjunction with the rapid acceleration of pCRH, CRH-binding protein levels decline sharply toward the end of gestation, increasing the availability of bioactive pCRH (McLean et al., 1995). Levels of binding protein have been associated with birth outcome (Hobel et al., 1999) and may moderate the activity and the effects of pCRH on the fetal nervous system. Maternal production of cortisol binding globulin (CBG) is stimulated by estrogen and thus levels increase progressively with advancing gestation, but then decline before term increasing fetal exposure to maternal cortisol in late gestation (Ho et al., 2007). Variations in CBG may contribute to individual differences in developmental outcomes because levels have been shown to be lower in women with growth-restricted fetuses (Ho et al., 2007). Another gestational timing effect may relate to the activity of the placental enzyme 11β-HSD2. This enzyme oxidizes maternal cortisol into cortisone, inactivating it and protecting the fetus from its direct and sometimes harmful effects during critical periods of development. The levels of placental 11β-HSD2 rise as gestation progresses before falling precipitously near term ensuring maturation of the fetal lungs, CNS and other organ systems in full-term births (Murphy and Clifton, 2003). Despite the presence of this protective enzyme early in gestation, maternal cortisol does reach the fetus and the amount varies with circulating maternal levels (Gitau et al., 1998). In addition, maternal stress down-regulates 11β-HSD2 activity in the placenta allowing a greater proportion of maternal cortisol to cross the placenta to reach the fetus (Mairesse et al., 2007; O'Donnell et al., 2011) with negative consequences for fetal growth and development (Kajantie et al., 2003). This is another mechanism whereby the consequences of maternal stress for the developing fetus may be amplified. Because of the timetable of fetal development and the changes in maternal and placental physiology, the consequences of stress exposures will vary on the basis of the gestational period of exposure.

It is critical to acknowledge that the differences in reproductive and stress physiology, even in very closely related species such as humans and nonhuman primates, limit the validity of generalizing from animal models (Power et al., 2006). For these reasons, this review focuses on studies of gestational stress in humans.

25.3.3　Fetal Adrenal Development

The fetal adrenals make unique contributions to both the regulation of fetal development and the timing of parturition. Cortisol is thought to play critical roles in the promotion of fetal maturation in preparation for extrauterine life. Further, DHEA sulfate produced by the fetal adrenal is an obligate precursor for placental estrogen and is thought to contribute to the initiation of parturition.

Morphologically, the fetal adrenal gland is comprised of two zones: the outer, definitive zone, and the large, inner fetal zone. Between these two zones is the transitional zone. The fetal and definitive zones can be recognized after the eighth gestational week. The fetal adrenals grow rapidly until the third trimester so that at term the fetal adrenals are significantly larger, relative to body weight, than the adult adrenals. At the end of human pregnancy, the fetal zone begins to atrophy. The human fetal adrenal has steroidogenic enzymes as early as the seventh gestational week and cortisol secretion can be detected as early as the eighth week. Cortisol production from the fetal adrenal is regulated by ACTH and ACTH-containing cells can be seen in the pituitary by eight gestational weeks (Jaffe et al., 1998; Kempna and Fluck, 2008). There is evidence that the fetus responds to pain with an increase in cortisol during the latter half of gestation (Gitau et al., 2001).

25.3.3.1 *Fetal Brain Development and Susceptibility to Stress and Stress Hormones*

The rapid changes in the developing fetal brain render it particularly susceptible to organizing and disorganizing influences of stress-responsive hormones such as cortisol and pCRH. Between 8 and 16 gestational weeks migrating neurons form the subplate zone, awaiting connections from afferent neurons originating in the thalamus, basal forebrain, and brainstem. Once neurons reach their final destination, they arborize and branch in an attempt to establish appropriate connections (Sidman and Rakic, 1973). Cessation of neuronal proliferation and migration to the cortical plate occurs around 20–24 weeks. However, dramatic changes in the organization of the cerebral cortex continue through term (Bourgeois, 1997; Volpe, 2008). During the last third of human pregnancy the fetal brain is forming secondary and tertiary gyri, and exhibiting neuronal differentiation, dendritic arborization, axonal elongation, synapse formation and collateralization, and myelination (Bourgeois, 1997; Volpe, 2008). Synapse formation during this period accelerates to a rate of approximately 40 000 synapses per minute (Bourgeois, 1997). Linear increases in total gray matter volume of 1.4% per week are seen from 29 to 41 gestational weeks and approximately 50% of the increase in cortical volume occurs between 34 and 40 gestational weeks (Kinney, 2006).

Regions of the brain that are both integral to the regulation of stress responses and vulnerable to exposure to stress hormones, including the hippocampus and amygdala, develop rapidly throughout gestation. Both are identifiable between 6 and 8 gestational weeks and by term the basic neuroanatomical architecture of these regions is present. Limited information exists regarding the time course of prenatal development of cortisol receptors in humans. There is evidence that both types

of cortisol receptors are present in the human hippocampus by 24 gestational weeks (Noorlander et al., 2006).

Data from animal models have documented that exposure to prenatal maternal biological and psychosocial stress influences the developing fetal brain and endocrine systems producing long-term effects on cognition, emotion, and physiology in the offspring (Kapoor et al., 2006). Evidence for persistent organizational changes or programming influences on the nervous system has been growing and may include changes in neurotransmitter levels, cell growth and survival, and adult neurogenesis. For instance, at high concentrations, the CRH and cortisol may inhibit growth and differentiation of the developing nervous system. Considerable evidence indicates that glucocorticoids, such as cortisol, are neurotoxic to hippocampal CA3 pyramidal cells, and fetal exposure to high levels of glucocorticoids produces irreversible damage to the hippocampus. Similar neurotoxic effects are observed with exposure to high levels of CRH. Exogenously administered CRH increases neuronal excitation leading to seizures in limbic areas associated with learning and memory and may participate in the mechanisms of neuronal injury. These data from animal models suggest mechanisms by which early-life stress may provoke long-term effects on stress, emotional regulation, and cognition (see Joels and Baram, 2009; Seckl, 2008 for reviews).

25.3.4 Gestational Stress Influences the Human Fetus

In humans, a compelling body of work has documented that both maternal report of elevated maternal stress or anxiety and exposure to traumatic events during pregnancy are associated with increased risk for preterm birth. These associations are independent of sociodemographic and obstetric risk factors. There are several pathways by which maternal stress may lead to preterm birth including accelerated production of pCRH and altered vascular and immune functioning (see Dunkel Schetter and Glynn, 2011 for review). Preterm birth is associated with pervasive developmental delays (Aarnoudse-Moens et al., 2009). It is believed, however, that intrauterine exposures, including stress, contribute to these impairments independently from birth outcomes. The study of human fetal development is important because it provides a direct test of the fetal programming hypothesis with the opportunity to assess the effects of gestational stress on development before the effects of external forces, such as birth outcome, parenting, and socialization, are exerted.

The idea that the fetus is responsive to maternal distress is not new and there are a number of reports suggesting that maternal exposure to trauma (e.g., earthquake) during pregnancy has direct and profound implications for fetal physiology and behavior (Ianniruberto and Tajani, 1981). More direct tests of the effect of maternal psychological state on the fetus come from studies manipulating maternal stress or anxiety and evaluating the consequences for fetal behavior. Fetuses display a consistent response profile in response to maternal exposures to moderate laboratory challenges such as the Stroop color-word test or viewing videos of labor and delivery (DiPietro, 2004). The nature of these responses appears to be moderated by maternal psychological state (Kinsella and Monk, 2009). These laboratory studies provide compelling evidence that fetuses are responsive to maternal stress and anxiety and they raise the possibility that repeated exposures over the course of gestation may influence the developing fetal nervous system.

Direct measures of fetal responses to external stimulation provide an index of fetal nervous system development and have been used to assess the developmental consequences of exposure to physical or maternal psychological stress. The response to a vibroacoustic stimulus (VAS) is an indication of fetal maturity reflecting maturation and integrity of neural pathways through the cerebral cortex, midbrain, brainstem, vagus nerve, and the cardiac conduction system. Using the fetal response to VAS, it has been shown that stress signals, most clearly pCRH trajectories, influence the developing fetal nervous system. Low pCRH is associated with more mature or earlier development of the fetus' ability to mount a response to the VAS and with a more mature profile to a classic habituation/dishabituation paradigm (Class et al., 2008; Sandman et al., 1999). Other maternal stress signals including overexpression of beta-endorphin and under-expression of ACTH have additionally been linked to the fetal response to VAS (Sandman et al., 2003).

These studies provide evidence that signals of maternal stress during gestation exert programming influences on the nervous system that cannot be explained by postnatal experiences. Continuity between the fetal and infant periods in assessments of movement and heart rate indicate that maternal influences that shape developmental trajectories during the prenatal period will continue to influence functioning postnatally (DiPietro, 2004). It is additionally clear that consideration of sexually dimorphic profiles of fetal development is necessary, as discussed later in this chapter.

25.3.5 Prenatal Maternal Psychosocial Stress and Infant and Child Development

25.3.5.1 Socioemotional Development

A growing literature indicates that prenatal exposure to elevated levels of maternal psychosocial stress is associated with behavioral and emotional disturbances during infancy and childhood among healthy full-term infants that is independent of postpartum maternal psychosocial stress (Bergman et al., 2007; Blair et al., 2011; Davis et al., 2004, 2007). Both maternal report of psychosocial stress and report of stressful life events are associated with more fearful and reactive behaviors during infancy and toddlerhood. Effects on social and emotional development continue to be observed during childhood and adolescence. Maternal antenatal anxiety and depression predict reports of childhood behavioral and emotional problems, including attention deficit hyperactivity disorder and anxiety problems (Davis and Sandman, 2012; O'Connor et al., 2002; Van den Bergh and Marcoen, 2004). These associations remain significant after controlling for birth outcomes and postnatal maternal psychological state. This suggests that signals of maternal psychosocial stress influence the fetal developmental trajectory with implications for children's functioning after birth.

25.3.5.2 HPA Axis Functioning

Alterations to the fetal HPA axis are frequently proposed as the primary biological pathway underlying fetal programming of later health and development. Animal studies suggest that the fetal HPA axis may be particularly vulnerable to prenatal exposure to maternal stress (Kapoor et al., 2006); however, relatively little is known about the consequences of prenatal maternal stress for HPA axis functioning in humans. There is evidence for higher cortisol levels, during a laboratory challenge among infants and toddlers exposed prenatally to elevated maternal cortisol (Grant et al., 2009). Several studies have suggested that prenatal maternal psychosocial stress is associated with altered circadian regulation during childhood and adolescence, both on typical days (O'Connor et al., 2005; Van den Bergh et al., 2007) and on days when stressors such as the first day of school are experienced (Gutteling et al., 2005). These studies suggest that prenatal exposure to maternal psychosocial stress may influence the developing fetal HPA axis and that this has implications for the regulation of cortisol production during infancy, childhood, and adolescence.

The mechanism by which maternal psychosocial stress is communicated to the fetus is unknown. It is unlikely that maternal cortisol mediates the effect of maternal report of psychosocial stress on child outcomes, as the majority of published studies find that during gestation maternal cortisol and psychosocial distress are not correlated and they exert independent effects on child outcomes (e.g., Davis and Sandman, 2010; de Weerth and Buitelaar, 2005; Harville et al., 2009). Data from human and animal models indicate that an epigenetic mechanism may underlie both the maternal communication of adversity to the fetus and the persistent influence of the exposure (Champagne and Curley, 2009; Oberlander et al., 2008; Szyf, 2011). Further, maternal psychosocial stress exerts widespread influences on a number of stress-sensitive systems other than the HPA axis including the immune and vascular systems (Dunkel Schetter and Glynn, 2011) indicating alternative pathways by which maternal psychological distress signals may impact the developing fetus.

25.3.5.3 Neurodevelopment

The influence of gestational exposure to maternal psychosocial stress on cognitive and motor development is less clear. There is evidence that maternal self-report of elevated stress and anxiety as well as exposure to traumatic life events, such as severe ice storms, during pregnancy are associated with delayed infant and child cognitive, language, and neuromotor development, and that these deficits may persist into adolescence. However, not all studies have demonstrated such associations and there is evidence that modest elevations in psychosocial stress during late gestation may actually increase cognitive maturation (Davis and Sandman, 2010; DiPietro et al., 2006).

Inconsistencies across studies may suggest that moderate exposure to prenatal maternal psychosocial stress does not negatively affect neurodevelopment. It is plausible, however, that generalized self-report measures of psychological distress do not adequately characterize stress that is unique during pregnancy. As reviewed in Davis and Sandman (2010), evidence is emerging that measures of pregnancy-specific stress (e.g., "I am fearful regarding the health of my baby," "I am concerned or worried about losing my baby") are better than measures of generalized psychological distress for predicting neurodevelopmental outcomes including fetal behavior, infant neurodevelopment, infant and toddler cognitive and motor development, and child brain development. It is important to note that these associations are not explained by actual medical risk associated with pregnancy and birth outcome. Support for the importance of pregnancy-specific stress for developmental outcomes comes from a recent study documenting associations between elevated pregnancy-specific anxiety and decreased gray matter density at 6–10 years of age (Buss et al., 2010). Elevations in pregnancy-specific anxiety, particularly early in gestation, are associated with gray matter volume reductions in the prefrontal cortex, the premotor cortex, the medial temporal lobe, the lateral temporal cortex, the postcentral gyrus as well as the cerebellum, extending to the middle occipital gyrus and the fusiform gyrus. These brain regions are associated with a variety of cognitive processes including reasoning, planning, attention, memory, and language, and raise the possibility that developmental

alterations to these regions may underlie associations between elevated pregnancy-specific anxiety and cognitive performance observed in prior studies (Buss et al., 2011).

25.3.6 Prenatal Maternal Biological Stress Signals and Infant and Child Development

Alterations to the maternal HPA and placental axis are most frequently cited as the mechanisms that underly fetal programming of later health and developmental outcomes. Although robust evidence exists from animal models, there are only a handful of studies that have evaluated the influence of biological stress signals during gestation on human development.

25.3.6.1 Social/emotional Development

Prenatal exposure to elevated maternal cortisol predicts increased fussiness, negative behavior, and fearfulness during infancy (Blair et al., 2011; Davis et al., 2007; de Weerth et al., 2003) and toddlerhood (Bergman et al., 2007). Further, there is evidence that elevations in pCRH contribute to the increased fearfulness observed during infancy (Davis et al., 2005). These studies suggest that maternal HPA and placental axis hormones contribute to the development of a fearful or reactive temperament. These data indicate that maternal and placental hormones contribute to the development of fearful temperament. New data from our prospective cohort suggest that this effect persists. Elevated gestational cortisol levels are associated with an increase in anxiety risk among preadolescent children (Davis and Sandman, 2012).

25.3.6.2 HPA Axis Functioning

Few studies have evaluated the consequences of prenatal maternal cortisol on HPA axis regulation in the offspring. In a prospective study with multiple prenatal assessments, it was documented that prenatal exposure to elevated levels of maternal cortisol is associated with a larger and more prolonged infant response to stress (Davis et al., 2011b). There is emerging, but as yet limited, evidence that this effect persists. In samples of 24 and 29 children, respectively, Gutteling et al. (2004) found that elevated maternal cortisol stress measured at one time during gestation (15–17 gestational weeks) independently predicted children's higher cortisol levels on the day of an inoculation and on the first day of a new school year. These studies provide evidence for effects of prenatal maternal cortisol on HPA axis functioning. Further prospective studies with multiple prenatal and postnatal measures are needed to address this question.

25.3.6.3 *Neurodevelopment*

Although little is known about the role of biological stress signals in shaping infant and child cognitive development, it is apparent that timing of exposure will be critical to understanding the developmental effects on outcome. Evidence suggests that the trajectory of maternal cortisol across gestation is the strongest predictor of child neurodevelopment. Elevated maternal cortisol during early and mid gestation has been associated with decreased neuromuscular maturity in the newborn (Ellman et al., 2008) and delayed cognitive development during toddlerhood (Bergman et al., 2010; Davis and Sandman, 2010). Conversely, elevated maternal cortisol late in gestation has been associated with significantly higher scores on measures of mental development at 1 year (Davis and Sandman, 2010).

These findings linking cortisol to neurodevelopment are remarkably consistent with its function in the maturation of the human fetus. Early in pregnancy, the fetus is protected from the naturally occurring increases in maternal cortisol by 11β-HSD2 (see Figure 25.3). However, because 11β-HSD2 is only a partial barrier, excessive increases in maternal cortisol early in gestation will expose the fetus to toxic levels with potentially detrimental consequences. In contrast, as pregnancy advances toward term, exposure to cortisol is necessary and beneficial for fetal maturation and exposure to increased cortisol is facilitated by the sharp drop in 11β-HSD2 activity, allowing a greater proportion of maternal cortisol to cross the placental barrier (Murphy and Clifton, 2003). The beneficial effects of modestly elevated cortisol during late gestation are consistent with animal models demonstrating that modest cortisol increases during the early postnatal period are associated with persisting beneficial effects for the developing brain (Catalani et al., 2000).

25.3.7 Is This Fetal Programming?

One concern that challenges research with humans is whether associations between maternal stress and anxiety and fetal outcomes should be interpreted as fetal programming, or alternatively, as a reflection of shared genetic factors. In the studies of naturally occurring variations in maternal cortisol or maternal self-reported stress, it is difficult to differentiate between these alternative explanations. The programming findings reported here, however, are consistent with animal models where random assignment is possible (Kapoor et al., 2006) and with human studies that evaluated the consequences of randomly occurring traumatic events, such as natural disasters (LaPlante et al., 2004; Yehuda et al., 2005) and with prenatal exposure to synthetic glucocorticoids (Davis et al., 2006, 2011a; French et al., 1999). More

convincingly, in a recent human study, similar effects of prenatal stress on child outcomes were documented among children conceived by *in vitro* fertilization in a model where mother and fetus were genetically unrelated (Rice et al., 2009). Thus, while in most human studies of prenatal stress, genetic mechanisms cannot be ruled out as a possible explanation, there is reasonable evidence to warrant the conclusion that maternal stress has effects on the neurodevelopment of the fetus.

25.3.8 Summary

Both psychosocial and biological maternal stress signals are associated with developmental consequences for the fetus. Further, these effects cannot be accounted for by birth outcome or postnatal maternal psychological distress. However, it is important to acknowledge that biological and psychosocial stress signals tend not to be correlated during pregnancy and have been shown to exert independent influences on developmental outcomes (Sandman and Davis, 2012). Future research will have to examine vascular or immune pathways that could be mechanisms by which increases in maternal psychosocial stress might also affect the fetus.

These studies emphasize the importance of performing prospective longitudinal studies in order to evaluate the trajectory of maternal stress signals across gestation and its association with infant and child developmental outcomes. Data indicate that the trajectory or profile biological and psychosocial stress signals may be more critical for determining developmental outcomes as compared to level at a given gestational interval (Davis and Sandman, 2010; Glynn et al., 2008). Both severity and timing of exposure must be considered in order to evaluate associations between prenatal stress measures and infant outcomes. The adaptive significance of these associations is yet to be determined and requires long-term follow-up evaluating the interaction between the prenatal and the postnatal environments.

25.4 POSTNATAL DEVELOPMENT

25.4.1 Postnatal Development of the HPA Axis

At birth, the HPA axis is not morphologically mature (Gunnar and Vazquez, 2006). The adrenal cortex still includes the fetal zone that involutes over the first six postnatal months. As the fetal zone involutes, the structure of the mature gland becomes more distinct. CBG is low in the neonate, increasing gradually over the initial postnatal months. As a result, small increases in total (bound and unbound) plasma cortisol concentrations in response to stressors in the neonatal period may index large increases in the biologically active unbound fraction of the hormone. There is also some indication that adrenal sensitivity to ACTH decreases over the first postnatal months (Forest, 1978), being higher in months 1–4 than later in development. Thus for several months following birth, the human HPA axis is more highly reactive to stimulation than it is later in development. This is consistent with evidence that during the first 3 or 4 months even mild stimulation (e.g., being undressed, weighed, and measured) provokes significant elevations in cortisol, while by 4 months of age the same stimulation no longer elevates cortisol in most typically developing infants.

The pattern of diurnal cortisol production also changes over the first few years of life (Gunnar and Vazquez, 2006). At birth, basal levels of cortisol bear no relation to time-of-day, although they are associated with levels of behavioral arousal. By as early as 6 weeks of age, an early morning peak and evening nadir can be observed; however, because of marked day-to-day variability, observing these aspects of the diurnal rhythm requires analytical methods that isolate average patterns from variability. In the next months, the peak and nadir become more distinct; but, it is not until about 4 years of age as children give up their daily naps that a more fully mature pattern is noted from mid-morning to late afternoon. Throughout the first few years of life, napping is associated with decreases in cortisol over the nap periods and rebounds to prenap levels by about 30–45 min after the child awakens. Other normal daily events have also been associated with decreases and then rebounds in cortisol levels; the rebounds can be misinterpreted as a stress response. Such normal events include the car trip the child takes on the way to the laboratory for testing. Thus accurate assessment of HPA activity during infancy requires control of more than time-of-day.

Following infancy, there is little developmental change in cortisol levels or reactivity until puberty. Over the pubertal transition, basal cortisol levels increase, with the sharpest increase around Tanner stage 3 (Netherton et al., 2004). Reactivity of the axis to psychosocial stressors also appears to increase with puberty (Stroud et al., 2009). As there is evidence that pubertal maturation is associated with increased reactivity of emotion systems, researchers have speculated that increased HPA activity at puberty may contribute to the psychiatric vulnerability associated with this period of development (Spear, 2000). It may also be that the impacts of early experience on the development of HPA reactivity and regulation are not fully realized until the neurobiological changes associated with puberty and adolescent development have occurred.

Because the HPA axis is under multifactorial regulation (Ulrich-Lai and Herman, 2009), developmental changes in HPA reactivity and regulation are also dependent on the development of the many systems whose activity ultimately impinges on CRH-producing neurons in the hypothalamus. With regard to psychological stressors, the amygdala and BNST are critical in activation of the axis, while the hippocampus and medial prefrontal cortex (mPFC) are involved in constraining and terminating its response (see Figure 25.2). There is no information on how development of these extrahypothalamic structures affects reactivity and regulation of the human HPA response to stressors during childhood and adolescence. Furthermore, there is little information on the developmental changes in GRs in extrahypothalamic regions in humans. There is evidence that, unlike in the rodent, there is no developmental increase in GR in the hippocampus over development, suggesting that the capacity of the hippocampus to contain and terminate HPA responses to stressors is relatively mature at birth (Pryce, 2008). In contrast, GR mRNA expression levels do increase into adolescence in regions of the prefrontal cortex, consistent with evidence that GR expression is as high or higher in the neocortex in humans as it is in the hippocampus (Pryce, 2008). This latter finding suggests that in human development, activity of the HPA axis both impacts and is importantly affected by the ontogeny of prefrontal circuits. The protracted development of neural systems involved in and affected by activity of the HPA axis further suggests that postnatal experiences will shape the complex circuitry of what Joels and Baram (2009) termed the 'neurosymphony of stress.'

25.4.2 Social Regulation of the HPA Axis in Human Development

In rodents and nonhuman primates, there is considerable evidence that proximity and contact with the caregiver is critical to regulation of the HPA axis (Sanchez et al., 2001). This is also true in human development. As early as the first months of life, infants cared for by more sensitive and responsive caregivers show a more rapid return to baseline following activation of the axis (Blair et al., 2006), while those receiving insensitive care exhibit increases in cortisol during parent–infant play sessions (Spangler et al., 1994). The role of sensitive and responsive care in regulating the axis extends to infants' experiences with surrogate caregivers, including child care providers (Vermeer and van IJzendoorn, 2006). Indeed, throughout the preschool period there is evidence that, for children in out-of-home child care, elevations in cortisol over the child care day are larger for those who receive less sensitive and/or more intrusive/overcontrolling care than for those who are receiving more sensitive care.

Sensitive responsive caregiving is an important feature in the development of secure attachment relationships. Attachment security is not a trait of the child but characterizes the relationship between a child and their adult caregiver. Children can be securely attached to one parent and insecurely attached to the other. Notably, all studies of the relations between attachment security and regulation of the HPA axis have been done examining mother–child relationships. Overall, children are able to use the presence and availability of their mother more effectively to regulate reactivity of the HPA axis if their relationship is secure rather than insecure (Gunnar and Donzella, 2002). This has been shown in studies using brief separations followed by reunions as the stressor, as well as in studies using physical stressors (i.e., inoculations). In studies of psychosocial threat (e.g., approach by strange objects, individuals), whether or not the effect is observed depends on whether the child exhibits fear behavior or not. For those exhibiting fear, the presence and availability of the attachment figure in secure relationships blocks elevations in cortisol, while in insecure relationships elevations are observed. What this suggests is that access to the mother in secure attachment relationships prevents fear and other negative emotions from stimulating elevations in cortisol. However, this is not the case when the mother is present but the attachment relationship is insecure. There is no understanding of the neurobiological processes underlying maternal buffering of the axis for frightened/distressed infants. One possibility is that secure relationships stimulate larger increases in oxytocin, a presumed antistress neuropeptide (Gouin et al., 2010). Among adults, intranasal infusions of oxytocin increase the stress-buffering effects of social support (Heinrichs et al., 2003). Thus greater productions of oxytocin among fearful infants when the mother is present in secure relationships may help to buffer or block cortisol increases.

While sensitive, responsive care and secure attachment relationships are expected to shape more regulated anxiety and stress responses over time, it is clear that early in life the actual presence of the attachment figure is needed to buffer activity of the HPA axis. A study of toddlers entering a new child care arrangement has shown that while toddlers in secure attachment relationships had lower cortisol levels during days when their mother accompanied them to child care (adaptation), as soon as she was no longer present, no differences in cortisol levels were observed as a function of attachment security (Ahnert et al., 2004). Furthermore, marked elevations in cortisol, relative to home levels, were noted for the first several weeks of child care, with these elevations decreasing but not eliminated after 5 months of experience in the care setting. Thus, however the buffering effect of attachment security is operating early in life, it is a characteristic of the relationship and requires the presence of the adult figure with whom the child is securely attached.

25.4.3 Early Care and HPA Axis Functioning

Given the powerful role of social relationships in regulating reactivity of the HPA axis early in life, one might expect that a child's early care history will shape the development of behavioral and HPA reactivity and regulation. That is, as in the animal literature, programming of threat reactive systems will extend into the postnatal period (Sanchez et al., 2001). While there is no question that early pathogenic care increases the risk of a variety of poor outcomes in human development, in every domain studied individual differences in susceptibility are substantial and numerous factors beyond early exposure to adverse care appear to be involved in predicting whether significant early experience effects are observed or not (Cicchetti and Rogosch, 1996).

The range of early care conditions examined in human studies is large, including not only normal variations in parental sensitivity and responsiveness but also neglect, physical and sexual abuse, exposure to interparental conflict and violence, global deprivation (as in orphanages/institutional care), parental loss, and conditions that often interfere with adequate caregiving (e.g., maternal depression, extreme poverty, homelessness). Some of the conditions involve repeated, acute experiences of threat (i.e., physical abuse), others more chronic experiences of the absence of supportive care (e.g., institutional rearing, neglect). Typically, these experiences do not occur in isolation from one another, and often their effects appear to be cumulative (Brown et al., 2009). With rare exceptions (e.g., children adopted from orphanages), care conditions early in life are more or less predictive of care conditions throughout childhood. Thus it is often challenging to isolate the impacts of early versus later experiences on activity of the HPA axis. Adding to the challenge of understanding the role of early care qualities is the fact that few studies of vulnerable children have examined reactivity and regulation of the HPA axis under conditions known to elicit stress responses. Most studies have either examined patterns of diurnal cortisol production and/or exposed children to mild challenges that do not activate the axis in most individuals (i.e., cognitive testing without feedback; Dickerson and Kemeny, 2004).

25.4.3.1 Diurnal Cortisol Patterns

As noted, the diurnal rhythm of the HPA axis is still maturing during the first years of life. It appears to be sensitive to adverse care, although alterations in diurnal patterns do not appear to be permanent, but rather associated with the more immediate contexts of early care. Effects of care context on diurnal patterns were first noted in research on 2- and 3-year-olds living in a Romanian institution. Of the 46 children studied, not one exhibited a typical diurnal pattern as sampled soon after wakeup, at noon, and in the early evening. Compared to home-reared Romanian children, average levels were significantly lower in the early morning and tended to be higher, but not significantly so, in the early evening hours (Carlson and Earls, 1997). Similar results have been obtained in studies of preschool-aged children recently removed from maltreating homes and placed in foster homes (Bruce et al., 2009). It is not clear whether the child's age or type of adverse care influences associations with diurnal patterns. There is some evidence that among preschool-aged children emotional maltreatment may be associated with elevated early morning cortisol levels, while neglect is associated with abnormally low levels (Bruce et al., 2009). Studies of children adopted from institutions have noted similar low early morning levels, and hence a relatively flat pattern of cortisol production over the daytime hours. Nonetheless, when post-institutionalized children are studied several years after adoption they all exhibit the typical pattern of high morning and low evening cortisol production (e.g., Kertes et al., 2008). In addition, for children studied monthly for a year in foster care, increasingly flat diurnal patterns have been observed in the absence of interventions designed to help foster parents provide more supportive care, with lower early morning levels being associated with the degree of parenting stress reported by the foster parent (Fisher and Stoolmiller, 2008). Thus, it seems likely that the degree, and possibly the type of alteration in the diurnal rhythm in young children indexes stress in relation to the current context and is not a reflection of permanent alterations in the set point or regulation of the diurnal rhythm.

There are several caveats worth noting. First, although the results have focused on average patterns in cortisol production over the day, day-to-day variability may be just as important. Indeed, one of the impacts of providing support and training to foster parents is that over the course of a year, evening cortisol levels in the children become increasingly stable and low from day to day, while this is not the case for children in regular foster care. Second, in most studies, early morning samples were taken 30 or more minutes after awakening, and thus it is not clear whether the magnitude of the cortisol awakening response is what is reflected in the findings and not wakeup cortisol, which reflects the diurnal rhythm. Thus far, to our knowledge, there are no published studies of the cortisol awakening response during early childhood in children experiencing early adverse care.

25.4.4 Effects of Early Care on Cortisol Set Points and Reactivity

Animal models of early care lead to the prediction of unidirectional effects. That is, poorer quality care early in life shapes the set point of the HPA axis and increases behavioral and physiological reactivity to threat. Recently, Boyce and Ellis (2005) have argued for a U-shaped pattern whereby moderate parental support early in life reduces, while both extremely high and low supportive care results in heightened behavioral and physiological reactivity. Thus far, there have been few tests of the argument that extremely supportive care is associated with heightened behavioral and physiological reactivity to threat. There is more support for the argument that poorer care and/or conditions associated with poorer early-life care predict increases in HPA set points and reactivity or lability. Thus, children whose mothers reported significant economic stress and depressive symptoms during the children's infancy exhibited higher afternoon cortisol levels at age four if the family was still under economic stress at that time, and levels of cortisol at this age predicted heightened internalizing problems in first and second grade (Essex et al., 2002; Smider et al., 2002). Among extremely poor families in rural Mexico, a cash-transfer program instituted during the children's infancy was associated with lower cortisol levels, but not less HPA reactivity, when the children were 2 to 6 years old (Fernald and Gunnar, 2009). Likewise, after controlling for later depressive symptoms in the mother, offspring of women who suffered postnatal depression were found to exhibit higher and more labile early morning cortisol concentrations at age 13, which mediated the expression of depression in the children by age 16 (Halligan et al., 2007).

In contrast to these findings, studies of internationally adopted children who experienced severe neglect before adoption challenge the expectation of elevated set points and reactivity following lack of supportive care early in life. Among a small sample of preschool-aged children studied 3 years post adoption, severe preadoption neglect was associated with higher basal urinary cortisol levels and poorer cortisol regulation following a mother–child, but not stranger–child stressor task (Wismer Fries et al., 2008). Among a larger group of children studied in middle childhood, higher early morning levels were noted, but only as a function of physical growth delay at adoption (Kertes et al., 2008). Consistent with the possibility of continued programming of the axis during the postnatal period, this latter finding is very reminiscent of programming effects that have associated poorer intra-uterine growth with activity of the HPA axis later in life (Phillips et al., 2000).

Thus far, work on postinstitutionalized children has not revealed evidence of hyperresponsivity of the axis to psychosocial stress in the form of the Trier Social Stress Test (TSST), a public speaking stressor (Gunnar et al., 2009), although only one study has been reported thus far and the children were largely prepubertal. Notably, a recent study of teenage girls in child protective service examined cortisol and cardiac reactivity to the TSST and noted blunted cortisol but not cardiac responses in these maltreated girls relative to nonmaltreated controls (MacMillan et al., 2009). It is not clear whether the difference between findings for postinstitutionalized and maltreated children reflect differences in the type and timing of early adversity, the timing of assessment (pre- versus postpubertal) or the concurrent life circumstances of the children (stable families versus foster care).

Adult studies of individuals maltreated as children have shown that patterns of HPA reactivity to psychosocial stressors depend on psychiatric diagnosis. For those suffering major depression, childhood maltreatment is associated with hyperreactivity and poor regulation of the axis, a pattern not seen among depressed adults without an early history of maltreatment (Heim et al., 2008). Conversely, among emotionally healthy adults, histories of more adverse family life conditions in childhood (i.e., maltreatment, interparental conflict, rejection) have been associated with a lower set point of the HPA axis and smaller cortisol responses to psychosocial stressor tasks (Carpenter et al., 2007). It has also been noted that among school-aged children, whether those who were sexually abused before age 5 exhibited altered cortisol production during a 5-day camp for maltreated children depended on whether the child exhibited internalizing behavior problems or not (Cicchetti et al., 2010). The confluence of internalizing problems and early sexual abuse predicted altered cortisol production, while neither factor alone appeared to have any association with cortisol levels (although for contradictory findings, see MacMillan et al., 2009).

The effects of trauma on the HPA axis also appear to depend on whether the individual develops posttraumatic stress disorder (PTSD) or not. Studied among adults, individuals with PTSD exhibit normal to low basal levels of cortisol; with varying evidence of hyper- or hyporeactivity to stressors, perhaps related to the extent that the stressor is reminicient of the original traumatizing experience (Yehuda, 2000). Among prepubertal children, PTSD appears to be associated with elevated levels of cortisol production (see review, Tarullo and Gunnar, 2006). Among sexually abused girls with PTSD assessed over time, elevated cortisol levels have been noted in childhood, with decreasing levels beginning in adolescence and levels lower than normal by adulthood (Trickett et al., 2010). It is still not clear whether puberty shifts the relation between PTSD and set points of the axis from hyper to hypo, or whether it is the time since trauma, which gradually alters the axis from hyper- to hypofunctioning.

25.4.5 Individual Differences

Numerous factors likely influence susceptibility to both pre- and postnatal programming of fear- and stress-response systems. Three factors are briefly mentioned here: sex differences, temperamental fearfulness or inhibition, and genes.

There is a developing literature suggesting that there are sexually dimorphic profiles of fetal development (Clifton, 2010). Compelling data from animal models document sex-specific consequences of stress hormone exposure (Goel and Bale, 2009). Further, there is growing appreciation that human fetal neurological development is different for males and females (Bernardes et al., 2008; Buss et al., 2009; DiPietro et al., 1998) and that gestational influences on fetal neurodevelopment may be associated with sex-specific postnatal developmental trajectories including risk for psychiatric disorders (Costello et al., 2007; Goel and Bale, 2009). The role of sex as a moderator of the effects of early postnatal care on fear- and stress-response systems is not clear (Gunnar and Vazquez, 2006). While there are some studies that indicate larger impacts of early parental care for boys than girls, many other studies have not indicated differential sensitivity by sex to variations in supportive early care.

Behavioral inhibition describes a temperament dimension defined by greater behavioral reactivity and inhibition of approach to novel, arousing, or unpredictable stimulation. Behaviorally inhibited infants often develop into children who are shy and prone to social anxiety (Chronis-Tuscano et al., 2009). While there is some evidence that extremely inhibited children have higher cortisol levels and reactivity (Kagan et al., 1987; Schmidt et al., 1997), this is not always found and likely depends on contextual factors, such as the degree of support the child has from attachment figures and whether inhibiting approach is an effective option for regulating stress (Gunnar, 2001; Gunnar and Donzella, 2002). The larger cortisol responses that fearful, inhibited children experience in the context of less supportive care may help mediate the greater vulnerability of these children to adverse early-life care (Phillips et al., 2011). However, their sensitivity may also allow them to excel in higher-quality care environments; a possibility proposed by Boyce and Ellis in their biological sensitivity to context theory (Boyce and Ellis, 2005).

As suggested by the work on behavioral inhibition, sensitivity to variations in the quality of early-life care may be influenced by genetic inheritance. For example, the type 1 CRH receptor mediates behavioral and physiological reactions to threat-provoking stimuli and there is now evidence that variants in the CRHR1 gene interact with early abuse histories to increase the risk of depression (Gillespie et al., 2009). A number of genetic polymorphisms have also been noted in genes that play key roles in the regulation of GR and thus the expression of glucocorticoid-responsive genes (Derijk and de Kloet, 2008). These too may influence risk and resilience to trauma and variations in the quality of early care environments (Derijk and de Kloet, 2008; Gillespie et al., 2009).

In the studies of children, variations in other genes, including those regulating major neurotransmitter systems, have been found to moderate relations between care experiences and reactivity of the HPA and sympathetic nervous system (Frigerio et al., 2009). Notably, although from a stress-diathesis perspective one might view genetic factors as increasing vulnerability to adverse early care, they may instead produce differential susceptibility to both the positive and negative qualities of early care (Belsky and Pluess, 2009).

25.4.6 Summary

Research on infants and children conducted over the last few decades has begun to illuminate the impact of postnatal experiences on the development of stress-mediating neurobiological systems. The human HPA axis continues to develop postnatally with significant changes in functioning over the first six postnatal months. Throughout infancy and early childhood the axis is under strong social regulation. Sensitive, responsive care maintains the axis in a relatively quiescent state and permits rapid returns to basal functioning following perturbations. There is increasing evidence that adverse care early in life impacts neurobehavioral development and impacts diurnal patterns of cortisol production assessed concurrently. There is emerging, but as yet sparse and sometimes conflicting, evidence that early neglect and abuse may be associated with long-term changes in the reactivity and regulation of the HPA axis and that effects on this neuroendocrine system may mediate some of the poor mental and physical outcomes of early maltreatment. In part, some of the conflicting findings appear related to the age at assessment as there may be changes in HPA functioning following early maltreatment or trauma that differ as a function of whether assessment is conducted before or after puberty. Some of the conflicting findings also appear to be related to whether or not the individual is suffering from an affective disorder or not, and if so, the nature of the disorder (i.e., depression, PTSD). Genetic variation likely contributes to the heterogeneity of outcomes noted both with regards to impacts on the HPA axis and on emotional functioning. Of particular note, however, are recent arguments that the same genes and/or emotional temperament that may result in heightened vulnerability to poor outcomes following early adverse care may also result in more optimal functioning, given positive early care experiences. Thus, while postnatal experiences appear to influence the developing neurobiology of stress reactivity and regulation, the role of individual differences and developmental timing are critical issues whose effects are still in need of a more complete explication.

25.5 FUTURE DIRECTIONS

The study of pre- and postnatal stress and its impact on neurodevelopment and health has thus far proceeded largely independently. It is time for this work to become integrated, both empirically and theoretically. While researchers studying prenatal stress have sometimes obtained measures of outcomes later in life, it is rare that postnatal experiences are examined as anything other than potential confounds to be controlled statistically. Yet, postnatal experiences have the potential to either ameliorate or exacerbate prenatal effects. In addition, alterations in infant functioning related to prenatal experiences may result in differential sensitivity to variations in early care experiences and/or behaviors that elicit different responses from caregivers. As noted, some researchers are beginning to address how the postnatal care interacts with prenatal stress exposure to influence cognitive, behavioral, and health outcomes. More of this work is needed.

From a fetal programming perspective, it is especially critical that conceptual models are examined via longitudinal studies that track postnatal development. Thus, if as some models posit, fetal programming via stress mediators prepares the fetus to survive in a harsh postnatal world (Gluckman and Hanson, 2004), then evidence is needed to support such a hypothesis. There is some evidence that this may be the case with regard to nutrition such that concordance in pre- and postnatal nutrition leads to more functional health outcomes than discordance (Armitage et al., 2005; Cleal et al., 2007). However, it is not known whether a similar adaptive advantage exists for other types of prenatal stressors and/or for other kinds of postnatal outcomes.

Consistent with early literature on outcomes for premature infants, it is also possible that a supportive postnatal environment may ameliorate and a harsh environment exacerbates the neurobehavioral sequelae of prenatal stress. In this case, discordance in the harshness of pre- and postnatal experience may predict better outcomes. Recent human studies have provided support for this possibility demonstrating that high-quality maternal care can compensate for the negative effects of prenatal stress exposures (Bergman et al., 2008, 2010). It remains likely that the effects of stress during the prenatal and postnatal periods will differ by developmental outcome and there is a clear need for prospective studies with multiple prenatal and postnatal assessments.

As noted earlier, there is an emerging literature that suggests that prenatal stress may operate on circuits that support enhanced behavioral inhibition or anxiety. If so, then an intriguing possibility is that prenatal stress may enhance the young child's sensitivity to its rearing context, resulting in more optimal outcomes under supportive conditions and poorer outcomes under harsh conditions than may be noted for infants from nonstressed pregnancies. Whether this is the case has not been explored, but is an example of the need to integrate pre- and postnatal conceptual models.

Thus far, only these future direction comments on the value of incorporating research and theory on postnatal stress into models and research on prenatal stress have been focused upon. Perhaps a more critical need is for those studying postnatal stress to consider the role that prenatal stress may be playing in their findings. Certainly it seems likely that children who are abused, neglected, or abandoned to orphanage care may be the product of stressed pregnancies. It also is likely that they are the products of pregnancies complicated by poor nutrition and exposure to alcohol and drugs. Unfortunately, in many studies of postnatal stress there is meager information about prenatal conditions or even birth outcomes. Retrospective reports obtained from parents in studies, for example, of child maltreatment must be suspect. For children abandoned to orphanage care, often even the child's age at abandonment is unknown, let alone their gestational age and health at birth and any record of prenatal conditions. Nonetheless, despite the challenge of obtaining accurate information on prenatal conditions for children identified because of their poor postnatal care, studies are needed where such information can and has been obtained. This is particularly important in studies examining interventions to improve outcomes as prenatal conditions may moderate how the child responds.

Finally, one area that would seem ripe for integration into research on pre- and postnatal stress is the role that the dramatic hormonal changes that accompany pregnancy contribute to the quality of postnatal caregiving. Stress and reproductive hormones during pregnancy are associated with maternal cognitive functioning (Glynn, 2010), the development of postpartum depression (Yim et al., 2009), and the quality of maternal care (Feldman et al., 2007). The maternal hormonal milieu also impacts the developing fetus, as was discussed. Thus, to close the loop on the understanding of the ways in which stress impacts the developing fetus and young child, studies that incorporate the impacts of stress and maternal hormonal changes on the mother, her caregiving, and her response to her infant are needed.

SEE ALSO

Cognitive Development: The Neural Correlates of Cognitive Control and the Development of Social Behavior.

References

Aarnoudse-Moens, C.S., Weisglas-Kuperus, N., van Goudoever, J.B., Oosterlaan, J., 2009. Meta-analysis of neurobehavioral outcomes in very preterm and/or very low birth weight children. Pediatrics 124 (2), 717–728.

Ahnert, L., Gunnar, M.R., Lamb, M.E., Barthel, M., 2004. Transition to child care: Associations with infant–mother attachment infant negative emotion and cortisol elevations. Child Development 75, 639–650.

Armitage, J.A., Taylor, P.D., Poston, L., 2005. Experimental models of developmental programming: Consequences of exposure to an energy rich diet during development. The Journal of Physiology 565 (Pt 1), 3–8.

Barker, D.J., 1998a. In utero programming of chronic disease. Clinical Science 95 (2), 115–128.

Barker, D.J.P., 1998b. Mothers, Babies and Health in Later Life. Churchill Livingstone, Edinburgh.

Belsky, J., Pluess, M., 2009. Beyond diathesis stress: Differential susceptibility to environmental influences. Psychological Bulletin 135, 885–908.

Bergman, K., Sarkar, P., O'Connor, T.G., Modi, N., Glover, V., 2007. Maternal stress during pregnancy predicts cognitive ability and fearfulness in infancy. Journal of the American Academy of Child and Adolescent Psychiatry 46 (11), 1454–1463.

Bergman, K., Sarkar, P., Glover, V., O'Connor, T.G., 2008. Quality of child–parent attachment moderates the impact of antenatal stress on child fearfulness. Journal of Child Psychology and Psychiatry 49 (10), 1089–1098.

Bergman, K., Sarkar, P., Glover, V., O'Connor, T.G., 2010. Maternal prenatal cortisol and infant cognitive development: Moderation by infant–mother attachment. Biological Psychiatry 67, 1026–1032.

Bernardes, J., Goncalves, H., Ayres-de-Campos, D., Rocha, A.P., 2008. Linear and complex heart rate dynamics vary with sex in relation to fetal behavioural states. Early Human Development 84 (7), 433–439.

Blair, C., Granger, D., Willoughby, M., Kivlighan, K., 2006. Maternal sensitivity is related to hypothalamic-pituitary-adrenal axis stress reactivity and regulation in response to emotion challenge in 6-month-old infants. Annals of the New York Academy of Sciences 1094, 263–267.

Blair, M.M., Glynn, L.M., Sandman, C.A., Davis, E.P., 2011. Prenatal maternal anxiety and early childhood temperament. Stress 14 (6), 644–651.

Bourgeois, J.P., 1997. Synaptogenesis, heterochrony and epigenesis in the mammalian neocortex. Acta Paediatrica. Supplement 422, 27–33.

Boyce, W.T., Ellis, B.J., 2005. Biological sensitivity to context: I. An evolutionary-developmental theory of the origins and functions of stress reactivity. Devel Psychopathology 17, 271–301.

Brown, D.W., Anda, R.F., Tiemeier, H., et al., 2009. Adverse childhood experiences and the risk of premature mortality. American Journal of Preventive Medicine 37, 389–396.

Bruce, J., Fisher, P.A., Pears, K.C., Levine, S., 2009. Morning cortisol levels in preschool-aged foster children: Differential effects of maltreatment type. Developmental Psychobiology 51, 14–23.

Buss, C., Davis, E.P., Class, Q.A., et al., 2009. Maturation of the human fetal startle response: Evidence for sex-specific maturation of the human fetus. Early Human Development 85 (10), 633–638.

Buss, C., Davis, E.P., Muftuler, L.T., Head, K., Sandman, C.A., 2010. High pregnancy anxiety during mid-gestation is associated with decreased gray matter density in 6–9-year-old children. Psychoneuroendocrinology 35 (1), 141–453.

Buss, C., Davis, E.P., Hobel, C.J., Sandman, C.A., 2011. Maternal pregnancy-specific anxiety is associated with child executive function at 6–9 years age. Stress 14 (6), 665–676. http://dx.doi.org/10.3109/10253890.2011.623250.

Carlson, M., Earls, F., 1997. Psychological and neuroendocrinological sequelae of early social deprivation in institutionalized children in Romania. Annals of the New York Academy of Sciences 807, 419–428.

Carpenter, L.L., Carvalho, J.P., Tyrka, A.R., et al., 2007. Decreased adrenocorticotropic hormone and cortisol responses to stress in healthy adults reporting significant childhood maltreatment. Biological Psychiatry 62, 1080–1087.

Catalani, A., Casolini, P., Scaccianoce, S., Patacchioli, F.R., Spinozzi, P., Angelucci, L., 2000. Maternal corticosterone during lactation permanently affects brain corticosteroid receptors, stress response and behaviour in rat progeny. Neuroscience 100, 319–325.

Champagne, F.A., Curley, J.P., 2009. Epigenetic mechanisms mediating the long-term effects of maternal care on development. Neuroscience and Biobehavioral Reviews 33 (4), 593–600.

Chronis-Tuscano, A., Degnan, K.A., Pine, D.S., et al., 2009. Stable early maternal report of behavioral inhibition predicts lifetime social anxiety disorder in adolescence. Journal of the American Academy of Child and Adolescent Psychiatry 48, 928–935.

Cicchetti, D., Rogosch, F.A., 1996. Equifinality and multifinality in developmental psychopathology. Development and Psychopathology 8, 597–600.

Cicchetti, D., Rogosch, F.A., Gunnar, M.R., Toth, L., 2010. The differential impacts of early abuse on internalizing problems and diurnal cortisol activity in school-aged children. Child Development 81 (1), 252–269.

Class, Q.A., Buss, C., Davis, E.P., et al., 2008. Low levels of corticotropin-releasing hormone during early pregnancy are associated with precocious maturation of the human fetus. Developmental Neuroscience 30 (6), 419–426.

Cleal, J.K., Poore, K.R., Boullin, J.P., et al., 2007. Mismatched pre- and postnatal nutrition leads to cardiovascular dysfunction and altered renal function in adulthood. Proceedings of the National Academy of Sciences of the United States of America 104 (22), 9529–9533.

Clifton, V.L., 2010. Review: Sex and the human placenta: Mediating differential strategies of fetal growth and survival. Placenta 31 (supplement), S33–S39.

Costello, E.J., Worthman, C., Erkanli, A., Angold, A., 2007. Prediction from low birth weight to female adolescent depression: A test of competing hypotheses. Archives of General Psychiatry 64 (3), 338–344.

Davis, E.P., Sandman, C.A., 2010. The timing of prenatal exposure to maternal cortisol and psychosocial stress is associated with human infant cognitive development. Child Development 81 (1), 131–148.

Davis, E.P., Sandman, C.A., 2012. Prenatal psychobiological predictors of anxiety risk in preadolescent children. Psychoneuroendocrinology http://dx.doi.org/10.1016/j.psyneuen.2011.12.016.

Davis, E.P., Waffam, F., Sandman, C.A., 2011a. Prenatal treatment with glucocorticoids sensitizes the hpa axis response to stress among full-term infants. Developmental Psychobiology 53 (2), 175–183.

Davis, E.P., Glynn, L.M., Waffarn, F., Sandman, C.A., 2011b. Prenatal maternal stress programs infant stress regulation. Journal of Child Psychology and Psychiatry 52 (2), 119–129.

Davis, E.P., Snidman, N., Wadhwa, P.D., Dunkel Schetter, C., Glynn, L., Sandman, C.A., 2004. Prenatal maternal anxiety and depression predict negative behavioral reactivity in infancy. Infancy 6 (3), 319–331.

Davis, E.P., Glynn, L.M., Dunkel Schetter, C., Hobel, C., Chicz-De Met, A., Sandman, C.A., 2005. Maternal plasma corticotropin-releasing hormone levels during pregnancy are associated with infant temperament. Developmental Neuroscience 27 (5), 299–305.

Davis, E.P., Townsend, E.L., Gunnar, M.R., et al., 2006. Antenatal betamethasone treatment has a persisting influence on infant HPA axis regulation. Journal of Perinatology 26 (3), 147–153.

Davis, E.P., Glynn, L.M., Dunkel Schetter, C., Hobel, C., Chicz-DeMet, A., Sandman, C.A., 2007. Prenatal exposure to maternal depression and cortisol influences infant temperament. Journal of the American Academy of Child and Adolescent Psychiatry 46 (6), 737–746.

de Weerth, C., Buitelaar, J.K., 2005. Physiological stress reactivity in human pregnancy - a review. Neuroscience & Biobehavioral Reviews 29 (2), 295–312.

de Weerth, C., van Hees, Y., Buitelaar, J., 2003. Prenatal maternal cortisol levels and infant behavior during the first 5 months. Early Human Development 74, 139–151.

Denver, R.J., 1997. Environmental stress as a developmental cue: Corticotropin-releasing hormone is a proximate mediator of adaptive phenotypic plasticity in amphibian metamorphosis. Hormones and Behavior 31, 169–179.

Derijk, R.H., de Kloet, E.R., 2008. Corticosteroid receptor polymorphisms: Determinants of vulnerability and resilience. European Journal of Pharmacology 583, 303–311.

Dickerson, S.S., Kemeny, M.E., 2004. Acute stressors and cortisol responses: A theoretical integration and synthesis of laboratory research. Psychological Bulletin 130, 355–391.

DiPietro, J.A., 2004. The role of prenatal maternal stress in child development. Current Directions in Psychological Science 13 (2), 71–74.

DiPietro, J.A., Costigan, K.A., Shupe, A.K., Pressman, E.K., Johnson, T.R., 1998. Fetal neurobehavioral development: Associations with socioeconomic class and fetal sex. Developmental Psychobiology 33 (1), 79–91.

DiPietro, J.A., Novak, M.F.S.X., Costigan, K.A., Atela, L.D., Ruesing, S.P., 2006. Maternal psychological distress during pregnancy in relation to child development at age two. Child Development 77 (3), 573–587.

Dunkel Schetter, C., Glynn, L., 2011. Stress in pregnancy: Empirical evidence and theoretical issues to guide interdisciplinary research. In: Contrada, R., Baum, A. (Eds.), Handbook of Stress Science. Springer Publishing Company, LLC., New York, NY.

Ellman, L.M., Dunkel-Schetter, C., Hobel, C.J., Chicz-DeMet, A., Glynn, L.M., Sandman, C.A., 2008. Timing of fetal exposure to stress hormones: Effects on newborn physical and neuromuscular maturation. Developmental Psychobiology 50, 232–241.

Essex, M.J., Klein, M., Cho, E., Kalin, N.H., 2002. Maternal stress beginning in infancy may sensitize children to later stress exposure: Effects on cortisol and behavior. Biological Psychiatry 52, 776–784.

Feldman, R., Weller, A., Zagoory-Sharon, O., Levine, A., 2007. Evidence for a neuroendocrinological foundation of human affiliation: plasma oxytocin levels across pregnancy and the postpartum period predict mother-infant bonding. Psychol Sci 18 (11), 965–970.

Fernald, L.C., Gunnar, M.R., 2009. Poverty-alleviation program participation and salivary cortisol in very low-income children. Social Science & Medicine 68, 2180–2189.

Fisher, P.A., Stoolmiller, M., 2008. Intervention effects on foster parent stress: Associations with child cortisol levels. Development and Psychopathology 20, 1003–1021.

Forest, M.G., 1978. Age-related response of plasma testosterone D4-androstenedione and cortisol to adrenocorticotropin in infants children and adults. Journal of Clinical Endocrinology and Metabolism 47, 931–937.

French, N., Hagan, R., Evans, S.F., Godfrey, M., Newnham, J., 1999. Repeated antenatal corticosteroids: Size at birth and subsequent development. American Journal of Obstetrics and Gynecology 180 (114), 121.

Frigerio, A., Ceppi, E., Rusconi, M., Giorda, R., Raggi, M.E., Fearon, P., 2009. The role played by the interaction between genetic factors and attachment in the stress response in infancy. Journal of Child Psychology and Psychiatry 50, 1513–1522.

Gillespie, C.F., Phifer, J., Bradley, B., Ressler, K.J., 2009. Risk and resilience: Genetic and environmental influences on development of the stress response. Depression and Anxiety 26, 984–989.

Gitau, R., Cameron, A., Fisk, N.M., Glover, V., 1998. Fetal exposure to maternal cortisol. Lancet 352 (9129), 707–708.

Gitau, R., Fisk, N.M., Teixeira, J.M., Cameron, A., Glover, V., 2001. Fetal hypothalamic pituitary-adrenal stress responses to invasive procedures are independent of maternal responses. Journal of Clinical Endocrinology and Metabolism 86 (1), 104–109.

Gluckman, P.D., Hanson, M.A., 2004. Living with the past: Evolution, development, and patterns of disease. Science 305 (5691), 1733–1736.

Glynn, L.M., 2010. Giving birth to a new brain: Hormone exposures of pregnancy influence human memory. Psychoneuroendocrinology 35, 1148–1155.

Glynn, L.M., Dunkel Schetter, C., Hobel, C., Sandman, C.A., 2008. Pattern of perceived stress and anxiety in pregnancy predict preterm birth. Health Psychology 27 (1), 42–51.

Goel, N., Bale, T.L., 2009. Examining the intersection of sex and stress in modelling neuropsychiatric disorders. Journal of Neuroendocrinology 21 (4), 415–420.

Gouin, J.P., Carter, C.S., Pournajafi-Nazarloo, H., et al., 2010. Marital behavior oxytocin vasopressin and wound healing. Psychoneuroendocrinology 35 (7), 1082–1090.

Grant, K.A., McMahon, C., Austin, M.P., Reilly, N., Leader, L., Ali, S., 2009. Maternal prenatal anxiety, postnatal caregiving and infants' cortisol responses to the still-face procedure. Developmental Psychobiology 51 (8), 625–637.

Gunnar, M.R., 2001. The role of glucocorticoids in anxiety disorders: A critical analysis. In: Vasey, M.W., Dadds, M. (Eds.), The Developmental Psychopathology of Anxiety. Oxford University Press, New York, pp. 143–159.

Gunnar, M.R., Donzella, B., 2002. Social regulation of the cortisol levels in early human development. Psychoneuroendocrinology 27, 199–220.

Gunnar, M.R., Vazquez, D., 2006. Stress neurobiology and developmental psychopathology. In: 2nd edn In: Cicchetti, D., Cohen, D. (Eds.), Developmental Psychopathology: Developmental Neuroscience, vol. 2. Wiley, New York, pp. 533–577.

Gunnar, M.R., Frenn, K., Wewerka, S., Van Ryzin, M.J., 2009. Moderate versus severe early life stress: Associations with stress reactivity and regulation in 10- to 12-year old children. Psychoneuroendocrinology 34, 62–75.

Gutteling, B.M., De Weerth, C., Buitelaar, J.K., 2004. Maternal prenatal stress and 4–6 year old children's salivary cortisol concentrations pre- and post-vaccination. Stress 7 (4), 257–260.

Gutteling, B.M., de Weerth, C., Buitelaar, J.K., 2005. Prenatal stress and children's cortisol reaction to the first day of school. Psychoneuroendocrinology 30 (6), 541–549.

Halligan, S.L., Herbert, J., Goodyer, I., Murray, L., 2007. Disturbances in morning cortisol excretion in association with maternal postnatal depression predict subsequent depressive symptomatology in adolescents. Biological Psychiatry 62, 40–46.

Harville, E.W., Savitz, D.A., Dole, N., Herring, A.H., Thorp, J.M., 2009. Stress questionnaires and stress biomarkers during pregnancy. Journal of Women's Health (2002) 18 (9), 1425–1433.

Heim, C., Newport, D.J., Mletzko, T., Miller, A.H., Nemeroff, C.B., 2008. The link between childhood trauma and depression: Insights from HPA axis studies in humans. Psychoneuroendocrinology 33, 693–710.

Heinrichs, M., Baumgartner, T., Kirschbaum, C., Ehlert, U., 2003. Social support and oxytocin interact to suppress cortisol and subjective responses to psychosocial stress. Biological Psychiatry 54, 1389–1398.

Ho, J.T., Lewis, J.G., O'Loughlin, P., et al., 2007. Reduced maternal corticosteroid-binding globulin and cortisol levels in pre-eclampsia and gamete recipient pregnancies. Clinical Endocrinology 66 (6), 869–877.

Hobel, C., Arora, C.P., Korst, L.M., 1999. Corticotrophin-releasing hormone and CRH binding protein. Differences between patients at risk for preterm birth and hypertension. Annals of the New York Academy of Sciences 897, 54–65.

Ianniruberto, A., Tajani, E., 1981. Ultrasonographic study of fetal movements. Seminars in Perinatology 5 (2), 175–181.

Jaffe, R.B., Mesiano, S., Smith, R., Coulter, C.L., Spencer, S.J., Chakravorty, A., 1998. The regulation and role of fetal adrenal development in human pregnancy. Endocrine Research 24 (34), 919–926.

Joels, M., Baram, T.Z., 2009. The neuro-symphony of stress. Nature Reviews Neuroscience 10 (6), 459–466.

Kagan, J., Reznick, J.S., Snidman, N., 1987. The physiology and psychology of behavioral inhibition in children. Child Development 58, 1459–1473.

Kajantie, E., Dunkel, L., Turpeinen, U., et al., 2003. Placental 11 beta-hydroxysteroid dehydrogenase-2 and fetal cortisol/cortisone shuttle in small preterm infants. Journal of Clinical Endocrinology and Metabolism 88 (1), 493–500.

Kapoor, A., Dunn, E., Kostaki, A., Andrews, M.H., Matthews, S.G., 2006. Fetal programming of the hypothalamic-pituitary-adrenal function: Prenatal stress and glucocorticoids. The Journal of Physiology 572, 31–44.

Kempna, P., Fluck, C.E., 2008. Adrenal gland development and defects. Best Practice & Research. Clinical Endocrinology & Metabolism 22 (1), 77–93.

Kertes, D.A., Gunnar, M.R., Madsen, N.J., Long, J., 2008. Early deprivation and home basal cortisol levels: A study of internationally-adopted children. Development and Psychopathology 20, 473–491.

Kinney, H.C., 2006. The near-term (late preterm) human brain and risk for periventricular leukomalacia: A review. Seminars in Perinatology 30 (2), 81–88.

Kinsella, M.T., Monk, C., 2009. Impact of maternal stress, depression and anxiety on fetal neurobehavioral development. Clinical Obstetrics and Gynecology 52 (3), 425–440.

Korosi, A., Baram, T.Z., 2008. The central corticotropin releasing factor system during development and adulthood. European Journal of Pharmacology 583, 204–214.

Kuzawa, C., 2005. The origins of the developmental origins hypothesis and the role of postnatal environments: Response to Koletzko. American Journal of Human Biology 17 (5), 662–664.

LaPlante, D.P., Barr, R.G., Brunet, A., et al., 2004. Stress during pregnancy affects general intellectual and language functioning in human toddlers. Pediatric Research 56 (3), 400–410.

Levine, S., 1957. Infantile experience and resistance to physiological stress. Science 126, 405–406.

MacMillan, H.L., Georgiades, K., Duku, E.K., et al., 2009. Cortisol response to stress in female youths exposed to childhood maltreatment: Results of the youth mood project. Biological Psychiatry 66, 62–68.

Mairesse, J., Lesage, J., Breton, C., et al., 2007. Maternal stress alters endocrine function of the feto-placental unit in rats. American Journal of Physiology, Endocrinology and Metabolism 292 (6), E1526–E1533.

McLean, M., Bisits, A., Davies, J., Woods, R., Lowry, P., Smith, R., 1995. A placental clock controlling the length of human pregnancy. Nature Medicine 1, 460–463.

Murphy, V.E., Clifton, V.L., 2003. Alterations in human placental 11beta-hydroxysteroid dehydrogenase type 1 and 2 with gestational age and labour. Placenta 24 (7), 739–744.

Nathanielsz, P.W., 1999. Life in the Womb: The Origin of Health and Disease. Promethean Press, Ithaca, NY.

Netherton, C., Goodyer, I., Tamplin, A., Herbert, J., 2004. Salivary cortisol and dehydroepiandrosterone in relation to puberty and gender. Psychoneuroendocrinology 29, 125–140.

Noorlander, C.W., De Graan, P.N., Middeldorp, J., Van Beers, J.J., Visser, G.H., 2006. Ontogeny of hippocampal corticosteroid receptors: Effects of antenatal glucocorticoids in human and mouse. The Journal of Comparative Neurology 499 (6), 924–932.

O'Connor, T.G., Heron, J., Glover, V., 2002. Antenatal anxiety predicts child behavioral/emotional problems independently of postnatal depression. Journal of the American Academy of Child and Adolescent Psychiatry 41 (12), 1470–1477.

O'Connor, T.G., Ben-Shlomo, Y., Heron, J., Golding, J., Adams, D., Glover, V., 2005. Prenatal anxiety predicts individual differences in cortisol in pre-adolescent children. Biological Psychiatry 58, 211–217.

O'Donnell, K.J., Bugge Jensen, A., Freeman, L., Khalife, N., O'Connor, T.G., Glover, V., 2011. Maternal prenatal anxiety and downregulation of placental 11beta-HSD2. Psychoneuroendocrinology http://dx.doi.org/10.1016/j.psyneuen.2011.09.014.

Oberlander, T.F., Weinberg, J., Papsdorf, M., Grunau, R., Misri, S., Devlin, A.M., 2008. Prenatal exposure to maternal depression, neonatal methylation of human glucocorticoid receptor gene (NR3C1) and infant cortisol stress responses. Epigenetics 3 (2), 97–106.

Phillips, D.A., Fox, N.A., Gunnar, M.R., 2011. Same place different experiences: Bringing individual differences to research in child care. Child Development Perspectives 5 (1), 44–49.

Phillips, D.I., Walker, B.R., Reynolds, R.M., et al., 2000. Low birth weight predicts elevated plasma cortisol concentrations in adults from 3 populations. Hypertension 35, 1301–1306.

Power, M.L., Bowman, M.E., Smith, R., et al., 2006. Pattern of maternal serum corticotropin-releasing hormone concentration during pregnancy in the common marmoset (*Callithrix jacchus*). American Journal of Primatology 68 (2), 181–188.

Pryce, C.R., 2008. Postnatal ontogeny of expression of the corticosteroid receptor genes in mammalian brains: Inter-species and intra-species differences. Brain Research Reviews 57, 596–605.

Rice, F., Harold, G.T., Boivin, J., van den Bree, M., Hay, D.F., Thapar, A., 2009. The links between prenatal stress and offspring development and psychopathology: Disentangling environmental and inherited influences. Psychological Medicine 1–11.

Rosen, J.B., Schulkin, J., 1998. From normal fear to pathological anxiety. Psychological Review 105, 325–350.

Sanchez, M.M., Ladd, C.O., Plotsky, P.M., 2001. Early adverse experience as a developmental risk factor for later psychopathology: Evidence from rodent and primate models. Development and Psychopathology 13, 419–450.

Sandman, C.A., Davis, E.P., 2010. Gestational stress influences cognition and behavior. Future Neurology 5 (5), 675–690.

Sandman, C.A., Davis, E.P., 2012. Neurobehavioural risk is associated with gestational exposure to stress hormones. Expert Review of Endocrinology and Metabolism 7 (4), 445–459.

Sandman, C.A., Wadhwa, P.D., Chicz-DeMet, A., Garite, T.J., Porto, M., 1999. Maternal corticotropin-releasing hormone and habituation in the human fetus. Developmental Psychobiology 34 (3), 163–173.

Sandman, C.A., Glynn, L., Wadhwa, P.D., Chicz-DeMet, A., Porto, M., Garite, T., 2003. Maternal hypothalamic-pituitary-adrenal dysregulation during the third trimester influences human fetal responses. Developmental Neuroscience 25 (1), 41–49.

Sandman, C.A., Glynn, L.M., Dunkel Schetter, C., Wadwha, P.D., Garite, T., Hobel, C., 2006. Elevated maternal cortisol early in pregnancy predicts third trimester levels of placental corticotropin releasing hormone (CRH): Priming the placental clock. Peptides 27, 1457–1463.

Sapolsky, R.M., Romero, L.M., Munck, A.U., 2000. How do glucocorticoids influence stress responses? Integrating permissive suppressive stimulatory and preparative actions. Endocrine Reviews 21, 55–89.

Schmidt, L.A., Fox, N.A., Rubin, K.H., et al., 1997. Behavioral and neuroendocrine responses in shy children. Developmental Psychobiology 30, 127–140.

Seckl, J.R., 2008. Glucocorticoids, developmental 'programming' and the risk of affective dysfunction. Progress in Brain Research 167, 17–34.

Shonkoff, J.P., Boyce, W.T., McEwen, B.S., 2009. Neuroscience molecular biology and the childhood roots of health disparities: Building a new framework for health promotion and disease prevention. Journal of the American Medical Association 301, 2252–2259.

Sidman, R.L., Rakic, P., 1973. Neuronal migration, with special reference to developing human brain: A review. Brain Research 62 (1), 1–35.

Smider, N.A., Essex, M.J., Kalin, N.H., et al., 2002. Salivary cortisol as a predictor of socioemotional adjustment during kindergarten: A prospective study. Child Development 73, 75–92.

Spangler, G., Schieche, M., Ilg, U., Maier, U., Ackermann, C., 1994. Maternal sensitivity as an external organizer for biobehavioral regulation in infancy. Developmental Psychobiology 27, 425–437.

Spear, L.P., 2000. The adolescent brain and age-related behavioral manifestations. Neuroscience and Biobehavioral Reviews 24, 417–463.

Stroud, L., Foster, E., Handwerger, K., et al., 2009. Stress response and the adolescent transition: Performance versus peer rejection stress. Development and Psychopathology 21, 47–68.

Szyf, M., 2011. The early life social environment and DNA methylation; DNA methylation mediating the long-term impact of social environments early in life. Epigenetics 6 (8).

Tarullo, A.R., Gunnar, M.R., 2006. Child maltreatment and the developing HPA axis. Hormones and Behavior 50, 632–639.

Trickett, P.K., Noll, J.G., Susman, E.J., Shenk, C.E., Putnam, F.W., 2010. Attenuation of cortisol across development for victims of sexual abuse. Development and Psychopathology 22, 165–175.

Ulrich-Lai, Y.M., Herman, J.P., 2009. Neural regulation of endocrine and autonomicstress responses. Nature Reviews Neuroscience 10, 397–409.

Van den Bergh, B.R.H., Marcoen, A., 2004. High antenatal maternal anxiety is related to ADHD symptoms, externalizing problems, and anxiety in 8/9-year-olds. Child Development 75 (4), 1085–1097.

Van den Bergh, B.R.H., Van Calster, B., Smits, T., Van Huffel, S., Lagae, L., 2007. Antenatal maternal anxiety is related to HPA-axis dysregulation and self-reported depressive symptoms in adolescence: A prospective study on the fetal origins of depressed mood. Neuropsychopharmacology 1–10.

Vermeer, H.J., van IJzendoorn, M.H., 2006. Children's elevated cortisol levels at daycare: A review and meta-analysis. Early Childhood Research Quarterly 21, 390–401.

Volpe, J.J., 2008. Neurology of the Newborn, 5th edn. Elsevier, Philadelphia, PA.

Wismer Fries, A.B., Shirtcliff, E.A., Pollak, S.D., 2008. Neuroendocrine dysregulation following early social deprivation in children. Developmental Psychobiology 50, 588–599.

Yehuda, R., 2000. Biology of posttraumatic stress disorder. The Journal of Clinical Psychiatry 61 (supplement 7), 15–21.

Yehuda, R., Engel, S.M., Brand, S.R., Seckle, J., Marcus, S.M., Berkowitz, G.S., 2005. Transgenerational effects of posttraumatic stress disorder in babies of mothers exposed to the World Trade Center attacks during pregnancy. Journal of Clinical Endocrinology and Metabolism 90 (7), 4115–4118.

Yim, I.S., Glynn, L.M., Dunkel-Schetter, C., Hobel, C.J., Chicz-DeMet, A., Sandman, C.A., 2009. Risk of postpartum depressive symptoms with elevated corticotropin-releasing hormone in human pregnancy. Archives of General Psychiatry 66 (2), 162–169.

Sex Differences in Brain and Behavioral Development

A.M. Beltz[1], J.E.O. Blakemore[2], S.A. Berenbaum[1]

[1]The Pennsylvania State University, University Park, PA, USA; [2]Indiana University – Purdue University Fort Wayne, Fort Wayne, IN, USA

O U T L I N E

Neural Circuit Development and Function in the Brain: Comprehensive Developmental Neuroscience, Volume 3 http://dx.doi.org/10.1016/B978-0-12-397267-5.00064-9

© 2013 Elsevier Inc. All rights reserved.

26.1 INTRODUCTION

Psychological and neural sex differences are a topic of great interest for their relevance to both applied and basic science questions. From an applied perspective, they figure into discussions ranging from women's underrepresentation in science and math careers (Ceci and Williams, 2010; Halpern et al., 2007) to sex differences in incidences and forms of psychopathology (Hartung and Widiger, 1998; Zahn-Waxler et al., 2008). From a basic science perspective, sex represents an important dimension of individual difference, and factors that lead to differences between the sexes can help us to understand variation generally. The focus of this chapter is sex differences in the human brain and behavior, with an emphasis on cognitive abilities; our goal is to describe what is known and hypothesized about the differences, their development, and their etiology, and to begin to link brain differences to cognitive differences.

The chapter is organized into several sections. First, we summarize the evidence concerning psychological sex differences in human beings and the ways in which they develop across age, focusing on changes from infancy through young adulthood. Second, we consider the main theories that have been offered to explain the differences and the evidence that supports those perspectives. Third, we summarize the evidence concerning sex differences in the human brain and what (little) is known about their development. Fourth, we review the small body of research about the etiology and psychological correlates of brain sex differences. Fifth, we emphasize the gaps in our knowledge and provide suggestions for further research.

26.2 PSYCHOLOGICAL SEX DIFFERENCES: NATURE AND DEVELOPMENT

Sex differences are found in many psychological domains, from sensory thresholds to career choices. We focus here on cognitive sex differences but note other significant psychological domains in which the sexes differ. Our goal is not to provide an exhaustive review of the extensive literature in this area, but to provide a summary of the findings, highlighting the key differences and, in later sections, explanations of these differences, as well as the links between cognitive and brain sex differences. Detailed reviews, with supporting references, are available elsewhere, as cited throughout the chapter. Consistent with others, we describe sex differences in terms of *effect size*, d (Cohen, 1988): small ($d \sim 0.2$, 85% overlap in distributions of the sexes), moderate ($d \sim 0.5$, 67% overlap, probably noticeable), and large ($d \sim 0.8$, 53% overlap, very noticeable).

As a prelude to a description of the differences, we note that there is some controversy over the size of these differences, whether they have declined over time, and whether they even exist (Hyde, 2005). Some of the controversy relates to what is a meaningful difference, with some arguing that sex differences are too small to be worthwhile. But sex differences are among the largest effects in psychology; compare the effects described below with typical effects in psychology and medicine described elsewhere (e.g., Meyer et al., 2001). This is illustrated in Figure 26.1. Furthermore, there is not always a perfect mapping between the size of a difference and its significance.

26.2.1 Cognitive Abilities

The sexes do not differ in overall intelligence (discussed in Camarata and Woodcock, 2006; Halpern, 2012; Roivainen, 2011), but, as detailed below, they do differ in their pattern of abilities: on average, boys and men are better than girls and women in spatial and some mathematical skills, whereas girls and women are better than boys and men in many verbal abilities, memory, and processing speed (Halpern, 2012). One analysis of the structure of intelligence (Johnson and Bouchard, 2007) concludes that it consists of three dimensions: a rotation–verbal dimension (at one end of the dimension, tasks measuring mental rotation and other spatial skills, and at the other end, verbal tasks); a focus–diffusion dimension (at one end, attention focused on one main stimulus in the environment, and at the other end, attention focused diffusely on several cues simultaneously); and memory. Typically, men were found

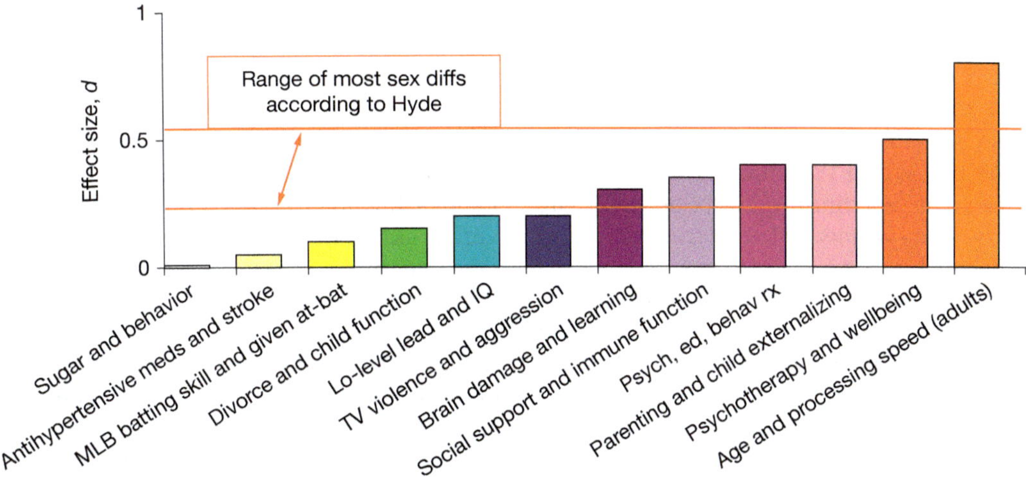

FIGURE 26.1   Psychological sex differences in behavior compared to other health and psychological effects, described in standard deviation units, *d. (data from Hyde, 2005; Meyer et al., 2001)*

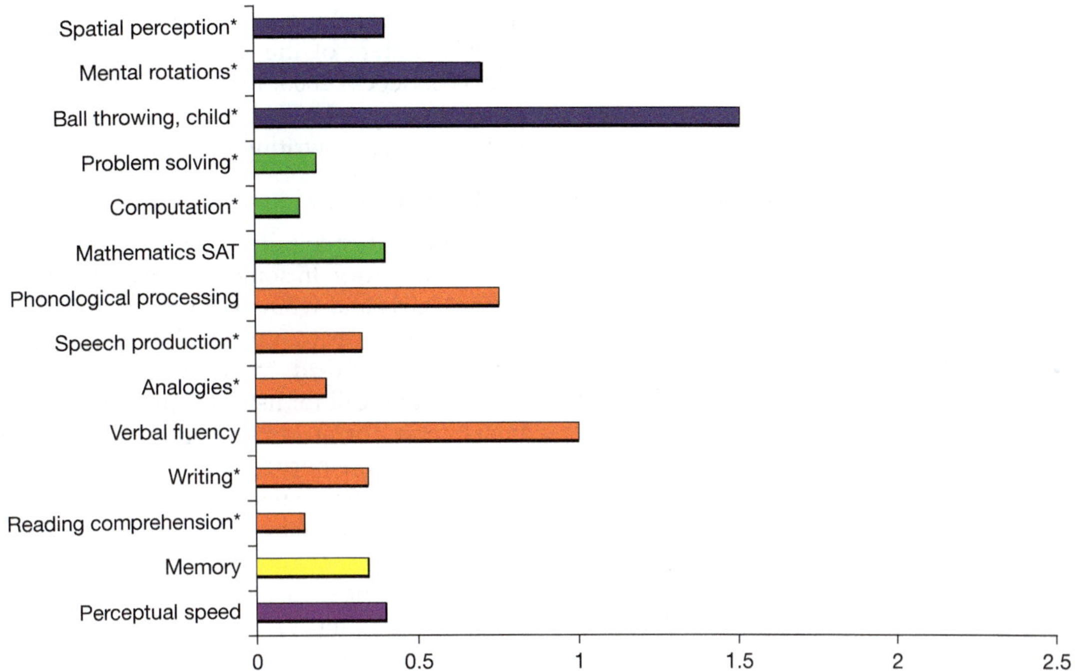

FIGURE 26.2   Sample sex differences in cognitive abilities: absolute value of the differences in standard deviation units, *d (data from Blakemore et al., 2009)*; asterisks (*) indicate results from meta-analyses.

to be nearer the rotation and focus ends of those two dimensions, and women at the verbal and diffusion ends, and women to have better memory. Figure 26.2 provides a summary of the cognitive sex differences expressed in standard deviation units, *d.*

26.2.1.1 *Spatial Abilities*

One widely studied cognitive sex difference is spatial ability. There are several ways to parse the domain, but boys and men outperform girls and women in most aspects of spatial ability, with the size of the difference

varying across abilities (Halpern, 2012; Lawton, 2010). The largest sex difference is in mental rotation, especially rotation of objects in three dimensions. This sex difference is apparent beginning in infancy (Moore and Johnson, 2008; Quinn and Liben, 2008), with boys and men having better ability than girls and women and with the difference increasing slightly from childhood to adulthood (Geiser et al., 2008).

There are moderately sized sex differences in spatial perception and the ability to identify spatial relations with respect to one's body in relation to external space or to identify the true vertical or horizontal (Halpern, 2012). This ability is measured by tasks such as the 'rod and frame task' (Voyer and Bryden, 1993) and 'water level task' (Vasta and Liben, 1996).

There are also sex differences in abilities related to navigating in the real world (Lawton, 2010), an ability typically referred to as 'wayfinding.' Boys and men are better than girls and women at remembering and navigating to distant locations in large spaces; some of the difference results from sex differences in strategy, with men relying on cardinal (north, south, east, and west) directions and women relying on landmarks. Considering larger issues of the spatial environment, there is a huge sex disparity among National Geography Bee winners (despite equal participation from boys and girls); the sex ratio increases at each level of competition, so that in many years, all ten finalists are boys (Liben, 1995).

There is one aspect of spatial ability on which the sex difference is reversed: memory for spatial location. Girls and women are better than boys and men in remembering the location of objects (Voyer et al., 2007).

26.2.1.2 Mathematical Abilities

The sexes differ in quantitative abilities, with the differences again varying by type of ability and age (Halpern, 2012). In school, girls get better grades than do boys in arithmetic and mathematics classes (as they do in all classes), but they do not perform as well on standardized tests of math knowledge and skills (e.g., the SAT). The main math sex difference is in problem-solving tasks, with boys outperforming girls, especially at older ages. Recent data show the sex difference to be small (Lindberg et al., 2010), although the size of the difference is larger in some countries than in others, in ways that are sometimes, but not always, related to indices of gender equality in those countries (Else-Quest et al., 2010; Guiso et al., 2008). There are no sex differences in understanding mathematical concepts, and girls outperform boys in computation, especially before puberty.

The sexes also differ in their variability in spatial and mathematical abilities (Hedges and Nowell, 1995). For example, for the past 20 years, about four times as many seventh-grade boys as girls have scored in the high ranges of the SAT-Mathematics (Wai et al., 2010); before that time, there were even higher ratios of boys to girls at the top end.

26.2.1.3 Verbal Abilities

There are many different kinds of verbal and language-related skills, such as vocabulary size, use of correct grammar, reading, doing anagrams, and following verbal instructions. The sexes are similar on some skills. When there are differences, they are generally in the direction of girls and women having better skills than boys and men, although the size of the difference varies with ability and age (Halpern, 2012).

Language learning occurs earlier in girls than in boys, with boys catching up by age 6 (Bornstein et al., 2004; Wallentin, 2009). Language disorders are more common in boys than in girls (Wallentin, 2009). In terms of specific verbal skills, females have a small to moderate advantage over males in several skills, most prominently reading comprehension, verbal fluency, and phonological processing, but males have a small edge in analogies (Halpern, 2012). In terms of verbal abilities that are particularly important educationally, there is great concern about boys' lag in reading performance (Chudowsky and Chudowsky, 2010). Boys also write less well than girls, a difference that is especially notable at the highest ability levels (Hedges and Nowell, 1995; Wai et al., 2010).

Females also process verbal materials more rapidly than males, including intelligence scale subtests, verbal fluency tests asking for many words to be generated quickly, letters of the alphabet, and digits (Camarata and Woodcock, 2006; Roivainen, 2011). Females' faster processing speed may be one contributor to their superior reading and writing abilities.

26.2.1.4 Memory

As with other cognitive domains, there are several aspects of memory. In some, the sexes do not differ. But in several, girls and women are better than boys and men: they show more accurate recall for learning facts or material that they read, more readily learn lists of words, have better recall for lists of common objects such as animals, food, furniture, and appliances, and have better recognition memory (Halpern, 2012; Johnson and Bouchard, 2007). Women have better verbal memory in part because they use more efficient clustering strategies.

Sex differences in memory are especially consistent for episodic memory, that is, memory of specific events and episodes. For example, women are better than men at recognizing faces that they have seen before. Many episodic memories are verbally based, but women have better episodic memories even when the tasks are not verbal (e.g., face recognition) except when the memory tasks are clearly spatial, in which case men do better (Herlitz et al., 2010). Women also recall the identities and locations of objects better than men, although it has been suggested that this might be due to women's better memories overall (Voyer et al., 2007). Sex differences in memory are generally small to moderate in size.

26.2.1.5 Perceptual Speed

In addition to faster processing of verbally based cognitive tasks, girls and women are faster than boys and men in another type of speeded task, called perceptual speed (Burns and Nettelbeck, 2005). Most intelligence tests have some subscales that measure this ability. In general, these tests involve the ability to perceive details and shift attention quickly, often while using fine motor skills such as finger movements (Halpern, 2012). The difference ranges from small to large, depending on the particular measure used.

26.2.2 Noncognitive Sex Differences

There are many other psychological domains in which the sexes differ, but we focus here on those that have received the most attention, that are relatively large in size, or that illustrate important points about studying sex differences (e.g., the importance of developmental status, measurement, and social context). Figure 26.3 provides a summary of the noncognitive sex differences expressed in standard deviation units, d.

26.2.2.1 Physical and Motor Skills

Sex differences in physical and motor skills vary by aspect and age (Blakemore et al., 2009). Boys are more active than girls (Eaton and Enns, 1986; Else-Quest et al., 2006), and activity level becomes increasingly sex-differentiated with age, with the differences largest in familiar, nonthreatening settings and when peers are present. There are few sex differences in early milestones of reaching, sitting, crawling, and walking, but differences in motor skills begin to appear in the second year (Mondschein et al., 2000). The earlier neurological development of girls compared to boys results in

early development of some abilities, such as eye–hand coordination. Overall, girls develop sooner than and are superior to boys with respect to fine motor skills, whereas boys have better gross motor skills and are stronger than girls (Largo et al., 2001a,b), which combine with boys' advantage in spatial skills to result in large sex differences in targeting (hitting a target with a ball) (Kimura, 1999). Sex differences in many (but not all) physical and motor abilities increase with age, although the patterns vary considerably depending on the particular skill (Dorfberger et al., 2009; Thomas and French, 1985).

26.2.2.2 Activity Interests

One of the largest and most important sex differences concerns interests (Blakemore et al., 2009; Ruble et al., 2006). In childhood, boys and girls prefer and engage with different toys (e.g., trucks and dolls) and participate in different activities (e.g., playing sports and playing dress-up). In adolescence, boys and girls continue to prefer and participate in different leisure activities (e.g., building things and dance), household chores (e.g., taking out the garbage and preparing food), and academic pursuits (e.g., math and language arts). In adulthood, men and women continue to prefer and participate in different activities, and are differentially represented in different occupations; for example, men prefer occupations that involve working with things, and women prefer occupations that involve working with people (Su et al., 2009). Although the sexes overlap in their interests and activity participation, the average differences

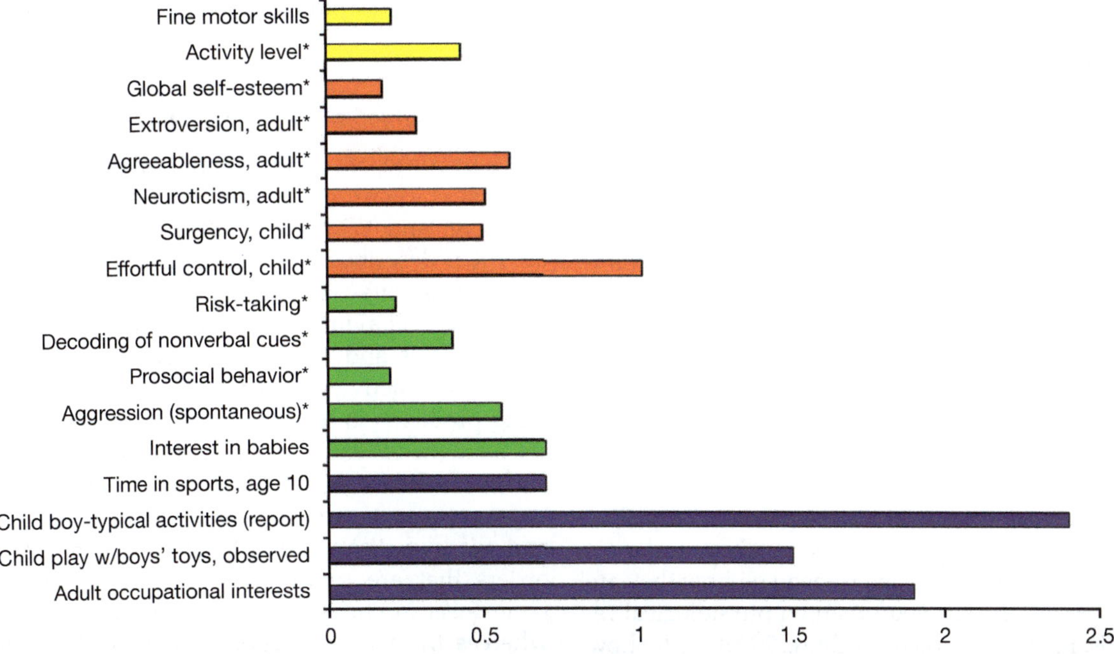

FIGURE 26.3 Sample sex differences in noncognitive characteristics (not including play or sex partners): absolute value of the differences in standard deviation units, d (data from Blakemore et al., 2009); asterisks (*) indicate results from meta-analyses.

are large to very large. Interests are important because they are reasonably stable, and shape future career choices, especially with respect to domains such as mathematics and the physical sciences (Diekman et al., 2010).

26.2.2.3 Temperament and Personality

Overall, most aspects of temperament show small or no sex differences (Else-Quest et al., 2006). The main exception is effortful control, which shows a moderate advantage for girls, and has been suggested to be consistent with the greater incidence of externalizing disorders in boys, which are characterized by inattentive, antisocial, and aggressive behaviors. In addition, boys have higher levels of surgency than girls, reflecting their greater activity and high-intensity pleasure.

Research on 'the big five' personality factors shows small differences between adult men and women. Costa and colleagues (2001) reported that women were higher than men in neuroticism, agreeableness, and extroversion. They also found cross-cultural similarities in the patterns of sex differences, but the differences between men and women were found to be largest in European and North American countries, and smallest in African and Asian countries.

There are also small sex differences favoring boys and men in global self-esteem, especially after childhood (Kling et al., 1999). Sex differences vary across the different domains of self-esteem, with males having consistently higher appearance-related, personal self, self-satisfaction, and athletic self-esteem, and females having higher ethical and behavioral conduct self-esteem (Gentile et al., 2009).

26.2.2.4 Social Behaviors

The stereotypes about sex differences in social behavior – that girls and women are emotionally expressive and perceptive, and more attuned to babies, and that boys and men take risks and are aggressive – have some basis in evidence, but the differences are not large and depend on context. Consider some examples (for details, see Blakemore et al., 2009). Regarding emotional expression, there is some suggestion that boys and men express more anger and less fear and sadness than girls, but more convincing is evidence that boys come to hide emotions like sadness and fear purposefully, especially in the presence of peers (Kyratzis, 2001). Regarding perceiving emotion in others, girls and women are somewhat more sympathetic, empathic, and accurate at decoding emotions than are boys and men, but the differences are small and depend on how they are measured (larger on self-report than on physiological response) (Eisenberg et al., 2006; McClure, 2000). Girls show more interest in and nurturant interaction with babies than boys do, although the size of the sex difference varies

across the particular situations measured in various studies (summarized in Blakemore et al., 2009). Regarding risk-taking, boys are more likely than girls to take risks and to be injured (Byrnes et al., 1999; Morrongiello and Matheis, 2007).

Aggression is one social behavior where sex differences are well known. Aggression is defined as behavior intended to harm others and includes physical, verbal, and indirect aggression. Males are more physically and verbally aggressive than females (Card et al., 2008; Dodge et al., 2006) from childhood through adulthood. Although the social context affects aggression, and the size of the difference varies across age and culture, there are no instances in which females are more directly aggressive than males.

Sometimes physical aggression moves into the domain of seriously antisocial or criminal behavior. Although few people show the type of high levels of physical aggression that would be called violent or antisocial, or at the very extreme, commit murder, the majority of those who do are male, in childhood, adolescence, and adulthood (Archer, 2004; Dodge et al., 2006; Moffitt et al., 2001; Rutter et al., 1998).

Another form of aggression is often called indirect, social, or relational aggression (e.g., Crick, 1995). This form of aggression involves such things as manipulating social relationships or purposefully excluding others. It has been seen as the 'feminine' form of aggression and often (but not always) reported to be more common in girls (see discussion in Blakemore et al., 2009). However, a recent meta-analysis (Card et al., 2008) found no significant differences between boys and girls in indirect aggression from childhood through adolescence.

26.2.2.5 Psychological Disorders

Many forms of serious behavioral problems and mental illness occur at different rates in the two sexes (Hartung and Widiger, 1998). Psychological problems are often described as being internalizing (e.g., anxiety and depression) or externalizing (e.g., conduct disorder, antisocial behavior, including, among other things, criminal acts and excessive aggression, and attention-deficit hyperactivity disorder). The incidence of internalizing problems is higher in girls and women, whereas the incidence of externalizing problems is higher in boys and men (Zahn-Waxler, 1993; Zahn-Waxler et al., 2008); boys are also more likely than girls to have learning and reading disabilities (Blakemore et al., 2009). Disorders that are predominant in males tend to have their onset in childhood (Hartung and Widiger, 1998), whereas the female preponderance of depression and eating disorders begins at puberty (Nolen-Hoeksema and Hilt, 2009).

26.3 EXPLANATIONS FOR PSYCHOLOGICAL SEX DIFFERENCES

Hypothesized causes of psychological sex differences tend to focus on either genetic and biological factors or social and cultural factors. We highlight the key theoretical perspectives focusing on proximal processes; we refer readers elsewhere for discussion and critiques of evolutionary explanations of cognitive and neural sex differences (e.g., Eagly and Wood, 1999; Geary, 1998; Hannagan, 2008; Wood and Eagly, 2002).

26.3.1 Socialization Perspectives

Most work on psychological sex differences comes from a socialization perspective, that is, that sex differences develop as an individual navigates, observes others, is socialized, and internalizes information about the social world. There are several types of socialization theories, differing in the extent to which they emphasize the role of basic social learning mechanisms, subtle socialization practices, social identity (as male or female), cognitive schemas derived from gender identity that guide behavior, and the role of sex-typed social roles and resulting stereotypes and expectancies.

26.3.1.1 Socialization of Noncognitive Sex Differences

Most studies of gender socialization have focused on social behaviors. This literature has been reviewed elsewhere (Blakemore et al., 2009; Martin and Ruble, 2010; Ruble et al., 2006), so we provide a brief summary. In essence, boys and girls are socialized differently in ways that affect a variety of psychological outcomes, with this gendered socialization coming from a variety of sources. Much of the focus has been on socialization by peers and parents, but there are powerful influences from other social forces, including other adults such as teachers, coaches, and clergy, and information received via the many forms of media.

Peers are a key enforcer of sex-typing. Children have strong preferences for interaction with members of their own sex, with these preferences maintained by children themselves and resistant to change by adults (Maccoby, 1998; Ruble et al., 2006). The more children play with others of the same sex, the more they engage in sex-typed activities and play styles (Martin and Fabes, 2001).

Parents also shape sex-typing, as seen in two examples. First, parents influence the career choices of their offspring in several ways: by differential encouragement of sex-appropriate activities, by the attitudes they espouse regarding what is appropriate for boys versus girls, and by the resources they provide to their children (e.g., paying for and providing computer-related materials and encouraging extracurricular involvement in math and science for sons, but not for daughters) (Simpkins et al., 2005). Second, parents socialize emotion differently in their sons and their daughters, through their conversations (e.g., Fivush, 1998; Fivush and Buckner, 2000): With daughters as compared to with sons, parents use more emotion-related words, elaborate their discussion of emotion more extensively, and focus these conversations on the emotional aspects of interpersonal relationships; over time, girls' own discussions include more extensive focus on emotion and emotional issues than do boys' own discussions.

Children also help to socialize themselves through their use of gender schemas. Children are motivated to be like others of their own sex and form cognitive constructions or networks of associations about the sexes that influence their behavior and thinking (Martin et al., 2002). These gender schemas direct children's attention, influence how information is interpreted, organized, and remembered, and guide behavior with objects and people; specifically, children selectively attend to and remember sex-typed information and show biases towards members of their own group (for review, see Martin et al., 2002). Many studies confirm the power of gender schemas to influence many aspects of behavior and thinking, including children's toy play, play partners, the ability to learn about activities that are traditionally sex-typed for the other sex, and impressions of others (reviewed in Blakemore et al., 2009; Martin et al., 2002; Ruble et al., 2006).

26.3.1.2 Socialization of Cognitive Sex Differences

Socialization perspectives have been applied less to cognitive abilities than to other sex-typed characteristics. Nevertheless, there are several types of evidence confirming the importance of social influences on sex-typed abilities.

Sex-typed academic skills, such as math and language, are influenced by family socialization. For example, although parents think that academic achievement is equally important for sons and daughters, they provide support for extracurricular involvement in math and science more for sons than for daughters (Simpkins et al., 2005), and parents' subtle beliefs about the inherent superiority of boys in such domains appears to undermine girls' subsequent academic performance, especially in math (Eccles et al., 2000).

Sex differences in spatial ability have been seen to depend on socioeconomic status (SES), with differences apparent in children from middle and high SES backgrounds, but not in children from low SES backgrounds (Levine et al., 2005). Such SES effects were suggested to result in part from access to experiences that facilitate spatial ability.

Spatial abilities have been suggested to develop from childhood sex-typed activities (e.g., Connor and Serbin, 1977, reviewed in Lawton, 2010). In particular, boy-typed toys and activities (such as building with Lego) are seen to encourage manipulation and exploration of the environment, and some have claimed that sex differences in spatial abilities would be reduced or even eliminated if girls were encouraged to play more with boys' toys (Halpern, 1986; Tracy, 1990). Evidence does show a weak-to-moderate link between spatial ability and aspects of sex-typed activities (e.g., Newcombe et al., 1983), but there is some variability and inconsistency that likely reflects methodological and conceptual issues (Baenninger and Newcombe, 1989; Voyer et al., 2000). It is important to note that these associations are not evidence of causation: engagement in boy-typed activities might enhance spatial ability or instead reflect that ability, that is, children with high spatial ability might be attracted to toys that allow manipulation and exploration. Some longitudinal data suggest that the causal path is from abilities to activities rather than the reverse (Newcombe and Dubas, 1992).

It is therefore important to note direct experimental evidence that spatial ability can be enhanced by experience. In particular, spatial ability can be improved through practice and training, with generalization beyond training stimuli. For example, playing an action video game was seen to improve both spatial attention and mental rotation ability (Feng et al., 2007). Training benefits both sexes, with women sometimes benefiting more than men, so that training may eliminate a sex difference in this domain (Lawton, 2010; Uttal et al., 2012).

Finally, stereotypes that emphasize women's cognitive inferiority appear to impair their performance (Steele, 1997). This has been demonstrated in studies that involve experimental manipulations, as illustrated for both math and spatial ability. Women who were told that sex differences in math have genetic causes performed worse on math tests than those who were told that the differences have experiential causes (Dar-Nimrod and Heine, 2006). Women who were told that men outperform women on spatial tasks performed worse on a mental rotations test than women who received neutral information, and the poorer performance of the group given negative stereotypes appeared to reflect increased emotional load (Wraga et al., 2007).

26.3.2 Genetic Perspectives

It is reasonable to expect that sex-related characteristics would be influenced by genes on the sex chromosomes, reflecting sex differences in sex chromosome composition (XX for typical females, XY for typical males). Work in individuals with sex chromosome abnormalities using sophisticated approaches has provided intriguing evidence that genes on the X chromosome

may influence aspects of cognition, including aspects of spatial ability (Ross et al., 2006). Evidence in rodents suggests that genes on the Y chromosome influence sex-typed behavior, including spatial ability and parenting (Arnold, 2009; Arnold and Chen, 2009). There are very limited human data on this topic, but they do not currently show direct effects of genes on the Y chromosome: Women with complete androgen insensitivity syndrome, who have a Y chromosome but no effective androgen exposure, reported similar childhood and adulthood gender role behavior to unaffected women (Hines et al., 2003a).

26.3.3 Hormonal Perspectives

26.3.3.1 Principles

Most of the research on biological mechanisms underlying gendered characteristics has focused on sex hormones, primarily androgens (including testosterone) and estrogens. This research is rooted in the work of Phoenix, Goy, and colleagues (Gibber and Goy, 1985; Phoenix et al., 1959, 1973), showing the long-lasting effects of early sex hormones on sex differences in behavior in rodents and primates, and in the subsequent work of Money and Ehrhardt (1972) in human beings. Hormones affect behavior in two ways (Becker et al., 2008; Goy and McEwen, 1980). First, sex hormones produce permanent changes to brain structures and the behaviors they subserve ('organizational' effects). Such effects usually occur early in life (during the prenatal period in human beings), although adolescence may also be an important organizational period (Schulz et al., 2009; Sisk and Zehr, 2005). Second, sex hormones produce temporary alterations to the brain and behavior (through ongoing changes to neural circuitry) as the hormones circulate in the body throughout adolescence and adulthood ('activational' effects). The main distinctions between organizational and activational effects concern timing and permanence, although these distinctions are not absolute (Arnold and Breedlove, 1985).

26.3.3.2 Evidence for Hormonal Influences on Nonhuman Sex-typed Behavior

Studies in many nonhuman animal species show that sex hormones are crucial for behavioral sex differences. Much work has confirmed and extended the early work of Phoenix, Goy, and colleagues, showing that hormones present during early life organize the brain so that they have long-lasting effects (Becker et al., 2002, 2008; Ryan and Vandenbergh, 2002; Wallen, 2005, 2009). These studies generally involve experimental manipulations of hormones (e.g., injecting females with testosterone, castrating males), but behavior has also been seen to be influenced by naturally occurring variations in hormones, such as those that result from an animal's position in the uterus, particularly the sex of its littermates (for reviews, see Clark and Galef, 1998; Ryan and Vandenbergh, 2002).

Even in nonhuman animals, however, the effects are complex. For example, studies in monkeys highlight two aspects of this complexity: (a) there are several sensitive periods for androgen effects on behavior, even during the prenatal period, with some behaviors masculinized by exposure early (but not late) in gestation, and other behaviors masculinized by exposure late (but not early) in gestation (Goy et al., 1988); and (b) environmental context modifies behavioral effects of hormones (Wallen, 1996).

Further, sex hormones may continue to exert organizational effects well beyond early life. Evidence from rodents indicates that puberty is another organizational period, with sex hormones at puberty producing permanent changes to the brain and behavior (Schulz et al., 2009; Sisk and Zehr, 2005).

There is also an extensive literature showing that sex hormones are necessary for the expression of sex-typed behaviors in adulthood (activational effects). Much work has focused on the importance for sexual behavior of testosterone in males and estradiol in females, but these hormones also play a role in the expression of nonsexual behaviors in adult animals, such as maternal behavior and aggression (Becker et al., 2002, 2008).

26.3.3.3 *Methods for Studying Early Hormonal Influences on Human Behavior*

Studies in people cannot involve experimental manipulations of hormones, but have taken considerable advantage of natural experiments, individuals whose hormone levels were atypical for their sex during early development as a result of a genetic disease or maternal ingestion of drugs during pregnancy to prevent miscarriage (for details, see Blakemore et al., 2009). The most extensively studied natural experiment is congenital adrenal hyperplasia (CAH), a genetic disease resulting in exposure to high levels of androgens beginning early in gestation because of an enzyme defect affecting cortisol production. If human psychological sex differences are affected by androgens present during critical periods of development (as occurs in nonhuman animals), then females with CAH should be behaviorally more male-typed and less female-typed than a comparison group of females without CAH.

Because CAH is not a perfect experiment (e.g., high levels of prenatal androgen also lead to masculinized genitalia, which might affect socialization, and CAH requires lifelong treatment with cortisone), it is important to seek converging evidence from other sources. Such evidence has been obtained for a number of behavioral domains from other natural experiments and from normal

individuals with typical variations in hormones. The latter includes those whose hormones have been directly measured (in amniotic fluid) and those whose hormones have been inferred (by virtue of sharing a uterine environment with an opposite-sex twin).

We briefly summarize the evidence that prenatal hormones – particularly androgens – influence human psychological sex differences and highlight key findings and issues. This evidence has been discussed in detail elsewhere (Berenbaum, 2006; Berenbaum and Beltz, 2011; Blakemore et al., 2009; Cohen-Bendahan et al., 2005; Hines, 2010).

26.3.3.4 *Prenatal Hormone Influences on Human Behavior: Noncognitive Sex Differences*

Data from several groups and countries with a variety of sound methods (including observations, tests, self-reports, and parent-reports) make clear that girls and women with CAH are more male-typed and less female-typed in aspects of their feelings, preferences, and behavior than are girls and women without CAH (in most studies, their unaffected sisters; reviewed in Berenbaum and Beltz, 2011; Blakemore et al., 2009; Hines, 2010). The largest difference between females with and without CAH is in sex-typed activity interests and engagement: girls and women with CAH prefer and are more likely to participate in male-typed activities from childhood through adulthood. They also have male-typed occupational interests; for example, females with CAH reported more interest in occupations that involve working with things (versus people) than unaffected female siblings (Beltz et al., 2011). The male-typed activity preferences and engagement in girls with CAH are directly associated with prenatal androgen and have not been shown to be influenced by parents' behavior (Meyer-Bahlburg et al., 2006; Nordenström et al., 2002; Pasterski et al., 2005). Females with CAH are sex-atypical in other domains (reviewed in Berenbaum and Beltz, 2011; Blakemore et al., 2009; Hines, 2010). Compared to typical females, females with CAH are more aggressive and less interested in babies and are more likely to have bisexual or homosexual orientation (although most are exclusively heterosexual). These masculinized characteristics stand in contrast to female-typical identity in the overwhelming majority of females with CAH (Dessens et al., 2005).

Data from other natural experiments also show that male-typical prenatal androgen levels are associated with male-typed activity interests and nonheterosexual orientation (e.g., Meyer-Bahlburg, 2005, reviewed in Berenbaum and Beltz, 2011; Blakemore et al., 2009). But the most compelling converging evidence comes from a study linking amniotic testosterone levels to activity interests in boys and girls at ages 6–10 (Auyeung et al., 2009): the most male-typed interests and characteristics (as reported by parents on a widely used measure) were found in children who had had high levels (for their sex) of naturally occurring prenatal testosterone. Unfortunately, there is little other good evidence about links between amniotic testosterone and other sex-typed characteristics in typical samples; the few existing studies had small samples and other methodological limitations.

There has also been interest in effects of early hormones on sex differences in psychopathology. Early organizational androgens have been hypothesized to contribute to the male vulnerability for childhood disorders (e.g., Baron-Cohen et al., 2004; Martel et al., 2009); however, there has been little consideration of the ways that different disorders would be affected by hormones. For example, autism has been claimed to reflect 'the extreme male brain' and therefore result from exposure to high prenatal androgens (Baron-Cohen et al., 2004), but many other disorders show similar male predominance and have been suggested by others to also result from high prenatal androgens (e.g., Martel et al., 2009).

26.3.3.5 *Prenatal Hormone Influences on Human Behavior: Cognitive Sex Differences*

Evidence from multiple sources provides moderate support for the notion that prenatal androgens influence later spatial and related abilities. Nothing is known about prenatal hormonal effects on mathematical abilities or on abilities that show a female advantage, because the sex differences are small to moderate, and it has been hard to accrue large enough samples.

Females with CAH have been found to have higher spatial ability than their sisters in childhood, adolescence, and adulthood (Berenbaum et al., 2012; Hampson et al., 1998; Hines et al., 2003b; Mueller et al., 2008; Resnick et al., 1986), with a meta-analysis suggesting that the effect is small to moderate (Puts et al., 2008). The relatively small size of the effect may explain why the effect is not always seen (e.g., Hines et al., 2003b; Malouf et al., 2006).

Confirming evidence for the effects of androgen on spatial ability comes from individuals at the other end of androgen levels: males with low early androgen levels due to idiopathic hypogonadotropic hypogonadism (IHH) have lower spatial ability than controls (Hier and Crowley, 1982). Importantly, the external genitals of males with IHH appear typical, suggesting that enhanced spatial ability in females with CAH is not due to social responses to their genitals.

There are now several studies of androgen effects on spatial ability in typical samples that provide converging evidence for prenatal androgen effects on spatial ability. In three studies of opposite-sex twins, females with a male co-twin (who are thought to have above-average prenatal exposure to testosterone) have been shown to have higher spatial ability than females with a female co-twin (Cole-Harding et al., 1988; Heil et al., 2011; Vuoksimaa et al., 2010). But, prenatal hormone influences are confounded with postnatal socialization in some of these findings because females with a male co-twin are reared with a male sibling of the same age. Evidence against socialization effects comes from women with slightly older siblings: those with a brother did not have better spatial ability than those with a sister (Heil et al., 2011). In one study of amniotic hormones, testosterone was associated with some indices of spatial ability in girls at age 7: compared to girls who had low amniotic testosterone, girls with high levels showed faster, but not more accurate, mental rotation; the effect was seen only in girls who showed evidence of using a mental rotation strategy (Grimshaw et al., 1995b).

26.3.3.6 *Adolescent Hormone Influences on Human Behavior*

In light of the recent animal evidence on the behavioral importance of organizational hormones at puberty, there has been increased attention to the effects of pubertal hormones on the human brain and behavior (see also Section 26.5). With respect to behavior, there is considerable speculation – but not a lot of data – about the ways in which changes at adolescence are triggered by the surge in sex hormones at that time, including girls' increased vulnerability to depression and eating disorders (Crick and Zahn-Waxler, 2003; Martel et al., 2009) and boys' increased risk taking and substance use (Forbes and Dahl, 2010; Steinberg, 2008). This is an area of active investigation, thus it is likely we will know much more about pubertal hormone effects on the brain and behavior within the next decade.

26.3.3.7 *Circulating Hormone Influences on Human Behavior*

There is an extensive literature (primarily in adolescents and adults) linking circulating levels of sex hormones (especially estradiol and testosterone) to sex-typed characteristics, primarily aggression, mood, and cognitive abilities (reviewed in Buchanan et al., 1992; Hampson, 2007; Maki and Sundermann, 2009; Puts et al., 2010). This work focuses on activational (transient) effects of hormones, so that behavior changes when hormone levels change. Findings are complex and difficult to summarize briefly (see reviews above for details). Much of the complexity reflects small effects, reliance on observational studies in adults, and bidirectional effects of behavior and hormones (e.g., aggressive behavior can increase testosterone). Hormones do not produce simple changes in behavior, and the most valuable studies are those that examine the ways in which hormones act indirectly and interact with social factors to change sex-related characteristics.

There is some consistency in the findings regarding cognitive effects of circulating sex hormones, with data coming from studies of natural variations in hormones across individuals and within individuals (e.g., in association with the menstrual cycle) and effects of hormone replacement (in association with aging or surgical removal of the ovaries). Findings are complex (for reviews, see Hampson, 2007; Maki and Sundermann, 2009), but generally suggest that spatial ability is facilitated by testosterone in the moderate range (levels that are high for females and low for males) and verbal memory is facilitated by estradiol, especially in postmenopausal women who are relatively young or recently menopausal. Nevertheless, links between cognition and hormones are not always found, probably due to factors that modify the effects of both hormones (e.g., diet) and cognition (e.g., experiences).

26.3.4 Integrated Perspectives

It is clear that sex-typed characteristics are influenced by multiple factors, and it is unfortunate that most studies focus on a single set of factors, rather than examining them in concert. As we have argued elsewhere (Berenbaum et al., 2011), sex differences can best be understood by integrating the different perspectives; focusing on only one set of causes (either social or biological) can lead to a distorted or even misleading understanding of sex-typed processes.

Consider spatial ability as an example of the ways in which an understanding of sex differences could be enhanced by attention to both biology *and* social experiences (for detailed discussion and other examples, see Berenbaum et al., 2011). As discussed above, evidence makes it clear that spatial ability depends on social experiences, genes on the X chromosome, and sex hormones during prenatal development and again in adult life. But spatial ability almost certainly develops from *joint* influences of biology and social experiences, as illustrated in two ways. First, biological influences on spatial ability are likely mediated through experience. As noted above, girls who are exposed to high levels of androgen during prenatal life (because of normal variation or CAH) are more likely than girls with low levels to play with boys' toys, and those toys may facilitate the development of spatial ability (e.g., Newcombe et al., 1983). Preliminary evidence from females with CAH does indeed indicate that their enhanced spatial abilities are in part mediated by their masculinized activity interests (Berenbaum et al., 2012). Second, biological predispositions likely facilitate learning. Although existing training studies are not compelling in this regard (men, who have high androgen levels, do not appear to be more likely to benefit from practice than women, who have low androgen

levels), the situation might differ in childhood, when abilities are developing, and on tests that do not show ceiling effects. This question can be studied in girls with CAH and in typical children, by examining the effects of practice at varying ages and with varying environmental experiences (e.g., as a function of SES).

26.4 BRAIN SEX DIFFERENCES: NATURE, DEVELOPMENT, AND CONSEQUENCES

The brains of men and women, and of boys and girls, are similar in many ways, but there are some systematic differences in anatomical structure, physiological functioning, and development. An understanding of these differences – and what causes them – should provide insight into the mechanisms underlying sex differences in human health, disease, and behavior, including the characteristics discussed above (Cosgrove et al., 2007; McCarthy and Arnold, 2011).

In this section, we review the work on human brain sex differences. We focus on topics where there is converging evidence for brain sex differences, but we also present some exciting new findings that will likely spark future research. Most of the work reviewed comes from studies using magnetic resonance imaging (MRI), as it is the dominant research tool in human developmental neuroscience (Casey et al., 2005; Luna et al., 2010), but we also consider work using other techniques, such as positron emission tomography (PET), perceptual asymmetries, and post mortem examinations.

26.4.1 Issues in Studying Brain Sex Differences

MRI is widely used as a measure of brain anatomy and physiology. Structural MRI (sMRI) provides measures of brain morphology and architecture (e.g., volumes of gray matter, white matter, and subcortical structures). Blood oxygen level-dependent (BOLD) functional MRI (fMRI) is a measure of change in blood oxygenation thought to reflect neural activity. Diffusion tensor imaging (DTI) is the most frequently used assessment of water diffusion in the brain, thought to reflect white matter microstructure and thus information transmission among brain regions. All MRI measures – sMRI, fMRI, and DTI – are indirect measures of brain structure and function, involving multiple assumptions and inferences, and each reflects multiple dynamic processes occurring at cellular and subcelluar levels (for discussion relevant to developmental science, see Casey et al., 2005; Paus, 2010).

The analysis of MRI data, particularly functional data, is complex and based on multiple assumptions, partly because assessments are indirect. The brain is typically partitioned into volumetric pixels, or voxels, and statistics are conducted on each voxel as if it were independent of all others. This 'mass univariate approach' can result in a score of false positives unless appropriate corrections to *p*-values are made for multiple comparisons (Friston, 2004). In studies linking behavior to brain activation, findings may be spurious if only voxels exceeding some preset threshold are examined (Vul et al., 2009). This is essentially a selection bias: only data from voxels reflecting significant brain activation are examined in relation to behavior. This problem can be avoided through *a priori* selection of brain regions of interest (ROIs) for analyses (Poldrack, 2000; Vul et al., 2009).

Concerns have been expressed about the interpretations of findings of brain sex differences (Fine, 2010; Jordan-Young, 2010). Such problems appear to result from the inappropriate use of data, rather than from the data themselves. First, people may over-interpret brain images because they are compelling; research containing irrelevant neuroscience explanations and brain images is rated by nonexperts as more satisfying or reasonable than accurate work without such information (Beck, 2010; McCabe and Castel, 2008; Weisberg et al., 2008). Second, brain sex differences are often interpreted in simplistic ways. For example, a bigger brain is not necessarily a better brain (see Section 26.4.2.1), and brain sex differences are not indicators of predetermined behavioral inequalities (see also McCarthy and Arnold, 2011; Poldrack, 2000). Third, brain sex differences need to be considered in conjunction with behavioral sex differences (see Sections 26.4.2.5 and 26.4.3).

As we discuss below, there is much careful and important work demonstrating brain sex differences. Research questions developed from a clear conceptual framework are often investigated using sound methodology and a careful approach to examining brain–behavior relations. Inferences are often balanced and appropriate, and there is considerable convergence of evidence. Of course, as in any area of science, not all studies are perfect, but limitations in methodology and inference are no more prevalent or problematic in this field than elsewhere (Fiedler, 2011).

26.4.2　Sex Differences in Brain Structure and Their Development

An understanding of brain sex differences requires consideration of development. Some of the most compelling work on brain–behavior relations has shown both a change in structural sex differences across development and the importance of considering developmental trajectories (pattern of brain changes over time), not just a single measurement (see Rubenstein and Rakic, 2013). For

example, general intelligence has been shown to relate to changes in cortical thickness across childhood and adolescence, but not to absolute measures of cortical thickness at a given age (Shaw et al., 2006); compared to typical children, children with attention-deficit hyperactivity disorder have delayed trajectories of cortical thickness maturation (Shaw et al., 2010). Thus, sex differences in brain development are best examined in longitudinal neuroimaging studies; however, there are many more studies of sex differences at a single point in time, so we review those studies (typically done in adults) as well as the longitudinal neuroimaging data. We first review findings of sex differences in brain structure and then examine how structural differences relate to behavioral sex differences.

26.4.2.1 *Brain Volume*

There are sex differences in intracranial and cerebral volume, with both approximately 10% larger in males than in females. These differences are seen in measurements made on postmortem brains (Holloway, 1980) and in live ones (using MRI) (De Bellis et al., 2001; Giedd and Rapoport, 2010; Giedd et al., 1997; Goldstein et al., 2001; Lenroot and Giedd, 2010; Lenroot et al., 2007; Nopoulos et al., 2000; Sowell et al., 2002); they are also seen in several nonhuman primate species (Falk et al., 1999; Holloway, 1980). The human differences appear to be primarily due to the larger occipital and frontal poles of men compared to women (Sowell et al., 2007). The differences in brain size are due in part to sex differences in body size, reflecting overall growth differences (Halpern, 2012; Holloway, 1980; Peters, 1991; Peters et al., 1998).

The brain reaches about 95% of its adult size by age 6, but the relative proportions of gray matter and white matter continue to change throughout childhood and adolescence. The peak cerebral volume occurs about 4 years earlier in girls than in boys, but boys have a larger absolute cerebrum size than girls throughout development (Giedd and Rapoport, 2010; Lenroot et al., 2007).

The implications of brain size sex differences for brain function or behavior, however, are not clear (see also Section 26.4.2.5). Sex differences in the size of other body structures (e.g., men's larger hearts and noses) do not translate into sex differences in function. Further, a larger brain does not necessarily mean a smarter brain. Although there is a moderate association between brain size and intelligence (Flashman et al., 1997; Luders et al., 2009b), there are no sex differences in general intelligence, and neural connectivity (rather than size) is key to understanding the brain substrates of intelligence (Song et al., 2008; van den Heuvel et al., 2009). Interpretations of structural size differences are complicated for several reasons: normal brain maturation involves cell death and synaptic pruning, reducing brain size; brain size may be larger in individuals with developmental disorders, such as autism (Shaw et al., 2010; Sowell et al., 2001, 2002); there are dynamic relations between experiences and brain anatomy (e.g., taxi driving experience is associated with hippocampal shape) (Maguire et al., 2000, 2003).

The sex difference in body size creates difficulties in studying sex differences in brain size, which, in turn, creates difficulties in studying other aspects of brain sex differences, leading to controversies over whether to correct for body size and overall brain size. Some have argued that sex differences in brain size are best corrected by body weight or height (Halpern, 2012; Holloway, 1980). Others have claimed that the relation between body height or weight and brain size is weak (Peters, 1991), particularly because there is great variation in human body size and type (Peters et al., 1998). Further, the relation between brain and body size changes across development (Giedd and Rapoport, 2010; Lenroot and Giedd, 2010), suggesting that brain–body size corrections should only be made in adult samples.

There are several ways to correct for sex differences in intracranial or cerebral volume in order to examine sex differences in specific brain structures (see Bishop and Wahlsten, 1997). First, brain volume can be covaried. This approach is most appropriate if there is a linear relationship between brain size and the size of the brain structure or region being investigated, and it can misrepresent sex differences if the relation is present in only one sex. Second, a ratio can be computed, reflecting the volume of the structure or regions being investigated as a proportion of brain size; this approach is easy to understand, but its anatomical interpretation is unclear. Finally, recent innovative approaches have been suggested, including the use of different numerical scaling factors for different characteristics of brain anatomy because larger brains are not uniform expansions of smaller brains; for example, there is greater relative increase in cortical surface area than cortical thickness with brain size increases (Im et al., 2008). Longitudinal studies overcome the problem by allowing examination of within-individual change across time and subsequent comparison of rates of change between the sexes (e.g., Thambisetty et al., 2010).

26.4.2.2 *Regional Structure Volume*

There are several consistently replicated sex differences in the size of human regional brain structures. But these effects are not always detected, likely because they are small and approaches to correcting for sex differences in brain size vary across studies. Generally, as detailed below, some key regions of the brain implicated in interhemispheric communication and memory (e.g.,

hippocampus and caudate nuclei) are larger in women than men, and other subcortical regions implicated in affective and sexual behaviors (e.g., amygdala and hypothalamus) are larger in men than women.

26.4.2.2.1 SEX DIFFERENCES IN INTERHEMISPHERIC COMMISSURES

Several studies have investigated sex differences in interhemispheric commissures (fiber bundles spanning the two halves of the brain). Commissures are important for facilitating information flow across hemispheres and thus for promoting cognitive function (Bryden, 1982; Kimura, 1999).

There has been particular focus on sex differences in the most posterior portion of the corpus callosum (CC), the splenium, following an early histological report of a sex difference in this region (De Lacoste-Utamsing and Holloway, 1982). A meta-analysis of 49 studies concluded that there are no systematic sex differences in this structure when appropriate corrections are made for the sex difference in brain size (Bishop and Wahlsten, 1997), and recent MRI studies support this finding (Leonard et al., 2008; Luders et al., 2006b). Nonetheless, research with large, adult samples using a sophisticated analysis technique (in which derivations of an individual's CC from a template CC are captured at many points), indicates that women do, in fact, have larger splenia than men (Davatzikos and Resnick, 1998; Dubb et al., 2003). Other research has positively linked the size of the splenium in women (but not men) to cognitive abilities, such as verbal fluency, two-dimensional mental rotations, and memory (Davatzikos and Resnick, 1998; Hines et al., 1992); however, these links are not consistently seen in children and adolescents (e.g., Luders et al., 2011).

The CC develops in a rostral-to-caudal direction throughout childhood and early adolescence, and the rate of growth may be greater in girls than in boys (Luders et al., 2010; Thompson et al., 2000). This sex difference is particularly striking because boys have faster overall rates of white matter development than girls (Section 26.4.2.4). Even from late adolescence through late adulthood, the splenium appears to increase in size in women more than in men (Dubb et al., 2003).

There are sex differences in the anterior commissure (connecting the right and left temporal lobes) and massa intermedia (connecting the right and left thalamus), but less is known about their functional significance than that of the CC. Both structures are generally found to be larger in women than men (Allen and Gorski, 1991, 1992; Kimura, 1999). The massa intermedia is also more frequently absent in men than in women (Rabl, 1958), and absence of this structure is linked to greater performance IQ scores for men, but not for women (Lansdell and Davie, 1972).

26.4.2.2.2 SEX DIFFERENCES IN STRUCTURES INVOLVED IN LEARNING AND MEMORY

Several brain structures implicated in learning and memory are also larger in females than males, consistent with the female advantage in memory (Section 26.2.1.4). The hippocampus, which is important for memory formation, retention, and recall, is larger in women than men (Cahill, 2005; Filipek et al., 1994; Goldstein et al., 2001; Halpern, 2012; Lenroot and Giedd, 2010), though this difference is most consistently reported in children and adolescents, perhaps because the hippocampus is generally found to grow at a faster rate and thus to mature earlier in girls than boys (Giedd et al., 1997; Lenroot and Giedd, 2010). The caudate nuclei, which are part of the basal ganglia and implicated in learning and memory, are also larger in females than males; this finding has been reported in both children (Giedd et al., 1997; Sowell et al., 2002) and adults (Filipek et al., 1994; Goldstein et al., 2001). Caution must be used when interpreting the implications of these findings because the sex difference in verbal memory remained in a sample of men and women who had left anterior temporal lobectomy, suggesting that other brain regions and sex differences in strategy are also important for sex differences in memory (Berenbaum et al., 1997).

26.4.2.2.3 SEX DIFFERENCES IN SUBCORTICAL STRUCTURES INVOLVED IN AFFECTIVE AND SEXUAL BEHAVIORS

Brain structures consistently found to be larger in men than women include regions of the hypothalamus and the amygdala. Sex differences in the human hypothalamus parallel early studies showing a very large sex difference in the rodent preoptic hypothalamus (Gorski et al., 1978). In particular, one of the four interstitial nuclei of the anterior hypothalamus (INAH-3) is larger in men than women (reviewed in Halpern, 2012; Kimura, 1999; LeVay, 1991). There are also reports that some regions of the bed nucleus of the stria terminalis (BNST), the central portion of the connection between the hypothalamus and amygdala, are larger in men than women (Allen and Gorski, 1990; Zhou et al., 1995). Although the functional implications of these sex differences are unclear, there is some suggestion that the INAH and BNST play a role in gender identity and sexual orientation (for discussion, see Hines, 2011; Savic et al., 2010).

The amygdalae, a pair of bilateral structures in the medial temporal lobe, are also larger in boys and men than in girls and women (Cahill, 2005; Goldstein et al., 2001; Halpern, 2012; Lenroot and Giedd, 2010) and develop faster in boys than in girls (Giedd et al., 1997; Lenroot and Giedd, 2010). They are thought to be important for the detection of and behavioral response to affective visual cues, perhaps because of their connectivity to many cortical and subcortical regions (Pessoa and Adolphs, 2010).

26.4.2.3 *Gray Matter*

There are sex differences in gray matter, with girls and women having relatively more gray matter than boys and men across the lifespan, particularly in parts of the frontal, temporal, and parietal lobes (Gur et al., 1999; Lenroot et al., 2007; Leonard et al., 2008; Luders and Toga, 2010; Luders et al., 2002; Paus, 2010). Relatedly, women have greater cortical thickness than do men in some areas of the brain, particularly in the frontal and parietal lobes. Supporting evidence comes from several different samples and research methodologies, including MRI combined with surface morphometry (Im et al., 2006; Lv et al., 2010), pattern algorithms (Luders et al., 2006a; Sowell et al., 2007), and tissue segmentation (Koscik et al., 2009). Some histological findings are inconsistent, but their interpretation is complicated by small sample sizes (e.g., Rabinowicz et al., 1999).

There are parallel differences in cortical complexity, or patterning of cerebral convolutions, with women showing greater complexity than men, particularly in the frontal and parietal lobes (Luders et al., 2004, 2006c). Findings of sex differences appear to vary with method: differences favoring females have been found using a new mesh-based approach (in which the convolutions at several thousand surface points are estimated from three-dimensional brain scans), but not when using the gyrification index (ratio of cortical surface to visible gyral surface, calculated from two-dimensional postmortem brain slices) (Zilles et al., 1988).

Gray matter develops in an inverted U-shaped trajectory across childhood and adolescence, increasing until puberty and then decreasing through early adulthood (Giedd and Rapoport, 2010; Giedd et al., 1999; Lenroot and Giedd, 2010; Lenroot et al., 2007). The shape and timing of the peak in gray matter trajectories differ across brain regions: parietal lobe volumes peak first, followed by frontal and temporal lobe volumes. Relatedly, cortical maturity, marked by decreases in cortical density, occurs in a specified pattern: the medial sensorimotor cortex matures first, with development proceeding in rostral and lateral directions, such that the prefrontal and lateral temporal cortices are the last regions to mature (Gogtay et al., 2004). This reduction in gray matter density, which occurs after pubertal onset, likely reflects the elimination of irrelevant brain connections and strengthening of relevant ones (i.e., pruning). The sexes have similarly shaped trajectories of gray matter development across all brain regions, but girls have an earlier average peak than boys, consistent with their earlier pubertal maturation. Generally, girls show a peak in regional gray matter between 8 and 10 years of age, whereas boys show a peak between 9 and 11 years of age (Lenroot et al., 2007).

26.4.2.4 *White Matter*

There is not clear consistency from MRI studies regarding sex differences in overall brain proportions of white matter. Some evidence indicates that men have a greater proportion of white matter than do women (Allen et al., 2003; Gur et al., 1999), but the difference is not always found and inconsistencies are not explained by differential corrections for the sex difference in brain size (Luders et al., 2002; Nopoulos et al., 2000). Women are rarely found to have greater white matter volume than men, suggesting that small samples make it difficult to consistently detect the increased white matter in men compared to women. Inconsistencies may also reflect age differences; sex differences in the proportion of brain white matter generally emerge in adolescence and persist through adulthood (Paus, 2010).

Converging evidence from MRI also suggests no sex differences in fractional anisotropy (FA), which is thought to reflect directed information transmission along white matter paths. This evidence comes from large, cross-sectional DTI studies (Eluvathingal et al., 2007; Giorgio et al., 2008; Lebel et al., 2008).

White matter volume increases linearly through childhood and into adulthood, and this pattern is consistent across brain regions (Giedd et al., 1999; Lenroot et al., 2007). Girls and boys both experience linear increases in white matter volume, but boys have a more rapid rate of increase compared to girls, particularly in adolescence (Lenroot et al., 2007). FA similarly increases across early childhood and into adulthood in an inferior-to-superior and posterior-to-anterior fashion (Colby et al., 2011); white matter connections around subcortical structures and in the CC undergo the greatest change across development, and frontal–temporal tracts are the last to mature (Eluvathingal et al., 2007; Giorgio et al., 2008; Lebel et al., 2008). Increases in FA may mark a transition from functional localization to distributed neural network functioning in the developing brain. For example, adolescents show a decrease in frontal gray matter density as FA increases in pathways connecting frontal regions to other brain areas (Giorgio et al., 2008). The ways in which sex interacts with age changes in FA are still unclear (for a review, see Schmithorst and Yuan, 2010): there are few patterns among the observed sex differences, and inconsistent findings likely reflect samples of varying age, the sex difference in white matter development, differences in FA measurement across studies (e.g., analyses at the level of voxels versus ROIs), and problems inherent in FA methodology (e.g., measurement of perpendicular fiber tracts within a voxel).

26.4.2.5 *Implications of Sex Differences in Brain Structure*

Some – but not all (e.g., Luders et al., 2009a) – of the sex differences in brain morphology might result from the different shaping of larger versus smaller brains (Allen et al., 2003; Im et al., 2008; Luders et al., 2002, 2010; Seldon, 2005; Zhang and Sejnowski, 2000). In larger as compared to smaller brains, the cortex tends to be flatter and thinner because it fills a larger intracranial space; this is consistent with reports of greater cortical thickness and complexity in women than men. There is also more white matter in larger brains because longer axonal connections are made to cortical regions that are farther apart than in smaller brains. This is consistent with evidence that men have greater white matter volumes than women.

What do sex differences in brain structure mean for cognitive sex differences? Compared to women, men appear to have larger regions of the parietal lobe, which is the primary brain area implicated in spatial ability (Brun et al., 2009). Parietal lobe surface area is, in fact, positively related to mental rotations performance in men, whereas parietal lobe gray matter volume is negatively related to mental rotations performance in women (Koscik et al., 2009); this suggests that different parietal lobe morphology subserves mental rotations performance in men and women, but it does not indicate whether men and women equally engage parietal regions during mental rotations tasks (this requires data on brain activation). There is also evidence from structural studies that some brain regions implicated in language, especially the temporal lobe, are larger in women than men (Brun et al., 2009; Harasty et al., 1997). But, these studies are difficult to interpret because verbal ability was not assessed. Sex differences in brain activation during spatial and language tasks are reviewed below (Sections 26.4.3.1 and 26.4.3.2, respectively).

It is probable that not all structural sex differences have behavioral significance, as there are likely multiple paths to the same outcome (De Vries, 2004; McCarthy and Arnold, 2011); this also appears to be the case for other bodily organs (e.g., the heart). Despite large variations in structure, brain function is remarkably similar and stable across people (Sporns, 2011). If this 'functional homeostasis' is maintained, then links between brain structure and behavior will not always exist.

26.4.3 Sex Differences in Brain Function (Activation)

There is a considerable literature examining sex differences in brain function as measured by task-specific activation (fMRI). The most meaningful work relates brain activation to task performance. An important issue concerns the sex difference on the behavioral outcome of interest. If men and women perform at different levels, brain activation differences might tell us about the neural substrates of the performance difference. If men and women perform at similar levels, brain activation patterns might tell us about the (potentially different) processes or strategies they use to arrive at the same outcome.

26.4.3.1 Spatial Abilities

There has been interest in finding the neural correlates of the male advantage in spatial ability described in Section 26.2.1.1. Generally, in both sexes, spatial processing is associated with activation of the parietal lobes as well as temporal, premotor, and extrastriate areas (Butler et al., 2006; Christova et al., 2008; Halari et al., 2006; Hugdahl et al., 2006; Thomsen et al., 2000; Weiss et al., 2003a; see also Chapter 15). There is also some evidence that right hemisphere regions are more engaged in spatial tasks than left hemisphere regions, particularly for men (see Section 26.4.3.4).

The brain regions activated by men and women during mental rotations tasks are largely overlapping when there are no sex differences in behavioral performance, but some brain differences have been reported. Women engage portions of the frontal lobe, in particular the right inferior frontal gyrus, that men generally do not (Hugdahl et al., 2006; Jordan et al., 2002; Weiss et al., 2003a). For women, it is the activation of these frontal regions (as well as some temporal and parietal regions) that positively predicts task performance accuracy, whereas activation of parietal regions (e.g., postcentral gyrus and precuneus) positively predicts accuracy for men (Butler et al., 2006). The different patterns of task-related brain activation for women and men might reflect women's use of top-down processing and men's use of bottom-up processing to complete mental rotations tasks (Butler et al., 2006).

Functional brain sex differences during mental rotations tasks when the sexes differ in performance may tell us about the neural substrates of the cognitive difference, although the performance differences may complicate interpretation. Most behavioral tests of mental rotations involve comparisons among multiple figures, whereas most neuroimaging studies of mental rotations involve a pairwise comparison because of scanner space constraints; the sex difference in mental rotations performance is likely reduced in the latter compared to the former (Peters and Battista, 2008). In one fMRI study in which men outperformed women, there were no sex differences in brain regions activated by the task, but women had significantly greater activation of those regions than men (Halari et al., 2006); this is consistent with the notion that increased activation reflects the need to work harder (e.g., Gur et al., 2000).

There are also sex differences in brain activation during other visuospatial tasks. When matching the orientation of angled lines, men were reported to have greater activation than women of left occipital and cingulate regions, and this effect increased with age (Clements

et al., 2006; Clements-Stephens et al., 2009); in contrast, no regional brain sex differences were found in a similar task in which a joystick was used to move a cursor in a specified angle away from a central stimulus (Christova et al., 2008). During spatial navigation, men were seen to engage more left temporal regions than women, and women to engage more right frontal, parietal, and hippocampal regions than men (Grön et al., 2000), although sex differences in related tasks have not always been found (Ohnishi et al., 2006). During spatial attention tasks, men appear to activate left parietal regions more than women, whereas women engage more right frontal regions than men (Rubia et al., 2010), although this difference is not always found (Bell et al., 2006). These studies converge to indicate that visuospatial tasks engage different brain systems in the two sexes: men recruit left hemisphere regions to a greater degree than women, and women recruit frontal lobe regions that men generally do not. Unfortunately, it is difficult to review these findings in relation to behavioral performance because of the limited number of studies available and differences across studies in the tasks used.

26.4.3.2 Language

Several studies have focused on finding the neural substrates of the female advantage in verbal ability described in Section 26.2.1.3. Generally, in both sexes, activity in regions of the left hemisphere, including the temporal lobe, prefrontal cortex, inferior frontal gyrus, cingulate, and regions of the parietal lobe is associated with the performance of most language tasks, including verbal fluency, rhyming, and comprehension (Allendorfer et al., 2012; Buckner et al., 1995; Burman et al., 2008; Clements et al., 2006; Frost et al., 1999; Gauthier et al., 2009; Halari et al., 2006; Plante et al., 2006; Shaywitz et al., 1995; Weiss et al., 2003b).

As for spatial ability, findings of sex differences in brain activation for language tasks using fMRI depend upon whether the sexes are matched on behavioral performance. When they are matched on task performance, most studies do not find differences in brain activation (Allendorfer et al., 2012; Clements et al., 2006; Donnelly et al., 2011; Frost et al., 1999; Weiss et al., 2003b). Nonetheless, differences favoring both sexes have also been reported: men have been seen to have greater activation than women in language-related regions (Buckner et al., 1995; Gauthier et al., 2009); the reverse has also been found, with women having greater activation than men, particularly in right hemisphere regions (Plante et al., 2006; Shaywitz et al., 1995). When the sexes are not matched on task performance, women behaviorally outperform and display greater brain activation than men, especially in right hemisphere regions (Burman et al., 2008; Halari et al., 2006; Plante et al., 2006), consistent with other data discussed in Section 26.4.3.4 suggesting bilateral representation of language in women. (This argument might appear inconsistent with that presented in Section 26.4.3.1, that is, that the greater frontal lobe activity of women during spatial tasks reflected their need to work harder than men. Issues pertaining to the interpretation of sex differences in brain function are complicated, however, as we discuss in Section 26.4.3.6.) Furthermore, the link between performance accuracy and brain activation differs for men and women: in women, accuracy is strongly linked to primary language areas in the bilateral frontal and left temporal lobes; in men, accuracy is weakly linked to primary left hemisphere language regions and also to secondary language areas, such as the left caudate and cingulate and right parietal lobe (Allendorfer et al., 2012; Burman et al., 2008; Donnelly et al., 2011).

26.4.3.3 *Emotional Recognition*

Much imaging work has concerned the neural substrates of sex differences in the processing of emotions, enough for reviews of sex differences in brain activity during the processing of emotional information, with a focus on amygdala engagement during the viewing of emotional human faces (Fusar-Poli et al., 2009; Sergerie et al., 2008). (It is interesting that most studies on emotion have focused on faces, rather than voices or prosody. See Chapter 18 for further reading on the neurobiology of face processing.) The viewing of emotional faces generally elicits activation in several brain regions, including the amygdala, frontal and prefrontal cortices, anterior cingulate cortex, insula, as well as areas in the temporal, parietal, and occipital lobes (Fusar-Poli et al., 2009; Sergerie et al., 2008). The amygdala is typically engaged in the viewing of all emotions, in particular happy, sad, and fearful faces (Fusar-Poli et al., 2009); this is likely a reflection of its central connectivity to visual, subcortical, and cortical regions of the brain, all of which are engaged in the detection of emotional cues and subsequent planning of behavioral responses (Pessoa and Adolphs, 2010).

There are sex differences in amygdala volume (larger in men than in women) and amygdala activation during the viewing of emotional faces (men have greater activation than do women). Two meta-analyses of over 100 empirical studies each have confirmed the sex differences in brain activation during emotion recognition tasks: men have greater bilateral activity than women in limbic areas, including the amygdala, and in prefrontal regions (Fusar-Poli et al., 2009; Sergerie et al., 2008), perhaps reflecting their need to work harder on the task, since women typically outperform men in the decoding of emotions (Section 26.2.2.4). Task performance was not explicitly considered in the meta-analyses, and studies varied in their measurement and reporting of behavioral sex differences. Few studies have, in fact, investigated sex differences in the link between amygdala

activation and performance on emotion recognition tasks: the frontal lobe inhibits amygdala activation during explicit identification of emotions versus the passive viewing of emotions (Critchley et al., 2000; Hariri et al., 2000), so most emotion recognition tasks do not require explicit behavioral responses. The limited available data suggest that bilateral amygdala activity is positively linked to emotion recognition accuracy for both men and women (Derntl et al., 2009; Habel et al., 2007).

There are also sex differences in amygdala activation during other emotion-related tasks. In a manner consistent with findings on facial emotion recognition, men show more amygdala activity than women during the viewing of sexual stimuli (reviewed in Hamann, 2005). During emotional memory tasks, the amygdala is activated in both sexes, but activation is greater in the left amygdala in women and in the right amygdala in men. This finding is not easy to interpret, but it has been suggested that sex differences in internalizing psychopathology arise in part through differences in memory for the gist of (right hemisphere) versus the details of (left hemisphere) negative emotional episodes (reviewed in Cahill, 2010; Hamann, 2005) (See Chapter 16 for further reading on the neurobiology of memory development.) Compared to emotional memory tasks, which emphasize emotions with a negative valence and have dominated the literature, tasks that emphasize emotions with a positive valence elicit different patterns of brain sex differences, with women showing more frontal and temporal activity than men, and men showing more left amygdala activity than women (Stevens and Hamann, 2012).

26.4.3.4 *Lateralized Functions*

Lateralization, also referred to as hemispheric specialization or functional asymmetry, has been a popular explanation for cognitive sex differences, particularly for language and spatial abilities. Typically, the left hemisphere, which houses both Broca's and Wernicke's areas, is dominant for sequential processing, including language, and the right hemisphere is dominant for simultaneous processing, including spatial abilities (for reviews, see Bryden, 1982; Hall et al., 2008; Kansaku and Kitazawa, 2001; Kimura, 1999). Data going back several decades, derived from patients with brain damage and behavioral tasks in typical individuals (such as split-field visual tasks, dichotic listening, and electroencephalogram asymmetry), suggest that women are less lateralized than men, especially for language (Bryden, 1982; McGlone, 1980). The topic has also been studied using fMRI, but results are often task- and method-dependent and subject to misinterpretation. For example, evidence reviewed above suggests that women have more right hemisphere activation than men for language tasks, and men have more left hemisphere activation than women for spatial tasks, but this does not mean that men are left-lateralized for language and women are right-lateralized for spatial tasks. Lateralization must be explicitly tested by comparing activation in the two hemispheres; it is not sufficient to show that the activity of only one hemisphere is above baseline or that there are significant sex differences in specific regions of a single hemisphere. When the correct analyses are conducted, women show less lateralization than men. Women's reduced lateralization has been suggested to be linked to greater interhemispheric communication facilitated by their larger interhemispheric commissures (e.g., CC and anterior commissure) (Halpern, 2012; Hines et al., 1992).

There is some evidence for sex differences in functional lateralization of spatial ability, although findings from neuroimaging (PET and fMRI) are not always consistent with those from perceptual asymmetry assessments (e.g., split visual field and dichotic listening tasks), suggesting that the methods are measuring different aspects of lateralization. A meta-analysis of sex differences in lateralization of spatial tasks measured by a variety of methods (e.g., fMRI, PET, split visual field, brain damage) found that men primarily engage the right hemisphere and women engage both hemispheres when solving spatial tasks (Vogel et al., 2003). However, the findings from the fMRI studies on spatial ability reviewed above (Section 26.4.2.1) in which lateralization was explicitly examined are mixed: some reported greater right than left hemisphere activity for both men and women (Halari et al., 2006; Hugdahl et al., 2006), others reported more right-lateralized activity in men than women (e.g., Christova et al., 2008) or in women than men (e.g., Clements et al., 2006), and still others reported no lateralization effects (e.g., Weiss et al., 2003a).

Women have somewhat greater bilateral representation for language than men. Two recent meta-analyses of fMRI and PET studies on language lateralization reported weak support for left-lateralization in men but not women in several domains (Sommer et al., 2004, 2008), and meta-analyses using other behavioral methods (e.g., dichotic listening) reveal similar differences (Bryden, 1982; Sommer et al., 2008; Voyer, 2011). Findings from the fMRI studies on language reviewed above (Section 26.4.3.2) are consistent with these meta-analyses, with some reporting greater left than right hemisphere activity in both sexes (Frost et al., 1999; Plante et al., 2006; Weiss et al., 2003b), and others reporting this effect only for men (Clements et al., 2006; Shaywitz et al., 1995).

Sex differences in amygdala lateralization during emotion-related tasks are task-dependent. Studies on emotion recognition reviewed above (Section 26.4.3.3), in which task performance and lateralization effects were explicitly examined, found no support for sex differences (Derntl et al., 2009; Habel et al., 2007). This bilateral pattern of activation may depend on development, though: recent work shows right-lateralized amygdala

activity during facial emotion recognition for adolescent boys (but not girls) (Schneider et al., 2011). There is also evidence for sex differences in amygdala lateralization during other emotion tasks in adults: men have right-lateralized and women have left-lateralized amygdala activity for emotion-related memories (Cahill, 2010; Hamann, 2005). A meta-analysis of fMRI and PET studies on a combination of emotion-related tasks found that men had more right-lateralized activity than women (Wager et al., 2003), but other work suggests that sex differences disappear with improved meta-analytic methodology (e.g., incorporating effect sizes of original findings into the analysis) (Sergerie et al., 2008).

There are few studies directly examining the link between sex differences in lateralized activation and sex differences in cognitive performance. For spatial ability, greater (usually right) lateralization is thought to be linked to better performance because the processing of spatial information is unhindered by language circuitry. For language, the lateralization-performance link is unclear: greater lateralization has been associated with poorer language skills in typical males (compared to females), and incomplete lateralization has been hypothesized to be a risk factor for language disorders (Hall et al., 2008). Interestingly, the limited data on the association between lateralization and performance concern language tasks. As expected, results are largely inconsistent, likely due to study variations in task designs and the poor temporal resolution of PET and fMRI – a general limitation of the techniques that is particularly critical in examinations of the speed and timing of interhemispheric processing (for discussion, see Hall et al., 2008; Kansaku and Kitazawa, 2001; Kitazawa and Kansaku, 2005; Ortigue et al., 2005). It is important to note that rarely are women reported to be more lateralized than men, so inconsistencies likely reflect differential measurement sensitivity and small sex differences. Given the relatively small sex differences in brain lateralization and the relatively large sex differences in some abilities, it is very unlikely that lateralization differences completely account for cognitive sex differences.

26.4.3.5 Development of Sex Differences in Brain Function

There are fewer studies of functional brain development than structural brain development, although structure and function are certainly interdependent. Typically, brain activity related to cognitive task performance becomes less diffuse and more fine-tuned with age, probably related to increasing connectivity among brain regions (Casey et al., 2005; Giorgio et al., 2008). For example, brain function in children is driven by short-range connections among anatomically close regions; in adults, many local connections are replaced by long-range ones, reflecting greater neural integration with development (Fair et al., 2007, 2009).

Brain function in adolescence is likely affected by the different trajectories of gray and white matter development, and these brain changes may underlie the increased risk-taking behavior characteristic of this developmental period. Limbic regions, which are implicated in affective processing, mature in early-to-mid adolescence, whereas the prefrontal cortex, which is thought to subserve cognitive control functions, is among the last brain region to undergo cortical thinning and myelination. The discordant maturational timing between these affective and cognitive neural processing networks is thought to result in limited top-down control of responses to appetitive stimuli such as peers and addictive substances (Casey et al., 2011; Steinberg, 2008). Because girls have faster brain development compared to boys, they are thought to experience a shorter period of discordance between limbic and prefrontal regions and therefore to be less likely to engage in the type of risk-taking behavior that emerges in adolescence (e.g., risky sex, substance use).

26.4.3.6 *Implications of Sex Differences in Brain Function*

There are several conclusions that emerge from the studies reviewed; a simplified summary of sex differences in neural activation to spatial, language, and emotion-related tasks is provided in Figure 26.4. First, with respect to visuospatial tasks, particularly mental rotations, both men and women engage parietal regions, but women engage frontal regions that men typically do not, and men engage some left-hemisphere regions that women typically do not (Figure 26.4(a)). Second, with respect to language tasks, women recruit more bilateral frontal and temporal regions than do men (Figure 26.4(b)). Third, with respect to the processing of emotional faces, men show greater bilateral amygdala activity than do women (Figure 26.4(c), right), perhaps reflecting their need to work harder than women to decode emotions. Fourth, with respect to negative emotional memory, women show greater left amygdala activity than men, and men show greater right amygdala activity than women (Figure 26.4(c), left); this difference is considered to have implications for sex differences in psychopathology (Cahill, 2010; Hamann, 2005; Stevens and Hamann, 2012).

We can also draw some general conclusions regarding sex differences in the lateralization of brain function for cognition and the link between brain activation and performance on cognitive tasks. With respect to lateralization, although the evidence is mixed, it appears that women are more likely than men to process material in both hemispheres; the lateralized activity of men appears to contribute to some cognitive advantages (e.g., in spatial ability) and disadvantages (e.g. in language).

With respect to brain activation–task performance links, performance accuracy is associated with activation of different brain regions for men and women for both visuospatial and language tasks, perhaps reflecting the use of sex-specific strategies.

Thus, the sexes engage similar brain regions to solve cognitive tasks, but they also appear to diverge in some notable ways. The sex differences primarily reflect two brain processes: efficiency and integration. Greater brain activity in the sex that typically performs worse on a behavioral task than the sex that performs better can be understood in terms of neural efficiency. When performance is poor, the task is seen to be difficult, and thus more neural resources are recruited to solve it. This interpretation is consistent with findings from studies in which task difficulty was manipulated: the neural substrates of a task become more widely distributed as task difficulty increases (e.g., Gur et al., 2000). Greater brain activity in the sex that typically performs better on a behavioral task than the sex that performs worse can be understood in terms of neural integration. When performance is good, the task elicits a diverse set of associations, subserved by greater neural engagement. This interpretation is consistent with findings on the neural systems underlying intelligence: functional connectivity among brain regions is positively associated with intelligence (Song et al., 2008). Whether efficiency or integration is invoked as an explanation for functional brain sex differences is likely dependent upon the nature of the task (e.g., complexity) and methodology (e.g., sample characteristics, task design, data analysis procedures); they are not mutually exclusive explanations, as evidenced by recent work on the efficiency of small-world networks (Fair et al., 2009; Sporns, 2011).

It is important to note that evidence regarding sex differences in brain function during performance of spatial, language, and emotion-related tasks is not completely consistent, for several reasons. First, results regarding sex differences in brain activation depend upon whether or not sex differences in task performance occurred and how they were considered. Second, samples are small in many neuroimaging studies, leading to low statistical power (and difficulty in detecting sex differences). Third, tasks and stimulus presentation methods vary across studies, so brain activation patterns are difficult to compare across studies. Fourth, participants vary across studies in sample characteristics, such as age; given sex differences in developmental trajectories (McClure et al., 2004; Plante et al., 2006; Rubia et al., 2010), it is important to be sure that findings reflect sex rather than developmental status. Fifth, the scanning environment constrains behavioral assessments in several ways (e.g., space, stimulus presentation), so the behavior measured in the scanner likely differs from the behavior that shows a sex difference outside the scanner (e.g., Peters and Battista, 2008).

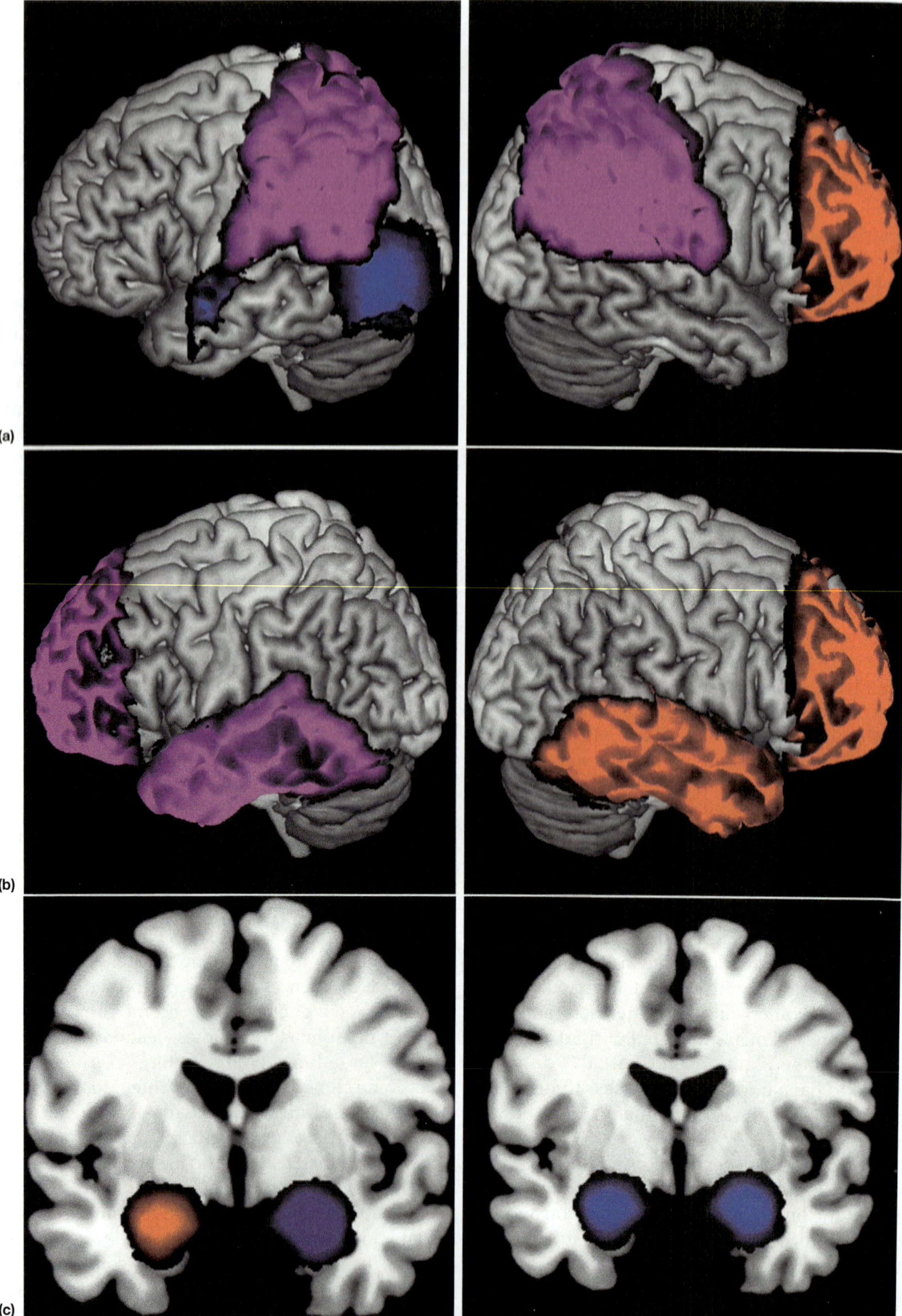

FIGURE 26.4 Summary of sex differences in neural activation to sex-typed tasks. Colors indicate areas of brain activation: Men in blue, women in red, overlap in purple. (a) Spatial tasks; right: right hemisphere, left: left hemisphere. (b) Language tasks; right: right hemisphere, left: left hemisphere. (c) Amygdala during emotion-related tasks depicted on coronal slice; right: facial recognition, left: emotional memories (see text for details).

26.5 CAUSES OF BRAIN SEX DIFFERENCES

The brain is plastic, and its development is influenced by a combination of genetic, epigenetic, hormonal, and experiential effects throughout the lifespan. Most aspects of brain development reflect a dynamic interplay among these factors and are therefore difficult to isolate and investigate (Cosgrove et al., 2007; McCarthy and Arnold, 2011; Poldrack, 2000).

Animal models are important for informing examinations of human neural processes. Many brain regions that show sex differences in human beings correspond to sexually dimorphic brain regions in nonhuman animals (Goldstein et al., 2001), and animal studies have helped to guide many human studies; however, the degree to which animal findings generalize to human beings is not always clear, particularly with respect to cognitive domains that are species-specific, such as language.

Research on the causes of human brain sex differences is relatively new; thus, the studies we review below must be interpreted critically and require replication. Parallel to our discussion of the causes of psychological sex differences (Section 26.3), we consider influences on brain sex differences of socialization, genes, and focus on hormones (prenatal, adolescent, and circulating).

26.5.1 Socialization Contributions to Brain Sex Differences

Brain sex differences are usually considered to be innate, but the brain changes in response to experiences, so it is important to consider socialization (Cosgrove et al., 2007). Although sex-differential experiences were not measured in most studies reviewed above (Section 26.4), socialization effects could account for variability in findings and inconsistencies across reports. For example, studies failing to find sex differences in white matter volume (as discussed in Section 26.4.2.4) could reflect effects of sex-differential experiences; perhaps certain male-typed experiences (e.g., engaging in spatial activities) facilitate white matter growth, so girls who have these experiences have more 'male-typical' white matter than girls who do not have these experiences.

There is much interesting work to be done regarding socialization influences on brain sex differences, particularly examining the ways in which training on cognitive tasks might differentially mediate (or be mediated by) brain sex differences. For example, we suggest above (Section 26.4.3.1) that sex differences in brain activation during the performance of spatial tasks might reflect sex-differential strategy use during task completion (see also Butler et al., 2006; Clements-Stephens et al., 2009; Jordan et al., 2002; Thomsen et al., 2000). We also review above evidence for the potential elimination of the behavioral sex difference in spatial tasks through training (Section 26.3.1.2). Thus, the sex difference in brain activation during spatial task performance might be partially explained by sex-differential experiences with spatial stimuli, and consequently, learned strategies for manipulation of those stimuli. Investigations of cognitive training effects on brain function of males and females are warranted.

26.5.2 Genetic Contributions to Brain Sex Differences

There is little conclusive evidence regarding sex chromosome contributions to human brain development and the ways in which genes and the environment interact to influence the brain in sex-differential ways. Recent twin data show greater genetic influences on white matter development in boys than girls (Chiang et al., 2011). Consistent with this, intriguing preliminary evidence indicates that several genes on the Y chromosome are expressed in the human male prenatal brain at midgestation (Reinius and Jazin, 2009) and the adult male brain (Vawter et al., 2004); however, development of the external genitalia is underway by midgestation and complete maturation has occurred by adulthood, so these findings might also reflect hormone (e.g., androgens from the testes) or experiential effects on the brain.

Some insight into the role of the X chromosome in brain sex differences is also provided by studies of girls and women with Turner syndrome (TS), who have a single (X) sex chromosome. Compared to unaffected controls, girls and women with TS have decreased parietal lobe volume and decreased activation in parietal lobe regions during a variety of visuospatial tasks; they also have increased amygdala volume and increased amygdala activation during the recognition of fearful faces. These findings are consistent with their cognitive profile, which is characterized, in part, by visuospatial and emotion recognition deficits (reviewed in Knickmeyer and Davenport, 2011). The implications of these findings for brain sex differences in the unaffected population are unclear for several reasons. First, brain differences are confounded by cognitive performance differences: TS and unaffected control groups generally differ on task performance in functional imaging studies. Second, sex hormone production is low in individuals with TS; thus, differential patterns of brain structure and function in individuals with TS as compared to those in unaffected girls and women might reflect sex hormone influences instead of X-chromosome influences on the brain. This is consistent with evidence showing that estradiol replacement in adolescents with TS reduces differences between girls with TS and control girls (Lepage et al., 2012). Third, it is unclear whether findings in individuals with TS reflect effects of the X chromosome, having a single (be it X or Y) sex chromosome, or X-chromosome dosage. Effects of X-chromosome dosage have been investigated by comparing brain structure in men with Klinefelter Syndrome (XXY) to unaffected men and women. Unfortunately, results from these studies are not easy to interpret because (as in studies in women with TS) X-dosage effects are confounded with effects of sex hormones (for discussion, see Bryant et al., 2011; Lentini et al., 2012). There is clearly much opportunity for future work on sex chromosome influences on brain sex differences.

26.5.3 Hormonal Contributions to Brain Sex Differences

Sex hormones, particularly androgens, are the most investigated cause of sex differences in the human brain. There is evidence from clinical and typical samples for prenatal, adolescent, and circulating sex hormone influences on the brain, but this research area is relatively new, and conclusions must be tentative.

26.5.3.1 *Prenatal Hormone Links to Brain Sex Differences*

The influence of androgens present during early development on structural brain sex differences has been examined in individuals with CAH and in typical boys whose amniotic testosterone levels were assessed. As discussed above (Section 26.3.4), androgens masculinize a variety of characteristics in girls and women with CAH (including activity interests, some social behaviors, and spatial ability). Structural imaging studies reveal effects that are generally more consistent with the disease process than with prenatal hormone effects on the brain; smaller amygdala volumes were reported in both males and females with CAH than in unaffected males and females (Merke et al., 2003), likely reflecting effects of postnatal cortisone treatment in individuals with CAH. In typical boys aged 8 to 11 years, testosterone levels from amniotic fluid have been linked to gray matter volumes in brain regions that show sex differences. Specifically, testosterone was positively linked to gray matter volumes in temporal regions, but negatively linked to volumes in occipital and frontal regions (Lombardo et al., 2012). More research is necessary to determine if this pattern of results is also seen in girls and if it changes with development.

Findings regarding prenatal androgen influences on brain function suggest altered amygdala and hippocampal activity in individuals with CAH, but interpretation is not straightforward. Consider findings from one set of studies in which images of emotional faces were viewed, rated, and recalled (Ernst et al., 2007; Mazzone et al., 2011). Brain activation findings showed that girls with CAH had greater amygdala activation while viewing negative facial emotions and less hippocampal activation while recalling emotional faces compared to unaffected girls; however, there were also group differences in behavioral ratings and task performance, complicating interpretation of brain activation differences. In a separate PET functional imaging study, women with CAH were not seen to differ from unaffected women in their neural response to olfactory stimuli: both groups displayed increased amygdala activation to a masculine pheromone and increased hypothalamus activation to a feminine pheromone, compared to men who showed the reciprocal pattern of brain activity (Ciumas et al., 2009).

Information about prenatal androgen effects on brain function also comes from typical samples, in which testosterone levels have been measured in amniotic fluid and linked to performance on dichotic listening tasks in childhood. Amniotic testosterone levels were positively associated with left-lateralized language processing in one sample of 6-year-old girls and boys (Lust et al., 2010) and another sample of 10-year-old girls (Grimshaw et al., 1995a). This is consistent with the sex difference in language lateralization reported in adults and children (Section 26.4.3.4). Amniotic testosterone levels have also been associated with right-lateralized emotion processing (measured with an emotion–word dichotic listening task) in 10-year-old boys (Grimshaw et al., 1995a), consistent with meta-analytic findings on right-lateralized amygdala activation in men (but not women) during emotion-related tasks (Wager et al., 2003).

26.5.3.2 *Adolescent Hormone Links to Brain Sex Differences*

There is considerable interest in the ways that brain changes in adolescence reflect the direct effects of pubertal sex hormones, although most work has not been able to differentiate permanent (organizational) effects of hormones from transient (activational) effects. Generally, pubertal increases in sex-specific gonadal hormones (estrogen in girls and testosterone in boys) are associated with decreases in cortical gray matter volume (reviewed in Peper et al., 2011). When boys and girls are matched on pubertal development, however, sex differences in links between testosterone and regional gray matter volumes are more difficult to interpret, with boys showing fewer significant associations than girls (Bramen et al., 2012). This apparent discrepancy may be explained by androgen sensitivity. The efficiency of the androgen receptor gene has been suggested to moderate the influence of sex hormones on cortical thinning: fewer repeats of a functional polymorphism within the gene predicted more 'masculinized' patterns of maturation (Paus et al., 2010; Raznahan et al., 2010).

There is stronger evidence for a link between pubertal hormones and white matter development in boys than in girls. In boys, increases in testosterone at puberty have been linked to increases in white matter volume (reviewed in Peper et al., 2011), with the efficiency of the androgen receptor gene moderating this association (Paus et al., 2010; Perrin et al., 2008). Testosterone levels – but not pubertal stage – have also been linked to FA increases in boys (Bava et al., 2011; Herting et al., 2012). Findings are inconsistent for girls: advanced pubertal stage has been positively linked to increases in white matter volume (Peper et al., 2011), but estradiol levels have been negatively associated with FA in girls (Herting et al., 2012). Differences in brain links with hormone levels versus pubertal stage may reflect limitations in self-reports of pubertal stage for both sexes (discussed in Dorn et al., 2006).

Pubertal hormone effects on brain function have been studied less than effects on brain structure, but some insight is provided by a study on the neural substrates of a monetary reward task (Forbes et al., 2010). Girls and boys with more advanced pubertal development (compared to same-age peers with less advanced development) showed less striatum and more prefrontal cortex activation at reward *presentation*, consistent with evidence for the discordant maturation of affective and cognitive neural processing networks in adolescence (Section 26.4.3.5). For boys only, testosterone levels were positively linked with striatum activity during reward *anticipation*, perhaps reflecting a sex difference in the neural substrates of consummatory anticipation that emerges at puberty. Estradiol links to brain function were not examined in this study, presumably because the measurement of this hormone is problematic, particularly during puberty (Dorn and Biro, 2011).

The influence of pubertal hormones on brain sex differences has also been investigated in boys with extremely early (disordered) puberty (Mueller et al., 2009, 2011a,b). Results are difficult to interpret, however, because of age and performance differences between clinical and control groups. Nonetheless, this research illustrates a valuable approach (i.e., neuroimaging individuals with atypical pubertal timing) to understanding hormonal influences on brain development.

26.5.3.3 *Circulating Hormone Links to Brain Sex Differences*

Relations between regional volumes and circulating hormone levels in adults provide evidence for transient (activational) effects of sex hormones on sex-related brain structure, as seen in both clinical and typical samples. Changes in overall brain volume were observed in a sample of transsexuals after 4 months of hormone treatment (antiandrogens + estrogens for male-to-female, and androgens for female-to-male) (Hulshoff Pol et al., 2006). Compared to untreated controls, male-to-female transsexuals showed decreases in overall brain volume, and female-to-male transsexuals showed increases in overall brain and hypothalamus volumes. These findings are consistent with the sex differences reported above (Sections 26.4.2.1. and 26.4.2.2.3) and suggest that androgens increase overall and some regional brain volumes. Nonetheless, the results regarding circulating androgen influences on overall brain volume are surprising, given that the sex difference in brain volume is apparent in childhood (Section 26.4.2.1). Regional specificity of hormone effects has been seen in samples of typical adults, but results differ across studies, likely due to differences in sample characteristics (e.g., age) and methods (e.g., how sex differences were accounted for in brain-hormone links). Circulating estrogen levels were positively linked to gray matter volume in the left superior parietal gyrus, but results differ across studies, likely due to differences in sample characteristics (e.g., age) and methods (e.g., how sex differences were accounted for in brain-hormone links). Circulating testosterone levels were negatively linked to gray matter volume in the left inferior frontal gyrus (Witte et al., 2010) and positively linked to gray matter volumes in the hippocampus, amygdala, insula, and occipital regions (Lentini et al., 2012)

Examinations of structural and functional brain changes at menopause and during different phases of the menstrual cycle provide indirect evidence for activational effects of estrogens on the brain. Generally, the hippocampus and frontal lobes are implicated as the primary sites for estrogen effects on brain structure in humans; thus, estrogen effects on brain function are most evident in verbal memory tasks, which show a sex difference that favors females (Section 26.2.1.4), and are subserved by the hippocampus and frontal lobes (Maki, 2005; Maki and Resnick, 2001). At menopause, there is evidence for decreases in whole brain, frontal lobe, and hippocampal volumes (Goto et al., 2011; Robertson et al., 2009). Further, estrogen therapy begun around the onset of menopause and of a relatively short duration (about 5 years) offsets these volume reductions (Erickson et al., 2010; Lord et al., 2008; Resnick et al., 2009). Postmenopausal women undergoing estrogen therapy have also been shown to have greater activation of frontal, parietal, and hippocampal regions during verbal and working memory tasks in comparison to postmenopausal women who never used estrogen therapy (Berent-Spillson et al., 2010; Dumas et al., 2010; Persad et al., 2009; Shaywitz et al., 1999). Interestingly, the increased activation with hormone therapy did not predict better memory in these studies; there were no performance differences between estrogen therapy users and nonusers on the memory tasks. Estrogen therapy users, however, had greater cerebral blood flow and better memory task performance than nonusers in research using PET; thus, cerebral blow flood may be one mechanism through which estrogen influences cognition (and is best measured by PET; Maki and Resnick, 2000; Resnick et al., 1998). This is a promising area for future research.

Indirect evidence for estrogen influences on the structure and function of the brain is also provided by research with women at different phases of their menstrual cycle. In high-estrogen phases of the menstrual cycle as compared to low-estrogen phases, women generally show more hippocampal volume as well as left frontal and temporal activation during the completion of verbal and mental rotation tasks (Dietrich et al., 2001; Fernández et al., 2003; Konrad et al., 2008; Protopopescu et al., 2008; Schöning et al., 2007). As with findings from studies on estrogen therapy at menopause, the implications of this research are not clear, because there is currently little evidence for a relation between estrogen-influenced brain activation and task performance: in the functional studies reviewed above, increased left frontal lobe activation in the high-estrogen phase of the cycle did not predict differential task performance (Dietrich et al., 2001; Fernández et al., 2003; Konrad et al., 2008). Similarly, during the high-estrogen phase of the menstrual cycle as compared to the low-estrogen phase, estradiol levels predicted greater left frontal lobe activation during a verbal memory task, but they did not predict better task performance (Craig et al., 2008).

In summary, there is little conclusive evidence regarding social, genetic, or hormonal contributions to sex differences in human brain structure and function, but this research area is growing. Most of the available data concern hormones. Preliminary findings suggest that prenatal androgen influences the activity of the hippocampus, amygdala, and lateralization of the brain for language- and emotion-related tasks. There is also some indication that sex-specific hormones at puberty are linked to the reduction of gray matter volume in adolescence, and that testosterone in boys is linked to the adolescent increase in white matter. Finally, there is evidence for activational effects of sex hormones on the adult brain, particularly for estrogen influences on hippocampal and frontal lobe function. Unfortunately, there are few data linking sex hormone influences on the brain with behavior. There are also few data regarding sex chromosome expression in the human brain, and no data regarding socialization influences on brain sex differences. Thus, there is much opportunity for future research.

26.6 CONCLUSIONS AND FUTURE DIRECTIONS

As reviewed above, the sexes differ in significant ways in both behavior and the brain (including structure and function), although there is still much to be learned regarding sex differences and their development and causes. It is important to be aware of the ways in which age and sample size can affect conclusions regarding the existence – or not – of sex differences.

There is good evidence that both sex hormones and social factors influence the development of human sex-related psychological characteristics, although little is currently known about the pathways by which socialization acts on and modifies biological predispositions. Determinants of brain sex differences are still largely unknown, but it is important to remember that the brain is plastic and that sex differences likely emerge from the interplay of genes, sex hormones, and social experiences. Brain sex differences cannot be assumed to be innate.

There are several opportunities to study the etiology of sex differences in brain and behavior. First, sex differences in early brain development can be studied in natural experiments in which prenatal hormone exposure is sex-atypical, such as CAH. These conditions help us to separate the relative influences of early (prenatal and neonatal) hormones and postnatal socialization, and, more important, allow the study of their interplay. Examples of the information that can be gained from these conditions have been reviewed in this chapter and elsewhere (Berenbaum and Beltz, 2011; Blakemore et al., 2009). Second, sex differences in adolescent brain development can be studied in individuals with disordered pubertal development. This provides an opportunity to test hypotheses about permanent changes to the brain induced by sex hormones (Sisk and Zehr, 2005, reviewed in Berenbaum and Beltz, 2011; Giedd et al., 2006). Third, sex differences in adolescent brain development can also be studied in typical individuals, by linking variations in pubertal development to changes in behavior and in the brain. There is some intriguing recent cross-sectional work in this area (reviewed in Berenbaum and Beltz, 2011; Peper et al., 2011), but longitudinal studies are optimal for learning how hormones and socialization at puberty influence brain changes and for examining the permanence of the changes.

The links between neural and behavioral sex differences are beginning to be understood, but they must be directly tested. It is not sufficient to find sex differences in regions of the brain known to subserve specific sex-related behavioral characteristics; sex differences in those regions of the brain must be shown to *explain* the behavioral sex differences.

There are several opportunities to understand direct links between sex differences in the brain and behavior. First, the meaning of sex differences in brain activation – when they reflect strategy use (when the sexes perform similarly) or behavioral differences (when the sexes differ in performance) – can be understood by manipulating task difficulty during functional brain scanning. Such types of designs are rare in examinations of sex differences, but they are typical in other research areas (e.g., Poldrack, 2000). Second, causal inferences about links between brain and behavior can be strengthened by using effective connectivity mapping which takes advantage of the time series nature of functional neuroimaging data. Work on functional and structural connectivity reviewed in this chapter has identified brain regions with related functional activity (e.g., Fair et al., 2007, 2009) and structural connections (e.g., Schmithorst and Yuan, 2010), but effective connectivity mapping would allow prediction of regional brain activity from activity in other brain regions or task performance (e.g., Gates et al., 2010, 2011; Kim et al., 2007). For example, effective connectivity mapping enables inferences regarding the direction of a connection between two brain regions: the connection might go from region A to region B for women, but from region B to region A for men. Third, neural effects of sex-differential experiences can be studied using longitudinal neuroimaging designs. For example, studies of cognitive training effects on changes in brain structure or function can reveal the plasticity of the brain and indicate how sex differences arise and how they might be modified (and the within-subject design avoids the confounding influence of sex differences in brain size) (e.g., Thambisetty et al., 2010).

In sum, the sexes are similar in many ways, but there are notable differences in their brain and behavior (including cognitive abilities). An understanding of sex differences has implications for discussions about women's underrepresentation in science and mathematics careers as well as for the differences in the etiology, appearance, and treatment of psychopathology between boys and men as compared to girls and women. Studying sex differences can also tell us about individual differences generally. By identifying the differences, delineating their development, and investigating their causes, we begin to uncover the mechanisms underlying variation in human health, disease, and behavior.

References

Allen, L.S., Gorski, R.A., 1990. Sex difference in the bed nucleus of the stria terminalis of the human brain. The Journal of Comparative Neurology 302, 697–706.

Allen, L.S., Gorski, R.A., 1991. Sexual dimorphism of the anterior commisure and massa intermedia of the human brain. The Journal of Comparative Neurology 312, 97–104.

Allen, L.S., Gorski, R.A., 1992. Sexual orientation and the size of the anterior commissure in the human brain. Proceedings of the National Academy of Sciences of the United States of America 89, 7199–7202.

Allen, J.S., Damasio, H., Grabowski, T.J., Bruss, J., Zhang, W., 2003. Sexual dimorphism and asymmetries in the gray-white composition of the human cerebrum. NeuroImage 18, 880–894.

Allendorfer, J.B., Lindsell, C.J., Siegel, M., et al., 2012. Females and males are highly similar in language performance and cortical activation patterns during verb generation. Cortex 48, 1218–1233.

Archer, J., 2004. Sex differences in aggression in real-world settings: A meta-analytic review. Review of General Psychology 8, 291–322.

Arnold, A.P., 2009. The organizational–activational hypothesis as the foundation for a unified theory of sexual differentiation of all mammalian tissues. Hormones and Behavior 55, 570–578.

Arnold, A.P., Breedlove, S.M., 1985. Organizational and activational effects of sex steroids on brain and behavior: A reanalysis. Hormones and Behavior 19, 469–498.

Arnold, A.P., Chen, X., 2009. What does the 'four core genotypes' mouse model tell us about sex differences in the brain and other tissues? Frontiers in Neuroendocrinology 30, 1–9.

Auyeung, B., Baron-Cohen, S., Ashwin, E., et al., 2009. Fetal testosterone predicts sexually differentiated childhood behavior in girls and in boys. Psychological Science 20, 144–148.

Baenninger, M., Newcombe, N., 1989. The role of experience in spatial test performance: A meta-anaylsis. Sex Roles 20, 327–344.

Baron-Cohen, S., Lutchmaya, S., Knickmeyer, R., 2004. Prenatal Testosterone in Mind. MIT Press, Cambridge, MA.

Bauer, P., 2013. Memory development. Neural Circuit Development and Function in the Healthy and Diseased Brain: Comprehensive Developmental Neuroscience, Volume 3. Elsevier, Oxford.

Bava, S., Boucquey, V., Goldenberg, D., et al., 2011. Sex differences in adolescent white matter architecture. Brain Research 1375, 41–48.

Beck, D.M., 2010. The appeal of the brain in the popular press. Perspectives on Psychological Science 5, 762–766.

Becker, J.B., Breedlove, S.M., Crews, D., McCarthy, M.M. (Eds.), 2002. Behavioral Endocrinology. MIT Press, Cambridge, MA.

Becker, J.B., Berkley, K.J., Geary, N., Hampson, E., Herman, J., Young, E. (Eds.), 2008. Sex Differences in the Brain: From Genes to Behavior. Oxford University Press, New York.

Bell, E.C., Willson, M.C., Wilman, A.H., Dave, S., Silverstone, P.H., 2006. Males and females differ in brain activation during cognitive tasks. NeuroImage 30, 529–538.

Beltz, A.M., Swanson, J.L., Berenbaum, S.A., 2011. Gendered occupational interests: Prenatal androgen effects on psychological orientation to things versus people. Hormones and Behavior 60, 313–317.

Berenbaum, S.A., 2006. Psychological outcome in children with disorders of sex development: Implications for treatment and understanding typical development. Annual Review of Sex Research 17, 1–38.

Berenbaum, S.A., Beltz, A.M., 2011. Sexual differentiation of human behavior: Effects of prenatal and pubertal organizational hormones. Frontiers in Neuroendocrinology 32, 183–200.

Berenbaum, S.A., Baxter, L., Seidenberg, M., Hermann, B., 1997. Role of the hippocampus in sex differences in verbal memory: Memory outcome following left anterior temporal lobectomy. Neuropsychology 11, 585–591.

Berenbaum, S.A., Blakemore, J.E.O., Beltz, A.M., 2011. A role for biology in gender-related behavior. Sex Roles 64, 804–825.

Berenbaum, S.A., Bryk, K.L.K., Beltz, A.M., 2012. Early androgen effects on spatial and mechanical abilities: Evidence from congenital adrenal hyperplasia. Behavioral Neuroscience 126, 86–96.

Berent-Spillson, A., Persad, C.C., Love, T., et al., 2010. Early menopausal hormone use influences brain regions used for visual working memory. Menopause 17, 692–699.

Bishop, K.M., Wahlsten, D., 1997. Sex differences in the human corpus callosum: Myth or reality? Neuroscience and Biobehavioral Reviews 21, 581–601.

Blakemore, J.E.O., Berenbaum, S.A., Liben, L.S., 2009. Gender Development. Psychology Press/Taylor & Francis, New York.

Bornstein, M.H., Hahn, C.-S., Haynes, O.M., 2004. Specific and general language performance across early childhood: Stability and gender considerations. First Language 24, 267–304.

Bramen, J.E., Hranilovich, J.A., Dahl, R.E., et al., 2012. Sex matters during adolescence: Testosterone-related cortical thickness maturation differs between boys and girls. PloS One 7, e33850.

Brun, C.C., Leporé, N., Luders, E., et al., 2009. Sex differences in brain structure in auditory and cingulate regions. NeuroReport 20, 930–935.

Bryant, D.M., Hoeft, F., Lai, S., et al., 2011. Neuroanatomical phenotype of Klinefelter Syndrome in childhood: A voxel-based morphometry study. Journal of Neuroscience 31, 6654–6660.

Bryden, M.P., 1982. Laterality: Functional Asymmetry in the Intact Brain. Academic Press, New York.

Buchanan, C.M., Eccles, J.S., Becker, J.B., 1992. Are adolescents the victims of raging hormones: Evidence for activational effects of hormones on moods and behavior at adolescence. Psychological Bulletin 111, 62–107.

Buckner, R.L., Raichle, M.E., Petersen, S.E., 1995. Dissociation of human prefrontal cortical areas across different speech production tasks and gender groups. Journal of Neurophysiology 74, 2163–2173.

Burman, D.D., Bitan, T., Booth, J.R., 2008. Sex differences in neural processing of language among children. Neuropsychologia 46, 1349–1362.

Burns, N.R., Nettelbeck, T., 2005. Inspection time and speed of processing: Sex differences on perceptual speed but not it. Personality and Individual Differences 39, 439–446.

Butler, T., Imperato-McGinley, J., Pan, H., et al., 2006. Sex differences in mental rotation: Top-down versus bottom-up processing. NeuroImage 32, 445–456.

Byrnes, J.P., Miller, D.C., Schafer, W.D., 1999. Gender differences in risk taking: A meta-analysis. Psychological Bulletin 125, 367–383.

Cahill, L., 2005. His brain, her brain. Scientific American 292, 40–47.

Cahill, L., 2010. Sex influences on brain and emotional memory: The burden of proof has shifted. Progress in Brain Research 186, 29–40.

Camarata, S., Woodcock, R.W., 2006. Sex differences in processing speed: Developmental effects in males and females. Intelligence 34, 231–252.

Card, N.A., Stucky, B.D., Sawalani, G.M., Little, T.D., 2008. Direct and indirect aggression during childhood and adolescence: A meta-analytic review of gender differences, intercorrelations, and relations to maladjustment. Child Development 79, 1185–1229.

Casey, B.J., Tottenham, N., Liston, C., Durston, S., 2005. Imaging the developing brain: What have we learned about cognitive development? Trends in Cognitive Sciences 9, 104–110.

Casey, B.J., Jones, R.M., Somerville, L.H., 2011. Braking and accelerating of the adolescent brain. Journal of Research on Adolescence 21, 21–33.

Ceci, S.J., Williams, W.M., 2010. Sex differences in math-intensive fields. Current Directions in Psychological Science 19, 275–279.

Chiang, M.C., McMahon, K.L., de Zubicaray, G.I., et al., 2011. Genetics of white matter development: A DTI study of 705 twins and their siblings aged 12 to 29. NeuroImage 54, 2308–2317.

Christova, P.S., Lewis, S.M., Tagaris, G.A., Uğurbil, K., Georgopoulos, A.P., 2008. A voxel-by-voxel parametric fMRI study of motor mental rotation: Hemispheric specialization and gender differences in neural processing efficiency. Experimental Brain Research 189, 79–90.

Chudowsky, N., Chudowsky, V., 2010. State Test Score Trends Through 2007–08, Part 5: Are There Differences in Achievement Between Boys and Girls? Center on Education Policy, Washington, DC.

Ciumas, C., Hirschberg, A.L., Savic, I., 2009. High fetal testosterone and sexually dimorphic cerebral networks in females. Cerebral Cortex 19, 1167–1174.

Clark, M.M., Galef, B.G., 1998. Effects of intrauterine position on the behavior and genital morphology of litter-bearing rodents. Developmental Neuropsychology 14, 197–211.

Clements, A.M., Rimrodt, S.L., Abel, J.R., et al., 2006. Sex differences in cerebral laterality of language and visuospatial processing. Brain and Language 98, 150–158.

Clements-Stephens, A.M., Rimrodt, S.L., Cutting, L.E., 2009. Developmental sex differences in basic visuospatial processing: Differences in strategy use? Neuroscience Letters 449, 155–160.

Cohen, J., 1988. Statistical Power Analysis for the Behavioral Sciences. Academic Press, New York.

Cohen-Bendahan, C.C.C., van de Beek, C., Berenbaum, S.A., 2005. Prenatal sex hormone effects on child and adult sex-typed behavior: Methods and findings. Neuroscience and Biobehavioral Reviews 29, 353–384.

Colby, J.B., Van Horn, J.D., Sowell, E.R., 2011. Quantitative in vivo evidence for broad regional gradients in the timing of white matter maturation during adolescence. NeuroImage 54, 25–31.

Cole-Harding, S., Morstad, A.L., Wilson, J.R., 1988. Spatial ability in members of opposite-sex twin pairs (abstract). Behavior Genetics 18, 710.

Connor, J.M., Serbin, L.A., 1977. Behaviorally based masculine and feminine activity-preference scales for preschoolers: Correlates with other classroom behaviors and cognitive tests. Child Development 48, 1411–1416.

Cosgrove, K.P., Mazure, C.M., Staley, J.K., 2007. Evolving knowledge of sex differences in brain structure, function, and chemistry. Biological Psychiatry 62, 847–855.

Costa Jr., P., Terracciano, A., McCrae, R.R., 2001. Gender differences in personality traits across cultures: Robust and surprising findings. Journal of Personality and Social Psychology 81, 322–331.

Craig, M.C., Fletcher, P.C., Daly, E.M., et al., 2008. Physiological variation in estradiol and brain function: A functional magnetic resonance imaging study of verbal memory across the follicular phase of the menstrual cycle. Hormones and Behavior 53, 503–508.

Crick, N.R., 1995. Relational aggression: The role of intent attributions, feelings of distress, and provocation type. Development and Psychopathology 7, 313–322.

Crick, N.R., Zahn-Waxler, C., 2003. The development of psychopathology in females and males: Current progress and future challenges. Development and Psychopathology 15, 719–742.

Critchley, H., Daly, E., Phillips, M., et al., 2000. Explicit and implicit neural mechanisms for processing of social information from facial expressions: A functional magnetic resonance imaging study. Human Brain Mapping 9, 93–105.

Dar-Nimrod, I., Heine, S.J., 2006. Exposure to scientific theories affects women's math performance. Science 314, 435.

Davatzikos, C., Resnick, S.M., 1998. Sex differences in anatomic measures of interhemispheric connectivity: Correlations with cognition in women but not men. Cerebral Cortex 8, 635–640.

De Bellis, M.D., Keshavan, M.S., Beers, S.R., et al., 2001. Sex differences in brain maturation during childhood and adolescence. Cerebral Cortex 11, 552–557.

De Lacoste-Utamsing, C., Holloway, R.L., 1982. Sexual dimorphism in the human corpus callosum. Science 216, 1431–1432.

De Vries, G.J., 2004. Minireview: Sex differences in adult and developing brains: Compensation, compensation, compensation. Endocrinology 145, 1063–1068.

Derntl, B., Habel, U., Windischberger, C., et al., 2009. General and specific responsiveness of the amygdala during explicit emotion recognition in females and males. BMC Neuroscience 10, 91.

Dessens, A.B., Slijper, F.M.E., Drop, S.L.S., 2005. Gender dysphoria and gender change in chromosomal females with congenital adrenal hyperplasia. Archives of Sexual Behavior 34, 389–397.

Diekman, A.B., Brown, E.R., Johnston, A.M., Clark, E.K., 2010. Seeking congruity between goals and roles: A new look at why women opt out of science, technology, engineering, and mathematics careers. Psychological Science 21, 1051–1057.

Dietrich, T., Krings, T., Neulen, J., et al., 2001. Effects of blood estrogen level on cortical activation patterns during cognitive activation as measured by functional MRI. NeuroImage 13, 425–432.

Dodge, K.A., Coie, J.D., Lynam, D., 2006. Aggression and antisocial behavior in youth. In: Eisenberg, N. (Ed.), Handbook of Child Psychology 6th edn. Social, Emotional, and Personality Development, vol. 3. Wiley, Hoboken, NJ.

Donnelly, K.M., Allendorfer, J.B., Szaflarski, J.P., 2011. Right hemispheric participation in semantic decision improves performance. Brain Research 1419, 105–116.

Dorfberger, S., Adi-Japhab, E., Karnia, A., 2009. Sex differences in motor performance and motor learning in children and adolescents: An increasing male advantage in motor learning and consolidation phase gains. Behavioural Brain Research 198, 165–171.

Dorn, L.D., Biro, F.M., 2011. Puberty and its measurement: A decade in review. Journal of Research on Adolescence 21, 180–195.

Dorn, L.D., Dahl, R.E., Woodward, H.R., Biro, F., 2006. Defining the boundaries of early adolescence: A user's guide to assessing pubertal status and pubertal timing in research with adolescents. Applied Developmental Science 10, 30–56.

Dubb, A., Gur, R., Avants, B., Gee, J., 2003. Characterization of sexual dimorphism in the human corpus callosum. NeuroImage 20, 512–519.

Dumas, J.A., Kutz, A.M., Naylor, M.R., Johnson, J.V., Newhouse, P.A., 2010. Increased memory load-related frontal activation after estradiol treatment in postmenopausal women. Hormones and Behavior 58, 929–935.

Eagly, A.H., Wood, W., 1999. The origins of sex differences in human behavior: Evolved dispositions versus social roles. American Psychologist 54, 408–423.

Eaton, W.O., Enns, L.R., 1986. Sex differences in human motor activity level. Psychological Bulletin 100, 19–28.

Eccles, J.S., Freedman-Doan, C., Frome, P., Jacobs, J., Yoon, K.S., 2000. Gender-role socialization in the family: A longitudinal approach. In: Eckes, T., Trautner, H.M. (Eds.), The Developmental Social Psychology of Gender. Erlbaum, Mahwah, NJ.

Eisenberg, N., Spinrad, T.L., Sadovsky, A., 2006. Empathy-related responding in children. In: Killen, M., Smetana, J.G. (Eds.), Handbook of Moral Development. Lawrence Erlbaum Associates, Mahwah, NJ.

Else-Quest, N.M., Hyde, J.S., Goldsmith, H.H., Van Hulle, C.A., 2006. Gender differences in temperament: A meta-analysis. Psychological Bulletin 132, 33–72.

Else-Quest, N.M., Hyde, J.S., Linn, M.C., 2010. Cross-national patterns of gender differences in mathematics: A meta-analysis. Psychological Bulletin 136, 103–127.

Eluvathingal, T.J., Hasan, K.M., Kramer, L., Fletcher, J.M., Ewing-Cobbs, L., 2007. Quantitative diffusion tensor tractography of association and projection fibers in normally developing children and adolescents. Cerebral Cortex 17, 2760–2768.

Erickson, K.I., Voss, M.W., Prakash, R.S., Chaddock, L., Kramer, A.F., 2010. A cross-sectional study of hormone treatment and hippocampal volume in postmenopausal women: Evidence for a limited window of opportunity. Neuropsychology 24, 68–76.

Ernst, M., Maheu, F.S., Schroth, E., et al., 2007. Amygdala function in adolescents with congenital adrenal hyperplasia: A model for the study of early steroid abnormalities. Neuropsychologia 45, 2104–2113.

Fair, D.A., Dosenbach, N.U.F., Church, J.A., et al., 2007. Development of distinct control networks through segregation and integration. Proceedings of the National Academy of Sciences of the United States of America 104, 13507–13512.

Fair, D.A., Cohen, A.L., Power, J.D., et al., 2009. Functional brain networks develop from a 'local to distributed' organization. PLoS Computational Biology 5, e1000381.

Falk, D., Froese, N., Sade, D.S., Dudek, B.G., 1999. Sex differences in brain/body relationships of rhesus monkeys and humans. Journal of Human Evolution 36, 233–238.

Feng, J., Spence, I., Pratt, J., 2007. Playing an action video game reduces gender differences in spatial cognition. Psychological Science 18, 850–855.

Fernández, G., Weis, S., Stoffel-Wagner, B., et al., 2003. Menstrual cycle-dependent neural plasticity in the adult human brain is hormone, task, and region specific. Journal of Neuroscience 23, 3790–3795.

Fiedler, K., 2011. Voodoo correlations are everywhere-not only in neuroscience. Perspectives on Psychological Science 6, 163–171.

Filipek, P.A., Richelme, C., Kennedy, D.N., Caviness, V.S., 1994. The young adult human brain: An MRI-based morphometric analysis. Cerebral Cortex 4, 344–360.

Fine, C., 2010. Delusions of Gender: How Our Minds, Society, and Neurosexism Create Difference. W.W. Norton, New York.

Fivush, R., 1998. Gendered narratives: Elaboration, structure, and emotion in parent-child reminiscing across the preschool years. In: Thompson, C.P., Herrmann, D.J. (Eds.), Autobiographical Memory: Theoretical and Applied Perspectives. Lawrence Erlbaum Associates, Mahwah, NJ.

Fivush, R., Buckner, J.P., 2000. Gender, sadness, and depression: The development of emotional focus through gendered discourse. In: Fischer, A.H. (Ed.), Gender and Emotion: Social Psychological Perspectives. Studies in Emotion and Social InteractionCambridge University Press, New York.

Flashman, L.A., Andreasen, N.C., Flaum, M., Swayze, V.W., 1997. Intelligence and regional brain volumes in normal controls. Intelligence 25, 149–160.

Forbes, E.E., Dahl, R.E., 2010. Pubertal development and behavior: Hormonal activation of social and motivational tendencies. Brain and Cognition 72, 66–72.

Forbes, E.E., Ryan, N.D., Phillips, M.L., et al., 2010. Healthy adolescents' neural response to reward: Associations with puberty, positive affect, and depressive symptoms. Journal of the American Academy of Child and Adolescent Psychiatry 49, 162–172.

Friston, K.J., 2004. Human Brain Function. Elsevier, San Diego, CA.

Frost, J.A., Binder, J.R., Springer, J.A., et al., 1999. Language processing is strongly left lateralized in both sexes: Evidence from functional MRI. Brain 122, 199–208.

Fusar-Poli, P., Placentino, A., Carletti, F., et al., 2009. Functional atlas of emotional faces processing: A voxel-based meta-analysis of 105 functional magnetic resonance imaging studies. Journal of Psychiatry & Neuroscience 34, 418–432.

Gates, K.M., Molenaar, P.C.M., Hillary, F.G., Ram, N., Rovine, M.J., 2010. Automatic search for fMRI connectivity mapping: An alternative to granger causality testing using formal equivalences among SEM path modeling, VAR, and unified SEM. NeuroImage 50, 1118–1125.

Gates, K.M., Molenaar, P.C.M., Hillary, F.G., Slobounov, S., 2011. Extended unified SEM approach for modeling event-related fMRI data. NeuroImage 54, 1151–1158.

Gauthier, C.T., Duyme, M., Zanca, M., Capron, C., 2009. Sex and performance level effects on brain activation during a verbal fluency task: A functional magnetic resonance imaging study. Cortex 45, 164–176.

Geary, D.C., 1998. Male, Female: The Evolution of Human Sex Differences. American Psychological Association, Washington, DC.

Geiser, C., Lehmann, W., Eid, M., 2008. A note on sex differences in mental rotation in different age groups. Intelligence 36, 556–563.

Gentile, B., Grabe, S., Dolan-Pascoe, B., Twenge, J.M., Wells, B.E., Maitino, A., 2009. Gender differences in domain-specific self-esteem: A meta-analysis. Review of General Psychology 13, 34–45.

Gibber, J., Goy, R.W., 1985. Infant directed behavior in young rhesus monkey: Sex difference and effects of prenatal androgen. American Journal of Primatology 8, 235–237.

Giedd, J.N., Castellanos, F.X., Rajapakse, J.C., Vaituzis, A.C., Rapoport, J.L., 1997. Sexual dimorphism of the developing human brain. Progress in Neuro-Psychopharmacology & Biological Psychiatry 21, 1185–1201.

Giedd, J.N., Blumenthal, J., Jeffries, N.O., et al., 1999. Brain development during childhood and adolescence: A longitudinal MRI study. Nature Neuroscience 2, 861–863.

Giedd, J.N., Clasen, L.S., Lenroot, R., et al., 2006. Puberty-related influences on brain development. Molecular and Cellular Endocrinology 254, 154–162.

Giedd, J.N., Rapoport, J.L., 2010. Structural MRI of pediatric brain development: What have we learned and where are we going? Neuron 67, 728–734.

Giorgio, A., Watkins, K.E., Douaud, G., et al., 2008. Changes in white matter microstructure during adolescence. NeuroImage 39, 52–61.

Gogtay, N., Giedd, J.N., Lusk, L., et al., 2004. Dynamic mapping of human cortical development during childhood through early adulthood. Proceedings of the National Academy of Sciences of the United States of America 101, 8174–8179.

Goldstein, J.M., Seidman, L.J., Horton, N.J., et al., 2001. Normal sexual dimorphism of the adult human brain assessed by in vivo magnetic resonance imaging. Cerebral Cortex 11, 490–497.

Gorski, R.A., Gordon, J.H., Shryne, J.E., Southam, A.M., 1978. Evidence for a morphological sex difference within the medial preoptic area of the rat brain. Brain Research 148, 333–346.

Goto, M., Abe, O., Miyati, T., et al., 2011. 3 tesla MRI detects accelerated hippocampal volume reduction in postmenopausal women. Journal of Magnetic Resonance Imaging 33, 48–53.

Goy, R.W., McEwen, B.S., 1980. Sexual Differentiation of the Brain. MIT Press, Cambridge.

Goy, R.W., Bercovitch, F.B., McBrair, M.C., 1988. Behavioral masculinization is independent of genital masculinization in prenatally androgenized female rhesus macaques. Hormones and Behavior 22, 552–571.

Grimshaw, G.M., Bryden, M.P., Finegan, J.A.K., 1995a. Relations between prenatal testosterone and cerebral lateralization in children. Neuropsychology 9, 68–79.

Grimshaw, G.M., Sitarenios, G., Finegan, J.A., 1995b. Mental rotation at 7 years: Relations with prenatal testosterone levels and spatial play experience. Brain and Cognition 29, 85–100.

Grön, G., Wunderlich, A.P., Spitzer, M., Tomczak, R., Riepe, M.W., 2000. Brain activation during human navigation: Gender-different neural networks as substrate of performance. Nature Neuroscience 3, 404–408.

Guiso, L., Monte, F., Sapienza, P., Zingales, L., 2008. Culture, gender, and math. Science 320, 1164–1165.

Gur, R.C., Turetsky, B.I., Matsui, M., et al., 1999. Sex differences in brain gray and white matter in healthy young adults: Correlations with cognitive performance. Journal of Neuroscience 19, 4065–4072.

Gur, R.C., Alsop, D., Glahn, D., et al., 2000. An fMRI study of sex differences in regional activation to a verbal and a spatial task. Brain and Language 74, 157–170.

Habel, U., Windischberger, C., Derntl, B., et al., 2007. Amygdala activation and facial expressions: Explicit emotion discrimination versus implicit emotion processing. Neuropsychologia 45, 2369–2377.

Halari, R., Sharma, T., Hines, M., Andrew, C., Simmons, A., Kumari, V., 2006. Comparable fMRI activity with differential behavioural performance on mental rotation and overt verbal fluency tasks in healthy men and women. Experimental Brain Research 169, 1–14.

Hall, J.J., Neal, T.J., Dean, R.S., 2008. Lateralization of cerebral functions. In: Horton, A.M., Wedding, D. (Eds.), The Neuropsychology Handbook. 3rd edn. Springer, New York.

Halpern, D.F., 1986. A different answer to the question, 'do sex-related differences in spatial abilities exist?'. American Psychologist 41, 1014–1015.

Halpern, D.F., 2012. Sex Differences in Cognitive Abilities. Psychology Press, New York.

Halpern, D.F., Benbow, C.P., Geary, D.C., Gur, R.C., Hyde, J.S., Gernsbacher, M.A., 2007. The science of sex differences in science and mathematics. Psychological Science in the Public Interest 8, 1–51.

Hamann, S., 2005. Sex differences in the responses of the human amygdala. The Neuroscientist 11, 288–293.

Hampson, E., 2007. Endocrine contributions to sex differences in visuospatial perception and cognition. In: Becker, J.B., Berkley, K.J., Geary, N., Hampson, E., Herman, J., Young, E. (Eds.), Sex Differences in the Brain: From Genes to Behavior. Oxford University Press, New York.

Hampson, E., Rovet, J.F., Altmann, D., 1998. Spatial reasoning in children with congenital adrenal hyperplasia due to 21-hydroxylase deficiency. Developmental Neuropsychology 14, 299–320.

Hannagan, R.J., 2008. Gendered political behavior: A Darwinian feminist approach. Sex Roles 59, 465–475.

Harasty, J., Double, K.L., Halliday, G.M., Kril, J.J., McRitchie, D.A., 1997. Language-associated cortical regions are proportionally larger in the female brain. Archives of Neurology 54, 171–176.

Hariri, A.R., Bookheimer, S.Y., Mazziotta, J.C., 2000. Modulating emotional responses: Effects of a neocortical network on the limbic system. NeuroReport 11, 43–48.

Hartung, C.M., Widiger, T.A., 1998. Gender differences in the diagnosis of mental disorders: Conclusions and controversies of the DSM-IV. Psychological Bulletin 123, 260–278.

Hedges, L.V., Nowell, A., 1995. Sex differences in mental test scores, variability, and numbers of high-scoring individuals. Science 269, 41–45.

Heil, M., Kavsek, M., Rolke, B., Beste, C., Jansen, P., 2011. Mental rotation in female fraternal twins: Evidence for intra-uterine hormone transfer? Biological Psychology 86, 90–93.

Herlitz, A., Lovén, J., Thilers, P., Rehnman, J., 2010. Sex differences in episodic memory: The where but not the why. In: Bäckman, L., Nyberg, L. (Eds.), Memory, Aging and the Brain. Psychology Press, New York.

Herting, M.M., Maxwell, E.C., Irvine, C., Nagel, B.J., 2012. The impact of sex, puberty, and hormones on white matter microstructure in adolescents. Cerebral Cortex 22, 1979–1992.

Hier, D.B., Crowley, W.F., 1982. Spatial ability in androgen-deficient men. The New England Journal of Medicine 302, 1202–1205.

Hines, M., 2010. Sex-related variation in human behavior and the brain. Trends in Cognitive Sciences 14, 448–456.

Hines, M., 2011. Gender development and the human brain. Annual Review of Neuroscience 34, 69–88.

Hines, M., McAdams, L.A., Chiu, L., Bentler, P.M., Lipcamon, J., 1992. Cognition and the corpus callosum: Verbal fluency, visuospatial ability, and language lateralization related to midsaggital surface areas of callosal subregions. Behavioral Neuroscience 106, 3–14.

Hines, M., Ahmed, F., Hughes, I.A., 2003a. Psychological outcomes and gender-related development in complete androgen insensitivity syndrome. Archives of Sexual Behavior 32, 93–101.

Hines, M., Fane, B.A., Pasterski, V.L., Mathews, G.A., Conway, G.S., Brook, C., 2003b. Spatial abilities following prenatal androgen abnormality: Targeting and mental rotations performance in individuals with congenital adrenal hyperplasia. Psychoneuroendocrinology 28, 1010–1026.

Holloway, R.L., 1980. Within-species brain-body weight variability: A reexamination of the Danish data and other primate species. American Journal of Physical Anthropology 53, 109–121.

Hugdahl, K., Thomsen, T., Ersland, L., 2006. Sex differences in visuospatial processing: An fMRI study of mental rotation. Neuropsychologia 44, 1575–1583.

Hulshoff Pol, H.E., Cohen-Kettenis, P.T., Van Haren, N.E.M., et al., 2006. Changing your sex changes your brain: Influences of testosterone and estrogen on adult human brain structure. European Journal of Endocrinology 155, S107–S114.

Hyde, J.S., 2005. The gender similarities hypothesis. American Psychologist 60, 581–592.

Im, K., Lee, J.M., Lee, J., et al., 2006. Gender difference analysis of cortical thickness in healthy young adults with surface-based methods. NeuroImage 31, 31–38.

Im, K., Lee, J.M., Lyttelton, O., Kim, S.H., Evans, A.C., Kim, S.I., 2008. Brain size and cortical structure in the adult human brain. Cerebral Cortex 18, 2181–2191.

Johnson, W., Bouchard, T.J., 2007. Sex differences in mental abilities: G masks the dimensions on which they lie. Intelligence 35, 23–39.

Jordan, K., Wurstenberg, T., Heinze, H.J., Peters, M., Jancke, L., 2002. Women and men exhibit different cortical activation patterns during mental rotation tasks. Neuropsychologia 40, 2397–2408.

Jordan-Young, R.M., 2010. Brain Storm: The Flaws in the Science of Sex Differences. Harvard University Press, Cambridge, MA.

Kansaku, K., Kitazawa, S., 2001. Imaging studies on sex differences in the lateralization of language. Neuroscience Research 41, 333–337.

Kim, J., Zhu, W., Chang, L., Bentler, P.M., Ernst, T., 2007. Unified structural equation modeling approach for the analysis of multisubject, multivariate functional MRI data. Human Brain Mapping 28, 85–93.

Kimura, D., 1999. Sex and Cognition. MIT Press, Cambridge, MA.

Kitazawa, S., Kansaku, K., 2005. Sex difference in language lateralization may be task-dependent. Brain 128, E30.

Kling, K.C., Hyde, J.S., Showers, C.J., Buswell, B.N., 1999. Gender differences in self-esteem: A meta-analysis. Psychological Bulletin 125, 470–500.

Knickmeyer, R.C., Davenport, M., 2011. Turner syndrome and sexual differentiation of the brain: Implications for understanding male-biased neurodevelopmental disorders. Journal of Neurodevelopmental Disorders 3, 293–306.

Konrad, C., Engelien, A., Schöning, S., et al., 2008. The functional anatomy of semantic retrieval is influenced by gender, menstrual cycle, and sex hormones. Journal of Neural Transmission 115, 1327–1337.

Koscik, T., O'Leary, D., Moser, D.J., Andreasen, N.C., Nopoulos, P., 2009. Sex differences in parietal lobe morphology: Relationship to mental rotation performance. Brain and Cognition 69, 451–459.

Kyratzis, A., 2001. Emotion talk in preschool same-sex friendship groups: Fluidity over time and context. Early Education and Development 12, 359–392.

Lansdell, H., Davie, J.C., 1972. Massa intermedia: Possible relation to intelligence. Neuropsychologia 10, 207–210.

Largo, R.H., Caflisch, J.A., Hug, F., Muggli, K., Molnar, A.A., Molinari, L., 2001a. Neuromotor development from 5 to 18 years. Part 2: Associated movements. Developmental Medicine and Child Neurology 43, 444–453.

Largo, R.H., Caflisch, J.A., Hug, F., et al., 2001b. Neuromotor development from 5 to 18 years. Part 1: Timed performance. Developmental Medicine and Child Neurology 43, 436–443.

Lawton, C.A., 2010. Gender, spatial abilities, and wayfinding. In: Chrisler, J.C., McCreary, D.R. (Eds.), Handbook of Gender Research in Psychology. Gender Research in General and Experimental Psychology, vol. 1. Springer Science + Business Media, New York.

Lebel, C., Walker, L., Leemans, A., Phillips, L., Beaulieu, C., 2008. Microstructural maturation of the human brain from childhood to adulthood. NeuroImage 40, 1044–1055.

Lenroot, R.K., Giedd, J.N., 2010. Sex differences in the adolescent brain. Brain and Cognition 72, 46–55.

Lenroot, R.K., Gogtay, N., Greenstein, D.K., et al., 2007. Sexual dimorphism of brain developmental trajectories during childhood and adolescence. NeuroImage 36, 1065–1073.

Lentini, E., Kasahara, M., Arver, S., Savic, I., 2012. Sex differences in the human brain and the impact of sex chromosomes and sex hormones. Cerebral Cortex.

Leonard, C.M., Towler, S., Welcome, S., et al., 2008. Size matters: Cerebral volume influences sex differences in neuroanatomy. Cerebral Cortex 18, 2920–2931.

Lepage, J.F., Mazaika, P.K., Hong, D.S., Raman, M., Reiss, A.L., 2012. Cortical brain morphology in young, estrogen-naive, and adolescent, estrogen-treated girls with Turner Syndrome. Cerebral Cortex.

LeVay, S., 1991. A difference in hypothalamic structure between heterosexual and homosexual men. Science 253, 1034–1037.

Levine, S.C., Vasilyeva, M., Lourenco, S.F., Newcombe, N.S., Huttenlocher, J., 2005. Socioeconomic status modifies the sex difference in spatial skill. Psychological Science 16, 841–845.

Liben, L.S., 1995. Psychology meets geography: Exploring the gender gap on the national geography bee. Psychological Science Agenda 8, 8–9.

Lindberg, S.M., Hyde, J.S., Petersen, J.L., Linn, M.C., 2010. New trends in gender and mathematics performance: A meta-analysis. Psychological Bulletin 136, 1123–1135.

Lombardo, M.V., Ashwin, E., Auyeung, B., et al., 2012. Fetal testosterone influences sexually dimorphic gray matter in the human brain. Journal of Neuroscience 32, 674–680.

Lord, C., Buss, C., Lupien, S.J., Pruessner, J.C., 2008. Hippocampal volumes are larger in postmenopausal women using estrogen therapy compared to past users, never users and men: A possible window of opportunity effect. Neurobiology of Aging 29, 95–101.

Luders, E., Steinmetz, H., Jancke, L., 2002. Brain size and grey matter volume in the healthy human brain. NeuroReport 13, 2371–2374.

Luders, E., Narr, K.I., Thompson, P.M., et al., 2004. Gender differences in cortical complexity. Nature Neuroscience 7, 799–800.

Luders, E., Harr, K.L., Thompson, P.M., et al., 2006a. Gender effects on cortical thickness and the influence of scaling. Human Brain Mapping 27, 314–324.

Luders, E., Narr, K.L., Zaidel, E., Thompson, P.M., Toga, A.W., 2006b. Gender effects on callosal thickness in scaled and unscaled space. NeuroReport 17, 1103–1106.

Luders, E., Thompson, P.M., Narr, K.L., Toga, A.W., Jancke, L., Gaser, C., 2006c. A curvature-based approach to estimate local gyrification on the cortical surface. NeuroImage 29, 1224–1230.

Luders, E., Gaser, C., Narr, K.L., Toga, A.W., 2009a. Why sex matters: Brain size independent differences in gray matter distributions between men and women. Journal of Neuroscience 29, 14265–14270.

Luders, E., Narr, K.L., Thompson, P.M., Toga, A.W., 2009b. Neuroanatomical correlates of intelligence. Intelligence 37, 156–163.

Luders, E., Thompson, P.M., Toga, A.W., 2010. The development of the corpus callosum in the healthy human brain. Journal of Neuroscience 30, 10985–10990.

Luders, E., Toga, A.W., 2010. Sex differences in brain anatomy. Progress in Brain Research 186, 3–12.

Luders, E., Thompson, P.M., Narr, K.L., et al., 2011. The link between callosal thickness and intelligence in healthy children and adolescents. NeuroImage 54, 1823–1830.

Luna, B., Velanova, K., Geier, C.F., 2010. Methodological approaches in developmental neuroimaging studies. Human Brain Mapping 31, 863–871.

Lust, J.M., Geuze, R.H., Van de Beek, C., Cohen-Kettenis, P.T., Groothuis, A.G.G., Bouma, A., 2010. Sex specific effect of prenatal testosterone on language lateralization in children. Neuropsychologia 48, 536–540.

Lv, B., Li, J., He, H.G., et al., 2010. Gender consistency and difference in healthy adults revealed by cortical thickness. NeuroImage 53, 373–382.

Maccoby, E.E., 1998. The Two Sexes: Growing Up Apart, Coming Together. Harvard University Press, Cambridge, MA.

Maguire, E.A., Gadian, D.G., Johnsrude, I.S., et al., 2000. Navigation-related structural change in the hippocampi of taxi drivers. Proceedings of the National Academy of Sciences of the United States of America 97, 4398–4403.

Maguire, E.A., Spiers, H.J., Good, C.D., Hartley, T., Frackowiak, R.S.J., Burgess, N., 2003. Navigation expertise and the human hippocampus: A structural brain imaging analysis. Hippocampus 13, 250–259.

Maki, P.M., 2005. Estrogen effects on the hippocampus and frontal lobes. International Journal of Fertility and Women's Medicine 50, 67–71.

Maki, P.M., Resnick, S.M., 2000. Longitudinal effects of estrogen replacement therapy on PET cerebral blood flow and cognition. Neurobiology of Aging 21, 373–383.

Maki, P.M., Resnick, S.M., 2001. Effects of estrogen on patterns of brain activity at rest and during cognitive activity: A review of neuroimaging studies. NeuroImage 14, 789–801.

Maki, P.M., Sundermann, E., 2009. Hormone therapy and cognitive function. Human Reproduction Update 15, 667–681.

Malouf, M.A., Migeon, C.J., Carson, K.A., Petrucci, L., Wisniewski, A.B., 2006. Cognitive outcome in adult women affected by congenital adrenal hyperplasia due to 21-hydroxylase deficiency. Hormone Research 65, 142–150.

Martel, M.M., Klump, K., Nigg, J.T., Breedlove, S.M., Sisk, C.L., 2009. Potential hormonal mechanisms of attention-deficit/hyperactivity disorder and major depressive disorder: A new perspective. Hormones and Behavior 55, 465–479.

Martin, C.L., Fabes, R.A., 2001. The stability and consequences of young children's same-sex peer interactions. Developmental Psychology 37, 431–446.

Martin, C.L., Ruble, D.N., 2010. Patterns of gender development. Annual Review of Psychology 61, 353–381.

Martin, C.L., Ruble, D.N., Szkrybalo, J., 2002. Cognitive theories of early gender development. Psychological Bulletin 128, 903–933.

Mazzone, L., Mueller, S.C., Maheu, F., VanRyzin, C., Merke, D.P., Ernst, M., 2011. Emotional memory in early steroid abnormalities: An fMRI study of adolescents with congenital adrenal hyperplasia. Developmental Neuropsychology 36, 473–492.

McCabe, D.P., Castel, A.D., 2008. Seeing is believing: The effect of brain images on judgments of scientific reasoning. Cognition 107, 343–352.

McCarthy, M.M., Arnold, A.P., 2011. Reframing sexual differentiation of the brain. Nature Neuroscience 14, 677–683.

McClure, E.B., 2000. A meta-analytic review of sex differences in facial expression processing and their development in infants, children, and adolescents. Psychological Bulletin 126, 424–453.

McClure, E.B., Monk, C.S., Nelson, E.E., et al., 2004. A developmental examination of gender differences in brain engagement during evaluation of threat. Biological Psychiatry 55, 1047–1055.

McGlone, J., 1980. Sex differences in human brain asymmetry: A critical survey. The Behavioral and Brain Sciences 3, 215–227.

Merke, D.P., Fields, J.D., Keil, M.F., Vaituzis, A.C., Chrousos, G.P., Giedd, J.N., 2003. Children with classic congenital adrenal hyperplasia have decreased amygdala volume: Potential prenatal and postnatal hormonal effects. Journal of Clinical Endocrinology and Metabolism 88, 1760–1765.

Meyer, G.J., Finn, S.E., Eyde, L.D., et al., 2001. Psychological testing and psychological assessment: A review of evidence and issues. American Psychologist 56, 128–165.

Meyer-Bahlburg, H.F.L., 2005. Gender identity outcome in female-raised 46, XY persons with penile agenesis, cloacal exstrophy of the bladder, or penile ablation. Archives of Sexual Behavior 34, 423–438.

Meyer-Bahlburg, H.F.L., Dolezal, C., Baker, S.W., Ehrhardt, A.A., New, M.I., 2006. Gender development in women with congenital adrenal hyperplasia as a function of disorder severity. Archives of Sexual Behavior 35, 667–684.

Moffitt, T.E., Caspi, A., Rutter, M., Silva, P.A., 2001. Sex Differences in Antisocial Behaviour: Conduct Disorder, Delinquency, and Violence in the Dunedin Longitudinal Study. Cambridge University Press, New York.

Mondschein, E.R., Adolph, K.E., Tamis-LeMonda, C.S., 2000. Gender bias in mothers' expectations about infant crawling. Journal of Experimental Child Psychology 77, 304–316.

Money, J., Ehrhardt, A.A., 1972. Man and Woman, Boy and Girl. Johns Hopkins University Press, Baltimore, MD.

Moore, D.S., Johnson, S.P., 2008. Mental rotation in human infants: A sex difference. Psychological Science 19, 1063–1066.

Morrongiello, B.A., Matheis, S., 2007. Understanding children's injury-risk behaviors: The independent contributions of cognitions and emotions. Journal of Pediatric Psychology 32, 926–937.

Mueller, S.C., Temple, V., Oh, E., et al., 2008. Early androgen exposure modulates spatial cognition in congenital adrenal hyperplasia (CAH). Psychoneuroendocrinology 33, 973–980.

Mueller, S.C., Mandell, D., Leschek, E.W., Pine, D.S., Merke, D.P., Ernst, M., 2009. Early hyperandrogenism affects the development of hippocampal function: Preliminary evidence from a functional magnetic resonance imaging study of boys with familial male precocious puberty. Journal of Child and Adolescent Psychopharmacology 19, 41–50.

Mueller, S.C., Merke, D.P., Leschek, E.W., et al., 2011a. Grey matter volume correlates with virtual water maze task performance in boys with androgen excess. Neuroscience 197, 225–232.

Mueller, S.C., Merke, D.P., Leschek, E.W., Fromm, S., VanRyzin, C., Ernst, M., 2011b. Increased medial temporal lobe and striatal grey-matter volume in a rare disorder of androgen excess: A voxel-based morphometry (VBM) study. The International Journal of Neuropsychopharmacology 14, 445–457.

Nelson, C., Righi, G., 2013. Face processing. Neural Circuit Development and Function in the Healthy and Diseased Brain: Comprehensive Developmental Neuroscience, Volume 3. Elsevier, Oxford.

Newcombe, N., Dubas, J.S., 1992. A longitudinal study of predictors of spatial ability in adolescent females. Child Development 63, 37–46.

Newcombe, N., Bandura, M.M., Taylor, D.G., 1983. Sex differences in spatial ability and spatial activities. Sex Roles 9, 377–386.

Nolen-Hoeksema, S., Hilt, L., 2009. Gender differences in depression. In: Gotlieb, I.H., Hammen, C.L. (Eds.), Handbook of Depression. 2nd edn. Guilford, New York.

Nopoulos, P., Flaum, M., O'Leary, D., Andreasen, N.C., 2000. Sexual dimorphism in the human brain: Evaluation of tissue volume, tissue composition and surface anatomy using magnetic resonance imaging. Psychiatry Research: Neuroimaging 98, 1–13.

Nordenström, A., Servin, A., Bohlin, G., Larsson, A., Wedell, A., 2002. Sex-typed toy play behavior correlates with the degree of prenatal androgen exposure assessed by CYP21 genotype in girls with congenital adrenal hyperplasia. Journal of Clinical Endocrinology and Metabolism 87, 5119–5124.

Ohnishi, T., Matsuda, H., Hirakata, M., Ugawa, Y., 2006. Navigation ability dependent neural activation in the human brain: An fMRI study. Neuroscience Research 55, 361–369.

Ortigue, S., Thut, G., Landis, T., Michel, C.M., 2005. Time-resolved sex differences in language lateralization. Brain 128, E28.

Pasterski, V.L., Geffner, M.E., Brain, C., Hindmarsh, P., Brook, C., Hines, M., 2005. Prenatal hormones and postnatal socialization by parents as determinants of male-typical toy play in girls with congenital adrenal hyperplasia. Child Development 76, 264–278.

Paus, T., 2010. Sex differences in the human brain: A developmental perspective. Progress in Brain Research 186, 13–28.

Paus, T., Nawaz-Khan, I., Leonard, G., et al., 2010. Sexual dimorphism in the adolescent brain: Role of testosterone and androgen receptor in global and local volumes of grey and white matter. Hormones and Behavior 57, 63–75.

Peper, J.S., Hulshoff Pol, H.E., Crone, E.A., van Honk, J., 2011. Sex steroids and brain structure in pubertal boys and girls: A mini-review of neuroimaging studies. Neuroscience 191, 28–37.

Perrin, J.S., Hervé, P.-Y., Leonard, G., et al., 2008. Growth of white matter in the adolescent brain: Role of testosterone and androgen receptor. Journal of Neuroscience 28, 9519–9524.

Persad, C.C., Zubieta, J.K., Love, T., Wang, H., Tkaczyk, A., Smith, Y.R., 2009. Enhanced neuroactivation during verbal memory processing in postmenopausal women receiving short-term hormone therapy. Fertility and Sterility 92, 197–204.

Pessoa, L., Adolphs, R., 2010. Emotion processing and the amygdala: From a 'low road' to 'many roads' of evaluating biological significance. Nature Reviews Neuroscience 11, 773–782.

Peters, M., 1991. Sex differences in human brain size and the general meaning of differences in brain size. Canadian Journal of Psychology 45, 507–522.

Peters, M., Battista, C., 2008. Applications of mental rotation figures of the shepard and metzler type and description of a mental rotation stimulus library. Brain and Cognition 66, 260–264.

Peters, M., Jäncke, L., Staiger, J.F., Schlaug, G., Huang, Y., Steinmetz, H., 1998. Unsolved problems in comparing brain sizes in homo sapiens. Brain and Cognition 37, 254–285.

Phoenix, C.H., Goy, R.W., Gerall, A.A., Young, W.C., 1959. Organizing action of prenatally administered testosterone propionate on the tissues mediating mating behavior in the female guinea pig. Endocrinology 65, 369–382.

Phoenix, C.H., Slob, A.K., Goy, R.W., 1973. Effects of castration and replacement therapy on sexual behavior of adult male rhesuses. Journal of Comparative and Physiological Psychology 84, 472–481.

Plante, E., Schmithorst, V.J., Holland, S.K., Byars, A.W., 2006. Sex differences in the activation of language cortex during childhood. Neuropsychologia 44, 1210–1221.

Poldrack, R.A., 2000. Imaging brain plasticity: Conceptual and methodological issues – A theoretical review. NeuroImage 12, 1–13.

Protopopescu, X., Butler, T., Pan, H., et al., 2008. Hippocampal structural changes across the menstrual cycle. Hippocampus 18, 985–988.

Puts, D.A., McDaniel, M.A., Jordan, C.L., Breedlove, N.J., 2008. Spatial ability and prenatal androgens: Meta-analyses of congenital adrenal hyperplasia and digit ratio (2D:4D) studies. Archives of Sexual Behavior 37, 100–111.

Puts, D.A., Cárdenas, R.A., Bailey, D.H., Burriss, R.P., Jordan, C.L., Breedlove, S.M., 2010. Salivary testosterone does not predict mental rotation performance in men or women. Hormones and Behavior 58, 282–289.

Quinn, P.C., Liben, L.S., 2008. A sex difference in mental rotation in young infants. Psychological Science 19, 1067–1070.

Rabinowicz, T., Dean, D.E., Petetot, J.M.C., de Courten-Myers, G.M., 1999. Gender differences in the human cerebral cortex: More neurons in males; more processes in females. Journal of Child Neurology 14, 98–107.

Rabl, R., 1958. Strukturstudien an der aassa intermedia des thalamus opticus. Journal für Hirnforschung 4, 78–112.

Raznahan, A., Lee, Y., Stidd, R., et al., 2010. Longitudinally mapping the influence of sex and androgen signaling on the dynamics of human cortical maturation in adolescence. Proceedings of the National Academy of Sciences of the United States of America 107, 16988–16993.

Reinius, B., Jazin, E., 2009. Prenatal sex differences in the human brain. Molecular Psychiatry 14, 988–989.

Resnick, S.M., Berenbaum, S.A., Gottesman, I.I., Bouchard, T.J., 1986. Early hormonal influences on cognitive functioning in congenital adrenal hyperplasia. Developmental Psychology 22, 191–198.

Resnick, S.M., Maki, P.M., Golski, S., Kraut, M.A., Zonderman, A.B., 1998. Effects of estrogen replacement therapy on pet cerebral blood flow and neuropsychological performance. Hormones and Behavior 34, 171–182.

Resnick, S.M., Espeland, M.A., Jaramillo, S.A., et al., 2009. Postmenopausal hormone therapy and regional brain volumes the WHIMS-MRI study. Neurology 72, 135–142.

Robertson, D., Craig, M., van Amelsvoort, T., et al., 2009. Effects of estrogen therapy on age-related differences in gray matter concentration. Climacteric 12, 301–309.

Roivainen, E., 2011. Gender differences in processing speed: A review of recent research. Learning and Individual Differences 21, 145–149.

Ross, J., Roeltgen, D., Zinn, A., 2006. Cognition and the sex chromosomes: Studies in Turner syndrome. Hormone Research 65, 47–56.

Rubenstein, J.L.R., Rakic, P., 2013. Patterning and Cell Types Specification in the Developing CNS and PNS.

Rubia, K., Hyde, Z., Halari, R., Giampietro, V., Smith, A., 2010. Effects of age and sex on developmental neural networks of visual-spatial attention allocation. NeuroImage 51, 817–827.

Ruble, D.N., Martin, C.L., Berenbaum, S.A., 2006. Gender development. In: Eisenberg, N. (Ed.), Handbook of Child Psychology. 6th ed Social, Emotional, and Personality Development, vol. 3. Wiley, Hoboken, NJ.

Rutter, M., Giller, H., Hagell, A., 1998. Antisocial Behavior by Young People. Cambridge University Press, New York.

Ryan, B.C., Vandenbergh, J.G., 2002. Intrauterine position effects. Neuroscience and Biobehavioral Reviews 26, 665–678.

Savic, I., Garcia-Falgueras, A., Swaab, D.F., 2010. Sexual differentiation of the human brain in relation to gender identity and sexual orientation. Progress in Brain Research 186, 41–62.

Schmithorst, V.J., Yuan, W.H., 2010. White matter development during adolescence as shown by diffusion MRI. Brain and Cognition 72, 16–25.

Schneider, S., Peters, J., Bromberg, U., et al., 2011. Boys do it the right way: Sex-dependent amygdala lateralization during face processing in adolescents. NeuroImage 56, 1847–1853.

Schöning, S., Engelien, A., Kugel, H., et al., 2007. Functional anatomy of visuo-spatial working memory during mental rotation is influenced by sex, menstrual cycle, and sex steroid hormones. Neuropsychologia 45, 3203–3214.

Schulz, K.M., Zehr, J.L., Salas-Ramirez, K.Y., Sisk, C.L., 2009. Testosterone programs adult social behavior before and during, but not after, adolescence. Endocrinology 150, 3690–3698.

Seldon, H.L., 2005. Does brain white matter growth expand the cortex like a balloon? Hypothesis and consequences. Laterality 10, 81–95.

Sergerie, K., Chochol, C., Armony, J.L., 2008. The role of the amygdala in emotional processing: A quantitative meta-analysis of functional neuroimaging studies. Neuroscience and Biobehavioral Reviews 32, 811–830.

Shaw, P., Greenstein, D., Lerch, J., et al., 2006. Intellectual ability and cortical development in children and adolescents. Nature 440, 676–679.

Shaw, P., Gogtay, N., Rapoport, J., 2010. Childhood psychiatric disorders as anomalies in neurodevelopmental trajectories. Human Brain Mapping 31, 917–925.

Shaywitz, B.A., Shaywitz, S.E., Pugh, K.R., et al., 1995. Sex differences in the functional organization of the brain for language. Nature 373, 607–609.

Shaywitz, S.E., Shaywitz, B.A., Pugh, K.R., et al., 1999. Effect of estrogen on brain activation patterns in postmenopausal women during working memory tasks. Journal of the American Medical Association 281, 1197–1202.

Simpkins, S.D., Davis-Kean, P.E., Eccles, J.S., 2005. Parents' socializing behavior and children's participation in math, science, and computer out-of-school activities. Applied Developmental Science 9, 14–30.

Sisk, C.L., Zehr, J.L., 2005. Pubertal hormones organize the adolescent brain and behavior. Frontiers in Neuroendocrinology 26, 163–174.

Sneider, J.T., Sava, S., Rogowska, J., Yurgelun-Todd, D.A., 2011. A preliminary study of sex differences in brain activation during a spatial navigation task in healthy adults. Perceptual and Motor Skills 113, 461–480.

Sommer, I.E.C., Aleman, A., Bouma, A., Kahn, R.S., 2004. Do women really have more bilateral language representation than men? A meta-analysis of functional imaging studies. Brain 127, 1845–1852.

Sommer, I.E., Aleman, A., Somers, M., Boks, M.P., Kahn, R.S., 2008. Sex differences in handedness, asymmetry of the planum temporale and functional language lateralization. Brain Research 1206, 76–88.

Song, M., Zhou, Y., Li, J., et al., 2008. Brain spontaneous functional connectivity and intelligence. NeuroImage 41, 1168–1176.

Sowell, E.R., Thompson, P.M., Tessner, K.D., Toga, A.W., 2001. Mapping continued brain growth and gray matter density reduction in dorsal frontal cortex: Inverse relationships during postadolescent brain maturation. Journal of Neuroscience 21, 8819–8829.

Sowell, E.R., Trauner, D.A., Gamst, A., Jernigan, T.L., 2002. Development of cortical and subcortical brain structures in childhood and adolescence: A structural MRI study. Developmental Medicine and Child Neurology 44, 4–16.

Sowell, E.R., Peterson, B.S., Kan, E., et al., 2007. Sex differences in cortical thickness mapped in 176 healthy individuals between 7 and 87 years of age. Cerebral Cortex 17, 1550–1560.

Sowell, E., Colby, J., O'Hare, E.D., 2013. Brain development: Birth through adolescence. Neural Circuit Development and Function in the Healthy and Diseased Brain: Comprehensive Developmental Neuroscience, Volume 3. Elsevier, Oxford.

Sporns, O., 2011. Networks of the Brain. MIT Press, Cambridge, MA.

Steele, C., 1997. A threat in the air. How stereotypes shape intellectual identity and performance. American Psychologist 52, 613–629.

Steinberg, L., 2008. A social neuroscience perspective on adolescent risk-taking. Developmental Review 28, 78–106.

Stevens, J.S., Hamann, S., 2012. Sex differences in brain activation to emotional stimuli: A meta-analysis of neuroimaging studies. Neuropsychologia 50, 1578–1593.

Stiles, J., Akshoomoff, N., Haist, F., 2013. Spatial cognition. Neural Circuit Development and Function in the Healthy and Diseased Brain: Comprehensive Developmental Neuroscience, Volume 3. Elsevier, Oxford.

Su, R., Rounds, J., Armstrong, P.I., 2009. Men and things, women and people: A meta-analysis of sex differences in interests. Psychological Bulletin 135, 859–884.

Thambisetty, M., Wan, J., Carass, A., An, Y., Prince, J.L., Resnick, S.M., 2010. Longitudinal changes in cortical thickness associated with normal aging. NeuroImage 52, 1215–1223.

Thomas, J.R., French, K.E., 1985. Gender differences across age in motor performance: A meta-analysis. Psychological Bulletin 98, 260–282.

Thompson, P.M., Giedd, J.N., Woods, R.P., MacDonald, D., Evans, A.C., Toga, A.W., 2000. Growth patterns in the developing brain detected by using continuum mechanical tensor maps. Nature 404, 190–193.

Thomsen, T., Hugdahl, K., Ersland, L., et al., 2000. Functional magnetic resonance imaging (fMRI) study of sex differences in a mental rotation task. Medical Science Monitor 60, 1186–1196.

Tracy, D.M., 1990. Toy-playing behavior, sex-role orientation, spatial ability, and science achievement. Journal of Research in Science Teaching 27, 637–649.

Uttal, D.H., Meadow, N.G., Tipton, E., et al., 2012. The malleability of spatial skills: A meta-analysis of training studies. Psychological Bulletin.

van den Heuvel, M.P., Stam, C.J., Kahn, R.S., Pol, H.E.H., 2009. Efficiency of functional brain networks and intellectual performance. Journal of Neuroscience 29, 7619–7624.

Vasta, R., Liben, L.S., 1996. The water-level task: An intriguing puzzle. Current Directions in Psychological Science 5, 171–177.

Vawter, M.P., Evans, S., Choudary, P., et al., 2004. Gender-specific gene expression in post-mortem human brain: Localization to sex chromosomes. Neuropsychopharmacology 29, 373–384.

Vogel, J.J., Bowers, C.A., Vogel, D.S., 2003. Cerebral lateralization of spatial abilities: A meta-analysis. Brain and Cognition 52, 197–204.

Voyer, D., 2011. Sex differences in dichotic listening. Brain and Cognition 76, 245–255.

Voyer, D., Bryden, M.P., 1993. Masking and visual field effects on a lateralized rod-and-frame test. Canadian Journal of Experimental Psychology 47, 26–37.

Voyer, D., Nolan, C.L., Voyer, S., 2000. The relation between experience and spatial performance in men and women. Sex Roles 43, 891–915.

Voyer, D., Postma, A., Brake, B., Imperato-McGinley, J., 2007. Gender differences in object location memory: A meta-analysis. Psychonomic Bulletin and Review 14, 23–38.

Vul, E., Harris, C., Winkielman, P., Pashler, H., 2009. Puzzlingly high correlations in fMRI studies of emotion, personality, and social cognition. Perspectives on Psychological Science 4, 274–290.

Vuoksimaa, E., Kaprio, J., Kremen, W.S., et al., 2010. Having a male co-twin masculinizes mental rotation performance in females. Psychological Science 21, 1069–1071.

Wager, T.D., Phan, K.L., Liberzon, I., Taylor, S.F., 2003. Valence, gender, and lateralization of functional brain anatomy in emotion: A meta-analysis of findings from neuroimaging. NeuroImage 19, 513–531.

Wai, J., Cacchio, M., Putallaz, M., Makel, M.C., 2010. Sex differences in the right tail of cognitive abilities: A 30 year examination. Intelligence 38, 412–423.

Wallen, K., 1996. Nature needs nurture: The interaction of hormonal and social influences on the development of behavioral sex differences in rhesus monkeys. Hormones and Behavior 30, 364–378.

Wallen, K., 2005. Hormonal influences on sexually differentiated behavior in nonhuman primates. Frontiers in Neuroendocrinology 26, 7–26.

Wallen, K., 2009. The organizational hypothesis: Reflections on the 50th anniversary of the publication of Phoenix, Goy, Gerall, and Young (1959). Hormones and Behavior 55, 561–565.

Wallentin, M., 2009. Putative sex differences in verbal abilities and language cortex: A critical review. Brain and Language 108, 175–183.

Weisberg, D.S., Keil, F.C., Goodstein, J., Rawson, E., Gray, J.R., 2008. The seductive allure of neuroscience explanations. Journal of Cognitive Neuroscience 20, 470–477.

Weiss, E., Siedentopf, C.M., Hofer, A., et al., 2003a. Sex differences in brain activation pattern during a visuospatial cognitive task: A functional magnetic resonance imaging study in healthy volunteers. Neuroscience Letters 344, 169–172.

Weiss, E.M., Siedentopf, C., Hofer, A., et al., 2003b. Brain activation pattern during a verbal fluency test in healthy male and female volunteers: A functional magnetic resonance imaging study. Neuroscience Letters 352, 191–194.

Witte, A.V., Savli, M., Holik, A., Kasper, S., Lanzenberger, R., 2010. Regional sex differences in grey matter volume are associated with sex hormones in the young adult human brain. NeuroImage 49, 1205–1212.

Wood, W., Eagly, A.H., 2002. A cross-cultural analysis of the behavior of women and men: Implications for the origins of sex differences. Psychological Bulletin 128, 699–727.

Wraga, M., Helt, M., Jacobs, E., Sullivan, K., 2007. Neural basis of stereotype-induced shifts in women's mental rotation performance. Social Cognitive and Affective Neuroscience 2, 12–19.

Zahn-Waxler, C., 1993. Warriors and worriers: Gender and psychopathology. Development and Psychopathology 5, 79–89.

Zahn-Waxler, C., Shirtcliff, E.A., Marceau, K., 2008. Disorders of childhood and adolescence: Gender and psychopathology. Annual Review of Clinical Psychology 4, 275–303.

Zhang, K., Sejnowski, T.J., 2000. A universal scaling law between gray matter and white matter of cerebral cortex. Proceedings of the National Academy of Sciences of the United States of America 97, 5621–5626.

Zhou, J.N., Hofman, M.A., Gooren, L.J.G., Swaab, D.F., 1995. A sex difference in the human brain and its relation to transsexuality. Nature 378, 68–70.

Zilles, K., Armstrong, E., Schleicher, A., Kretschmann, H.J., 1988. The human pattern of gyrification in the cerebral cortex. Anatomy and Embryology 179, 173–179.

SECTION III

DISEASES

Neural-Tube Defects

C. Pyrgaki[1], L. Niswander[1,2]

[1]University of Colorado Denver, Aurora, CO, USA; [2]Howard Hughes Medical Institute, Aurora, CO, USA

27.1 NEURAL-TUBE DEFECTS

After heart defects, neural-tube defects (NTDs) are the second most common birth defect, occurring in approximately 300 000 newborns worldwide ($\sim$1:1000 live births) (Botto et al., 1999). The prevalence of NTDs varies considerably between different racial and ethnic groups (Carmichael et al., 2004; Feldman et al., 1982; Feuchtbaum et al., 1999). In the United States, the two most common types of NTDs, spina bifida and anencephaly, are estimated to affect $\sim$3000 pregnancies each year, with caudal NTDs occurring at a higher rate than cranial NTDs (Canfield et al., 2009; CDC, 2004). Both cranial and caudal NTDs can be associated with craniofacial defects (Rittler et al., 2008), but the majority of NTDs are nonsyndromic (Hall et al., 1988). NTDs, especially those of the cranial region, seem to have a higher prevalence among females (Rogers and Morris, 1973; Seller, 1987), although the reason for this sex difference is not well understood.

27.1.1 Types of NTDs

NTDs are divided into two major groups, cranial NTDs and spinal NTDs, and the nomenclature and classification of NTDs in humans are based on position and severity of the NTDs. Cranial NTDs result from failure of the neural tube to close in the cranial region and are classified as follows:

1. Encephalocele, defined as herniation of the cranial vault. Encephalocele can be either a meningocele,

© 2013 Elsevier Inc. All rights reserved.

if the herniation contains only cerebral spinal fluid (CSF) and meninges, or a meningoencephalocele, if it also contains neural tissue (McComb, 1997).

2. Anencephaly, where the cranial vault is absent and the neural tissue degenerates by week 8 of gestation (Calzolari et al., 2004).

Spinal NTDs also are referred to as spinal dysraphisms. This term was coined by Lichtenstein in 1940 to describe incomplete fusion or malformations of structures in the dorsal midline of the back, particularly congenital abnormalities of the vertebral column and spinal cord (Tavafoghi et al., 1978). These can be further divided into three groups:

1. Spina bifida occulta, which in its strict definition refers only to bone-fusion defects in the spine.
2. Spina bifida cystica, which refers to meningocele and myelomeningocele, where the herniation contains not only CSF, but also neuronal tissue.
3. Spina bifida aperta (SBA), in which the neural tissue is exposed to the environment (Kaufman, 2004).

A relatively rare form of dysraphism that is open or exposed results from the failure of closure of the neural tube throughout its entire length and is called craniorachischisis (Coskun et al., 2009). The term *spina bifida* has become commonly associated with the open spinal dysraphism of a myelomeningocele.

27.1.2 Clinical Aspects and Prognosis of NTDs

NTDs can be fatal; all exencephalic embryos are stillborn or die shortly after birth, whereas the mortality rate of babies with spina bifida is especially high over the first year of life. Individuals with less severe myelomeningocele suffer from lifelong disabilities, including reduced mobility, little or no bowel and bladder control, and urological infections, and they often require surgical interventions to control the effect of hydrocephalus, the build-up of fluid inside the skull caused by obstruction of CSF circulation (Simeonsson et al., 2002). NTDs pose a considerable monetary burden on the health care system (Kinsman and Doehring, 1996), as well as a significant emotional burden on the affected individual and his/her family. Therefore, understanding neural-tube development and the causes of NTDs are among the most important health-related studies today.

27.1.3 Etiology of NTDs

The etiology of NTDs is complex and involves both genetic and environmental factors, which makes NTDs a classic example of a multifactorial disorder.

27.1.3.1 Genetics of NTDs

The genetic risk of a recurrent NTD in a family with one child with an NTD is 2–5%, a 50-fold increase compared with the rest of the population (Forrester and Merz, 2005). Despite the evidence that genetics plays a pivotal role in the occurrence of NTDs, there are few data on single-gene defects directly associated with NTDs in humans. A number of chromosome rearrangements, such as trisomies 13 and 18, and known genetic syndromes, for example, Meckel syndrome, are associated with NTDs (Lynch, 2005). Chromosomal abnormalities, such as aneuploidy, are present in 5–17% of cases with NTDs (Kennedy et al., 1998), and a 13q deletion, with a critical region at 13q33-34, is strongly correlated with the occurrence of NTDs (Luo et al., 2000). These data can be used to extract clues in an effort to determine the identity of genes involved in neural-tube closure. Clues as to candidate genes for association studies of human NTDs also have been derived from research in animal models of NTD, as described below and as outlined in comprehensive reviews by Harris and Juriloff in 2007 and 2010.

27.1.3.1.1 ENVIRONMENT AND NTDS

Maternal diabetes during pregnancy is a key environmental factor, as there is a greater than tenfold increase in NTD frequency in the offspring of diabetic mothers as compared to the general population (Milunsky et al., 1982). The mechanism by which diabetes contributes to failure of neural-tube closure in humans remains unclear. However, a mouse model of diabetes showed a correlation between elevated blood glucose in the mother, increased apoptosis of neural cells, and reduced levels of Pax3, a gene required for neural-tube (NT) formation (Fine et al., 1999; Phelan et al., 1997). Genomics studies show that maternal diabetes can alter transcriptional programs in the developing mouse embryo, potentially affecting genes that directly and indirectly control NT formation (Pavlinkova et al., 2009). Maternal hyperthermia during pregnancy can increase the risk of NTDs up to twofold, although the mechanism responsible for this increase is unknown (Lynberg et al., 1994). Pharmacological agents, among them valproic acid (Nau et al., 1991) and antiepileptic drugs, are associated with an increased incidence of NTDs. For example, antiepileptic drugs can increase NTD risk by 10- to 20-fold (Lindhout et al., 1992). Lifestyle choices such as drinking, smoking, and recreational drugs also increase the risk of NTDs through unknown mechanisms (Suarez et al., 2008). Finally, studies have provided a link between dietary choices and risk for NTDs. In general, vitamin intake is not correlated with reduced risk for NTDs (Carmichael et al., 2003); however, there is one dietary supplement, folic acid, which can reduce the risk for NTDs when added to the maternal diet periconceptionally. Folic acid is the environmental factor that has attracted the most attention from developmental biologists, epidemiologists, and the general public due to the view that it has a protective effect against NTDs (Honein et al., 2001). Epidemiological studies started in the early 1980s showed that maternal folic-acid supplementation led to a significant reduction in NTD incidence (Laurence, 1985). This evidence dramatically influenced public-health policies and led to food-fortification programs in the United States and a number of other countries. The mechanism by which folic acid affects the incidence of NTDs remains largely unknown, despite the investigative efforts of a number of laboratories worldwide concerning its effect on the developing embryo.

As noted, there is a gap in knowledge of the mechanisms that lead to NTDs with respect to both genetic and environmental causes. This lack of knowledge significantly limits efforts of the scientific and medical communities toward the prevention of NTDs, both through genetic counseling and through control of the embryonic environment before and during NT formation.

27.2 VERTEBRATE NEURULATION

27.2.1 Animal Models for Vertebrate Neurulation Studies

Neural-tube development requires the coordination of multiple tissues (neural ectoderm, the neighboring mesenchyme, and the overlying ectoderm) in both space and time. For over 50 years, a number of animal systems have been used to study NT development. Model systems such as birds (Averbuch-Heller et al., 1994; Bel-Vialar et al., 2002; Schoenwolf et al., 1989) and amphibians (Clarke et al., 1991; Davidson and Keller, 1999; Roffers-Agarwal et al., 2008) and zebrafish (Nyholm et al., 2009; Puschel et al., 1992; Sumanas

et al., 2005) have contributed considerable insight into neurulation. The mouse is the most extensively studied mammalian experimental model for neurulation and NTD, and it most closely recapitulates human neurulation. However, even mice have limitations as a model of human NT development. For example, there are thought to be differences between mice and humans in the number of closure initiation sites, the sites where the neural folds first meet. Moreover, very few mouse models of NTD are representative of nonsyndromic human NTDs (Juriloff and Harris, 2000). Despite these differences, the mouse undergoes neurulation in a manner that is closer to human than to other animal models used to study neurulation to date. Moreover, similar genes are likely responsible for NT closure in humans as in mice (Harris and Juriloff, 2007, 2010). Finally, the great number of mouse NTD mutants provides an invaluable tool for understanding the underlying genetic, molecular, and cellular basis of NTD.

27.2.2 Primary and Secondary Neurulation

The process that forms the primordium of the central nervous system, the neural tube, is called neurulation. Modes of neurulation vary across vertebrate species; however, cellular comparisons reveal conserved mechanisms. This reinforces the idea that study of neurulation in different animal systems can contribute to our understanding of the molecular basis of NT formation and NTDs in humans.

There are two different modes of neurulation: primary and secondary neurulation (Figure 27.1). During primary neurulation, the flat and thickened epithelial layer on the dorsal surface of the embryo called the neural plate bends, and the neural folds seal together to form a hollow neural tube. Secondary neurulation occurs in the most caudal region of the embryo. During secondary neurulation, a cluster of mesenchymal cells condenses below the surface ectoderm to form a cord-like structure. A central lumen forms in this compact structure through

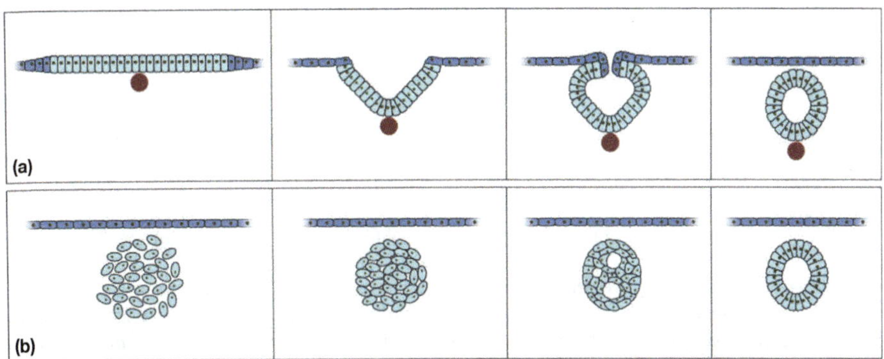

FIGURE 27.1 Primary and secondary neurulation. (a) Primary neurulation and (b) secondary neurulation in amniotes. Organization of cells from onset (*left*) to completion (*right*) of neurulation.

a process known as cavitation, giving rise to the NT (Griffith et al., 1992; Schoenwolf, 1979; Schoenwolf and Delongo, 1980). All amniotes studied to date, including humans (Saitsu et al., 2004), undergo both primary and secondary neurulation.

Although anamniotes have provided considerable information on how neurulation occurs and have furthered our understanding of neurulation in mammals, for the purposes of this chapter, we will focus on amniotes and more specifically mice due to their similarity in development and genetics to human neurulation.

27.2.3 Neurulation in Amniote Model Systems

Amniote neurulation occurs in at least four stages: (i) the acquisition of a neural fate, (ii) the elevation of the neural folds, (iii) the bending of the neural plate, and (iv) the meeting of the neural folds at the midline and fusion that forms the closed neural tube (Figure 27.2).

These four stages occur concurrently along the rostral–caudal axis and in coordination with movements of the primitive streak. Following neural induction, the neuroectoderm becomes morphologically distinct from the non-neural ectoderm through apicobasal thickening of the neuroectoderm to form a tall and thick epithelial sheet that is relatively wide along the medial–lateral axis and short along the anterior–posterior (AP) axis (Figure 27.2(a)). Subsequently, the neural plate undergoes convergent extension (see below) to narrow along the medial–lateral axis and elongate along the AP axis.

Bending of the neural plate at the lateral edges forms the neural folds, which continue to elevate and converge to meet in the midline and form a closed tube. Forces from the neural ectoderm, the mesenchyme and the adjacent non-neural ectoderm promote elevation of the neural folds (Moury and Schoenwolf, 1995). Further bending of the neural folds is driven by formation of "hinge" points involving apical constriction of a limited number of cells in three positions along the neural plate. The first hinge point, called the floor plate or medial hinge point (MHP), forms in the middle of the neural plate above the notochord (posterior) or prechordal mesoderm (anterior). The next two hinge points form in pairs on the dorsolateral sides of the elevating neural folds and are called dorsolateral hinge points (DLHPs; Figure 27.2(b) and 27.2(c)) (Colas and Schoenwolf, 2001; Shum and Copp, 1996). The inductive signal for MHP formation comes from the notochord/prechordal mesoderm and is the secreted protein Sonic Hedgehog (Shh) (Smith and Schoenwolf, 1989; van Straaten et al., 1988). Shh, in a dose-dependent manner, also participates in positioning of the DLHPs (Ybot-Gonzalez et al., 2002). Bending the neural folds is critically important in bringing the neural folds in close proximity to the midline and thus facilitates closure.

The final step of neurulation is when the neural folds meet in the midline, the non-neural ectoderm separates from the neural ectoderm, and each tissue layer seals together to form a continuous sheet of ectoderm overlying the closed neural tube (Figure 27.2(d)). This closure is traditionally called fusion, although the cells do not actually fuse. Very little is known about this last step of neurulation in any organism. For instance, in the mammalian embryo, it has been debated whether the neural or non-neural ectoderm initiates closure. Recent live imaging of the mouse embryo during neurulation showed that the non-neural ectoderm cells are highly dynamic and extend thin, long, and fast-moving cellular extensions that connect the two folds as they approach each other, indicating that the non-neural ectoderm initiates and possibly drives neural-fold fusion (Pyrgaki, 2010b). A role for adhesion molecules is hypothesized in fusion. For example, mutations in adhesion molecules, such as ephrinA5 and ephrinA7, can cause NTDs in mice (Holmberg et al., 2000). However, a direct involvement of these molecules in neural-fold fusion has not been demonstrated.

The fusion of the neural folds starts from one or more initial closure points, depending on the species examined. The classical view has been of a zipper-like closure that proceeds rostrally and caudally from these initial closure points. Recent live imaging of the mouse embryo cranial

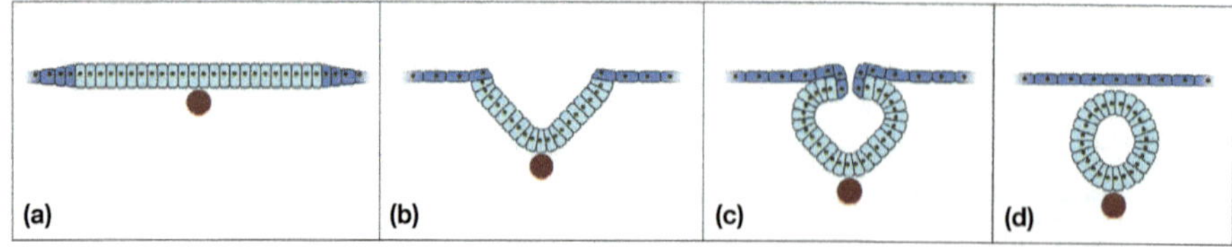

FIGURE 27.2　Four distinct steps of neural-tube formation in amniotes. (a) Acquisition of a neural fate and shaping of the neural plate. (b) Elevation of the neural folds. The non-neural ectoderm exerts force on the neural folds to contribute to neural-fold elevation. (c) Bending of the neural plate and formation of hinge points. In three positions, the neural ectoderm cells change their shape from cuboidal to wedge-shaped to form the medial hinge point and dorsolateral hinge points. These cellular changes facilitate further bending of neuroectoderm to bring the neural folds close together along the midline. (d) Meeting of the neural folds at the midline and fusion to form the closed neural tube and the overlying ectodermal sheet.

region supports a zipper-like closure in the hindbrain. However, closure in the mouse midbrain suggests a different mechanism in which intermediate closure points facilitate closure of the folds in a "buttoning-up" manner (Pyrgaki et al., 2010b).

Migration of the neural crest cells, which form at the boundary between the neural and non-neural ectoderm and then migrate laterally to different parts of the body, occurs concurrently with neural-tube closure in some rostral–caudal regions or after neural-tube closure in other regions.

27.2.4 Neurulation in Humans

Practical and ethical considerations limit our ability to study neurulation in humans, and animal models are therefore needed to provide insight into the cellular and molecular bases of human neurulation. However, there is sufficient information regarding neurulation in humans to suggest that some of the principles of neurulation in amniotes also apply to the human embryo.

Neural-tube formation and closure in humans occurs within the first month of pregnancy, between weeks 3–4 of gestation. In humans, primary neurulation is responsible for generating most of the neural tube. The caudal eminence of the human embryo is formed via secondary neurulation, which will be the point of origin of the terminal parts of the spinal cord, notochord, somites, vertebrae, hindgut, nerves, and blood vessels. This ectoderm-covered mass of pluripotent tissue is a continuation of the primitive streak (at stages 9–12), and it gradually replaces the streak (at stages 12 and 13 of the development of the human embryo, at about 4 weeks post fertilization) and forms a solid cellular mass called the neural cord. The neural cord forms the lower sacral and coccygeal parts of the spinal cord (Muller and O'Rahilly, 1987). The rostral and caudal neuropores close when approximately 20 and 25 pairs of somites are visible, respectively.

One potential difference between human and mouse neural-tube formation may be the number of initial closure points. The mouse has three initial fusion points. In humans, there is an ongoing debate about the number of fusion points. Clinical studies of human embryos with NTDs have argued for 3–5 closure points (Martinez-Frias et al., 1996; Nakatsu et al., 2000; Srinivas et al., 2008). These data are disputed in a study published in 2002 that supports the idea that the human embryo has only two initial closure sites and two neuropores (O'Rahilly and Muller, 2002). It is unclear whether the question of the number of closure points in the human embryo can be answered by case reports. Therefore, one cannot be sure that animal models fully represent the human condition, and it is very likely they do not. We need to be aware that, despite their usefulness, animal models do have limitations, and one needs to be careful when using them to infer analogies to human neurulation.

27.3 GENETIC APPROACHES USED TO UNCOVER REGULATORS OF NEURAL-TUBE CLOSURE IN MICE

Genetic approaches in model organisms are powerful tools for elucidating the biological bases of NTDs and NT development. The mouse provides an excellent model for studies of mammalian neurulation, as the technology of genetic manipulation is very advanced, allowing for a specific gene to be altered and studied with respect to its effect on the organism (Sedivy and Joyner, 1992).

Targeted mutagenesis approaches in the mouse have dramatically assisted efforts to understand better the processes underlying human disease. A great number of mouse models of human conditions are used in research to provide insight into the mechanisms of disease and to explore novel treatments for these conditions.

Phenotype-driven approaches (also referred to as forward genetic screens) start from animals that have been selected for a phenotype of interest and then move to identifying the gene that has been mutated. Many of the forward genetic screens utilize mutagenic agents, ENU being the agent most widely used, that create point mutations and hence can result not only in the inactivation of genes, but also in the generation of novel alleles (i.e., neomorphic, hypomorphic, and alteration of a conserved protein domain). These novel alleles may better represent a disease phenotype than a complete loss-of-function allele.

Over the past decade, forward genetic screens in mice from a number of laboratories worldwide have provided a new resource for the study of multiple biological systems, including neural-tube closure (Nishimura et al., 2003; Zohn et al., 2005). Additional studies to determine the molecular and cellular processes regulated by these key developmental genes have promoted new levels of understanding of the mechanisms that underlie embryonic development. In terms of neural-tube development, forward genetic approaches in the mouse have revealed a number of novel genes that oftentimes had not been functionally analyzed in any system, thus highlighting the effectiveness of phenotype-based screens in elucidating novel aspects of NT development.

27.4 MOLECULAR BASIS OF NTDs

The following sections will describe the molecular mechanisms that control normal neurulation and how these can be disturbed and subsequently lead to NTDs, with the focus on mice and humans.

Following neural induction, the early neural plate is initially specified to give rise to rostral brain tissues (i.e., the forebrain). Posterior character is then imparted to the neural tissue such that it acquires more caudal fates (midbrain, hindbrain, and spinal cord). Further rostrocaudal patterning of the neural plate then begins to take place, as well as patterning in the dorsoventral axis. As the process of neural induction, patterning, and formation of the neural tube is covered in detail in Rubenstein and Rakic, 2013. We will focus here on how disruptions of genes that regulate these processes lead to NTDs both in mice and humans.

27.4.1 Genes Controlling Convergent Extension/PCP Pathway and NTDs

After neural induction, the neural plate narrows and elongates along the embryo's medial–lateral and AP axes, respectively. This shaping of the neural plate is achieved mainly via the morphogenetic process called convergent extension: a medially directed movement and intercalation of the neural-plate cells toward the midline resulting in the lengthening and narrowing of the neural plate (Figure 27.3(a) and 27.3(b)).

It is well established that the planar cell polarity (PCP) pathway controls convergent extension of both neural and mesodermal tissues in vertebrates (reviewed in Wallingford et al., 2002). Vertebrate cognates of the PCP cascade are required for neural-tube closure in frog, zebrafish, and mouse embryos. In the mouse embryo, mutations in five different genes of the PCP pathway lead to the most severe form of NTD, in which the NT remains open throughout its length and mimics the human condition known as craniorachischisis. These embryos have an abnormally wide neural plate, and the two folds are physically too far apart to be joined to each other, due to defects in convergent extension.

The *loop tail* (*Lp*) mouse is the oldest mouse model of craniorachischisis, first described in 1949 by Strong and Hollander. Heterozygous *Lp* mice are viable with a characteristic looped tail, whereas homozygotes exhibit craniorachischisis (Strong and Hollander, 1949). The neural folds of *Lp* homozygous embryos at 8.5 days after fertilization (E8.5) are elevated, but the MHP is absent and the midline region assumes a U shape, in contrast to the normal "V" shape. The gene mutated in *Lp* is homologous to *Drosophila strabismus* and is called *Vangl2* (also *Lpp1* or *Lptap*) (Murdoch

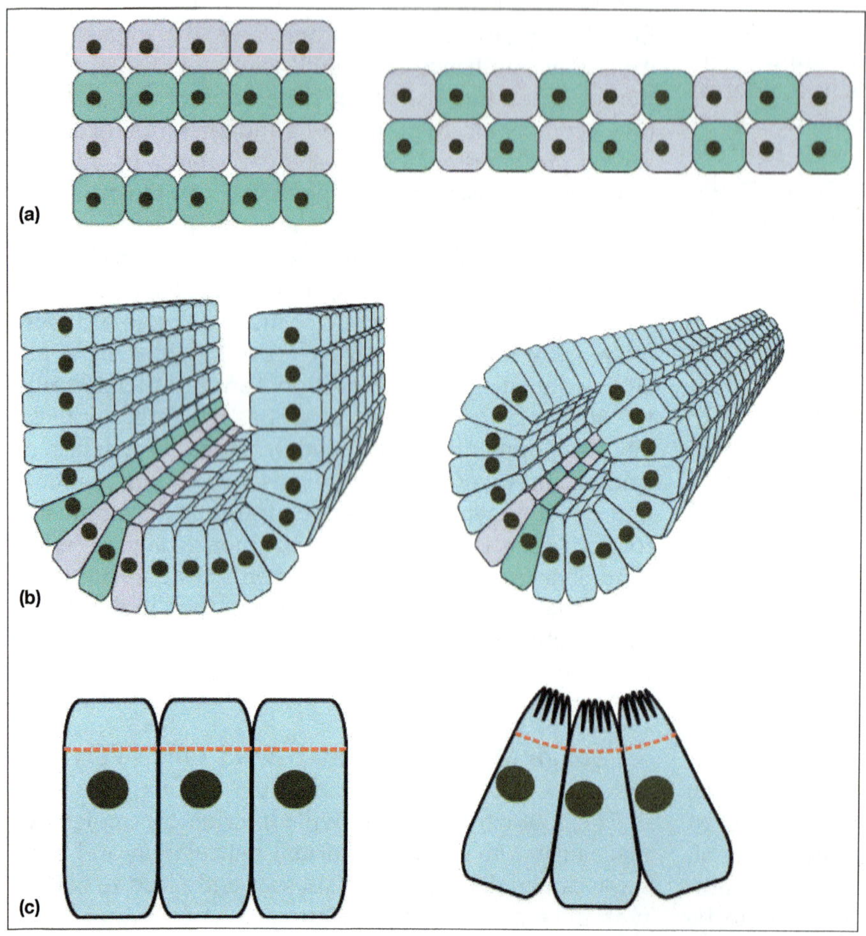

FIGURE 27.3 Convergent extension and apical constriction. (a) Convergent extension cell movements result in tissue elongation via intercalation of adjacent cells (depicted by *dark* and *light squares*) in an epithelial sheet to form a narrower, longer strip of tissue. (b) Convergent extension facilitates the narrowing of the neural plate and brings the neural folds in closer proximity. (c) Apical constriction. Contraction on the apical side of neural cells leads to formation of hinge points and bending of the neural tube.

et al., 2001a). *Vangl2* encodes a transmembrane protein with a C-terminal PDZ-binding motif and is expressed broadly in the early neuroectoderm (Jessen and Solnica-Krezel, 2004). Mice null for *Vangl1*, another Vangl family member in mammals, show subtle alterations in the polarity of inner-hair cells of the cochlea (another manifestation of a PCP defect), but trans-heterozygotes for *Vangl1* and *Vangl2* exhibit craniorachischisis, similar to *Vangl2* null mice (Torban et al., 2008). Genetic interactions have been also observed between *Lp* and *circletail* (*Crc*) mice (Murdoch et al., 2001b). Mice heterozygous for both *Lp* and *Crc*, though not allelic, display the same phenotype of craniorachichis that each of these mice displays when homozygous for each individual gene. The *Crc* mouse harbors a stop mutation in the cell polarity gene *Scribble*, which is the mouse orthologue of *Scribble* in flies (Murdoch et al., 2003). *Lp* also genetically interacts with a regulator of the PCP pathway, PTK7. PTK7 is highly conserved among species and encodes a transmembrane tyrosine kinase protein that, when mutated, leads to NTD and defects in stereociliary bundle orientation in mice (Lu et al., 2004). Two other mouse mutant lines, *Crash* and *Spin cycle*, display NTD and inner-ear defects that result from disruption of the PCP gene *Celsr1*, a large protocadherin seven-pass transmembrane molecule, orthologous to *Flamingo*, a *Drosophila* gene also known as *Starry night* (Curtin et al., 2003). Finally, double mutants of mouse *Disheveled1* and *Disheveled2* (*Dsh1* and *Dsh2*) exhibit NTDs very similar to the craniorachischisis observed in other PCP mutants. *Dsh1* and *Dsh2* share 65% sequence identity, and the fact that *Dsh1* null mice have a far less severe phenotype (Hamblet et al., 2002) than the double mutant indicates functional redundancy between the two proteins.

Recent studies indicate another role for the PCP proteins in cilia biogenesis. In *Xenopus*, loss of function of *Fuzzy* or *Inturned*, two orthologues of fly PCP effector proteins, results in NTD and failure of ciliogenesis due to incorrect microtubule orientation (Park et al., 2006). Disruption of *Fuz* in mice also causes NTDs, apparently due to disrupted ciliogenesis that in turn leads to defective Shh signaling (Gray et al., 2009). Furthermore, *inturned* in mammals is an important regulator of cilia formation, NT closure, and embryonic development, and *inturned* mutations affect Shh signaling (Zeng et al., 2010). A connection between ciliary function and NT closure has also been identified in humans. Meckel syndrome, the most frequently identified syndromic NTD, is associated with disruption of *MKS1* and *MKS3* (Kyttala et al., 2006; Smith et al., 2006). Mks1 is required for ciliogenesis and Hedgehog signaling in mice (Weatherbee et al., 2009). Genes involved in Bardet–Biedl syndrome (BBS), a human ciliary defect, have been disrupted in mice, leading to phenotypes shared with PCP mutants, including NTDs (exencephaly), open

eyelids, and disrupted cochlear stereociliary bundles. BBS genes interact with *Vangl2* in mice and zebrafish and Vangl2, like BBS proteins, is localized to the basal body and axoneme of the cilia (Ross et al., 2005). Within the PCP pathway genes, the coding region and splice sites of *Vangl1* and 2 have been analyzed in a population of 66 human patients with NTDs, and only one variant in *Vangl1* was identified, which is predicted to result in a nonsynonymous change that is unlikely to have functional significance (Doudney et al., 2005). More recently, however, three patients with NTD were found to be heterozygotes for mutated VANGL1 alleles and five novel mutations in VANGL1 have also been found in patients with NTDs (Kibar et al., 2007, 2009). Moreover in a 2010 study, three novel missense mutations in VANGL2 were found in three different human fetuses with severe NTDs (Lei et al., 2010). Recent studies took advantage of this PCP pathway-specific knowledge generated from animal models to interrogate a larger set of PCP genes for mutations in samples from patients with craniorachischisis. This identified eight potentially causative mutations in CELSR1 and SCRIB. Overall, these studies make a strong case for a connection between PCP genes and NTDs.

27.4.2 Neural-Tube Patterning

Patterning of the neural tube along the AP and dorsal–ventral (DV) axes relies on tightly regulated molecular cascades. The patterning centers of the early embryo as well as their mechanisms of actions are described in Rubenstein and Rakic, 2013. Here, we will describe how disruption of patterning-related genes alters NT development and leads to NTDs in humans and mice.

27.4.2.1 Anterior–posterior Patterning Genes and NTDs

The major players of AP patterning include graded signals from fibroblast growth factors (FGFs), Wnt, and retinoic acid (RA). FGFs are key players in caudal transformation of anterior neural tissue to a more posterior fate (Kudoh et al., 2004; Rentzsch et al., 2004). FGFs are expressed in the regressing primitive streak, which continues to promote the progressive caudal patterning of the neural tissue (Dasen et al., 2003; Liu et al., 2001).

Along with FGFs, a number of other molecules, such as Wnts, Nodal, and RA, are present in the posterior of the embryo, and they refine the AP fate of the neural tissue (reviewed in Gamse and Sive, 2000). RA signaling acts to promote the neural character of the hindbrain and the anterior spinal cord (Bel-Vialar et al., 2002; Liu et al., 2001). RA and FGF act in opposition to each other to impart AP fates along the spinal cord. Studies in *Xenopus* embryos indicate a requirement for RA signaling for expression of posterior markers (Hoxb9, N-tubulin, and Xlim1) and for establishment of posterior fate (Blumberg et al., 1996), and overexpression of Cyp26, an enzyme that mediates the degradation of RA, causes anterior structures to develop where normally there should be posterior structures (de Roos et al., 1999; Hollemann et al., 1998). The mammalian homologue, Cyp26A1, when ablated in mice, causes embryonic lethality, spina bifida, and truncation of the tail and lumbrosacral region. A higher risk for NTDs in humans has been associated with polymorphisms in RALDH2, an enzyme in the biosynthetic pathway of RA. Other enzymes which function in RA synthesis or metabolism, such as ALDH1, CYP26A1, CYP26B1, and CYP26C1, as well as cellular RA-binding proteins, such as CRABP1 and CRABP2, have been examined as potential NTD risk factors, but none of these genes has so far shown significant correlation with NTD risk in humans (Deak et al., 2005b).

Graded activity of Wnt signaling is also required for proper positioning of the midbrain–hindbrain border. The mechanism by which Wnt leads to posteriorization of the neural tissue is indirect and requires FGF signaling (Domingos et al., 2001). Mutations in components of the Wnt pathway can lead to NTDs. For example, a missense mutation of a highly conserved amino acid in the low-density, lipoprotein-receptor-related protein 6 (Lrp6), a co-receptor required for Wnt canonical signaling, causes exencephaly in homozygous mutant mice. Mice heterozygous for this mutation display the characteristic phenotype of a crooked tail (Carter et al., 2005).

27.4.2.2 Dorsal–ventral Patterning Genes and NTDs

Dorsoventral patterning largely utilizes a common mechanism throughout the length of the embryonic body. Four different classes of secreted factors influence DV patterning: Shh, bone morphogenetic proteins (BMPs), FGFs, and RA.

Correct patterning of the ventral half of the NT largely relies on the activity of Shh, whereas the dorsal half requires the inductive activities of the transforming growth factor-beta superfamily, which includes the BMPs, expressed in the overlying ectoderm and the roof plate. The roof plate is the dorsal equivalent of the floor plate, and it expresses BMP4, 5, and 7, which in turn induce dorsal genes, such as the transcription factors Pax3 and Msx (Liem et al., 1995, 1997). BMPs also act to set the borders of expression of bHLH (*Math*, *Mash*, and *Ngn*) and LIM (*Lbx* and *Lmx*) transcription factors (Helms and Johnson, 2003).

Tissues from humans with SBA have been evaluated for mutations in Bmp4 or the BMP inhibitor Noggin. Four heterozygous missense mutations in Bmp4 inherited from unaffected heterozygous parents in four different patients have been identified (Felder et al., 2002). Three out of the four missense mutations in Bmp4 (S91C, T225A, and R226W) are in the protein region responsible for dimerization, and the sequence alterations in this region have been proposed to influence the stability of the mature protein. The fourth missense mutation (S367T) was found within the C-terminal region that is physiologically important and hence might disrupt the function of the mature BMP4 protein. Moreover, one missense mutation (G92E) was found in Noggin, which replaces a highly conserved glycine in a region essential for proper structure and function of the Noggin protein. The sequence variants in these SBA patients may cause changes in protein stability and/or activity and therefore contribute to the failure of NT development (Felder et al., 2002).

Shh signaling, from the notochord and floor plate, acts on the surrounding neural cells to regulate the expression of transcription factors that control the formation of ventral cell types, such as motor neurons and ventral interneurons. As described above, genes involved in ciliogenesis are also needed for Hedgehog signaling and disruptions in this process are associated with NTD in mice and humans.

The Pax family of patterning molecules has been extensively studied for association with NTD. Many Pax genes are tightly regulated in both space and time during NT development (Stoykova and Gruss, 1994), and BMPs act in conjunction with Shh to set the correct expression boundaries of Pax genes. Mutations of *Pax3* in mice cause the *splotch* (*sp*) phenotype, which includes NTD (exencephaly and spina bifida) and dysgenesis of spinal ganglia, limb, and heart structures (reviewed in Gruss and Walther, 1992). *Pax3* mutations are associated with the Waardenburg syndrome in humans (pigmentation, limb, and enteric nervous-system defects and deafness) (Baldwin et al., 1994; Tassabehji et al., 1993). Attempts have been made to associate NTD in humans with mutations in *PAX 1*, *3*, *7*, or *9*. In two studies of 79 and 38 familial cases of NTDs, one mutation in the paired domain of *Pax1* was found in a patient with spina bifida (Hol et al., 1996); however, no further Pax mutations have been detected. Thus, the data to date do not show a strong association between Pax gene mutations and human NTDs.

27.4.2.3 Bending of the Neural Plate and NTDs

Another crucial step during NT formation is the bending of the neural plate in order to elevate and bring the two folds into close proximity along the midline. The MHP and two DLHPs are differentially used at different axial levels. Moreover, their formation is differentially controlled at a molecular level. For example, lack of *Zic2* function in mice causes loss of DLHP, although the MHP forms properly (Ybot-Gonzalez et al., 2007), and this results in extensive spina bifida (Elms et al.,

2003; Nagai et al., 2000). In mouse, the posterior neuropore completes NT formation, although it lacks a MHP; thus, the presence of the DLHP is sufficient for closure. The MHP may be dispensable for spinal neural-tube closure, as mice that lack Shh, HNF3β, Gli2, or Gli1 and Gli2 do not develop spinal NTD, despite failure to form the MHP, although less-acute bending still occurs by an unknown mechanism (Ang and Rossant, 1994; Chiang et al., 1996; Ding et al., 1998; Matise et al., 1998; Park et al., 2000; Weinstein et al., 1994).

The Shh pathway is critical in the bending of the neural plate. High Shh levels suppress DLHP formation (Ybot-Gonzalez et al., 2002), and low Shh levels induce their formation (Echelard et al., 1993), thus helping to set the position of DLHP formation. Loss of Patched, the Shh receptor that keeps the Shh pathway in an "off" state when a ligand is absent, causes both cranial and spinal NTDs in mice (Goodrich et al., 1997). One aspect of the phenotype could be that loss of Ptch constitutively actives the pathway and, like Shh overexpression, prevents DLHP formation. Loss of PKA, a protein kinase that downregulates Shh signaling, leads to spinal NTD (Huang et al., 2002). Mutation of Rab23, another negative regulator of the Shh pathway, disrupts DLHP formation and neural patterning leading to cranial NTD (Gunther et al., 1994).

The cellular basis for NT bending is apical constriction. The neuroepithelium, a pseudostratified epithelium of a single-layer thickness, begins to change its shape from columnar to wedge, first in the medial region and then at dorsolateral positions, where the hinge points will form (Figures 27.2(b), 27.2(c), and 27.3(c)). Our understanding of the cellular basis of this highly regulated constriction is limited. Two molecules, Shroom3 and p190RhoGAP, regulate apical constriction in the NT. Shroom3 belongs in the Shroom family but to date only Shroom3 has been associated with NTDs in mice. Shroom3 controls apical constriction via binding of filamentous actin. Mice null for *Shroom3* display exencephaly and loss of actin that would normally be localized on the apical-cell surface (Hildebrand and Soriano, 1999). P190RhoGAP is a negative regulator of Rho GTPase, which regulates actin microfilaments. Mice lacking p190RhoGAP have a number of defects, including NTD, and an excessive accumulation of polymerized actin in cells of the neural plate (Brouns et al., 2000). In addition to apical constriction, basal expansion, a process that is potentially also relying on cytoskeletal rearrangements, has been thought to assist NT bending (Smith and Schoenwolf, 1987); however, no mutants that disrupt this process have been identified.

Like *Shroom3*, other genes that regulate actin rearrangement at the cell surface are associated with NTDs. For example, cranial NTDs result from the disruption of protein kinase C, which targets a number of cytoskeletal regulators including the actin-binding protein vinculin, the cytoskeleton-related genes *mena* (*Enah*)/*profilin1* (*pfn1*), and *Macs* and *Mlp* (Koleske et al., 1998; Lanier et al., 1999; Stumpo et al., 1995; Xu et al., 1998). These data support the involvement of actin in NT formation. However, both older (Schoenwolf et al., 1988) and more recent (Ybot-Gonzalez and Copp, 1999) studies indicate that the MHP and DLHP are not affected by cytochalasin treatment, which blocks actin polymerization and elongation. This suggests that actin microfilaments may act to stabilize the hinge points after their formation, rather than to drive their formation.

Some of the genes involved in apical constriction and bending of the neural plate have been examined in human NTDs. In the Shh pathway, GLI3 mutations cause two different syndromes, the Greig cephalopolydactyly syndrome and the Pallister–Hall syndrome (Johnston et al., 2005, Radhakrishna et al., 1997), but neither syndrome presents with NTD. The lack of NTD is perhaps not surprising, given that the *Gli3* mutant mouse (*extratoes*) shows cranial NTDs with low penetrance and only in homozygous mutant embryos, and this also results in perinatal lethality (Winter and Huson, 1988). Mutations in *Shh* (Nanni et al., 1999) and in its receptor, *patched-1*, are associated with holoprosencephaly (Ming et al., 2002), a structural forebrain anomaly found with high frequency in humans. In studies conducted on human embryos with craniorachischisis, abnormal *Shh* expression was found (Kirillova et al., 2000), similar to altered expression of *Shh* in *Lp* mice (Murdoch et al., 2001a). However, the primary event leading to abnormal *Shh* expression was not identified. Variations in the *Shh* gene have not been correlated with other human NTDs, although these studies have largely been limited to the N-terminus part of the Shh protein, but not the C-terminus, which contains the autocatalytic cleavage site (Zhu et al., 2003). Variations in genes such as *PKA*, *Zic1, 2, 3* also failed to show correlation with human NTDs. In terms of apical constriction, only *MLP* and *MACS* have been examined as potential candidates for NTDs, but no correlation was found at least in 43 Caucasian simplex families in which the affected child had a lumbrosacral myelomeningocele (Stumpo et al., 1998).

27.4.2.4 NTDs due to Disruption of Neural-fold Fusion

The last step of NT closure, the joining of the two folds along the midline, also called fusion, is the least studied and least understood. It has been hypothesized that adhesion molecules facilitate neural-fold joining. However, whether such molecules play a direct role in fusion is unknown, although as outlined below, a few adhesion molecules are required for NT closure. In addition, it has been suggested that components of the extracellular matrix (ECM) might play a role in fusion (Moran and

Rice, 1975; Sadler, 1978), although it is unclear which ECM components may be involved.

Defects in a few adhesion molecules have been linked to NTDs. For example, in mice, inactivation of the ephrin receptor subfamily members *Eph-A5* or *Eph-A7* causes NTDs (Holmberg et al., 2000). Both ephrin receptors are expressed in the cranial, but not spinal, neuroepithelium. Ephrin signaling is thought to act through components of the PCP pathway. According to Lee et al. (2006), ephrinB1, a cell-surface-ephrin ligand, controls migration of cells during frog development by acting through PCP-pathway components and clustering of Dishevelled proteins. Two members of the cadherin family of adhesion molecules, N-cadherin and E-cadherin, have complementary expression patterns during neurulation, with N-cadherin expressed in the neural plate and E-cadherin expressed in the non-neural ectoderm (Detrick et al., 1990). Mice null for *N-cadherin* (*Cdh2*) undergo normal neurulation, showing that N-cadherin is not essential for NT formation (Radice et al., 1997). A role for E-cadherin has been more difficult to discern, as loss of E-cadherin results in embryonic lethality prior to neurulation. However, two pieces of evidence support a role for E-cadherin in NT closure. First, the culture of rat embryos with an antisense oligonucleotide against E-cadherin leads to cranial NT malformations (Chen and Hales, 1995). Secondly, when E-cadherin is downregulated in the non-neural ectoderm as occurs in Grhl2 mutants, the neural folds, face, and body wall fail to fuse (Pyrgaki et al., 2010a; Werth et al., 2010). NCAM1 is another adhesion molecule expressed in the neuroectoderm. Although *NCAM1* null mice do not display NTD (Cremer et al., 1994), NCAM1 has been studied in 132 families with spina bifida, and variations in this molecule may contribute to NTD risk (Deak et al., 2005a). Finally, NTD is observed in mouse embryos lacking protease activator receptor 1 and 2 (PAR1 and PAR2), which are expressed in the non-neural ectoderm during NT formation (Camerer et al., 2010). Thus, both cellular and molecular evidence support the idea that the non-neural ectoderm is involved in initiating and perhaps driving neural-fold fusion.

27.4.2.5 NTD due to Disruption of Apoptosis

During NT development, as during embryonic development in general, apoptosis and cell proliferation must be tightly regulated for development to proceed normally. Apoptosis or programmed cell death occurs in the neuroepithelium during neurulation. This was first documented by ultrastructural studies (Geelen and Langman, 1979; Schluter, 1973) and subsequently shown to be apoptotic cell death (Lawson et al., 1999). Either an increased or decreased rate of apoptosis has been associated with NTDs in mice. For example, NTDs in

embryos lacking *ApoB* (Homanics et al., 1995), *Bcl10* (Ruland et al., 2001) or *Mdm4* are associated with increased apoptosis. In these cases, the phenotype is attributed to insufficient numbers of cells participating in neurulation.

Conversely, mutations in genes such as *Trp-53* (Sah et al., 1995), *Casp3* (Kuida et al., 1998), and *Apaf1* (Cecconi et al., 1998) lead to NTDs associated with reduced levels of apoptosis. In these cases, excessive numbers of cells that would otherwise be reduced by apoptosis are thought to prevent the normal morphogenesis of the neural plate. Despite the dramatic phenotypes caused by misregulation of apoptosis, experimental studies to block apoptosis with pharmacological agents during mouse neurulation indicates that apoptosis facilitates NT closure but is not necessary for its completion (Massa et al., 2009; Yamaguchi et al., 2011). To date, there have been no studies that attempt to correlate the risk for human NTDs with molecules of the apoptotic pathway.

27.4.2.6 NTD due to Disruption of Proliferation

Proliferation must be tightly controlled, as either increased or decreased rates of proliferation can lead to NTDs. During cranial neural closure, proliferation is differentially regulated along the dorsoventral axis of the NT, with the dorsal half displaying more proliferation than the ventral half (Copp et al., 2003). As proliferation and differentiation are inversely correlated, this difference is reflected by a greater number of differentiated cells in the ventral NT at this early stage of neurulation.

Changes in expression of genes that control proliferation often lead to abnormal NT development. For example, overexpression of *Notch3* results in exencephaly, and these mice have increased numbers of neuronal progenitors (Lardelli et al., 1996). Tumor-suppressor genes, when mutated in mice, can cause abnormal neural-cell proliferation and lead to exencephaly. Examples include loss of function of *p53* (Sah et al., 1995) and a hypomorphic mutation in *Brca1* (Gowen et al., 1996). A study of an Irish population showed that two noncoding variants of *p53* are associated with NTD risk and two different variants with maternal risk (Pangilinan et al., 2008). In addition, variations in *Brca1* in a family-based association study correlate with spina bifida, although this may not be causative (King et al., 2007). Mutations in *Tsc2*, another tumor-suppressor gene, cause exencephaly and embryonic lethality in mice (Kobayashi et al., 1999). Mice with heterozygous mutations for genes of the chromatin-remodeling complex, such as *Brg1* and *Srg1*, are predisposed to exencephaly, and they often get tumors of the nervous system (Kim et al., 2001). Hyperproliferation also causes exencephaly in the mouse *Phactr4^{humdy}* mutant, which indirectly regulates the tumor-suppressor protein Rb (Kim et al., 2007). Hyperproliferation of the neuroepithelium before and during closure creates excess tissue, which inhibits bending of the NT and/or prevents the two folds from reaching each other. Reduced proliferation, on the other hand, leads to a number of cells inadequate to complete NT closure. Loss of mLin41 causes cranial NTD due to reduced proliferation and premature differentiation of the neuroectoderm, thus acting as a temporal regulator of neural progenitor maintenance (Chen et al., 2012).

Changes in cell number in tissues that surround the neuroepithelium can indirectly have a deleterious effect on NT formation. One example is reduced proliferation in the hindgut of *curly tail* mice, which likely carry a mutation in *Grhl3* (reviewed in Brouns et al., 2005). This hypoproliferation causes growth imbalance between the ventral and dorsal tissue, which in turn leads to excessive curvature within the caudal region and the failure of the posterior neuropore to close (Gustavsson et al., 2007; van Straaten and Copp, 2001). To date, the connection between the disruption of *Grhl3* and the proliferation defect remains elusive. Another example is the *open mind* mutation, which disrupts Hectd1 ubiquitin ligase, and results in inappropriate numbers of head-mesoderm cells and exencephaly (Zohn et al., 2007). Finally, *Twist1* mutant mice display exencephaly and severe craniofacial defects (Soo et al., 2002). *Twist1* is expressed in the head mesenchyme, where it regulates both proliferation (Ota et al., 2004) and apoptosis (Hjiantoniou et al., 2003), as well as neural-crest delamination and migration (Vincentz et al., 2008). The Saethre–Chotzen syndrome (SCS) in humans is due to TWIST1 mutations (Elanko et al., 2001; Kress et al., 2006). The SCS phenotype includes craniofacial defects but not NTD, although these phenotypes are thought to be due to haploinsufficiency rather than complete loss of TWIST1 function (el Ghouzzi et al., 1997).

27.5 FUTURE DIRECTIONS

In mice, there are currently >200 genes that are known to be required for neural-tube closure. However, surprisingly few of these models have been used to go beyond the question of the function of an individual gene. An interesting next level of exploration will be to determine the interplay between these genes. NTD in humans is largely thought to be a multigenic disease in which small changes in multiple genes may combine to lead to NTDs. Although the analysis of model organisms with homozygous mutations or severe loss-of-function phenotypes has been extremely useful, these models do not adequately reflect the complex genetics of human NTD. Therefore, it will be of interest to probe the phenotypic interplay between different gene

mutations further, including that between genes that do not obviously fit within a particular pathway or cellular process.

The genetic complexity of human NTD leads to a cautionary note with respect to genetic counseling. Based on the already large number of gene candidates identified from animal NTD models and the fact that a significant portion of the genome is still not functionally analyzed, there will likely be on the order of 500–1000 genes required for NT closure and which, when mutated, can lead to NTD. This highlights the difficulty in trying to identify causative single-gene mutations in human NTD. Moreover, it becomes a daunting task to conceive of a comprehensive genetic screen. Ultimately, advances in personalized medicine and whole-genome analysis will open this field of research to a more comprehensive overview. Studies are underway to determine whether genome copy-number variations are associated with NTD, but these data are currently lacking. The potential to identify a large number of polymorphisms leads to the next question, namely, how to determine whether a polymorphism identified from association studies in humans may create a meaningful change in gene expression or protein function. This question can now be quite readily addressed by techniques that allow sophisticated manipulation of the mouse genome such that specific polymorphisms can be created and tested *in vivo* for changes in phenotype and/or protein function.

Another area ripe for exploration lies in determining the interplay between mouse NTD models and known environmental factors. For example, would NTD risk increase in mice heterozygous for a mutation in a gene required for NT closure when gestation occurs in a mother that is a model for diabetes or obesity? If so, this would greatly aid research in determining the molecular basis of NTD risk defined by epidemiological studies.

Epidemiological research lies at the heart of one of the most profoundly influential studies that has changed our approach to NTD risk. Epidemiological studies have clearly shown a decreased incidence of NTD as diets across the world have improved, and this positive response has been strongly correlated with a higher level of folic acid in the diet. Although the preventative effect of folic acid on NTD incidence has been recognized since the 1980s, surprisingly little basic research has been done to understand the mechanistic basis for this effect. To date, only 18 mouse NTD models have been tested for their responsiveness to folic acid. Of these, FA supplementation reduced the incidence of NTDs in six cases, was not beneficial in nine cases, and, contrary to expectations, exacerbated NTDs in three cases (Gray et al., 2010; Harris 2009; Marean et al., 2011). Testing of many more of these mouse models will provide critical information for understanding the molecular logic that may underlie responsiveness to folic-acid supplementation. For example, are there particular pathways or cellular processes that are associated with responsiveness to folic-acid supplementation? Is the response allele-specific, or can a null allele be rescued, the latter perhaps suggesting that folic acid may bypass the defect? Folic acid provides the building blocks for many cellular reactions, including cell proliferation, gene expression, and genome stability, through its effects on nucleotide biosynthesis and as a methyl donor. Mouse NTD models can be used to tease out the complex function of folic acid, for instance, by targeting one arm of the biosynthetic pathway (i.e., supplementation with purines or biasing the diet toward methyl donors) in order to determine the mechanisms of folic-acid responsiveness better. Furthermore, there is a need to reconsider the treatment traditionally done in the mouse studies, which consists of short-term (days) exposure to folic-acid supplementation. This is not reflective of the trend in long-term folic-acid exposure in the U.S. population, as mandatory FA fortification began in 1998 and periconceptual use has been strongly recommended since the early 1990s. Indeed, recent long-term and multigenerational studies in mice indicate that the incidence of NTD for a given genetic allele can differ between short- and long-term folic acid diets and that the epigenetic profile can vary widely in wild-type mice fed a methyl diet for six generations (Li et al., 2011; Marean et al., 2011). These unexpected findings highlight the need to understand how FA influences NT closure and the mechanisms and genetics underlying the response to FA. Finally, in both mice and humans, there are a relatively large proportion of folate-resistant NTDs. One goal will be to utilize the knowledge gained from understanding the normal function of the mutated gene and its downstream pathways to determine whether other periconceptual therapies can be designed based on this insight. The types of studies outlined above will be critical to a better understanding of the complex genetic interplay that likely represents non-syndromic NTD in humans, as well as the intriguing but poorly understood intersection between genes and environmental influences in the highly complicated process of neural-tube closure.

References

Ang, S.L., Rossant, J., 1994. HNF-3 beta is essential for node and notochord formation in mouse development. Cell 78, 561–574.

Averbuch-Heller, L., Pruginin, M., Kahane, N., Tsoulfas, P., Parada, L., Rosenthal, A., et al., 1994. Neurotrophin 3 stimulates the differentiation of motoneurons from avian neural tube progenitor cells. Proceedings of the National Academy of Sciences of the United States of America 91, 3247–3251.

Baldwin, C.T., Lipsky, N.R., Hoth, C.F., Cohen, T., Mamuya, W., Milunsky, A., 1994. Mutations in PAX3 associated with Waardenburg syndrome type I. Human Mutation 3, 205–211.

Bel-Vialar, S., Itasaki, N., Krumlauf, R., 2002. Initiating Hox gene expression: In the early chick neural tube differential sensitivity to FGF and RA signaling subdivides the HoxB genes in two distinct groups. Development 129, 5103–5115.

Blumberg, B., Bolado Jr., J., Derguini, F., Craig, A.G., Moreno, T.A., Chakravarti, D., et al., 1996. Novel retinoic acid receptor ligands in *Xenopus* embryos. Proceedings of the National Academy of Sciences of the United States of America 93, 4873–4878.

Botto, L.D., Moore, C.A., Khoury, M.J., Erickson, J.D., 1999. Neural-tube defects. The New England Journal of Medicine 341, 1509–1519.

Brouns, M.R., Matheson, S.F., Hu, K.Q., Delalle, I., Caviness, V.S., Silver, J., et al., 2000. The adhesion signaling molecule p190 Rho-GAP is required for morphogenetic processes in neural development. Development 127, 4891–4903.

Brouns, M.R., Peeters, M.C., Geurts, J.M., Merckx, D.M., Engelen, J.J., Hekking, J.W., et al., 2005. Toward positional cloning of the curly tail gene. Birth Defects Research. Part A, Clinical and Molecular Teratology 73, 154–161.

Calzolari, F., Gambi, B., Garani, G., Tamisari, L., 2004. Anencephaly: MRI findings and pathogenetic theories. Pediatric Radiology 34, 1012–1016.

Camerer, E., Barker, A., Duong, D.N., Ganesan, R., Kataoka, H., Cornelissen, I., et al., 2010. Local protease signaling contributes to neural tube closure in the mouse embryo. Developmental Cell 18, 25–38.

Canfield, M.A., Marengo, L., Ramadhani, T.A., Suarez, L., Brender, J.D., Scheuerle, A., 2009. The prevalence and predictors of anencephaly and spina bifida in Texas. Paediatric and Perinatal Epidemiology 23, 41–50.

Carmichael, S.L., Shaw, G.M., Selvin, S., Schaffer, D.M., 2003. Diet quality and risk of neural tube defects. Medical Hypotheses 60, 351–355.

Carmichael, S.L., Shaw, G.M., Kaidarova, Z., 2004. Congenital malformations in offspring of Hispanic and African-American women in California, 1989-1997. Birth Defects Research. Part A, Clinical and Molecular Teratology 70, 382–388.

Carter, M., Chen, X., Slowinska, B., Minnerath, S., Glickstein, S., Shi, L., et al., 2005. Crooked tail (Cd) model of human folate-responsive neural tube defects is mutated in Wnt coreceptor lipoprotein receptor-related protein 6. Proceedings of the National Academy of Sciences of the United States of America 102, 12843–12848.

CDC, 2004. Spina bifida and anencephaly before and after folic acid mandate—United States, 1995–1996 and 1999–2000. MMWR. Morbidity and Mortality Weekly Report 53, 362–365.

Cecconi, F., Alvarez-Bolado, G., Meyer, B.I., Roth, K.A., Gruss, P., 1998. Apaf1 (CED-4 homolog) regulates programmed cell death in mammalian development. Cell 94, 727–737.

Chen, B., Hales, B.F., 1995. Antisense oligonucleotide down-regulation of E-cadherin in the yolk sac and cranial neural tube malformations. Biology of Reproduction 53, 1229–1238.

Chen, J., Lai, F., Niswander, L., 2012. The ubiquitin ligase mLin41 temporally promotes neural progenitor cell maintenance through FGF signaling. Genes Dev 26, 803–815.

Chiang, C., Litingtung, Y., Lee, E., Young, K.E., Corden, J.L., Westphal, H., et al., 1996. Cyclopia and defective axial patterning in mice lacking Sonic hedgehog gene function. Nature 383, 407–413.

Clarke, J.D., Holder, N., Soffe, S.R., Storm-Mathisen, J., 1991. Neuroanatomical and functional analysis of neural tube formation in notochordless *Xenopus* embryos; laterality of the ventral spinal cord is lost. Development 112, 499–516.

Colas, J.F., Schoenwolf, G.C., 2001. Towards a cellular and molecular understanding of neurulation. Developmental Dynamics 221, 117–145.

Copp, A.J., Greene, N.D., Murdoch, J.N., 2003. The genetic basis of mammalian neurulation. Nature Reviews Genetics 4, 784–793.

Coskun, A., Kiran, G., Ozdemir, O., 2009. Craniorachischisis totalis: A case report and review of the literature. Fetal Diagnosis and Therapy 25, 21–25.

Cremer, H., Lange, R., Christoph, A., Plomann, M., Vopper, G., Roes, J., et al., 1994. Inactivation of the N-CAM gene in mice results in size reduction of the olfactory bulb and deficits in spatial learning. Nature 367, 455–459.

Curtin, J.A., Quint, E., Tsipouri, V., Arkell, R.M., Cattanach, B., Copp, A.J., et al., 2003. Mutation of Celsr1 disrupts planar polarity of inner ear hair cells and causes severe neural tube defects in the mouse. Current Biology 13, 1129–1133.

Dasen, J.S., Liu, J.P., Jessell, T.M., 2003. Motor neuron columnar fate imposed by sequential phases of Hox-c activity. Nature 425, 926–933.

Davidson, L.A., Keller, R.E., 1999. Neural tube closure in Xenopus laevis involves medial migration, directed protrusive activity, cell intercalation and convergent extension. Development 126, 4547–4556.

de Roos, K., Sonneveld, E., Compaan, B., ten Berge, D., Durston, A.J., van der Saag, P.T., 1999. Expression of retinoic acid 4-hydroxylase (CYP26) during mouse and Xenopus laevis embryogenesis. Mechanisms of Development 82, 205–211.

Deak, K.L., Boyles, A.L., Etchevers, H.C., Melvin, E.C., Siegel, D.G., Graham, F.L., et al., 2005a. SNPs in the neural cell adhesion molecule 1 gene (NCAM1) may be associated with human neural tube defects. Human Genetics 117, 133–142.

Deak, K.L., Dickerson, M.E., Linney, E., Enterline, D.S., George, T.M., Melvin, E.C., et al., 2005b. Analysis of ALDH1A2, CYP26A1, CYP26B1, CRABP1, and CRABP2 in human neural tube defects suggests a possible association with alleles in ALDH1A2. Birth Defects Research. Part A, Clinical and Molecular Teratology 73, 868–875.

Detrick, R.J., Dickey, D., Kintner, C.R., 1990. The effects of N-cadherin misexpression on morphogenesis in Xenopus embryos. Neuron 4, 493–506.

Ding, Q., Motoyama, J., Gasca, S., Mo, R., Sasaki, H., Rossant, J., et al., 1998. Diminished Sonic hedgehog signaling and lack of floor plate differentiation in Gli2 mutant mice. Development 125, 2533–2543.

Domingos, P.M., Itasaki, N., Jones, C.M., Mercurio, S., Sargent, M.G., Smith, J.C., et al., 2001. The Wnt/beta-catenin pathway posteriorizes neural tissue in Xenopus by an indirect mechanism requiring FGF signalling. Developmental Biology 239, 148–160.

Doudney, K., Ybot-Gonzalez, P., Paternotte, C., Stevenson, R.E., Greene, N.D., Moore, G.E., et al., 2005. Analysis of the planar cell polarity gene Vangl2 and its co-expressed paralogue Vangl1 in neural tube defect patients. American Journal of Medical Genetics. Part A 136, 90–92.

Echelard, Y., Epstein, D.J., St-Jacques, B., Shen, L., Mohler, J., McMahon, J.A., et al., 1993. Sonic hedgehog, a member of a family of putative signaling molecules, is implicated in the regulation of CNS polarity. Cell 75, 1417–1430.

el Ghouzzi, V., Le Merrer, M., Perrin-Schmitt, F., Lajeunie, E., Benit, P., Renier, D., et al., 1997. Mutations of the TWIST gene in the Saethre-Chotzen syndrome. Nature Genetics 15, 42–46.

Elanko, N., Sibbring, J.S., Metcalfe, K.A., Clayton-Smith, J., Donnai, D., Temple, I.K., et al., 2001. A survey of TWIST for mutations in craniosynostosis reveals a variable length polyglycine tract in asymptomatic individuals. Human Mutation 18, 535–541.

Elms, P., Siggers, P., Napper, D., Greenfield, A., Arkell, R., 2003. Zic2 is required for neural crest formation and hindbrain patterning during mouse development. Developmental Biology 264, 391–406.

Felder, B., Stegmann, K., Schultealbert, A., Geller, F., Strehl, E., Ermert, A., et al., 2002. Evaluation of BMP4 and its specific inhibitor NOG as candidates in human neural tube defects (NTDs). European Journal of Human Genetics 10, 753–756.

Feldman, J.G., Stein, S.C., Klein, R.J., Kohl, S., Casey, G., 1982. The prevalence of neural tube defects among ethnic groups in Brooklyn, New York. Journal of Chronic Diseases 35, 53–60.

Feuchtbaum, L.B., Currier, R.J., Riggle, S., Roberson, M., Lorey, F.W., Cunningham, G.C., 1999. Neural tube defect prevalence in California (1990-1994): Eliciting patterns by type of defect and maternal race/ethnicity. Genetic Testing 3, 265–272.

Fine, E.L., Horal, M., Chang, T.I., Fortin, G., Loeken, M.R., 1999. Evidence that elevated glucose causes altered gene expression, apoptosis, and neural tube defects in a mouse model of diabetic pregnancy. Diabetes 48, 2454–2462.

Forrester, M.B., Merz, R.D., 2005. Precurrence risk of neural tube defects in siblings of infants with lipomyelomeningocele. Genetics in Medicine 7, 457.

Gamse, J., Sive, H., 2000. Vertebrate anteroposterior patterning: The Xenopus neurectoderm as a paradigm. Bioessays 22, 976–986.

Geelen, J.A., Langman, J., 1979. Ultrastructural observations on closure of the neural tube in the mouse. Anatomy and Embryology 156, 73–88.

Goodrich, L.V., Milenkovic, L., Higgins, K.M., Scott, M.P., 1997. Altered neural cell fates and medulloblastoma in mouse patched mutants. Science 277, 1109–1113.

Gowen, L.C., Johnson, B.L., Latour, A.M., Sulik, K.K., Koller, B.H., 1996. Brca1 deficiency results in early embryonic lethality characterized by neuroepithelial abnormalities. Nature Genetics 12, 191–194.

Gray, R.S., Abitua, P.B., Wlodarczyk, B.J., Szabo-Rogers, H.L., Blanchard, O., Lee, I., et al., 2009. The planar cell polarity effector Fuz is essential for targeted membrane trafficking, ciliogenesis and mouse embryonic development. Nature Cell Biology 11, 1225–1232.

Gray, J.D., Nakouzi, G., Slowinska-Castaldo, B., Dazard, J.E., Rao, J.S., Nadeau, J.H., et al., 2010. Functional interactions between the LRP6 WNT co-receptor and folate supplementation. Human Molecular Genetics 19, 4560–4572.

Griffith, C.M., Wiley, M.J., Sanders, E.J., 1992. The vertebrate tail bud: Three germ layers from one tissue. Anatomy and Embryology 185, 101–113.

Gruss, P., Walther, C., 1992. Pax in development. Cell 69, 719–722.

Gunther, T., Struwe, M., Aguzzi, A., Schughart, K., 1994. Open brain, a new mouse mutant with severe neural tube defects, shows altered gene expression patterns in the developing spinal cord. Development 120, 3119–3130.

Gustavsson, P., Greene, N.D., Lad, D., Pauws, E., de Castro, S.C., Stanier, P., et al., 2007. Increased expression of Grainyhead-like-3 rescues spina bifida in a folate-resistant mouse model. Human Molecular Genetics 16, 2640–2646.

Hall, J.G., Friedman, J.M., Kenna, B.A., Popkin, J., Jawanda, M., Arnold, W., 1988. Clinical, genetic, and epidemiological factors in neural tube defects. American Journal of Human Genetics 43, 827–837.

Hamblet, N.S., Lijam, N., Ruiz-Lozano, P., Wang, J., Yang, Y., Luo, Z., et al., 2002. Dishevelled 2 is essential for cardiac outflow tract development, somite segmentation and neural tube closure. Development 129, 5827–5838.

Harris, M.J., 2009. Insights into prevention of human neural tube defects by folic acid arising from consideration of mouse mutants. Birth Defects Research. Part A, Clinical and Molecular Teratology 85, 331–339.

Harris, M.J., Juriloff, D.M., 2007. Mouse mutants with neural tube closure defects and their role in understanding human neural tube defects. Birth Defects Research. Part A, Clinical and Molecular Teratology 79, 187–210.

Harris, M.J., Juriloff, D.M., 2010. An update to the list of mouse mutants with neural tube closure defects and advances toward a complete genetic perspective of neural tube closure. Birth Defects Research (Part A) 88, 653–669.

Helms, A.W., Johnson, J.E., 2003. Specification of dorsal spinal cord interneurons. Current Opinion in Neurobiology 13, 42–49.

Hildebrand, J.D., Soriano, P., 1999. Shroom, a PDZ domain-containing actin-binding protein, is required for neural tube morphogenesis in mice. Cell 99, 485–497.

Hjiantoniou, E., Iseki, S., Uney, J.B., Phylactou, L.A., 2003. DNazyme-mediated cleavage of Twist transcripts and increase in cellular apoptosis. Biochemical and Biophysical Research Communications 300, 178–181.

Hol, F.A., Geurds, M.P., Chatkupt, S., Shugart, Y.Y., Balling, R., Schrander-Stumpel, C.T., et al., 1996. PAX genes and human neural tube defects: An amino acid substitution in PAX1 in a patient with spina bifida. Journal of Medical Genetics 33, 655–660.

Hollemann, T., Chen, Y., Grunz, H., Pieler, T., 1998. Regionalized metabolic activity establishes boundaries of retinoic acid signalling. European Molecular Biology Organization Journal 17, 7361–7372.

Holmberg, J., Clarke, D.L., Frisen, J., 2000. Regulation of repulsion versus adhesion by different splice forms of an Eph receptor. Nature 408, 203–206.

Homanics, G.E., Maeda, N., Traber, M.G., Kayden, H.J., Dehart, D.B., Sulik, K.K., 1995. Exencephaly and hydrocephaly in mice with targeted modification of the apolipoprotein B (Apob) gene. Teratology 51, 1–10.

Honein, M.A., Paulozzi, L.J., Mathews, T.J., Erickson, J.D., Wong, L.Y., 2001. Impact of folic acid fortification of the US food supply on the occurrence of neural tube defects. Journal of the American Medical Association 285, 2981–2986.

Huang, Y., Roelink, H., McKnight, G.S., 2002. Protein kinase A deficiency causes axially localized neural tube defects in mice. The Journal of Biological Chemistry 277, 19889–19896.

Jessen, J.R., Solnica-Krezel, L., 2004. Identification and developmental expression pattern of van gogh-like 1, a second zebrafish strabismus homologue. Gene Expression Patterns 4, 339–344.

Johnston, J.J., Olivos-Glander, I., Killoran, C., Elson, E., Turner, J.T., Peters, K.F., et al., 2005. Molecular and clinical analyses of Greig cephalopolysyndactyly and Pallister-Hall syndromes: robust phenotype prediction from the type and position of GLI3 mutations. American Journal of Human Genetics 76, 609–622.

Juriloff, D.M., Harris, M.J., 2000. Mouse models for neural tube closure defects. Human Molecular Genetics 9, 993–1000.

Kaufman, B.A., 2004. Neural tube defects. Pediatric Clinics of North America 51, 389–419.

Kennedy, D., Chitayat, D., Winsor, E.J., Silver, M., Toi, A., 1998. Prenatally diagnosed neural tube defects: Ultrasound, chromosome, and autopsy or postnatal findings in 212 cases. American Journal of Medical Genetics 77, 317–321.

Kibar, Z., Torban, E., McDearmid, J.R., Reynolds, A., Berghout, J., Mathieu, M., et al., 2007. Mutations in VANGL1 associated with neural-tube defects. The New England Journal of Medicine 356, 1432–1437.

Kibar, Z., Bosoi, C.M., Kooistra, M., Salem, S., Finnell, R.H., De Marco, P., et al., 2009. Novel mutations in VANGL1 in neural tube defects. Human Mutation 30, E706–E715.

Kim, J.K., Huh, S.O., Choi, H., Lee, K.S., Shin, D., Lee, C., et al., 2001. Srg3, a mouse homolog of yeast SWI3, is essential for early embryogenesis and involved in brain development. Molecular and Cellular Biology 21, 7787–7795.

Kim, T.H., Goodman, J., Anderson, K.V., Niswander, L., 2007. Phactr4 regulates neural tube and optic fissure closure by controlling PP1-, Rb-, and E2F1-regulated cell-cycle progression. Developmental Cell 13, 87–102.

King, T.M., Au, K.S., Kirkpatrick, T.J., Davidson, C., Fletcher, J.M., Townsend, I., et al., 2007. The impact of BRCA1 on spina bifida meningomyelocele lesions. Annals of Human Genetics 71, 719–728.

Kinsman, S.L., Doehring, M.C., 1996. The cost of preventable conditions in adults with spina bifida. European Journal of Pediatric Surgery 6 (supplement 1), 17–20.

Kirillova, I., Novikova, I., Auge, J., Audollent, S., Esnault, D., Encha-Razavi, F., et al., 2000. Expression of the sonic hedgehog gene in human embryos with neural tube defects. Teratology 61, 347–354.

Kobayashi, T., Minowa, O., Kuno, J., Mitani, H., Hino, O., Noda, T., 1999. Renal carcinogenesis, hepatic hemangiomatosis, and embryonic lethality caused by a germ-line Tsc2 mutation in mice. Cancer Research 59, 1206–1211.

Koleske, A.J., Gifford, A.M., Scott, M.L., Nee, M., Bronson, R.T., Miczek, K.A., et al., 1998. Essential roles for the Abl and Arg tyrosine kinases in neurulation. Neuron 21, 1259–1272.

Kress, W., Schropp, C., Lieb, G., Petersen, B., Busse-Ratzka, M., Kunz, J., et al., 2006. Saethre-Chotzen syndrome caused by TWIST 1 gene mutations: Functional differentiation from Muenke coronal synostosis syndrome. European Journal of Human Genetics 14, 39–48.

Kudoh, T., Concha, M.L., Houart, C., Dawid, I.B., Wilson, S.W., 2004. Combinatorial Fgf and Bmp signalling patterns the gastrula ectoderm into prospective neural and epidermal domains. Development 131, 3581–3592.

Kuida, K., Haydar, T.F., Kuan, C.Y., Gu, Y., Taya, C., Karasuyama, H., et al., 1998. Reduced apoptosis and cytochrome c-mediated caspase activation in mice lacking caspase 9. Cell 94, 325–337.

Kyttala, M., Tallila, J., Salonen, R., Kopra, O., Kohlschmidt, N., Paavola-Sakki, P., et al., 2006. MKS1, encoding a component of the flagellar apparatus basal body proteome, is mutated in Meckel syndrome. Nature Genetics 38, 155–157.

Lanier, L.M., Gates, M.A., Witke, W., Menzies, A.S., Wehman, A.M., Macklis, J.D., et al., 1999. Mena is required for neurulation and commissure formation. Neuron 22, 313–325.

Lardelli, M., Williams, R., Mitsiadis, T., Lendahl, U., 1996. Expression of the Notch 3 intracellular domain in mouse central nervous system progenitor cells is lethal and leads to disturbed neural tube development. Mechanisms of Development 59, 177–190.

Laurence, K.M., 1985. Prevention of neural tube defects by improvement in maternal diet and preconceptional folic acid supplementation. Progress in Clinical and Biological Research 163B, 383–388.

Lawson, A., Schoenwolf, G.C., England, M.A., Addai, F.K., Ahima, R.S., 1999. Programmed cell death and the morphogenesis of the hindbrain roof plate in the chick embryo. Anatomy and Embryology 200, 509–519.

Lee, H.S., Bong, Y.S., Moore, K.B., Soria, K., Moody, S.A., Daar, I.O., 2006.Dishevelled mediates ephrinB1 signalling in the eye field through the planar cell polarity pathway. Nat Cell Biol 8(1), 55-63. Erratum in: Nat Cell Biol 2006;8(2):202.

Lei, Y.P., Zhang, T., Li, H., Wu, B.L., Jin, L., Wang, H.Y., 2010. VANGL2 mutations in human cranial neural-tube defects. The New England Journal of Medicine 362, 2232–2235.

Li, C.C., Cropley, J.E., Cowley, M.J., Preiss, T., Martin, D.I., Suter, C.M., 2011. A sustained dietary change increases epigenetic variation in isogenic mice. PLoS Genetics 7, e1001380.

Liem Jr., K.F., Tremml, G., Roelink, H., Jessell, T.M., 1995. Dorsal differentiation of neural plate cells induced by BMP-mediated signals from epidermal ectoderm. Cell 82, 969–979.

Liem Jr., K.F., Tremml, G., Jessell, T.M., 1997. A role for the roof plate and its resident TGFbeta-related proteins in neuronal patterning in the dorsal spinal cord. Cell 91, 127–138.

Lindhout, D., Omtzigt, J.G., Cornel, M.C., 1992. Spectrum of neuraltube defects in 34 infants prenatally exposed to antiepileptic drugs. Neurology 42, 111–118.

Liu, J.P., Laufer, E., Jessell, T.M., 2001. Assigning the positional identity of spinal motor neurons: Rostrocaudal patterning of Hox-c expression by FGFs, Gdf11, and retinoids. Neuron 32, 997–1012.

Lu, X., Borchers, A.G., Jolicoeur, C., Rayburn, H., Baker, J.C., Tessier-Lavigne, M., 2004. PTK7/CCK-4 is a novel regulator of planar cell polarity in vertebrates. Nature 430, 93–98.

Luo, J., Balkin, N., Stewart, J.F., Sarwark, J.F., Charrow, J., Nye, J.S., 2000. Neural tube defects and the 13q deletion syndrome: Evidence for a critical region in 13q33-34. American Journal of Medical Genetics 91, 227–230.

Lynberg, M.C., Khoury, M.J., Lu, X., Cocian, T., 1994. Maternal flu, fever, and the risk of neural tube defects: A population-based case-control study. American Journal of Epidemiology 140, 244–255.

Lynch, S.A., 2005. Non-multifactorial neural tube defects. American Journal of Medical Genetics. Part C, Seminars in Medical Genetics 135C, 69–76.

Marean, A., Graf, A., Zhang, Y., Niswander, L., 2011. Folic acid supplementation can adversely affect murine neural tube closure and embryonic survival. Human Molecular Genetics 20, 678–683.

Martinez-Frias, M.L., Urioste, M., Bermejo, E., Sanchis, A., Rodriguez-Pinilla, E., 1996. Epidemiological analysis of multi-site closure failure of neural tube in humans. American Journal of Medical Genetics 66, 64–68.

Massa, V., Savery, D., Ybot-Gonzalez, P., Ferraro, E., Rongvaux, A., Cecconi, F., et al., 2009. Apoptosis is not required for mammalian neural tube closure. Proceedings of the National Academy of Sciences of the United States of America 106, 8233–8238.

Matise, M.P., Epstein, D.J., Park, H.L., Platt, K.A., Joyner, A.L., 1998. Gli2 is required for induction of floor plate and adjacent cells, but not most ventral neurons in the mouse central nervous system. Development 125, 2759–2770.

McComb, J.G., 1997. Spinal and cranial neural tube defects. Seminars in Pediatric Neurology 4, 156–166.

Milunsky, A., Alpert, E., Kitzmiller, J.L., Younger, M.D., Neff, R.K., 1982. Prenatal diagnosis of neural tube defects VIII. The importance of serum alpha-fetoprotein screening in diabetic pregnant women. American Journal of Obstetrics and Gynecology 142, 1030–1032.

Ming, J.E., Kaupas, M.E., Roessler, E., Brunner, H.G., Golabi, M., Tekin, M., et al., 2002. Mutations in PATCHED-1, the receptor for SONIC HEDGEHOG, are associated with holoprosencephaly. Human Genetics 110, 297–301.

Moran, D., Rice, R.W., 1975. An ultrastructural examination of the role of cell membrane surface coat material during neurulation. The Journal of Cell Biology 64, 172–181.

Moury, J.D., Schoenwolf, G.C., 1995. Cooperative model of epithelial shaping and bending during avian neurulation: Autonomous movements of the neural plate, autonomous movements of the epidermis, and interactions in the neural plate/epidermis transition zone. Developmental Dynamics 204, 323–337.

Muller, F., O'Rahilly, R., 1987. The development of the human brain, the closure of the caudal neuropore, and the beginning of secondary neurulation at stage 12. Anatomy and Embryology 176, 413–430.

Murdoch, J.N., Doudney, K., Paternotte, C., Copp, A.J., Stanier, P., 2001a. Severe neural tube defects in the loop-tail mouse result from mutation of Lpp 1, a novel gene involved in floor plate specification. Human Molecular Genetics 10, 2593–2601.

Murdoch, J.N., Rachel, R.A., Shah, S., Beermann, F., Stanier, P., Mason, C.A., et al., 2001b. Circletail, a new mouse mutant with severe neural tube defects: Chromosomal localization and interaction with the loop-tail mutation. Genomics 78, 55–63.

Murdoch, J.N., Henderson, D.J., Doudney, K., Gaston-Massuet, C., Phillips, H.M., Paternotte, C., et al., 2003. Disruption of scribble (Scrb1) causes severe neural tube defects in the circletail mouse. Human Molecular Genetics 12, 87–98.

Nagai, T., Aruga, J., Minowa, O., Sugimoto, T., Ohno, Y., Noda, T., et al., 2000. Zic2 regulates the kinetics of neurulation. Proceedings of the National Academy of Sciences of the United States of America 97, 1618–1623.

Nakatsu, T., Uwabe, C., Shiota, K., 2000. Neural tube closure in humans initiates at multiple sites: Evidence from human embryos and implications for the pathogenesis of neural tube defects. Anatomy and Embryology 201, 455–466.

Nanni, L., Ming, J.E., Bocian, M., Steinhaus, K., Bianchi, D.W., Die-Smulders, C., et al., 1999. The mutational spectrum of the sonic hedgehog gene in holoprosencephaly: SHH mutations cause a significant proportion of autosomal dominant holoprosencephaly. Human Molecular Genetics 8, 2479–2488.

Nau, H., Hauck, R.S., Ehlers, K., 1991. Valproic acid-induced neural tube defects in mouse and human: Aspects of chirality, alternative drug development, pharmacokinetics and possible mechanisms. Pharmacology and Toxicology 69, 310–321.

Nishimura, I., Drake, T.A., Lusis, A.J., Lyons, K.M., Nadeau, J.H., Zernik, J., 2003. ENU large-scale mutagenesis and quantitative trait linkage (QTL) analysis in mice: Novel technologies for searching polygenetic determinants of craniofacial abnormalities. Critical Reviews in Oral Biology and Medicine 14, 320–330.

Nyholm, M.K., Abdelilah-Seyfried, S., Grinblat, Y., 2009. A novel genetic mechanism regulates dorsolateral hinge-point formation during zebrafish cranial neurulation. Journal of Cell Science 122, 2137–2148.

O'Rahilly, R., Muller, F., 2002. The two sites of fusion of the neural folds and the two neuropores in the human embryo. Teratology 65, 162–170.

Ota, M.S., Loebel, D.A., O'Rourke, M.P., Wong, N., Tsoi, B., Tam, P.P., 2004. Twist is required for patterning the cranial nerves and maintaining the viability of mesodermal cells. Developmental Dynamics 230, 216–228.

Pangilinan, F., Geiler, K., Dolle, J., Troendle, J., Swanson, D.A., Molloy, A.M., et al., 2008. Construction of a high resolution linkage disequilibrium map to evaluate common genetic variation in TP53 and neural tube defect risk in an Irish population. American Journal of Medical Genetics. Part A 146A, 2617–2625.

Park, H.L., Bai, C., Platt, K.A., Matise, M.P., Beeghly, A., Hui, C.C., et al., 2000. Mouse Gli1 mutants are viable but have defects in SHH signaling in combination with a Gli2 mutation. Development 127, 1593–1605.

Park, T.J., Haigo, S.L., Wallingford, J.B., 2006. Ciliogenesis defects in embryos lacking inturned or fuzzy function are associated with failure of planar cell polarity and Hedgehog signaling. Nature Genetics 38, 303–311.

Pavlinkova, G., Salbaum, J.M., Kappen, C., 2009. Maternal diabetes alters transcriptional programs in the developing embryo. BMC Genomics 10, 274.

Phelan, S.A., Ito, M., Loeken, M.R., 1997. Neural tube defects in embryos of diabetic mice: Role of the Pax-3 gene and apoptosis. Diabetes 46, 1189–1197.

Puschel, A.W., Westerfield, M., Dressler, G.R., 1992. Comparative analysis of Pax-2 protein distributions during neurulation in mice and zebrafish. Mechanisms of Development 38, 197–208.

Pyrgaki, C., Liu, A., Niswander, L., 2010a. Grainyhead-like 2 regulates neural tube closure and adhesion molecule expression during neural fold fusion. Developmental Biology 353, 38–49.

Pyrgaki, C., Trainor, P., Hadjantonakis, A.K., Niswander, L., 2010b. Dynamic imaging of neural tube closure. Developmental Biology 344, 941–947.

Radhakrishna, U., Wild, A., Grzeschik, K.H., Antonarakis, S.E., 1997. Mutation in GLI3 in postaxial polydactyly type A. Nature Genetics 17, 269–271.

Radice, G.L., Rayburn, H., Matsunami, H., Knudsen, K.A., Takeichi, M., Hynes, R.O., 1997. Developmental defects in mouse embryos lacking N-cadherin. Developmental Biology 181, 64–78.

Rentzsch, F., Bakkers, J., Kramer, C., Hammerschmidt, M., 2004. Fgf signaling induces posterior neuroectoderm independently of Bmp signaling inhibition. Developmental Dynamics 231, 750–757.

Rittler, M., Lopez-Camelo, J.S., Castilla, E.E., Bermejo, E., Cocchi, G., Correa, A., et al., 2008. Preferential associations between oral clefts and other major congenital anomalies. The Cleft Palate-Craniofacial Journal 45, 525–532.

Robinson, et al. 2012. Mutations in the planar cell polarity genes CELSR1 and SCRIB are associated with the severe neural tube defect craniorachischisis. Human Mutation 33, 440–447.

Roffers-Agarwal, J., Xanthos, J.B., Kragtorp, K.A., Miller, J.R., 2008. Enabled (Xena) regulates neural plate morphogenesis, apical constriction, and cellular adhesion required for neural tube closure in *Xenopus*. Developmental Biology 314, 393–403.

Rogers, S.C., Morris, M., 1973. Anencephalus: A changing sex ratio. British Journal of Preventive & Social Medicine 27, 81–84.

Ross, A.J., May-Simera, H., Eichers, E.R., Kai, M., Hill, J., Jagger, D.J., et al., 2005. Disruption of Bardet-Biedl syndrome ciliary proteins perturbs planar cell polarity in vertebrates. Nature Genetics 37, 1135–1140.

Rubenstein, J.L.R., Rakic, P., 2013. Patterning and Cell Types Specification in the Developing CNS and PNS.

Ruland, J., Duncan, G.S., Elia, A., del Barco Barrantes, I., Nguyen, L., Plyte, S., et al., 2001. Bcl10 is a positive regulator of antigen receptor-induced activation of NF-kappaB and neural tube closure. Cell 104, 33–42.

Sadler, T.W., 1978. Distribution of surface coat material on fusing neural folds of mouse embryos during neurulation. Anatomical Record 191, 345–349.

Sah, V.P., Attardi, L.D., Mulligan, G.J., Williams, B.O., Bronson, R.T., Jacks, T., 1995. A subset of p53-deficient embryos exhibit exencephaly. Nature Genetics 10, 175–180.

Saitsu, H., Yamada, S., Uwabe, C., Ishibashi, M., Shiota, K., 2004. Development of the posterior neural tube in human embryos. Anatomy and Embryology 209, 107–117.

Schluter, G., 1973. Ultrastructural observations on cell necrosis during formation of the neural tube in mouse embryos. Zeitschrift für Anatomie und Entwicklungsgeschichte 141, 251–264.

Schoenwolf, G.C., 1979. Histological and ultrastructural observations of tail bud formation in the chick embryo. Anatomical Record 193, 131–147.

Schoenwolf, G.C., Delongo, J., 1980. Ultrastructure of secondary neurulation in the chick embryo. The American Journal of Anatomy 158, 43–63.

Schoenwolf, G.C., Folsom, D., Moe, A., 1988. A reexamination of the role of microfilaments in neurulation in the chick embryo. Anatomical Record 220, 87–102.

Schoenwolf, G.C., Everaert, S., Bortier, H., Vakaet, L., 1989. Neural plate- and neural tube-forming potential of isolated epiblast areas in avian embryos. Anatomy and Embryology 179, 541–549.

Sedivy, J.M., Joyner, A.L., 1992. Gene Targeting. Oxford University Press, New York.

Seller, M.J., 1987. Neural tube defects and sex ratios. American Journal of Medical Genetics 26, 699–707.

Shum, A.S., Copp, A.J., 1996. Regional differences in morphogenesis of the neuroepithelium suggest multiple mechanisms of spinal neurulation in the mouse. Anatomy and Embryology 194, 65–73.

Simeonsson, R.J., McMillen, J.S., Huntington, G.S., 2002. Secondary conditions in children with disabilities: Spina bifida as a case example. Mental Retardation and Developmental Disabilities Research Reviews 8, 198–205.

Smith, J.L., Schoenwolf, G.C., 1987. Cell cycle and neuroepithelial cell shape during bending of the chick neural plate. Anatomical Record 218, 196–206.

Smith, J.L., Schoenwolf, G.C., 1989. Notochordal induction of cell wedging in the chick neural plate and its role in neural tube formation. The Journal of Experimental Zoology 250, 49–62.

Smith, U.M., Consugar, M., Tee, L.J., McKee, B.M., Maina, E.N., Whelan, S., et al., 2006. The transmembrane protein meckelin (MKS3) is mutated in Meckel-Gruber syndrome and the wpk rat. Nature Genetics 38, 191–196.

Soo, K., O'Rourke, M.P., Khoo, P.L., Steiner, K.A., Wong, N., Behringer, R.R., et al., 2002. Twist function is required for the morphogenesis of the cephalic neural tube and the differentiation of the cranial neural crest cells in the mouse embryo. Developmental Biology 247, 251–270.

Srinivas, D., Sharma, B.S., Mahapatra, A.K., 2008. Triple neural tube defect and the multisite closure theory for neural tube defects: Is there an additional site? Case report. Journal of Neurosurgery. Pediatrics 1, 160–163.

Stoykova, A., Gruss, P., 1994. Roles of Pax-genes in developing and adult brain as suggested by expression patterns. Journal of Neuroscience 14, 1395–1412.

Strong, L.C., Hollander, W.F., 1949. Hereditary loop-tail in the house mouse. Journal of Heredity 40, 329–334.

Stumpo, D.J., Bock, C.B., Tuttle, J.S., Blackshear, P.J., 1995. MARCKS deficiency in mice leads to abnormal brain development and perinatal death. Proceedings of the National Academy of Sciences of the United States of America 92, 944–948.

Stumpo, D.J., Eddy Jr., R.L., Haley, L.L., Sait, S., Shows, T.B., Lai, W.S., et al., 1998. Promoter sequence, expression, and fine chromosomal mapping of the human gene (MLP) encoding the MARCKS-like protein: Identification of neighboring and linked polymorphic loci for MLP and MACS and use in the evaluation of human neural tube defects. Genomics 49, 253–264.

Suarez, L., Felkner, M., Brender, J.D., Canfield, M., Hendricks, K., 2008. Maternal exposures to cigarette smoke, alcohol, and street drugs and neural tube defect occurrence in offspring. Maternal and Child Health Journal 12, 394–401.

Sumanas, S., Zhang, B., Dai, R., Lin, S., 2005. 15-zinc finger protein Bloody Fingers is required for zebrafish morphogenetic movements during neurulation. Developmental Biology 283, 85–96.

Tassabehji, M., Read, A.P., Newton, V.E., Patton, M., Gruss, P., Harris, R., et al., 1993. Mutations in the PAX3 gene causing Waardenburg syndrome type 1 and type 2. Nature Genetics 3, 26–30.

Tavafoghi, V., Ghandchi, A., Hambrick Jr., G.W., Udverhelyi, G.B., 1978. Cutaneous signs of spinal dysraphism. Report of a patient with a tail-like lipoma and review of 200 cases in the literature. Archives of Dermatology 114, 573–577.

Torban, E., Patenaude, A.M., Leclerc, S., Rakowiecki, S., Gauthier, S., Andelfinger, G., et al., 2008. Genetic interaction between members of the Vangl family causes neural tube defects in mice. Proceedings of the National Academy of Sciences of the United States of America 105, 3449–3454.

van Straaten, H.W., Copp, A.J., 2001. Curly tail: A 50-year history of the mouse spina bifida model. Anatomy and Embryology 203, 225–237.

van Straaten, H.W., Hekking, J.W., Wiertz-Hoessels, E.J., Thors, F., Drukker, J., 1988. Effect of the notochord on the differentiation of a floor plate area in the neural tube of the chick embryo. Anatomy and Embryology 177, 317–324.

Vincentz, J.W., Barnes, R.M., Rodgers, R., Firulli, B.A., Conway, S.J., Firulli, A.B., 2008. An absence of Twist1 results in aberrant cardiac neural crest morphogenesis. Developmental Biology 320, 131–139.

Wallingford, J.B., Fraser, S.E., Harland, R.M., 2002. Convergent extension: The molecular control of polarized cell movement during embryonic development. Developmental Cell 2, 695–706.

Weatherbee, S.D., Niswander, L.A., Anderson, K.V., 2009. A mouse model for Meckel syndrome reveals Mks1 is required for ciliogenesis and Hedgehog signaling. Human Molecular Genetics 18, 4565–4575.

Weinstein, D.C., Ruiz, A., Altaba, I., Chen, W.S., Hoodless, P., Prezioso, V.R., et al., 1994. The winged-helix transcription factor HNF-3 beta is required for notochord development in the mouse embryo. Cell 78, 575–588.

Werth, M., Walentin, K., Aue, A., Schonheit, J., Wuebken, A., Pode-Shakked, N., et al., 2010. The transcription factor grainyhead-like 2 regulates the molecular composition of the epithelial apical junctional complex. Development 137, 3835–3845.

Winter, R.M., Huson, S.M., 1988. Greig cephalopolysyndactyly syndrome: A possible mouse homologue (Xt-extra toes). American Journal of Medical Genetics 31, 793–798.

Xu, W., Baribault, H., Adamson, E.D., 1998. Vinculin knockout results in heart and brain defects during embryonic development. Development 125, 327–337.

Yamaguchi, Y., Shinotsuka, N., Nonomura, K., Takemoto, K., et al., 2011. Live imaging of apoptosis in a novel transgenic mouse highlights its role in neural tube closure. J Cell Biol 195 (6), 1047–1060.

Ybot-Gonzalez, P., Copp, A.J., 1999. Bending of the neural plate during mouse spinal neurulation is independent of actin microfilaments. Developmental Dynamics 215, 273–283.

Ybot-Gonzalez, P., Cogram, P., Gerrelli, D., Copp, A.J., 2002. Sonic hedgehog and the molecular regulation of mouse neural tube closure. Development 129, 2507–2517.

Ybot-Gonzalez, P., Gaston-Massuet, C., Girdler, G., Klingensmith, J., Arkell, R., Greene, N.D., et al., 2007. Neural plate morphogenesis during mouse neurulation is regulated by antagonism of Bmp signalling. Development 134, 3203–3211.

Zeng, H., Hoover, A.N., Liu, A., 2010. PCP effector gene Inturned is an important regulator of cilia formation and embryonic development in mammals. Developmental Biology 339, 418–428.

Zhu, H., Barber, R., Shaw, G.M., Lammer, E.J., Finnell, R.H., 2003. Is Sonic hedgehog (SHH) a candidate gene for spina bifida? A pilot study. American Journal of Medical Genetics. Part A 117A, 87–88.

Zohn, I.E., Anderson, K.V., Niswander, L., 2005. Using genomewide mutagenesis screens to identify the genes required for neural tube closure in the mouse. Birth Defects Research. Part A, Clinical and Molecular Teratology 73, 583–590.

Zohn, I.E., Anderson, K.V., Niswander, L., 2007. The Hectd1 ubiquitin ligase is required for development of the head mesenchyme and neural tube closure. Developmental Biology 306, 208–221.

Fetal Alcohol Spectrum Disorder
Targeted Effects of Ethanol on Cell Proliferation and Survival

S.M. Mooney[1,2], P.J. Lein[3], M.W. Miller[2,4]

[1]University of Maryland, Baltimore, MD, USA; [2]State University of New York Upstate Medical University, Syracuse, NY, USA; [3]University of California at Davis School of Veterinary Medicine, Davis, CA, USA; [4]Veterans Affairs Medical Center, Syracuse, NY, USA

28.1 INTRODUCTION

Developmental disorders arise from genetic and environmental causes. Often, either the genetic or environmental factor takes precedence and defines the disorder, for example, as in Huntington's chorea or mercury poisoning, respectively. In other situations, the effects of the two contributors are not mutually exclusive. In fact, the effect of a genetic or environmental alteration alone may be masked, and a disorder may result only when a genetically susceptible organism is exposed to an environmental factor. One such developmental disorder is fetal alcohol spectrum disorder (FASD).

Besides genetic and environmental contributions, another major factor defining developmental disorders such as FASD is fetal programming. On the basis of this concept, nutrition or exposure to a substance during fetal development can cause alterations over the lifespan. These consequences often can be quiescent for years and then have powerful effects that shape the behavior of adolescent or adult humans; for example, fetal exposure to substances can shape lifelong mental health status (Salisbury et al., 2009; Schlotz and Phillips, 2009). Taken more broadly, fetal programming can be part of a cycle that promotes the continued pathological situation. In the case of alcohol use, this has been referred to as the "alcoholism generator" (Miller and Spear, 2006). Accordingly, fetal exposure increases the likelihood of alcohol use in adolescents and reduces the age of initiation of alcohol consumption. These behaviors contribute to an increased incidence of alcoholism/alcohol abuse in adults, which in turn begets children who are exposed to alcohol *in utero*.

© 2013 Elsevier Inc. All rights reserved.

28.2 CLINICAL CONSEQUENCES OF ALCOHOL EXPOSURE

The effects of ethanol on a fetus are extensive, devastating, often permanent, and when clustered, are referred to as FASD. Depending upon the population, 2% or more of all neonates have FASD (Centers for Disease Control and Prevention, 2009). Defects can include a stereotypical set of craniofacial malformations such as narrow palpebral fissures, a deficient philtrum, and a flattened nasal bridge (Jones et al., 1973; Lemoine et al., 1968). Based on rodent studies, these features occur when the exposure is restricted to or includes the period of gastrulation (Sulik, 2005; Sulik et al., 1981).

Prenatal alcohol exposure has profound effects on the nervous system in humans, including learning/memory deficits and hyperactivity (Coles, 2006; Fryer et al., 2006). In fact, prenatal alcohol exposure is the leading known cause of mental retardation (Abel and Sokol, 1992; Stratton et al., 1996). Careful examination of the brains of children with FASD shows profound changes. Postmortem and imaging studies show that their brains are smaller, can be covered with sheets of neuroglial heterotopias, and exhibit abnormal gyral patterns and dysmorphic corpora callosa (e.g., Bookstein et al., 2002; Clarren et al., 1978; Jellinger et al., 1981; Pfeiffer et al., 1979; Riley et al., 1995; Swayze et al., 1997; Wisniewski et al., 1983). Quantitative magnetic resonance imaging studies show that ethanol induces microencephaly, as evidenced by the reduced size of specific nuclei, for example, the basal ganglia (Mattson et al., 1996) and cerebellum (Sowell et al., 1996).

28.3 ANIMAL MODELS OF ETHANOL INTAKE

When interpreting the data on fetal ethanol effects, it is important to appreciate the various animal models of FASD because each has strengths and weaknesses, as discussed in a number of excellent reviews (e.g., Cudd, 2005; Kelly et al., 2009; Schneider et al., 2011). For the purposes of this review, it is important to consider three features of a chosen model when determining its utility and appropriateness: the timing, duration, and amount of ethanol exposure. That is, the method of choice is based on the focus of the study and on determining the variables that require controls.

The timing and duration of ethanol exposure are critical because neural development is a constantly changing and asynchronous process among the multitude of brain components. Grossly, the brain increases in size steadily during gestation and then transiently exhibits a relative burst during what is described as the brain growth spurt (Dobbing and Sands, 1979). In primates, this spurt occurs during the third trimester of gestation, and in rodents, it occurs during the first 2 postnatal weeks. Much of this growth results from the early morphogenesis of neuronal processes and the production of glia. Another major event that contributes to early brain growth and can shape the response to ethanol neurotoxicity is the timing and duration of neuronal production. Neurons in most components of the rodent brain are generated prenatally, but in some brain regions (e.g., the hippocampus, thalamus, and cerebellum), some neurons are produced postnatally (e.g., Altman and Bayer, 1990; Altman and Das, 1965, 1966; Mooney and Miller, 2007a; Rao and Jacobson, 2005). The complexity of the effect of ethanol on select brain structures results from the incidence and different length periods of this neuronogenesis. These periods may be discrete or may overlap; thus, individual or multiple brain structures may be vulnerable to ethanol at a particular time during gestation.

The amount of exposure, that is, the dose of ethanol, is also a critical variable (Bonthius and West, 1988a,b, 1990). Dose can be defined by the peak exposure and the time over which this exposure is maintained. For example, the size of the cerebellum and the density of Purkinje neurons are adversely affected by ethanol when ethanol is administered in a larger bolus over a short period of time than when it is delivered more slowly over a more protracted period, even if the total amount of ethanol exposure is equalized.

Prenatal administration of ethanol has generally relied on one of three methods: (1) pair-feeding of a liquid diet, (2) intragastric gavage of the dam, and (3) delivery via intraperitoneal injection of the dam. Integral to these models is the use of controls that account for caloric and nutritional intake and for stress. (1) In the pair-feeding approach, animals in the ethanol-exposed group are given an ethanol-containing liquid diet at the same time each day, and the same volume as an isocaloric, non-ethanol-containing diet is given to the pair-fed control animal. The pair-feeding model is the most biologically relevant to the human situation; however, it leads to variability in blood ethanol concentrations over the diurnal cycle (Miller, 1992). Often, changes in pair-fed controls have been compared with animals fed chow and water *ad libitum*, but this diet differs from the liquid ethanol-containing and control diets. Therefore, it is advisable to provide the controls fed *ad libitum* free access to the liquid control diets (Eade et al., 2010; Youngentob et al., 2007). (2) In the gavage model, an ethanol-containing liquid is delivered to pregnant dams directly into their stomachs (Kelly and Lawrence, 2008). This can lead to stress, the effects of which can be minimized by frequent and equivalent handling of all the subjects. (3) In the injection model (usually used with mice), pregnant dams are administered a diluted solution of ethanol or an equivalent volume of saline. This can produce quick increases in peak ethanol exposure; however, it also has the complication of stress. The latter two models should include careful monitoring of nutritional intake.

Administration of ethanol in the early postnatal period is a bit more complicated. Using one approach, referred to as the pup-in-a-cup method (Diaz and Samson, 1980; West et al., 1981), each 4-day-old rat pup has a permanent gavage tube implanted directly into its stomach and an ethanol-containing solution of artificial milk or a non-ethanol control solution is delivered directly. The liquid is usually provided every 2 h, but this schedule can be manipulated. This method allows for careful delivery and monitoring of the ethanol, but the pups are stressed and do not have maternal interactions. Comparisons to suckling control pups can partially obviate these confounding variables. An alternative method is to provide the ethanol or control solution via intragastric gavage (Kelly and Lawrence, 2008). Following gavage, the pups are returned to their dam. This method minimizes the stress inherent in the pup-in-a-cup method, but it does have some inherent stress that can affect maternal–pup interactions, and it is labor intensive.

28.4 CELL NUMBERS

Research on the effects of ethanol on the rodent brain, specifically the cerebral cortex, is more advanced than similar research on humans (Miller, 2006a). The rat cerebral cortex has been extensively studied and is profoundly affected by prenatal ethanol exposure. It is smaller and exhibits multiple abnormalities such as neurons that are in the wrong place (i.e., ectopic neurons), dysmorphic and dysfunctional neurons, and alterations in synaptogenesis and myelinogenesis. These developmental problems lead to permanent alterations. For example, (1) the numbers of neurons in cortex and other structures in the mature brain are reduced (e.g., Bonthius et al., 1992; Marcussen et al., 1994; Miller, 1995a,b, 1999; Miller and Muller, 1989; Miller and Potempa, 1990; Mooney and Miller, 2007a), (2) surviving neurons form aberrant connections (e.g., Al-Rabiai and Miller, 1989; Clamp and Lindsley, 1998; Miller, 1987, 1997; Miller and Al-Rabiai, 1994; Miller et al., 1990; West et al., 1981), and (3) cortical metabolism is reduced (e.g., Miller and Dow-Edwards, 1988, 1993; Vingan et al., 1986). These changes are downstream from primary defects in early development, for example, in neuronal production and survival. This review focuses on these two processes, which are primary determinants of the numbers of neurons in a specific brain structure.

28.4.1 Cell Proliferation

In the rodent, the proliferation of most central nervous system (CNS) neurons occurs prenatally and, in some structures (e.g., the cerebellum, thalamus, and hippocampus), it extends into the first couple of weeks postnatally. In most cases, this proliferation takes place within zones that line or are proximal to the ventricles. The principal exceptions to this pattern are the external granule layer (EGL) of the cerebellum and the subgranular zone (SGZ) of the dentate gyrus, which are seeded by cells lining the fourth and lateral ventricles, respectively (Chizhikov et al., 2006; Rao and Jacobson, 2005). Cell proliferation is defined by the cycling behavior of the cells (i.e., how quickly cells pass through the four phases of the cell cycle) and the numbers of cells that are actively cycling, which is known as the growth fraction.

28.4.1.1 Cell Cycle

28.4.1.1.1 CEREBRAL CORTEX

Ethanol has a profound effect on the cell cycle kinetics and growth fraction for proliferating populations. *Ad libitum* consumption of ethanol by pregnant rat dams during the last 2 weeks of gestation (achieving peak blood ethanol concentrations of $\sim$150 mg dl^{-1}) affects cell proliferation in the brains of the developing fetuses. The cycling of cells in the cortical ventricular zone (VZ) is retarded (Kennedy and Elliott, 1985; Miller, 1989; Miller and Nowakowski, 1991). On the other hand, ethanol has no effect on the numbers of cells that are actively cycling. Ethanol has a similar effect on VZ cells in organotypic slices of the dorsal telencephalon (Siegenthaler and Miller, 2005a) and cultures of dissociated neuroblastoma cells (Luo and Miller, 1999a) and neural stem cells (Hicks et al., 2010). In each of these cases, ethanol has a consistent effect of prolonging the length of the cell cycle. This occurs principally through lengthening of the G1 phase of the cell cycle, though other phases, notably S, are also vulnerable.

The changes in the cell cycle kinetics appear to result from activation of specific cell cycle checkpoints. Ethanol induces a strong genomic response, which leads to the down- or upregulation of many transcripts associated with passage through G1 (Hicks et al., 2010). Inhibition of the G1/S checkpoint is a consequence of the silencing of genes that are necessary for the progression of cells through G2 and M. Methylation and another epigenetic event, acetylation, are key mechanisms underlying gene silencing and are targets of ethanol (Haycock, 2009; Hicks et al., 2010; Liu et al., 2009; Moonat et al., 2010; Oberlander et al., 2008; Pandey et al., 2008; Zhou et al., 2011). The occurrence of methylation is environmentally affected by the presence of an ambient growth factor (Hicks et al., 2010). Such findings are consistent with evidence that ethanol toxicity can be offset by choline supplementation (Thomas et al., 2004, 2010).

The response of VZ cells to ethanol differs from that of cycling cells in non-VZ proliferative populations, for example, the subventricular zone (SZ). The SZ gives rise to cortical neurons largely in the superficial cortex (Miller, 1989; Nieto et al., 2004; Pontious et al., 2008; Tarabykin et al., 2001) and neurons in the olfactory bulb (Lledo et al., 2008; Whitman and Greer, 2009). Exposure to moderate amounts of ethanol *in vivo* (wherein the blood ethanol concentration is ~150 mg dl^{-1}) increases SZ cell proliferation (Miller and Nowakowski, 1991). This results from an increase in the growth fraction and not from a change in the cell cycle kinetics of SZ cells. The net outcome is that there is a latent surge in the generation of cortical neurons, particularly those that are normally destined for the superficial cortex (Miller, 1986, 1988a, 1997). A similar growth-promoting effect is also evident for neural progenitors in the SGZ of the dentate gyrus; however, this effect is dose-dependent, with high doses of ethanol depressing neural production (Miller, 1995a).

28.4.1.1.2 THALAMUS

The ventrobasal (VB) nucleus of the thalamus is a special CNS site; it has two nonoverlapping periods of neuronal generation (Altman and Bayer, 1979, 1989; Mooney and Miller, 2007b). The second (postnatal) period of neuronogenesis uniquely occurs within the brain parenchyma *per se.* This second period is even more intriguing because it occurs concurrent with a number of desynchronous developmental events within the VB. Such events include the elaboration of neurites particularly by the prenatally generated neurons, the formation of synapses, the generation of projections to the cortex, and the innervation of the VB by cortical axons.

Prenatal exposure to ethanol has a latent effect on the proliferating cells in the VB of young pups (Mooney and Miller, 2010). Specifically, the length of the cell cycle is shorter in ethanol-exposed animals than in controls, and this has a direct, albeit transient, effect on the number of neurons generated daily. This change is consistent with the notion that ethanol exposure initiates a sequence of fetal programming, that is, ethanol causes effects in the fetus that have long-term consequences, which become evident in the more mature animal. The proliferation and survival of the postnatally generated VB neurons are regulated by neurotrophins, for example, nerve growth factor (NGF) and brain-derived neurotrophic factor (BDNF; Mooney and Miller, 2011). Prenatal exposure to ethanol affects the expression of NGF, and it only alters postnatal VB neuronogenesis in the presence of the dyadic effector NGF.

28.4.1.1.3 BRAINSTEM

Early exposure to ethanol also affects the generation of neurons in the brainstem. This includes neurons in the trigeminal brainstem nuclear complex (Miller, 1999; Miller and Muller, 1989), which are derived from cells lining the fourth ventricle. This effect has been refined by restricting the exposure to ethanol to a single episode on a single day. As might be predicted, exposure on a day of neuronogenesis (when neuronal precursors pass through their last mitotic division) is sufficient to reduce the neuronal number in some cranial nerve nuclei (Mooney and Miller, 2007c). It is surprising, however, that these nuclei are equally vulnerable to exposure on a day of gastrulation [e.g., gestational day (G) 7 or G8] and refractile to ethanol toxicity during the 3–4 days between gastrulation and neuronogenesis. These patterns are consistent with the notion that cycling stem cells are vulnerable to ethanol and that these effects are programmed at an early time.

Most cerebellar neurons are derived from an auxiliary proliferative zone, EGL (Rao and Jacobson, 2005). The EGL gives rise to the granule neurons that populate the internal granular layer. The generation of cerebellar granule neurons is negatively affected by ethanol (Bauer-Moffett and Altman, 1977; Kornguth et al., 1979). This alteration results from an ethanol-induced lengthening of the cell cycle of their precursors in the external granule layer (Borges and Lewis, 1983) and by the dysregulation of cyclins and cyclin-dependent kinases (Li et al., 2002).

28.4.1.2 Fetal Programming

Prenatal exposure to ethanol has many long-term consequences, and at least two of them can be attributed to fetal programming. They relate to the two systems that exhibit postnatal neurogenesis: (1) the SGZ of the dentate gyrus and (2) the anterior SZ and olfactory system. Postnatal neurogenesis is highly responsive to environmental perturbations.

28.4.1.2.1 DENTATE GYRUS

The dentate gyrus has captured the interest of neuroscientists because of its apparent role in learning and memory, notably spatial navigation. Increased neuronogenesis (the production of new neurons) correlates with improved learning and memory (Deng et al., 2010; Eisch et al., 2008; Leuner and Gould, 2010). For example, rats that exercise on a running wheel have increased neuronal production and improved learning in a spatial learning task in a Morris water maze (van Praag et al., 2005). The numbers of neurons generated and the behavioral improvement are positively correlated to the amount of running. Furthermore, training on hippocampal-dependent associative learning tasks doubles the number of neurons in the rat dentate gyrus (Gould et al., 1999). This contrasts with training on hippocampal-independent tasks wherein there is no effect on the number of neurons in the dentate gyrus. Some pathological situations, such as traumatic brain injury (e.g., stroke), can stimulate neuronal production (Dash et al., 2001). This can lead to improved recovery and learning outcome (Chang et al., 2005; Chou et al., 2006; Kleindienst et al., 2005; Lu et al., 2005; Nakatomi et al., 2002; Wang et al., 2004).

SGZ neuronogenesis is reduced in a number of situations. The antiproliferative agent methylazoxymethanol (MAM) can eliminate the postnatal generation of neural cells in the dentate gyrus. Animals treated with MAM have impaired performance in hippocampal-dependent memory tasks, for example, conditioned fear (Shors et al., 2001). Recovery of neuronogenesis correlates with a return in ability to acquire trace memories. Postnatal neuronogenesis is highly responsive to environmental perturbations. Inflammation and stress can cause a decrease in neuronogenesis and degradation in memory (Ekdahl et al., 2003; Monje et al., 2002, 2003). Different psychiatric states can affect neural stem cell proliferation in the adult dentate gyrus. Postmortem samples from schizophrenic patients reveal depressed proliferation, whereas proliferation in patients with bipolar disorder is unaffected (Reif et al., 2006). Furthermore, SGZ neurogenesis decreases with aging (Jin et al., 2005; Kuhn et al., 1996) as does the ability to learn new tasks and to recall memories. Ablation of proliferating SGZ cells by irradiation compromises spatial discrimination, navigation, and learning (Clelland et al., 2009).

Ethanol exposure has fascinating effects on neuronal generation in the dentate gyrus in rats of all ages. Exposure of adolescents and adults to high doses of ethanol reduces SGZ neurogenesis (Crews et al., 2006; Nixon and Crews, 2002). This corresponds to ethanol-induced reductions in learning and memory. Moreover, at least in nursing rats, the effects of ethanol are dose-dependent (Miller, 1995a). Whereas exposure to high doses reduces neurogenesis, low doses can promote neuronal production. Exposure to other substances of abuse (e.g., opiates and cocaine) also depresses neuronal generation and reduces the ability to learn and remember new behaviors (Domínguez-Escribà et al., 2006; Eisch et al., 2000). Note that all of these substances of abuse appear to target the proliferation of new cells rather than their differentiation and survival. Possibly the most intriguing findings are that prenatal exposure to ethanol alters SGZ neurogenesis in adolescents and adults (Domínguez-Escribà et al., 2006; Klintsova et al., 2007; Redila et al., 2006). The implication is that such fetal programming is transmitted via proliferating neural stem cells.

28.4.1.2.2 OLFACTORY SYSTEM

The olfactory system is affected by ethanol. Exposure to ethanol during the first 2 postnatal weeks leads to a reduction in the numbers of granule and mitral cells in the olfactory bulb (Bonthius et al., 1992). This reduction persists into adulthood despite replacement from cells generated in the anterior SZ. There is no direct evidence that ethanol affects the proliferation of cells in the anterior SZ; however, studies do show that prenatal exposure to ethanol can shape the olfactory behavior of adolescents (Eade et al., 2009, 2010; Eade and Youngentob, 2009; Youngentob et al., 2007). Not only is there an enhanced neurophysiological response to ethanol odor among rats exposed to ethanol *in utero*, but such animals are also more likely to follow an ethanol-exposed peer than a water-exposed rat. In addition, ethanol-exposed animals show an increased avidity for bitter-tasting substances including ethanol and quinine (Youngentob and Glendinning, 2009; Youngentob et al., 2007). Together, these findings show that animals exposed to ethanol prenatally are programmed to prefer the odor and taste of ethanol. Human studies also show a link between prenatal exposure to ethanol and later use/abuse of the drug (Molina et al., 2007; Pepino and Mennella, 2007). Indeed, this is the basis for the 'alcoholism generator' (Miller and Spear, 2006).

28.4.1.3 Cell Fate

Cell fate is defined by many events; one appears to be cell migration. Prenatal exposure to ethanol disrupts the migration of cortical neurons *in vivo* (Miller, 1986, 1988a, 1993) and *in situ* (Siegenthaler and Miller, 2004). *In vivo* studies show that late-generated neurons that normally reside in layer II/III of the cortex often terminate their migration in layers V and VI after ethanol exposure (Miller, 1986, 1988a, 1997). Some early-generated neurons destined for layer V end up in the superficial cortex (Miller, 1986, 1987, 1988a; Miller et al., 1990). Despite these defects, the neurons retain their connectional phenotype. That is, many of the late- and early-generated neurons are callosal or corticospinal projection neurons, respectively, regardless of whether they are distributed in their correct position or in an ectopic location. Interestingly, this disruption is remarkably similar to the pattern that occurs when cortical precursors from a later time in cortical neuronogenesis are transplanted into the proliferative zones of younger fetuses (McConnell, 1988). The implication from this heterochronic transplantation experiment is that cells that are generated late in cortical histogenesis are predominantly neuronal progenitors, and the fates of these cells are largely immutable regardless of their ultimate laminar residence. By extension, it also appears that ethanol does not affect cell fate when the exposure includes the time when cells pass through their final mitotic cycle, that is, their birth date (Miller, 1987, 1997). Note that a study of cultured neural stem cells also shows that ethanol has no effect on the diversity or numbers of progeny (Hicks et al., 2010).

Some *in vivo* data support the notion that cell fate can be altered by ethanol exposure. Prenatal exposure to ethanol alters the fates of hematopoietic progenitors in the bone marrow of mouse neonates, and lymphocyte development is delayed (Wang et al., 2006, 2009). Presumably, this contributes to the immunosuppression and vulnerability of children with FASD (Sliwowska et al., 2006; Zhang et al., 2005).

Numerous studies of cultured precursors concur that ethanol can affect cell fate. For example, the diversity of cells generated by precursors in neurospheres can be reduced by ethanol (Santillano et al., 2005), and the differentiation of cultured neural stem cells is affected by ethanol (Tateno et al., 2005). Accordingly, ethanol induces precursors to become glia (astrocytes and oligodendrocytes) and reduces neuronal differentiation. This occurs in the absence of an effect on cell viability. A study of human brain-derived neural stem and progenitor cells shows that ethanol alters the expression profile of glia- and neuron-committed precursors (Vangipuram and Lyman, 2010). The concept of ethanol-induced cell fate switching is also addressed by the cell and *in ovo* culture studies by Vernadakis and colleagues. The embryonic chick telencephalic wall contains proliferating pluripotential cells, that is, neural stem cells (Kentroti and Vernadakis, 1992, 1995). Ethanol can cause the selective elimination of cells with a particular lineage (i.e., after lineages are determined Kentroti and Vernadakis, 1996); however, other evidence shows that ethanol causes cells to switch their neurochemical phenotypic lineage (Brodie and Vernadakis, 1992; Kentroti and Vernadakis, 1992).

Ethanol can affect the differentiation of cycling and recently postmitotic cells via targeted alterations of genetic expression (Hashimoto-Torii et al., 2011; Hicks et al., 2010; Liu et al., 2009; Miller et al., 2006; Zhou et al., 2011). This is exemplified by altered expression of genes associated with cell proliferation (e.g., cyclins and cyclin-dependent kinases), growth factor function (e.g., transforming growth factor (TGF) β1, insulin-like growth factor (IGF) I, epidermal growth factor (EGF) receptor), and extracellular matrix molecules (e.g., integrins, L1, and neural cell adhesion molecule), as well as mRNAs underlying cell determination and morphogenesis such as *Wasf1*, *SatB2*, *Bhlhb5*, *ID2*, *NR4A3*, *FoxP1*, neurogenin, *Sox5*, and *Bhlhe22*. One mechanism by which the profile of expressed transcripts is altered is the ethanol-induced selective hyper- and hypomethylation of CpG islands of genes associated with neural development such as *Bub1*; cyclins A2, B1, and F; securin; IGF-I; and EGF-containing fibulin-like extracellular matrix protein I (Hicks et al., 2010; Liu et al., 2009; Zhou et al., 2011). The changes in methylation are particularly notable for cells treated with TGFβ1.

28.4.1.4 Growth Factor Regulation of Cell Proliferation

28.4.1.4.1 INSULIN-LIKE GROWTH FACTOR I

The behavior of neural stem cells can be affected by pro- and antimitogenic growth factors. IGF-I is a key promitogenic player in brain development (Rubin and Baserga, 1995). Reduction or elimination of IGF-I (e.g., via pharmacological blockade or gene knockout or knockdown) leads to smaller fetuses and microencephaly. One of the contributing effects of IGF-I toward brain growth is its ability to promote cell proliferation as exemplified by a shortened doubling time for cultured neural cells (Resnicoff et al., 1994). IGF-I initiates its action by binding to and activating specific membrane-bound receptors that sequentially lead to the activation of extracellular signal-regulated kinase (ERK) 1/2.

Microencephaly and reductions in overall body growth caused by prenatal exposure to ethanol correlate with reductions in plasma IGF-I (Breese et al., 1993; de la Monte et al., 2005; Lynch et al., 2001; Mauceri et al., 1993; Singh et al., 1994; Soscia et al., 2006). Some studies also show that IGF-2 is altered (Singh et al., 1994), whereas others show that IGF-2 expression is unchanged (Breese and Sonntag, 1995). A further contributor to the brain and body growth reduction is a decrease in IGF-I transcript and protein in pregnant dams. The acute changes during gestation have long-term consequences (Breese et al., 1993). Despite being normal during the first 3 postnatal weeks, IGF-I concentrations eventually fall in weanling and adolescent rats. The dynamism of ethanol-induced changes has also been examined in the chick (Lynch et al., 2001). IGF-I expression is unaffected before day 6, drops transiently on day 6, and then rises 2 days later. This increase appears to be a response to a reduction in the availability of IGF-binding protein. Supplementation of IGF-I in the rat can partially offset ethanol-induced alterations (McGough et al., 2009) and alleviate the behavioral effects of ethanol such as on motor coordination. On the other hand, IGF-I does not mitigate ethanol-induced hyperactivity and spatial learning deficits.

Ethanol inhibits the effects of IGF-I to promote cell proliferation (Resnicoff et al., 1994). This is associated with a reduction in receptor phosphorylation and the association of phosphatidylinositol-3 kinase (PI3K) with insulin receptor substrate 1. Thus, central IGF signaling mechanisms are altered by ethanol. Apparently, these changes lead to altered feedback regulation. Impaired insulin and IGF-I signaling leads to a general depression of the transcription of genes for insulin, IGF-I and IGF receptors (de la Monte et al., 2005). The outcome of these changes is the inhibition of glucose transport and the associated production of ATP.

28.4.1.4.2 PLATELET-DERIVED GROWTH FACTOR

Like IGF, platelet-derived growth factor (PDGF) is a potent promitogenic factor. PDGF ligands and receptors are expressed by cells in the immature brain (Reddy and Pleasure, 1992; Valenzuela et al., 1997) and cycling neural cells (Luo and Miller, 1997, 1999b). Moreover, PDGF ligands affect the behavior of cycling neural cells (Luo and Miller, 1997, 1999b). This is mediated by an acceleration of the cell cycle, presumably by shortening the G1 phase.

Of the two high-affinity receptors for PDGF, ethanol targets the α isoform (Luo and Miller, 1999b). It upregulates the expression and inhibits the activation of the PDGFα receptor. PDGF signals through a receptor-activated Ras-Raf-ERK1/2 pathway in proliferating neural cells. Ethanol affects the PDGF-initiated activation of each mediator in the Ras-Raf-ERK1/2 cascade with the ultimate effect being a change in the pattern of ERK1/2 phosphorylation. Interestingly, ethanol causes the upregulation of ERK1/2, which changes the PDGF-promoted phasic stimulation into a tonic activation.

28.4.1.4.3 TRANSFORMING GROWTH FACTOR β1

Antimitogenic factors are a counterbalancing set of proteins. They are critical for limiting cell proliferation and restraining the expansion of neural precursor populations. A prime example of an antimitogenic factor is TGFβ1. TGFβ1 reduces neural generation, not by slowing the cell cycle, but by moving cells out of a proliferative population (Hicks et al., 2010; Siegenthaler and Miller, 2005b). That is, TGFβ1 reduces the growth fraction by promoting cell cycle exit. This is transduced through a p21-mediated mechanism. Furthermore, TGFβ1 facilitates this cell cycle exit and promotes the migration of postmitotic cells away from the proliferative populations (Siegenthaler and Miller, 2004).

TGFβ1 binds to a heterodimerized receptor with serine/threonine kinase activity (Danielpour and Song, 2006; ten Dijke and Hill, 2004). When activated, the receptor phosphorylates Smad2/3, which translocates to the nucleus and promotes transcription. In cortical proliferative zones, TGFβ1 also activates ERK1/2 either directly or through crosstalk with activated Smad2/3 (Powrozek and Miller, 2009). TGFβ1 activation of ERK1/2 is a sustained (tonic) response, not unlike that initiated by ethanol (Luo and Miller, 1999a).

Ethanol inhibits the TGFβ1-mediated inhibition of cell proliferation in various populations of neural precursors: astrocytes and C6 glioma cells (Miller and Luo, 2002a), B104 neuroblastoma cells (Luo and Miller, 1999a), and neuronal progenitors (Miller and Luo, 2002b). The principal mechanism involves ethanol-mediated interference

with the TGFβ1-induced reduction in the growth fraction. Concomitantly, it may cause the death of neural cells through a TGFβ1-mediated mechanism (Hicks and Miller, 2011; Kuhn and Sarkar, 2008). Prenatal exposure to ethanol affects the expression of TGFβ receptors in the fetal cerebral wall (Miller, 2003), which has downstream effects on the two signaling pathways triggered by TGFβ1 (Powrozek and Miller, 2009). These two pathways rely on Smad2/3 and ERK1/2. Not only do these pathways interact, but ethanol can affect this interplay. In fact, it appears that ethanol mimics, and presumably acts through, TGFβ1.

28.4.2 Neuronal Death/Survival

Neuronal death is a normal process of neural development. Neurons can die via a number of processes: apoptosis, necrosis, excitotoxicity, and autophagy. Apoptosis is the most common mode during development; it is characterized by morphological and biochemical changes (Danial and Korsmeyer, 2004; Kerr, 2002; Kerr et al., 1972; Mooney and Henderson, 2006; Wyllie, 1997). Morphological changes include chromatin condensation, membrane blebbing, endonucleolytic DNA cleavage, and formation of apoptotic bodies. Biochemical changes include activation of caspase 3, fragmentation of the nuclear DNA, and the consequent generation of polyadenylated strands of DNA.

Neuronal death is time dependent, and it can affect proliferating and postproliferative cells. Though the proliferative behavior of both stem and progenitor cells can be affected by ethanol (Miller, 2006b; Zawada and Das, 2006), there is an apparent difference in the susceptibility of these two types of precursors to ethanol-induced death. Stem cells (which are commonly in the VZ) may be vulnerable to ethanol; however, neural progenitors (e.g., those in the SZ) appear to be impervious to ethanol-induced death (Camarillo and Miranda, 2008; Hicks and Miller, 2011; Prock and Miranda, 2007; Santillano et al., 2005). Interestingly, these data run counter to studies showing that ethanol has no apparent effect on the survival of stem cells (Tateno et al., 2005). Such findings are in accord with the evidence that hypoxic ischemia has little effect on stem cells but compromises the viability of progenitors (Romanko et al., 2004).

The survival of postproliferative cells has most thoroughly been studied in the cerebral cortex. In the cortex, the period of naturally occurring neuronal death (NOND) takes place primarily during the second postnatal week (e.g., Ferrer et al., 1990; Finlay and Slattery, 1983; Heumann and Leuba, 1983; Heumann et al., 1978; Miller, 1995c). Indeed, the pattern of NOND follows the inside-to-outside sequence of cortical neuronal generation. That is, it is defined by the neuronal time of origin so that neurons in the deep cortex (e.g., layer V) die before those in the superficial cortex (layer II/III; Miller, 1988b, 1995c).

Developmental exposure to ethanol can induce neuronal death in various brain regions, and this death appears to be generalized. For example, in the principal sensory nucleus of the trigeminal nerve, all constituent neurons appear to be equally vulnerable (Miller, 1995b, 1999). Likewise, there is no discernible pattern to the incidence of death among Purkinje and granule neurons within a cerebellar lobule (Pierce et al., 1999). Patterns of biochemical changes indicate that ethanol-induced neuronal death also occurs in neocortex (Kuhn and Miller, 1998; Miller, 1996; Mooney and Miller, 2001; Olney et al., 2002a,b). Based on anatomical studies of the expression of 'death markers' (e.g., caspase 3 immunolabeling and terminal uridylated nick-end labeling), it appears that the cerebral cortex is different in that select subpopulations are particularly vulnerable to ethanol exposure during the period of NOND, notably neurons in layers II/III and V (Ikonomidou et al., 2000; Olney et al., 2002a,b; Young et al., 2003).

Ethanol can affect multiple mechanisms of neuronal death (Mooney and Henderson, 2006). Three are described here: (1) intrinsic, (2) extrinsic, and (3) caspase 3-independent pathways.

28.4.2.1 Intrinsic Pathway

The intrinsic pathway is a mitochondrial-dependent pathway that is typically activated in response to an apoptotic signal such as DNA damage or reactive oxygen species (ROS; Green and Reed, 1998; Miller et al., 2000; Mooney and Henderson, 2006; Soengas et al., 1999). Proapoptotic proteins are released from the mitochondrial intermembrane space. Permeabilization of the mitochondrial outer membrane is mediated by Bcl proteins and promotes binding of p53 to proapoptotic proteins, for example, Bcl-XS or Bax (Chipuk et al., 2004). Bax upregulation may allow insertion of Bax homodimers into the mitochondrial membrane, thereby altering its permeability and permitting intermembrane substances to leak into the cytoplasm. These substances cause activation of caspase 3, which represses DNA repair and initiates DNA fragmentation and cell death.

Exposure to ethanol alters the *in vivo* expression of Bcl proteins (Mooney and Miller, 2003). Changes may be rapid, as in the case of the transcripts (Inoue et al., 2002; Moore et al., 1999), or delayed, as detected by changes in protein expression (Mooney and Miller, 2001; Heaton et al., 2003b). Ethanol also increases the expression of active caspase 3 (Carloni et al., 2004; Han et al., 2005; Ikonomidou et al., 2000; Mooney and Miller, 2003; Nowoslawski et al., 2005; Olney et al., 2002a,b) and induces production of ROS, which then

causes DNA damage (Heaton et al., 2003a,b; Kotch et al., 1995; Maffi et al., 2008; Ramachandran et al., 2001, 2003). Interestingly, Bax is apparently required for ethanol-induced cell death, but caspase 3 is not (Young et al., 2003, 2005). Mice deficient in *Bax* do not exhibit argyrophilic (degenerating) cells in response to acute ethanol exposure, whereas caspase 3-null animals do.

28.4.2.2 Extrinsic Pathway

The extrinsic pathway is activated by binding the Fas ligand (FasL) to its cell surface receptor Fas. This binding causes receptor oligomerization, and the recruitment of the Fas-associated death domain (Fadd) and its association with procaspases 8 and 10 (Benn and Wolff, 2004). The Fadd-caspase 8/10 complex forms the death-inducing signaling complex, which cleaves and activates caspase 3. As with the intrinsic pathway, active caspase 3 inactivates poly-ADP-ribose polymerase (PARP) and allows DNA fragmentation. In addition, active caspase 8 cleaves the Bcl family protein, Bid. Truncated Bid (tBid) translocates to the mitochondria where it can activate the intrinsic pathway by promoting insertion of Bax homodimers into the mitochondrial membrane. Ethanol alters expression of FasL and Fas (Cheema et al., 2000; de la Monte and Wands, 2002; Hicks and Miller, 2011) and can increase caspase 8 activity (Vaudry et al., 2002).

28.4.2.3 Caspase 3-Independent Pathway

As with the intrinsic pathway, the caspase 3-independent pathway is mitochondria-dependent. Following its release from the mitochondrial intermembrane space, apoptosis-inducing factor can either directly upregulate DNase activity or can cleave and inactivate PARP. The effect of this is the repression of DNA repair and promotion of cell degeneration. Although there is little direct evidence that ethanol activates the caspase 3-independent pathway, inhibiting caspase activity does not prevent ethanol-induced cell death (D'Mello et al., 2000; Keramaris et al., 2000; Miller et al., 1997; Selznick et al., 2000; Stefanis et al., 1999). This implies that a caspase 3-independent pathway is able to be activated, regardless of whether it is the main pathway affected by ethanol. It is noteworthy that both the intrinsic and extrinsic pathways are upstream of p53 activation (Mooney and Henderson, 2006); ethanol affects p53 expression (Kuhn and Miller, 1998) and p53-mediated cell death (Miller et al., 2003).

28.4.2.4 Growth Factor Targets

Neurotrophins play critical roles in normal brain development (e.g., Bibel and Barde, 2000). Developmental expression of neurotrophins is necessary for neuronal survival, process outgrowth, and synaptogenesis. Two neurotrophins particularly important for brain development are NGF and BDNF. Although NGF is not expressed by cortical progenitors *in vitro* or in the VZ *in vivo*, which suggests that it is not required for cell proliferation or the initiation of migration (Barnabé-Heider and Miller, 2003; Fukumitsu et al., 1998; Maisonpierre et al., 1990), it is highly expressed in the early postnatal period, suggesting a role in postmitotic development (Das et al., 2001; Heaton et al., 2003b).

BDNF is present in cortical progenitor cells *in vivo* and *in vitro* (Barnabé-Heider and Miller, 2003; Fukumitsu et al., 1998; Maisonpierre et al., 1990) and remains evident in the cerebral cortex through adulthood (Climent et al., 2002; Das et al., 2001; Heaton et al., 2003b; Itami et al., 2000; Pitts and Miller, 2000; Vitalis et al., 2002).

Developmental exposure to ethanol alters the expression of neurotrophins and their receptors, although there is no consensus as to the effect (e.g., Climent et al., 2002; Heaton et al., 1999, 2000b, 2003a,b; Light et al., 2001; Seabold et al., 1998). Prenatal and postnatal exposure to ethanol increases NGF expression in the cortex and striatum (Heaton et al., 2000a, 2003a,b). Cortical BDNF expression (Climent et al., 2002) is reduced during the first 2 postnatal weeks by prenatal exposure to ethanol, whereas postnatal exposure to ethanol increases cortical BDNF expression (Heaton et al., 2003b).

For neurotrophins to have an effect on brain development, they must bind to and activate receptors. Thus, the expression, both the amount and location, of receptors may provide greater insight into the role of the neurotrophin systems in development. Cortical expression of trkA and trkB is upregulated following prenatal exposure to ethanol (Climent et al., 2002; Gottesfeld et al., 1990; Valles et al., 1994). Exposure to ethanol in the early postnatal period also upregulates cortical p75 expression (Toesca et al., 2003) as does ethanol treatment of cultured cortical neurons (Seabold et al., 1998). Evidence suggests that p75 can mediate apoptotic death or mediate the protective effect of a neurotrophin (Blochl and Blochl, 2007; Casaccia-Bonnefil et al., 1996, 1999; Seabold et al., 1998). Thus, the changes in receptor expression may be in response to ethanol-induced reductions in neurotrophin concentrations and subserve a protective mechanism.

Exogenous neurotrophins can ameliorate ethanol-induced cell death. Animals that overexpress NGF are less vulnerable to ethanol-induced neurotoxicity (Heaton et al., 2000a). The addition of NGF or BDNF to cultured cells reduces ethanol-induced cell death (Bhave et al., 1999; de la Monte et al., 2000, Heaton et al., 1992, 1993, 1994; Miller et al., 2003; Seabold et al., 1998). The ability of NGF to protect cortical neurons against ethanol-induced death is both site-specific and age-dependent. NGF fails to protect against ethanol-induced death in cultures of cortical neurons (Seabold et al., 1998), or in slice cultures from 16-day-old rat fetuses (Mooney and Miller, 2007a). In contrast, NGF is neuroprotective in

slice cultures taken from a 3-day-old rat brain. This neuroprotection is seen only in the more mature lower cortical plate and not in the upper cortical plate, which largely contains migrating neurons. The implication is that the neuroprotection depends upon the maturity of the neurons.

NGF-induced neuroprotection is correlated with receptor activity. Ethanol inhibits the NGF-induced phosphorylation of trkA in the primary neurons taken from 16-day-old fetuses but does not affect p-trk expression in slices from 3-day-old pups. Activation, that is, phosphorylation, of trkA is generally associated with neuronal survival. Thus, an ethanol-induced reduction in p-trk expression may be predictive of an inability of NGF to protect against ethanol-induced death.

In addition to altering the expression of neurotrophin ligands and receptors, exposure to ethanol alters downstream signaling. Trk receptors can signal survival via the ERK1/2 and PI3K pathways, and p75 may also signal survival via PI3K. Activation of both pathways is age-dependent. Growth factor-dependent activation is downregulated in the presence of ethanol (Climent et al., 2002; Kalluri and Ticku, 2002). Ethanol inhibits endogenous phospho-ERK1/2 expression (Kalluri and Ticku, 2002). Ethanol also alters growth factor-induced phosphorylation of the ERK1/2 pathway in cortical cells (Climent et al., 2002; Luo and Miller, 1999a,b) and PI3K/Akt in cerebellar cells (de la Monte and Wands, 2002; Li et al., 2004). Interestingly, the ethanol-induced reduction in growth factor-stimulated PI3K/Akt activation in cerebellar neurons occurs in the absence of a change in receptor or ligand expression (de la Monte and Wands, 2002). The effect of ethanol on pathway activation is cell type dependent. For example, ethanol inhibits the BDNF-mediated increase in expression of phosphorylated jun-*N*-terminal kinase (p-JNK) and phosphorylated Akt (p-Akt) in cultured mouse cerebellar granule cells (Li et al., 2004). In contrast, in a human neuroblastoma cell line (TB8 cells), ethanol inhibits the BDNF-mediated increase in p-Akt but not p-JNK expression, suggesting that different pathways are activated by BDNF in these cells.

28.5 BEHAVIORAL CONSEQUENCES

Social interactions are crucial for the survival of humans and other mammalian species. In humans, peer relationships are sources of knowledge about behavioral patterns, attitudes, values, and consequences in different situations (Deutsch and Gerard, 1955). In the same way, peer-directed social activity of rodents seems crucial for establishing social organization in a group or between partners and for developing the ability to express and understand intraspecific communication signals (Vanderschuren et al., 1997).

A secure and consistent social milieu is important not only for humans, but for laboratory rodents as well, with social deprivation being stressful (Hall, 1998). Rats exhibit numerous social behaviors including play fighting, contact behavior, and social investigation. Abnormal social behaviors have been reported in rats following a number of developmental insults, including neonatal lesions of the amygdala (Daenen et al., 2002), neonatal exposure to Borna disease virus (Lancaster et al., 2007), and ethanol exposure during development (Kelly et al., 2000; Thomas et al., 1998). Developmental exposure to ethanol alters play-fighting behavior (Lawrence et al., 2008; Lugli et al., 2003; Meyer and Riley, 1986; Royalty, 1990), as well as the amount and type of active social interactions in adolescents and adults (Kelly and Dillingham, 1994; Lugli et al., 2003). Prenatal exposure to ethanol alters social behavior in a sex-dependent fashion (Meyer and Riley, 1986), with males showing less play behavior and females demonstrating more play fighting. Interestingly, a single exposure to ethanol on G12 alters the social behavior of adolescent and adult rats, and many of the changes are sex-specific, that is, they are detected only in males (Mooney and Varlinskaya, 2011; Mooney et al., 2009).

28.6 SUMMARY

Early developmental exposure to ethanol has multiple consequences including targeted effects on cell proliferation and survival. In general, ethanol reduces cell proliferation and compromises neuronal survival, but the scope of the changes are defined by the site, the exposure, the maturity of the cells, and the growth factor availability and receptivity. The culmination of these changes affects the total numbers of neurons in the brain. Interestingly, the numbers of neurons in structures within a particular system (e.g., the somatosensory system) may be differentially altered. For example, the numbers of neurons in the rat trigeminal brainstem nuclei (Miller, 1999; Miller and Muller, 1989) and the somatosensory cortex (Miller and Potempa, 1990; Powrozek and Zhou, 2005) are 30–50% fewer in ethanol-exposed animals; however, no detectable effect is evident in the thalamus (Livy et al., 2001; Mooney and Miller, 1999, 2010). This breakdown in the numerical matching of neuronal populations within a system of interconnected structures implies that the assemblage of neurons within each structure occurs either autonomously or asynchronously. The consequences of such differential effects are functional changes (Miller and Dow-Edwards, 1993; Vingan et al., 1986; Xie et al., 2010) that likely underlie ethanol-induced dysfunction such as motor incoordination and cognitive deficits (Coles, 2006; Fryer et al., 2006).

Two intriguing contributors to postnatal responses and responsivity are genetic–environmental interactions and fetal programming. It is becoming clearer that ethanol has targeted effects on genomic expression and epigenetic modifications (e.g., Haycock, 2009; Hicks et al., 2010; Pandey et al., 2008). Overlying these interactions, and possibly a result of them, is the effect of early exposure to ethanol on fetal programming (Miller and Spear, 2006). Such programming may be mediated through neural stem cells. The mechanism and the scope of this programming are at present unknown. Understanding them will not only provide key insights into normal brain ontogeny, but it is also likely critical for developing strategies to modify neural development and outcome and ultimately preventing or offsetting developmental disorders such as FASD.

Acknowledgments

The authors thank the National Institutes of Health (AA06916, AA18693, and AA178231, SMM; ES14901, PJL; AA06916, AA07568, and AA178231, MW Miller), Autism Speaks (SM Mooney), University of California at Davis MIND Institute (PJ Lein), and the Department of Veterans Affairs (MW Miller) for their support.

References

Abel, E.L., Sokol, R.J., 1992. A revised conservative estimate of the incidence of FAS and economic impact. Alcoholism, Clinical and Experimental Research 15, 514–524.

Al-Rabiai, S., Miller, M.W., 1989. Effects of prenatal exposure to ethanol on the ultrastructure of layer V in somatosensory cortex of mature rats. Journal of Neurocytology 18, 711–729.

Altman, J., Bayer, S.A., 1979. Development of the diencephalon in the rat. VI. Re-evaluation of the embryonic development of the thalamus on the basis of thymidine-radiographic datings. The Journal of Comparative Neurology 188, 501–524.

Altman, J., Bayer, S.A., 1989. Development of the rat thalamus: IV. The intermediate lobule of the thalamic neuroepithelium, and the time and site of origin and settling pattern of neurons of the ventral nuclear complex. The Journal of Comparative Neurology 284, 534–566.

Altman, J., Bayer, S.A., 1990. Migration and distribution of two populations of hippocampal progenitors during the perinatal and postnatal periods. The Journal of Comparative Neurology 124, 319–335.

Altman, J., Das, G.P., 1965. Autoradiographic and histological evidence of postnatal hippocampal neurogenesis in rats. The Journal of Comparative Neurology 124, 319–336.

Altman, J., Das, G.P., 1966. Autoradiographic and histological studies of postnatal neurogenesis. I. A longitudinal investigation of the kinetics, migration and transformation of cells incorporating tritiated thymidine in neonate rats, with special reference to postnatal neurogenesis in some brain regions. The Journal of Comparative Neurology 126, 337–389.

Barnabé-Heider, F., Miller, F.D., 2003. Endogenously produced neurotrophins regulate survival and differentiation of cortical progenitors via distinct signaling pathways. Journal of Neuroscience 23, 5149–5160.

Bauer-Moffett, C., Altman, J., 1977. The effect of ethanol chronically administered to preweanling rats on cerebellar development: A morphological study. Brain Research 119, 249–268.

Benn, S.C., Wolff, C.J., 2004. Adult neuron survival strategies – Slamming on the brakes. Nature Reviews Neuroscience 5, 686–700.

Bhave, S.V., Ghoda, L., Hoffman, P.L., 1999. Brain-derived neurotrophic factor mediates the anti-apoptotic effect of NMDA in cerebellar granule neurons: Signal transduction cascades and site of ethanol action. Journal of Neuroscience 19, 3277–3286.

Bibel, M., Barde, Y.A., 2000. Neurotrophins: Key regulators of cell fate and cell shape in the vertebrate nervous system. Genes & Development 14, 2919–2937.

Blochl, A., Blochl, R., 2007. A cell-biological model of p75NTR signaling. Journal of Neurochemistry 102, 289–305.

Bonthius, D.J., West, J.R., 1990. Alcohol-induced neuronal loss in developing rats: increased brain damage with binge exposure. Alcoholism: Clinical and Experimental Research 14, 107–118.

Bonthius, D.J., Bonthius, N.E., Napper, R.M., West, J.R., 1992. Early postnatal alcohol exposure acutely and permanently reduces the number of granule cells and mitral cells in the rat olfactory bulb: A stereological study. The Journal of Comparative Neurology 324, 557–566.

Bonthius, D.J., West, J.R., 1988a. Blood alcohol concentration and microencephaly: A dose–response study in the neonatal rat. Teratology 37, 223–231.

Bonthius, D.J., West, J.R., 1988b. Alcohol-induced neuronal loss in developing rats: Increased brain damage with binge exposure. Alcoholism, Clinical and Experimental Research 14, 107–118.

Bookstein, F.L., Streissguth, A.P., Sampson, P.D., Connor, P.D., Barr, H.M., 2002. Corpus callosum shape and neuropsychological deficits in adult males with heavy fetal alcohol exposure. Neuro-Image 15, 233–251.

Borges, S., Lewis, P.D., 1983. Effects of ethanol on postnatal cell acquisition in the rat cerebellum. Brain Research 271, 388–391.

Breese, C.R., Sonntag, W.E., 1995. Effect of ethanol on plasma and hepatic insulin-like growth factor regulation in pregnant rats. Alcoholism, Clinical and Experimental Research 19, 867–873.

Breese, C.R., D'Costa, A., Ingram, R.L., Lenham, J., Sonntag, W.E., 1993. Long-term suppression of insulin-like growth factor-1 in rats after in utero ethanol exposure: Relationship to somatic growth. Journal of Pharmacology and Experimental Therapeutics 264, 448–456.

Brodie, C., Vernadakis, A., 1992. Ethanol increases cholinergic and decreases GABAergic neuronal expression in cultures derived from 8-day-old chick embryo cerebral hemispheres: Interaction of ethanol and growth factors. Developmental Brain Research 65, 253–257.

Camarillo, C., Miranda, R.C., 2008. Ethanol exposure during neurogenesis induces persistent effects on neural maturation: Evidence from an ex vivo model of fetal cerebral cortical neuroepithelial progenitor maturation. Gene Expression 14, 159–171.

Carloni, S., Mazzoni, E., Balduini, W., 2004. Caspase-3 and calpain activities after acute and repeated ethanol administration during the rat brain growth spurt. Journal of Neurochemistry 89, 197–203.

Casaccia-Bonnefil, P., Carter, B.D., Dobrowsky, R.T., Chao, M.V., 1996. Death of oligodendrocytes mediated by the interaction of nerve growth factor with its receptor p75. Nature 383, 716–719.

Casaccia-Bonnefil, P., Gu, C., Khursigara, G., Chao, M.V., 1999. p75 neurotrophin receptor as a modulator of survival and death decisions. Microscopy Research and Technique 45, 217–224.

Centers for Disease Control and Prevention (2009) Reducing alcohol-exposed pregnancies. A Report of the National Task Force on Fetal Alcohol Syndrome and Fetal Alcohol Effect. US Dept Human Hlth Serv. http://www.cdc.gov/ncbddd/fasd/documents/121972RedAlcohPreg+Cov.pdf.

Chang, Y.S., Mu, D., Wendland, M., et al., 2005. Erythropoietin improves functional and histological outcome in neonatal stroke. Pediatric Research 58, 106–111.

Cheema, Z.F., West, J.R., Miranda, R.C., 2000. Ethanol induces Fas/Apo [Apoptosis]-1 mRNA and cell suicide in the developing cerebral cortex. Alcoholism, Clinical and Experimental Research 24, 535–543.

Chipuk, J.E., Kuwana, T., Bouchier-Hayes, L., et al., 2004. Direct activation of Bax by p53 mediates mitochondrial membrane permeabilization and apoptosis. Science 303, 1010–1014.

Chizhikov, V.V., Lindgren, A.G., Currle, D.S., Rose, M.F., Monuki, E.S., Millen, K.J., 2006. The roof plate regulates cerebellar cell-type specification and proliferation. Development 133, 2793–2804.

Chou, J., Harvey, B.K., Chang, C.F., Shen, H., Morales, M., Wang, Y., 2006. Neuroregenerative effects of BMP7 after stroke in rats. Journal of Neurological Sciences 240, 21–29.

Clamp, P.A., Lindsley, T.A., 1998. Early events in the development of neuronal polarity in vitro are altered by ethanol. Alcoholism, Clinical and Experimental Research 22, 1277–1284.

Clarren, S.K., Alvord, E.C., Sumi, S.M., Streissguth, A.P., Smith, D.W., 1978. Brain malformations related to prenatal exposure to ethanol. Journal of Pediatrics 92, 64–67.

Clelland, C.D., Choi, M., Romberg, C., et al., 2009. A functional role for adult hippocampal neurogenesis in spatial pattern separation. Science 325, 210–213.

Climent, E., Pascual, M., Renau-Piqueras, J., Guerri, C., 2002. Ethanol exposure enhances cell death in the developing cerebral cortex: Role of brain-derived neurotrophic factor and its signaling pathways. Journal of Neuroscience Research 68, 213–225.

Coles, C.D., 2006. Prenatal alcohol exposure and human development. In: Miller, M.W. (Ed.), Brain Development. Normal Processes and the Effects of Alcohol and Nicotine. Oxford Univ Press, New York, pp. 123–142.

Crews, F.T., Mdzinarishvili, A., Kim, D., He, J., Nixon, K., 2006. Neurogenesis in adolescent brain is potently inhibited by ethanol. Neuroscience 137, 437–445.

Cudd, T.A., 2005. Animal model systems for the study of alcohol teratology. Exp Biol Med (Maywood) 230, 389–393.

Daenen, E.W., Wolterink, G., Gerrits, M.A., Van Ree, J.M., 2002. The effects of neonatal lesions in the amygdala or ventral hippocampus on social behaviour later in life. Behavioural Brain Research 136, 571–582.

Danial, N.N., Korsmeyer, S.J., 2004. Cell death: Critical control points. Cell 116, 205–219.

Danielpour, D., Song, K., 2006. Cross-talk between IGF-I and TGFβ signaling pathways. Cytokine & Growth Factor Reviews 17, 59–74.

Das, K.P., Chao, S.L., White, L.D., et al., 2001. Differential patterns of nerve growth factor, brain-derived neurotrophic factor and neurotrophin-3 mRNA and protein levels in developing regions of rat brain. Neuroscience 103, 739–761.

Dash, P.K., Mach, S.A., Moore, A.N., 2001. Enhanced neurogenesis in the rodent hippocampus following traumatic brain injury. Journal of Neuroscience Research 63, 313–319.

de la Monte, S.M., Wands, J.R., 2002. Chronic gestational exposure to ethanol impairs insulin-stimulated survival and mitochondrial function in cerebellar neurons. Cellular and Molecular Life Sciences 59, 882–893.

de la Monte, S.M., Ganju, N., Banerjee, K., Brown, N.V., Luong, T., Wands, J.R., 2000. Partial rescue of ethanol-induced neuronal apoptosis by growth factor activation of phosphoinositol-3-kinase. Alcoholism: Clinical and Experimental Research 24, 716–726.

de la Monte, S.M., Lahousse, S.A., Carter, J., Wands, J.R., 2002. ATP luminescence-based motility-invasion assay. Biotechniques 33, 98–100.

de la Monte, S.M., Xu, X.J., Wands, J.R., 2005. Ethanol inhibits insulin expression and actions in the developing brain. Cellular and Molecular Life Sciences 62, 1131–1145.

Deng, W., Aimone, J.B., Gage, F.H., 2010. New neurons and new memories: How does adult hippocampal neurogenesis affect learning and memory? Nature Reviews Neuroscience 11, 339–350.

Deutsch, M., Gerard, H.B., 1955. A study of normative and informational social influences upon individual judgement. Journal of Abnormal Psychology 51, 629–636.

Diaz, J., Samson, H.H., 1980. Impaired brain growth in neonatal rats exposed to ethanol. Science 208, 751–753.

D'Mello, S.R., Kuan, C.Y., Flavell, R.A., Rakic, P., 2000. Caspase-3 is required for apoptosis-associated DNA fragmentation but not for cell death in neurons deprived of potassium. Journal of Neuroscience Research 59, 24–31.

Dobbing, J., Sands, J.D., 1979. Comparative aspects of the brain growth spurt. Early Human Development 3, 79–83.

Domínguez-Escribà, L., Hernández-Rabaza, V., Soriano-Navarro, M., et al., 2006. Chronic cocaine exposure impairs progenitor proliferation but spares survival and maturation of neural precursors in adult rat dentate gyrus. European Journal of Neuroscience 24, 586–594.

Eade, A.M., Youngentob, S.L., 2009. Adolescent ethanol experience alters immediate and long-term behavioral responses to ethanol odor in observer and demonstrator rats. Behavioral and Brain Functions 5, 23.

Eade, A.M., Sheehe, P.R., Molina, J.C., Spear, N.E., Youngentob, L.M., Youngentob, S.L., 2009. The consequence of fetal ethanol exposure and adolescent odor re-exposure on the response to ethanol odor in adolescent and adult rats. Behavioral and Brain Functions 5, 3.

Eade, A.M., Sheehe, P.R., Youngentob, S.L., 2010. Ontogeny of the enhanced fetal-ethanol-induced behavioral and neurophysiologic olfactory response to ethanol odor. Alcoholism, Clinical and Experimental Research 34, 206–213.

Eisch, A.J., Barrot, M., Schad, C.A., Self, D.W., Nestler, E.J., 2000. Opiates inhibit neurogenesis in the adult rat hippocampus. Proceedings of the National Academy of Sciences of the United States of America 97, 7579–7584.

Eisch, A.J., Cameron, H.A., Encinas, J.M., Meltzer, L.A., Ming, G.L., Overstreet-Wadiche, L.S., 2008. Adult neurogenesis, mental health, and mental illness: Hope or hype? Journal of Neuroscience 28, 11785–11791.

Ekdahl, C.T., Claesen, J.H., Bonde, S., Kokaia, Z., Lindvall, O., 2003. Inflammation is detrimental for neurogenesis in adult brain. Proceedings of the National Academy of Sciences of the United States of America 100, 13632–13637.

Ferrer, I., Bernet, E., Soriano, E., Del Rio, T., Fonseca, M., 1990. Naturally occurring cell death in the cerebral cortex of the rat and removal of dead cells by transitory phagocytes. Neuroscience 39, 451–458.

Finlay, B.L., Slattery, M., 1983. Local differences in the amount of early cell death in neocortex predict adult local specializations. Science 219, 1349–1351.

Fryer, S.L., McGee, C.L., Spadoni, A.D., Riley, R.P., 2006. Influence of alcohol on the structure of the developing human brain. In: Miller, M.W. (Ed.), Brain Development. Normal Processes and the Effects of Alcohol and Nicotine. Oxford University Press, New York, pp. 143–152.

Fukumitsu, H., Furukawa, Y., Tsusaka, M., et al., 1998. Simultaneous expression of brain-derived neurotrophic factor and neurotrophin-3 in Cajal-Retzius, subplate and ventricular progenitor cells during early development stages of the rat cerebral cortex. Neuroscience 84, 115–127.

Gottesfeld, Z., Morgan, B., Perez-Polo, J.R., 1990. Prenatal alcohol exposure alters the development of sympathetic synaptic components and of nerve growth factor receptor expression selectivity in lymphoid organs. Journal of Neuroscience Research 26, 308–316.

Gould, E., Beylin, A., Tanapat, P., Reeves, A., Shors, T.J., 1999. Learning enhances adult neurogenesis in the hippocampal formation. Nature Neuroscience 2, 260–265.

Green, D.R., Reed, J.C., 1998. Mitochondria and apoptosis. Science 281, 1309–1312.

Hall, F.S., 1998. Social deprivation of neonatal, adolescent, and adult rats has distinct neurochemical and behavioral consequences. Critical Reviews in Neurobiology 12, 129–162.

Han, J.Y., Joo, Y., Kim, Y.S., et al., 2005. Ethanol induces cell death by activating caspase-3 in the rat cerebral cortex. Molecules and Cells 20, 189–195.

Hashimoto-Torii, K., Kawasawa, Y.I., Kuhn, A., Rakic, P., 2011. Combined transcriptome analysis of fetal human and mouse cerebral cortex exposed to alcohol. Proceedings of the National Academy of Sciences of the United States of America 108, 4212–4217.

Haycock, P.C., 2009. Fetal alcohol spectrum disorders: The epigenetic perspective. Biology of Reproduction 81, 607–617.

Heaton, M.B., Swanson, D.J., Paiva, M., Walker, D.W., 1992. Ethanol exposure affects trophic factor activity and responsiveness in chick embryo. Alcohol 9, 161–166.

Heaton, M.B., Paiva, M., Swanson, D.J., Walker, D.W., 1993. Modulation of ethanol neurotoxicity by nerve growth factor. Brain Research 620, 78–85.

Heaton, M.B., Paiva, M., Swanson, D.J., Walker, D.W., 1994. Responsiveness of cultured septal and hippocampal neurons to ethanol and neurotrophic substances. Journal of Neuroscience Research 39, 305–318.

Heaton, M.B., Mitchell, J.J., Paiva, M., 1999. Ethanol-induced alterations in neurotrophin expression in developing cerebellum: Relationship to periods of temporal susceptibility. Alcoholism, Clinical and Experimental Research 23, 1637–1642.

Heaton, M.B., Mitchell, J.J., Paiva, M., 2000a. Overexpression of NGF ameliorates ethanol neurotoxicity in the developing cerebellum. Journal of Neurobiology 45, 95–104.

Heaton, M.B., Mitchell, J.J., Paiva, M., Walker, D.W., 2000b. Ethanol-induced alterations in the expression of neurotrophic factors in the developing rat central nervous system. Developmental Brain Research 121, 97–107.

Heaton, M.B., Paiva, M., Madorsky, I., Mayer, J., Moore, D.B., 2003a. Effects of ethanol on neurotrophic factors, apoptosis-related proteins, endogenous antioxidants, and reactive oxygen species in neonatal striatum: Relationship to periods of vulnerability. Developmental Brain Research 140, 237–252.

Heaton, M.B., Paiva, M., Madorsky, I., Shaw, G., 2003b. Ethanol effects on neonatal rat cortex: Comparative analyses of neurotrophic factors, apoptosis-related proteins, and oxidative processes during vulnerable and resistant periods. Developmental Brain Research 145, 249–262.

Heumann, D., Leuba, G., 1983. Neuronal death in the development and aging of the cerebral cortex of the mouse. Neuropathology and Applied Neurobiology 9, 297–311.

Heumann, D., Leuba, G., Rabinowicz, T., 1978. Postnatal development of the mouse cerebral neocortex. IV. Evolution of the total cortical volume of the population of neurons and glial cells. Journal für Hirnforschung 19, 385–393.

Hicks, S.D., Miller, M.W., 2011. Ethanol causes the death of neural stem cells via transforming growth factor β1-dependent and independent mechanisms. Experimental Neurology 229, 372–380.

Hicks, S.D., Middleton, F.A., Miller, M.W., 2010. Ethanol-induced methylation of cell cycle genes in neural stem cells. Journal of Neurochemistry 114, 1767–1780.

Ikonomidou, C., Bittigau, P., Ishimaru, M.J., et al., 2000. Ethanol-induced apoptotic neurodegeneration and fetal alcohol syndrome. Science 287, 1056–1060.

Inoue, M., Nakamura, K., Iwahashi, K., Ameno, K., Itoh, M., Suwaki, H., 2002. Changes of bcl-2 and bax mRNA expressions in the ethanol-treated mouse brain. Nihon Arukōru Yakubutsu Igakkai Zasshi 37, 120–129.

Itami, C., Mizuno, K., Kohno, T., Nakamura, S., 2000. Brain-derived neurotrophic factor requirement for activity-dependent maturation of glutamatergic synapse in developing mouse somatosensory cortex. Brain Research 857, 141–150.

Jellinger, K., Gross, H., Kaltenback, E., Grisold, W., 1981. Holoprosencephaly and agenesis of the corpus callosum: Frequency of associated malformations. Acta Neuropathologica 55, 1–10.

Jin, K., Minami, M., Xie, L., et al., 2005. Ischemia-induced neurogenesis is preserved but reduced after traumatic brain injury. Journal of Neurotrauma 22, 1011–1017.

Jones, K.L., Smith, D.W., Ulleland, C.N., Streissguth, P., 1973. Pattern of malformation in offspring of chronic alcoholic mothers. Lancet 1, 1267–1271.

Kalluri, H.S., Ticku, M.K., 2002. Ethanol-mediated inhibition of mitogen-activated protein kinase phosphorylation in mouse brain. European Journal of Pharmacology 439, 53–58.

Kelly, S.J., Dillingham, R.R., 1994. Sexually dimorphic effects of perinatal alcohol exposure on social interactions and amygdala DNA and DOPAC concentrations. Neurotoxicology and Teratology 16, 377–384.

Kelly, S.J., Lawrence, C.R., 2008. Intragastric intubation of alcohol during the perinatal period. Alcohol 447, 102–110.

Kelly, S.J., Day, N., Streissguth, A.P., 2000. Effects of prenatal alcohol exposure on social behavior in humans and other species. Neurotoxicology and Teratology 22, 143–149.

Kelly, S.J., Goodlett, C.R., Hannigan, J.H., 2009. Animal models of fetal alcohol spectrum disorders: impact of the social environment. Developmental Disabilities Research Reviews 15, 200–208.

Kennedy, L.A., Elliott, M.J., 1985. Cell proliferation in the embryonic mouse neocortex following acute maternal alcohol intoxication. International Journal of Developmental Neuroscience 3, 311–315.

Kentroti, S., Vernadakis, A., 1992. Ethanol administration during early embryogenesis affects neuronal phenotypes at a time when neuroblasts are pluripotential. Journal of Neuroscience Research 33, 617–625.

Kentroti, S., Vernadakis, A., 1995. Early neuroblasts are pluripotential: Colocalization of neurotransmitters and neuropeptides. Journal of Neuroscience Research 41, 696–707.

Kentroti, S., Vernadakis, A., 1996. Ethanol neurotoxicity in culture: Selective loss of cholinergic neurons. Journal of Neuroscience Research 44, 577–585.

Keramaris, E., Stefanis, L., MacLaurin, J., et al., 2000. Involvement of caspase 3 in apoptotic death of cortical neurons evoked by DNA damage. Molecular and Cellular Neurosciences 15, 368–379.

Kerr, J.F., 2002. History of the events leading to the formulation of the apoptosis concept. Toxicology 181–182, 471–474.

Kerr, J.F., Wyllie, A.H., Currie, A.R., 1972. Apoptosis: A basic biological phenomenon with wide-ranging implications in tissue kinetics. British Journal of Cancer 26, 239–257.

Kleindienst, A., McGinn, M.J., Harvey, H.B., Colello, R.J., Hamm, R.J., Bullock, M.R., 2005. Enhanced hippocampal neurogenesis by intraventricular S100B infusion is associated with improved cognitive recovery after traumatic brain injury. Journal of Neurotrauma 22, 645–655.

Klintsova, A.Y., Helfer, J.L., Calizo, L.H., Dong, W.K., Goodlett, C.R., Greenough, W.T., 2007. Persistent impairment of hippocampal neurogenesis in young adult rats following early postnatal alcohol exposure. Alcoholism, Clinical and Experimental Research 31, 2073–2082.

Kornguth, S.E., Rutledge, J.J., Sunderland, E., et al., 1979. Impeded cerebellar development and reduced serum thyroxine levels associated with fetal alcohol intoxication. Brain Research 177, 347–360.

Kotch, L., Chen, S.Y., Sulik, K., 1995. Ethanol-induced teratogenesis: Free radical damage as a possible mechanism. Teratology 52, 128–136 in mental retardation. The International Journal of Biochemistry and Cell Biology 41:96–107.

Kuhn, H.G., Dickinson-Anson, H., Gage, F.H., 1996. Neurogenesis in the dentate gyrus of the adult rat: Age-related decrease of neuronal progenitor population. Journal of Neuroscience 16, 2027–2033.

Kuhn, P.E., Miller, M.W., 1998. Expression of p53 and ALZ-50 immunoreactivity in rat cortex: Effect of prenatal exposure to ethanol. Experimental Neurology 154, 418–429.

Kuhn, P., Sarkar, D., 2008. Ethanol induces apoptotic death of β-endorphin neurons in the rat hypothalamus by a TGFβ1-dependent mechanism. Alcoholism, Clinical and Experimental Research 32, 707–714.

Lancaster, K., Dietz, D.M., Moran, T.H., Pletnikov, M.V., 2007. Abnormal social behaviors in young and adult rats neonatally infected with Borna disease virus. Behavioural Brain Research 176, 141–148.

Lawrence, C.R., Bonner, H.C., Newsom, R.J., Kelly, S.J., 2008. Effects of alcohol exposure during development on play behavior and c-Fos expression in response to play behavior. Behavioural Brain Research 188, 209–218.

Lemoine, P., Harousseau, H., Borteyru, J.-P., Menuet, J.-C., 1968. Les enfants de parents alcooliques: Anomalies observées à propos de 127 cas. Ouest-Médical 21, 476–482.

Leuner, B., Gould, E., 2010. Structural plasticity and hippocampal function. Annual Review of Psychology 61, 111–140.

Li, Z., Miller, M.W., Luo, J., 2002. Effects of prenatal exposure to ethanol on the cyclin-dependent kinase system in the developing rat cerebellum. Developmental Brain Research 139, 237–245.

Li, Z., Ding, M., Thiele, C.J., Luo, J., 2004. Ethanol inhibits brain-derived neurotrophic factor-mediated intracellular signaling and activator protein-1 activation in cerebellar granule neurons. Neuroscience 126, 149–162.

Light, K.E., Ge, Y., Belcher, S.M., 2001. Early postnatal ethanol exposure selectively decreases BDNF and truncated TrkB-T2 receptor mRNA expression in the rat cerebellum. Molecular Brain Research 93, 46–55.

Liu, Y., Balaraman, Y., Wang, G., Nephew, K.P., Zhou, F.C., 2009. Alcohol exposure alters DNA methylation profiles in mouse embryos at early neurulation. Epigenetics 4, 500–511.

Livy, D.J., Maier, S.E., West, J.R., 2001. Fetal alcohol exposure and temporal vulnerability: Effects of binge-like alcohol exposure on the ventrolateral nucleus of the thalamus. Alcoholism, Clinical and Experimental Research 25, 774–780.

Lledo, P.M., Merkle, F.T., Alvarez-Buylla, A., 2008. Origin and function of olfactory bulb interneuron diversity. Trends in Neurosciences 31, 392–400.

Lu, J., Wu, Y., Sousa, N., Almeida, O.F., 2005. SMAD pathway mediation of BDNF and TGFβ2 regulation of proliferation and differentiation of hippocampal granule neurons. Development 132, 3231–3242.

Lugli, G., Krueger, J.M., Davis, J.M., Persico, A.M., Keller, F., Smalheiser, N.R., 2003. Methodological factors influencing measurement and processing of plasma reelin in humans. BMC Biochemistry 4, 9.

Luo, J., Miller, M.W., 1997. Basic fibroblast growth factor- and platelet-derived growth factor-mediated cell proliferation in B104 neuroblastoma cells: Effect of ethanol on cell cycle kinetics. Brain Research 770, 139–150.

Luo, J., Miller, M.W., 1999a. Transforming growth factor β1 (TGFβ1) mediated inhibition of B104 neuroblastoma cell proliferation and neural cell adhesion molecule expression: Effect of ethanol. Journal of Neurochemistry 72, 2286–2293.

Luo, J., Miller, M.W., 1999b. Platelet-derived growth factor (PDGF) mediated signal transduction underlying astrocyte proliferation: Site of ethanol action. Journal of Neuroscience 19, 10014–10025.

Lynch, S.A., Elton, C.W., Melinda Carver, F., Pennington, S.N., 2001. Alcohol-induced modulation of the insulin-like growth factor system in early chick embryo cranial tissue. Alcoholism, Clinical and Experimental Research 25, 755–763.

Maffi, S.K., Rathinam, M.L., Cherian, P.P., et al., 2008. Glutathione content as a potential mediator of the vulnerability of cultured fetal cortical neurons to ethanol-induced apoptosis. Journal of Neuroscience Research 86, 1064–1076.

Maisonpierre, P.C., Belluscio, L., Friedman, B., et al., 1990. NT-3, BDNF, and NGF in the developing rat nervous system: Parallel as well as reciprocal patterns of expression. Neuron 5, 501–509.

Marcussen, B.L., Goodlett, C.R., Mahoney, J.C., West, J.R., 1994. Developing rat Purkinje cells are more vulnerable to alcohol-induced depletion during differentiation than during neurogenesis. Alcohol 11, 147–156.

Mattson, S.N., Riley, E.P., Sowell, E.R., Jernigan, T.L., Sobel, D.F., Jones, K.L., 1996. A decrease in the size of the basal ganglia in children with fetal alcohol syndrome. Alcoholism, Clinical and Experimental Research 20, 1088–1093.

Mauceri, H.J., Unterman, T., Dempsey, S., Lee, W.H., Conway, S., 1993. Effect of ethanol exposure on circulating levels of insulin-like growth factor I and II, and insulin-like growth factor binding proteins in fetal rats. Alcoholism, Clinical and Experimental Research 17, 1201–1206.

McConnell, S.K., 1988. Fates of visual cortical neurons in the ferret after isochronic and heterochronic transplantation. Journal of Neuroscience 8, 945–974.

McGough, N.N., Thomas, J.D., Dominguez, H.D., Riley, E.P., 2009. Insulin-like growth factor-I mitigates motor coordination deficits associated with neonatal alcohol exposure in rats. Neurotoxicology and Teratology 31, 40–48.

Meyer, L.S., Riley, E.P., 1986. Social play in juvenile rats prenatally exposed to alcohol. Teratology 34, 1–7.

Miller, M.W., 1986. Effects of alcohol on the generation and migration of cerebral cortical neurons. Science 233, 1308–1311.

Miller, M.W., 1987. Effect of prenatal exposure to ethanol on the distribution and time of origin of corticospinal neurons in the rat. The Journal of Comparative Neurology 257, 372–382.

Miller, M.W., 1988a. Effect of prenatal exposure to ethanol on the development of cerebral cortex: I. Neuronal generation. Alcoholism, Clinical and Experimental Research 12, 440–449.

Miller, M.W., 1988b. Development of projection and local circuit neurons in cerebral cortex. In: Peters, A., Jones, E.G. (Eds.), Cerebral Cortex. Development and Maturation of Cerebral Cortex7, Plenum, New York, pp. 133–175.

Miller, M.W., 1989. Effect of prenatal exposure to ethanol on the development of cerebral cortex: II. Cell proliferation in the ventricular and subventricular zones of the rat. The Journal of Comparative Neurology 287, 326–338.

Miller, M.W., 1992. Circadian rhythm of cell proliferation in the telencephalic ventricular zone: Effect of *in utero* exposure to ethanol. Brain Research 595, 17–24.

Miller, M.W., 1993. Migration of cortical neurons is altered by gestational exposure to ethanol. Alcoholism, Clinical and Experimental Research 17, 304–314.

Miller, M.W., 1995a. Generation of neurons in the rat dentate gyrus and hippocampus: Effects of prenatal and postnatal treatment with ethanol. Alcoholism, Clinical and Experimental Research 19, 1500–1509.

Miller, M.W., 1995b. Effect of pre- or postnatal exposure to ethanol on the total number of neurons in the principal sensory nucleus of the trigeminal nerve: Cell proliferation versus neuronal death. Alcoholism, Clinical and Experimental Research 19, 1359–1364.

Miller, M.W., 1995c. Relationship of time of origin and death of neurons in rat somatosensory cortex: Barrel versus septal cortex and projection versus local circuit neurons. The Journal of Comparative Neurology 355, 6–14.

Miller, M.W., 1996. Limited ethanol exposure selectively alters the proliferation of precursors cells in cerebral cortex. Alcoholism, Clinical and Experimental Research 20, 139–143.

Miller, M.W., 1997. Effects of prenatal exposure to ethanol on callosal projection neurons in rat somatosensory cortex. Brain Research 766, 121–128.

Miller, M.W., 1999. Kinetics of the migration of neurons to rat somatosensory cortex. Developmental Brain Research 115, 111–122.

Miller, M.W., 2003. Expression of transforming growth factor β (TGFβ) in developing rat cerebral cortex: Effects of prenatal exposure to ethanol. The Journal of Comparative Neurology 460, 410–424.

Miller, M.W., 2006a. Brain Development. Normal Processes and the Effects of Alcohol and Nicotine. Oxford University Press, New York.

Miller, M.W., 2006b. Early exposure to ethanol affects the proliferation of neuronal precursors. In: Miller, M.W. (Ed.), Normal Processes and the Effects of Alcohol and Nicotine. Oxford University Press, New York, pp. 182–198.

Miller, M.W., Al-Rabiai, S., 1994. Effects of prenatal exposure to ethanol on the number of axons in the pyramidal tract of the rat. Alcoholism, Clinical and Experimental Research 18, 346–354.

Miller, M.W., Dow-Edwards, D.L., 1988. Structural and metabolic alterations in rat cerebral cortex induced by prenatal exposure to ethanol. Brain Research 474, 316–326.

Miller, M.W., Dow-Edwards, D.L., 1993. Vibrissal stimulation affects glucose utilization in the rat trigeminal/somatosensory system both in normal rats and in rats prenatally exposed to ethanol. The Journal of Comparative Neurology 335, 283–294.

Miller, M.W., Luo, J., 2002a. Effects of ethanol and basic fibroblast growth factor (bFGF) on the transforming growth factor β1 (TGFβ1) regulated proliferation of cortical astrocytes and C6 astrocytoma cells. Alcoholism, Clinical and Experimental Research 26, 671–675.

Miller, M.W., Luo, J., 2002b. Effects of ethanol and transforming growth factor β (TGFβ) on neuronal proliferation and nCAM expression. Alcoholism, Clinical and Experimental Research 26, 1073–1079.

Miller, M.W., Muller, S.J., 1989. Structure and histogenesis of the principal sensory nucleus of the trigeminal nerve: Effects of prenatal exposure to ethanol. The Journal of Comparative Neurology 282, 570–580.

Miller, M.W., Nowakowski, R.S., 1991. Effect of prenatal exposure to ethanol on the cell cycle kinetics and growth fraction in the proliferative zones of fetal rat cerebral cortex. Alcoholism, Clinical and Experimental Research 15, 229–232.

Miller, M.W., Potempa, G., 1990. Numbers of neurons and glia in mature rat somatosensory cortex: Effects of prenatal exposure to ethanol. The Journal of Comparative Neurology 293, 92–102.

Miller, M.W., Chiaia, N.L., Rhoades, R.W., 1990. Intracellular recording and injection study of corticospinal neurons in the rat somatosensory cortex: Effect of prenatal exposure to ethanol. The Journal of Comparative Neurology 297, 91–105.

Miller, T.M., Moulder, K.L., Knudson, C.M., et al., 1997. Bax deletion further orders the cell death pathway in cerebellar granule cells and suggests a caspase-independent pathway to cell death. J Cell Biol 139, 205–217.

Miller, F.D., Pozniak, C.D., Walsh, G.S., 2000. Neuronal life and death: An essential role for the p53 family. Cell Death and Differentiation 7, 880–888.

Miller, M.W., Peters, A., Wharton, S.B., Wyllie, A.H., 2003. Effects of p53 on the proliferation and survival of conditionally immortalized murine neural cells. Brain Research 965, 57–66.

Miller, M.W., Mooney, S.M., Middleton, F.A., 2006. Transforming growth factor β1 and ethanol affect transcription of genes for cell adhesion proteins in B104 neuroblastoma cells. Journal of Neurochemistry 97, 1182–1190.

Miller, M.W., Spear, L.P., 2006. The alcoholism generator. Alcoholism, Clinical and Experimental Research 30, 1466–1469.

Molina, J.C., Spear, N.E., Spear, L.P., Mennella, J.A., Lewis, M.J., 2007. The international society for developmental psychobiology 39th annual meeting symposium: Alcohol and development: Beyond fetal alcohol syndrome. Developmental Psychobiology 49, 227–242.

Monje, M.L., Mizumatsu, S., Fike, J.R., Palmer, T.D., 2002. Irradiation induces neural precursor-cell dysfunction. Nature Medicine 8, 955–962.

Monje, M.L., Toda, H., Palmer, T.D., 2003. Inflammatory blockade restores adult hippocampal neurogenesis. Science 302, 1760–1765.

Moonat, S., Starkman, B.G., Sakharkar, A., Pandey, S.C., 2010. Neuroscience of alcoholism: Molecular and cellular mechanisms. Cellular and Molecular Life Sciences 67, 73–88.

Mooney, S.M., Henderson, G.I., 2006. Intracellular pathways of neuronal death. In: Miller, M.W. (Ed.), Brain Development. Normal Processes and the Effects of Alcohol and Nicotine. Oxford University Press, New York, pp. 91–103.

Mooney, S.M., Miller, M.W., 1999. Effects of prenatal exposure to ethanol on systems matching: The number of neurons in the ventrobasal thalamic nucleus of the mature rat thalamus. Developmental Brain Research 117, 121–125.

Mooney, S.M., Miller, M.W., 2001. Effects of prenatal exposure to ethanol on the expression of bcl-2, bax and caspase 3 in the developing rat cerebral cortex and thalamus. Brain Research 911, 71–81.

Mooney, S.M., Miller, M.W., 2003. Ethanol-induced neuronal death in organotypic cultures of rat cerebral cortex. Developmental Brain Research 147, 135–141.

Mooney, S.M., Miller, M.W., 2007a. Postnatal neuronogenesis in the thalamic ventrobasal nucleus. Journal of Neuroscience 27, 5023–5032.

Mooney, S.M., Miller, M.W., 2007b. Nerve growth factor protects against ethanol-induced neuronal death in organotypic slice cultures of rat cerebral cortex. Neuroscience 149, 372–381.

Mooney, S.M., Miller, M.W., 2007c. Time-specific effects of ethanol exposure on cranial nerve nuclei: gastrulation and neuronogenesis. Exp Neurol 205, 56–63.

Mooney, S.M., Miller, M.W., 2010. Effect of ethanol on postnatal neuronogenesis in the ventrobasal thalamus. Experimental Neurology 223, 566–573.

Mooney, S.M., Miller, M.W., 2011. Role of neurotrophins in postnatal neurogenesis in thalamus. Neuroscience 179, 256–266.

Mooney, S.M., Varlinskaya, E.I., 2011. Acute prenatal exposure to ethanol and social behavior: Effects of age, sex, and timing of exposure. Behavioural Brain Research 216, 358–364.

Mooney, S.M., Youngentob, S.L., Varlinskaya, E.I., 2009. Behavioral effects of acute ethanol exposure to ethanol are time-limited. Alcoholism, Clinical and Experimental Research 33, 35A.

Moore, D.B., Walker, D.W., Heaton, M.B., 1999. Neonatal ethanol exposure alters bcl-2 family mRNA levels in the rat cerebellar vermis. Alcoholism: Clinical and Experimental Research 23, 1251–1261.

Moore, S.J., Turnpenny, P., Quinn, A., et al., 2000. A clinical study of 57 children with fetal anticonvulsant syndromes. Journal of Medical Genetics 37, 489–497.

Nakatomi, H., Kuriu, T., Okabe, S., et al., 2002. Regeneration of hippocampal pyramidal neurons after ischemic brain injury by recruitment of endogenous neural progenitors. Cell 110, 429–441.

Nieto, M., Monuki, E.S., Tang, H., et al., 2004. Expression of Cux-1 and Cux-2 in the subventricular zone and upper layers II-IV of the cerebral cortex. The Journal of Comparative Neurology 479, 168–180.

Nixon, K., Crews, F.T., 2002. Binge ethanol exposure decreases neurogenesis in adult rat hippocampus. Journal of Neurochemistry 83, 1087–1093.

Nowoslawski, L., Klocke, B.J., Roth, K.A., 2005. Molecular regulation of acute ethanol-induced neuron apoptosis. Journal of Neuropathology and Experimental Neurology 64, 490–497.

Oberlander, T.F., Weinberg, J., Papsdorf, M., Grunau, R., Misri, S., Devlin, A.M., 2008. Prenatal exposure to maternal depression, neonatal methylation of human glucocorticoid receptor gene (NR3C1) and infant cortisol stress responses. Epigenetics 3, 97–106.

Olney, J.W., Tenkova, T., Dikranian, K., Qin, Y.Q., Labruyere, J., Ikonomidou, C., 2002a. Ethanol-induced apoptotic neurodegeneration in the developing C57BL/6 mouse brain. Developmental Brain Research 133, 115–126.

Olney, J.W., Tenkova, T., Dikranian, K., et al., 2002b. Ethanol-induced caspase-3 activation in the in vivo developing mouse brain. Neurobiology of Disease 9, 205–219.

Pandey, S.C., Ugale, R., Zhang, H., Tang, L., Prakash, A., 2008. Brain chromatin remodeling: A novel mechanism of alcoholism. Journal of Neuroscience 28, 3729–3737.

Pepino, M.Y., Mennella, J.A., 2007. Effects of cigarette smoking and family history of alcoholism on sweet taste perception and food cravings in women. Alcoholism, Clinical and Experimental Research 31, 1891–1899.

Pfeiffer, J., Majewski, F., Fischbach, H., Bierich, J.R., Volk, B., 1979. Alcohol embryo- and fetopathy. Neuropathology of 3 children and 3 fetuses. Journal of Neurological Sciences 41, 125–137.

Pierce, D.R., Williams, D.K., Light, K.E., 1999. Purkinje cell vulnerability to developmental ethanol exposure in the rat cerebellum. Alcoholism, Clinical and Experimental Research 23, 1650–1659.

Pitts, A.F., Miller, M.W., 2000. Expression of nerve growth factor, brain-derived neurotrophic factor, and neurotrophin-3 in the somatosensory cortex of the rat: Co-expression with high affinity neurotrophin receptors. The Journal of Comparative Neurology 418, 241–254.

Pontious, A., Kowalczyk, T., Englund, C., Hevner, R.F., 2008. Role of intermediate progenitor cells in cerebral cortex development. Developmental Neuroscience 30, 24–32.

Powrozek, T.A., Miller, M.W., 2009. Ethanol affects transforming growth factor beta 1-initiated signals: Cross-talking pathways in the developing rat cerebral wall. Journal of Neuroscience 29, 9521–9533.

Powrozek, T.A., Zhou, F.C., 2005. Effects of prenatal alcohol exposure on the development of the vibrissal somatosensory cortical barrel network. Developmental Brain Research 155, 135–146.

Prock, T.L., Miranda, R.C., 2007. Embryonic cerebral cortical progenitors are resistant to apoptosis, but increase expression of suicide receptor DISC-complex genes and suppress autophagy following ethanol exposure. Alcoholism, Clinical and Experimental Research 31, 694–703.

Ramachandran, V., Perez, A., Chen, J., Senthil, D., Schenker, S., Henderson, G.I., 2001. In utero ethanol exposure causes mitochondrial dysfunction, which can result in apoptotic cell death in fetal brain: A potential role for 4-hydroxynonenal. Alcoholism, Clinical and Experimental Research 25, 862–871.

Ramachandran, V., Talley Watts, L., Maffi, S.K., Chen, J.J., Schenker, S., Henderson, G.I., 2003. Ethanol-induced oxidative stress precedes mitochondrially mediated apoptotic death of cultured fetal cortical neurons. Journal of Neuroscience Research 74, 577–588.

Rao, M., Jacobson, M.S., 2005. Developmental Neurobiology. Kluwer-Plenum, New York.

Reddy, U.R., Pleasure, D., 1992. Expression of platelet-derived growth factor (PDGF) and PDGF receptor genes in the developing nervous rat brain. Journal of Neuroscience Research 31, 670–677.

Redila, V.A., Olson, A.K., Swann, S.E., et al., 2006. Hippocampal cell proliferation is reduced following prenatal ethanol exposure but can be rescued with voluntary exercise. Hippocampus 16, 305–311.

Reif, A., Fritzen, S., Finger, M., et al., 2006. Neural stem cell proliferation is decreased, but not in depression. Molecular Psychiatry 11, 514–522.

Resnicoff, M., Rubini, M., Baserga, R., Rubin, R., 1994. Ethanol inhibits insulin-like growth factor-1-mediated signalling and proliferation of C6 rat glioblastoma cells. Laboratory Investigation 71, 657–662.

Riley, E.P., Mattson, S.N., Sowell, E.R., Jernigan, T.L., Sobel, D.F., Jones, K.L., 1995. Abnormalities of the corpus callosum in children prenatally exposed to alcohol. Alcoholism, Clinical and Experimental Research 19, 1198–1202.

Romanko, M.J., Rothstein, R.P., Levison, S.W., 2004. Neural stem cells in the subventricular zone are resilient to hypoxia/ischemia whereas progenitors are vulnerable. Journal of Cerebral Blood Flow and Metabolism 24, 814–824.

Royalty, J., 1990. Effects of prenatal ethanol exposure on juvenile play-fighting and postpubertal aggression in rats. Psychology Reports 66, 551–560.

Rubin, R., Baserga, R., 1995. Insulin-like growth factor-I receptor. Its role in cell proliferation, apoptosis, and tumorigenicity. Laboratory Investigation 73, 311–331.

Salisbury, A.L., Ponder, K.L., Padbury, J.F., Lester, B.M., 2009. Fetal effects of psychoactive drugs. Clinics in Perinatology 36, 595–619.

Santillano, D.R., Kumar, L.S., Prock, T.L., Camarillo, C., Tingling, J.D., Miranda, R.C., 2005. Ethanol induces cell-cycle activity and reduces stem cell diversity to alter both regenerative capacity and differentiation potential of cerebral cortical neuroepithelial precursors. BMC Neuroscience 6, 59.

Schlotz, W., Phillips, D.I., 2009. Fetal origins of mental health: Evidence and mechanisms. Brain, Behavior, and Immunity 23, 905–916.

Schneider, M.L., Moore, C.F., Adkins, M.M., 2011. The effects of prenatal alcohol exposure on behavior: rodent and primate studies. Neuropsychol Rev 21, 186–203.

Seabold, G., Luo, J., Miller, M.W., 1998. Effect of ethanol on neurotrophin-mediated cell survival and receptor expression in cortical neuronal cultures. Brain Research 128, 139–145.

Selznick, L.A., Zheng, T.S., Flavell, R.A., Rakic, P., Roth, K.A., 2000. Amyloid beta-induced neuronal death is bax-dependent but caspase-independent. Journal of Neuropathology and Experimental Neurology 59, 271–279.

Shors, T.J., Miesegaes, G., Beylin, A., Zhao, M., Rydel, T., Gould, E., 2001. Neurogenesis in the adult is involved in the formation of trace memories. Nature 410, 372–376.

Siegenthaler, J.A., Miller, M.W., 2004. Transforming growth factor β1 regulates cell migration in rat cortex: Effects of ethanol. Cerebral Cortex 14, 602–613.

Siegenthaler, J.A., Miller, M.W., 2005a. Transforming growth factor β1 promotes cell cycle exit through the cyclin-dependent kinase inhibitor p21 in the developing cerebral cortex. Journal of Neuroscience 21, 8627–8636.

Siegenthaler, J.A., Miller, M.W., 2005b. Ethanol disrupts cell cycle regulation in developing rat cortex: Interaction with transforming growth factor β1. Journal of Neurochemistry 95, 902–912.

Singh, S.P., Srivenugopal, K.S., Ehmann, S., Yuan, X.H., Snyder, A.K., 1994. Insulin-like growth factors (IGF-I and IGF-II), IGF-binding proteins, and IGF gene expression in the offspring of ethanol-fed rats. The Journal of Laboratory and Clinical Medicine 124, 183–192.

Sliwowska, J.H., Zhang, X., Weinberg, J., 2006. Prenatal ethanol exposure and fetal programming: Implications for endocrine and immune development and long-term health. In: Miller, M.W. (Ed.), Normal Processes and the Effects of Alcohol and Nicotine. Oxford University Press, New York, pp. 153–181.

Soengas, M.S., Alarcon, R.M., Yoshids, H., et al., 1999. Apaf-1 and caspase-9 in p53-dependent apoptosis and tumor inhibition. Science 284, 156–159.

Soscia, S.J., Tong, M., Xu, X.J., et al., 2006. Chronic gestational exposure to ethanol causes insulin and IGF resistance and impairs acetylcholine homeostasis in the brain. Cellular and Molecular Life Sciences 63, 2039–2056.

Sowell, E.R., Jernigan, T.L., Mattson, S.N., Riley, E.P., Sobel, D.F., Jones, K.L., 1996. Abnormal development of the cerebellar vermis in children prenatally exposed to alcohol: Size reduction in lobules I–V. Alcoholism, Clinical and Experimental Research 20, 31–34.

Stefanis, L., Parks, D.S., Friedman, W.J., Green, L.A., 1999. Caspase-dependent and independent death of camptothecin-treated embryonic cortical neurons. Journal of Neuroscience 19, 6235–6247.

Stratton, K., Howe, C., Battaglia, F., 1996. Fetal Alcohol Syndrome. Diagnosis, Epidemiology, Prevention, and Treatment. National Academy, Washington, DC.

Sulik, K.K., 2005. Genesis of alcohol-induced craniofacial dysmorphism. Experimental Biology and Medicine 230, 366–375.

Sulik, K.K., Johnston, M.C., Webb, M.A., 1981. Fetal alcohol syndrome: Embryogenesis in a mouse model. Science 214, 936–938.

Swayze 2nd, V.W., Johnson, V.P., Hanson, J.W., et al., 1997. Magnetic resonance imaging of brain anomalies in fetal alcohol syndrome. Pediatrics 99, 232–240.

Tarabykin, V., Stoykova, A., Usman, N., Gruss, P., 2001. Cortical upper layer neurons derive from the subventricular zone as indicated by Svet1 gene expression. Development 128, 1983–1993.

Tateno, M., Ukai, W., Yamamoto, M., Hashimoto, E., Ikeda, H., Saito, T., 2005. The effect of cell fate determination of neural stem cells. Alcoholism, Clinical and Experimental Research 29, 225S–229S.

ten Dijke, P., Hill, C.S., 2004. Emerging roles for TGFβ1 in nervous system development. Trends in Biochemical Sciences 29, 265–273.

Thomas, J.D., Garrison, M., O'Neill, T.M., 2004. Perinatal choline supplementation attenuates behavioral alterations associated with neonatal alcohol exposure in rats. Neurotoxicology and Teratology 26, 35–45.

Thomas, J.D., Idrus, N.M., Monk, B.R., Dominguez, H.D., 2010. Prenatal choline supplementation mitigates behavioral alterations associated with prenatal alcohol exposure in rats. Birth Defects Research 88, 827–837.

Thomas, S.E., Kelly, S.J., Mattson, S.N., Riley, E.P., 1998. Comparison of social abilities of children with fetal alcohol syndrome to those of children with similar IQ scores and normal controls. Alcoholism, Clinical and Experimental Research 22, 528–533.

Toesca, A., Giannetti, S., Granato, A., 2003. Overexpression of the p75 neurotrophin receptor in the sensori-motor cortex of rats exposed to ethanol during early postnatal life. Neuroscience Letters 342, 89–92.

Valenzuela, C.F., Kazlauskas, A., Weiner, J.L., 1997. Roles of platelet-derived growth factor and its receptor in developing and mature nervous system. Brain Research Reviews 24, 77–89.

Valles, S., Lindo, L., Montoliu, C., Renau-Piqueras, J., Guerri, C., 1994. Prenatal exposure to ethanol induces changes in the nerve growth factor and its receptor in proliferating astrocytes in primary culture. Brain Research 656, 281–286.

van Praag, H., Shubert, T., Zhao, C., Gage, F.H., 2005. Exercise enhances learning and hippocampal neurogenesis in aged mice. Journal of Neuroscience 25, 8680–8685.

Vanderschuren, L.J., Niesink, R.J., Van Ree, J.M., 1997. The neurobiology of social play behavior in rats. Neuroscience and Biobehavioral Reviews 21, 309–326.

Vangipuram, S.D., Lyman, W.D., 2010. Ethanol alters cell fate of fetal human brain-derived stem and progenitor cells. Alcoholism, Clinical and Experimental Research 34, 1574–1583.

Vaudry, D., Rousselle, C., Basille, M., et al., 2002. Pituitary adenylate cyclase-activating polypeptide protects rat cerebellar granule neurons against ethanol-induced apoptotic cell death. Proceedings of the National Academy of Sciences of the United States of America 99, 6398–6403.

Vingan, R.D., Dow-Edwards, D.L., Riley, E.P., 1986. Cerebral metabolic alterations in rats following prenatal alcohol exposure: A deoxyglucose study. Alcoholism, Clinical and Experimental Research 10, 22–26.

Vitalis, T., Cases, O., Gillies, K., et al., 2002. Interactions between TrkB signaling and serotonin excess in the developing murine somatosensory cortex: A role in tangential and radial organization of thalamocortical axons. Journal of Neuroscience 22, 4987–5000.

Wang, L., Zhang, Z., Wang, Y., Zhang, R., Chopp, M., 2004. Treatment of stroke with erythropoietin enhances neurogenesis and angiogenesis and improves neurological function in rats. Stroke 35, 1732–1737.

Wang, H., Zhou, H., Moscatello, K.M., et al., 2006. In utero exposure to alcohol alters cell fate decisions by hematopoietic progenitors in the bone marrow of offspring mice during neonatal development. Cellular Immunology 239, 75–85.

Wang, H., Zhou, H., Chervenak, R., et al., 2009. Ethanol exhibits specificity in its effects on differentiation of hematopoietic progenitors. Cellular Immunology 255, 1–7.

West, J.R., Hodges, C.A., Black Jr., A.C., 1981. Distal infrapyramidal granule cell axons possess typical mossy fiber morphology. Brain Research Bulletin 6, 119–124.

Whitman, M.C., Greer, C.A., 2009. Adult neurogenesis and the olfactory system. Progress in Neurobiology 89, 162–175.

Wisniewski, K., Dambska, M., Sher, J.H., Qazi, Q., 1983. A clinical neuropathological study of the fetal alcohol syndrome. Neuropediatrics 14, 197–201.

Wyllie, A.H., 1997. Apoptosis: An overview. British Medical Bulletin 53, 451–465.

Xie, N., Yang, Q., Chappell, T.D., Cx, Li, Waters, R.S., 2010. Prenatal alcohol exposure reduces the size of the forelimb representation in motor cortex in rat: An intracortical microstimulation (ICMS) mapping study. Alcohol 44, 185–194.

Young, C., Klocke, B.J., Tenkova, T., et al., 2003. Ethanol-induced neuronal apoptosis *in vivo* requires Bax in the developing mouse brain. Cell Death and Differentiation 10, 1148–1155.

Young, C., Roth, K.A., Klocke, B.J., et al., 2005. Role of caspase-3 in ethanol-induced developmental neurodegeneration. Neurobiology of Disease 20, 608–614.

Youngentob, S.L., Glendinning, J.I., 2009. Fetal ethanol exposure increases ethanol intake by making it smell and taste better. Proceedings of the National Academy of Sciences of the United States of America 106, 5359–5364.

Youngentob, S.L., Kent, P.F., Sheehe, P.R., Molina, J.C., Spear, N.E., Youngentob, L.M., 2007. Experience-induced fetal plasticity: The effect of gestational ethanol exposure on the behavioral and neurophysiologic olfactory response to ethanol odor in early postnatal and adult rats. Behavioral Neuroscience 121, 1293–1305.

Zawada, W.M., Das, M., 2006. Effects of ethanol on the regulation if cell cycle in neural stem cells. In: Miller, M.W. (Ed.), Normal Processes and the Effects of Alcohol and Nicotine. Oxford University Press, New York, pp. 199–215.

Zhang, X., Sliwowska, J.H., Weinberg, J., 2005. Prenatal alcohol exposure and fetal programming: Effects on neuroendocrine and immune function. Experimental Biology and Medicine 230, 376–388.

Zhou, F.C., Balaraman, Y., Teng, M.X., Liu, Y., Singh, R.P., Nephew, K.P., 2011. Alcohol alters DNA methylation patterns and inhibits neural stem cell differentiation. Alcoholism, Clinical and Experimental Research 35, 735–746.

Azetidine-2-Carboxylic Acid and Other Nonprotein Amino Acids in the Pathogenesis of Neurodevelopmental Disorders

E. Rubenstein

Stanford University School of Medicine, Stanford, CA, USA

29.1 INTRODUCTION

There are approximately 900 naturally occurring amino acids, of which only 22 have been evolutionarily selected for inclusion as subunits in the structure of proteins (Rosenthal and Bell, 1979; Rubenstein, 2000). The others, nonprotein amino acids, are found principally in plants and lower marine forms in which they participate in metabolism and provide a means of nitrogen storage. In addition, many are deployed as toxic weapons, deterring intruding vegetation and marauding predators.

Some amino acids have entered the food chain. Notable among this group is azetidine-2-carboxylic acid (Aze), which is the lower homolog of the protein amino acid proline. It readily escapes the gate-keeping function of prolyl aminoacyl-tRNA synthetases and replaces proline in proteins, corrupting their structure, function, and antigenicity. Aze may be an environmental agent that contributes to susceptibility to a wide range of disorders, especially those involving the central nervous system (Rubenstein, 2008).

Proline is an anomalous amino acid (Figure 29.1(c)). Its nitrogen atom is linked to only one hydrogen atom instead of two. It cannot act as a hydrogen bond donor but can serve as an acceptor. The nitrogen is covalently bound to the ring structure, constraining its mobility in peptide bond formation. The φ backbone dihedral angle is about $-75°$, and proline displays conformational rigidity. Unlike other amino acids, which prefer the *trans* configuration of the hydrogen and oxygen atoms in the peptide bond, proline readily shifts to the *cis* architecture in response to regional changes in spatial charge distribution. Lacking a hydrogen atom, the proline molecule isomerizes by swinging its entire ring into the *cis* position, a conformational change that realigns the local peptide sequence, resulting in folding of the protein (Baldwin, 2008; Smith et al., 2012).

Neural Circuit Development and Function in the Brain: Comprehensive Developmental Neuroscience, Volume 3 http://dx.doi.org/10.1016/B978-0-12-397267-5.00220-X

© 2013 Elsevier Inc. All rights reserved.

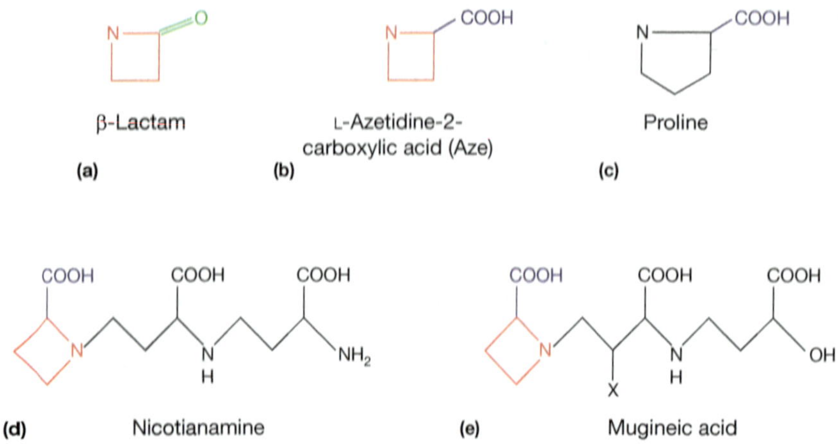

FIGURE 29.1   Small-ring molecules. This figure shows the structures of four-membered rings as free molecules: β-lactam (a), L-azetidine-2-carboxylic acid (Aze) (b), and as constituents of the two plant chelator molecules, nicotianamine (d) and mugineic acid (e). Note the similarity of β-lactam (a) and L-azetidine-2-carboxylic acid (Aze) (b). Proline (c) is the five-membered ring higher homologue of Aze (b). The four-membered rings are red, the ring acyl group is green, and the ring carboxyl groups are blue.

Because of its idiosyncrasies as an amino acid, proline is not found within alpha helices or beta sheets. It can take a position at the ends of these structures and in turns. Repeating prolyl residues may fold into a left-handed helix (PPII), with backbone dihedral angles of about $-75°$, $150°$, and a *trans* configuration of their peptide bonds. A right-handed helix (PPI) forms with dihedral angles of about $-75°$ and $160°$ when there is a *cis* configuration of their peptide bonds.

Aze, bearing a ring that contains four members instead of five, is otherwise nearly identical to proline (Figure 29.1(b)). There are, of course, subtle differences in bond angles, including those that determine the degree of ring puckering. Azetidine changes its *cis–trans* configuration under slightly different conditions of local charge, and its presence can thus cause misfolding (Tsai et al., 1990).

29.1.1 Ring Strain and Reactivity

Biomolecules comprise atoms in chains that tend to adopt a linear configuration in which their mutually repelling electron clouds maintain maximal distance from each other. This is not possible in small-ring compounds. The conformation of such cyclic molecules subjects them to various forms of strain or structural stress. The increased internal energy acquired during synthesis is released during chemical transactions (Hassner, 1983). Crowding can result in narrowing of preferred bond angles, compression of electron orbitals, and malpositioning of ring members. For these reasons, strain energy is exceedingly high in small cyclic molecules, increasing their chemical reactivity.

Molecules with five or more rings abound in biochemistry, but there appear to be only two four-membered ring compounds that have entered human life, and both have done so as exogenous substances: beta-lactam and Aze.

Beta-lactam closely resembles Aze. It is a cyclic amide with an acyl group on a carbon adjacent to the nitrogen (Figure 29.1(a)). It is the central component of the penicillins. The penicillin molecule binds to and inactivates the bacterial cell wall transpeptidase enzyme that forms an acyl intermediate with a D-alanyl-D-alanyl terminal peptide residue. Penicillin intrudes into the active site of the enzyme, where its beta-lactam impersonates the structure of the normal substrate. It irreversibly inhibits the enzyme, thus preventing cross-linking of the bacterial cell wall (Stryer, 1988). The ring strain of the four-membered beta-lactam contributes to its remarkable reactivity (Hassner, 1983).

Aze is strikingly similar to beta-lactam. Instead of an acyl group, it contains a carboxyl group on ring atom number 2. Like beta-lactam, Aze is highly strained and is thus highly reactive.

29.1.2 The Aminoacyl-tRNA Prolyl Synthetases

There are two aminoacyl-tRNA synthetases that recognize proline, EPRS (cytoplasmic) and PARS2 (nucleus encoded but mitochondrial). EPRS is multifunctional, recognizing glutamic acid as well as proline. The human forms of these synthetases lack editing domains. The chromosomal location of the gene *EPRS* is 1q41-q42 and that of *PARS2* is 1p32.2 (Antonellis and Green, 2008). Certain nonprotein amino acids escape the gate-keeping function of aminoacyl-tRNA synthetases and gain entry into proteins, replacing their homologs and changing the structure, function, and antigenicity of the misassembled molecule. The large number of disorders associated with synthetase dysfunction include the Charcot-Marie-Tooth disease, amyotrophic lateral sclerosis (ALS), leukoencephalopathy, Parkinson's disease, various neoplasms, autoimmune polymyositis, dermatomyositis, rheumatoid and erosive arthritis, and type 2 diabetes (Antonellis and Green, 2008; Beebe et al., 2008; Park et al., 2008).

29.1.3 Azetidine-2-Carboxylic Acid

Notable among nonprotein amino acids with disease-causing potential is azetidine-2-carboxylic acid (Aze), the lower homolog of proline that is identical to proline except in that it contains four instead of five members in its ring (Fowden, 1956). Aze is ubiquitously present in low concentrations in vegetation, where it is a central component of the chelating molecules, nicotianamine and mugineic acid, which capture metals from the soil and transfer them to various parts of the plants (Figure 29.1(d) and 29.1(e)) (Curie et al., 2009; von Wiren et al., 1999).

In certain species, Aze accumulates to exceedingly high levels, notably in the bulbous roots of table beets and sugar beets (*Beta vulgaris*). Such plants deploy the toxic Aze to deter the intrusion of competing vegetation and to poison infecting microorganisms and predating animals. Thus, Aze is present in the food chain via small concentrations in vegetables. However, it is in high concentrations in certain foods, especially in dairy products derived from livestock fed sugar beet by-products (Rubenstein, 2008).

29.1.4 Aze in the Food Chain

The entry of Aze into the food chain is a consequence of geopolitical events of the 19th and 20th centuries (Rubenstein, 2008). Previously sugar beets were not cultivated or used for human consumption. They grew wildly and were foraged by animals, notably red deer in Europe. The sugar beet is an unattractive, large, bulbous plant with a flavor offensive to humans. The root was occasionally used as a therapeutic agent for various illnesses. In 1747, the German apothecary chemist Andreas Margraff discovered low concentrations of sucrose in the root. During the following 50 years, his student and research associate, Franz Achard, succeeded in increasing the concentration of sugar several times over, and at present the sucrose concentration can be as high as 18%. The Napoleonic wars resulted in the blockades of Europe, which severed the supply of Caribbean sugar. The entire continent was deprived of its principal source of sweets for nearly a decade.

To quell unrest, Napoleon issued a decree in 1810 allotting 8000 acres of land near Paris to sugar beet production. He also ordered the construction of facilities to process sugar beets and organized schools to teach sugar beet agronomy. After a slow start, the sugar beet industry grew rapidly, and by the middle of the century it was flourishing. Since then, it has become the fourth leading agricultural enterprise in Europe and a major farming activity in both Canada and the United States. Sugar beets are now grown and processed virtually worldwide, including in Asia, the Middle East, and the Far East. At present, about one-third of the world's sucrose supply comes from sugar beets.

Sugar beet processing involves washing the plants, dicing them, and then boiling them in alkali. The sucrose is extracted after several rounds of distillation. The residue, consisting of plant parts and thick molasses and rich in Aze, is fed to farm animals, especially to dairy cattle, which savor its unusually sweet flavor. A dairy cow may be fed as much as 10 pounds of sugar beet molasses per day. This increases its consumption of grasses and grains so that its daily food intake supports the required volume of its milk production. The health and milk production of dairy cattle begin to falter after about 5 years, at which time they are slaughtered. It is this association between sugar beets and milk and meat production that has led to the introduction of Aze into milk, other dairy products, and, notably, gelatin (Rubenstein et al., 2009).

Aze is also present in high concentrations in table (garden) beets, which are a staple food in Central and Eastern Europe. They are the principal constituent of beet soup (borscht). Sliced beets are also a popular offering in salad bars in the United States (Rubenstein et al., 2006).

29.1.5 The Multiple Sclerosis/Aze Hypothesis

Five independent lines of evidence suggest an association between Aze consumption and the pathogenesis of multiple sclerosis. The central concept is that Aze intrusion into the proline-laden consensual epitope of myelin basic protein leads to the molecule's architectural modification, functional impairment, and eventual emergence as an antigen. Thereafter, immune mechanisms are triggered and progressively damage the myelin sheath.

Correlation does not establish causation; however, the chronologic, geographic, migrational, enzootic, biochemical, and genetic associations warrant further investigation (Bessonov et al., 2010a,b; Rubenstein, 2008; Sobel, 2008).

29.1.6 Chronologic and Geographic Evidence

The emergence of dietary Aze in high concentrations closely preceded in time and place Charcot's description of what appeared to be a new disease, multiple sclerosis. This was in Paris in 1868 (Talley, 2005). The geographic distribution of the disease thereafter coincided with that of sugar beet agriculture, extending at higher latitudes across Europe, North America, and then virtually worldwide, including Sardinia, the Middle East, Scandinavia, and the Orkney and Faroe Islands. Micro hot-spots, such as those in Alberta, Canada, and the Tokachi Province in Japan, add support to the geographical evidence.

Alberta is the epicenter of sugar beet production in Canada and has the highest prevalence of multiple sclerosis in that country. Tokachi province, a small region in northern Japan, produces half of that nation's sugar beet crop and has the highest prevalence of multiple sclerosis among any Asian population (Rubenstein, 2008).

29.1.7 Migrational Evidence

Myelination of the central nervous system begins in early life and continues through adolescence, a fact pertinent to the effect of population migration on the occurrence of multiple sclerosis (Kennedy et al., 2006). An example is the abrupt appearance of the disease in the French West Indies during the 1990s among returning West Indians who had emigrated to France during the 1950s and 1960s (Cabre et al., 2005).

The risk was highest for those who arrived in France before the age of 15. Their mean duration of stay was 12.3 years, and the mean interval between their arrival in France and the appearance of the disease was 19.1 years. This epidemiology recapitulates in mirror image the original emergence and spread of the disease (Rubenstein, 2008).

29.1.8 Enzootic Evidence

Swayback is an enzootic disease closely mimicking multiple sclerosis in clinical and histological characteristics. An outbreak in Alberta in 1972 killed 60 of 100 newborn lambs. An investigation into the circumstances surrounding this event led to the conclusion that the only unique feature was that the ewes and their offspring were fed an unusual diet containing large amounts of sugar beet silage (Cancilla and Barlow, 1969; Chalmers, 1974).

29.1.9 Biochemical Evidence

As already described, the misincorporation of Aze into a protein can result in alteration of protein structure, function, and immunogenicity (Bessonov et al., 2010a,b; Rubenstein, 2008). In the case of multiple sclerosis, the putative site of the misincorporation lies in the myelin basic protein (Boggs, 2006; Harauz et al., 2009). Specifically, the misincorporation would occur in the position of one or more of the four prolyls in the consensual epitopic sequence, which embraces the hexapeptide stretch PRTPPP (residues 96–101, human numbering). This highly conserved proline-rich motif is stochastically vulnerable to the deleterious effects of Aze replacement of proline.

Aze may intrude within other proline-rich molecules involved in myelination. An enormous body of meticulous research has contributed to our knowledge of myelin basic protein, its posttranslational modification, and its roles in signaling and protein–protein binding (Boggs, 2006; Harauz et al., 2009).

The random incursion of Aze into myelin basic protein and related ensemble myelin proteins may account for some of the anatomic, pathologic, and clinical heterogeneity of the disorder. Such misassembly could ultimately ignite or contribute to the progression of immune responses, which can have dire clinical consequences.

The metabolic burdens caused by the inherently error-prone processes of translation are compounded by exposure to an alien amino acid and by an impaired aminoacyl-tRNA synthetase mechanism. Defective editing causes the myriad pathologic effects already listed. The presence of extensive oligodendrocyte apoptosis, together with the absence of lymphocyte and myelin phagocytosis in newly formed multiple sclerosis lesions, supports the concept that autoimmunity plays a critical role in demyelination.

29.1.10 Genetic Evidence

The important role of genetics in the pathogenesis of multiple sclerosis has been authoritatively reviewed by Oksenberg et al. (2008). These considerations have led to a targeted genomic search for SNPs involving genes related to Aze. The work, conducted in collaboration with the laboratory of Dr. Ronald Davis at Stanford, is currently in progress.

29.1.11 Alternative Splicing and Amplification

Alien amino acids ordinarily affect only the molecule that has been misincorporated; however, the misassembly of molecules involved in alternative splicing can alter hundreds of different kinds of 'downstream' proteins that undergo the splicing process. Therefore, the misassembly of alternative splicing proteins can greatly amplify the molecular pathology induced by Aze.

A growing body of evidence has implicated the role of disordered alternative splicing in disease pathogenesis. This mechanism has been identified in a number of developmental disorders, especially those involving the nervous system. Recent reports have focused on the A2BP1 protein, which binds to the C-terminus of ataxin-2.

Ataxin-2 has been found to act as a coregulator of zinc-finger transcriptional activity (Hallen et al., 2011). Defective function of this molecular mechanism can lead to amplified abnormal splicing of a large number of proteins and thus to an exceptionally diverse group of disorders.

Prominent among the diseases associated with abnormal splicing are multiple abnormalities caused by defective neurodevelopment, including various forms of autism (Abrahams and Geschwind, 2008; Geschwind and Levitt, 2007; Hammock and Levitt, 2011; Martin et al., 2007; Sakai et al., 2011; Voineagu et al., 2011). Exposure to Aze-containing milk and other dairy products early in life may be of relevance, serving as an environmental susceptibility factor in autism spectrum disorders (see Chapter 34 and Rubenstein and Rakic, 2013).

Another neurodegenerative disorder that may be related to ataxin-2 and A2BP1 is ALS (Bonini and Gitler, 2011; Daoud et al., 2011; Lee et al., 2011; Van Damme et al., 2011). It is relevant to point out that the nonprotein amino acid β-*N*-methyl-amino-L-alanine (BMAA) has long been suspected to be an environmental susceptibility factor in ALS, initially in several Western Pacific islands. β-*N*-oxalylamino-L-alanine (BOAA) has been implicated in neurolathyrism endemic in certain regions of East Africa and Southern Asia. Canavanine, a homolog of arginine, is present in hundreds of legumes and in high concentrations in alfalfa seeds and sprouts. It has been suspected of playing a role in the pathogenesis of autoimmune disorders, including systemic lupus erythematosus (Cox et al., 2005; Rubenstein, 2000).

Other nonneurologic disorders associated with ataxin-2/A2BP1 include type 2 diabetes (Lehtinen et al., 2011). Similarly, obesity has been linked to the alternative splicing proteins (Ma et al., 2010). Biliary cirrhosis has also been associated with the same protein (Joshita et al., 2010).

29.1.12 Toxicity of Aze

A large number of early reports confirmed that Aze causes biochemical, morphological, and clinical defects. Among the biochemical abnormalities are misincorporation into vasopressin, hemoglobin, ovalbumin, and proline peptides. Inadvertent administration of Aze has resulted in butyric aminoaciduria. Morphologic abnormalities include abnormalities of somite formation, Golgi structure, osteoblasts, epithelial buds, cartilage, thyroid tissue, cleft palate, vertebral ossification, testes, keratin, hair, collagen, teeth, renal ducts, and mammary epithelium (Rubenstein, 2000; Tsai et al., 1990). Aze has recently been shown to be toxic to neurons and astrocytes in cell cultures, with associated increases in oxidized and ubiquitinated proteins (Dasuri et al., 2011).

Recent studies conducted by the Azetidine Group of collaborating investigators at the Stanford Medical Center have focused on the molecular details of misincorporation. This group includes M. Albertelli, M.J. Butte, R. Davis, J.E. Elias, M. Fontaine, K.V. Grimes, S. Krishnakumar, M. Mindrinos, G. Sonderstrup, R. Sobel, and E. Rubenstein. Some of this work has been summarized in the report by R.A. Sobel et al.

Aze has been found in milk samples from Boulder, CO. The use of sugar beets for dairy cattle feeding ceased in California about 3 years ago, and milk samples from California are currently free of Aze.

The administration (by oral gavage or intraperitoneal injection) of Aze in doses of 300 or 600 mg/kg to newborn CD-1 mice resulted in oligodendrocyte nuclear swelling and hepatic injury. Three of the five mice fed the larger dose by gavage were found dead or euthanized at 12–15 days. Three of the five mice in the 600-mg intraperitoneal group were euthanized or found dead after 20–24 days of Aze administration. The remaining mice in the 600-mg group and all the mice in the 300-mg group and control groups remained healthy. Mice that were euthanized early or found dead showed lethargy and weight loss; two also had tail paresis or paralysis. Some of the livers displayed mottled pale areas of hepatocyte vacuolization (Sobel et al., 2011).

Our collaborator in Canada, G. Harauz, has reported studies of myelin basic protein in recombinant *Escherichia coli*. These showed that Aze resulted in severe diminution in growth rate and that Aze was misincorporated in 3 of 11 possible sites in myelin basic protein. Molecular modeling led to the conclusion that Aze could cause a severe bend in the polypeptide chain and could disrupt a polyproline II structure (Bessonov et al., 2010a,b).

29.1.13 ATAXIN-2 and A2BP1 as Proline-Rich Proteins

These considerations have led to an analysis of compositional bias among the amino acids within ataxin-2 and A2BP1. The study revealed that ataxin-2 is exceptionally proline rich. Of its 1313 residues, a total of 177 are prolyls (13.4%). Positions 47–158 contain 30 prolyls among its 112 amino acids (26.8%), and within this sequence, prolyls comprise 7 of the 10 amino acids in positions 55–64. Among the 184 amino acids in positions 561–734, there are 44 prolyls (23.9%). Finally, there are 34 prolyls (21.7%) among the 157 amino acids in the sequence 929–1085 (UniProt ID Q 99700).

A2BP1 is less proline-rich, containing 40 prolyls among its 370 amino acids (10.8%). Noteworthy is the sequence PPPPIP, which occupies positions 281–286 (GenBank AA143817.1).

29.2 DISCUSSION

It is clear that Aze can displace proline in proteins and that such misassembly can have deleterious effects on protein structure, function, and immunogenicity. Whether Aze, introduced into the diet or into medicinal agents, has contributed to the shifting prevalence and incidence of clinical disorders is unknown. The complexities associated with an ever-changing biosphere and with population shifts constitute nearly chaotic conditions that challenge epidemiology.

Caution is required before inferences are drawn.

Is it possible that a simple compound, a small and apparently innocent amino acid found in plants, has become a susceptibility factor in the pathogenesis of a broad spectrum of diseases in which genetic predisposition plays a critically important role? Well-planned studies will be required to address this question. These must take into account such confounding factors as dose and time and duration of exposure, from early embryonic life through adulthood.

Some supporting evidence can be accumulated by demonstrating that Aze is in the environment of those affected and that Aze is in their tissues and in the specific cells that are damaged, for instance, in central nervous system cells in the case of neurologic disease. In most instances, this would require examination of autopsy material from patients and from controls. If a suitable animal model exists, the administration of Aze in appropriate doses at appropriate states of development should reproduce illness. Finally, the removal of Aze, perhaps by washout with pure proline, might prove beneficial.

The group at Stanford is testing the reactivity of fresh T cells from patients with various diseases to synthetic proteins in which Aze has replaced proline in differing combinations and permutations.

It is important to proceed with great caution in forming conclusions. It is essential to avoid the needlessly damaging consequences of inappropriate disclosures disseminated by the popular media. The public needs to be reminded periodically about the significance of environmental hazards and that even sunlight, oxygen, and water can be detrimental. It will require ingenuity, collaborative effort, and persistence to identify the environmental agent that may underlie the suffering experienced by millions.

Acknowledgments

This work has been supported by The Stanford Spark Program. Dairy samples from Colorado have been provided by M. Feder.

References

Abrahams, B.S., Geschwind, D.H., 2008. Advances in autism genetics; on the threshold of a new neurobiology. Nature Reviews. Genetics 9, 341–355.

Antonellis, A., Green, E.D., 2008. The role of aminoacyl-tRNA synthetases in genetic disease. Annual Review of Genomics and Human Genetics 9, 87–107.

Baldwin, R.L., 2008. The search for folding intermediates and the mechanism of protein folding. Annual Review of Biophysics 37, 1–21.

Beebe, K., Mock, M., Merriman, E., Schimmel, P., 2008. Distinct domains of tRNA synthetase recognize the same base pair. Nature 451, 90–93.

Bessonov, K., Bamm, V.V., Harauz, G., 2010a. Misincorporation of the proline homologue Aze (azetidine-2-carboxylic acid) into recombinant myelin basic protein. Phytochemistry 71, 502–507.

Bessonov, K., Bamm, V.V., Harauz, G., 2010b. Misincorporation of the proline homologue Aze (azetidine-2-carboxylic acid) into recombinant myelin basic protein. Phytochemistry 71, 1032–1034.

Boggs, J.M., 2006. Myelin basic protein; a multifunctional protein. Cellular and Molecular Life Sciences 63, 1945–1961.

Bonini, N.M., Gitler, A.D., 2011. Model organisms reveal insight into human neurodegenerative disease: Ataxin-2 intermediate-length polyglutamine expansions are a risk for ALS. Journal of Molecular Neuroscience 45, 676–683.

Cabre, P., Signate, A., Olindo, S., et al., 2005. Role of return migration in the emergence of multiple sclerosis in the French West Indies. Brain 128, 2899–2910.

Cancilla, P.A., Barlow, R.M., 1969. Structural changes of the central nervous system in swayback (enzootic ataxia) in lambs. IV. Electron microscopy of the white matter of the spinal cord. Acta Neuropathologica (Berlin) 12, 307–317.

Chalmers, G.A., 1974. Swayback (enzootic ataxia) in Alberta lambs. Canadian Journal of Comparative Medicine 38, 111–117.

Cox, P.A., Banack, S.A., Murch, S.J., et al., 2005. Diverse taxa of cyanobacteria produce beta-N-methylamino-L-alanine, a neurotoxic amino acid. Proceedings of the National Academy of Sciences of the United States of America 102, 5074–5078.

Curie, C., Cassin, G., Couch, D., et al., 2009. Metal movement within the plant: Contribution of nicotianamine and yellow stripe 1-like transporters. Annals of Botany 103, 1–11.

Daoud, H., Balzil, V., Martins, S., et al., 2011. Association of long ATXN2 CAG repeat sizes with increased risk of amyotrophic lateral sclerosis. Archives of Neurology 6, 739–742.

Dasuri, K., Ebenezer, P.J., Uranga, R.M., et al., 2011. Amino acid analog toxicity in primary rat neuronal and astrocyte cultures: Implications for protein misfolding and TDP-43 regulation. Journal of Neuroscience Research 89, 1471–1477.

Fowden, L., 1956. Azetidine-2-carboxylic acid: a new cyclic amino acid occurring in plants. Biochemical Journal 64, 323–332.

Geschwind, D.H., Levitt, P.P., 2007. Autism spectrum disorders: Developmental disconnection syndromes. Current Opinion in Neurobiology 17, 103–111.

Hallen, L., Klein, H., Stoschek, C., et al., 2011. The KRAB-containing zinc-finger transcriptional regulator ZBRK1 activates SCA2 gene through directed interaction with its gene product, ataxin-2. Human Molecular Genetics 20, 104–114.

Hammock, E.A., Levitt, P., 2011. Developmental expressional of a gene implicated in multiple neurodevelopmental disorders A2BP1 (Fox1). Developmental Neuroscience 33, 64–74.

Harauz, G., Ladizhansky, V., Boggs, J.M., 2009. Structural polymorphism and multifunctionality of myelin basic protein. Biochemistry 48, 8094–9104.

Hassner, A., 1983. Small Ring Heterocycles – Part 2. John Wiley, New York p. ix.

Joshita, S., Umemura, T., Yoshizawa, K., Katsuyama, Y., Tanaka, E., Ota, M., 2010. Shinshu PBC Study Group. A2BP1 as a novel susceptible gene for primary biliary cirrhosis in Japanese patients. Human Immunology 71, 520–524.

Kennedy, J., O'Connor, P., Sadovnick, A.D., Perara, M., Yee, I., Banwell, B., 2006. Age of onset of multiple sclerosis may be influenced by place of residence during childhood rather than ancestry. Neuroepidemiology 26, 162–167.

Lee, T., Li, Y.R., Chesi, A., et al., 2011. Evaluating the prevalence of polyglutamine repeat expansions in amyotrophic lateral sclerosis. Neurology 76, 2062–2065.

Lehtinen, A.B., Cox, A.J., Ziegler, J.T., et al., 2011. Genetic mapping of vascular calcified plaque loci on chromosome 16p in European

Americans from the diabetes heart study. Annals of Human Genetics 75, 222–235.

Ma, L., Hanson, R.L., Traurig, M.T., et al., 2010. Evaluation of A2BP1 as an obesity gene. Diabetes 59, 2837–2845.

Martin, C.L., Duvall, J.A., Ilkin, Y., et al., 2007. Cytogenetic and molecular characterization of A2B1/FOX1 as a candidate gene for autism. American Journal of Medical Genetics. Part B, Neuropsychiatric Genetics 144B, 869–876.

Oksenberg, J.R., Barazini, S.E., Sawcer, S., Hauser, S.L., 2008. SNPs to pathways to pathogenesis. Nature Review Genetics 9, 516–526.

Park, S.G., Schimmel, P., Kim, S., 2008. Aminoacyl trNA synthetases and their connection to disease. Proceedings of the National Academy of Sciences of the United States of America 105, 11043–11049.

Rosenthal, G.A., Bell, E.H., 1979. Naturally occurring toxic protein amino acids. In: Rosenthal, G.A., Janzen, D.H. (Eds.), Herbivores: Their Interaction with Secondary Plant Metabolites. Academic Press, New York, pp. 361–363.

Rubenstein, E., 2000. Biologic effects of and clinical disorders caused by nonprotein amino acids. Medicine (Baltimore) 79, 80–89.

Rubenstein, E., 2008. Misincorporation of the proline analogue; azetidine-2-carboxylic acid in the pathogenesis of multiple sclerosis: A hypothesis. Journal of Neuropathology and Experimental Neurology 67, 1035–1040.

Rubenstein, J.L.R., Rakic, P., 2013. Patterning and Cell Types Specification in the Developing CNS and PNS.

Rubenstein, E., Zhou, H., Krasinska, K.M., Chien, A., Becker, C.H., 2006. Azetidine-2-carboxylic acid in garden beets (*Beta vulgaris*). Phytochemistry 67, 898–903.

Rubenstein, E., McLaughlin, T., Winant, R.C., et al., 2009. Azetidine-2-carboxylic acid in the food chain. Phytochemistry 70, 100–104.

Sakai, Y., Shaw, C.A., Dawson, D.C., et al., 2011. Protein interactome reveals converging molecular pathways among autism disorders. Science Translational Medicine 3 (86), ra49.

Smith, G.S.T., De Avila, M., Paez, P.M., et al., 2012. Proline substitutions and threonine pseudophosphorylation of the SH3 ligand of 18.5-kDa myelin basic protein decrease its affinity for the Fyn-SH3 domain and alter process development and protein localization in oligodendrocytes. Journal of Neuroscience Research 90, 28–47.

Sobel, R.A., 2008. A novel unifying hypothesis of multiple sclerosis. Journal of Neuropathology and Experimental Neurology 67, 1032–1034.

Sobel, R.A., Albertelli, M.A., Butte, M.J., Elias, J.E., Grimes, K.V., Rubenstein, E., 2011. Studies on the AZE (Azetidine-2-carboxylic acid) hypothesis of MS pathogenesis. In: Keystone Symposium on Immunology, Genetics and Repair of Multiple Sclerosis, Taos, NM.

Stryer, L., 1988. Biochemistry, 3rd edn. W.H. Freeman, New York p. 201.

Talley, C.L., 2005. The emergence of multiple sclerosis, 1870-1950: A puzzle of historical epidemiology. Perspectives in Biology and Medicine 48, 383–395.

Tsai, F.H., Overberger, C.G., Zand, R., 1990. Synthesis and peptide bond orientation in tetrapeptides containing L-azetidine-2-carboxylic acid and L-proline. Biopolymers 30, 1039–1049.

Van Damme, P., Veldinik, J.H., van Bitterswijk, M., et al., 2011. Expanded ATNX2 CAG repeat size in ALS identifies genetic overlap between ALS and SCA2. Neurology 76, 2066–2072.

Voineagu, I., Wang, X., Johnston, P., et al., 2011. Transcriptomic analysis of autistic brain reveals convergent molecular pathology. Nature 474, 380–384.

von Wiren, N., Klair, S., Bansal, S., et al., 1999. Nicotianamine chelates both FeIII and FeII. Implications for metal transport in plants. Plant Physiology 119, 1107–1114.

30

Down Syndrome

A.B. Bowman[1], K.C. Ess[1], K.K. Kumar[1], K.L. Summar[2]

[1]Vanderbilt University, Nashville, TN, USA; [2]George Washington University, Washington, DC, USA

OUTLINE

30.1 INTRODUCTION

Trisomy of human chromosome 21 (HSA21) is the genetic basis of Down syndrome (DS). DS is associated with a spectrum of developmental disabilities and physiological and health disturbances of varying penetrance. Despite the variability of DS-associated phenotypes, all individuals express some degree of intellectual disability.

Triplication of a large number of genes on one human chromosome has made basic research into the molecular genetics and the cellular and neuropathological basis quite challenging. Yet advancing our understanding of DS pathophysiology as well as identifying new interventional strategies requires living animal and cellular models. Herein lies the challenge; how can one most accurately model a disease caused by trisomy of an entire

Neural Circuit Development and Function in the Brain: Comprehensive Developmental Neuroscience, Volume 3 http://dx.doi.org/10.1016/B978-0-12-397267-5.00130-8

© 2013 Elsevier Inc. All rights reserved.

human chromosome containing more than 400 individual genes? The best solution to date has been to duplicate large segments of syntenic regions of mouse chromosomes that carry subsets of genes located on HSA21. Alternatively, the insertion of segments of HSA21 into the mouse has also been employed to define specific genes that underlie aspects of the DS phenotype. Mouse models are, therefore, immensely valuable in deciphering genetic interactions of conserved human disease genes. But as the number of genes being studied grows, so does the risk of species-specific differences confounding the application of what can be learned in the mouse to the human disease. The recent advent of induced pluripotent stem cell (iPSC) technology offers the promise of human cellular models of DS that have full trisomy 21.

This chapter begins in Section 30.2 with a description of the DS genetic and clinical phenotypes, with an emphasis on neurological features. Section 30.3 details the experimental models of DS, with an emphasis on their utility and accuracy for translation to the human condition. Section 30.4 describes our current understanding of the neuroanatomical and functional correlates of the DS phenotype, with an emphasis on comparing research from experimental models with what is known about the human phenotype. These DS endophenotypes span both time (from embryonic neurodevelopment to adult aging) and scale (from molecular and cellular aspects up to local circuits and systems neuroscience). Section 30.5 outlines genotype–phenotype correlations in DS and the contribution of specific HSA21 genes as well as other genetic interactions with the DS endophenotypes. The chapter concludes with a discussion of directions for future research using iPSCs and consideration of potential therapeutic interventions.

30.2 DS MOLECULAR GENETICS AND CLINICAL PHENOTYPE

30.2.1 Trisomy of Human Chromosome 21 and the Molecular Genetics of DS

Euploid human cells contain 46 chromosomes (2N), including 22 pairs of autosomes and one pair of sex chromosomes (46,XX [female] and 46,XY [male]). Haploid human cells contain 23 chromosomes (N), which is characteristic of normal ovum and spermatozoa. Typical development is dependent upon the gene content of these chromosomes as well as chromosomal balance. Cells that do not contain an exact multiple of the haploid number of chromosomes are termed *aneuploid cells*. This abnormal chromosomal distribution is the most common and clinically significant type of human chromosome abnormality, occurring in 3–4% of all pregnancies (Lee and Summar, 2009). Human aneuploidies include

monosomies (one copy of a particular chromosome) and polysomies (more than two copies of a particular chromosome). The majority of aneuploid conceptuses are spontaneously lost during the first trimester of pregnancy. Nondisjunction is the most common mechanism of aneuploidy. When nondisjunction occurs, during meiosis I or meiosis II, gametes are created that either contain an extra chromosome or lack a copy of a chromosome. During conception, these abnormal cells combine with haploid gametes and result in monosomic or trisomic zygotes (Figure 30.1). Nondisjunction occurring later during mitosis results in a mosaic cell line. In addition to nondisjunction, chromosomal rearrangement such as translocation, partial duplication, and ring chromosome formation can result in aneuploidy.

Trisomy 21 is the quintessential example of aneuploidy in humans and results in the clinical phenotype of DS. This mechanism is responsible for the most common genetic cause of intellectual disability, occurring in 1 in 833 live births and making DS an excellent model for studying the relationship of genetics to neural function (http://www.cdc.gov/features/dsdownsyndrome). Individuals with DS have variations in neurodevelopment and neurodegeneration as well as disorders involving the central nervous system and other systems of the body. To further explore the complex relationship between aneuploidy, human development, and the clinical phenotype of DS, many researchers from diverse disciplines, including biologists, educators, physicians, and psychologists, have investigated pieces of the puzzle. A comprehensive review of all of these areas is beyond the scope of this text. The focus of this discussion is the influence of molecular genetics on the neurological phenotype of DS.

To elucidate the origins of the phenotypic features, scientists have approached aneuploidy and DS with the assumption that if a gene is present in three copies, the expression of that gene will be increased. The gene dosage hypothesis states that the increased expression of multiple trisomic genes directly leads to the features of DS (Antonarakis et al., 2004; Korbel et al., 2009). The amplified developmental instability hypothesis states that the phenotype is not directly due to specific genes on chromosome 21 but rather to a change in genetic balance and regulation (Patterson and Costa, 2005). Evidence of globally increased expression of genes on chromosome 21 in human fetal material has been documented by measuring mRNA levels (Mao et al., 2003). These same transcriptome studies failed to document the increased expression of genes not found on chromosome 21 in the same tissues. Proteomic studies have documented that mRNA from genes mapped to chromosome 21 translates into increased protein production in transgenic mice as well as in human tissues (Gulesserian et al., 2001; Tan et al., 1973; Wang et al.,

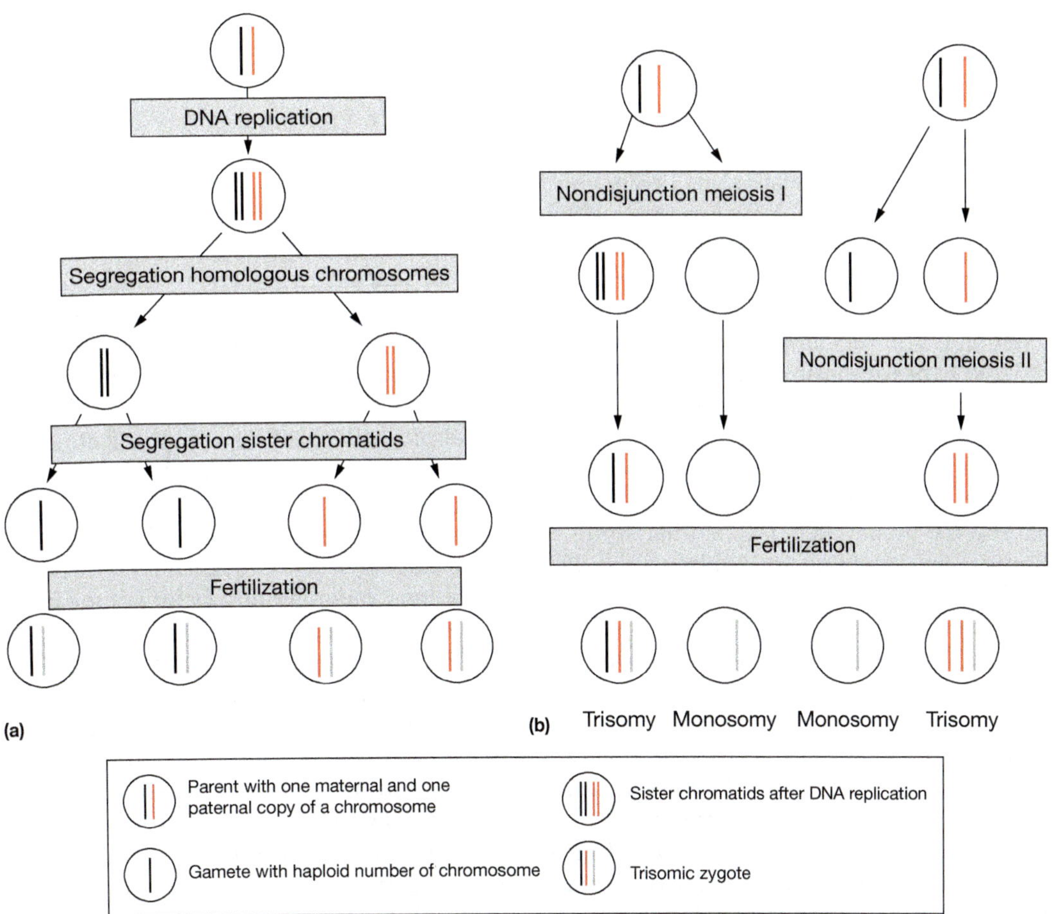

FIGURE 30.1 Meiotic nondisjunction. (a) Normal meiosis results in gametes with a haploid number of chromosomes. When fertilized with other normal gametes, a normal diploid state is reestablished. (b) In meiotic nondisjunction, two chromosomes migrate to the same daughter cell resulting in gametes with a diploid (or null) number of chromosomes. When fertilized with other normal gametes, abnormal trisomic or monosomic states are established. (a, b) Symbol key is provided in the box at the bottom of the figure.

1998). Recently, posttranscriptional regulators, including miRNAs, have been found to be overexpressed in DS brain and heart tissue specimens, resulting in a decreased expression of target genes, which very likely contributes to the DS phenotype (Kuhn et al., 2008).

Not only gene dosage, a factor in the phenotype of DS, but also gene content plays a crucial role. In 2000, through the collaborative work of many investigators, the complete DNA sequence of human chromosome 21 was published. This landmark work showed that human chromosome 21 is a relatively gene-poor chromosome, with 225 genes embedded in 33.8 Mb of genomic DNA compared with the 545 genes embedded in 33.4 MB of chromosome 22 (Hattori et al., 2000). With new sequencing technology, the number of genes on the human chromosome is now thought to be between 300 and 500 (http://www.ensembl.org). This finding of comparatively few genes on chromosome 21 offers a possible explanation of why trisomy 21 is compatible with postnatal survival and longevity whereas trisomy of HSA13, HSA15, and HSA22 are not. Finally, allelic variation of

chromosome 21, epigenetic factors, mosaic status, and environmental effects must be taken into consideration when studying the correlations between the genotype and phenotype of DS. There are some aspects of the phenotype of trisomy 21 that appear consistently in every individual with DS (i.e., facial appearance, intellectual disability, and generalized hypotonia). However, each of these common phenotypic features exhibits a great deal of variability in expression from one individual to another (e.g., intellectual disability ranges from mild to moderate cognitive impairment). Clearly, more work is needed to fully understand all the factors involved in these variations.

30.2.2 Neurodevelopmental and Neurodegenerative Features of the DS Phenotype

Historically, DS has been the prototypical intellectual disability disorder in research as well as in societal portrayals and thus implies a nonspecific alteration in brain development and pathology. However, DS is associated with distinct neuroanatomic and neuropathologic findings that in turn present with specific cognitive deficits. Postmortem observations (Crome et al., 1966; Davidson, 1928; Wisniewski et al., 1985) and volumetric magnetic resonance imaging (MRI) studies (Aylward et al., 1997b, 1999; Emerson et al., 1995; Pearlson et al., 1998; Raz et al., 1995) indicate that people with DS have reduced total brain volumes, most notably in the frontal and temporal areas as well as the cerebellum. Altered temporal volume, architecture, and function in DS (particularly of the hippocampal area) are considered to be responsible, in part, for the cognitive deficits associated with DS (Pennington et al., 2003). Disturbances in cerebellar volume and architecture are associated with the generalized hypotonia of DS as well as the inability to learn new skills (Daum and Ackermann, 1995; Konczak and Timmann, 2007). A similar finding of reduced hippocampal volume in children with DS suggests that these are neurodevelopmental findings rather than neurodegenerative findings of an aging brain (Pinter et al., 2001a). By contrast, people with DS have normal volumes of subcortical areas, including the lenticular nuclei and posterior parietal and occipital gray matter.

Interest in the neuropathology of DS existed long before the genetic etiology was delineated. Pathologic findings characteristic of those seen in dementia of the Alzheimer's type were described in the brains of individuals with DS long before the normal number of human chromosomes was even determined (Jervis, 1948; Struwe, 1929). It is now known that virtually all adults with DS who have come to autopsy have the hallmark neuropathology of Alzheimer disease (AD), with prominent β-amyloid plaques and neurofibrillary tangles (NFTs). Furthermore, decreased numbers of apical dendritic spines of pyramidal neurons in the hippocampus and cingulate gyrus were noted when comparing the brains of people with DS to those of normal controls (Becker et al., 1986; Takashima et al., 1994). Significantly reduced dendritic spines, increased β-amyloid plaques, and increased NFTss in DS are considered to be specific findings that are not seen in other types of intellectual disability (Suetsugu and Mehraein, 1980).

Like the gross anatomical and histological findings, the cognitive phenotype found in individuals with DS is specific (Silverman, 2007). Cognitive deficits in DS are not uniform across all domains. Particular deficits exist in learning, memory, and language that lead to difficulty with intellectual functioning. To further complicate this issue, a great deal of variability exists between individuals. People with DS have difficulty with verbal short-term memory and explicit (episodic or conscious) long-term memory (Lott and Dierssen, 2010; Silverman, 2007). The function of explicit memory has been linked to the hippocampus (Squire, 1992), which has been shown to function abnormally in DS (Pennington et al., 2003).

Implicit memory – the ability to use previous experience to assist with the performance of a task without conscious awareness – is preserved in DS (Carlesimo et al., 1997). Prefrontal cortex function also affects explicit and working memory and has been shown to have deficits in DS (Nadel, 1999). This impairment seems to involve verbal information with sparing of visuospatial domains, offering a potential explanation of why people with DS learn better in an environment that includes visual materials in addition to verbal presentation (Frenkel and Bourdin, 2009; Lanfranchi et al., 2009). Morphology of grammar (internal structure of words) and syntax (how words are put together in phrases) are impaired in people with DS. The communication difficulties are specific to DS when compared to others with intellectual disabilities and mental age-matched controls. Communication is further complicated for people with DS because intelligibility is negatively affected by poor oral motor tone and anatomic factors of the mouth and pharynx (Abbeduto et al., 2001). New technologies, including functional imaging techniques such as functional MRI (fMRI) and spectroscopy, should allow for further studies of correlations between described cognitive and language deficits and brain function in humans with DS.

Longitudinal studies with adults with DS reveal the onset of cognitive decline (from baseline) in middle age similar to that seen in much older individuals without intellectual disability (Devenny et al., 1996, 2000). These observations along with the knowledge of the ubiquitous presence of amyloid plaques and NFTs in postmortem trisomy 21 specimens have led to many investigations into the risk of Alzheimer's type dementia in people with DS. Multiple studies of varying design have established that approximately 50–70% of adults with DS will develop dementia by age 60–70 years. Importantly, an alternative statement of these data is that some individuals with DS will survive dementia-free into old age (Devenny et al., 1996).

Understanding the genotype–phenotype relationship in DS may lead to a greater understanding of neurodevelopment and neurodegeneration through mechanisms involved in the gene dosage effect as well as gene balance and regulation. This has obvious application to the continued improvement of the lives of people with DS, but it may also lead to novel therapies for the general population that suffers from neurodevelopmental as well as neurodegenerative disorders of the central nervous system.

30.2.3 Other CNS Features of DS

As noted in Table 30.1, people with DS have increased risk for seizures, psychiatric disorders, and Alzheimer dementia. Alzheimer dementia was discussed in the

TABLE 30.1 Common medical comorbidities in children with Down syndrome

System	Finding
Ears, eyes, nose	Hearing loss (conductive and sensorineural), frequent and chronic otitis media, obstructive sleep apnea, glaucoma, cataracts (congenital and acquired), refractive errors, strabismus, nystagmus, amblyopia
Cardiovascular	Atrioventricular septal defect, ventricular septal defect, atrial septal defect, patent ductus arteriosus, tetralogy of Fallot, and persistent pulmonary hypertension; adolescents and young adults – mitral valve prolapse and aortic regurgitation
Endocrine	Diabetes mellitus, hypothyroidism, hyperthyroidism
Gastrointestinal	Constipation, celiac sprue (25%)
Hematological	Transient myeloproliferative disorder, leukemia
Neurological	Intellectual disability (mental retardation); hypotonia, seizures; psychiatric disorders, autism spectrum disorder
Orthopedic	Atlantooccipital instability, pes planus, subluxation of patella, developmental hip dysplasia
Dermatological	Hyperkeratosis, seborrhea, cutis marmorata, xerosis, folliculitis

previous section and will not be addressed here. Seizures have a bimodal distribution in DS with 40% of seizure disorders presenting in the first year of life and another 40% present in the third decade or later (Pueschel et al., 1991). Epilepsy in the very young can manifest as infantile spasms or generalized tonic clonic ('grand mal') seizures. Infantile spasms due to any etiology are associated with hypsarrhythmia, a specific electroencephalographic finding, and infants with DS and infantile spasms have more frequent diagnoses of an autism spectrum disorder later in life (Eisermann et al., 2003). Seizures presenting in adolescents and adults can be generalized tonic clonic or partial ('petit mal'). Both types of seizures are usually accompanied by some alteration of mental status that can be brief. This has led to missed diagnoses in people with DS as the seizures may be ascribed to a lack of focus or attention. Finally, age of seizure onset appears to be important in DS as adults older than 45 years are more likely to develop dementia, whereas early onset of seizures is not associated with dementia (Lott and Dierssen, 2010; Puri et al., 2001).

Behavioral and psychiatric disorders are more common in people with all intellectual disabilities when compared to the general population. However, people with DS score significantly lower on maladaptive behaviors when compared to people with other cognitive disabilities (Dykens and Kasari, 1997). Children with DS, for example, are frequently diagnosed with disruptive behaviors including attention deficit hyperactivity disorder, oppositional defiant disorder, or aggression (Myers and Pueschel, 1991). Less is known about behavioral and psychiatric problems in adolescents with DS. Preliminary data show that they may have fewer externalizing (disruptive) behaviors than their younger cohorts (Dykens et al., 2002). Adults with DS are more prone to depression when compared to the typical population and to others with intellectual disabilities (6–11% compared to 1–2%; Collacott et al., 1992). Historically, 'diagnostic overshadowing' has led to a failure to recognize co-occurrence of psychiatric illness in individuals with intellectual disability (Reiss et al., 1982). With improved recognition of these secondary disorders in this population, prompt treatment and eventually even more effective therapies may be discovered.

30.2.4 Non-CNS Features of DS

Some medical comorbidities occur more often in people with DS as compared to their typically developing peers, including congenital heart disease (100-fold increase; Ferencz et al., 1985), leukemia (30-fold; Hasle et al., 2000), and Alzheimer dementia (10-fold increase in DS; Zigman and Lott, 2007). Of the congenital heart lesions, the most common lesions are atrioventricular septal defects (45% of newborns with DS), ventricular septal defects (35%), atrial septal defects (8%), patent ductus arteriosus (7%), tetralogy of Fallot (4%), and others (1%) (Roizen and Patterson, 2003). Due to the prevalence of congenital heart disease in DS, the American Academy of Pediatrics recommends an echocardiogram and pediatric cardiology evaluation in every newborn with DS regardless of the presence of symptoms or physical findings (American Academy of Pediatrics and Committee on Genetics, 2011). Hearing loss (both conductive and sensorineural) and visual disturbances (glaucoma, cataracts, strabismus, myopia, astigmatism) occur with much more frequency in people with DS and can present in infancy through adulthood. For these reasons, vigilant screening of vision and hearing is recommended to prevent further deleterious effects on intellectual and adaptive functioning (American Academy of Pediatrics and Committee on Genetics, 2011; Roizen, 2002; Smith, 2001).

Advances in medical care have improved survival for people with DS (increasing from an average of 12 years in the 1940s to an average of 57.8 years for women and 61.1 years for men) (Bittles et al., 2007; Glasson et al., 2003). Despite this improved survival, there are many physical problems involving every organ system of the body that can affect people with DS. Additional studies are needed to better understand the impact of these comorbidities on the intellectual and adaptive functioning of people with DS.

Just as secondary medical conditions can occur with increased frequency in DS, there are comorbidities that appear to be less common in DS than in the general population. For example, analysis of the causes of deaths between these groups revealed that mortality due to ischemic heart disease is less than one half and deaths due to solid tumors less than one tenth in DS patients compared to the general population (standardized mortality odds ratio of 0.42 and 0.07, respectively, for causes of death; Yang et al., 2002). Review of cancer registries has confirmed that people with DS have fewer solid tumors even though they have an increased risk of leukemia (Hasle et al., 2000). Two candidate genes from HSA21 have provided potential mechanisms for this 'protective factor' against solid tumors. The first gene is the Down syndrome candidate region 1 (DSCR1/RCAN1), which is a calcineurin inhibitor. DSCR1 is modestly overexpressed in transgenic mice, and this overexpression is sufficient to provide suppression of tumor growth by inhibiting angiogenesis by attenuation of calcineurin activity (Baek et al., 2009). Other mouse studies have implicated the upregulation of the oncogene Ets 2 as a possible factor in the prophylactic effect of trisomy 21 on solid tumor development (Sussan et al., 2008). Further studies need to be done to clarify this apparent advantage in cancer among people with DS and distinguish genetic from environmental differences. Potentially, novel therapeutic interventions for cancer in the general population could be designed by a better understanding of DS.

30.3 EXPERIMENTAL MODELS OF DS

30.3.1 Overview of DS Mouse Models

The need for animal models to study human development and disease has been recognized for millennia dating back to at least the time of Aristotle and Galen (Guerrini, 2003). The incredible pace of research during the last half century has allowed the creation of increasingly sophisticated and relevant animal models of many diseases. The mouse has become the favored model system given the well-defined anatomy and physiology of this species; completely sequenced genome; availability of robust quantitative assays of learning, memory, and anxiety; as well as a huge inventory of available tools including specific antibodies and genetic engineering to modify the mouse genome. This latter development continues to improve with refined and subtle manipulations now routine. However, modeling the complexity of a chromosomal human disease such as DS in the mouse has proven to be very difficult. This is due largely to

evolutionary differences between mice and humans in the sequence, expression patterns, and function of many genes. Furthermore, HSA21 is not present in a direct orthologous fashion in mice as genetic material corresponding to that found on HSA21 is found mainly on mouse chromosomes 16 (Mmu16) but also on Mmu10 and Mmu17. Conversely, additional genes on Mmu16, Mmu10, and Mmu17 are not found on HSA21 and thus are likely not relevant to the study of DS. These evolutionary changes have made a 'simple' mouse model of DS very challenging. However, recent advances to allow complex genomic engineering have recently been achieved, leading to the recent creation of a mouse with trisomy of almost all the genetic information that is normally found on HSA21 (Yu et al., 2010). These models will likely be most informative when combined with single-gene knockouts to test whether three copies of specific genes contribute to the pathogenesis of DS and whether decreasing gene dosage can reverse the clinical features. Such findings would provide a clear rationale to selectively target identified pathways pharmacologically (Table 30.2).

30.3.2 DS Mouse Models with Mouse Additional Segments

Initial attempts to model DS reasoned that most HSA21 genes are found on a syntenic region of mouse chromosome 16. Animals with balanced chromosomal translocations were identified and then selectively bred to generate an Mmu16 trisomy mouse. These mice died *in utero*, likely because of cardiac and placental abnormalities (Miyabara et al., 1982). Additional relevant phenotypes include generalized edema and delayed development of the cerebral cortex. While these results strongly support that genes present on a syntenic region of HSA21 in the mouse are required for normal development, their lack of viability severely hampered further progress and determination of the relevance to DS. Furthermore, Mmu16 trisomy is unable to provide any information about the role of HSA21 orthologs found on Mmu10 and Mmu17. Finally, as Mmu16 contains multiple genes not found on HSA21, any results from these mice have to be interpreted with caution.

A major advance in DS research was the generation of a mouse with a balanced translocation of Mmu16 that corresponded closely to the syntenic region of HSA21 (Figure 30.2). Selective breeding of these animals allowed the generation of Ts65Dn mice that are trisomic for about 14 Mb of mouse DNA and contain the majority (at least 132) of the genes on HSA21 (Holtzman et al., 1996). Ts65Dn mice, however, have significant limitations including some mortality during early postnatal life and phenotypic variability. In addition, males are infertile, complicating the maintenance of this mouse

TABLE 30.2 Mouse models of DS

	Triplication	Abnormalities	Limitations	Reference
Mouse chromosome				
Mmu16	Entire Mmu16	Edema, abnormal cerebral cortex, heart defects, thymic hypoplasia	Lethal *in utero*, no Mmu10 and Mmu17 genes	Miyabara et al. (1982)
Ts65Dn	Partial Mmu16	Some perinatal death, cerebellar hypoplasia, cortical and hippocampal spine alterations, craniofacial dysmorphology	Perinatal death, no cardiac abnormalities, hyperactive, no Mmu10 and Mmu17 genes	Holtzman et al. (1996)
Ts1Cje	Partial Mmu16	Cerebellar hypoplasia, craniofacial dysmorphology, hippocampal spine alterations	No cardiac abnormalities, hypoactive, no Mmu10 and Mmu17 genes	Sago et al. (1998)
Ts1Rhr	Partial Mmu16	Increased size of mandible	Lack of CNS and cardiac phenotypes??	Olson et al. (2004)
Ms1Rhr	Deletion partial Mmu16	None	Lack of CNS phenotype	Olson et al. (2007)
Duplication of human chromosome				
Tc1	Extra copy of HSA21	Decreased density of cerebellar granule neurons, cardiac defects	Variable penetrance likely due to mosaicism	O'Doherty et al. (2005)
Chromosomal engineering				
Ts1Yah	Mmu17	Impaired working memory, enhanced spatial memory	No Mmu16 genes	Pereira et al. (2009)
Dp(10, 16, 17)	Mmu10, Mmu16, Mmu17	Impaired cognition, hippocampal-mediated deficits, hydrocephalus	Only initial characterization to date	Yu et al. (2010)

colony. While Ts65Dn mice exhibit no cardiac defects, over the years they have proven invaluable because of CNS and craniofacial abnormalities (Kurt et al., 2000; Roper et al., 2006b). The brain abnormalities include a small cerebellum with decreased numbers of granule neurons and Purkinje cells (Baxter et al., 2000). In addition, decreased spine density was seen on dendrites from hippocampal dentate granule neurons (Belichenko et al., 2004). The failure of Ts65Dn mice to more closely phenocopy DS should not be too surprising given the complexity of this syndrome and the lack of triplication of many additional orthologous genes that reside on Mmu10 and Mmu17. The Ts1Cje model (Figure 30.2) has also been used by many investigators and features a partial trisomy of Mmu16 though the region is smaller than in Ts65Dn containing approximately 85 orthologous genes (Sago et al., 1998). These mice have CNS abnormalities similar to those seen in Ts65Dn, including small cerebellums and abnormalities in the number of cerebellar granule cells, but not Purkinje neurons. Similar to Ts65Dn, decreased spine density with dendritic spine enlargement was seen in this model (Belichenko et al., 2007). Multiple histogenic and physiological processes have been examined in these mouse models, revealing important insights into a spectrum of DS-relevant phenotypes (see Sections 30.3.3 and 30.4 for more details).

While models with triplication of Mmu16 continue to provide important information about global DS phenotypes and preclinical testing, additional mouse models are clearly needed to more finely define genetic requirements for the pathogenesis of DS. Also, other models were needed to test the emerging hypothesis that trisomy of a 'critical region' of HSA21 (Down syndrome critical region, DSCR) is sufficient or required for the phenotypes seen in DS. This hypothesis was based on the identification of rare patients with DS that appeared to have trisomy of a relatively discrete region of HSA21. This hypothesis has been more rigorously tested in mice using Cre/LoxP technology to duplicate (Ts1Rhr) or delete (Ms1Rhr) an approximately 4-Mb region of Mmu16 containing the approximately 33 orthologous genes found in the DSCR (Olson et al., 2004). Interbreeding with wild-type mice then allows the production of animals with one or three copies of the mouse DSCR. A particularly elegant application of this technique was to breed Ms1Rhr mice to Ts65Dn or Ts1Cje mice. This allowed the generation of mice that were trisomic for genes within Ts65Dn or Ts1Cje but disomic within the DSCR. These Ms1Rhr/Ts65Dn mice did not reverse abnormalities of cranial morphology, and Ts1Rhr mice that were trisomic for just the DSCR also did not exhibit cranial abnormalities (Olson et al., 2004). CNS

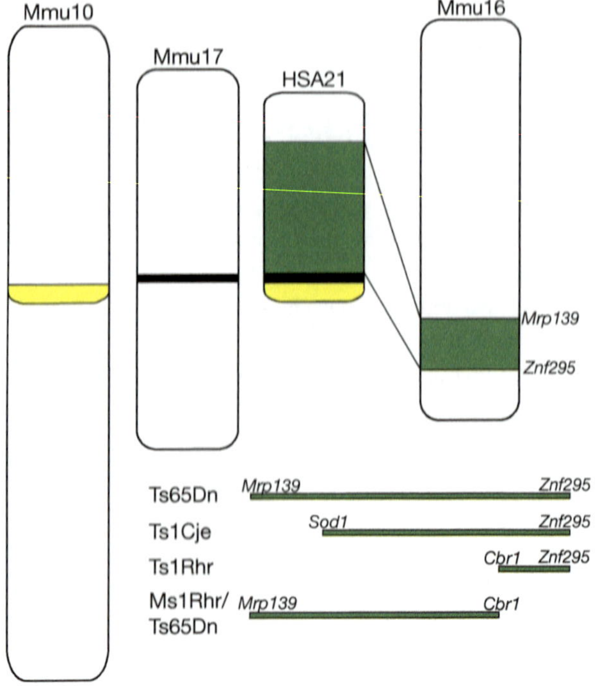

FIGURE 30.2 Schematic of human and mouse chromosomes relevant to DS and comparison of the most commonly used mouse models. The majority of HSA21 orthologous genes are located on Mmu16 but they can also be found on Mmu10 and Mmu17. The commonly used Ts65Dn, Ts1Cje, Ts1Rhr, and Ms1Rhr/T65Dn mouse models of DS feature triplication of variable segments of Mmu16 as shown. The Ts1Rhr region contains the 'Down syndrome critical region' (DSCR). The names of specific genes are listed to denote the approximate beginning and end of each Mmu16 segment.

manifestations have not been extensively studied though trisomy of the DSCR in Ts1Rhr mice was insufficient to produce abnormalities of spatial learning (Olson et al., 2007). However, removing the DSCR region with Ms1Rhr/Ts65Dn mice reversed hippocampal-dependent memory deficits. In addition, detailed analyses of Ts1Rhr mice revealed an increased size of dendritic spines and decreased spine density as seen in Ts65Dn or Ts1Cje mice (Belichenko et al., 2009a,b). These models demonstrate that despite its name, genes within the DSCR are not sufficient for the craniofacial abnormalities and some of the CNS manifestations seen in DS. This approach again underscores a complex interplay of genes in the DSCR with non-DSCR genes located on Mmu16 as well as Mmu10 and Mmu17. A further study of the DSCR was recently reported with overexpression of *RCAN1* (Dierssen et al., 2011). These mice have disruptions in visual–spatial learning similar to that seen in people with DS.

A similar technique of chromosomal engineering was used to make Ts1Yah mice that are trisomic for the syntenic region of Mmu17 including 12 orthologous genes found on HSA21 (Pereira et al., 2009). These mice have learning and behavioral abnormalities but enhanced spatial learning, which is typically a hippocampal-dependent process. These somewhat contradictory results firmly support the need for animal models that have genetic alterations of genes outside of Mmu16 and further underscore the complexity inherent to DS.

The most exciting recent advance using chromosomal engineering was the creation of a mouse model that is trisomic for essentially all the relevant portions of Mmu16, Mmu10, and Mmu17 that correspond to HSA21 (Yu et al., 2010). These Dp(10)1Yey/+, Dp(16)1Yey/+, and Dp(17)1Yey/+ mice (Dp(10, 16, 17)) exhibit impaired hippocampal-dependent learning and hippocampal long-term potentiation (LTP). In addition, a few Dp(10, 16, 17) mice had aqueductal stenosis and hydrocephalus. No potential abnormalities of the cerebral cortex or craniofacial or cardiac defects relevant to DS have been reported to date.

30.3.3 Transchromosomic DS Mouse Model Using Human DNA

While the new Dp(10, 16, 17) mouse described previously is very promising, alterations in DS may require increased expression of *human* genes and/or functional changes in *human* proteins. To address this possibility, mice have been generated using microcell-mediated chromosomal transfer (Shinohara et al., 2001). These Tc1 animals contain an approximately 42-Mb portion of HSA21 (O'Doherty et al., 2005). However, these animals are limited as a DS model because of loss of the human chromosome in a variable fashion in different organs and tissues. This mosaicism may be broadly similar to that seen in people with DS but significantly hampers the use of Tc1 mice for biomedical research.

30.3.4 Nonmouse Models of DS

While mouse models have dominated the field of DS research, additional models have been developed in *Drosophila*, *Caenorhabditis elegans*, and even yeast by modifying orthologous genes found within HSA21 (Guipponi et al., 2000; Moldrich, 2007; Tejedor et al., 1995). As discussed below, these additional model systems have allowed the testing of individual genes relevant to DS and the study of conserved gene expression and protein function (Chang and Min, 2005; Raich et al., 2003; Yu et al., 2009). *Drosophila* has been a particularly valuable and experimentally tractable system. For example, overexpression of any of the three HSA21 orthologs *dap160* (*ITSN1*), *synj* (*SYNJ1*), and *nla* (*DSCR1*) produced abnormalities in brain synaptic connections, but overexpression of all three genes impaired synaptic vesicle recycling and motor function (Chang and Min, 2009). Such 'simpler' model systems can complement existing

TABLE 30.3 Nonmouse models of DS

Organism	Gene alteration	Abnormalities	Limitations	Reference
Drosophila melanogaster	Overexpression of *dap160* or *synj* or *nla1*	Reduced recycling of neurotransmitter vesicles	Divergent evolution, lack of interactions with other DS-related genes	Chang and Min (2009)
Drosophila melanogaster	Overexpression of *dap160* and *synj* and *nla1*	Reduced recycling of neurotransmitter vesicles and impaired locomotion	Divergent evolution, studies limited to three overexpressed genes	Chang and Min (2009)
Drosophila melanogaster	Loss of *Dscam* exon 19	Impaired neural development and axonogenesis	Divergent evolution, unable to study interactions with other DS-related genes, unclear relevance to DS	Yu et al. (2009)
Caenorhabditis elegans	Overexpression of *mbk-1 (DYRK1A)*	Impaired odortaxis	Divergent evolution, unable to study interactions with other DS-related genes	Raich et al. (2003)
Homo sapiens	iPSC with trisomy 21	None reported	*In vitro* studies only	Park et al. (2008)

mouse models and test single and multiple gene function and interaction within the developing brain. Furthermore, these additional model systems may be well positioned to identify and test new therapies to ameliorate various brain phenotypes relevant to DS (Table 30.3).

30.3.5 Human Cellular Models of DS: Stem Cells and Patient-Specific Models of Disease

Mouse models of DS described above were generated with the express intent of identifying biological pathways altered by the trisomy of genes found on HSA21. As described above, these models include Ts65Dn, Ts1Cje, Ts1Rhr, Ms1Rhr, Tc1, Ts1Yah, and Dp(10, 16, 17). They have all contributed to our knowledge in defining genomic regions required for subsets of DS phenotypes including craniofacial and age-related neurodegeneration, cardiovascular phenotypes, learning and memory deficits, cerebellar hypoplasia, and partial penetrance neonatal lethality. These models may have served best by reaffirming the complexity of DS and the ongoing challenges in modeling a 'simple' trisomic disease. In addition, these mouse models have focused attention on a subset of HSA21 genes (e.g., *DSCR1*, *DYRK1A*, *APP*, and *SOD1*) and biological pathways (nuclear factor of activated T cells (NFAT) activation, Hedgehog signaling, and microtubule-based intracellular transport) that appear to play an important role in the generation of these phenotypes in the mouse models (see Section 30.4). The existing mouse models and other model organisms as described earlier suffer from limitations that are important to consider when applying research findings to DS. First, most of the models are trisomic for only a subset (one or more) of the homologous genes on HSA21 and as such fail to account for the impact of these other genes in the processes and phenotypes being studied. The generation of

the Tc1 model carrying an almost intact HSA21 largely overcame this problem. However, as described above, the mosaic loss of HSA21 from a significant portion of cells in these animals complicates the interpretation of the phenotypes in this model. Second, regulatory control of exogenous human genes may differ from that of their endogenous mouse counterparts; thus, the degree of overexpression of trisomic genes may not accurately model the expression of these genes in patients. Indeed, the lack of DS-like anatomical and morphological phenotypes in the Tc1 model may be due to differences in expression of human genes in the mouse. Third, genetic interactions with genes not found on HSA21 may show species-specific differences; thus, pathways identified in the mouse may play a stronger or weaker role in the human condition. Finally, there are several discrepancies between the mouse models and human patients including differences in skeletal modifications, absence of comparable gastrointestinal abnormalities, failure to observe cortical lamination defects, and hypoplasia of the cerebrum in the mouse models (Delabar et al., 2006). The recent creation of Dp(10, 16, 17) mice may circumvent many of these concerns though it remains possible that human genes and proteins have different pathological roles leading to the phenotypes seen in people with DS. Much more extensive evaluations of Dp(10, 16, 17) mice will be required to see if they will become the preeminent animal model for DS. Despite these shortcomings, trisomy mouse models are currently the best available research tool for studying the molecular pathophysiology of DS in a whole animal. However, a human-based neuronal/glial model of DS would address many of the caveats discussed here, facilitate the translation of research findings between human and animal research, and speed the discovery and testing of therapeutic targets.

At the end of 2007, three independent laboratories, headed by Shinya Yamanaka (Kyoto University), George Daley (Harvard Medical School), and James Thomson

(University of Wisconsin–Madison), generated pluripotent stem cells from adult human dermal cells with the developmental potential seemingly equivalent to that of human embryonic stem cells (Park et al., 2008; Takahashi et al., 2007; Yu et al., 2007). These iPSCs were found to be competent to generate all three major cell type lineages and resemble embryonic stem cells in their pluripotency and other important characteristics. Human iPSCs can be differentiated into embryoid bodies as has been done with human embryonic stem cells (Takahashi et al., 2007; Yu et al., 2007). Neural progenitor cells can be harvested from embryoid bodies and cultured prior to differentiation into neurons and glia (Yeo et al., 2007, 2008). The ability to generate iPSCs from human patients is anticipated to open up a new frontier of research into human disease. The study of the pathogenesis of DS represents an almost ideal application of this technology given the challenges of modeling trisomy of HSA21 in other model systems, as discussed earlier. Furthermore, the ability to differentiate neurons and glia from DS iPSCs offers a chance to examine in detail changes in both the development and maintenance of neural function caused by trisomy HSA21 in the context of human cells. Finally, the relative ease to expand, maintain, culture, and differentiate iPSCs will enable this resource to be utilized by a broader range of research laboratories around the world, expanding both the scope and depth of DS research being pursued. To date, there has been a single report of iPSCs being generated from trisomy 21 fibroblasts (Park et al., 2008). This is an important 'proof of principle' and establishes that trisomy of HSA21 does not prevent the reprogramming process required to make iPSCs and that this exciting new technology can be effectively applied to the study of DS.

30.4 NEUROLOGICAL AND FUNCTIONAL CORRELATES OF THE DS PHENOTYPE

Research efforts to better understand the cellular and functional basis of the DS phenotype have followed a two-pronged strategy. The first strategy is to characterize and define the neurological, functional, and neuroanatomical correlates of the DS genotype in human patients. This enables understanding of the specific neurological phenotypes associated with DS. The second strategy seeks to refine this knowledge by a detailed study of the molecular, cellular, and functional correlates of the DS genotype using experimental model systems. Under this approach, researchers have sought to understand the pathophysiological basis of the DS phenotype by characterizing and identifying the myriad observable differences between DS and control groups and developing a hierarchy of potential cause–effect

relationships between biomarkers of the disease and the behavioral, cognitive, and neurological challenges facing patients. Thus, phenotypes in DS animal and cellular models are consistently compared and contrasted to the clinical phenotypes of the DS patient. This approach seeks to increase scientific confidence in the accuracy of DS model systems to recapitulate the pathophysiology of trisomy HSA21. Here, we examine known DS model phenotypes and their possible relationships with known DS phenotypes. Studies on phenotypic correlates of the DS genotype fall under two broad categories: (1) studies examining the influence of specific trisomic genes to particular DS phenotypes and (2) studies using models with multiple trisomic genes (from several genes up to near-complete representation of HSA21 genes) to better understand the cellular, behavioral, and functional correlates of the DS genotype.

30.4.1 Behavioral, Motor, and Cognitive Alterations

Mouse models of DS have been carefully examined for cognitive, motor, and behavioral deficits consistent with the intellectual disability, motor, and behavioral features associated with DS. The specific nature of the DS phenotype has been extensively characterized as well as compared and contrasted with other developmental disorders. The intellectual quotient of individuals with DS averages about 50 (Contestabile et al., 2010; Vicari, 2004; Vicari et al., 2000, 2005). However, mental retardation is insufficient to explain the specific pattern of neuronal functional domains affected in DS. For example, despite similar or even less severe mental retardation, the distinctive characteristics of children with DS, compared to children with other developmental and intellectual disabilities, have been recognized as advantageous for their rearing and social integration (Corrice and Glidden, 2009). Individuals with DS exhibit a decreased risk for maladaptive behaviors, psychopathology, or emotional disturbance compared to others having a similar degree of mental retardation (Dykens, 2007). However, DS individuals carry a tenfold larger risk for a comorbid diagnosis of autism than the general population (Molloy et al., 2009). Despite this, social interactions and repetitive movements are not considered a part of the core DS phenotype (Corrice and Glidden, 2009; Dykens, 2007; Molloy et al., 2009; Zigler and Hodapp, 1991). Instead, speech and language difficulties, along with learning and memory deficits, are the principal phenotypic features of DS (Contestabile et al., 2010). In addition, individuals with DS show elevated gaze fixation and decreased scanning for novel stimuli in the environment, as well as subdued emotional responsiveness (Zigler and Hodapp, 1991). Also, significant delays in the development of cognitive and neurological milestones are reported for DS (Nadel, 2003). Hippocampal and prefrontal cortical dysfunction appears to be at the core of these deficits (Nadel, 2003). Declines in both short- and long-term memory as well as in memory consolidation have been reported, with deficits in both visual–spatial as well as verbal memory. Finally, a large portion of patients with DS exhibit a progressive cognitive decline with aging and the pathological features of AD.

Several recent reviews summarize the behavioral phenotypes of DS mouse models (Contestabile et al., 2010; Rachidi and Lopes, 2007). Common behavioral abnormalities in these models include decreased fear conditioning, decreased novel object recognition and long-term memory, decreased spatial learning and memory, hypoactivity and decreased exploratory behavior, and delayed acquisition of sensorimotor skills. These behavioral phenotypes are consistent with the human DS phenotype and arguably validate these models for mechanistic study and detailed characterization. By far the most widely used and extensively characterized of these models is the Ts65Dn segmental trisomy mouse model (Figure 30.2). Characterization of these mice revealed a significant developmental delay in acquisition of sensorimotor behaviors (Holtzman et al., 1996). Of particular note was a delay in the pattern of ultrasonic vocalizations, possibly similar to the delay in speech seen in human DS children (Holtzman et al., 1996). The newest and most complete mouse segmental trisomic DS genetic model, Dp(10, 16, 17), exhibits deficits in hippocampal-dependent learning and memory by the Morris water maze test, in which trisomic animals spent significantly less time searching for a hidden platform in the target quadrant as well as the contextual fear conditioning test (Yu et al., 2010). Analysis of the HSA21 'transchromosomic' mouse model, Tc1, also revealed spatial learning and memory deficits by Morris water maze and novel object memory deficits (Morice et al., 2008; O'Doherty et al., 2005). However, while defects in short-term spatial and novel object memory were detected, long-term memory appeared to be intact. This is in contrast to the Ts65Dn model, which exhibits deficits in long-term memory of novel objects (Fernandez and Garner, 2008), or the Ts1Cje model, which has normal performance in short- and long-term novel object memory tests (Fernandez and Garner, 2007). The basis for these discrepancies in DS mouse models is unresolved. Possible explanations include (1) the inclusion of different partially overlapping sets of HSA21 gene or related genes in the various models, (2) difficulties in modeling the influence of human trisomy in a rodent model, and (3) environmental factors or experimental differences that may influence mouse behavior experiments.

Genetic dissection of DS mouse phenotypes, to attribute specific HSA21 genes or chromosomal regions to the DS phenotype, has received considerable attention in the field. Approaches to asking questions on the role of genes specific to the DS phenotype take advantage of various mouse models with partially overlapping sets of triplicated genes, for example, the analysis of novel object memory in the Ts65Dn and Ts1Cje lines mentioned earlier (Fernandez and Garner, 2007), as the Ts65Dn line contains additional genetic material homologous to HSA21 not found in the Ts1Cje line. Comparison of partially overlapping segmental trisomic mouse models can provide insight into the genetic basis of the DS behavioral phenotypes. For example, the work of Epstein and colleagues with the Ts65Dn, Ts1Cje, and Ms1Ts65 lines examined correlations between motor activity and learning and memory by Morris water maze between the *App* to *Sod1*, *Sod1* to *Mx11*, and *App* to *Mx1* genetic intervals (Sago et al., 2000). However, complexities and differences in phenotypic penetrance have made clear genotype–phenotype correlations challenging. This complexity is further exemplified by work on a segmental trisomic model (Ts1Yah) that contains genes found in the subtelomeric region of HSA21 (*Abcg1* to *U2af1*) from Mmu17 and genes that are excluded from the widely used Ts65Dn, Ts1Cje, and Ms1Ts65 mice, which include only HSA21 homologous genes found on Mmu16 (Pereira et al., 2009). The Ts1Yah line exhibits deficits in novel object recognition, but improvements above control lines for spatial memory (Pereira et al., 2009). This work again revealed additional complexity in the combinatorial genotype–phenotype correlations that could contribute to the overall DS phenotype.

A complementary approach to genetic dissection of HSA21 genes and their contribution to the DS phenotype is to perform detailed phenotypic characterization in single-locus genetic models. For example, a BAC transgenic model of the *Abcg1* locus found no alterations in behavioral tests for anxiety or working memory (Parkinson et al., 2009). While these results suggest that the *Abcg1* locus does not contribute to the cognitive deficits in DS, caution must be taken in drawing this conclusion, given that the potential influence of other HSA21 loci cannot be ruled out. A *Dyrk1a* BAC transgenic model, in contrast, exhibited profound learning and memory deficits on the Morris water maze test (Ahn et al., 2006). While these results strongly suggest that *Dyrk1a* contributes to learning and memory deficits in DS, caution must also be taken concerning this conclusion as its relative contribution in the context of full trisomy HSA21 is unknown. Another example is the *Sim2* transgenic lines that display anxiety-related behaviors similar to those found in segmental trisomy models that include *Sim2* such as Ts65Dn (Chrast et al., 2000). In addition to overexpression models, genetic knockout of

HSA21 homologous genes may also inform us about the role of genes in the DS phenotype. For example, the *RCAN1* null mouse exhibits deficits in spatial learning and memory as well as contextual learning (Hoeffer et al., 2007). The phenotype is similar to lines with elevated calcineurin activity. Closer study revealed the *RCAN1* null mouse has elevated activity in this pathway (Hoeffer et al., 2007). Functional analysis of HSA21 genes can provide mechanistic insight into the neural pathways underlying behavioral phenotypes.

30.4.2 Neurogenesis

DS is associated with reduced brain size (~80% of typically developing humans) from very early in development (~5-month fetuses) through adulthood (Pinter et al., 2001b; Schmidt-Sidor et al., 1990; Winter et al., 2000). Both hypocellularity and hypoplasia likely contribute to this phenotype (Contestabile et al., 2010). Several regions of the brain are affected including the neocortex (entorhinal, frontal, prefrontal, and temporal cortices), hippocampus, amygdala, cerebellum, hypothalamus, and brainstem (Aylward et al., 1997a, 1999; Contestabile et al., 2010; Kesslak et al., 1994; Pine et al., 1997; Raz et al., 1995; Sylvester, 1983; Teipel and Hampel, 2006; Teipel et al., 2003, 2004). Several studies converge upon particularly substantial decreases in brain volume for the hippocampus and temporal lobe. Indeed, behavioral- and neural activity-based animal studies have focused upon phenotypic abnormalities of the hippocampus in part for this reason. Neurogenesis defects, reductions in the numbers of neurons generated in development, are likely to substantially contribute to reduced brain size in DS and have been observed in both DS fetuses and children (Contestabile et al., 2010; Guidi et al., 2008; Larsen et al., 2008; Wisniewski, 1990; Wisniewski et al., 1984). Schmidt-Sidor and colleagues examined brains of fetuses of 15–22 weeks gestational age and from birth to 60 months of age and reported that initial defects occur sometime between 22 weeks of age and birth (Schmidt-Sidor et al., 1990). Thus neurogenesis defects likely represent the earliest neurological phenotypic evidence of the DS genotype. Indeed, some of the earliest descriptions of neurological phenotypes in mouse models of DS (e.g., trisomy Mmu16) are reports of neurogenesis defects in these models (Haydar et al., 1996; Sweeney et al., 1989).

Detailed characterization and mechanistic studies in animal models of DS, such as the Ts65Dn and Ts1Cje, have confirmed defects in cell proliferation and neurogenesis (Chakrabarti et al., 2007; Contestabile et al., 2009a,b; Moldrich et al., 2009). In agreement with rodent models, a human neuroprogenitor model of DS revealed that despite similar initial production of neurons, progenitors with trisomy HSA21 produced fewer neurons when cultured for over 10 weeks (Bhattacharyya et al., 2009). Examination of granule cell precursors from the Ts65Dn mouse revealed that prolonged G1 and G2 phases of the cell cycle contributed to decreased proliferation (Contestabile et al., 2009a). Examination of cortical and hippocampal neuroprogenitors in the Ts65Dn model also revealed evidence of a longer cell cycle contributing to reduced neurogenesis (Chakrabarti et al., 2007; Contestabile et al., 2007). Furthermore, reductions in neurogenesis in the Ts65Dn mouse model are associated with decreased synaptogenesis in the first postnatal week (Chakrabarti et al., 2007). Multiple neuronal cell types are potentially influenced by alterations in neurogenesis and neurodevelopment, including the serotonergic, noradrenergic, GABAergic, and cholinergic neurons (Bar-Peled et al., 1996; Dierssen et al., 1997; Fiedler et al., 1994; Granholm et al., 2000; Kleschevnikov et al., 2004; Whittle et al., 2007). Finally, the neurogenesis defect may not occur only in development, as the Ts1Cje line exhibits a reduction in adult neurogenesis with a concomitant rise in production of astrocytes (Hewitt et al., 2010).

DYRK1A is a member of the dual-specificity tyrosine-regulated kinase family that maps to HSA21 21q22.2. The *Drosophila* homolog of *DYRK1A*, *minibrain* (*mnb*), is required for postembryonic neurogenesis. Loss of *mnb* causes reduced optic lobes and central brain likely caused at least in part by abnormal spacing of neuroprogenitors in the larval brain (Tejedor et al., 1995). This gene family shares homology with other kinases involved in regulation of cell division. The DYRK kinase family is so named because of its autophosphorylation of tyrosine residues in its YXY activation loop and serine/threonine phosphorylation of its protein substrates. The subsequent mapping to HSA21, findings of early neural expression in humans and rodents, and overexpression in DS brain and the Ts65Dn mouse model drew attention to this gene as a candidate for contributing to the neurogenesis and neurodevelopment phenotypes in DS patients (Guimerá et al., 1996; Guimera et al., 1999; Shindoh et al., 1996; Song et al., 1996). Decreased expression of the mouse homolog causes developmental delay and alteration in brain morphology (Fotaki et al., 2002). To more specifically examine the influence of this gene on neurogenesis in DS, D'Arcangelo and colleagues overexpressed mouse *Dyrk1A* in the cortex by *in utero* electroporation (Yabut et al., 2010). They report that overexpression promotes neuronal maturation while inhibiting proliferation of neural progenitors (Yabut et al., 2010). Minimal influence of cell fate was observed, suggesting that *Dyrk1A* acted predominantly by influencing production of neurons rather than alterations in the neurodevelopmental program itself. Furthermore, these phenotypes were dependent on kinase activity and nuclear export and protein degradation of cyclin D1 (Yabut et al., 2010). In further support of a role of this gene in regulation of neurogenesis and brain size, patients with a loss of the *DYRK1A* gene present with microcephaly (Fujita et al., 2010; Moller et al., 2008).

Other genes likely also contribute to defects in neurogenesis and neurodevelopment. For example, the *Olig1* and *Olig2* genes contribute to an imbalance in excitatory and inhibitory neuronal systems (Chakrabarti et al., 2010). While many neuronal populations exhibit decreased numbers, examination of the Ts65Dn model revealed an increase in the parvalbumin and somatostatin subclasses of inhibitory neurons, with no change in the number of calretinin inhibitory neurons (Chakrabarti et al., 2010). Restoration of disomy to the *Olig1* and *Olig2* genes was found to rescue this phenotype. Other genes that may contribute include *TTC3*, which has been shown to inhibit neuronal differentiation in a cell culture model (Berto et al., 2007). *Sim2* plays a critical role in *Drosophila* neurogenesis (Chrast et al., 2000), and the HSA21 gene *Sim2* mouse homolog exhibits CNS-specific expression patterns consistent with a role in mammalian neurogenesis (Rachidi et al., 2005). Finally, altered responses to key neuronal proliferative signaling systems may also underlie alterations in neurogenesis. Cerebellar granule neuron precursors normally exhibit a strong mitogenic response to the morphogen Sonic hedgehog (SHH). Purified granule neuron precursors from the Ts65Dn model were found to have a significantly reduced proliferative response to SHH (Roper et al., 2006a). Interestingly, administration of a small molecule agonist of the SHH signaling pathway (Smoothed Agonist) was able to rescue the decreased proliferative phenotype of these cells (Roper et al., 2006a). In summary, a complex interplay of developmental signaling and altered expression of multiple HSA21 genes likely contributes to the DS neurogenesis phenotype.

30.4.3 Regional Connectivity and Development of Neural Circuits

One plausible explanation for the learning and memory deficits and other psychiatric and cognitive phenotypes associated with DS is a failure to set up the normal neural circuits and appropriate brain regional connectivity in development. Thus, in addition to alterations in the numbers and types of neurons generated during development (discussed earlier in Section 30.4.2), the establishment of abnormal brain circuitry during development is also hypothesized to contribute to the DS phenotype. However, less work has been done in this area relative to studies examining the contribution of alterations in neurogenesis and/or changes in neuronal activity at the cellular level. Indeed,

reported alterations in brain morphology are often attributed to alterations in regional neurogenesis rather than changes in the neural circuits themselves (Contestabile et al., 2010; Rachidi and Lopes, 2007). MRI studies of people with DS indicate the potential of structural changes (e.g., disproportionate loss of volume in the cerebellum, hippocampus, and a subregion of the superior temporal gyrus and increased subcortical gray matter) to underlie DS cognitive and behavioral phenotypes (Frangou et al., 1997; Pinter et al., 2001a,b; Raz et al., 1995). A recent fMRI study of children with DS during a story-listening task revealed reduced activation of receptive language areas (superior and temporal gyri; Losin et al., 2009). Likewise, a magnetoencephalography study has suggested that cortical activation patterns during motor tasks and motor observational tasks are less coherent in the DS brain, showing broader activation patterns and reduced lateralization (Virji-Babul et al., 2010). Finally, changes in amino acid and monoamine neurotransmitters (e.g., serotonin, dopamine, and GABA) in the DS fetal brain also suggest defects in the development of neural circuits (Whittle et al., 2007). Despite evidence that alterations in brain structure and connectivity underlie mental retardation in DS and other developmental disorders, there is growing recognition that ameliorative therapy may still be possible (Dierssen and Ramakers, 2006).

As mentioned previously, a prominent feature of the DS phenotype is reduced brain size. Detailed three-dimensional quantification of brain morphology by high-resolution MRI in the Ts65Dn and Ts1Rhr models has examined the relative contribution of changes in brain volume and brain shape to this phenotype (Aldridge et al., 2007). Such analysis demonstrates severe reductions in cerebellar volume, with minimal volume loss in other brain regions (Aldridge et al., 2007), confirming early observations of reduced cerebellar volume in the Ts65Dn model as well as DS patients (Baxter et al., 2000; Scott et al., 1983). Interestingly, the Ts1Rhr model shows a subtle decrease in overall brain volume, with a relative sparing of the cerebellum (Aldridge et al., 2007). This suggests that genes triplicated only in the large segmental trisomic Ts65Dn model may be responsible for the reduced cerebellar volume phenotype. Furthermore, it is noteworthy that overall brain shape, measured using 29 independent morphological landmarks, is unchanged in the Ts65Dn model (Aldridge et al., 2007). One interpretation of these findings is that alterations in the cell number and perhaps degree of connectivity (see below) may be more important for the DS phenotype than dysmorphic or improper/aberrant neural connections across different brain regions.

Specific HSA21 genes have been implicated in the connectivity and neural circuitry phenotypes in DS. The triplicated *Dscam* gene functions in neuronal morphogenesis and connectivity by contributing to establishment of diverse neural connectivity patterns in development (Gao, 2007; Schmucker, 2007; Yu et al., 2010). In addition, mosaic spacing, axonal targeting, and appropriate arborization of the neuronal pathways in the mouse and *Drosophila* brain require *Dscam* (Fuerst et al., 2008; Hattori et al., 2007; Hummel et al., 2003). Additional evidence for altered connectivity was acquired indirectly by measures of cellular transport between the hippocampal to forebrain circuit. That the circuit appears to be intact, and indeed more robust than wild type, suggests that this circuit is not functioning properly in Ts65Dn mice (Bearer et al., 2007). Lastly, alterations in the structural features on synapses likely contribute to the DS phenotype. Lucifer yellow injections into neurons of the hippocampus and cortex of the Ts65Dn model revealed enlarged dendritic spines and other changes in synaptic architecture (Belichenko et al., 2004). The *Dyrk1a* gene (trisomic in both Ts65Dn and Ts1Rhr models) has been strongly implicated in defects in both neuronal morphology and connectivity by altering synapse formation and dendritic architecture (Contestabile et al., 2010). Heterozygosity (loss of a single allele) of *Dyrk1a* in the mouse alters cortical circuitry by decreasing dendritic branching and spines (Benavides-Piccione et al., 2005), perhaps via alterations in cAMP-responsive element-binding protein or NFAT activity (Arron et al., 2006; Hammerle et al., 2003). Furthermore, alterations in the Notch signaling pathway due to increased Dyrk1a activity could contribute to alterations in neural developmental pathways in DS (Fernandez-Martinez et al., 2009). Additional evidence for alterations in neural circuitry comes from analysis of electrophysiological alterations in the hippocampus revealing increased connectivity despite reduced excitatory and inhibitory inputs (Hanson et al., 2007). An important question for potential therapeutic intervention is whether these and other changes in neuronal morphology can be altered once established. Evidence in favor of this view is emerging, based in part on improvements in cortical pyramidal cell dendritic and spine phenotypes associated with environmental enrichment in Ts65Dn models (Dierssen et al., 2003). Interestingly, such environmental enrichment-related alterations in brain structure correlate with amelioration of behavioral, cell signal transduction, and neurogenesis phenotypes in the Ts65Dn model (Baamonde et al., 2011; Chakrabarti et al., 2011; Martínez-Cué et al., 2005). These data add to a growing awareness that even neuronal structure-related phenotypes in DS may be amenable to clinical intervention (Dierssen and Ramakers, 2006).

30.4.4 Neuronal Dysfunction in DS

Alterations in neuronal function are diverse, as expected given the DS genotype. Our discussion of a key subset of altered cellular function is split into four broad categories below.

30.4.4.1 Plasticity and Synaptic Function

There is evidence for alterations in both excitatory and inhibitory circuits in DS. Loss of asymmetric synapses (predominantly excitatory), without changes in the number of symmetric synapses (predominantly inhibitory), in the cortex of the Ts65Dn model suggested that the excitatory circuits may be the first to be affected (Kurt et al., 2000). However, detailed examination revealed alterations in the size of symmetric synapses (Belichenko et al., 2009a,b). Furthermore, immunostaining for markers of inhibitory neural circuits revealed increased calretinin GABAergic synapses in the cortex of the Ts65Dn model (Pérez-Cremades et al., 2010). Modulatory circuits may also be affected, as suggested by the loss of striatal interneuron LTP in the Ts65Dn model (Di Filippo et al., 2010). Indeed, changes in synaptic plasticity have been observed throughout the brain including decreased LTP in the hippocampus (Hanson et al., 2007; Kleschevnikov et al., 2004; Siarey et al., 1997). LTP and LTD deficits have also been reported in the Ts1Cje DS mouse model (Siarey et al., 2005). Importantly, the most complete of the segmental trisomic mouse models, the Dp(10, 16, 17) mouse model, validated electrophysiological and corresponding behavioral deficits (Yu et al., 2010). Changes in a pair associative stimulation (PAS) paradigm confirmed these findings of LTP deficits in DS patients (Battaglia et al., 2008). Reduced activation of NMDA receptors, increased GABA$_B$ potassium currents, and elevated rates of mESPCs in the Ts65Dn model likely contribute to the changes in synaptic plasticity (Best et al., 2007, 2008; Kleschevnikov et al., 2004). Depressed hippocampal LTP was found to be suppressed by pharmacological interventions that increase inhibitory tone (Kleschevnikov et al., 2004). The GIRK2 potassium channel, encoded by an HSA21 gene, has also been implicated in changes in the balance of excitatory and inhibitory transmission in DS (Best et al., 2007; Harashima et al., 2006). Another HSA21 gene, *RCAN1*, regulates LTP via phosphatase signaling (Hoeffer et al., 2007). Additionally, LTP and LTD are altered in a *DYRK1A* BAC transgenic model overexpressing this single HSA21 gene (Ahn et al., 2006).

Careful expression analysis of functional components has revealed subtle defects in synaptic functional components in DS. Proteomic analysis of synaptosomal fractions in Ts65Dn mice suggested that the overall composition of synaptic proteins is similar to that in wild-type animals (Fernandez et al., 2009), though careful quantitative assessment of key components (e.g., NR2A, GAD65/67, VGAT, GluR2, Cdk5, neurotrophin-3, and GABA$_A$ receptor) has revealed subtle changes (Altafaj et al., 2008; Belichenko et al., 2009a,b; Pérez-Cremades et al., 2010; Pollonini et al., 2008). One contributing genetic factor is likely to be trisomic expression of *Dyrk1a*, which has been linked to changes in

NR2A expression and calcium transients (Altafaj et al., 2008). The HSA21 gene *SYNJ1* encoding Synaptojanin 1 is a key regulator of the phosphatidylinositol signaling and metabolism that is required for normal neuronal transmission (Voronov et al., 2008).

30.4.4.2 Genetic and Proteomic Dysregulation

The most obvious of the cellular changes expected in DS are gene expression differences due to the extra copy of all HSA21 genes. However, likely due to the complexities of gene regulation, not all HSA21 genes are overexpressed by the expected 1.5-fold (Ait Yahya-Graison et al., 2007; FitzPatrick et al., 2002; Mao et al., 2003, 2005). Indeed, these studies revealed genes on HSA21 whose expression was either unchanged from diploid, overexpressed beyond what is expected from the dose alone, or even downregulated (FitzPatrick, 2005; Rachidi and Lopes, 2008). Furthermore, genes located on other chromosomes also have been found to exhibit profound differences in expression. These results from human tissues have been corroborated across DS mouse models including the Ts1Cje and Ts65Dn (Amano et al., 2004; Hewitt et al., 2010; Kahlem et al., 2004; Moldrich et al., 2009; Wang et al., 2004).

Clearly, the magnitude of expression differences dictates the phenotype, and the temporal and physical pattern of expression contributes as well to the phenotypic consequences of DS. HSA21 gene and HSA21 gene homolog expression maps have been generated to identify tissues and brain regions where and when HSA21 genes are expressed and misexpressed in DS (Gitton et al., 2002; Kahlem et al., 2004; Reymond et al., 2002). The influence of trisomy HSA21 is further complicated by the finding that overexpression of the HSA21 homolog Dyrk1a is sufficient to elicit global changes in cortical gene splicing that are also seen in DS fetal brain samples (Toiber et al., 2010). Thus, even a single gene whose expression is altered by trisomy can influence gene expression throughout the genome. Finally, changes in gene expression do not consistently result in matching changes in protein levels. Thus recent analysis of the DS proteome will also be necessary to achieve an accurate model of the cellular consequences of the DS genotype (Delom et al., 2009; Patterson, 2009; Shin et al., 2006, 2007).

30.4.4.3 Dysregulation of Vital Cellular Processes

The transcriptome and proteome alterations in DS potentially lead to diverse dysregulation of vital cellular processes. For example, several HSA21 genes have been functionally linked to membrane and vesicular trafficking. The HSA21 homologous gene, *Itsn1*, plays a role in endocytic processing and vesicular trafficking (Yu et al., 2010). Misexpression of the HSA21 homolog of β-amyloid precursor protein (β-APP) contributes to

alterations in nerve growth factor (NGF) transport (Salehi et al., 2006), while the *DSCR1/RCAN1* homologue regulates vesicle exocytosis (Keating, 2008). Other changes include alterations in the levels of myo-inositol in the brain, altered proteolytic processing of APP, and increased sensitivity to genotoxic stress (Beacher et al., 2005; Micali et al., 2010; Tansley et al., 2007).

30.4.5 Aging and Neurodegeneration

Neurodegeneration in DS is known to produce significant cognitive deficits and declines in patients afflicted with this illness. The pathogenesis of the disease is thought to be based on several major age-related processes: the formation of amyloid plaques, the degeneration of basal forebrain cholinergic neurons (BFCNs), and increased production of reactive oxygen species (ROS). Neurodegenerative mechanisms contribute to DS cognitive impairment given that early-onset AD occurs with an extremely high prevalence in these individuals (Contestabile et al., 2010). These neurodegenerative processes are known to have a strong genetic component, as several genes promote amyloidosis and respond to oxidative stress in the pathophysiology of DS. As the lifespan of DS individuals has increased significantly in recent years, obtaining greater understanding of pathways critical to the neuropathology of DS has been an active area of research that may yield future therapeutic targets to halt the progression of neurological symptoms.

30.4.5.1 DS and the Formation of Amyloid Plaques and NFTs

Individuals with DS present in the clinic with formations of amyloid plaques and progressive degeneration of BFCN (Casanova et al., 1985; Lockrow et al., 2009; Mufson et al., 2003). The presentations of senile plaques generally occur during middle age and are highly similar to that observed in the pathology of AD. The formation of these AD-type brain lesions occurs by the fourth decade of life, which is 20–30 years earlier in onset than patients suffering from AD (Iqbal and Grundke-Iqbal, 2010; Iqbal et al., 2010). These plaques are causally related to overexpression of β-APP whose gene is triplicated in DS (Contestabile et al., 2010). Full-length β-APP is proteolytically cleaved by β- and γ-secretase generating Aβ peptides, which then form plaques in the DS brain (Contestabile et al., 2010). Aβ lesions in the form of this 'preamyloid' appear by around 12 years of age (though they have been described in the DS neonatal brain as well) and are described as amorphous non-fibrillar aggregates with few dystrophic neurites (Contestabile et al., 2010; Giaccone et al., 1989; Kida et al., 1995; Lemere et al., 1996; Mann and Esiri, 1989; Motte and Williams, 1989; Wisniewski et al., 1994). These preamyloid plaques progress to mature Aβ plaques that are associated with neuronal damage and generally appear during the third decade of life (Contestabile et al., 2010; Lemere et al., 1996; Wisniewski et al., 1994). Current research suggests that in addition to amyloidosis, overproduction of Aβ can trigger early cognitive impairment in dementia by modulating synaptic activity (Contestabile et al., 2010; Conti and Cattaneo, 2005; Gasparini and Dityatev, 2008). Aβ aggregates block LTP in excitatory synapses and enhance LTD by inhibiting glutamate receptors and downregulating N-methyl-D-aspartate receptors (NMDARs; Contestabile et al., 2010; Lambert et al., 1998; Li et al., 2009; Shankar et al., 2008; Townsend et al., 2006; Walsh et al., 2002; Wang et al., 2002).

In addition to Aβ, intraneuronal inclusions of hyperphosphorylated tau, a microtubule-stabilizing protein, develop in the proximal dendrites and cell bodies of neurons and within dystrophic neurites (Contestabile et al., 2010; Gasparini et al., 2007). Abnormal phosphorylation of tau causes the formation of NFTs, neuropil threads, and plaque dystrophic neurites associated with various neocortical tauopathies that cause dementia (Grundke-Iqbal et al., 1986a,b, 1988; Iqbal et al., 1986, 1989, 2009; Lee et al., 1991). In addition, Aβ and tau lesions are known to impact many brain regions in DS such as the prefrontal cortex, hippocampus, basal ganglia, thalamus, hypothalamus, and midbrain (Contestabile et al., 2010; Wisniewski et al., 1985). These lesions are believed to mediate the cognitive decline and dementia associated with DS.

The age-dependent DS phenotypes in the central nervous system are not restricted to cognitive processes but they also impact the visual system. DS patients present with distinctive early-onset cerulean cataracts, 'blue dot' cataracts that emerge at the equatorial periphery of the lens (Moncaster et al., 2010). The lenses of DS individuals have a characteristic pattern of supranuclear opacification accompanied by accumulation of Aβ, amyloid pathology, and fiber cell Aβ aggregates (Moncaster et al., 2010). Laboratory experiments have demonstrated that incubation of human lens protein in synthetic Aβ promoted light scattering, protein aggregation, and amyloid formation suggesting that the DS lens phenotype is a genetic cataract (Moncaster et al., 2010).

30.4.5.2 Degeneration of BFCNs in DS

In addition to amyloid plaque formation, degeneration of BFCNs and decreased activity of choline acetyltransferase with increasing age has been observed in DS. One of the most significant symptoms associated with death of cholinergic neurons in DS patients is memory loss (Ginsberg et al., 2006; Lockrow et al., 2009). In addition, BFCNs are responsible for providing the majority of cholinergic innervations to the hippocampus

and cortex and play a significant role in attention and cognition in both humans and mouse models (Lockrow et al., 2009; Perry et al., 1977; Whitehouse et al., 1982). In particular, DS-related cognitive declines and adult-onset degeneration of BFCNs are observed in Ts65Dn mice (Chen et al., 2008; Cooper et al., 2001; Granholm et al., 2000; Holtzman et al., 1996; Lockrow et al., 2009). These results are not unexpected, as there is a triplication of amyloid precursor protein (*App*) gene in the Ts65Dn mouse (Cataldo et al., 2003; Lockrow et al., 2009). *App* triplication in the Ts65Dn mouse also influences early endosomal processes, NGF trafficking, and loss of BFCNs (Lockrow et al., 2009; Salehi et al., 2006). Interestingly, these declines occur despite absence of amyloid plaques in the Ts65dn mouse model (Seo and Isacson, 2005).

30.4.5.3 ROS, Oxidative Stress, and Neurodegeneration in DS

Based on the close mirroring of the neuropathological findings between DS and AD, the role of mitochondria, specifically oxidative stress, has been highly implicated in the pathogenesis of DS. After development of a prooxidant state early in the progression of the disease, individuals suffering from DS demonstrate elevated levels of DNA damage and lipid peroxidation (Jovanovic et al., 1998; Lockrow et al., 2009; Pallardo et al., 2006). Furthermore, cortical neurons in DS individuals show elevation in ROS, which contributes to neuronal degeneration and induction of apoptosis (Busciglio and Yankner, 1995; Lockrow et al., 2009). Decreased levels of endogenous antioxidants such as α-tocopherol, an inhibitor of lipid peroxidation, have been demonstrated in patients with AD; however, no evidence for a protective effect has been reported for DS individuals (Azzi et al., 2003; Lockrow et al., 2009). ROS elevation does not determine the fate of cortical neurons, as administration of antioxidants such as vitamin E has been shown to be restorative (Behar and Colton, 2003; Lockrow et al., 2009; Schuchmann and Heinemann, 2000). This may provide a therapeutic window into improvement of learning and memory in DS patients.

30.4.5.4 Genetic Component of Neurodegeneration in DS

Genetics also plays a profound role in the development of plaques and NFTs. The protein dual-specificity tyrosine phosphorylation-regulated kinase 1A (DYRK1A) is a serine/threonine kinase that has activity both in neurodevelopment as well as in neurodegeneration (Contestabile et al., 2010). Upregulation of DYRK1A has been demonstrated in DS. This enhancement of expression may result in hyperphosphorylation of tau that further phosphorylates glycogen synthase-3β causing tau self-aggregation into NFTs (Contestabile et al., 2010; Liu et al., 2008). DS does not solely associate with upregulation of DYRK1A, but its kinase activity is also enhanced, further promoting assembly of tau into filaments (Iqbal et al., 2009; Liu et al., 2008). Similarly, Down syndrome critical region 1 (*DSCR1*), a calcineurin inhibitor, is thought to play a role in DS given its important role regulating mitochondrial function and oxidative stress. *DSCR1* functions as a stress response element as it is upregulated in reaction to elevated hydrogen peroxide (Contestabile et al., 2010; Crawford et al., 1997; Lin et al., 2003). Despite this critical role in reducing oxidative damage to cells, chronic overexpression of *DSCR1* in the context of DS has a number of negative side effects consistent with an AD-type degenerative process (Contestabile et al., 2010). DSCR1 inhibits the Ca2+/calmodulin-dependent protein serine/threonine phosphatase calcineurin and regulates synaptic activity (Contestabile et al., 2010; Ermak et al., 2002; Fuentes et al., 2000; Kingsbury and Cunningham, 2000). DSCR1-mediated inhibition of calcineurin can lead to increased NMDAR mean open time and opening probability, which alters neuronal excitability in the brain (Contestabile et al., 2010; Lieberman and Mody, 1994). In addition to modulating synaptic excitability, calcineurin acts to dephosphorylate tau, and inhibition of this pathway leads to a pathological hyperphosphorylation of tau that correlates with cognitive impairment (Contestabile et al., 2010; Gong et al., 1996; Luo et al., 2008; Yu et al., 2006).

30.5 GENETIC MECHANISMS UNDERLYING DS

The prevailing model of the genetic mechanism underlying the DS phenotype is the 'gene dosage effect' model (Contestabile et al., 2010; Patterson, 2007; Rachidi and Lopes, 2007; Roper and Reeves, 2006; Salehi et al., 2007). This model suggests that individual genes along HSA21 each contribute to specific endophenotypes that culminate in the overall DS phenotype. This is likely an oversimplification, as it does not take into account genetic interactions among HSA21 genes. Thus, the emergent influence of interactions between multiple trisomic genes has also been proposed as a genetic mechanism in DS. However, this model can be viewed as just an elaboration of the 'gene dosage' model that accounts for genetic interaction effects. Indeed, strong evidence exists to support such a role of genetic interactions between HSA21 genes in the DS phenotype. For example, overexpression of homologs for three HSA21 genes (*ITSN1*, *SYNJ1*, and *DSCR1*) alters synaptic function in *Drosophila*, but restoring expression of just one ameliorates this phenotype (Chang and Min, 2009). Likewise, substantial evidence exists supporting a role for functional interactions between the HSA21 genes *DSCR1* and *DYRK1A* (de la Luna and Estivill, 2006). The concept

of the DSCR has been put forward to support the hypothesis that a subset of genes on HSA21 contributes to many of the major DS traits (Contestabile et al., 2010; Patterson, 2009). Extensive phenotyping of mouse models that contain portions of the DSCR has elucidated genetic intervals that contribute to specific brain-related and other phenotypes (Belichenko et al., 2009a,b; Olson et al., 2007). In addition, case reports of patients with microdeletions and/or small segmental duplications have furthered the correlation of gene interval with specific DS endophenotypes; some recent notable examples are cited here (Eggermann et al., 2010; Fujita et al., 2010; Kondo et al., 2006; Lyle et al., 2009; Ronan et al., 2007; Sato et al., 2008). Lastly, nonspecific effects of the genetic imbalance caused by trisomy have the potential to modify global expression patterns to contribute to the DS phenotype (Contestabile et al., 2010; Patterson, 2007; Rachidi and Lopes, 2007; Roper and Reeves, 2006; Salehi et al., 2007). Together, these three genetic mechanisms are thought to underlie the complex genotype–phenotype correlations that exist in DS.

30.6 TRANSLATIONAL AND THERAPEUTIC STRATEGIES IN DS

As detailed above, owing to the strength of evidence that mouse models of DS display clinically relevant phenotypes, researchers have begun to explore pharmacological and other intervention strategies for therapeutic benefit. Therapeutic strategies range from very early phenotypes (e.g., decreased neurogenesis) to behavioral phenotypes (e.g., learning and memory deficits, anxiety) and neurodegenerative phenotypes (e.g., cognitive decline). For example, given the reduced levels of serotonin in the DS fetal brain, the selective serotonin reuptake inhibitor fluoxetine was tested for rescue of the reduced neurogenesis phenotype in the Ts65Dn model (Clark et al., 2006; Whittle et al., 2007). Indeed, 2- to 3-week treatment paradigms increased neurogenesis in the hippocampus, subventricular zone, cortex, and striatum (Bianchi et al., 2010; Clark et al., 2006). Fluoxetine treatment was also found to restore expression levels of the serotonin 1A receptor (5-HT1A) and brain-derived neurotrophic factor (BDNF) as well as to show improvement in a hippocampal-dependent memory task (Bianchi et al., 2010). The NMDA receptor uncompetitive antagonist memantine is FDA-approved for treatment of moderate to severe AD. Memantine was subsequently shown to rescue memory task deficits, reduced neuronal number in the hippocampus, and decreased vesicular glutamate transporter-1 (VGAT1) expression (Costa et al., 2008; Rueda et al., 2010). More recent studies in the Ts65Dn mouse model suggest therapeutic potential for GABA$_A$ receptor inverse agonists or β1-adrenergic receptor agonists in amelioration of cognitive phenotypes (Braudeau et al., 2011; Faizi et al., 2011). In addition, nutritional manipulation has also shown therapeutic promise, for example, perinatal choline, vitamin E, or green tea extract supplementation (Guedj et al., 2009; Lockrow et al., 2009; Moon et al., 2010). Future studies aimed at counteracting the influence of trisomy HSA21 genes and dysregulated gene networks provide promise of therapeutic intervention.

30.7 SUMMARY

Trisomy of HSA21 perhaps yields a remarkably mild clinical phenotype in the face of the profound genetic insult expected from triplication of over 400 genes. The complexity of the genetics raises an incredible challenge to elucidate genotype–phenotype correlations across the constellation of endophenotypes and incomplete penetrance that contribute to the clinical presentation of this disorder. Experimental models of DS range in complexity, model organism, and genetic completeness. But without doubt, the mainstay of DS basic research has been mouse genetic models with trisomy of mouse chromosomal regions that are syntenic to HSA21 (e.g., Ts65Dn and Dp(10, 16, 17)). However, *Drosophila, C. elegans*, yeast, other model organisms and human cell lines also provide important tools in dissecting the function and interactions of HSA21 genes. The recent generation of DS iPSC lines yields the promise of human neuronal models of disease for mechanistic and translational studies. The major neurological phenotypes associated with DS include (1) behavioral, motor, psychiatric, and cognitive alteration; (2) deficient neurogenesis during development and perhaps in adulthood; (3) alterations in neural connectivity and synaptogenesis; (4) neuronal dysfunction including deficits in plasticity; and (5) neurodegenerative features similar to AD. Analysis of the genetic mechanisms underlying DS has revealed a prominent role for a subset of HSA21 genes. This provides hope that therapeutic targeting of specific functional deficits can ameliorate DS endophenotypes and provide clinical and behavioral benefits.

References

Abbeduto, L., Pavetto, M., Kesin, E., et al., 2001. The linguistic and cognitive profile of Down syndrome: Evidence from a comparison with fragile X syndrome. Down Syndrome Research and Practice 7, 9–15.

Amano, K., Sago, H., Uchikawa, C., et al., 2004. Dosage-dependent overexpression of genes in the trisomic region of Ts1Cje mouse model for Down syndrome. Human Molecular Genetics 13, 1333–1340.

Ahn, K., Jeong, H., Choi, H., et al., 2006. DYRK1A BAC transgenic mice show altered synaptic plasticity with learning and memory defects. Neurobiology of Disease 22, 463–472.

Ait Yahya-Graison, E., Aubert, J., Dauphinot, L., et al., 2007. Classification of human chromosome 21 gene-expression variations in Down syndrome: Impact on disease phenotypes. American Journal of Human Genetics 81, 475–491.

Aldridge, K., Reeves, R.H., Olson, L.E., Richtsmeier, J.T., 2007. Differential effects of trisomy on brain shape and volume in related aneuploid mouse models. American Journal of Medical Genetics. Part A 143A, 1060–1070.

Altafaj, X., Ortiz-Abalia, J., Fernández, M., et al., 2008. Increased NR2A expression and prolonged decay of NMDA-induced calcium transient in cerebellum of TgDyrk1A mice, a mouse model of Down syndrome. Neurobiology of Disease 32, 377–384.

American Academy of Pediatrics, Committee on Genetics, 2011. American Academy of Pediatrics: Health supervision for children with Down syndrome. Pediatrics 128, 393–407.

Antonarakis, S.E., Lyle, R., Dermitzakis, E.T., Reymond, A., Deutsch, S., 2004. Chromosome 21 and Down syndrome: From genomics to pathophysiology. Nature Reviews Genetics 5, 725–738.

Arron, J.R., Winslow, M.M., Polleri, A., et al., 2006. NFAT dysregulation by increased dosage of DSCR1 and DYRK1A on chromosome 21. Nature 441, 595–600.

Aylward, E.H., Burt, D.B., Thorpe, L.U., Lai, F., Dalton, A., 1997a. Diagnosis of dementia in individuals with intellectual disability. Journal of Intellectual Disability Research 41 (Pt 2), 152–164.

Aylward, E.H., Habbak, R., Warren, A.C., et al., 1997b. Cerebellar volume in adults with Down syndrome. Archives of Neurology 54, 209–212.

Aylward, E.H., Li, Q., Honeycutt, N.A., et al., 1999. MRI volumes of the hippocampus and amygdala in adults with Down's syndrome with and without dementia. The American Journal of Psychiatry 156, 564–568.

Azzi, A., Gysin, R., Kempná, P., et al., 2003. The role of alpha-tocopherol in preventing disease: From epidemiology to molecular events. Molecular Aspects of Medicine 24, 325–336.

Baamonde, C., Martínez-Cué, C., Flórez, J., Dierssen, M., 2011. G-protein-associated signal transduction processes are restored after postweaning environmental enrichment in Ts65Dn, a Down syndrome mouse model. Developmental Neuroscience http://dx.doi.org/10.1159/000329425.

Baek, K., Zaslavsky, A., Lynch, R., et al., 2009. Down syndrome suppression of tumor growth and the role of the calcineurin inhibitor DSCR1. Nature 459, 1126–1130.

Balme, D.M., Gotthelf, A. (Eds.), 2002. Aristotle. Historia Animalium. Books I-X. Cambridge University Press, Cambridge, p. v.

Bar-Peled, O., Korkotian, E., Segal, M., Groner, Y., 1996. Constitutive overexpression of Cu/Zn superoxide dismutase exacerbates kainic acid-induced apoptosis of transgenic-Cu/Zn superoxide dismutase neurons. Proceedings of the National Academy of Sciences of the United States of America 93, 8530–8535.

Battaglia, F., Quartarone, A., Rizzo, V., et al., 2008. Early impairment of synaptic plasticity in patients with Down's syndrome. Neurobiology of Aging 29, 1272–1275.

Baxter, L.L., Moran, T.H., Richtsmeier, J.T., Troncoso, J., Reeves, R.H., 2000. Discovery and genetic localization of Down syndrome cerebellar phenotypes using the Ts65Dn mouse. Human Molecular Genetics 9, 195–202.

Beacher, F., Simmons, A., Daly, E., et al., 2005. Hippocampal myoinositol and cognitive ability in adults with Down syndrome: An in vivo proton magnetic resonance spectroscopy study. Archives of General Psychiatry 62, 1360–1365.

Bearer, E.L., Zhang, X., Jacobs, R.E., 2007. Live imaging of neuronal connections by magnetic resonance: Robust transport in the hippocampal-septal memory circuit in a mouse model of Down syndrome. NeuroImage 37, 230–242.

Becker, L.E., Armstrong, D.L., Chan, F., 1986. Dendritic atrophy in children with Down's syndrome. Annals of Neurology 20, 520–526.

Behar, T., Colton, C., 2003. Redox regulation of neuronal migration in a Down syndrome model. Free Radical Biology & Medicine 35, 566–575.

Belichenko, P., Masliah, E., Kleschevnikov, A., et al., 2004. Synaptic structural abnormalities in the Ts65Dn mouse model of Down syndrome. The Journal of Comparative Neurology 480, 281–298.

Belichenko, P., Kleschevnikov, A., Salehi, A., Epstein, C., Mobley, W., 2007. Synaptic and cognitive abnormalities in mouse models of Down syndrome: Exploring genotype-phenotype relationships. The Journal of Comparative Neurology 504, 329–345.

Belichenko, N.P., Belichenko, P.V., Kleschevnikov, A.M., Salehi, A., Reeves, R.H., Mobley, W.C., 2009a. The 'Down syndrome critical region' is sufficient in the mouse model to confer behavioral, neurophysiological, and synaptic phenotypes characteristic of Down syndrome. Journal of Neuroscience 29, 5938–5948.

Belichenko, P.V., Kleschevnikov, A.M., Masliah, E., et al., 2009b. Excitatory-inhibitory relationship in the fascia dentata in the Ts65Dn mouse model of Down syndrome. The Journal of Comparative Neurology 512, 453–466.

Benavides-Piccione, R., Dierssen, M., Ballesteros-Yanez, I., et al., 2005. Alterations in the phenotype of neocortical pyramidal cells in the Dyrk1A+/- mouse. Neurobiology of Disease 20, 115–122.

Berto, G., Camera, P., Fusco, C., et al., 2007. The Down syndrome critical region protein TTC3 inhibits neuronal differentiation via RhoA and Citron kinase. Journal of Cell Science 120, 1859–1867.

Best, T.K., Siarey, R.J., Galdzicki, Z., 2007. Ts65Dn, a mouse model of Down syndrome, exhibits increased GABA(B)-induced potassium current. Journal of Neurophysiology 97, 892–900.

Best, T.K., Cho-Clark, M., Siarey, R.J., Galdzicki, Z., 2008. Speeding of miniature excitatory post-synaptic currents in Ts65Dn cultured hippocampal neurons. Neuroscience Letters 438, 356–361.

Bhattacharyya, A., Mcmillan, E., Chen, S.I., Wallace, K., Svendsen, C.N., 2009. A critical period in cortical interneuron neurogenesis in Down syndrome revealed by human neural progenitor cells. Developmental Neuroscience 31, 497–510.

Bianchi, P., Ciani, E., Guidi, S., et al., 2010. Early pharmacotherapy restores neurogenesis and cognitive performance in the Ts65Dn mouse model for Down syndrome. Journal of Neuroscience 30, 8769–8779.

Bittles, A.H., Bower, C., Hussain, R., Glasson, E.J., 2007. The four ages of Down syndrome. European Journal of Public Health 17, 221–225.

Braudeau, J., Delatour, B., Duchon, A., et al., 2011. Specific targeting of the GABA-A receptor {alpha}5 subtype by a selective inverse agonist restores cognitive deficits in Down syndrome mice. Journal of Psychopharmacology 8, 1030–1042.

Busciglio, J., Yankner, B., 1995. Apoptosis and increased generation of reactive oxygen species in Down's syndrome neurons in vitro. Nature 378, 776–779.

Carlesimo, G.A., Marotta, L., Vicari, S., 1997. Long-term memory in mental retardation: Evidence for a specific impairment in subjects with Down's syndrome. Neuropsychologia 35, 71–79.

Casanova, M.F., Walker, L.C., Whitehouse, P.J., Price, D.L., 1985. Abnormalities of the nucleus basalis in Down's syndrome. Annals of Neurology 18, 310–313.

Cataldo, A., Petanceska, S., Peterhoff, C., et al., 2003. App gene dosage modulates endosomal abnormalities of Alzheimer's disease in a segmental trisomy 16 mouse model of Down syndrome. Journal of Neuroscience 23, 6788–6792.

Chakrabarti, L., Galdzicki, Z., Haydar, T.F., 2007. Defects in embryonic neurogenesis and initial synapse formation in the forebrain of the Ts65Dn mouse model of Down syndrome. Journal of Neuroscience 27, 11483–11495.

Chakrabarti, L., Best, T.K., Cramer, N.P., et al., 2010. Olig1 and Olig2 triplication causes developmental brain defects in Down syndrome. Nature Neuroscience 13, 927–934.

Chakrabarti, L., Scafidi, J., Gallo, V., Haydar, T.F., 2011. Environmental enrichment rescues postnatal neurogenesis defect in the male and female Ts65Dn mouse model of Down syndrome. Developmental Neuroscience 33, 428–441. http://dx.doi.org/10.1159/000329423.

Chang, K.T., Min, K.-T., 2005. Drosophila melanogaster homolog of Down syndrome critical region 1 is critical for mitochondrial function. Nature Neuroscience 8, 1577–1585.

Chang, K.T., Min, K.-T., 2009. Upregulation of three Drosophila homologs of human chromosome 21 genes alters synaptic function: Implications for Down syndrome. Proceedings of the National Academy of Sciences of the United States of America 106, 17117–17122.

Chen, Y., Dyakin, V., Branch, C., et al., 2008. In vivo MRI identifies cholinergic circuitry deficits in a Down syndrome model. Neurobiology of Aging 30, 1453–1465.

Chrast, R., Scott, H.S., Madani, R., et al., 2000. Mice trisomic for a bacterial artificial chromosome with the single-minded 2 gene (Sim2) show phenotypes similar to some of those present in the partial trisomy 16 mouse models of Down syndrome. Human Molecular Genetics 9, 1853–1864.

Clark, S., Schwalbe, J., Stasko, M.R., Yarowsky, P.J., Costa, A.C.S., 2006. Fluoxetine rescues deficient neurogenesis in hippocampus of the Ts65Dn mouse model for Down syndrome. Experimental Neurology 200, 256–261.

Collacott, R.A., Cooper, S.A., Mcgrother, C., 1992. Differential rates of psychiatric disorders in adults with Down's syndrome compared with other mentally handicapped adults. The British Journal of Psychiatry 161, 671–674.

Contestabile, A., Fila, T., Ceccarelli, C., et al., 2007. Cell cycle alteration and decreased cell proliferation in the hippocampal dentate gyrus and in the neocortical germinal matrix of fetuses with Down syndrome and in Ts65Dn mice. Hippocampus 17, 665–678.

Contestabile, A., Fila, T., Bartesaghi, R., Ciani, E., 2009a. Cell cycle elongation impairs proliferation of cerebellar granule cell precursors in the Ts65Dn mouse, an animal model for Down syndrome. Brain Pathology 19, 224–237.

Contestabile, A., Fila, T., Cappellini, A., Bartesaghi, R., Ciani, E., 2009b. Widespread impairment of cell proliferation in the neonate Ts65Dn mouse, a model for Down syndrome. Cell Proliferation 42, 171–181.

Contestabile, A., Benfenati, F., Gasparini, L., 2010. Communication breaks-down: From neurodevelopment defects to cognitive disabilities in Down syndrome. Progress in Neurobiology 91, 1–22.

Conti, L., Cattaneo, E., 2005. Controlling neural stem cell division within the adult subventricular zone: An APPealing job. Trends in Neurosciences 28, 57–59.

Cooper, J., Salehi, A., Delcroix, J., et al., 2001. Failed retrograde transport of NGF in a mouse model of Down's syndrome: Reversal of cholinergic neurodegenerative phenotypes following NGF infusion. Proceedings of the National Academy of Sciences of the United States of America 98, 10439–10444.

Corrice, A.M., Glidden, L.M., 2009. The Down syndrome advantage: Fact or fiction? American Journal on Intellectual and Developmental Disabilities 114, 254–268.

Costa, A., Scott-Mckean, J., Stasko, M., 2008. Acute injections of the NMDA receptor antagonist memantine rescue performance deficits of the Ts65Dn mouse model of Down syndrome on a fear conditioning test. Neuropsychopharmacology 33, 1624–1632.

Crawford, D., Leahy, K., Abramova, N., Lan, L., Wang, Y., Davies, K., 1997. Hamster adapt78 mRNA is a Down syndrome critical region homologue that is inducible by oxidative stress. Archives of Biochemistry and Biophysics 342, 6–12.

Crome, L., Cowie, V., Slater, E., 1966. A statistical note on cerebellar and brain stem weight in Mongolism. Journal of Mental Deficiency Research 10, 69–72.

Daum, I., Ackermann, H., 1995. Cerebellar contributions to cognition. Behavioural Brain Research 67, 201–210.

Davidson, L.M., 1928. The brain in Mongolian idiocy. Archives of Neurology and Psychiatry 20, 1229–1257.

De La Luna, S., Estivill, X., 2006. Cooperation to amplify gene-dosage-imbalance effects. Trends in Molecular Medicine 12, 451–454.

Delabar, J., Aflalo-Rattenbac, R., Creau, N., 2006. Developmental defects in trisomy 21 and mouse models. Scientific World Journal 6, 1945–1964.

Delom, F., Burt, E., Hoischen, A., et al., 2009. Transchromosomic cell model of Down syndrome shows aberrant migration, adhesion and proteome response to extracellular matrix. Proteome Science 7, 31.

Devenny, D.A., Silverman, W.P., Hill, A.L., Jenkins, E., Sersen, E.A., Wisniewski, K.E., 1996. Normal ageing in adults with Down's syndrome: A longitudinal study. Journal of Intellectual Disability Research 40 (Pt 3), 208–221.

Devenny, D., Krinsky-Mchale, S., Sersen, G., Silverman, W., 2000. Sequence of cognitive decline in dementia in adults with Down's syndrome. Journal of Intellectual Disability Research 44 (Pt 6), 654–665.

Di Filippo, M., Tozzi, A., Ghiglieri, V., et al., 2010. Impaired plasticity at specific subset of striatal synapses in the Ts65Dn mouse model of Down syndrome. Biological Psychiatry 67, 666–671.

Dierssen, M., Vallina, I.F., Baamonde, C., Garcia-Calatayud, S., Lumbreras, M.A., Florez, J., 1997. Alterations of central noradrenergic transmission in Ts65Dn mouse, a model for Down syndrome. Brain Research 749, 238–244.

Dierssen, M., Benavides-Piccione, R., Martínez-Cué, C., et al., 2003. Alterations of neocortical pyramidal cell phenotype in the Ts65Dn mouse model of Down syndrome: Effects of environmental enrichment. Cerebral Cortex 13, 758–764.

Dierssen, M., Ramakers, G.J.A., 2006. Dendritic pathology in mental retardation: From molecular genetics to neurobiology. Genes, Brain, and Behavior 5 (Suppl 2), 48–60.

Dierssen, M., Arque, G., McDonald, J., et al., 2011. Behavioral characterization of a mouse model overexpressing DSCR1/ RCAN1. PLoS One 6, e17010.

Dykens, E., 2007. Psychiatric and behavioral disorders in persons with Down syndrome. Mental Retardation and Developmental Disabilities Research Reviews 13, 272–278.

Dykens, E., Kasari, C., 1997. Maladaptive behavior in children with Prader-Willi syndrome, Down syndrome, and nonspecific mental retardation. American Journal on Mental Retardation 102, 228–237.

Dykens, E.M., Shah, B., Sagun, J., Beck, T., King, B.H., 2002. Maladaptive behaviour in children and adolescents with Down's syndrome. Journal of Intellectual Disability Research 46, 484–492.

Eggermann, T., Schönherr, N., Spengler, S., et al., 2010. Identification of a 21q22 duplication in a Silver-Russell syndrome patient further narrows down the Down syndrome critical region. American Journal of Medical Genetics. Part A 152A, 356–359.

Eisermann, M.M., Delaraillere, A., Dellatolas, G., et al., 2003. Infantile spasms in Down syndrome – Effects of delayed anticonvulsive treatment. Epilepsy Research 55, 21–27.

Emerson, J.F., Kesslak, J.P., Chen, P.C., Lott, I.T., 1995. Magnetic resonance imaging of the aging brain in Down syndrome. Progress in Clinical and Biological Research 393, 123–138.

Ermak, G., Harris, C., Davies, K., 2002. The DSCR1 (Adapt78) isoform 1 protein calcipressin 1 inhibits calcineurin and protects against acute calcium-mediated stress damage, including transient oxidative stress. The FASEB Journal 16, 814–824.

Faizi, M., Bader, P.L., Tun, C., et al., 2011. Comprehensive behavioral phenotyping of Ts65Dn mouse model of Down syndrome: Activation of β1-adrenergic receptor by xamoterol as a potential cognitive enhancer. Neurobiology of Disease 2, 397–413.

Ferencz, C., Rubin, J.D., Mccarter, R.J., et al., 1985. Congenital heart disease: Prevalence at livebirth. The Baltimore-Washington Infant Study. American Journal of Epidemiology 121, 31–36.

Fernandez, F., Garner, C.C., 2007. Object recognition memory is conserved in Ts1Cje, a mouse model of Down syndrome. Neuroscience Letters 421, 137–141.

Fernandez, F., Garner, C.C., 2008. Episodic-like memory in Ts65Dn, a mouse model of Down syndrome. Behavioural Brain Research 188, 233–237.

Fernandez, F., Trinidad, J.C., Blank, M., Feng, D.-D., Burlingame, A.L., Garner, C.C., 2009. Normal protein composition of synapses in Ts65Dn mice: A mouse model of Down syndrome. Journal of Neurochemistry 110, 157–169.

Fernandez-Martinez, J., Vela, E.M., Tora-Ponsioen, M., Ocaña, O.H., Nieto, M.A., Galceran, J., 2009. Attenuation of Notch signalling by the Down-syndrome-associated kinase DYRK1A. Journal of Cell Science 122, 1574–1583.

Fiedler, J.L., Epstein, C.J., Rapoport, S.I., Caviedes, R., Caviedes, P., 1994. Regional alteration of cholinergic function in central neurons of trisomy 16 mouse fetuses, an animal model of human trisomy 21 (Down syndrome). Brain Research 658, 27–32.

Fitzpatrick, D.R., 2005. Transcriptional consequences of autosomal trisomy: Primary gene dosage with complex downstream effects. Trends in Genetics 21, 249–253.

Fitzpatrick, D.R., Ramsay, J., Mcgill, N.I., Shade, M., Carothers, A.D., Hastie, N.D., 2002. Transcriptome analysis of human autosomal trisomy. Human Molecular Genetics 11, 3249–3256.

Fotaki, V., Dierssen, M., Alcantara, S., et al., 2002. Dyrk1A haploinsufficiency affects viability and causes developmental delay and abnormal brain morphology in mice. Molecular and Cellular Biology 22, 6636–6647.

Frangou, S., Aylward, E., Warren, A., Sharma, T., Barta, P., Pearlson, G., 1997. Small planum temporale volume in Down's syndrome: A volumetric MRI study. The American Journal of Psychiatry 154, 1424–1429.

Frenkel, S., Bourdin, B., 2009. Verbal, visual, and spatio-sequential short-term memory: Assessment of the storage capacities of children and teenagers with Down's syndrome. Journal of Intellectual Disability Research 53, 152–160.

Fuentes, J., Genescà, L., Kingsbury, T., et al., 2000. DSCR1, overexpressed in Down syndrome, is an inhibitor of calcineurin-mediated signaling pathways. Human Molecular Genetics 9, 1681–1690.

Fuerst, P.G., Koizumi, A., Masland, R.H., Burgess, R.W., 2008. Neurite arborization and mosaic spacing in the mouse retina require DSCAM. Nature 451, 470–474.

Fujita, H., Torii, C., Kosaki, R., et al., 2010. Microdeletion of the Down syndrome critical region at 21q22. American Journal of Medical Genetics. Part A 152A, 950–953.

Gao, F.-B., 2007. Molecular and cellular mechanisms of dendritic morphogenesis. Current Opinion in Neurobiology 17, 525–532.

Gasparini, L., Dityatev, A., 2008. [beta]-amyloid and glutamate receptors. Experimental Neurology 212, 1–4.

Gasparini, L., Terni, B., Spillantini, M., 2007. Frontotemporal dementia with tau pathology. Neurodegenerative Diseases 4, 236–253.

Giaccone, G., Tagliavini, F., Linoli, G., et al., 1989. Down patients: Extracellular preamyloid deposits precede neuritic degeneration and senile plaques. Neuroscience Letters 97, 232–238.

Ginsberg, S., Che, S., Wuu, J., Counts, S., Mufson, E., 2006. Down regulation of trk but not p75NTR gene expression in single cholinergic basal forebrain neurons mark the progression of Alzheimer's disease. Journal of Neurochemistry 97, 475–487.

Gitton, Y., Dahmane, N., Baik, S., et al., 2002. A gene expression map of human chromosome 21 orthologues in the mouse. Nature 420, 586–590.

Glasson, E.J., Sullivan, S.G., Hussain, R., Petterson, B.A., Montgomery, P.D., Bittles, A.H., 2003. Comparative survival advantage of males with Down syndrome. American Journal of Human Biology 15, 192–195.

Gong, C., Shaikh, S., Grundke-Iqbal, I., Iqbal, K., 1996. Inhibition of protein phosphatase-2B (calcineurin) activity towards Alzheimer abnormally phosphorylated tau by neuroleptics. Brain Research 741, 95–102.

Granholm, A.C., Sanders, L.A., Crnic, L.S., 2000. Loss of cholinergic phenotype in basal forebrain coincides with cognitive decline in a mouse model of Down's syndrome. Experimental Neurology 161, 647–663.

Grundke-Iqbal, I., Iqbal, K., Quinlan, M., Tung, Y., Zaidi, M., Wisniewski, H., 1986a. Microtubule-associated protein tau. A component of Alzheimer paired helical filaments. Journal of Biological Chemistry 261, 6084–6089.

Grundke-Iqbal, I., Iqbal, K., Tung, Y., Quinlan, M., Wisniewski, H., Binder, L., 1986b. Abnormal phosphorylation of the microtubule-associated protein tau (tau) in Alzheimer cytoskeletal pathology. Proceedings of the National Academy of Sciences of the United States of America 83, 4913–4917.

Grundke-Iqbal, I., Vorbrodt, A.W., Iqbal, K., Tung, Y.C., Wang, G.P., Wisniewski, H.M., 1988. Microtubule-associated polypeptides tau are altered in Alzheimer paired helical filaments. Brain Research 464, 43–52.

Guedj, F., Sébrié, C., Rivals, I., et al., 2009. Green tea polyphenols rescue of brain defects induced by overexpression of DYRK1A. PLoS One 4, e4606.

Guerrini, A., 2003. Experimenting with Humans and Animals: From Galen to Animal Rights. The Johns Hopkins University Press, Baltimore.

Guidi, S., Bonasoni, P., Ceccarelli, C., et al., 2008. Neurogenesis impairment and increased cell death reduce total neuron number in the hippocampal region of fetuses with Down syndrome. Brain Pathology 18, 180–197.

Guimerá, J., Casas, C., Pucharcòs, C., et al., 1996. A human homologue of Drosophila minibrain (MNB) is expressed in the neuronal regions affected in Down syndrome and maps to the critical region. Human Molecular Genetics 5, 1305–1310.

Guimera, J., Casas, C., Estivill, X., Pritchard, M., 1999. Human minibrain homologue (MNBH/DYRK1): Characterization, alternative splicing, differential tissue expression, and overexpression in Down syndrome. Genomics 57, 407–418.

Guipponi, M., Brunschwig, K., Chamoun, Z., et al., 2000. C21orf5, a novel human chromosome 21 gene, has a Caenorhabditis elegans ortholog (pad-1) required for embryonic patterning. Genomics 68, 30–40.

Gulesserian, T., Seidl, R., Hardmeier, R., Cairns, N., Lubec, G., 2001. Superoxide dismutase SOD1, encoded on chromosome 21, but not SOD2 is overexpressed in brains of patients with Down syndrome. Journal of Investigative Medicine 49, 41–46.

Hammerle, B., Carnicero, A., Elizalde, C., Ceron, J., Martinez, S., Tejedor, F., 2003. Expression patterns and subcellular localization of the Down syndrome candidate protein MNB/DYRK1A suggest a role in late neuronal differentiation. European Journal of Neuroscience 17, 2277–2286.

Hanson, J.E., Blank, M., Valenzuela, R.A., Garner, C.C., Madison, D.V., 2007. The functional nature of synaptic circuitry is altered in area CA3 of the hippocampus in a mouse model of Down's syndrome. Journal of Physiology (London) 579, 53–67.

Harashima, C., Jacobowitz, D.M., Witta, J., et al., 2006. Abnormal expression of the G-protein-activated inwardly rectifying potassium channel 2 (GIRK2) in hippocampus, frontal cortex, and substantia nigra of Ts65Dn mouse: A model of Down syndrome. The Journal of Comparative Neurology 494, 815–833.

Hasle, H., Clemmensen, I.H., Mikkelsen, M., 2000. Risks of leukaemia and solid tumours in individuals with Down's syndrome. Lancet 355, 165–169.

Hattori, M., Fujiyama, A., Taylor, T., et al., 2000. The DNA sequence of human chromosome 21. Nature 405, 311–319.

Hattori, D., Demir, E., Kim, H.W., Viragh, E., Zipursky, S.L., Dickson, B.J., 2007. Dscam diversity is essential for neuronal wiring and self-recognition. Nature 449, 223–227.

Haydar, T.F., Blue, M.E., Molliver, M.E., Krueger, B.K., Yarowsky, P.J., 1996. Consequences of trisomy 16 for mouse brain development: Corticogenesis in a model of Down syndrome. Journal of Neuroscience 16, 6175–6182.

Hewitt, C.A., Ling, K.-H., Merson, T.D., et al., 2010. Gene network disruptions and neurogenesis defects in the adult Ts1Cje mouse model of Down syndrome. PLoS One 5, e11561.

Hoeffer, C., Dey, A., Sachan, N., et al., 2007. The Down syndrome critical region protein RCAN1 regulates long-term potentiation and memory via inhibition of phosphatase signaling. Journal of Neuroscience 27, 13161–13172.

Holtzman, D., Santucci, D., Kilbridge, J., et al., 1996. Developmental abnormalities and age-related neurodegeneration in a mouse model of Down syndrome. Proceedings of the National Academy of Sciences of the United States of America 93, 13333–13338.

Hummel, T., Vasconcelos, M.L., Clemens, J.C., Fishilevich, Y., Vosshall, L.B., Zipursky, S.L., 2003. Axonal targeting of olfactory receptor neurons in Drosophila is controlled by Dscam. Neuron 37, 221–231.

Iqbal, K., Grundke-Iqbal, I., 2010. Alzheimer's disease, a multifactorial disorder seeking multitherapies. Alzheimer's & Dementia 6, 420–424.

Iqbal, K., Grundke-Iqbal, I., Zaidi, T., et al., 1986. Defective brain microtubule assembly in Alzheimer's disease. Lancet 2, 421–426.

Iqbal, K., Grundke-Iqbal, I., Smith, A., George, L., Tung, Y., Zaidi, T., 1989. Identification and localization of a tau peptide to paired helical filaments of Alzheimer disease. Proceedings of the National Academy of Sciences of the United States of America 86, 5646–5650.

Iqbal, K., Liu, F., Gong, C., Alonso, A.C., Grundke-Iqbal, I., 2009. Mechanisms of tau-induced neurodegeneration. Acta Neuropathologica 118, 53–69.

Iqbal, K., Wang, X., Blanchard, J., Liu, F., Gong, C.X., Grundke-Iqbal, I., 2010. Alzheimer's disease neurofibrillary degeneration: Pivotal and multifactorial. Biochemical Society Transactions 38, 962–966.

Jervis, G.A., 1948. Early senile dementia in mongoloid idiocy. The American Journal of Psychiatry 105, 102–106.

Jovanovic, S., Clements, D., Macleod, K., 1998. Biomarkers of oxidative stress are significantly elevated in Down syndrome. Free Radical Biology & Medicine 25, 1044–1048.

Kahlem, P., Sultan, M., Herwig, R., et al., 2004. Transcript level alterations reflect gene dosage effects across multiple tissues in a mouse model of Down syndrome. Genome Research 14, 1258–1267.

Keating, D.J., Dubach, D., Zanin, M.P., et al., 2008. DSCR1/RCAN1 regulates vesicle exocytosis and fusion pore kinetics: Implicatons for Down syndrome and Alzheimer's disease. Human Molecular Genetics 17, 1020–1030.

Kesslak, J.P., Nagata, S.F., Lott, I., Nalcioglu, O., 1994. Magnetic resonance imaging analysis of age-related changes in the brains of individuals with Down's syndrome. Neurology 44, 1039–1045.

Kida, E., Wisniewski, K.E., Wisniewski, H.M., 1995. Early amyloid-beta deposits show different immunoreactivity to the amino- and carboxy-terminal regions of beta-peptide in Alzheimer's disease and Down's syndrome brain. Neuroscience Letters 193, 105–108.

Kingsbury, T., Cunningham, K., 2000. A conserved family of calcineurin regulators. Genes & Development 14, 1595–1604.

Kleschevnikov, A.M., Belichenko, P.V., Villar, A.J., Epstein, C.J., Malenka, R.C., Mobley, W.C., 2004. Hippocampal long-term potentiation suppressed by increased inhibition in the Ts65Dn mouse, a genetic model of Down syndrome. Journal of Neuroscience 24, 8153–8160.

Konczak, J., Timmann, D., 2007. The effect of damage to the cerebellum on sensorimotor and cognitive function in children and adolescents. Neuroscience and Biobehavioral Reviews 31, 1101–1113.

Kondo, Y., Mizuno, S., Ohara, K., et al., 2006. Two cases of partial trisomy 21 (pter-q22.1) without the major features of Down syndrome. American Journal of Medical Genetics. Part A 140, 227–232.

Korbel, J.O., Tirosh-Wagner, T., Urban, A.E., et al., 2009. The genetic architecture of Down syndrome phenotypes revealed by high-resolution analysis of human segmental trisomies. Proceedings of the National Academy of Sciences of the United States of America 106, 12031–12036.

Kuhn, D.E., Nuovo, G.J., Martin, M.M., et al., 2008. Human chromosome 21-derived miRNAs are overexpressed in Down syndrome brains and hearts. Biochemical and Biophysical Research Communications 370, 473–477.

Kurt, M.A., Davies, D.C., Kidd, M., Dierssen, M., Flórez, J., 2000. Synaptic deficit in the temporal cortex of partial trisomy 16 (Ts65Dn) mice. Brain Research 858, 191–197.

Lambert, M.P., Barlow, A.K., Chromy, B.A., et al., 1998. Diffusible, non-fibrillar ligands derived from Abeta1-42 are potent central nervous system neurotoxins. Proceedings of the National Academy of Sciences of the United States of America 95, 6448–6453.

Lanfranchi, S., Carretti, B., Spano, G., Cornoldi, C., 2009. A specific deficit in visuospatial simultaneous working memory in Down syndrome. Journal of Intellectual Disability Research 53, 474–483.

Larsen, K.B., Laursen, H., Graem, N., Samuelsen, G.B., Bogdanovic, N., Pakkenberg, B., 2008. Reduced cell number in the neocortical part of the human fetal brain in Down syndrome. Annals of Anatomy 190, 421–427.

Lee, B., Summar, K., 2009. Down syndrome and other abnormalities of chromosome number. In: Berman, R.E., Vaughan, V.C. (Eds.), Nelson Textbook of Pediatrics. WB Saunders, Philadelphia, PA.

Lee, V., Balin, B., Otvos, L.J., Trojanowski, J., 1991. A68: A major subunit of paired helical filaments and derivatized forms of normal Tau. Science 251, 675–678.

Lemere, C.A., Blusztajn, J.K., Yamaguchi, H., Wisniewski, T., Saido, T.C., Selkoe, D.J., 1996. Sequence of deposition of heterogeneous amyloid [beta]-peptides and APO E in Down syndrome: Implications for initial events in amyloid plaque formation. Neurobiology of Disease 3, 16–32.

Li, S., Hong, S., Shepardson, N.E., Walsh, D.M., Shankar, G.M., Selkoe, D., 2009. Soluble oligomers of amyloid Beta protein facilitate hippocampal long-term depression by disrupting neuronal glutamate uptake. Neuron 62, 788–801.

Lieberman, D., Mody, I., 1994. Regulation of NMDA channel function by endogenous Ca(2+)-dependent phosphatase. Nature 369, 235–239.

Lin, H., Michtalik, H., Zhang, S., et al., 2003. Oxidative and calcium stress regulate DSCR1 (Adapt78/MCIP1) protein. Free Radical Biology & Medicine 35, 528–539.

Liu, F., Liang, Z., Wegiel, J., et al., 2008. Overexpression of Dyrk1A contributes to neurofibrillary degeneration in Down syndrome. Federation of American Societies for Experimental Biology Journal 22 (9), 3224–3233.

Lockrow, J., Prakasam, A., Huang, P., Bimonte-Nelson, H., Sambamurti, K., Granholm, A.-C., 2009. Cholinergic degeneration and memory loss delayed by vitamin E in a Down syndrome mouse model. Experimental Neurology 216, 278–289.

Losin, E.A.R., Rivera, S.M., O'hare, E.D., Sowell, E.R., Pinter, J.D., 2009. Abnormal fMRI activation pattern during story listening in individuals with Down syndrome. American Journal on Intellectual and Developmental Disabilities 114, 369–380.

Lott, I.T., Dierssen, M., 2010. Cognitive deficits and associated neurological complications in individuals with Down's syndrome. Lancet Neurology 9, 623–633.

Luo, J., Ma, J., Yu, D., et al., 2008. Infusion of FK506, a specific inhibitor of calcineurin, induces potent tau hyperphosphorylation in mouse brain. Brain Research Bulletin 76, 464–468.

Lyle, R., Béna, F., Gagos, S., et al., 2009. Genotype-phenotype correlations in Down syndrome identified by array CGH in 30 cases of partial trisomy and partial monosomy chromosome 21. European Journal of Human Genetics 17, 454–466.

Mann, D.M.A., Esiri, M.M., 1989. The pattern of acquisition of plaques and tangles in the brains of patients under 50 years of age with Down's syndrome. Journal of Neurological Sciences 89, 169–179.

Mao, R., Zielke, C.L., Zielke, H.R., Pevsner, J., 2003. Global up-regulation of chromosome 21 gene expression in the developing Down syndrome brain. Genomics 81, 457–467.

Mao, R., Wang, X., Spitznagel, E.L., et al., 2005. Primary and secondary transcriptional effects in the developing human Down syndrome brain and heart. Genome Biology 6, R107.

Martínez-Cué, C., Rueda, N., García, E., Davisson, M.T., Schmidt, C., Flórez, J., 2005. Behavioral, cognitive and biochemical responses to different environmental conditions in male Ts65Dn mice, a model of Down syndrome. Behavioural Brain Research 163, 174–185.

Micali, N., Longobardi, E., Iotti, G., et al., 2010. Down syndrome fibroblasts and mouse Prep1-overexpressing cells display increased sensitivity to genotoxic stress. Nucleic Acids Research 38, 3595–3604.

Miyabara, S., Gropp, A., Winking, H., 1982. Trisomy 16 in the mouse fetus associated with generalized edema and cardiovascular and urinary tract anomalies. Teratology 25, 369–380.

Moldrich, R.X., 2007. A yeast model of Down syndrome. International Journal of Developmental Neuroscience 25, 539–543.

Moldrich, R.X., Dauphinot, L., Laffaire, J., et al., 2009. Proliferation deficits and gene expression dysregulation in Down's syndrome (Ts1Cje) neural progenitor cells cultured from neurospheres. Journal of Neuroscience Research 87, 3143–3152.

Moller, R., Kubart, S., Hoeltzenbein, M., et al., 2008. Truncation of the Down syndrome candidate gene DYRK1A in two unrelated patients with microcephaly. American Journal of Human Genetics 82, 1165–1170.

Molloy, C.A., Murray, D.S., Kinsman, A., et al., 2009. Differences in the clinical presentation of Trisomy 21 with and without autism. Journal of Intellectual Disability Research 53, 143–151.

Moncaster, J., Pineda, R., Moir, R., et al., 2010. Alzheimer's disease amyloid-beta links lens and brain pathology in Down syndrome. PLoS One 5, e10659.

Moon, J., Chen, M., Gandhy, S.U., et al., 2010. Perinatal choline supplementation improves cognitive functioning and emotion regulation in the Ts65Dn mouse model of Down syndrome. Behavioral Neuroscience 124, 346–361.

Morice, E., Andreae, L.C., Cooke, S.F., et al., 2008. Preservation of long-term memory and synaptic plasticity despite short-term impairments in the Tc1 mouse model of Down syndrome. Learning and Memory 15, 492–500.

Motte, J., Williams, R.S., 1989. Age-related changes in the density and morphology of plaques and neurofibrillary tangles in Down syndrome brain. Acta Neuropathologica 77, 535–546.

Mufson, E.J., Ginsberg, S.D., Ikonomovic, M.D., Dekosky, S.T., 2003. Human cholinergic basal forebrain: Chemoanatomy and neurologic dysfunction. Journal of Chemical Neuroanatomy 26, 233–242.

Myers, B.A., Pueschel, S.M., 1991. Psychiatric disorders in persons with Down syndrome. The Journal of Nervous and Mental Disease 179, 609–613.

Nadel, L., 1999. Down syndrome in cognitive neuroscience perspective. In: Tager-Flusberg, H. (Ed.), Neurodevelopmental Disorders. MIT Press, Cambridge, MA.

Nadel, L., 2003. Down's syndrome: A genetic disorder in biobehavioral perspective. Genes, Brain, and Behavior 2, 156–166.

O'Doherty, A., Ruf, S., Mulligan, C., et al., 2005. An aneuploid mouse strain carrying human chromosome 21 with Down syndrome phenotypes. Science 309, 2033–2037.

Olson, L., Roper, R., Baxter, L., Carlson, E., Epstein, C., Reeves, R., 2004. Down syndrome mouse models Ts65Dn, Ts1Cje, and Ms1Cje/Ts65Dn exhibit variable severity of cerebellar phenotypes. Developmental Dynamics 230, 581–589.

Olson, L., Roper, R., Sengstaken, C., et al., 2007. Trisomy for the Down syndrome 'critical region' is necessary but not sufficient for brain phenotypes of trisomic mice. Human Molecular Genetics 16, 774–782.

Pallardo, F., Degan, P., D'ischia, M., et al., 2006. Multiple evidence for an early age pro-oxidant state in Down Syndrome patients. Biogerontology 7, 211–220.

Park, I., Arora, N., Huo, H., et al., 2008. Disease-specific induced pluripotent stem cells. Cell 134, 877–886.

Parkinson, P.F., Kannangara, T.S., Eadie, B.D., Burgess, B.L., Wellington, C.L., Christie, B.R., 2009. Cognition, learning behaviour and hippocampal synaptic plasticity are not disrupted in mice overexpressing the cholesterol transporter ABCG1. Lipids in Health and Disease 8, 5.

Patterson, D., 2007. Genetic mechanisms involved in the phenotype of Down syndrome. Mental Retardation and Developmental Disabilities Research Reviews 13, 199–206.

Patterson, D., 2009. Molecular genetic analysis of Down syndrome. Human Genetics 126, 195–214.

Patterson, D., Costa, A.C., 2005. Down syndrome and genetics – A case of linked histories. Nature Reviews Genetics 6, 137–147.

Pearlson, G.D., Breiter, S.N., Aylward, E.H., et al., 1998. MRI brain changes in subjects with Down syndrome with and without dementia. Developmental Medicine and Child Neurology 40, 326–334.

Pennington, B.F., Moon, J., Edgin, J., Stedron, J., Nadel, L., 2003. The neuropsychology of Down syndrome: Evidence for hippocampal dysfunction. Child Development 74, 75–93.

Pereira, P.L., Magnol, L., Sahun, I., et al., 2009. A new mouse model for the trisomy of the Abcg1-U2af1 region reveals the complexity of the combinatorial genetic code of Down syndrome. Human Molecular Genetics 18, 4756–4769.

Pérez-Cremades, D., Hernández, S., Blasco-Ibáñez, J.M., Crespo, C., Nacher, J., Varea, E., 2010. Alteration of inhibitory circuits in the somatosensory cortex of Ts65Dn mice, a model for Down's syndrome. Journal of Neural Transmission 117, 445–455.

Perry, E., Perry, R., Blessed, G., Tomlinson, B., 1977. Necropsy evidence of central cholinergic deficits in senile dementia. Lancet 1, 189.

Pine, S.S., Landing, B.H., Shankle, W.R., 1997. Reduced inferior olivary neuron number in early Down syndrome. Pediatric Pathology & Laboratory Medicine 17, 537–545.

Pinter, J.D., Brown, W.E., Eliez, S., Schmitt, J.E., Capone, G.T., Reiss, A.L., 2001a. Amygdala and hippocampal volumes in children with Down syndrome: A high-resolution MRI study. Neurology 56, 972–974.

Pinter, J.D., Eliez, S., Schmitt, J.E., Capone, G.T., Reiss, A.L., 2001b. Neuroanatomy of Down's syndrome: A high-resolution MRI study. The American Journal of Psychiatry 158, 1659–1665.

Pollonini, G., Gao, V., Rabe, A., Palminiello, S., Albertini, G., Alberini, C.M., 2008. Abnormal expression of synaptic proteins and neurotrophin-3 in the Down syndrome mouse model Ts65Dn. Neuroscience 156, 99–106.

Pueschel, S.M., Louis, S., Mcknight, P., 1991. Seizure disorders in Down syndrome. Archives of Neurology 48, 318–320.

Puri, B.K., Ho, K.W., Singh, I., 2001. Age of seizure onset in adults with Down's syndrome. International Journal of Clinical Practice 55, 442–444.

Rachidi, M., Lopes, C., 2007. Mental retardation in Down syndrome: From gene dosage imbalance to molecular and cellular mechanisms. Neuroscience Research 59, 349–369.

Rachidi, M., Lopes, C., 2008. Mental retardation and associated neurological dysfunctions in Down syndrome: A consequence of dysregulation in critical chromosome 21 genes and associated molecular pathways. European Journal of Paediatric Neurology 12, 168–182.

Rachidi, M., Lopes, C., Charron, G., et al., 2005. Spatial and temporal localization during embryonic and fetal human development of the transcription factor SIM2 in brain regions altered in Down syndrome. International Journal of Developmental Neuroscience 23, 475–484.

Raich, W., Moorman, C., Lacefield, C., et al., 2003. Characterization of Caenorhabditis elegans homologs of the Down syndrome candidate gene DYRK1A. Genetics 163, 571–580.

Raz, N., Torres, I.J., Briggs, S.D., et al., 1995. Selective neuroanatomic abnormalities in Down's syndrome and their cognitive correlates: Evidence from MRI morphometry. Neurology 45, 356–366.

Reiss, S., Levitan, G.W., Szyszko, J., 1982. Emotional disturbance and mental retardation: Diagnostic overshadowing. American Journal of Mental Deficiency 86, 567–574.

Reymond, A., Marigo, V., Yaylaoglu, M.B., et al., 2002. Human chromosome 21 gene expression atlas in the mouse. Nature 420, 582–586.

Roizen, N.J., 2002. Medical care and monitoring for the adolescent with Down syndrome. Adolescent Medicine 13, 345–358, vii.

Roizen, N.J., Patterson, D., 2003. Down's syndrome. Lancet 361, 1281–1289.

Ronan, A., Fagan, K., Christie, L., Conroy, J., Nowak, N.J., Turner, G., 2007. Familial 4.3 Mb duplication of 21q22 sheds new light on the Down syndrome critical region. Journal of Medical Genetics 44, 448–451.

Roper, R., Reeves, R., 2006. Understanding the basis for Down syndrome phenotypes. PLoS Genetics 2, e50.

Roper, R., Baxter, L., Saran, N., Klinedinst, D., Beachy, P., Reeves, R., 2006a. Defective cerebellar response to mitogenic Hedgehog signaling in Down syndrome mice. Proceedings of the National Academy of Sciences of the United States of America 103, 1452–1456.

Roper, R., St John, H., Philip, J., Lawler, A., Reeves, R., 2006b. Perinatal loss of Ts65Dn Down syndrome mice. Genetics 172, 437–443.

Rueda, N., Llorens-Martín, M., Flórez, J., et al., 2010. Memantine normalizes several phenotypic features in the Ts65Dn mouse model of Down syndrome. Journal of Alzheimer's Disease 21, 277–290.

Sago, H., Carlson, E., Smith, D., et al., 1998. Ts1Cje, a partial trisomy 16 mouse model for Down syndrome, exhibits learning and behavioral abnormalities. Proceedings of the National Academy of Sciences of the United States of America 95, 6256–6261.

Sago, H., Carlson, E., Smith, D., et al., 2000. Genetic dissection of region associated with behavioral abnormalities in mouse models for Down syndrome. Pediatric Research 48, 606–613.

Salehi, A., Delcroix, J., Belichenko, P., et al., 2006. Increased App expression in a mouse model of Down's syndrome disrupts NGF transport and causes cholinergic neuron degeneration. Neuron 51, 29–42.

Salehi, A., Faizi, M., Belichenko, P., Mobley, W., 2007. Using mouse models to explore genotype-phenotype relationship in Down syndrome. Mental Retardation and Developmental Disabilities Research Reviews 13, 207–214.

Sato, D., Kawara, H., Shimokawa, O., et al., 2008. A girl with Down syndrome and partial trisomy for 21pter-q22.13: A clue to narrow the Down syndrome critical region. American Journal of Medical Genetics. Part A 146A, 124–127.

Schmidt-Sidor, B., Wisniewski, K.E., Shepard, T.H., Sersen, E.A., 1990. Brain growth in Down syndrome subjects 15 to 22 weeks of gestational age and birth to 60 months. Clinical Neuropathology 9, 181–190.

Schmucker, D., 2007. Molecular diversity of Dscam: Recognition of molecular identity in neuronal wiring. Nature Reviews Neuroscience 8, 915–920.

Schuchmann, S., Heinemann, U., 2000. Diminished glutathione levels cause spontaneous and mitochondria-mediated cell death in neurons from trisomy 16 mice: A model of Down's syndrome. Journal of Neurochemistry 74, 1205–1214.

Scott, B.S., Becker, L.E., Petit, T.L., 1983. Neurobiology of Down's syndrome. Progress in Neurobiology 21, 199–237.

Seo, H., Isacson, O., 2005. Abnormal APP, cholinergic and cognitive function in Ts65Dn Down's model mice. Experimental Neurology 193, 469–480.

Shankar, G.M., Li, S., Mehta, T.H., et al., 2008. Amyloid-[beta] protein dimers isolated directly from Alzheimer's brains impair synaptic plasticity and memory. Nature Medicine 14, 837–842.

Shin, J., Gulesserian, T., Verger, E., Delabar, J., Lubec, G., 2006. Protein dysregulation in mouse hippocampus polytransgenic for chromosome 21 structures in the Down syndrome critical region. Journal of Proteome Research 5, 44–53.

Shin, J., Guedj, F., Delabar, J., Lubec, G., 2007. Dysregulation of growth factor receptor-bound protein 2 and fascin in hippocampus of mice polytransgenic for chromosome 21 structures. Hippocampus 17, 1180–1192.

Shindoh, N., Kudoh, J., Maeda, H., et al., 1996. Cloning of a human homolog of the Drosophila minibrain/rat Dyrk gene from 'the Down syndrome critical region' of chromosome 21. Biochemical and Biophysical Research Communications 225, 92–99.

Shinohara, T., Tomizuka, K., Miyabara, S., et al., 2001. Mice containing a human chromosome 21 model behavioral impairment and cardiac anomalies of Down's syndrome. Human Molecular Genetics 10, 1163–1175.

Siarey, R.J., Stoll, J., Rapoport, S.I., Galdzicki, Z., 1997. Altered long-term potentiation in the young and old Ts65Dn mouse, a model for Down syndrome. Neuropharmacology 36, 1549–1554.

Siarey, R.J., Villar, A.J., Epstein, C.J., Galdzicki, Z., 2005. Abnormal synaptic plasticity in the Ts1Cje segmental trisomy 16 mouse model of Down syndrome. Neuropharmacology 49, 122–128.

Silverman, W., 2007. Down syndrome: Cognitive phenotype. Mental Retardation and Developmental Disabilities Research Reviews 13, 228–236.

Smith, D.S., 2001. Health care management of adults with Down syndrome. American Family Physician 64, 1031–1038.

Song, W.J., Sternberg, L.R., Kasten-Sportès, C., et al., 1996. Isolation of human and murine homologues of the Drosophila minibrain gene: Human homologue maps to 21q22.2 in the Down syndrome 'critical region'. Genomics 38, 331–339.

Squire, L.R., 1992. Memory and the hippocampus: A synthesis from findings with rats, monkeys, and humans. Psychological Review 99, 195–231.

Struwe, F., 1929. Histopathological studies on the origin and nature of senile plaques. Neurologie et Psychiatrie 122, 291.

Suetsugu, M., Mehraein, P., 1980. Spine distribution along the apical dendrites of the pyramidal neurons in Down's syndrome. A quantitative Golgi study. Acta Neuropathologica 50, 207–210.

Sussan, T., Yang, A., Li, F., Ostrowski, M., Reeves, R., 2008. Trisomy represses Apc^{Min} mediated tumours in mouse models of Down's syndrome. Nature 451, 73–75.

Sweeney, J.E., Höhmann, C.F., Oster-Granite, M.L., Coyle, J.T., 1989. Neurogenesis of the basal forebrain in euploid and trisomy 16 mice: An animal model for developmental disorders in Down syndrome. Neuroscience 31, 413–425.

Sylvester, P.E., 1983. The hippocampus in Down's syndrome. Journal of Mental Deficiency Research 27 (Pt 3), 227–236.

Takahashi, K., Tanabe, K., Ohnuki, M., et al., 2007. Induction of pluripotent stem cells from adult human fibroblasts by defined factors. Cell 131, 861–872.

Takashima, S., Iida, K., Mito, T., Arima, M., 1994. Dendritic and histochemical development and ageing in patients with Down's syndrome. Journal of Intellectual Disability Research 38 (Pt 3), 265–273.

Tan, Y.H., Tischfield, J., Ruddle, F.H., 1973. The linkage of genes for the human interferon-induced antiviral protein and indophenol oxidase-B traits to chromosome G-21. The Journal of Experimental Medicine 137, 317–330.

Tansley, G., Burgess, B., Bryan, M., et al., 2007. The cholesterol transporter ABCG1 modulates the subcellular distribution and proteolytic processing of beta-amyloid precursor protein. Journal of Lipid Research 48, 1022–1034.

Teipel, S.J., Schapiro, M.B., Alexander, G.E., et al., 2003. Relation of corpus callosum and hippocampal size to age in nondemented adults with Down's syndrome. The American Journal of Psychiatry 160, 1870–1878.

Teipel, S.J., Alexander, G.E., Schapiro, M.B., Moller, H.J., Rapoport, S.I., Hampel, H., 2004. Age-related cortical grey matter reductions in non-demented Down's syndrome adults determined by MRI with voxel-based morphometry. Brain 127, 811–824.

Teipel, S.J., Hampel, H., 2006. Neuroanatomy of Down syndrome in vivo: A model of preclinical Alzheimer's disease. Behavior Genetics 36, 405–415.

Tejedor, F., Zhu, X.R., Kaltenbach, E., et al., 1995. Minibrain: A new protein kinase family involved in postembryonic neurogenesis in Drosophila. Neuron 14, 287–301.

Toiber, D., Azkona, G., Ben-Ari, S., Torán, N., Soreq, H., Dierssen, M., 2010. Engineering DYRK1A over-dosage yields Down syndrome-characteristic cortical splicing aberrations. Neurobiology of Disease 40, 348–359.

Townsend, M., Shankar, G., Mehta, T., Walsh, D., Selkoe, D., 2006. Effects of secreted oligomers of amyloid beta-protein on hippocampal synaptic plasticity: A potent role for trimers. The Journal of Physiology 572, 477–492.

Vicari, S., 2004. Memory development and intellectual disabilities. Acta Paediatrica. Supplement 93 (445), 60–63 discussion 63-64.

Vicari, S., Bellucci, S., Carlesimo, G.A., 2000. Implicit and explicit memory: A functional dissociation in persons with Down syndrome. Neuropsychologia 38, 240–251.

Vicari, S., Bellucci, S., Carlesimo, G.A., 2005. Visual and spatial long-term memory: Differential pattern of impairments in Williams and Down syndromes. Developmental Medicine and Child Neurology 47, 305–311.

Virji-Babul, N., Moiseev, A., Cheung, T., Weeks, D.J., Cheyne, D., Ribary, U., 2010. Neural mechanisms underlying action observation in adults with Down syndrome. American Journal on Intellectual and Developmental Disabilities 115, 113–127.

Voronov, S.V., Frere, S.G., Giovedi, S., et al., 2008. Synaptojanin 1-linked phosphoinositide dyshomeostasis and cognitive deficits in mouse models of Down's syndrome. Proceedings of the National Academy of Sciences of the United States of America 105, 9415–9420.

Walsh, D.M., Klyubin, I., Fadeeva, J.V., et al., 2002. Naturally secreted oligomers of amyloid [beta] protein potently inhibit hippocampal long-term potentiation in vivo. Nature 416, 535–539.

Wang, P., Chen, H., Qin, H., et al., 1998. Overexpression of human copper, zinc-superoxide dismutase (SOD1) prevents postischemic injury. Proceedings of the National Academy of Sciences of the United States of America 95, 4556–4560.

Wang, H.-W., Pasternak, J.F., Kuo, H., et al., 2002. Soluble oligomers of [beta] amyloid (1-42) inhibit long-term potentiation but not long-term depression in rat dentate gyrus. Brain Research 924, 133–140.

Wang, C.C., Kadota, M., Nishigaki, R., et al., 2004. Molecular hierarchy in neurons differentiated from mouse ES cells containing a single human chromosome 21. Biochemical and Biophysical Research Communications 314, 335–350.

Whitehouse, P., Struble, R., Clark, A., Price, D., 1982. Alzheimer disease: Plaques, tangles, and the basal forebrain. Annals of Neurology 12, 494.

Whittle, N., Sartori, S.B., Dierssen, M., Lubec, G., Singewald, N., 2007. Fetal Down syndrome brains exhibit aberrant levels of neurotransmitters critical for normal brain development. Pediatrics 120, e1465–e1471.

Winter, T.C., Ostrovsky, A.A., Komarniski, C.A., Uhrich, S.B., 2000. Cerebellar and frontal lobe hypoplasia in fetuses with trisomy 21: Usefulness as combined US markers. Radiology 214, 533–538.

Wisniewski, K.E., 1990. Down syndrome children often have brain with maturation delay, retardation of growth, and cortical dysgenesis. American Journal of Medical Genetics. Supplement 7, 274–281.

Wisniewski, K.E., Laure-Kamionowska, M., Wisniewski, H.M., 1984. Evidence of arrest of neurogenesis and synaptogenesis in brains of patients with Down's syndrome. The New England Journal of Medicine 311, 1187–1188.

Wisniewski, K.E., Wisniewski, H.M., Wen, G.Y., 1985. Occurrence of neuropathological changes and dementia of Alzheimer's disease in Down's syndrome. Annals of Neurology 17, 278–282.

Wisniewski, H., Silverman, W., Wegiel, J., 1994. Ageing, Alzheimer disease and mental retardation. Journal of Intellectual Disability Research 38 (Pt 3), 233–239.

Yabut, O., Domogauer, J., D'arcangelo, G., 2010. Dyrk1A overexpression inhibits proliferation and induces premature neuronal differentiation of neural progenitor cells. Journal of Neuroscience 30, 4004–4014.

Yang, Q., Rasmussen, S.A., Friedman, J.M., 2002. Mortality associated with Down's syndrome in the USA from 1983 to 1997: A population-based study. Lancet 359, 1019–1025.

Yeo, G.W., Xu, X., Liang, T.Y., et al., 2007. Alternative splicing events identified in human embryonic stem cells and neural progenitors. PLoS Computational Biology 3, 1951–1967.

Yeo, G.W., Coufal, N., Aigner, S., et al., 2008. Multiple layers of molecular controls modulate self-renewal and neuronal lineage specification of embryonic stem cells. Human Molecular Genetics 17, R67–R75.

Yu, D., Luo, J., Bu, F., Song, G., Zhang, L., Wei, Q., 2006. Inhibition of calcineurin by infusion of CsA causes hyperphosphorylation of tau and is accompanied by abnormal behavior in mice. Biological Chemistry 387, 977–983.

Yu, J., Vodyanik, M.A., Smuga-Otto, K., et al., 2007. Induced pluripotent stem cell lines derived from human somatic cells. Science 318, 1917–1920.

Yu, H.-H., Yang, J.S., Wang, J., Huang, Y., Lee, T., 2009. Endodomain diversity in the Drosophila Dscam and its roles in neuronal morphogenesis. Journal of Neuroscience 29, 1904–1914.

Yu, T., Li, Z., Jia, Z., et al., 2010. A mouse model of Down syndrome trisomic for all human chromosome 21 syntenic regions. Human Molecular Genetics 19, 2780–2791.

Zigler, E., Hodapp, R.M., 1991. Behavioral functioning in individuals with mental retardation. Annual Review of Psychology 42, 29–50.

Zigman, W., Lott, I., 2007. Alzheimer's disease in Down syndrome: Neurobiology and risk. Mental Retardation and Developmental Disabilities Research Reviews 13, 237–246.

Further Reading

http://www.cdc.gov/features/dsdownsyndrome. CDC Data of Statistics – Feature: Down Syndrome Cases at Birth Increased.

http://www.ensembl.org. Ensembl Genome Browser.

Lissencephalies and Axon Guidance Disorders

E.H. Sherr, L. Fernandez
University of California, San Francisco, CA, USA

OUTLINE

Neural Circuit Development and Function in the Brain: Comprehensive Developmental Neuroscience, Volume 3 http://dx.doi.org/10.1016/B978-0-12-397267-5.00128-X

© 2013 Elsevier Inc. All rights reserved.

31.1 INTRODUCTION: DISORDERS OF AXON GUIDANCE

During development, neurons extend their axons over long distances to form connections with synaptic targets (Lin et al., 2009; Schmidt et al., 2009). For this process to occur, a choreographed sequence of events must take place (Garbe and Bashaw, 2004; Izzi and Charron, 2011; Lin et al., 2009; O'Donnell et al., 2009). First, neurons and their surrounding target tissues must be specified to express the correct complement of receptors and guidance cues, respectively (Garbe and Bashaw, 2004; Izzi and Charron, 2011; O'Donnell et al., 2009). Second, receptors must be assembled into the appropriate complexes and localized to the axonal or dendritic growth cones, whereas guidance cues must be correctly trafficked to and localized within the extracellular environment (Garbe and Bashaw, 2004; Izzi and Charron, 2011; O'Donnell et al., 2009). Third, signaling mechanisms must be in place to integrate and transmit signals from the surface receptors to changes in the growth cone actin cytoskeleton, resulting in stereotyped steering decisions (Garbe and Bashaw, 2004; Izzi and Charron, 2011; O'Donnell et al., 2009). Each of these steps provides many potential levels for the regulation of axon guidance decisions (Garbe and Bashaw, 2004). These processes of axon guidance are controlled by the so-called axon guidance proteins (Schmidt et al., 2009), and remarkably, only a handful of human disorders resulting from primary errors in these processes have been identified (Engle, 2010).

The physician's ability to detect disorders of axon guidance has been augmented by classic, pathological, radiological, and electrophysiological techniques (Engle, 2010). Diagnostic radiological and postmortem neuropathological studies detect overall changes in white matter volume and major abnormalities of axon tracts demarcated from the background, such as the corpus callosum, anterior and posterior commissures, optic chiasm, and cerebellar peduncles (Engle, 2010). Neuropathological studies can also detect the absence of axons that normally cross the midline at many points in the brain stem and spinal cord, which are more difficult to visualize by standard magnetic resonance imaging (MRI) (Engle, 2010). Electrophysiological studies such as evoked potentials can reveal aberrant central connections of peripheral sensory or motor nerves (Engle, 2010). Exciting advances in neuroimaging and genetics are revolutionizing the ability to define axon guidance disorders (Engle, 2010). Detailed fiber tract anatomy can now be visualized using noninvasive tractography such as diffusion tensor imaging (DTI) and diffusion spectrum imaging (Engle, 2010). These techniques provide tract orientation by determining the anisotropic properties of water diffusion and can be used to reconstruct the trajectories of fiber systems in three-dimensional space (Tovar-Moll et al., 2007; Wahl et al., 2009). Human genetics also is providing unbiased approaches to identify the etiologies of disorders with aberrant axon tracts (Engle, 2010). For some syndromes, animal and *in vitro* studies have confirmed that the encoded protein has a primary role in axon guidance (Engle, 2010). For others, such studies reveal a primary role in neuronal specification and migration rather than, or in addition to, a role in axon guidance (Engle, 2010). Some neurodevelopmental disorders without clinical, pathological, or radiological evidence of aberrant axon tracts have been found to result from mutations in genes that contribute to axon guidance in animal models (Engle, 2010).

The major human genetic disorders that result, or are proposed to result, from defective axon guidance include genetic mutations that alter axon growth cone ligands and receptors, downstream signaling molecules, and axon transport, as well as proteins without currently recognized roles in axon guidance (Engle, 2010). For example, mutations in the β-tubulin isotype III (*TUBB3*) gene have been identified in a series of autosomal dominant disorders of axon guidance, which are known as the *TUBB3* syndromes (Tischfield et al., 2010). Understanding mechanisms of axon growth and axon guidance are critical to elucidating developmental disorders (Yu et al., 2010).

31.2 LIGAND/RECEPTOR SYSTEMS MEDIATING AXON GUIDANCE

During development, neuronal growth cones, the highly motile, specialized structures at the tips of extending axons, follow specific pathways and navigate series of intermediate choice points to find their correct targets (Allen and Chilton, 2009; Bush and Soriano, 2009; Cooper, 2002; Izzi and Charron, 2011; Kaprielian et al., 2001; Nie et al., 2010; O'Donnell et al., 2009; Richards et al., 2004; Schmidt et al., 2009). Pathway selection by axons is oriented by a large variety of short- and long-range guidance cues distributed along the entire pathway, to which different axons respond differently (Barkovich et al., 2009; Bush and Soriano, 2009; Cooper, 2002; Izzi and Charron, 2011; Kaprielian et al., 2001; Richards et al., 2004; Schmidt et al., 2009). Axon guidance proteins can be either secreted or associated with membranes (Kaprielian et al., 2001; Schmidt et al., 2009). In the case of secreted proteins, gradients are formed that enable them to signal over long distances, whereas membrane-bound molecules are responsible for local signaling (Schmidt et al., 2009). Generating precise patterns of connectivity depends on the regulated action of conserved families of guidance cues and their neuronal receptors (Bashaw and Klein, 2010;

Kaprielian et al., 2001; Schmidt et al., 2009). Cell surface receptors residing on growth cones and their associated axons interpret these signals as positive/attractive or negative/repulsive forces (Bush and Soriano, 2009; Izzi and Charron, 2011; Kaprielian et al., 2001; Schmidt et al., 2009). Local regulation of protein synthesis and degradation in the axon also contribute to the rapid changes in growth cone dynamics that occur during axonal navigation (Nie et al., 2010).

Identifying the molecules responsible for guiding growing axons to their target is only the first step; it is crucial to know how they are localized within the neuron itself (Allen and Chilton, 2009). The intracellular signaling cascade initiated on detection of the guidance cue by the axon-bound receptor triggers dynamic rearrangements of the actin cytoskeleton within the growth cone, promoting cycles of extension and retraction of filopodia at the leading edge (Bashaw and Klein, 2010; Cooper, 2002). This allows continual reassessment of the immediate environment by the growth cone (Cooper, 2002). In the case of chemoattraction, movement along the desired trajectory is achieved by elongation of the actin cytoskeleton, leading to the promotion of filopodia extension toward the source of the guidance cue (Cooper, 2002). In contrast, chemorepulsion promotes actin depolymerization and filopodia retraction, resulting in growth cone collapse and ultimately migration away from the ligand source (Cooper, 2002).

Several phylogenetically conserved families of guidance cues and receptors have been discovered (Chiappedi and Bejor, 2010; Koeberle and Bahr, 2004; O'Donnell et al., 2009; Schmidt et al., 2009). The four classic ligand/receptor systems mediating axon guidance are (1) semaphorins (sema) and their plexin (plex) and neuropilin receptors, (2) netrins and their DCC and UNC5 receptors, (3) slits and their roundabout (Robo) receptors, and (4) ephrins and their Eph receptors (Koeberle and Bahr, 2004; Lin et al., 2009; Nie et al., 2010; O'Donnell et al., 2009; Schmidt et al., 2009). It is the type of receptor, or receptor complex, expressed on the growth cone's surface, rather than a given guidance cue, that determines the direction of axon growth (Cooper, 2002; O'Donnell et al., 2009). Additional protein families previously recognized for other developmental functions have been implicated in growth cone guidance, including sonic hedgehog (SHH), bone morphogenetic proteins (BMPs), and wingless-type (WNT) proteins (Chiappedi and Bejor, 2010; Giger et al., 2010; O'Donnell et al., 2009; Schmidt et al., 2009). For example, WNT proteins have been identified as axon guidance proteins that play a role in guiding commissural axons from the spinal cord to the brain (Giger et al., 2010; Schmidt et al., 2009).

The expression of guidance cues and receptors is exquisitely tailored to allow growth cones to make appropriate path-finding decisions at specific times and places throughout development (O'Donnell et al., 2009). A wide variety of mechanisms are in place to ensure the correct presentation and receipt of guidance signals, ranging from spatially and temporally restricted transcriptional regulation of cues and receptors to their specific posttranslational trafficking (Chiappedi and Bejor, 2010; O'Donnell et al., 2009).

Secreted guidance cues, such as netrins and slits, have been shown to act as long-range cues, secreted from intermediate or final targets, and are presumed to form a chemotactic gradient along the pathway of the exploring growth cone (Cooper, 2002). In other instances, these factors can behave as short-range guidance cues, where they act over a distance of only a few cell diameters to affect changes in the direction of growth cone migration at specific choice points (Cooper, 2002). The molecular mechanisms that can influence the spatial distribution of guidance cues are the interaction with heparin sulfate proteoglycans, protein components of the extracellular matrix or axon-bound proteins, and selective proteolytic cleavage of the guidance protein (Cooper, 2002). Guidance receptors transport their ligands along axons to new locations distant from their points of synthesis, thereby determining their spatial distribution (Cooper, 2002).

Regulating the delivery of guidance receptors to the growth cone plasma membrane can have profound influences on axon growth and guidance; the regulation of receptor expression at the cellular level is not only strictly confined to surface expression but also includes regulated removal by endocytosis (Bashaw and Klein, 2010; O'Donnell et al., 2009). Endocytosis may be a necessary aspect of guidance receptor activation and signaling (Bashaw and Klein, 2010). In several cases, receptor endocytosis seems to be an obligate step in receptor activation that is evoked by ligand binding, whereas other examples point to the modulation of guidance responses by receptor endocytosis that is triggered by an independent pathway (O'Donnell et al., 2009). Endocytosis of guidance receptors also is influenced by other membrane receptor systems and can lead to changes in responsiveness to the guidance cue (Bashaw and Klein, 2010). In addition to contributing to receptor signaling, endocytosis can also modulate axon responses by regulating which receptors are expressed at the surface of the growth cone (O'Donnell et al., 2009).

Similar to endocytosis, proteolytic processing contributes to receptor activation and modulates guidance responses (Bashaw and Klein, 2010; O'Donnell et al., 2009). A role for proteolysis in axon guidance was supported by a number of early studies demonstrating that growth cones secrete proteases, and investigators proposed that cleavage of extracellular matrix components is required to advance through the extracellular environment (O'Donnell et al., 2009). Later, genetic screens for defects in axonal navigation at the midline in Drosophila implicated the Kuzbanian ADAM (a disintegrin and metalloprotease) family transmembrane metalloprotease in the regulation of axon extension and guidance at the midline (O'Donnell et al., 2009). Several additional studies have implicated ADAM metalloproteases and matrix metalloproteases in contributing to axon guidance *in vivo* in both invertebrate and vertebrate nervous systems (O'Donnell et al., 2009).

ADAM10 forms a stable complex with ephrin-A2, and when EphR interacts with ephrin-A2, the resulting ligand–receptor complex is clipped by selective ADAM10-dependent cleavage of ephrin-A2; cleavage events are restricted to only those ephrin ligands that are engaged by receptors (Bashaw and Klein, 2010; O'Donnell et al., 2009). Ligand/receptor binding and formation of an active complex expose a new recognition sequence for ADAM10, resulting in the optimal positioning of the protease domain with respect to the substrate (Bashaw and Klein, 2010; O'Donnell et al., 2009). The ligand dependence of the cleavage event provides an elegant explanation for how an initially adhesive interaction can be converted to repulsion and offers an efficient strategy for axon detachment and attenuation of signaling (Bashaw and Klein, 2010; O'Donnell et al., 2009). The matrix metalloprotease family can play a similar role in converting ephrinB/EphB adhesion into axon retraction by specific cleavage of the EphB2 receptor (Bashaw and Klein, 2010; O'Donnell et al., 2009). Thus, both ephrin ligands and Eph receptors can be substrates for regulated proteolysis, and these proteolytic events seem to be critical in mediating axon retraction (Bashaw and Klein, 2010; O'Donnell et al., 2009). There also seems to be a common regulatory mechanism for the DCC and a number of ephrin ligands in which metalloprotease-mediated ectodomain shedding is followed by intramembranous gamma-secretase cleavage (Bashaw and Klein, 2010; O'Donnell et al., 2009).

31.2.1 Ephrins and Their Eph Receptors

Ephrins are members of a family of guidance molecules that are either anchored to the cell membrane by a glycosylphosphatidylinositol (GPI) linkage or have a transmembrane domain (Bush and Soriano, 2009; Kaprielian et al., 2001; Koeberle and Bahr, 2004). The ephrin/eph family has a variety of important functions, including axonal outgrowth and pruning, neuronal connectivity, synaptic maturation and plasticity, and neuronal apoptosis (Bush and Soriano, 2009; Lin et al., 2009). The Eph receptors represent the largest family of receptor tyrosine kinases in the mammalian genome and regulate various signaling pathways through a number of downstream effectors, including guanine nucleotide

exchange factors (GEFs), guanosine triphosphatase (GTPase)–activating proteins, tyrosine kinases, phosphatases, and adaptor proteins (Nie et al., 2010). Eph/ephrin signaling is important in the formation of retinal connections, other axonal projections, and the corpus callosum (Nie et al., 2010).

The interaction of the Eph receptor tyrosine kinase family with their membrane-bound ligands, the ephrins, drives axon path-finding throughout the developing central and peripheral nervous systems via a chemorepulsive mechanism (Cooper, 2002; Giger et al., 2010). However, it has been shown that chemoattractive functions for ephrins also exist (Koeberle and Bahr, 2004). Eph receptors and their ephrin ligands both are capable of transmitting signals in the cell in which they are expressed: Eph receptor signaling is termed *forward signaling* and ephrin ligand signaling is termed *reversed signaling* (Bashaw and Klein, 2010; O'Donnell et al., 2009). The capacity for bidirectional signaling is a hallmark of the Eph/ephrin signaling system. In addition to their ability to signal through cognate Eph tyrosine kinase receptors, ephrins can also transduce a reverse signal into the cell in which they are expressed (Bush and Soriano, 2009). Binding of ephrin ligands triggers Eph receptor clustering, autophosphorylation, and downstream signaling cascades, which cause cytoskeletal rearrangements and changes in cell adhesion (Nie et al., 2010). Through these mechanisms, Eph receptors control axon turning, retraction, and branching (Nie et al., 2010).

Ephrin ligands and Eph receptors contribute to the guidance of retinal ganglion cell (RGC) axons in the visual system (Koeberle and Bahr, 2004; Nie et al., 2010; O'Donnell et al., 2009). Endocytosis of activated Eph receptors at the growth cone is necessary to allow for proper forward signaling, leading to growth cone retraction (Allen and Chilton, 2009; O'Donnell et al., 2009). In the case of membrane-associated ephrins, endocytosis of the ephrin–Eph complex is required for efficient cell detachment (parallel to proteolytic cleavage; Bashaw and Klein, 2010). Vav family GEFs have been implicated as regulators of Eph receptor endocytosis, signaling, or both (Bashaw and Klein, 2010). Vav proteins may trigger Eph internalization into signaling endosomes from where Eph receptors mediate dynamic changes of the actin cytoskeleton underlying growth cone collapse (Bashaw and Klein, 2010). Alternatively, Vav proteins may act in concert with other regulators of Rho GTPases to regulate Eph repulsive signaling, independent of endocytosis (Bashaw and Klein, 2010). For ephrinB-EphB–induced repulsive guidance, efficient cell detachment requires bidirectional endocytosis (Bashaw and Klein, 2010). Vav proteins may primarily promote cell detachment by mediating local Rac-dependent endocytosis of the ephrin–Eph complex and membrane (Bashaw and Klein, 2010).

Ephrin-A5 has been shown to act as a repulsive cue for somatosensory thalamocortical (TC) axons expressing EphA receptors in the TC system (Torii and Levitt, 2005). It seems to control the topography of TC projections within cortical areas (Torii and Levitt, 2005). Analysis of EphA/ephrin-A mutants suggests that ephrin-A5 signaling may also control the specificity of TC projections into individual cortical areas by regulating the positioning of TC axons within the subcortical telencephalon (ST) (Torii and Levitt, 2005). Gradients of EphA7 and ephrin-A5 exhibit complementary expression patterns from embryonic to postnatal ages, consistent with putative molecular interactions during development (Torii and Levitt, 2005). It has been hypothesized that ephrin-A5 serves as a dominant ligand *in vivo* for EphA7 among combinatorial gradients of ephrin-As (Torii and Levitt, 2005). The gradient of EphA7 receptor levels by neocortical neurons is a critical regulator of the topographic targeting of corticothalamic (CT) axons through local interactions within thalamic nuclei, and this regulation is independent of the relative positioning of CT axons within the ST (Torii and Levitt, 2005). The topography of CT projections can be disrupted while the inter- and intra-areal topography of TC projections develops normally (Torii and Levitt, 2005).

A-type Eph receptors and their ligands, ephrin-As, also have been implicated in controlling cell positioning in a variety of developmental contexts by sorting cell types, restricting their intermingling, or regulating their migration (Torii et al., 2009). EphA and ephrin-A signaling has a proapoptotic effect in the embryonic cortex without affecting the proliferation or cell cycle progression (Torii et al., 2009). EphA and ephrin-A signaling regulates lateral neuronal dispersion and intermingling during the multipolar stage of radial migration, and this mechanism is required to generate cortical columns with appropriate cellular components (Torii et al., 2009).

31.2.2 Semaphorins (Semas) and Their Plexin (Plex) and Neuropilin Receptors

Semaphorins are a large family of secreted and membrane-associated proteins most commonly associated with a role in axon guidance (Parrinello et al., 2008). The most extensively studied biological function of semaphorins is their role in guiding axons to their targets in the developing nervous system by providing repulsive signals (Cooper, 2002; Giger et al., 2010; Koeberle and Bahr, 2004; Parrinello et al., 2008). However, more recently, it has been reported that semaphorins can also act as axonal attractants and mediate adhesive signals in a variety of tissues, both in development and adulthood (Koeberle and Bahr, 2004; Parrinello et al., 2008). For example, semaphorins have also been found to act as chemoattractive guidance cues for cortical dendrites and olfactory bulb axons (Cooper, 2002).

The first semaphorin receptors to be identified were the neuropilins (Np-1 and -2), which recognize only the secreted semaphorins (Cooper, 2002; Koeberle and Bahr, 2004). Semaphorin 3A–Np-1 interactions are required for the fasciculation of the peripheral fibers of the trigeminal and vagal projections (Cooper, 2002). Np-2 is required for the organization and fasciculation of several cranial and spinal nerves (Cooper, 2002). Receptors for membrane-bound semaphorins are members of the family of plexins (Koeberle and Bahr, 2004). Plexins can interact with Np-1, conferring different semaphoring-binding specificities (Koeberle and Bahr, 2004).

Activation of the semaphorin receptor complex can lead to chemoattractive or chemorepulsive responses depending on the molecular composition of the receptor complex (Cooper, 2002). The growth cone response to semaphorins can be modulated by the direct physical interaction between neuropilins and members of the receptor families, highlighting the importance of receptor cross-talk in determining growth cone responses to guidance cues (Cooper, 2002).

The guidance factor semaphorin 3C, which is expressed by corpus callosum neurons, acts through the Np-1 receptor to orient axons crossing through the corpus callosum; transient neurons work together with their glial partners in guiding callosal axons (Niquille et al., 2009). The transducer that mediates the semaphoring 3C attractive response remains so far undefined (Niquille et al., 2009). Midline glial cells are the principal corpus callosum guidepost cells and secrete guidance factors that channel the callosal axons into the correct path (Niquille et al., 2009). These guidance signaling factors include netrin 1/DCC, Slit2/Robo1, ephrins/Eph, semaphorin/Np-1, and WNT (Niquille et al., 2009). Mutant mice for these guidance cues and their receptors exhibit callosal defects that range from minor, with few axons leaving the callosal track, to severe, with complete agenesis of the corpus callosum (ACC) (Niquille et al., 2009). Transmembrane proteins, including the tyrosine kinase receptors MET, ERBB2, OTK, and VEGFR2, participate in semaphorin responses by regulating diverse intracellular signaling events and functional outcomes (Niquille et al., 2009).

The cell adhesion molecule L1 has also been shown to act as a receptor for semaphorins (Koeberle and Bahr, 2004). A dependency on endocytosis to trigger axon retraction is observed in neurons responding to Sema3A, where the L1 IgCAM (immunoglobulin cell adhesion molecules), a component of the sema receptor complex, mediates endocytosis of the Sema3A holoreceptor in response to ligand binding (Bashaw and Klein, 2010; O'Donnell et al., 2009). The cell adhesion molecules L1 and TAG-1 (transient axonal glycoprotein-1) promote Sema3A activity through interaction and coendocytosis with its receptor neurolipin-1 (Bashaw and Klein, 2010).

Semaphorin 5A (Sema5A) is a membrane-bound protein and interacts with the receptor of plexin 3B (Lin et al., 2009). During development, Sema5A functions as an attractive and repulsive guidance molecule (RGM), and altered expression has been linked with aberrant development of axonal connections in the forebrain (Lin et al., 2009).

Semaphorin 4D (Sema4D) and B-type plexins demonstrate distinct expression patterns over critical time windows during the development of the murine cortex (Hirschberg et al., 2010). Sema4D–plexin-B2 interactions regulate mechanisms underlying cell specification, differentiation, and migration during corticogenesis (Hirschberg et al., 2010).

31.2.3 Netrins and Their DCC and UNC5 Guidance System

The DCC axon guidance receptor and its ligands, the netrins, have been shown to play a pivotal role in the guidance of axonal projections toward the ventral midline throughout the developing nervous system (Cooper, 2002; O'Donnell et al., 2009). The DCC belongs to a family of transmembrane proteins that possess four immunoglobulin domains and six fibronectin type III repeats (Kaprielian et al., 2001). Netrins are diffusible bifunctional molecules that can act as chemoattractants or chemorepellents for developing axons (Koeberle and Bahr, 2004). Secreted netrins and their receptors are one of the well-characterized axon guidance pathway families (Lin et al., 2009). The interaction of netrin-1 with DCC results in a chemoattractive response while interaction with the UNC5 family of netrin receptors results in chemorepulsion (Cooper, 2002; Lin et al., 2009). Netrin-1 functions as a floor plate-derived chemoattractant, which directs the pathfinding of commissural axons (Kaprielian et al., 2001). Netrin-1 binds to DCC that is expressed on commissural axons and their associated growth cones (Kaprielian et al., 2001). Mice lacking DCC or netrin-1 exhibit severe defects in commissural axon extension toward the floor plate and also lack several major commissures within the forebrain, including the corpus callosum and the hippocampal commissure (Cooper, 2002; Kaprielian et al., 2001). Studies have also revealed that DCC is crucial for the migration of some neuronal populations (Cooper, 2002).

In the developing mammalian neural tube, the DCC protein is present on the surface of commissural axons, as they migrate toward the floorplate, the source of the netrin gradient (Cooper, 2002). Once these axons have crossed the floorplate, they no longer respond to netrin

despite the fact that they still retain expression of DCC on the axonal membrane (Cooper, 2002). Instead, they become responsive to the chemorepellents Slit2 and class 3 semaphorins, which are produced by the floorplate and the ventral neural tube, respectively (Cooper, 2002). This switch in responsiveness to chemorepulsive cues once having crossed the midline is believed to propel axons away from the midline and explains why axons are never seen to recross the midline after reaching the contralateral side (Schmidt et al., 2009). The key to the silencing of the chemoattractive response of the netrin-1-DCC interaction in this context lies in the absence or presence of the Slit receptor, Robo (Cooper, 2002). Axons projecting toward the midline express DCC but not Robo on their surface (Cooper, 2002). When on the ipsilateral side, netrin engagement by DCC homodimers triggers a chemoattractive response (Cooper, 2002). The direct coupling of the DCC and Robo receptors provides a precise temporal and spatial mechanism that accurately controls growth cone responses at a given choice point comprising conflicting directional information (Cooper, 2002). Slit–Robo chemorepulsion overrides netrin–DCC chemoattraction, thus becoming the driving force for that growth cone (Cooper, 2002).

The DCC is subservient to another chemorepulsive guidance receptor, UNC5 (Cooper, 2002). The chemoattractive response of DCC–netrin interactions is converted to chemorepulsion by the direct interaction between the cytoplasmic domains of DCC and UNC5 in the presence of netrin-1 (Cooper, 2002). In the presence of UNC5, the DCC is forced to change its polarity and act as a chemorepulsive netrin receptor (Cooper, 2002). In the presence of netrin-1, the affinity of the DCC is higher for Robo and UNC5 than it is for another DCC receptor (Cooper, 2002).

Modifying serotonin (5-HT) abundance in the embryonic mouse brain disrupts the precision of sensory maps formed by TC axons (TCAs), suggesting that 5-HT influences TCAs during development (Bonnin et al., 2007). 5-HT 1B and 5-HT 1D receptor expression in the fetal forebrain overlaps with that of the DCC and UNC5c expression. 5-HT coverts the attraction exerted by netrin-1 on posterior TCAs to repulsion (Bonnin et al., 2007).

Protein kinase A (PKA) has a role in regulating translocation of the DCC receptor to the growth cone plasma membrane (O'Donnell et al., 2009). Netrin-dependent inhibition of Rho activity also contributes to the DCC mobilization (O'Donnell et al., 2009). Rho GTPases constitute a major signaling output of guidance receptor activation (O'Donnell et al., 2009). The trafficking and polarized localization of netrin and Slit receptors are critical for proper direction of axon outgrowth (O'Donnell et al., 2009). Specifically, mutations in the UNC-73 Trio-family RacGEF or the VAB-8 kinesin-related protein disrupt the normal localization of the SAX-3 (Robo) and UNC-40 (DCC) receptors, and in the case of UNC-40, regulation of localization also requires the MIG-2 Rac small GTPase (O'Donnell et al., 2009). These perturbations in normal receptor localization lead to significant defects in Slit and netrin-dependent posterior oriented cell and growth cone migration and further emphasize important regulatory roles for Rho GtPases in the control of axon guidance receptor localization (O'Donnell et al., 2009). In addition to these positive regulatory mechanisms, the trafficking of SAX-3 (Robo) and UNC-5 can also be negatively regulated with important outcomes for axon growth (O'Donnell et al., 2009).

In the regulated endocytosis of the repulsive netrin receptor UNC5 in vertebrate neurons, activation of PKC triggers the formation of a protein complex, including the cytoplasmic domain of UNC5H1, proteins interacting with C-kinase 1 (Pick1), and PKC and leads to specific removal of UNC5H1 from the growth cone surface (Bashaw and Klein, 2010; O'Donnell et al., 2009). Furthermore, the PKC activation leads to colocalization of UNC5A with early endosomal markers (Bashaw and Klein, 2010; O'Donnell et al., 2009). Thus, PKC-mediated removal of surface UNC5 provides a means to switch netrin responses from repulsion, mediated by either UNC5 alone or an UNC5-DCC complex, to attraction mediated by DCC (Bashaw and Klein, 2010; O'Donnell et al., 2009). G-protein-coupled adenosine 2b (A2b) receptor is a likely mediator of PKC activation because activation of A2b leads to the PKC-dependent endocytosis of UNC5 (Bashaw and Klein, 2010; O'Donnell et al., 2009). A2b is a netrin receptor that, together with DCC, appears to be required to mediate axon attraction, although this proposal has been controversial, and other evidence indicates that either A2b plays no role in netrin signaling or its role in netrin signaling is to modulate netrin responses (Bashaw and Klein, 2010; O'Donnell et al., 2009). In the context of UNC5 regulation, A2b acts independently of netrin, and its ability to regulate UNC5 surface levels supports its role as a potent modulator of netrin responses (Bashaw and Klein, 2010; O'Donnell et al., 2009).

31.2.4 Slits and Their Robo Guidance System

Slits are secreted proteins that control midline repulsion during development in vertebrates, signaling through receptors that belong to the Robo family (Allen and Chilton, 2009; Cooper, 2002; Kaprielian et al., 2001; Koeberle and Bahr, 2004). Slit can have dual roles, also promoting axonal growth (Koeberle and Bahr, 2004). Slit is a large extracellular matrix protein that is secreted by midline glia and is associated with the surfaces of axons (Kaprielian et al., 2001).

The Robo receptor is the best demonstration of both the importance of receptor trafficking for mediating axon guidance and the complexities yet to be unraveled (Allen and Chilton, 2009). This system is a striking example of the need for the growth cone to change its response as it grows toward, through, and beyond a guidance cue and the fundamental role played by the surface expression of receptors and their associated signaling components (Allen and Chilton, 2009). In both vertebrates and invertebrates, surface expression of the Robo receptor on axons of longitudinally projecting neurons and on precrossed commissural ones prevents them from approaching the embryonic midline (Allen and Chilton, 2009). Robo is the receptor for the chemorepellent protein Slit, which emanates from the midline (Allen and Chilton, 2009). Downregulation of Robo on commissural axons during midline crossing abrogates the repulsive effect of Slit so that a contralateral projection can be formed (Allen and Chilton, 2009). Following crossing, Robo is then restored to the axonal growth cone, which is not proven to prevent recrossing (Allen and Chilton, 2009).

In *Drosophila*, Slit is expressed by glia at the ventral midline, where it acts as a chemorepulsive guidance cue (Allen and Chilton, 2009; Cooper, 2002). The Slit receptor, Robo, is expressed at high levels on those axons that never cross the midline (Allen and Chilton, 2009; Cooper, 2002). In contrast, axons destined to cross the midline express very low levels of Robo when projecting on the ipsilateral side (Allen and Chilton, 2009; Cooper, 2002). Once on the contralateral side, Robo protein is greatly upregulated on the axonal membrane, and these axons never cross the midline again (Allen and Chilton, 2009; Cooper, 2002). Robo loss-of-function mutations result in both the commissural and noncomissural axons crossing the midline multiple times (Cooper, 2002).

Three Slit and three Robo orthologs have been identified in mammals (Cooper, 2002). The ability of mammalian Slits to act as chemorepulsive guidance cues has been demonstrated for a variety of axon populations, including olfactory bulb, hippocampal, and spinal motor axons (Cooper, 2002). The chemorepulsive activity of Slits has also been implicated in the targeted migration of neuroblasts within the rostral migratory stream toward the olfactory bulb and GABAergic neurons from the ganglionic eminence into the cortex (Cooper, 2002). Slit2 has also been shown to induce axon branching in sensory neurons (Cooper, 2002).

Heparan sulfate proteoglycans are essential for Slit-driven chemorepulsion (Cooper, 2002). These proteoglycans may be responsible for establishing effective local Slit concentrations and/or presenting Slit to the receptor in an appropriate format (Cooper, 2002).

Slit2 binds laminin-1, suggesting that the localization of Slit2 to precise choice points may be due to direct interactions with the laminin isoforms within the surrounding extracellular matrix (Cooper, 2002). Slit2 also interacts directly with netrin-1 (Cooper, 2002). Both Slit and netrin-1 are coexpressed in many regions of the embryonic brain, including the floorplate of the neural tube (Cooper, 2002). Netrin attracts commissural axons toward the floorplate, while Slit acts to repel axons from the floorplate (Cooper, 2002). Once at the floorplate, the chemoattractive response to netrin-1 is silenced by the direct coupling of the netrin receptor, DCC, with the chemorepulsive Slit receptor, Robo, allowing the growth cones to escape the attractive forces of netrin-1 and move away from the floor plate (Cooper, 2002).

31.3 DOWNSTREAM SIGNALING MECHANISMS AND OTHER PROTEINS INVOLVED IN AXON GUIDANCE

On binding of axon guidance proteins to their growth cone receptors, intracellular signaling cascades are activated, resulting in extensive remodeling of the cytoskeleton and subsequent steering of the growth cone (Izzi and Charron, 2011; O'Donnell et al., 2009; Schmidt et al., 2009). Axon guidance is not a one-way process in which the growth cone passively receives information through its growth cone receptor (Schmidt et al., 2009). Instead, neurons can regulate the expression of receptors on the growth cone surface and modulate their response to axon guidance cues through various mechanisms (Schmidt et al., 2009). Also, different axon guidance cues can influence each other by binding to the same binding partner and/or through interaction of different signaling cascades (Schmidt et al., 2009). Furthermore, axon guidance receptor activity can be regulated by physical interactions between different guidance receptors (Schmidt et al., 2009).

Activation of specific signaling pathways can promote attraction or repulsion, result in growth cone collapse, or affect the rate of axon extension (Izzi and Charron, 2011; O'Donnell et al., 2009). The way a given guidance signal is interpreted also depends on the activities of a number of second-messenger pathways within the cell, and these pathways are potent modulators of axon responses *in vivo* (O'Donnell et al., 2009).

Calcium and cyclic nucleotides (cAMP and cGMP) can act *in vitro* to directly mediate guidance receptor signaling and also can modulate the strength and valence of guidance responses (Bashaw and Klein, 2010; O'Donnell et al., 2009). Disruption of calcium and cyclic nucleotide signaling leads to guidance defects in many systems, and in some cases, direct links have been made to specific guidance receptor signaling pathways (Bashaw and Klein, 2010). These two signaling systems show extensive cross talk in the regulation of growth cone guidance:

Calcium signaling can promote the production of cyclic nucleotides through activation of soluble adenylyl cyclases and nitric oxide synthase, and cyclic nucleotides can regulate cellular calcium concentration by controlling the activity of plasma membrane calcium channels as well as through the regulation of calcium-induced calcium release (CICR) from internal stores (Bashaw and Klein, 2010). This positive feedback could potentially play a role in amplifying responses to shallow gradients of guidance cues (Bashaw and Klein, 2010). High cyclic nucleotide levels favor attraction, whereas low levels favor repulsion (O'Donnell et al., 2009). It is the ratio of cAMP/cGMP that is important in determining the polarity of the guidance response and has implicated calcium channel modulation in the control of guidance responses (O'Donnell et al., 2009). A clear demonstration that cyclic nucleotides and their downstream effectors can convert responses from attraction to repulsion and vice versa during axon guidance *in vivo* is lacking; however, a growing body of evidence supports a potent role for cyclic nucleotide signaling in modulating the strength of receptor responses (O'Donnell et al., 2009).

The signals mediating changes in cyclic nucleotide levels are thought to be independent of the guidance receptors whose responses they modulate; however, a more direct role of guidance receptor signaling in activating cAMP signaling has been suggested in the case of netrin (Ming et al., 1997; O'Donnell et al., 2009). Netrin acting through DCC (or A2b) leads to the elevation of cAMP and activation of PKA, and these events have been proposed to be instrumental in promoting netrin-mediated axon outgrowth and attraction (O'Donnell et al., 2009). There is an important role for cyclic nucleotides in modulating the strength of guidance responses *in vivo* rather than switching the polarity of responses (O'Donnell et al., 2009). The challenge now is to define the signals and receptors that regulate cyclic nucleotide signaling *in vivo* and to define specific contexts where their modulatory effects influence axon guidance (O'Donnell et al., 2009).

Exposure of growth cones to *in vitro* gradients of guidance cues can induce a corresponding gradient of calcium elevation (Bashaw and Klein, 2010). These asymmetric changes in calcium concentrations appear to be instructive signals to direct growth cone turning, because focal elevation of calcium is sufficient to induce turning responses (Bashaw and Klein, 2010). Increases in calcium influx and CICR can be triggered by guidance cues, and the outcome for growth cone behavior (either attraction or repulsion) can be influenced by the magnitude of the calcium elevation, the slope of the calcium gradient, and potentially the specific source of the calcium as well (Bashaw and Klein, 2010). In general, moderate amplitude increases in calcium (often involving CICR) favor attraction, whereas high- or low-amplitude increases favor repulsion, although differences in neuron type, growth substrate, and resting calcium concentrations can affect growth cone responses (Bashaw and Klein, 2010).

Electrophysiological recordings from growth cones indicate that attractive and repulsive guidance cues trigger rapid and reciprocal changes in membrane potential; attractants such as brain-derived neutrophic factor (BDNF) and netrin lead to membrane depolarization and repellents, such as Slit and semaphorin, lead to hyperpolarization (Bashaw and Klein, 2010). Moreover, the polarity of the change in membrane potential determines attraction versus repulsion (Bashaw and Klein, 2010). For netrin- and BDNF-mediated attraction, transient receptor potential (TRP) calcium channels contribute to membrane depolarization, and calcium influx through these channels is required for chemoattraction (Bashaw and Klein, 2010). BDNF and netrin through engagement of their respective TrkB and DCC receptors lead to calcium release from internal stores and activation of TRP channels: The subsequent TRP channel-dependent membrane depolarization is sufficient to activate voltage-dependent calcium channels (VDCCs), and the resulting calcium influx is essential for attractive turning (Bashaw and Klein, 2010). A role for semaphorins in the activation of cyclic nucleotide gated (CNG) calcium channels strengthens the case for the specific regulation of calcium influx through plasma membrane channels and points to the importance of cross regulation of cyclic nucleotide and calcium signaling (Bashaw and Klein, 2010). Here, Sema signaling through plexin receptors stimulates the production of cGMP, which in turn is required for membrane hyperpolarization, activation of CNG channels, and growth cone repulsion (Bashaw and Klein, 2010).

Similar to calcium signaling, cyclic nucleotides (cAMP or cGMP) can have profound effects on growth cone responses to guidance cues (Bashaw and Klein, 2010). The levels of cyclic nucleotides, specifically the ratio of cAMP to cGMP, can determine whether the response to a guidance cue will be attractive or repulsive, with high cyclic nucleotide levels (or high cAMP/cGMP ratios) favoring attraction and low levels (or low cAMP/cGMP ratios) favoring repulsion; cyclic nucleotide signaling can clearly modulate the strength of receptor responses (Bashaw and Klein, 2010).

Sema–plexin signaling leads to the production of cGMP, and cGMP plays an instructive role in promoting repulsion by regulating membrane hyperpolarization and the influx of calcium through CNG channels (Bashaw and Klein, 2010). Netrin outgrowth and attraction require DCC (or A2b)-mediated elevation of cAMP and activation of PKA (Bashaw and Klein, 2010).

Rho-family GTPases, a subgroup of the Ras superfamily of small GTPases, have been extensively studied for their role in cell motility and regulation of cytoskeletal structures (O'Donnell et al., 2009). Members of the Rho (Rac homology) family include Rac, Cdc42, and RhoA (O'Donnell et al., 2009). Rho-family GTPases catalyze the hydrolysis of bound GTP to GDP, switching from active (GTP-bound) and inactive (GDP-bound) states (O'Donnell et al., 2009). The activity of these GTPases has profound effects on actin cytoskeletal and microtubule dynamics (O'Donnell et al., 2009). Rho-family GTPases are very important in mediating axon guidance receptor signaling output (O'Donnell et al., 2009). Guidance cues including slits, netrins, ephrins, and semaphorins can all influence the activity of Rho-family GTPases (Bashaw and Klein, 2010; O'Donnell et al., 2009). Slits, acting through Robo receptors, lead to decreased levels of active Cdc42 and increased RhoA and Rac activity (Bashaw and Klein, 2010; O'Donnell et al., 2009). Ephrins, through Eph receptors, result in increased RhoA activity as well, but they can also cause transient, decreased Rac activity in RGCs, whereas Eph–ephrin reverse signaling activates Rac and Cdc43 to direct repulsive axon pruning (Bashaw and Klein, 2010; O'Donnell et al., 2009). There is no clear consensus for how Rho GTPases mediate repulsion, because these repulsive guidance pathways each influence RhoA, Rac, and Cdc42 activity in distinct ways (Bashaw and Klein, 2010; O'Donnell et al., 2009).

The primary regulators of Rho GTPase cycling and activity are the Rho-family GEFs and GAPs (GTPase-activating proteins) (O'Donnell et al., 2009). Upstream regulation is likely to provide tissue specific as well as temporal control of Rho GTPase signaling during growth cone guidance (O'Donnell et al., 2009). Guidance receptors can directly regulate Rho GTPases (O'Donnell et al., 2009). Identifying the GEFs and GAPs that function downstream of a given guidance receptor is critical to understanding the mechanism of guidance receptor signal transduction (O'Donnell et al., 2009).

Identification of individual Rho GTPase regulators that are essential mediators of guidance receptor signaling pathways is complicated by at least three major factors: (1) redundancy can obscure important functions, (2) individual GEFs and GAPs can act in multiple signaling pathways, (3) GEFs and GAPs often contribute to only part of any given signaling output (O'Donnell et al., 2009).

Ras-GAP α-chimaerin is an essential mediator of the ephrinB3/EphA4 guidance pathway (Bashaw and Klein, 2010; O'Donnell et al., 2009). The α2-chimaerin isoform contains two interaction domains for EphA4, the N-terminal SH2 domain, which can interact with phosphorylated juxtamembrane tyrosines of EphA4, and a second region in the C-terminus that constitutively interacts with the kinase domain of EphA4 (O'Donnell et al., 2009). The association of α2-chimaerin with Eph receptors appears to be direct or mediated by the Nck2 (Grb4) adaptor protein (Bashaw and Klein, 2010). EphA4-dependent tyrosine phosphorylation of α2-chimaerin occurs in response to ephrinB3, and this treatment increases the Ras-GAP activity of α-chimaerin (O'Donnell et al., 2009). The interaction with EphA4 activates the intrinsic GAP activity of α2-chimaerin, and this leads to inactivation of Rac1 (Bashaw and Klein, 2010). The cooperative action of α2-chimaerin in reducing Rac1-mediated actin polymerization and ephexin1 in enhancing RhoA-mediated actin depolymerization appears to induce efficient axon retraction (Bashaw and Klein, 2010). In addition, α-chimaerin's diacylglyceron (DAG)-binding C1 domain is very likely to regulate the GAP activity of α2-chimaerin (O'Donnell et al., 2009). The GAP domain in β2-chimaerin is occluded by the N-terminal SH2 motif, mediated by intramolecular interactions with the C1 domain, and ligand binding to the C1 domain is predicted to result in exposure of the Rac-GAP domain (O'Donnell et al., 2009). Increases in DAG production would be expected to increase the Rac-GAP activity of α2-chimaerin (O'Donnell et al., 2009). Similarly, SH2-mediated interactions with receptors may free the GAP domain for Rac inhibition (O'Donnell et al., 2009). Although a reduction in Rac activity is required to mediate ephrin-A-induced collapse, Rac activation also appears to be necessary for responses to ephrins (O'Donnell et al., 2009). Interference with Rac signaling blocks growth cone collapse in response to both semaphorins and ephrins (O'Donnell et al., 2009). Although decreases in Rac activity are observed following ephrin stimulation, reactivation of Rac is temporally correlated with growth cone collapse (O'Donnell et al., 2009).

Rac activity appears to be required for endocytosis; semaphorin 3A or ephrin treatment of retinal growth cones results in Rac-dependent endocytosis, which appears to mediate contact repulsion (O'Donnell et al., 2009). Specifically, for class B Eph/ephrins, bidirectional endocytosis occurs as the ephrin ligand and the Eph receptor are each internalized in trans to neighboring cells in a process that depends on their cytoplasmic domains and Rac activity (O'Donnell et al., 2009). The conserved Vav subfamily of Dbl GEFs plays a central role in this process, which appears to be instrumental for growth cone retraction (O'Donnell et al., 2009). Vav2/3 and α2-chimaerin have opposing effects on Rac1, and yet both are mediators of EphA forward signaling (Bashaw and Klein, 2010). Unlike Vav2/3, α2-chimaerin does not influence Eph receptor endocytosis (Bashaw and Klein, 2010). It is possible that the activated Eph receptor first activates α2-chimaerin to induce axon retraction and then activates Vav2/3 to locally activate Rac1-dependent endocytosis to allow cell detachment (Bashaw and Klein, 2010). Ephrins also function through

Rho activation, and this activation appears to be mediated by the Dbl-family Rho GEF ephexin (O'Donnell et al., 2009). Ephexin activates RhoA, Rac, and Cdc42, but activation of the EphA receptor results in preferential activity toward RhoA (O'Donnell et al., 2009).

Genetic analysis of motor axon guidance in *Drosophila* indicates that plexin-B mediates repulsion in part by binding to active Rac (Bashaw and Klein, 2010). In contrast to plexin-Bs, plexin-A-induced growth cone repulsion requires the activation of Rac (Bashaw and Klein, 2010). The FERM domain-containing GEF protein, FARP2, associates with the plexin-A1/Npn-1 complex (Bashaw and Klein, 2010). Sema3A binding to Npn-1 causes FARP2 to dissociate from plexin-A1, activating FARP2's GEF activity and raising the levels of Rac-GTP in the cell (Bashaw and Klein, 2010). Activation of Rac triggers Rnd1 binding to plexin-A1, thereby activating plexin-A1's intrinsic Ras-GAP activity (Bashaw and Klein, 2010). Activated plexin-A1 downregulates R-Ras activity, which may lead to inhibition of integrin function and growth cone repulsion (Bashaw and Klein, 2010).

In contrast to the mechanism of plexin-B1 activation via Rac sequestration and RhoA activation, growth cone collapse induced by Sema3A requires activation of Rac (O'Donnell et al., 2009). Plexin-A1, together with neuropilin, transduces guidance signals from class 3 semaphorins, leading to Rac activation, Rnd1 recruitment, and reduction in R-Ras activity (O'Donnell et al., 2009).

In the absence of ephrin stimulation, nonphosphorylated ephexin1 is bound to EphA4 and activates RhoA, Rac1, and Cdc42, leading to a balance of GTPase activation that promotes axonal growth (Bashaw and Klein, 2010). Eph tyrosine kinase activity is required, but not sufficient to promote ephexin1 phosphorylation; instead, ephrin-induced clustering of Ephs appears to promote ephexin1 phosphorylation, probably involving Src tyrosine kinase (Bashaw and Klein, 2010). Tyrosine phosphorylation of ephexin1 shifts its exchange activity toward RhoA, thereby causing growth cone collapse *in vitro* (Bashaw and Klein, 2010). When Ephs are activated in a portion of the growth cone, tyrosine phosphorylated ephexin1 may tip the local balance of GTPase activity toward RhoA, thereby causing actin depolymerization and local retraction (Bashaw and Klein, 2010). In other regions of the growth cone that have not made contact with ephrins, ephexin1 promotes growth by activating RhoA, Rac1, and Cdc42 (Bashaw and Klein, 2010).

In motile cells, activation of Rho GTPases results in modulation of cytoskeletal dynamics via effector proteins, and one of the best characterized of these is the dual Cdc42/Rac effector, p21-activated kinase (O'Donnell et al., 2009). A well-established pathway of PAK activation via Cdc42 or Rac results in inhibition of the actin depolymerizing factor cofilin by activating its inhibitor, LIM kinase (O'Donnell et al., 2009). Other notable targets of PAK include the myosin activator, myosin light chain (MLC) kinase, and the microtubule destabilizing protein, Op18/stathmin, which are each inhibited by PAK phosphorylation (O'Donnell et al., 2009). GTP-bound Rac and Cdc42 regulate Pak activity through binding to its Cdc42/Rac interactive binding domain, relieving autoinhibition of Pak by its N-terminal domain (O'Donnell et al., 2009). Pak likely functions downstream of Rac/Cdc42 in axon guidance (O'Donnell et al., 2009). *Drosophila pak, dock*, and *rac* each functions in midline axon repulsion and interacts genetically with the Slit/Robo pathway (O'Donnell et al., 2009). Expression of a constitutively membrane-targeted Pak suppresses defects caused by *rac* loss of function, which suggests that these Rac-dependent defects likely occur through loss of PAK regulation (O'Donnell et al., 2009). Both Cdc42 and Rac likely also function through pathways independently of PAK, particularly in axon growth (O'Donnell et al., 2009). Regulation of outgrowth via Rac can occur through a PAK-independent mechanism; however, guidance mediated through Rac and Cdc42 at least partly involves PAK function (O'Donnell et al., 2009).

Regulation of Pak leads to modulation of actin dynamics in axon growth and guidance by regulating the actin depolymerizing factor, cofilin, by modulating the activity of the serine/threonine kinase, LIMK (lin-11 and Mec-3 kinase) (O'Donnell et al., 2009). Cofilin destabilizes F-actin through pointed-end severing of actin filaments, although this activity may be necessary to maintain the supply of monomeric G-actin, thus promoting actin polymerization (O'Donnell et al., 2009). This activity is inhibited by phosphorylation at the N-terminal Ser3: Phosphorylation at this site is reciprocally regulated by the LIM and TES (testis-derived transcript) kinases and by the Slingshot phosphatase (Shh) (O'Donnell et al., 2009). In some cases, the rate of cycling between phosphorylated and nonphosphorylated states, rather than the absolute level of either species, can determine the influence of cofilin on actin dynamics (O'Donnell et al., 2009). How the LIM kinase and slingshot function in concert to regulate growth cone dynamics by regulating cofilin is of great interest in understanding receptor-mediated guidance (O'Donnell et al., 2009). Cycling of cofilin is required for promoting axon growth (O'Donnell et al., 2009). The LIMK also appears to mediate both axon outgrowth and attraction in certain contexts (O'Donnell et al., 2009). A gradient of phosphorylated cofilin accompanies the attractive response to BMP7, and repulsive responses from the same ligand are mediated by Ssh activity, demonstrating that distinct responses can be generated through activities converging on a single actin regulator (O'Donnell et al., 2009). The LIMK is required for growth cone attraction in certain contexts (O'Donnell et al., 2009).

Reverse signaling by receptorlike ephrinB proteins has been implicated in axon guidance (Bashaw and Klein, 2010). Following interactions with cognate Ephs, ephrinB proteins become clustered, and signaling is initiated either by Src-mediated tyrosine phosphorylation of the ephrinB cytoplasmic tail or by recruitment of PDZ domain-containing effectors (Bashaw and Klein, 2010). The NCK2 (NCK adaptor protein 2) is recruited to the phosphorylated ephrinB protein and is essential for several ephrinB-mediated processes, including spine formation (Bashaw and Klein, 2010). Tyrosine phosphorylation-dependent ephrin-B3 reverse signaling controls the stereotyped pruning of exuberant mossy fiber axons in the hippocampus, and NCK2 has been implicated in this process (Bashaw and Klein, 2010). The NCK2 appears to couple ephrinB3 with DOCK180 and PAK, leading to the activation of RAC1 (ras-related C3 botulinum toxin substrate 1) and CDC42 (cell division cycle 42) and downstream signaling that results in axon retraction/pruning (Bashaw and Klein, 2010).

Although inhibition of RAC can accompany repulsive guidance decisions, activation of RAC may also be involved in mediating responses to repulsive cues exemplified by Eph-dependent growth cone retraction and Ephrin-dependent axon pruning (Bashaw and Klein, 2010; O'Donnell et al., 2009). In the context of Slit–Robo-mediated repulsion, activation of Robo receptors by Slit leads to the activation of RAC, and limiting RAC function reduces the efficiency of Slit–Robo signaling (Bashaw and Klein, 2010; O'Donnell et al., 2009). The conserved RAC GAP, Vilse/CrGAP, contributes to Slit-dependent guidance decisions and antagonizes RAC function (Bashaw and Klein, 2010; O'Donnell et al., 2009). In axons, high levels of Vilse overexpression causes similar defects to Robo loss of function, and low levels of overexpression cause dosage-dependent defects in Slit–Robo repulsion similar to loss of function of Vilse (Bashaw and Klein, 2010; O'Donnell et al., 2009). The consequences of increasing or decreasing Vilse function are similar: Both lead to a compromise in the efficiency of Slit–Robo midline repulsion, suggesting that the precise levels of spatial distribution of Vilse RAC GAP activity may be instructive for Robo repulsion (Bashaw and Klein, 2010; O'Donnell et al., 2009). The interaction of Vilse/crGAP with Robo may be regulated in different subcellular contexts or during distinct stages of Slit–Robo repulsion (O'Donnell et al., 2009). Both the Slit–Robo and the forward Eph pathway use RAC GAP (Vilse for Robo and α-chimaerin for Ephs) and a Rac GEF (Sos for Robo and Vav for Ephs) to mediate repulsion (Bashaw and Klein, 2010; O'Donnell et al., 2009). A complex of proteins, including the adapters Nick, Pak, and Rac, are recruited to the receptors to mediate repulsive signaling (Bashaw and Klein, 2010). Coordinated action of GEFs and GAPs may promote the cycling of Rac activity, which may be more important than the overall levels of Rac-GTP in a responding growth cone (Bashaw and Klein, 2010; O'Donnell et al., 2009). Alternatively, these GAPs and GEFs may represent distinct steps in the repulsive signal transduction output (Bashaw and Klein, 2010; O'Donnell et al., 2009). Rac activation in the case of Robo receptors may precede internalization as well (Bashaw and Klein, 2010; O'Donnell et al., 2009). Parallel mechanisms of repulsion may exist in these distinct ligand–receptor systems (O'Donnell et al., 2009).

Src family kinases (Src, Fyn, Yes, and others; collectively known as SFKs) are nonreceptor protein tyrosine kinases that have emerged as essential mediators of various guidance receptors (Bashaw and Klein, 2010). SFKs appear to be required in motor (LMC; lateral motor column) axons for limb trajectory selection and are critical for Eph forward signaling (Bashaw and Klein, 2010). Recruitment and activation of SFKs have been documented downstream of reverse signaling via GPI-anchored ephrinA ligands (Bashaw and Klein, 2010). The link between GPI-anchored ephrinAs and SFKs may be provided by transmembrane proteins such as p75 and TrkB (Bashaw and Klein, 2010). SFKs and focal adhesion kinase (FAK) are also essential mediators of netrin signaling (Bashaw and Klein, 2010).

The morphogen SHH mediates cell fate and axon guidance in the developing nervous system by two distinct pathways (Bashaw and Klein, 2010). Cell fate specification by Shh is mediated by the receptor Patched (Ptc) via the canonical pathway requiring the Gli family of transcription factors (Bashaw and Klein, 2010). In contrast, axon guidance by Shh is mediated by SFK in a smoothened-dependent manner via a rapidly acting, noncanonical signaling pathway not requiring transcription (Bashaw and Klein, 2010).

In addition to their critical contributions to downstream signaling, second messengers and Rho GTPases can also influence guidance responses by regulating the surface localization and activation of guidance receptors (Bashaw and Klein, 2010). Netrins induce outgrowth and attractive turning via the DCC family of receptors at least in part by regulating Rho GTPases (O'Donnell et al., 2009). Outgrowth of commissural axons in response to netrin requires Rho GTPase activity (O'Donnell et al., 2009). Netrin induces rapid activation of RAC1, CDC42, and PAK1 (p21 protein-activated kinase 1), which may occur in a complex containing the constitutive components DCC and NCK-1, as well as netrin-induced components, RAC1, CDC42, PAK1, and N-WASP (O'Donnell et al., 2009). Activation of this complex by netrin causes profound changes in growth cone morphology (O'Donnell et al., 2009). Netrin stimulation leads to an increase in the DCC surface levels, and this effect is enhanced by PKA activation (Bashaw and

Klein, 2010). Blocking adenylate cyclase, PKA activity, or exocytosis prevents the increase in the DCC surface levels and blunts netrin-induced axon outgrowth (Bashaw and Klein, 2010). In addition to PKA's role in regulating DCC, netrin-dependent inhibition of Rho activity also contributes to DCC membrane localization (Bashaw and Klein, 2010). The trafficking and polarized localization of netrin and Slit receptors are critical for proper direction of axon outgrowth (Bashaw and Klein, 2010). There are important upstream regulatory roles of Rho GTPases in the control of axon guidance receptor localization (Bashaw and Klein, 2010). Trio is an important regulator of axon guidance decisions in several contexts (O'Donnell et al., 2009). Trio contains two Rho GEF domains, one with specificity for Rac and RhoG and the other that activates RhoA (O'Donnell et al., 2009).

Stimulation of RhoA results in activation of Rho kinase (O'Donnell et al., 2009). Rho kinases (ROCK or Rok) are serine/threonine kinases that, similar to PAK, regulate LIMK (O'Donnell et al., 2009). In addition, Rho kinases can regulate myosin activity through the phosphorylation of MLC, which results in activation and increases actin–myosin contractility (O'Donnell et al., 2009). Rho kinase also indirectly regulates myosin activity by phosphorylating and inhibiting MLC phosphatase (O'Donnell et al., 2009). ROCK activity is necessary for RhoA-induced retraction, likely through regulation of myosin II (O'Donnell et al., 2009). Inhibition of ROCK prevents the stability and contraction of actin arcs, which are filamentous actin structures that form in the transition zone of growth cones and affect microtubule bundling and dynamics (O'Donnell et al., 2009). ROCK can phosphorylate LIMK to regulate cofilin activity (O'Donnell et al., 2009). ROCK also phosphorylates the colapsin response mediator protein-2 (CRMP-2) after LPA (lipoprotein, Lp) or ephrin-A5 stimulation inhibits its ability to bind tubulin heterodimers (O'Donnell et al., 2009). CRMP-2 normally promotes axon outgrowth and branching, presumably by nucleating and promoting microtubule assembly (O'Donnell et al., 2009). The same residue in CRMP-2 that is targeted by ROCK downstream of LPA and ephrinA5 is phosphorylated by Cdk5 downstream of Sema3a-induced growth cone collapse, suggesting that independent signaling pathways can converge on the regulation of CRMP-2 phosphorylation (O'Donnell et al., 2009).

The RGM gene family consists of three members, RGMa, RGMb, and RGMc (Severyn et al., 2009). Each gene encodes a protein whose expression is restricted to a small number of tissues and is hypothesized to be involved in distinct biological functions ranging from control of iron metabolism to regulation of axonal guidance and neuronal survival in the developing nervous system (Severyn et al., 2009). RGMa is a cell membrane-associated GPI-linked two-chain axonal guidance protein found primarily in the developing and adult central nervous system (Severyn et al., 2009). It has been shown that RGMa regulates repulsive guidance of retinal axons via binding to neogenin, a transmembrane protein that is also a receptor for netrins (Severyn et al., 2009).

Nervous system development is highly dependent on the microtubule cytoskeleton (Tischfield et al., 2010). Microtubules are copolymers assembled from tubulin heterodimers, which contain several different α- and β-isotypes (Tischfield et al., 2010). Each isotype may have properties necessary for specific cellular functions (Tischfield et al., 2010). TUBB3, one of at least six β-tubulins found in mammals, is distinct because purified microtubules enriched in TUBB3 are considerably more dynamic than those composed from other β-tubulin isotypes and because its expression is primarily limited to neurons (Tischfield et al., 2010). TUBB3 expression is greatest during periods of axon guidance and maturation; the level decreases in the adult central nervous system but remains high in the peripheral nervous system (Tischfield et al., 2010). The neuronal β-tubulin isotype is required in axon guidance and normal brain development (Tischfield et al., 2010).

During neurite extension, various families of microtubule-associated proteins (MAPs) bind to microtubules to regulate their stability, and hence the rate and direction of process outgrowth (Allen and Chilton, 2009). Some of these proteins appear to be specifically localized in either axons or dendrites (Allen and Chilton, 2009). Understanding how these MAPs become recruited to different compartments could shed light on how this occurs with axon guidance molecules (Allen and Chilton, 2009). The stability of microtubules will directly affect the facility with which motor proteins can convey their cargoes to the growing axon tip (Allen and Chilton, 2009).

Following axogenesis, the proximal section of the axon forms a diffusion impermeable barrier termed the axon initial segment (AIS) (Allen and Chilton, 2009). The AIS, microtubule polarity, and motor protein affinities collectively provide a series of filters to selectively transport necessary proteins to the axonal tip (Allen and Chilton, 2009). The growth cone autonomously executes intrinsic mechanisms to sort and retain from these deliveries those that are required (Allen and Chilton, 2009). The interplay between the cell adhesion molecule L1 and the neuropilin–semaphorin signaling axis provides the best current insight into how the turnover of axon guidance receptors is regulated at the growth cone (Allen and Chilton, 2009). L1 and other members of its family of cell adhesion molecules are important for the overall adhesion of a developing axon to its substrate (Allen and Chilton, 2009). By regulation, the

distribution and turnover of semaphorin signaling complexes, L1 and maybe other cell adhesion molecules such as neuronal cell adhesion molecule, required for the signaling of Sema3B and Sema3F through Np-2, may play a key role in determining which receptors reach and stay at the developing axon tip (Allen and Chilton, 2009).

L1 is expressed along the length of the axon, but it is preferentially inserted into the growth cone membrane (Allen and Chilton, 2009). It functions as a critical component of the Sema3A receptor complex by binding Np-1 and determining whether a repulsive or attractive response is generated (Allen and Chilton, 2009). L1 also regulates the turnover of neuropilin receptors by controlling its endocytosis following semaphoring binding (Allen and Chilton, 2009). At the growth cone, L1 interacts with Ezrin–Radixin–Moesin proteins to influence the actin cytoskeleton and axon outgrowth (Allen and Chilton, 2009). L1 is anchored to the axonal membrane via ankyrinB, which is colocalized with L1 along the axonal shaft (Allen and Chilton, 2009). A mechanism is suggested whereby L1 could carry neuropilin to the growth cone and modulate its signaling once there (Allen and Chilton, 2009). Given the large turnover of plasma membrane occurring in developing growth cones, association with L1 in its cycle of surface renewal would be a very efficient means of locally recycling neuropilin, obviating the need to continually transport fresh receptor down the growing axon (Allen and Chilton, 2009). Following a change in its phosphorylation, L1 can subsequently associate with different cytoskeletal components and assume a more passive, adhesive role in the axon (Allen and Chilton, 2009). The interaction between L1 and neuropilin provides a means for both selective targeting of the latter to the developing growth cone and regulation of its endocytic turnover once there (Allen and Chilton, 2009).

There is another mechanism to retain guidance receptors at the axonal tip (Allen and Chilton, 2009). For neuropilin and the receptors for ephrin and netrin guidance molecules, local variations in the composition of the plasma membrane play an important role (Allen and Chilton, 2009). In the axon, glycolipid-enriched complexes, also known as lipid rafts, mediate the sorting of GPI-anchored proteins (Allen and Chilton, 2009). The responses induced by semaphorin and netrin are both dependent on the integrity of lipid rafts within the growth cone (Allen and Chilton, 2009). The GPI-anchored ephrinA and the transmembrane ephrinB families of molecules are also present in lipid rafts (Allen and Chilton, 2009).

Palmitoylation of proteins is a posttranslational modification, which can enhance the association of proteins with lipid rafts, and is a critical component of the nervous system development (Allen and Chilton, 2009). This modification occurs with the netrin receptor, DCC

(Allen and Chilton, 2009). This is required for its proper function in promoting both axonal growth and turning (Allen and Chilton, 2009). Lipid modification and microenvironments could be crucial for getting receptors to the developing growth cone (Allen and Chilton, 2009). Restriction within the growth cone of enzymes, which regulate palmitoylation, could then locally modulate retention of guidance receptors and associated signaling components (Allen and Chilton, 2009).

31.3.1 Agenesis of the Corpus Callosum

The corpus callosum is the largest fiber tract in the central nervous system and the major interhemispheric fiber bundle in the brain (http://www.ncbi.nlm.nih.gov/omim) (Bloom and Hynd, 2005; Donahoo and Richards, 2009; Paul et al., 2007; Richards et al., 2004). It consists of over 190 million axons, connects neurons in the left and right cerebral hemispheres, and is essential for the coordinated transfer of information between them (Bloom and Hynd, 2005; Chiappedi and Bejor, 2010; Donahoo and Richards, 2009; Paul et al., 2007; Richards et al., 2004). The corpus callosum contains homotopic and heterotopic interhemispheric connections (Paul et al., 2007).

Formation of the corpus callosum begins as early as 6 weeks of gestation, with the first fibers crossing the midline at 11–12 weeks of gestation and completion of the basic shape by 18–20 weeks (http://www.ncbi.nlm.nih.gov/omim) (Bloom and Hynd, 2005; Chiappedi and Bejor, 2010). The corpus callosum first enlarges caudally then develops rostrally (Bloom and Hynd, 2005). Myelination occurs slowly over the lifespan, with the process completing in puberty (Bloom and Hynd, 2005). Myelination progresses caudally to rostrally, much as the corpus callosum develops, from the splenium to the genu and rostrum (Bloom and Hynd, 2005; Richards et al., 2004). The most anterior portion of the callosum is the genu, which connects the prefrontal cortices on either hemisphere (Bloom and Hynd, 2005; Richards et al., 2004). The middle portions of the corpus callosum connect the motor and somatosensory regions (Bloom and Hynd, 2005). The caudal part of the body of the corpus callosum connects the cortex from the temporoparietal–occipital junction, as do portions of the splenium, the most posterior section of the corpus callosum (Bloom and Hynd, 2005; Richards et al., 2004). The splenium also connects the dorsal parietal and occipital regions (Bloom and Hynd, 2005).

For the corpus callosum to form, several critical developmental events must occur in sequence (Donahoo and Richards, 2009; Paul et al., 2007). These include correct midline patterning, formation of telencephalic hemispheres, birth and specification of commissural neurons,

and axon guidance across the midline to the final target in the contralateral hemisphere (Donahoo and Richards, 2009; Paul et al., 2007; Richards et al., 2004).

Np-1 has been shown to be involved in corpus callosum formation (Donahoo and Richards, 2009). The first axons to cross the midline arise from neurons in the cingulated cortex (Paul et al., 2007). Cingulate pioneering axons of the corpus callosum express Np-1 at a crucial temporal stage for callosal development (Donahoo and Richards, 2009; Paul et al., 2007). Callosal axons are guided medially to the midline, where they are channeled into the contralateral hemisphere (Donahoo and Richards, 2009). Midline cellular populations that have been shown to assist in the formation of the corpus callosum include the midline zipper glia, the glial wedge, the indusium griseum glia, and the subcallosal sling (Bush and Soriano, 2009; Donahoo and Richards, 2009; Paul et al., 2007; Richards et al., 2004). The glial wedge has been shown to express molecules such as Slit2 required for callosal axons to cross the midline (Richards et al., 2004). The indusium griseum also expresses the guidance molecule Slit2 and resides directly above the corpus callosum (Richards et al., 2004).

After crossing the midline, callosal axons grow into the contralateral hemisphere toward their designated target region, usually homotopic to their region of origin, and then innervate the appropriate cortical layer (Paul et al., 2007). Such processes probably involve both molecular recognition of the appropriate target region and activity-dependent mechanisms that regulate axon targeting to the correct layer and the subsequent refinement of the projection (Paul et al., 2007).

ACC encompasses a broad range of diagnoses and is one of the most frequent malformations in the brain with a reported incidence ranging between 0.5 and 70 in 10 000 births (http://www.ncbi.nlm.nih.gov/omim) (Chiappedi and Bejor, 2010; Richards et al., 2004). Both complete and partial ACC can result from disruption in any one of the multiple steps of callosal development, such as cellular proliferation and migration, axonal growth, or glial patterning at the midline (Bush and Soriano, 2009; Engle, 2010; Paul et al., 2007; Richards et al., 2004). ACC is a clinically and genetically heterogeneous condition, which can be observed as either an isolated condition or a manifestation in the context of a congenital syndrome (http://www.ncbi.nlm.nih.gov/omim. Further heterogeneity in ACC can arise from concomitant abnormalities in the anterior commissure (Paul et al., 2007). Malformation of the corpus callosum can result in either the complete absence of the corpus callosum or partial absence (hypogenesis) or thinning of the corpus callosum (Donahoo and Richards, 2009; Paul et al., 2007; Richards et al., 2004).

The wide range of disorders in which callosal abnormalities are found underscores the importance of understanding the basic nature of the development and function of the corpus callosum (Bloom and Hynd, 2005).

31.3.1.1 Clinical Characteristics

Corpus callosum dysgenesis accompanies a multitude of inherited disorders and results in a clinical spectrum ranging from normal to severe mental retardation (Engle, 2010; Richards et al., 2004). The most frequent clinical findings in patients with ACC are mental retardation (60%), visual problems (33%), speech delay (29%), seizures (25%), and feeding problems (20%) (http://www.ncbi.nlm.nih.gov/omim).

Isolated ACC, even when not ascertained clinically, still causes behavioral and cognitive impairment (Paul et al., 2007). As more individuals with primary ACC are identified and assessed with sensitive standardized neuropsychological measures, a pattern of deficits in higher order cognition and social skills has become apparent (Chiappedi and Bejor, 2010; Paul et al., 2007). The major anatomical feature of primary ACC is the absence of the corpus callosum, and it is presumed to be the cause of the cognitive and behavioral changes in these individuals (Paul et al., 2007). However, colpocephaly (abnormal enlargement of the occipital horns of the brain) and Probst bundles (large, bilateral, barrel-shaped axonal structures that do not cross the midline) are common in people with primary ACC and together with other subtle anatomical changes probably also affect behavior (Paul et al., 2007).

The functional consequences of structural changes in brain connectivity contribute to cognitive impairment (Paul et al., 2007). Primary ACC has a limited impact on general cognitive ability (Paul et al., 2007). Although the full-scale intelligence quotient (IQ) can be lower than expected based on family history, scores frequently remain within the average range (Paul et al., 2007). In an unexpectedly large number of persons with primary ACC (as many as 60%), performance IQ and verbal IQ are significantly different (Paul et al., 2007). However, there is no consistency with respect to which of the two is more affected (Paul et al., 2007). Impairments in abstract reasoning, problem solving, generalization, and category fluency have all been consistently observed in patients with primary ACC (Chiappedi and Bejor, 2010; Paul et al., 2007). Data from large sample sizes suggest that problem-solving abilities become more impaired as task complexity increases (Paul et al., 2007).

The most comprehensively examined higher cognitive domain in patients with ACC is language (Paul et al., 2007). Overall, individuals with primary ACC have intact general naming, receptive language, and lexical reading skills (Paul et al., 2007). However, impairments have been reported in the comprehension of syntax and linguistic pragmatics, and in phonological processing and rhyming (Paul et al., 2007). With respect to linguistic pragmatics, persons with primary ACC are impaired in the comprehension of idioms, proverbs, vocal prosody, and narrative humor (Paul et al., 2007). Patients with primary ACC also show marked difficulty with expressive language (Paul et al., 2007).

Parents of individuals with primary ACC consistently describe impaired social skills and poor personal insight as the features that interfere most prominently with the daily lives of their children (Chiappedi and Bejor, 2010; Paul et al., 2007). Also, parents frequently report health concerns such as feeding or sleeping issues, elimination problems, and unusual tolerance for pain (Chiappedi and Bejor, 2010). Specific traits include emotional immaturity, lack of introspection, impaired social competence, general deficits in social judgment and planning, and poor communication of emotions (Chiappedi and Bejor, 2010; Paul et al., 2007). Consequently, patients with primary ACC often have impoverished and superficial relationships, suffer social isolation, and have interpersonal conflict both at home and at work because of misinterpretation of social cues (Chiappedi and Bejor, 2010; Paul et al., 2007).

Responses of adults with primary ACC on self-report measures also suggest diminished self-awareness (Paul et al., 2007). The patients' self-reports are often in direct conflict with observations from friends and family (Paul et al., 2007). One potential factor contributing to poor self-awareness may be a more general impairment in comprehension and description of social situations (Paul et al., 2007).

Taken together, the neuropsychological findings in primary ACC highlight a pattern of deficits in problem solving, social pragmatics of language and communication, and processing emotion (Chiappedi and Bejor, 2010; Paul et al., 2007). Behavioral and emotional factors are frequently associated: A tendency for deficient social cognition in individuals with ACC seems to stem from a combination of difficulty integrating information from multiple sources, using paralinguistic cues for emotion, and understanding nonliteral speech (Chiappedi and Bejor, 2010).

The deficits in social communication and social interaction in patients with primary ACC overlap with the diagnostic criteria for autism (Chiappedi and Bejor, 2010; Paul et al., 2007). Furthermore, people with primary ACC may display a variety of other social, attentional, and behavioral symptoms that can resemble those of certain psychiatric disorders (Paul et al., 2007). Examination of symptom overlap between psychiatric disorders and ACC may help to isolate the symptoms that are directly caused by a dysfunction in corticocortical connectivity (Paul et al., 2007).

There are also structural similarities between ACC and some psychiatric disorders (Paul et al., 2007). Structural correlates of abnormal brain connectivity are evident in essentially every psychiatric disorder that has been examined (Paul et al., 2007). For example, several studies have found altered morphology of the corpus callosum in schizophrenia patients, including changes in size and shape, as well as microstructural changes in callosal regions that are revealed by diffusion MRI (Bloom and Hynd, 2005; Paul et al., 2007). There are also a number of reports of complete ACC in patients with schizophrenia, underscoring a direct connection between ACC and schizophrenia (Paul et al., 2007). Corpus callosum size, especially its anterior sectors, is also decreased in some cases of autism (Bloom and Hynd, 2005; Chiappedi and Bejor, 2010; Paul et al., 2007). Microstructural changes in the corpus callosum have also been found in patients with Tourette's syndrome (a disorder characterized by repetitive, stereotyped, involuntary movements and vocalizations called tics) and attention deficit hyperactivity disorder (Bloom and Hynd, 2005; Paul et al., 2007). Abnormalities in the size of the corpus callosum have also been found in patients diagnosed with mental retardation, Down syndrome, developmental dyslexia, and developmental language disorders (Bloom and Hynd, 2005). Deviant asymmetry of cortical areas, possibly related to callosal abnormalities, has been found in developmental dyslexia and specific language impairment (Bloom and Hynd, 2005).

31.3.1.2 Genetics

The genetics of ACC is variable and reflects the underlying complexity of callosal development (Bush and Soriano, 2009; Chiappedi and Bejor, 2010; Paul et al., 2007). A combination of genetic mechanisms, including single-gene Mendelian mutations, single-gene sporadic mutations, and complex genetics, might have a role in the etiology of ACC (Paul et al., 2007). Retrospective chart reviews and cross-sectional cohort studies report that 30–45% of cases of ACC have identifiable causes (Chiappedi and Bejor, 2010; Paul et al., 2007). Approximately 10% have chromosomal anomalies, and the remaining 20–35% have recognizable genetic syndromes (Paul et al., 2007). However, if only individuals with complete ACC are considered, then the percentage of patients with recognizable syndromes drops to 10–15%, and thus 75% of cases of complete ACC do not have an identified cause (Chiappedi and Bejor, 2010; Paul et al., 2007).

One example of ACC associated with a Mendelian disorder is X-linked lissencephaly with ACC and ambiguous genitalia (XLAG), which results from a mutation in the aristaless-related homeobox gene (ARX) (Chiappedi and Bejor, 2010; Paul et al., 2007).

Another syndrome caused by a single-gene mutation is CRASH syndrome (corpus callosum agenesis, retardation, adducted thumbs, spastic paraplegia, and hydrocephalus), which is accompanied by diminutive corticospinal tracts within the brainstem (Paul et al., 2007). CRASH is caused by mutations in the L1 cell adhesion molecule (L1CAM) gene that codes for a transmembrane cell adhesion protein broadly expressed in the central nervous system (Paul et al., 2007).

Andermann syndrome, an autosomal recessive condition prevalent in the Saguenay-Lac-St-Jean region of Quebec, presents with callosal hypoplasia or ACC, cognitive impairment, episodes of psychosis, and a progressive central and peripheral neuropathy (Chiappedi and Bejor, 2010; Paul et al., 2007). It is caused by mutation of the KCl cotransporter KCC3 (potassium chloride cotransporters 3) (Paul et al., 2007).

Mowat–Wilson syndrome (MWS) in addition to ACC causes Hirschsprung disease, congenital heart disease, genitourinary anomalies, microcephaly, epilepsy, and severe cognitive impairment (Chiappedi and Bejor, 2010; Paul et al., 2007). The MWS is caused by heterozygous inactivating mutations in the gene zinc finger homeobox 1b on chromosome 2q22, which codes for SMAD-interacting protein 1 (Paul et al., 2007). ACC is not observed in all MWS cases (Paul et al., 2007).

Aicardi syndrome is another disorder probably caused by sporadic mutations on the X chromosome (Chiappedi and Bejor, 2010; Paul et al., 2007). It is only observed in females and XXY males with Klinefelter syndrome (Paul et al., 2007). It consists of ACC, infantile spasms and chorioretinal lacunae, and additional cerebral and ophthalmological abnormalities (Paul et al., 2007).

ACC is a notable facet of craniofrontonasal syndrome (CFNS) (Bush and Soriano, 2009). Mutations in the EPHRIN-B1 gene result in a wide spectrum of developmental abnormalities that constitute this syndrome (Bush and Soriano, 2009). This syndrome includes cleft palate, craniofrontonasal dysplasia, craniosynostosis, axial skeletal defects such as asymmetry of the thoracic skeleton and limb abnormalities, and neurological defects such as mental retardation (Bush and Soriano, 2009). It is an X-linked condition (Bush and Soriano, 2009).

Some other syndromes associated with ACC are ACC with fatal lactic acidosis, HSAS/MASA syndromes (X-linked hydrocephalus) (L1CAM), acrocallosal syndrome, Chudley–MCCullough syndrome, Donnai–Barrow syndrome, FG syndrome, gentiopatellar syndrome, Temtamy syndrome, Toriello–Carey syndrome, Vici syndrome, ACC with spastic paraparesis (SPG11), CFNS, Fryns syndrome, Marden–Walker syndrome, Meckel–Gruber syndrome, microphthalmia with linear skin defects, Opitz G syndrome, orofaciodigital syndrome, pyruvate decarboxylase deficiency, Rubinstein–Taybi syndrome, septo-optic dysplasia (DeMorsier syndrome), Sotos syndrome, Warburg micro syndrome, and Wolf–Hirschhorn syndrome (Bush and Soriano, 2009; Chiappedi and Bejor, 2010; Paul et al., 2007).

In most individuals with ACC, there is no clearly inherited cause or a recognized genetic syndrome, suggesting that ACC can be caused by sporadic (*de novo*) genetic events (Paul et al., 2007). It is likely that some cases of ACC are caused by haploinsufficiency at other genetic loci (Paul et al., 2007). This is supported by many reports of patients with ACC who have sporadic chromosomal mutations with particular loci identified repeatedly (Paul et al., 2007).

The number of patients with ACC in which chromosomal rearrangements are found has increased following technical improvements, from conventional karyotyping to subtelomeric and array-CGH (comparative genomic hybridization) analysis (Chiappedi and Bejor, 2010). Candidate genes have been located especially on chromosome 1 and also on 3, 7, 8, 13, 15, 18, and 21 (Chiappedi and Bejor, 2010). Data obtained using microarray-based comparative genomic hybridization demonstrate that patients with ACC have chromosomal deletions or duplications that are smaller than those that can be detected using conventional cytogenetics (Paul et al., 2007). The risk of having a child with ACC is nearly threefold higher for mothers aged 40 and above, which is consistent with causal sporadic chromosomal changes (Paul et al., 2007).

Many cases of ACC might be caused by polygenic and other complex interactions (Paul et al., 2007). The abundance of case reports of ACC associated with specific diseases probably also reflects complex underlying mechanisms (Paul et al., 2007).

Environmental factors might contribute to ACC (Chiappedi and Bejor, 2010; Paul et al., 2007). One example of environmental influences on callosal development is provided by fetal alcohol syndrome (FAS) (Chiappedi and Bejor, 2010; Paul et al., 2007). Alcohol exposure *in utero* decreases gliogenesis and glial–neuronal interactions, processes that are vital for corpus callosum development (Paul et al., 2007). Ethanol disrupts the transcription and biochemical function of L1CAM, implicating an interplay of environment and genetics in ACC (Paul et al., 2007). The incidence of ACC in FAS is approximately 6.8% (Paul et al., 2007). In many FAS cases, the corpus callosum is hypoplastic; this may result not only from the disruption of early events in callosal formation but also from later dysregulation of axonal pruning (Paul et al., 2007). Such mechanisms might also cause callosal hypoplasia in other disorders such as schizophrenia and autism (Paul et al., 2007). Other environmental factors may also influence postnatal and prenatal callosal development, including musical training, hypothyroidism, and enrichment or deprivation of experience (Paul et al., 2007).

31.3.1.3 Neuroradiological Findings

Consistent with the broad range of genetic factors involved in ACC, the cognitive, behavioral, and neurological consequences of ACC are wide ranging (Paul et al., 2007; Richards et al., 2004). One approach to defining clinical subsets of the ACC patient population is to categorize individuals according to specific neuroanatomical findings and subsequently relate these to the behavioral symptoms within these groups (Paul et al., 2007). The presence of polymicrogyria (PMG) (an excessive number of gyri on the surface of the brain), pachygyria (too few gyri on the surface of the brain), and heterotopia, detected using MRI, correlate with moderate to severe developmental delay (Chiappedi and Bejor, 2010; Paul et al., 2007).

In ACC, the superomedial aspects of the lateral ventricles are deformed by the fibers of the cerebral hemispheres that were destined to cross in the corpus callosum and that, with agenesis, course instead longitudinally as the bundles of Probst (Chiappedi and Bejor, 2010). Crescentic lateral ventricles result from the impression of medial ventricular wall by these bundles (Chiappedi and Bejor, 2010). The other relevant neuroradiological sign is the eversion of the cingulated gyri which can be seen in coronal scans (Chiappedi and Bejor, 2010).

Ultrasonography can be helpful, even if MRI is thought to be far superior at least for partial agenesis (Chiappedi and Bejor, 2010). Morphologically, two types of ACC can be distinguished: In type 1, axons are present but unable to cross the midline, forming large aberrant fiber bundles (Probst bundles), while in the less frequent type 2, axons fail to form (Chiappedi and Bejor, 2010).

The most relevant sonographic sign in sagittal views is the superior displacement of the third ventricle, while parasagittal views show that the medial cortical sulci radiates superiorly instead of horizontally and the absence of the normally echogenic pericallosal sulcus (Chiappedi and Bejor, 2010). Coronal scans show absence of callosum and Probst longitudinal bundles indenting the dorsomedial aspect of lateral ventricles (Chiappedi and Bejor, 2010).

With partial agenesis, the posterior portion is nearly always affected, with the notable exception of the anterior involvement that occurs when partial agenesis is associated with the holoprosencephalies (Chiappedi and Bejor, 2010).

In MRI, the four components of the corpus callosum are best viewed on sagittal imaging, although its relationship to the cerebral hemispheres is best shown on coronal images (Chiappedi and Bejor, 2010). The corpus callosum is a densely packed white matter structure, with high signal on T_1-weighted and low signal in T_2-weighted images after the age of 24 months (Chiappedi and Bejor, 2010). MRI imaging shows that myelination is more advanced in the posterior parts of the corpus callosum when compared to the anterior regions (Chiappedi and Bejor, 2010).

Prenatal diagnosis of complete callosal agenesis is feasible from the midtrimester onward by expert sonography (Chiappedi and Bejor, 2010). In the axial view, suspicious findings are absent cavum septi pellucid and teardrop configuration of the lateral ventricles with possible ventriculomegaly; the nonvisualization of the corpus callosum at transfontanellar ultrasound in either the sagittal or coronal plane is diagnostic (Chiappedi and Bejor, 2010). More subtle findings, such as hypoplasia and partial ACC, may also be recognized antenatally (Chiappedi and Bejor, 2010). Fetal MRI is worth doing in order to reinforce a difficult sonographic diagnosis and, at the same time, to exclude possible additional cerebral anomalies, which may be overlooked at ultrasound (Chiappedi and Bejor, 2010).

Malformations of the brainstem and cerebellum have been increasingly recognized in patients with malformations of the cerebrum, including callosal anomalies (Barkovich et al., 2009).

31.3.2 L1 Syndrome

The L1 syndrome is a highly variable X-linked neurological disorder (Bertolin et al., 2010; Engle, 2010; Schafer et al., 2010). It results from mutations in the *L1CAM* gene (Bertolin et al., 2010; Engle, 2010; Schafer et al., 2010; Vos et al., 2010). L1 is a transmembrane neural adhesion molecule that acts as a short-range axon guidance cue and is highly expressed in developing axons and apical dendrites of cortical neurons and within migratory axons of the corpus callosum and corticospinal tract (Bertolin et al., 2010; Engle, 2010; Schrander-Stumpel and Vos, 1993).

The L1CAM is one of a subgroup of structurally related integral membrane glycoproteins belonging to a large class of immunoglobulin superfamily cell adhesion molecules that mediate cell-to-cell adhesion at the cell surface (http://www.ncbi.nlm.nih.gov/omim) (Bertolin et al., 2010; Schrander-Stumpel and Vos, 1993; Vos et al., 2010). The various functions of L1CAM include guidance of neurite outgrowth in development, neuronal cell migration, axon bundling, synaptogenesis, myelination, neuronal cell survival, and hippocampal long-term potentiation (http://www.ncbi.nlm.nih.gov/omim) (Bertolin et al., 2010; Schafer et al., 2010; Schrander-Stumpel and Vos, 1993).

L1 has multiple extracellular binding partners, including $\beta 1$ integrins, neuronal cell adhesion molecule, TAG-1/axonin-1, contactin, Np-1, and L1 itself, through which it potentiates cell adhesion, provides a mechanical link to the actin cytoskeleton, and serves as a coreceptor to assist in intracellular signal transduction (Engle, 2010). L1 homophilic binding increases cell adhesion and enhances neuronal migration and neurite outgrowth, whereas binding to Np-1 mediates Sema3A-induced growth cone collapse and axon repulsion (Engle, 2010).

L1 has also multiple intracellular binding partners; L1 links to the actin cytoskeleton through interactions with ankyrin or FERM domain-containing proteins, and the interaction of L1 with adaptor protein 2 is required for sorting of L1 to the axonal growth cone (Engle, 2010).

31.3.2.1 Clinical Characteristics

The L1 syndrome was originally recognized as four distinct entities: X-linked hydrocephalus due to stenosis of the aqueduct of Sylvius, MASA syndrome (mental retardation, aphasia, shuffling gait, adducted thumbs), X-linked complicated spastic paraplegia type I, and X-linked corpus callosum agenesis (Bertolin et al., 2010; Engle, 2010; Vos et al., 2010). On the basis of their genetic homogeneity and phenotypic overlap, these disorders are considered a single entity (Engle, 2010; Schafer et al., 2010).

This X-linked syndrome comprises a broad phenotypic spectrum, including hydrocephalus, mental retardation, aphasia, spastic paraplegia, and adducted thumbs (Engle, 2010; Schafer et al., 2010; Schrander-Stumpel and Vos, 1993).

X-linked hydrocephalus with stenosis of the aqueduct of Sylvius is the most common genetic form of congenital hydrocephalus, with a prevalence of approximately 1 in 30 000 (Schrander-Stumpel and Vos, 1993). This accounts for approximately 5–10% of males with nonsyndromic hydrocephalus (Schrander-Stumpel and Vos, 1993). Hydrocephalus may be present prenatally and result in stillbirth or death in early infancy (Schrander-Stumpel and Vos, 1993). Males with hydrocephalus with stenosis of the aqueduct of Sylvius are born with severe hydrocephalus and adducted thumbs (Schrander-Stumpel and Vos, 1993). Seizures may occur (Schrander-Stumpel and Vos, 1993). In less severely affected males, hydrocephalus may be subclinically present and documented only because of developmental delay (Schrander-Stumpel and Vos, 1993). Mild-to-moderate ventricular enlargement is compatible with long survival (Schrander-Stumpel and Vos, 1993).

In hydrocephalus with stenosis of the aqueduct of Sylvius, intellectual disability is usually severe and is independent of shunting procedures in individuals with severe hydrocephalus (Schrander-Stumpel and Vos, 1993). In MASA syndrome, intellectual disability ranges from mild (IQ of 50–70) to moderate (IQ of 30–50) (Schrander-Stumpel and Vos, 1993).

Boys initially exhibit hypotonia of the legs, which evolves into spasticity during the first years of life (Schrander-Stumpel and Vos, 1993). In adult males, the spasticity tends to be somewhat progressive (Schrander-Stumpel and Vos, 1993). Spasticity usually results in atrophy of the muscles of the legs and contractures that together cause the shuffling gait (Schrander-Stumpel and Vos, 1993).

Carrier females may manifest minor features such as adducted thumbs and/or subnormal intelligence (Schrander-Stumpel and Vos, 1993). Rarely do females manifest the complete L1 syndrome phenotype (Schrander-Stumpel and Vos, 1993).

31.3.2.2 Genetics

The *L1CAM* gene is located on the long arm of X-chromosome (Xq28) and is comprised of 29 exons, the first being noncoding, spanning about 16 kb (Bertolin et al., 2010; Schrander-Stumpel and Vos, 1993; Vos et al., 2010). The L1 syndrome results from missense, nonsense, splice site, and frameshift mutations scattered throughout the exons and intron–exon boundaries of the *L1CAM* gene (Bertolin et al., 2010; Engle, 2010; Schrander-Stumpel and Vos, 1993). Since the known *L1CAM* mutations are scattered over the whole coding region, the entire gene has to be sequenced in order to achieve molecular diagnosis (Bertolin et al., 2010). To date, more than 240 different mutations have been reported in the *L1CAM* gene (Bertolin et al., 2010). Most *L1CAM* mutations are unique to each family, that is, they appear to be private mutations (Vos et al., 2010). Only a few families harbor the same recurrent mutation (Vos et al., 2010).

Missense mutations account for over one-third of pathological L1 mutations described, and those affecting extracellular domains result in more severe clinical consequences than those affecting the cytoplasmic part of L1 (Schafer et al., 2010). Some missense mutations affect structurally important amino acid sites in the extracellular domains, which may cause protein misfolding (Schafer et al., 2010). Such missense mutations interfere with homo- and heterophilic ligand binding, intracellular trafficking, neurite growth, and neurite branching (Bertolin et al., 2010; Engle, 2010; Schafer et al., 2010). Children with a truncating mutation are more likely to die before the age of 3 than children with a missense mutation (Vos et al., 2010).

Axon guidance defects occur with both extra- and intracellular mutations (Engle, 2010). The role of L1CAM in neuronal migration and survival, synaptogenesis, and long-term potentiation may also contribute to the phenotype (Engle, 2010).

L1CAM mutation analysis is offered to patients suspected of having L1 syndrome (Vos et al., 2010). Once a mutation has been established, prenatal testing can be performed in subsequent pregnancies, and carriership testing can be carried out to determine the potential presence of an *L1CAM* mutation in female relatives (Vos et al., 2010).

31.3.2.3 Neuroradiological Findings

Neuroimaging reveals hydrocephalus with or without stenosis of the aqueduct of Sylvius in combination with corpus callosum agenesis/hypogenesis and/or cerebellar hypoplasia, small brainstem, and agenesis of the pyramids (corticospinal tracts) (Engle, 2010; Schrander-Stumpel and Vos, 1993). Bilateral absence of the pyramids detected by MRI is an almost pathognomonic inding (Schrander-Stumpel and Vos, 1993). Aqueductal stenosis is not a constant feature of L1 syndrome (Schrander-Stumpel and Vos, 1993).

L1 syndrome cannot be reliably diagnosed on the basis of prenatal ultrasound only (Schrander-Stumpel and Vos, 1993). A diagnosis of hydrocephalus often requires serial ultrasound examination and cannot be guaranteed before 20–24 weeks of gestation or even the third trimester of pregnancy (Schrander-Stumpel and Vos, 1993). Apparently, normal ultrasound findings in a pregnancy with a priori increased risk are not reliable in ruling out L1 syndrome in the fetus (Schrander-Stumpel and Vos, 1993).

31.3.3 Joubert Syndrome and Related Disorders

Joubert syndrome (JS) is an autosomal recessive and genetically heterogeneous trait characterized by combinations of congenital hypotonia, ataxia, abnormal respiratory patterns, mental retardation, social disabilities including autism, and synkinetic mirror movements (Engle, 2010; Parisi, 2009; Valente et al., 2008). The JS can also cosegregate with retinopathy, kidney disease, liver disease, polydactyly, obesity, and/or situs inversus (Engle, 2010; Valente et al., 2008). This spectrum is now called JS and related disorders (JSRD) (Engle, 2010).

The prevalence of JSRD has been estimated as approximately 1:100 000 in the United States, but this is likely an underestimate given by the broad spectrum of features particularly in those with milder manifestations (Parisi, 2009).

JS and JSRD are classified as ciliopathies because the mutated genes encode signal transduction and scaffolding proteins implicated in the function of the primary cilium or its anchoring structure, the basal body (Engle, 2010; Valente et al., 2008). Cilia sense environmental cues and mediate signals through receptor-dependent pathways such as SHH, noncanonical Wnt, and platelet-derived growth factor (PDGF) receptor (Engle, 2010; Parisi, 2009).

The JSRD, similar to many of the disorders considered ciliopathies, shows considerable heterogeneity in clinical features and on a molecular basis (Parisi, 2009; Valente et al., 2008). The clinical features of JSRD are shared by many ciliary disorders and typically involve the renal epithelium, retinal photoreceptor cells, central nervous system, body axis, sensory organs, and others (Parisi, 2009; Valente et al., 2008).

31.3.3.1 Clinical Characteristics

The features necessary for a diagnosis of classic JB include the following: The molar tooth sign on axial views from cranial MRI studies comprised of these three findings: cerebellar vermis hypoplasia, deepened interpendicular fossa, and thick, elongated superior cerebellar peduncles; intellectual impairment/developmental delay of variable degree; hypotonia in infancy; one or both of the following (not required but supportive of the diagnosis): irregular breathing pattern in infancy (episodic apnea and/or tachypnea, sometimes alternating) and abnormal eye movements (nystagmus and/or oculomotor apraxia (OMA)) (Parisi, 2009; Valente et al., 2008).

Many children with JS exhibit dysmorphic facial features that include a broad forehead, arched eyebrows, eyelid ptosis, wide-spaced eyes, open mouth configuration, and facial hypotonia (Parisi, 2009). Some individuals also have polydactyly of the hands and/or feet, which can take many forms (Parisi, 2009).

The JSRD encompasses classic JS as described previously, as well as conditions with other features such as central nervous system anomalies (including occipital encephalocele, corpus callosal agenesis), ocular coloboma, retinal dystrophy, renal disease (including cystic dysplasia or nephronophthisis (NPHP) (cystic kidney disease), and hepatic fibrosis (Parisi, 2009; Valente et al., 2008). When the ocular and renal systems are involved, the syndromes are sometimes described as cerebello-oculo-renal syndromes (Parisi, 2009; Valente et al., 2008). An association between kidney disease and retinal involvement has been observed, with specific findings of NPHP plus retinal dystrophy known as Senior–Loken syndrome (Parisi, 2009; Valente et al., 2008).

One JSRD is the condition known as COACH syndrome (coloboma, oligophrenia/developmental delay, ataxia, cerebellar vermis hypoplasia, hepatic fibrosis) (Parisi, 2009; Valente et al., 2008). Liver involvement, when coupled with renal cystic disease, has prompted the inclusion of JSRD as a congenital hepatorenal fibrocystic disease (Parisi, 2009; Valente et al., 2008).

Cognitive impairment in JSRD is highly variable, with many children exhibiting moderately severe disability (Parisi, 2009). The average age of independent sitting has been reported to be 19 months, and the average of walking is reported to be 4 years for those who developed these skills (Parisi, 2009).

The clinical features related to the complex hindbrain malformation include ataxia, which typically becomes apparent as children develop ambulation, and ocular, oral-motor, and speech dyspraxia (Parisi, 2009). Some children require assistive devices or use sign language to communicate given expressive language impairment (Parisi, 2009).

Seizures have been reported in some children with JSRD (Parisi, 2009). Some children with JSRD and autistic features have been described (Parisi, 2009). Behavioral problems, typically impulsivity, perseveration, and temper tantrums, appear to be relatively common particularly with increasing age (Parisi, 2009).

There is a broad spectrum of ocular findings in JSRD (Parisi, 2009; Valente et al., 2008). Abnormalities of ocular motility are very common, particularly nystagmus, which can be horizontal, vertical, and/or torsional, and typically has a pendular or sometimes see-saw pattern, and OMA, which is characterized by difficulty in smooth visual tracking, dysconjugate eye movements, and head thrusting to compensate for poor saccade initiation (Parisi, 2009; Valente et al., 2008). Nystagmus and OMA are often present at birth and may improve with age (Parisi, 2009; Valente et al., 2008).

Other common ocular anomalies that may require medical or surgical treatment include strabismus, amblyopia, and ptosis (Parisi, 2009). Third nerve palsy, Duane anomaly (unilateral or bilateral restriction in the ability to move the affected eye outward, and when attempting to move the affected eye inward, it retracts into the orbit, resulting in narrowing of the width of the palpebral fissure) and optic disc drusen have also been observed (Parisi, 2009).

Coloboma, a congenital ocular developmental defect, is present in a subset of individuals with JSRD and typically involves the choroids and retina, but rarely the iris (Parisi, 2009). Many children with unilateral or bilateral colobomas also develop liver disease, as in COACH syndrome, but colobomas are not a necessary feature of this disorder (Parisi, 2009). However, retinal dystrophy is not typical in COACH syndrome, in contrast to other JSRDs (Parisi, 2009).

Kidney disease is relatively common on JSRD, with a prevalence of up to 30% of subjects in early surveys; this estimate may be even higher with long-term follow-up given its age-dependent penetrance (Parisi, 2009; Valente et al., 2008). Two different forms of kidney disease have been described: cystic dysplasia and juvenile NPHP (Parisi, 2009).

Cystic dysplasia may be identified prenatally or congenitally by ultrasound findings of multiple cysts of many different sizes in immature kidneys with fetal lobulations (Parisi, 2009; Valente et al., 2008). This finding is the characteristic of Dekaban-Arima syndrome, a JSRD that includes congenital blindness and occasional encephalocele (Parisi, 2009; Valente et al., 2008).

The other, more common renal disorder in JSRD is juvenile NPHP, characterized by tubulointerstitial nephritis and cysts concentrated at the corticomedullary junction (Parisi, 2009; Valente et al., 2008). Most children present with urine-concentrating defects in the first or second decade of life as manifested by polydipsia, polyuria, anemia, and growth failure, with a rise in serum creatinine around 9 years and progression to the end-stage renal disease by approximately 13 years of age (Parisi, 2009; Valente et al., 2008). Mutations in at least nine ciliary genes have been identified in individuals with NPHP, about 20% of whom have extrarenal manifestations, including cerebellar malformations, OMA, and retinal dystrophy (Vos et al., 2010). It is possible that the renal disease in JSRD is part of a continuum of findings with the common etiology involving abnormal ciliary proteins leading to tubular dysfunction (Parisi, 2009).

In rare cases, the congenital renal disease in JSRD consists of enlarged kidneys, microscopic cysts distributed throughout the cortex and medulla, and infantile hypertension, similar to the renal disease of autosomal recessive polycystic kidney disease (Parisi, 2009; Valente et al., 2008). Several of these patients have *MKS3* (Meckel–Gruber syndrome Gene 3) mutations similar to those with COACH syndrome (Parisi, 2009).

Hepatic involvement in JSRD is likely underreported, as manifestations of liver disease are usually not apparent at birth (Parisi, 2009). However, since current management guidelines recommend routine screening for liver dysfunction in all children with JSRD, hepatic involvement is being identified presymptomatically (Parisi, 2009). The liver disease in COACH syndrome has demonstrated variable progression (Parisi, 2009; Valente et al., 2008). Some JSRD/COACH patients present with evidence of portal hypertension, including hematemesis, esophageal varices or portosystematic shunting, and occasionally life-threatening bleeding events (Parisi, 2009; Valente et al., 2008); others present with elevated or fluctuating levels of serum transaminases (ALT or AST) or gamma-glutamyl transferase (Parisi, 2009). Physical examination findings may include hepatomegaly with or without splenomegaly (Parisi, 2009).

The skeletal findings in JSRD include coneshaped epiphyses and polydactyly (Parisi, 2009). Coneshaped epiphyses have been most observed in children with Mainzer–Saldino syndrome (cerebellar ataxia with NPHP and retinal dystrophy) (Parisi, 2009). Polydactyly is often postaxial, although preaxial polydactyly of the hands or great toes has been observed (Parisi, 2009). Mesaxial polydactyly has been described in individuals with the oral-facial-digital type VI syndrome, a JSRD with oral frenulae, lingual tumors or hamartomas, and craniofacial findings that include wide-spaced eyes and a midline lip groove (Parisi, 2009). With age, some children with JSRD develop scoliosis related to abnormal tone (Parisi, 2009).

Endocrine abnormalities are not uncommon in JSRD, and some children exhibit pituitary hormone dysfunction such as isolated growth hormone or thyroid hormone deficiency, or even more extensive panhypopituitarism, with some males demonstrating micropenis (Parisi, 2009).

The vast majority of infants and children diagnosed with JSRD survive the neonatal period and many demonstrate improvement with time in their tone,

respiratory function, and feeding behaviors (Parisi, 2009). Because of the risk of later development of retinal, renal, and hepatic complications, ongoing monitoring is essential (Parisi, 2009).

31.3.3.2 Genetics

The JSRD is genetically heterogeneous, and at least nine loci and eight genes *INPP5E* (inositol polyphosphate-5-phosphatase), *AHI1* (Abelson helper integration site 1), *NPHP1*, *CEP290* (centrosomal protein 290 kDa), *TMEM67* (transmembrane protein 67), *RPGRIP1L* (RPGR-interacting protein-1-like protein), *ARL13B* (ADP-ribosylation factorlike 13B), and *CC2D2A* (coiled-coil and C2 domain containing 2A) have been identified (Engle, 2010).

The *INPP5E* gene was identified as causative, and the retinal phenotype is predominant (Parisi, 2009). Also, it is possible that this gene represents another COACH gene (Parisi, 2009). This gene encodes an inositol polyphosphatase-5-phosphatase E necessary for cilia stability and indicates a link between phosphotidylinoditol signaling and ciliary function (Parisi, 2009).

Mutations in the 29-exon gene, *AHI1*, have been identified in JSRD (Valente et al., 2008). This gene encodes a protein of unknown function named jouberin, containing several protein–protein interaction domains (Valente et al., 2008). The most common clinical association in *AHI1*-related JSRD is retinal dystrophy, occurring in ~80% of those with mutations (Parisi, 2009; Valente et al., 2008). Renal disease has been observed in some subjects (Parisi, 2009). However, no subjects with *AHI1* mutations have had features of encephalocele, polydactyly, or liver fibrosis (Parisi, 2009). Other central nervous system anomalies, including PMG, corpus callosum anomalies, and frontal lobe atrophy, have been described in some individuals with *AHI1* mutations (Parisi, 2009). Single heterozygous mutations in the *AHI1* gene have been identified in a number of JSRD patients within the reported screenings (Valente et al., 2008). The vast majority of identified mutations are nonsense (truncating, frameshift, or splice-site mutations) that cluster mainly in the first half of the gene (exons 7–16) and in exons encoding the functional domains (Valente et al., 2008). Missense mutations are rare and are predicted to affect amino acid residues that are crucial for the correct functioning of such domains (Valente et al., 2008). No correlation can be drawn between the type or site of mutations and the associated phenotype (Valente et al., 2008).

The first gene associated with JSRD, the 30-exon *NPHP1* gene, was identified as causing juvenile NPHP (Parisi, 2009). *NPHP* resides within a ~290 kb region of genomic DNA flanked by large inverted repeat elements on chromosome 2q13 that is homozygously deleted in JSRD or NPHP; a few individuals are compound heterozygotes for a deletion and a point mutation in *NPHP1* (Parisi, 2009; Valente et al., 2008). The *NPHP1* mutation detection rate for the purely renal disorder is ~20–30%, whereas the mutation rate is only about 1–3% in individuals with JSRD (Parisi, 2009). Some individuals with the common deletion have congenital OMA known as Cogan syndrome, and others have Senior–Loken syndrome with retinal impairment, but in general, the neurologic symptoms tend to be milder than in many children with JSRD (Parisi, 2009; Valente et al., 2008).

The large, 54-exon *CEP290* gene, mapping to the long arm of chromosome 12, has been associated with multiple clinical disorders ranging from isolated Leber congenital amaurosis to JSRD, MKS, and BBS (Parisi, 2009; Valente et al., 2008). Most subjects with JSRD due to *CEP290* mutations have retinal dystrophy or congenital blindness, and many also develop renal disease consistent with NPHP or renal cortical cysts (Parisi, 2009; Valente et al., 2008). Findings in some affected individuals have included ocular colobomas, encephaloceles, septal heart defects, hepatic disease, and situs anomalies (Parisi, 2009). The most frequent *CEP290* mutation, and the one that holds the strongest genotype–phenotype correlation, is the intronic mutation c.2991 + 1655>G that creates a splice-donor site and inserts a cryptic exon in the *CEP290* messenger RNA between exons 26 and 27, introducing a stop codon immediately downstream of exon 26 (Valente et al., 2008).

The 28-exon *TMEM67/MKS3* gene, mapping to chromosome 8q22, encodes a 995-amino acid protein called meckelin, which plays a role in primary cilium formation and interacts with other known ciliary proteins (Parisi, 2009; Valente et al., 2008). Originally identified as a causative for MKS, mutations in this gene have been reported in JSRD (Parisi, 2009; Valente et al., 2008). The *MKS3* mutations identified in MKS are typically compound heterozygous missense and truncating mutations or homozygous splice-site mutations that are found across the whole gene length, whereas the disease-associated mutations in JSRD/COACH tend to be missense mutations or the combination of a missense mutation and a splice-site or nonsense mutation, with very few mutations overlapping with those seen in MKS (Parisi, 2009; Valente et al., 2008).

Mutations in the *RPGRIP1L* gene were first identified in patients with the renal form of JSRD (Parisi, 2009). This gene has 26 exons encoding for a 1315 amino acid protein containing several coiled-coil domains, required for protein–protein interactions (Valente et al., 2008). The phenotype spectrum includes renal disease with some affected individuals manifesting occipital encephaloceles and polydactyly, and rarely, retinal disease or colobomas; a few have scoliosis, clubfoot, or pituitary hormone deficiency (Parisi, 2009). Hepatic fibrosis and

COACH syndrome have been described in individuals with JSRD due to *RPGRIP1L* (Parisi, 2009). Overall, estimates of the prevalence of *RPGRIP1L* mutations in the cerebello-renal form of JSDR range from ~9 to 12% (Parisi, 2009).

ARL13B, a 10-exon gene, encodes a protein that is a member of the Ras GTPase family and localizes to the primary cilia of cerebellar neurons, kidney, and retina (Parisi, 2009).

The phenotype in patients with mutations in the 38-exon *CC2D2A* gene has ranged from classic JS to JS with encephalocele to the COACH phenotype with coloboma, liver, and kidney involvement (Parisi, 2009). *CC2D2A* has been shown to interact with *CEP290* (Parisi, 2009). This gene is estimated to cause almost 10% of JSRD (Parisi, 2009).

The genes mentioned previously account for an estimated 50% of causative mutations in JSRD (Parisi, 2009). The same JSRD gene can cause multiple different phenotypes, and several different genes can be associated with the same clinical features (Parisi, 2009; Valente et al., 2008). In addition, clinical features can vary between affected siblings within the same family (Parisi, 2009). This intrafamilial variability supports the existence of genetic modifiers and epistatic effects (Parisi, 2009; Valente et al., 2008). Commercial clinical testing is available for many of the JSRD genes (Parisi, 2009). Additional JSRD genes remain to be identified (Parisi, 2009).

The gene products associated with JSRD are known to localize to the primary cilium and/or basal body and the centrosome apparatus, which has been identified in almost all cell types, and many of these proteins are important for the structure, function, and/or stability of this organelle and its related structures (Parisi, 2009; Valente et al., 2008).

Primary cilia play a role in intraflagellar transport, cell division, tissue differentiation, establishment of body axis, growth, and mechanosensation involved in cellular signaling processes and are essential for signal transduction processes that underlie many aspects of vertebrate development and morphogenesis, including the SHH, Wnt/β-catenin, PDGF receptor alpha signaling pathways, Ras-GTP, and phosphotidylinositol signaling (Parisi, 2009; Valente et al., 2008).

In the developing brain, primary cilia have been involved in regulating some of the most powerful pathways active in the early embryo, such as those of Wnt and Sonic hedgehog (Valente et al., 2008). It appears that cilia can function as a sort of cellular sensor, picking up environmental signals and transducing them to the nucleus, regulating cell cycle and proliferation (Valente et al., 2008).

Also, the proper cell polarity and orientation of tubular structures in tissues require normal ciliary function

(Parisi, 2009). The expanding group of human disorders known as ciliopathies shares overlapping clinical manifestations that reflect the critical role cilia play in the growth and differentiation of various tissues (Parisi, 2009).

31.3.3.3 Neuroradiological Findings

The diagnosis of JSRD is dependent on the presence in MRI of the 'molar tooth' sign, a toothlike shape on axial images at the level of the midbrain–hindbrain junction that reflects cerebellar vermian hypoplasia, a deepened interpeduncular fossa, and horizontally oriented and thickened superior cerebellar peduncle (SCP) (Engle, 2010; Parisi, 2009; Valente et al., 2008). When a diagnosis of JSRD is suspected, a detailed cranial MRI to evaluate the molar tooth sign is essential (Parisi, 2009).

The peculiar neuroradiological malformations seen in JS have their correspondence at the neuropathological level, with various abnormalities of the midbrain and hindbrain being consistently identified in brains from JS patients (Valente et al., 2008). Postmortem studies of individuals with genetically undefined JS have revealed severe cerebellar vermian hypoplasia or absence, with midline clefting, dysplasia of the deep cerebellar and inferior olivary nuclei, elongation of the caudal midbrain tegmentum, reduction in pontine neurons, hypoplasia of the solitary, trigeminal and dorsal column nuclei and tracts, and occasional heterotopia of Purkinje-like neurons (Engle, 2010; Valente et al., 2008).

Approximately 10% of individuals with JSRD demonstrate fluid collections in the posterior fossa resembling the Dandy–Walker malformation (agenesis of the cerebellar vermis, cystic dilation of the fourth ventricle, and enlargement of the posterior fossa (Parisi, 2009). Although hydrocephalus is uncommon in JSRD, rare patients have required a shunt for symptomatic elevations of intracranial pressure (Parisi, 2009).

In addition, occipital encephaloceles or meningoceles have been observed, suggesting overlap with Meckel syndrome, a typical prenatal or perinatal lethal ciliopathy characterized by brain anomalies (especially encephalocele), cystic renal dysplasia, and the hepatic ductal plate malformation (Parisi, 2009). Other brain anomalies in JSRD have included PMG, ACC, and cerebellar heterotopias (Parisi, 2009).

Prenatal imaging via ultrasound and/or fetal MRI is the best and most practical diagnostic option (Parisi, 2009). Extracranial anomalies such as polydactyly or renal cysts and major structural brain malformations such as encephalocele may facilitate prenatal diagnosis of JSRD as early as the first trimester or may suggest the diagnosis (Parisi, 2009). Early diagnosis is more difficult when extracranial findings are not present and because the molar tooth sign has not been reported before 27 weeks of gestation (Parisi, 2009).

31.3.4 Horizontal Gaze Palsy with Progressive Scoliosis

Horizontal gaze palsy with progressive scoliosis (HGPPS) is a rare, clinically and genetically homogeneous disorder in which hindbrain axons fail to cross the midline (Amouri et al., 2009; Avadhani et al., 2010; Bomfim et al., 2009; Engle, 2010). This disorder is characterized by congenital absence of conjugate horizontal eye movements and preservation of vertical gaze and convergence, which is associated with progressive scoliosis developing in childhood and adolescence (Amouri et al., 2009; Avadhani et al., 2010; Bomfim et al., 2009; Otaduy et al., 2009).

31.3.4.1 Clinical Characteristics

Clinical findings characteristic of HGPPS include absence of all horizontal gaze reflexes, conjugate pendular nystagmus, and progressive scoliosis (Bomfim et al., 2009; Engle, 2010; Otaduy et al., 2009). Affected individuals are born with restricted horizontal gaze and develop scoliosis within the first decade of life (Engle, 2010; Otaduy et al., 2009).

The gaze palsy may result from errors in axon connectivity into and out of the abducens nucleus (Bomfim et al., 2009; Engle, 2010). The normal contralateral inputs onto the abducens nucleus from the pontine paramedian reticular formation and vestibular nuclei are predicted to be ipsilateral in HGPPS, and this would likely alter the firing patterns of motor and internuclear neurons (Bomfim et al., 2009; Engle, 2010). Axons of the abducens internuclear neurons would also fail to cross the midline via the medial longitudinal fasciculus to synapse on medial rectus motor neurons in the contralateral oculomotor nucleus, further perturbing horizontal gaze (Bomfim et al., 2009; Engle, 2010).

Although the etiology of scoliosis is also speculative, HGPPS provides the first genetic evidence of a neurogenic cause for this disability (Bomfim et al., 2009; Engle, 2010). It has been suggested that the pathogenesis of progressive idiopathic scoliosis involves a primary neurological dysfunction involving the proprioceptive inputs mediated by the posterior column pathways of the spinal cord and medial lemniscus (Bomfim et al., 2009).

Individuals with HGPPS perform normally on neuropsychological testing and have normal fine motor control without mirror movements, suggesting that the pathologically ipsilateral corticospinal axons find their appropriate target, albeit on the wrong side (Engle, 2010).

The differential diagnosis of HGPPS embraces several genetic disorders of eye movement, such as Duane retraction syndrome (DRS), Möbius syndrome, and others (Bomfim et al., 2009; Otaduy et al., 2009). Clinical and neuroimaging findings can differentiate these entities from HGPPS (Bomfim et al., 2009).

31.3.4.2 Genetics

HGPPS is an autosomal recessive trait and results from mutations in the *ROBO3* gene (http://www.ncbi.nlm.nih.gov/omim) (Amouri et al., 2009; Avadhani et al., 2010; Bomfim et al., 2009; Engle, 2010; Otaduy et al., 2009). *ROBO3* encodes a transmembrane receptor, and it is a divergent member of the Robo family of axon guidance molecules (Amouri et al., 2009; Engle, 2010; Otaduy et al., 2009). Robo 3 is essential for midline crossing of hindbrain and spinal cord commissural and precerebellar axons (Avadhani et al., 2010; Engle, 2010). Robo 3 is also necessary for midline crossing of precerebellar neurons, and defects in neuronal migration may also contribute to the HGPPS phenotype (Amouri et al., 2009; Avadhani et al., 2010; Engle, 2010).

ROBO3 alternative splicing produces two functionally antagonistic isoforms with distinct carboxy termini (Engle, 2010). *ROBO3.1* inhibits the responsiveness of commissural axons to Slit repellents and is present on commissural axons before and during midline crossing, whereas *ROBO3.3* is Slit responsive and appears on the growth cone postcrossing to block the recrossing (Engle, 2010). HGPPS mutations reported to date alter nucleotides common to both isoforms (Engle, 2010).

Indistinguishable phenotypes result from *ROBO3* nonsense, frameshift, splice-site, or missense mutations spread across the gene, supporting a complete loss of *ROBO3* function (Engle, 2010). Over ten different mutations located in different domains of the encoded protein have been identified and are thought to diminish the function of this receptor (Amouri et al., 2009).

31.3.4.3 Neuroradiological Findings

Electrophysiological and neuroimaging studies in HGPPS support the absence of decussating axons in the pons and medulla (Engle, 2010). MRI reveals ventral flattening and hypoplasia of the hindbrain, absence of facial colliculi, and a butterfly-shaped medulla with a midline pontine cleft (Avadhani et al., 2010; Bomfim et al., 2009; Engle, 2010; Otaduy et al., 2009). The unusual appearance of the medulla and abnormal functional results suggest that sensorimotor projections do not cross the midline in HGPPS (Amouri et al., 2009). Functional MRI reveals ipsilateral rather than the normal contralateral activation in the primary motor cortex following motor tasks (Engle, 2010).

The split pons sign is attributable to abnormal development of the abducens nuclei and medial longitudinal fasciculus occurring between 5 and 8 weeks of gestation (Bomfim et al., 2009). Hypoplasia of the medial lemniscus, which is located posterior to the pyramids, is thought to explain why the inferior olivary nuclei are unusually more prominent than the pyramids (Bomfim et al., 2009). The deep midline cleft along the ventral aspect of the medulla oblongata has been described as the result of uncrossed corticospinal tracts (Bomfim et al., 2009).

DTI, with its ability to demonstrate white matter tracts, is a very suitable technique to further evaluate the abnormalities underlying this disease (Otaduy et al., 2009). Previous DTI studies have described the absences of superior cerebellar and pyramidal decussations, major pontine fibers, and decussation of the superior cerebellar peduncles, with fMRI combined study confirming the ipsilateral sensorimotor findings (Avadhani et al., 2010; Engle, 2010; Otaduy et al., 2009). The cortex, corpus callosum, and exiting cranial nerves appear structurally normal (Engle, 2010).

31.3.5 Kallmann Syndrome

Individuals with Kallman syndrome (KS) have congenital anosmia and hypogonadotropic hypogonadism (HH) (http://www.ncbi.nlm.nih.gov/omim) (Engle, 2010; Fechner et al., 2008; Hardelin and Dode, 2008; Kaplan et al., 2010; Kim et al., 2008). Anosmia is related to the absence or hypoplasia of the olfactory bulb and tracts (Hardelin and Dode, 2008). Hypogonadism is due to gonadotropin-releasing hormone (GnRH) deficiency, which presumably results from a failure of the embryonic migration of neuroendocrine GnRH cells from the olfactory epithelium to the forebrain (Engle, 2010; Fechner et al., 2008; Hardelin and Dode, 2008; Kaplan et al., 2010; Kim et al., 2008). This failure could be a consequence of the early degeneration of olfactory nerve and terminal nerve fibers, because the latter normally act as guiding cues for the migration of GnRH cells (Engle, 2010; Hardelin and Dode, 2008). Defects in GnRH cell fate specification, differentiation, axon elongation, or axon targeting to the hypothalamus median eminence may, however, also contribute to GnRH deficiency, at least in some genetic forms of the disease (Hardelin and Dode, 2008). Both olfactory sensory neurons and GnRH are born in the olfactory placode of the developing nose (Engle, 2010; Fechner et al., 2008). It is proposed that errors in growth and guidance of olfactory axons can result in KS (Engle, 2010).

The prevalence of the disease has been estimated at 1 out of 8000 in boys (Hardelin and Dode, 2008). In girls, the prevalence might be five times lower (Hardelin and Dode, 2008).

31.3.5.1 Clinical Characteristics

Transmitting females have partial or complete anosmia (http://www.ncbi.nlm.nih.gov/omim). Often, the lack of smell goes unnoticed, and individuals with KS are not diagnosed until they fail to undergo secondary sexual development during the teenage years (Engle, 2010; Fechner et al., 2008; Kaplan et al., 2010; Kim et al., 2008). The KS may also be suspected as early as in infancy in boys, in the presence of cryptorchidism or a micropenis, combined with subnormal LH (luteinizing hormone) and FSH (follicle-stimulating hormone) concentrations (Hardelin and Dode, 2008; Kaplan et al., 2010). Microphallus has been noted in up to 65% of individuals with KS, and cryptorchidism has been reported in up to 73% of males with KS (Kaplan et al., 2010). The postnatal surge in FSH, LH, and testosterone in the male infant, as a consequence of the continued function of the fetal GnRH pulse generator, provides a 6-month window of opportunity to establish the diagnosis of HH (Hardelin and Dode, 2008).

Other frequently occurring features in this syndrome include characteristic face and hand dysmorphia, hypotonia, arhinencephaly, semicircular canal agenesis or hypoplasia, deafness, urinary tract anomalies, orofacial clefting, dysphagia, and tracheo-esophagial anomalies (Fechner et al., 2008; Hardelin and Dode, 2008; Kim et al., 2008).

The renal abnormality most frequently associated with KS is unilateral renal aplasia, but other anomalies such as renal diverticulum, horseshoe kidney, malrotated kidney, multicystic dysplastic kidney, and vesiculoureteral reflux have been reported (Kaplan et al., 2010; Kim et al., 2008). Unilateral renal aplasia is more often right sided in KS patients, while renal agenesis in individuals without KS is more likely to be left sided (Kaplan et al., 2010). All patients with suspected KS should have a renal ultrasound to rule out a renal anomaly (Kaplan et al., 2010; Kim et al., 2008).

Cleft lip ± palate has been noted in up to 13–14% of individuals with KS (Kaplan et al., 2010; Kim et al., 2008). The incidence of clefting in KS patients is significant, as the incidence of cleft lip ± palate in the general population is only 0.1–0.2% (Kaplan et al., 2010). In addition to cleft lip ± palate, cleft palate alone and dental agenesis have been reported in several patients with KS (Kaplan et al., 2010).

The incidence of hearing loss in patients with KS has been reported to be as high as 28% (Kaplan et al., 2010; Kim et al., 2008). Both sensorineural and conductive hearing loss have been described (Kaplan et al., 2010). Hearing loss is commonly unilateral (Kaplan et al., 2010). Individuals in whom KS is suspected should undergo formal auditory evaluation (Kaplan et al., 2010). Abnormalities of the inner ear have been noted on CT scan (Kaplan et al., 2010).

Other less-common findings associated with KS include musculoskeletal anomalies, such as clinodactyly, camptodactyly, and fusion of the fourth and fifth metacarpal bones; oculomotor anomalies, such as ptosis, iris coloboma, and nystagmus; high-arched palate, unilateral nasal cartilage agenesis; malrotation of the gut; visual attention defects; and cardiac defects (Kaplan et al., 2010; Kim et al., 2008). Cardiac defects reported include atrial septal defect, ventricular septal defect, right-sided aortic arch, double-outlet right ventricle, transposition of the great arteries, and arrhythmias (Kaplan et al., 2010). X-linked ichthyosis, mental retardation, chondrodysplasia punctata, and short stature can also occur, usually as part of a contiguous gene syndrome (Kim et al., 2008).

Studies of X-linked Kallmann syndrome have found instances of renal agenesis and also pointed to mirror movements of the hands (synkinesia), pes cavus, high-arched palate, and cerebellar ataxia (http://www.ncbi.nlm.nih.gov/omim) (Fechner et al., 2008; Kim et al., 2008).

The KS is diagnosed when low-serum gonadotropins and gonadal steroids are coupled with a compromised sense of smell (Fechner et al., 2008; Hardelin and Dode, 2008; Kim et al., 2008). The latter should be ascertained by the means of detailed questioning and olfactory screening tests (Fechner et al., 2008; Hardelin and Dode, 2008; Kim et al., 2008).

The treatment of KS is that of the hypogonadism (Fechner et al., 2008; Hardelin and Dode, 2008; Kim et al., 2008). It aims first to initiate virilization or breast development and second to develop fertility (Fechner et al., 2008; Hardelin and Dode, 2008; Kim et al., 2008). Hormone replacement therapy, with testosterone for males and combined estrogen and progesterone for females, is the treatment to stimulate the development of secondary sexual characteristics (Fechner et al., 2008; Hardelin and Dode, 2008; Kim et al., 2008).

31.3.5.2 Genetics

The KS is genetically heterogeneous (Engle, 2010). Most KS patients present as sporadic cases (two thirds of cases), but many cases are clearly familial and can be inherited as an X-linked, autosomal dominant, and autosomal recessive trait (Engle, 2010; Fechner et al., 2008; Hardelin and Dode, 2008; Kaplan et al., 2010; Kim et al., 2008). In the autosomal dominant form, incomplete penetrance has been emphasized, which makes it difficult to identify affected patients (Fechner et al., 2008; Hardelin and Dode, 2008; Kim et al., 2008). Overall, the autosomal dominant mode of inheritance seems to account for 86% of inherited cases of KS (Fechner et al., 2008).

Six genes have been reported, accounting for approximately 30% of cases: *KAL1* (Kallmann syndrome 1 sequence), *FGFR1* (fibroblast growth factor receptor 1), *PROK2* (prokineticin 2), *PROKR2* (prokineticin receptor 2), *FGF8* (fibroblast growth factor 8), and *CHD7* (chromodomain helicase DNA-binding protein-7) (Engle, 2010;

Kaplan et al., 2010; Kim et al., 2008). These KS genes encode transmembrane receptors and ligands that may be important for growth cone guidance (Engle, 2010). Some KS proteins also interact with one another and with heparan sulfate proteoglycans to amplify downstream signaling pathways (Engle, 2010).

The KS can be oligogenic, resulting from combinations of mutations in more than one KS gene (Engle, 2010; Hardelin and Dode, 2008; Kim et al., 2008). It is possible that oligogenic inheritance accounts in part for the long recognized incomplete penetrance of the disease, at least in some cases (Hardelin and Dode, 2008; Kim et al., 2008).

X-linked KS is caused by loss-of-function mutations in *KAL1*, which is expressed in developing olfactory bulb (http://www.ncbi.nlm.nih.gov/omim) (Engle, 2010; Hardelin and Dode, 2008; Kaplan et al., 2010; Kim et al., 2008). Two-thirds of males harboring *KAL1* mutations also have mirror movements and enlarged, aberrant ipsilateral CSTs, supporting a role of *KAL1* in the guidance of CST and olfactory axons (http://www.ncbi.nlm.nih.gov/omim) (Engle, 2010; Hardelin and Dode, 2008; Kaplan et al., 2010).

The *KAL1* gene is found in the pseudoautosomal region of the X-chromosome and encodes the secreted glycoprotein anosmin-1, which has cell adhesion, neurite outgrowth, and axon guidance and branch-promoting activities (http://www.ncbi.nlm.nih.gov/omim) (Engle, 2010; Fechner et al., 2008; Hardelin and Dode, 2008; Kaplan et al., 2010; Kim et al., 2008). Anosmin-1 requires heparin sulfate for its functions (Kim et al., 2008). Anosmin is required for the normal migration of olfactory and GnRh neurons from the olfactory placode to the hypothalamus (Fechner et al., 2008; Kim et al., 2008).

The vast majority of *KAL1* mutations reported so far are nonsense mutations, frameshift mutations, or large gene deletions, which are expected to inactivate protein synthesis, and are apparently sufficient to produce the abnormal phenotype in males (Fechner et al., 2008; Hardelin and Dode, 2008; Kim et al., 2008). The *KAL1* gene is also expressed in developing Purkinje cells of the cerebellum, meso- and metanephros, oculomotor nucleus, and facial mesenchyme (Fechner et al., 2008).

Mutations in *FGFR1*, also known as *KAL2*, underlie an autosomal dominant form of the disease (Fechner et al., 2008; Hardelin and Dode, 2008; Kaplan et al., 2010; Kim et al., 2008). This gene is localized on chromosome 8p11.2–p12 (Fechner et al., 2008; Kim et al., 2008). *FGFR1* is a member of the receptor tyrosine kinase superfamily (Hardelin and Dode, 2008). FGF signaling controls cell proliferation, migration, differentiation, and survival and thus, plays essential roles in various processes of embryonic development (Hardelin and Dode, 2008; Kim et al., 2008). Mutations in this gene are associated

with failed morphogenesis of the olfactory bulbs (Fechner et al., 2008). Mutations of *KAL1* and *KAL2* account for less than 20% of clinical cases (Kim et al., 2008). Anosmin-1 and *FGFR1* are involved in the same cellular signaling pathway (Kim et al., 2008).

Mutations in *PROKR2* and *PROK2* have been found in heterozygous, homozygous, or compound heterozygous states (Hardelin and Dode, 2008; Kaplan et al., 2010; Kim et al., 2008). Most of these mutations are missense mutations, and many are also present in apparently unaffected individuals (Hardelin and Dode, 2008). For most of the mutations, however, deleterious effects on prokineticin signaling have been shown (Hardelin and Dode, 2008). These two genes are likely to be involved in both monogenic recessive and digenic or oligogenic KS transmission modes (Hardelin and Dode, 2008). *PROKR2* belongs to the G-protein-coupled seven transmembrane domain receptor family (Hardelin and Dode, 2008; Kim et al., 2008).

Other patients carrying heterozygous mutations in *PROKR2*, *PROK2*, or hypomorphic mutations in *KAL1* are expected to carry additional mutations in other, as yet unknown, KS genes (Hardelin and Dode, 2008). For each genetic form of KS identified so far, the clinical heterogeneity of the disease within affected families clearly indicates that the manifestation of KS phenotypes is dependent on factors other than the mutated gene itself, probably factors such as modifier genes and epigenetic factors (Hardelin and Dode, 2008; Kim et al., 2008).

Some of the possible modifier genes and candidate genes that may account for the remaining KS cases are the nasal embryonic LHRH factor (*NELF*), *CHD7*, and early B-cell factor 2 (Kim et al., 2008). *NELF* has been shown to serve as a common guidance molecule for the olfactory axon and GnRH neurons across the nasal region during mouse embryonic development (Kim et al., 2008). *CHD7* is the only known locus associated with CHARGE syndrome (Kim et al., 2008). CHARGE syndrome is a developmental disorder defined by iris coloboma, congenital heart disease, choanal atresia, mental and growth retardation, genital hypoplasia, and ear malformations and/or deafness; it sometimes includes HH associated with a defective sense of smell and abnormal olfactory bulbs development (Kim et al., 2008). *EBF* genes encode a family of transcription factors, and they have been implicated in various neural developmental processes. Their function includes axon navigation and migration (Kim et al., 2008).

A greater variability in the degree of hypogonadism has been observed in patients carrying mutations in *FGFR1*, *FGF8*, *PROKR2*, or *PROK2* than in *KAL1* patients (Hardelin and Dode, 2008). Unilateral renal agenesis occurs in approximately 30% of *KAL1* patients but has not been reported in patients with *FGFR1*, *FGF8*, *PROKR2*, or *PROK2* mutations (Hardelin and Dode, 2008; Kaplan et al., 2010). If a renal anomaly is present, FISH analysis and/or sequencing of *KAL1* should be the first line genetic test (Kaplan et al., 2010).

Tooth agenesis and hearing impairment are common to several genetic forms of KS, although the mechanism of the hearing impairment could vary between different genetic forms (Hardelin and Dode, 2008; Kaplan et al., 2010). Palate defects should also be considered as one of these shared traits, even though the severity differs between *KAL1* and other genetic forms (Hardelin and Dode, 2008; Kaplan et al., 2010; Kim et al., 2008). Clefting has been associated with mutations in *FGFR1*, *FGF8*, and *CHD7* but not with *KAL1*, *PROK2*, or *PROKR2* (Kaplan et al., 2010). Microphallus and cryptorchidism are more common in patients with *KAL1* mutations versus *FGFR1* mutations (Kaplan et al., 2010). Hearing loss has been reported in patients with *KAL1*, *FGFR1*, *FGF8*, *PKOKR2*, and *CHD7* mutations (Kaplan et al., 2010).

31.3.5.3 Neuroradiological Findings

MRI of the forebrain can be carried out to show hypoplasia or aplasia of the olfactory bulbs and tracts (Fechner et al., 2008; Hardelin and Dode, 2008).

A hypoplastic olfactory bulb seen on cerebral MRI does not always correlate with the degree of olfactory deficit, and normal images can also be found in KS (Kim et al., 2008).

31.3.6 Albinism

Albinism is an autosomal recessive inherited condition present at birth (Renugadevi et al., 2010; Summers, 2009). The phenotype ranges from a complete lack of pigmentation in the skin, hair, and iris, called oculocutaneous albinism (OCA), or a lack of pigmentation in the iris alone, termed ocular albinism (OA) (Gronskov et al., 2007; Renugadevi et al., 2010; Summers, 2009). Several defects can cause albinism, including a complete lack of melanocytes or few pigment cells, interference in the migration of cells to their proper location during embryo development, and failure of the cells to produce melanin because of a lack of tyrosinase or abnormalities within the cells (Renugadevi et al., 2010; Summers, 2009).

The prevalence of albinism in the United States is estimated to be 1 in 18 000 (Summers, 2009). Albinism can affect people of all ethnic backgrounds (Gronskov et al., 2007). Prevalence of the different forms of albinism varies considerably worldwide, partly explained by the different founder mutations in different genes and the fact that it can be difficult clinically to distinguish between the different subtypes of albinism among the large normal spectrum of pigmentation (Gronskov et al., 2007).

31.3.6.1 Clinical Characteristics

OCA is a group of four autosomal recessive disorders caused by either a complete lack or a reduction of melanin biosynthesis in the melanocytes resulting in hypopigmentation of the hair, skin, and eyes (Gronskov et al., 2007; Summers, 2009). Reduction of melanin in the eyes results in reduced visual acuity caused by foveal hypoplasia and misrouting of the optic nerve fibers (Gronskov et al., 2007; Summers, 2009). The clinical spectrum of OCA varies, with OCA1A being the most severe type characterized by a complete lack of melanin production throughout life, while the milder forms OCA1B, OCA2, OCA3, and OCA4 show some pigment accumulation over time (Gronskov et al., 2007). The different types of OCA are caused by mutations in different genes, but the clinical phenotype is not always distinguishable, making molecular diagnosis a useful tool and essential for genetic counseling (Gronskov et al., 2007).

All types of OCA and OA have similar findings, including various degrees of congenital nystagmus, hypopigmentation of the iris leading to iris translucency, reduced pigmentation of the retinal pigment epithelium, foveal hypoplasia, reduced visual acuity and refractive errors, and sometimes a degree of color vision impairment (Gronskov et al., 2007; Renugadevi et al., 2010; Summers, 2009). Photophobia may be prominent (Gronskov et al., 2007; Renugadevi et al., 2010). A characteristic finding is misrouting of the optic nerves, consisting of an excessive crossing of the fibers in the optic chiasma, which can result in strabismus and reduced stereoscopic vision (Gronskov et al., 2007; Summers, 2009). Absence of misrouting excludes the diagnosis of albinism (Gronskov et al., 2007).

A few patients with albinism who have vision 20/50 or better have some rudimentary foveal development, and some thinning of the retina in the foveal area has been demonstrated with optical coherence tomography (Summers, 2009). Careful inspection can show granular melanin pigment in the macula in a few patients with albinism and occasionally finely granular pigment beyond the macula (Summers, 2009). The presence of melanin pigment in the macula correlates with better visual acuity (Summers, 2009). Recognition of visual acuity among persons with albinism varies from 20/20 to 20/400 but is commonly close to 20/80 (Gronskov et al., 2007; Summers, 2009).

Nystagmus typically develops by 6–8 weeks of age (Summers, 2009). Nystagmus is initially slow and has a large amplitude, but the amplitude typically decreases within the first year of life (Summers, 2009). Delayed visual maturation has been reported in albinism (Summers, 2009). Parents with an infant with albinism have noted poor fixation on faces and objects and a delay in visual development (Summers, 2009).

Pattern visual-evoked potentials performed with monocular visual stimulation demonstrate the excessive retinostriate decussation that is characteristic of albinism (Gronskov et al., 2007; Summers, 2009). In individuals with a questionable phenotype for albinism, the visual-evoked potentials can be useful in identifying those with the disorder (Summers, 2009). This abnormal decussation may account for absent stereoacuity that is often found in albinism (Summers, 2009).

Disorders in which albinism is part of a larger syndrome include Hermansky–Pudlak syndrome (HPS), Chediak–Higashi syndrome, Griscelli syndrome, and Waardenburg syndrome type 2 (WS2) (Gronskov et al., 2007; Summers, 2009). All, except WS2, are inherited as autosomal recessive traits and can be distinguished on the basis of clinical and biochemical criteria (Gronskov et al., 2007). Also, an association of hypopigmentation in Prader–Willi syndrome and Angelman disease with a deletion on 15q11 has been found (Gronskov et al., 2007).

Lifespan in patients with OCA is not limited, and medical problems are generally not increased compared to those in the general population (Gronskov et al., 2007). Skin cancers may occur, and regular skin checks should be offered (Gronskov et al., 2007). Development and intelligence are normal (Gronskov et al., 2007).

31.3.6.2 Genetics

The OCA is a group of congenital heterogeneous disorders of melanin biosynthesis in the melanocytes (Gronskov et al., 2007). At least four genes are responsible for different types of OCA (Gronskov et al., 2007; Summers, 2009). The current classification of albinism is determined by the affected gene (Summers, 2009). Most patients are compound heterozygotes (Gronskov et al., 2007). Siblings with albinism can show variable expression in visual function and clinical phenotype, suggesting that other genes modify the classical phenotype (Summers, 2009).

The OCA1 is caused by mutations in the tyrosinase gene (*TYR*) on chromosome 11q14–21, encoding the enzyme tyrosinase that catalyzes rate-limiting steps in the melanin biosynthetic pathway (Engle, 2010; Gronskov et al., 2007; Renugadevi et al., 2010; Summers, 2009). Mutations completely abolishing tyrosinase activity result in OCA1A, while mutations rending some enzyme activity result in OCA1B, allowing some accumulation of melanin pigment over time (Engle, 2010; Gronskov et al., 2007; Summers, 2009). Almost 200 mutations in *TYR* are known (Gronskov et al., 2007).

Mutations in the *OCA2* gene (formerly known as the P-gene) cause the OCA2 phenotype (Gronskov et al., 2007; Renugadevi et al., 2010; Summers, 2009). This gene maps to chromosome 15q11.2 and encodes an integral melanosomal protein (Engle, 2010; Gronskov et al.,

2007; Summers, 2009). OCA2 protein is important for normal biogenesis of melanosomes and for normal processing and transport of melanosomal proteins such as TYR and tyrosinase-related protein1 (TYRP1) (Gronskov et al., 2007; Renugadevi et al., 2010). Seventy-two mutations in *OCA2* are listed as causes of OCA (Gronskov et al., 2007).

The OCA3 is caused by mutations in *TYRP1* (Gronskov et al., 2007; Renugadevi et al., 2010). It maps to chromosome 9p23 (Renugadevi et al., 2010). TYRP1 is an enzyme in the melanin biosynthesis pathway, catalyzing the oxidation of 5,6-dihydroxyindole-2-carboxylic acid monomers into melanin (Gronskov et al., 2007). Tyrp1 functions to stabilize Tyr (Gronskov et al., 2007).

Mutations in the membrane-associated transporter protein gene (*MATP*, also known as *SCL45A2*) cause OCA4 (Gronskov et al., 2007; Renugadevi et al., 2010; Summers, 2009). *MATP* maps to 5p13.3 and encodes a protein expressed in the melanosomal membrane; it may function as a membrane transporter directing melanosomal traffic and other substances to melanosomes (Gronskov et al., 2007; Renugadevi et al., 2010).

Another type of albinism caused by mutations on *OA1* (Xp22.3), OA (OA1), affects males because of X-linked inheritance (Engle, 2010; Gronskov et al., 2007; Renugadevi et al., 2010; Summers, 2009). *OA1* encodes a G-protein-coupled receptor on the melanosome membrane (Engle, 2010; Renugadevi et al., 2010).

31.3.6.3 Neuroradiological Findings

Among disorders where albinism is part of a larger syndrome, such as HPS, cerebral atrophy, most marked in the occipital lobes, can be seen (Budisteanu et al., 2010; Gronskov et al., 2007).

31.3.7 Congenital Fibrosis of the Extraocular Muscles Type I

Congenital fibrosis of the extraocular muscles type 1 (CFEOM1) is a complex strabismus syndrome categorized as one of the congenital cranial dysinnervation disorders (Andrews et al., 1993; Engle, 2010; Heidary et al., 2008). It is the most common form of CFEOM (Andrews et al., 1993; Yamada et al., 2005). The minimum prevalence of CFEOM1 has been estimated to be 1/230 000 (Andrews et al., 1993; Heidary et al., 2008).

31.3.7.1 Clinical Characteristics

Affected individuals are born with congenital bilateral nonprogressive blepharoptosis and strabismus as well as congenital bilateral nonprogressive external ophthalmoplegia with the eyes fixed in an infraducted position approximately 20–30° below the horizontal midline (Andrews et al., 1993; Engle, 2010; Heidary et al., 2008; Kakinuma and Kiyama, 2009; Khan et al., 2010; Yamada et al., 2005). The eyes look down at rest and cannot be elevated, whereas horizontal movement can range from absence to full (Andrews et al., 1993; Engle, 2010). These patients lack binocular vision (Andrews et al., 1993; Yamada et al., 2005). Forced duction testing is positive for marked restriction of extraocular motility (Andrews et al., 1993; Heidary et al., 2008). Patients typically assume a compensatory 'chin up' head posture to fixate on objects (Heidary et al., 2008).

Affected individuals often have ocular synkinesis (aberrant patterns of eye movement), including synergistic convergence, synergistic divergence, Marcus Gunn jaw-winking phenomenon (congenital synkinetic movement due to synkinesis between the upper eyelid and the pterygoids), and no pupillary involvement (Engle, 2010; Heidary et al., 2008; Khan et al., 2010; Yamada et al., 2005). Synkinetic eye movements are thought to result from aberrant axonal routing (Heidary et al., 2008).

The Marcus Gunn jaw-winking phenomenon is categorized clinically as an ocular miswiring syndrome (Yamada et al., 2005). Affected individuals have ptosis accompanied by elevation of the ptotic eyelid on movement of the lower jaw (Andrews et al., 1993; Yamada et al., 2005). It is first noted in young infants when they are being fed (Andrews et al., 1993). This syndrome is proposed to result from misdirection of axons intended to travel within the motor branch of the trigeminal nerve to innervate the ipsilateral pterygoid muscle (Andrews et al., 1993; Yamada et al., 2005). Instead, these axons aberrantly innervate myofibers of the levator palpebrae superioris muscle, which is normally innervated by a branch of the oculomotor nerve (Andrews et al., 1993; Yamada et al., 2005).

There is phenotypic heterogeneity with respect to involvement of the horizontal extraocular muscles (Heidary et al., 2008). Horizontal movements may be severely restricted or absent, and horizontal strabismus may be present with an increased incidence of exotropia versus esotropia (Heidary et al., 2008). As a consequence of the marked limitation of eye movements and the blepharoptosis, many CFEOM patients develop strabismic and deprivation amblyopia (Andrews et al., 1993; Heidary et al., 2008; Yamada et al., 2005). Among families with CFEOM1, the vertical strabismus is quite uniform, but the horizontal strabismus can vary (Andrews et al., 1993). CFEOM1 patients generally show normal cognitive and physical development (Heidary et al., 2008).

31.3.7.2 *Genetics*

CFEOM1 is inherited as an autosomal dominant trait with complete penetrance and minimal variation in expression (Andrews et al., 1993; Khan et al., 2010; Yamada et al., 2005). It results from heterozygous mutations in *KIF21A* (kinesin family member 21A), which encodes a kinesin motor and maps to 12q12 (http://www.ncbi.nlm.nih.gov/omim) (Andrews et al., 1993; Engle, 2010; Heidary et al., 2008; Kakinuma and Kiyama, 2009; Khan et al., 2010; Yamada et al., 2005). Eighty mutation-positive patients of multiple ethnicities reported to date harbor only 11 unique missense mutations, which are often *de novo*, and 75% harbor 2860C>T (R954W) (Andrews et al., 1993; Engle, 2010). These mutations alter only 7 of the 1675 amino acids in *KIF21A* (Engle, 2010).

The KIF21A is a neuronally expressed protein, which is important in axonal maintenance (Heidary et al., 2008; Yamada et al., 2005). It has been characterized as a member of the Kif4-class superfamily of kinesin motors and acts as a plus-end kinesin motor (Kakinuma and Kiyama, 2009). This gene encodes a developmental motor kinesin responsible for anterograde axonal transport of cargo along neurons such as that of the superior division of cranial nerve III (Engle, 2010; Heidary et al., 2008; Khan et al., 2010).

31.3.7.3 *Neuroradiological Findings*

Central nervous system maldevelopment, including cortical dysplasia and basal ganglia abnormalities, has been reported (Heidary et al., 2008). Hypoplasia of the superior rectus and levator palpebrae superioris is a common feature of CFEOM1 patients (Heidary et al., 2008). Orbital MRI has demonstrated an absent or severely hypoplastic superior division of the oculomotor nerve (Heidary et al., 2008). Imaging of the skull base has confirmed hypoplasia of the oculomotor nerve as it exited the brain stem (Heidary et al., 2008).

31.3.8 Duane Retraction Syndrome

The DRS is a unilateral or bilateral congenital anomaly of the sixth cranial nerve nuclei with aberrant innervations by supply from the third cranial nerve (Gabay et al., 2010; Yuksel et al., 2010). This syndrome accounts for 1–5% of all cases of strabismus (Zanin et al., 2010). Unilateral Duane syndrome, which accounts for 85% of all cases of DRS, is predominantly sporadic (90%), more prevalent in females (60%), and mainly affects the left eye (Gutowski, 2000; Yuksel et al., 2010; Zanin et al., 2010). It is characterized by marked limitation or absence of abduction, variable limitation of adduction, palpebral fissure narrowing, and globe retraction on attempted adduction (Gabay et al., 2010; Zanin et al., 2010).

Disturbance between the fourth and tenth weeks of embryogenesis seems most obvious and could explain the various nonocular and ocular abnormalities in combination with the Duane's syndrome (Yuksel et al., 2010).

A teratogenic event during the second month of gestation seems to cause most ocular and extraocular abnormalities observed in combination with DRS (Yuksel et al., 2010). Thalidomide has been reported as having a teratogenic effect (Yuksel et al., 2010).

31.3.8.1 Clinical Characteristics

Affected individuals have restricted horizontal gaze greatest with attempted abduction and ocular synkinesis resulting in globe retraction with attempted adduction (Engle, 2010; Gutowski, 2000). When an affected individual attempts to adduct his/her eyes, both the intended medial rectus and the pathologically innervated lateral rectus muscle contract, resulting in retraction of the eyeball into the orbit (Engle, 2010; Gutowski, 2000). Electromyography shows increased electrical activity in a paretic lateral rectus muscle (Engle, 2010; Gutowski, 2000).

Binocular vision is preserved in DRS (Yuksel et al., 2010). The compensation by abnormal head posture allows binocularity in one field of gaze despite the severe eye motility deficit in the other field of gaze (Yuksel et al., 2010). The degree of sensorial binocular status plays an important role in the conjugacy of saccades (Yuksel et al., 2010).

The DRS has been associated with several other conditions where anomalous axonal guidance occurs, such as OCA (Gutowski, 2000; Yuksel et al., 2010). Some other syndromes associated with DRS are Okihiro syndrome (forearm malformation and hearing loss), Wildervanck syndrome (fusion of neck vertebrae and hearing loss), Holt–Oram syndrome (abnormalities of the upper limbs and heart), morning-glory syndrome (abnormalities of the optic disc or blind spot), and Goldenhar syndrome (malformations of the jaw, cheek, and ear, usually on one side of the face) (Yuksel et al., 2010).

Three types of DRS are recognized, depending on the amount of aberrant innervation present (Gutowski, 2000). Palpebral fissure narrowing and retraction of the affected eyeball on adduction tend to be constant findings (Gutowski, 2000). Palpebral fissure narrowing is due to recti contraction and mechanical factors (Gutowski, 2000).

DRS type I consists of defective abduction with normal or minimally defective adduction (Gutowski, 2000; Yuksel et al., 2010). In DRS type II, adduction is defective, and there is exotropia of the affected eye and normal or minimally defective abduction (Gutowski, 2000; Yuksel et al., 2010). Both adduction and abduction are defective in type III (Gutowski, 2000; Yuksel et al., 2010). All three types of DRS frequently produce additional vertical eye movement anomalies, which are characterized by changes in the ocular axes (Gutowski, 2000).

Eye movement recordings are an additional tool for understanding the underlying pathogenesis of DRS (Yuksel et al., 2010). This is a noninvasive technique that has given valuable information about the neural control of movement (Yuksel et al., 2010). Most DRS patients compensate well for the disorder and do not require further management (Yuksel et al., 2010). Standard management of DRS may, in some cases, involve eye muscle surgery (Yuksel et al., 2010).

31.3.8.2 Genetics

The majority of DRS cases are sporadic, with only 2–5% of patients showing a familial pattern (Gutowski, 2000; Yuksel et al., 2010). A high prevalence of DRS has been noted in individuals with thalidomide embryopathy (Gutowski, 2000; Yuksel et al., 2010). Both genetic and environmental factors are likely to play a role in the development of DRS (Yuksel et al., 2010; Zanin et al., 2010).

Studies of sporadic forms of DRS showed 10–20 times greater risk for having other congenital malformations divided in mainly four categories: skeletal, auricular, ocular, and neural (Yuksel et al., 2010). The skeletal abnormalities involved the palate and vertebral column (Yuksel et al., 2010). The auricular malformations included the external ear, the external auditory meatus, and the semicircular canals (Yuksel et al., 2010). Ocular defects concerned the extraocular muscles and the eyelids, including ocular dermoids (Yuksel et al., 2010). Neural defects involved the third, fourth, and sixth cranial nerves (Yuksel et al., 2010).

Most cases of familial DRS are autosomal dominant without associated abnormalities (Gutowski, 2000; Yuksel et al., 2010). There are also several autosomal dominant syndromes in which DRS is a recognized feature (Gutowski, 2000). In some families with dominant DRS, the disease shows reduced penetrance and variable expressivity (Yuksel et al., 2010). An autosomal recessive pattern of inheritance has also been suggested in several reports of DRS (Gutowski, 2000; Yuksel et al., 2010). Autosomal recessive DRS can occur either in isolation or in association with other abnormalities (Gutowski, 2000).

Several chromosomal loci for genes contributing to DRS have been suggested (Gutowski, 2000; Yuksel et al., 2010; Zanin et al., 2010). CHN1 (chimerin 1) is one of the genes responsible for DRS (Engle, 2010; Zanin et al., 2010). Individuals harboring CHN1 mutations have a higher incidence of vertical movement abnormalities and bilateral eye involvement when compared to individuals with nonfamilial DRS (Engle, 2010). CHN1 mutations alter the development of abducens and, to a lesser extent, oculomotor axons (Engle, 2010).

Deletions of chromosomal material on chromosomes 4 and 8 and the presence of an extra marker chromosome, thought to be derived from chromosome 22, has been documented in DRS individuals (Yuksel et al., 2010).

SALL4 (sal-like 4) and *HOXA1* (*homeobox A1*) have been found to be associated also with syndromic forms (Zanin et al., 2010). However, patients with isolated DRS have not been found to carry these mutations (Zanin et al., 2010).

31.3.8.3 *Neuroradiological Findings*

Motion-encoded MRI has been used for the study of human extraocular muscle function; local physiological contraction and elongation have been quantified (Yuksel et al., 2010). The visualization of the abducens nucleus itself at a neural level remains unfeasible, but the nerve can be explored at the pontomedullar level (Yuksel et al., 2010).

MRI of individuals harboring *CHN1* mutations can reveal hypoplasia of the oculomotor nerve and oculomotor-innervated muscles in addition to the expected abducens nerve hypoplasia and aberrant lateral rectus innervation (Engle, 2010).

MRI has confirmed the maldevelopment of the abducens nerve in DRS and has showed the compensatory innervation by the third nerve at a peripheral level (Yuksel et al., 2010). MRI in cases of DRS type I has demonstrated the absence of the abducens nerve (Yuksel et al., 2010).

31.3.9 Pontine Tegmental Cap Dysplasia

Pontine tegmental cap dysplasia (PTCD) is a cerebellar, brain stem, and cranial nerve malformation syndrome (Engle, 2010). It was first reported in four patients, in 2007 (Barth et al., 2007; Macferran et al., 2010). Pathomechanism of the condition involves a defect in the migration or navigation of axons of rhombencephalic neurons (Szczaluba et al., 2010).

PTCD belongs to a group of related conditions that share in common malformations of the midbrain hindbrain (Macferran et al., 2010). Under this category are those conditions that have in common the molar tooth sign (Macferran et al., 2010).

Conditions described in this group include the following syndromes: Joubert, Senior–Loken, COACH, Dekaban-Arima, oro-facial-digital type VI, and encephalocele with renal cysts (Engle, 2010; Macferran et al., 2010).

It is suggested that many cases of pontocerebellar hypoplasia or Moebius syndrome should be revised for the features of PTCD (Szczaluba et al., 2010). To date, no known etiology for PTCD has been identified (Macferran et al., 2010).

31.3.9.1 *Clinical Characteristics*

The affected children described to date have mild to severe developmental delay, ataxia, and a combination of restricted horizontal eye movements, ocular apraxia,

facial weakness, deafness, and swallowing and feeding impairments (Engle, 2010; Macferran et al., 2010). Also, they have shown bilateral trigeminal nerve dysfunction, seizures, and central ventilation abnormalities (Macferran et al., 2010). Commonly reported somatic abnormalities include vertebral anomalies and craniofacial dysmorphism (Macferran et al., 2010).

31.3.9.2 *Genetics*

The reported children had neither a positive family history nor consanguineous parents, so it remains to be proved that PTCD is genetic (Engle, 2010). It is plausible that it results from *de novo* dominant mutations or recessive mutations in an unidentified gene (Engle, 2010).

A very small (96 kb) 2q13 deletion has been identified as the first case of a molecular genetic abnormality as the identified cause of the condition (Macferran et al., 2010). The deleted region encompasses the *NPHP1* gene (Macferran et al., 2010). This gene has been reported in association with JS (Macferran et al., 2010).

31.3.9.3 *Neuradiological Findings*

Neuroimaging reveals pontine hypoplasia with ventral flattening and dorsal protrusion of tissue into the fourth ventricle ('tegmental cap') (Engle, 2010; Macferran et al., 2010). Cerebellar vermian hypoplasia and elongated and laterally misplaced SCP result in a modified molar tooth sign (Engle, 2010; Macferran et al., 2010). The middle and inferior cerebellar peduncles and cranial nerves VII and VIII are small (Engle, 2010; Macferran et al., 2010). DTI reveals failure of the SCP, MCP, and axons of the pontine nuclei to decussate and defines the tegmental cap as an ectopic dorsal transverse fiber bundle (Engle, 2010).

31.4 INTRODUCTION: NEURONAL MIGRATION

Development of central nervous system is a highly complicated process, and it is organized in the following steps (Spalice et al., 2009; Verrotti et al., 2010):

1. Primary neurulation (3–4 weeks of gestation), beginning of neuronal migration (fifth week of gestation)
2. Prosencephalic development (2–3 months of gestation)
3. Neuronal proliferation (3–4 months of gestation)
4. Neuronal migration (1–5 months of gestation)
5. Organization (5 months of gestation to after birth)
6. Myelination (after the birth)

Cell migration has an essential role in the developing cerebral cortex because all neurons that eventually populate the six-layered cerebral cortex and other brain regions undergo mitosis in distant compartments and then migrate great distances to achieve final positioning (Kerjan and Gleeson, 2007).

Neuronal migration consists of nerve cells moving from their original site in the ventricular and subventricular zones to their final location (Kerjan and Gleeson, 2007; Pang et al., 2008; Spalice et al., 2009; Verrotti et al., 2010). Regulation of timing and direction of these simultaneous migrations are highly ordered (Spalice et al., 2009; Verrotti et al., 2010).

Instructed by extracellular cues, the activation of guidance receptors and their downstream signaling pathways enable newborn neurons to migrate through the developing nervous system until they reach their destination (Valiente and Marin, 2010). On arrival to their final destination, neurons cancel their migratory program and continue their differentiation into mature neurons (Valiente and Marin, 2010). It has been suggested that early patterns of activity generated in the target region may influence this process (Valiente and Marin, 2010).

Neurons originating in the cortical ventricular zone migrate radially to form the cortical plate (CP) and mainly become projecting neurons (Spalice et al., 2009; Verrotti et al., 2010).

Migration of neocortical neurons occurs mostly between the 12th and the 24th weeks of gestation (Spalice et al., 2009; Verrotti et al., 2010). The first postmitotic neurons produced in the periventricular germinative neuroepithelium will migrate to form a subpial preplate or primitive plexiform zone (Spalice et al., 2009). Subsequently, produced neurons, which will form the CP, migrate into the preplate and split it into the superficial molecular layer (or layer I or the marginal zone containing Cajal–Retzius neurons) and the deep subplate (Spalice et al., 2009).

Schematically, the successive waves of migratory neurons will pass the subplate neurons and finish their migratory pathway below layer I, forming successively (but with substantial overlap) cortical layers VI, V, IV, III, and II (Spalice et al., 2009). This means that the neurons that migrate first will stop in the deepest cortical layers, and those that migrate afterward pass through the layers formed previously to form the outer cortical layers according to a migration scheme defined 'inside-out' (Pang et al., 2008; Spalice et al., 2009; Verrotti et al., 2010).

Neocortical migrating neurons can adopt different types of trajectories: A large proportion of neurons migrate radially, along radial glial guides, from the germinative zone to the CP (Spalice et al., 2009; Verrotti et al., 2010). Radial glial cells are specialized glial cells present in the neocortex during neuronal migration (Spalice et al., 2009; Verrotti et al., 2010).

Another important group of neuronal precursors initially adopts a tangential trajectory at the level of the ventricular or subventricular germinative zones before adopting a classic radial migrating pathway along radial glia (Spalice et al., 2009; Verrotti et al., 2010). Tangentially, migrating neurons have also been located at the intermediate zone level (prospective white matter) (Spalice et al., 2009; Verrotti et al., 2010).

The phenotype of radial glial seems to be determined by both migrating neurons and intrinsic factors expressed by glial cells (Spalice et al., 2009; Verrotti et al., 2010). Among the latter, the transcription factor Paired Box Gene (*PAX6*), which is specifically localized in radial glia during cortical development, is critical for the morphology, number, function, and cell cycle of radial glia (Spalice et al., 2009; Verrotti et al., 2010).

There are several molecules involved in the control of neuronal migration and in targeting the exact destination of the neurons (Spalice et al., 2009; Verrotti et al., 2010). These molecules can be divided into four broad categories: molecules of the cytoskeleton, which play an important role in the initiation and ongoing progression of neuronal migration, such as Filamin A, ARF-GEF2, doublecortin, LIS1, TUBA1A; signaling molecules playing a role in lamination, such as reelin and some reelin receptors such as p35, cdk5, and Brn1/Brn2; molecules modulating glycosylation, which seem to provide stop signs for migrating neurons, such as, POMT1, POMGnT1, fukutin, and FAK; and other factors including neurotransmitters such as glutamate and GABA, trophic factors such as brain-derived neurotrophic factor and thyroid hormones, molecules deriving from peroxisomal metabolism, and environmental factors such as ethanol and cocaine (Spalice et al., 2009; Verrotti et al., 2010).

Reelin is crucial for the lamination of cortical structures, but the molecular mechanisms underlying its action remain unclear (Valiente and Marin, 2010). Genetic studies have positioned Reelin, apolipoprotein E receptor 2 (ApoER2), Vldlr, and Dab1 into a common signaling pathway that leads to the phosphorylation of Dab1 in migrating neurons, an event that is required for normal layering of the cortex (Valiente and Marin, 2010). Both structural barriers at the pial surface of the brain and molecular stop signals are involved in mediating neuronal migration arrest (Pang et al., 2008).

Cell migration requires the dynamic regulation of adhesion complexes between migrating cells and the surrounding extracellular matrix proteins (Valiente and Marin, 2010). In many cell types, this process involves integrin-mediated adhesion, but the function of this signaling system in neuronal migration has remained controversial (Valiente and Marin, 2010). The striking morphology of neurons, as they migrate, extends dendrites and axons and connects with other cells, implying a strictly regulated program of cytoskeletal organization (Kerjan and Gleeson, 2007).

31.5 OVERVIEW OF NEURONAL MIGRATION DISORDERS

Malformations of cortical development are an important cause of epilepsy and development delay (Pang et al., 2008; Spalice et al., 2009; Verrotti et al., 2010). It is estimated that up to 40% of children with refractory epilepsy have a cortical malformation (Pang et al., 2008; Spalice et al., 2009; Verrotti et al., 2010). These malformations have also been associated with mental retardation and cerebral palsy (Mochida, 2009). Malformations of cortical development encompass a large spectrum of disorders related to abnormal cortical development with varied genetic etiologies, anatomic abnormalities, and clinical manifestations (Mochida, 2009; Pang et al., 2008).

Cerebral cortical development requires orchestrated movement of cells arising from different regions within the brain, and born at different times, to achieve specific laminar position, orientation, and connections with other cells (Kerjan and Gleeson, 2007; Pang et al., 2008). Disruptions at these various stages may result in malformations of cortical development (Pang et al., 2008). Although cortical development has been separated into various stages, there is a significant overlap between the stages, and many abnormalities may cause dysfunction at more than one level (Pang et al., 2008).

Malformation syndromes are typically classified based on the earliest disruption of development (Barkovich et al., 2005; Pang et al., 2008). This classification system divides brain malformations into disorders of cell proliferation, neuronal migration, and cortical organization (Barkovich et al., 2005; Mochida, 2009). Neuronal migration disorders include lissencephaly, heterotopia, focal cortical dysgenesis, PMG, and schizencephaly (SCZ) (Verrotti et al., 2010).

The pathogenesis of these malformations is multifactorial: Genetic mutations or environmental insults, whether acquired *in utero* at different stages of brain development, or during the perinatal or postnatal period after corticogenesis, may all contribute to the development of these disorders (Jaglin and Chelly, 2009; Pang et al., 2008). The timing, severity, and type of environmental influences, as well as genetic factors, will ultimately determine the type and extent of the malformation (Pang et al., 2008).

Genetic studies in humans and mice have identified a spectrum of mutations in genes involved in a large array of crucial processes such as cell proliferation, cell adhesion, cell migration, chemoattraction and repulsion, posttranslational modifications, and dynamics of the cytoskeleton that often disrupts the development of the cerebral cortex and can lead to severe cortical malformations (Jaglin and Chelly, 2009).

Following neurogenesis, the disruption of neuronal migration resulting from genetic mutations represents a major cause of cortical dysgenesis and encompasses a large variety of malformations (Jaglin and Chelly, 2009). Many of these genes encode important effectors that modulate cytoskeletal dynamics during the migration of neuronal cells (Jaglin and Chelly, 2009). Mutations in *DCX* (*doublecortin*) and *LIS1* genes, which encode MAPs, have been shown to be associated with a large spectrum of neuronal migration disorders (Jaglin and Chelly, 2009).

Despite the significant progress over the past few years, many cases of cortical dysgenesis are still unexplained (Jaglin and Chelly, 2009). The identification of further genes is important for the transfer to the clinic and genetic counseling, as well as to have a better understanding of the physiopathology of human cortical dysgenesis (Jaglin and Chelly, 2009).

31.5.1 Lissencephaly

Lissencephaly is a group of disorders that is characterized by an abnormally smooth surface of the cerebral cortex (Mochida, 2009; Pang et al., 2008; Vallee and Tsai, 2006; Verrotti et al., 2010). It is a severe brain malformation characterized by agyria (absence of gyri) and pachygyria (reduced number of broadened gyri), thickened cortex, abnormal cortical layering, enlarged ventricles, and neuronal heterotopias (abnormal positioning of neurons) (Ghai et al., 2006; Pang et al., 2008; Reiner et al., 2006; Verrotti et al., 2010). The lifespan of patients with these disorders is short and most of them die within the first year of life, usually because of aspiration pneumonia and sepsis (Reiner et al., 2006).

Lissencephaly is a neuronal migration disorder that results from impaired migration of postmitotic neurons from the ventricular zone to the developing CP (Ghai et al., 2006; Reiner et al., 2006).

On the basis of etiologies and associated malformations, five groups of lissencephaly can be identified: classical lissencephaly, cobblestone lissencephaly, X-linked lissencephaly with ACC, lissencephaly with cerebellar hypoplasia (LCH), and microlissencephaly (Verrotti et al., 2010). The onset of lissencephaly is considered to occur no later than the 12th–16th week of gestation (Verrotti et al., 2010).

Functions of some lissencephaly genes are closely related to microtubules (Mochida, 2009). The network of microtubules and molecular motor, dynein, are critical to this movement of the centrosome and nucleus in migrating neurons (Mochida, 2009). Some lissencephaly genes are associated with specific neuropathology of the cerebral cortex (Mochida, 2009). Mutations of six genes have been associated with lissencephaly, including *LIS1, DCX, TUBA1A, RELN*, very low-density lipoprotein receptor (*VLDLR*), and *ARX*, whereas codeletion of α*WHAE* with *LIS1* appears to act as a modifier locus (Spalice et al., 2009).

31.5.1.1 *Classical Lissencephaly*

Classical lissencephaly, previously known as type 1 lissencephaly, causes a combination of agyria and pachygyria (Kerjan and Gleeson, 2007; Verrotti et al., 2010). This disorder represents one of the most severe disorders of neocortical neuronal migration (Kerjan and Gleeson, 2007). It is distinguished from the other forms of lissencephaly based on the absence of additional characteristic features (Kerjan and Gleeson, 2007). This type of lissencephaly occurs in Miller–Dieker syndrome (MDS) and in isolated lissencephaly sequence (ILS) (Reiner et al., 2006). The incidence of classical lissencephaly has been estimated to be 1.2 in 100 000 births (Verrotti et al., 2010).

Microscopically, the cortex appears poorly structured with only four immature layers of neurons instead of the normal six highly organized layers present in a well-developed brain (Kerjan and Gleeson, 2007; Pang et al., 2008; Verrotti et al., 2010). This disorder is derived from both tangential and radial migration disorders of neurons (Spalice et al., 2009). The pathogenesis of classic lissencephaly is unlikely to be due to defective neuronal migration alone but may include an aspect of abnormal proliferation (Mochida, 2009).

31.5.1.1.1 CLINICAL CHARACTERISTICS

Children with classical lissencephaly are diagnosed in the first few months of life (Kerjan and Gleeson, 2007). ILS is clinically characterized by early hypotonia, which may evolve to limb spasticity, seizures, psychomotor retardation, and often microcephaly (Jaglin and Chelly, 2009; Pang et al., 2008; Spalice et al., 2009; Verrotti et al., 2010). Seizures are present in almost the totality of children with onset in early age, mostly in the first 6–12 months (Spalice et al., 2009; Verrotti et al., 2010). High prevalence (80%) of infantile spasms with or without typical hypsarrhythmia on EEG has been reported (Spalice et al., 2009; Verrotti et al., 2010). Later, most children have a more complex epileptic syndrome, including atypical absences, drop attacks, and myoclonic, partial complex, tonic, and tonic–clonic seizures (Spalice et al., 2009; Verrotti et al., 2010). The EEG demonstrated diffused fast rhythms with high amplitude, considered peculiar of this condition (Spalice et al., 2009).

Children with MDS have a severe form of ILS with facial dysmorphisms, including a prominent forehead, bitemporal hollowing, a short nose with upturned nares, and a long and protuberant upper lip with thin vermillion border and small jaw (Mochida, 2009; Verrotti et al., 2010). Other children with MDS might also present cardiac and renal abnormalities, cryptorchidism, sacral dimple, omphaloceles, and clinodactyly (Verrotti et al., 2010).

31.5.1.1.2 GENETICS

Lissencephaly associated with *RELN, VLDLR*, and *ARX* are pathologically and radiologically distinct from lissencephaly because of *LIS1, DCX*, and *TUBA1A* (Mochida, 2009). The LIS1 gene is the first gene that was correlated with human lissencephaly (Jaglin and Chelly, 2009; Mochida, 2009; Reiner et al., 2006; Verrotti et al., 2010). LIS1 localizes to chromosome 17p13.3 (Mochida, 2009). LIS1 is a MAP that localizes primarily to the centromere in migrating neurons (Kerjan and Gleeson, 2007; Vallee and Tsai, 2006). LIS1 is highly conserved in evolution both in sequence and in multiple functional aspects (Reiner et al., 2006).

A tight relationship between LIS1, microtubule regulation, and microtubule-based motor proteins has been suggested for many organisms (Reiner et al., 2006). LIS1 controls mitotic spindle orientation in both the neuroepithelial stem cells and the radial glial progenitor cells (Verrotti et al., 2010). It has a role in preserving the normal microtubule network organization (Reiner et al., 2006; Vallee and Tsai, 2006).

In addition to a direct role for LIS1 in regulating tubulin dynamics, LIS1 interacts with a plethora of MAP (Reiner et al., 2006). This includes interactions with DCX, CLIP-170, and MAP1b (Reiner et al., 2006). LIS1 deletion causes dysfunction of the dynein, a microtubular cytoplasmic protein involved in neuronal migration processes (Jaglin and Chelly, 2009; Mochida, 2009; Verrotti et al., 2010). LIS1 interacts with several subunits of the retrograde, microtubule-based motor complex dynein/dynactin (Reiner et al., 2006; Vallee and Tsai, 2006). LIS1 regulates cytoplasmic dynein activity and participates in several dynein-mediated activities such as intracellular transport and mitosis (Reiner et al., 2006; Vallee and Tsai, 2006). In migrating cells, the presence of dynein, dynactin, and LIS1 at the leading cell cortex is essential for directed cell motility (Reiner et al., 2006).

ILS is caused by intragenic mutations or deletions of the LIS1 gene or by small deletions involving 17p13.3 (Ghai et al., 2006; Kerjan and Gleeson, 2007; Mochida, 2009; Verrotti et al., 2010). Complete deletion of both *LIS1* and 14-3-3 α*WHAE* genes on chromosome 17p13 causes MDS (Ghai et al., 2006; Kerjan and Gleeson, 2007; Mochida, 2009; Pang et al., 2008; Verrotti et al., 2010). MDS has been considered a contiguous gene deletion syndrome (Kerjan and Gleeson, 2007). α*WHAE* belongs to the 14-3-3 family of proteins that can have many effects on phosphoproteins, including protection from dephosphorylation (Verrotti et al., 2010). 14-3-3 binds to CDK5/p35-phosphorylated NUDEL, and this binding maintains NUDEL phosphorylation (Verrotti et al., 2010). NUDEL is a LIS1-binding protein that, together with LIS1, regulates the cytoplasmic dynein heavy chain function through phosphorylation by CDK5/p35, a complex known to be essential for neuronal migration (Jaglin and Chelly, 2009; Mochida, 2009; Reiner et al., 2006; Verrotti et al., 2010). A smaller deletion, encompassing the region of LIS1 gene but not the 14-3-3 gene, has been associated with a milder phenotype of isolated LIS (Verrotti et al., 2010). αWHAE is suggested to be a dosage-dependent modifier of the severity of classical lissencephaly (Kerjan and Gleeson, 2007).

Lissencephaly due to mutations or deletions of LIS1 is a dominant trait (Mochida, 2009). Most LIS1 mutations are *de novo*, and therefore the recurrence risk is generally low (Mochida, 2009). However, in some cases, a parent harbors a balanced translocation involving the LIS1 gene, and so their risk of recurrence could be much higher (Mochida, 2009). Most cases of MDS and ILS are sporadic (Ghai et al., 2006). However, approximately 20% of patients with MDS inherited a genetic deletion from a parent (Ghai et al., 2006).

Mutations of the doublecortin (*DCX*) gene on chromosome Xq22.3 are also known to cause classical lissencephaly in males while heterozygous mutations in females are associated with subcortical band heterotopia (SBH) (Jaglin and Chelly, 2009; Mochida, 2009; Reiner et al., 2006; Verrotti et al., 2010). Males with SBH and *DCX* mutations have rarely been reported (Verrotti et al., 2010). SBH is also known as 'double cortex' syndrome as a band of heterotopic neurons is found within the cerebral white matter between a normal-appearing cortex and the ventricular surface higher (Mochida, 2009). About 20% of patients with classical ILS or SBH have mutations of the DCX gene, resulting in X-linked lissencephaly (LISX1) or double cortex syndrome (DC) (Spalice et al., 2009). Mutations in DCX may be inherited from a mother with SBH to her son, causing lissencephaly, or to her daughter, causing SBH (Mochida, 2009). Doublecortin is expressed in postmitotic neurons, but neither in proliferating cells of the ventricular zone during the development period nor in mature neurons of the adult brain (Verrotti et al., 2010). DCX is a cytoplasmic protein that appears to direct neuronal migration by regulating the organization and stability of microtubules (Jaglin and Chelly, 2009; Mochida, 2009; Verrotti et al., 2010). Nearly all mutations identified to date are premature protein truncations or missense mutations that cluster within the repeated tubulin-binding domain of DCX and inactivate its effects on microtubules (Kerjan and Gleeson, 2007).

Mutations in the alpha tubulinic complex (*TUBA1A*) gene located on chromosome 12q12–q14 have been correlated with the agyria–pachygyria-band spectrum of phenotype (Jaglin and Chelly, 2009; Kerjan and Gleeson, 2007; Mochida, 2009; Verrotti et al., 2010). This gene encodes for brain-specific alpha tubulin protein that represents one of the major component of microtubule complex required for cell movement (Mochida, 2009; Verrotti et al., 2010). Mutations in *TUBA1A* are considered to affect the folding of tubulin heterodimers and influence interactions with microtubule-binding proteins (doublecortin and kinesin KIF1A), resulting in disorders of microtubular function and deficits in the motility of neuronal progenitor cells (Verrotti et al., 2010). Congenital microcephaly, spastic diplegia or quadriplegia, and mental retardation are common clinical features seen in patients with the *TUBA1A* mutation phenotype (Mochida, 2009; Verrotti et al., 2010). In addition to microcephaly, rare occurrences of agyria and subcortical heterotopia (SBH) demonstrate that TUBA1A-related lissencephaly could encompass a large spectrum of cortical abnormalities (Jaglin and Chelly, 2009). This abnormal gyral pattern is combined with dysgenesis of the anterior limb of the internal capsule to give a dysmorphic aspect to the basal ganglia (Jaglin and Chelly, 2009). Moreover, other extracortical defects often include complete to partial ACC and mild to severe cerebellar hypoplasia (Jaglin and Chelly, 2009).

Lissencephaly due to *TUBA1A* mutations manifests as a dominant trait, and therefore, the mutations are generally *de novo*, as seen with *LIS1* (Mochida, 2009). Neuronal axonal guidance and/or growth defects, in addition to early neuronal differentiation abnormalities, are likely to be involved in the pathogeny of *TUBA1A*-related cortical dysgenesis (Jaglin and Chelly, 2009).

LIS1 and *DCX* collectively account for about three-quarters of isolated classic lissencephaly, and *TUBA1A* is estimated to account for about 4% of cases (Jaglin and Chelly, 2009; Mochida, 2009). Mutations in these three genes generally cause similar clinical phenotypes, including microcephaly, mental retardation, with or without epilepsy, and motor deficits (Mochida, 2009). Mutations of these three genes lead to this form of lissencephaly in which cortical thickness is increased fourfold and produce a recognizable gradient in which the malformation is more severe anteriorly (*DCX*) or posteriorly (*LIS1* and *TUBA1A*) (Jaglin and Chelly, 2009; Mochida, 2009).

ARX is a homeobox gene that is expressed in the ganglionic eminences and the neocortical ventricular zone (Pang et al., 2008). This gene plays an important role in the proliferation of neuronal precursors and differentiation of the forebrain (Pang et al., 2008). Mutations of ARX are a rare cause of lissencephaly, although less severe mutations result in a more common developmental disorder, cryptogenic infantile spasms (Spalice et al., 2009). Mutations in this gene cause the X-linked lissencephaly syndrome with ambiguous genitalia (Mochida, 2009; Pang et al., 2008). These patients have neonatal-onset epilepsy, hypothalamic dysfunction causing temperature dysregulation, chronic diarrhea, and ambiguous genitalia (micropenis and cryptorchidism) (Pang et al., 2008). This gene is a transcription factor expressed in the forebrain that regulates nonradial migration of interneurons from ventral regions to the developing cortex (Spalice et al., 2009). Severe seizures are presumably related to a severe deficiency of inhibitory interneurons (Spalice et al., 2009).

Patients with *ARX* mutations have abnormalities of the basal ganglia and absence of the corpus callosum, whereas those with *RELN* and *VLDLR* mutations have less cortical thickening, absence of a cell-sparse zone, and profound cerebellar hypoplasia (Spalice et al., 2009).

Reelin (*RELN*) is a signaling glycoprotein secreted by the early neurons on the surface of the cerebral cortex known as the Cajal–Retzius cells (Pang et al., 2008). This is a large extracellular matrix protein, which, when absent, causes reversal of cortical layers with deeper layers being made up of younger rather than older born neurons (Reiner et al., 2006). Activation of the Reelin signaling pathway is thought to be essential for proper positioning of migratory neurons into the appropriate lamina of the cortex (Pang et al., 2008). Mutations in *RELN* give rise to seizures, developmental delay, and hypotonia (Pang et al., 2008). Also, the loss of cerebellar organization likely contributes to ataxia (Pang et al., 2008). Only a few patients have been described with mutations in this gene, and generalized pachygyria, severe cerebellar hypoplasia, and hippocampal abnormalities seem to be the common features (Mochida, 2009).

VLDLR mutations cause similar abnormalities to *RELN*, with severe cerebellar hypoplasia, but the simplification of gyri may be milder features (Mochida, 2009). The product of the *VLDLR* gene belongs to the same biological pathway as *RELN* (Mochida, 2009). VLDLR, along with APOER2, acts as a receptor for the RELN protein in migrating neurons and transmits the extracellular RELN signal to the intracellular signaling pathway (Mochida, 2009). There is also notable evidence of interaction between RELN signaling and LIS1 (Mochida, 2009).

In some forms of classic lissencephaly, defects in GABAergic inhibitory interneurons have been suggested (Mochida, 2009). GABAergic interneurons of the cerebral cortex are derived in the ganglionic eminence (which develops into basal ganglia) and migrate tangentially into the cerebral cortex (Mochida, 2009). The best known examples of defects in GABAergic interneurons are due to *ARX* mutations (Mochida, 2009). Also, the number of inhibitory neurons is greatly diminished in the brain of patients with a LIS1 (Mochida, 2009).

Lissencephaly is often associated with severe, intractable epilepsy, and defects in interneurons, in addition to abnormal cortical lamination, may be in part responsible for this (Mochida, 2009). Mutations in *TUBA1A* seem to be associated with a lower incidence of epilepsy compared with *LIS1* (Mochida, 2009).

31.5.1.1.3 NEURORADIOLOGICAL FINDINGS

MRI of the brain in classic lissencephaly demonstrates an hour-glass configuration with areas of pachygyria and agyria, poorly developed sylvian and rolandic fissures, and failure of opercularization of the insular areas (Pang et al., 2008). It shows some degree of abnormality in the spacing of the gyri and sulci (Kerjan and Gleeson, 2007). The cortex is moderately thickened (5–10 mm) with white matter signal abnormalities (Kerjan and Gleeson, 2007; Verrotti et al., 2010). Associated findings may include dilatation of lateral ventricles, mild hypoplasia of the corpus callosum, and persistent cavum septum pellucidum (Verrotti et al., 2010).

Mutations in *LIS1* are often associated with abnormalities prevalent in the parietal and occipital cortex, whereas DCX lissencephaly is more pronounced in the frontal and temporal cortex (Pang et al., 2008; Verrotti et al., 2010). Mutations in *TUBA1A* have led to gyral malformations that are more severe in posterior than in anterior regions of the brain, often combined with dysgenesis of the corpus callosum, cerebellar and brainstem hypoplasia, and variable cortical malformation, including subtle SBH, ventricular dilatation, and absence or hypoplasia of the anterior limb of the internal capsule (Pang et al., 2008; Verrotti et al., 2010).

In individuals harboring *ARX* mutations, the lissencephaly is worse posteriorly than anteriorly, and there is absence of the corpus callosum (Pang et al., 2008). The cortex is moderately thickened (5–10 mm) with white matter signal abnormalities as well as cystic or fragmented basal ganglia (Pang et al., 2008). In *RELN* mutations, the lissencephaly is associated with cerebellar hypoplasia and hippocampal and brainstem abnormalities (Pang et al., 2008).

31.5.1.2 Cobblestone Lissencephaly

Cobblestone lissencephaly, previously type II, is a complex brain malformation characterized by global disorganization of cerebral organogenesis (Verrotti et al., 2010). It is characterized by a defective basement membrane in which breaches are formed (Jaglin and Chelly, 2009). It refers to the nodular appearance of the cerebral cortex caused by disorganization of the cortical layers and overmigration of neurons through the pial surface of the brain into the leptomeninges (Jaglin and Chelly, 2009). The cortex displays irregular grooves imparting a cobblestone pattern and consists of cluster and circular arrays of neurons, with no recognizable layers, separated by glial and vascular septa (Verrotti et al., 2010). The brain phenotype includes multiple anomalies such as hydrocephalus and neuronal overmigration, causing a cobblestone cortex, lissencephaly, and ACC (Reiner et al., 2006). Brainstem abnormalities also may be present (Ghai et al., 2006).

All patients with cobblestone lissencephaly show defects in the O-linked glycosylation of the glycoprotein α-dystroglycan, a protein that bridges the actin cytoskeleton of cells and the extracellular matrix component, laminin callosum (Reiner et al., 2006).

31.5.1.2.1 CLINICAL CHARACTERISTICS

It is associated with various eye abnormalities and congenital muscular dystrophies (Reiner et al., 2006). Cobblestone lissencephaly has been described in three syndromes: the Walker–Warburg syndrome (WWS), muscle–eye–brain disease (MEB), and Fukuyama-type congenital muscular dystrophy (FCMD) (Ghai et al., 2006). WWS is the most severe of these small groups of syndromes (Verrotti et al., 2010). It has a worldwide distribution, while FCMD has been found in Japan and MEB primarily in Finland (Verrotti et al., 2010).

The major clinical features of WWS are macrocephaly, cerebellar malformation, ventricular dilation/hydrocephalus, retinal malformation, anterior chamber abnormality, and congenital muscular dystrophy (Verrotti et al., 2010).

MEB disease results in a severe form of congenital muscular dystrophy with mental retardation and myoclonic jerks (Verrotti et al., 2010). Ocular disorders include progressive myopia, retinal dystrophy, glaucoma, and optic atrophy (Verrotti et al., 2010).

FCMD is the mild form of cobblestone lissencephaly, and it is characterized by severe hypotonia, progressive weakness, and developmental delay (Pang et al., 2008; Verrotti et al., 2010). The majority of patients are unable to walk unsupported (Pang et al., 2008). Mental retardation is a universal finding (Pang et al., 2008). The association of epilepsy and seizure-related disorders in FCMD is widely accepted: Febrile seizures and epilepsy with generalized tonic convulsions appear in about 50% of children, but they are usually not severe (Pang et al., 2008).

31.5.1.2.2 GENETICS

Cobblestone lissencephaly follows an autosomal recessive inheritance pattern (Pang et al., 2008). Several genes have been implicated in the etiology of WWS (Verrotti et al., 2010). Different mutations have been found in the proteins O-mannosyltransferase 1 and 2 (*POMT1* at 9q34 and *POMT2* genes) and also in each of the fukutin (*FKTN* at 9q31–33) and fukutin-related 9

protein (*FKRP* at 19q13–32) genes (Pang et al., 2008; Verrotti et al., 2010).

The MEB gene has been localized on the chromosome 1p32–34 (*POMGnT1* gene for protein O mannose β-1,2-N-acetylglucosaminyltransferase) (Pang et al., 2008; Verrotti et al., 2010). FCMD is associated with mutations of the gene *FKTN* on chromosome 9q31, which encodes a novel 461-amino acid protein termed 'fukutin' (Pang et al., 2008; Verrotti et al., 2010).

All these genes are involved in the glycoylation of α-dystroglycan, an extracellular protein capable of binding to components of extracellular matrix such as laminin, agrin, neurexin, and perlecan (Pang et al., 2008; Verrotti et al., 2010). Mutations in these genes compromise the integrity of the superficial marginal zone of the cortex, so that neurons overmigrate beyond this structure into the pial surface, forming the cobblestone (Pang et al., 2008; Verrotti et al., 2010).

31.5.1.2.3 NEURORADIOLOGICAL FINDINGS

Brain MRI demonstrates the typical cobblestone lissencephaly with varying degrees of severity (Pang et al., 2008; Verrotti et al., 2010). MRI in WWS and MEB reveals pontine hypogenesis with a distinct dorsal 'kink' at the mesencephalic-pontine junction; a 'Z-shaped' hypoplastic brainstem is considered a key feature (Pang et al., 2008; Verrotti et al., 2010).

31.5.1.3 *Lissencephaly X-linked with ACC*

Lissencephaly X-linked with ACC (XLAG) includes a thickened cortex and three-layered cortex with gyral malformations that are more severe in posterior than anterior brain regions, atrophic striatal and thalamic nuclei, poorly myelinated white matter, ACC, and ambiguous genitalia (Jagla et al., 2008; Miyata et al., 2009; Okazaki et al., 2008).

31.5.1.3.1 CLINICAL CHARACTERISTICS

XLGA is associated with intractable neonatal onset epilepsy, temperature instability, probably due to a hypothalamic dysfunction, severe diarrhea, postnatal microcephaly, abnormal genitalia with micropenis and cryptorchidism, and early death (Jagla et al., 2008; Miyata et al., 2009; Okazaki et al., 2008).

31.5.1.3.2 GENETICS

XLAG results from defects in the *ARX* (aristaless-related homeobox) gene located at Xp22.13 (Okazaki et al., 2008). The *ARX* gene product has two functional domains, *prd*-like homeodomain, and *aristaless* domain (Okazaki et al., 2008). Disruption of the *prd*-like homeodomain leads to XLAG (Okazaki et al., 2008). The functional domain, *prd*-like homeodomain, has very important functions in the formation of the normal brain in early development (Okazaki et al., 2008).

The major function of ARX protein is thought to be not only the regulation of proliferation and tangential migration of GABAergic interneurons but also the radial migration of pyramidal neurons death (Okazaki et al., 2008).

31.5.1.3.3 NEURORADIOLOGICAL FINDINGS

Imaging studies show a thick cerebral cortex (5–6 mm) with anterior pachygyria and posterior agyria (Okazaki et al., 2008). Other findings include abnormal signal of white matter, absence of corpus callosum, and cystic or fragmental basal ganglia (Okazaki et al., 2008).

31.5.1.4 *Lissencephaly with Cerebellar Hypoplasia*

LCH has been recently defined as a different group of lissencephaly, which is neither classical nor cobblestone type (Verrotti et al., 2010). It is associated with severe abnormalities of the cerebellum, ranging from vermian hypoplasia to total aplasia with either classical or cobblestone lissencephaly, and abnormalities in the hippocampus and brainstem (Hong et al., 2000; Verrotti et al., 2010).

31.5.1.4.1 CLINICAL CHARACTERISTICS

Affected children show a motor, language, and cognitive delay; they neither sit and stand unsupported nor develop linguistic skills (Verrotti et al., 2010). Hypotonia and severe ataxia are frequent; in addition, generalized epilepsy begins at an early age (Verrotti et al., 2010).

31.5.1.4.2 GENETICS

It is an autosomal recessive disorder (Hong et al., 2000) that maps to chromosome 7q22 and is associated with mutations in the gene encoding reelin (*RELN*) (Hong et al., 2000; Verrotti et al., 2010). The mutations disrupt splicing of RELN cDNA, resulting in low or undetectable amounts of reelin protein (Hong et al., 2000). RELN encodes a large secreted protein that acts on migrating cortical neurons by binding to the VLDLR, the APOER2, $\alpha 3 \beta 1$ integrin, and protocadherins (Hong et al., 2000). LCH is also correlated with mutations of the gene, which encodes for VLDLR (Verrotti et al., 2010).

31.5.1.4.3 NEURORADIOLOGICAL FINDINGS

MRI demonstrates diffuse pachygyria, hippocampal dysplasia, and hypoplastic brainstem (Verrotti et al., 2010). Cerebellar manifestations range from midline hypoplasia to diffuse volume reduction and disturbed foliation (Verrotti et al., 2010).

31.5.1.5 *Microlissencephaly*

Microlissencephaly differs from other variants of LIS1 by the presence of a severe microcephaly (Verrotti et al., 2010). It is caused by an abnormal neuronal proliferation or survival combined with neuronal migration disorders (Verrotti et al., 2010). Two main types of microlissencephaly are recognizable: type A (Norman-Roberts syndrome) with no infratentorial anomalies and type B (or Barth syndrome), which is associated with a severe hypoplasia of the cerebellum and corpus callosum (Verrotti et al., 2010). A recent form has been reported in which primordial osteodysplastic dwarfism is associated with severe microcephaly (Verrotti et al., 2010).

31.5.1.5.1 CLINICAL CHARACTERISTICS

Norman-Roberts syndrome presents with a wide phenotypic heterogeneity (Natacci et al., 2007). Clinical characteristics include microcephaly, bitemporal hollowing, a low sloping forehead, slightly prominent occiput, widely set eyes, a broad and prominent nasal bridge, and severe postnatal growth deficiency (Verrotti et al., 2010). Neurological features include hypertonia, hyperreflexia, seizures, and severe mental retardation (Verrotti et al., 2010).

31.5.1.5.2 GENETICS

Norman-Roberts syndrome is an autosomal recessive disorder (Natacci et al., 2007).

31.5.1.5.3 NEURORADIOLOGICAL FINDINGS

The brain MRI has shown changes consistent with lissencephaly type I (Verrotti et al., 2010).

31.5.2 Heterotopia

Heterotopia is a neuronal migration disorder characterized by a cluster of disorganized neurons in abnormal locations, and it includes three main groups: periventricular nodular heterotopia (PNH), SBH, and marginal glioneural heterotopia (Spalice et al., 2009; Verrotti et al., 2010).

PNH is a rare malformation in which primary neuronal cells never begin migration and remain adjacent to the lateral ventricles (Pang et al., 2008; Spalice et al., 2009; Verrotti et al., 2010). PNH may involve both sides of the brain or, less frequently, be restricted to a single hemisphere (Spalice et al., 2009; Verrotti et al., 2010). Apparently, the cerebral cortex is normal (Pang et al., 2008).

SBH or 'double cortex' syndrome is characterized by a diffuse laminar band of gray matter located below the cerebral cortex and separated from it by a thin band of white matter (Pang et al., 2008; Spalice et al., 2009; Verrotti et al., 2010).

Marginal zone heterotopias or leptomeningeal glioneuronal heterotopias are one form of dysplasia in which ectopic nests of glial and neuronal cells are observed in the cortical MZ or overlying leptomeninges, respectively (Verrotti et al., 2010).

31.5.2.1 Clinical Characteristics

The spectrum of clinical presentation of PNH is wide (Spalice et al., 2009). Epilepsy is the main aspect (Pang et al., 2008; Spalice et al., 2009). About 90% of patients with PNH have epilepsy that can begin at any age, and it is usually intractable (Pang et al., 2008; Spalice et al., 2009; Verrotti et al., 2010). There is no clear relationship between the epilepsy severity and extent of nodular heterotopia (Pang et al., 2008). Surgical removal of the heterotopia cortex is generally successful (Spalice et al., 2009; Verrotti et al., 2010).

Other symptoms of PNH include severe developmental delay, microcephaly, and infantile spasms (Verrotti et al., 2010). However, in general, these patients have normal intelligence. Some patients may have learning problems such as impaired reading fluency (Pang et al., 2008). Some patients with PNH have also been described with Chiari I and amniotic band syndrome (Spalice et al., 2009; Verrotti et al., 2010).

The main clinical manifestation of SBH is epilepsy (Spalice et al., 2009). Individuals with SBH have variable degrees of mental retardation and intractable epilepsy, which seem to correlate with the thickness of the band and the overlying cortex (Spalice et al., 2009; Verrotti et al., 2010).

The cortex may be normal or associated with pachygyria (Verrotti et al., 2010). Lennox–Gastaut syndrome is another potential presentation (Spalice et al., 2009). Epilepsy surgery for focal seizures yields poor results, while callosotomy has been associated with improvement in drop attacks (Spalice et al., 2009). The clinical spectrum of SBH in male subjects overlaps with that in females in terms of seizure type representation, epilepsy syndromes, and response to antiepileptic therapy (Verrotti et al., 2010). However, there is increased heterogeneity with respect to cognitive function, neuroimaging, and molecular genetic data in males compared with females (Verrotti et al., 2010).

31.5.2.2 Genetics

PNH can be caused by genetic mutations or extrinsic factors, such as infections or prenatal injuries (Pang et al., 2008; Verrotti et al., 2010). Mutations of the *FLNA* gene (Xq28) cause bilateral PNH in the majority of patients; this form is often fatal for males, therefore explaining the female preponderance (Pang et al., 2008; Spalice et al., 2009; Verrotti et al., 2010). This gene encodes for Filamin A, an actin-binding protein that stabilizes the cytoskeleton to generate the forces necessary for cell mobility and mediates focal adhesions along the ventricular epithelium (Pang et al., 2008; Spalice et al., 2009; Verrotti et al., 2010). There is enrichment of FLNa (filamin A, alpha) in postmitotic migrating neurons, an expression pattern that might be maintained in part by FILIP (filamin-A-interacting protein), a potent degrader of FLNa (Spalice et al., 2009). FLNa mutations resulting in PH often involve truncation or disruption of the actin-binding domain, indicating that FLNa's ability to cross-link actin may be necessary for migration (Spalice et al., 2009).

The autosomal recessive form of PNH is caused by mutations in the *ARFGEF2* (ADP-ribosylation factor guanine exchange factor 2) gene localized at 20q13.13, which encodes for the protein brefeldin-inhibited GEF2 (Pang et al., 2008; Spalice et al., 2009; Verrotti et al., 2010). Mutations in ARF-GEF may impair the targeted transport of FLNA to the cell surface within neural progenitors along the neuroependyma (Pang et al., 2008; Verrotti et al., 2010). The disruption of these cells could contribute to PNH formation with microcephaly (Verrotti et al., 2010).

PNH has also been associated with copy number variations, including duplication 5p15.1 or 5p15.33 and deletion 6q26–q27 or 7q11.33 (Spalice et al., 2009; Verrotti et al., 2010). At least 15 distinct PNH syndromes have been described (Spalice et al., 2009).

SBH is caused by alterations in two genes: *LIS1* at 17p13.32 and *DCX* at Xq22.3–q23.3 (Pang et al., 2008; Verrotti et al., 2010). Mutations of the *DCX* gene have been found in all familial cases and in 53–84% of patients with sporadic, diffuse, or anteriorly predominant band heterotopia, which represent the most common forms of SBH (Verrotti et al., 2010). Other genetic causes of SBH remain unexplained (Verrotti et al., 2010). The DCX protein is thought to direct neuronal migration by regulating the organization and stability of microtubules necessary for neuronal motility (Pang et al., 2008).

31.5.2.3 Neuroradiological Findings

MRI patients with X-linked dominant mutation show bilateral symmetric nodules lying adjacent to the lateral ventricular walls; additional findings include hypoplasia of the corpus callosum and posterior fossa abnormalities such as cerebellar hypoplasia and enlarged cisterna magna (Pang et al., 2008; Verrotti et al., 2010). Unilateral PNH is commonly located in the posterior paratrigonal zone of the lateral ventricles and may involve the adjacent white matter (Verrotti et al., 2010).

Patients with autosomal recessive mutations of the *ARFGEF2* gene present with microcephaly, slightly enlarged ventricles, and delayed myelination (Pang et al., 2008; Verrotti et al., 2010). Symmetrical nodular heterotopia lining the ventricles is also seen, and the overlying cortex may be thinned with abnormal gyri (Pang et al., 2008).

MRI of the brain in SBH demonstrates two parallel layers of gray matter, a thin outer ribbon and a thick inner band, separated by a very thin layer of white matter (Pang et al., 2008; Verrotti et al., 2010).

31.5.3 Polymicrogyria

PMG is a cortical malformation characterized by an irregular brain surface with an excessive number of small and partly fused gyri separated by shallow sulci, giving the surface of the cortex its characteristic lumpy appearance (Pang et al., 2008; Spalice et al., 2009; Verrotti et al., 2010).

PMG can be focal or diffused, unilateral or bilateral (Pang et al., 2008; Verrotti et al., 2010). It is a very common cortical malformation and can be an isolated lesion associated with other brain malformations such as heterotopia, white matter lesions, or a part of several multiple congenital anomaly/mental retardation syndromes (Spalice et al., 2009; Verrotti et al., 2010). Bilateral involvement of the cortex is frequently seen, with a symmetric or asymmetric distribution, affecting the frontal, frontoparietal, parieto-occipital, perisylvian, and mesial occipital regions (Pang et al., 2008; Spalice et al., 2009).

PMG pathogenesis is not understood; brain pathology demonstrates abnormal development or loss of neurons in middle and deep cortical layers, variably associated with an unlayered cortical structure (Spalice et al., 2009). Two types of PMG can be identified histopathologically: a simplified four-layered form (in which there is a layer of intracortical laminar necrosis with subsequent alterations of late migration and postmigratory disruption of cortical organization) and an unlayered form (in which the molecular layers are continuous and do not follow the borders of convolutions and the neurons below have radial distribution while laminar organization is absent) (Verrotti et al., 2010). The incidence of PMG is unknown because of its clinical and etiological heterogeneity (Verrotti et al., 2010).

31.5.3.1 Clinical Characteristics

The wide spectrum of clinical manifestations is related to the extension of PMG, which varies greatly (Spalice et al., 2009). Almost all children with PMG have a high risk of developing epilepsy (Verrotti et al., 2010). Seizures usually begin between 4 and 12 years of age, and they are drug resistant in approximately 65% of patients (Spalice et al., 2009; Verrotti et al., 2010). A small number of children present with focal epilepsy, while the most frequent seizure types are atypical absences, tonic or atonic drop attacks, or tonic–clonic convulsions (Spalice et al., 2009; Verrotti et al., 2010).

In the bilateral frontal type, the most common symptoms include delayed motor and language milestones, spastic hemiparesis or quadriparesis, and mild to moderate mental retardation (Spalice et al., 2009; Verrotti et al., 2010). In the bilateral frontoparietal form, the clinical presentation is characterized by global developmental delay, esotropia, and pyramidal and cerebellar signs

and seizures, which occur in 94% of patients and are mostly generalized (Verrotti et al., 2010).

Bilateral perisylvian PMG-affected patients can present pseudobulbar palsy with diplegia of the facial, pharyngeal, and masticatory muscles and pyramidal signs and seizures (Verrotti et al., 2010). Infantile spasms may be the presenting seizure type even if seizures develop only before the end or after the first decade (Verrotti et al., 2010). Most patients develop multiple seizure types, and seizure control is poor in more than half of the cases (Verrotti et al., 2010).

Other forms of PMG, such as bilateral parasagittal parietooccipital, bilateral generalized, and unilateral ones, can produce various kinds of seizures and EEG abnormalities (including status epilepticus during sleep), cognitive slowing, motor delay, and cerebral palsy (Verrotti et al., 2010).

31.5.3.2 Genetics

PMG has been associated with mutations of a few genes, including *SRPX2* (sushi-repeat-containing protein, X-linked 2), *PAX6* (paired box 6), *TBR2* (T-box-brain2), *GPR56* (G-protein-coupled receptor 56), *KIAA1279*, *RAB3GAP1* (RAB3 GTPase-activating protein subunit 1), and *COL18A1* (collagen, type XVIII, alpha 1), with all but *SRPX2* found in rare syndromes (Spalice et al., 2009).

The genetic role in the etiopathogenesis of PMG is also supported by its association with Aicardi syndrome, Zellweger syndrome, and WWS or with chromosomal abnormalities such as 22q11 deletion, 1p36 monosomy, and trisomy of chromosome 13 (Pang et al., 2008; Verrotti et al., 2010).

The familial transmission of PMG has been identified in bilateral frontoparietal, bilateral perysylvian, and bilateral generalized forms (Pang et al., 2008; Verrotti et al., 2010). Bilateral frontoparietal PMG seems to be related to the mutation of the gene *GPR56* on chromosome 16q12.2–21 with an autosomic recessive inheritance (Pang et al., 2008; Verrotti et al., 2010). This gene encodes for a G-protein-coupled receptor, a regulator of cell cycle signaling in neuronal progenitor cells at all ages, and plays an essential role in the regional patterning of the human cerebral cortex (Spalice et al., 2009; Verrotti et al., 2010).

Bilateral perysylvian PMG is mainly attributed to different patterns of inheritance, including X-linked dominant, X-linked recessive, autosomal recessive, autosomal dominant with reduced penetrance, autosomal recessive with pseudodominance, and autosomal dominant (Verrotti et al., 2010). A locus for X-linked bilateral perisylvian PMG maps on the distal long arm of the X chromosome (Xq28), but the linkage has not been confirmed and no gene has been identified (Verrotti et al., 2010).

TBR2 is a gene involved in the genesis of PMG, associated with microcephaly and corpus callosum agenesis (Verrotti et al., 2010). It maps to chromosome 3p (Verrotti et al., 2010). This gene encodes a transcription factor, a member of the T-box family, critical in invertebrate embryonic development of the central nervous system and mesoderm (Verrotti et al., 2010). It may also be involved in neuronal division and migration (Verrotti et al., 2010).

31.5.3.3 *Neuroradiological findings*

MRI in PMG demonstrates small irregular gyri and an indistinct gray and white matter junction (Verrotti et al., 2010). The polymicrogyric cortex often appears mildly thickened (6–10 mm) because of cortical overfolding (Verrotti et al., 2010). T2 signals within the cortex are usually normal, although there may be delayed myelination (Verrotti et al., 2010).

31.5.4 Schizencephaly

SCZ is a structural abnormality of the brain, characterized by congenital clefts spanning the cerebral hemisphere from the pial surface to the lateral ventricle and lined by cortical gray matter with communication between the ventricles and extra-axial subarachnoid space (Pang et al., 2008; Spalice et al., 2009; Verrotti et al., 2010). Cleft localization varies widely, but the perisylvian region is more frequently involved (Spalice et al., 2009; Verrotti et al., 2010). The cortex overlying the cleft is often polymicrogyric (Verrotti et al., 2010). Actually, SCZ is classified within the same group as PMG (Pang et al., 2008; Verrotti et al., 2010).

This malformation can be unilateral or bilateral, symmetric or asymmetric, and can be divided into two subtypes: 'closed or fused lips' or type I (if the cleft walls are in apposition) and 'open lips' or type II (if the cleft walls are separated) (Pang et al., 2008; Spalice et al., 2009; Verrotti et al., 2010). Type II SCZ is more common than type I (Barth et al., 2007). SCZ tends to involve the insular, precentral, and postcentral regions (Pang et al., 2008). In addition, SCZ is often associated with septo-optic dysplasia (Spalice et al., 2009). The etiology of this disorder has not been clearly established, and several causes, including genetic, vascular, toxic, metabolic, and infectious factors, may be involved (Spalice et al., 2009; Verrotti et al., 2010).

31.5.4.1 *Clinical characteristics*

Patients with unilateral closed-lipped SCZ generally have mild hemiparesis and seizures but no impairment of normal developmental milestones (Pang et al., 2008; Spalice et al., 2009; Verrotti et al., 2010). When the cleft is open, patients have mild to moderate developmental delay, microcephaly, seizures, spasticity, and hemiparesis (Pang et al., 2008; Verrotti et al., 2010). Patients with bilateral clefts show severe mental deficits and severe motor abnormalities, including spastic quadriparesis (Pang et al., 2008; Spalice et al., 2009; Verrotti et al., 2010). Blindness due to optic nerve hypoplasia can be common (Verrotti et al., 2010). Language development is more likely to be normal in patients with unilateral SCZ compared to patients with bilateral clefts (Spalice et al., 2009; Verrotti et al., 2010). Noncentral nervous system manifestations have also been reported, such as gastroschisis and bowel atresias (Pang et al., 2008).

Several types of seizure have been reported, including generalized tonic–clonic, partial motor, and sensorial (Spalice et al., 2009; Verrotti et al., 2010). Infantile spasms have been described in a few children (Verrotti et al., 2010). Seizures are usually resistant to medical therapy, and stabilization may be achieved through surgery (Verrotti et al., 2010).

31.5.4.2 *Genetics*

EMX2 (empty spiracles homeobox 2) gene mutations may be correlated with type II SCZ (Pang et al., 2008; Spalice et al., 2009; Verrotti et al., 2010). *EMX2* is a homeotic gene expressed in proliferating neuroblasts and is probably involved in controlling cortical migration and structural patterning of the developing rostral brain (Pang et al., 2008; Spalice et al., 2009; Verrotti et al., 2010). However, recent studies criticize the true role of *EMX2* in SCZ (Pang et al., 2008; Spalice et al., 2009; Verrotti et al., 2010).

31.5.4.3 *Neuroradiological findings*

In addition to PMG, MRI may demonstrate agenesis of the septum pellucidum and agenesis or thinning of the corpus callosum, hippocampal malformations, posterior fossa abnormalities, ventricular diverticula and arachnoid cysts, and multiple calcifications (Pang et al., 2008; Verrotti et al., 2010). Periventricular heterotopic nodules have also been found in a minority of cases (Verrotti et al., 2010). The malformations associated with SCZ may also involve extracerebral structures (Verrotti et al., 2010).

References

Allen, J., Chilton, J.K., 2009. The specific targeting of guidance receptors within neurons: Who directs the directors? Developmental Biology 327 (1), 4–11.

Amouri, R., Nehdi, H., et al., 2009. Allelic ROBO3 heterogeneity in Tunisian patients with horizontal gaze palsy with progressive scoliosis. Journal of Molecular Neuroscience 39 (3), 337–341.

Andrews, C.V., Hunter, D.G., et al., 1993. Congenital fibrosis of the extraocular muscles. In: Pagon, R.A., Bird, T.D., Dolan, C.R., Stephens, K. (Eds.), GeneReviews. University of Washington, Seattle, WA.

Avadhani, A., Ilayaraja, V., et al., 2010. Diffusion tensor imaging in horizontal gaze palsy with progressive scoliosis. Magnetic Resonance Imaging 28 (2), 212–216.

Barkovich, A.J., Kuzniecky, R.I., et al., 2005. A developmental and genetic classification for malformations of cortical development. Neurology 65 (12), 1873–1887.

Barkovich, A.J., Millen, K.J., Dobyns, W.B., 2009. A developmental and genetic classification for midbrain–hindbrain malformations. Brain 132 (Pt 12), 3199–3230.

Barth, P.G., Majoie, C.B., et al., 2007. Pontine tegmental cap dysplasia: A novel brain malformation with a defect in axonal guidance. Brain 130 (Pt 9), 2258–2266.

Bashaw, G.J., Klein, R., 2010. Signaling from axon guidance receptors. Cold Spring Harbor Perspectives in Biology 2 (5), a001941.

Bertolin, C., Boaretto, F., et al., 2010. Novel mutations in the L1CAM gene support the complexity of L1 syndrome. Journal of Neurological Sciences 294 (1–2), 124–126.

Bloom, J.S., Hynd, G.W., 2005. The role of the corpus callosum in interhemispheric transfer of information: Excitation or inhibition? Neuropsychology Review 15 (2), 59–71.

Bomfim, R.C., Tavora, D.G., et al., 2009. Horizontal gaze palsy with progressive scoliosis: CT and MR findings. Pediatric Radiology 39 (2), 184–187.

Bonnin, A., Torii, M., et al., 2007. Serotonin modulates the response of embryonic thalamocortical axons to netrin-1. Nature Neuroscience 10 (5), 588–597.

Budisteanu, M., Arghir, A., et al., 2010. Oculocutaneous albinism associated with multiple malformations and psychomotor retardation. Pediatric Dermatology 27 (2), 212–214.

Bush, J.O., Soriano, P., 2009. Ephrin-B1 regulates axon guidance by reverse signaling through a PDZ-dependent mechanism. Genes and Development 23 (13), 1586–1599.

Chiappedi, M., Bejor, M., 2010. Corpus callosum agenesis and rehabilitative treatment. Italian Journal of Pediatrics 36, 64.

Cooper, H.M., 2002. Axon guidance receptors direct growth cone pathfinding: Rivalry at the leading edge. International Journal of Developmental Biology 46 (4), 621–631.

Donahoo, A.L., Richards, L.J., 2009. Understanding the mechanisms of callosal development through the use of transgenic mouse models. Seminars in Pediatric Neurology 16 (3), 127–142.

Engle, E.C., 2010. Human genetic disorders of axon guidance. Cold Spring Harbor Perspectives in Biology 2 (3), a001784.

Fechner, A., Fong, S., McGovern, P., 2008. A review of Kallmann syndrome: Genetics, pathophysiology, and clinical management. Obstetrical and Gynecological Survey 63 (3), 189–194.

Gabay, S., Henik, A., et al., 2010. Ocular motor ability and covert attention in patients with Duane retraction syndrome. Neuropsychologia 48 (10), 3102–3109.

Garbe, D.S., Bashaw, G.J., 2004. Axon guidance at the midline: From mutants to mechanisms. Critical Reviews in Biochemistry and Molecular Biology 39 (5–6), 319–341.

Ghai, S., Fong, K.W., et al., 2006. Prenatal US and MR imaging findings of lissencephaly: Review of fetal cerebral sulcal development. Radiographics 26 (2), 389–405.

Giger, R.J., Hollis II, E.R., Tuszynski, M.H., 2010. Guidance molecules in axon regeneration. Cold Spring Harbor Perspectives in Biology 2 (7), a001867.

Gronskov, K., Ek, J., Brondum-Nielsen, K., 2007. Oculocutaneous albinism. Orphanet Journal of Rare Diseases 2, 43.

Gutowski, N.J., 2000. Duane's syndrome. European Journal of Neurology 7 (2), 145–149.

Hardelin, J.P., Dode, C., 2008. The complex genetics of Kallmann syndrome: KAL1, FGFR1, FGF8, PROKR2, PROK2, et al. Sexual Development 2 (4–5), 181–193.

Heidary, G., Engle, E.C., Hunter, D.G., 2008. Congenital fibrosis of the extraocular muscles. Seminars in Ophthalmology 23 (1), 3–8.

Hirschberg, A., Deng, S., et al., 2010. Gene deletion mutants reveal a role for semaphorin receptors of the plexin-B family in mechanisms underlying corticogenesis. Molecular and Cellular Biology 30 (3), 764–780.

Hong, S.E., Shugart, Y.Y., et al., 2000. Autosomal recessive lissencephaly with cerebellar hypoplasia is associated with human RELN mutations. Nature Genetics 26 (1), 93–96.

Izzi, L., Charron, F., 2011. Midline axon guidance and human genetic disorders. Clinical Genetics 80 (3), 226–234.

Jagla, M., Kruczek, P., Kwinta, P., 2008. Association between X-linked lissencephaly with ambiguous genitalia syndrome and lenticulostriate vasculopathy in neonate. Journal of Clinical Ultrasound 36 (6), 387–390.

Jaglin, X.H., Chelly, J., 2009. Tubulin-related cortical dysgeneses: Microtubule dysfunction underlying neuronal migration defects. Trends in Genetics 25 (12), 555–566.

Kakinuma, N., Kiyama, R., 2009. A major mutation of KIF21A associated with congenital fibrosis of the extraocular muscles type 1 (CFEOM1) enhances translocation of Kank1 to the membrane. Biochemical and Biophysical Research Communications 386 (4), 639–644.

Kaplan, J.D., Bernstein, J.A., et al., 2010. Clues to an early diagnosis of Kallmann syndrome. American Journal of Medical Genetics Part A 152A (11), 2796–2801.

Kaprielian, Z., Runko, E., Imondi, R., 2001. Axon guidance at the midline choice point. Developmental Dynamics 221 (2), 154–181.

Kerjan, G., Gleeson, J.G., 2007. Genetic mechanisms underlying abnormal neuronal migration in classical lissencephaly. Trends in Genetics 23 (12), 623–630.

Khan, A.O., Khalil, D.S., et al., 2010. Germline mosaicism for KIF21A mutation (p.R954L) mimicking recessive inheritance for congenital fibrosis of the extraocular muscles. Ophthalmology 117 (1), 154–158.

Kim, S.H., Hu, Y., et al., 2008. Diversity in fibroblast growth factor receptor 1 regulation: Learning from the investigation of Kallmann syndrome. Journal of Neuroendocrinology 20 (2), 141–163.

Koeberle, P.D., Bahr, M., 2004. Growth and guidance cues for regenerating axons: Where have they gone? Journal of Neurobiology 59 (1), 162–180.

Lin, L., Lesnick, T.G., et al., 2009. Axon guidance and synaptic maintenance: Preclinical markers for neurodegenerative disease and therapeutics. Trends in Neurosciences 32 (3), 142–149.

Macferran, K.M., Buchmann, R.F., et al., 2010. Pontine tegmental cap dysplasia with a 2q13 microdeletion involving the NPHP1 gene: Insights into malformations of the mid-hindbrain. Seminars in Pediatric Neurology 17 (1), 69–74.

Ming, G.L., Song, H.J., et al., 1997. cAMP-dependent growth cone guidance by netrin-1. Neuron 19 (6), 1225–1235.

Miyata, R., Hayashi, M., et al., 2009. Analysis of the hypothalamus in a case of X-linked lissencephaly with abnormal genitalia (XLAG). Brain and Development 31 (6), 456–460.

Mochida, G.H., 2009. Genetics and biology of microcephaly and lissencephaly. Seminars in Pediatric Neurology 16 (3), 120–126.

Natacci, F., Bedeschi, M.F., et al., 2007. Norman-Roberts syndrome: Characterization of the phenotype in early fetal life. Prenatal Diagnosis 27 (6), 568–572.

Nie, D., Di Nardo, A., et al., 2010. Tsc2-Rheb signaling regulates EphA-mediated axon guidance. Nature Neuroscience 13 (2), 163–172.

Niquille, M., Garel, S., et al., 2009. Transient neuronal populations are required to guide callosal axons: A role for semaphorin 3C. PLoS Biology 7 (10), e1000230.

O'Donnell, M., Chance, R.K., Bashaw, G.J., 2009. Axon growth and guidance: Receptor regulation and signal transduction. Annual Review of Neuroscience 32, 383–412.

Okazaki, S., Ohsawa, M., et al., 2008. Aristaless-related homeobox gene disruption leads to abnormal distribution of GABAergic interneurons in human neocortex: Evidence based on a case of X-linked lissencephaly with abnormal genitalia (XLAG). Acta Neuropathologica 116 (4), 453–462.

Otaduy, M.C., Leite Cda, C., et al., 2009. Further diffusion tensor imaging contribution in horizontal gaze palsy and progressive scoliosis. Arquivos de Neuro-Psiquiatria 67 (4), 1054–1056.

Pang, T., Atefy, R., Sheen, V., 2008. Malformations of cortical development. The Neurologist 14 (3), 181–191.

Parisi, M.A., 2009. Clinical and molecular features of Joubert syndrome and related disorders. American Journal of Medical Genetics Part C: Seminars in Medical Genetics 151C (4), 326–340.

Parrinello, S., Noon, L.A., et al., 2008. NF1 loss disrupts Schwann cell-axonal interactions: A novel role for semaphorin 4F. Genes and Development 22 (23), 3335–3348.

Paul, L.K., Brown, W.S., et al., 2007. Agenesis of the corpus callosum: Genetic, developmental and functional aspects of connectivity. Nature Reviews Neuroscience 8 (4), 287–299.

Reiner, O., Sapoznik, S., Sapir, T., 2006. Lissencephaly 1 linking to multiple diseases: Mental retardation, neurodegeneration, schizophrenia, male sterility, and more. Neuromolecular Medicine 8 (4), 547–565.

Renugadevi, K., Sil, A.K., et al., 2010. Spectrum of candidate gene mutations associated with Indian familial oculocutaneous and ocular albinism. Molecular Vision 16, 1514–1524.

Richards, L.J., Plachez, C., Ren, T., 2004. Mechanisms regulating the development of the corpus callosum and its agenesis in mouse and human. Clinical Genetics 66 (4), 276–289.

Schafer, M.K., Nam, Y.C., et al., 2010. L1 syndrome mutations impair neuronal L1 function at different levels by divergent mechanisms. Neurobiology of Disease 40 (1), 222–237.

Schmidt, E.R., Pasterkamp, R.J., van den Berg, L.H., 2009. Axon guidance proteins: Novel therapeutic targets for ALS? Progress in Neurobiology 88 (4), 286–301.

Schrander-Stumpel, C., Vos, Y.J., 1993. L1 syndrome. In: Pagon, R.A., Bird, T.D., Dolan, C.R., Stephens, K. (Eds.), GeneReviews. University of Washington, Seattle, WA.

Severyn, C.J., Shinde, U., Rotwein, P., 2009. Molecular biology, genetics and biochemistry of the repulsive guidance molecule family. Biochemistry Journal 422 (3), 393–403.

Spalice, A., Parisi, P., et al., 2009. Neuronal migration disorders: Clinical, neuroradiologic and genetics aspects. Acta Paediatrica 98 (3), 421–433.

Summers, C.G., 2009. Albinism: Classification, clinical characteristics, and recent findings. Optometry and Vision Science 86 (6), 659–662.

Szczaluba, K., Szymanska, K., et al., 2010. Pontine tegmental cap dysplasia: A hindbrain malformation caused by defective neuronal migration. Neurology 74 (22), 1835.

Tischfield, M.A., Baris, H.N., et al., 2010. Human TUBB3 mutations perturb microtubule dynamics, kinesin interactions, and axon guidance. Cell 140 (1), 74–87.

Torii, M., Levitt, P., 2005. Dissociation of corticothalamic and thalamocortical axon targeting by an EphA7-mediated mechanism. Neuron 48 (4), 563–575.

Torii, M., Hashimoto-Torii, K., et al., 2009. Integration of neuronal clones in the radial cortical columns by EphA and ephrin-A signalling. Nature 461 (7263), 524–528.

Tovar-Moll, F., Moll, J., et al., 2007. Neuroplasticity in human callosal dysgenesis: A diffusion tensor imaging study. Cerebral Cortex 17 (3), 531–541.

Valente, E.M., Brancati, F., Dallapiccola, B., 2008. Genotypes and phenotypes of Joubert syndrome and related disorders. European Journal of Medical Genetics 51 (1), 1–23.

Valiente, M., Marin, O., 2010. Neuronal migration mechanisms in development and disease. Current Opinion in Neurobiology 20 (1), 68–78.

Vallee, R.B., Tsai, J.W., 2006. The cellular roles of the lissencephaly gene LIS1, and what they tell us about brain development. Genes and Development 20 (11), 1384–1393.

Verrotti, A., Spalice, A., et al., 2010. New trends in neuronal migration disorders. European Journal of Paediatric Neurology 14 (1), 1–12.

Vos, Y.J., de Walle, H.E., et al., 2010. Genotype–phenotype correlations in L1 syndrome: A guide for genetic counselling and mutation analysis. Journal of Medical Genetics 47 (3), 169–175.

Wahl, M., Strominger, Z., et al., 2009. Variability of homotopic and heterotopic callosal connectivity in partial agenesis of the corpus callosum: A 3T diffusion tensor imaging and Q-ball tractography study. American Journal of Neuroradiology 30 (2), 282–289.

Yamada, K., Hunter, D.G., et al., 2005. A novel KIF21A mutation in a patient with congenital fibrosis of the extraocular muscles and Marcus Gunn jaw-winking phenomenon. Archives of Ophthalmology 123 (9), 1254–1259.

Yu, L.M., Miller, F.D., Stoichet, M.S., 2010. The use of immobilized neurotrophins to support neuron survival and guide nerve fiber growth in compartmentalized chambers. Biomaterials 31 (27), 6987–6999.

Yuksel, D., Orban de Xivry, J.J., Lefevre, P., 2010. Review of the major findings about Duane retraction syndrome (DRS) leading to an updated form of classification. Vision Research 50 (23), 2334–2347.

Zanin, E., Gambarelli, N., Denis, D., 2010. Distinctive clinical features of bilateral Duane retraction syndrome. Journal of AAPOS 14 (4), 293–297.

Developmental Disabilities, Autism, and Schizophrenia at a Single Locus:
Complex Gene Regulation and Genomic Instability of 15q11–q13 Cause a Range of Neurodevelopmental Disorders

N. Urraca, L.T. Reiter
University of Tennessee Health Science Center, Memphis, TN, USA

Nomenclature

ARHGAP11B Rho GTPase activating protein 11B
ATP10A ATPase, class V, type 10A
C15orf2 Chromosome 15 open reading frame 2
CHRNA7 Cholinergic receptor, nicotinic, alpha 7
CTCF CCCTC-binding factor
CYFIP1 Cytoplasmic FMR1 interacting protein 1
Dube3a Drosophila ube3a
ECT2 Epithelial cell transforming sequence 2 oncogene
GABRA5 Gamma-aminobutyric acid (GABA) A receptor, alpha 5

Gabrb3 Mice GABA A receptor, subunit beta 3 gene
GABRB3 Gamma-aminobutyric acid (GABA) A receptor, beta 3 gene
GABRG3 Gamma-aminobutyric acid (GABA) A receptor, gamma 3
HBII52 Small nucleolar RNA, C/D box 115 cluster
Idic (15) Isodicentric chromosome 15
Int dup (15) Interstitial duplication chromosome 15
KLF13 Kruppel-like factor 13
MAGEL2 MAGE-like 2
MECP2 Methyl CpG binding protein 2
MKRN3 Makorin ring finger protein 3
MTMR10 Myotubularin-related protein 10

Neural Circuit Development and Function in the Brain: Comprehensive Developmental Neuroscience, Volume 3 http://dx.doi.org/10.1016/B978-0-12-397267-5.00017-0

© 2013 Elsevier Inc. All rights reserved.

MTMR15 Myotubularin-related protein 15

NDN Necdin homolog

NIPA1 Nonimprinted gene in Prader–Willi syndrome/Angelman syndrome chromosome region 1

NIPA2 Nonimprinted gene in Prader–Willi syndrome/Angelman syndrome chromosome region 2

OTUD7A OTU domain-containing 7A

PWRN1 Prader–Willi region non-protein-coding RNA 1

SmN Survival of motor neuron 1

snoRNAs Small nucleolar RNAs

SNRPN Small nuclear ribonucleoprotein polypeptide N

SNURF SNRPN upstream reading frame

TCF4 Transcription factor 4

TRPM1 Transient receptor potential cation channel, subfamily M, member 1

TUBGCP5 Tubulin, gamma complex associated protein 5

UBE3A Ubiquitin protein ligase E3A

Ube3a Mice Ube3a

32.1 COMPLEX GENOMIC CHARACTERISTICS AND GENE REGULATION AT THE HUMAN 15q LOCUS

The chromosome 15q locus is an enormously complex region of the human genome responsible for several devastating neuropsychological syndromes as well as a major causative locus in autism spectrum disorder (ASD) (see Chapter 34). Gene regulation in this region is extremely complex and is governed by a variety of molecular changes, including DNA methylation, antisense transcripts, microRNAs, and genomic rearrangements. The purpose of this chapter is to provide a broad introduction to some of the neurological disorders and causative genes in the 15q11–q13 locus.

Chromosome 15 is acrocentric; thus, it essentially has one long coding arm (q) and a short arm (p) containing mostly heterochromatin arrays of ribosomal RNA genes, satellite sequences, and other repetitive DNA sequences. The long q arm contains 2.9% of the coding genes in the human genome (Zody et al., 2006). A key structural feature of chromosome 15 is the high number of low-copy segmental duplications, with 8.8% of its euchromatin being inundated with low-copy repeats (LCRs) (Zody et al., 2006). These LCRs predispose genes in the region to dosage changes due to both deletion and duplication events mediated by nonallelic homologous recombination (NAHR) events and result in a spectrum of neurodevelopmental disorders (mechanism reviewed in Shaw and Lupski, 2004). This locus is complicated even further by a 2-Mb cluster domain of genes preferentially expressed from one parental allele or imprinted genes.

Imprinted genes are epigenetically marked in gametes to drive expression of that gene from a single parental allele in the offspring. An epigenetic mark (or imprint) is set depending on the gene and on the germ cell. In fertilization, somatic cells of the developing embryo contain both maternally and paternally inherited alleles; this epigenetic mark ensures that just one allele is expressed (Weaver et al., 2009). In the embryonic germ cells, the original imprint is erased and a new one is added depending on the sex and the gene. This parent-specific expression is directly due to these epigenetic modifications of the DNA (DNA methylation, histone modifications, and noncoding RNAs) and not due to changes in the primary sequence, thus altering the conformation of chromatin fibers and therefore the regulation of the expression of the nearby genes. Although these modifications can have dramatic effects on the phenotype, they do not alter the primary DNA sequence. The monoallelic expression of an imprinted gene may occur only in one of possibly several isoforms, or in particular tissues, or at particular stages of development. The untranslated mRNA H19 was the first gene shown to be imprinted in humans (Zhang and Tycko, 1992); as many as 40 such genes have now been identified, and another 156 new genes have been predicted by both computational and experimental approaches to be regulated in a parent-specific manner through epigenetic modification (Luedi et al., 2007).

The 15q11–q13 region contains several genes that show parent-of-origin-specific expression and others that show biallelic expression in the brain. Genes expressed exclusively from the paternal chromosome in the brain are *MKRN3*, *MAGEL2*, *NDN*, *SNRPN*, and a cluster of *snoRNAs*. *C15orf2* and *PWRN1* are genes that have shown monoallelic expression in human fetal brain (Buiting et al., 2007). Two genes show maternal-specific expression in the brain: *UBE3A*, an E3 ubiquitin ligase responsible for Angelman syndrome (AS) phenotypes that shows biallelic expression in most tissues except for a preferential expression of the maternal allele in human brains, and *ATP10A*, which also is preferentially expressed in the maternal chromosome in human brain as well as in fibroblasts (Herzing et al., 2001; Meguro et al., 2001). Recently it has been suggested that the *ATP10A* may be monoallelic in expression, dependent on sex and common genetic variation and not regulated by an imprinting mechanism (Hogart et al., 2008).

Recently, a large region between *Snrpn* and *Ndn* was found to be imprinted in the mouse brain, with several paternally expressed ncRNAs and a paternally expressed gene called *DOKist4* (Gregg et al., 2010). Although the phenotypic outcome of these additional paternally expressed genes and ncRNAs is still unknown, it clearly speaks to the complexity of this locus in terms of genotype to phenotype correlations.

This region is one of the most unstable regions in the human genome (Knoll et al., 1993) because of a number of complex LCRs that predispose the region to misalignment during meiosis I, which leads to unequal NAHR events involving both sister chromatid and interchromosomal exchanges (Lupski, 1998; Robinson et al., 1998).

In addition, there is a high rate of recombination in women in this locus (Robinson and Lalande, 1995). 15q11–q13 homologous pairing is an epigenetic regulatory mechanism disrupted by genomic rearrangements such as duplications and deletions. CCCTC-binding factor (*CTCF*) controls the process of X chromosome pairing as well as pairing at the 15q11–q13 locus (Meguro-Horike et al., 2011). At least five recurrent breakpoints (BPs) have been identified containing LCRs involved in microdeletions, microduplications, inversions, as well as isodicentric chromosomes (see Figure 32.1).

Severe phenotypic consequences at the 15q locus can be caused by deletions, duplications, mutations, or defects on a single active allele, be it maternal or paternal. In rare instances, two copies of chromosome 15 can be inherited from a single parent (uniparental disomy), causing the same severe phenotypic consequences as loss of function for one 15q allele. The two primary genomic disorders that cause neuropathology in this region are the Prader–Willi syndrome (PWS) and Angelman syndrome (AS), whose phenotypes result from loss of the paternal or maternal contribution of the 15q11–q13 genomic region, respectively. The first nine individuals with PWS were described in 1956 by the endocrinologists Prader, Labhart, and Willi (Clarke and Boer, 1995). By 1976, it

was apparent that a recurrent microdeletion could be visualized by high-resolution chromosome analysis in patients with PWS and AS (Ledbetter et al., 1981). In 1989, it was established that AS is a direct result of maternal deletion, whereas PWS results from a paternal deletion of the same region on chromosome 15 (Knoll et al., 1989). This 2-Mb domain is often referred to as the PWS/AS critical region. In 95% of PWS/AS patients, there are two main types of deletions: class I deletions, with breakpoints at BP1 (proximal) to BP3 (distal), and class II deletions, with breakpoints from BP2 (proximal) to BP3 (distal) (see Figure 32.1). The remaining 5% of PWS/AS patients have the distal breakpoint at BP4. Patients with smaller deletions from BP1 to BP2 have been identified, but these individuals do not present with classic PWS or AS and may represent a distinct neurological syndrome unrelated to the primary causative genes for PWS and AS (Doornbos et al., 2009). As with most genomic disorders, these LCRs mediate not only deletion events, but also reciprocal duplication events. The same LCRs and therefore the same breakpoints in addition to BP5 can mediate inverted dup(15) marker chromosomes and some cases of interstitial duplications and triplications of chromosome 15q11–q13 (Roberts et al., 2002). The clinical presentation associated with deletions tends

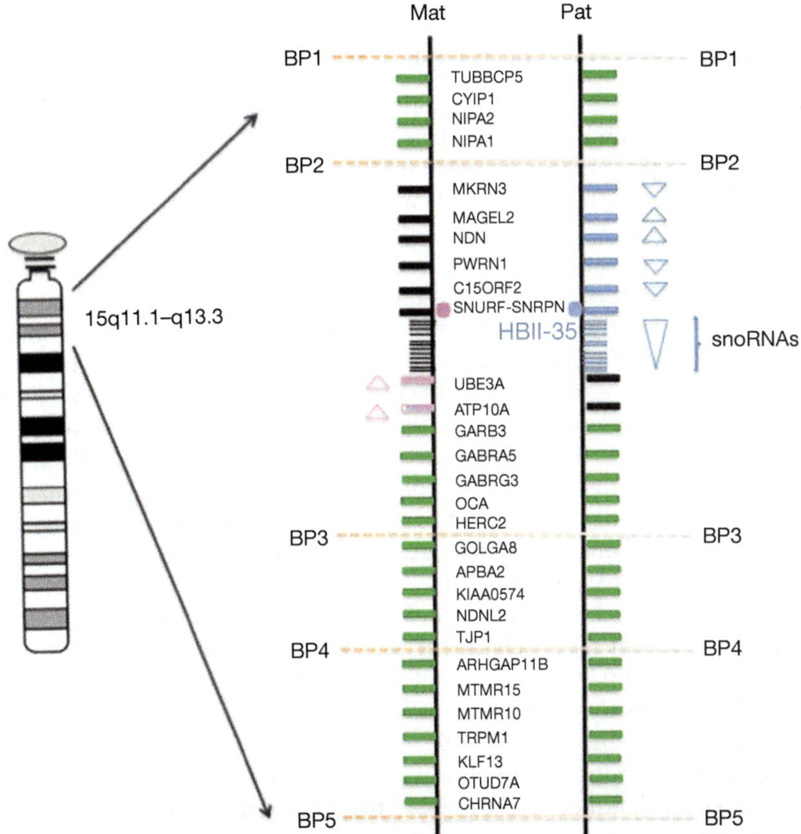

FIGURE 32.1 15q11.1–q13.3 region. Maternal and paternal alleles are shown separately. Maternally expressed genes are in pink, paternally expressed in *blue*, biallelically expressed genes in green, and the genes silenced on a particular allele are in *black*. *ATP10A* varies among populations, so its pink/blue. Dots pink and blue refer to AC-IC and PW-IC, respectively. The five different common breakpoints for both duplications and deletions in this region appear as dotted lines.

to be more uniform, and commonly includes dysmorphology, which facilitates diagnosis, while duplications present with a more subtle phenotype that often includes autism when maternally inherited, but may also include anxiety and other neurological effects in paternally inherited duplications (Hogart et al., 2010). The widespread use of array comparative genomic hybridization (array CGH) has increased the frequency of detection of submicroscopic interstitial 15q duplications, as well as smaller, possibly pathogenic deletions in this region. Smaller deletions and duplications found in neurodevelopmental disorders such as schizophrenia and autism will be described later in this chapter.

32.2 EPIGENETIC CONTROL OF GENE EXPRESSION ACROSS 15q11–q13

Imprinted gene expression on 15q is coordinately controlled in *cis* by an imprinting center (IC), a genetic element functional in germline and/or early postzygotic development that regulates the establishment of parental specific allelic differences in replication timing, DNA methylation, and chromatin structure. In other words, the IC's function is to reset the maternal and paternal imprints. A mutation at the IC sets the imprint permanently, and the parental imprint will be fixed on the chromosome on which the mutation arose, resulting in a heritable form of the imprinting disorder (Saitoh et al., 1996). The IC regulates imprinting in the whole region and is located in the 5′ untranslated region of the small nuclear ribonucleoprotein polypeptide N (*SNRPN*) gene. The PWS/AS IC has two functional components: a 4.3-kb DNA segment at the promoter of *SNRPN* involved in PWS (PWS-IC) and a 0.9-kb element ∼35 kb upstream of *SNRPN* which regulates the gene responsible for AS (AS-IC) (Buiting et al., 1995). AS-IC acquires a primary imprint during gametogenesis, thus establishing the maternal epigenotype. The inactive maternal IC is marked by CpG methylation and by histone H3 Lys 9 di-methylation, while the active paternal IC is marked by histone H3 Lys4 methylation (Soejima and Wagstaff, 2005). The unmethylated maternal allele of the AS-IC most likely binds a *trans*-acting factor that confers methylation on the PWS-IC maternal allele after fertilization. It is assumed that, once established, the PWS-IC paternal epigenotype spreads across the entire PWS/AS domain in the soma.

32.2.1 Epigenetic Control Across the 15q Region

Histone acetylation and chromatin structure are influenced by DNA methylation, resulting in the production of compacted chromatin that is refractory to transcription. In the PWS/AS region, *MKRN3*, *NDN*, and *SNRPN* are silenced by maternal-specific CpG methylation of the promoter/exon I region (Driscoll et al., 1992; Glenn et al., 1996; Jay et al., 1997).

The *SNRPN* gene is a bicistronic-imprinted gene that encodes two polypeptides, the SNURF (SNRPN upstream reading frame) and the SmN (splicing factor involved in RNA processing) (Gray et al., 1999). This gene has 10 exons: exons 1–3 encode the SNURF protein, and exons 4–10 encode the SmN. This SNRP–SmN transcript unit also hosts the genes to encode multiple noncoding RNAs (snoRNAs), and it serves as the start site for the *UBE3A* antisense transcript (Runte et al., 2001). Because the CpG island of the promoter is hypermethylated on the maternal chromosome and hypomethylated on the paternal chromosome, the gene is transcribed exclusively from the parentally inherited chromosome and shows a high level of expression in the brain and the heart. Different methods have taken advantage of the differentially methylated *SNRPN* promoter for the diagnosis of PWS and AS. Our group recently utilized this differentially methylated region to develop a high-resolution melting curve assay to determine the parent of origin of the duplicated chromosome in individuals with 15q duplication (Urraca et al., 2010).

32.3 GENES KNOWN TO CAUSE DISEASE IN THE REGION

32.3.1 The E3 Ubiquitin Ligase Gene (*UBE3A*)

The *UBE3A* gene product is the canonical HECT domain E3 ubiquitin ligase enzyme known as E6-associated protein (E6-AP). Although originally identified as a host protein that interacts with viral E6 protein during papillomavirus infection (Huibregtse et al., 1993), the primary role of this enzyme is thought to be the monoubiquitination of protein substrates that are destined for recycling by the ubiquitin proteasome system or are tagged for trafficking to various cellular compartments. In addition to this enzymatic function, E6-AP can also function as a transcriptional co-activator of steroid hormone receptor genes (Nawaz et al., 1999). Maternal deficiency for *UBE3A* results in the neurodevelopmental disorder Angelman syndrome (Kishino et al., 1997). This gene exhibits maternal allele-specific expression in most regions of the brain in mice and humans, with the strongest expression restricted to hippocampal and cerebellar neurons (Albrecht et al., 1997; Dindot et al., 2008). The discovery of maternally inherited loss-of-function mutations in *UBE3A* has clearly established *UBE3A* as the causative gene in AS (Kishino et al., 1997; Matsuura et al., 1997). Although a handful of putative UBE3A substrates have now been identified, most notably the Arc protein which regulates synaptic AMPA receptors (Greer et al., 2010) and ECT2 protein which is a key regulator of the actin cytoskeleton (Reiter et al., 2006), it is still unclear if the key proteins responsible for the AS phenotype are regulated by UBE3A at the protein or transcriptional level, or both.

Cytogenetic abnormalities in the PWS/AS region have been described in 1–3% of autistic individuals. Consistent with a role for *UBE3A* in autism are studies that have demonstrated copy number variants not only at the *UBE3A* locus, but also at other loci regulating the ubiquitin–proteasome pathway in autistic individuals (Glessner et al., 2009). The *UBE3A* gene itself has been proposed as a candidate gene for autism susceptibility, although linkage and association studies have been somewhat inconsistent (Cook et al., 1998; Guffanti et al., 2010; Nurmi et al., 2001).

32.3.2 SnoRNA Cluster

Small nucleolar RNAs (snoRNAs) are noncoding RNAs located in the nucleolus which are involved in rRNA modifications. Pre-rRNA maturation includes endonucleolytic and exonucleolytic cleavages plus modifications such as methylation or pseudouridylation. There are two main classes of snoRNA: the C/D box snoRNAs, which are associated with methylation, and the H/ACA box snoRNAs, which are associated with pseudouridylation (Bachellerie et al., 2002). There is a cluster of C/D box snoRNA genes encoded in the 15q11–q13 region that are processed from introns of the paternally expressed *SNURF-SNRPN* sense *UBE3A* antisense transcript (see Figure 32.1). *HBII52* has 47 copies, *HBII85* has 24 and the others genes (*HBII-36, HBII-13, HBII-437, HBII-238A,* and *HBII-438B*) one copy each. These snoRNAS are highly expressed in the brain (Cavaille et al., 2000). Recent studies demonstrate that chromatin de-condensation of MBII/HBII snoRNAs plays a significant role in the regulation of nucleolar size during neuronal maturation (Leung et al., 2009). Reports on balanced translocations affecting the paternal copy of 15q11–q13 and three microdeletions in the snoRNA region (15q11.2) have now implicated the *HBII-85* snoRNA as the key paternally expressed causative gene(s) for the PWS phenotype (De Smith et al., 2009; Duker et al., 2010; Gallagher et al., 2002; Sahoo et al., 2008).

32.3.3 Gamma-Aminobutyric Acid Genes

The 15q11–q13 region contains a cluster of GABA$_A$ receptor subunits genes: *GABRB3, GABRA5,* and *GABRG3,* which encode for the gamma-aminobutyric acid (GABA) receptor subunits β3, α5, and γ3, respectively. During development, GABA$_A$ receptors play a role in proliferation, migration, and differentiation of precursor cells that direct the development of the embryonic brain (Owens and Kriegstein, 2002). GABAergic dysfunction in the brains

of autistic individuals has been reported and particular attention has been devoted to the GABRB3 subunit (Pizzarelli and Cherubini, 2011). It has been observed that this gene is normally biallelically expressed in the brain, but in autistic individuals monoallelic expression has been found in combination with a reduction in GABA receptor protein (Hogart et al., 2007). Additional studies indicate that *GABRB3* expression is reduced in the parietal cortex cerebellum of subjects with autism (Fatemi et al., 2009). It has been shown that a rare coding variant of the GABRB3 gene is associated with autism when transmitted maternally (Delahanty et al., 2011). This same variant was identified in two independent families segregating with childhood absence epilepsy (Tanaka et al., 2008).

32.3.4 CYFIP1 and Other Autism-Related Genes

More severe behavioral problems (autism, ADHD, and obsessive–compulsive disorder) appear in patients with type I deletions (BP1–BP3) than in patients with type II deletions (BP2–BP3); this could be influenced by the genes between BP1 and BP2: *TUBGCP5, NIPA1, NIPA2,* and *CYFIP1*. The latter three of these are widely expressed in the central nervous system, while TUBGCP5 is expressed in the subthalamic nuclei. Patients with a microdeletion at 15q11.2 between BP1 and BP2, which includes the four genes, present delayed motor and speech development, dysmorphisms, and behavioral problems (Doornbos et al., 2009). It seems that the haploinsufficiency of *NIPA1* does not cause any disease, as PWS/AS patients do not exhibit progressive spastic paraplegia compared to autosomal dominant hereditary spastic paraplegia. In terms of autism, more attention has been focused on *CYFIP1* since Prader–Willi phenotypes have been observed in fragile X syndrome (FXS) patients (see Chapter 33), and it was found that the fragile X associated protein (FMRP) acts in concert with CYFIP1 protein to regulate mRNA translation in neurons (Napoli et al., 2008). In patients with this FXS sub-phenotype that show PWS overlap, *CYFIP1* mRNA levels are generally reduced by two- to fourfold as compared to normal controls or individuals with FXS without PWS features (Nowicki et al., 2007).

32.4 CONTIGUOUS GENE DELETION/DUPLICATION SYNDROMES ON 15q

32.4.1 Deletion Syndromes

32.4.1.1 PWS (OMIM #176270)

PWS is a contiguous gene syndrome caused by the loss of function in those genes situated within the 15q11–q13 region. The syndrome is the direct result of loss of a paternally expressed gene or genes in the 15q region. The molecular events resulting in PWS include interstitial deletions (70%), uniparental disomy (UPD) (25%), imprinting center defects (<5%), and, on rare occasions, chromosomal translocations (<1%).

The syndrome has a prevalence of 1/15,000–1/30,000, and the main clinical features include an initial failure to thrive, followed later in life by an obsession with food and other behavioral problems, including intellectual disability (for a detailed review of phenotypic features and management, see Cassidy and Driscoll, 2009; Goldstone et al., 2008, respectively). There are also significant phenotypic differences between patients with 15q deletions as opposed to maternal UPD. In UPD patients, the facial features and the hypopigmentation are less frequent than in patients with the deletion; in contrast, sleep disorders and behavioral abnormalities such as psychosis and autism are more common (Cassidy, 1997). Seizures are present in half of the individuals with a deletion and in less than 10% in UPD cases (Varela et al., 2005). Individuals with class I deletions have a more severe phenotype than those with class II deletions, including self-injurious behavior, deficits in adaptive behavior, obsessive–compulsive behavior, and difficulties in visual-motor integration (Butler et al., 2004). Cardiovascular and respiratory disorders related to the obesity that results from obsessive food compulsion are the most frequent causes of death in these patients. In 1993, a group of clinicians established the clinical diagnostic criteria for PWS (Holm et al., 1993). With the availability now of molecular laboratory testing, when there are clinical findings pointing to the disorder the revised criteria recommends to test for PWS at different ages depending on the phenotypes observed (Gunay-Aygun et al., 2001). The most widely used laboratory test is based on the paternal differences in DNA methylation from the 5′ end of the *SNRPN* gene. PWS deletion patients have only a maternal methylated allele. Since the phenotypic outcomes can be somewhat variable depending on the size and character of a genomic defect, it is important to determine the molecular class of the loss of function for genetic counseling purposes. Chromosomal analysis can identify large deletions and translocations, *SNRPN* fluorescence *in situ* hybridization (FISH) probes detect deletions, and DNA polymorphisms from the patient and parents are useful in the diagnosis of UPD cases. MS-HRM has also been shown to be an efficient and cost-effective method for the identification of PWS or AS deletions (Hung et al., 2011; White et al., 2007).

32.4.1.2 *Angelman syndrome (OMIM 105830)*

Angelman syndrome (AS) was first recognized in 1965 by physician Harry Angelman (pronounced as if a "male angel"), who described three unrelated subjects all presenting with a similar curious phenotype, including inappropriate laughter, happy demeanor, and a "puppet-like" dangling of the arms (Angelman, 1965). The estimated prevalence of AS is between 1:1000 and 1:20,000 (Williams, 2005). AS is a neurogenetic disorder characterized by severe intellectual disability, ataxia, seizures, EEG abnormalities, bouts of inappropriate laughter, absent speech, autistic-like behavior, sleep disorder and postnatal microcephaly, macrostomia, maxillary hypoplasia, and prognathia (Williams et al., 2006). AS is difficult to detect in infancy, as hypotonia and developmental delay are fairly nonspecific features that can occur in infancy for a number of reasons. The characteristic phenotypes become clearer after age one. Seizure onset is often between 1 and 3 years. Many seizure types have been reported, but atypical absence and myoclonic seizures are most frequent. The most common EEG finding is the presence of triphasic delta activity with a maximum over the formal regions (Conant et al., 2009). The Angelman phenotype becomes less striking in adulthood, the sleep problems improve, but most patients continue to have seizures (for further review and diagnostic approaches, see Van Buggenhout and Fryns, 2009). It has been proposed that, to rule out AS, *SNRPN* methylation assay tests should be performed on any patient with developmental delay, severe intellectual disability, speech impairment, and happy disposition (Varela et al., 2004).

The molecular causes of AS include interstitial deletions of the maternal chromosome (70%), paternal UPD in 3–5% of the cases, imprinting center mutations in 10%, and maternally inherited *UBE3A* mutations in 5–10%, leading to the absence of the gene product. Of the ~10% of AS patients who do not have chromosome 15q defects, it appears that at least some of these individuals may have mutations in potential UBE3A pathway genes such as *MECP2* and the transcription factor *TCF4* (Takano et al., 2010; Watson et al., 2001). Maternally derived deletions and *UBE3A* mutations result in a more severe phenotype than uniparental disomy or imprinting defects, both of which still retain two intact copies of *UBE3A*. These maternal loss-of-function individuals often have severe microcephaly, greater delay in developmental milestones, more severely impaired communication skills, and more severe seizures, in addition to hypopigmentation in the deletion cases (Lossie et al., 2001). The nonimprinted gene responsible for this pigmentation loss is the *OCA* gene, which also causes type II oculocutaneous albinism and is located in the distal portion of 15q11–q13 within the common class I and class II deletion boundaries (Fridman et al., 2003). Comparing the main deletion classes, there are no major phenotypic differences, except for the absence of speech in class I patients (Varela et al., 2004) and perhaps an increased risk for ASD in class I individuals (Peters et al., 2004).

32.4.2 Duplications Syndromes

As many as 3–5% of individuals with broad ASD have copy number changes at the 15q locus (Hogart et al., 2010). Two types of 15q duplications have been identified that result in an ASD phenotype. These duplications can occur as either simple interstitial duplications containing fewer than 18 genes or more complex extrachromosomal isodicentric duplications containing >60 genes.

32.4.3 Isodicentric Chromosome 15q Duplications

A marker chromosome is a structurally abnormal, unidentified extra piece of chromosomal material. Marker chromosomes usually occur in addition to the normal chromosome complement and are thus also referred to as supernumerary chromosomes. When a marker chromosome occurs in the PWS/AS region, the isodicentric chromosome 15 [idic(15)], it is formed by the inverted duplication of proximal 15q resulting in an additional small di-centromeric chromosome detectable by cytogenetic and FISH analysis (Battaglia, 2005). This [idic(15)] is one of the most common supernumerary marker chromosomes in humans (Webb, 1994). Idic (15) marker chromosomes that do not include the PWS/AS critical region have no obvious clinical effect (Huang et al., 1997), whereas most of those including this region are maternally derived, arise *de novo*, and lead to a neurobehavioral phenotype (Battaglia, 2008). Most idic (15) chromosomes arise through BP3:BP3 or BP4:BP5 recombination events (see Figure 32.1). The clinical features of idic (15) syndrome include moderate to profound developmental delay/intellectual disability, autism or autistic-like behavior, central hypotonia, minor dysmorphisms, and seizures (Battaglia, 2008). Tetrasomy of this region is associated with a more severe phenotype than trisomy, suggesting that there is a dosage effect for a gene or genes duplicated in this region. In particular, there is an apparent correlation between the frequency and severity of seizures in these individuals and the increase in copy number, clearly implicating the duplicated GABA receptor gene cluster as a contributor to the increased seizure risk and the difficult seizure management.

32.4.4 Interstitial 15q11–q13 Duplication

This submicroscopic duplication is almost impossible to detect by traditional karyotype analysis; however, with the widespread use of array CGH in the clinic, it is now becoming a frequent cause of ASD where chromosomal rearrangements are involved. Most int dup (15) patients with an interstitial duplication 15q11–q13 share the common deletion breakpoints of PWS/AS and are the result of the reciprocal NAHR event that forms the common PWS/AS deletions. In most cases, the duplication is *de novo*, but there are a few familial cases, which have been critical to understanding the difference in phenotype associated with maternal versus paternal inheritance of the duplication. There is only one report of an interstitial duplication that expanded during meiosis when transmitted from a carrier mother to her son; otherwise, the duplication is stable (Gurrieri et al., 1999). There is a wide range of severity in the developmental disabilities; however, most individuals with this smaller duplication present with a mild, somewhat sociable form of autism combined with moderate-to-severe anxiety disorder. The severity of the phenotype does not show an obvious correlation between the duplicated region sizes. There has been one report of an adult with a maternally derived int dup (15) who had a severe phenotype characterized by intractable epilepsy (Orrico et al., 2009). However, in the authors' own cohort of 15 interstitial duplication 15q subjects, just two individuals present with severe seizures (manuscript in review).

Like the deletions, there is also a parent-of-origin effect for the duplications containing the PWS/AS critical region. Maternally transmitted 15q duplication is most frequently detected, because it is consistently associated with autistic features with variable degrees of developmental delay. Unfortunately, genotype–phenotype correlations have not consistently demonstrated that these duplications are the sole cause of ASD in these cases. Boyar et al. reviewed the literature and described the phenotype in 31 confirmed maternal cases, with intellectual disability, developmental delay, learning disabilities, language impairment, and autism/autistic-like behavior as the most common finding, followed by hypotonia, repetitive behavior, hyperactivity, poor attention, and clumsiness. Less frequent are seizures, echolalia, aggression, and hypopigmentation (Boyar et al., 2001).

On the other hand, very few cases of paternally transmitted 15q duplication have been reported, and most of these individuals appear neurotypical, as one can see from familial cases in which unaffected mothers have transmitted paternally derived duplication chromosomes to their affected children (Cook et al., 1997; Roberts et al., 2002). However, there are a few paternal cases reported with speech delay and behavior problems. Mohandas et al. reported a patient with nonspecific developmental delay and partial agenesis of rostral corpus callosum; Mao et al. described a patient with global developmental delay, depression, obesity, food-seeking behavior, and self-injurious tendencies; and Roberts et al. presented a series of 16 cases in which only one patient had paternal duplication, and his phenotype included developmental delay and behavioral disorder (Mao et al., 2000; Mohandas et al., 1999; Roberts et al., 2002).

The maternal bias of duplication-associated risk in autism suggests the requirement of a maternally expressed transcript for greatest ASD risk, clearly implicating the maternally expressed *UBE3A* gene in the ASD phenotype in 15q duplication individuals.

32.5 COMPLEX DISEASES

32.5.1 Microdeletion 15q13.3

The use of array CGH in patients with developmental delay, intellectual disability, and/or dysmorphic features has resulted in the recognition of a recurrent 15q13.3 microdeletion (Ben-Shachar et al., 2009; Sharp et al., 2008). The critical deletion spans 1.5 Mb and arises through NAHR between LCR sequences on BP4 and BP5, telomeric to PWS/AS region.

The first description of recurrent 15q13.3 microdeletion was by Sharp et al., who screened for idiopathic intellectual disability using whole-genome array CGH or quantitative polymerase reaction (PCR) and found the microdeletion with a frequency of 1/350. Interestingly, in their series 7 out of 9 had epilepsy and/or abnormal EEG findings (Sharp et al., 2008). Incomplete penetrance (65–70%) has been reported and some relatives of 15q13.3 microdeletion individuals were reported as neurotypical, while others had intellectual disability, epilepsy, and/or neuropsychiatric disorders (Ben-Shachar et al., 2009; Shinawi et al., 2009).

Based on the findings of Sharp et al., a case-control study was conducted in patients with idiopathic generalized epilepsy, and there were 12 cases with the 15q13.3 microdeletion (12/1223) and none in the controls (0/3699). However, in these 12 cases, 9 cases did not have dysmorphic features or intellectual disability. Dibbens et al. proposed the segregation of the microdeletion as a susceptibility variant, because they analyzed families and they found that the microdeletion did not account for all the epilepsy risk in their families (i.e., there were multiple affected individuals without the microdeletion as well as unaffected individuals with the deletion (Dibbens et al., 2009)). Finally, there is a report of a patient carrying a homozygous microdeletion, inherited from both parents, with severe epileptic encephalopathy, retinopathy, autistic features, and choreoathetosis (Masurel-Paulet et al., 2010).

The chromosomal region for the 15q13.3 microdeletion encompasses seven genes: *ARHGAP11B, MTMR15, MTMR10, TRPM1, KLF13, OTUD7A,* and *CHRNA7* (see Figure 32.1). The PWS-IC seems to regulate *CHRNA7* levels with MECP2 as the link between epigenetic mark and transcriptional regulation (Yasui et al., 2011). The *CHRNA7* gene is highly expressed in the brain and is considered a good candidate gene, at least for the

epilepsy phenotypes in 15q13.3 deletion individuals. *CHRNA7* encodes a synaptic ion channel protein that mediates neuronal signal transmission, and mutations in other members of the nicotine receptor subunit gene family cause the autosomal dominant nocturnal frontal lobe epilepsy.

A smaller deletion ($\sim$680 kb) that includes the *CHRNA7* gene has been identified with a frequency of 1 in 2960, but the phenotype is quite similar to the larger deletion of 1.5 Mb (Shinawi et al., 2009). In some ethnicities, the BP4–BP5 region can be inverted (Sharp et al., 2008), which predisposes this region to the smaller deletion by NAHR between CHRNA7-LCR copies on the normal and inverted chromosomes (Shinawi et al., 2009).

32.5.2 Schizophrenia and Behavioral Abnormalities

Schizophrenia is a severe mental disorder characterized by hallucinations, delusions, cognitive deficits, and apathy. Like other common diseases, it is a complex genetic disorder with high heritability (Lichtenstein et al., 2009). There is no doubt that genetic factors play a role in the pathogenesis of the disorder, but there is no single major gene that confers the risk for the whole phenotype. It is likely that common alleles of small effect and some rare alleles with relatively large effects are combining to increase susceptibility to the development of schizophrenia (Wang et al., 2005). Nowadays, it is known that there are copy number variations (microdeletions/microduplications) that contribute to the etiology of schizophrenia. Genome-wide scans for structural variants have identified deletions on chromosomes 1q21.1, 3q29, 15q11.2, 15q13.3, and 22q11.2 and duplications on chromosomes 16p11.2 and 16p13.1 that increase the risk of schizophrenia (Consortium, 2008; Ingason et al., 2011; Mccarthy et al., 2009; Mulle et al., 2010; Stefansson et al., 2008). Maternal 15q11–q13 duplication has also been found to be a risk factor for schizophrenia, with an odds ratio of 7.3 (Ingason et al., 2011).

Interestingly, a genome-wide linkage analysis showed that an endophenotype found in schizophrenia, P50 auditory gating deficit, is linked (LOD score = 5.3) to chromosome 15q13–q14, the locus for the *CHRNA7* gene (Freedman et al., 1997). This gene is considered a candidate gene for schizophrenia based on positive association studies (Freedman et al., 2001; Stephens et al., 2009).

32.5.3 Microduplication 15q13.3

Recently, a few individuals with the reciprocal duplication between BP4 and BP5 have been reported. This duplication originates by NAHR between BP4 and BP5 LCRs and presents with no recognizable phenotype as in the microdeletion cases with developmental delay/intellectual disability and psychiatric disease, but not epilepsy. There is also inter- and intrafamilial variability suggesting incomplete penetrance, since other family members with the duplication appear clinically neurotypical (Van Bon et al., 2009). Interestingly, a smaller duplication (358–680 kb) that includes the *CHRNA7* gene has been reported with the same clinical outcomes as the larger BP4–BP5 duplication. These cases are all inherited and some family members also present with a psychiatric disease that can be due to a very low penetrance or could be explained as a risk factor for alcoholism and psychiatric or behavioral disorders (Szafranski et al., 2010). Thus, the only reports of this smaller duplication differ in the duplication size, but all include the *CHRNA7* gene (Szafranski et al., 2010). Therefore, studies to date show a remarkably variable expressivity for this 1.5-Mb deletion/duplication on chromosome 15q13.3, which includes neurotypical phenotype, intellectual disability, autism, seizures, bipolar disorder, and schizophrenia.

32.6 ANIMAL MODELS OF 15q REGION DISORDERS

Perhaps no other single region of the human genome has been the inspiration for such a variety of transgenic mouse models as the 15q region. One clear advantage of studying the 15q region in mice is that the entire 15q11.2–q13 locus from BP1–BP5 is syntenic to the mouse chromosome 7qC region, making it feasible to study not only single-gene disorders, but also the effects of multiple-gene deletions and duplications in the genomic disorders that occur at 15q in humans.

Several mouse models have now been produced for the single-gene imprinting disorder AS through both knock-out and knock-in strategies (Jiang et al., 1998; Miura et al., 2002). As in the human conditions, these mice only exhibit phenotypes when the *Ube3a* deficiency is passed though the maternal germline. These models have provided a system for the study of ataxia, motor control, and long-term potentiation (LTP) defects observed in AS patients. They have also been used to identify novel phenotypes due to loss of *Ube3a* related to cerebellar controlled lick rhythm (Heck et al., 2008) and experience dependent maturation of cortical neurons of the visual cortex (Kashiro et al., 2009; Yashiro et al., 2009). In addition, the recent construction of a YFP–*Ube3a* fusion mouse has revealed not only that *Ube3a* is maternally expressed in most regions of the brain, but also that the paternally inherited allele is not completely silent, as previously thought (Dindot et al., 2008). In addition, GFAP-positive neuronal precursors in the vermus are not subject to imprinted regulation and show bi-allelic expression of *Ube3a* (Dindot et al., 2008). Mice with a maternal deletion from *Ube3a* to *Gabrb3* had increased seizure activity, altered ultrasonic vocalization, and impairment in learning and memory tasks (Jiang et al., 2010).

Two recently developed mouse models have revealed details about the function of the PWS/AS imprinting center as well as the molecular mechanism of PWS phenotypes. HBII-85 snoRNAs sequences are highly conserved between humans and mice. Mice lacking the MBII-85 snoRNA ortholog show postnatal growth retardation, delayed sexual maturation, motor learning deficit, and hyperphagia, as well as other features seen in PWS patients (Ding et al., 2008). The snoRNAs at this locus and a neuron-specific long antisense UBE3A transcript (Rougeulle et al., 1998) that initiates at the *SNRPN* promoter (see Figure 32.1) are both involved in the complex orchestration of allele-specific repression that results in the *UBE3A* gene being expressed from the maternal allele and the MBII-85 snoRNA from the paternal allele. The development of a mouse with a 35-kb targeted deletion of the PWS-IC has greatly accelerated the understanding of the epigenetic and transcriptional changes that must take place to cause allele-specific expression (Yang et al., 1998). This mouse model of PWS accurately recapitulates neonatal phenotypes including feeding difficulties, failure to thrive, and small size. However, it has proven difficult to utilize this model for behavioral studies in adult mice due to its early lethality in a C57BL/6 genetic background. Chamberlain et al. were able to solve this dilemma by moving the PWS-IC^{del} mutation to a variety of genetic backgrounds until they were able to rescue this lethality, which appears to be unrelated to the mutation at the PWS-IC (Chamberlain et al., 2004). This adult viable PWS-IC^{del} mouse was used to identify cognitive abnormalities as well as to assay imprinted expression for genes in the 15q/7C region, with particular emphasis on *Ube3a*, which is known to be maternally expressed in neurons, as well as for two genes that have been the subject of some debate in the literature, *ATP10A* and the *Gabrb3* cluster, which were both shown to be biallelically expressed in neurons in this mouse model of PWS (Relkovic et al., 2010). In an effort to understand how the snoRNA cluster, and specifically the PWS-associated *HBII-85* gene, is regulated in neurons, Leung et al. looked for changes in chromatin decondensation and nucleolar size in the brains of PWS-IC^{del} animals and compared them to brain tissue from both AS and PWS *SNRPN>UBE3A* deletion individuals (Leung et al., 2009). In both the animal model and the human brain samples, they identified a large region (~888 kb) from the *Snrpn* gene to the *Ube3a* gene (see Figure 32.1) that undergoes dramatic chromatin decondensation in neurons on the paternal allele only, thus providing an open chromatin conformation and presumably transcriptionally active zone around the paternal *Snrpn* promoter, which transcribes the *HBII-85* snoRNA gene as well as the antisense-*Ube3a* transcript (Leung et al., 2009). Most individuals with PWS are also deleted for additional 15q genes or have maternal uniparental disomy for this region, resulting in overexpression of the maternal *UBE3A* gene. Although loss of paternal *HBII-85* transcription alone may not account for all aspects of the Prader–Willi phenotype, the use of this imprinting center mouse model to reveal the complex mechanisms of imprinted expression and regional epigenetic regulation on 15q is just now becoming evident.

32.6.1 Additional Mouse Models for Genes Deleted or Duplicated in the 15q11–q13 Region

One complication in the phenotypic analysis of AS, PWS, and duplication 15q autism is that most of these individuals are deleted or duplicated for a region encompassing some 18 genes, including several genes that, when deleted in mouse models, cause phenotypes on their own. The *Gabrb3* null mice, for example, display epilepsy, learning and memory deficits, as well as defects in social behavior (Delorey et al., 1998). There is also an interesting relationship between the *p* gene and the *GABRA5* and *GABRB3* subunits genes. Deletion of both *p* alleles causes rearrangements of Gabaα5 and Gabaβ3 receptors, producing a peculiar phenotype characterized by ataxia, jerky gait, and seizures (Nakatsu et al., 1993). Obviously, the duplication of this cluster in humans contributes to the difference in severity of phenotypes observed between interstitial duplication 15q individuals and the more severely affected isodicentric 15q duplication cases, where as many as 6 copies of the *GABRB3* cluster may be present (Hogart et al., 2007).

A mouse model of paternal uniparental disomy showed high incidence of failure to thrive with spontaneous death in the first month, similar to both the *HBII-85*-deficient and PWS-IC^{del} models of PWS. Survivors developed obesity, hyperactive behavior, ataxic gait, and electroencephalograph with high-amplitude delta rhythmic activity (Cattanach et al., 1997).

In an effort to reflect the interstitial duplication 15q syndrome in mice more accurately, Nakatani et al. used chromosomal engineering techniques to duplicate the entire 7qC region syntenic to typical human class II duplication breakpoints in interstitial duplication 15q cases (Nakatani et al., 2009). It was somewhat surprising that the animals with paternally but not maternally inherited duplications of 7qC showed poor social interaction, behavioral inflexibility, abnormal ultrasonic vocalizations, and anxiety. However, the animals with maternal duplication did trend toward significance on several tests, and, more importantly, an accurate test for autism assessment in mice has not yet been developed; so, in the future, these animals may exhibit subtle autistic features, like their human counterparts. That being said, as more interstitial duplication 15q cases are identified using array-CGH methods, there has been an increase in the number of paternal 15q duplication cases, who typically present with both anxiety and sleep problems but generally do not have cognitive defects (L. Reiter and N.C. Schanen, unpublished observations). In the end, this mouse model of int dup(15) may be a better reflection of the actual phenotypes observed in a large cohort of int dup(15) from both maternal and paternal origins.

32.6.2 *Drosophila* Models of 15q Syndromes

A recent avenue of research has been the construction of human disease models in the well-studied and easily manipulated genetic model organism *Drosophila melanogaster* (Doronkin and Reiter, 2008; Pfleger and Reiter, 2008). A model for Angelman syndrome revealed motor defects, LTP deficits, and circadian rhythm abnormalities in flies lacking the *UBE3A* ortholog *Dube3a* (Wu et al., 2008). Flies lacking Dube3a or even overexpressing Dube3a in the nervous system did not show any gross morphological changes in brain structure of neuronal connections as predicted by phenotype-observed mouse models of AS. Additional studies revealed that both overexpression and loss of Dube3a can affect the number and complexity of dendritic arbors in the fly peripheral nervous system (Lu et al., 2009), suggesting that Dube3a may control more subtle aspects of synaptic connectivity and is perhaps not involved in nervous system development, *per se*.

Our group has capitalized on the mis-expression methodologies available in the *Drosophila* system, primarily the GAL4/UAS system, which allows for overexpression of a given construct in a variety of specific tissue types or under the control of temporal regulators such as heatshock induction (Duffy, 2002). Using a proteomics approach, the authors of this study have identified over 80 unique protein or transcription targets of Dube3a (manuscript in revision). The first is a Rho-GEF involved in actin cytoskeletal remodeling, which the authors identified as a protein that decreased in intensity when they overexpressed human UBE3A in fly heads (Reiter et al., 2006). This protein, known as Pebble in flies and ECT2 in mammals, physically interacts with both fly and human UBE3A proteins in 293T cells. In addition, in a mouse *Ube3a* loss-of-function model of AS changes in Ect2 expression is observed in both the hippocampus and cerebellar regions of the brain (Reiter et al., 2006). More recently, the authors identified a protein called Punch in flies and GTP cyclohydrolase I in mammals using overexpression of fly Dube3a in heads. The regulation of Punch, however, appears to be through the transcription co-activation function of Dube3a, as the ubiquitin ligase function is not required for increased Punch transcript levels, increased dopamine levels, or hyperactivity observed in these flies (Ferdousy et al., 2011).

32.7 CONCLUSIONS

Human chromosome 15q11–q13 is a complex genomic region subject to regulation by parental imprinting. Several LCRs in this region predispose 15q to genomic rearrangements, causing parent-specific dosage changes that lead to a variety of neurodevelopmental disorders, including autism. The widespread use of array-CGH technology to detect CNVs in the 15q region has made it a focal point for the study of PWS/AS autism and even certain forms of schizophrenia. In addition, animal models of 15q disorders run the gamut from flies to mice and include not only single gene knock-outs, but also chromosomally engineered models duplicating or deleting the entire region of the mouse genome syntentic to 15q11.2–q13. With such a wide array of resources available in both human and model organism genetics, it is only a matter of time before someone unlocks the mechanisms of gene regulation, methylation, imprinting, and genomic recombination that cause these neurological disorders with the hope of eventually designing therapeutics for the treatment of these syndromes.

Glossary

Array-CGH: A microarray method commonly used to identify deleted and duplicated segments of the genome.

C57BL/6 C57 black 6: A common inbred strain of laboratory mouse.

Endophenotype: Psychiatric concept that refers to a biomarker (neuro-physiologic, biochemical, endocrinological, neuroanatomical, cognitive, or neuropsychological) that is associated with the phenotype in the populations, is heritable, is illness independent (manifests in individuals whether or not illness is active), co-segregates with illness, and is found in non affected family members at a higher rate than in the general population (Gottesman and Gould, 2003).

Epigenetic: Refers to DNA and chromatin modifications that can affect gene function without change in the genotype.

Epigenotype: DNA-exclusive states of gene expression and epigenetic modification; the maternal state and the paternal state.

HBII: Nomenclature used for human genes of the snoRNA cluster; there are different genes (as HBII-13, HB-36, HBII-85, etc.).

HRM: High-resolution melting curve analysis; a method that can quickly distinguish short DNA fragments which differ by as little as a single nucleotide without the need for DNA sequencing.

GAL4/UAS: Bipartite expression system used in *Drosophila* that was derived from the yeast GAL4 gene promoter.

Imprinting: A difference in gene expression that depends on the parent of origin of the allele.

Imprinting center (IC): A region of the DNA usually marked by differential methylation that regulates the imprinted expression of a gene or genes in the region.

Knock-in mouse: Insertion of a protein-coding cDNA sequence at a particular locus in the mouse within the gene of interest.

Knock-out mouse: One or more genes turned off through a targeted mutation in the mouse.

Ortholog: Genes in different species that are similar in DNA sequence and also encode proteins with the same function.

Penetrance: The proportion of individuals with a genotype known to cause disease and present clinical symptoms.

Segmental duplication: Regions bigger than 1 kb that are not high-copy repeats and have more than 90% identity to another region in the genome.

Syntenic: Gene loci that are in sequence on a chromosome between two species.

Uniparental disomy: Two copies of a specific chromosome, both inherited from one parent.

References

Albrecht, U., et al., 1997. Imprinted expression of the murine Angelman syndrome gene, Ube3a, in hippocampal and Purkinje neurons. Nature Genetics 17 (1), 175–178.

Angelman, H., 1965. 'Pupper' children a report on three cases. Developmental Medicine and Child Neurology 7 (6), 681–688.

Bachellerie, J.P., Cavaille, J., Huttenhofer, A., 2002. The expanding snoRNA world. Biochimie 84 (8), 775–790.

Battaglia, A., 2005. The inv dup(15) or idic(15) syndrome: A clinically recognisable neurogenetic disorder. Brain & Development 27 (5), 365–369.

Battaglia, A., 2008. The inv dup (15) or idic (15) syndrome (Tetrasomy 15q). Orphanet Journal of Rare Diseases 19 (3), 30.

Ben-Shachar, S., et al., 2009. Microdeletion 15q13.3: A locus with incomplete penetrance for autism, mental retardation, and psychiatric disorders. Journal of Medical Genetics 46 (6), 382–388.

Boyar, F.Z., et al., 2001. A family with a grand-maternally derived interstitial duplication of proximal 15q. Clinical Genetics 60 (6), 421–430.

Buiting, K., et al., 1995. Inherited microdeletions in the Angelman and Prader-Willi sydromes define an imprinting center on human chromosome 15. Nature Genetics 9 (4), 395–400.

Buiting, K., et al., 2007. C15orf2 and a novel noncoding transcript from the Prader-Willi/Angelman syndrome region show monoallelic expression in fetal brain. Genomics 89 (5), 588–595.

Butler, M.G., Bittel, D.C., Kibiryeva, N., Talebizadeh, Z., Thompson, T., 2004. Behavioral differences among subjects with Prader-Willi syndrome and type I or type II deletion and maternal disomy. Pediatrics 113 (3), 565–573.

Cassidy, S.B., 1997. Prader-Willi syndrome. Journal of Medical Genetics 34 (11), 917–923.

Cassidy, S.B., Driscoll, D.J., 2009. Prader-Willi syndrome. European Journal of Human Genetics 17 (1), 3–13.

Cattanach, B.M., et al., 1997. A candidate model for Angelman syndrome in the mouse. Mammalian Genome 8 (7), 472–478.

Cavaille, J., et al., 2000. Identification of brain-specific and imprinted small nucleolar RNA genes exhibiting an unusual genomic organization. Proceedings of the National Academy of Sciences of the United States of America 97 (26), 14311–14316.

Chamberlain, S.J., et al., 2004. Evidence for genetic modifiers of postnatal lethality in PWS-IC deletion mice. Human Molecular Genetics 13 (23), 2971–2977.

Clarke, D.J., Boer, H., 1995. Genetic and behavioural aspects of Prader-Willi syndrome: A review with a translation of the original paper. Mental Handicap Research 8 (1), 38–53.

Conant, K.D., Thibert, R.L., Thiele, E.A., 2009. Epilepsy and the sleep-wake patterns found in Angelman syndrome. Epilepsia 50 (11), 2497–2500.

Consortium, I.S., 2008. Rare chromosomal deletions and duplications increase risk of schizophrenia. Nature 455 (7210), 237–241.

Cook Jr., E.H., et al., 1997. Autism or atypical autism in maternally but not paternally derived proximal 15q duplication. The American Journal of Human Genetics 60 (4), 928–934.

Cook Jr., E.H., et al., 1998. Linkage-disequilibrium mapping of autistic disorder, with 15q11-13 markers. The American Journal of Human Genetics 62 (5), 1077–1083.

De Smith, A.J., et al., 2009. A deletion of the HBII-85 class of small nucleolar RNAs (snoRNAs) is associated with hyperphagia, obesity and hypogonadism. Human Molecular Genetics 28 (17), 3257–3265.

Delahanty, R.J., et al., 2011. Maternal transmission of a rare GABRB3 signal peptide variant is associated with autism. Molecular Psychiatry 16, 86–96.

Delorey, T.M., et al., 1998. Mice lacking the B3 subunit of the GABAA receptor have the epilepy phenotype and many of the behavioral characteristics of Angelman syndrome. Journal of Neuroscience 18 (20), 8505–8514.

Dibbens, L.M., et al., 2009. Familial and sporadic 15q13.3 microdeletions in idiopathic generalized epilepsy: Precedent for disorders with complex inheritance. Human Molecular Genetics 18 (19), 3626–3631.

Dindot, S.V., Antalffy, B.A., Bhattacharjee, M.B., Beaudet, A.L., 2008. The Angelman syndrome ubiquitin ligase localizes to the synapse and nucleus, and maternal deficiency results in abnormal dendritic spine morphology. Human Molecular Genetics 17 (1), 111–118.

Ding, F., et al., 2008. SnoRNA Snord116 (Pwcr1/MBII-85) deletion causes growth deficiency and hyperphagia in mice. PLoS One 3 (3), e1709.

Doornbos, M., et al., 2009. Nine patients with a microdeletion 15q11.2 between breakpoints 1 and 2 of the Prader-Willi critical region, possibly associated with behavioural disturbances. European Journal of Human Genetics 52 (2–3), 108–115.

Doronkin, S., Reiter, L.T., 2008. *Drosophila* orthologues to human disease genes: An update on progress. Progress in Nucleic Acid Research and Molecular Biology 82, 1–32.

Driscoll, D.J., et al., 1992. A DNA methylation imprint, determined by the sex of the parent, distinguishes the Angelman and Prader-Willi syndromes. Genomics 13 (4), 917–924.

Duffy, J.B., 2002. GAL4 system in *Drosophila*: A fly geneticist's Swiss army knife. Genesis 34 (1–2), 1–15.

Duker, A.L., et al., 2010. Paternally inherited microdeletion at 15q11.2 confirms a significant role for the SNORD116 C/D box snoRNA cluster in Prader-Willi syndrome. European Journal of Human Genetics 18 (11), 1196–1201.

Fatemi, S.H., Reutiman, T.J., Folsom, T.D., Thuras, P.D., 2009. GABAA receptor downregulation in brains of subjects with autism. Journal of Autism and Developmental Disorders 39 (2), 223–230.

Ferdousy, F., et al., 2011. *Drosophila* Ube3a regulates monoamine synthesis by increasing GTP cyclohydrolase I activity via a non-ubiquitin ligase mechanism. Neurobiology of Disease 41 (3), 669–677.

Freedman, R., et al., 1997. Linkage of a neurophysiological deficit in schizophrenia to a chromosome 15 locus. Proceedings of the National Academy of Sciences of the United States of America 94 (2), 587–592.

Freedman, R., et al., 2001. Linkage disequilibrium for schizophrenia at the chromosome 15q13-14 locus of the alpha7-nicotinic acetylcholine receptor subunit gene (CHRNA7). American Journal of Medical Genetics 105 (1), 20–22.

Fridman, C., et al., 2003. Angelman syndrome associated with oculocutaneous albinism due to an intragenic deletion of the P gene. American Journal of Medical Genetics 119A (2), 180–183.

Gallagher, R., Pils, B., Albalwi, M., Francke, U., 2002. Evidence for the role of PWCR1/HBII-85 C/D box small nucleolar RNAs in Prader-Willi syndrome. The American Journal of Human Genetics 71 (3), 669–678.

Glenn, C.C., et al., 1996. Gene structure, DNA methylation, and imprinted expression of the human *SNRPN* gene. The American Journal of Human Genetics 58 (2), 335–346.

Glessner, J.T., et al., 2009. Autism genome-wide copy number variation reveals ubiquitin and neuronal genes. Nature 459 (7246), 569–573.

Goldstone, A.P., Holand, A.J., Hauffa, B.P., Hokken-Koelega, A.C., Tauber, M., 2008. Recommendations for the diagnosis and management of Prader-Willi syndrome. Journal of Clinical Endocrinology and Metabolism 93 (11), 4183–4197.

Gottesman, I.I., Gould, T.D., 2003. The endophenotype concept in psychiatry: Etymology and strategic intentions. The American Journal of Psychiatry 160 (4), 636–645.

Gray, T.A., Saitoh, S., Nicholls, R.D., 1999. An imprinted, mammalian bicistronic transcript encodes two independent proteins. Proceedings of the National Academy of Sciences of the United States of America 96 (10), 5616–5621.

Greer, P.L., et al., 2010. The Angelmans yndrome protein Ube3A regulates synapse development by ubiquitinating arc. Cell 140 (5), 704–716.

Gregg, C., et al., 2010. High-resolution analysis of parent-of-origin allelic expression in the mouse brain. Science 329 (5992), 643–648.

Guffanti, G., et al., 2010. Role of UBE3A and ATP10A genes in autism susceptibility region 15q11-q13 in an Italian population: A positive replication for UBE3A. Psychiatry Research 185 (1–2), 33–38.

Gunay-Aygun, M., Schwartz, S., Heeger, S., O'riordan, M.A., Cassidy, S., 2001. The changing purpose of Prader-Willi syndrome clinical diagnostic criteria and proposed revised criteria. Pediatrics 108 (5), 1–5.

Gurrieri, F., et al., 1999. Pervasive developmental disorder and epilepsy due to maternally derived duplication of 15q11-q13. Neurology 52 (8), 1694–1697.

Heck, D.H., Zhao, Y., Roy, S., Ledoux, M.S., Reiter, L.T., 2008. Analysis of cerebellar function in Ube3a-deficient mice reveals novel genotype-specific behaviors. Human Molecular Genetics 17 (14), 2181–2189.

Herzing, L.B., Kim, S.J., Cook Jr., E.H., Ledbetter, D.H., 2001. The human aminophospholipid-transporting ATPase gene ATP10C maps adjacent to UBE3A and exhibits similar imprinted expression. The American Journal of Human Genetics 68 (6), 1501–1505.

Hogart, A., Nagarajan, R.P., Patzel, K.A., Yasui, D.H., Lasalle, J.M., 2007. 15q11-13 GABAA receptor genes are normally biallelically expressed in brain yet are subject to epigenetic dysregulation in autism-spectrum disorders. Human Molecular Genetics 16 (6), 691–703.

Hogart, A., Patzel, K.A., Lasalle, J.M., 2008. Gender influences monoallelic expression of ATP10A in human brain. Human Genetics 124 (3), 235–242.

Hogart, A., Wu, D., Lasalle, J.M., Schanen, N.C., 2010. The comorbidity of autism with the genomic disorders of chromosome 15q11.2–q13. Neurobiology of Disease 38 (2), 181–191.

Holm, V.A., et al., 1993. Prader-Willi syndrome: Consensus diagnostic criteria. Pediatrics 91 (2), 398–402.

Huang, B., et al., 1997. Refined molecular characterization of the breakpoints in small inv dup(15) chromosomes. Human Genetics 99 (1), 11–17.

Huibregtse, J.M., Scheffner, M., Howley, P.M., 1993. Cloning and expression of the cDNA for E6-AP, a protein that mediates the interaction of the human papillomavirus E6 oncoprotein with p53. Journal of Molecular Cell Biology 13 (2), 775–784.

Hung, C.C., et al., 2011. Quantitative and qualitative analyses of the SNRPN gene using real-time PCR with melting curve analysis. Journal of Molecular Diagnostics 13, 609–613.

Ingason, A., et al., 2011. Maternally derived microduplications at 15q11-q13: Implication of imprinted genes in psychotic illness. The American Journal of Psychiatry 168 (4), 408–417.

Jay, P., et al., 1997. The human necdin gene, NDN, is maternally imprinted and located in the Prader-Willi syndrome chromosomal region. Nature Genetics 17 (3), 357–361.

Jiang, Y.H., et al., 1998. Mutation of the Angelman ubiquitin ligase in mice causes increased cytoplasmic p53 and deficits of contextual learning and long-term potentiation. Neuron 21 (4), 799–811.

Jiang, Y.H., et al., 2010. Altered ultrasonic vocalization and impaired learning and memory in Angelman syndrome mouse model with a large maternal deletion from Ube3a to Gabrb3. PLoS One 5 (8), e12278.

Kashiro, K., et al., 2009. Ube3a is required for experience-dependent maturation of the neocortex. Nature Neuroscience 12 (6), 777–783.

Kishino, T., Lalande, M., Wagstaff, J., 1997. UBE3A/E6-AP mutations cause Angelman syndrome. Nature Genetics 15 (1), 70–73.

Knoll, J.H., et al., 1989. Angelman and Prader-Willi syndromes share a common chromosome 15 deletion but differ in parental origin of the deletion. American Journal of Medical Genetics 32 (2), 285–290.

Knoll, J.H., Wagstaff, J., Lalande, M., 1993. Cytogenetic and molecular studies in the Prader-Willi and Angelman syndromes: An overview. American Journal of Medical Genetics 46 (1), 2–6.

Ledbetter, D.H., et al., 1981. Deletions of chromosome 15 as a cause of the Prader-Willi syndrome. The New England Journal of Medicine 304 (6), 325–329.

Leung, K.N., Vallero, R.O., Dubose, A.J., Resnick, J.L., Lasalle, J.M., 2009. Imprinting regulates mammalian snoRNA-encoding chromatin decondensation and neuronal nucleolar size. Human Molecular Genetics 18 (22), 4227–4238.

Lichtenstein, P., et al., 2009. Common genetics determinants of schizophrenia and bipolar disorder in Swedish families: A population-based study. Lancet 372 (9659), 234–239.

Lossie, A.C., et al., 2001. Distinct phenotypes distinguish the molecular classes of Angelman syndrome. Journal of Medical Genetics 38 (12), 834–845.

Lu, Y., et al., 2009. The *Drosophila* homologue of the Angelman syndrome ubiquitin ligase regulates the formation of terminal dendritic branches. Human Molecular Genetics 18 (3), 454–462.

Luedi, P.P., et al., 2007. Computational and experimental identification of novel human imprinted genes. Genome Research 17 (12), 1723–1730.

Lupski, J.R., 1998. Genomic disorders: Structural features of the genome can lead to DNA rearrangements and human disease traits. Trends in Genetics 14 (10), 417–422.

Mao, R., et al., 2000. Characteristics of two cases with dup(15) (q11.2-q12): One of maternal and one of paternal origin. Genetics in Medicine 2 (2), 131–135.

Masurel-Paulet, A., et al., 2010. Delineation of 15q13.3 microdeletions. Clinical Genetics 72 (8), 149–161.

Matsuura, T., et al., 1997. De novo truncating mutations in E6-AP ubiquitin-protein ligase gene (UBE3A) in Angelman syndrome. Nature Genetics 15 (1), 74–77.

Mccarthy, S.E., et al., 2009. Microduplications of 16p11.2 are associated with schizophrenia. Nature Genetics 41 (11), 1223–1227.

Meguro, M., et al., 2001. A novel maternally-expressed gene, ATP10C, encodes a putative aminophospholipid translocase associated with Angelman syndrome. Nature Genetics 28 (1), 19–20.

Meguro-Horike, M., et al., 2011. Neuron-specific impairment of inter-chromosomal pairing and transcription in a novel model of human 15q-duplication syndrome. Human Molecular Genetics 20 (19), 3798–3810.

Miura, K., et al., 2002. Neurobehavioral and electroencephalographic abnormalities in Ube3a maternal-deficient mice. Neurobiology of Disease 9 (2), 149–159.

Mohandas, T.K., et al., 1999. Paternally derived de novo interstitial duplication of proximal 15q in a patient with developmental delay. American Journal of Medical Genetics 82 (4), 294–300.

Mulle, J.G., et al., 2010. Microdeletions of 3q29 confer high risk for schizophrenia. The American Journal of Human Genetics 87 (2), 229–236.

Nakatani, J., et al., 2009. Abnormal behavior in a chromosome-engineered mouse model for human 15q11-13 duplication seen in autism. Cell 137 (7), 1235–1246.

Nakatsu, Y., et al., 1993. A cluster of three GABAA receptor subunit genes is deleted in a neurological mutant of the mouse p locus. Nature 364 (6436), 448–450.

Napoli, I., et al., 2008. The fragile X syndrome protein represses activity-dependent translation through CYFIP1, a new 4E-BP. Cell 134 (6), 1042–1054.

Nawaz, Z., et al., 1999. The Angelman syndrome-associated protein, E6-AP, is a coactivator for the nuclear hormone receptor superfamily. Molecular and Cellular Biology 19 (2), 1182–1189.

Nowicki, S.T., et al., 2007. The Prader-Willi phenotype of fragile X syndrome. Journal of Developmental and Behavioral Pediatrics 28 (2), 133–138.

Nurmi, E.L., et al., 2001. Linkage disequilibrium at the Angelman syndrome gene UBE3A in autism families. Genomics 77 (1–2), 105–113.

Orrico, A., et al., 2009. Late-onset Lennox-Gastaut syndrome in a patient with 15q11.2-q13.1 duplication. American Journal of Medical Genetics. Part A 149 (5), 1033–1035.

Owens, D.F., Kriegstein, A.R., 2002. Is there more to GABA than synaptic inhibition? Nature Reviews Neuroscience 3 (9), 715–727.

Peters, S.U., Beaudet, A.L., Madduri, N., Bacino, C.A., 2004. Autism in Angelman syndrome: Implications for autism research. Clinical Genetics 66 (6), 530–536.

Pfleger, C.M., Reiter, L.T., 2008. Recent efforts to model human diseases in vivo in *Drosophila*. Fly (Austin) 2 (3), 129–132.

Pizzarelli, R., Cherubini, E., 2011. Alterations of GABAergic signaling in autism spectrum disorders. Neural Plasticity 297153.

Reiter, L.T., Seagroves, T.N., Bowers, M., Bier, E., 2006. Expression of the Rho-GEF Pbl/ECT2 is regulated by the UBE3A E3 ubiquitin ligase. Human Molecular Genetics 15 (18), 2825–2835.

Relkovic, D., et al., 2010. Behavioural and cognitive abnormalities in an imprinting centre deletion mouse model for Prader-Willi syndrome. European Journal of Neuroscience 31 (1), 156–164.

Roberts, S.E., Ne, Dennis, Ce, Browne, et al., 2002. Characterization of interstitial duplications and triplications of chromosome 15q11-q13. Human Genetics 110 (3), 227–234.

Robinson, W.P., Lalande, M., 1995. Sex-specific meiotic recombination in the Prader-Willi/Angelman syndrome imprinted region. Human Molecular Genetics 4 (5), 801–806.

Robinson, W.P., et al., 1998. The mechanisms involved in formation of deletions and duplications of 15q11-q13. Journal of Medical Genetics 35 (2), 130–136.

Rougeulle, C., Cardoso, C., Fontes, M., Colleaux, L., Lalande, M., 1998. An imprinted antisense RNA overlaps UBE3A and a second maternally expressed transcript. Nature Genetics 19 (1), 15–16.

Runte, M., et al., 2001. The IC-SNURF-SNRPN transcript serves as a host for multiple small nucleolar RNA species and as an antisense RNA for UBE3A. Human Molecular Genetics 10 (23), 2687–2700.

Sahoo, T., et al., 2008. Prader-Willi phenotype caused by paternal deficiency for the HBII-85 C/D box small nucleolar RNA cluster. Nature Genetics 40 (6), 719–721.

Saitoh, S., et al., 1996. Minimal definition of the imprinting center and fixation of chromosome 15q11-q13 epigenotype by imprinting mutations. Proceedings of the National Academy of Sciences of the United States of America 93 (15), 7811–7815.

Sharp, A.J., et al., 2008. A recurrent 15q13.3 microdeletion syndrome associated with mental retardation and seizures. Nature Genetics 40 (3), 322–328.

Shaw, C.J., Lupski, J.R., 2004. Implications of human genome architecture for rearrangement-based disorders: The genomic basis of disease. Human Molecular Genetics 13 (1), R57–R64.

Shinawi, M., et al., 2009. A small recurrent deletion within 15q13.3 is associated with a range of neurodevelopmental phenotypes. Nature Genetics 41 (12), 1269–1271.

Soejima, H., Wagstaff, J., 2005. Imprinting centers, chromatin structure, and disease. Journal of Cellular Biochemistry 95 (2), 226–233.

Stefansson, H., et al., 2008. Large recurrent microdeletions associated with schizophrenia. Nature 455 (7210), 232–236.

Stephens, S.H., et al., 2009. Association of the 5'-upstream regulatory region of the alpha7 nicotinic acetylcholine receptor subunit gene (CHRNA7) with schizophrenia. Schizophrenia Research 109 (1–3), 102–112.

Szafranski, P., et al., 2010. Structures and molecular mechanisms for common 15q13.3 microduplications involving CHRNA7: Benign or pathological? Human Mutation 31 (7), 840–850.

Takano, K., Lyons, M., Moyes, C., Jones, J., Schwartz, C., 2010. Two percent of patients suspected of having Angelman syndrome have TCF4 mutations. Clinical Genetics 78 (3), 282–288.

Tanaka, M., et al., 2008. Hyperglycosylation and reduced GABA currents of mutated GABRB3 polypeptide in remitting childhood absence epilepsy. The American Journal of Human Genetics 82 (6), 1249–1261.

Urraca, N., Davis, L., Cook Jr., E.H., Schanen, N.C., Reiter, L.T., 2010. A single-tube quantitative high-resolution melting curve method for parent-of-origin determination of 15q duplications. Genetic Testing and Molecular Biomarkers 14 (4), 571–576.

Van Bon, B.W., et al., 2009. Further delineation of the 15q13 microdeletion and duplication syndromes: A clinical spectrum varying from non-pathogenic to a severe outcome. Journal of Medical Genetics 46 (8), 511–523.

Van Buggenhout, G., Fryns, J.P., 2009. Angelman syndrome (AS, MIM 105830). European Journal of Human Genetics 17 (11), 1367–1373.

Varela, M.C., Kok, F., Oto, P.A., Koiffmann, C.P., 2004. Phenotypic variability in Angelman syndrome: Comparison among different deletion classes and between deletion and UPD subjects. European Journal of Human Genetics 12 (12), 987–992.

Varela, M.C., Kok, F., Setian, N., Kim, C.A., Koiffmann, C.P., 2005. Impact of molecular mechanisms, including deletion size, on Prader-Willi syndrome phenotype: Study of 75 patients. Clinical Genetics 67 (1), 47–52.

Wang, W.Y., Barratt, B.J., Clayton, D.G., Todd, J.A., 2005. Genome-wide association studies: Theoretical and practical concerns. Nature Reviews Genetics 6 (2), 109–118.

Watson, P., et al., 2001. Angelman syndrome phenotype associated with mutations in MECP2, a gene encoding a methyl CpG binding protein. Journal of Medical Genetics 38 (4), 224–228.

Weaver, J.R., Susiarjo, M., Bartolomei, M.S., 2009. Imprinting and epigenetic changes in the early embryo. Mammalian Genome 20 (9–10), 532–543.

Webb, T., 1994. Inv dup(15) supernumerary marker chromosomes. Journal of Medical Genetics 31 (8), 585–594.

White, H.E., Hall, V.J., Cross, N.C., 2007. Methylation-sensitive high-resolution melting-curve analysis of the SNRPN gene as a diagnostic screen for Prader-Willi and Angelman syndromes. Clinical Chemistry 53 (11), 1960–1962.

Williams, C.A., 2005. Neurological aspects of the Angelman syndrome. Brain & Development 27 (2), 88–94.

Williams, C.A., et al., 2006. Angelman syndrome 2005: Updated consensus for diagnostic criteria. American Journal of Medical Genetics 140 (5), 413–418.

Wu, Y., et al., 2008. A Drosophila model for Angelman syndrome. Proceedings of the National Academy of Sciences of the United States of America 105 (34), 12399–12404.

Yang, T., et al., 1998. A mouse model for Prader-Willi syndrome imprinting-centre mutations. Nature Genetics 19 (1), 25–31.

Yashiro, K., et al., 2009. Ube3a is required for experience-dependent maturation of the neocortex. Nature Neuroscience 12 (6), 777–783.

Yasui, D.H., et al., 2011. 15q11.2-13.3 chromatin analysis reveals epigenetic regulation of CHRNA7 with deficiencies in Rett and autism brain. Human Molecular Genetics 3 (4), 356–364.

Zhang, Y., Tycko, B., 1992. Monoallelic expression of the human H19 gene. Nature Genetics 1 (1), 40–44.

Zody, M., et al., 2006. Analysis of the DNA sequence and duplication history of human chromosome 15. Nature 440 (7084), 671–675.

Fragile X Clinical Features and Neurobiology

M.J. Leigh, R.J. Hagerman

University of California at Davis, M.I.N.D. Institute, Sacramento, CA, USA

OUTLINE

Abbreviations

ADHD Attention deficit hyperactivity disorder
AMPA Alpha-amino-3-hydroxy-5-methyl-4-isoxazole-propionic acid
ASD Autism spectrum disorders
CGG Cytosine–guanine–guanine
DD Developmental disabilities

DSM-IV-TR Diagnostic and Statistical Manual, fourth edition, text revision
FMR1 Fragile X mental retardation 1
FMRP Fragile X mental retardation protein
FSH Follicle stimulation hormone
FXPOI Fragile X-associated primary ovarian insufficiency
FXS Fragile X syndrome

Neural Circuit Development and Function in the Brain: Comprehensive Developmental Neuroscience, Volume 3 http://dx.doi.org/10.1016/B978-0-12-397267-5.00044-3

© 2013 Elsevier Inc. All rights reserved.

FXTAS Fragile X-associated tremor/ataxia syndrome
ID Intellectual disability
KO Knock out
MCP Middle cerebellar peduncle
MDD Major depressive disorder
mGluR Metabotropic glutamate receptor
MMP9 Matrix metalloproteinase 9
MP Metacarpal-phalangeal
MPEP 2-Methyl-6-(phenylethynyl)-pyridine hydrochloride
MS Multiple sclerosis
MVP Mitral valve prolapse
NCS-R National Comorbidity Survey Replication
PCR Polymerase chain reaction
PECS Picture exchange communication system
POF Premature ovarian failure
POI Primary ovarian insufficiency
PWP Prader–Willi phenotype
SCID Structured Clinical Interview for DSM IV
SCL-90 Symptom Checklist-90
SSRIs Selective serotonin reuptake inhibitors

33.1 INTRODUCTION

33.1.1 Fragile X Syndrome and Fragile X-Associated Disorders

Fragile X syndrome (FXS) and the fragile X family of disorders may affect generations of a family in a variety of ways. It is essential for readers to understand the spectrum of involvement from FXS and premutation disorders, as significant advances in the knowledge of these disorders have occurred in recent years. As one enters into a new stage of targeted treatments, identification and understanding of the fragile X-associated disorders are even more important. The study of fragile X-associated disorders serves as a pathway for gaining insight into the molecular pathogenesis of other neurodevelopmental disorders.

FXS is the most common inherited cause of intellectual disability (ID) and the most common single genetic cause of autism. It is also known as Martin–Bell syndrome and, in South America, as Escalante syndrome. The syndrome was first described in 1943 by Martin and Bell, two British physicians who reported a sex-linked ID syndrome in a large family without definitive physical characteristics (Martin, 1943). Sex-linked ID in the sons of otherwise unaffected daughters of a Saskatchewan family was described by Renpenning in 1962 (Renpenning et al., 1962). In 1969, Lubs noted an unusual secondary constriction of the long arm of the X chromosome in a family with four intellectually disabled males and three carrier females (Lubs, 1969). Further investigations were aided by studies showing that observation of these fragile sites on the X chromosome depended on what type of tissue culture medium was used, allowing increased frequency of observation (Sutherland, 1977). The fragile X chromosome and ID were further linked to macroorchidism by Turner

(Turner et al., 1978) and Sutherland (Sutherland and Ashforth, 1979). In 1981, Richards et al. reexamined the pedigree described by Martin and Bell in 1943, which did not appear to have physical anomalies. Seven members of the original family reported by Martin and Bell were evaluated, and five were found to carry the fragile X chromosome. Unusual physical features were also noted upon further examination, including large ears, prognathism, and macroorchidism (Richards et al., 1981). Of note, the family described by Renpenning in 1962 was reevaluated by Fox and colleagues, and members of the family were found to have microcephaly and no macroorchidism, both of which are atypical of FXS. It has since been suggested that Renpenning syndrome refers to the condition of mental retardation, microcephaly, and sex-linked inheritance, but without an associated fragile X chromosome (Fox et al., 1980).

The molecular etiology of the FXS, an unstable cytosine–guanine–guanine (CGG) repetitive sequence region of DNA on the X chromosome, was discovered in 1991 (Verkerk et al., 1991; Yu et al., 1991). A mutation in the *fragile X mental retardation-1* (*FMR1*) gene leads to expanded CGG repeats on the 5′ end of the *FMR1* gene in successive generations (Fu et al., 1991). The number of CGG repeats determines whether an individual is categorized as having a full mutation, or a premutation, or is in the gray or intermediate zone (Figure 33.1).

33.1.1.1 Full Mutation

In the general population, the number of CGG repeats ranges from 5 to 44 repeats. Greater than 200 CGG repeats confers the full mutation (Maddalena et al., 2001). In the full mutation, the *FMR1* gene is usually completely methylated and transcription of *FMR1* mRNA is silenced, so that very little fragile X mental

FIGURE 33.1   A family affected by fragile X. *Left*, a young man with fragile X syndrome; *middle*, his mother who is a premutation carrier; *right*: his maternal grandmother who is a premutation carrier with FXTAS. Notice the young man's gaze away from the camera.

retardation protein (FMRP) is produced (Bell et al., 1991; Pieretti et al., 1991). It is a lack of this protein that leads to the syndrome (Loesch et al., 2004). FXS is characterized by features including developmental delay, deficits in short-term memory, and speech delays. Affected individuals may have hyperactivity, attention difficulties, and an autism spectrum disorder (ASD). A sometimes subtle physical phenotype can be associated with the full mutation, including macroorchidism (from puberty onward) and prominent ears. These are discussed in detail later in the chapter.

Studies of the prevalence of FXS have focused on individuals with significant cognitive impairment. These studies resulted in a prevalence of 1 in 3,600 to 1 in 4,000 (Crawford et al., 2002; Turner et al., 1996). The allele frequency in the general population is 1 in 2,500 for both males and females identified in newborn screening (Fernandez-Carvajal et al., 2009b; Hagerman, 2008). Most affected people have the full mutation with >200 CGG repeats; however, in rare cases, the disorder can be caused by a deletion or point mutation in the *FMR1* gene or by an FMRP deficit in the upper range of the premutation.

33.1.1.2 *Premutation Carriers*

Carriers of fragile X, also known as premutation carriers, are defined as having 55–200 CGG repeats (Maddalena et al., 2001). The prevalence of the premutation ranges from 1 in 250 to 810 in males and 1 in 130 to 260 in females (Dombrowski et al., 2002; Fernandez-Carvajal et al., 2009b; Hagerman, 2008). The premutation was previously thought to have no implications for disease. Most individuals with the premutation are cognitively unaffected. However, research has shown effects of the premutation, first in reproduction with the discovery of fragile X-associated primary ovarian insufficiency (FXPOI). Primary ovarian insufficiency (POI) refers to menopause before the age of 40. Neurological symptoms including tremor and ataxia were reported in 2001 in an older male subgroup of fragile X premutation carriers (Hagerman et al., 2001). In children with the premutation, most are unaffected, but a subgroup demonstrates evidence of difficulties including attention deficits and social deficits (Farzin et al., 2006). A pattern of psychiatric problems in a subgroup of premutation carriers is also emerging, which is discussed later in the chapter.

33.1.1.3 *Gray or Intermediate Zone*

Among the general population, noncarriers have 5–54 CGG repeats. However, those with 45–54 repeats have been discovered to have unique features as well. These individuals are known as being in the gray or intermediate zone (Maddalena et al., 2001). They are usually phenotypically unaffected, although there can be elevated *FMR1* mRNA and twice the rate of POI in the general

population (Bodega et al., 2006; Bretherick et al., 2005; Loesch et al., 2007). On occasion, alleles in the gray zone can be unstable, and, therefore, this category requires further study for both phenotypic involvement and allelic stability. There have been rare reports of males with gray zone alleles who have had some fragile X features, but whether the features are due to the gray zone allele is not certain (Aziz et al., 2003).

33.2 CLINICAL DESCRIPTION

33.2.1 Clinical Description of FXS

The phenotype of FXS can be quite broad, with a mix of physical, cognitive, and behavioral features. Typically, children with FXS are not diagnosed until 3 years of age, when their behaviors bring them to the attention of their physician (Bailey et al., 2003). ID is a key feature in FXS, as it is the most common inherited cause of ID. Eighty-five percent of males and 25–30% of females are found to have an IQ <70 (de Vries et al., 1996; Hagerman et al., 1992, 2008b; Loesch et al., 2004). Lower FMRP levels are associated with more severe cognitive deficits. Females with FXS usually present with learning disorders and a normal or borderline IQ (Chonchaiya et al., 2009a).

33.2.1.1 *Clinical Phenotype*

The classical physical phenotype associated with males with FXS is of a person with macroorchidism, large and prominent ears, and a long, narrow face. This physical picture is typical for adult males with FXS. The physical features are often not present in the prepubertal child with FXS (Chudley and Hagerman, 1987).

Macroorchidism was the first physical feature identified in FXS. It is present in more than 90% of adult males (Merenstein et al., 1996). The macroorchidism does not appear to affect fertility as males with FXS are fertile (Willems et al., 1992). The sperm of full-mutation males alone contains the premutation (Reyniers et al., 1993). Due to this, fathers with FXS pass on the premutation to all of their daughters, and as they pass on their Y chromosome to their sons, the latter are not affected. Using an orchidometer, Butler et al. compiled growth charts for 185 males with FXS including height, weight, and testicular volume measurements. In the study, the average testicular volume in an adult male with FXS was 45 ml with a 95% confidence interval of 25–70 ml (Butler et al., 1992). The normal testicular volume is around 25–30 ml (Prader, 1966), and a volume greater than 30 ml is considered to be macroorchidism.

Prominent ears were described in 78% of 97 prepubertal boys with the full mutation (Merenstein et al., 1996). The ears can be long and wide and may demonstrate loss

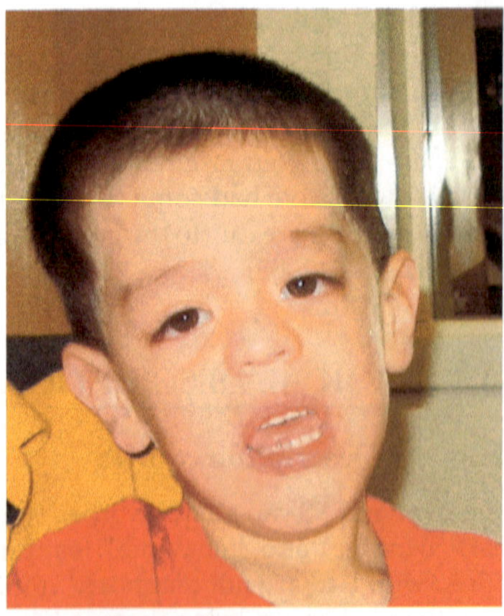

FIGURE 33.2   Image of a young boy with fragile X syndrome. Note the broad forehead, epicanthal folds, and mildly prominent ears with cupping of the pinnae bilaterally.

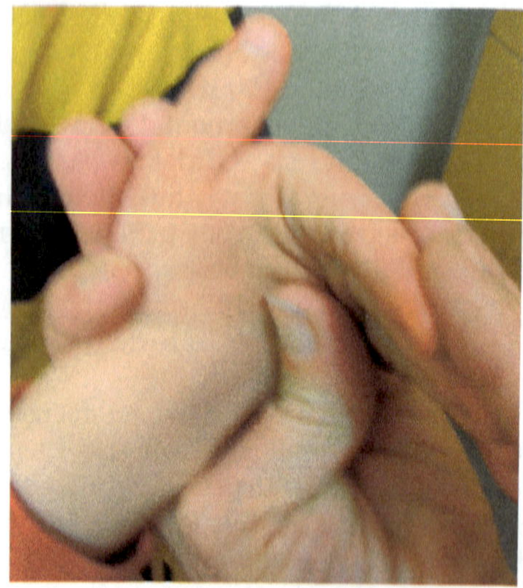

FIGURE 33.3   Hyperextensibility of MP joints with extension to 90°.

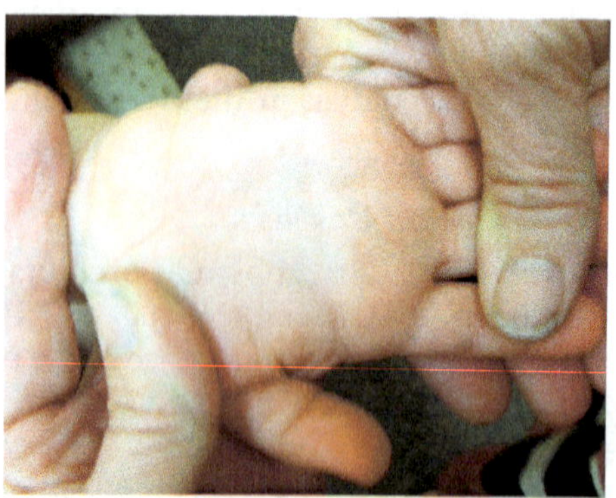

FIGURE 33.4   Single palmar crease in a boy with FXS.

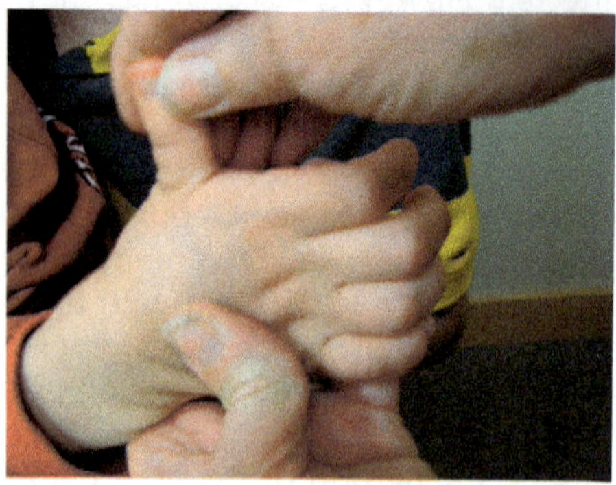

FIGURE 33.5   Double-jointed thumb in a boy with FXS.

of the antihelical fold, leading to 'cupping' of the upper pinna (Hagerman, 2002b). If the prominent ears are undesired, a surgical pinning procedure can be done (Figure 33.2).

Facial features were characterized in a retrospective study of children with FXS by Hockey and Crowhurst. Common characteristics in prepubertal children included puffiness around the eyes, strabismus, and hypotonia (Hockey and Crowhurst, 1988). Epicanthal folds were described by Simko et al. (1989). A high palate is a common finding, and often the palate is narrow as well (Simko et al., 1989). Jaw length increases disproportionately to body height in patients with FXS (Loesch and Sampson, 1993), and jaw prominence can be seen in adults, in addition to a long face after puberty. Dental maturity may be advanced in FXS, more so in younger children. Dental maturity was advanced in girls with the premutation as well (Kotilainen and Pirinen, 1999).

Connective tissue abnormalities are seen in FXS, including pes plenus, joint laxity, scoliosis, and hyperextensibility of the finger joints (Hagerman et al., 1984). Davids et al. examined 150 males with FXS and found pes planus in 50% of them. Hyperextensibility of the metacarpal–phalangeal (MP) joints was observed in 73% of patients younger than 11 years of age, in 56% of those 11–19 years old, and in 30% of those older than 20 years of age. Hyperextensibility is defined as an MP joint angle greater than or equal to 90° (Davids et al., 1990). Double-jointed thumbs may also be a physical finding. The skin is soft and 'velvety' in texture in FXS. A single or bridged palmar crease and calluses from hand-biting behaviors may be present. Hypotonia is described in young patients (Hagerman et al., 1983), and it appears to be due to a general effect of CNS dysfunction. Clonus may be present in adult FXS individuals more frequently than in children with FXS. It should be noted that approximately 30% of individuals with FXS do not have the typical physical features (Figures 33.3, 33.4, and 33.5).

In female individuals with FXS, the spectrum of physical involvement is tempered by the unaffected X chromosome. Phenotypic involvement is closely associated with the activation ratio (Abrams et al., 1994; Sobesky et al., 1996). The activation ratio refers to the fraction of normal *FMR1* alleles on the active X chromosome. Females with FXS were described in 1971 by Escalante, who noted flat feet and a high palate (Escalante, 1971). Cronister et al. looked at 105 full-mutation females and found an increased incidence of voluntary thumb dislocation and hyperextensible MP joints compared to 90 unaffected controls (Cronister et al., 1991). Mildly prominent ears are a common finding in prepubertal girls. Female individuals with FXS may have the full spectrum of involvement including ID and physical features or, on the other side of the spectrum, have no ID. Angkustsiri et al. reported a girl with the full mutation who had not only gifted intellectual abilities but also mild auditory processing problems and significant anxiety (Angkustsiri et al., 2008).

33.2.1.2 *Behavioral Phenotype*

Young children with FXS may present with language delays and are diagnosed with developmental delays at an average age of 21 months (Bailey et al., 2003). Hyperactivity is commonly reported, as are irritability, tantrums, perseveration, self-injurious behavior, mood instability, hand flapping, and hand biting (Hagerman, 2002b).

Patients with FXS often seem to have hyperarousability to auditory, visual, or tactile stimuli. This may be due to autonomic dysregulation with both sympathetic hyperarousal to sensory stimuli (Miller et al., 1999) and decreased parasympathetic activity compared to controls (Boccia and Roberts, 2000). Hyperactivity was noted in 47% of boys with FXS by Finelli et al. (1985). Bregman et al. found attention problems in all of 14 boys studied, but only 71% met the criteria for attention deficit hyperactivity disorder (ADHD; Bregman et al., 1988). ADHD may be the initially noted complaint in higher-functioning boys with FXS (Hagerman et al., 1985). Impulsivity is evident in boys with FXS when compared to boys with Down syndrome and unaffected control boys (Munir et al., 2000). Sullivan and colleagues have documented attention deficits in the majority of boys with FXS studied (Sullivan et al., 2006).

Shyness and social anxiety are commonly seen in both males and females with FXS and may improve with age. It may be more of a problem in females with FXS because ADHD symptoms are often more severe in boys, and ADHD may serve to temper the shyness (Merenstein et al., 1996). Psychiatric problems may also occur in individuals with FXS. Franke et al. evaluated psychiatric problems in mothers with the full mutation and in siblings with both the premutation and without the premutation. Fifty-four percent of full-mutation mothers had a psychiatric problem. Anxiety disorders were seen in 46% of those mothers with the full mutation, and social

phobia was seen in 31%. Bipolar disorder was present in 15% and major depressive disorder (MDD) in 15%. Schizotypal and schizoid personality disorder was seen in 23% and avoidant personality disorder in 23% (Franke et al., 1998).

Executive function defects and difficulties with math are commonly reported in females with FXS. Attention problems are less frequent in females in comparison to males, with around 33% having ADHD (Hagerman et al., 1992). Females tend to have less hyperactivity but impulsivity is usually evident. Selective mutism may be seen in females as well (Hagerman et al., 1999).

33.2.1.3 *Autism*

Autism is a disorder characterized by deficits in social communication, language, and repetitive movements. It is closely related to FXS, as approximately 2–7% of individuals with autism will be positive for the fragile X mutation (Brown et al., 1986; Reddy, 2005; Wassink et al., 2001). Once a child receives the diagnosis of autism or ASD, fragile X DNA testing is recommended. Around 25% of males with FXS were diagnosed with autism in the 1990s using the Childhood Autism Rating Scale (Bailey et al., 1998). When evaluated with current gold standard diagnostic tools such as the Autism Diagnostic Observation Schedule and the Autism Diagnostic Interview, 30–35% of males meet the diagnostic criteria for autism (Harris et al., 2008; Rogers et al., 2001). Autistic-like features such as hand biting, hand flapping, perseveration, shyness, and poor eye contact have been noted in FXS, but it is the core social and communication deficits that lead to a diagnosis of autism (Baumgardner et al., 1995; Hagerman et al., 1986; Hatton et al., 2006; Kaufmann et al., 2004; Kerby and Dawson, 1994; Lewis et al., 2006). Social communication is not typically a weak point in the majority of patients, except in those with autism. Individuals with FXS are typically sensitive to social cues (Simon and Finucane, 1996).

In females with FXS as well, autism has been diagnosed, but on a less frequent basis than in males. Several studies have been conducted that evaluated the number of females with FXS and autism, and the rate is approximately 4–10% (Dissanayake et al., 2009; Hagerman, 2002a; Leigh et al., 2010).

33.2.1.4 *Associated Medical Conditions*

Seizures are the most important medical problem associated with FXS. They occur in approximately 20% of individuals and are more common in early childhood (Berry-Kravis, 2002; Hagerman and Stafstrom, 2009; Musumeci et al., 1999). The seizures are usually well controlled by anticonvulsants if treated early, but, in some cases, multiple anticonvulsants are necessary (Berry-Kravis, 2002). All types of seizures can occur, but partial complex seizures are perhaps the most common.

Strabismus has been noted in 8–40% of patients with FXS (Hatton et al., 1998; King et al., 1995; Maino et al., 1991; Storm et al., 1987). The type of strabismus reported has included exotropia, esotropia, and hyperdeviations. Surgery or patching may be necessary for correction, and, if found on physical examination, a referral to an ophthalmologist is recommended.

Otitis media has been reported by Hagerman and colleagues in 63% of 30 boys with FXS in comparison to 15% of their typical siblings and 38% of developmentally disabled children without FXS. Forty-three percent were treated with polyethylene (PE) tubes or prophylactic antibiotics for persistent middle ear effusions (Hagerman et al., 1987). Monitoring for otitis media should be done, as the resulting effects such as hearing loss may worsen language and cognitive defects. The facial structure in FXS is thought to contribute to the propensity for otitis media. Sinusitis was noted in 23% of 43 fragile X full-mutation males (Hagerman, 2002b).

Mitral valve prolapse (MVP) has been observed in FXS and is likely due to connective tissue abnormalities. Loehr et al. found MVP by echocardiogram in 55% of 40 male and female full-mutation individuals, and the prolapse was associated with a click or murmur on physical examination (Loehr et al., 1986). A study of 13 males and females using standard ECG, Holter ECG, and echocardiography found MVP in 77%, tricuspid prolapse in 15%, and mild pulmonary artery dilation in 23%. Other defects were noted including posterior aortic leaflet prolapse, mild aortic regurgitation, and mild pulmonary artery dilation (Puzzo et al., 1990). Any signs or physical examination symptoms suggestive of a cardiac problem warrant a referral to a cardiologist for further evaluation.

In FXS, abnormal growth patterns have been noted, including macrocephaly (Meryash et al., 1984). Short stature is common in both males and females (Loesch et al., 1987). There is evidence of hypothalamic–pituitary dysfunction with enhanced cortisol levels after stress and a blunted thyroid stimulation hormone response to thyrotropin-releasing hormone (Hessl et al., 2004; Wilson et al., 1988; Wisbeck et al., 2000).

The Prader–Willi phenotype (PWP) of FXS is characterized by extreme obesity with a full, round face; small, broad hands and feet; a small penis; and hyperphagia (de Vries et al., 1993; Nowicki et al., 2007). This PWP is phenotypically similar to Prader–Willi syndrome, but occurs without cytogenetic or methylation abnormalities at 15q11–13. It is estimated to occur in <10% of males with FXS, and occasionally occurs in females. CYFIP1 mRNA levels are generally reduced by two- to fourfold in the patients with the PWP compared to individuals with FXS alone and controls (Nowicki et al., 2007).

Children with FXS typically have disordered sleep early in childhood, and this usually improves with time (Kronk et al., 2009). Treatment with melatonin can improve this wakefulness in the middle of the night (Wirojanan et al., 2009).

33.2.2 Premutation Involvement

The authors' understanding regarding involvement in the premutation has evolved over the years: from considering those with the premutation as completely unaffected or 'nonpenetrant' in the 1980s to identification of premature ovarian failure (POF) in approximately 20% of women with the premutation in 1991 (Cronister et al., 1991) and now to the presence of a large spectrum of psychiatric, medical, and neurological problems in a limited number of carriers (Chonchaiya et al., 2009a). Over the last decade, the understanding of the molecular mechanism leading to premutation involvement through RNA toxicity related to elevated levels of FMR1 mRNA first identified by Tassone et al. (2000a) has grown. Below each type of premutation, the involvement is described in detail.

33.2.2.1 Developmental Involvement in Children with the Premutation

Involvement in children with the premutation was first identified in 1995 with a report of ADHD and social difficulties in high-end premutation carriers (Hagerman et al., 1996). Further reports of developmental problems followed (Aziz et al., 2003; Goodlin-Jones et al., 2004; Tassone et al., 2000b), and it was the problems seen in some young carriers that led to the discovery of elevated mRNA in carriers (Tassone et al., 2000a). Although the original cases were biased by clinic referral, Farzin et al. carried out a controlled study comparing the developmental problems of boys with the premutation who presented clinically as the proband to those who were identified through cascade testing once the proband had been diagnosed and to brothers who did not have the premutation. High rates of ASD (73%) and ADHD (90%) were found in those that were the probands compared to a rate of 8% of ASD and 38% of ADHD in the non-probands, and no ASD and 13% of ADHD in the typical brothers without the premutation (Farzin et al., 2006). Although the rate of ASD was not significantly different in the nonproband boys with the premutation compared to the non-carrier brothers, social deficits as measured by the Social Communication Questionnaire were significantly higher than in the unaffected brothers. Therefore, a subgroup of boys with the premutation is at a higher risk for ASD, social anxiety, and social interactional difficulties, in addition to ADHD.

Similar findings of premutation involvement were seen in the family comorbidity study of Bailey et al. where 1276 families completed an online survey about the children affected by fragile X in the family. Although most of the patients evaluated by this survey had the full mutation, the survey also included 57 boys and 199 girls with the premutation. Developmental delay was found in 32% of males and 6% of females with the premutation, attention problems in 45% of males and 14% of females, hyperactivity in 30% of males and 3% of females, autism in 19% of males and 1% of females, and anxiety in 36% of males and 31% of females (Bailey et al., 2008). However, this survey may also be biased toward clinical involvement in those with the premutation. Determining the predictors of those at risk with the premutation and the percentage who will be affected by developmental problems will require larger studies of unselected populations, such as those who are identified by newborn screening. Such studies are currently underway with longitudinal follow-up of those in whom the premutation was identified at birth.

Recent studies by Chen and colleagues of premutation neuronal cell cultures from the *FMR1* knock-in (KI) mouse have demonstrated a deficit of neuronal cell branching and larger synaptic connections compared to neurons without the premutation. In addition, enhanced neuronal cell death in culture was seen by 21 days of division compared to neurons without the premutation (Chen et al., 2010). These findings suggest that premutation neurons may be more vulnerable to early cell death, which could lead to developmental problems of connectivity, such as ASD in early development, or perhaps neurodegeneration in later life. Additionally, neurons with the premutation may be a population genetically vulnerable to environmental toxins that could increase neuronal cell death. The authors have recently published four cases of carriers who were exposed to environmental neurotoxins from chemical plants in close proximity to their home and whose neurological symptoms began early in life (Paul et al., 2010). These neurological problems included multiple sclerosis (MS) symptoms in early adulthood in two women and fragile X-associated tremor/ataxia syndrome (FXTAS) symptoms beginning in their 40s to early 50s. This is earlier than the mean age of onset of FXTAS symptoms at age 60 as reported by Leehey et al. (2007) and described below.

33.2.2.2 *Fragile X-Associated Tremor/Ataxia Syndrome*

The FXTAS was first reported in five cases of older males with the premutation described in 2001 (Hagerman et al., 2001). All these patients had an intention or action tremor combined with ataxia leading to frequent falls. They progressed to needing a cane and eventually a wheelchair as weakness, autonomic dysfunction, and fatigue developed. The brain imaging demonstrated brain atrophy and white matter disease, usually in the periventricular region and in the middle cerebellar peduncles (MCP) sign of the cerebellum

(Brunberg et al., 2002). Subsequently, criteria for the diagnosis of FXTAS were outlined in 2003 (Jacquemont et al., 2003) and then modified in 2004 (Hagerman and Hagerman, 2004). The modifications included the presence of eosinophilic intranuclear inclusions in neurons and astrocytes reported by Greco et al. (2002, 2006) that are unique to FXTAS. These inclusions occur throughout the central and the peripheral nervous system with the highest rate in the hippocampus and limbic system (Greco et al., 2006; Hunsaker et al., 2011). Molecular studies of the inclusions have demonstrated the presence of elevated *FMR1* mRNA in addition to other proteins (Iwahashi et al., 2006). The inclusions may help to arrest the cellular dysregulation that occurs in the presence of elevated mRNA, and they are also seen in the KI premutation mouse (Brouwer et al., 2008; Wenzel et al., 2010).

Recently, inclusions have been seen in peripheral tissue including the myenteric plexi of the gastrointestinal system, in the thyroid gland, in the adrenal gland, and in the testicles, particularly in the Leydig cells that make testosterone (Gokden et al., 2009; Greco et al., 2007; Louis et al., 2006). Impotence and testosterone deficiency are common and often occur even before the onset of FXTAS (Hagerman et al., 2008a). A report by Jacquemont et al. (2004) surveyed all the positive families in California, and a study was carried out on all the premutation carriers. It was found that in male carriers, 17% in their 50s were affected by tremor and ataxia, 38% in their 60s, 42% in their 70s, and 75% in their 80s. These rates were significantly different compared to age-matched male relatives who did not have the premutation. The female carriers were not significantly different from controls in this study.

Subsequently, the authors and others have reported on older female carriers who have developed FXTAS (Berry-Kravis et al., 2007; Coffey et al., 2008; Hagerman and Hagerman, 2004; O'dwyer et al., 2005; Rodriguez-Revenga et al., 2009). Typically, fewer females with the premutation develop FXTAS (8–16%) compared to males (Coffey et al., 2008; Rodriguez-Revenga et al., 2009), and when they do develop FXTAS, they have less brain atrophy, less white matter disease, and only 13% have the MCP sign (Adams et al., 2007). Seritan and colleagues reported the presence of dementia in approximately 50% of male carriers with FXTAS, but it is rare in female carriers with FXTAS (Seritan et al., 2008).

Cognitive problems usually begin with memory deficits and then progress to executive function deficits by the time the patients present with tremor and ataxia problems (Brega et al., 2008; Grigsby et al., 2006, 2007, 2008; Koldewyn et al., 2008; Loesch et al., 2003). Many patients may have a disinhibited sense of humor, leading to problems in public, which can be a burden to families. The dementia that develops is a frontal subcortical dementia initially, and, on occasion, the initial symptom of FXTAS can be cognitive disorientation (Bourgeois et al., 2009). Treatment of the cognitive loss includes the use of medications important for Alzheimer disease such as memantine, since glutamate toxicity may be a factor in the pathophysiology of this disease. Treatment of early depression or anxiety as described below is also important.

33.2.2.3 Primary Ovarian Insufficiency

Approximately 16–20% of carriers can experience POI, and in carriers this is called FXPOI. As the CGG repeat number increases, the prevalence of POI is increased until about 120 repeats, and then the prevalence will drop to less than 16% (Sullivan et al., 2005). There is evidence that RNA toxicity affects the viability of the granulosa cells that support the ovum, although RNA toxicity may also decrease the viability of the ovum directly. Women with the premutation may also demonstrate mild elevations of follicle stimulation hormone (FSH) compared to controls even before they go into FXPOI (Welt et al., 2004). Of note, POI was previously referred to as premature ovarian failure (POF), but was renamed POI because a limited number of patients may become pregnant after experiencing the lack of menses (Welt, 2008). The American College of Obstetrics and Gynecology has recommended that all women who present with POI should be screened by fragile X DNA testing to see if they have the premutation (ACOG Committee Opinion, 2006). Approximately 1–7.5% of those with spontaneous POI have the premutation, and 13% of those who have a familial history of POI turn out to have the premutation (Welt, 2008). Once FXPOI is diagnosed, there are treatments including hormonal support and treatment of the emotional problems that often accompany this diagnosis (Nelson, 2009; Wittenberger et al., 2007).

33.2.2.4 Psychiatric Manifestations of Premutation Carriers

Emotional problems are common in carriers of the premutation both with and without FXTAS. A study carried out by Rodriquez-Ravenga et al. evaluated 34 women with the premutation and a child with FXS compared to 39 women without the premutation but with a child with developmental disabilities (DD) compared to age-matched control women without the premutation and without a child with DD. They used psychiatric questionnaires including the Symptom Checklist-90 (SCL-90) and the Beck Inventory to assess psychiatric problems. They found higher rates of psychiatric problems in women with the premutation and in women with a child with DD compared to controls, and this likely reflects the increased emotional stress that occurs in raising a child with FXS or DD. However, the women with the premutation had higher scores on the depression subtest than the other two groups of women (Rodriguez-Revenga et al., 2008). Similar results were seen utilizing the SCL-90 in 144 women and 68 men with the premutation (Hessl et al., 2005). Compared to published normative data, there was a higher rate of obsessive–compulsive symptoms in carriers of both sexes compared to controls. In addition, elevations in the *FMR1* mRNA levels correlated with increased symptoms on the obsessive–compulsive scale and on the psychoticism scale in males both with and without FXTAS (Hessl et al., 2005).

Roberts et al. utilized the Structured Clinical Interview for DSM IV (SCID) for mood and anxiety disorders in 93 women with the premutation. They found a significant increase in the rate of lifetime major depressive disorder (MDD) in premutation women (43%) that was significantly different from the rate of MDD in the National Comorbidity Survey Replication (NCS-R) data set. Forty-eight percent of these women had their first episode before the birth of their child with FXS, and the mean age of the first episode was 27 years. In the overall group, 31% had sought help from a professional, and 35% were on psychiatric medication. The premutation group also had four times the rate of lifetime panic disorder without agoraphobia and current agoraphobia without depressive disorder compared to the NCS-R. There was an inverse correlation between the CGG repeat number and the prevalence of depression, suggesting that the lowered FMRP levels at the upper end of the premutation range was protective for MDD (Roberts et al., 2009). Sobesky and colleagues in their studies of women with the full mutation found a protective benefit of lowered FMRP in that executive function deficits in the full mutation interfered with their understanding of psychiatric problems (Sobesky et al., 1996). The same issue may occur at the upper end of the premutation range.

Bourgeois et al. carried out an SCID on 85 individuals with the fragile X premutation, 47 with the FXTAS (33 males, 14 females, mean age 66) and 38 without FXTAS (16 males, 22 females, mean age 52). They found that in FXTAS, 30 cases (65%) met the lifetime Diagnostic and Statistical Manual, fourth edition, text revision (DSM-IV-TR) criteria for mood disorder, and 24 cases (52%) met the lifetime DSM-IV-TR criteria for anxiety disorder. Among the non-FXTAS subjects, there were 15 cases (42%) of lifetime mood disorder and 18 cases (47%) of lifetime anxiety disorder. When compared to age-specific NCS-R data, the lifetime prevalence of any mood disorder, MDD, any anxiety disorder, panic disorder, specific phobia, and post-traumatic stress disorder (PTSD) was significantly higher in subjects with FXTAS. The lifetime rates of social phobia in individuals with the premutation without FXTAS were significantly higher than NCS-R data (Bourgeois et al., 2011). Therefore, individuals with the premutation may manifest psychiatric problems well before the onset of neurological problems. The psychiatric problems are likely related to the RNA toxicity that is apparent in the limbic system before the onset of FXTAS. Hormonal difficulties may also exacerbate the psychiatric problems that are common in carriers, as noted earlier.

33.2.2.5 *Associated Medical Conditions in Carriers*

A recent study of 146 women with the premutation demonstrated that approximately 8% who were not initially referred for this problem suffered from FXTAS (Coffey et al., 2008). In addition, in those with FXTAS, about 61% had hypertension, 50% had thyroid problems, usually hypothyroidism, 22% had seizures, and 43% had fibromyalgia; all these problems were higher than that found in age-matched controls without the premutation. Neuropathy problems were diagnosed in 53% in those with FXTAS, which is expected, but many women without FXTAS but with the premutation had neurological symptoms. These problems include symptoms of numbness and tingling in 34%, muscle pain in 20%, and intermittent tremor in 9% of women without FXTAS. These problems were more frequent than in age-matched controls without the premutation. MS occurs in 2–3% of carriers, which is a higher prevalence than what is seen in the general population of women (Zhang et al., 2009). One carrier who experienced a rapid decline over 15 years of MS demonstrated both active MS lesions in the CNS with inflammation and inclusions of FXTAS (Greco et al., 2008). There appears to be more autoimmune disease in carriers, and this is currently being investigated.

33.3 GENETICS AND NEUROBIOLOGY

33.3.1 Genetics – *FMR1* Gene

The genetic cause of FXS, the *FMR1* gene mutation, was elucidated in 1991 (Verkerk et al., 1991). The gene is 38 kB in length (Penagarikano et al., 2007). It has an unstable CGG region at the 5′ promoter region, and the fragile site is located at band Xq27.3 (Harrison et al., 1983). The fragile site was named FRAXA and was the first fragile site described on the X chromosome.

33.3.1.1 *Trinucleotide Repeat Disorder*

FXS was one of the first disorders associated with a trinucleotide repeat expansion. These trinucleotide repeat expansions may occur in coding regions in syndromes such as in Kennedy disease (spinal–bulbar muscular atrophy). The expansion can also occur in noncoding regions, as is the case in FXS. The reason for the trinucleotide repeat expansion is unclear. One theory is that it occurs during an unequal exchange of genetic

material during meiosis or mitosis (Penagarikano et al., 2007).

Premutation alleles are unmethylated. The premutation alleles are unstable and may undergo expansions during oogenesis (Fu et al., 1991; Oberle et al., 1991). This leads to the phenomenon of premutation carrier progeny having either the premutation or the full mutation. It is known that all mothers of children with the full mutation either are premutation carriers or have the full mutation themselves. In general, premutation sizes of 90–100 CGG repeats usually expand to the full mutation (Nolin et al., 2003). There has been no documentation of a normal-sized allele expanding to a full mutation in one generation. However, there has been one instance reported of a gray zone allele expanding to a full mutation in two generations. The maternal grandfather had 52 CGG repeats (gray zone), his daughter had 56 CGG repeats (premutation), and his grandson had the full mutation with 538 CGG repeats (Fernandez-Carvajal et al., 2009a).

As stated previously, having greater than 200 CGG repeats is considered a full mutation. It is sometimes reported as a single number or may be reported as a range of repeat sizes. The *FMR1* gene, its upstream CpG island, and surrounding sequence are hypermethylated, which inactivates the gene (Sutcliffe et al., 1992). With complete methylation, there is little or no *FMR1* mRNA produced. Methylation of the gene works to inactivate transcription directly by inhibiting the binding of transcription factors and also works indirectly by inducing condensation of the chromosome, which then prevents transcription factors from binding (Penagarikano et al., 2007). This leads to the halting of production of *FMR1* mRNA and the gene's product, FMRP. It is the loss of this protein that leads to the phenotype in FXS (Pieretti et al., 1991).

Mosaicism may refer to the size of the allele or the pattern of methylation. Approximately 12% of individuals with the full mutation are considered to be size mosaics, with a combination of premutation or full-size alleles. Six percent of full-mutation individuals are methylation mosaics, with both methylated and unmethylated alleles present on Southern blot (Rousseau et al., 1994). The percentage of methylation can have significant effects on the phenotype. Merenstein and colleagues examined methylation in 218 full-mutation male patients, and those with complete methylation had the lowest IQ scores and greatest physical involvement (Merenstein et al., 1996).

There are additional ways to cause the fragile X phenotype besides the trinucleotide expansion. More than 15 different deletions affecting the *FMR1* gene and point mutations that can lead to the phenotype have been reported (De Boulle et al., 1993; Gedeon et al., 1992; Hammond et al., 1997). The phenotype of FXS can also occur in an individual with the premutation who has significantly low levels of FMRP.

Contraction of the allele occurs when an individual transmits a smaller-sized allele to offspring. This has been documented in mother-to-daughter transmission of the *FMR1* gene (Brown et al., 1996) and also in approximately one-third of father-to-daughter transmissions (Fisch et al., 1995; Nolin et al., 1996).

33.3.1.2 FMRP and Upregulation of Other Proteins

The FMRP has a maximum length of 632 amino acids and a molecular mass of 80 kDa (Ashley et al., 1993). It is involved in a variety of cellular processes and is expressed in different types of tissues but primarily in neurons and in the testes (Devys et al., 1993; Khandjian et al., 1995). FMRP has been shown to interact as a selective RNA-binding protein (Bassell and Warren, 2008). FMRP shuttles between the nucleus and cytoplasm and is localized to dendrites (Feng et al., 1997). It is involved in regulating the translation of a subset of mRNAs at synapses. It has been hypothesized that FMRP is involved in chromatin remodeling in the nucleus as well (Krueger and Bear, 2011). Usually, FMRP inhibits translation, but this inhibition is released with activation of metabotropic glutamate receptors (mGluRs). Activation of the mGluR5 pathway leads to long-term depression and weak synaptic connections (Bear et al., 2004). In the absence of FMRP, there is upregulation of the mGluR5 pathway leading to a reduction of alpha-amino-3-hydroxy-5-methyl-4-isoxazole-propionic acid (AMPA) receptors in the dendrite. Without FMRP, there is dysregulated translation of other mRNAs, which leads to altered synaptic plasticity (Brown et al., 2001).

Individuals with FXS have more dendritic spines, which are thinner and longer compared to those in unaffected individuals. Bagni and Greenough presented a description of how FMRP may lead to the synaptic alterations, including regulation of translation and protein interactions including Cytoplasmic FMR1-interacting protein 1 (CYFIP1), Cytoplasmic FMR1-interacting protein 2 (CYFIP2), myosin VA, and nucleolin (Bagni and Greenough, 2005). The effect of the FMRP defect on synapses has been demonstrated in a *Drosophila* model of FXS, where the loss-of-function models show defects in synaptic structure and disturbed neurotransmission. The *dfxr* (*Drosophila* fragile X-related) gene was shown to inversely regulate expression of a microtubule-associated protein, Futsch. Therefore, dFXR was proposed to function as a translational repressor of Futsch (Zhang et al., 2001). Subsequent work has led to the understanding of FMRP as a master protein that regulates hundreds of mRNAs usually through inhibition of translation (Darnell et al., 2011). The synaptic structural defect can be seen in the fragile X mouse model as well. Cruz-Martin and colleagues observed a delayed downregulation in dendritic spine turnover from immature filopodia to mature spines during the early postnatal period compared to wild-type mice. This suggests a developmental delay in the maturation of dendritic spines in FXS (Cruz-Martin et al., 2010).

The decrease in FMRP may explain the increased prevalence of developmental delays and autism. FMRP regulates the translation of proteins associated with autism including neuroligins, neurorexins, the SHANK family of proteins, PTEN, and CYFIP (Bassell and Warren, 2008; Darnell et al., 2005, 2011).

The extracellular signal-related kinase (ERK) pathway seems to be affected in the fragile X mouse model, and the ERK1/2 pathway activation is delayed in individuals with FXS (Weng et al., 2008). Osterweil and colleagues found that basal protein synthesis in the hippocampus of the knockout (KO) mouse was reduced to wild-type levels when treated with an mGluR5 inhibitor as well as an ERK 1/2 inhibitor (Osterweil et al., 2010). The ERK pathway may also be a target for therapies in FXS.

Besides upregulation of the mGluR5 pathway, the lack of FMRP is associated with upregulation of the mammalian target of rapamycin (mTOR) pathway and downregulation of the PTEN pathway, the GABA receptors, and the dopamine pathway (Bassell and Warren, 2008; D'hulst and Kooy, 2007; Wang et al., 2008). These pathways may be the targets of treatment for FXS, which is discussed later. The review by Krueger and Bear can be referred to for further details regarding milestones in the pathophysiology of FXS and implications for treatment approaches as well (Krueger and Bear, 2011).

33.3.1.3 Inheritance

The inheritance pattern of FXS is X-linked with variable penetrance. Since it is an X chromosome-linked disorder, it affects more males than females.

33.4 TESTING

33.4.1 DNA Testing

The first method for testing for FXS was cytogenetic testing. The newest standard method of screening for FXS is through DNA testing, which includes a polymerase chain reaction (PCR), and Southern blot. The advantage of PCR is that it uses smaller amounts of DNA, is less expensive, and has a faster result time in comparison to the Southern blot. Its drawback is that it cannot detect longer DNA sequences, which is why both PCR and Southern blots are typically used. The number of CGG repeats is typically reported, as is methylation status in the full mutation. Using both methods, testing is 99% sensitive (McConkie-Rosell et al., 2005). Testing methods do not typically look for conventional mutations.

Guidelines for testing have been issued by the American College of Medical Genetics (Sherman et al., 2005; Table 33.1).

TABLE 33.1 Individuals for Whom Fragile X Testing Should be Considered

Individuals with mental retardation, developmental delay, or autism, especially if they have any physical or behavioral characteristics of fragile X syndrome, a family history of fragile X syndrome, or male or female relatives with undiagnosed mental retardation

Individuals seeking reproductive counseling who have a family history of fragile X syndrome or a family history of undiagnosed mental retardation

Fetuses of known carrier mothers

Affected individuals or their relatives in the context of a positive cytogenetic fragile X test result who are seeking further counseling related to the risk of carrier status among themselves or their relatives

Source: Sherman S, Pletcher BA, and Driscoll DA (2005) Fragile X syndrome: Diagnostic and carrier testing. Genetics in Medicine 7: 584–587.

Other conditions that may be related to the fragile X family of disorders should also be evaluated with fragile X testing. For women experiencing ovarian dysfunction, the American College of Medical Genetics recommends testing with elevated FSH levels, especially if they have a family history of POF, or FXS, or relatives with undiagnosed mental retardation. Testing for the fragile X premutation and FXTAS is recommended for those individuals who are experiencing late-onset intention tremor and cerebellar ataxia of unknown origin, especially if they have a family history of movement disorders or FXS, or relatives with undiagnosed mental retardation (Sherman et al., 2005).

A new method for quantifying FMRP has been discovered, which uses an ELISA assay (Iwahashi et al., 2009). It is now possible to look more quantitatively at exactly how much protein a particular individual with FXS is able to make.

33.4.2 Newborn Screening/Blood Spot

Population-based screening was examined by Palomaki (1994), and subsequent studies showed the efficacy of screening programs in identifying carriers and affected fetuses (Pesso et al., 2000; Toledano-Alhadef et al., 2001). Fetal testing through amniotic cells or chorionic villi is also possible when one of the parents is known to be a carrier.

Different methods for screening large populations for FXS have been developed using newborn blood spots and PCR. Coffee et al. described screening 36 124 de-identified newborn males with an assay. The assay had 100% specificity and sensitivity for detecting *FMR1* methylation in males and detected excess *FMR1* methylation in 82% of females with full mutations. They identified 64 males with *FMR1* methylation and found 7 to have the full mutation (Coffee et al., 2009). Another method utilizing nested PCR was developed by Tassone

et al., and this can identify both males and females with the full mutation or the premutation (Tassone et al., 2008). This methodology was implemented in Spain where a total of 10 000 newborn blood spots were screened. So far, this study has reported on 5267 male blood spots where 199 gray zone alleles, 21 premutation alleles, and two full-mutation alleles (1 in 2633) have been identified (Fernandez-Carvajal et al., 2009b). Modifications to the Tassone technique have been carried out and recently reported (Filipovic-Sadic et al., 2010). Newborn screening is currently occurring in numerous centers around the United States.

33.4.3 Cascade Testing

Identifying affected members of a family is important, as changes in the fragile X gene can affect multiple family members through its unstable trinucleotide expansion. Usually, a proband is identified, which is often an affected child with FXS. Other members of the family should be identified and counseled about being tested.

33.4.4 Genetic Counseling

Genetic counselors are important members of the treatment team, particularly in helping patients and their families understand the diagnosis. Counseling sessions provide key educational opportunities. Important issues to be discussed should include discussion of the clinical presentation and inheritance patterns, treatments, therapies, follow-up recommendations, referrals for medical, educational, and mental health treatments, and contact information for support groups (McConkie-Rosell et al., 2005). Degree of involvement in both the premutation and the full mutation should be reviewed by the genetic counselor for the whole family tree. There are typically multiple family members involved whenever a proband is identified (Chonchaiya et al., 2009b; McConkie-Rosell et al., 2007).

33.5 TARGETED TREATMENTS

With better understanding of the molecular pathways affected by the lack of FMRP, targeted treatments are currently being studied as described below.

33.5.1 mGluR5 Antagonists

FMRP normally modulates dendritic maturation involving inhibition of the mGluR system 1- and 5-mediated translation in neurons (Aschrafi et al., 2005; Weiler et al., 2004). The mGluR5 system has been found to be upregulated in the fragile X KO mouse, with a resulting enhanced long-term depression in the hippocampus (Huber et al., 2002). Long-term depression is the phenomenon of a decrease in synaptic effectiveness, thus leading to weakened synaptic connections (Bear and Abraham, 1996).

Lack of inhibition of the mGluR5 system also results in increased internalization of AMPA receptors (Snyder et al., 2001). AMPA receptors in the synapse have been correlated to synaptic protrusion. The mGluR systems are excitatory systems involved in many different areas, and the resulting upregulation may explain some of the physical and neurological aspects of FXS including seizures, electroencephalographic abnormalities including spike wave discharges, anxiety, tactile defensiveness, and difficulty with coordination (Bear et al., 2004).

Features of the fragile X KO mouse have been thought to be due to the increased activity of the mGluR5 system. The effects of decreasing mGluR5 receptors have been evaluated in the KO mouse model of fragile X. In 2007, Dolen et al. created a fragile X KO mouse that also had a 50% decrease in the number of mGluR5 receptors. The mouse demonstrated rescue of phenotypic features such as dendritic spines and a normalization in audiogenic seizures. Macroorchidism was the one feature studied that was not rescued by decreasing mGlur5 receptors (Dolen et al., 2007). mGluR5 antagonists have also been shown to rescue the *Drosophila* model of fragile X (McBride et al., 2005).

mGluR5 antagonists are being studied as a targeted treatment for FXS. In the *FMR1* KO mouse, treatment with MPEP (2-methyl-6-(phenylethynyl)-pyridine hydrochloride), an mGluR5 antagonist, was shown to decrease the startle response and rescue the protrusion phenotype of hippocampal neurons (de Vrij et al., 2008). Trials of mGluR5 antagonists in patients with FXS have also begun. One example is fenobam, a highly potent and selective mGluR5 antagonist. In 2009, Berry-Kravis and colleagues completed a pilot single-dose trial of fenobam and observed calm behavior in 9 out of 12 patients studied with no significant adverse effects. There was an improvement in 6 out of 12 patients in prepulse inhibition testing, which is a measure of impulse control and sensorimotor gating (Berry-Kravis et al., 2009). The use of the mGluR5 antagonists is likely to be beneficial in treating seizures because seizures are ameliorated in the KO animal model with mGluR5 antagonists (Hagerman and Stafstrom, 2009; Yan et al., 2005). AFQ056, an mGluR5 antagonist, was studied in 30 male individuals with FXS aged 18–35 in a double-blind, crossover study. Improvements were seen in seven patients who had fully methylated *FMR1* promoter regions and no response in patients with partial promoter methylation. The most common adverse event was mild to moderately severe fatigue or headache (Jacquemont et al., 2011).

33.5.2 GABA Agonists

GABA dysregulation has been observed in the fragile X KO mouse, and amygdala hyperexcitability has been normalized in the knockout mouse with treatment using the GABA agonist, gaboxadol (Olmos-Serrano et al., 2010). Studies of the GABA B agonist arbaclofen, which is the R isomer of baclofen, are currently being carried out, including a double-blind crossover study in children and adults with FXS from 5 years of age and older. Arbaclofen works at a presynaptic receptor and lowers the level of glutamate at the synapse. It also lowers the level of excess protein produced at the synapse in FXS. A trial of arbaclofen in 63 patients with FXS has shown efficacy in those with autism or significant social deficits (Berry-Kravis et al., 2010).

33.5.3 Minocycline

Minocycline is a second-generation semisynthetic tetracycline derivative (Shetty, 2002). It was first introduced in 1967 and is generally well tolerated (Smith and Leyden, 2005). Minocycline is commonly used as an antibiotic in the treatment of acne vulgaris, and its class of medications, tetracyclines, are the treatment of choice for certain bacterial infections such as Rocky Mountain spotted fever and brucellosis. It has been investigated as a neuroprotective agent in neurological diseases including Huntington's disease, amyotrophic lateral sclerosis, and MS (Kim and Suh, 2009). The mechanism of action by which minocycline exerts its neuroprotective effects is not precisely known, but likely has to do with its anti-inflammatory and anti-apoptotic effects. Minocycline's effects are reported to be through inhibition of cytochrome C, caspases 1 and 3, cytokines, and the suppression of metalloproteinase activity (Chen et al., 2000; Stirling et al., 2005). The latter effect is the key to minocycline's efficacy in the treatment of FXS.

As stated before, the lack of FMRP in FXS leads to upregulation of other proteins, one of these being matrix metalloproteinase 9 (MMP9). Matrix metalloproteinases are involved in extracellular degradation of proteins and are important for synaptic structure and plasticity (Sternlicht and Werb, 2001). Elevated MMP9 activity has been proposed as one mechanism for impaired dendritic spine maturation in FXS. Bilousova et al. demonstrated that minocycline treatment lowers MMP9 levels and matures synaptic connections in cultured hippocampal cells and that minocycline treatment for 1 month rescued synaptic abnormalities in the *FMR1* KO mouse. The treated mice showed improvements in anxiety on elevated plus maze and exploratory behavior on Y maze (Bilousova et al., 2009).

Minocycline treatment trials for FXS are currently underway. Utari and colleagues described 50 males and females with FXS who were treated clinically with minocycline for 2 weeks or more. Using parent impressions, behavioral and cognitive changes were noted including improvements in language (54%), attention (50%), social communication (44%), and anxiety (30%). The side effects described included pigmentation of the nails (in one patient), loss of appetite, gastrointestinal upset, loose stools, and headache (Utari et al., 2010). An open-label study by Paribello and colleagues showed significant improvement in the Aberrant Behavior Checklist irritability subscale scores, Clinical Global Impression Scale – Improvement scores, and visual analog for behavior scores after 8 weeks of treatment with minocycline (Paribello et al., 2010). Currently, trials of minocycline are underway including a double-blind controlled trial in children and adolescents with FXS.

Side effects of minocycline include graying of teeth and gums, particularly in those younger than 8 years of age, gastrointestinal upset, sun sensitivity, and, in rare cases, pseudotumor cerebri and a lupus-like syndrome. These side effects should be monitored in any child or adult undergoing minocycline treatment. For children treated for longer than 3–6 months, an antinuclear antibody blood test should be done and, if positive, consideration should be given as to whether to continue with the medication in the long term.

33.5.4 Riluzole

Riluzole is a medication that is considered to have inhibitory effects on the glutamatergic system and is thought to potentiate the GABA system. It is approved for use in amyotrophic lateral sclerosis in adults. It has been associated with improvements in treatment-resistant depression (Zarate et al., 2004) and, in combination with other medications, was helpful in treatment-resistant obsessive–compulsive disorder (OCD) (Grant et al., 2007). In a 6-week, open-label study in six adults with FXS, riluzole was well tolerated; although it was associated with corrected peripheral ERK activation, riluzole was associated with a clinical response in only one of the six participants (Erickson et al., 2011).

33.5.5 Symptomatic Treatments

A number of currently available medications are useful in the treatment for individuals with FXS and are discussed in detail below.

33.5.5.1 *For Mood Stability*

Mood instability is seen in the majority of boys with FXS, and perhaps the best medication for this problem is a low dose of aripiprazole (Chonchaiya et al., 2009a; Hagerman et al., 2009). Aripiprazole is an atypical antipsychotic that has fewer problems with weight gain than risperidone, although most children do gain weight on this medication. In the authors' experience, aripiprazole is helpful for anxiety and for ADHD symptoms in addition to tantrums. Usually, just 0.5–2 mg works best when given at bedtime to younger children. Rarely does the dose need to go over 5 mg, and sometimes a higher dose leads to an increase in irritability or activation.

Additional mood stabilizers include anticonvulsants such as valproate or lamotrigine, and these medications can be helpful for aggression. Lithium may also be considered a targeted treatment for FXS because it works by lowering the mGluR5 system. An open trial of lithium was helpful in a small cohort of children and adults with FXS who had a poor response to other medications (Berry-Kravis et al., 2008).

33.5.5.2 *For ADHD Symptoms*

ADHD symptoms respond well to stimulants with approximately 66% demonstrating a positive response if the medications are given after 5 years of age (Hagerman et al., 2009). The alpha agonists including clonidine and guanfacine can also be helpful for calming the hyperarousal and tantrum behavior (Hagerman et al., 1998, 2009). These medications are often more helpful than stimulants for the child under 5 years of age. Clonidine can also be helpful for sleep disturbance, although melatonin should be tried first for sleep problems (Wirojanan et al., 2009).

33.5.5.3 *For Anxiety*

Anxiety is almost a universal problem for individuals with FXS and for many carriers as described above. The use of selective serotonin reuptake inhibitors (SSRIs) is very helpful for these problems. Fluoxetine is the most activating of the SSRIs, but that activation may be helpful to boost language in those with selective mutism. Sertraline has had the most common use in young children, and it deserves further study regarding its ability to enhance language in toddlers and to decrease the rate of autism (Hagerman et al., 2009). Many patients may benefit from a combination of a stimulant for ADHD and an SSRI for anxiety.

33.6 BEHAVIORAL INTERVENTIONS

33.6.1 Educational

Effective treatment for FXS consists of a combination of medication therapies as discussed previously and behavioral interventions. With the variable phenotype of FXS and premutation involvement, individualized therapy recommendations are essential. There have been very few studies examining educational and behavioral therapies in FXS (Hall, 2009). Formal testing for language abilities and cognitive measures may be helpful in formulating an educational and therapy plan. Children with FXS will qualify for specialized educational methods including an individualized education plan.

Work in related disorders such as autism has shown that behavioral interventions can help, especially when started at a young age (Dawson et al., 2010). For those with FXS plus autism, well-established treatment models such as the Denver model (Dawson et al., 2010) and discrete trial training (Applied Behavior Analysis) (Smith et al., 2007) are recommended. Early identification of children with FXS is important, as therapies may be implemented sooner. Additionally, for patients with social deficits such as ASD or social anxiety, social skills groups may be helpful.

Behavior problems have been one of the greatest concerns in parent surveys (Bailey et al., 2000). Different types of therapies have been initiated, including sensory integration therapy and psychological counseling for behavioral interventions. Heightened sensitivity to sensory stimuli is related to hyperarousal and the autonomic dysfunction previously described. Occupational therapists trained in sensory integration therapy as well as fine motor therapy can be a valuable resource for the patients and their families for behavioral problems (Scharfenaker et al., 2002). Eye contact may be improved through behavioral intervention and was shown to be enhanced using a method of percentile scheduling with reinforcement in three out of six fragile X males studied (Hall et al., 2009). However, therapy to improve eye contact is controversial because it may worsen anxiety. A parent-initiated sleep therapy was evaluated in a population of five autistic children and seven children with FXS. The program reduced settling down to sleep and night awakenings (Weiskop et al., 2005).

Speech and language therapy can be essential to the development of a child with FXS as most can have speech delays. An audiology evaluation may be helpful, especially if hearing loss is suspected or recurrent otitis media is a problem. A conductive hearing loss due to fluid accumulating behind the ear is common in FXS. If this occurs as a child during the particularly crucial formative years of speech, it can further delay language development, so aggressive treatment with ear, nose and throat (ENT, also known as otolaryngology) evaluation and PE tubes is recommended (Hagerman, 2002a). Physical therapy can be used to help with the coordination difficulties in FXS and in premutation carriers with FXTAS.

For all of the above therapies, it must be emphasized that while the hours spent working with a therapist are fundamental, a home therapy program that families can continue to implement is needed to achieve the most benefit from the therapies.

33.6.2 Computer Work

Technological advances have been very helpful in providing a way for those with FXS to improve deficiencies. Assistive technology refers to technology that can help enhance functional abilities. For example, computerized learning systems have been created such as Writing with Symbols, which is a word prediction program. Assistive technology may also refer to picture cards, such as those used in the Picture Exchange Communication System. These can enhance communication, particularly for those who have speech delays (Hess et al., 2009). These devices can be obtained through the school or through private purchase. Children with FXS are often adept at computer use, making these interventions enjoyable as well as educational. There are also assistive technology specialists who can provide support with these interventions. There are current trials to assess the utility of assistive technology alone and in combination with medications.

For patients with FXTAS, using computer-assisted devices such as video games or the Nintendo Wii system with a balance board may be helpful to improve balance. Studies examining this are beginning, and the Wii system was found to be useful in improving balance and lower limb muscle strength in a pilot study of unaffected women (Nitz et al., 2010).

33.7 SUMMARY AND FUTURE PERSPECTIVES

33.7.1 Summary

In summary, FXS and its associated disorders have a rich history as a group of disorders caused by a CGG trinucleotide repeat sequence in a single gene, *FMR1*. Although the fragile X family of disorders has an identified single gene mutation cause, the resulting clinical phenotype can be quite diverse. Individuals with FXS may display ID, attention problems, autistic features or autism, as well as physical features including macroorchidism and prominent ears. Females with FXS may have learning disabilities, social anxiety, or impulsivity, in addition to variable physical features. Premutation carriers may develop FXTAS, with cognitive problems, tremor, and difficulties with balance. FXPOI may also occur in female premutation carriers in addition to psychiatric problems, hypertension, autoimmune dysfunction, and mild neurological problems such as neuropathy. Testing is widely available using PCR and Southern blotting in blood and also in amniotic cells or other tissue. Newborn screening is being studied around the world to increase early identification and treatment. Genetic counseling is important, given the possible implications of a diagnosis for both patients and their extended families.

A greater understanding of how the lack or deficiency of FMRP brings about the phenotype of FXS is developing, including the involvement of pathways such as the mGluR5 pathway, GABA pathways, AMPA receptors, PTEN pathway, and mTOR pathway. Treatments have been developed that use the understanding of these pathways to provide reversal of the neurobiological deficits in FXS. mGluR5 antagonists, GABA agonists, and minocycline as well as medications to help anxiety, aggression, and mood stabilization are currently being employed. Educational and behavioral interventions are also important components of a treatment regimen.

33.7.2 Future Perspectives

Treatment for FXS will likely combine both targeted treatments to reverse the neurobiological abnormalities and educational and learning paradigms to strengthen the synaptic connections that are facilitated by the targeted treatments. New treatments are being assessed now in cultured neurons for both premutation and full-mutation disorders. The use of mGluR5 antagonists has been shown to be helpful in a model of autism in the mouse (Silverman et al., 2010), and this suggests that these targeted treatments for FXS will also be helpful for autism. Currently, studies of arbaclofen are being conducted in autism, and trials of mGluR5 antagonists will be initiated in autism in the near future. Fragile X is leading the way for new targeted treatments in autism and perhaps for other neurodevelopmental disorders that have similar molecular dysfunctions. The future is bright for reversing the intellectual disabilities and the autism in FXS.

SEE ALSO

Cognitive Development: Developing Attention and Self Regulation in Infancy and Childhood; Statistical Learning Mechanisms in Infancy; The Development of Visuospatial Processing; The Effects of Stress on Early Brain and Behavioral Development; The Neural Architecture and Developmental Course of Face Processing; The Neural Correlates of Cognitive Control and the Development of Social Behavior. **Diseases**: Autisms; Excitation-Inhibition | Epilepsies; Language Impairment; The Developmental Neurobiology of Repetitive Behavior.

Acknowledgments

This work was supported by the National Fragile X Foundation and the MIND Institute, National Institute of Health grants HD036071 and HD055510; Neurotherapeutic Research Institute (NTRI) grants DE019583, and DA024854; National Institute on Aging grants AG032119 and AG032115; National Center for Research Resources UL1 RR024146; support from the Health and Human Services Administration of Developmental Disabilities grant 90DD05969.

References

Abrams, M.T., Reiss, A.L., Freund, L.S., et al., 1994. Molecular-neurobehavioral associations in females with the fragile X full mutation. American Journal of Medical Genetics 51, 317–327.

ACOG Committee Opinion, 2006. Screening for fragile X syndrome. Obstetrics and Gynecology 107, 1483–1485.

Adams, J.S., Adams, P.E., Nguyen, D., et al., 2007. Volumetric brain changes in females with fragile X-associated tremor/ataxia syndrome (FXTAS). Neurology 69, 851–859.

Angkustsiri, K., Wirojanan, J., Deprey, L.J., Gane, L.W., Hagerman, R.J., 2008. Fragile X syndrome with anxiety disorder and exceptional verbal intelligence. American Journal of Medical Genetics. Part A 146, 376–379.

Aschrafi, A., Cunningham, B.A., Edelman, G.M., Vanderklish, P.W., 2005. The fragile X mental retardation protein and group I metabotropic glutamate receptors regulate levels of mRNA granules in brain. Proceedings of the National Academy of Sciences of the United States of America 102, 2180–2185.

Ashley Jr., C.T., Wilkinson, K.D., Reines, D., Warren, S.T., 1993. FMR1 protein: Conserved RNP family domains and selective RNA binding. Science 262, 563–566.

Aziz, M., Stathopulu, E., Callias, M., et al., 2003. Clinical features of boys with fragile X premutations and intermediate alleles. American Journal of Medical Genetics 121B, 119–127.

Bagni, C., Greenough, W.T., 2005. From mRNP trafficking to spine dysmorphogenesis: The roots of fragile X syndrome. Nature Reviews Neuroscience 6, 376–387.

Bailey Jr., D.B., Mesibov, G.B., Hatton, D.D., et al., 1998. Autistic behavior in young boys with fragile X syndrome. Journal of Autism and Developmental Disorders 28, 499–508.

Bailey, D.B., Skinner, D., Hatton, D., Roberts, J., 2000. Family experiences and factors associated with the diagnosis of fragile X syndrome. Journal of Developmental and Behavioral Pediatrics 21, 315–321.

Bailey Jr., D.B., Skinner, D., Sparkman, K.L., 2003. Discovering fragile X syndrome: Family experiences and perceptions. Pediatrics 111, 407–416.

Bailey Jr., D.B., Raspa, M., Olmsted, M., Holiday, D.B., 2008. Co-occurring conditions associated with *FMR1* gene variations: Findings from a national parent survey. American Journal of Medical Genetics. Part A 146A, 2060–2069.

Bassell, G.J., Warren, S.T., 2008. Fragile X syndrome: Loss of local mRNA regulation alters synaptic development and function. Neuron 60, 201–214.

Baumgardner, T.L., Reiss, A.L., Freund, L.S., Abrams, M.T., 1995. Specification of the neurobehavioral phenotype in males with fragile X syndrome. Pediatrics 95, 744–752.

Bear, M.F., Abraham, W.C., 1996. Long-term depression in hippocampus. Annual Review of Neuroscience 19, 437–462.

Bear, M.F., Huber, K.M., Warren, S.T., 2004. The mGluR theory of fragile X mental retardation. Trends in Neurosciences 27, 370–377.

Bell, M.V., Hirst, M.C., Nakahori, Y., et al., 1991. Physical mapping across the fragile X: Hypermethylation and clinical expression of the fragile X syndrome. Cell 64, 861–866.

Berry-Kravis, E., 2002. Epilepsy in fragile X syndrome. Developmental Medicine and Child Neurology 44, 724–728.

Berry-Kravis, E., Abrams, L., Coffey, S.M., et al., 2007. Fragile X-associated tremor/ataxia syndrome: Clinical features, genetics, and testing guidelines. Movement Disorders 22, 2018–2030.

Berry-Kravis, E., Sumis, A., Hervey, C., et al., 2008. Open-label treatment trial of lithium to target the underlying defect in fragile X syndrome. Journal of Developmental and Behavioral Pediatrics 29, 293–302.

Berry-Kravis, E., Hessl, D., Coffey, S., et al., 2009. A pilot open label, single dose trial of fenobam in adults with fragile X syndrome. Journal of Medical Genetics 46, 266–271.

Berry-Kravis, E., Cherubini, M., Zarevics, P., et al., 2010. Arbaclofen for the treatment of children and adults with fragile X syndrome: Results of a phase 2, randomized, double-blind, placebo-controlled, crossover study [Abstract]. In: International Meeting for Autism Research, Philadelphia, PA, p. 741.

Bilousova, T.V., Dansie, L., Ngo, M., et al., 2009. Minocycline promotes dendritic spine maturation and improves behavioural performance in the fragile X mouse model. Journal of Medical Genetics 46, 94–102.

Boccia, M.L., Roberts, J.E., 2000. Behavior and autonomic nervous system function assessed via heart period measures: The case of hyperarousal in boys with fragile X syndrome. Behavior Research Methods, Instruments, and Computers 32, 5–10.

Bodega, B., Bione, S., Dalpra, L., et al., 2006. Influence of intermediate and uninterrupted FMR1 CGG expansions in premature ovarian failure manifestation. Human Reproduction 21, 952–957.

Bourgeois, J.A., Coffey, S.M., Rivera, S.M., et al., 2009. A review of fragile X premutation disorders: Expanding the psychiatric perspective. Journal of Clinical Psychiatry 70, 852–862.

Bourgeois, J.A., Seritan, A.L., Casillas, E.M., et al., 2011. Lifetime prevalence of mood and anxiety disorders in fragile X premutation carriers. Journal of Clinical Psychiatry 72, 175–182.

Brega, A.G., Goodrich, G., Bennett, R.E., et al., 2008. The primary cognitive deficit among males with fragile X-associated tremor/ataxia syndrome (FXTAS) is a dysexecutive syndrome. Journal of Clinical and Experimental Neuropsychology 30 (8), 853–869.

Bregman, J.D., Leckman, J.F., Ort, S.I., 1988. Fragile X syndrome: Genetic predisposition to psychopathology. Journal of Autism and Developmental Disorders 18, 343–354.

Bretherick, K.L., Fluker, M.R., Robinson, W.P., 2005. FMR1 repeat sizes in the gray zone and high end of the normal range are associated with premature ovarian failure. Human Genetics 117, 376–382.

Brouwer, J.R., Huizer, K., Severijnen, L.A., et al., 2008. CGG-repeat length and neuropathological and molecular correlates in a mouse model for fragile X-associated tremor/ataxia syndrome. Journal of Neurochemistry 107, 1671–1682.

Brown, W.T., Jenkins, E.C., Cohen, I.L., et al., 1986. Fragile X and autism: A multicenter survey. American Journal of Medical Genetics 23, 341–352.

Brown, W.T., Houck Jr., G.E., Ding, X., et al., 1996. Reverse mutations in the fragile X syndrome. American Journal of Medical Genetics 64, 287–292.

Brown, V., Jin, P., Ceman, S., et al., 2001. Microarray identification of FMRP-associated brain mRNAs and altered mRNA translational profiles in fragile X syndrome. Cell 107, 477–487.

Brunberg, J.A., Jacquemont, S., Hagerman, R.J., et al., 2002. Fragile X premutation carriers: Characteristic MR imaging findings in adult males with progressive cerebellar and cognitive dysfunction. American Journal of Neuroradiology 23, 1757–1766.

Butler, M.G., Brunschwig, A., Miller, L.K., Hagerman, R.J., 1992. Standards for selected anthropometric measurements in males with the fragile X syndrome. Pediatrics 89, 1059–1062.

Chen, M., Ona, V.O., Li, M., et al., 2000. Minocycline inhibits caspase-1 and caspase-3 expression and delays mortality in a transgenic mouse model of Huntington disease. Nature Medicine 6, 797–801.

Chen, Y., Tassone, F., Berman, R.F., et al., 2010. Murine hippocampal neurons expressing *Fmr1* gene premutations show early developmental deficits and late degeneration. Human Molecular Genetics 19, 196–208.

Chonchaiya, W., Schneider, A., Hagerman, R.J., 2009a. Fragile X: A family of disorders. Advances in Pediatrics 56, 165–186.

Chonchaiya, W., Utari, A., Pereira, G.M., et al., 2009b. Broad clinical involvement in a family affected by the fragile X premutation. Journal of Developmental and Behavioral Pediatrics 30, 544–551.

Chudley, A.E., Hagerman, R.J., 1987. Fragile X syndrome. Journal of Pediatrics 110, 821–831.

Coffee, B., Keith, K., Albizua, I., et al., 2009. Incidence of fragile X syndrome by newborn screening for methylated FMR1 DNA. American Journal of Human Genetics 85, 503–514.

Coffey, S.M., Cook, K., Tartaglia, N., et al., 2008. Expanded clinical phenotype of women with the FMR1 premutation. American Journal of Medical Genetics. Part A 146A, 1009–1016.

Crawford, D.C., Meadows, K.L., Newman, J.L., et al., 2002. Prevalence of the fragile X syndrome in African-Americans. American Journal of Medical Genetics 110, 226–233.

Cronister, A., Schreiner, R., Wittenberger, M., et al., 1991. Heterozygous fragile X female: Historical, physical, cognitive, and cytogenetic features. American Journal of Medical Genetics 38, 269–274.

Cruz-Martin, A., Crespo, M., Portera-Cailliau, C., 2010. Delayed stabilization of dendritic spines in fragile X mice. Journal of Neuroscience 30, 7793–7803.

Darnell, J.C., Mostovetsky, O., Darnell, R.B., 2005. FMRP RNA targets: Identification and validation. Genes, Brain, and Behavior 4, 341–349.

Darnell, J.C., Van Driesche, S.J., Zhang, C., et al., 2011. FMRP stalls ribosomal translocation on mRNAs linked to synaptic function and autism. Cell 146, 247–261.

Davids, J.R., Hagerman, R.J., Eilert, R.E., 1990. Orthopaedic aspects of fragile-X syndrome. Journal of Bone and Joint Surgery. American Volume 72, 889–896.

Dawson, G., Rogers, S., Munson, J., et al., 2010. Randomized, controlled trial of an intervention for toddlers with autism: The Early Start Denver Model. Pediatrics 125, e17–e23.

De Boulle, K., Verkerk, A.J., Reyniers, E., et al., 1993. A point mutation in the FMR-1 gene associated with fragile X mental retardation. Nature Genetics 3, 31–35.

De Vries, B.B., Fryns, J.P., Butler, M.G., et al., 1993. Clinical and molecular studies in fragile X patients with a Prader–Willi-like phenotype. Journal of Medical Genetics 30, 761–766.

De Vries, B.B., Wiegers, A.M., Smits, A.P., et al., 1996. Mental status of females with an FMR1 gene full mutation. American Journal of Human Genetics 58, 1025–1032.

De Vrij, F.M., Levenga, J., Van Der Linde, H.C., et al., 2008. Rescue of behavioral phenotype and neuronal protrusion morphology in Fmr1 KO mice. Neurobiology of Disease 31, 127–132.

Devys, D., Lutz, Y., Rouyer, N., Bellocq, J.P., Mandel, J.L., 1993. The FMR-1 protein is cytoplasmic, most abundant in neurons and appears normal in carriers of a fragile X premutation. Nature Genetics 4, 335–340.

D'hulst, C., Kooy, R.F., 2007. The GABAA receptor: A novel target for treatment of fragile X? Trends in Neurosciences 30, 425–431.

Dissanayake, C., Bui, Q., Bulhak-Paterson, D., Huggins, R., Loesch, D.Z., 2009. Behavioural and cognitive phenotypes in idiopathic autism versus autism associated with fragile X syndrome. Journal of Child Psychology and Psychiatry 50, 290–299.

Dolen, G., Osterweil, E., Rao, B.S., et al., 2007. Correction of fragile X syndrome in mice. Neuron 56, 955–962.

Dombrowski, C., Levesque, M.L., Morel, M.L., et al., 2002. Premutation and intermediate-size FMR1 alleles in 10 572 males from the general population: Loss of an AGG interruption is a late event in the generation of fragile X syndrome alleles. Human Molecular Genetics 11, 371–378.

Erickson, C.A., Weng, N., Weiler, I.J., et al., 2011. Open-label riluzole in fragile X syndrome. Brain Research 1380, 264–270.

Escalante, J., 1971. Estudo genetic da deficicncia mental. University of Sao Paulo; PhD.

Farzin, F., Perry, H., Hessl, D., et al., 2006. Autism spectrum disorders and attention-deficit/hyperactivity disorder in boys with the fragile X premutation. Journal of Developmental and Behavioral Pediatrics 27, S137–S144.

Feng, Y., Gutekunst, C.A., Eberhart, D.E., et al., 1997. Fragile X mental retardation protein: Nucleocytoplasmic shuttling and association with somatodendritic ribosomes. Journal of Neuroscience 17, 1539–1547.

Fernandez-Carvajal, I., Lopez Posadas, B., Pan, R., et al., 2009a. Expansion of an FMR1 grey-zone allele to a full mutation in two generations. Journal of Molecular Diagnostics 11, 306–310.

Fernandez-Carvajal, I., Walichiewicz, P., Xiaosen, X., et al., 2009b. Screening for expanded alleles of the FMR1 gene in blood spots from newborn males in a Spanish population. Journal of Molecular Diagnostics 11, 324–329.

Filipovic-Sadic, S., Sah, S., Chen, L., et al., 2010. A novel FMR1 PCR method for the routine detection of low abundance expanded alleles and full mutations in fragile X syndrome. Clinical Chemistry 56, 399–408.

Finelli, P.F., Pueschel, S.M., Padre-Mendoza, T., O'brien, M.M., 1985. Neurological findings in patients with the fragile-X syndrome. Journal of Neurology, Neurosurgery, and Psychiatry 48, 150–153.

Fisch, G.S., Snow, K., Thibodeau, S.N., et al., 1995. The fragile X premutation in carriers and its effect on mutation size in offspring. American Journal of Human Genetics 56, 1147–1155.

Fox, P., Fox, D., Gerrard, J.W., 1980. X-linked mental retardation: Renpenning revisited. American Journal of Medical Genetics 7, 491–495.

Franke, P., Leboyer, M., Gansicke, M., et al., 1998. Genotype–phenotype relationship in female carriers of the premutation and full mutation of FMR-1. Psychiatry Research 80, 113–127.

Fu, Y.H., Kuhl, D.P., Pizzuti, A., et al., 1991. Variation of the CGG repeat at the fragile X site results in genetic instability: Resolution of the Sherman paradox. Cell 67, 1047–1058.

Gedeon, A.K., Baker, E., Robinson, H., et al., 1992. Fragile X syndrome without CCG amplification has an FMR1 deletion. Nature Genetics 1, 341–344.

Gokden, M., Al-Hinti, J.T., Harik, S.I., 2009. Peripheral nervous system pathology in fragile X tremor/ataxia syndrome (FXTAS). Neuropathology 29, 280–284.

Goodlin-Jones, B., Tassone, F., Gane, L.W., Hagerman, R.J., 2004. Autistic spectrum disorder and the fragile X premutation. Journal of Developmental and Behavioral Pediatrics 25, 392–398.

Grant, P., Lougee, L., Hirschtritt, M., Swedo, S.E., 2007. An open-label trial of riluzole, a glutamate antagonist, in children with treatment-resistant obsessive-compulsive disorder. Journal of Child and Adolescent Psychopharmacology 17, 761–767.

Greco, C.M., Hagerman, R.J., Tassone, F., et al., 2002. Neuronal intranuclear inclusions in a new cerebellar tremor/ataxia syndrome among fragile X carriers. Brain 125, 1760–1771.

Greco, C.M., Berman, R.F., Martin, R.M., et al., 2006. Neuropathology of fragile X-associated tremor/ataxia syndrome (FXTAS). Brain 129, 243–255.

Greco, C.M., Soontarapornchai, K., Wirojanan, J., et al., 2007. Testicular and pituitary inclusion formation in fragile X associated tremor/ataxia syndrome. Journal of Urology 177, 1434–1437.

Greco, C.M., Tassone, F., Garcia-Arocena, D., et al., 2008. Clinical and neuropathologic findings in a woman with the FMR1 premutation and multiple sclerosis. Archives of Neurology 65, 1114–1116.

Grigsby, J., Brega, A.G., Jacquemont, S., et al., 2006. Impairment in the cognitive functioning of men with fragile X-associated tremor/

ataxia syndrome (FXTAS). Journal of Neurological Sciences 248, 227–233.

Grigsby, J., Brega, A.G., Leehey, M.A., et al., 2007. Impairment of executive cognitive functioning in males with fragile X-associated tremor/ataxia syndrome. Movement Disorders 22, 645–650.

Grigsby, J., Brega, A.G., Engle, K., et al., 2008. Cognitive profile of fragile X premutation carriers with and without fragile X-associated tremor/ataxia syndrome. Neuropsychology 22, 48–60.

Hagerman, R.J., 2002a. Medical follow-up and pharmacotherapy. In: Hagerman, R.J., Hagerman, P.J. (Eds.), Fragile X Syndrome: Diagnosis, Treatment, and Research. Johns Hopkins University Press, Baltimore, MD, pp. 287–338.

Hagerman, R.J., 2002b. Physical and behavioral phenotype. In: Hagerman, R.J., Hagerman, P.J. (Eds.), Fragile X Syndrome: Diagnosis, Treatment and Research. Johns Hopkins University Press, Baltimore, MD, pp. 3–109.

Hagerman, P.J., 2008. The fragile X prevalence paradox. Journal of Medical Genetics 45, 498–499.

Hagerman, R.J., Altshul-Stark, D., Mcbogg, P., 1987. Recurrent otitis media in the fragile X syndrome. American Journal of Diseases of Children 141, 184–187.

Hagerman, R.J., Bregman, J.D., Tirosh, E., 1998. Clonidine. In: Reiss, S., Aman, M.G. (Eds.), Psychotropic Medication and Developmental Disabilities: The International Consensus Handbook. The Ohio State University Nisonger Center, Columbus, OH, pp. 259–269.

Hagerman, P.J., Hagerman, R.J., 2004. The fragile-X premutation: A maturing perspective. American Journal of Human Genetics 74, 805–816.

Hagerman, R.J., Hills, J., Scharfenaker, S., Lewis, H., 1999. Fragile X syndrome and selective mutism. American Journal of Medical Genetics 83, 313–317.

Hagerman, R.J., Jackson, A.W., Levitas, A., Rimland, B., Braden, M., 1986. An analysis of autism in fifty males with the fragile X syndrome. American Journal of Medical Genetics 23, 359–374.

Hagerman, R., Kemper, M., Hudson, M., 1985. Learning disabilities and attentional problems in boys with the fragile X syndrome. American Journal of Diseases of Children 139, 674–678.

Hagerman, R.J., Mcbogg, P., Hagerman, P.J., 1983. The fragile X syndrome: History, diagnosis, and treatment. Journal of Developmental and Behavioral Pediatrics 4, 122–130.

Hagerman, R.J., Rivera, S.M., Hagerman, P.J., 2008b. The fragile X family of disorders: A model for autism and targeted treatments. Current Pediatric Reviews 4, 40–52.

Hagerman, P.J., Stafstrom, C.E., 2009. Origins of epilepsy in fragile x syndrome. Epilepsy Currents 9, 108–112.

Hagerman, R.J., Van Housen, K., Smith, A.C., Mcgavran, L., 1984. Consideration of connective tissue dysfunction in the fragile X syndrome. American Journal of Medical Genetics 17, 111–121.

Hagerman, R.J., Jackson, C., Amiri, K., et al., 1992. Girls with fragile X syndrome: Physical and neurocognitive status and outcome. Pediatrics 89, 395–400.

Hagerman, R.J., Staley, L.W., O'conner, R., et al., 1996. Learning-disabled males with a fragile X CGG expansion in the upper premutation size range. Pediatrics 97, 122–126.

Hagerman, R.J., Leehey, M., Heinrichs, W., et al., 2001. Intention tremor, parkinsonism, and generalized brain atrophy in male carriers of fragile X. Neurology 57, 127–130.

Hagerman, R.J., Hall, D.A., Coffey, S., et al., 2008a. Treatment of fragile X-associated tremor ataxia syndrome (FXTAS) and related neurological problems. Clinical Interventions in Aging 3, 251–262.

Hagerman, R.J., Berry-Kravis, E., Kaufmann, W.E., et al., 2009. Advances in the treatment of fragile X syndrome. Pediatrics 123, 378–390.

Hall, S.S., 2009. Treatments for fragile X syndrome: A closer look at the data. Developmental Disabilities Research Reviews 15, 353–360.

Hall, S.S., Maynes, N.P., Reiss, A.L., 2009. Using percentile schedules to increase eye contact in children with Fragile X syndrome. Journal of Applied Behavior Analysis 42, 171–176.

Hammond, L.S., Macias, M.M., Tarleton, J.C., Shashidhar Pai, G., 1997. Fragile X syndrome and deletions in FMR1: New case and review of the literature. American Journal of Medical Genetics 72, 430–434.

Harris, S.W., Hessl, D., Goodlin-Jones, B., et al., 2008. Autism profiles of males with fragile X syndrome. American Journal of Mental Retardation 113, 427–438.

Harrison, C.J., Jack, E.M., Allen, T.D., Harris, R., 1983. The fragile X: A scanning electron microscope study. Journal of Medical Genetics 20, 280–285.

Hatton, D.D., Buckley, E.G., Lachiewicz, A., Roberts, J.E., 1998. Ocular status of young boys with fragile X syndrome: A prospective study. Journal of American Association for Pediatric Ophthalmology and Strabismus 2, 298–301.

Hatton, D.D., Sideris, J., Skinner, M., et al., 2006. Autistic behavior in children with fragile X syndrome: Prevalence, stability, and the impact of FMRP. American Journal of Medical Genetics. Part A 140, 1804–1813.

Hess, L., Chitwood, K., Harris, S., 2009. Assistive technology use by persons with fragile X syndrome: Three case reports. In: I. The American Occupational Therapy Association, Special Interest Section Quarterly: Technology. American Occupational Therapy Association Inc., Bethesda, MD.

Hessl, D., Rivera, S.M., Reiss, A.L., 2004. The neuroanatomy and neuroendocrinology of fragile X syndrome. Mental Retardation and Developmental Disabilities Research Reviews 10, 17–24.

Hessl, D., Tassone, F., Loesch, D.Z., et al., 2005. Abnormal elevation of FMR1 mRNA is associated with psychological symptoms in individuals with the fragile X premutation. American Journal of Medical Genetics. Part B, Neuropsychiatric Genetics 139, 115–121.

Hockey, A., Crowhurst, J., 1988. Early manifestations of the Martin–Bell syndrome based on a series of both sexes from infancy. American Journal of Medical Genetics 30, 61–71.

Huber, K.M., Gallagher, S.M., Warren, S.T., Bear, M.F., 2002. Altered synaptic plasticity in a mouse model of fragile X mental retardation. Proceedings of the National Academy of Sciences of the United States of America 99, 7746–7750.

Hunsaker, M.R., Greco, C.M., Spath, M.A., et al., 2011. Widespread non-central nervous system organ pathology in fragile X premutation carriers with fragile X-associated tremor/ataxia syndrome and CGG knock-in mice. Acta Neuropathologica 122 (4), 467–479.

Iwahashi, C.K., Yasui, D.H., An, H.J., et al., 2006. Protein composition of the intranuclear inclusions of FXTAS. Brain 129, 256–271.

Iwahashi, C., Tassone, F., Hagerman, R.J., et al., 2009. A quantitative ELISA assay for the fragile x mental retardation 1 protein. Journal of Molecular Diagnostics 11, 281–289.

Jacquemont, S., Hagerman, R.J., Leehey, M., et al., 2003. Fragile X premutation tremor/ataxia syndrome: Molecular, clinical, and neuroimaging correlates. American Journal of Human Genetics 72, 869–878.

Jacquemont, S., Hagerman, R.J., Leehey, M.A., et al., 2004. Penetrance of the fragile X-associated tremor/ataxia syndrome in a premutation carrier population. Journal of the American Medical Association 291, 460–469.

Jacquemont, S., Curie, A., Des Portes, V., et al., 2011. Epigenetic modification of the FMR1 gene in fragile X syndrome is associated with differential response to the mGluR5 antagonist AFQ056. Science Translational Medicine 3, 64ra1.

Kaufmann, W.E., Cortell, R., Kau, A.S., et al., 2004. Autism spectrum disorder in fragile X syndrome: Communication, social interaction, and specific behaviors. American Journal of Medical Genetics 129A, 225–234.

Kerby, D.S., Dawson, B.L., 1994. Autistic features, personality, and adaptive behavior in males with the fragile X syndrome and no autism. American Journal of Mental Retardation 98, 455–462.

Khandjian, E.W., Fortin, A., Thibodeau, A., et al., 1995. A heterogeneous set of FMR1 proteins is widely distributed in mouse tissues and is modulated in cell culture. Human Molecular Genetics 4, 783–789.

Kim, H.S., Suh, Y.H., 2009. Minocycline and neurodegenerative diseases. Behavioural Brain Research 196, 168–179.

King, R.A., Hagerman, R.J., Houghton, M., 1995. Ocular findings in fragile X syndrome. Developmental Brain Dysfunction 8, 223–229.

Koldewyn, K., Hessl, D., Adams, J., et al., 2008. Reduced hippocampal activation during recall is associated with elevated FMR1 mRNA and psychiatric symptoms in men with the fragile X premutation. Brain Imaging and Behavior 2, 105–116.

Kotilainen, J., Pirinen, S., 1999. Dental maturity is advanced in fragile X syndrome. American Journal of Medical Genetics 83, 298–301.

Kronk, R., Dahl, R., Noll, R., 2009. Caregiver reports of sleep problems on a convenience sample of children with fragile X syndrome. American Journal on Intellectual and Developmental Disabilities 114, 383–392.

Krueger, D.D., Bear, M.F., 2011. Toward fulfilling the promise of molecular medicine in fragile X syndrome. Annual Review of Medicine 62, 411–429.

Leehey, M.A., Berry-Kravis, E., Min, S.J., et al., 2007. Progression of tremor and ataxia in male carriers of the FMR1 premutation. Movement Disorders 22, 203–206.

Leigh, M., Tassone, F., Mendoza-Morales, G., et al., 2010. Evaluation of autism spectrum disorders in females with Fragile X syndrome [Abstract]. In: International Meeting for Autism Research, Philadelphia, PA.

Lewis, P., Abbeduto, L., Murphy, M., et al., 2006. Cognitive, language and social-cognitive skills of individuals with fragile X syndrome with and without autism. Journal of Intellectual Disability Research 50, 532–545.

Loehr, J.P., Synhorst, D.P., Wolfe, R.R., Hagerman, R.J., 1986. Aortic root dilatation and mitral valve prolapse in the fragile X syndrome. American Journal of Medical Genetics 23, 189–194.

Loesch, D.Z., Huggins, R.M., Hagerman, R.J., 2004. Phenotypic variation and FMRP levels in fragile X. Mental Retardation and Developmental Disabilities Research Reviews 10, 31–41.

Loesch, D.Z., Hay, D.A., Sutherland, G.R., et al., 1987. Phenotypic variation in male-transmitted fragile X: Genetic inferences. American Journal of Medical Genetics 27, 401–417.

Loesch, D.Z., Sampson, M.L., 1993. Effect of the fragile X anomaly on body proportions estimated by pedigree analysis. Clinical Genetics 44, 82–88.

Loesch, D.Z., Bui, Q.M., Grigsby, J., et al., 2003. Effect of the fragile X status categories and the fragile X mental retardation protein levels on executive functioning in males and females with fragile X. Neuropsychology 17, 646–657.

Loesch, D.Z., Bui, Q.M., Huggins, R.M., et al., 2007. Transcript levels of the intermediate size or grey zone fragile X mental retardation 1 alleles are raised, and correlate with the number of CGG repeats. Journal of Medical Genetics 44, 200–204.

Louis, E., Moskowitz, C., Friez, M., Amaya, M., Vonsattel, J.P., 2006. Parkinsonism, dysautonomia, and intranuclear inclusions in a fragile X carrier: A clinical–pathological study. Movement Disorders 21, 420–425.

Lubs, H., 1969. A marker X chromosome. American Journal of Human Genetics 23, 189–194.

Maddalena, A., Richards, C.S., Mcginniss, M.J., et al., 2001. Technical standards and guidelines for fragile X: The first of a series of disease-specific supplements to the Standards and Guidelines for Clinical Genetics Laboratories of the American College of Medical

Genetics. Quality Assurance Subcommittee of the Laboratory Practice Committee. Genetics in Medicine 3, 200–205.

Maino, D.M., Wesson, M., Schlange, D., Cibis, G., Maino, J.H., 1991. Optometric findings in the fragile X syndrome. Optometry and Vision Science 68, 634–640.

Martin, J.B.J., 1943. A pedigree of mental defect showing sex linkage. Archives of Neurology and Psychiatry 6, 154–157.

Mcbride, S.M., Choi, C.H., Wang, Y., et al., 2005. Pharmacological rescue of synaptic plasticity, courtship behavior, and mushroom body defects in a Drosophila model of fragile X syndrome. Neuron 45, 753–764.

Mcconkie-Rosell, A., Finucane, B., Cronister, A., et al., 2005. Genetic counseling for fragile x syndrome: Updated recommendations of the national society of genetic counselors. Journal of Genetic Counseling 14, 249–270.

Mcconkie-Rosell, A., Abrams, L., Finucane, B., et al., 2007. Recommendations from multi-disciplinary focus groups on cascade testing and genetic counseling for fragile X-associated disorders. Journal of Genetic Counseling 16, 593–606.

Merenstein, S.A., Sobesky, W.E., Taylor, A.K., et al., 1996. Molecular-clinical correlations in males with an expanded FMR1 mutation. American Journal of Medical Genetics 64, 388–394.

Meryash, D.L., Cronk, C.E., Sachs, B., Gerald, P.S., 1984. An anthropometric study of males with the fragile-X syndrome. American Journal of Medical Genetics 17, 159–174.

Miller, L.J., Mcintosh, D.N., Mcgrath, J., et al., 1999. Electrodermal responses to sensory stimuli in individuals with fragile X syndrome: A preliminary report. American Journal of Medical Genetics 83, 268–279.

Munir, F., Cornish, K.M., Wilding, J., 2000. A neuropsychological profile of attention deficits in young males with fragile X syndrome. Neuropsychologia 38, 1261–1270.

Musumeci, S.A., Hagerman, R.J., Ferri, R., et al., 1999. Epilepsy and EEG findings in males with fragile X syndrome. Epilepsia 40, 1092–1099.

Nelson, L.M., 2009. Clinical practice. Primary ovarian insufficiency. The New England Journal of Medicine 360, 606–614.

Nitz, J.C., Kuys, S., Isles, R., Fu, S., 2010. Is the Wii Fit a new-generation tool for improving balance, health and well-being? A pilot study. Climacteric 13 (5), 487–491.

Nolin, S.L., Lewis 3rd, F.A., Ye, L.L., et al., 1996. Familial transmission of the FMR1 CGG repeat. American Journal of Human Genetics 59, 1252–1261.

Nolin, S.L., Brown, W.T., Glicksman, A., et al., 2003. Expansion of the fragile X CGG repeat in females with premutation or intermediate alleles. American Journal of Human Genetics 72, 454–464.

Nowicki, S.T., Tassone, F., Ono, M.Y., et al., 2007. The Prader–Willi phenotype of fragile X syndrome. Journal of Developmental and Behavioral Pediatrics 28, 133–138.

Oberle, I., Rousseau, F., Heitz, D., et al., 1991. Instability of a 550-base pair DNA segment and abnormal methylation in fragile X syndrome. Science 252, 1097–1102.

O'dwyer, J.P., Clabby, C., Crown, J., Barton, D.E., Hutchinson, M., 2005. Fragile X-associated tremor/ataxia syndrome presenting in a woman after chemotherapy. Neurology 65, 331–332.

Olmos-Serrano, J.L., Paluszkiewicz, S.M., Martin, B.S., et al., 2010. Defective GABAergic neurotransmission and pharmacological rescue of neuronal hyperexcitability in the amygdala in a mouse model of fragile X syndrome. Journal of Neuroscience 30, 9929–9938.

Osterweil, E.K., Krueger, D.D., Reinhold, K., Bear, M.F., 2010. Hypersensitivity to mGluR5 and ERK1/2 leads to excessive protein synthesis in the hippocampus of a mouse model of fragile X syndrome. Journal of Neuroscience 30, 15616–15627.

Palomaki, G.E., 1994. Population based prenatal screening for the fragile X syndrome. Journal of Medical Screening 1, 65–72.

Paribello, C., Tao, L., Folino, A., et al., 2010. Open-label add-on treatment trial of minocycline in fragile X syndrome. BMC Neurology 10, 91.

Paul, R., Pessah, I.N., Gane, L., et al., 2010. Early onset of neurological symptoms in fragile X premutation carriers exposed to neurotoxins. Neurotoxicology 31, 399–402.

Penagarikano, O., Mulle, J.G., Warren, S.T., 2007. The pathophysiology of fragile x syndrome. Annual Review of Genomics and Human Genetics 8, 109–129.

Pesso, R., Berkenstadt, M., Cuckle, H., et al., 2000. Screening for fragile X syndrome in women of reproductive age. Prenatal Diagnosis 20, 611–614.

Pieretti, M., Zhang, F.P., Fu, Y.H., et al., 1991. Absence of expression of the FMR-1 gene in fragile X syndrome. Cell 66, 817–822.

Prader, A., 1966. Testicular size: Assessment and clinical importance. Triangle 7, 240–243.

Puzzo, A., Fiamma, G., Rubino, V.E., et al., 1990. Cardiovascular aspects of Martin–Bell syndrome. Cardiologia 35, 857–862.

Reddy, K.S., 2005. Cytogenetic abnormalities and fragile-X syndrome in Autism Spectrum Disorder. BMC Medical Genetics 6, 3.

Renpenning, H., Gerrard, J.W., Zaleski, W.A., Tabata, T., 1962. Familial sex-linked mental retardation. Canadian Medical Association Journal 87, 954–956.

Reyniers, E., Vits, L., De Boulle, K., et al., 1993. The full mutation in the FMR-1 gene of male fragile X patients is absent in their sperm. Nature Genetics 4, 143–146.

Richards, B.W., Sylvester, P.E., Brooker, C., 1981. Fragile X-linked mental retardation: The Martin–Bell syndrome. Journal of Mental Deficiency Research 25 (Pt 4), 253–256.

Roberts, J.E., Bailey Jr., D.B., Mankowski, J., et al., 2009. Mood and anxiety disorders in females with the FMR1 premutation. American Journal of Medical Genetics. Part B, Neuropsychiatric Genetics 150B, 130–139.

Rodriguez-Revenga, L., Madrigal, I., Alegret, M., Santos, M., Mila, M., 2008. Evidence of depressive symptoms in fragile-X syndrome premutated females. Psychiatric Genetics 18, 153–155.

Rodriguez-Revenga, L., Madrigal, I., Pagonabarraga, J., et al., 2009. Penetrance of FMR1 premutation associated pathologies in fragile X syndrome families. European Journal of Human Genetics 17, 1359–1362.

Rogers, S.J., Wehner, D.E., Hagerman, R., 2001. The behavioral phenotype in fragile X: Symptoms of autism in very young children with fragile X syndrome, idiopathic autism, and other developmental disorders. Journal of Developmental and Behavioral Pediatrics 22, 409–417.

Rousseau, F., Heitz, D., Tarleton, J., et al., 1994. A multicenter study on genotype-phenotype correlations in the fragile X syndrome, using direct diagnosis with probe StB12.3: The first 2,253 cases. American Journal of Human Genetics 55, 225–237.

Scharfenaker, S., O'connor, R., Stackhouse, T., Braden, M., Gray, K., 2002. An integrated approach to intervention. In: Hagerman, P.J., Hagerman, R.J. (Eds.), Fragile X Syndrome Diagnosis, Treatment and Research. Johns Hopkins University Press, Baltimore, MD, pp. 363–427.

Seritan, A.L., Nguyen, D.V., Farias, S.T., et al., 2008. Dementia in fragile X-associated tremor/ataxia syndrome (FXTAS): Comparison with Alzheimer's disease. American Journal of Medical Genetics. Part B, Neuropsychiatric Genetics 147B, 1138–1144.

Sherman, S., Pletcher, B.A., Driscoll, D.A., 2005. Fragile X syndrome: Diagnostic and carrier testing. Genetics in Medicine 7, 584–587.

Shetty, A.K., 2002. Tetracyclines in pediatrics revisited. Clinical Pediatrics (Philadelphia) 41, 203–209.

Silverman, J.L., Tolu, S.S., Barkan, C.L., Crawley, J.N., 2010. Repetitive self-grooming behavior in the BTBR mouse model of autism is blocked by the mGluR5 antagonist MPEP. Neuropsychopharmacology 35, 976–989.

Simko, A., Hornstein, L., Soukup, S., Bagamery, N., 1989. Fragile X syndrome: Recognition in young children. Pediatrics 83, 547–552.

Simon, E.W., Finucane, B.M., 1996. Facial emotion identification in males with fragile X syndrome. American Journal of Medical Genetics 67, 77–80.

Smith, K., Leyden, J.J., 2005. Safety of doxycycline and minocycline: A systematic review. Clinical Therapeutics 27, 1329–1342.

Smith, T., Mozingo, D., Mruzek, D.W., Zarcone, J.R., 2007. Applied behavior analysis in the treatment of autism. In: Hollander, E., Anagnostou, E. (Eds.), Clinical Manual for the Treatment of Autism. American Psychiatric Publishing, Washington, DC, pp. 153–177.

Snyder, E.M., Philpot, B.D., Huber, K.M., et al., 2001. Internalization of ionotropic glutamate receptors in response to mGluR activation. Nature Neuroscience 4, 1079–1085.

Sobesky, W.E., Taylor, A.K., Pennington, B.F., et al., 1996. Molecular/ clinical correlations in females with fragile X. American Journal of Medical Genetics 64, 340–345.

Sternlicht, M.D., Werb, Z., 2001. How matrix metalloproteinases regulate cell behavior. Annual Review of Cell and Developmental Biology 17, 463–516.

Stirling, D.P., Koochesfahani, K.M., Steeves, J.D., Tetzlaff, W., 2005. Minocycline as a neuroprotective agent. The Neuroscientist 11, 308–322.

Storm, R.L., Pebenito, R., Ferretti, C., 1987. Ophthalmologic findings in the fragile X syndrome. Archives of Ophthalmology 105, 1099–1102.

Sullivan, A.K., Marcus, M., Epstein, M.P., et al., 2005. Association of FMR1 repeat size with ovarian dysfunction. Human Reproduction 20, 402–412.

Sullivan, K., Hatton, D., Hammer, J., et al., 2006. ADHD symptoms in children with FXS. American Journal of Medical Genetics. Part A 140, 2275–2288.

Sutcliffe, J.S., Nelson, D.L., Zhang, F., et al., 1992. DNA methylation represses FMR-1 transcription in fragile X syndrome. Human Molecular Genetics 1, 397–400.

Sutherland, G.R., 1977. Fragile sites on human chromosomes: Demonstration of their dependence on the type of tissue culture medium. Science 197, 265–266.

Sutherland, G.R., Ashforth, P.L., 1979. X-linked mental retardation with macro-orchidism and the fragile site at Xq 27 or 28. Human Genetics 48, 117–120.

Tassone, F., Hagerman, R.J., Taylor, A.K., et al., 2000a. Elevated levels of FMR1 mRNA in carrier males: A new mechanism of involvement in the fragile-X syndrome. American Journal of Human Genetics 66, 6–15.

Tassone, F., Hagerman, R.J., Taylor, A.K., et al., 2000b. Clinical involvement and protein expression in individuals with the FMR1 premutation. American Journal of Medical Genetics 91, 144–152.

Tassone, F., Pan, R., Amiri, K., Taylor, A.K., Hagerman, P.J., 2008. A rapid polymerase chain reaction-based screening method for identification of all expanded alleles of the fragile X (FMR1) gene in newborn and high-risk populations. Journal of Molecular Diagnostics 10, 43–49.

Toledano-Alhadef, H., Basel-Vanagaite, L., Magal, N., et al., 2001. Fragile-X carrier screening and the prevalence of premutation and full-mutation carriers in Israel. American Journal of Human Genetics 69, 351–360.

Turner, G., Till, R., Daniel, A., 1978. Marker X chromosomes, mental retardation and macro-orchidism. The New England Journal of Medicine 299, 1472.

Turner, G., Webb, T., Wake, S., Robinson, H., 1996. Prevalence of fragile X syndrome. American Journal of Medical Genetics 64, 196–197.

Utari, A., Chonchaiya, W., Rivera, S.M., et al., 2010. Side effects of minocycline treatment in patients with fragile X syndrome and exploration of outcome measures. American Journal on Intellectual and Developmental Disabilities 115, 433–443.

Verkerk, A.J., Pieretti, M., Sutcliffe, J.S., et al., 1991. Identification of a gene (FMR-1) containing a CGG repeat coincident with a breakpoint cluster region exhibiting length variation in fragile X syndrome. Cell 65, 905–914.

Wang, H., Wu, L.J., Kim, S.S., et al., 2008. FMRP acts as a key messenger for dopamine modulation in the forebrain. Neuron 59, 634–647.

Wassink, T.H., Piven, J., Patil, S.R., 2001. Chromosomal abnormalities in a clinic sample of individuals with autistic disorder. Psychiatric Genetics 11, 57–63.

Weiler, I.J., Spangler, C.C., Klintsova, A.Y., et al., 2004. Fragile X mental retardation protein is necessary for neurotransmitter-activated protein translation at synapses. Proceedings of the National Academy of Sciences of the United States of America 101, 17504–17509.

Weiskop, S., Richdale, A., Matthews, J., 2005. Behavioural treatment to reduce sleep problems in children with autism or fragile X syndrome. Developmental Medicine and Child Neurology 47, 94–104.

Welt, C.K., 2008. Primary ovarian insufficiency: A more accurate term for premature ovarian failure. Clinical Endocrinology 68, 499–509.

Welt, C.K., Smith, P.C., Taylor, A.E., 2004. Evidence of early ovarian aging in fragile X premutation carriers. Journal of Clinical Endocrinology and Metabolism 89, 4569–4574.

Weng, N., Weiler, I.J., Sumis, A., Berry-Kravis, E., Greenough, W.T., 2008. Early-phase ERK activation as a biomarker for metabolic status in fragile X syndrome. American Journal of Medical Genetics. Part B, Neuropsychiatric Genetics 147B, 1253–1257.

Wenzel, H.J., Hunsaker, M.R., Greco, C.M., Willemsen, R., Berman, R.F., 2010. Ubiquitin-positive intranuclear inclusions in neuronal and glial cells in a mouse model of the fragile X premutation. Brain Research 1318, 155–166.

Willems, P.J., Van Roy, B., De Boulle, K., et al., 1992. Segregation of the fragile X mutation from an affected male to his normal daughter. Human Molecular Genetics 1, 511–515.

Wilson, D.P., Carpenter, N.J., Berkovitz, G., 1988. Thyroid function in men with fragile X-linked MR. American Journal of Medical Genetics 31, 733–734.

Wirojanan, J., Jacquemont, S., Diaz, R., et al., 2009. The efficacy of melatonin for sleep problems in children with autism, fragile X syndrome, or autism and fragile X syndrome. Journal of Clinical Sleep Medicine 5, 145–150.

Wisbeck, J.M., Huffman, L.C., Freund, L., et al., 2000. Cortisol and social stressors in children with fragile X: A pilot study. Journal of Developmental and Behavioral Pediatrics 21, 278–282.

Wittenberger, M.D., Hagerman, R.J., Sherman, S.L., et al., 2007. The FMR1 premutation and reproduction. Fertility and Sterility 87, 456–465.

Yan, Q.J., Rammal, M., Tranfaglia, M., Bauchwitz, R.P., 2005. Suppression of two major fragile X syndrome mouse model phenotypes by the mGluR5 antagonist MPEP. Neuropharmacology 49, 1053–1066.

Yu, S., Pritchard, M., Kremer, E., et al., 1991. Fragile X genotype characterized by an unstable region of DNA. Science 252, 1179–1181.

Zarate Jr., C.A., Payne, J.L., Quiroz, J., et al., 2004. An open-label trial of riluzole in patients with treatment-resistant major depression. American Journal of Psychiatry 161, 171–174.

Zhang, Y.Q., Bailey, A.M., Matthies, H.J., et al., 2001. Drosophila fragile X-related gene regulates the MAP1B homolog Futsch to control synaptic structure and function. Cell 107, 591–603.

Zhang, L., Coffey, S., Lua, L.L., et al., 2009. FMR1 premutation in females diagnosed with multiple sclerosis. Journal of Neurology, Neurosurgery, and Psychiatry 80, 812–814.

Relevant Websites

http://www.fraxa.org/ – FRAXA Research Foundation.

http://www.fragilex.org/ – National Fragile X Foundation.

http://www.ucdmc.ucdavis.edu/mindinstitute/research/fxrtc.html – UC Davis Medical Center MIND Institute Fragile X Research and Treatment Center.

Autisms

A.M. Persico

University Campus Bio-Medico, Rome, Italy; I.R.C.C.S. Fondazione Santa Lucia, Rome, Italy

OUTLINE

© 2013 Elsevier Inc. All rights reserved.

34.1 INTRODUCTION

34.1.1 The History of Autism

The term *autism* was coined in 1911 by Swiss psychiatrist Eugen Bleuler to designate one of the hallmarks of schizophrenia, namely the social withdrawal resulting in enclosure in one's self (self = α'υτός autòs, in ancient Greek) (Bleuler, 1911). During the following three decades, this term reached an ever broader audience in psychiatry, mainly through the work of Eugène Minkowski (1927), who addressed schizophrenic autism in great detail in his famous text 'La Schizophrénie'. However, schizophrenic autism must not be confused with autism spectrum disorder (ASD), which defines an independent nosological entity and not a mere symptom. This disorder was first described in 1943 by Leo Kanner in a cohort of 11 children, who essentially shared an 'enclosure in one's self' as their distinctive trait (Kanner, 1943). Only 1 year later, in 1944, the Austrian pediatrician Hans Asperger described four boys displaying some, but not all, of the behavioral symptoms present in Kanner's patients (Asperger, 1944a). Asperger's work, written in German, reached a wider audience after it was publicized in 1981 by Lorna Wing, who described 34 individuals, ranging from 5 to 35 years of age, whose clinical picture was closer to Asperger's cases than to Kanner's (Wing, 1981). Thereafter, it was translated into English by Uta Frith in 1991 (Asperger, 1944b). Hence, the existence of clinical heterogeneity in autism is by no means a recent acquisition; it was recognized from the beginning that autistic patients do indeed share some common features, primarily an enclosure in one's self, and display an impressive variability in symptom patterns, developmental trajectories, disease course, and severity of impairment, spread along a dimensional continuum that was later designated as the 'autism spectrum' (Piven et al., 1997). This impressive clinical variability is underscored by an equally impressive degree of etiological heterogeneity, which has led the term *autisms* to designate a set of neurodevelopmental disorders with early onset in life that share autism as a common feature but that are produced through distinct processes. This chapter will summarize the current state of knowledge regarding these 'autisms'; readers interested in 'schizophrenic autism' are referred to the excellent review by Parnas et al. (2002).

34.1.2 Definition and Epidemiology of Autism Spectrum Disorder

ASD is characterized by deficits in social interaction and communication, as well as by stereotyped behaviors and insistence on sameness (i.e., restricted patterns of interest and activities) (American Psychiatric Association, 1994). Its onset occurs in early childhood, before 3 years of age (American Psychiatric Association, 1994). ASD essentially encompasses three different pervasive developmental disorders listed in the fourth edition of the *Diagnostic and Statistical Manual of Mental Disorders* (DSM-IV) (American Psychiatric Association, 1994), namely autistic disorder, Asperger disorder, and pervasive developmental disorder not otherwise specified (PDD-NOS). These separate diagnostic categories likely will merge into a single comprehensive ASD category in the upcoming DSM-V (American Psychiatric Association, 2012) following the ever-swinging logic behind categorical diagnosis in psychiatry, which historically alternates between analysis and synthesis.

The incidence of ASD has risen dramatically during the last two decades, from 2–5 in 10 000 to approximately 1–2 in 1000 children; broader diagnostic criteria and increased awareness in the medical community certainly have contributed to this trend, but a real increase in incidence, possibly due to gene–environment interactions, is also likely (Fombonne, 2005; Persico and Bourgeron, 2006; Rutter, 2005). Males are particularly susceptible, with male-to-female ratios ranging from approximately 4:1 to 8:1, depending on disease severity and recruiting context (Fombonne, 2005; Rutter, 2005). An additional layer of complexity stems from comorbidity with seizures and mental retardation (MR) present in up to 30% and 65% of cases, respectively (Fombonne, 2005; Tuchman and Rapin, 2002).

34.1.3 Neuropathological and Systemic Abnormalities in Autism Spectrum Disorder

Altered neurodevelopment occurring during the first and second trimesters of prenatal life is now widely recognized as the underlying neuropathological cause of ASD (DiCicco-Bloom et al., 2006). Postmortem studies of autistic brains have uncovered important neuroanatomical abnormalities in the central nervous system (CNS) of ASD patients, generally resulting from reduced programmed cell death and/or increased cell proliferation, altered cell migration, and abnormal cell differentiation with reduced neuronal size and abnormal wiring (Bauman and Kemper, 2005). These neuropathological anomalies, especially the patchy cytoarchitectonic abnormalities present in the cerebral and cerebellar cortex, would seemingly explain the imbalance in local vs long-distance connectivity on the one hand and excitatory vs inhibitory connectivity on the other currently believed to underlie disrupted sensory integration, altered social information processing, and frequent comorbidity with epilepsy (Courchesne and Pierce, 2005; Geschwind and Levitt, 2007; Rubenstein and Merzenich, 2003). All these neurodevelopmental processes physiologically occur during the first and second trimesters of pregnancy (Rice and Barone, 2000). Hence, despite some methodological limitations and predictable brain-to-brain variability, neuropathological studies collectively have been instrumental in indicating a prenatal origin for autism. Further support comes from behavioral analyses demonstrating a delayed appearance or inhibition of specific motor reflexes already on the day of birth or early on in neonates later diagnosed with an ASD (Teitelbaum et al., 2004). An additional confirmation of the existence of a prenatal time window for autism vulnerability comes from studies of teratogenic drugs and congenital infections known to cause autism in some cases (see Section 34.4). This temporal framework ought to be considered when attempting to incorporate potential environmental factors into realistic pathogenetic models, which should not necessarily exclude modulatory roles for early postnatal exposures but *must* incorporate this crucial prenatal component.

Viewing autism exclusively as a brain disease would be an oversimplification; ASD patients also display variable degrees of systemic involvement, with signs and symptoms frequently including macrosomy (Sacco et al., 2007a), gastrointestinal disorders (Buie et al., 2010), and immune dysreactivity (Ashwood et al., 2006; Jyonouchi et al., 2005). In summary, autism should be viewed as a multiorgan systemic disorder, primarily involving but not restricted to the nervous system, with prenatal onset and postnatal clinical expression.

34.1.4 Toward a Classification of the Autisms

To address the great heterogeneity present in ASD, investigators have aimed at identifying subgroups of patients who at least partly share common pathophysiological underpinnings. These attempts essentially have followed two complementary strategies, namely the study of endophenotypes and the use of genetic approaches:

1. An endophenotype can be best described as a familial and heritable quantitative trait associated with a complex disease (Gottesman and Gould, 2003). The most important endophenotypes reported to date in autism research are summarized in Table 34.1. A detailed discussion of endophenotypes will be provided elsewhere (Persico and Sacco, 2013). The study of endophenotypes in complex disorders, such as autism, provides several advantages: (a) the lesser complexity of an endophenotype and its greater proximity to the genetic level, as compared with clinical affection status and behavioral symptoms, facilitates the interpretation of the results; (b) a continuous measure reflects more faithfully the existence of a continuum of signs and symptoms in the autism spectrum compared with a categorical

'case versus control' distinction; and (c) standardized and automated procedures are used to measure biological parameters (Sacco et al., 2010).
2. For more than two decades, autism has been identified as 'the most genetic' neuropsychiatric disorder because of the monozygotic twin concordance rate as high as 73–95%, impressive heritability (>90%, as estimated by twin studies), and

TABLE 34.1 Endophenotypes in the Autism Spectrum

Behavioral/neurodevelopmental

- Delayed expressive speech (Alarcón et al., 2008; Spence et al., 2006)
- ADI-R domains: social interaction domain; restricted and repetitive behaviors (Liu et al., 2008; Sakurai et al., 2006)
- Savant skills: absolute pitch, calendar calculations, etc. (Wallace et al., 2009)
- Social Responsiveness Scale scores (Duvall et al., 2007)

Neuropsychological

- Pattern of face processing (Adolphs et al., 2008; Hernandez et al., 2009; Klin et al., 2002)
- Executive functions (Delorme et al., 2007)

Neurophysiological

- Reduced cingulate self-response in a visual imagery task, when playing with a human partner (Chiu et al., 2008)
- Abnormal patterns of cortical auditory activation (Boddaert et al., 2003; Bonnel et al., 2010; Bruneau et al., 2003; Gomot et al., 2008)
- Dysfunctional mirror neuron systems (Cattaneo et al., 2007; Dapretto et al., 2006; Martineau et al., 2010)
- Centroparietal and temporal EEG related to autistic behaviors and intellectual impairment (Roux et al., 1997)
- Blunted or delayed frontal activation during visual attention tasks (Belmonte et al., 2010)

Morphological

- Macrocephaly (Fombonne et al., 1999; Lainhart et al., 1997; Miles et al., 2000; Sacco et al., 2007a; Stevenson et al., 1997; Woodhouse et al., 1996)
- Macrosomy (Bigler et al., 2010; Dissanayake et al., 2006; Lainhart et al., 2006; Sacco et al., 2007a; van Daalen et al., 2007)
- Minor physical anomalies (Hammond et al., 2008; Miles et al., 2008; Tripi et al., 2008)

Biochemical

- Hyperserotoninemia (Hérault et al., 1996; McBride et al., 1998; Mulder et al., 2004; Piven et al., 1991)
- Oligopeptiduria (Reichelt et al., 1981; Sacco et al., 2010)
- Urinary dopamine and HVA levels (Hameury et al., 1995)
- Decreased plasma fatty acids (Vancassel et al., 2001)

Endocrine

- Decreased melatonin plasma levels (Melke et al., 2008)
- Decreased oxytocin plasma levels (Modahl et al., 1998)

Immunological

- Increased proinflammatory and IL-10-producing immune cells, decreased CD4+ T lymphocytes, increased naive and effector memory CD8+ T lymphocytes (Saresella et al., 2009) a noticeable sibling recurrence risk (5–6% for full-blown autistic disorder, approximately 15% for broad ASD) (for reviews of autism genetics, see Abrahams and Geschwind, 2008; Freitag, 2007; Geschwind, 2011; Muhle et al., 2004; Persico and Bourgeron, 2006). These heritability estimates, obtained primarily in the UK and in Northern Europe in the early 1990s, were not replicated by a more recent California-based twin study that supported a larger proportion of variance explained by shared environmental factors as opposed to genetic heritability (55% vs. 37% for strict autism, respectively) (Hallmayer et al., 2011). Conceivably, the relative weight of genetic and environmental factors may be region-specific and change over time. Nonetheless, the parallel increase in sibling recurrence risk, estimated by recent baby sibling studies at 18.7% (26.2% for males and 9.1% for females) (Ozonoff et al., 2011), and the presence of mild autistic traits in many first-degree relatives of autistic patients (Piven et al., 1997) still point toward a strong genetic component in ASD playing a sizable permissive role at a minimum. Linkage and association studies have identified numerous susceptibility genes located on various chromosomes, especially 2q, 7q, 15q, and on the X chromosome. The clinical heterogeneity of ASD is believed at least partly to reflect the complexity of its genetic underpinnings, the general underlying mechanisms of which are summarized in Table 34.2.

34.2 'CLASSIC' SYNDROMIC AUTISMS

34.2.1 General Description

In approximately 10% of ASD cases, autistic symptoms are part of a broader syndrome due to a known medical cause. These syndromes can stem from (a) genomic DNA mutations, triplet repeat expansions, or cytogenetic abnormalities visible by classical G band karyotyping, conditions summarized in Table 34.3; (b) mitochondrial DNA (mtDNA) mutations or gene dosage abnormalities, which are listed in Table 34.4; or (c) copy number variants (CNVs), genomic DNA microdeletions/microduplications detectable only using microarray technologies. Genetic and genomic forms have been reviewed by Gillberg (1998), Cohen et al. (2005), Feinstein and Singh (2007), Zafeiriou et al. (2007), and Benvenuto et al. (2009); autism linked to mitochondrial disease and mtDNA abnormalities has been reviewed recently by Palmieri and Persico (2010) and by Rossignol and Frye (2011); CNVs have been reviewed by Merikangas et al. (2009), Guilmatre et al. (2009), Weiss (2009), and Carvalho et al. (2010).

In general, malformations and/or facial dysmorphisms, moderate-to-profound mental retardation, severe epilepsy, neurological signs, and symptoms are largely more frequent in syndromic autism than in idiopathic forms. Overall, the M:F gender ratio is close to 1, although males are particularly prone to suffer from specific syndromes. Abnormal growth in the form

---

TABLE 34.2   Mechanisms Underlying the Complexity of Autism Genetics

1. Genetic heterogeneity	Different contributing genes cause the disease in distinct patients, who may display similar clinical phenotypes
2. Different modes of inheritance	
(a) Polygenic or oligogenic	Several functional polymorphisms located in different genes and widely distributed in the general population ('common variants'), each conferring a small risk, are collectively required for an individual to develop the disease
(b) Monogenic	Genetic mutations or genomic rearrangements affecting a single gene cause the disease, typically in a single or in very few patients ('private' or 'rare variants,' respectively)
(c) Combined genetic and genomic quasi-recessive mode	Convergence onto the same individual of one allele carrying a null mutation inherited from one parent and the other allele carrying a genomic rearrangement (typically a microdeletion) inherited from the other parent. Both mutation and microdeletion are recessive, and neither by itself is pathogenic in either parent; they may even be present at low frequency in the general population
Phenocopies	Cases exclusively due to environmental factors and clinically indistinguishable from genetic cases
Variable penetrance and expressivity	Variable degrees of phenotypic expression of genetic variants: generally high level of pathology caused by rare variants, lower degree of expression for common variants
Epistasis	Gene–gene interactions, with permissive and blunting effects exerted by common variants ('modifier genes'). Phenotypic expression is complex, and not a mere summation of single-gene effects. By this mechanism, an identical mutation can produce different phenotypes in different individuals or mouse inbred strains
Gene–environment interactions	Genes may confer vulnerability to or protection from the disease by lowering or raising the threshold of sensitivity to pathogenic environmental factors

TABLE 34.3   Syndromic Autisms Due to Known Mutations, Triplet Repeat Expansions, or Cytogenenetic Abnormalities Visible by G Band Karyotyping

	Gene/ch. region	Prevalence	Percentage of autistic patients with the syndrome	Percentage of patients with the syndrome who are autistic	Signs and symptoms
Fragile X syndrome	FMR1	1/3500–1/9000	2.1	25–33	Facial dysmorphisms, macroorchidism, poor eye contact, social anxiety, language impairment, stereotypies, hyperactivity, sensory hyper-reactivity
Tuberous sclerosis	TSC1 TSC2	1–1.7/10 000	1–4% (8–14% if seizures present)	16–65	Hamartomas in skin, CNS, kidney, heart, lungs, retina; autism, mental retardation, learning disability, epilepsy (infantile spasms)
Neurofibromatosis type 1	NF1	1/3000–1/4000	≤1.4	?	Café-au-lait macules, neurofibromas, axillary or groin frecklings, optic pathway tumors, bone dysplasias
Untreated phenylketonuria	PAH	1/10 000–1/15 000	–	5.7	Microcephaly, hypertonia, mental retardation, language impairment, psychomotor agitation, autism, seizures
Adenylosuccinate lyase deficiency	ADSL	?	≤1	80–100	Mental retardation and severe autism, seizures, psychomotor regression
Smith–Lemli–Opitz syndrome	DHCR7	1/10 000–1/60 000	≤1	46–53	Microcephaly, facial dysmorphism, malformations (sometimes lethal, usually cleft palate, cardiac m., hypospadia), short stature, variable mental retardation, sensory hyper-reactivity, language impairment, self-injurious behavior, sleep disturbance, opisthokinesis and other stereotypies
Cohen syndrome	COH-1 ?	1/105 000	≤1	48	Microcephaly, facial dysmorphism, truncal obesity, hematologic and eye abnormalities, mental retardation, motor clumsiness, hypotonia, language impairment, autism
Cornelia de Lange syndrome	NIPBL SMC1A SMC3 ?	1/10 000	≤1	35–50	Facial dysmorphism, growth deficiency with short stature, major malformations (especially cardiac, gastrointestinal, musculoskeletal), developmental delay, mental retardation, feeding difficulties, extreme shyness, self-injurious behavior, hyperactivity with attention deficit, aggression, obsessive-compulsive behavior, depression
Sotos syndrome	NSD1	1/10 000–1/50 000 (?)	≤1	?	Macrocephaly, pre- and postnatal overgrowth, facial dysmorphism, developmental delay
Cole–Hughes macrocephaly	?	?	≤1	?	Macrocephaly, mental retardation, attention deficit and hyperactivity, developmental delay, autism, language impairment, obesity, delayed bone age, facial dysmorphism
Lujan–Fryns syndrome	UPF3B MED12	?	≤1	80	X-linked mental retardation with marfanoid habitus (tall stature, facial dysmorphism), hypotonia, mild-to-moderate mental retardation, ascending aortic aneurysm, autism, aggression, hyperactivity, emotional instability
San Filippo syndromes: A	SGSH	0.3–1.6/100 000	≤1	?	Prominent regression or developmental delay, autism, motor and verbal stereotypies, hyperactivity, aggression, sleep disturbance,

Continued

TABLE 34.3 Syndromic Autisms Due to Known Mutations, Triplet Repeat Expansions, or Cytogenenetic Abnormalities Visible by G Band Karyotyping—cont'd

	Gene/ch. region	Prevalence	Percentage of autistic patients with the syndrome	Percentage of patients with the syndrome who are autistic	Signs and symptoms
B	NAGLU				inappropriate effect, variable malformations (visceromegaly, facial, skeletal, etc.). Onset, usually (but not always) beyond age 3, qualifies for DSMIV disintegrative disorder
C	HGSNAT				
D	GNS				
ARX syndrome	ARX	?	≤1	?	X-linked mental retardation with or without autism, and X-linked infantile spasms for insertion/missense mutations; X-linked lissencephaly with agenesis of the corpus callosum and ambiguous genitalia for truncating mutations (death due to neurodevelopmental delay and intractable seizures)
Ch 2q37 deletion syndrome	2q37	?	≤1	?	Brachymetaphalangism, mental retardation, autism
Williams–Beuren syndrome	7q11.23 del	2–5/ 100000	≤1	?	Elfin face, stenosis of the aorta and other arteries, short stature, dental malformations, hypercalcemia, loquaciousness, sociability, autism (rare), attention deficit, hyperactivity, anxiety, visuocognitive deficits
Williams–Beuren region duplication syndrome	7q11.23 dup	?	≤1	?	Growth delay, facial and dental dysmorphisms, autism, mental retardation, developmental delay, impaired expressive language, seizures
Ch 13 deletion syndrome	13q	?	≤1	?	Mental retardation, language impairment, retinoblastoma, growth retardation, various malformations (cardiac, craniofacial, gastrointestinal, renal, limbs and digits)

15Q CHROMOSOMAL SYNDROMES

	Gene/ch. region	Prevalence	Percentage of autistic patients with the syndrome	Percentage of patients with the syndrome who are autistic	Signs and symptoms
Angelman syndrome	Del or mutation in maternal UBE3A	1/10000– 1/12000	≤1	42	Facial dysmorphism, developmental delay, speech impairment, stereotypies, mental retardation, gait ataxia, 'happy puppet' attitude, hyperactivity with attention deficit, temper tantrums, frequently microcephaly and seizures
Prader–Willi syndrome	Del of paternal allele at 15q11–q13	1/10000– 1/15000	?	25.3	Developmental delay, short stature, mental retardation, hyperphagia, obesity, hypotonia, hypogonadism, obsessive-compulsive behavior
Isodicentric 15q	Dup 15q11– q13, GABRB3	1/30000 (?)	≤1	70	Short stature, diabetes, anal and jejunal atresias, acanthosis nigrans, severe autism, developmental delay, mental retardation, hypotonia, seizures
Hypomelanosis of Ito	Chr dup/ dels, often 15q11–q13	1/10000	≤1	10	Hypopigmented macules, neurological deficits, variable mental retardation and seizures, multiple malformations (brain, ocular, musculoskeletal)
Smith–Magenis syndrome	17p11.2 del	1/25000	≤1	93	Mental retardation, developmental delay, self-injurious behavior, facial dysmorphisms, hearing impairment; skeletal, renal, cardiac, and eye abnormalities

TABLE 34.3 Syndromic Autisms Due to Known Mutations, Triplet Repeat Expansions, or Cytogenenetic Abnormalities Visible by G Band Karyotyping—cont'd

	Gene/ch. region	Prevalence	Percentage of autistic patients with the syndrome	Percentage of patients with the syndrome who are autistic	Signs and symptoms
Potocki–Lupsky syndrome	17p11.2 dup	?	≤1	?	Hyperactivity, attention deficit, autism, mental retardation, developmental delay, short stature, hypotonia, mild dysmorphism, cardiac and dental abnormalities
Down syndrome	Trisomy of ch. 21	1/1000	≤2.5	≤10	Facial dysmorphism, cardiac and intestinal malformations, variable degree of mental retardation, severe autism (when present)
Velofaciocardial/ Di George syndrome	22q11.2 del	1/4000	≤1	20–31	Facial dysmorphism, cleft palate, cardiac malformations, hypoplasia of the thymus, hypoparathyroidism, autism, mental retardation, developmental delay, attention deficit, hyperactivity, psychosis, seizures
Ch 22q11 duplication syndrome	22q11.2 dup	?	≤1	?	Facial dysmorphism, velopharyngeal insufficiency, autism, mental retardation, developmental delay
Ch 22q13.3 deletion syndrome	22q13.3 del	?	≤1	?	Mild dysmorphisms, severe hypotonia, mental retardation, developmental delay, impaired language development

?, no data available.

TABLE 34.4 Syndromic Autisms Due to mtDNA Mutations or Rearrangements

References	Mutation	mtDNA gene	Number of patients	Signs and symptoms
Graf et al. (2000)	8363G>A	tRNA^{Lys}	Two siblings	Brother: autism, behavioral regression, extreme hyperactivity, lack of attention, mild fine and gross motor dyspraxia
				Sister: partial complex seizures, unsteady gait, myoclonus, swallowing dysfunction, moderate mental retardation
Fillano et al. (2002)	Large mtDNA deletions		Five ASD patients	Autism, ataxia, cardiomyopathy
Pons et al. (2004)	3243A>G	tRNA$^{Leu(UUR)}$	Two ASD and their two mothers	Highly heterogeneous: typically mitochondrial encephalopathy with lactic acidosis and stroke-like episodes (MELAS), or maternally inherited progressive external ophthalmoplegia. In these two patients, autism with developmental delay, clumsiness, attention deficit, neurologic deterioration in the presence of fever, microcephaly or macrocephaly
	?	mtDNA? genomic DNA?	One ASD with mtDNA depletion	Autism, muscle hypotonia, seizures, myoclonus, and developmental delay
Weissman et al. (2008)	3397A>G	ND1 subunit of complex I	25 ASD patients with primary mit. disorder (3/25 mtDNA mutation carriers)	Autism, excessive fatigability and/or exercise intolerance, gastrointestinal dysfunction, cardiovascular abnormalities, facial dysmorphisms, microcephaly or macrocephaly, developmental gross motor delays, growth retardation
	4295A>G	tRNA^{Ileu}		
	11984T>C	ND4 subunit of complex I		

of microsomy, or less frequently macrosomy, is not unusual. In many syndromes, clinical manifestations of autism can be highly heterogeneous, even in the presence of the same well-characterized mutation or genomic rearrangement.

34.2.2 Mitochondrial Autisms

Biochemical parameters linked to mitochondrial function are frequently abnormal in ASD (Giulivi et al., 2010; Palmieri and Persico, 2010; Rossignol and Frye, 2011).

As many as 5% of autistic children even satisfy diagnostic criteria for a full-blown mitochondrial disease (Rossignol and Frye, 2011). Yet, mutations or chromosomal rearrangements in mtDNA or nuclear DNA (nDNA) are detected only in approximately 20% of children with ASD and mitochondrial disease (i.e., ≤1% of all ASD children), and each mtDNA mutation or chromosomal rearrangement listed in Table 34.4 is detected in ≤0.1% of all cases. Hence, mitochondrial dysfunction appears to be secondary in the vast majority of patients, that is, downstream of other pathophysiological abnormalities such as excessive oxidative stress (Palmieri and Persico, 2010). Importantly, since mitochondrial function requires approximately 1500 nuclear genes and oxidative phosphorylation involves at least 80 proteins, only 13 encoded by mtDNA, mutations, and chromosomal rearrangements should be sought both in nDNA and in mtDNA (Shadel, 2008; Zeviani and Di Donato, 2004). Indeed, chromosomal rearrangements, which could affect mitochondrial functions, include deletions in 15q11–q13 (cytochrome C oxidase subunit 5A, COX5A), 13q13–q14.1 (mitochondrial ribosomal protein 31, MRPS31), 4q32–q34.68 (electron-transferring-flavoprotein dehydrogenase, ETFDH), and 2q37.3 (NADH dehydrogenase ubiquinone 1 alpha subcomplex 10, NDUFA10), as recently reviewed by Smith et al. (2009).

Mitochondrial autism, despite an even more prominent clinical heterogeneity, often displays some peculiarities which should prompt clinicians to request molecular investigations (Palmieri and Persico, 2010; Rossignol and Frye, 2011). Its neurological signs and symptoms, such as oculomotor abnormalities, dysarthria, ptosis, hearing deficits, hypertonia, and movement disorders, are generally atypical for autism. Behavioral regression, especially in concomitance with fever, is frequently reported by parents (Shoffner et al., 2010; Weissman et al., 2008). Except in the case of mitochondrial depletion, family history is generally positive for mitochondrial diseases in the maternal lineage. At least one biochemical parameter among several typically assessed to screen for mitochondrial disorders is usually abnormal in children. The incidence of microcephaly and microsomy is unusually high, reaching approximately 20% of all cases. Neuroanatomical abnormalities are relatively frequent, although highly variable in nature (Nissenkorn et al., 2000; Shoffner et al., 2010; Weissman et al., 2008). 'Ragged red fibers,' characterized by a segmental proliferation and accumulation of abnormal mitochondria under the sarcolemmal membrane, are usually visible in muscle biopsies of adults, but in most affected children muscle tissue histology will be negative.

34.2.3 Copy Number Variants

The recent advent of microarray-based high-resolution genome analysis has dramatically increased our ability to detect genomic deletions and duplications. CNVs are deletions and duplications of at least 1 kb in size, undetectable by standard chromosomal banding and karyotyping techniques. They are, however, discernible using microarray-based approaches, such as array-comparative genome hybridization (CGH) techniques employing either bacterial artificial chromosome (BAC) or single nucleotide polymorphism (SNP) arrays, whereby signal intensity is used to estimate the number of alleles. Initial genome-wide studies reported enhanced frequencies of CNVs in autistic patients compared to controls (on average 6–10% vs. 1–3%, respectively). In particular, Jacquemont et al. (2006) found 8 of 29 (27.5%) autistic patients carrying deletions or duplications between 1.4 and 16.0 Mb in size, including six *de novo* chromosomal rearrangements. Sebat et al. (2007) found *de novo* CNVs in 12/118 (10%) autistic children from simplex families (i.e., families with only one autistic child) and in 2/196 (2%) normal trios. Marshall et al. (2008) found 27/427 (6.3%) autistic patients carrying *de novo* CNVs, which were more common in simplex (4/56 = 7.1%) than in multiplex (1/49 = 2.0%) families. Christian et al. (2008) reported the presence of seven *de novo* and 44 inherited CNVs in 397 ASD patients. Collectively, these results were compatible with the existence of genomic instability in a sizable subgroup of autistic patients. However, later studies have not replicated genome-wide differences in CNV frequency between ASD patients and controls using genomic DNA extracted from leukocytes or lymphoblastoid cell lines (Bucan et al., 2009; Glessner et al., 2009; Pinto et al., 2010). We have recently performed a small-scale study using genomic DNA extracted from neocortical post-mortem specimens, finding increased genomic instability in only one out of ten autistic brains compared to ten matched controls (Roberto Sacco, Antonio M., Persico, Shawn Levy, and colleagues, unpublished observation). Therefore, excessive genomic instability may characterize some families with autistic patients, but it does not represent a widespread hallmark of autism either in the CNS or in peripheral tissues. CNV location may instead play a more relevant role compared to CNV frequency and mean size. Rare or even private CNVs seemingly affect the coding region of functionally important genes more often among ASD patients than in controls: disrupted loci belong to gene families involved in synaptogenesis, cell proliferation and migration, ubiquitination, and GTPase/Ras signaling (Bucan et al., 2009; Glessner et al., 2009; Pinto et al., 2010). This conclusion has been further strengthened by two large data sets that have recently uncovered highly heterogeneous *de novo* copy-number variants which collectively affect several hundred loci and presumably account for 5–8% of cases of simplex forms of ASD (Levy et al., 2011; Sanders et al., 2011; for comment see Schaaf and Zoghbi, 2011). Network-based functional analysis of these rare CNVs confirms the involvement of these loci in synapse development, axon targeting, and neuron motility (Gilman et al., 2011). We shall encounter again many of these genes in surveying monogenic forms of autism (Section 34.3).

Most CNVs are unique to any given patient, both in size and genomic distribution. However, recurrent microdeletion syndromes have also been identified: their chromosomal location and associated clinical features are listed in Table 34.5 (Fernandez et al., 2010; Kumar et al., 2008; Liang et al., 2009; Rajcan-Separovic et al., 2007; Weiss et al., 2008). In general, CNVs can be associated with a variety of clinical features, including major or minor malformations, facial dysmorphisms, severe neurological symptoms, full-blown autism, milder autism-spectrum traits, or even behavioral disorders outside of the autism spectrum (frequently seen in siblings carrying the same CNV as their autistic sib). Variable penetrance and great phenotypic heterogeneity thus characterize CNV expressivity to the same extent as we have seen occur in many 'classical' syndromic forms listed in Table 34.3. This is true to the point that it is often difficult to determine whether in a given patient a CNV is the sole cause of autism, confers vulnerability to the disease, or represents a chance finding. Indeed, the majority of CNVs are inherited from either one of the parents, who may show some autism spectrum traits, but certainly do not satisfy criteria for autistic disorder. Also, a sizable percentage of population controls carries CNVs, available in public databases (Iafrate et al., 2004). Finally, many CNVs found in ASD patients are not autism specific, but are found also in patients with mental retardation, schizophrenia, or other psychopathologies.

34.3 NOVEL SYNDROMIC FORMS OF MONOGENIC AUTISMS

34.3.1 General Description

In recent years, several monogenic forms of autism have been uncovered (see review by Lintas and Persico, 2009). Each is present in a small number of patients (i.e., <1%) and can result from mutations or cytogenetic anomalies proved to be absent from large pools of control chromosomes. These findings have led to the proposal that most autisms may represent a collection of syndromes due to rare, if not even, private mutations or CNVs (Buxbaum, 2009). However, causal mutations and chromosomal rearrangements should ideally appear *de novo*, but they are more often segregating in the family, which again underscores their variable degree of penetrance and heterogeneous expressivity. We shall now review the characteristics and neurobiological bases of the most important monogenic forms recently discovered, which are listed in Table 34.6.

34.3.2 Synaptic Genes

Several genes involved in monogenic autisms are known to play a role in synapse formation, maturation, and stabilization. This functional role in the establishment and fine-tuning of neuronal connections is pathophysiologically appealing, when considering autism as a 'dysconnection syndrome' (Courchesne and Pierce, 2005; Geschwind and Levitt, 2007; Rubenstein

TABLE 34.5 Syndromic Autisms Due to Recurrent CNVs

Ch region	Del/Dup	Neurodevelopmental signs and symptoms	Other signs and symptoms
1q21	Del	Autism, attention deficit, hyperactivity, antisocial behavior, anxiety, epilepsy, mental retardation, developmental delay, depression, hallucinations, schizophrenia	Minor dysmorphisms, cardiac defects, cataracts, multiple congenital malformations
	Dup	Autism, attention deficit, hyperactivity, epilepsy, mental retardation, developmental delay, impaired language, learning disability	Minor dysmorphisms, multiple congenital malformations
2p15–2p16.1	Del	Autism, developmental delay	Microsomy, microcephaly, dysmorphic features
15q13	Del	Autism, attention deficit, hyperactivity, aggression, anxiety, epilepsy, mental retardation, developmental delay, impaired language, schizophrenia	Minor dysmorphisms, cardiac defects
	Dup	Autism, anxiety, bipolar disorder, mental retardation, developmental delay, obsessive-compulsive disorder, language delay	Minor dysmorphisms, hypotonia, obesity, recurrent ear infections
16p11.2	Del	Autism, Asperger syndrome, attention deficit, hyperactivity, dyslexia, bipolar disorder, anxiety, epilepsy, mental retardation, developmental delay, language impairment, schizophrenia	Minor dysmorphisms, hypotonia, multiple congenital malformations
	Dup	Autism, attention deficit, hyperactivity, anxiety, epilepsy, mental retardation, developmental delay, obsessive-compulsive disorder	

and Merzenich, 2003). However, there are also issues raising caution in interpreting autistic signs and symptoms as merely due to reduced synaptogenesis, as briefly addressed in Section 34.6.

34.3.2.1 The Neuroligin Genes (NLGN3, NLGN4, and NLGN4Y)

The *NLGN3*, *NLGN4*, and possibly *NLGN4Y* genes, located in human ch Xq13, Xq22.33, and Yq11.2, respectively, have been found to host mutations seemingly responsible for behavioral phenotypes, including autism (Table 34.6). Neuroligin genes encode for cell adhesion molecules located postsynaptically in glutamatergic (*NLGN1, NLGN3, NLGN4, NLGN4Y*) and GABAergic (NLGN2) synapses (for reviews, see Betancur et al., 2009; Craig and Kang, 2007; Südhof, 2008). At the extracellular level, postsynaptic neuroligins interact with presynaptic α- or β-neurexins (see Section 34.3.2.3); at the intracellular level, neuroligins associate with postsynaptic scaffolding proteins, such as SHANK3 (see Section 34.3.2.2). This network of synaptic proteins appears to play a critical role in synapse generation,

TABLE 34.6 Mutations and Cytogenetic Abnormalities, Either *de novo* or Segregating, Affecting NLGN, SHANK3, NRXN1, MECP2, HOXA1, PTEN, and Calcium-Related Genes, their Incidence in Samples of ASD Individuals, and Associated Clinical Phenotypes

Gene	References	Mutations/del	Incidence	Clinical phenotype
NLGN3	Jamain et al. (2003)	R451C	1/158 (0.6%)	Autism, Asperger's syndrome
NLGN4	Jamain et al. (2003)	D396X	1/158 (0.6%)	Autism, Asperger's syndrome
	Laumonnier et al. (2004)	D429X	One family with 13 affected males	Autism, MR, PDD-NOS
	Yan et al. (2005)	G99S	1/148 (0.7%)	Severe autism, MR, language disability
		K378R	1/148 (0.7%)	Autism
		V403M	1/148 (0.7%)	PDD-NOS
		R704C	1/148 (0.7%)	Autism
	Lawson-Yuen et al. (2008)	del exons 4,5,6	One family with one affected male[c]	Autism with motor tics
	Daoud et al. (2009)	−355G>A	1/96 (1.0%)	Autism with severe MR
	Pampanos et al. (2009)	K378R	1/169 (0.6%)	Mild autism with normal IQ
NLGN4Y	Yan et al. (2008a)	I679V	1/290 (0.3%)	Autism
SHANK3	Durand et al. (2007)	142 kb del	1/227 (0.4%)	Autism with severe language deficits and MR
		E409X	1/227 (0.4%)	Autism with severe language deficits and MR
		800 kb del	1/227 (0.4%)	Autism with severe language deficits and MR (The trisomic brother has Asperger's syndrome)
	Moessner et al. (2007)	277 kb del, 3.2 Mb del, 4.36 Mb del	1/400 (0.25%) each, 3/400 (0.75%) total	Autism with severe language deficits and MR
		Q321R	1/400 (0.25%)	PDD-NOS with regression of spoken words
	Gauthier et al. (2009)	L68P	1/427 (0.23%)	PDDNOS with regression of spoken words
		c. 2265C +1delG	1/427 (0.23%)	Autism
NRXN1	Feng et al. (2006)	S14L	3/264 (1.1%)	Autism, MR, seizures, mild facial dysmorphism
		T40S	1/264 (0.4%)	Autism, MR, mild facial dysmorphism
	The Autism Genome Project Consortium (2007)	300 kb del at 2p16	2/196 (0.5%)	Autism, MR, mild to severe spoken language deficits
	Kim et al. (2008)	ins(16;2)(q22.1; p16.1p16.3)[f]	case report	Autism, MR

TABLE 34.6 Mutations and Cytogenetic Abnormalities, Either *de novo* or Segregating, Affecting NLGN, SHANK3, NRXN1, MECP2, HOXA1, PTEN, and Calcium-Related Genes, their Incidence in Samples of ASD Individuals, and Associated Clinical Phenotypes—cont'd

Gene	References	Mutations/del	Incidence	Clinical phenotype
		t(1;2)(q31.3;p16.3)	case report	PDD-NOS, ADHD, conduct disorder, intermittent explosive disorder
		L18Q	1/57 + 0/53 (0.9%)	Autism (?)
		L748I	1/57 + 2/53 (2.7%)	Autism (?) with incomplete penetrance
	Yan et al. (2008a,b)	R8P, L13F, c1024 +1 G>A, T665I, E715K	1/116 (0.9%) each 5/116 (4.3%) total	Autism (?)
	Zweier et al. (2009)	180 kb del + p. S979X	1/179 (0.6%)	Autism, severe MR, lack of spoken language
MECP2	Lam et al. (2000)	IVS2+2delTAAG	1/21F (4.8%)	Autism, MR. No regression, epilepsy, or microcephaly
	Vourc'h et al. (2001)	–	0/59 (42M,17F)	
	Beyer et al. (2002)	–	0/202 (154M,48F)	
	Carney et al. (2003)	1157del41, R294X	2/69F (2.9%)	Autism, MR, history of regression. No stereotypies, epilepsy, or microcephaly
	Zappella et al. (2003)	R133C, R453X	2/19F (4.7%)	Preserved speech variant of Rett syndrome
	Shibayama et al. (2004)	c.1638 G>C, c. 6809 T>C, P376R	1/24 (4.1%) each 3/24 (12.5%) total	Autism (?)
	Lobo-Menendez et al. (2004)	–	0/99 (58M,41F)	
	Li et al. (2005)	–	0/65 (49M,16F)	
	Xi et al. (2007)	c.1461 G>A	1/31M (3.2%)	Autism (?)
	Harvey et al. (2007)	–	0/401 (266M,135F)	
	Coutinho et al. (2007a,b)	G206A	1/172 (0.6%) (141M, 31F)	Autistic male with severe MR and lack of spoken language
		Twelve 3'UTR variants, c.27-55G>A, c.377+18C>G	1/172 (0.6%) each	Autism (?)
HoxA1	Tischfield et al. (2005) and Bosley et al. (2007)	c.84 C>G (Y28X)	One patient from a Turkish consanguineous family	BSAS with variable degrees of horizontal gaze abnormalities, deafness, focal weakness, hypoventilation, vascular malformations of the internal carotid arteries and cardiac outflow tract, MR and autism (present in 3/9 Saudi Arabian patients)
		175-176insG	Nine patients from 5 Saudi Arabian consanguineous fam.	
PTEN	Goffin et al. (2001)	Y178X	Case report	Cowden syndrome with autism and progressive macrocephaly
	Butler et al. (2005)	H93R, D252G, F241S	3/18 (16.6%) all macrocephalic	Extreme macrocephaly and macrosomy
	Boccone et al. (2006)	I135R	1 (case report)	Bannayan–Riley–Ruvalcaba syndrome with reactive nuclear lymphoid hyperplasia and autism
	Buxbaum et al. (2007)	D326N	1/88 (1.1%) all macrocephalic	Macrocephaly (+9.6 SD), polydactily at both feet, autism, MR, language delay
	Orrico et al. (2009)	Y176C, N276S	1/40 (2.5%) each, all macrocephalic	Autism

Continued

TABLE 34.6 Mutations and Cytogenetic Abnormalities, Either *de novo* or Segregating, Affecting NLGN, SHANK3, NRXN1, MECP2, HOXA1, PTEN, and Calcium-Related Genes, their Incidence in Samples of ASD Individuals, and Associated Clinical Phenotypes—cont'd

Gene	References	Mutations/del	Incidence	Clinical phenotype
		H118P	1/40 (2.5%), all macrocephalic	Developmental delay without autism
	Herman et al. (2007a) and Varga et al. (2009)	520insT, R130X, E157G, L139X, IVS6-3C>G	5/60 (8.3%) total, 5/27 (18.5%) macrocephalic	Macrocephaly, autism or PDD-NOS, developmental delay, MR, language delay
EIF4E	Neves-Pereira et al. (2009)	46,XY,t(4,5)(q23; q31.3)	Case report	Autism with regression (loss of spoken words and social interaction at age 2)
		C_8-4EBE	2/120 (1.6%) multiplex families ($N=4$ subjects)	Severe autism, language delay
CACNA1C	Splawski et al. (2004)	G406R	13 patients with Timothy syndrome	Timothy syndrome: lethal arrhythmias, webbing of fingers and toes, congenital heart disease, immune deficiency, intermittent hypoglycemia, autism and MR
CACNA1F	Hemara-Wahanui et al. (2005), Hope et al. (2005)	I745T	One pedigree with 3 ASD males out of 10 mutation carriers	5/10 mutation carriers have MR, and 3 of these 5 individuals has MR+autism
CACNA1H	Splawski et al. (2006)	R212C, R902W, W962C	1/491 (0.2%) each, 3/491 (0.6%) total	Autism (?)
		R1871Q+A1874V	3/491 (0.6%)	Autism (?)
BKCa	Laumonnier et al. (2006)	46,XY, t(9,10)(q23, q22)	Case report	Autism, lack of spoken language
		Ala138Val	1/116 (0.9%)	Autism
SCN2A	Weiss et al. (2003)	R1902C	1/229 (0.4%) families	Autism (AGRE family AU0247, only in one of two affected children)

Variants are not listed if also present in control samples.

MR, mental retardation; PDD-NOS, pervasive developmental disorder not otherwise specified; (?), clinical phenotype not described.

maturation, stabilization, and maintenance (Betancur et al., 2009; Craig and Kang, 2007; Südhof, 2008). *In vitro* assays initially suggested that the interaction between neuroligins and neurexins may trigger the formation of functional presynaptic boutons in contacting neurites, both in neuronal and even in non-neuronal cells (Dean et al., 2003; Scheiffele et al., 2000). Later studies of triple knockout mice lacking neuroligins 1, 2, and 3 showed that their absolute numbers of synapses are unchanged, whereas synaptic transmission is severely hampered, leading to respiratory failure and death on the day of birth (Varoqueaux et al., 2006). These results underscore the role of neuroligins as critical to synaptic function, more than to synaptogenesis *per se*.

Jamain et al. (2003) reported one frameshift (D396X in *NLGN4*) and one missense (R451C in *NLGN3*) mutation in two unrelated Swedish families, both inherited from apparently unaffected mothers. Mouse mutants carrying the human R451C mutation show a mild behavioral phenotype, described by Tabuchi et al. (2007) and by Chadman et al. (2008). Reduced ultrasonic vocalizations in males represent the most consistent behavioral abnormality, followed by impaired social novelty preference (Tabuchi et al., 2007, but see Chadman et al., 2008). Surprisingly, enhanced spatial learning abilities and increased inhibitory synaptic transmission were also recorded (Tabuchi et al., 2007). Functional *in vitro* studies of the R451C mutation show defective vesicle trafficking with partial retention of NLGN3 in the endoplasmic reticulum. Reduced synapse induction properties are due to blunted NLGN3 delivery to the cell membrane, as mutated NLGN3 retains synaptogenetic properties in the minority of cells where delivery to the membrane does occur (Chih et al., 2004; Chubykin et al., 2005; Comoletti et al., 2004). These abnormalities lead to reduced glutamate-driven excitation in the neocortex, while AMPA-driven excitation, NR2B subunit delivery, and long-term potentiation are all enhanced in the hippocampus (Etherton et al., 2011a). Instead, the R704C mutation initially reported by Yan et al. (2005) causes a major and selective decrease in AMPA receptor-mediated synaptic transmission, leaving the number of synapses unchanged (Etherton et al., 2011b).

Multiple studies collectively confirm the low frequency of *NLGN* gene mutations among idiopathic ASD patients (Table 34.6) (Lintas and Persico, 2009). The clinical phenotype of patients carrying *NLGN* mutations is highly heterogeneous, ranging from severe autistic disorder to Asperger's syndrome (the 'speech-preserved' variant of autism), PDD-NOS (the autism variant satisfying only some, but not all diagnostic criteria), nonspecific X-linked mental retardation, specific language impairment, and Tourette syndrome (Table 34.6). Disease onset may be slow and insidious or sudden and regressive (see Section 34.6). Mutation carriers typically display no dysmorphic feature and are phenotypically indistinguishable (Lintas and Persico, 2009).

34.3.2.2 *The SH3 and Multiple Ankyrin Repeat Domains 3 Gene (SHANK3)*

The *SHANK3* gene, located in chromosome 22q13.3, encodes for a scaffolding protein found in the postsynaptic density complex of excitatory synapses, where it binds to neuroligins and to actin, affecting actin polymerization, growth cone motility, dendritic spine morphology, and synaptic transmission (Durand et al., 2011). Several recent studies have described rare mutations or small cytogenetic rearrangements affecting *SHANK3* in patients with an autistic phenotype mainly characterized by severe expressive language impairment (Table 34.6). Similarly, 13 patients carrying deletions of the terminal 22q13 region encompassing or breaking the *SHANK3* locus all share mental retardation and severe delay in or absence of expressive speech (Dhar et al., 2010). As for neuroligins, also *SHANK3* mutations or deletions/duplications represent rare events, affecting only 9/1054 (0.85%) ASD individuals (see Table 34.6). No evidence of association was found in large samples (Qin et al., 2009; Sykes et al., 2009), demonstrating that the *SHANK3* gene hosts rare variants, but not common variants. With the possible exception of language impairment, mutations and cytogenetic abnormalities affecting *SHANK3* display highly variable phenotypic expression: (a) they are most often inherited from parents described as either healthy or epileptic; (b) in some families, they are present also in unaffected siblings of the autistic proband; (c) two *de novo* mutations, R536W and R1117X, different from those found in ASD patients, were detected in patients with borderline or moderate mental retardation and either schizoaffective disorder, attention-deficit/hyperactivity disorder (ADHD), or schizophrenia/atypical chronic psychosis (Gauthier et al., 2010). Importantly, autistic individuals carrying inherited 22q13 deletions involving *SHANK3* (800 kb in Durand et al. (2007) and 3.2 Mb at 22q13

in Moessner et al. (2007)) due to a paternal balanced translocation have siblings with partial 22q13 trisomy diagnosed with Asperger syndrome, showing early language development and ADHD. Hence, a physiological window for *SHANK3* may be functionally crucial to cognitive development in humans.

34.3.2.3 *The Neurexin 1 Gene (NRXN1)*

Presynaptic neurexins are able to induce postsynaptic differentiation by interacting with postsynaptic neuroligins (Betancur et al., 2009; Craig and Kang, 2007; Südhof, 2008). In addition, α neurexins are involved in neurotransmitter release, as they link calcium (Ca^{2+}) channels to synaptic vesicle exocytosis (Missler et al., 2003; Zhang et al., 2005). The three neurexin genes (*NRXN1*, *NRXN2*, and *NRXN3*, located on human ch 2p16.3, 11q13, and 14q24.3–q31.1, respectively) have two independent promoters, yielding long and short mRNAs which encode for α and β neurexins, respectively (Ichtchenko et al., 1995). Several studies have reported rare sequence variants or CNVs affecting the *NRXN1* locus, as summarized in Table 34.6. However, an exact definition of *NRXN1* roles in autism is complicated by an extreme interindividual variability in genotype–phenotype correlations. An initial *NRXN1*β screening conducted by Feng et al. (2006) identified two heterozygous missense mutations (S14L and T40S) present in 4/264 (1.5%) ASD patients and not in 729 controls (Table 34.6). These 'mutations' are actually rare segregating polymorphisms found also in first-degree relatives, who clinically range from apparently normal behavior to hyperactivity, depression, and/or learning problems. Similarly, one of two chromosomal rearrangements affecting the *NRXN1* gene was also shown to be paternally inherited in one patient (Kim et al., 2008). In another study, a *de novo* heterozygous 300 kb deletion in the coding exons of the *NRXN1* gene was found in two autistic sisters (Autism Genome Project Consortium, 2007); interestingly, one girl was reported to be nonverbal, whereas the other only had mild language regression. CNVs disrupting the *NRXN1* coding sequence can result in schizophrenia and not in ASD (Kirov et al., 2009; Rujescu et al., 2009). This phenotypic variability is further underscored by a large retrospective study involving 3540 individuals, identifying in 12 of them exonic *NRXN1* microdeletions causing very heterogeneous clinical phenotypes including ASD, mental retardation, language delay, and muscle hypotonia to a variable degree (Ching et al., 2010). Finally, a recent study by Zweier et al. (2009) concerning a girl with severe mental retardation and autism demonstrated the inheritance in compound heterozygosity of a 180 kb deletion from her unaffected mother and a stop mutation in exon 15 from her healthy father. These genetic abnormalities are predicted to deprive this patient of the longer alpha *NRXN1* isoform, whose lack cannot be functionally complemented by the shorter beta isoform, leading to significantly decreased numbers of synapses both in alpha *NRXN1* knockout mice and in *Drosophila* (Etherton et al., 2009; Li et al., 2007a; Zeng et al., 2007). In Section 34.6, we shall return to compound heterozygosity as a viable mechanism able to explain the complexities of rare variant contributions to autism pathogenesis in some families.

34.3.3 Chromatin Architecture Genes

Rett syndrome is a peculiar PDD initially described by the Austrian pediatrician Andreas Rett in 1966. This severe neurodevelopmental disorder is characterized by regression, autism, microcephaly, stereotyped behaviors, epilepsy, and breathing problems (Chahrour and Zoghbi, 2007; Rett, 1966). The discovery that approximately 80% of females with Rett syndrome carry *de novo* mutations in the methyl-CpG-binding protein 2 (MeCP2) gene, which plays a critical role in determining chromatin structure by modulating DNA methylation (Amir et al., 1999), spurred great interest in the role of this gene in other PDDs and more broadly in the role of epigenetic mechanisms in autism.

34.3.3.1 *The Methyl-CpG-Binding Protein 2 Gene*

The methyl-CpG-binding protein 2 (MeCP2) binds to methylated CpG dinucleotides, recruiting histone deacetylase 1 (HDAC1) and other proteins involved in chromatin repression at specific gene promoters. It thus acts as a transcriptional repressor (Chahrour and Zoghbi, 2007), not only during development but throughout adult life (McGraw et al., 2011). *De novo* mutations of the *MECP2* gene located on chromosome Xq28, in addition to classical Rett syndrome, can also result in asymptomatic phenotypes, mild mental retardation, and verbal Rett variants, depending upon the specific mutation, the genetic background of the patient, and the X-inactivation pattern, which is highly skewed in the presence of mutations affecting X-linked genes, such as *NLGN3* and *MECP2*, albeit not being skewed in ASD families altogether (Gong et al., 2008). Instead, *MECP2* mutations are generally lethal in males (Amir et al., 1999; Chahrour and Zoghbi, 2007).

Several groups have screened the *MECP2* gene for mutations in nonsyndromic ASD patients, finding positives in 5 of the 397 females (1.3%) and none of the 741 males assessed to date (Lintas and Persico, 2009) (Table 34.6). Three *de novo MECP2* mutations have been found in two out of eleven studies assessing female ASD patients: the IVS2+2delTAAG splice variant in intron 2, the frameshift mutation 1157del41, and the nonsense mutation R294X (Carney et al., 2003; Lam et al., 2000). Two additional *de novo* mutations, R133C and R453X, were identified in two autism-spectrum girls fulfilling

criteria for the 'preserved speech' variant of Rett syndrome (Zappella et al., 2003). A few other missense, intronic, or 3'-UTR variants listed in Table 34.6 either are inherited from one of the parents, or it is not specified whether they are inherited or occurring *de novo*. Importantly, young girls carrying *MECP2* mutations appear autistic and mentally retarded, but display none of the symptoms typical of Rett syndrome (epilepsy, microcephaly, stereotypies, and breathing problems). Also, regression is not consistently reported by parents (Carney et al., 2003). Signs and symptoms more typical of Rett syndrome may appear when they grow older (Young et al., 2008). There is thus a rationale for *MECP2* gene screenings in female autistic patients and for follow-up programs to monitor these patients clinically over time (Lintas and Persico, 2009).

34.3.4 Morphogenetic and Growth-Regulating Genes

Many syndromic patients display facial dysmorphisms, minor or major malformations, microcephaly or macrocephaly either in isolation or as part of a broader microsomy or macrosomy, respectively (Tables 34.3 and 34.5). Also, idiopathic autistic children sometimes display minor facial dysmorphisms, which are beginning to be characterized using standardized or even automated methods for consistency (Hammond et al., 2004, 2008; Miles et al., 2008; Tripi et al., 2008). In addition, head and body growth rates are often abnormal. Macrocephaly has been consistently found in approximately 20% of autistic patients (Fombonne et al., 1999; Lainhart et al., 1997; Miles et al., 2000; Sacco et al., 2007a; Stevenson et al., 1997; Woodhouse et al., 1996). Head circumference in these ASD patients is typically normal at birth, and an overgrowth seemingly develops over the first few years of life (Courchesne et al., 2007). This macrocephaly is part of a broader macrosomy in most, though not all, patients (Bigler et al., 2010; Dissanayake et al., 2006; Lainhart et al., 2006; Sacco et al., 2007a). On the contrary, a small subset of idiopathic autistic patients is instead microcephalic and usually also microsomic (see Figures 1 and 2 in Sacco et al., 2007a).

34.3.4.1 *The Homeobox A1 Gene (HOXA1)*

HOXA1 is a homeobox gene located on chromosome 7p15.3. It is critically involved in the development of head and neck structures directly or indirectly derived from the distal part of rhombomere 4 and from rhombomere 5 during embryogenesis (Chisaka et al., 1992; Mark et al., 1993; Rossell and Capecchi, 1999): these include brainstem, cerebellum, several cranial nerves, medium and internal ear, and occipital and hyoid bones. Both common and rare *HOXA1* gene variants have been sought. The common A218G polymorphism exerts a sizable effect on head growth rates both in autistic and in typically developing children, with the G allele yielding faster head growth and smaller cerebellar volumes (Canu et al., 2009; Conciatori et al., 2004; Muscarella et al., 2007, 2010). This measurable effect on the growth of regions known to be involved in autism, such as the cerebellum (Courchesne, 1997), is intriguing, despite the nonreplication of an initial report suggesting that this common *HOXA1* variant could possibly contribute to autism (Ingram et al., 2000b; but see Gallagher et al., 2004; Li et al., 2002; Romano et al., 2003; Sen et al., 2007; Talebizadeh et al., 2002). The G allele leads to the substitution of the second of ten contiguous histidines by an arginine (His73Arg). This stretch of ten histidines underlies protein–protein interactions, which could be modulated by this gene variant, although direct experimental evidence is still lacking.

In reference to rare *HOXA1* gene variants possibly causing autism, the study of five consaguineous families from Saudi Arabia and of one from Turkey disclosed homozygous stop codon mutations in ten affected individuals (Bosley et al., 2007; Tischfield et al., 2005). Two different mutations were identified: a c.84 C>G mutation, which results in the introduction of a stop codon (Y28X) in the Turkish patient, and a 175–176insG, which causes a reading frame shift and a premature protein truncation in nine Saudi Arabian patients (Bosley et al., 2007; Tischfield et al., 2005). Mutation carriers show some phenotypic similarities, including horizontal gaze abnormalities, deafness, focal weakness, hypoventilation, vascular malformations of the internal carotid arteries and cardiac outflow tract, mental retardation, and autism: this set of clinical symptoms and malformations was named Bosley–Salih–Alorainy syndrome (BSAS) (Bosley et al., 2007). Importantly, all signs and symptoms (even vascular malformations) display a significant degree of interindividual variability in both presence and severity, with developmental delay and autism reported only in a subset of patients (see Section 34.6). Many signs and symptoms of BSAS overlap with those present in the Athabascan brainstem dysgenesis syndrome found in Native American children carrying a distinct *HOXA1* R26X mutation (Bosley et al., 2008). Since these causal mutations are recessive, similar phenotypes should be sought only in areas where inbreeding is substantial. In other geographical areas, oligogenic heterozygosity also involving rare *HOXA1* variants may play a role in idiopathic autism (Schaaf et al., 2011).

34.3.4.2 *The Phosphatease and Tensin Homolog Gene (PTEN)*

PTEN is a tumor suppressor gene located on human chromosome 10q23, which favors cell-cycle arrest in G1 and apoptosis. In conjunction with other tumor-suppressor genes, such as *TSC1*, *TSC2*, and *NF1*, it balances the stimulation physiologically exerted on cell proliferation and body growth by nutrient availability (sugars and proteins), insulin release, and pro-inflammatory cytokines, through the ERK/PI3K/mTOR pathway (Figure 34.1) (Ma and Blenis, 2009). Mutations inactivating these tumor-suppressor genes cause diseases often associated with syndromic autisms (see Table 34.3). *PTEN* knockout mice indeed display macrosomy, macrocephaly, CNS overgrowth with thickening of the neocortex and cytoarchitectonic abnormalities in the hippocampus, excessive dendritic and axonal growth, and increased numbers of synapses (Kwon et al., 2006). Furthermore, in humans, germline mutations resulting in *PTEN* haploinsufficiency facilitate cell-cycle progression and oncogenesis, leading to macrocephaly/macrosomy and to cancer development, respectively (Eng, 2003). Germline *PTEN* mutations have been documented in the vast majority of patients diagnosed with Cowden syndrome, which carries enhanced risk for breast, endometrial, and thyroid cancers (Eng, 2003). Also, individuals suffering from other related hamartoma disorders, such as Bannayan–Riley–Ruvalcaba syndrome, Proteus syndrome, and Proteus-like syndromes, display germline *PTEN* mutations in 60%, 20%, and 50% of cases, respectively (Eng, 2003). Interestingly, genetic syndromes due to *PTEN* germline haploinsufficiency are also frequently associated with autism or mental retardation, as initially reported by Goffin et al. (2001) (e.g., see Boccone et al., 2006). Instead, several missense mutations affecting evolutionarily conserved amino acid residues have been detected in macrocephalic individuals affected by idiopathic autism (Table 34.6). ASD patients carrying *PTEN* mutations are invariably characterized by severe to extreme macrocephaly (i.e., cranial circumference >97th percentile or +2 SD, but most *PTEN* mutation carriers typically display ≥+3 SD). In some cases, the overgrowth starts prenatally, whereas other *PTEN* mutation carriers display a normal head circumference at birth and macrocephaly develops only postnatally, as generally occurs in macrocephalic autistic children (Courchesne et al., 2007). In addition, the majority of macrocephalic autistic patients are actually macrosomic, underscoring a systemic disruption of body growth control mechanisms (Sacco et al., 2007a). Although all mutation carriers share macrocephaly as a unifying feature, behavioral phenotypes can again differ significantly between patients, and some mutations are inherited from apparently normal fathers (Varga et al., 2009). The incidence of *PTEN de novo* mutations can be estimated at 4.7% (6/126) among macrocephalic/macrosomic ASD patients, who in turn are approximately 20% of all ASD patients (Lintas and Persico, 2009). The percentage of PTEN mutation carriers may be even higher in selected clinical populations (Butler et al.,

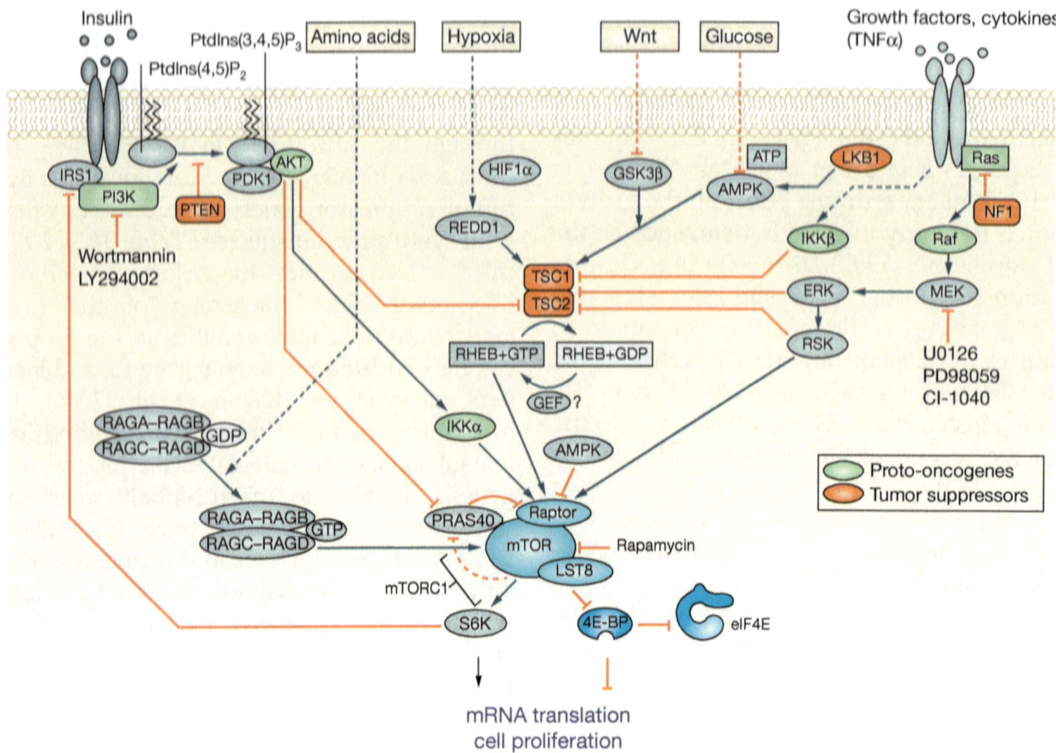

FIGURE 34.1   The ERK/PI3K/mTOR pathway (with permission from Ma and Blenis (2009), modified).

2005; Varga et al., 2009). This high genetic yield and the predisposition toward malignancies underscore the importance of screening for *PTEN* mutations in macrocephalic/macrosomic autistic children (Herman et al., 2007a,b; Lintas and Persico, 2009; McBride et al., 2010; Schaefer and Mendelson, 2008; Varga et al., 2009).

34.3.4.3 *The Eukaryotic Translation Initiation Factor 4E Gene (EIF4E)*

The *EIF4E* gene, located on human chromosome 4q21–q25, encodes the rate-limiting component of eukaryotic translation initiation and the downstream effector in the mTOR pathway (Figure 34.1). Recently, a balanced translocation disrupting the *EIF4E* locus was found in an autistic boy with loss of initial language and social interactions at 2 years of age (Neves-Pereira et al., 2009) (Table 34.6). In the same study, both affected children of 2/120 multiplex families were found to inherit a C insertion from an apparently unaffected father, extending to eight a stretch of seven cytosines located in the basal promoter element of the *EIF4E* gene (4-EBE) (Neves-Pereira et al., 2009). The C$_8$-4EBE allele seemingly binds with much higher affinity an abundant nuclear protein (probably hnRNPK), resulting in a twofold increase in gene expression compared to the C$_7$-4EBE allele (Neves-Pereira et al., 2009). Interestingly, despite carrying the same paternally inherited C$_8$-4EBE allele, only one of these four autistic children underwent

behavioral regression, as had occurred to the proband carrying the *de novo* translocation (Neves-Pereira et al., 2009).

34.3.5 Calcium-Related Genes

Many lines of evidence indicate that excessive Ca^{2+} signaling is pivotal in the pathophysiological processes leading to autism, as reviewed by Krey and Dolmetsch (2007). Excessive intracellular Ca^{2+} spikes can modulate the aspartate/glutamate mitochondrial carrier AGC1, leading to abnormal energy metabolism and enhanced oxidative stress (for review, see Palmieri and Persico, 2010). Rare gain-of-function mutations causing autism or multisystem disorders encompassing autistic behaviors have been detected in genes encoding ion channels either directly conducting Ca^{2+} or indirectly prolonging the opening time of voltage-gated Ca^{2+} channels.

34.3.5.1 *The Ion Channel-encoding Genes CACNA1C, CACNA1F, CACNA1H, BKCa, and SCN2A*

Gain-of-function mutations in the gene encoding for the L-type voltage-gated Ca^{2+} channel Ca$_v$1.2 (*CACNA1C*) cause Timothy syndrome, a multisystem disorder characterized by lethal arrhythmias, webbing of fingers and toes, congenital heart disease, immune deficiency, intermittent hypoglycemia, autism, and mental retardation (Splawski et al., 2004). Similarly, mutations in the L-type voltage-gated Ca^{2+} channel $Ca_v1.4$ (*CACNA1F*) cause an incomplete form of X-linked congenital stationary night blindness (CSNB2), frequently accompanied by cognitive impairment and autism or epilepsy, but only with gain-of-function and never with loss-of-function *CACNA1F* mutations (Hemara-Wahanui et al., 2005; Hope et al., 2005). Surprisingly, *CACNA1F* is not expressed in the CNS, except for the pineal gland (Hemara-Wahanui et al., 2005). In general, these gain-of-function mutations reduce or block voltage-dependent channel inactivation, resulting in excessive Ca^{2+} influx (Hemara-Wahanui et al., 2005; Splawski et al., 2004). Also, mutations and chromosomal abnormalities indirectly boosting cytosolic Ca^{2+} levels or amplifying intracellular Ca^{2+} signaling by hampering Ca^{2+}-activated negative feedback mechanisms have been found associated with autism. An autistic boy was found to carry a balanced translocation disrupting one copy of the *KCNMA1* gene, which encodes the α subunit of the large conductance Ca^{2+}-activated K^+ (BK_{Ca}) channel: the inactivation of one copy of this gene results in a more depolarized resting membrane potential and in a relatively less efficient control of neuronal excitability (Laumonnier et al., 2006). The R1902C variant located in the *SCN2A* gene decreases the affinity of this voltage-gated sodium channel for Ca^{2+}-bound calmodulin, which would stabilize the inactivation gate and minimize sustained channel activity during depolarization (Kim et al., 2004; Weiss et al., 2003). The situation is less clear with mutations affecting the *CACNA1H* gene, which encodes for the T-type voltage-gated Ca^{2+} channel $Ca_v3.2$ (Splawski et al., 2006). These mutations are found in ASD families and not in controls, but do not fully segregate with an autistic phenotype; furthermore, mutant channels require greater depolarizations to activate conductance and overall conduct substantially less than wild-type channels (Splawski et al., 2006). However, in the case of these autism-predisposing mutations, more depolarized potentials are also necessary to produce channel inactivation, indicating that longer-lasting calcium influx will be generated by small perturbations from the resting membrane potential (Splawski et al., 2006). Therefore, studies of patients carrying mutations in calcium-related genes collectively support excessive Ca^{2+} signaling as a critical player in the pathophysiological processes leading to autism.

34.4 NONSYNDROMIC AUTISMS: THE ROLE OF COMMON VARIANTS

34.4.1 General Description

Functional polymorphisms widely distributed in the general population can indeed confer vulnerability or protection in complex disorders, such as autism. Conceivably, a host of unfavorable common variants could even cause a disease phenotype, either directly or by lowering the sensitivity threshold to widespread environmental agents. Common genetic variants are typically sought by applying a candidate gene or, more recently, a genome-wide association approach. Although several common variants have been found associated with autism, as reviewed in detail elsewhere (Abrahams and Geschwind, 2008; Freitag, 2007; Muhle et al., 2004; Persico and Bourgeron, 2006), the evidence from independent replications and from functional studies is not equally strong for all of them (see Tables 2 and 4 in Abrahams and Geschwind, 2008). We shall now briefly summarize the results and pathophysiological bases concerning some of the most consistently replicated genes.

34.4.2 Reelin (RELN)

The *RELN* gene encodes for reelin, a critical stop signal for migrating neurons in several CNS districts, including the neocortex, the cerebellum, and the hindbrain (Herz and Chen, 2006; Rice and Curran, 2001). *RELN* maps to human chromosome 7q22, in a region linked with autism in several studies (Muhle et al., 2004; Persico and Bourgeron, 2006). Reelin acts by binding to a variety of receptors, including the VLDL receptors, APOE-R2, and $\alpha3\beta1$ integrins, and by exerting a proteolytic activity on extracellular matrix proteins (D'Arcangelo et al., 1999; Hiesberger et al., 1999; Quattrocchi et al., 2002). *Reeler* mice lack reelin protein due to spontaneous deletions of the *RELN* gene (D'Arcangelo et al., 1995). Their brains display major cytoarchitectonic abnormalities (Goffinet, 1984), the distribution of which largely overlaps with regions of altered neuronal migration in autistic brains, as reviewed by Persico and Bourgeron (2006). Importantly, post-mortem studies have documented reductions in *RELN* and *DAB1* gene expression, as well as elevations in *VLDLR* mRNA, in the cerebral and cerebellar cortex of autistic individuals compared to controls (Fatemi et al., 2005; Lintas and Persico, 2010). Similar reductions have been found *in vivo* when measuring reelin plasma levels in ASD patients compared to controls (Fatemi et al., 2002; Lugli et al., 2003).

RELN gene mutations resulting in a lack of reelin protein yield the Norman–Roberts syndrome, a severe autosomal recessive disease characterized by lissencephaly and cerebellar hypoplasia, with severe mental retardation, abnormal neuromuscular connectivity, and congenital lymphoedema (Hong et al., 2000). *RELN* gene variants with a less dramatic functional impact have been found to confer liability to neuropsychiatric disorders, such as autism and schizophrenia (for review, see Lintas and Persico, 2008). Genetic association studies on *RELN* gene variants and autism are listed

TABLE 34.7 Genetic Association Studies on *RELN* Gene Variants and Autism

References	Polymorphisms	Experimental design	Race and ethnicity	Outcome
Persico et al. (2001)	5′UTR: GGC repeat Intron 5: rs607755 Exon 50: rs2229864	Case–control Family-based	Italians; U.S.-Caucasians	Association with GGC repeat and with haplotypes formed by GGC + rs607755 + rs2229864
Zhang et al. (2002)	5′UTR: GGC repeat	Case–control Family-based	Not specified (families from Canada and AGRE)	Association with GGC repeat
Krebs et al. (2002)	5′UTR: GGC repeat	Family-based	Mostly (94%) EU-Caucasians	No association with GGC repeat
Bonora et al. (2003)	5′UTR: GGC repeat intron 5: rs607755 exon 22: rs362691 intron 31: rs362726 exon 50: rs2229864	Family-based	EU-Caucasians: IMGSAC and German families	No association with any common variant; rare missense variants are present (see Table 34.1)
Li et al. (2004)	5′UTR: GGC repeat	Family-based	Not specified	No association with GGC repeat
Devlin et al. (2004)	5′UTR: GGC repeat	Family-based	Not specified (NIH CPEA families)	No association with GGC repeat
Skaar et al. (2005)	5′UTR: GGC repeat Intron 5: rs607755 exon 44: rs2075043 exon 45: rs362746 exon 50: rs2229864 intron 59: rs736707	Family-based	U.S.-Caucasians from Duke Univ., AGRE, and Tufts University	Association with GGC triplet and with specific haplotypes
Serajee et al. (2006)	Exon 22: rs362691[a] intron 59: rs736707[a]	Family-based	U.S.-Caucasians from AGRE	Association with rs362691 and rs736707
Dutta et al. (2007), Dutta et al. (2008)	5′UTR: GGC repeat intron 5: rs607755 intron 12: rs727531 exon 15: rs2072403 intron 15: rs2072402 exon 22: rs362691 intron 41: rs362719 exon 50: rs2229864 intron 59: rs736707	Case–control Family-based	Indian from West Bengal and Assam	No association with any common variant; possible paternal transmission of GGC 10-repeat allele
Li et al. (2008)	Intron 59: rs736707[b]	Case–control	Eastern China	Association with rs736707
Kelemenova et al. (2010)	5′UTR: GGC repeat	Case–control	Slovaks	Association with GGC triplet

[a]*Only the two SNPs found associated with autism are listed here, out of 34 SNPs assessed.*
[b]*Only one SNP found associated with autism is listed here, out of 12 SNPs assessed.*

in Table 34.7. A polymorphic GGC repeat located immediately 5′ of the AUG translation initiation codon and ranging from 4 to 23 repeats in our sample (Persico et al., 2001, 2006) has been found associated with autism in several, though not all, studies. In particular, 'long' GGC alleles (i.e., >10 GGC repeats) were shown to decrease *RELN* gene expression in neuronal (SN56 and N2A) and non-neuronal (CHO) cell lines (Persico et al., 2006). These alleles are present in approximately 20% of autistic individuals, compared with 10% of population controls. Other studies have pointed toward more 3′ regions of this large gene as possibly hosting functional variants. Some studies reporting no association with the GGC triplet repeat have nonetheless found rare variants of potential interest (Bonora et al., 2003). *RELN* variants different from those involved in autism may possibly contribute to schizophrenia: the GGC variant does not seem to confer vulnerability to schizophrenia, which has instead been found to be associated with SNP rs7341475 located in intron 4 (Shifman et al., 2008). However, this SNP does not appear to influence brain structure, working memory, or *RELN* gene expression, so the functional correlates of this association remain unclear (Tost et al., 2010). Gene–gene interactions with other schizophrenia liability genes (Hall et al., 2007) and epigenetic control of *RELN* gene expression (Grayson et al., 2006; Lintas and Persico, 2010) may perhaps play more prominent roles than single functional variants in conferring vulnerability to schizophrenia, especially after puberty.

Possible gene–gene and gene–environment interactions involving *RELN* have been previously presented, especially in reference to PON1 gene polymorphisms and prenatal exposure to organophosphate pesticides (see Section 34.4 and Persico and Bourgeron, 2006). Briefly, reelin's proteolytic activity, crucial for neuronal migration, is inhibited by organophosphates (Quattrocchi et al., 2002). Furthermore, the *PON1* gene, encoding for the organophosphate-detoxifying enzyme present in human serum, is also associated with autism and provides evidence of gene–gene interactions with *RELN* alleles (D'Amelio et al., 2005). We thus proposed a gene–gene–environment interaction model, whereby individuals carrying genetic or epigenetic variants resulting in reduced *RELN* gene expression and in less active paraoxonase isoforms, if exposed prenatally to organophosphates during critical periods in neurodevelopment, will more likely suffer from altered neuronal migration resulting in autistic disorder (Persico and Bourgeron, 2006). We have recently measured two different *PON1* enzymatic activities in the serum of 174 ASD patients, 144 controls, and 175 first-degree relatives finding significant reductions in arylesterase, but not in diazoxonase activity, primarily due to a functional inhibition of this enzymatic activity and not due to quantitative decreases in PON1 protein levels (Gaita et al., 2010). These results were unexpected, because diazoxon is one of the most widespread organophosphates in the United States, whereas decreases in arylesterase activity have so far been recorded in the presence of enhanced oxidative stress and/or immune activation, as during viral hepatitis C (Ferré et al., 2005; Kilic et al., 2005), influenza (van Lenten et al., 2001), and HIV infections (Parra et al., 2007). Recent neuroanatomical, gene expression, and brain imaging studies strongly support an abnormal activation of the immune system in autism (see Sections 34.4.5.3 and 34.6). Within this framework, *RELN–PON1* interactions may be better explained by a joint modulation of inflammatory processes, especially monocyte recruitment and migration into the CNS (Ahmed et al., 2003; Cameron and Landreth, 2010; Gaita et al., 2010; van Lenten et al., 2002).

34.4.3 The MET Protooncogene

The MET receptor tyrosine kinase, encoded by the *MET* protooncogene located in human ch 7q31, plays an important role in modulating cell proliferation and migration, as reviewed by Levitt et al. (2004) and by Levitt and Campbell (2009). Briefly, the MET receptor binds hepatocyte growth factor (HGF), which is translated as an inactive precursor and activated by proteolytic cleavage to yield the MET receptor ligand: this cleavage is achieved by the protease plasminogen activator, urokinase-type (uPA), when uPA binds to its receptor, the urokinase plasminogen activator receptor (uPAR) (see Figure 3 in Levitt et al., 2004). The cleavage-mediated activation of HGF can instead be suppressed by the plasminogen activator inhibitor-1 (PAI-1). *MET* gene variants have been found to be associated with autism in four independent studies involving at least seven distinct family samples (Campbell et al., 2006, 2008; Jackson et al., 2009; Sousa et al., 2009). The *MET* gene variant conferring autism vulnerability in the initial study was the C allele at rs1858830, located in the *MET* gene promoter (Campbell et al., 2006). This allele dramatically reduces the binding of transcription factors SP1 and PC4, thereby decreasing *MET* transcription assessed in neuronal (SN56 and N2A) and non-neuronal (HEK) cell lines using luciferase-expressing reporter constructs (Campbell et al., 2006). The association between autism and the C allele at rs1858830 was replicated by the same group in an independent sample (Campbell et al., 2008), whereas a study from the IMGSAC Consortium found autism associated with SNP rs38845, located in intron 1 of the *MET* gene (Sousa et al., 2009). A fourth study, employing only case–control contrasts, confirmed an association with the C allele at rs1858830 in a South Carolina, and not in an Italian cohort (but the latter was deeply underpowered: South Carolina sample = 174 ASD patients vs. 369 controls; Italian sample = 65 ASD patients vs. 126 controls) (Jackson et al., 2009).

Analyses of postmortem tissue from the superior temporal gyrus (Brodmann area 41/42) confirmed an approximate twofold decrease in *MET* transcript and protein expression in ASD patients compared to matched controls (Campbell et al., 2007). Moreover, significantly lower MET protein levels were found among controls carrying the C/C, as compared to the G/G genotype at rs1858830 (Campbell et al., 2007). The same tissues unveiled increased expression of the *HGF, PLAUR,* and *SERPINE1* genes, which encode for HGF, uPAR, and PAI-1, respectively (Campbell et al., 2007). Conceivably, decreased *MET* gene expression at the neocortical level triggers compensatory increases in the expression of other molecules belonging to the same pathway, such as *HGF, PLAUR,* and *SERPINE1*. The latter two genes apparently also host common genetic variants independently contributing to autism vulnerability: a *SERPINE1* haplotype and the *PLAUR* promoter T allele at rs344781 are both associated with autism (Campbell et al., 2008). Significant gene–gene interactions have also been shown for *MET* and *PLAUR* (Campbell et al., 2008).

The functional correlates of this genetic predisposition are perhaps more interesting than its behavioral correlates, which appear rather nonspecific (Campbell et al., 2010). The MET/HGF pathway is known to play an important role in the developing CNS, in the immune system, and in gastrointestinal repair, all strongly linked to the pathophysiology of autism. Blunted MET/HGF signaling negatively affects interneuron migration from the ganglionic eminence into the cerebral cortex and granule cell proliferation in the cerebellum (Ieraci et al., 2002; Levitt et al., 2004): reduced numbers of neocortical GABAergic interneurons and a reduction in cerebellar size, especially in the vermis, may be particularly relevant to the elevated comorbidity between autism and epilepsy, as well as to brain imaging findings of reduced cerebellar vermis size in ASD patients (Courchesne, 1997). The same also occurs in zebrafish, where *MET* is critically involved in cerebellar development and, interestingly, in the migration of cells forming the facial motor nucleus (Elsen et al., 2009; also see Section 34.4.2, Rodier et al. (1996), and Rodier (2002)). In the mouse, *MET* expression is especially pronounced in cortical projection neurons between P7 and P14, when long-range cortical connections are wired through neuronal sprouting and active synaptogenesis (Judson et al., 2009). Emx1-Cre-driven deletion of *MET* in dorsal pallially derived forebrain neurons affects dendritic development both in pyramidal cells (decreased apical and increased basal dendritic harbor length) and in medium spiny neurons (increased dendritic harbor length), the latter postsynaptic to *MET*-expressing corticostriatal afferents during development. The number of dendritic spines is unchanged, but spine head volume is significantly enlarged (Judson et al., 2010). These same animals show a twofold stronger connectivity between cortical layers 2/3 and corticostriatal, but not corticopontine, layer 5 pyramidal neurons (Qiu et al., 2011). Although human genetic variants modulate *MET* gene expression to a moderate extent, compared to these experimental manipulations, these studies clearly implicate excessive local and decreased long-range connectivity at the neocortical level as the most likely mechanism underlying the ASD risk conferred by *MET* gene alleles.

Additional evidence converging on the MET/HGF pathway also comes from *Plaur*-deficient mice, which show disrupted forebrain interneuron development, increased susceptibility to seizures, anxiety, and abnormal social behavior (Levitt et al., 2004). These same *Plaur* knockout mice also display severely impaired granulocyte and monocyte migration toward inflammatory foci (Allgayer, 2006). The MET/HGF pathway is indeed known to play both 'pro-inflammatory' roles (stimulating leukocyte adhesion and migration, migration of dendritic cells, antagonizing the effects of TGF-β) and anti-inflammatory roles (suppression of the antigen-presenting function of dendritic cells; blunting of eosinophila and airway hyperresponsiveness in animal models of asthma) (Beilman et al., 2000; Okunishi et al., 2005). Finally, the C allele at rs1858830 has been found to be associated primarily with autistic syndromes encompassing gastrointestinal symptoms (Campbell et al., 2009), which are frequently encountered in autistic patients (Buie et al., 2010). Autistic children with gastrointestinal symptoms may also display decreased serum levels of HGF (Russo et al., 2009; Sugihara et al., 2007), again pointing toward the translation of impaired MET/HGF signaling into inefficient gastrointestinal repair mechanisms.

Collectively, current evidence points toward multiple common gene variants promoting a dysregulation of the MET/HGF pathway, which represents a significant contributor to neurodevelopmental, immune, and gastrointestinal abnormalities in autism.

34.4.4 The Oxytocin Receptor Gene

The oxytocin receptor gene (OXTR) is a high-affinity G-protein-coupled receptor encoded by the *OXTR* gene located on human ch 3p26.2. It binds oxytocin (OXT), a nine-amino-acid neurohypophyseal hormone encoded by the *OXT* gene, which also encodes for neurophysin I and is located on human ch 20p13. This hormone and neuromodulator, largely distributed in limbic areas such as the nucleus accumbens and the amygdala, in addition to well-established roles in parturition and breast feeding, physiologically influences social cognition in a relatively species- and sex-specific manner (for review, see Carter, 2007; Carter et al., 2008). OXT or OXTR knockout mice display impaired social memory, while parturition is largely unaffected (Ferguson et al., 2000; Takayanagi et al., 2005). Interestingly, only male mice with a targeted forebrain OXTR knockout fail to recognize individuals of their own species, suggesting the existence of compensatory mechanisms in females (Sun et al., 2008). Heterozygous *reeler* mice display reduced neocortical *OXTR* gene expression, suggesting an intriguing crosstalk between the RELN and OXT pathways (Liu et al., 2005). Several human studies employing an intranasal administration paradigm demonstrate that OXT stimulates affiliative behaviors, subjective feelings of trustworthiness, facial recognition, and in general all social cognitive functions evolutionarily involved in the establishment of a strong emotional bond between parents and neonate (Ebstein et al., 2009).

Genetic studies of the OXT and arginine vasopressin (AVP) systems were undertaken under the assumption that a disruption of these hormonal/neurochemical systems could underlie the deficits in social cognition which characterize ASD patients (for review, see Ebstein et al., 2009). Indeed, at least six out of eight genetic studies performed to date have reported a positive association between ASD and the *OXTR* gene (Table 34.8). This consistency is surprising, especially when considering that many studies were severely underpowered. Single marker and haplotype association analyses primarily point toward a haplotype block encompassing exon 3 and the beginning of intron 3 as possibly hosting a functional variant conferring autism liability, although some evidence also points toward the 3'UTR of the gene encoded by exon 4 (Table 34.8). Findings on the *OXT* gene have been less consistent, if not entirely negative. Some studies have reported positive findings for the AVP receptor gene *AVP1a* (Kim et al.,

TABLE 34.8 Genetic Association Studies on *OXTR* Gene Variants and Autism

References	Polymorphisms	Experimental design	Race and ethnicity	Outcome
Wu et al. (2005)	Eight SNPs	Case–control Family-based	Chinese Han	Association with rs2254298 and rs53576, and with haplotypes involving rs53576
Jacob et al. (2007)	Intron 3: rs2254298 Intron 3: rs53576	Family-based	Caucasian-Americans	Association with rs2254298
Lerer et al. (2008)	16 SNPs	Family-based	Israelis	Association with rs2268494 and rs1042778, and with a 5-SNP haplotype involving rs2254298 and rs2268494
Yrigollen et al. (2008)	Intron 3: rs237885, rs2268493, rs237898	Family-based	Americans (Caucasians = 93%)	Association with rs2268493
Wermter et al. (2010)	22 SNPs	Family-based	Germans	Association with rs2270465 and with a four SNP haplotype spanning the entire locus
Tansey et al. (2010)	18 SNPs	Family-based Gene expression in LCLs and amygdala	Caucasians (Irish + Portuguese + UK)	No association with ASD; three SNPs show association with gene expression
Kelemenova et al. (2010)	Exon 3: rs2228485	Case–control	Slovaks	No association with ASD
Liu et al. (2010)	11 SNPs	Family-based Case–control	Japanese	Association with rs2254298 (case–control only), and with a 5-SNP haplotype involving rs2254298

2002; Wassink et al., 2004; Yirmiya et al., 2006) and for the gene encoding CD38, a transmembrane glycoprotein mainly expressed in immune cells (NK, T, B cells and macrophages) and responsible for triggering the release of OXT in neurons (Jin et al., 2007; Munesue et al., 2010). Interestingly, another study (Gregory et al., 2009) described a genomic deletion encompassing the *OXTR* gene in an autistic proband belonging to a multiplex family and in his mother, who displayed relevant obsessive–compulsive traits. The other affected sibling did not carry the deletion, but instead his *OXTR* promoter was hypermethylated at CpG islands located at −934, −924, and −901 bp from the translation start site. Hypermethylation of the *OXTR* promoter at several CpG islands located between −860 and −959 was also demonstrated in genomic DNA extracted from leukocytes and from post-mortem brain tissue of autistic patients, compared to matched controls (Gregory et al., 2009). As predicted, enhanced methylation was correlated with decreased *OXTR* transcript levels in temporocortical post-mortem brain tissue (Gregory et al., 2009). Hence, predisposition to autism can be apparently conferred by the *OXTR* locus through distinct genomic, genetic, and epigenetic mechanisms, all resulting in hampered OXT signaling (Gurrieri and Neri, 2009). Abnormal neuropeptide processing in autistic children, yielding reduced OXT blood levels despite enhanced concentrations of OXT precursor (Green et al., 2001; Modahl et al., 1998), may further exacerbate this deficit, bringing OXT signaling below a critical threshold necessary for the physiological development of social behavior. Based on the influence of *OXTR* gene variants on amygdalar volume, which is bilaterally smaller in healthy adults carrying the G allele at rs2254298 (Furman et al., 2011; Inoue et al., 2010), blunted OXT signaling can be predicted to have a negative impact on the development and function of specific cortical and limbic regions critical to social cognition.

34.4.5 The Contacting-Associated Protein-Like 2 Gene (CNTNAP2)

The *CNTNAP2* gene, located on human ch 7q35–q36, encodes for the contacting-associated protein (CASPR2), a member of the Neurexin family which also includes Neurexin1 (see Section 34.3.2.3). This locus was originally identified by two groups applying linkage analysis both on affection status and on language delay, used as a quantitative trait locus (QTL) (Alarcón et al., 2008; Arking et al., 2008). Follow-up association analyses carried out on potential candidate genes supported *CNTNAP2* as being solely responsible for the linkage peak (Alarcón et al., 2008; Arking et al., 2008), although negative association findings were also published (Li et al., 2010). The *CNTNAP2* allele appears to confer autism vulnerability primarily in males (Alarcón et al., 2008), and possibly if inherited from the maternal side (Arking et al., 2008). *CNTNAP2* was shown to be highly expressed in frontal and temporal regions of the human fetus, as well as striatum, amygdala, and thalamus, all areas strongly involved in linguistic functions and emotional information processing (Abrahams et al., 2007; Alarcón et al., 2008). T1-weighted anatomical MRI scans performed in 314 healthy volunteers revealed that the autism-associated allele seemingly yields reduced gray and white matter volumes and fractional anisotropy, following sex-specific distributions involving several autism-related brain regions, such as frontal cortex, fusiform gyrus, and cerebellum (Tan et al., 2010). Further analyses by fMRI revealed that carriers of the *CNTNAP2* risk allele have widespread and bilateral connectivity distributed throughout the frontal cortex and anterior temporal poles, whereas the protective allele is associated with a left-lateralized network composed of left inferior frontal gyrus, insula, anterior temporal pole, superior temporal gyrus, and angular gyrus (Scott-Van Zeeland et al., 2010). The latter results point toward *CNTNAP2* alleles as conferring autism vulnerability by affecting the lateralization and possibly the extent of long-range connectivity. At the cellular level, the *Drosophila* orthologs of *CASPR2* and *NRXN1* have been shown to colocalize partly at synaptic active sites, and overexpression of either gene increases the density of active zones and modulates the shape of synapses (Zweier et al., 2009). These 'synaptic' roles for *CNTNAP2*, especially if applied to long-range neural pathways connecting language-related cortical regions, would indeed fit with the converging evidence on synaptic roles summarized here for several other genes (see Section 34.3.2).

Multiple lines of evidence point toward the relative nonspecificity of many 'autism' genes, which may play cognitive roles that, if deranged, can translate into different human disorders: this seems to apply even more to *CNTNAP2*. Several rare genetic variants and *de novo* cytogenetic abnormalities in *CNTNAP2* have been described in autistic probands, which oftentimes present also with seizures and regression (Bakkaloglu et al., 2008; Jackman et al., 2009; Poot et al., 2010; Rossi et al., 2008). However, *CNTNAP2* was also identified by genome-wide CNV analysis as relevant to the development of idiopathic generalized and focal epilepsies (Mefford et al., 2010). Further evidence linking *CNTNAP2* perhaps more directly with language development than with autism *per se* comes from the functional connection between *CNTNAP2* and *FOXP2*, a transcription factor critically involved in the development of expressive language (Lai et al., 2001): *FOXP2* binds to the promoter of *CNTNAP2* and dramatically downregulates its expression (Vernes et al., 2008). *CNTNAP2* gene variants have also been found to be associated with specific language impairment (Vernes et al., 2008). Moreover, common *CNTNAP2* variants

have been found to confer vulnerability to schizophrenia and bipolar disorder (O'Dushlaine et al., 2010), whereas rare variants have been described in ADHD patients (Elia et al., 2010). Meanwhile, although several rare variants present in autistic probands were not found in large numbers of control chromosomes, the vast majority are inherited from an unaffected parent and many of them are transmitted to some, but not all, affected siblings in multiplex families, suggesting that they may enhance autism risk, but are not sufficient to cause the disease (Bakkaloglu et al., 2008). Hence, *CNTNAP2* may play a broader role in shaping the autistic phenotype toward language deficit and possibly epilepsy, rather than strictly conferring autism vulnerability.

34.4.6 The Engrailed 2 Gene

The Engrailed genes play an important role in the patterning of the midbrain/hindbrain region, the only CNS area where they are actively expressed during development (Davis et al., 1998). In particular, engrailed 2 (*EN2*) is expressed in cerebellum, pons, periaqueductal gray, and colliculi. *EN2* knockout mice display decreased seizure threshold to kainic acid (Tripathi et al., 2009) and a relatively subtle behavioral phenotype, with abnormalities in developmental motor, social, and memory tasks (Cheh et al., 2006). Their cerebellar size is reduced and their compartmentalization is interestingly disrupted both in the vermis and in cerebellar hemispheres (Kuemerle et al., 1997; Millen et al., 1994, 1995). In addition, 5-HT and 5-hydroxy-indolacetic acid (5-HIAA) levels are doubled in the cerebellum only (Cheh et al., 2006). The role of *EN2* in cerebellar development and the frequency of cerebellar abnormalities reported in neuroanatomical and brain imaging studies of ASD patients (Courchesne, 1997) spurred interest in genetic studies of *EN2* as early as 1995, when a significant association with autism was first reported by Petit et al. Nine studies followed this initial report, and at least seven of them replicated the initial association in various racial and ethnic groups (Table 34.9). The *EN2* gene, located in human chromosome 7q36.3, encompasses two exons and one intron. The most replicated association was found with the A–C haplotype at SNPs rs1861972 and rs1861973, embedded into intron 1. Following transfection with constructs encompassing the luciferase reporter gene, this haplotype consistently yields approximately 20% higher expression levels in neuronal PC12 cells, non-neuronal HEK293T cells, and in primary cultures of mouse cerebellar granule cells harvested on postnatal day 6 (P6) (Benayed et al., 2009). This difference in gene expression is not due to cryptic splicing, but rather to allele-specific transcription factor binding. Increased expression of *EN2* can be predicted to result in faster differentiation of the midbrain/hindbrain region and of cerebellar circuits (Benayed et al., 2009). This should occur at the expense of the proliferating and migrating pools, thus yielding decreased Purkinje cell numbers and cytoarchitectonic abnormalities in the cerebellar cortex.

34.4.7 Gamma-Aminobutyric Acid Receptor β3 (GABRB3)

Several lines of investigation indicate the existence of abnormalities in the brain gamma-aminobutyric acid (GABA) system of autistic children. The frequent co-morbidity with epilepsy and the morphogenetic roles of GABA, an inhibitory neurotransmitter in adult brain but an excitatory neurotransmitter during prenatal neuro-development due to high intracellular chloride concen-trations in immature neurons (see review by Jentsch et al., 2002), have spurred interest into GABA receptor (GABAR) subunit genes as potential candidates for autism (Blatt et al., 2001; Hussman, 2001). In addition, the 15q11–q13 region deleted/duplicated in 1–4% of autistic patients (see Tables 34.3 and 34.5) (McCauley et al., 2004; Schroer et al., 1998) encompasses the GABAA receptor gene cluster, which consists of three GABAR genes, namely *GABRB3*, *GABRA5*, and *GABRG3*. Inves-tigations of these genes have provided some support especially to *GABRB3*. A significant association between autism and markers located within or nearby *GABRB3* has been found in most studies (Buxbaum et al., 2002; Cook et al., 1998; Curran et al., 2005; Kim et al., 2006; Martin et al., 2000; McCauley et al., 2004; Yoo et al., 2009), although negative reports have also appeared (Ma et al., 2005; Maestrini et al., 1999; Menold et al., 2001; Salmon et al., 1999; Tochigi et al., 2007). Some have pro-posed that behavioral traits, such as savant skills and insis-tence on sameness, may be especially linked to genes located in the 15q11–q13 region (Nurmi et al., 2003; Shao et al., 2003). Maternal transmission of a *GABRB3* signal peptide variant (P11S), previously implicated in childhood absence epilepsy, is associated with autism (Delahanty et al., 2011). This rare variant, present in 17 (1.47%) of

TABLE 34.9 Genetic Association Studies on *EN2* Gene Variants and Autism

References	Polymorphisms	Experimental design	Race and ethnicity	Outcome
Petit et al. (1995)	Two RFLPs (MP4 and MP5 probes)	Case–control	French (all Caucasians)	Association with the MP4 probe and PvuII
Zhong et al. (2003)	Exon 1: rs3735653	Family-based	AGRE families (race not specified)	No association with rs3735653
Gharani et al. (2004)	Exon 1: rs3735653 intron 1: rs1861972 intron 1: rs1861973 exon 2: rs2361689	Family-based	Caucasian-Americans	Association with rs1861972 and rs1861973
Benayed et al. (2005)	Intron 1: rs1861972, intron 1: rs1861973, PvuII/MP4, and 13 other SNPs	Family-based	Two samples: AGRE and NIMH (race not specified)	Association with rs1861972 and rs1861973 (intronic haplotype)
Brune et al. (2008)	Intron 1: rs1861972	Family-based	Not specified (NIH CPEA families)	Association only with broad autism and under a recessive model
Wang et al. (2008)	Eight SNPs	Family-based	Chinese Han	Association with haplotypes involving rs3824068 (intron 1)
Yang et al. (2008)	Intron 1: rs1861972, intron 1: rs1861973	Case–control	Chinese Han	Association with single markers; 'protective' haplotype
Benayed et al. (2009)	16 SNPs	Family-based	Three samples: AGRE I, AGRE II, NIMH (Caucasian non-Hispanic subset for association)	Maximum association with rs1861972 and rs1861973 (intron 1); AC haplotype yields increased expression
Sen et al. (2010)	Exon 1: rs3735653 promoter: rs34808376 promoter: rs6150410 intron 1: rs1861972, intron 1: rs1861973	Family-based	Indian from West Bengal and Assam	Association with rs1861973
Yang et al. (2010)	Five SNPs in intron 1: rs3824068, rs3824067, rs1861972, rs1861973 and rs3830031	Case–control	Chinese Han	Association with the A–C haplotype at rs1861972 and rs1861973 (intron 1)

1152 simplex and multiplex families, yields reduced whole-cell current and decreased β3 subunit protein on the cell surface due to impaired intracellular β3 subunit processing (Delahanty et al., 2011). *GABRB3* gene variants must also be viewed within the framework of the entire set of GABAR-encoding genes, as several gene–gene interactions between them have been detected (Ashley-Koch et al., 2006; Ma et al., 2005).

In addition to genetic variants, epigenetic dysregulation of the *GABRB3* locus may also contribute to autism. *GABRB3* expression is reduced on average by as much as 50% in several neocortical and cerebellar regions (Fatemi et al., 2009). Interestingly, a sizable subset of ASD brains displays either monoallelic or abnormally downregulated *GABRB3* expression instead of the normal levels of biallelic expression present in controls (Hogart et al., 2007, 2009). Interestingly, GABRB3-deficient mice exhibit impaired social and exploratory behaviors, deficits in nonselective attention, and hypoplasia of the cerebellar vermis, all features relevant to autism (DeLorey et al., 2008). In addition, mice deficient in MeCP2 display reductions in GABRB3 protein, as MeCP2 acts as a positive regulator of *GABRB3* gene expression (Samaco et al., 2005). Collectively, these results suggest the existence of genetic and epigenetic influences leading to a behaviorally relevant downregulation of *GABRB3* in autistic brains.

34.4.8 The Serotonin Transporter (SLC6A4) and Integrin β3 Subunit Genes

Elevated whole-blood serotonin (5-HT), present in about one-third of ASD patients, represents one of the most consistent biological endophenotypes in autism research (Table 34.1). Hyperserotoninemia appears to be a genetically determined familial trait, as first-degree relatives display mean 5-HT blood levels intermediate between those of their autistic family members and of population controls (Abramson et al., 1989; Cook et al., 1990; Leventhal et al., 1990; Piven et al., 1991). In most patients, elevated 5-HT blood levels in autism seemingly stem from accumulation of 5-HT in platelets due to increased densities of functionally active serotonin transporter (5-HTT) molecules on platelet membranes, with no change in 5-HTT affinity for 5-HT and no elevation in free 5-HT plasma level (Cook et al., 1988; Katsui et al., 1986; Marazziti et al., 2000). Autism-associated hyperserotonemia has been the object of intense investigation, because either it could play a role in the etiological processes leading to the disease, or it could at least mark a relatively homogeneous subgroup of ASD patients. Genes encoding proteins involved in 5-HT metabolism and neurotransmission include, among others, the 5-HT transporter gene (*SLC6A4*) and the integrin β3 subunit gene (*ITGB3*). The serotonin transporter (5-HTT) responsible for platelet 5-HT uptake is identical in its primary sequence to the 5-HTT expressed in serotoninergic neurons: both are indeed produced by a single *SLC6A4* gene, located on chromosome 17q12 (Lesch et al., 1993). In reference to functional common variants, this gene contains two variable number tandem repeats (VNTRs) affecting expression levels: (a) the 5-HTT gene-linked polymorphic region (5-HTTLPR), located in the promoter, encompasses a 'long' 16-repeat allele, yielding approximately 50% higher 5-HTT gene expression and tritiated 5-HT uptake compared with homozygosity for the 'short' 14-repeat allele or with heterozygosity (Lesch et al., 1996); (b) the serotonin transporter intron 2 (STin2) VNTR includes 9, 10, and 12 repeat alleles, with the latter acting as a transcriptional enhancer (MacKenzie and Quinn, 1999). Overall, meta-analyses of association studies between autism and these two VNTRs have been negative, although there may be some association with the 5-HTTLPR 'short' allele in North-American families only (see review Huang and Santangelo, 2008). Furthermore, contributions of these VNTRs to enhanced 5-HT blood levels appear marginal at best (Anderson et al., 2002; Betancur et al., 2002; Coutinho et al., 2004, 2007a,b; Cross et al., 2008; Persico et al., 2002). Hence, immune factors, such as TNFα and other proinflammatory cytokines, which are known to activate 5-HTT transport activity (Zhu et al., 2006), as well as common variants in other genes, such as *ITGB3* (see below), may influence 5-HT blood levels to a larger extent.

In addition to these two common VNTRs, several rare *SLC6A4* variants have been identified as significantly enhancing autism risk. In particular, four coding substitutions located at highly conserved positions and 15 other variants located in 5′ noncoding and other intronic regions are transmitted to autistic probands exhibiting rigid-compulsive behaviors (Sutcliffe et al., 2005). Two of these variants, Phe465Leu and Leu550Val, confer elevated 5-HTT surface density (V_{max}), while retaining a capacity for acute protein kinase G (PKG) and p38 mitogen-activated protein kinase (MAPK) regulation; five other variants (Thr4Ala, Gly56Ala, Glu215Lys, Lys605Asn, and Pro612Ser) demonstrate no change in V_{max}, but show a complete loss of 5-HT uptake stimulation after acute PKG and p38 MAPK activation (Prasad et al., 2005). Finally, two other variants, Gly56Ala and Ile425Leu, show markedly reduced 5-HTT association with protein phosphatase 2A (PP2A), leading to profound and long-lasting 5-HTT internalization following phosphorylation by PKC (Prasad et al., 2005, 2009). When expressed stably in CHO cells, both Gly56Ala and Ile425Leu display a striking loss of 5-HTT protein following catalytic activation (Prasad et al., 2009). Since the Gly56Ala variant is less rare than the other variants, it will be interesting to see the results of studies contrasting wild-type Gly56 versus mutated Ala56 mice (Veenstra-Vanderweele et al., 2009). Collectively, despite showing a complex array of effects, rare *SLC6A4* variants conferring autism vulnerability (a) display enhanced 5-HT transport activity, presumably leading to decreased extracellular 5-HT levels and/or shorter exposure of 5-HT receptors to their ligand, or (b) lose the plastic regulation normally mediated by intracellular kinases and able to adapt 5-HTT activity to the functional needs of local circuits (Prasad et al., 2009).

The *ITGB3* gene was first identified as a QTL for 5-HT blood levels, initially in the Hutterites (Weiss et al., 2004, 2005a) and then in the general population (Weiss et al., 2005b). *ITGB3* maps in ch 17q21.32, under a replicated linkage peak for autism (Cantor et al., 2005; Stone et al., 2004), and *ITGB3* alleles, either alone or in interaction with *SLC6A4*, have been found at least nominally associated with autism in all five studies performed to date (Coutinho et al., 2007a,b; Ma et al., 2010; Mei et al., 2007; Weiss et al., 2006a,b). Several lines of evidence support functional interactions between *ITGB3* and *SLC6A4*: (a) the integrin receptor composed of an αIIb subunit and of the *ITGB3*-encoded β3 subunit was recently identified as a novel component of the *SLC6A4* regulatory protein complex (Carneiro et al., 2008; Weiss et al., 2006a); (b) the *ITGB3* SNP rs5918 (Leu33Pro) modulates *SLC6A4* trafficking and transport activity (Carneiro et al., 2008); (c) *ITGB3* and *SLC6A4* gene expression levels are correlated in human and mouse tissues (Weiss et al., 2006a); and finally, (d) several published studies have described significant *SLC6A4* and *ITGB3* interactions for both autism risk and 5-HT blood levels, with a male-specific effect (Coutinho et al., 2007a,b; Ma et al., 2010; Mei et al., 2007; Weiss et al., 2004, 2005b, 2006a,b). Recent results from our laboratory point toward the existence of at least two distinct functional genetic *ITGB3* variants, located at opposite ends of the *ITGB3* gene: the 5′ variants significantly influence 5-HT blood levels in ours and in other studies (Weiss et al., 2006b), whereas autism-associated SNPs cluster toward the 3′ end of the gene (Napolioni et al., 2011). Interestingly, these results closely resemble the association patterns previously reported for asthma and wheezing versus allergies and IgE levels, also associated with 5′ and 3′ markers, respectively (Thompson et al., 2007; Weiss et al., 2005c).

Finally, mice carrying a combined haploinsufficiency at *PTEN* and *SLC6A4* (PTEN+/−;SLC6A4+/−) develop larger head sizes and greater sex-specific behavioral abnormalities, compared to PTEN+/−;SLC6A4+/+ or PTEN+/+;SLC6A4+/− mice (Page et al., 2009). This influence of 5-HT on the penetrance of other morphogenetic and neurodevelopmental genes is not entirely surprising in view of the many neurotrophic roles exerted by 5-HT during development (for review, see Di Pino et al., 2004; Persico, 2009). Yet, this study underscores the urgent need to move from assessments of 5-HTT and ITGB3 contributions to the 'presence/absence' of autism, toward less reductionist and more biologically meaningful approaches, addressing 5-HT modulation of the disease phenotype in the context of gene–gene interactions.

34.5 NONSYNDROMIC AUTISMS: ENVIRONMENTAL FORMS

34.5.1 General Description

Certain environmental factors have been shown to enhance the risk of developing autism significantly, to the point that at least in some patients they can be regarded as 'the primary cause' of the disease (for review, see Landrigan, 2010). These environmental agents include teratogenic drugs, namely valproic acid, misoprostol, and thalidomide, as well as prenatal rubella and cytomegalovirus (CMV) infections. Cases conclusively shown to derive from exposure to these environmental agents are relatively rare, and the role of genetic vulnerability conferred by common variants at the individual level cannot be overlooked. Nonetheless, these cases possess high heuristic potential, as the time window of exposure is usually known and often narrow, typically occurring during early prenatal neurodevelopment.

34.5.2 The Fetal Anticonvulsant Syndrome

Prenatal exposure to phenytoin, sodium valproate, and carbamazepine, either alone or in combination, causes the 'fetal anticonvulsant syndrome.' In the largest population-based study reported to date (Rasalam et al., 2005), strict diagnostic criteria for autistic disorder were met by 5/56 (8.9%) and 2/80 (2.5%) children exposed to valproate and carbamazepine alone, respectively; including also patients treated with drug combinations, cumulative percentages rise to 9/77 (11.7%) and 5/110 (4.5%) for valproate and carbamazepine, respectively. Differently from idiopathic autism, the M:F ratio here is close to 1 (Moore et al., 2000; Rasalam et al., 2005). Clinical descriptions of these children are provided both in larger cohort studies (Moore et al., 2000; Rasalam et al., 2005) and in a number of case reports (Christianson et al., 1994; Williams and Hersh, 1997; Williams et al., 2001): patients almost invariantly display speech delay, whereas motor delay is much less prevalent; cognitive impairment is typically mild or absent, and there is no history of regression or loss of skills; major malformations are sometimes present, but minor facial dysmorphology is certainly more prevalent (notice the hypertelorism, frontal bossing, and other dysmorphic features in Figures 1 and 2 of Moore et al., 2000).

Sodium valproate is believed to exert its teratogenic effects mainly by inhibiting histone deacetylase (Phiel et al., 2001): the persistent acetylation of histones and demethylation of cytosine at the promoters of neurodevelopmentally relevant genes would then lead to a dysregulation in excess of their gene expression. This pharmacological effect impinges on epigenetic mechanisms partly overlapping with those involved in *MECP2* hemizygosity, as summarized above, yielding a broad range of neurodevelopmental and behavioral abnormalities. As predicted by its pharmacological action, valproic acid increases the expression of several neurally relevant genes, such as *HoxA1* (Stodgell et al., 2006), GATA-3 (Rout and Clausen, 2009), WNT (Wiltse, 2005), and RELN and GAD67 (Dong et al., 2007). Curiously, reduced expression has been reported for *NLGN3* (Kolozsi et al., 2009) and for genes involved in the differentiation of serotoninergic neurons, including Sonic hedgehog, its receptor Patched, and the transcription factors Gli1 and Pet-1 (Miyazaki et al., 2005). The latter effect delays differentiation, resulting in enhanced growth and abnormal distribution of serotoninergic pathways (Miyazaki et al., 2005; Tsujino et al., 2007). Neuroanatomical anomalies primarily include abnormal cerebellar cytoarchitectonics and brainstem nuclei formation (Ingram et al., 2000a; Rodier, 2002; Rodier et al., 1996). Particularly reminiscent of human autism are the local hyperconnectivity present in the neocortex and the abnormal morphology of motor cortical neurons, accompanied by delayed motor development (Rinaldi et al., 2008; Schneider and Przewłocki, 2005; Snow et al., 2008). Other neurochemical and behavioral abnormalities include increased monoamine concentrations due to enhanced expression of tyrosine hydroxylase (D'Souza et al., 2009; Narita et al., 2002), altered circadian rhythms (Tsujino et al., 2007), elevated nociceptive threshold and enkephalinergic tone (Schneider et al., 2001, 2007), abnormal fear conditioning and amygdala processing (Markram et al., 2008), increased LTP due to enhanced expression of NMDA receptors (Rinaldi et al., 2007), and a hyperactive mesocortical dopaminergic pathway (Nakasato et al., 2008). These human and animal data, in conjunction with data from Rett patients and MECP2 inactivation, strongly underscore the importance of epigenetic control over gene expression as an important player in autism pathogenesis (LaSalle, 2007; Schanen, 2006).

34.5.3 Other Teratogenic Agents: Thalidomide and Misoprostol

Thalidomide and misoprostol are two teratogenic drugs, known to induce a variety of systemic malformations. Thalidomide was commercialized as a sedative drug in the late 1950s before being withdrawn from the market in 1961. Teratogenicity is due to its angiogenesis inhibiting activity, which causes multiple systemic malformations, as well as abnormal cortical development and neuronal hyperexcitability (Hallene et al., 2006). Misoprostol is a methyl ester derivative of prostaglandin E1, used especially in Central and South America to treat gastric ulcers, but also a popular abortion inducer due to its powerful stimulatory effect on uterine contractions: the teratogenic effects of misoprostol have been studied in children born after unsuccessful abortion attempts (Bandim et al., 2003). These two drugs display several interesting parallels: both hamper fetal blood perfusion either directly (thalidomide) or indirectly (misoprostol); both produce systemic and especially ophthalmologic malformations, primarily coloboma and microphtalmos (Miller et al., 2004, 2005); both frequently cause prenatally exposed children to develop signs of Moebius sequence, including horizontal strabismus (Duane syndrome) and facial nerve palsy due to the involvement of the VI and VII cranial nerves (Bandim et al., 2003; Miller et al., 2005); both are associated with enhanced risk of autism and/or mental retardation, provided exposure occurs early in development (Miller et al., 2005; Strömland et al., 1994). The critical period for teratogenetic induction of autism has been defined in great detail for thalidomide, where it appears to be restricted to as early as 4–6 weeks into gestation (i.e., 6–8 weeks since the last menstrual cycle) (Miller et al., 2005; Strömland et al., 1994). The critical period for misoprostol has not been defined with the same precision, but it is known that maximum fetal vulnerability occurs during the first 2 months of pregnancy, and possible 5–6 weeks after fertilization (i.e., 7–8 weeks since the last menstrual cycle) (Bandim et al., 2003). Patients with idiopathic Moebius sequence and with no history of prenatal exposure to thalidomide or misoprostol have been found at enhanced risk of autism by some (Gillberg and Steffenburg, 1989), but not by others (Briegel et al., 2009). It will be important to determine conclusively whether there is a significant association between Moebius sequence and autism, because this would demonstrate that autism specificity is conferred more by a sensitive time window during development, rather than by the specific nature of prenatal insults or teratogenic mechanisms involved.

34.5.4 Environmental Pollutants as Potential Teratogens

The number of potentially teratogenic chemicals to which pregnant women may be exposed is theoretically elevated. In practice, prolonged and/or intensive exposure at critical times would be necessary to negatively influence development in any meaningful way (Rice and Barone, 2000). Such an exposure may occur in some geographical areas, primarily for ambient and indoor air pollutants (exterminators, can sprays, and pest bombs), and for pesticides routinely used in agriculture (Landrigan, 2010; Zhang and Smith, 2003). Compounds for which preliminary evidence supports possible roles in enhancing autism risk include organochlorine pesticides, organophosphates (most clearly chlorpyrifos), heavy metals, and chlorinated solvents (Engel et al., 2007; Roberts et al., 2007; Whyatt and Barr, 2001; Whyatt et al., 2003; Windham et al., 2006). Prenatal exposure to organophosphates, such as chlorpyrifos, has been found to be associated with lower IQ, developmental delay, ADHD, and autism-spectrum traits defined as PDD-NOS (Eskenazi et al., 2008; Rauh et al., 2006). Some individuals may be genetically vulnerable to suffer from the consequences of prenatal organophosphate exposure, depending on functional genetic variants at loci such as *PON1*, the gene encoding for paraoxonase, and the HDL-associated serum enzyme responsible for organophosphate detoxification in humans (D'Amelio et al., 2005; Gaita et al., 2010). More definitive evidence linking autism, genetic vulnerability, and prenatal exposure to toxic agents is being sought through various efforts, including large epidemiological studies, such as the 'Childhood Autism Risks from Genetics and the Environment' (CHARGE) study (Hertz-Picciotto et al., 2006).

34.5.5 Congenital Viral Infections

Rubella and CMV represent the two infectious agents best known to enhance autism risk following a congenital infection (for review, see Libbey et al., 2005; van den Pol, 2006). Autistic children prenatally infected with these viruses generally present severe mental retardation and physical anomalies, such as ophthalmologic malformations, deafness, and cardiac malformations. Brain imaging findings are highly variable, ranging from cortical malformations (polymicrogyria, pachygyria, heterotopias) indicative of migration defects to abnormal intensity of the periventricular white matter suggestive of abnormal myelination in the absence of any cortical malformation. Epilepsy and cerebral palsy are also frequent.

34.5.5.1 *Congenital Rubella*

The largest longitudinal study involving several hundred children prenatally exposed to rubella virus estimates at 7.4% the rate of autism in this group; risk appears especially high if the infection occurs during the first 8 weeks postconception (Chess, 1971, 1977; Chess et al., 1978). Congenital rubella symptoms often change over time: some neurodevelopmental symptoms undergo remission, others are permanent, others may progressively worsen or even appear in late childhood or adolescence (Banatvala and Brown, 2004).

Interestingly, mental retardation and autism do not covary over time in these children, but seemingly follow independent trajectories (Chess, 1977). The occurrence of 'late-onset' autism (i.e., onset later than 3 years of age) following congenital rubella has also been reported (Chess et al., 1978).

These data should be viewed with some caution, because: (a) the incidence of ASD among 'rubella children' was estimated well before the establishment of current standards for a clinical diagnosis of ASD and (b) these variable clinical courses should be confirmed by applying current diagnostic criteria and modern tools for clinical follow-up. However, the former limitation can be predicted to lead to an underestimation of ASD incidence following congenital rubella, since ASD diagnostic criteria have now become overinclusive, as compared to Kanner's classical criteria which would have been applied in the seventies (Berger et al., 2010). Secondly, not only psychiatric, but also physical signs and symptoms of congenital rubella change significantly over time (the 'late manifestations' can even appear during adolescence or adulthood). Furthermore, even in idiopathic ASD, a spontaneous remission by age 3 of autistic behaviors diagnosed at a younger age is not an entirely unusual event (Turner and Stone, 2007; van Daalen et al., 2009). Finally, the partial spontaneous improvement of severe autistic behaviors rapidly developed by some children following postsurgical cerebellar vermal lesions without any specific rehabilitation (Riva and Giorgi, 2000) suggests that environmental etiologies can produce clinical courses more unstable than those seen in the majority of idiopathic ASD children.

34.5.5.2 *Congenital Cytomegalovirus Infection*

Evidence linking prenatal CMV infection to autism is more circumstantial. Several case reports have been published (Ivarsson et al., 1990; Kawatani et al., 2010; López-Pisón et al., 2005; Markowitz, 1983; Stubbs, 1978; Stubbs et al., 1984; Sweeten et al., 2004; Yamashita et al., 2003), but risk estimates are essentially based on a small cohort of seven prenatally CMV-infected children, two of whom displayed autistic features (2/7 = 28.6%) (Yamashita et al., 2003). The presence of normal cortical gyri, indicating a substantial sparing of neuronal migration even in the presence of periventricular white matter abnormalities, led the authors to point toward the third trimester of pregnancy as the critical time window for autism-causing CMV infections (Yamashita et al., 2003). It remains to be determined to what extent autism ensues from direct viral damage, from the strong immune response driven by herpes viruses, such as CMV, or from the nature and location of cerebral malformations which are particularly frequent in congenital CMV infection (Engman et al., 2010).

34.5.5.3 *Future Perspectives: Possible Novel Roles for Congenital Viral Infections*

In addition to congenital rubella and CMV infections, our group is currently exploring vertical viral transmission as a novel mechanism potentially able to explain high 'heritability' (i.e., parent-to-offspring transmission) in the presence of relatively low rates of disease-specific genetic abnormalities (Maher, 2008). Viral genomes present in parental gametes (egg and/or sperm cells) could be passed onto the offspring already at the time of fertilization, and start being actively transcribed in permissive cells of the fetus only at a later stage during development (Lintas et al., 2010; Persico, 2010). Gamete-mediated vertical viral transmission has been well documented for several viruses, including human immunodeficiency virus, hepatitis B virus, and hepatitis C virus (Englert et al., 2004). Alternatively, seminal fluids could act as vehicles for viral transfer from father to offspring, passing horizontally through the mother. In either case, damage would be due to direct viral interference with cellular functions in permissive fetal cells, to maternal immune response prenatally, and to the patient's immune response in the late prenatal and postnatal periods.

We have recently found the genomes of polyomaviruses (BKV, JCV, and SV40) in *post-mortem* temporocortical tissue (Brodmann areas 41/42) belonging to 10/15 autistic patients and 3/13 controls ($P<0.05$) (Lintas et al., 2010). Also, a trend toward poly-viral infections, including multiple polyoma and/or other neurotropic viruses, was recorded (40% vs. 7.7%, respectively; $P=0.08$). Congenital polyomavirus infections, either alone or in synergy with other viruses, could conceivably explain several puzzling features of autistic disorder, as discussed in Lintas et al. (2010). Briefly, (a) converging experimental approaches have demonstrated an inappropriate and persistent activation of the innate immune system, compatible with an unresolved, early-onset viral infection accompanied by autoimmune phenomena (Garbett et al., 2008; Lintas et al., 2009; Vargas et al., 2005); (b) polyomaviruses can cause autoimmune disorders (Rekvig et al., 2006), which are also frequently encountered in first-degree relatives of autistic patients (Comi et al., 1999); (c) polyomaviruses can produce genomic instability through the activity of their early gene product large-T antigen (LTAg) (Frisque et al., 2006); (d) polyomavirus replication is more active in males, as witnessed by viruria consistently higher in males compared to females (Knowles, 2006); (e) JCV can indeed infect cultured neural progenitor cells, oligodendrocytes and astrocytes, whereas neurons are nonsusceptible to JCV infection (Hou et al., 2006); (f) following transformation of canine MDCK cells and human mesothelial cells, the LTAg of SV40 polyomaviruses has been shown to induce the production and secretion of HGF, which in turn activates by phosphorylating its receptor encoded by the *MET* gene (Cacciotti et al., 2001), which hosts some of the most consistently replicated common variants conferring vulnerability to autism (see Section 34.5.3); and (g) a recent MRI study documented for the first time the presence of temporal lobe and/or white matter abnormalities similar to those produced by viral infections, in as many as 36% of autistic children (Boddaert et al., 2009).

These preliminary results should be viewed with caution, because polyomavirus infections in autistic brains could be the consequence of immunosuppression or tissue susceptibility rather than the cause of autism (Persico, 2010). Furthermore, viral infections may have been active only prenatally and during early infancy, making it difficult to assess viral roles using biomaterials collected later in life. We are currently undertaking a thorough search of polyomaviruses in male gametes of fathers of autistic children and controls, which has already confirmed that a sizable percentage of mobile sperm samples from ASD fathers host polyomavirus genomes (Lintas and Persico, unpublished results), as previously shown by others for control samples (Martini et al., 1996). It will be very interesting to assess whether polyomavirus genomes extracted from mobile sperm cells are able to develop a cytopathic effect in permissive cells following transfection.

34.6 CONCLUSIONS: WHERE AND HOW DO COMMON VARIANTS MEET WITH RARE VARIANTS AND/OR WITH THE ENVIRONMENT?

A single pathophysiological scenario is clearly not compatible with the diversity of nonsyndromic autisms. Yet, the data summarized in this survey allow one to reach some firm conclusions and foster evidence-based speculation.

(1) Specific rare genetic variants have been convincingly shown to cause autism, at least in some cases; however, genotype–phenotype correlations are extremely labile. Not only can mutations located in the same gene result in very different clinical phenotypes, as repeatedly described in Tables 34.5–34.9 for multiple genes: the very same mutation can cause behavioral and morphological phenotypes displaying a surprising degree of variability in different patients, even in affected members of the same extended family. A clear example, introduced in Section 34.3.4.1, is provided by interindividual differences in cerebrovascular malformations seen in individuals from consanguineous families homozygous for

truncating *HOXA1* mutations, each resulting in HoxA1 protein isoforms lacking all functional domains (Bosley et al., 2007; Tischfield et al., 2005). This phenotypic variation is not at all novel in human genetics (Wolf, 1997): a similar degree of interindividual phenotypic variability occurs in syndromic forms due to well-characterized mutations, triplet repeat expansions, or genomic rearrangements, such as fragile-X syndrome or tuberous sclerosis (Table 34.3). This phenotypic variability closely mimics the impressive phenotypic variability seen when a gene inactivated by homologous recombination is backcrossed onto the genetic backgrounds of different mouse inbred strains (Doetschman, 2009). These phenotypic differences clearly emphasize the importance of common genetic variants ('genetic background,' 'modifier genes') in determining the penetrance and expressivity of rare variants.

(2) CNV studies performed to date provide several indications. Briefly: (a) estimates of the percentage of ASD patients and population controls carrying CNVs are likely to increase with the improvement of available technologies; (b) if there truly is a subgroup of ASD patients with excessive genomic instability, its size is relatively small ($\leq$10%) and thus subject to high stochastic variability in independent samples; (c) CNVs *per se* may be more immediately related to evolution than to health and disease: their presence in population controls is physiological, subject to high stochastic variability in independent human samples and may be related to increasing paternal age according to rodent models (Flatscher-Bader et al., 2011); (d) when pathogenic, CNVs seemingly act as rare variants with variable penetrance and expressivity: some *de novo* CNVs may act dominantly and even display complete penetrance, while other CNVs may follow a 'quasi-recessive' mechanism, as described by Zweier et al. (2009) for *NRXN1* (see Section 34.3.2.3) and in a recent report by Vorstman et al. (2010), who identified an autistic individual carrying a maternally inherited deletion and a paternally inherited nonsynonymous amino acid substitution, both affecting the *DIAPH3* locus in human ch. 13q21.2; (e) the genomic location of a CNV is more critical to its pathogenic potential than the total number or mean size of CNVs present in a given individual. CNVs most frequently encountered in ASD patients often encompass genes which, when mutated, are responsible for monogenic forms of autism, such as *NLGN4*, *SHANK3*, and *NRXN1*; (f) the progressive transfer of array-based approaches from the laboratory into clinical practice will indeed enhance the ability of clinicians to detect an increasing number of submicroscopic *de novo* chromosomal abnormalities; (g) deletions and duplications spanning entire genes affect expression only in a minority of 'gene dosage-sensitive loci,' as only approximately 29% of genes duplicated in trisomy 21 are actually overexpressed at or above RNA levels predicted on the basis of allele copy number (Aït Yahya-Graison et al., 2007; Lockstone et al., 2007). Homeostatic mechanisms regulating gene expression through *trans*-acting elements and noncoding RNAs can exert negative feedbacks able to establish close-to-normal gene expression levels. Hence, the conclusive definition of a given CNV as the primary cause of ASD in a given patient cannot be exclusively based on genomic data, as currently proposed (Kaminsky et al., 2011; Miller et al., 2010), but requires functional demonstration of abnormal gene expression in patient-derived cells or cell lines. Genomic evidence of *de novo* status or the absence of a patient's CNV in a very large sample of control chromosomes should be regarded as highly suggestive, but not as conclusive evidence of pathogenicity until gene expression correlates are demonstrated.

(3) Environmental factors can represent the primary cause of autism in some cases. Toxic and viral agents generally also produce major/minor malformations and neurological signs, essentially due to brainstem damage. However, no environmental teratogen or congenital infection causes autism in every single exposed subject, as summarized in Section 34.5. This again underscores the permissive role of common genetic variants, which determine the sensitivity threshold to environmental teratogens and infectious agents. Given the prenatal timing of autism-inducing teratology, more attention should be paid to common genetic variants characterizing the 'feto-maternal' unit, rather than merely the offspring.

(4) Stochastic events typically represent an overlooked nuisance to scientists, but they have been shown to provide significant contributions to deranged developmental processes (Kurnit et al., 1987). Part of the variance in affection status, symptom pattern, and disease severity currently attributed to common genetic variants may actually depend on stochastic events, possibly 'personalizing' genotype–phenotype correlations on top of differences in genetic background.

(5) A history of clinical regression, even when documented by home videos, does not necessarily imply the existence of environmental factors striking at the time when behavioral abnormalities

become manifest. Several children carrying pathogenic mutations in *NLGN* genes or in *EIF4E* ever since conception undergo apparently normal development until a severe regression occurs at approximately 2 years of age, with loss of initially acquired social and verbal milestones (see Sections 34.3.2.1 and 34.3.4.3. Regression may, thus, simply stem from a functional collapse of neural networks, occurring at the time when either they should come 'online' to support the harmonious expansion of social cognition or they are overwhelmed by pathological levels of oxidative stress or other dysfunctional processes (see below).

(6) Environmental forms indicate that pathogenic processes responsible for autism must act early on in neurodevelopment, possible as early as weeks 4–8 post-conception. An exclusively postnatal exposure of a non-genetically-vulnerable individual to prolonged psychological traumas, toxic chemicals, infectious microorganisms, and pathological reactions to vaccines, may in some cases produce psychopathology, but this is clearly distinguishable from autism (the closest example being the 'quasi-autism' of Romanian adoptees institutionalized until adoption and grown in a state of early deprivation of interpersonal contacts, which has a distinctive set of symptoms and a generally more favorable prognostic outcome, as described by Rutter et al. (2007)). In addition, autism induction may be more related to the prenatal timing of the pathogenic insult than to the nature of the insulting agent *per se*.

(7) *SHANK3*, *NRXN1*, *NLGN3*, and *NLGN4* are commonly addressed as 'synaptic genes' and their role in synaptogenesis is often depicted as critical in determining the functional disconnection of distant cortical and subcortical regions, which seemingly characterizes the CNS of autistic patients (Belmonte et al., 2010; Courchesne and Pierce, 2005; Geschwind and Levitt, 2007; Rubenstein and Merzenich, 2003). The major limit of this interpretation is that the peak of synaptogenesis in association cortices occurs late in human neurodevelopment, namely around 2 years of age. This timing is incompatible with abnormalities in cell proliferation and migration clearly documented by neuroanatomical studies of post-mortem brains and with the early prenatal timing of environmental insult in teratological forms. Furthermore, *NLGN* knockout mice and mice carrying human mutations, such as R451C, display relatively modest behavioral phenotypes overall in comparison to the severity of human autism (Chadman et al., 2008; Radyushkin et al., 2009). Finally, *NLGN* gene inactivation in *C. elegans*

surprisingly yields increased sensitivity to oxidative stress as one of its main biochemical features (Hunter et al., 2010). Hence, the critical pathogenetic step may not consist in reduced synapse formation, which has never been convincingly described in ASD brains; it may also not consist in structural abnormalities of long-range neural pathways, which have been assumed to exist on the basis of neurophysiological data more than being demonstrated neuroanatomically; instead, it could consist in the excessive energy requirements imposed on brain cells by malformed and malfunctioning synapses, leading to excessive oxidative stress and consequent functional disconnections (Chauhan and Chauhan, 2006; Palmieri and Persico, 2010; Palmieri et al., 2010).

(8) Abnormal growth rates, either of the head alone or more often of the whole body, represent a frequent feature in autistic children, especially during early infancy (Courchesne et al., 2007). Converging evidence from several syndromic forms strongly points toward the ERK/PI3K/mTOR pathway as playing a pivotal role in autism (Figure 34.1) (Levitt and Campbell, 2009; Ma and Blenis, 2009). A thorough understanding of intracellular pathways involved in autism pathogenesis will be critical, as novel treatments are beginning to show promise of reversing genetically determined abnormalities even in adult mouse models of PTEN haploinsufficiency, fragile-X, and Rett syndrome (Dölen et al., 2007; Tropea et al., 2009; Zhou et al., 2009). Different autisms converging downstream on an hyperactivation of the ERK/PI3K/mTOR pathway may clinically improve using pharmacological inhibitors of this pathway, regardless of the upstream genetic background responsible for generating this biochemical imbalance.

(9) Common genetic variants are typically considered as conferring autism vulnerability. However, genetic variants conferring protection from autism could be equally important. As a concrete example, subjects VII16, 17, 47, and 48 in Hope et al. (2005), despite carrying the very same I745T mutation in the *CACNA1F* gene yielding congenital stationary night blindness type-2 in 16 members of this extended family (see Section 34.3.5.1), do not suffer from autism or profound mental retardation, presumably through the action of protective gene variants. In families with an autistic proband, protective gene variants are preferentially transmitted from parents to unaffected siblings. Electrophysiological data indicate that 'unaffected' family members suffer from disconnections between distant cortical regions

similar to those affecting their autistic siblings, but can implement compensatory circuits which are apparently not available to affected family members (Belmonte et al., 2010). Protective genetic variants may consist in the 'non-predisposing' allele at some loci, but conceivably there should be instances where one allele can be pathogenically neutral and the other can be exquisitely protective, as occurs with the sickle cell anemia allele conferring protection from malaria (Allison, 2009). *SLC25A12* and *GLOI* are two examples of genes possibly hosting common protective variants (Palmieri et al., 2010; Sacco et al., 2007b).

(10) Neuroanatomical, genome-wide expression, and brain imaging studies provide converging evidence of an abnormal activation of the immune system in autism, and particularly of its innate components (Garbett et al., 2008; Laurence and Fatemi, 2005; Petropoulos et al., 2006; Vargas et al., 2005; for review of genome-wide expression studies, see Sacco et al., 2011). This could be due to abnormal synaptic function and/or molecular processing leading to proinflammatory cytokine production, as occurs in Alzheimer disease (Meda et al., 1999). However, temporal lobe abnormalities reminiscent of virally generated lesions have been detected in 48% of autistic children in a recent brain imaging study (Boddaert et al., 2009). These results are compatible with the presence of a persistent, virally triggered immune reaction in a subgroup of genetically predisposed autistic children. Studies of vertically transmitted viruses are thus justified, as they may thus contribute to solve the mystery of the 'missing heritability' (Maher, 2008) in autism research.

(11) In a translational perspective, it may be initially easier to estimate autism risk using a limited set of the most influential common variants, than to search for rare or private variants by sequencing a large enough panel of candidate genes, if not the entire genome. A first example of a test providing combinatorial autism risk estimates from four loci, each hosting one common biallelic SNP, has recently been published (Carayol et al., 2010). As the number of common variants will increase, current 18% sensitivity will hopefully rise to match an already satisfactory 92% specificity (Carayol et al., 2010). In general, this diagnostic approach will be most useful in estimating the risk of autism in (a) newborn siblings of autistic children and (b) sporadic cases displaying initial behavioral abnormalities at 1–2 years of age and potentially evolving by age 3 toward normal behavior, or into full-blown autism, softer ASD traits, specific language impairment, ADHD, or other behavioral syndromes. However, the latter use will require testing the specificity of common variants conferring autism vulnerability, which may not reliably separate behavioral syndromes given the labile genotype–phenotype correlations presented throughout this survey. Tests of this sort will be increasingly sought, as early behavioral intervention programs, targeted to address ASD signs and symptoms much earlier than age 3, have begun showing significant efficacy in controlled trials (Dawson et al., 2010; for review, see Rogers and Vismara, 2008; Howlin et al., 2009) and are being actively pursued in many clinical centers. Interestingly, treatment response is also not a uniform dimension, as post-treatment measures always display larger dispersion compared to pretreatment, indicating the presence of treatment 'responders' and 'nonresponders' (Dawson et al., 2010). Therefore, genetic and biochemical markers may also be sought to predict treatment response. Finally, the short-term efficacy of early intervention programs is comforting, though not entirely unexpected when considering that environmental enrichment largely reverses behavioral abnormalities in an animal model as 'organic' as rats prenatally exposed to valproic acid (Schneider et al., 2006), and in *MECP2* knockout mice (Lonetti et al., 2010). Early interventions most likely yield better outcomes because they act during critical periods of greater plasticity in postnatal brain development. It will be important to see whether behavioral improvements are permanent, or whether periodic/continuous maintenance treatment will be required, most likely through parent training strategies.

Acknowledgments

The author's work is supported by the Italian Ministry for University, Scientific Research and Technology (Programmi di Ricerca di Interesse Nazionale, prot. n.2006058195), the Italian Ministry of Health (RFPS-2007-5-640174), Autism Speaks (Princeton, NJ), the Autism Research Institute (San Diego, CA), the Fondazione Gaetano e Mafalda Luce (Milan, Italy), and Autism Aids Onlus (Naples, Italy). The author is grateful to the NICHD Brain & Tissue Bank for Developmental Disorders (Baltimore, MD), the Harvard Brain Tissue Resource Center (Belmont, MA), and the Autism Tissue Program (Princeton, NJ), for coordinating and kindly providing post-mortem brain tissues essential to the author's work. The author would also like to acknowledge the members of the author's laboratory, Roberto Sacco, Carla Lintas, Laura Altieri, Federica Lombardi and Valerio Napolioni, for their commitment to autism research.

This text is dedicated to the author's father, Andrea, who passed away 1 week before its completion: "He was a great father, a devoted husband, and a proud Navy Officer: I owe to his care and to his example much of what I am today".

References

Abrahams, B.S., Geschwind, D.H., 2008. Advances in autism genetics: On the threshold of a new neurobiology. Nature Reviews Genetics 9, 341–355.

Abrahams, B.S., Tentler, D., Perederiy, J.V., Oldham, M.C., Coppola, G., Geschwind, D.H., 2007. Genome-wide analyses of human perisylvian cerebral cortical patterning. Proceedings of the National Academy of Science of the United States of America 104, 17849–17854.

Abramson, R.K., Wright, H.H., Carpenter, R., et al., 1989. Elevated blood serotonin in autistic probands and their first-degree relatives. Journal of Autism and Developmental Disorders 19, 397–407.

Adolphs, R., Spezio, M.L., Parlier, M., Piven, J., 2008. Distinct face-processing strategies in parents of autistic children. Current Biology 18, 1090–1093.

Ahmed, Z., Babaei, S., Maguire, G.F., et al., 2003. Paraoxonase-1 reduces monocyte chemotaxis and adhesion to endothelial cells due to oxidation of palmitoyl, linoleoyl glycerophosphorylcholine. Cardiovascular Research 57, 225–231.

Aït Yahya-Graison, E., Aubert, J., Dauphinot, L., et al., 2007. Classification of human chromosome 21 gene-expression variations in Down syndrome: Impact on disease phenotypes. American Journal of Human Genetics 81, 475–491.

Alarcón, M., Abrahams, B.S., Stone, J.L., et al., 2008. Linkage, association, and gene-expression analyses identify CNTNAP2 as an autism-susceptibility gene. American Journal of Human Genetics 82, 150–159.

Allgayer, H., 2006. Regulation and clinical significance of urokinase-receptor (u-PAR), an invasion-related molecule. Gastroenterology 44, 503–511.

Allison, A.C., 2009. Genetic control of resistance to human malaria. Current Opinion in Immunology 21, 499–505.

American Psychiatric Association, 1994. Diagnostic and Statistical Manual of Mental Disorders, 4th edn. American Psychiatric Association, Washington, DC.

American Psychiatric Association, 2010. http://www.dsm5.org/Pages/Default.aspx (accessed September 2011).

American Psychiatric Association, 2012. DSM-5 development. Available at http://www.dsm5.org/Pages/Default.aspx (accessed October 2012).

Amir, R.E., Van den Veyver, I.B., Wan, M., Tran, C.Q., Francke, U., Zoghbi, H.Y., 1999. Rett syndrome is caused by mutations in X-linked MECP2, encoding methyl-CpG-binding protein 2. Nature Genetics 23, 185–188.

Anderson, G.M., Gutknecht, L., Cohen, D.J., et al., 2002. Serotonin transporter promoter variants in autism: Functional effects and relationship to platelet hyperserotonemia. Molecular Psychiatry 7, 831–836.

Arking, D.E., Cutler, D.J., Brune, C.W., et al., 2008. A common genetic variant in the neurexin superfamily member CNTNAP2 increases familial risk of autism. American Journal of Human Genetics 82, 160–164.

Ashley-Koch, A.E., Mei, H., Jaworski, J., et al., 2006. An analysis paradigm for investigating multi-locus effects in complex disease: Examination of three GABA receptor subunit genes on 15q11-q13 as risk factors for autistic disorder. Annals of Human Genetics 70, 281–292.

Ashwood, P., Wills, S., Van de Water, J., 2006. The immune response in autism: A new frontier for autism research. Journal of Leukocyte Biology 80, 1–15.

Asperger, H., 1944a. Die 'Autistischen Psychopathen' im Kindesalter. Archiv fur Psychiatrie und Nervenkrankheiten 117, 76–136.

Asperger, H., 1944b. Autistic Psychopathy in Childhood. In: Translated from Frith U (1991) Autism and Asperger's Syndrome. Cambridge University Press, Cambridge, UK.

Autism Genome Project Consortium, 2007. Mapping autism risk loci using genetic linkage and chromosomal rearrangements. Nature Genetics 39, 319–328.

Bakkaloglu, B., O'Roak, B.J., Louvi, A., et al., 2008. Molecular cytogenetic analysis and resequencing of contactin associated protein-like 2 in autism spectrum disorders. American Journal of Human Genetics 82, 165–173.

Banatvala, J.E., Brown, D.W., 2004. Rubella. Lancet 363, 1127–1137.

Bandim, J.M., Ventura, L.O., Miller, M.T., Almeida, H.C., Costa, A.E., 2003. Autism and Möbius sequence: An exploratory study of children in northeastern Brazil. Arquivos de Neuro-Psiquiatria 61, 181–185.

Bauman, M.L., Kemper, T.L., 2005. Neuroanatomic observations of the brain in autism: A review and future directions. International Journal of Developmental Neuroscience 23, 183–187.

Beilman, M., Vande Woude, G.F., Dienes, H.P., Schirmacher, P., 2000. Hepatocyte growth factor-stimulated invasiveness of monocytes. Blood 95, 3964–3969.

Belmonte, M.K., Gomot, M., Baron-Cohen, S., 2010. Visual attention in autism families: 'Unaffected' sibs share atypical frontal activation. Journal of Child Psychology and Psychiatry 51, 259–276.

Benayed, R., Choi, J., Matteson, P.G., et al., 2009. Autism-associated haplotype affects the regulation of the homeobox gene, EN-GRAILED 2. Biological Psychiatry 66, 911–917.

Benayed, R., Gharani, N., Rossman, I., et al., 2005. Support for the homeobox transcription factor gene ENGRAILED 2 as an autism spectrum disorder susceptibility locus. American Journal of Human Genetics 77, 851–868.

Benvenuto, A., Moavero, R., Alessandrelli, R., Manzi, B., Curatolo, P., 2009. Syndromic autism: Causes and pathogenetic pathways. World Journal of Pediatrics 5, 169–176.

Berger, B.E., Navar-Boggan, A.M., Omer, S.B., 2010. Congenital rubella syndrome and autism spectrum disorder prevented by rubella vaccination–United States, 2001–2010. BMC Public Health 11, 340.

Betancur, C., Corbex, M., Spielewoy, C., et al., 2002. Serotonin transporter gene polymorphisms and hyperserotonemia in autistic disorder. Molecular Psychiatry 7, 67–71.

Betancur, C., Sakurai, T., Buxbaum, J.D., 2009. The emerging role of synaptic cell-adhesion pathways in the pathogenesis of autism spectrum disorders. Trends in Neurosciences 32, 402–412.

Beyer, K.S., Blasi, F., Bacchelli, E., et al., 2002. Mutation analysis of the coding sequence of the MECP2 gene in infantile autism. Human Genetics 111, 305–309.

Bigler, E.D., Abildskov, T.J., Petrie, J.A., et al., 2010. Volumetric and voxel-based morphometry findings in autism subjects with and without macrocephaly. Developmental Neuropsychology 35, 278–295.

Blatt, G.J., Fitzgerald, C.M., Guptill, J.T., Booker, A.B., Kemper, T.L., Bauman, M.L., 2001. Density and distribution of hippocampal neurotransmitter receptors in autism: An autoradiographic study. Journal of Autism and Developmental Disorders 31, 537–543.

Bleuler, E., 1911. Dementia Praecox oder Gruppe der Schizophrenien. In: Aschaffenburg, G. (Ed.), Handbuch der Psychiatrie. Deuticke, Leipzig.

Boccone, L., Dessì, V., Zappu, A., et al., 2006. Bannayan-Riley-Ruvalcaba syndrome with reactive nodular lymphoid hyperplasia and autism and a PTEN mutation. American Journal of Medical Genetics Part A 140A, 1965–1969.

Boddaert, N., Belin, P., Chabane, N., et al., 2003. Perception of complex sounds: Abnormal pattern of cortical activation in autism. The American Journal of Psychiatry 160, 2057–2060.

Boddaert, N., Zilbovicius, M., Philipe, A., et al., 2009. MRI findings in 77 children with non-syndromic autistic disorder. PLoS One 4, e4415.

Bonnel, A., McAdams, S., Smith, B., et al., 2010. Enhanced pure-tone pitch discrimination among persons with autism but not Asperger syndrome. Neuropsychologia 48, 2465–2475.

Bonora, E., Beyer, K.S., Lamb, J.A., et al., 2003. Analysis of reelin as a candidate gene for autism. Molecular Psychiatry 8, 885–892.

Bosley, T.M., Salih, M.A., Alorainy, I.A., et al., 2007. Clinical characterization of the HOXA1 syndrome BSAS variant. Neurology 69, 1245–1253.

Bosley, T.M., Alorainy, I.A., Salih, M.A., et al., 2008. The clinical spectrum of homozygous HOXA1 mutations. American Journal of Medical Genetics Part A 146A, 1235–1240.

Briegel, W., Schimek, M., Kamp-Becker, I., Hofmann, C., Schwab, K.O., 2009. Autism spectrum disorders in children and adolescents with Moebius sequence. European Child and Adolescent Psychiatry 18, 515–519.

Brune, C.W., Korvatska, E., Allen-Brady, K., et al., 2008. Heterogeneous association between engrailed-2 and autism in the CPEA network. American Journal of Medical Genetics Part B Neuropsychiatric Genetics 147B, 187–193.

Bruneau, N., Bonnet-Brilhault, F., Gomot, M., Adrien, J.L., Barthélémy, C., 2003. Cortical auditory processing and communication in children with autism: Electrophysiological/behavioral relations. International Journal of Psychophysiology 51, 17–25.

Bucan, M., Abrahams, B.S., Wang, K., et al., 2009. Genome-wide analyses of exonic copy number variants in a family-based study point to novel autism susceptibility genes. PLoS Genetics 5, e1000536.

Buie, T., Campbell, D.B., Fuchs III, G.J., et al., 2010. Evaluation, diagnosis, and treatment of gastrointestinal disorders in individuals with ASDs: A consensus report. Pediatrics 125, S1–S18.

Butler, M.G., Dasouki, M.J., Zhou, X.P., et al., 2005. Subset of individuals with autism spectrum disorders and extreme macrocephaly associated with germline PTEN tumour suppressor gene mutations. Journal of Medical Genetics 42, 318–321.

Buxbaum, J.D., 2009. Multiple rare variants in the etiology of autism spectrum disorders. Dialogues in Clinical Neurosciences 11, 35–43.

Buxbaum, J.D., Silverman, J.M., Smith, C.J., et al., 2002. Association between a GABRB3 polymorphism and autism. Molecular Psychiatry 7, 311–316.

Buxbaum, J.D., Cai, G., Chaste, P., et al., 2007. Mutation screening of the PTEN gene in patients with autism spectrum disorders and macrocephaly. American Journal of Medical Genetics Part B Neuropsychiatric Genetics 144B, 484–491.

Cacciotti, P., Libener, R., Betta, P., et al., 2001. SV40 replication in human mesothelial cells induces HGF/Met receptor activation: A model for viral-related carcinogenesis of human malignant mesothelioma. Proceedings of the National Academy of Science of the United States of America 98, 12032–12037.

Cameron, B., Landreth, G.E., 2010. Inflammation, microglia, and Alzheimer's disease. Neurobiology of Disease 37, 503–509.

Campbell, D.B., Sutcliffe, J.S., Ebert, P.J., et al., 2006. A genetic variant that disrupts *MET* transcription is associated with autism. Proceedings of the National Academy of Science of the United States of America 103, 16834–16839.

Campbell, D.B., D'Oronzio, R., Garbett, K., et al., 2007. Disruption of cerebral cortex MET signaling in autism spectrum disorder. Annals of Neurology 62, 243–250.

Campbell, D.B., Li, C., Sutcliffe, J.S., Persico, A.M., Levitt, P., 2008. Genetic evidence implicating multiple genes in the MET receptor tyrosine kinase pathway in autism spectrum disorder. Autism Research 1, 159–168.

Campbell, D.B., Buie, T.M., Winter, H., et al., 2009. Distinct genetic risk based on association of MET in families with co-occurring autism and gastrointestinal conditions. Pediatrics 123, 1018–1024.

Campbell, D.B., Warren, D., Sutcliffe, J.S., Lee, E.B., Levitt, P., 2010. Association of *MET* with social and communication phenotypes in individuals with autism spectrum disorder. American Journal of Medical Genetics Part B Neuropsychiatric Genetics 153B, 438–446.

Cantor, R.M., Kono, N., Duvall, J.A., et al., 2005. Replication of autism linkage: Fine-mapping peak at 17q21. American Journal of Human Genetics 76, 1050–1056.

Canu, E., Boccardi, M., Ghidoni, R., et al., 2009. HOXA1 A218G polymorphism is associated with smaller cerebellar volume in healthy humans. Journal of Neuroimaging 19, 353–358.

Carayol, J., Schellenberg, G.D., Tores, F., Hager, J., Ziegler, A., Dawson, G., 2010. Assessing the impact of a combined analysis of four common low-risk genetic variants on autism risk. Molecular Autism 1, 4.

Carneiro, A.M., Cook, E.H., Murphy, D.L., Blakely, R.D., 2008. Interactions between integrin alphaIIbbeta3 and the serotonin transporter regulate serotonin transport and platelet aggregation in mice and humans. Journal of Clinical Investigation 118, 1544–1552.

Carney, R.M., Wolpert, C.M., Ravan, S.A., et al., 2003. Identification of MeCP2 mutations in a series of females with autistic disorder. Pediatric Neurology 28, 205–211.

Carter, C.S., 2007. Sex differences in oxytocin and vasopressin: Implications for autism spectrum disorders? Behavioral Brain Research 176, 170–186.

Carter, C.S., Grippo, A.J., Pournajafi-Nazarloo, H., Ruscio, M.G., Porges, S.W., 2008. Oxytocin, vasopressin and sociality. Progress in Brain Research 170, 331–336.

Carvalho, C.M., Zhang, F., Lupski, J.R., 2010. Evolution in health and medicine Sackler colloquium: Genomic disorders: A window into human gene and genome evolution. Proceedings of the National Academy of Science of the United States of America 107 (supplement 1), 1765–1771.

Cattaneo, L., Fabbri-Destro, M., Boria, S., et al., 2007. Impairment of actions chains in autism and its possible role in intention understanding. Proceedings of the National Academy of Science of the United States of America 104, 17825–17830.

Chadman, K.K., Gong, S., Scattoni, M.L., et al., 2008. Minimal aberrant behavioral phenotypes of neuroligin-3 R451C knockin mice. Autism Research 1, 147–158.

Chahrour, M., Zoghbi, H.Y., 2007. The story of Rett syndrome: From clinic to neurobiology. Neuron 56, 422–437.

Chauhan, A., Chauhan, V., 2006. Oxidative stress in autism. Pathophysiology 13, 171–181.

Cheh, M.A., Millonig, J.H., Roselli, L.M., et al., 2006. En2 knockout mice display neurobehavioral and neurochemical alterations relevant to autism spectrum disorder. Brain Research 1116, 166–176.

Chess, S., 1971. Autism in children with congenital rubella. Journal of Autism & Childhood Schizophrenia 1, 33–47.

Chess, S., 1977. Follow-up report on autism in congenital rubella. Journal of Autism & Childhood Schizophrenia 7, 69–81.

Chess, S., Fernandez, P., Korn, S., 1978. Behavioral consequences of congenital rubella. Journal of Pediatrics 93, 699–703.

Chih, B., Afridi, S.K., Clark, L., Scheiffele, P., 2004. Disorder-associated mutations lead to functional inactivation of neuroligins. Human Molecular Genetics 13, 1471–1477.

Ching, M.S., Shen, Y., Tan, W.H., et al., 2010. Deletions of NRXN1 (neurexin-1) predispose to a wide spectrum of developmental disorders. American Journal of Medical Genetics Part B Neuropsychiatric Genetics 153B, 937–994.

Chisaka, O., Musci, T.S., Capecchi, M.R., 1992. Developmental defects of the ear, cranial nerves, and hindbrain resulting from targeted disruption of the mouse homeobox gene *Hox-1.6*. Nature 355, 516–520.

Chiu, P.H., Kayali, M.A., Kishida, K.T., et al., 2008. Self responses along cingulate cortex reveal quantitative neural phenotype for high-functioning autism. Neuron 57, 463–473.

Christian, S.L., Brune, C.W., Sudi, J., et al., 2008. Novel submicroscopic chromosomal abnormalities detected in autism spectrum disorder. Biological Psychiatry 63, 1111–1117.

Christianson, A.L., Chesler, N., Kromberg, J.G., 1994. Fetal valproate syndrome: Clinical and neuro-developmental features in two sibling pairs. Developmental Medicine & Child Neurology 36, 361–369.

Chubykin, A.A., Liu, X., Comoletti, D., Tsigelny, I., Taylor, P., Südhof, T.C., 2005. Dissection of synapse induction by neuroligins: Effect of a neuroligin mutation associated with autism. Journal of Biological Chemistry 280, 22365–22374.

Cohen, D., Pichard, N., Tordjman, S., et al., 2005. Specific genetic disorders and autism: Clinical contribution towards their identification. Journal of Autism and Developmental Disorders 35, 103–116.

Comi, A.M., Zimmerman, A.W., Frye, V.H., Law, P.A., Peeden, J.N., 1999. Familial clustering of autoimmune disorders and evaluation of medical risk factors in autism. Journal of Child Neurology 14, 388–394.

Comoletti, D., De Jaco, A., Jennings, L.L., et al., 2004. The Arg451Cys-neuroligin-3 mutation associated with autism reveals a defect in protein processing. The Journal of Neuroscience 24, 4889–4893.

Conciatori, M., Stodgell, C.J., Hyman, S.L., et al., 2004. Association between the HOXA1 A218G polymorphism and increased head circumference in patients with autism. Biological Psychiatry 55, 413–419.

Cook, E.H., Leventhal, B.L., Heller, W., Metz, J., Wainwright, M., Freedman, D.X., 1990. Autistic children and their first-degree relatives: Relationships between serotonin and norepinephrine levels and intelligence. The Journal of Neuropsychiatry & Clinical Neuroscience 2, 268–274.

Cook Jr., E.H., Courchesne, R.Y., Cox, N.J., et al., 1998. Linkage-disequilibrium mapping of autistic disorder, with 15q11-13 markers. American Journal of Human Genetics 62, 1077–1083.

Cook Jr., E.H., Leventhal, B.L., Freedman, D.X., 1988. Free serotonin in plasma: Autistic children and their first-degree relatives. Biological Psychiatry 24, 488–491.

Courchesne, E., 1997. Brainstem, cerebellar and limbic neuroanatomical abnormalities in autism. Current Opinion in Neurobiology 7, 269–278.

Courchesne, E., Pierce, K., 2005. Why the frontal cortex in autism might be talking only to itself: Local over-connectivity but long-distance disconnection. Current Opinions in Neurobiology 15, 225–230.

Courchesne, E., Pierce, K., Schumann, C.M., et al., 2007. Mapping early brain development in autism. Neuron 56, 399–413.

Coutinho, A.M., Oliveira, G., Katz, C., et al., 2007. MECP2 coding sequence and 3'UTR variation in 172 unrelated autistic patients. American Journal of Medical Genetics Part B Neuropsychiatric Genetics 144B, 475–483.

Coutinho, A.M., Oliveira, G., Morgadinho, T., et al., 2004. Variants of the serotonin transporter gene (SLC6A4) significantly contribute to hyperserotonemia in autism. Molecular Psychiatry 9, 264–271.

Coutinho, A.M., Sousa, I., Martins, M., et al., 2007. Evidence for epistasis between SLC6A4 and ITGB3 in autism etiology and in the determination of platelet serotonin levels. Human Genetics 121, 243–256.

Craig, A.M., Kang, Y., 2007. Neurexin-neuroligin signaling in synapse development. Current Opinion in Neurobiology 17, 43–52.

Cross, S., Kim, S.J., Weiss, L.A., et al., 2008. Molecular genetics of the platelet serotonin system in first-degree relatives of patients with autism. Neuropsychopharmacology 33, 353–360.

Curran, S., Roberts, S., Thomas, S., et al., 2005. An association analysis of microsatellite markers across the Prader-Willi/Angelman critical region on chromosome 15 (q11-13) and autism spectrum disorder. American Journal of Medical Genetics Part B Neuropsychiatric Genetics 137B, 25–28.

D'Amelio, M., Ricci, I., Sacco, R., et al., 2005. Paraoxonase gene variants are associated with autism in North America, but not in Italy: Possible regional specificity in gene-environment interactions. Molecular Psychiatry 10, 1006–1016.

D'Arcangelo, G., Miao, G.G., Chen, S.C., Soares, H.D., Morgan, J.I., Curran, T., 1995. A protein related to extracellular matrix proteins deleted in the mouse mutant reeler. Nature 374, 719–723.

D'Arcangelo, G., Homayouni, R., Keshvara, L., Rice, D.S., Sheldon, M., Curran, T., 1999. Reelin is a ligand for lipoprotein receptors. Neuron 24, 471–479.

Daoud, H., Bonnet-Brilhault, F., Védrine, S., et al., 2009. Autism and nonsyndromic mental retardation associated with a *de novo* mutation in the NLGN4X gene promoter causing an increased expression level. Biological Psychiatry 66, 906–910.

Dapretto, M., Davies, M.S., Pfeifer, J.H., et al., 2006. Understanding emotions in others: Mirror neuron dysfunction in children with autism spectrum disorders. Nature Neuroscience 9, 28–30.

Davis, C.A., Noble-Topham, S.E., Rossant, J., Joyner, A.L., 1998. Expression of the homeobox-containing gene En-2 delineates a specific region of the developing mouse brain. Genes and Development 2, 361–371.

Dawson, G., Rogers, S., Munson, J., et al., 2010. Randomized, controlled trial of an intervention for toddlers with autism: The Early Start Denver Model. Pediatrics 125, e17–e23.

Dean, C., Scholl, F.G., Choih, J., et al., 2003. Neurexin mediates the assembly of presynaptic terminals. Nature Neuroscience 6, 708–716.

Delahanty, R.J., Kang, J.Q., Brune, C.W., et al., 2011. Maternal transmission of a rare GABRB3 signal peptide variant is associated with autism. Molecular Psychiatry 16, 86–96.

DeLorey, T.M., Sahbaie, P., Hashemi, E., Homanics, G.E., Clark, J.D., 2008. Gabrb3 gene deficient mice exhibit impaired social and exploratory behaviors, deficits in non-selective attention and hypoplasia of cerebellar vermal lobules: A potential model of autism spectrum disorder. Behavioral Brain Research 187, 207–220.

Delorme, R., Goussé, V., Roy, I., et al., 2007. Shared executive dysfunctions in unaffected relatives of patients with autism and obsessive-compulsive disorder. European Psychiatry 22, 32–38.

Devlin, B., Bennett, P., Dawson, G., et al., 2004. Alleles of a reelin CGG repeat do not convey liability to autism in a sample from the CPEA network. American Journal of Medical Genetics Part B Neuropsychiatric Genetics 126, 46–50.

Dhar, S.U., del Gaudio, D., German, J.R., et al., 2010. 22q13.3 deletion syndrome: Clinical and molecular analysis using array CGH. American Journal of Medical Genetics part A 152A, 573–581.

Di Pino, G., Moessner, R., Lesch, K.P., Lauder, J.M., Persico, A.M., 2004. Roles for serotonin in neurodevelopment: More than just neural transmission. Current Neuropharmacology 2, 403–418.

DiCicco-Bloom, E., Lord, C., Zwaigenbaum, L., et al., 2006. The developmental neurobiology of autism spectrum disorder. The Journal of Neuroscience 26, 6897–6906.

Dissanayake, C., Bui, Q.M., Huggins, R., Loesch, D.Z., 2006. Growth in stature and head circumference in high-functioning autism and Asperger disorder during the first 3 years of life. Developmental Psychopathology 18, 381–393.

Doetschman, T., 2009. Influence of genetic background on genetically engineered mouse phenotypes. Methods in Molecular Biology 530, 423–433.

Dölen, G., Osterweil, E., Rao, B.S., et al., 2007. Correction of fragile X syndrome in mice. Neuron 56, 955–962.

Dong, E., Guidotti, A., Grayson, D.R., Costa, E., 2007. Histone hyperacetylation induces demethylation of reelin and 67-kDa glutamic acid decarboxylase promoters. Proceedings of the National Academy of Science of the United States of America 104, 4676–4681.

D'Souza, A., Onem, E., Patel, P., La Gamma, E.F., Nankova, B.B., 2009. Valproic acid regulates catecholaminergic pathways by concentration-dependent threshold effects on TH mRNA synthesis and degradation. Brain Research 1247, 1–10.

Durand, C.M., Betancur, C., Boeckers, T.M., et al., 2007. Mutations in the gene encoding the synaptic scaffolding protein SHANK3 are associated with autism spectrum disorders. Nature Genetics 39, 25–27.

Durand, C.M., Perroy, J., Loll, F., et al., 2011. SHANK3 mutations identified in autism lead to modification of dendritic spine morphology via an actin-dependent mechanism. Molecular Psychiatry 17, 71–84.

Dutta, S., Guhathakurta, S., Sinha, S., et al., 2007. Reelin gene polymorphisms in the Indian population: A possible paternal 5'UTR-CGG-repeat-allele effect on autism. American Journal of Medical Genetics Part B Neuropsychiatric Genetics 144B, 106–112.

Dutta, S., Sinha, S., Ghosh, S., Chatterjee, A., Ahmed, S., Usha, R., 2008. Genetic analysis of reelin gene (RELN) SNPs: No association with autism spectrum disorder in the Indian population. Neuroscience Letters 441, 56–60.

Duvall, J.A., Lu, A., Cantor, R.M., Todd, R.D., Constantino, J.N., Geschwind, D.H., 2007. A quantitative trait locus analysis of social responsiveness in multiplex autism families. American Journal of Psychiatry 164, 656–662.

Ebstein, R.P., Israel, S., Lerer, E., et al., 2009. Arginine vasopressin and oxytocin modulate human social behavior. Annals of the New York Academy of Science 1167, 87–102.

Elia, J., Gai, X., Xie, H.M., et al., 2010. Rare structural variants found in attention-deficit hyperactivity disorder are preferentially associated with neurodevelopmental genes. Molecular Psychiatry 15, 637–646.

Elsen, G.E., Choi, L.Y., Prince, V.E., Ho, R.K., 2009. The autism susceptibility gene met regulates zebrafish cerebellar development and facial motor neuron migration. Developmental Biology 335, 78–92.

Eng, C., 2003. PTEN: One gene, many syndromes. Human Mutation 22, 183–198.

Engel, S.M., Berkowitz, G.S., Barr, D.B., et al., 2007. Prenatal organophosphate metabolite and organochlorine levels and performance on the Brazelton Neonatal Behavioral Assessment Scale in a multiethnic pregnancy cohort. American Journal of Epidemiology 165, 1397–1404.

Englert, Y., Lesage, B., Van Vooren, J.P., et al., 2004. Medically assisted reproduction in the presence of chronic viral diseases. Human Reproduction Update 10, 149–162.

Engman, M.L., Lewensohn-Fuchs, I., Mosskin, M., Malm, G., 2010. Congenital cytomegalovirus infection; the impact of cerebral cortical malformations. Acta Paediatrica 99, 1344–1349.

Eskenazi, B., Rosas, L.G., Marks, A.R., et al., 2008. Pesticide toxicity and the developing brain. Basic and Clinical Pharmacology & Toxicology 102, 228–236.

Etherton, M.R., Blaiss, C.A., Powell, C.M., Südhof, T.C., 2009. Mouse neurexin-1alpha deletion causes correlated electrophysiological and behavioral changes consistent with cognitive impairments. Proceedings of the National Academy of Science of the United States ofAmerica 106, 17998–18003.

Etherton, M., Földy, C., Sharma, M., et al., 2011a. Autism-linked neuroligin-3 R451C mutation differentially alters hippocampal and cortical synaptic function. Proceedings of the National Academy of Science of the United States ofAmerica 108, 13764–13769.

Etherton, M.R., Tabuchi, K., Sharma, M., Ko, J., Südhof, T.C., 2011b. An autism-associated point mutation in the neuroligin cytoplasmic tail selectively impairs AMPA receptor-mediated synaptic transmission in hippocampus. EMBO Journal 30, 2908–2919.

Fatemi, S.H., Stary, J.M., Egan, E.A., 2002. Reduced blood levels of reelin as a vulnerability factor in pathophysiology of autistic disorder. Cellular and Molecular Neurobiology 22, 139–152.

Fatemi, S.H., Snow, A.V., Stary, J.M., et al., 2005. Reelin signaling is impaired in autism. Biological Psychiatry 57, 777–787.

Fatemi, S.H., Reutiman, T.J., Folsom, T.D., Thuras, P.D., 2009. GABA(A) receptor downregulation in brains of subjects with autism. Journal of Autism and Developmental Disorders 39, 223–230.

Feinstein, C., Singh, S., 2007. Social phenotypes in neurogenetic syndromes. Child & Adolescent Psychiatric Clinics of North America 16, 631–647.

Feng, J., Schroer, R., Yan, J., et al., 2006. High frequency of neurexin 1beta signal peptide structural variants in patients with autism. Neuroscience Letters 409, 10–13.

Ferguson, J.N., Young, L.J., Hearn, E.F., Matzuk, M.M., Insel, T.R., Winslow, J.T., 2000. Social amnesia in mice lacking the oxytocin gene. Nature Genetics 25, 284–288.

Fernandez, B.A., Roberts, W., Chung, B., et al., 2010. Phenotypic spectrum associated with de novo and inherited deletions and duplications at 16p11.2 in individuals ascertained for diagnosis of autism spectrum disorder. Journal of Medical Genetics 47, 195–203.

Ferré, N., Marsillach, J., Camps, J., et al., 2005. Genetic association of paraoxonase-1 polymorphisms and chronic hepatitis C virus infection. Clinica Chimica Acta 361, 206–210.

Fillano, J., Goldenthal, M.J., Rhodes, C.H., Marín-García, J., 2002. Mitochondrial dysfunction in patients with hypotonia, epilepsy, autism, and developmental delay: HEADD syndrome. Journal of Child Neurology 17, 435–439.

Flatscher-Bader, T., Foldi, C.J., Chong, S., et al., 2011. Increased *de novo* copy number variants in the offspring of older males. Translational Psychiatry 1, e34.

Fombonne, E., 2005. Epidemiology of autistic disorder and other pervasive developmental disorders. The Journal of Clinical Psychiatry 66 (supplement 10), 3–8.

Fombonne, E., Rogé, B., Claverie, J., Courty, S., Frémolle, J., 1999. Microcephaly and macrocephaly in autism. Journal of Autism and Developmental Disorders 29, 113–119.

Freitag, C.M., 2007. The genetics of autistic disorders and its clinical relevance: A review of the literature. Molecular Psychiatry 12, 2–22.

Frisque, R.J., Hofstetter, C., Tyagarajan, S.K., 2006. Transforming activities of JC virus early proteins. In: Ahsan, N. (Ed.), Polyomaviruses and Human Diseases Advances in Experimental Medicine and Biology 577, Eureka.com/Landes Bioscience, Georgetown, TX, pp. 288–309.

Furman, D.J., Chen, M.C., Gotlib, I.H., 2011. Variant in oxytocin receptor gene is associated with amygdala volume. Psychoneuroendocrinology 36, 891–897.

Gaita, L., Manzi, B., Sacco, R., et al., 2010. Decreased serum arylesterase activity in autism spectrum disorders. Psychiatry Research 180, 105–113.

Gallagher, L., Hawi, Z., Kearney, G., Fitzgerald, M., Gill, M., 2004. No association between allelic variants of HOXA1/HOXB1 and autism. American Journal of Medical Genetics Part B Neuropsychiatric Genetics 124B, 64–67.

Garbett, K., Ebert, P.J., Mitchell, A., et al., 2008. Immune transcriptome alterations in the temporal cortex of subjects with autism. Neurobiology of Disease 30, 303–311.

Gauthier, J., Spiegelman, D., Piton, A., et al., 2009. Novel de novo SHANK3 mitation in autistic patients. American Journal of Medical Genetics Part B Neuropsychiatric Genetics 150B, 421–424.

Gauthier, J., Champagne, N., Lafrenière, R.G., et al., 2010. De novo mutations in the gene encoding the synaptic scaffolding protein SHANK3 in patients ascertained for schizophrenia. Proceedings of the National Academy of Science of the United States of America 107, 7863–7868.

Geschwind, D.H., 2011. Genetics of autism spectrum disorders. Trends in Cognitive Sciences 15, 409–416.

Geschwind, D.H., Levitt, P., 2007. Autism spectrum disorders: Developmental disconnection syndromes. Current Opinion in Neurobiology 17, 103–111.

Gharani, N., Benayed, R., Mancuso, V., Brzustowicz, L.M., Millonig, J.H., 2004. Association of the homeobox transcription factor, ENGRAILED 2, 3, with autism spectrum disorder. Molecular Psychiatry 9, 474–484.

Gillberg, C., 1998. Chromosomal disorders and autism. Journal of Autism and Developmental Disorders 28, 415–425.

Gillberg, C., Steffenburg, S., 1989. Autistic behaviour in Moebius syndrome. Acta Paediatrica Scandinavica 78, 314–316.

Gilman, S.R., Iossifov, I., Levy, D., Ronemus, M., Wigler, M., Vitkup, D., 2011. Rare de novo variants associated with autism implicate a large functional network of genes involved in formation and function of synapses. Neuron 70, 898–907.

Giulivi, C., Zhang, Y.F., Omanska-Klusek, A., et al., 2010. Mitochondrial dysfunction in autism. Journal of the American Medical Association 304, 2389–2396.

Glessner, J.T., Wang, K., Cai, G., et al., 2009. Autism genome-wide copy number variation reveals ubiquitin and neuronal genes. Nature 459, 569–573.

Goffin, A., Hoefsloot, L.H., Bosgoed, E., Swillen, A., Fryns, J.P., 2001. PTEN mutation in a family with Cowden syndrome and autism. American Journal of Medical Genetics 105, 521–524.

Goffinet, A.M., 1984. Events governing organization of postmigratory neurons: Studies on brain development in normal and reeler mice. Brain Research Reviews 7, 261–296.

Gomot, M., Belmonte, M.K., Bullmore, E.T., Bernard, F.A., Baron-Cohen, S., 2008. Brain hyper-reactivity to auditory novel targets in children with high-functioning autism. Brain 131, 2479–2488.

Gong, X., Bacchelli, E., Blasi, F., et al., 2008. Analysis of X chromosome inactivation in autism spectrum disorders. American Journal of Medical Genetics Part B Neuropsychiatric Genetics 147B, 830–835.

Gottesman, I.I., Gould, T.D., 2003. The endophenotype concept in psychiatry: Etymology and strategic intentions. The American Journal of Psychiatry 160, 636–645.

Graf, W., Marin-Garcia, J., Gao, H.G., et al., 2000. Autism associated with the mitochondrial DNA G8363A transfer RNA(Lys) mutation. Journal of Child Neurology 15, 357–361.

Grayson, D.R., Chen, Y., Costa, E., et al., 2006. The human reelin gene: Transcription factors (+), repressors (−) and the methylation switch (+/−) in schizophrenia. Pharmacology and Therapeutics 111, 272–286.

Green, L., Fein, D., Modahl, C., Feinstein, C., Waterhouse, L., Morris, M., 2001. Oxytocin and autistic disorder: Alterations in peptide forms. Biological Psychiatry 50, 609–613.

Gregory, S.G., Connelly, J.J., Towers, A.J., et al., 2009. Genomic and epigenetic evidence for oxytocin receptor deficiency in autism. BMC Medicine 7, 62.

Guilmatre, A., Dubourg, C., Mosca, A.L., et al., 2009. Recurrent rearrangements in synaptic and neurodevelopmental genes and shared biologic pathways in schizophrenia, autism, and mental retardation. Archives of General Psychiatry 66, 947–956.

Gurrieri, F., Neri, G., 2009. Defective oxytocin function: A clue to understanding the cause of autism? BMC Medicine 7, 63.

Hall, H., Lawyer, G., Sillén, A., et al., 2007. Potential genetic variants in schizophrenia: A Bayesian analysis. World Journal of Biological Psychiatry 8, 12–22.

Hallene, K.L., Oby, E., Lee, B.J., et al., 2006. Prenatal exposure to thalidomide, altered vasculogenesis, and CNS malformations. Neuroscience 142, 267–283.

Hallmayer, J., Cleveland, S., Torres, A., et al., 2011. Genetic heritability and shared environmental factors among twin pairs with autism. Archives of General Psychiatry 68, 1095–1102.

Hameury, L., Roux, S., Barthélémy, C., et al., 1995. Quantified multidimensional assessment of autism and other pervasive developmental disorders. Application for bioclinical research. European Child and Adolescent Psychiatry 4, 123–135.

Hammond, P., Hutton, T.J., Allanson, J.E., et al., 2004. 3D analysis of facial morphology. American Journal of Medical Genetics Part A 126A, 339–348.

Hammond, P., Forster-Gibson, C., Chudley, A.E., et al., 2008. Face-brain asymmetry in autism spectrum disorders. Molecular Psychiatry 13, 614–623.

Harvey, C.G., Menon, S.D., Stachowiak, B., et al., 2007. Sequence variants within exon 1 of MECP2 occur in females with mental retardation. American Journal of Medical Genetics Part B Neuropsychiatric Genetics 144B, 355–360.

Hemara-Wahanui, A., Berjukow, S., Hope, C.I., et al., 2005. A *CACNA1F* mutation identified in an X-linked retinal disorder shifts the voltage dependence of the $Ca_v1.4$ channel activation. Proceedings of the National Academy of Science of the United States of America 102, 7553–7558.

Hérault, J., Petit, E., Martineau, J., et al., 1996. Serotonin and autism: Biochemical and molecular biology features. Psychiatry Research 65, 33–43.

Herman, G.E., Butter, E., Enrile, B., Pastore, M., Prior, T.W., Sommer, A., 2007a. Increasing knowledge of PTEN germline mutations: Two additional patients with autism and macrocephaly. American Journal of Medical Genetics Part A 143, 589–593.

Herman, G.E., Henninger, N., Ratliff-Schaub, K., Pastore, M., Fitzgerald, S., McBride, K.L., 2007b. Genetic testing in autism: How much is enough? Genetic in Medicine 9, 268–273.

Hernandez, N., Metzger, A., Magné, R., et al., 2009. Exploration of core features of a human face by healthy and autistic adults analyzed by visual scanning. Neuropsychologia 47, 1004–1012.

Hertz-Picciotto, I., Croen LA, L.A., Hansen, R., Jones, C.R., van de Water, J., Pessah, I.N., 2006. The CHARGE study: An epidemiologic investigation of genetic and environmental factors contributing to autism. Environmental Health Perspectives 114, 1119–1125.

Herz, J., Chen, Y., 2006. Reelin, lipoprotein receptors and synaptic plasticity. Nature Reviews Neuroscience 7, 850–859.

Hiesberger, T., Trommsdorff, M., Howell, B.W., et al., 1999. Direct binding of reelin to VLDL receptor and APOE receptor 2 induces tyrosine phosphorylation of disabled-1 and modulated tau phosphorylation. Neuron 24, 481–489.

Hogart, A., Nagarajan, R.P., Patzel, K.A., Yasui, D.H., Lasalle, J.M., 2007. 15q11-13 GABAA receptor genes are normally biallelically expressed in brain yet are subject to epigenetic dysregulation in autism-spectrum disorders. Human Molecular Genetics 16, 691–703.

Hogart, A., Leung, K.N., Wang, N.J., et al., 2009. Chromosome 15q11-13 duplication syndrome brain reveals epigenetic alterations in gene expression not predicted from copy number. Journal of Medical Genetics 46, 86–93.

Hong, S.E., Shugart, Y.Y., Huang, D.T., et al., 2000. Autosomal recessive lissencephaly with cerebellar hypoplasia is associated with RELN mutations. Nature Genetics 26, 93–96.

Hope, C.I., Sharp, D.M., Hemara-Wahanui, A., et al., 2005. Clinical manifestations of a unique X-linked retinal disorder in a large New Zealand family with a novel mutation in *CACNA1F*, the gene responsible for CSNB2. Clinical and Experimental Ophthalmology 33, 129–136.

Hou, J., Seth, P., Major, E.O., 2006. JC virus can infect human immune and nervous system progenitor cells: Implications for pathogenesis. In: Ahsan, N. (Ed.), Polyomaviruses and Human Diseases Advances in Experimental Medicine and Biology 577, Eureka.com/Landes Bioscience, Georgetown, TX, pp. 266–273.

Howlin, P., Magiati, I., Charman, T., 2009. Systematic review of early intensive behavioral interventions for children with autism. American Journal on Intellectual and Developmental Disabilities 114, 23–41.

Huang, C.H., Santangelo, S.L., 2008. Autism and serotonin transporter gene polymorphisms: A systematic review and meta-analysis. American Journal of Medical Genetics Part B Neuropsychiatric Genetics 147B, 903–913.

Hunter, J.W., Mullen, G.P., McManus, J.R., et al., 2010. Neuroligin-deficient mutants of *C. elegans* have sensory processing deficits and are hypersensitive to oxidative stress and mercury toxicity. Disease Models & Mechanisms 3, 366–376.

Hussman, J.P., 2001. Suppressed GABAergic inhibition as a common factor in suspected etiologies of autism. Journal of Autism and Developmental Disorders 31, 247–248.

Iafrate, A.J., Feuk, L., Rivera, M.N., et al., 2004. Detection of large-scale variation in the human genome. Nature Genetics 36, 949–951. See Database of Genomic Variants at http://projects.tcag.ca/variation/ (accessed September 2011).

Ichtchenko, K., Hata, Y., Nguyen, T., et al., 1995. Neuroligin 1: A splice site-specific ligand for beta-neurexins. Cell 81, 435–443.

Ieraci, A., Forni, P.E., Ponzetto, C., 2002. Viable hypomorphic signaling mutant of the Met receptor reveals a role for hepatocyte growth factor in postnatal cerebellar development. Proceedings of the National Academy of Science of the United States of America 99, 15200–15205.

Ingram, J.L., Peckhan, S.M., Tisdale, B., Rodier, P.M., 2000a. Prenatal exposure of rats to valproic acid reproduces the cerebellar anomalies associated with autism. Neurotoxicology and Teratology 22, 319–324.

Ingram, J.L., Stodgell, C.J., Hyman, S.L., Figlewicz, D.A., Weitkamp, L.R., Rodier, P.M., 2000b. Discovery of allelic variants of *HOXA1* and *HOXB1*: Genetic susceptibility to autism spectrum disorders. Teratology 62, 393–405.

Inoue, H., Yamasue, H., Tochigi, M., et al., 2010. Association between the oxytocin receptor gene and amygdalar volume in healthy adults. Biological Psychiatry 68, 1066–1072.

Ivarsson, S.A., Bjerre, I., Vegfors, P., Ahlfors, K., 1990. Autism as one of several disabilities in two children with congenital cytomegalovirus infection. Neuropediatrics 21, 102–103.

Jackman, C., Horn, N.D., Molleston, J.P., Sokol, D.K., 2009. Gene associated with seizures, autism, and hepatomegaly in an Amish girl. Pediatric Neurology 40, 310–313.

Jackson, P.B., Boccuto, L., Skinner, C., et al., 2009. Further evidence that the rs1858830 C variant in the promoter region of the MET gene is associated with autistic disorder. Autism Research 2, 232–236.

Jacob, S., Brune, C.W., Carter, C.S., Leventhal, B.L., Lord, C., Cook Jr., E.H., 2007. Association of the oxytocin receptor gene (OXTR) in Caucasian children and adolescents with autism. Neuroscience Letters 417, 6–9.

Jacquemont, M.L., Sanlaville, D., Redon, R., et al., 2006. Array-based comparative genomic hybridisation identifies high frequency of cryptic chromosomal rearrangements in patients with syndromic autism spectrum disorders. Journal of Medical Genetics 43, 843–849.

Jamain, S., Quach, H., Betancur, C., et al., 2003. Mutations of the X-linked genes encoding neuroligins NLGN3 and NLGN4 are associated with autism. Nature Genetics 34, 27–29.

Jentsch, T.J., Stein, V., Weinreich, F., Zdebik, A.A., 2002. Molecular structure and physiological function of chloride channels. Physiological Reviews 82, 503–568.

Jin, D., Liu, H.X., Hirai, H., et al., 2007. CD38 is critical for social behaviour by regulating oxytocin secretion. Nature 446, 41–45.

Judson, M.C., Bergman, M.Y., Campbell, D.B., Eagleson, K.L., Levitt, P., 2009. Dynamic gene and protein expression patterns of the autism-associated met receptor tyrosine kinase in the developing mouse forebrain. Journal of Comparative Neurology 513, 511–531.

Judson, M.C., Eagleson, K.L., Wang, L., Levitt, P., 2010. Evidence of cell-nonautonomous changes in dendrite and dendritic spine morphology in the met-signaling-deficient mouse forebrain. Journal of Comparative Neurology 518, 4463–4478.

Jyonouchi, H., Geng, L., Ruby, A., Zimmerman-Bier, B., 2005. Dysregulated innate immune responses in young children with autism spectrum disorders: Their relationship to gastrointestinal symptoms and dietary intervention. Neuropsychobiology 51, 77–85.

Kaminsky, E.B., Kaul, V., Paschall, J., et al., 2011. An evidence-based approach to establish the functional and clinical significance of copy number variants in intellectual and developmental disabilities. Genetic Medicine 13, 777–784.

Kanner, L., 1943. Autistic disturbances of affective contact. Nervous Child 2, 217–250.

Katsui, T., Okuda, M., Usuda, S., Koizumi, T., 1986. Kinetics of ^{3}H-serotonin uptake by platelets in infantile autism and developmental language disorder (including five pairs of twins). Journal of Autism and Developmental Disorders 16, 69–76.

Kawatani, M., Nakai, A., Okuno, T., et al., 2010. Detection of cytomegalovirus in preserved umbilical cord from a boy with autistic disorder. Pediatrics International 52, 304–307.

Kelemenova, S., Schmidtova, E., Ficek, A., Celec, P., Kubranska, A., Ostatnikova, D., 2010. Polymorphisms of candidate genes in Slovak autistic patients. Psychiatric Genetics 20, 137–139.

Kilic, S.S., Aydin, S., Kilic, N., Erman, F., Aydin, S., Celik, I., 2005. Serum arylesterase and paraoxonase activity in patients with chronic hepatitis. World Journal of Gastroenterology 11, 7351–7354.

Kim, S.J., Young, L.J., Gonen, D., et al., 2002. Transmission disequilibrium testing of arginine vasopressin receptor 1A (AVPR1A) polymorphism in autism. Molecular Psychiatry 7, 503–507.

Kim, J., Ghosh, S., Liu, H., Tateyama, M., Kass, R.S., Pitt, G.S., 2004. Calmodulin mediates Ca^{2+} sensitivity of sodium channels. The Journal of Biological Chemistry 43, 45004–45012.

Kim, S.A., Kim, J.H., Park, M., Cho, I.H., Yoo, H.J., 2006. Association of GABRB3 polymorphisms with autism spectrum disorders in Korean trios. Neuropsychobiology 54, 160–165.

Kim, H.G., Kishikawa, S., Higgins, A.W., et al., 2008. Disruption of neurexin 1 associated with autism spectrum disorder. American Journal of Human Genetics 82, 199–207.

Kirov, G., Rujescu, D., Ingason, A., Collier, D.A., O'Donovan, M.C., Owen, M.J., 2009. Neurexin 1 (*NRXN1*) deletions in schizophrenia. Schizophrenia Bulletin 35, 851–854.

Klin, A., Jones, W., Schultz, R., Volkmar, F., Cohen, D., 2002. Visual fixation patterns during viewing of naturalistic social situations as predictors of social competence in individuals with autism. Archives of General Psychiatry 59, 809–816.

Knowles, W.A., 2006. Discovery and epidemiology of the human polyomaviruses BK virus (BKV) and JC virus (JCV). In: Ahsan, N. (Ed.), Polyomaviruses and Human Diseases. Advances in Experimental Medicine and Biology 577, Eureka.com/Landes Bioscience, Georgetown, TX, pp. 19–45.

Kolozsi, E., Mackenzie, R.N., Roullet, F.I., deCatanzaro, D., Foster, J.A., 2009. Prenatal exposure to valproic acid leads to reduced expression of synaptic adhesion molecule neuroligin 3 in mice. Neuroscience 163, 1201–1210.

Krebs, M.O., Betancur, C., Leroy, S., et al., 2002. Absence of association between a polymorphic GGC repeat in the 5′ untranslated region of the reelin gene and autism. Molecular Psychiatry 7, 801–804.

Krey, J., Dolmetsch, R., 2007. Molecular mechanisms of autism: A possible role for Ca^{2+} signaling. Current Opinion in Neurobiology 17, 112–119.

Kuemerle, B., Gulden, F., Cherosky, N., Williams, E., Herrup, K., 1997. Pattern deformities and cell loss in Engrailed-2 mutant mice suggest two separate patterning events during cerebellar development. The Journal of Neuroscience 17, 7881–7889.

Kumar, R.A., KaraMohamed, S., Sudi, J., et al., 2008. Recurrent 16p11.2 microdeletions in autism. Human Molecular Genetics 17, 628–638.

Kurnit, D.M., Layton, W.M., Matthysse, S., 1987. Genetics, chance, and morphogenesis. American Journal of Human Genetics 41, 979–995.

Kwon, C.H., Luikart, B.W., Powell, C.M., et al., 2006. Pten regulates neuronal arborization and social interaction in mice. Neuron 50, 377–388.

Lai, C.S.L., Fisher, S.E., Hurst, J.A., Vargha-Khadem, F., Monaco, A.P., 2001. A forkhead-domain gene is mutated in a severe speech and language disorder. Nature 413, 519–523.

Lainhart, J.E., Piven, J., Wzorek, M., et al., 1997. Macrocephaly in children and adults with autism. Journal of the American Academy of Child and Adolescent Psychiatry 36, 282–290.

Lainhart, J.E., Bigler, E.D., Bocian, M., et al., 2006. Head circumference and height in autism: A study by the Collaborative Program of Excellence in Autism. American Journal of Medical Genetics Part A 140, 2257–2274.

Lam, C.W., Yeung, W.L., Ko, C.H., et al., 2000. Spectrum of mutations in the MECP2 gene in patients with infantile autism and Rett syndrome. Journal of Medical Genetics 37, e41.

Landrigan, P.J., 2010. What causes autism? Exploring the environmental contribution. Current Opinion in Pediatrics 22, 219–225.

LaSalle, J.M., 2007. The Odyssey of MeCP2 and parental imprinting. Epigenetics 2, 5–10.

Laumonnier, F., Bonnet-Brilhault, F., Gomot, M., et al., 2004. X-linked mental retardation and autism are associated with a mutation in the NLGN4 gene, a member of the neuroligin family. American Journal of Human Genetics 74, 552–557.

Laumonnier, F., Roger, S., Guérin, P., et al., 2006. Association of a functional deficit of the BKCa channel, a synaptic regulator of neuronal excitability, with autism and mental retardation. American Journal of Psychiatry 163, 1622–1629.

Laurence, J.A., Fatemi, S.H., 2005. Glial fibrillary acidic protein is elevated in superior frontal, parietal and cerebellar cortices of autistic subjects. Cerebellum 4, 206–210.

Lawson-Yuen, A., Saldivar, J.S., Sommer, S., Picker, J., 2008. Familial deletion within NLGN4 associated with autism and Tourette syndrome. European Journal of Human Genetics 16, 614–618.

Lerer, E., Levi, S., Salomon, S., Darvasi, A., Yirmiya, N., Ebstein, R.P., 2008. Association between the oxytocin receptor (OXTR) gene and autism: Relationship to Vineland Adaptive Behavior Scales and cognition. Molecular Psychiatry 13, 980–988.

Lesch, K.P., Bengel, D., Heils, A., et al., 1996. Association of anxiety-related traits with a polymorphism in the serotonin transporter gene regulatory region. Science 274, 1527–1531.

Lesch, K.P., Wolozin, B.L., Murphy, D.L., Riederer, P., 1993. Primary structure of the human platelet serotonin (5-HT) uptake site: Identity with the brain 5-HT transporter. Journal of Neurochemistry 60, 2319–2322.

Leventhal, B.L., Cook Jr., E.H., Morford, M., Ravitz, A., Freedman, D.X., 1990. Relationships of whole blood serotonin and plasma norepinephrine within families. Journal of Autism and Developmental Disorders 20, 499–511.

Levitt, P., Campbell, D.B., 2009. The genetic and neurobiologic compass points toward common signaling dysfunctions in autism spectrum disorders. The Journal of Clinical Investigation 119, 747–754.

Levitt, P., Eagleson, K.L., Powell, E.M., 2004. Regulation of neocortical interneuron development and the implications for neurodevelopmental disorders. Trends in Neuroscience 27, 400–406.

Levy, D., Ronemus, M., Yamrom, B., et al., 2011. Rare de novo and transmitted copy-number variation in autistic spectrum disorders. Neuron 70, 886–897.

Li, J., Tabor, H.K., Nguyen, L., et al., 2002. Lack of association between HoxA1 and HoxB1 gene variants and autism in 110 multiplex families. American Journal of Medical Genetics Part B Neuropsychiatric Genetics 114, 24–30.

Li, J., Nguyen, L., Gleason, C., et al., 2004. Lack of evidence for an association between WNT2 and RELN polimorphisms and autism. American Journal of Medical Genetics Part B Neuropsychiatric Genetics 126, 51–57.

Li, H., Yamagata, T., Mori, M., Yasuhara, A., Momoi, M.Y., 2005. Mutation analysis of methyl-CpG binding protein family genes in autistic patients. Brain Development 27, 321–325.

Li, J., Ashley, J., Budnik, V., Bhat, M.A., 2007. Crucial role of *Drosophila* neurexin in proper active zone apposition to postsynaptic densities, synaptic growth, and synaptic transmission. Neuron 55, 741–755.

Li, H., Li, Y., Shao, J., et al., 2008. The association analysis of RELN and GRM8 genes with autistic spectrum disorder in Chinese Han population. American Journal of Medical Genetics part B Neuropsychiatric Genetics 147B, 194–200.

Li, X., Hu, Z., He, Y., et al., 2010. Association analysis of CNTNAP2 polymorphisms with autism in the Chinese Han population. Psychiatric Genetics 20, 113–117.

Liang, J.S., Shimojima, K., Ohno, K., et al., 2009. A newly recognised microdeletion syndrome of 2p15-16.1 manifesting moderate developmental delay, autistic behaviour, short stature, microcephaly, and dysmorphic features: A new patient with 3.2 Mb deletion. Journal of Medical Genetics 46, 645–647.

Libbey, J.E., Sweeten, T.L., McMahon, W.M., Fujinami, R.S., 2005. Autistic disorder and viral infections. Journal of Neurovirology 11, 1–10.

Lintas, C., Persico, A.M., 2008. Reelin gene polymorphisms in autistic disorder. In: Fatemi, S.H. (Ed.), Reelin Glycoprotein, Biology, Structure and Roles in Health and Disease. Springer, New York, pp. 385–400.

Lintas, C., Persico, A.M., 2009. Autistic phenotypes and genetic testing: State-of-the-art for the clinical geneticist. Journal of Medical Genetics 46, 1–8.

Lintas, C., Persico, A.M., 2010. Neocortical RELN promoter methylation increases significantly after puberty. NeuroReport 21, 114–118.

Lintas, C., Sacco, R., Garbett, K., et al., 2009. Involvement of the PRKCB1 gene in autistic disorder: Significant genetic association and reduced neocortical gene expression. Molecular Psychiatry 14, 705–718.

Lintas, C., Altieri, L., Lombardi, F., Sacco, R., Persico, A.M., 2010. Association of autism with polyomavirus infection in postmortem brains. Journal of Neurovirology 16, 141–149.

Liu, W., Pappas, G.D., Carter, C.S., 2005. Oxytocin receptors in brain cortical regions are reduced in haploinsufficient (+/−) reeler mice. Neurological Research 27, 339–345.

Liu, X.Q., Paterson, A.D., Szatmari, P., Autism Genome Project Consortium, 2008. Genome-wide linkage analyses of quantitative and categorical autism subphenotypes. Biological Psychiatry 64, 561–570.

Liu, X., Kawamura, Y., Shimada, T., et al., 2010. Association of the oxytocin receptor (OXTR) gene polymorphisms with autism spectrum disorder (ASD) in the Japanese population. Journal of Human Genetics 55, 137–141.

Lobo-Menendez, F., Sossey-Alaoui, K., Bell, J.M., et al., 2004. Absence of MeCP2 mutations in patients from the South Carolina autism project. American Journal of Medical Genetics Part B Neuropsychiatric Genetics 117B, 97–101.

Lockstone, H.E., Harris, L.W., Swatton, J.E., Wayland, M.T., Holland, A.J., Bahn, S., 2007. Gene expression profiling in the adult Down syndrome brain. Genomics 90, 647–660.

Lonetti, G., Angelucci, A., Morando, L., Boggio, E.M., Giustetto, M., Pizzorusso, T., 2010. Early environmental enrichment moderates the behavioral and synaptic phenotype of MeCP2 null mice. Biological Psychiatry 67, 657–665.

López-Pisón, J., Rubio-Rubio, R., Ureña-Hornos, T., et al., 2005. Diagnóstico retrospectivo de infección congénita por citomegalovirus en un caso clínico infantil. Revista de Neurología 40, 733–736.

Lugli, G., Krueger, J.M., Davis, J.M., Persico, A.M., Keller, F., Smalheiser, N.R., 2003. Methodological factors influencing measurement and processing of plasma reelin in humans. BMC Biochemistry 4, 9.

Ma, X.M., Blenis, J., 2009. Molecular mechanisms of mTOR-mediated translational control. Nature Reviews Molecular and Cellular Biology 10, 307–318.

Ma, D.Q., Whitehead, P.L., Menold, M.M., et al., 2005. Identification of significant association and gene-gene interaction of GABA receptor subunit genes in autism. American Journal of Human Genetics 77, 377–388.

Ma, D.Q., Rabionet, R., Konidari, I., et al., 2010. Association and gene-gene interaction of SLC6A4 and ITGB3 in autism. American Journal of Medical Genetics Part B Neuropsychiatric Genetics 153, 477–483.

MacKenzie, A., Quinn, J., 1999. A serotonin transporter gene intron 2 polymorphic region, correlated with affective disorders, has allele-dependent differential enhancer-like properties in the mouse embryo. Proceedings of the National Academy of Science of the United States of America 96, 15251–15255.

Maestrini, E., Lai, C., Marlow, A., et al., 1999. Serotonin transporter (5-HTT) and gamma-aminobutyric acid receptor subunit beta3 (GABRB3) gene polymorphisms are not associated with autism in the IMGSA families. The International Molecular Genetic Study of Autism Consortium. American Journal of Medical Genetics 88, 492–496.

Maher, B., 2008. Personal genomes: The case of the missing heritability. Nature 456, 18–21.

Marazziti, D., Muratori, F., Cesari, A., et al., 2000. Increased density of the platelet serotonin transporter in autism. Pharmacopsychiatry 33, 165–168.

Mark, M., Lufkin, T., Vonesch, J.L., et al., 1993. Two rhombomeres are altered in *Hoxa-1* mutant mice. Development 119, 319–338.

Markowitz, P.I., 1983. Autism in a child with congenital cytomegalovirus infection. Journal of Autism and Developmental Disorders 13, 249–253.

Markram, K., Rinaldi, T., La Mendola, D., Sandi, C., Markram, H., 2008. Abnormal fear conditioning and amygdala processing in an animal model of autism. Neuropsychopharmacology 33, 901–912.

Marshall, C.R., Noor, A., Vincent, J.B., et al., 2008. Structural variation of chromosomes in autism spectrum disorder. American Journal of Human Genetics 82, 477–488.

Martin, E.R., Menold, M.M., Wolpert, C.M., et al., 2000. Analysis of linkage disequilibrium in gamma-aminobutyric acid receptor subunit genes in autistic disorder. American Journal of Medical Genetics 96, 43–48.

Martineau, J., Andersson, F., Barthélémy, C., Cottier, J.P., Destrieux, C., 2010. Atypical activation of the mirror neuron system during perception of hand motion in autism. Brain Research 1320, 168–175.

Martini, F., Iaccheri, L., Lazzarin, L., et al., 1996. SV40 early region and large T antigen in human brain tumors, peripheral blood cells, and sperm fluids from healthy individuals. Cancer Research 56, 4820–4825.

McBride, P.A., Anderson, G.M., Hertzig, M.E., et al., 1998. Effects of diagnosis, race, and puberty on platelet serotonin levels in autism and mental retardation. Journal of the American Academy of Child and Adolescent Psychiatry 37, 767–776.

McBride, K.L., Varga, E.A., Pastore, M.T., et al., 2010. Confirmation study of PTEN mutations among individuals with autism or developmental delays/mental retardation and macrocephaly. Autism Research 3, 137–141.

McCauley, J.L., Olson, L.M., Delahanty, R., et al., 2004. A linkage disequilibrium map of the 1-Mb 15q12 GABA(A) receptor subunit cluster and association to autism. American Journal of Medical Genetics Part B Neuropsychiatric Genetics 131B, 51–59.

McGraw, C.M., Samaco, R.C., Zoghbi, H.Y., 2011. Adult neural function requires MeCP2. Science 333, 186.

Meda, L., Baron, P., Prat, E., et al., 1999. Proinflammatory profile of cytokine production by human monocytes and murine microglia stimulated with beta-amyloid. Journal of Neuroimmunology 93, 45–52.

Mefford, H.C., Muhle, H., Ostertag, P., et al., 2010. Genome-wide copy number variation in epilepsy: Novel susceptibility loci in idiopathic generalized and focal epilepsies. PLoS Genetics 6, e1000962.

Mei, H., Cuccaro, M.L., Martin, E.R., 2007. Multifactor dimensionality reduction-phenomics: A novel method to capture genetic heterogeneity with use of phenotypic variables. American Journal of Human Genetics 81, 1251–1261.

Melke, J., Goubran Botros, H., Chaste, P., Betancur, C., et al., 2008. Abnormal melatonin synthesis in autism spectrum disorders. Molecular Psychiatry 13, 90–98.

Menold, M.M., Shao, Y., Wolpert, C.M., et al., 2001. Association analysis of chromosome 15 gabaa receptor subunit genes in autistic disorder. Journal of Neurogenetics 15, 245–259.

Merikangas, A.K., Corvin, A.P., Gallagher, L., 2009. Copy-number variants in neurodevelopmental disorders: Promises and challenges. Trends in Genetics 25, 536–544.

Miles, J.H., Hadden, L.L., Takahashi, T.N., Hillman, R.E., 2000. Head circumference is an independent clinical finding associated with autism. American Journal of Medical Genetics Part B Neuropsychiatric Genetics 95, 339–350.

Miles, J.H., Takahashi, T.N., Hong, J., et al., 2008. Development and validation of a measure of dysmorphology: Useful for autism subgroup classification. American Journal of Medical Genetics Part A 146A, 1101–1116.

Millen, K.J., Wurst, W., Herrup, K., Joyner, A.L., 1994. Abnormal embryonic development and patterning of postnatal foliation in two mouse Engrailed-2 mutants. Development 120, 695–706.

Millen, K.J., Hui, C.C., Joyner, A.L., 1995. A role for En-2 and other murine homologues of *Drosophila* segment polarity genes in regulating positional information in the developing cerebellum. Development 121, 3935–3945.

Miller, M.T., Strömland, K., Ventura, L., Johansson, M., Bandim, J.M., Gillberg, C., 2004. Autism with ophthalmologic malformations: The plot thickens. Transactions of the American Ophthalmological Society 102, 107–120.

Miller, M.T., Strömland, K., Ventura, L., Johansson, M., Bandim, J.M., Gillberg, C., 2005. Autism associated with conditions characterized by developmental errors in early embryogenesis: A mini review. International Journal of Developmental Neuroscience 23, 201–219.

Miller, M.T., Ventura, L., Strömland, K., 2009. Thalidomide and misoprostol: Ophthalmologic manifestations and associations both expected and unexpected. Birth Defects Research. Part A Clinical and Molecular Teratology 85, 667–676.

Miller, D.T., Adam, M.P., Aradhya, S., et al., 2010. Consensus statement: Chromosomal microarray is a first-tier clinical diagnostic test for individuals with developmental disabilities or congenital anomalies. American Journal of Human Genetics 86, 749–764.

Minkowski, E., 1927. La schizophrénie. Psychopathologie des schizoïdes et des schizophrènes, 1st edn. Payot, Paris.

Missler, M., Zhang, W., Rohlmann, A., et al., 2003. Alpha-neurexins couple Ca^{2+} channels to synaptic vesicle exocytosis. Nature 423, 939–948.

Miyazaki, K., Narita, N., Narita, M., 2005. Maternal administration of thalidomide or valproic acid causes abnormal serotonergic neurons in the offspring: Implication for pathogenesis of autism. International Journal of Developmental Neuroscience 23, 287–297.

Modahl, C., Green, L., Fein, D., et al., 1998. Plasma oxytocin levels in autistic children. Biological Psychiatry 43, 270–277.

Moessner, R., Marshall, C.R., Sutcliffe, J.S., et al., 2007. Contribution of SHANK3 mutations to autism spectrum disorder. American Journal of Human Genetics 81, 1289–1297.

Moore, S.J., Turnpenny, P., Quinn, A., et al., 2000. A clinical study of 57 children with fetal anticonvulsant syndromes. Journal of Medical Genetics 37, 489–497.

Muhle, R., Trentacoste, S.V., Rapin, I., 2004. The genetics of autism. Pediatrics 113, e472–e486.

Mulder, E.J., Anderson, G.M., Kema, I.P., et al., 2004. Platelet serotonin levels in pervasive developmental disorders and mental retardation: Diagnostic group differences, within-group distribution, and behavioral correlates. Journal of the American Academy of Child and Adolescent Psychiatry 43, 491–499.

Munesue, T., Yokoyama, S., Nakamura, K., et al., 2010. Two genetic variants of CD38 in subjects with autism spectrum disorder and controls. Neuroscience Research 67, 181–191.

Muscarella, L.A., Guarnieri, V., Sacco, R., et al., 2007. HOXA1 gene variants influence head growth rates in humans. American Journal of Medical Genetics Part B Neuropsychiatric Genetics 144B, 388–390.

Muscarella, L.A., Guarnieri, V., Sacco, R., et al., 2010. Candidate gene study of HOXB1 in autism spectrum disorder. Molecular Autism 1, 9.

Nakasato, A., Nakatani, Y., Seki, Y., Tsujino, N., Umino, M., Arita, H., 2008. Swim stress exaggerates the hyperactive mesocortical dopamine system in a rodent model of autism. Brain Research 1193, 128–135.

Napolioni, V., Lombardi, F., Sacco, R., et al., 2011. Family-based association study of ITGB3 in autism spectrum disorder and its endophenotypes. European Journal of Human Genetics 19, 353–359.

Narita, N., Kato, M., Tazoe, M., Miyazaki, K., Narita, M., Okado, N., 2002. Increased monoamine concentration in the brain and blood of fetal thalidomide- and valproic acid-exposed rat: Putative animal models for autism. Pediatric Research 52, 576–579.

Neves-Pereira, M., Müller, B., Massie, D., et al., 2009. Deregulation of EIF4E: A novel mechanism for autism. Journal of Medical Genetics 46, 759–765.

Nissenkorn, A., Zeharia, A., Lev, D., et al., 2000. Neurologic presentations of mitochondrial disorders. Journal of Child Neurology 15, 44–48.

Nurmi, E.L., Dowd, M., Tadevosyan-Leyfer, O., Haines, J.L., Folstein, S.E., Sutcliffe, J.S., 2003. Exploratory subsetting of autism families based on savant skills improves evidence of genetic linkage to 15q11-q13. The Journal of the American Academy of Child and Adolescent Psychiatry 42, 856–863.

O'Dushlaine, C., Kenny, E., Heron, E., et al., 2010. Molecular pathways involved in neuronal cell adhesion and membrane scaffolding contribute to schizophrenia and bipolar disorder susceptibility. Molecular Psychiatry 16, 286–292.

Okunishi, K., Dohi, M., Nakagome, K., et al., 2005. A novel role of hepatocyte growth factor as an immune regulator through suppressing dendritic cell function. The Journal of Immunology 175, 4745–4753.

Orrico, A., Galli, L., Buoni, S., Orsi, A., Vonella, G., Sorrentino, V., 2009. Novel PTEN mutations in neurodevelopmental disorders and macrocephaly. Clinical Genetics 75, 195–198.

Ozonoff, S., Young, G.S., Carter, A., et al., 2011. Recurrence risk for autism spectrum disorders: A Baby Siblings Research Consortium Study. Pediatrics 128, e488–495.

Page, D.T., Kuti, O.J., Prestia, C., Sur, M., 2009. Haploinsufficiency for Pten and Serotonin transporter cooperatively influences brain size and social behavior. Proceedings of the National Academy of Science of the United States of America 106, 1989–1994.

Palmieri, L., Persico, A.M., 2010. Mitochondrial dysfunction in autism spectrum disorders: Cause or effect? Biochimica et Biophysica Acta 1797, 1130–1137.

Palmieri, L., Papaleo, V., Porcelli, V., et al., 2010. Altered calcium homeostasis in autism-spectrum disorders: Evidence from biochemical and genetic studies of the mitochondrial aspartate/glutamate carrier AGC1. Molecular Psychiatry 15, 38–52.

Pampanos, A., Volaki, K., Kanavakis, E., et al., 2009. A substitution involving the NLGN4 gene associated with autistic behavior in the Greek population. Genetic Testing and Molecular Biomarkers 13, 611–615.

Parnas, J., Bovet, P., Zahavi, D., 2002. Schizophrenic autism: Clinical phenomenology and pathogenic implications. World Psychiatry 1, 131–136.

Parra, S., Alonso-Villaverde, C., Coll, B., et al., 2007. Serum paraoxonase-1 activity and concentration are influenced by human immunodeficiency virus infection. Atherosclerosis 194, 175–181.

Persico, A.M., 2009. Developmental roles of the serotonin transporter. In: Kalueff, A.V. (Ed.), Experimental Models in Serotonin Transporter Research. Cambridge University Press, Cambridge, UK, pp. 78–104.

Persico, A.M., 2010. Polyomaviruses and autism: More than simple association? Journal of Neurovirology 16, 332–333.

Persico, A.M., Bourgeron, T., 2006. Searching for ways out of the autism maze: Genetic, epigenetic and environmental clues. Trends in Neuroscience 29, 349–358.

Persico, A.M., Sacco, R., 2013. Endophenotypes in Autism Spectrum Disorder. In: Patel, V.B., Preedy, V.R., Martin, C. (Eds.) The Comprehensive Guide to Autism. Springer Science + Business Media B.V., Dordrecht, Netherlands, in press.

Persico, A.M., D'Agruma, L., Maiorano, N., et al., 2001. Reelin gene alleles and haplotypes as a factor predisposing to autistic disorder. Molecular Psychiatry 6, 150–159.

Persico, A.M., Pascucci, T., Puglisi-Allegra, S., et al., 2002. Serotonin transporter gene promoter variants do not explain the hyperserotoninemia in autistic children. Molecular Psychiatry 7, 795–800.

Persico, A.M., Levitt, P., Pimenta, A., 2006. Polymorphic GGC repeat differentially regulates human reelin gene expression levels. Journal of Neural Transmission 113, 1373–1382.

Petit, E., Hérault, J., Martineau, J., et al., 1995. Association study with two markers of a human homeogene in infantile autism. Journal of Medical Genetics 32, 269–274.

Petropoulos, H., Friedman, S.D., Shaw, D.W., Artru, A.A., Dawson, G., Dager, S.R., 2006. Gray matter abnormalities in autism spectrum disorder revealed by T2 relaxation. Neurology 67, 632–636.

Phiel, C.J., Zhang, F., Huang, E.Y., Guenther, M.G., Lazar, M.A., Klein, P.S., 2001. Histone deacetylase is a direct target of valproic acid, a potent anticonvulsant, mood stabilizer, and teratogen. Journal of Biological Chemistry 276, 36734–36741.

Pinto, D., Pagnamenta, A.T., Klei, L., et al., 2010. Functional impact of global rare copy number variation in autism spectrum disorders. Nature 466, 368–372.

Piven, J., Palmer, P., Jacobi, D., Childress, D., Arndt, S., 1997. Broader autism phenotype: Evidence from a family history study of multiple-incidence autism families. American Journal of Psychiatry 154, 185–190.

Piven, J., Tsai, G., Nehme, E., Coyle, J.T., Chase, G.A., Folstein, S.E., 1991. Platelet serotonin, a possible marker for familial autism. Journal of Autism and Developmental Disorders 21, 51–59.

Pons, R., Andreu, A.L., Checcarelli, N., et al., 2004. Mitochondrial DNA abnormalities and autistic spectrum disorders. Journal of Pediatrics 144, 81–85.

Poot, M., Beyer, V., Schwaab, I., et al., 2010. Disruption of CNTNAP2 and additional structural genome changes in a boy with speech delay and autism spectrum disorder. Neurogenetics 11, 81–89.

Prasad, H.C., Zhu, C.B., McCauley, J.L., et al., 2005. Human serotonin transporter variants display altered sensitivity to protein kinase G and p38 mitogen-activated protein kinase. Proceedings of the National Academy of Science of the United States of America 102, 11545–11550.

Prasad, H.C., Steiner, J.A., Sutcliffe, J.S., Blakely, R.D., 2009. Enhanced activity of human serotonin transporter variants associated with autism. Philosophical Transactions of the Royal Society B: Biological Sciences 364, 163–173.

Qin, J., Jia, M., Wang, L., et al., 2009. Association study of SHANK3 gene polymorphisms with autism in Chinese Han population. BMC Medical Genetics 10, 61.

Qiu, S., Anderson, C.T., Levitt, P., Shepherd, G.M., 2011. Circuit-specific intracortical hyperconnectivity in mice with deletion of the autism-associated Met receptor tyrosine kinase. The Journal of Neuroscience 31, 5855–5864.

Quattrocchi, C.C., Wannenes, F., Persico, A.M., et al., 2002. Reelin is a serine protease of the extracellular matrix. Journal of Biological Chemistry 277, 303–309.

Radyushkin, K., Hammerschmidt, K., Boretius, S., et al., 2009. Neuroligin-3-deficient mice: Model of a monogenic heritable form of autism with an olfactory deficit. Genes, Brain and Behavior 8, 416–425.

Rajcan-Separovic, E., Harvard, C., Liu, X., et al., 2007. Clinical and molecular cytogenetic characterisation of a newly recognised microdeletion syndrome involving 2p15-16.1. Journal of Medical Genetics 44, 269–276.

Rasalam, A.D., Hailey, H., Williams, J.H., et al., 2005. Characteristics of fetal anticonvulsant syndrome associated autistic disorder. Developmental Medicine & Child Neurology 47, 551–555.

Rauh, V.A., Garfinkel, R., Perera, F.P., et al., 2006. Impact of prenatal chlorpyrifos exposure on neurodevelopment in the first 3 years of life among inner-city children. Pediatrics 118, e1845–e1859.

Reichelt, K.L., Hole, K., Hamberger, A., et al., 1981. Biologically active peptide-containing fractions in schizophrenia and childhood autism. Advances in Biochemical Psychopharmacology 28, 627–643.

Rekvig, O.P., Bendiksen, S., Moens, U., 2006. Immunity and autoimmunity induced by polyomaviruses: Clinical, experimental and theoretical aspects. In: Ahsan, N. (Ed.), Polyomaviruses and Human Diseases. Advances in Experimental Medicine and Biology 577, Eureka.com/Landes Bioscience, Georgetown, TX, pp. 117–147.

Rett, A., 1966. Über ein zerebral-atrophisches Syndrom bei Hyperammonämie. Wiener medizinische Wochenschrift 116, 723–726.

Rice, D., Barone Jr., S., 2000. Critical periods of vulnerability for the developing nervous system: Evidence from humans and animal models. Environmental Health Perspectives 108 (supplement 3), 511–533.

Rice, D.S., Curran, T., 2001. Role of the reelin signaling pathway in central nervous system development. Annual Review of Neuroscience 24, 1005–1039.

Rinaldi, T., Kulangara, K., Antoniello, K., Markram, H., 2007. Elevated NMDA receptor levels and enhanced postsynaptic long-term potentiation induced by prenatal exposure to valproic acid. Proceedings of the National Academy of Science of the United States of America 104, 13501–13506.

Rinaldi, T., Silberberg, G., Markram, H., 2008. Hyperconnectivity of local neocortical microcircuitry induced by prenatal exposure to valproic acid. Cerebral Cortex 18, 763–770.

Riva, D., Giorgi, C., 2000. The cerebellum contributes to higher functions during development: Evidence from a series of children surgically treated for posterior fossa tumours. Brain 123, 1051–1061.

Roberts, E.M., English, P.B., Grether, J.K., Windham, G.C., Somberg, L., Wolff, C., 2007. Maternal residence near agricultural pesticide applications and autism spectrum disorders among children in the California Central Valley. Environmental Health Perspectives 115, 1482–1489.

Rodier, P.M., 2002. Converging evidence for brain stem injury in autism. Developmental Psychopathology 14, 537–557.

Rodier, P.M., Ingram, J.L., Tisdale, B., Nelson, S., Romano, J., 1996. Embryological origin for autism: Developmental anomalies of the cranial nerve motor nuclei. Journal of Comparative Neurology 370, 247–261.

Rogers, S.J., Vismara, L.A., 2008. Evidence-based comprehensive treatments for early autism. Journal of Clinical Child Adolescent Psychology 37, 8–38.

Romano, V., Calì, F., Mirisola, M., et al., 2003. Lack of association of HOXA1 and HOXB1 mutations and autism in Sicilian (Italian) patients. Molecular Psychiatry 8, 716–717.

Rossell, M., Capecchi, M.R., 1999. Mice mutant for both *Hoxa1* and *Hoxb1* show extensive remodeling of the hindbrain and defects in craniofacial development. Development 126, 5027–5040.

Rossi, E., Verri, A.P., Patricelli, M.G., et al., 2008. A 12Mb deletion at 7q33-q35 associated with autism spectrum disorders and primary amenorrhea. European Journal of Medical Genetics 51, 631–638.

Rossignol, D.A., Frye, R.E., 2011. Mitochondrial dysfunction in autism spectrum disorders: A systematic review and meta-analysis. Molecular Psychiatry 69, 41R–47R.

Rout, U.K., Clausen, P., 2009. Common increase of GATA-3 level in PC-12 cells by three teratogens causing autism spectrum disorders. Neuroscience Research 64, 162–169.

Roux, S., Bruneau, N., Garreau, B., et al., 1997. Bioclinical profiles of autism and other developmental disorders using a multivariate statistical approach. Biological Psychiatry 42, 1148–1156.

Rubenstein, J.L., Merzenich, M.M., 2003. Model of autism: Increased ratio of excitation/inhibition in key neural systems. Genes, Brain and Behavior 2, 255–267.

Rujescu, D., Ingason, A., Cichon, S., et al., 2009. Disruption of the neurexin 1 gene is associated with schizophrenia. Human Molecular Genetics 18, 988–996.

Russo, A.J., Krigsman, A., Jepson, B., Wakefield, A., 2009. Decreased serum hepatocyte growth factor (HGF) in autistic children with severe gastrointestinal disease. Biomarker Insights 2, 181–190.

Rutter, M., 2005. Incidence of autism spectrum disorders: Changes over time and their meaning. Acta Paediatrica 94, 2–15.

Rutter, M., Kreppner, J., Croft, C., et al., 2007. Early adolescent outcomes of institutionally deprived and non-deprived adoptees III. Quasi-autism. Journal of Child Psychology and Psychiatry 48, 1200–1207.

Sacco, R., Militerni, R., Frolli, A., et al., 2007a. Clinical, morphological, and biochemical correlates of head circumference in autism. Biological Psychiatry 62, 1038–1047.

Sacco, R., Papaleo, V., Hager, J., et al., 2007b. Case–control and family-based association studies of candidate genes in autistic disorder and its endophenotypes: TPH2 and GLO1. BMC Medical Genetics 8, 11.

Sacco, R., Curatolo, P., Manzi, B., et al., 2011. Principal pathogenetic components and biological endophenotypes in autism spectrum disorders. Autism Research 3, 237–252.

Sacco, R., Persico, A.M., Garbett, K.A., Mirnics, K., 2011. Genome-wide expression studies in autism-spectrum disorders: Moving from neurodevelopment to neuroimmunology. In: Clelland, J.D. (Ed.), Advances in Neurobiology 2 – Genomics, Proteomics, and the Nervous System. Springer, New York, pp. 469–487.

Sakurai, T., Ramoz, N., Reichert, J.G., et al., 2006. Association analysis of the NrCAM gene in autism and in subsets of families with severe obsessive-compulsive or self-stimulatory behaviors. Psychiatric Genetics 16, 251–257.

Salmon, B., Hallmayer, J., Rogers, T., et al., 1999. Absence of linkage and linkage disequilibrium to chromosome 15q11-q13 markers in 139 multiplex families with autism. American Journal of Medical Genetics 88, 551–556.

Samaco, R.C., Hogart, A., LaSalle, J.M., 2005. Epigenetic overlap in autism-spectrum neurodevelopmental disorders: MECP2 deficiency causes reduced expression of UBE3A and GABRB3. Human Molecular Genetics 14, 483–492.

Sanders, S.J., Ercan-Sencicek, A.G., Hus, V., et al., 2011. Multiple recurrent de novo CNVs, including duplications of the 7q11.23 Williams syndrome region, are strongly associated with autism. Neuron 70, 863–885.

Saresella, M., Marventano, I., Guerini, F.R., et al., 2009. An autistic endophenotype results in complex immune dysfunction in healthy siblings of autistic children. Biological Psychiatry 66, 978–984.

Schaaf, C.P., Zoghbi, H.Y., 2011. Solving the autism puzzle a few pieces at a time. Neuron 70, 806–808.

Schaaf, C.P., Sabo, A., Sakai, Y., et al., 2011. Oligogenic heterozygosity in individuals with high-functioning autism spectrum disorders. Human Molecular Genetics 20, 3366–3375.

Schaefer, G.B., Mendelson, N.J., 2008. Genetic evaluation for the etiologic diagnosis of autism spectrum disorders. Genetics in Medicine 10, 4–12.

Schanen, N.C., 2006. Epigenetics of autism spectrum disorders. Human Molecular Genetics 15 (Spec No 2), R138–R150.

Scheiffele, P., Fan, J., Choih, J., Fetter, R., Serafini, T., 2000. Neuroligin expressed in nonneuronal cells triggers presynaptic development in contacting axons. Cell 101, 657–669.

Schneider, T., Przewłocki, R., 2005. Behavioral alterations in rats prenatally exposed to valproic acid: Animal model of autism. Neuropsychopharmacology 30, 80–89.

Schneider, T., Labuz, D., Przewłocki, R., 2001. Nociceptive changes in rats after prenatal exposure to valproic acid. Polish Journal of Pharmacology 53, 531–534.

Schneider, T., Turczak, J., Przewłocki, R., 2006. Environmental enrichment reverses behavioral alterations in rats prenatally exposed to valproic acid: Issues for a therapeutic approach in autism. Neuropsychopharmacology 31, 36–46.

Schneider, T., Ziòłkowska, B., Gieryk, A., Tyminska, A., Przewłocki, R., 2007. Prenatal exposure to valproic acid disturbs the enkephalinergic system functioning, basal hedonic tone, and emotional responses in an animal model of autism. Psychopharmacology (Berl) 193, 547–555.

Schroer, R.J., Phelan, M.C., Michaelis, R.C., et al., 1998. Autism and maternally derived aberrations of chromosome 15q. American Journal of Medical Genetics 76, 327–336.

Scott-Van Zeeland, A.A., Abrahams, B.S., Alvarez-Retuerto, A.I., et al., 2010. Altered functional connectivity in frontal lobe circuits is associated with variation in the autism risk gene CNTNAP2. Science Translational Medicine 2, 56ra80.

Sebat, J., Lakshmi, B., Malhotra, D., et al., 2007. Strong association of de novo copy number mutations with autism. Science 316, 445–449.

Sen, B., Sinha, S., Ahmed, S., Ghosh, S., Gangopadhyay, P.K., Usha, R., 2007. Lack of association of HOXA1 and HOXB1 variants with autism in the Indian population. Psychiatric Genetics 17, 1.

Sen, B., Singh, A.S., Sinha, S., et al., 2010. Family-based studies indicate association of Engrailed 2 gene with autism in an Indian population. Genes Brain & Behavior 9, 248–255.

Serajee, F.J., Zhong, H., Huq, A.H., 2006. Association of reelin gene polymorphisms with autism. Genomics 87, 75–83.

Shadel, G., 2008. Expression and maintenance of mitochondrial DNA: New insights into human disease pathology. American Journal of Pathology 172, 1445–1456.

Shao, Y., Cuccaro, M.L., Hauser, E.R., et al., 2003. Fine mapping of autistic disorder to chromosome 15q11-q13 by use of phenotypic subtypes. American Journal of Human Genetics 72, 539–548.

Shibayama, A., Cook Jr., E.H., Feng, J., et al., 2004. MECP2 structural and 3′-UTR variants in schizophrenia, autism and other psychiatric diseases: A possible association with autism. American Journal of Medical Genetics Part B Neuropsychiatric Genetics 128B, 50–53.

Shifman, S., Johannesson, M., Bronstein, M., et al., 2008. Genome-wide association identifies a common variant in the reelin gene that increases the risk of schizophrenia only in women. PLoS Genetics 4, e28.

Shoffner, J., Hyams, L., Langley, G.N., et al., 2010. Fever plus mitochondrial disease could be risk factors for autistic regression. Journal of Child Neurology 25, 429–434.

Skaar, D.A., Shao, Y., Haines, J.L., et al., 2005. Analysis of the RELN gene as a genetic risk factor for autism. Molecular Psychiatry 10, 563–571.

Smith, M., Spence, M.A., Flodman, P., 2009. Nuclear and mitochondrial genome defects in autisms. Annals of the New York Academy of Science 1151, 102–132.

Snow, W.M., Hartle, K., Ivanco, T.L., 2008. Altered morphology of motor cortex neurons in the VPA rat model of autism. Developmental Psychobiology 50, 633–639.

Sousa, I., Clark, T.G., Toma, C., et al., 2009. MET and autism susceptibility: Family and case–control studies. European Journal of Human Genetics 17, 749–758.

Spence, S.J., Cantor, R.M., Chung, L., Kim, S., Geschwind, D.H., Alarcón, M., 2006. Stratification based on language related endophenotypes in autism: Attempt to replicate reported linkage.

American Journal of Medical Genetics, Part B Neuropsychiatric Genetics 141B, 591–598.

Splawski, I., Timothy, K.W., Sharpe, L.M., et al., 2004. Ca$_V$1.2 calcium channel dysfunction causes a multisystem disorder including arrhythmia and autism. Cell 119, 19–31.

Splawski, I., Yoo, D.S., Stotz, S.C., Cherry, A., Clapham, D.E., Keating, M.T., 2006. CACNA1H mutations in autism spectrum disorders. The Journal of Biological Chemistry 281, 22085–22091.

Stevenson, R.E., Schroer, R.J., Skinner, C., Fender, D., Simensen, R.J., 1997. Autism and macrocephaly. Lancet 349, 1744–1745.

Stodgell, C.J., Ingram, J.L., O'Bara, M., Tisdale, B.K., Nau, H., Rodier, P.M., 2006. Induction of the homeotic gene Hoxa1 through valproic acid's teratogenic mechanism of action. Neurotoxicology and Teratology 28, 617–624.

Stone, J.L., Merriman, B., Cantor, R.M., et al., 2004. Evidence for sex-specific risk alleles in autism spectrum disorder. American Journal of Human Genetics 75, 1117–1123.

Strömland, K., Nordin, V., Miller, M., Akerström, B., Gillberg, C., 1994. Autism in thalidomide embryopathy: A population study. Developmental Medicine and Child Neurology 36, 351–356.

Stubbs, E.G., 1978. Autistic symptoms in a child with congenital cytomegalovirus infection. Journal of Autism and Childhood Schizophrenia 8, 37–43.

Stubbs, E.G., Ash, E., Williams, C.P., 1984. Autism and congenital cytomegalovirus. Journal of Autism and Developmental Disorders 14, 183–189.

Südhof, T.C., 2008. Neuroligins and neurexins link synaptic function to cognitive disease. Nature 455, 903–911.

Sugihara, G., Hashimoto, K., Iwata, Y., et al., 2007. Decreased serum levels of hepatocyte growth factor in male adults with high-functioning autism. Progress in Neuro-Psychopharmacology and Biological Psychiatry 31, 412–415.

Sun, L., Huang, L., Nguyen, P., et al., 2008. DNA methyltransferase 1 and 3B activate BAG-1 expression via recruitment of CTCFL/BORIS and modulation of promoter histone methylation. Cancer Research 68, 2726–2735.

Sutcliffe, J.S., Delahanty, R.J., Prasad, H.C., et al., 2005. Allelic heterogeneity at the serotonin transporter locus (SLC6A4) confers susceptibility to autism and rigid-compulsive behaviors. American Journal of Human Genetics 77, 265–279.

Sweeten, T.L., Posey, D.J., McDougle, C.J., 2004. Brief report: Autistic disorder in three children with cytomegalovirus infection. Journal of Autism and Developmental Disorders 34, 583–586.

Sykes, N.H., Toma, C., Wilson, N., et al., 2009. Copy number variation and association analysis of SHANK3 as a candidate gene for autism in the IMGSAC collection. European Journal of Human Genetics 17, 1347–1353.

Tabuchi, K., Blundell, J., Etherton, M.R., et al., 2007. A neuroligin-3 mutation implicated in autism increases inhibitory synaptic transmission in mice. Science 318, 71–76.

Takayanagi, Y., Yoshida, M., Bielsky, I.F., et al., 2005. Pervasive social deficits, but normal parturition, in oxytocin receptor-deficient mice. Proceedings of the National Academy of Science of the United States of America 102, 16096–16101.

Talebizadeh, Z., Bittel, D.C., Miles, J.H., et al., 2002. No association between HOXA1 and HOXB1 genes and autism spectrum disorders (ASD). Journal of Medical Genetics 39, e70.

Tan, G.C., Doke, T.F., Ashburner, J., Wood, N.W., Frackowiak, R.S., 2010. Normal variation in fronto-occipital circuitry and cerebellar structure with an autism-associated polymorphism of CNTNAP2. Neuroimage 53, 1030–1042.

Tansey, K.E., Brookes, K.J., Hill, M.J., et al., 2010. Oxytocin receptor (OXTR) does not play a major role in the aetiology of autism: Genetic and molecular studies. Neuroscience Letters 474, 163–167.

Teitelbaum, O., Benton, T., Shah, P.K., Prince, A., Kelly, J.L., Teitelbaum, P., 2004. Eshkol-Wachman movement notation in diagnosis: The early detection of Asperger's syndrome. Proceedings of the National Academy of Science of the United States of America 101, 11909–11914.

Thompson, E.E., Pan, L., Ostrovnaya, I., et al., 2007. Integrin beta 3 genotype influences asthma and allergy phenotypes in the first 6 years of life. Journal of Allergy and Clinical Immunology 119, 1423–1429.

Tischfield, M.A., Bosley, T.M., Salih, M.A., et al., 2005. Homozygous HOXA1 mutations disrupt human brainstem, inner ear, cardiovascular and cognitive development. Nature Genetics 37, 1035–1037.

Tochigi, M., Kato, C., Koishi, S., et al., 2007. No evidence for significant association between GABA receptor genes in chromosome 15q11-q13 and autism in a Japanese population. Journal of Human Genetics 52, 985–989.

Tost, H., Lipska, B.K., Vakkalanka, R., et al., 2010. No effect of a common allelic variant in the reelin gene on intermediate phenotype measures of brain structure, brain function, and gene expression. Biological Psychiatry 68, 105–107.

Tripathi, P.P., Sgadò, P., Scali, M., et al., 2009. Increased susceptibility to kainic acid-induced seizures in Engrailed-2 knockout mice. Neuroscience 159, 842–849.

Tripi, G., Roux, S., Canziani, T., Bonnet Brilhault, F., Barthélémy, C., Canziani, F., 2008. Minor physical anomalies in children with autism spectrum disorder. Early Human Development 84, 217–223.

Tropea, D., Giacometti, E., Wilson, N.R., et al., 2009. Partial reversal of Rett syndrome-like symptoms in MeCP2 mutant mice. Proceedings of the National Academy of Science of the United States of America 106, 2029–2034.

Tsujino, N., Nakatani, Y., Seki, Y., et al., 2007. Abnormality of circadian rhythm accompanied by an increase in frontal cortex serotonin in animal model of autism. Neuroscience Research 57, 289–295.

Tuchman, R., Rapin, I., 2002. Epilepsy in autism. Lancet Neurology 1, 352–358.

Turner, L.M., Stone, W.L., 2007. Variability in outcome for children with an ASD diagnosis at age 2. Journal of Child Psychology and Psychiatry 48, 793–802.

van Daalen, E., Swinkels, S.H., Dietz, C., van Engeland, H., Buitelaar, J.K., 2007. Body length and head growth in the first year of life in autism. Pediatric Neurology 37, 324–330.

van Daalen, E., Kemner, C., Dietz, C., Swinkels, S.H., Buitelaar, J.K., van Engeland, H., 2009. Inter-rater reliability and stability of diagnoses of autism spectrum disorder in children identified through screening at a very young age. European Child & Adolescent Psychiatry 18, 663–674.

van den Pol, A.N., 2006. Viral infections in the developing and mature brain. Trends in Neurosciences 29, 398–406.

Van Lenten, B.J., Wagner, A.C., Nayak, D.P., Hama, S., Navab, M., Fogelman, A.M., 2001. High-density lipoprotein loses its anti-inflammatory properties during acute influenza a infection. Circulation 103, 2283–2288.

Van Lenten, B.J., Wagner, A.C., Anantharamaiah, G.M., et al., 2002. Influenza infection promotes macrophage traffic into arteries of mice that is prevented by D-4F, an apolipoprotein A-I mimetic peptide. Circulation 106, 1127–1132.

Vancassel, S., Durand, G., Barthélémy, C., et al., 2001. Plasma fatty acid levels in autistic children. Prostaglandins, Leukotrienes and Essential Fatty Acids 65, 1–7.

Varga, E.A., Pastore, M., Prior, T., Herman, G.E., McBride, K.L., 2009. The prevalence of PTEN mutations in a clinical pediatric cohort with autism spectrum disorders, developmental delay, and macrocephaly. Genetics in Medicine 11, 111–117.

Vargas, D.L., Nascimbene, C., Krishnan, C., Zimmerman, A.W., Pardo, C.A., 2005. Neuroglial activation and neuroinflammation in the brain of patients with autism. Annals of Neurology 57, 67–81.

Varoqueaux, F., Aramuni, G., Rawson, R.L., et al., 2006. Neuroligins determine synapse maturation and function. Neuron 51, 741–754.

Veenstra-Vanderweele, J., Jessen, T.N., Thompson, B.J., et al., 2009. Modeling rare gene variation to gain insight into the oldest biomarker in autism: Construction of the serotonin transporter Gly56Ala knock-in mouse. Journal of Neurodevelopmental Disorders 1, 158–171.

Vernes, S.C., Newbury, D.F., Abrahams, B.S., et al., 2008. A functional genetic link between distinct developmental language disorders. The New England Journal of Medicine 359, 2337–2345.

Vorstman, J.A., van Daalen, E., Jalali, G.R., et al., 2010. A double hit implicates DIAPH3 as an autism risk gene. Molecular Psychiatry 16, 442–451.

Vourc'h, P., Bienvenu, T., Beldjord, C., et al., 2001. No mutations in the coding region of the Rett syndrome gene MECP2 in 59 autistic patients. European Journal of Human Genetics 9, 556–558.

Wallace, G.L., Happé, F., Giedd, J.N., 2009. A case study of a multiply talented savant with an autism spectrum disorder: Neuropsychological functioning and brain morphometry. Philosophical Transactions of the Royal Society B: Biological Sciences 364, 1425–1432.

Wang, L., Jia, M., Yue, W., et al., 2008. Association of the ENGRAILED 2 (EN2) gene with autism in Chinese Han population. American Journal of Medical Genetics Part B Neuropsychiatric Genetics 147B, 434–438.

Wassink, T.H., Piven, J., Vieland, V.J., et al., 2004. Examination of AVPR1a as an autism susceptibility gene. Molecular Psychiatry 9, 968–972.

Weiss, L.A., 2009. Autism genetics: Emerging data from genome-wide copy-number and single nucleotide polymorphism scans. Expert Review of Molecular Diagnostics 9, 795–803.

Weiss, L.A., Escayg, A., Kearney, J.A., et al., 2003. Sodium channels SCN1A, SCN2A and SCN3A in familial autism. Molecular Psychiatry 8, 186–194.

Weiss, L.A., Veenstra-Vanderweele, J., Newman, D.L., et al., 2004. Genome-wide association study identifies ITGB3 as a QTL for whole blood serotonin. European Journal of Human Genetics 12, 949–954.

Weiss, L.A., Abney, M., Cook Jr., E.H., Ober, C., 2005a. Sex-specific genetic architecture of whole blood serotonin levels. American Journal of Human Genetics 76, 33–41.

Weiss, L.A., Abney, M., Parry, R., Scanu, A.M., Cook Jr., E.H., Ober, C., 2005b. Variation in ITGB3 has sex-specific associations with plasma lipoprotein(a) and whole blood serotonin levels in a population-based sample. Human Genetics 117, 81–87.

Weiss, L.A., Lester, L.A., Lester, L.A., et al., 2005c. Variation in ITGB3 is associated with asthma and sensitization to mold allergen in four populations. American Journal of Respiratory and Critical Care Medicine 172, 67–73.

Weiss, L.A., Ober, C., Cook Jr., E.H., 2006. ITGB3 shows genetic and expression interaction with SLC6A4. Human Genetics 120, 93–100.

Weiss, L.A., Kosova, G., Delahanty, R.J., et al., 2006a. Variation in ITGB3 is associated with whole-blood serotonin level and autism susceptibility. European Journal Human Genetics 14, 923–931.

Weiss, L.A., Shen, Y., Korn, J.M., et al., 2008. Association between microdeletion and microduplication at 16p11.2 and autism. New England Journal of Medicine 358, 667–675.

Weissman, J.R., Kelley, R.I., Bauman, M.L., et al., 2008. Mitochondrial disease in autism spectrum disorder patients: A cohort analysis. PLoS One 3, e3815.

Wermter, A.K., Kamp-Becker, I., Hesse, P., Schulte-Körne, G., Strauch, K., Remschmidt, H., 2010. Evidence for the involvement of genetic variation in the oxytocin receptor gene (OXTR) in the etiology of autistic disorders on high-functioning level. American Journal of Medical Genetics Part B Neuropsychiatric Genetics 153B, 629–639.

Whyatt, R.M., Barr, D.B., 2001. Measurement of organophosphate metabolites in postpartum meconium as a potential biomarker of prenatal exposure: A validation study. Environmental Health Perspectives 109, 417–420.

Whyatt, R.M., Barr, D.B., Camann, D.E., et al., 2003. Contemporary-use pesticides in personal air samples during pregnancy and blood samples at delivery among urban minority mothers and newborns. Environmental Health Perspectives 111, 749–756.

Williams, P.G., Hersh, J.H., 1997. A male with fetal valproate syndrome and autism. Developmental Medicine & Child Neurology 39, 632–634.

Williams, G., King, J., Cunningham, M., Stephan, M., Kerr, B., Hersh, J.H., 2001. Fetal valproate syndrome and autism: Additional evidence of an association. Developmental Medicine & Child Neurology 43, 202–206.

Wiltse, J., 2005. Mode of action: Inhibition of histone deacetylase, altering WNT-dependent gene expression, and regulation of beta-catenin—developmental effects of valproic acid. Critical Reviews in Toxicology 35, 727–738.

Windham, G.C., Zhang, L., Gunier, R., Croen, L.A., Grether, J.K., 2006. Autism spectrum disorders in relation to distribution of hazardous air pollutants in the san francisco bay area. Environmental Health Perspectives 114, 1438–1444.

Wing, L., 1981. Asperger's Syndrome: A clinical account. Psychological Medicine 11, 115–130.

Wolf, U., 1997. Identical mutations and phenotypic variation. Human Genetics 100, 305–321.

Woodhouse, W., Bailey, A., Rutter, M., Bolton, P., Baird, G., Le Couteur, A., 1996. Head circumference in autism and other pervasive developmental disorders. Journal of Child Psychology and Psychiatry 37, 665–671.

Wu, S., Jia, M., Ruan, Y., et al., 2005. Positive association of the oxytocin receptor gene (OXTR) with autism in the Chinese Han population. Biological Psychiatry 58, 74–77.

Xi, C.Y., Ma, H.W., Lu, Y., et al., 2007. MeCP2 gene mutation analysis in autistic boys with developmental regression. Psychiatric Genetics 17, 113–116.

Yamashita, Y., Fujimoto, C., Nakajima, E., Isagai, T., Matsuishi, T., 2003. Possible association between congenital cytomegalovirus infection and autistic disorder. Journal of Autism and Developmental Disorders 33, 455–459.

Yan, J., Oliveira, G., Coutinho, A., et al., 2005. Analysis of the neuroligin 3 and 4 genes in autism and other neuropsychiatric patients. Molecular Psychiatry 10, 329–332.

Yan, J., Feng, J., Schroer, R., et al., 2008a. Analysis of the neuroligin 4Y gene in patients with autism. Psychiatric Genetics 18, 204–207.

Yan, J., Noltner, K., Feng, J., et al., 2008b. Neurexin 1alpha structural variants associated with autism. Neuroscience Letters 438, 368–370.

Yang, P., Lung, F.W., Jong, Y.J., Hsieh, H.Y., Liang, C.L., Juo, S.H., 2008. Association of the homeobox transcription factor gene ENGRAILED 2 with autistic disorder in Chinese children. Neuropsychobiology 57, 3–8.

Yang, P., Shu, B.C., Hallmayer, J.F., Lung, F.W., 2010. Intronic single nucleotide polymorphisms of engrailed homeobox 2 modulate the disease vulnerability of autism in a Han Chinese population. Neuropsychobiology 62, 104–115.

Yirmiya, N., Rosenberg, C., Levi, S., et al., 2006. Association between arginine vasopressin receptor 1A (AVPR1a) gene and autism in a family-based study: Mediation by social skills. Molecular Psychiatry 11, 488–494.

Yoo, H.K., Chung, S., Hong, J.P., Kim, B.N., Cho, S.C., 2009. Microsatellite marker in gamma-aminobutyric acid-a receptor beta 3 subunit gene and autism spectrum disorders in Korean trios. Yonsei Medical Journal 50, 304–306.

Young, D.J., Bebbington, A., Anderson, A., et al., 2008. The diagnosis of autism in a female: Could it be Rett syndrome? European Journal of Pediatrics 167, 661–669.

Yrigollen, C.M., Han, S.S., Kochetkova, A., et al., 2008. Genes controlling affiliative behavior as candidate genes for autism. Biological Psychiatry 63, 911–916.

Zafeiriou, D.I., Ververi, A., Vargiami, E., 2007. Childhood autism and associated comorbidities. Brain and Development 29, 257–272.

Zappella, M., Meloni, I., Longo, I., et al., 2003. Study of MECP2 gene in Rett syndrome variants and autistic girls. American Journal of Medical Genetics Part B Neuropsychiatric Genetics 119B, 102–107.

Zeng, X., Sun, M., Liu, L., Chen, F., Wei, L., Xie, W., 2007. Neurexin-1 is required for synapse formation and larvae associative learning in *Drosophila*. FEBS Letters 581, 2509–2516.

Zeviani, M., Di Donato, S., 2004. Mitochondrial disorders. Brain 127, 2153–2172.

Zhang, J., Smith, K.R., 2003. Indoor air pollution: A global health concern. British Medical Bulletin 68, 209–225.

Zhang, H., Liu, X., Zhang, C., et al., 2002. Reelin gene alleles and susceptibility to autism spectrum disorders. Molecular Psychiatry 7, 1012–1017.

Zhang, W., Rohlmann, A., Sargsyan, V., et al., 2005. Extracellular domains of alpha-neurexins participate in regulating synaptic transmission by selectively affecting N- and P/Q-type Ca^{2+} channels. The Journal of Neuroscience 25, 4330–4342.

Zhong, H., Serajee, F.J., Nabi, R., Huq, A.H., 2003. No association between the EN2 gene and autistic disorder. Journal of Medical Genetics 40, e4.

Zhou, J., Blundell, J., Ogawa, S., et al., 2009. Pharmacological inhibition of mTORC1 suppresses anatomical, cellular, and behavioral abnormalities in neural-specific Pten knock-out mice. The Journal of Neuroscience 29, 1773–1783.

Zhu, C.B., Blakely, R.D., Hewlett, W.A., 2006. The proinflammatory cytokines interleukin-1beta and tumor necrosis factor-alpha activate serotonin transporters. Neuropsychopharmacology 31, 2121–2131.

Zweier, C., de Jong, E.K., Zweier, M., et al., 2009. CNTNAP2 and NRXN1 are mutated in autosomal-recessive Pitt-Hopkins-like mental retardation and determine the level of a common synaptic protein in *Drosophila*. American Journal of Human Genetics 85, 655–666.

Neurodevelopmental Genomics of Autism, Schizophrenia, and Related Disorders

J.F. Cubells, D. Moreno-De-Luca

Emory University School of Medicine, Atlanta, GA, USA

35.1 INTRODUCTION AND HISTORICAL OVERVIEW

In 1943, Leo Kanner first described 'infantile autism' as "... children's *inability to relate themselves* in the ordinary way to people and situations from the beginning of life" (page 242, emphasis in original; Kanner, 1943). The modern Diagnostic and Statistical Manual of Mental Disorders, 4th edition text revision (DSM-IV TR; American Psychiatric Association, 2000), categorizes that classical syndrome as autistic disorder, which is thought of as a part of a heterogeneous spectrum of behaviorally defined disorders, the autism spectrum disorders (ASDs). By definition, the ASDs manifest clinically within the first 3 years of life; impair language, communication, and social interaction; and are typified by idiosyncratic interests and repetitive behaviors. In the DSM-IV TR, ASDs also include Asperger disorder, defined as the presence of autism symptoms in the absence of language

delay and intellectual disability (ID), and pervasive developmental disorder not otherwise specified, in which autistic features are present but do not meet full criteria for a diagnosis of autistic disorder (see Chapter 34).

Kanner explicitly borrowed his diagnostic term from Eugen Bleuler (1950), who used 'autism' to describe the process of withdrawal into one's own world, which he identified as a hallmark of a heterogeneous set of adult behavioral disorders in his classic book, *Dementia Praecox or the Group of Schizophrenias*. Today, 'schizophrenia' refers to a set of behaviorally defined syndromes, the diagnostic features of which include 'positive symptoms' such as hallucinations, delusions (firmly fixed false beliefs not explained by cultural context), thought disorder, and disorganized or bizarre behavior and 'negative symptoms' such as apathy, anhedonia (inability to feel pleasure), and social withdrawal. Schizophrenia is one of several 'schizophrenia spectrum disorders' (SSDs) that include schizoaffective disorder, in which

© 2013 Elsevier Inc. All rights reserved.

cycles of depression and/or mania occur together with chronic psychosis, and psychotic disorder not otherwise specified, which includes individuals with psychosis who do not meet full criteria for schizophrenia. Family studies also suggest that a nonpsychotic syndrome, called schizotypal personality disorder, is part of the schizophrenia spectrum (Kety et al., 1971, 1994; Tienari et al., 2003). To avoid the misconception that either autism or schizophrenia is a unitary disorder, we will use the terms ASD and SSD throughout this chapter.

35.2 PHENOTYPIC SIMILARITIES AND DIFFERENCES BETWEEN ASD AND SSD

ASD and SSD both produce lifelong disability and, as we have seen from the history of their nosology, share as hallmark manifestations, impaired abilities to function in human social groups. ASD and SSD share many other phenotypic characteristics, but there are also some important differences. ASD and SSD patients share higher rates of ID (Bhaumik et al., 2008; Hemmings, 2006; Matson and Shoemaker, 2009; Morgan et al., 2008), epilepsy (Ep; Gaitatzis et al., 2004), and motor disturbances (Welham et al., 2009) compared with the general population. To make a diagnosis of an ASD according to the criteria of the DSM-IV-TR, behavioral signs of the disorder must be evident by the age of 3 years. In contrast, SSDs are usually first diagnosed during late adolescence or early adulthood. However, cases meeting diagnostic criteria for SSD clearly occur during childhood (Rapoport, 2009) and mounting evidence suggests that neurobehavioral impairments often precede the onset of full-blown SSD, a clinical phenomenon known as the schizophrenia prodrome (Thomas and Woods, 2006). Manifestations of the schizophrenia prodrome include unusual beliefs and patterns of thought, learning disabilities, minor neurological abnormalities, social withdrawal, and problems with peer relationships. Such differences are readily detectable in many cases, such as in a classic study in which raters, unaware of current diagnosis, observed childhood home movies of patients' families and were able reliably to identify the child destined to develop SSD years later (Walker and Lewine, 1990). Thus, as with ASD, neurodevelopmental dysfunction is often evident early in the lives of patients with SSD.

ASDs are approximately fourfold more common in males than females (Control, 2009). In contrast, there appears to be little to no difference in the prevalence of SSD in males versus females (Saha et al., 2005). However, good evidence supports the conclusion that SSD in males tends to be earlier in onset with a less favorable course over the lifespan than in females (Angermeyer et al., 1990; Hafner et al., 1993), so there

is some degree of sexual dimorphism in SSD. The range of impairment in psychosocial function across affected individuals is broad for both ASD and SSD, with some patients able to integrate into society and function independently, most exhibiting substantial and lifelong need for help in coping with being part of society, and severely affected individuals exhibiting profound impairment that destroys their abilities to care for themselves or function even minimally in society. From the information just reviewed, it is reasonable to view ASD and SSD as partially distinct sets of neurodevelopmental disorders (NDDs) that share a variety of phenotypic features.

35.3 GENETIC MECHANISMS IN ASD AND SSD

35.3.1 Heritability of Risk for ASD and SSD

ASD and SSD exhibit similar degrees of risk heritability. Family studies show that the recurrence risk for ASD and SSD in siblings of affected individuals is approximately 5–10%, which is substantially greater than the ~1% prevalence of either set of disorders in the general population. Additionally, twin studies show concordance rates in monozygotic twins to be substantially higher than in dizygotic twins for both ASD (up to 90% vs. ~10%) and SSD (up to 80% vs. ~10%), leading to heritability estimates of 0.8–0.9 (Folstein and Rosen-Sheidley, 2001; Losh et al., 2008; Riley and Kendler, 2004).

35.3.2 Overlapping Molecular Genetic Associations in ASD and SSD: Single Genes

In some cases, SSD and ASD appear to share common, or at least substantially overlapping, genetic etiologies. Table 35.1 shows a list of genes in which mono- or biallelic mutations or loss-of-function, single-gene copy number variants (CNVs) have been involved in the etiology of both ASD and SSD. Interestingly, these genes encode synaptic or cell adhesion molecules, which play important roles in neural development and therefore appear to be excellent candidates as key players in the pathophysiology of ASD and SSD (Guilmatre et al., 2009).

At the moment, however, the greatest apparent overlap of genetic factors in ASD and SSD comes from recent discoveries regarding the role of genomic CNV (defined below) in both sets of disorders as well as in other neurodevelopmental conditions including Ep and ID. Although CNVs have thus far been associated with ASD and SSD only in small proportions of patients, the number of CNVs associated with both sets of disorders continues to grow. Even if they account only for a minority

TABLE 35.1 Genes Associated with Both ASD and SSZ Through Loss-of-Function Single-Gene CNVs and/or Functional Mutations

Gene symbol	Chromosomal location	Gene name	OMIM ID	ASD references	SSD references
SHANK3	22q13.3	SH3 and multiple ankyrin repeat domains 3	606230	Durand et al. (2007) and Gauthier et al. (2009)	Gauthier et al. (2010)
CNTNAP2	7q35	Contactin-associated protein-like 2	604569	Strauss et al. (2006) and Zweier et al. (2009)	Friedman et al. (2008)
NRXN1	2p16.3	Neurexin 1	600565	Ching et al. (2010), Wisniowiecka-Kowalnik et al. (2010), and Zweier et al. (2009)	Ikeda et al. (2010) and Rujescu et al. (2009)

of cases, CNV-associated cases of ASDs and SSDs promise to be invaluable for understanding specific relationships between genetic differences and disorder-related phenotypes, by virtue of the clarity with which associations can be established between genetic differences (i.e., presence vs. absence of a CNV) and individual patients. Thus, by understanding similarities and differences among patients carrying a common CNV but diagnosed clinically with an ASD, an SSD, or both, we may be able to elucidate meaningful differences, and commonalities, in developmental and pathophysiological processes leading to one set of disorders or the other.

35.4 CNV IN THE GENOME: KNOWN FOR DECADES, UNDERAPPRECIATED UNTIL RECENTLY

Large-scale variation in the genome, such as chromosomal rearrangements, duplications, and deletions, has been known to cytogeneticists for many decades. However, the recent advent of molecular methods allowing high-resolution examination of the entire genome, such as array comparative hybridization (aCGH; Cowell, 2004) and genome-wide single-nucleotide polymorphism (SNP)-genotyping platforms (Ding and Jin, 2009), has led to a much fuller appreciation of how common and how relevant to complex neurobehavioral disorders such large-scale variation actually is. CNVs comprise a class of genomic variants in which long stretches of DNA, ranging in size from thousands to millions of base pairs (bp), occur in variable numbers of contiguous copies on chromosomes from different individuals (Zhang et al., 2009). While some CNV appears to be 'private,' occurring only within a single individual or family, recurrent CNV can be identified at specific locations in the genome in unrelated individuals.

35.4.1 Genomic Architecture Predisposing to Recurrent CNV

The most frequent overlapping rearrangements are usually recurrent and arise from nonallelic homologous recombination (NAHR), where a segment of unique DNA sequence (50 kb–10 Mb) is lost or duplicated due to the presence of flanking segmental duplications, or low-copy repeats (LCR) – large (>10 kb), highly repetitive (>95% homology), DNA sequences that predispose to genomic instability (Dibbens et al., 2009). When two of these paired segmental duplications are found in the same orientation along the chromosome, they can be improperly aligned during meiosis, leading to the duplication of the intervening sequence in one allele and the deletion of this same region in the other; this is the mechanism behind NAHR. Interestingly, some of the regions in which these recurrent deletions occur also harbor inversion polymorphisms which change the orientation of the paired LCR from inverted, or facing on different directions along the chromosome, to aligned, hence greatly increasing the chance of NAHR occurring. In some instances, the presence of the appropriate inversion allele is necessary for the CNV to occur, as is the case for the 17q21.31 region (Alkan et al., 2009; Sharp et al., 2006).

LCR-mediated NAHR thus repeatedly generates within the population recurrent CNV that is identical or nearly so in size and location. Recognition of recurrent CNV in research and clinical settings has increased tremendously over the past few years, leading to the identification of several previously unknown genomic disorders in which autism and/or schizophrenia is part of the phenotype. As the number of array-based research studies and diagnostic procedures increases, and collaborative pooling of information across research and clinical laboratories grows, more recurrent the CNV that causes or predisposes to NDD.

35.5 RECURRENT CNV ASSOCIATED WITH ASD, SSD, AND OTHER NDDS

The remainder of this chapter will focus on recurrent CNVs and their associations to ASD, SSD, and often other NDDs such as Ep (see Chapter 36) and ID. Emerging evidence implicates a growing list of CNVs that associate with ASD, SSD, and other NDDs. This emerging pattern of common CNVs playing a presumably causal role in diverse phenotypic outcomes suggests that subgroups of patients with clinically different disorders manifest variable phenotypes arising from common underlying genetic disturbances. It appears likely that at least some of the commonality of risk elevation in ASD, SSD, and other NDDs reflects disruption in fundamental brain developmental processes, some of which might be responsive to specific environmental conditions, epigenetic events, or influences of specific loci distant from the CNV that could influence the trajectory of phenotypic outcome. Identifying and disentangling factors associated with variation in phenotypic outcomes among carriers of specific recurrent CNVs are a major challenge, yet meeting that challenge promises to shed light on the basis for ASD, SSD, and other NDDs. Phenotypic studies of participants specifically ascertained by CNV status will be an important first step toward understanding the phenotypic variability associated with recurrent CNVs. We propose that studies ascertaining participants according to the presence or absence of specific sets of CNVs and focusing on a broad variety of behavioral, cognitive, physiological, metabolic, cellular, and molecular phenotypes are likely to be productive strategies for understanding the relationship of ASDs to SSDs and other NDDs.

In other words, if the goal is to organize developmental and neurobiological analysis of NDDs according to biologically meaningful criteria, then direct ascertainment of cases by CNV status will be superior to doing so by phenomenology. The former approach offers the opportunity to ascertain cases on the basis of likely molecular etiology, while the latter has proved unreliable for classifying cases according to similar mechanisms of clinical risk. Where cost, ethical concerns, and territoriality in scientific funding priorities are not important constraints, CNV would ideally be ascertained in population-based samples of infants at birth, and cases defined by CNV status would be followed prospectively together with demographically matched noncarrier controls. Although in the short-term ascertainment of CNV within large phenotypically identified collections, such as the AGRE resource (Lajonchere, 2010), is the most practical approach to studying CNV and NDD, the ascertainment bias inherent in identifying cases by nonspecific behavioral phenotypes will plague the field until unbiased molecular ascertainment of CNV carriers (some of whom will have only very mild, or absent, clinical phenotypes) becomes a reality in neurodevelopmental epidemiology.

Table 35.2 lists recurrent CNV associated with ASD and SSD, as well as other NDDs (most commonly ID or Ep). Note that most of these pathogenic recurrent CNVs also associate with a large variety of medical and anatomic disorders (the list of such associated disorders probably remains incomplete). Although a detailed discussion of the medical complications of CNV disorders is beyond the scope of this chapter, their presence is very important because it emphasizes the urgent need to begin training clinicians and educators who evaluate ASD, SSD, and other NDDs to include CNV disorders in the differential diagnosis so that appropriate referral for cytogenetic evaluation can be made, and medical as well as behavioral interventions can be instituted. Recently, Miller and colleagues (Miller et al., 2010) suggested that testing for CNV by microarray be made standard of care for evaluation of ASD and ID. No such suggestion has been published for SSD, although in our view, such testing should at least be considered when there is clinical evidence suggesting CNV (e.g., the simultaneous presentation of psychosis and ID). Bassett and Chow (1999)

TABLE 35.2   Recurrent CNVs Identified Across ASD and Schizophrenia

Genomic region	Position (Mb)[a]	Size (Mb)[a]	Number of genes[b]	CNV	References
1q21.1	chr1:144 963 732–145 864 377	0.9	7	del	Brunetti-Pierri et al. (2008) and Consortium (2008)
3q29	chr3:197 244 288–198 830 238	1.6	21	del	Mulle et al. (2010) and Willatt et al. (2005)
15q13.3	chr15:28 698 632–30 234 007	1.5	6	del	Ben-Shachar et al. (2009), Consortium (2008), and Stefansson et al. (2008) (see Chapter 32)
16p11.2	chr16:29 557 553–30 107 434	0.5	25	dup	McCarthy et al. (2009) and Weiss et al. (2008)
16p13.11	chr16:15 421 876–16 200 195	0.8	7	dup	Ingason et al. (2009) and Ullmann et al. (2007)
17q12	chr17:31 893 783–33 277 865	1.4	15	del	Loirat et al. (2010) and Moreno-De-Luca et al. (2010)
22q11.2	chr22:17 412 646–19 797 314	2.4	41	del	Antshel et al. (2007), Consortium (2008), Karayiorgou et al. (1995), Pulver et al. (1994), and Vorstman et al. (2006)

[a] Size and position are calculated in the hg18 genome assembly and exclude the DNA sequence from flanking segmental duplications.
[b] The number of genes in each region is based on RefSeq coding genes.

have elaborated specific clinical criteria indicating testing by fluorescent *in situ* hybridization (FISH) for 22q11 deletion syndrome (22q11DS) in patients with SSD. Such criteria need reevaluation and expansion as microarrays are rapidly supplanting FISH as the primary molecular–cytogenetic diagnostic procedure (FISH remains essential for confirmation of positive microarray results). The need to consider genome-wide testing for recurrent and rare CNV during the work-up of SSD is, in our view, a critical area for future translational research.

In the following sections, we briefly review current knowledge regarding the relationship of specific recurrent CNV to ASD and SSD. The discussion begins with a review of the 22q11 deletion syndrome (22q11DS) because the molecular basis of 22q11DS has been known for 2 decades, and its associations to NDDs have been most extensively studied. We regard 22q11DS as a prototypic CNV disorder, as it exhibits the phenotypic heterogeneity (i.e., variable expressivity) and variable penetrance that appear to be common to all SSD- and ASD-related CNV disorders thus far described. Following discussion of the 22q11DS, we will review the other disorders listed in Table 35.3 in order of the chromosomes they affect, and then will conclude with remarks on implications and future directions for neurodevelopmental research.

35.5.1 22q11.2 Deletion Syndrome: The Prototypic CNV Disorder

First described in an autopsy series of four infants (Kirkpatrick and DiGeorge, 1968), DiGeorge syndrome (DGS) referred to a constellation of severe immune deficiency and findings suggestive of maldevelopment of the third and fourth pharyngeal arches, including thymic aplasia, parathyroid hypoplasia, and abnormalities

of the aortic arch. Additional syndromes, called conotruncal anomaly face (CTAF) syndrome in a case series from Japan (Kinouchi et al., 1976) and velocardiofacial syndrome (VCFS) in another series from the United States (Shprintzen et al., 1978), consisted of velopharyngeal anomalies, cardiac anomalies, typical facial appearance, learning disabilities, and speech and language problems. In 1991, Scambler and colleagues reported that microdeletions at 22q11.2 associated with sporadic and familial DGS, and subsequent studies soon showed the majority of cases of DGS, CTAF, and VCFS were all associated with similar deletions (Burn et al., 1993; Carey et al., 1992; Scambler et al., 1992). Although all the foregoing syndrome designations are still in use, it is clear that the term 22q11DS subsumes almost all of the cases meeting the various phenotype definitions just reviewed and is therefore the most appropriate designation for this common CNV disorder (estimated at 1/4000 live births; Botto et al., 2003).

Behavioral manifestations in 22q11DS vary widely (Ousley et al., 2007) but are common in children and adults. By the mid-1980s, behavioral difficulties in children with VCFS had been described (Golding-Kushner et al., 1985), and psychosis in adolescents with the disorder was reported in 1992 (Shprintzen et al., 1992). Pulver and colleagues confirmed that schizophrenia was common in patients with 22q11DS (Pulver et al., 1994), and Karayiorgou and colleagues found several previously undiagnosed cases of 22q11DS in a series of patients with schizophrenia diagnosed solely on the basis of clinical presentation (Karayiorgou et al., 1995). The latter study was a landmark because it raised the prospect that a small but clinically and epidemiologically significant proportion of the SSD population carried undiagnosed 22q11 deletions. It is now clear that 22q11DS occurs at a low but nontrivial rate (~0.75%, about 30-fold more

TABLE 35.3   Candidate Genes in the 22q11 Deletion Region Plausibly Contributing to Risk for SSD or ASD

Locus	Gene product and function	References
COMT	Catechol-*O*-methyltransferase catalyzes catabolism of neurotransmitters dopamine and norepinephrine	Abdolmaleky et al. (2006), Bassett et al. (2007), Lachman et al. (1996), Munafo et al. (2005), and Shifman et al. (2002)
PRODH	Proline dehydrogenase participates in synthetic pathway for excitatory neurotransmitter, glutamate	Gogos et al. (1999), Jacquet et al. (2002), and Meechan et al. (2009)
ZDHHC8	Zinc finger, DHHC-type containing 8, likely a transmembrane palmitoyl transferase, which posttranslationally modifies proteins involved in intracellular trafficking and synaptic function. Variants associated with abnormal smooth-pursuit eye movements (SPEM) in SSD	Shin et al. (2010)
RANBP1	Variants associated with abnormal SPEM might be due to linkage disequilibrium with *ZDHHC8*	Cheong et al. (2011)
RTN4R	No Go-66 receptor, a key protein in axonal pathfinding during development	Sinibaldi et al. (2004)

frequently than in the population at large; Hoogendoorn et al., 2008) in clinically diagnosed SSD patients. Selecting specific phenotypic characteristics prior to molecular testing (e.g., facial dysmorphology, conotruncal heart defects, high-arched palate or cleft palate, ID) can substantially increase the diagnostic yield for the deletion in cohorts of patients with SSD (Bassett and Chow, 1999).

The most common 22q11 deletion (~80% of cases) is approximately 3 megabases (Mb) long, occurring between two LCRs flanking the deletion (Edelmann et al., 1999a, 1999b). Two additional LCRs lie within the 3 Mb deletion region and account for the majority of remaining deletions (~10% of 1.5 Mb and the rest of varied size). The 3-Mb region encompasses 40 genes, and several of these appear to be compelling candidates as genes contributing to SSD risk. However, none of them has yet been confirmed with sufficient confidence to call them 'schizophrenia genes.' Table 35.3 summarizes candidate genes residing within the 22q11DS region for which there is some evidence (usually mixed positive and negative results) supporting associations to risk for SSD.

It is worth noting that even accepting only the positive evidence supporting associations of individual loci within the 22q11DS region as the truth, the effect sizes of those associations are much smaller than the magnitude of the association of the 22q11 deletion itself with SSD (OR <1.5 for any given locus vs. OR ~20 for 22q11DS). It is thus possible that hemizygosity of multiple genes within the 22q11DS region, each of small individual effect, somehow synergizes to produce a more substantial influence on brain development when risk genes in the region are affected simultaneously. Alternatively, it is possible that deletion of single copies of multiple genes creates numerous opportunities for deleterious effects of risk loci elsewhere in the genome. While speculative, such hypotheses are useful because they suggest specific strategies for examining how 22q11DS elevates risk for SSD, ASD, or other NDDs. For example, a genome-wide analysis of variants in 22q11DS patients affected or unaffected by ASD, SSD, or both could identify specific loci elsewhere in the genome that elevate developmental sensitivity to the effects of hemizygosity of loci at 22q11.

The above summary shows that exciting progress has been made in understanding potential genetic mechanisms in SSD related to 22q11DS. Several genes involved either in neurotransmission, neuronal function, or development may be contributing to risk for SSD. The association of 22q11DS to ASD was described only relatively recently (Antshel et al., 2007; Vorstman et al., 2006), so fewer studies of ASD and individual genes within the deletion region have been published.

However, a recent study comparing symptom profiles among ASD patients with 22q11DS, those with Klinefelter syndrome (KS: karyotype 47 XXY), and ASD patients with no specifically defined genetic syndrome suggested that the ranges of overall ASD symptoms in both 22q11DS and KS were narrower than in the idiopathic group and differed from each other (Bruining et al., 2010). Discriminant function analysis of symptom scores (derived from the Autism Diagnostic Interview, Revised) showed clear distinctions in the profiles of each group. While the study can be criticized for several methodological difficulties (e.g., the patient samples were ascertained separately, thus introducing the possibility of selection bias), it is an important step forward because the results suggest that specific aspects of the broad ASD phenotype may specifically associate with definable genetic differences among patients. If true, such associations could form the basis for dissection of genetically regulated developmental pathways leading to specific 'points' in the 'phenotypic space' of ASD and may also provide a basis for comparing pathways leading to ASD versus SSD.

Finally, 22q11DS may serve as an important prototype for demonstrating the clinical importance of diagnosing specific CNV disorders in patients whose clinical presentation is entirely or predominantly behavioral. From the earliest days, DGS was known to associate with hypocalcemia that in its most severe form could lead to fatal status epilepticus. Hypocalcemia in 22q11DS results from the variable degree of parathyroid hypoplasia that occurs in those afflicted and can vary over an individual patient's lifetime. The condition can usually be corrected by dietary supplementation with vitamin D and calcium, but to establish the diagnosis, clinicians must test ionized serum calcium levels, rather than rely on the total calcium levels available in standard metabolic panels. The failure to test specifically for ionized calcium in patients with 22q11DS can lead to poor clinical outcomes. For example, SSD patients are almost always treated with antipsychotic medications. As a class, those agents tend to lower the seizure threshold. Clozapine is an antipsychotic medication that is among the worst offenders in terms of its effect on seizure threshold but remains a uniquely effective treatment in some patients whose psychotic symptoms do not respond well to other antipsychotic medications (Kane et al., 1988). Patients with SSD, 22q11DS, and untreated hypocalcemia who are given clozapine will be at extremely high risk for seizures owing to the synergistic effects of low calcium and the medication. Clinicians unaware of the patient's 22q11DS diagnosis would almost certainly discontinue clozapine when the patient has a seizure after starting clozapine. In that scenario, prophylactic administration of vitamin D and calcium, rather than discontinuation of clozapine, would have been an unexplored therapeutic alternative. Failure to diagnose 22q11DS in the ~0.75% of SSD patients who carry such deletions can thereby deprive them of potentially effective treatments. Such cases have been documented, so this matter is of more than theoretical interest (Caluseriu et al., 2007). As we learn more about the large variety of clinical management issues associated with each CNV disorder, moving from phenomenological diagnosis to molecular diagnosis of NDD promises to become increasingly valuable for patient care.

35.5.2 1q21 Deletions

With the advent of cytogenomic array testing, recurrent deletions of 1q21, as well as several other genomic disorders that went previously undetected, have been identified in the last few years. This CNV spans 1.35 Mb and was originally identified in a large cohort of patients with schizophrenia (Stefansson et al., 2008). The 1q21 deletion was detected in 11 out of 4718 cases with schizophrenia (0.23%), compared to 8 of 41 199 controls (0.02%), providing compelling evidence of the association of this CNV and schizophrenia. Almost simultaneously, the identical CNV was identified in patients referred for clinical testing with a wide array of phenotypic features, which include isolated heart defects, cataracts, Müllerian aplasia, and microcephaly (Mefford et al., 2008). It is noteworthy that patients with the reciprocal 1q21 duplication have macrocephaly, pointing toward a dosage-sensitive gene involved in head growth. Additionally, several of these patients had a behavioral phenotype manifesting as ASD (Brunetti-Pierri et al., 2008; Mefford et al., 2008). Differing from other well-known microdeletion syndromes, such as Angelman and Prader-Willi (15q11.q13 deletions), but similar to 22q11DS, deletions in 1q21 lead to wide phenotypic variability and are sometimes found in apparently unaffected parents of affected individuals. However, a thorough phenotypic assessment in these apparently unaffected parents frequently reveals a subclinical neurocognitive or behavioral phenotype (Girirajan and Eichler, 2010).

There are at least seven genes in the 1q21 deleted interval, although the one responsible for the behavioral phenotype of patients with this CNV has not been identified. *GJA5* and *GJA8* are both part of the connexin family of genes and play an important role in membrane junctions. Mutations in *GJA5* are known to cause atrial fibrillation, whereas mutations in *GJA8* give rise to cataracts. Interestingly, GJA8 has been previously associated with schizophrenia (Ni et al., 2007). However, the gene responsible for the neurobehavioral phenotype of these patients remains unknown.

35.5.3 3q29 Deletions

Causing another novel genomic disorder, microdeletions of 3q29 have been associated with a variable array of phenotypes, including mild-to-moderate mental retardation, slightly dysmorphic facial features, gate ataxia, and long tapering fingers (Willatt et al., 2005).

Also mediated by segmental duplications, this recurrent deletion is 1.6 Mb in size and contains 21 genes, several of which are interesting candidates for NDD. *DLG1* and *PAK2* are interesting candidates, as they are both autosomal homologs of well-described X-linked ID genes *DLG3* and *PAK3*. *DLG1*, also known as synapse-associated protein 97, also interacts directly with *PTEN* to inhibit axonal stimulation of myelination. This molecular brake is important in maintaining proper myelin thickness and, when removed, produces myelin outfoldings and demyelination. In fact, this brake ceases to function in peripheral neuropathies such as Charcot–Marie–Tooth (Cotter et al., 2010). Additionally, and perhaps most importantly, DLG1 interacts directly with AMPA and NMDA receptors, both key components of the glutamatergic synapse (Howard et al., 2010), in line with recent research showing the role of glutamatergic dysfunction in schizophrenia (Gaspar et al., 2009). *PAK2* also appears as an interesting candidate, as it regulates cytoskeleton dynamics, consequently regulating the morphology of the synapse and glutamate receptor complexes in the process (Kreis and Barnier, 2009).

35.5.4 15q13.2–13.3 Deletions

The proximal region of the long arm of chromosome 15 contains a series of five highly similar LCRs, which predispose this region of the genome to a variety of rearrangements (Mignon-Ravix et al., 2007: see Chapter 32). The most well known of such rearrangements consist of deletions between the two most proximal LCRs, giving rise to the imprinted disorders Prader-Willi syndrome or Angelman syndrome, depending on whether the deletion is paternally or maternally inherited, respectively. More distal deletions on chromosome 15, involving loss of DNA between the third and fifth or fourth and fifth LCRs, give rise to 15q13.2–13.3 deletion syndrome (15q13DS). Such deletions have consistently been identified in genome-wide association study (GWAS) of ASD (Miller et al., 2009; Pagnamenta et al., 2009) and SSD (Consortium, 2008; Stefansson et al., 2008). ID is common in 15q13.2–13.3 DS, as are difficulties with aggressive behavior and rage outbursts (Ben-Shachar et al., 2009; Miller et al., 2009; Sharp et al., 2008). Ep is also common in 15q13DS, with one study estimating this single set of CNVs to account for ~1% of idiopathic cases of Ep (Dibbens et al., 2009).

Among the loci deleted in 15q13DS is *CHRNA7*, encoding the α7 nicotinic acetylcholine receptor (α7nAChR). This observation is of great interest for at least two reasons. First, linkage and association studies have implicated *CHRNA7* as an important genetic modifier of an SSD-related physiological phenotype known as gating of the P50 auditory-evoked potential (P50-AEP). The P50-AEP is a positive deflection on EEG that occurs ~50 ms after a brief auditory stimulus. Gating of the P50-AEP refers to the phenomenon in which an auditory stimulus shortly before the index stimulus attenuates the P50-AEP. Such gating is often impaired in persons with SSD, as well as in their unaffected relatives. Freedman and colleagues reported linkage between markers on 15q13 and P50-AEP gating in families segregating SSD (Freedman et al., 1997), and subsequent association studies suggested that variation at *CHRNA7* accounts for this linkage (Leonard et al., 1998). Together with the foregoing results, the association of 15q13DS with ASD and SSD suggests *CHRNA7* as a prime candidate gene relevant to altered development and function of the brain in ASD and SSD.

The second reason for specific interest in *CHRNA7* is that the pharmacology of the α7nAChR is well developed, with many compounds available for use in animal models and in some cases humans, which could be used as probes with which to examine the role of the receptor in development and possibly even for therapeutics. We recently described an adult male patient with 15q13DS, SSD, rage outbursts, and Ep, whose aggressive behavior was substantially attenuated by treatment with an acetylcholinesterase-inhibiting positive allosteric regulator of the α7nAChR, galantamine. The case provides at least a single example in which diagnosis of a CNV disorder led directly to altered pharmacotherapy in a clinical situation (Cubells et al., 2011).

35.5.5 16p11.2 Duplications

Weiss and colleagues (Weiss et al., 2008) performed a genome-wide search for recurrent CNV associated with ASD, using data from SNP-genotyping arrays from several large GWAS of ASD. They noted significant associations of *both* deletions and duplications in a ~590-kb region flanked by LCRs, located on chromosome 16p11.2. The finding was even more noteworthy because macrocephaly (enlarged head circumference), an endophenotype observed in a substantial minority of children with ASD, was also associated with the deletion at 16p11.2.

Simultaneously with the report of the Weiss et al. group, two other research teams reported associations between ASD and CNV at 16p11.2 (Kumar et al., 2008; Marshall et al., 2008). Importantly, one of those studies (Kumar et al., 2008) confirmed an association between 16p11.2 deletions and ASD using an independent molecular approach: array comparative hybridization. Although those investigators also found a single case of 16p11.2 duplication in their sample, they observed the duplication in two of their control subjects and therefore did not conclude the duplication associated with ASD. However, other studies have confirmed both the deletion and duplication as clearly associated with ASD

(Fernandez et al., 2010; Marshall et al., 2008), despite the variable expressivity highlighted by observations of apparently unaffected individuals occasionally carrying the duplication. Detailed examination of ASD probands carrying CNV at 16p11.2 revealed that the heterogeneous phenotypic spectra associated with these genomic variants include ID and variable facial dysmorphology (Fernandez et al., 2010). That study also found evidence suggesting that 16p11.2 deletions may be more penetrant with regard to ASD than are duplications. Another study confirmed the phenotypic heterogeneity associated with CNV at 16p11.2, adding Ep and motor delay to the manifestations associated with either the deletion or duplication, and attention deficit hyperactivity disorder to those associated with the duplication (Shinawi et al., 2010). Interestingly, that same study found macrocephaly to associate with the deletion and microcephaly (diminished head circumference) with the duplication.

McCarthy and colleagues reported that SSD are also associated with duplications (but not deletions) at 16p11.2 (McCarthy et al., 2009). Interestingly, consistent with earlier results in ASD, this group also observed an association of the 16p11.2 duplication with head circumference in the SSD sample, suggesting this endophenotype may not be specific to ASD but rather associated with the duplication and a broader risk for NDDs. If the 'specificity' of the association between SSD and only the duplication at 16p11.2 withstands more extensive study in additional cohorts of patients, such an observation could help distinguish genes within the CNV region that might more specifically predispose to ASD (when haploinsufficient) rather than SSD (when present in excess). However, more data are necessary before it is clear that only the duplication associates with SSD.

35.5.6 16p13.11 Deletions and Duplications

The short arm of chromosome 16 is particularly rich in LCRs (Martin et al., 2004), with the result that nonhomologous recombination events in the region are common. Thus, another set of recurrent CNVs distal to the 16p11.2 region just discussed occurs on 16p13.11. These CNVs vary somewhat in length, due to the complexity of the region, but most are ~1.4–1.65 Mb in length. Both duplications and deletions were originally described as associated with ASD and ID (Pinto et al., 2010; Ullmann et al., 2007), although apparently unaffected carriers of the duplications were identified in several families with affected members. However, a study that screened a large cohort of patients with ID or multiple congenital anomalies (MCA), as well as two cohorts of European-ancestry adults with no known evidence of NDD (but who were not specifically evaluated) found only the deletions to occur significantly more frequently in patients than in the control individuals. Those observations led the authors of that study to suggest that duplications at 16p13.1–13.2 might be nonpathogenic variants. A more recent study, of >4300 patients with SSD and >35 000 controls ascertained in several European countries and evaluated using SNP-genotyping arrays, found an overall association between duplications at 16p13.1 and SSD, with approximately a threefold excess observed in the patient group. When the investigators classified the duplications according to their positions across three subregions of 16p13.1, they found a stronger association with duplications residing in the proximal two subregions (with respective odds ratios increasing from 3.27 to 7.27). The authors of the latter study, while acknowledging the difficulty of declaring duplications at 16p13.1 to be pathogenic, given heterogeneity in duplication size and the prior inconclusive results with regard to ID and MCA, argue that additional factors add to evidence that such duplications are pathogenic. They point out that the subregion analysis just summarized, in addition to strengthening the statistical association also identifies a strong candidate locus, *NDE1*. That gene encodes a protein that interacts with *DISC1*, which itself is the product of a strongly supported 'schizophrenia gene.' Another binding partner of *NDE1* is the gene product of *LIS1*, which is strongly associated with the severe NDD, lissencephaly.

35.5.7 17q12 Deletions

Deletions in 17q12 were until recently believed to be one of the few recurrent genomic disorders that spared the central nervous system. This 1.4-Mb recurrent deletion harbors the *HNF1B* gene, which is responsible for the renal cysts and diabetes (RCAD, MIM ID #137920) syndrome. Generally, affected patients have various degrees of renal compromise, including renal cysts and hyperechogenic kidneys that might progress to renal failure (Sovik et al., 2002). Additionally, maturity-onset diabetes of the young type 5 is usually seen by early adulthood. Affected females may have uterine malformations such as bicornuate uterus and Müllerian aplasia (Bellanne-Chantelot et al., 2004). This clinical presentation is often accompanied by a characteristic facial gestalt that includes macrocephaly and prominent forehead, downslanting palpebral fissures, depressed nasal bridge, and protruding maxilla in adulthood (Moreno-De-Luca et al., 2010). However, patients with a milder phenotype, even without these core clinical features and hence without a diagnosis of RCAD, are not infrequent (Moreno-De-Luca et al., 2010).

As alluded to previously, until recently, there was no evidence that a central nervous system phenotype is associated with 17q12 deletion syndrome. However, recent studies have shown that ID, ranging from mild to moderate, is common in patients with 17q12DS. More interestingly, behavioral anomalies have been identified in small cohorts of these patients, particularly involving problems in social interactions reminiscent of ASD. A recent report (Moreno-De-Luca et al., 2010) found that six of nine patients with 17q12DS met DSM-IV TR criteria for ASD. This finding was then confirmed in larger cohorts of patients with ASD (Moreno-De-Luca et al., 2010). Additionally, given the clinical and genetic overlap between ASD and SSD, large cohorts of patients with schizophrenia were investigated to assess the frequency of the deletion. A strong association was identified between 17q12DS and SSD. This CNV was absent from a very large sample of control individuals (52 448), which could be interpreted as a strong impact of this CNV over the phenotype of affected individuals, albeit with variable expressivity (Moreno-De-Luca et al., 2010).

Interestingly, the 17q12 region overlaps with a replicated linkage and association peak found in families with ASD (IMGSAC, 2001; McCauley et al., 2005; Stone et al., 2004, 2007; Yonan et al., 2003). It is tempting to hypothesize that one of the genes within this region is responsible for that linkage signal. However, the frequency of the deletion across different populations is 1 in 900 on average, and the frequency of yet-undiscovered mutations in one of the genes, which would account for a similar phenotype, is very likely rare as well and might not fully explain the linkage signal.

The 17q12DS region harbors 15 genes, and haploinsufficiency in one or more of these likely accounts for the neurocognitive phenotypes observed in these patients. *HNF1B* is responsible for the core features of RCAD (Bellanne-Chantelot et al., 2004); however, patients with point mutations or single-gene deletions do not appear to have a behavioral phenotype, which would mean that one of the other genes within the region might be responsible for the central nervous system findings. *LHX1* is a transcription factor involved in brain development and axonal guidance (Avraham et al., 2009) and appears as an interesting candidate. Knockout mice lack proper patterning of the midbrain–hindbrain barrier (Shawlot and Behringer, 1995). Nevertheless, no mutations in humans have been documented, so the pathogenic role of haploinsufficiency cannot be clearly established. More studies are needed to clarify this issue.

35.6 CONCLUSIONS AND FUTURE DIRECTIONS

The information just summarized illustrates exciting progress in understanding genomic mechanisms contributing to ASD, SSD, and other NDDs. While our current, very incomplete, knowledge of the roles of CNV in ASD and SSD suggests common pathways of risk for

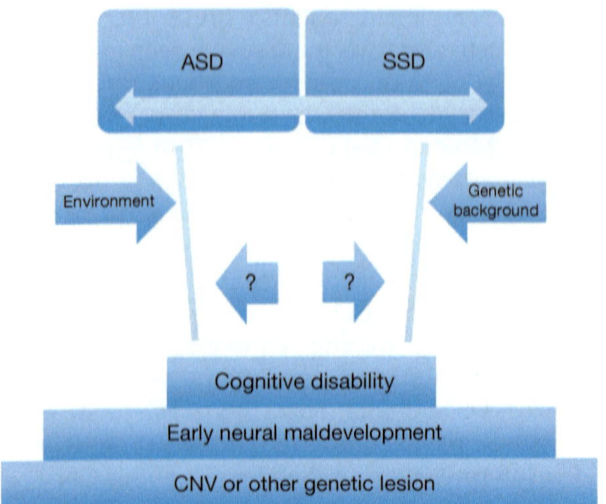

FIGURE 35.1 A heuristic model relating CNV to risk for ASD and SSD. The model is based on the premises that (i) genes within CNV participate in gene networks regulating neural development; (ii) specific factors, including environmental events, epigenetic factors, and stochastic processes can alter developmental trajectories, thereby influencing phenotypic outcomes; and (iii) timing of such events may be critical, especially relative to key developmental epochs such as the prenatal, early childhood, and adolescent periods. Note that the model posits ID as an outcome that precedes ASD or SSD. That component of the model reflects the virtual universal association of some form or degree of ID (or at least learning disorders) with CNV. Thus, we hypothesize that impaired abilities to engage in social learning, probably beginning at birth or possibly earlier, contribute cumulatively to the development of ASD and SSD. Fundamentally, however, such learning difficulties arise from suboptimal function at the cellular and molecular levels.

these two sets of disorders, a major challenge is to understand how similar or identical CNV results in widely different clinical outcomes in different individuals. In Figure 35.1, we summarize a heuristic model for generating testable hypotheses regarding the developmental impact of CNV on risk for ASD, SSD, and other NDDs (although for simplicity, we include only ASD and SSD in the diagram). The model incorporates the following three hypotheses:

1. CNV predisposing to ASD, SSD, and other NDDs do so by altering the function of gene-regulatory networks in which loci at or near the particular CNV participate. Note that a variety of mechanisms could alter such networks, including under- or overexpression of the products of dosage-sensitive genes within a CNV region; unmasking (in the case of deletion) or enhancement (in duplications) of expression of recessive deleterious alleles within the CNV region; effects of CNV on chromatin structure, leading to 'spreading' effects on regulation of genes in *cis* near the CNV or altered *trans* regulation via mechanisms such as chromatin looping (Miele and Dekker, 2009); or the presence of risk alleles at loci elsewhere in the genome but involved in the same regulatory networks as those within the CNV.

2. Specific environmental factors might interact with CNV-associated gene-regulatory networks to alter developmental trajectories. A well-known (and therapeutically modifiable) environmental factor affecting neural development, for example, is maternal intake of folic acid during pregnancy, which strongly impacts the risk of open neural tube defects in offspring (Berry and Li, 2002; see Chapter 27). Understanding how such factors interact with genetic networks at the cellular level during development of the brain could lead to effective preventive or ameliorative strategies in patients with CNV disorders.

3. Variability in the timing or magnitude of specific environmental exposures in the context of specific CNVs might alter developmental trajectories. As noted above, ASD and SSD differ in their typical timing of clinical presentation. However, the typical epochs of presentation (early childhood for ASD, or during or just after puberty for SSD) are periods of profound developmental changes in brain structure or function. It is possible that specific CNVs set up carriers for vulnerability to specific risk factors occurring during these critical periods.

Future research is needed at all levels on CNVs and their association to risk for NDD. To date, patients with CNV have been ascertained almost entirely within clinical contexts or in case–control studies where cases have been selected based on phenomenology. 'Unaffected' (or more likely, mildly enough affected to escape clinical notice) carriers have generally been discovered upon family testing once a proband has been identified, or in the context of case–control studies where 'controls' may not represent the general population, but rather are ascertained for absence of a particular set of syndromes. Current literature, while extremely valuable, is therefore almost certainly laden with biased ascertainment and other difficulties that preclude rigorous epidemiological delineation of the role of CNV in public health. The availability of relatively low-cost methods for scanning the genome for CNV (e.g., aCGH) should support the development of sampling strategies in which molecular rather than phenomenological criteria drive case and control ascertainment. Such studies would be particularly valuable for clarifying factors that distinguish clinical subtypes of specific CNV disorders from each other. As alluded to in examples from 22q11DS and 15q13.3 DS, the clear fact that subsets of patients diagnosed according to phenomenological schema such as the DSM-IVTR are at elevated risk for carrying specific CNV has potentially profound implications for diagnosis and treatment of behavioral disorders.

Another exciting direction enabled by expanding knowledge of CNV-associated behavioral disorders is that credible animal models for fundamentally human disorders become possible. Thus, while nobody will ever be able to guess what constitutes psychosis or autism in a mouse, experiments on rodents carrying engineered syntenic CNVs are shedding light on specific neural and developmental mechanisms likely to be relevant to behavioral deficits in SSD and ASD (Meechan et al., 2009; Sigurdsson et al., 2010). Animal models of neuro-developmentally important CNV promise to introduce novel strategies for understanding pathological neural development as well as new platforms for testing therapeutic drugs.

While political and economic barriers, such as insurance discrimination against behavioral disorders and (at least in the United States) inadequate or absent health coverage for large proportions of the population, will continue to impede progress in the diagnosis and treatment of behavioral disorders, the emerging literature on CNV as causative factors in mental illness adds to overwhelming evidence that NDDs are every bit as 'biological' and therefore medically 'real' as other complex disorders. Hopefully, the mountain of evidence supporting such a proposition will eventually get large enough that even the American Congress will be unable to continue ignoring it, thus leading to the elimination of barriers to healthcare access that currently severely impact patients with ASD and SSD. In this regard, the on-going explosion of knowledge on CNV disorders will benefit patients with NDD whether or not they carry associated CNV.

References

Abdolmaleky, H.M., Cheng, K.H., Faraone, S.V., et al., 2006. Hypo-methylation of MB-COMT promoter is a major risk factor for schizophrenia and bipolar disorder. Human Molecular Genetics 15, 3132–3145.

Alkan, C., Kidd, J.M., Marques-Bonet, T., et al., 2009. Personalized copy number and segmental duplication maps using next-generation sequencing. Nature Genetics 41, 1061–1067.

American Psychiatric Association, 2000. Diagnostic and Statistical Manual of Mental Disorders, Text Revision: DSM-IV-TR. American Psychiatric Association, Washington, DC.

Angermeyer, M.C., Kuhn, L., Goldstein, J.M., 1990. Gender and the course of schizophrenia: Differences in treated outcomes. Schizophrenia Bulletin 16, 293–307.

Antshel, K.M., Aneja, A., Strunge, L., et al., 2007. Autistic spectrum disorders in velo-cardio facial syndrome (22q11.2 deletion). Journal of Autism and Developmental Disorders 37, 1776–1786.

Avraham, O., Hadas, Y., Vald, L., et al., 2009. Transcriptional control of axonal guidance and sorting in dorsal interneurons by the Lim-HD proteins Lhx9 and Lhx1. Neural Development 4, 21.

Bassett, A.S., Chow, E.W., 1999. 22q11 deletion syndrome: A genetic subtype of schizophrenia. Biological Psychiatry 46, 882–891.

Bassett, A.S., Caluseriu, O., Weksberg, R., Young, D.A., Chow, E.W., 2007. Catechol-O-methyl transferase and expression of schizophrenia in 73 adults with 22q11 deletion syndrome. Biological Psychiatry 61, 1135–1140.

Bellanne-Chantelot, C., Chauveau, D., Gautier, J.F., et al., 2004. Clinical spectrum associated with hepatocyte nuclear factor-1beta mutations. Annals of Internal Medicine 140, 510–517.

Ben-Shachar, S., Lanpher, B., German, J.R., et al., 2009. Microdeletion 15q13.3: A locus with incomplete penetrance for autism, mental retardation, and psychiatric disorders. Journal of Medical Genetics 46, 382–388.

Berry, R.J., Li, Z., 2002. Folic acid alone prevents neural tube defects: Evidence from the China study. Epidemiology 13, 114–116.

Bhaumik, S., Tyrer, F.C., McGrother, C., Ganghadaran, S.K., 2008. Psychiatric service use and psychiatric disorders in adults with intellectual disability. Journal of Intellectual Disability Research 52, 986–995.

Bleuler, E., 1950. Dementia Praecox or The Group of Schizophrenias (Translated by J. Zinkin). International Universities Press, Madison, CT.

Botto, L.D., May, K., Fernhoff, P.M., et al., 2003. A population-based study of the 22q11.2 deletion: Phenotype, incidence, and contribution to major birth defects in the population. Pediatrics 112, 101–107.

Bruining, H., de Sonneville, L., Swaab, H., et al., 2010. Dissecting the clinical heterogeneity of autism spectrum disorders through defined genotypes. PLoS One 5, e10887.

Brunetti-Pierri, N., Berg, J.S., Scaglia, F., et al., 2008. Recurrent reciprocal 1q21.1 deletions and duplications associated with microcephaly or macrocephaly and developmental and behavioral abnormalities. Nature Genetics 40, 1466–1471.

Burn, J., Takao, A., Wilson, D., et al., 1993. Conotruncal anomaly face syndrome is associated with a deletion within chromosome 22q11. Journal of Medical Genetics 30, 822–824.

Caluseriu, O., Tayyeb, T., Chow, E., Bassett, A.S., 2007. Clozapine-associated seizures in a 22q deletion syndrome subtype of schizophrenia. Schizophrenia Research 60, 70.

Carey, A.H., Kelly, D., Halford, S., et al., 1992. Molecular genetic study of the frequency of monosomy 22q11 in DiGeorge syndrome. American Journal of Human Genetics 51, 964–970.

Cheong, H.S., Park, B.L., Kim, E.M., et al., 2011. Association of RANBP1 haplotype with smooth pursuit eye movement abnormality. American Journal of Medical Genetics. Part B, Neuropsychiatric Genetics 156B, 67–71.

Ching, M.S., Shen, Y., Tan, W.H., et al., 2010. Deletions of NRXN1 (neurexin-1) predispose to a wide spectrum of developmental disorders. American Journal of Medical Genetics. Part B, Neuropsychiatric Genetics 153B, 937–947.

Consortium, I.S., 2008. Rare chromosomal deletions and duplications increase risk of schizophrenia. Nature 455, 237–241.

Cotter, L., Ozcelik, M., Jacob, C., et al., 2010. Dlg1-PTEN interaction regulates myelin thickness to prevent damaging peripheral nerve overmyelination. Science 328, 1415–1418.

Cowell, J.K., 2004. High throughput determination of gains and losses of genetic material using high resolution BAC arrays and comparative genomic hybridization. Combinatorial Chemistry and High Throughput Screening 7, 587–596.

Cubells, J., Deoreo, E., Harvey, P., et al., 2011. Pharmaco-genetically guided treatment of recurrent rage outbursts in an adult male with 15q13.3 deletion syndrome. American Journal of Medical Genetics. Part A 155, 805–810.

Dibbens, L.M., Mullen, S., Helbig, I., et al., 2009. Familial and sporadic 15q13.3 microdeletions in idiopathic generalized epilepsy: Precedent for disorders with complex inheritance. Human Molecular Genetics 18, 3626–3631.

Ding, C., Jin, S., 2009. High-throughput methods for SNP genotyping. Methods in Molecular Biology 578, 245–254.

Durand, C.M., Betancur, C., Boeckers, T.M., et al., 2007. Mutations in the gene encoding the synaptic scaffolding protein SHANK3 are associated with autism spectrum disorders. Nature Genetics 39, 25–27.

Edelmann, L., Pandita, R.K., Morrow, B.E., 1999a. Low-copy repeats mediate the common 3-Mb deletion in patients with velo-cardio-facial syndrome. American Journal of Human Genetics 64, 1076–1086.

Edelmann, L., Pandita, R.K., Spiteri, E., et al., 1999b. A common molecular basis for rearrangement disorders on chromosome 22q11. Human Molecular Genetics 8, 1157–1167.

Fernandez, B.A., Roberts, W., Chung, B., et al., 2010. Phenotypic spectrum associated with de novo and inherited deletions and duplications at 16p11.2 in individuals ascertained for diagnosis of autism spectrum disorder. Journal of Medical Genetics 47, 195–203.

Folstein, S.E., Rosen-Sheidley, B., 2001. Genetics of autism: Complex aetiology for a heterogeneous disorder. Nature Reviews Genetics 2, 943–955.

Freedman, R., Coon, H., Myles-Worsley, M., et al., 1997. Linkage of a neurophysiological deficit in schizophrenia to a chromosome 15 locus. Proceedings of the National Academy of Sciences of the United States of America 94, 587–592.

Friedman, J.I., Vrijenhoek, T., Markx, S., et al., 2008. CNTNAP2 gene dosage variation is associated with schizophrenia and epilepsy. Molecular Psychiatry 13, 261–266.

Gaitatzis, A., Trimble, M.R., Sander, J.W., 2004. The psychiatric comorbidity of epilepsy. Acta Neurologica Scandinavica 110, 207–220.

Gaspar, P.A., Bustamante, M.L., Silva, H., Aboitiz, F., 2009. Molecular mechanisms underlying glutamatergic dysfunction in schizophrenia: Therapeutic implications. Journal of Neurochemistry 111, 891–900.

Gauthier, J., Spiegelman, D., Piton, A., et al., 2009. Novel de novo SHANK3 mutation in autistic patients. American Journal of Medical Genetics. Part B, Neuropsychiatric Genetics 150B, 421–424.

Gauthier, J., Champagne, N., Lafreniere, R.G., et al., 2010. De novo mutations in the gene encoding the synaptic scaffolding protein SHANK3 in patients ascertained for schizophrenia. Proceedings of the National Academy of Sciences of the United States of America 107, 7863–7868.

Girirajan, S., Eichler, E.E., 2010. Phenotypic variability and genetic susceptibility to genomic disorders. Human Molecular Genetics 19, R176–R187.

Gogos, J.A., Santha, M., Takacs, Z., et al., 1999. The gene encoding proline dehydrogenase modulates sensorimotor gating in mice. Nature Genetics 21, 434–439.

Golding-Kushner, K.J., Weller, G., Shprintzen, R.J., 1985. Velo-cardio-facial syndrome: Language and psychological profiles. Journal of Craniofacial Genetics and Developmental Biology 5, 259–266.

Guilmatre, A., Dubourg, C., Mosca, A.L., et al., 2009. Recurrent rearrangements in synaptic and neurodevelopmental genes and shared biologic pathways in schizophrenia, autism, and mental retardation. Archives of General Psychiatry 66, 947–956.

Hafner, H., Maurer, K., Loffler, W., Riecher-Rossler, A., 1993. The influence of age and sex on the onset and early course of schizophrenia. British Journal of Psychiatry 162, 80–86.

Hemmings, C.P., 2006. Schizophrenia spectrum disorders in people with intellectual disabilities. Current Opinion in Psychiatry 19, 470–474.

Hoogendoorn, M.L., Vorstman, J.A., Jalali, G.R., et al., 2008. Prevalence of 22q11.2 deletions in 311 Dutch patients with schizophrenia. Schizophrenia Research 98, 84–88.

Howard, M.A., Elias, G.M., Elias, L.A., Swat, W., Nicoll, R.A., 2010. The role of SAP97 in synaptic glutamate receptor dynamics. Proceedings of the National Academy of Sciences of the United States of America 107, 3805–3810.

Ikeda, M., Aleksic, B., Kirov, G., et al., 2010. Copy number variation in schizophrenia in the Japanese population. Biological Psychiatry 67, 283–286.

IMGSAC, 2001. A genomewide screen for autism: Strong evidence for linkage to chromosomes 2q, 7q, and 16p. American Journal of Human Genetics 69, 570–581.

Ingason, A., Rujescu, D., Cichon, S., et al., 2009. Copy number variations of chromosome 16p131 region associated with schizophrenia. Molecular Psychiatry 16, 17–25.

Jacquet, H., Raux, G., Thibaut, F., et al., 2002. PRODH mutations and hyperprolinemia in a subset of schizophrenic patients. Human Molecular Genetics 11, 2243–2249.

Kane, J.M., Honigfeld, G., Singer, J., Meltzer, H., 1988. Clozapine in treatment-resistant schizophrenics. Psychopharmacology Bulletin 24, 62–67.

Kanner, L., 1943. Autistic disturbances of affective contact. The Nervous Child 2, 217–250.

Karayiorgou, M., Morris, M.A., Morrow, B., et al., 1995. Schizophrenia susceptibility associated with interstitial deletions of chromosome 22q11. Proceedings of the National Academy of Sciences of the United States of America 92, 7612–7616.

Kety, S.S., Rosenthal, D., Wender, P.H., Schulsinger, F., 1971. Mental illness in the biological and adoptive families of adopted schizophrenics. American Journal of Psychiatry 128, 302–306.

Kety, S.S., Wender, P.H., Jacobsen, B., et al., 1994. Mental illness in the biological and adoptive relatives of schizophrenic adoptees: Replication of the Copenhagen study in the rest of Denmark. Archives of General Psychiatry 51, 442–455.

Kinouchi, A., Mori, K., Ando, M., Takao, A., 1976. Facial appearance with conotruncal anomalies. Pediatrics in Japan 84, 84–89.

Kirkpatrick Jr., J.A., Digeorge, A.M., 1968. Congenital absence of the thymus. American Journal of Roentgenology, Radium Therapy, and Nuclear Medicine 103, 32–37.

Kreis, P., Barnier, J.V., 2009. PAK signalling in neuronal physiology. Cellular Signalling 21, 384–393.

Kumar, R.A., Karamohamed, S., Sudi, J., et al., 2008. Recurrent 16p11.2 microdeletions in autism. Human Molecular Genetics 17, 628–638.

Lachman, H.M., Papolos, D.F., Saito, T., Yu, Y.M., Szumlanski, C.L., Weinshilboum, R.M., 1996. Human catechol-O-methyltransferase pharmacogenetics: Description of a functional polymorphism and its potential application to neuropsychiatric disorders. Pharmacogenetics 6, 243–250.

Lajonchere, C.M., 2010. Changing the landscape of autism research: The autism genetic resource exchange. Neuron 68, 187–191.

Leonard, S., Gault, J., Moore, T., et al., 1998. Further investigation of a chromosome 15 locus in schizophrenia: Analysis of affected sibpairs from the NIMH Genetics Initiative. American Journal of Medical Genetics 81, 308–312.

Loirat, C., Bellanne-Chantelot, C., Husson, I., Deschenes, G., Guigonis, V., Chabane, N., 2010. Autism in three patients with cystic or hyperechogenic kidneys and chromosome 17q12 deletion. Nephrology, Dialysis, Transplantation 25, 3430–3433.

Losh, M., Sullivan, P.F., Trembath, D., Piven, J., 2008. Current developments in the genetics of autism: From phenome to genome. Journal of Neuropathology and Experimental Neurology 67, 829–837.

Marshall, C.R., Noor, A., Vincent, J.B., et al., 2008. Structural variation of chromosomes in autism spectrum disorder. American Journal of Human Genetics 82, 477–488.

Martin, J., Han, C., Gordon, L.A., et al., 2004. The sequence and analysis of duplication-rich human chromosome 16. Nature 432, 988–994.

Matson, J.L., Shoemaker, M., 2009. Intellectual disability and its relationship to autism spectrum disorders. Research in Developmental Disabilities 30, 1107–1114.

McCarthy, S.E., Makarov, V., Kirov, G., et al., 2009. Microduplications of 16p11.2 are associated with schizophrenia. Nature Genetics 41, 1223–1227.

McCauley, J.L., Li, C., Jiang, L., et al., 2005. Genome-wide and Ordered-Subset linkage analyses provide support for autism loci on 17q and 19p with evidence of phenotypic and interlocus genetic correlates. BMC Medical Genetics 6, 1.

Meechan, D.W., Tucker, E.S., Maynard, T.M., Lamantia, A.S., 2009. Diminished dosage of 22q11 genes disrupts neurogenesis and cortical development in a mouse model of 22q11 deletion/DiGeorge syndrome. Proceedings of the National Academy of Sciences of the United States of America 106, 16434–16445.

Mefford, H.C., Sharp, A.J., Baker, C., et al., 2008. Recurrent rearrangements of chromosome 1q21.1 and variable pediatric phenotypes. New England Journal of Medicine 359, 1685–1699.

Miele, A., Dekker, J., 2009. Mapping *cis*- and *trans*- chromatin interaction networks using chromosome conformation capture (3C). Methods in Molecular Biology 464, 105–121.

Mignon-Ravix, C., Depetris, D., Luciani, J.J., et al., 2007. Recurrent rearrangements in the proximal 15q11-q14 region: A new breakpoint cluster specific to unbalanced translocations. European Journal of Human Genetics 15, 432–440.

Miller, D.T., Shen, Y., Weiss, L.A., et al., 2009. Microdeletion/duplication at 15q13.2q133 among individuals with features of autism and other neuropsychiatric disorders. Journal of Medical Genetics 46, 242–248.

Miller, D.T., Adam, M., Aradhya, S., et al., 2010. Consensus statement: Chromosomal microarray Is a first-tier clinical diagnostic test for individuals with developmental disabilities or congenital anomalies. American Journal of Human Genetics 86, 749–764.

Moreno-De-luca, D., Mulle, J.G., Kaminsky, E.B., et al., 2010. Deletion 17q12 is a recurrent copy number variant that confers high risk of autism and schizophrenia. American Journal of Human Genetics 87, 618–630.

Morgan, V.A., Leonard, H., Bourke, J., Jablensky, A., 2008. Intellectual disability co-occurring with schizophrenia and other psychiatric illness: Population-based study. British Journal of Psychiatry 193, 364–372.

Mulle, J.G., Dodd, A.F., McGrath, J.A., et al., 2010. Microdeletions of 3q29 confer high risk for schizophrenia. American Journal of Human Genetics 87, 229–236.

Munafo, M.R., Bowes, L., Clark, T.G., Flint, J., 2005. Lack of association of the COMT (Val158/108 Met) gene and schizophrenia: A meta-analysis of case-control studies. Molecular Psychiatry 10, 765–770.

Ni, X., Valente, J., Azevedo, M.H., Pato, M.T., Pato, C.N., Kennedy, J.L., 2007. Connexin 50 gene on human chromosome 1q21 is associated with schizophrenia in matched case control and family-based studies. Journal of Medical Genetics 44, 532–536.

Ousley, O., Rockers, K., Dell, M.L., Coleman, K., Cubells, J.F., 2007. A review of neurocognitive and behavioral profiles associated with 22q11 deletion syndrome: Implications for clinical evaluation and treatment. Current Psychiatry Reports 9, 148–158.

Pagnamenta, A.T., Wing, K., Akha, E.S., et al., 2009. A 15q13.3 microdeletion segregating with autism. European Journal of Human Genetics 17, 687–692.

Pinto, D., Pagnamenta, A.T., Klei, L., et al., 2010. Functional impact of global rare copy number variation in autism spectrum disorders. Nature 466, 368–372.

Pulver, A.E., Nestadt, G., Goldberg, R., et al., 1994. Psychotic illness in patients diagnosed with velo-cardio-facial syndrome and their relatives. Journal of Nervous and Mental Disease 182, 476–478.

Rapoport, J.L., 2009. Personal reflections on observational and experimental research approaches to childhood psychopathology. Journal of Child Psychology and Psychiatry 50, 36–43.

Riley, B.P., Kendler, K.S., 2004. Schizophrenia: Genetics, Comprehensive Textbook of Psychiatry, 8th edn. Lippincott Williams & Wilkins, Philadelphia, PA.

Rujescu, D., Ingason, A., Cichon, S., et al., 2009. Disruption of the neurexin 1 gene is associated with schizophrenia. Human Molecular Genetics 18, 988–996.

Saha, S., Chant, D., Welham, J., McGrath, J., 2005. A systematic review of the prevalence of schizophrenia. PLoS Medicine 2, e141.

Scambler, P.J., Kelly, D., Lindsay, E., et al., 1992. Velo-cardio-facial syndrome associated with chromosome 22 deletions encompassing the DiGeorge locus. Lancet 339, 1138–1139.

Sharp, A.J., Cheng, Z., Eichler, E.E., 2006. Structural variation of the human genome. Annual Review of Genomics and Human Genetics 7, 407–442.

Sharp, A.J., Mefford, H.C., Li, K., et al., 2008. A recurrent 15q13.3 microdeletion syndrome associated with mental retardation and seizures. Nature Genetics 40, 322–328.

Shawlot, W., Behringer, R.R., 1995. Requirement for Lim1 in head-organizer function. Nature 374, 425–430.

Shifman, S., Bronstein, M., Sternfeld, M., et al., 2002. A highly significant association between a COMT haplotype and schizophrenia. American Journal of Human Genetics 71, 1296–1302.

Shin, H.D., Park, B.L., Bae, J.S., et al., 2010. Association of ZDHHC8 polymorphisms with smooth pursuit eye movement abnormality. American Journal of Medical Genetics. Part B, Neuropsychiatric Genetics 153B, 1167–1172.

Shinawi, M., Liu, P., Kang, S.H., et al., 2010. Recurrent reciprocal 16p11.2 rearrangements associated with global developmental delay, behavioural problems, dysmorphism, epilepsy, and abnormal head size. Journal of Medical Genetics 47, 332–341.

Shprintzen, R.J., Goldberg, R.B., Lewin, M.L., et al., 1978. A new syndrome involving cleft palate, cardiac anomalies, typical facies, and learning disabilities: Velo-cardio-facial syndrome. Cleft Palate Journal 15, 56–62.

Shprintzen, R.J., Goldberg, R., Golding-Kushner, K.J., Marion, R.W., 1992. Late-onset psychosis in the velo-cardio-facial syndrome. American Journal of Medical Genetics 42, 141–142.

Sigurdsson, T., Stark, K.L., Karayiorgou, M., Gogos, J.A., Gordon, J.A., 2010. Impaired hippocampal-prefrontal synchrony in a genetic mouse model of schizophrenia. Nature 464, 763–767.

Sinibaldi, L., de Luca, A., Bellacchio, E., et al., 2004. Mutations of the Nogo-66 receptor (RTN4R) gene in schizophrenia. Human Mutation 24, 534–535.

Sovik, O., Sagen, J., Njolstad, P.R., Nyland, H., Myhr, K.M., 2002. Contributions to the MODY5 phenotype. Journal of Inherited Metabolic Disease 25, 597–598.

Stefansson, H., Rujescu, D., Cichon, S., et al., 2008. Large recurrent microdeletions associated with schizophrenia. Nature 455, 232–236.

Stone, J.L., Merriman, B., Cantor, R.M., et al., 2004. Evidence for sex-specific risk alleles in autism spectrum disorder. American Journal of Human Genetics 75, 1117–1123.

Stone, J.L., Merriman, B., Cantor, R.M., Geschwind, D.H., Nelson, S.F., 2007. High density SNP association study of a major autism linkage region on chromosome 17. Human Molecular Genetics 16, 704–715.

Strauss, K.A., Puffenberger, E.G., Huentelman, M.J., et al., 2006. Recessive symptomatic focal epilepsy and mutant contactin-associated protein-like 2. New England Journal of Medicine 354, 1370–1377.

Thomas, L.E., Woods, S.W., 2006. The schizophrenia prodrome: A developmentally informed review and update for psychopharmacologic treatment. Child and Adolescent Psychiatric Clinics of North America 15, 109–133.

Tienari, P., Wynne, L.C., Laksy, K., et al., 2003. Genetic boundaries of the schizophrenia spectrum: Evidence from the Finnish Adoptive Family Study of Schizophrenia. American Journal of Psychiatry 160, 1587–1594.

Ullmann, R., Turner, G., Kirchhoff, M., et al., 2007. Array CGH identifies reciprocal 16p13.1 duplications and deletions that predispose to autism and/or mental retardation. Human Mutation 28, 674–682.

U.S.C.D.C., 2009. Prevalence of autism spectrum disorders – Autism and Developmental Disabilities Monitoring Network, United States, 2006. Morbidity and Mortality Weekly Report. Surveillance Summaries 58, 1–20.

Vorstman, J.A., Morcus, M.E., Duijff, S.N., et al., 2006. The 22q11.2 deletion in children: High rate of autistic disorders and early onset of psychotic symptoms. Journal of the American Academy of Child and Adolescent Psychiatry 45, 1104–1113.

Walker, E., Lewine, R.J., 1990. Prediction of adult-onset schizophrenia from childhood home movies of the patients. American Journal of Psychiatry 147, 1052–1056.

Weiss, L.A., Shen, Y., Korn, J.M., et al., 2008. Association between microdeletion and microduplication at 16p11.2 and autism. New England Journal of Medicine 358, 667–675.

Welham, J., Isohanni, M., Jones, P., McGrath, J., 2009. The antecedents of schizophrenia: A review of birth cohort studies. Schizophrenia Bulletin 35, 603–623.

Willatt, L., Cox, J., Barber, J., et al., 2005. 3q29 microdeletion syndrome: Clinical and molecular characterization of a new syndrome. American Journal of Human Genetics 77, 154–160.

Wisniowiecka-Kowalnik, B., Nesteruk, M., Peters, S.U., et al., 2010. Intragenic rearrangements in NRXN1 in three families with autism spectrum disorder, developmental delay, and speech delay. American Journal of Medical Genetics. Part B, Neuropsychiatric Genetics 153B, 983–993.

Yonan, A.L., Alarcon, M., Cheng, R., et al., 2003. A genomewide screen of 345 families for autism-susceptibility loci. American Journal of Human Genetics 73, 886–897.

Zhang, F., Gu, W., Hurles, M.E., Lupski, J.R., 2009. Copy number variation in human health, disease, and evolution. Annual Review of Genomics and Human Genetics 10, 451–481.

Zweier, C., de Jong, E.K., Zweier, M., et al., 2009. CNTNAP2 and NRXN1 are mutated in autosomal-recessive Pitt-Hopkins-like mental retardation and determine the level of a common synaptic protein in Drosophila. American Journal of Human Genetics 85, 655–666.

Excitation–Inhibition Epilepsies

A.X. Thomas, A.R. Brooks-Kayal

University of Colorado, Aurora, CO, USA

OUTLINE

36.1 INTRODUCTION

Epilepsy affects more than 50 million people across the world, making it the second most common neurological disorder (Hauser et al., 1993). Children in particular are at high risk for seizures, and 5 out of every 1000 children will develop epilepsy in any given year. Known risk factors can account for only 25–45% of cases of early-life epilepsy, including congenital malformations of the central nervous system (CNS), moderate to severe head trauma, CNS infections, hypoxic–ischemic encephalopathy, inherited metabolic conditions, and genetic predisposition (Cowan, 2002).

To begin the discussion of epilepsy research, it is important to define the terms *epilepsy* and *epileptogenesis*. *Epilepsy*

Neural Circuit Development and Function in the Brain: Comprehensive Developmental Neuroscience, Volume 3 http://dx.doi.org/10.1016/B978-0-12-397267-5.00042-X

© 2013 Elsevier Inc. All rights reserved.

is a neurological disorder that is characterized by recurrent spontaneous seizures. *Epileptogenesis* is defined as the process during which changes occur in the brain after a precipitating injury or insult that results in the development of spontaneous recurrent seizure activity or epilepsy. The time period between the initial insult and the onset of epilepsy is referred to as the latent period. Much of the research in epilepsy is geared toward understanding the underlying molecular mechanisms associated with epileptogenesis. It is important to elucidate the underlying mechanisms of epileptogenesis both to identify better treatments for existing epilepsy and to develop new therapies to prevent epilepsy, as well as the associated neurocognitive disorders that afflict many individuals with epilepsy (LaFrance et al., 2008).

Although epilepsy is defined by the presence of recurrent seizures, it is often associated with cognitive or behavioral dysfunction, particularly in children, among whom 30% of epilepsy patients will have comorbid autism, intellectual or developmental disabilities, or both (Nolan et al., 2003; Tuchman et al., 2009). Children with epilepsy also show higher rates of attention deficit/hyperactivity disorder, learning disorders, and behavioral problems, as well as depression and anxiety (Beghi et al., 2006; Dunn et al., 2002; Gonzalez-Heydrich et al., 2007; Jones et al., 2007; Laurent and Arzimanoglou, 2006). A number of well-known genetic disorders share epilepsy, intellectual disability, and autism as prominent phenotypic features, including tuberous sclerosis, Rett syndrome, fragile X, Angelman syndrome, and genetic disorders of cortical development/migration with mutations such as aristaless-related homeobox (ARX), DCX, and Lis1 (reviewed in Brooks-Kayal, 2011; Deonna and Roulet, 2006; Guzzetta, 2006; Wolff et al., 2006). Seizures themselves, particularly when occurring during early life, may also produce a variety of cellular and molecular changes in the hippocampus that may contribute to an enhanced risk of cognitive or behavioral dysfunction (reviewed in Brooks-Kayal, 2011). Comorbid changes can be progressive, both over periods of years and over the course of the lifespan, and their severity has been correlated with age of onset, seizure frequency, total number of seizures, and increasing age (Dam, 1990; Hermann et al., 2002, 2003).

36.2 DIFFERENCES BETWEEN THE DEVELOPING AND MATURE BRAIN

To understand the complex effects of seizures on the developing brain, it is imperative to understand the differences between the developing and the adult brain. The functional development of inhibitory and excitatory neurons is an important part of the physiology of the normal developing neonatal brain. Three different types of pyramidal neuronal populations have been identified in the newborn rat hippocampus (Tyzio et al., 1999). These early neurons are described by their responsiveness to the neurotransmitters GABA and glutamate, and the populations consist of silent neurons, neurons that respond to only GABA, and neurons that respond to GABA and glutamate. The silent neurons lack spontaneous or evoked postsynaptic currents (PSCs), even in the presence of toxins that would normally stimulate transmitter release. These neurons have functional $GABA_A$ and glutamate receptors. The neurons that respond to only GABA demonstrate GABA-mediated but not glutamate-mediated PSCs. The neurons that respond to GABA and glutamate demonstrate both GABA-mediated and glutamate-mediated PSCs. These three populations of pyramidal neurons also differ morphologically. The silent neurons have a small soma and an axon with no apical dendrites. The neurons that respond to only GABA are more differentiated and have a larger soma and a small apical dendrite with no basal dendrite. The neurons that respond to GABA and glutamate are even more differentiated and have a basal dendrite and an apical dendrite that reaches the distal part of the stratum lacunosum moleculare (Tyzio et al., 1999).

Even more pertinent to human development is a similar observation made with respect to macaque monkey embryos in utero (Khazipov et al., 2001). In macaque monkeys, neurogenesis is complete in the hippocampus at E165 and morphological differentiation of the hippocampal pyramidal neurons occurs during the second half of gestation. The neurons are highly mature at the time of birth. Embryonic recordings conducted in the second half of gestation from embryonic day (E) 85 to E154 demonstrated that at E85 the CA1 pyramidal neurons were silent, with some neurons that respond to only GABA starting to form. At E105, the proportion of these neurons that respond to only GABA had increased and penetrated deeper into the stratum radiatum with a small proportion of neurons that respond to GABA and glutamate. From E119 and onward, the recorded pyramidal neurons exhibited both GABAergic and glutamatergic PSCs. The ability of $GABA_A$ receptor antagonist bicuculline to initiate epileptiform activity during various stages of development also was studied. At mid-gestation, epileptiform activity was not seen after the administration of the $GABA_A$ antagonists bicuculline but was observed at E105–109, which coincides with the appearance of neurons that respond to only GABA with axonal collaterals and the appearance of spines and glutamatergic PSCs. The occurrence and severity of the epileptiform events coincided with the development of the pyramidal neurons and the increase in the number of spines. Therefore, in primates, maturation of GABAergic and glutamatergic neurotransmission emerges early in utero. The CA1 pyramidal neurons are initially silent and then mature and differentiate to acquire morphological and physiological properties needed to generate network-driven epileptiform activities well before birth. Early hippocampal development is conserved throughout mammalian development. The steps in hippocampal development seem to be similar in rodents and primates but occur at a faster rate in primates (Khazipov et al., 2001).

The first synapses of the principal neurons in the hippocampus are formed on the apical dendrites (Dupuy and Houser, 1997; Rozenberg et al., 1989). Glutamatergic synapses are formed after the maturity of their postsynaptic target, whereas GABAergic synapses are formed between the axons of GABA neurons and the dendrites of pyramidal neurons. Inhibitory neurons play an important role in normal development of the neocortex and in regulating the critical period. The critical period is defined as a window of time during which experience provides input that is essential for normal development and permanently alters function (Hensch, 2005). During the critical period of rodent visual cortex development, GABA release is important for ocular dominance plasticity (Fagiolini and Hensch, 2000; Hensch et al., 1998). Enhanced inhibition with administration of benzodiazepines just after eye opening is able to accelerate the onset of the critical period (Fagiolini and Hensch, 2000; Fagiolini et al., 2004; Iwai et al., 2003). Tonic GABA

release is sufficient to trigger the completion of the critical period even with the complete absence of visual input (Hensch, 2005).

36.2.1 GABAergic Neurotransmission

GABA and glutamate receptors undergo dramatic changes over the course of postnatal development (Figure 36.1). GABAergic neurotransmission provides very important functions in the brain that differ between immature and mature neurons (Ben-Ari, 2002). GABA is a neurotransmitter that acts in an excitatory manner in the immature brain, while acting in an inhibitory manner in the mature brain (Ben-Ari et al., 1989) and is tied to the temporally regulated expression of K^+–Cl^--coupled cotransporter (KCC2) (Rivera et al., 1999). There is also a developmentally regulated shift in the expression of GABA receptor subunits (Brooks-Kayal and Pritchett, 1993; Brooks-Kayal et al., 2001; Fritschy et al., 1999; Wisden et al., 1992).

Ben-Ari et al. first reported that GABA functioned as a depolarizing, or excitatory, neurotransmitter in young neurons from neonatal hippocampal slices, unlike those of adults (Ben-Ari et al., 1989). In disassociated

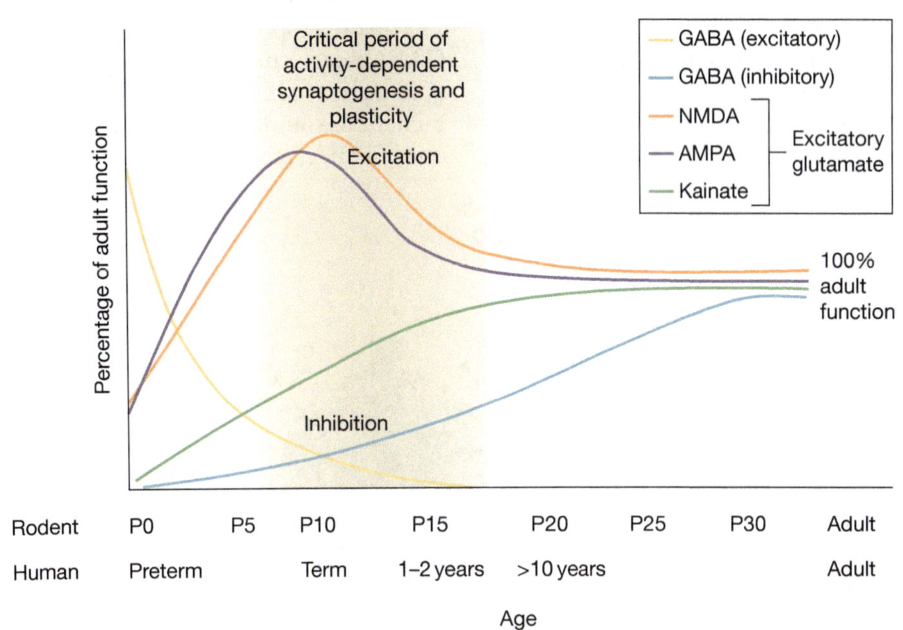

FIGURE 36.1 Equivalent developmental periods are displayed for rats and humans on the top and bottom x-axes, respectively. Activation of GABA receptors is depolarizing in rats early in the first postnatal week and in humans up to and including the neonatal period. Functional inhibition, however, is gradually reached over development in rats and humans. Before full maturation of GABA-mediated inhibition, the NMDA and AMPA subtypes of glutamate receptors peak between the first and second postnatal weeks in rats and in the neonatal period in humans. Kainate receptor binding is initially low and gradually rises to adult levels by the fourth postnatal week. Abbreviations: AMPA, -amino-3-hydroxy-5-methyl-4-isoxazole propionate; GABA, γ-aminobutyric acid; NMDA, N-methyl-D-aspartate; P, postnatal day. *Reprinted with permission of John Wiley & Sons, Inc. Neonatal seizures.* Annals of Neurology 62(2): 2007, 112–120 © 2007 Wiley-Liss, Inc., A Wiley Company.

embryonic spinal cord neurons, the opening of a single GABA channel is enough to trigger a sodium action potential (Serafini et al., 1995). The activation of GABA receptors also has the ability to generate calcium currents by directly activating voltage-dependent calcium channels (Fukuda et al., 1998; Leinekugel et al., 1995, 1997). The developmental shift in GABA function has been observed in a wide range of species and has been accepted as a general rule that has been conserved throughout vertebrate evolution. Glutamatergic neurons, which are the main excitatory neurons in the adult brain, are formed after GABAergic neurons (Ben-Ari et al., 1989). Thus, in the immature brain, GABA release and activation of GABA$_A$ receptors initiate depolarization and increased concentration of calcium, whereas glutamate serves this function in the adult brain (Ben-Ari, 2002).

GABA$_A$ receptor activity is sufficient to drive synchronous neuronal activity and can be inhibited by GABA$_A$ receptor blockade (Ben-Ari et al., 1989). In cultured hippocampal slices from young rats, elevated extracellular potassium-induced ictal-like epileptiform activity was blocked by GABA$_A$ receptor antagonists, bicuculline and gabazine, and increased in frequency and duration by GABA$_A$ receptor agonists, isoguvacine and muscimol (Dzhala and Staley, 2003). This exacerbation of epileptiform activity with GABA$_A$ receptor agonists and blockade with GABA$_A$ receptor antagonists is opposite to that seen in adult rats.

36.2.1.1 Chloride Gradient

To understand the implications of the chloride gradient on GABAergic transmission, it is important to compare the influx and efflux of chloride ions in neonatal and mature neurons. In neonatal hippocampal slices,

Ben-Ari et al. used electrochemical methods to observe an increase in chloride ions in neonatal hippocampal neurons as compared to the adult (Ben-Ari et al., 1989). The increase in chloride concentration by 20–40 mM is sufficient to shift the function of GABA from inhibition in the adult to excitation in the neonate (Ben-Ari, 2002). There are two main families of chloride transporters responsible for influx and efflux of chloride ions in neurons. The chloride gradient is driven by the Na$^+$–K$^+$–2Cl$^-$ cotransporter (NKCC1), which imports two ions of chloride for every ion of sodium and potassium exported, and KCC2, which exports one ion of potassium and chloride. Therefore, the activation of NKCC1 allows for the accumulation of intracellular chloride and the activation of KCC2 decreases the concentration of intracellular chloride. NKCC1 is expressed at early developmental stages and is responsible for the high intracellular chloride concentration inside immature neurons (Fukuda et al., 1998). Expression of KCC2 mRNA levels slowly increases to adult levels at 2 weeks postnatally in various brain regions, which coincides with the functional maturation of these areas. The transfection of immature hippocampal neurons with KCC2 caused an early switch of GABA function from excitatory to inhibitory. GABA also remains excitatory in mice that lack KCC2 expression (Rivera et al., 1999). Thus, the key transporter implicated in the developmentally regulated switch from the excitatory to the inhibitory effects of GABA is KCC2 (Figure 36.2).

The developmental shift from excitatory to inhibitory GABAergic currents is regulated by GABA itself (Ganguly et al., 2001). Blocking GABA$_A$ receptors with bicuculline and picrotoxin prevented KCC2 increases and GABA remained excitatory. The developmental

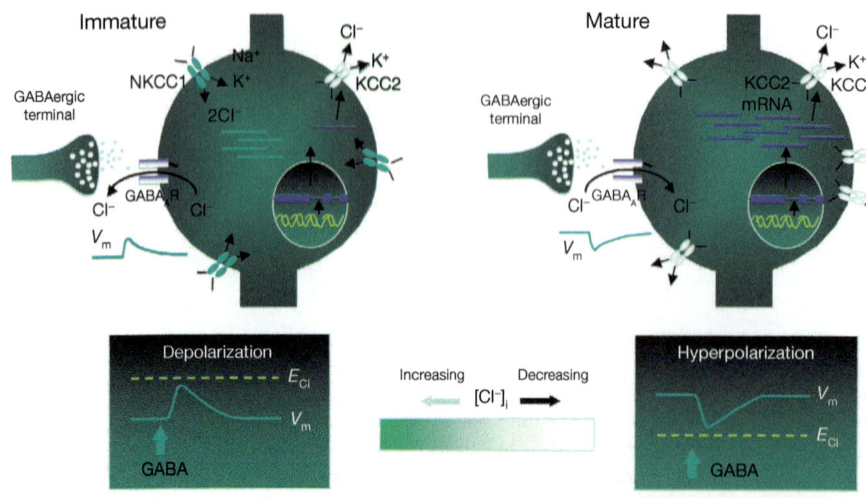

FIGURE 36.2 This schematic illustrates how the developmental shift in the chloride gradient (Cl$^-$) affects GABAergic transmission in the immature and mature brain. GABA$_{AR}$, GABA$_A$ receptor; KCC2, transports Cl$^-$ out of the cell; NKCC1, transports Cl$^-$ into the cell; mRNA, messenger RNA; V_m, membrane voltage; E_{Cl}, chloride equilibrium potential. *Reprinted from Brooks-Kayal AR (2005) Rearranging receptors. Epilepsia 46(supplement 7): 29–38, with permission.*

shift is independent of action potentials, since tetrodotoxin, a sodium channel blocker, had no effect on the shift from excitation to inhibition. Therefore, an action potential-independent quantal release of GABA resulting in miniature PSCs is sufficient to cause the expression of KCC2 and reduce $[Ca^{2+}]_i$ and appears to be required. Blocking glutamate activity had no effect on the developmental shift of GABA (Ganguly et al., 2001). Taken together, the studies demonstrate that the shift in GABA function from excitatory to inhibitory is primarily caused by the temporally regulated expression of KCC2 which is downstream of GABA$_A$ receptor activation.

36.2.1.2 GABA Receptor Subunit Changes

There are three main GABA subtypes: GABA$_A$, GABA$_B$, and GABA$_C$. GABA$_A$ receptors are ligand-gated ion channels composed of heterogeneous pentameric protein complexes. The pentameric receptors are formed by various combinations of different subunits: α, β, γ, δ, ε, π, and θ. The various combinations of GABA subunits bestow certain physiological characteristics as well as pharmacological properties. GABA$_A$ receptors are typically located postsynaptically and mediate fast synaptic inhibition in the adult brain and excitation in the developing brain. The GABA$_A$ receptors are primarily anion selective, and typically gate chloride ions but, in certain conditions, gate bicarbonate ions as well. GABA$_A$ receptors are typically composed of two α subunits, two β subunits, and one γ subunit. The γ subunit can be replaced by a δ, an ε, a θ, or a π subunit, depending on the neuronal subtype and subcellular localization of the receptor (McKernan and Whiting, 1996; Rudolph and Möhler, 2004). GABA$_B$ receptors are G-protein-coupled metabotropic receptors that are located both presynaptically and postsynaptically and exhibit longer-duration slow inhibitory currents through K^+ and Ca^{2+} channels. GABA$_B$ receptors are composed of a heterodimer consisting of GABA$_{B1}$ and GABA$_{B2}$. The molecular diversity seen in the GABA$_B$ receptors arises from the splice variants of the GABA$_{B1a}$ and GABA$_{B1b}$ subunit isoforms (Bettler and Tiao, 2006; Couve et al., 2000). GABA$_A$ and GABA$_B$ receptors are distributed across the CNS (Bowery et al., 1987).

There are significant spatiotemporal differences in the subunit expression of GABA$_A$ receptors. Alpha$_2$, α_3, and α_5 subunits peak early in development and stabilize or decline, whereas the expression of α_1 and γ_2 is at low levels early in development and increases to adult levels (Fritschy et al., 1994; Wisden et al., 1992). In rats, there is a greater than twofold increase in mRNA expression of α_1, α_4, and γ_2 from postnatal day (P)7 to adulthood and a tenfold decrease in the expression of α_5 mRNA during the same period (Brooks-Kayal et al., 2001). Developmentally regulated changes in GABA$_A$ receptor

subunit composition have also been observed in humans. Between the 36th week of gestation and adulthood, there is a threefold increase in the expression of α_1 mRNA levels in the cortex and cerebellum (Brooks-Kayal and Pritchett, 1993). There are also spatiotemporal differences in the subunit expression of GABA$_B$ receptors. In rodents, the presynaptic subunit of GABA$_B$, GABA$_{B1a}$, is high at birth and progressively declines to adult levels by P14 (Fritschy et al., 1999). However, the postsynaptic GABA$_B$ subunit, GABA$_{B1b}$, is low at birth, progressively increases in the first two postnatal weeks, and then declines to adult levels (McLean et al., 1996). There is some presynaptic GABA$_B$ receptor activity at birth, but no postsynaptic activity until the second or third postnatal week (Fukuda et al., 1993; Gaiarsa et al., 1995).

36.2.2 Glutamatergic Neurotransmission

Excitation outweighs inhibition during the first few years of life in the cerebral cortex and limbic structures in humans and rodents (Fox et al., 1996; Huttenlocher et al., 1982). Glutamate receptors are the main excitatory receptors in the adult brain and are divided into two main classes: ionotropic and metabotropic. Most pertinent to the discussion on developmental epilepsies is the ionotropic class of glutamate receptors. These ionotropic glutamate receptors activate cation-selective ion channels permeable to Na^+, K^+, and Ca^{2+} possessing varying degrees of permeability and blockade by the divalent cation Mg^{2+} (Mayer and Westbrook, 1987). There are three main types of ionotropic glutamate receptors: α-amino-3-hydroxyl-5-methyl-4-isoxazolepropionate (AMPA), kainate, and N-methyl-D-aspartic acid (NMDA) (Mayer, 2005). NMDA receptors are mainly located postsynaptically, require glycine as a coagonist, and have a voltage-dependent magnesium blockade that must be released before agonist binding can open the channel to allow gating of calcium and sodium. NMDA receptors are composed of an obligatory NR1 subunit with a combination of NR2A, NR2B, NR2C, and/or NR3A subunits (Lau and Zukin, 2007). AMPA receptors are typically located on the postsynaptic membrane and are composed of various combinations of GluR1, GluR2, GluR3, and/or GluR4 (Shepherd and Huganir, 2007). Kainate receptors are located presynaptically and postsynaptically on both neurons and glia. Kainate receptors are almost exclusively composed of various combinations of GluR5, GluR6, GluR7, KA1, and/or KA2 and gate sodium (Pinheiro and Mulle, 2006). NMDA receptors are always permeable to calcium ions, whereas AMPA and kainate receptors are selectively permeable to divalent cations depending on their subunit compositions. When AMPA receptors contain low levels of GluR2 or lack it entirely, the receptors are calcium-permeable (Shepherd and Huganir, 2007). When kainate receptors have low levels of GluR5 or GluR6 or lack these subunits altogether, they allow increased gating of calcium ions (Pinheiro and Mulle, 2006).

The expression of glutamate receptor subunits is spatially and developmentally regulated and has important functional implications. In the mature brain, NR2A is present in almost all areas of the brain, while NR2B is located in the forebrain, NR2C is limited to the cerebellum, and NR2D is restricted to the thalamus and subthalamus (Monyer et al., 1994). In the immature brain, NMDA receptors have high levels of NR2B, NR2D, and NR3A subunits. Elevated levels of NR2B subunits extend current decay times, and the high levels of NR2D and NR3A reduce NMDA receptor sensitivity to magnesium ions (Lau and Zukin, 2007). The increased expression of the NR2B, NR2D, and NR3A subunits is implicated in increased NMDA receptor-mediated calcium influx, lower threshold for seizures, and excitotoxic hypoxic–ischemic injury during development (Rakhade and Jensen, 2009). AMPA receptor subunit composition is also developmentally regulated. The immature rat brain has lower GluR2 expression and a higher prevalence of GluR2 subunit-deficient receptors, which causes an increased calcium and sodium influx. When GluR2 is present in the receptor, AMPA receptors primarily gate sodium (Kumar et al., 2002; Sanchez and Jensen, 2001). In the developing human brain, there is a relative deficiency in GluR2 subunit-expressing AMPA receptors in cortical neurons during term and the early postnatal period. There is a direct correlation between the expression of GluR2-deficient AMPA receptors and vulnerability to hypoxic/ischemic injury (Talos et al., 2006a,b).

36.2.2.1 *GABA, NMDA, and AMPA*

In the developing brain, GABAergic and glutamatergic synapses work in conjunction with each other. Early glutamatergic neurotransmission is solely NMDA-receptor mediated, without any significant contribution of AMPA receptors (Durand et al., 1996; Isaac et al., 1997; Wu et al., 1996). Voltage-dependent Mg^{2+} blockade of NMDA receptors takes place in both neonates and adults (Khazipov et al., 1995; Swann et al., 1999); however, the degree varies across development, since NMDA receptors containing NR2D and NR3A are expressed in immature neurons and are less sensitive to Mg^{2+} blockade (Lau and Zukin, 2007). In the CA1 region of the hippocampus, the first glutamatergic synapses formed at P2 are silent at resting potential; however, the proportion of silent synapses progressively decreases from P2 to P5 (Durand et al., 1996; Liao and Malinow, 1996). Typically, the activation of AMPA receptors is sufficient to remove the Mg^{2+} blockade of NMDA channels in adult neurons,

resulting in their activation. However, in neonatal hippocampal slices, AMPA activation is not sufficient to remove Mg^{2+} blockade of NMDA channels (Liao and Malinow, 1996); instead, GABA$_A$ receptor activation serves this role (Leinekugel et al., 1995). The opposite is true in the adult brain, where GABA$_A$ receptor activation prevents the activation of NMDA receptors, thereby inhibiting NMDA-dependent forms of synaptic plasticity (Artola and Singer, 1987; Wigström and Gustafsson, 1983). GABA-dependent activation of P2–P5 CA3 pyramidal neurons by the GABA$_A$ receptor agonist, isoguvacine, is sufficient to remove the voltage-dependent Mg^{2+} blockade of NMDA receptors, thereby increasing the affinity of NMDA channels for magnesium and increasing calcium ion influx (Leinekugel et al., 1995). Therefore, in the developing brain, GABA$_A$ receptor activation acts synergistically with NMDA receptor activation, providing the same function that AMPA receptors offer in the adult brain.

36.3 EXPERIMENTAL MODELS OF DEVELOPMENTAL EPILEPSY

Epilepsy is a very complex disease with many different causes and pathologies; therefore, it is important to develop research models that try to mimic the various types of human epilepsies. Many of the research models of seizures and epilepsy have been thoroughly reviewed (Pitkänen et al., 2005; Table 36.1).

36.3.1 Environmental/Perfusion

36.3.1.1 *Hypoxia*

The most common cause of neonatal seizures is hypoxic encephalopathy (Aicardi and Chevrie, 1970; Hauser et al., 1993). The neonatal rodent hypoxia model has been utilized to elucidate how global hypoxia in neonates leads to increased susceptibility to seizures later in life. Humans and rodents have a similar age-dependent susceptibility to hypoxia-induced seizures early in life, which can cause long-term susceptibility to seizure and neuronal death. P9–P12 Long–Evans rats demonstrate a high susceptibility to hypoxia-induced seizures, with a peak in P10 pups. Sprague–Dawley rats exhibit peak seizures at P8–P9 (Owens et al., 1997). Similar windows of susceptibility are also seen in mice. The hypoxia model uses a brief 15-min exposure to a graded global hypoxia (7–4% O_2) by altering the levels of O_2 and N_2 in an airtight chamber. The acute hypoxia-induced seizures lead to a long-term increase in the susceptibility to seizures, which is usually ascertained through the use of threshold doses of various chemoconvulsants,

TABLE 36.1 Animal Models of Neonatal Seizures and Epilepsy

Age	Experimental model	Human condition	
		Acute model	Recurrent spontaneous seizures (epilepsy)
Neonatal			
P0–P5	Flurothyl	Neonatal seizures	No
P3–P7	Hypoxic ischemia	Hypoxic–ischemic encephalopathy and perinatal stroke	Yes
Adolescent			
P8–12	Hypoxia	Hypoxic encephalopathy	Yes
P7–P14	Hypoxic ischemia	Hypoxic–ischemic encephalopathy and perinatal stroke	Yes
P10–P11	Hyperthermia	Febrile seizures	Yes
P10	Tetanus toxin single dose	Complex partial seizures	Yes
P10–P12	Tetanus toxin continuous infusion	Infantile spasms	Yes
P7–P10	Lithium–pilocarpine	Neonatal seizures	No
P12–P20	Lithium–pilocarpine	Neonatal seizures	Yes*
P7–P10	Kainate	Neonatal seizures	No
P20–Adult	Kainate	Neonatal seizures	Yes[a]

[a] *The increase in rate of epilepsy development directly correlates with the age.*

including pentylenetetrazol (PTZ), flurothyl, and kainate (Jensen et al., 1992; Rakhade et al., 2008b), as well as to spontaneous epileptiform/ictal EEG discharges in a subset of exposed animals (Rakhade et al., 2008a).

36.3.1.2 Hypoxic Ischemia

Neonatal seizures also occur after hypoxic–ischemic encephalopathy and perinatal stroke modeled by the hypoxic–ischemic rodent model of neonatal seizures. Typically, this method is used with P3–P14 rodents by ligating a cerebral artery, usually a unilateral carotid or middle cerebral artery. After ligation, the rodents are allowed to recover and are then subjected to 15–30 min of hypoxia (8% O_2) (Jensen et al., 1994; Rice et al., 1981; Vannucci et al., 1999). Similar to the hypoxia model, the hypoxic–ischemic model results in long-term increases in seizure susceptibility as assessed by injections of chemoconvulsants, as well as emergence of epilepsy later in life (Wirrell et al., 2001; Yager et al., 2002).

36.3.1.3 Hyperthermia

The most common type of seizure in infants and young children is febrile seizures, with a prevalence of 2.3–4%. Simple febrile seizures are short and generalized and occur with high fevers in infants and young children from ~3 months to 5 years of age, peaking between 6 months and 2 years of age. Complex febrile seizures are prolonged or focal seizures or seizures that recur within a single febrile episode. Although the majority of simple febrile seizures are benign, complex prolonged febrile seizures have been associated with an increased risk of later epilepsy development (Cowan, 2002; Hauser, 1994). The model for prolonged febrile seizures was first developed in the immature rat (Baram et al., 1997; Toth et al., 1998) and then adapted for mice (Dubé et al., 2005). The febrile seizure model induces hyperthermia to produce prolonged seizures. The core temperature of P10–11 rats or P14–15 mice is increased gradually and maintained at hyperthermia (40–42 °C) for 30 min, with the animal's core temperature being tightly regulated. The rodents are then allowed to recover for 1 h. In this model, approximately 24% of rats go on to develop spontaneous temporal lobe seizures in adulthood. If the duration of hyperthermia is increased to 64 min, then 45% of rats develop spontaneous seizures in adulthood (Dubé et al., 2010).

36.3.2 Toxin

36.3.2.1 Tetanus Toxin

The ability of tetanus toxin to induce chronic epilepsy was first described by Roux and Borrel (1898). The modern form of the tetanus toxin model of epilepsy involves stereotaxic injection of minute amounts of tetanus toxin into the brain to create an epileptic focus that spontaneously discharges. After interhippocampal infusion of tetanus toxin in P10 rats, anywhere from one to seven seizures are typically observed during the first week, with each seizure lasting from a few seconds to several minutes. The seizure frequency usually peaks on day two and declines over the week following infusion (Benke and Swann, 2004). When the immature rats become adults, they can exhibit unprovoked seizures, typically with a low prevalence (Lee et al., 1995).

36.3.3 Chemoconvulsant

36.3.3.1 Lithium–Pilocarpine

The lithium–pilocarpine model is a variation of the pilocarpine model used in adults. Honchar et al. determined that systemic activation of the cholinergic system in lithium-treated rats induced seizures (Honchar et al., 1983). Typically, lithium is administered 24 h before administration of pilocarpine. The pretreatment with lithium greatly decreases the amount of pilocarpine needed to induce status epilepticus (SE). There are two dosing regimens, the first with a high dose of pilocarpine (25–60 mg kg^{-1}) 24 h after lithium treatment and the other with divided doses (10 mg kg^{-1}) administered 24 h after lithium treatment every 30 min until SE is induced (Glien et al., 2001). In P12 rats, 25% of the rats experience spontaneous electrographic seizures of limbic onset (i.e., temporal lobe epilepsy (TLE)) 3 months after SE (Kubová et al., 2004). By P20, this percentage increases to 67% (Raol et al., 2003).

36.3.3.2 Kainate

The seizure-inducing properties of kainate were first reported by Nadler et al. (1978). The kainate model is another useful model of TLE, because the rats experience an episode of SE on the order of hours after injection with kainate, followed by days to weeks of a seizure-free latent period, finally ending with progressive recurrent spontaneous seizures, typically when SE occurs in rodents $\geq$P20. Spontaneous recurrent seizures do not develop in P0 and P5 kainate-induced rats (Stafstrom et al., 1992). There are two main dosing regimens for kainate-induced SE, one using a single high dose (Meier et al., 1992) and the other using repetitive low doses (Meier and Dudek, 1996).

36.3.3.3 Flurothyl

Flurothyl is an inhaled GABAergic antagonist that induces seizures within minutes (Truitt et al., 1960). This method can be used to induce seizures up to five times per day (Cha et al., 2002; Holmes et al., 1998). Seizures are arrested within 30 s of returning the rats to room air; therefore, it is a useful model to study the effect of recurrent seizures. Continuous exposure to flurothyl is able to produce SE (Sperber et al., 1999). Immature rats (P0–P5) that receive recurrent flurothyl seizures do not go on to develop spontaneous recurrent seizures; however, the rats do have a reduced seizure threshold later in life (Holmes et al., 1998; Sogawa et al., 2001).

36.3.4 Kindling

The phenomenon of kindling was first described by Graham Goddard (1967). Goddard studied the effect of repeated hippocampal electrical stimulation on learning in rats. The small electrical insults led progressively to greater severity of seizures and resulted in a permanent increase in seizure susceptibility (Goddard et al., 1969). Kindling is initiated by periodic insults that result in network synchronization that are accompanied by behavioral seizures. There are two main methods of kindling, electrical and chemical. The electrical model of kindling uses repetitive electrical stimulation to various brain centers including the amygdala, perforant

pathway, dorsal hippocampus, olfactory bulb, and perirhinal cortex. Chemical kindling utilizes repetitive administration of chemical agents that evoke repetitive epileptic spiking such as carbacol, acetylcholine, bicuculline, lidocaine, cocaine, and PTZ.

36.3.5 Infantile Spasms Models

Infantile spasms are a severe form of developmental epilepsy that begins between 3 and 12 months of age and has numerous etiologies, including a wide range of acquired and congenital causes (Frost and Hrachovy, 2005). Infantile spasms are associated with specific and unique electroencephalogram (EEG) findings, consisting of chaotic, high-voltage slow background activity intermixed with multifocal spikes and generalized slow waves (hypsarrhythmia), followed by generalized voltage attenuation during the spasm seizures. Only a few medications are effective for infantile spasms (glucocorticoids, adrenocorticotropic hormone, and vigabatrin), and the outcome of infantile spasms is usually poor, with other seizure types occurring after the first year of life and severely abnormal neurological development. The prognosis is especially poor in cases where delay is noted prior to the onset of spasms and the spasms do not disappear with therapy (Stafstrom, 2009). Recently, several animal models have been developed that recapitulate many of the features of infantile spasms, including the NMDA model (Kábová et al., 1999), the tetrodotoxin infusion model (Lee et al., 2008), and the multiple injury model (Scantlebury and Moshé, 2006). In addition, two genetic models of infantile spasms have been developed, the ARX model and the Ts65Dn Down syndrome model, which are discussed later. These new models should provide insights that will improve the understanding, treatment, and prevention of this devastating early-childhood epilepsy syndrome.

36.3.6 Genetic

In humans, mutations in the X-linked ARX gene are linked to structural brain abnormalities, including lissencephaly and abnormal migration of inhibitory interneurons, and neurological deficits, including severe intellectual disability and early-life epilepsy (known as an infantile epileptic encephalopathy). Conditional ARX knockouts exhibit behavioral and electrographic early-life seizures resembling infantile spasms (Marsh et al., 2009; Price et al., 2009). The administration of GABA$_B$ agonists to a mouse model of Down syndrome Ts65Dn mice between 1 week and 2 months old causes them to develop clusters of extensor spasms, which resemble infantile spasms, accompanied by polyspike-wave bursts and electrodecremental responses on EEG (Cortez et al., 2009).

Transgenic mice with mutations in KCNQ3 are known to develop seizures within the first 2 weeks of life. Mutations of KCNQ2 and KCNQ3 are linked to benign familial neonatal convulsions in humans (Castaldo et al., 2002). $K_v1.1$ knockouts exhibit a seizure-sensitive predisposition at P10, demonstrating an increase in early-life seizure susceptibility. Mutations in KCNA1, the human homolog of $K_v1.1$, can lead to episodic ataxia and seizures in humans (Rho et al., 1999). There are many other genetic mouse models with genes that are implicated in human early-life seizures, including SCN2A, SCN1B, and KCNQ2, to name a few. However, these mouse models do not demonstrate early-life seizures but rather develop seizures later in life (Chen et al., 2004; Kearney et al., 2001; Peters et al., 2005).

36.4 SEIZURE-INDUCED CHANGES IN THE BRAIN

The changes in the brain due to seizures can be divided into three overlapping temporal groups: acute, subacute, and chronic. Acute changes are those changes that occur in the brain within the first few minutes to days after a seizure. The subacute changes take place from hours to weeks with the chronic changes taking place over weeks to months. The acute changes that occur are activation of immediate early genes (IEGs), post-translational protein modifications, and changes to ion channel activity. The subacute changes in the brain include neuronal death, activation of neurotrophic factors, inflammation, neurogenesis, and alterations in transcription factors. The chronic changes include mossy fiber sprouting, dendritic plasticity, and increased susceptibility to recurrent seizures (Rakhade and Jensen, 2009).

36.4.1 Acute

36.4.1.1 *Immediate Early Genes*

IEGs are genes that are rapidly transcribed in response to cellular stimuli, such as neuronal activity, and are implicated in synaptic plasticity and synaptogenesis. Many IEGs are transcription factors and DNA-binding proteins and have the ability to activate specialized signaling cascades. Repeated synaptic activity induces IEG activation which leads to depolarization and opening of NMDA receptor channels. The calcium influx following NMDA receptor activation can cause activation of kinase cascades that result in phosphorylation of specific transcription factors such as cyclic-AMP response element-binding protein (CREB) and CREB-binding protein (CBP) (Greer and Greenberg, 2008). CREB is a bZIP transcription factor that is activated when phosphorylated at its Ser133 site. The phosphorylated CREB (pCREB) then translocates into the nucleus and dimerizes and binds to the promoter consensus sequence TGACGTCA, which is known as the cAMP response element (CRE) motif. pCREB along with CBP, a chromatin regulator, has the ability to upregulate transcription of target genes (Lonze and Ginty, 2002). The upregulated IEGs can activate secondary response genes that have the ability to modulate synaptic activity (Greer and Greenberg, 2008). IEGs, including *Fos, Jun, EGr1, Egr4, Homer1, Nurr77, Arc,* and *CREB,* have been identified as activated in adult animal models of epilepsy (Herdegen and Leah, 1998), some as early as 30 min after seizure induction (Honkaniemi and Sharp, 1999). IEG activation has also been seen in the developing brain, but to a lesser degree. *Fos* and *Jun* activation has been seen in hypoxia-induced seizures and lithium–pilocarpine SE models (Dubé et al., 1998; Jensen et al., 1993). Activation of *Fos* differs based on age and mode of seizure induction. PTZ- and flurothyl-induced seizures show similar patterns of *Fos* activation in the amygdala, piriform cortex, and hypothalamus, but patterns in the cortex vary based on the age of the rat. *Fos* activation in adults is seen in the superficial layers of the cortex, whereas in P10 rats the activation is seen in the deep layers of the neocortex. Hypoxia-induced seizures demonstrated that *Fos* activation is confined to layer VI of the neocortex and is rarely involved in the limbic structures as seen with chemoconvulsant models (Jensen et al., 1993).

36.4.1.2 *Ion Channel and Receptor Posttranscriptional Regulation*

The calcium influx after a seizure has the ability to activate phosphatases and kinases that alter ion channel and neurotransmitter receptor function. The calcium-calmodulin-activated phosphatase, calcineurin, can lead to $GABA_A$ receptor endocytosis, decreased inhibitory postsynaptic potential frequency, and reduced network inhibition after early-life hypoxia-induced seizures (Sanchez et al., 2005) and adult pilocarpine-induced SE *in vivo* (Kurz et al., 2001). The same observation of calcineurin-regulated $GABA_A$ receptor-mediated synaptic inhibition has also been seen in neuronal culture and whole hippocampal slices (Khalilov et al., 2003). Seizure-induced calcineurin activation causes dephosphorylation and endocytosis of $GABA_A$ receptors (Blair et al., 2004; Sanchez et al., 2005) and $K_v2.1$ channels (Bernard et al., 2004). Pretreatment with FK506, a calcineurin inhibitor, reversed SE-induced dephosphorylation of $GABA_A$ receptor 2/3 subunits. Calcineurin-mediated dephosphorylation and endocytosis of $GABA_A$ receptors reduce their inhibitory function and may contribute to neuronal excitability in the hippocampus after seizures and epileptogenesis (Wang et al., 2009).

Glutamate receptors contain kinase phosphorylation sites that can be activated by seizures in the developing brain. Within minutes after a seizure in juvenile rats, protein kinase C activity and calcium–calmodulin-dependent kinase II activity cause an increase in phosphorylation of Ser831 on GluR1 and Ser880 on GluR2 subunits of the AMPA receptor. Protein kinase A activity and phosphorylation of Ser845 on GluR1 are also observed (Rakhade et al., 2008a,b). The phosphorylation of Ser831 on GluR1 is known to increase channel conductance and phosphorylation of Ser845 on GluR1 increases open-channel probability (Shepherd and Huganir, 2007), which can lead to AMPA receptor-mediated potentiation (Rakhade et al., 2008a,b). The administration of AMPA receptor antagonists 48 h after a seizure is sufficient to prevent kinase activation, phosphorylation of Ser831 and Ser845 on GluR1, altered AMPA receptor activity, and increased seizure susceptibility later in life (Rakhade et al., 2008a,b). Increased phosphorylation of Ser880 on GluR2 is known to cause endocytosis of GluR2 subunit and increased Ca^{2+} permeability (Shepherd and Huganir, 2007). Activation of cellular sarcoma kinases can induce phosphorylation of NR2A and NR2B receptor subunits in NMDA receptors in the hypoxic–ischemic model of developmental epilepsy (Jiang et al., 2008).

36.4.2 Subacute

36.4.2.1 Neuronal Death

Adult models of epilepsy have shown progressive loss of CA3 and CA1 neurons via necrosis and apoptosis in the hippocampus in response to electrical or chemoconvulsant stimulation (Henshall and Murphy, 2008; Henshall and Simon, 2005). Models of early-life seizures have failed to demonstrate neuronal loss in the amygdala, hippocampus, or temporal cortical regions of animals younger than 2 weeks of age (Holopainen, 2008). There is limited neuronal death seen in the hippocampal neurons of immature brains in the flurothyl-induced seizure model (Wasterlain et al., 2002) and in the febrile seizure model (Toth et al., 1998). In the lithium–pilocarpine model, regional and age-related differences in neuronal cell loss have been demonstrated. In P14 and P21 rats exposed to lithium–pilocarpine-induced seizures, few damaged neurons were seen in the CA1 region of the hippocampus, whereas some neuronal damage was observed in the hilar and CA3 regions at P14, with extensive damage at P21 in some, but not all, animals (Raol et al., 2003; Sankar et al., 1998). The damage is thought to be mediated via necrosis, as demonstrated by eosinophilic cell infiltration, and apoptosis, as demonstrated by terminal deoxynucleotidyl transferase-mediated biotinylated UTP nick end labeling (TUNEL) stain. A direct relationship between age and vulnerability of neurons in the amygdala and dentate gyrus was also observed (Sankar et al., 1998). Interestingly, the degree of cell loss did not correlate with the risk

of later development of epilepsy (Raol et al., 2003). Some variability in neuronal death in chemoconvulsant models has been seen with different routes of administration. Intracerebroventricular (i.c.v.) delivery of kainic acid causes more severe acute and progressive damage than an intraperitoneal (i.p.) route of delivery. Administration of kainic acid in P7 rats i.c.v. causes a dose-dependent acute neuronal loss in the CA3 region of the hippocampus with apoptosis seen in CA3 and CA1 as demonstrated by electron microscopy and TUNEL staining (Humphrey et al., 2002). However, kainic-acid-induced SE in P9 rats delivered by i.p. did not produce neuronal damage (Rizzi et al., 2003).

There is another population of neurons, the subplate neurons, that is susceptible to neuronal loss in the hypoxic–ischemic model of developmental epilepsies (McQuillen et al., 2003). The subplate neurons are located in the deep gray matter proper of the neocortex and become interstitial neurons of the subcortical white matter during the prenatal and neonatal periods. They are important for normal maturation of cortical networks, such as the visual cortex (Kanold et al., 2003; Kostovic and Rakic, 1980; Rakic, 1977). Subplate neurons also possess high levels of NMDA and AMPA receptors (Talos et al., 2006a,b) but lack oxidative stress defense mechanisms, making them especially susceptible to hypoxic–ischemic insults (McQuillen et al., 2003). Administration of antioxidants, such as erythropoietin, after acute hypoxia increased the latency of seizures, reduced the duration of seizures, protected against hippocampal cell loss, and decreased apoptosis in P10 rats (Mikati et al., 2007).

36.4.2.2 Neurotrophic Factors

Neurotrophic factors are important mediators of normal synaptogenesis and are expressed at higher amounts during postnatal development than in the adult (Greer and Greenberg, 2008). Early-life seizures in animal models have been shown to increase neurotrophic factor expression (Tandon et al., 1999). Chronic intrahippocampal injection of brain-derived neurotrophic factor (BDNF) results in spontaneous limbic seizures through TrkB signaling in adult animals, suggesting that BDNF is sufficient to produce epileptogenesis (Scharfman et al., 2002). Mice with a conditional knockout of TrkB in neurons have been shown to be protected from epileptogenesis (McNamara et al., 2006). In the adult rat kindling model, BDNF/TrkB activation suppresses the surface expression of KCC2, thereby suppressing chloride-dependent fast GABAergic inhibition (Rivera et al., 2002). However, BDNF has also been demonstrated to play a neuroprotective role during seizures in immature brains. During kainate-induced seizures in P19 rats, BDNF concentrations increase twofold and the administration of antisense oligodeoxynucleotides specific to BNDF increased seizure duration and loss of CA1 and CA3 pyramidal cells and hilar interneurons. The neuronal loss was determined not to be dependent on seizure duration (Tandon et al., 1999).

36.4.2.3 Inflammation

The inflammatory processes that occur after seizures in the adult brain are important contributors to neuronal cell death (Kunz and Oliw, 2001); however, the role of inflammatory processes in the developing brain is less well understood. SE-induced glial activation and cytokine transcription are age-dependent. At P9, there is very little increase in the activation of glia and cytokine induction after kainate-induced seizures. At P15, immunostaining of microglia and astrocytes increased, as did IL-1β mRNA expression and CA3 neuronal injury. At P21, the immunostaining of microglia and astrocytes was significantly increased, as were IL-1β, TNF-α, IL-6, and Ra (endogenous IL-1 receptor antagonist). The neuronal injury in CA1 and CA3 was also significantly increased (Rizzi et al., 2003). The precise mechanism of neuronal death is poorly understood; however, increase in cytokine expression may play a role. One possible mechanism may involve IL-1β, as IL-1β is known to initiate phosphorylation of NMDA receptors (Viviani et al., 2003), thereby altering the receptor channel-gating properties to favor Ca^{2+} influx (Ali and Salter, 2001). In the rodent model of prolonged febrile seizure, hippocampal levels of IL-1β were significantly elevated after prolonged febrile seizures for a period of over 24 h. Chronically, IL-1β levels were elevated only in the subset of rats that developed spontaneous limbic seizures after febrile SE, consistent with a role for this inflammatory mediator in epileptogenesis (Dubé et al., 2010).

Microglia are a glial cell type that reside in the brain and spinal cord and function, in a manner similar to macrophages, as active immune defense for the CNS. Microglia reach maximal density in the brain during early development in humans (Billiards et al., 2006) and rodents (Dalmau et al., 2003). In addition to the phagocytic role of microglia, they are able to produce neurotrophic molecules, cytokines, and chemokines (Kim and de Vellis, 2005). Microglia are also involved in classical complement cascade-mediated CNS synapse elimination in the developing brain (Stevens et al., 2007). Seizures can directly activate microglia that trigger a cytokine-mediated inflammatory response, as well as complement factors and major histocompatibility class factors (Vezzani et al., 2008). In adult rats, the administration of the microglial inactivators minocyclin and doxycycline has been shown to protect against neuronal death in the kainate-induced SE model (Heo et al., 2006). This observation could not be reproduced in juvenile rat kainate-induced seizures because of the lack of neuronal cell loss during the second postnatal week in this model (Holopainen, 2008).

Cytokine activation can induce two main developmentally regulated inflammatory pathways: inducible nitric oxide synthesis (iNOS) (Romero et al., 1996) and/or the cyclooxygenase (COX) pathway (Tocco et al., 1997). COX-2 is the rate-limiting enzyme for conversion of arachidonic acid to prostaglandins. The enzyme is upregulated in response to seizures and may contribute to SE-induced CA3 hippocampal neuron damage in adult rats (Kawaguchi et al., 2005; Tu and Bazan, 2003). The expression of COX-2 markedly increases between P7 and P14 and reaches adult levels at P21 (Tocco et al., 1997). The developmentally regulated nature of COX-2 expression may in part explain the age-dependent effect of inflammation and neuronal death. In P9 rats, iNOS and COX-2 were upregulated after NMDA injection-induced excitotoxic shock. iNOS was observed in infiltrated neutrophils and in ramified protoplasmic astrocytes closely associated with blood vessels, whereas COX-2 was observed in active microglial and neuronal cells (Acarin et al., 2002). In P21 KA-induced rats, two phases were seen after seizure induction. The first phase occurs within 30 min, is localized to the hippocampus, and is caused by kainic acid receptor activation. Pretreatment with a selective COX-2 inhibitor, NS398, is able to inhibit the inflammatory process almost entirely. The late phase of the inflammatory process seems to be due to prolonged COX-2 expression localized to the hippocampus (Yoshikawa et al., 2006). Pretreatment with celecoxib, a selective COX-2 inhibitor, in flurothyl-induced P7 rats is able to delay signs of seizure and attenuate COX-2 expression (Kim and Jang, 2006). Interestingly, inflammatory mediators can affect neuronal function by altering activity-dependent long-term synaptic plasticity, neuronal excitability, and synaptic transmission in CA1 pyramidal neurons (Chen and Bazan, 2005).

36.4.2.4 Alteration in Transcription of Receptors

In addition to acute changes in posttranslational regulation of ion channels and neurotransmitter receptors, seizures can induce transcriptional changes in GABA and glutamate receptors (Holopainen, 2008). Kainate-induced seizures in P9 rats are sufficient to alter the normal maturation of $GABA_A$ receptor expression by altering region-selective expression of $α_1$, $α_2$ $β_3$, and $γ_2$ subunit mRNAs in the hippocampus that can last up to a week (Laurén et al., 2005). In addition to acute and subacute changes in $GABA_A$ receptor composition in young rats, long-term alterations in $GABA_A$ receptor $α_1$ subunit mRNA and protein expression persisted up to 3 months in the dentate gyrus of the hippocampus in SE-induced P10 and P20 rats (Raol et al., 2006; Zhang et al., 2004). The shifting role of $GABA_A$ receptors from excitatory to inhibitory in the developing brain is demonstrated when seizures generated by functional excitatory GABAergic synapses cause fast oscillations that are necessary to transform normal network activity to an epileptic network (Khazipov et al., 2004). Interestingly, in the immature brain, inhibition of GABA$_A$ receptors can prevent long-lasting sequelae of seizures, whereas in the adult brain GABA$_A$ receptor inhibition leads to high-frequency seizures (Khalilov et al., 2005).

Seizures also affect the expression of both ionotropic and metabotropic glutamate receptor subunits. Downregulation of kainate receptors in the CA3 and dentate gyrus was observed in recurrent kainate-induced seizures in P12 rats (Tandon et al., 2002). In addition, elevation of mGluR1$_\alpha$ protein expression was seen in the inhibitory interneurons of the CA1 stratum oriens–alveus, amygdala, and piriform cortex of the same kainate-induced model. This change in mGluR1$_\alpha$ expression may induce synchrony of the limbic network suppression that can prevent further seizure propagation (Avallone et al., 2006). There is a decrease in GluR2 protein levels at P10 after lithium–pilocarpine (Zhang et al., 2004) and hypoxia-induced seizures (Sanchez et al., 2001). The decrease in GluR2 levels could be explained by excitotoxicity, which activates the repressor element 1-silencing transcription factor that suppresses GluR2 promoter activity, leading to AMPA receptor-mediated neuronal death (Calderone et al., 2003). Ionotropic glutamate receptors such as NMDA receptors are responsible for the excitotoxicity seen with prolonged receptor activation and Ca^{2+} influx in seizure-related neuronal death in adult rats (Furukawa et al., 1997).

Pilocarpine-induced SE in P14 rats showed increased AMPA GluR2 and kainate KA2 subunit mRNAs with decreases in AMPA GluR3 and kainate GluR6 mRNAs, but only in mature dentate granule cell neurons. In the study, immature dentate granule cells showed a decrease only in kainate GluR6 mRNA levels (Porter et al., 2006). These changes to kainate receptor subunits may play a role in the altered kainate receptor conductance and dentate granule cell excitability seen in chronic epilepsy (Epsztein et al., 2005). At P7, a single kainate-induced seizure has the ability to cause a long-term decrease in expression of GluR1 and NR2A subunits and an increase in PSD-95, a primary subsynaptic scaffold, in CA1 (Cornejo et al., 2007).

36.4.2.5 *Neurogenesis*

Increased neurogenesis has been seen in autopsy specimens and tissue biopsies from pediatric epilepsy surgery patients (Takei et al., 2007). Timing of the seizure insult plays a very important role in seizure-induced neurogenesis. There is a significant decrease in the neurogenesis observed in the granule cell layer of the dentate gyrus following a series of 25 flurothyl-induced

seizures at P0–P4 as determined via 5-bromo-2′-deoxyuridine-5′-monophosphate (BrdU) and NeuN co-labeling (McCabe et al., 2001; Schmid et al., 1999). The decrease in BrdU labeling continued for 6 days after the last seizure. A single flurothyl-induced seizure showed no difference in BrdU-labeled cells when compared to controls. Adult rats that were subjected to a series of 25 flurothyl-induced seizures showed a marked increase in dentate gyrus neurogenesis when compared to controls (McCabe et al., 2001). Between 1 and 4 weeks of age, chemoconvulsant-induced seizures cause long-term increases in dentate granule cells after flurothyl-induced seizures (Holmes et al., 1998) and lithium–pilocarpine-induced seizures (Porter et al., 2004; Sankar et al., 2000). Lithium- and pilocarpine-induced SE rats at P20 showed a sixfold increase in BrdU labeling 8 days after SE induction in the dentate gyrus, which decreased to a threefold increase 3 weeks after induction. In addition to increased neurogenesis, an increase in apoptosis was observed in the same samples as determined by a threefold increase in TUNEL staining 8 days after induction as compared to controls. Only a subset of the newborn cells actually went on to become mature neurons as demonstrated by NeuN staining (Porter et al., 2004). The number of episodes of SE and type of chemoconvulsant also have important implications for neurogenesis. One or two episodes of kainate-induced seizures showed no difference at P6 and P9 in BrdU labeling in the dentate gyrus when compared to controls. However, three episodes of kainate-induced seizures demonstrated a decrease in BrdU labeling in the dentate gyrus at P6, P9, and P13 at 48 h after seizure. No difference in cell death or apoptosis was observed in the kainate-induced SE rats when compared to controls; however, the newly born cells demonstrate irregular morphology (Liu, 2003). Alterations in neurogenesis may have a significant functional impact on normal brain development since 80% of granule cells develop after birth, and about half develop after P5 in normal rat brain development (Bayer, 1980).

36.4.3 Chronic

36.4.3.1 *Sprouting*

In adult rodent models of epilepsy and chronic epileptic foci removed from adult and older pediatric patients, hippocampal granule cell mossy fibers 'sprout' aberrant collaterals to the inner molecular layer of the dentate gyrus, forming monosynaptic connections and a positive feedback loop that may contribute to the seizure focus (Williams et al., 2007). The degree and localization of mossy fiber sprouting differ greatly with age (Holmes et al., 1998, 1999; Liu et al., 1999). In neonatal seizures, much of the sprouting is minimal and located in the CA3 (Holmes et al., 1998, 1999). The adult has a greater amount of sprouting, which is localized to the supragranular region and results from cell loss in the hilus and CA3 (Liu et al., 1999). The pattern of mossy fiber sprouting in early-life seizures is distinct from adult patterns in that the mossy fiber synapses terminate on the basal dendrites in the CA3 region and the stratum oriens (Holmes et al., 1999). There is some controversy regarding the effect of sprouting on epileptogenesis. Early studies have demonstrated that direct infusion of the protein inhibitor cyclohexamide was able to block pilocarpine- and kainate-induced mossy fiber sprouting in adult rats but not epileptogenesis (Longo and Mello, 1997). However, more recent studies have demonstrated that direct infusion of cyclohexamide to the dentate gyrus of adult rats spanning the period of pilocarpine treatment was unable to block mossy fiber sprouting or epileptogenesis (Toyoda and Buckmaster, 2005; Williams et al., 2002). The mechanism of mossy fiber sprouting is still unclear. Some hypothesize that the sprouting is caused by hyperexcitability; however, recent studies do not support this hypothesis. In adults, blocking L-type calcium channels and sodium channels is unable to stop mossy fiber sprouting in the pilocarpine model (Buckmaster, 2004; Ingram et al., 2009). However, in P14 mice, pilocarpine-induced L-type calcium channel blockade was able to inhibit mossy fiber sprouting (Ikegaya et al., 2000). Early treatment with rapamycin, an inhibitor of the mammalian target of rapamycin, was able to decrease seizure susceptibility in a transgenic mouse model of tuberous sclerosis (Meikle et al., 2008) and block mossy fiber sprouting in the pilocarpine model only with constant infusion, as the sprouting developed again after the cessation of infusion (Buckmaster et al., 2009).

36.5 TREATMENTS

The most challenging aspect of pediatric epilepsy treatment is to identify drugs that can protect against seizures while allowing normal functioning and development of the nervous system. Bromide was the first effective epilepsy treatment. Bromides act by augmenting $GABA_A$ receptor-mediated inhibition by increasing $GABA_A$-mediated currents due to a threefold increase in receptor permeability to Br^- over Cl^- (Bormann et al., 1987; Gallagher et al., 1978). Bromide has been replaced with less toxic antiepileptic drugs (AEDs), some of which target the same GABAergic pathway. Many newer AEDs work through non-GABAergic pathways or possess mixed mechanisms of action. The three main mechanisms of AEDs include enhancement of synaptic inhibition, reduction of synaptic excitation, or modulation of voltage-gated ion channels. Molecular targets of AEDs include voltage-gated sodium channels, voltage-gated calcium channels, GABAergic neurotransmission, and glutamatergic neurotransmission (Rogawski and Löscher, 2004; Table 36.2).

Methods of AED discovery have been recently reviewed by Bialer and White (2010). It is important to note that AED drug testing is typically studied in the adult rat brain. Drugs that pass this first level of screening are then studied in models of epilepsy, typically first in adult models and only later (if at all) in developmental epilepsies. There are three main types of drug-testing modalities used currently: maximal electroshock, subcutaneous PTZ injection, and electric kindling. The maximal electroshock modality utilizes electrical pulses delivered via ear clips or corneal electrodes to cause tonic extension of the hind limbs. If the administration of the drug inhibits the extension of the hind limbs, then the potential AED is thought to prevent the spread of seizure activity through neural tissue and is useful in screening AEDs effective against generalized tonic–clonic seizures. In the subcutaneous PTZ injection, animals are evaluated for clonic seizures. The effectiveness of a drug is determined by its ability to suppress these clonic seizures, making the model useful for finding AEDs effective against generalized myoclonic seizures. The drug discovery model of electrical kindling uses corneal electrodes to deliver 6 Hz electrical signals to induce complex partial seizures. Animals that do not display a clonic phase and automatisms are thought to be protected, making this method useful in screening drugs effective against partial seizures (Bialer et al., 2002; Holmes and Zhao, 2008). Unfortunately, these methods are not always accurate in their predictive efficacy. Levitacetam failed the subcutaneous PTZ and maximal electroshock models, but has proven to be a very effective AED (Klitgaard et al., 1998; Löscher et al., 1998). Therefore, new models of drug screening will be important in the future discovery of effective AEDs, especially for pediatric epilepsies.

36.5.1 GABA

36.5.1.1 Phenobarbital

Phenobarbital is a positive allosteric modulator of $GABA_A$ receptors that works by increasing channel-opening probability (Macdonald and Olsen, 1994). In addition to its primary GABAergic effect, phenobarbital can also block sodium channels and T-type calcium channels (Ffrench-Mullen et al., 1993). Phenobarbital is currently used as a first-line therapy for pediatric seizures but is mostly ineffective due to the shift in chloride gradient previously described. The effects of prenatal exposure to phenobarbital on CNS development have been studied extensively (Fishman and Yanai, 1983; Yanai et al., 1979) and have been reviewed (Kaindl et al., 2006). Perinatal and early-life exposure to phenobarbital can reduce brain

TABLE 36.2   Antiepileptic Drugs Used to Treat Neonatal Seizures and Pediatric Epilepsies

Structure	Drug	Primary effect	Site of action
	Phenobarbital	Enhances inhibition	GABA(A) – Mature brain
	Benzodiazepines	Enhances inhibition	GABA(A) – Mature brain
	Topiramate	Decreases excitation and enhances inhibition	Voltage-dependent sodium channels, Calcium channels, AMPA, and Kainate receptors/ GABA(A)
	Levetiracetam	Decreases excitation and enhances inhibition	Calcium channels, and SV2A/ GABA(A)
	Bumetanide	Enhances inhibition	NKCC1 (chloride gradient)

TABLE 36.2   Antiepileptic Drugs Used to Treat Neonatal Seizures and Pediatric Epilepsies—cont'd

Structure	Drug	Primary effect	Site of action
	Talampanel	Decreases excitation	AMPA receptors
	Flupirtine	Decreases excitation and enhances inhibition	NMDA receptors/KCNQ channels

weight in infant rats, with reductions in DNA, RNA, protein, cholesterol, neuronal number, and neurogenesis occurring, as well as enhanced apoptosis and alterations in gene expression (Bittigau et al., 2002; Diaz and Schain, 1978; Pick and Yanai, 1985; Raol et al., 2005; Stefovska et al., 2008; Yanai et al., 1989). Behavioral studies evaluating the morphological and neurochemical effects of perinatal and early-life exposure to phenobarbital, including deficits in various spatial-learning tasks and memory (Mikati et al., 1994; Pick and Yanai, 1985; Rogel-Fuchs et al., 1992; Yanai et al., 1989). Perinatal exposure to phenobarbital also increased aggression and activity levels in neonatal rats (Diaz and Schain, 1978; File and Wilks, 1990). Phenobarbital also has effects on synaptogenesis and myelination. The administration of phenobarbital to neonatal rats resulted in persistent abnormalities in mitochondria, myelin sheaths, and lamellar inclusion bodies in the cerebellum when compared to controls (Fishman et al., 1989).

36.5.1.2 Benzodiazepines

Benzodiazepines are first-line drugs in the treatment of SE. The actions of benzodiazepines are mediated by allosterically modulating $GABA_A$ receptors that contain the γ_2 with the $\alpha1$, $\alpha2$, $\alpha3$, and $\alpha5$ subunits (Rudolph et al., 1999; Smith, 2001), increasing $GABA_A$ receptor currents

(Bai et al., 2001). Behavioral deficits have been attributed to the effects of benzodiazepines on memory, possibly due to an alteration in attention (Pereira et al., 1989). The administration of a single dose of diazepam to P11 rats caused a significant decrease in cell proliferation (Pawlikowski et al., 1987). Diazepam and clonazepam administration to P7 rats caused widespread apoptotic neurodegeneration, which is associated with reduced expression of neurotrophins and other prosurvival proteins (Bittigau et al., 2002). Diazepam administration from P10 to P40 in rats permanently altered the expression of genes for GABA receptor subunits, GABA transporters, and GABA-synthesizing enzymes in hippocampal dentate granule neurons (Raol et al., 2005). Diazepam also affects cell proliferation, cell differentiation, and myelination making it less than ideal for treating pediatric epilepsies (Stefovska et al., 2008).

36.5.2 Non-GABA

36.5.2.1 Topiramate

Topiramate works through many pharmacological mechanisms including modulation of voltage-dependent sodium channels, modulation of calcium channels, enhancement of the effect of GABA on $GABA_A$ receptors, and modulation of AMPA and kainate receptors, all of which have been demonstrated to possess possible neuroprotective qualities. The administration of topiramate before hypoxia-induced seizures suppressed acute seizures in a dose-dependent manner. In addition to blocking acute seizures, topiramate administration was also able to eliminate the long-term susceptibility to kainate-induced seizures and the seizure-induced neuronal injury observed after acute hypoxia (Koh and Jensen, 2001). In lithium–pilocarpine-induced seizures in P20 rats, topiramate treatment resulted in an improved visuospatial performance in the water maze test when compared to saline-treated controls; however, no differences were found in histological examination of the hippocampus. The administration of topiramate in seizure-free neonatal rats resulted in no difference in water maze or histological studies (Cha et al., 2002). Topiramate demonstrates no neurotoxicity to the developing brain at anticonvulsant doses in P7 rats, in stark contrast to phenobarbital (Glier et al., 2004). The administration of topiramate to cultured fetal rat hippocampal and cortical neurons showed an increase in neurite outgrowth (Smith-Swintosky et al., 2001).

36.5.2.2 Levetiracetam

The mechanism of levetiracetam is not fully understood. The drug can block N-type calcium channels and reverse the inhibition by negative allosteric modulators such as zinc and beta-carbolines on neuronal GABA and glycine-gated currents (Rigo et al., 2002). Another pharmacological function distinct to levetiracetam is its binding to synaptic protein SV2A (Lynch et al., 2004). Recent studies have determined that SV2A is an important broad-spectrum anticonvulsant target (Kaminski et al., 2008). More studies are needed to elucidate the role of SV2A in ictal events. The administration of levetiracetam resulted in no deficits in visuospatial memory as observed via water maze testing in amygdala-kindled rats (Lamberty et al., 2000). In neonatal rats, levetiracetam resulted in no neurotoxicity when administered from P0 to P7 and analyzed from 2 to 5 days after administration (Manthey et al., 2005). Levetiracetam is effective in treating tonic convulsions and absence-like seizures in spontaneously epileptic rats during drug administration for up to 8 days after final administration, possibly demonstrating an antiepileptogenic effect (Ji-qun et al., 2005). Levetiracetam administration resulted in no alteration in hippocampal cell proliferation; however, significant effects in protein expression associated with the cytoskeleton, energy metabolism, neurotransmission, signal transduction, myelination, and stress response were observed (Paulson et al., 2010).

36.5.2.3 Bumetanide

Bumetanide is a well-known loop diuretic that is a specific inhibitor of NKCC1. Since enhanced expression of NKCC1 relative to KCC2 is responsible for the excitatory actions of GABA in the developing brain, it is easy to hypothesize that inhibiting NKCC1 would result in decreased excitatory actions of GABA. Dzahla et al. demonstrated the ability of bumetanide to shift the chloride gradient in immature neurons, suppress the epileptiform activity in hippocampal slices, and attenuate electrographic seizures in neonatal rats *in vivo* (Dzhala et al., 2005). Recent *in vitro* studies have shown that bumetanide can enhance the efficacy of phenobarbital in a neonatal seizure model. In the study, phenobarbital alone was able to abolish recurrent seizures in only 30% of the hippocampal slices, whereas the administration of bumetanide with phenobarbital was able to abolish recurrent seizures in 70% of the hippocampal slices. In addition to preventing recurrent seizures, the coadministration of bumetanide and phenobarbital also reduced the frequency, duration, and power of seizures in the remaining 30% (Dzhala et al., 2008). Bumetanide is able to block focal epileptic seizures in immature rat hippocampus, but is not able to prevent the formation of an epileptogenic mirror focus *in vitro* (Nardou et al., 2009).

36.5.2.4 Talampanel

Talampanel is a noncompetitive antagonist of the AMPA receptor and possesses anticonvulsant and neuroprotective properties. Talampanel administration to P10 rats in the hypoxia-induced seizure model suppressed seizures in a dose-dependent manner. Pretreatment of P10 rats before hypoxia prevented increases in seizure-induced neuronal injury later in life (Aujla et al., 2009). Talampanel is also neuroprotective in the CA1 region of the hippocampus in the fluid percussion traumatic brain injury model when administered within 30 min of injury but not when administered at 3 h (Belayev et al., 2001).

36.5.2.5 Flupirtine

Flupirtine, a selective neuronal potassium channel opener, activates KCNQ channels (G-protein-regulated inward rectifying K^+ channels) and stabilizes the resting membrane potential. Flupirtine also works via NMDA receptor antagonism by enhancing Mg^{2+} blockade (Kornhuber et al., 1999). The anticonvulsant properties of flupirtine have only recently been studied. Flupirtine proved successful in completely suppressing seizures in the flurothyl model of neonatal seizures and in preventing electrographic and behavioral seizures in the kainic-acid-induced SE model in P10 rats (Raol et al., 2009). Further studies are necessary to determine the long-term effects of flupirtine treatment of neonatal seizures.

36.6 CONCLUSION

Epilepsy is a very complex disease that affects millions of people worldwide. There are many different types of epilepsy, with various animal models used to elucidate the molecular mechanisms of epileptogenesis. The developing brain is especially sensitive to seizures and epilepsy due to the early enhancement of excitation in the immature brain. The acute and chronic effects of SE are quite complicated and include alterations in ion channels and neurotransmitter receptors, neuronal death, neurotrophin expression, inflammation, and mossy fiber sprouting. Since the development of the human brain is quite dynamic, the same therapeutic modalities used in adults have proved unsuccessful in controlling neonatal seizures. Newer therapeutics that act via novel mechanisms and are more effective in the developing brain are needed in order to treat those suffering from neonatal seizures and pediatric epilepsies adequately.

References

Acarin, L., et al., 2002. Expression of inducible nitric oxide synthase and cyclooxygenase-2 after excitotoxic damage to the immature rat brain. Journal of Neuroscience Research 68 (6), 745–754.

Aicardi, J., Chevrie, J.J., 1970. Convulsive status epilepticus in infants and children: A study of 239 cases. Epilepsia 11 (2), 187–197.

Ali, D.W., Salter, M.W., 2001. NMDA receptor regulation by Src kinase signalling in excitatory synaptic transmission and plasticity. Current Opinion in Neurobiology 11 (3), 336–342.

Artola, A., Singer, W., 1987. Long-term potentiation and NMDA receptors in rat visual cortex. Nature 330 (6149), 649–652.

Aujla, P.K., Fetell, M.R., Jensen, F.E., 2009. Talampanel suppresses the acute and chronic effects of seizures in a rodent neonatal seizure model. Epilepsia 50 (4), 694–701.

Avallone, J., et al., 2006. Distinct regulation of metabotropic glutamate receptor (mGluR1 alpha) in the developing limbic system following multiple early-life seizures. Experimental Neurology 202, 100–111.

Bai, D., et al., 2001. Distinct functional and pharmacological properties of tonic and quantal inhibitory postsynaptic currents mediated by gamma-aminobutyric acid(A) receptors in hippocampal neurons. Molecular Pharmacology 59 (4), 814–824.

Baram, T.Z., Gerth, A., Schultz, L., 1997. Febrile seizures: An appropriate-aged model suitable for long-term studies. Brain Research. Developmental Brain Research 98 (2), 265–270.

Bayer, S.A., 1980. Development of the hippocampal region in the rat. II. Morphogenesis during embryonic and early postnatal life. Journal of Comparative Neurology 190 (1), 115–134.

Beghi, M., et al., 2006. Learning disorders in epilepsy. Epilepsia 47 (supplement 2), 14–18.

Belayev, L., et al., 2001. Talampanel, a novel noncompetitive AMPA antagonist, is neuroprotective after traumatic brain injury in rats. Journal of Neurotrauma 18 (10), 1031–1038.

Ben-Ari, Y., 2002. Excitatory actions of gaba during development: The nature of the nurture. Nature Reviews Neuroscience 3 (9), 728–739.

Ben-Ari, Y., et al., 1989. Giant synaptic potentials in immature rat CA3 hippocampal neurones. Journal of Physiology 416, 303–325.

Benke, T.A., Swann, J., 2004. The tetanus toxin model of chronic epilepsy. Advances in Experimental Medicine and Biology 548, 226–238.

Bernard, C., et al., 2004. Acquired dendritic channelopathy in temporal lobe epilepsy. Science 305 (5683), 532–535.

Bettler, B., Tiao, J.Y.-H., 2006. Molecular diversity, trafficking and subcellular localization of GABA$_B$ receptors. Pharmacology and Therapeutics 110 (3), 533–543.

Bialer, M., White, H.S., 2010. Key factors in the discovery and development of new antiepileptic drugs. Nature Reviews. Drug Discovery 9 (1), 68–82.

Bialer, M., et al., 2002. Progress report on new antiepileptic drugs: A summary of the Sixth Eilat Conference (EILAT VI). Epilepsy Research 51 (1–2), 31–71.

Billiards, S.S., et al., 2006. Development of microglia in the cerebral white matter of the human fetus and infant. Journal of Comparative Neurology 497 (2), 199–208.

Bittigau, P., et al., 2002. Antiepileptic drugs and apoptotic neurodegeneration in the developing brain. Proceedings of the National Academy of Sciences of the United States of America 99 (23), 15089–15094.

Blair, R.E., et al., 2004. Epileptogenesis causes acute and chronic increases in GABA$_A$ receptor endocytosis that contributes to the induction and maintenance of seizures in the hippocampal culture model of acquired epilepsy. Journal of Pharmacology and Experimental Therapeutics 310 (3), 871–880.

Bormann, J., Hamill, O.P., Sakmann, B., 1987. Mechanism of anion permeation through channels gated by glycine and gamma-aminobutyric acid in mouse cultured spinal neurons. Journal of Physiology 385, 243–286.

Bowery, N.G., Hudson, A.L., Price, G.W., 1987. GABA$_A$ and GABA$_B$ receptor site distribution in the rat central nervous system. Neuroscience 20 (2), 365–383.

Brooks-Kayal, A., 2011. Molecular mechanisms of cognitive and behavioral comorbidities of epilepsy in children. Epilepsia 52 (supplement 1), 13–20.

Brooks-Kayal, A.R., Pritchett, D.B., 1993. Developmental changes in human gamma-aminobutyric acidA receptor subunit composition. Annals of Neurology 34 (5), 687–693.

Brooks-Kayal, A.R., et al., 2001. Gamma-aminobutyric acid(A) receptor subunit expression predicts functional changes in hippocampal dentate granule cells during postnatal development. Journal of Neurochemistry 77 (5), 1266–1278.

Buckmaster, P.S., 2004. Prolonged infusion of tetrodotoxin does not block mossy fiber sprouting in pilocarpine-treated rats. Epilepsia 45 (5), 452–458.

Buckmaster, P.S., Ingram, E.A., Wen, X., 2009. Inhibition of the mammalian target of rapamycin signaling pathway suppresses dentate granule cell axon sprouting in a rodent model of temporal lobe epilepsy. Journal of Neuroscience 29 (25), 8259–8269.

Calderone, A., et al., 2003. Ischemic insults derepress the gene silencer REST in neurons destined to die. Journal of Neuroscience 23 (6), 2112–2121.

Castaldo, P., et al., 2002. Benign familial neonatal convulsions caused by altered gating of KCNQ2/KCNQ3 potassium channels. Journal of Neuroscience 22 (2) RC199.

Cha, B.H., et al., 2002. Effect of topiramate following recurrent and prolonged seizures during early development. Epilepsy Research 51 (3), 217–232.

Chen, C., Bazan, N.G., 2005. Endogenous PGE2 regulates membrane excitability and synaptic transmission in hippocampal CA1 pyramidal neurons. Journal of Neurophysiology 93 (2), 929–941.

Chen, C., et al., 2004. Mice lacking sodium channel beta1 subunits display defects in neuronal excitability, sodium channel expression, and nodal architecture. Journal of Neuroscience 24 (16), 4030–4042.

Cornejo, B.J., et al., 2007. A single episode of neonatal seizures permanently alters glutamatergic synapses. Annals of Neurology 61 (5), 411–426.

Cortez, M.A., et al., 2009. Infantile spasms and Down syndrome: A new animal model. Pediatric Research 65 (5), 499–503.

Couve, A., Moss, S.J., Pangalos, M.N., 2000. GABA$_B$ receptors: A new paradigm in G protein signaling. Molecular and Cellular Neurosciences 16 (4), 296–312.

Cowan, L.D., 2002. The epidemiology of the epilepsies in children. Mental Retardation and Developmental Disabilities Research Reviews 8 (3), 171–181.

Dalmau, I., et al., 2003. Dynamics of microglia in the developing rat brain. Journal of Comparative Neurology 458 (2), 144–157.

Dam, M., 1990. Children with epilepsy: The effect of seizures, syndromes, and etiological factors on cognitive functioning. Epilepsia 31 (supplement 4), S26–S29.

Deonna, T., Roulet, E., 2006. Autistic spectrum disorder: Evaluating a possible contributing or causal role of epilepsy. Epilepsia 47 (supplement 2), 79–82.

Diaz, J., Schain, R.J., 1978. Phenobarbital: Effects of long-term administration on behavior and brain of artificially reared rats. Science 199 (4324), 90–91.

Dubé, C.M., et al., 1998. C-Fos, Jun D and HSP72 immunoreactivity, and neuronal injury following lithium-pilocarpine induced status epilepticus in immature and adult rats. Brain Research. Molecular Brain Research 63 (1), 139–154.

Dubé, C.M., et al., 2005. Interleukin-1beta contributes to the generation of experimental febrile seizures. Annals of Neurology 57 (1), 152–155.

Dubé, C.M., et al., 2010. Epileptogenesis provoked by prolonged experimental febrile seizures: Mechanisms and biomarkers. Journal of Neuroscience 30 (22), 7484–7494.

Dunn, D.W., et al., 2002. Teacher assessment of behaviour in children with new-onset seizures. Seizure 11 (3), 169–175.

Dupuy, S.T., Houser, C.R., 1997. Developmental changes in GABA neurons of the rat dentate gyrus: An in situ hybridization and birthdating study. Journal of Comparative Neurology 389 (3), 402–418.

Durand, G.M., Kovalchuk, Y., Konnerth, A., 1996. Long-term potentiation and functional synapse induction in developing hippocampus. Nature 381 (6577), 71–75.

Dzhala, V.I., Staley, K.J., 2003. Excitatory actions of endogenously released GABA contribute to initiation of ictal epileptiform activity in the developing hippocampus. Journal of Neuroscience 23 (5), 1840–1846.

Dzhala, V.I., et al., 2005. NKCC1 transporter facilitates seizures in the developing brain. Nature Medicine 11 (11), 1205–1213.

Dzhala, V.I., Brumback, A.C., Staley, K.J., 2008. Bumetanide enhances phenobarbital efficacy in a neonatal seizure model. Annals of Neurology 63 (2), 222–235.

Epsztein, J., et al., 2005. Recurrent mossy fibers establish aberrant kainate receptor-operated synapses on granule cells from epileptic rats. Journal of Neuroscience 25, 8229–8239.

Fagiolini, M., Hensch, T.K., 2000. Inhibitory threshold for critical-period activation in primary visual cortex. Nature 404 (6774), 183–186.

Fagiolini, M., et al., 2004. Specific GABA$_A$ circuits for visual cortical plasticity. Science 303 (5664), 1681–1683.

Ffrench-Mullen, J.M., Barker, J.L., Rogawski, M.A., 1993. Calcium current block by (−)-pentobarbital, phenobarbital, and CHEB but not (+)-pentobarbital in acutely isolated hippocampal CA1 neurons: Comparison with effects on GABA-activated Cl⁻ current. Journal of Neuroscience 13 (8), 3211–3221.

File, S.E., Wilks, L.J., 1990. Changes in seizure threshold and aggression during chronic treatment with three anticonvulsants and on drug withdrawal. Psychopharmacology 100 (2), 237–242.

Fishman, R.H., Yanai, J., 1983. Long-lasting effects of early barbiturates on central nervous system and behavior. Neuroscience and Biobehavioral Reviews 7 (1), 19–28.

Fishman, R.H., Ornoy, A., Yanai, J., 1989. Correlated ultrastructural damage between cerebellum cells after early anticonvulsant treatment in mice. International Journal of Developmental Neuroscience 7 (1), 15–26.

Fox, K., et al., 1996. Glutamate receptor blockade at cortical synapses disrupts development of thalamocortical and columnar organization in somatosensory cortex. Proceedings of the National Academy of Sciences of the United States of America 93 (11), 5584–5589.

Fritschy, J.M., et al., 1994. Switch in the expression of rat GABA$_A$-receptor subtypes during postnatal development: An immunohistochemical study. Journal of Neuroscience 14 (9), 5302–5324.

Fritschy, J.M., et al., 1999. GABAergic neurons and GABA$_A$-receptors in temporal lobe epilepsy. Neurochemistry International 34 (5), 435–445.

Frost, J.D., Hrachovy, R.A., 2005. Pathogenesis of infantile spasms: A model based on developmental desynchronization. Journal of Clinical Neurophysiology 22 (1), 25–36.

Fukuda, A., Mody, I., Prince, D., 1993. Differential ontogenesis of presynaptic and postsynaptic GABA$_B$ inhibition in rat somatosensory cortex. Journal of Neurophysiology 70 (1), 448–452.

Fukuda, A., et al., 1998. Changes in intracellular Ca^{2+} induced by GABA$_A$ receptor activation and reduction in Cl⁻ gradient in neonatal rat neocortex. Journal of Neurophysiology 79 (1), 439–446.

Furukawa, K., et al., 1997. The actin-severing protein gelsolin modulates calcium channel and NMDA receptor activities and vulnerability to excitotoxicity in hippocampal neurons. Journal of Neuroscience 17 (21), 8178–8186.

Gaiarsa, J., Tseeb, V., Ben-Ari, Y., 1995. Postnatal development of pre- and postsynaptic GABA$_B$-mediated inhibitions in the CA3 hippocampal region of the rat. Journal of Neurophysiology 73 (1), 246–255.

Gallagher, J.P., Higashi, H., Nishi, S., 1978. Characterization and ionic basis of GABA-induced depolarizations recorded *in vitro* from cat primary afferent neurones. Journal of Physiology 275, 263–282.

Ganguly, K., et al., 2001. GABA itself promotes the developmental switch of neuronal GABAergic responses from excitation to inhibition. Cell 105 (4), 521–532.

Glien, M., et al., 2001. Repeated low-dose treatment of rats with pilocarpine: Low mortality but high proportion of rats developing epilepsy. Epilepsy Research 46 (2), 111–119.

Glier, C., et al., 2004. Therapeutic doses of topiramate are not toxic to the developing rat brain. Experimental Neurology 187 (2), 403–409.

Goddard, G.V., 1967. Development of epileptic seizures through brain stimulation at low intensity. Nature 214 (5092), 1020–1021.

Goddard, G.V., McIntyre, D.C., Leech, C.K., 1969. A permanent change in brain function resulting from daily electrical stimulation. Experimental Neurology 25 (3), 295–330.

Gonzalez-Heydrich, J., et al., 2007. Psychiatric disorders and behavioral characteristics of pediatric patients with both epilepsy and attention-deficit hyperactivity disorder. Epilepsy and Behavior 10 (3), 384–388.

Greer, P.L., Greenberg, M.E., 2008. From synapse to nucleus: Calcium-dependent gene transcription in the control of synapse development and function. Neuron 59 (6), 846–860.

Guzzetta, F., 2006. Cognitive and behavioral outcome in West syndrome. Epilepsia 47 (supplement 2), 49–52.

Hauser, W.A., 1994. The prevalence and incidence of convulsive disorders in children. Epilepsia 35 (supplement 2), S1–S6.

Hauser, W.A., Annegers, J.F., Kurland, L.T., 1993. Incidence of epilepsy and unprovoked seizures in Rochester, Minnesota: 1935–1984. Epilepsia 34 (3), 453–468.

Hensch, T.K., 2005. Critical period plasticity in local cortical circuits. Nature Reviews Neuroscience 6 (11), 877–888.

Hensch, T.K., et al., 1998. Local GABA circuit control of experience-dependent plasticity in developing visual cortex. Science 282 (5393), 1504–1508.

Henshall, D.C., Murphy, B.M., 2008. Modulators of neuronal cell death in epilepsy. Current Opinion in Pharmacology 8 (1), 75–81.

Henshall, D.C., Simon, R.P., 2005. Epilepsy and apoptosis pathways. Journal of Cerebral Blood Flow and Metabolism 25 (12), 1557–1572.

Heo, K., et al., 2006. Minocycline inhibits caspase-dependent and -independent cell death pathways and is neuroprotective against hippocampal damage after treatment with kainic acid in mice. Neuroscience Letters 398 (3), 195–200.

Herdegen, T., Leah, J.D., 1998. Inducible and constitutive transcription factors in the mammalian nervous system: Control of gene expression by Jun, Fos and Krox, and CREB/ATF proteins. Brain Research. Brain Research Reviews 28 (3), 370–490.

Hermann, B., et al., 2002. The neurodevelopmental impact of childhood-onset temporal lobe epilepsy on brain structure and function. Epilepsia 43 (9), 1062–1071.

Hermann, B., et al., 2003. Neurodevelopmental vulnerability of the corpus callosum to childhood onset localization-related epilepsy. NeuroImage 18 (2), 284–292.

Holmes, G.L., Zhao, Q., 2008. Choosing the correct antiepileptic drugs: From animal studies to the clinic. Pediatric Neurology 38 (3), 151–162.

Holmes, G.L., et al., 1998. Consequences of neonatal seizures in the rat: Morphological and behavioral effects. Annals of Neurology 44 (6), 845–857.

Holmes, G.L., et al., 1999. Mossy fiber sprouting after recurrent seizures during early development in rats. Journal of Comparative Neurology 404 (4), 537–553.

Holopainen, I.E., 2008. Seizures in the developing brain: Cellular and molecular mechanisms of neuronal damage, neurogenesis and cellular reorganization. Neurochemistry International 52, 935–947.

Honchar, M.P., Olney, J.W., Sherman, W.R., 1983. Systemic cholinergic agents induce seizures and brain damage in lithium-treated rats. Science 220 (4594), 323–325.

Honkaniemi, J., Sharp, F.R., 1999. Prolonged expression of zinc finger immediate-early gene mRNAs and decreased protein synthesis following kainic acid induced seizures. European Journal of Neuroscience 11 (1), 10–17.

Humphrey, W.M., et al., 2002. Immediate and delayed hippocampal neuronal loss induced by kainic acid during early postnatal development in the rat. Brain Research. Developmental Brain Research 137 (1), 1–12.

Huttenlocher, P.R., et al., 1982. Synaptogenesis in human visual cortex – Evidence for synapse elimination during normal development. Neuroscience Letters 33 (3), 247–252.

Ikegaya, Y., Nishiyama, N., Matsuki, N., 2000. L-type Ca(2+) channel blocker inhibits mossy fiber sprouting and cognitive deficits following pilocarpine seizures in immature mice. Neuroscience 98 (4), 647–659.

Ingram, E.A., et al., 2009. Prolonged infusion of inhibitors of calcineurin or L-type calcium channels does not block mossy fiber sprouting in a model of temporal lobe epilepsy. Epilepsia 50 (1), 56–64.

Isaac, J.T., et al., 1997. Silent synapses during development of thalamocortical inputs. Neuron 18 (2), 269–280.

Iwai, Y., et al., 2003. Rapid critical period induction by tonic inhibition in visual cortex. Journal of Neuroscience 23 (17), 6695–6702.

Jensen, F.E., et al., 1992. Age-dependent changes in long-term seizure susceptibility and behavior after hypoxia in rats. Epilepsia 33 (6), 971–980.

Jensen, F.E., Firkusny, I.R., Mower, G.D., 1993. Differences in c-fos immunoreactivity due to age and mode of seizure induction. Brain Research. Molecular Brain Research 17 (3–4), 185–193.

Jensen, F.E., et al., 1994. The putative essential nutrient pyrroloquinoline quinone is neuroprotective in a rodent model of hypoxic/ischemic brain injury. Neuroscience 62 (2), 399–406.

Jiang, X., et al., 2008. Activated Src kinases interact with the N-methyl-D-aspartate receptor after neonatal brain ischemia. Annals of Neurology 63 (5), 632–641.

Ji-qun, C., et al., 2005. Long-lasting antiepileptic effects of levetiracetam against epileptic seizures in the spontaneously epileptic rat (SER): Differentiation of levetiracetam from conventional antiepileptic drugs. Epilepsia 46 (9), 1362–1370.

Jones, J.E., et al., 2007. Psychiatric comorbidity in children with new onset epilepsy. Developmental Medicine and Child Neurology 49 (7), 493–497.

Kábová, R., et al., 1999. Age-specific N-methyl-D-aspartate-induced seizures: Perspectives for the West syndrome model. Epilepsia 40 (10), 1357–1369.

Kaindl, A.M., et al., 2006. Antiepileptic drugs and the developing brain. Cellular and Molecular Life Sciences 63 (4), 399–413.

Kaminski, R.M., et al., 2008. SV2A protein is a broad-spectrum anticonvulsant target: Functional correlation between protein binding and seizure protection in models of both partial and generalized epilepsy. Neuropharmacology 54 (4), 715–720.

Kanold, P.O., et al., 2003. Role of subplate neurons in functional maturation of visual cortical columns. Science 301 (5632), 521–525.

Kawaguchi, K., et al., 2005. Cyclooxygenase-2 expression is induced in rat brain after kainate-induced seizures and promotes neuronal death in CA3 hippocampus. Brain Research 1050 (1–2), 130–137.

Kearney, J.A., et al., 2001. A gain-of-function mutation in the sodium channel gene Scn2a results in seizures and behavioral abnormalities. Neuroscience 102 (2), 307–317.

Khalilov, I., Holmes, G.L., Ben-Ari, Y., 2003. *In vitro* formation of a secondary epileptogenic mirror focus by interhippocampal propagation of seizures. Nature Neuroscience 6 (10), 1079–1085.

Khalilov, I., et al., 2005. Epileptogenic actions of GABA and fast oscillations in the developing hippocampus. Neuron 48, 787–796.

Khazipov, R., Ragozzino, D., Bregestovski, P., 1995. Kinetics and Mg^{2+} block of N-methyl-D-aspartate receptor channels during postnatal development of hippocampal CA3 pyramidal neurons. Neuroscience 69 (4), 1057–1065.

Khazipov, R., et al., 2001. Early development of neuronal activity in the primate hippocampus in utero. Journal of Neuroscience 21, 9770–9781.

Khazipov, R., et al., 2004. Developmental changes in GABAergic actions and seizure susceptibility in the rat hippocampus. European Journal of Neuroscience 19 (3), 590–600.

Kim, S.U., de Vellis, J., 2005. Microglia in health and disease. Journal of Neuroscience Research 81 (3), 302–313.

Kim, D.K., Jang, T.J., 2006. Cyclooxygenase-2 expression and effect of celecoxib in flurothyl-induced neonatal seizure. International Journal of Experimental Pathology 87 (1), 73–78.

Klitgaard, H., et al., 1998. Evidence for a unique profile of levetiracetam in rodent models of seizures and epilepsy. European Journal of Pharmacology 353 (2–3), 191–206.

Koh, S., Jensen, F.E., 2001. Topiramate blocks perinatal hypoxia-induced seizures in rat pups. Annals of Neurology 50 (3), 366–372.

Kornhuber, J., et al., 1999. Flupirtine shows functional NMDA receptor antagonism by enhancing Mg^{2+} block via activation of voltage independent potassium channels: Rapid communication. Journal of Neural Transmission 106 (9–10), 857–867.

Kostovic, I., Rakic, P., 1980. Cytology and time of origin of interstitial neurons in the white matter in infant and adult human and monkey telencephalon. Journal of Neurocytology 9 (2), 219–242.

Kubová, H., et al., 2004. Status epilepticus in immature rats leads to behavioural and cognitive impairment and epileptogenesis. European Journal of Neuroscience 19 (12), 3255–3265.

Kumar, S.S., et al., 2002. A developmental switch of AMPA receptor subunits in neocortical pyramidal neurons. Journal of Neuroscience 22 (8), 3005–3015.

Kunz, T., Oliw, E.H., 2001. The selective cyclooxygenase-2 inhibitor rofecoxib reduces kainate-induced cell death in the rat hippocampus. European Journal of Neuroscience 13 (3), 569–575.

Kurz, J.E., et al., 2001. A significant increase in both basal and maximal calcineurin activity in the rat pilocarpine model of status epilepticus. Journal of Neurochemistry 78 (2), 304–315.

LaFrance, W.C., Kanner, A.M., Hermann, B., 2008. Psychiatric comorbidities in epilepsy. International Review of Neurobiology 83, 347–383.

Lamberty, Y., Margineanu, D.G., Klitgaard, H., 2000. Absence of negative impact of levetiracetam on cognitive function and memory in normal and amygdala-kindled rats. Epilepsy and Behavior 1 (5), 333–342.

Lau, C.G., Zukin, R.S., 2007. NMDA receptor trafficking in synaptic plasticity and neuropsychiatric disorders. Nature Reviews Neuroscience 8 (6), 413–426.

Laurén, H.B., et al., 2005. Kainic acid-induced status epilepticus alters GABA receptor subunit mRNA and protein expression in the developing rat hippocampus. Journal of Neurochemistry 94 (5), 1384–1394.

Laurent, A., Arzimanoglou, A., 2006. Cognitive impairments in children with nonidiopathic temporal lobe epilepsy. Epilepsia 47 (supplement 2), 99–102.

Lee, C.L., et al., 1995. Tetanus toxin-induced seizures in infant rats and their effects on hippocampal excitability in adulthood. Brain Research 677 (1), 97–109.

Lee, C.L., et al., 2008. A new animal model of infantile spasms with unprovoked persistent seizures. Epilepsia 49 (2), 298–307.

Leinekugel, X., et al., 1995. Synaptic $GABA_A$ activation induces Ca^{2+} rise in pyramidal cells and interneurons from rat neonatal hippocampal slices. Journal of Physiology 487 (Pt 2), 319–329.

Leinekugel, X., et al., 1997. Ca^{2+} oscillations mediated by the synergistic excitatory actions of GABA(A) and NMDA receptors in the neonatal hippocampus. Neuron 18 (2), 243–255.

Liao, D., Malinow, R., 1996. Deficiency in induction but not expression of LTP in hippocampal slices from young rats. Learning and Memory 3 (2–3), 138–149.

Liu, H., 2003. Suppression of hippocampal neurogenesis is associated with developmental stage, number of perinatal seizure episodes, and glucocorticosteroid level. Experimental Neurology 184 (1), 196–213.

Liu, Z., et al., 1999. Consequences of recurrent seizures during early brain development. Neuroscience 92 (4), 1443–1454.

Longo, B.M., Mello, L.E., 1997. Blockade of pilocarpine- or kainate-induced mossy fiber sprouting by cycloheximide does not prevent subsequent epileptogenesis in rats. Neuroscience Letters 226 (3), 163–166.

Lonze, B.E., Ginty, D.D., 2002. Function and regulation of CREB family transcription factors in the nervous system. Neuron 35 (4), 605–623.

Löscher, W., Hönack, D., Rundfeldt, C., 1998. Antiepileptogenic effects of the novel anticonvulsant levetiracetam (ucb L059) in the kindling model of temporal lobe epilepsy. Journal of Pharmacology and Experimental Therapeutics 284 (2), 474–479.

Lynch, B.A., et al., 2004. The synaptic vesicle protein SV2A is the binding site for the antiepileptic drug levetiracetam. Proceedings of the National Academy of Sciences of the United States of America 101 (26), 9861–9866.

Macdonald, R.L., Olsen, R.W., 1994. $GABA_A$ receptor channels. Annual Review of Neuroscience 17, 569–602.

Manthey, D., et al., 2005. Sulthiame but not levetiracetam exerts neurotoxic effect in the developing rat brain. Experimental Neurology 193 (2), 497–503.

Marsh, E., et al., 2009. Targeted loss of Arx results in a developmental epilepsy mouse model and recapitulates the human phenotype in heterozygous females. Brain 132 (Pt 6), 1563–1576.

Mayer, M.L., 2005. Glutamate receptor ion channels. Current Opinion in Neurobiology 15 (3), 282–288.

Mayer, M.L., Westbrook, G.L., 1987. The physiology of excitatory amino acids in the vertebrate central nervous system. Progress in Neurobiology 28 (3), 197–276.

McCabe, B.K., et al., 2001. Reduced neurogenesis after neonatal seizures. Journal of Neuroscience 21 (6), 2094–2103.

McKernan, R.M., Whiting, P.J., 1996. Which $GABA_A$-receptor subtypes really occur in the brain? Trends in Neurosciences 19 (4), 139–143.

McLean, H.A., et al., 1996. Spontaneous release of GABA activates $GABA_B$ receptors and controls network activity in the neonatal rat hippocampus. Journal of Neurophysiology 76 (2), 1036–1046.

McNamara, J.O., Huang, Y.Z., Leonard, A.S., 2006. Molecular signaling mechanisms underlying epileptogenesis. Science's Signal Transduction Knowledge Environment 2006 (356), re12.

McQuillen, P.S., et al., 2003. Selective vulnerability of subplate neurons after early neonatal hypoxia-ischemia. Journal of Neuroscience 23 (8), 3308–3315.

Meier, C.L., Dudek, F.E., 1996. Spontaneous and stimulation-induced synchronized burst afterdischarges in the isolated CA1 of kainate-treated rats. Journal of Neurophysiology 76 (4), 2231–2239.

Meier, C.L., Obenaus, A., Dudek, F.E., 1992. Persistent hyperexcitability in isolated hippocampal CA1 of kainate-lesioned rats. Journal of Neurophysiology 68 (6), 2120–2127.

Meikle, L., et al., 2008. Response of a neuronal model of tuberous sclerosis to mammalian target of rapamycin (mTOR) inhibitors: Effects on mTORC1 and Akt signaling lead to improved survival and function. Journal of Neuroscience 28 (21), 5422–5432.

Mikati, M.A., et al., 1994. Phenobarbital modifies seizure-related brain injury in the developing brain. Annals of Neurology 36 (3), 425–433.

Mikati, M.A., El Hokayem, J.A., El Sabban, M.E., 2007. Effects of a single dose of erythropoietin on subsequent seizure susceptibility in rats exposed to acute hypoxia at P10. Epilepsia 48 (1), 175–181.

Monyer, H., et al., 1994. Developmental and regional expression in the rat brain and functional properties of four NMDA receptors. Neuron 12 (3), 529–540.

Nadler, J.V., Perry, B.W., Cotman, C.W., 1978. Intraventricular kainic acid preferentially destroys hippocampal pyramidal cells. Nature 271 (5646), 676–677.

Nardou, R., Ben-Ari, Y.l., Khalilov, I., 2009. Bumetanide, an NKCC1 antagonist, does not prevent formation of epileptogenic focus but blocks epileptic focus seizures in immature rat hippocampus. Journal of Neurophysiology 2878–2888.

Nolan, M.A., et al., 2003. Intelligence in childhood epilepsy syndromes. Epilepsy Research 53 (1–2), 139–150.

Owens, J., et al., 1997. Acute and chronic effects of hypoxia on the developing hippocampus. Annals of Neurology 41 (2), 187–199.

Paulson, L., et al., 2010. Effect of levetiracetam on hippocampal protein expression and cell proliferation in rats. Epilepsy Research 90 (1–2), 110–120.

Pawlikowski, M., et al., 1987. Effects of diazepam on cell proliferation in cerebral cortex, anterior pituitary and thymus of developing rats. Life Sciences 40 (11), 1131–1135.

Pereira, M.E., et al., 1989. Inhibition by diazepam of the effect of additional training and of extinction on the retention of shuttle avoidance behavior in rats. Behavioral Neuroscience 103 (1), 202–205.

Peters, H.C., et al., 2005. Conditional transgenic suppression of M channels in mouse brain reveals functions in neuronal excitability, resonance and behavior. Nature Neuroscience 8 (1), 51–60.

Pick, C.G., Yanai, J., 1985. Long term reduction in eight arm maze performance after early exposure to phenobarbital. International Journal of Developmental Neuroscience 3 (3), 223–227.

Pinheiro, P., Mulle, C., 2006. Kainate receptors. Cell and Tissue Research 326 (2), 457–482.

Pitkänen, A., Schwartzkroin, P.A., Moshé, S.L., 2005. Models of Seizures and Epilepsy, 1st edn Elsevier Academic, New York.

Porter, B.E., Maronski, M., Brooks-Kayal, A.R., 2004. Fate of newborn dentate granule cells after early life status epilepticus. Epilepsia 45 (1), 13–19.

Porter, B.E., Cui, X.-N., Brooks-Kayal, A.R., 2006. Status epilepticus differentially alters AMPA and kainate receptor subunit expression in mature and immature dentate granule neurons. European Journal of Neuroscience 23 (11), 2857–2863.

Price, M.G., et al., 2009. A triplet repeat expansion genetic mouse model of infantile spasms syndrome, Arx$^{(GCG)10+7}$, with interneuronopathy, spasms in infancy, persistent seizures, and adult cognitive and behavioral impairment. Journal of Neuroscience 29 (27), 8752–8763.

Rakhade, S.N., Huynh, T., et al., 2008a. Early-life hypoxia-induced seizures in the rat lead to spontaneous electrographic seizures in later life, as well as altered hippocampal neurotransmitter receptor expression. Amer Epil Soc Abstr.

Rakhade, S.N., Zhou, C., et al., 2008b. Early alterations of AMPA receptors mediate synaptic potentiation induced by neonatal seizures. Journal of Neuroscience 28 (32), 7979–7990.

Rakhade, S.N., Jensen, F.E., 2009. Epileptogenesis in the immature brain: Emerging mechanisms. Neurology 5, 380–391.

Rakic, P., 1977. Prenatal development of the visual system in rhesus monkey. Philosophical Transactions of the Royal Society of London. Series B, Biological Sciences 278 (961), 245–260.

Raol, Y.H., Budreck, E.C., Brooks-Kayal, A.R., 2003. Epilepsy after early-life seizures can be independent of hippocampal injury. Annals of Neurology 53 (4), 503–511.

Raol, Y.H., et al., 2005. Long-term effects of diazepam and phenobarbital treatment during development on GABA receptors, transporters and glutamic acid decarboxylase. Neuroscience 132 (2), 399–407.

Raol, Y.H., et al., 2006. Increased GABA(A)-receptor alpha1-subunit expression in hippocampal dentate gyrus after early-life status epilepticus. Epilepsia 47, 1665–1673.

Raol, Y.H., et al., 2009. A KCNQ channel opener for experimental neonatal seizures and status epilepticus. Annals of Neurology 65 (3), 326–336.

Rho, J.M., et al., 1999. Developmental seizure susceptibility of kv1.1 potassium channel knockout mice. Developmental Neuroscience 21 (3–5), 320–327.

Rice, J.E., Vannucci, R.C., Brierley, J.B., 1981. The influence of immaturity on hypoxic-ischemic brain damage in the rat. Annals of Neurology 9 (2), 131–141.

Rigo, J.-M., et al., 2002. The anti-epileptic drug levetiracetam reverses the inhibition by negative allosteric modulators of neuronal GABA- and glycine-gated currents. British Journal of Pharmacology 136 (5), 659–672.

Rivera, C., et al., 1999. The K$^+$/Cl$^-$ co-transporter KCC2 renders GABA hyperpolarizing during neuronal maturation. Nature 397 (6716), 251–255.

Rivera, C., et al., 2002. BDNF-induced TrkB activation down-regulates the K$^+$–Cl$^-$ cotransporter KCC2 and impairs neuronal Cl$^-$ extrusion. Journal of Cell Biology 159 (5), 747–752.

Rizzi, M., et al., 2003. Glia activation and cytokine increase in rat hippocampus by kainic acid-induced status epilepticus during postnatal development. Neurobiology of Disease 14 (3), 494–503.

Rogawski, M.A., Löscher, W., 2004. The neurobiology of antiepileptic drugs. Nature Reviews Neuroscience 5 (7), 553–564.

Rogel-Fuchs, Y., et al., 1992. Hippocampal cholinergic alterations and related behavioral deficits after early exposure to phenobarbital. Brain Research Bulletin 29 (1), 1–6.

Romero, L.I., et al., 1996. Roles of IL-1 and TNF-alpha in endotoxin-induced activation of nitric oxide synthase in cultured rat brain cells. American Journal of Physiology 270 (2(Pt 2)), R326–R332.

Roux, E., Borrel, A., 1898. Tétanos cérébral et immunité contre le tétanos. Annales de l'Institut Pasteur 4, 225–239.

Rozenberg, F., et al., 1989. Distribution of GABAergic neurons in late fetal and early postnatal rat hippocampus. Brain Research. Developmental Brain Research 50 (2), 177–187.

Rudolph, U., et al., 1999. Benzodiazepine actions mediated by specific gamma-aminobutyric acid(A) receptor subtypes. Nature 401 (6755), 796–800.

Rudolph, U., Möhler, H., 2004. Analysis of GABA$_A$ receptor function and dissection of the pharmacology of benzodiazepines and general anesthetics through mouse genetics. Annual Review of Pharmacology and Toxicology 44 (14), 475–498.

Sanchez, R.M., Jensen, F.E., 2001. Maturational aspects of epilepsy mechanisms and consequences for the immature brain. Epilepsia 42 (5), 577–585.

Sanchez, R.M., et al., 2001. Decreased glutamate receptor 2 expression and enhanced epileptogenesis in immature rat hippocampus after perinatal hypoxia-induced seizures. Journal of Neuroscience 21, 8154–8163.

Sanchez, R.M., et al., 2005. AMPA/kainate receptor-mediated downregulation of GABAergic synaptic transmission by calcineurin after seizures in the developing rat brain. Journal of Neuroscience 25 (13), 3442–3451.

Sankar, R., et al., 1998. Patterns of status epilepticus-induced neuronal injury during development and long-term consequences. Journal of Neuroscience 18 (20), 8382–8393.

Sankar, R., et al., 2000. Granule cell neurogenesis after status epilepticus in the immature rat brain. Epilepsia 41 (supplement 6), S53–S56.

Scantlebury, M.H., Moshé, S.L., 2006. A new model of infantile spasms. American Epilepsy Society Abstract 6111.

Scharfman, H.E., et al., 2002. Spontaneous limbic seizures after intrahippocampal infusion of brain-derived neurotrophic factor. Experimental Neurology 174 (2), 201–214.

Schmid, R., et al., 1999. Effects of neonatal seizures on subsequent seizure-induced brain injury. Neurology 53 (8), 1754–1761.

Serafini, R., et al., 1995. Depolarizing GABA-activated Cl$^-$ channels in embryonic rat spinal and olfactory bulb cells. Journal of Physiology 488 (Pt 2), 371–386.

Shepherd, J.D., Huganir, R.L., 2007. The cell biology of synaptic plasticity: AMPA receptor trafficking. Annual Review of Cell and Developmental Biology 23, 613–643.

Smith, T.A., 2001. Type A gamma-aminobutyric acid (GABA$_A$) receptor subunits and benzodiazepine binding: Significance to clinical syndromes and their treatment. British Journal of Biomedical Science 58 (2), 111–121.

Smith-Swintosky, V.L., et al., 2001. Topiramate promotes neurite outgrowth and recovery of function after nerve injury. Neuroreport 12 (5), 1031–1034.

Sogawa, Y., et al., 2001. Timing of cognitive deficits following neonatal seizures: Relationship to histological changes in the hippocampus. Brain Research. Developmental Brain Research 131 (1–2), 73–83.

Sperber, E.F., et al., 1999. Flurothyl status epilepticus in developing rats: Behavioral, electrographic histological and electrophysiological studies. Brain Research. Developmental Brain Research 116 (1), 59–68.

Stafstrom, C.E., 2009. Infantile spasms: A critical review of emerging animal models. Epilepsy Currents – American Epilepsy Society 9 (3), 75–81.

Stafstrom, C.E., Thompson, J.L., Holmes, G.L., 1992. Kainic acid seizures in the developing brain: Status epilepticus and spontaneous recurrent seizures. Brain Research. Developmental Brain Research 65 (2), 227–236.

Stefovska, V.G., et al., 2008. Sedative and anticonvulsant drugs suppress postnatal neurogenesis. Annals of Neurology 64 (4), 434–445.

Stevens, B., et al., 2007. The classical complement cascade mediates CNS synapse elimination. Cell 131 (6), 1164–1178.

Swann, J.W., et al., 1999. Developmental neuroplasticity: Roles in early life seizures and chronic epilepsy. Advances in Neurology 79, 203–216.

Takei, H., et al., 2007. Evidence of increased cell proliferation in the hippocampus in children with Ammon's horn sclerosis. Pathology International 57 (2), 76–81.

Talos, D.M., Fishman, R.E., et al., 2006a. Developmental regulation of alpha-amino-3-hydroxy-5-methyl-4-isoxazole-propionic acid receptor subunit expression in forebrain and relationship to regional susceptibility to hypoxic/ischemic injury I. Rodent cerebral white matter and cortex. Journal of Comparative Neurology 497 (1), 42–60.

Talos, D.M., Follett, P.L., et al., 2006b. Developmental regulation of alpha-amino-3-hydroxy-5-methyl-4-isoxazole-propionic acid receptor subunit expression in forebrain and relationship to regional susceptibility to hypoxic/ischemic injury II. Human cerebral white matter and cortex. Journal of Comparative Neurology 497 (1), 61–77.

Tandon, P., et al., 1999. Neuroprotective effects of brain-derived neurotrophic factor in seizures during development. Neuroscience 91 (1), 293–303.

Tandon, P., et al., 2002. Downregulation of kainate receptors in the hippocampus following repeated seizures in immature rats. Brain Research. Developmental Brain Research 136 (2), 145–150.

Tocco, G., et al., 1997. Maturational regulation and regional induction of cyclooxygenase-2 in rat brain: Implications for Alzheimer's disease. Experimental Neurology 144 (2), 339–349.

Toth, Z., et al., 1998. Seizure-induced neuronal injury: Vulnerability to febrile seizures in an immature rat model. Journal of Neuroscience 18 (11), 4285–4294.

Toyoda, I., Buckmaster, P.S., 2005. Prolonged infusion of cycloheximide does not block mossy fiber sprouting in a model of temporal lobe epilepsy. Epilepsia 46 (7), 1017–1020.

Truitt, E.B., Ebersberger, E.M., Ling, A.S.C., 1960. Measurement of brain excitability by use of hexaflurodiethyl ether (Indoklon). Journal of Pharmacology and Experimental Therapeutics 129, 445–453.

Tu, B., Bazan, N.G., 2003. Hippocampal kindling epileptogenesis upregulates neuronal cyclooxygenase-2 expression in neocortex. Experimental Neurology 179 (2), 167–175.

Tuchman, R., Moshé, S.L., Rapin, I., 2009. Convulsing toward the pathophysiology of autism. Brain and Development 31 (2), 95–103.

Tyzio, R., et al., 1999. The establishment of GABAergic and glutamatergic synapses on CA1 pyramidal neurons is sequential and correlates with the development of the apical dendrite. Journal of Neuroscience 19 (23), 10372–10382.

Vannucci, R.C., et al., 1999. Mini-review rat model of perinatal hypoxic-ischemic brain damage. Journal of Neuroscience Research 163, 158–163.

Vezzani, A., Balosso, S., Ravizza, T., 2008. The role of cytokines in the pathophysiology of epilepsy. Brain, Behavior, and Immunity 22 (6), 797–803.

Viviani, B., et al., 2003. Interleukin-1beta enhances NMDA receptor-mediated intracellular calcium increase through activation of the Src family of kinases. Journal of Neuroscience 23 (25), 8692–8700.

Wang, A., et al., 2009. Calcineurin-mediated GABA(A) receptor dephosphorylation in rats after kainic acid-induced status epilepticus. Seizure 18 (7), 519–523.

Wasterlain, C.G., et al., 2002. Seizure-induced neuronal death in the immature brain. Progress in Brain Research 135, 335–353.

Wigström, H., Gustafsson, B., 1983. Facilitated induction of hippocampal long-lasting potentiation during blockade of inhibition. Nature 301 (5901), 603–604.

Williams, P.A., et al., 2002. Reassessment of the effects of cycloheximide on mossy fiber sprouting and epileptogenesis in the pilocarpine model of temporal lobe epilepsy. Journal of Neurophysiology 88 (4), 2075–2087.

Williams, P.A., et al., 2007. Development of spontaneous seizures after experimental status epilepticus: Implications for understanding epileptogenesis. Epilepsia 48 (supplement 5), 157–163.

Wirrell, E.C., et al., 2001. Prolonged seizures exacerbate perinatal hypoxic-ischemic brain damage. Pediatric Research 50 (4), 445–454.

Wisden, W., et al., 1992. The distribution of 13 $GABA_A$ receptor subunit mRNAs in the rat brain. I. Telencephalon, diencephalon, mesencephalon. Journal of Neuroscience 12 (3), 1040–1062.

Wolff, M., Cassé-Perrot, C., Dravet, C., 2006. Severe myoclonic epilepsy of infants (Dravet syndrome): Natural history and neuropsychological findings. Epilepsia 47 (Suppl 2), 45–48.

Wu, G., Malinow, R., Cline, H.T., 1996. Maturation of a central glutamatergic synapse. Science 274 (5289), 972–976.

Yager, J.Y., et al., 2002. Prolonged neonatal seizures exacerbate hypoxic-ischemic brain damage: Correlation with cerebral energy metabolism and excitatory amino acid release. Developmental Neuroscience 24 (5), 367–381.

Yanai, J., Rosselli-Austin, L., Tabakoff, B., 1979. Neuronal deficits in mice following prenatal exposure to phenobarbital. Experimental Neurology 64 (2), 237–244.

Yanai, J., et al., 1989. Neural and behavioral alterations after early exposure to phenobarbital. Neurotoxicology 10 (3), 543–554.

Yoshikawa, K., et al., 2006. Profiling of eicosanoid production in the rat hippocampus during kainic acid-induced seizure: dual phase regulation and differential involvement of COX-1 and COX-2. Journal of Biological Chemistry 281 (21), 14663–14669.

Zhang, G., et al., 2004. Effects of status epilepticus on hippocampal $GABA_A$ receptors are age-dependent. Neuroscience 125 (2), 299–303.

The growing number of animal models for developmental sensory disorders has been tremendously useful for understanding both normal and abnormal development and sensory processing.

37.2 HEARING IMPAIRMENT

Developmental auditory disorders may involve defects at any stage from the moment sound enters the external ear to the sensorineural synapse to the downstream processing pathways of the central nervous system. Disorders of the auditory organ are thus roughly classified as *conductive*, affecting primarily the mechanical structures of the external and middle ear; *sensorineural*, affecting primarily the sensory, neural, and supporting structures of the inner ear and central auditory pathway; or *mixed*. Because hearing impairment that precedes language acquisition can be especially significant, hearing loss is often characterized as *prelingual*, occurring before 2 years of age, or *postlingual*. Severity of hearing loss may be mild (<40 dB), moderate (40–70 dB), severe (70–90 dB), or profound (>90 dB).

37.2.1 Acquired Hearing Loss

Many forms of hearing loss are preventable. The hair cells, for example, are particularly vulnerable to acoustic trauma and to a wide range of ototoxic drugs (e.g., cisplatin, isoretinoin, alcohol, aminoglyocoside antibiotics, loop-diuretics), and as mammalian hair cells do not regenerate, the inner hair cells constitute a scarce resource at only about 3500 cells per human cochlea. Infection has long been a major cause of hearing loss. Both acquired and congenital syphilis can produce otosyphilis, which if untreated causes endolymphatic hydrops (fluid accumulation) and subsequent degeneration accompanied by hearing loss and vestibular dysfunction. Although reports of congenital otosyphilis are rare in the current literature, there is concern that this disorder could rise with the reemergence of syphilis. Historically, rubella (German measles) was a major cause of hearing loss, and over half of all cases of congenital rubella syndrome, acquired through maternal infection, still lead to hearing loss. Vaccination programs have largely eliminated rubella from high-income countries, though the disease and its consequent hearing loss are still common in many low-income countries (Tucci et al., 2010). Hearing loss as a result of anoxia, maternal diabetes, and maternal herpes simplex virus are also still seen.

The infectious agent with greatest relevance to childhood hearing loss is perhaps the herpes virus cytomegalovirus (CMV), which is now considered a leading cause of congenital developmental disorders. CMV is widespread in humans, with CMV infection rates at birth estimated at 0.2–1%, and sensorineural hearing loss may develop in children who appear healthy at birth. Congenital CMV is estimated to cause 6% of hearing impairment in newborns and 15–25% of sensorineural hearing loss in children in the United States, and is now the primary cause of noninherited sensorineural hearing loss in children. Despite the incidence of this disorder, and the recent introduction of mouse models for CMV-induced hearing loss, the pathology of this CMV-induced hearing loss is still not well understood,

and control remains focused on hygiene and on development of a vaccine or antiviral.

37.2.2 Inherited Hearing Loss

While a significant proportion of human congenital hearing loss was once due to disorders associated with maternal infection, in middle- and high-income countries today, the majority of congenital hearing loss has a genetic component. The most common form of sensory impairment, hearing loss, is also one of the most common birth abnormalities. Clinically, the various sensory disorders are typically categorized as nonsyndromic or primary (e.g., isolated deafness not accompanied by other apparent abnormalities) and syndromic (e.g., deafness accompanied by other sensory, morphological, or physiological abnormalities). Nonsyndromic disorders, though less numerous, account for a majority of cases of deafness of genetic origin; the various nonsyndromic disorders are identified by inheritance pattern, with DFN, DFNA, and DFNB referring to X-linked, autosomal-dominant, and autosomal-recessive deafness disorders (see Table 37.1). In practice, this increasingly long list can be confusing due to genetic heterogeneity and to overlap with some of the syndromic forms of deafness. As the list of mapped loci and identified genes for inherited deafness disorders is still growing, any attempt to encompass the known hereditary hearing disorders is quickly made obsolete by the description of new disorders in mouse and human. A selection from the wide range of auditory disorders will be briefly described here, and the reader is encouraged to consult updated websites and reviews for the most current, in-depth information (Dror and Avraham, 2009; Petit and Richardson, 2009).

37.2.2.1 Hereditary Conductive or Mixed Hearing Loss

In normal hearing, sound is collected at the external ear and focused into the external ear canal. The sound waves set up vibrations of the tympanic membrane that are transmitted by the three small bones of the ossicular chain through the air-filled cavity of the middle ear to the membranous oval window at the border between the middle and inner ear. The hinged lever-like mechanical advantage system of the interconnected ossicles, together with the reduction in surface area between the tympanic membrane and the oval window, allow pressure at the tympanic membrane to be amplified sufficiently to drive compressive waves in the fluid-filled cochlea of the inner ear. When conductive hearing loss presents in children, it is most often associated with otitis media, or middle ear infection, and its sequelae, which can include chronic perforation, mastoiditis, cholesteatoma, ossicular fixation, or tympanic membrane

TABLE 37.1 Disorders of the ear

Name	Gene	Gene product	Reference
Nonsyndromic			
X-linked			
DFN3	POU3F4	POU domain, class 3, transcription factor 4	de Kok et al. (1995)
Autosomal-dominant			
DFNA1	DIAPH1	Protein diaphanous homolog 1	Lynch et al. (1997)
DFNA2A DFNA2B	KCNQ4 GJB3	Voltage-gated K^+ channel subfamily KQT member 4; Kv7.4 Gap–junction beta-3 protein; connexin 31	Kubisch et al. (1999) Xia et al. (1998)
DFNA3A DFNA3B	GJB2 GJB6	Gap–junction beta-2 protein; connexin 26 Gap–junction beta-6 protein; connexin 30	Kelsell et al. (1997) Grifa et al. (1999)
DFNA4	MYH14	Myosin-14	Donaudy et al. (2004)
DFNA5	DFNA5	Nonsyndromic hearing impairment protein 5	Van Laer et al. (1998)
DFNA6/14/38	WFS1	Wolframin	Bespalova et al. (2001) Young et al. (2001)
DFNA8/12	TECTA	Tectorin alpha	Verhoeven et al. (1998)
DFNA9	COCH	Cochlin	Robertson et al. (1998)
DFNA10	EYA4	Eyes-absent homolog 4	Wayne et al. (2001)
DFNA11	MYO7A	Myosin-7a	Liu et al. (1997)
DFNA13	COL11A2	Collagen, type XI, alpha 2	McGuirt et al. (1999)
DFNA15	POU4F3	POU class 4 homeobox 3	Vahava et al. (1998)
DFNA17	MYH9	Myosin-9	Lalwani et al. (2000)
DFNA20/26	ACTG1	Gamma actin	Zhu et al. (2003) van Wijk et al. (2003)
DFNA22	MYO6	Myosin-6	Melchionda et al. (2001)
DFNA23	SIX1	Sine-oculis homeobox homolog 1	Ruf et al. (2004)
DFNA25	SLC17A8	Solute carrier family 17, member 8; vesicular glutamate transporter 3	Ruel et al. (2008)
DFNA28	GRHL2	Grainyhead-like protein 2 homolog	Peters et al. (2002)
DFNA36	TMC1	Transmembrane channel-like protein 1	Kurima et al. (2002)
DFNA39	DSPP	Dentin sialophosphoprotein	Xiao et al. (2001)
DFNA44	CCDC50	Coiled-coil domain containing protein 50; Ymer	Modamio-Hoybjor et al. (2007)
DFNA48	MYO1A	Myosin-1a	Donaudy et al. (2003)
	CRYM	Mu-crystallin homolog	Abe et al. (2003)
Autosomal-recessive			
DFNB1A DFNB1B	GJB2 GJB6	Gap–junction beta-2 protein; connexin 26 Gap–junction beta-6 protein; connexin 30	Kelsell et al. (1997) del Castillo et al. (2002)
DFNB2	MYO7A	Myosin-7a	Liu et al. (1997) Weil et al. (1997)
DFNB3	MYO15	Myosin-15	Wang et al. (1998)
DFNB4	SLC26A4	Pendrin	Li et al. (1998)

Continued

TABLE 37.1   Disorders of the ear—cont'd

Name	Gene	Gene product	Reference
DFNB6	TMIE	Transmembrane inner ear expressed protein	Naz et al. (2002)
DFNB7/11	TMC1	Transmembrane channel-like protein 1	Kurima et al. (2002)
DFNB8/10	TMPRSS3	Transmembrane protease serine 3	Scott et al. (2001)
DFNB9	OTOF	Otoferlin	Yasunaga et al. (1999)
DFNB12	CDH23	Cadherin-23	Bork et al. (2001)
DFNB16	STRC	Stereocilin	Verpy et al. (2001)
DFNB18	USH1C	Harmonin	Ouyang et al. (2002) Ahmed et al. (2002)
DFNB21	TECTA	Tectorin alpha	Mustapha et al. (1999)
DFNB22	OTOA	Otoancorin	Zwaenepoel et al. (2002)
DFNB23	PCDH15	Protocadherin-15	Ahmed et al. (2003b)
DFNB24	RDX	Radixin	Khan et al. (2007)
DFNB28	TRIOBP	TRIO and F-actin-binding protein	Shahin et al. (2006) Riazuddin et al. (2006b)
DFNB29	CLDN14	Claudin-14	Wilcox et al. (2001)
DFNB30	MYO3A	Myosin-3a	Walsh et al. (2002)
DFNB31	WHRN	Whirlin	Mburu et al. (2003)
DFNB36	ESPN	Espin	Naz et al. (2004)
DFNB37	MYO6	Myosin-6	Ahmed et al. (2003a)
DFNB49	MARVELD2	MARVEL domain-containing protein 2; tricellulin	Riazuddin et al. (2006a)
DFNB53	COL11A2	Collagen, type XI, alpha 2	Chen et al. (2005)
DFNB59	PJVK	Deafness, autosomal-recessive 59; pejvakin	Delmaghani et al. (2006)
	SLC26A5	Prestin	Liu et al. (2003)
Syndromic			
Alport syndrome			
X-linked	COL4A5	Type IV collagen alpha-5	Barker et al. (1990)
Autosomal-recessive	COL4A3	Collagen alpha-3(IV) chain	Mochizuki et al. (1994)
Autosomal-recessive	COL4A4	Collagen alpha-4(IV) chain	Mochizuki et al. (1994)
Branchio–oto–renal syndrome			
BOR1	EYA1	Eyes-absent homolog 1	Abdelhak et al. (1997)
BOR2	SIX5	Homeobox protein SIX5	Hoskins et al. (2007)
BOS3	SIX1	Sine-oculis homeobox homolog 1	Ruf et al. (2004)
Keratitis–ichthyosis–deafness syndrome (KID)	GJB2	Gap–junction beta-2 protein; connexin 26	Richard et al. (2002)
Jervell and Lange-Nielsen syndrome			
JLNS1	KCNQ1	Voltage-gated K^+ channel subfamily KQT member 1	Neyroud et al. (1997)
JLNS2	KCNE1	Voltage-gated K^+ channel, Isk-related family, member 1	Tyson et al. (1997) Schulze-Bahr et al. (1997)
Norrie disease	NDP	Norrin	Berger et al. (1992)

TABLE 37.1 Disorders of the ear—cont'd

Name	Gene	Gene product	Reference
Pendred syndrome			
	SLC26A4	Pendrin	Everett et al. (1997)
	FOXI1	Forkhead box protein I1	Yang et al. (2007)
Stickler syndrome			
STL1	COL2A1	Collagen alpha-1(II) chain	Ahmad et al. (1991)
STL2	COL11A1	Collagen alpha-1(XI) chain	Richards et al. (1996)
STL3	COL11A2	collagen, type XI, alpha-2	Vikkula et al. (1995)
STL4	COL9A1	Collagen alpha-1(IX) chain	Van Camp et al. (2006)
Treacher–Collins syndrome	TCOF1	Treacle protein	TCSCG (1996)
Usher syndrome			
USH1B	MYO7A	Myosin-7a	Weil et al. (1995)
USH1C	USH1C	Harmonin	Verpy et al. (2000)
USH1D	CDH23	Cadherin-like 23	Bolz et al. (2001)
USH1F	PCDH15	Protocadherin-15	Ahmed et al. (2001)
USH1G	SANS	Usher syndrome type-1 G protein	Mustapha et al. (2002)
USH2A	USH2A	Usherin	Eudy et al. (1998)
USH2C	VLGR1	G protein-coupled receptor 98	Weston et al. (2004)
USH2D	WHRN	Whirlin	Ebermann et al. (2007)
USH3	USH3	Clarin 1	Joensuu et al. (2001)
Vohwinkel syndrome	GJB2	Gap–junction beta-2 protein; connexin 26	Maestrini et al. (1999)
Waardenburg syndrome			
WS1/3	PAX3	Paired box protein Pax-3	Tassabehji et al. (1992) Baldwin et al. (1992) Hoth et al. (1993)
WS2A	MITF	Microphthalmia-associated transcription factor	Tassabehji et al. (1994)
WS2D	SNAI2	Zinc finger protein SNAI2; slug	Sanchez-Martin et al. (2002)
WS2E	SOX10	SRY-BOX 10; SOX10 transcription factor	Bondurand et al. (1999)
WS4A	EDNRB	Endothelin B receptor	Puffenberger et al. (1994)
WS4B	EDN3	Endothelin-3	Edery et al. (1996)
WS4C	SOX10	SRY-BOX 10	Pingault et al. (1998)
Muckle–Wells syndrome	NLRP3	NACHT, LRR, and PYD domain-containing protein 3	Hoffman et al. (2001)
Bartter syndrome and deafness (DFNB73)			
Type 4A	BSND	Barttin	(Birkenhager et al. (2001)
Type 4b	CLCNKA CLCNKB	Chloride channel ClC-Ka; CLCK1 Chloride channel ClC-Kb; CLCKB	Schlingmann et al. (2004) Schlingmann et al. (2004)

Continued

TABLE 37.1　Disorders of the ear—cont'd

Name	Gene	Gene product	Reference
Hypoparathyroidism, deafness, and renal dysplasia	GATA3	Trans-acting T-cell-specific transcription factor GATA-3	Van Esch et al. (2000)
Renal tubular acidosis with deafness	ATP6B1	V-type proton ATPase subunit B	Karet et al. (1999)
Microtia, hearing impairment, and cleft palate	HOXA2	Homeobox A2	Alasti et al. (2008)
Otopalatodigital syndrome (OPD1)	FLNA	Filamin A	(Robertson et al. (2003)

Compiled in part from Online Mendelian Inheritance in Man (OMIM); OMIM gene symbols are used throughout

retraction, among others. Congenital conductive hearing loss most often results from malformations of the middle ear that cause fixation or disruption of the ossicular chain. As a group, the conductive losses are relatively tractable, as the various forms can often be successfully treated with surgery.

Deafness with perilymphatic Gusher is an X-linked disorder that involves congenital mixed conductive and sensorineural hearing loss and is caused by mutation of Brain 4 (Brn4, or POU3F4), a POU-domain transcription factor expressed in the developing otic mesenchyme. The conductive hearing loss results from stapes fixation to the oval window (with the name of the disorder arising from perilymph that gushes out from the cochlea during surgery to free the stapes). The sensorineural hearing loss in the mouse model likely results from a critical role of *Brn4* in establishing the endocochlear potential. Female carriers have been reported to have milder hearing loss that may be conductive or sensorineural. Mutation of the T-box transcription factor gene TBX1 is associated with velocardiofacial syndrome/DiGeorge syndrome, which is often accompanied by chronic otitis media. In mice, *Brn4* has been shown to interact with *Tbx1* to specify cochlear structure (Braunstein et al., 2008).

Several syndromes involving joint disorders, such as Cushing symphalangism, multiple synostoses, and the recently described Teunissen–Cremers syndrome ('stapes ankylosis with broad thumbs and toes'), result in moderate conductive hearing loss. These disorders can be caused by mutations affecting noggin (NOG) or growth and differentiation factor 5 (GDF5), both of which interact with bone morphogenetic proteins (BMP) during development. The secreted signaling proteins BMPs and GDFs induce bone formation, whereas the BMP antagonist NOG halts cartilage production at the site of joint initiation and is thus a critical player in joint development. GDF5 gain of function mutations and NOG loss of function mutations are both associated with ossicular fixation. In heterozygous $Nog^{+/-}$ mice

with varying degrees of hearing loss, impairment is associated with the presence of an extra bone fragment of the stapes that interferes with normal stapedial movement, and complete loss of *Nog* results in a more extreme phenotype with fused ossicles (Hwang and Wu, 2008). Thus, in the developing middle ear, *Bmp4*, *Gdf5*, and *Nog* may cooperate to establish the stapes and to separate it from the styloid process during ossicular development.

As much as 2% of profound deafness in childhood may be caused by the autosomal-dominant disorder branchio–oto–renal syndrome (BOR), which is characterized by branchial arch and renal abnormalities, together with variable hearing loss of sensorineural, conductive, or mixed origin. BOR can include structural anomalies of the external, middle, and inner ear and is associated with mutations of *Drosophila* eyes-absent homolog 1 (EYA1). In mouse, *Eya1* is expressed in adult hair cells, as well as in the developing mouse otocyst, and deletion of *Eya1* halts development at the otocyst stage. The *Eya1* gene product is a transcriptional coregulator that translocates to the nucleus together with sine-oculis *Six1*, and mutations in SIX1 result in both BOR and the closely related branchiootic syndrome 3 (BOS3) (Kochhar et al., 2007).

Cholesteatoma occurs when keratin-producing squamous epithelium develops within the temporal bone or middle ear; as this mass expands, it impinges on nearby structures, causing conductive hearing loss. This disorder, which can present in childhood or adulthood, most often occurs after otitis media or injury to the tympanic membrane, occurring as a uni- or bilateral conductive hearing loss, and may result from a genetic predisposition to otitis media. A rare congenital cholesteatoma has also been reported in cases of BOR and in adenomatous polyposis coli, a disorder associated with mutations of the adenomatous polyposis coli (APC) gene, multiple adenomatous polyps, and colorectal cancer. Several forms of nonsyndromic and syndromic hearing loss may involve chronic otitis media. For example, sensorineural hearing loss DFNA10 is a postlingual progressive hearing loss that typically presents late, in at least the second decade. This disorder involves the eyes-absent homolog EYA4, and the $Eya4^{-/-}$ mouse model displays a predisposition to otitis media. Several additional, as yet unidentified genes that function during development to direct middle ear anatomy likely also result in a susceptibility to otitis media (Rye et al., 2011). The rare autosomal-dominant craniofacial disorder, Treacher–Collins syndrome, is often accompanied by conductive hearing loss due to abnormalities of the ossicular chain. The syndrome is caused by a mutation affecting the nucleolar phosphoprotein treacle (TCOF1), which controls ribosome production and is required for proliferation of neural crest cells.

Semicircular canal dehiscence is a rare condition in which an opening in the bone of the semicircular canal effectively creates a third window in the inner ear. This extra outlet acts as a relief valve, such that oval window vibrations that would normally drive only the cochlear perilymph toward the round window are transmitted in part toward this third window, stimulating the vestibular organs. In patients with this condition, acoustic stimuli or fullness of the middle ear can produce dizziness and nystagmus. Otopalatodigital syndrome and frontometaphyseal dysplasia are X-linked skeletal dysplasias that also result in moderate to profound conductive deafness and are linked to the actin-binding protein filamin A (FLNA). Finally, congenital atresia of the oval window and perilymph fistula (loss of perilymph into the middle ear) are rare conductive disorders of unknown etiology.

37.2.2.2 Hereditary Sensorineural Hearing Loss

Sensorineural hearing loss is most commonly related to defects within the cochlea, though a small percentage of hearing loss is more central. The mammalian inner ear, encased in the temporal bone, comprises six sensory organs: five belonging to the vestibular system and one, the cochlea, to the auditory system. The cochlea coils about 2.5 times in humans (less in certain disorders), narrowing toward its apex, and in cross section is divided into three fluid-filled ducts that run in parallel nearly the length of the coiled cochlea. Two of these ducts, the perilymph-filled scala vestibuli and scala tympani, join at the apical end of the cochlea to form a continuous chamber with the oval window at one end and the membranous round window at the other. The endolymph-filled scala media is sandwiched between the scalae vestibuli and tympani and contains the receptor organ itself, the organ of Corti, which comprises the mechanosensitive hair cells – one row of inner hair cells, the actual sensory receptor cells along the inside of the cochlear spiral, and three rows of outer hair cells that provide cochlear amplification – as well as several types of supporting cells. Each hair cell is a polarized neuroepithelial cell whose apical, mechanotransducing pole contains the stereocilia hair bundle, and whose basal, synaptic pole is surrounded by supporting cells of the basilar membrane. The stereocilia project into the endolymph, and the tallest stereocilia of the outer hair cells are inserted in the tectorial membrane, a gelatinous extracellular matrix. At the inner, coiled, core of the cochlea lies the spiral ganglion, containing the cell bodies that give rise to the afferent fibers of the auditory nerve.

Vibrations of the oval window are transmitted to the perilymph in the scala vestibuli, setting up a traveling wave that propagates along the basilar membrane, which is tuned by mechanical properties along its length to map low-frequency sounds near the apex and high-frequency sounds near the base of the cochlea. The traveling wave at the basilar membrane induces movement relative to the hair cell bundle. Tethered only at one end, the tectorial membrane can move in part independently of the organ of Corti, and the shearing motion of the tectorial membrane against the organ of Corti causes deflection of the hair bundles. Outer hair cells actively amplify the cochlear mechanics through a mechanism that is still controversial.

Inward current through the mechanoelectrical transduction channels produces a receptor potential that alters the tonic release of glutamate at the hair cell–ganglion cell synapse; the majority of hair cells innervated by ganglion cells of the cochlear nerve are inner hair cells. The active mechanical properties of the cochlea cause the healthy cochlea to emit faint sounds, both spontaneously and in response to auditory stimuli. These otoacoustic emissions (OAEs) can be recorded and used clinically to assess outer hair cell motility and/or to identify whether the site of hearing loss is cochlear. As a noninvasive test that does not require a behavioral response, OAE recording is also commonly used to screen newborns for hearing loss. Central dysfunction is indicated by abnormal auditory brainstem responses (ABRs), auditory evoked potentials recorded at surface electrodes. Auditory neuropathy refers to any one of several forms of nonsyndromic sensorineural hearing loss in which the deficit is localized to the inner hair cell–ganglion cell synapse, or to neurons of the auditory nerve and auditory brain stem.

37.2.2.3 Hearing Loss and the Tectorial Membrane

The longest stereocilia of each outer hair cell contact the tectorial membrane, a specialized extracellular matrix consisting of collagen fibril bundles, together with three non-collagenous glycoproteins – otogelin, α-tectorin, and β-tectorin – that are expressed only in the inner ear. As mRNA for both tectorins is only transiently expressed during cochlear development, whatever form the tectorial membrane takes during that period likely persists for the lifetime of the organism. Though loss of otogelin in mouse causes early vestibular dysfunction, followed by a variable, progressive hearing loss and subtle abnormalities of the tectorial membrane, no human deafness associated with otogelin mutations has yet been identified (Simmler et al., 2000). In contrast to the subtle changes caused by the lack of otogelin, the absence of α-tectorin results in a failure to form the striated sheet matrix, the non-collagenous portion of the tectorial membrane. Mutations affecting α-tectorin (TECTA) are found in prelingual mid-frequency hearing loss of mild to profound severity, in both autosomal-dominant (DFNA8/12) and autosomal-recessive (DFNB21) forms of nonsyndromic hearing loss. In animal models lacking α-tectorin, otogelin and β-tectorin are not incorporated in the tectorial membrane and the tectorial membrane fails to associate normally with the spiral limbus and the organ of Corti, with the result that basilar membrane sensitivity is reduced and OAEs are abolished (Legan et al., 2000). To date, no mutations affecting β-tectorin (TECTB) have been associated with human hearing loss, though loss of *Tectb* in mouse results in a low-frequency hearing loss.

Two additional genes expressed uniquely in the inner ear are associated with sensorineural hearing loss. Otoancorin (OTOA) mutations associate with moderate to severe prelingual autosomal-recessive hearing loss DFNB22. Otoancorin is expressed at the apical surface of supporting cells adjacent to hair cells, but its role at the interface between the tectorial membrane and the sensory epithelium remains to be identified. Stereocilin (STRC) localizes to the distal portion of outer hair cell stereocilia. In humans, STRC mutations are associated with the early postlingual hearing loss DFNB16. In the mouse model $Strc^{-/-}$, which also experiences early hearing loss, the stereocilia array is established, but without the apical connectors between stereocilia or the links attaching stereocilia to the tectorial membrane, the stereocilia array rapidly deteriorates (Verpy et al., 2011).

37.2.2.4 *Hypothyroidism and Hearing Loss*

Congenital hypothyroidism in humans has been associated with a range of conductive and sensorineural hearing loss. Though none of these pathologies is entirely clear, animal studies point to a few possible pathways for hypothyroidism-induced hearing loss. The transcription factors thyroid receptor TRα1 (*Thra*) and T3 thyroid receptor TRβ (*Thrb*) are variably expressed during embryogenesis in the cochlea, *Thra* throughout the cochlea, and *Thrb* at high levels in the organ of Corti and at lower levels in the stria vascularis. Embryonic hypothyroidism delays development of the organ of Corti and prevents the normal transient developmental upregulation of β-tectorin mRNA. The Snell dwarf $Pou1f^{dw}$ mouse model highlights several effects of congenital hypothyroidism: (1) the tectorial membrane is abnormal, possibly due to altered β-tectorin expression; (2) KCNQ4 expression and function in outer hair cells is reduced; and (3) the endocochlear potential is decreased due to reduced expression of the inwardly rectifying K^+ channel KCNJ10 in intermediate cells of the stria vascularis (Mustapha et al., 2009).

37.2.2.5 *Pendred Syndrome*

Pendred syndrome (deafness with goiter) is the most common syndromic form of deafness, accounting for up to 7.8% of cases of congenital deafness and occurring in an estimated 7.5 of 100,000 births. Patients inheriting this autosomal-recessive disorder have variable degrees of deafness at birth and typically develop goiter in the second decade. The syndrome is accompanied by structural defects of the temporal bone such as Mondini dysplasia and enlarged vestibular aqueduct (EVA) that are likely driven in part by accumulation of endolymph. The vestibular aqueduct, embedded within the temporal bone, is a small canal containing the endolymphatic duct and extending from the vestibule between the cochlea and the labyrinth to the endolymphatic sac. In Mondini dysplasia, the apical turn of the cochlea fails to form, and patients are profoundly deaf at birth. In EVA, vestibular dysfunction may be present and the hearing loss is variable. The mutated Pendred syndrome gene PDS (SLC26A4) is a member of the solute carrier protein 26 anion transporter family, and the gene product pendrin is a transmembrane $Cl^-/I^-/HCO_3^-$ transporter. Allelic heterogeneity produces some SLC26A4 variants with Pendred syndrome and others with non-syndromic deafness with EVA (DFNB4).

Pendrin is expressed in the inner ear and kidney, and in the thyroid, where it mediates apical iodide transport in thyroid follicular cells (Kopp et al., 2008). Although Pendred syndrome is sometimes accompanied by hypothyroidism that could itself contribute to hearing loss, the distribution of the mouse pendrin throughout the endolymphatic duct and sac, and in specific areas of utricle, saccule, and external sulcus, points to a specific role of pendrin in fluid resorption and in regulating the ionic composition of the cochlear endolymph. The cochlear endolymph of the scala media is a unique extracellular fluid of high K^+ concentration and low Na^+ and Ca^{2+} concentrations. The unusual ionic makeup of the cochlear endolymph gives rise to a positive 'endocochlear potential' of about +80 mV and provides the driving force responsible for the large receptor potentials recorded from inner hair cells. Maintaining the ionic composition of the endolymph that underpins the endocochlear potential is critical to the generation of receptor potentials; this is a primary role of the stria vascularis, the extensively vascularized epithelium in the lateral wall of the scala media (Hibino et al., 2010).

Loss of pendrin in the knockout $Pds^{-/-}$ mouse leads to a thinned stria vascularis and a markedly reduced endocochlear potential. The Pendred syndrome alleles cause a complete loss of function, due to the retention of pendrin protein in the endoplasmic reticulum, whereas the nonsyndromic EVA alleles result in reduced anion transport and a less severe phenotype. In the absence of pendrin, both $Kcnj10$ expression in the stria vascularis and the endocochlear potential are lost in early postnatal life. The decreased K^+ concentration of $Pds^{-/-}$ endolymph may be explained by the loss of $Kcnj10$ expression, while the increased Ca^{2+} concentration results from disruption of pendrin-mediated transport of HCO_3^- into the endolymph. Loss of HCO_3^- exchange lowers the pH of the endolymph and can inhibit TRPV5/6 Ca^{2+} channels, increasing the endolymphic Ca^{2+} concentration (Nakaya et al., 2007). The winged helix/forkhead protein *Foxil* has similar expression patterns to pendrin and may be an upstream regulator, as $Foxil^{-/-}$ mice lack pendrin transcripts and display an enlarged endolymphatic chamber. Mutations of FOXI1 are also found in human patients with Pendred syndrome or nonsyndromic enlarged vestibular aqueduct.

To maintain the endolymphic concentration necessary for transduction, K^+ must be actively recycled; K^+ exits the hair cells and diffuses through the perilymph to the lateral wall, where it is transported by N^+–K^+–ATPase and/or Na^+–K^+–Cl^- cotransporter (NKCCl) into the stria vascularis, from which it is reintroduced into the endolymph by the voltage-gated K^+ channel KCNQ1. NKCCl is expressed during development in the stria vascularis and in immature epithelial cells lining the developing scala media, and its removal causes the collapse of the cochlear ducts. Mutations of KCNQ1 are also associated with the cardio-auditory Jervell and Lange-Nielsen syndrome, an autosomal-recessive disorder characterized by congenital deafness, prolonged QT syndrome, ventricular arrhythmia, and sudden death (Wangemann, 2006). Mouse models of this disorder display similar, though somewhat variable, defects.

Autosomal-dominant deafness DFNA2A – mild to severe progressive hearing loss with widely variable age of onset (early childhood to sixth decade) and occasional tinnitus – is associated with mutations affecting the voltage-gated potassium channel KCNQ4. In mouse, the native $Kcnq4$ is expressed at high levels at the base of outer hair cells, where it forms a depolarization-activated K^+ channel with relatively slow kinetics. The slow progressive hearing loss seen in $Kcnq4^{-/-}$ mice may be due to degeneration of outer hair cells that are chronically depolarized (Kharkovets et al., 2006).

Mutations involving the connexins Cx26 (GJB2) and Cx30 (GJB6) may account for half of all cases of prelingual autosomal-recessive hearing loss, with loss of Cx26/30 leading to hair cell degeneration and subsequent degeneration of the spiral ganglion cells. The precise functions of Cx26 and 30 in the inner ear are unknown, though they are important for maintaining the high K^+ concentration of the endolymph. Cx26 and Cx30 are both highly expressed in adults in the supporting cells of the organ of Corti and in the stria vascularis, and though the connexins are not found in the developing hair cells themselves, both connexins are required for the survival of hair cells and the organ of Corti. Extensive gap junction coupling in the neonatal organ of Corti is likely supported by Cx26/30 gap junctions; the function of this coupling is unknown. In mice lacking Cx30, the capillary endothelial barrier in the stria vascularis breaks down, the endocochlear potential disappears, and the hair cells degenerate (Cohen-Salmon et al., 2007).

The SLC26 family protein prestin (SLC26A5), which is closely related to pendrin, is linked to mild to profound hearing loss DFNB61, presenting from birth to the fourth decade. Prestin is expressed specifically in cochlear outer hair cells, where it functions as an anion transporter in the lateral membrane of outer hair cells. Electromotility, the hair cells' ability to change shape in an applied electric field, is driven by a physical change in the lateral membrane of outer hair cells and underlies cochlear amplification. Prestin is likely the critical voltage-driven motor protein for electromotility, and the loss of prestin and hair cell electromotility may yield a loss in cochlear amplification of 40–60 dB (Liu et al., 2003).

37.2.2.6 Other Hearing Disorders

Although hereditary hearing loss affecting primarily low frequencies (<2 kHz) is rare, a few distinct forms have been reported. The nonsyndromic hearing loss DFNA1 is an autosomal-dominant sensorineural disorder that manifests at about 10 years with low-frequency hearing loss and progresses to entail profound deafness at all frequencies by age 30. The disorder results from mutations affecting the homolog of *Drosophila* diaphanous (DIAPH1), which is widely expressed in many tissues in addition to the cochlea. As diaphanous is a member of the family of highly conserved formin proteins, which stabilize microtubules and regulate actin nucleation and elongation (Chesarone et al., 2010), the disorder likely affects the actin skeleton of hair cells or stereocilia. A second form of low-frequency sensorineural deafness, DFN6/14/38, results in moderate hearing loss with occasional tinnitus. Little is known of the pathology of this disorder, which involves wolframin (WFS1), a membrane glycoprotein found at the endoplasmic reticulum.

Several forms of inherited hearing loss are associated with mutations in mitochondrial DNA and can present as prelingual or postlingual hearing loss. In particular, Mohr–Tranebjaerg syndrome (MTS), or dystonia–deafness syndrome, is an X-linked progressive syndrome (DFN1) involving early postlingual hearing loss and dystonia that can also be accompanied by myopia, cortical blindness, and mental impairment. The mutation is due to a loss of function in the TIMM8A gene (translocase of inner mitochondrial membrane 8), but the mechanism of hearing loss is unknown. Additional disorders linked to mutations in mitochondrial genes include diabetes–deafness syndrome, Ballinger–Wallace syndrome, multisystem disorder, and a number of nonsyndromic sensorineural deafness disorders.

37.2.2.7 Auditory Neuropathy

Auditory neuropathy accounts for about 10% of profound childhood hearing loss. Some of the major preventable causes of auditory neuropathy include anoxia, certain infectious diseases, and hyperbilirubinemia. Neonatal hyperbilirubinemia is easily treatable, but it is also common, and bilirubin neurotoxicity may account for 50% of all cases of auditory neuropathy. Various genetic mutations underlying auditory neuropathies of mild to profound hearing loss have also been identified.

One of the first auditory neuropathy genes identified was found in a candidate gene search for the profound prelingual hearing loss DFNB9 (Yasunaga et al., 1999). The membrane-anchored Ca^{2+}-binding protein otoferlin (OTOF) is expressed in hair cells and in the spiral ganglion during development in the mouse, and in adult localizes to vesicles of the ribbon synapses at the basal pole of the inner hair cells. Though otoferlin is not required for the development of ribbon synapses, it is required both for Ca^{2+}-dependent exocytosis at ribbon synapses of inner hair cells and at ribbon synapses that are transiently expressed during development at the outer hair cells. Otoferlin mutations may also result in minor defects in hair bundle structure.

The postlingual progressive hearing loss of variable severity DFNA15 is linked to mutations in POU4F3. Brn-3.1, a member of the Class IV POU-domain transcription factor family, is expressed exclusively in hair cells of the inner ear; $Brn-3.1^{-/-}$ mice show severe balance deficits and are completely deaf at a young age. In $Brn-3.1^{-/-}$ mice, the overall structure of the cochlea and spiral ganglion is normal at birth, but the hair cells do not differentiate and extend stereocilia; within a short period, the hair cells and supporting cells of the cochlea and the neurons of the spiral ganglion, degenerate (Erkman et al., 1996). Brn-3.1 likely mediates terminal differentiation of the hair cells, and the failure of hair cell differentiation may lead to secondary degeneration of supporting cells and postsynaptic neurons.

Glutamate release at the synapse between inner hair cell and spiral ganglion cell requires the vesicular glutamate transporter VGLUT3 (SLC17A8), and mutation of the SLC17A8 gene in humans is associated with the progressive sensorineural hearing loss DFNA25. Loss of SLC17A8 in mouse results in deafness, seizing, and deficits in central auditory circuit refinement; the central deficits may be due to a role for VGLUT3 in inhibitory synapse maturation (Seal et al., 2008). The gene for SLC17A8 is one of a few known to have effects both early in the cochlea and more centrally in the auditory pathway. Another is pejvakin (Persian for "echo"), which is associated with recessive auditory neuropathy DFNB59, causing prelingual severe hearing impairment. Pejvakin is expressed in cell bodies in the cochlea, spiral ganglion, and neuronal subpopulations in the auditory brain stem and midbrain in mouse, and though the function is unknown, the pathology likely affects both outer hair cells and downstream auditory pathways.

Charcot–Marie–Tooth disease describes a diverse group of inherited peripheral neuropathies, with onset in the first or second decade, of which the most common is the demyelinating form CMT1. About 5% of patients with CMT exhibit sensorineural hearing loss, with a likely locus in the cochlear nerve, specifically those with point mutations or deletions in the genes for peripheral myelin protein 22 (PMP22), myelin protein zero (MPZ), or connexin-32 (GJB1).

37.3 MIXED AUDITORY AND VISUAL IMPAIRMENT – USHER SYNDROME

Hereditary deafness–blindness in its most common form, Usher syndrome, is an autosomal-recessive disorder that involves bilateral sensorineural hearing loss and retinitis pigmentosa, in addition to vestibular deficits in some cases. Over half of all cases of hereditary deafness–blindness are associated with Usher syndrome, which has an estimated global prevalence of 1:16,000–1:50,000. The three clinically recognized subtypes of the syndrome are distinguished from one another by severity of hearing loss, age of onset of retinitis pigmentosa and the presence or absence of vestibular dysfunction. Usher syndrome types 1 and 2 are most common, type 1 being more severe and type 2 more frequent. Patients with the rarer form of type 3 Usher syndrome exhibit progressive hearing loss of intermediate severity, together with vestibular dysfunction and retinitis pigmentosa of varying severity and onset. Examination of the hair cells in Usher syndrome reveals significant disorder of the hair bundles, and the syndrome is considered a ciliopathy, as the primary deficit is at hair cell stereocilia in the cochlea and at the ciliated stalk connecting inner and outer photoreceptor segments in the retina.

Usher syndrome can result from mutations in any one of at least nine genes, all of which are involved in the organization of the stereocilia during development of the hair bundle. The stereocilia are not true cilia, but rather are specialized microvilli formed of tightly packed cross-linked actin filaments. During development, however, each hair bundle expresses a single true cilium, the kinocilium, which in mammalian cochlear hair cells eventually degenerates. During early development of the hair bundle, a central kinocilium, surrounded by microvilli at apparently random locations, is first extended at the apical pole of the hair cell. The kinocilium migrates to one side of the hair bundle, and then the kinocilia of all hair cells align to the same side of their hair cells. A height-ordered array of stereocilia begins to form, with those stereocilia nearest the kinocilium elongating first, followed sequentially by stereocilia farther removed from the kinocilium, and an organized, hexagonal array of stereocilia forms, in stepwise order of increasing height. Stereocilia are continually extended from the apical surface during this period, many to be later resorbed. Once elongation is complete, the addition of actin filaments to the stereocilia increases stereocilia width, the stereocilia extend actin rootlets into the cytoplasm of the hair cell, and the basal ends of the stereocilia taper to a point (Goodyear et al., 2006; Tilney et al., 1992).

Mature stereocilia are linked at their apical tips by tip links, filamentous strands oriented in parallel with the direction of hair bundle deflection that connect the tip of each stereocilium to the side of its taller, adjacent stereocilium. In addition to the tip links, several additional filamentous links between stereocilia are also expressed, especially during development. In the E17.5 mouse, lateral links, tip links, and some kinociliary links are present. Ankle links appear at around P2 and remain until about P9. The different links can be distinguished by their position along the stereocilia, from apical tip to basal ankle, and by their distinct antibody binding.

37.3.1 Usher Syndrome Type 1

Individuals with USH1 typically exhibit severe hearing loss ($\geq$90 dB) and vestibular dysfunction at birth. Balance problems contribute to an average 12-month delay in walking, and onset of retinitis pigmentosa occurs before puberty. Although Usher type 1 is genetically heterogeneous, all identified relevant gene products are expressed in cochlear hair cells, where they function during development to organize and hold together the hair bundle. The five known gene products are myosin VIIa (USH1B), an actin-based motor protein, harmonin (USH1C), cadherin 23 (USH1D), protocadherin-15 (USH1F), and the putative scaffolding protein Sans (USH1G). Harmonin can bind to F-actin, as well as to the four other USH1 proteins. Harmonin, cadherin 23, and protocadherin-15 are expressed at the hair bundle as soon as stereocilia are extended, and cadherin 23 and protocadherin-15 together

form the tip link connecting the tip of one stereocilia to its taller neighbor. The various *Pcdh15* isoforms differ in sites of expression along the stereocilia and in timing of expression during development, and may direct distinct aspects of hair bundle development. Myosin VIIa is expressed throughout the hair cell, and likely plays several roles in development of the hair bundle. One function involves directing harmonin to its appropriate position on the stereocilia, as harmonin remains at the base of the hair bundle in the absence of *Myo7a* (Boeda et al., 2002). Sans function is less well understood; this transiently expressed protein can interact with myosin VIIa or with harmonin and may be a component of the developmentally expressed lateral links.

37.3.2 Usher Syndrome Type 2

Usher syndrome type 2, which accounts for 3–6% of congenital hearing loss, is characterized by mild congenital nonprogressive hearing loss, onset of retinitis pigmentosa in the first to second decades, and a lack of vestibular dysfunction. Three causative genes have been identified for USH2; all proteins are expressed in the cochlea during development. Type 2A, the most common form of Usher syndrome, results from mutations in usherin (USH2A), type 2C from mutations in GPR98, and type 2D from mutations in whirlin (WHRN).

The highly conserved 600 kDa long usherin variant is transiently associated with the growing stereocilia of inner and outer hair cells. The long isoform, found in both retina and inner ear, contains a PDZ-binding motif on its cytoplasmic tail, as do harmonin b and whirlin, which are also present in the differentiating hair bundle. GPR98 (VLGR1) is a transmembrane, calcium-binding G protein-coupled receptor that is transiently expressed at the base of the hair bundle during development. Both usherin and Vlgr1 bind to whirlin and myo7a, and association of Vlgr with usherin allows the formation of the ankle links at the base of the hair bundle. Ankle links are required for controlling the shape of the hair bundle, and loss of Vlgr during development leads to reduced transduction currents in the outer hair cells, alteration of the overall hair bundle structure, and deafness. Whirlin localizes to the stereocilia tips, where it interacts with myo15a or with myo7a in the control of stereocilia length during the growth of the hair bundle. In mice, targeted N-terminus whirlin mutants exhibit loss of both vision and hearing, whereas C-terminus mutants exhibit an isolated auditory disorder (Yang et al., 2010).

37.3.3 Usher Syndrome Type 3

Type 3 (USH3) is a relatively rare form of Usher syndrome reported primarily in Finland and among Ashkenazi Jewish populations, usually diagnosed in the first decade. The disorder results from mutations in the transmembrane protein clarin-1 (CLRN1), most instances occurring due to a single amino acid substitution that affects glycosylation and membrane insertion of the protein. *Clarin-1* is expressed transiently in the cochlea during development, and in Muller cells of the retina. In the Clrn1 knockout mouse, loss of clarin-1 results in disorganized stereocilia of the outer hair cells, followed by slow progressive disorganization of the hair bundle and age-related stereocilia lengthening and eventual hair cell degeneration, leading to vestibular dysfunction and complete loss of hearing (Geller et al., 2009). The protein's ability to interact with cell adhesion molecules and induce actin filament reorganization suggests that the pathology results from disruption of interactions between cell-adhesion molecules and the actin cytoskeleton.

37.3.4 Retinitis Pigmentosa in Usher Syndrome

As this disease is characterized by the progressive degeneration of the rod photoreceptors, the cone photoreceptors, and finally retinal pigment epithelium (RPE) cells, patients present first with night blindness (nyctalopia). Progressive loss of peripheral vision leads to tunnel vision, whereas central vision (color and high acuity) may be compromised after a variable period. The pigment deposits for which the disorder is named appear late in the disease, after photoreceptors degenerate and cells of the RPE detach and migrate through the eye (Milam et al., 1998).

The ciliary defect in hearing loss in Usher syndrome, and the existence of a ciliary stalk between the inner and outer segments of retinal photoreceptors, points to photoreceptors as the primary site of the visual deficit in Usher syndrome, and the first measurable retinal loss in humans with USH 1B, 1 F, 2A, and 2C is of photoreceptors. Despite the existence of various animal models for Usher syndrome, an ongoing challenge has been that affected animals exhibit hearing loss and vestibular dysfunction but most mouse models fail to develop the retinitis pigmentosa characteristic of Usher syndrome in humans. Additionally, many well-known mouse models of deafness (e.g., waltzer, shaker-1, deaf circler, Ames waltzer, and Jackson shaker) possess mutated USH1 genes and exhibit vestibular but not visual impairment. In the retina, usherin is expressed at all ages at the connecting cilium in photoreceptors, and one promising development is the mouse model for USH2A, which loses large numbers of photoreceptors (Liu et al., 2007).

37.4 VISUAL IMPAIRMENT

Many historical forms of childhood blindness have been largely eradicated in middle- and high-income countries, and are being eliminated from low-income countries, where programs combating measles and vitamin A deficiency have been especially successful. Corneal disorders still account for 19% of childhood blindness worldwide, however, and are disproportionately common in the developing world, amounting to perhaps 1% of all cases of blindness in the highest income regions and as much as 36% of all cases in sub-Saharan Africa (Gilbert, 2009). Most of these cases are preventable, resulting from vitamin A (retinol) deficiency, measles, and/or use of traditional treatments. Prenatal or neonatal lack of dietary vitamin A, increased demand for vitamin A during infection, and malabsorption of vitamin A due to illness can singly or together lead to retinol deficiency, a risk factor for xerophthalmia (dry eye), which commonly leads to corneal ulceration and scarring. Corneal ulceration and scarring can also result from measles infection, an effect that is potentiated if vitamin A deficiency is already present.

Microphthalmia, an abnormally small eye with normal structure, develops in response to genetic or environmental factors, and may be present in as much as one-tenth of all childhood blindness. Microphthalmia most often occurs bilaterally as part of a heritable syndrome, and most of the genes linked to this disorder code for transcription factors and homeobox genes involved in early development of the eye, for example in neuronal differentiation, proliferation of retinal progenitor cells, or formation of the lens placode. Some of the known genes include transcription factor SOX-2 (SOX2), retina homeobox protein Rx (RAX), and visual system homeobox 2 (VSX, also CHX10), among others (see Table 37.2).

37.4.1 Disorders of the Anterior Segment

The anterior segment of the eye includes the cornea, iris, and lens, together with their supporting structures. As the first structure through which light passes, and as part of the interface with the most refractive power, the cornea occupies a special position; hence, environmental insults or genetic disorders that affect the cornea can profoundly affect visual sensation. Transparency of the lens also plays a critical role in conducting light to the retina. Disorders of the anterior segment can commonly include glaucoma of unknown etiology. Aniridia, for example, the complete or partial absence of the iris, may occur with glaucoma, as well as with cataracts or photophobia. Aniridia is associated with over 100 different mutations of the transcription factor paired box 6 (PAX6), which is linked to several additional disorders of abnormal eye development, such as morning glory disc (a dysplasia of the optic disc), ectopic pupil, optic nerve hypoplasia, and foveal hypoplasia. Corneal dystrophy refers to a group of disorders that lead to accumulation of deposits in the corneal layers and to a loss of corneal transparency; depending on the specific type of dystrophy, the

TABLE 37.2 Disorders of the eye

Name	Gene	Gene product	Reference
Microphthalmia			
MCOPCT2	SIX6	Homeobox protein SIX6	Gallardo et al. (2004)
MCOP3	RAX	Retina homeobox protein Rx	Voronina et al. (2004)
MCOP5	MFRP	Membrane frizzled-related protein	Sundin et al. (2005)
MCOPS2	BCOR	BCL-6 corepressor	Ng et al. (2004)
MCOPS3	SOX2	Transcription factor SOX-2	Fantes et al. (2003)
MCOPS5	OTX2	Orthodenticle homeobox 2	Ragge et al. (2005)
MCOPS6	BMP4	Bone-morphogenic protein 4	Bakrania et al. (2008)
MCOPS7	HCCS	Holocytochrome c synthase	Wimplinger et al. (2006)
MCOPS9	STRA6	Stimulated by retinoic acid gene 6 protein	Pasutto et al. (2007)
Corneal dystrophy			
CHED2	SLC4A11	Sodium bicarbonate transporter-like protein 11	Vithana et al. (2006) Desir et al. (2007)
CDGDL	TACSTD2	Tumor-associated calcium signal transducer 2	Tsujikawa et al. (1999)
MECD	KRT3 KRT12	Keratin 3 Keratin 12	Irvine et al. (1997)
Congenital cataract			
	CRYAA	Alpha crystallin A	Litt et al. (1998)
	CRYAB	Alpha crystallin B	Berry et al. (2001)
	CRYBA1	Beta crystallin A1	Kannabiran et al. (1998)
	CRYBB1	Beta crystallin B1	Cohen et al. (2007)
	CRYBB2	Beta crystallin B2	Litt et al. (1997)
	CRYGC	Gamma C crystallin	Heon et al. (1999)
	CRYGD	Gamma D crystallin	Nandrot et al. (2003)
	CRYGS	Gamma S crystallin	Sun et al. (2005)
Retinitis pigmentosa			
RP1	RP1	Retinitis pigmentosa RP1 protein	Pierce et al. (1999) Sullivan et al. (1999)
RP2	RPGR	Retinitis pigmentosa 2	Schwahn et al. (1998)
RP3	RPGR	Retinitis pigmentosa GTPase regulator	Meindl et al. (1996)
RP4	RHO	Rhodopsin	Dryja et al. (1990) Humphries et al. (1997)
RP7	PRPH2 ROM1	Peripherin-2 (retinal degeneration slow) Rod outer segment protein 1	Farrar et al. (1991) Kajiwara et al. (1994)
RP8	MTTS2	tRNA, mitochondrial, serine, 2	Mansergh et al. (1999)
RP9	RP9	Retinitis pigmentosa 9 protein	Keen et al. (2002)
RP10	IMPDH1	Inosine-5′-monophosphate dehydrogenase 1	Bowne et al. (2006)
RP11	PRPF31	Pre-mRNA processing factor 31	Vithana et al. (2001)
RP12	CRB1	Crumbs homolog 1	den Hollander et al. (1999)

Continued

TABLE 37.2 Disorders of the eye—cont'd

Name	Gene	Gene product	Reference
RP13	PRPF8	Pre-mRNA-processing factor 8	McKie et al. (2001)
RP14	TULP1	Tubby-related protein 1	Hagstrom et al. (1998)
RP17	CA4	Carbonic anhydrase IV	Rebello et al. (2004)
RP18	PRPF3	Pre-mRNA-processing factor 3	Chakarova et al. (2002)
RP19	ABCA4	Retina-specific ATP-binding cassette transporter	Rozet et al. (1999)
RP20	RPE65	Retinoid isomerohydrolase	Gu et al. (1997)
RP25	EYS	Eyes-shut homolog	Abd El-Aziz et al. (2008)
RP26	CERKL	Ceramide kinase-like	Tuson et al. (2004)
RP27	NRL	Neural retina-specific leucine zipper protein	Bessant et al. (1999)
RP30	FSCN2	Fascin 2	Wada et al. (2001)
RP31	TOPORS	E3 ubiquitin–protein ligase Topors	Chakarova et al. (2007)
RP33	SNRNP200	Small nuclear ribonucleoprotein 200 kDa	Zhao et al. (2009)
RP35	SEMA4A	Semaphorin-4A	Abid et al. (2006)
RP36	PRCD	Progressive rod–cone degeneration, dog, homolog	Zangerl et al. (2006)
RP37	NR2E3	Photoreceptor-specific nuclear receptor	Coppieters et al. (2007)
RP38	MERTK	MER tyrosine kinase	Gal et al. (2000)
RP39	USH2A	Usherin	Rivolta et al. (2000)
RP40	PDE6B	Rod cGMP-specific cyclic phosphodiesterase, beta subunit	McLaughlin et al. (1993)
RP41	PROM1	Prominin 1	Maw et al. (2000)
RP42	KLHL7	Kelch-like protein 7	Friedman et al. (2009)
RP43	PDE6A	Rod cGMP-specific cyclic phosphodiesterase alpha subunit	Huang et al. (1995)
RP44	RGR	Retinal G-protein coupled receptor	Morimura et al. (1999)
RP45	CNGB1	Cyclic nucleotide-gated channel beta 1	Bareil et al. (2001)
RP46	IDH3B	Isocitrate dehydrogenase 3 beta subunit	Hartong et al. (2008)
RP47	SAG	S-antigen; S-arrestin	Nakazawa et al. (1998)
RP48	GUCA1B	Guanylate cyclase activator 1B	Sato et al. (2005)
RP49	CNGA1	Cyclic nucleotide gated channel alpha 1	Dryja et al. (1995)
RP50	BEST1	Bestrophin 1	Davidson et al. (2009)
Juvenile, AR	SPATA7	Spermatogenesis associated 7	Wang et al. (2009)
Newfoundland rod–cone	RLBP1	Retinaldehyde-binding protein 1	Eichers et al. (2002)
Cone–rod dystrophy			
CORD2	CRX	Cone–rod homeobox protein	Freund et al. (1997)
CORD3	ABCA4	Retina-specific ATP-binding cassette transporter Rim)	(Cremers et al. (1998)
CORD5	PITPNM3	Phosphatidylinositol transfer protein, membrane-associated 3	Kohn et al. (2007)
CORD6	GUCY2D	Retinal guanylyl cyclase 1	Kelsell et al. (1998)
CORD7	RIMS1	Protein-regulating synaptic membrane exocytosis 1	Johnson et al. (2003)
CORD9	ADAM9	A disintegrin and metalloproteinase domain 9	Parry et al. (2009)

TABLE 37.2 Disorders of the eye—cont'd

Name	Gene	Gene product	Reference
CORD10	SEMA4A	Semaphorin-4A	Abid et al. (2006)
CORD12	PROM1	Prominin 1	Yang et al. (2008)
CORD13	RPGRIP1	Retinitis pigmentosa GTPase regulator-interacting protein	Hameed et al. (2003)
CORD14	GUCA1A	Guanylate cyclase activator 1A	Payne et al. (1998)
CORD15	CDHR1	Cadherin-related family member 1; photoreceptor cadherin	
CORDX1	RPGR	Retinitis pigmentosa GTPase regulator	Demirci et al. (2002)
CORDX3	CACNA1F	Voltage-dependent L-type Ca^{2+} channel, alpha-1 F; Cav1.4	Jalkanen et al. (2006)
Early-onset severe	LRAT	Lecithin retinol acyltransferase	Thompson et al. (2001)
Stargardt disease			
STGD1	ABCA4 CNGB3	Retina-specific ATP-binding cassette transporter Cyclic nucleotide-gated cation channel beta-3	Allikmets et al. (1997) (Nishiguchi et al., 2005)
STGD3	ELOVL4	Elongation of very long chain fatty acids protein 4	Zhang et al. (2001)
STGD4	PROM1	Prominin 1	Yang et al. (2008)
Leber congenital amaurosis			
LCA1	GUCY2D	Retinal guanylate cyclase	Perrault et al. (1996)
LCA2	RPE65	Retinoid isomerohydrolase	Marlhens et al. (1997)
LCA3	SPATA7	Spermatogenesis-associated protein 7	Wang et al. (2009)
LCA4	AIPL1	Aryl–hydrocarbon-interacting protein-like 1	Sohocki et al. (2000)
LCA6	RPGRIP1	Retinitis pigmentosa GTPase regulator-interacting protein	Dryja et al. (2001)
LCA7	CRX	Cone–rod homeobox protein	Freund et al. (1997)
LCA8	CRB1	Crumbs homolog 1	Abouzeid et al. (2006)
LCA10	CEP290	Centrosomal protein of 290 kD	den Hollander et al. (2006)
LCA11	IMPDH1	Inosine-5′-monophosphate dehydrogenase 1	Bowne et al. (2006)
LCA12	RD3	Retinal degeneration 3	Friedman et al. (2006)
LCA13	RDH12	Retinol dehydrogenase 12	Janecke et al. (2004)
LCA15	TULP1	Tubby-like protein 1	Hagstrom et al. (1998)
LCA16	KCNJ13	K channel, inward rectifier, Kir7.1	Sergouniotis et al. (2011)
Night blindness			
CSNBAD1	RHO	Rhodopsin	Dryja et al. (1993)
CSNBAD2	PDE6B	Rod cGMP-specific cyclic phosphodiesterase beta subunit	Gal et al. (1994)
CSNBAD3	GNAT1	G protein, alpha transducing 1 (transducin, alpha subunit)	Dryja et al. (1996)
CSNB1A	NYX	Nyctalopin	Bech-Hansen et al. (2000)
CSNB1B	GRM6	Metabotropic glutamate receptor 6	Dryja et al. (2005)

Continued

TABLE 37.2   Disorders of the eye—cont'd

Name	Gene	Gene product	Reference
CSNB1C	TRPM1	Transient receptor potential cation channel subfamily M member 1	Bellone et al. (2008) Li et al. (2009)
CSNB1D	SLC24A1	Sodium/potassium/calcium exchanger	Riazuddin et al. (2010)
CSNB2A	CACNA1F	Voltage-dependent L-type Ca^{2+} channel, alpha-1 F; Cav1.4	Bech-Hansen et al. (1998) Strom et al. (1998)
CSNB2B	CABP4	Calcium-binding protein-4	Zeitz et al. (2006)
Oguchi disease 1	SAG	S-arrestin	Fuchs et al. (1995)
Oguchi disease 2	GRK1	Rhodopsin kinase	Yamamoto et al. (1997)
Joubert syndrome			
JBTS5	CEP290	Centrosomal protein of 290 kD	Sayer et al. (2006) Valente et al. (2006)
JBTS7	RPGRIP1L	Retinitis pigmentosa GTPase regulator-interacting protein	Delous et al. (2007)
JBTS8	ARL13B	ADP-ribosylation factor-like protein 13B	Cantagrel et al. (2008)
JBTS9	CC2D2A	Coiled-coil and C2 domain-containing protein 2A	Gorden et al. (2008) Noor et al. (2008)
JBTS10	CXORF5	Oral–facial–digital syndrome 1	Coene et al. (2009)
Peters' anomaly	PAX6	Paired box 6 protein Pax-6	Hanson et al. (1994) Mirzayans et al. (1995)
	PITX2	Pituitary homeobox 2	Doward et al. (1999)
	CYP1B1	Cytochrome P450 1B1	Stoilov et al. (1998)
	FOXC1	Forkhead box protein C1	Honkanen et al. (2003)
Axenfeld–Rieger syndrome	PITX2	Pituitary homeobox 2	Semina et al. (1996)
Iridogoniodysgenesis			
IRID1	FOXC1	Forkhead box protein C1	Nishimura et al. (1998)
IRID2	PITX2	Pituitary homeobox 2	Semina et al. (1996)
Posterior embryotoxon	JAG1	Jagged 1	Le Caignec et al. (2002)
Aniridia	PAX6	Paired box 6 protein Pax-6	Jordan et al. (1992)
Anterior segment mesenchymal dysgenesis	PITX3	Pituitary homeobox 3	Semina et al. (1998)
	FOXE3	Forkhead box protein E3	Semina et al. (2001)
Ectopia pupillae	PAX6	Paired box 6 protein Pax-6	Hanson et al. (1999)
Optic nerve hypoplasia	PAX6	Paired box 6 protein Pax-6	Azuma et al. (2003)
Foveal hypoplasia	PAX6	Paired box 6 protein Pax-6	Azuma et al. (1996) Hanson et al. (1999)
Coloboma of the optic nerve (morning glory disc)	PAX6	Paired box 6 protein Pax-6	Azuma et al. (2003)
Krause–Kivlin syndrome			
Cataract–microcornea syndrome	GJA8	Gap–junction alpha-8 protein	Devi and Vijayalakshmi (2006)
Bardet–Biedl syndrome			
	ARL6	ADP-ribosylation factor-like protein 6	Chiang et al. (2004)

TABLE 37.2   Disorders of the eye—cont'd

Name	Gene	Gene product	Reference
	MKKS	McKusick–Kaufman/Bardet–Biedl syndromes putative chaperonin	Katsanis et al. (2000) Slavotinek et al. (2000)
	MKS1	Meckel syndrome type 1 protein	Leitch et al. (2008)
	CEP290	Centrosomal protein of 290 kD	Leitch et al. (2008)
	CCDC28B	Coiled-coil domain containing 28B	Badano et al. (2006)

Compiled in part from OMIM; OMIM gene symbols are used throughout.

disorder can manifest as early as the first decade. Genes encoding a sodium borate cotransporter (SLC4A11) and transforming growth factor, beta-induced (TGFBI) have been implicated in multiple corneal dystrophies; however, relatively few animal models exist, corneal transparency itself is not well understood, and both the normal functions and the pathologies of these gene products are largely unknown (Klintworth, 2009).

Cataract, an opacification of the lens, occurs congenitally in 1–6 of 10,000 births. Worldwide, congenital cataract may account for 15% of childhood blindness, and if left untreated, this peripheral disorder can induce significant central visual deficits such as amblyopia. Inherited congenital cataract in humans is commonly autosomal-dominant and phenotypically variable, as cataracts appear with varying densities, colors, and locations within the lens.

The lens itself consists primarily of lens fiber cells surrounded by an epithelial cell layer on the outer surface; this structure continues to grow throughout life from the inside out, adding new lens fiber cells, formed from the outer epithelial cell layer to the body of the lens underneath the epithelial cell layer. Lens transparency is ensured in part by the loss of all intracellular organelles in lens fiber cells as they differentiate from the epithelial cell layer. In the avascular lens, homeostasis is maintained through a gap–junction coupled network. The cytosolic crystallins form the principal protein component of lens fiber cells, and the inside–out growth process of the lens dictates that the crystallins be extremely stable, as they cannot be replaced during the life of the lens. Furthermore, the crystallins play a role in setting refractive index and maintaining transparency of the lens. The α-crystallins are heat-shock proteins and chaperones, with α-crystallin making up almost half of the total lens protein by weight. Both αA-crystallin and αB-crystallin are found in the developing lens, and β- and γ-crystallins are also present in the adult lens (Andley, 2007; Bassnett, 2009).

Mutations of several of the crystallins, including αA, αB, βA1, βB1, βB2, γC, γD, and γS, have been associated with hereditary congenital cataracts as well as with a developmental cataract that is absent at birth but develops in early childhood. The pathologies of the crystallin mutations are incompletely understood; they may involve several distinct pathways that ultimately affect solubility and transparency of the crystallins. For example, levels of the chaperone αA-crystallin may control the ratio of soluble to insoluble αB protein, as the loss of αA-crystallin in mice leads to higher levels of insoluble αB-crystallin and reduced lens transparency (Brady et al., 1997). Additional mutations may alter protein transparency by actions on protein folding or protein solubility, or through effects on crystallin–crystallin interactions.

Connexins expressed in the lens, Cx46 (GJA3) and Cx50 (GJA8), allow gap–junction coupling of both lens fiber cells and lens epithelial cells. Although it is still not precisely understood how gap–junction coupling maintains transparency, gap–junction coupling provides a circulation system for the avascular lens, promoting homeostasis among the anuclear lens fiber cells and promoting crystalline solubility, and mutations of Cx46 and Cx50 are associated with congenital cataract. In mice, knockouts for Cx46 and for Cx50 exhibit different cataract phenotypes; nevertheless, it appears that Cx50 and Cx46 are functionally redundant and that the number of gap junctions is more critical than the specific connexin components for preserving transparency (White, 2002).

37.4.2 Disorders of the Retina

37.4.2.1 Retinopathy of Prematurity

Retinopathy of prematurity (ROP) is a disorder of retinal vascularization that first appeared in high-income countries in the 1940s, when advances in neonatal care led to increased survival rates for premature infants. Today, global incidence of this disorder describes an inverse U-shape as a function of economic context. Low-income countries typically have high infant mortality rates, which are associated with very low ROP incidence because premature infants do not survive. High-income countries also have low rates of ROP, and it is middle-income countries with variable quality in neonatal care that have the highest rates of ROP. Oxygen exposure was identified as a risk factor early on, but the survival of increasingly smaller infants in the developed world has kept ROP incidence high. Although genetic factors may play a role in developing ROP, the primary risk factors remain oxygen exposure, degree of prematurity, and low birth weight. By current estimates, a majority (68%) of premature infants born in first-world countries and weighing 1250 g or less at birth will develop ROP (Good et al., 2005).

Two phases of ROP are seen, the first characterized by a loss of vascularization and the second by a pathological angiogenesis. In normal development, the retinal vasculature develops radially, reaching the peripheral extent of the retina at about 36 weeks gestation. Thus, in premature infants, an avascular area, with extent determined by gestational age at birth, surrounds the central vascularized retina. At birth, retinal vascularization slows, possibly in response to the relatively hyperoxic environment outside the uterus. During the first phase of ROP, the avascularized retina, unable to meet metabolic demand, becomes hypoxic until about 32 weeks gestational age. The second phase begins at 32–34 weeks gestation and involves neovascularization in response to hypoxia. In more extreme cases, the pathological blood vessel growth can pull the retina away from the RPE, leading to retinal detachment and blindness; prevention consists of careful monitoring and laser photocoagulation to prevent retinal detachment. Less extreme forms of ROP that resolve spontaneously are associated with abnormally thick retinae with broad, shallow foveal pits, and with altered rod photoreceptor development. Vascular endothelial growth factor (VEGF), which can be regulated by hypoxia, plays a critical role in vascularization of the retina. The proximal cause of the first phase of ROP may be hyperoxic downregulation of VEGF-induced vascular growth, whereas the second stage is likely caused by hypoxic upregulation of VGEF to stimulate angiogenesis. VEGF activity requires insulin-like growth factor (IGF-1), in a permissive or added manner. Thus, the loss of maternally supplied IGF-1 at birth in premature infants may abolish VEGF activity (Smith, 2008). Potential preventative treatments for ROP include supplemental IGF-1 in the preterm infant and the development of small-molecule antagonists to VEGF receptors.

The primary ROP risk factors of gestational age and oxygen exposure can be potentiated by genetic background; in particular, several mutations in the Wnt–beta-catenin pathway affect development of retinal vasculature and may also increase risk for developing ROP. These include the Norrie disease protein norrin, the canonical Wnt receptor *Frizzled-4* (*Fz4*; gene FZD4), and the Wnt coreceptor Lrp5. Norrin, normally produced in the Muller glia, is a high-affinity ligand for the canonical Wnt receptor *Fz4* at retinal epithelial cells. Additional inherited pathologies of retinal

vascularization that are associated with this group of proteins include X-linked familial exudative vitreoretinopathy (FEVR), Coats' disease, and persistent fetal vasculature syndrome (PFVS).

37.4.2.2 *Retinitis Pigmentosa*

Retinitis pigmentosa occurs frequently in isolation; the term refers to a large group of genetically heterogeneous degenerative disorders with a prevalence of 2–3 in 10,000. The disease may be inherited in X-linked, autosomal-dominant, or autosomal-recessive mode, and age of onset is variable after the first year. Diagnosis typically occurs in young adulthood, but early-onset forms may manifest in the first 5 years. This disorder is essentially a rod–cone dystrophy, in that rod photoreceptors degenerate before cone photoreceptors, resulting in a loss of peripheral and night vision.

Over 10% of retinitis pigmentosa patients have the most severe, X-linked form (XLRP), which is associated with mutations in retinitis pigmentosa 2 (RP2) or in retinitis pigmentosa GTPase regulator (RPGR). RP2 is a membrane-associated protein that is found throughout the retina and in the RPE. RP2 shares sequence homology with tubulin-specific cofactor C, a GTPase activator for tubulin, and regulates the GTP-bound form of the microtubule-associated protein ADP ribosylation factor-like 3 (Arl3). RP2 and Arl3 are localized to the connecting cilium between inner and outer photoreceptor segments, where they mediate trafficking of vesicles from the Golgi (Evans et al., 2010).

RPGR interacts with the delta subunit of rod cyclic phosphodiesterase (PDE), which mediates transport of certain proteins to photoreceptor outer segments, and retinitis pigmentosa alleles of RPGR exhibit reduced binding to PDE/delta. RPGR also interacts with a protein localized to the connecting cilia of photoreceptors: PPGR-interacting protein 1 (RPGRIP1), and its expression overlaps that of usherin.

The photopigment of the photoreceptor outer segment consists of a transmembrane opsin (in cones) or rhodopsin (in rods) G-protein-coupled receptor and a covalently attached small organic moiety, 11-*cis*-retinal, that determines the photoreceptor's wavelength sensitivity. Absorption of a photon isomerizes the 11-*cis*-retinal, causing a conformational change in the opsin that leads to activation of the membrane-bound G protein transducin. Transducin activation turns on a cGMP PDE that hydrolyzes cGMP, causing the normally open cGMP-gated cation channels in the outer segment plasma membrane to close. Closure of the cation channel hyperpolarizes the photoreceptor and leads to a decrease in tonic neurotransmitter release at the photoreceptor–bipolar cell synapse. Inactivation of opsin is accomplished in part by S-arrestin, while 11-*cis*-retinal is regenerated through the retinoid cycle. Well over 100 mutant rhodopsin (RHO) alleles, some with mouse models, can cause retinitis pigmentosa. These alleles may cause protein misfolding that prevents formation of a functional photopigment, interfere with trafficking to the outer segment, affect activation of transducin, and/or initiate the formation of stable rhodopsin that is constitutively active in the dark. Protein misfolding can underlie certain forms of retinitis pigmentosa, though the mechanism of degeneration is unknown. Mutations of S-arrestin (SAG) are also associated with retinitis pigmentosa, probably due to constitutive rhodopsin activation. The cyclic nucleotide channels in photoreceptor outer segment membrane are heteromeric tetramers composed of A (CNGA1) and B (CNGB1) subunits; each subunit has been linked to retinitis pigmentosa (Wright et al., 2010).

The RPE plays a critical role in retinal health, phagocytosing about 10% of the photoreceptor outer segments daily and performing many of the steps necessary in the retinoid cycle in which 11-*cis*-retinal is regenerated from the all-*trans*-retinal product of photoisomerization in the outer segments, and mutations affecting the retinoid cycle are associated with many retinal disorders. One such mutation in retinitis pigmentosa affects the retinoid isomerase RPE65, which is normally expressed in RPE and in cone photoreceptors and is necessary for the isomerization of all-*trans*-retinyl to 11-*cis*-retinal. Loss of RPE65 results in a gradual disorganization of the outer segments and a slow degeneration of photoreceptors, loss of the rod ERG and of rhodopsin, and an accumulation in the RPE of all-*trans*-retinyl esters. The rhodopsin homology RGR (retinal G-protein-coupled receptor) is another RPE protein involved in the retinoid cycle that has been implicated in retinitis pigmentosa. The retina-specific ATP-binding cassette transporter ABCA4 gene encoding the photoreceptor Rim protein also functions in the retinoid cycle, though it is expressed in the disk membrane of photoreceptor outer segments, where it is involved in active transport of retinoids from the lumen of the disk to the cytoplasm. The all-*trans*-retinal product of photoactivation is a precursor for potentially toxic diretinal compounds. Loss of ABCA4 activity allows accumulation of diretinal compounds that are then incorporated in the RPE during outer segment phagocytosis. Finally, mutations of the anion channel bestrophin 1 (BEST1), which is also expressed in RPE cells, lead to a retinitis pigmentosa of unknown pathology.

The tubby-related protein TULP1 is expressed in the photoreceptor inner segments where it can bind dynamin, a critical endocytosis protein. In mouse, loss of Tulp1 results in the loss of ribbon synapses at photoreceptor terminals, the mislocalization of the visual pigment rhodopsin to the inner segment, and photoreceptor degeneration. The transcription factor neural retina leucine zipper (NRL) is expressed primarily in rod photoreceptors. The retinae of mice lacking *Nrl* develop with no rod photoreceptors and with more short-wavelength cones than normal, a phenotype similar to that seen in humans with retinitis pigmentosa caused by mutations of NRL (Mears et al., 2001).

Retinitis pigmentosa is distinguished from other retinal dystrophies in that cone photoreceptor death follows the main phase of rod photoreceptor death, and little is currently known about the events that initiate cone photoreceptor death. Several models of retinitis pigmentosa share a common metabolic pathway for cone death, however, suggesting that in diverse cases the late loss of foveal vision may be caused by cell starvation.

37.4.2.3 Cone–Rod Dystrophy

Cone–rod dystrophy is distinguished from retinitis pigmentosa by the order in which photoreceptors are lost. The cone–rod dystrophies are less prevalent than retinitis pigmentosa, but can be severe because the loss of high-acuity, color vision precedes night blindness and the loss of peripheral vision. The transcription factor cone–rod homeobox (CRX) is a member of the orthodenticle family that is expressed primarily in photoreceptor inner segments. Crx transactivates photoreceptor-specific genes including rhodopsin and arrestin, and was the earliest gene identified in cone–rod dystrophy. Additional causes of cone–rod dystrophy are similar to those discussed earlier with retinitis pigmentosa.

37.4.2.4 Leber Congenital Amaurosis

The most severe of the retinal diseases, Leber congenital amaurosis (LCA), underlies about 5% of inherited retinopathies; it includes a heterogeneous group of severe inherited disorders in which nystagmus and retinitis pigmentosa are present, the ERG response is absent, and pupils are amaurotic (i.e., they do not respond to light shining directly in the same eye) (Walia et al., 2010). Identified mutations include the photoreceptor transcription factor Crx, which leads to a failure to form photoreceptor outer segments and the formation of abnormal photoreceptor synapses, and the Crumbs protein (CRB1), a regulator of cell polarity and morphology whose loss leads to dysmorphic Muller glia cells, retinal disorganization, and subsequent degeneration.

Another subgroup of LCA cases results from mutations of visual cycle proteins, either RPE65 or retinol dehydrogenase 12 (RDH12). The role of RDH12 in LCA is still unclear, however, as the Rdh12 deficiency appears to be compensated for in the mouse knockout. Maintenance of the cGMP-gated current in the photoreceptor outer segments requires local cGMP synthesis, and the membrane-bound retinal guanylate cyclase GUC2D has been associated with LCA. *In vitro* results suggest that a GUC2D mutation commonly found in LCA impairs cyclase activity and causes the constitutive closure of the cGMP-gated channels.

37.4.2.5 *Stargardt's Disease*

Stargardt's disease, an autosomal-recessive juvenile-onset macular dystrophy, presents with loss of central vision and progressive atrophy of the RPE at the fovea. Known causes include mutations of ABCA4 and of the cyclic-nucleotide-gated channel, discussed earlier under retinitis pigmentosa.

37.4.2.6 *Night Blindness*

The inherited night blindnesses typically cause myopia, reduced visual acuity, and nystagmus in addition to reduced vision in dim light (nyctalopia). Complete congenital stationary night blindness (cCSNB), or type 1 CSNB, is a nonprogressive X-linked or autosomal-recessive disorder that affects the photoreceptor–bipolar cell synapse and can be seen on electroretinogram (ERG). Different forms of cCSNB have been linked with mutations to genes encoding a leucine-rich proteoglycan, nyctalopin (NYX), the metabotropic glutamate receptor mGluR6 (GRM6), and the transient receptor potential channel melastatin (TRPM1). In addition to numerous transgenic mouse models for night blindness, naturally occurring animal models exist in the Nob mouse ("no b-wave," a Nyx mutant) and certain colorations of appaloosa horse (Trpm1 mutant).

In the normal retina at rest, glutamate tonically released from the photoreceptor activates the retina-specific Group III metabotropic glutamate receptor mGluR6, which is expressed specifically on the dendrites of an ON bipolar cell, to cause the closure of a nonspecific cation channel. The reduction in glutamate release with light stimulation causes the mGluR6-coupled cation channels to open, depolarizing the ON bipolar cell. Though the function of nyctalopin is unknown, it is expressed in close proximity to mGluR6 in dendrites of rod bipolar cells, TRPM1 is likely regulated by mGluR6 activity, and loss of *Trpm* causes an abnormal ERG characteristic of ON bipolar dysfunction; in fact, depolarizing bipolar cells from Nob mouse retina show no response to glutamate application. Thus, the cCSNB defect appears to be localized to a site on the ON bipolar cell postsynaptic membrane where nyctalopin, mGluR6, and TRPM1 are likely functionally linked.

Congenital stationary night blindness type-2 (CSNB2) is an X-linked disorder caused by a presynaptic channelopathy at the rod–bipolar synapse. CSNB2 is caused by mutations that affect the $\alpha 1$ subunit of the retinal L-type calcium channel $Ca_v 1.4$ ($Ca_v 1.4\alpha 1$; gene CACNA1F) or the calmodulin-like calcium-binding protein-4 (CABP4), which modulates $Ca_v 1.4$ function. The $Ca_v 1.4$ channel inactivates slowly in response to depolarization and it lacks calcium-dependent inactivation; both qualities contribute to the photoreceptor's ability to release glutamate tonically, and both the $\alpha 1$ subunit and CABP4 are necessary for transmitter release at the photoreceptor–bipolar synapse.

37.5 DISORDERS OF CHEMICAL SENSATION

Recognized disorders or chemical sensation usually present after infection or after use of certain drugs, and developmental disorders of the chemical senses are rare.

37.5.1 Taste

One well-studied taste 'disorder' is the inability to taste the synthetic compound phenylthiocarbamide (PTC). PTC and 6-n-propylthiouracil (PROP) are members of a class of thiourea compounds that are perceived bimodally, a large number of individuals tasting these compounds as very bitter, and somewhat fewer individuals failing to taste them at all; PTC nontasters exist in nearly every population tested. Bitter tastes are transduced by receptors in the Taste Receptor 2 (TAS2R) family of 7-transmembrane, G-protein-coupled receptors, which are coexpressed with the G protein α subunit gustducin in a subset of taste receptor cells within the taste buds. The TAS2R38 gene is not highly conserved across species, and like other members of the TAS2R gene family, the human PTC (TAS2R38) gene shows broad allelic heterogeneity. Though the lack of PTC conservation across species and the broad distribution of non-tasting PTC alleles might suggest that PTC sensing is less a disorder than an interesting genetic oddity, the PTC-sensing phenotype may confer certain benefits. It has been suggested, for example, that PTC tasters might enjoy an evolutionary advantage by avoiding bitter-tasting toxic compounds, and also that a genetically encoded ability to taste bitter compounds could be protective against nicotine and alcohol consumption that might otherwise lead to addiction. Indeed, specific single-nucleotide polymorphisms in TAS2R38 are associated with a higher incidence of nicotine use in certain populations (Mangold et al., 2008). Interestingly, TAS2R38 and several other bitter-taste receptors are also expressed in ciliated cells of the airway epithelium. When stimulated with bitter compounds, these cells show an increase in ciliary beating (Shah et al., 2009), a response that may play a role in protecting the lungs from exposure to noxious compounds. Variability is not restricted to the bitter receptors, and altered taste perception of sucrose correlates with single-nucleotide polymorphisms in the promoter region of the TAS1R3 gene. Taste anomalies or deficits are exhibited by many of the knockout mice for gustducin, various members of the TAS1R (sweet) and TAS2R (bitter) families, the taste tissue-specific phospholipase Cβ2 (Plcb2), and the TRP channel Trpm5, and several human taste preferences correlate with polymorphisms in various TAS2Rs (Hayes et al., 2011). Whether or not genetic variability in taste should be considered a disorder is unclear; nevertheless, the broad variability of genetically encoded taste perception offers opportunities for further study.

37.5.2 Olfaction

37.5.2.1 *Kallmann Syndrome*

Kallmann syndrome encompasses a heterogeneous group of disorders that include severe hypogonadotropic hypogonadism (impaired or absent function of the testes or ovaries) and hyposmia or anosmia. Diagnosis is typically made when patients fail to enter puberty due to a deficiency of gonadotropin-releasing hormone (GnRH). Olfactory impairment, anosmia or hyposmia, is typically discovered only at this time. The sensory impairment results from the absence or hypoplasia of the olfactory bulbs, which are the principal target of the olfactory epithelium, and of the lateral olfactory tracts. A broad range of additional, variable neurological deficits may also be present. Kallmann syndrome is not precisely a disorder of the olfactory organ, as the deficit is one of axon migration and neurogenesis in central pathways, but is worth mention here as one of very few identified olfactory disorders. Much about this disorder is still unclear, and in fact the genes that have been identified to date represent a minority of all cases (Dode and Hardelin, 2009).

The principal known genes for Kallmann syndrome include KAL1, which shows X-linked inheritance and codes for anosmin-1; KAL2, an autosomal-dominant form associated with mutations of fibroblast growth factor receptor 1 (FGFR1); KAL3 and KAL4, encoding prokineticin receptor-2 (PROKR2) and prokineticin-2 (PROK2); and KAL6, encoding fibroblast growth factor 8 (FGF8). Anosmin-1 is a secreted glycoprotein of embryonic olfactory bulb, tract, and cortex that acts as a chemoattractant and promotes neurite branching of olfactory neurons, possibly signaling through FGFR1, which is co-expressed with anosmin-1 in the olfactory bulbs during development. PROK2 is a secreted ligand for PROKR2, a G protein-coupled membrane receptor necessary for differentiation and migration of neural progenitor cells to the olfactory bulb. It is likely that many of the still unidentified gene products underlying Kallmann syndrome will be proteins that interact with FGFR1 – possibly FGF2 – or with PROKR2.

37.5.2.2 *Other Congenital Olfactory Disorders*

Hyposmia has also been reported in a few additional syndromic disorders. CHARGE (coloboma, heart disease, atresia choanae, retardation, genital and ear anomalies) syndrome, with an estimated incidence on the order of 10^{-4}, is related to the less severe Kallmann syndrome, type 5. CHARGE syndrome is typically diagnosed in infancy, may be accompanied by deaf–blindness, and nearly always involves olfactory deficits. Over 50% of CHARGE cases are caused by mutations in chromodomain helicase DNA-binding protein 7 (CHD7), and several mouse models of this disorder are associated with *Chd7* mutations. *Chd7* and the transcription factor *Sox2* form a transcriptional network for regulating several target genes in neural stem cells (Engelen et al., 2011).

In both humans and mice, PAX6 mutations and aniridia may be accompanied by olfactory bulb hypoplasia

TABLE 37.3 Disorders of chemical sensation

Name	Gene	Gene product	Reference
PTC non-taster			
PTC	TAS2R38	Taste receptor, type 2, member 38	Kim et al. (2003)
Kallmann syndrome			
KAL1	KAL1	Kallmann syndrome 1 protein; anosmin-1	Franco et al. (1991) Legouis et al. (1991)
KAL2	FGFR1	Fibroblast growth factor 1	Dode et al. (2003)
KAL3	PROKR2	Prokineticin receptor 2	Dode et al. (2006)
KAL4	PROK2	Prokineticin-2	Dode et al. (2006)
KAL5	CHD7	Chromodomain helicase DNA-binding protein 7	Kim et al. (2008)
KAL6	FGF8	Fibroblast growth factor 8	Falardeau et al. (2008)
Aniridia and hyposmia	PAX6	Paired box 6 protein Pax-6	Sisodiya et al. (2001)

OMIM gene symbols are used throughout.

and hyposmia, whereas mutation of SOX 9 (campomelic dysplasia) or SOX10 (Waardenburg syndrome) may be accompanied by a complete absence of the olfactory bulbs. Finally, hyposmia has been reported in some cases of pseudohypoparathyroidism, Bardet–Biedl syndrome, and Leber congenital amaurosis (Table 37.3).

37.6 CONCLUSION

Congenital sensory impairment is widespread. With the success of vaccination, maternal health, and perinatal care programs worldwide, the incidence of congenital sensory impairment has decreased greatly, and inherited disorders are now the most common form of congenital sensory loss. This is both a challenge and a boon: a challenge because these disorders can be quite heterogeneous; a boon because the growth of animal models for specific genetic disorders offers rich opportunities for research into normal and pathological sensory development.

References

Abd El-Aziz, M.M., Barragan, I., O'Driscoll, C.A., et al., 2008. EYS, encoding an ortholog of Drosophila spacemaker, is mutated in autosomal recessive retinitis pigmentosa. Nature Genetics 40, 1285–1287.

Abdelhak, S., Kalatzis, V., Heilig, R., et al., 1997. A human homologue of the Drosophila eyes absent gene underlies branchio-oto-renal (BOR) syndrome and identifies a novel gene family. Nature Genetics 15, 157–164.

Abe, S., Katagiri, T., Saito-Hisaminato, A., et al., 2003. Identification of CRYM as a candidate responsible for nonsyndromic deafness, through cDNA microarray analysis of human cochlear and vestibular tissues. American Journal of Human Genetics 72, 73–82.

Abid, A., Ismail, M., Mehdi, S.Q., Khaliq, S., 2006. Identification of novel mutations in the SEMA4A gene associated with retinal degenerative diseases. Journal of Medical Genetics 43, 378–381.

Abouzeid, H., Li, Y., Maumenee, I.H., Dharmaraj, S., Sundin, O., 2006. A G1103R mutation in CRB1 is co-inherited with high hyperopia and Leber congenital amaurosis. Ophthalmic Genetics 27, 15–20.

Ahmad, N.N., Ala-Kokko, L., Knowlton, R.G., et al., 1991. Stop codon in the procollagen II gene (COL2A1) in a family with the Stickler syndrome (arthro-ophthalmopathy). Proceedings of the National Academy of Sciences of the United States of America 88, 6624–6627.

Ahmed, Z.M., Riazuddin, S., Bernstein, S.L., et al., 2001. Mutations of the protocadherin gene PCDH15 cause Usher syndrome type 1F. American Journal of Human Genetics 69, 25–34.

Ahmed, Z.M., Smith, T.N., Riazuddin, S., et al., 2002. Nonsyndromic recessive deafness DFNB18 and Usher syndrome type IC are allelic mutations of USHIC. Human Genetics 110, 527–531.

Ahmed, Z.M., Morell, R.J., Riazuddin, S., et al., 2003. Mutations of MYO6 are associated with recessive deafness, DFNB37. American Journal of Human Genetics 72, 1315–1322.

Ahmed, Z.M., Riazuddin, S., Ahmad, J., et al., 2003. PCDH15 is expressed in the neurosensory epithelium of the eye and ear and mutant alleles are responsible for both USH1F and DFNB23. Human Molecular Genetics 12, 3215–3223.

Alasti, F., Sadeghi, A., Sanati, M.H., et al., 2008. A mutation in HOXA2 is responsible for autosomal-recessive microtia in an Iranian family. American Journal of Human Genetics 82, 982–991.

Allikmets, R., Singh, N., Sun, H., et al., 1997. A photoreceptor cell-specific ATP-binding transporter gene (ABCR) is mutated in recessive Stargardt macular dystrophy. Nature Genetics 15, 236–246.

Andley, U.P., 2007. Crystallins in the eye: Function and pathology. Progress in Retinal and Eye Research 26, 78–98.

Azuma, N., Nishina, S., Yanagisawa, H., Okuyama, T., Yamada, M., 1996. PAX6 missense mutation in isolated foveal hypoplasia. Nature Genetics 13, 141–142.

Azuma, N., Yamaguchi, Y., Handa, H., et al., 2003. Mutations of the PAX6 gene detected in patients with a variety of optic-nerve malformations. American Journal of Human Genetics 72, 1565–1570.

Badano, J.L., Leitch, C.C., Ansley, S.J., et al., 2006. Dissection of epistasis in oligogenic Bardet-Biedl syndrome. Nature 439, 326–330.

Bakrania, P., Efthymiou, M., Klein, J.C., et al., 2008. Mutations in BMP4 cause eye, brain, and digit developmental anomalies: Overlap between the BMP4 and hedgehog signaling pathways. American Journal of Human Genetics 82, 304–319.

Baldwin, C.T., Hoth, C.F., Amos, J.A., da Silva, E.O., Milunsky, A., 1992. An exonic mutation in the HuP2 paired domain gene causes Waardenburg's syndrome. Nature 355, 637–638.

Bareil, C., Hamel, C.P., Delague, V., Arnaud, B., Demaille, J., Claustres, M., 2001. Segregation of a mutation in CNGB1 encoding the beta-subunit of the rod cGMP-gated channel in a family with autosomal recessive retinitis pigmentosa. Human Genetics 108, 328–334.

Barker, D.F., Hostikka, S.L., Zhou, J., et al., 1990. Identification of mutations in the COL4A5 collagen gene in Alport syndrome. Science 248, 1224–1227.

Bassnett, S., 2009. On the mechanism of organelle degradation in the vertebrate lens. Experimental Eye Research 88, 133–139.

Bech-Hansen, N.T., Naylor, M.J., Maybaum, T.A., et al., 1998. Loss-of-function mutations in a calcium-channel alpha1-subunit gene in Xp11.23 cause incomplete X-linked congenital stationary night blindness. Nature Genetics 19, 264–267.

Bech-Hansen, N.T., Naylor, M.J., Maybaum, T.A., et al., 2000. Mutations in NYX, encoding the leucine-rich proteoglycan nyctalopin, cause X-linked complete congenital stationary night blindness. Nature Genetics 26, 319–323.

Bellone, R.R., Brooks, S.A., Sandmeyer, L., et al., 2008. Differential gene expression of TRPM1, the potential cause of congenital stationary night blindness and coat spotting patterns (LP) in the Appaloosa horse (Equus caballus). Genetics 179, 1861–1870.

Berger, W., Meindl, A., van de Pol, T.J., et al., 1992. Isolation of a candidate gene for Norrie disease by positional cloning. Nature Genetics 2, 84.

Berry, V., Francis, P., Reddy, M.A., et al., 2001. Alpha-B crystallin gene (CRYAB) mutation causes dominant congenital posterior polar cataract in humans. American Journal of Human Genetics 69, 1141–1145.

Bespalova, I.N., Van Camp, G., Bom, S.J., et al., 2001. Mutations in the Wolfram syndrome 1 gene (WFS1) are a common cause of low frequency sensorineural hearing loss. Human Molecular Genetics 10, 2501–2508.

Bessant, D.A., Payne, A.M., Mitton, K.P., et al., 1999. A mutation in NRL is associated with autosomal dominant retinitis pigmentosa. Nature Genetics 21, 355–356.

Birkenhager, R., Otto, E., Schurmann, M.J., et al., 2001. Mutation of BSND causes Bartter syndrome with sensorineural deafness and kidney failure. Nature Genetics 29, 310–314.

Boeda, B., El-Amraoui, A., Bahloul, A., et al., 2002. Myosin VIIa, harmonin and cadherin 23, three Usher I gene products that cooperate to shape the sensory hair cell bundle. The EMBO Journal 21, 6689–6699.

Bolz, H., von Brederlow, B., Ramírez, A., et al., 2001. Mutation of CDH23, encoding a new member of the cadherin gene family, causes Usher syndrome type 1D. Nature Genetics 27, 108–112.

Bondurand, N., Kuhlbrodt, K., Pingault, V., et al., 1999. A molecular analysis of the yemenite deaf-blind hypopigmentation syndrome: SOX10 dysfunction causes different neurocristopathies. Human Molecular Genetics 8, 1785–1789.

Bork, J.M., Peters, L.M., Riazuddin, S., et al., 2001. Usher syndrome 1D and nonsyndromic autosomal recessive deafness DFNB12 are caused by allelic mutations of the novel cadherin-like gene CDH23. American Journal of Human Genetics 68, 26–37.

Bowne, S.J., Sullivan, L.S., Mortimer, S.E., et al., 2006. Spectrum and frequency of mutations in IMPDH1 associated with autosomal dominant retinitis pigmentosa and leber congenital amaurosis. Investigative Ophthalmology & Visual Science 47, 34–42.

Brady, J.P., Garland, D., Duglas-Tabor, Y., Robison Jr., W.G., Groome, A., Wawrousek, E.F., 1997. Targeted disruption of the mouse alpha A-crystallin gene induces cataract and cytoplasmic inclusion bodies containing the small heat shock protein alpha B-crystallin. Proceedings of the National Academy of Sciences of the United States of America 94, 884–889.

Braunstein, E.M., Crenshaw 3rd, E.B., Morrow, B.E., Adams, J.C., 2008. Cooperative function of Tbx1 and Brn4 in the periotic mesenchyme is necessary for cochlea formation. Journal of the Association for Research in Otolaryngology 9, 33–43.

Cantagrel, V., Silhavy, J.L., Bielas, S.L., et al., 2008. Mutations in the cilia gene ARL13B lead to the classical form of Joubert syndrome. American Journal of Human Genetics 83, 170–179.

Chakarova, C.F., Hims, M.M., Bolz, H., et al., 2002. Mutations in HPRP3, a third member of pre-mRNA splicing factor genes, implicated in autosomal dominant retinitis pigmentosa. Human Molecular Genetics 11, 87–92.

Chakarova, C.F., Papaioannou, M.G., Khanna, H., et al., 2007. Mutations in TOPORS cause autosomal dominant retinitis pigmentosa with perivascular retinal pigment epithelium atrophy. American Journal of Human Genetics 81, 1098–1103.

Chen, W., Kahrizi, K., Meyer, N.C., et al., 2005. Mutation of COL11A2 causes autosomal recessive non-syndromic hearing loss at the DFNB53 locus. Journal of Medical Genetics 42, e61.

Chesarone, M.A., DuPage, A.G., Goode, B.L., 2010. Unleashing formins to remodel the actin and microtubule cytoskeletons. Nature Reviews Molecular Cell Biology 11, 62–74.

Chiang, A.P., Nishimura, D., Searby, C., et al., 2004. Comparative genomic analysis identifies an ADP-ribosylation factor-like gene as the cause of Bardet-Biedl syndrome (BBS3). American Journal of Human Genetics 75, 475–484.

Coene, K.L., Roepman, R., Doherty, D., et al., 2009. OFD1 is mutated in X-linked Joubert syndrome and interacts with LCA5-encoded lebercilin. American Journal of Human Genetics 85, 465–481.

Cohen, D., Bar-Yosef, U., Levy, J., et al., 2007. Homozygous CRYBB1 deletion mutation underlies autosomal recessive congenital cataract. Investigative Ophthalmology & Visual Science 48, 2208–2213.

Cohen-Salmon, M., Regnault, B., Cayet, N., et al., 2007. Connexin30 deficiency causes instrastrial fluid-blood barrier disruption within the cochlear stria vascularis. Proceedings of the National Academy of Sciences of the United States of America 104, 6229–6234.

Coppieters, F., Leroy, B.P., Beysen, D., et al., 2007. Recurrent mutation in the first zinc finger of the orphan nuclear receptor NR2E3 causes autosomal dominant retinitis pigmentosa. American Journal of Human Genetics 81, 147–157.

Cremers, F.P., van de Pol, D.J., van Driel, M., et al., 1998. Autosomal recessive retinitis pigmentosa and cone-rod dystrophy caused by splice site mutations in the Stargardt's disease gene ABCR. Human Molecular Genetics 7, 355–362.

Davidson, A.E., Millar, I.D., Urquhart, J.E., et al., 2009. Missense mutations in a retinal pigment epithelium protein, bestrophin-1, cause retinitis pigmentosa. American Journal of Human Genetics 85, 581–592.

de Kok, Y.J., van der Maarel, S.M., Bitner-Glindzicz, M., et al., 1995. Association between X-linked mixed deafness and mutations in the POU domain gene POU3F4. Science 267, 685–688.

del Castillo, I., Villamar, M., Moreno-Pelayo, M.A., et al., 2002. A deletion involving the connexin 30 gene in nonsyndromic hearing impairment. The New England Journal of Medicine 346, 243–249.

Delmaghani, S., del Castillo, F.J., Michel, V., et al., 2006. Mutations in the gene encoding pejvakin, a newly identified protein of the afferent auditory pathway, cause DFNB59 auditory neuropathy. Nature Genetics 38, 770–778.

Delous, M., Baala, L., Salomon, R., et al., 2007. The ciliary gene RPGRIP1L is mutated in cerebello-oculo-renal syndrome (Joubert syndrome type B) and Meckel syndrome. Nature Genetics 39, 875–881.

Demirci, F.Y., Rigatti, B.W., Wen, G., et al., 2002. X-linked cone-rod dystrophy (locus COD1): Identification of mutations in RPGR exon ORF15. American Journal of Human Genetics 70, 1049–1053.

den Hollander, A.I., ten Brink, J.B., de Kok, Y.J., et al., 1999. Mutations in a human homologue of Drosophila crumbs cause retinitis pigmentosa (RP12). Nature Genetics 23, 217–221.

den Hollander, A.I., Koenekoop, R.K., Yzer, S., et al., 2006. Mutations in the CEP290 (NPHP6) gene are a frequent cause of Leber congenital amaurosis. American Journal of Human Genetics 79, 556–561.

Desir, J., Moya, G., Reish, O., et al., 2007. Borate transporter SLC4A11 mutations cause both Harboyan syndrome and non-syndromic corneal endothelial dystrophy. Journal of Medical Genetics 44, 322–326.

Devi, R.R., Vijayalakshmi, P., 2006. Novel mutations in GJA8 associated with autosomal dominant congenital cataract and microcornea. Molecular Vision 12, 190–195.

Dode, C., Hardelin, J.P., 2009. Kallmann syndrome. European Journal of Human Genetics 17, 139–146.

Dode, C., Levilliers, J., Dupont, J.M., et al., 2003. Loss-of-function mutations in FGFR1 cause autosomal dominant Kallmann syndrome. Nature Genetics 33, 463–465.

Dode, C., Teixeira, L., Levilliers, J., et al., 2006. Kallmann syndrome: Mutations in the genes encoding prokineticin-2 and prokineticin receptor-2. PLoS Genetics 2, e175.

Donaudy, F., Ferrara, A., Esposito, L., et al., 2003. Multiple mutations of MYO1A, a cochlear-expressed gene, in sensorineural hearing loss. American Journal of Human Genetics 72, 1571–1577.

Donaudy, F., Snoeckx, R., Pfister, M., et al., 2004. Nonmuscle myosin heavy-chain gene MYH14 is expressed in cochlea and mutated in patients affected by autosomal dominant hearing impairment (DFNA4). American Journal of Human Genetics 74, 770–776.

Doward, W., Perveen, R., Lloyd, I.C., Ridgway, A.E., Wilson, L., Black, G.C., 1999. A mutation in the RIEG1 gene associated with Peters' anomaly. Journal of Medical Genetics 36, 152–155.

Dror, A.A., Avraham, K.B., 2009. Hearing loss: Mechanisms revealed by genetics and cell biology. Annual Review of Genetics 43, 411–437.

Dryja, T.P., McGee, T.L., Reichel, E., et al., 1990. A point mutation of the rhodopsin gene in one form of retinitis pigmentosa. Nature 343, 364–366.

Dryja, T.P., Berson, E.L., Rao, V.R., Oprian, D.D., 1993. Heterozygous missense mutation in the rhodopsin gene as a cause of congenital stationary night blindness. Nature Genetics 4, 280–283.

Dryja, T.P., Finn, J.T., Peng, Y.W., McGee, T.L., Berson, E.L., Yau, K.W., 1995. Mutations in the gene encoding the alpha subunit of the rod cGMP-gated channel in autosomal recessive retinitis pigmentosa. Proceedings of the National Academy of Sciences of the United States of America 92, 10177–10181.

Dryja, T.P., Hahn, L.B., Reboul, T., Arnaud, B., 1996. Missense mutation in the gene encoding the alpha subunit of rod transducin in the Nougaret form of congenital stationary night blindness. Nature Genetics 13, 358–360.

Dryja, T.P., Adams, S.M., Grimsby, J.L., et al., 2001. Null RPGRIP1 alleles in patients with Leber congenital amaurosis. American Journal of Human Genetics 68, 1295–1298.

Dryja, T.P., McGee, T.L., Berson, E.L., et al., 2005. Night blindness and abnormal cone electroretinogram ON responses in patients with mutations in the GRM6 gene encoding mGluR6. Proceedings of the National Academy of Sciences of the United States of America 102, 4884–4889.

Ebermann, I., Scholl, H.P., Charbel Issa, P., et al., 2007. A novel gene for Usher syndrome type 2: Mutations in the long isoform of whirlin are associated with retinitis pigmentosa and sensorineural hearing loss. Human Genetics 121, 203–211.

Edery, P., Attie, T., Amiel, J., et al., 1996. Mutation of the endothelin-3 gene in the Waardenburg-Hirschsprung disease (Shah-Waardenburg syndrome). Nature Genetics 12, 442–444.

Eichers, E.R., Green, J.S., Stockton, D.W., et al., 2002. Newfoundland rod-cone dystrophy, an early-onset retinal dystrophy, is caused by splice-junction mutations in RLBP1. American Journal of Human Genetics 70, 955–964.

Engelen, E., Akinci, U., Bryne, J.C., et al., 2011. Sox2 cooperates with Chd7 to regulate genes that are mutated in human syndromes. Nature Genetics 43, 607–611.

Erkman, L., McEvilly, R.J., Luo, L., et al., 1996. Role of transcription factors Brn-3.1 and Brn-3.2 in auditory and visual system development. Nature 381, 603–606.

Eudy, J.D., Weston, M.D., Yao, S., et al., 1998. Mutation of a gene encoding a protein with extracellular matrix motifs in Usher syndrome type IIa. Science 280, 1753–1757.

Evans, R.J., Schwarz, N., Nagel-Wolfrum, K., Wolfrum, U., Hardcastle, A.J., Cheetham, M.E., 2010. The retinitis pigmentosa protein RP2 links pericentriolar vesicle transport between the Golgi and the primary cilium. Human Molecular Genetics 19, 1358–1367.

Everett, L.A., Glaser, B., Beck, J.C., et al., 1997. Pendred syndrome is caused by mutations in a putative sulphate transporter gene (PDS). Nature Genetics 17, 411–422.

Falardeau, J., Chung, W.C., Beenken, A., et al., 2008. Decreased FGF8 signaling causes deficiency of gonadotropin-releasing hormone in humans and mice. The Journal of Clinical Investigation 118, 2822–2831.

Fantes, J., Ragge, N.K., Lynch, S.A., et al., 2003. Mutations in SOX2 cause anophthalmia. Nature Genetics 33, 461–463.

Farrar, G.J., Kenna, P., Jordan, S.A., et al., 1991. A three-base-pair deletion in the peripherin-RDS gene in one form of retinitis pigmentosa. Nature 354, 478–480.

Franco, B., Guioli, S., Pragliola, A., et al., 1991. A Gene Deleted in Kallmanns Syndrome Shares Homology with Neural Cell-Adhesion and Axonal Path-Finding Molecules. Nature 353, 529–536.

Freund, C.L., Gregory-Evans, C.Y., Furukawa, T., et al., 1997. Cone-rod dystrophy due to mutations in a novel photoreceptor-specific homeobox gene (CRX) essential for maintenance of the photoreceptor. Cell 91, 543–553.

Friedman, J.S., Chang, B., Kannabiran, C., et al., 2006. Premature truncation of a novel protein, RD3, exhibiting subnuclear localization is associated with retinal degeneration. American Journal of Human Genetics 79, 1059–1070.

Friedman, J.S., Ray, J.W., Waseem, N., et al., 2009. Mutations in a BTB-Kelch protein, KLHL7, cause autosomal-dominant retinitis pigmentosa. American Journal of Human Genetics 84, 792–800.

Fuchs, S., Nakazawa, M., Maw, M., Tamai, M., Oguchi, Y., Gal, A., 1995. A homozygous 1-base pair deletion in the arrestin gene is a frequent cause of Oguchi disease in Japanese. Nature Genetics 10, 360–362.

Gal, A., Orth, U., Baehr, W., Schwinger, E., Rosenberg, T., 1994. Heterozygous missense mutation in the rod cGMP phosphodiesterase beta-subunit gene in autosomal dominant stationary night blindness. Nature Genetics 7, 64–68.

Gal, A., Li, Y., Thompson, D.A., et al., 2000. Mutations in MERTK, the human orthologue of the RCS rat retinal dystrophy gene, cause retinitis pigmentosa. Nature Genetics 26, 270–271.

Gallardo, M.E., Rodriguez De Cordoba, S., Schneider, A.S., Dwyer, M.A., Ayuso, C., Bovolenta, P., 2004. Analysis of the developmental SIX6 homeobox gene in patients with anophthalmia/microphthalmia. American Journal of Medical Genetics. Part A 129A, 92–94.

Geller, S.F., Guerin, K.I., Visel, M., et al., 2009. CLRN1 is nonessential in the mouse retina but is required for cochlear hair cell development. PLoS Genetics 5, e1000607.

Gilbert, C., 2009. Worldwide causes of blindness in children. In: Wilson, M.E., Saunders, R.A., Trivedi, R.H. (Eds.), Pediatric Ophthalmology. Springer, Berlin.

Good, W.V., Hardy, R.J., Dobson, V., et al., 2005. The incidence and course of retinopathy of prematurity: Findings from the early treatment for retinopathy of prematurity study. Pediatrics 116, 15–23.

Goodyear, J.R., Kros, C.J., Richardson, G.P., 2006. The development of hair cells in the inner ear. In: Eatock, R.A., Fay, R.R., Popper, A.N. (Eds.), Vertebrate Hair Cells. Springer, New York.

Gorden, N.T., Arts, H.H., Parisi, M.A., et al., 2008. CC2D2A is mutated in Joubert syndrome and interacts with the ciliopathy-associated basal body protein CEP290. American Journal of Human Genetics 83, 559–571.

Grifa, A., Wagner, C.A., D'Ambrosio, L., et al., 1999. Mutations in GJB6 cause nonsyndromic autosomal dominant deafness at DFNA3 locus. Nature Genetics 23, 16–18.

Gu, S.M., Thompson, D.A., Srikumari, C.R., et al., 1997. Mutations in RPE65 cause autosomal recessive childhood-onset severe retinal dystrophy. Nature Genetics 17, 194–197.

Hagstrom, S.A., North, M.A., Nishina, P.L., Berson, E.L., Dryja, T.P., 1998. Recessive mutations in the gene encoding the tubby-like protein TULP1 in patients with retinitis pigmentosa. Nature Genetics 18, 174–176.

Hameed, A., Abid, A., Aziz, A., Ismail, M., Mehdi, S.Q., Khaliq, S., 2003. Evidence of RPGRIP1 gene mutations associated with recessive cone-rod dystrophy. Journal of Medical Genetics 40, 616–619.

Hanson, I.M., Fletcher, J.M., Jordan, T., et al., 1994. Mutations at the PAX6 locus are found in heterogeneous anterior segment malformations including Peters' anomaly. Nature Genetics 6, 168–173.

Hanson, I., Churchill, A., Love, J., et al., 1999. Missense mutations in the most ancient residues of the PAX6 paired domain underlie a spectrum of human congenital eye malformations. Human Molecular Genetics 8, 165–172.

Hartong, D.T., Dange, M., McGee, T.L., Berson, E.L., Dryja, T.P., Colman, R.F., 2008. Insights from retinitis pigmentosa into the roles of isocitrate dehydrogenases in the Krebs cycle. Nature Genetics 40, 1230–1234.

Hayes, J.E., Wallace, M.R., Knopik, V.S., Herbstman, D.M., Bartoshuk, L.M., Duffy, V.B., 2011. Allelic variation in TAS2R bitter receptor genes associates with variation in sensations from and ingestive behaviors toward common bitter beverages in adults. Chemical Senses 36, 311–319.

Heon, E., Priston, M., Schorderet, D.F., et al., 1999. The gamma-crystallins and human cataracts: A puzzle made clearer. American Journal of Human Genetics 65, 1261–1267.

Hibino, H., Nin, F., Tsuzuki, C., Kurachi, Y., 2010. How is the highly positive endocochlear potential formed? The specific architecture of the stria vascularis and the roles of the ion-transport apparatus. Pflügers Archiv 459, 521–533.

Hoffman, H.M., Mueller, J.L., Broide, D.H., Wanderer, A.A., Kolodner, R.D., 2001. Mutation of a new gene encoding a putative pyrin-like protein causes familial cold autoinflammatory syndrome and Muckle-Wells syndrome. Nature Genetics 29, 301–305.

Honkanen, R.A., Nishimura, D.Y., Swiderski, R.E., et al., 2003. A family with Axenfeld-Rieger syndrome and Peters Anomaly caused by a point mutation (Phe112Ser) in the FOXC1 gene. American Journal of Ophthalmology 135, 368–375.

Hoskins, B.E., Cramer, C.H., Silvius, D., et al., 2007. Transcription factor SIX5 is mutated in patients with branchio-oto-renal syndrome. American Journal of Human Genetics 80, 800–804.

Hoth, C.F., Milunsky, A., Lipsky, N., Sheffer, R., Clarren, S.K., Baldwin, C.T., 1993. Mutations in the paired domain of the human PAX3 gene cause Klein-Waardenburg syndrome (WS-III) as well as Waardenburg syndrome type I (WS-I). American Journal of Human Genetics 52, 455–462.

Huang, S.H., Pittler, S.J., Huang, X., Oliveira, L., Berson, E.L., Dryja, T.P., 1995. Autosomal recessive retinitis pigmentosa caused by mutations in the alpha subunit of rod cGMP phosphodiesterase. Nature Genetics 11, 468–471.

Humphries, M.M., Rancourt, D., Farrar, G.J., et al., 1997. Retinopathy induced in mice by targeted disruption of the rhodopsin gene. Nature Genetics 15, 216–219.

Hwang, C.H., Wu, D.K., 2008. Noggin heterozygous mice: An animal model for congenital conductive hearing loss in humans. Human Molecular Genetics 17, 844–853.

Irvine, A.D., Corden, L.D., Swensson, O., et al., 1997. Mutations in cornea-specific keratin K3 or K12 genes cause Meesmann's corneal dystrophy. Nature Genetics 16, 184–187.

Jalkanen, R., Mantyjarvi, M., Tobias, R., et al., 2006. X linked cone-rod dystrophy, CORDX3, is caused by a mutation in the CACNA1F gene. Journal of Medical Genetics 43, 699–704.

Janecke, A.R., Thompson, D.A., Utermann, G., et al., 2004. Mutations in RDH12 encoding a photoreceptor cell retinol dehydrogenase cause childhood-onset severe retinal dystrophy. Nature Genetics 36, 850–854.

Joensuu, T., Hamalainen, R., Yuan, B., et al., 2001. Mutations in a novel gene with transmembrane domains underlie Usher syndrome type 3. American Journal of Human Genetics 69, 673–684.

Johnson, S., Halford, S., Morris, A.G., et al., 2003. Genomic organisation and alternative splicing of human RIM1, a gene implicated in autosomal dominant cone-rod dystrophy (CORD7). Genomics 81, 304–314.

Jordan, T., Hanson, I., Zaletayev, D., et al., 1992. The human PAX6 gene is mutated in two patients with aniridia. Nature Genetics 1, 328–332.

Kajiwara, K., Berson, E.L., Dryja, T.P., 1994. Digenic retinitis pigmentosa due to mutations at the unlinked peripherin/RDS and ROM1 loci. Science 264, 1604–1608.

Kannabiran, C., Rogan, P.K., Olmos, L., et al., 1998. Autosomal dominant zonular cataract with sutural opacities is associated with a splice mutation in the betaA3/A1-crystallin gene. Molecular Vision 4, 21.

Karet, F.E., Finberg, K.E., Nelson, R.D., et al., 1999. Mutations in the gene encoding B1 subunit of H+–ATPase cause renal tubular acidosis with sensorineural deafness. Nature Genetics 21, 84–90.

Katsanis, N., Beales, P.L., Woods, M.O., et al., 2000. Mutations in MKKS cause obesity, retinal dystrophy and renal malformations associated with Bardet-Biedl syndrome. Nature Genetics 26, 67–70.

Keen, T.J., Hims, M.M., McKie, A.B., et al., 2002. Mutations in a protein target of the Pim-1 kinase associated with the RP9 form of autosomal dominant retinitis pigmentosa. European Journal of Human Genetics 10, 245–249.

Kelsell, D.P., Dunlop, J., Stevens, H.P., et al., 1997. Connexin 26 mutations in hereditary non-syndromic sensorineural deafness. Nature 387, 80–83.

Kelsell, R.E., Gregory-Evans, K., Payne, A.M., et al., 1998. Mutations in the retinal guanylate cyclase (RETGC-1) gene in dominant cone-rod dystrophy. Human Molecular Genetics 7, 1179–1184.

Khan, S.Y., Ahmed, Z.M., Shabbir, M.I., et al., 2007. Mutations of the RDX gene cause nonsyndromic hearing loss at the DFNB24 locus. Human Mutation 28, 417–423.

Kharkovets, T., Dedek, K., Maier, H., et al., 2006. Mice with altered KCNQ4 K^+ channels implicate sensory outer hair cells in human progressive deafness. The EMBO Journal 25, 642–652.

Kim, U.K., Jorgenson, E., Coon, H., Leppert, M., Risch, N., Drayna, D., 2003. Positional cloning of the human quantitative trait locus underlying taste sensitivity to phenylthiocarbamide. Science 299, 1221–1225.

Kim, H.G., Kurth, I., Lan, F., et al., 2008. Mutations in CHD7, encoding a chromatin-remodeling protein, cause idiopathic hypogonadotropic hypogonadism and Kallmann syndrome. American Journal of Human Genetics 83, 511–519.

Klintworth, G.K., 2009. Corneal dystrophies. Orphanet Journal of Rare Diseases 4, 7.

Kochhar, A., Fischer, S.M., Kimberling, W.J., Smith, R.J., 2007. Branchio-oto-renal syndrome. American Journal of Medical Genetics. Part A 143A, 1671–1678.

Kohn, L., Kadzhaev, K., Burstedt, M.S., et al., 2007. Mutation in the PYK2-binding domain of PITPNM3 causes autosomal dominant cone dystrophy (CORD5) in two Swedish families. European Journal of Human Genetics 15, 664–671.

Kopp, P., Pesce, L., Solis, S.J., 2008. Pendred syndrome and iodide transport in the thyroid. Trends in Endocrinology and Metabolism 19, 260–268.

Kubisch, C., Schroeder, B.C., Friedrich, T., et al., 1999. KCNQ4, a novel potassium channel expressed in sensory outer hair cells, is mutated in dominant deafness. Cell 96, 437–446.

Kurima, K., Peters, L.M., Yang, Y., et al., 2002. Dominant and recessive deafness caused by mutations of a novel gene, TMC1, required for cochlear hair-cell function. Nature Genetics 30, 277–284.

Lalwani, A.K., Goldstein, J.A., Kelley, M.J., Luxford, W., Castelein, C.M., Mhatre, A.N., 2000. Human nonsyndromic hereditary deafness DFNA17 is due to a mutation in nonmuscle myosin MYH9. American Journal of Human Genetics 67, 1121–1128.

Le Caignec, C., Lefevre, M., Schott, J.J., et al., 2002. Familial deafness, congenital heart defects, and posterior embryotoxon caused by cysteine substitution in the first epidermal-growth-factor-like domain of jagged 1. American Journal of Human Genetics 71, 180–186.

Legan, P.K., Lukashkina, V.A., Goodyear, R.J., Kossi, M., Russell, I.J., Richardson, G.P., 2000. A targeted deletion in alpha-tectorin reveals that the tectorial membrane is required for the gain and timing of cochlear feedback. Neuron 28, 273–285.

Legouis, R., Hardelin, J.P., Levilliers, J., et al., 1991. The Candidate Gene for the X-Linked Kallmann Syndrome Encodes a Protein Related to Adhesion Molecules. Cell 67, 423–435.

Leitch, C.C., Zaghloul, N.A., Davis, E.E., et al., 2008. Hypomorphic mutations in syndromic encephalocele genes are associated with Bardet-Biedl syndrome. Nature Genetics 40, 443–448.

Li, X.C., Everett, L.A., Lalwani, A.K., et al., 1998. A mutation in PDS causes non-syndromic recessive deafness. Nature Genetics 18, 215–217.

Li, Z., Sergouniotis, P.I., Michaelides, M., et al., 2009. Recessive mutations of the gene TRPM1 abrogate ON bipolar cell function and cause complete congenital stationary night blindness in humans. American Journal of Human Genetics 85, 711–719.

Litt, M., Carrero-Valenzuela, R., LaMorticella, D.M., et al., 1997. Autosomal dominant cerulean cataract is associated with a chain termination mutation in the human beta-crystallin gene CRYBB2. Human Molecular Genetics 6, 665–668.

Litt, M., Kramer, P., LaMorticella, D.M., Murphey, W., Lovrien, E.W., Weleber, R.G., 1998. Autosomal dominant congenital cataract associated with a missense mutation in the human alpha crystallin gene CRYAA. Human Molecular Genetics 7, 471–474.

Liu, X.Z., Walsh, J., Tamagawa, Y., et al., 1997. Autosomal dominant non-syndromic deafness caused by a mutation in the myosin VIIA gene. Nature Genetics 17, 268–269.

Liu, X.Z., Ouyang, X.M., Xia, X.J., et al., 2003. Prestin, a cochlear motor protein, is defective in non-syndromic hearing loss. Human Molecular Genetics 12, 1155–1162.

Liu, X., Bulgakov, O.V., Darrow, K.N., et al., 2007. Usherin is required for maintenance of retinal photoreceptors and normal development of cochlear hair cells. Proceedings of the National Academy of Sciences of the United States of America 104, 4413–4418.

Lynch, E.D., Lee, M.K., Morrow, J.E., Welcsh, P.L., Leon, P.E., King, M.C., 1997. Nonsyndromic deafness DFNA1 associated with mutation of a human homolog of the Drosophila gene diaphanous. Science 278, 1315–1318.

Maestrini, E., Korge, B.P., Ocana-Sierra, J., et al., 1999. A missense mutation in connexin26, D66H, causes mutilating keratoderma with sensorineural deafness (Vohwinkel's syndrome) in three unrelated families. Human Molecular Genetics 8, 1237–1243.

Mangold, J.E., Payne, T.J., Ma, J.Z., Chen, G., Li, M.D., 2008. Bitter taste receptor gene polymorphisms are an important factor in the development of nicotine dependence in African Americans. Journal of Medical Genetics 45, 578–582.

Mansergh, F.C., Millington-Ward, S., Kennan, A., et al., 1999. Retinitis pigmentosa and progressive sensorineural hearing loss caused by a C12258A mutation in the mitochondrial MTTS2 gene. American Journal of Human Genetics 64, 971–985.

Marlhens, F., Bareil, C., Griffoin, J.M., et al., 1997. Mutations in RPE65 cause Leber's congenital amaurosis. Nature Genetics 17, 139–141.

Maw, M.A., Corbeil, D., Koch, J., et al., 2000. A frameshift mutation in prominin (mouse)-like 1 causes human retinal degeneration. Human Molecular Genetics 9, 27–34.

Mburu, P., Mustapha, M., Varela, A., et al., 2003. Defects in whirlin, a PDZ domain molecule involved in stereocilia elongation, cause deafness in the whirler mouse and families with DFNB31. Nature Genetics 34, 421–428.

McGuirt, W.T., Prasad, S.D., Griffith, A.J., et al., 1999. Mutations in COL11A2 cause non-syndromic hearing loss (DFNA13). Nature Genetics 23, 413–419.

McKie, A.B., McHale, J.C., Keen, T.J., et al., 2001. Mutations in the pre-mRNA splicing factor gene PRPC8 in autosomal dominant retinitis pigmentosa (RP13). Human Molecular Genetics 10, 1555–1562.

McLaughlin, M.E., Sandberg, M.A., Berson, E.L., Dryja, T.P., 1993. Recessive mutations in the gene encoding the beta-subunit of rod phosphodiesterase in patients with retinitis pigmentosa. Nature Genetics 4, 130–134.

Mears, A.J., Kondo, M., Swain, P.K., et al., 2001. Nrl is required for rod photoreceptor development. Nature Genetics 29, 447–452.

Meindl, A., Dry, K., Herrmann, K., et al., 1996. A gene (RPGR) with homology to the RCC1 guanine nucleotide exchange factor is mutated in X-linked retinitis pigmentosa (RP3). Nature Genetics 13, 35–42.

Melchionda, S., Ahituv, N., Bisceglia, L., et al., 2001. MYO6, the human homologue of the gene responsible for deafness in Snell's waltzer mice, is mutated in autosomal dominant nonsyndromic hearing loss. American Journal of Human Genetics 69, 635–640.

Milam, A.H., Li, Z.Y., Fariss, R.N., 1998. Histopathology of the human retina in retinitis pigmentosa. Progress in Retinal and Eye Research 17, 175–205.

Mirzayans, F., Pearce, W.G., MacDonald, I.M., Walter, M.A., 1995. Mutation of the PAX6 gene in patients with autosomal dominant keratitis. American Journal of Human Genetics 57, 539–548.

Mochizuki, T., Lemmink, H.H., Mariyama, M., et al., 1994. Identification of mutations in the alpha 3(IV) and alpha 4(IV) collagen genes in autosomal recessive Alport syndrome. Nature Genetics 8, 77–81.

Modamio-Hoybjor, S., Mencia, A., Goodyear, R., et al., 2007. A mutation in CCDC50, a gene encoding an effector of epidermal growth factor-mediated cell signaling, causes progressive hearing loss. American Journal of Human Genetics 80, 1076–1089.

Morimura, H., Saindelle-Ribeaudeau, F., Berson, E.L., Dryja, T.P., 1999. Mutations in RGR, encoding a light-sensitive opsin homologue, in patients with retinitis pigmentosa. Nature Genetics 23, 393–394.

Mustapha, M., Weil, D., Chardenoux, S., et al., 1999. An alpha-tectorin gene defect causes a newly identified autosomal recessive form of sensorineural pre-lingual non-syndromic deafness, DFNB21. Human Molecular Genetics 8, 409–412.

Mustapha, M., Chouery, E., Torchard-Pagnez, D., et al., 2002. A novel locus for Usher syndrome type I, USH1G, maps to chromosome 17q24-25. Human Genetics 110, 348–350.

Mustapha, M., Fang, Q., Gong, T.W., et al., 2009. Deafness and permanently reduced potassium channel gene expression and function in hypothyroid Pit1dw mutants. Journal of Neuroscience 29, 1212–1223.

Nakaya, K., Harbidge, D.G., Wangemann, P., et al., 2007. Lack of pendrin HCO_3^- transport elevates vestibular endolymphatic [Ca^{2+}] by inhibition of acid-sensitive TRPV5 and TRPV6 channels. American Journal of Physiology. Renal Physiology 292, F1314–F1321.

Nakazawa, M., Wada, Y., Tamai, M., 1998. Arrestin gene mutations in autosomal recessive retinitis pigmentosa. Archives of Ophthalmology 116, 498–501.

Nandrot, E., Slingsby, C., Basak, A., et al., 2003. Gamma-D crystallin gene (CRYGD) mutation causes autosomal dominant congenital cerulean cataracts. Journal of Medical Genetics 40, 262–267.

Naz, S., Giguere, C.M., Kohrman, D.C., et al., 2002. Mutations in a novel gene, TMIE, are associated with hearing loss linked to the DFNB6 locus. American Journal of Human Genetics 71, 632–636.

Naz, S., Griffith, A.J., Riazuddin, S., et al., 2004. Mutations of ESPN cause autosomal recessive deafness and vestibular dysfunction. Journal of Medical Genetics 41, 591–595.

Neyroud, N., Tesson, F., Denjoy, I., et al., 1997. A novel mutation in the potassium channel gene KVLQT1 causes the Jervell and Lange-Nielsen cardioauditory syndrome. Nature Genetics 15, 186–189.

Ng, D., Thakker, N., Corcoran, C.M., et al., 2004. Oculofaciocardiodental and Lenz microphthalmia syndromes result from distinct classes of mutations in BCOR. Nature Genetics 36, 411–416.

Nishiguchi, K.M., Sandberg, M.A., Gorji, N., Berson, E.L., Dryja, T.P., 2005. Cone cGMP-gated channel mutations and clinical findings in patients with achromatopsia, macular degeneration, and other hereditary cone diseases. Human Mutation 25, 248–258.

Nishimura, D.Y., Swiderski, R.E., Alward, W.L., et al., 1998. The forkhead transcription factor gene FKHL7 is responsible for glaucoma phenotypes which map to 6p25. Nature Genetics 19, 140–147.

Noor, A., Windpassinger, C., Patel, M., et al., 2008. CC2D2A, encoding a coiled-coil and C2 domain protein, causes autosomal-recessive mental retardation with retinitis pigmentosa. American Journal of Human Genetics 82, 1011–1018.

Ouyang, X.M., Xia, X.J., Verpy, E., et al., 2002. Mutations in the alternatively spliced exons of USH1C cause non-syndromic recessive deafness. Human Genetics 111, 26–30.

Parry, D.A., Toomes, C., Bida, L., et al., 2009. Loss of the metalloprotease ADAM9 leads to cone-rod dystrophy in humans and retinal degeneration in mice. American Journal of Human Genetics 84, 683–691.

Pasutto, F., Sticht, H., Hammersen, G., et al., 2007. Mutations in STRA6 cause a broad spectrum of malformations including anophthalmia, congenital heart defects, diaphragmatic hernia, alveolar capillary dysplasia, lung hypoplasia, and mental retardation. American Journal of Human Genetics 80, 550–560.

Payne, A.M., Downes, S.M., Bessant, D.A., et al., 1998. A mutation in guanylate cyclase activator 1A (GUCA1A) in an autosomal dominant cone dystrophy pedigree mapping to a new locus on chromosome 6p21.1. Human Molecular Genetics 7, 273–277.

Perrault, I., Rozet, J.M., Calvas, P., et al., 1996. Retinal-specific guanylate cyclase gene mutations in Leber's congenital amaurosis. Nature Genetics 14, 461–464.

Peters, L.M., Anderson, D.W., Griffith, A.J., et al., 2002. Mutation of a transcription factor, TFCP2L3, causes progressive autosomal dominant hearing loss, DFNA28. Human Molecular Genetics 11, 2877–2885.

Petit, C., Richardson, G.P., 2009. Linking genes underlying deafness to hair-bundle development and function. Nature Neuroscience 12, 703–710.

Pierce, E.A., Quinn, T., Meehan, T., McGee, T.L., Berson, E.L., Dryja, T.P., 1999. Mutations in a gene encoding a new oxygen-regulated photoreceptor protein cause dominant retinitis pigmentosa. Nature Genetics 22, 248–254.

Pingault, V., Bondurand, N., Kuhlbrodt, K., et al., 1998. SOX10 mutations in patients with Waardenburg-Hirschsprung disease. Nature Genetics 18, 171–173.

Puffenberger, E.G., Hosoda, K., Washington, S.S., et al., 1994. A missense mutation of the endothelin-B receptor gene in multigenic Hirschsprung's disease. Cell 79, 1257–1266.

Ragge, N.K., Brown, A.G., Poloschek, C.M., et al., 2005. Heterozygous mutations of OTX2 cause severe ocular malformations. American Journal of Human Genetics 76, 1008–1022.

Rebello, G., Ramesar, R., Vorster, A., et al., 2004. Apoptosis-inducing signal sequence mutation in carbonic anhydrase IV identified in patients with the RP17 form of retinitis pigmentosa. Proceedings of the National Academy of Sciences of the United States of America 101, 6617–6622.

Riazuddin, S., Ahmed, Z.M., Fanning, A.S., et al., 2006. Tricellulin is a tight-junction protein necessary for hearing. American Journal of Human Genetics 79, 1040–1051.

Riazuddin, S., Khan, S.N., Ahmed, Z.M., et al., 2006. Mutations in TRIOBP, which encodes a putative cytoskeletal-organizing protein, are associated with nonsyndromic recessive deafness. American Journal of Human Genetics 78, 137–143.

Riazuddin, S.A., Shahzadi, A., Zeitz, C., et al., 2010. A mutation in SLC24A1 implicated in autosomal-recessive congenital stationary night blindness. American Journal of Human Genetics 87, 523–531.

Richard, G., Rouan, F., Willoughby, C.E., et al., 2002. Missense mutations in GJB2 encoding connexin-26 cause the ectodermal dysplasia keratitis-ichthyosis-deafness syndrome. American Journal of Human Genetics 70, 1341–1348.

Richards, A.J., Yates, J.R., Williams, R., et al., 1996. A family with Stickler syndrome type 2 has a mutation in the COL11A1 gene resulting in the substitution of glycine 97 by valine in alpha 1 (XI) collagen. Human Molecular Genetics 5, 1339–1343.

Rivolta, C., Sweklo, E.A., Berson, E.L., Dryja, T.P., 2000. Missense mutation in the USH2A gene: Association with recessive retinitis pigmentosa without hearing loss. American Journal of Human Genetics 66, 1975–1978.

Robertson, N.G., Lu, L., Heller, S., et al., 1998. Mutations in a novel cochlear gene cause DFNA9, a human nonsyndromic deafness with vestibular dysfunction. Nature Genetics 20, 299–303.

Robertson, S.P., Twigg, S.R., Sutherland-Smith, A.J., et al., 2003. Localized mutations in the gene encoding the cytoskeletal protein filamin A cause diverse malformations in humans. Nature Genetics 33, 487–491.

Rozet, J.M., Gerber, S., Ghazi, I., et al., 1999. Mutations of the retinal specific ATP binding transporter gene (ABCR) in a single family segregating both autosomal recessive retinitis pigmentosa RP19 and Stargardt disease: Evidence of clinical heterogeneity at this locus. Journal of Medical Genetics 36, 447–451.

Ruel, J., Emery, S., Nouvian, R., et al., 2008. Impairment of SLC17A8 encoding vesicular glutamate transporter-3, VGLUT3, underlies nonsyndromic deafness DFNA25 and inner hair cell dysfunction in null mice. American Journal of Human Genetics 83, 278–292.

Ruf, R.G., Xu, P.X., Silvius, D., et al., 2004. SIX1 mutations cause branchio-oto-renal syndrome by disruption of EYA1-SIX1-DNA complexes. Proceedings of the National Academy of Sciences of the United States of America 101, 8090–8095.

Rye, M.S., Bhutta, M.F., Cheeseman, M.T., et al., 2011. Unraveling the genetics of otitis media: From mouse to human and back again. Mammalian Genome 22, 66–82.

Sanchez-Martin, M., Rodriguez-Garcia, A., Perez-Losada, J., Sagrera, A., Read, A.P., Sanchez-Garcia, I., 2002. SLUG (SNAI2) deletions in patients with Waardenburg disease. Human Molecular Genetics 11, 3231–3236.

Sato, M., Nakazawa, M., Usui, T., Tanimoto, N., Abe, H., Ohguro, H., 2005. Mutations in the gene coding for guanylate cyclase-activating protein 2 (GUCA1B gene) in patients with autosomal dominant retinal dystrophies. Graefe's Archive for Clinical and Experimental Ophthalmology 243, 235–242.

Sayer, J.A., Otto, E.A., O'Toole, J.F., et al., 2006. The centrosomal protein nephrocystin-6 is mutated in Joubert syndrome and activates transcription factor ATF4. Nature Genetics 38, 674–681.

Schlingmann, K.P., Konrad, M., Jeck, N., et al., 2004. Salt wasting and deafness resulting from mutations in two chloride channels. The New England Journal of Medicine 350, 1314–1319.

Schulze-Bahr, E., Haverkamp, W., Wedekind, H., et al., 1997. Autosomal recessive long-QT syndrome (Jervell Lange-Nielsen syndrome) is genetically heterogeneous. Human Genetics 100, 573–576.

Schwahn, U., Lenzner, S., Dong, J., et al., 1998. Positional cloning of the gene for X-linked retinitis pigmentosa 2. Nature Genetics 19, 327–332.

Scott, H.S., Kudoh, J., Wattenhofer, M., et al., 2001. Insertion of beta-satellite repeats identifies a transmembrane protease causing both congenital and childhood onset autosomal recessive deafness. Nature Genetics 27, 59–63.

Seal, R.P., Akil, O., Yi, E., et al., 2008. Sensorineural deafness and seizures in mice lacking vesicular glutamate transporter 3. Neuron 57, 263–275.

Semina, E.V., Reiter, R., Leysens, N.J., et al., 1996. Cloning and characterization of a novel bicoid-related homeobox transcription factor gene, RIEG, involved in Rieger syndrome. Nature Genetics 14, 392–399.

Semina, E.V., Ferrell, R.E., Mintz-Hittner, H.A., et al., 1998. A novel homeobox gene PITX3 is mutated in families with autosomal-dominant cataracts and ASMD. Nature Genetics 19, 167–170.

Semina, E.V., Brownell, I., Mintz-Hittner, H.A., Murray, J.C., Jamrich, M., 2001. Mutations in the human forkhead transcription factor FOXE3 associated with anterior segment ocular dysgenesis and cataracts. Human Molecular Genetics 10, 231–236.

Sergouniotis, P.I., Davidson, A.E., Mackay, D.S., et al., 2011. Recessive Mutations in KCNJ13, Encoding an Inwardly Rectifying Potassium Channel Subunit, Cause Leber Congenital Amaurosis. American Journal of Human Genetics 89, 183–190.

Shah, A.S., Ben-Shahar, Y., Moninger, T.O., Kline, J.N., Welsh, M.J., 2009. Motile cilia of human airway epithelia are chemosensory. Science 325, 1131–1134.

Shahin, H., Walsh, T., Sobe, T., et al., 2006. Mutations in a novel isoform of TRIOBP that encodes a filamentous-actin binding protein are responsible for DFNB28 recessive nonsyndromic hearing loss. American Journal of Human Genetics 78, 144–152.

Simmler, M.C., Cohen-Salmon, M., El-Amraoui, A., et al., 2000. Targeted disruption of otog results in deafness and severe imbalance. Nature Genetics 24, 139–143.

Sisodiya, S.M., Free, S.L., Williamson, K.A., et al., 2001. PAX6 haploinsufficiency causes cerebral malformation and olfactory dysfunction in humans. Nature Genetics 28, 214–216.

Slavotinek, A.M., Stone, E.M., Mykytyn, K., et al., 2000. Mutations in MKKS cause Bardet-Biedl syndrome. Nature Genetics 26, 15–16.

Smith, L.E., 2008. Through the eyes of a child: Understanding retinopathy through ROP the Friedenwald lecture. Investigative Ophthalmology & Visual Science 49, 5177–5182.

Sohocki, M.M., Bowne, S.J., Sullivan, L.S., et al., 2000. Mutations in a new photoreceptor-pineal gene on 17p cause Leber congenital amaurosis. Nature Genetics 24, 79–83.

Stoilov, I., Akarsu, A.N., Alozie, I., et al., 1998. Sequence analysis and homology modeling suggest that primary congenital glaucoma on 2p21 results from mutations disrupting either the hinge region or the conserved core structures of cytochrome P4501B1. American Journal of Human Genetics 62, 573–584.

Strom, T.M., Nyakatura, G., Apfelstedt-Sylla, E., et al., 1998. An L-type calcium-channel gene mutated in incomplete X-linked congenital stationary night blindness. Nature Genetics 19, 260–263.

Sullivan, L.S., Heckenlively, J.R., Bowne, S.J., et al., 1999. Mutations in a novel retina-specific gene cause autosomal dominant retinitis pigmentosa. Nature Genetics 22, 255–259.

Sun, H., Ma, Z., Li, Y., et al., 2005. Gamma-S crystallin gene (CRYGS) mutation causes dominant progressive cortical cataract in humans. Journal of Medical Genetics 42, 706–710.

Sundin, O.H., Leppert, G.S., Silva, E.D., et al., 2005. Extreme hyperopia is the result of null mutations in MFRP, which encodes a Frizzled-related protein. Proceedings of the National Academy of Sciences of the United States of America 102, 9553–9558.

Tassabehji, M., Newton, V.E., Read, A.P., 1994. Waardenburg syndrome type 2 caused by mutations in the human microphthalmia (MITF) gene. Nature Genetics 8, 251–255.

Tassabehji, M., Read, A.P., Newton, V.E., et al., 1992. Waardenburg's syndrome patients have mutations in the human homologue of the Pax-3 paired box gene. Nature 355, 635–636.

TCSCG (The Treacher Collins Syndrome Collaborative Group), 1996. Positional cloning of a gene involved in the pathogenesis of Treacher Collins syndrome. Nature Genetics 12, 130–136.

Thompson, D.A., Li, Y., McHenry, C.L., et al., 2001. Mutations in the gene encoding lecithin retinol acyltransferase are associated with early-onset severe retinal dystrophy. Nature Genetics 28, 123–124.

Tilney, L.G., Tilney, M.S., DeRosier, D.J., 1992. Actin filaments, stereocilia, and hair cells: How cells count and measure. Annual Review of Cell Biology 8, 257–274.

Tsujikawa, M., Kurahashi, H., Tanaka, T., et al., 1999. Identification of the gene responsible for gelatinous drop-like corneal dystrophy. Nature Genetics 21, 420–423.

Tucci, D., Merson, M.H., Wilson, B.S., 2010. A summary of the literature on global hearing impairment: Current status and priorities for action. Otology & Neurotology 31, 31–41.

Tuson, M., Marfany, G., Gonzalez-Duarte, R., 2004. Mutation of CERKL, a novel human ceramide kinase gene, causes autosomal recessive retinitis pigmentosa (RP26). American Journal of Human Genetics 74, 128–138.

Tyson, J., Tranebjaerg, L., Bellman, S., et al., 1997. IsK and KvLQT1: Mutation in either of the two subunits of the slow component of the delayed rectifier potassium channel can cause Jervell and Lange-Nielsen syndrome. Human Molecular Genetics 6, 2179–2185.

Vahava, O., Morell, R., Lynch, E.D., et al., 1998. Mutation in transcription factor POU4F3 associated with inherited progressive hearing loss in humans. Science 279, 1950–1954.

Valente, E.M., Silhavy, J.L., Brancati, F., et al., 2006. Mutations in CEP290, which encodes a centrosomal protein, cause pleiotropic forms of Joubert syndrome. Nature Genetics 38, 623–625.

Van Camp, G., Snoeckx, R.L., Hilgert, N., et al., 2006. A new autosomal recessive form of Stickler syndrome is caused by a mutation in the COL9A1 gene. American Journal of Human Genetics 79, 449–457.

Van Esch, H., Groenen, P., Nesbit, M.A., et al., 2000. GATA3 haploinsufficiency causes human HDR syndrome. Nature 406, 419–422.

Van Laer, L., Huizing, E.H., Verstreken, M., et al., 1998. Nonsyndromic hearing impairment is associated with a mutation in DFNA5. Nature Genetics 20, 194–197.

van Wijk, E., Krieger, E., Kemperman, M.H., et al., 2003. A mutation in the gamma actin 1 (ACTG1) gene causes autosomal dominant hearing loss (DFNA20/26). Journal of Medical Genetics 40, 879–884.

Verhoeven, K., Van Laer, L., Kirschhofer, K., et al., 1998. Mutations in the human alpha-tectorin gene cause autosomal dominant non-syndromic hearing impairment. Nature Genetics 19, 60–62.

Verpy, E., Leibovici, M., Zwaenepoel, I., et al., 2000. A defect in harmonin, a PDZ domain-containing protein expressed in the inner ear sensory hair cells, underlies Usher syndrome type 1C. Nature Genetics 26, 51–55.

Verpy, E., Masmoudi, S., Zwaenepoel, I., et al., 2001. Mutations in a new gene encoding a protein of the hair bundle cause non-syndromic deafness at the DFNB16 locus. Nature Genetics 29, 345–349.

Verpy, E., Leibovici, M., Michalski, N., et al., 2011. Stereocilin connects outer hair cell stereocilia to one another and to the tectorial membrane. The Journal of Comparative Neurology 519, 194–210.

Vikkula, M., Mariman, E.C., Lui, V.C., et al., 1995. Autosomal dominant and recessive osteochondrodysplasias associated with the COL11A2 locus. Cell 80, 431–437.

Vithana, E.N., Abu-Safieh, L., Allen, M.J., et al., 2001. A human homolog of yeast pre-mRNA splicing gene, PRP31, underlies autosomal dominant retinitis pigmentosa on chromosome 19q13.4 (RP11). Molecular Cell 8, 375–381.

Vithana, E.N., Morgan, P., Sundaresan, P., et al., 2006. Mutations in sodium-borate cotransporter SLC4A11 cause recessive congenital hereditary endothelial dystrophy (CHED2). Nature Genetics 38, 755–757.

Voronina, V.A., Kozhemyakina, E.A., O'Kernick, C.M., et al., 2004. Mutations in the human RAX homeobox gene in a patient with anophthalmia and sclerocornea. Human Molecular Genetics 13, 315–322.

Wada, Y., Abe, T., Takeshita, T., Sato, H., Yanashima, K., Tamai, M., 2001. Mutation of human retinal fascin gene (FSCN2) causes autosomal dominant retinitis pigmentosa. Investigative Ophthalmology & Visual Science 42, 2395–2400.

Walia, S., Fishman, G.A., Jacobson, S.G., et al., 2010. Visual acuity in patients with Leber's congenital amaurosis and early childhood-onset retinitis pigmentosa. Ophthalmology 117, 1190–1198.

Walsh, T., Walsh, V., Vreugde, S., et al., 2002. From flies' eyes to our ears: Mutations in a human class III myosin cause progressive non-syndromic hearing loss DFNB30. Proceedings of the National Academy of Sciences of the United States of America 99, 7518–7523.

Wang, A., Liang, Y., Fridell, R.A., et al., 1998. Association of unconventional myosin MYO15 mutations with human nonsyndromic deafness DFNB3. Science 280, 1447–1451.

Wang, H., den Hollander, A.I., Moayedi, Y., et al., 2009. Mutations in SPATA7 cause Leber congenital amaurosis and juvenile retinitis pigmentosa. American Journal of Human Genetics 84, 380–387.

Wangemann, P., 2006. Supporting sensory transduction: Cochlear fluid homeostasis and the endocochlear potential. The Journal of Physiology 576, 11–21.

Wayne, S., Robertson, N.G., DeClau, F., et al., 2001. Mutations in the transcriptional activator EYA4 cause late-onset deafness at the DFNA10 locus. Human Molecular Genetics 10, 195–200.

Weil, D., Blanchard, S., Kaplan, J., et al., 1995. Defective myosin VIIA gene responsible for Usher syndrome type 1B. Nature 374, 60–61.

Weil, D., Kussel, P., Blanchard, S., et al., 1997. The autosomal recessive isolated deafness, DFNB2, and the Usher 1B syndrome are allelic defects of the myosin-VIIA gene. Nature Genetics 16, 191–193.

Weston, M.D., Luijendijk, M.W., Humphrey, K.D., Moller, C., Kimberling, W.J., 2004. Mutations in the VLGR1 gene implicate G-protein signaling in the pathogenesis of Usher syndrome type II. American Journal of Human Genetics 74, 357–366.

White, T.W., 2002. Unique and redundant connexin contributions to lens development. Science 295, 319–320.

Wilcox, E.R., Burton, Q.L., Naz, S., et al., 2001. Mutations in the gene encoding tight junction claudin-14 cause autosomal recessive deafness DFNB29. Cell 104, 165–172.

Wimplinger, I., Morleo, M., Rosenberger, G., et al., 2006. Mutations of the mitochondrial holocytochrome c-type synthase in X-linked dominant microphthalmia with linear skin defects syndrome. American Journal of Human Genetics 79, 878–889.

Wright, A.F., Chakarova, C.F., Abd El-Aziz, M.M., Bhattacharya, S.S., 2010. Photoreceptor degeneration: Genetic and mechanistic dissection of a complex trait. Nature Reviews Genetics 11, 273–284.

Xia, J.H., Liu, C.Y., Tang, B.S., et al., 1998. Mutations in the gene encoding gap junction protein beta-3 associated with autosomal dominant hearing impairment. Nature Genetics 20, 370–373.

Xiao, S., Yu, C., Chou, X., et al., 2001. Dentinogenesis imperfecta 1 with or without progressive hearing loss is associated with distinct mutations in DSPP. Nature Genetics 27, 201–204.

Yamamoto, S., Sippel, K.C., Berson, E.L., Dryja, T.P., 1997. Defects in the rhodopsin kinase gene in the Oguchi form of stationary night blindness. Nature Genetics 15, 175–178.

Yang, T., Vidarsson, H., Rodrigo-Blomqvist, S., Rosengren, S.S., Enerback, S., Smith, R.J., 2007. Transcriptional control of SLC26A4 is involved in Pendred syndrome and nonsyndromic enlargement of vestibular aqueduct (DFNB4). American Journal of Human Genetics 80, 1055–1063.

Yang, Z., Chen, Y., Lillo, C., et al., 2008. Mutant prominin 1 found in patients with macular degeneration disrupts photoreceptor disk morphogenesis in mice. The Journal of Clinical Investigation 118, 2908–2916.

Yang, J., Liu, X., Zhao, Y., et al., 2010. Ablation of whirlin long isoform disrupts the USH2 protein complex and causes vision and hearing loss. PLoS Genetics 6, e1000955.

Yasunaga, S., Grati, M., Cohen-Salmon, M., et al., 1999. A mutation in OTOF, encoding otoferlin, a FER-1-like protein, causes DFNB9, a nonsyndromic form of deafness. Nature Genetics 21, 363–369.

Young, T.L., Ives, E., Lynch, E., et al., 2001. Non-syndromic progressive hearing loss DFNA38 is caused by heterozygous missense mutation in the Wolfram syndrome gene WFS1. Human Molecular Genetics 10, 2509–2514.

Zangerl, B., Goldstein, O., Philp, A.R., et al., 2006. Identical mutation in a novel retinal gene causes progressive rod-cone degeneration in dogs and retinitis pigmentosa in humans. Genomics 88, 551–563.

Zeitz, C., Kloeckener-Gruissem, B., Forster, U., et al., 2006. Mutations in CABP4, the gene encoding the Ca2+–binding protein 4, cause autosomal recessive night blindness. American Journal of Human Genetics 79, 657–667.

Zhang, K., Kniazeva, M., Han, M., et al., 2001. A 5-bp deletion in ELOVL4 is associated with two related forms of autosomal dominant macular dystrophy. Nature Genetics 27, 89–93.

Zhao, C., Bellur, D.L., Lu, S., et al., 2009. Autosomal-dominant retinitis pigmentosa caused by a mutation in SNRNP200, a gene required for unwinding of U4/U6 snRNAs. American Journal of Human Genetics 85, 617–627.

Zhu, M., Yang, T., Wei, S., et al., 2003. Mutations in the gamma-actin gene (ACTG1) are associated with dominant progressive deafness (DFNA20/26). American Journal of Human Genetics 73, 1082–1091.

Zwaenepoel, I., Mustapha, M., Leibovici, M., et al., 2002. Otoancorin, an inner ear protein restricted to the interface between the apical surface of sensory epithelia and their overlying acellular gels, is defective in autosomal recessive deafness DFNB22. Proceedings of the National Academy of Sciences of the United States of America 99, 6240–6245.

Relevant Websites

http://hereditaryhearingloss.org/ – Hereditary Hearing Loss Homepage.

http://www.sph.uth.tmc.edu/RetNet/ – Retinal Information Network.

The Developmental Neurobiology
of Repetitive Behavior

S.-J. Kim[1], M. Lewis[2], J. Veenstra-VanderWeele[3]

[1]University of Washington, Seattle, WA, USA; Center for Integrative Brain Research, Seattle Children's
Research Institute, Seattle, WA, USA; [2]University of Florida, Gainesville, FL, USA; [3]Vanderbilt University,
Nashville, TN, USA

O U T L I N E

Neural Circuit Development and Function in the Brain: Comprehensive Developmental
Neuroscience, Volume 3 http://dx.doi.org/10.1016/B978-0-12-397267-5.00039-X

© 2013 Elsevier Inc. All rights reserved.

38.1 INTRODUCTION

Repetitive behavior can be expressed as a variety of topographies, ranging from very simple motor behaviors to extremely complex rituals. The most succinct terminology for these related behaviors comes from the autism literature, where restricted, repetitive behavior (RRB) refers to a broad class of behaviors linked by repetition, inflexibility, and lack of obvious purpose or function. As detailed later, RRB is characteristic of a number of neurodevelopmental and neuropsychiatric disorders, but various forms of aberrant RRB also occur in a significant percentage of individuals without cognitive or neuropsychiatric impairment (Castellanos et al., 1996; Rafaeli-Mor et al., 1999; Singer, 2009). A number of neurobiology research tools have improved the understanding of the mediation of RRB, including molecular genetics, neuroimaging, and animal models. Data from these different methodological domains are beginning to coalesce, resulting in the identification of key brain regions, circuits, and neurotransmitter systems.

38.2 PHENOMENOLOGY OF REPETITIVE BEHAVIOR

38.2.1 Normative Developmental Pattern of Repetitive Behavior

Repetitive motor behavior, compulsions, and rituals are well documented in normative development, and some attempts have been made to chart the developmental trajectory of such behaviors. For example, Thelen (1980) demonstrated that specific rhythmical, stereotyped movements in infants have a clear onset, peak (at about 2 years of age), and decline coincident with emerging voluntary motor control. Whereas very young children engage in a number of repetitive motor behaviors (e.g., swaying, rocking, and flapping), older children display a variety of behaviors that are compulsive and ritualistic (e.g., insistence on certain clothing or foods and bedtime rituals), reflecting an insistence on sameness (IS), rigidity, and ritualization of daily activities (Thelen, 1980). Attachment to a favorite object, perseverating on certain thoughts and topics, and having intense, restricted interests are also common in preschoolers. Evans et al. (1997) have shown that such behaviors peak at about 2–3 years of age and begin to decline after about age 5.

38.2.2 Phenomenology and Treatment of Repetitive Behavior in Tourette Syndrome

Tourette syndrome (TS) is defined by repetitive behaviors that include both motor and vocal tics (American Psychiatric Association, 2000). Motor tics are stereotyped movements that range from very simple, such as an eye blink, to more complex, such as a full body salute. Vocal tics similarly range from very simple, such as throat clearing, to more complex, including complete phrases. Although patients often describe a premonitory urge or sensory experience preceding a tic, no description of an internal experience is necessary to make the diagnosis. Men are more likely to be affected, and TS follows a developmental time course, with onset typically in middle childhood and slow improvement of behavior in many patients as they enter adulthood (Bloch and Leckman, 2009).

Many people with TS do not require medical care, but treatment is often helpful. Behavioral therapy, which includes monitoring of urges and the use of voluntary replacement behaviors, has been shown to be effective in many patients (Piacentini et al., 2010). A variety of medications have been tested in patients with TS, with most data favoring dopamine receptor D2 antagonist drugs that were developed as antipsychotics (reviewed in Swain et al., 2007). Data also support the use of norepinephrine alpha-2 agonist drugs, such as clonidine, in the treatment of TS (Leckman et al., 1991). Importantly, patients with TS are often more severely impaired by comorbid conditions, such as attention deficit hyperactivity disorder (ADHD) or obsessive–compulsive disorder (OCD), than by the tics themselves.

38.2.3 Phenomenology and Treatment of Repetitive Behavior in OCD

OCD is characterized by repetitive, distressing thoughts and accompanying repetitive behaviors that relieve distress (American Psychiatric Association, 2000). OCD can emerge during childhood or adulthood, with men more likely to have earlier onset and an associated tic disorder or TS. Most people with OCD experience both obsessive thoughts and compulsive behaviors, but some, particularly children, manifest only compulsions. In turn, some, particularly adults, manifest only obsessions, often with compulsive thoughts but no outward compulsive behavior. Unlike the repetitive tics of TS, repetitive behavior in OCD is typically purposeful, such as hand washing to eliminate germs. Obsessions and compulsions can be divided into a variety of semi-independent symptom dimensions, including forbidden thoughts, symmetry, cleaning, and hoarding (Bloch et al., 2008).

Treatment for OCD is often partially effective but rarely relieves symptoms completely. Cognitive behavioral therapy (CBT), which involves exposure to the obsessive stimulus without engaging in the compulsive response, has been shown to be helpful in many patients (Foa et al., 2005). Serotonin reuptake inhibitors (SRIs), but not other monoamine reuptake inhibitors (Leonard et al., 1989), are also helpful, particularly when combined with CBT (Pediatric OCD Treatment Study Team, 2004). Dopamine receptor D2 antagonist drugs can also be helpful when added to SRIs (Bloch et al., 2006). More recent data suggest that medications acting on the glutamate system may also be helpful for some people with OCD (Pittenger et al., 2006). Severe cases of OCD have also been treated successfully with a variety of neurosurgical approaches, including cingulotomy and deep brain stimulation (DBS) in the ventral internal capsule/ventral striatum region (Greenberg et al., 2008).

38.2.4 Phenomenology and Treatment of Repetitive Behavior in Autism Spectrum Disorder

As a cluster of positive symptoms, RRB rather than general developmental delay is often the first clue that a child has an autism spectrum disorder (ASD) (Pediatric OCD Treatment Study Team, 2004). The category of RRB in ASD is broad and heterogeneous, including (1) encompassing preoccupation; (2) apparently inflexible adherence to specific, nonfunctional routines or rituals; (3) stereotyped and repetitive motor mannerisms; and (4) persistent preoccupation with parts of objects (American Psychiatric Association, 2000). Turner (1999) conceptualized repetitive behaviors as falling into two clusters: 'lower order' motor actions that are characterized by repetition of movement, and more complex or 'higher order' behaviors that have a distinct cognitive component. Since then, several groups (Cuccaro et al., 2003; Lam et al., 2008; Szatmari et al., 2006) have examined the structure of RRB as measured on the Autism Diagnostic Interview-Revised (ADI-R) (Lord et al., 1994), the gold standard instrument for autism ascertainment. These studies have identified a three-factor model of RRB: repetitive sensory and motor behavior (RSMB), which corresponds to 'lower order' RRB, IS, and circumscribed interest (CI).

Further support for a developmental pattern of repetitive behavior emergence and resolution can be found in studies of children with developmental disorders. For example, Esbensen et al. (2009) assessed individuals with autism across a wide range of ages (2–62 years) using a modified cross-sectional design. They found that repetitive behaviors decreased across age regardless of subtype of repetitive behavior. Somewhat different results were obtained by Richler et al. (2010), who assessed the development of RRB longitudinally at 2, 3, 5, and 9 years of age in children with autism and developmental delay. They measured two major subtypes of repetitive behaviors (repetitive sensorimotor behavior and IS), and reported decreased sensorimotor repetitive behaviors and increased IS behaviors with increasing age. Further research is needed to understand the transition between normative developmental RRB and pathological RRB that exceeds the expected level of RRB at a particular developmental stage. Such information would allow for identification of environmental and neurobiological mechanisms that mediate this transition and the persistence of repetitive behavior.

The occurrence of RRB in normative development, as well as in TS and OCD, raises the question of whether there is a unique pattern of behavior in ASD. Most data suggest that RRB is quantitatively increased at each developmental stage in ASD, but without a unique topography that distinguishes this behavior from normative development or OCD (Bartak and Rutter, 1976; Bodfish et al., 2000; Freeman et al., 1981; Hermelin and O'Connor, 1963; Lord, 1995; Lord and Pickles, 1996; Russell et al., 2005; Szatmari et al., 1989; Watt et al., 2008; Wing and Gould, 1979). Instead, an elevated pattern of occurrence and severity of RRB, particularly rituals and restricted interests, appears to distinguish autism from other disorders (Bartak and Rutter, 1976; Bodfish et al., 2000; Lam et al., 2008). Most attempts to treat RRB in ASD center on the same systems and approaches used in OCD and TS, with most data favoring the use of dopamine D2 antagonist drugs (McDougle et al., 2005) and not so much supporting SRIs (King et al., 2009).

38.2.5 Cognitive, Sensory, and Motor Correlates of RRB in Typical and Atypical Development

Deficits in executive function are often reported in individuals with RRB (Turner, 1997). Executive function is defined as a broad category of cognitive processes involved in the planning and execution of flexible, goal-directed behavior (O'Hearn et al., 2008). Evans et al. (2004) demonstrated that some executive function tasks, such as set shifting and response inhibition/motor suppression, were related to the frequency and intensity of compulsive behaviors in typically developing children. In individuals with ASD, positive correlations are seen between executive function deficits and RRB symptoms (Lopez et al., 2005; Turner, 1997). Furthermore, Lopez et al. (2005) reported that the degree of RRB in individuals with autism was positively correlated with deficits in cognitive flexibility, even after controlling for level of cognitive function. However, other groups have had mixed results in ASD (Joseph and Tager-Flusberg, 2004; South et al., 2005). Consistent with results in typical development and in ASD, individuals with OCD also show deficits in executive function, including response inhibition, set shifting, and reversal learning (Menzies et al., 2008). Some executive function deficits are also reported in TS, particularly in response inhibition, although these studies may be complicated by the common comorbidity with ADHD (Eddy et al., 2009).

38.2.6 Medication-Induced Repetitive Behavior

A number of medications have been observed to cause or exacerbate repetitive behavior in clinical populations. This is most commonly seen in the context of an existing neuropsychiatric disorder that has led to medication treatment in the first place. One obvious appeal to these observations is the opportunity to translate these induced repetitive behavior symptoms into animal models that could probe the underlying mechanisms of action, as well as potential treatments.

The most common scenario is the child with ADHD who develops a motor tic after starting a dopamine reuptake-inhibiting stimulant medication, such as methylphenidate, or a dopamine-releasing medication, such as amphetamine or dextroamphetamine. Given the frequency of comorbid tic disorder or TS in ADHD, many instances of tics emerging during stimulant treatment may not be due to the medication. Overall, the data suggest that it is no more common for tics to worsen than to improve on methylphenidate, and that high doses of dextroamphetamine appear to worsen tic severity (Bloch et al., 2009).

In addition, chronic dopaminergic replacements, such as L-DOPA (L-3,4-dihydroxyphenylalanine) treatment for Parkinson's disease (PD) or high-dose psychostimulants (e.g., cocaine and amphetamine), have been shown to cause dyskinesias, compulsions, and punding (i.e., abnormal repetitive nongoal-oriented behavior) (Fasano and Petrovic, 2010; O'Sullivan et al., 2007; Voon et al., 2009). Treatment strategies include a gradual reduction in dopamine dosage and N-methyl-D-aspartate (NMDA) blockers (e.g., amantadine) in PD patients.

Two other medications have been reported to either trigger or increase OCD symptoms in some, but not all, studies. At high doses, clozapine and, to a much lesser extent, other atypical antipsychotics can lead to repetitive behaviors or OCD symptoms (Sa et al., 2009). The complex pharmacology of clozapine, with actions on multiple receptors in the serotonin, dopamine, and histamine systems, makes it difficult to assess what particular receptors are responsible for this effect. Susceptibility to clozapine treatment-emergent OCD symptoms was associated in one study with polymorphisms in the neuronal glutamate transporter SLC1A1 gene (Kwon et al., 2009), which is also associated with idiopathic OCD, as reviewed later. Finally, serotonin receptor 5-HT$_{1B}$ agonists, including sumatriptan, have been reported to cause worsened OCD symptoms in some patients with existing OCD (Gross-Isseroff et al., 2004), but reports are mixed across studies.

38.2.7 Autoimmune-Mediated Repetitive Behavior

Interest in the relationship between repetitive behavior and immune system response to infection first emerged from the observation of tics and compulsions in some children with Sydenham's chorea (SC), a movement disorder that emerges during poststreptococcal infection rheumatic fever (RF). Swedo and colleagues assessed whether some children may have OCD or tic symptoms without manifesting the full symptoms of SC or RF, identifying a subset of children who follow a pattern that they termed postinfections autoimmune neuropsychiatric disorders associated with group A streptococcal (PANDAS) infection (Swedo et al., 1998). In addition to tic or compulsive symptom onset soon after a strep infection, individuals who fit the PANDAS pattern are typically prepubertal, often show additional comorbidities such as hyperactivity, and are more likely to show a variable, intermittent pattern of symptoms (Murphy et al., 2010). Plasma exchange and intravenous immunoglobulin showed promise in an initial randomized, controlled trial in the PANDAS population (Perlmutter et al., 1999). Similarly, one study shows successful prevention of symptom exacerbations with antibiotic treatment (Snider et al., 2005). While these preliminary treatment studies are encouraging, it has been more difficult to connect symptom exacerbations to strep infections or to connect PANDAS pattern to a specific autoimmune mechanism. While some studies show increased antistreptolysin O titers in children with tics (Cardona and Orefici, 2001), prospective data are limited and do not yet show a clear relationship between strep infections and symptom exacerbations (Leckman et al., 2011). Considerable effort has gone into pursuing the specific antibody or immune cell response that may mediate PANDAS (reviewed in Murphy et al., 2010), with some promising, but not yet consistent, findings.

38.3 GENETICS OF REPETITIVE BEHAVIOR

A variety of approaches has been brought to bear on understanding the genetic susceptibility to repetitive behavior across neurodevelopmental and neuropsychiatric disorders. Twin studies are the gold standard for evaluating the genetic contribution to susceptibility. Family studies can offer further support for genetic susceptibility and allow for modeling of inheritance patterns, but environmental effects within families can confound results. A variety of molecular approaches is applied to try to locate genes that may contribute to susceptibility. Chromosomal rearrangements, deletions, or duplications that are detected by karyotyping can point to regions of interest or even implicate single genes that are disrupted. Linkage mapping within extended families or sibling pairs can identify chromosomal regions that are shared more often than chance among affected family members. Association studies can identify particular alleles that are more common among people with a disorder than in the general population. Each of these approaches has strengths and weaknesses, and multiple approaches are often required to identify a gene of interest. Technology is moving rapidly in this arena, with the promise of whole-genome sequencing in the near future. The limiting factor in gene identification may be collecting samples of sufficient size to apply the latest technologies while correcting for the huge number of statistical tests performed across the genome. Given the rapid movement in this area, only those studies with the highest impact or replication across multiple samples are reviewed below.

38.3.1 Genetic Susceptibility to TS

TS affects approximately 1% of the population (Robertson, 2008), with isolated motor and vocal tic disorders affecting a larger number. The largest twin study suggests that TS is strongly heritable, with 53% of monozygotic twins (MZ) sharing the diagnosis compared to only 8% of dizygotic twins (DZ) (Price et al., 1985). These numbers rise when chronic tics are included, suggesting that isolated motor tics or, less commonly, vocal tics share a genetic susceptibility with TS. Family studies also support a significant role for heritability in TS (reviewed in O'Rourke et al., 2009). In addition to chronic tic disorders, OCD is also more common in the relatives of probands with TS (Pauls et al., 1986). While early segregation analysis suggested dominant inheritance of TS (Pauls and Leckman, 1986), genetic linkage studies have not yet revealed a major susceptibility gene.

The initial molecular genetic studies in TS used candidate gene association approaches with little success. Several small- to medium-sized genetic linkage studies have been conducted, with only one achieving genome-wide statistical significance on chromosome 2p23 (Tourette Syndrome Association International Consortium for Genetics, 2007). Linkage mapping in a single family followed by full exon sequencing of 51 genes on chromosome 15q led to the identification of a nonsense mutation in the L-histidine decarboxylase gene (*HDC*) in all affected family members (Ercan-Sencicek et al., 2010). Expression of this HDC nonsense variant interfered with the activity of the wild-type allele, suggesting that it acts in a dominant negative fashion to decrease synthesis of histamine (Ercan-Sencicek et al., 2010). Recurrent chromosomal rearrangements have been reported at a number of sites, and copy number variants may also play a role in TS susceptibility (Sundaram et al., 2010). A chromosome 13q inversion in one patient with TS pointed toward the slit- and trk-like 1 neuronal trophic factor (*SLITRK1*) as a positional candidate gene (Abelson et al., 2005). One frameshift mutation and a recurrent rare variant at a microRNA-binding site were identified in a group of TS patients, although subsequent studies have provided only mixed support (O'Roak et al., 2010). Despite their rarity, the promising findings at *HDC* and *SLITRK1* may point toward molecular mechanisms involved in the larger group of patients with TS.

38.3.2 Genetic Susceptibility to OCD

OCD affects approximately 2% of the population, with similar numbers of females and males, but with an earlier onset in males. Twin studies with modern diagnostic criteria are lacking, but more recent twin studies based upon OCD symptoms rather than diagnosis suggest a strong heritability (Hudziak et al., 2004). Family studies strongly favor genetic susceptibility, with most evidence coming from families of probands with childhood-onset OCD (Nestadt et al., 2000). TS and chronic tic disorders are also more common in first-degree relatives of probands with OCD (Grados et al., 2001), paralleling what is seen in probands with TS.

Most molecular genetics studies in OCD have focused on candidate genes, with mixed success. Based on response to the SRIs, the most studied candidate gene is the serotonin transporter (SERT, *SLC6A4*). Studies of common *SLC6A4* polymorphisms also favor a role for the serotonin transporter gene (SERT, *SLC6A4*) in OCD, but different polymorphisms have been associated across studies, with some suggestion of sex differences (Voyiaziakis et al., 2011; Wendland et al., 2008). Ozaki et al. (2003) identified a rare, hyperfunctioning Ile425Val variant of *SLC6A4* in two families with OCD and other psychiatric disorders, although subsequent studies have provided mixed support (Voyiaziakis et al., 2011). Only three small- to medium-sized genome-wide linkage studies have been carried out, with no statistically significant findings in any study (Hanna et al., 2002, 2007; Shugart et al., 2006). A follow-up linkage study for hoarding symptoms in OCD families yielded a significant linkage finding on chromosome 14q (Samuels et al., 2007). Additionally, one suggestive linkage finding on chromosome 9p24 (Hanna et al., 2002) was replicated in an independent linkage sample (Willour et al., 2004). Follow-up studies of the neuronal glutamate transporter EAAC1 gene (*SLC1A1*) have revealed a significant association in multiple samples, although different polymorphisms in the 3' region of the gene have been associated across studies (reviewed in Wendland et al., 2009). Single candidate gene association studies have pointed to other genes in the glutamate system, converging on the finding that cerebrospinal fluid glutamate levels are elevated in OCD patients (Chakrabarty et al., 2005).

38.3.3 Genetic Susceptibility to Repetitive Behavior in ASD

ASD is a highly heritable complex genetic disorder with much higher concordance rates in monozygotic twins (64–91%) than in dizygotic twins and siblings

(0–9%) (Bailey et al., 1995; Bolton et al., 1994; Steffe nburg et al., 1989). The ADI-R and Autism Diagnostic Observation Schedule (ADOS) (Lord et al., 1999, 2000), the gold standard instruments for ASD ascertainment, have been used extensively to reduce phenotypic heterogeneity in ASD research (Hus et al., 2007; Veenstra-VanderWeele and Cook, 2004). In addition, researchers have applied a stratification strategy by using 'subphenotypes' to enhance the ability to dissect the genetic etiology of ASD (Buxbaum et al., 2001; Cuccaro et al., 2003; Lam et al., 2008; Shao et al., 2002). Since RRB is one of the core domain features of ASD, specific forms of RRB can be used as a 'subphenotype' to identify a common genetic etiology in ASD. For example, Shao et al. (2003) have found increased linkage evidence at the *GABRB3* locus in the15q11-q13 region in families sharing high IS factor scores. Brune et al. also reported an association between the 5-HTTLPR long/long genotype of the serotonin transporter gene (*SLC6A4*) and RSMB (Brune et al., 2006).

Several groups have reported that RRB has a strong tendency to run within ASD families, indicating separate genetic factors for RRB. For instance, Silverman and colleagues examined the variance of the RRB subdomain scores on the ADI-R in 212 ASD sibling pairs (Silverman et al., 2002). They found statistically significant familiality in two ADI-R subdomains that captures the RRB similar to the CI and IS factors. In line with the Silverman study, Tadevosyan-Leyfer and colleagues also reported the familiality of the 'compulsion' factor derived from the ADI-R that captures both CI and IS factors (Tadevosyan-Leyfer et al., 2003). However, the RSMB factor did not appear to be familial in these studies. Instead, the RSMB factor was associated with individual characteristics, such as IQ, age, social/communication impairments, and the presence of regression (Lam et al., 2008). In summary, 'higher order' RRBs, such as IS and CI, appear to be under genetic control, whereas the 'lower order' RSMB factor reflects variation in developmental levels. Moreover, genetic factors for RRB are likely independent of those that influence the social or communication deficits in ASD (Mandy and Skuse, 2008; Ronald et al., 2006; Silverman et al., 2002).

38.3.4 Simple Genetic Disorders with Repetitive Behavior Phenotypes

RRB is also common in specific, relatively rare, genetic syndromes including but not limited to the genetic syndromes detailed below. A few example syndromes are reviewed briefly here, but further details can be found on the Online Mendelian Inheritance in Man (OMIM), which provides a comprehensive review of individual genetic syndromes. These individual syndromes may provide insight into the molecular pathways and brain systems involved in RRB in the larger group of patients with developmental or neuropsychiatric disorders. Syndromes other than those highlighted below, including Angelman syndrome, Cornelia de Lange syndrome, Down syndrome (DS), and Cri-du-Chat syndrome, also include RRB as a significant part of their clinical profile. Each of these syndromes includes some degree of intellectual impairment, which raises the issue of interpretation of RRB in the developmental context. Importantly, individuals with each of these syndromes appear to have more RRB than IQ-matched peers.

Prader–Willi syndrome (PWS, OMIM #176270) is a rare genomic imprinting disorder with an estimated incidence rate of ~1 in 15000 (Butler, 1990). PWS occurs when the *paternal* contribution is absent in the 15q11-q13 region. Interestingly, studies have reported a higher rate of ASD among individuals with PWS (Veltman et al., 2005). In addition, PWS has been associated with clinically significant RRB (Bittel and Butler, 2005; Dykens and Shah, 2003; Dykens et al., 1999; State and Dykens, 2000). For example, skin picking is reported in most individuals with PWS (Dykens et al., 1999; Thompson and Gray, 1994; Torrado et al., 2006; Veltman et al., 2004; Webb et al., 2002; Whitman and Accardo, 1987). A sizable group of individuals with PWS has prominent OCD symptoms, such as hoarding, ordering/arranging, concerns with symmetry/exactness, rewriting, and need to tell/know/ask (Dykens et al., 1996).

Rett syndrome (RS, OMIM #312750) is classified as one of the pervasive developmental disorders in the Diagnostic and Statistical Manual of Mental Disorders, 4th Edition (DSM-IV) (American Psychiatric Association, 2000). RS is caused by mutations in *MECP2*, located on the X chromosome (Xq28), and occurs almost exclusively in females with an estimated prevalence of ~1 in 20000 females (Kozinetz et al., 1993; Leonard et al., 1997). The affected infants show normal prenatal and postnatal development for the first 5 months, followed by a deceleration of head growth rate, loss of acquired skills, impairments in social communication, and characteristic stereotypic repetitive hand movements such as mouthing or hand-wringing (Ben Zeev Ghidoni, 2007). Interestingly, an increase in *MECP2* gene dosage also appears to be pathological. Ramocki and colleagues described clinical characteristics of nine boys with duplication of the chromosomal interval including *MECP2* (Ramocki et al., 2009). This *MECP2* duplication syndrome is associated with severe to profound mental retardation, as well as repetitive behaviors including hand stereotypies and resistance to changes (Ramocki et al., 2010).

Smith–Magenis syndrome (SMS, OMIM #182290) is most commonly caused by a 3.7-Mb interstitial deletion in chromosome 17p11.2. SMS can also be caused by mutations in *RAI1* within the chromosome 17p11.2 region. The individuals with SMS show brachycephaly, midface hypoplasia, prognathism, hoarse voice, and speech delay with or without hearing loss, psychomotor and growth retardation, as well as RRB, such as stereotypic body movements (e.g., mouthing objects, hand biting, teeth grinding, body rocking, spinning/twirling, and lick and flip), restricted interests, obsessions, repetitive speech, ritualistic behavior, and attachment to people (preference for adults) (Clarke and Boer, 1998; Dykens and Smith, 1998; Dykens et al., 1997; Moss et al., 2008).

Smith–Lemli–Opitz syndrome (SLOS, OMIM #270400) is an autosomal recessive genetic disorder caused by a defect in cholesterol biosynthesis due to mutations of *DHCR7*. The estimated incidence of SLOS varies, ranging from 1 in 20 000 to 1 in 80 000 births (Kelley and Hennekam, 2000; Lowry and Yong, 1980; Ryan et al., 1998). The SLOS phenotypes vary but include microcephaly, mental retardation, hypotonia, facial dysmorphism, digital anomalies, and ambiguous genitalia. In addition, various forms of RRBs are common in SLOS. For instance, Tierney and colleagues reported repetitive forceful and rapid backward head/trunk arching and backward thrusting (opisthokinesis) in 50% of their study subjects, stereotypic stretching with brief and rapid hand movements, as well as self-injurious behavior, such as biting themselves or banging their heads on objects (Tierney et al., 1999, 2000).

38.4 NEUROPATHOLOGY AND NEUROIMAGING OF REPETITIVE BEHAVIOR

Neuropathology studies of RRB in human samples have been limited by sample availability. Even in TS, where a number of studies have been published, there are not sufficient samples to allow for replication of key findings. Neuroimaging techniques are a promising avenue and are yielding some consistent findings across studies. A variety of structural and functional neuroimaging techniques have been applied to the study of repetitive behavior in defined neurodevelopmental or neuropsychiatric syndromes. The structural techniques include variations on magnetic resonance imaging (MRI), including structural MRI, magnetic resonance spectroscopy (MRS), and diffusion tensor imaging. A variety of techniques is also used to evaluate functional activity of particular brain regions, including positron emission tomography (PET), which is typically used as a measure of brain metabolism, as well as single-photon emission computed tomography (SPECT) and functional MRI, which are used as measures of regional blood flow. Additionally, each of these functional measures can be applied in the resting state or during the

performance of particular cognitive tasks, making the resulting findings quite difficult to evaluate for consistency. Finally, various PET ligands can be used to evaluate the availability or binding of specific receptor or transporter proteins. The summaries that follow reflect findings that have emerged in large sample sizes or across multiple studies. Except where noted, most of these studies have been conducted in older children, adolescents, and adults because of the difficulties in imaging younger children. Importantly, data across these disorders point to common brain regions, including the basal ganglia, but findings are sometimes in opposite directions in different disorders, making it difficult to draw clear lines between RRB and specific regional alterations.

38.4.1 Neuropathology and Neuroimaging in TS

A limited number of postmortem cases are available in TS. In small sample studies, two findings have emerged with evidence across at least two studies or different brain regions. First, dopamine transporter binding density is decreased in both the basal ganglia and the frontal cortex (Singer et al., 1991; Yoon et al., 2007). Although this finding provides further support for the involvement of the dopaminergic system in TS and RRB, it is difficult to evaluate whether it is a primary or secondary change due to exposure to medications acting on the dopamine system as well as chronic TS. Second, two studies have found a decrease in parvalbumin-positive interneurons in the caudate (Kalanithi et al., 2005; Kataoka et al., 2010). As detailed below, this may point to a more specific population of neurons than can be identified in structural neuroimaging studies.

The presence of frequent motor tics may limit the ability of patients with TS to remain in the scanner. Perhaps as a result, relatively few structural neuroimaging studies have been performed in TS, with little consistency among studies to date (Plessen et al., 2009). The largest structural MRI study pointed to decreases in caudate volume in children and adults with TS (Peterson et al., 2003). A follow-up study provided further support for the importance of caudate volume, observing that lower caudate volume during childhood predicted higher tic severity in young adulthood (Bloch et al., 2005). The same large study found increased volumes in the orbitofrontal and parietal cortex of children with TS but decreased volumes in adults (Peterson et al., 2001). Furthermore, cortical volumes correlated inversely with symptom severity, suggesting that these alterations may serve to compensate for the primary deficit in TS (Peterson et al., 2001). Abnormalities have also been reported in other brain regions, including the corpus callosum, the limbic system, and the thalamus, but consistent findings have yet to emerge (Plessen et al., 2009). A variety of PET ligands has been used to examine monoamine receptors and transporters in TS, with a particular focus on the basal ganglia. The most consistent finding to emerge from these studies is an increase in amphetamine-induced release of dopamine, as assessed by a change in receptor binding after drug administration (Singer et al., 2002; Steeves et al., 2010; Wong et al., 2008).

Functional neuroimaging studies in TS have shown little consistency, likely because of variation in methodologies and subject populations (Rickards, 2009). If subjects are able to remain still in the scanner to allow adequate image acquisition, they will, as a result, be scanned in the act of tic suppression, regardless of what other cognitive task may be assigned. Perhaps as a result, PET, SPECT, and fMRI studies show considerable variability, with some support for decreased metabolism or blood flow in the basal ganglia (Braun et al., 1993; Eidelberg et al., 1997; Klieger et al., 1997; Peterson et al., 1998; Riddle et al., 1992). When considering the neuroimaging results as a whole, then, there is reasonable evidence favoring the involvement of the basal ganglia and the dopamine system, but much less consistency regarding the importance of other brain regions.

38.4.2 Neuroimaging of OCD

Postmortem studies have not been published to date in OCD. Individual structural MRI studies vary, but the aggregate data provide significant support for cortical–striatal pathology in OCD. Radua and Mataix-Cols (2009) performed a voxel-wise meta-analysis to identify individual regions with consistent findings across 12 MRI studies, including 3 studies of children and adolescents, and 9 of adults. Increased gray matter volume was found in the basal ganglia, including a large portion of the putamen and extending into the caudate. The severity of OCD in the patients included in each study correlated with the severity of this increased basal ganglia volume. Additionally, decreased gray matter volume was found in a continuous region encompassing the dorsal mediofrontal cortex and anterior cingulate cortex. No consistent evidence was obtained for changes in the orbitofrontal cortex (OFC), although individual structural neuroimaging studies do report changes in this region (Pujol et al., 2004; Szeszko et al., 2008), as do the functional neuroimaging studies reviewed later.

MRS has also been applied in OCD to understand the chemical composition of brain regions implicated in functional studies. A couple of studies have reported decreased N-acetylasparate, a putative marker of neuronal integrity, in the anterior cingulate cortex (Jang et al., 2006; Yucel et al., 2007). The Rosenberg group has also reported increased glutamatergic signal, comprising both glutamate and GABA, in the caudate that normalizes with treatment with SRIs in pediatric OCD patients (Rosenberg et al., 2000). This may match one study that found increased glutamate levels in cerebrospinal fluid from patients with OCD (Chakrabarty et al., 2005). PET and SPECT studies have also been used to evaluate ligand binding to particular receptor and transporter proteins in OCD. Despite some inconsistency, a few studies report decreased binding to the serotonin transporter in a number of regions, including the thalamus (Matsumoto et al., 2010; Reimold et al., 2007; Zitterl et al., 2007). Other studies implicate the dopamine system, including decreased ligand binding to the dopamine D2 receptor, suggesting increased dopamine in the synapse (Moresco et al., 2007; Perani et al., 2008).

Functional neuroimaging studies provide further evidence of the involvement of the basal ganglia in OCD and also point to the OFC. Different methodologies have been used to examine functional activation of brain regions in OCD, including PET, SPECT, and fMRI, which are detailed elsewhere in this volume. PET studies emerged first, typically using [18]fluorodeoxyglucose to assess regional brain metabolism. SPECT and fMRI use different measures of regional blood flow, which may correspond to regional activation. Measures of metabolism and blood flow may provide different results, although a few consistent findings have emerged across different modalities. These methods can also be applied in different settings, either to measure baseline activity at rest or to measure changes in activity during particular cognitive tasks. Resting PET studies favor increased brain metabolism in the caudate and the OFC, with some evidence also favoring the thalamus (Baxter et al., 1987, 1988; Whiteside et al., 2004), although findings are not always consistent (Whiteside et al., 2004). fMRI, SPECT, and PET studies have also used symptom provocation or executive function tasks to evaluate differences in brain region activation in particular contexts, with the data again highlighting the importance of the basal ganglia and OFC (Chamberlain et al., 2008; Menzies et al., 2008).

The coupling of functional neuroimaging methodology with successful treatment provides the best evidence for involvement of the corticostriatal pathway in OCD. The first reports of increases of caudate and OFC glucose metabolism by PET were followed quickly by reports of decreased caudate and OFC metabolic activity by PET after successful treatment with SRIs (Baxter et al., 1992; Benkelfat et al., 1990). OCD is not the only disorder to be treated with SRIs; they can also be helpful for depression and anxiety disorders, raising the possibility that changes in regional glucose metabolism could be nonspecific. Saxena et al. (2002) tested this possibility by comparing changes in regional glucose metabolism with an SRI in patients with OCD compared to patients with major depression, finding that the decreased metabolic activity in the caudate and OFC was specific to OCD. Adding further support to the specificity of this change to OCD treatment response, a decreased caudate glucose metabolism was also seen after successful CBT for OCD (Baxter et al., 1992; Schwartz et al., 1996). Taken as a whole, the structural and functional neuroimaging data support the concept of an OCD circuit including the OFC, the anterior cingulate cortex, the basal ganglia, and the thalamus.

38.4.3 Neuroimaging of Repetitive Behavior in ASD

Little, if any, information is available associating postmortem findings with RRB in individuals with neurodevelopmental disorders (Amaral et al., 2008). There are some associations, however, between structural neuroimaging findings (i.e., regional volumetric measurements) and repetitive behavior. For example, Sears et al. (1999) reported a significant negative association between caudate volume and three ADI-R repetitive behavior items: difficulties with minor changes in routine, compulsions/rituals, and complex mannerisms. A directionally different result was obtained by Hollander et al. (2005), who found increased right caudate volumes in individuals with autism and a positive correlation between right caudate volumes and ADI-R repetitive behavior domain total scores. Interestingly, no relationship was found between right caudate volume and RSMB. Instead, the association was due to the correlation between right caudate volume and IS/resistance to change factor scores. The same pattern was observed when putamen volumes were correlated with repetitive behavior scores. These findings were largely replicated by Rojas et al. (2006). In addition, significant positive partial correlations with the ADI-R repetitive and stereotyped behavior domain were also found in the left inferior frontal gyrus and right amygdala. Smaller volumes of the superior temporal gyri, left postcentral gyrus, and cerebellar regions were associated with worse repetitive behavior domain scores. Pierce and Courchesne (2001) found a positive correlation between repetitive behavior exhibited in the experimental setting of an exploration task and frontal lobe volume in a study of young children with autism. This measure of repetitive behavior was negatively correlated with cerebellar vermis volume. These neuroanatomic measures were not associated with ADI-R or ADOS repetitive behavior scores, however, Kates et al. (2005) compared boys with stereotypies who had no other known developmental or neurological disorder with matched controls. In this study, decreases in frontal white matter were found even after total white matter volume was taken into account. Caudate volumes did not differ between groups when expressed relative to total brain volume.

There has been little utilization of functional MRI to determine the neurobiological basis of repetitive behavior in neurodevelopmental disorders. One notable exception has been the work of Thakkar et al. (2008) who made use of an antisaccade task, which involves suppression of the prepotent response of looking toward rather than away from a stimulus. High-functioning individuals with ASD exhibited significantly higher error rates in the antisaccade condition and significantly increased anterior cingulate cortex (ACC) activation during correct trials. Moreover, higher ADI-R repetitive behavior scores were associated with greater ACC activation during correct trials, with repetitive sensorimotor behavior scores more strongly related to ACC activation than resistance to change/IS factor scores. An exaggerated ACC response to correct trials has also been observed with OCD subjects. An association between repetitive behavior and ACC activation has also been shown by Shafritz et al. (2008). These investigators used both a response-shifting and a set-shifting cognitive task in individuals with high-functioning autism. Individuals with autism showed deficits in response shifting but, surprisingly, not cognitive set shifting when compared to controls. Reduced activation in frontal, striatal, and parietal regions was observed during these trials in the ASD group. The severity of repetitive behavior was negatively correlated with activation in anterior cingulate and posterior parietal regions.

Despite the paucity of imaging studies related to repetitive behavior, the extant studies, taken together, highlight the importance of corticostriatal-thalamocortical circuitry in the mediation of repetitive behavior, as in TS and OCD. As will be seen in later sections, findings from animal models of repetitive behavior strongly support the involvement of this circuitry. Of particular value for the future would be studies such as the work of Langen et al. (2009), examining trajectories of development of key brain structures such as the caudate. Such work would need, however, to track coincident changes in the expression of RRB (Table 38.1).

38.5 MODELING REPETITIVE BEHAVIOR IN ANIMALS

As reviewed by the authors recently (Lewis et al., 2007), animal models of repetitive behavior that are relevant to clinical disorders can be categorized as follows: repetitive behavior associated with targeted insults to the CNS, repetitive behavior induced by pharmacological agents, and repetitive behavior associated with restricted environments and experience. Since that review, there have been reports confirmed by the authors' own observations of spontaneous repetitive behavior in two inbred mouse strains. As reviewed above, repetitive behavior occurs in multiple clinical populations, and animal models of repetitive behavior are therefore unlikely to be specific to a particular disorder. Even without achieving specificity,

TABLE 38.1   Summary of Clinical Research Areas and Major Findings

Research area	Twin study	Molecular genetic study	Neuropathology and neuroimaging study
TS	53% monozygotic (MZ) versus 8% dizygotic (DZ)	*HDC*	↓ DAT binding density in basal ganglia and frontal cortex
		SLITRK1	↓ Parvalbumin-positive neuron in caudate
		2p23	↓ Caudate volume
			↓ Caudate volume during childhood ↑tic severity in adulthood
			↑ Orbitofrontal and parietal cortex volume in children versus ↓ in adults
			↓ Cortical volume ↑tic severity
			↓ Metabolism in basal ganglia
OCD	None using DSM criteria	*SLC6A4*	↑ Glutamate in CSF
		SLC1A1	↓ Binding to 5-HT transporter
		14q	↓ Ligand binding to D2 receptor
		9p24	↑ Gray matter volume in basal ganglia
			↓ Gray matter volume in dorsal mediofrontal cortex and anterior cingulate cortex
			↑ Metabolism in caudate, thalamus, OFC
			↓ Metabolism in caudate and OFC after successful treatment with SRIs
ASD	~91% MZ versus ~10% DZ	*GABRB3* – IS	↑ ADI-R RRB item score ↓ caudate volume
		SLC6A4 – RSMB	↑ ADI-R IS score ↑ right caudate and putamen volume
			↑ ADI-R RSMB score ↑ left inferior frontal gyrus and right amygdala volume
			↑ RRB ↑ frontal lobe and ↓ cerebellar vermis volume

however, these models can be very useful in deciphering the circuit or the pharmacological contributions to repetitive behavior across clinical and nonclinical populations.

38.5.1 Repetitive Behavior in Animal Models of Targeted CNS Insult

There are a small number of mouse models that involve mutations of specific genes or chromosomal regions that have been reported to result in specific forms of repetitive behavior as part of the phenotype. For example, compulsive grooming resulting in hair removal and self-inflicted wounds has been observed in the *Hoxb8* homozygous mutant mouse (Greer and Capecchi, 2002) and the *Sapap3* KO mouse (Welch et al., 2007). In the latter model, the SAPAP3 protein is expressed selectively in glutamate synapses in striatum, whereas high levels of expression of *Hoxb8* were observed in brain regions known to comprise circuitry mediating OCD symptoms in patients. These models may have specific relevance to OC spectrum disorders such as trichotillomania or self-injurious behaviors common in individuals with severe neurodevelopmental

disorders. Mice expressing truncated MeCP2 protein, which serve to model RS, exhibit repetitive forelimb movements resembling the distinctive hand stereotypies (e.g., hand wringing, waving, and clapping) observed in patients with this syndrome (Moretti et al., 2005; Shahbazian et al., 2002). PWS patients exhibit a variety of compulsive behaviors including skin picking associated with deletions of the q11–13 region of chromosome 15 (Dykens, 2004). Among the genes that lie within this region is the *GABRB3* gene, which codes for the β3 subunit of the GABAA receptor. The *Gabrb3* homozygous knockout mouse exhibits stereotyped behavior including intense circling or 'tail-chasing' (DeLorey et al., 1998; Homanics et al., 1997). Ts65Dn mice have segmental trisomy for orthologs of a number of genes on human chromosome 21 and thus serve as a model of DS. The authors have shown that such mice exhibit repetitive hindlimb jumping and cage-top twirling (Turner et al., 2001). Similar stereotypies (jumping and cage top circling) have been reported in the amyloid precursor protein transgenic mouse model of Alzheimer disease (TgCRND8) (Ambree et al., 2006). Alterations in the neurexins, neuroligins, and associated proteins including SHANK3 have been implicated in the etiology of autism. Transgenic animals overexpressing neuroligin 2 (TgNL2) have altered synapse development and neuronal excitability and behaviorally exhibit limb clasping similar to *MECP2* KO mice and stereotyped vertical jumping (Hines et al., 2008). Neurexin 1α (*Nrxn1*), neuroligin 1 (*Nl1*), and *Shank3* KO mice each show increased grooming behavior, potentially pointing to a common repetitive behavior outcome for disruption of this autism-associated synaptic protein system.

CNS insult leading to repetitive behaviors in animals has also included nongenomic factors. Perhaps the most striking demonstration of such factors comes from Martin et al. (2008). These investigators purified antibodies from women who had at least two children with ASD and injected them into pregnant rhesus macaques. The offspring of these macaques were observed to engage in spontaneous whole-body stereotypies that persisted in the 6 months following weaning and were observed in multiple test conditions. Exposure to maternally derived IgGs that cross the placenta has been implicated in other disorders involving tics and compulsive disorders. Prenatal exposure to valproic acid has been linked to autism susceptibility. In rats, exposure to valproic acid on embryonic day 12.5 induces stereotypic activity (Ingram et al., 2000; Rodier et al., 1997; Schneider and Przewlocki, 2005). Repetitive behavior can also be induced by exposure of newborn rats to Borna disease virus (Hornig et al., 1999). In nonhuman primates, early lesions encompassing the amygdala, hippocampal formation, and adjacent temporal cortex result in repetitive behavior (Bachevalier and Loveland, 2006). A delayed (after year 1 of life) emergence of repetitive motor behavior following amygdala or hippocampal lesions in macaque infants has also been reported by Bauman et al. (2008), who also showed that amygdala damage induced self-directed behaviors, whereas hippocampal lesions induced repetitive head twisting.

Although these studies provide potentially valuable models, repetitive behavior has generally not been the focus or rationale for the work. Thus, the repetitive behavior observed has often not been well characterized and little additional work has investigated the specific neurobiological mechanisms associated with the expression of the repetitive behavior.

38.5.2 Animal Models of Drug-Induced Repetitive Behavior

As early as the 1960s, it was known that specific pharmacological agents such as amphetamine and apomorphine can induce repetitive behavior in animals. As these drugs act on dopamine receptors or uptake sites that are enriched in striatum, these findings pointed to the importance of the basal ganglia in the mediation of repetitive behaviors. Confirmation came from experiments showing that dopamine or a dopamine agonist injected directly into the corpus striatum induced stereotyped behavior in rats (e.g., Ernst and Smelik, 1966). Similarly, intrastriatal administration of the glutamate receptor ligand, NMDA, also induced stereotyped behavior (Karler et al., 1997). Intracortical manipulations enhancing the activity of excitatory corticostriatal projections exacerbate the expression of stereotypy. For instance, administration of either the D_2 antagonist sulpiride or the GABA antagonist bicuculline into the frontal cortex enhances the motor stimulatory effects of amphetamine (Karler et al., 1998; Kiyatkin and Rebec, 1999). Conversely, amphetamine-induced stereotypy can be attenuated via intracortical infusion of DA or GABAergic agonists (Karler et al., 1998). Experiments in which the expression of drug-induced stereotypy was shown to be sensitive to manipulations in the substantia nigra pars reticulata (SNpr) and the subthalamic nucleus (STN) also support the hypothesized role of cortical–basal ganglia circuitry in repetitive behaviors (Barwick et al., 2000; Scheel-Kruger et al., 1978). These, and many other relevant findings, provide clear evidence of the preeminent role played by the cortical–basal ganglia circuitry in the expression of drug-induced repetitive motor behaviors.

38.5.3 Repetitive Behavior and Environmental Restriction

Abnormal repetitive behaviors are considered sentinel behaviors by applied ethologists signaling poor animal welfare. This is not surprising as a wide variety of species of animals housed in restricted or impoverished environments (e.g., zoo, farm, and laboratory) exhibit abnormal repetitive behavior (Mason and Rushen, 2006). In fact, as Wurbel (2001) has pointed out, repetitive behaviors are the most common category of abnormal behavior observed in confined animals. Demonstrations of the attenuation or prevention of repetitive behavior by rearing animals in larger, more complex environments (environmental enrichment) provide strong evidence for the role of environmental restriction in the induction of repetitive behavior. Early social deprivation as a special case of environmental restriction has been shown to have powerful deleterious effects on humans and nonhuman primates including the induction of abnormal repetitive behavior (Carlson and Earls, 1997; Mason and Rushen, 2006).

The authors' own work has involved an animal model that falls under the category of repetitive behavior associated with environmental restriction. In this model, deer mice (*Peromyscus maniculatus*) exhibit repetitive hindlimb jumping and backward somersaulting as a consequence of being reared in standard laboratory caging. These behaviors occur at a high rate, persist across much of the life of the animal, and appear relatively early in development, sometimes as early as weaning. The authors have shown in several studies that environmental enrichment markedly attenuates the development and expression of the repetitive behavior. This outcome was associated with biochemical and morphological changes in basal ganglia circuitry (Lewis, 2004).

38.5.4 Repetitive Behavior in Inbred Mouse Strains

Examination of inbred mouse strains for autistic-like behavioral traits led to the observation that C58/J mice displayed stereotyped jumping and backward flipping behaviors not observed in other strains (Moy et al., 2008a,b). A similar behavioral phenotype has been reported in the C57BL/10 strain (Deacon et al., 2007), with mice of this strain exhibiting spontaneous repetitive vertical jumping with no such behavior observed in the closely related C57BL/6 strain. The authors have confirmed both sets of observations (unpublished findings). In addition, Crawley and her colleagues have described an excessive grooming phenotype in BTBR (BTBRT+tf/J) mice (Silverman et al., 2010). Thus, several inbred mouse strains display a repetitive behavior phenotype, which should be highly advantageous for addressing the issue of the genetics of repetitive behavior.

38.5.5 Resistance to Change/IS in Animal Models

As the findings in the previous sections make clear, it is the repetitive sensory–motor factor of RRB that is most frequently modeled in animals. This cluster of behaviors is easier to model than behaviors related to the IS or resistance to change factor. Nevertheless, some animal work has addressed cognitive flexibility or resistance to change. This work has entailed a range of behavioral tests from response extinction to reversal learning to intra- and extradimensional set shifting (e.g., Colacicco et al., 2002). In some cases, there has been an attempt to correlate repetitive motor behaviors or stereotypies with measures of cognitive flexibility. For example, the amount of environmental restriction-induced stereotypy observed in bank voles and bears was significantly inversely correlated with extinction learning (Garner and Mason, 2002; Vickery and Mason, 2005). Similar findings were obtained with Orange-wing Amazon parrots using performance on a variation of a gambling task that indexed the tendency to repeat responses or perseverate. Birds with higher stereotypy scores exhibited greater sequential dependency in their responses on this task (Garner et al., 2003). In their own work, the authors have examined the performance of deer mice in a procedural T-maze learning task. Their results indicate

that high levels of stereotypy in deer mice were associated with deficits in reversal learning in the T-maze (Tanimura et al., 2008b). The relationship between cognitive rigidity (deficits in set shifting, extinction, and reversal learning) and motor stereotypy is perhaps not surprising given the common mediation by cortical–basal ganglia pathways. Much greater emphasis needs to be placed on modeling 'higher order' repetitive behaviors in animals in future studies, however.

38.6 NEUROCIRCUITRY OF REPETITIVE BEHAVIOR

38.6.1 Basal Ganglia Circuitry and Repetitive Behavior

As reviewed recently by the authors (Lewis and Kim, 2009), neural mediation of the expression of repetitive behavior rests on pathways that project from select areas of the cortex to the striatum and then onto other basal ganglia, then the thalamus, and finally back to the cortex. Medium spiny GABAergic striatal projection neurons receive input from sensory–motor and associative areas of cortex, and, in turn, give rise to the so-called direct and indirect pathways that constitute corticostriato-thalamocortical loops. GABAergic medium spiny neurons in the striatum that express the neuropeptides dynorphin and substance P as well as D_1 dopamine receptors and A_1 adenosine receptors constitute striatonigral or direct-pathway neurons. These neurons send projections from the striatum to the internal segment of the globus pallidus (GPi) and SNpr. Striatal medium spiny neurons that express the neuropeptide enkephalin as well as D_2 dopamine receptors and A_2 adenosine receptors constitute striatopallidal or indirect-pathway neurons. Indirect-pathway neurons project to the external segment of the globus pallidus (GPe) and then to STN before projecting to GPi and SNpr. Output from the GPi/SNpr goes to the thalamus and then on to the cortex to complete the circuitry (Gerfen, 2000; Olanow et al., 2000; Steiner and Gerfen, 1998). The classic view has been that the direct pathway facilitates movement via disinhibition of glutamatergic thalamocortical firing, whereas the indirect pathway inhibits ongoing movement via inhibition of thalamocortical afferents (Gerfen et al., 1990).

The medium spiny cells that give rise to either the direct or the indirect pathways constitute about 85% of projection neurons and the matrix compartment of the striatum. The remaining projection neurons form patchy areas or striosomes that are distributed throughout the extrastriosomal matrix. Striosomal projection neurons

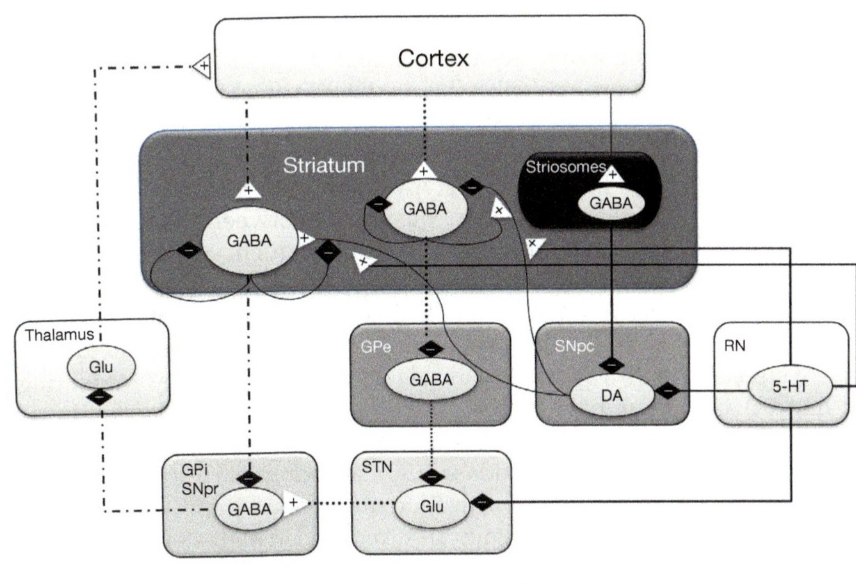

FIGURE 38.1 Cortical–basal ganglia circuitry. The direct and indirect pathways through the basal ganglia are shown, along with other neurotransmitter systems that impact basal ganglia circuitry. +, excitatory input; −, inhibitory input; Glu, glutamate; 5-HT, serotonin; DA, dopamine; GPe, globus pallidus externa; GPi, globus pallidus interna; RN, raphe nucleus; SNpc, substantia nigra pars compacta; SNpr, substantia nigra pars riticulata; STN, subthalamic nucleus.

receive input preferentially from limbic cortical areas (e.g., OFC, anterior cingulate/posterior medial prefrontal cortex [PFC]) and, in turn, project to the substantia nigra pars compacta (SNpc) (Canales and Graybiel, 2000a,b). Striosomal projections can, therefore, directly mediate nigrostriatal dopamine pathway activity, which, in turn, will strongly influence the activity of direct- and indirect-pathway neurons in the striatum. Moreover, the activity of striosomal projections should strongly impact reward through regulation of ascending DA projections. White and Hiroi (White and Hiroi, 1998) provided some support for this idea by showing that high rates of intracranial self-stimulation were associated with electrode placement either in or next to striosomes. Conversely, 'normal' sensory–motor function (e.g., grooming and locomotion) in rats appears to be mediated by the extrastriosomal matrix (Brown et al., 2002). Findings from nonhuman primates have suggested that striosomal output innervates SNpr and GP as well as SNpc (Levesque and Parent, 2005), so these pathways also may not be as segregated as once believed (Figure 38.1).

38.6.2 Cortical–Basal Ganglia Circuitry and Repetitive Behavior

As many as five parallel information-processing circuits have been identified that make up the cortical–basal ganglia circuitry. These five circuits have been labeled as the motor, oculomotor, dorsolateral prefrontal, lateral orbitofrontal, and anterior cingulate circuits (Alexander et al., 1986). These circuits, though anatomically distinct, are nevertheless not segregated. Of these, the motor circuit has been the most studied and emerges as the best candidate for mediation of repetitive behavior.

Direct evidence for the role of such circuitry in repetitive behavior is limited. The authors have shown that stereotyped behavior in early socially deprived rhesus macaques was associated with dopamine receptor supersensitivity (Lewis et al., 1990), loss of dopamine innervation in striatum and dopamine cells in substantia nigra, and decreases in medium spiny striatal projection neurons as indexed by neuropeptide staining (Martin et al., 1991).

The importance of the corticobasal ganglia circuitry in repetitive behavior is further highlighted by the behavioral phenotype of the *Sapap3* knockout mouse (Welch et al., 2007). *Sapap3* encodes a postsynaptic scaffolding protein, highly expressed in the striatum and important in regulating glutamatergic corticostriatal synapses. Mice homozygous for the gene deletion exhibited excessive grooming leading to lesions of the head, neck, and snout. In addition, these animals exhibited alterations in α-amino-3-hydroxy-5-methyl-4-isoxazoleproprionic acid (AMPA) and NMDA-receptor-dependent transmission at corticostriatal synapses. Interestingly, administration of the serotonin uptake inhibitor fluoxetine given systemically for 6 days reversed the compulsive grooming. Finally, the behavioral phenotype was rescued by transduction of the *Sapap3* gene into preweaning mice. This work shows that deletion of even a single protein that functions to maintain the activity of the cortical–basal ganglia circuitry can result in a robust repetitive behavior phenotype. Moreover, an SRI can reverse the effects of loss of a glutamate synapse protein. Further support for the involvement of corticostriatal signaling is evidenced by mice lacking the *Slitrk5* gene, which encodes another member of the protein family that includes *Slitrk1*, implicated in TS. Like the Sapap3 knockout mouse, the *Slitrk5* knockout mouse also shows increased grooming and anxiety-like behavior that improves with fluoxetine (Shmelkov et al., 2010). It also shows abnormalities in corticostriatal neurotransmission, perhaps due to altered striatal cellular composition, and increased OFC activity. These emerging genetic models, while not yet mapping onto known RRB susceptibility genes, suggest that the complex circuitry underlying RRB can be effectively modulated in multiple ways.

Grabli et al. (2004) have reported induction of stereotyped behavior (e.g., licking and biting of fingers) in monkeys by the GABA antagonist bicuculline microinjected into the limbic aspect of the GPe (part of the indirect pathway). In a follow-up study (Baup et al., 2008), this group showed that DBS applied to the STN dramatically reduced these drug-induced repetitive behaviors. The importance of the STN, and thus the indirect pathway, was also highlighted by Winter et al. (2008). In this study, rats that sustained ibotenic acid lesions to the STN exhibited an increase in compulsive lever pressing in the signal attenuation model of OCD. This same research group has also shown that bilateral high-frequency stimulation of the STN as well as its pharmacological inactivation reduced compulsive checking in rats induced by the dopamine agonist quinpirole (Winter et al., 2008). This latter finding is consistent with clinical observations that DBS applied to the STN reduced the severity of symptoms in previously treatment-refractory OCD patients (Mallet et al., 2008).

The authors have shown, in multiple studies, that early environmental enrichment markedly attenuated the development of repetitive behavior in deer mice. In addition, brain changes associated with this attenuation occurred in brain areas associated with the cortical–basal ganglia circuitry but not in other brain regions (e.g., hippocampus; see Lewis, 2004). In other experiments with deer mice, antagonism of corticostriatal glutamatergic projections or nigrostriatal dopaminergic projections by intrastriatal administration of selective pharmacological agents selectively reduced stereotyped behavior (Presti et al., 2003). Interestingly, the D_1 dopamine receptor antagonist SCH23390 exhibited such effects, whereas no such attenuation was observed following intrastriatal administration of the D_2 dopamine receptor antagonist raclopride (Presti et al., 2004).

Dysregulation of corticostriato-thalamocortical circuitry associated with motor disorders is thought to be due to an imbalance between the direct and indirect pathways comprising this circuit. Because dynorphin and enkephalin serve as markers for direct- and indirect-pathway neurons, respectively, concentrations of these striatal neuropeptides were measured to index the relative activation of these basal ganglia pathways in stereotypic deer mice (Presti and Lewis, 2005). Measurements were made in dorsolateral striatum using

deer mice exhibiting different levels of spontaneous stereotypy. Results indicated significantly increased dynorphin/enkephalin content ratios in high-stereotypy mice relative to low-stereotypy mice. This ratio difference was due to significantly lower leu-enkephalin content in high-stereotypy mice. Moreover, a significant positive correlation was found between the dynorphin/enkephalin content ratio and frequency of stereotypy in these mice, whereas a significant negative correlation was found for enkephalin content and stereotypy.

To extend these findings, the authors assessed indirect-pathway activation relative to stereotypy by measuring neuronal metabolic activation of the STN, a key brain region in the indirect pathway (Tanimura et al., 2008a). Using cytochrome oxidase (CO) histochemistry to index long-term neuronal activation, they found that CO staining in the STN was significantly reduced in high-stereotypy mice. Further, CO staining was significantly negatively correlated with the frequency of stereotypy. Thus, higher rates of spontaneous stereotypy were associated with reduced neuronal activation of the indirect pathway.

The authors hypothesized that if high rates of spontaneous stereotypy were associated with decreases in indirect-pathway activation, then stimulation of this pathway by a selective pharmacological agent should attenuate repetitive behavior. As A2A receptors are enriched in striatum, expressed on striatopallidal neurons and activate Gs/olf proteins upon stimulation, activation of these receptors should attenuate stereotypy. When administered alone, however, the selective A2A receptor agonist CGS21680 failed to reduce stereotypy. The addition of the selective A1 agonist N6-cyclopentyladine (CPA) to CGS21680 did selectively attenuate stereotypy in a dose-dependent manner without adverse suppression of general motor activity. The relative efficacy of the combined stimulation of A2A and A1 receptors compared to A2A alone may be explained by the results reported by Karcz-Kubicha et al. (2006). In this work, administration of an A1 or A2A receptor agonist alone did not induce striatal c-Fos expression. Stimulation of both receptor subtypes, however, did induce striatal c-Fos expression and in a selective fashion with activation seen in striatopallidal, but not striatonigral, neurons. This combined treatment of A1 and A2A receptor agonists also increased striatal enkephalin expression. The attenuation by repetitive behavior of this drug combination provides additional evidence for the importance of the indirect pathway and also highlights potential novel therapeutic targets.

38.6.3 Long-Term Neuroadaptations and Repetitive Behavior

The development and persistence of repetitive behavior in neurodevelopmental disorders presumably involve long-term, experience-dependent plasticity in the cortical–basal ganglia pathways. As yet, there is little information available as to the nature of these neuroadaptations or the mechanisms that mediate such plasticity. There are other models of such long-term basal ganglia neuroadaptations, however, that may be highly informative for the understanding of pathway changes that may mediate the development and expression of repetitive behavior associated with clinical disorders.

One model for such neuroadaptation would be habit learning or habit formation. Habits and repetitive behaviors share a number of important similarities; indeed, as Graybiel (2008) has suggested, repetitive behaviors can be thought of as 'extreme' habits. Habit formation, typically examined in the context of procedural learning, involves adaptations of cortical–basal ganglia loops. Discrete shifts in neural activity patterns associated with the transition from a goal-directed to a habit-driven behavior have been identified using chronic electrophysiological monitoring of ensembles of neurons in rodents and nonhuman primates (reviewed in Graybiel, 2008). Amphetamine sensitization, another model of dopamine-dependent striatal plasticity, involves long-term neuronal changes following repeated, intermittent drug exposure. Amphetamine sensitization accelerates the development of habit learning or formation (Nelson and Killcross, 2006), and also results in significantly increased levels of repetitive motor behavior (Canales and Graybiel, 2000b).

Experience-dependent neuroadaptations are generally thought to be driven by differential gene expression mediated by transcription factors. A leading candidate for mediating long-term striatal plasticity is the transcription factor ΔFosB (Nestler et al., 1999, 2001). ΔFosB undergoes posttranslational modifications that result in highly stable isoforms, which heterodimerize with Jun proteins and bind to AP-1 sites expressed in the promoter regions of genes encoding key striatal proteins (e.g., AMPA glutamate receptor subunit, GluR2, and dynorphin; Bibb et al., 2001; Chen et al., 1997; Kelz et al., 1999). ΔFosB is induced after chronic exposure to stimuli relevant to repetitive behavior (e.g., stress, drugs of abuse, and chronic wheel running) and persists in the brain for long periods of time (McClung et al., 2004). Thus, ΔFosB might have a more general role in the development of repetitive behavior induced by a wide range of stimuli.

Repetitive behavior including compulsions and dyskinesias can be induced by chronic L-DOPA administration to individuals with PD. In rats, L-DOPA-induced dyskinesias have been shown to be associated with increased striosomal FosB relative to matrix FosB (Andersson et al., 1999; Cenci et al., 1999). Pulsatile administration of a D_1 dopamine agonist to parkinsonian nonhuman primates markedly elevated striatal ΔFosB but only in those animals that developed dyskinesias (Doucet et al., 1996). These FosB-related proteins appear to be expressed preferentially in direct-pathway neurons (Andersson et al., 1999). Similarly, in mouse models of L-DOPA-induced dyskinesias, activation of extracellular-signal-regulated kinase (ERK), the extracellular-regulated kinases that mediate downstream transcription, was restricted to direct-pathway neurons (Santini et al., 2009). Selective induction of ΔFosB in striatal direct-pathway neurons is associated with compulsive wheel running in rodents. Transgenic mice that selectively overexpress ΔFosB in these projection neurons display compulsive wheel running, whereas this behavior is significantly inhibited in animals that overexpress the gene in enkephalin-containing or indirect-pathway neurons (Werme et al., 2002).

The various models described in the previous sections provide important candidate mechanisms that may explain the transition from normative behavior to aberrant repetitive behavior. Identification of such mechanisms would provide novel potential therapeutic targets for drug development.

38.7 SUMMARY

Repetitive behavior has been studied using a variety of tools, from molecular genetics to neuroimaging to model organisms, but the wide variety of repetitive behaviors observed across normative development and human disorders presents a substantial challenge. The phenomenology of repetitive behavior reveals a continuum beginning with very simple repetitive motor sequences, such as hand flapping or grunting, that are typically seen in early childhood and also in ASD and TS. At the other end of the continuum are IS and compulsive rituals, which typically emerge in the preschool years and decline thereafter except when seen in ASD and OCD.

Within the human repetitive behavior disorders, some consistent findings have emerged across cognitive, treatment, genetic, and neuroimaging studies. Neuropsychological testing reveals deficits in executive functioning across disorders. Treatment studies point to behavioral therapies that stress response inhibition in each study, although approaches vary quite widely depending upon the RRB target. Pharmacology studies point to dopamine D2 receptor antagonists across disorders, but there is less consistency in response to SRIs, which are useful in OCD but show mixed data in ASD. Genetic studies support substantial heritability for repetitive behavior, both within and across disorders. Initial genetic findings do not show clear consistency across studies, with the exception of the serotonin transporter gene, which has been implicated in OCD and in RRB within ASD. Neuroimaging studies have coalesced to some degree around the corticostriatal circuit, although the strongest structural findings in TS and

OCD are in opposite directions in the caudate. Other inconsistencies have also emerged across disorders. Multiple studies, from genetics to MRS to CSF, point to the glutamate system in OCD. In contrast, postmortem findings in TS point to decreased GABAergic neurons in the caudate.

Ideally, animal models should be used to dissect the ways that genetic and environmental influences impact brain circuits to feed RRB. Initial work using genetic models points to the same corticostriatal circuits that are implicated by neuroimaging studies. Some of the best current models are combinations of environmental deprivation in species or inbred strains that show some baseline repetitive behavior. These naturalistic models have significant potential to translate back to human disorders. For example, the social disconnectedness that is diagnostic in autism may actually parallel the environmental deprivation that triggers RRB in animals. Research in naturalistic and pharmacological animal models of RRB favors imbalance in the direct and indirect pathway of the basal ganglia. As molecular genetic studies yield more RRB susceptibility genes, the different ways in which this imbalance may be triggered could be better understood. With the rapid progress possible using multiple different research approaches, the emerging understanding of the circuits underlying RRB may translate to potential treatments for TS, OCD, or RRB within neurodevelopmental disorders.

References

Abelson, J.F., Kwan, K.Y., O'roak, B.J., et al., 2005. Sequence variants in SLITRK1 are associated with Tourette's syndrome. Science 310, 317–320.

Alexander, G.E., Delong, M.R., Strick, P.L., 1986. Parallel organization of functionally segregated circuits linking basal ganglia and cortex. Annual Review of Neuroscience 9, 357–381.

Amaral, D.G., Schumann, C.M., Nordahl, C.W., 2008. Neuroanatomy of autism. Trends in Neurosciences 31, 137–145.

Ambree, O., Touma, C., Gortz, N., et al., 2006. Activity changes and marked stereotypic behavior precede Abeta pathology in TgCRND8 Alzheimer mice. Neurobiology of Aging 27, 955–964.

American Psychiatric Association, 2000. Diagnostic and Statistical Manual of Mental Disorders-Text Revision (DSM-IV-TR), 4th edn. American Psychiatric Association Press, Washington, DC.

Andersson, M., Hilbertson, A., Cenci, M.A., 1999. Striatal fosB expression is causally linked with L-DOPA-induced abnormal involuntary movements and the associated upregulation of striatal prodynorphin mRNA in a rat model of Parkinson's disease. Neurobiology of Disease 6, 461–474.

Bachevalier, J., Loveland, K.A., 2006. The orbitofrontal–amygdala circuit and self-regulation of social–emotional behavior in autism. Neuroscience and Biobehavioral Reviews 30, 97–117.

Bailey, A., Le Couteur, A., Gottesman, I., et al., 1995. Autism as a strongly genetic disorder: Evidence from a British twin study. Psychological Medicine 25, 63–78.

Bartak, L., Rutter, M., 1976. Differences between mentally retarded and normally intelligent autistic children. Journal of Autism and Developmental Disorders 6, 106–120.

Barwick, V.S., Jones, D.H., Richter, J.T., Hicks, P.B., Young, K.A., 2000. Subthalamic nucleus microinjections of 5-HT2 receptor antagonists suppress stereotypy in rats. Neuroreport 11, 267–270.

Bauman, M.D., Toscano, J.E., Babineau, B.A., Mason, W.A., Amaral, D.G., 2008. Emergence of stereotypies in juvenile monkeys (Macaca mulatta) with neonatal amygdala or hippocampus lesions. Behavioral Neuroscience 122, 1005–1015.

Baup, N., Grabli, D., Karachi, C., et al., 2008. High-frequency stimulation of the anterior subthalamic nucleus reduces stereotyped behaviors in primates. Journal of Neuroscience 28, 8785–8788.

Baxter, L.R., Phelps, M.E., Mazziotta, J.C., Guze, B.H., Schwartz, J.M., Selin, C.E., 1987. Local cerebral metabolic rates in obsessive-compulsive disorder: A comparison with rates in unipolar depression and in normal controls. Archives of General Psychiatry 44, 211–218.

Baxter Jr., L.R., Schwartz, J.M., Mazziotta, J.C., et al., 1988. Cerebral glucose metabolic rates in nondepressed patients with obsessive-compulsive disorder. American Journal of Psychiatry 145, 1560–1563.

Baxter, L., Schwartz, J., Bergman, K., et al., 1992. Caudate glucose metabolic rate changes with both drug and behavior therapy for obsessive–compulsive disorder. Archives of General Psychiatry 49, 681–689.

Ben Zeev Ghidoni, B., 2007. Rett syndrome. Child and Adolescent Psychiatric Clinics of North America 16, 723–743.

Benkelfat, C., Nordahl, T., Semple, W., King, A., Murphy, D., Cohen, R., 1990. Local cerebral glucose metabolic rates in obsessive–compulsive disorder: Patients treated with clomipramine. Archives of General Psychiatry 47, 840–848.

Bibb, J.A., Nishi, A., O'callaghan, J.P., et al., 2001. Phosphorylation of protein phosphatase inhibitor-1 by Cdk5. Journal of Biological Chemistry 276, 14490–14497.

Bittel, D.C., Butler, M.G., 2005. Prader–Willi syndrome: Clinical genetics, cytogenetics and molecular biology. Expert Reviews in Molecular Medicine 7, 1–20.

Bloch, M.H., Leckman, J.F., Zhu, H., Peterson, B.S., 2005. Caudate volumes in childhood predict symptom severity in adults with Tourette syndrome. Neurology 65, 1253–1258.

Bloch, M.H., Landeros-Weisenberger, A., Kelmendi, B., Coric, V., Bracken, M.B., Leckman, J.F., 2006. A systematic review: Antipsychotic augmentation with treatment refractory obsessive–compulsive disorder. Molecular Psychiatry 11, 622–632.

Bloch, M.H., Landeros-Weisenberger, A., Rosario, M.C., Pittenger, C., Leckman, J.F., 2008. Meta-analysis of the symptom structure of obsessive–compulsive disorder. American Journal of Psychiatry 165, 1532–1542.

Bloch, M.H., Leckman, J.F., 2009. Clinical course of Tourette syndrome. Journal of Psychosomatic Research 67, 497–501.

Bloch, M.H., Panza, K.E., Landeros-Weisenberger, A., Leckman, J.F., 2009. Meta-analysis: Treatment of attention-deficit/hyperactivity disorder in children with comorbid tic disorders. Journal of the American Academy of Child and Adolescent Psychiatry 48, 884–893.

Bodfish, J.W., Symons, F.J., Parker, D.E., Lewis, M.H., 2000. Varieties of repetitive behavior in autism: Comparisons to mental retardation. Journal of Autism and Developmental Disorders 30, 237–243.

Bolton, P., Macdonald, H., Pickles, A., et al., 1994. A case–control family history study of autism. Journal of Child Psychology and Psychiatry 35, 877–900.

Braun, A.R., Stoetter, B., Randolph, C., et al., 1993. The functional neuroanatomy of Tourette's syndrome: An FDG-PET study. I. Regional changes in cerebral glucose metabolism differentiating patients and controls. Neuropsychopharmacology 9, 277–291.

Brown, L.L., Feldman, S.M., Smith, D.M., Cavanaugh, J.R., Ackermann, R.F., Graybiel, A.M., 2002. Differential metabolic activity in the striosome and matrix compartments of the rat striatum during natural behaviors. Journal of Neuroscience 22, 305–314.

Brune, C.W., Kim, S.J., Salt, J., Leventhal, B.L., Lord, C., Cook Jr., E.H., 2006. 5-HTTLPR genotype-specific phenotype in children and adolescents with autism. American Journal of Psychiatry 163, 2148–2156.

Butler, M.G., 1990. Prader–Willi syndrome: Current understanding of cause and diagnosis. American Journal of Medical Genetics 35, 319–332.

Buxbaum, J.D., Silverman, J.M., Smith, C.J., et al., 2001. Evidence for a susceptibility gene for autism on chromosome 2 and for genetic heterogeneity. American Journal of Human Genetics 68, 1514–1520.

Canales, J.J., Graybiel, A.M., 2000a. A measure of striatal function predicts motor stereotypy. Nature Neuroscience 3, 377–383.

Canales, J.J., Graybiel, A.M., 2000b. Patterns of gene expression and behavior induced by chronic dopamine treatments. Annals of Neurology 47, S53–S59.

Cardona, F., Orefici, G., 2001. Group A streptococcal infections and tic disorders in an Italian pediatric population. Journal of Pediatrics 138, 71–75.

Carlson, M., Earls, F., 1997. Psychological and neuroendocrinological sequelae of early social deprivation in institutionalized children in Romania. Annals of the New York Academy of Sciences 807, 419–428.

Castellanos, F.X., Ritchie, G.F., Marsh, W.L., Rapoport, J.L., 1996. DSM-IV stereotypic movement disorder: Persistence of stereotypies of infancy in intellectually normal adolescents and adults. Journal of Clinical Psychiatry 57, 116–122.

Cenci, M.A., Tranberg, A., Andersson, M., Hilbertson, A., 1999. Changes in the regional and compartmental distribution of FosB- and JunB-like immunoreactivity induced in the dopamine-denervated rat striatum by acute or chronic L-dopa treatment. Neuroscience 94, 515–527.

Chakrabarty, K., Bhattacharyya, S., Christopher, R., Khanna, S., 2005. Glutamatergic dysfunction in OCD. Neuropsychopharmacology 30, 1735–1740.

Chamberlain, S.R., Menzies, L., Hampshire, A., et al., 2008. Orbitofrontal dysfunction in patients with obsessive–compulsive disorder and their unaffected relatives. Science 321, 421–422.

Chen, J., Kelz, M.B., Hope, B.T., Nakabeppu, Y., Nestler, E.J., 1997. Chronic Fos-related antigens: Stable variants of deltaFosB induced in brain by chronic treatments. Journal of Neuroscience 17, 4933–4941.

Clarke, D.J., Boer, H., 1998. Problem behaviors associated with deletion Prader–Willi, Smith–Magenis, and cri du chat syndromes. American Journal of Mental Retardation 103, 264–271.

Colacicco, G., Welzl, H., Lipp, H.P., Wurbel, H., 2002. Attentional set-shifting in mice: Modification of a rat paradigm, and evidence for strain-dependent variation. Behavioural Brain Research 132, 95–102.

Cuccaro, M.L., Shao, Y., Grubber, J., et al., 2003. Factor analysis of restricted and repetitive behaviors in autism using the Autism Diagnostic Interview-R. Child Psychiatry and Human Development 34, 3–17.

Deacon, R.M., Thomas, C.L., Rawlins, J.N., Morley, B.J., 2007. A comparison of the behavior of C57BL/6 and C57BL/10 mice. Behavioural Brain Research 179, 239–247.

Delorey, T.M., Handforth, A., Anagnostaras, S.G., et al., 1998. Mice lacking the beta3 subunit of the GABAA receptor have the epilepsy phenotype and many of the behavioral characteristics of Angelman syndrome. Journal of Neuroscience 18, 8505–8514.

Doucet, J.P., Nakabeppu, Y., Bedard, P.J., et al., 1996. Chronic alterations in dopaminergic neurotransmission produce a persistent elevation of deltaFosB-like protein(s) in both the rodent and primate striatum. European Journal of Neuroscience 8, 365–381.

Dykens, E.M., 2004. Maladaptive and compulsive behavior in Prader–Willi syndrome: New insights from older adults. American Journal of Mental Retardation 109, 142–153.

Dykens, E.M., Smith, A.C., 1998. Distinctiveness and correlates of maladaptive behaviour in children and adolescents with Smith–Magenis syndrome. Journal of Intellectual Disability Research 42 (Pt 6), 481–489.

Dykens, E., Shah, B., 2003. Psychiatric disorders in Prader–Willi syndrome: Epidemiology and management. CNS Drugs 17, 167–178.

Dykens, E.M., Leckman, J.F., Cassidy, S.B., 1996. Obsessions and compulsions in Prader–Willi syndrome. Journal of Child Psychology and Psychiatry 37, 995–1002.

Dykens, E.M., Finucane, B.M., Gayley, C., 1997. Brief report: Cognitive and behavioral profiles in persons with Smith–Magenis syndrome. Journal of Autism and Developmental Disorders 27, 203–211.

Dykens, E.M., Cassidy, S.B., King, B.H., 1999. Maladaptive behavior differences in Prader–Willi syndrome due to paternal deletion versus maternal uniparental disomy. American Journal of Mental Retardation 104, 67–77.

Eddy, C.M., Rizzo, R., Cavanna, A.E., 2009. Neuropsychological aspects of Tourette syndrome: A review. Journal of Psychosomatic Research 67, 503–513.

Eidelberg, D., Moeller, J.R., Antonini, A., et al., 1997. The metabolic anatomy of Tourette's syndrome. Neurology 48, 927–934.

Ercan-Sencicek, A.G., Stillman, A.A., Ghosh, A.K., et al., 2010. L-histidine decarboxylase and Tourette's syndrome. New England Journal of Medicine 362, 1901–1908.

Ernst, A.M., Smelik, P.G., 1966. Site of action of dopamine and apomorphine on compulsive gnawing behaviour in rats. Experientia 22, 837–838.

Esbensen, A.J., Seltzer, M.M., Lam, K.S., Bodfish, J.W., 2009. Age-related differences in restricted repetitive behaviors in autism spectrum disorders. Journal of Autism and Developmental Disorders 39, 57–66.

Evans, D.W., Leckman, J.F., Carter, A., et al., 1997. Ritual, habit, and perfectionism: The prevalence and development of compulsive-like behavior in normal young children. Child Development 68, 58–68.

Evans, D.W., Lewis, M.D., Iobst, E., 2004. The role of the orbitofrontal cortex in normally developing compulsive-like behaviors and obsessive–compulsive disorder. Brain and Cognition 55, 220–234.

Fasano, A., Petrovic, I., 2010. Insights into pathophysiology of punding reveal possible treatment strategies. Molecular Psychiatry 15, 560–573.

Foa, E.B., Liebowitz, M.R., Kozak, M.J., et al., 2005. Randomized, placebo-controlled trial of exposure and ritual prevention, clomipramine, and their combination in the treatment of obsessive–compulsive disorder. American Journal of Psychiatry 162, 151–161.

Freeman, B.J., Ritvo, E.R., Schroth, P.C., Tonick, I., Gurhrie, D., Wake, L., 1981. Behavioral characteristics of high-and-low-IQ autistic children. American Journal of Psychiatry 138, 25–29.

Garner, J.P., Mason, G.J., 2002. Evidence for a relationship between cage stereotypies and behavioural disinhibition in laboratory rodents. Behavioural Brain Research 136, 83–92.

Garner, J.P., Meehan, C.L., Mench, J.A., 2003. Stereotypies in caged parrots, schizophrenia and autism: Evidence for a common mechanism. Behavioural Brain Research 145, 125–134.

Gerfen, C.R., 2000. Molecular effects of dopamine on striatal-projection pathways. Trends in Neurosciences 23 (supplement 10), S64–S70.

Gerfen, C., Engber, T., Mahan, L., et al., 1990. D1 and D2 dopamine receptor-regulated gene expression of striatonigral and striatopallidal neurons. Science 250, 1429–1431.

Grabli, D., Mccairn, K., Hirsch, E.C., et al., 2004. Behavioural disorders induced by external globus pallidus dysfunction in primates: I. Behavioural study. Brain 127, 2039–2054.

Grados, M.A., Riddle, M.A., Samuels, J.F., et al., 2001. The familial phenotype of obsessive–compulsive disorder in relation to tic disorders: The Hopkins OCD family study. Biological Psychiatry 50, 559–565.

Graybiel, A.M., 2008. Habits, rituals, and the evaluative brain. Annual Review of Neuroscience 31, 359–387.

Greenberg, B.D., Gabriels, L.A., Malone Jr., D.A., et al., 2008. Deep brain stimulation of the ventral internal capsule/ventral striatum for obsessive–compulsive disorder: Worldwide experience. Molecular Psychiatry 15, 64–79.

Greer, J.M., Capecchi, M.R., 2002. Hoxb8 is required for normal grooming behavior in mice. Neuron 33, 23–34.

Gross-Isseroff, R., Cohen, R., Sasson, Y., Voet, H., Zohar, J., 2004. Serotonergic dissection of obsessive–compulsive symptoms: A challenge study with m-chlorophenylpiperazine and sumatriptan. Neuropsychobiology 50, 200–205.

Hanna, G., Veenstra-Vander Weele, J., Cox, N., et al., 2002. Genome-wide linkage analysis of families with obsessive–compulsive disorder ascertained through pediatric probands. American Journal of Medical Genetics 114, 541–552.

Hanna, G.L., Veenstra-Vanderweele, J., Cox, N.J., et al., 2007. Evidence for a susceptibility locus on chromosome 10p15 in early-onset obsessive–compulsive disorder. Biological Psychiatry 62, 856–862.

Hermelin, B., O'connor, N., 1963. The response of self-generated behaviour of severely disturbed children and severely subnormal controls. British Journal of Social and Clinical Psychology 2, 37–43.

Hines, R.M., Wu, L., Hines, D.J., et al., 2008. Synaptic imbalance, stereotypies, and impaired social interactions in mice with altered neuroligin 2 expression. Journal of Neuroscience 28, 6055–6067.

Hollander, E., Anagnostou, E., Chaplin, W., et al., 2005. Striatal volume on magnetic resonance imaging and repetitive behaviors in autism. Biological Psychiatry 58, 226–232.

Homanics, G.E., Delorey, T.M., Firestone, L.L., et al., 1997. Mice devoid of gamma-aminobutyrate type A receptor beta3 subunit have epilepsy, cleft palate, and hypersensitive behavior. Proceedings of the National Academy of Sciences of the United States of America 94, 4143–4148.

Hornig, M., Weissenbock, H., Horscroft, N., Lipkin, W.I., 1999. An infection-based model of neurodevelopmental damage. Proceedings of the National Academy of Sciences of the United States of America 96, 12102–12107.

Hudziak, J.J., Van Beijsterveldt, C.E., Althoff, R.R., et al., 2004. Genetic and environmental contributions to the Child Behavior Checklist Obsessive–Compulsive Scale: A cross-cultural twin study. Archives of General Psychiatry 61, 608–616.

Hus, V., Pickles, A., Cook Jr., E.H., Risi, S., Lord, C., 2007. Using the autism diagnostic interview – Revised to increase phenotypic homogeneity in genetic studies of autism. Biological Psychiatry 61, 438–448.

Ingram, J.L., Peckham, S.M., Tisdale, B., Rodier, P.M., 2000. Prenatal exposure of rats to valproic acid reproduces the cerebellar anomalies associated with autism. Neurotoxicology and Teratology 22, 319–324.

Jang, J.H., Kwon, J.S., Jang, D.P., et al., 2006. A proton MRSI study of brain N-acetylaspartate level after 12 weeks of citalopram treatment in drug-naive patients with obsessive–compulsive disorder. American Journal of Psychiatry 163, 1202–1207.

Joseph, R.M., Tager-Flusberg, H., 2004. The relationship of theory of mind and executive functions to symptom type and severity in children with autism. Development and Psychopathology 16, 137–155.

Kalanithi, P.S., Zheng, W., Kataoka, Y., et al., 2005. Altered parvalbumin-positive neuron distribution in basal ganglia of individuals with Tourette syndrome. Proceedings of the National Academy of Sciences of the United States of America 102, 13307–13312.

Karcz-Kubicha, M., Ferre, S., Diaz-Ruiz, O., et al., 2006. Stimulation of adenosine receptors selectively activates gene expression in striatal enkephalinergic neurons. Neuropsychopharmacology 31, 2173–2179.

Karler, R., Bedingfield, J.B., Thai, D.K., Calder, L.D., 1997. The role of the frontal cortex in the mouse in behavioral sensitization to amphetamine. Brain Research 757, 228–235.

Karler, R., Calder, L.D., Thai, D.K., Bedingfield, J.B., 1998. The role of dopamine in the mouse frontal cortex: A new hypothesis of behavioral sensitization to amphetamine and cocaine. Pharmacology, Biochemistry, and Behavior 61, 435–443.

Kataoka, Y., Kalanithi, P.S., Grantz, H., et al., 2010. Decreased number of parvalbumin and cholinergic interneurons in the striatum of individuals with Tourette syndrome. Journal of Comparative Neurology 518, 277–291.

Kates, W.R., Lanham, D.C., Singer, H.S., 2005. Frontal white matter reductions in healthy males with complex stereotypies. Pediatric Neurology 32, 109–112.

Kelley, R.I., Hennekam, R.C., 2000. The Smith–Lemli–Opitz syndrome. Journal of Medical Genetics 37, 321–335.

Kelz, M.B., Chen, J., Carlezon Jr., W.A., et al., 1999. Expression of the transcription factor deltaFosB in the brain controls sensitivity to cocaine. Nature 401, 272–276.

King, B.H., Hollander, E., Sikich, L., et al., 2009. Lack of efficacy of citalopram in children with autism spectrum disorders and high levels of repetitive behavior: Citalopram ineffective in children with autism. Archives of General Psychiatry 66, 583–590.

Kiyatkin, E.A., Rebec, G.V., 1999. Striatal neuronal activity and responsiveness to dopamine and glutamate after selective blockade of D1 and D2 dopamine receptors in freely moving rats. Journal of Neuroscience 19, 3594–3609.

Klieger, P.S., Fett, K.A., Dimitsopulos, T., Kurlan, R., 1997. Asymmetry of basal ganglia perfusion in Tourette's syndrome shown by technetium-99m-HMPAO SPECT. Journal of Nuclear Medicine 38, 188–191.

Kozinetz, C.A., Skender, M.L., Macnaughton, N., et al., 1993. Epidemiology of Rett syndrome: A population-based registry. Pediatrics 91, 445–450.

Kwon, J.S., Joo, Y.H., Nam, H.J., et al., 2009. Association of the glutamate transporter gene SLC1A1 with atypical antipsychotics-induced obsessive–compulsive symptoms. Archives of General Psychiatry 66, 1233–1241.

Lam, K.S., Bodfish, J.W., Piven, J., 2008. Evidence for three subtypes of repetitive behavior in autism that differ in familiality and association with other symptoms. Journal of Child Psychology and Psychiatry 49, 1193–1200.

Langen, M., Schnack, H.G., Nederveen, H., et al., 2009. Changes in the developmental trajectories of striatum in autism. Biological Psychiatry 66, 327–333.

Leckman, J.F., Hardin, M.T., Riddle, M.A., Stevenson, J., Ort, S.I., Cohen, D.J., 1991. Clonidine treatment of Gilles de la Tourette's syndrome. Archives of General Psychiatry 48, 324–328.

Leckman, J.F., King, R.A., Gilbert, D.L., et al., 2011. Streptococcal upper respiratory tract infections and exacerbations of tic and obsessive–compulsive symptoms: A prospective longitudinal study. Journal of the American Academy of Child and Adolescent Psychiatry 50, 108–118.e3.

Leonard, H.L., Swedo, S.E., Rapoport, J.L., et al., 1989. Treatment of obsessive–compulsive disorder with clomipramine and desipramine in children and adolescents: A double-blind crossover comparison. Archives of General Psychiatry 46, 1088–1092.

Leonard, H., Bower, C., English, D., 1997. The prevalence and incidence of Rett syndrome in Australia. European Child and Adolescent Psychiatry 6 (supplement 1), 8–10.

Levesque, M., Parent, A., 2005. The striatofugal fiber system in primates: A reevaluation of its organization based on single-axon tracing studies. Proceedings of the National Academy of Sciences of the United States of America 102, 11888–11893.

Lewis, M.H., 2004. Environmental complexity and central nervous system development and function. Mental Retardation and Developmental Disabilities Research Reviews 10, 91–95.

Lewis, M., Kim, S.-J., 2009. The pathophysiology of repetitive behavior. Journal of Neurodevelopmental Disorders 1, 114–132.

Lewis, M.H., Gluck, J.P., Beauchamp, A.J., Keresztury, M.F., Mailman, R.B., 1990. Long-term effects of early social isolation in Macaca mulatta: Changes in dopamine receptor function following apomorphine challenge. Brain Research. Developmental Brain Research 513, 67–73.

Lewis, M.H., Tanimura, Y., Lee, L.W., Bodfish, J.W., 2007. Animal models of restricted repetitive behavior in autism. Behavioural Brain Research 176, 66–74.

Lopez, B.R., Lincoln, A.J., Ozonoff, S., Lai, Z., 2005. Examining the relationship between executive functions and restricted, repetitive symptoms of autistic disorder. Journal of Autism and Developmental Disorders 35, 445–460.

Lord, C., 1995. Follow-up of two-year-olds referred for possible autism. Journal of Child Psychology and Psychiatry 36, 1365–1382.

Lord, C., Pickles, A., 1996. Language level and nonverbal social-communicative behaviors in autistic and language-delayed children. Journal of the American Academy of Child and Adolescent Psychiatry 35, 1542–1550.

Lord, C., Rutter, M., Le Couteur, A., 1994. Autism Diagnostic Interview – Revised: A revised version of a diagnostic interview for caregivers of individuals with possible pervasive developmental disorders. Journal of Autism and Developmental Disorders 24, 659–685.

Lord, C., Rutter, M., Dilavore, P.C., Risi, S., 1999. The ADOS-G (Autism Diagnostic Observation Schedule-Generic). Western Psychological Services, Los Angeles.

Lord, C., Risi, S., Lambrecht, L., et al., 2000. The autism diagnostic observation schedule-generic: A standard measure of social and communication deficits associated with the spectrum of autism. Journal of Autism and Developmental Disorders 30, 205–223.

Lowry, R.B., Yong, S.L., 1980. Borderline normal intelligence in the Smith-Lemli-Opitz (RSH) syndrome. American Journal of Medical Genetics 5, 137–143.

Mallet, L., Polosan, M., Jaafari, N., et al., 2008. Subthalamic nucleus stimulation in severe obsessive–compulsive disorder. New England Journal of Medicine 359, 2121–2134.

Mandy, W.P., Skuse, D.H., 2008. Research review: What is the association between the social-communication element of autism and repetitive interests, behaviours and activities? Journal of Child Psychology and Psychiatry 49, 795–808.

Martin, L.J., Spicer, D.M., Lewis, M.H., Gluck, J.P., Cork, L.C., 1991. Social deprivation of infant rhesus monkeys alters the chemoarchitecture of the brain: I. Subcortical regions. Journal of Neuroscience 11, 3344–3358.

Martin, L.A., Ashwood, P., Braunschweig, D., Cabanlit, M., Van De Water, J., Amaral, D.G., 2008. Stereotypies and hyperactivity in rhesus monkeys exposed to IgG from mothers of children with autism. Brain, Behavior, and Immunity 22, 806–816.

Mason, G., Rushen, J., 2006. Stereotypies in Captive Animals: Fundamentals and Implications for Welfare. CAB International, Wallingford.

Matsumoto, R., Ichise, M., Ito, H., et al., 2010. Reduced serotonin transporter binding in the insular cortex in patients with obsessive–compulsive disorder: A [11C]DASB PET study. NeuroImage 49, 121–126.

McClung, C.A., Ulery, P.G., Perrotti, L.I., Zachariou, V., Berton, O., Nestler, E.J., 2004. DeltaFosB: A molecular switch for long-term adaptation in the brain. Brain Research – Molecular Brain Research 132, 146–154.

McDougle, C.J., Scahill, L., Aman, M.G., et al., 2005. Risperidone for the core symptom domains of autism: Results from the study by the autism network of the research units on pediatric psychopharmacology. American Journal of Psychiatry 162, 1142–1148.

Menzies, L., Chamberlain, S.R., Laird, A.R., Thelen, S.M., Sahakian, B.J., Bullmore, E.T., 2008. Integrating evidence from neuroimaging and neuropsychological studies of obsessive–compulsive disorder: The orbitofronto-striatal model revisited. Neuroscience and Biobehavioral Reviews 32, 525–549.

Moresco, R.M., Pietra, L., Henin, M., et al., 2007. Fluvoxamine treatment and D2 receptors: A pet study on OCD drug-naive patients. Neuropsychopharmacology 32, 197–205.

Moretti, P., Bouwknecht, J.A., Teague, R., Paylor, R., Zoghbi, H.Y., 2005. Abnormalities of social interactions and home-cage behavior in a mouse model of Rett syndrome. Human Molecular Genetics 14, 205–220.

Moss, J., Oliver, C., Arron, K., Burbidge, C., Berg, K., 2008. The prevalence and phenomenology of repetitive behavior in genetic syndromes. Journal of Autism and Developmental Disorders 39, 572–588.

Moy, S.S., Nadler, J.J., Poe, M.D., et al., 2008a. Development of a mouse test for repetitive, restricted behaviors: Relevance to autism. Behavioural Brain Research 188, 178–194.

Moy, S.S., Nadler, J.J., Young, N.B., et al., 2008b. Social approach and repetitive behavior in eleven inbred mouse strains. Behavioural Brain Research 191, 118–129.

Murphy, T.K., Kurlan, R., Leckman, J., 2010. The immunobiology of Tourette's disorder, pediatric autoimmune neuropsychiatric disorders associated with Streptococcus, and related disorders: A way forward. Journal of Child and Adolescent Psychopharmacology 20, 317–331.

Nelson, A., Killcross, S., 2006. Amphetamine exposure enhances habit formation. Journal of Neuroscience 26, 3805–3812.

Nestadt, G., Samuels, J., Riddle, M., et al., 2000. A family study of obsessive–compulsive disorder. Archives of General Psychiatry 57, 358–363.

Nestler, E.J., Kelz, M.B., Chen, J., 1999. DeltaFosB: A molecular mediator of long-term neural and behavioral plasticity. Brain Research 835, 10–17.

Nestler, E.J., Barrot, M., Self, D.W., 2001. DeltaFosB: A sustained molecular switch for addiction. Proceedings of the National Academy of Sciences of the United States of America 98, 11042–11046.

O'Hearn, K., Asato, M., Ordaz, S., Luna, B., 2008. Neurodevelopment and executive function in autism. Development and Psychopathology 20, 1103–1132.

O'Roak, B.J., Morgan, T.M., Fishman, D.O., et al., 2010. Additional support for the association of SLITRK1 var321 and Tourette syndrome. Molecular Psychiatry 15, 447–450.

O'Rourke, J.A., Scharf, J.M., Yu, D., Pauls, D.L., 2009. The genetics of Tourette syndrome: A review. Journal of Psychosomatic Research 67, 533–545.

O'Sullivan, S.S., Evans, A.H., Lees, A.J., 2007. Punding in Parkinson's disease. Practical Neurology 7, 397–399.

Olanow, C.W., Schapira, A.H., Roth, T., 2000. Waking up to sleep episodes in Parkinson's disease. Movement Disorders 15, 212–215.

Ozaki, N., Goldman, D., Kaye, W., et al., 2003. A missense mutation in the serotonin transporter is associated with a complex neuropsychiatric phenotype. Molecular Psychiatry 8, 933–936.

Pauls, D.L., Leckman, J.F., 1986. The inheritance of Gilles de la Tourette's syndrome and associated behaviors. New England Journal of Medicine 315, 993–997.

Pauls, D.L., Towbin, K.E., Leckman, J.F., Zahner, G.E., Cohen, D.J., 1986. Gilles de la Tourette's syndrome and obsessive–compulsive disorder. Evidence supporting a genetic relationship. Archives of General Psychiatry 43, 1180–1182.

Pediatric OCD Treatment Study Team, 2004. Cognitive-behavior therapy, sertraline, and their combination for children and adolescents with obsessive–compulsive disorder: The Pediatric OCD Treatment Study (POTS) randomized controlled trial. Journal of the American Medical Association 292, 1969–1976.

Perani, D., Garibotto, V., Gorini, A., et al., 2008. In vivo PET study of 5HT(2A) serotonin and D(2) dopamine dysfunction in drug-naive obsessive–compulsive disorder. NeuroImage 42, 306–314.

Perlmutter, S.J., Leitman, S.F., Garvey, M.A., et al., 1999. Therapeutic plasma exchange and intravenous immunoglobulin for obsessive–compulsive disorder and tic disorders in childhood. Lancet 354, 1153–1158.

Peterson, B.S., Skudlarski, P., Anderson, A.W., et al., 1998. A functional magnetic resonance imaging study of tic suppression in Tourette syndrome. Archives of General Psychiatry 55, 326–333.

Peterson, B.S., Staib, L., Scahill, L., et al., 2001. Regional brain and ventricular volumes in Tourette syndrome. Archives of General Psychiatry 58, 427–440.

Peterson, B.S., Thomas, P., Kane, M.J., et al., 2003. Basal Ganglia volumes in patients with Gilles de la Tourette syndrome. Archives of General Psychiatry 60, 415–424.

Piacentini, J., Woods, D.W., Scahill, L., et al., 2010. Behavior therapy for children with Tourette disorder: A randomized controlled trial. Journal of the American Medical Association 303, 1929–1937.

Pierce, K., Courchesne, E., 2001. Evidence for a cerebellar role in reduced exploration and stereotyped behavior in autism. Biological Psychiatry 49, 655–664.

Pittenger, C., Krystal, J.H., Coric, V., 2006. Glutamate-modulating drugs as novel pharmacotherapeutic agents in the treatment of obsessive–compulsive disorder. NeuroRx 3, 69–81.

Plessen, K.J., Bansal, R., Peterson, B.S., 2009. Imaging evidence for anatomical disturbances and neuroplastic compensation in persons with Tourette syndrome. Journal of Psychosomatic Research 67, 559–573.

Presti, M.F., Lewis, M.H., 2005. Striatal opioid peptide content in an animal model of spontaneous stereotypic behavior. Behavioural Brain Research 157, 363–368.

Presti, M.F., Mikes, H.M., Lewis, M.H., 2003. Selective blockade of spontaneous motor stereotypy via intrastriatal pharmacological manipulation. Pharmacology, Biochemistry, and Behavior 74, 833–839.

Presti, M.F., Gibney, B.C., Lewis, M.H., 2004. Effects of intrastriatal administration of selective dopaminergic ligands on spontaneous stereotypy in mice. Physiology and Behavior 80, 433–439.

Price, R.A., Kidd, K.K., Cohen, D.J., Pauls, D.L., Leckman, J.F., 1985. A twin study of Tourette syndrome. Archives of General Psychiatry 42, 815–820.

Pujol, J., Soriano-Mas, C., Alonso, P., et al., 2004. Mapping structural brain alterations in obsessive–compulsive disorder. Archives of General Psychiatry 61, 720–730.

Radua, J., Mataix-Cols, D., 2009. Voxel-wise meta-analysis of grey matter changes in obsessive–compulsive disorder. British Journal of Psychiatry 195, 393–402.

Rafaeli-Mor, N., Foster, L., Berkson, G., 1999. Self-reported body-rocking and other habits in college students. American Journal of Mental Retardation 104, 1–10.

Ramocki, M.B., Peters, S.U., Tavyev, Y.J., et al., 2009. Autism and other neuropsychiatric symptoms are prevalent in individuals with MeCP2 duplication syndrome. Annals of Neurology 66, 771–782.

Ramocki, M.B., Tavyev, Y.J., Peters, S.U., 2010. The MECP2 duplication syndrome. American Journal of Medical Genetics Part A 152A, 1079–1088.

Reimold, M., Smolka, M.N., Zimmer, A., et al., 2007. Reduced availability of serotonin transporters in obsessive–compulsive disorder correlates with symptom severity – A [11C]DASB PET study. Journal of Neural Transmission 114, 1603–1609.

Richler, J., Huerta, M., Bishop, S.L., Lord, C., 2010. Developmental trajectories of restricted and repetitive behaviors and interests in children with autism spectrum disorders. Development and Psychopathology 22, 55–69.

Rickards, H., 2009. Functional neuroimaging in Tourette syndrome. Journal of Psychosomatic Research 67, 575–584.

Riddle, M.A., Rasmusson, A.M., Woods, S.W., Hoffer, P.B., 1992. SPECT imaging of cerebral blood flow in Tourette syndrome. Advances in Neurology 58, 207–211.

Robertson, M.M., 2008. The prevalence and epidemiology of Gilles de la Tourette syndrome. Part 1: The epidemiological and prevalence studies. Journal of Psychosomatic Research 65, 461–472.

Rodier, P.M., Ingram, J.L., Tisdale, B., Croog, V.J., 1997. Linking etiologies in humans and animal models: Studies of autism. Reproductive Toxicology 11, 417–422.

Rojas, D.C., Peterson, E., Winterrowd, E., Reite, M.L., Rogers, S.J., Tregellas, J.R., 2006. Regional gray matter volumetric changes in autism associated with social and repetitive behavior symptoms. BMC Psychiatry 6, 56.

Ronald, A., Happe, F., Bolton, P., et al., 2006. Genetic heterogeneity between the three components of the autism spectrum: A twin study. Journal of the American Academy of Child and Adolescent Psychiatry 45, 691–699.

Rosenberg, D.R., Macmaster, F.P., Keshavan, M.S., Fitzgerald, K.D., Stewart, C.M., Moore, G.J., 2000. Decrease in caudate glutamatergic concentrations in pediatric obsessive–compulsive disorder patients taking paroxetine. Journal of the American Academy of Child and Adolescent Psychiatry 39, 1096–1103.

Russell, A.J., Mataix-Cols, D., Anson, M., Murphy, D.G., 2005. Obsessions and compulsions in Asperger syndrome and high-functioning autism. British Journal of Psychiatry 186, 525–528.

Ryan, A.K., Bartlett, K., Clayton, P., et al., 1998. Smith-Lemli-Opitz syndrome: A variable clinical and biochemical phenotype. Journal of Medical Genetics 35, 558–565.

Sa, A.R., Hounie, A.G., Sampaio, A.S., Arrais, J., Miguel, E.C., Elkis, H., 2009. Obsessive–compulsive symptoms and disorder in patients with schizophrenia treated with clozapine or haloperidol. Comprehensive Psychiatry 50, 437–442.

Samuels, J., Shugart, Y.Y., Grados, M.A., et al., 2007. Significant linkage to compulsive hoarding on chromosome 14 in families with obsessive–compulsive disorder: Results from the OCD Collaborative Genetics Study. American Journal of Psychiatry 164, 493–499.

Santini, E., Alcacer, C., Cacciatore, S., et al., 2009. L-DOPA activates ERK signaling and phosphorylates histone H3 in the striatonigral medium spiny neurons of hemiparkinsonian mice. Journal of Neurochemistry 108, 621–633.

Saxena, S., Brody, A.L., Ho, M.L., et al., 2002. Differential cerebral metabolic changes with paroxetine treatment of obsessive–compulsive disorder vs major depression. Archives of General Psychiatry 59, 250–261.

Scheel-Kruger, J., Arnt, J., Braestrup, C., Christensen, A.V., Cools, A.R., Magelund, G., 1978. GABA-dopamine interaction in substantia nigra and nucleus accumbens – Relevance to behavioral stimulation and stereotyped behavior. Advances in Biochemical Psychopharmacology 19, 343–346.

Schneider, T., Przewlocki, R., 2005. Behavioral alterations in rats prenatally exposed to valproic acid: Animal model of autism. Neuropsychopharmacology 30, 80–89.

Schwartz, J.M., Stoessel, P.W., Baxter Jr., L.R., Martin, K.M., Phelps, M.E., 1996. Systematic changes in cerebral glucose metabolic rate after successful behavior modification treatment of obsessive–compulsive disorder. Archives of General Psychiatry 53, 109–113.

Sears, L.L., Vest, C., Mohamed, S., Bailey, J., Ranson, B.J., Piven, J., 1999. An MRI study of the basal ganglia in autism. Progress in Neuro-Psychopharmacology and Biological Psychiatry 23, 613–624.

Shafritz, K.M., Dichter, G.S., Baranek, G.T., Belger, A., 2008. The neural circuitry mediating shifts in behavioral response and cognitive set in autism. Biological Psychiatry 63, 974–980.

Shahbazian, M., Young, J., Yuva-Paylor, L., et al., 2002. Mice with truncated MeCP2 recapitulate many Rett syndrome features and display hyperacetylation of histone H3. Neuron 35, 243.

Shao, Y., Raiford, K.L., Wolpert, C.M., et al., 2002. Phenotypic homogeneity provides increased support for linkage on chromosome 2 in autistic disorder. American Journal of Human Genetics 70, 1058–1061.

Shao, Y., Cuccaro, M.L., Hauser, E.R., et al., 2003. Fine mapping of autistic disorder to chromosome 15q11-q13 by use of phenotypic subtypes. American Journal of Human Genetics 72, 539–548.

Shmelkov, S.V., Hormigo, A., Jing, D., et al., 2010. Slitrk5 deficiency impairs corticostriatal circuitry and leads to obsessive–compulsive-like behaviors in mice. Nature Medicine 16, 598–602.

Shugart, Y.Y., Samuels, J., Willour, V.L., et al., 2006. Genomewide linkage scan for obsessive–compulsive disorder: Evidence for susceptibility loci on chromosomes 3q, 7p, 1q, 15q, and 6q. Molecular Psychiatry 11, 763–770.

Silverman, J., Smith, C., Schmeidler, J., et al., 2002. Symptom domains in autism and related conditions: Evidence for familiality. American Journal of Medical Genetics 114, 64–73.

Silverman, J.L., Tolu, S.S., Barkan, C.L., Crawley, J.N., 2010. Repetitive self-grooming behavior in the BTBR mouse model of autism is blocked by the mGluR5 antagonist MPEP. Neuropsychopharmacology 35, 976–989.

Singer, H.S., 2009. Motor stereotypies. Seminars in Pediatric Neurology 16, 77–81.

Singer, H.S., Hahn, I.H., Moran, T.H., 1991. Abnormal dopamine uptake sites in postmortem striatum from patients with Tourette's syndrome. Annals of Neurology 30, 558–562.

Singer, H.S., Szymanski, S., Giuliano, J., et al., 2002. Elevated intrasynaptic dopamine release in Tourette's syndrome measured by PET. American Journal of Psychiatry 159, 1329–1336.

Snider, L.A., Lougee, L., Slattery, M., Grant, P., Swedo, S.E., 2005. Antibiotic prophylaxis with azithromycin or penicillin for childhood-onset neuropsychiatric disorders. Biological Psychiatry 57, 788–792.

South, M., Ozonoff, S., Mcmahon, W.M., 2005. Repetitive behavior profiles in Asperger syndrome and high-functioning autism. Journal of Autism and Developmental Disorders 35, 145–158.

State, M.W., Dykens, E.M., 2000. Genetics of childhood disorders: XV. Prader–Willi syndrome: Genes, brain, and behavior. Journal of the American Academy of Child and Adolescent Psychiatry 39, 797–800.

Steeves, T.D., Ko, J.H., Kideckel, D.M., et al., 2010. Extrastriatal dopaminergic dysfunction in Tourette syndrome. Annals of Neurology 67, 170–181.

Steffenburg, S., Gillberg, C., Hellgren, L., et al., 1989. A twin study of autism in Denmark, Finland, Iceland, Norway and Sweden. Journal of Child Psychology and Psychiatry 30, 405–416.

Steiner, H., Gerfen, C.R., 1998. Role of dynorphin and enkephalin in the regulation of striatal output pathways and behavior. Experimental Brain Research 123, 60–76.

Sundaram, S.K., Huq, A.M., Wilson, B.J., Chugani, H.T., 2010. Tourette syndrome is associated with recurrent exonic copy number variants. Neurology 74, 1583–1590.

Swain, J.E., Scahill, L., Lombroso, P.J., King, R.A., Leckman, J.F., 2007. Tourette syndrome and tic disorders: A decade of progress. Journal of the American Academy of Child and Adolescent Psychiatry 46, 947–968.

Swedo, S.E., Leonard, H.L., Garvey, M., et al., 1998. Pediatric autoimmune neuropsychiatric disorders associated with streptococcal infections: Clinical description of the first 50 cases. American Journal of Psychiatry 155, 264–271.

Szatmari, P., Bartolucci, G., Bremner, R., 1989. Asperger's syndrome and autism: Comparison of early history and outcome. Developmental Medicine and Child Neurology 31, 709–720.

Szatmari, P., Georgiades, S., Bryson, S., et al., 2006. Investigating the structure of the restricted, repetitive behaviours and interests domain of autism. Journal of Child Psychology and Psychiatry 47, 582–590.

Szeszko, P.R., Christian, C., Macmaster, F., et al., 2008. Gray matter structural alterations in psychotropic drug-naive pediatric obsessive-compulsive disorder: An optimized voxel-based morphometry study. American Journal of Psychiatry 165, 1299–1307.

Tadevosyan-Leyfer, O., Dowd, M., Mankoski, R., et al., 2003. A principal components analysis of the autism diagnostic interview-revised. Journal of the American Academy of Child and Adolescent Psychiatry 42, 864–872.

Tanimura, Y., Yang, M.C., Lewis, M.H., 2008a. Procedural learning and cognitive flexibility in a mouse model of restricted, repetitive behaviour. Behavioural Brain Research 189, 250–256.

Tanimura, Y., Yang, M.C., Lewis, M.H., 2008b. Procedural learning and cognitive flexibility in a mouse model of restricted, repetitive behaviour. Behavioural Brain Research 189(2), 250-256.

Thakkar, K.N., Polli, F.E., Joseph, R.M., et al., 2008. Response monitoring, repetitive behaviour and anterior cingulate abnormalities in autism spectrum disorders (ASD). Brain 131, 2464–2478.

Thelen, E., 1980. Determinants of amounts of stereotyped behavior in normal human infants. Ethology and Sociobiology 1, 141–150.

Thompson, T., Gray, D.B. (Eds.), 1994. Destructive Behavior in Developmental Disabilities: Diagnosis and Treatment. Sage, Thousand Oaks, CA.

Tierney, E., Nwokoro, N.A., Porter, F., 1999. The behavioral phenotype of Smith-Lemli-Opitz syndrome. In: 46th Annual Meeting of the American Academy of Child and Adolescent Psychiatry, Chicago.

Tierney, E., Nwokoro, N.A., Kelley, R.I., 2000. Behavioral phenotype of RSH/Smith-Lemli-Opitz syndrome. Mental Retardation and Developmental Disabilities Research Reviews 6, 131–134.

Torrado M, Araoz V, Baialardo E, et al. (2006) Clinical-etiologic correlation in children with Prader–Willi syndrome (PWS): An interdisciplinary study. American Journal of Medical Genetics Part A 143(5), 460–468.

Tourette Syndrome Association International Consortium for Genetics, 2007. Genome scan for Tourette disorder in affected-sibling-pair and multigenerational families. American Journal of Human Genetics 80, 265–272.

Turner, M., 1997. Towards an executive dysfunction account of repetitive behaviour in autism. In: Russell, J. (Ed.), Autism as an Executive Disorder. Oxford University Press, Oxford.

Turner, M., 1999. Annotation: Repetitive behaviour in autism: a review of psychological research. Journal of Child Psychology and Psychiatry 40, 839–849.

Turner, C.A., Presti, M.F., Newman, H.A., Bugenhagen, P., Crnic, L., Lewis, M.H., 2001. Spontaneous stereotypy in an animal model of Down syndrome: Ts65Dn mice. Behavior Genetics 31, 393–400.

Veenstra-Vanderweele, J., Cook, E.H., 2004. Molecular genetics of autism spectrum disorder. Molecular Psychiatry 9, 819–832.

Veltman, M.W., Thompson, R.J., Roberts, S.E., Thomas, N.S., Whittington, J., Bolton, P.F., 2004. Prader–Willi syndrome – A study comparing deletion and uniparental disomy cases with reference to autism spectrum disorders. European Child and Adolescent Psychiatry 13, 42–50.

Veltman, M.W., Craig, E.E., Bolton, P.F., 2005. Autism spectrum disorders in Prader–Willi and Angelman syndromes: A systematic review. Psychiatric Genetics 15, 243–254.

Vickery, S., Mason, G.J., 2005. Behavioural persistence in captive bears: A reply to Criswell and Galbreath. Ursus 16, 274–279.

Voon, V., Fernagut, P.O., Wickens, J., et al., 2009. Chronic dopaminergic stimulation in Parkinson's disease: From dyskinesias to impulse control disorders. Lancet Neurology 8, 1140–1149.

Voyiaziakis E, Evgrafov O, Li, D, et al., 2011. Association of SLC6A4 variants with obsessive–compulsive disorder in a large multicenter US family study. Molecular Psychiatry 16(1), 108–120.

Watt, N., Wetherby, A.M., Barber, A., Morgan, L., 2008. Repetitive and stereotyped behaviors in children with autism spectrum disorders in the second year of life. Journal of Autism and Developmental Disorders 38, 1518–1533.

Webb, T., Whittington, J., Clarke, D., Boer, H., Butler, J., Holland, A., 2002. A study of the influence of different genotypes on the physical and behavioral phenotypes of children and adults ascertained clinically as having PWS. Clinical Genetics 62, 273–281.

Welch, J.M., Lu, J., Rodriguiz, R.M., et al., 2007. Cortico-striatal synaptic defects and OCD-like behaviours in Sapap3-mutant mice. Nature 448, 894–900.

Wendland, J.R., Moya, P.R., Kruse, M.R., et al., 2008. A novel, putative gain-of-function haplotype at SLC6A4 associates with obsessive–compulsive disorder. Human Molecular Genetics 17, 717–723.

Wendland, J.R., Moya, P.R., Timpano, K.R., et al., 2009. A haplotype containing quantitative trait loci for SLC1A1 gene expression and its association with obsessive–compulsive disorder. Archives of General Psychiatry 66, 408–416.

Werme, M., Messer, C., Olson, L., et al., 2002. Delta FosB regulates wheel running. Journal of Neuroscience 22, 8133–8138.

White, N.M., Hiroi, N., 1998. Preferential localization of self-stimulation sites in striosomes/patches in the rat striatum. Proceedings of the National Academy of Sciences of the United States of America 95, 6486–6491.

Whiteside, S.P., Port, J.D., Abramowitz, J.S., 2004. A meta-analysis of functional neuroimaging in obsessive–compulsive disorder. Psychiatry Research 132, 69–79.

Whitman, B.Y., Accardo, P., 1987. Emotional symptoms in Prader–Willi syndrome adolescents. American Journal of Medical Genetics 28, 897–905.

Willour, V.L., Yao Shugart, Y., Samuels, J., et al., 2004. Replication study supports evidence for linkage to 9p24 in obsessive–compulsive disorder. American Journal of Human Genetics 75, 508–513.

Wing, L., Gould, J., 1979. Severe impairments of social interaction and associated abnormalities in children: Epidemiology and classification. Journal of Autism and Developmental Disorders 9, 11–29.

Winter, C., Mundt, A., Jalali, R., et al., 2008. High frequency stimulation and temporary inactivation of the subthalamic nucleus reduce quinpirole-induced compulsive checking behavior in rats. Experimental Neurology 210, 217–228.

Wong, D.F., Brasic, J.R., Singer, H.S., et al., 2008. Mechanisms of dopaminergic and serotonergic neurotransmission in Tourette syndrome: Clues from an in vivo neurochemistry study with PET. Neuropsychopharmacology 33, 1239–1251.

Wurbel, H., 2001. Ideal homes? Housing effects on rodent brain and behaviour. Trends in Neurosciences 24, 207–211.

Yoon, D.Y., Gause, C.D., Leckman, J.F., Singer, H.S., 2007. Frontal dopaminergic abnormality in Tourette syndrome: A postmortem analysis. Journal of Neurological Sciences 255, 50–56.

Yucel, M., Harrison, B.J., Wood, S.J., et al., 2007. Functional and biochemical alterations of the medial frontal cortex in obsessive–compulsive disorder. Archives of General Psychiatry 64, 946–955.

Zitterl, W., Aigner, M., Stompe, T., et al., 2007. [123I]-beta-CIT SPECT imaging shows reduced thalamus-hypothalamus serotonin transporter availability in 24 drug-free obsessive–compulsive checkers. Neuropsychopharmacology 32, 1661–1668.

Disorders of Cognitive Control

B.J. Casey, N. Franklin, M.M. Cohen

Sackler Institute for Developmental Psychobiology, New York, NY, USA

39.1 INTRODUCTION

A core feature of several neurodevelopmental disorders is the difficulty in overriding or suppressing inappropriate thoughts and behaviors in favor of appropriate ones. This ability is referred to as cognitive control. Examples of disorders of cognitive control include attention deficit hyperactivity disorder (ADHD), which is characterized by both distractibility and impulsivity; Tourette syndrome, which is characterized by difficulty suppressing repetitive movements and vocalizations that may be complex, emotionally provocative, and exacerbated by stressful situations; obsessive–compulsive disorder (OCD), which is characterized by intrusive thoughts and ritualistic behaviors; and schizophrenia, which involves disorganized thoughts, delusions, or hallucinations and difficulty suppressing them. Cognitive control problems are found in a range of disabilities with limited ability to regulate attention, thought, behavior, or emotions. The number of disorders with cognitive control problems underscores the need for a clearer understanding of the development and neurobiological bases of cognitive control.

Most theoretical and neuroanatomical accounts of cognitive control have focused on the role of the prefrontal cortex in this ability. This focus is based largely on the protracted development of the prefrontal cortex that coincides with cognitive maturity and the neuropsychological literature showing that frontal lobe damage impairs the ability to regulate behavior and suppress inappropriate thoughts or actions (e.g., Phineas Gage, as cited in Harlow, 1869). Although the approach has been important in describing problems in cognitive control, it has not captured the biological basis of how behavior is regulated or how it breaks down in disorders of cognitive control.

This chapter integrates findings on the neurobiological basis of disorders of cognitive control, highlighting the role of cortical and subcortical brain regions in signaling and implementing control. At first glance, the involvement of a number of the same brain regions across many disorders would seem too disparate to be informative. However, considering these findings in the context of cognitive control theory and development paints a clearer picture of each region's contribution to the regulation of behavior and potential disruption in different developmental disorders.

Neural Circuit Development and Function in the Brain: Comprehensive Developmental Neuroscience, Volume 3 http://dx.doi.org/10.1016/B978-0-12-397267-5.00119-9

© 2013 Elsevier Inc. All rights reserved.

39.2 DEVELOPMENT OF COGNITIVE CONTROL

A key feature of cognitive development is a steady increase in the ability to suppress irrelevant information and inappropriate actions in favor of appropriate ones. This ability becomes more efficient throughout childhood and adolescence. Failure to develop this ability results in cognition that is susceptible to interference from competing external or internal information without resolution.

A classic example of the child's developing ability to resolve behavioral conflict is demonstrated by the Piagetian A-not-B task. In this task, the child reaches for a hidden toy in one location (a covered well) and is then required to find the hidden toy in a new location. The infant continues to reach in the old well even though the toy is hidden in full view of the infant and even though the child may look in the correct new direction. This failure has been interpreted as reflecting inhibitory inefficiency rather than a lack of object permanence, as originally proposed by Piaget (Diamond, 1988). As such, obtaining the goal of reaching and finding the toy in the original location biases motor systems in that direction that compete with the new goal and the means to reach in the new opposing direction. Development of the prefrontal cortex has been implicated in the development of this goal-directed behavior.

The development of the ability to override inappropriate actions in favor of appropriate ones has a protracted course of development. In older children, this ability is measured by developmentally appropriate versions of adult neuropsychological tasks including go/no-go, working memory, and attention-switching tasks. In all cases, children have a more difficult time ignoring or suppressing irrelevant salient information or responses in favor of the relevant ones than adults. Performance on these tasks shows a developmental trend over the ages of 4–12 years, approximating adult levels by 12–13 years of age, as indexed by mean reaction times and accuracy rates (Enns and Cameron, 1987).

The age-related differences in performance of cognitive tasks are not observed on tasks in the absence of competing information. For example, tasks that measure the ability to detect and predict statistical regularities in the environment are mastered by infancy (Saffran et al., 1996). This ability is indicated by anticipatory saccades in the direction of the expected stimulus or by a longer duration of looking in the direction of an expected stimulus, when it is not presented (violation in expectation). Learning about one's environment is essential for adaptive functioning, as well as for the development of neural specialization and regulatory ability during toddler, childhood, and adolescent years. Knowing when or what or in which context to expect an event is critical for planning and maintaining appropriate actions in different contexts over time. Adjusting behavior when these expectations are violated is an essential element of cognitive control and an aspect of cognition that shows more protracted development (Mayr et al., 2005).

If this ability develops differently, or along a different trajectory, it may contribute to difficulties in the maturation of self-control abilities (or the concomitant neural systems in which these abilities are instantiated). Thus, with development, these cognitive systems become more differentiated as they are modulated both by experience and by the top-down cortical projections from the prefrontal cortex that help the organism alter behavior when these predictions are violated (Casey, 2005; Casey et al., 2006).

39.3 BRAIN DEVELOPMENT AND COGNITIVE CONTROL

A significant amount of brain development occurs *in utero*, but changes continue postnatally. These postnatal changes coincide with changes in cognitive control. This period is characterized by rapid synapse formation that begins well before birth in nonhuman primates (Rakic, 1974) and results in overproduction of synapses relative to its adult state. This process of synaptogenesis appears to occur concurrently across diverse regions of the nonhuman primate cerebral cortex (Rakic et al., 1986). In both human and nonhuman primate studies, the early synaptic density peaks are followed by a plateau phase that decreases during childhood and into adulthood. The plateau and pruning phases of some cortical regions (e.g., prefrontal cortex) in primates are relatively protracted in comparison to others (e.g., sensorimotor and subcortical regions; Bourgeois et al., 1994; Huttenlocher and Dabholkar, 1997). Positron emission tomography (PET) studies of glucose metabolism suggest that maturation of local metabolic rates parallel the time course of overproduction and subsequent pruning of synapses (Chugani et al., 1987). These studies imply different time courses in regional brain development.

Magnetic resonance imaging (MRI) technologies have introduced a new set of tools for capturing features of brain development in living, developing humans. MRI is particularly well suited to the study of children, as it provides exquisitely accurate anatomical images without the use of ionizing radiation. However, these methods lack the resolution to definitively characterize the mechanism of change with development (e.g., dendritic arborization, synaptic pruning, and myelination). The most informative studies to date are those based on volumetric measures and large sample sizes (Giedd et al., 1999a; Sowell et al., 2003). These studies have yielded three consistent findings. First, total cerebral volume shows little significant change after 6 years of age. Second, there is a significant decrease in cortical gray matter by approximately 12 years of age in prefrontal and association cortices that is preceded by earlier maturing cortical development in sensorimotor regions. Finally, there is an increase in cerebral white matter throughout childhood and young adulthood, especially in prefrontal white matter tracts (Klingberg et al., 1999).

The protracted development of prefrontal and association cortices, along with white matter fiber tract development in this circuitry, contributes to children's developing capacity for cognitive control, as shown by diffusion tensor imaging (DTI). DTI is a measure sensitive to myelination and neuroanatomical changes in white matter microstructure (Liston et al., 2006). Further, variability in the myelination and regularity of prefrontal white matter fibers contribute to individual differences in cognitive control and have been linked to disorders of cognitive control such as ADHD (Casey et al., 2007a).

To investigate neural circuits underlying disruption of cognitive control in developmental disorders more directly, an *in vivo* assessment of the physiological time-course of behavior is needed. Functional magnetic resonance imaging (fMRI) provides this ability (Logothetis et al., 2001). These studies show that children recruit distinct but often larger, more diffuse brain regions when performing cognitive control tasks than do adults. The pattern of activity within brain regions central to cognitive control performance, such as prefrontal cortex, becomes more focal or fine-tuned, based on cross-sectional (Brown et al., 2005) and longitudinal studies (Durston et al., 2004). This pattern of activity is suggestive of development within, and refinement of, projections to and from the prefrontal cortex with maturation. Recent developmental functional connectivity data (Kelly et al., 2009) are consistent with this observation of diffuse correlations among frontal brain regions in children, whereas adults exhibit more focal connections with distal regions.

39.4 DOPAMINE AND COGNITIVE CONTROL

Structural and functional changes in the brain with development occur together with changes in neurotransmitter systems. For example, significant changes are observed in the dopamine system that innervates prefrontal circuitry. The development of the dopaminergic system parallels the development of performance on cognitive control tasks; postnatal innervation of dopaminergic neurons to cortical and subcortical targets peaks at times coincident with the development of cognitive control. Such age-dependent effects of dopamine have been best studied in humans using delayed response

tasks and Piaget's A not B task in typically developing and phenylketonuria (PKU)-affected children (reviewed in Diamond, 1998). The genetic mutation that causes PKU results in decreased levels of tyrosine, a precursor of dopamine, in the central nervous system. In behavioral tasks such as the A not B task, typically developing children show sharp improvements in the first year of life and gradual improvements in age-appropriate versions of the task until age 10; however, in PKU children, the sharp increase in performance is not observed, and only mild improvements occur. In typically developing children, this task is highly dependent on prefrontal cortex activation, a brain area that shows extremely high dopamine turnover, and is thus impaired in PKU children.

In nonhuman primates, the A not B task also shows a sharp increase in performance during the first 6 months of life. Behavioral performance is paralleled by an observed postnatal increase in dopamine levels and in dopamine receptor gene expression (Goldman-Rakic and Brown, 1982; Lidow et al., 1991). An examination of the laminar distribution patterns of tyrosine hydroxylase (TH)-positive processes in prefrontal regions has revealed that a gradual maturation of axons and varicosities occurs up until 2–3 years of age in the monkey (Rosenberg and Lewis, 1995), which is roughly 10–12 years in the human.

In rats with a pharmacologically induced form of PKU, performance is also altered in delayed alternation tasks (Diamond et al., 1994). Neuroanatomical studies in rats have shown that dopaminergic cells in the ventral tegmental area (VTA) and substantia nigra (SN) are fully differentiated and have axons projecting into the prefrontal cortex by embryonic day 16. Only after birth, however, does a dramatic increase in arborization and innervation into deeper cortical layers occur. This postnatal innervation is also paralleled by an increase in dopamine receptor gene expression and in dopamine levels coincident with a postnatal period of rapid synaptogenesis in dendritic spines followed by a slower plateau phase of growth until adolescence (Granger et al., 1995).

Several dopamine-related hypotheses have emerged in an attempt to explain developmental disorders of cognitive control such as ADHD (Swanson et al., 2007) and schizophrenia (Snyder, 1976) based on pharmacological treatments for these disorders. Computational models of dopamine based on animal and human imaging studies suggest that dopamine controls the flow of information from other areas of the brain to the prefrontal cortex in gating and maintaining information (Braver and Cohen, 2000). Both gating and maintenance are important in overriding or suppressing inappropriate thoughts and behaviors in favor of appropriate ones (i.e., cognitive control). Thus, dysfunction in this system can result in dysregulated attention, behavior, and thoughts.

39.5 BRAIN CIRCUITRY IMPLICATED IN COGNITIVE CONTROL

Most functional and developmental studies have focused on the role of the prefrontal cortex in cognitive control. However, both cortical and deep subcortical structures have been implicated in disorders of cognitive control. Considering these findings in the context of development and cognitive control theory may paint a clearer picture of each region's contribution in regulating behavior and its disruption in developmental disorders.

39.5.1 Disorders of Cognitive Control

As depicted in Figure 39.1, clinical neuroimaging studies have identified several brain regions that appear to be altered in childhood disorders of cognitive control. These regions include the prefrontal cortex, anterior cingulate cortex, posterior parietal cortex, basal ganglia, and cerebellum. Abnormalities in these structures have been reported in ADHD (Bush et al., 1999; Castellanos et al., 1996; Durston et al., 2003; Tamm et al., 2006), Tourette syndrome (Leckman et al., 2010), OCD (Baxter et al., 1988; Fitzgerald et al., 2005; Swedo et al., 1992), and childhood-onset and adult schizophrenia (Carter et al., 2001; Frazier et al., 1996).

Abnormalities in size, asymmetry, function, or glucose metabolism are typically reported. For example, MRI volumetric studies of ADHD have revealed abnormalities in the size of the prefrontal cortex, basal ganglia, and cerebellum, showing decreased volumes in each. Decreased activity in these regions as well as in the anterior cingulate and parietal cortices during performance of cognitive control tasks has been reported (Bush et al., 1999; Tamm et al., 2006). Activity in these regions, and cognitive control, is largely normalized with stimulant medications (Epstein et al., 2007; Vaidya et al., 1998), presumably due to increases in the availability of dopamine (Volkow et al., 2001).

PET studies of OCD have revealed hypermetabolic activity in these regions, particularly in the caudate nucleus, anterior cingulate cortex, and orbitofrontal cortex (Baxter et al., 1988; Swedo et al., 1989). Both pharmacological and behavioral treatments have been shown to normalize these patterns of activity. Abnormalities in the basal ganglia, specifically the striatum, in children with Tourette syndrome have been reported in fMRI studies during provocation of symptoms (Peterson et al., 1998). Structural imaging studies have shown cortical thinning in the frontal and parietal cortices in individuals with Tourette syndrome relative to typically developing children (Sowell et al., 2008).

Individuals with schizophrenia show suboptimal levels of activity in prefrontal and parietal regions when performing cognitive control tasks such as the Wisconsin Card Sorting Task or the n-back working memory tasks (Barch and Csernansky, 2007; Berman et al., 1988). MRI-based decreases in volume of the basal ganglia and cerebellum have been reported in this disorder too, especially in those individuals with childhood onset (Frazier et al., 1996; Giedd et al., 1999b). Therefore, a number of cortical and subcortical regions appear to be significantly involved in a range of disorders that have as a key symptom a problem overriding inappropriate actions (i.e., cognitive control).

The presence of common disturbances in cognitive and neural systems across discrete syndromes may be surprising at first. However, there has been increasing concern regarding the validity of the boundaries between discrete syndromes as well as the underlying dimensional nature of specific functional systems underlying these disorders (Frances et al., 1990). Specific disorders may be associated with differential salience, valence, or combinations of the core underlying functional systems. The identification of core processes involved in a disorder can move a field from a disparate set of data-driven findings to a more theoretically coherent collection of findings. Theoretical understanding and identification of specific neurophysiologic function may provide valuable information for validating the core features of and distinctions between psychiatric disorders.

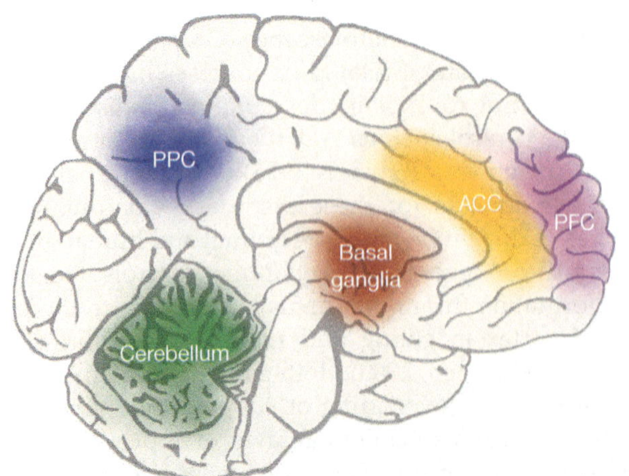

FIGURE 39.1   Brain regions implicated in disorders of cognitive control.

39.5.2 Theoretical Framework: Signaling and Implementation of Control

Cognitive control processes such as the ability to suppress or override competing attentional and behavioral responses have been included in a number of cognitive theories (Cohen and Servan-Schreiber, 1992; Desimone and Duncan, 1995; Shallice, 1988). For example, Shallice (1988) proposed a "supervisory attention system" as a system for inhibiting or replacing routine, reflexive behaviors with more appropriate behaviors. Desimone and Duncan (1995) describe top-down biasing signals as important in attending to relevant information by virtue of mutual inhibition or suppression of irrelevant information. Finally, Miller and Cohen (2001) proposed a model of cognitive control based largely on their respective nonhuman primate and computational modeling studies of prefrontal function. According to this theory, the function of the prefrontal cortex in cognitive control is active maintenance of patterns of activity that represent goals and the means to achieve them. The prefrontal cortex biases relevant sensory and motor systems for goal-directed behavior, through the buildup and integration of rules learned throughout development (Bunge and Zelazo, 2006). Functional imaging studies of cognitive control have focused largely on this form of top-down cortical control in trying to understand the breakdown in regulating actions and biasing of attention in favor of relevant information.

The focus on prefrontal brain regions is based on the neuropsychological literature showing that frontal lobe damage impairs the ability to regulate behavior. Although this approach has been important in describing problems in cognitive control, it has not captured the biological basis of how this ability breaks down in such a broad range of clinical disorders. Figure 39.2 illustrates a theoretical framework that includes the popular construct of top-down implementation of control by the prefrontal cortex. In addition, it provides mechanisms by which control is called. Specifically, different brain regions can signal the prefrontal cortex for control. Each of the regions identified is part of unique circuitry that project both to and from the prefrontal cortex, thus providing a means for signaling prefrontal regions to

help impose top-down control of behavior by biasing signals relevant to goal-related behavior.

Immaturity, developmental delay, or dysfunction within these circuits can lead to cognitive control problems. Ineffective signaling of control systems by any one of these regions could lead to poor regulation of attention or behavior, though with subtle differences, depending on the system impacted. Likewise, intact signaling of prefrontal systems in the presence of inefficient top-down control could result in poor regulation of attention or behavior, though presumably in a more general way. This theoretical framework of cognitive control moves findings of a variety of disparate brain regions' involvement in several disorders toward a cohesive understanding of where cognitive control can break down in these disorders.

The conditions in which these regions may signal the prefrontal cortex are summarized below in terms of frontocortical and frontosubcortical circuitry. Each circuit is generally described. Examples are provided for how each circuit might break down and lead to what appears to be an implementation of control problem, but which could simply be a signaling problem. For example, prefrontal dysfunction could result in a failure of top-down biasing of attention or behavior. Dysfunction in corticocortical signaling from posterior parietal cortex or from anterior cingulate cortex could result in a failure to resolve perceptual conflict among competing inputs or response conflict among competing outputs, respectively. Dysfunction in corticosubcortical signaling from the cerebellum or from the basal ganglia could result in a failure to detect and alter responses to violations in predicted timing or frequency of events, respectively. Finally, a detailed description of one of the frontosubcortical circuits is provided as an example of how a signaling or implementation problem could occur at one of many different locations within a given circuit. This

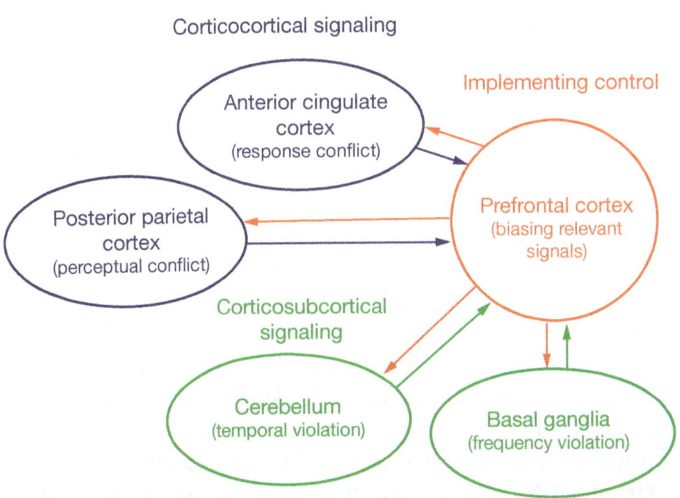

FIGURE 39.2 Locations within prefrontal circuitry where control might break down in disorders of cognitive control. Each region is part of unique circuits that project both to and from the prefrontal cortex, thus providing a means for signaling prefrontal regions to help impose top-down control of behavior and a location where control could break down.

example emphasizes the importance of circuit development and function rather than a homunculus approach of emphasizing a single brain region such as the frontal lobes to explain disorders of cognitive control.

39.5.2.1 Frontal Cortical Control Circuitry

There is an expansive literature implicating the prefrontal cortex in cognitive control from imaging studies (Cohen et al., 1994; D'Espostio et al., 1995; Duncan and Owen, 2000; Smith and Jonides, 1999) to studies of patients with frontal lobe lesions (e.g., Milner et al., 1985; Stuss et al., 1982) to electrophysiological and lesion studies in animals (Fuster, 1980; Mishkin and Pribram, 1955) to computational models (Braver and Cohen, 2000). Each of these approaches consistently supports the involvement of the prefrontal cortex in implementation of cognitive control.

Anterior cingulate and posterior parietal cortices have been shown to be involved in aspects of cognitive control too, especially in overcoming prepotent response tendencies and switching attentional sets (Barber and Carter, 2005; Liston et al., 2006). Activity in both regions has been shown to predict increased prefrontal activity and subsequently enhance behavioral performance in switching behavioral and attentional sets (Hedden and Gabrieli, 2010; Liston et al., 2006). These results are consistent with a network of cortical structures that regulate prefrontal activity by signaling the need for greater control (Birrell and Brown, 2000; Blais and Bunge, 2010; Dias et al., 1996a; Fox et al., 2003; McAlonan and Brown, 2003; O'Reilly et al., 2002).

Each of these brain regions has functionally been dissociated. There is a growing consensus that the prefrontal cortex acts to support task-relevant representations of stimulus information and stimulus–response mappings, thus favoring them in competitions with task-inappropriate representations in posterior cortex (Desimone and Duncan, 1995; Miller and Cohen, 2001). The posterior parietal cortex has been implicated in the generation of motor plans via transformations of sensory inputs from multiple modalities (Andersen and Buneo, 2002) in the service of perceptual decision-making (Gold and Shadlen, 2001; Platt and Glimcher, 1999). As such, this region helps to detect and resolve perceptual conflict. The anterior cingulate cortex, in contrast, has been shown to be involved in the response to conflict detection and its resolution (Botvinick et al., 2001; Posner and Petersen, 1990). One influential theory, known as the conflict-monitoring hypothesis, provides a plausible account of how anterior cingulate and dorsolateral prefrontal cortices act in concert to detect conflict and implement control to resolve it. Accordingly, the anterior cingulate cortex monitors conflicts in information processing and recruits lateral portions of the prefrontal cortex to resolve competition as needed (Botvinick et al., 2001).

While the prefrontal cortex, anterior cingulate, and posterior parietal cortex respond to manipulations of conflict, the role of the anterior cingulate is limited to conflict at the level of the response and not at the level of the stimulus representation (Bhanji et al., 2010; Milham and Banich, 2005). Just as the anterior cingulate is anatomically well situated to detect conflicts at the level of a motor response and signal these to the lateral prefrontal cortex (Barbas and Pandya, 1989; Bates and Goldman-Rakic, 1993), several studies suggest that the posterior parietal cortex is anatomically well suited to detect stimulus conflict and signal this to the prefrontal cortex. The primate posterior parietal cortex receives ample, direct input from the extrastriate visual cortex and sends direct projections to the lateral prefrontal cortex (Wise et al., 1997). Previous studies have emphasized a role for the posterior parietal cortex in detecting unexpected or behaviorally relevant stimuli and facilitating goal-directed attention to task-relevant aspects of a visual stimulus (Corbetta and Shulman, 2002; Corbetta et al., 2000). Together, these findings suggest one mechanism by which these processes may be mediated: detection of conflicts in information processing at the level of the stimulus representation may signal to the prefrontal cortex the need for enhanced top-down control (Casey et al., 2000; Desimone and Duncan, 1995; Dias et al., 1996b; O'Reilly et al., 2002).

These findings are consistent with the anterior cingulate and parietal cortex being involved in signaling the prefrontal cortex in the presence of competing inputs and outputs (perceptual or response conflict) as depicted in Figure 39.2. The ability to detect conflict is necessary for triggering cognitive control to bias the relevant input or output being promoted to resolve the conflict (Miller and Cohen, 2001). If conflict is not detected, then the control system is not signaled and competition between inputs or outputs may not be resolved in the goal-oriented direction (e.g., the schizophrenic child attends to intrusive thoughts over appropriate ones).

39.5.2.2 Frontal Subcortical Control Circuitry

Subcortical regions implicated in disorders of cognitive control include the basal ganglia and cerebellum. The basal ganglia and cerebellum have been implicated in learning about the frequency and the timing of events (i.e., learning what to expect and when). These regions make up frontostriatal and frontocerebellar loops that have similar features. For example, both the cerebellum and the basal ganglia project to the prefrontal cortex via the thalamus. The primary neurotransmitter in both the basal ganglia and the cerebellum is GABA, an inhibitory neurotransmitter. Glutamate is found in the prefrontal cortex and thalamus, which is an excitatory neurotransmitter and dopamine is a critical neuromodulator of both circuits (Braver and Barch, 2002; Cohen et al., 1992;

Montague et al., 1996; Schultz et al., 1997) that is expressed preferentially in portions of the prefrontal cortex, basal ganglia, and dentate nucleus of the cerebellum, all regions implicated in disorders of cognitive control.

These circuits have been shown to support both motor and cognitive behavior with cognitive-related actions being driven by projections from the prefrontal cortex and modulated by input from the dentate nucleus of the cerebellum and from the dorsal and ventral striatum of the basal ganglia (caudate and nucleus accumbens). This circuitry is perhaps most well described within the basal ganglia thalamocortical circuitry for which at least five circuits have been identified (Alexander et al., 1986). The basal ganglia thalamocortical circuits include a motor, oculomotor, prefrontal (dorsolateral and lateral orbital), and limbic circuit. These circuits involve the same projection regions (basal ganglia, thalamus, and cortex), but differ in the exact projection zone within each region and in the set of thoughts and actions they support.

The general organization of these circuits involves prefrontal projections to different portions of the striatum (i.e., putamen or caudate nuclei), which then project to either a direct or an indirect pathway (see Figure 39.3). The direct pathway involves an inhibitory projection to the internal capsule of the globus pallidus (GPi) and substantia nigra (SNr) resulting in the dampening of an inhibitory projection to the thalamus (i.e., disinhibition). The indirect pathway consists of an inhibitory GABA projection to the external capsule of the globus pallidus (GPe) that dampens the inhibitory projection to the subthalamic nuclei (STN) resulting in excitation of the internal capsule of the globus pallidus and substantia nigra. This in turn leads to inhibition of the thalamus. The direct pathway presumably facilitates thalamo-cortically mediated behavior, while the indirect pathway is thought to inhibit thalamo-cortically mediated behavior.

As the prefrontal cortex projects directly to the basal ganglia and cerebellum and as both project back to the prefrontal cortex via the thalamus, an account of suppression of competing actions may be described here in somewhat more mechanistic terms. That is, the basal ganglia and cerebellum have been implicated in monitoring the frequency and/or timing of events (Davidson et al., 2003; Hayes et al., 1998; Ivry and Keele, 1989; McClure et al., 2003; Spencer et al., 2003; Van Mier and Petersen, 2002). The ability to predict what and when an event will occur is an essential component of cognitive control, in planning and maintaining appropriate thoughts and actions in different contexts over time. Maintaining representations of such events and information is critical in suppressing competing ones.

Thus, frontocerebellar and frontostriatal circuits may provide neural mechanisms for the maintenance of representations of events over time. In contrast, detecting violations in such predictions (which presumably allows the system to attend to and learn new information) may be linked to intrinsic inhibitory functions of GABA-related functions of the basal ganglia and cerebellum in the absence of frontally driven planned thoughts and actions. Accordingly, the basal ganglia and cerebellum do not generate planned or voluntary movements or behaviors *per se*, but rather the prefrontal cortex generates these voluntary actions. They do, however, detect violations in the timing and nature of events, providing the system with a way to shift out of prefrontally driven behavior when highly salient events occur. The basal ganglia and cerebellum then act broadly to inhibit competing thoughts and behaviors that would otherwise interfere with the prefrontally driven goal or behavior (Casey, 2000; Mink, 1996).

39.6 DISRUPTION OF COGNITIVE CONTROL CIRCUITRY

How does the previously described circuitry contribute to the symptoms and behaviors observed in the childhood disorders of cognitive control? First, the typical assumption made when a child presents with a cognitive control problem is that it is due to top-down prefrontal dysfunction or inefficiency. For example, disruption of the prefrontally mediated direct pathway within the basal ganglia corticosubcortical circuit described can result in constantly interrupted behaviors such as those observed in ADHD or constantly interrupted thoughts observed in schizophrenia. In contrast, if the basal ganglia and cerebellum are involved in shifting out of prefrontally driven thoughts or behaviors (Redgrave et al., 1999), then their disruption may result in irrepressible repetitive behaviors and thoughts similar to those observed in OCD and Tourette syndrome. Likewise, disruption in cortical signaling from the posterior parietal and anterior cingulate cortex of perceptual or

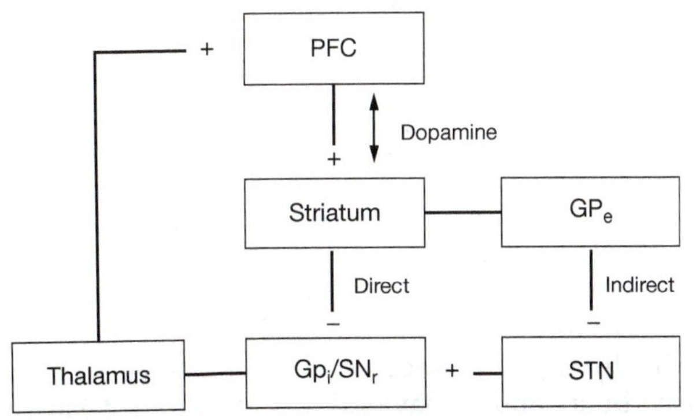

FIGURE 39.3 Direct and indirect pathways of the basal ganglia corticosubcortical circuit involved in cognitive control.

response conflict may result in failure of prefrontal systems to bias appropriate signals in favor of goal-directed behavior. As such, inappropriate attentional or behavioral responses (i.e., distractibility and impulsivity) as observed in ADHD may result (Bush et al., 1999; Tamm et al., 2006). Finally, neuromodulatory (e.g., dopamine) imbalances or deficiency in prefrontal circuitry can lead to problems in cognitive control resulting in constantly interrupted behaviors and thoughts as seen in ADHD and schizophrenia. Several psychiatric and neurologic disorders have been linked to disruptions in specific frontostriatal (Alexander et al., 1986, 1991), frontocerebellar (Dum and Strick, 2003; Middleton and Strick, 2002), and corticocortical loops (Bush et al., 2005; Casey et al., 2007b).

The theoretical framework distinguishing between signaling and implementation of control suggests that information is maintained in an active state over time in the prefrontal cortex by means of recurrent excitatory connectivity (Cohen et al., 1992). The prefrontal cortex, which consists primarily of excitatory projections (glutamate), is thus involved in maintenance of relevant information for action and disruption of this brain region results in deficits in the ability to carry out the relevant actions, as evidenced in ADHD. The basal ganglia and cerebellum, which consist primarily of inhibitory projections (GABA), are involved in switching or shifting attention elsewhere when there is a lack of sufficient prefrontal input to drive the behavior in an organized way. Disruption to these brain regions (basal ganglia and cerebellum) or their development may therefore result in cognitive control deficits related to an inability to shift out of particular behavioral sets (Hayes et al., 1998), as evidenced in OCD and schizophrenia. Likewise, disruption in cortical signaling from the posterior parietal and anterior cingulate cortex of perceptual or response conflict may result in failure of prefrontal systems to bias appropriate signals in favor of goal-directed behavior, as observed in ADHD (Bush et al., 1999; Tamm et al., 2006).

39.7 SUMMARY

A key feature of cognitive development is the gradual increase in the ability to suppress competing thoughts and actions in favor of goal-oriented ones, referred to as cognitive control. This ability is disrupted in several neurodevelopmental disorders including ADHD, OCD, Tourette syndrome, and schizophrenia. The protracted development of the prefrontal cortex has made this region a primary candidate in the study of the development of cognitive control. Further, the neuropsychological literature showing that frontal lobe damage impairs the ability to regulate behavior and suppress inappropriate thoughts or actions has driven the focus of clinical studies.

Recent theoretical and empirical studies suggest that the function of the prefrontal cortex in cognitive control relates to the active maintenance of patterns of activity that represent goals and the means to achieve them. The implementation of control requires signaling of the prefrontal cortex to bias relevant sensory and motor systems for goal-directed behavior. These signals may arise from cortical or subcortical regions. Cortical regions include the posterior parietal cortex and anterior cingulate cortex, which have been implicated in detecting perceptual and response conflict and signaling the prefrontal cortex to implement control. Subcortical signaling regions include the basal ganglia and cerebellum. These regions have been implicated in detecting violations in the expected nature and timing of events, providing the system with a way to shift out of prefrontal-driven behavior when highly salient or novel events occur.

Each region is thus unique and part of an interactive circuitry that project both to and from the prefrontal cortex, thus providing a means for signaling prefrontal regions to help impose top-down control of attention and behavior by biasing signals relevant to goal-related behavior. Immaturity, developmental delay, or dysfunction within these circuits can lead to cognitive control problems. Ineffective signaling of control systems by any one of these regions can lead to poor regulation of attention or behavior, but with subtle differences depending on the system impacted. Ineffective biasing of circuits by control systems can likewise lead to poor regulation of attention or behavior, though in a more general way.

In sum, basic learning and attention systems are important in signaling top-down control systems to adjust attention and behavior when predicted outcomes are violated (Botvinick et al., 1999; Casey et al., 2000). The basic assumption is that learning where, when, or what contexts to expect an event is critical for goal-directed behavior across different contexts over time. Deficits in learning to detect regularities in the environment can lead to less signaling of control systems to help alter or adjust behavior when these expectations are violated. Such deficits can mimic those observed when top-down control systems themselves are impaired.

SEE ALSO

Circuit Development: Cerebellar Circuits. **Cognitive Development**: Developing Attention and Self Regulation in Infancy and Childhood; Statistical Learning Mechanisms in Infancy; Structural Brain Development: Birth Through Adolescence; The Effects of Stress on Early Brain and Behavioral Development. **Diseases**: Neurodevelopmental Genomics of Autism, Schizophrenia and Related Disorders.

Glossary

ADHD Characterized by both distractibility and impulsivity, with frequent overlap of these behaviors; symptoms usually start before 7 years of age.

Anterior cingulate cortex The frontal part of the cingulate cortex with connections to the prefrontal cortex; implicated in response conflict detection and its resolution.

Basal ganglia A group of nuclei situated at the base of the forebrain with strong connections to the prefrontal cortex and thalamus; implicated in learning about the frequency of events.

Cerebellum It is tucked underneath the cerebral hemispheres at the posterior ventral portion of the brain with connections to the prefrontal cortex and thalamus; implicated in learning about the timing of events.

Cognitive control The ability to override or suppress inappropriate thoughts and behaviors in favor of appropriate ones.

Diffusion tensor imaging (DTI) An imaging technique that enables the measurement of the restricted diffusion of water in tissue in order to produce images of neural tracts in the brain.

Direct pathway Involves an inhibitory projection from the striatum to the internal capsule of the globus pallidus (GPi) and substantia nigra (SNr) resulting in the dampening of an inhibitory projection to the thalamus (i.e., disinhibition); implicated in facilitation of thalamo-cortically mediated behavior.

Dopamine A catecholamine neurotransmitter produced in the substantia nigra and ventral tegmental area; acts as a neuromodulator of the frontostriatal and frontocerebellar loops and is implicated in controlling the flow of information from other areas of the brain to the prefrontal cortex in the service of gating and maintaining information.

Functional magnetic resonance imaging (fMRI) A specialized type of MRI that measures the hemodynamic response (change in blood flow) related to neural activity in the brain or spinal cord of humans or other animals.

Glutamate The most abundant excitatory neurotransmitter in the nervous system with high concentrations in the prefrontal cortex and thalamus; implicated in maintenance of information for action.

Go/No-go task A task assessing cognitive control in which a subject presses a button to each target stimulus and must withhold a response when a rare nontarget stimulus appears on screen.

Indirect pathway Consists of an inhibitory GABA projection from the striatum to the external capsule of the globus pallidus (GPe) that dampens the inhibitory projection to the subthalamic nuclei (STN) resulting in excitation of the internal capsule of the globus pallidus and substantia nigra, leading to inhibition of the thalamus; implicated in the inhibition of thalamo-cortically mediated behavior.

Magnetic resonance imaging (MRI) A noninvasive medical imaging technique used in radiology to visualize the detailed internal structure of the body and brain.

N-back working memory task A task assessing cognitive control in which the participant sees stimuli one at a time and must respond whether the stimulus they see is the same stimulus as they saw a predetermined number of trials back.

OCD An anxiety disorder characterized by intrusive thoughts and ritualistic behaviors; age of onset is seen earlier in men, ranges from childhood to young adulthood, and is often before 15 years of age.

Phenylketonuria (PKU) An autosomal recessive metabolic genetic disorder characterized by a deficiency in the hepatic enzyme phenylalanine hydroxylase, which ultimately results in decreased levels of tyrosine, a precursor of dopamine, in the central nervous system.

Piaget's A-not-B task A classic task demonstrating a child's developing ability to resolve behavioral conflict in which an infant reaches for a hidden toy in one location and is then required to find the toy in a new location hidden in full view.

Positron emission tomography (PET) A nuclear medicine imaging technique which produces a three-dimensional image of functional processes in the body through the detection of gamma rays emitted indirectly by a positron-emitting tracer, which is introduced into the body on a biologically active molecule.

Posterior parietal cortex The posterior section of the parietal cortex with connections to the prefrontal cortex; implicated in the generation of motor plans via transformations of sensory inputs from multiple modalities in the service of perceptual decision-making and detecting and resolving perceptual conflict.

Prefrontal cortex Located in the anterior part of the frontal lobes of the brain with widespread connections to many areas of the brain, including the anterior cingulate cortex, the posterior parietal cortex, the basal ganglia, and the cerebellum; implicated in cognitive control capabilities and biasing relevant signals; development is protracted and maturity is not reached until young adulthood.

Schizophrenia A mental illness characterized by disorganized thoughts, delusions, or hallucinations and difficulty suppressing them; symptom onset typically occurs in young adulthood.

Tourette syndrome Characterized by difficulty suppressing repetitive movements and vocalizations that may be complex, emotionally provocative, and exacerbated by stressful situations; disease onset is usually seen in childhood.

Wisconsin card sorting task A task assessing cognitive control in which the participant sorts cards according to rules that change over the course of the experiment.

References

Alexander, G.E., DeLong, M.R., Strick, P.L., 1986. Parallel organization of functionally segregated circuits linking basal ganglia and cortex. Annual Review of Neuroscience 9, 357–381.

Alexander, G.E., Crutcher, M.D., DeLong, M.R., 1991. Basal gangliathalmocortical circuits: Parallel substrates for motor oculomotor, prefrontal and limbic functions. Progress in Brain Research 85, 119–145.

Andersen, R.A., Buneo, C.A., 2002. Intentional maps in posterior parietal cortex. Annual Review of Neuroscience 25, 189–220.

Barbas, H., Pandya, D.N., 1989. Architecture and intrinsic connections of the prefrontal cortex in the rhesus monkey. The Journal of Comparative Neurology 286, 353–375.

Barber, A.D., Carter, C.S., 2005. Cognitive control involved in overcoming prepotent response tendencies and switching between tasks. Cerebral Cortex 15, 899–912.

Barch, D.M., Csernansky, J.G., 2007. Abnormal parietal cortex activation during working memory in schizophrenia: Verbal phonological coding disturbances versus domain-general executive dysfunction. The American Journal of Psychiatry 164, 1090–1098.

Bates, J.F., Goldman-Rakic, P.S., 1993. Prefrontal connections of medial motor areas in the rhesus monkey. The Journal of Comparative Neurology 336, 211–228.

Baxter, L.R., Schwartz, J.M., Mazziotta, J.C., et al., 1988. Cerebral glucose metabolic rates in nondepressed patients with obsessive-compulsive disorder. The American Journal of Psychiatry 145, 1560–1563.

Berman, K.F., Illowsky, B.P., Weinberger, D.R., 1988. Physiological dysfunction of dorsolateral prefrontal cortex in schizophrenia: IV. Further evidence for regional and behavioral specificity. Archives of General Psychiatry 45, 616–622.

Bhanji, J.P., Beer, J.S., Bunge, S.A., 2010. Taking a gamble or playing by the rules: Dissociable prefrontal systems implicated in probabilistic versus deterministic rule-based decisions. NeuroImage 49, 1810–1819.

Birrell, J.M., Brown, V.J., 2000. Medial frontal cortex mediates perceptual attentional set shifting in the rat. Journal of Neuroscience 20, 4320–4324.

Blais, C., Bunge, S.A., 2010. Behavioral and neural evidence for item-specific performance monitoring. Journal of Cognitive Neuroscience 22, 2758–2767.

Botvinick, M., Nystrom, L.E., Fissell, K., Carter, C.S., Cohen, J.D., 1999. Conflict monitoring versus selection-for-action in anterior cingulate cortex. Nature 402, 179–181.

Botvinick, M.M., Braver, T.S., Barch, D.M., Carter, C.S., Cohen, J.D., 2001. Conflict monitoring and cognitive control. Psychology Review 108, 624–652.

Bourgeois, J.P., Goldman-Rakic, P.S., Rakic, P., 1994. Synaptogenesis in the prefrontal cortex of rhesus monkeys. Cerebral Cortex 4, 78–96.

Braver, T.S., Barch, D.M., 2002. A theory of cognitive control, aging cognition, and neuromodulation. Neuroscience and Biobehavioral Reviews 26, 809–817.

Braver, T.S., Cohen, J.D., 2000. On the control of control: The role of dopamine regulating prefrontal function and working memory. In: Monsell, S., Driver, J. (Eds.), Control of Cognitive Processes Attention and Performance XVIII. MIT Press, Cambridge, MA.

Brown, T.T., Lugar, H.M., Coalson, R.S., Miezin, F.M., Petersen, S.E., Schlaggar, B.L., 2005. Developmental changes in human cerebral functional organization for word generation. Cerebral Cortex 15, 275–290.

Bunge, S.A., Zelazo, P.D., 2006. A brain-based account of the development of rule use in childhood. Current Directions in Psychological Science 15, 118–121.

Bush, G., Frazier, J.A., Rauch, S.L., et al., 1999. Anterior cingulate cortex dysfunction in ADHD revealed by fMRI and the counting Stroop. Biological Psychiatry 45, 1542–1552.

Bush, G., Valera, E.M., Seidman, L.J., 2005. Functional neuroimaging of attention-deficit/hyperactivity disorder: A review and suggested future directions. Biological Psychiatry 57, 1273–1284.

Carter, C.S., MacDonald 3rd, A.W., Ross, L.L., Stenger, V.A., 2001. Anterior cingulate cortex activity and impaired self-monitoring of performance in patients with schizophrenia: An event-related fMRI study. The American Journal of Psychiatry 158, 1423–1428.

Casey, B.J., 2000. Disruption of inhibitory control in developmental disorders: A mechanistic model of implicated frontostriatal circuitry. In: Siegler, R.S., McClelland, J.L. (Eds.), Mechanisms of Cognitive Development: The Carnegie Symposium on Cognition. Erlbaum, Hillsdale, NJ.

Casey, B.J., 2005. Frotostriatal and frontocerebellar circuitry underlying cognitive control. In: Mayr, U., Owh, E., Keele, S.W. (Eds.), Developing Individuality in the Human Brain. American Psychological Association, Washington, DC.

Casey, B.J., Thomas, K.M., Welsh, T.F., et al., 2000. Dissociation of response conflict, attentional control, and expectancy with functional magnetic resonance imaging (fMRI). Proceedings of the National Academy of Sciences of the United States of America 97, 8727–8733.

Casey, B.J., Amso, D., Davidson, M.C., 2006. Learning about learning and development with neuroimaging. In: Johnson, M., Munakata, Y. (Eds.), Attention and Performance XXI: Processes of Change in Brain and Cognitive Development. Oxford University Press, New York.

Casey, B.J., Epstein, J.N., Buhle, J., et al., 2007a. Frontostriatal connectivity and its role in cognitive control in parent-child dyads with ADHD. The American Journal of Psychiatry 164, 1729–1736.

Casey, B.J., Nigg, J.T., Durston, S., 2007b. New potential leads in the biology and treatment of attention deficit-hyperactivity disorder. Current Opinion in Neurology 20, 119–124.

Castellanos, F.X., Geidd, J.N., Marsh, W.L., et al., 1996. Quantative brain magnetic resonance imaging in attention-deficit hyperactivity disorder. Archives of General Psychiatry 53, 607–616.

Chugani, H.T., Phelps, M.E., Mazziotta, J.C., 1987. Positron emission tomography study of human brain functional development. Annals of Neurology 22, 487–497.

Cohen, J.D., Servan-Schreiber, D., 1992. Context, cortex, and dopamine: A connectionist approach to behavior and biology in schizophrenia. Psychology Review 99, 45–77.

Cohen, J.D., Servan-Schreiber, D., McClelland, J.L., 1992. A parallel distributed processing approach to automaticity. The American Journal of Psychology 105, 239–269.

Cohen, J.D., Forman, S.D., Braver, T.S., Casey, B.J., Servan-Schreiber, D., Noll, D.C., 1994. Activation of prefrontal cortex in a non spatial working memory task with Functional MRI. Human Brain Mapping 1, 293–304.

Corbetta, M., Shulman, G.L., 2002. Control of goal-directed and stimulus-driven attention in the brain. Nature Reviews Neuroscience 3, 201–215.

Corbetta, M., Kincade, J.M., Ollinger, J.M., McAvoy, M.P., Shulman, G.L., 2000. Voluntary orienting is dissociated from target detection in human posterior parietal cortex. Nature Neuroscience 3, 292–297.

D'Espostio, M., Detre, J.A., Alsop, D.C., Shin, R.K., Atlas, S., Grossman, M., 1995. The neural basis of the central executive system of working memory. Nature 378, 279–281.

Davidson, M.C., Thomas, K.M., Casey, B.J., 2003. Imaging the developing brain with fMRI. Mental Retardation and Developmental Disabilities Research Reviews 9, 161–167.

Desimone, R., Duncan, J., 1995. Neural mechanisms of selective visual attention. Annual Reviews in Neuroscience 18, 193–222.

Diamond, A., 1988. Abilities and neural mechanisms underlying AB performance. Child Development 59, 523–527.

Diamond, A., 1998. Evidence for the importance of dopamine for prefrontal cortex functions early in life. In: Robert, A.C., Robbins, T.W., Weiskrantz, L. (Eds.), The Prefrontal Cortex. Oxford University Press, Oxford, pp. 144–164.

Diamond, A., Ciaramitaro, V., Donner, E., Djali, S., Robinson, M.B., 1994. An animal model of early-treated PKU. Journal of Neuroscience 14, 3072–3082.

Dias, R., Robbins, T.W., Roberts, A.C., 1996a. Dissociation in prefrontal cortex of affective and attentional shifts. Nature 380, 69–72.

Dias, R., Robbins, T.W., Roberts, A.C., 1996b. Primate analogue of the Wisconsin Card Sorting Test: Effects of excitotoxic lesions of the prefrontal cortex in the marmoset. Behavioral Neuroscience 110, 872–886.

Dum, R.P., Strick, P.L., 2003. An unfolded map of the cerebellar dentate nucleus and its projections to the cerebral cortex. Journal of Neurophysiology 89, 634–639.

Duncan, J., Owen, A.M., 2000. Common regions of the human frontal lobe recruited by diverse cognitive demands. Trends in Neuroscience 23, 475–483.

Durston, S., Tottenham, N.T., Thomas, K.M., et al., 2003. Differential patterns of striatal activation in young children with and without ADHD. Biological Psychiatry 53, 871–878.

Durston, S., Hulshoff Pol, H.E., Schnack, H.G., et al., 2004. Magnetic resonance imaging of boys with attention-deficit/hyperactivity disorder and their unaffected siblings. Journal of the American Academy of Child and Adolescent Psychiatry 43, 332–340.

Enns, J.T., Cameron, S., 1987. Selective attention in young children: The relations between visual search, filtering, and priming. Journal of Experimental Child Psychology 44, 38–63.

Epstein, J.N., Casey, B.J., Tonev, S.T., et al., 2007. ADHD- and medication-related brain activation effects in concordantly affected parent-child dyads with ADHD. Journal of Child Psychology and Psychiatry 48, 899–913.

Fitzgerald, K.D., Welsh, R.C., Gehring, W.J., et al., 2005. Error-related hyperactivity of the anterior cingulate cortex in obsessive-compulsive disorder. Biological Psychiatry 57, 287–294.

Fox, M.T., Barense, M.D., Baxter, M.G., 2003. Perceptual attentional set-shifting is impaired in rats with neurotoxic lesions of posterior parietal cortex. Journal of Neuroscience 23, 676–681.

Frances, A., Pincus, H.A., Widiger, T.A., Davis, W.W., First, M.B., 1990. DSM-IV: Work in progress. The American Journal of Psychiatry 147, 1439–1448.

Frazier, J.A., Geidd, J.N., Hamburger, S.D., et al., 1996. Brain magnetic resonance imaging in childhood-onset schizophrenia. Archives of General Psychiatry 53, 617–624.

Fuster, J.M., 1980. The Prefrontal Cortex: Anatomy, Physiology, and Neuropsychology of the Frontal Lobe. Raven Press, New York.

Giedd, J.N., Blumenthal, J., Jeffries, N.O., et al., 1999a. Brain development during childhood and adolescence: A logitudinal MRI study. Nature Neuroscience 10, 861–863.

Giedd, J.N., Jeffries, N.O., Blumenthal, J., et al., 1999b. Childhood-onset schizophrenia: Progressive brain changes during adolescence. Biological Psychiatry 46, 892–898.

Gold, J.I., Shadlen, M.N., 2001. Neural computations that underlie decisions about sensory stimuli. Trends in Cognitive Science 5, 10–16.

Goldman-Rakic, P.S., Brown, R.M., 1982. Postnatal development of monoamine content and synthesis in the cerebral cortex of rhesus monkeys. Brain Research 256, 339–349.

Granger, B., Tekaia, F., LeSourd, A.M., Rakic, P., Bourgeois, J.P., 1995. Tempo of neurogenesis and synaptogenesis in the primate cingulate mesocortex: Comparison with the neocortex. The Journal of Comparative Neurology 360, 363–376.

Harlow, J.M., 1869. Recovery from the passage of an iron bar through the head. Boston Medical and Surgical Journal 3, 116–117.

Hayes, A.E., Davidson, M.C., Keele, S.W., Rafal, R.D., 1998. Toward a functional analysis of the basal ganglia. Journal of Cognitive Neuroscience 10, 178–198.

Hedden, T., Gabrieli, J.D., 2010. Shared and selective neural correlates of inhibition, facilitation, and shifting processes during executive control. NeuroImage 51, 421–431.

Huttenlocher, P.R., Dabholkar, A.S., 1997. Regional differences in synaptogenesis in human cerebral cortex. The Journal of Comparative Neurology 387, 167–178.

Ivry, R.B., Keele, S.W., 1989. Timing functions of the cerebellum. Journal of Cognitive Neuroscience 1, 136–152.

Kelly, A.M., Di Martino, A., Uddin, L.Q., et al., 2009. Development of anterior cingulate functional connectivity from late childhood to early adulthood. Cerebral Cortex 19, 640–657.

Klingberg, T., Vaidya, C.J., Gabrieli, J.D., Moseley, M.E., Hedehus, M., 1999. Myelination and organization of the frontal white matter in children: A diffusion tensor MRI study. Neuroreport 10, 2817–2821.

Leckman, J.F., Bloch, M.H., Smith, M.E., Larabi, D., Hampson, M., 2010. Neurobiological substrates of Tourette's disorder. Journal of Child and Adolescent Psychopharmacology 20, 237–247.

Lidow, M.S., Goldman-Rakic, P.S., Gallager, D.W., Rakic, P., 1991. Distribution of dopaminergic receptors in the primate cerebral cortex: Quantitative autoradiographic analysis using [3H]raclopride, [3H]spiperone and [3H]SCH23390. Neuroscience 40, 657–671.

Liston, C., Watts, R., Tottenham, N., et al., 2006. Frontostriatal microstructure modulates efficient recruitment of cognitive control. Cerebral Cortex 16, 553–560.

Logothetis, N.K., Pauls, J., Augath, M., Trinath, T., Oeltermann, A., 2001. Neurophysiological investigation of the basis of the fMRI signal. Nature 412, 150–157.

Mayr, U., Awh, E., Keele, S. (Eds.), 2005. Developing Individuality in the Human Brain: A Tribute to Michael I. Posner. American Psychological Association, Washington.

McAlonan, K., Brown, V.J., 2003. Orbital prefrontal cortex mediates reversal learning and not attentional set shifting in the rat. Behavioural Brain Research 146, 97–103.

McClure, S.M., Berns, G.S., Montague, P.R., 2003. Temporal prediction errors in a passive learning task activate human striatum. Neuron 38, 339–346.

Middleton, F.A., Strick, P.L., 2002. Basal-ganglia 'projections' to the prefrontal cortex of the primate. Cerebral Cortex 12, 926–935.

Milham, M.P., Banich, M.T., 2005. Anterior cingulate cortex: An fMRI analysis of conflict specificity and functional differentiation. Human Brain Mapping 25, 328–335.

Miller, E.K., Cohen, J.D., 2001. An integrative theory of prefrontal cortex function. Annual Review of Neuroscience 24, 167–202.

Milner, B., Petrides, M., Smith, M.L., 1985. Frontal lobes and the temporal organization of memory. Human Neurobiology 4, 137–142.

Mink, J.W., 1996. The basal ganglia: Focused selectio oand inhibition of competing motor programs. Progress in Neurobiology 50, 381–425.

Mishkin, M., Pribram, K.H., 1955. Analysis of the effects of frontal lesions in monkey. I. Variations of delayed alternation. Journal of Comparative and Physiological Psychology 48, 492–495.

Montague, P.R., Dayan, P., Sejnowski, T.J., 1996. A framework for mesencephalic dopamine systems based on predictive Hebbian learning. Journal of Neuroscience 16, 1936–1947.

O'Reilly, R.C., Noelle, D.C., Braver, T.S., Cohen, J.D., 2002. Prefrontal cortex and dynamic categorization tasks: Representational organization and neuromodulatory control. Cerebral Cortex 12, 246–257.

Peterson, B.S., Skudlarski, P., Anderson, A.W., et al., 1998. A functional magnetic resonance imaging study of tic suppression in Tourette syndrome. Archives of General Psychiatry 55, 326–333.

Platt, M.L., Glimcher, P.W., 1999. Neural correlates of decision variables in parietal cortex. Nature 400, 233–238.

Posner, M.I., Petersen, S.E., 1990. The attention system of the human brain. Annual Review of Neuroscience 13, 25–42.

Rakic, P., 1974. Neurons in rhesus monkey visual cortex: Systematic relation between time of origin and eventual disposition. Science 183, 425–427.

Rakic, P., Bourgeois, J.P., Eckenhoff, M.F., Zecevic, N., Goldman-Rakic, P.S., 1986. Concurrent overproduction of synapses in diverse regions of the primate cerebral cortex. Science 232, 232–235.

Redgrave, P., Prescott, T.J., Gurney, K., 1999. The basal ganglia: A vertebrate solution to the selection problem? Neuroscience 89, 1009–1023.

Rosenberg, D.R., Lewis, D.A., 1995. Postnatal maturation of the dopaminergic innervation of monkey prefrontal and motor cortices: A tyrosine hydroxylase immunohistochemical analysis. The Journal of Comparative Neurology 358, 383–400.

Saffran, J.R., Aslin, R.N., Newport, E.L., 1996. Statistical learning by 8-month-old infants. Science 274, 1926–1928.

Schultz, W., Dayan, P., Montague, P.R., 1997. A neural substrate of prediction and reward. Science 275, 1593–1599.

Shallice, T., 1988. From Neuropsychology to Mental Structure. Cambridge University Press, New York.

Smith, E.E., Jonides, J., 1999. Storage and executive processes in the frontal lobes. Science 283, 1657–1661.

Snyder, S.H., 1976. The dopamine hypothesis of schizophrenia: Focus on the dopamine receptor. The American Journal of Psychiatry 133, 197–202.

Sowell, E.R., Peterson, B.S., Thompson, P.M., Welcome, S.E., Henkenius, A.L., Toga, A.W., 2003. Mapping cortical change across the human life span. Nature Neuroscience 6, 309–315.

Sowell, E.R., Kan, E., Yoshii, J., et al., 2008. Thinning of sensorimotor cortices in children with Tourette syndrome. Nature Neuroscience 11, 637–639.

Spencer, R.M., Zelaznik, H.N., Diedrichsen, J., Ivry, R.B., 2003. Disrupted timing of discontinuous but not continuous movements by cerebellar lesions. Science 300, 1437–1439.

Stuss, D.T., Kaplan, E.F., Benson, D.F., Weir, W.S., Chiulli, S., Sarazin, F.F., 1982. Evidence for the involvement of orbitofrontal cortex in memory functions: An interference effect. Journal of Comparative and Physiological Psychology 96, 913–925.

Swanson, J.M., Kinsbourne, M., Nigg, J., et al., 2007. Etiologic subtypes of attention-deficit/hyperactivity disorder: Brain imaging, molecular genetic and environmental factors and the dopamine hypothesis. Neuropsychology Review 17, 39–59.

Swedo, S.E., Schapiro, M.B., Grady, C.L., et al., 1989. Cerebral glucose metabolism in childhood-onset obsessive-compulsive disorder. Archives of General Psychiatry 46, 518–523.

Swedo, S.E., Pietrini, P., Leonard, H.L., et al., 1992. Cerebral glucose metabolism in childhood-onset obsessive-compulsive disorder. Revisualization during pharmacotherapy. Archives of General Psychiatry 49, 690–694.

Tamm, L., Menon, V., Reiss, A.L., 2006. Parietal attentional system aberrations during target detection in adolescents with attention deficit hyperactivity disorder: Event-related fMRI evidence. The American Journal of Psychiatry 163, 1033–1043.

Vaidya, C.J., Austin, G., Kirkorian, G., et al., 1998. Selective effects of methylphenidate in attention deficit hyperactivity disorder: A functional magnetic resonance study. Proceedings of the National Academy of Sciences of the United States of America 95, 14494–14499.

Van Mier, H.I., Petersen, S.E., 2002. Role of the cerebellum in motor cognition. Annals of the New York Academy of Sciences 978, 334–353.

Volkow, N.D., Wang, G., Fowler, J.S., et al., 2001. Therapeutic doses of oral methylphenidate significantly increase extracellular dopamine in the human brain. Journal of Neuroscience 21, RC121.

Wise, S.P., Boussaoud, D., Johnson, P.B., Caminiti, R., 1997. Premotor and parietal cortex: Corticocortical connectivity and combinatorial computations. Annual Review of Neuroscience 20, 25–42.

Language Impairment

R.H. Fitch

University of Connecticut, Storrs, CT, USA

O U T L I N E

40.1 INTRODUCTION

Language represents a complex and unique aspect of human neurocognitive processing and, as such, is vulnerable to disturbance through disruption of or damage to underlying neural systems. Such disruption is generally divided into two types: developmental (encompassing pre- and perinatal brain injury) and acquired.

Acquired disturbances of language generally present as deficits in the comprehension (reception) and/or production (expression) of language following traumatic injury to *established* neural language systems, generally in the left cortical hemisphere (although a minority of individuals seem to have language functions in the right cortical hemisphere; Figure 40.1). These language regions include interconnected areas, such as a left perisylvian posterior lateral temporal (auditory) region called Wernicke's area. This region is believed to mediate the perception and processing of spoken language. Another key language area is Broca's area, found in the left inferior frontal/motor cortex, which generally is thought to mediate production of oral speech. Additional areas routinely activated during neuroimaging measures of language tasks include left parietotemporal and occipitotemporal areas implicated in reading. Damage to these regions results in a range of language deficits collectively termed *aphasias*, with symptoms reflecting the functions mediated by the damaged regions. For example, Wernicke's aphasia presents as a loss in receptive language processing, whereas Broca's aphasia presents as a general deficit in language production (but not comprehension). More recent evidence suggests that these disorders are not as circumscribed as once believed (e.g., Blank et al., 2002), but the terms continue to be used in clinical diagnosis. More global forms of aphasia, encompassing both expressive and receptive language deficits, may also be seen after more extensive damage affecting multiple brain regions. Acquired language disorders may also include deficits in reading skills, termed *acquired dyslexia* or *alexia*. Importantly, all of these disorders are – by definition – confined to the assessment of acquired injury in previously language-competent individuals (older children and adults) and reflect a loss in language skills due to damage of *already-developed* neural systems subserving these processes (see Jordan and Hillis, 2006, for review). Interestingly, language functions do seem to reorganize to the right hemisphere after left-hemisphere damage in young children, evidencing a

© 2013 Elsevier Inc. All rights reserved.

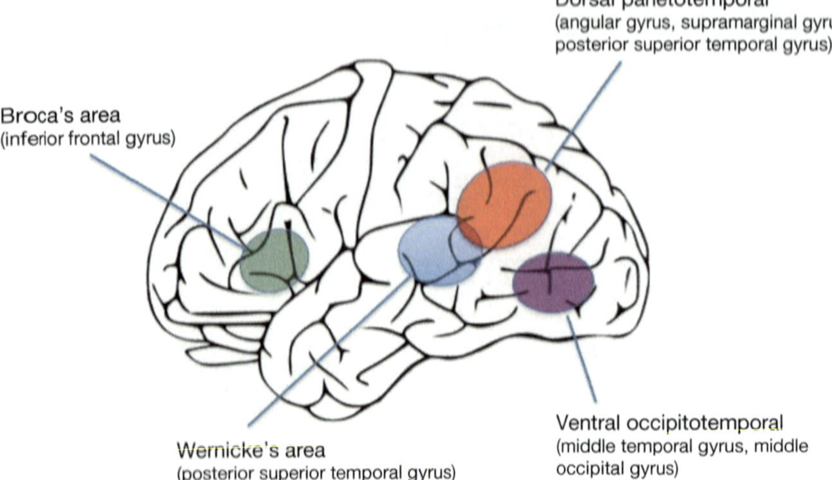

FIGURE 40.1 Left-hemisphere cortical regions implicated in language processing.

high degree of plasticity in language systems during early development. Indeed, young children who undergo left temporal lobectomy because of intractable epilepsy show a surprisingly normal course of language development (Vargha-Khadem et al., 1997). However, comparable injuries in older children and adults with established language functions lead to more permanent language function loss and substantially less reorganization (less plasticity). These early plasticity effects parallel evidence that a second language is acquired more quickly and effortlessly in young children compared with adults.

In contrast to acquired aphasias, *developmental* disruptions to language systems occur as a result of early factors influencing the *initial* establishment of neural systems needed to support emergent language. These disruptions may occur as one component of a more global neural disruption, resulting in overall degradation of cognitive processing (e.g., mental retardation, learning disorders, or pervasive developmental disorders; Broman and Grafman, 1994; Pennington, 1991), or may be *specific* to language systems (meaning overall intellect or IQ is within the normal range). The latter condition is most commonly associated with the term *language impairment* (LI), and therefore the rest of this chapter addresses behavioral features and underlying factors implicated in the developmental disruption of language systems.

40.2 SPECIFIC LANGUAGE IMPAIRMENT AND RELATED DISABILITIES

The ability to understand and produce spoken words represents a profoundly complex process that most young children acquire with remarkable ease, despite a lack of formal instruction (i.e., most young children are not explicitly taught how to speak). Such evidence of innate predisposition toward language skill has fueled debates about intrinsic 'preadaptations' in the human brain for language, particularly compared with neural substrates underlying other species-specific use of communicative vocalizations or even learned communicative systems (such as sign or computer use in nonhuman primates; see Fischer and Marcus, 2006; Hauser, 1996). Despite the language-ready predisposition seen in most typically developing human children, a subset nonetheless experiences delays in the achievement of language milestones in the absence of known causal factors. That is, even when potential underlying impairments such as epilepsy or other neurologic abnormalities, disorders of vision or hearing, psychiatric impairments, or environmental deprivation are excluded, about 5–10% of children exhibit unexplained deficits/delays in language development (Beitchman et al., 1986; Leonard, 1998; see Heim and Benasich, 2006, for review). These problems are specific to language and occur despite a nonverbal IQ in the normal range. Such language-specific developmental disorders are termed either (LI) or *specific language impairment* (SLI), and deficits associated with SLI can be further subdivided into receptive (*SLI-R*, comprehension), expressive (*SLI-E*, production), or combined deficits.

What is implied by SLI being defined, in part, by the *exclusion* of other impairments and disabilities? One example of this 'exclusionary criteria' would be applied when fetal lead or fetal alcohol exposure (FAS) results in generalized cognitive impairments that include impaired language skills. A young child with language delays related to FAS would *not* likely be diagnosed as having SLI, based on the more generalized underlying cognitive deficit. A diagnosis of SLI also requires the exclusion of other related disorders with similar or overlapping symptoms. For example, a subset of exclusively *expressive* language difficulties specific to articulation, phonology, and/or oral-motor skills are thought to represent a related but separate disability termed *speech-sound disorder* (SSD; Pennington and Bishop, 2009; Peterson et al., 2007). *Receptive* deficits in language processing may also overlap with central auditory processing disorder (CAPD or APD). However, CAPD generally presents as a broad-based difficulty in listening to and processing sounds that include, but are not limited to, speech sounds. Moreover, CAPD has been characterized by the American Speech & Hearing Association to include deficiencies in 'blocking out' noise and acoustic distracters (Bamiou et al., 2001), which are not diagnostic features of SLI. Therefore, CAPD may be a contributing factor in the development of SLI, but the two disorders are, at least at the present time, considered clinically different.

SLI can also be diagnosed as *comorbid* with other co-occurring developmental disabilities. For example, SLI might co-occur with attention difficulties, as seen in attention deficit disorders with or without hyperactivity (ADD/ADHD) (see Chapter 22). SLI might also be comorbid with SSD or CAPD. From a diagnostic perspective, SLI might further 'overlap' with characteristics of related disorders. For example, autism spectrum disorder (ASD; see Chapter 34) encompasses substantial communicative disabilities, and ASD individuals presenting with significant LI have even been designated as ASD-LI (e.g., Williams et al., 2008). While these overlaps may lead to complications and difficulties in individual diagnosis, when taken from a research perspective, the existence of overlapping phenotypic deficits across SLI and ASD-LI populations presents an opportunity to study potential commonalities in underlying etiology. Such studies have revealed some evidence of phenotypic similarity between ASD-LI and SLI populations (e.g., evidence of common neuroanatomic abnormalities, and similarities in anomalous information processing; Herbert and Kenet, 2007), but others suggest that the underlying etiology (i.e., genetics) leading to language disruptions in these two disorders may follow different neurodevelopmental routes and that ASD-LI versus SLI are in fact likely to stem from very different causes (Williams et al., 2008). Ongoing research in this area will be critical in ascertaining whether common mechanisms across these disorders may account for the similarities in phenotypic language disturbance or whether perhaps differing vectors of disruption to early brain development might exert such fundamental (core) effects on information processing that they ultimately lead to seemingly similar phenotypes in higher-order language processing ('phenocopies').

Overall, the most common comorbidity seen with SLI is the subsequent diagnosis of *dyslexia* or *reading disability* (RD, also called *specific reading disability* or *developmental dyslexia*). Dyslexia is defined by exclusionary criteria (much like SLI), but in this case represents an unexpected delay or deficit in the acquisition and performance of reading, despite an overall nonverbal IQ in the normal range. This prevalent comorbidity between SLI and dyslexia likely reflects the fact that, while most children with SLI do eventually reach normal language milestones (i.e., they learn to understand and produce speech), the subsequent milestones in reading require the translation of learned phonemes (letter sounds) onto orthographic representation (written letters). Since deficits in phonemic representation and/or phonologic processing comprise a core and persisting component of SLI (see below for more discussion), it is not surprising that a large portion (>50%) of children with SLI go on to be diagnosed as dyslexic/RD (Bishop and Snowling, 2004; Catts et al., 2005; Schuele, 2004; Sices et al., 2007). It is a matter of ongoing debate whether those children who are initially diagnosed as SLI but overcome their disorder without further reading difficulties may, in fact, comprise a different subtype of the SLI population as compared to those who go on to be comorbid for dyslexia/RD. Moreover, another subgroup may be comprised of older children/adults diagnosed with dyslexia/RD in the *absence* of any prior history of SLI (see Bishop and Snowling, 2004; Pennington and Bishop, 2009, for discussion). In fact, it has been suggested that the latter subset of dyslexics, generally thought to correspond to a subcategory termed orthographic, or surface, dyslexics, exhibit primarily visual and higher-order reading deficits but lack the core phonological deficits characteristic of SLI. Conversely, the remaining dyslexics do show core phonological deficits and are accordingly designated phonologic dyslexics. Many or most of the dyslexics in this category do have some history of SLI or language-related difficulties.

Ongoing assessment of these various subgroups appears likely to support the existence of core functional components that characterize both SLI and dyslexia/RD, and may speak to the fact that some of these core deficits show a large amount of overlap across the two disorders (Figure 40.2). For this reason – and recognizing the obvious underlying heterogeneity inherent to the term – some researchers have begun to use combined terms such as *language disabilities* (LD), *language-learning disabilities*, or *language-learning impairments* to refer to this set of language-related developmental disabilities (Peterson et al., 2007; Tallal et al., 1993; but see Bishop and Snowling, 2004, for an opposing view). The remainder of this chapter will discuss the phenotypic features and underlying genetics for this *common* category of LD, inclusive of both SLI and dyslexia/RD. This grouping reflects the fact that much of the pertinent neuroimaging and genetic research has been performed using adult dyslexic and/or combined LD populations, with

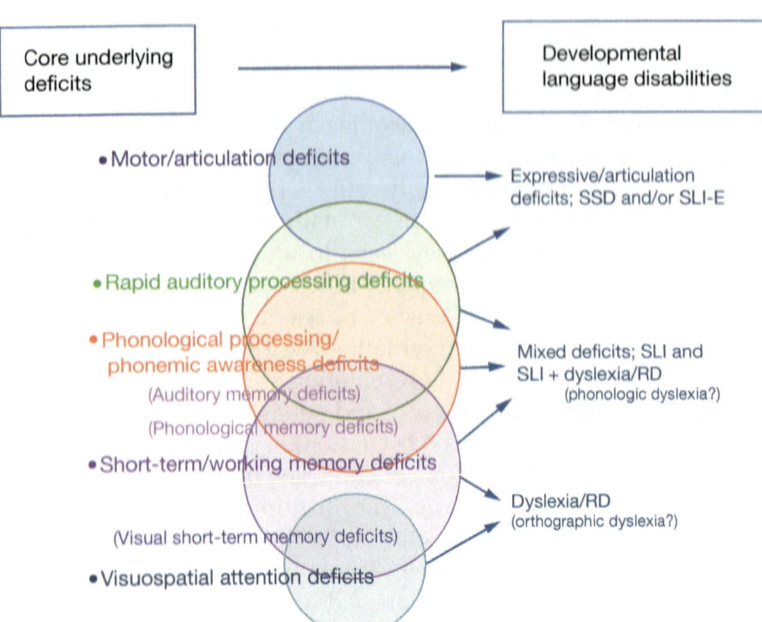

FIGURE 40.2 A proposed schematic diagram relating identified core function deficits associated with developmental language disability, and clinically defined populations evidencing those underlying deficits. SSD, speech sound disorder; SLI, specific language impairment; SLI-E, specific expressive language impairment; RD, reading disability.

a comparatively limited amount of research performed exclusively on SLI children (but see Webster and Shevell, 2004 for data specific to SLI).

40.3 BEHAVIORAL FEATURES OF LD

Behavioral profiles in clinically defined language-disabled groups vary, but are typically characterized by deficits in both lower-order (e.g., sensory/motor) and higher-order (e.g., grammatical, syntactic, and verbal memory) processes (e.g., Menghini et al., 2010). In fact, the LD population appears to be a heterogeneous mix of individuals with an assortment of deficits corresponding to some subset of the following functional core areas (along with other possible deficits not listed here). (1) Oral-motor skills and articulation, as required for tasks such as rapid naming. (2) Rapid auditory processing, as required to discriminate rapidly changing acoustic sounds such as consonant–vowel syllables. (3a) Phonemic awareness, as required to distinguish and identify phonemes, or subunits of words, and (3b) phonological processing, as required to manipulate phonemes within words and perform mapping to-and-from sounds and letters, as required for non-word reading. (4) Short-term and/or working verbal memory and the related auditory and phonological memory, required for nonword repetition. (5) Visuospatial processing, visual memory, and/or visuospatial attention, required for specific reading tasks. Evidence further suggests that unique patterns of these core deficits may characterize specific subsets of individuals (i.e., SLI/SSD only, SLI+dyslexic/RD, and dyslexic/RD only; see Figure 40.2). Thus, dyslexic/RD-only individuals (orthographic dyslexics) are less likely to show motor/articulation deficits. Conversely, SLI-E populations are less likely to show visual-processing and memory deficits. However, both groups, as well as the comorbid overlapping population, share evidence of *common* deficits in rapid auditory processing (Farmer and Klein, 1995; Fitch and Tallal, 2003), phonemic and phonological processing (Pennington and Bishop, 2009; Schuele, 2004; Shankweiler et al., 1995), and aspects of short-term/working memory (Briscoe and Rankin, 2009; Montgomery et al., 2010). Each of these core functional features of LD (including SLI and dyslexia/RD) is discussed in greater detail below.

40.3.1 Motor/Articulation Deficits, SSD, and the KE family

As young children learn to speak, it is not unusual to hear substitutions of letter sounds (e.g., /w/ for /r/, as in 'wabbit'), lisps, or stutters. Normally, children outgrow these charming but immature articulation patterns. Some children, however, persist with articulation difficulties (lisp, stutter, and other oral mispronunciations) that may reflect chronic defects in oral musculature, vocal apparatus, palate formation, or general oral-motor 'weakness.' When evidenced in the absence of any delays in language comprehension or reductions in vocabulary, these articulation deficits tend to be less predictive of long-term language and literacy problems, although the chronic incorrect encoding of certain letter sounds may, in fact, lead to some difficulties in the orthographic translation of those same sounds. On the other hand, children with expressive language difficulties characterized by *multiple* speech-sound errors (phonemic omissions, substitutions, and distortions characterized as phonological processing difficulties) exhibit more marked impairment of articulation and expressive language. (In a phonological (or phonemic) disorder (PD), a child has difficulty learning the sound system of the language, and, specifically, fails to recognize and identify specific sounds as unique. For example, the sounds /k/ and /t/ may not be recognized as different and may therefore be substituted for each other). Children exhibiting language difficulties characterized by such deficits in the oral production or expression of language are, depending on other deficits, likely to be diagnosed with *specific expressive language impairment (SLI-E)*, *PD*, *SSD*, or some combination of these (Fischer and Scharff, 2009; Pennington and Bishop, 2009). Concurrent research also aims to examine a putative role for more generalized motor deficits associated with SLI, and to separate these features from related overlapping disorders such as developmental coordination disorder (e.g., Archibald and Alloway, 2008).

An intriguing series of studies have been performed on a single extended family with an inherited form of expressive LI (also called 'verbal dyspraxia'; Alcock et al., 2000). Half the members of the patient KE's family (so named for anonymity) show this disorder, exhibiting severe impairments in phonology and syntax, as well as oral praxis. Ongoing studies of this family have led to the isolation of a gene termed *FOXP2* that may be associated with the expressive and oral language deficits evidenced in the KE family. Interestingly, animal studies of the *Foxp2* homolog have shown that disruptions of this gene lead to defects in motor learning, including the acquisition of birdsong and the production of vocalizations in mice (reviewed in Fischer and Scharff, 2009; see below for further discussion).

Ongoing research will continue to assess the specific role of articulation and motor-specific deficits in defining subtypes within the heterogeneous LD population.

40.3.2 Rapid Auditory Processing Deficits

A key series of studies by Tallal and colleagues showed that SLI children were unable to discriminate two sequential tones when the interval between the tones fell below a threshold interval and were also impaired in the identification and discrimination of consonant–vowel syllables characterized by short, rapidly changing formant transitions (e.g., /ba/, /da/, /pa/, /ta/; see Tallal, 1977, 1980; Tallal and Newcombe,

1978; Tallal and Piercy, 1973a,b, 1975; Tallal and Stark, 1981; reviewed in Fitch and Tallal, 2003). Additional studies showed that performance on these auditory tasks was correlated not only with speech perception indices, but also with nonword reading scores (Tallal, 1980). Thus, it was posited that basic defects in low-level auditory processing may be associated with deficits in speech perception and phonology and thus could represent a core factor in the development of LD.

This suggestion of a low-level auditory processing deficit as a possible causal factor in developmental LD has elicited some controversy (e.g., Mody et al., 1997; Ramus, 2003; Rosen and Manganari, 2001). However, ongoing behavioral and psychophysical studies of language-disabled subjects appear to support the notion of an auditory processing deficit that may underlie (or at least be related to) some deficits in speech perception and subsequent language development (Au and Lovegrove, 2007; Cardy et al., 2005; Cohen-Mimran and Sapir, 2007; Corbera et al., 2006; Edwards et al., 2004; Farmer and Klein, 1995; Gaab et al., 2007; Hari and Kiesla, 1996; King et al., 2007; Kraus et al., 1996; McAnally and Stein, 1996, 1997; McCrosky and Kidder, 1980; Neville et al., 1993; Reed, 1989; Renvall and Hari, 2002; Robin et al., 1989; Sutter et al., 2000; Watson, 1992; Witton et al., 1998; Wright et al., 1997). Again, while it remains to be determined whether these auditory deficits represent causal or comorbid (parallel but noncausal) deficits (see McArthur and Bishop, 2001; Ramus, 2003; Rosen and Manganari, 2001), an early deficit in auditory processing could reasonably precede observable LI, in which case assessment of auditory processing in at-risk infants would be extremely useful in identifying those likely to develop later language problems. In fact, Benasich et al. (2002) found that infants with a family history of LI or dyslexia (and thus at an elevated risk of developing language problems themselves; Tallal et al., 1991) were impaired relative to controls in their ability to discriminate two-tone sequences when there was a short interval between the tones, but not with a longer interval. Prospective follow-up of these children revealed a predictive relationship between the threshold at which rapidly presented auditory stimuli could be processed in infancy and language outcomes at 12–24 months (Benasich et al., 2006; Choudhury et al., 2007). More recently, a comparable relationship was seen between early auditory-evoked response potential and electroencephalogram (AERP/EEG) scores using these same stimuli and later language outcomes (Choudhury et al., 2007). In addition, predictive associations between early auditory processing skills have been related to language performance later in life in normally developing samples. Trehub and Henderson (1996) found that children who had performed above the median on a variety of acoustic gap detection tasks at 6 or 12 months were found to have larger productive vocabularies, use longer, more complex sentences, and produce more irregular words compared with children who had scored below the median. Such findings are also supported by evidence from studies recording evoked response potentials (ERPs) to auditory stimuli in infancy. Molfese and Molfese (1997) found that ERPs to consonant–vowel syllables recorded from infants within 36 h of birth differed between children whose verbal IQ was above, versus below, the norm at 5 years of age. Similarly, infants with a family history of dyslexia showed different patterns of ERPs to consonant–vowel stimuli as compared with matched controls when recorded at 1 week and at 6 months (Leppänen and Lyytinen, 1997; Leppänen et al., 1999; Pihko et al., 1999). Importantly, evidence indicates that these group differences do in fact relate to emergent language skills, based on longitudinal analysis (Benasich et al., 2006; Choudhury et al., 2007).

Collective data thus support the notion that the ability to make fine auditory discriminations (rapid auditory processing) is strongly correlated with later language development and that deficits in this basic function may impair subsequent language development, with ultimate implications for higher-order processes seemingly distal to basic acoustic processing (e.g., reading). Ongoing research will continue to evaluate the role of basic auditory processing deficits in the later emergence of language deficits.

40.3.3 Phonemic and Phonological Processing Deficits

A wide variety of language-based developmental disorders appear to encompass a central deficit in the ability to learn to identify and discriminate individual speech sounds as unique (phonemic awareness). Studies of both SLI and dyslexic/RD populations have consistently shown deficits in phonemic awareness, as well as phonological processing (i.e., using and manipulating phonemes and translating them to and from print) within these populations. Tasks that tap such processes include asking individuals to remove phonemes from words (e.g., /slid/ to /lid/), substitute phonemes (e.g., /bat/ to /hat/), map phonemes from or onto letters (phoneme to grapheme), and/or decode phonemes during nonword reading tasks. The consistency of profound deficits observed in such tasks within both SLI and dyslexic populations has led to some speculation that phonological deficits may form a critical core deficit spanning diverse language disorders (e.g., Bradley and Bryant, 1983; Catts et al., 2005; reviewed in Vellutino et al., 2004). However, speculation still continues regarding the putative etiological underpinning of core phonemic and

phonological difficulties. For example, many studies suggest that difficulties with phonological processing in LD populations may reflect underlying deficits in sound processing and identification (or rapid auditory processing deficits, as described above), while others suggest that phonological deficits may arise in parallel with separate auditory problems, but remain specific to linguistic systems (e.g., Ramus, 2003). Still others ascribe core phonological deficits in LD to primary difficulties with phonological memory (e.g., Gathercole and Baddeley, 1990). In fact, a plausible explanation would suggest that the core demands of phonological processing in language development may be such that disruption from a variety of *different* routes or causes (bottom-up or top-down) might lead to impairments in this key process of language development and reading and lead in turn to shared or overlapping phenotypes of language disruption.

Again, ongoing research into the role of core deficits in phonemic awareness and phonological processing will continue to inform the diagnosis, treatment, and etiologic understanding of developmental LD as a whole.

40.3.4 Short-Term/Working Memory Deficits

Evidence has shown consistent evidence of deficits in language processing using working and/or verbal *short-term memory* (STM) to process sentence morphology, syntax, and/or semantics in LD populations (Brady et al., 1983; Shankweiler and Crain, 1986; Shankweiler et al., 1995; Smith et al., 1989). Clearly, the effective use of language requires the use of STM to hold letter sounds in memory during word processing (prior to combining the sounds into a meaningful word), as well as in processing complex semantic meaning within sentences, paragraphs, and narratives (e.g., the meaning of later parts of a sentence may be modified by early sentence structure that must be held in STM to fully interpret the complete sentence). Tasks that tap these underlying capacities include digit-span recall tasks, word-memory tasks, and more complex assessments of active narrative comprehension during reading (see Montgomery et al., 2010, for review). Evidence for STM deficits specific to processing phonological information, such as is required for repeating nonwords, has also been reviewed (Gathercole and Baddeley, 1990; see also Catts et al., 2005). In a recent review by Briscoe and Rankin (2009), the authors discuss data dissociating deficits specific to the 'phonological memory loop' versus more generalized deficits in executive working memory systems in SLI subjects, concluding that evidence more strongly supports a deficit in core phonological memory processes (phonological loop) as opposed to overall executive memory. However, these assertions are counted by findings such as those of

Smith-Spark and Fisk (2007), who demonstrated deficits in cross-modal executive working memory, as well as phonological memory, in dyslexics.

An intriguing family-based genetic association study has also shown a genetic–behavioral linkage within a subset of dyslexic individuals, with deficits in STM appearing to correspond to variations in a segment of the dyslexia-risk gene *DYX1C1* (Marino et al., 2007). Specifically, a significant linkage was observed between single-letter backward-span scores and a genetic variant within this segment of the *DYX1C1* gene. Further evidence suggests associations between memory and/or attentional difficulties and anomalies in other dyslexia-risk genes such as *DCDC2* (Berninger et al., 2008). These collective results support the view that ongoing genetic research into LD may eventually reveal a correspondence between specific genes (out of multiple LD risk genes that have been, and are likely yet to be, identified) and specific core functional deficits contributing to LD (e.g., STM).

40.3.5 Visuospatial Processing, Visuospatial Memory, and/or Visuospatial Attention Deficits

A series of studies in the 1990s suggested that, in addition to rapid auditory processing deficits, adult dyslexics might also be characterized by deficits in processing rapidly changing, but not slowly changing (or static), visual information. This assertion was based on evidence of impairments in processing visual rapid change in human dyslexics and led to suggestions that LD may include a core deficit in 'magnocellular' system processing, since rapidly changing visual information is processed in the magnocellular (rather than parvocellular) subsystem of the visual thalamic nucleus (Lehmkuhle et al., 1993; Livingstone et al., 1991; Lovegrove et al., 1990; Slaghuis et al., 1992). Evidence of a 'magnocellular theory' of dyslexia continues to be explored (e.g., Stein, 2001). With regard to visual memory, a recent study of adult dyslexics and age/IQ-matched controls reported evidence of working memory deficits in dyslexics for both verbal and nonverbal (visuospatial) working memory span tasks that encompassed simple, complex, and dynamic span assessments (Smith-Spark and Fisk, 2007). These latter studies were of particular interest in showing that STM deficits in dyslexic populations are *not* seen exclusively in the phonological domain and may therefore reflect a more fundamental defect in neurological memory systems that in turn impacts on language processing (as well as other functions; but see Briscoe and Rankin, 2009). Ongoing research also continues to explore the role of visual memory and/or attention as a feature of LD populations. For example, Shaywitz and Shaywitz (2008) have shown that attentional processing plays a key role in reading for dyslexic individuals as measured by fMRI, further suggesting that this component of LD may even be amenable to pharmacological remediation, similar to treatment for ADHD.

Clearly, research has shown a range of core functional deficits associated with SLI and/or dyslexia/RD (Figure 40.2), and research continues to refine these behaviorally established criteria for reliable subtyping within the LD population. Thus, a very strong impetus exists to establish markers of genetic or neural features that could be used to segregate this heterogeneous LD population. Unfortunately, what are needed first are reliable markers by which to identify homogeneous subgroups for testing, which will in turn yield significance in genetic linkage and neuroimaging studies. This makes the task circular and very difficult. Recognizing these limits, studies have proceeded nonetheless by using heterogeneous populations (i.e., both SLI and dyslexic/RD) with *post hoc* analysis of subgroups, by separating groups based on core behavioral features (e.g., those with/without phonologic deficits), by correlating one putative marker with another (e.g., memory scores and genetic mapping), by using longitudinal analyses of emergent language measures to segregate groups, and by assessing distinctions and overlaps with related disorders (such as ASD-LI and SLI). Each of these approaches (as well as others) moves us forward in understanding the underlying etiology of LD. The following sections will review some of the anatomic, neuroimaging and genetic data that have resulted from these approaches.

40.4 NEUROPATHOLOGY OF LD

To date, structural neuroimaging and *postmortem* studies performed on clinically defined childhood and adult LD populations (SLI and dyslexic/RD) have, by and large, failed to reveal consistent neurological factors as markers for developmental language and reading disabilities. Some evidence of reduced cortical and subcortical volume in children with LD (e.g., Jernigan et al., 1991) has been reported. Others have reported evidence of anomalous hemispheric asymmetry in language-disabled subjects as measured by *postmortem* anatomical analysis (e.g., Galaburda, 1991; Galaburda et al., 1985; Humphreys et al., 1990) and MRI (Hynd et al., 1990, 1991; Jancke et al., 1994; Larsen et al., 1990; Leonard et al., 1993, 2006; Robichon et al., 2000; Schultz et al., 1994). Generally, these studies report greater evidence of abnormal (increased) symmetry in affected populations, specifically in the *planum temporale*, an area between Heschl's gyrus and the sylvian fissure and including Wernicke's area, and, more recently, in the inferior frontal gyrus (with both regions showing a left greater than right asymmetry in typical populations). Interestingly, these effects appear to be larger and more reliable for SLI subjects, as compared to dyslexics. In fact, recent work by Leonard et al. (1993, 2006) specifically compared cortical asymmetry for SLI and dyslexic populations and found evidence for smaller cerebral volume and greater temporal symmetry only in SLI children. Some evidence also shows reductions in cerebellar volume, as well as in the cortex and caudate nucleus, in individuals with a speech/language disorder (Watkins et al., 2002). The cerebellar deficits are postulated to relate more specifically to motor component deficits in LD. Clearly, more research is needed using homogeneous subtypes of the LD population in order to gain information about the biological underpinnings of identified functional deficits associated with LD.

Notably, failure to identify a consistently observable, gross anatomical neural feature of LD does *not* imply a normal brain, which has led some to investigate more subtle, but no less deleterious, mechanisms of neural disruption in developmental disabilities of language. For example, key neurodevelopmental processes may be disrupted early in life, leading to anomalous neurocircuitry not necessarily evident at the gross anatomical level. Such complex and subtle effects, however, could have potentially severe behavioral consequences. One scenario that could lead to such effects involves a confluence of genetic and/or environmental factors that occurs in the perinatal period, during key periods of neural development that disrupt the cascade of normal developmental events. One such example is neuronal migration disorders (Barth, 1987). Neuronal migration disorders occur when control mechanisms regulating the final positioning of newly generated migrating neural cells are disrupted and can result in permanent cellular/structural anomalies including agyria/pachygyria, microgyria, dysplasia, and neuronal and leptomeningeal heterotopias and focal ectopias (see Barth, 1987 for review). In fact, Galaburda and Kemper (1979) reported the presence of such focal cellular anomalies in the *postmortem* brain of an adult dyslexic male. This subject was characterized by delayed speech development in the preschool period, and difficulties with reading and spelling soon after entering elementary school, despite normal intelligence and special tutoring. These neuroanatomical findings, including the presence of cortical dysplasias and polymicrogyria in the left temporal lobe and cortical dysplasias throughout the left hemisphere, provided one of the first suggestions that language and reading impairments may represent behavioral manifestations of an underlying neuroanatomical disorder. Subsequently, Galaburda et al. (1985) published an account of three additional brains obtained at *post mortem* from adult male dyslexics. One of these subjects exhibited

delayed speech development followed by reading difficulties in school, and the others exhibited early childhood learning disabilities followed by reading difficulties. Again, numerous neuronal ectopias and cortical dysplasias were observed, primarily in left perisylvian/temporal (language) regions. The authors ascribed these cellular anomalies to focal disruption of neocortical neuronal migration, which probably occurred during the prenatal period (see also Chang et al., 2005, 2007; Preis et al., 1998; Rocha de Vasconcelos Hage et al., 2006). This assertion is supported by developmental studies of animal models exhibiting similar neuronal migration disorders.

In addition to neocortical neuronal migration disorders, fMRI studies have revealed clear evidence of abnormal cortical activation in LD subjects during language tasks, which is thought to reveal abnormal cortical connectivity and/or circuitry in these subjects (e.g., Hoeft et al., 2006; Shaywitz et al., 1998, 2002; Temple et al., 2001). In general, these results correspond to anatomic findings, with evidence of reduced activation in left perisylvian, occipitotemporal, and temperoparietal regions during phonological and reading tasks in dyslexic/RD individuals (see Figure 40.1). Abnormalities in white matter tracts connecting anterior and posterior perisylvian regions have also been reported in subjects with impaired reading ability (Klingberg et al., 2000). Anomalous patterns of activation could in turn stem from a common underlying etiology relating to processes involved in neuronal migration, *or* they could reflect aspects of LD largely orthogonal to focal anomalies. An improved understanding of how these disruptions in neural functioning may underlie, or relate to, expressed behavioral deficits would further our understanding of the relationships among and between the behavioral deficits that characterize LD overall (as well as within various subtypes). In other words, the ambiguity in our understanding of the etiology and interrelation of *neural* anomalies characterizing LD parallels clinical controversies regarding the role of core characteristic *behavioral* deficits that define human LD (as discussed above). Are neural anomalies – perhaps with a common underlying etiology – expressed in a variety of interrelated structures and in turn expressed as a cascade of interwoven functional deficits defining the fabric of 'language disability'? Alternatively, can multiple and semi-orthogonal anomalies simultaneously arise in neural development (both cortical and subcortical), leading in turn to semi-orthogonal functional deficits that contribute (in varying degrees) to a heterogeneous behavioral expression of LD (with some features more pronounced in certain subtypes than others)?

As an interesting aside, histopathological analysis of sections from the same *postmortem* brains that had previously shown focal cortical anomalies (Galaburda et al., 1985) also subsequently revealed abnormalities at the thalamic level, specifically, in the lateral geniculate (visual) nucleus (LGN; Livingstone et al., 1991). The dyslexic brains showed significantly smaller magnocellular LGN cells (28% smaller in surface area), but no size differences in parvocellular LGN neurons, as compared to controls. Concurrent electrophysiological evidence showed that healthy adult dyslexics exhibited anomalies in neural activation during performance of visual tasks known to depend on the magnocellular system. Livingstone and colleagues suggested that the focal cortical anomalies seen in dyslexics were linked to disruptions of thalamic development, including the visual pathways of the LGN responsible for transmission of low spatial frequency, low luminance contrast, and high temporal rate of change of information (magnocellular pathways; see also Lehmkuhle et al., 1993; Lovegrove et al., 1990; Slaghuis et al., 1992). Subsequent analysis of the same brains showed a similar type of anatomical change in the auditory (medial geniculate) nucleus (MGN). Specifically, dyslexics exhibited a significant shift toward more small and fewer large cells in the left MGN as compared to controls (Galaburda et al., 1994). The latter findings have been viewed in light of concurrent data demonstrating that language-disabled subjects also exhibit fundamental defects in the processing of rapidly changing auditory information (discussed above).

40.5 CANDIDATE LD AND DYSLEXIA SUSCEPTIBILITY GENES

Another approach to understanding the basic etiology of language disorders derives from epidemiological and family studies designed to draw linkages between the incidence of specific genetic variants and the incidence of language disorders (including dyslexia/RD). Studies have examined the incidence of LD in families in order to determine estimates of heritability, used linkage analysis to assess genetic 'markers' shared at above-chance levels among affected individuals from the same family, and used genetic association studies to examine correspondence between variations in regions of genes and specific phenotypes within a group (generally a clinically defined group, such as dyslexics). Such studies have revealed that language disorders are characterized by a degree of heritability, but do not conform to any single-gene models. Moreover, a degree of environmental influence seems to exist as well, since affected individuals can appear in families with no history of LD. The etiology of these 'nonheritable' incidences *could* reflect factors such as chronic ear infections (which impair hearing and thus may disrupt language acquisition), perinatal birth incidents, or the cumulative effects

multiple 'risk' genes inherited from parents with subthreshold expression, but expressed in offspring due to recombination or environmental exacerbation.

Cumulative studies appear to implicate a role for regions on chromosomes 1, 3, 6, and 15 for SSD; 13, 16, and 19 for SLI; and 1, 2, 3, 6, 15, 18, and X for dyslexia/RD (reviewed in Bishop, 2009; Gibson and Gruen, 2008; Pennington and Bishop, 2009). Studies honing in on specific genes have further revealed a role for the *FOXP2* gene in the KE family, as associated with the incidence of an expressive language disorder (Fischer and Scharff, 2009). Researchers have also reported on evidence for six candidate dyslexia susceptibility genes. These include *DYX1C1*, *KIAA0319*, *DCDC2*, *ROBO1*, *MRPL19*, and *C2ORF3* (see Anthoni et al., 2007; Brkanac et al., 2007; Cope et al., 2005; Francks et al., 2004; Hannula-Jouppi et al., 2005; Harold et al., 2006; McGrath et al., 2006; Meng et al., 2005; Paracchini et al., 2006; Schumacher et al., 2006; Taipale et al., 2003, see Fischer and Francks, 2006, for review).

Of these genes, *KIAA0319* and *DCDC2* are located on chromosome 6, *FOXP2* is located on chromosome 7, *ROBO1* is located on chromosome 13, and *DYX1C1* is located on chromosome 15. *MRPL19* and *C2ORF3* are coregulated genes on chromosome 2, the function of which is not yet known. Research suggests that each of the remaining genes appears to be involved in early cortical development, including neuronal migration, axon growth, and synaptic plasticity (Wang et al., 2006). Since neurodevelopmental processes such as neuronal migration and axon growth share several common features and requirements, including dependence upon coordinated changes in cell adhesion and cytoskeletal restructuring, the overlapping functions implicated for these genes is not surprising. For example, *ROBO1* has well-understood roles in axon growth and neuronal migration. The proteins in the *DCX* family, of which *DCDC2* is a member, play well-documented roles in neuronal migration to neocortex and may also play a role in the development of the corpus callosum (LoTurco et al., 2006; Miller, 2005; Rosen et al., 2007; Wang et al., 2006). In addition, *KIAA0319* appears to play a role in cortical cell adhesion and migration, as does *FOXP2* (possibly via regulation of contactin-associated protein). Moreover, *FOXP2* appears to specifically modulate cortical synaptic plasticity, and acts as a transcriptional factor (transcriptional repressor) that may further regulate the expression of other genes important to brain development (Fischer and Scharff, 2009). Indeed, research has revealed very minor changes to a promoter sequence in the *FOXP2* gene when comparing humans and nonhuman primates, and this evolutionary change may modulate some aspects of the development of unique human language-processing capabilities (Konopka et al., 2009).

In terms of functional implications, studies using animal models have provided a unique insight into the role of these various genes by linking individual gene actions with core functional deficits characteristic of human LD populations. For example, studies have demonstrated that knockdown of the avian ortholog to *FOXP2* in birds leads to an impairment in the motor learning of birdsong, and also that *Foxp2* knockout mice show anomalous vocalizations (Gaub et al., 2010; Shu et al., 2005). Similarly, recent research has used rodent models to knock down the actions of identified dyslexia-risk genes transiently, specifically through the transfection of 'interference RNA' (RNAi) into the cerebral ventricles of fetal rats. This RNAi is taken up by new neurons in the ventricular zone, thus deactivating the target genes in these specific cells. Behavioral testing of rodents treated using this technology shows that male rats with embryonic cortical RNAi of the homolog *Dyx1c1* exhibit deficits in complex auditory processing later in life (Threlkeld et al., 2007). These results have intriguing implications for the possible role of *DYX1C1* in modulating auditory processing deficits, and perhaps even associated phonemic/phonological deficits. More recent research has extended these findings to rodent models using another dyslexia-risk gene, *KIAA0319*. Specifically, rats with embryonic RNAi for the homolog *Kiaa0319* also showed significant deficits in rapid auditory processing for fast frequency-modulated (FM) sweep stimuli as compared to sham controls (Szalkowski et al., 2012). Again, these effects are interpreted to parallel evidence of rapid auditory processing deficits in LD populations. As an aside, it is important to note that neither group of RNAi animals was categorically impaired in sound processing tasks, and both groups performed quite normally on discrimination tasks with easier/slower stimuli (e.g., single tones). This point is important when developing animal models for clinical conditions, since a genetic defect resulting in comprehensive learning and cognitive impairments would provide a poor model for the relatively specific pattern(s) of deficits seen within the LD phenotype.

More recently, another study examined STM performance on a delayed match to sample radial arm maze task in male rats with or without embryonic RNAi of *Dyx1c1* (Szalkowski et al., 2011). In this study, results showed persistent STM deficits in *Dyx1c1* knockdown rats, as indicated by higher numbers of errors, paralleling evidence of cross-modal STM deficits in human dyslexics (Smith-Spark and Fisk, 2007). These effects were seen despite parallel evidence that *Dyx1c1* RNAi rats do *not* show deficits on basic spatial maze learning (Morris maze) unless migrational anomalies specifically extended into the hippocampus (as was seen in a small subset of subjects; Threlkeld et al., 2007). Interestingly, similar working memory deficits were *not* seen in a sample of *Kiaa0319* knockout rats, suggesting that these genetic risk factors may exert dissociable effects on core behavioral deficits related to dyslexia (Szalkowski et al., 2012). Again, this pattern of behavioral–genetic associations provides a relatively specific profile to model deficits seen in LD (i.e., deficits in STM). Future work will continue to explore a role for animal models in evaluating these gene–brain–behavior relationships.

40.6 CONCLUSIONS

As will by now be readily apparent from the current chapter, research in the field of language-related disabilities represents one of the most comprehensive and complex cross-disciplinary research fields currently confronting the scientific and clinical community. On the one hand, the breadth of expertise and approaches brought to bear on the issue has led to a vast database concerning the behavioral, neural, and genetic features of LD. On the other hand, the field also suffers from the challenge inherent to processing and integrating the enormous amount of information that now exists. These difficulties are reflected in the fact that different researchers have created conflicting acronyms for what appear to be the same terms, that substantial disagreements exist regarding the distinctions and overlap between subgroups as a function of behavioral criteria (leading to studies that use differently defined populations as subjects for assessment and experimentation), and that it remains quite nearly impossible to achieve expertise in fields as diverse as molecular genetics to clinical behavioral evaluation (thus constraining efforts to integrate data on LD populations fully). Nonetheless, researchers continue to move forward in refining and reshaping our conceptions of language-related developmental disorders from a behavioral perspective and in adding new information to the database of neural and genetic features that characterize these populations. Finally, animal studies provide a unique opportunity to test and assess some of the putative genetic–neural–behavioral links in an experimental manner that is virtually impossible to employ with human subjects.

With regard to future directions, two of the most promising goals toward which research efforts continue to strive are (1) the development and implementation of effective early diagnostic criteria that might detect infants at risk for later language problems, and provide an opportunity for very early intervention (while brain systems are still highly plastic), and (2) the development of effective remediation strategies tailored to unique patterns of deficits in subpopulations of LD (Hoeft et al., 2006). Promising efforts currently continue in both respects. For example, studies have shown that measures of acoustic processing obtained from infants as young as 6 months of age may predict their later risk for language difficulties (Benasich et al., 2006; Choudhury et al., 2007). In addition, intriguing neuroimaging studies have shown that effective intervention strategies can actually shift the abnormal patterns of brain activity in affected subjects observed during language tasks toward more normative patterns following training (e.g., Gaab et al., 2007). All of these promising routes of investigation will continue in the next decade, as we continue to make inroads in our diagnosis, understanding, and treatment of developmental LD.

References

Alcock, K.J., Passingham, R.E., Watkins, K.E., Vargha-Khadem, F., 2000. Oral dyspraxia in inherited speech and language impairment and acquired dysphasia. Brain and Language 75, 17–33.

Anthoni, H., Zucchelli, M., Matsson, H., et al., 2007. A locus on 2p12 containing the co-regulated MRPL19 and C2ORF3 genes is associated to dyslexia. Human Molecular Genetics 16, 667–677.

Archibald, L.M., Alloway, T.P., 2008. Comparing language profiles: Children with specific language impairment and developmental coordination disorder. International Journal of Language and Communication Disorders 43, 165–180.

Au, A., Lovegrove, B., 2007. The contribution of rapid visual and auditory processing to the reading of irregular words and pseudowords presented singly and in contiguity. Perception and Psychophysics 69, 1344–1359.

Bamiou, D.D., Musiek, F.E., Luxon, L.M., 2001. Aetiology and clinical presentations of auditory processing disorders – A review. Archives of Diseases in Childhood 85, 361–365.

Barth, P.G., 1987. Disorders of neuronal migration. Canadian Journal of Neurological Science 14, 1–16.

Beitchman, J.H., Nair, R., Patel, P.G., 1986. Prevalence of speech and language disorders in 5-year old kindergarten children in the Ottowa-Carlton region. Journal of Speech and Hearing Disorders 51, 98–110.

Benasich, A.A., Thomas, J.J., Choudhury, N., Leppanen, P.H.T., 2002. The importance of rapid auditory processing abilities to early language development: Evidence from converging methodologies. Developmental Psychobiology 40, 278–292.

Benasich, A.A., Choudhury, N., Friedman, J.T., Realpe-Bonilla, T., Chojnowska, C., Gou, Z., 2006. The infant as a prelinguistic model for language learning impairments: Predicting from event-related potentials to behavior. Neuropsychologia 44, 396–411.

Berninger, V.W., Raskind, W., Riachards, T., Abbott, R., Stock, P., 2008. A multi-disciplinary approach to understanding developmental dyslexia within working-memory architecture: Genotypes, phenotypes, brain, and instruction. Developmental Neuropsychology 33, 707–744.

Bishop, D.D.M., 2009. Genes, cognition and communication: Insights from neurodevelopmental disorders. Annals of the New York Academy of Sciences 1156, 1–18.

Bishop, D.V.M., Snowling, M.J., 2004. Developmental dyslexia and specific language impairment: Same or different? Psychological Bulletin 130, 858–886.

Blank, S.C., Scott, S.K., Murphy, K., Warburton, E., Wise, R.J., 2002. Speech production: Wernicke, Broca and beyond. Brain 125, 1829–1838.

Bradley, L., Bryant, P., 1983. Categorizing sounds and learning to read – A causal connection. Nature 301, 419–421.

Brady, S., Shankweiler, D., Mann, V., 1983. Speech perception and memory coding in relation to reading ability. Journal of Experimental Child Psychology 35, 345–367.

Briscoe, J., Rankin, P.M., 2009. Exploration of a 'double-jeopardy' hypothesis within working memory profiles for children with specific language impairment. International Journal of Language and Communication Disorders 44, 236–250.

Brkanac, Z., Chapman, N.H., Matsushita, M.M., et al., 2007. Evaluation of candidate genes for DYX1 and DYX2 in families with dyslexia. American Journal of Medical Genetics 144, 556–560.

Broman, S.H., Grafman, J., 1994. Atypical Cognitive Deficits in Developmental Disorders: Implications for Brain Function. Lawrence Earlbaum Associates, Hillsdale, NJ.

Cardy, J.E.O., Flagg, E.J., Roberts, W., Brian, J., Roberts, T.P.L., 2005. Magnetoencephalography identifies rapid temporal processing deficit in autism and language impairment. NeuroReport 16, 329–332.

Catts, H.W., Adlof, S.M., Hogan, T.P., Weismer, S.M., 2005. Are specific language impairment and dyslexia distinct disorders? Journal of Speech, Language, and Hearing Research 48, 1378–1396.

Chang, B.S., Ly, J., Appignani, B., et al., 2005. Reading impairment in the neural migration disorder of periventricular nodular heterotopia. Neurology 64, 799–803.

Chang, B.S., Katzir, T., Liu, T., et al., 2007. A structural basis for reading fluency: White matter defects in a genetic brain malformation. Neurology 69, 2146–2154.

Choudhury, N., Leppannen, P.H.T., Leevers, H.J., Benasich, A.A., 2007. Infant information processing and family history of specific language impairment: Converging evidence for RAP deficits from two paradigms. Developmental Science 10, 213–236.

Cohen-Mimran, R., Sapir, S., 2007. Auditory temporal processing deficits in children with reading disabilities. Dyslexia 13, 175–192.

Cope, N., Harold, D., Hill, G., et al., 2005. Strong evidence that *KIAA0319* on chromosome 6p is a susceptibility gene for developmental dyslexia. American Journal of Human Genetics 76, 581–591.

Corbera, S., Escera, C., Artigas, J., 2006. Impaired duration mismatch negativity in developmental dyslexia. NeuroReport 17, 10051–10055.

Edwards, V.T., Giaschi, D.E., Dougherty, R.F., et al., 2004. Psychophysical indexes of temporal processing abnormalities in children with developmental dyslexia. Developmental Neuropsychology 25, 321–354.

Farmer, M.E., Klein, R.M., 1995. The evidence for a temporal processing deficit linked to dyslexia. Psychonomic Bulletin and Review 2, 460–493.

Fischer, S.E., Francks, C., 2006. Genes, cognition and dyslexia: Learning to read the genome. Trends in Cognitive Sciences 10, 250–257.

Fischer, S.E., Marcus, G.F., 2006. The eloquent ape: Genes, brains and the evolution of language. Nature Reviews Genetics 7, 9–20.

Fischer, S.E., Scharff, C., 2009. FOXP2 as a molecular window into speech and language. Trends in Genetics 25, 166–177.

Fitch, R.H., Tallal, P., 2003. Neural mechanisms of language based learning impairments: Insights from human populations and animal models. Behavioral and Cognitive Neuroscience Reviews 2 (3), 155–178.

Francks, C., Paracchini, S., Smith, S.D., et al., 2004. A 77-kilobase region of chromosome 6p22.2 is associated with dyslexia in families in the United Kingdom and from the United States. American Journal of Human Genetics 75, 1046–1058.

Gaab, N., Gabrieli, J.D.E., Deutsch, G.K., Tallal, P., Temple, E., 2007. Neural correlates of rapid auditory processing are disrupted in children with developmental dyslexia and ameliorated with training: An fMRI study. Restorative Neurology and Neuroscience 25, 295–310.

Galaburda, A.M., 1991. Anatomy of dyslexia: Argument against phrenology. In: Duane, D.D. (Ed.), The Reading Brain: The Biological Basis of Dyslexia. York Press, Parkton, MD, pp. 119–131.

Galaburda, A.M., Kemper, T.L., 1979. Cytoarchitectonic abnormalities in developmental dyslexia: A case study. Annals of Neurology 6, 94–100.

Galaburda, A.M., Sherman, G.F., Rosen, G.D., Aboitiz, F., Geschwind, N., 1985. Developmental dyslexia: Four consecutive cases with cortical anomalies. Annals of Neurology 18, 222–233.

Galaburda, A.M., Menard, M.T., Rosen, G.D., Livingstone, M.S., 1994. Evidence for aberrant auditory anatomy in developmental dyslexia. Proceedings of the National Academy of Sciences 91, 8010–8013.

Galaburda, A.M., Rosen, G.D., Denenberg, V.H., Fitch, R.H., LoTurco, J.J., Sherman, G.F., 2001. Models of temporal processing and language development. Clinical Neuroscience Research 1 (3), 230–237.

Gathercole, S.E., Baddeley, A.D., 1990. Phonological memory deficits in language impaired children: Is there a causal connection? Journal of Memory and Language 29, 336–360.

Gaub, S., Groszer, M., Fischer, S.E., Ehret, G., 2010. The structure of vocalizations in Foxp2-deficient mouse pups. Genes, Brain, and Behavior 9 (4), 390–401.

Gibson, C.J., Gruen, J.R., 2008. The human lexinome: Genes of language and reading. Journal of Communication Disorders 41, 409–420.

Goldman-Rakic, P.S., Rakic, P., 1978. Prenatal removal of the frontal association cortex in the fetal rhesus monkey: Anatomical and functional consequences in postnatal life. Brain Research 152, 451–485.

Hannula-Jouppi, K., Kaminen-Ahola, N., Taipale, M., et al., 2005. The axon guidance receptor gene ROBO1 is a candidate gene for developmental dyslexia. PLoS Genetics 1, e50.

Hari, R., Kiesla, P., 1996. Deficit of temporal auditory processing in dyslexic adults. Neuroscience Letters 205, 138–140.

Harold, D., Paracchini, S., Scerri, T., et al., 2006. Further evidence that the KIAA0319 gene confers susceptibility to developmental dyslexia. Molecular Psychiatry 11, 1085–1091, 1061.

Hauser, M.D., 1996. The Evolution of Communication. MIT Press, Cambridge, MA.

Hautus, M.J., Setchell, G.J., Waldie, K.E., Kirk, I.J., 2003. Age-related improvements in auditory temporal resolution in reading-impaired children. Dyslexia 9, 37–45.

Heim, S., Benasich, A.A., 2006. Developmental disorders of language. In: Ciccheti, D., Cohen, D. (Eds.), Developmental Psychopathology. Risk, Disorder and Adaptation, vol. 3. Wiley, New York, pp. 268–316, 2nd edn.

Herbert, M.R., Kenet, T., 2007. Brain abnormalities in language disorders and autism. Pediatric Clinics of North America 54 (3), 563–583.

Herman, A., Galaburda, A., Fitch, R.H., Carter, A.R., Rosen, G., 1997. Cerebral microgyria, thalamic cell size and auditory temporal processing in male and female rats. Cerebral Cortex 7, 453–464.

Hoeft, F., Hernandez, A., McMillon, G., et al., 2006. The neural basis of dyslexia: A comparison between dyslexic and nondyslexic children equated for reading ability. Journal of Neuroscience 26, 10700–10708.

Humphreys, P., Kaufmann, W.E., Galaburda, A.M., 1990. Developmental dyslexia in women: Neuropathological findings in three cases. Annals of Neurology 28, 727–738.

Hynd, G.W., Semrud-Clikeman, M., Lorys, A.R., Novey, E.S., Eliopulos, D., 1990. Brain morphology in developmental dyslexia and attention deficit disorder/hyperactivity. Archives of Neurology 47, 919–926.

Hynd, G.W., Marshall, R.M., Semrud-Clikeman, M., 1991. Developmental dyslexia, neurolinguistic theory and deviations in brain morphology. Reading and Writing 3, 345–362.

Jancke, L., Schlaug, G., Huang, Y., Steinmetz, H., 1994. Asymmetry of the planum temporale. NeuroReport 5, 1161–1163.

Jernigan, T.L., Hesselin, J.R., Sowell, E., Tallal, P.A., 1991. Cerebral structure on magnetic resonance imaging in language- and learning-impaired children. Archives of Neurology 48, 539–545.

Jordan, L.C., Hillis, A.E., 2006. Disorders of speech and language: Aphasia, apraxia & dysarthria. Current Opinions in Neurology 19 (6), 580–585.

King, B., Wood, C., Faulkner, D., 2007. Sensitivity to auditory stimuli in children with developmental dyslexia. Dyslexia 14, 116–141.

Klingberg, T., Hedehus, M., Temple, E., et al., 2000. Microstructure of tempero-parietal white matter as a basis for reading ability: Evidence from diffusion tensor magnetic resonance imaging. Neuron 23, 493–500.

Konopka, G., Bomar, J.M., Winden, K., et al., 2009. Human-specific transcriptional regulation of CNS development genes by FOXP2. Nature 462, 213–217.

Kraus, N., McGee, T.J., Carrell, T.D., Zecker, S.G., Nicol, T.G., Koch, D.B., 1996. Auditory neurophysiologic responses and discrimination deficits in children with learning problems. Science 273, 971–973.

Larsen, J., Hoien, T., Lundberg, I., Odegaard, H., 1990. MRI evaluation of the size and symmetry of the planum temporale in adolescents with developmental dyslexia. Brain and Language 39, 289–301.

Lehmkuhle, S., Garzia, R.P., Turner, L., Hash, T., Baro, J.A., 1993. A defective visual pathway in children with reading disability. New England Journal of Medicine 328, 989–996.

Leonard, L.B., 1998. Children with Specific Language Impairment. MIT Press, Cambridge, MA.

Leonard, C., Voeller, K.K.S., Lombardino, L.J., et al., 1993. Anomalous cerebral structure in dyslexia revealed with magnetic resonance imaging. Archives of Neurology 50, 461–469.

Leonard, C., Eckert, M., Given, B., Virginia, B., Eden, G., 2006. Individual differences in anatomy predict reading and oral language impairments in children. Brain 129, 3329–3342.

Leppänen, P.H.T., Lyytinen, H., 1997. Auditory event-related potentials in the study of developmental language-related disorders. Audiology and Neuro-Otology 2, 308–340.

Leppänen, P.H.T., Pihko, E., Eklund, K.M., Lyytinen, H., 1999. Cortical responses of infants with and without a genetic risk for dyslexia: II. Group effects. NeuroReport 10, 969–973.

Livingstone, M.S., Rosen, G.D., Drislane, F.W., Galaburda, A.M., 1991. Physiological and anatomical evidence for a magnocellular defect in developmental dyslexia. Proceedings of the National Academy of Sciences of the United States of America 88, 7943–7947.

LoTurco, J., Wang, Y., Paramasivam, M., 2006. Neuronal migration and dyslexia susceptibility. In: Rosen, G. (Ed.), The Dyslexic Brain: New Pathways in Neuroscience Discovery. Lawrence Erlbaum Associates, Mahway, NJ, pp. 119–128.

Lovegrove, W., Garzia, R., Nicholson, S., 1990. Experimental evidence for a transient system deficit in specific reading disability. Journal of the American Optometric Association 2, 137–146.

Marino, C., Citterio, A., Giorda, R., et al., 2007. Association of short-term memory with a variant within DYX1C1 in developmental dyslexia. Genes, Brain, and Behavior 6, 640–646.

McAnally, K.I., Stein, J.F., 1996. Auditory temporal coding in dyslexia. Proceedings of the Royal Society of London B: Biological Sciences 263, 961–965.

McAnally, K.I., Stein, J.F., 1997. Scalp potentials evoked by amplitude modulated tones in dyslexia. Journal of Speech, Language, and Hearing Research 40, 939–945.

McArthur, G.M., Bishop, D.V.M., 2001. Auditory perceptual processing in people with reading and oral language impairments: Current issues and recommendations. Dyslexia 7, 150–170.

McCrosky, R., Kidder, H., 1980. Auditory fusion among learning disabled, reading disabled and normal children. Journal of Learning Disabilities 13, 69–76.

McGrath, L.M., Smith, S.D., Pennington, B.F., 2006. Breakthroughs in the search for dyslexia candidate genes. Trends in Molecular Medicine 112, 333–341.

Meng, H., Smith, S.D., Hager, K., et al., 2005. *DCDC2* is associated with reading disability and modulates neuronal development in the brain. Proceedings of the National Academy of Sciences of the United States of America 102, 17053–17058.

Menghini, D., Finzi, A., Benassi, M., et al., 2010. Different underlying neurocognitive deficits in developmental dyslexia: A comparative study. Neuropsychologia 48, 863–872.

Miller, G., 2005. Genes that guide brain development linked to dyslexia. Science 310, 759.

Mody, M., Studdert-Kennedy, M., Brady, S., 1997. Speech perception deficits in poor readers: Auditory processing or phonological coding? Journal of Experimental Child Psychology 64, 199–231.

Molfese, D.L., Molfese, V.J., 1997. Discrimination of language skills at five years of age using event-related potentials recorded at birth. Developmental Neuropsychology 13, 135–156.

Montgomery, J.W., Magimairaj, B.M., Finney, M.C., 2010. Working memory and specific language impairment: An update on the relation and perspectives on assessment and treatment. American Journal of Speech-Language Pathology 19, 78–94.

Neville, H.J., Coffey, S.A., Holcomb, P.J., Tallal, P., 1993. The neurobiology of sensory and language processing in language-impaired children. Journal of Cognitive Neuroscience 5 (2), 235–253.

Paracchini, V., Seia, M., Coviello, D., et al., 2006. The chromosome 6p22 haplotype associated with dyslexia reduces the expression of KIAA0319, a novel gene involved in neuronal migration. Human Molecular Genetics 73, 346–352.

Pennington, B.F., 1991. Diagnosing Learning Disorders: A Neuropsychological Framework. The Guildford Press, New York.

Pennington, B.F., Bishop, D.V.M., 2009. Relations among speech, language, and reading disorders. Annual Review of Psychology 60, 283–306.

Peterson, R.L., McGrath, L.M., Smith, S.D., Pennington, B.F., 2007. Neuropsychology and genetics of speech, language, and literacy disorders. Pediatric Clinics of North America 54, 543–561.

Pihko, E., Leppänen, P.H.T., Eklund, K.M., Cheour, M., Guttorm, T.K., Lyytinen, H., 1999. Cortical responses of infants with and without a genetic risk for dyslexia: I. Age effects. NeuroReport 10, 901–905.

Preis, S., Engelbrecht, V., Huang, Y., Steinmetz, H., 1998. Focal grey matter heterotopias in monozygotic twins with developmental language disorder. European Journal of Pediatrics 157, 849–852.

Ramus, F., 2003. Developmental dyslexia: Specific phonological deficit or general sensorimotor dysfunction? Current Opinion in Neurobiology 13, 212–218.

Reed, M.A., 1989. Speech perception and the discrimination of brief auditory cues in reading disabled children. Journal of Experimental Child Psychology 48, 270–292.

Renvall, H., Hari, R., 2002. Auditory cortical responses to speech-like stimuli in dyslexic adults. Journal of Cognitive Neuroscience 14, 757–768.

Robichon, F., Levriere, O., Farnarier, P., Habib, M., 2000. Developmental dyslexia: Atypical cortical asymmetries and functional significance. European Journal of Neurology 7, 35–46.

Robin, D., Tomblin, J.B., Kearney, A., Hug, L., 1989. Auditory temporal pattern learning in children with severe speech and language impairment. Brain and Language 36, 604–613.

Rocha de Vasconcelos Hage, S., Cendes, F., Augusta Montenegro, M., Abramides, D.V., Giumaraes, C.A., Guerreio, M.M., 2006. Specific language impairment: Linguistic and neurobiological aspects. Arquivos de Neuro-Psiquiatria 64, 173–180.

Rosen, G.D., Bai, J., Wang, Y., et al., 2007. Disruption of neuronal migration by RNAi of Dyx1c1 results in neocortical and hippocampal malformations. Cerebral Cortex 17, 2562–2572.

Rosen, S., Manganari, E., 2001. Is there a relationship between speech and non-speech auditory processing in children with dyslexia? Journal of Speech, Language, and Hearing Research 44, 720–736.

Schuele, C.M., 2004. The impact of developmental speech and language impairments on the acquisition of literacy skills. Mental Retardation and Developmental Disabilities Research Reviews 10 (3), 176–183.

Schultz, R.T., Cho, N.K., Staib, L.H., et al., 1994. Brain morphology in normal and dyslexic children: The influence of sex and age. Annals of Neurology 35, 732–742.

Schumacher, J., Anthoni, H., Dahdouh, F., et al., 2006. Strong genetic evidence of DCDC2 as a susceptibility gene for dyslexia. American Journal of Human Genetics 78, 52–62.

Shankweiler, D., Crain, S., 1986. Language mechanisms and reading disorder: A modular approach. Cognition 24, 139–168.

Shankweiler, D., Crain, S., Katz, L., et al., 1995. Cognitive profiles of reading-disabled children: Comparison of language skills in phonology, morphology and syntax. Psychological Science 6, 149–156.

Shaywitz, S.E., Shaywitz, B.A., 2008. Paying attention to reading: The neurobiology of reading and dyslexia. Development and Psychopathology 20, 1329–1349.

Shaywitz, B.A., Shaywitz, S.E., Pugh, K.R., et al., 1995. Sex differences in the functional organization of the brain for language. Nature 373, 607–609.

Shaywitz, S.E., Shaywitz, B.A., Pugh, K., et al., 1998. Functional disruption in the organization of the brain for reading in dyslexia. Proceedings of the National Academy of Sciences 95, 2636–2641.

Shaywitz, B.A., Shaywitz, S.E., Pugh, K.R., et al., 2002. Disruption of posterior brain systems for reading in children with developmental dyslexia. Biological Psychiatry 52, 101–110.

Shu, W., Cho, J.Y., Jiang, Y., et al., 2005. Altered ultrasonic vocalization in mice with a disruption in the Foxp2 gene. Proceedings of the National Academy of Sciences 102, 9643–9648.

Sices, L., Taylor, H.G., Freebairn, L., Hansen, A., Lewis, B., 2007. Relationship between speech-sound disorders and early literacy skills in preschool-age children: Impact of comorbid language impairment. Journal of Developmental and Behavioral Pediatrics 28 (6), 438–447.

Slaghuis, W.L., Lovegrove, W.J., Freestun, J., 1992. Letter recognition in peripheral vision and metacontrast masking in dyslexic and normal readers. Clinical Visual Sciences 7, 53–65.

Smith, S.T., Macaruso, P., Shankweiler, D., Crain, S., 1989. Syntactic comprehension in young poor readers. Applied Psycholinguistics 10, 429–454.

Smith-Spark, J.H., Fisk, J.E., 2007. Working memory functioning in developmental dyslexia. Memory 15, 34–56.

Stein, J., 2001. The magnocellular theory of developmental dyslexia. Dyslexia 7, 12–36.

Sutter, M.L., Petkov, C., Baynes, K., O'Connor, K.N., 2000. Auditory scene analysis in dyslexics. NeuroReport 11, 1967–1971.

Szalkowski, C.E., Hinman, J., Threlkeld, S.W., et al., 2011. Persistent spatial working memory deficits in rats following in utero RNAi of Dyx1c1. Genes, Brain, and Behavior 10, 244–252.

Szalkowski, C.E., Fiondella, C.G., Galaburda, A.M., Rosen, G.D., LoTurco, J.J., Fitch, R.H., 2012. Neocortical disruption and behavioral impairments in rats following in utero RNAi of candidate dyslexia risk gene Kiaa0319. International Journal of Developmental Neuroscience 30, 290–302.

Taipale, M., Kaminen, N., Nopola-Hemmi, J., et al., 2003. A candidate gene for developmental dyslexia encodes a nuclear tetratricopeptide repeat domain protein dynamically regulated in brain. Proceedings of the National Academy of Sciences of the United States of America 100, 11553–11558.

Tallal, P., 1977. Auditory perception, phonics and reading disabilities in children. Journal of the Acoustical Society of America 62, S100.

Tallal, P., 1980. Auditory temporal perception, phonics and reading disabilities in children. Brain and Language 9, 182–198.

Tallal, P., 2004. Improving language and literacy is a matter of time. Nature Neuroscience Reviews 5, 721–728.

Tallal, P., Newcombe, F., 1978. Impairment of auditory perception and language comprehension in dysphasia. Brain and Language 5, 13–24.

Tallal, P., Piercy, M., 1973a. Developmental aphasia: Impaired rate of non-verbal processing as a function of sensory modality. Neuropsychologia 11, 389–398.

Tallal, P., Piercy, M., 1973b. Defects of non-verbal auditory perception in children with developmental aphasia. Nature 241, 468–469.

Tallal, P., Piercy, M., 1975. Developmental aphasia: The perception of brief vowels and extended stop consonants. Neuropsychologia 13, 69–74.

Tallal, P., Stark, R.E., 1981. Speech acoustic-cue discrimination abilities of normally developing and language-impaired children. Journal of the Acoustical Society of America 69 (2), 568–574.

Tallal, P., Townsend, J., Curtiss, J., Wulfeck, B., 1991. Phenotypic profiles of language impaired children based on genetic/family history. Brain and Language 41, 81–95.

Tallal, P., Miller, S., Fitch, R.H., 1993. Neurobiological basis of speech: A case for the preeminence of temporal processing. In: Tallal, P., Galaburda, A.M., Llinas, R., von Euler, C. (Eds.), Temporal Information Processing in the Nervous System, with Special Reference to Dyslexia and Dysphasia. Annals of the New York Academy of Sciences, vol. 682. New York Academy of Sciences, New York, pp. 27–47.

Temple, E., Poldrack, R.A., Salidis, J., et al., 2001. Disrupted neural responses to phonological and orthographic processing in dyslexic children: An fMRI study. NeuroReport 12, 299–307.

Temple, E., Deutsch, G.K., Poldrack, R.A., et al., 2003. Neural deficits in children with dyslexia ameliorated by behavioral remediation: Evidence from functional MRI. Proceedings of the National Academy of Sciences 100, 2860–2865.

Threlkeld, S.W., McClure, M.M., Bai, J., et al., 2007. Developmental disruptions and behavioral impairments in rats following *in utero* RNAi of *Dyx1c1*. Brain Research Bulletin 71, 508–514.

Trehub, S.E., Henderson, J.L., 1996. Temporal resolution in infancy and subsequent language development. Journal of Speech and Hearing Research 39 (6), 1315–1320.

Vargha-Khadem, F., Carr, L.J., Isaacs, E., Brett, E., Adams, C., Mishkin, M., 1997. Onset of speech after left hemispherectomy in a nine-year-old boy. Brain 120, 159–182.

Vellutino, F.R., Fletcher, J.M., Snowling, M.J., Scanlon, D.M., 2004. Specific reading disability (dyslexia): What have we learned in the past four decades? Journal of Child Psychology and Psychiatry 45, 2–40.

Wang, Y., Paramasivam, M., Thomas, A., et al., 2006. *Dyx1c1* functions in neuronal migration in developing neocortex. Neuroscience 143, 515–522.

Watkins, K.E., Vargha-Khadem, F., Ashburner, J., et al., 2002. MRI analysis of an inherited speech and language disorder: Structural brain abnormalities. Brain 125, 465–478.

Watson, B.U., 1992. Auditory temporal acuity in normally achieving and learning-disabled college students. Journal of Speech and Hearing Research 35, 148–156.

Webster, R.I., Shevell, M.I., 2004. Neurobiology of specific language impairment. Journal of Child Neurology 19, 471–481.

Williams, D., Botting, N., Boucher, J., 2008. Language in autism and specific language impairment: Where are the links? Psychological Bulletin 134, 944–963.

Witton, C., Talcott, J., Hansen, P., et al., 1998. Sensitivity to dynamic auditory and visual stimuli predicts nonword reading ability in both dyslexic and normal readers. Current Biology 8, 791–797.

Wright, B., Lombardino, L., King, W., Puranik, C., Leonard, C., Merzenich, M., 1997. Deficits in auditory temporal and spectral resolution in language-impaired children. Nature 387, 176–178.

Index

Note: Page numbers followed by *b* indicate boxes, *f* indicate figures and *t* indicate tables. **Bold** locators refer to articles/chapters.